Anatomy & Physiology

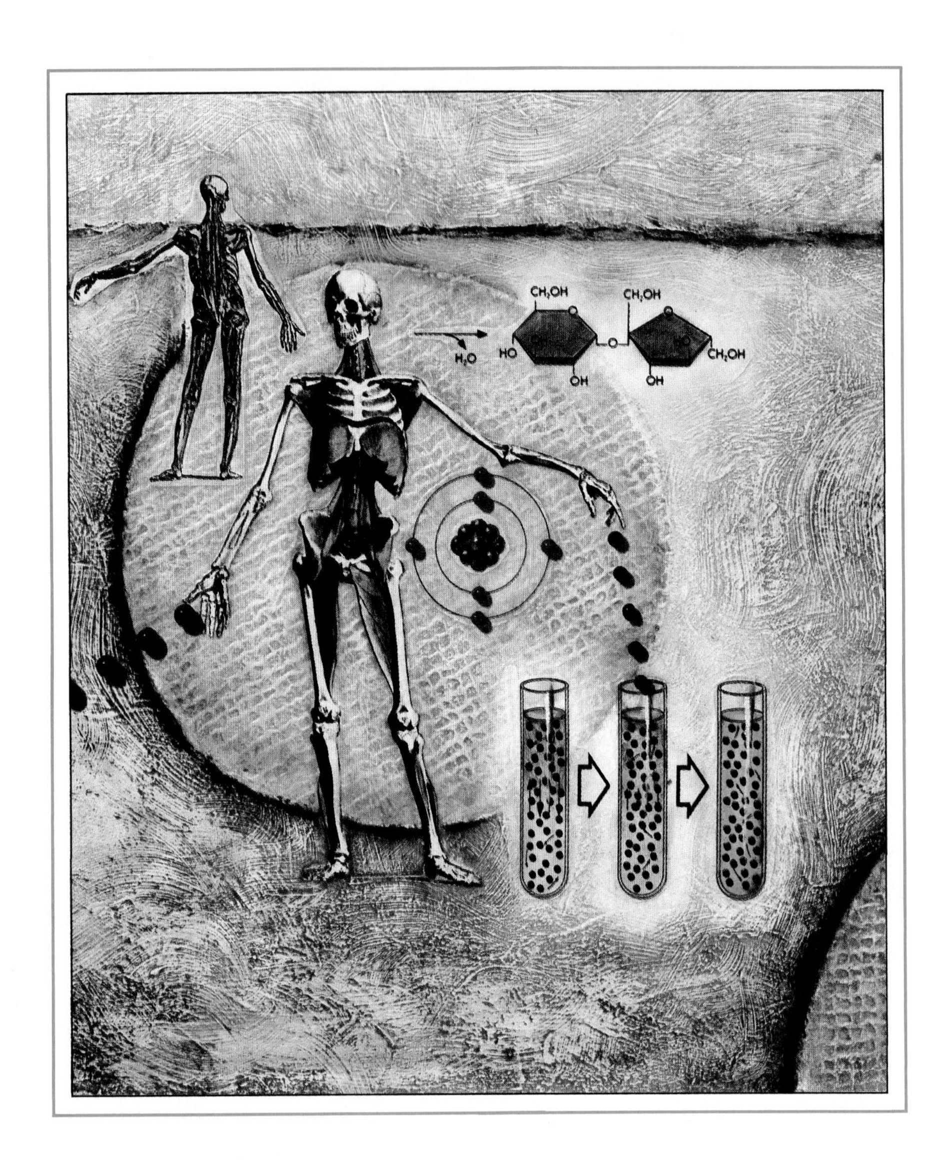

The art of collage (from French, "to glue") can show many aspects
or layers of an idea.
Our opener collages were created by a medical illustrator, and
designed to be unique partners in your discovery of
anatomy and physiology.
You are invited to explore each collage and discover the functions,
relationships, and applications it represents.
As your knowledge of anatomy and physiology increases, you will
become aware of a new image, or perhaps find that
an old image has new significance.

EXPLORE AND ENJOY!

ANATOMY & PHYSIOLOGY

ROD R. SEELEY, *Ph.D.*
Professor of Physiology
Idaho State University

TRENT D. STEPHENS, *Ph.D.*
Professor of Anatomy and Embryology
Idaho State University

PHILIP TATE, *D.A. (Biological Education)*
Instructor of Anatomy and Physiology
Phoenix College
Maricopa Community College District

SECOND EDITION

St. Louis Baltimore Boston Chicago London Philadelphia Sydney Toronto

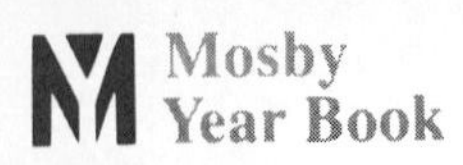

Dedicated to Publishing Excellence

Editor: Deborah Allen
Developmental editor: Robert J. Callanan
Project manager: Patricia E. Tannian
Production editor: Teresa Breckwoldt
Book design and collage development: Susan Lane
Illustration coordinator: Elizabeth Rohne Rudder
Photo researcher: Mary Temperelli

Cover and part opener collages by William B. Westwood, M.S.

Anatomical Plates provided by R.T. Hutchings from *Color Atlas of Human Anatomy*, edition 2, by R.M.H. McMinn and R.T. Hutchings, Year Book Medical Publishers, Inc., 1988.

Credits for all materials used by permission appear after the index.

SECOND EDITION

Printed in the United States of America.

Mosby–Year Book, Inc.
11830 Westline Industrial Drive
St. Louis, Missouri 63146

Library of Congress Cataloging in Publication Data

Seeley, Rod R.
Anatomy and physiology / Rod R. Seeley, Trent D. Stephens, Philip Tate.—Ed. 2.
p. cm.
Includes bibliographical references and index.
ISBN 0-8016-4832-7
1. Human physiology. 2. Human anatomy. I. Stephens, Trent D. II. Tate, Philip. III. Title.
[DNLM: 1. Anatomy. 2. Physiology. QT 104 S452a]
QP34.5.S4 1992
612—dc20
DNLM/DLC
for Library of Congress
91-25661
CIP

92 93 94 95 96 GW/CX/VH 9 8 7 6 5 4 3 2 1

Preface

HUMAN ANATOMY AND PHYSIOLOGY COURSES PRESENT A TREMENDOUS CHALLENGE TO BOTH STUDENTS AND TEACHERS. Acquisition of knowledge is an essential beginning to the study of anatomy and physiology, but it is also important for students to develop the ability to solve practical, real-life problems related to the principles that they have learned. It is impossible to memorize appropriate responses to all of the situations to which one is exposed throughout life. Students who are prepared to respond reasonably and effectively in new situations are better prepared to be effective health care professionals and citizens.

Like the first edition, the second edition of *Anatomy and Physiology* is designed to help students learn basic anatomy and physiology. Information is presented in a readable form that seeks to explain concepts so that they may be truly understood rather than simply memorized. When teaching beginning students, it is important not to obscure the "big picture" with an overwhelming deluge of detail. It is also important to provide enough pieces of information to allow the students to solve basic problems. It was our goal to present basic content in such a way that it could also be used as a basis for developing problem-solving skills.

Anatomy and Physiology compares very favorably to other excellent texts in respect to presentation of content, but it is unique in its approach to the development of problem-solving skills. The second edition provides the instructor with a great deal of flexibility in that it can be used very successfully in the numerous courses that focus solely on content and learning of new vocabulary. Or it can provide a gentle introduction to problem-solving techniques that can be emphasized to a greater extent as students progress through the course. Finally, it can be used in courses that concentrate on developing the problem-solving skills that will eventually be required of future professionals.

Themes

We have chosen to emphasize two major themes throughout this text: *The Relationship Between Structure and Function* and *Homeostasis*.

Just as the structure of a hammer makes it well suited for the function of pounding nails, the forms of specific cells, tissues, and organs within the body allow them to perform specific functions effectively. For example, muscle cells contain proteins that make contraction possible, and bone cells surround themselves with a mineralized matrix that provides strength and support. Knowledge of structure and function relationships makes it easier to understand anatomy and physiology and greatly enhances appreciation for the subject.

Homeostasis, the maintenance of an internal environment within an acceptably narrow range of values, is necessary for the survival of the human body. For instance, if the blood delivers inadequate amounts of oxygen to the cells of the body, heart and respiration rates increase until oxygen delivery becomes adequate. The emphasis in this book is on how mechanisms operate to maintain homeostasis. However, because failure of these mechanisms also illustrates how they work, pathological conditions that result in dysfunction, diseases, and possibly death are also presented. A consideration of pathology adds relevance and interest that makes the material more meaningful. These two themes—the relationship between structure and function, and homeostasis—combined with the book's strong problem-solving orientation make this text quite *unique* among anatomy and physiology texts.

General Features

Four general features distinguish *Anatomy and Physiology:*

1. Systematic presentation of content
2. Balanced coverage of anatomy and physiology
3. Relevant clinical examples used to encourage problem solving
4. Systematic presentation of questions that require solution of practical problems

Systematic presentation of content—Explanations are based on a conceptual framework that allows the student to tie together individual pieces of information. Simple facts are presented first and explanations develop in a logical sequence.

Balanced coverage of anatomy and physiology—Many texts emphasize the anatomy content at the expense of the physiology content. As a result, when health professionals return to school for further training, it is invariably because they need a better understanding of physiology. Although this text provides a solid foundation in anatomy, it also provides a thorough coverage of physiology. Two chapters in this text are particularly illustrative of the emphasis we put on providing adequate coverage of physiology. These are Chapter 9, "Membrane Potentials," and Chapter 17, "Functional Organization of the Endocrine System." Furthermore, the relationship between anatomy (structure) and physiology (function) is constantly stressed, because this further enables students to solve problems. For example, if the structure and the location of a tissue are known, it is also possible to predict the tissue's function.

Relevant clinical examples used to encourage problem solving—Clinical information should never be an end in itself. In some texts, mere clinical description or medical terminology represents a significant portion of the material. This text provides clinical examples to both promote interest and demonstrate relevance, but clinical information is used primarily to illustrate the application of basic knowledge. The ability to apply information is a skill that will always be an asset for students, even after knowledge learned today is no longer current. Students using *Anatomy and Physiology* are encouraged in their professional or private lives to effectively use the knowledge they have gained through comprehending and solving basic clinical problems.

Systematic presentation of questions that require the solution of practical problems—At best, some anatomy and physiology texts include a few "thought" questions that, for the most part, involve a restatement or a summary of content. Yet once students understand the material well enough to state it in their own words, it only seems logical for them to proceed to the next step—that is, to apply the knowledge to hypothetical situations. This text features two sets of problem-solving questions in every chapter, **Predict Questions** and **Concept Questions** (to be highlighted in more detail later in this preface), which provide students with that opportunity and challenge because we believe that practice in solving problems greatly enhances problem-solving skills.

A brief example will demonstrate the difference between knowing content and being able to use that content. Suppose the following information has been given:

1. Within cells there are structures called mitochondria.
2. Mitochondria can transfer the energy in food molecules to adenosine triphosphate (ATP), an energy storage molecule that is used by cells to perform the activities necessary for life.
3. The transfer of energy to ATP requires oxygen: the more oxygen that is available, the more ATP that is produced.

Typical questions requiring students to remember or understand the information might include the following:

1. In what structures is ATP produced?
2. When oxygen consumption increases, what happens to the number of ATPs produced?

On the other hand, it is quite possible to design questions that require the students to apply newly learned information to solve problems. In this way, material is not merely learned and regurgitated. Instead, that material is utilized in a practical scenario, a scenario that will certainly stay with a student much longer than a review question. For example:

1. Given two different tissue types, predict, based on oxygen consumption rates, which tissue has the greatest number of mitochondria.
2. Given two different tissue types, propose (synthesize) an experiment that will determine which tissue has the greatest number of mitochondria.

This example illustrates that it is possible to do more with content than repeat or rephrase the information.

This text helps to develop problem-solving skills in several ways. *First,* all the information necessary to solve a problem is presented at a level that is sufficiently simple to avoid unnecessary confusion. *Second,* the opportunity to practice problem solving is made available through **Predict Questions** and **Concept Questions,** one group embedded within the chapter material, and the other group found at the end of each chapter. *Third,* answers and explanations for these kinds of problems are provided. The explanations illustrate the methods used to solve problems and provide a model for the development of problem-solving skills. When students are exposed to the reasoning used to correctly solve a problem, they are more likely to be able to successfully apply that reasoning to future problems. The acquisition of problem-solving skills is necessary for a complete understanding of anatomy and physiology; it is fun; and it makes it possible for the student to deal with the many problems that occur as a part of professional and everyday life.

Illustration Program

The statement, "A single picture is worth ten thousand words," is especially true in anatomy and physiology. Structure-function relationships become immediately apparent in the well-designed, accurate illustrations in this text. To maximize the effectiveness of the illustrations, they have been placed as close as possible to the narrative where they are cited, and special attention has been devoted to the figure legends, which summarize or emphasize the important features of each illustration. Although the anatomical drawings are accurate and the physiological flow diagrams and graphs are conceptually clear, the illustrations accomplish more than just presentation of important information. They have been designed to be non-intimidating and esthetically pleasing, features that encourage the student to spend time with the illustration for maximum learning and pleasure. All of the artwork in this textbook is in full color, making the illustrations artistically attractive while emphasizing the important structures. In addition to the illustrations, numerous photographs bring a dimension of realism to the text. In many cases, photographs are accompanied by line drawings that emphasize important features of the photograph. Another feature is the *Mini-Atlas of Human Anatomy*, containing human cadaver dissections, which is found at the end of the text as one of the appendixes. Cadaver dissections are also used in many places within the actual chapters to visually reinforce a concept or description. These atlas-quality photographs provide a visual overview of the anatomy, helping to bridge the gap between illustrations and actual structures, and to stimulate student interest in the subject.

Developmental Story

No matter how innovative our original vision for this text may have been, there is no doubt that, without the help of numerous instructors who were willing to help us implement our ideas and hone the results to near-perfection, we would have not been able to produce this text. It was our goal to produce a text embodying our unique ideas, which would also be judged suitable for widespread classroom use. Fortunately, many of our reviewers were in agreement with our goals, for they too had often experienced frustrations with their existing texts.

The actual development of this text began with a detailed survey of the anatomy and physiology market, in which we learned a great deal about the relative strengths and weaknesses of various anatomy and physiology texts. We also learned a lot about the challenges facing instructors in these courses. It was undoubtedly this market survey that convinced us of the validity and viability of our then proposed text.

We began preparing for the second edition of *Anatomy and Physiology* immediately after the first edition was published. We began keeping a detailed catalog of comments submitted to us by instructors using our text. We also invited 16 instructors to keep detailed daily diaries of how our text was working for them and their students on each subject they covered in their course. These comments formed the basis of the first draft of our second edition.

We also held a focus group prior to beginning our text revision. Authors, reviewers, and editors met for several days to review the first edition of *Anatomy and Physiology* in detail. A similar focus group including our artists was held to evaluate the illustration program in the first edition and to initiate revisions for the second edition. Finally, the entire manuscript was submitted to a reviewer panel after the first draft was prepared and again after the second draft was completed. Our reviewers were extremely valuable in helping us to update the text and to improve the educational value of the illustrations.

New To This Edition

It is impossible to list every change that has taken place in the second edition of *Anatomy and Physiology*, but the major highlights of the revision are listed below.

1. Chapter 2 has been renamed "The Chemical Basis of Life" and has been extensively rewritten with the introductory student in mind. Chemistry is now presented as an overview rather than as a review. This allows the student who has never taken a course in biology or chemistry to fully understand the material presented.
2. Chapter 3 has also been extensively revised to delete excess detail, while ensuring that beginning students have the information that they will need later in the course. A brief introduction to the overall process of metabolism has been added to this chapter. The full coverage of metabolism remains in Chapter 25.
3. The Nervous System (Part III) has been reorganized. All senses are now covered in Chapter 15, "The Senses." General senses—touch, vibration, etc.—have been added to this chapter to compare them to the special senses of taste, sight, hearing, balance, and smell. This chapter is

placed directly following the coverage of the peripheral nervous system. Chapter 16 is now "Autonomic Nervous System" and directly precedes coverage of the endocrine system. This move enables the two major control systems of internal homeostasis to be covered in succession.

4. Chapter 21, "Cardiovascular System: Peripheral Circulation and Regulation," now includes coverage of the lymphatic vessels, previously covered in Chapter 22. This move provides students with a complete picture of circulation—from arteries to veins to lymphatics—and enables them to get a clearer picture of fluid dynamics. Lymphatic organs and immunity are now covered in Chapter 22 giving a focused view of immunity—perhaps the hottest topic in anatomy and physiology today.
5. Chapter 24, "Digestive System," now includes information about the digestion and absorption of macromolecules. This was previously covered in Chapter 25 with metabolism. This move allows the students to learn the complete physiology of the digestive tract in one chapter—from the mechanical breakdown of food through the chemical breakdown and absorption of food.
6. Chapter 25 has been renamed "Nutrition and Metabolism" to reflect the increased emphasis on nutrition in this edition. Information about nutritional requirements and sources of nutrients makes the biochemical events of digestion and metabolism more relevant to the students.
7. Approximately 300 new illustrations have been prepared for this edition of *Anatomy and Physiology*. Many of these replace flow charts based totally on "boxes and arrows" with anatomical renderings placed within feedback pathways. Feedback pathways have been consistently color coded so that students may understand them readily. Green arrows are always excitatory and red arrows are always inhibitory. Further, many anatomical renderings are now accompanied by small orientation drawings which allow the student to see where the structure is located relative to the whole body.
8. Many more tables appear in this edition. A number of these tables that did not appear in the first edition organize additional anatomical information for the student. Some of these tables are accompanied by illustrations that will help the student visualize material as he organizes it for himself. Additional new tables summarize key processes in physiology. These enable the student to see the stepwise progression of a process in a clear, concise format and therefore will facilitate his learning of physiological processes.
9. Simple summaries have been added for each system of the body. These enable the student to review at a glance the major structures of each system, the major functions of that system, and the major role it plays in maintaining homeostasis.
10. New predict and concept questions have been added or changed to make the problems more relevant and consistent with real-life situations.

Learning Aids

As the amount of information in a textbook increases, it becomes more and more difficult for students to organize the material in their minds, determine the main points, and evaluate the progress of their learning. Above all, the text must be an effective teaching tool. Because each student may learn best in a different way, a variety of teaching and learning aids are provided.

Objectives. Each chapter begins with a series of learning objectives. The objectives are not a detailed cataloging of everything to be learned in the chapter. Rather, they emphasize the important facts, topics, and concepts to be covered. The chapter objectives are a conceptual framework to which additional material will be added as the chapter is read in detail.

Vocabulary aids. Learning anatomy and physiology is, in many ways, like learning a new language. To communicate effectively, a basic terminology, dealing with important or commonly used facts and concepts, must be mastered. At the beginning of each chapter are the **key terms,** a list of some of the more important new words to be learned. Throughout the text, these and additional terms are presented in **boldface print.** In cases where it is instructionally valuable, the **derivation** or **origin** of the word is given. In their original language, words are often descriptive, and knowing the original meaning can enhance understanding and make it easier to remember the definition of the word. Common prefixes, suffixes, and combining forms of many biological terms appear on the inside covers of the text and provide additional information on the derivation of words. When pronunciation of the word is complex, a **pronunciation key** is presented. Often simply being able to pronounce a word correctly is the key to remembering it. The **glossary,** which collects the most important terms into one location for easy reference, also has a pronunciation guide.

Related topics. Knowledge of anatomy and physiology is interrelated and cumulative, with new information building on previous information. It is difficult to understand advanced material without mastering the basics. This section at the beginning of each chapter points out material that should be understood or reviewed before proceeding with the new chapter material.

Asides. The aside is a brief statement following the discussion of an important concept. It clarifies the concept by presenting an example of the concept in action. For example, the aside may illustrate the normal response of a system to exercise, or it might describe a pathological condition that shows how a system responds to an abnormal situation. The advantage of the aside is that it appears right after the concept is presented. In this way the relevance of the concept is immediately apparent, helping the student to better appreciate and understand it.

Boxed essays. The boxed essays are expanded versions of the asides that permit a more detailed or complete coverage of a topic. Subjects covered include pathologies, current research, sports medicine, exercise physiology, pharmacology, and clinical applications. They are designed not only to illustrate the chapter content but to stimulate interest as well.

Predict questions. While the aside or boxed essay can illustrate how a concept works, the predict question requires the application of the concept. When reading a text, it is very easy to become a passive learner; everything seems very clear to passive learners until they attempt to use the information. The predict question converts the passive learner into an active learner who must use new information to solve a problem. The answer to this kind of question is not a mere restatement of fact, but rather a prediction, an analysis of the data, the synthesis of an experiment, or the evaluation and weighing of the important variables of the problem. For example, "Given a stimulus, predict how a system will respond." Or, "Given a clinical condition, explain why the observed symptoms occurred." Predict questions are practice problems that help to develop the skills necessary to solve the concept questions at the end of the chapter. In this regard, not only are possible answers given for the questions, but explanations are provided that demonstrate the process of problem solving.

Tables. The book contains many tables that have several uses. They provide more specific information than that included in the text discussion, allowing the text to concentrate on the general or main points of a topic. The tables also summarize some aspect of the chapter's content, providing a convenient way to find information quickly. Often, a table is designed to accompany an illustration, so that a written description and a visual presentation are combined to communicate information effectively.

Chapter summary. As the student reads the chapter, details may obscure the overall picture. The chapter summary is an outline that briefly states the important facts and concepts and provides a perspective of the "big picture."

Content review questions. The content review questions are another method used in this text to transform the passive learner into an active learner. The questions systematically cover the content and require students to summarize and restate the content in their own words.

Concept questions. Following mastery of the content questions, and therefore chapter content, the concept questions require the application of that content to new situations. These are not essay questions that involve the restatement or summarization of chapter content. Instead, they provide additional practice in problem solving and promote the development and acquisition of problem-solving skills.

Appendixes. Appendix A is a full-color mini-atlas of human anatomy that illustrates the integrated relationships between the structures of different systems. These photographs of actual human dissections are taken from *Color Atlas of Human Anatomy*, edition 2, by R.M.H. McMinn and R.T. Hutchings (Year Book Medical Publishers, Inc., 1988). Reference to these photographs will help students bridge the gap between the idealized illustrations of human anatomy in the textbook and the actual anatomical relationships they will see in the laboratory. Appendix B is a table of measurements that will help the student relate the metric system to the more familiar English system when determining the size or weight of a structure. Appendix C will help the student understand the shorthand of scientific notation. Appendix D explains the rationale behind how various solutions may be described. Appendix E explains the concept of pH and how it is measured. Appendix F contains tables of routine clinical tests along with their normal values and clinical significance. Reference to this appendix will provide students with the homeostatic values of many common substances in the blood and urine. Also, the importance of laboratory testing in the diagnosis and/or treatment of illnesses becomes readily apparent to the students.

Supplements

Any textbook can be used alone, but thoughtfully developed supplements increase its effectiveness for both student and instructor. The supplements prepared to accompany the second edition of *Anatomy and Physiology* have been designed and written to firmly support the pedagogical model developed in the text. The authors who joined us in producing this supplement package are committed to quality education and have eagerly shared their expertise in producing an outstanding array of support materials for both the student and the teacher.

Study Guide. The second edition of the study guide by Philip Tate and James Kennedy of Phoenix College and Rod Seeley of Idaho State University continues to support the text. It introduces the student to the content of anatomy and physiology using matching, labeling, and completion exercises. A Mastery Learning Activity consisting of multiple choice questions emphasizes comprehension of the material, evaluates progress, and prepares the student for classroom testing. In addition, a Final Challenges section consisting of essay questions provides practice with questions similar to the predict and concept questions of the textbook. Answers are given for all exercises, and explanations are furnished for the Mastery Learning Activity and the Final Challenges. Explanations are provided for higher cognitive level questions to help students understand the thought processes essential to arriving at a correct answer by bringing together information and creating an appropriate solution to the problem. The intent is to help students develop their problem-solving ability. Carefully reviewed by experienced instructors who currently teach anatomy and physiology, the study guide provides the reinforcement and practice so essential to the student's success in the course.

Laboratory Manual. The second edition of *Anatomy and Physiology Laboratory Manual* by Jay Templin of Widener University has been extensively revised to include a number of interactive laboratory exercises that support the active learning model encouraged by the textbook. Consultants Patricia Fink from the University of Arkansas at Little Rock and Leslie Wiemerslage from Belleville Area Community College have provided sound advice on developing creative learning experiences for the laboratory. The revision includes additional full-color sections on both human and cat anatomy as well as the full-color reference section on histology. The laboratory report format has been significantly improved for facilitation of learning by the student and ease of grading by the instructor. Once again, the suggestions and corrections of reviewers with wide experience in teaching anatomy and physiology have been incorporated into this revision.

The accompanying Preparator's Manual provides detailed information necessary for the instructor to prepare for the lab. Answers to all questions on the lab reports in the Lab Manual are also included.

Instructor's Resource Manual. This invaluable resource, written by Margaret Weck of Lake Superior State University, facilitates development and presentation of a well-integrated course. It suggests ways to organize the material and is keyed to relevant transparencies, illustrations, and laboratory exercises. Major points that deserve emphasis are included, hints on how to reinforce concepts are given, typical problem areas are noted along with ways to deal with the problems, and possible topics for discussion are considered. Answers for the concept questions at the end of each chapter in the text and for the essay questions in the test bank are found in the *Instructor's Resource Manual.* There are also 100 transparency masters including key diagrams and tables from the text and additional material useful for student handouts. Each chapter also includes a list of relevant audiovisual and software sources; the manual also contains an extensive list of laboratory supply houses.

Special features of the *Instructor's Resource Manual* include "conversion notes" that detail the difference in terms of organization and coverage between our second edition and several other leading anatomy and physiology texts. This is an ideal tool to assist you in converting your lecture notes from your current text to *Anatomy and Physiology*, second edition. Also included is a "teaching survival guide" that provides practical tips for novice instructors and teaching assistants. We thank Linda Van Thiel of Wayne State University for this very helpful essay. Finally, Margaret Weck has added tips for instructors to help them take advantage of the exercises in the text that foster the development of problem-solving skills and to help them assist their students to become more active learners.

Test Bank. There must be consistency between the material presented to students and what they actually see on the test. Written by Dorothy Martin of Black Hawk College, the test bank has been carefully integrated to complement both the text and the study guide. The test bank contains over 2200 items including multiple choice, completion, matching, and essay questions. A number of new test items have been added which allow the instructor to evaluate the students' developing problem-solving skills. Each question is classified according to the knowledge level it requires and all answers are provided. As with all other materials that students will use, the test questions have been thoroughly reviewed by anatomy and physiology instructors and painstakingly polished to offer the best possible teaching tool.

Test-generating system. Qualified adopters of this text may request a computerized Diploma test bank package, compatible for use on IBM PC, Apple IIc, or Apple IIe computers. This software package is a unique combination of user-friendly computerized aids for the instructor:

- **Testing:** A test generator allows the user to select items from the test bank either manually or randomly; to add, edit, or delete test items through a preset format that includes multiple choice, true-false, short answer, or matching options; and to print exams with or without saving them for future use.
- **Grading:** A computerized record keeper saves student names (up to 250), assignments (up to 50), and related grades in a format similar to that used in manual grade books. Statistics on individual or class performance, test weighting, and push-button grade curving are features of this software.
- **Tutoring:** A tutorial package uses items from the test bank for student review. Student scores can be merged into the grading records.
- **Scheduling:** A computerized datebook makes class planning and schedule management quick and convenient.

Computerized test banks are also available for the Macintosh computer. This convenient program allows instructors to select, edit, or delete items from the test bank of questions provided. This testing system is packed with features such as the ability to place graphics with test questions and a pop-up menu to select question type and level. "Windows" allow instructors to reference test questions simultaneously with their notes.

Transparency Acetates. Full color transparencies with large, easy-to-read labels emphasize the major anatomical structures and physiological processes covered in the text. These images provide a common vehicle for communication between the instructor and the student as they enrich and further clarify lecture presentations.

Human Body Systems Software. Available in an IBM format, this interactive software program, written by Kevin Patton and Kathryn Baalman of St. Charles County Community College, helps beginning students achieve success in anatomy and physiology. Individual modules introduce each of the eleven body systems. Each module contains an introduction, a tutorial with practice review questions, practical applications, and a final quiz.

Human Cadaver Dissection Video. This 60-minute video, narrated by Trent Stephens and produced at Logan College of Chiropractic, takes the student through a dissection of the musculature of the human body as well as the internal organs of the thorax and abdomen. Presented at an introductory level, this video provides vivid dissection close-ups with clear, precise commentary.

The Human Body Videodisc. Compact, versatile, and easy to use, this new resource provides unprecedented visual reinforcement for teaching anatomy and physiology. Containing over 1000 still images and over 50 minutes of moving images and animations, the human body videodisc provides an overview of the structure and function of each system in the human body. For each system, it includes anatomical artwork, photographs of gross anatomy, micrographs of relevant tissues, and moving sequences to show the system in action or how a process works. Animations were prepared specifically for use on this videodisc to enable students to visualize complex processes such as movement across a cell membrane, muscle contraction, the mechanics of breathing, and the formation of urine.

A Print Directory accompanies the videodisc which lists all of the images available with their frame references and provides a copy of the narration for each of the moving sequences. The Lecture Guide, prepared by Frank Peek of the State University of New York in Morristown, contains lecture outlines with suggestions for using the videodisc in teaching anatomy and physiology. Barcodes to images and sequences on the disc allow immediate access.

Acknowledgments

No modern textbook is solely the work of the authors. To adequately acknowledge the support of loved ones is not possible. They have had the patience and understanding to tolerate our frustration and absence and they have been willing to provide undying encouragement. We also wish to express our gratitude to the staff of Mosby–Year Book, Inc., for their steadfast help and encouragement. We sincerely thank Deborah Allen and Bob Callanan for their hours of work, suggestions, and tremendous patience and encouragement. We also thank our production editor Teresa Breckwoldt who spent many hours turning our manuscript into a book. The Mosby–Year Book employees with whom we have worked are excellent professionals. They have been consistently helpful and their efforts are appreciated. Their commitment to this project has clearly been more than a job to them.

We also thank Liz Rudder and Christine Oleksyk who worked on the development and execution of the illustration program for the second edition of *Anatomy and Physiology*. Their efforts have helped us to build on the strong art program coordinated by Tom and Elizabeth Sims in the first edition. The art program for this text represents a monumental effort and we appreciate their contribution to the overall appearance and pedagogical value of the illustrations.

Finally, we sincerely thank the reviewers and the instructors who have provided us with excellent constructive criticism. The remuneration they received represents only a token payment for their efforts. To conscientiously review a textbook requires a true commitment and dedication to excellence in teaching. Their helpful criticisms and suggestions for improvement were a significant contribution that we greatly appreciate.

Rod R. Seeley
Trent D. Stephens
Philip Tate

Focus Group Participants

James Aldridge
Palm Beach Community College

John Conroy
University of Winnipeg

Patricia O'Mahoney-Damon
University of Southern Maine

Betsy Ott
Tyler Junior College

Richard Satterlie
Arizona State University

Willard Woodward
Parkland College

Donna Van Wynesberghe
University of Wisconsin at Milwaukee

Second Edition Reviewers

Marshall Anderson
Rockhurst College

Cynthia Bottrell
Scott Community College

Spencer Bowes
Oakton Community College

Anthony Chee
Houston Community College

Salvatore Drogo
Mohawk Valley Community College

Douglas Eder
Southern Illinois University at Edwardsville

Betty Elder
Central Community College

James Ezell
J. Sargeant Reynolds Community College

Craig Gundy
Weber State College

Linden Haynes
Hinds Community College

Vickie Hennessey
Sinclair Community College

James Larsen
University of Southern Mississippi

Kathleen Lee
University of Vermont

Alexsandria Manrov
Tidewater Community College

Elden Martin
Bowling Green State University

Randall Oelerich
Duluth Community College

Glenn Powell
Bellevue Community College

Ralph Stevens, III
Old Dominion University

Cynthia Surmacz
Bloomsburg University

Cecilia Thomas
Hinds Community College

Carl Thurman
University of Missouri at St. Louis

William Weaver
Miami-Dade Community College

James White
Prairie State College

About the Authors

Rod R. Seeley
Professor of Physiology, Idaho State University

With a B.S. in zoology from Idaho State University and an M.S. and Ph.D. in zoology from Utah State University, Rod Seeley has built a solid reputation as a widely published author of journal and feature articles, a popular public lecturer, and an award-winning instructor. Very much involved in the methods and mechanisms that help students learn, he contributes to this text his teaching expertise and proven ability to communicate effectively in any medium.

Trent D. Stephens
Professor of Anatomy and Embryology, Idaho State University

A versatile educator, Trent Stephens teaches such courses as human anatomy, neuroanatomy, and embryology. His skill as a biological illustrator has greatly influenced every illustration in this text. With B.S. and M.S. degrees in zoology from Brigham Young University and a Ph.D. in anatomy from the University of Pennsylvania, Trent Stephens uses his background to reach diverse audiences. His students continually rate him highly on their evaluations—you will too!

Philip Tate
Instructor of Anatomy and Physiology, Phoenix College

From the community college to the private 4-year college, Phil Tate has taught anatomy and physiology to all levels of students: nursing and allied health, physical education, and biology majors. At San Diego State University, Phil earned B.S. degrees in both mathematics and zoology and a M.S. in ecology. He earned his doctorate in biological education from Idaho State University.

Contents in Brief

Contents

I ORGANIZATION OF THE HUMAN BODY

II SUPPORT AND MOVEMENT

III INTEGRATION AND CONTROL SYSTEMS

17 *Functional Organization of the Endocrine System* 525

18 *Endocrine Glands* 545

IV REGULATION AND MAINTENANCE

21 Cardiovascular System: Peripheral Circulation and Regulation 645

22 Lymphatic Organs and Immunity 699

23 *Respiratory System* 735

24 *Digestive System* 771

25 Nutrition and Metabolism 819

26 Urinary System 849

27 Water, Electrolytes, and Acid-Base Balance 881

V REPRODUCTION AND DEVELOPMENT

PART I

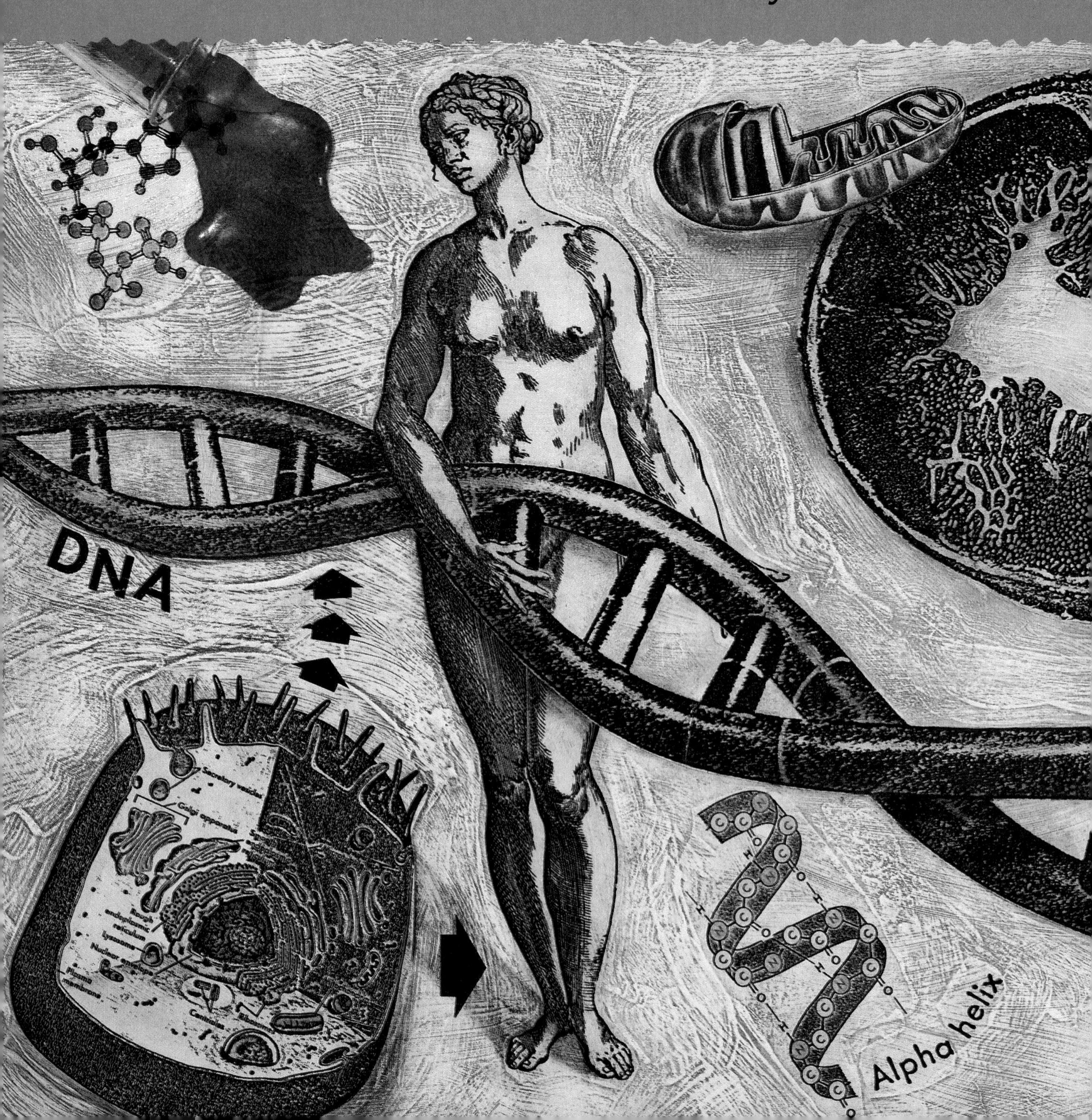

ORGANIZATION OF THE HUMAN BODY

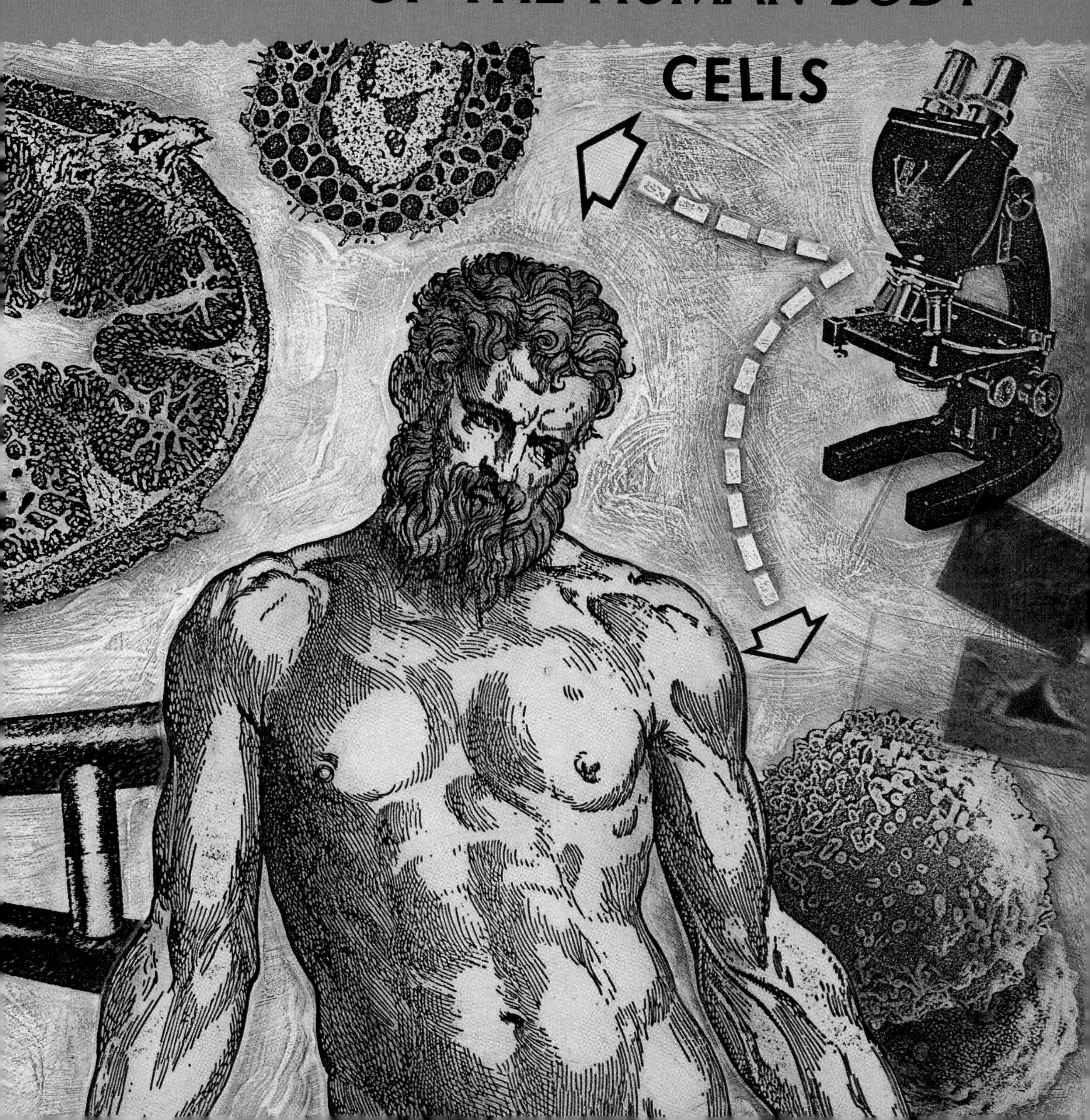

CHAPTER 1 OBJECTIVES

After reading this chapter you should be able to

1. Explain the importance of understanding the relationship between structure and function.
2. Define the terms anatomy and physiology and identify the modern frontiers of each.
3. Describe the molecular, organelle, cellular, tissue, organ, organ system, and whole organism levels of organization.
4. Define the term tissue and name the four primary tissue types.
5. List the 11 organ systems and indicate the major functions of each.
6. Explain the importance of studying other animals to help understand human anatomy and physiology.
7. Define homeostasis.
8. Diagram a negative-feedback system and a positive-feedback system and describe the relationship of each to homeostasis.
9. Describe the anatomical position.
10. Use the directional terms in Table 1-2 to describe specific body structures.
11. Name and describe the three major planes of the body or of an organ.
12. Define the axial and appendicular regions of the body and describe the subdivisions of each region.
13. Describe two ways to subdivide the abdominal region and explain the importance of the divisions.
14. Define the terms thoracic cavity, abdominal cavity, pelvic cavity, and mediastinum.
15. Define a serous membrane and explain the relationship between parietal and visceral serous membranes.
16. Name the membranes that line the walls and cover the organs of each body cavity and name the fluid found inside each cavity.
17. Define mesentery and describe its function.
18. Define the term retroperitoneal and list examples of retroperitoneal organs.

The Human Organism

KEY TERMS

anatomical position

anatomy

cell

homeostasis
(ho′me-o-sta′sis)

mesentery
(mes′en-tĕr′e)

negative feedback

organ

organ system

organelle

organism

parietal (pă-ri′ĕ-tal)

physiology

plane

positive feedback

retroperitoneal
(rĕ′tro-pĕr′ĭ-to-ne′al)

tissue

visceral (vis′er-al)

HUMAN ANATOMY AND PHYSIOLOGY COMPRISE THE STUDY OF HOW THE HUMAN BODY IS ORGANIZED AND HOW IT FUNCTIONS. Knowledge from such a study makes it possible to predict how a cell, organ, or organ system will respond to various stimuli and how this response will affect the whole person. For example, knowledge of the structure and function of the circulatory and nervous systems will help you understand why it is important to control high blood pressure and arteriosclerosis to prevent stroke. This knowledge will also help you understand the symptoms exhibited when a person has a stroke. The goal of this text is to develop in the reader a sound functional knowledge of anatomy and physiology, an appreciation for the intricate processes on which humans depend for survival, and the ability to apply this knowledge in problem-solving situations.

The study of anatomy and physiology is an essential prerequisite for those who plan to pursue the health sciences since a firm knowledge of the body's structure and function is essential for health professionals to perform their duties adequately. A sound background in anatomy and physiology is also an advantage to those who are not in the health professions. The ability to evaluate one's own physiological activities, understand recommended treatments, critically evaluate advertisements and reports in popular literature, and interact with health professionals is improved with this background.

ANATOMY

Anatomy is the scientific discipline that investigates the body's structure. Modern anatomy is a dynamic discipline that covers a wide range of anatomical studies, including the functions of anatomical structures, their microscopic organization, and the processes by which anatomical structures develop. **Microscopic anatomy** includes cytology and histology. **Cytology** (si-tol'o-je) examines the structural features of cells, and **histology** (his-tol'o-je) is the study of tissues. The light microscope, the scanning electron microscope, and the transmission electron microscope are important tools used in both fields of investigation.

Gross anatomy, the study of structures that can be examined without the aid of a microscope, is approached from either a systemic or regional perspective. **Systemic anatomy** is the study of the body by systems and is the approach taken in this and most other undergraduate textbooks. A system is a group of structures that have one or more common functions. Examples are the circulatory, nervous, respiratory, skeletal, and muscular systems. **Regional anatomy** is the study of the body's organization by areas and is the approach taken in most graduate programs at medical and dental schools. Within each region such as the head, abdomen, or arm, all systems are studied simultaneously.

Neuroanatomy and developmental anatomy are among the major frontiers of modern anatomy. **Neuroanatomy** deals mainly with the structural and functional organization of the brain and spinal cord. **Developmental anatomy** emphasizes the structural changes that occur in an individual from the time of fertilization to adulthood. Developmental anatomists are concerned especially with those changes that occur before birth, and they attempt to discover factors that control those changes.

Surface anatomy is the study of the external form of the body and its relation to deeper structures. For example, the sternum (breastbone) and parts of the ribs can be seen and palpated (felt) on the front of the chest. These structures can be used as landmarks to identify regions of the heart and points on the chest where certain heart sounds can best be heard. **Anatomic imaging** involves the use of x-rays, ultrasound, nuclear magnetic resonance, and other technologies to create pictures of internal structures. Anatomic imaging is a rapidly advancing discipline, and much of the future work in anatomy and surgery will depend on computer-assisted anatomic imaging. Both surface anatomy and anatomic imaging provide important information in diagnosing disease.

PHYSIOLOGY

Physiology is the scientific discipline that deals with the vital processes or functions of living things. Understanding, analyzing, and predicting responses of cells, organ systems, and organisms to stimuli are major goals of physiology. Physiology is studied by systems rather than by regions since most physiological processes involve the interactions of one or more systems. It is also important when studying physiology to recognize structures as dynamic rather than static entities. Physiological processes allow the organism to maintain relatively constant conditions in the presence of a continually changing environment.

Physiology can be subdivided based on the organism involved, the levels of organization within a given organism, or the specific system studied. Molecular, cellular, and systemic physiology are examples of physiology that emphasize specific organizational levels. Human physiology refers to the study of a specific organism, the human, as opposed, for example, to plant physiology or insect physiology. Neurophysiology, like neuroanatomy, is one of the major modern frontiers of science. Neurophysiologists address questions such as how normal and abnormal brain functions take place, what controls them, and how drugs affect those functions.

STRUCTURAL AND FUNCTIONAL ORGANIZATION

The body can be considered conceptually at seven structural levels: the molecular, organelle, cellular, tissue, organ, organ system, and complete organism (Figure 1-1). Although the boundaries between each of the levels are not always clear, the categories are useful in making the entire body easier to comprehend.

Molecular

The structural and functional characteristics of all organisms are determined by their chemical characteristics. (A brief overview of basic chemistry is presented in Chapter 2.) The **molecular,** or chemical, level of organization involves interactions between elements and their combinations into molecules. The function of molecules is related intimately to their structure. For example, collagen molecules are strong, ropelike fibers that provide the skin with structural strength. If collagen loses this ropelike structure, the skin becomes fragile and is easily torn.

Organelle

An **organelle** is a "small organ" or structure contained within a cell that performs one or more specific functions. Mitochondria are organelles in which the cell's energy is packaged. (Organelles are discussed more fully in Chapter 3.)

Cellular

Cells are the basic living units of all plants and animals. Cell types differ in their structure and function but have many characteristics in common. Knowledge of these characteristics and of the variations in them is essential to a basic understanding of anatomy and physiology. (The cell is discussed in Chapter 3.)

Tissue

A group of cells with similar structure and function plus the extracellular substances located between them constitute a **tissue.** The large number of tissues that comprise the body is classified into four primary tissue types: epithelial, connective, muscle, and nervous. Each primary tissue type has several characteristics that distinguish it from the other primary types, and each has several subcategories. (Tissues are discussed in Chapter 4.)

Organ

Most **organs** are composed of two or more tissue types that perform one or more common functions. Examples of organs are the skin, liver, pancreas, stomach, eye, tongue, and heart. Each tissue in an organ plays a role in that organ's function.

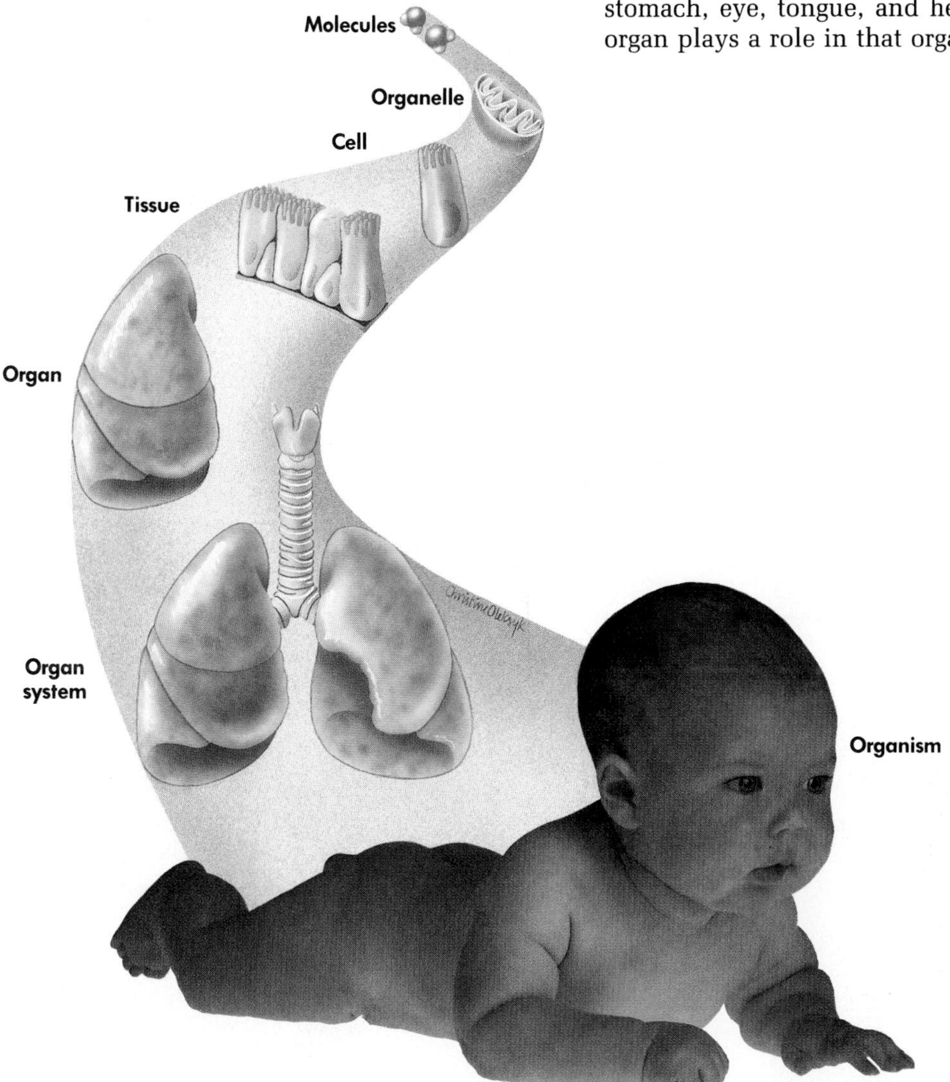

FIGURE 1-1 Seven levels of organization: molecular, organelle, cellular, tissue, organ, organ system, and organism.

Organ System

An **organ system** is a group of organs classified as a unit because of a common function or set of functions. In this text the body is divided into 11 major organ systems: the integumentary, skeletal, muscular, nervous, endocrine, cardiovascular, lymphatic, respiratory, digestive, urinary, and reproductive systems (Table 1-1). The classification of organ systems is somewhat arbitrary. For example, the muscular and skeletal systems can be combined and studied as the musculoskeletal system, or the nervous system can be subdivided into the peripheral and central nervous systems.

Organism

An **organism** is any living thing considered as a whole, whether composed of one cell or many. The human organism is a complex of mutually dependent organ systems. The survival of the individual organism depends on the effective operation and coordination of the organ systems. For example, the cardiovascular system transports oxygen that enters the blood within the respiratory system and carries cellular waste products that are eliminated from the blood by the urinary system. The coordination of these systems is accomplished by the nervous and endocrine systems.

Text continued on p. 11.

Table 1-1

Organ systems of the body

SYSTEM	MAJOR COMPONENTS	FUNCTIONS
Integumentary system	Skin, hair, nails, and sweat glands	Protects, regulates temperature, prevents water loss, and produces vitamin D precursors
Skeletal system	Bones, associated cartilage, and joints	Protects, supports, and allows body movement, produces blood cells, and stores minerals
Muscular system	Muscles attached to the skeleton	Allows body movement, maintains posture, and produces body heat

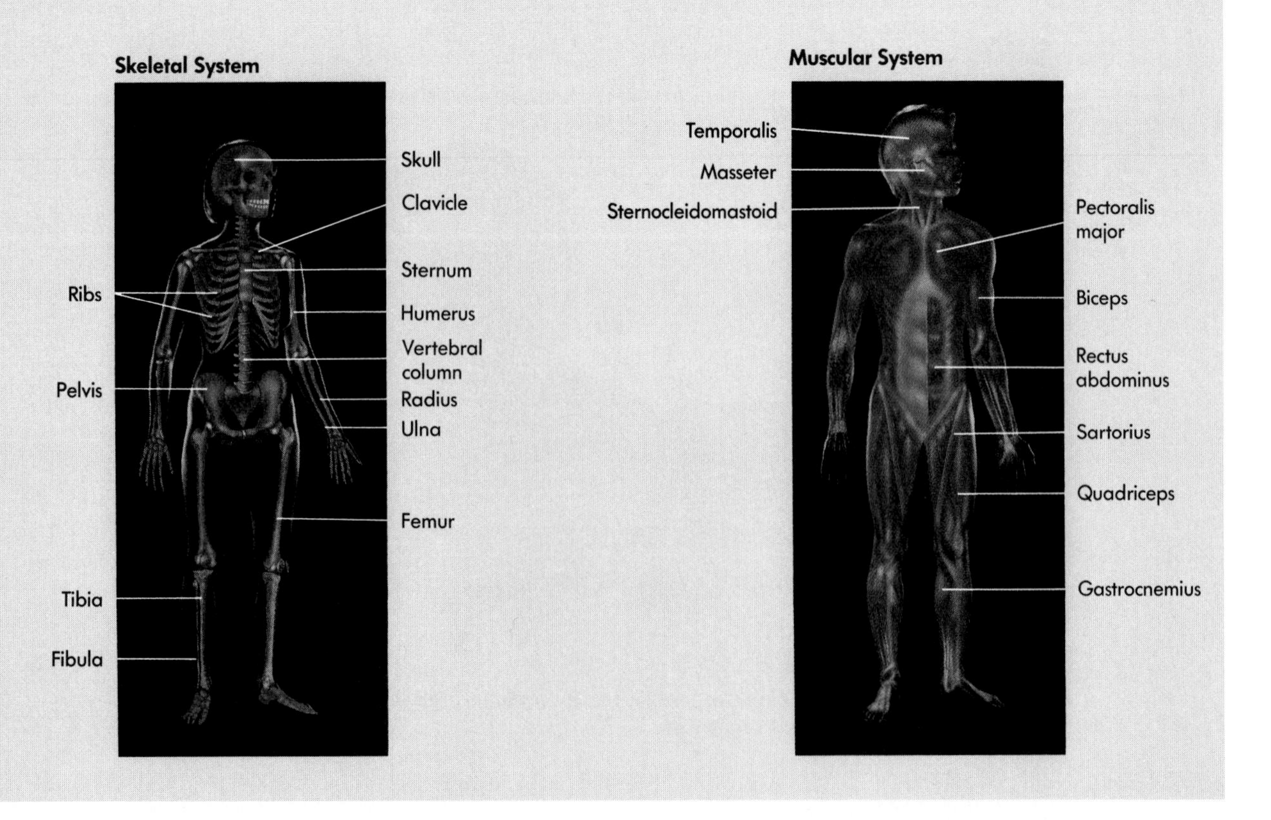

Table 1-1

Organ systems of the body—cont'd

SYSTEM	MAJOR COMPONENTS	FUNCTIONS
Nervous system	Brain, spinal cord, nerves, and sensory receptors	A major regulatory system: detects sensation, controls movements, controls physiological and intellectual functions
Endocrine system	Endocrine glands such as the pituitary, thyroid, and adrenal glands	A major regulatory system: participates in the regulation of metabolism, reproduction, and many other functions
Cardiovascular system	Heart, blood vessels, and blood	Transports nutrients, waste products, gases, and hormones throughout the body; plays a role in the immune response and the regulation of body temperature

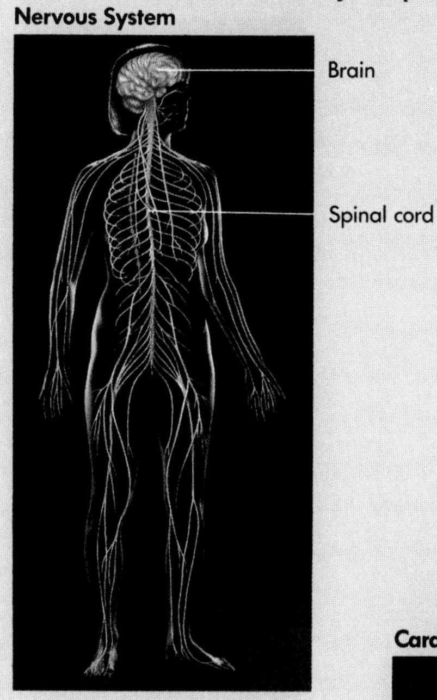

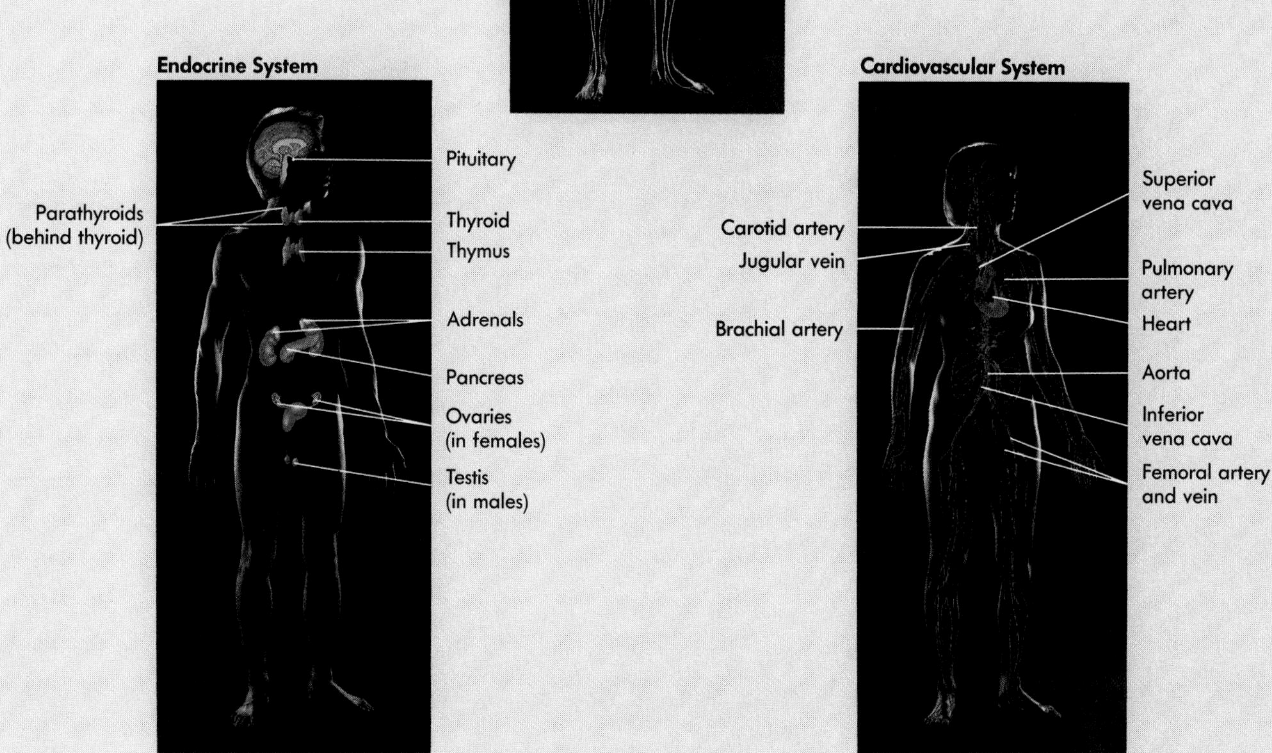

Continued.

Table 1-1

Organ systems of the body—cont'd

SYSTEM	MAJOR COMPONENTS	FUNCTIONS
Lymphatic system	Lymph vessels, lymph nodes, and other lymph organs	Removes foreign substances from the blood and lymph, combats disease, maintains tissue fluid balance, and absorbs fats
Respiratory system	Lungs and respiratory passages	Exchanges gases (oxygen and carbon dioxide) between the blood and the air and regulates blood pH
Digestive system	Mouth, esophagus, stomach, intestines, and accessory structures	Performs the mechanical and chemical processes of digestion, absorption of nutrients, and elimination of wastes

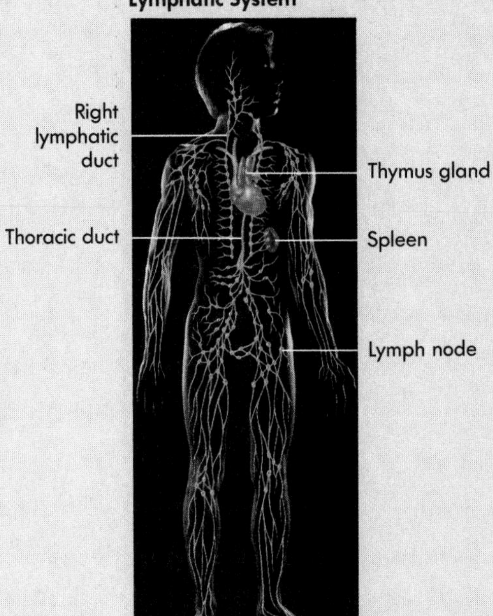

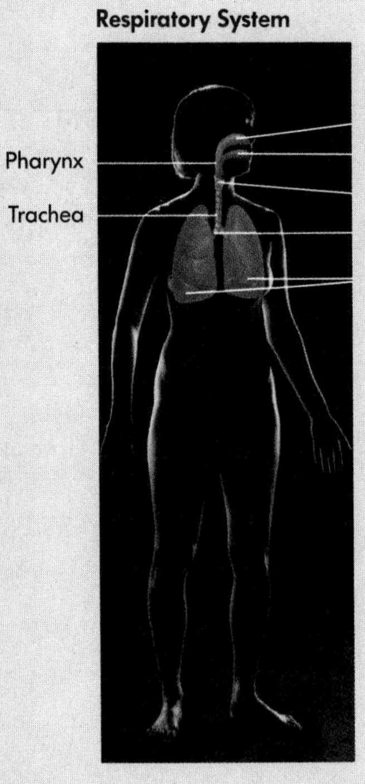

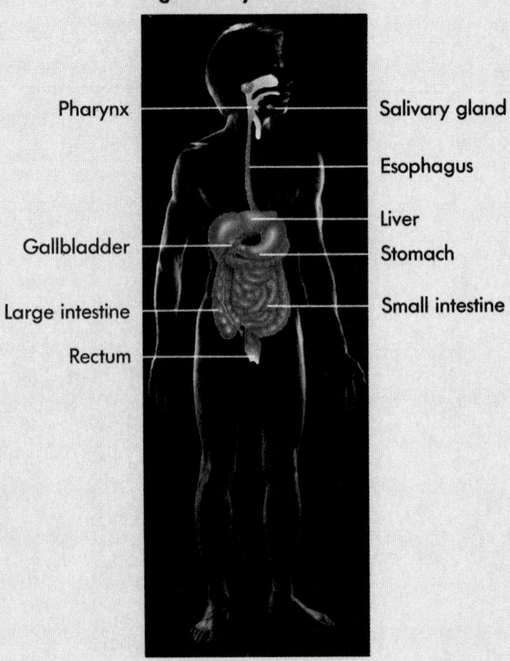

Table 1-1

Organ systems of the body—cont'd

SYSTEM	MAJOR COMPONENTS	FUNCTIONS
Urinary system	Kidneys, urinary bladder, and ducts that carry urine	Removes waste products from the circulatory system; regulates blood pH, ion balance, and water balance
Reproductive system	Gonads, accessory structures, and genitals of males and females	Performs the processes of reproduction and controls sexual functions and behaviors

Urinary System

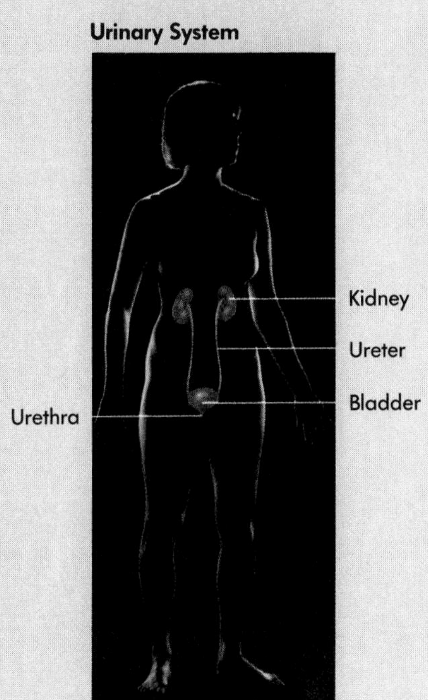

Reproductive System—Male

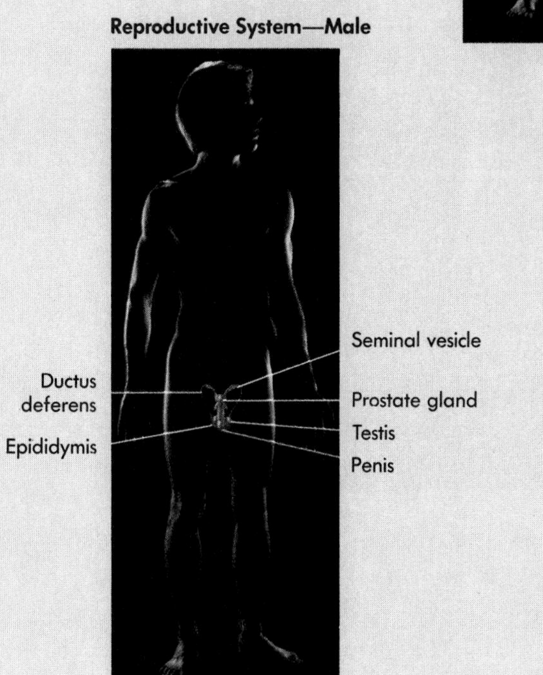

Reproductive System—Female

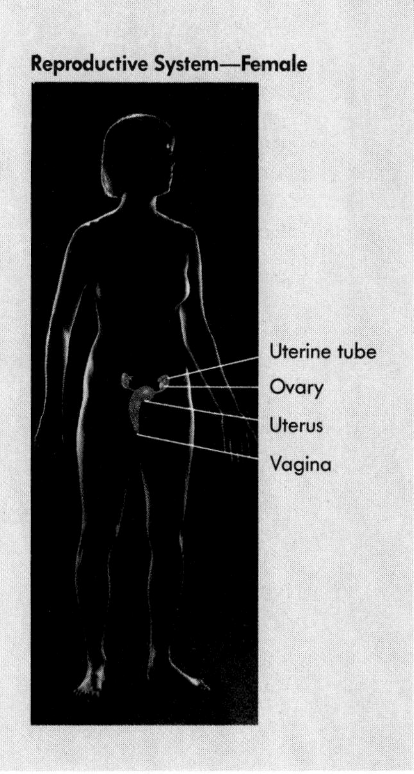

Continued.

Table 1-1

Organ systems of the body—cont'd

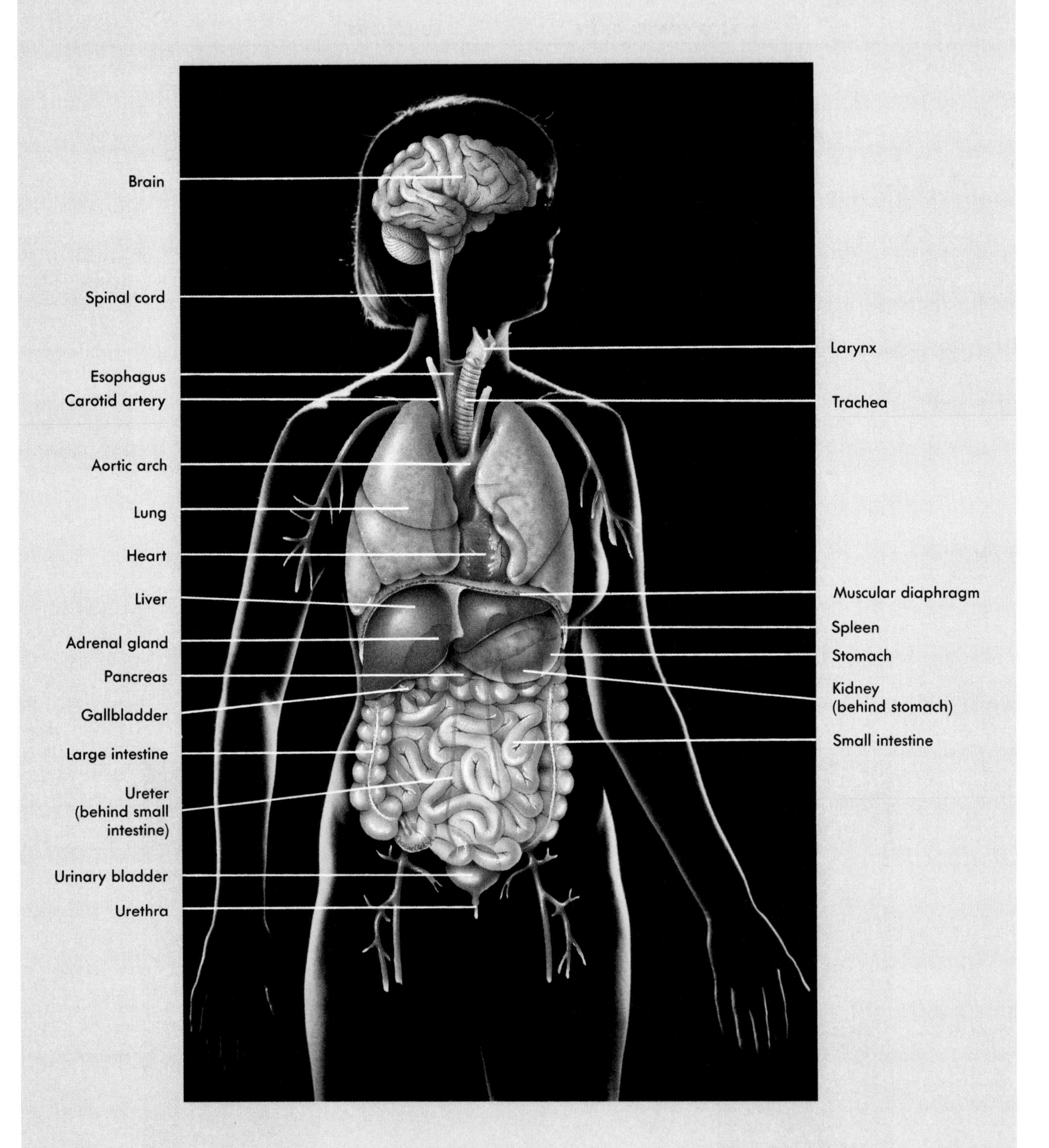

THE HUMAN ORGANISM

Humans are organisms and have many characteristics in common with other organisms. The most important feature common to all organisms is that they are alive. The characteristics of life are organization, metabolism, responsiveness, growth and development, and reproduction.

Characteristics of Life

Organization. Living things are highly organized. All living things are composed of cells or, in the case of viruses, must be contained within cells to perform living functions. Cells in turn are composed of highly organized organelles, which depend on the extremely accurate organization of large molecules. Many of the large molecules quickly lose their structure if they are removed from the cell (as a result of cell rupture, for example), and the cell quickly loses its function if its internal organization is disrupted.

Metabolism. Metabolism is the ability to assimilate food (e.g., sugars and proteins) and/or energy (e.g., sunlight in plants) and to use them to perform vital functions such as growth, movement, and reproduction. The intake of energy through metabolism is absolutely vital to everything else that happens in the organism.

Responsiveness. A major characteristic of living things is that they are able to respond to the environment. The ability to respond requires that the organism can sense changes in the environment and can make adjustments that help maintain its life. Responses include movement away from danger or poor environmental conditions toward a more suitable environment, food, or water or adjustment of the internal environment to maintain a constant internal balance in the presence of a changing external environment.

Growth and development. A common feature of most living creatures is the ability to grow and develop. Growth is the ability to take in food, break it down into basic building blocks, and from those building blocks produce new cells and tissues, resulting in the overall enlargement of all or part of the organism. Development includes the changes an organism undergoes through time. Development usually involves growth, but it also involves changes in cell function from generalized to specialized (differentiation) and changes in the shape of tissues, organs, and the entire organism (morphogenesis). Development of the individual begins with fertilization and ends at death. The greatest developmental changes occur before birth, but many changes continue after birth, and some continue throughout life.

Reproduction. A very important characteristic of living things is their ability to reproduce. Without reproduction of cells there would be no growth and development, and a living thing would remain as a single cell, without the development of complexity in size, shape, or function. Without reproduction of the species there is only extinction. The reproduction of all organisms is based on the possession of the genetic molecule DNA.

Biomedical Research

Much of what is known about the human organism has come from the study of other organisms. The basic processes of life and the basic nature of DNA are similar in all organisms from viruses to humans. Much knowledge about the functions of DNA and its importance to life has come from the study of bacteria and viruses.

Biomedical research uses organisms other than humans for several reasons. The use of organisms whose cells have less genetic information than do human cells allows the acquisition of knowledge about genetic mechanisms without the complications present in studying the more complex human systems. The tremendous strides in molecular biology have been possible because bacteria and viruses are relatively simple. The knowledge thus gained can be applied to human cells.

Some biomedical research requires the use of complete, intact multicellular organisms rather than isolated cells or single-celled organisms; however, this research may also require either great risk to the organism or the sacrifice of the organism and must therefore be conducted on animals other than humans. The great strides made during the past few years in open heart surgery and heart transplantation, for example, required that the experimental techniques be perfected on other mammals before being attempted on humans. Strict laws govern the use of animals in biomedical research. These laws are designed to assure minimum suffering on the part of

Failure to appreciate the differences between humans and other animals led to many misconceptions by early scientists. One of the first great anatomists was a Roman physician, Claudius Galen (130-201 AD). Galen described a large number of anatomical structures supposedly present in humans but observed only in other animals. The errors introduced by Galen persisted for more than 1300 years until a Flemish anatomist, Andreas Vesalius (1514-1564), who is considered the first modern anatomist, carefully examined human cadavers and began to correct the textbooks. This experience should serve as a word of caution: some current knowledge in molecular biology and even physiology has not been confirmed in humans.

the animal and to discourage unnecessary experimentation.

Other organisms are obtained much more easily for study than are humans. This is especially true in the study of anatomy. Few undergraduate institutions can obtain human cadavers; consequently, cats, fetal pigs, or plastic models often are used.

However, no matter how useful other organisms are in helping answer questions about humans, the final answers ultimately must come from the study of ourselves. In anatomy, for example, no matter how closely related an animal may be to us, some features remain unique to humans.

HOMEOSTASIS

Homeostasis (ho′me-o-sta′sis) is the maintenance of a relatively constant internal environment (within the body) in the presence of a constantly and sometimes dramatically changing external environment (outside the body). Under normal conditions the body responds to changes in the external environment by making adjustments to maintain the internal environment within specific limits. Each cell of the body is surrounded by a small amount of fluid, and the normal function of that cell depends on the maintenance of its fluid environment within a narrow range of conditions, including volume, temperature, and chemical content. If the composition of the fluid surrounding cells deviates from its normal range, the cells and possibly the individual may die.

All the organ systems of the body contribute to the cellular environment and are controlled so that this environment remains relatively constant. For example, the amount of water taken in through the digestive system and the amount expelled by the digestive, respiratory, integumentary, and urinary systems are controlled so that the body's fluid content remains relatively constant. The digestive, respiratory, circulatory, integumentary, and urinary systems are regulated so that each cell receives adequate oxygen and nutrients and so that waste products do not accumulate to a toxic level. Many of the organ systems involved in homeostasis are regulated by the nervous and endocrine systems.

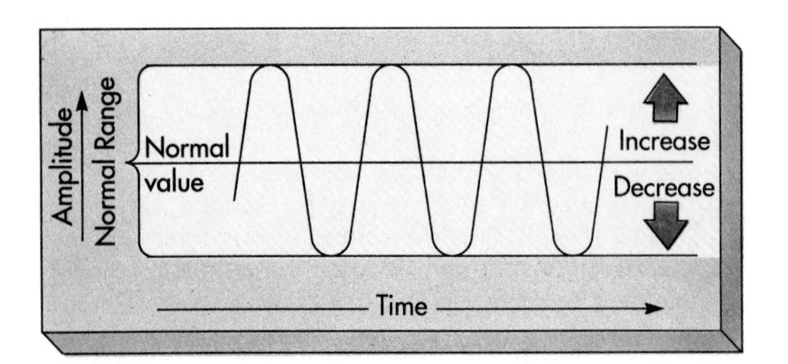

FIGURE 1-2 Negative feedback. The normal range is the range over which a given homeostatic condition is maintained. Deviations (increases or decreases) from the normal value are restricted by negative-feedback mechanisms, maintaining the values within a normal range.

Negative Feedback

Most systems of the body are regulated by **negative-feedback** mechanisms that function to maintain homeostasis. "Negative" means that any deviation from an ideal normal value is resisted or negated. Negative feedback does not prevent variation but maintains that variation within a normal range (Figure 1-2). Deviation from the ideal normal value (i.e., at the center of the normal range) initiates negative-feedback responses that tend to return the value back to the center of the range (Figure 1-3).

An example of negative feedback occurring in the body is the maintenance of normal blood pressure (Figure 1-4). If blood pressure decreases slightly from the ideal normal value, negative-feedback mechanisms increase blood pressure and return it toward the normal value; or if blood pressure increases slightly above the ideal normal value, negative-feedback mechanisms decrease blood pressure. As a result, blood pressure constantly rises and falls around the ideal normal value, establishing a normal range of values for blood pressure.

1

Describe the consequences when a negative-feedback mechanism cannot bring the value of some parameter such as blood pressure back to its normal level.

? ? ? ? ? ? ? ? ? ? ?

Although negative-feedback control mechanisms tend to promote homeostasis, there are conditions during which a deviation from the normal range of values benefits the individual, for homeostasis is not simply the maintenance of all conditions within the same narrow range of values at all times. For example, during exercise the normal ranges for blood pressure and heart rate are significantly elevated (Figure 1-5). The elevated blood pressure is required to supply muscle cells with the extra food and oxygen they need to maintain their increased rate of activity.

2

Explain how negative-feedback mechanisms work to control respiratory rate and heart rate when a person is at rest and when the person is exercising.

? ? ? ? ? ? ? ? ? ? ?

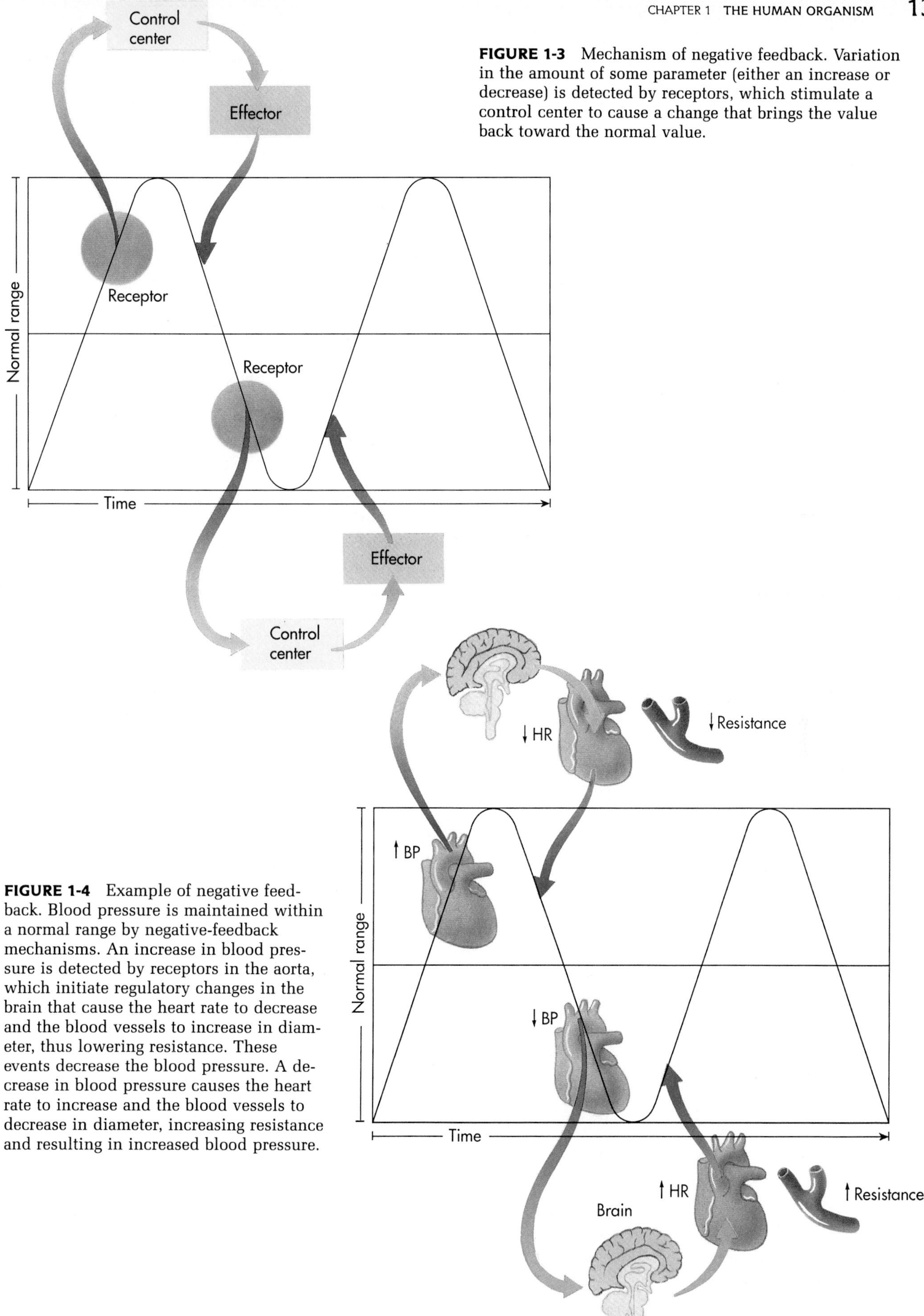

FIGURE 1-3 Mechanism of negative feedback. Variation in the amount of some parameter (either an increase or decrease) is detected by receptors, which stimulate a control center to cause a change that brings the value back toward the normal value.

FIGURE 1-4 Example of negative feedback. Blood pressure is maintained within a normal range by negative-feedback mechanisms. An increase in blood pressure is detected by receptors in the aorta, which initiate regulatory changes in the brain that cause the heart rate to decrease and the blood vessels to increase in diameter, thus lowering resistance. These events decrease the blood pressure. A decrease in blood pressure causes the heart rate to increase and the blood vessels to decrease in diameter, increasing resistance and resulting in increased blood pressure.

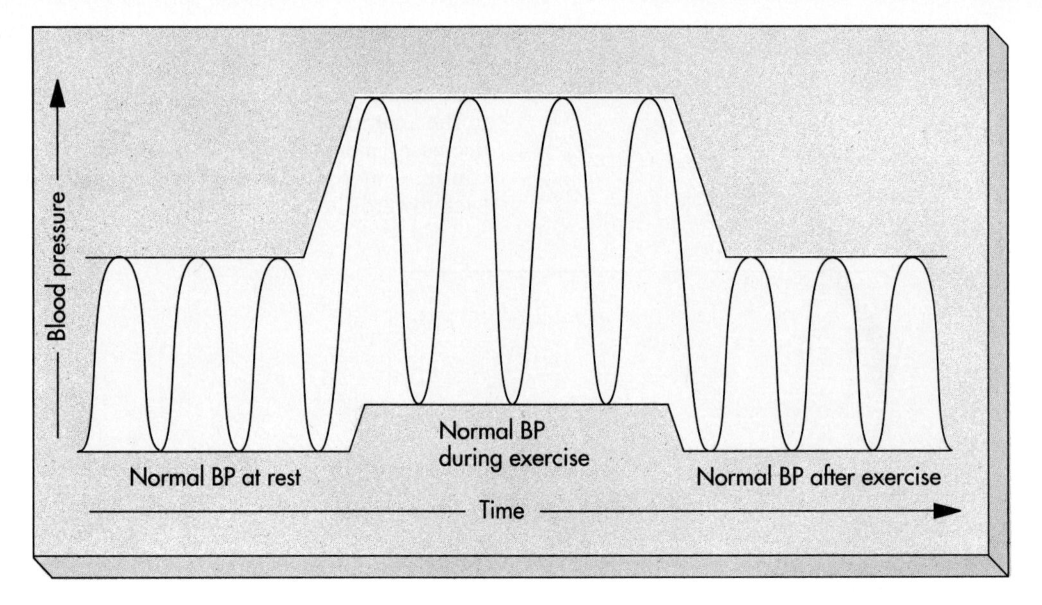

FIGURE 1-5 Change in blood pressure during exercise. During exercise the demand for oxygen by muscle tissues increases. This demand is met by an increase in blood flow to the tissues caused by an increase in blood pressure. The increased blood pressure is not an abnormal or nonhomeostatic condition but is simply a resetting of the normal homeostatic range to meet the increased demand. The reset range is higher and broader than the resting range. After exercise ceases, the range returns to that of the resting condition.

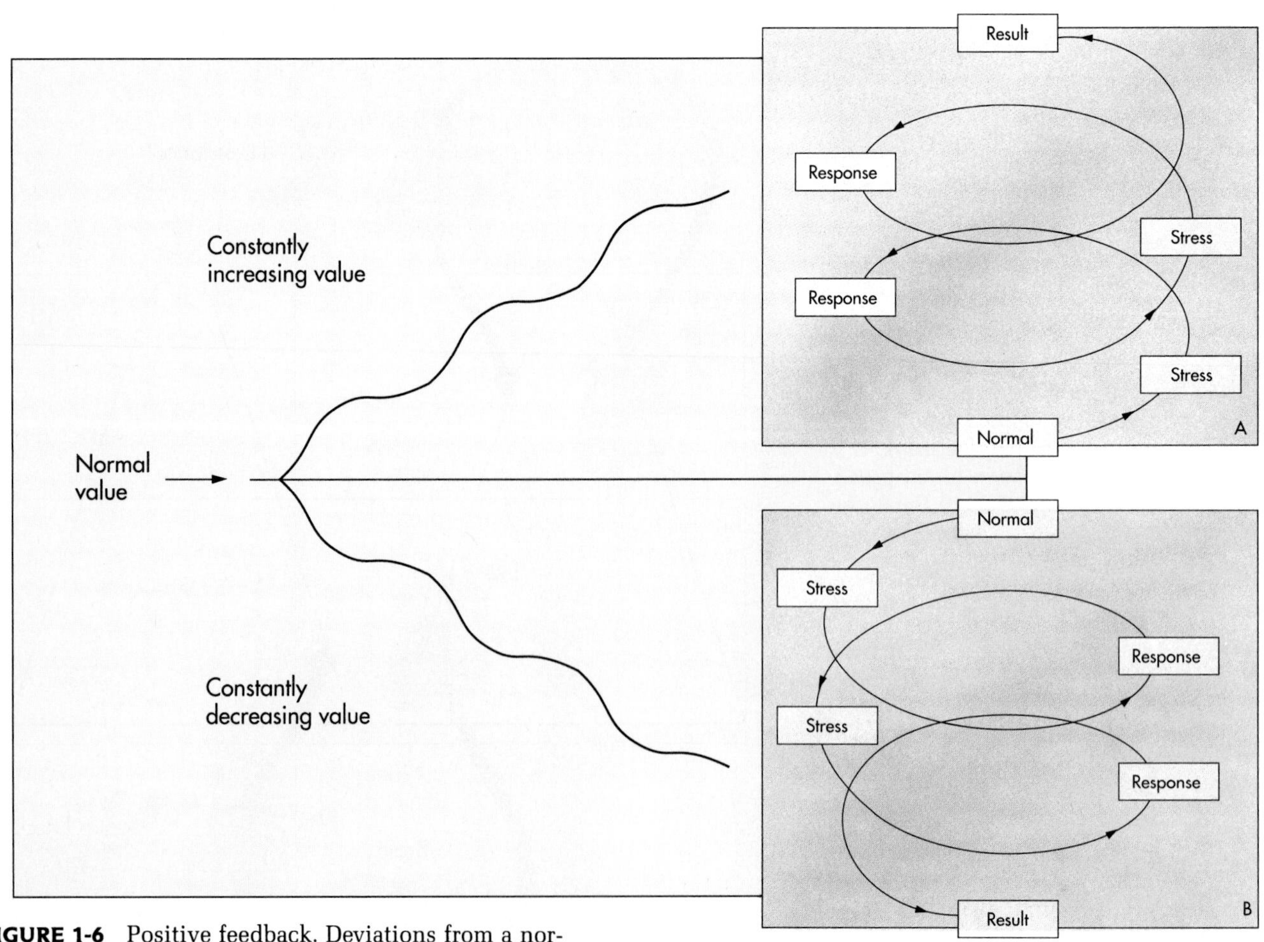

FIGURE 1-6 Positive feedback. Deviations from a normal value cause an additional deviation away from the normal value in either a positive or negative direction.

Since good health requires the maintenance of homeostasis, which is regulated by negative feedback, illness often occurs when negative-feedback mechanisms are disrupted. Medical therapy frequently is designed to overcome illness by aiding the negative-feedback process. One example is the introduction of fluid into the blood or the administration of blood, which can reverse a constantly decreasing blood pressure and thus restore homeostasis.

Positive Feedback

Positive-feedback responses are not homeostatic in nature, yield a different result than negative-feedback responses, and are far less common in healthy individuals. "Positive" implies that when a deviation from a normal value occurs, the response of the system is to make the deviation larger (Figure 1-6). Therefore positive feedback usually creates a "vicious cycle," leading away from homeostasis and possibly leading to death. For example, if blood pressure declines to a sufficiently low value (e.g., after extreme blood loss), too little blood flows back to the heart. Consequently, the heart pumps less blood, and the blood pressure drops further. The additional decrease in blood pressure causes less blood to be returned to the heart and causes the heart to pump even less blood, again decreasing the blood pressure. The process continues until the blood pressure is too low to sustain life or until the positive-feedback cycle is interrupted by medical intervention such as a blood transfusion.

A few positive-feedback mechanisms do operate in the body under normal conditions, but in all cases they eventually are limited in some way. The birth process is an example of a normally occurring positive-feedback mechanism (Figure 1-7). During labor the fetus' head pushes against the opening of the uterus at the cervix. The resulting stretch of the cervix stimulates contractions of the uterine wall muscles. The uterine contractions push the fetus against the cervix, further stretching it and stimulating additional contractions, which result in additional stretching. This positive-feedback sequence is terminated only when the fetus is expelled from the uterus and the stretching stimulus is eliminated.

3

A hormone from the pituitary gland causes the ovary (the female reproductive organ) to secrete estrogen. Estrogen, in turn, causes an increase in the rate of pituitary hormone secretion. That process, once it begins, continues until ovulation (release of the egg from the ovary) occurs. After ovulation the ovary secretes progesterone, which causes the rate of pituitary hormone secretion to decline toward its previous "normal" level. Is the system just described a negative-feedback system, a positive-feedback system, or a system involving both negative and positive feedback? Explain.

? ? ? ? ? ? ? ? ? ? ?

A

Birth
Increased contraction
Increased stretch
Contraction
Stretch
Normal

B

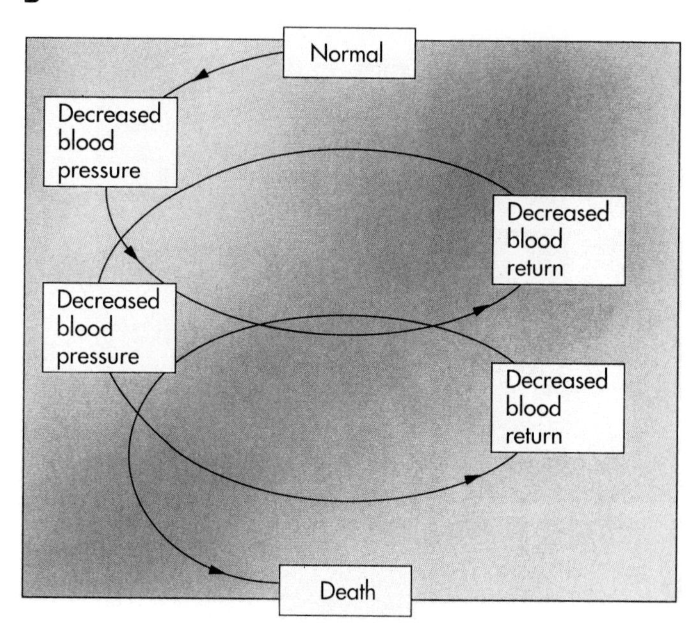

FIGURE 1-7 Example of positive feedback. **A** Childbirth (parturition). Stretching of the uterus as the developing fetus grows eventually stimulates the start of contractions. The contractions push the fetus toward the opening of the uterus, causing it to stretch even more, and the additional stretching initiates more contractions. This positive-feedback cycle continues until the fetus is delivered and the stretching stimulation is eliminated. **B** As blood pressure decreases less blood returns to the heart. With a decreased blood return, the blood pressure decreases even more, until death ultimately occurs.

TERMINOLOGY AND THE BODY PLAN

When first studying anatomy and physiology, the number of new words can seem overwhelming. Nonetheless, it is necessary to learn this new vocabulary. When writing reports or talking with colleagues, you must use correct terminology to avoid confusion and errors. Learning is less difficult and much more interesting if attention is paid to the derivation, or etymology, of new words. Most of the terms in these fields are derived from Latin or Greek, and in the original languages the terms are descriptive. For example, *foramen* is a Latin word meaning a hole, and *magnum* means large. Therefore the "foramen magnum" is a large hole, which is located at the base of the skull. Words are often modified by adding a prefix or suffix. The suffix *-itis* means an inflammation, so "appendicitis" is an inflammation of the appendix. As new terms are introduced, their meanings are often explained. A list of important terms associated with the text is in the glossary at the end of the text.

FIGURE 1-8 Anatomical position. A human in anatomical position is standing with the feet and palms of the hands facing forward with the thumbs to the outside.

Directional Terms

When describing parts of the body, it is often important to refer to their relative positions. Directional terms have been developed to facilitate such references. Directional terms always refer to the body in the anatomical position (Figure 1-8), regardless of its actual position. The **anatomical position** refers to a person standing erect with the feet facing forward, arms hanging to the sides, and palms of the hands facing forward with the thumbs to the outside. Directions such as up, down, in front of, and in back of are not usually used by anatomists because they can be confusing when the body is in various positions such as standing, sitting, or lying down. Of the three general directional concepts (right or left, up or down, front or back), only **right** and **left** are retained as directional terms in anatomical terminology. In human anatomy up and down are replaced by **superior** and **inferior,** respectively, and front and back are replaced by **anterior** and **posterior,** respectively. A series of important directional terms is presented in Table 1-2. It is important to become familiar with these terms as soon as possible since they will be used repeatedly throughout the text.

The word anterior means that which goes before or first, and the anterior surface of the human body is the **ventral,** or belly, surface because the belly "goes first" when we are walking. The posterior (that which follows) surface of the body is the **dorsal** surface or the back. Superior (higher) is equal to **cephalic** (toward the head) since, when we are in the standing position, the head (cephalic end) is the highest point. Inferior (lower) is usually equal to **caudal** (toward the tail) in humans. Since the term caudal means toward the tail, which would be located at the end of the vertebral column if humans had tails, in the lower limbs the term caudal does not apply and is not equal to inferior.

4

The anatomical position of a cat refers to the animal standing erect on all four limbs and facing forward. Based on the etymology of the directional terms, what two terms would indicate movement toward the head? What two terms would mean movement toward the animal's back? Compare these terms with those referring to a human in the anatomical position.

? ? ? ? ? ? ? ? ? ? ?

Proximal means nearest, whereas **distal** means to be distant. These terms are used to refer to linear structures in which one end is near some other structure (its attachment point) and the other end is far off. These terms are commonly used in reference to structures in the limbs; each limb is attached at its

Table 1-2

Directional terms for humans

TERMS	ETYMOLOGY*	DEFINITION
Left		Toward the left side
Right		Toward the right side
Superior	L., higher	A structure higher than another (usually synonymous with cephalic)
Inferior	L., lower	A structure lower than another (usually synonymous with caudal)
Cephalic	G. *kephale,* head	Closer to the head than another structure (usually synonymous with superior)
Caudal	L. *cauda,* a tail	Closer to the tail than another structure (usually synonymous with inferior)
Proximal	L. *proximus,* nearest	Closer than another structure to the point of attachment to the trunk
Distal	L. *di-* plus *sto,* to stand apart or to be distant	Farther than another structure from the point of attachment to the trunk
Medial	L. *medialis,* middle	Toward the middle or the midline of the body
Lateral	L. *latus,* side	Away from the middle or midline of the body
Anterior	L., before	The front of the body (synonymous with ventral)
Posterior	L. *posterus,* following	The back of the body (synonymous with dorsal)
Ventral	L. *venter* or *ventr-,* belly	Toward the belly (synonymous with anterior)
Dorsal	L. *dorsum,* back	Toward the back (synonymous with posterior)
Superficial	L. *superficialis*	Toward or on the surface (not shown in figure)
Deep	O.E. *deop,* deep	Away from the surface, internal (not shown in figure)

*Origin and meaning of the word—
L., Latin; G., Greek; O.E., Old English.

Left
Right
Medial
Proximal
Superior
(Cephalic)
Distal
Lateral
Proximal
Distal
Inferior
(Caudal)

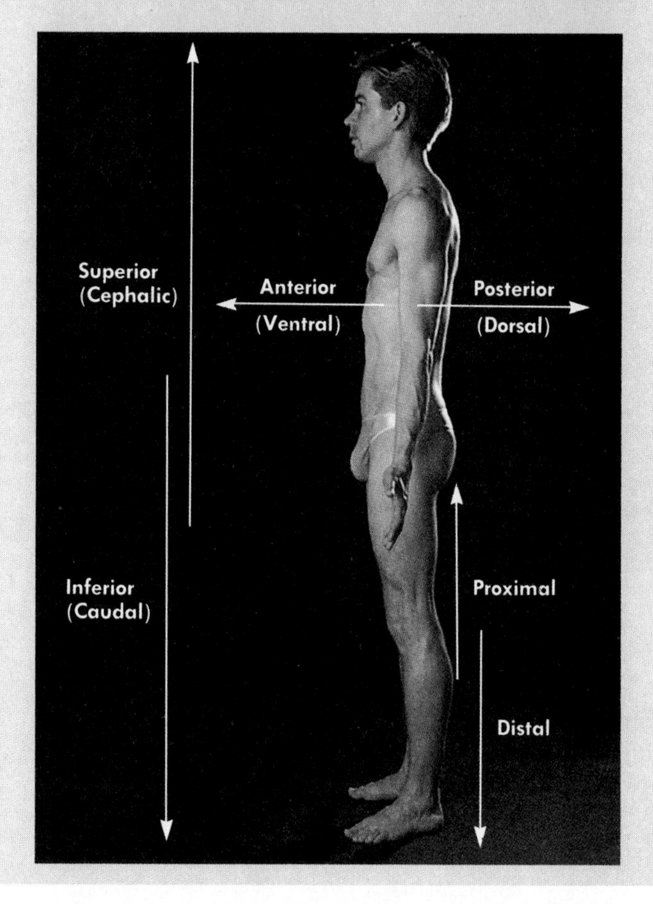

proximal end to the body, and the other end, the distal end (the hand or foot), is some distance away. Proximal and distal may also be used in reference to tubular systems such as the kidney or the digestive tract. The proximal end of the small intestine is attached to the stomach, and the distal end of the small intestine connects to the large intestine.

Medial and **lateral** mean toward the midline and away from the midline, respectively. The nose is located in a medial position in the face, and the eyes are lateral to the nose. The terms **superficial** and **deep** refer to a structure close to the surface of the body and more toward the interior of the body, respectively. The skin is superficial to muscle and bone.

Two positional terms used in anatomy and related to body movement also should be mentioned. **Prone** means to lie or be placed with the anterior surface down, and **supine** means to lie or be placed so that the anterior surface faces up.

It often is not possible to describe the relationship between two structures using only one directional term any more than it is possible to describe directions on the Earth's surface using only the terms north, south, east, and west. Just as we combine north and east to describe a town that lies northeast of us, we can also combine terms such as anterior and lateral to describe a structure that lies anterolateral to another structure.

5

Describe in as many directional terms as you can the relation between your kneecap and your heel.

? ? ? ? ? ? ? ? ? ? ?

Planes

At times it is conceptually useful to describe the body as having a series of **planes** (imaginary flat surfaces) passing through it (Figure 1-9). A **sagittal** (saj′ĭ-tal) plane runs vertically through the body and sep-

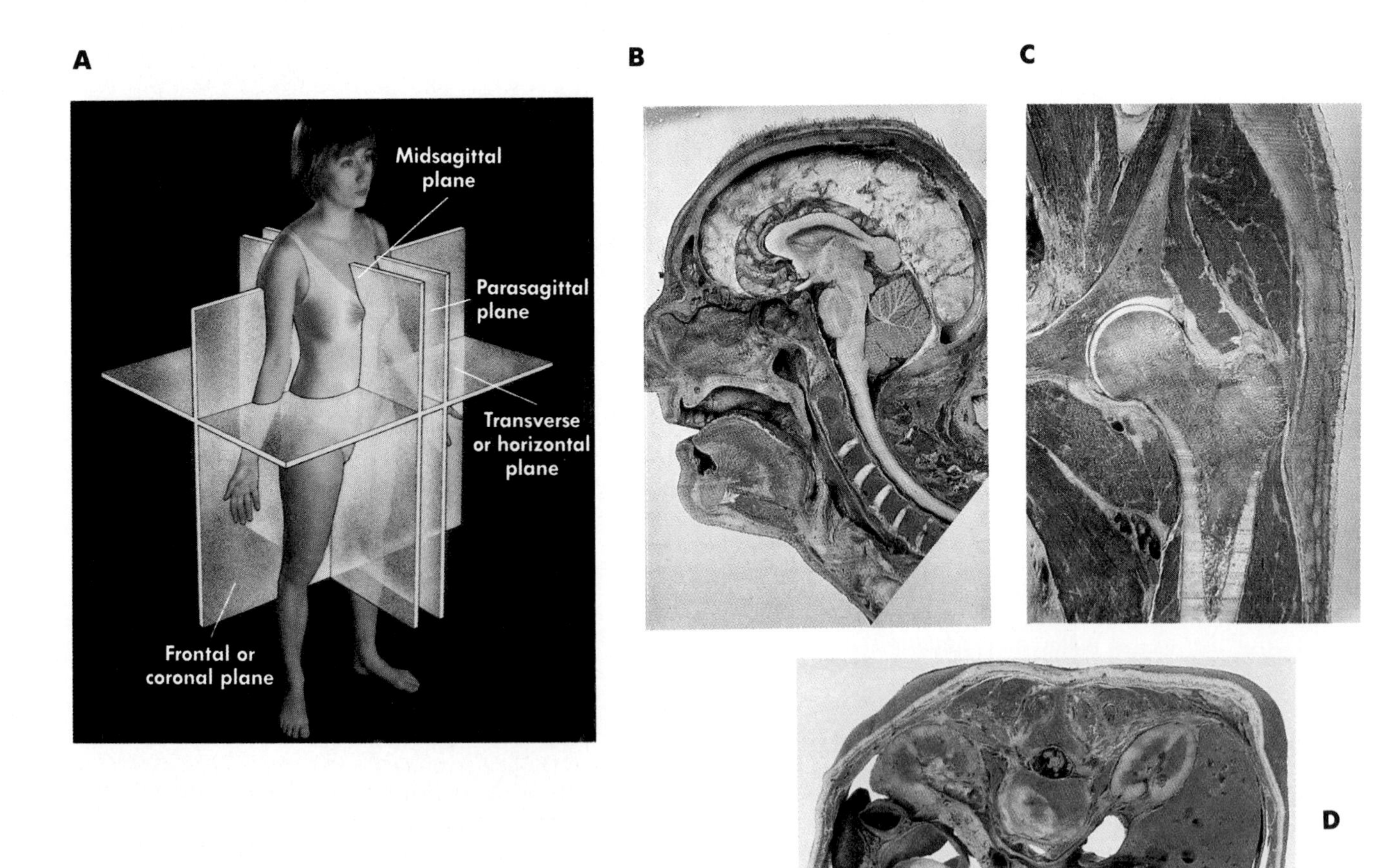

FIGURE 1-9 Body planes or planes of section. **A** Planes are indicated by "glass" sheets. **B** Human head as seen in midsagittal section. **C** Frontal section through the left hip joint. **D** Transverse section through the abdomen.

arates it into right and left portions. The word sagittal literally means "the flight of an arrow" and refers to the way the body would be split by an arrow passing anterior to posterior through the body. If the plane divides the body into equal right and left halves, it is a **midsagittal,** or a **median,** section, and if the plane is to one side of the midline, it is **parasagittal** (*para-* means alongside of). A **transverse,** or **horizontal,** sectional plane divides the body into superior and inferior portions and runs parallel to the surface of the ground. A **frontal,** or **coronal** (ko-ro'nal), plane runs vertically and divides the body into anterior and posterior portions.

Organs are often sectioned to reveal their internal structure (Figure 1-10). A cut through the long axis of the organ is a **longitudinal** section, and a cut at right angles to the long axis is a **cross,** or **transverse,** section. If a cut is made across the long axis at other than a right angle, it is called an **oblique** section.

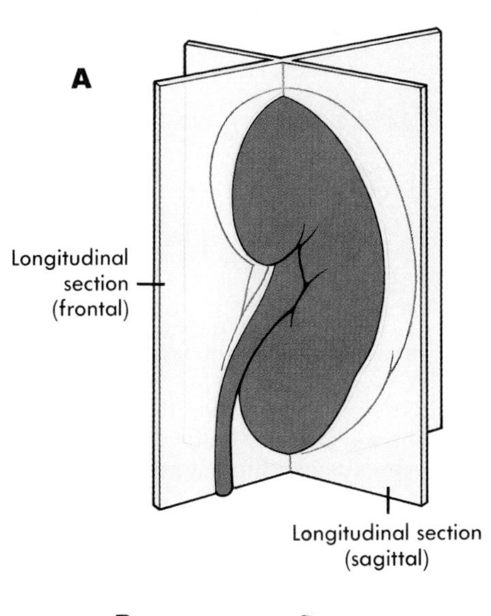

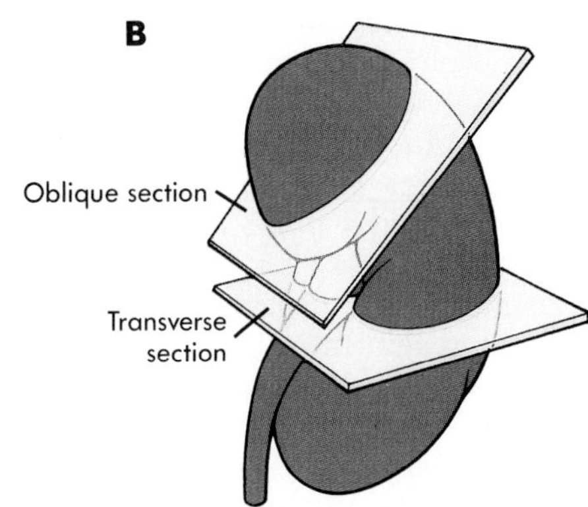

FIGURE 1-10 Planes of section through an organ. **A** Longitudinal frontal and sagittal sections. **B** Oblique and transverse sections.

Body Regions

The body is commonly divided into several regions (Figure 1-11). The first major division is between the **appendicular** (ap'pen-dik'u-lar) and **axial** (ak'se-al) regions. The appendicular region includes the limbs or extremities and their associated girdles (the bony structures by which the limbs are attached to the body). The upper limb is divided into the arm (brachial region), forearm, wrist, and hand. The **arm** extends from the shoulder to the elbow, and the **forearm** extends from the elbow to the wrist. The upper limb is attached to the body by the **shoulder,** or **pectoral** (pek'to-ral) **girdle.** The lower extremity is divided into the thigh (femoral region), leg, ankle, and foot. The **thigh** extends from the hip to the knee, and the **leg** extends from the knee to the ankle. The lower limb is attached to the body by the **hip,** or **pelvic** (pel'vik) **girdle.** Note that the terms arm and leg, contrary to popular usage, refer only to a portion of the respective limb.

The axial portion of the body consists of the **head** (cephalic region), **neck** (cervical region), and **trunk** (the body excluding the head and limbs). The trunk can be divided into the **thorax** (chest), **abdomen** (region between the thorax and pelvis), and **pelvis** (the inferior end of the trunk, associated with the hips).

The abdomen is often subdivided superficially into four **quadrants** (Figure 1-12, *A*). They include the upper right, upper left, lower right, and lower left quadrants. The dividing lines consist of two imaginary lines on the surface of the abdomen—one horizontal and the other vertical—that intersect at the umbilicus (navel). The four-quadrant approach is commonly used by clinicians to describe the location of underlying organs or of a clinical problem such as a pain or a tumor. For example, the pain of acute appendicitis usually is located in the lower-right quadrant. When more precision in locating abdominal structures is necessary (e.g., in abdominal surgery), the abdominal area can be subdivided into nine **regions** by four imaginary lines—two horizontal and two vertical. These four lines create an imaginary tic-tac-toe board on the abdomen. The names of these nine regions are listed in Figure 1-12, *B*.

Body Cavities

The body contains many cavities such as the nasal, cranial, abdominal, and bone marrow cavities. Some of these cavities open to the outside of the body, and some do not. Discussion in this chapter is limited to the major trunk cavities that do not open to the outside. Traditionally, anatomy and physiology textbooks describe a dorsal body cavity, which contains the brain and spinal cord, and a ventral body cavity, which includes all the trunk cavities. However, this concept is not discussed in standard works

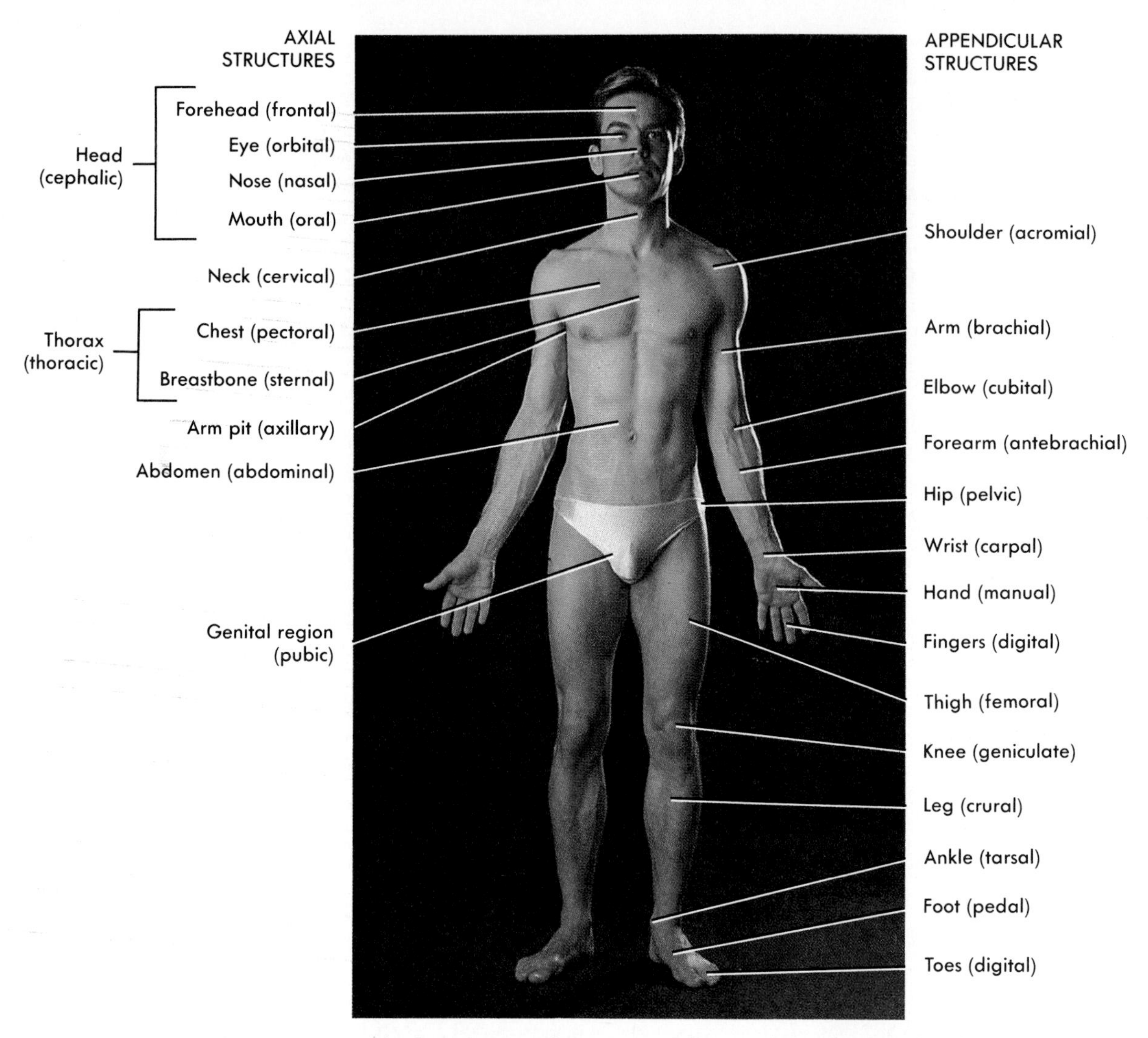

FIGURE 1-11 Major body regions and structures. The axial region includes the head, neck, and trunk. The appendicular region includes the limbs, or extremities, and their associated girdles. The arm is that part of the upper limb between the shoulder and elbow. The forearm is between the elbow and wrist. In the lower limb the thigh is between the hip and the knee, whereas the leg is between the knee and the ankle.

on anatomy. There are no parallels, neither embryonic, anatomic, nor histologic, between the fluid-filled space around the central nervous system and the trunk cavities. For these reasons the concept of a dorsal body cavity is not included in this book. A clear concept of the large cavities in the trunk is helpful in understanding how organs are arranged in the body.

The trunk contains three large cavities, the **thoracic cavity,** the **abdominal cavity,** and the **pelvic cavity** (Figure 1-13). The thoracic cavity is divided into two portions by a midline structure called the **mediastinum** (me'de-as-ti'num; a middle wall). The mediastinum is that portion of the thorax containing the trachea, esophagus, thymus, heart, and associated structures. The organs that comprise the mediastinum form a wall that completely separates the right and left lungs. The thoracic cavity is surrounded by the rib cage and is separated from the abdominal cavity by the muscular diaphragm.

There is no physical separation between the abdominal and pelvic cavities, which sometimes are called the abdominopelvic cavity. The division between these cavities is an imaginary plane drawn between the symphysis pubis and the sacral promontory (Figure 1-13). The abdominal cavity is bounded primarily by the abdominal muscles and contains the stomach, intestines, liver, spleen, pancreas, and kidneys. The pelvic cavity is a small space enclosed by the bones of the pelvis and contains the urinary bladder, part of the large intestine, and the internal reproductive organs.

The trunk cavities and their organs are lined with serous membranes (membranes that produce a watery substance). To understand the serous membranes, imagine an inflated balloon into which a fist

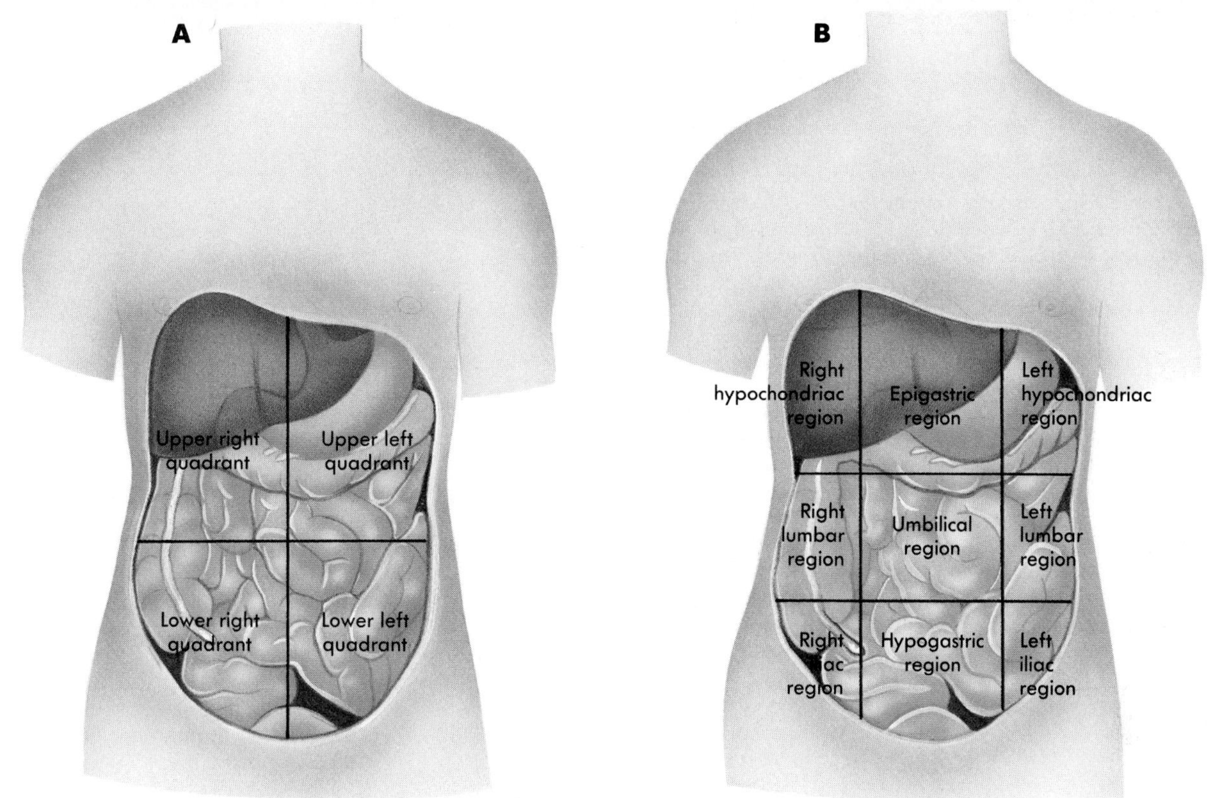

FIGURE 1-12 **A** Abdominal quadrants. Lines are superimposed over internal organs to demonstrate the relationship of the organs to the quadrants. **B** Abdominal regions. The abdomen can also be divided into nine regions. Lines are superimposed over internal organs.

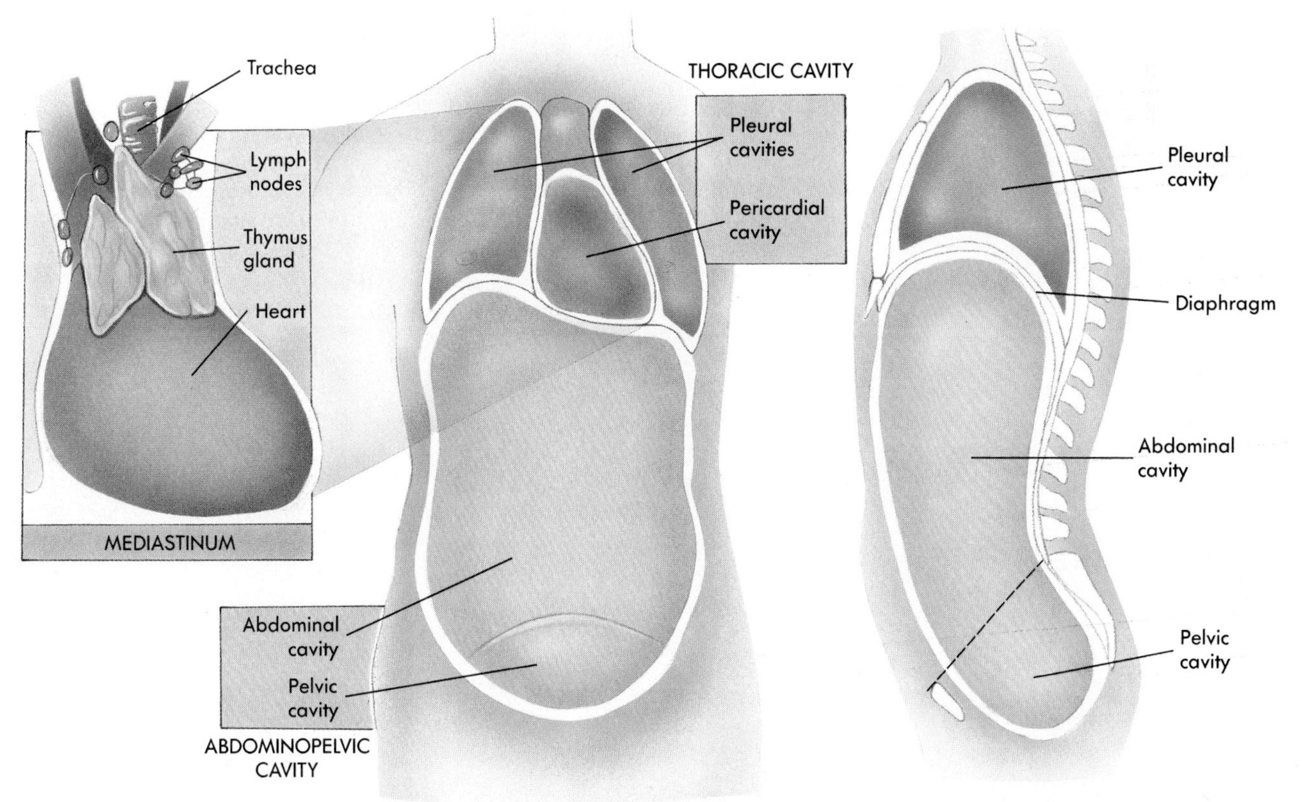

FIGURE 1-13 Body cavities. The thoracic cavity includes the two pleural cavities and the pericardial cavity. Some of the contents of the mediastinum are shown on the left. The abdominopelvic cavity contains the abdominal cavity and the pelvic cavity.

Anatomical Imaging

Anatomical imaging has revolutionized medical science. It has been estimated that during the past 20 years there has been as much progress in clinical medicine as in all the previous history of medicine combined, and anatomical imaging has made a major contribution to that progress. Anatomical imaging allows medical personnel to look inside the body with amazing accuracy and without the trauma and risk of exploratory surgery. Although most of the technology of anatomical imaging is very new, the concept and earliest technology are quite old.

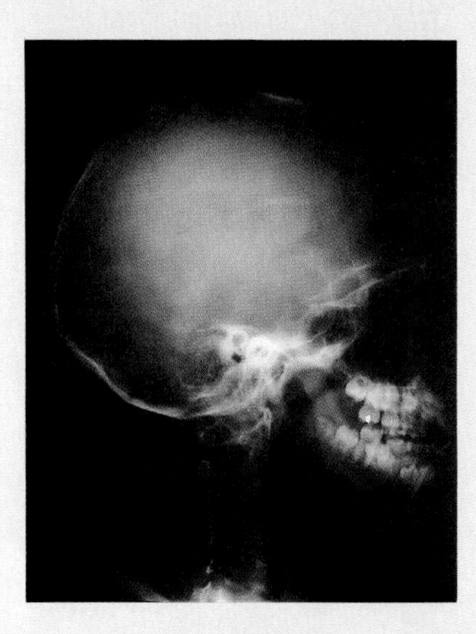

FIGURE 1-A

X-rays, the magical, mysterious rays that can see inside the body, were first used in medicine by Wilhelm Roentgen in 1885. They were called x-rays because no one knew what they were. The rays, extremely shortwave electromagnetic radiation (see Chapter 2), can pass through most tissues of the body and expose a photographic plate; but bones and radiopaque dyes absorb the rays and create underexposed areas that appear white on the photographic film (Figure 1-A). X-rays have been in common use for many years and have numerous uses. A major limitation of x-rays is that they give only a flat, two-dimensional (2-D) image of the body, which is a three-dimensional (3-D) structure. Almost everyone has had an x-ray, either to visualize a broken bone or to check for a cavity in a tooth.

Ultrasound (sonography) is the second oldest imaging technique. First developed in the early 1950s as an extension of World War II sonar technology, it uses high-frequency sound waves. The sound waves are emitted from a transmitter-receiver placed on the skin over the area that is scanned. The sound waves strike internal organs and are reflected back to the receiver on the skin. Even though the basic technology is fairly old, the most important advances in the field occurred only after it became possible to analyze the reflected sound waves by computer. Once the computer has analyzed the pattern of sound waves, the information is transferred to a monitor in which the result is visualized as an ultrasound image (Figure 1-B). One of the more recent advances in ultrasound technology is the ability of the more advanced computers to analyze changes in position through time and to display those changes as "real-time" movements. Among other medical uses, ultrasound commonly is used to evaluate the condition of the fetus during pregnancy.

Computer analysis was also the basis of another major medical breakthrough in imaging. **Computed tomographic (CT)** scans, developed in 1972 and originally called **computerized axial tomographic (CAT)** scans, are computer-analyzed x-ray images. A low-intensity x-ray tube is rotated through a 360-degree arc around the patient, and the images are fed into a computer. The computer then constructs the image of a "slice" through the body at the point where the x-ray beam was focused and rotated (Figure 1-C). It is also possible with some computers to take several scans short distances apart and stack the slices to produce a 3-D image of a part of the body (Figure 1-D).

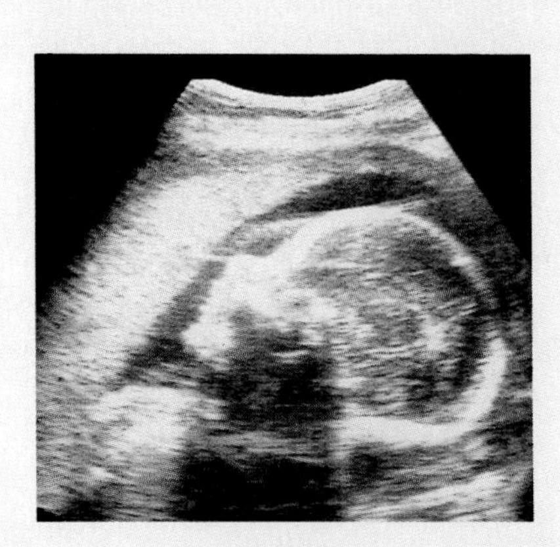

FIGURE 1-B

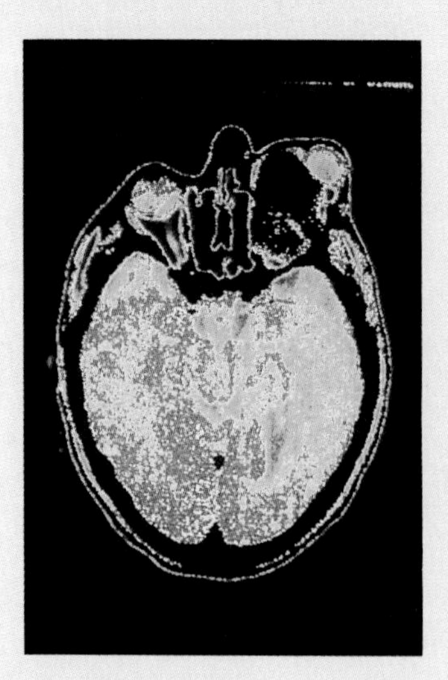

FIGURE 1-C

FIGURE 1-D

Dynamic spatial reconstruction (DSR) takes CT one step farther. Instead of using a single rotating x-ray machine to take single slices and add them together, DSR uses numerous (approximately 30) x-ray tubes. The images from all the tubes are compiled simultaneously, rapidly producing a 3-D image. Because of the speed of the process, multiple images can be compiled to show changes through time, giving the system a dynamic quality. This system allows us to move away from seeing only static structure and toward seeing dynamic structure and function.

Digital subtraction angiography (DSA) is also one step beyond CT scans. A 3-D x-ray image of an organ such as the heart is made and stored in a computer. A radiopaque dye is injected into the circulation, and a second x-ray computer image is made. Then the first image is subtracted from the second one, greatly enhancing the differences, with the primary difference the presence of the injected dye (Figure 1-E). These computer images can be dynamic and can be used, for example, to guide a catheter into a coronary artery during angioplasty (the insertion of a tiny balloon into a coronary artery to compress material clogging the artery).

Magnetic resonance imaging (MRI), which subjects a person to a large electromagnetic field and certain radio waves, also is based on principles that have been known for years but have been applied only recently to medicine. The magnetic field causes the protons of various atoms to align (see Chapter 2). Because of the large amounts of water in the body, the alignment of hydrogen atom protons is at present most important in this imaging system. Radio waves of certain frequencies, which change the alignment of the hydrogen atoms, then are directed at the patient. When the radio waves are turned off, the hydrogen atoms realign in accordance with the magnetic field. The time it takes the hydrogen atoms to realign is different for various tissues of the body. These differences can be analyzed by computer to produce very clear sections through the body (Figure 1-F). The technique is also very sensitive in detecting some forms of cancer and can detect a tumor far more readily than can a CT scan.

Positron emission tomographic (PET) scans are able to identify the metabolic states of various tissues. This technique is particularly useful in analyzing the brain. When cells are active, they are using energy. The energy they need is supplied by the breakdown of glucose (blood sugar). If radioactively labeled glucose is given to a patient, the active cells take up the labeled glucose. As the radioactivity in the glucose decays, positrons (positively charged subatomic particles) are emitted. When the positrons collide with electrons, the two particles annihilate each other, and gamma rays are given off that can be detected, pinpointing the cells that are metabolically active (Figure 1-G).

Any time the human body is exposed to x-rays, ultrasound, electromagnetic fields, or radioactively labeled substances, there is potential risk. In the medical application of anatomical imaging, the risk must be weighed against the benefit, and the risk is minimized by using the lowest possible doses that will still provide the necessary information. Numerous studies have been conducted and are still being conducted to determine the outcomes of diagnostic and therapeutic exposures. It is well known that x-rays can cause cell damage, particularly to the reproductive cells. As a result of this knowledge, the number of x-rays and the level of exposure are kept to a minimum, the x-ray beam is focused as closely as possible to avoid scattering of the rays, areas of the body not being x-rayed are shielded, and x-ray personnel are shielded. There are no known risks from ultrasound or electromagnetic fields at the levels used in diagnosis.

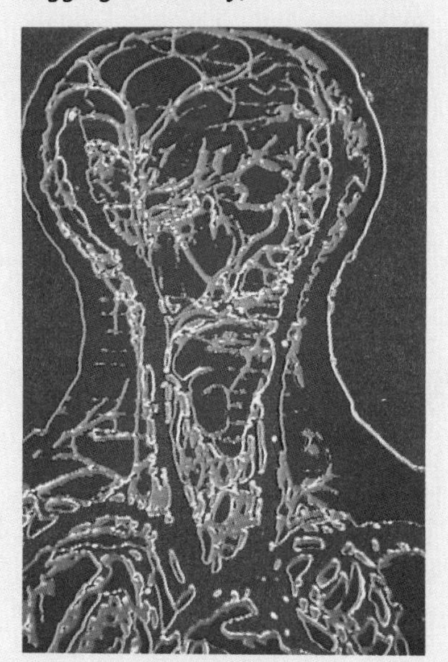

FIGURE 1-E

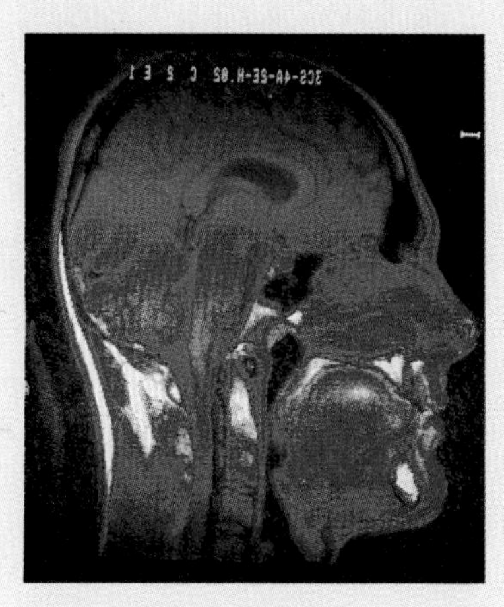

FIGURE 1-F

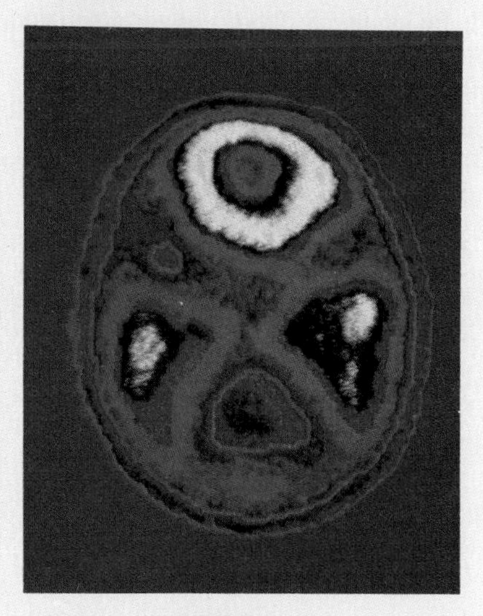

FIGURE 1-G

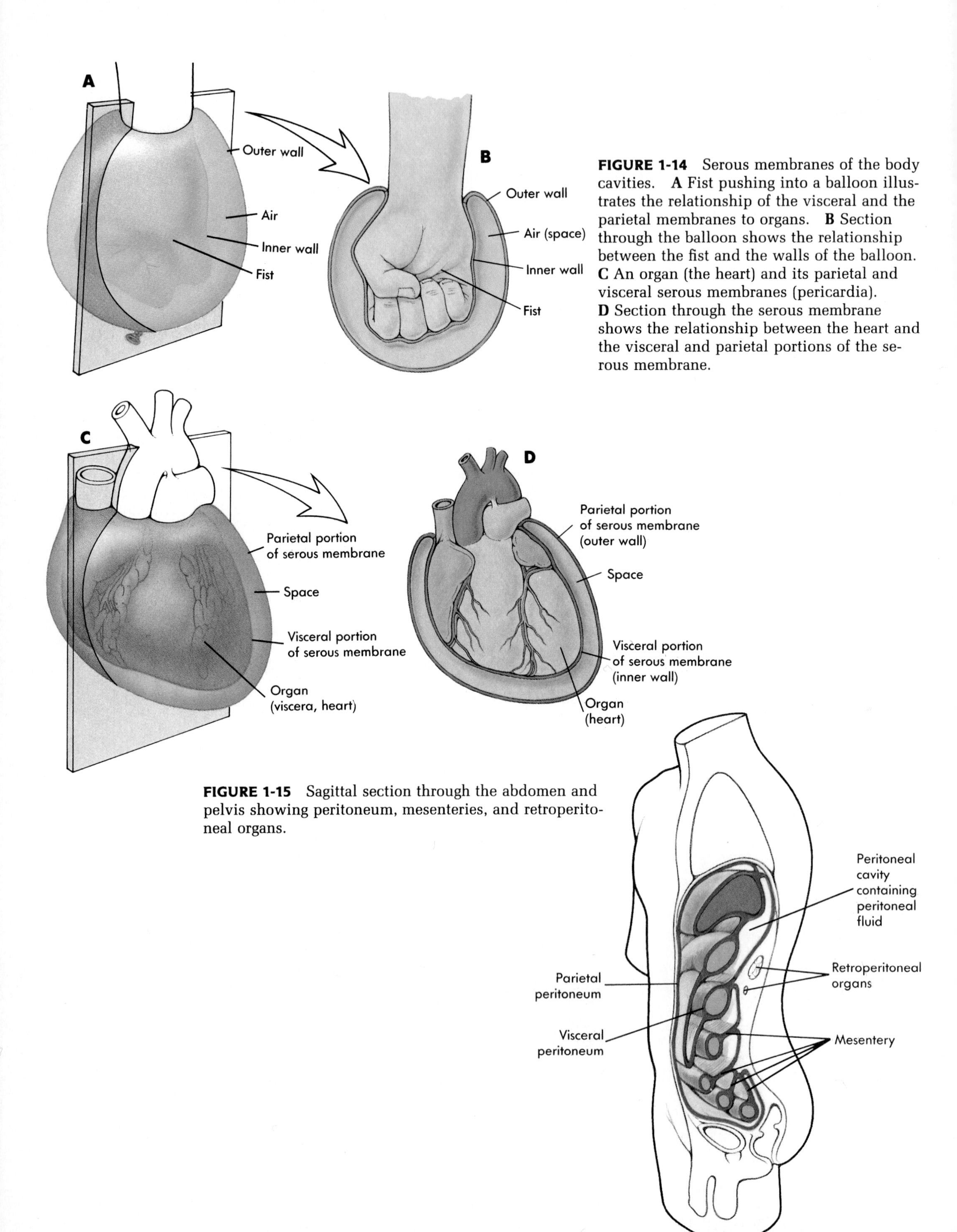

FIGURE 1-14 Serous membranes of the body cavities. **A** Fist pushing into a balloon illustrates the relationship of the visceral and the parietal membranes to organs. **B** Section through the balloon shows the relationship between the fist and the walls of the balloon. **C** An organ (the heart) and its parietal and visceral serous membranes (pericardia). **D** Section through the serous membrane shows the relationship between the heart and the visceral and parietal portions of the serous membrane.

FIGURE 1-15 Sagittal section through the abdomen and pelvis showing peritoneum, mesenteries, and retroperitoneal organs.

has been pushed (Figure 1-14, *A* and *B*). The walls of the balloon represent the serous membranes, and the fist is analogous to the internal organs or **viscera** (vis'er-ah) (Figure 1-14, *C* and *D*). Many internal organs extend into the serous membranes in a manner similar to the way the fist pushes into the balloon. The portion of the serous membrane in contact with the organ is referred to as **visceral** (vis'er-al; organ), and the part of the membrane in contact with the wall of the cavity is referred to as **parietal** (pă-ri'ĕ-tal; wall).

A potential cavity or space is located between the visceral and parietal membranes and normally is filled with a thin, lubricating film of serous fluid produced by the membranes. If an organ rubs against another organ or against the body wall, the serous fluid and smooth serous membranes function to reduce friction.

The thorax contains three serous membrane–lined cavities: two **pleural** (ploor'al; associated with the ribs) **cavities** and a **pericardial** (pĕr-ĭ-kar'de-al; around the heart) **cavity.** Each lung is covered by a **visceral pleura,** and the interior of the lateral walls of the thorax and the lateral surfaces of the mediastinum are lined with **parietal pleura.** These membranes are continuous where the bronchi and vessels enter and leave each lung. The pleural cavity is located between the visceral and parietal pleura and contains pleural fluid.

The heart is covered by the **visceral pericardium** and is contained within a connective tissue sac, the **pericardial sac,** which is lined with the **parietal pericardium.** The visceral and parietal pericardia are continuous with each other where vessels enter and exit the superior portion of the heart. The pericardial cavity, which contains pericardial fluid, is located between the visceral and parietal pericardium.

The abdominal and pelvic cavities are lined by the **parietal peritoneum** (pĕr'ĭ-to-ne'um; to stretch over), and many of the organs contained in the cavity are covered by the **visceral peritoneum** (Figure 1-15). The space between the two membranes contains peritoneal fluid and is called the **peritoneal cavity.**

The serous membranes can become inflamed—often as a result of bacterial infection. Peritonitis is an inflammation of the peritoneum, pericarditis is an inflammation of the pericardium, and pleurisy (a term more commonly used than pleuritis) is an inflammation of the pleura.

The parietal peritoneum is continuous with a double-layered membrane called a **mesentery** (mes'en-tĕr'e) (see Figure 1-15), which anchors some of the abdominal organs to the body wall, and provides a pathway for nerves and vessels to reach the organs. Some of the abdominal and pelvic organs closely associated with the body wall are covered only by parietal peritoneum, and do not have mesenteries. These organs are said to be **retroperitoneal** (rĕ'tro-pĕr'ĭ-to-ne'al; behind the peritoneum) (see Figure 1-15). The retroperitoneal organs include the kidneys, adrenal glands, pancreas, portions of the intestine, and the urinary bladder.

6

Explain how an organ can be located within the abdominal cavity but not within the peritoneal cavity.

? ? ? ? ? ? ? ? ? ? ?

SUMMARY

A FUNCTIONAL KNOWLEDGE OF ANATOMY AND PHYSIOLOGY CAN BE USED TO SOLVE PROBLEMS CONCERNING THE BODY WHEN HEALTHY OR DISEASED.

ANATOMY p. 4

1. Anatomy is the study of the body's structures.
   - Microscopic anatomy examines the small details of cells (cytology) and tissues (histology).
   - Gross anatomy emphasizes organs from a systemic or regional perspective.
2. Neuroanatomy deals with the structure and function of the brain and spinal cord, and developmental anatomy is concerned with changes that occur between fertilization and adulthood.
3. Surface anatomy uses superficial structures to locate deeper structures, and anatomic imaging is a noninvasive technique for identifying deep structures.

PHYSIOLOGY p. 4

1. Physiology is the study of the body's functions.
2. The discipline can be approached from an organismal point of view (e.g., human) or an organizational perspective (e.g., cellular or systemic).

STRUCTURAL AND FUNCTIONAL ORGANIZATION p. 4

1. Basic chemical characteristics are responsible for the structure and functions of life.
2. Organelles are small structures within cells that perform specific functions.
3. Cells are the basic living units of plants and animals and have many common characteristics.
4. Tissues are groups of cells of similar structure and function and their associated extracellular substances. The four primary tissue types are epithelial, connective, muscle, and nervous tissues.
5. Most organs are structures composed of two or more tissues that perform specific functions.
6. Organs are arranged into the 11 organ systems of the human body (see Table 1-1).
7. Organ systems interact to form a whole, functioning organism.

THE HUMAN ORGANISM *p. 11*

Characteristics of Life

Humans have many characteristics such as organization, metabolism, responsiveness, growth and development, and reproduction in common with other organisms.

Biomedical Research

Much of what is known about humans is derived from research on other organisms.

HOMEOSTASIS *p. 12*

Homeostasis is a state of equilibrium in which body functions, fluids, and other factors of the internal environment are maintained at levels suitable to support life.

Negative Feedback

Negative-feedback mechanisms operate to restore homeostasis. Positive-feedback mechanisms usually increase deviations from normal.

Positive Feedback

Although a few positive-feedback mechanisms normally exist in the body, most positive-feedback mechanisms are harmful.

TERMINOLOGY AND THE BODY PLAN *p. 16*

Directional Terms

1. A human standing erect with the feet facing forward, the arms hanging to the side, and the palms facing forward is in the anatomical position.
2. Directional terms always refer to the anatomical position no matter what the actual position of the body (see Table 1-2).

Planes

1. Planes of the body
   - A midsagittal or median section divides the body into equal left and right parts. A parasagittal section produces unequal left and right parts.
   - A transverse (horizontal) plane divides the body into superior and inferior portions.
   - A frontal (coronal) plane divides the body into anterior and posterior parts.
2. Sections of an organ
   - A longitudinal section of an organ divides it along the long axis.
   - A transverse (cross) section cuts at a right angle to the long axis of an organ.
   - An oblique section cuts across the long axis of an organ at an angle other than a right angle.

Body Regions

1. The body can be divided into appendicular (limbs and girdles) and axial (head, neck, and trunk) regions.
2. Superficially the abdomen can be divided into four quadrants or nine regions. These divisions are useful for locating internal organs or describing the location of a pain or tumor.

Body Cavities

1. The trunk contains the thoracic, abdominal, and pelvic cavities.
   - The thoracic cavity is subdivided by the mediastinum.
   - The diaphragm separates the thoracic and abdominal cavities.
   - The pelvic cavity is surrounded by the pelvic bones.
2. The trunk cavities are lined by serous membranes.
   - The parietal portion of a serous membrane lines the wall of the cavity, and the visceral portion is in contact with the internal organs.
   - The serous membranes secrete fluid that fills the space between the visceral and parietal membranes. The serous membranes protect organs from friction.
   - The pleural membranes surround the lungs, the pericardial membranes surround the heart, and the peritoneal membranes line the abdominal and pelvic cavities and surround their organs.
3. Mesenteries are parts of the peritoneum that hold the abdominal organs in place and provide a passageway for blood vessels and nerves to the organs.
4. Retroperitoneal organs are located "behind" the parietal peritoneum.

CONTENT REVIEW

1. What is meant by a functional knowledge of anatomy and physiology?
2. Define anatomy. Contrast microscopic anatomy, gross anatomy, neuroanatomy, and developmental anatomy.
3. Define physiology. What are two ways in which physiology can be studied?
4. List seven structural levels at which the body can be considered conceptually.
5. Define a tissue. What are the four primary tissue types?
6. Define an organ and an organ system. What are the 11 organ systems of the body and their functions?
7. Why is it important to realize that humans share many characteristics with other animals?
8. What is meant by the term homeostasis? If a deviation from homeostasis occurs, what mechanism restores it?
9. Define positive feedback. Why are positive-feedback mechanisms often harmful?
10. Why is knowledge of the etymology of anatomical and physiological terms useful?
11. What is the anatomical position?
12. List two terms that in humans would indicate toward the head. Name two terms that would mean the opposite.
13. List two terms that would indicate the back in humans. What two terms would mean the front?
14. Define the following terms, and give the word that means the opposite: proximal, prone, lateral, and superficial.
15. Define the three planes of the body. What is the difference between a parasagittal section and a midsagittal section?
16. What are the two major body regions?
17. Describe the four-quadrant and the nine-region methods of subdividing the abdominal region. What is the purpose of these divisions?
18. Define the thoracic, abdominal, and pelvic cavities. What is the mediastinum?
19. Differentiate between parietal and visceral serous membranes. What is the function of the serous membranes?
20. Name the serous membranes lining each of the body cavities.
21. What are mesenteries? Explain their function.
22. What are retroperitoneal organs? List four examples.

CONCEPT QUESTIONS

1. Diabetes is a disorder in which the pancreas (an organ) fails to produce insulin (a chemical that is normally made by the pancreas and released into the circulation). List as many levels of organization as you can in which this disorder could be corrected.
2. During physical exercise respiration rate increases. Two anatomy and physiology students are arguing about the mechanisms involved: Student A claims that they are positive feedback, and Student B claims they are negative feedback. Do you agree with Student A or Student B and why?
3. The following observations were made on a patient who had suffered from a bullet wound:
   - Considerable blood loss
   - Blood pressure very low and dropping
   - Heart rate elevated and rising, but a continually dropping blood pressure
   - Increase in blood pressure after a transfusion

   Which of the following statements are consistent with the observations?

   A A positive-feedback mechanism was interrupted by the transfusion.
   B A negative-feedback mechanism was interrupted by the transfusion.
   C Negative-feedback mechanisms are occasionally inadequate without medical intervention.
   D The transfusion was not necessary.
   E A and C.
4. Which of the following manipulations should be accomplished to expose the anterior surface of a patient's heart?
   1. Make an opening into the mediastinum.
   2. Lay the patient in a supine position.
   3. Lay the patient in a prone position.
   4. Make an incision through the pericardial sac.
   5. Make an opening into the abdomen.

   A 1, 2, 4 **D** 3, 4, 5
   B 1, 3, 4 **E** 2 and 4 only
   C 2, 4, 5
5. During pregnancy which of the mother's body cavities will increase most in size?
   A Cranial
   B Thoracic
   C Abdominal
   D Pelvic
6. A bullet enters the left side of a man's chest, passes through the left lung, and lodges in the heart. Name in order the serous membranes through which the bullet passes.
7. A woman is stabbed in the abdomen with a fencing foil. The foil passes through the abdominal body wall and into and through the stomach, pierces the diaphragm, and finally stops in the left lung. List in order the serous membranes the foil pierces.

? ?

ANSWERS TO PREDICT QUESTIONS

1 p. 12. When a negative-feedback mechanism fails to bring a value back to its normal level, the value will continue to deviate from normal. Homeostasis is not maintained in this situation, and the health of the individual is compromised.

2 p. 12. Negative-feedback mechanisms work to control respiratory rate and heart rate when a person is at rest by controlling both rates within ranges that support normal life functions such as delivering food and oxygen to the tissues of the body. If one or the other rate begins to decline or increase, negative feedback-mechanisms return the rate to its normal levels. When a person is exercising, the active tissues require more food and oxygen. The rate of respiration and the heart rate must increase to keep pace with the tissue's demands for food and oxygen. During exercise the overall ranges are reset to higher levels, and the rates are controlled within these new, higher ranges by negative-feedback mechanisms.

3 p. 15. The system described is a system involving both negative and positive feedback. The pituitary hormone causes the ovary to secrete more estrogen, which, in turn, causes an increase in the rate of pituitary hormone secretion, which in turn causes the ovary to secrete more estrogen, and so on. This is an example of a positive-feedback system. The positive-feedback system continues until the process ends with ovulation. After ovulation the ovary secretes progesterone, which causes the rate of pituitary hormone secretion to decline toward its previous "normal" level. This is a negative-feedback mechanism, which tends to counteract deviation (increase) from the normal levels of pituitary hormone secretion and force pituitary hormone secretion levels back to their normal range.

4 p. 16. In the cat cephalic and anterior are toward the head; dorsal and superior are toward the back. In humans cephalic and superior are toward the head; dorsal and posterior are toward the back.

5 p. 18. Your kneecap is both proximal and superior to the heel. It is also anterior to the heel since it is on the anterior side of the lower limb whereas the heel is on the posterior side.

6 p. 25. The abdominal cavity contains the abdominal organs and the peritoneal cavity. The peritoneal cavity is located between the organs or between the organs and the wall of the abdominal cavity. It is bounded by the parietal peritoneum and the visceral peritoneum and contains no organs, only serous fluid. Because the peritoneal cavity contains no organs, an organ can be located within the abdominal cavity but not within the peritoneal cavity.

CHAPTER 2 OBJECTIVES

After reading this chapter you should be able to

1. Define the terms matter, energy, element, and atom.
2. Describe, using diagrams, the structure of several common atoms. Include a description of the subatomic particles.
3. Explain ionic bonds, covalent bonds, polar covalent bonds, and hydrogen bonds.
4. Describe and give an example of each of the following reaction types: synthesis, decomposition, exchange, dehydration, hydrolysis, and reversible.
5. List the factors that affect the rate of chemical reactions.
6. Define potential and kinetic energy. Describe electrical, electromagnetic, chemical, and heat energy.
7. Distinguish between exergonic and endergonic reactions.
8. List the properties of water that make it important for living organisms.
9. Define solution, solvent, and solute. Describe two ways the concentration of a solute in a solvent can be expressed.
10. Define acid and base and differentiate between a strong acid or base and a weak acid or base.
11. Describe the pH scale and its relationship to acidity and alkalinity.
12. Define a salt and a buffer and state why they are important.
13. Explain the importance of oxygen and carbon dioxide to living organisms.
14. Describe the chemical structure of carbohydrates and state the role of carbohydrates in the body.
15. List and describe the importance of the major types of lipids.
16. Describe the different structural levels of proteins.
17. Define enzymes and state their functions.
18. Contrast the structure and function of deoxyribonucleic acid (DNA) and ribonucleic acid (RNA).
19. Explain the function of adenosine triphosphate (ATP).

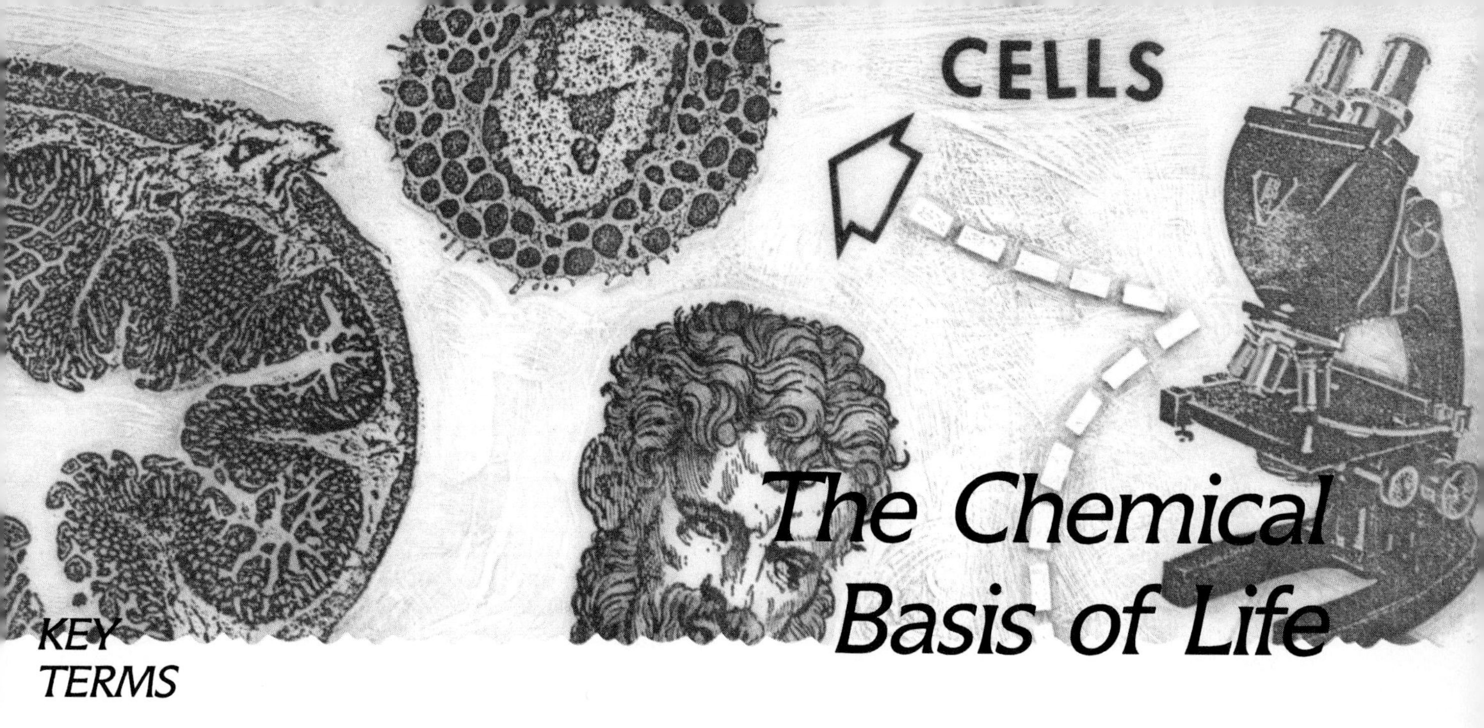

The Chemical Basis of Life

KEY TERMS

acid

atom

base

buffer

carbohydrate

covalent bond

endergonic (en-der-gon′ik)

enzyme (en′zīm)

exergonic (ek′ser-gon′ik)

ion

ionic bond

lipid

molecule

nucleic acid

protein

solution

RELATED TOPICS

The following concept is important for a good understanding of this chapter. If you are not familiar with it, you should review it before proceeding.

Levels of organization (Chapter 1)

CHEMISTRY IS THE SCIENTIFIC DISCIPLINE THAT DEALS WITH THE COMPOSITION AND STRUCTURE OF SUBSTANCES AND WITH THE REACTIONS THEY UNDERGO. A basic knowledge of chemical principles is essential for understanding anatomy and physiology. For example, the physiological processes of digestion, muscle contraction, and metabolism and the generation of nerve impulses can all be described in chemical terms. In addition, many abnormal conditions and their treatments can be explained in chemical terms, even though their symptoms are exhibited as malfunctions in organ systems. For example, Parkinson's disease, which has the symptom of uncontrolled shaking movements, results from the lack of an adequate amount of a chemical called dopamine in certain nerve cells of the brain. It is treated by giving patients another chemical that is converted to dopamine by brain cells.

This chapter outlines some basic chemical principles and emphasizes the relationship of these principles to living organisms. It is not a comprehensive review of chemistry, but it does review some of the basic chemical principles that make anatomy and physiology more understandable. You should refer to this chapter when chemical phenomena are discussed later in the text.

BASIC CHEMISTRY
Matter and Elements

All living and nonliving things are composed of **matter,** which is anything that occupies space. **Mass** is the amount of matter in an object, and **weight** is the gravitational force acting on an object of a given mass. For example, the weight of an apple results from the force of gravity "pulling" on the apple's mass.

1

The difference between mass and weight can be illustrated by considering an astronaut. How would an astronaut's mass and weight in outer space compare to his mass and weight on the earth's surface?

? ? ? ? ? ? ? ? ? ? ?

Matter is composed of **elements,** which are substances that cannot be broken down into simpler substances by ordinary chemical reactions. A list of the elements commonly found in the human body is found in Table 2-1. Approximately 96% of the weight of the body is due to the elements oxygen, carbon, hydrogen, and nitrogen.

There are 92 naturally occurring elements. Additional elements have been made artificially in the laboratory by smashing one element (or part of an element) into another element at very high speeds (near the speed of light). To date, 17 such elements are known, and the number can be expected to increase.

Atoms

The basic building blocks of elements are atoms. An **atom** is the smallest piece of an element that has the chemical characteristics of that element. An element is composed of atoms of only one kind. For example, the element carbon is composed of only carbon atoms, and the element oxygen is composed of only oxygen atoms.

An element, or an atom of that element, often is represented by a symbol. Usually the first letter or letters of the element's name are used—for example, C for carbon, H for hydrogen, Ca for calcium, and Cl for chlorine. Occasionally the symbol is taken from

Table 2-1

Some common elements

ELEMENT	SYMBOL	ATOMIC NUMBER	ATOMIC MASS	ATOMIC WEIGHT	AMOUNT IN HUMAN BODY BY WEIGHT (%)	AMOUNT IN HUMAN BODY BY NUMBER OF ATOMS (%)
Hydrogen	H	1	1	1.008	9.5	63.0
Carbon	C	6	12	12.011	18.5	9.5
Nitrogen	N	7	14	14.0067	3.3	1.4
Oxygen	O	8	16	15.999	65.0	25.5
Fluorine	F	9	19	18.9984	Trace	Trace
Sodium	Na	11	23	22.9898	0.2	0.3
Magnesium	Mg	12	24	24.305	0.1	0.1
Phosphorus	P	15	31	30.9738	1.0	0.22
Sulfur	S	16	32	32.06	0.3	0.05
Chlorine	Cl	17	35	35.453	0.2	0.03
Potassium	K	19	39	39.10	0.4	0.06
Calcium	Ca	20	40	40.08	1.5	0.31
Chromium	Cr	24	52	51.996	Trace	Trace
Manganese	Mn	25	55	54.9380	Trace	Trace
Iron	Fe	26	56	55.84	Trace	Trace
Cobalt	Co	27	59	58.9332	Trace	Trace
Copper	Cu	29	63	63.54	Trace	Trace
Zinc	Zn	30	64	65.37	Trace	Trace
Selenium	Se	34	80	78.96	Trace	Trace
Molybdenum	Mo	42	98	95.94	Trace	Trace
Iodine	I	53	127	126.9045	Trace	Trace

the Latin name for the element—for example, Na from the Latin word *natrium* is the symbol for sodium.

Atomic structure

The characteristics exhibited by living and nonliving matter result from the structure, organization, and behavior of its atoms. The three major types of subatomic particles that constitute atoms are **neutrons, protons,** and **electrons.** Neutrons have no electrical charge. Protons have positive charges, and electrons have negative charges equal in magnitude but opposite in polarity (charge). Because there are equal numbers of protons and electrons in an atom, the individual charges cancel each other, and the atom is electrically neutral.

Protons and neutrons organize in atoms to form a central **nucleus,** and the electrons move around it (Figure 2-1). Each electron constantly orbits the nucleus. It is impossible, however, to know precisely where any given electron is located at any particular moment, although the region in which it is most likely to be found can be specified. This region, known as the electron's **orbital,** is designated by its shape and its distance from the nucleus. Orbitals are grouped together to form **electron shells,** which can contain more than one electron.

Atoms can be represented in several ways (Figure 2-2). Commonly, atoms are diagrammed as a round nucleus surrounded by concentric circles that represent electron shells. Hydrogen has a single electron in the first shell, carbon has two electrons in the first shell and four electrons in the second shell, and oxygen has two electrons in the first shell and six electrons in the second shell.

Atomic number and mass number

The **atomic number** of an element is equal to the number of protons in each atom, and because the number of electrons and protons is equal, the atomic number also indicates the number of electrons. Each element is uniquely defined by the number of protons in the atoms of that element. For example, only hydrogen atoms have one proton, and only carbon atoms have six protons (see Table 2-1).

Protons and neutrons have approximately the same mass, and they are responsible for most of the mass of atoms; electrons, on the other hand, have very little mass. The **mass number** of an element equals the number of protons plus the number of

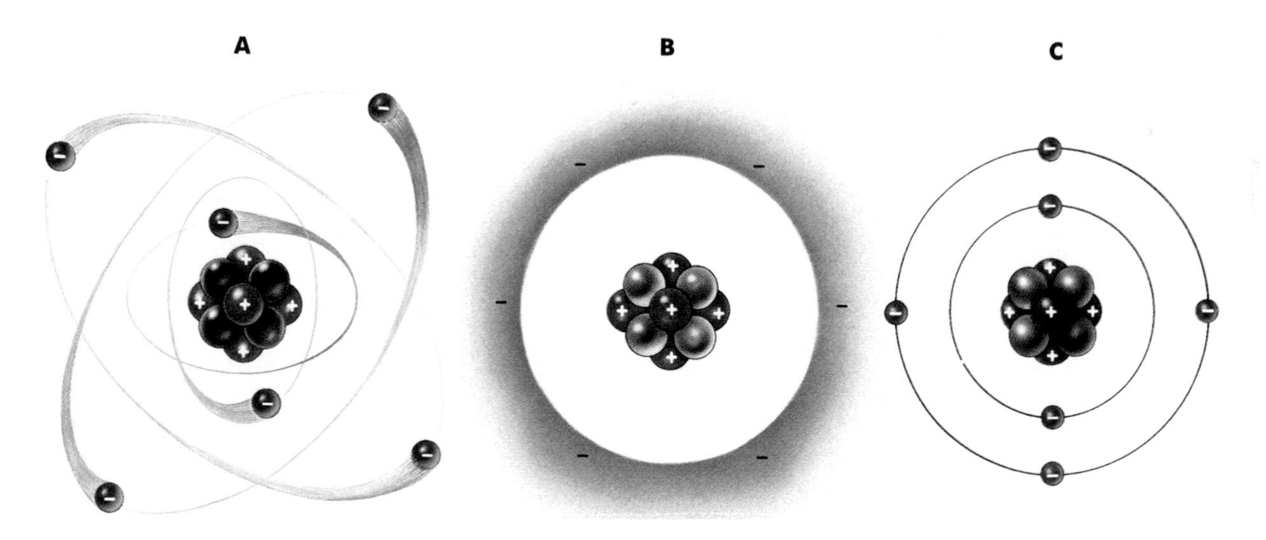

FIGURE 2-1 Three models of an atom (carbon). **A** Historic model of electrons orbiting the nucleus like a minute solar system. **B** More modern model demonstrating that the location of an electron at any point in time is a probability estimate. The frequency of the dots represents the relative probability that the electron is in a given location. **C** Model of concentric circles showing relative distances (orbitals) of the electrons from the nucleus.

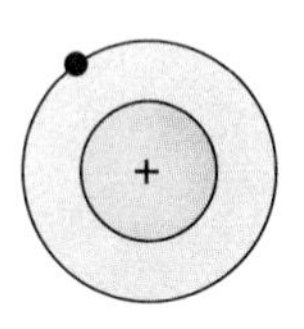

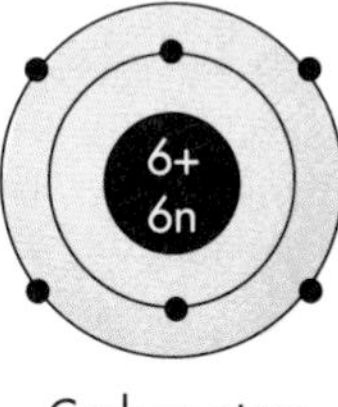

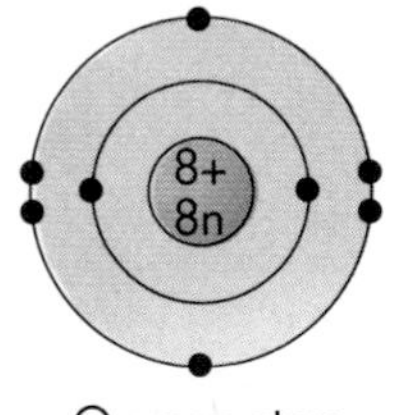

FIGURE 2-2 Hydrogen, carbon, and oxygen atoms. Within the nucleus, the number of protons (+) and neutrons (*n*) is indicated. The electrons are shown as small black dots. Note that the number of protons and the number of electrons within an atom are equal to each other.

neutrons in each atom. For example, the mass number for carbon is 12 because it has six protons and six neutrons. The number of protons, neutrons, and electrons in an atom determine the chemical characteristics of each atom and the kinds of chemical bonds the atoms form with other atoms.

2

The atomic number of potassium is 19, and the mass number is 39. What is the number of protons, neutrons, and electrons in an atom of potassium?

? ? ? ? ? ? ? ? ? ? ?

Atomic weight

Individual atoms have very little weight, and it takes a large number of atoms to produce a weight that is easily measured. The number of atoms used, 6.023×10^{23}, is called **Avogadro's number** (see Appendix C for an explanation of the scientific notation of numbers). The **atomic weight** of an element is the weight in grams of Avogadro's number of atoms of that element. For example, the atomic weight of Avogadro's number of hydrogen atoms is 1.008 g, and the atomic weight of Avogadro's number of carbon atoms is 12.011 g (see Table 2-1).

A **mole** is defined as Avogadro's number of atoms. Therefore 1 mole of an element is the atomic weight of the atom in grams. Just as a grocer sells eggs in lots of 12, that is, a dozen, a chemist groups atoms in lots of 6.023×10^{23} (Avogadro's number), that is, a mole. The mole is used as a convenient way to determine the number of atoms in a sample of an element. For example, 1.008 g of hydrogen (1 mole) has the same number of atoms as 12.011 g of carbon (1 mole).

3

Are there more or fewer atoms in 12 g of carbon than in 12 g of magnesium? (HINT: Use Table 2-1.)

? ? ? ? ? ? ? ? ? ? ?

Electrons and Chemical Bonds

The chemical behavior of an atom is determined largely by the electrons in its outermost electron shell. **Chemical bonds** are formed when the outermost electrons are transferred or shared between atoms. The resulting combination of atoms is called a **molecule.** If a molecule has two or more different kinds of atoms, it may be referred to as a **compound.** A molecule can be represented by a **molecular formula,** which consists of the symbols of the atoms in the molecule plus a subscript denoting the number of each type of atom. For example, the molecular formula for glucose (a sugar) is $C_6H_{12}O_6$, indicating that glucose has 6 carbon, 12 hydrogen, and 6 oxygen atoms.

When atoms combine into molecules, the outermost electron shell of each atom becomes **stable,** or **complete.** The first shell is complete when it contains two electrons, and the second is complete when it has eight electrons. Although the third electron shell has the capacity of holding 18 electrons, it is most stable when it contains eight. For larger atoms additional electron shells exist, each of which contains an additional number of electrons.

Chemical bonds can be grouped into four categories: ionic, covalent, polar covalent, and hydrogen bonds. Table 2-2 summarizes the important characteristics of these bonds.

Ionic bonds

Ionic bonds result when one atom loses an electron and another atom accepts that electron. For example, sodium can lose an electron that can be accepted by chlorine. Sodium has one electron in its outermost (third) shell and eight electrons in its second shell (Figure 2-3). If sodium could gain seven electrons, its outermost shell would be complete. However, it is easier for sodium to lose its outermost electron, in which case its second shell with eight electrons would become its complete outermost shell. Chlorine has seven electrons in its outermost third shell, and if it would accept the electron from sodium, chlorine's outermost shell would be complete with eight electrons. Thus by transferring an electron from sodium to chlorine, both atoms achieve complete outermost electron shells.

An atom is electrically neutral because it has an equal number of protons and electrons. After an atom donates an electron, it has one more proton than it has electrons and is positively charged. After an atom accepts a donated electron, it has one more electron than it has protons and is negatively charged. These charged atoms are called **ions** (i'ons). Positively charged ions are called **cations** (kat'i-ons), and negatively charged ions are called **anions** (an'i-ons). Because oppositely charged ions are attracted to each other, cations and anions tend to remain close together. The bonds that result from this close attraction are called ionic bonds. Some important ions are listed in Table 2-3.

4

Calcium atoms have two electrons in their outermost shell. Explain how calcium atoms form ionic bonds with chlorine atoms, which have seven electrons in their outermost shell.

? ? ? ? ? ? ? ? ? ? ?

Table 2-2

Comparison of chemical bonds

DEFINITION	CHARGE DISTRIBUTION	EXAMPLE
Ionic Bond Complete transfer of electrons between atoms	Separate positively charged and negatively charged ions	Na^+Cl^- Sodium chloride
Nonpolar Covalent Bond Equal sharing of electrons between atoms	Charge evenly distributed among the atoms of the molecule	H—C—H (with H above and below C) Methane
Polar Covalent Bond Unequal sharing of electrons between atoms	Slight positive charge on one end of the molecule and a slight negative charge on the other end of the molecule	δ^+H—$O\delta^-$—δ^+H Water
Hydrogen Bond Attraction of oppositely charged ends of one polar molecule to another polar molecule	Charge distribution within the polar molecules is due to polar covalent bonds	δ^+H, δ^+H—$O\delta^-$ - - - - δ^+H—$O\delta$—δ^+H Water molecules

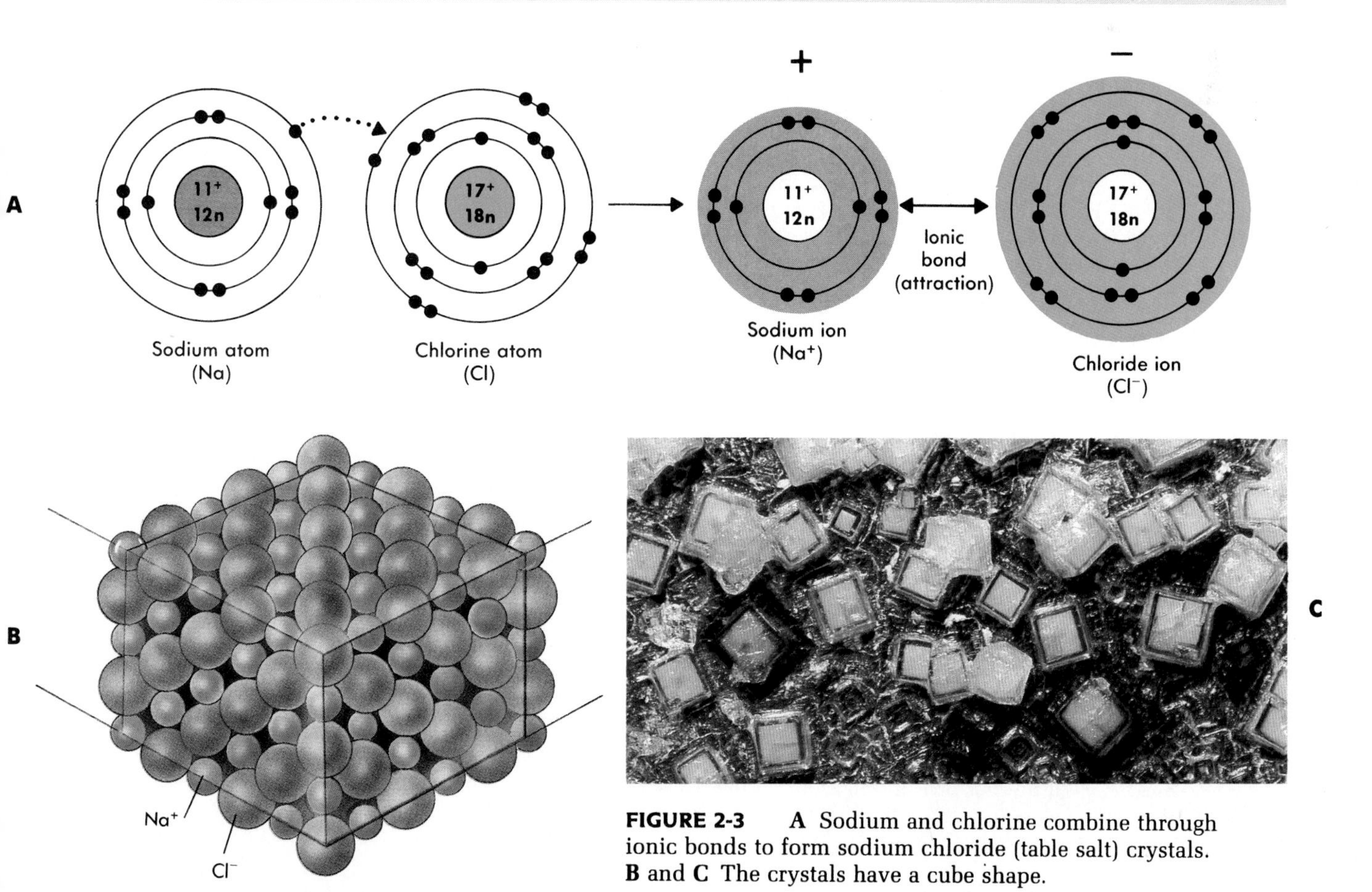

FIGURE 2-3 **A** Sodium and chlorine combine through ionic bonds to form sodium chloride (table salt) crystals. **B** and **C** The crystals have a cube shape.

Radioactive Isotopes and X-rays

Isotopes are elements having the same atomic numbers but different mass numbers; thus the number of protons and electrons are equal, but the number of neutrons in the nucleus is different for each isotope of a given element. For example, there are three isotopes of hydrogen: hydrogen, deuterium, and tritium. All three isotopes have one proton and one electron, but hydrogen has no neutrons in its nucleus, deuterium has one neutron, and tritium has two neutrons (Figure 2-A). Isotopes can be denoted using the symbol of the element preceded by the isotope's mass number (number of protons and neutrons). Thus hydrogen is ^{1}H, deutrium is ^{2}H, and tritium is ^{3}H.

Radioactive isotopes have unstable nuclei that spontaneously change to form more stable nuclei. As a result, either new isotopes or new elements are produced. In the process of nuclear change three kinds of rays, called alpha, beta, and gamma rays, are emitted from the nuclei of radioactive isotopes. Alpha rays are positively charged helium ions (He^{2+}) with a mass of 4 and a charge of $+2$ (i.e., two protons and two neutrons). Beta rays are electrons formed as neutrons change into protons. The electrons are ejected from the nuclei, and the protons remain in the nuclei. Gamma rays are a form of electromagnetic radiation released from nuclei as they lose energy.

Because all isotopes of an element have the same atomic number, their chemical behavior is very similar. For example, tritium can substitute for hydrogen, and either ^{125}iodine or ^{131}iodine can substitute for ^{126}iodine in chemical reactions.

Radioactive isotopes are commonly used by clinicians and researchers because sensitive measuring devices can detect the radioactive rays emitted from isotopes even when they are present in very small amounts. Several procedures used to determine the concentration of substances such as hormones depend on the incorporation of small amounts of radioactive isotopes such as ^{125}iodine into the substances being measured. Disorders of the thyroid gland, the adrenal gland, and the reproductive organs can be diagnosed more accurately using these procedures.

Radioactive isotopes are also used to treat cancer. Some of the particles released from isotopes have a very high energy content and can penetrate and destroy tissues; thus radioactive isotopes can be used to destroy rapidly growing tumors, which are more sensitive to radiation than healthy cells. Radiation can also be used to sterilize materials that cannot be exposed to high temperature. Some fabric and plastic items used during surgical procedures, for example, are sterilized by exposure to radioactive isotopes. In addition, radioactive emissions can provide a convenient and safe method of sterilizing food and other items.

X-rays are electromagnetic radiations with a much shorter wavelength than visible light. When electric current is used to heat a filament to very high temperatures, the energy of the electrons becomes so great that some electrons are emitted from the hot filament. When these electrons strike a positive electrode, they release some of their energy in the form of x-rays.

X-rays do not penetrate dense material as readily as they penetrate less dense material, and x-rays will expose photographic film. Consequently, an x-ray beam can pass through a person and onto photographic film. Dense tissues of the body will absorb the x-rays; in these areas the film will be underexposed and will appear white or light in color on the developed film. The x-rays will pass readily through less dense tissue, on the other hand, and the film in these areas will be overexposed and will appear black or dark in color. For example, in an x-ray of the skeletal system the dense bones are white, and the less dense soft tissues are dark, often so dark that no details can be seen. Because the dense bone material is clearly visible, x-rays can be used to determine if bones are broken or have other abnormalities.

Photographing soft tissues can be accomplished using low-energy x-rays and radiopaque substances, which are dense materials that absorb x-rays. Mammograms are low-energy x-rays of the breast that can be used to detect tumors because tumors are slightly more dense than normal tissue. If a radiopaque liquid is given to a patient, the liquid will assume the shape of the organ into which it is placed. For example, if a barium solution is swallowed, the outline of the upper digestive tract can be photographed using x-rays to detect abnormalities such as ulcers.

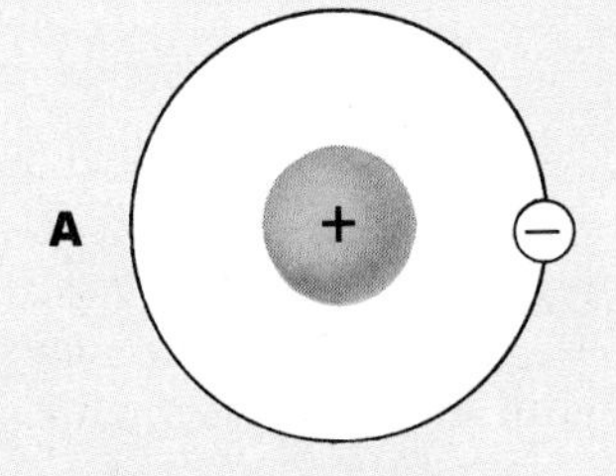

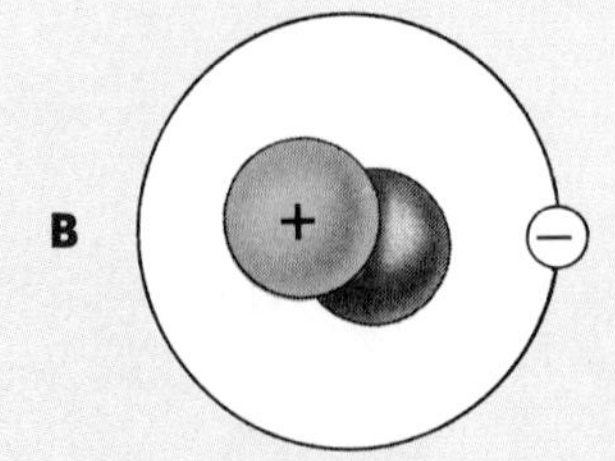

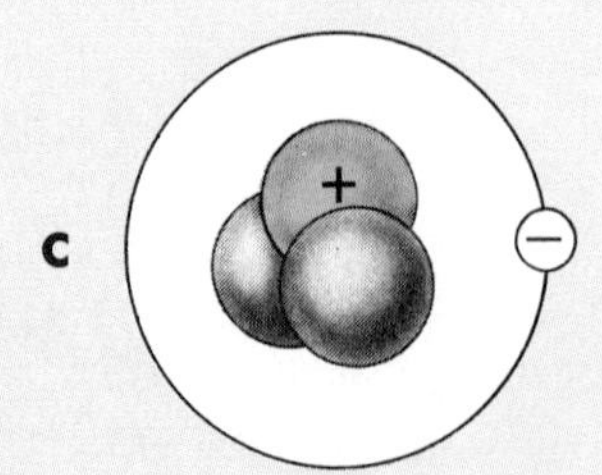

FIGURE 2-A Isotopes of hydrogen. **A** Hydrogen has one proton and no neutrons in the nucleus. **B** Deuterium has one proton and one neutron in its nucleus. **C** Tritium has one proton and two neutrons in its nucleus.

Table 2-3

Important ions

COMMON IONS	SYMBOL	FUNCTION
Calcium	Ca^{2+}	Bones, teeth, blood clotting, muscle contraction
Sodium	Na^{+}	Membrane potentials, water balance
Potassium	K^{+}	Membrane potentials
Hydrogen	H^{+}	Acid-base balance
Hydroxide	OH^{-}	Acid-base balance
Chloride	Cl^{-}	Acid-base balance
Bicarbonate	HCO_3^{-}	Acid-base balance
Ammonium	NH_4^{+}	Acid-base balance
Phosphate	PO_4^{3-}	Bone, teeth, energy exchange, acid-base balance
Iron	Fe^{2+}	Red blood cell formation
Magnesium	Mg^{2+}	Necessary for enzymes
Iodide	I^{-}	Present in thyroid hormones

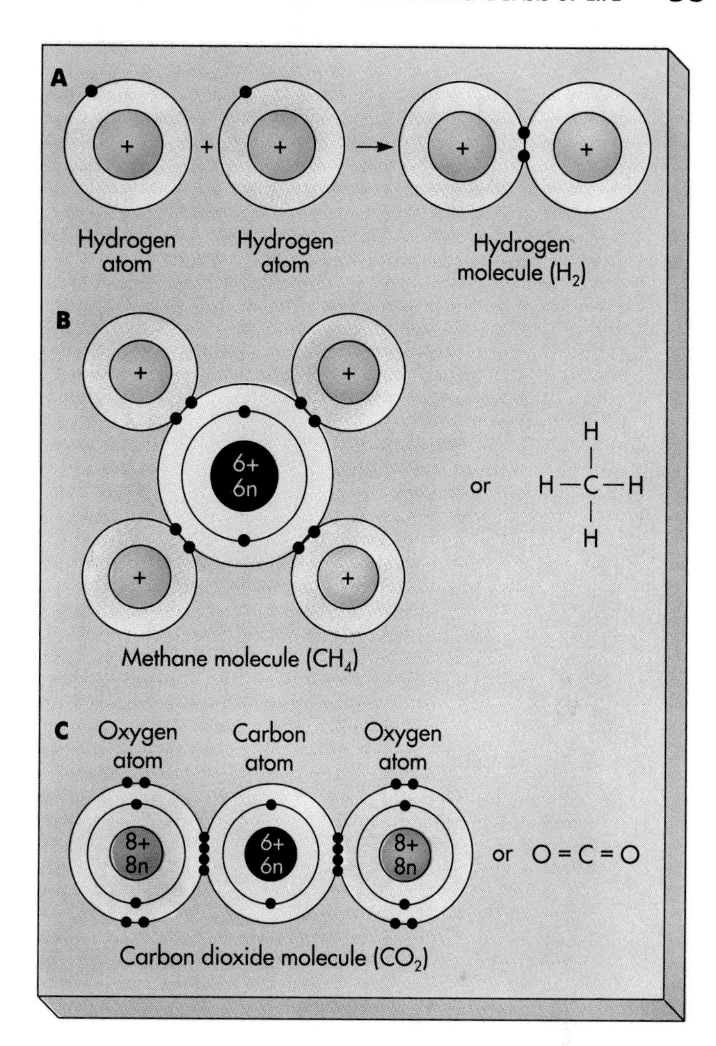

FIGURE 2-4 **A** Single covalent bond between two hydrogen atoms results in a hydrogen molecule (hydrogen gas). Each atom shares two electrons. **B** Single covalent bonds between four hydrogen atoms and one carbon atom form methane (a highly flammable gas). The four electrons in the outer shell of the carbon atom and the single electron of each hydrogen atom are shared so that the carbon atom has eight electrons in its outer shell and each hydrogen atom has two electrons in its outer shell. **C** Double covalent bonds between a carbon atom and two oxygen atoms to produce the gas carbon dioxide. Carbon shares four electrons with each of the oxygen atoms.

Cations and anions are sometimes referred to as **electrolytes** because they have the capacity to conduct electrical current when they are in solution. An electrocardiogram (ECG) is a recording of the electrical currents produced by the heart. These currents can be detected by electrodes on the surface of the body because the ions in the body fluids conduct the electrical currents.

Covalent bonds

Covalent bonds result when two or more atoms complete their electron shells by sharing electrons. When an electron pair is shared between two atoms, a **single covalent bond** results. An example is the bond between two hydrogen atoms to form a hydrogen molecule. Each hydrogen atom has one electron and requires one more electron to complete its outer electron shell. Hydrogen atoms can form a single covalent bond by sharing their electrons, allowing the two electrons to orbit around both of the nuclei (Figure 2-4, *A*). A single covalent bond can be represented by a single line between the symbols of the atoms involved (e.g., H—H).

Coavalent bonds are very common in the molecules that constitute living matter. Carbon readily forms single covalent bonds with hydrogen. Carbon has four electrons in its outer shell, requiring an additional four electrons to complete it. A single carbon atom can share electrons with four hydrogen atoms to complete the second electron shell of the carbon atom and the first electron shell of the four hydrogen atoms (Figure 2-4, *B*). A carbon atom can also form single covalent bonds with other atoms such as carbon, oxygen, nitrogen, and sulfur.

A **double covalent bond** results when two atoms share two electron pairs. When carbon combines with oxygen atoms to form carbon dioxide, two double covalent bonds are formed (Figure 2-4, *C*). Double covalent bonds are indicated by a double line between the atoms (e.g., O = C = O).

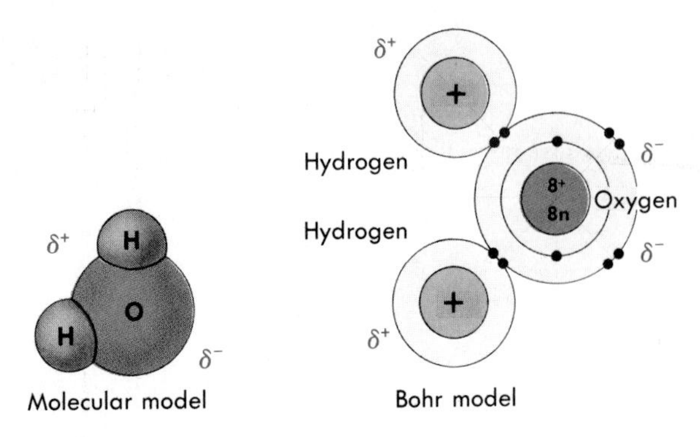

FIGURE 2-5 Two diagrams of a polar covalently bound molecule (water). The two hydrogen atoms are nearer one end of the molecule, giving that end a partial positive charge (indicated by δ^+). The opposite end of the molecule has a partial negative charge (indicated by δ^-).

Complex molecules result from the ability of carbon to form four covalent bonds. A series of carbon atoms bound together constitute the "backbone" of many large molecules. Variation in the length of the carbon chains and the combination of atoms bound to the carbon backbone allow the formation of a wide variety of molecules. For example, some protein molecules have thousands of carbon atoms bound by covalent bonds to one another or to other atoms such as nitrogen, sulfur, hydrogen, and oxygen. Without the ability of carbon to form covalent bonds, the complex molecules that are common to living organisms could not exist.

When electrons are shared equally between atoms, as in carbon dioxide, the bonds are called **nonpolar covalent bonds.** Atoms bound to one another by a covalent bond, however, do not always share their electrons equally. Bonds of this type are called **polar covalent bonds** and are common in both living and nonliving matter. For example, oxygen atoms attract electrons more strongly than do hydrogen atoms. In water an oxygen atom and two hydrogen atoms are bound by covalent bonds, but the electrons orbit in the vicinity of the oxygen nucleus more than in the vicinity of the hydrogen nuclei. Because electrons have a negative charge, the area around the oxygen nucleus is slightly more negative than the area around the hydrogen nuclei (Figure 2-5 and Table 2-2).

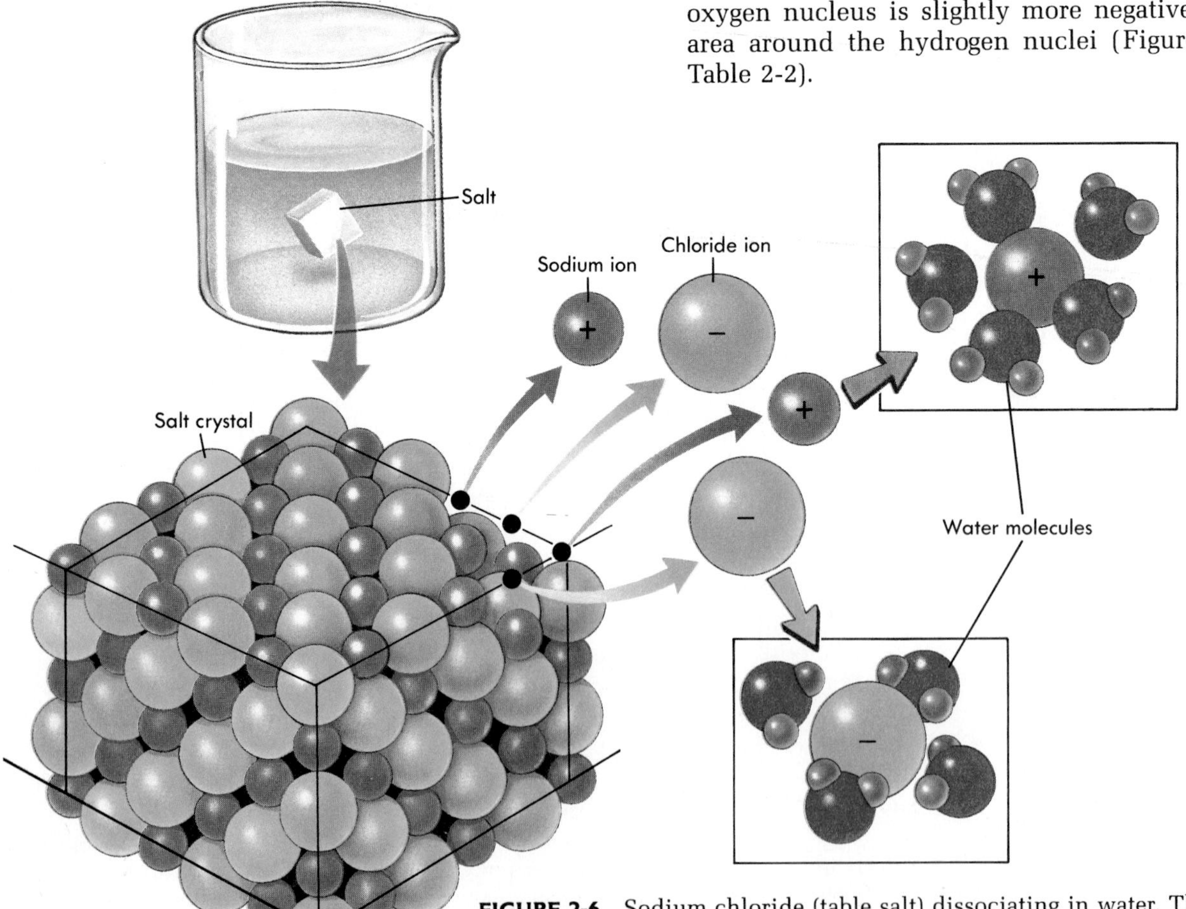

FIGURE 2-6 Sodium chloride (table salt) dissociating in water. The positively charged sodium ions (Na^+) are attracted to the negative oxygen *(red)* end of the water molecule, and the negatively charged chlorine ions (Cl^-) are attracted to the positively charged hydrogen *(blue)* end of the water molecule.

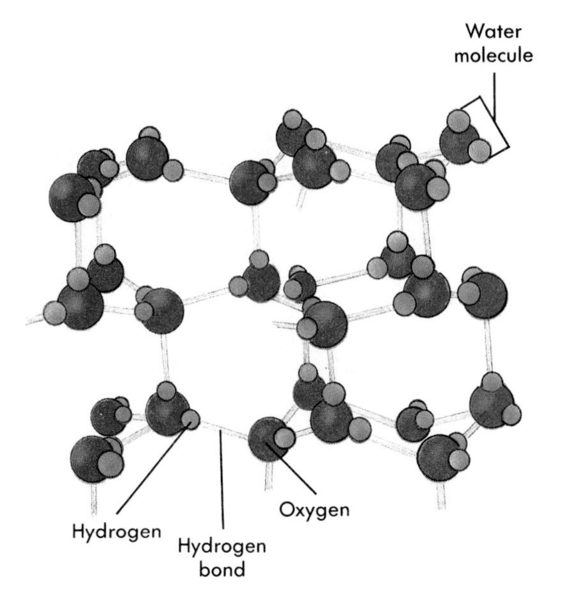

FIGURE 2-7 Hydrogen bonds between water molecules. The positive hydrogen part of one water molecule forms a hydrogen bond with the negative oxygen part of another water molecule.

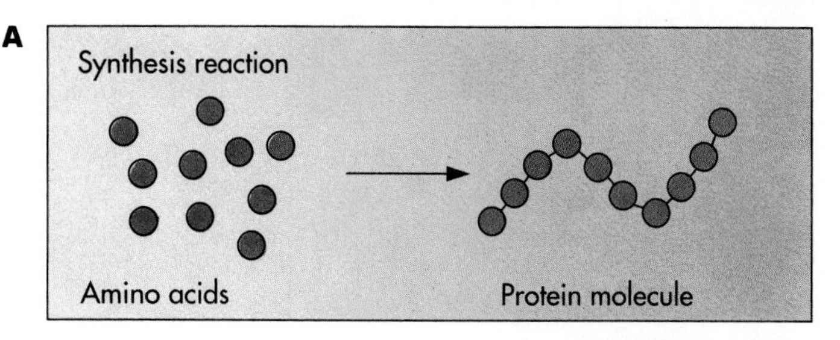

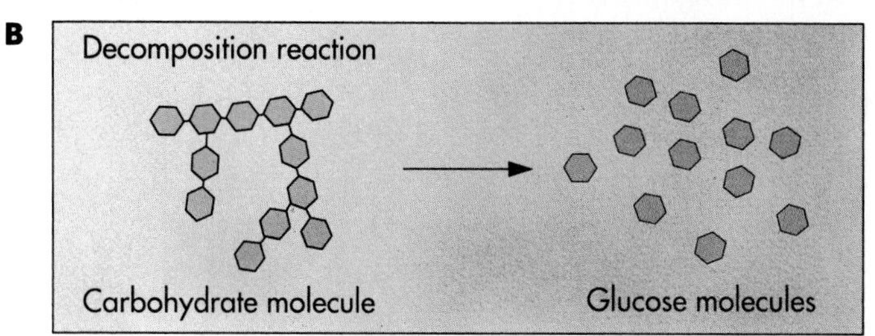

FIGURE 2-8 **A** Synthesis reaction in which amino acids combine to form a protein molecule. **B** Decomposition reaction in which a carbohydrate molecule breaks down into smaller glucose molecules.

Polar and charged substances dissolve in water readily, and nonpolar substances such as oils do not. Because water molecules contain both a partially negative region and a partially positive region, they attract and surround ions and other polar molecules. When ionic substances dissolve in water, the cations and anions separate, or **dissociate,** and water molecules surround and isolate the ions, keeping them in solution (Figure 2-6).

Hydrogen bonds

Molecules with polar covalent bonds exhibit weak attractions for each other. The slight positive charge on one molecule is weakly attracted to the slight negative charge on another molecule. For example, hydrogen atoms that are bound covalently to either oxygen or nitrogen atoms have a small positive charge and are weakly attracted to the small negative charge associated with the atoms (e.g., oxygen or nitrogen) of other molecules. These attractions are referred to as **hydrogen bonds** (Figure 2-7). Hydrogen bonds play an important role in determining the shape of complex molecules (see sections about proteins and nucleic acids later in this chapter) because the hydrogen bonds between different polar parts of the molecule function to hold the molecule together.

CHEMICAL REACTIONS

A **chemical reaction** is the process by which atoms, ions, or molecules interact either to form or to break chemical bonds. The substances that enter into a chemical reaction are called the **reactants,** and the substances that result from the chemical reaction are called the **products.**

Classification of Chemical Reactions

Chemical reactions are classified as synthesis, decomposition, or exchange reactions. When two or more atoms, ions, or molecules combine to form a new and larger molecule, the process is called a **synthesis reaction.** The molecules characteristic of life, such as proteins, carbohydrates, lipids, and nucleic acids, are produced by synthesis reactions. For example, smaller molecules called amino acids combine to form a larger protein molecule (Figure 2-8, *A*). All of the synthesis reactions that occur within the body are referred to collectively as **anabolism.** The growth, maintenance, and repair of the body could not take place without anabolic reactions.

Decompose means to break down into smaller parts. A **decomposition reaction** is the reverse of a synthesis reaction—larger molecules are broken down to form smaller molecules, ions, or atoms. The breakdown of carbohydrates into smaller molecules called glucose is an example (Figure 2-8, *B*). All of the decomposition reactions that occur in the body are collectively called **catabolism.** They include the digestion of food molecules in the intestine and within cells, the breakdown of fat stores, and the breakdown of foreign matter and microorganisms in certain blood cells that function to protect the body. All of the anabolic and catabolic reactions in the body are collectively defined as **metabolism.**

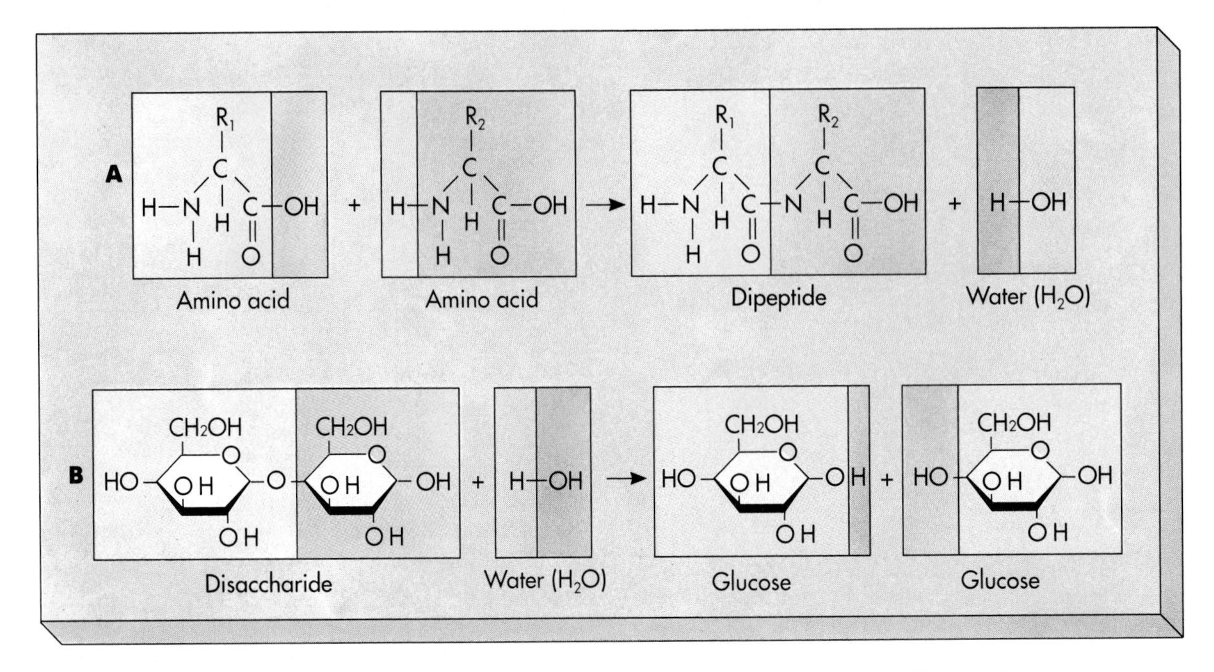

FIGURE 2-9 **A** A synthesis reaction in which two amino acids combine to form a dipeptide. It is also an exchange reaction because one amino acid exchanges parts with another amino acid. The reaction also results in the removal of a water molecule from the amino acids, and it is therefore a dehydration reaction. **B** Decomposition reaction in which a disaccharide breaks apart to form glucose molecules. It is also an exchange reaction because the disaccharide exchanges parts with the water molecule. Because the reaction uses a water molecule, it is also a hydrolysis reaction.

Atoms usually are bound chemically to other atoms to form molecules. Consequently, to form different molecules, part of one molecule is exchanged for part of another molecule. This is called an **exchange reaction,** which can be represented symbolically as follows:

$$AB + CD \rightarrow AD + CB$$

where parts B and D of the molecules are exchanged.

Some synthesis and decomposition reactions can also be exchange reactions. Two amino acids joined together to form a dipeptide are an example (Figure 2-9, *A*). In this particular synthesis reaction water is removed from the amino acids as they are bound together. Synthesis reactions in which water is a product are called **dehydration** (water out) **reactions.**

An example of a decomposition reaction that is also an exchange reaction is the separation of a disaccharide (a type of carbohydrate) into glucose molecules (Figure 2-9, *B*). Note that this particular decomposition reaction requires that water be split into two parts. Consequently, decomposition reactions that require water are called **hydrolysis** (water dissolution) **reactions.** Figure 2-9 illustrates that a single chemical reaction can be defined in several different ways.

Reversible Reactions

Some synthesis, decomposition, or exchange reactions are reversible; the reaction can proceed from reactants to products and from products to reactants. When the rate of product formation is equal to the reverse reaction, the relationship is said to be at equilibrium. At **equilibrium** the ratio of the products to the reactants tends to remain constant. Thus if additional reactants are added to a reaction mixture, some will form product until the original ratio of products to reactants is reestablished. For example, the reaction between carbon dioxide (CO_2) and water (H_2O) to form carbonic acid (H_2CO_3) is reversible:

$$CO_2 + H_2O \rightleftharpoons H_2CO_3$$

Carbonic acid then separates by a reversible reaction to form hydrogen ions (H^+) and bicarbonate ions (HCO_3^-):

$$H_2CO_3 \rightleftharpoons H^+ + HCO_3^-$$

If carbon dioxide is added to water, additional carbonic acid forms, which, in turn, causes more hydrogen ions and bicarbonate ions to form. Therefore the ratio of hydrogen and bicarbonate ions to carbon dioxide remains constant. Maintaining a constant level of hydrogen ions is necessary for proper functioning of the nervous system. This can be achieved, in part, by regulating blood carbon dioxide levels. For example, slowing down the respiration rate would cause blood carbon dioxide levels to increase.

5

If the respiration rate increases, carbon dioxide is eliminated from the blood. What effect would this change have on blood hydrogen ion levels?

? ? ? ? ? ? ? ? ? ? ?

Rate of Chemical Reactions

The rate at which a chemical reaction proceeds is influenced by several factors, including how easily the substances react with one another, their concentration, the temperature, and the presence of a catalyst.

Substances differ in their ability to react with other substances. Iron, for example, reacts slowly with oxygen to form rust. The components of dynamite, on the other hand, react violently with each other in a fraction of a second.

Within limits, the greater the concentration of the reactants, the greater is the rate at which a given chemical reaction will proceed. This occurs because as the concentration of molecules increases, they are more likely to come into contact with one another. For example, the normal concentration of oxygen inside cells enables oxygen to come into contact with other molecules, producing the chemical reactions necessary for life. If the oxygen concentration decreases, the rate of chemical reactions decreases. This decrease can impair cell function and even result in death.

The speed of chemical reactions also increases when the temperature is increased. As temperature increases, molecules move at greater speeds, and they collide with each other more frequently and with greater force, increasing the likelihood of a chemical reaction. When a person has a fever of only a few degrees, reactions occur throughout the body at an accelerated rate, resulting in increased activity in the organ systems (e.g., increased heart and respiratory rates). When body temperature drops, various metabolic processes slow. The sluggish movement of very cold fingers results largely from the reduced rate of chemical reactions in cold muscle tissue.

At normal body temperatures most chemical reactions would proceed very slowly if it were not for the action of the body's enzymes. **Enzymes** (en'zīmz) are protein molecules in the body that act as catalysts. A **catalyst** (ka'tă-list) is a substance that increases the rate at which a chemical reaction proceeds without itself being permanently changed or depleted. Many of the chemical reactions that occur in the body require enzymes, and the regulation of chemical events in cells is due primarily to mechanisms that control either the concentration or the activity of enzymes. Enzymes are considered in greater detail later in this chapter.

ENERGY

Energy, unlike matter, does not occupy space and has no mass. **Energy** is defined as the capacity to do **work,** that is, to move matter. Energy can be subdivided into potential energy and kinetic energy. **Potential energy** is stored energy that could do work but is not doing so. For example, a coiled spring stores potential energy. It could push against an object and move the object, but as long as the spring does not uncoil, no work is accomplished. **Kinetic energy** is energy caused by an object's moving and is the form of energy that actually does work. An uncoiling spring pushing an object causing it to move is an example. Thus when potential energy is released, it becomes kinetic energy, which does work.

Potential and kinetic energy can be found in many different forms. Of particular interest to the study of human organisms are electrical, electromagnetic, chemical, and heat energy.

Electrical Energy

Electrical energy involves the movement of ions or electrons. Examples are nerve impulses and the electric current supplying a light bulb. A nerve impulse, which functions to carry messages from one part of the body to another, results from the movement of ions across cell membranes. Nerve impulses are discussed in Chapter 9.

Electromagnetic Energy

Electromagnetic energy is energy that moves in waves. Energy waves of different wavelength comprise the electromagnetic spectrum. The parts of the electromagnetic spectrum, listed from the shortest to the longest wavelengths, and some of their important uses include the following: gamma rays (radiation therapy), x-rays (medical examination), ultraviolet light (stimulation of vitamin D production, responsible for sunburn), visible light (activation of chemicals in the eye, resulting in vision), infrared radiation (loss or gain of heat), microwaves (electrical appliances, radar), and radio waves (radio, television, nuclear magnetic resonance for medical examination).

Chemical Energy

Chemical energy is a form of potential energy within the electrons forming a chemical bond. Electrons within an atom have potential energy caused by the attraction of the negatively charged electrons to the positively charged protons; the greater the distance of an electron from the nucleus, the greater is its potential energy. This is analogous to an object's potential energy at different distances from the earth's surface. Consider the effects of dropping a glass from a table to the ground vs. dropping a glass from a 15-story building. The potential energy in the glass on the table, which is close to the ground, is small. When the glass falls to the ground and that potential energy is released, the glass may or may not break. On the other hand, the potential energy in the glass on top of the 15-story building is much larger, as can be demonstrated by dropping the glass and releasing that energy. The glass surely will be smashed to pieces.

The electron shells of an atom are located at increasing distances from the nucleus, and the amount of potential energy in an electron is determined by the electron shell in which the electron is found. Electrons within the same electron shell have virtually the same amount of potential energy. Electrons with higher potential energy levels are found in shells located farther from the nucleus than electrons with lower potential energy levels, which are found in electron shells located closer to the nucleus. The electrons in the outermost shell have the most potential energy, and these electrons are the ones most commonly involved in chemical reactions.

A chemical bond is a form of potential energy. **Exergonic** (ek'ser-gon'ik) **reactions** release energy because the products of the chemical reactions contain less potential energy than the reactants (Figure 2-10, *A*). Some of the energy released is used to drive processes such as muscle contraction or to produce new molecules, and some of the energy is released as heat.

In **endergonic** (en-der-gon'ik) **reactions** the products contain more potential energy than the reactants (Figure 2-10, *B*). Therefore these reactions require the input of energy from another source. The energy released by exergonic reactions during the breakdown of food molecules is the energy source for endergonic reactions in the body. The synthesis of the molecules found in living organisms is the result of endergonic reactions. In general, catabolism (decomposition) involves exergonic reactions that provide the energy to drive the endergonic reactions of anabolism (synthesis).

The energy that makes almost all life on earth possible ultimately comes from the sun. Plants, in the process of photosynthesis, capture the energy in sunlight. Through endergonic reactions, plants convert the sun's energy into chemical bonds in the sugar glucose. Plants and organisms that eat plants break down glucose through exergonic reactions. The released energy fuels the chemical reactions of life.

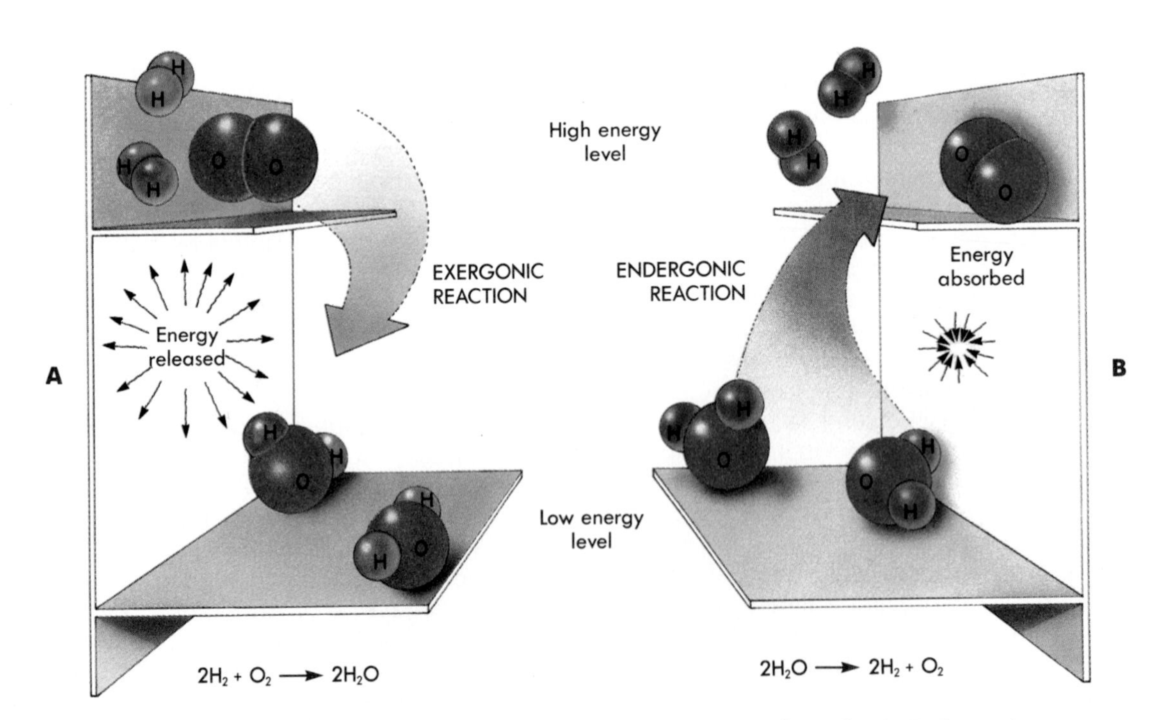

FIGURE 2-10 **A** Exergonic reaction in which energy is released. **B** Endergonic reaction in which energy is required for the reaction to proceed. In each figure the upper shelf represents a higher energy state, and the lower shelf represents a lower energy state.

Heat Energy

Heat energy results from the random movement of atoms, ions, or molecules; the faster the movement, the greater is the heat energy. The temperature of an object is a measurement of how much heat energy it contains; the higher the temperature, the greater is the heat energy.

All other forms of energy can be converted into heat energy. For example, when a moving object comes to rest, its kinetic energy is converted into heat energy by friction. The potential energy in chemical bonds can also be released as heat energy during exergonic reactions. The body temperature of humans is maintained by heat produced in this fashion.

6

Why does body temperature increase during exercise?

? ? ? ? ? ? ? ? ? ? ?

INORGANIC MOLECULES

Originally it was believed that inorganic molecules were those that came from nonliving sources and organic molecules were those extracted from living organisms. However, as the science of chemistry developed, it became apparent that organic molecules could be manufactured in the laboratory so that the original definitions were no longer valid. Currently, **inorganic molecules** are defined as those that do not contain carbon, and **organic molecules** are molecules that do contain carbon. These definitions have exceptions. For example, carbon dioxide and carbon monoxide are classified as inorganic molecules.

Water

Approximately 60% to 80% of most cells' volume is water; and plasma, the liquid portion of blood, is 92% water. A molecule of **water** is composed of one atom of oxygen joined to two atoms of hydrogen by covalent bonds. Because the oxygen atom attracts electrons more strongly than do the hydrogen atoms, water molecules are polar, with a partial positive charge on the hydrogen atoms and a partial negative charge on the oxygen atom. Hydrogen bonds form between the positively charged hydrogen atoms of one water molecule and the negatively charged oxygen atoms of another water molecule. These hydrogen bonds constitute the major force that holds water molecules together as a liquid (see Figures 2-5 and 2-7).

Water has physical and chemical properties well suited for the many purposes it serves in living organisms.

Stabilizing body temperature. Water has a high **specific heat,** meaning that a relatively large amount of heat is required to raise its temperature; therefore it tends to resist large temperature fluctuations. When water evaporates, it changes from a liquid to a gas, and because heat is required for that process, the evaporation of water from the surface of the body rids the body of excess heat.

Protection. Water is an effective lubricant that provides protection against damage resulting from friction. For example, tears protect the eye's surface from the rubbing of the eyelids. In addition, water forms a fluid cushion around organs and within joints that helps to protect them from trauma. Cerebrospinal fluid around the brain is an example.

Chemical reactions. The chemical reactions necessary for life do not take place unless the reacting molecules are dissolved in water. For example, sodium chloride must dissociate in water into sodium and chloride ions before they can react with other ions. Water also directly participates in many chemical reactions. A dehydration reaction is a synthesis reaction in which water is produced, and a hydrolysis reaction is a decomposition reaction in which a water molecule is depleted (see Figure 2-9).

Mixing medium. Water can mix with other substances to form solutions, suspensions, and colloids. A **solution** is any liquid that contains dissolved substances. For example, sweat is a solution in which sodium chloride and other substances are dissolved in water. A **suspension** is a liquid that contains nondissolved materials that will settle out of the liquid unless it is continually shaken. Red blood cells in plasma provide an example of a suspension. A **colloid** (kol'loyd) is liquid that contains nondissolved materials that do not settle out of the liquid. Water and proteins inside cells form a colloid.

In living organisms the fluids inside and outside cells are complex, consisting of solutions, suspensions, and colloids. The ability of water to mix with other substances enables it to act as a medium for transport; nutrients, gases, and waste products are transported from one point in the body to another in body fluids such as plasma.

Solution Concentrations

The liquid portion of a solution is the **solvent,** and the substances dissolved in the solvent are **solutes.** For example, water is the solvent and sodium chloride is the solute in a sodium chloride solution. The concentration of solute particles dissolved in solvents can be expressed in several ways. One common way is to indicate the percent of solute by weight. For example, a 10% solution of sodium chloride can be made by dissolving 10 g of sodium chloride into enough water to make 100 ml of solution.

Physiologists determine concentrations in **osmoles,** which express the number of particles in a solution. A particle can be an atom, ion, or molecule. An osmole is 1 mole (Avogadro's number) of particles in 1 kilogram (kg) of water. The **osmolality** of a solution is a reflection of the number, not the type, of particles in a solution. Thus a 1-osmolal solution contains 1 osmole of particles per kilogram of solution, but the particles may be all one type or a complex mixture of different types.

The concentration of particles in body fluids is so low that the measurement **milliosmole** (mOsm), 1/1000 of an osmole, is used. Most body fluids have an osmotic concentration of approximately 300 mOsm and contain many different ions and molecules. The osmotic concentration of body fluids is important because it influences the movement of water into or out of cells (see Chapter 3). Appendix D contains more information on calculating concentrations.

Acids and Bases

Many molecules are classified as acids or bases. For most purposes an **acid** is defined as a proton donor. Because a hydrogen atom without its electron is a proton (H^+), any substance that releases hydrogen ions is an acid. For example, hydrochloric acid (HCl) forms hydrogen and chloride ions (H^+ and Cl^-) in solution and therefore is an acid.

$$HCl \rightarrow H^+ + Cl^-$$

A **base** is defined as a proton acceptor, and any substance that binds to (accepts) hydrogen ions is a base. Many bases function as proton acceptors by releasing hydroxide ions (OH^-) when they dissociate. For example, the base sodium hydroxide (NaOH) dissociates to form sodium and hydroxide ions:

$$NaOH \rightarrow Na^+ + OH^-$$

The hydroxide ions are proton acceptors that combine with hydrogen ions to form water:

$$OH^- + H^+ \rightarrow H_2O$$

Acids and bases are classified as strong or weak. Strong acids or bases dissociate almost completely when dissolved in water. Consequently, they release almost all of their hydrogen or hydroxide ions. The more completely the acid or base dissociates the stronger it is. Hydrochloric acid, for example, is a strong acid because it completely dissociates in water.

$$HCl \rightarrow H^+ + Cl^-$$

Not freely reversible

Weak acids or bases only partially dissociate in water. Consequently, they only release some of their hydrogen or hydroxide ions. For example, when acetic acid (CH_3COOH) is dissolved in water, some of it dissociates, but some of it remains in the undissociated form. An equilibrium is established between the ions and the undissociated weak acid.

$$CH_3COOH \leftrightarrow CH_3COO^- + H^+$$

Freely reversible

For a given weak acid or base, the ratio of the dissociated ions to the weak acid is a constant.

The pH scale

The pH scale, which runs from 0 to 14, is a means of referring to the hydrogen ion concentration in a solution (Figure 2-11). Pure water is defined as a **neutral solution** and has a pH of 7. A neutral solution has equal concentrations of hydrogen and hydroxide ions. Solutions with a pH less than 7 are **acidic** and have a greater concentration of hydrogen ions than hydroxide ions. **Alkaline,** or **basic,** solutions have a pH greater than 7 and have fewer hydrogen ions than hydroxide ions.

The symbol pH stands for power (p) of hydrogen ion (H^+) concentration. The power is a factor of 10, which means that a change in the pH of a solution by one pH unit represents a tenfold change in the hydrogen ion concentration. For example, a solution of pH 6 has a hydrogen ion concentration 10 times greater than a solution of pH 7 and 100 times greater than a solution of pH 8. As the pH value becomes smaller, the solution has more hydrogen ions and is more acidic, and as the pH value becomes larger, the solution has fewer hydrogen ions and is more basic. Appendix E considers pH in greater detail.

The normal pH range for human blood is 7.35 to 7.45. The condition of acidosis results if blood pH drops below 7.35, in which case the nervous system becomes depressed, and the individual becomes disoriented and possibly comatose. Alkalosis results if blood pH rises above 7.45. The nervous system becomes overexcitable, and the individual may be extremely nervous or have convulsions. Both acidosis and alkalosis can result in death.

Salts

A **salt** is a molecule consisting of a cation other than hydrogen and an anion other than hydroxide. Salts are formed by the interaction of an acid and a base in which the hydrogen ions of the acid are replaced by the positive ions of the base. For example, when hydrochloric acid reacts with the base sodium hydroxide (NaOH), the salt sodium chloride (NaCl) is formed.

$$HCl + NaOH \rightarrow NaCl + H_2O$$

Salts dissociate to form positively and negatively charged ions when dissolved in water.

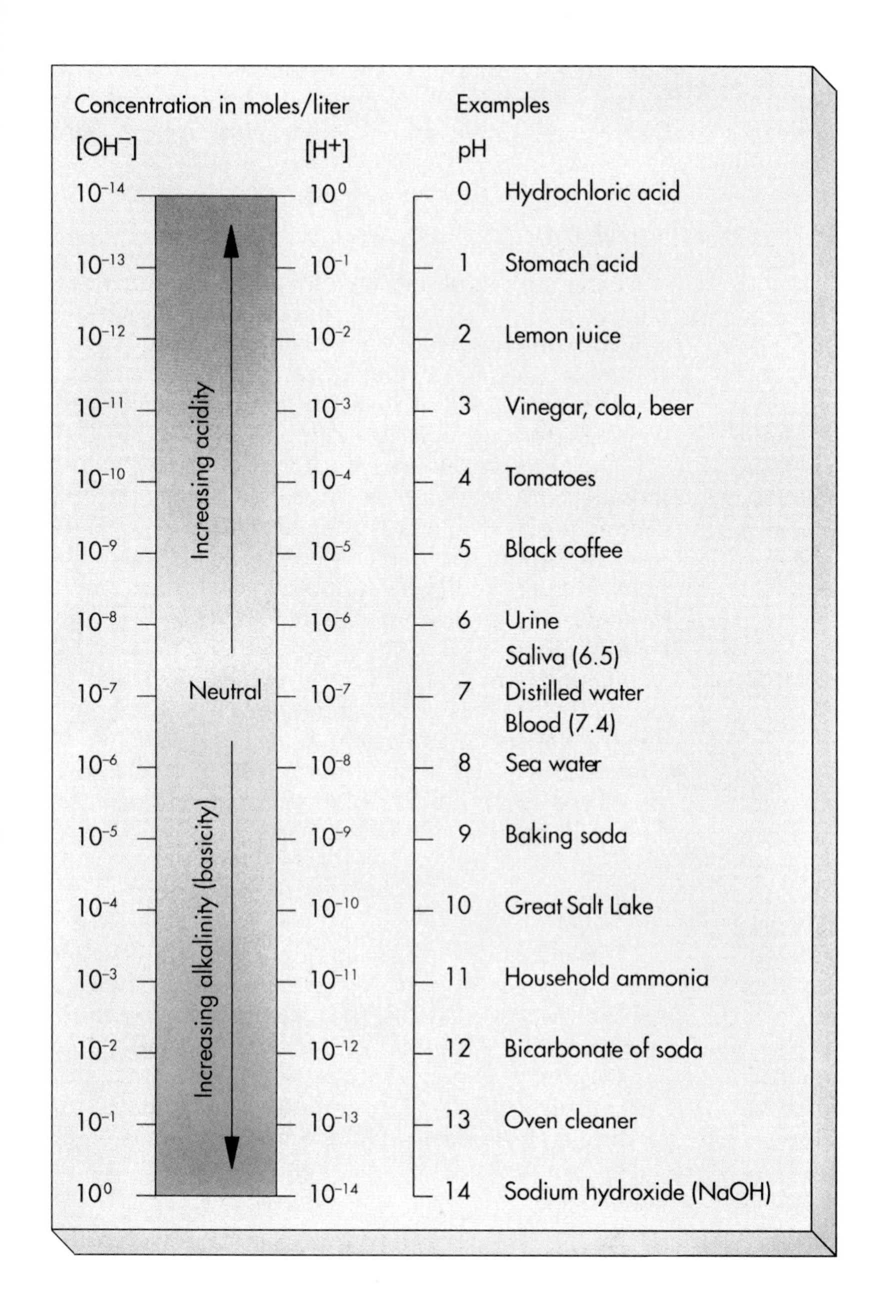

FIGURE 2-11 The pH scale. A pH of 7 is considered neutral. Values less than 7 are acidic (the lower the number, the more acidic). Values greater than 7 are basic (the higher the number, the more basic). Representative fluids and their approximate pH values are listed.

Buffers

The chemical behavior of many molecules changes as the pH of the solution in which they are dissolved changes. For example, many enzymes work optimally within narrow ranges of pH. An organism's survival depends on its ability to regulate body fluid pH within a narrow range. Deviations from the normal pH range for human blood (7.35 to 7.45) are life threatening. One way body fluid pH is regulated involves the action of **buffers,** chemicals that resist changes in solution pH when either acids or bases are added.

Weak acids and weak bases are effective buffers because of the equilibrium that exists between the acid or base and the ions into which it dissociates. Carbonic acid is an example:

$$\underset{\text{Carbonic acid}}{H_2CO_3} \leftrightarrow \underset{\text{Hydrogen ion}}{H^+} + \underset{\text{Bicarbonate ion}}{HCO_3^-}$$

If hydrogen ions are added to a solution of carbonic acid, many of the hydrogen ions and bicarbonate ions combine to form carbonic acid; thus the concentration of hydrogen ions is not increased as much as it would be without this reaction. If hydroxide ions are added to the solution, they combine with hydrogen ions to form water. Then additional carbonic acid molecules dissociate to form hydrogen and bicarbonate ions, maintaining the hydrogen ion concentration (pH) within physiological limits.

The greater the buffer concentration, the more effective it is in resisting a change in pH, but buffers cannot prevent some change in the pH of a solution. For example, when an acid is added to a solution containing a buffer, the pH will decrease but not to the extent it would have without the buffer. Several very important buffers are found in living systems and include bicarbonate, phosphates, amino acids, and proteins as components.

Oxygen

Oxygen (O_2) is an inorganic molecule consisting of two oxygen atoms bound together by a double covalent bond. Approximately 21% of the gas in the atmosphere is oxygen, and it is essential for most living organisms. Oxygen is required by humans in the final step of a series of reactions in which energy is extracted from food molecules (see Chapters 3 and 25).

Carbon Dioxide

Carbon dioxide (CO_2) consists of one carbon atom bound by double covalent bonds to two oxygen atoms. Carbon dioxide is produced when organic molecules such as glucose (a simple sugar) are metabolized within the cells of the body (see Chapters 3 and 25). Much of the energy stored in the covalent bonds of glucose is transferred to other organic molecules when the bonds are broken and carbon dioxide is released. Once carbon dioxide is produced, it is eliminated from the cell as a metabolic by-product, is transferred to the lungs by blood, and is exhaled during respiration. If carbon dioxide is allowed to accumulate within cells, it becomes toxic.

ORGANIC MOLECULES

Carbon's capacity to form four covalent bonds allows it to form a wide variety of complex molecules. The four major groups of organic molecules essential to living organisms are carbohydrates, lipids, proteins, and nucleic acids. Each of these groups has specific structural and functional characteristics.

Carbohydrates

Carbohydrates are primarily composed of carbon, hydrogen, and oxygen atoms and range in size from small to very large. In most carbohydrates, for each carbon atom there are two hydrogen atoms and one oxygen atom. The general molecular formula is

$$C_nH_{2n}O_n \quad \text{or} \quad (CH_2O)_n$$

where *n* may be 1, 2, 3, etc. For example, for the sugar glucose *n* is 6 and the molecular formula is $C_6H_{12}O_6$. The large number of oxygen atoms in carbohydrate molecules makes them relatively polar molecules. Consequently, they are soluble in polar solutes such as water.

Monosaccharides

Large carbohydrates are composed of numerous, relatively simple building blocks called **monosaccharides** (mono- means one; saccharide means sugar) or simple sugars. Some monosaccharides contain as few as three carbons (trioses); others contain five carbons (pentoses), six carbons (hexoses), or more.

The monosaccharides most important to humans include both pentoses and hexoses. Common six-carbon sugars, such as glucose, fructose, and galactose, are **isomers,** which are molecules that have the same number and types of atoms but differ in their three-dimensional arrangement (Figure 2-12). Glucose, or blood sugar, is the major carbohydrate found in the blood and is a major nutrient for most cells of the body. Fructose and galactose are also important dietary nutrients. Important five-carbon sugars include ribose and deoxyribose (see Figure 2-23), which are components of ribonucleic acid (RNA) and deoxyribonucleic acid (DNA), respectively.

Disaccharides

Disaccharides (di- means two) are composed of two simple sugars bonded together through a dehydration reaction. Glucose and fructose, for example, combine to form a disaccharide called **sucrose** (table sugar) plus a molecule of water (Figure 2-13, *A*). Several disaccharides are important to humans, including sucrose; lactose (glucose plus galactose), or milk sugar; and maltose (two glucose molecules), or malt sugar.

FIGURE 2-12 Sugars: glucose, fructose, and galactose. Fructose is a structural isomer of glucose because it has identical chemical groups bonded in a different arrangement in the molecule *(purple shading)*. Galactose is a stereoisomer of glucose because it has exactly the same groups bonded to each carbon atom but located in a different three-dimensional orientation *(yellow shading)*. The sugars are represented as linear models to illustrate more readily the relationships between the atoms of the molecules. In actuality, these sugar molecules almost always are found in the indicated ring form.

FIGURE 2-13 **A** Formation of sucrose, a disaccharide, by a dehydration reaction involving glucose and fructose (monosaccharides). **B** Glycogen is a polysaccharide formed by combining many glucose molecules. The photo shows glycogen granules in a liver cell.

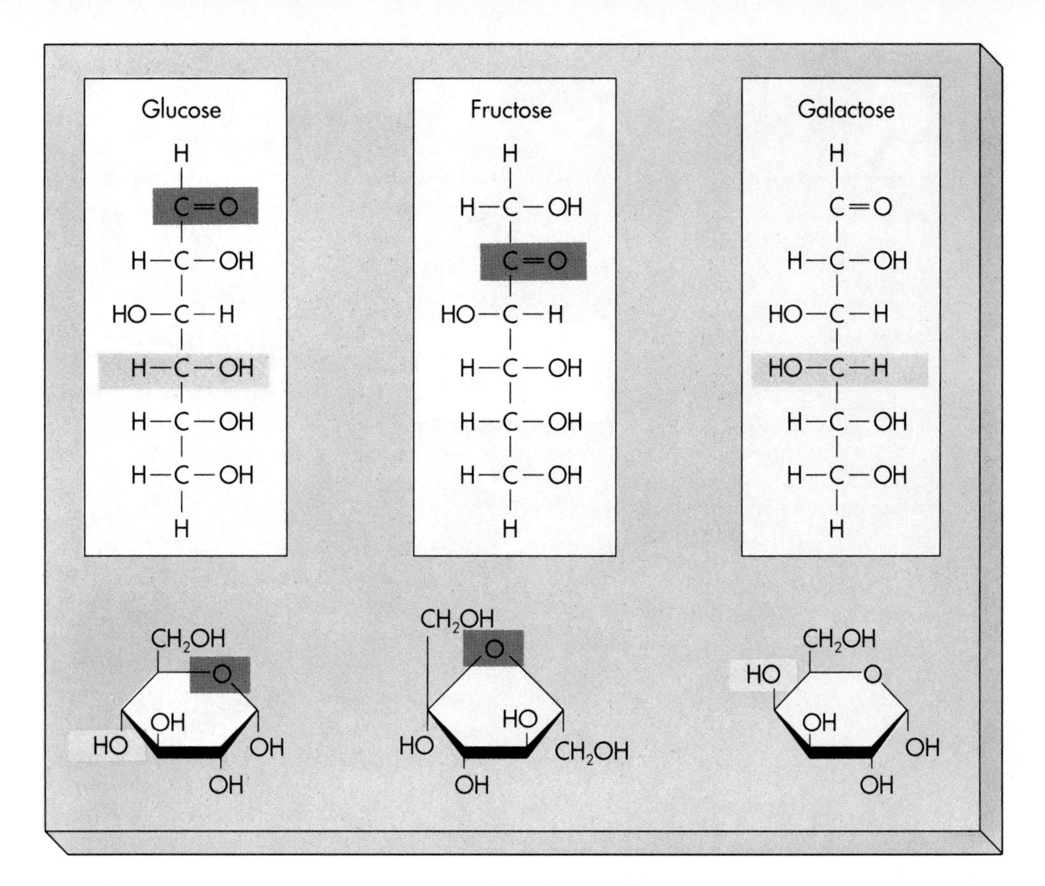

FIGURE 2-12 For legend see opposite page.

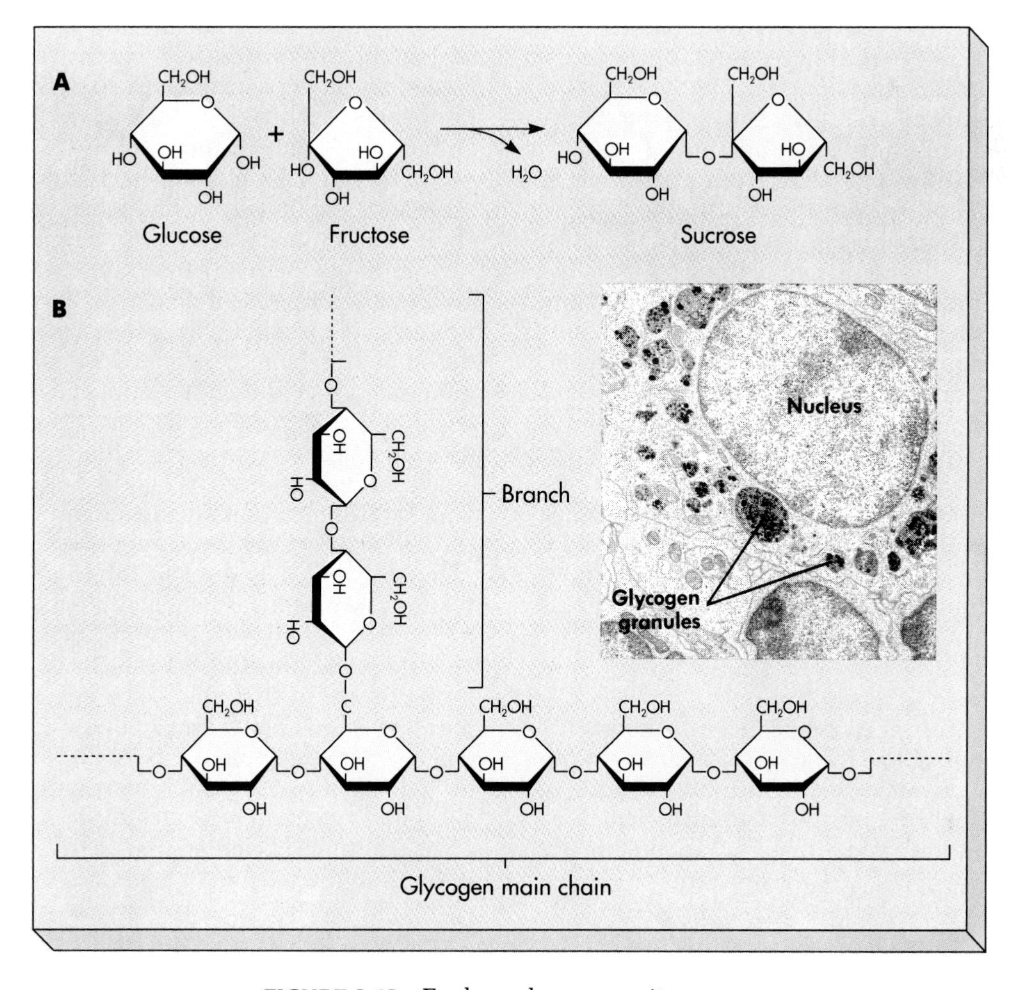

FIGURE 2-13 For legend see opposite page.

Table 2-4

Role of carbohydrates in the body

ROLE	EXAMPLE
Structure	Ribose forms part of RNA and ATP molecules, and deoxyribose forms part of DNA.
Energy	Monosaccharides (glucose, fructose, galactose) can be used as energy sources. Disaccharides (sucrose, lactose, maltose) and polysaccharides (starch, glycogen) must be broken down to monosaccharides before they can be used for energy. Glycogen is an important energy storage molecule in muscles and in the liver.
Bulk	Cellulose forms bulk in the feces.

Polysaccharides

Polysaccharides (*poly-* means many) consist of many monosaccharides bound together to form long chains that are either straight or branched. Glycogen, or animal starch, is a polysaccharide composed of many glucose molecules (Figure 2-13, *B*). Because glucose can be metabolized rapidly and the resulting energy can be used by cells, glycogen is an important storage molecule. A substantial amount of the glucose that is metabolized to produce energy for muscle contraction during exercise is stored in the form of glycogen in the cells of the liver and skeletal muscles.

Starch and cellulose are important polysaccharides found in plants, and both are composed of long chains of glucose molecules. Starch is used as a storage molecule in plants in the same way that glycogen is used in animals, and cellulose is an important structural component of plant cell walls. When humans ingest plants, the starch can be broken down and used as an energy source. The cellulose, however, is not digestible and is eliminated in the feces in which it provides bulk. Table 2-4 summarizes the role of carbohydrates in the body.

Lipids

Lipids constitute a second major group of organic molecules common to living systems. Like carbohydrates, they are composed principally of carbon, hydrogen, and oxygen, but other elements, such as phosphorus and nitrogen, are minor components of some lipids. Lipids contain a lower ratio of oxygen to carbon than do carbohydrates, which makes them less polar. Consequently, lipids can be dissolved in nonpolar organic solvents, such as alcohol or acetone, but they are relatively insoluble in water. The definition of lipids is so general that several different kinds of molecules, such as the fats, phospholipids, steroids, and prostaglandins, fit into this category.

Fats are a major type of lipid, and, like carbohydrates, fats are ingested and broken down by hydrolysis reactions in cells to release energy for use by those cells. Conversely, if intake exceeds need, excess chemical energy from any source can be stored in the body as fat for later use as energy is needed. Fats also provide protection by surrounding and padding organs, and under the skin fats act as an insulator to prevent heat loss.

Triglycerides, which comprise 95% of the fats in the human body, consist of two different types of building blocks: glycerol and fatty acids. **Glycerol** is a three-carbon molecule with a hydroxyl group attached to each carbon atom, and **fatty acids** consist of a straight chain of carbon atoms with a carboxyl group attached at one end (Figure 2-14, *A*) A **carboxyl group** (—COOH) consists of both an oxygen atom and a hydroxyl group attached to a carbon atom. The carboxyl group is responsible for the acidic nature of the molecule because it releases hydrogen ions into solution. Glycerides can be described according to the number and kinds of fatty acids that combine with glycerol through dehydration reactions. Monoglycerides have one fatty acid, diglycerides have two fatty acids, and triglycerides have three fatty acids bound to a glycerol molecule (Figure 2-14, *B* and *C*).

Fatty acids differ from each other according to the length and the degree of saturation of their carbon chains. Most naturally occurring fatty acids contain an even number of carbon atoms with 14- to 18-carbon chains most common. A fatty acid is **saturated** if it contains only single covalent bonds between the carbon atoms. Sources of saturated fats include beef, pork, whole milk, cheese, butter, eggs, coconut oil, and palm oil. The carbon chain is **unsaturated** if it has one or more double covalent bonds between carbon atoms (Figure 2-15). Because the double covalent bonds can occur anywhere along the carbon chain, many types of unsaturated fatty acids with an equal degree of unsaturation are possible. **Monounsaturated** fats, such as olive and peanut oils, have one double covalent bond between carbon atoms. **Polyunsaturated** fats, such as safflower, sunflower, corn, or fish oils, have two or more double covalent bonds between carbon atoms. Unsaturated fats are the best type of fats in the diet because saturated fats apparently are more conducive to the development of cardiovascular disease.

Phospholipids are similar to triglycerides except that one of the fatty acids bound to the glycerol mol-

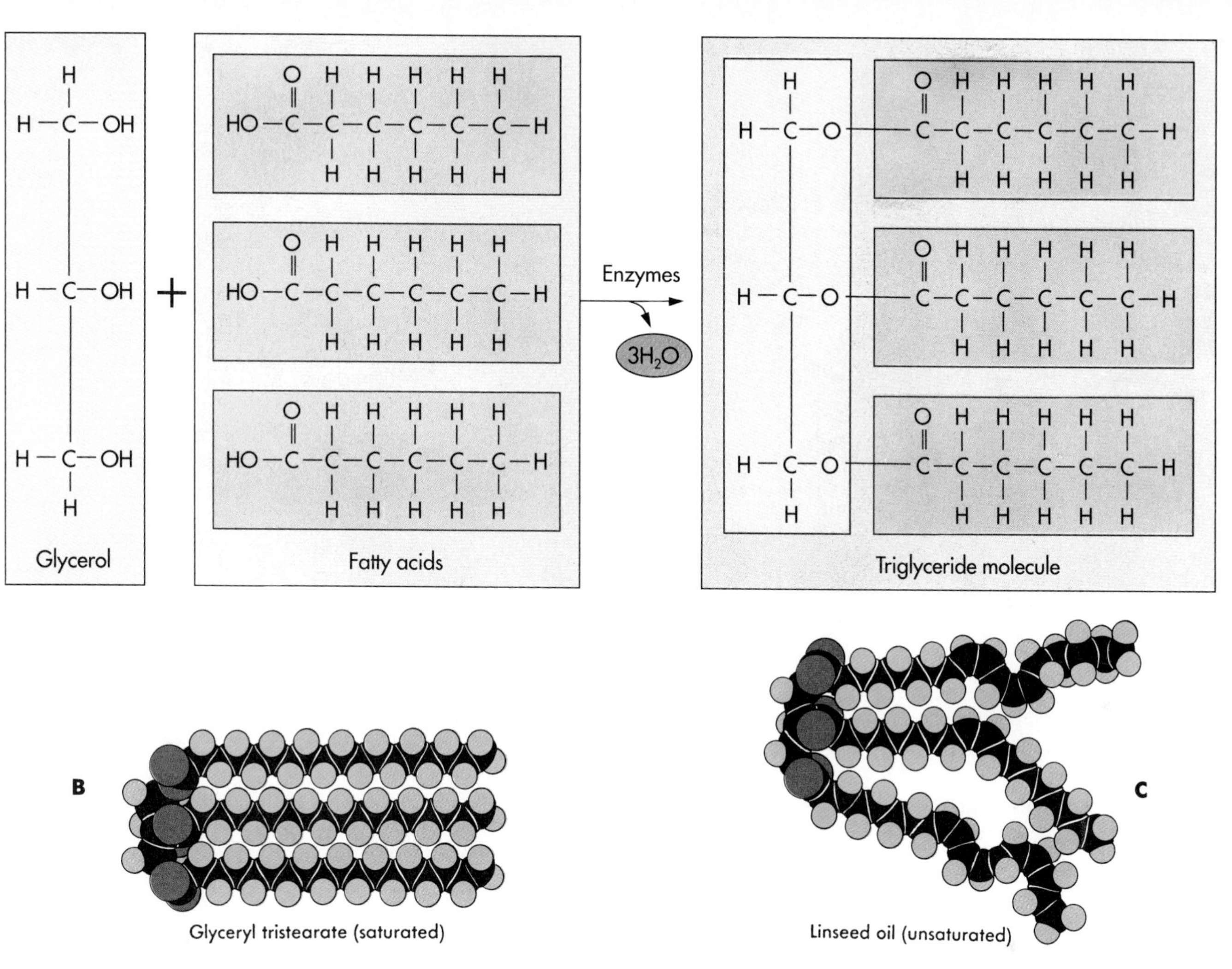

FIGURE 2-14 Triglycerides. **A** Production of a triglyceride from one glycerol molecule and three fatty acids. **B** Glycerol tristearate (saturated). The chains are all straight because each carbon is bonded to the next by a single bond. **C** Linseed oil (unsaturated). The chains are bent at each point where the chain is unsaturated because of double bonds between the carbon atoms.

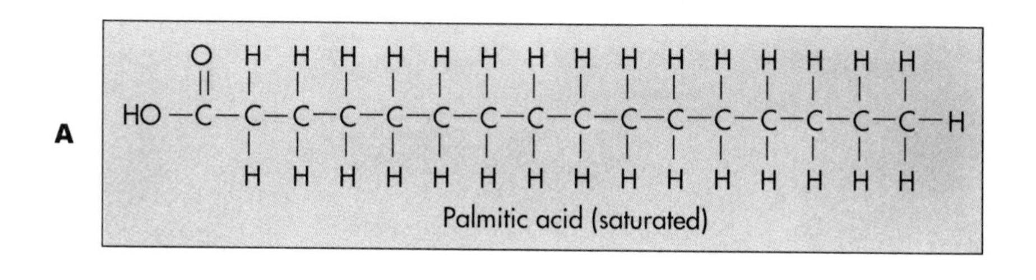

B

Linolenic acid (unsaturated)

FIGURE 2-15 Fatty acids. **A** Palmitic acid (saturated with no double bonds between the carbons). **B** Linolenic acid (unsaturated with double bonds between the carbons).

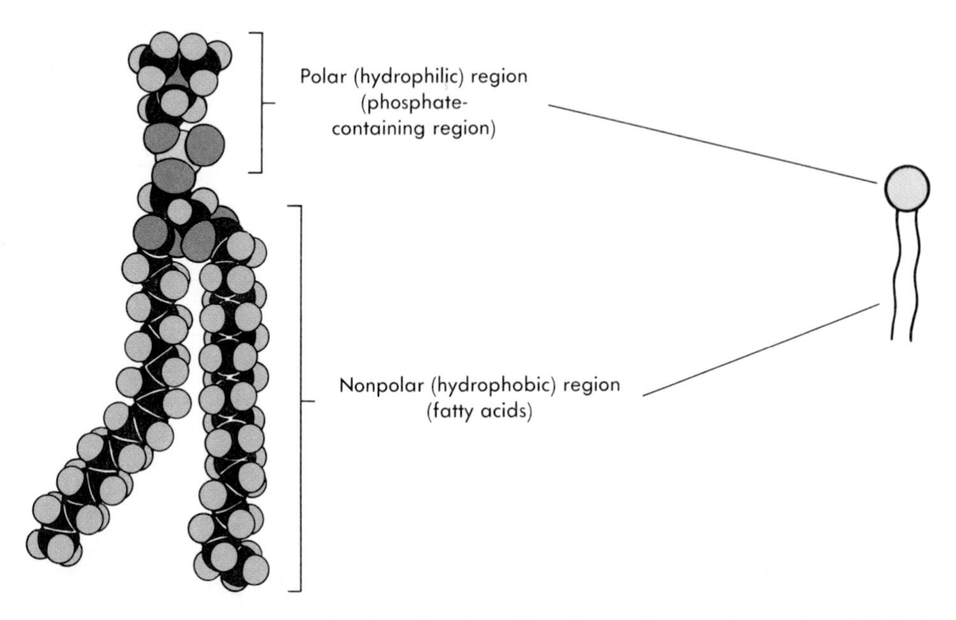

FIGURE 2-16 *Left*, Phospholipid molecule with polar and nonpolar regions labeled. *Right*, The way phospholipids are often depicted.

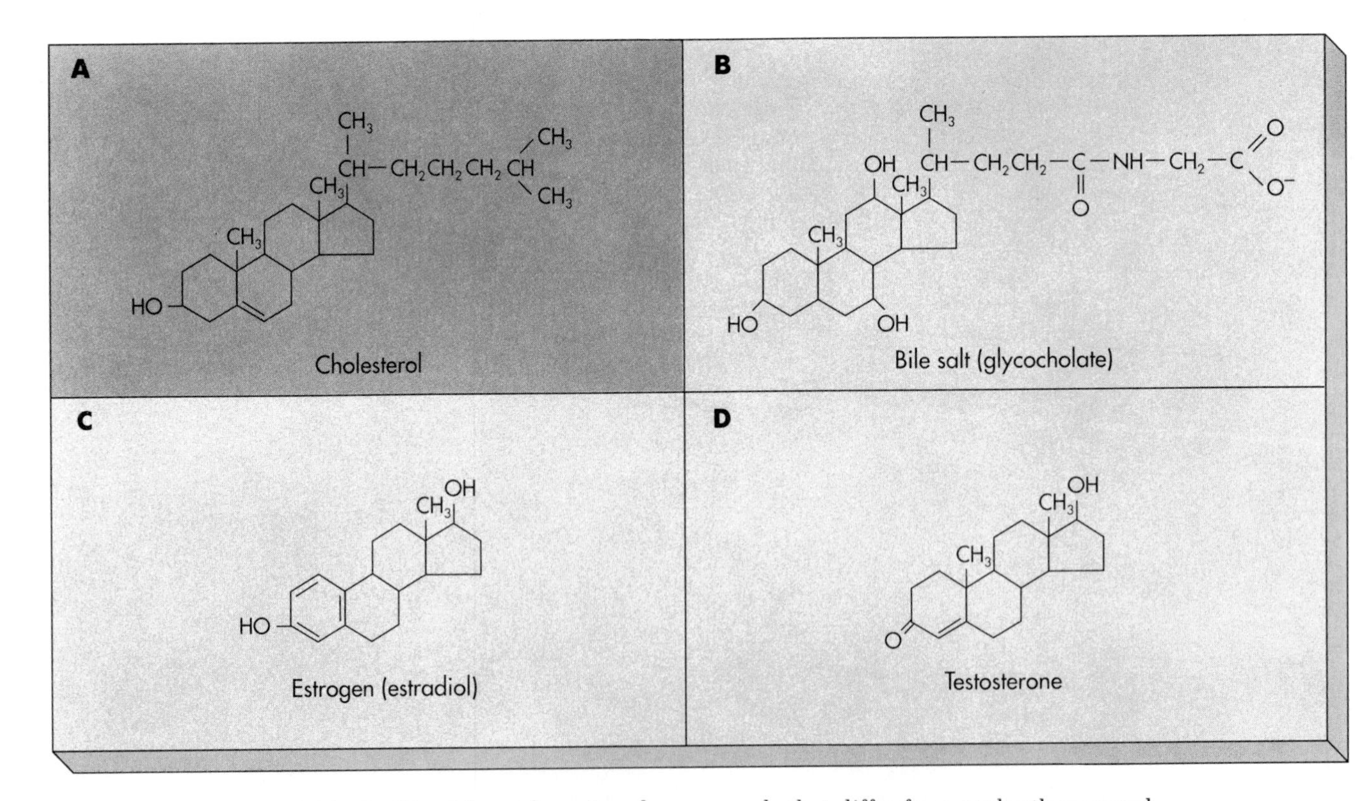

FIGURE 2-17 Steroids are four-ringed compounds that differ from each other according to the groups attached to the rings. **A** Cholesterol, the most common steroid, can be modified to produce other steroids. **B** The major bile salt, glycocholate. **C** Estrogen. **D** Testosterone.

Table 2-5

Role of lipids in the body

ROLE	EXAMPLE
Protection	Fat surrounds and pads organs.
Insulation	Fat under the skin prevents heat loss. Myelin surrounds nerve cells and electrically insulates the cells from each other.
Regulation	Steroid hormones regulate many physiological processes. For example, estrogen and testosterone are sex hormones responsible for many of the differences between males and females. Prostaglandins help regulate tissue inflammation and repair.
Vitamins	Fat-soluble vitamins perform a variety of functions. Vitamin A forms retinol, which is necessary for seeing in the dark; active vitamin D promotes calcium uptake by the small intestine; vitamin E promotes wound healing; and vitamin K is necessary for the synthesis of proteins responsible for blood clotting.
Structure	Phospholipids and cholesterol are important components of cell membranes.
Energy	Lipids can be stored and broken down later for energy; per unit of weight they yield more energy than carbohydrates or proteins.

ecule is replaced by a molecule containing phosphate and, usually, nitrogen (Figure 2-16). They are polar at the end of the molecule to which the phosphate is bound and nonpolar at the other end. The polar end of the molecule is hydrophilic (attracted to water), and the nonpolar end is hydrophobic (repelled by water). Phospholipids are important components of cell membranes (see Chapter 3).

Prostaglandins, thromboxanes, and **leukotrienes** are lipids derived from fatty acids. They are made in most cells and are important regulatory molecules. Among their numerous effects is their role in the response of tissues to injuries. Prostaglandins have been implicated in regulating the secretion of some hormones, blood clotting, some reproductive functions, and many other processes. Many of the therapeutic effects of aspirin and other anti-inflammatory drugs are due to their ability to inhibit prostaglandin synthesis.

Steroids differ in chemical structure from other lipid molecules, but their solubility characteristics are similar. All steroid molecules, composed of carbon atoms bound together into four ringlike structures, are structurally similar (Figure 2-17), but their functions are diverse. Important steroid molecules include cholesterol, bile salts, estrogen, progesterone, and testosterone. Cholesterol is an important steroid because other molecules are synthesized from it. For example, bile salts, which increase fat absorption in the intestines, are derived from cholesterol, as are the reproductive hormones estrogen, progesterone, and testosterone. In addition, cholesterol is an important component of cell membranes. Although high levels of cholesterol in the blood increase the risk of cardiovascular disease, normal amounts of cholesterol are vital for normal function.

Another class of lipids is the **fat-soluble vitamins.** Their structures are not closely related to one another, but they are nonpolar molecules essential for many normal functions of the body (see Chapter 25). Table 2-5 lists the functions of lipids in the body.

Proteins

All **proteins** contain carbon, hydrogen, oxygen, and nitrogen bound together by covalent bonds, and most proteins contain some sulfur. In addition, some proteins contain small amounts of phosphorus, iron, and iodine. The molecular weights of proteins can be very large. For the purpose of comparison, the molecular weight of water is 18, sodium chloride, approximately 58, and glucose, approximately 180; but the molecular weights of proteins range from approximately 1000 to several million.

Protein structure

The basic building blocks for proteins are the 20 **amino acid** molecules. Each amino acid has a carboxyl group (—COOH), an amine group (—NH_2), a hydrogen atom, and a group called "R" attached to the same carbon atom. R represents a variety of chemical structures, and the differences in R groups make the amino acids different from one another (Figure 2-18).

Covalent bonds formed between amino acid molecules during protein synthesis are called **peptide bonds** (Figure 2-19). A dipeptide is two amino acids bound together by a peptide bond, a tripeptide is three amino acids bound together by peptide bonds, and a polypeptide is many amino acids bound to-

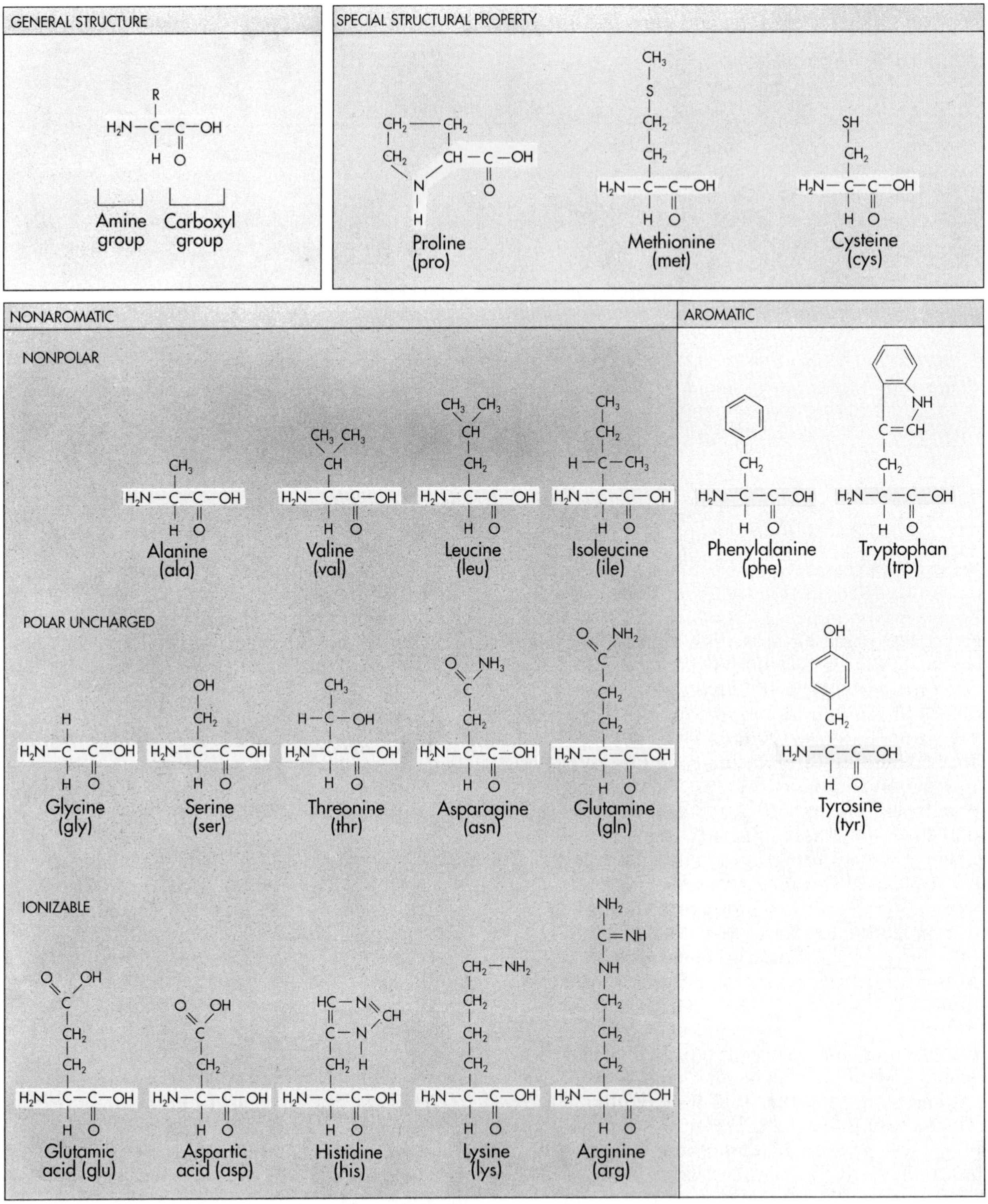

GENERAL STRUCTURE

Amine group — Carboxyl group

SPECIAL STRUCTURAL PROPERTY

Proline (pro) — Methionine (met) — Cysteine (cys)

NONAROMATIC

NONPOLAR

Alanine (ala) — Valine (val) — Leucine (leu) — Isoleucine (ile)

POLAR UNCHARGED

Glycine (gly) — Serine (ser) — Threonine (thr) — Asparagine (asn) — Glutamine (gln)

IONIZABLE

Glutamic acid (glu) — Aspartic acid (asp) — Histidine (his) — Lysine (lys) — Arginine (arg)

AROMATIC

Phenylalanine (phe) — Tryptophan (trp) — Tyrosine (tyr)

FIGURE 2-18 The general structure of an amino acid *(upper left)* shows the amine group (—NH_2) and the carboxyl group (—COOH). The R group is the part of an amino acid that makes it different from other amino acids. Twenty principal amino acids are divided into subgroups according to various chemical characteristics such as aromatic (those with a carbon ring) vs. nonaromatic (those without a carbon ring), polarity, and ionizability. Amino acids with structures that do not fit these classes are classified as "special." Glycine is the simplest amino acid; tyrosine is an important component of thyroid hormones; improper metabolism of phenylalanine in the genetic disease phenylketonuria (PKU) can cause mental retardation; aspartic acid combined with phenylalanine forms the artificial sweetener aspartame (marketed as Nutrasweet and Equal).

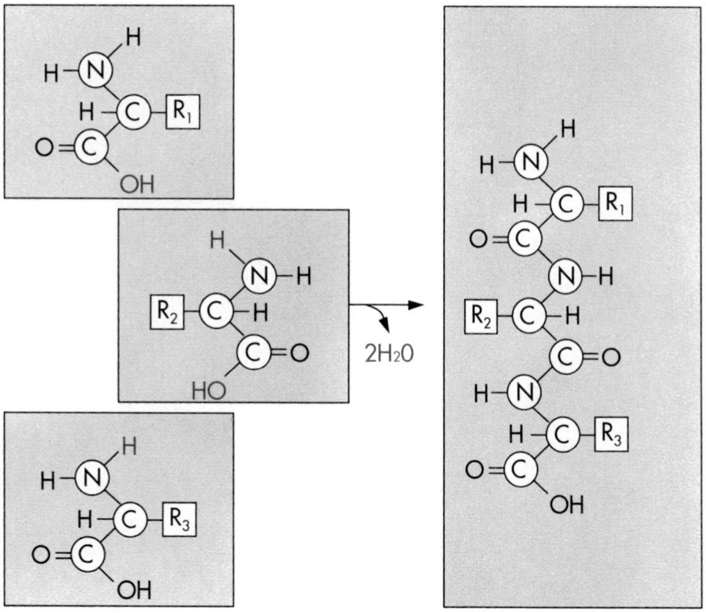

FIGURE 2-19 *Left,* A dehydration reaction between three amino acids to form, *right,* a tripeptide. One water molecule (H_2O) is given off for each peptide bond formed (shaded yellow).

gether by peptide bonds. Proteins are polypeptides composed of hundreds of amino acids. Because there are 20 different amino acids and because each amino acid may be located at any position along a polypeptide chain, the potential number of different protein molecules is enormous.

The **primary structure** (Figure 2-20, *A*) of a protein is determined by the sequence of the amino acids bound by peptide bonds. The **secondary structure** (Figure 2-20, *B*) of a protein is determined by the hydrogen bonds between amino acids that cause the protein to coil into helices, or pleated sheets. The ability of proteins to perform their functions depends on their shape. If the hydrogen bonds that maintain the protein's shape are broken, the protein becomes nonfunctional. This change in shape is called **denaturation,** and it can be caused by abnormally high temperatures or changes in pH of body fluids. An everyday example of denaturation is the change in the proteins of egg whites when they are cooked.

The **tertiary structure** (Figure 2-20, *C*) results from the folding of the helices, or pleated sheets. It can be caused by the formation of covalent bonds between sulfur atoms of one amino acid and sulfur atoms in another amino acid located at a different place in the sequence of amino acids. Some amino acids are polar and are therefore hydrophilic, and other amino acids are less polar and more hydrophobic. Hydrophobic regions of proteins tend to fold into a globular shape to minimize the contact of that region with water, and hydrophilic portions remain unfolded to maximize their contact with water. Changes in only a few amino acids in the sequence can markedly change the protein's tertiary structure. Because the function of a protein is influenced by its three-dimensional structure, changes that influence the tertiary structure of the protein may also affect the function of the protein.

If two or more proteins associate to form a functional unit, the individual proteins are called subunits. The **quaternary structure** refers to the spatial relationships between the individual subunits (Figure 2-20, *D*).

Enzymes

Proteins perform many roles in the body (Table 2-6), including acting as enzymes. An **enzyme** is a protein catalyst that increases the rate at which a chemical reaction proceeds without the enzyme being permanently changed. For a chemical reaction to begin, energy must be provided. For example, heat in the form of a spark is required to start the reaction between oxygen and gasoline. Once some oxygen molecules react with gasoline, the energy released can start additional reactions.

Activation energy is the energy required to start a chemical reaction. For most of the chemical reactions in the body, the activation energy is so high that few chemical reactions can occur naturally without help. Enzymes provide that help by lowering the activation energy (Figure 2-21). With an enzyme, the rate of a chemical reaction can take place more than a million times faster than without the enzyme.

The three-dimensional shape of enzymes is critical for their normal function because it determines

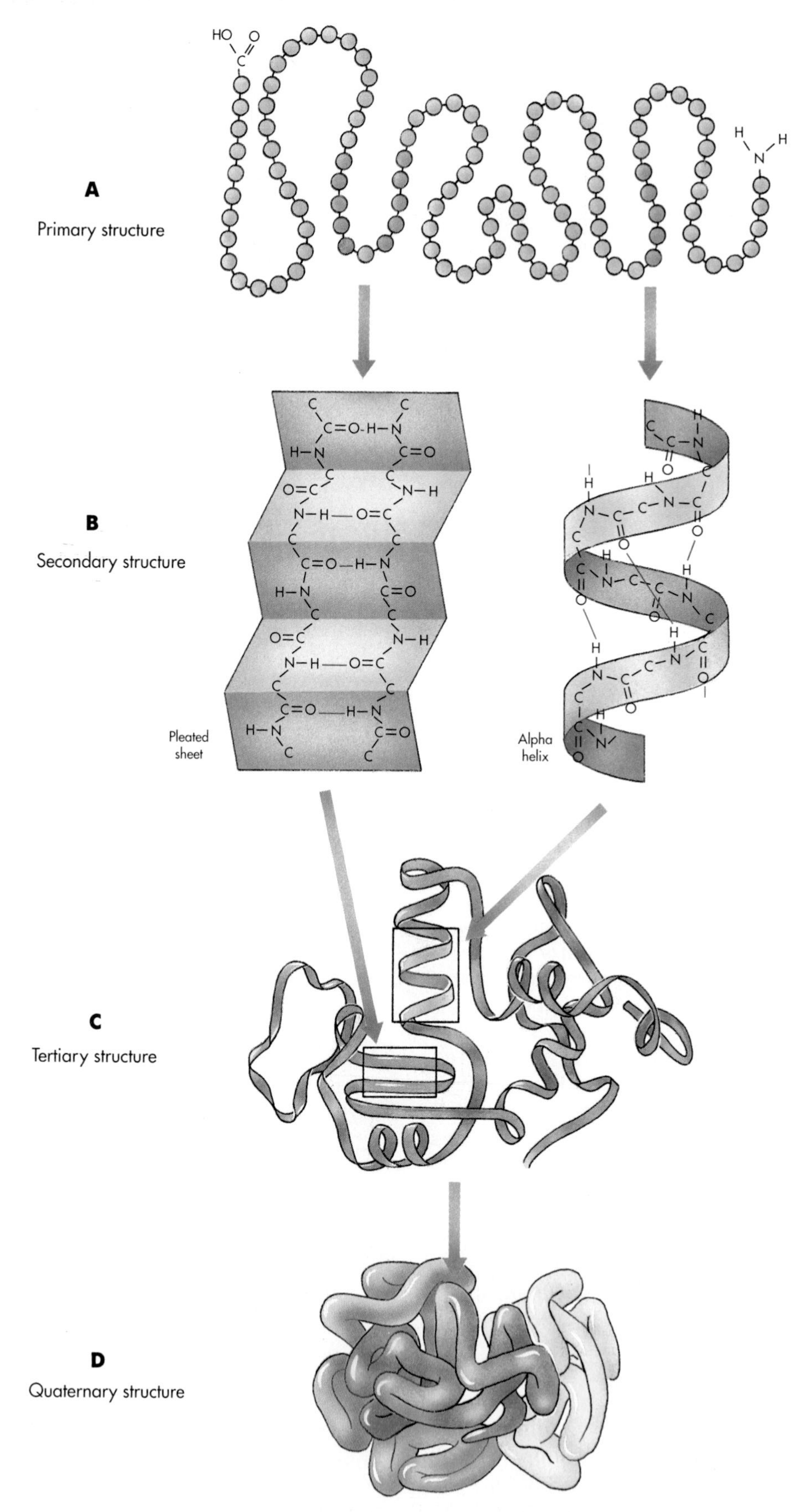

FIGURE 2-20 Protein structure. **A** Primary structure—its amino acid sequence. **B** Secondary structure, with folding as a result of hydrogen bonding *(red lines)*. **C** Tertiary structure with secondary folding caused by interactions within the polypeptide and its immediate environment. **D** Quaternary structure refers to the relationships between individual subunits.

Table 2-6

Role of proteins in the body

ROLE	EXAMPLE
Regulation	Enzymes control chemical reactions. Hormones regulate many physiological processes; for example, thyroid hormone affects metabolic rate.
Transport	Hemoglobin transports oxygen and carbon dioxide in the blood. Plasma proteins transport many substances in the blood. Proteins in cell membranes control the movement of materials into and out of the cell.
Protection	Antibodies and complement protect against microorganisms and other foreign substances.
Contraction	Actin and myosin in muscle are responsible for muscle contraction.
Structure	Collagen fibers form a structural framework in many parts of the body. Keratin adds strength to skin, hair, and nails.
Energy	Proteins can be broken down for energy; per unit of weight, they yield as much energy as carbohydrates.

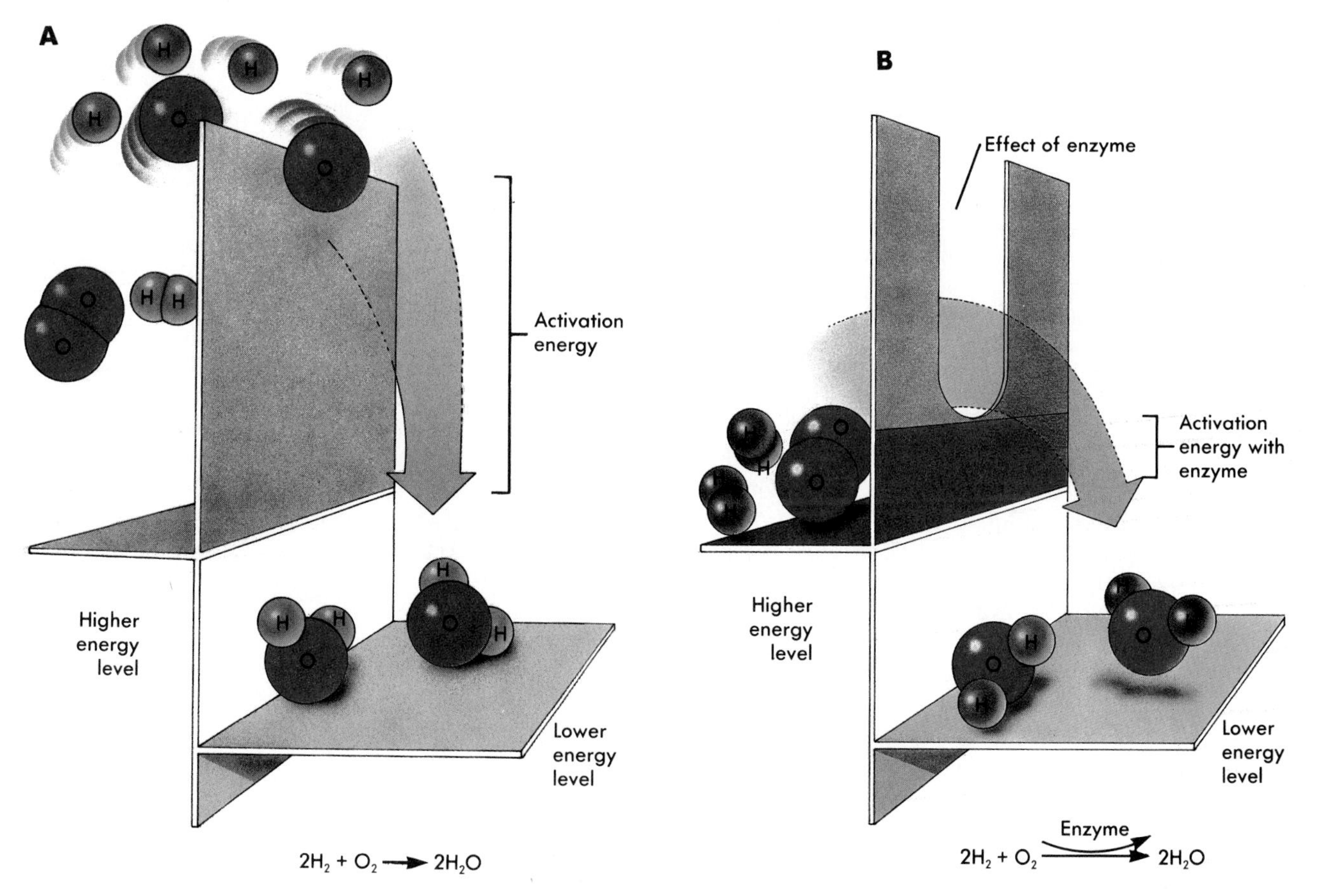

FIGURE 2-21 **A** Activation energy needed to change hydrogen and oxygen to water. The upper shelf represents a higher energy state, and the lower shelf represents a lower energy state. The "wall" extending above the upper shelf represents the activation energy. Even though energy is given up moving from the upper to the lower shelf, the activation energy wall must be overcome before the reaction can proceed. **B** The enzyme lowers the activation energy, making it easier for the reaction to proceed.

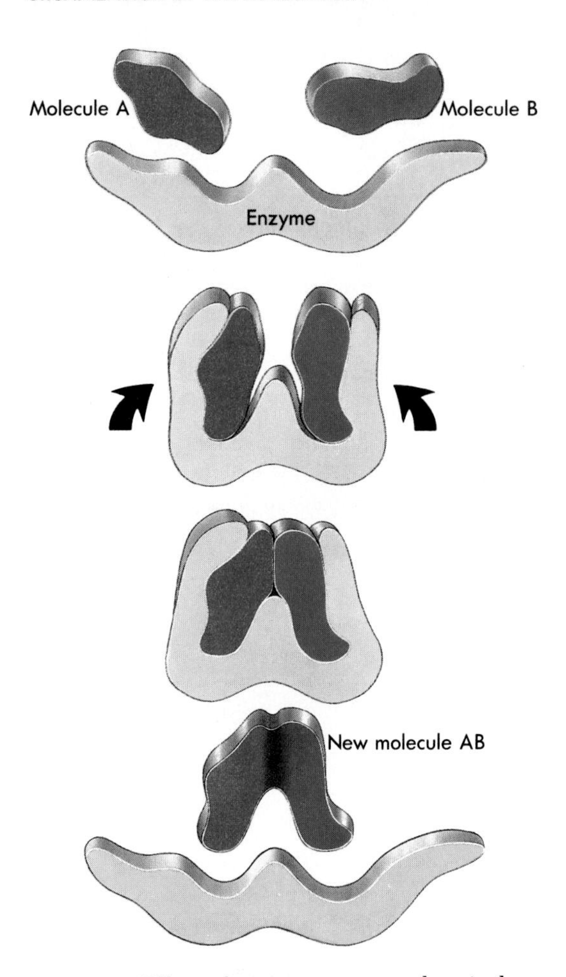

FIGURE 2-22 Effect of an enzyme on a chemical reaction. The enzyme brings the two reacting molecules (*A* and *B*) together at its active site so that the reactive regions between the two molecules come close together. The enzymes may also participate in some reactions but are not permanently altered by the reaction.

the structure of the enzyme's **active site.** According to the **lock-and-key model** of enzyme action, reactants must bind to a specific active site on the enzyme. At the active site reactants are brought into close proximity (Figure 2-22), reducing the activation energy of the reaction. After the reactants combine, they are released from the enzyme's active site, and the enzyme is capable of catalyzing additional reactions. Slight changes in the structure of an enzyme can destroy the ability of the active site to function. Enzymes are very sensitive to changes in temperature or pH, which may influence their structure.

7

Describe how changing one amino acid in an enzyme could affect the function of an enzyme.

? ? ? ? ? ? ? ? ? ? ?

Some enzymes require additional, nonprotein substances to be functional. The nonprotein component of an enzyme, called a **cofactor,** can be an ion or a complex organic molecule. Cofactors normally form part of the enzyme's active site and are required to make the enzyme functional. Some vitamins function as cofactors for certain enzymes.

Enzymes are highly specific, and each enzyme catalyzes a specific chemical reaction involving only certain reactants and products and no others. Therefore many different enzymes are needed to catalyze the chemical reactions of the body. Enzymes often are named by adding the suffix *-ase* to the name of the molecules on which they act. For example, an enzyme that catalyzes the breakdown of lipids is a **lipase** (li'pās), and an enzyme that breaks down proteins is called a **protease** (pro'te-ās).

Enzymes control the rate at which most chemical reactions proceed; consequently, they control essentially all cellular activities. The activity of enzymes also is regulated by several mechanisms that exist within the cells. Some mechanisms control the enzyme concentration by influencing the rate at which the enzymes are synthesized, and others alter the activity of existing enzymes. Much of what is known about the regulation of cellular activity involves knowledge of how enzyme activity is controlled.

Nucleic Acids: DNA and RNA

The **nucleic acids** constitute another group of very important organic molecules. **Deoxyribonucleic acid (DNA)** is the genetic material of cells. The information directing the chemical processes that occur in organisms and therefore determine their characteristics is contained in DNA. **Ribonucleic acid (RNA)** is structurally related to DNA, and there are three types of RNA that play important roles in protein synthesis. In Chapter 3 the means by which DNA and RNA direct the function of the cell are described.

The nucleic acids are large molecules composed of carbon, hydrogen, oxygen, nitrogen, and phosphorus. Both DNA and RNA consist of basic building blocks called **nucleotides.** Each nucleotide is composed of a monosaccharide to which a nitrogenous organic base and a phosphate group are attached (Figure 2-23). The monosaccharide is deoxyribose for DNA and ribose for RNA. The organic bases are thymine and cytosine, which are single-ringed pyrimidines, and adenine, guanine, and uracil, which are double-ringed purines (Figure 2-24). The nucleotides are joined together in a chain by covalent bonds to form the nucleic acids.

DNA is composed of two strands of nucleotides (Figure 2-25). Each nucleotide of DNA contains one of the organic bases: adenine, thymine, cytosine, or guanine. The organic bases of one strand are bound to the organic bases of the other strand by hydrogen bonds to produce a ladderlike structure that is shaped like a helix. Adenine binds only to thymine because the structure of these organic bases allows two hydrogen bonds to form between them. Cytosine binds only to guanine because the structure of these organic bases allows three hydrogen bonds to form between them.

DNA molecules are associated with globular histone proteins to form **chromatin** (kro'mah-tin). The histone proteins provide a site of attachment for the DNA and are involved with regulating DNA function. For most of a cell's life, chromatin is organized as a string with beads. During cell division, however, the chromatin condenses into structures called **chromosomes** (kro'mo-somz). DNA, chromatin, and chromosomes are considered in greater detail in Chapter 3.

RNA has a structure similar to a single strand of DNA. Like DNA, four different nucleotides comprise the RNA molecule, and the organic bases are the same except uracil substitutes for thymine (see Figure 2-24). Uracil can bind only to adenine.

The sequence of organic bases in DNA molecules stores genetic information. Each DNA molecule con-

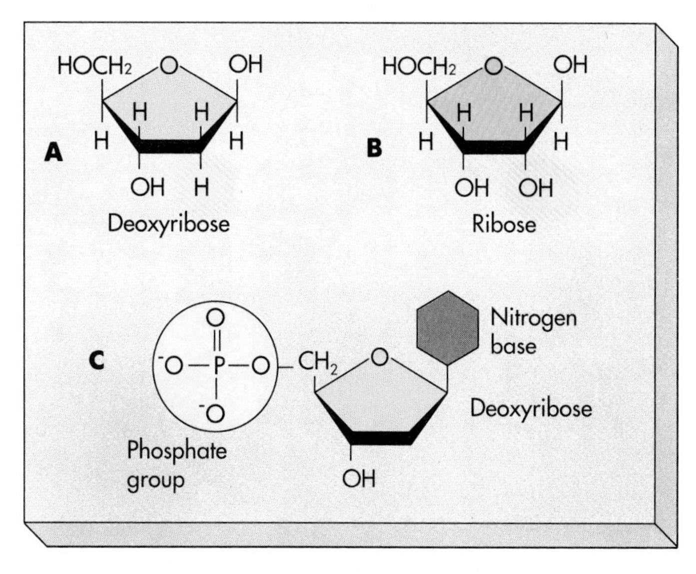

FIGURE 2-23 Components of nucleotides. **A** Deoxyribose sugar that forms nucleotides used in DNA production. **B** Ribose sugar that forms nucleotides used in RNA production. Note that deoxyribose is ribose minus an oxygen atom. **C** Deoxyribonucleotide consisting of deoxyribose, a nitrogen base, and a phosphate group.

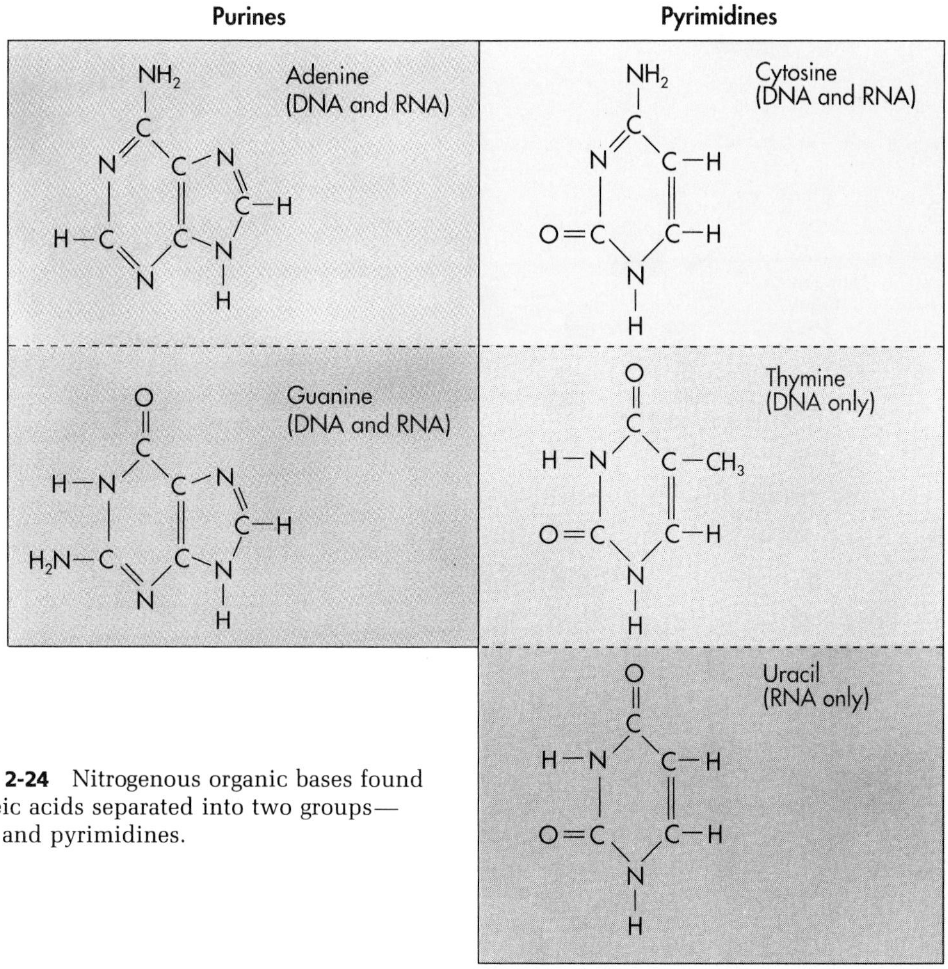

Purines
Pyrimidines
Adenine (DNA and RNA)
Guanine (DNA and RNA)
Cytosine (DNA and RNA)
Thymine (DNA only)
Uracil (RNA only)

FIGURE 2-24 Nitrogenous organic bases found in nucleic acids separated into two groups—purines and pyrimidines.

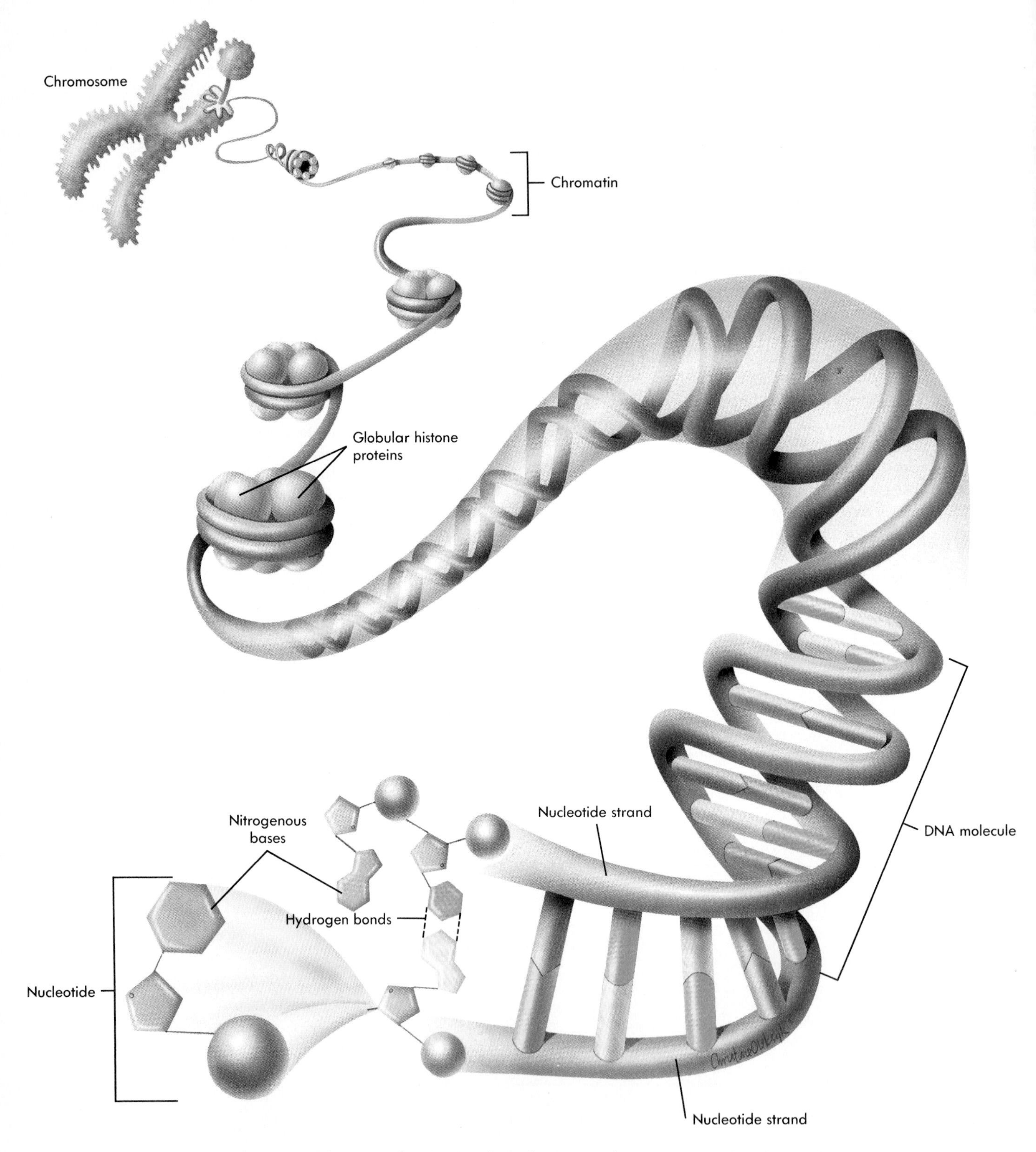

FIGURE 2-25 Structure of DNA. Nucleotides join to form two strands. The nucleotides of one strand are joined by hydrogen bonds to the nucleotides of the other strand to form a DNA molecule. Associated with the DNA molecule are globular histone proteins. Usually the DNA molecule is stretched out, resembling a string of beads, and is called chromatin. During cell division, however, the chromatin condenses to form bodies called chromosomes.

sists of millions of organic bases, and their sequence ultimately determines the type and sequence of amino acids found in protein molecules. Because enzymes are proteins, DNA structure determines the rate and type of chemical reactions that occur in cells by controlling enzyme structure. Therefore it is the information contained in DNA that ultimately defines all cellular activities. Other proteins coded by DNA such as collagen determine many of the structural features of humans.

Adenosine Triphosphate

Adenosine triphosphate (ATP) is an especially important organic molecule found in all living organisms. It consists of the organic base adenine, the sugar ribose, and three phosphate groups (Figure 2-26). The potential energy stored in the covalent bond between the second and third phosphate groups is important to living organisms because it provides the energy used in nearly all of the endergonic reactions within cells.

The catabolism of glucose and other nutrient molecules results in exergonic reactions that release energy. Some of that energy is used to synthesize ATP from adenosine diphosphate (ADP) and a phosphate group (P):

$$\text{ADP} + \text{P} + \text{Energy (from catabolism)} \rightarrow \text{ATP}$$

The transfer of energy from nutrient molecules to ATP involves a series of chemical reactions in which a high energy electron is transferred from one molecule to the next molecule in the series. The transfer of the electron can be complete, resulting in an ionic bond, or there can be a partial shift of the electron, resulting in a covalent bond. The loss of an electron by an atom is called **oxidation,** and the gain of an electron by an atom is called **reduction.** Because the loss of an electron by one atom is accompanied by the gain of that electron by another atom, these reactions are called **oxidation-reduction reactions.** In Chapter 25 the oxidation-reduction reactions of metabolism are considered in greater detail.

Once produced, ATP is used to provide energy for other chemical reactions (anabolism) or to drive cell processes such as muscle contraction. In the process ATP is converted back to ADP and a phosphate group.

$$\text{ATP} \rightarrow \text{ADP} + \text{P} + \text{Energy (for anabolism and other cell processes)}$$

ATP is often called the energy currency of cells because ATP is capable of both storing and providing energy. The concentration of ATP is maintained within a narrow range of values, and essentially all endergonic chemical reactions stop when there is inadequate ATP.

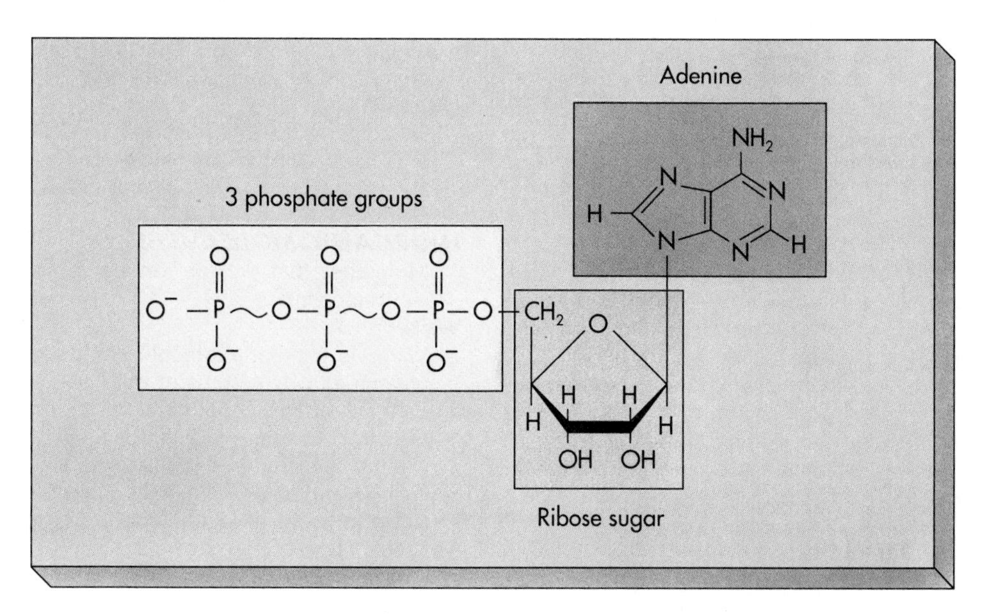

FIGURE 2-26 Adenosine triphosphate (ATP) molecule.

SUMMARY

CHEMISTRY IS THE STUDY OF THE COMPOSITION, STRUCTURE, AND PROPERTIES OF SUBSTANCES AND THE REACTIONS THEY UNDERGO. MUCH OF THE STRUCTURE AND FUNCTION OF HEALTHY OR DISEASED ORGANISMS CAN BE UNDERSTOOD AT THE CHEMICAL LEVEL.

BASIC CHEMISTRY *p. 30*

Matter and Elements

1. Matter is anything that occupies space. Mass is the amount of matter in an object, and weight is due to the gravitational attraction of the earth for matter.
2. Matter is composed of elements.

Atoms

1. The atom is the smallest unit of matter that cannot be altered by chemical means. An element is matter composed of only one kind of atom.
2. Atoms consist of protons, neutrons, and electrons.
   - Protons are positively charged, electrons are negatively charged, and neutrons have no charge.
   - Protons and neutrons are found in the center, or nucleus, of the atom, and electrons move around the nucleus in orbitals. An electron shell is an orbital or a group of orbitals.
3. The atomic number is the number of protons in an atom. The mass number is the sum of the protons and the neutrons.
4. The atomic weight is the weight in grams of Avogadro's number of atoms. A mole is Avogadro's number of atoms.

Electrons and Chemical Bonds

1. The chemical behavior of atoms is determined mainly by the electrons in their outermost electron shell.
   - A chemical bond occurs when atoms share or transfer electrons to achieve a stable outer electron shell.
   - A molecule is two or more atoms joined together by chemical bonds.
2. Ions (electrolytes) are atoms that have gained or lost electrons.
   - An atom that loses an electron becomes positively charged and is called a cation. An anion is an atom that becomes negatively charged after accepting an electron.
   - The oppositely charged cation and anion attract each other, forming an ionic bond.
3. A covalent bond is the sharing of electron pairs between atoms. A polar covalent bond results when the sharing of electrons is unequal and produces a molecule that is negative on one end and positive on the other end.
4. A hydrogen bond is the weak attraction that occurs between the oppositely charged regions of polar molecules. Hydrogen bonds are important in determining the three-dimensional structure of large molecules.

CHEMICAL REACTIONS *p. 37*

Classification of Chemical Reactions

1. Synthesis reactions are the combining of two or more atoms, ions, or molecules to form a new or larger molecule. Anabolism is the sum of all the synthesis reactions in the body.
2. Decomposition reactions are the breaking of the chemical bonds in a large molecule to produce smaller molecules, ions, or atoms. All of the decomposition reactions in the body are called catabolism.
3. Exchange reactions occur when part of a molecule is replaced by another element or molecule.
4. Dehydration reactions are synthesis reactions in which water is produced. Hydrolysis reactions are decomposition reactions in which water is depleted.

Reversible Reactions

Freely reversible reactions produce an equilibrium condition in which a certain amount of product and of reactant are always present.

Rate of Chemical Reactions

The rate of chemical reactions can be affected by the nature of the reactants, the concentration of the reactants, temperature, and enzymes (catalysts).

ENERGY *p. 39*

Energy is the ability to do work. Potential energy is stored energy, and kinetic energy is energy resulting from an object's moving.

Electrical Energy

Electrical energy involves the movements of ions or electrons and is responsible for nerve impulses.

Electromagnetic Energy

Electromagnetic energy moves in waves.

Chemical Energy

1. Chemical bonds are a form of potential energy.
2. Exergonic reactions are chemical reactions in which the reactants have less potential energy than the products. The decrease in energy may be lost as heat or be transferred to other products. Exergonic reactions are typical of catabolic reactions.
3. In endergonic reactions the products contain more potential energy than the reactants. Endergonic reactions are typical of anabolic reactions.

Heat Energy

Heat energy results from the random movement of atoms, ions, or molecules. Heat energy is released in exergonic reactions and is responsible for body temperature.

INORGANIC MOLECULES *p. 41*

Molecules that do not contain carbon atoms are inorganic molecules.

Water

1. Water is a polar molecule composed of one atom of oxygen and two atoms of hydrogen.
2. Water stabilizes body temperature, protects against friction and trauma, makes chemical reactions possible, directly participates in chemical reactions (e.g., dehydration and hydrolysis reactions), and is a mixing medium (e.g., solutions, suspensions, and colloids).

Solution Concentrations

1. A liquid (solvent) containing a dissolved substance (solute) is a solution.
2. An osmole contains 1 mole (Avogadro's number) of particles (i.e., atoms, ions, or molecules) in 1 kg water. A milliosmole is 1/1000 of an osmole.

Acids and Bases

1. Acids are proton (i.e., hydrogen ions) donors, and bases (e.g., hydroxide ions) are proton acceptors.
2. A strong acid or base almost completely dissociates in water. A weak acid or base partially dissociates.

The pH Scale

1. A neutral solution has an equal number of hydrogen ions and hydroxide ions and is assigned a pH of 7.
2. Acid solutions, in which the number of hydrogen ions is greater than the number of hydroxide ions, have pH values less than 7.
3. Basic or alkaline solutions have more hydroxide ions than hydrogen ions and a pH greater than 7.

Salts

A salt is a molecule consisting of a cation other than hydrogen and an anion other than hydroxide. Salts are formed when acids react with bases.

Buffers

Chemicals (weak acids or bases) that resist changes in pH are buffers.

Oxygen

Oxygen is necessary in the reactions that extract energy from food molecules in living organisms.

Carbon Dioxide

During metabolism when the chemical bonds of organic molecules are broken down, carbon dioxide and energy are released.

ORGANIC MOLECULES p. 44

Organic molecules contain carbon atoms bound together by covalent bonds (except carbon dioxide and carbon monoxide).

Carbohydrates

1. Monosaccharides are the basic building blocks of other carbohydrates. They, especially glucose, are important sources of energy. Examples are ribose, deoxyribose, glucose, fructose, and galactose.
2. Disaccharide molecules are formed by dehydration reactions between two monosaccharides. They are broken apart into monosaccharides by hydrolysis reactions. Examples of disaccharides are sucrose, lactose, and maltose.
3. Polysaccharides are many monosaccharides bound together to form long chains. Examples include cellulose, starch, and glycogen.

Lipids

1. Triglycerides are composed of glycerol and fatty acids. One, two, or three fatty acids can attach to the glycerol molecule.
   - Fatty acids are straight chains of carbon molecules of varying lengths, which may be saturated (only single covalent bonds between carbon atoms) or unsaturated (one or more double covalent bonds between carbon atoms).
   - Energy is stored in fats.
2. Phospholipids are lipids in which a fatty acid is replaced by a phosphate-containing molecule. Phospholipids are a major structural component of cell membranes.
3. Steroids are lipids composed of four interconnected ring molecules. Examples include cholesterol, bile salts, and sex hormones.
4. Other lipids include fat-soluble vitamins, prostaglandins, thromboxanes, and leukotrienes.

Proteins

1. The building blocks of protein are amino acids, which are joined by peptide bonds.
2. The number, kinds, and arrangement of amino acids determine a protein's primary structure. Hydrogen bonds between amino acids determine secondary structure, and hydrogen bonds between amino acids and water determine tertiary structure. Interactions between different protein subunits determine quaternary structure.
3. Enzymes are specialized protein catalysts that lower the activation energy for chemical reactions. Enzymes speed up chemical reactions but are not consumed or altered in the process.
4. Activation energy is the energy required to start exergonic or endergonic reactions. Enzymes lower activation energy.
5. Enzymes are specific and operate according to the lock-and-key model.
6. Cofactors are ions or organic molecules such as vitamins that are required for some enzymes to function.

Nucleic Acids: DNA and RNA

1. The basic unit of nucleic acids is the nucleotide, which is a monosaccharide with an attached phosphate and organic base.
2. Deoxyribonucleic acid (DNA) nucleotides contain the monosaccharide deoxyribose and the organic bases adenine, thymine, guanine, or cytosine. DNA occurs as a double strand of joined nucleotides and is the hereditary material of cells.
3. Ribonucleic acid (RNA) nucleotides are composed of the monosaccharide ribose. The organic bases are the same as for DNA except that thymine is replaced with uracil.

Adenosine Triphosphate

Adenosine triphosphate (ATP) stores energy derived from catabolism. The energy is released from ATP and is used in anabolism and other cell processes.

CONTENT REVIEW

1. Define chemistry. Why is an understanding of chemistry important for studying human anatomy and physiology?
2. Define matter, mass, and weight.
3. Describe or diagram the structure of a single atom. Contrast the charge and the weight of the subatomic particles.
4. Define atomic number, atomic mass, atomic weight, and mole.
5. Distinguish between ionic, covalent, polar covalent, and hydrogen bonds. Define cation and anion.
6. Define a chemical reaction. Contrast what occurs in synthesis and decomposition reactions. How do anabolism, catabolism, and metabolism relate to synthesis and decomposition reactions?
7. What is an exchange reaction? How can an exchange reaction result in a dehydration or hydrolysis reaction?
8. Describe reversible reactions. What is meant by the equilibrium condition in freely reversible reactions?
9. List four factors that affect the rate of chemical reactions. How must each factor change to increase the rate of reaction?
10. Define energy. How are potential and kinetic energy different from each other?
11. Describe electrical energy, electromagnetic energy, chemical energy, and heat energy.
12. Contrast exergonic and endergonic reactions. Which is typical of catabolism? Of anabolism?
13. Define an inorganic and an organic molecule.
14. List four functions that water performs in living systems.
15. Define solution, solute, and solvent.
16. What is the osmolality of a solution?
17. Define acid and base. Describe the pH scale. What is the difference between a strong acid or base and a weak acid or base?
18. What is a salt? What is a buffer, and why are buffers important to organisms?
19. What are the functions of oxygen and carbon dioxide in living systems?
20. Name the four major types of organic molecules important to life.
21. Name the basic building blocks of carbohydrates, fats, proteins, and nucleic acids.
22. Distinguish between fats, phospholipids, and steroids. Name an example of each.
23. What makes proteins different from each other? Define a peptide bond.
24. Describe the primary, secondary, tertiary, and quaternary structures of proteins.
25. Chemically, what type of organic molecule is an enzyme? What do enzymes do, and how do they work? Define cofactor.
26. What are the structural and functional differences between DNA and RNA?
27. Describe the structure of ATP. What role does this molecule play in energy exchange?

CONCEPT QUESTIONS

1. Iron has an atomic number of 26 and a mass number of 56. How many protons, neutrons, and electrons are in an atom of iron? If an atom of iron lost three electrons, what would be the charge of the resulting ion? Write the correct symbol for this ion.
2. A mixture of chemicals was warmed slightly. As a consequence, although no more heat was added, the solution became very hot. Explain what occurred to make the solution so hot.
3. Two solutions, A and B, when mixed together at room temperature, produced a chemical reaction. However, when the solutions were boiled and allowed to cool to room temperature before mixing, no chemical reaction took place. Explain.
4. In terms of exergonic and endergonic reactions, explain why eating food is necessary for increasing muscle mass.
5. Solution A has a pH of 2, and solution B has a pH of 8. If equal amounts of solutions A and B are mixed, will the resulting solution be acidic or basic?
6. Given a buffered solution that is based on the following equilibrium:

$$CO_2 + H_2O \leftrightarrow H_2CO_3 \leftrightarrow H^+ + HCO_3^-$$

what will happen to the pH of the solution if $NaHCO_3$ is added to the solution?

7. An enzyme, E, catalyzes the following reaction:

$$A + B \xrightarrow{E} C$$

However, the product, C, binds to the active site of the enzyme in a reversible fashion. What will happen if A and B are continually added to a solution that contains a fixed amount of the enzyme?

8. Given the materials commonly found in a kitchen, explain how to distinguish between a protein and a lipid.
9. A student was given two unlabeled substances, one a typical phospholipid and one a typical protein. She was asked to determine which substance was the protein and which was the phospholipid. The available techniques allowed her to determine the elements in each sample. How could she identify each substance?

? ?

ANSWERS TO PREDICT QUESTIONS

1 p. 30. The mass (amount of matter) of the astronaut on the earth's surface and in outer space would not change. In outer space where the force of gravity from the earth is very small, the astronaut would be "weightless" compared to his weight on the earth's surface.

2 p. 32. Potassium has 19 protons (the atomic number), 20 neutrons (the mass number minus the atomic number), and 19 electrons (because the number of electrons equals the number of protons).

3 p. 32. In 12 g of carbon there is Avogadro's number of atoms. In 12 g of magnesium, which is approximately half a mole of magnesium, there is approximately one half of Avogadro's number of atoms.

4 p. 32. Calcium chloride's chemical formula is $CaCl_2$. The two electrons from the outer shell of the calcium atom are donated to two chlorine atoms, filling the outer shell of each with eight electrons and eliminating the partial outer shell of calcium. As a result, the calcium atom becomes a calcium ion with a double positive charge (because it is missing two electrons), and the chlorine atoms become chloride ions with one negative charge each (because they have gained one electron each). The positive calcium ion attracts two negative chloride ions to it to form calcium chloride through ionic bonding.

5 p. 39. A decrease in blood carbon dioxide would decrease the amount of carbonic acid and therefore the blood hydrogen ion level. Because carbon dioxide and water are in equilibrium with hydrogen ions and bicarbonate ions, with carbonic acid as an intermediate, a decrease in carbon dioxide causes some hydrogen ions and bicarbonate ions to join together to form carbonic acid, which dissociates to form carbon dioxide and water. Consequently, the hydrogen ion concentration decreases.

6 p. 41. During exercise muscle contractions increase. This increase requires the release of stored energy in exergonic reactions. Some of the energy is used to drive muscle contractions, and some of it is released as heat. Because the rate of these reactions increases during exercise, more heat is produced than when at rest, and body temperature increases.

7 p. 54. Changing one amino acid in a protein chain may alter the three-dimensional structure of the protein chain. If the three-dimensional structure of an enzyme is changed, the function of the enzyme can decrease.

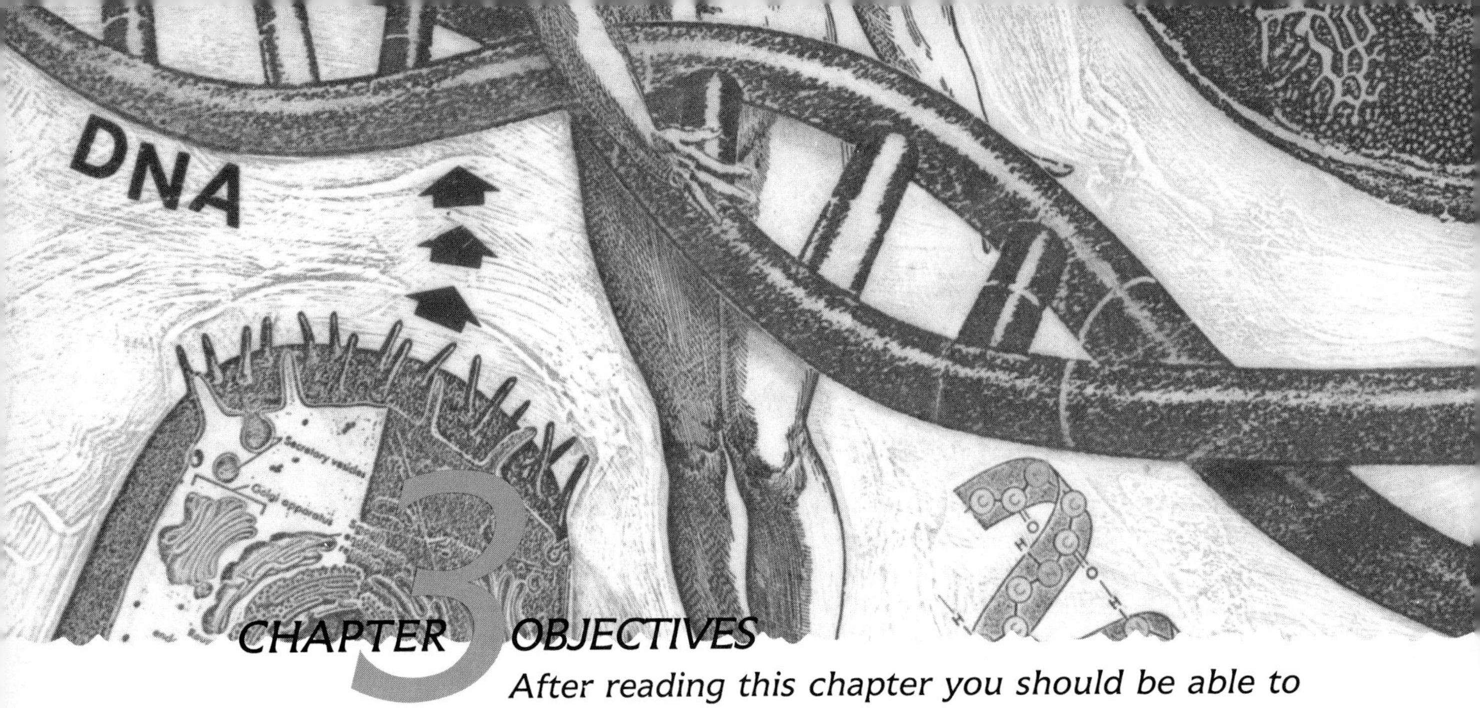

CHAPTER 3 OBJECTIVES

After reading this chapter you should be able to

1. Describe the structure of the plasma membrane. Explain why the plasma membrane is more permeable to lipid-soluble substances and small molecules than to large water-soluble substances.
2. Describe the factors that affect the rate and the direction of diffusion of a solute in a solvent.
3. Explain the role of osmosis and of osmotic pressure in controlling the movement of water across the plasma membrane. Compare isotonic, hypertonic, and hypotonic solutions with isosmotic, hyperosmotic, and hyposmotic solutions.
4. Describe mediated transport and explain the characteristics of specificity, competition, and saturation.
5. Describe the processes of facilitated diffusion, active transport, phagocytosis, pinocytosis, and exocytosis.
6. Describe the structure and function of the nucleus and nucleoli.
7. Define cytoplasm, cytosol, and organelle.
8. Contrast microtubules, microfilaments, and intermediate filaments.
9. Compare the structure and function of rough and smooth endoplasmic reticulum.
10. Explain the role in secretion of the Golgi apparatus and secretory vesicles.
11. Distinguish between lysosomes and peroxisomes.
12. Describe the structure and function of mitochondria.
13. Describe centrioles, spindle fibers, cilia, flagella, and microvilli.
14. Define cell metabolism and contrast aerobic and anaerobic respiration.
15. Describe the process of protein synthesis.
16. Explain what is accomplished during mitosis and cytokinesis.
17. Describe the events of meiosis and explain how they result in the production of genetically unique individuals.

Structure and Function of the Cell

KEY TERMS

active transport

cytoplasm (si'to-plazm)

diffusion

endoplasmic reticulum (en'do-plaz'mik rĕ-tik'u-lum)

facilitated diffusion

Golgi (gōl'je) apparatus

meiosis (mi-o'sis)

mitochondria (mi'to-kon'dre-ah)

mitosis (mi-to'sis)

nucleus (nu'kle-us)

osmosis (os-mo'sis)

plasma membrane

ribosome

transcription

translation

RELATED TOPICS

The following concepts are important for a good understanding of this chapter. If you are not familiar with them, you should review them before proceeding.

Ions, molecules, classes of organic molecules, and solutions (Chapter 2)

A **CELL** IS THE STRUCTURAL AND FUNCTIONAL UNIT OF LIVING ORGANISMS. Trillions of cells and the substances between them compose the human body. All human cells originate from a single fertilized egg, and as **differentiation** (cells becoming different from each other) proceeds during embryonic development, cells specialize and give rise to a wide variety of cell types such as nerve, muscle, bone, fat, and blood cells.

Although cells eventually may have quite different structures and functions, all cells initially share some common characteristics (Figure 3-1). The **plasma** (plaz'mah) or **cell membrane** forms the outer boundary of the cell through which the cell interacts with its external environment. The **nucleus** (nu'kle-us) usually is located centrally and functions to direct cell activities, most of which take place in the **cytoplasm** (si'to-plazm), located between the plasma membrane and the nucleus.

This chapter presents the structure and function of the cell's components. It is a brief overview of cell biology and provides the reader with adequate background information for the remainder of this text.

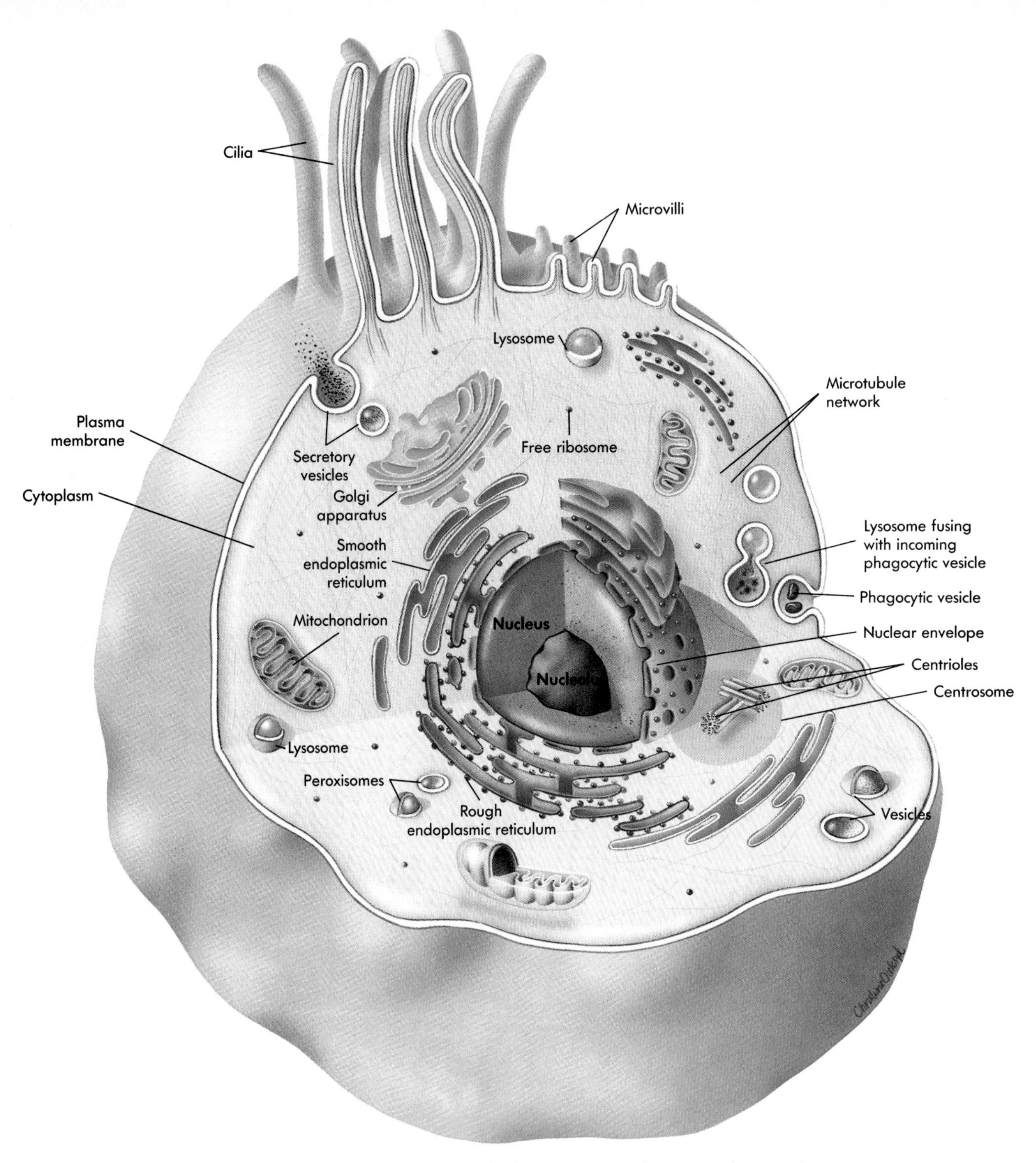

FIGURE 3-1 Generalized cell showing the plasma membrane, nucleus, and cytoplasm with its organelles. Although no single cell contains all these organelles, many cells contain a large number of them.

STRUCTURE OF THE PLASMA MEMBRANE

The plasma membrane is the outermost component of a cell. Substances outside the plasma membrane are **extracellular** or **intercellular**, and substances inside it are **intracellular.** The functions of the plasma membrane are to enclose and support the cell contents and to determine what moves into and out of the cell.

The plasma membrane consists primarily of lipids and proteins and small amounts of carbohydrates (Figure 3-2). The predominant lipids are phospholipids and cholesterol. **Phospholipids** readily assemble to form a **lipid bilayer,** a double layer of lipid molecules, because they have a polar (charged) head and a nonpolar (uncharged) tail (see Chapter 2). The polar **hydrophilic** (water-loving) heads are exposed to water inside and outside the cell, whereas the nonpolar **hydrophobic** (water-fearing) tails face each other in the interior of the plasma membrane. The other major lipid in the plasma membrane is **cholesterol** (see Chapter 2), which is interspersed between the phospholipids. Cholesterol is believed to increase the mechanical stability and flexibility of the plasma membrane.

Although the basic structure of the plasma membrane is determined mainly by its lipids, the functions of the plasma membrane are determined mainly by its proteins. Some protein molecules penetrate the lipid bilayer from one surface to the other, whereas others are attached to either the inner or outer surfaces of the lipid bilayer. Some of the proteins function in membrane transport by forming membrane channels or by acting as carrier molecules. Other proteins function as receptors, markers, enzymes, or structural supports in the membrane.

Receptor molecules are part of an intercellular communication system that enables coordination of the activities of the cells. For example, a nerve cell can release a chemical messenger that moves to a muscle cell and binds to its receptor. The binding acts as a signal that triggers a response such as con-

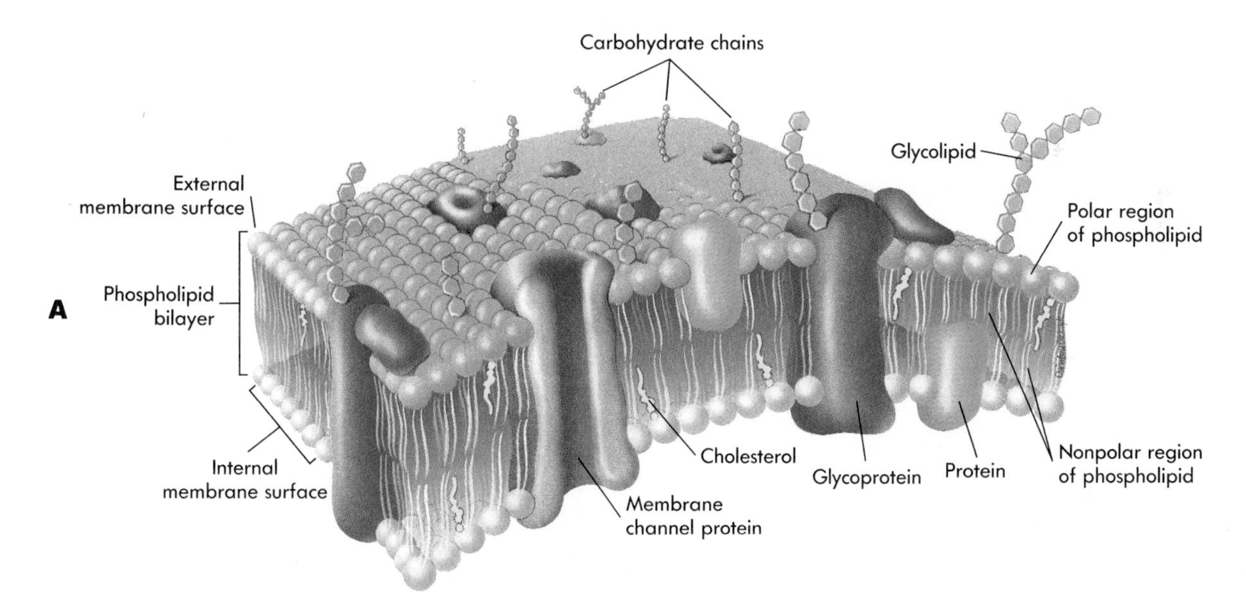

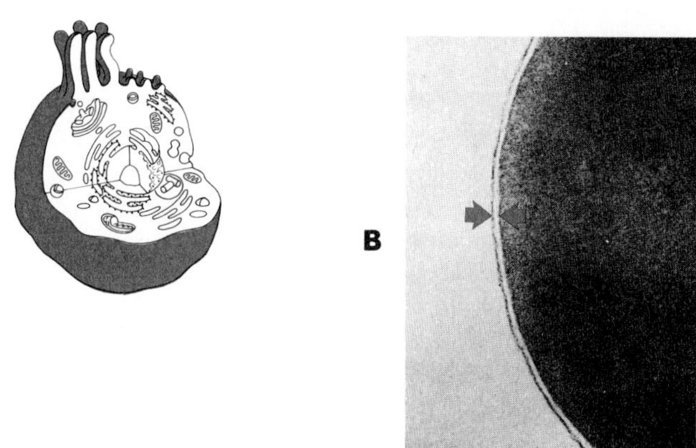

FIGURE 3-2 **A** Fluid mosaic model of the plasma membrane composed of a lipid bilayer of phospholipids and cholesterol with proteins "floating" in the membrane. The nonpolar hydrophobic portion of each phospholipid molecule is directed toward the center of the membrane, and the polar hydrophilic portion is directed toward the water environment either outside or inside the cell. **B** Electron micrograph of human red blood cell with membrane indicated by the arrow. Proteins at either surface of the lipid bilayer stain more readily than the lipid bilayer and give the membrane the appearance of consisting of three parts: the two dark outer parts are proteins and the phospholipid heads, and the lighter central part is the phospholipid tails.

traction in the muscle cell. The same chemical messenger would have no effect on another cell that lacks the receptor molecule. **Marker molecules** allow cells to identify each other. Examples include recognition of the egg by the sperm and the ability of the immune system to distinguish between self-cells and foreign cells such as bacteria or donor cells in an organ transplant. Intercellular communication and recognition are important because cells are not isolated entities, and they must work together to ensure normal body functions.

The ability of receptor and marker molecules to function depends on their three-dimensional shape. This is similar to the effect of an enzyme's shape (lock-and-key model; see Chapter 2) on the ability of the enzyme to function. The shape of receptor and marker molecules is affected by carbohydrates, most of which are chains of monosaccharides (sugars) that attach to proteins and lipids to form **glycoproteins** and **glycolipids** on the outer surface of the plasma membrane.

The modern concept of the plasma membrane, the **fluid mosaic model,** suggests that the plasma membrane is neither rigid nor static in structure but is highly flexible and may change its shape and composition through time. Each half of the lipid bilayer functions as a liquid in which other molecules such as proteins "float." The fluid nature of the lipid bilayer has several important consequences. It provides an important means of distributing molecules within the plasma membrane, and slight damage to the membrane can be repaired because the phospholipids tend to reassemble (flow) around the damaged site and seal it closed. In addition, the fluid nature of the lipid bilayer enables membranes to fuse with one another.

MOVEMENT THROUGH THE PLASMA MEMBRANE

The plasma membrane separates the extracellular material from the intracellular material and is **selectively permeable,** allowing some substances but not others to pass through it. The intracellular material has a different composition from that of the extracellular material, and the survival of the cell depends on the maintenance of those differences. Enzymes, glycogen, and potassium ions are found in higher concentrations intracellularly, and sodium, calcium, and chloride ions are found in greater concentrations extracellularly. In addition, nutrients must continually enter the cell, and waste products must exit while the volume of the cell remains unchanged. Because of the plasma membrane's permeability characteristics and its ability to transport molecules selectively, the cell is able to maintain homeostasis. Rupture of the membrane, alteration of its permeability characteristics, or inhibition of transport processes can disrupt the normal concentration differences across the plasma membrane and lead to cell death.

Substances move across the plasma membrane in four ways: (1) directly through the lipid bilayer, (2) through membrane channels, (3) with carrier molecules in the membrane, or (4) in vesicles. Molecules that are soluble in lipids, such as oxygen, carbon dioxide, and steroids, pass through the plasma membrane readily by dissolving in the lipid bilayer. The lipid bilayer acts as a barrier to most polar substances, which are not soluble in lipids.

Membrane channels are composed of large protein molecules that extend from one surface of the plasma membrane to the other (see Figure 3-2, *A*). Several channel types exist, and each type allows molecules of only a certain size range to pass through it. In addition, most channels are positively charged, and because like charges repel one another, positive ions pass through the channels less readily than neutral or negatively charged molecules of the same size. Water and chloride ions (Cl^-) pass through the membrane channels relatively easily, but sodium ions (Na^+) and potassium ions (K^+) pass through more slowly.

Large polar substances such as glucose and amino acids cannot pass through the plasma membrane in significant amounts because they are too large to move through membrane channels and their polar nature prevents their dissolving in the lipid bilayer. Protein molecules within the membrane function as carrier molecules, which combine with large polar substances on one side of the membrane and transport them to the other side of the membrane.

Large polar molecules, small pieces of matter, and even whole cells can be transported across the plasma membrane in a **vesicle** (ves'ĭ-kl), which is a membrane-bound sac. Because of the fluid nature of membranes, the vesicle and the plasma membrane fuse, allowing the contents of the vesicle to cross the plasma membrane.

Diffusion

A solution is either a liquid or a gas and consists of one or more substances called **solutes** dissolved in the predominant liquid or gas, which is called the **solvent. Diffusion** can be viewed as the tendency for solute molecules to move from an area of high concentration to an area of low concentration in solution (Figure 3-3, *A* and *B*). Diffusion is a product of the constant random motion of all atoms, molecules, or ions in a solution. Because there are more solute particles in an area of high concentration than in an area of low concentration and because the particles move randomly, the chances are greater that solute particles will move from the higher to the lower concentration than in the opposite direction. Thus the

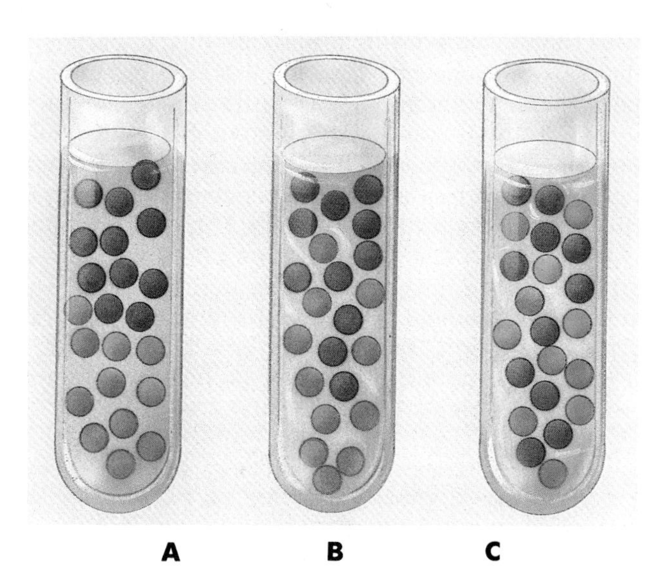

FIGURE 3-3 Diffusion. **A** One solution (*red*, representing one type of molecule) is layered onto a second solution (*blue*, representing a second type of molecule). There is a concentration gradient for the red molecules from the red solution into the blue solution because there are no red molecules in the blue solution. There is also a concentration gradient for the blue molecules from the blue solution into the red solution because there are no blue molecules in the red solution. **B** Red molecules move with their concentration gradient into the blue solution, and the blue molecules move with their concentration gradient into the red solution. **C** Red and blue molecules are distributed evenly throughout the solution. Even though the red and blue molecules continue to move randomly, an equilibrium exists, and no net movement occurs because no concentration gradient exists.

overall or net movement is from the high concentration to the area of low concentration. At equilibrium the net movement of solutes stops, although the random molecular motion continues, and the movement of solutes in any one direction is balanced by an equal movement in the opposite direction (Figure 3-3, C). The movement and distribution of smoke or perfume throughout a room in which there are no air currents or of a dye throughout a beaker of still water are examples of diffusion.

A concentration difference exists when the concentration of a solute is greater at one point than at another point in a solvent. The concentration difference between two points divided by the distance between those two points is called the **concentration gradient.** Solutes diffuse with their concentration gradients (from a high to a low concentration) until an equilibrium is achieved. For a given concentration difference between two points in a solution the concentration gradient is larger if the distance between the two points is small, and the concentration gradient is smaller if the distance between the two points is large.

The rate of diffusion is influenced by the magnitude of the concentration gradient, the temperature of the solution, the size of the diffusing molecules, and the viscosity of the solvent. The greater the concentration gradient, the greater is the number of solute particles moving from high to low concentration. As the temperature of a solution increases, the speed at which all molecules move increases, resulting in a greater diffusion rate. Small molecules diffuse through a solution more readily than do large molecules. **Viscosity** is a measure of how easily a liquid flows; thick solutions such as syrup are more viscous than water. Diffusion occurs more slowly in viscous solvents than in thin, watery solvents.

Diffusion of molecules is an important means by which substances move between the extracellular and intracellular fluids in the body. Substances that can diffuse through either the lipid bilayer or the membrane channels can pass through the plasma membrane. Thus some nutrients enter and some waste products leave the cell by diffusion, and maintenance of the appropriate intracellular concentration of these substances depends to a large degree on the process of diffusion. For example, if the extracellular concentration of oxygen is reduced, inadequate oxygen will diffuse into the cell, and normal cell function cannot occur.

1

Urea is a toxic waste produced inside cells. It diffuses from the cells into the blood and is eliminated from the body by the kidneys. What would happen to the intracellular and extracellular concentration of urea if the kidneys stopped functioning?

? ? ? ? ? ? ? ? ? ? ?

Osmosis

Osmosis (os-mo′sis) is the diffusion of water (solvent) across a selectively permeable membrane (e.g., a plasma membrane) which allows water but not all solutes to diffuse through it. Water diffuses from a solution with proportionately more water across a selectively permeable membrane into a solution with proportionately less water. Because solution concentrations are defined in terms of solute concentrations (see Chapter 2), water diffuses from the less concentrated solution (fewer solutes, more water) into the more concentrated solution (more solutes, less water). Osmosis is important to cells because large volume changes caused by water movement disrupt normal cell function.

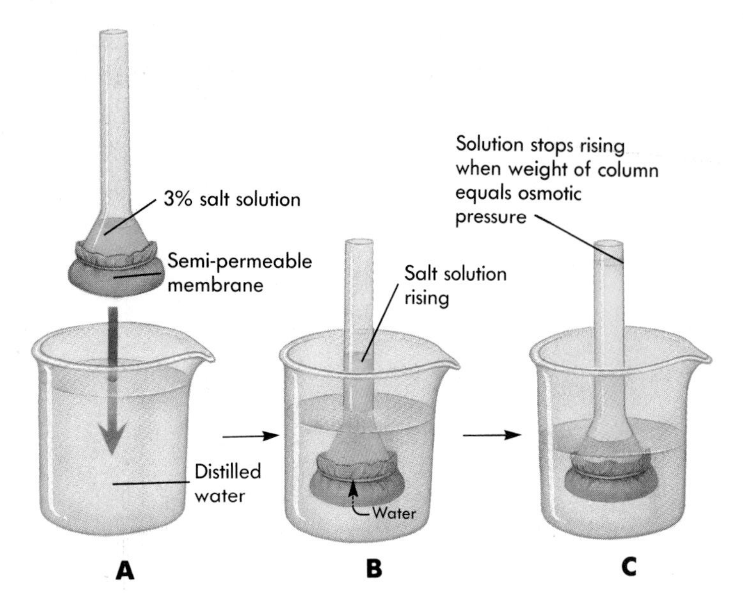

FIGURE 3-4 Osmosis. **A** The end of a tube containing a 3% salt solution is closed at one end with a semipermeable membrane that allows water molecules to pass through it but retains the salt molecules within the tube. **B** Tube is immersed in distilled water. Because the tube contains salt as well as water molecules, the tube has proportionately less water than the beaker, which contains only water. The water molecules diffuse with their concentration gradient into the tube. Because the salt molecules cannot leave the tube, the total fluid volume inside the tube increases, and fluid moves up the glass tube as a result of osmosis. **C** Water will continue to move into the tube until the weight of the column of water in the tube (hydrostatic pressure) exerts a downward force equal to the osmotic force moving water molecules into the tube. The hydrostatic pressure that prevents net movement of water into the tube is termed the osmotic pressure of the solution in the tube.

Osmotic pressure is the force required to prevent the movement of water by osmosis across a selectively permeable membrane. The osmotic pressure of a solution can be determined by placing the solution into a tube that is closed at one end by a selectively permeable membrane (Figure 3-4). The tube is then immersed in distilled water. Water molecules move by osmosis through the membrane into the tube, forcing the solution to move up the tube. As the solution rises into the tube, its weight produces hydrostatic pressure that moves water out of the tube back into the distilled water surrounding the tube. At equilibrium net movement of water stops, which means the movement of water into the tube by osmosis is equal to the movement of water out of the tube caused by hydrostatic pressure. The osmotic pressure of the solution in the tube is equal to the hydrostatic pressure that prevents net movement of water into the tube.

Osmotic pressure is a measure of the tendency for water to move by osmosis across a selectively permeable membrane into a solution. Because water moves from less concentrated solutions (fewer solutes, more water) into more concentrated solutions (more solutes, less water), the greater the concentration of a solution (the less water it has), the greater is the tendency for water to move into the solution, and the greater the osmotic pressure must be to prevent that movement. Thus the greater the concentration of a solution, the greater is the osmotic pressure of the solution, and the greater is the tendency for water to move into the solution.

2

Given the experiment in Figure 3-4, what would happen to osmotic pressure if the membrane were not selectively permeable but instead allowed all solutes and water to pass through it?

? ? ? ? ? ? ? ? ? ? ?

Three terms frequently are used to compare the osmotic pressure of solutions. They are useful because solutions of differing concentrations often are separated by plasma membranes. Solutions with the same concentration of solute particles (see Chapter 2), even though their types of solute particles may differ, have the same osmotic pressure and are **isosmotic** (i'sos-mot'ik). If a solution has a greater concentration of solute particles and therefore a greater osmotic pressure than another solution, that solution is **hyperosmotic** (hi'per-os-mot'ik) with respect to the more dilute solution. The more dilute solution with the lower osmotic pressure is **hyposmotic** (hi'pos-mot'ik) with respect to the more concentrated solution.

Three additional terms describe the tendency of cells to shrink or swell when placed in a solution. If a cell is placed in a solution and the cell neither shrinks nor swells, the solution is **isotonic** (i'so-ton'ik). If the cell shrinks, the solution is **hypertonic** (hi'per-ton'ik); and if the cell swells, the solution is **hypotonic** (hi'po-ton'ik) (Figure 3-5).

An isotonic solution can be isosmotic. Because isosmotic solutions have the same concentration of solutes and water as the cell's cytoplasm, there is no net movement of water, and the cell neither swells nor shrinks (see Figure 3-5, *B*). Hypertonic solutions can be hyperosmotic and have a greater concentration of solute molecules and a lower concentration of water than the cell's cytoplasm. Therefore water moves by osmosis from the cell into the hypertonic solution, causing the cell to shrink, a process called **crenation** (kre-na'shun) (see Figure 3-5, *C*). Hypotonic solutions can be hyposmotic and have a smaller concentration of solute molecules and a greater concentration of water than the cell's cytoplasm. Therefore water moves by osmosis into the cell, causing it

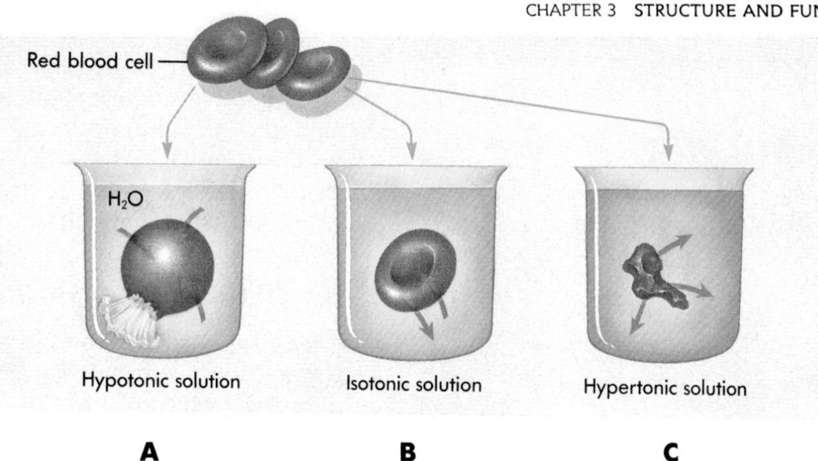

FIGURE 3-5 Effects of hypotonic, isotonic, and hypertonic solutions on red blood cells. **A** Hypotonic solutions with low ion concentrations result in swelling and lysis of cells. **B** Isotonic solutions with normal ion concentrations result in normal-shaped cells. **C** Hypertonic solutions with high ion concentrations result in shrinkage (crenation) of the cell.

to swell. If the cell swells enough, it can rupture, a process called **lysis** (see Figure 3-5, *A*). Solutions injected into the circulatory system or the tissues must be isotonic because crenation or swelling of cells disrupts their normal function and can lead to cell death.

The "osmotic" terms refer to solution concentration, and the "tonic" terms refer to the tendency of cells to swell or shrink. These terms should not be used interchangeably. For example, not all isosmotic solutions are isotonic. It is possible to prepare a solution of glycerol and a solution of mannitol that are isosmotic to the cell's cytoplasm. Because the solutions are isosmotic, they have the same concentration of solutes and water as the cell's cytoplasm. However, glycerol can diffuse across the plasma membrane, but mannitol cannot. When glycerol diffuses into the cell, the solute concentration of the cell's cytoplasm increases, and its water concentration decreases. Therefore water moves by osmosis into the cell, causing it to swell, and the glycerol solution is both isosmotic and hypotonic. In contrast, mannitol (and therefore water) cannot enter the cell, and the isosmotic mannitol solution is also isotonic.

Filtration

Filtration results when a partition containing small holes is placed in a stream of moving liquid. Particles small enough to pass through holes move through the partition with the liquid, but particles larger than the holes are prevented from moving beyond the partition. In contrast to diffusion, filtration depends on a pressure difference on either side of the partition. The liquid moves from the side of the partition with the greater pressure to the side with the lower pressure.

Filtration occurs in the kidneys as a step in urine formation. Blood pressure moves fluid from the blood through a partition, or filtration membrane. Ions and small molecules pass through the partition, whereas most proteins and blood cells remain in the blood.

Mediated Transport Mechanisms

Many essential molecules (e.g., amino acids and glucose) cannot enter the cell by diffusion, and many products (e.g., some proteins) cannot exit the cell by diffusion. **Mediated transport mechanisms** involve carrier molecules within the plasma membrane that function to move large, water-soluble molecules or electrically charged molecules across the plasma membrane. The carrier molecules are proteins that extend across the plasma membrane. Once a molecule to be transported binds to the carrier molecule on one side of the membrane, the three-dimensional shape of the carrier molecule changes, and the transported molecule is moved to the opposite side of the membrane. The carrier molecule then resumes its original shape and is available to transport other molecules (Figure 3-6).

Mediated transport mechanisms exhibit three characteristics: specificity, competition, and saturation. **Specificity** means that each carrier molecule binds to and transports only a single type of molecule. For example, the carrier molecule that transports glucose will not bind to amino acids or ions. The site on the carrier molecule that binds to and

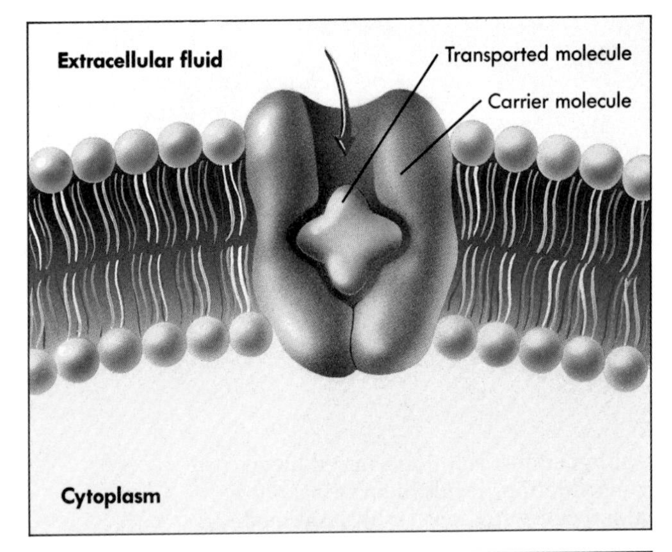

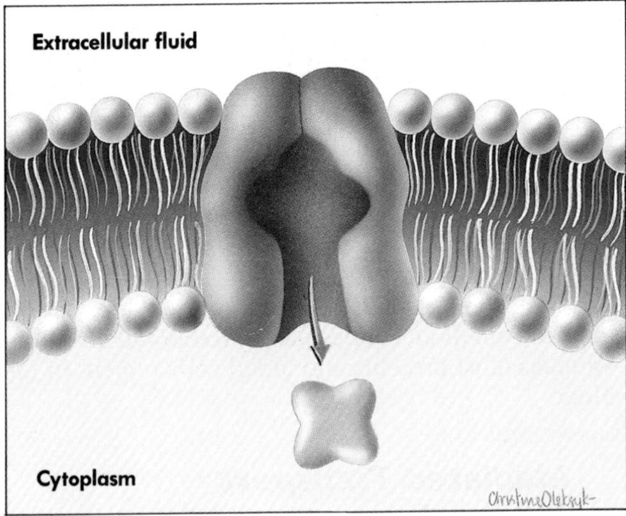

FIGURE 3-6 Mediated transport by a carrier molecule. The carrier molecule binds with a molecule on one side of the plasma membrane, changes shape, and releases the molecule on the other side of the plasma membrane.

transports molecules is the **active site,** and the chemical structure of the active site determines the specificity of the carrier molecule (Figure 3-7, *A*). **Competition** is the result of similar molecules binding to the carrier molecule. Although the active sites of carrier molecules exhibit specificity, closely related substances may bind to the same active site. The substance in the greater concentration or the substance that binds to the active site more readily will be transported across the plasma membrane at the greater rate (Figure 3-7, *B*). **Saturation** means that the rate of transport of molecules across the membrane is limited by the number of available carrier molecules. As the concentration of a transported substance increases, more carrier molecules have their active sites occupied. The rate at which the substance is transported increases, but once the concentration of the substance is increased so that all the active sites are occupied, the rate of transport remains constant even though the concentration of the substance is increased further (Figure 3-7, *C*).

There are two kinds of mediated transport: facilitated diffusion and active transport.

Facilitated diffusion

Facilitated diffusion is a carrier-mediated process that moves substances into or out of cells from a high to a low concentration. Facilitated diffusion does not require metabolic energy to transport substances across the plasma membrane. The rate at which molecules are transported is directly proportional to their concentration gradient up to the point of saturation when all the carrier molecules are being used. Then the rate of transport remains constant at its maximum rate.

3

The transport of glucose into and out of most cells occurs by facilitated diffusion. Once glucose enters a cell, it is rapidly converted to other molecules such as glucose 6-phosphate or glycogen. What effect does this conversion have on the ability of the cell to acquire glucose? Explain.

? ? ? ? ? ? ? ? ? ? ?

Active transport

Active transport is a carrier-mediated process that requires energy provided by adenosine triphosphate (ATP). Movement of the transported substance to the opposite side of the membrane and its subsequent release from the carrier molecule is fueled by the breakdown of ATP. The maximum rate at which active transport proceeds depends on the number of carrier molecules in the plasma membrane and the availability of adequate ATP.

Active transport processes are important because they can move substances against their concentration gradient from a low concentration to a high concentration. Consequently, they have the ability to accumulate substances on one side of the plasma membrane at concentrations many times greater than those on the other side. For example, glucose and amino acids are transported actively from the small intestine into the blood. In other cases the active transport mechanism may exchange one substance for another. For example, the sodium-potassium exchange pump moves sodium out of cells and potassium into cells (Figure 3-8). The result is a high concentration of sodium outside the cell and a high concentration of potassium inside the cell (see Chapter 9).

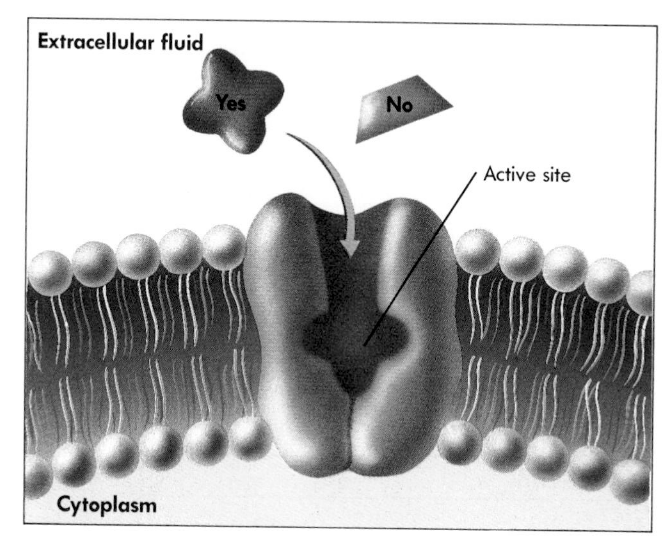

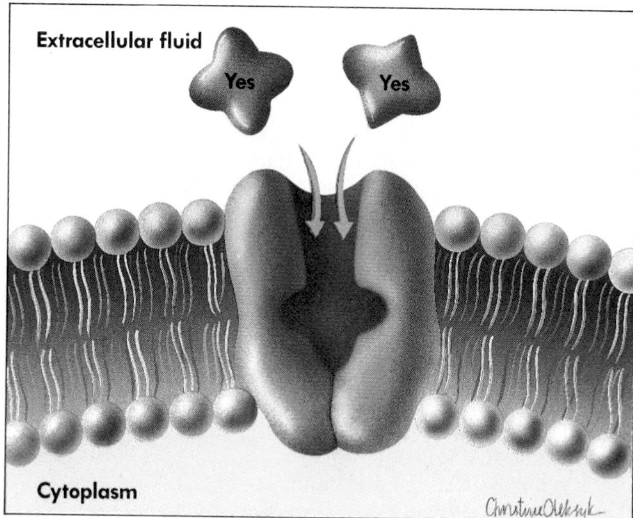

C

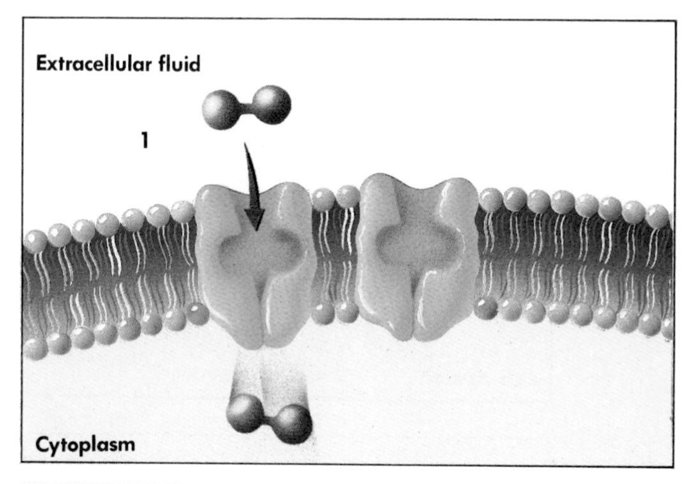

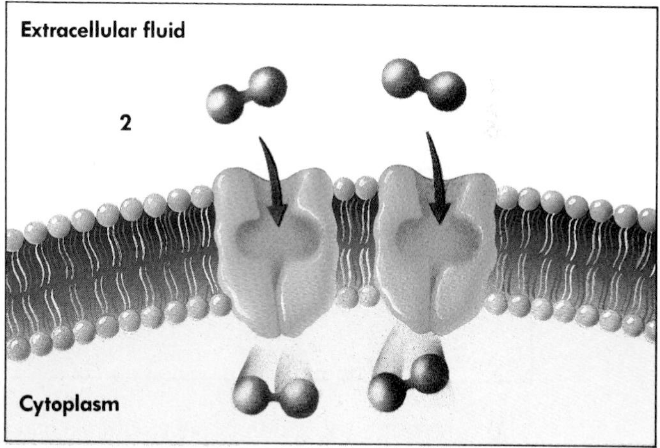

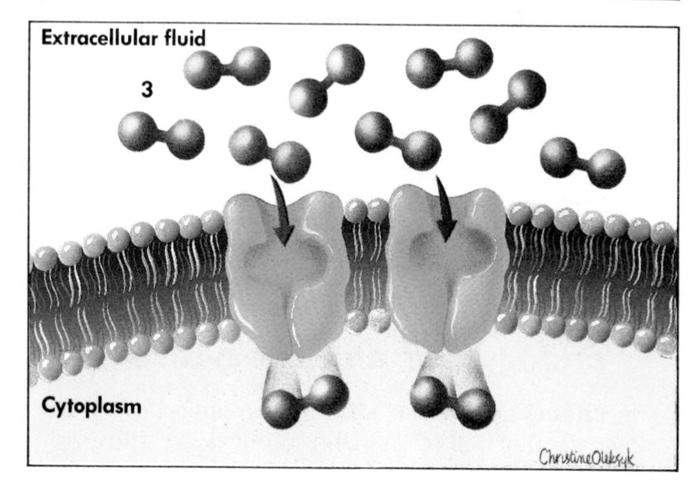

FIGURE 3-7 Characteristics of mediated transport.
A Specificity. Only molecules capable of binding to the active site are transported. **B** Competition. Similar molecules compete for the same active site.
C Saturation. At *1* there are more carrier molecules than molecules for transport. At *2* the concentration of molecules available for transport increases until all the carrier molecules are involved in transporting molecules (i.e., the system is saturated). At *3* even though the concentration of molecules for transport has increased, the rate of transport is limited by the number of carrier molecules. The rate of transport is the same as at *2*.

If movement of a substance is from a lower to higher concentration and ATP is used, the process is active transport. Active transport, however, can also move substances from higher to lower concentrations. For example, immediately after a meal rich in glucose, the concentration of glucose in the small intestine is greater than in the blood. Active transport uses ATP to move glucose into the blood, even though the movement is from higher to lower concentration.

4

GIVEN: A transport process exhibits saturation; poisons that block ATP production do not affect the transport process; and movement is always from a high to a low concentration.
Is the transport process simple diffusion, facilitated diffusion, or active transport? Explain.

? ? ? ? ? ? ? ? ? ? ?

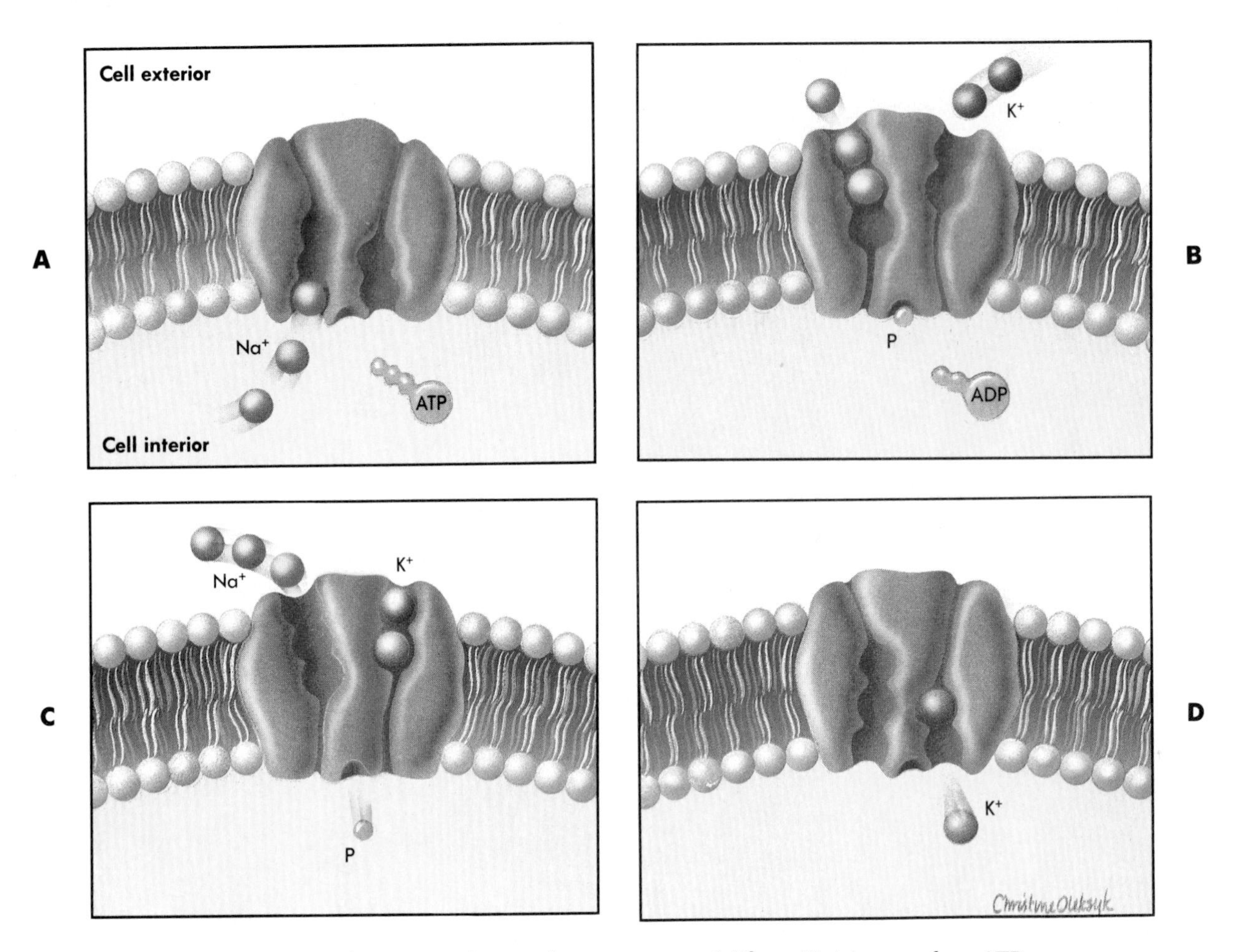

FIGURE 3-8 Sodium-potassium exchange pump. **A** Three Na^+ ions and an ATP bind to the carrier molecule. **B** The ATP breaks down to ADP and the phosphate attaches to the carrier molecule. The carrier molecule changes shape and the Na^+ ions are transported across the membrane. **C** The Na^+ ions diffuse away from the carrier molecule, two K^+ ions bind to the carrier molecule, and the phosphate is released. **D** The carrier molecule changes shape, transporting K^+ ions across the membrane, and the K^+ ions diffuse away from the carrier molecule. The carrier molecule can bind again to Na^+ ions and ATP.

Endocytosis and Exocytosis

Endocytosis (en′do-si′to-sis) includes both phagocytosis (fag′o-si-to′sis) and pinocytosis (pin′o-si-to′sis) and refers to the bulk uptake of material through the plasma membrane by the formation of a vesicle. A vesicle is a membrane-bound droplet found within the cytoplasm of a cell. A portion of the plasma membrane wraps around a particle or droplet and fuses so that the particle or droplet is surrounded by a membrane. That portion of the membrane then "pinches off" so that the particle or droplet, surrounded by a membrane, is within the cytoplasm of the cell, and the plasma membrane is left intact.

Phagocytosis means cell eating (Figure 3-9, *A*) and applies to endocytosis when solid particles are ingested and phagocytic vesicles are formed. White blood cells and some other cell types phagocytize bacteria, cell debris, and foreign particles. Therefore phagocytosis is important in the elimination of harmful substances from the body.

Pinocytosis means cell drinking and is distinguished from phagocytosis in that smaller vesicles are formed and they contain molecules dissolved in liquid rather than particles (Figure 3-9, *B*). Pinocytosis often forms vesicles near the tips of deep invaginations of the plasma membrane. It is a common transport phenomenon in a variety of cell types and occurs in certain cells of the kidney, epithelial cells

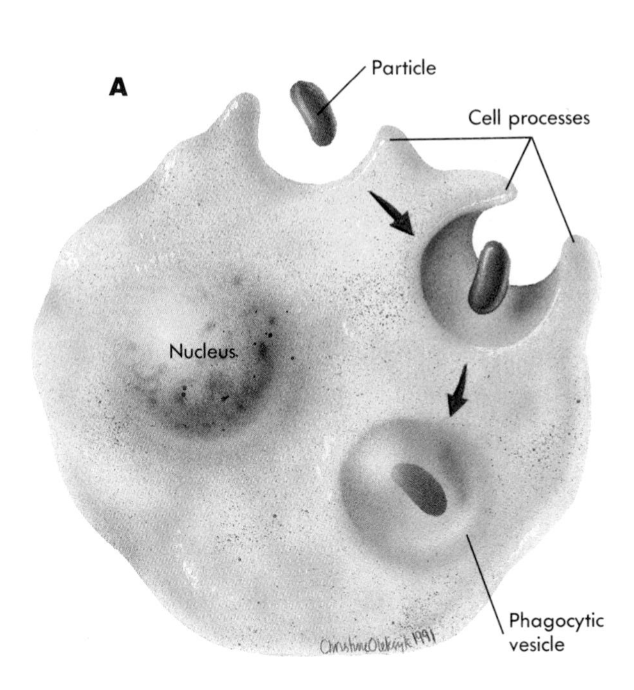

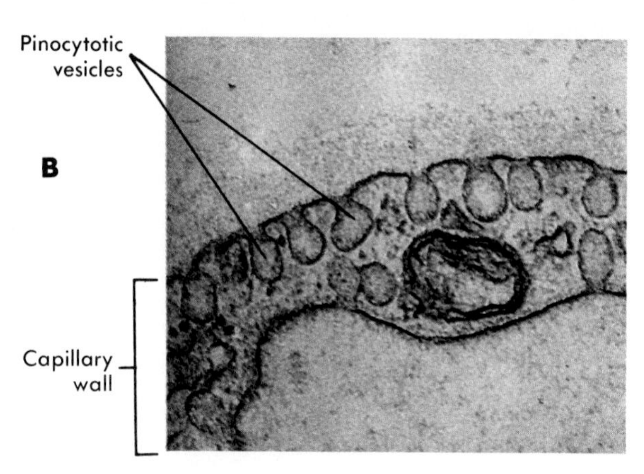

FIGURE 3-9 **A** Phagocytosis. Cell processes extend from the cell and surround the particle to be taken into the cell by phagocytosis. The cell processes have surrounded the particle, and they fuse to form a vesicle that contains the particle. The vesicle is internalized within the cell. **B** Pinocytosis is much like phagocytosis except the cell processes and therefore the vesicles formed are much smaller and the material inside the vesicle is liquid rather than particulate.

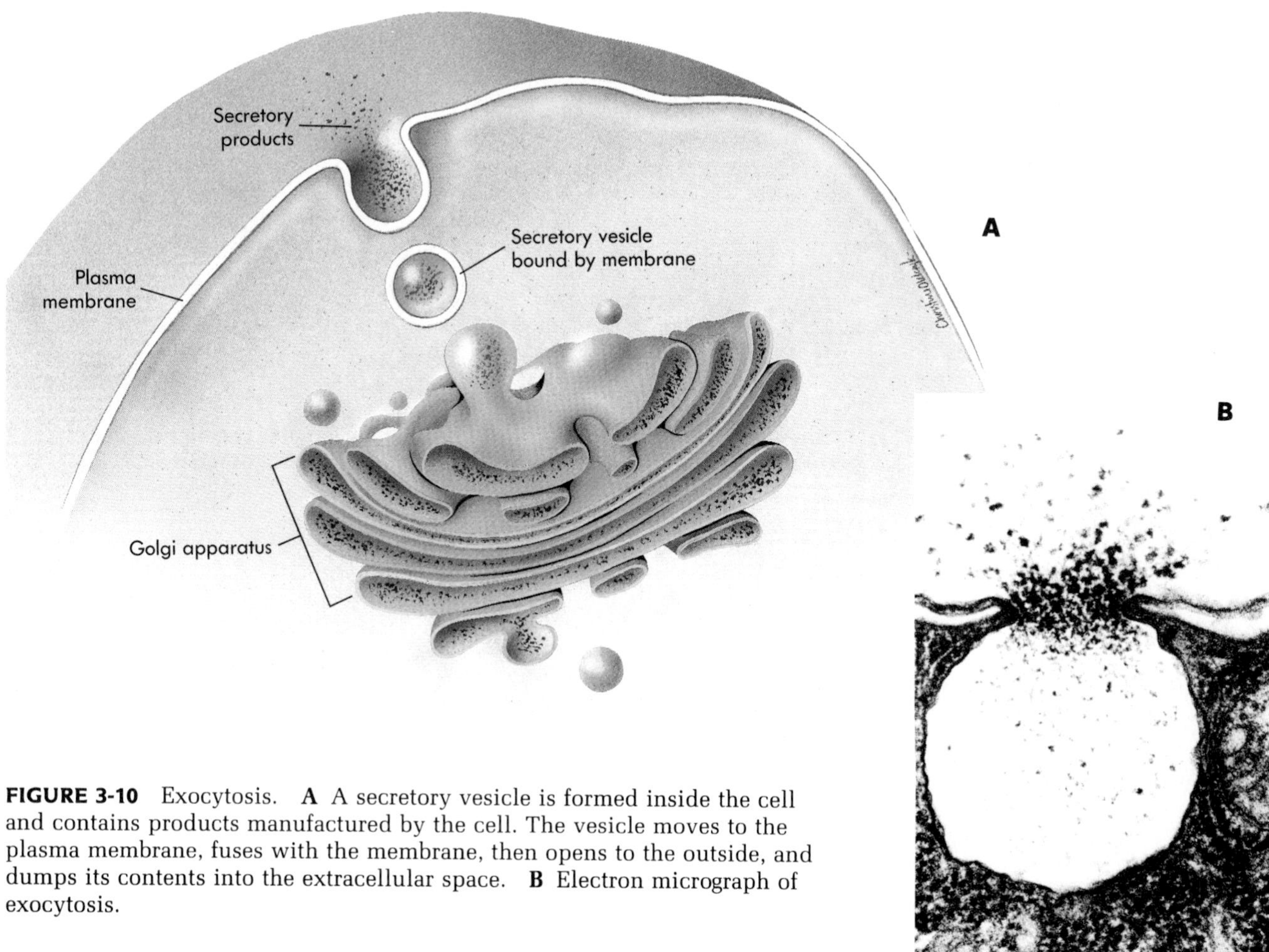

FIGURE 3-10 Exocytosis. **A** A secretory vesicle is formed inside the cell and contains products manufactured by the cell. The vesicle moves to the plasma membrane, fuses with the membrane, then opens to the outside, and dumps its contents into the extracellular space. **B** Electron micrograph of exocytosis.

Table 3-1

Comparison of membrane transport mechanisms

TRANSPORT MECHANISM	DESCRIPTION	SUBSTANCES TRANSPORTED	EXAMPLE
Diffusion	Random movement of molecules results in net movement from areas of higher to lower concentration.	Lipid-soluble molecules dissolve in the lipid bilayer and diffuse through it; ions and small molecules diffuse through membrane channels.	Oxygen, carbon dioxide, and lipids such as steroid hormones dissolve in the lipid bilayer; Cl^- ions and urea move through membrane channels.
Osmosis	Water diffuses across a selectively permeable membrane.	Water diffuses through membrane channels.	Water moves from the stomach into the blood.
Filtration	Liquid moves through a partition that allows some, but not all, of the substances in the liquid to pass through it; movement is due to a pressure difference across the partition.	Liquid and substances pass through holes in the partition.	Filtration in the kidneys allows filtration of everything in blood except proteins and blood cells.
Facilitated diffusion	Carrier molecules combine with substances and move them across the plasma membrane; no ATP is used; substances always are moved from areas of higher to lower concentration; it exhibits the characteristics of specificity, saturation, and competition.	Substances too large to pass through membrane channels and too polar to dissolve in the lipid bilayer are transported.	Glucose (most cells) and fructose are moved.
Active transport	Carrier molecules combine with substances and move them across the plasma membrane; ATP is used; substances can be moved from areas of lower to higher concentration; it exhibits the characteristics of specificity, saturation, and competition.	Substances too large to pass through membrane channels and too polar to dissolve in the lipid bilayer; substances that are accumulated in higher concentrations on one side of the membrane than the other are transported.	Glucose (intestines and kidneys), amino acids, and ions such as Na^+, K^+, Ca^{2+}, and H^+ are transported.
Endocytosis	The plasma membrane forms a vesicle around the substances to be transported, and the vesicle is taken into the cell; it requires ATP; in receptor-mediated endocytosis specific substances are ingested.	Phagocytosis takes in cells and solid particles; pinocytosis takes in molecules dissolved in liquid.	Immune system cells called phagocytes ingest bacteria and cellular debris; most cells take in substances through pinocytosis.
Exocytosis	Materials manufactured by the cell are packaged in secretory vesicles that fuse with the plasma membrane and release their contents to the outside of the cell; it requires ATP.	Secreted proteins and lipids are transported.	Digestive enzymes, hormones, neurotransmitters, and glandular secretions are transported and cell waste products are eliminated.

of the intestine, cells of the liver, and cells that line capillaries.

Phagocytosis and pinocytosis can exhibit specificity. For example, cells that phagocytize bacteria and necrotic tissue do not phagocytize healthy cells. Certain molecules of the plasma membrane, called receptor sites, recognize substances to be transported by phagocytosis or pinocytosis. This is called **receptor-mediated endocytosis,** and the receptor sites combine only with certain molecules. This mechanism increases the rate at which specific substances are taken up by the cells. Both phagocytosis and pinocytosis require energy in the form of ATP and are therefore active processes. However, because they involve the bulk movement of material into the cell, phagocytosis and pinocytosis do not exhibit either the degree of specificity or saturation that active transport exhibits.

In some cells secretions accumulate within vesicles. These secretory vesicles then move to the plasma membrane where the membrane of the vesicle fuses with the plasma membrane and the content of the vesicle is expelled from the cell. This process is called **exocytosis** (eks-o-si-to'sis) (Figure 3-10). Secretion of digestive enzymes by the pancreas, mucus by the salivary glands, and milk from the mammary glands are examples of exocytosis. In some respects the process is similar to phagocytosis and pinocytosis but occurs in the opposite direction.

Table 3-1 summarizes and compares the mechanisms by which different kinds of molecules are transported across the plasma membrane.

NUCLEUS

The nucleus is a large membrane-bound structure usually located near the center of the cell. It may be spherical, elongated, or lobed, depending on the cell type. All cells of the body have a nucleus at some point in their life cycle, although some cells such as red blood cells lose their nuclei as they develop. Other cells, such as osteoclasts (a type of bone cell) and skeletal muscle cells, contain more than one nucleus.

The nucleus is surrounded by a **nuclear envelope** (Figure 3-11) composed of two membranes separated by a space. At many points on the surface of the nuclear envelope the inner and outer membranes fuse to form porelike structures, the **nuclear pores.** Evidence indicates that molecules move between the nucleus and the cytoplasm through these nuclear pores.

Deoxyribonucleic acid (DNA) and associated proteins are dispersed throughout the nucleus as thin strands approximately 4 to 5 nm in diameter (see Appendix B). The proteins include **histones** (his'tōnz) and other proteins that play a role in the regulation of DNA function. The DNA and protein strands can be stained with dyes and are called **chromatin** (kro'mah-tin; colored material). Chromatin is distributed throughout the nucleus but is more condensed and more readily stained in some areas than in others. The more highly condensed chromatin apparently is less functional than the more evenly distributed chromatin, which stains lighter. During cell division the chromatin condenses to form the more solid bodies called **chromosomes** (colored bodies).

In the nucleus DNA determines the structure of **messenger RNA (mRNA),** which moves out of the nucleus through the nuclear pores into the cytoplasm where mRNA is involved in the synthesis of proteins (see the discussion of protein synthesis on page 87). Because all enzymes are proteins and because most chemical reactions within the cell are regulated by enzymes, the information contained within nuclear DNA determines most of the chemical events that occur within the cell.

Because mRNA synthesis occurs within the nucleus, cells without nuclei accomplish protein synthesis only as long as mRNA, produced before the

A

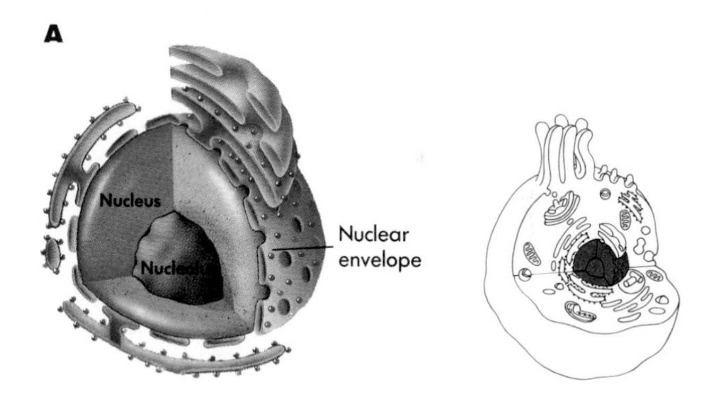

B

FIGURE 3-11 Nuclear envelope. **A** The nuclear envelope consists of inner and outer membranes that become fused at the nuclear pores. **B** Micrograph showing the inner and outer membranes of the nuclear envelope and the nuclear pores.

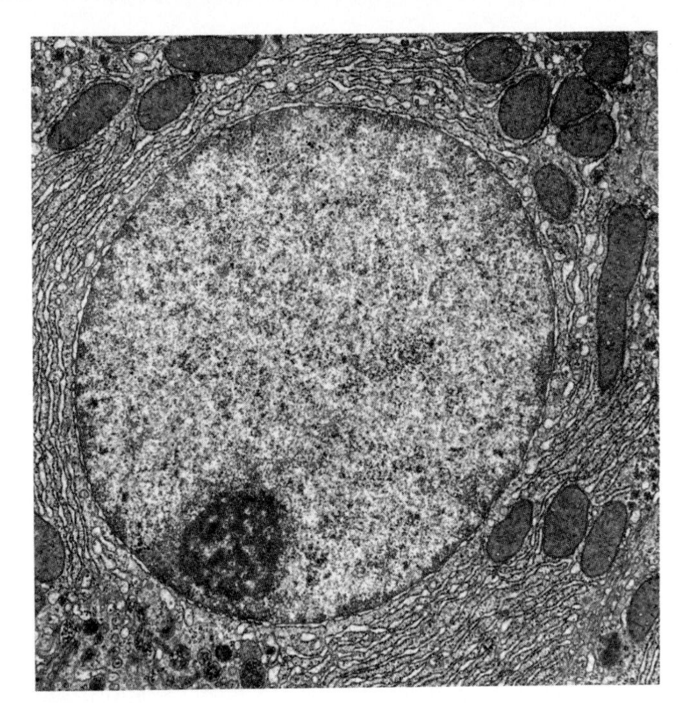

FIGURE 3-12 Electron micrograph of the nucleolus within the nucleus.

nuclear degeneration, remains functional. The nuclei of developing red blood cells are extruded before the red blood cells enter the blood where they survive without a nucleus for approximately 120 days. In comparison many cells with nuclei such as nerve and skeletal muscle cells survive as long as the individual survives.

A **nucleolus** (nu-kle′o-lus) is a somewhat rounded, dense, well-defined nuclear body with no surrounding membrane (Figure 3-12). There are one to four nucleoli per nucleus, depending on the cell. Within the nucleolus the subunits of ribosomes are manufactured (see the discussion of ribosomes on page 77).

CYTOPLASM

Cytoplasm, the cellular material outside the nucleus but inside the plasma membrane, is approximately half cytosol and half organelles.

Cytosol

Cytosol consists of a fluid portion, a cytoskeleton, and cytoplasmic inclusions. The fluid portion of cytoplasm is a solution with dissolved ions and molecules and a colloid with suspended molecules, especially proteins. Many of these proteins are enzymes that catalyze the breakdown of molecules for energy or catalyze the synthesis of sugars, fatty acids, nucleotides, amino acids, and other molecules.

Cytoskeleton

The **cytoskeleton** functions to support the cell and hold the nucleus and organelles in place. It is also responsible for cell movements, that is, changes in cell shape or movement of cell organelles. The cytoskeleton consists of three groups of proteins: microtubules, actin filaments, and intermediate filaments (Figure 3-13).

Microtubules are hollow tubules composed primarily of protein units called **tubulin.** The microtubules are approximately 25 nm in diameter with walls that are approximately 5 nm thick. Microtubules vary in length but are normally several micrometers (μm) long. Microtubules play a variety of roles within cells. They help provide support to the cytoplasm of the cell. They are involved in the process of cell division and form essential components of certain cell organelles such as centrioles, spindle fibers, cilia, and flagella.

Actin filaments, or **microfilaments,** are small fibrils approximately 8 nm in diameter that form bundles, sheets, or networks in the cytoplasm of cells. Actin filaments provide structure to the cytoplasm and mechanical support for microvilli. Some actin filaments are involved with cell movement. For example, muscle cells contain a large number of highly organized actin filaments responsible for the muscle's contractile capabilities (see Chapter 10).

Intermediate filaments are protein fibers approximately 10 nm in diameter. They provide mechanical strengths to cells. For example, intermediate filaments support the extensions of nerve cells, which have a very small diameter but may be a meter in length.

Cytoplasmic inclusions

Also part of the cytosol are **cytoplasmic inclusions,** which are aggregates of chemicals either produced by the cell or taken in by the cell. For example, lipid droplets or glycogen granules store energy-rich molecules; hemoglobin in red blood cells transports oxygen; melanin is a pigment that colors the skin, hair, and eyes; and **lipochromes** (lip′o-krōmz) are pigments that increase in amount with age. Dust, minerals, and dyes (as in a tattoo) can also accumulate in the cytoplasm.

Organelles

Organelles are small structures within cells and are specialized for particular functions such as manufacturing proteins or producing ATP. Most organelles have membranes that are similar to the plasma membrane. The membranes separate the organelles from the rest of the cytoplasm, creating a subcellular compartment with its own enzymes and capable of

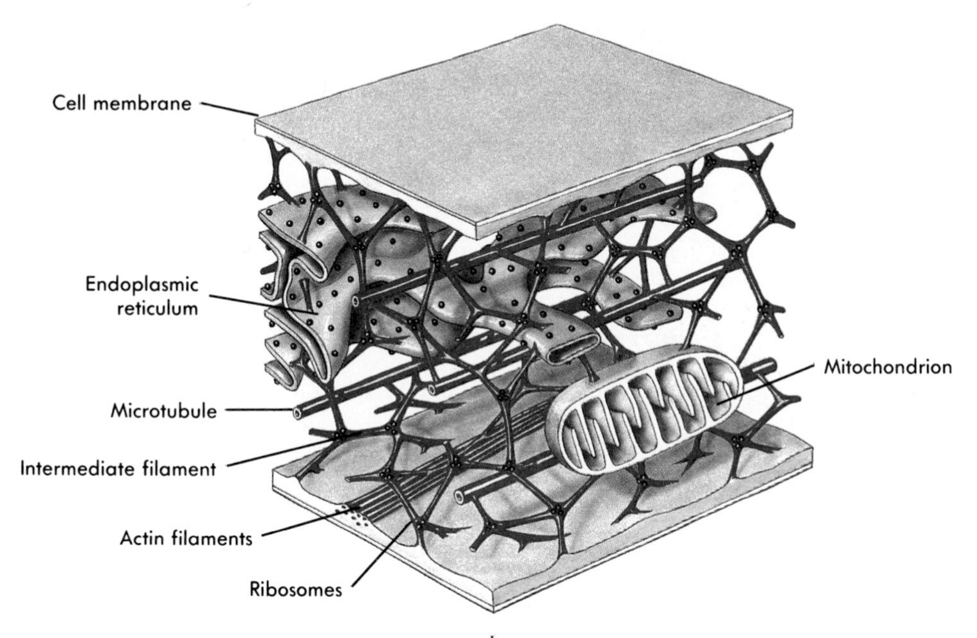

FIGURE 3-13 Cytoskeleton showing microtubules, actin filaments, and intermediate filaments.

carrying out its own unique chemical reactions. The nucleus is an example of an organelle.

The number and type of cytoplasmic organelles within each cell are related to the cell's specific structure and function. Cells secreting large amounts of protein contain well-developed organelles that synthesize and secrete protein, whereas cells actively transporting substances such as sodium ions across their plasma membrane contain highly developed organelles that produce ATP. The following sections describe the structure and main functions of the major cytoplasmic organelles found in cells.

Ribosomes

Ribosomes are the site of protein synthesis. The ribosome is composed of a large subunit and a smaller subunit joined together. The ribosomal subunits, which consist of **ribosomal RNA (rRNA)** and proteins, are assembled separately in the nucleolus of the nucleus. The ribosomal subunits then move through the nuclear pores into the cytoplasm in which they join together to form the functional ribosome (Figure 3-14). Ribosomes can be found free in the cytoplasm or associated with a membrane called the endoplasmic reticulum. **Free ribosomes** primarily synthesize proteins used inside the cell, whereas endoplasmic reticulum ribosomes can produce proteins that are secreted from the cell.

Endoplasmic reticulum

The outer membrane of the nuclear envelope is continuous with a series of membranes distributed

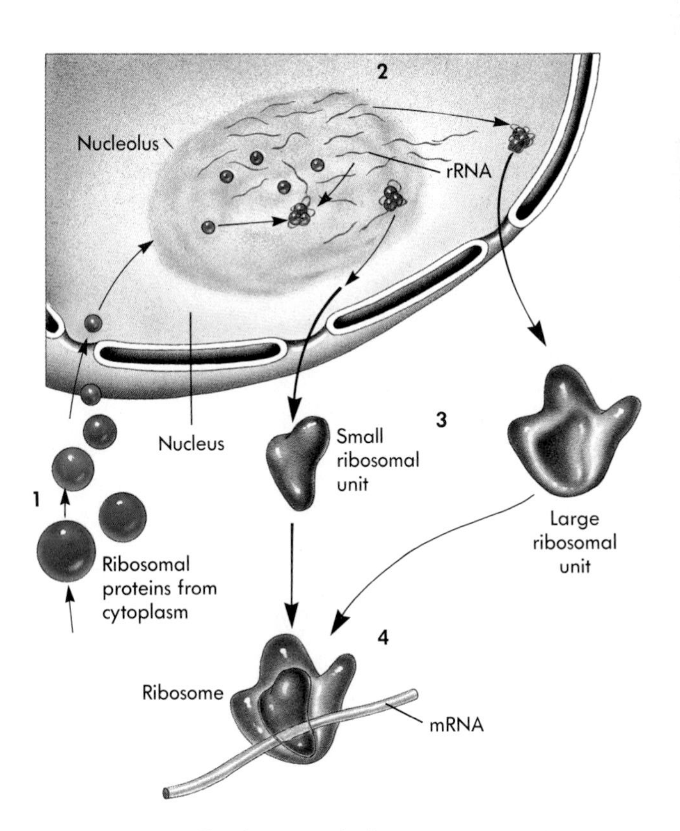

FIGURE 3-14 Production of ribosomes. At *1* ribosomal proteins, produced in the cytoplasm, are transported through nuclear pores into the nucleolus. AT *2* rRNA, most of which is produced in the nucleolus, is assembled with ribosomal proteins to form small and large subunits. At *3* the small and large ribosomal subunits leave the nucleolus and the nucleus through nuclear pores. At *4* the small and large subunits, now in the cytoplasm, combine with each other and with mRNA.

throughout the cytoplasm of the cell, collectively referred to as the **endoplasmic reticulum** (en′do-plaz′mik rĕ-tik′u-lum; network inside the cytoplasm) (see Figure 3-15). The endoplasmic reticulum forms broad, flattened sacs and tubules that interconnect. The containers formed by these tubules and flattened sacs, which are separated from the rest of the cytoplasm, are called **cisternae** (sis-ter′ne).

Rough endoplasmic reticulum, which is endoplasmic reticulum with attached ribosomes, produces proteins for secretion and for internal use. The amount and the configuration of the endoplasmic reticulum within the cytoplasm depend on the cell type and function. Cells with abundant rough endoplasmic reticulum synthesize large amounts of protein that are secreted for use outside of the cell.

Smooth endoplasmic reticulum, which is endoplasmic reticulum without attached ribosomes, manufactures lipids (e.g., phospholipids, cholesterol, steroid hormones) and carbohydrates (e.g., glycogen). Dense accumulations of smooth endoplasmic reticulum are found in cells that synthesize large amounts of lipid. Enzymes required for lipid synthesis are associated with the membranes of the smooth endoplasmic reticulum. Smooth endoplasmic reticulum also functions in detoxification processes in which chemicals and drugs are acted on by enzymes to change their structure and reduce their toxicity. The smooth endoplasmic reticulum of skeletal muscle stores calcium ions that function in muscle contraction.

Golgi apparatus

The **Golgi** (gōl′je) **apparatus** (see Figure 3-15) is composed of flattened membranous sacs stacked on

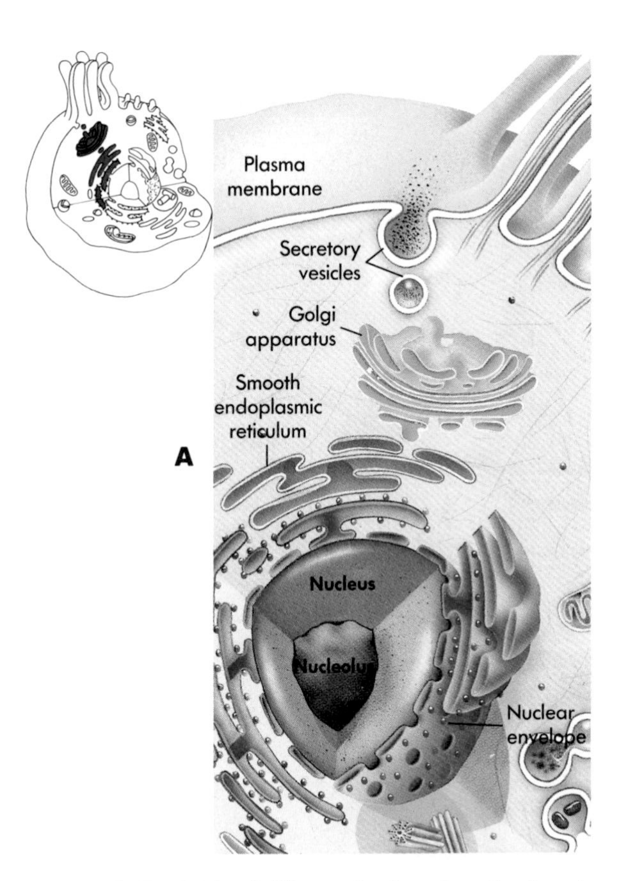

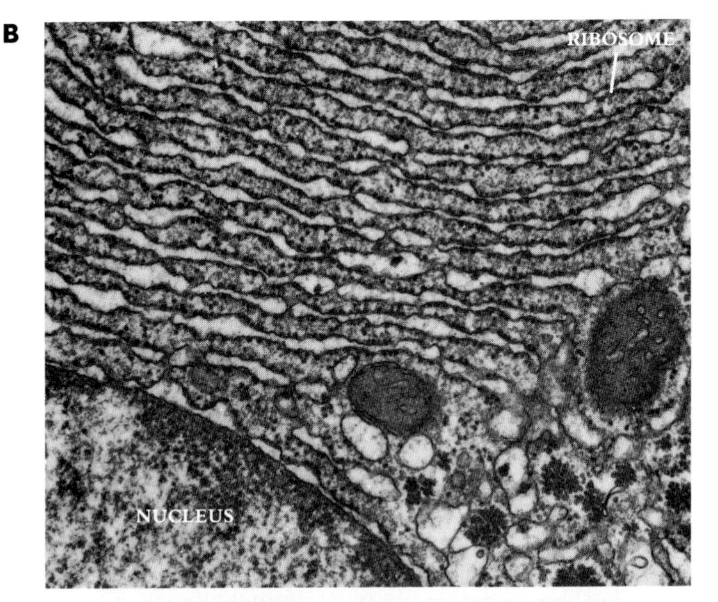

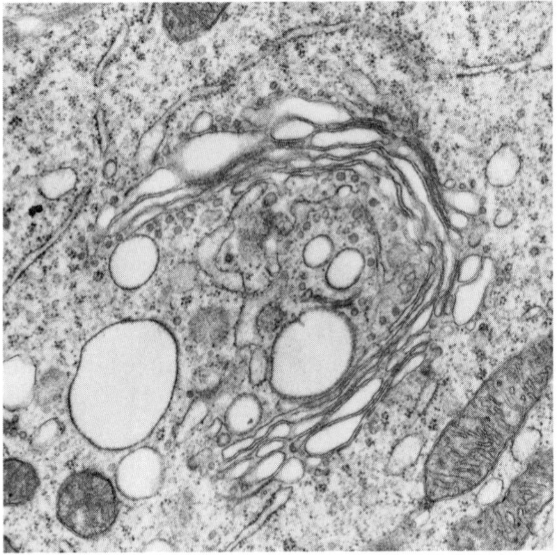

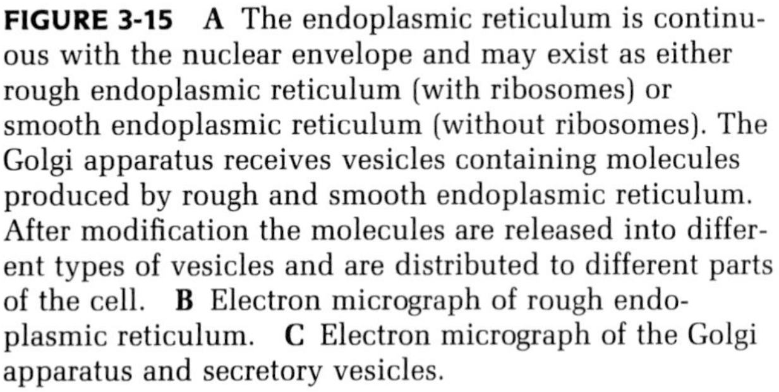

FIGURE 3-15 **A** The endoplasmic reticulum is continuous with the nuclear envelope and may exist as either rough endoplasmic reticulum (with ribosomes) or smooth endoplasmic reticulum (without ribosomes). The Golgi apparatus receives vesicles containing molecules produced by rough and smooth endoplasmic reticulum. After modification the molecules are released into different types of vesicles and are distributed to different parts of the cell. **B** Electron micrograph of rough endoplasmic reticulum. **C** Electron micrograph of the Golgi apparatus and secretory vesicles.

each other like dinner plates. The Golgi apparatus modifies, packages, and distributes proteins and lipids manufactured by the rough and smooth endoplasmic reticulum. Proteins produced at the ribosomes of the rough endoplasmic reticulum are surrounded by a vesicle that forms from the membrane of the endoplasmic reticulum. The vesicle moves to the Golgi apparatus, fuses with the membrane of the Golgi apparatus, and releases the protein into the cisterna of the Golgi apparatus. The Golgi apparatus concentrates and in some cases chemically modifies the proteins by synthesizing and attaching carbohydrate molecules to the proteins to form glycoproteins or lipids to proteins to form **lipoproteins.** The proteins then are packaged into vesicles that pinch off from the margins of the Golgi apparatus and are distributed to various locations. Some vesicles carry proteins to the plasma membrane where the proteins are secreted from the cell by exocytosis; other vesicles contain proteins that become part of the plasma membrane; and still other vesicles contain enzymes that are used within the cell.

The Golgi apparatuses are most numerous and most highly developed in cells that secrete large amounts of protein or glycoproteins such as cells in the salivary glands and the pancreas.

Secretory vesicles

The membrane-bound **secretory vesicles** (see Figure 3-15) that pinch off from the Golgi apparatus move to the surface of the cell, their membranes fuse with the plasma membrane, and the contents of the vesicle are released to the exterior by exocytosis. The membranes of the vesicles then are incorporated into the plasma membrane.

Secretory vesicles accumulate in many cells, but their contents frequently are not released to the exterior until a signal is received by the cell. For example, secretory vesicles that contain the hormone insulin do not release it until the concentration of glucose in the blood increases and acts as a signal for the secretion of insulin from the cells.

Lysosomes

Lysosomes (li′so-sōmz) are membrane-bound vesicles that pinch off from the Golgi apparatus (Figure 3-15). They contain a variety of hydrolytic enzymes that function as intracellular digestive systems. Endocytic vesicles fuse with the lysosomes to form one vesicle and to expose the phagocytized materials to hydrolytic enzymes (Figure 3-16). Enzymes

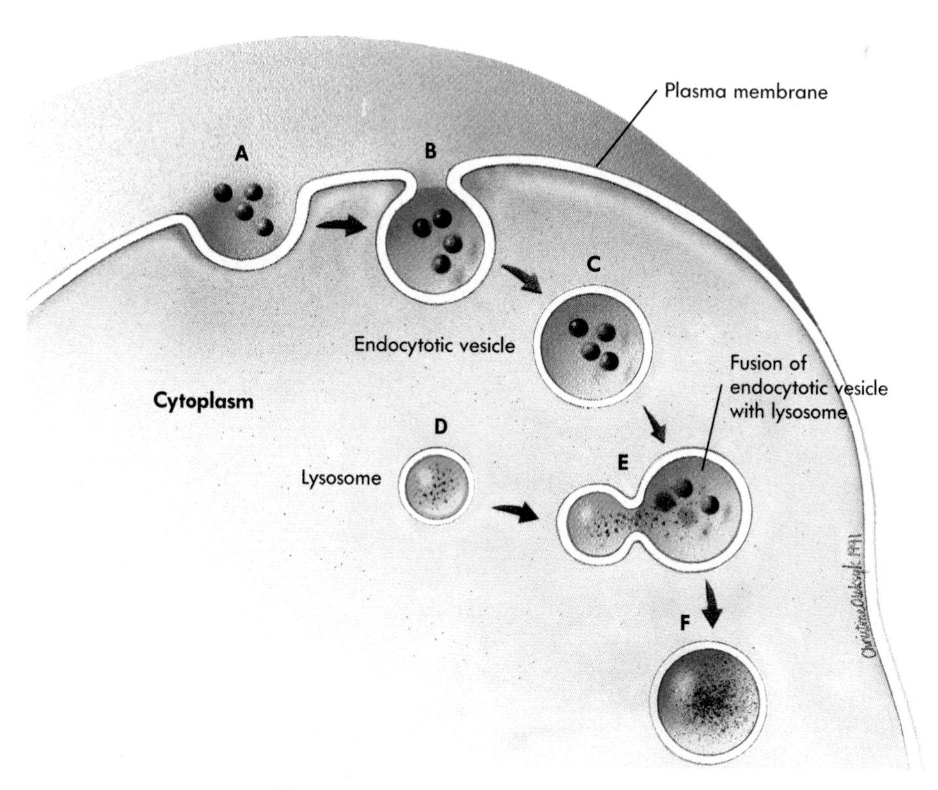

FIGURE 3-16 Action of lysosomes. **A** Material is taken into the cell by endocytosis. **B** The endocytotic vesicle is pinched off from the plasma membrane and, **C**, becomes a separate vesicle. **D** A lysosomal vesicle approaches the endocytotic vesicle. **E** The two vesicles merge. **F** The enzymes from the lysosome mix with the endocytotic material, and the enzymes digest the material.

within lysosomes include ones that digest nucleic acids, proteins, polysaccharides, and lipids. Certain white blood cells have large numbers of lysosomes that contain enzymes to digest phagocytized bacteria. Lysosomes also digest organelles of the cell that are no longer functional in a process called **autophagia** (aw'to-fa'je-ah, self-eating). Furthermore, when tissues are damaged, ruptured lysosomes within the damaged cells release their enzymes, which digest both damaged and healthy cells. In other cells the lysosomes move to the plasma membrane, and the enzymes are secreted by exocytosis. For example, the normal process of bone growth involves the breakdown of material outside the cell. Enzymes responsible for that degradation are released into the extracellular fluid from lysosomes.

Some diseases result from nonfunctional lysosomal enzymes. For example, Pompe's disease results from the inability of lysosomal enzymes to break down glycogen. The glycogen accumulates in large amounts in the heart, liver, and skeletal muscles, an accumulation that often leads to heart failure. Lipid storage disorders, which often are hereditary, are characterized by the accumulation of large amounts of lipids in phagocytic cells that lack the normal enzymes required to break down the lipid droplets that are taken into the cells by phagocytosis. Symptoms include enlargement of the spleen and the liver and replacement of bone marrow by the affected cells.

Peroxisomes

Peroxisomes (per-oks'ĭ-sōmz) are membrane-bound vesicles that are smaller than lysosomes. Peroxisomes generate hydrogen peroxide (H_2O_2) that can be used to detoxify potentially harmful molecules such as alcohol or formaldehyde. Excess hydrogen peroxide, however, can also be toxic, so peroxisomes contain the enzyme catalase, which breaks down hydrogen peroxide to water and oxygen. Cells that are active in detoxification, such as liver and kidney cells, have many peroxisomes.

Mitochondria

Mitochondria (mi'to-kon'dre-ah) usually are illustrated as small rod-shaped structures (Figure 3-17). In living cells time-lapse photography reveals that mitochondria constantly change shape from spherical to rod shaped or even to long threadlike structures. Mitochondria are the major sites of ATP production, which is the major energy source for most endergonic chemical reactions within the cell. Each mitochondrion has an inner and outer membrane separated by an intermembranous space. The outer membrane has a smooth contour, but the inner membrane has numerous infoldings called **cristae** (kris'te) that project like shelves into the interior of the mitochondria.

A complex series of mitochondrial enzymes forms two major enzyme systems that are responsible for oxidative metabolism and most ATP synthesis (see Chapter 25). The enzymes of the citric acid (or Kreb's) cycle are found in the **matrix,** which is the substance located in the space formed by the inner membrane. The enzymes of the electron transport chain are embedded within the inner membrane. Cells with a greater energy requirement have more mitochondria with more cristae than cells with lower energy requirements. Within the cytoplasm of a given cell the mitochondria are more numerous in areas where ATP is utilized. For example, mitochondria are numerous in cells that perform active transport and are packed near the membrane where active transport occurs.

Increases in the number of mitochondria result from the division of preexisting mitochondria. When muscles enlarge as a result of exercise, the number of mitochondria within the muscle cells increases to provide the additional ATP required for muscle contraction.

The structure of some mitochondrial proteins is determined by DNA contained within the mitochondria themselves, and the proteins are synthesized on ribosome-like structures within the mitochondria. However, the structure of many other mitochondrial proteins is determined by nuclear DNA, and those proteins are synthesized on ribosomes.

5

Describe the structural characteristics of cells that are highly specialized to do the following: (A) synthesize and secrete proteins; (B) actively transport substances into the cell; (C) synthesize lipids; and (D) phagocytize foreign substances.

? ? ? ? ? ? ? ? ? ? ?

Centrioles and spindle fibers

The **centrosome** (sen'tro-sōm) is a specialized zone of cytoplasm close to the nucleus that contains two **centrioles** (sen'tre-ōls). Each centriole is a small cylindrical organelle (0.3 to 0.5 μm in length and approximately 0.15 μm in diameter), and the two centrioles normally are oriented perpendicular to each other within the centrosome (see Figure 3-1). The wall of the centriole is composed of nine evenly spaced, longitudinally oriented, parallel units, or triplets. Each unit consists of three parallel microtubules joined together (Figure 3-18).

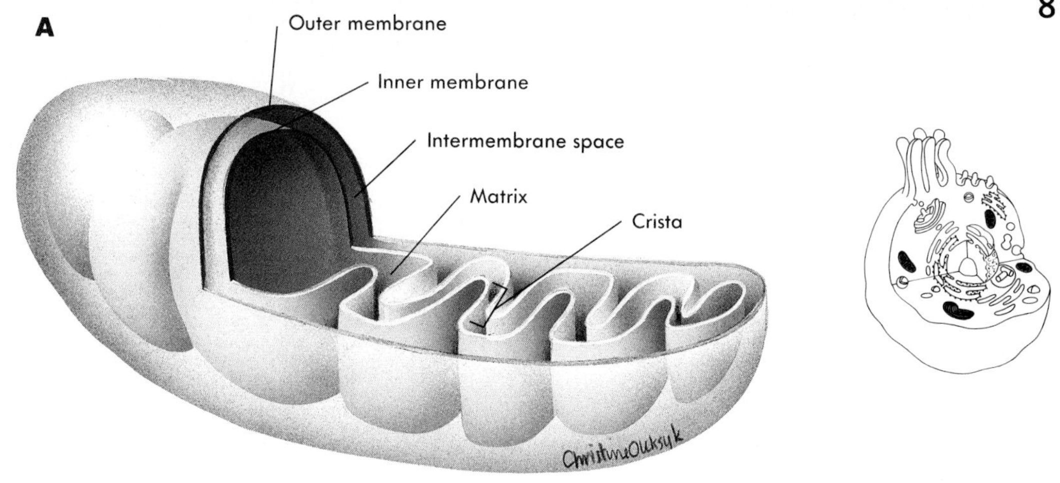

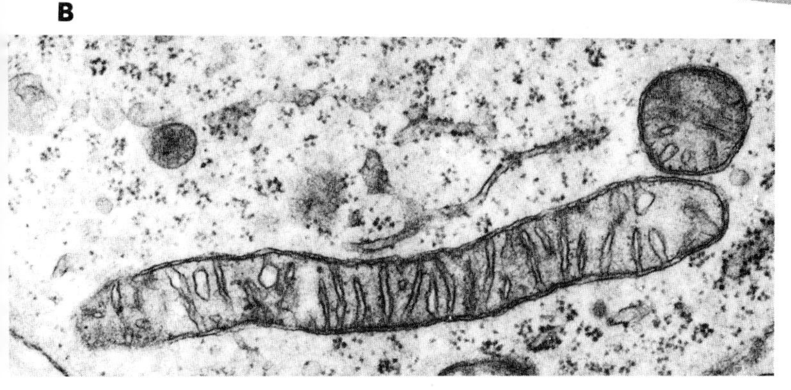

FIGURE 3-17 Mitochondria.
A Typical mitochondrion structure.
B Electron micrograph of a mitochondrion in longitudinal and cross section.

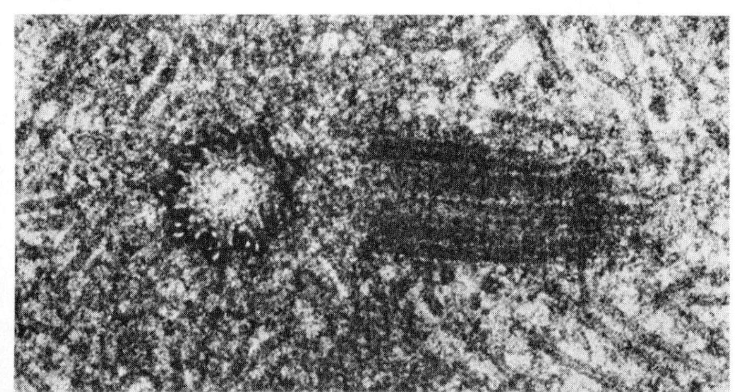

FIGURE 3-18 Centrioles. **A** Structure of a centriole, which comprises nine triplets of microtubules. Each triplet contains one complete microtubule fused to two incomplete microtubules. **B** Electron micrograph of a pair of centrioles, which normally are located together near the nucleus. One is cut in cross section and one in longitudinal section.

Before cell division, centrioles double in number, and one pair moves to each side of the nucleus. The centrioles apparently function as organizers that determine the poles or ends of the cell. Special microtubules called **spindle fibers** develop from the area of the centrosome and extend toward the chromosomes. Spindle fibers attach to the double-stranded chromosomes and aid in the normal separation of the chromosomes during cell division (see Cell Division in this chapter).

Cilia and flagella

Cilia (sil'e-ah) are appendages that project from the surface of cells and are capable of moving. They usually are limited to one surface of a given cell and vary in number from one to thousands per cell. Cilia are cylindrical in shape (approximately 10 μm in length and 0.2 μm in diameter), and the shaft of each cilium is enclosed by the plasma membrane. Two centrally located microtubules and nine peripheral

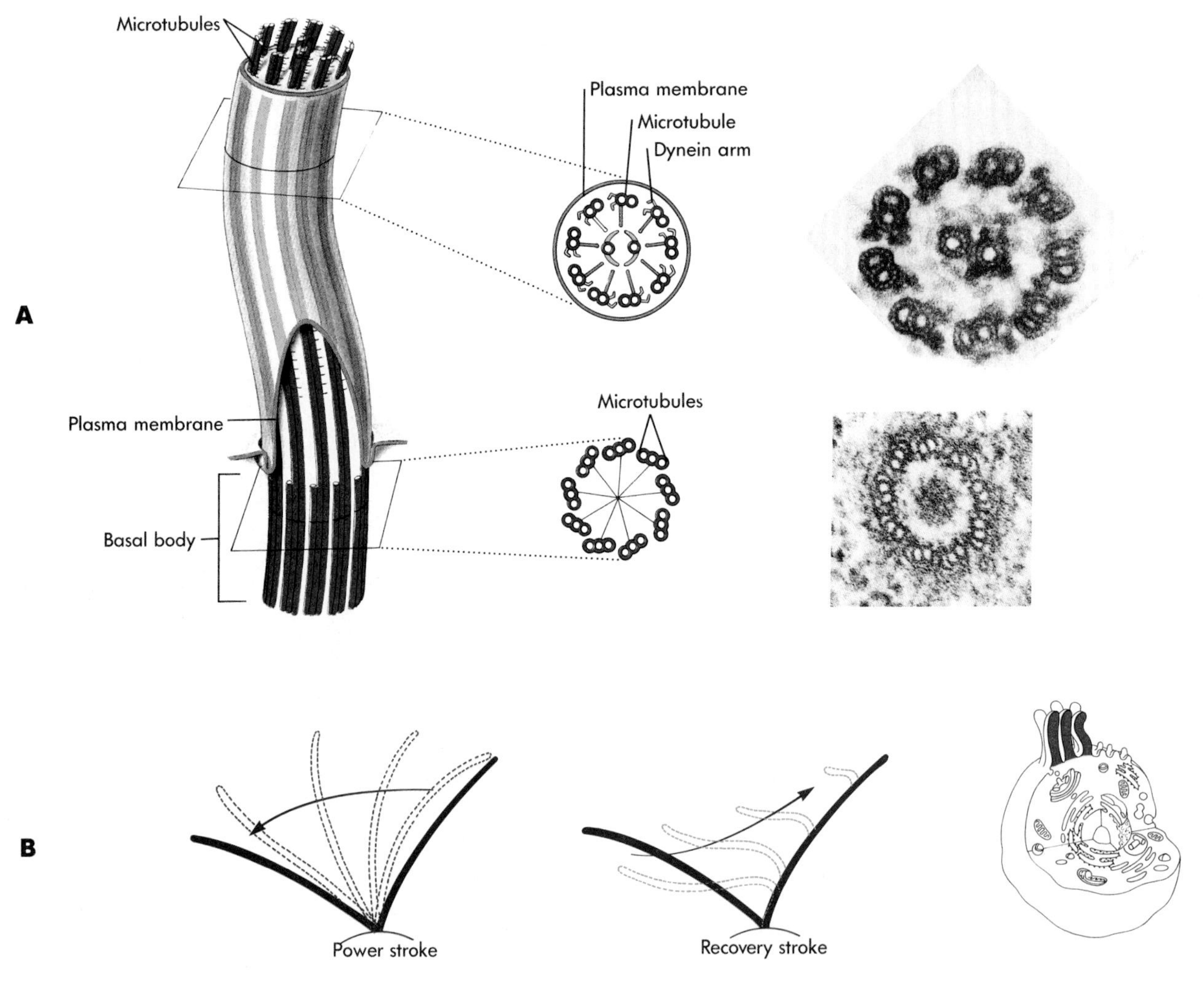

FIGURE 3-19 **A** Ciliary/flagellar structures. The shaft is composed of nine microtubule doublets around its periphery and two in the center. Dynein arms are proteins that connect one pair of microtubules to another pair. Dynein arm movement, which requires ATP, causes the microtubules to slide past each other, resulting in bending or movement of the cilium/flagellum. A basal body attaches the cilium/flagellum to the plasma membrane. **B** Ciliary movement, showing power and recovery strokes.

pairs of fused microtubules extend from the base to the tip of each cilium (Figure 3-19, *A*). Movement of the microtubules past each other, a process that requires energy from ATP, is responsible for movement of the cilia. A **basal body,** which is a modified centriole, is located in the cytoplasm at the base of the cilium.

Cilia are numerous on surface cells that line the respiratory tract and the female reproductive tract. In these regions cilia move in a coordinated fashion with a power stroke in one direction and a recovery stroke in the opposite direction (Figure 3-19, *B*). Their motion moves materials over the surface of the cells. For example, cilia in the trachea move mucus embedded with dust particles upward and away from the lungs. This action helps keep the lungs clear of debris.

Flagella (flă-jel'ah) have a structure similar to cilia but are longer (55 μm), and there is usually only one per cell. For example, each spermatozoon (sperm cell) is propelled by a single flagellum. Rather than having a power stroke and a recovery stroke like cilia, flagella move in a whiplike fashion.

Microvilli

Microvilli (mi'kro-vil'i) (Figure 3-20) are cylindrically shaped extensions of the plasma membrane (approximately 0.5 to 1 μm in length and 90 nm in diameter). Normally many microvilli are on each cell, and they function to increase the cell surface area. Microvilli can be confused with cilia, but they do not move, and they are supported with actin filaments, not microtubules. Microvilli are found in the intestine, kidney, and other areas in which absorption is an important function. In certain locations of the body microvilli are highly modified to function as sensory receptors. For example, elongated microvilli in hair cells of the inner ear respond to sound.

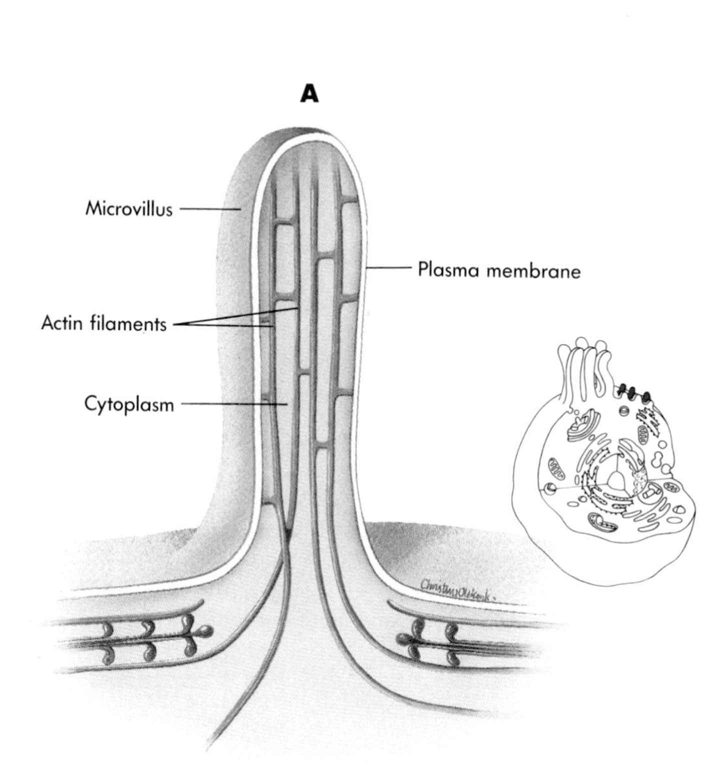

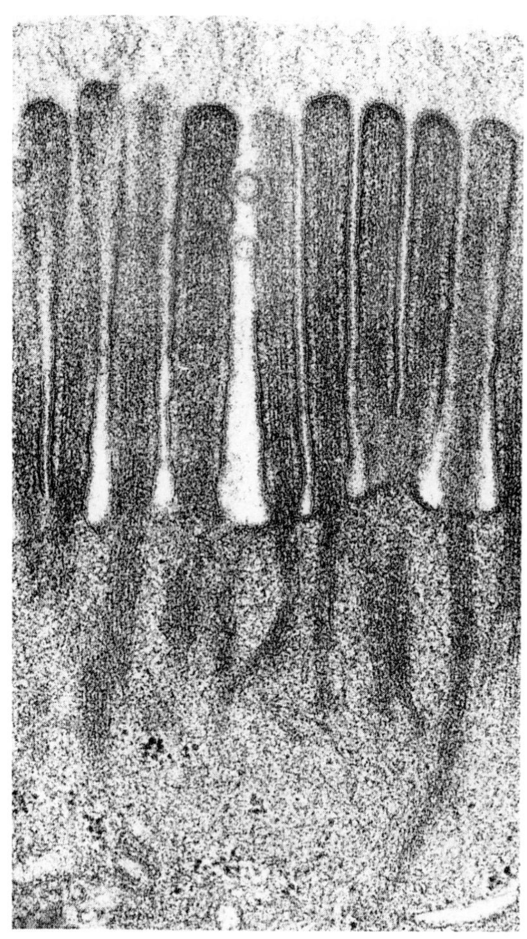

FIGURE 3-20 **A** A microvillus is a tiny tubular extension of the cell and contains cytoplasm and some actin filaments (microfilaments). **B** Electron micrograph of microvilli.

Table 3-2

Summary of cell parts

CELL PART	STRUCTURE	FUNCTION
Plasma Membrane	Lipid bilayer composed of phospholipids and cholesterol and proteins that extend across or attach to either surface of the lipid bilayer	Outer boundary of cells that controls entry and exit of substances; receptor molecules function in intercellular communication; marker molecules enable cells to recognize each other
Nucleus		
Nuclear envelope	Double membrane enclosing the nucleus; the outer membrane is continuous with the endoplasmic reticulum; nuclear pores extend through the nuclear envelope	Separates nucleus from cytoplasm and regulates movement of materials into and out of the nucleus
Chromatin	Dispersed thin strands of DNA, histones, and other proteins; condenses to form chromosomes during cell division	DNA regulates protein (e.g., enzyme) synthesis and therefore the chemical reactions of the cell; DNA is the genetic or hereditary material
Nucleolus	One to four dense bodies consisting of ribosomal RNA and proteins	Assembly site of large and small ribosomal subunits
Cytoplasm: Cytosol		
Fluid portion	Water with dissolved ions and molecules; colloid with suspended proteins	Contains enzymes that catalyze decomposition and synthesis reactions; ATP is produced in glycolysis reactions
Cytoskeleton		
Microtubules	Hollow cylinders composed of the protein tubulin; 25 nm in diameter	Support the cytoplasm and form centrioles, spindle fibers, cilia, and flagella; responsible for cell movements
Actin filaments	Small fibrils of protein; 8 nm in diameter	Support the cytoplasm, form microvilli, responsible for cell movements
Intermediate fibers	Protein fibers; 10 nm in diameter	Support the cytoplasm
Cytoplasmic inclusions	Aggregates of molecules manufactured or ingested by the cell; may or may not be membrane bound	Function depends on the molecules: energy storage (lipids, glycogen), oxygen transport (hemoglobin), skin color (melanin), and others
Cytoplasm: Organelles		
Ribosome	Ribosomal RNA and proteins form large and small subunits; attached to endoplasmic reticulum or free	Site of protein synthesis
Rough endoplasmic reticulum	Membranous tubules and flattened sacs with attached ribosomes	Protein synthesis and transport to Golgi apparatus
Smooth endoplasmic reticulum	Membranous tubules and flattened sacs with no attached ribosomes	Manufactures lipids and carbohydrates; detoxifies harmful chemicals; stores calcium
Golgi apparatus	Flattened membrane sacs stacked on each other	Modification, packaging, and distribution of proteins and lipids for secretion or internal use
Secretory vesicle	Membrane-bound sac pinched off Golgi apparatus	Carries proteins and lipids to cell surface for secretion
Lysosome	Membrane-bound vesicle pinched off Golgi apparatus	Contains digestive enzymes
Peroxisome	Membrane-bound vesicle	Detoxifies harmful molecules and breaks down hydrogen peroxide

Table 3-2

Summary of cell parts—cont'd

CELL PART	STRUCTURE	FUNCTION
Cytoplasm: Organelles—cont'd		
Mitochondria	Spherical, rod-shaped, or threadlike structures; enclosed by double membrane; inner membrane forms projections called cristae	Major site of ATP synthesis when oxygen is available
Centrioles	Pair of cylindrical organelles consisting of triplets of parallel microtubules	Organizers that determine the polarity (ends) of the cell; form the basal bodies of cilia and flagella
Spindle fibers	Microtubules extending from the centrosome (area around centrioles) to chromosomes and other parts of the cell (i.e., aster fibers)	Assist in the separation of chromosomes during cell division
Cilia	Extensions of the plasma membrane containing doublets of parallel microtubules; 10 μm in length	Move materials over the surface of cells
Flagellum	Extension of the plasma membrane containing doublets of parallel microtubules; 55 μm in length	In humans responsible for movement of spermatozoa
Microvilli	Extensions of the plasma membrane containing microfilaments	Increases surface area of the plasma membrane for absorption and secretion; modified to form sensory receptors

WHOLE CELL ACTIVITY

Although the structure and functions of individual cell components can be described (Table 3-2), to understand how a cell functions the interactions between the parts must be considered. For example, the active transport of many food molecules into the cell by the plasma membrane requires proteins (i.e., carrier molecules) and ATP manufactured within the cell. The ATP is produced in the cytosol and in mitochondria, and the proteins are produced on ribosomes. The production of proteins requires amino acids transported into the cell by the plasma membrane, ATP produced by the mitochondria, and assembly instructions provided by the nucleus. Many of the proteins produced by ribosomes also function as enzymes that control chemical reactions in the plasma membrane, mitochondria, ribosomes, nucleus, and other parts of the cell. Thus a picture of mutual interdependence of cell parts emerges when the whole cell is examined. The next topics, cell metabolism, protein synthesis, the cell life cycle, and meiosis, illustrate the interactions of cell parts that result in a functioning cell.

CELL METABOLISM

Cell metabolism is the sum of all the catabolic (decomposition) and anabolic (synthesis) reactions in the cell. The breakdown of food molecules such as carbohydrates, lipids, and proteins releases energy that is used to synthesize ATP. Each ATP molecule contains a portion of the energy originally stored in the chemical bonds of the food molecules. The ATP molecules are smaller "packets" of energy that can be used to drive other chemical reactions or processes such as active transport.

The production of ATP takes place in the cytosol and in mitochondria through a series of chemical reactions (see Chapter 25 for details). Energy from food molecules is transferred to ATP in a controlled fashion. If the energy in food molecules were released all at once, the cell literally would burn up.

The breakdown of the sugar glucose (e.g., that found in a candy bar) is used to illustrate the production of ATP from food molecules. Once glucose is transported into a cell, a series of reactions takes place within the cytosol. These chemical reactions, collectively called **glycolysis,** convert the glucose to

Solving the Problems of Life

A single cell and the human body must solve similar problems to survive. Both must interact with the external environment and be able to maintain homeostasis of their internal environment. They must take in, distribute, and break down food molecules and oxygen; synthesize new parts; eliminate waste products; and so on. It may help to understand the structures and functions of the cell by comparing the parts of a cell with the parts of the body that address the same basic needs (Table 3-A). Which is the more amazing solution for dealing with the problems of life—a single cell or the specialization and interactions of the trillions of cells that comprise the human body?

Table 3-A
Comparing a cell to the human body

FUNCTION	CELL PART	BODY PART
Regulation of homeostasis	Nucleus	Nervous and endocrine systems
Support	Cytoskeleton	Skeletal system
Movement	Cytoskeleton	Muscular system
Protective outer boundary capable of interacting with external environment	Plasma membrane	Integumentary system
Intake of food molecules and gases	Plasma membrane, microvilli	Digestive and respiratory systems
Distribution of molecules	Golgi apparatus, vesicles, membrane transport mechanisms	Circulatory system
Synthesis of new molecules and parts	Cytosol, ribosomes, endoplasmic reticulum, nucleoli, mitochondria	All body cells
Energy storage	Cytoplasmic inclusions	Adipose (fat) tissue, glycogen in skeletal muscle and the liver
Breakdown of molecules	Cytosol, mitochondria, lysosomes, peroxisomes	Digestive system and all body cells
Elimination of wastes	Plasma membrane, lysosomes	Urinary, digestive, and respiratory systems
Reproduction	Chromosomes, spindle fibers, centrioles, other organelles	Reproductive system

pyruvic acid. Pyruvic acid can enter different biochemical pathways, depending on oxygen availability (Figure 3-21).

Aerobic respiration occurs when oxygen is available. The pyruvic acid molecules enter mitochondria and through another series of chemical reactions—the citric acid cycle and the electron transport chain—are converted to carbon dioxide and water. Aerobic respiration can produce 38 ATP molecules from the energy contained in each glucose molecule.

There are several important points to note about aerobic respiration. First, the quantities of ATP produced through aerobic respiration are absolutely necessary to maintain the energy-requiring chemical reactions of life in human cells. Second, aerobic respiration requires oxygen because the last chemical reaction that takes place in aerobic respiration is the combination of oxygen with hydrogen to form water. If this reaction does not take place, the reactions immediately preceding it do not occur either. This explains why breathing oxygen is necessary for life: without oxygen aerobic respiration is inhibited, and the cells do not produce enough ATP to sustain life. Finally, during aerobic respiration the carbon atoms of food molecules are separated from each other to form carbon dioxide. Thus the carbon dioxide humans breathe out comes from the food they eat.

Anaerobic respiration occurs without oxygen and includes the conversion of pyruvic acid to lactic acid. There is a net production of two ATP molecules for each glucose molecule utilized. Anaerobic res-

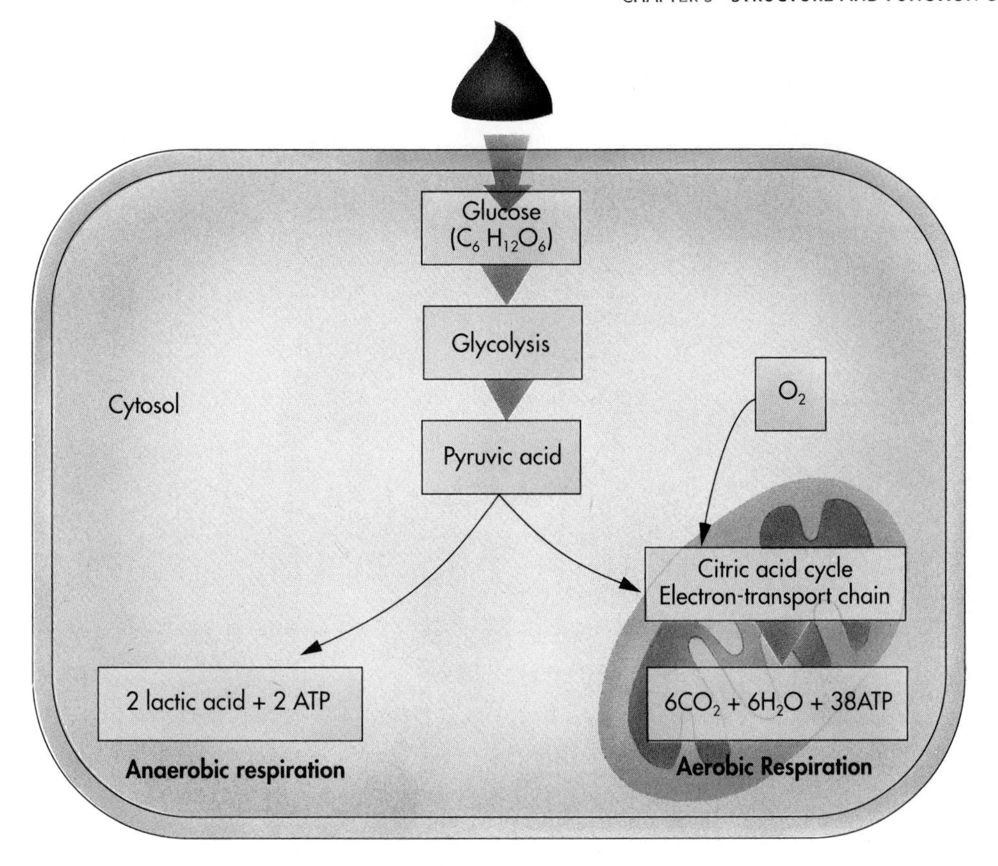

FIGURE 3-21 Overview of cell metabolism. Aerobic respiration requires oxygen and produces more ATP per glucose molecule than anaerobic respiration.

piration is not as productive as aerobic respiration in producing ATP, but it does allow the cells to function for short periods when oxygen levels are too low for aerobic respiration to provide all the needed ATP. For example, during intense exercise when aerobic respiration has depleted the oxygen supply, anaerobic respiration can provide additional ATP.

PROTEIN SYNTHESIS

Normal cell structure and function would not be possible without proteins, which form the cytoskeleton and other structural components of cells and function as transport molecules, receptors, and enzymes. In addition, proteins secreted from cells perform vital functions: collagen is a structural protein that gives tissues flexibility and strength; enzymes control the chemical reactions of food digestion in the intestines; and protein hormones regulate the activities of many tissues.

Ultimately, the production of all the proteins in the body is under the control of DNA. Recall from Chapter 2 that the building blocks of DNA are nucleotides containing adenine (A), cytosine (C), guanine (G), and thymine (T). The nucleotides form two parallel strands of nucleic acids, and the sequence of the nucleotides is a method of storing information. Every three nucleotides, called a **triplet,** codes for an amino acid, and amino acids are the building blocks of proteins. All of the triplets required to code for the synthesis of a specific protein are called a **gene.**

The production of proteins from the stored information in DNA involves two steps: transcription and translation, which can be illustrated with an analogy. Suppose a cook wants a recipe that is found only in a reference book in the library. Because the book cannot be checked out, the cook makes a handwritten copy, or **transcription,** of the recipe. Later, in the kitchen the information contained in the copied recipe is used to prepare the meal. The changing of something from one form to another (from recipe to meal) is called **translation.**

In terms of this analogy, DNA is the reference book that contains many recipes for making different proteins. DNA, however, is too large a molecule to pass through the nuclear envelope to go to the ribosomes (the kitchen) where the proteins are prepared. Just as the reference book stays in the library, DNA remains in the nucleus. Therefore through transcription the cell makes a copy of the information in DNA necessary to make a particular protein (the recipe). The copy, which is called **messenger RNA (mRNA),** travels from the nucleus to ribosomes in the cytoplasm where the information in the copy is used to construct a protein, that is, translation. Of

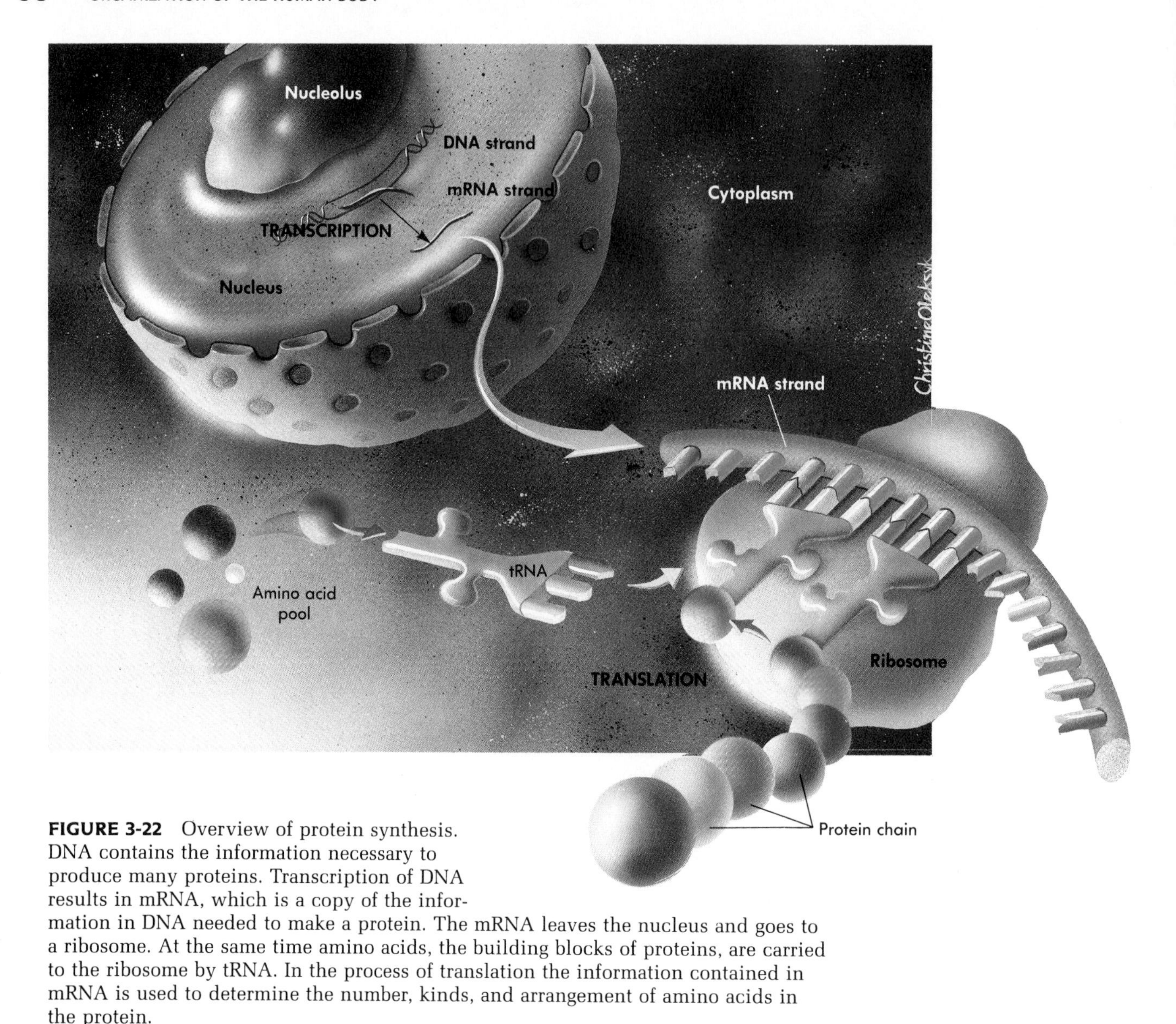

FIGURE 3-22 Overview of protein synthesis. DNA contains the information necessary to produce many proteins. Transcription of DNA results in mRNA, which is a copy of the information in DNA needed to make a protein. The mRNA leaves the nucleus and goes to a ribosome. At the same time amino acids, the building blocks of proteins, are carried to the ribosome by tRNA. In the process of translation the information contained in mRNA is used to determine the number, kinds, and arrangement of amino acids in the protein.

course, to turn a recipe into a meal the actual ingredients are needed. The ingredients necessary to synthesize a protein are amino acids. Specialized transport molecules, called **transfer RNA (tRNA),** carry the amino acids to the ribosome (Figure 3-22).

In summary, the synthesis of proteins involves transcription, making a copy of part of the stored information in DNA, and translation, converting that copied information into a protein. The details of transcription and translation are considered next.

Transcription

Transcription is the synthesis of mRNA using the sequence of nucleotides in DNA. It occurs when the double strands of a DNA segment separate and RNA nucleotides pair with DNA nucleotides (Figure 3-23). Nucleotides pair with each other according to the following rule: adenine pairs with thymine or uracil, and cytosine pairs with guanine. Because thymine is only in DNA and uracil is only in RNA, adenine, thymine, cytosine, and guanine nucleotides of DNA pair with uracil, adenine, guanine, and cytosine nucleotides of mRNA, respectively. This pairing relationship between nucleotides ensures that the information in DNA is transcribed correctly to mRNA. The RNA nucleotides combine through dehydration reactions catalyzed by RNA polymerase enzymes to form a long mRNA segment. After an mRNA segment has been produced, portions of the mRNA molecule may be removed, or two or more mRNA molecules may be combined.

The mRNA molecule contains the information required to determine the sequence of amino acids in a protein. The information, called the **genetic code,** is carried in groups of three nucleotides called **codons.** The number and sequence of codons in the mRNA are determined by the number and sequence of triplets in the segments of DNA that were transcribed. For example, the triplet code of CTA in DNA results in the codon GAU in mRNA, which codes for aspartic acid. Each codon codes for a specific amino acid. There are 64 possible mRNA codons, but only 20 amino acids are in proteins. As a result, the genetic code is redundant because more than one codon may code for the same amino acid. For example, CGA, CGG, CGT, and CGC all code for the amino acid alanine, and UUU and UAC both code for phenylalanine. Some codons do not code for amino acids but perform other functions (e.g., UAA acts as a signal for stopping the production of a protein).

Translation

The synthesis of a protein at the ribosome in response to the codons of mRNA is called translation. In addition to mRNA, translation requires tRNA and ribosomes, which consist of **ribosomal RNA (rRNA)** and proteins. Like mRNA, tRNA and rRNA are produced in the nucleus by transcription.

The function of tRNA is to match a specific amino acid to a specific codon of mRNA. To accomplish this task, one end of each kind of tRNA combines with a specific amino acid. The other end of the tRNA has an **anticodon,** which consists of three nucleotides. Based on the pairing relationships between nucleotides, the anticodon can combine only with its matched codon. For example, the tRNA that binds to aspartic acid has the anticodon CUA, which combines with the codon GAU of mRNA. Therefore the codon GAU codes for aspartic acid.

Ribosomes align the codons of the mRNA with the anticodons of tRNA and then join the amino acids of adjacent tRNA molecules. As the amino acids are joined together, a protein (i.e., polypeptide, a chain of amino acids) is formed. The step-by-step process of protein synthesis at the ribosome is described in detail in Figure 3-24. Some proteins are composed of two or more amino acid chains that are joined after each chain is produced on separate ribosomes.

After the initial portion of mRNA is used by a ribosome, another ribosome can attach to the mRNA and begin to make a protein. The resulting cluster of ribosomes attached to the mRNA is called a **polyribosome.** Each ribosome in a polyribosome produces an identical protein, and polyribosomes are an efficient way to use a single mRNA molecule to produce many copies of the same protein.

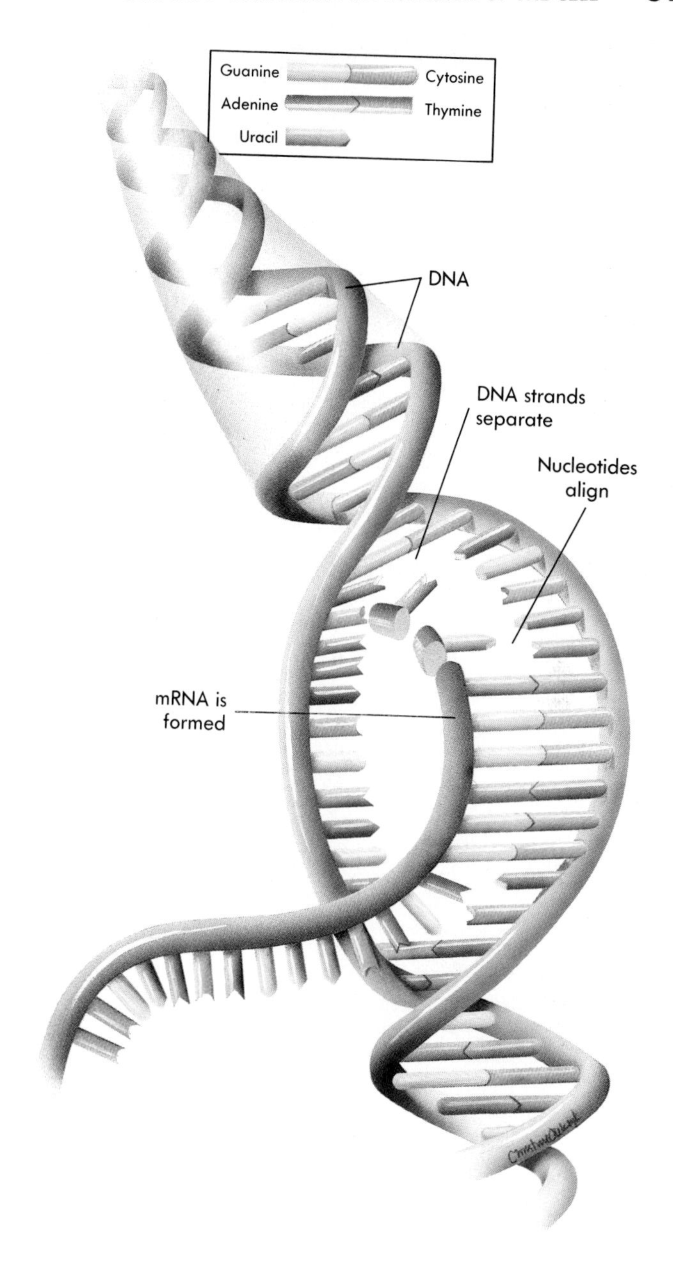

FIGURE 3-23 Formation of mRNA by transcription of DNA strands in the cell nucleus. A segment of the DNA molecule is opened, and RNA polymerase (an enzyme that is not shown) assembles nucleotides into mRNA according to the base pair combinations shown in the inset. Thus the sequence of nucleotides in DNA determines the sequence of nucleotides in mRNA. As nucleotides are added, an mRNA molecule is formed.

6

Explain how changing one nucleotide within a DNA molecule of a cell could change the structure of a protein produced by that cell. What effect would this change have on the protein's function?

? ? ? ? ? ? ? ? ? ? ?

FIGURE 3-24 Translation of mRNA to produce a protein

A To start protein synthesis a ribosome binds to mRNA. The ribosome also has two binding sites for tRNA, one of which is occupied by a tRNA with its amino acid. Note that the codon of mRNA and the anticodon of tRNA are aligned and joined. The other tRNA binding site is open.

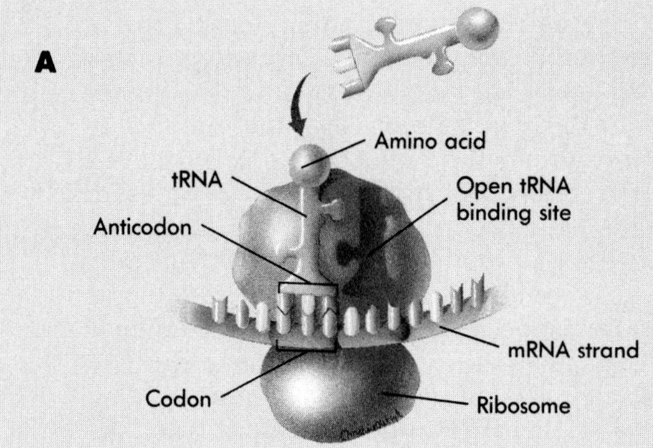

B By occupying the open tRNA binding site the next tRNA is properly alilgned with mRNA and with the other tRNA.

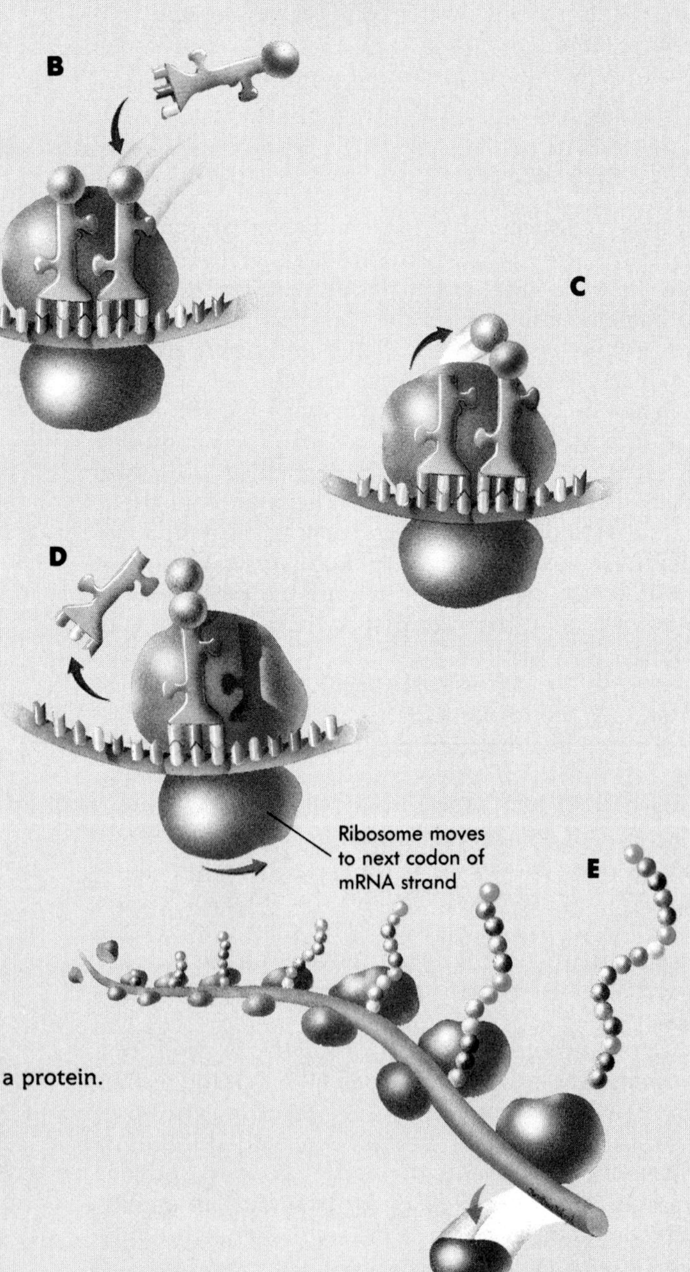

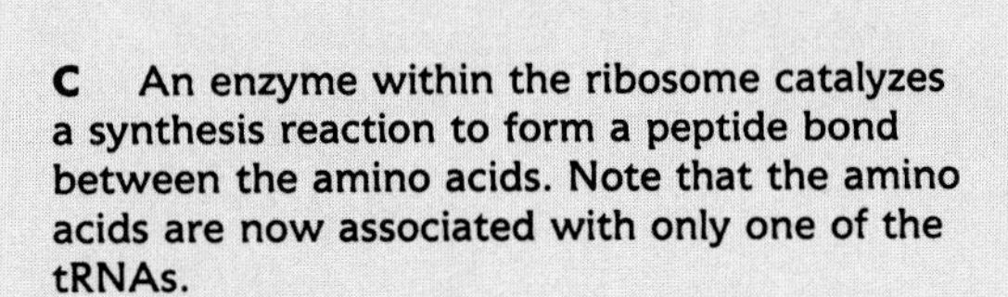
C An enzyme within the ribosome catalyzes a synthesis reaction to form a peptide bond between the amino acids. Note that the amino acids are now associated with only one of the tRNAs.

D The ribosome shifts potision by three nucleotides. The tRNA without the amino acid is released from the ribosome, and the tRNA with the amino acids takes its position. A tRNA binding site is left open by the shift. Additional amino acids can be added by repeating steps **B-D.** Eventually a stop codon in the mRNA ends the production of the protein, which is released from the ribosome.

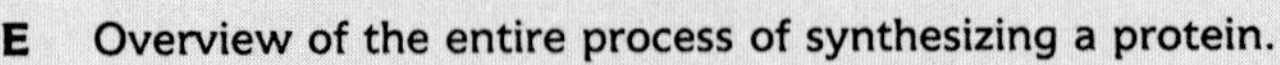
E Overview of the entire process of synthesizing a protein.

Regulation of Protein Synthesis

All of the cells in the body except for sex cells have the same DNA. The transcription of mRNA in cells, however, is regulated so that all portions of all DNA molecules are not continually transcribed. The proteins associated with DNA in the nucleus play a role in regulating the transcription. As cells differentiate and become specialized for specific functions during development, part of the DNA becomes nonfunctional (i.e., it is not transcribed), whereas other segments of DNA remain very active. For example, in most cells little if any hemoglobin is synthesized. However, in developing red blood cells hemoglobin synthesis occurs rapidly.

Within a single cell protein synthesis normally is not constant but occurs more rapidly at some times than others. Regulatory molecules that interact with the nuclear proteins can either increase or decrease the transcription rate of specific DNA segments. For example, thyroxine, a hormone released by cells of the thyroid gland, enters cells such as skeletal muscle cells and increases specific types of mRNA transcription and consequently the production of certain proteins. As a result, an increase in the number of mitochondria and an increase in metabolism occur in these cells.

CELL LIFE CYCLE

The **cell life cycle** includes the changes a cell undergoes from the time it is formed until it divides to produce two new cells. The life cycle of a cell can be divided into interphase and cell division.

Interphase

Interphase is the time between cell divisions. Ninety percent or more of a typical cell's life cycle is spent in interphase. During this time the cell carries out the metabolic activities necessary for life and performs its specialized functions such as secreting digestive enzymes. In addition, the cell prepares for cell division. This preparation includes an increase in cell size as many of the cell's components double in quantity and a **replication** (duplication) of the cell's DNA. Consequently, when the cell divides, each new cell receives the organelles and DNA necessary for continued functioning.

During interphase DNA and its associated proteins appear as dispersed chromatin threads within the nucleus. When DNA replication begins, the two strands of each DNA molecule separate from one another. Each strand then serves as a template for the production of a new strand of DNA, which is formed as new nucleotides pair with the existing nucleotides of each strand of the separated DNA molecule. As a result, two identical DNA molecules are produced. Each of the two new DNA molecules has one strand of nucleotides derived from the original DNA molecule and one newly synthesized strand (Figure 3-25). During interphase the centrioles within the centrosome also are duplicated.

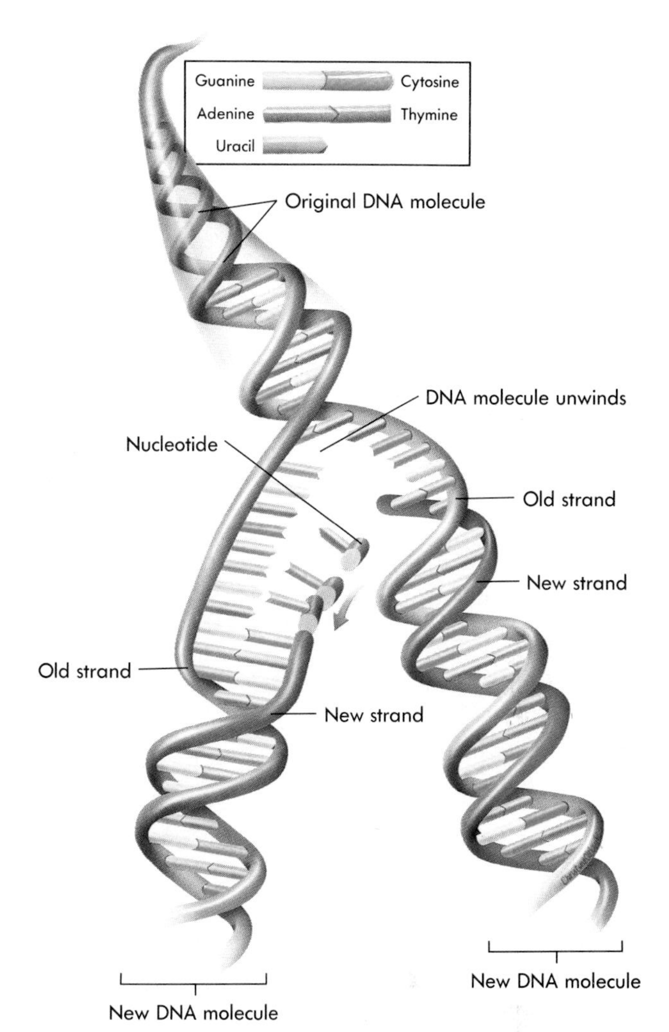

FIGURE 3-25 Replication of DNA during interphase produces two identical molecules of DNA. The strands of the DNA molecule separate from each other, and each strand functions as a template on which another strand is formed. The base pairing relationship between nucleotides determines the sequence of nucleotides in the newly formed strand.

7

Suppose a molecule of DNA separates, forming strands one and two. Part of the nucleotide sequence in strand one is ATGCTA. From this template, what would be the sequence of nucleotides in the DNA replicated from strand one and strand two?

? ? ? ? ? ? ? ? ? ? ?

FIGURE 3-26 *Interphase, mitosis, and cytokinesis*

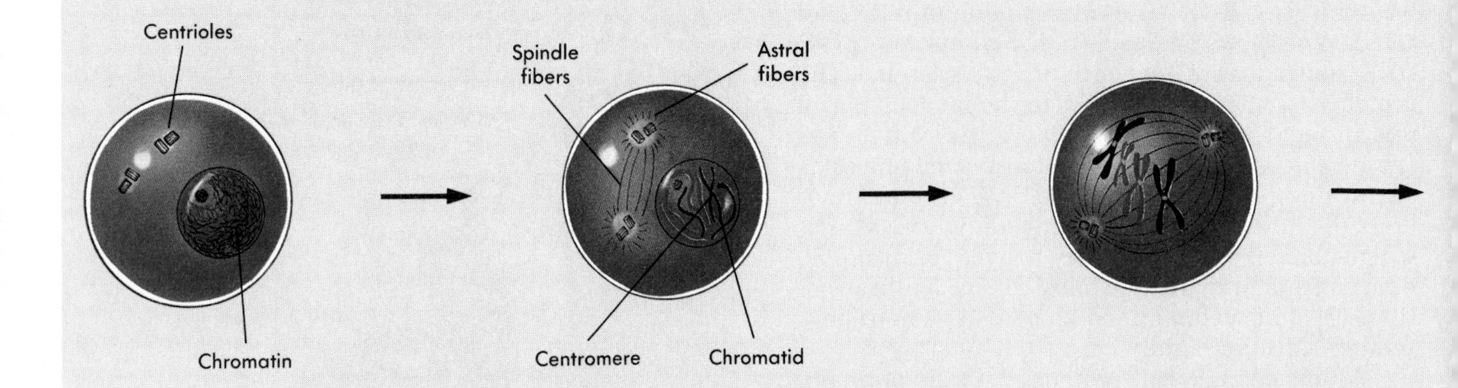

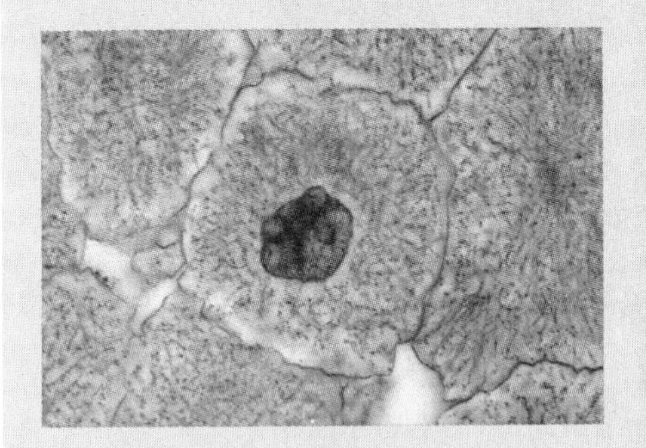

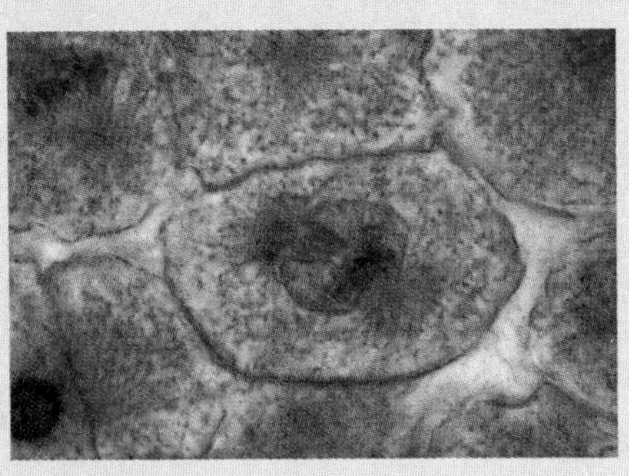

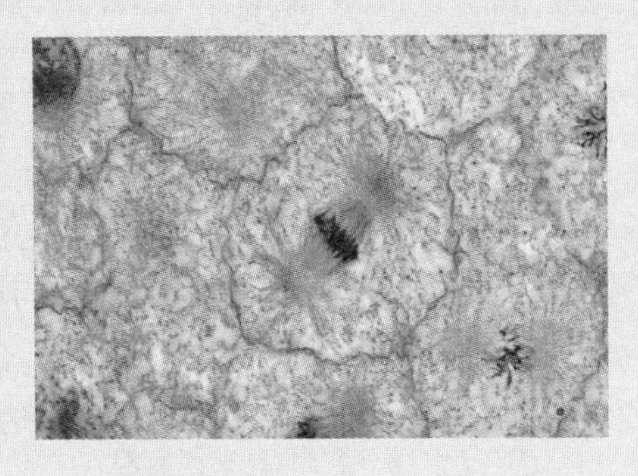

A Interphase
DNA, which is dispersed as chromatin, replicates. The two strands of each DNA molecule separate, and a copy of each strand is made.
Consequently, two identical DNA molecules are produced. The pair of centrioles replicate to produce two pairs of centrioles.

B Prophase
Chromatin strands condense to form chromosomes. Each chromosome is composed of two identical strands called **chromatids,** which are joined together at one point by a specialized region called the **centromere.** Each chromatid contains one of the DNA molecules replicated during interphase.

One pair of centrioles moves to each side, or pole, of the cell. Microtubules form near the centrioles and project in all directions. Some of the microtubules end blindly and are called **astral fibers.** Others, known as **spindle fibers** project toward an invisible line, called the **equator,** and either overlap with fibers from other centrioles or attach to the centromeres of the chromosomes. At the end of prophase the nuclear envelope degenerates, and the nucleoli disappear.

C Metaphase
The chromosomoes align along the equator with spindle fibers from each pair of centrioles attached to their centromeres.

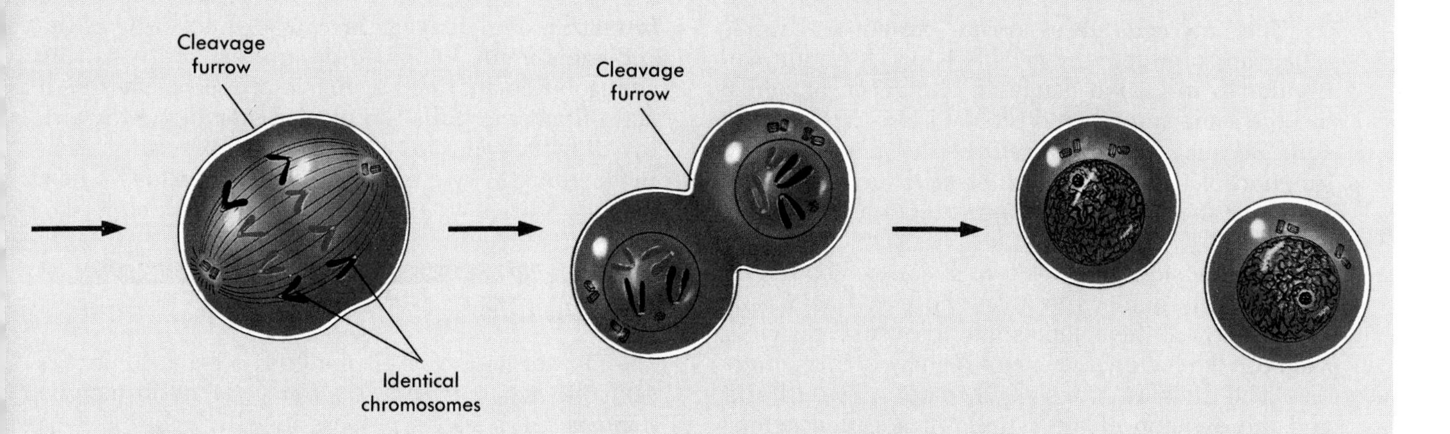

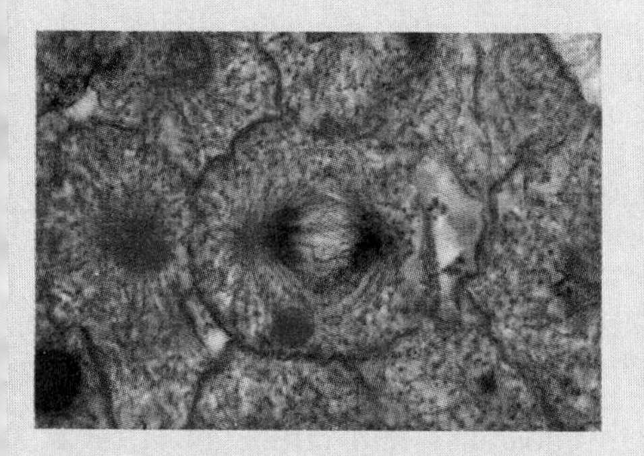

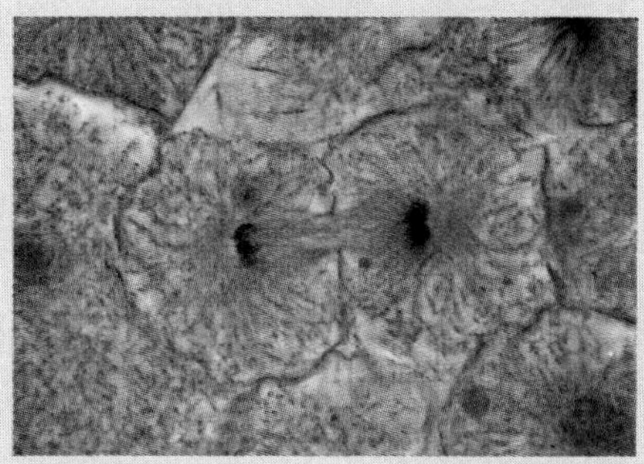

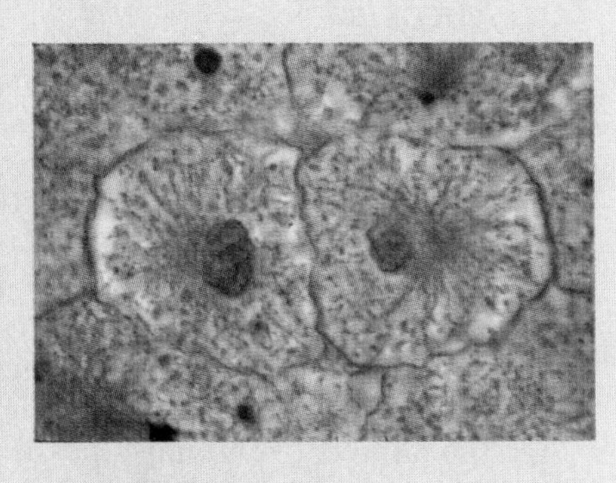

D Anaphase
The centromeres separate, and each chromatid is then referred to as a chromosome. Thus, when the centromeres divide, the chromosome number doubles and there are two identical sets of chromosomes. The two sets of chromosomes are pulled by the spindle fibers toward the poles of the cell. Separation of the chromatids signals the beginning of anaphase, and by the time anaphase has ended, the chromosomes have reached the poles of the cell. The beginning of cytokinesis is evident during anaphase; along the equator of the cell the cytoplasm becomes narrower as the plasma membrane pinches inward.

E Telophase
The migration of each set of chromosomes is complete. A new nuclear envelope develops from the endoplasmic reticulum, and the nucleoli reappear. During the latter portion of telophase the spindle fibers disappear, and the chromosomes unravel to become less distinct chromatin threads. The nuclei of the two daughter cells assume the appearance of interphase nuclei, and the process of mitosis is complete.

F Interphase
Cytokinesis, which continues from anaphase through telophase, becomes complete when the plasma membranes move close enough together at the equator of the cell to fuse, completely separating the two new daughter cells, each of which now has a complete set of chromosomes (a diploid number of chromosomes) identical to the parent cell.

Cell Division

The new cells necessary for growth and tissue repair are produced by cell division. A parent cell divides to form two daughter cells, each of which has the same amount and type of DNA as the parent cell. Because DNA determines the structure and function of cells, the daughter cells have the same structure and perform the same functions as the parent cell.

Cell division involves two major events: the division of the nucleus to form two new nuclei, and the division of the cytoplasm to form two new cells, each of which contains one of the newly formed nuclei. The division of the nucleus occurs by mitosis, and the division of the cytoplasm is called cytokinesis.

Mitosis

Mitosis (mi-to'sis) is the division of the nucleus into two nuclei, each of which has the same amount and type of DNA as the original nucleus. The DNA, which was dispersed as chromatin in interphase, condenses in mitosis to form chromosomes. In each of the human **somatic** (so-mat'ik) cells, which include all cells except the sex cells, there are 46 chromosomes, which are referred to as a **diploid** (dip'loyd) number of chromosomes. Sex cells have half the number of chromosomes as somatic cells (see Meiosis in this chapter). The 46 chromosomes in somatic cells are organized into 23 pairs of chromosomes. Twenty-two of these pairs are called **autosomes.** Each member of an autosomal pair of chromosomes looks structurally alike, and together they are called a **homologous** (ho-mol'o-gus) pair of chromosomes. One member of each autosomal pair is derived from the person's father, and the other is derived from the mother. The remaining pair of chromosomes comprises the sex chromosomes. In females sex chromosomes look alike, and each is called an **X chromosome.** In males the sex chromosomes do not look alike; one is an X chromosome, and the other is smaller and is called a **Y chromosome.**

For the convenience of description, mitosis is divided into four stages: **prophase, metaphase, anaphase,** and **telophase.** Although each stage represents major events, mitosis is a continuous process, and there are no discrete jumps from one stage to another. Learning the characteristics associated with each stage is helpful, but a more important concept is how each daughter cell obtains the same number and type of chromosomes as the parent cell. The major events of mitosis are summarized in Figure 3-26.

Cytokinesis

Cytokinesis (si-to-kin-e'sis) is the division of the cell's cytoplasm to produce two new cells. Cytokinesis begins in anaphase, continues through telophase, and ends in the following interphase (see Figure 3-26, *D-F*). The first sign of cytokinesis is the formation of a **cleavage furrow,** or puckering of the cell membrane, which forms midway between the centrioles. A contractile ring composed primarily of actin filaments pulls the plasma membrane inward, dividing the cell into two halves. Cytokinesis is complete when the two halves separate to form two new cells.

MEIOSIS

The formation of all the body's cells except for sex cells occurs by mitosis. Sex cells are formed by **meiosis** (mi-o'sis), a process in which the nucleus undergoes two divisions resulting in four nuclei, each containing half as many chromosomes as the parent cell. The daughter cells that are produced by cytokinesis differentiate into **gametes** (gam'ētz), or sex cells. The gametes are reproductive cells—spermatozoa in males and oocytes in females. Each gamete not only has half the number of chromosomes found in a somatic cell but also has one chromosome from each homologous pair found in the parent cell. The complement of chromosomes in a gamete is referred to as a **haploid** number. Oocytes contain 22 autosomal chromosomes (one from each homologous pair) and an X chromosome, and spermatozoa have 22 autosomal chromosomes and either an X or Y chromosome. During fertilization when a spermatozoon fuses with an oocyte, the normal number of 46 chromosomes in 23 pairs is reestablished.

The first division during meiosis is divided into four stages: prophase I, metaphase I, anaphase I, and telophase I (Figure 3-27). As in prophase of mitosis, the nuclear envelope degenerates, spindle fibers form, and the already duplicated chromosomes become visible. Each chromosome consists of two chromatids joined by a centromere. However, in prophase I the four chromatids of a homologous pair of chromosomes join together, or **synapse,** to form a **tetrad.** In metaphase I the tetrads align at the equatorial plane, and in anaphase I each pair of homologous chromosomes separates and moves toward opposite poles of the cell. For each pair of homologous chromosomes, one daughter cell receives one member of the pair, and the other daughter cell receives the other member. Thus each daughter cell has 23 chromosomes, each of which is composed of two chromatids. Telophase I, with cytokinesis, is similar to mitosis, and two daughter cells are produced.

Interkinesis is the time between the formation of the daughter cells and the second meiotic division. There is no duplication of DNA during interkinesis.

The second division of meiosis has four stages: prophase II, metaphase II, anaphase II, and telophase II. These stages occur much as they do in mitosis except that there are 23 chromosomes instead of 46. The chromosomes align at the equatorial plane in

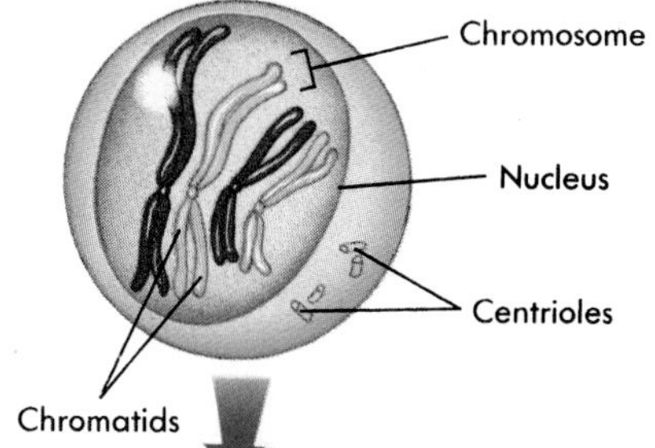

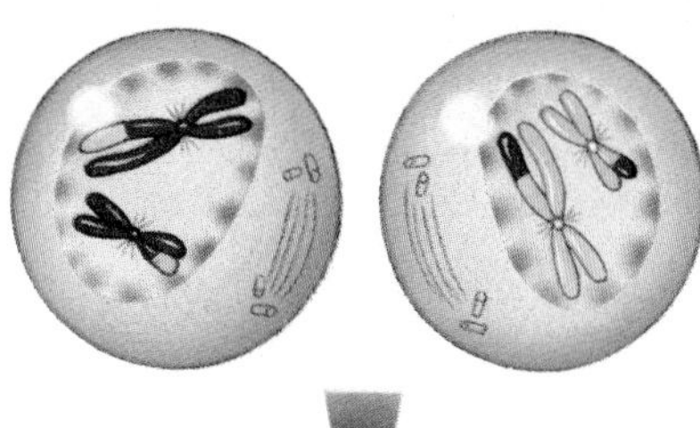

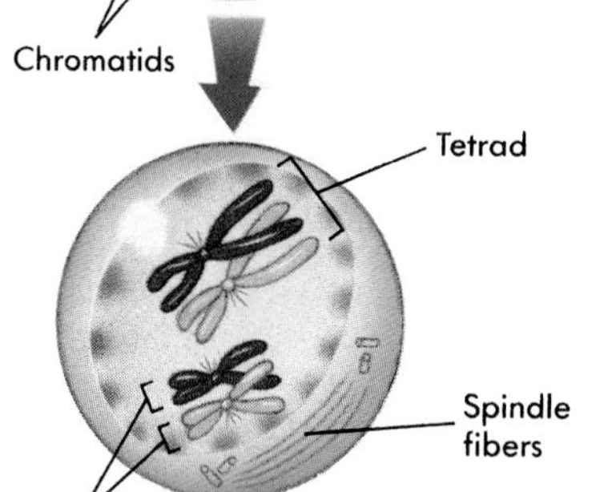

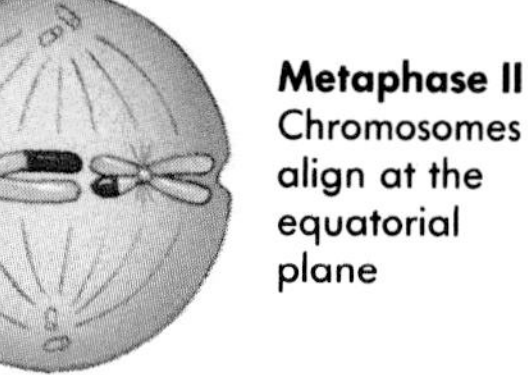

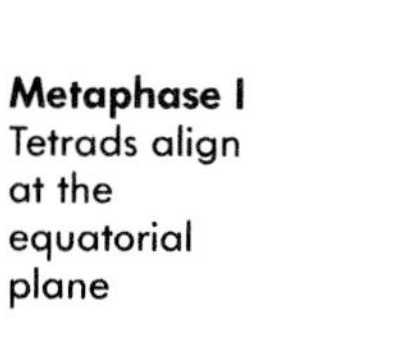
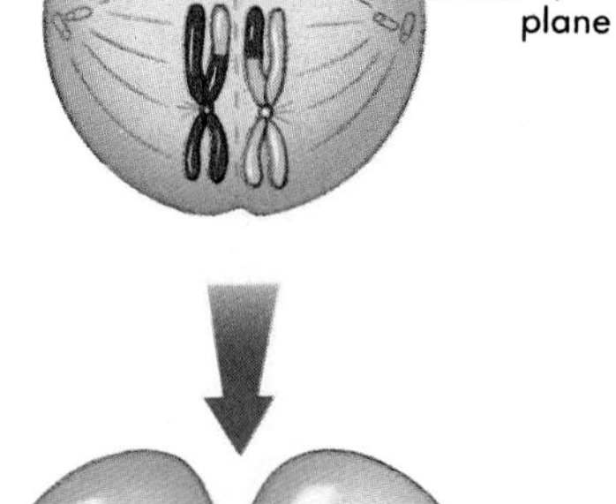

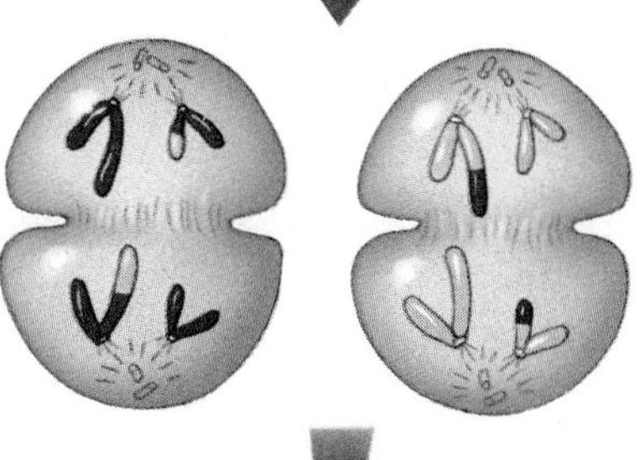

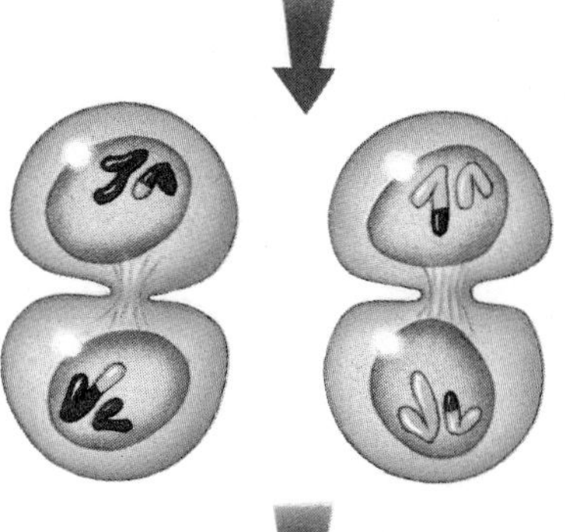

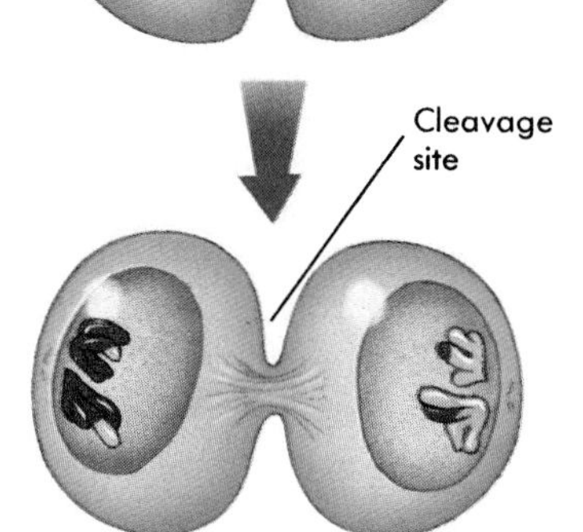

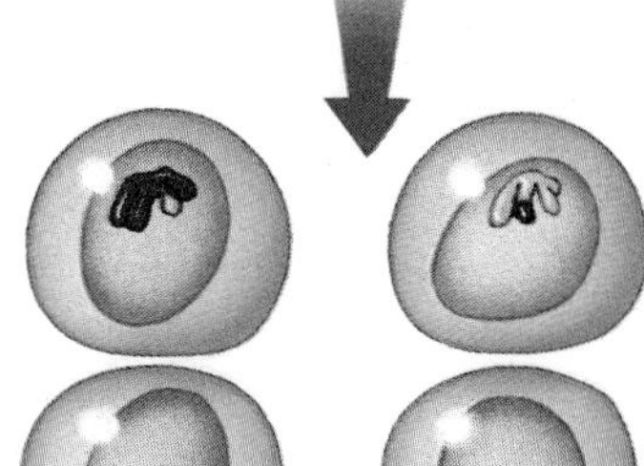
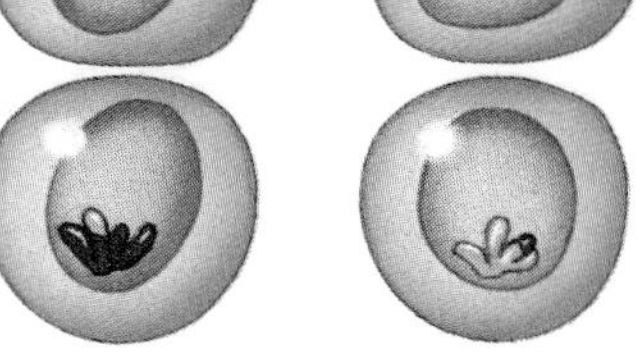

FIGURE 3-27 Meiosis.

Table 3-3

Comparison of mitosis and meiosis

FEATURE	MITOSIS	MEIOSIS
Time of DNA replication	Interphase	Interphase
Number of cell divisions	One	Two; there is no replication of DNA in between meiotic divisions
Cells produced	Two daughter cells genetically identical to the parent cell; each daughter cell has the diploid number of chromosomes	Four gametes, each different from the parent cell and each other; the gametes have the haploid number of chromosomes
Function	New cells are formed during growth or tissue repair; new cells have identical DNA and can perform the same functions as the parent cells	Gametes are produced for reproduction; during fertilization the haploid number of chromosomes in each gamete unite to restore the diploid number typical of somatic cells; genetic variability is increased because of crossing over and random assortment of homologous chromosomes during meiosis

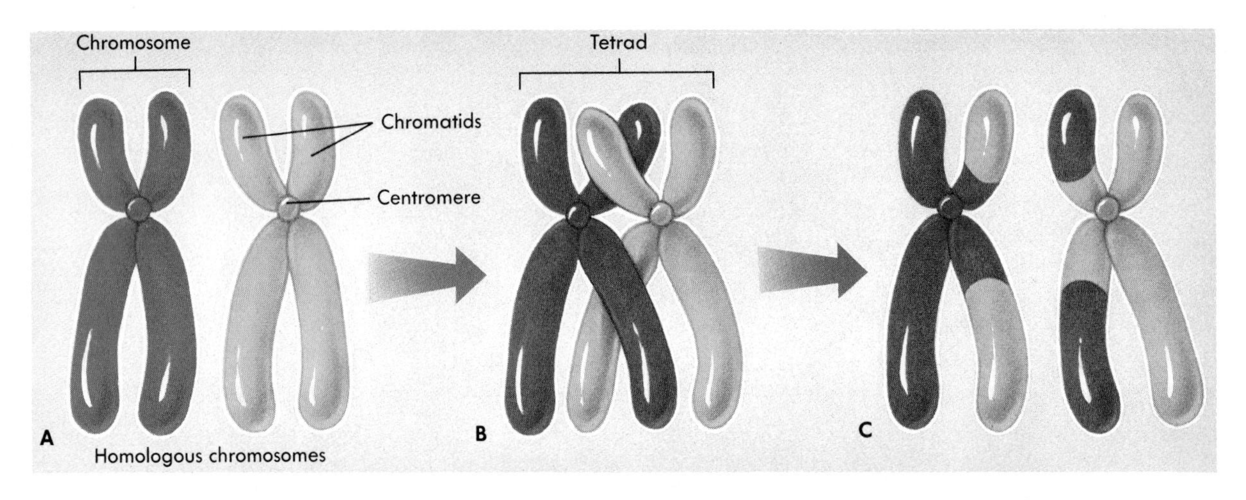

FIGURE 3-28 Crossing over may occur during prophase I of meiosis. **A** Pair of replicated homologous chromosomes. **B** Chromatids of the homologous chromosomes form a tetrad. **C** Genetic material is exchanged.

metaphase II, and their chromatids split apart in anaphase II. The chromatids then are called chromosomes, and each new cell receives 23 chromosomes. Table 3-3 compares mitosis and meiosis.

In addition to reducing the number of chromosomes in a cell from 46 to 23, meiosis is also responsible for genetic diversity for two reasons. First, when tetrads are formed, some of the chromatids break apart, and part of one chromatid from one homologous pair is exchanged for part of another chromatid from the other homologous pair (Figure 3-28). This exchange is called **crossing over;** as a result, chromatids with different DNA content are formed.

Second, there is a random distribution of the genetic material received from each parent. One member of each homologous pair of chromosomes was derived from the person's father and the other member from the person's mother. The homologous chromosomes align randomly during metaphase I; when they split apart, each daughter cell receives some of the father's and some of the mother's genetic material. How much of the father's or mother's genetic material, however, is determined by chance.

With crossing over and random assortment of homologous chromosomes, the possible number of gametes with different genetic makeup is practically unlimited. When the different gametes of two individuals unite, it is virtually certain that the resulting genetic makeup never has occurred before and never will occur again.

SUMMARY

A CELL IS THE STRUCTURAL AND FUNCTIONAL UNIT OF LIVING ORGANISMS.

1. The plasma membrane forms the outer boundary of the cell.
2. The nucleus directs the cell's activities.
3. Cytoplasm, between the nucleus and plasma membrane, is where most cell activities take place.

STRUCTURE OF THE PLASMA MEMBRANE *p. 65*

1. The plasma membrane passively or actively regulates what enters or leaves the cell.
2. The plasma membrane is composed of a phospholipid bilayer in which proteins float (fluid mosaic model). The proteins function as membrane channels, carrier molecules, receptor molecules, marker molecules, enzymes, and structural components of the membrane.

MOVEMENT THROUGH THE PLASMA MEMBRANE *p. 66*

1. Lipid-soluble molecules pass through the plasma membrane readily by dissolving in the lipid bilayer.
2. Small molecules pass through membrane channels. The channels are positively charged, allowing negatively charged ions and neutral molecules to pass through more readily than positively charged ions.
3. Large polar substances (e.g., glucose and amino acids) are transported through the membrane by carrier molecules.

Diffusion

1. Diffusion is the movement of a substance from an area of high concentration to an area of low concentration (with a concentration gradient).
2. The concentration gradient is the difference in solute concentration between two points divided by the distance separating the points.
3. The rate of diffusion increases with an increase in the concentration gradient, an increase in temperature, a decrease in molecular size, and a decrease in viscosity.
4. The end result of diffusion is a uniform distribution of molecules.
5. Diffusion requires no expenditure of energy.

Osmosis

1. Osmosis is the diffusion of water (solvent) across a selectively permeable membrane.
2. Osmotic pressure is a measure of the tendency of water to move across the selectively permeable membrane.
3. Isosmotic solutions have the same concentration of solute particles. Hyperosmotic solutions have a greater concentration, and hyposmotic solutions have a smaller concentration of solute particles than a reference solution.
4. Cells placed in an isotonic solution neither swell nor shrink. In a hypertonic solution they shrink (crenate), and in a hypotonic solution they swell and may lyse.

Filtration

1. Filtration is the movement of a liquid through a partition with holes that allow the liquid, but not everything in the liquid, to pass through them.
2. Liquid movement is due to a pressure difference across the partition.

Mediated Transport Mechanisms

1. Mediated transport is the movement of a substance across a membrane by means of a carrier molecule. The substances transported tend to be large, water-soluble molecules.
   - The carrier molecules have active sites that bind with either a single transport molecule or a group of similar transport molecules. This selectiveness is called specificity.
   - Similar molecules can compete for carrier molecules, with each reducing the rate of transport of the other.
   - Once all the carrier molecules are in use, the rate of transport cannot increase further (saturation).
2. There are two kinds of mediated transport.
   - Facilitated diffusion moves substances with their concentration gradient and does not require energy expenditure (ATP).
   - Active transport can move substances against their concentration gradient and requires ATP. An exchange pump is an active transport mechanism that simultaneously moves two substances in opposite directions across the plasma membrane.

Endocytosis and Exocytosis

1. Endocytosis is the bulk movement of materials into cells.
   - Phagocytosis is the bulk movement of solid material into cells by the formation of a vesicle.
   - Pinocytosis is similar to phagocytosis except that the ingested material is much smaller or is in solution.
2. Exocytosis is the secretion of materials from cells by vesicle formation.
3. Endocytosis and exocytosis use vesicles, can be specific for the substance transported, and require energy.

NUCLEUS *p. 75*

1. The nuclear envelope consists of two separate membranes with nuclear pores.
2. DNA and associated proteins are found inside the nucleus as chromatin. DNA is the hereditary material of the cell and controls the activities of the cell through the production of proteins through RNA.
3. Proteins (histones) play a role in the regulation of DNA activity.
4. Nucleoli consist of RNA and proteins and are the sites of ribosomal subunit assembly.

CYTOPLASM *p. 76*

Cytosol

1. Cytosol consists of a fluid portion (the site of chemical reactions), the cytoskeleton, and cytoplasmic inclusions.
2. The cytoskeleton supports the cell and enables cell movements. It consists of protein fibers.
   - Microtubules are hollow tubes composed of the protein tubulin. They form spindle fibers and are components of centrioles, cilia, and flagella.
   - Actin filaments are small protein fibrils that provide structure to the cytoplasm or cause cell movements.
   - Intermediate filaments are protein fibers that provide structural strength to cells.

Organelles

Organelles are subcellular structures specialized for specific functions.

Ribosomes

1. Ribosomes consist of small and large subunits manufactured in the nucleolus and assembled in the cytoplasm.
2. Ribosomes are the sites of protein synthesis.
3. Ribosomes can be free or associated with the endoplasmic reticulum.

Endoplasmic Reticulum

1. The endoplasmic reticulum is an extension of the outer membrane of the nuclear envelope and forms tubules or sacs (cisternae) throughout the cell.
2. Rough endoplasmic reticulum has ribosomes and is a site of protein synthesis.
3. Smooth endoplasmic reticulum lacks ribosomes and is involved in lipid production, detoxification, and calcium storage.

Golgi Apparatus

The Golgi apparatus is a series of closely packed, modified cisternae that function to modify, package, and distribute lipids and proteins produced by the endoplasmic reticulum.

Secretory Vesicles

Secretory vesicles are membrane-bound sacs that carry substances from the Golgi apparatus to the plasma membrane where the vesicle's contents are released by exocytosis.

Lysosomes

1. Lysosomes are membrane-bound sacs containing hydrolytic enzymes. Within the cell the enzymes break down phagocytized material and nonfunctional organelles (autophagia).
2. Enzymes released from the cell by lysis or enzymes secreted from the cell can digest extracellular material.

Peroxisomes

Peroxisomes are membrane-bound sacs containing enzymes that detoxify molecules and catalyze the breakdown of hydrogen peroxide.

Mitochondria

1. Mitochondria are the major sites of the production of ATP, which is used as an energy source by cells.
2. The mitochondria have a smooth outer membrane and an inner membrane that is infolded to produce cristae.
3. Mitochondria contain their own DNA, can produce some of their own proteins, and can replicate independently of the cell.

Centrioles and Spindle Fibers

1. Centrioles, cylindrical organelles, are located in the centrosome, a specialized zone of the cytoplasm. Centrioles determine the poles of cells during cell division.
2. Spindle fibers are involved in the separation of chromosomes during cell division.

Cilia and Flagella

1. Movement of materials over the surface of the cell is facilitated by cilia.
2. Flagella, much longer than cilia, propel spermatozoa.

Microvilli

Microvilli increase the surface area of the plasma membrane for absorption or secretion.

WHOLE CELL ACTIVITIES *p. 85*

Understanding the interactions between cell parts is necessary to an understanding of the functions of whole cells.

CELL METABOLISM *p. 85*

1. Aerobic respiration requires oxygen and produces carbon dioxide, water, and 38 ATP from a molecule of glucose.
2. Anaerobic respiration does not require oxygen and produces lactic acid and two ATP from a molecule of glucose.

PROTEIN SYNTHESIS *p. 87*

1. Information stored in DNA is copied to mRNA.
2. The mRNA goes to ribosomes in which it directs the synthesis of proteins.

Transcription

1. DNA unwinds and, through nucleotide pairing, produces mRNA (transcription).
2. The genetic code, which codes for amino acids, consists of codons, which are a sequence of three nucleotides in mRNA.

Translation

1. The mRNA moves through the nuclear pores to ribosomes.
2. Transfer RNA that carries amino acids interacts at the ribosome with mRNA. The anticodons of tRNA bind to the codons of mRNA, and the amino acids are joined to form a protein (translation).

Regulation of Protein Synthesis

1. Cells become specialized because of inactivation of certain parts of the DNA molecule and activation of other parts.
2. The rate of DNA activity and thus protein production can be controlled internally or can be affected by regulatory substances secreted by other cells.

CELL LIFE CYCLE *p. 91*

Interphase

Interphase is the period between cell divisions. DNA unwinds, and each strand produces a new DNA molecule during interphase.

Cell Division

1. Mitosis is the replication of the cell's nucleus, and cytokinesis is division of the cell's cytoplasm.
2. Humans have 22 pairs of homologous chromosomes called autosomes. Females also have two X chromosomes, and males also have an X chromosome and a Y chromosome.
3. Mitosis is a continuous process divided into four stages.
   - Prophase. Chromatin condenses to become visible as chromosomes. Each chromosome consists of two chromatids joined at the centromere. Centrioles move to opposite poles of the cell, and astral fibers and spindle fibers form. Nucleoli disappear, and the nuclear envelope degenerates.
   - Metaphase. Chromosomes align at the equatorial plane.
   - Anaphase. The chromatids of each chromosome separate at the centromere. Each chromatid then is called a chromosome. The chromosomes migrate to opposite poles.
   - Telophase. Chromosomes unravel to become chromatin. The nuclear envelope and nucleoli reappear.
4. Cytokinesis begins with the formation of the cleavage furrow during anaphase. It is complete when the plasma membrane comes together at the equator, producing two new daughter cells.

***MEIOSIS** p. 94*

1. Meiosis results in the production of gametes (oocytes or spermatozoa).
2. All gametes receive one half of the homologous autosomes (one from each homologous pair). Oocytes also receive an X chromosome. Spermatozoa have an X or a Y chromosome.
3. There are two cell divisions in meiosis. Each division has four stages (prophase, metaphase, anaphase, and telophase) similar to those in mitosis.
   - In the first division tetrads form, crossing over occurs, and homologous chromosomes are distributed randomly. Two cells are formed, each with 23 chromosomes. Each chromosome has two chromatids.
   - In the second division the chromatids of each chromosome separate, and each cell receives 23 chromatids, which then are called chromosomes.
   - Genetic variability is increased by crossing over and random assortment of chromosomes.

CONTENT REVIEW

1. Define plasma membrane, nucleus, and cytoplasm.
2. What is the function of the plasma membrane? How do phospholipids, cholesterol, and proteins contribute to that function? Describe the fluid mosaic model of the plasma membrane.
3. How do large lipid-soluble molecules move across the plasma membrane?
4. How do small molecules, water- or lipid-soluble, pass through the membrane? What effect does the electrical charge of molecules have on the ease of passage?
5. Define diffusion. How do the concentration gradient, temperature, molecule size, and viscosity affect the rate of diffusion?
6. Define osmosis and osmotic pressure.
7. What happens to a cell that is placed in an isotonic solution? In a hypertonic or hypotonic solution? What are crenation and lysis?
8. Define filtration.
9. What is mediated transport? Explain the basis for specificity, competition, and saturation of transport mechanisms.
10. Contrast active transport and facilitated diffusion in relationship to energy expenditure and movement of substances with or against their concentration gradients.
11. Name three ways in which phagocytosis, pinocytosis, and exocytosis are similar. How do they differ?
12. List and describe the parts of cytosol.
13. What is the function of the cytoskeleton? Name the three groups of proteins of which it is made.
14. Describe the structure of the nuclear envelope.
15. What is the difference between chromatin and chromosomes? Name the two components of chromatin, and explain their functions.
16. Where are ribosomes assembled, and what kinds of molecules are found in them? Give the function of ribosomes.
17. What is endoplasmic reticulum? Contrast rough and smooth endoplasmic reticulum according to structure and function.
18. Describe the Golgi apparatus, and state its function.
19. Where are secretory vesicles produced? What are their contents, and how are they released?
20. What are a lysosome and a peroxisome? Explain the function of lysosomes and peroxisomes.
21. Describe the structure of mitochondria. Name the important molecule produced by mitochondria. For what is this molecule used?
22. Describe the structure and function of centrioles, spindle fibers, cilia, flagella, and microvilli.
23. Define glycolysis. Where does it take place?
24. Contrast the production of ATP by aerobic and anaerobic respiration. What happens to the oxygen we breathe in? The carbon dioxide we breathe out comes from where?
25. Explain what happens during transcription. How does mRNA get to the ribosomes?
26. Discuss translation. What kinds of RNA are involved? How are codons and anticodons involved in the synthesis of proteins?
27. Distinguish between mitosis and cytokinesis.
28. Discuss the events that occur during interphase, prophase, metaphase, anaphase, and telophase of mitosis.
29. Define meiosis, and describe the events that occur during meiosis. What happens to the number of chromosomes during meiosis?
30. How are the chromosomes of males and females different?
31. What are two ways in which genetic variability is increased?

CONCEPT QUESTIONS

1. A man's body was found floating in the salt water of San Francisco Bay. The salt water is more concentrated than the fluids of the body. When seen during an autopsy, the cells in his lung tissues were clearly swollen. Choose the following most logical conclusion.
   A He probably drowned in the bay.
   B He was probably murdered.
   C He did not drown.
2. Solution A is hyperosmotic to solution B. If solution A were separated from solution B by a selectively permeable membrane, would water move from solution A into solution B or vice versa? Explain.
3. A dialysis membrane is selectively permeable, and substances smaller than proteins are able to pass through it. If you wanted to use a dialysis machine to remove only urea (a small molecule) from blood, what could you use for the dialysis fluid?
   A A solution that is isotonic and contains only protein
   B A solution that is isotonic and contains the same concentration of all substances except that it has no urea
   C Distilled water
   D Blood
4. For a carrier-mediated mechanism, Figure 3-7, *D*, illustrates the rate of transport of a molecule across a plasma membrane vs. the concentration of the molecule to be transported. For simple diffusion, which is not carrier mediated, draw a graph for the rate of movement across a plasma membrane vs. the concentration.
5. A researcher wanted to determine the nature of the transport mechanism that moved substance X into a cell. She could measure only the concentration of substance X in the extracellular fluid and within the cell, so she did a series of experiments and gathered the data presented below.

   At A the extracellular concentration of substance X was equal to the intracellular concentration of substance X. Choose the transport process that is consistent with the data.
   A Diffusion
   B Active transport
   C Facilitated diffusion
   D Not enough information to make a judgment
6. Given the following data from electron micrographs of a cell, predict the major function of the cell:
   - Moderate number of mitochondria
   - Well-developed rough endoplasmic reticulum
   - Moderate number of lysosomes
   - Well-developed Golgi apparatuses
   - Dense nuclear chromatin
   - Numerous vesicles
7. If you had the ability to inhibit mRNA synthesis with a drug, explain how you could distinguish between proteins released from secretory vesicles in which they had been stored and proteins released from cells in which they have been newly synthesized.

? ?

ANSWERS TO PREDICT QUESTIONS

1 p. 67. Urea is continually produced by metabolizing cells and diffuses from the cells into the interstitial spaces and from the interstitial spaces into the blood. If the kidneys stop eliminating urea, it begins to accumulate in the blood. Because the concentration of urea increases in the blood, urea cannot diffuse from the interstitial spaces. As urea accumulates in the interstitial spaces, the rate of diffusion from cells into the interstitial spaces slows because the urea must pass from a higher to a lower concentration by the process of diffusion. The urea finally reaches concentrations high enough to be toxic to cells, causing cell damage followed by cell death.

2 p. 68. If the membrane were freely permeable, the solutes in the tube would diffuse from the tube (higher concentration of solutes) into the beaker (lower concentration of solutes) until there were equal amounts of solutes inside the tube and beaker (i.e., equilibrium). In a similar fashion, water in the beaker would diffuse from the beaker (higher concentration of water) into the tube (lower concentration of water) until there were equal amounts of water inside the tube and beaker. Consequently, the solution concentrations inside the tube and beaker would be the same because they would both contain the same amounts of solutes and water. Under these conditions there would be no net movement of water into the tube. Thus osmotic pressure would be zero. This simple experiment demonstrates that osmosis and osmotic pressure require a membrane that is selectively permeable.

3 p. 70. Glucose transported by facilitated diffusion across the plasma membrane will move from a high to a low concentration gradient. If glucose molecules are quickly converted to some other molecule as they enter the cell, a steep concentration gradient is maintained. The rate of glucose transport into the cell is directly proportional to the magnitude of the concentration gradient.

4 p. 71. The transport process is facilitated diffusion. Active transport and facilitated diffusion both exhibit saturation. It is not active transport because the inhibitors of metabolism and therefore ATP production do not inhibit the transport process. ATP is required for active transport but not for facilitated diffusion.

5 p. 80.

A. Cells highly specialized to synthesize and secrete proteins would have large amounts of rough endoplasmic reticulum (ribosomes attached to endoplasmic reticulum) because these organelles are important for protein synthesis. Golgi apparatuses would be well developed because they package materials for release in secretory vesicles. Also there would be numerous secretory vesicles in the cytoplasm.

B. Cells highly specialized to actively transport substances into the cell would have a large surface area exposed to the fluid from which substances are actively transported, and numerous mitochondria would be present near the membrane across which active transport occurs.

C. Cells highly specialized to synthesize lipids would have large amounts of smooth endoplasmic reticulum. Depending on the kind of lipid produced, lipid droplets may accumulate in the cytoplasm.

D. Cells highly specialized to phagocytize foreign substance would have numerous lysosomes in their cytoplasm and evidence of phagocytic vesicles.

6 p. 89. By changing a single nucleotide within a DNA molecule, a change in the nucleotide of messenger RNA produced from that segment of DNA also would occur, and a different amino acid would be placed in the amino acid chain for which the messenger RNA provides direction. Because a change in the amino acid sequence of a protein could change its structure, one substitution of a nucleotide in a DNA chain could result in altered protein structure and function.

7 p. 91. Because adenine pairs with thymine (there is no uracil in DNA) and cytosine pairs with guanine, the sequence of DNA replicated from strand one would be TACGAT. This sequence would also be the sequence of DNA in the original strand two. A replicate of strand two, therefore, would be ATGCTA, which is the same as the original strand one.

4 CHAPTER OBJECTIVES

After reading this chapter you should be able to

1. List the characteristics used to classify tissues into one of the four major tissue types.
2. List the features that characterize epithelium.
3. Describe the characteristics that are used to classify the various epithelial types.
4. List for each epithelial type its number of cell layers, cell shapes, major cellular organelles, surface specializations, and the functions to which it is adapted.
5. Explain why junctional complexes between cells are important to the normal function of epithelium.
6. Define the term gland and describe the two major categories of glands.
7. List the features that characterize connective tissue.
8. List the major large molecules of the connective tissue matrix and explain their functions in the matrix.
9. List the major categories of connective tissue and describe the characteristics of each.
10. List the general characteristics of muscle.
11. Name the main types of muscles and list their major characteristics.
12. Describe the characteristics of nervous tissue.
13. Name the three embryonic germ layers.
14. List the functional and structural characteristics of serous and mucous membranes.
15. Describe the process of inflammation and explain why inflammation is protective to the body.
16. Describe the major events involved in tissue repair.

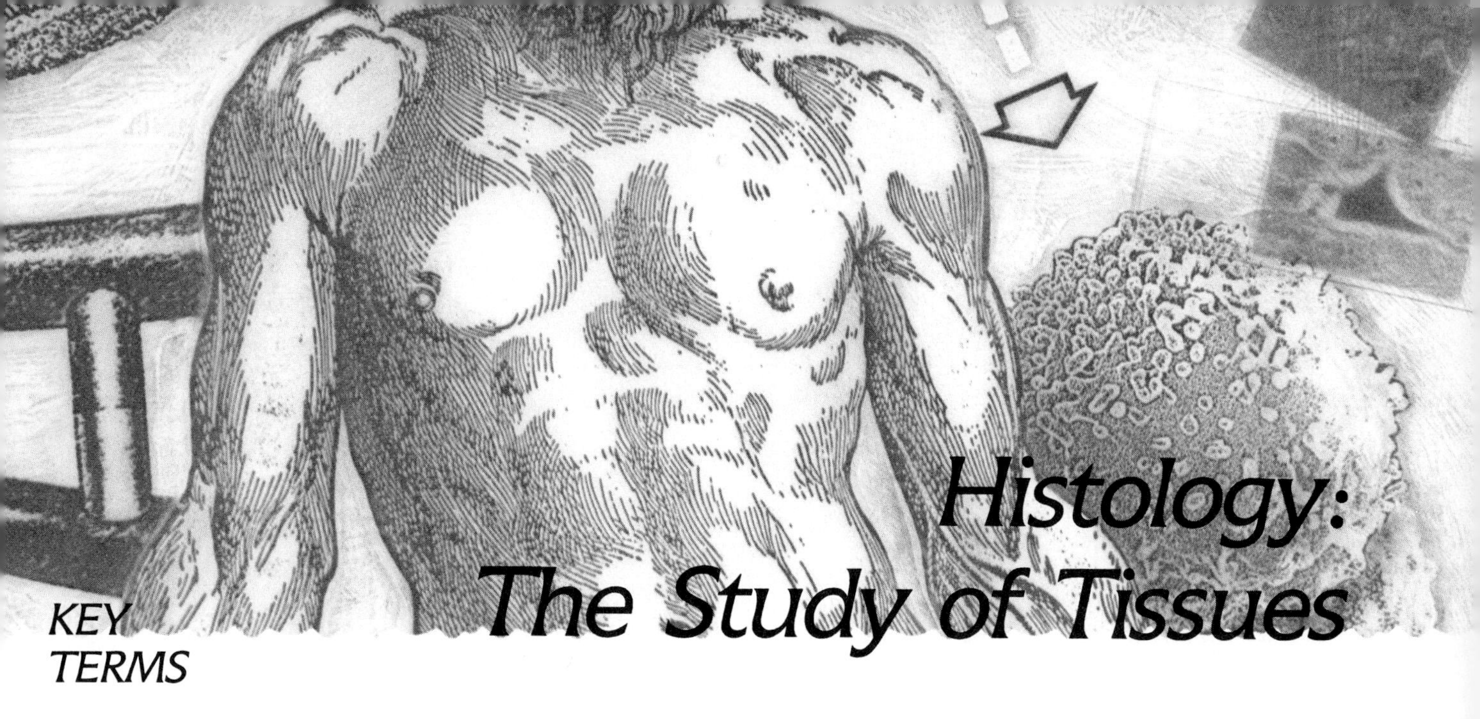

Histology: The Study of Tissues

KEY TERMS

basement membrane

connective tissue

desmosome (dez′mo-sōm)

epithelium

exocrine (ek′so-krin) gland

extracellular matrix

germ layer

histology

inflammation

mucous membrane

muscular tissue

nervous tissue

serous membrane

squamous (skwa′mus)

tissue repair

RELATED TOPICS

The following terms or concepts are important for a good understanding of this chapter. If you are not familiar with them, you should review them before proceeding.

Body cavities and the membranes lining them
(Chapter 1)

Structure and function of the cellular organelles
(Chapter 3)

CELLS ARE GROUPED INTO TISSUES, AND EACH TISSUE TYPE IS HIGHLY SPECIALIZED TO PERFORM SPECIFIC FUNCTIONS. Knowledge of tissue structure and function is important in understanding how inidvidual cells are organized to form tissues, organs, organ systems, and the complete organism. The microscopic study of tissues is **histology**. This chapter discusses the structure of the major tissue types and their functional characteristics. Structure and function are so closely related in tissues that a student should be able to predict a tissue's function when given its structure and vice versa.

The structure of cells and the composition of the **extracellular matrix** (the noncellular substances surrounding cells) are characteristics used to classify tissue types. The four basic tissue types are (1) **epithelial,** (2) **connective,** (3) **muscular,** and (4) **nervous tissue.** Epithelial and connective tissues are the most diverse of the four tissue types and are components of every organ. They are classified according to structure (i.e., how closely the cells of each tissue are packed and what the position of the tissue is relative to other tissues). Muscular and nervous tissues, on the other hand, are defined mainly according to their functions.

EPITHELIAL TISSUE

A major characteristic of **epithelium** (plural: epithelia) is that it consists almost entirely of cells that have very little extracellular material between them. Epithelium covers surfaces (e.g., the outside of the body and the lining of the digestive tract, the vessels, and many body cavities) or forms structures (e.g., glands) that are derived developmentally from the body surfaces. Therefore most epithelial tissues have one free surface that is not associated with other cells. Most epithelia also have a **basement membrane** (exceptions include the lymph vessels and liver sinusoids). The basement membrane is a specialized type of extracellular material that is secreted primarily by the epithelial cells on the side opposite their free surface and helps attach the epithelial cells to the underlying tissues. Blood vessels do not penetrate the basement membrane to reach the epithelium; thus all gases and nutrients carried in the blood must reach the epithelium by diffusing across the basement membrane from blood vessels in the underlying connective tissue. Thus the most metabolically active cells are close to the basement membrane. In stratified epithelia, cells die as they move farther away from the basement membrane.

Classification of Epithelium

The major types of epithelia and their distributions are illustrated in Figure 4-1. Epithelia are classified primarily according to the number of cell layers and the shape of the cells. **Simple epithelium** consists of a single layer of cells, with each cell extending from the basement membrane to the free surface. **Stratified epithelium** consists of more than one layer of cells, and only one of the layers is adjacent to the basement membrane. **Pseudostratified epithelium** consists of a combination of cells, with all the cells attached to the basement membrane but only some of them reaching the free surface. This epithelium is called pseudostratified because, although it consists of a single cell layer, it appears multilayered. The arrangement of the nuclei gives a stratified appearance.

Categories of epithelium based on cell shape include **squamous** (skwa'mus; flat), **cuboidal** (cubelike), and **columnar** (tall and thin, similar to a column). In most cases an epithelium is given two names (e.g., simple squamous, stratified squamous, simple columnar, pseudostratified columnar), with the first name indicating the number of layers and the second indicating the shape of the cells (Table 4-1).

Stratified squamous epithelium can be classified further as either moist or keratinized, according to the condition of the outermost layer of cells. In both types the deepest layers are composed of living cells. In moist stratified squamous epithelium, found in areas such as the mouth and vagina, the outermost layers also consist of living cells. On the other hand, keratinized stratified squamous epithelium, found in the skin (see Chapter 5), has outer layers composed of dead cells, which give the tissue a very tough, moisture-resistant character.

Table 4-1

Classification of epithelium

TYPES OF EPITHELIUM	SHAPE OF CELLS
Simple (single layer of cells)	Squamous Cuboidal Columnar
Stratified (more than one layer of cells)	Squamous Moist Keratinized Cuboidal (very rare) Columnar (very rare)
Pseudostratified (modification of simple epithelium)	Columnar
Transitional (modification of stratified epithelium)	Roughly cuboidal or many surfaced

1

Explain why it would be a disadvantage to have skin composed of moist, stratified epithelium or simple epithelium.

? ? ? ? ? ? ? ? ? ? ?

One type of epithelium is specialized to line the urinary bladder and ureters in which considerable expansion may occur. This epithelium is stratified, but the number of cells that comprise the stratified layer is variable, depending on whether it is stretched or not. The surface cells are roughly cuboidal but become more squamouslike when the organ is stretched. This type of tissue is called **transitional epithelium.**

Functional Characteristics

Epithelial tissues perform many functions (Table 4-2, p. 108-109), including formation of a barrier between a free surface and the underlying delicate tissues, secretion, transportation, and absorption of selected molecules. The type and arrangement of organelles within each cell (see Chapter 3), the shape

Text continued on p. 110.

FIGURE 4-1 *Epithelial tissues*

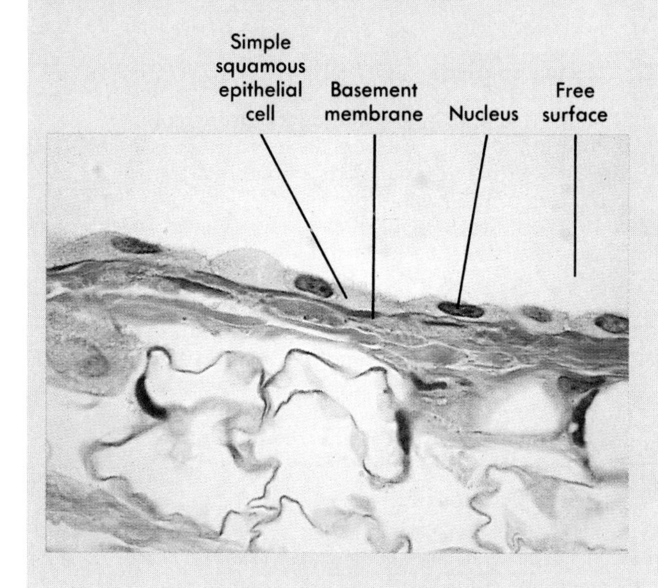

A Simple squamous epithelium

LOCATION: Lining of blood and lymph vessels (endothelium) and small ducts, alveoli of the lungs, loop of Henle in kidney tubules, lining of serous membranes (mesothelium), and inner surface of the eardrum.
STRUCTURE: Single layer of flat, often hexagonal cells. Since the cells are so flat, the nuclei appear as bumps when viewed as a cross section.
FUNCTION: Diffusion, filtration, some protection against friction, secretion, and absorption.

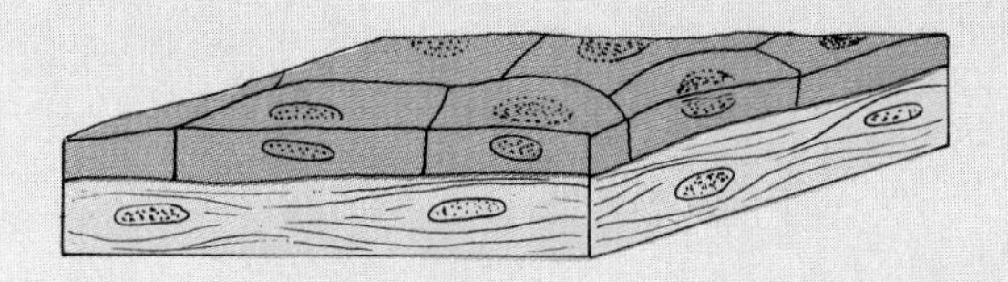

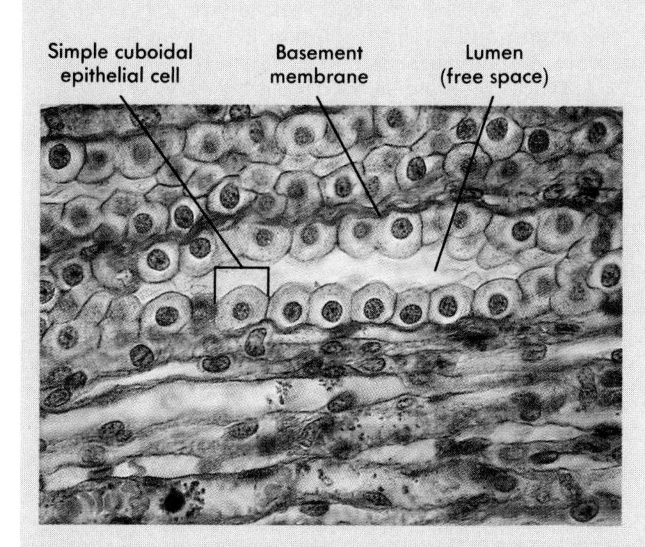

B Simple cuboidal epithelium

LOCATION: Glands and their ducts, terminal bronchioles of lungs, kidney tubules, choroid plexus of the brain, and surface of the ovaries.
STRUCTURE: Single layer of cube-shaped cells. Some cells have cilia (terminal bronchioles) or microvilli (kidney tubules).
FUNCTION: Movement of mucus-containing particles out of the terminal bronchioles by ciliated cells. Absorption and secretion by cells of the kidney tubules. Secretion by cells of the choroid plexus and glands.

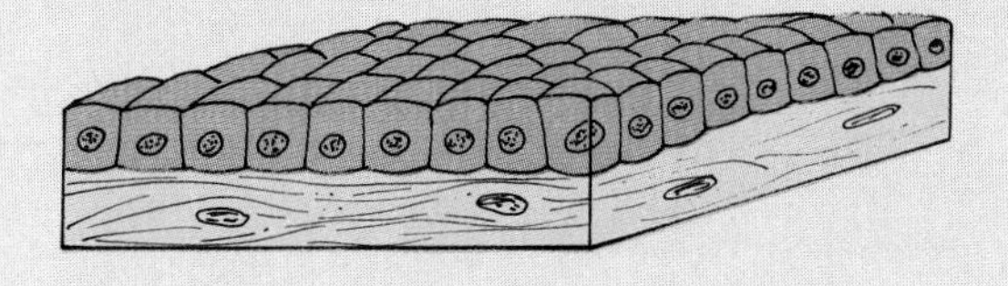

Continued.

FIGURE 4-1—cont'd

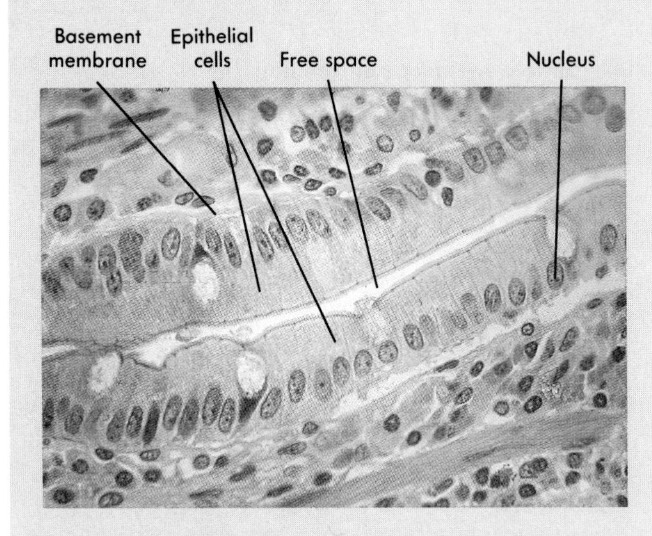

C Simple columnar epithelium

LOCATION: Glands and some ducts, bronchioles of lungs, auditory tube, uterus, uterine tubes, stomach, intestines, gallbladder, bile ducts, and ventricles of the brain.

STRUCTURE: Single layer of tall, narrow cells. Some cells have cilia (bronchioles of lungs, auditory tubes, uterine tubes, and uterus) or microvilli (intestines).

FUNCTION: Movement of particles out of the bronchioles of the lungs and partially responsible for the movement of the egg through the uterine tubes by ciliated cells. Secretion by cells of the glands, the stomach, and the intestine. Absorption by cells of the intestine.

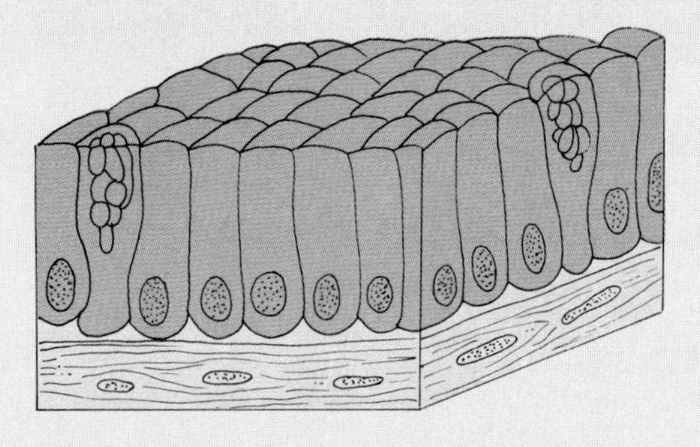

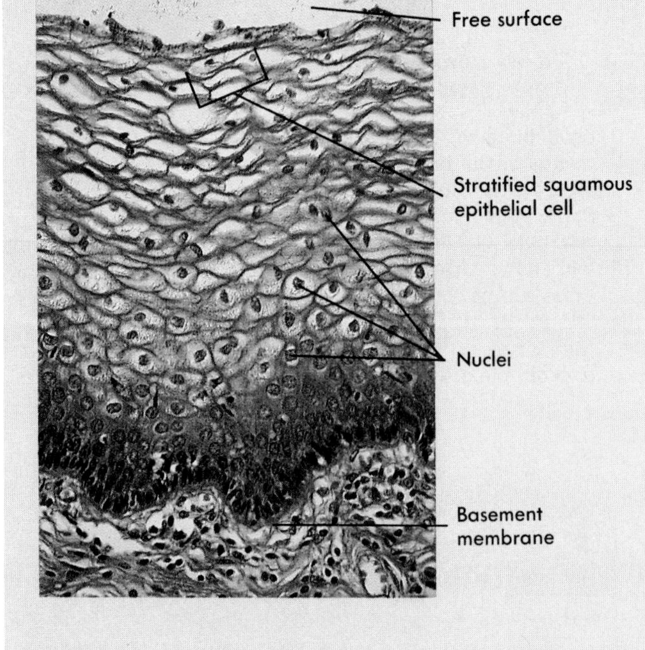

D Stratified squamous epithelium

LOCATION: Moist—mouth, throat, larynx, esophagus, anus, vagina, inferior urethra, and cornea. Keratinized—skin.

STRUCTURE: Multiple layers of cells that are cuboidal in the basal layer and progressively flattened toward the surface. The epithelium may be moist or keratinized. In moist stratified squamous epithelium the surface cells retain a nucleus and cytoplasm. In keratinized cells the cytoplasm is replaced by keratin, and the cells are dead.

FUNCTION: Protection against abrasion and infection.

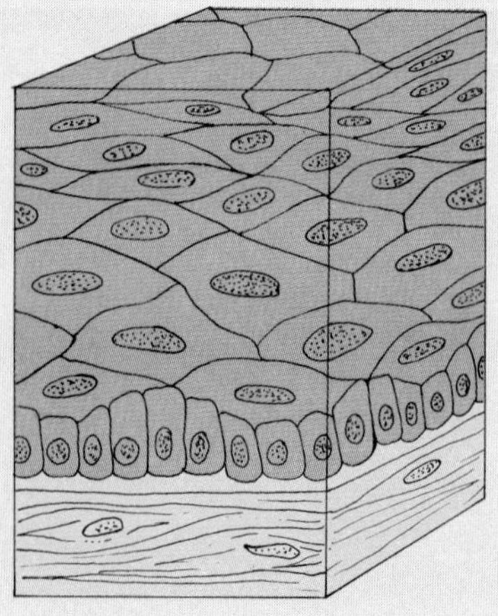

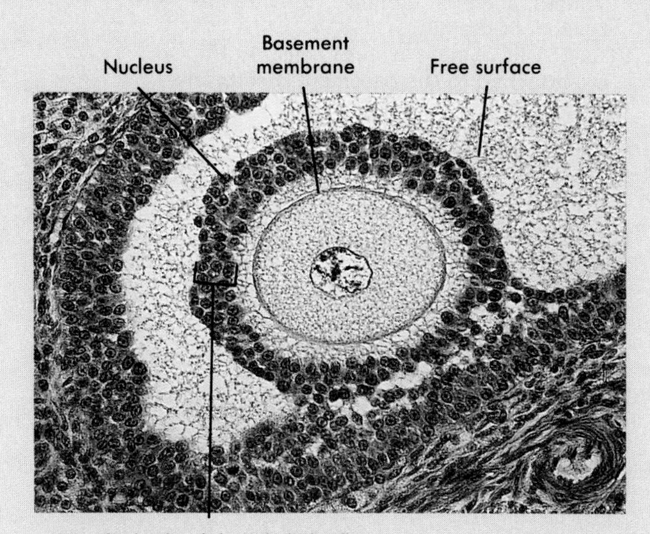

E Stratified cuboidal epithelium

LOCATION: Sweat gland ducts and ovarian follicular cells.
STRUCTURE: Multiple layers of somewhat cube-shaped cells.
FUNCTION: Secretion, absorption, and protection against infection.

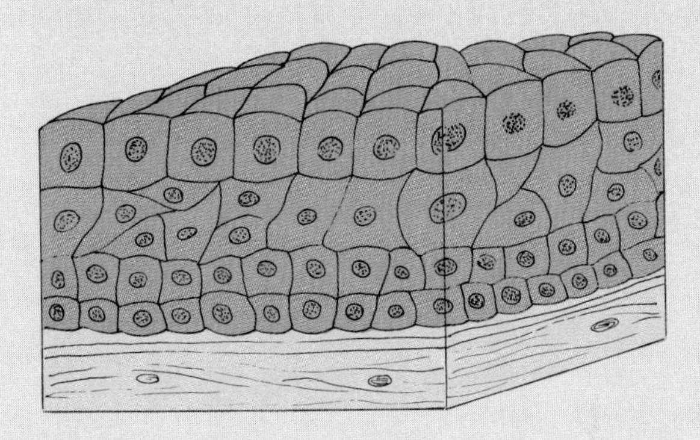

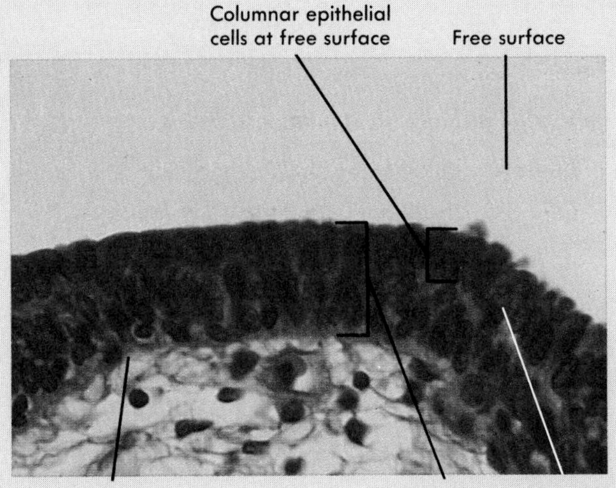

F Stratified columnar epithelium

LOCATION: Mammary gland duct, larynx, and a portion of the male urethra.
STRUCTURE: Multiple layers of cells, with tall, thin cells resting on layers of more cuboidal cells. The cells are ciliated in the larynx.
FUNCTION: Protection and secretion.

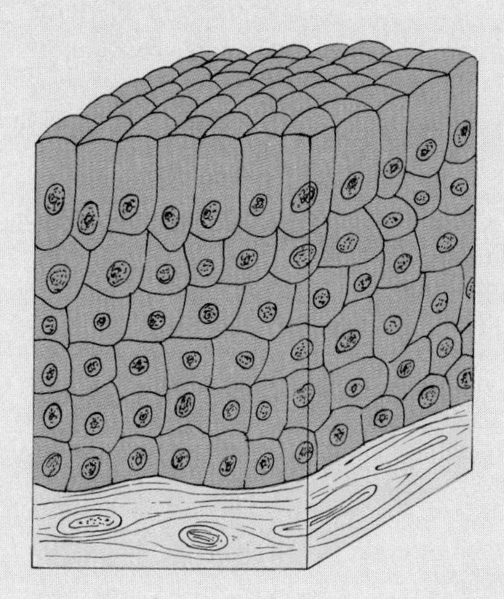

Continued.

FIGURE 4-1—cont'd

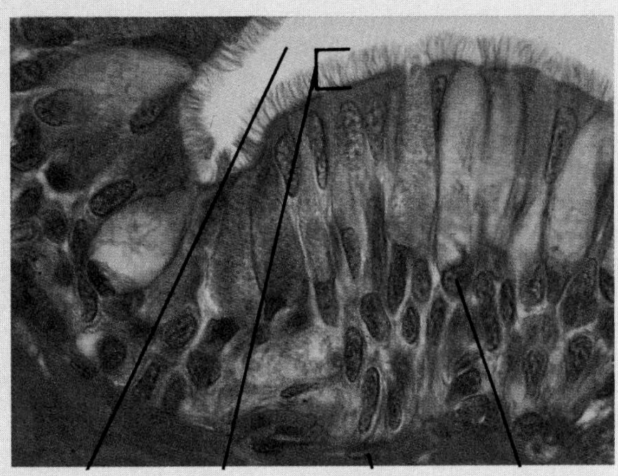

G Pseudostratified columnar epithelium

LOCATION: Larynx, nasal cavity, paranasal sinuses, pharynx, auditory tube, trachea, and bronchi of the lungs.

STRUCTURE: Single layer of cells. All the cells are attached to the basement membrane. Some cells are tall and thin and reach the free surface, and others do not. The nuclei of these cells are at different levels and appear stratified. The cells are almost always ciliated and are associated with goblet cells.

FUNCTION: Movement of fluid (often mucus) that contains foreign particles.

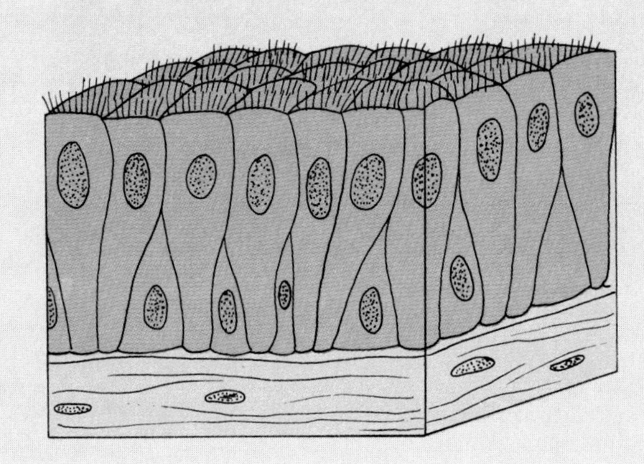

Table 4-2

Function and location of epithelial tissue

FUNCTION	SIMPLE SQUAMOUS	SIMPLE CUBOIDAL	SIMPLE COLUMNAR
Diffusion	Blood and lymph capillaries, alveoli of lungs, thin segment of loop of Henle		
Filtration	Bowman's capsule of kidney		
Secretion or absorption	Mesothelium (serous fluid)	Choroid plexus (cerebrospinal fluid), part of kidney tubule, many glands	Stomach, small intestine, large intestine, uterus, many glands
Protection (against friction and abrasion)	Endothelium, mesothelium		
Movement of mucus (ciliated)		Terminal bronchioles of lungs	Bronchioles of lungs, auditory tubes, uterine tubes, uterus
Capable of great stretching			
Miscellaneous	Lines the inner part of the eardrum, smallest ducts of glands	Pancreatic duct, surface of ovary, inside lining of eye (pigmented epithelium of retina), ducts of glands	Bile duct, gallbladder, ependyma (lining of brain ventricles and central canal of spinal cord), ducts of glands

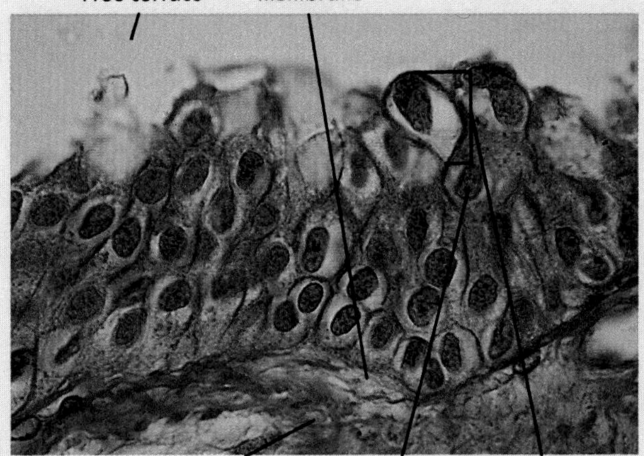

H Transitional epithelium

LOCATION: Urinary bladder, ureters, and superior urethra.

STRUCTURE: Stratified cells that appear cubelike when the organ or tube is relaxed and appear squamous when the organ or tube is distended by fluid.

FUNCTION: Formation of a permeability barrier and protection against the caustic effect of urine. Accommodation of fluid-content fluctuations in organ or tube.

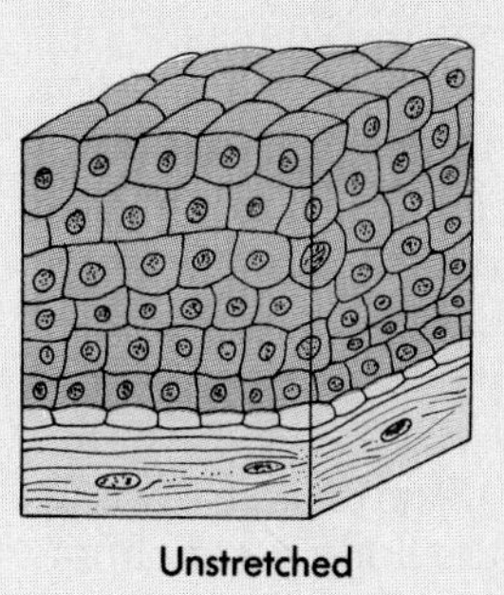

Unstretched

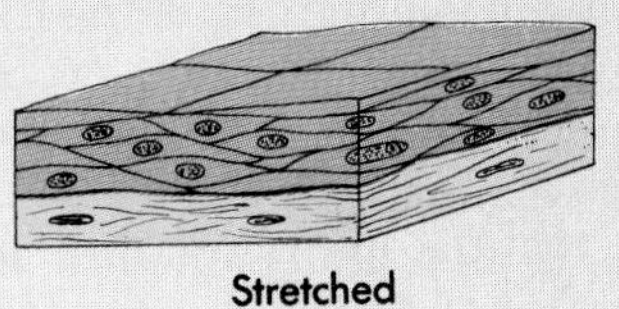

Stretched

STRATIFIED SQUAMOUS	STRATIFIED CUBOIDAL	STRATIFIED COLUMNAR	PSEUDOSTRATIFIED COLUMNAR	TRANSITIONAL
Skin (epidermis), cornea, mouth and throat, epiglottis, larynx, esophagus, anus, vagina				
			Larynx, nasal cavity, paranasal sinus, nasopharynx, auditory tube, trachea, bronchi of lungs	
				Urinary bladder, ureter, upper part of urethra
Lower part of urethra, sebaceous gland duct	Sweat gland duct	Part of male urethra, epididymis, ductus deferens, mammary gland duct	Part of male urethra, salivary gland duct	

of cells, and the organization of cells within each epithelial type reflect these functional characteristics. Accordingly, epithelial types are specialized, and this specialization can be understood best in terms of the functions they perform.

Cell layers and cell shapes

Simple epithelium with its single layer of cells is found in organs in which the principal functions are diffusion (lungs), filtration (kidneys), secretion (glands), or absorption (intestines). The selective movement of materials through epithelium would be hindered by a stratified epithelium, which is found in areas where protection is a major function. The multiple layers of cells in stratified epithelium are well adapted for a protective role because as the outer cells are damaged, they are replaced by cells from deeper layers and a continuous barrier of epithelial cells is maintained in the tissue. Stratified squamous epithelium is found in areas of the body where abrasion can occur such as the skin, mouth, throat, esophagus, anus, and vagina.

Differential function is also reflected in cell shape. Cells involved in diffusion and filtration are normally flat and thin. For example, simple squamous epithelium forms blood and lymph capillaries, the alveoli (air sacs) of the lungs, and parts of the kidney tubules. Cells with the major function of secretion or absorption are usually cuboidal or columnar. Their greater cytoplasmic volume, as compared to that of squamous epithelium, is a result of the presence of the organelles responsible for the tissue's function. For example, pseudostratified columnar epithelium, which secretes large amounts of mucus, lines the respiratory tract (see Chapter 23) and contains large, mucus-filled goblet cells, which are specialized columnar epithelial cells.

Cell surfaces

The surface of epithelial cells can be divided into three categories: (1) a free surface that faces away from underlying tissue, (2) a surface that faces other cells, and (3) a surface that faces the basement membrane, the basilar surface. Types of free surfaces include smooth, microvillar, and ciliated. Smooth surfaces reduce friction. Simple squamous epithelium with a smooth surface forms the covering (the mesothelium) of serous membranes. The lining of blood vessels (endothelium) is a simple squamous epithelium that reduces friction as blood flows through the vessels (see Chapter 21).

Microvilli and cilia were described in Chapter 3. Microvilli greatly increase surface area and are found in cells involved in absorption or secretion (e.g., the lining of the small intestine; see Chapter 24). Cilia propel materials across the surface of the cell. Simple ciliated cuboidal, simple ciliated columnar, and pseudostratified ciliated columnar epithelia are in the respiratory tract (see Chapter 23) where mucus that contains foreign substances (e.g., dust particles) is removed from the respiratory passages by the ciliary movements of these tissues.

Transitional epithelium has a rather unusual cell membrane specialization in which rigid sections of membrane are separated by rather flexible regions. This differentiation allows the cell membrane of transitional epithelium to fold like an accordion. When transitional epithelium is stretched, the cell surface can unfold.

Cell connections

Cell surfaces other than free surfaces have modifications that serve to hold cells to each other or to the basement membrane (Figure 4-2). These modi-

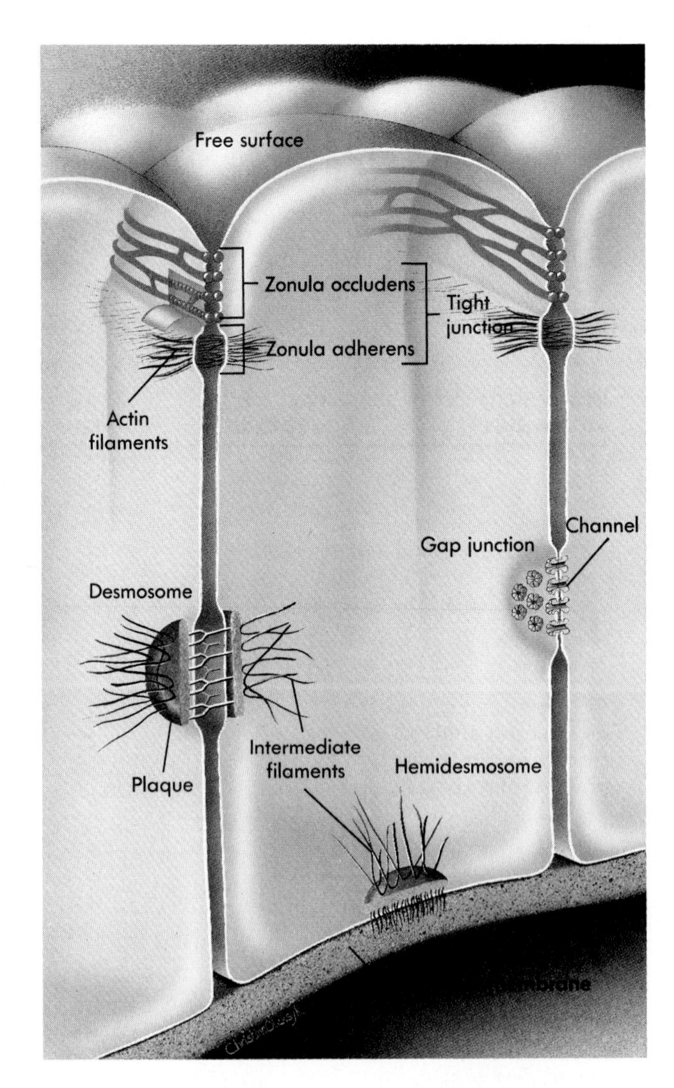

FIGURE 4-2 Cell connections. Zonula adherens and zonula occludens form a junctional complex called a "tight" junction. Desmosomes, hemidesmosomes, and gap junctions are also shown, although no cell will have all of these different connections.

Microscopic Imaging

We see objects because light either passes through the object or is reflected off the object and enters our eye where it is translated into an image (see Chapter 15). However, we are quite limited in what we can see with the unaided eye. Without the aid of magnifying lenses, the smallest object we can resolve is approximately 100 μm in diameter (0.1 mm; approximately the size of a fine pencil dot). To resolve (distinguish between two points) smaller structures, we must use a microscope.

There are two basic types of microscopes: light microscopes and electron microscopes. As their names imply, light microscopes use light to produce an image, and electron microscopes use beams of electrons. **Light microscopes** usually use transmitted light (light passing through the object), but some light microscopes are equipped to use reflected light. Glass lenses are used in the microscope to magnify the image, and the image either can be observed directly by looking into the microscope or the light from the image can be used to expose photographic film to make a photograph of the image. The resolution of light microscopes is limited by the wavelength of light, and the lower limit is approximately 0.1 μm (near the size of a small bacterium).

Light microscopy is used on a regular basis to examine cells and tissues. Because the image usually is generated using transmitted light, tissues to be examined must be cut very thinly to allow the light to pass through them. To make thin enough sections for light microscopy (sections usually are cut 1 to 20 μm thick), the tissue must be fixed (a process that preserves the tissue and makes it more rigid) and embedded in some material (wax or plastic) that will make the tissue rigid enough for cutting into sections. Since most tissues are colorless and transparent when thinly sectioned, the tissue must be stained with a colored dye so the structural detail can be seen. As a result, the colors seen in color photomicrographs are not the true colors of the tissue but are the colors of the stains used. Often the color of the stain, once applied to the tissue, can provide specific information about the tissue.

To observe objects much smaller than a cell such as cell organelles, the light microscope will not do, and we must use an **electron microscope,** whose limit of resolution is approximately 0.1 nm (the size of large molecules). In electron microscopy a beam of electrons either is passed through an object (transmission electron microscopy [TEM]) or is reflected off the surface of an object (scanning electron microscopy [SEM]). The electron beam is magnified with electromagnets. For both processes the specimen must be fixed, and for TEM the specimen must be imbedded in plastic and thinly sectioned (0.01 μm to 0.15 μm thick). When a focused electron beam strikes most tissues, it causes them to quickly disintegrate, so the tissue must be coated with some material such as metal that will not disintegrate in the electron beam (this is true for both TEM and SEM). Furthermore, the electron beam is not visible to the human eye, so it must be directed onto a fluorescent or photographic plate on which the electron beam is converted into a visible image. Since the electron beam does not transmit color information, electron micrographs are black and white unless color enhancement has been added.

The magnification ability of SEM is not as great as that of TEM. However, depth of focus of SEM is much greater, allowing for the production of a clear three-dimensional image of the tissue structure.

fications accomplish three tasks: (1) they mechanically bind the cells together; (2) they form a permeability barrier; and (3) they provide a mechanism for intercellular communication. Epithelial cells secrete glycoproteins that attach the cells to the basement membrane and to each other. This relatively weak binding between cells is reinforced by **desmosomes** (dez'mo-sōmz), disk-shaped structures with especially adhesive glycoproteins that bind cells to each other (see Figure 4-2). Many desmosomes are found in epithelia that are subjected to stress such as the stratified squamous epithelium of the skin. **Hemidesmosomes,** similar to one half of a desmosome, attach epithelial cells to the basement membrane.

The **zonula adherens** (zo'nu-lah ad-hĕ'renz) is located between the cell membranes of adjacent cells and acts like a weak glue that holds cells together by forming a girdle of adhesive glycoprotein around each cell. These attachments are not as strong as those involving desmosomes.

The **zonula occludens** (o-klood'enz) forms a permeability barrier. The cell membranes of adjacent cells join each other in a jigsaw fashion to form a tight seal (see Figure 4-2). Near the free surface the zonulae occludens completely surround the cell and prevent the passage of materials between cells. Thus material must move through the cells, which can actively regulate what is absorbed or secreted. Zonulae occludens are found in areas where a permeability barrier exists such as the lining of the intestines. The zonula adherens and zonula occludens, taken together, form what is known as a **tight junction** (see Figure 4-2).

A **gap junction** is a small protein channel that allows the passage of ions and small molecules between cells to provide a means of intercellular communication (see Figure 4-2). The exact function of gap junctions in epithelium is not clear, but in cardiac and smooth muscle gap junctions are important in coordinating muscle function between cells. The gap junctions between cardiac muscle cells are called intercalated disks.

Glands

Glands are secretory organs. Most glands are composed primarily of epithelium, with a supporting network of connective tissue. They develop from an infolding or outfolding of epithelium in the embryo. If the gland maintains an open contact with the epithelium from which it developed, then a duct is present. Glands with ducts are called **exocrine** (ek′so-krin) **glands;** the ducts are lined with epithelium. Alternatively, the gland may become separate from the epithelium of its origin. In this case the cellular products (hormones) are secreted into the bloodstream and are carried throughout the body. These glands have no ducts and are called **endocrine** (en′do-krin) **glands.**

Although most exocrine glands are composed of many cells (multicellular glands), some exocrine glands can be composed of a single cell (unicellular glands) (Figure 4-3, *A*) such as the goblet cells of the respiratory system. Multicellular glands can be classified further according to the structure of their ducts (Figure 4-3, *B-G*). Glands that have ducts with few branches are called **simple,** and glands with ducts that branch repeatedly are called **compound.** Further classification is based on whether the ducts end in **tubules** (small tubes) or saclike structures called **acini** (as′ĭ-ne; meaning grapes and suggesting a cluster of grapes or small sacs) or **alveoli** (al-ve′o-le; a hollow sac). Tubular glands can be classified as **straight** or **coiled.** Tubular glands can be simple and straight, simple and coiled, compound and straight, or compound and coiled. Acinar glands can be simple or compound.

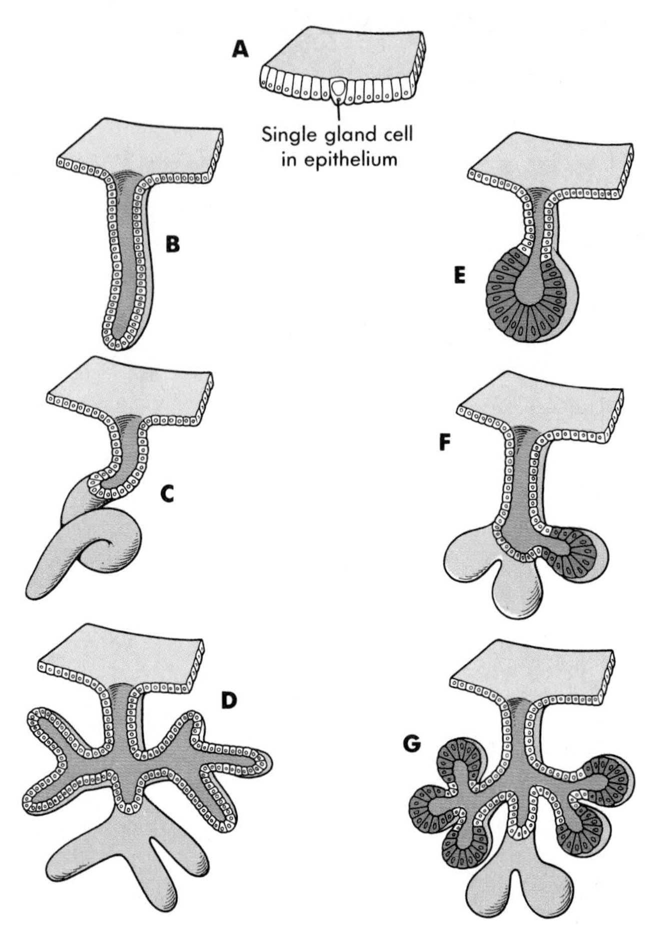

FIGURE 4-3 Structure of exocrine glands. **A** Unicellular. **B** Simple straight tubular. **C** Simple coiled tubular. **D** Compound tubular. **E** Simple acinar. **F** Simple branched acinar. **G** Compound acinar.

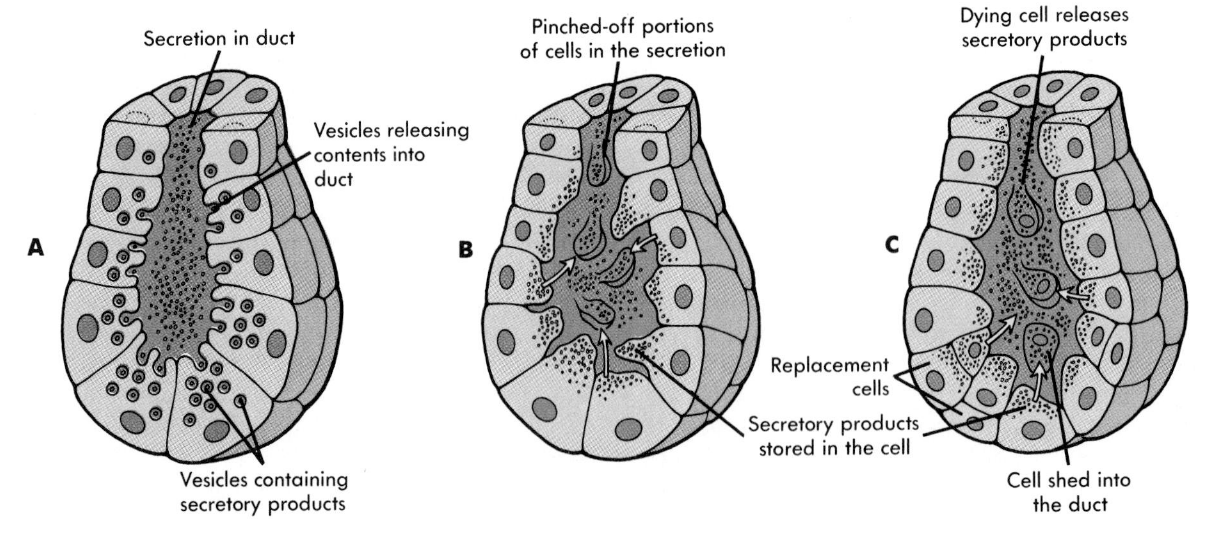

FIGURE 4-4 Exocrine glands classified according to the type of secretion. **A** Merocrine. Cells of the gland produce vesicles that contain secretory products, and the vesicles empty their contents into the duct through exocytosis. **B** Apocrine. Secretory products are stored in the cell near the lumen of the duct. A portion of the cell near the duct that contains the secretory products is actually pinched off the cell and joins the secretion. **C** Holocrine. Secretory products are stored in the cells of the gland. Entire cells are shed by the gland and become part of the secretion. The lost cells are replaced by other cells deeper in the gland.

Exocrine glands can also be classified according to how products leave the cell. **Merocrine** (mĕr'o-krin) glands, such as water-producing sweat glands and the exocrine portion of the pancreas, secrete products with no loss of actual cellular material (Figure 4-4, *A*). Secretions either are actively transported or are packaged in vesicles and then released by the process of exocytosis at the free surface of the cell. **Apocrine** (ap'o-krin) glands such as the mammary glands discharge fragments of the gland's cells in the secretion (Figure 4-4, *B*). Products are retained within the cell, and large portions of the cell are pinched off to become part of the secretion. **Holocrine** (hōl'o-krin) glands, such as sebaceous (oil) glands of the skin, shed entire cells (Figure 4-4, *C*). Substances accumulate in the cytoplasm of each epithelial cell, the cell ruptures and dies, and the entire cell becomes part of the secretion.

Endocrine glands are so variable in their structure that they are not classified easily. They are described in Chapters 17 and 18.

CONNECTIVE TISSUE

The essential characteristic that distinguishes connective tissue from the other three tissue types is that it consists of cells separated from each other by extracellular matrix. This nonliving extracellular matrix gives most connective tissues their functional characteristics and is the basis for separation of connective tissues into subgroups.

Connective Tissue Cells

The cells of the various connective tissues are specialized to produce the extracellular matrix. The cells' names end with suffixes according to the cells' functions as blasts, cytes, or clasts. **Blasts** create the matrix, **cytes** maintain it, and **clasts** break it down for remodeling. For example, fibroblasts are cells that form fibrous connective tissue, and chondrocytes are cells that maintain cartilage (chondro- refers to cartilage). Osteoblasts form bone (osteo- means bone), osteocytes maintain it, and osteoclasts break it down (see Chapter 6).

The extracellular matrix has three major components: (1) protein fibers, (2) ground substance consisting of nonfibrous protein and other molecules, and (3) fluid. The relative amounts and types of these three components form the basis of connective tissue classification.

Protein Fibers of the Matrix

Three types of fibrous proteins—collagen, reticular fibers, and elastin—help form connective tissue.

Collagen (kol'lă-jen) is the most common protein in the body and accounts for one fourth to one third of the total body protein (approximately 6% of the total body weight). Each collagen molecule resembles a microscopic rope consisting of three polypeptide chains coiled together. Collagen is very strong and flexible but quite inelastic. There are at least 10 different types of collagen, many of which are specific to certain tissues.

Reticular (rĕ-tik'u-lar) **fibers** are actually very fine collagen fibers and therefore are not a chemically distinct entity. They are very short, thin fibers that branch to form a network (reticular means netlike) and appear different microscopically from other collagen fibers. Reticular fibers are not as strong as most collagen but are very effective at filling up loose space.

Another type of protein found in connective tissue is **elastin.** As the name suggests, this protein, with the ability to return to its original shape after being distended or compressed, gives the tissue in which it is found an elastic quality. The structure of an elastin molecule is similar to a coiled metal spring. The individual molecules are cross-linked to produce a large interwoven meshwork resembling a bedspring that extends through the entire tissue.

Other Matrix Molecules

Two types of large, nonprotein molecules are part of the extracellular matrix—hyaluronic acid and proteoglycans. These molecules constitute most of the **ground substance** of the matrix, the shapeless background against which the collagen fibers are seen in the microscope. However, the molecules themselves are not shapeless but are highly structured. **Hyaluronic** (hi'al-u-ron'ik; glassy appearance) **acid** is a long, unbranched polysaccharide chain composed of repeating disaccharide units. It provides a very slippery quality to the fluids that contain it and for that reason is a good lubricant for joint cavities (see Chapter 8). It also is found in large quantities in connective tissue and is the major component of the vitreous humor of the eye (see Chapter 16). A **proteoglycan** (pro'te-o-gli'kan; formed from proteins and polysaccharides) is a large molecule that consists of numerous polysaccharides, each attached at one end to a common protein core. These **proteoglycan monomers** resemble minute pine tree branches. The protein core is the branch of the tree, and the proteoglycans are the needles. The protein cores of proteoglycan monomers can attach to a molecule of hyaluronic acid to form a **proteoglycan aggregate.** The aggregate would resemble a complete pine tree, with hyaluronic acid represented by the tree trunk and the proteoglycan monomers forming the limbs. Proteoglycans tend to trap large quantities of water, which imparts considerable resiliency to tissues.

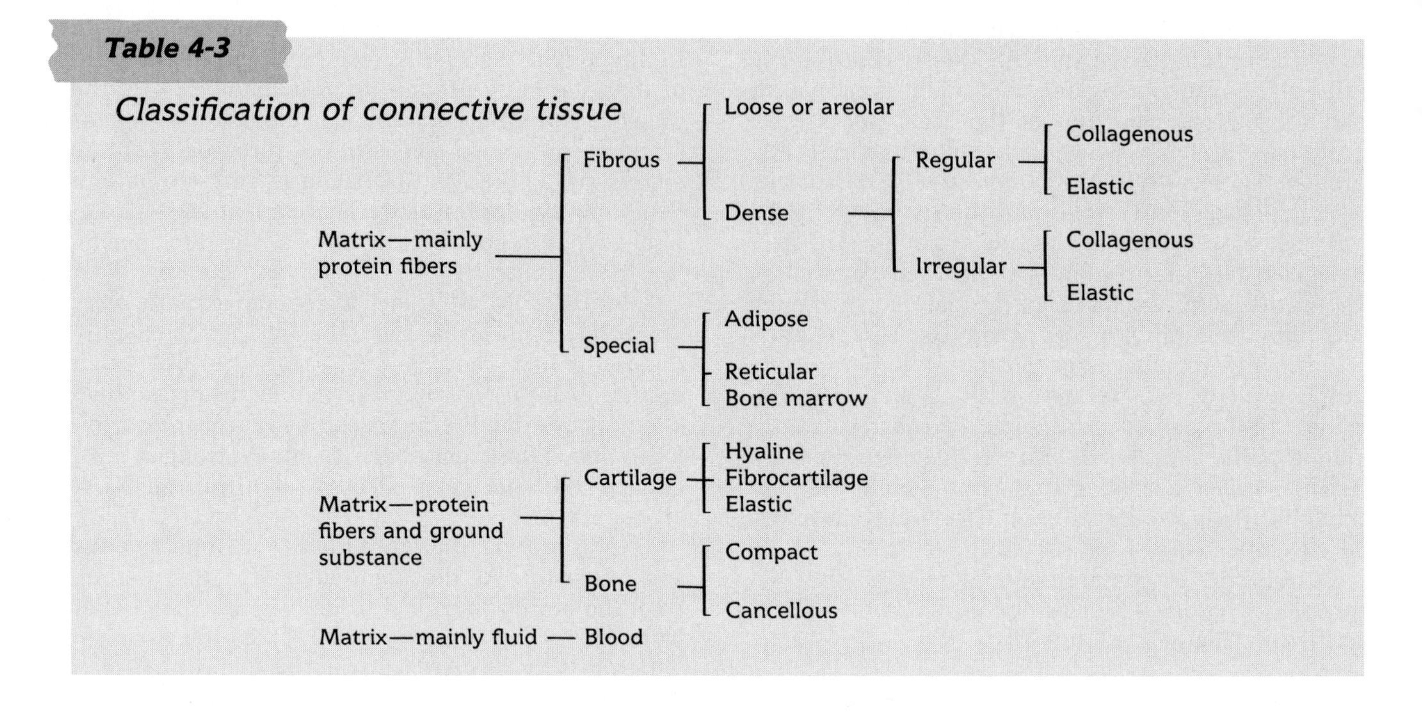

Classification of Connective Tissue

Connective tissue constitutes a continuum. Connective tissue types blend into one another, and the transition points cannot be defined precisely. As a result the classification scheme used to subdivide connective tissues into categories is somewhat arbitrary; but, roughly, the three major categories are (1) one in which the major feature of the extracellular matrix is the protein fiber, (2) one in which both protein fibers and ground substance are prominent in the matrix, and (3) one in which the matrix is mostly fluid (Table 4-3).

Matrix with Fibers as the Primary Feature

Connective tissue that has a matrix with fibers as the primary feature consists of two subtypes, fibrous and special.

Fibrous connective tissue

In **fibrous connective tissue** the protein fiber component of the matrix predominates. Fibrous connective tissue can be divided into two subtypes, loose and dense, depending on the amount of fibrous protein. In **loose connective tissue** (Figure 4-5, *A*) the protein fibers form a lacy network with numerous fluid-filled spaces. In dense connective tissue protein fibers fill nearly all the extracellular space.

Loose connective tissue sometimes is referred to as **areolar** (ah-re′o-lar; tissue with small spaces or areas between the fibers). Areolar tissue is the "loose packing" material of most organs and other tissues and attaches the skin to underlying tissues. It contains three major types of protein fibers—collagen, reticular fibers, and elastin—plus a variety of cells such as fibroblasts, which produce the fibrous matrix, macrophages, which move through the tissue engulfing bacteria and cell debris, and lymphocytes, which are involved in immunity. The loose packing of areolar tissue often is associated with other connective tissue types such as reticular tissue and fat (adipose tissue).

Dense connective tissue can be subdivided into two major groups, regular and irregular. The cells within dense connective tissue are spindle-shaped cells, collectively called **fibroblasts.** In **dense regular connective tissue** (Figure 4-5, *B* and *C*) the protein fibers of the extracellular matrix are oriented predominantly in one direction. Dense regular connective tissue forms structures such as tendons (fibers connecting muscle to bone; see Chapter 11) and ligaments (fibers connecting bone to bone; see Chapter 8). **Dense irregular connective tissue** (Figure 4-5, *D* and *E*) contains protein fibers arranged as a meshwork of randomly oriented fibers. Alternatively, the fibers within a given layer of irregular dense connective tissue may be oriented in one direction, whereas the fibers of adjacent layers are oriented at nearly right angles to that layer. Dense irregular connective tissue forms the dermis of the skin, the tough, inner portion of the skin from which leather is made.

Text continued on p. 119.

FIGURE 4-5 Connective tissue

A Areolar or loose

LOCATION: Widely distributed throughout the body; substance on which epithelial basement membranes rest; packing between glands, muscles, and nerves. Attaches the skin to underlying tissues.
STRUCTURE: Cells (e.g., fibroblasts, macrophages, and lymphocytes) within a fine network of mostly collagen fibers. Often merges with denser connective tissue.
FUNCTION: Loose packing, support, and nourishment for the structures with which it is associated.

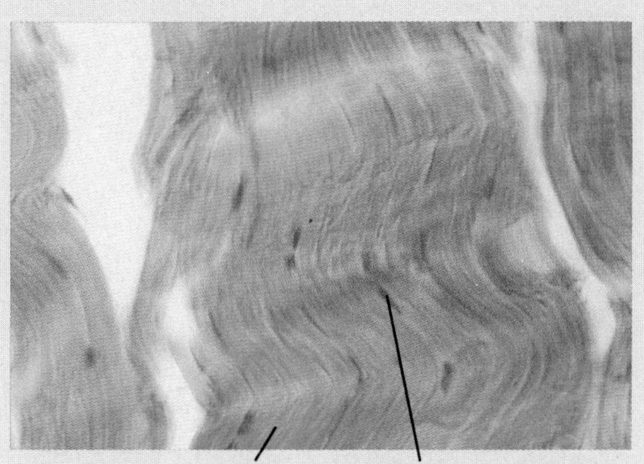

B Dense regular collagenous

LOCATION: Tendons (attach muscle to bone) and ligaments (attach bones to each other).
STRUCTURE: Matrix composed of collagen fibers running in somewhat the same direction.
FUNCTION: Ability to withstand great pulling forces exerted in the direction of fiber orientation, great tensile strength, and stretch resistance.

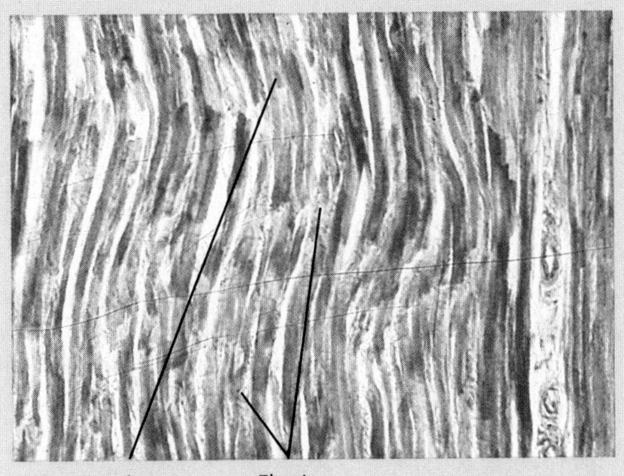

C Dense regular elastic

LOCATION: Ligaments between the vertebrae and along the dorsal aspect of the neck (nucha) and in the vocal cords.
STRUCTURE: Matrix composed of regularly arranged collagen fibers and elastin fibers.
FUNCTION: Capable of stretching and recoiling like a rubber band with strength in the direction of fiber orientation.

Continued.

FIGURE 4-5—cont'd

D Dense irregular collagenous

LOCATION: Aponeuroses and sheaths; dermis of the skin; organ capsules and septa; outer covering of body tubes.
STRUCTURE: Matrix composed of collagen fibers that run in all directions or in alternating planes of fibers oriented in a somewhat single direction.
FUNCTION: Tensile strength capable of withstanding stretching in all directions.

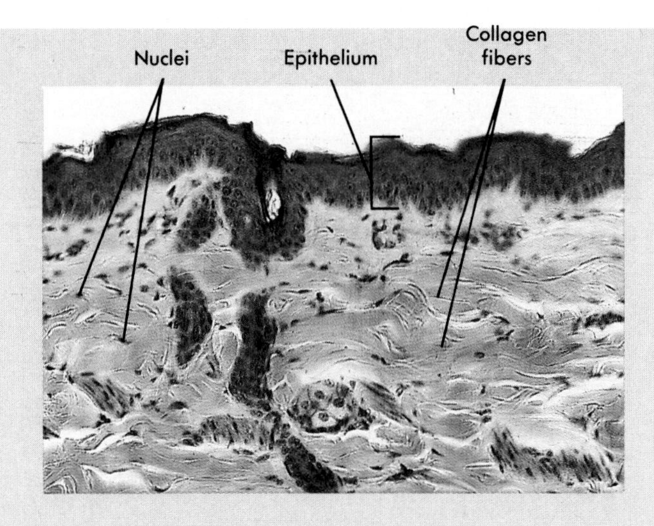

E Dense irregular elastic

LOCATION: Elastic arteries.
STRUCTURE: Matrix composed of bundles and sheets of collagenous and elastin fibers oriented in multiple directions.
FUNCTION: Capable of strength with stretching and recoil in several directions.

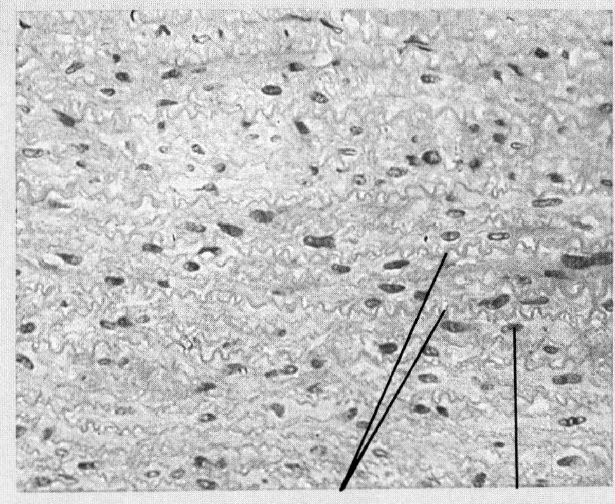

F Adipose tissue

LOCATION: Predominantly in subcutaneous areas, mesenteries, renal pelvis, around kidneys, and attached to the surface of colon. Where loose connective tissue penetrates into spaces and crevices.
STRUCTURE: Little extracellular material surrounding cells. The adipocytes, or fat cells, are so full of lipid that the cytoplasm is pushed to the periphery of the cell.
FUNCTION: Packing material, thermal insulator, energy storage, and protection of organs against injury from being bumped or jarred.

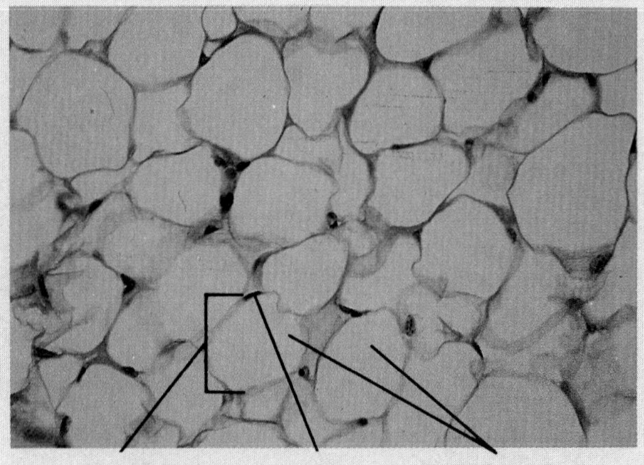

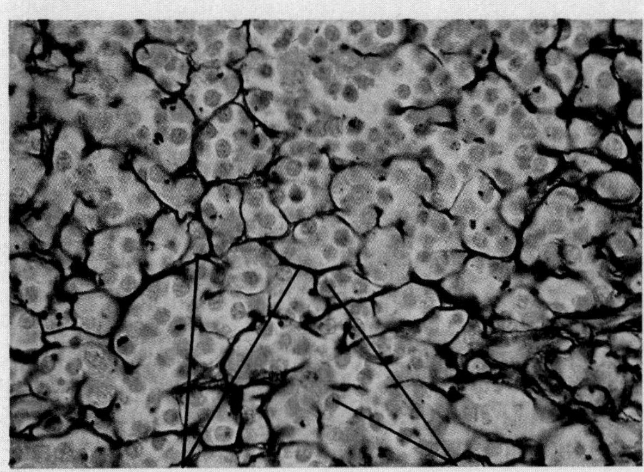

G Reticular

LOCATION: Within the lymph nodes, spleen, and bone marrow.
STRUCTURE: Fine network of reticular fibers irregularly arranged.
FUNCTION: Provides a superstructure for the lymphatic and hemopoietic tissues.

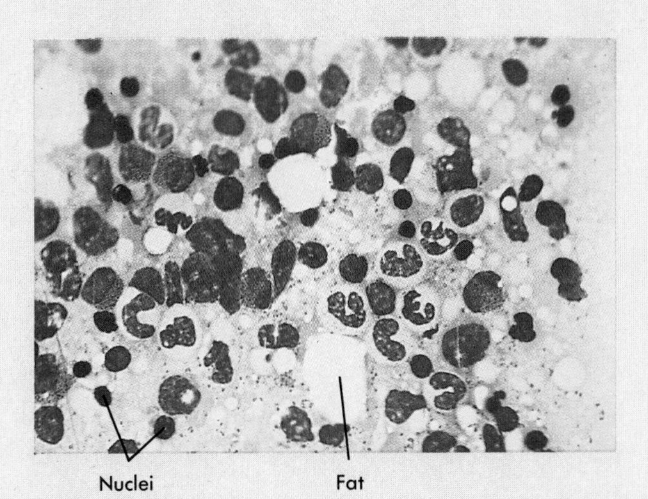

H Bone marrow

LOCATION: Within marrow cavities of bone. Two types: yellow marrow (mostly adipose tissue) in the shafts of long bones; and red marrow (hemopoietic or blood-forming tissue) in the ends of long bones and in short, flat, and irregularly shaped bones.
STRUCTURE: Reticular framework with numerous blood-forming cells (red marrow).
FUNCTION: Production of new blood cells (red marrow).

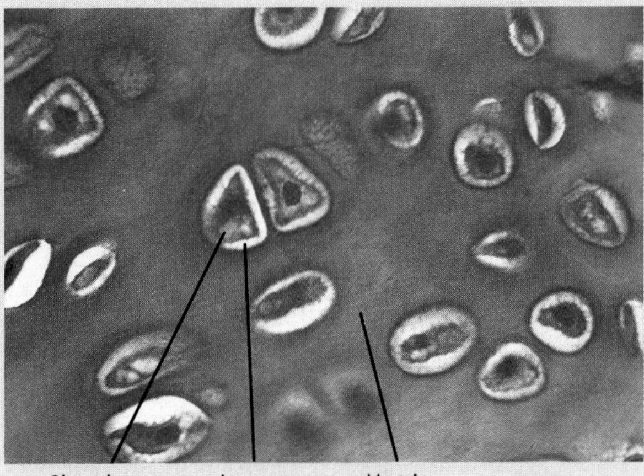

I Hyaline cartilage

LOCATION: Growing long bones, cartilage rings of the respiratory system, costal cartilage of ribs, nasal cartilage, articulating surface of bones, and the embryonic skeleton.
STRUCTURE: Collagen fibers of cartilage type that are small and evenly dispersed in the matrix, making the matrix appear transparent. The cartilage cells, or chondrocytes, are found in spaces, or lacunae, within the rigid matrix.
FUNCTION: Allows growth of long bones. Provides rigidity with some flexibility in the trachea, bronchi, ribs, and nose. Forms rugged, smooth, yet somewhat flexible articulating surfaces. Forms the embryonic skeleton.

Continued.

FIGURE 4-5—cont'd

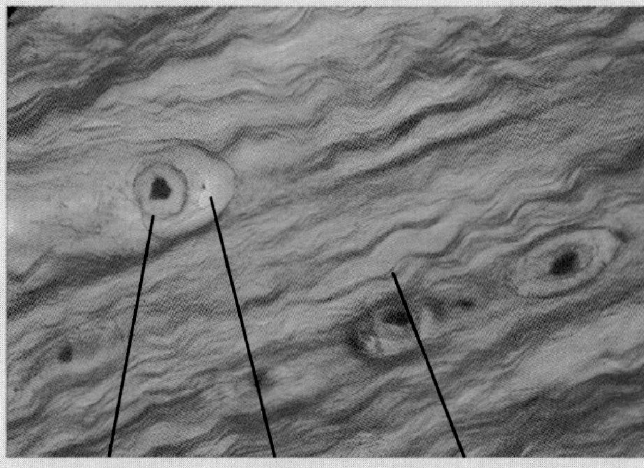

J Fibrocartilage

LOCATION: Intervertebral disks, symphysis pubis, articular disks (e.g., knee and temporomandibular [jaw] joints), and the round ligament.
STRUCTURE: Collagenous fibers similar to those in hyaline cartilages and the more general type of collagen fibers in other connective tissues. The fibers are more numerous than in other cartilages and are arranged in thick bundles.
FUNCTION: Somewhat flexible and capable of withstanding considerable pressure. Connects structures subjected to great pressure.

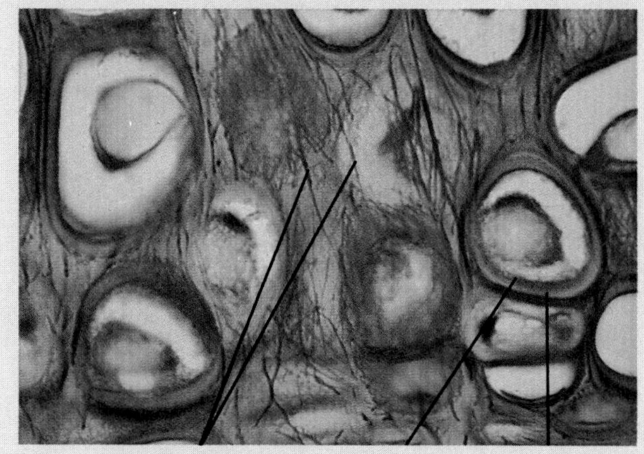

K Elastic cartilage

LOCATION: External ear, epiglottis, and auditory tube.
STRUCTURE: Similar to hyaline cartilage, but matrix also contains elastin fibers.
FUNCTION: Provides rigidity with even more flexibility than hyaline cartilage since elastic fibers return to their original shape after being stretched.

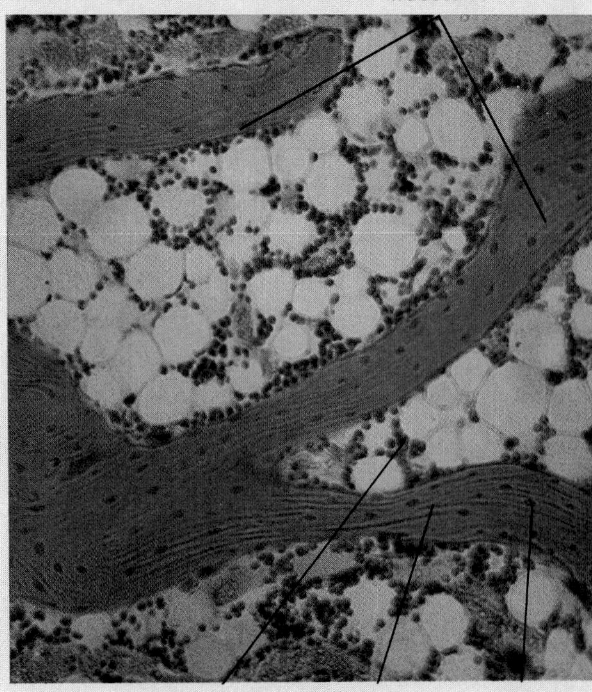

L Cancellous bone

LOCATION: In the interior of the bones of the skull, sternum, and pelvis. Also found in the ends of the long bones.
STRUCTURE: Latticelike network of scaffolding characterized by trabeculae with large spaces between them. The osteocytes, or bone cells, are located within lacunae in the trabeculae.
FUNCTION: Acts as a scaffolding to provide strength and support without the greater weight of solid bone.

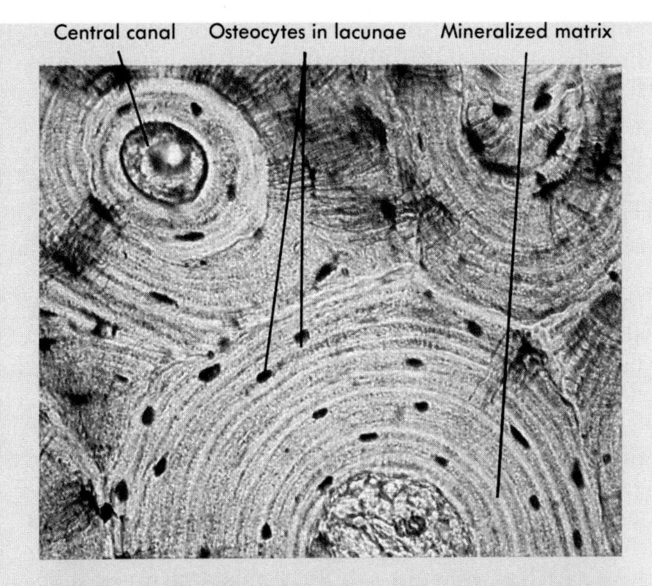

M Compact bone

LOCATION: Outer portions of all bones and the shafts of long bones.
STRUCTURE: Hard, bony matrix predominates. Many osteocytes are located within lacunae that are distributed in a circular fashion around the central canals. Small passageways connect adjacent lacunae.
FUNCTION: Provides great strength and support. Forms a solid outer shell on bones that keeps them from being easily broken or punctured.

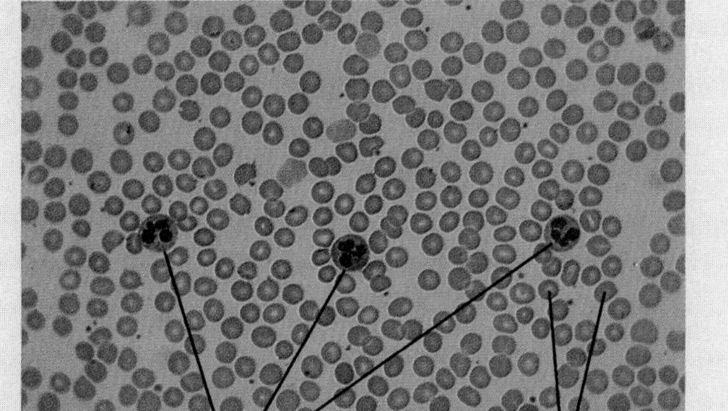

N Blood

LOCATION: Within the blood vessels. Produced by the hemopoietic tissues. White blood cells frequently leave the blood vessels and enter the interstitial spaces.
STRUCTURE: Blood cells and a fluid matrix.
FUNCTION: Transports oxygen, carbon dioxide, hormones, nutrients, waste products, and other substances. Protects the body from infections and is involved in temperature regulation.

The collagen fibers of dense connective tissue resist stretching and give the tissue considerable strength in the direction of the fiber orientation. Therefore dense regular connective tissue has considerable strength in one direction, and dense irregular connective tissue has less strength but in many directions.

2

Using tendons and skin as examples, explain why the differences in fiber orientation are functionally important.

? ? ? ? ? ? ? ? ? ? ?

The predominance in dense connective tissue of either collagen, which is quite flexible but inelastic, or elastin, which is flexible and elastic, is the basis of another dense connective tissue classification division. Dense connective tissue can be classified as **dense regular collagenous, dense regular elastic, dense irregular collagenous,** or **dense irregular elastic** (see Figure 4-5). Dense regular collagenous connective tissue forms tendons and most ligaments. Dense regular elastic connective tissue forms some elastic ligaments such as the **nuchal** (nu′kal) ligament, which lies along the posterior of the neck and helps hold the head upright.

Dense irregular collagenous connective tissue is characteristic of the dermis of the skin (see Chapter 5) and of the connective tissue capsules that surround organs such as the kidney and spleen. Dense irregular elastic connective tissue helps form the walls of large arteries.

3

Why would it be a disadvantage if tendons were elastic?

? ? ? ? ? ? ? ? ? ? ?

Special connective tissue

Adipose tissue and reticular tissue are special types of connective tissue. **Adipose** (ad'ĭ-pōs; fat) **tissue** (Figure 4-5, *F*, p. 116) consists of **adipocytes,** or fat cells, which contain large amounts of lipid. Unlike other connective tissue, adipose tissue is composed of large cells and a small amount of reticular matrix. Adipose tissue functions not only as an insulator and protective tissue but as a site of energy storage. Lipids take up less space per calorie than either carbohydrates or proteins and therefore are well adapted for energy storage.

Adipose tissue exists in both yellow (white) and brown forms, with yellow adipose tissue the more abundant of the two in adults. At birth a human's yellow adipose tissue is white but turns more yellow with age because of the accumulation of pigments such as carotene, a plant pigment that humans can metabolize as a source of vitamin A. Brown adipose tissue is found only in specific areas of the body such as the axillae (armpits), neck, and near the kidney. Its color results from the cytochrome pigments in the numerous mitochondria and its abundant blood supply. Brown fat is much more prevalent in babies in whom it may be difficult to distinguish from white fat. It is specialized to generate heat when lipid molecules are metabolized and may play a significant role in body temperature regulation in newborn babies.

Reticular tissue forms the framework of lymphatic tissue, bone marrow, and liver (Figure 4-5, *G*). It is characterized by a network of reticular fibers and a number of cell types. The reticular fibers are produced by **reticular cells,** which remain closely attached to the fibers. The spaces between the reticular fibers may contain a wide variety of cells such as **dendritic cells** (cells that look very much like reticular cells but are part of the immune system and do not produce reticular fibers; see Chapter 22), lymphocytes, macrophages, and other blood cells.

Another type of special connective tissue is **bone marrow** (Figure 4-5, *H*). There are two types of bone marrow, yellow marrow and red marrow (see Chapter 6). **Yellow marrow** consists of adipose tissue, and **red marrow** consists of hemopoietic (he'mo-poy-et'ik; blood forming) tissue. Hemopoietic tissue, which produces red and white blood cells, is described in detail in Chapter 19.

Matrix with Both Protein Fibers and Ground Substance

Cartilage

Cartilage (kar'tĭ-lij) is composed of cartilage cells, or **chondrocytes** (kon'dro-sītz), located in spaces called **lacunae** (lă-ku'ne) within an extensive and relatively rigid matrix. The matrix contains protein fibers that consist of collagen or in some cases collagen and elastin; ground substance consisting of proteoglycans and other organic molecules; and fluid. Most of the proteoglycans in the matrix form aggregates with hyaluronic acid. Within the cartilage matrix proteoglycan aggregates function as minute sponges capable of trapping large quantities of water. This trapped water allows cartilage to spring back after being compressed, and collagen gives cartilage considerable strength. Blood vessels do not penetrate the substance of cartilage; thus cartilage heals very slowly after an injury because the cells and nutrients necessary for tissue repair do not easily reach the damaged area.

There are three types of cartilage: (1) **hyaline** (hī'ă-lin) **cartilage** has large amounts of both collagen and proteoglycan; (2) **fibrocartilage** has more collagen than proteoglycan; and (3) **elastic cartilage** has elastin in addition to collagen and proteoglycan. Hyaline cartilage (Figure 4-5, *I*) consists of fine collagenous fibrils evenly dispersed throughout the ground substance, has a glassy, translucent matrix, and is extremely smooth. It is found in areas where strong support and some flexibility are needed (e.g., in the rib cage and in the rings within the trachea and bronchi; see Chapter 23). Hyaline cartilage covers the surfaces of bones that move against each other in joints. Hyaline cartilage also forms most of the skeleton before it is replaced by bone in the embryo, and it is involved in the growth of bones in length (see Chapter 6).

4

Describe the characteristics of hyaline cartilage that suit it to its function in joints.

? ? ? ? ? ? ? ? ? ? ?

Fibrocartilage (Figure 4-5, *J*) differs from hyaline cartilage in that it has much thicker bundles of collagen dispersed through the matrix. It is slightly compressible and very tough and is in areas of the body such as the knee, the jaw, and between vertebrae where a great deal of pressure is applied to joints.

Elastic cartilage (Figure 4-5, *K*) has numerous elastic fibers dispersed throughout the matrix and is found in areas such as the external ears that have rigid but elastic properties.

Bone

Bone is a hard connective tissue that consists of living cells and mineralized matrix. Bone matrix has an organic and an inorganic portion. The organic portion consists of protein fibers, primarily collagen, and other organic molecules. The mineral (inorganic) portion consists of calcium phosphate crystals called **hydroxyapatite** (hi-drok'se-ap-ah-tīt). The strength and ridigity of the mineralized matrix allow bones to support and protect other tissues and organs of the body. Bone cells, or **osteocytes,** are located within holes in the matrix, which are called lacunae and are similar to the lacunae of cartilage.

There are two types of bone, **cancellous** (kan'sĕ-lus), or **spongy, bone** (Figure 4-5, *L*) and **compact bone** (Figure 4-5, *M*). Spongy bone has spaces between the plates, or **trabeculae** (tră-bek'u-le; beams), of bone and therefore resembles a sponge, and compact bone essentially is solid. Bone, unlike cartilage, has a rich blood supply; for this reason bone can repair itself much more readily than can cartilage. Bone is described more fully in Chapter 6.

Predominantly Fluid Matrix

Blood is somewhat unique among the connective tissues because the matrix between the cells is liquid (Figure 4-5, *N*, p. 119). Thus, although the cells of most other connective tissues are more or less stationary within a relatively rigid matrix, blood cells are free to move within a fluid matrix. Some blood cells leave the bloodstream and wander through other tissues. The liquid matrix of blood allows it to flow rapidly through the body carrying food, oxygen, waste products, and other materials. The matrix of blood is also unique in that most of it is produced by cells contained in other tissues. Blood is discussed more fully in Chapter 19.

MUSCULAR TISSUE

The main characteristic of **muscular tissue** is that it is contractile and is therefore responsible for movement. Muscle contraction is accomplished by the interaction of contractile proteins, which are described in Chapter 10. Muscles contract to move the entire body, to pump blood through the heart and blood vessels, and to decrease the size of hollow organs such as the stomach.

The three types of muscular tissue—skeletal, cardiac, and smooth muscle—are classified according to both structure and function (Table 4-4). Muscular tissue grouped according to structure is either **striated,** in which microscopic bands or striations can be seen, or **nonstriated.** When classified according to function, a muscle is **voluntary,** consciously controlled, or **involuntary,** not normally consciously controlled. Thus the three muscular types are striated voluntary or **skeletal muscle** (Figure 4-6, *A*), striated involuntary or **cardiac muscle** (Figure 4-6, *B*), and nonstriated involuntary or **smooth muscle** (Figure 4-6, *C*).

Skeletal muscle is what normally is thought of as "muscle" (see Chapter 10). It constitutes the meat of animals and represents a large portion of the human body's total weight. Skeletal muscle, as the

Table 4-4

Comparison of muscle types

FEATURES	SKELETAL MUSCLE	CARDIAC MUSCLE	SMOOTH MUSCLE
Location	Attached to bones	Heart	Walls of hollow organs, blood vessels, eyes, glands, and skin
Cell shape	Very long, cylindrical cells (1-40 mm in length and may extend the entire length of the muscle; 10-100 μm in diameter)	Cylindrical cells that branch (100-500 μm in length; 100-200 μm in diameter)	Spindle-shaped cells (15-200 μm in length; 5-10 μm in diameter)
Nucleus	Multinucleated, peripherally located	Single, centrally located	Single, centrally located
Striations	Yes	Yes	No
Control	Voluntary	Involuntary	Involuntary
Ability to contract spontaneously	No	Yes	Yes
Function	Body movement	Pumps blood	Movement of food through the digestive tract, emptying of the urinary bladder, regulation of blood vessel diameter, change in pupil size, contraction of many gland ducts, movement of hair, and many more functions
Special features		Branching fibers, intercalated disks join the cells to each other (gap junctions)	Gap junctions

FIGURE 4-6 Muscular tissue

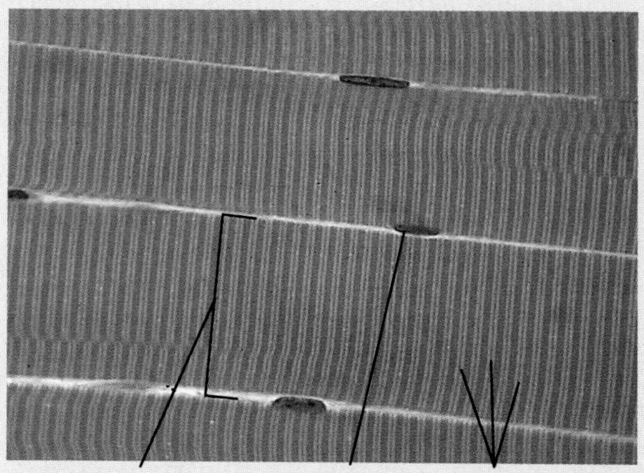

A Skeletal muscle

LOCATION: Attached to bone.
STRUCTURE: Appears striated. Cells are large, long, and cylindrical with several peripherally located nuclei in each cell.
FUNCTION: Movement of the body. Under voluntary control.

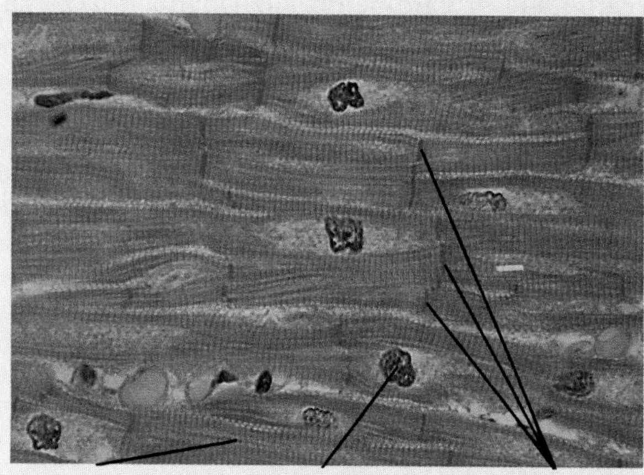

B Cardiac muscle

LOCATION: Heart.
STRUCTURE: Appears striated. Cells are cylindrical and branching with a single, centrally located nucleus. Cells are connected to each other by specialized junctions called intercalated disks.
FUNCTION: Pumps the blood. Under involuntary control.

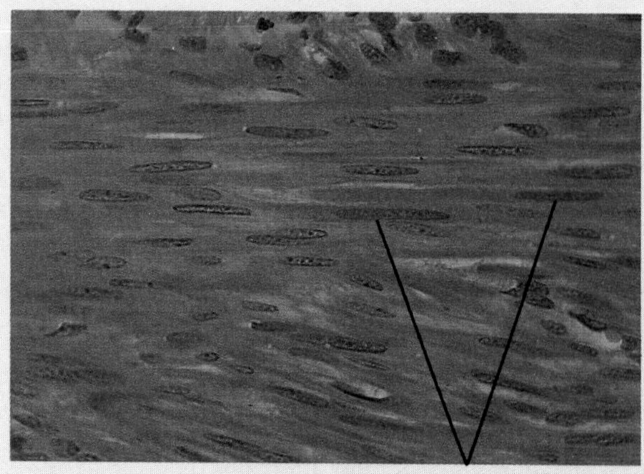

C Smooth muscle

LOCATION: In the walls of hollow organs, pupil of the eye, skin (attached to hair), and glands.
STRUCTURE: No striations. Cells are spindle shaped with a single, centrally located nucleus.
FUNCTION: Regulates the size of organs, forces fluid through tubes, controls the amount of light entering the eye, and produces "goose flesh" in the skin. Under involuntary control.

name implies, attaches to the skeleton and by contracting, causes the major body movements. Cardiac muscle is the muscle of the heart (see Chapter 20), and contraction of cardiac muscle is responsible for pumping blood. Smooth muscle is widespread throughout the body and is responsible for a wide range of functions such as movements in the digestive, urinary, and reproductive systems.

NERVOUS TISSUE

The fourth and final class of tissue is **nervous tissue** (see Chapter 12), which is characterized by the ability to conduct electrical signals called action potentials. It consists of neurons, which are responsible for this conductive ability, and support cells (neuroglia).

Neurons, or nerve cells (Figure 4-7), are the actual conducting cells of nervous tissue. They are composed of three major parts—cell body, dendrites, and axon. The **cell body** contains the nucleus and is the site of general cell functions. Dendrites and axons are two types of nerve cell processes (projections of cytoplasm surrounded by membrane). **Dendrites** (den′drītz) usually receive electrical impulses and conduct them toward the cell body, and the **axon** usually conducts impulses away from the cell body. Each neuron has only one axon but may have several dendrites.

Neurons that possess several dendrites and one axon are called **multipolar neurons** (Figure 4-7, *A*). Ones that possess a single dendrite and an axon are called **bipolar neurons.** Some very specialized neurons, **unipolar neurons** (Figure 4-7, *B*), have only one axon and no dendrites. Within each subgroup are many shapes and sizes of neurons, especially in the brain and the spinal cord.

Neuroglia (nu-rog′le-ah; nerve glue) are the support cells of the brain, spinal cord, and peripheral nerves (Figure 4-8). The term neuroglia originally referred only to the support cells of the central nervous system but now is applied also to cells in the peripheral nervous system. Neuroglia nourish, protect, and insulate neurons. Neurons and neuroglial cells are described in greater detail in Chapters 12 and 13.

FIGURE 4-7 *Nervous tissue*

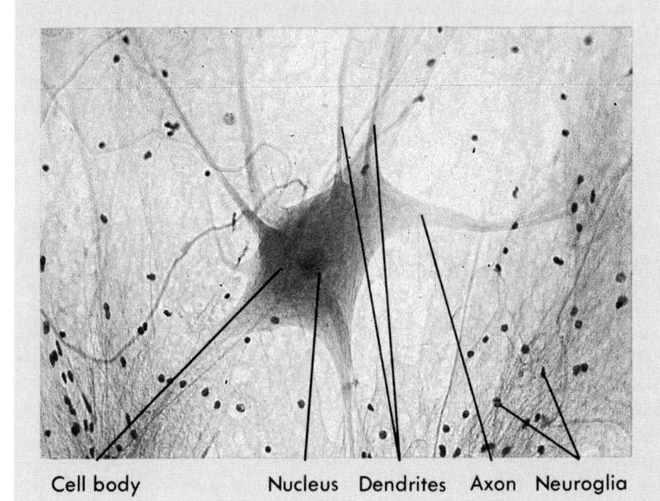

A Multipolar neuron

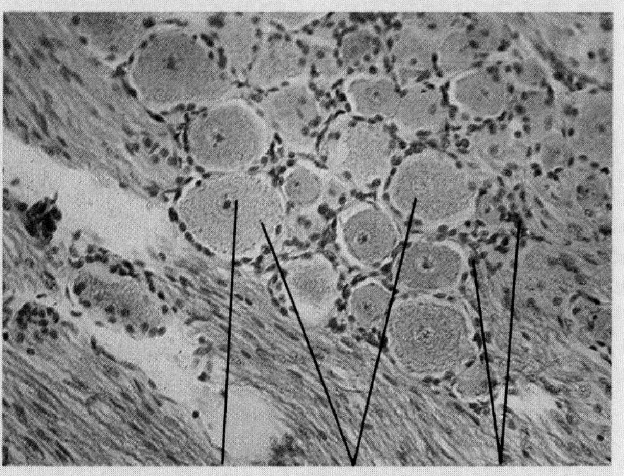

B Unipolar neuron

LOCATION: Cell bodies—in the brain, spinal cord, or ganglia; cell processes—all parts of the body.
STRUCTURE: Mainly relatively large cells with a variety of shapes. Characterized by cell processes, which vary in number from one (unipolar neuron) to two (bipolar neuron) to many (multipolar neuron).
FUNCTION: Conduct impulses, store "information," and in some way integrate and evaluate data.

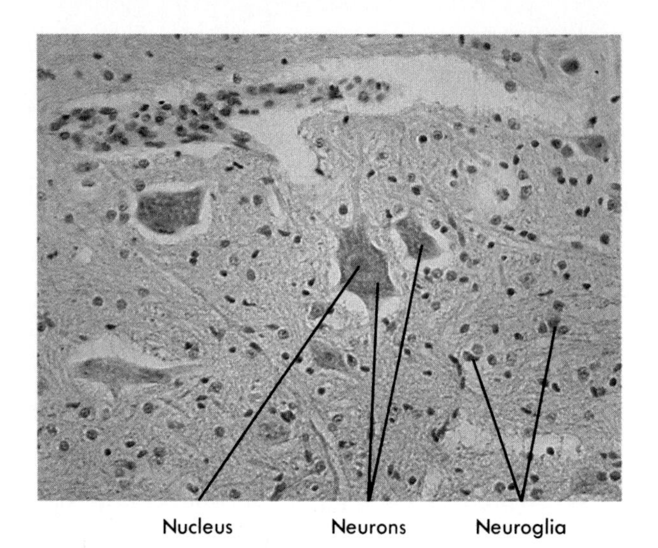

FIGURE 4-8 Neuroglia. (×400.)

LOCATION: In the brain, spinal cord, ganglia, and nerves.
STRUCTURE: Usually smaller and more numerous than neurons. Some surround blood vessels, some wrap around axons, others are phagocytic, and others wrap around peripheral nerve axons.
FUNCTION: Support neurons and form a selectively permeable barrier between neurons and other cell types, and function as insulators and protectors.

EMBRYONIC TISSUE DEVELOPMENT

All four tissue types are derived from three **germ layers,** the endoderm, the mesoderm, and the ectoderm, during the early development of the embryo (see Table 29-1). The **endoderm** (en'do-derm), the inner layer, forms the lining of the digestive tract and its derivatives. The **mesoderm** (mez'o-derm), the middle layer, forms tissues, such as muscle, bone, and blood vessels. The **ectoderm** (ek'to-derm), the outer layer, forms the skin. Some of the ectoderm, the **neuroectoderm,** becomes the central nervous system (see Chapter 13). Groups of cells that break away from the neuroectoderm during development, **neural crest cells,** give rise to portions of the peripheral and autonomic nerves (see Chapters 14 and 15) and to skin pigment (see Chapter 5) and many tissues of the face. The layers are named "germ" layers because the beginning of all the tissues of the adult can be traced back to one of them. The germ layers and their derivatives are described more fully in Chapter 29.

MEMBRANES

A membrane is a thin sheet or layer of tissue that covers a structure or lines a cavity. Most membranes are formed from epithelium and the connective tissue on which it rests. The three major categories of internal membranes are serous membranes, mucous membranes, and synovial membranes.

Serous membranes consist of simple squamous epithelium (mesothelium) and its basement membrane, which rest on a delicate layer of loose connective tissue. Serous membranes line cavities (pericardial, pleural, and peritoneal; see Chapter 1) that do not open to the exterior. Serous membranes do not contain glands but are moistened by a small amount of fluid similar to lymph called **serous fluid,** which they produce. Serous membranes protect the internal organs from friction, help hold them in place, and act as selectively permeable barriers that prevent the accumulation of large amounts of fluid within the serous cavities.

Mucous membranes consist of epithelial cells and their basement membrane, which rest on a thick layer of loose connective tissue called the **lamina propria.** They line cavities and canals that open to the outside of the body such as the digestive, respiratory, excretory, and reproductive passages. Many, but not all, mucous membranes contain mucous glands (some of which are goblet cells), which secrete a viscous substance called **mucus.** The functions of the mucous membranes vary, depending on their location, and include protection, absorption, and secretion.

Synovial membranes consist of modified connective tissue cells either intermixed with part of the dense connective tissue of the joint capsule or separated from the capsule by areolar or adipose tissue. Synovial membranes line freely movable joints (see Chapter 8). They produce a fluid rich in hyaluronic acid, which makes the joint fluid very slippery, facilitating smooth movement within the joint.

INFLAMMATION

The inflammatory response occurs when tissues are damaged (Figure 4-9) and/or in association with an immune response. Although there are many possible agents of injury (e.g., microorganisms, cold, heat, radiant energy, chemicals, electricity, or mechanical trauma), the inflammatory response to all causes is similar. The inflammatory response mobilizes the body's defenses, isolates and destroys microorganisms and other injurious agents, and removes foreign materials and damaged cells so that tissue repair can proceed. The details of the inflammatory response are presented in Chapter 22.

Inflammation produces five major signs: redness, heat, swelling, pain, and disturbance of function. Although unpleasant, these processes usually benefit recovery, and each of the symptoms can be understood in terms of events that occur during the inflammatory response.

After a person is injured, chemical substances called **mediators of inflammation** are released or ac-

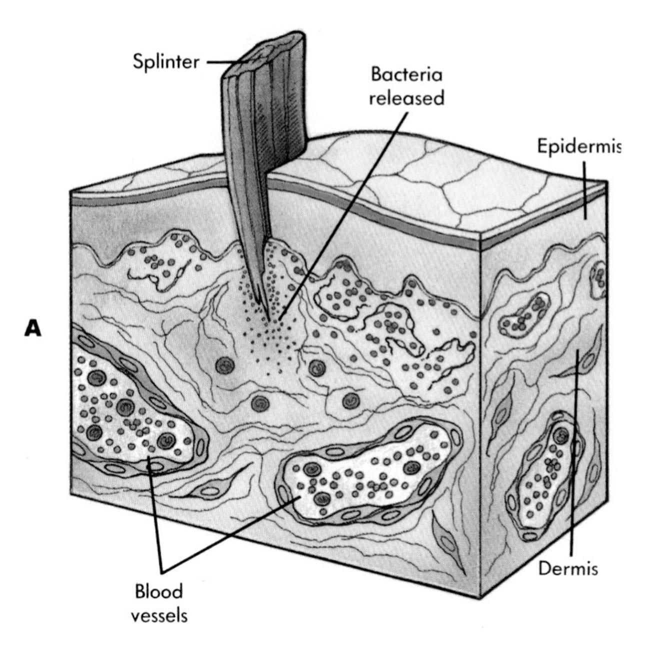

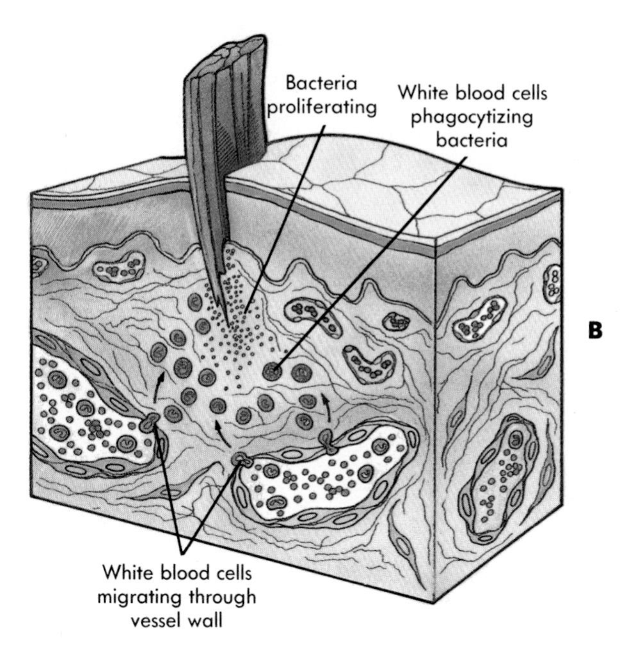

FIGURE 4-9 Inflammation. **A** A splinter in the skin causes tissue damage and releases bacteria. Dilated blood vessels (capillaries) cause skin to become red, and fluid from the tissues and vessels causes swelling. **B** White blood cells (e.g., neutrophils and macrophages) leave the blood vessels and arrive at the site of bacterial infection where they begin to phagocytize bacteria. Additional connective tissue fibers are formed to contain the infection and keep it from spreading.

tivated in the tissues and the adjacent blood vessels. The mediators include histamine, kinins, prostaglandins, leukotrienes, and others. Some mediators induce **vasodilation** (expansion) of blood vessels and produce the symptoms of redness and heat. Vasodilation is beneficial because it increases the speed with which white blood cells and other substances important for fighting infections and repairing the injury are brought to the site of injury.

Mediators also stimulate pain receptors and increase the permeability of blood vessels, allowing the movement of materials such as clotting proteins and white blood cells out of the blood vessels and into the tissue where they can deal directly with the injury. As proteins from the blood move into the tissue, they change the osmotic relationship between the blood and the tissue. Water follows the proteins by osmosis, and the tissue swells, producing **edema.** Edema increases the pressure in the tissue, which may also stimulate neurons and cause the sensation of pain.

Once the clotting proteins have diffused into the interstitial spaces, they tend to form a clot. Coagulation of blood also occurs in the more severely injured blood vessels. The effect of coagulation is to isolate the injurious agent and to separate it from the remainder of the body. Foreign particles and microorganisms that are present at the site of injury are "walled off" from tissues by the clotting process. Pain, limitation of movement resulting from edema, and tissue destruction all contribute to the disturbance of function. This disturbance can be valuable because it warns the person to protect the injured structure from further damage. Sometimes the inflammatory response lasts longer or is more intense than is desirable, and drugs are used to suppress the symptoms. Antihistamines block the effects of histamine, and aspirin prevents the synthesis of prostaglandins. On the other hand, the inflammatory response by itself may not be enough to fight off an infection, and antibiotics may be required.

TISSUE REPAIR

Tissue repair is the substitution of viable cells for dead cells, and it can occur by regeneration or replacement. In **regeneration** the new cells are the same type as those that were destroyed, and normal function is usually restored. In **replacement** a new type of tissue develops that eventually causes scar production and the loss of some tissue function. Most wounds heal through regeneration and replacement; which process dominates depends on the tissues involved and the nature or extent of the wound.

Cells can be classified into three groups—labile, stable, or permanent—according to their regenerative ability. **Labile** cells (e.g., the skin, mucous membranes, and hemopoietic and lymphoid tissues) continue to divide throughout life. Damage to these cells can be repaired completely by regeneration. **Stable** cells (e.g., connective tissues and glands, including

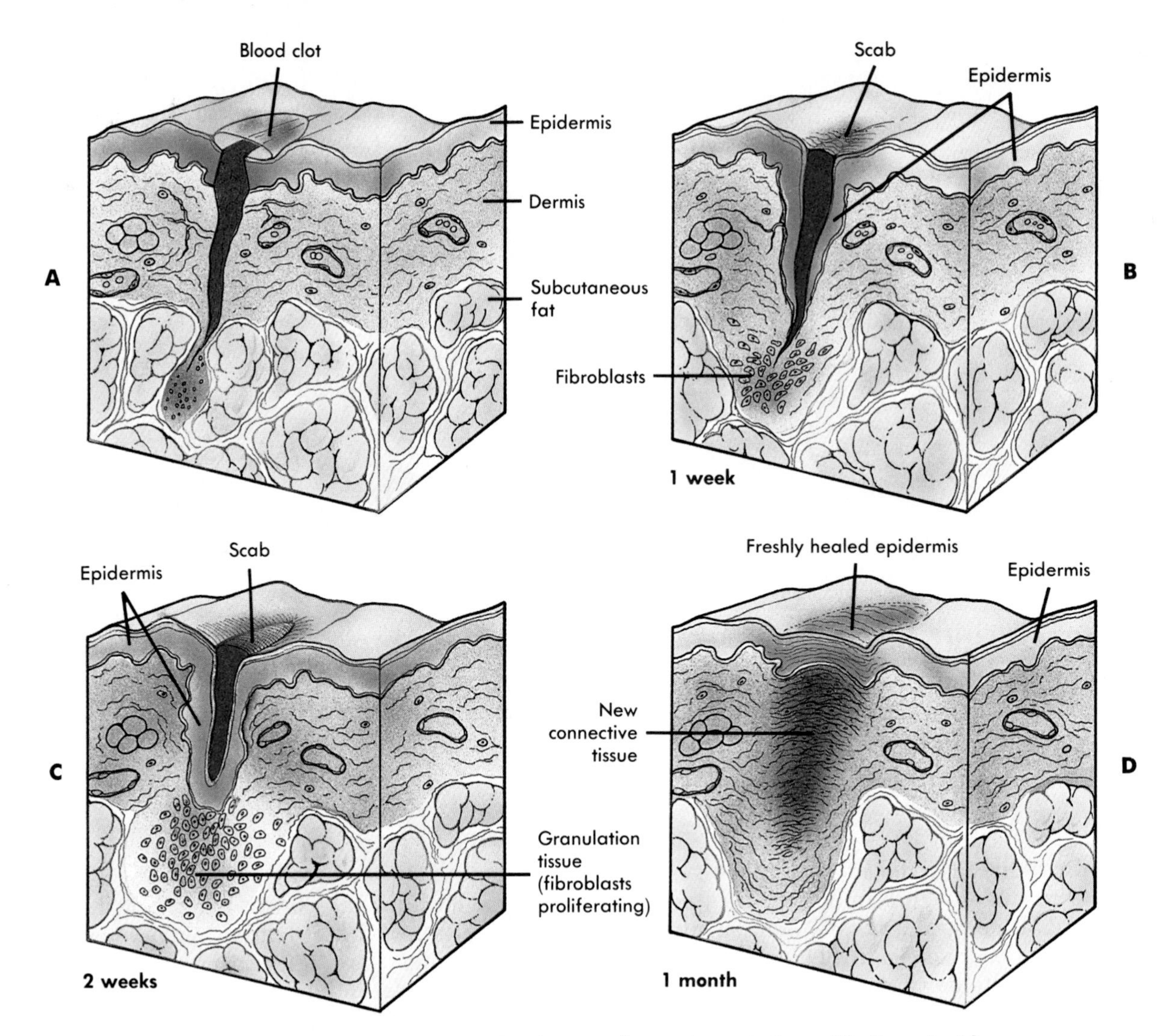

FIGURE 4-10 Tissue repair. **A** Fresh wound cuts through the epithelium (epidermis) and underlying connective tissue (dermis), and a clot forms. **B** Approximately 1 week after the injury a scab is present, and epithelium is growing into the wound. **C** Approximately 2 weeks after the injury the epithelium has grown completely into the wound, and granulation tissue has formed. **D** Approximately 1 month after the injury the wound has completely closed, the scab has been sloughed, and the granulation tissue is being replaced.

the liver, pancreas, and endocrine glands) do not actively replicate after growth ceases, but they do retain the ability to divide, if necessary, and are capable of regeneration. **Permanent** cells cannot replicate, and if killed, they are replaced by a different type of cell. Neurons fit this category, although neurons have some ability to recover from damage. If the cell body of the neuron is not destroyed, it is possible for the cell to replace a damaged axon or dendrite; but if the neuron's cell body is destroyed, the neuron cannot regenerate. Muscle cells also have little ability to regenerate, although they can repair themselves.

Repair of the skin provides a good example of wound repair (Figure 4-10). The basic pattern of the repair is the same for other tissues, especially ones covered by epithelium. If the edges of the wound are close together such as in a surgical incision, the wound heals by a process called primary union. If the edges are not close together or if there has been extensive loss of tissue, the process is called secondary union.

In **primary union** the wound fills with blood, and a clot forms (see Chapter 19). The clot contains a threadlike protein, fibrin, which binds the edges of the wound together. The surface of the clot dries to form a **scab,** which seals the wound and helps prevent infection. An inflammatory response induces vasodilation, bringing increased blood cells and

Cancer Tissue

Cancer refers to a malignant, spreading tumor and the illness that results from such a tumor. Tumor refers to any swelling, although modern usage has limited the term to swellings that involve neoplastic tissue. **Neoplasm** (ne'o-plazm) means "new growth" and refers to abnormal tissue growth that results from unusually rapid cellular proliferation that continues after normal growth of the tissue has stopped or slowed considerably. A neoplasm may be either **malignant** (with malice or intent to cause harm), able to spread and become worse, or **benign** (bĕ-nīn'; kind), not inclined to spread and not likely to become worse. Malignant tumors can spread by local growth and expansion or by **metastasis** (mĕ-tas'tă-sis; moving to another place), which results from tumor cells' separating from the main mass and being carried by the lymphatic or circulatory system to a new site where a second neoplasm is created. **Oncology** (ong-kol'o-je; tumor study) is the study of cancer and its associated problems.

A **carcinoma** is a malignant neoplasm derived from epithelial tissue. A **sarcoma** is a malignant neoplasm derived from connective tissue.

Malignant neoplasms lack the normal growth control that is exhibited by most other adult tissues, and in many ways they resemble embryonic tissue. Rapid growth is one characteristic of embryonic tissue, but as the tissue begins to reach its adult size and function, it slows or stops growing completely. This cessation of growth is controlled at the individual cell level; cancer results when a cell or group of cells for some reason breaks away from that control. This breaking loose involves the genetic machinery and can be induced by viruses, environmental toxins, and other causes. The illness associated with cancer usually occurs as the tumor invades and destroys the healthy surrounding tissue, eliminating its function.

Cancer therapy concentrates primarily on trying to confine and then kill the malignant cells. This goal is accomplished currently by killing the tissue with x-rays or lasers, by removing the tumor surgically, or by treating the patient with drugs that kill rapidly dividing cells. The major problem with current therapy is that some cancers cannot be removed completely by surgery or killed by x-rays or lasers. In addition, x-rays and lasers may also kill normal tissue adjacent to the tumor. Drugs used in cancer therapy do not kill only cancer tissue, but they kill any other rapidly growing tissue as well such as bone marrow (in which new blood cells are produced) and the lining of the intestinal tract. Loss of these tissues can result in anemia (caused by a lack of red blood cells) and nausea (caused by loss of the intestinal lining).

other substances to the area. Blood vessel permeability increases, resulting in edema. Fibrin and blood cells move into the wounded tissues because of the increased vascular permeability. The fibrin acts to isolate and wall off microorganisms and other foreign matter. Some of the white blood cells that move into the tissue are phagocytic cells called **neutrophils** (nu'tro-filz; see Figure 4-9, *B*). They ingest bacteria, thus helping to fight infection, and they also ingest tissue debris, clearing the area for repair. Neutrophils are killed in this process and may accumulate as a mixture of dead cells and fluid called **pus.**

While phagocytosis proceeds, the epithelium at the edge of the wound undergoes regeneration and migrates under the scab. After a few days the epithelial cells from the edges meet, forming a single layer of cells over the wound. The single layer proliferates and differentiates, restoring the original epithelium. The scab sloughs after the epithelium is repaired. As the new epithelium forms, a second type of phagocytic cell, a **macrophage,** removes the dead neutrophils and cellular debris.

Fibroblasts from surrounding connective tissue migrate into the clot, producing collagen and other extracellular matrix components. Capillaries grow from blood vessels at the edge of the wound and revascularize the area, and fibrin in the clot is decomposed (fibrinolysis) and removed. The result is the replacement of the clot by a delicate connective tissue, **granulation tissue,** that consists of fibroblasts, collagen and capillaries. A large amount of granulation tissue sometimes persists as a **scar,** which at first is bright red because of vascularization of the tissue. Later the scar will blanch and become white as collagen accumulates and the vascular channels are compressed.

Repair by **secondary union** proceeds in a fashion similar to healing by primary union, but there are some differences. Because the wound edges are far apart, the clot may not close the gap completely, and it will take the epithelial cells much longer to regenerate and cover the wound. With increased tissue damage the degree of the inflammatory response is greater, there is more cell debris for the phagocytes to remove, and the risk of infection is greater. Much more granulation tissue forms, and **wound contraction,** a result of the contraction of fibroblasts in the granulation tissue, reduces the size of the wound and speeds healing. Unfortunately, wound contraction can lead to disfiguring and debilitating scars. Thus it is advisable to suture a large wound so that it can heal by primary rather than secondary union; healing will be faster, the risk of infection will be lowered, and the degree of scarring will be reduced.

SUMMARY

STRUCTURE AND FUNCTION ARE CLOSELY RELATED IN TISSUES.
HISTOLOGY IS THE STUDY OF TISSUES.

EPITHELIAL TISSUE p. 104

1. Epithelium consists of cells with little extracellular matrix, covers surfaces, has a basement membrane, and has no blood vessels.
2. The basement membrane is secreted by the epithelial cells and attaches the epithelium to the underlying tissues.

Classification of Epithelium

1. Simple epithelium has a single layer of cells, stratified epithelium has two or more layers, and pseudostratified epithelium has a single layer that appears stratified.
2. Cells can be squamous (flat), cuboidal, or columnar.
3. Stratified squamous epithelium can be moist or keratinized.
4. Transitional epithelium is stratified with cells that can change shape from cuboidal to flattened.

Functional Characteristics

1. Simple epithelium generally is involved in diffusion, filtration, secretion, or absorption. Stratified epithelium serves a protective role. Squamous cells function in diffusion and filtration. Cuboidal or columnar cells, with a larger cell volume that contains many organelles, secrete or absorb.
2. A smooth free surface reduces friction (mesothelium and endothelium), microvilli increase absorption (intestines), and cilia move materials across the free surface (respiratory tract and uterine tubes). Transitional epithelium has a folded surface that allows the cell to change shape.
3. Cells are bound together mechanically by glycoproteins, desmosomes, and zonula adherens and to the basement membrane by hemidesmosomes. The zonula occludens and zonula adherens form a permeability barrier or tight junction, and gap junctions allow intercellular communication.

Glands

1. Glands are organs that secrete. Exocrine glands secrete through ducts, and endocrine glands release hormones that are absorbed directly into the blood.
2. Glands are classified as unicellular or multicellular. Multicellular exocrine glands have ducts, which are simple or compound (branched). The ducts can be tubular or end in small sacs (acini or alveoli). Tubular glands can be straight or coiled.
3. Glands are classified according to their mode of secretion. Merocrine glands (pancreas) secrete substances as they are produced, apocrine glands (mammary glands) accumulate secretions that are released when a portion of the cell pinches off, and holocrine glands (sebaceous glands) accumulate secretions that are released when the cell ruptures and dies.

CONNECTIVE TISSUE p. 113

1. Connective tissue is distinguished by its extracellular matrix.
2. The extracellular matrix results from the activity of specialized connective tissue cells; in general, blast cells form the matrix, cyte cells maintain it, and clast cells break it down.

Protein Fibers of the Matrix

1. Collagen fibers structurally resemble ropes. They are strong and flexible but resist stretching.
2. Reticular fibers are fine collagen fibers that form a branching network that supports other cells and tissues.
3. Elastin fibers have a structure similar to a bedspring. They can be stretched and then will revert back to their original shape.

Other Matrix Molecules

1. Hyaluronic acid and proteoglycans are important nonprotein molecules of the matrix.
2. Hyaluronic acid is a long, unbranched polysaccharide chain composed of repeating disaccharide units.
3. A proteoglycan is a large molecule that consists of numerous polysaccharide branches attached to a common protein core.

Classification of Connective Tissue

Connective tissue is classified according to the type of protein and the proportions of protein, ground substance, and fluid in the matrix.

Matrix with Fibers as the Primary Feature

1. Loose (areolar) connective tissue has many different cell types and a random arrangement of protein fibers with space between the fibers. This tissue fills spaces around the organs and attaches the skin to underlying tissues.
2. Dense regular connective tissue is composed of fibers arranged in one direction, providing strength in a direction parallel to the fiber orientation. There are two types of dense regular connective tissue, collagenous (tendons and ligaments) and elastic (ligaments of vertebrae).
3. Dense irregular connective tissue has fibers organized in many directions, producing strength in different directions. There are two types of dense irregular connective tissue, collagenous (capsules of organs and dermis of skin) and elastic (large arteries).
4. Adipose tissue has fat cells (adipocytes) filled with lipid and very little extracellular matrix (a few reticular fibers).
   - Adipose tissue functions as energy storage, insulation, and protection.
   - Adipose tissue may be yellow (white) or brown. Brown fat is specialized for generating heat.
5. Reticular tissue is a network of fine collagen and forms the framework of lymphoid tissue, bone marrow, and the liver.
6. Yellow bone marrow is a site of fat storage, and red bone marrow is the site of blood cell formation.

Matrix with Both Protein Fibers and Ground Substance

1. Cartilage has a relatively rigid matrix composed of protein fibers and proteoglycan aggregates. The major cell type is the chondrocyte, which is located within lacunae.
   - Hyaline cartilage has evenly dispersed collagen fibers that provide rigidity with some flexibility. Examples include the costal cartilage, the covering over the ends of bones in joints, the growing portion of long bones, and the embryonic skeleton.

- Fibrocartilage has collagen fibers arranged in thick bundles, can withstand great pressure, and is found between vertebrae, in the jaw, and in the knee.
- Elastic cartilage is similar to hyaline cartilage, but it has elastin fibers. It is more flexible than hyaline cartilage. It is found in the external ear.

2. Bone cells, or osteocytes, are located in lacunae that are surrounded by a mineralized matrix (hydroxyapatite) that makes bone very hard. Cancellous bone has spaces between bony trabeculae, and compact bone is more solid.

Predominantly Fluid Matrix

Blood cells are suspended in a fluid matrix.

MUSCULAR TISSUE *p. 121*

1. Muscular tissue has the ability to contract.
2. Skeletal (striated voluntary) muscle attaches to bone and is responsible for body movement. Skeletal muscle cells are long, cylindrical-shaped cells with several peripherally located nuclei.
3. Cardiac (striated involuntary) muscle cells are cylindrical, branching cells with a single, central nucleus. Cardiac muscle is responsible for pumping blood through the circulatory system.
4. Smooth (nonstriated involuntary) muscle forms the walls of hollow organs, the pupil of the eye, and other structures. Its cells are spindle shaped with a single, central nucleus.

NERVOUS TISSUE *p. 123*

1. Nervous tissue has the ability to conduct electrical impulses and is composed of neurons (conductive cells) and neuroglia (support cells).
2. Neurons have cell processes called dendrites and axons. The dendrites can receive electrical impulses, and the axons can conduct them. Neurons may be multipolar (several dendrites and an axon), bipolar (one dendrite and one axon), or unipolar (one axon).

EMBRYONIC TISSUE DEVELOPMENT *p. 124*

The endoderm, mesoderm, and ectoderm are the primary germ layers from which all adult structures arise.

MEMBRANES *p. 124*

1. Serous membranes line cavities that do not open to the exterior, do not contain glands, but do secrete serous fluid.
2. Mucous membranes line cavities that open to the outside and often contain mucous glands, which secrete mucus. The connective tissue on which mucous membranes rest is called the lamina propria.

INFLAMMATION *p. 124*

1. The function of the inflammatory response is to isolate injurious agents from the rest of the body and to attack and destroy the injurious agent.
2. The inflammatory response produces five symptoms: redness, heat, swelling, pain, and loss of function.

TISSUE REPAIR *p. 125*

1. Tissue repair is the substitution of viable cells for dead cells. Tissue repair occurs by regeneration or replacement.
   - Labile cells divide throughout life and can undergo regeneration.
   - Stable cells do not ordinarily replicate but can regenerate if necessary.
   - Permanent cells cannot replicate. If killed, permanent tissue is repaired by replacement.
2. Tissue repair by primary union occurs when the edges of the wound are close together. Secondary union occurs when the edges are far apart.

CONTENT REVIEW

1. Define histology and tissues. What two things distinguish one type of tissue from another?
2. Name the four primary tissue types, and give the general basis for defining them.
3. List four characteristics of epithelium.
4. What is the basement membrane, and how is it produced?
5. Name two general characteristics that are used to classify epithelium. Based on this classification scheme, describe the different kinds of epithelium.
6. Why is pseudostratified epithelium not really considered a stratified epithelium? Describe transitional epithelium.
7. What kind of functions would a single layer of epithelium be expected to perform? A stratified layer?
8. In locations where diffusion or filtration are occurring, what shape cells would be expected?
9. Why are cuboidal or columnar-shaped cells found where secretion or absorption is occurring?
10. What is the function of an epithelial free surface that is smooth, has cilia, has microvilli, or is folded? Give an example of an epithelium in which each surface type is found.
11. Name the ways in which epithelial cells are bound to each other and to the basement membrane.
12. Define the term gland. Distinguish between exocrine and endocrine glands. Describe the classification scheme for exocrine glands based on their duct systems.
13. Describe three different ways in which exocrine glands release their secretions. Give an example for each method.
14. What is the major characteristic that distinguishes connective tissue from other tissues?
15. Explain the difference between connective tissue cells that are termed blast, cyte, or clast cells.
16. Contrast the structure and characteristics of collagen fibers, reticular fibers, and elastin fibers.
17. What three components are found in the extracellular matrix of connective tissue? How are they used to classify connective tissue?
18. Describe the fiber arrangement in loose (areolar) connective tissue. What functions does this tissue accomplish?
19. What is the function of reticular tissue?
20. Structurally and functionally, what is the difference between dense regular connective tissue and dense irregular connective tissue?

21. Name the two kinds of dense regular connective tissue, and give an example of each. Do the same for dense irregular connective tissue.
22. What features of the extracellular matrix distinguish adipose tissue from other connective tissue? What is an adipocyte?
23. List the functions of adipose tissue. Name the two types of adipose tissue. Which one is important in generating heat?
24. What are red marrow and yellow marrow?
25. Describe the components of cartilage. How do hyaline cartilage, elastic cartilage, and fibrocartilage differ in structure and function? Give an example of each.
26. Describe the components of bone. Differentiate between cancellous and compact bone.
27. What characteristic separates blood from the other connective tissues?
28. Functionally, what is unique about muscle? Contrast the structure of skeletal, cardiac, and smooth muscle cells. Which of the muscle types is under voluntary control? What tasks does each type perform?
29. Functionally, what is unique about nervous tissue? What do neurons and neuroglia accomplish?
30. What is the difference between a dendrite and an axon? Describe the structure of multipolar, bipolar, and unipolar neurons.
31. What are the three embryonic germ layers, and what are the other two primary embryonic tissues?
32. Compare serous and mucous membranes according to the type of cavity they line and their secretions.
33. What is the function of the inflammatory response? Name the five symptoms of the inflammatory response, and explain how each is produced.
34. Define tissue repair. What is the difference between tissue repair that occurs by regeneration and by replacement?
35. Differentiate between labile cells, stable cells, and permanent cells. Give examples of each type. What is the significance of these cell types to tissue repair?
36. Describe the process of tissue repair. Contrast healing by primary union and by secondary union.

CONCEPT QUESTIONS

1. Given the observation that a tissue has more than one layer of cells lining a free surface, (1) list the possible tissue types that exhibit that characteristic and (2) explain what additional observations would have to be made to identify them as specific tissue types.
2. Compare the cell shapes and surface specializations of an epithelium that functions to resist abrasion with those of an epithelium that functions to carry out absorption of materials.
3. Tell how to distinguish between a gland that produces a merocrine secretion and a gland that produces a holocrine secretion. Assume that you have the ability to chemically analyze the composition of the secretions.
4. Given the following statement—if a tissue is capable of contracting, is under involuntary control, and has mononucleated cells, it is smooth muscle—indicate whether the statement is appropriate or not. Explain your answer.
5. Antihistamines block the effect of a chemical mediator, histamine, that is released during the inflammatory response. What effect would administering antihistamines have on the inflammatory response, and would doing so be beneficial?

? ?

ANSWERS TO PREDICT QUESTIONS

1 *p. 104.* Skin composed of moist, stratified epithelium or simple epithelium would provide no moisture barrier, and large amounts of fluid would be lost by evaporation through the skin. In addition, simple epithelium would provide no protection against abrasion since it is only one layer thick.

2 *p. 119.* Tendons function like cables in that they can tolerate tension applied to them in one direction with considerable force, and collagen fibers oriented in one direction are very important in providing strength. The forces pulling on skin are not as great as the forces on tendons, but the pull may occur in any direction. Therefore the skin's collagen fibers must be oriented in all directions to be oriented with the direction of pull.

3 *p. 119.* If tendons were elastic, there would be a reduction in the force delivered to the load being moved because a portion of the force of muscle contraction would be required to stretch the elastic tendon. Control would be lost because the elastic tendon would be stretched when the muscle contracted. Subsequently, the elastic tendon would recoil, producing movement over which there would be little control. It would be like lifting a weight with a rubber band rather than a string.

4 *p. 120.* Hyaline cartilage is well suited for its function in joints because it resists compression and has tensile strength, which resists stretching and bending, as a result of the presence of collagen fibers. In addition, the proteoglycans in the matrix make cartilage very smooth on its surface (reducing friction as bones move across the joint) and very resilient because of the presence of water, which creates a kind of shock absorber between the bones.

PART II

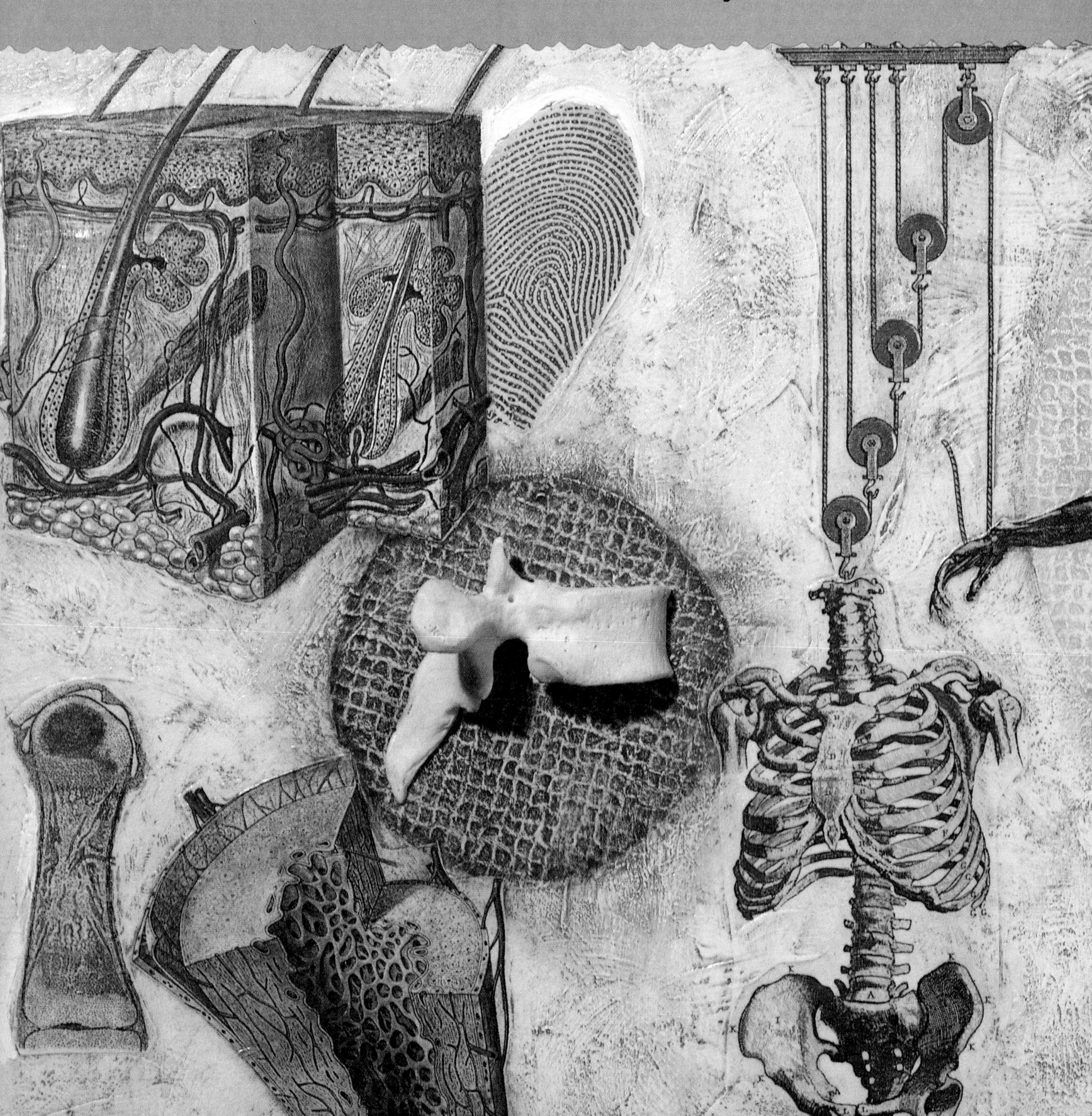

SUPPORT AND MOVEMENT

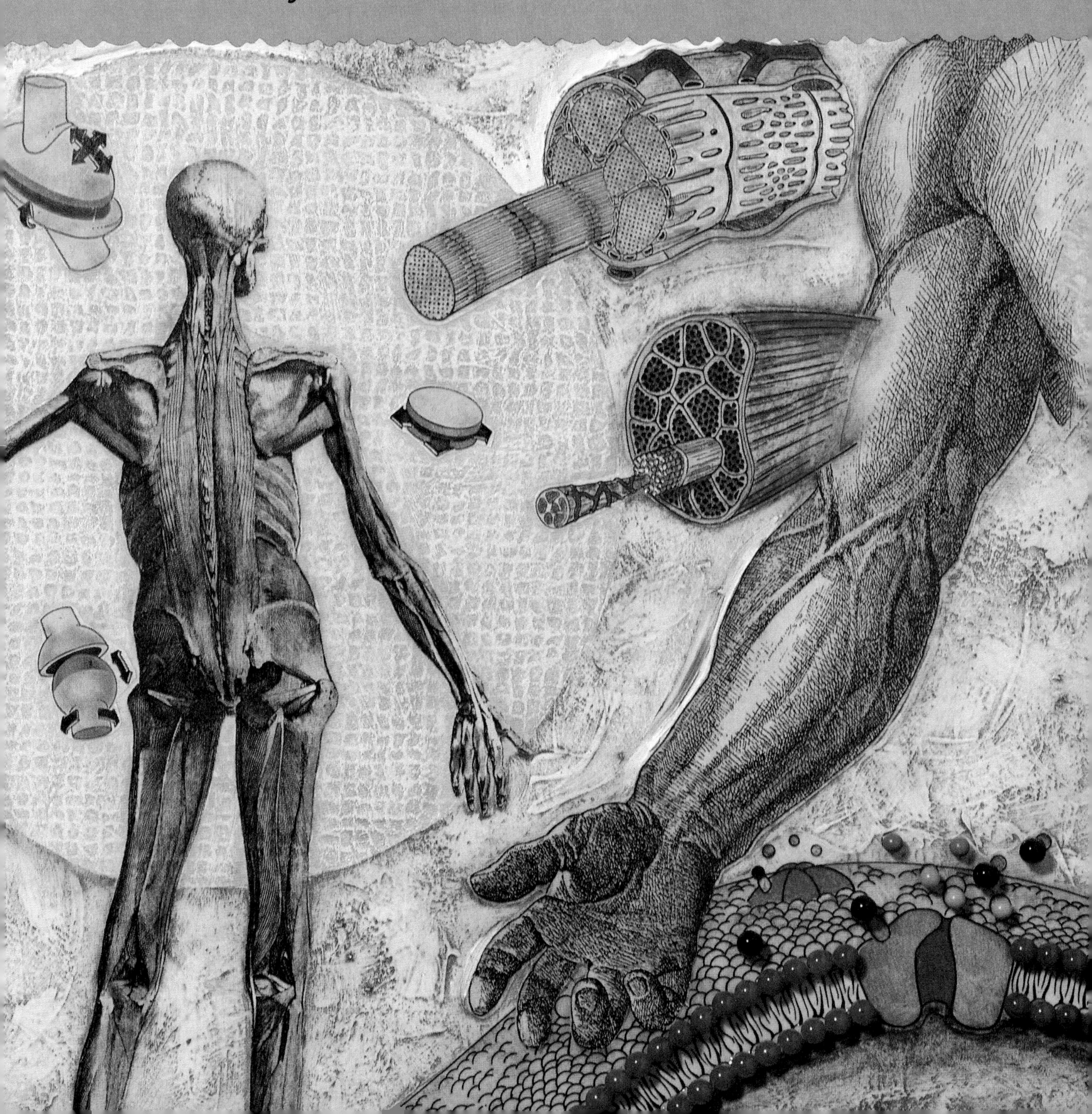

CHAPTER 5 OBJECTIVES

After reading this chapter you should be able to

1. Describe the structure and function of the hypodermis.
2. Name and describe the two layers of the dermis.
3. Explain the basis for dividing the epidermis into strata. List and describe each stratum.
4. Describe the events occurring during keratinization that produce a skin resistant to abrasion and water loss.
5. Contrast thick skin and thin skin.
6. Discuss melanocytes and the way they produce and transfer melanin.
7. Explain how melanin, carotene, and blood affect skin color.
8. Distinguish between lanugo, vellus, and terminal hair.
9. Describe the structure of a hair and the sheaths that surround the hair. Discuss the phases of hair growth.
10. Describe the production of "goose flesh."
11. Describe the glands of the skin and their secretions.
12. Describe the parts of a nail and explain how nails are produced.
13. Discuss the functions of the skin, hair, nails, and glands.
14. Describe the production of vitamin D by the body and the functions of vitamin D.
15. Describe the changes that occur in the integumentary system with age.

Integumentary System

KEY TERMS

arrector pili (ah-rek′tor pĭ′le)

carotene

dermis (der′mis)

epidermis (ep′ĭ-der′mis)

hair follicle

hypodermis

keratinization

keratinocyte (kĕ-rat′ĭ-no-sīt)

melanin

melanocyte (mel′ă-no-sīt)

nail matrix

sebaceous (se-ba′shus) gland

stratum

sweat gland

vitamin D

RELATED TOPICS

The following terms or concepts are important for a good understanding of this chapter. If you are not familiar with them, you should review them before proceeding.

Cellular organelles (Chapter 3)

Epithelial and connective tissue histology (Chapter 4)

Structure of glands (Chapter 4)

THE INTEGUMENTARY SYSTEM, CONSISTING OF THE SKIN AND ACCESSORY STRUCTURES SUCH AS HAIR, NAILS, AND A VARIETY OF GLANDS, IS THE LARGEST ORGAN SYSTEM IN THE BODY. The skin is composed of a layer of dense, irregular connective tissue called the **dermis** (der′mis) and is covered by a layer of epithelial tissue called the **epidermis** (ep′ĭ-der′mis) (Figure 5-1). The integumentary system is involved in a number of functions such as protecting internal structures from mechanical and chemical damage, preventing the entry of infectious agents, providing protection against ultraviolet radiation from the sun, preventing dehydration, regulating temperature, producing vitamin D, and detecting stimuli.

HYPODERMIS

Although not actually part of the skin, the **hypodermis** (see Figure 5-1) is included here because it attaches the skin to underlying bone and muscle and because blood vessels and nerves that supply the skin pass through it. The hypodermis is sometimes called **subcutaneous** (sub′ku-ta′ne-us) **tissue** or **superficial fascia** (fash′e-ah) and consists of loose connective tissue with collagen and elastin fibers. The main types of cells within the hypodermis are fibroblasts, fat cells, and macrophages.

Approximately half of the body's stored fat is in the hypodermis, although there is variation in amount and location with age, sex, and diet. For example, newborn infants have a large amount of fat, which accounts for their chubby appearance, and women have more fat than men, especially over the thighs, buttocks, and breasts. Fat in the hypodermis functions as padding and insulation and is responsible for some of the structural differences between men and women.

The hypodermis is used to estimate total body fat. The skin is pinched at selected locations, and the thickness of the fold of skin and underlying hypodermis is measured. The thicker the fold, the greater is the amount of total body fat. Clinically, the hypodermis is the site of subcutaneous injections.

SKIN

Dermis

The dermis is dense, irregular connective tissue with fibroblasts, a few fat cells, and macrophages. Collagen is the main connective tissue fiber, but elastin and reticular fibers are also present. The dermis is responsible for most of the structural strength of the skin. Compared to the hypodermis, there is a scarcity of fat cells and blood vessels in the dermis. Nerve endings, hair follicles, smooth muscles, glands, and lymphatics extend into the dermis (see Figure 5-1).

The dermis is divided into two indistinct layers (see Figure 5-1; Figure 5-2): the deeper **reticular** (rĕ-tik′u-lar) **layer** and the more superficial **papillary** (pap′ĭ-lĕr′e) **layer.** The reticular layer is the main fibrous layer of the dermis and blends into the hypodermis. It forms a mat of irregularly arranged fibers that are resistant to stretching in many directions. The elastin and collagen fibers are oriented more in some directions than in others and produce **cleavage,** or **tension lines,** in the skin. Knowledge of cleavage line directions is important to a surgeon because an incision made across the cleavage lines is likely to gap, producing considerable scar tissue, but an incision made parallel with the lines tends to gap less and produce less scar tissue. In addition, if the skin is overstretched, the dermis may rupture, leaving lines that are visible through the epidermis. These lines, called **striae** (strī′e), or **stretch marks,** may develop on the abdomen and breasts of a woman during pregnancy.

The dermis is that portion of an animal hide from which leather is made. The epidermis of the skin is removed, and the fibrous dermis is preserved by tanning. Clinically, the dermis in humans is sometimes the site of injections such as the tuberculin skin test.

The papillary layer derives its name from projections called **papillae** (pă-pil′a) that extend toward the epidermis. Compared to the reticular layer, the papillary layer has more cells and fewer fibers, and these fibers are finer and more loosely arranged. The papillary layer also contains a large number of blood vessels that supply the overlying avascular epidermis with nutrients, remove waste products, and aid in regulating body temperature.

Epidermis

The epidermis is stratified squamous epithelium separated from the dermis by a basement membrane. The epidermis is not as thick as the dermis. The epidermis contains no blood vessels and derives nourishment by diffusion from capillaries of the papillary layer. Most of the cells of the epidermis are **keratinocytes** (kĕ-rat′ĭ-no-sītz), which produce a protein mixture called **keratin** (kĕr′ah-tin). Keratinocytes are responsible for the structural strength and permeability characteristics of the epidermis. Other cells of the epidermis include **melanocytes** (mel′ă-no-sītz), which contribute to skin color, and **Langerhans cells,** which are part of the immune system (see Chapter 22).

Cells are produced in the deepest layers of the epidermis by mitosis. As new cells are formed, they push older cells to the surface where they slough or **desquamate** (des′kwă-māt). The outermost cells in this stratified arrangement protect the cells underneath, and the deeper replicating cells replace cells lost from the surface. During their movement from the deeper epidermal layers to the surface, the cells undergo **keratinization,** a process that involves

FIGURE 5-1 Skin and hypodermis. Figure represents a block of skin (dermis and epidermis), hypodermis, and accessory structures (hairs and glands).

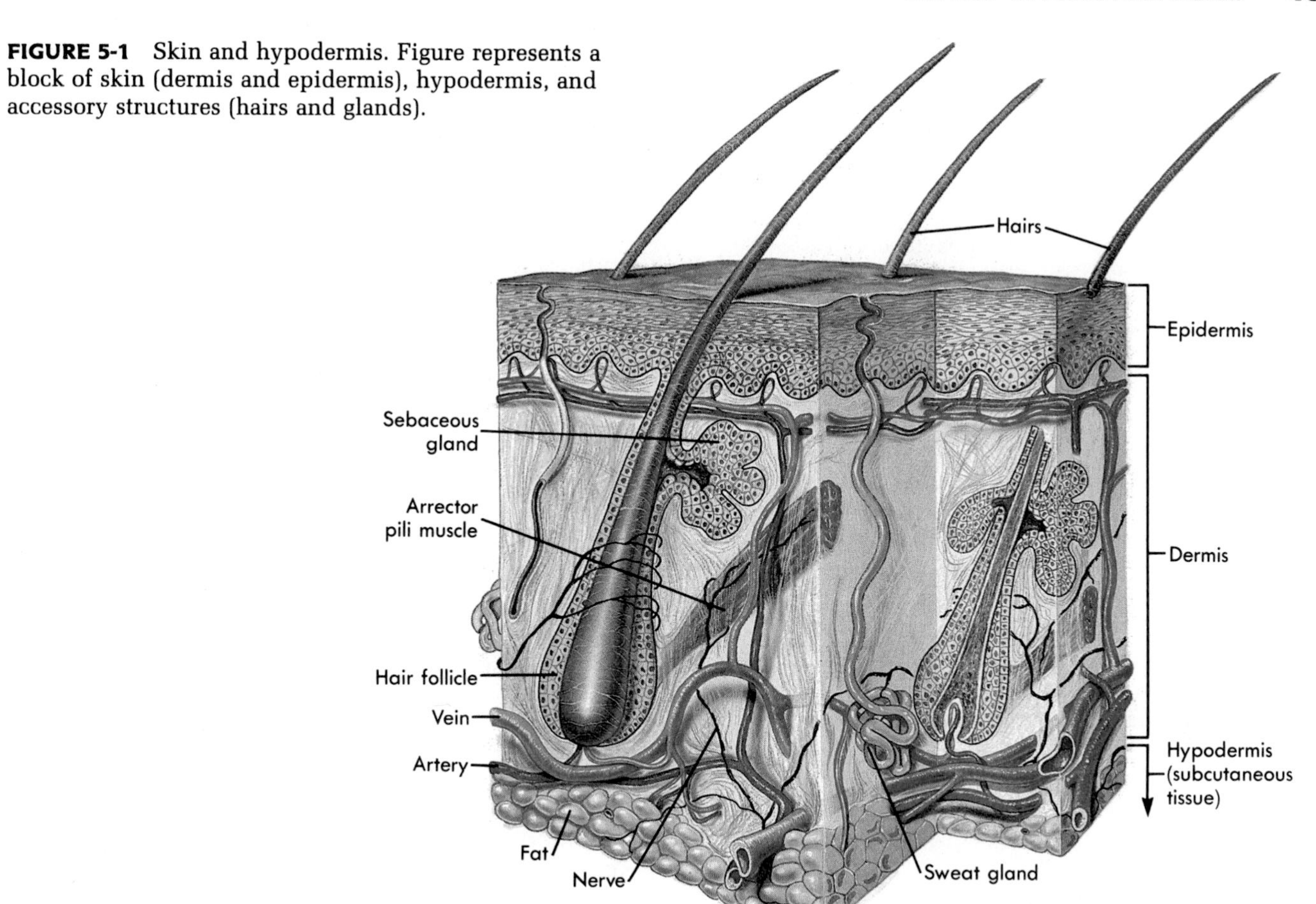

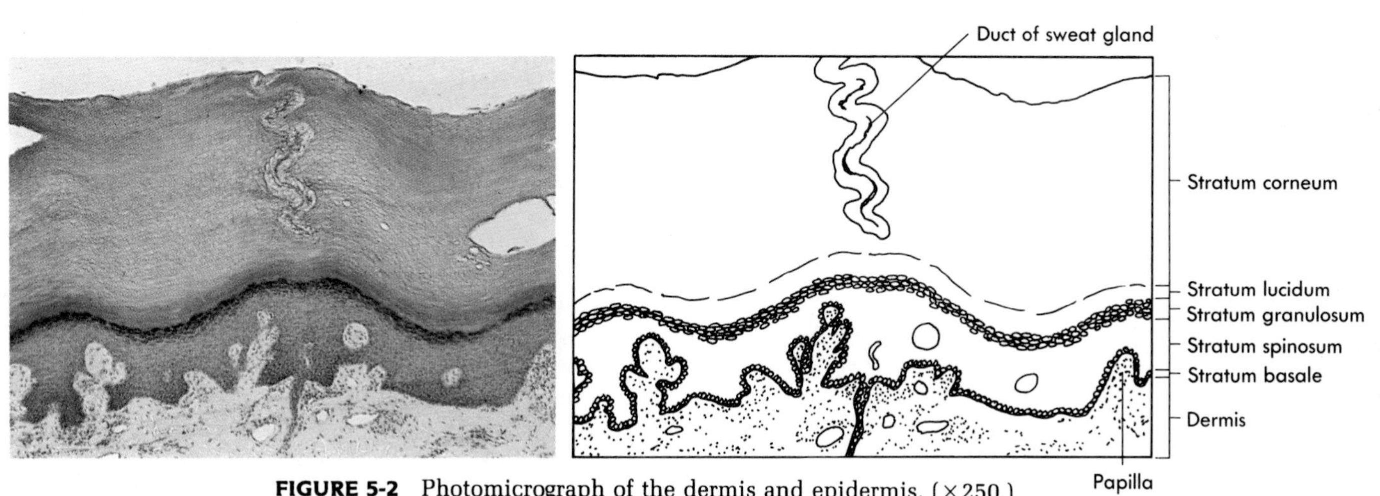

FIGURE 5-2 Photomicrograph of the dermis and epidermis. (×250.)

change in shape, structure, and chemical composition. These cells eventually die and produce an outer layer of cells that resists abrasion and forms a permeability barrier.

The study of keratinization is important because many skin diseases such as psoriasis result from malfunctions in this process. By comparing normal to abnormal keratinization, scientists may be able to develop effective therapies.

Although keratinization is a continual process, distinct transitional stages are recognized as the cells change. Based on these stages, the many layers of cells in the epidermis are divided into regions, or **strata** (Figure 5-3). From the deepest to the most superficial, five strata are observed: stratum basale, stratum spinosum, stratum granulosum, stratum lucidum, and stratum corneum. The number of cell layers in each stratum and even the number of strata in the skin vary, depending on their location in the body.

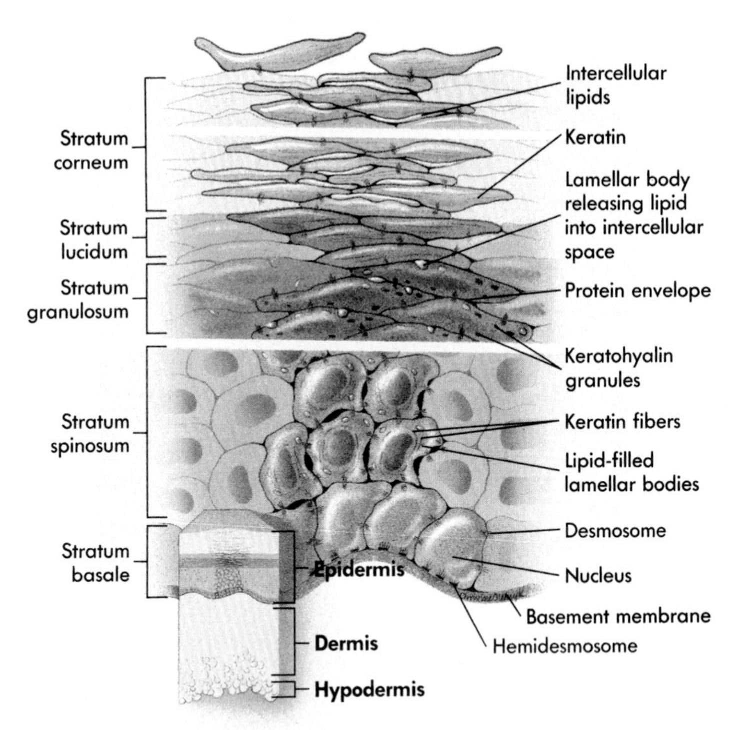

FIGURE 5-3 Epidermal strata. Stratum basale cells contain functional organelles and are capable of mitosis. Stratum spinosum cells accumulate keratin and lamellar bodies. Stratum granulosum cells contain keratohyalin and degenerating organelles. A hard protein envelope forms, and lamellar bodies release lipids. Stratum lucidum cells die and become transparent. Stratum corneum cells are dead, enveloped, and contain keratin. They are connected by many desmosomes and are surrounded by lipids.

Stratum basale

The deepest portion of the epidermis is a single layer of columnar cells, the **stratum basale** (bah-sal′e) (see Figures 5-2 and 5-3). Keratinocytes of the stratum basale secrete the basement membrane on which the epidermis rests. Structural strength is provided by hemidesmosomes, which anchor the epidermis to the basement membrane, and by desmosomes, which hold the keratinocytes together (see Chapter 4). Keratinocytes are strengthened internally by keratin fibers (intermediate filaments) that insert into the desmosomes. Keratinocytes undergo mitotic divisions approximately every 19 days. One daughter cell becomes a new stratum basale cell and divides again, but the other daughter cell is pushed toward the surface and becomes **keratinized.** It takes approximately 40 to 56 days for the cell to reach the epidermal surface and desquamate.

Stratum spinosum

Superficial to the stratum basale is the **stratum spinosum** (spi-no′sum), consisting of 8 to 10 layers of many-sided cells (see Figures 5-2 and 5-3). As the cells in this stratum are pushed to the surface, they flatten; desmosomes are broken apart, and new desmosomes are formed. During preparation for microscopic observation the cells usually shrink from each other except where attached by desmosomes, causing the cells to appear spiny—hence the name stratum spinosum. Additional keratin fibers and lipid-filled, membrane-bound organelles called lamellar bodies are formed inside the keratinocytes. A limited amount of cell division takes place in this stratum, and for this reason the stratum basale and stratum spinosum are sometimes considered a single stratum called the **stratum germinativum** (jer′mĭ-nă-tiv′um). Mitosis does not occur in the more superficial strata.

Stratum granulosum

The **stratum granulosum** (gran′u-lo′sum) consists of two to five layers of somewhat flattened, diamond-shaped cells that have their long axes oriented parallel to the surface of the skin (see Figures 5-2 and 5-3). This stratum derives its name from the non-membrane-bound protein granules of **keratohyalin** (kĕr′ă-to-hi′ă-lin), which accumulate in the cytoplasm of the cell. The lamellar bodies of these cells move to the cell membrane and release their lipid contents into the intercellular space; inside the cell, a protein envelope forms beneath the cell membrane. In the most superficial layers of the stratum granulosum the nucleus and other organelles degenerate, and the cell dies. Unlike the other organelles, however, the keratin fibers and keratohyalin granules do not degenerate.

Stratum lucidum

The **stratum lucidum** (lu'sĭ-dum) appears as a thin, clear zone above the stratum granulosum (see Figures 5-2 and 5-3) and consists of several layers of dead cells with indistinct boundaries. Keratin fibers are present, but the keratohyalin, which was evident as granules in the stratum granulosum, has dispersed around the keratin fibers, and the cells appear somewhat transparent. The stratum lucidum is absent in the skin of most areas of the body (see the discussion of thick and thin skin).

Stratum corneum

The last and most superficial stratum of the epidermis is the **stratum corneum** (kor'ne-um) (see Figures 5-2 and 5-3). This stratum is composed of as many as 25 or more layers of dead squamous cells joined by desmosomes. Eventually the desmosomes break apart, and the cells are desquamated from the surface of the skin. Dandruff is an example of desquamation of the stratum corneum. Less noticeably, cells are continually shed as clothes rub against the body or as the skin is washed.

The stratum corneum consists of **cornified cells** (i.e., dead cells), which have a hard protein envelope and are filled with the protein keratin (a mixture of keratin fibers and keratohyalin). The envelope and the keratin are responsible for the structural strength of the stratum corneum. The type of keratin found in the skin is soft keratin. Another type of keratin, hard keratin, is found in nails and the external parts of hair. Cells containing hard keratin are more durable than cells with soft keratin and do not desquamate.

Surrounding the cells are the lipids released from the lamellar bodies. The lipids are responsible for many of the permeability characteristics of the skin.

1

What kind of substances could pass easily through the skin by diffusion? What kinds would have difficulty?

? ? ? ? ? ? ? ? ? ? ?

Table 5-1 summarizes the structures and functions of the skin and hypodermis.

Thick and Thin Skin

Skin is classified as **thick** or **thin** based on the structure of the epidermis. Thick skin has all five epithelial strata, and the stratum corneum has many layers of cells. Thick skin is on the palms of the hands, the soles of the feet, and the fingertips (i.e., areas subject to pressure or friction). The papillae of the dermis underlying thick skin are in parallel, curving ridges that shape the overlying epidermis into fingerprints and footprints. The ridges increase friction and improve the grip of the hands and feet.

Fingerprints were first used in criminal investigation in 1880 by Henry Faulds, a Scottish medical missionary. The identity of a thief who had been drinking purified alcohol from the dispensary was confirmed by a greasy fingerprint on a bottle.

Thin skin is found over the rest of the body and is more flexible than thick skin. Each stratum contains fewer layers of cells than are found in thick skin; the stratum granulosum frequently consists of only one or two layers of cells, and the stratum lucidum generally is absent. The dermis under thin skin projects upward as separate papillae and does not produce the ridges seen in thick skin. Hair is found only in thin skin.

The entire skin, including both the epidermis and the dermis, varies in thickness from 0.5 mm in the eyelids to 5 mm for the back and shoulders. The terms thin and thick, which refer to the epidermis only, should not be used when total skin thickness is considered. Most of the difference in total skin thickness is due to variation in the thickness of the dermis. For example, the skin of the back is thin skin, whereas that of the palm is thick skin; however, the total skin thickness of the back is greater than that of the palm because there is more dermis in the skin of the back.

In skin subjected to friction or pressure the number of layers in the stratum corneum greatly increases, producing a thickened area called a **callus** (kal'us). The skin over bony prominences may develop a cone-shaped structure called a **corn.** The base of the cone is at the surface, but the apex extends deep into the epidermis, and pressure on the corn can be quite painful. Calluses and corns develop in both thin and thick skin.

Skin Color

Skin color is determined by pigments in the skin, by blood circulating through the skin, and by the thickness of the stratum corneum. **Melanin,** a brown-to-black pigment, is responsible for most skin color. Certain regions of the skin (e.g., freckles, moles, nipples and areolae of the breasts, the axillae, and the genitalia) have large amounts of melanin. Other areas of the body (e.g., the lips, the palms of the hands, and the soles of the feet) have less melanin.

Melanin is produced by **melanocytes,** irregularly shaped cells with many long processes that extend

Table 5-1

Comparison of the skin (epidermis and dermis) and hypodermis

PART	STRUCTURE	FUNCTION
Epidermis	Superficial part of skin; stratified squamous epithelium; composed of five strata	Barrier that prevents water loss and the entry of chemicals and microorganisms; protects against abrasion and ultraviolet light; produces vitamin D; gives rise to hair, nails, and glands
Stratum corneum	Most superficial strata of the epidermis; 25 or more layers of dead squamous cells	Provision of structural strength by keratin within cells; prevention of water loss by lipids surrounding cells; desquamation of most superficial cells resists abrasion
Stratum lucidum	Three to five layers of dead cells; appears transparent; present in thick skin, absent in most thin skin	Dispersion of keratohyalin around keratin fibers
Stratum granulosum	Two to five layers of flattened, diamond-shaped cells	Production of keratohyalin granules; lamellar bodies release lipids from cells; cells die
Stratum spinosum	Eight to 10 layers of many-sided cells	Production of keratin fibers; formation of lamellar bodies
Stratum basale	Deepest strata of the epidermis; single layer of columnar cells; basement membrane of the epidermis attaches to the dermis	Production of cells of the more superficial strata; melanocytes produce and distribute melanin, which protects against ultraviolet light
Dermis	Deep part of skin; dense irregular connective tissue composed of two layers	Responsible for the structural strength and flexibility of the skin; the epidermis exchanges gases, nutrients, and waste products with blood vessels in the dermis
Papillary layer	Projections toward the epidermis	Brings blood vessels close to the epidermis; forms fingerprints and footprints
Reticular layer	Mat of collagen and elastin fibers	Main fibrous layer of the dermis; strong in many directions; forms cleavage lines
Hypodermis	Not part of the skin; loose connective tissue with abundant fat deposits	Attaches the dermis to underlying structures; fat tissue provides energy storage, insulation, and padding; blood vessels and nerves from the hypodermis supply the dermis

between the keratinocytes of the stratum basale and the stratum spinosum. Melanin is packaged into vesicles called **melanosomes** (mel'ă-no-sōmz), which are released from the cell processes by exocytosis. Keratinocytes adjoining the melanocyte processes take up the melanin granules by phagocytosis and form their own melanosomes. Although all the keratinocytes may contain melanin, only the melanocytes produce it, and each melanocyte supplies melanin to approximately 36 keratinocytes.

Melanin production is determined by genetic factors, hormones, and exposure to light. Genetic factors are responsible for the amounts of melanin found in different races and for differences in amounts among people of the same race. Although many genes are responsible for skin color, a single mutation can prevent the manufacture of melanin, resulting in **albinism** (al'bĭ-nizm), which usually is a recessive genetic trait that causes a deficiency or absence of pigment in the skin, hair, and eyes. **Vitiligo** (vit-ĭ-li'go), the development of patches of white skin, occurs because the melanocytes in the affected area are destroyed, apparently by an autoimmune response (see Chapter 22). During pregnancy certain hormones cause an increase in melanin production in the mother, darkening the nipples, areolae, and genitalia. The cheekbones, forehead, and chest also may darken, resulting in the "mask of pregnancy," and a dark line of pigmentation may appear on the midline of the abdomen. Diseases such as Addison's disease that produce an imbalance of certain hormones also cause increased pigmentation. Exposure to ultraviolet light darkens melanin already present and stimulates melanin production, resulting in tanning of the skin.

Racial variations in skin color such as black, brown, yellow, and white are not determined by the number of melanocytes because all races have essentially the same number. Instead, variation in color is due to the amount of melanin produced by the melanocytes and to the size, number, and distribution of the melanosomes.

The location of pigments and other substances in the skin affects the color produced. As light passes through a substance, some wavelengths of light can be reflected or scattered to make a blue color. The color of the sky is an example of blue color produced by light reflecting from dust particles. In the skin collagen fibers of the dermis also scatter light to produce a blue color. Thus the deeper within the dermis or hypodermis any dark pigment is located, the bluer the pigment appears as a result of the light-scattering effect of the overlying tissue. The blue color of tattoos, bruises, and some superficial blood vessels is due to this effect.

Carotene is a yellow pigment found in plants such as carrots. Humans normally ingest carotene and use it as a source of vitamin A. Carotene is lipid soluble, and when large amounts of carotene are consumed, the excess carotene accumulates in the stratum corneum and in the fat cells of the dermis and hypodermis, causing the skin to develop a yellowish tint that disappears once carotene intake is reduced. Note, however, that the yellow hue of the skin of some Oriental races is due to variations in melanin and not to carotene.

Blood flowing through the skin imparts a reddish hue, and when blood flow increases (e.g., during blushing, anger, and the inflammatory response), the red color intensifies. A decrease in blood flow such as occurs in shock can make the skin appear pale, and a decrease in the blood oxygen content produces **cyanosis** (si-ă-no'sis), a bluish skin color.

2

Explain the differences in skin color between (A) palms of the hands and the lips; (B) palms of the hands of a person who does heavy manual labor and one who does not; (C) anterior and posterior surfaces of the forearm; and (D) genitals and the soles of the feet.

? ? ? ? ? ? ? ? ? ? ?

ACCESSORY SKIN STRUCTURES

Hair

The presence of **hair** is one of the characteristics common to all mammals; if the hair is thick and covers most of the body surface, it is called fur. In humans hair is found everywhere in the skin except the palms, soles, lips, nipples, parts of the external genitalia, and the distal segments of the fingers and toes.

By the fifth or sixth month of fetal development, delicate unpigmented hair called **lanugo** (lă-nu'go), which covers the fetus, is produced. Near the time of birth the lanugo of the scalp, eyelids, and eyebrows is replaced by **terminal hairs,** which are long, coarse, and pigmented. The lanugo on the rest of the body is shed and replaced by **vellus** (vel'us) **hairs,** which are short, fine, and usually unpigmented. At puberty much of the vellus hair is replaced by terminal hair, especially in the pubic and axillary regions. The hair of the chest, legs, and arms is approximately 90% terminal hair in males compared to approximately 35% in females. In males the vellus hairs of the face are replaced by terminal hairs to form the beard. The beard, pubic, and axillary hair are signs of sexual maturity. In addition, pubic and axillary hair may function as wicks for dispersing odors produced by secretions from specialized glands in the pubic and axillary regions. It also has been proposed that pubic hair provides protection against abrasion during intercourse, and axillary hair reduces friction when the arms move.

Text continued on p. 143.

Burns

Burns are classified according to the depth of the burn and the extent of surface area involved. Based on depth, burns are either partial-thickness or full-thickness burns (Figure 5-A). **Partial-thickness burns** are divided into **first-** and **second-degree burns.** First-degree burns involve only the epidermis and are red and painful, and slight edema (swelling) may be present. They can be caused by sunburn or brief exposure to hot or cold objects and heal in a week or so without scarring. Second-degree burns damage the epidermis and the dermis. If there is minimum dermal damage, symptoms include redness, pain, edema, and blisters. Healing takes approximately 2 weeks, and there is no scarring. If the burn goes deep into the dermis, however, the wound appears red, tan, or white, may take several months to heal, and might scar. In all second-degree burns the epidermis regenerates from epithelial tissue in hair follicles and sweat glands and from the edges of the wound.

Full-thickness burns are also termed **third-degree burns.** The epidermis and the dermis are completely destroyed, and deeper tissue may also be involved. Third-degree burns are often surrounded by first- and second-degree burns. Although the areas that have first- and second-degree burns are painful, the region of third-degree burn is usually painless because of destruction of sensory receptors. Third-degree burns appear white, tan, brown, black, or deep cherry red in color. Skin can regenerate in a third-degree burn only from the edges, and skin grafts are often necessary.

With destruction of the skin by a burn, fluid loss from the body increases greatly, possibly leading to shock and death. For proper fluid replacement it is necessary to anticipate the amount of fluid that will be lost. It can be calculated based on the body weight of the patient and the extent of the burn. The surface area that is burned can be estimated conveniently by the "rule of nines" in

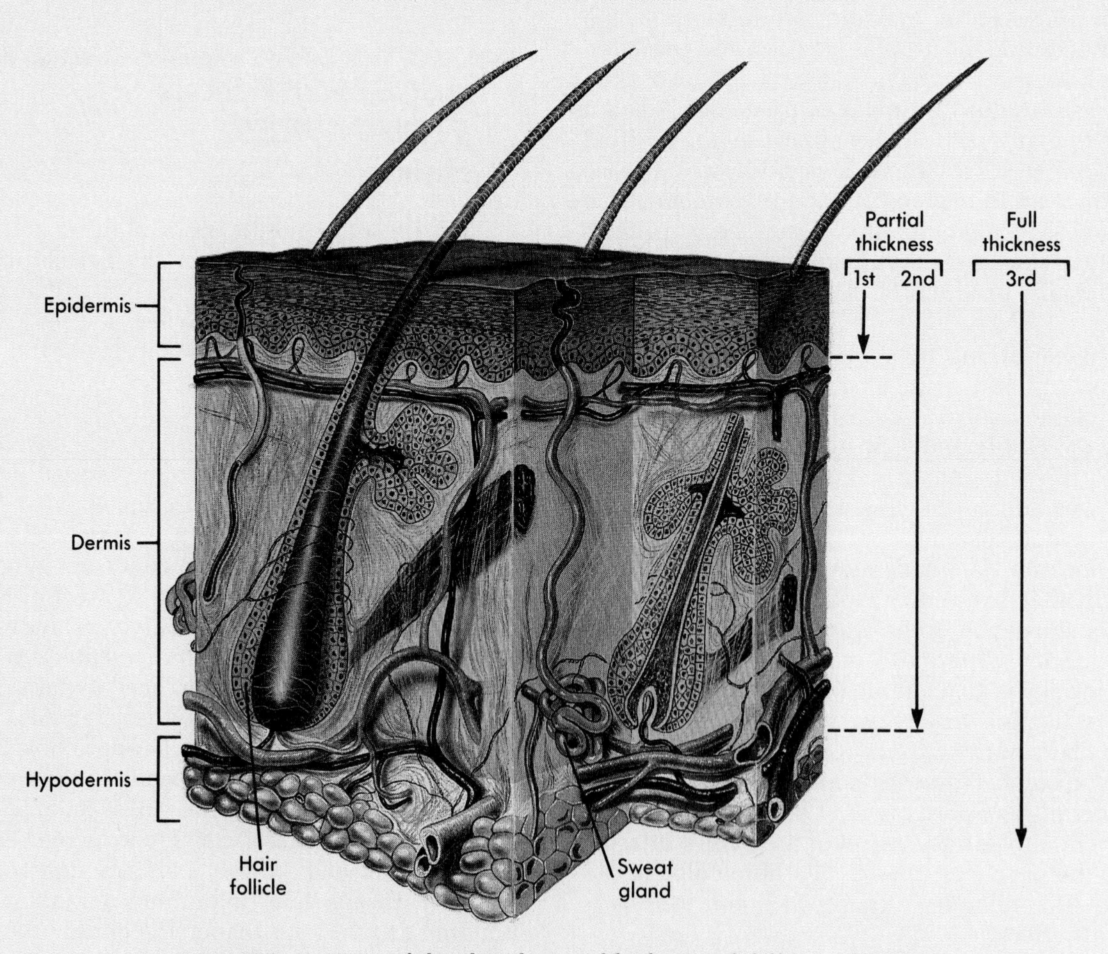

FIGURE 5-A Parts of the skin damaged by burns of different degrees.

which the body is divided into areas that are approximately 9% or multiples of 9% of the total body surface (Figure 5-B). This technique is adequate for adults, but surface area relationships are different in younger patients. For example, in an infant the head and neck comprise 21% of total surface area, whereas in an adult they comprise 9%. For burn victims less than 15 years of age, tables specifically developed for this age group should be consulted.

Deep partial-thickness and full-thickness burns are not only sites of fluid loss but also present an opportunity for the entry of microorganisms and infection. For this reason burn patients are maintained under aseptic conditions and are given antibiotics. Burns of these types also take a long time to heal and form scar tissue with disfiguring and debilitating wound contracture. To prevent these complications and to speed healing, skin grafts are performed. In a split skin graft the epidermis and part of the dermis are removed from another part of the body and are placed over the burn. Interstitial fluid from the burn nourishes the graft until it becomes vascularized. Meanwhile, the donor tissue produces new epidermis from epithelial tissue in the hair follicles and sweat glands such as occurs in patients with superficial second-degree burns. Other types of grafts are possible, and in cases in which a suitable donor site is not available, artificial skin or grafts from human cadavers or from pigs are used. These techniques are often unsatisfactory because the body's immune system recognizes the graft as a foreign substance and rejects it. A solution to this problem is laboratory-grown skin. A piece of healthy skin from the burn victim is removed and placed in a flask with nutrients and hormones that stimulate rapid growth. The skin that is produced consists only of epidermis and does not contain glands or hair.

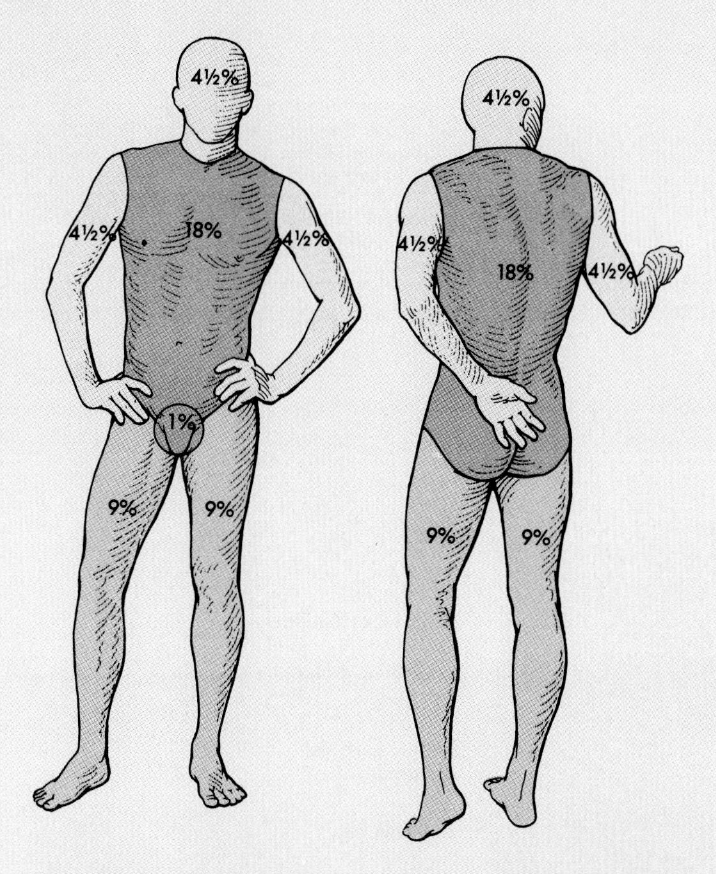

FIGURE 5-B Estimating surface areas with the rule of nines.

Hair structure

A hair is divided into the **shaft, root,** and **hair bulb** (Figure 5-4, *A*). The shaft protrudes above the surface of the skin, and the root and hair bulb are located below the surface. The root and the shaft of the hair are composed of columns of dead keratinized epithelial cells arranged in three concentric layers: the **medulla** (mĕ-dul'ah), the **cortex,** and the **cuticle** (ku'tĭ-kl). The medulla is the central axis of the hair and consists of two or three layers of cells containing soft keratin. The cortex forms the bulk of the hair and consists of cells containing hard keratin. The cuticle is a single layer of cells that forms the hair suface. The cuticle cells contain hard keratin, and the edges of the cuticle cells overlap like shingles on a roof.

Hard keratin contains more sulfur than does soft keratin. When hair burns, the sulfur combines with hydrogen to form hydrogen sulfide, which produces the unpleasant odor of rotten eggs. In some animals such as sheep the cuticle edges of the hair are raised and during textile manufacture catch each other and hold together to form threads.

As with skin color, varying amounts of melanin are responsible for different shades of color. An exception is red hair, which is due to a modified type of melanin containing iron. Hair color is controlled by genes, and dark hair color is not necessarily dominant over light. With age the amount of melanin in hair may decrease, causing the color of the hair to fade or become white (i.e., no melanin). Gray hair is usually a mixture of unfaded, faded, and white hairs.

The **hair follicle** consists of a **dermal root sheath** and an **epithelial root sheath** (Figure 5-4, *B*). The

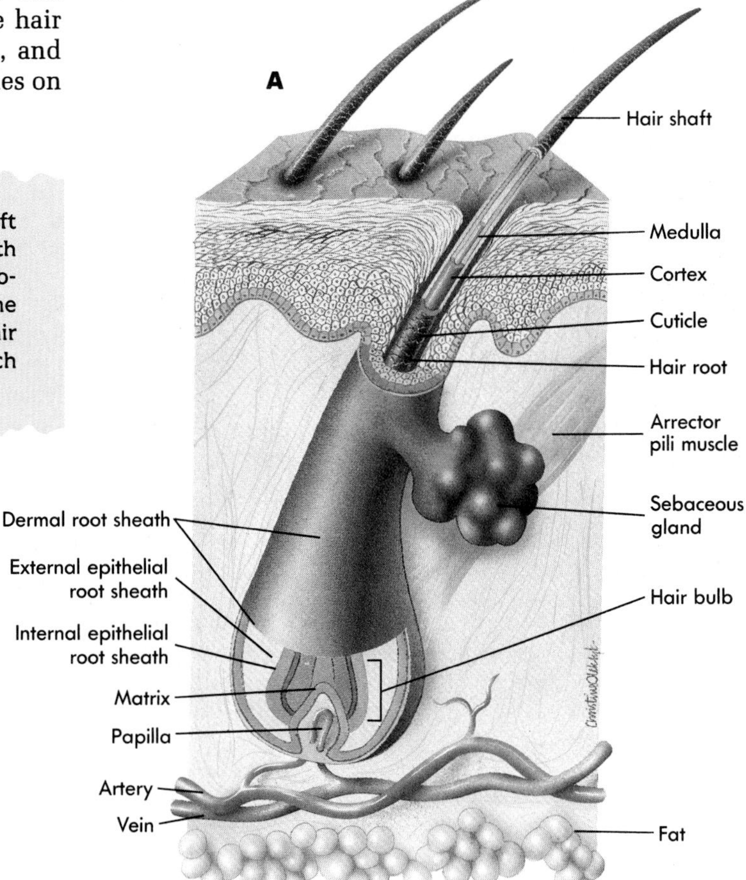

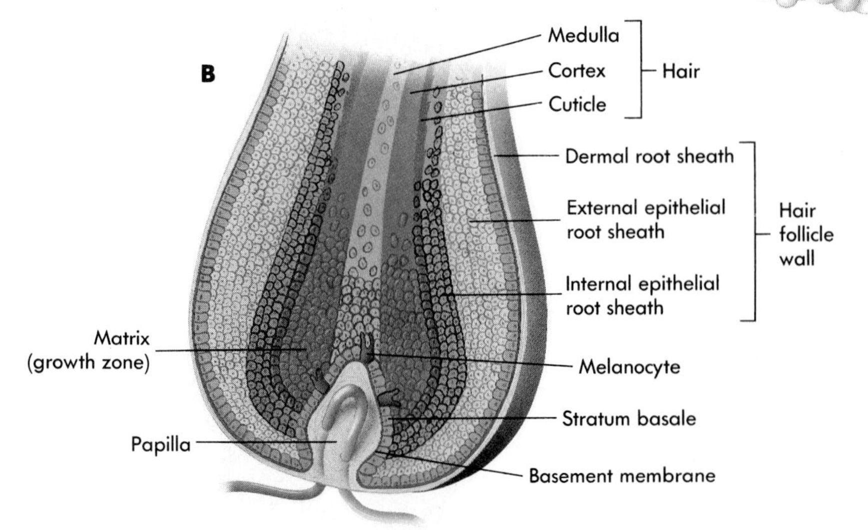

FIGURE 5-4 **A** Hair follicle. **B** Enlargement of hair follicle wall and hair bulb.

dermal root sheath is the portion of the dermis that surrounds the epithelial root sheath. The epithelial root sheath is divided into an external and an internal portion. At the opening of the follicle the external epithelial root sheath has all the strata found in thin skin, but deeper in the hair follicle the number of cells decreases until at the hair bulb only the stratum germinativum is present, having important consequences for the repair of the skin. If the epidermis and the superficial part of the dermis are damaged, the undamaged part of the hair follicle that lies deep in the dermis serves as a source of new epithelium. The internal epithelial root sheath has raised edges that mesh closely with the raised edges of the hair cuticle and hold the hair in place. When a hair is pulled out, the internal epithelial root sheath usually comes out as well and is plainly visible as whitish tissue around the root of the hair.

The hair bulb is an expanded knob at the base of the hair root (see Figure 5-4). Inside the hair bulb is a mass of undifferentiated epithelial cells, the **matrix,** which produces the hair and the internal epithelial root sheath. The dermis of the skin projects into the hair bulb as a papilla and contains blood vessels that provide nourishment to the cells of the matrix.

Hair growth

Hair is produced in cycles that involve a **growth stage** and a **resting stage.** During the growth stage hair is formed by cells of the matrix that differentiate, become keratinized, and die. The hair grows longer as cells are added at the base of the hair root. Eventually growth of the hair stops; the hair follicle shortens and holds the hair in place. After the resting period a new cycle begins, and a new hair replaces the old hair, which falls out of the hair follicle. Thus loss of hair normally means that the hair is being replaced. The length of each stage depends on the hair—eyelashes grow for approximately 30 days and rest for 105 days, whereas scalp hairs grow for a period of 3 years and rest for 1 to 2 years. At any given time an estimated 90% of the scalp hairs are in the growing stage, and there is a normal loss of approximately 100 scalp hairs per day.

The most common kind of permanent hair loss is "pattern baldness." It occurs when the hair follicle reverts to producing vellus hair, which is very short, transparent, and for practical purposes invisible. Although more common and more pronounced in certain men, baldness may also occur in women. Genetic factors and the hormone testosterone are involved in causing baldness.

The average rate of hair growth is approximately 0.3 mm per day, although hairs grow at different rates even in approximately the same location. Cutting, shaving, or plucking hair does not alter the growth rate or the character of the hair, but hair may feel coarse and bristly shortly after shaving because the short hairs are less flexible. Maximum hair length is determined by the rate of hair growth and the length of the growing phase. For example, scalp hair may become very long, but eyelashes are short.

3

Marie Antoinette's hair supposedly turned white overnight after she heard she would be sent to the guillotine. Explain why you believe or disbelieve this story.

? ? ? ? ? ? ? ? ? ? ?

Muscles

Associated with each hair follicle are smooth muscle cells, the **arrector pili** (ah-rek'tor pĭ'le), which extend from the dermal root sheath of the hair follicle to the papillary layer of the dermis (see Figure 5-4). Normally the hair follicle and the hair inside it are at an oblique angle to the surface of the skin. However, when the arrector pili contract, they pull the follicle into a position more perpendicular to the surface of the skin, causing the hair to "stand on end." Movement of the hair follicle produces a raised area called "goose flesh" or "goose bumps."

Contraction of the arrector pili occurs in response to cold or to frightening situations, and in animals with fur the response increases the thickness of the fur. When the response results from cold temperatures, it is beneficial because the fur traps more air and thus becomes a better insulator. In a frightening situation the animal appears larger and more ferocious, which might deter an attacker. It is unlikely that humans with their sparse amount of hair derive any important benefit from either response.

Glands

The major glands of the skin are the **sebaceous** (se-ba'shus) **glands** and the **sweat glands** (Figure 5-5). Sebaceous glands located in the dermis are simple or compound alveolar glands that produce **sebum** (se'bum), an oily, white substance rich in lipids. Because sebum is released by the lysis and death of the secretory cells, sebaceous glands are classified as holocrine glands (see Chapter 4). Most sebaceous glands are connected by a duct to the upper part of the hair follicles from which the sebum oils the hair and the skin surface, prevents drying, and provides protection against some bacteria. A few sebaceous glands located in the lips and the eyelids (meibomian glands) and on the genitalia are not associated with hairs but open directly onto the skin surface.

Sweat glands traditionally are classified as merocrine or apocrine, according to their mode of secre-

FIGURE 5-5 Glands of the skin.

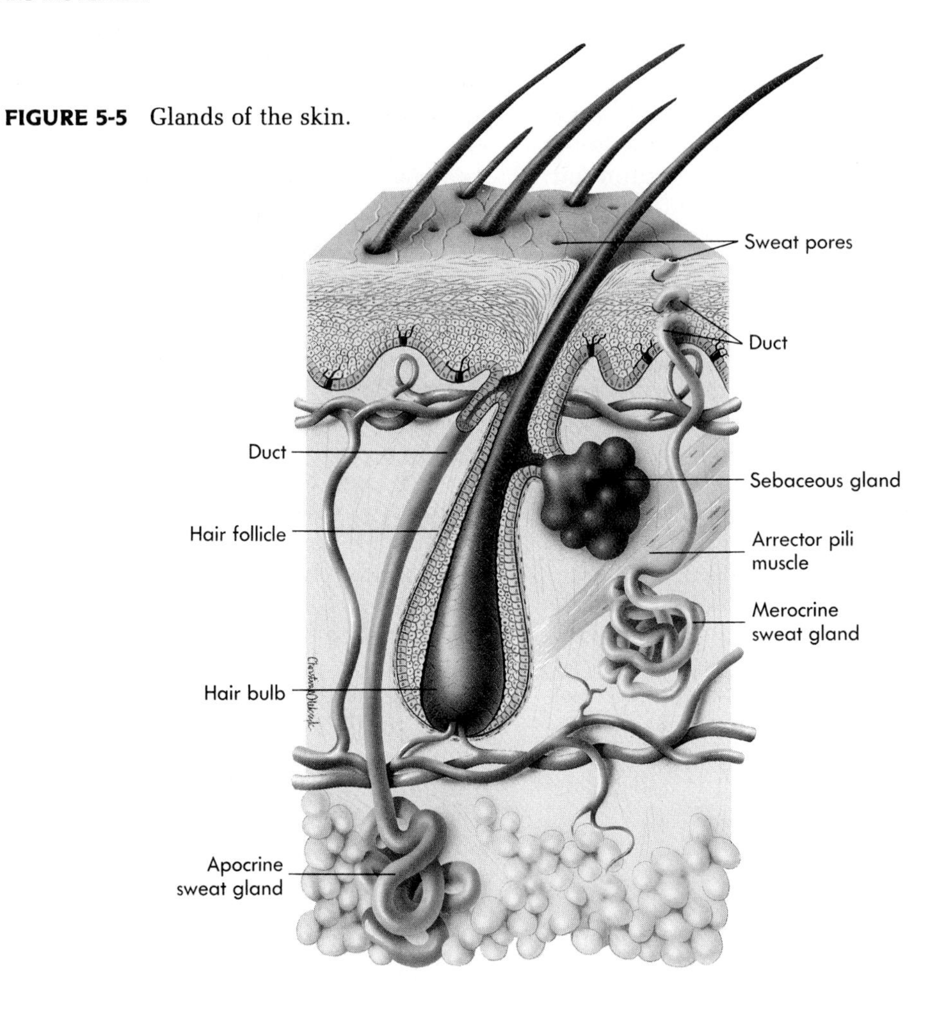

tion (see Chapter 4). Merocrine sweat glands, the most common type, are simple coiled tubular glands that open directly onto the surface of the skin through sweat pores (see Figure 5-5). Merocrine sweat glands can be divided into two parts: the deep coiled portion, which is located mostly in the dermis, and the duct, which passes to the surface of the skin. The coiled portion of the gland produces an isotonic fluid that is mostly water but also contains some salts (mainly sodium chloride) and small amounts of ammonia, urea, uric acid, and lactic acid. As this fluid moves through the duct, sodium chloride is removed by active transport, conserving salts and resulting in a hypotonic fluid called **sweat.** When the body temperature starts to rise above normal levels, the sweat glands produce sweat, which evaporates and cools the body. Sweat also may be released in the palms, soles, and axillae as a result of emotional stress.

Merocrine sweat glands are most numerous in the palms of the hands and the soles of the feet but are absent from the margin of the lips, the labia minora, and the tips of the penis and clitoris. Only a few mammals such as humans and horses have merocrine sweat glands in the hairy skin. Dogs, on the other hand, keep cool by water lost through panting instead of sweating.

Emotional sweating is used in lie detector (polygraph) tests because sweat gland activity may increase when a person tells a lie. The sweat produced, even in small amounts, can be detected because the salt solution conducts electricity and lowers the electrical resistance of the skin.

Apocrine sweat glands are compound coiled tubular glands that usually open into hair follicles superficial to the opening of the sebaceous glands (see Figure 5-5) but occasionally open directly onto the skin surface. They are larger than merocrine sweat glands and extend into the hypodermis. These glands are found in the axillae and genitalia (scrotum and labia majora) and around the anus. They become active at puberty as a result of the influence of sex hormones. Their secretion is an organic substance that is essentially odorless when first released but is quickly metabolized by bacteria to cause what commonly is known as body odor. (It is now known that apocrine glands are a type of merocrine gland; however, the apocrine glands retain their name.)

Other skin glands include the ceruminous glands of the external auditory meatus, which produce cerumen (earwax), and the mammary glands.

Nails

The distal joints of primate digits have nails, whereas reptiles, birds, and most mammals have claws or hooves. The nails protect the ends of the digits, aid in manipulation and grasping of small objects, and are used for scratching.

The **nail** consists of the proximal **nail root** and the distal **nail body** (Figure 5-6, *A*). The nail root is covered by skin, and the nail body is the visible portion of the nail. The lateral and proximal edges of the nail are covered by skin called the **nail fold,** and the edges are held in place by the **nail groove** (Figure 5-6, *B*). The stratum corneum of the nail fold grows onto the nail body as the **eponychium** (ep-on-nik′e-um), or cuticle. Beneath the free edge of the nail body is the **hyponychium** (hi-po-nik′e-um), a thickened region of the stratum corneum (Figure 5-6, *C*).

The nail root and the nail body attach to the **nail bed,** the proximal portion of which is the **nail matrix.** Only the stratum germinativum is present in the nail bed and nail matrix. The nail matrix is thicker than the nail bed and produces most of the nail, although the nail bed does contribute. The nail bed is visible through the clear nail and appears pink because of blood vessels in the dermis. A small part of the nail matrix, the **lunula** (lu′nu-lah), is seen through the nail body as a whitish, crescent-shaped area at the base of the nail. The lunula, seen best on the thumb, appears white because the blood vessels cannot be seen through the thicker nail matrix.

The nail is stratum corneum that contains hard keratin. The nail cells are produced in the nail matrix and are pushed distally over the nail bed. Nails grow at an average rate of 0.5 to 1.2 mm per day, and fingernails grow more rapidly than toenails. Nails, like hair, grow from the base. Unlike hair, they grow continuously throughout life and do not have a resting phase.

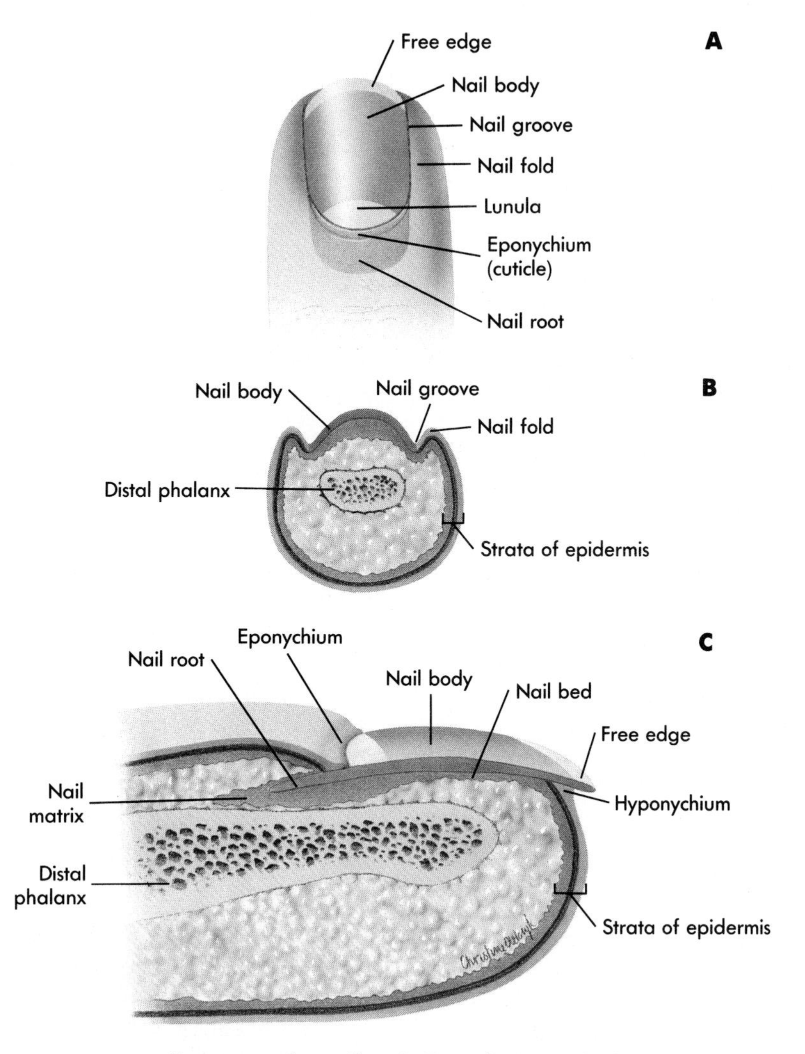

FIGURE 5-6 Structure of a nail. **A** Dorsal view. **B** Cross section. **C** Longitudinal section.

The Integumentary System as a Diagnostic Aid

The integumentary system is useful in diagnosis because it is easily observed and often reflects events occurring in other parts of the body. For example, cyanosis, a bluish color caused by decreased blood oxygen content, is an indication of impaired circulatory function or respiratory function. When red blood cells wear out, they are broken down, and part of their contents are excreted by the liver as bile pigments into the intestine. **Jaundice** (jawn'dis), a yellowish skin color, occurs when there are excess bile pigments in the blood. If the liver is damaged by a disease such as viral hepatitis, bile pigments are not excreted and accumulate in the blood.

Rashes and lesions in the skin can be symptomatic of problems elsewhere in the body. For example, scarlet fever results from a bacterial infection in the throat. The bacteria releases a toxin into the blood that causes the pink-red rash for which this disease was named. In allergic reactions (see Chapter 22) a release of histamine into the tissues produces swelling and reddening. The development of a rash (hives) in the skin can indicate an allergy to ingested foods or drugs such as penicillin.

The condition of the skin, hair, and nails is affected by nutritional status. In an individual with vitamin A deficiency the skin produces excess keratin and assumes a characteristic sandpaper texture, whereas in an individual with iron-deficiency anemia the nails lose their normal contour and become flat or concave (spoon shaped).

The hair concentrates many substances that can be detected by laboratory analysis, and comparison of a patient's hair to a "normal" hair can be useful in diagnosis. For example, lead poisoning results in high levels of lead in the hair. However, the use of hair analysis as a screening test to determine the health or nutritional status of an individual remains unreliable.

Bacterial Infections

Staphylococcus aureus is commonly found in pimples, boils, and carbuncles and causes impetigo, a disease of the skin that usually affects children and is characterized by small pus-containing blisters that easily rupture and form a thick yellowish crust. *Streptococcus pyogenes* causes erysipelas, swollen red patches in the skin. Burns often are infected by *Pseudomonas aeruginosa,* producing a characteristic blue-green pus caused by the bacterial pigment.

Acne is a disorder of the hair follicles and sebaceous glands that affects almost everyone at some time or another. Although the exact cause of acne is unknown, it is believed that four factors are involved: hormones, sebum, abnormal keratinization within the hair follicle, and the bacterium *Propionibacterium acnes*. The lesions apparently begin with hyperproliferation of the hair follicle epidermis, and many cells are desquamated. These cells are abnormally sticky and adhere to each other to form a mass of cells mixed with sebum that blocks the hair follicle. During puberty, hormones, especially testosterone, stimulate the sebaceous glands to increase sebum production. Because both the adrenal gland and the testes produce testosterone, the effect is seen in males and females. An accumulation of sebum behind the blockage produces a whitehead, which may continue to develop into a blackhead and/or a pimple. A blackhead results if the opening of the hair follicle is pushed open by the accumulating horny cells and sebum. Although there is general agreement that dirt is not responsible for the black color of blackheads, the exact cause of the black color is disputed. A pimple develops if the wall of the hair follicle ruptures. Once the wall of the follicle ruptures, *P. acnes* and other microorganisms stimulate an inflammatory response that results in the formation of a red pimple filled with pus. If tissue damage is extensive, scarring occurs.

Viral Infections

Some of the well-known viral infections of the skin include chickenpox, measles, German measles, and cold sores (herpes simplex). Warts, which are caused by a viral infection of the epidermis, are generally harmless and usually disappear without treatment.

Fungal Infections

Ringworm is a fungal infection that affects the keratinized portion of the skin, hair, and nails and produces patchy scaling and an inflammatory response. The lesions are often circular with a raised edge and in ancient times were thought caused by worms. Several species of fungus cause ringworm in humans, and the condition usually is described by its location on the body; in the scalp the condition is ringworm, in the groin it is jock itch, and in the feet it is athlete's foot.

Decubitus Ulcers

Decubitus (de-ku'bĭ-tus) **ulcers,** also known as bedsores or pressure sores, develop in patients who are immobile (e.g., bedridden or confined to a wheelchair). The weight of the body, especially in areas over bony projections such as the hip bones and heels, compresses tissue and causes **ischemia** (is-ke'me-ah), or reduced circulation. The consequence is destruction, or **necrosis** (nĕ-kro'sis), of the hypodermis and deeper tissues that is followed by death of the skin. Once the skin dies, microorganisms gain entry to produce an infected ulcer.

Bullae

Bullae (bul'e) are fluid-filled areas in the skin that develop when tissues are damaged, and the resultant inflammatory response produces edema. Infections or physical injuries can cause bullae or lesions in different layers of the skin.

Psoriasis

The cause of **psoriasis** (so-ri'ă-sis) is unknown, although there may be a genetic component. An increase in mitotic activity in the stratum basale, abnormal keratinization, and elongation of the dermal papillae toward the skin surface result in a thicker-than-normal stratum corneum that desquamates to produce large, silvery scales. If the scales are scraped away, bleeding occurs from the blood vessels at the top of the dermal papillae. Psoriasis is a chronic disease that can be controlled but as yet has no cure.

Eczema and Dermatitis

Eczema (ek'zĕ-mah) and **dermatitis** (der'mă-ti'tis) describe an inflammatory condition of the skin. Cause of the inflammation may be allergy, infection, poor circulation, or exposure to physical factors such as chemicals, heat, cold, or sunlight.

Birthmarks

Birthmarks are congenital (present at birth) disorders of the capillaries in the dermis of the skin. Usually they are only of concern for cosmetic reasons. A **strawberry birthmark** is a mass of soft, elevated tissue that appears bright red to deep purple in color. In 70% of the patients strawberry birthmarks disappear spontaneously by 7 years of age.

Port-wine stains appear as flat, dull, red or bluish patches that persist throughout life.

Moles

A **mole** is an elevation of the skin that is variable in size and often is pigmented and hairy. Histologically, a mole is an aggregation, or "nest," of melanocytes in the epidermis or dermis. They are a normal occurrence, and most people have 10 to 20 moles, which appear in childhood and enlarge until puberty.

Cancer

Skin cancer is the most common type of cancer. Although chemicals and radiation (x-rays) are known to induce cancer, the development of skin cancer is associated most often with exposure to ultraviolet radiation from the sun; consequently, most skin cancers develop on the face or neck. People most likely to have skin cancer are those who are fair skinned (i.e., have less protection from the sun) or those over the age of 50 (i.e., have had long exposure to the sun).

Basal cell carcinoma, the most frequent skin cancer, begins in the stratum basale and extends into the dermis to produce an open ulcer. Surgical removal or radiation therapy cures this type of cancer, and fortunately, there is little danger that the cancer will spread, or **metastasize** (me-tas'tă-siz), to other areas of the body if treated in time. **Squamous cell carcinoma** develops from stratum spinosum keratinocytes that continue to divide as they produce keratin. Typically the result is a nodular, keratinized tumor confined to the epidermis, but it can invade the dermis, metastasize, and cause death. **Malignant melanoma** is a rare form of skin cancer that arises from melanocytes, usually in a preexisting mole. The melanoma may appear as a large, flat, spreading lesion or as a deeply pigmented nodule. Metastasis is common, and unless diagnosed and treated early in development, this cancer is often fatal. Other types of skin cancer are possible (e.g., metastasis from other parts of the body to the skin).

A

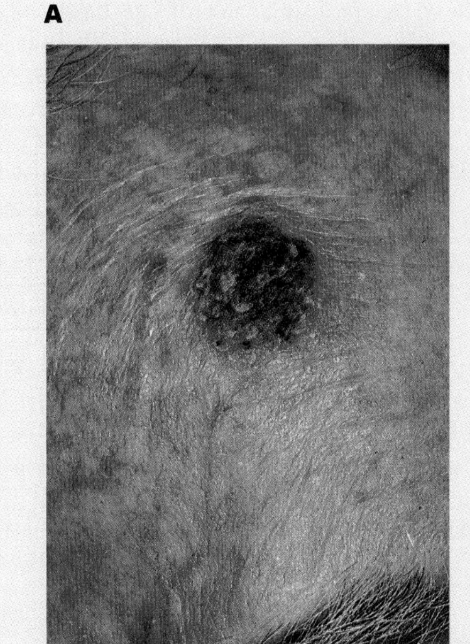

B

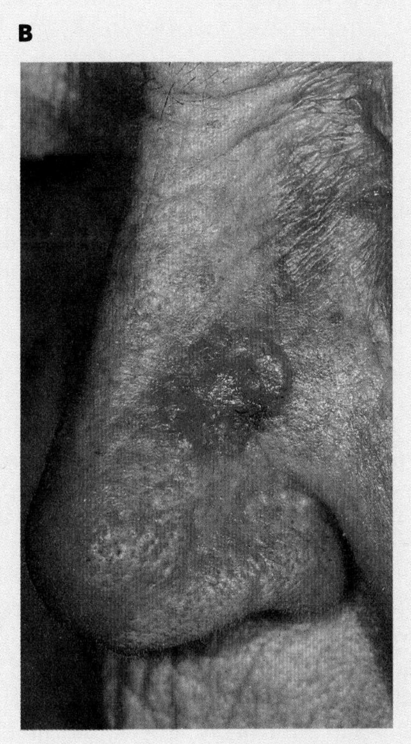

C

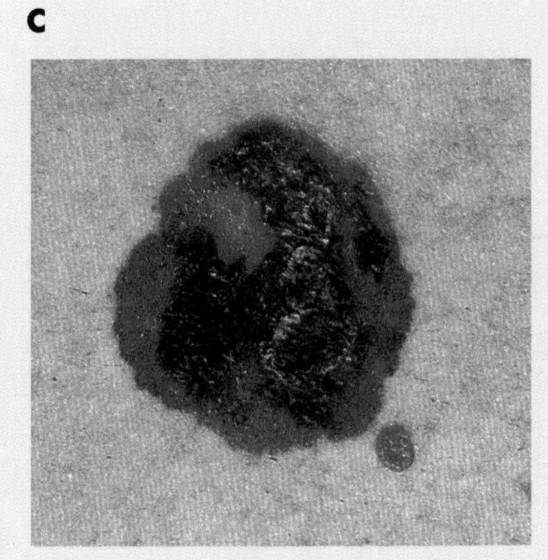

FIGURE 5-C Cancer of the skin.
A Basal cell carcinoma.
B Squamous cell carcinoma.
C Malignant melanoma.

FUNCTIONS OF THE INTEGUMENTARY SYSTEM

Protection

The integumentary system performs many protective functions. The intact skin forms a physical barrier that prevents the entry of microorganisms and other foreign substances into the body. Secretions from skin glands are slightly acidic and produce an environment unsuitable for some microorganisms; the skin also contains components of the immune system that act against microorganisms (see Chapter 22). The skin is a permeability barrier that determines what substances can diffuse into or out of the body surface, and it is especially important in preventing water loss.

Some lipid-soluble substances readily pass through the epidermis; thus lipid-soluble medications can be administered by applying them to the skin. For example, a medicine used to prevent sea sickness is impregnated into a patch, and the patch is placed behind the ear. The medication slowly diffuses through the skin into the blood.

The integumentary system provides protection against abrasion. As the outer cells of the stratum corneum are desquamated, they are replaced by cells from the stratum basale. Calluses develop in areas subject to heavy friction or pressure. Hair on the head and in the pubic and axillary regions also protects against abrasion. Hair and melanin in the skin protect against ultraviolet radiation, which can damage the DNA of cells. When exposure to ultraviolet light increases, the amount of melanin in the skin increases, providing additional protection. The eyebrows prevent sweat from entering the eyes; the eyelashes protect the eyes from foreign objects; hair in the nose and ears prevents the entry of dust and other foreign materials; and nails protect the ends of the digits from damage and can be used in defense.

Temperature Regulation

Body temperature tends to increase as a result of exercise, fever, or an increase in environmental temperature. Homeostasis is maintained by the loss of excess heat. The blood vessels (arterioles) in the dermis dilate and allow more blood to flow through the skin, thus transferring heat from deeper tissues to the skin. To counteract environmental heat gain or to get rid of excess heat produced by the body, sweat is produced. The sweat spreads over the surface of the skin and evaporates, carrying away heat in the process.

If body temperature begins to drop below normal, heat can be conserved by a decrease in the diameter of dermal blood vessels, thus reducing blood flow to the skin. However, with less warm blood flowing through the skin, the skin temperature decreases. If the skin temperature drops below approximately 15° C, blood vessels dilate as a protective mechanism to prevent tissue damage from the cold.

Contraction of the arrector pili causes hair to stand on end, but with the sparse amount of hair in humans this does not significantly reduce heat loss. Hair on the head, however, is an effective insulator. General temperature regulation is considered in Chapter 25.

4

You may have noticed that on very cold winter days people's noses and ears turn red. Can you explain why this happens?

? ? ? ? ? ? ? ? ? ? ?

Vitamin D Production

Vitamin D is synthesized in skin exposed to ultraviolet light, and humans can produce all the vitamin D they require by this process if enough ultraviolet light is available. However, because humans live indoors and wear clothing, their exposure to ultraviolet light may not be adequate for the manufacture of sufficient vitamin D. For example, people living in cold climates may have inadequate exposure to ultraviolet light because they remain indoors or are covered by warm clothing when outdoors. Fortunately, vitamin D can also be ingested and absorbed in the intestine. Natural sources of vitamin D are liver (especially fish liver), egg yolks, and dairy products (e.g., butter, cheese, and milk). In addition, the diet can be supplemented with vitamin D in fortified milk or vitamin pills.

Vitamin D synthesis begins when the precursor molecule, 7-dehydrocholesterol, is exposed to ultraviolet light and is converted into cholecalciferol. The cholecalciferol is released into the blood and is modified by hydroxylation (hydroxide ion is added) in the liver and kidneys to form active vitamin D (calciferol). Vitamin D functions as a hormone to stimulate uptake of calcium and phosphate from the intestines, to promote their release from bones, and to reduce their loss from the kidneys, resulting in increased blood calcium and phosphate levels. Adequate levels of these minerals are necessary for normal bone metabolism (see Chapter 6), and calcium is required for normal nerve and muscle function (see Chapters 9 and 10).

Sensation

The integumentary system is well supplied with sensory receptors, including touch receptors in the epidermis and in the dermal papillae; and pain, heat, cold, and pressure receptors in the dermis and deeper tissues. Hair follicles (but not the hair) are well innervated, and movement of the hair can be detected by sensory receptors surrounding the base of the hair follicle. Sensory receptors are discussed in more detail in Chapter 14.

Excretion

Excretion is the removal of waste products from the body. In addition to water and salts, sweat contains a small amount of waste products such as urea, uric acid, and ammonia. Large amounts of sweat can be lost from the body, especially during vigorous exercise in a hot environment, and lost water and salts must be replaced to restore fluid and electrolyte homeostasis. However, the sweat glands do not function in any significant way in excretion.

EFFECTS OF AGING ON THE INTEGUMENTARY SYSTEM

As the body ages, the blood flow to the skin is reduced, and the skin becomes thinner, is more easily damaged, and repairs more slowly. Elastic fibers in the dermis decrease in number and diameter, and the skin tends to sag. A loss of subcutaneous tissue, especially in the face, also causes sagging, wrinkled skin.

A decrease in the activity of sebaceous and sweat glands results in dry skin and poor thermoregulatory ability. The decrease in ability to sweat can contribute to death from heat prostration in elderly individuals who do not take proper precautions.

The number of functioning melanocytes generally decreases; but in some localized areas, especially on the hands and the face, melanocytes increase in number to produce age spots. (Age spots are different from freckles, which are caused by an increase in melanin production and not an increase in melanocyte numbers.) White or gray hairs also occur because of a decrease or lack of melanin production.

Skin that is exposed to sunlight appears to age more rapidly than nonexposed skin. This effect is observed on areas of the body such as the face and hands that receive sun exposure (Figure 5-7). The effects of chronic sunlight exposure on the skin, however, are different from the effects of normal aging. In skin exposed to sunlight normal elastic fibers are replaced by an interwoven mat of thick elastic-like material, the number of collagen fibers decreases, and the ability of keratinocytes to divide is impaired.

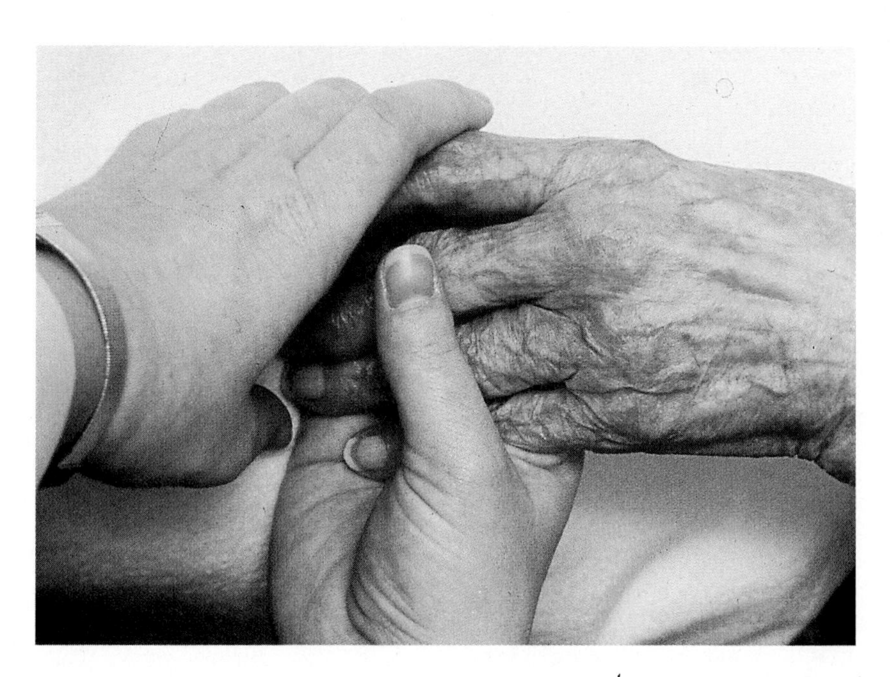

FIGURE 5-7 Skin changes with age. As we age, skin loses elasticity and wrinkles form.

The Integumentary System

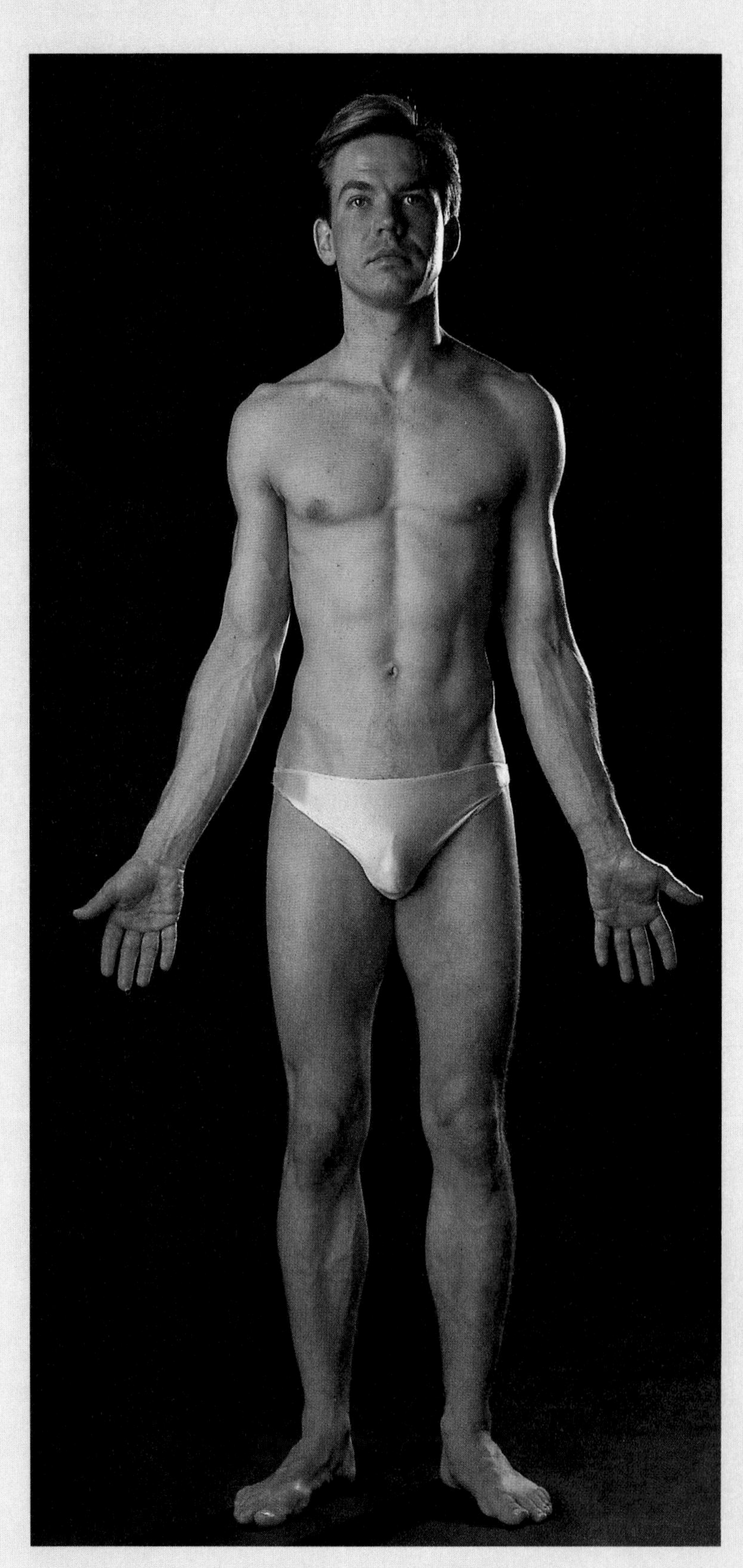

Major Components

Skin, hair, nails, and sweat glands

Major Functions

Protects against abrasions and ultraviolet light
Prevents the entry of microorganisms and harmful substances
Reduces water loss
Acts as a regulator of body temperature
Produces vitamin D precursors
Provides sensory information regarding heat, cold, pressure, and pain
Excretes small amounts of waste products in sweat

SUMMARY

THE INTEGUMENTARY SYSTEM CONSISTS OF THE SKIN, HAIR, NAILS, AND A VARIETY OF GLANDS. THE SKIN PROTECTS, HELPS REGULATE BODY TEMPERATURE, PRODUCES VITAMIN D, AND DETECTS STIMULI.

HYPODERMIS *p. 136*

1. Located beneath the dermis, the hypodermis is loose connective tissue that contains collagen and elastin fibers.
2. The hypodermis attaches the skin to underlying structures and is a site of fat storage.

SKIN *p. 136*

Dermis

1. The dermis, a dense, irregular connective tissue with few fat cells, is divided into two layers.
2. The reticular layer is the main fibrous layer and consists mostly of collagen.
3. The papillary layer is well supplied with capillaries.

Epidermis

1. The epidermis is stratified squamous epithelium divided into five strata.
2. The stratum basale consists of keratinocytes, which produce the cells of the more superficial strata.
3. The stratum spinosum consists of several layers of cells held together by many desmosomes. The stratum basale and the stratum spinosum are sometimes called the stratum germinativum.
4. The stratum granulosum consists of cells filled with granules of keratohyalin. Cell death occurs in this strata.
5. The stratum lucidum consists of a layer of dead transparent cells.
6. The stratum corneum consists of many layers of dead squamous cells. The most superficial cells are desquamated.
7. Keratinization is the transformation of the living cells of the stratum basale into the dead squamous cells of the stratum corneum.
   - Keratinized cells are filled with keratin and have a protein envelope, both of which contribute to structural strength. The cells are also held together by many desmosomes.
   - Intercellular spaces are filled with lipids that contribute to the impermeability of the epidermis to water.
8. Soft keratin is found in skin and the inside of hairs, whereas hard keratin occurs in nails and the outside of hairs. Hard keratin makes cells more durable, and these cells do not desquamate.

Thick and Thin Skin

1. Thick skin has all five epithelial strata. The dermis under thick skin produces fingerprints.
2. Thin skin contains fewer cell layers per strata, and the stratum lucidum is usually absent. Hair is found only in thin skin.

Skin Color

1. Melanocytes produce melanin inside melanosomes and then transfer the melanin to keratinocytes. The size and distribution of melanosomes determine skin color. Melanin production is determined genetically but can be influenced by hormones and ultraviolet light (tanning).
2. Carotene, an ingested plant pigment, can cause the skin to appear yellowish.
3. Increased blood flow produces a red skin color, whereas a decreased blood flow causes a pale skin. Decreased oxygen content in the blood results in a bluish color called cyanosis.

ACCESSORY SKIN STRUCTURES *p. 141*

Hair

1. Lanugo, fetal hair, is replaced near the time of birth by terminal hairs (eyelids, eyebrows, and scalp) and vellus hairs. At puberty vellus hairs can be replaced with terminal hairs.
2. Hair is dead keratinized epithelial cells consisting of a central axis of cells with soft keratin, the medulla, which is surrounded by a cortex of cells with hard keratin. The cortex is covered by the cuticle, a single layer of cells filled with hard keratin.
3. Hair color is determined by the amount and kind of melanin present.
4. A hair has three parts: the shaft, the root, and the hair bulb.
5. The hair bulb produces the hair in cycles involving a growth stage and a resting stage.

Muscles

Contraction of the arrector pili, which are smooth muscles, causes hair to "stand on end" and produces "goose flesh."

Glands

1. Sebaceous glands produce sebum, which oils the hair and the surface of the skin.
2. Merocrine sweat glands produce sweat that cools the body. Apocrine sweat glands produce an organic secretion that can be broken down by bacteria to cause body odor.
3. Other skin glands include ceruminous glands (ear) and the mammary glands.

Nails

1. The nail consists of a nail root and a nail body resting on the nail bed.
2. Part of the nail root, the nail matrix, produces the nail body, which is several layers of cells containing hard keratin.

FUNCTIONS OF THE INTEGUMENTARY SYSTEM *p. 150*

Protection

The skin prevents the entry of microorganisms, acts as a permeability barrier, and provides protection against abrasion and ultraviolet light.

Temperature Regulation

1. Through dilation and constriction of blood vessels, the skin controls heat loss from the body.
2. Sweat glands produce sweat that evaporates and lowers body temperature.

Vitamin D Production

1. Skin exposed to ultraviolet light produces cholecalciferol that is modified in the liver and then in the kidneys to form active vitamin D.
2. Vitamin D increases blood calcium levels by promoting calcium uptake from the intestine, release of calcium from bone, and reduction of calcium loss from the kidneys.

Sensation

The skin contains sensory receptors for pain, touch, hot, cold, and pressure that allow proper response to the environment.

Excretion

Skin glands remove small amounts of waste products (e.g., urea, uric acid, and ammonia) but are not important in excretion.

EFFECTS OF AGING ON THE INTEGUMENTARY SYSTEM *p. 151*

1. As the body ages, blood flow to the skin is reduced, the skin becomes thinner, and elasticity is lost.
2. Sweat and sebaceous glands are less active, and the number of melanocytes decreases.

CONTENT REVIEW

1. Name the components of the integumentary system.
2. List the functions of the integumentary system.
3. Describe the structure and the function of the hypodermis.
4. What type of tissue is the dermis? How is the dermis different from the hypodermis?
5. Name the two layers of the dermis, and contrast them to each other.
6. What are cleavage lines in the skin, and why are they important?
7. What kind of tissue is the epidermis? Name and describe the five strata of the epidermis. In which strata are new cells formed by mitosis? Which strata have live cells, and which have dead cells?
8. Define keratinization. Describe the structural features that result from keratinization and make the epidermis structurally strong and resistant to water loss.
9. Distinguish between soft and hard keratin, and state where each type of keratin is found.
10. Compare thick and thin skin. Is hair found in thick skin or thin skin?
11. What is a callus? A corn?
12. Which cells of the epidermis produce melanin? What happens to the melanin once it is produced? What factors determine the amount of melanin produced in the skin?
13. How do melanin, carotene, and blood affect skin color?
14. When and where are lanugo, vellus, and terminal hairs found in the skin?
15. Define the root, shaft, and hair bulb of a hair. Describe the three parts of the root or shaft seen in cross section.
16. What determines hair color?
17. Describe the parts of a hair follicle. Why is the epithelial root sheath important in the repair of the skin?
18. In what part of a hair does growth take place? What are the stages of hair growth?
19. What determines the length of hair? Why does baldness occur?
20. What happens when the arrector pili of the skin contract?
21. What secretion is produced by the sebaceous glands? What is the function of the secretion?
22. Which glands of the skin are responsible for cooling the body? Which glands are involved with the production of body odor?
23. Name the parts of a nail. Which part produces the nail? What is the lunula?
24. How does the skin provide protection?
25. How does the skin assist in the regulation of body temperature?
26. Where is cholecalciferol produced and then modified into vitamin D? What are the functions of the modified vitamin D?
27. What kind of sensory receptors are found in the skin, and why are they important?
28. What substances are excreted by skin glands? Is the skin an important site of excretion?
29. List the changes that occur in the skin with increasing age.

CONCEPT QUESTIONS

1. A woman has stretch marks on her abdomen, yet she states that she has never been pregnant. Is this possible?
2. The skin of infants is more easily penetrated and injured by abrasion than that of adults. Based on this fact, which stratum of the epidermis is probably much thinner in infants than in adults?
3. Melanocytes are found primarily in the stratum basale of the epidermis. In reference to their function, why does this location make sense?
4. Harry Fastfeet, a Caucasian, jogs on a cold day. What color would you expect his skin to be (1) just before starting to run, (2) during the run, and (3) 3 minutes after the run?
5. Why are your eyelashes not a foot long? Your fingernails?
6. Given what you know about the cause of acne, propose some ways to prevent or treat the disorder.
7. A patient has an ingrown toenail, a condition in which the nail grows into the nail fold. Would cutting the nail away from the nail fold permanently correct this condition? Why or why not?

ANSWERS TO PREDICT QUESTIONS

1 *p. 139*. Because the permeability barrier is composed mainly of lipids surrounding the epidermal cells, substances that are lipid soluble could pass through easily. Water-soluble substances would have difficulty.

2 *p. 141*.

A. The lips are pinker or redder than the palms of the hand. Several explanations for this difference are possible. There could be more blood vessels in the lips; there could be increased blood flow in the lips; or the blood vessels could be easier to see through the epidermis of the lips. The last possibility explains most of the difference in color between the lips and the palms. The epidermis of the lips is thinner and not as heavily keratinized as that of the palms. In addition, the papillae containing the blood vessels in the lips are "high" and closer to the surface.

B. A person who does manual labor has a thicker stratum corneum on the palms (and possibly calluses) than a person who does not perform manual labor. The thicker epidermis masks the underlying blood vessels, and the palms do not appear as pink. Additionally, carotene accumulating in the lipids of the stratum corneum might impart a yellowish cast to the palms.

C. The posterior surface of the forearm appears darker because of the tanning effect of ultraviolet light from the sun.

D. The genitals normally have more melanin and appear darker than the soles of the feet.

3 *p. 145*. The story is not true. Hair color is due to melanin that is added to the hair in the hair matrix as the hair grows. The hair itself is dead. To turn white, the hair must grow out without the addition of melanin.

4 *p. 150*. On cold days skin blood vessels of the ears and nose may dilate, bringing warm blood to the ears and nose and thus preventing tissue damage from the cold. The increased blood flow makes the ears and nose appear red.

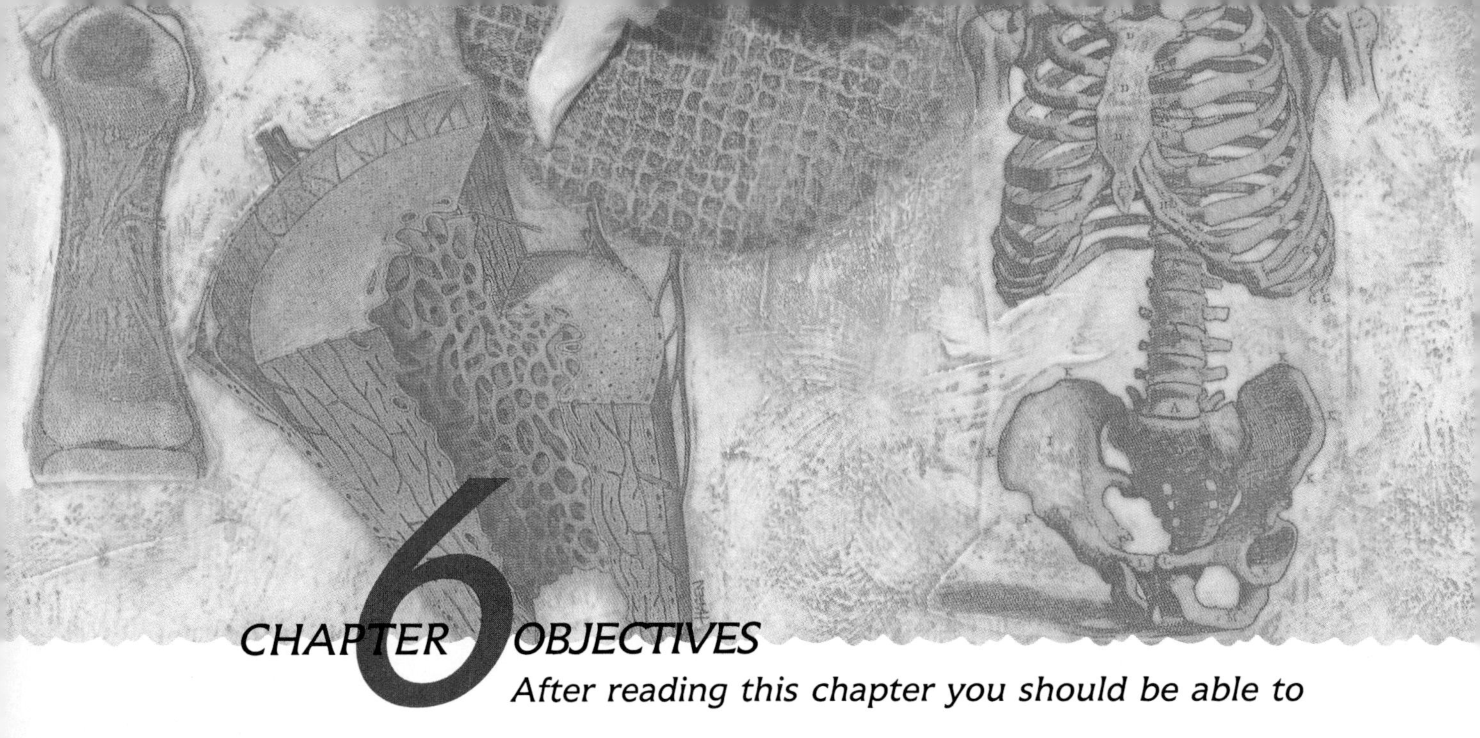

CHAPTER 6 OBJECTIVES

After reading this chapter you should be able to

1. List the major cellular components of tendons, ligaments, cartilage, and bone.
2. Describe the major components of the connective tissue matrix and indicate which features are most characteristic of tendons, ligaments, cartilage, and bone.
3. Describe the microscopic anatomy of tendons and ligaments and explain their functional characteristics.
4. Describe the microscopic anatomy of cartilage, including the perichondrium, and explain how its features influence its functional characteristics.
5. Explain how growth occurs in hyaline cartilage.
6. Name the major bone shapes and describe their anatomy.
7. Describe the composition and organization of bone matrix, list the three types of bone cells, and state the functions of each type of bone cell.
8. Describe the features that characterize cancellous and compact bone and explain how those features influence function in each case.
9. Name the two major types of ossification and describe the features of each.
10. Describe bone growth and explain how it differs from the growth of cartilage, tendons, and ligaments.
11. List the nutritional and hormonal requirements for bone growth.
12. Explain the role of bone in calcium homeostasis.
13. Explain how bone remodeling occurs and describe the conditions in which it occurs.
14. Describe the effects of mechanical strain and the effects of inadequate stress on bone.
15. Describe the process of bone repair, the cells involved, and the types of tissue produced.

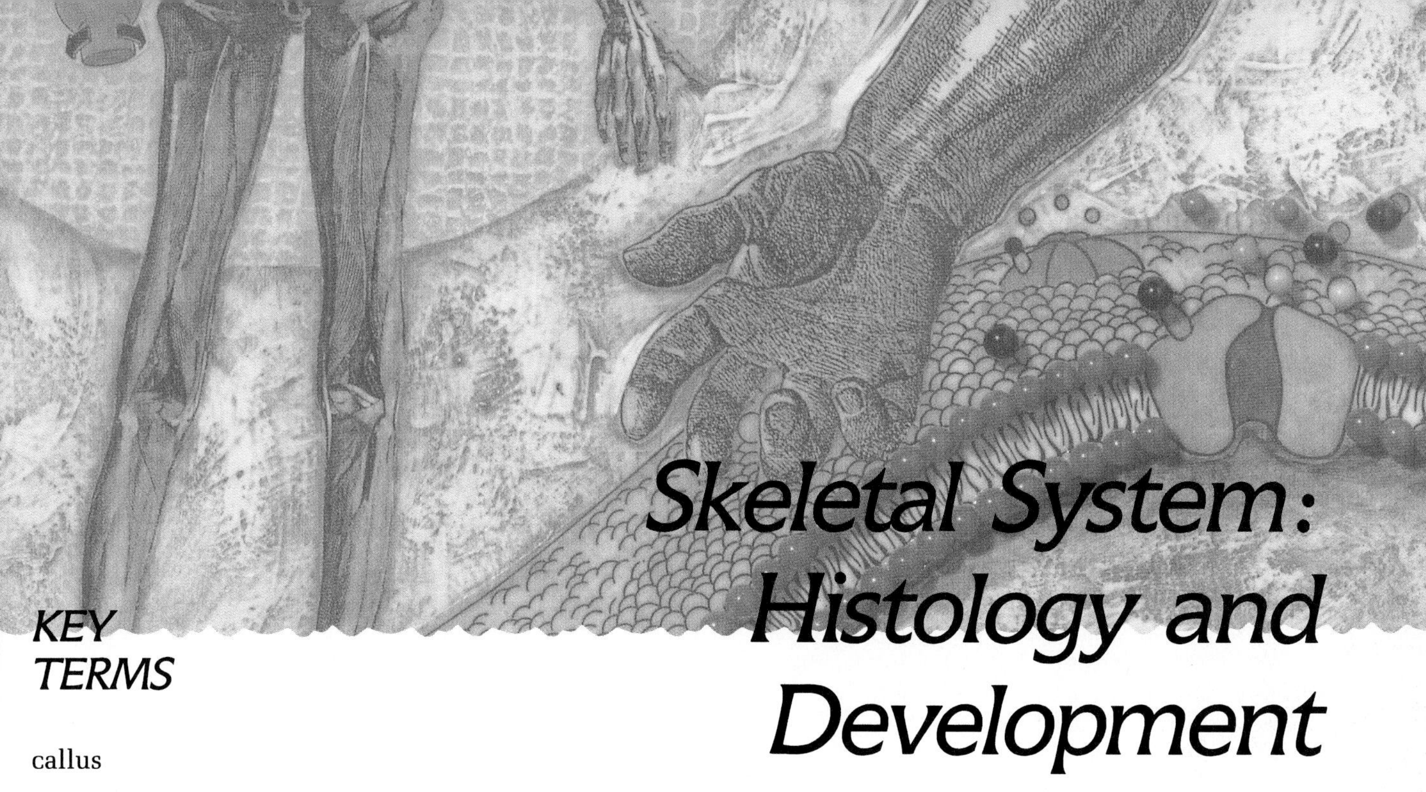

Skeletal System: Histology and Development

KEY TERMS

callus

cancellous bone

collagen (kol'lă-jen)

compact bone

endochondral ossification

epiphyseal (ep-ĭ-fiz'ĭ-al) plate

hydroxyapatite (hi-drok'se-ap'ĕ-tīt)

intramembranous ossification

lacuna (lă-ku'nah)

lamella (lă-mel'ah)

osteoblast (os'te-o-blast)

osteoclast (os'te-o-klast)

osteocyte (os'te-o-sīt)

osteon (os'te-on)

trabecula (tră-bek'u-lah)

RELATED TOPICS

The following term or concept is important for a good understanding of this chapter. If you are not familiar with it, you should review it before proceeding.

Histology of connective tissues (Chapter 4)

THE SKELETAL SYSTEM CONSISTS OF BONES AND THEIR ASSOCIATED CONNECTIVE TISSUES, INCLUDING CARTILAGE, TENDONS, AND LIGAMENTS. Because bone is very rigid, it is well adapted to help maintain the shape of the body. Cartilage, which is somewhat rigid but more flexible than bone, also provides support. Cartilage is abundant in the embryo and the fetus in which it provides a model for most of the adult bones and is a major site of skeletal growth in the embryo, fetus, and child. Tendons and ligaments are strong bands of fibrous connective tissue; tendons attach muscle to bone, and ligaments attach bone to bone.

This chapter considers the origin, structure, and functions of skeletal tissues. Chapter 7 deals with individual bones and Chapter 8 with the articulations or joints between bones.

FUNCTIONS OF THE SKELETAL SYSTEM

The skeletal system provides support and protection, allows body movements, stores minerals and fats, and is the site of blood cell production.

Support. Rigid, strong bone is well suited for bearing weight and is the major supporting element of the body. Cartilage provides a firm, yet flexible support within certain structures such as the nose, external ear, costal cartilages, and trachea.

Protection. Bone is hard and protects the organs it surrounds. For example, the brain is enclosed and protected by the skull, and the spinal cord is surrounded by the vertebrae. The heart, lungs, and other organs of the thorax are protected by the rib cage.

Movement. Skeletal muscles attach to bones by tendons, and contraction of the skeletal muscles causes the bones to move, producing body movements. Joints, which are formed where two or more bones come together, permit and control the movement between bones. Smooth cartilage covers the ends of bones within some joints, allowing the bones to move freely. Ligaments connect bones to each other and prevent excessive movements.

Storage. Excess minerals in the blood are taken into bone and stored. Should blood levels of the minerals decrease, the minerals are released from bone into the blood. The principal minerals stored are calcium and phosphorus. Bone also stores fat (adipose tissue) within cavities of the bone. If needed, the fats are released into the blood and are used by other tissues as a source of energy.

Blood cell production. The cavities of bone can also contain bone marrow that gives rise to blood cells and platelets (see Chapter 19) that leave the cavities and enter blood vessels.

TENDONS AND LIGAMENTS

Tendons attach muscles to bones, and **ligaments** attach bones to bones. Tendons and ligaments are dense, regular connective tissue, consisting almost entirely of thick bundles of densely packed parallel collagen fibers. The orientation of the collagen fibers in one direction makes the tendons and ligaments very strong in that direction. Because collagen is a white protein, most tendons and ligaments appear white; some ligaments, however, also contain elastin, which gives them a slightly yellow appearance.

Although their general structures are similar, the major histological differences between tendons and ligaments include the following: (1) collagen fibrils of ligaments are often less compact; (2) some fibrils of many ligaments are not parallel; and (3) ligaments usually are more flattened than tendons and form sheets or bands of tissue.

The cells of developing tendons and ligaments are spindle-shaped **fibroblasts.** Once a fibroblast becomes completely surrounded by matrix it is a **fibrocyte.**

Tendons and ligaments grow by two different processes. In the first process, called **appositional growth,** surface fibroblasts divide to produce additional fibroblasts, which secrete matrix to the outside of existing fibers. In the second process, **interstitial growth,** fibrocytes proliferate and secrete matrix inside the tissue.

1

Few nerves and blood vessels enter the substance of a tendon or a ligament. Explain why injured tendons take a long time to heal.

? ? ? ? ? ? ? ? ? ? ?

HYALINE CARTILAGE

Several types of **cartilage** are described in Chapter 4, but the discussion in this chapter is confined to **hyaline cartilage** since it is the type most intimately associated with bone function and development. The other types of cartilage are associated mainly with joints and are discussed in Chapter 8.

Collagen and proteoglycans create a supporting framework for the water-filled matrix of hyaline cartilage. Collagen is largely responsible for the extreme strength of cartilage. Proteoglycan aggregates provide sites for water entrapment and account for the water content of cartilage, which is higher than in almost any other connective tissue and is largely responsible for the resilient nature of cartilage (see Chapter 4).

Cells that produce new cartilage matrix on the outside of more mature cartilage are **chondroblasts** (kon'dro-blasts; *chondro* means cartilage). When a chondroblast is surrounded by matrix, it becomes a **chondrocyte,** which is a rounded cell that occupies a space within the matrix called a **lacuna** (lă-ku'nah) (Figure 6-1).

Cartilage is surrounded by a double-layered connective tissue sheath, the **perichondrium** (pĕr-e-kon'dre-um) (see Figure 6-1). The outer layer of the perichondrium is dense, irregular connective tissue containing fibroblasts. The inner, more delicate layer has fewer fibers and contains chondroblasts, which produce new cartilage. Blood vessels and nerves occupy the outer layer of the perichondrium but do not enter the cartilage matrix so that nutrients must diffuse through the matrix to reach the chondrocytes; as a result, cartilage heals very slowly after an injury. The articular (joint) surfaces of cartilage have no perichondrium, vessels, or nerves.

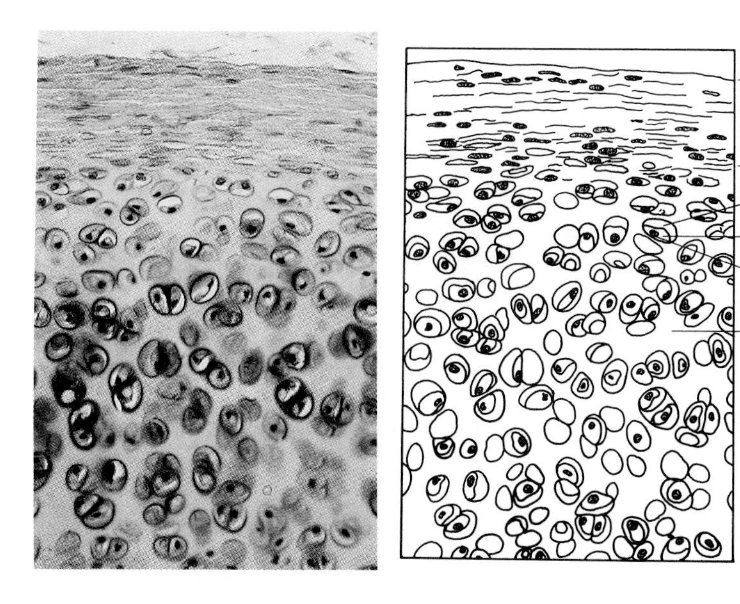

FIGURE 6-1 Photomicrograph of hyaline cartilage. Chondrocytes within lacunae are surrounded by cartilage matrix.

Like tendons and ligaments, cartilage grows by appositional and interstitial growth. Chondroblasts in the perichondrium lay down new matrix and add new chondrocytes to the outside of the tissue (appositional growth). Chondrocytes within the tissue divide and add more matrix from the inside (interstitial growth).

BONE

Bone Shape

Individual bones can be classified according to their shape as long, short, flat, or irregular (Figure 6-2). **Long bones** are longer than they are wide. Most of the bones of the upper and lower limbs are long bones. **Short bones** are approximately as broad as they are long. They are nearly cube shaped or round and are exemplified by the bones of the wrists (carpals) and ankle (tarsals). **Flat bones** have a relatively thin, flattened shape, and they usually are curved. Examples of flat bones are certain skull bones, ribs, the breastbone (sternum), and shoulder blades (scapulae). **Irregular bones** are ones such as the vertebrae and facial bones with shapes that do not fit readily into the other three categories.

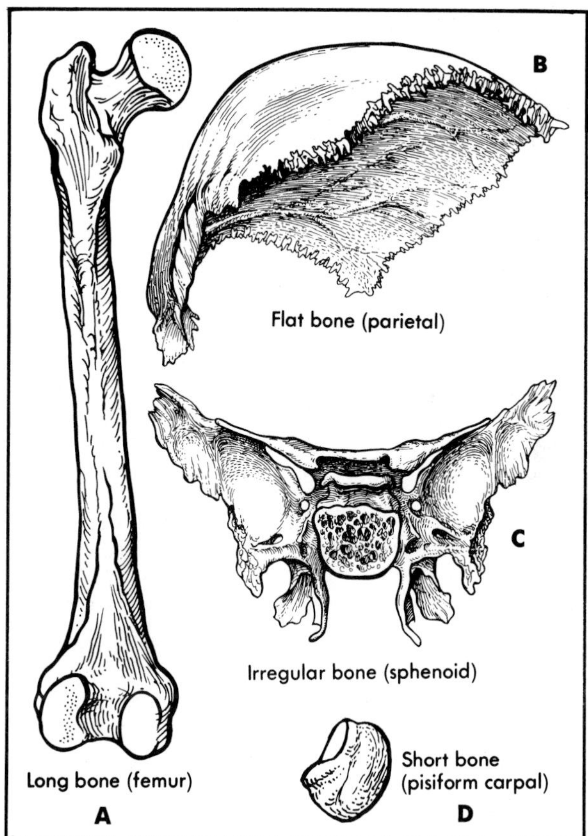

FIGURE 6-2 Various bone shapes. **A** Long bone (femur or thigh bone). **B** Flat bone (parietal bone from roof of skull). **C** Irregular bone (sphenoid bone from skull). **D** Short bone (carpal, or wrist, bone).

Bone Anatomy

Each growing long bone consists of three major components: (1) a shaft, the **diaphysis** (di-af'ĭ-sis); (2) an **epiphysis** (e-pif'ĭ-sis) at each end of the bone; and (3) an **epiphyseal,** or **growth, plate** (Figure 6-3, and Table 6-1). The epiphyseal plate is the site of major bone elongation (Figure 6-3, *A*). When bone growth stops, the epiphyseal plate becomes ossified and is called the **epiphyseal line** (Figure 6-3, *B*). The diaphysis is composed of **compact bone,** which is mostly bone matrix with few spaces. The epiphyses consist of **cancellous,** or **spongy, bone,** which has many small spaces or cavities within the bone matrix. The outer surface of the epiphyses consists of a layer of compact bone. Within joints the epiphyses are covered by **articular cartilage.**

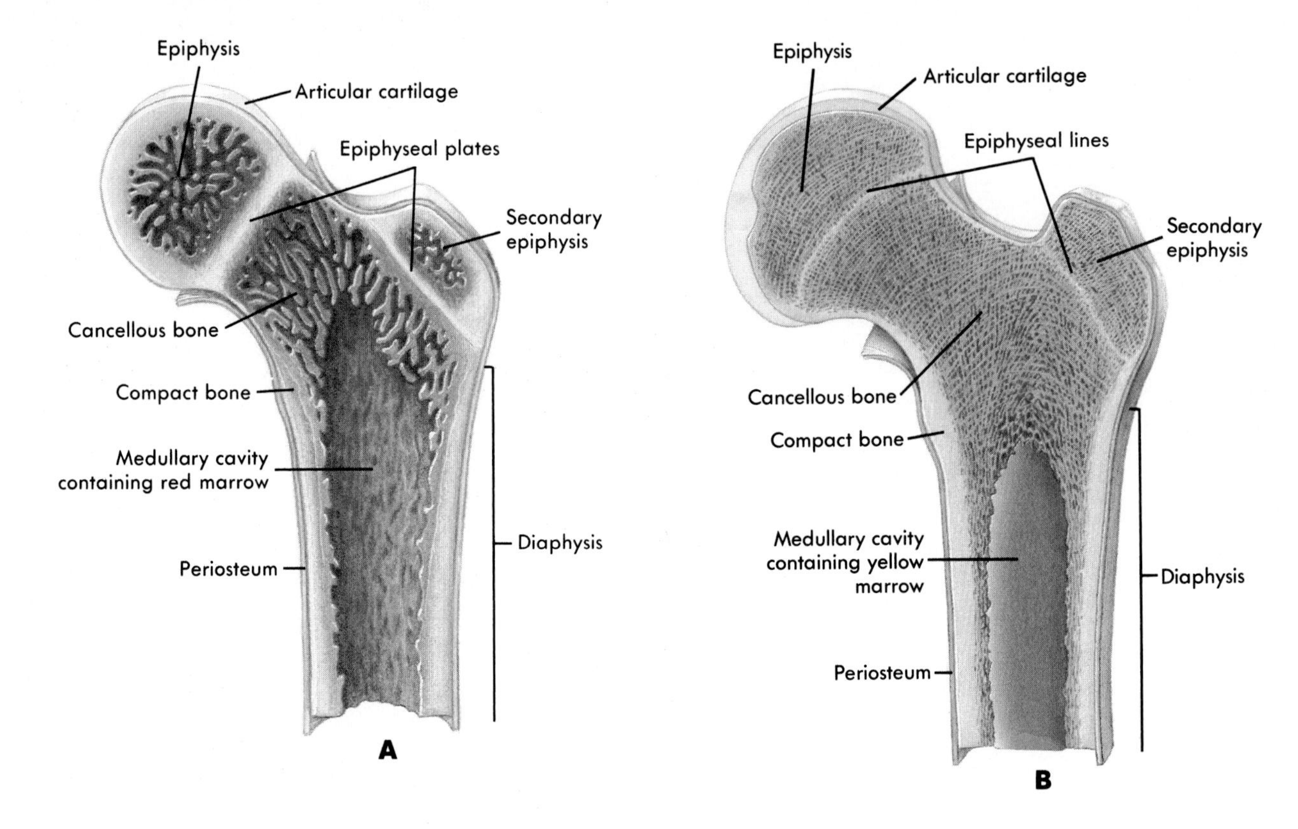

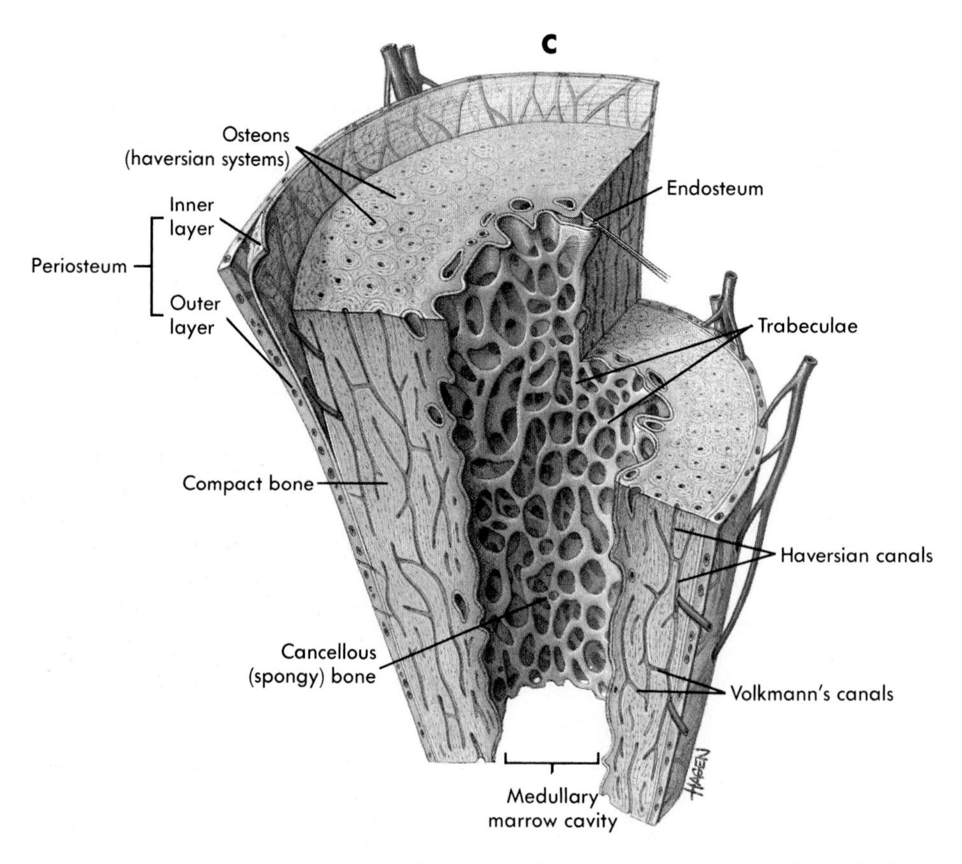

FIGURE 6-3 **A** Young long bone (the femur) showing epiphyses, epiphyseal plates, and diaphysis. **B** Adult long bone with epiphyseal lines. **C** Internal features of a portion of the long bone in **B.**

Table 6-1

Gross anatomy of long bone

PART	DESCRIPTION	PART	DESCRIPTION
Diaphysis	Shaft of the bone	Epiphyseal plate	Area of hyaline cartilage between the diaphysis and epiphysis; cartilage growth followed by endochondral ossification results in bone growth in length
Epiphyses	Ends of the bone	Cancellous (spongy) bone	Bone having many small spaces; found in the epiphysis; arranged into trabeculae
Periosteum	Double-layered connective tissue membrane covering the outer surface of bone except where there is articular cartilage; ligaments and tendons attach to bone through the periosteum; blood vessels and nerves from the periosteum supply the bone; the periosteum is the site of bone growth in diameter	Compact bone	Dense bone with few internal spaces organized into osteons; forms the diaphysis and covers the spongy bone of the epiphyses
Endosteum	Thin connective tissue membrane lining the inner cavities of bone	Medullary cavity	Large cavity within the diaphysis
Articular cartilage	Thin layer of hyaline cartilage covering a bone where it forms a joint (articulation) with another bone	Red marrow	Connective tissue in the spaces of spongy bone; the site of blood cell production
		Yellow marrow	Fat stored within the medullary cavity

In addition to the small spaces within cancellous bone and compact bone, some bones contain large cavities. The diaphyses of long bones have a large **medullary cavity,** and some of the skull bones have spaces called **sinuses.** The sinuses are filled with air (see Chapter 7), and the medullary cavity and the cavities of cancellous bone are filled with marrow. The medullary cavity of the adult diaphysis normally is filled with **yellow marrow,** which is mostly adipose tissue. The spaces in the proximal epiphyses of the larger adult long bones contain **red marrow,** which is the site of hematopoiesis (blood formation). In general, yellow marrow is associated with the long bones of the limbs, and red marrow is associated with the rest of the skeleton (Figure 6-4). Children's bones have more red marrow than do adult bones. Children even have red marrow located in the diaphyses of long bones. With a human's increasing age the red marrow in the limbs is replaced with yellow marrow.

The outer surface of bone consists of a **periosteum** (per′e-os′te-um), which, like the perichondrium of cartilage, has two layers (Figure 6-3, C). The outer fibrous layer is dense, fibrous, irregular collagenous connective tissue that contains blood vessels and nerves. The inner layer consists mostly of a single layer of osteoblasts with a few osteoclasts. **Osteoblasts** (os′te-o-blasts) are bone producing cells, and **osteoclasts** (os′te-o-klasts) are cells that break

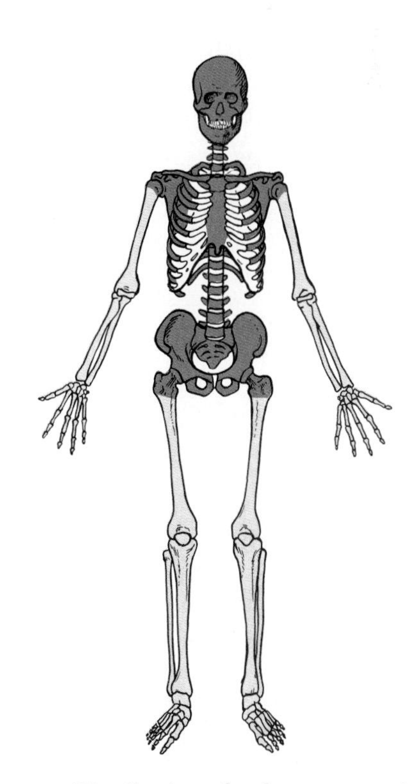

FIGURE 6-4 Distribution of red marrow and yellow marrow in an adult.

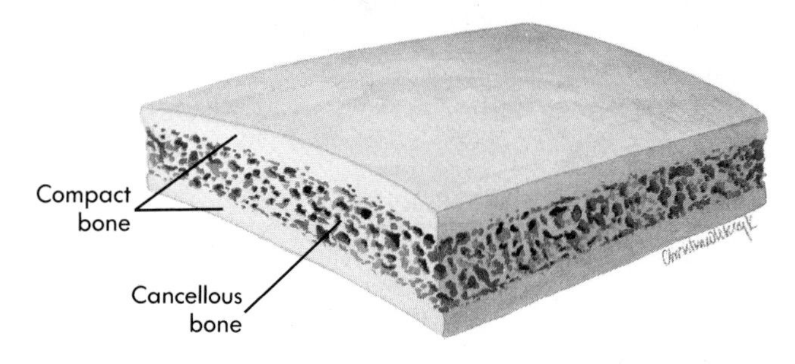

FIGURE 6-5 Structure of a flat bone. Outer layers of compact bone surround cancellous bone.

down bone. Where tendons and ligaments attach to bone, the collagen fibers of the tendon or ligament become continuous with those of the periosteum. Additional fibers, called **perforating,** or **Sharpey's, fibers,** penetrate the periosteum into the outer portion of the bone and help attach the tendons, ligaments, and periosteum to the bone.

The **endosteum** (en-dos′te-um) is a membrane that lines the inner surfaces of bone (i.e., the medullary cavity of the diaphysis and the smaller cavities of cancellous bone and compact bone). The endosteum consists mostly of a single layer of osteoblasts with some osteoclasts.

Flat bones usually have no diaphyses or epiphyses, and they contain an interior framework of cancellous bone sandwiched between two layers of compact bone (Figure 6-5). Short and irregular bones have a composition similar to the epiphyses of long bones. They have compact bone surfaces that surround a cancellous bone center with small spaces that usually are filled with marrow. Short and irregular bones are not elongated and have no diaphyses. However, certain regions of these bones (e.g., the processes of irregular bones) possess epiphyseal growth plates and therefore have small epiphyses.

Bone Histology

Bone matrix

By weight mature bone matrix normally is approximately 35% organic and 65% inorganic material. The major organic component is collagen, and the major inorganic components are the minerals calcium and phosphate. Most of the mineral in bone is in the form of calcium phosphate crystals called **hydroxyapatite** (hi-drok′se-ap′ĕ-tīt), which have the molecular formula $3Ca_3(PO_4)_2 \cdot Ca(OH)_2$. The collagen and mineral components are responsible for the major functional characteristics of bone. In terms of an analogy, bone matrix resembles reinforced concrete; collagen, like reinforcing steel bars, lends flexible strength to the matrix; and the mineral components, like concrete, give the matrix compression (weight-bearing) strength.

If all the mineral is removed from a long bone, collagen remains as the primary constituent, and the bone becomes very flexible. On the other hand, if the collagen is removed from the bone, the mineral component remains as the primary constituent, and the bone is very brittle and easily broken (Figure 6-6).

2

In elderly people the proportion of collagen to hydroxyapatite in bone decreases. Describe the effect of that decrease on the mechanical properties of bones in elderly people.

? ? ? ? ? ? ? ? ? ?

Bone matrix is produced by osteoblasts. Once an osteoblast becomes surrounded by matrix, it is an **osteocyte** (os′te-o-sīt). Bone matrix is broken down by osteoclasts, which are large cells with several nuclei. Secretions of osteoclasts include citric and lactic acids, which help dissolve bone mineral, and collagenase, which digests collagen. Osteoclasts play an important role in bone remodeling and mineral homeostasis.

Osteoblasts and osteoclasts have different origins. Osteoblasts are derived from uncommitted cells located in the periosteum and endosteum. Uncommitted cells have not become a specific cell type, but they can differentiate into more specialized cells. Osteoclasts are derived from uncommitted cells (related to monocytes, see Chapter 19) in red bone marrow.

Bone matrix is organized into thin sheets or layers approximately 3 to 7 μm thick called **lamellae** (lă-mel′e). Osteocytes are arranged in layers sandwiched between adjacent lamellae and occupy spaces in the matrix called **lacunae** (Figure 6-7). Osteocytes have long, thin cell processes that reach to similar processes of nearby osteocytes. The spaces occupied by these processes and extending between lacunae are called **canaliculi** (kan-ă-lik′u-le; little canals). Bone differs from tendons, ligaments, and cartilage in that bone cells are in contact with one another. This is a marked advantage in bone nutrition because nutrients can pass from cell to cell through the canaliculi rather than having to diffuse through the mineralized matrix.

The two major types of bone, based on their histological structure, are cancellous and compact bone. Cancellous bone consists of a lacy network of bony plates, whereas compact bone is mostly solid matrix and cells with few spaces.

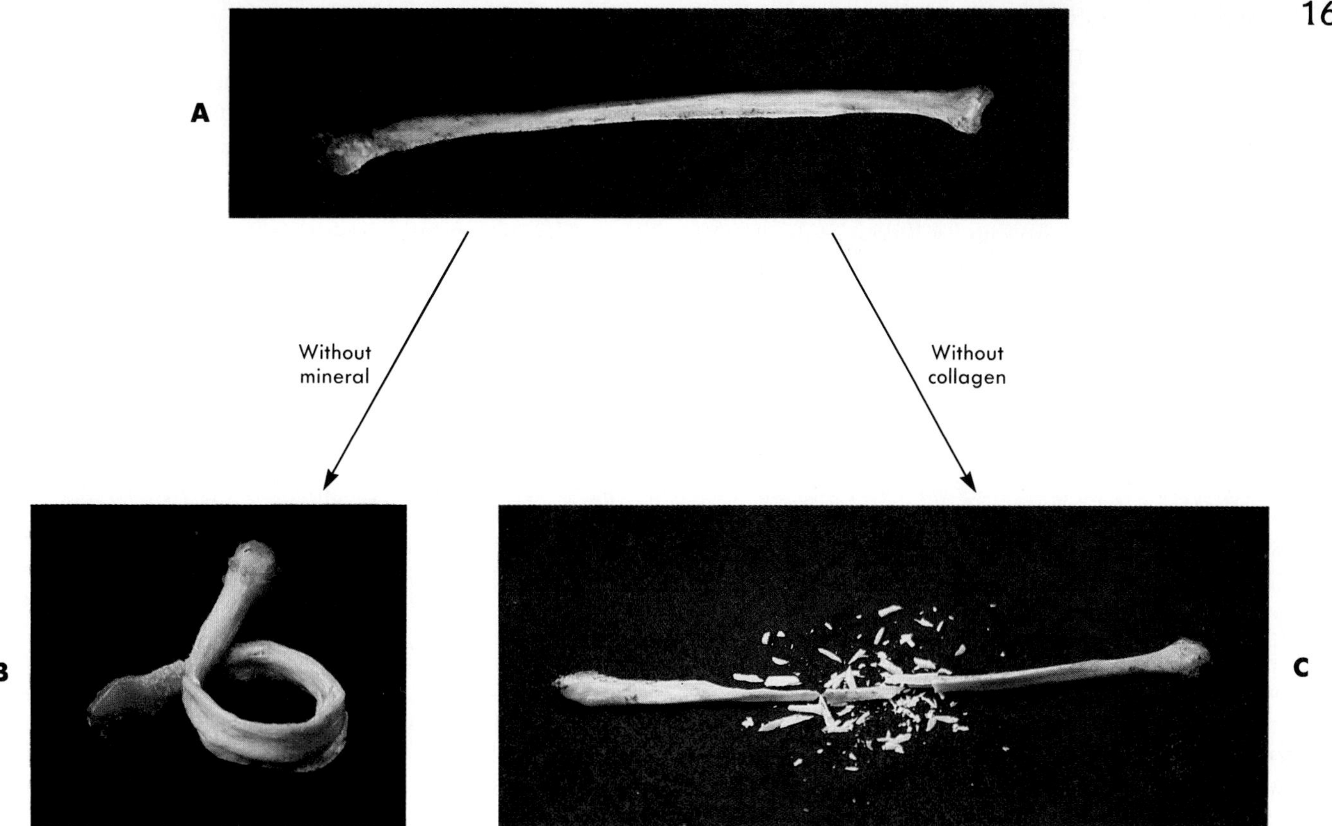

FIGURE 6-6 **A** A normal bone. **B** A demineralized bone, in which collagen is the primary remaining component, can be bent without breaking. **C** When collagen is removed, mineral is the primary remaining component, making the bone so brittle it is easily shattered.

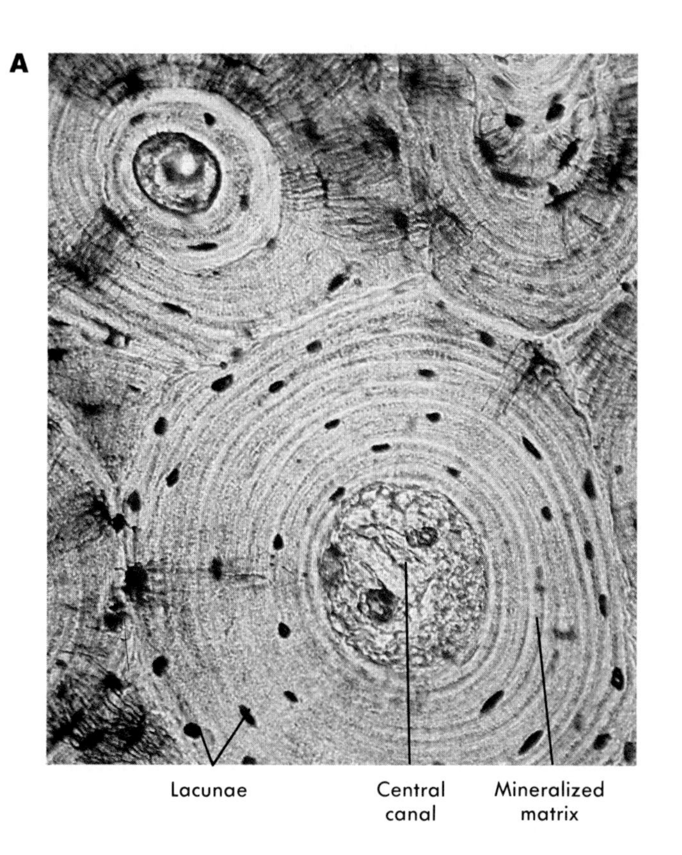

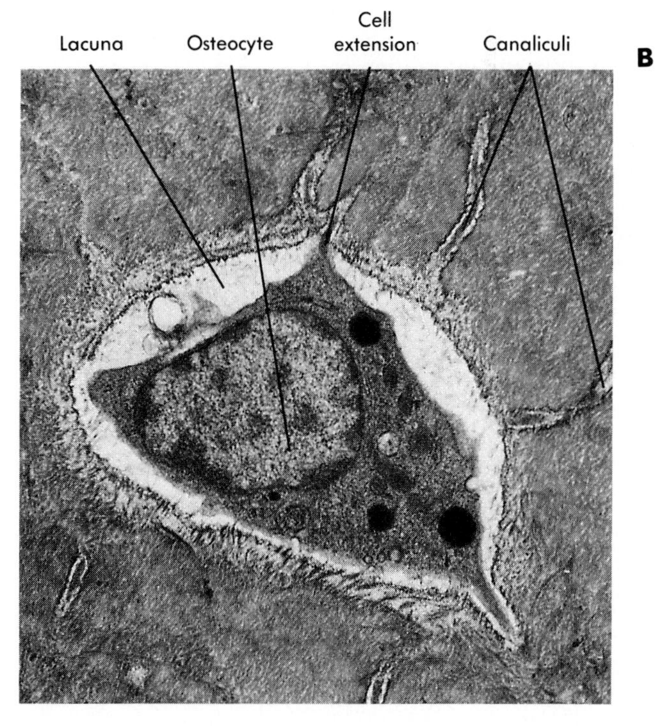

FIGURE 6-7 **A** Lamellae are thin layers of bone with osteocytes located between the layers. **B** Osteocyte within a lacuna. Cell extensions connect to other osteocytes through canaliculi.

Cancellous bone

Cancellous bone (Figure 6-8) consists of interconnecting rods or plates of bone called **trabeculae** (tră-bek'u-le; beam). The trabeculae consist of several lamellae with osteocytes that are located between the layers. Each osteocyte is associated with other osteocytes through canaliculi. The trabeculae have a layer of osteoblasts on their surfaces. Usually no blood vessels penetrate the trabeculae, so osteocytes must obtain nutrients through their canaliculi.

Trabeculae are oriented along the lines of stress within a bone (Figure 6-9). If the direction of weight-bearing stress is changed slightly because of a fracture that heals improperly, for example, the trabecular pattern realigns with the new lines of stress.

Compact bone

Compact bone (Figure 6-10) is more dense (with fewer spaces) than cancellous bone. Blood vessels enter the substance of the bone itself, and the osteocytes and lamellae of compact bone are primarily oriented around those blood vessels. Vessels that run parallel to the long axis of the bone are contained within **haversian** (hă-ver'shan), or **central, canals** surrounded by **concentric lamellae.** Haversian canals are lined with endosteum and contain blood vessels, nerves, and loose connective tissue. A **haversian system,** or **osteon** (os'te-on), consists of a single haversian canal, its contents, and associated concentric lamellae and osteocytes. In cross section the haversian system resembles a circular target; the "bull's-eye" of the target is the haversian canal, and 4 to 20 concentric lamellae form the rings. Osteocytes are located between the lamellar rings, and canaliculi radiate from each osteocyte across the lamellae to produce the appearance of minute cracks across the rings of the target.

The osteocytes receive nutrients and eliminate waste products through the canal system within compact bone. Blood vessels from the periosteum or endosteum enter the bone through **Volkmann's,** or **perforating, canals,** which run perpendicular to the long axis of the bone (see Figure 6-10). Volkmann's canals are not surrounded by concentric lamellae but cut across the grain of the concentric haversian lamellae. The haversian canals receive blood vessels from Volkmann's canals. Nutrients in the blood vessels enter the haversian canals, pass into the canaliculi, and are transported through the cytoplasm of the osteocytes that occupy the canaliculi and lacunae to the most peripheral cells within each osteon. Waste products are removed in the reverse direction.

The outer surfaces of compact bone are covered by **circumferential lamellae,** which are flat plates that extend around the bone (see Figure 6-10). In some bones such as certain bones of the face the layer of compact bone may be so thin that no osteons exist, and the compact bone is composed of only circumferential lamellae.

Trabeculae

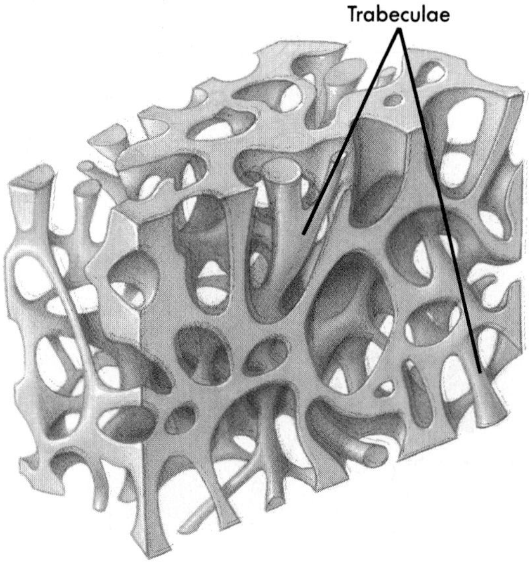

FIGURE 6-8 Fine structure of cancellous bone. Beams of bone, the trabeculae, surround spaces in the bone. In life the spaces are filled with red or yellow marrow.

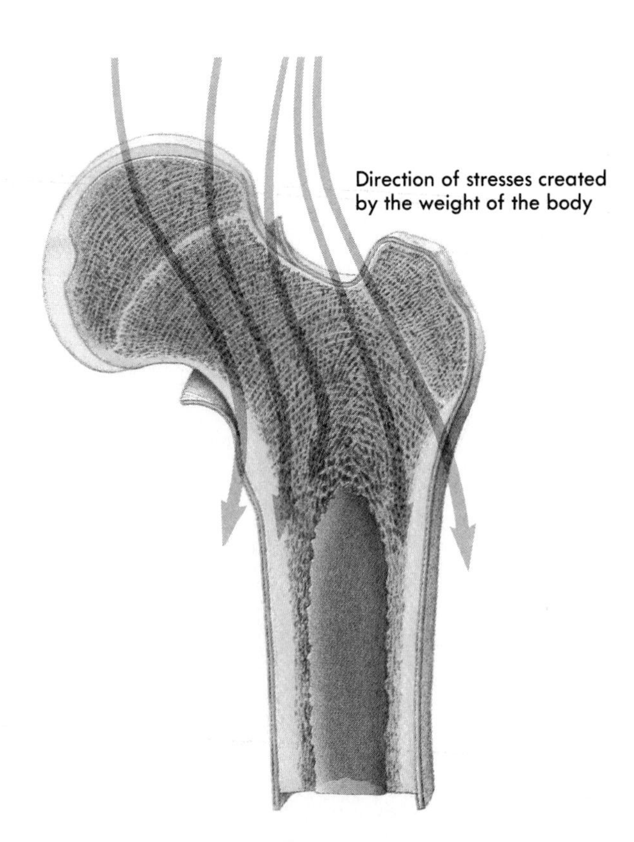

FIGURE 6-9 Long bone (femur) showing trabeculae oriented along lines of stress *(arrows)*.

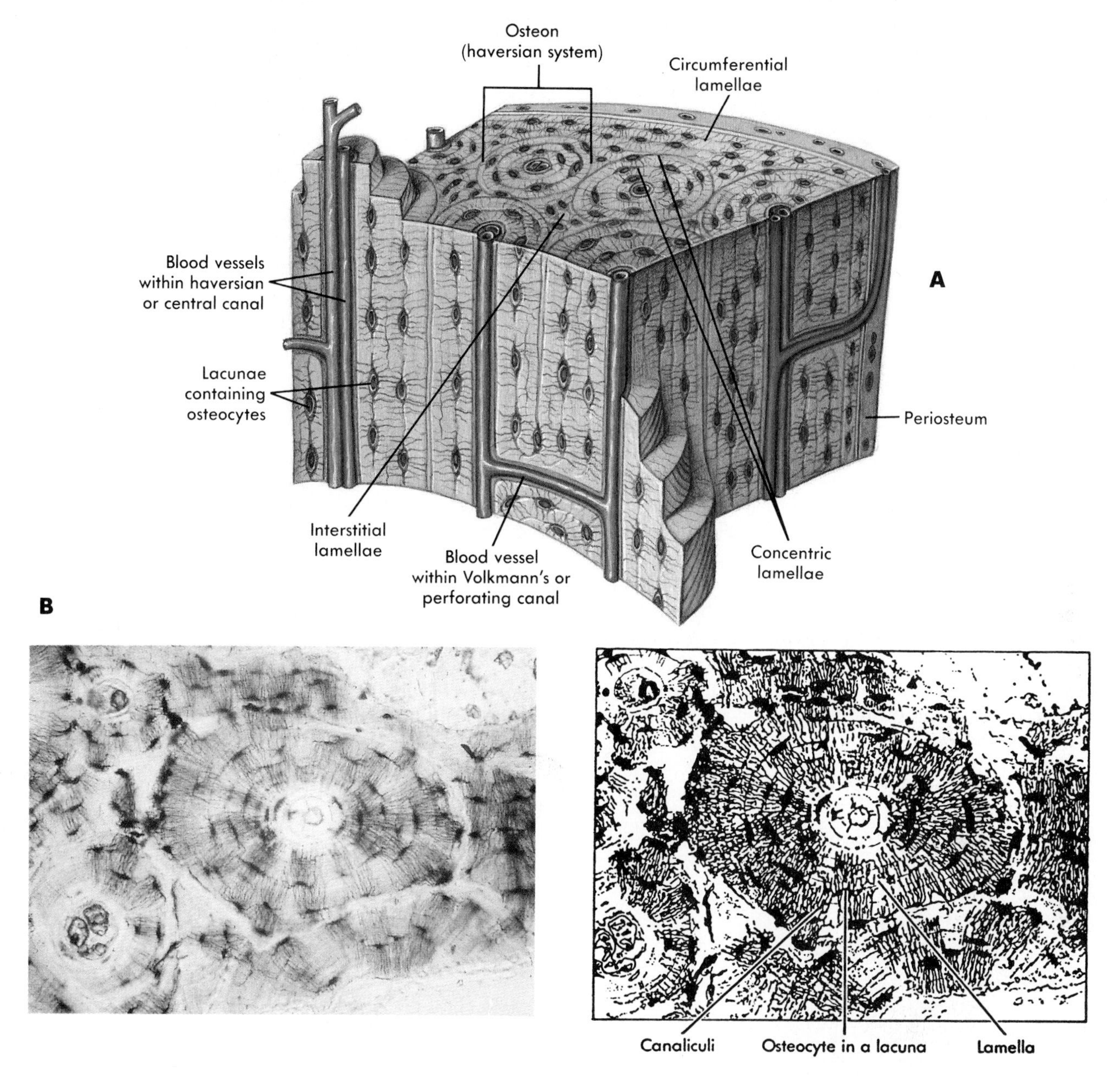

FIGURE 6-10 Fine structure of compact bone. **A** An osteon consists of concentric lamellae surrounding a blood vessel within the haversian canal. The outer surface of compact bone is circumferential lamellae. **B** Photomicrograph of an osteon. (×400.)

BONE OSSIFICATION

Ossification (os′ĭ-fĭ-ka′shun) is the formation of bone by osteoblasts and involves the synthesis of an organic extracellular matrix and the addition of minerals, mostly in the form of hydroxyapatite, to that matrix. Ossification occurs by two processes, each involving preexisting connective tissue. Bone formation that occurs within connective tissue membranes is **intramembranous** (within membranes), and bone formation in association with cartilage is **endochondral** (inside cartilage). Four points are important about these two types of ossification. First, both cancellous and compact bone result from each process. Second, the mechanism of bone deposition is essentially the same in both processes. Third, the bone matrix formed by one process is indistinguishable from that formed by the other. Fourth, although bone forms in association with previously existing connective tissue (either fibrous membranes or cartilage), it eventually replaces that tissue.

Intramembranous Ossification

Many skull bones and the clavicle (collarbone) develop in areas in which collagen membranes are produced by fibroblasts during embryonic development. **Osteoprogenitor cells** differentiate into osteoblasts and begin to produce bone in the connective tissue membranes. These areas of bone formation are **centers of ossification.** Thin bone trabeculae radiate out in many directions from each ossification center along the fibers of the membrane. Blood vessels and unspecialized cells invade the spaces within the bone, forming bone marrow, and the connective tissue surrounding the bone becomes the periosteum. Osteoblasts from the periosteum produce an outer layer of compact bone over an internal zone of bone trabeculae; thus the end products of intramembranous bone formation are bones with a cancellous center and a compact bone surface.

Usually two or more centers of ossification exist in each flat skull bone, and the skull bones result from the fusion of these centers as they enlarge (Figure 6-11). The flat bones of the skull are not completely formed at the time of birth, and the membranous gaps between the bones are called **fontanels,** or soft spots (see Chapter 8). The bones eventually grow together, closing the fontanels.

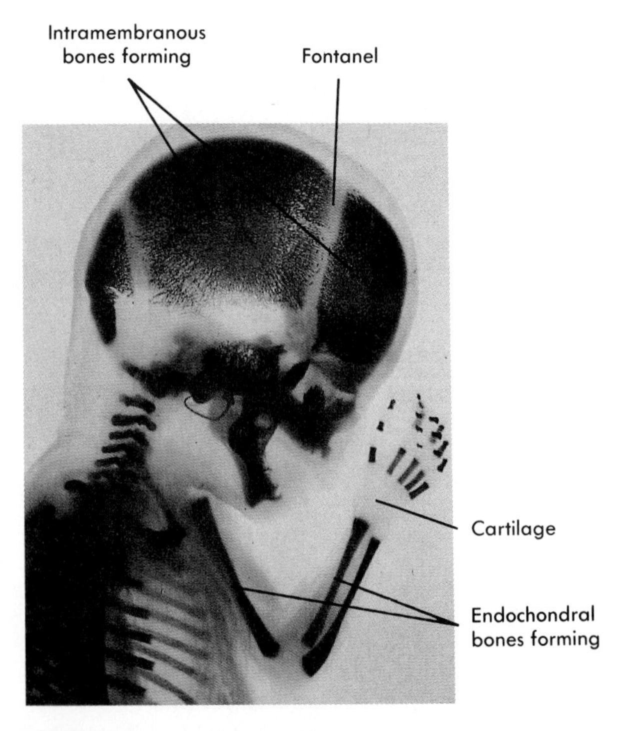

FIGURE 6-11 Bone formation in an 18-week-old fetus. Intramembranous ossification occurs at centers of ossification in the flat bones of the skull. Endochondral ossification has formed bones in the diaphyses of long bones. The epiphyses are still cartilage at this stage of development.

Endochondral Ossification

The bones at the base of the skull and most of the remaining skeletal system develop from hyaline cartilage templates through the process of endochondral ossification (see Figure 6-11). The first phase of endochondral ossification is the formation of a hyaline cartilage model by chondroblasts, which become chondrocytes as they are surrounded by cartilage matrix (Figure 6-12). The chondrocytes in the center of the future bone **hypertrophy** (hi-per'tro-fe; enlarge), and the matrix between the enlarged cells becomes mineralized with calcium carbonate. At this point the cartilage is referred to as calcified cartilage. The chondrocytes in this calcified area eventually die, leaving enlarged lacunae with thin walls of calcified matrix.

The second phase of endochondral ossification begins with osteoblasts. Apparently some of the cells of the perichondrium in these cartilage models remain uncommitted. As increasing numbers of capillaries invade the perichondrium, some of the uncommitted perichondrial cells become osteoblasts, and the perichondrium becomes the periosteum. The osteoblasts of this new periosteum form circumferential lamellae on the surface of the cartilage model, which becomes the compact bone of the diaphysis.

Blood vessels grow into the enlarged lacunae of the calcified cartilage at **primary ossification centers.** The connective tissue surrounding the blood vessels contains osteoblasts from the periosteum. The osteoblasts begin to produce bone on the surface of the calcified cartilage, forming bone trabeculae. Consequently, the cartilage of the diaphysis is changed into cancellous bone. As osteoblasts deposit additional bone on the trabeculae, the trabeculae become thicker, and the spaces between adjacent trabeculae become smaller. Eventually most of the spaces are filled in by the expanding trabeculae, and the diaphysis becomes compact bone. As the conversion of the diaphysis into compact bone proceeds, osteoclasts remove some of the bone to form the medullary cavity.

In long bones the diaphysis is the primary ossification center, and additional sites of ossification, called **secondary ossification centers,** appear in the epiphyses. Primary ossification centers appear during early fetal development, whereas secondary ossification centers appear in the proximal epiphysis of the femur, humerus, and tibia approximately 1 month before birth. A baby is considered full term if one of these three ossification centers can be seen on x-rays at the time of birth. At approximately 18 to 20 years of age the last secondary ossification center appears in the medial epiphysis of the clavicle.

Ossification and replacement of cartilage by bone continues in the cartilage model until all the cartilage, except that in the epiphyseal plate and on articular surfaces, has been replaced by bone. The

FIGURE 6-12 Endochondral Ossification

A A cartilage model is produced by chondroblasts that become chondrocytes enclosed by cartilage matrix.

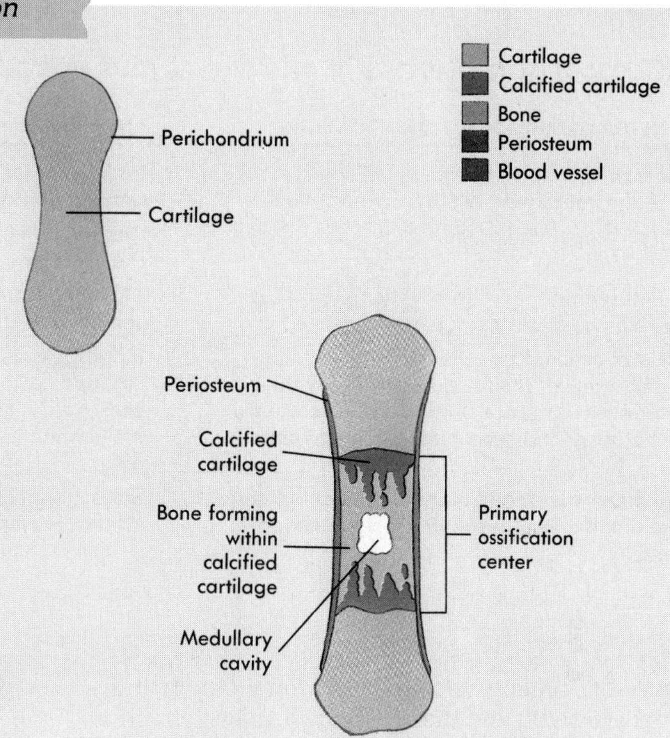

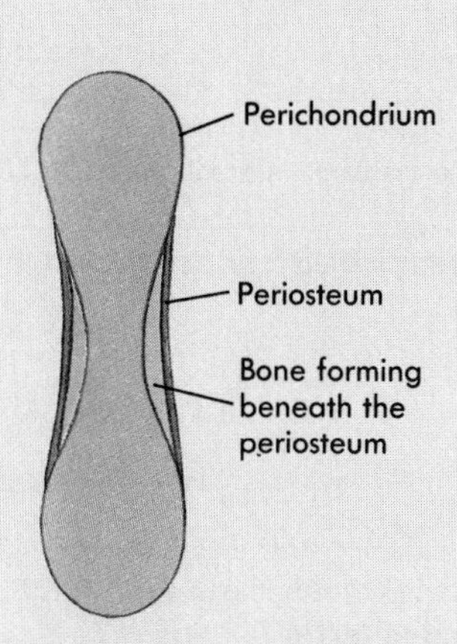

B Calcified cartilage is formed as chondrocytes hypertrophy and die and the cartilage matrix is mineralized. The perichondrium becomes the periosteum as osteoblasts begin producing bone.

C A primary ossification center forms as blood vessels and osteoblasts invade the calcified cartilage at the end of the second month of development. The osteoblasts produce bone that is later resorbed as osteoclasts form the medullary cavity.

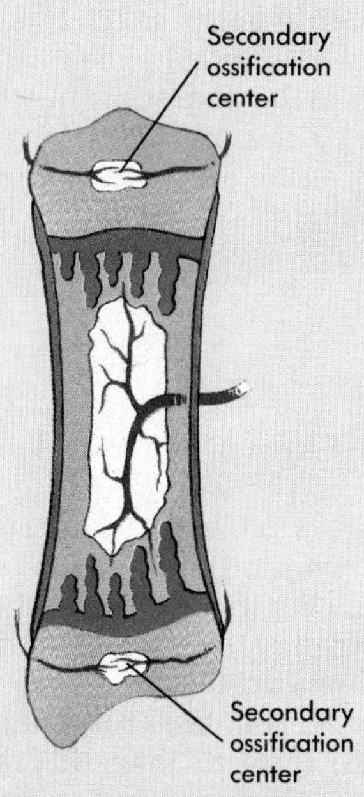

D Secondary ossification centers begin to form in some bones at about 8 months of development.

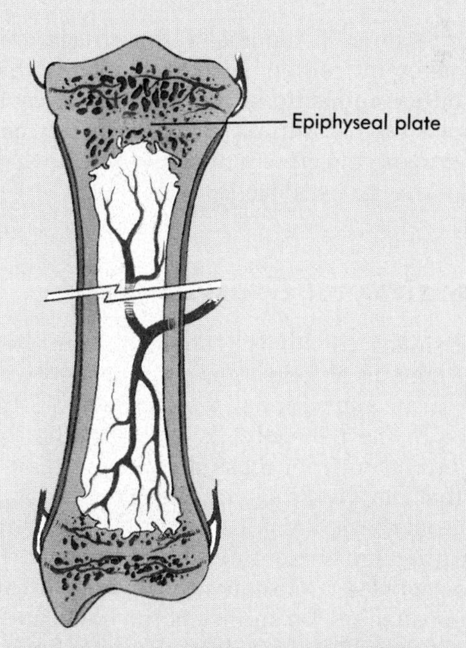

E The original cartilage model is ossified; the epiphyseal plate has become the epiphyseal line, and the articular cartilage remains over the articular surfaces at the ends of the bone. (The articular cartilage is not shown in this figure.)

Table 6-2

Comparison of intramembranous and endochondral ossification

INTRAMEMBRANOUS OSSIFICATION	ENDOCHONDRAL OSSIFICATION
Embryonic cells form a collagen membrane containing osteoprogenitor cells.	Embryonic cells become chondroblasts that produce a cartilage template, which is surrounded by the perichondrium.
No comparable stage.	Chondrocytes hypertrophy, the cartilage matrix is calcified, and the chondrocytes die.
Embryonic cells form the periosteum, which contains osteoblasts.	The perichondrium becomes the periosteum when uncommitted cells within the periosteum become osteoblasts.
Osteoprogenitor cells become osteoblasts at centers of ossification; internally the osteoblasts produce bone matrix; externally the periosteal osteoblasts produce bone.	Blood vessels and osteoblasts from the periosteum invade the calcified cartilage template; internally these osteoblasts produce bone matrix at primary ossification centers (and later at secondary ossification centers); externally the periosteal osteoblasts produce bone.
Intramembranous bone is remodeled and is indistinguishable from endochondral bone.	Endochondral bone is remodeled and is indistinguishable from intramembranous bone.

epiphyseal plate, which exists throughout an individual's growth, and the articular cartilage, which is a permanent structure, are derived from the original embryonic cartilage template. Table 6-2 compares the processes of intramembranous and endochondral ossification.

3

Explain why bones cannot undergo interstitial growth.

? ? ? ? ? ? ? ? ? ? ?

BONE GROWTH

Unlike tendons, ligaments, and cartilage, bones cannot grow by interstitial growth. Bone growth can occur by either **appositional growth,** the formation of new bone on the surface of bone, or by **endochondral growth,** the growth of cartilage followed by replacement of the cartilage by bone.

Appositional Growth

Appositional growth is responsible for the increase in diameter of long bones and most growth of other bones. In appositional growth osteoblasts on the surface of the bone divide. The superficial osteoblasts produced from these divisions remain osteoblasts that can divide again. The deep osteoblasts resulting from these divisions produce bone matrix, and when they are surrounded by the matrix, they become osteocytes. Consequently, a new layer of bone is deposited on the surface of the bone, and the bone has increased in size. In cancellous bone appositional growth adds bone matrix to the outer surface of trabeculae. In compact bone appositional growth is responsible for the formation of circumferential and concentric lamellae (see Bone Remodeling in this chapter).

Endochondral Growth

Endochondral growth is responsible for the increase in length of bones. In long bones endochondral growth at the epiphyseal plate results in an increase in the length of the diaphysis. In other bones growth at epiphyseal plates produces elongation of long projections such as the processes of vertebrae (see Chapter 7). Endochondral growth also occurs within articular cartilage and is responsible for growth of epiphyses.

Epiphyseal plate

In a long bone the epiphyseal plate separates the epiphysis from the diaphysis (Figure 6-13, *A*). The epiphyseal plate is highly organized, and its cells are oriented parallel to the direction of growth (Figure 6-13, *B*). The chondrocytes nearest the epiphysis are arranged randomly within the hyaline cartilage in the **zone of resting cartilage** and do not divide rapidly. The chondrocytes in the **proliferating zone** produce new cartilage through interstitial cartilage growth. The chondrocytes divide and are organized into columns resembling stacks of plates or coins. A maturation gradient exists in each column: cells near the epiphysis are younger and are actively proliferating; cells progressively nearer the diaphysis are older and are undergoing hypertrophy. As the matrix nearest

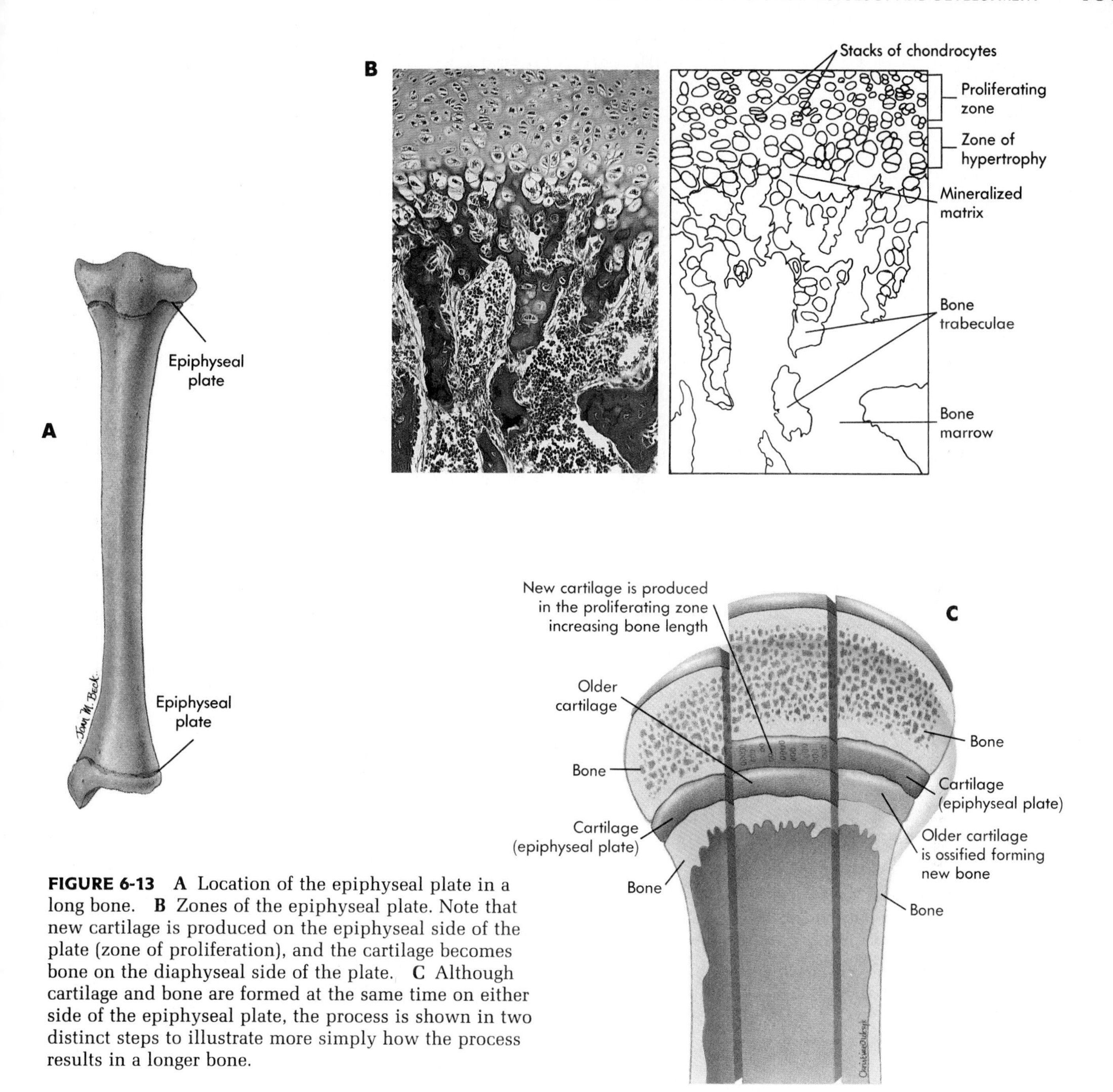

FIGURE 6-13 **A** Location of the epiphyseal plate in a long bone. **B** Zones of the epiphyseal plate. Note that new cartilage is produced on the epiphyseal side of the plate (zone of proliferation), and the cartilage becomes bone on the diaphyseal side of the plate. **C** Although cartilage and bone are formed at the same time on either side of the epiphyseal plate, the process is shown in two distinct steps to illustrate more simply how the process results in a longer bone.

the diaphysis becomes mineralized with calcium carbonate, the hypertrophied chondrocytes die. Blood vessels grow into the area, and the connective tissue surrounding the blood vessels contains osteoblasts from the endosteum. The osteoblasts line up on the surface of the calcified cartilage and deposit bone. Ossification and remodeling of the matrix then occurs (see Bone Remodeling in this chapter).

In summary, the diaphysis increases in length as a result of interstitial cartilage growth at the epiphyseal plate (Figure 6-13, C). The cartilage is calcified and dies, and osteoblasts invade the calcified cartilage, which is converted into bone. Basically, the production of bone by endochondral growth is the same process as the formation of bone in the fetus by endochondral ossification.

Since cartilage does not appear readily on x-ray film, growth plates appear as black areas between the white diaphysis and the epiphyses. The epiphyses fuse with the diaphysis between approximately 12 and 25 years of age, depending on the bone. This fusion results in ossification of the epiphyseal plate, which becomes the epiphyseal line, and growth in bone length ceases.

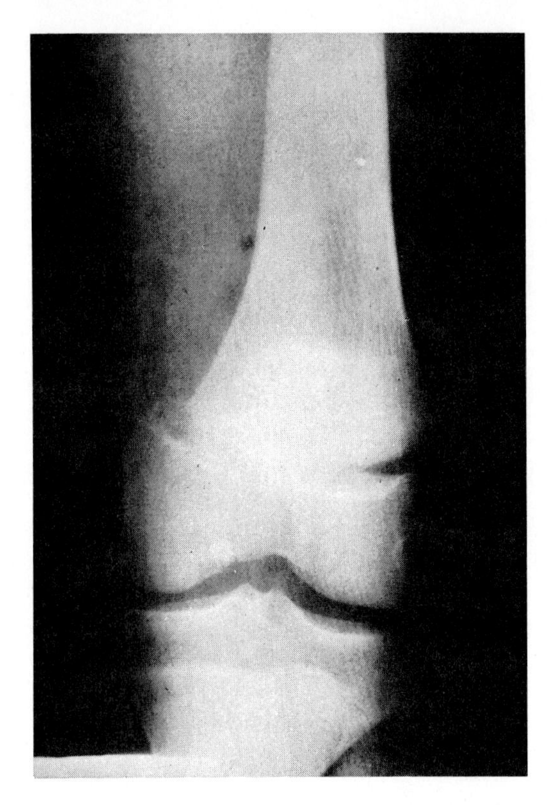

FIGURE 6-14 Fracture of the epiphyseal plate, separating the diaphysis from the epiphysis.

4

A 15-year-old football player is tackled during a game, and the epiphyseal plate of the left femur is damaged (Figure 6-14). What are the results of such an injury, and why is recovery difficult?

? ? ? ? ? ? ? ? ? ? ?

Articular cartilage

The secondary ossification centers in the epiphyses of long bones are surrounded by cartilage in which endochondral bone growth occurs. Bone growth within this epiphyseal cartilage is substantial after birth, resulting in expansion of the epiphyseal diameter. The endochondral growth in the epiphyseal cartilage is similar to that occurring in the epiphyseal plate except that the chondrocyte columns are not as pronounced and radiate out from the secondary ossification center. The chondrocytes near the surface of the epiphyseal cartilage are similar to those in the zone of resting cartilage of the epiphyseal plate. In the deepest part of the epiphyseal cartilage, nearest the secondary ossification center, the cartilage is calcified, dies, and is ossified to form new epiphyseal bone. The entire growth of short bones, which lack an epiphyseal plate, occurs by this same process.

Much of the epiphyseal cartilage is eventually replaced by cancellous bone with an outer surface of compact bone. However, the cartilage on the articular surface of the epiphysis is not replaced by bone. This **articular cartilage** loses its perichondrium and remains as a cartilage plate over the articular surface of the long bone.

Factors Affecting Bone Growth

Bones of an individual's skeleton usually reach a certain length, thickness, and shape through the processes described in the previous sections. The potential shape and size of a bone and an individual's final adult height are determined genetically, but factors such as nutrition and hormones may greatly modify the expression of those genetic factors.

Nutrition

Since bone growth requires chondroblast and osteoblast proliferation, any metabolic disorder that affects the rate of cell proliferation or the production of collagen and other matrix components affects bone growth, as does the availability of calcium or other minerals needed in the mineralization process.

The long bones of a child sometimes exhibit lines of arrested growth, which are transverse regions of greater bone density crossing an otherwise normal bone. These lines are caused by greater calcification below the epiphyseal plate of a bone where it has grown at a slower rate during an illness or severe nutritional deprivation. They demonstrate that illness or malnutrition during the time of bone growth can cause a person to be shorter than he or she would have been otherwise.

Certain vitamins are important in very specific ways to bone growth. **Vitamin D** is necessary for the normal absorption of calcium from the intestines (see Chapters 5 and 24). Vitamin D can be synthesized by the body or ingested orally. Its rate of synthesis is increased when the skin is exposed to sunlight.

Insufficient vitamin D in children causes **rickets,** a disease resulting from reduced mineralization of the organic matrix of bone. Children with rickets can have bowed bones and inflamed joints. During the winter in northern climates if children are not exposed to sufficient sunlight, vitamin D may be taken as a dietary supplement to prevent rickets. Vitamin D deficiency may also be caused by the body's inability to absorb fats (in which vitamin D is soluble). This condition can occur in adults who suffer from digestive disorders and can be one cause of "adult rickets," or **osteomalacia** (os'te-o-mă-la'shĭ-ah) (i.e., softening of the bones as a result of calcium depletion).

Vitamin C is necessary for normal collagen synthesis and matrix mineralization by osteoblasts. Vitamin C deficiency results in bones deficient in collagen and, to a certain extent, in mineral. In children vitamin C deficiency can cause growth retardation. In children and adults vitamin C deficiency can result in **scurvy**, which is marked by ulceration and hemorrhage in almost any area of the body because of the lack of normal collagen synthesis in connective tissues. Wound healing, which requires collagen synthesis, is hindered in patients with vitamin C deficiency, and in extreme cases the teeth may fall out because the ligaments that hold them in place break down.

Hormones

Hormones are very important in bone growth. **Growth hormone** from the anterior pituitary increases general tissue growth (see Chapters 17 and 18), including overall bone growth, by stimulating interstitial cartilage growth and appositional bone growth. **Thyroid hormone** is also required for normal growth of all tissues, including cartilage; therefore a decrease in this hormone can result in decreased size of the individual. **Sex hormones** also influence bone growth. Estrogen (a class of female sex hormones) and testosterone (a male sex hormone) initially stimulate bone growth, which accounts for the burst of growth at the time of puberty when production of these hormones increases. However, both hormones also stimulate ossification of epiphyseal plates and thus the cessation of growth. Because estrogens cause a quicker closure of the epiphyseal plate than does testosterone, females usually stop growing earlier than males, and since their entire growth period is somewhat shorter, females usually do not reach the same height as males. Decreased levels of testosterone or estrogen can prolong the growth phase of the epiphyseal plates, even though the bones grow more slowly. However, growth is very complex and is influenced by many factors in addition to sex hormones (e.g., other hormones, genetics, and nutrition).

5

A 12-year-old female has an adrenal tumor that produces large amounts of estrogen. If untreated, what effect will this condition have on her growth for the next 6 months? On her height when she is 18?

? ? ? ? ? ? ? ? ? ? ?

MAINTENANCE OF BLOOD CALCIUM LEVELS

Bones play an important role in regulating blood calcium levels, which must be maintained within narrow limits for functions such as muscle contraction and membrane potentials to occur normally (see Chapters 9, 10, and 12). Bone is the major storage site for calcium in the body, and movement of calcium into and out of bone helps to determine blood calcium levels. Calcium moves into bone as osteoblasts build new bone, and calcium moves out of bone as osteoclasts break down bone. When osteoblast and osteoclast activity is balanced, the movement of calcium into and out of a bone are equal. When blood calcium levels are too low, osteoclast activity increases. More calcium is released by osteoclasts from bone into the blood than is removed by osteoblasts from the blood to make new bone. Consequently, there is a net movement of calcium from bone into blood, and blood calcium levels increase. Conversely, if blood calcium levels are too high, osteoclast activity decreases. Less calcium is released by osteoclasts from bone into the blood than is taken from the blood by osteoblasts to produce new bone. As a result, there is a net movement of calcium from blood to bone, and blood calcium levels decrease.

Parathyroid hormone from the parathyroid glands is the major regulator of blood calcium levels. If blood calcium decreases, the secretion of parathyroid hormone increases, stimulating osteoclast activity, which results in increased bone breakdown and increased blood calcium levels. Tumors in the parathyroid glands that secrete large amounts of parathyroid hormone can cause so much bone breakdown that bones are weakened and fracture easily. On the other hand, an increase in blood calcium levels results in less parathyroid hormone secretion, decreased osteoclast activity, reduced calcium release from bone, and decreased blood calcium levels.

Parathyroid hormone also regulates blood calcium levels by increasing calcium uptake in the small intestine and decreasing calcium loss from the kidneys. Increased parathyroid hormone promotes the formation of vitamin D in the kidneys, and vitamin D increases the absorption of calcium from the small intestine. Parathyroid hormone also increases the reabsorption of calcium from urine in the kidneys; thus less calcium is lost in the urine.

Calcitonin (kal'sĭ-to'nin), secreted from the thyroid gland, also affects osteoclast activity. An increase in blood calcium levels stimulates the thyroid gland to secrete calcitonin, which inhibits osteoclast activity. Parathyroid hormone and calcitonin are described more fully in Chapters 18 and 27.

BONE REMODELING

Bone remodeling (i.e., the removal of old bone by osteoclasts and the deposition of new bone by osteoblasts) is involved in bone growth, changes in bone shape, the adjustment of the bone to stress, bone repair, and calcium ion regulation in the body. For example, as a long bone increases in length and in diameter, the size of the marrow cavity also increases (Figure 6-15). Otherwise, the bone would consist of nearly solid bone matrix and would be very heavy. When compared to a solid rod, a cylinder with the same height, weight, and composition but with a greater diameter can support much more weight without bending. Therefore bone has a mechanical advantage as a cylinder rather than as a rod. The relative thickness of compact bone is maintained by the removal of bone on the inside by osteoclasts and the addition of bone to the outside by osteoblasts.

Remodeling is also responsible for the formation of new osteons in compact bone. This process occurs in two ways (Figure 6-16). First, a few osteoclasts in the periosteum remove bone, resulting in groove formation along the surface of the bone. Periosteal capillaries lie within these grooves and become surrounded as the osteoblasts of the periosteum form new bone. Additional lamellae then are added around the capillary until an osteon results. Second, within already existing osteons osteoclasts enter a haversian canal through the blood vessels and begin to remove bone from the center of the osteon, resulting in an enlarged channel through the bone. New concentric lamellae then are formed around the vessels until the new osteon fills the area occupied by the old osteon.

Osteons and circumferential lamellae constantly are being removed by osteoclasts, and new osteons are being formed by osteoblasts. However, this process leaves portions of older osteons and circumferential lamellae, called **interstitial lamellae,** between the newly developed osteons (see Figure 6-10).

Remodeling, the formation of additional bone, alteration in trabecular alignment (to reinforce the scaffolding), or other changes can modify the strength of the bone in response to the amount of stress applied to it. Mechanical stress applied to bone increases osteoblast activity in bone tissue, and removal of mechanical stress decreases osteoblast activity. Under conditions of reduced stress such as being bedridden or paralyzed, osteoclasts continue to work at nearly their normal rate, but osteoblast activity is reduced, resulting in a decrease in bone density. In addition, pressure in bone causes an electrical change that decreases the bone resorption activity of osteoclasts. Therefore, by applying weight (pressure) to a broken bone, a person speeds the healing process. Weak pulses of electrical current applied to a broken bone sometimes are used clinically to speed the healing process.

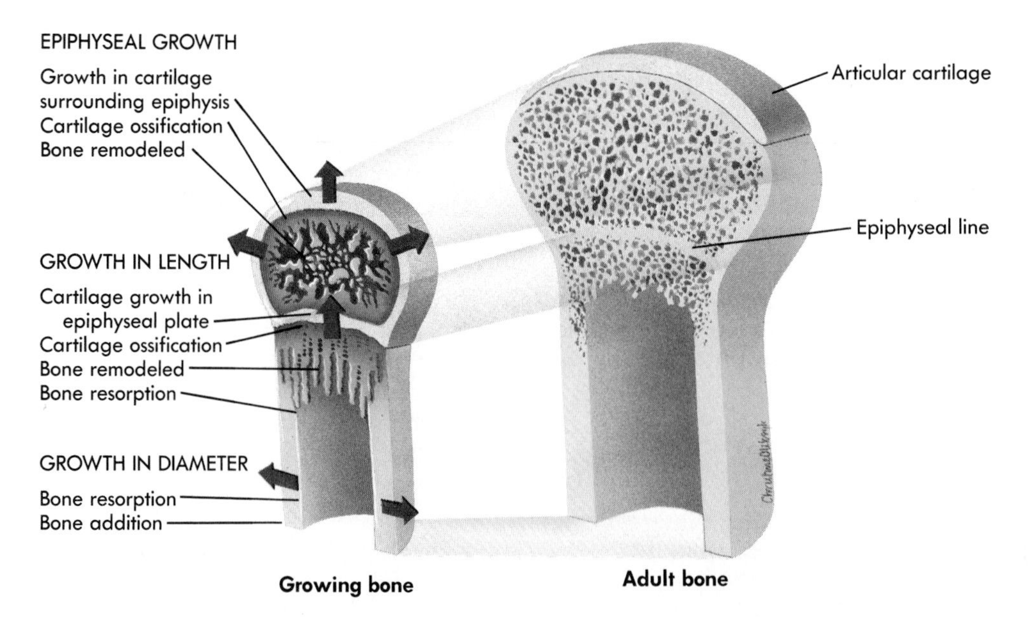

FIGURE 6-15 Remodeling of a long bone. Appositional bone growth on the outside of the shaft and bone resorption on the inside increase the diameter of the bone. Endochondral bone growth and subsequent bone remodeling cause the diaphysis to increase in length and the epiphysis to enlarge.

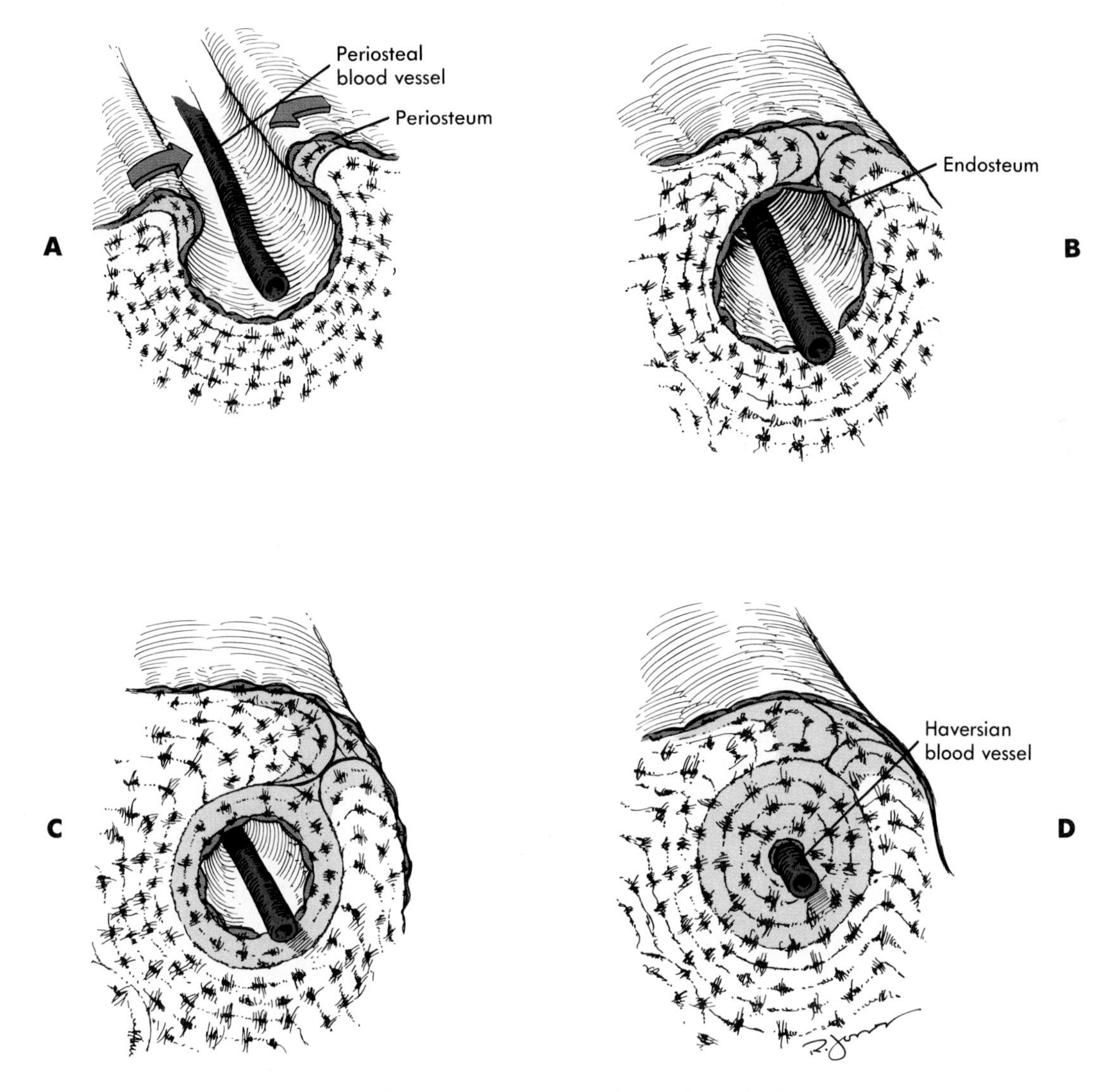

FIGURE 6-16 Formation of a new osteon. **A** The surface of a bone consists of grooves and ridges. Blood vessels in the periosteum tend to lie in the grooves. **B** New bone is added to the ridges, building them up. **C** When the bone built on adjacent ridges meets, the groove is transformed into a tunnel. The periosteum of the groove becomes the endosteum of the tunnel. **D** Bone deposition by this endosteum fills in the tunnel, thus completing the haversian system.

BONE REPAIR

When bone is damaged such as when fractured, the blood vessels in the area (mainly those of the periosteum) are also damaged. These vessels bleed, and a clot forms in the damaged area (Figure 6-17, *A*). Two to 3 days after the injury, blood vessels and uncommitted cells from surrounding tissues invade the area. Approximately 1 week after the injury, some of these uncommitted cells differentiate into fibroblasts, which produce a fibrous network between the broken bones. Other cells differentiate into chondroblasts, which produce islets of fibrocartilage in the fibrous network (Figure 6-17, *B*). This zone of tissue repair between the two bone fragments is a **fibrocartilage callus.** Osteoblasts from the periosteum and endosteum enter the fibrocartilage callus and convert it into a **bony callus.** The osteoblasts form bone through intramembranous bone formation in the fibrous network and by endochondral bone formation in the fibrocartilage islets. The bony callus has two portions: an external callus, located around the outside of the fracture, and an internal callus, located between the broken bone fragments (Figure 6-17, *C*).

Bone formation is usually complete by 4 to 6 weeks after the injury. Immobilization of the bone is critical during this time since movement can refracture the new matrix. Finally, the bone is slowly remodeled to form compact and cancellous bone, and the repair is complete (Figure 6-17, *D*). Total healing of the fracture may require several weeks or even months, depending on a number of conditions. If the fracture occurs in the diaphysis of a long bone, remodeling also restores the medullary cavity. This repair may be so finely remodeled that no evidence of the fracture remains on the bone, even when viewed on x-ray film. However, the repaired zone usually remains slightly thicker than the adjacent bone.

Bone Fractures

Bone fractures are classified in several ways. The most commonly used classification involves the severity of injury to the soft tissues surrounding the bone. An **open** fracture (formerly called compound) occurs when an open wound extends to the site of the fracture or when a fragment of bone protrudes through the skin. If the skin is not perforated, the fracture is **closed** (formerly called simple). If the soft tissues around a closed fracture are damaged, the fracture is complicated.

Two other terms to designate fractures are **complete,** in which the bone is broken into at least two fragments, and **incomplete,** in which the fracture does not extend completely across the bone (Figure 6-A, *A*). An incomplete fracture that occurs on the convex side of the bone's curve is a green-stick fracture. Hairline fractures are incomplete fractures in which the two sections of bone do not separate; they are common in skull fractures. Some hairline fractures are "occult" because there are clinical signs of the fracture but no x-ray study confirmation; however, x-ray studies 3 to 4 weeks later show evidence of bone repair.

Comminuted (kom-ĭ-nu'ted) fractures are ones in which the bone breaks into more than two pieces—usually two major fragments and a smaller fragment (Figure 6-A, *B*). **Impacted fractures** are those in which one fragment is driven into the cancellar portion of the other fragment (Figure 6-A, *C*).

Fractures are also classified according to the direction of the fracture within a bone. **Linear** fractures run parallel to the long axis of the bone, and **transverse** fractures are at right angles to the long axis (see Figure 6-A, *B*). **Oblique** fractures run obliquely in relation to the long axis, and **spiral** fractures have a helical course around the bone (Figure 6-A, *D*). **Dentate** fractures have rough, toothed, broken ends, and **stellate** fractures have breakage lines radiating from a central point.

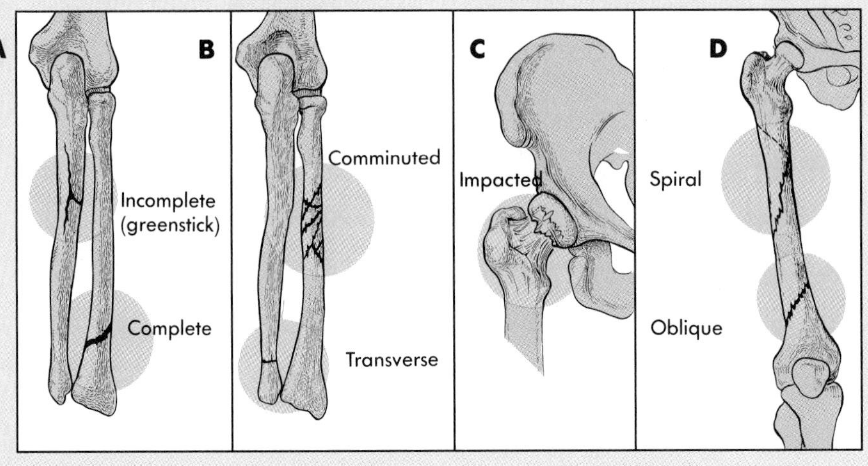

FIGURE 6-A Bone fractures.
A Complete and incomplete.
B Comminuted and transverse.
C Impacted.
D Oblique and spiral.

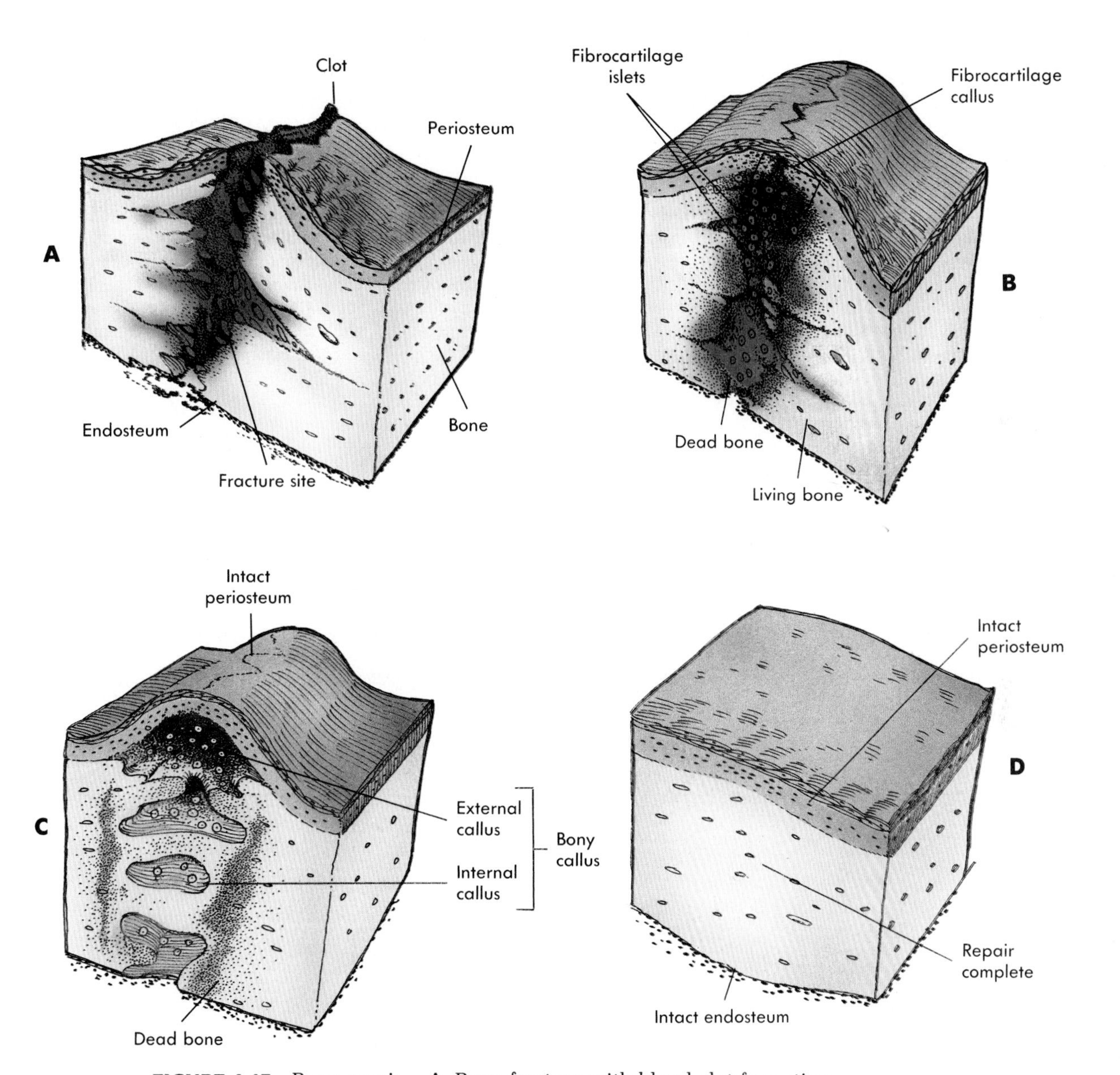

FIGURE 6-17 Bone repair. **A** Bone fracture with blood clot formation. **B** Fibrocartilage callus forms between bone fragments, closing the wound and attaching the fragments. **C** Cancellous bone replaces the fibrocartilage callus. **D** Bone is remodeled, and the cancellous bone is replaced by compact bone. Healing is complete.

Bone Disorders

Growth and Development Disorders

Giantism is a condition of abnormal size, or overgrowth, of the body (Figure 6-B, *A*). This abnormal increase in size usually is due to excessive cartilage and bone formation at the epiphyseal plates of long bones. There are several types of giantism. The most common type, a pituitary giant, results from excess secretion of pituitary growth hormone.

Acromegaly (ak'ro-meg'al-e) is also caused by excess pituitary growth hormone secretion. However, acromegaly involves growth of connective tissue, including bones, after the epiphyseal plates have ossified. The effect mainly involves increased diameter of all bones and is most strikingly apparent in the face and hands. Many pituitary giants also develop acromegaly later in life. However, the large stature of some individuals may result from genetic factors rather than from abnormal levels of growth hormone.

Dwarfism, the condition in which a person is abnormally small, is the opposite of giantism (see Figure 6-B, *A*). Pituitary dwarfism results when abnormally low levels of pituitary growth hormone affect the whole body, thus producing a small person who is proportioned normally. **Achondroplastic** (ă-kon'dro-plas'tik) dwarfism, involving a disproportionate shortening of the long bones, is more common than proportionate dwarfing and produces a person with a nearly normal–sized trunk and head but shorter-than-normal limbs. Most cases of achondroplastic dwarfism are the results of genetic defects that cause deficient or improper growth of the cartilage template, especially the epiphyseal plate, and often involve deficient collagen synthesis. Often the cartilage matrix does not have its normal integrity, and the chondrocytes of the epiphysis cannot form their normal columns, even though rates of cell proliferation may be normal.

A

FIGURE 6-B A Giant and dwarf.

Osteogenesis imperfecta, a group of genetic disorders producing very brittle bones that are fractured easily, occurs because insufficient collagen is formed to strengthen the bones properly. Intrauterine fractures of the extremities usually heal in poor alignment, causing the limbs to appear bent and short (Figure 6-B, *B*). Several other hereditary disorders of bone mineralization involve the enzymes responsible for normal phosphate or calcium metabolism. They closely resemble rickets and result in weak bones.

Bacterial Infections

Osteomyelitis is bone inflammation that often results from bacterial infection. It can lead to complete destruction of the bone. *Staphylococcus aureus,* often introduced into the body through wounds, is a common cause of osteomyelitis (Figure 6-B, *C*). Bone tuberculosis, a specific type of osteomyelitis, results from spread of the tubercular bacterium from the initial site of infection such as the lungs to the bones through the circulatory system.

Tumors

There are many types of tumors with a wide range of resultant bone defects and prognoses (Figure 6-B, *D*). Tumors may be benign or malignant. Malignant bone tumors may metastasize to other parts of the body or result from metastasizing tumors elsewhere.

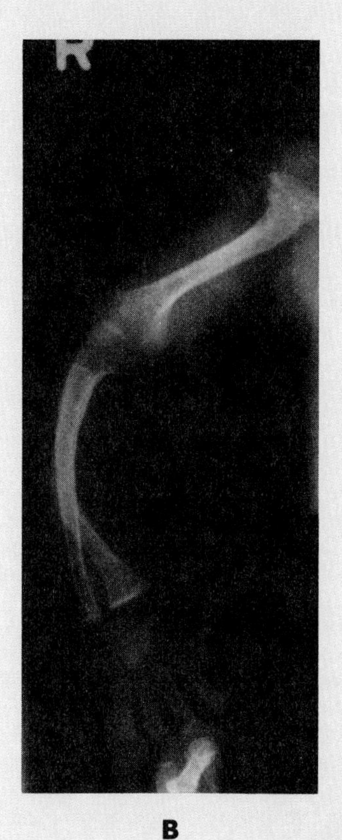

B

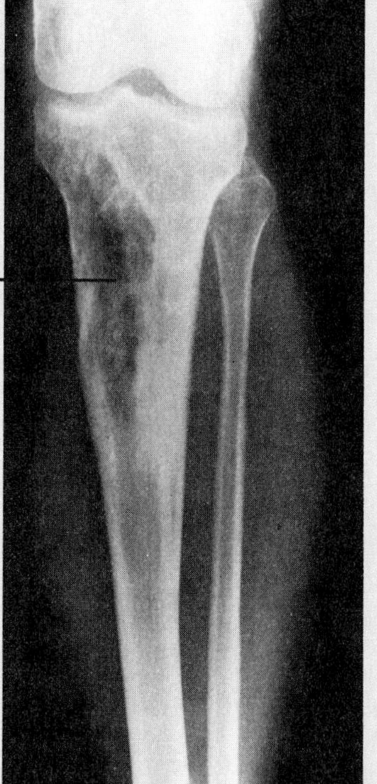

C

B Osteogenesis imperfecta.
C Osteomyelitis.
D Bone tumor.

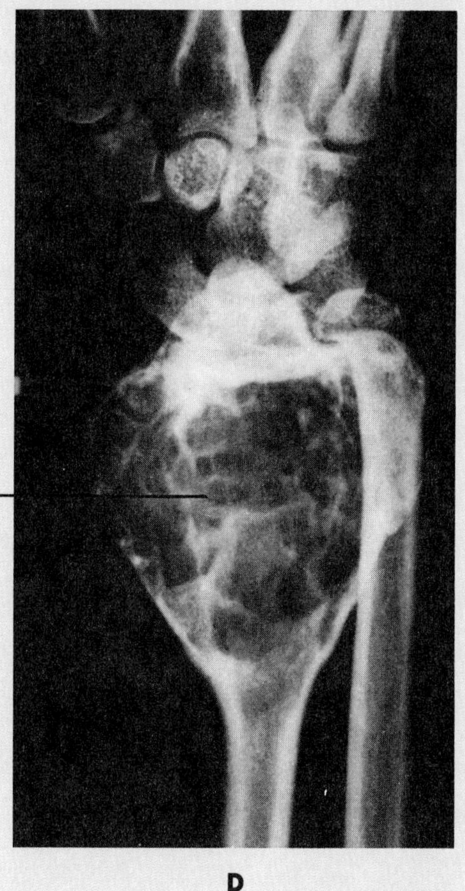

D

Continued.

Bone Disorders—cont'd

Decalcification

Osteomalacia, or the softening of bones, results from calcium depletion from bones. If the body has an unusual need for calcium (e.g., during pregnancy when growth of the fetus requires large amounts of calcium), the calcium may be removed from its bones, which consequently become soft and weakened.

Osteoporosis (os'te-o-po-ro'sis), or porous bone, results from reduction in the overall quantity of bone tissue (Figure 6-B, *E*). It occurs when the rate of bone resorption exceeds the rate of bone formation. The loss of bone mass makes bones so porous and weakened that they become deformed and prone to fracture. Osteoporosis is 2½ times more common in women than in men. The occurrence of osteoporosis also increases with age. In both men and women bone mass starts to decrease at approximately age 40, and continually decreases thereafter. Women eventually can lose approximately one half, and men one quarter, of their cancellous bone.

A common cause of osteoporosis is decreased production of the female sex hormone estrogen in postmenopausal women. Estrogen is secreted by the ovaries, and it normally contributes to the maintenance of normal bone mass by inhibiting the stimulatory effects of parathyroid hormone on osteoclasts. After menopause estrogen production decreases, resulting in degeneration of cancellous bone, especially in the vertebrae of the spine and the bones of the forearm. Collapse of the vertebrae can cause a decrease in height or in more severe cases produce kyphosis, or a "dowager's hump" in the upper back. Conditions other than menopause that result in decreased estrogen can also cause osteoporosis. Examples include removal of the ovaries before menopause, extreme exercise to the point of amenorrhea (lack of menstrual flow), anorexia nervosa (self starvation), and cigarette smoking.

Age-related osteoporosis affects both women and men, causing degeneration of cancellous and compact bone in the hips and legs. The cause of this type of osteoporosis is unclear.

Inadequate dietary intake or absorption of calcium can contribute to osteoporosis. Absorption of calcium from the small intestine decreases with age, and individuals with osteoporosis often have insufficient intake of calcium, vitamin D, and vitamin C.

Because hormones affect bone metabolism, hormonal imbalances can also cause osteoporosis. Overproduction of parathyroid hormone, which results in over-stimulation of osteoclasts, is the most common example.

Finally, osteoporosis can result from inadequate exercise or disuse caused by fractures or paralysis. Significant amounts of bone are lost after 8 weeks of immobilization.

Treatment of osteoporosis attempts to reduce bone loss and/or increase bone formation. Estrogen therapy has proven effective in reducing bone loss in postmenopausal women. Estrogen therapy, however, may have disadvantages such as increasing the risk of uterine or breast cancer. Increased dietary calcium and vitamin D can increase calcium uptake and reduce the amount of estrogen needed. Adequate calcium intake throughout life can also slow or prevent age-related osteoporosis. The greater the bone mass before the onset of osteoporosis, the greater is the tolerance for bone loss later in life. For this reason it is important for adults, especially women, in their twenties and thirties to ingest 1000 mg of calcium a day. Finally, exercise has proven effective not only in reducing bone loss, but in increasing bone mass.

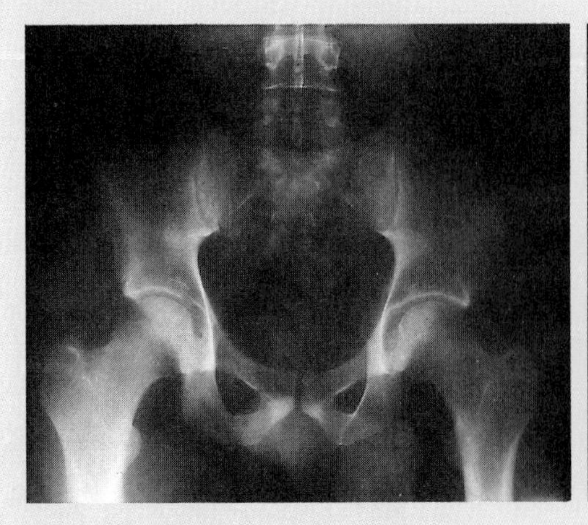

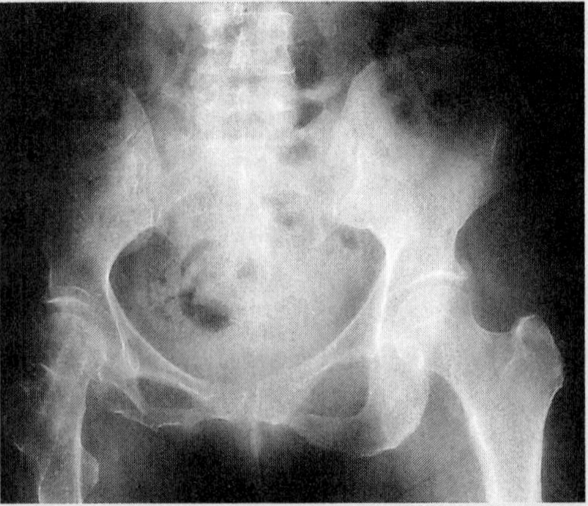

FIGURE 6-B, cont'd E Normal bone compared to bone with osteoporosis.

SUMMARY

THE SKELETAL SYSTEM CONSISTS OF BONES, CARTILAGE, TENDONS, AND LIGAMENTS.

FUNCTIONS OF THE SKELETAL SYSTEM *p. 158*

Bone supports the body, protects organs it surrounds, allows body movements, stores minerals and fats, and is the site of blood cell production.

TENDONS AND LIGAMENTS *p. 158*

1. Tendons attach muscles to bones, and ligaments attach bones to bones.
2. Fibroblasts produce the matrix of tendons and ligaments. Fibrocytes are fibroblasts completely surrounded by matrix.
3. Collagen fibers in tendons are arranged in parallel bundles. In ligaments not all the fibers are parallel, and some fibers contain elastin.
4. Growth
   - Appositional growth occurs when fibroblasts secrete matrix to the outside of existing fibers.
   - Interstitial growth occurs when fibrocytes produce new matrix inside the tissue.

HYALINE CARTILAGE *p. 158*

1. The matrix of cartilage contains large amounts of water, which makes the cartilage resilient.
2. Chondroblasts produce cartilage and become chondrocytes. Chondrocytes are located in lacunae surrounded by matrix.
3. The perichondrium surrounds cartilage.
   - The outer layer contains fibroblasts.
   - The inner layer contains chondroblasts.
4. Cartilage grows by appositional and interstitial growth.

BONE *p. 159*

Bone Shape

Individual bones can be classified as long, short, flat, or irregular.

Bone Anatomy

1. The diaphysis is the shaft of a long bone, and the epiphyses are the ends.
2. The epiphyseal plate is the site of bone growth in length.
3. The medullary cavity is a space within the diaphysis.
4. Red marrow is the site of blood cell production, and yellow marrow consists of fat.
5. The periosteum covers the outer surface of bone.
   - The outer layer contains blood vessels and nerves.
   - The inner layer contains osteoblasts and some osteoclasts.
   - Sharpey's fibers hold the periosteum, ligaments, and tendons in place.
6. The endosteum lines cavities inside bone and contains osteoblasts and osteoclasts.
7. Short, flat, and irregular bones have an outer covering of compact bone surrounding cancellous bone.

Bone Histology

1. Bone matrix
   - Bone is composed of an organic matrix (mostly collagen) that provides flexible strength and an inorganic matrix (hydroxyapatite) that provides compressional strength.
   - Osteoblasts produce bone matrix and become osteocytes. Osteocytes are located in lacunae and are connected to each other through canaliculi.
   - Osteoclasts break down bone.
   - Bone is arranged in thin layers called lamellae.
2. Cancellous bone
   - Lamellae combine to form trabeculae, beams of bone that interconnect to form a latticelike structure.
   - The trabeculae are oriented along lines of stress and provide structural strength.
3. Compact bone
   - Canals within compact bone provide a means for the exchange of gases, nutrients, and waste products: canaliculi connect osteocytes to each other and to haversian canals; haversian canals contain blood vessels that pass to Volkmann's canals; and Volkmann's canals carry blood vessels to and from the periosteum or endosteum.
   - Compact bone consists of highly organized lamellae: circumferential lamellae cover the outer surface of compact bones; concentric lamellae surround haversian canals forming haversian systems, or osteons; interstitial lamellae are remnants of the other lamellae left after bone remodeling.

BONE OSSIFICATION *p. 165*

Intramembranous Ossification

1. Some skull bones and the clavicle develop from membranes.
2. Within the membrane at ossification centers osteoblasts produce bone along the membrane fibers to form cancellous bone.
3. Beneath the periosteum osteoblasts lay down compact bone to form the outer surface of the bone.
4. Fontanels are areas of membrane that are not ossified at birth.

Endochondral Ossification

1. Most bones develop from a cartilage template.
2. The cartilage is calcified and dies. Osteoblasts form lamellae on the calcified cartilage, producing cancellous bone.
3. An outer surface of compact bone is formed beneath the periosteum or endosteum by osteoblasts.
4. Primary ossification centers form in the diaphysis during fetal development. Secondary ossification centers form in the epiphyses. Ossification is completed between birth and 25 years of age.
5. Articular cartilage on the ends of bones and the epiphyseal plate is cartilage that does not ossify.

BONE GROWTH *p. 168*

Appositional Growth

1. Long bones grow in width by bone apposition on the outer surface of the bone.
2. Short, flat, and irregular bones mostly increase in size by appositional growth.

Endochondral Growth

1. Endochondral growth involves the interstitial growth of cartilage followed by endochondral ossification of the cartilage. The result is an increase in bone length.
2. Endochondral growth at the epiphyseal plate results in increase in length of the diaphysis. Bone growth in length ceases when the epiphyseal plate becomes ossified and forms the epiphyseal line.
3. Endochondral growth at articular cartilage results in epiphyseal growth.

Factors Affecting Bone Growth

1. Genetic factors determine bone shape and size. The expression of genetic factors can be modified.
2. Factors such as deficiencies in vitamins D and C that alter the mineralization process or production of organic matrix can affect bone growth.
3. Growth hormone, thyroid hormone, estrogen, and testosterone stimulate bone growth.
4. Estrogen and testosterone cause increased bone growth and closure of the epiphyseal plate.

MAINTENANCE OF BLOOD CALCIUM LEVELS *p. 171*

Parathyroid hormone increases bone breakdown and thus increases blood calcium levels. Calcitonin has the opposite effect.

BONE REMODELING *p. 172*

1. Remodeling allows bone to change shape, adjust to stress, repair, and regulate body calcium levels.
2. Bone adjusts to stress by adding new bone and by realignment of bone through remodeling.

BONE REPAIR *p. 174*

1. Fracture repair begins with the formation of a blood clot.
2. The clot is replaced by a fiber network and by a fibrocartilage callus.
3. The fibrocartilage callus is ossified and remodeled.

CONTENT REVIEW

1. What are the components of the skeletal system?
2. List the functions of the skeletal system.
3. Describe the structure of tendons and ligaments. How do they differ?
4. Why do tendons and ligaments heal slowly?
5. How do tendons and ligaments grow?
6. Describe the structure of cartilage. Where are chondrocytes and chondroblasts found?
7. What is the perichondrium? Describe its structure.
8. Why is the size (thickness) of cartilage limited?
9. How does cartilage grow?
10. Describe the four basic shapes of individual bones.
11. Define diaphysis, epiphysis, epiphyseal plate, and epiphyseal line.
12. Define yellow and red marrow. Where are they located in a child and in an adult?
13. Where are the periosteum and the endosteum located?
14. Name the extracellular components of the bony matrix, and explain their contribution to the strength of bone.
15. What are osteoblasts, osteocytes, and osteoclasts?
16. Describe the structure of cancellous bone. What are trabeculae, and what is their function?
17. Describe the structure of compact bone. How does compact bone differ from cancellous bone?
18. Name the three types of lamellae found in compact bone.
19. Describe the canal system that brings nutrients to osteocytes. What is an osteon?
20. How are cancellous bone and compact bone formed during intramembranous and endochondral ossification?
21. When and where do primary ossification centers and secondary ossification centers appear?
22. How does a bone increase in width? Describe the increase in length of bone at the epiphyseal plate and articular cartilage.
23. Name the two basic components of bone that are affected by nutritional status. How do vitamin D and vitamin C affect bone growth?
24. Describe the conditions that result from too little or too much growth hormone in children and in adults.
25. What effects do estrogen and testosterone have on bone growth? How do these effects account for the average height difference observed in men and women?
26. Name the hormones that regulate calcium levels in the body. Describe the effect of these hormones on the activity of bone cells.
27. What cells are involved in bone remodeling? What is accomplished by remodeling bones?
28. How does bone adjust to stress? Describe the role of osteoblasts and osteoclasts in this process. What happens to bone that is not subjected to stress?
29. Describe the repair of a broken bone.

CONCEPT QUESTIONS

1. When a person develops Paget's disease, for unknown reasons the collagen fibers in the bone matrix run randomly in all directions. In addition, there is a reduction in the amount of trabecular bone. What symptoms would you expect to observe?
2. Why is a haversian system not necessary in the epiphyses of bone?
3. Assume that two patients have identical breaks in the femur (thigh bone). If one is bedridden and the other has a walking cast, which patient's fracture heals faster? Explain.
4. Explain why running helps prevent osteoporosis in the elderly. Does the benefit include all bones or mainly those of the legs?
5. Would a patient suffering from kidney failure be more likely to develop osteomalacia or osteoporosis? Explain.
6. In some cultures eunuchs were responsible for guarding harems (the collective wives of one male). Eunuchs were males who, as boys, were castrated, (i.e., the testes, the major site of testosterone production in males, were removed). Since testosterone is responsible for the sex drive in males, the reason for castration is obvious. As a side effect of this procedure, the eunuchs grew to above-normal heights. Can you explain why this growth happens?
7. A patient has hyperparathyroidism and produces excessive amounts of parathyroid hormone. What effect would this hormone have on bone? Would administration of large doses of vitamin D help the situation? Explain.

? ?

ANSWERS TO PREDICT QUESTIONS

1 *p. 158.* In the absence of a good blood supply, nutrients, chemicals, and cells involved in tissue repair enter the tissue of tendons and ligaments very slowly. As a result, the ability of these tissues to undergo repair is poor.

2 *p. 167.* Collagen provides bone with flexible strength, and a reduction in collagen results in brittle bones that are broken easily.

3 *p. 168.* Bones cannot undergo interstitial growth because bone matrix is rigid and cannot expand from within. Therefore new bone must be added to the surface by apposition.

4 *p. 170.* Damage to the epiphyseal plate interferes with bone elongation. As a result the bone and therefore the thigh will be shorter than normal. Recovery is difficult because cartilage repairs very slowly.

5 *p. 171.* Her growth for the next few months increases, but because the epiphyseal plates ossify earlier than normal, her total height (at age 18) is less than otherwise expected.

7 CHAPTER OBJECTIVES

After reading this chapter you should be able to

1. Name the major bony landmarks and explain the functional significance of each.
2. List the bones contributing to each major portion of the skull.
3. Describe the major features of the floor of the cranial vault, including the foramina and what passes through them.
4. Describe the major features of the skull as seen from the superior, posterior, lateral, inferior, and frontal views.
5. Describe the four major curvatures of the vertebral column, explain what causes them, and indicate when these curvatures develop.
6. List the features that characterize the vertebrae of the cervical, thoracic, lumbar, and sacral regions.
7. Give the number of and explain the difference between true, false, and floating ribs.
8. Describe the shape of the three portions of the sternum and their relationship to the ribs.
9. Describe the two bones of the pectoral girdle and point out the surface features that can be seen on a living human.
10. List the major features of the humerus and give the function of each.
11. Describe the major features of the ulna and radius and explain how those two bones interact when the radius is rotated around the ulna.
12. List the eight carpal bones and describe the carpal tunnel.
13. Indicate the skeletal differences between the thumb and the fingers.
14. Describe the coxa in relation to the three fused bones composing it.
15. List and explain the differences between the male pelvis and the female pelvis.
16. Describe the head and neck of the femur and compare them to those of the humerus.
17. Describe the relationship between the tibia, the fibula, and the femur.
18. List the tarsal bones and describe the relationship between the tibia, the fibula, the talus, and the calcaneus.

Skeletal System: Gross Anatomy

KEY TERMS

appendicular skeleton

axial skeleton

cranial vault

cranium (kra′ne-um)

false pelvis

false rib

foramen (fo-ra′men)

hard palate

nasal septum

orbit

ossicle

paranasal sinus

pectoral (pek′to-ral) girdle

pelvic girdle

vertebral column

RELATED TOPICS

The following terms or concepts are important for a good understanding of this chapter. If you are not familiar with them, you should review them before proceeding.

Directional terms and body regions (Chapter 1)

Bone histology (Chapter 6)

THE GROSS ANATOMY OF THE SKELETAL SYSTEM INCLUDES THOSE FEATURES OF BONES, CARTILAGES, TENDONS, AND LIGAMENTS THAT CAN BE SEEN WITHOUT THE AID OF THE MICROSCOPE. This chapter's content is confined almost entirely to the bones and some major cartilages. Tendons are described in Chapter 11 in relation to muscles, and ligaments are described in Chapter 8 in relation to joints.

Examination of skeletal gross anatomy uses dried, prepared bones. The advantage of this approach is that the major features of individual bones can be seen clearly without being obstructed by associated soft tissues such as muscles, tendons, ligaments, cartilage, nerves, and blood vessels. The disadvantage is that it is easy to ignore the important relationships between bones and soft tissues and the fact that bone itself has soft tissue (e.g., osteocytes and the periosteum) as components (see Chapter 6).

The named bones are divided into two categories: (1) the axial skeleton and (2) the appendicular skeleton. The **axial skeleton** consists of the skull, hyoid bone, vertebral column, and thoracic cage (rib cage). The **appendicular skeleton** consists of the limbs and their girdles. In this chapter bones are described according to these categories.

GENERAL CONSIDERATIONS

It is traditional to list 206 bones in the average adult skeleton (Table 7-1 and Figure 7-1), although the actual number varies from person to person and decreases with age as some bones become fused.

Many of the anatomical features of bones are listed in Table 7-2. Most of these features are based on the relationship between the bones and associated soft tissues. If a bone possesses a **tubercle** (lump) or **process** (projection), it is usually because something (e.g., a ligament or a tendon) was attached to that lump or projection during life. If a bone has a smooth, articular surface, that surface was part of a joint and was covered with articular cartilage. If the bone has a **foramen** (fo-ra'men; a hole) in it, that hole was occupied by something such as a nerve or blood vessel. Some bones contain epithelial-lined air spaces called **sinuses.** These bones are composed of paper-thin, translucent compact bone only and have little or no cancellous center (see Chapter 6).

Table 7-1

Number of named bones listed by category

BONES		NUMBER
Axial Skeleton		
Skull		
Cranial vault		
Paired	Parietal	2
	Temporal	2
Unpaired	Frontal	1
	Occipital	1
	Sphenoid	1
	Ethmoid	1
Face		
Paired	Maxilla	2
	Zygomatic	2
	Palatine	2
	Nasal	2
	Lacrimal	2
	Inferior nasal concha	2
Unpaired	Mandible	1
	Vomer	1
Auditory Ossicles		
Malleus		2
Incus		2
Stapes		2
	TOTAL SKULL	28
Hyoid		1
Vertebral column		
Cervical vertebrae		7
Thoracic vertebrae		12
Lumbar vertebrae		5
Sacrum		1
Coccyx		1
	TOTAL VERTEBRAL COLUMN	26
Thoracic cage (rib cage)		
Ribs		24
Sternum		1
	TOTAL THORACIC CAGE	25
	TOTAL AXIAL SKELETON	80

BONES		NUMBER
Appendicular Skeleton		
Pectoral girdle		
Scapula		2
Clavicle		2
Upper limb		
Humerus		2
Ulna		2
Radius		2
Carpals		16
Metacarpals		10
Phalanges		28
	TOTAL UPPER LIMB AND GIRDLE	64
Pelvic girdle		
Coxa		2
Lower limb		
Femur		2
Tibia		2
Fibula		2
Patella		2
Tarsals		14
Metatarsals		10
Phalanges		28
	TOTAL LOWER LIMB AND GIRDLE	62
	TOTAL APPENDICULAR SKELETON	126
	TOTAL AXIAL SKELETON	80
	TOTAL APPENDICULAR SKELETON	126
	TOTAL BONES	206

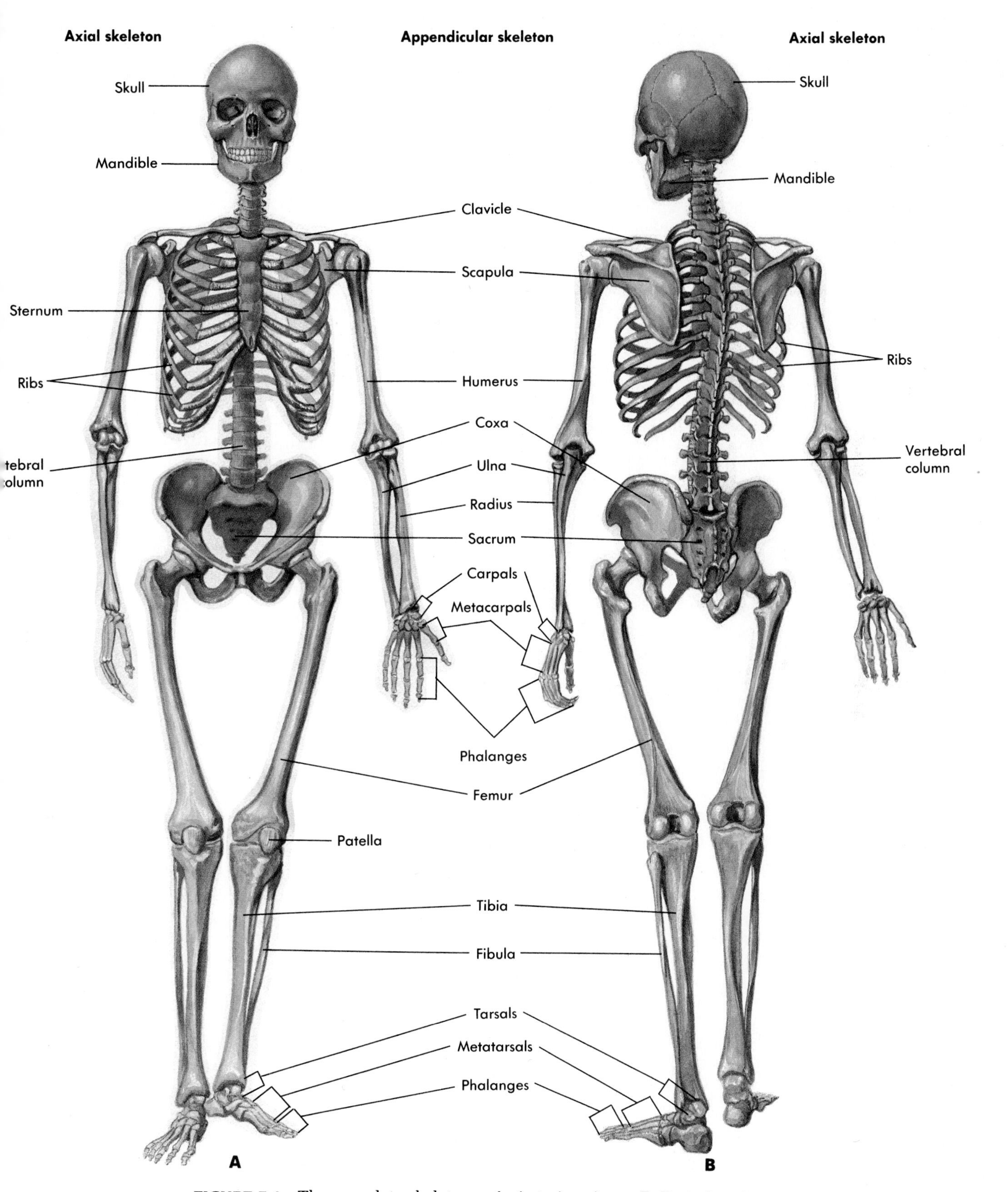

FIGURE 7-1 The complete skeleton. **A** Anterior view. **B** Posterior view.

Table 7-2

General anatomical terms for various features of bones

TERM	DESCRIPTION
Major Features	
Body	Main portion
Head	Enlarged (often rounded) end
Neck	Constricted area (between head and body)
Margin or border	Edge
Angle	Bend
Ramus	Branch off the body (beyond the angle)
Condyle	Smooth, rounded articular surface
Facet	Small, flattened articular surface
Ridges	
Line or linea	Low ridge
Crest or crista	Prominent ridge
Spine	Very high ridge
Projections	
Process	Prominent projection
Tubercle	Small, rounded process
Tuberosity or tuber	Knoblike process; usually larger than a tubercle
Trochanter	Large tuberosity found only on the proximal femur
Epicondyle	Near or above a condyle
Lingula	Flat, tongue-shaped process
Hamulus	Hook-shaped process
Cornu	Horn-shaped process
Openings	
Foramen	Hole
Nutrient foramen	Conveys blood vessels supplying the bone itself
Canal or meatus	Tunnel
Fissure	Cleft
Sinus or labyrinth	Cavity
Depressions	
Fossa	General term for a depression
Impression	Indentation made by a specific structure
Notch	Depression in the margin of a bone
Fovea	Little pit
Groove or sulcus	Deeper, narrow depression

AXIAL SKELETON

The axial skeleton is divided into the skull, hyoid bone, vertebral column, and thoracic cage. The axial skeleton forms the upright axis of the body. It also protects the brain, spinal cord, and the vital organs housed within the thorax.

Skull

The skull is composed of 28 separate bones (see Table 7-1) organized into the following groups: the auditory ossicles, the cranial vault, and the facial bones. The six **auditory ossicles** (two each of the malleus, incus, and stapes), which function in hearing (see Chapter 15), are located, three on each side of the head, inside cavities of the temporal bones and cannot be observed unless the temporal bones are cut open. The remaining 22 bones of the skull, or **cranium** (kra′ne-um), are roughly divided into two portions: the cranial vault and the face. The individual bones are illustrated in Figure 7-2. The **cranial vault,** or brain case, consists of eight bones that immediately surround and protect the brain. They include the paired parietal and temporal bones, and the unpaired frontal, occipital, sphenoid, and ethmoid bones.

The **facial bones** (14 in number) form the structure of the face in the anterior skull but do not contribute to the cranial vault (see Table 7-1). They are the maxilla (two), mandible (one), zygomatic (two), palatine (two), nasal (two), lacrimal (two), vomer (one), and inferior nasal concha (two) bones. The frontal and ethmoid bones, which are part of the cranial vault, also contribute to the face. The mandible is often listed as a facial bone, even though it is not part of the intact skull.

The facial bones provide protection for major sensory organs located in the face such as the eyes, nose, and tongue. The bones of the face also provide attachment points for muscles involved in **mastication** (mas′tĭ-ka-shun; chewing), facial expression, and eye movement. The jaws (mandible and maxillae) possess **alveolar** (al′ve-o′lar) processes with sockets for the attachment of the teeth. The bones of the face, along with their associated soft tissues, determine the unique facial features of each individual.

The **hyoid bone** (Figure 7-2, *N*), which is unpaired, is not actually part of the skull (see Table 7-1) and has no direct bony attachment to the skull. It is attached to the skull by muscles and ligaments, it "floats" in the superior aspect of the neck just below the mandible. The hyoid bone provides a base for the tongue muscles, and it also is an attachment point for important neck muscles that elevate the larynx during speech or swallowing.

Text continued on p. 193.

FIGURE 7-2 *Bones of the skull*

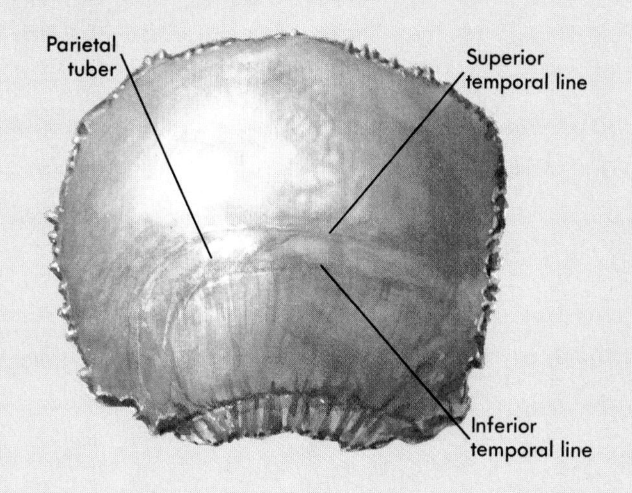

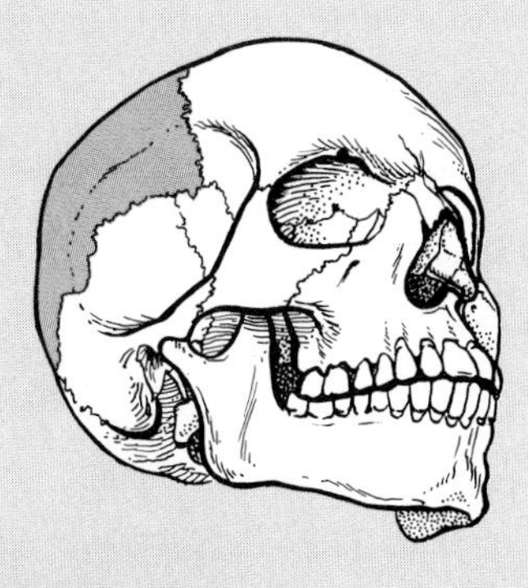

A Right parietal bone
(viewed from the lateral side)

SPECIAL FEATURE: Forms the lateral walls of the skull.

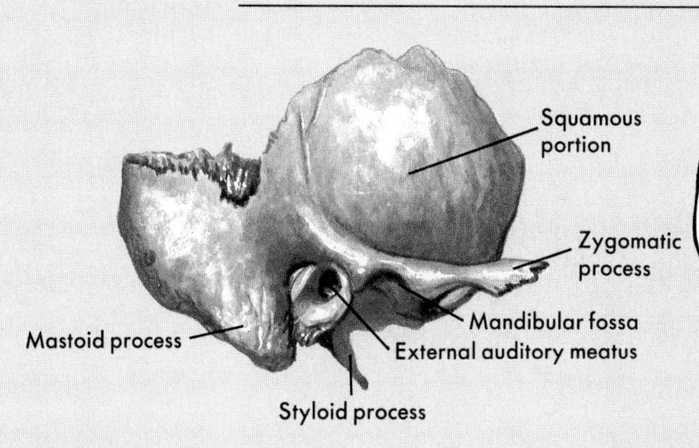

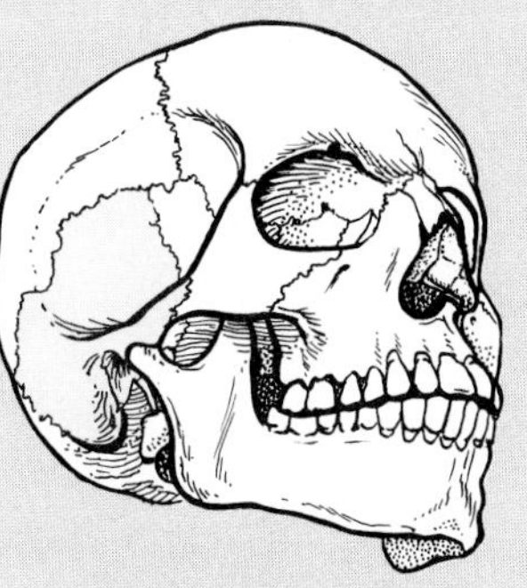

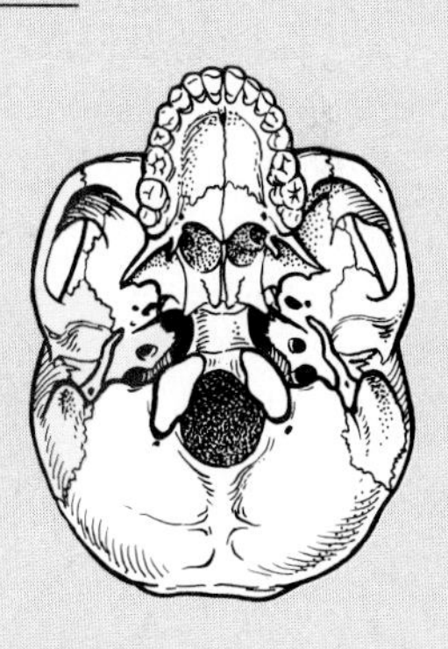

B Right temporal bone
(viewed from the lateral side)

SPECIAL FEATURES: Contains the external auditory meatus (external canal of the ear) and the middle and inner ears; has a mandibular fossa where the mandible articulates with the skull; has the styloid process for muscle attachment and the zygomatic process, which attaches to the zygomatic bone. Contains the mastoid air cells.

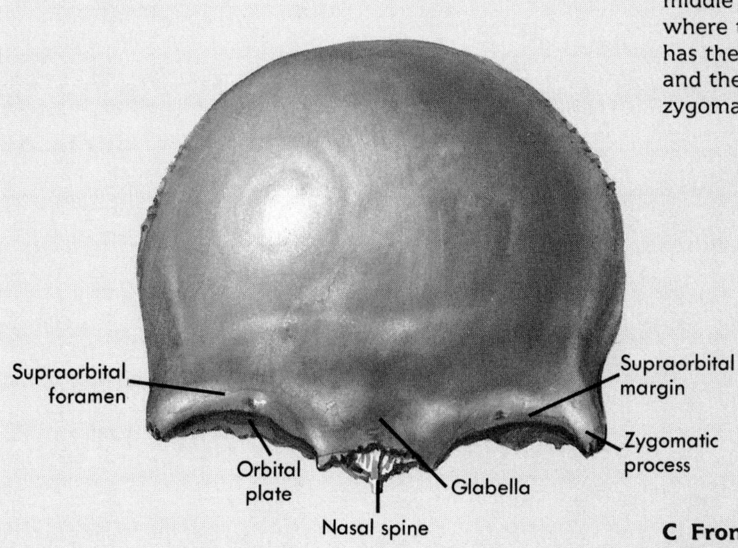

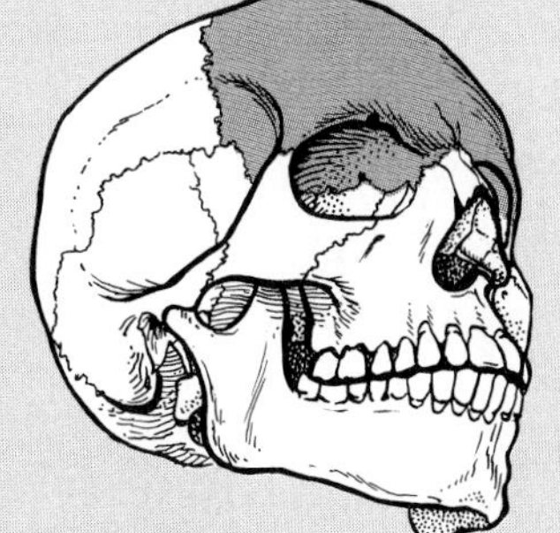

C Frontal bone
(viewed from in front and slightly above)

SPECIAL FEATURES: Forms the forehead and roof of the orbit. Contains the frontal sinus.

Continued.

FIGURE 7-2—cont'd

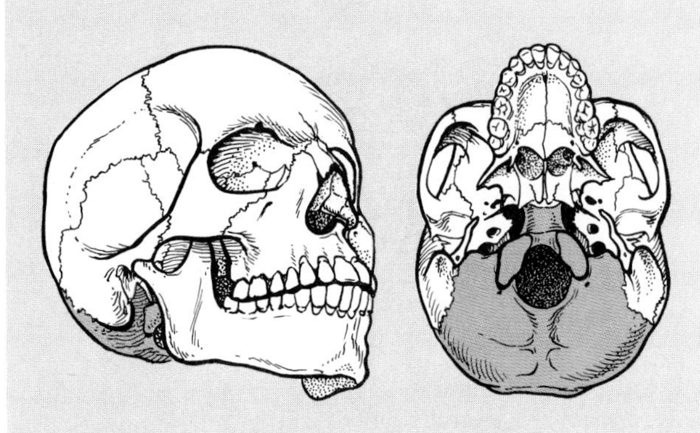

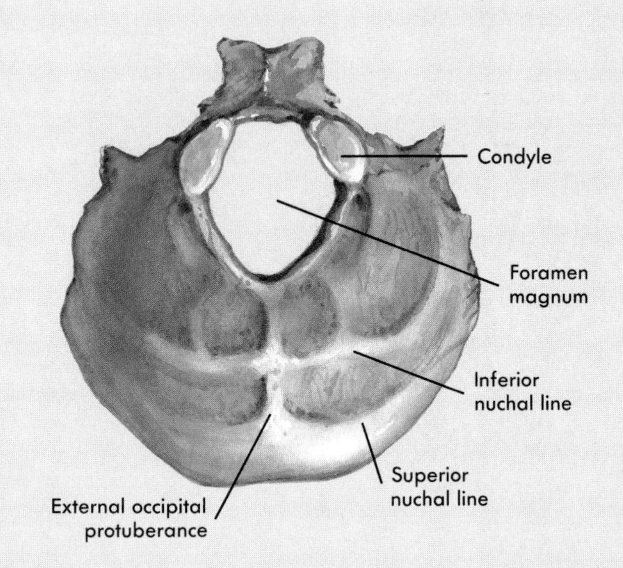

D Occipital bone (viewed from below)

SPECIAL FEATURES: Contains the foramen magnum through which the spinal cord passes and has ridges (nuchal lines) where neck muscles attach.

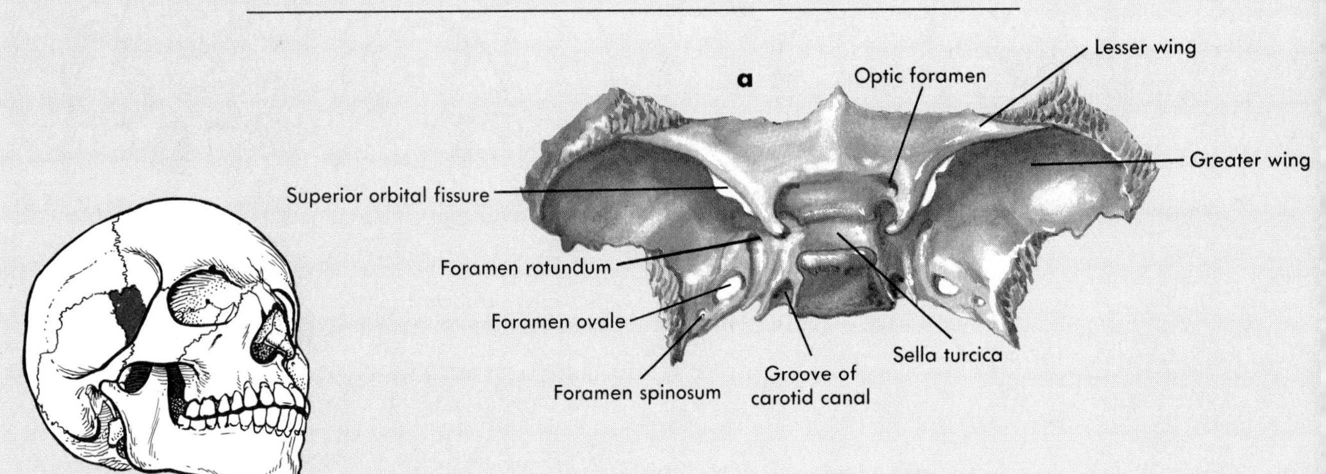

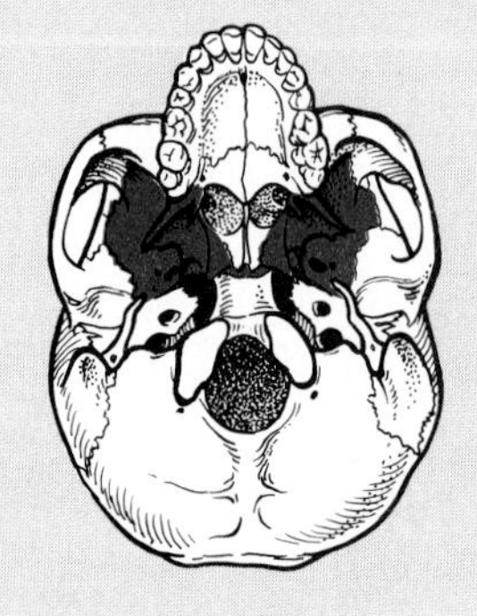

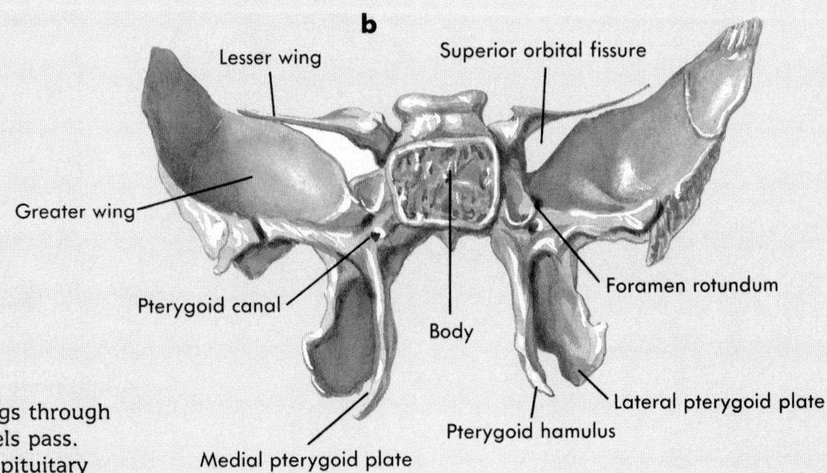

E Sphenoid bone
a, Superior view. *b,* Posterior view.

SPECIAL FEATURES: Several openings through which cranial nerves and blood vessels pass. Contains the sella turcica where the pituitary gland is located. Contains the sphenoidal sinus.

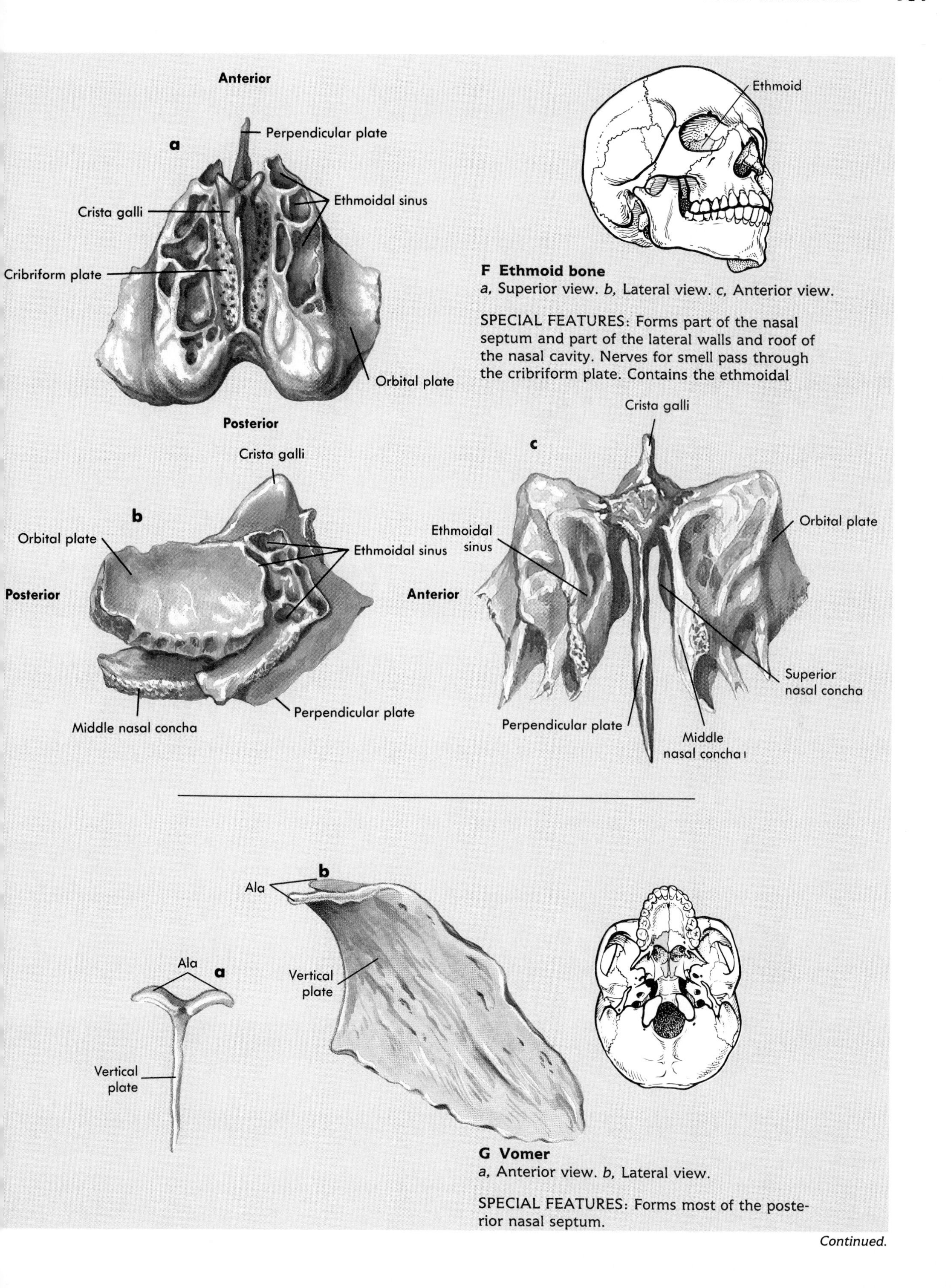

F Ethmoid bone
a, Superior view. *b,* Lateral view. *c,* Anterior view.

SPECIAL FEATURES: Forms part of the nasal septum and part of the lateral walls and roof of the nasal cavity. Nerves for smell pass through the cribriform plate. Contains the ethmoidal

G Vomer
a, Anterior view. *b,* Lateral view.

SPECIAL FEATURES: Forms most of the posterior nasal septum.

Continued.

FIGURE 7-2—cont'd

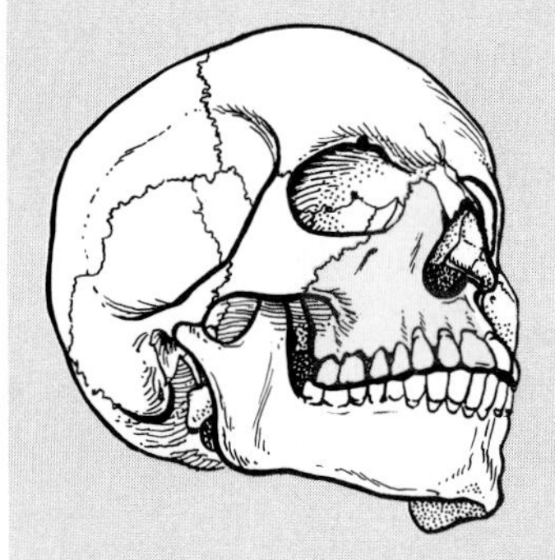

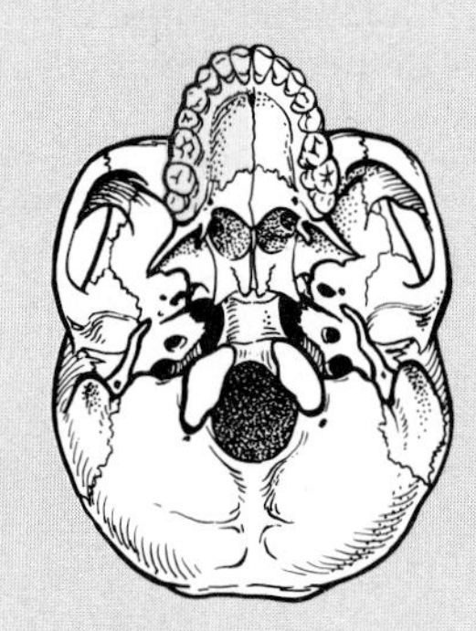

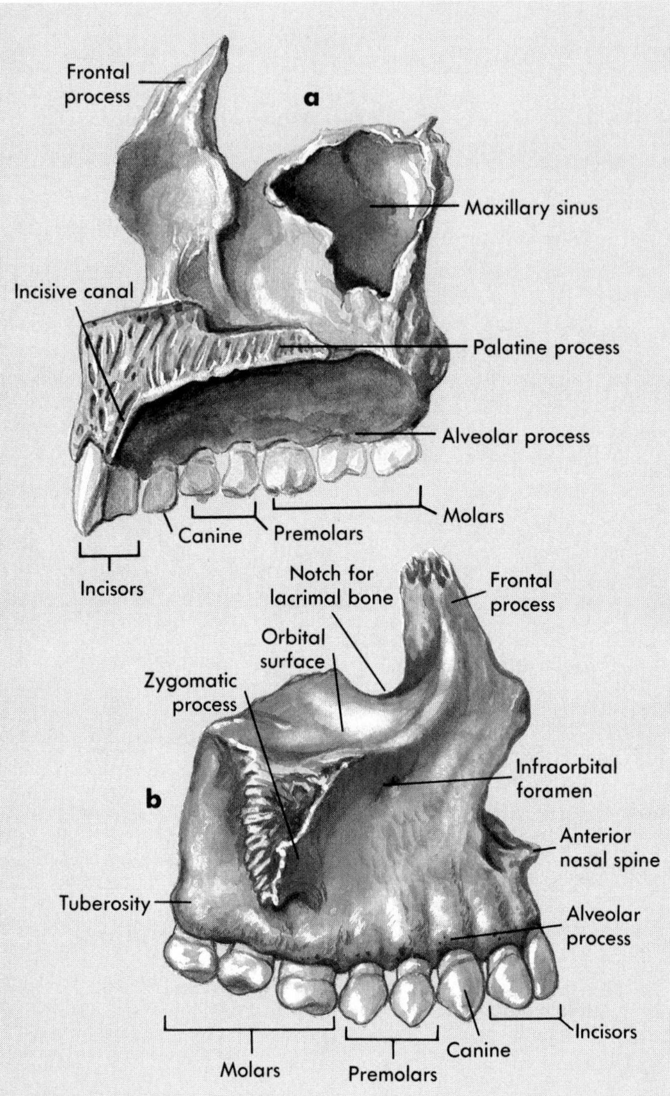

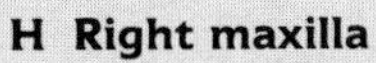

H Right maxilla
a, Medial view. *b,* Lateral view.

SPECIAL FEATURES: Helps form part of the hard palate, the floor of the orbit, and the lateral wall and floor of the nasal cavity. Holds the upper teeth. Contains the maxillary sinus.

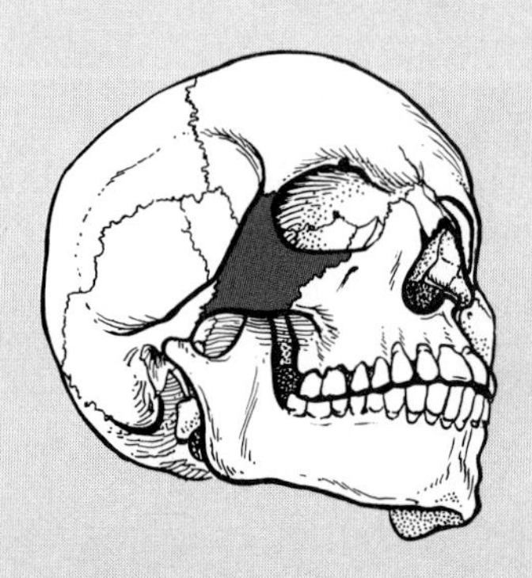

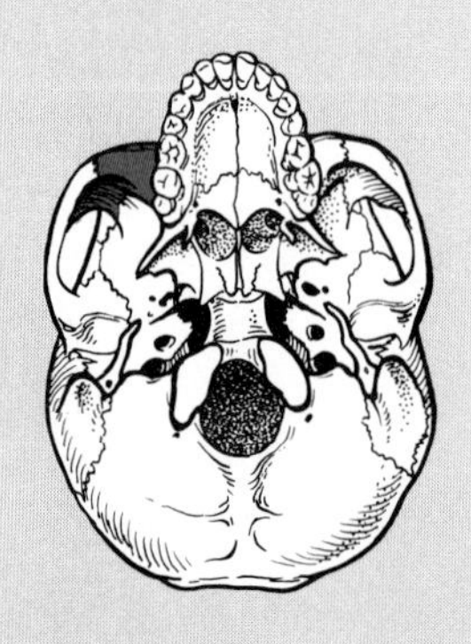

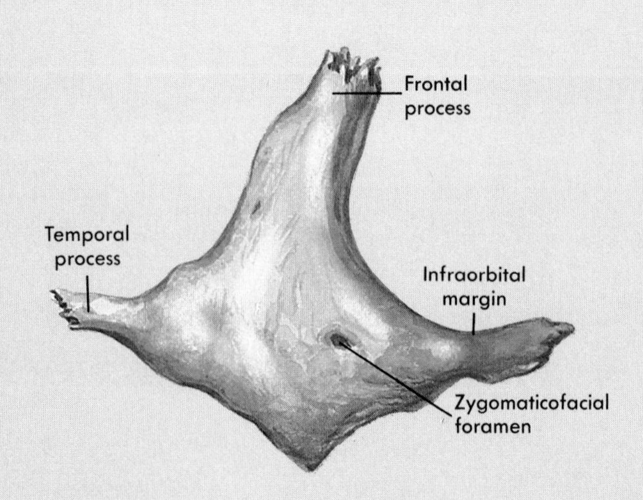

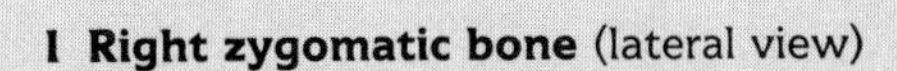

I Right zygomatic bone (lateral view)

SPECIAL FEATURES: Forms the prominence of the cheek. Forms the anterolateral wall of the orbit.

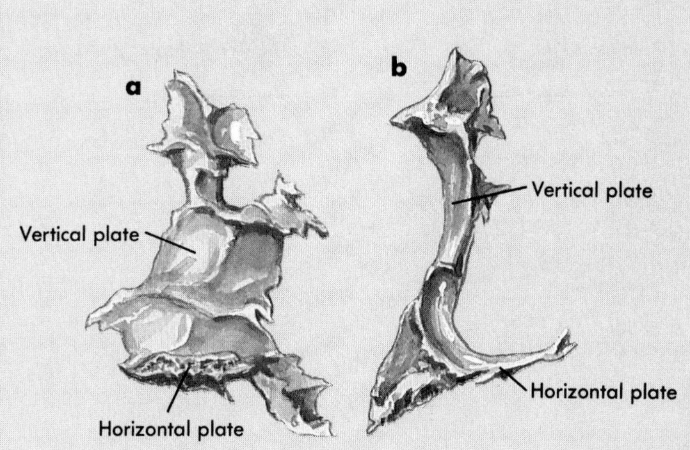

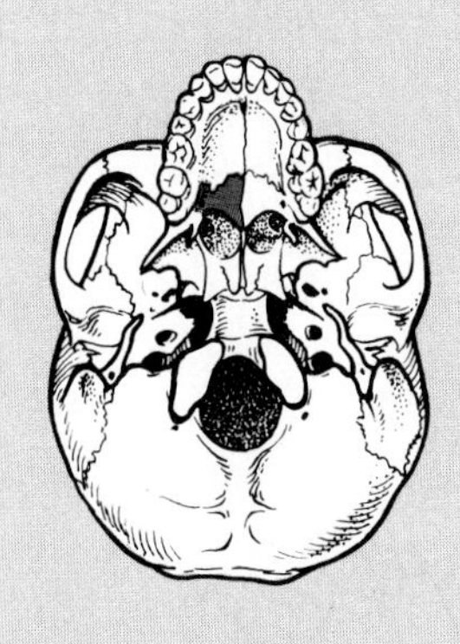

J Right palatine bone
a, Medial view. *b,* Anterior view.

SPECIAL FEATURES: Helps form part of the hard palate and a small part of the wall of the orbit.

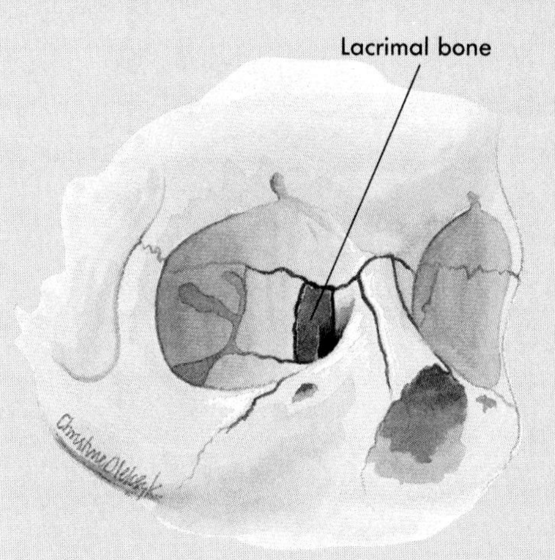

K Right lacrimal bone (lateral view)

SPECIAL FEATURE: Forms a small portion of the orbital wall.

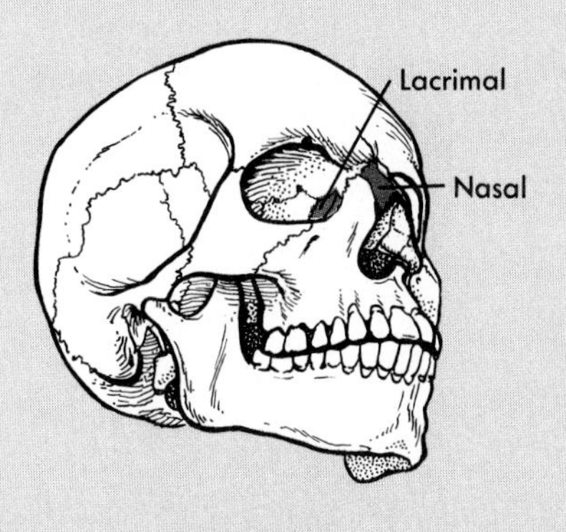

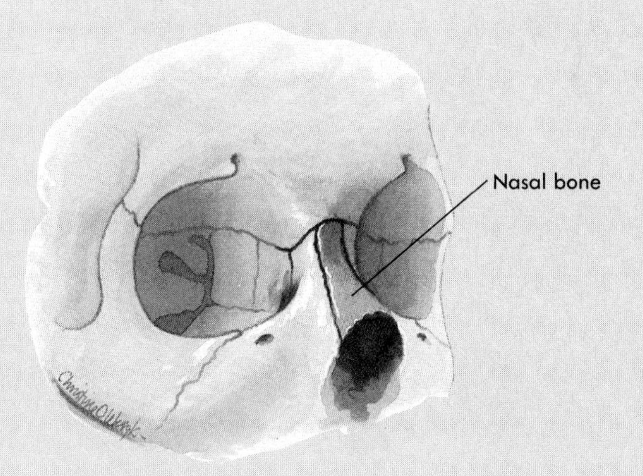

L Right nasal bone (lateral view)

SPECIAL FEATURE: Forms the bridge of the nose.

Continued.

FIGURE 7-2—cont'd

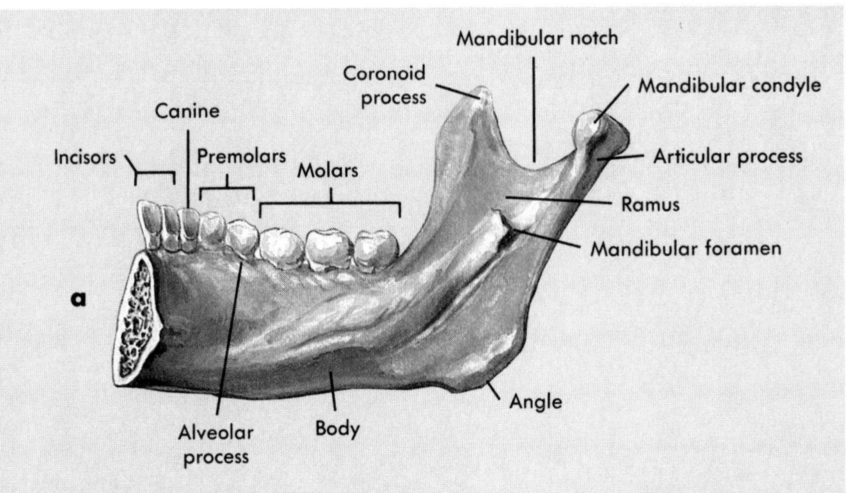

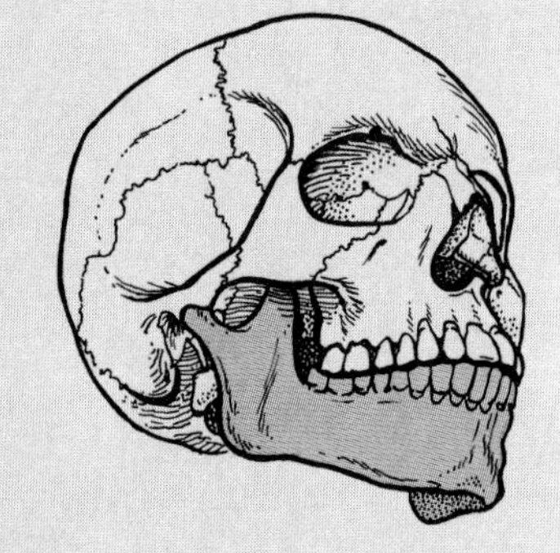

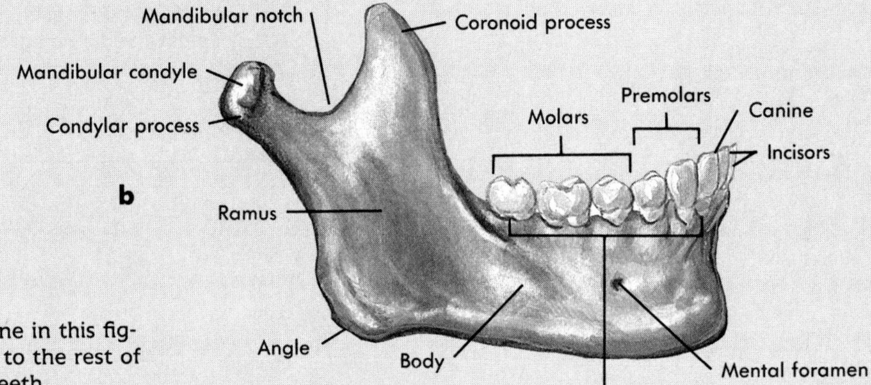

M Right half of mandible
a, Medial view. *b,* Lateral view.

SPECIAL FEATURES: The only bone in this figure that is freely movable relative to the rest of the skull bones. Holds the lower teeth.

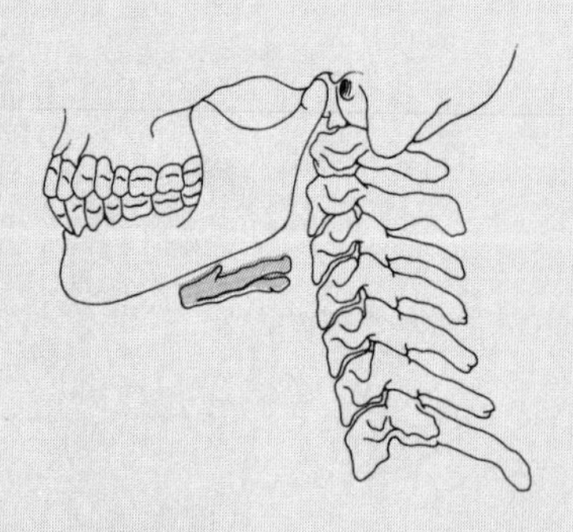

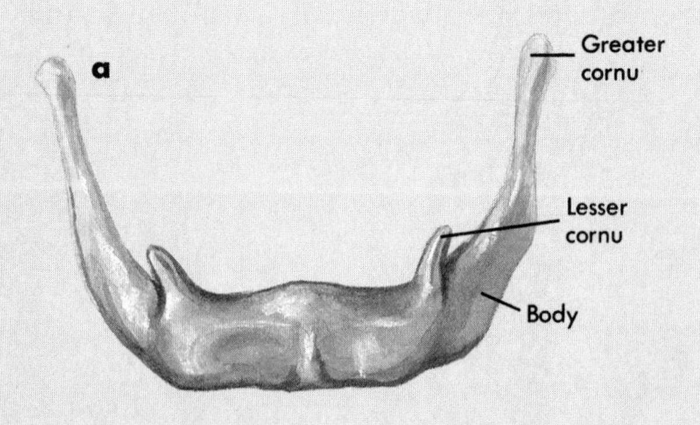

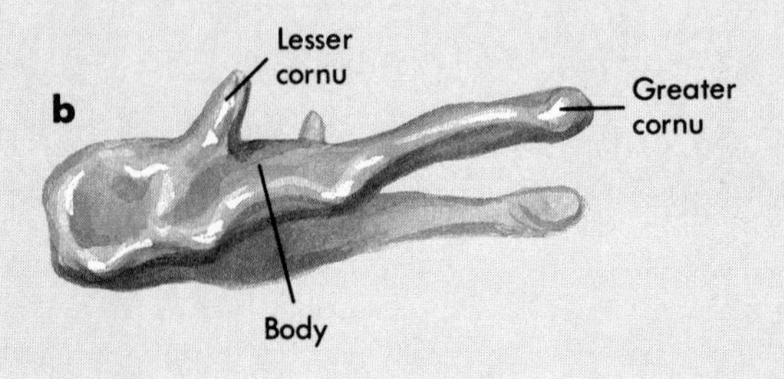

N Hyoid bone
a, Anterior view. *b,* Lateral view (from the left side).

SPECIAL FEATURES: Not actually part of the skull. One of the few bones of the body that does not articulate with another bone. It is attached to the skull by muscles and ligaments.

Unit skull

It is often convenient to think of the skull, excluding the mandible, as a single bone. The external features of the skull include ridges, lines, processes, and plates—important for the attachment of muscles or for articulations between the bones of the skull. There are also a number spaces and ridges that mark the internal aspect of the skull. Selected features of the intact skull are listed in Table 7-3.

Table 7-3

Processes and other features of the skull

FEATURE	BONE ON WHICH FEATURE IS FOUND	DESCRIPTION
External Features of Unit Skull		
Alveolar process	Maxilla	Ridge on maxilla containing the teeth
Horizontal plate	Palatine	Posterior one third of the hard palate
Mandibular fossa	Temporal	Depression where the mandible articulates with the skull
Mastoid process	Temporal	Enlargement posterior to the ear; attachment site for several muscles that move the head
Nuchal lines	Occipital	Attachment points for several posterior neck muscles
Occipital condyle	Occipital	Point of articulation between the skull and the vertebral column
Palatine process	Maxilla	Anterior two thirds of the hard palate
Pterygoid hamulus	Sphenoid	Hooked process on the inferior end of the medial pterygoid plate, around which the tendon of one palatine muscle passes; an important dental landmark
Pterygoid plates (medial and lateral)	Sphenoid	Bony plates on the inferior aspect of the sphenoid bone; sites of attachment for two muscles of mastication (chewing)
Styloid process	Temporal	Attachment site for three muscles (to the tongue, pharynx, and hyoid bone) and several ligaments
Temporal lines	Parietal	Where the temporalis muscle, which closes the jaw, attaches
Internal Features of Unit Skull		
Crista galli	Ethmoid	Process in the anterior part of the cranial vault to which one of the connective tissue coverings of the brain (dura mater) connects
Petrous portion	Temporal	Thick, interior portion of temporal bone; contains middle and inner ears and auditory ossicles
Sella turcica	Sphenoid	Bony structure resembling a saddle in which the pituitary gland is located
Mandible		
Alveolar process		Ridge on the mandible containing the teeth
Angle		Posterior, inferior corner of mandible
Coronoid process		Attachment point for the temporalis muscle
Genu		Chin (resembles a bent knee)
Mandibular condyle		Region where the mandible articulates with the skull
Ramus		Portion of the mandible superior to the angle

FIGURE 7-3
Skull as seen from a superior view.

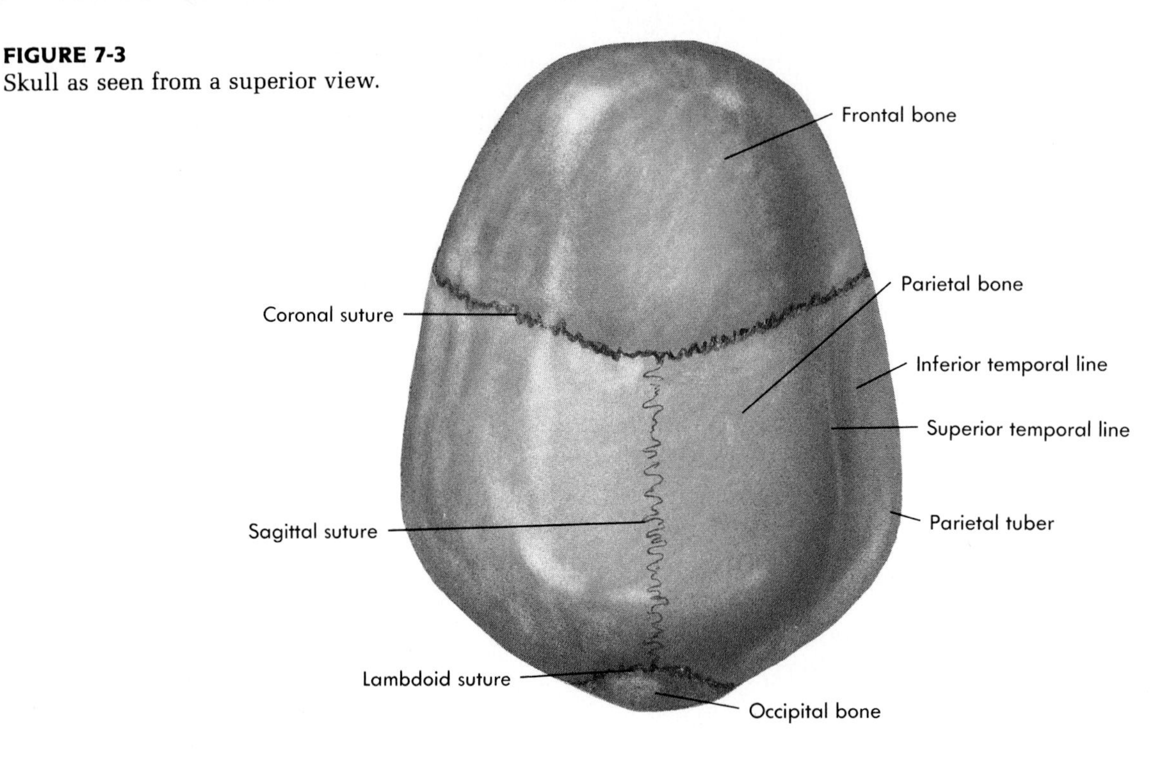

Superior view of the skull

The skull appears quite simple when viewed from above. Only four bones are seen from this view: the frontal bone, two parietal bones, and a small part of the occipital bone. The paired **parietal bones** are joined at the midline by the **sagittal suture,** and the parietal bones are connected anteriorly to the **frontal bone** by the **coronal suture** (Figure 7-3).

1

Explain the basis for the names of the sagittal and coronal sutures.

? ? ? ? ? ? ? ? ? ? ?

Posterior view of the skull

The parietal and occipital bones are the major structures seen from the posterior view (Figure 7-4). The parietal bones are joined to the occipital bone posteriorly by the **lambdoid** (lam′doyd; shape resembles the Greek letter lambda) **suture.** Occasionally small **sutural** (su′chur-ul; small, extra, irregular bones; also called wormian bones) **bones** form along the lambdoid suture.

An **external occipital protuberance** is present on the posterior surface of the occipital bone (see Figure 7-4). It can be felt through the scalp at the base of the head and varies considerably in size between individuals. The external occipital protuberance is the site of attachment of the **ligamentum nuchae** (nu′ke; nape of neck), an elastic ligament that extends down the neck and helps keep the head erect by pulling on the occipital region of the skull. **Nuchal lines** are a set of small ridges that extend laterally from the protuberance and are the points of attachment for several neck muscles.

The ligamentum nuchae and neck muscles in humans are not as strong as comparable structures in other animals; therefore the human bony prominence and lines of the posterior skull are not as well developed as in those animals. The location of the human foramen magnum allows the skull to balance above the vertebral column and allows an upright posture. Thus human skulls require less ligamental and muscular effort to balance the head on the spinal column than do the skulls of other animals (including other primates) whose skulls are not balanced over the vertebral column. The presence of small nuchal lines in hominids (animals with an upright stance like humans) reflects this decreased musculature and is one way anthropologists can establish probable upright posture in hominids.

Lateral view of the skull

The parietal bone and the squamous portion of the temporal bone (related to time; so named because the hair of the temples is often the first to turn white, indicating the passage of time) form a large portion of the side of the head (Figure 7-5). The **squamous suture** joins these bones. A prominent feature of the temporal bone is a large hole, the **external auditory**

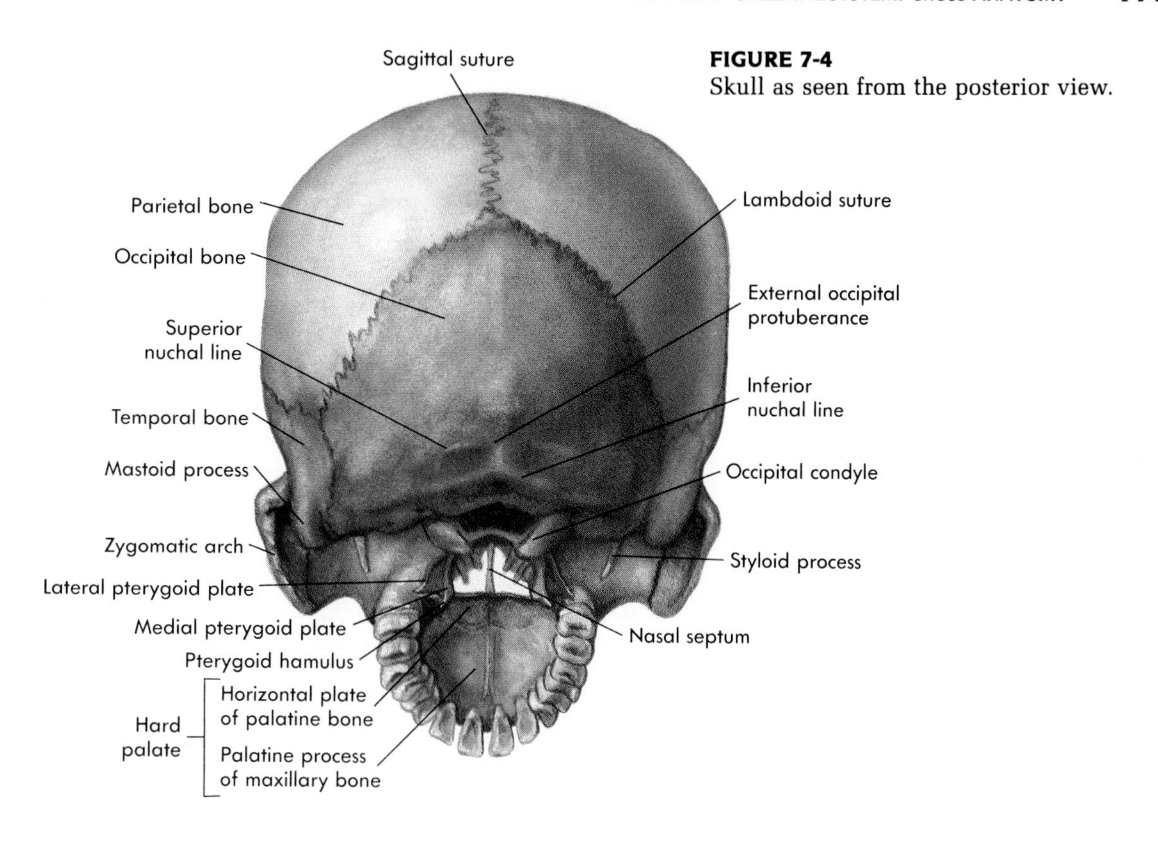

FIGURE 7-4
Skull as seen from the posterior view.

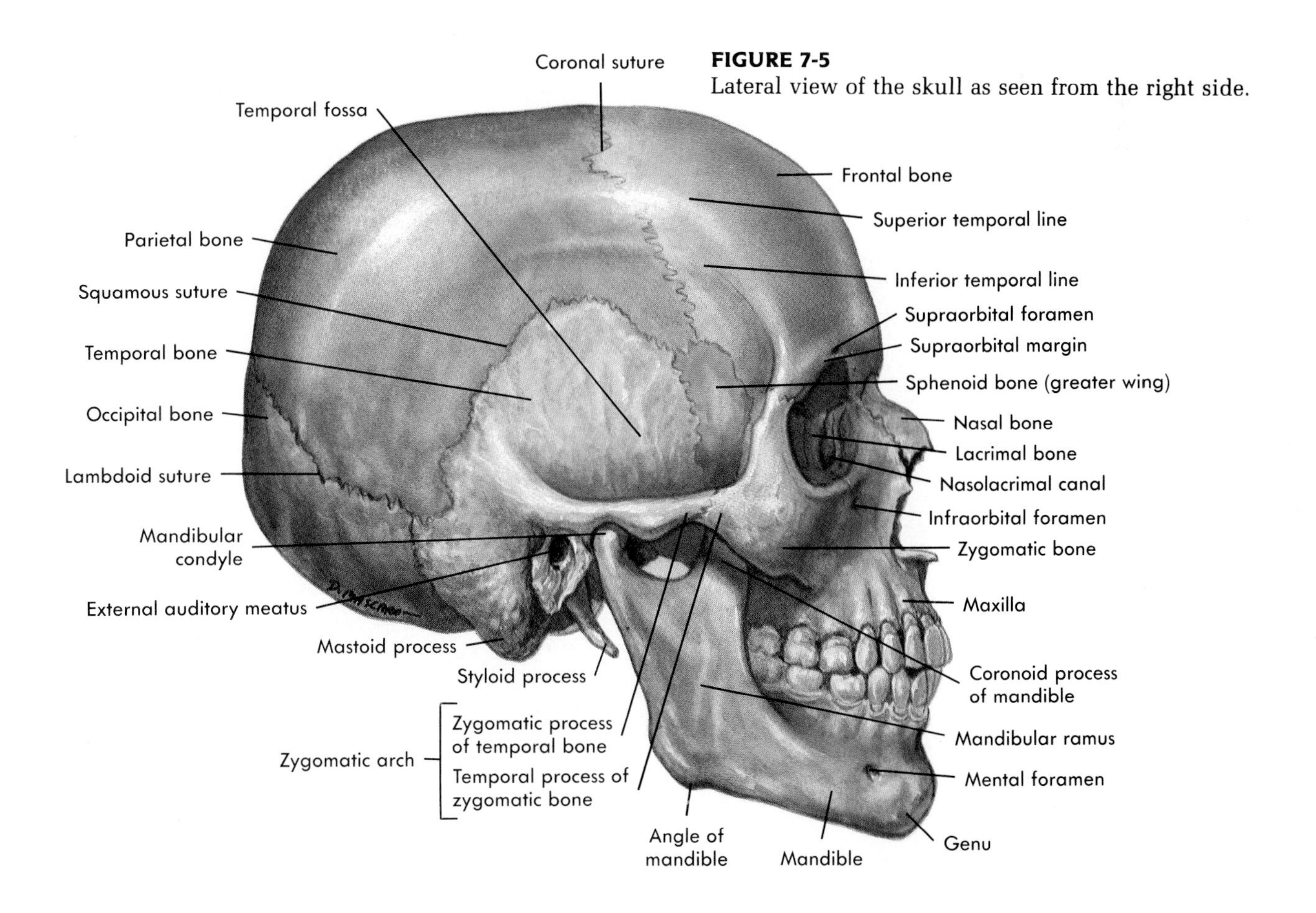

FIGURE 7-5
Lateral view of the skull as seen from the right side.

meatus (me-a'tus; passageway or tunnel), which transmits sound waves toward the eardrum. The external ear, or auricle, surrounds the meatus. Just posterior and inferior to the external auditory meatus is a large inferior projection, the **mastoid** (mas'toyd; resembling a breast) **process.** The process can be seen and felt as a prominent lump just posterior to the ear (see Figure 7-5). It is not solid bone but is filled with cavities called the mastoid air cells, which are connected to the middle ear. Important neck muscles involved in rotation of the head attach to the mastoid process.

The **temporal lines,** which are attachment points of the temporalis muscle (one of the major muscles of mastication), arch across the lateral surface of the parietal bone (see Figure 7-5). The **temporal fossa,** which extends inferior to the temporal lines and passes deep to the zygomatic arch (described below), is occupied by the temporalis muscle during life.

The temporal lines are important to anthropologists because a heavy temporal line suggests a strong temporalis muscle that supported a heavy jaw. In a male gorilla the temporalis muscles are so large that the temporal lines meet in the midline of the skull to form a heavy sagittal crest. In humans the temporal lines are much smaller.

The lateral surface of the **greater wing** of the **sphenoid** (sfē'noyd; wedge shaped) **bone** is immediately anterior to the temporal bone (see Figure 7-5). The sphenoid bone, although appearing to be two paired bones (one on each side of the skull), is actually a single bone that extends completely across the skull. Anterior to the sphenoid bone is the **zygomatic** (zi-go-mat'ik; a bar or yoke), or cheek, **bone** (see Figure 7-5), which can be easily seen and felt on the face (Figure 7-6).

The **zygomatic arch,** which consists of joined processes from the temporal and zygomatic bones, forms a bridge across the side of the skull, superficial to the temporal fossa (see Figure 7-5). The zygomatic arch is easily felt on the side of the face, and the muscles on either side of the arch can be felt as the jaws are opened and closed (see Figure 7-6).

The **maxilla** is anterior and inferior to the zygomatic bone to which it is joined (see Figure 7-5). The **mandible** is inferior to the maxilla and articulates posteriorly with the temporal bone. The maxilla contains the superior set of teeth, and the mandible contains the inferior teeth.

Frontal view of the skull

The major structures seen from the frontal view are the frontal bone (forehead), the zygomatic bones (cheeks), the maxillae (upper jaw), and the mandible (lower jaw) (Figure 7-7). The teeth, which are very prominent in this view, are discussed in Chapter 24. Many bones of the face may be easily felt through the skin of the face (Figure 7-8).

From this view the most prominent openings into the skull are the orbits and the nasal cavity. The **orbits** are cone-shaped fossae with their apices directed posteriorly (see Figures 7-7 and 7-9). They are called orbits because of the rotation of the eyes within the sockets. The bones of the orbits provide both protection for the eyes and attachment points for the

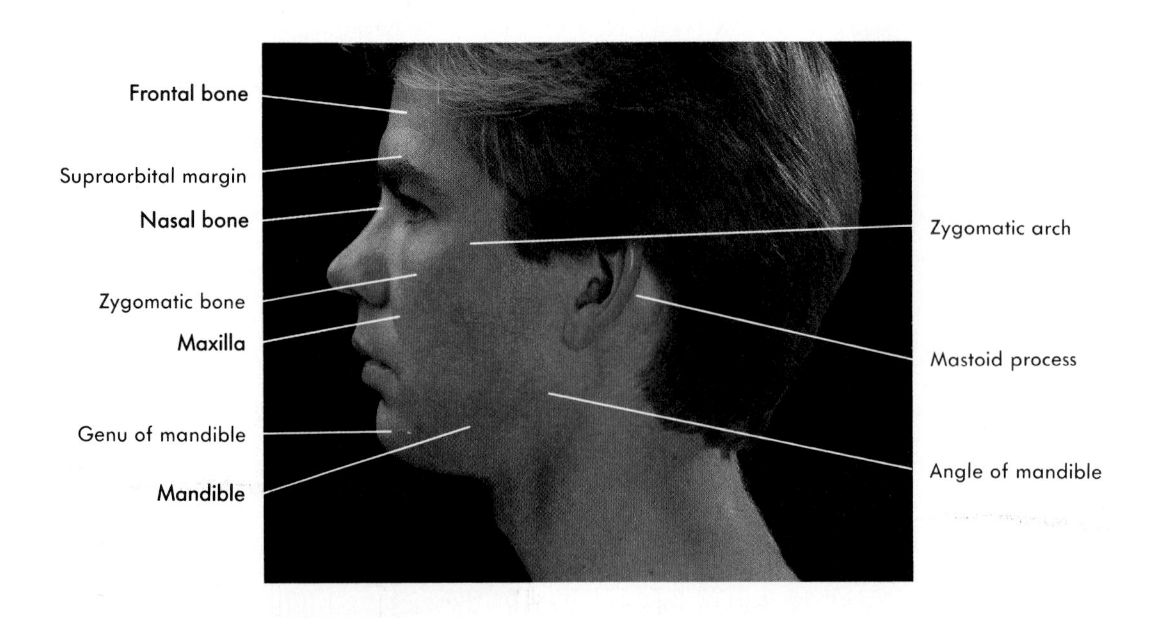

FIGURE 7-6 Lateral view of bony landmarks on the face.

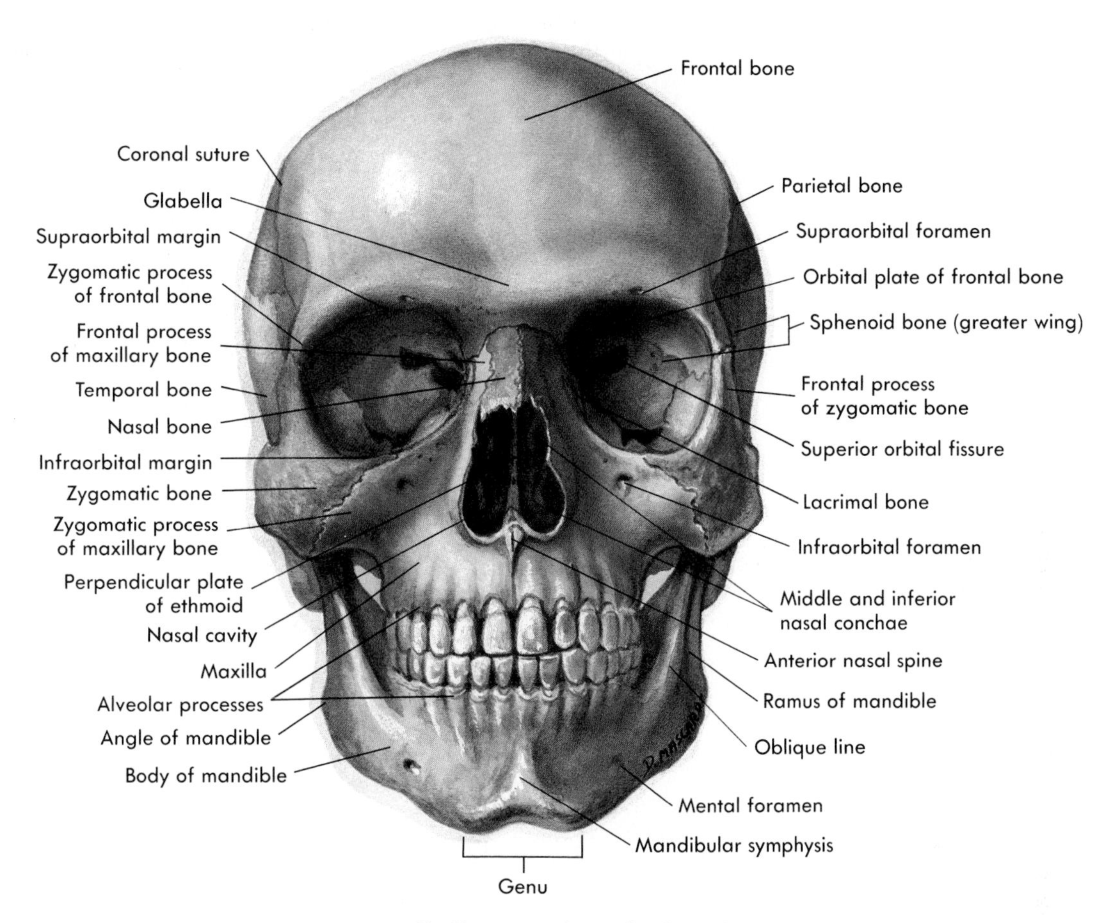

FIGURE 7-7 Skull as seen from the frontal view.

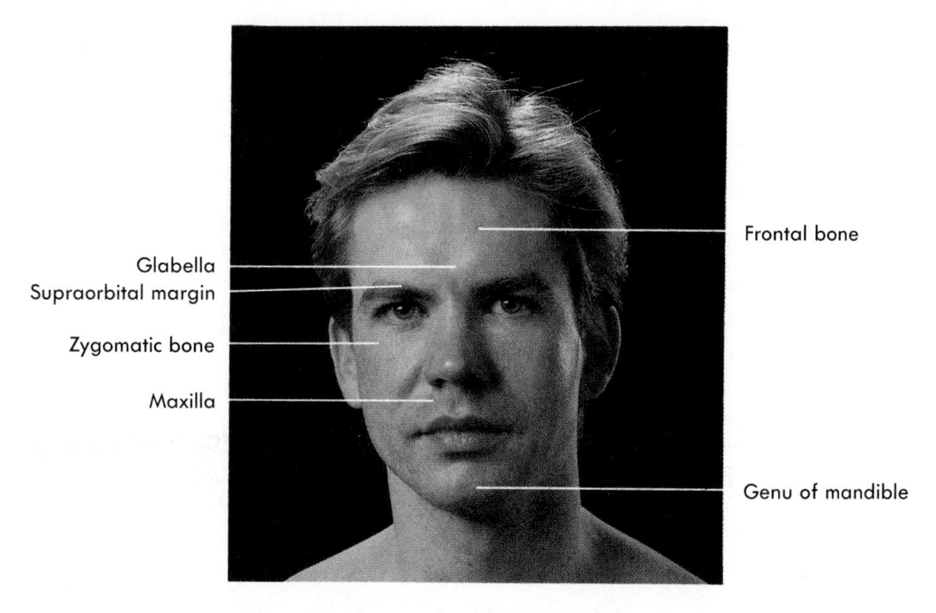

FIGURE 7-8 Anterior view of bony landmarks on the face.

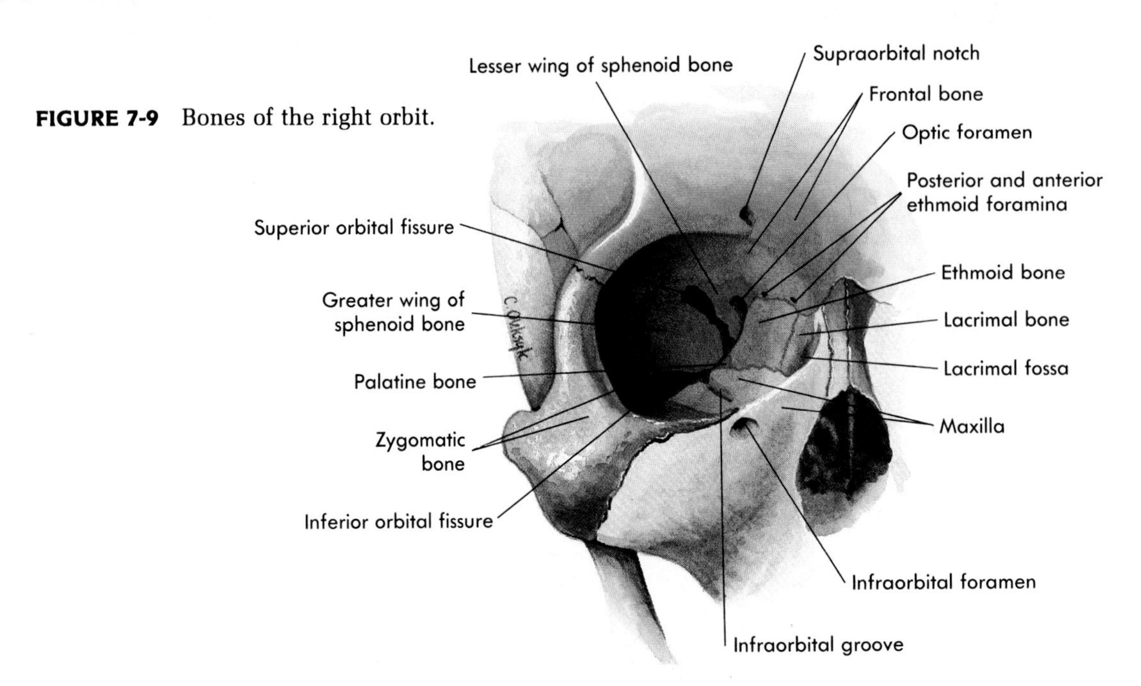

FIGURE 7-9 Bones of the right orbit.

Table 7-4

Bones forming the orbit (see Figures 7-7 and 7-9)

BONE	PORTION OF ORBIT
Frontal	Roof
Sphenoid	Roof and lateral wall
Zygomatic	Lateral wall
Maxilla	Floor
Lacrimal	Medial wall
Ethmoid	Medial wall
Palatine	Medial wall

muscles that move the eyes. The major portion of each eyeball is within the orbit, and the portion of the eye visible from the outside is relatively small. Each orbit contains blood vessels, nerves, and fat, as well as the eyeball and the muscles that move it. The bones forming the orbit are listed in Table 7-4.

The superolateral corner of the orbit (where the zygomatic and frontal bones join) is a weak point in the skull that is easily fractured by a severe blow to that region of the head. The bone tends to collapse into the orbit, resulting in an injury that is difficult to repair.

The orbit has several openings through which structures communicate between it and other cavities (see Table 7-6, p. 201). The nasolacrimal duct passes from the orbit into the nasal cavity through the **nasolacrimal canal** and carries tears from the eyes to the nasal cavity. The optic nerve for the sense of vision passes from the eye through the optic foramen at the posterior apex of the orbit and enters the cranial vault. Two fissures in the posterior region of the orbit provide openings through which nerves and vessels communicate with the orbit or pass to the face.

The **nasal cavity** (Table 7-5 and Figures 7-7 and 7-10) has a pear-shaped opening anteriorly and is divided into right and left halves by a **nasal septum** (sep′tum; wall). The bony part of the nasal septum consists primarily of the vomer and the perpendicular plate of the ethmoid bone. The anterior portion of the nasal septum is formed by hyaline cartilage.

The septum usually is located in the midsagittal plane until a person is 7 years old. Thereafter it tends to bulge slightly to one side or the other. The septum can deviate abnormally at birth or as a result of injury; the deviation can be severe enough to block the nasal passage on one side and interfere with normal breathing.

The external portion of the nose, formed mostly of hyaline cartilage, is almost entirely absent in the dried skeleton and is represented mainly by the nasal bones and the frontal processes of the maxillary bones, which form the bridge of the nose.

Table 7-5

Bones forming the nasal cavity (see Figures 7-7 and 7-10)

BONE	PORTION OF NASAL CAVITY	BONE	PORTION OF NASAL CAVITY
Frontal	Roof	Inferior nasal concha	Lateral wall
Nasal	Roof	Lacrimal	Lateral wall
Sphenoid	Roof	Maxilla	Floor
Ethmoid	Roof, septum, and lateral wall	Palatine	Floor and lateral wall
		Vomer	Septum

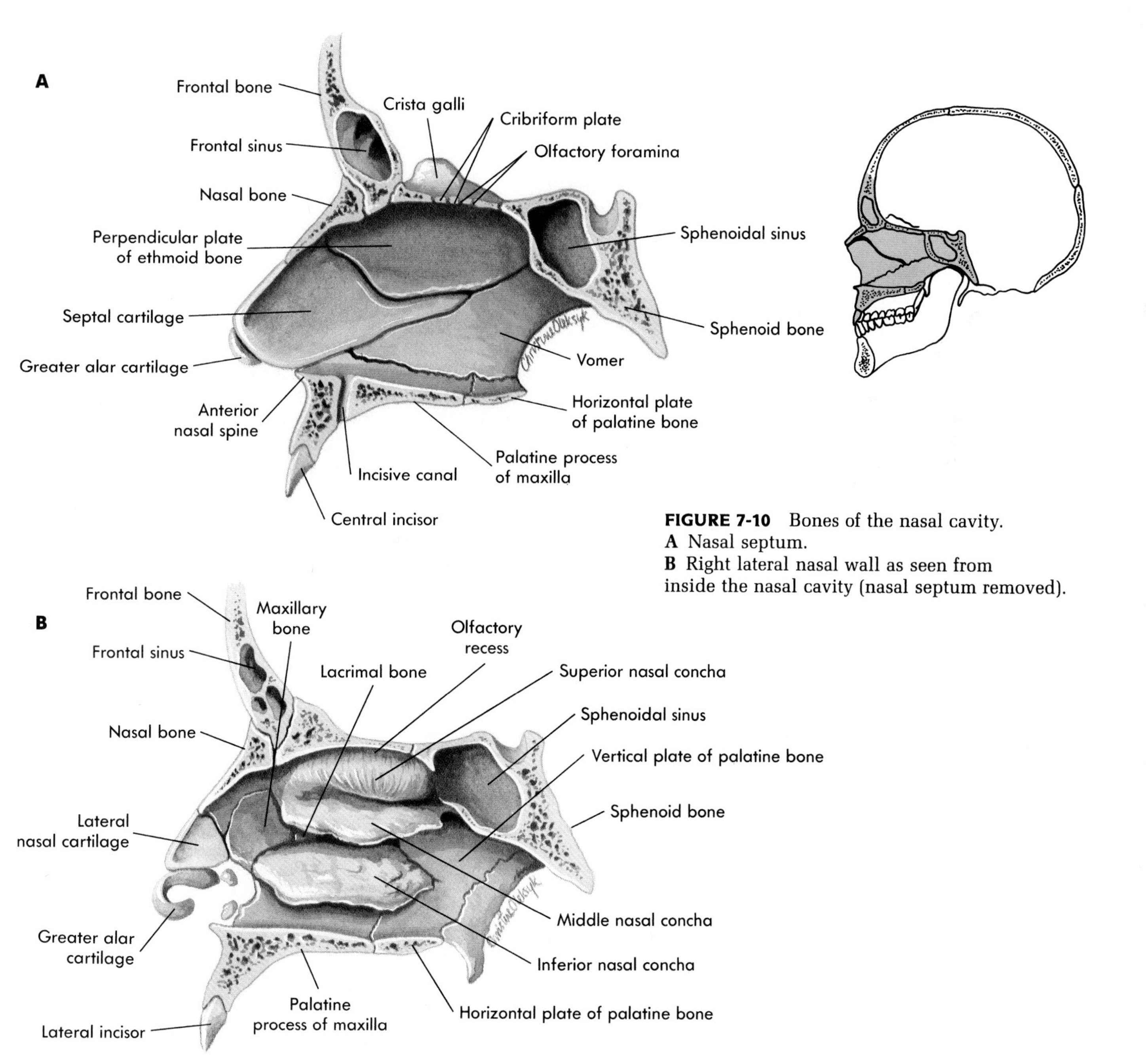

FIGURE 7-10 Bones of the nasal cavity. **A** Nasal septum. **B** Right lateral nasal wall as seen from inside the nasal cavity (nasal septum removed).

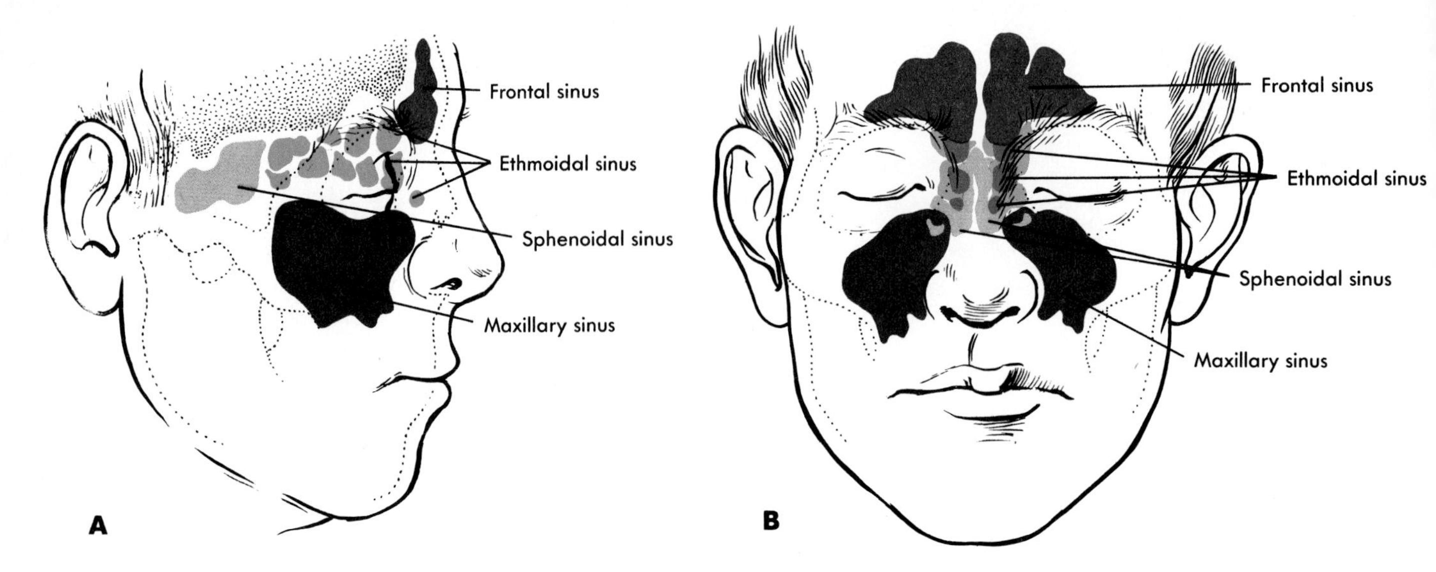

FIGURE 7-11 Paranasal sinuses. **A** Viewed from the side. **B** Viewed from in front.

2

A direct blow to the nose may result in a "broken nose." List at least three bones that may be broken.

? ? ? ? ? ? ? ? ? ? ?

The lateral wall of the nasal cavity has three bony shelves, the **nasal conchae** (kon'ke; resembling a conch shell), which are directed inferiorly. The inferior nasal concha is a separate bone, and the middle and superior nasal conchae are projections from the ethmoid bone. The conchae function to increase the surface area in the nasal cavity, facilitating moistening, removal of particles, and warming of the air inhaled through the nose.

Several of the bones associated with the nasal cavity have large cavities within them called the **paranasal sinuses,** which open into the nasal cavity (Figure 7-11). The sinuses decrease the weight of the skull and act as resonating chambers during voice production (compare the normal voice to the voice of a person who has a cold and whose sinuses are "stopped up"). The sinuses are named for the bones in which they are located and include the frontal, maxillary, ethmoidal, and sphenoidal sinuses.

Interior of the cranial vault

The floor of the cranial vault when viewed from above with the roof cut away can be divided roughly into three fossae (anterior, middle, and posterior), which are formed as the developing skull conforms to the shape of the brain (Figure 7-12, p. 202).

A prominent ridge, the **crista galli** (kris'tah gal'e; rooster's comb), is located in the center of the anterior fossa. The crista galli is a point of attachment for one of the **meninges** (mĕ-nin'jēz), a thick connective tissue membrane that supports and protects the brain (see Chapter 13). On either side of the crista galli is the fossa for the olfactory bulb, which receives the olfactory nerves for the sense of smell. The floor of each olfactory fossa is formed by the **cribriform** (krib'rĕ-form) **plate** of the ethmoid bone. The olfactory nerves extend from the cranial vault into the roof of the nasal cavity through sievelike perforations in the cribriform plate called olfactory foramina (see Chapter 15).

The cribriform plate may be fractured in an automobile accident (e.g., when the driver's nose strikes the steering wheel), in which case **cerebrospinal** (ser-e-bro-spi'nal) **fluid** from the cranial cavity may leak through the fracture into the nose. This leakage is a dangerous sign and requires immediate medical attention since risk of infection is very high.

One self-defense maneuver involves an open-handed upward chop with the base of the hand to an assailant's nose, a movement that may break the cribriform plate.

A central prominence located within the floor of the cranial vault is formed by the body of the sphenoid bone. This area is modified into a structure resembling a saddle, the **sella turcica** (sel'ah tur'sĭ-kah; Turkish saddle), which is occupied by the pituitary gland during life. The petrous portion of the temporal bone is on each side and is slightly posterior to the

Table 7-6

Skull foramina, fissures, and canals (see Figures 7-12 and 7-13)

OPENING	BONE CONTAINING THE OPENING	TRANSMITTED STRUCTURES
Carotid canal	Temporal	Carotid artery and carotid sympathetic nerve plexus
Ethmoid foramina, anterior and posterior	Between frontal and ethmoid	Anterior and posterior ethmoid nerves
External auditory meatus	Temporal	Sound waves enroute to eardrum
Foramen lacerum	Between temporal, occipital, and sphenoid	None because the foramen is filled with cartilage during life; carotid canal and pterygoid canal cross its superior part but do not actually pass through it
Foramen magnum	Occipital	Spinal cord, accessory nerves, and vertebral arteries
Foramen ovale	Sphenoid	Mandibular division of trigeminal nerve
Foramen rotundum	Sphenoid	Maxillary division of trigeminal nerve
Foramen spinosum	Sphenoid	Middle meningeal artery
Hypoglossal canal	Occipital	Hypoglossal nerve
Incisive foramen (canal)	Between maxillae	Incisive nerve
Inferior orbital fissure	Between sphenoid and maxilla	Infraorbital nerve and vessels and zygomatic nerve
Infraorbital foramen	Maxilla	Infraorbital nerve
Internal auditory meatus	Temporal	Facial nerve and vestibulocochlear nerve
Jugular foramen	Between temporal and occipital	Internal jugular vein, glossopharyngeal nerve, vagus nerve, and accessory nerve
Mandibular foramen	Mandible	Inferior alveolar nerve to mandibular teeth
Mental foramen	Mandible	Mental nerve
Nasolacrimal canal	Between lacrimal and maxilla	Nasolacrimal (tear) duct
Olfactory foramina	Ethmoid	Olfactory nerves
Optic foramen	Sphenoid	Optic nerve and ophthalmic artery
Palatine foramina, anterior and posterior	Palatine	Palatine nerves
Pterygoid canal	Sphenoid	Sympathetic and parasympathetic nerves to the face
Sphenopalatine foramen	Between palatine and sphenoid	Nasopalatine nerve and sphenopalatine vessels
Stylomastoid foramen	Temporal	Facial nerve
Superior orbital fissure	Sphenoid	Oculomotor nerve, trochlear nerve, ophthalmic division of trigeminal nerve, abducens nerve, and ophthalmic veins
Supraorbital foramen or notch	Frontal	Supraorbital nerve and vessels
Zygomaticofacial foramen	Zygomatic	Zygomaticofacial nerve
Zygomaticotemporal foramen	Zygomatic	Zygomaticotemporal nerve

sella turcica. This thick bony ridge is hollow and contains the middle and inner ears.

The prominent **foramen magnum,** through which the spinal cord and brain are connected, is located in the posterior fossa. The other foramina of the skull and the structures passing through them are listed in Table 7-6.

Inferior view of the skull

Seen from below with the mandible removed, the base of the skull is complex, with a number of foramina and specialized surfaces (Figure 7-13). The foramen magnum passes through the occipital bone just slightly posterior to the center of the skull base.

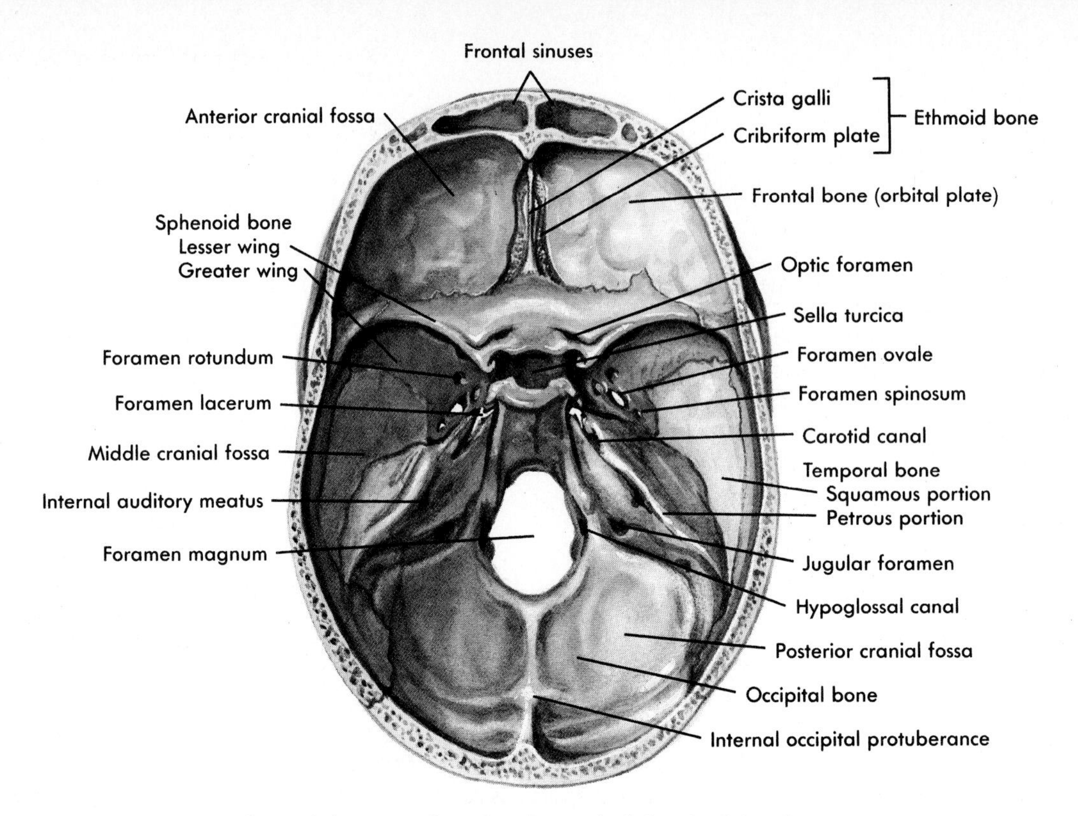

FIGURE 7-12 Floor of the cranial vault. The roof of the skull has been removed, and the floor is viewed from above.

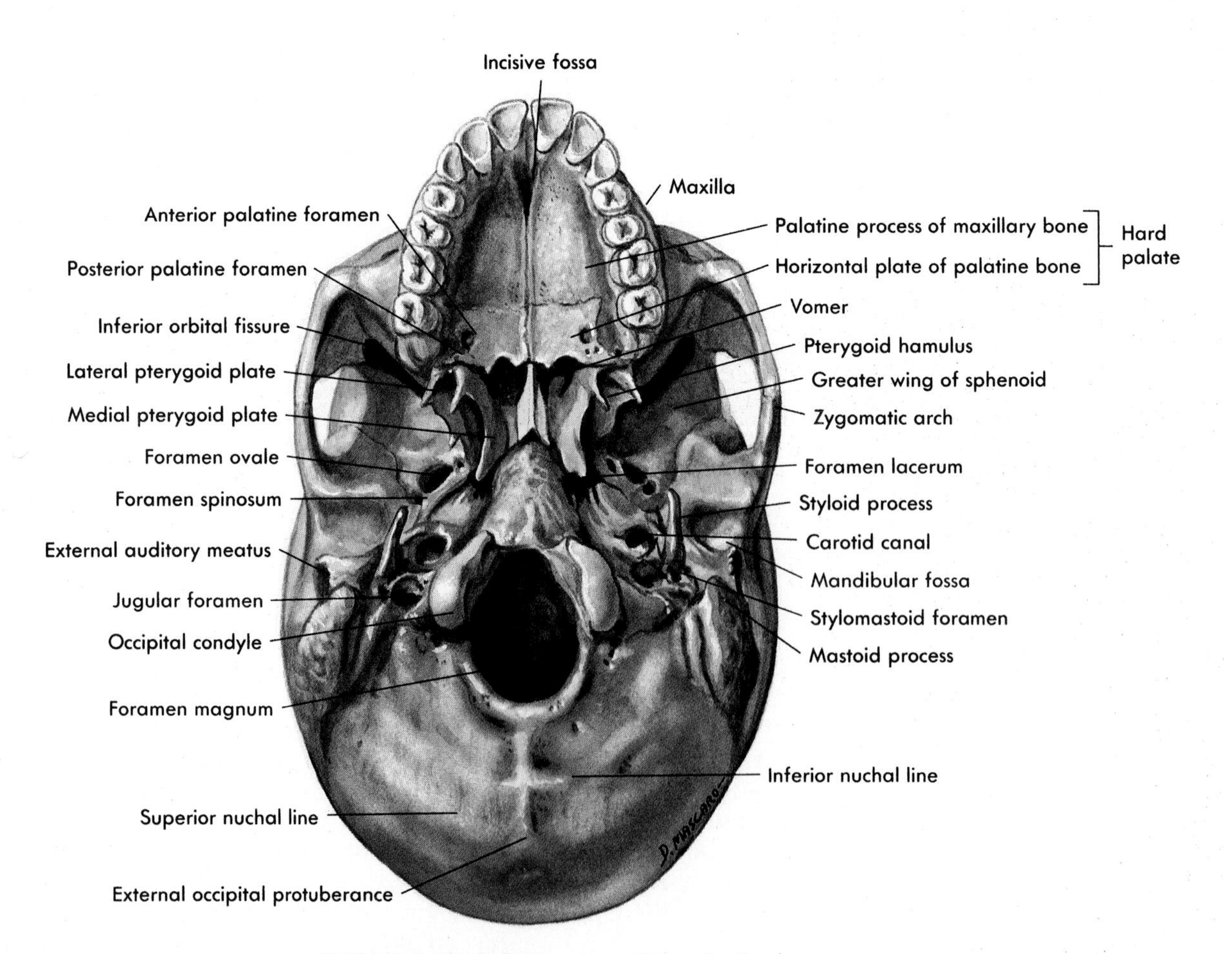

FIGURE 7-13 Inferior view of the skull.

Occipital condyles, the smooth points of articulation between the skull and the vertebral column, are located on the lateral and anterior margins of the foramen magnum.

The major entry and exit points for blood vessels that supply the brain can be seen from this view. Blood reaches the brain through the internal carotid arteries, which pass through the **carotid canals,** and the vertebral arteries, which pass through the foramen magnum. Immediately after the internal carotid artery enters the carotid canal, it turns medially almost 90 degrees, continues through the carotid canal, again turns almost 90 degrees, and enters the cranial cavity through the superior part of the **foramen lacerum** (lah-ser'um). A thin plate of bone separates the carotid canal from the middle ear. Thus it is possible for a person to hear his own heartbeat when, for example, he is frightened or after he has run. Most blood leaves the brain through the internal jugular veins, which exit through the **jugular foramina** located lateral to the occipital condyles.

Two long, pointed **styloid** (sti'loyd; stylus or pen shaped) **processes** project from the floor of the temporal bone (see Figures 7-5 and 7-13). Three muscles involved in movement of the tongue, hyoid bone, and pharynx attach to each process. The **mandibular fossa,** where the mandible articulates with the rest of the skull, is anterior to the mastoid process at the base of the zygomatic arch.

The posterior opening of the nasal cavity is bounded on each side by the vertical bony plates of the sphenoid bone, the **medial pterygoid** (tĕr'ĭ-goyd; wing shaped) **plate** and the **lateral pterygoid plate.** The medial and lateral pterygoid muscles, which help move the mandible, attach to the lateral plate (see Chapter 11). The **vomer** forms the posterior portion of the nasal septum and can be seen between the medial pterygoid plates in the center of the nasal cavity.

The **hard palate** forms the floor of the nasal cavity. Sutures join four bones to form the hard palate; the palatine processes of the two maxillary bones form the anterior two thirds of the palate, and the horizontal plates of the two palatine bones form the posterior one third of the palate. The tissues of the soft palate extend posteriorly from the hard or bony palate. The hard and soft palates separate the nasal cavity from the mouth, enabling humans to eat and breathe at the same time.

During development the facial bones sometimes fail to fuse with each other. A cleft lip results if the maxillae do not form normally; and a cleft palate occurs when the palatine processes of the maxillae do not fuse with each other. A cleft palate produces an opening between the nasal and oral cavities, making it difficult to eat or drink or to speak distinctly.

Vertebral Column

The **vertebral column** usually consists of 26 bones, which can be divided into five regions (Figure 7-14, *A*). There are seven **cervical** vertebrae, 12 **thoracic** vertebrae, five **lumbar** vertebrae, one **sacral** bone, and one **coccygeal** (kok-sij'e-al) bone. Originally approximately 34 vertebrae form during development, but the five sacral vertebrae fuse to form one bone, and the four or five coccygeal bones usually fuse into one bone.

The adult vertebral column has four major curvatures (see Figure 7-14, *A*). Two of the curves appear during embryonic development and reflect the C-shaped curve of the embryo and fetus within the uterus. When the infant raises its head in the first few months after birth, a secondary curve, which is convex anteriorly, develops in the neck. Later, when the infant learns to sit and then walk, the lumbar portion of the column also becomes convex anteriorly. Thus in the adult vertebral column the cervical region is convex anteriorly, the thoracic region is concave anteriorly, the lumbar region is convex anteriorly, and the sacral and coccygeal regions are, together, concave anteriorly.

If the convex curve of the lumbar region is exaggerated, the term **lordosis** (lor-do'sis; hollow back) is applied to the defect; and if the concave curve, especially in the thorax, is exaggerated, the term **kyphosis** (ki-fo'sis; hump back) is applied. **Scoliosis** (sko'le-o'sis) is an abnormal bending of the spine to the side and often is accompanied by secondary abnormal curvatures such as kyphosis (Figure 7-15).

General plan of the vertebrae

The vertebral column performs five major functions: (1) it supports the weight of the head and trunk; (2) it protects the spinal cord; (3) it allows spinal nerves to exit the spinal cord; (4) it provides a site for muscle attachment; and (5) it permits movement of the head and trunk. The general structure of a vertebra is outlined in Table 7-7. Each vertebra consists of a body, an arch, and various processes (see Figure 7-14, *B* and *C*). The weight-bearing portion of the vertebra is a bony disk called the **body.** During life **intervertebral disks** of fibrocartilage, which are located between the bodies of adjacent vertebrae (see Figure 7-14, *A*), provide additional support and prevent the vertebral bodies from rubbing against each other. The invertebral disks consist of an external **anulus fibrosus** (an'u-lus fi-bro'sus; fibrous ring) and an internal gelatinous **nucleus pulposus** (pul-po'sus; pulp). The disk becomes more compressed with increasing age so that the distance between vertebrae and therefore the overall height of the individual decreases. A ruptured, or herniated, disk results from the breakage or ballooning of the anulus fibrosus with a partial or complete release of the nucleus pulposus (Figure 7-16, *A*). The herniated portion of the disk

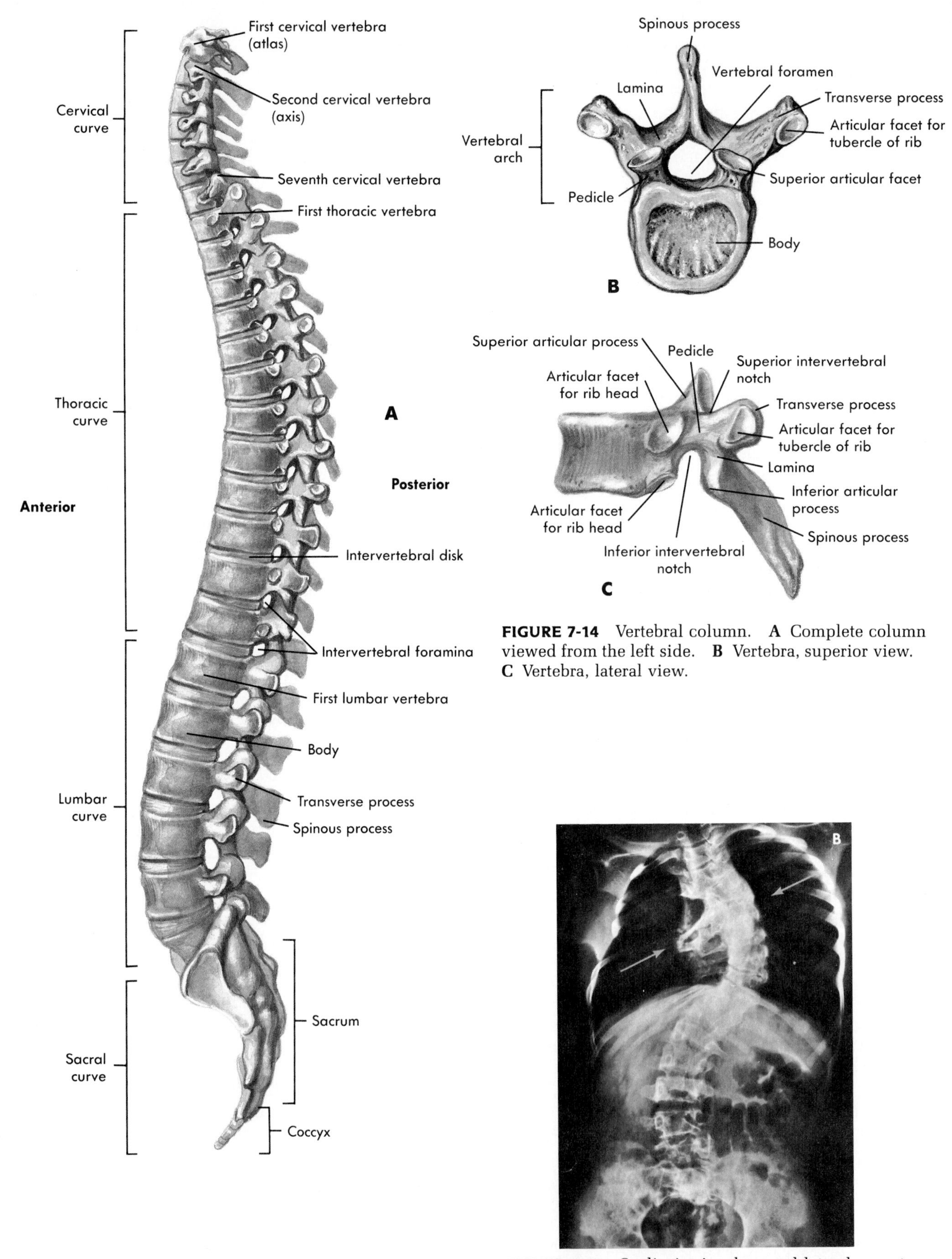

FIGURE 7-14 Vertebral column. **A** Complete column viewed from the left side. **B** Vertebra, superior view. **C** Vertebra, lateral view.

FIGURE 7-15 Scoliosis. An abnormal lateral curvature of the spine.

Table 7-7

General structure of a vertebra (see Figure 7-14, B and C)

FEATURE	DESCRIPTION
Body	Disk shaped; usually largest portion with flat surfaces directed superiorly and inferiorly; forms the anterior wall of vertebral foramen; intervertebral disks are located between the bodies
Arch	Portion surrounding the lateral and posterior portions of the vertebral foramen; possesses several processes and articular surfaces
Pedicle	Foot of the arch with one on each side; forms the lateral walls of the vertebral foramen
Lamina	Dorsal or posterior portion of the arch; forms the posterior wall of the vertebral foramen
Transverse process	Process projecting laterally from the junction of the lamina and pedicle; a site of muscle attachment
Spinous process	Process projecting posteriorly at the point where the two laminae join; a site of muscle attachment
Articular processes	Superior and inferior projections containing facets where vertebrae articulate with each other; strengthens the vertebral column and allows movements
Vertebral foramen	Hole in each vertebra through which the spinal cord passes; adjacent vertebral foramina form the vertebral canal
Intervertebral foramen	Opening between vertebrae through which spinal nerves exit the vertebral canal

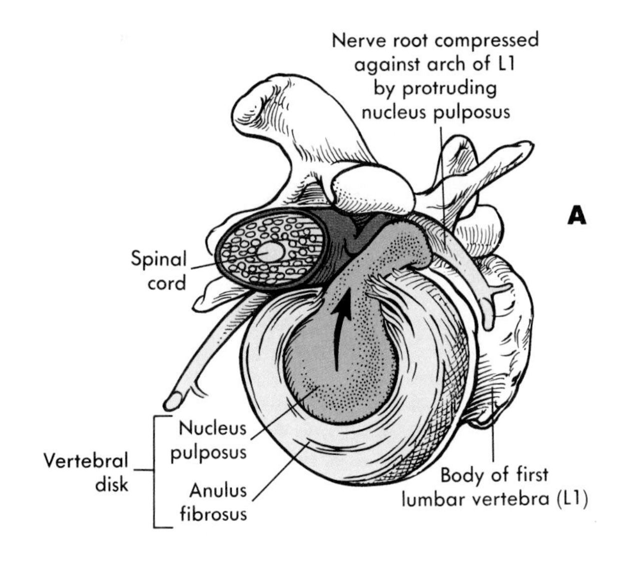

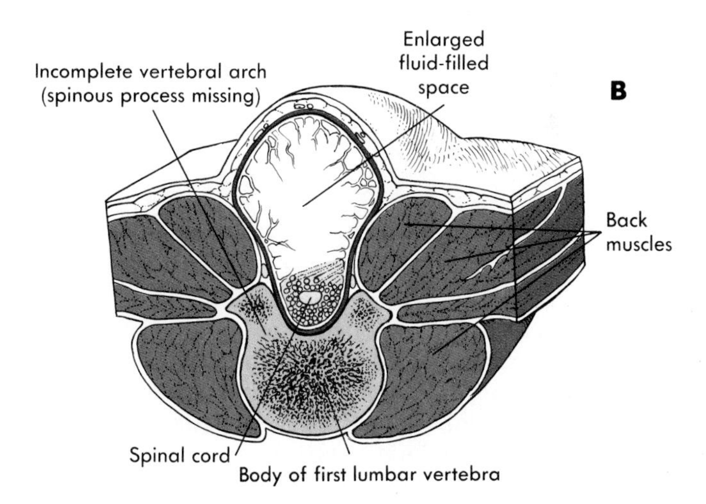

FIGURE 7-16 **A** Herniated disk. Part of the anulus fibrosus has been removed to reveal the nucleus pulposus in the center of the disk. **B** Spina bifida occurs when two vertebral laminae fail to fuse.

may push against the spinal cord or spinal nerves, compromising their normal function and producing pain.

Sometimes vertebral laminae may fail to fuse (or even fail to form) during development, resulting in a condition called **spina bifida** (spi'nah bif'ĭ-dah; split spine; Figure 7-16, *B*). This defect is most common in the lumbar region. If the defect is severe and involves the spinal cord, it may interfere with normal nervous function below the point of the defect. In some surgical procedures (such as removal of an intervertebral disk) the vertebrae are in the way. This problem can be solved by removing a lamina, a procedure called a **laminectomy.**

Herniated or ruptured disks can be repaired in one of several ways. One procedure uses prolonged bed rest in hopes that the herniated portion of the disk will recede and the anulus fibrosus will repair itself. In many cases, however, surgery is required, and the damaged disk is removed. To enhance the stability of the vertebral column, a piece of hipbone sometimes is inserted into the space previously occupied by the disk and the adjacent vertebrae are fused across the gap.

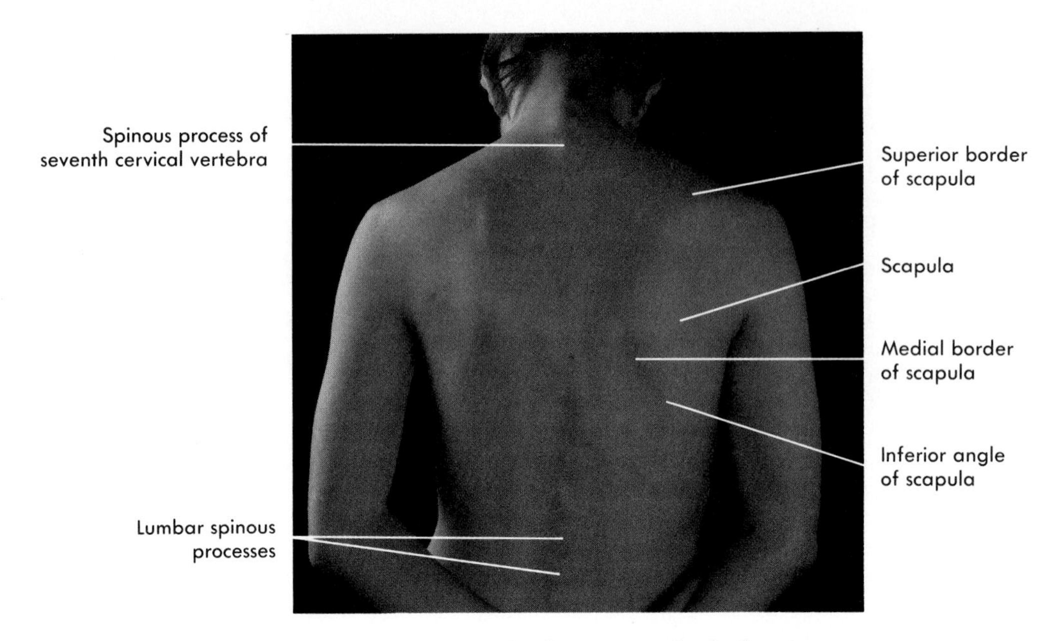

FIGURE 7-17 Photograph of a person's back showing the scapula and vertebral spinous processes.

Protection of the spinal cord is provided by the **vertebral arch** and the dorsal portion of the body, which surround a large opening, the **vertebral foramen.** The vertebral foramina of adjacent vertebrae combine to form the **vertebral canal,** which contains the spinal cord. The arch can be divided into left and right halves, and each half has two parts: the **pedicle** (ped'ĭ-kl; foot), which is attached to the body, and the **lamina** (lam'ĭ-nah; thin plate), which continues dorsally from the pedicle to join the lamina from the opposite half of the arch. Spinal nerves exit the spinal cord through the **intervertebral foramina** (see Figure 7-14, *A*), which are formed by notches in the pedicles of adjacent vertebrae.

Movement and additional support of the vertebral column are made possible by the vertebral processes. Each vertebra has a **superior** and an **inferior articular process,** with the superior process of one vertebra articulating with the inferior process of the next superior vertebra. Overlap of these processes increases the rigidity of the vertebral column. The region of overlap and articulation between the superior and inferior articular processes create a smooth "little face" on each articular process called an **articular facet** (fas'et). A **transverse process** extends laterally from each side of the arch, and a single **spinous process** is present at the point of junction between the two laminae. The spinous processes can be seen and felt as a series of lumps down the midline of the back (Figure 7-17). Much vertebral movement is accomplished by the contraction of skeletal muscles that are attached to the transverse and spinous processes (see Chapter 11).

Regional differences in vertebrae

The vertebrae of each region of the vertebral column have specific characteristics, which tend to blend at the boundaries between regions. The **cervical vertebrae** (see Figure 7-14, *A*; Figure 7-18, *A* to *C*) have very small bodies, partly **bifid** (bi'fid; split) spinous processes, and a **transverse foramen** in each transverse process through which the vertebral arteries extend toward the head. Only cervical vertebrae have transverse foramina.

Whiplash is a traumatic hyperextension of the cervical vertebrae. The head is a heavy object at the end of a flexible column, and it may become hyperextended as a result of a sudden acceleration of the body. This commonly occurs in "rear-end" automobile accidents in which the body is quickly forced forward while the head remains stationary. Common injuries resulting from whiplash are fracture of the spinous processes of the cervical vertebrae, ruptured disks (with an anterior tear of the anulus fibrosus), posterior pressure on the cord or spinal nerves, and strained or torn muscles, tendons, and ligaments.

The first cervical vertebra is called the **atlas** (see Figure 7-18, *A*) because it holds up the head, just as Atlas in classical mythology held up the world. The atlas has no body but has large superior articular

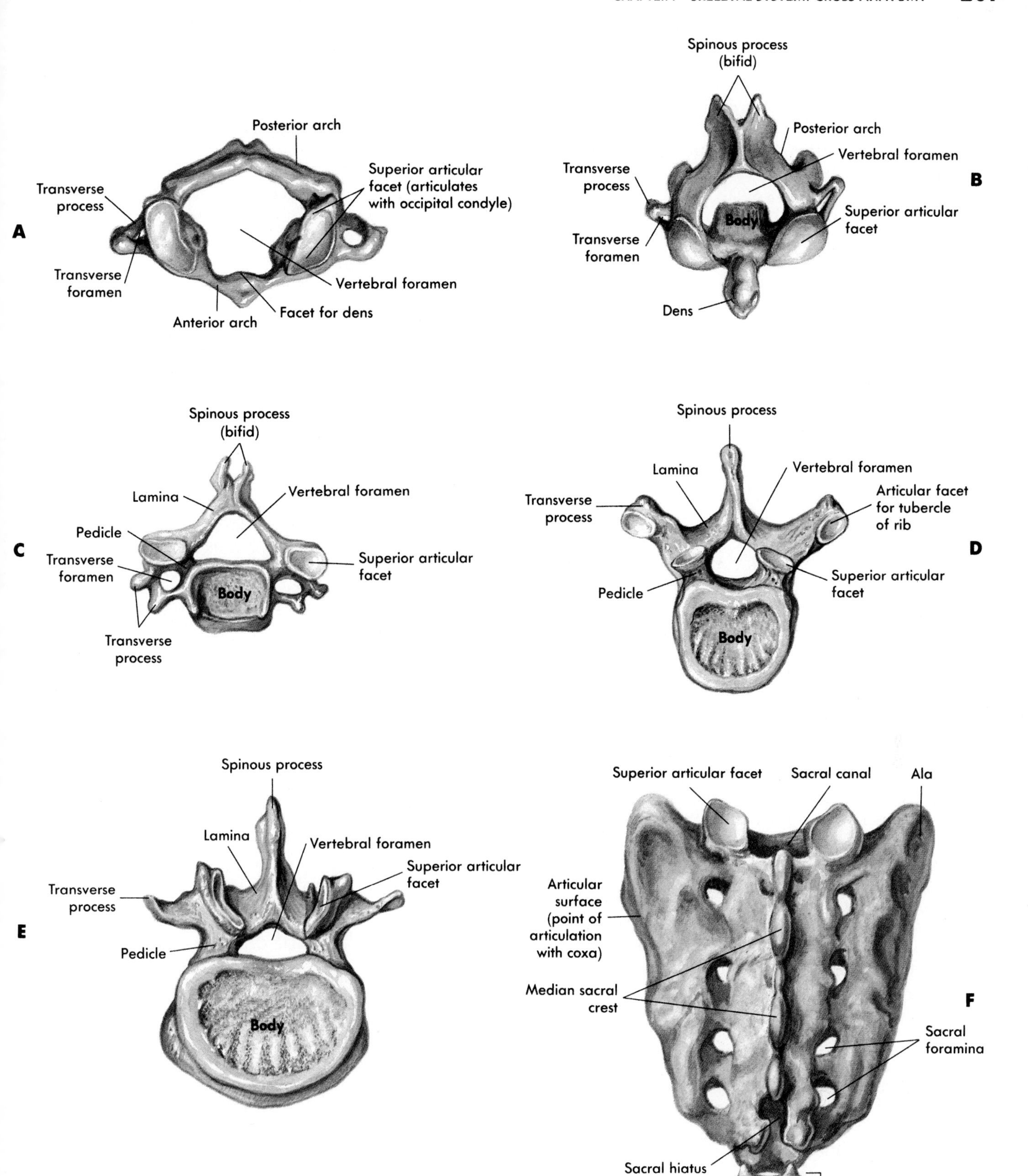

FIGURE 7-18 Vertebrae. **A** Atlas (first cervical vertebra), superior view. **B** Axis (second cervical vertebra), slightly posterior and superior view. **C** Fifth cervical vertebra, superior view. **D** Thoracic vertebra, superior view. **E** Lumbar vertebra, superior view. **F** Sacrum and coccyx, posterior view.

facets where it joins the occipital condyles on the base of the skull. This joint allows the head to move in a "yes" motion or to tilt from side to side. The second cervical vertebra is called the **axis** (see Figure 7-18, *B*) because a considerable amount of rotation occurs at this vertebra to produce a "no" motion of the head. The axis has a highly modified process on the superior side of its small body called the **dens,** or **odontoid process** (both dens and odontoid mean tooth shaped). The dens fits into the enlarged vertebral foramen of the atlas, and the latter rotates around this process. The spinous process of the seventh cervical vertebra, which is not bifid, is quite pronounced and often can be seen and felt as a lump between the shoulders (see Figure 7-17).

The **thoracic vertebrae** (see Figures 7-14 and 7-18, *D*) possess long, thin spinous processes, which are directed inferiorly, and they have relatively long transverse processes. The first 10 thoracic vertebrae have articular facets on their transverse processes where they articulate with the tubercles of the ribs. Additional facets are on the superior and inferior margins of the body where the heads of the ribs articulate. The head of most ribs articulates between two vertebrae so that each vertebra has half of a facet (sometimes called a demifacet) at the point where the rib head articulates.

The **lumbar vertebrae** (see Figures 7-14, *A*, and 7-18, *E*) have large, thick bodies and heavy, rectangular transverse and spinous processes. The superior articular processes face medially, and the inferior articular processes face laterally. When the superior articular surface of one lumbar vertebrae joins the inferior articulating surface of another lumbar vertebrae, an arrangement results that limits rotation of the lumbar vertebrae.

3

Why would the arrangement of the superior and inferior articulating processes in the lumbar vertebrae be beneficial? Why are the lumbar vertebrae more massive than the cervical vertebrae?

? ? ? ? ? ? ? ? ? ? ?

The **sacral vertebrae** (see Figure 7-14, *A*, and 7-18, *F*) are highly modified in comparison to the others. These five vertebrae are fused into a single bone called the **sacrum.** The transverse processes are fused to form the **alae** (ă'le; wings) that join the sacrum to the pelvic bones. The spinous processes of the first four sacral vertebrae are more or less separate projections on the dorsum of the bone called the **median sacral crest.** The spinous process of the fifth vertebra does not form, leaving a **sacral hiatus** at the inferior end of the sacrum, which is often the site of anesthetic injections. The intervertebral foramina are divided into dorsal and ventral foramina, the **sacral foramina,** which are lateral to the midline. The anterior edge of the body of the first sacral vertebra bulges to form the **sacral promontory** (see Figure 7-14, *A*), a landmark that separates the abdominal cavity from the pelvic cavity. The sacral promontory can be felt during a vaginal examination, and it is used as a reference point during measurement of the pelvic outlet.

The **coccyx** (kok'siks; shaped like a cuckoo bill; see Figures 7-14, *A* and 7-18, *F*), or tailbone, is the most inferior portion of the vertebral column and usually consists of four more-or-less fused vertebrae that form a triangle, with the apex directed inferiorly. The coccygeal vertebrae are greatly reduced in size compared to the other vertebrae and have neither vertebral foramina nor well-developed processes.

Because the cervical vertebrae are rather delicate and have small bodies, dislocations and fractures are more common in this area than in other regions of the column. Since the lumbar vertebrae have massive bodies and carry a large amount of weight, ruptured intervertebral disks are more common in this area than in other regions of the column. The coccyx is easily broken in falls during which a person sits down hard on a solid surface.

Thoracic Cage

The **thoracic cage,** or **rib cage,** protects the vital organs within the thorax and prevents the collapse of the lung during respiration. It consists of the thoracic vertebrae, the ribs with their associated costal (rib) cartilages, and the sternum (Figure 7-19, *A*).

Ribs and costal cartilages

The 12 pairs of **ribs** can be divided into the true and the false ribs. The superior seven pairs, called the **true,** or **vertebrosternal** (ver-te'bro-ster'nal), **ribs,** articulate with the thoracic vertebrae and attach directly through their **costal cartilages** to the sternum. The inferior five pairs, **false ribs,** articulate with the thoracic vertebrae but do not attach directly to the sternum. The false ribs consist of two groups. The eighth, ninth, and tenth ribs, the **vertebrochondral** (ver-te'bro-kon'dral) **ribs,** are joined to a common cartilage, which, in turn, is attached to the sternum. The eleventh and twelfth ribs, the **floating,** or **vertebral ribs,** have no attachment to the sternum. The costal cartilages are flexible and permit the thoracic cage to expand during respiration.

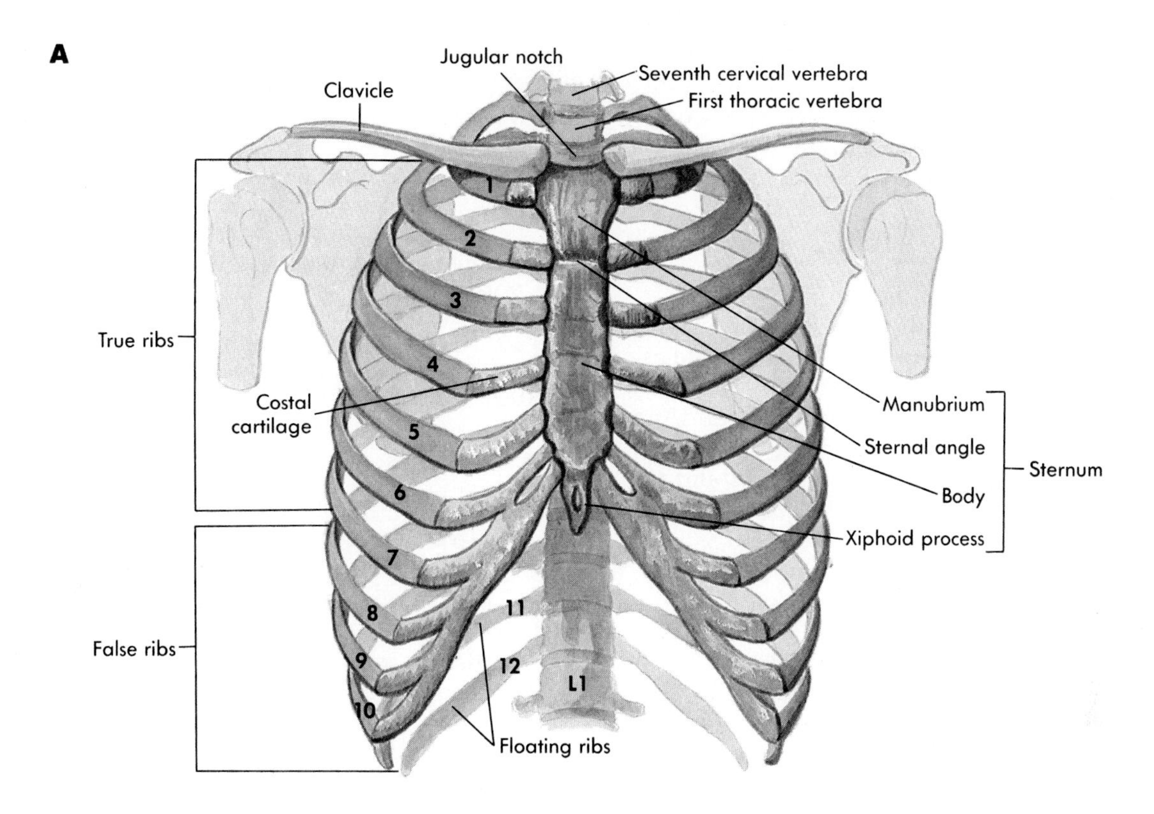

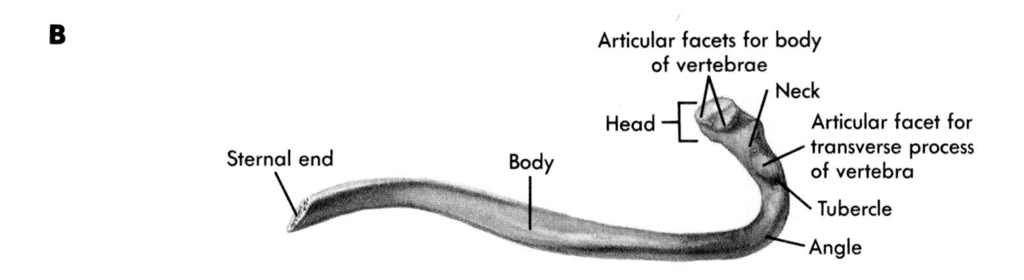

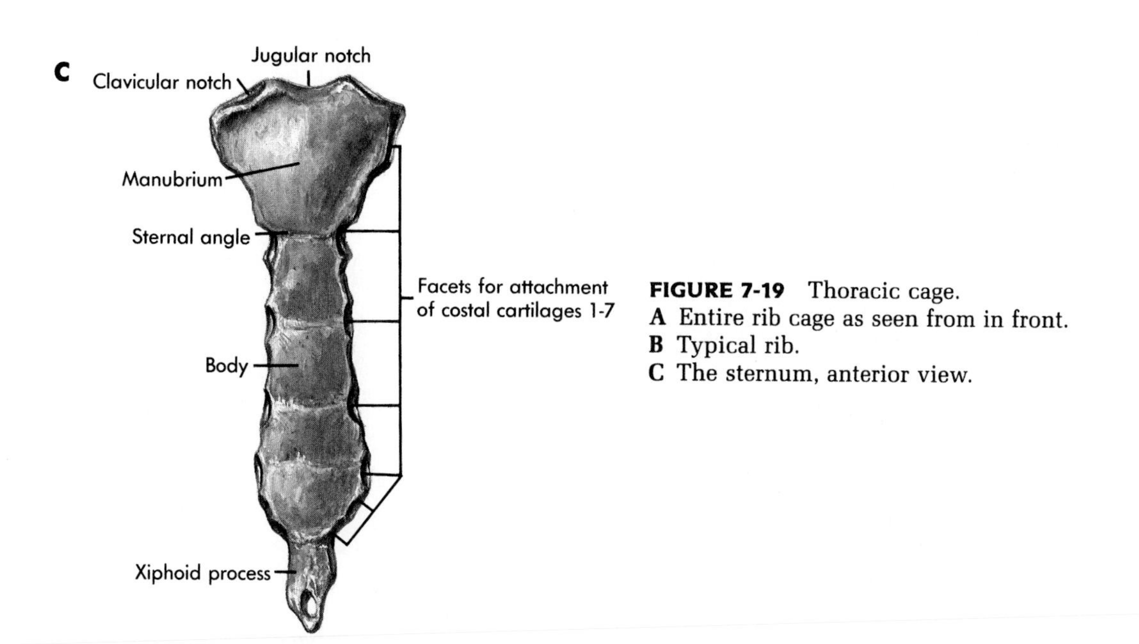

FIGURE 7-19 Thoracic cage.
A Entire rib cage as seen from in front.
B Typical rib.
C The sternum, anterior view.

A separated rib is a dislocation between a rib and its costal cartilage. As a result of the dislocation, the rib can move, override adjacent ribs, and cause pain. Separation of the tenth rib is the most common.

Most ribs have two points of articulation with the thoracic vertebrae (Figure 7-19, *B*). The **head** articulates with the bodies of two adjacent vertebrae and the intervertebral disk between them, and the tubercle articulates with the transverse process of one vertebra. The **neck** is between the head and tubercle, and the **body,** or shaft, is the main part of the rib.

Sternum

The **sternum,** or breastbone, is shaped like a sword (if you use your imagination) and is divided into three parts (Figure 7-19, C): the **manubrium** (mă-nu'bre-um; handle), the **body** (representing most of the sword blade; the old term for the body was gladiolus, meaning sword), and the **xiphoid** (zīf'oyd; meaning sword and representing the sword tip) **process.** The superior margin of the manubrium has a **jugular** (neck) **notch** in the midline, which can be easily felt at the anterior base of the neck (Figure 7-20). The first rib and the clavicle articulate with the manubrium. The point at which the manubrium joins the body of the sternum can be felt as a prominence on the anterior thorax called the **sternal angle.** The cartilage of the second rib attaches to the sternum at the sternal angle, the third through seventh ribs attach to the body of the sternum, and no ribs attach to the xiphoid process.

The sternal angle is important clinically because the second rib is found lateral to the sternal angle and can be used as a starting point for counting the other ribs. Counting ribs is important because they are landmarks that are used to locate structures in the thorax such as areas of the heart. The sternum often is used as a site for taking red bone marrow samples because it is readily accessible. Because the xiphoid process of the sternum is attached only at its superior end, it may be broken during cardiopulmonary resuscitation (CPR) and then may lacerate the liver.

APPENDICULAR SKELETON

The appendicular skeleton (see Figure 7-1) consists of the bones of the **upper** and **lower limbs** and the **girdles** by which they are attached to the body. Girdle means a belt or a zone and refers to the two zones, pectoral and pelvic, where the limbs are attached to the body.

Upper Limb

The human forelimb is capable of a wide range of movements, including lifting, grasping, pulling, and touching. Many structural characteristics of the upper limb reflect these functions. The upper limb and its girdle are attached rather loosely by muscles to the rest of the body, an arrangement that allows considerable freedom of movement of this extremity. This freedom of movement allows placement of the hand in a wide range of positions to accomplish its functions.

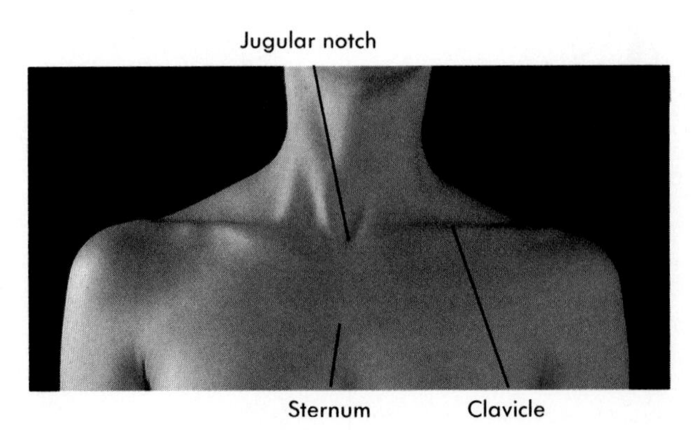

FIGURE 7-20 Surface anatomy showing bones of the upper thorax.

Pectoral girdle

The **pectoral** (pek'to-ral), or **shoulder, girdle** consists of two bones that attach the upper limb to the body: the **scapula** (skap'u-lah), or shoulder blade, and the **clavicle** (klav'ĭ-kl), or collar bone (Figure 7-21). The scapula (Figure 7-21, *A* and *B*) is a flat, triangular bone that can easily be seen and felt in a living person (see Figure 7-17). The base of the triangle, the superior border, faces superiorly, and the apex, the inferior angle, is directed inferiorly.

The large **acromion** (ak-ro'me-en; shoulder tip) **process,** which can be felt at the tip of the shoulder, has three functions: (1) to form a protective cover for the shoulder joint; (2) to form the attachment site for the clavicle; and (3) to provide attachment points for some of the shoulder muscles. The **scapular spine** extends from the acromion process across the posterior surface of the scapula into a small **supraspinous fossa** superior to the spine and a larger **infraspinous fossa** inferior to the spine. The deep, anterior surface of the scapula constitutes the **subscapular**

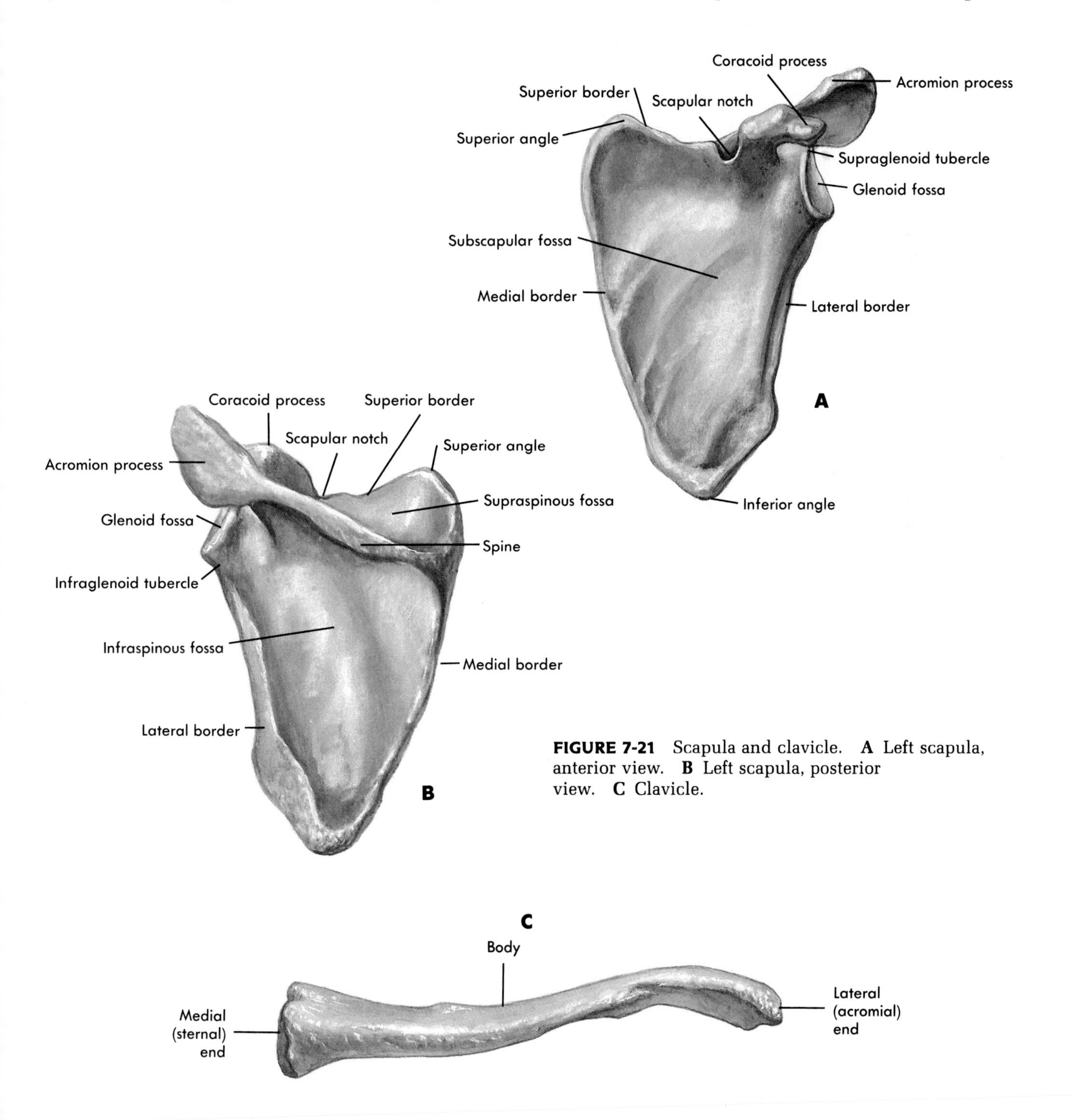

FIGURE 7-21 Scapula and clavicle. **A** Left scapula, anterior view. **B** Left scapula, posterior view. **C** Clavicle.

fossa. The smaller **coracoid** (shaped like a crow's beak) **process** provides attachments for some shoulder and arm muscles. A **glenoid** (glen'oyd) **fossa,** located in the superior lateral portion of the bone, articulates with the head of the humerus.

The clavicle (Figure 7-21, C) is a long bone with a slight sigmoid (S-shaped) curve and is easily seen and felt in the living human (see Figure 7-20). The lateral end of the clavicle articulates with the acromion process and its medial end articulates with the manubrium of the sternum. These articulations are the only bony connections between the pectoral girdle and the axial skeleton. The clavicle is important in holding the upper limb away from the body, facilitating the limb's mobility.

4

A broken clavicle would change the position of the upper limb in what way?

? ? ? ? ? ? ? ? ? ? ?

Arm

The arm (the portion of the upper limb from the shoulder to the elbow) contains only one bone, the **humerus** (Figure 7-22). The humeral **head** articulates with the glenoid fossa of the scapula. The **anatomical neck,** immediately distal to the head, is almost nonexistent, so a surgical neck is identified. The **surgical neck** is so named because it is a common fracture site that often requires surgical repair. If it becomes necessary to remove the humeral head because of disease or injury, it is removed down to the surgical neck. The **greater** and **lesser tubercles** are located on the lateral and anterior surfaces of the proximal end of the humerus where they function as sites of muscle attachment. The groove between the two tubercles contains one tendon of the biceps muscle (a muscle with "two heads") and is called the **intertubercular,** or **bicipital** (bi-sip'ĭ-tal), **groove.** The **deltoid tuberosity** is located on the lateral surface of the humerus approximately one third of the way along its length and is the attachment site for the deltoid muscle.

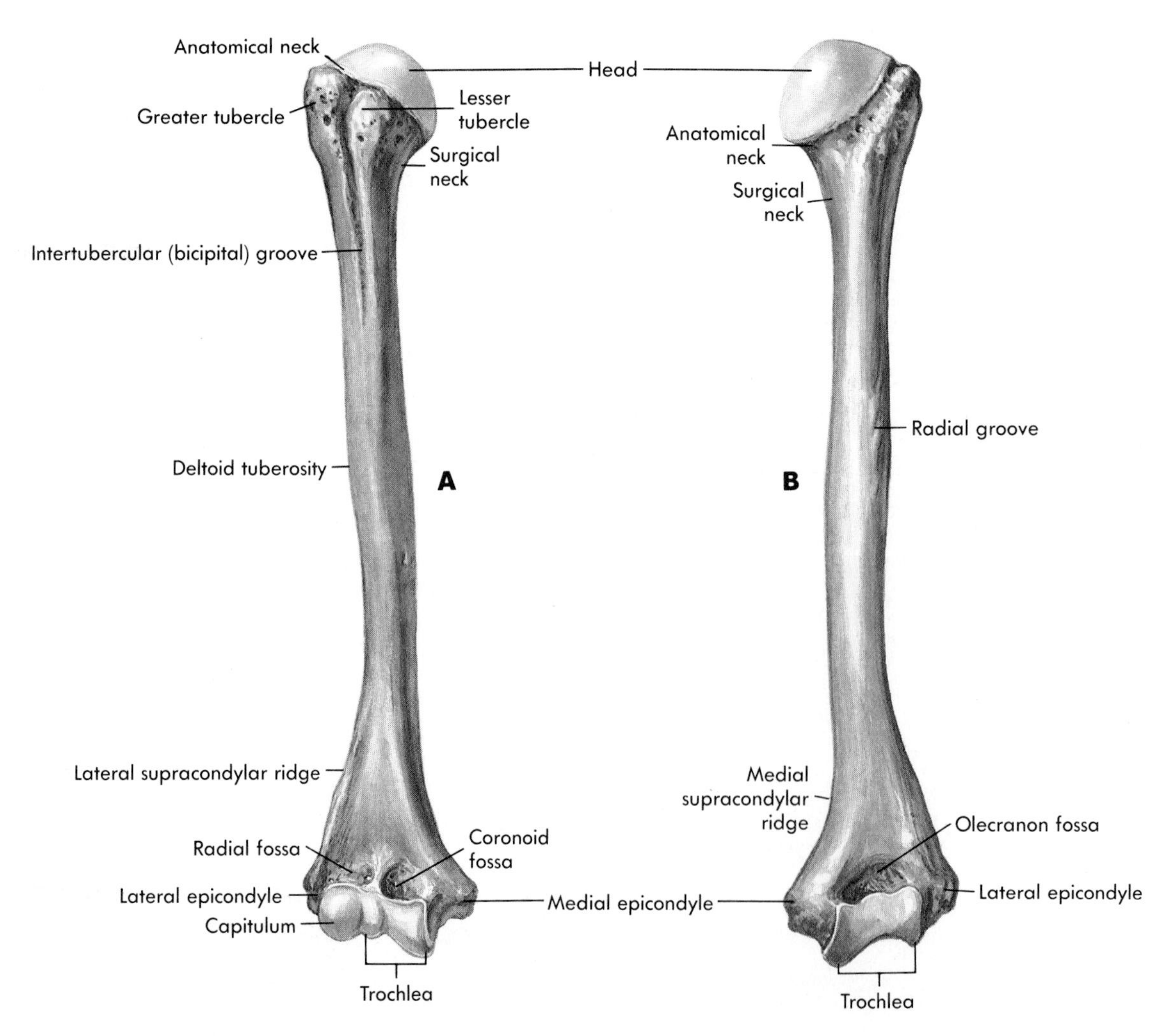

FIGURE 7-22 Right humerus. **A** Anterior view. **B** Posterior view.

Exaggerated exercise such as weight lifting increases the size of bony tubercles, including those of the humerus. Anthropologists use tubercular enlargements to estimate the occupation of the person whose bones are being examined. For example, a large deltoid tuberosity in a young skeleton suggests that the young person was required to lift heavier-than-normal weights. In some situations this information may indicate that the person was a slave required to carry heavy loads.

The articular surfaces of the distal end of the humerus exhibit some unusual features because the humerus articulates with the two forearm bones. The lateral portion of the articular surface is very rounded, articulates with the radius, and is called the **capitulum** (kă-pit′u-lum; head shaped). The medial portion somewhat resembles a spool (having a groove between two ridges), articulates with the ulna, and is called the **trochlea** (trōk′le-ah; spool). Proximal to the capitulum and the trochlea are the **medial** and **lateral epicondyles,** which function as points of muscle attachment for the muscles of the forearm.

Forearm

The forearm has two bones, the **ulna** on the medial side of the forearm, the side with the little finger, and the **radius** on the lateral, or thumb side, of the forearm (Figure 7-23).

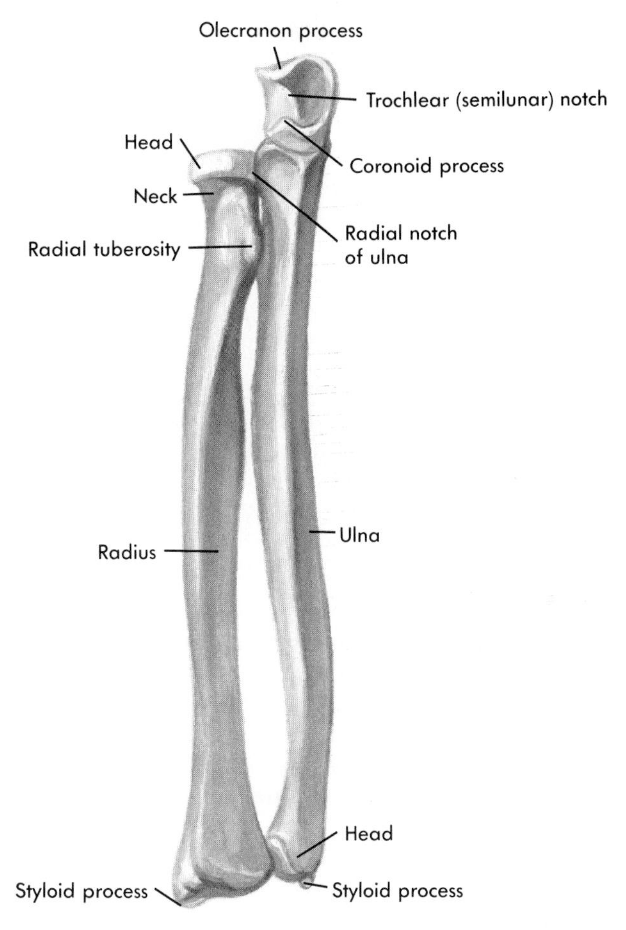

FIGURE 7-23 Ulna and radius of the right forearm.

FIGURE 7-24 Surface anatomy showing bones of the pectoral girdle and upper limb.

The proximal end of the ulna has a C-shaped articular surface that fits over the trochlea of the humerus. The larger, posterior process is the **olecranon** (o-lek'ră-non; meaning the point of the elbow) **process** and can be felt as the point of the elbow (Figure 7-24). Posterior arm muscles attach to the olecranon process. The smaller, anterior process is the **coronoid** (ko'ro-noyd; also means crow's beak) **process,** and the notch between the two where the ulna articulates with the humerus is the **trochlear,** or **semilunar, notch.**

5

Explain the function of the olecranon and coronoid fossae on the distal humerus (see Figure 7-22).

? ? ? ? ? ? ? ? ? ? ?

The distal end of the ulna has a small **head** that articulates with both the radius and the wrist bones (see Figures 7-23 and 7-24). The posteromedial side of the head exhibits a small **styloid** (sti'loyd; shaped like a stylus or writing instrument) **process** to which ligaments of the wrist are attached. This process can be felt on the medial (ulnar) side of the distal forearm.

The proximal end of the radius is the **head.** It is concave and articulates with the capitulum of the humerus, and the lateral surfaces of the head constitute a smooth cylinder where the radius rotates against the **radial notch** of the ulna. As the forearm is rotated (supination and pronation; see Chapter 8), the proximal end of the ulna tends to stay in place, and the radius tends to rotate. The **radial tuberosity** is the point at which a major anterior arm muscle attaches.

The distal end of the radius, which articulates with the ulna and the carpals, is somewhat broadened, and a **styloid process** to which wrist ligaments are attached is located on the lateral side of the distal radius.

Wrist

The wrist is a relatively short region between the forearm and hand and is composed of eight **carpal** (kar'pul) **bones,** which are arranged into two rows of four each (Figure 7-25). The eight carpals, taken together, are convex posteriorly and concave anteriorly. The anterior concavity of the carpals is accentuated by the tubercle of the trapezium at the base of the thumb and the hook of the hamate at the base of the little finger. A ligament stretches across the wrist

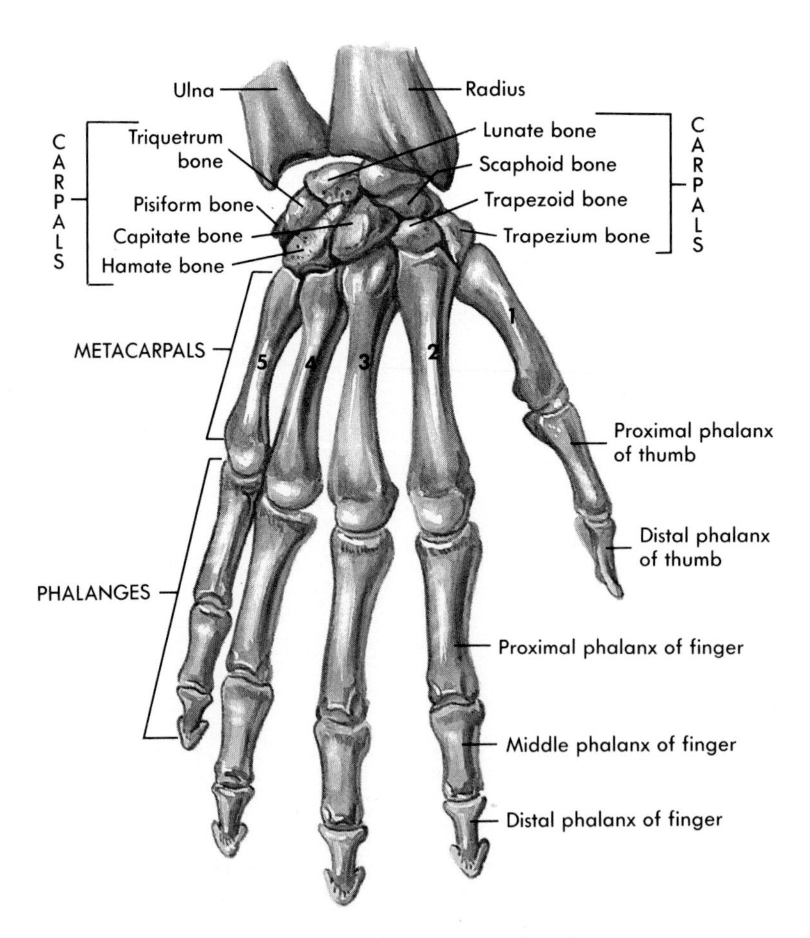

FIGURE 7-25 Bones of the right wrist and hand, posterior view.

from the tubercle of the trapezium to the hook of the hamate to form a tunnel on the anterior surface of the wrist called the **carpal tunnel.** Tendons, nerves, and blood vessels pass through this tunnel to enter the hand.

Because neither the bones nor the ligaments that form the walls of the carpal tunnel will stretch, edema (fluid buildup) or connective tissue deposition within the carpal tunnel that is caused by trauma or some other problem may apply pressure against the nerve and vessels passing through the tunnel. This pressure causes carpal tunnel syndrome, which consists of tingling, burning, and numbness in the hand.

Hand

Five **metacarpals** are attached to the carpal bones and constitute the bony framework of the hand (see Figure 7-25). The curvature of the carpal bones in forming the carpal tunnel causes the metacarpals also to form a curve so that, in the resting position, the palm of the hand is concave. The distal ends of the metacarpals help form the knuckles of the hand. The spaces between the metacarpals are occupied by soft tissue.

The five **digits** of each hand include one thumb and four fingers. Each digit consists of small long bones called **phalanges** (fă-lan'jēz; the singular term phalanx refers to the Greek *phalanx*—a line or wedge of soldiers holding their spears, tips outward, in front of them). The thumb has two phalanges, and each finger has three. One or two **sesamoid** (ses'ă-moyd; resembling a sesame seed) **bones** often form near the junction between the proximal phalanx and the metacarpal of the thumb. Sesamoid bones are small bones found within tendons.

6

Explain why the dried, articulated skeleton appears to have much longer "fingers" than are seen in the hand with the soft tissue intact.

? ? ? ? ? ? ? ? ? ? ?

Lower Limb

The general pattern of the lower limb is very similar to that of the upper limb except that the pelvic girdle is attached much more firmly to the body than is the pectoral girdle and the bones in general are thicker, heavier, and longer than those of the upper limb. These structures reflect the lower limb's function in support and locomotion of the body.

Pelvic girdle

The **pelvis** (pel'vis; basin), or **pelvic girdle,** is a ring of bones formed by the sacrum posteriorly and paired bones called the **coxae** (kok'se), or hip bones, laterally and anteriorly (Figure 7-26, *A*). Each coxa consists of a large, concave bony plate superiorly, a slightly narrower region in the center, and an expanded bony ring inferiorly, which surrounds a large **obturator** (ob'tur-a'tor; to occlude or close up, indicating that the foramen is occluded by soft tissue) **foramen** (Figures 7-26, *B* and *C*). A fossa called the **acetabulum** (a'sĕ-tab'u-lum; a shallow vinegar cup—a common household item in ancient times) is located on the lateral surface of each coxa and is the point of articulation of the lower limb with the girdle. The articular surface of the acetabulum is crescent shaped and occupies only the superior and lateral aspects of the fossa. The pelvic girdle is the place of attachment for the lower limbs, supports the weight of the body, and protects internal organs. In addition, the pelvic girdle protects the developing fetus and forms a passageway through which the fetus passes during delivery.

Each coxa is formed by the fusion of three bones during development: the **ilium** (il'e-um; groin), the **ischium** (ish'e-um; hip), and the **pubis** (pu'bis; refers to the genital hair). All three bones join near the center of the acetabulum (Figure 7-27). The superior portion of the ilium is called the **iliac crest.** The crest ends anteriorly as the **anterior superior iliac spine** and posteriorly as the **posterior superior iliac spine.** The crest and anterior spine can be felt and even seen in thin individuals (Figure 7-28). The anterior superior iliac spine is an important anatomical landmark that is used, for example, to find the correct location for giving injections in the hip muscle. A dimple overlies the posterior superior iliac spine just superior to the buttocks. The **auricular surface** of the ilium joins the sacrum to form the **sacroiliac joint.** The medial side of the ilium consists of a large depression called the **iliac fossa.**

The sacroiliac joint receives most of the weight of the upper body and is strongly supported by ligaments. Excessive strain on the joint, however, can cause slight movement of the joint and can stretch connective tissue and associated nerve endings in the area and cause pain. Thus is derived the expression, "My aching sacroiliac!" This problem sometimes develops in pregnant women because of the forward weight distribution of the fetus.

The ischium possesses a heavy **ischial tuberosity** where posterior thigh muscles attach and on which a person sits. A **greater sciatic** (si-at'ik) **notch** and

Text continued on p. 218.

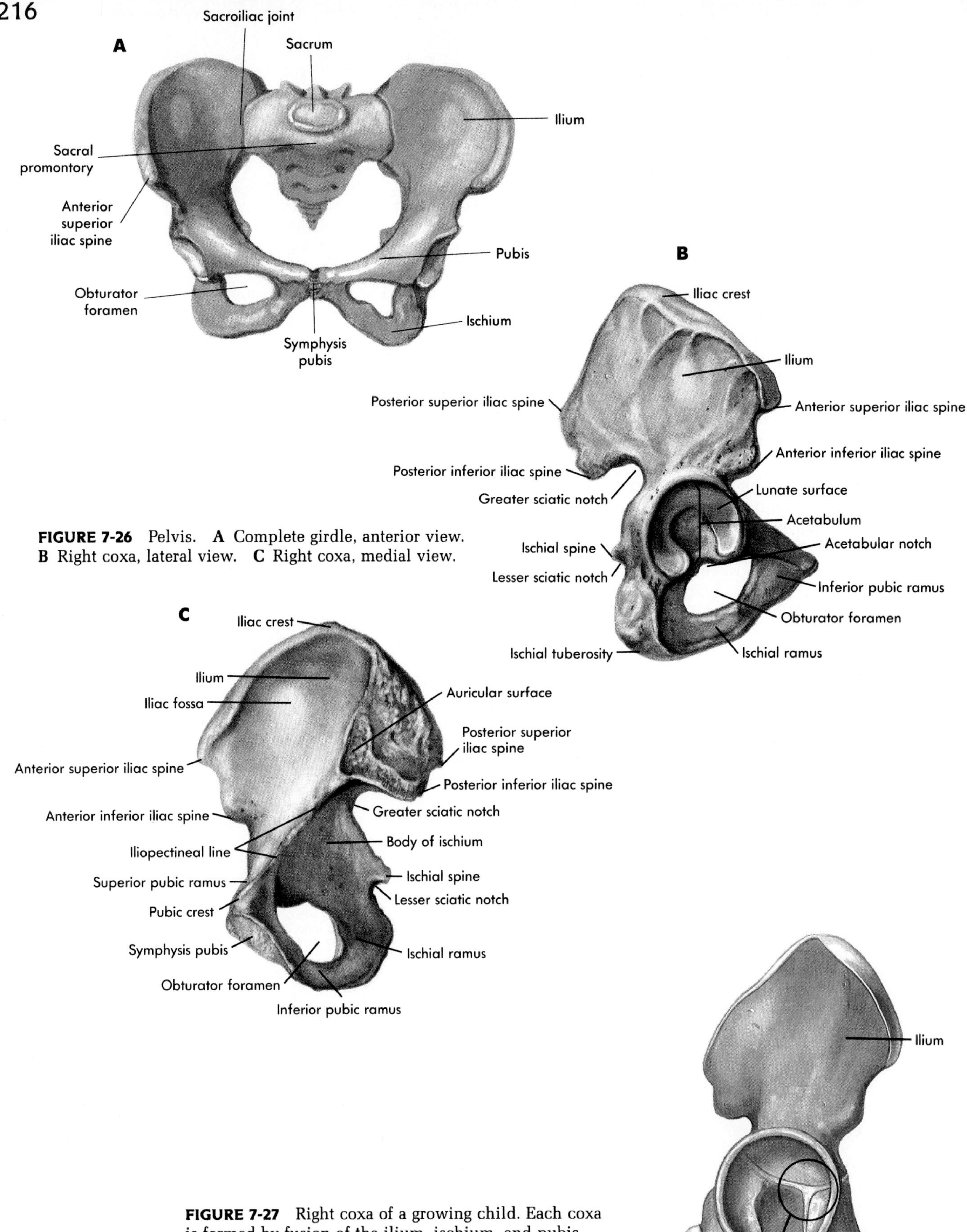

FIGURE 7-26 Pelvis. **A** Complete girdle, anterior view. **B** Right coxa, lateral view. **C** Right coxa, medial view.

FIGURE 7-27 Right coxa of a growing child. Each coxa is formed by fusion of the ilium, ischium, and pubis. The three bones can be seen joining near the center of the acetabulum, separated by lines of cartilage.

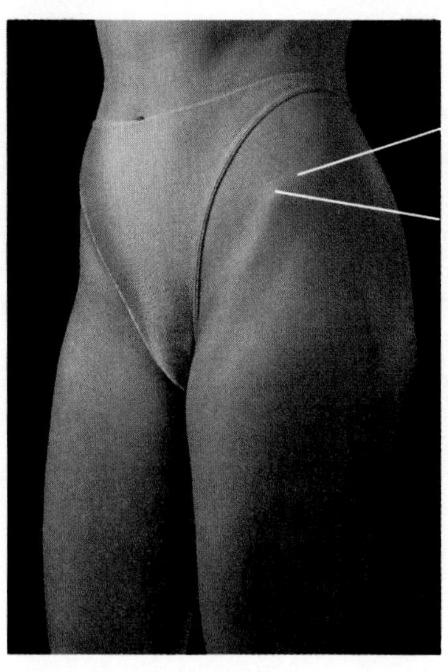

FIGURE 7-28 Surface anatomy showing an anterior view of the bones of the hips.

Table 7-8

Differences between male and female pelvis (see Figure 7-29)

AREA	DESCRIPTION
General	Female pelvis somewhat lighter in weight and wider laterally but shorter superiorly to inferiorly and less funnel shaped; less obvious muscle attachment points in female than in male
Sacrum	Broader in female with the inferior portion directed more posteriorly; the sacral promontory projects less far anteriorly in female
Pelvic inlet	Heart shaped in male; oval in female
Pelvic outlet	Broader and more shallow in female
Subpubic angle	Less than 90 degrees in male; 90 degrees or more in female
Ilium	More shallow and flared laterally in female
Ischial spines	Further apart in female
Ischial tuberosities	Turned laterally in female and medially in male

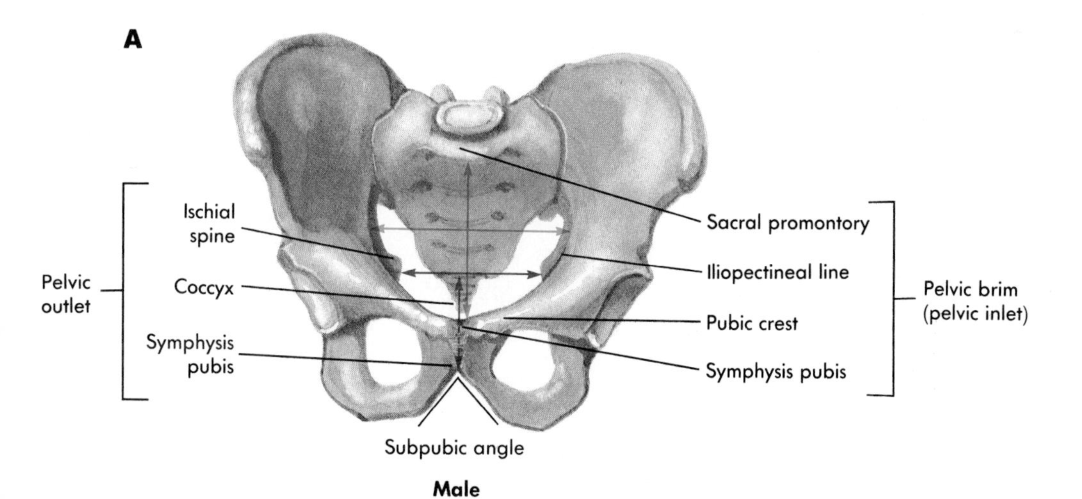

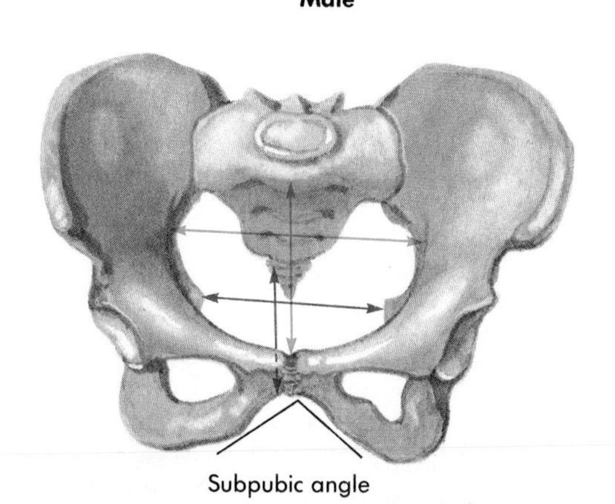

FIGURE 7-29 Comparison between male and female pelvis. **A** Male. Pelvic outlet *(red lines)* is small, and the subpubic angle is less than 90 degrees. The pelvic inlet *(blue lines)* is also shown. **B** Female. Pelvic outlet *(red lines)* is larger, and the subpubic angle is 90 degrees or greater. The pelvic inlet *(blue lines)* is also shown.

lesser sciatic notch are on the posterior side of the ischium. The sciatic nerve passes through the greater sciatic notch. The pubis possesses a **pubic crest** where abdominal muscles attach. The pubic crest can be felt anteriorly. Just inferior to the pubic crest is the point of junction, the **symphysis** (sim'fĭ-sis; a coming together) **pubis** (pubic symphysis) between the two coxae.

The pelvis can be divided into two parts by an imaginary plane passing from the sacral promontory along the **iliopectineal lines** of the ilium to the pubic crest. The bony boundary of this plane is the **pelvic brim.** The **greater,** or **false, pelvis** is superior to the pelvic brim and is partially surrounded by bone on the posterior and lateral sides. During life the abdominal muscles form the anterior wall of the false pelvis. The **true pelvis** is inferior to the pelvic brim and is completely surrounded by bone. The superior opening of the true pelvis, at the level of the pelvic brim, is the **pelvic inlet.** The inferior opening of the true pelvis (bordered by the inferior margin of the pubis, the ischial spines and tuberosities, and the coccyx) is the **pelvic outlet.**

The male pelvis usually is more massive than the female pelvis as a result of the greater weight and size of the male, and the female pelvis is broader and has a larger, more rounded pelvic inlet and outlet (Figure 7-29), reflecting the fact that the fetus must pass through these openings in the female pelvis during delivery. Table 7-8 lists additional differences between the male and female pelvis.

A wide circular pelvic inlet and a pelvic outlet with widely spaced ischial spines are ideal for delivery. Variations in the pelvic inlet and outlet can cause problems during delivery; thus the size of the pelvic inlet and outlet routinely is measured during pelvic examinations of pregnant women. If the pelvic outlet is too small for normal delivery, delivery can be accomplished by cesarean section, which is the surgical removal of the fetus through the abdominal wall.

Thigh

The thigh contains a single bone, the **femur,** which has a prominent rounded **head** where it articulates with the acetabulum and a well-defined **neck,** both located at an oblique angle to the shaft of the femur (Figure 7-30). The proximal shaft exhibits two tuberosities, a **greater trochanter** (tro'kan-ter;

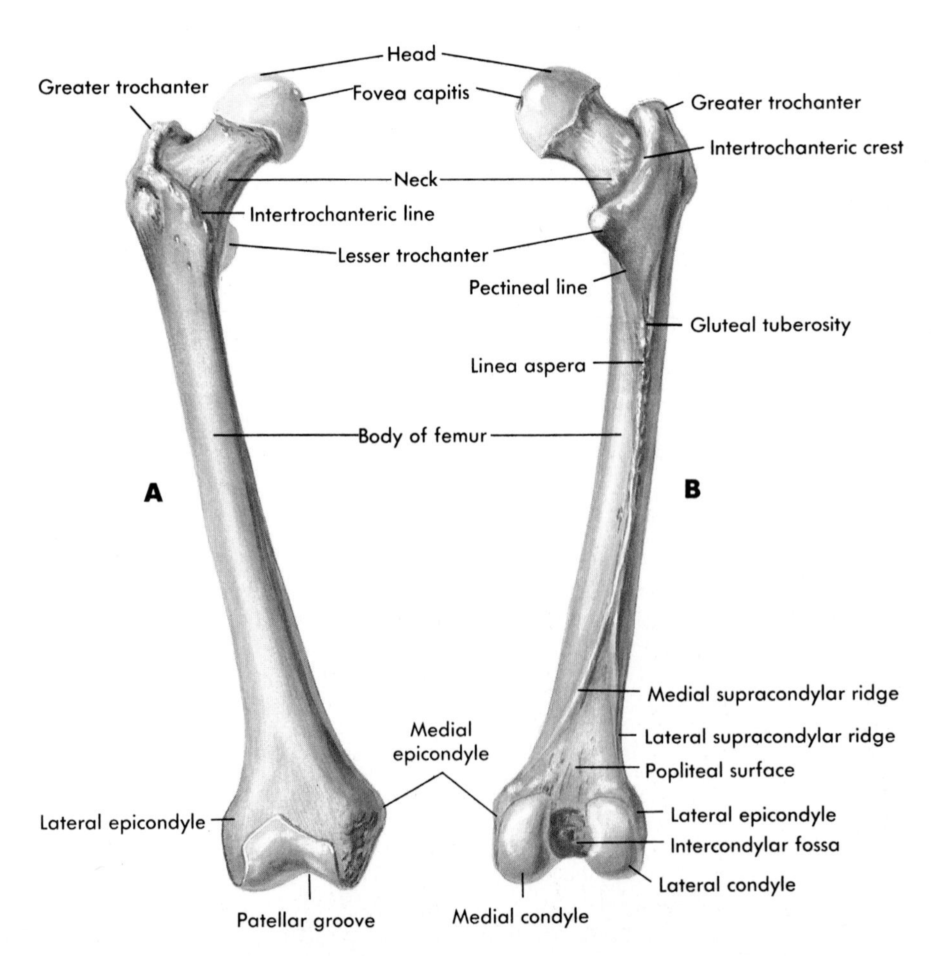

FIGURE 7-30 Right femur. **A** Anterior view. **B** Posterior view.

runner) lateral to the neck and a smaller or **lesser trochanter** inferior and posterior to the neck. Both trochanters are attachment sites for muscles that attach the hip to the thigh. The greater trochanter and its attached muscles form a bulge that can be seen as the widest part of the hips (see Figure 7-28). The distal end of the femur has **medial** and **lateral condyles,** smooth, rounded surfaces that articulate with the tibia. Located laterally and proximally to the condyles are the **medial** and **lateral epicondyles,** important sites of muscle and ligament attachment.

The **patella** is a large sesamoid bone located within the major anterior tendon of the thigh muscles (Figure 7-31). It articulates with the patellar groove of the femur to create a smooth articular surface over the anterior distal end of the femur. The patella increases the mechanical advantage of the tendon by changing the angle with which it passes over the knee.

7

Compare the following in terms of structure and function for the upper and lower limbs: depth of socket, size of bone, and size of tubercles. What is the significance of these differences?

? ? ? ? ? ? ? ? ? ? ?

Leg

The leg (the portion of the lower limb between the knee and the ankle) contains two bones, the **tibia** (tib'ĭ-ah; shin bone) and the **fibula** (fib'u-lah; resembling a clasp or buckle; Figure 7-32). The tibia is by far the larger of the two and supports most of the weight of the leg. A **tibial tuberosity,** the attachment point for the anterior thigh muscles, can easily be seen and felt just inferior to the patella (Figure 7-33). The **anterior crest** forms the shin. The proximal end of the tibia has flat medial and lateral condyles that

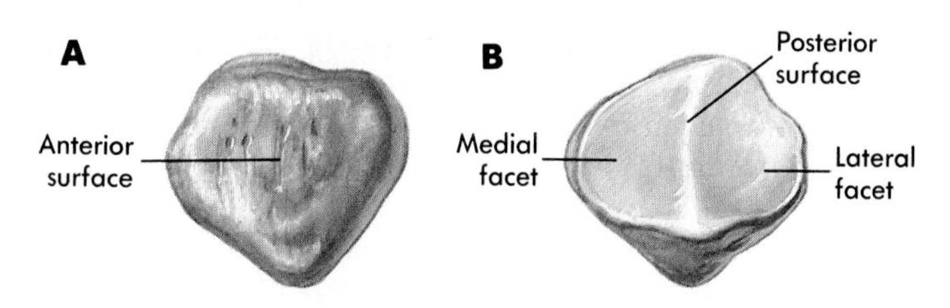

FIGURE 7-31 Right patella. **A** Anterior view. **B** Posterior view.

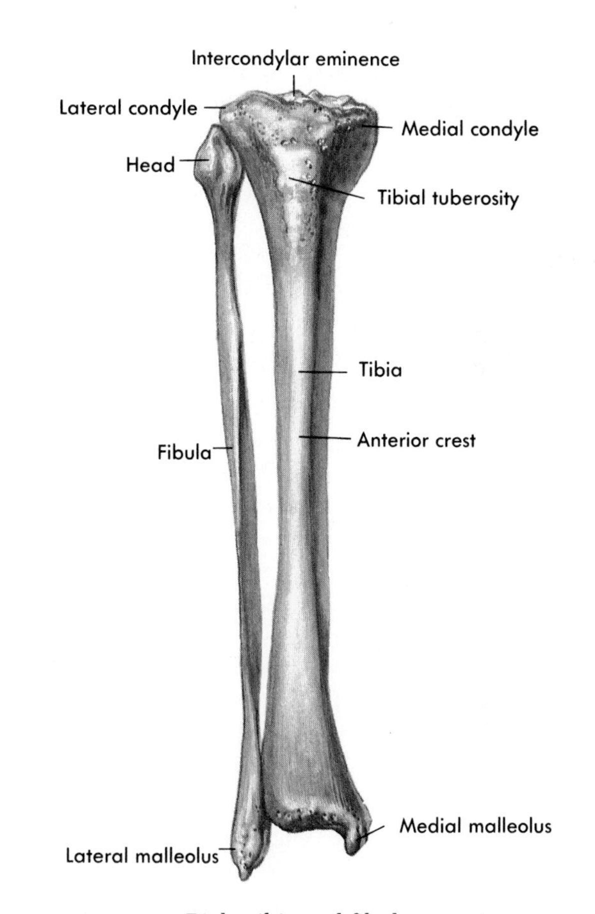

FIGURE 7-32 Right tibia and fibula, anterior view.

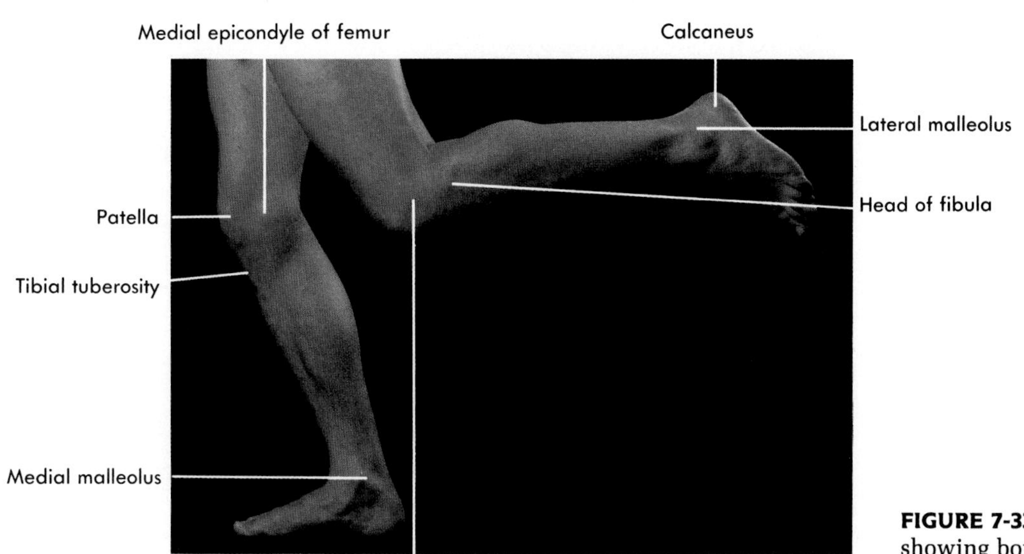

FIGURE 7-33 Surface anatomy showing bones of the lower limb.

articulate with the condyles of the femur. Located between the condyles is the **intercondylar eminence,** a site of ligament attachment. The knee is unusual in that it has ligaments within the joint (see Chapter 8). The distal end of the tibia is enlarged to form the **medial malleolus** (ma'le-o'lus; mallet shaped), which helps form the medial side of the ankle joint.

The fibula does not articulate with the femur but has a small proximal head where it articulates with the tibia. The distal end of the fibula is slightly enlarged as the **lateral malleolus** to create the lateral wall of the ankle joint. The lateral and medial malleoli can be felt and seen as prominent lumps on either side of the ankle (see Figure 7-33). The thinnest portion of the fibula is just proximal to the lateral malleolus.

8

Explain why modern ski boots are designed with high tops that extend partway up the leg.

? ? ? ? ? ? ? ? ? ? ?

Ankle

The ankle consists of seven **tarsal** (tar'sal; the sole of the foot) **bones,** which are depicted and named in Figure 7-34. The **talus** (tal'us; ankle bone) articulates with the tibia and the fibula to form the ankle joint. The **calcaneus** (kal-ka'ne-us; heel) is located inferior and just lateral to the talus and supports that bone. The calcaneus protrudes posteriorly where the calf muscles attach to it and where it can be easily felt as the heel. The ankle is relatively much larger than the wrist and includes the heel and the proximal one third of the foot.

Foot

The **metatarsals** and **phalanges** of the foot are arranged in a manner very similar to the metacarpals and phalanges of the hand, with the great toe analogous to the thumb (see Figure 7-34). Small sesamoid bones often form in the tendons of muscles attached to the great toe. As mentioned previously, the calcaneus bone of the ankle forms the heel of the foot. The ball of the foot is the junction between the metatarsals and phalanges. The foot as a unit is convex dorsally and concave ventrally to form the arches of the foot (described more fully in Chapter 8).

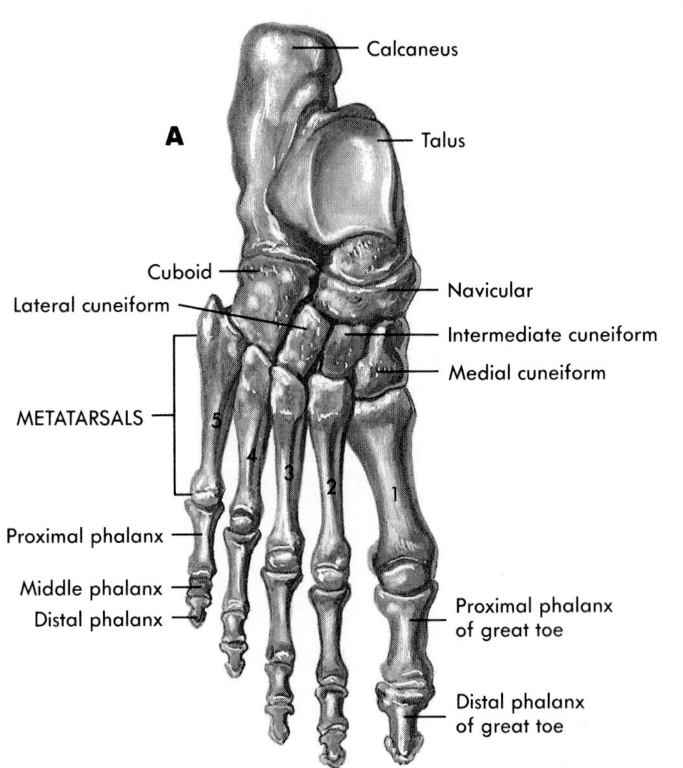

FIGURE 7-34 Bones of the right ankle and foot. **A** Dorsal view. **B** Medial view.

The Skeletal System

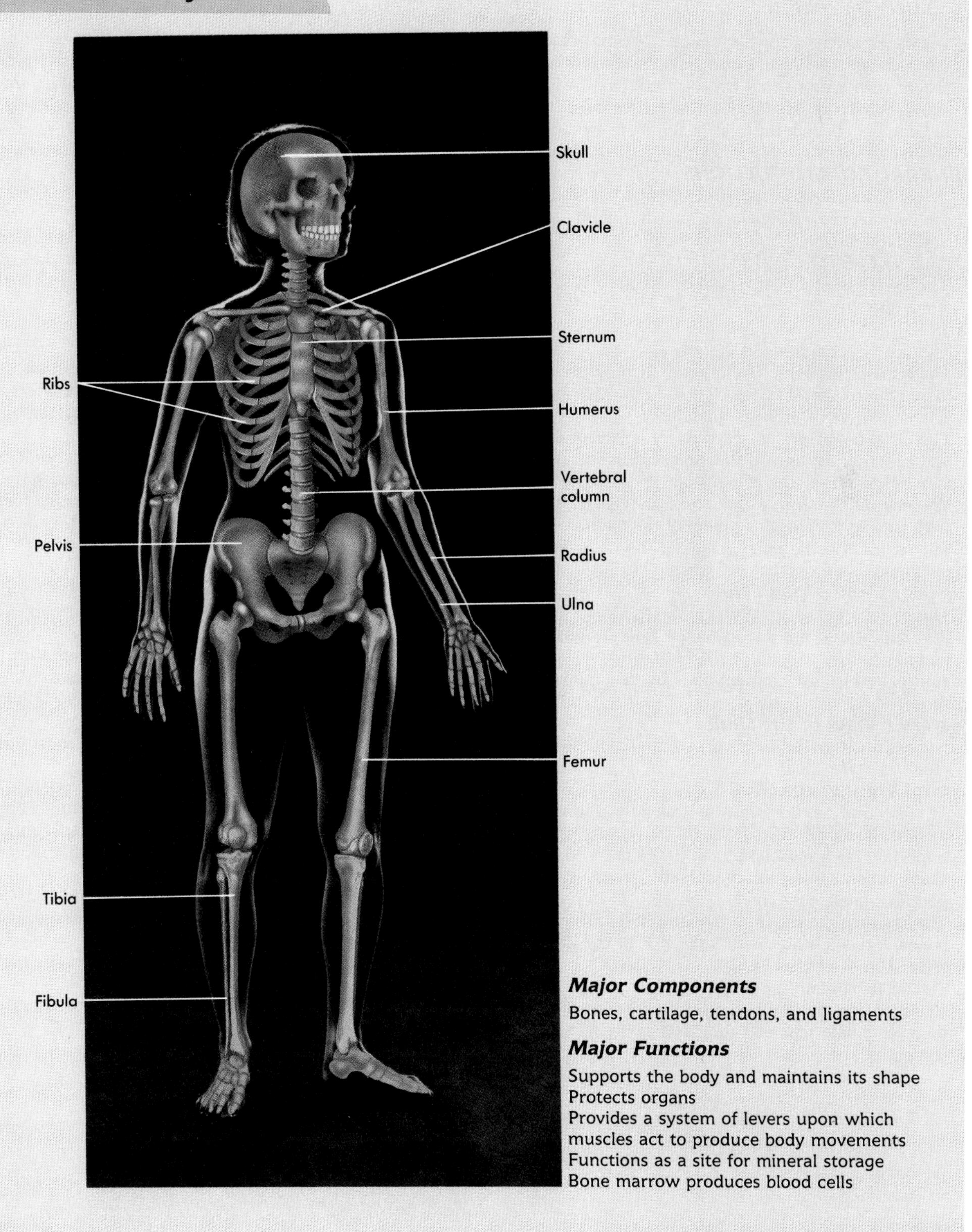

Major Components

Bones, cartilage, tendons, and ligaments

Major Functions

Supports the body and maintains its shape
Protects organs
Provides a system of levers upon which muscles act to produce body movements
Functions as a site for mineral storage
Bone marrow produces blood cells

SUMMARY

THE GROSS ANATOMY OF THE SKELETAL SYSTEM CONSIDERS THE FEATURES OF BONE, CARTILAGE, TENDONS, AND LIGAMENTS THAT CAN BE SEEN WITHOUT THE USE OF A MICROSCOPE. Dried, prepared bones display the major features of bone but obscure the relationship between bone and soft tissue.

GENERAL CONSIDERATIONS *p. 184*

Bones have processes, smooth surfaces, and holes that are associated with ligaments, muscles, joints, nerves, and blood vessels.

AXIAL SKELETON *p. 186*

The axial skeleton consists of the skull, vertebral column, and thoracic cage.

Skull

1. The skull is composed of 28 bones.
2. The ear ossicles, which function in hearing, are located inside the temporal bones.
3. The cranial vault protects the brain.
4. The facial bones protect the sensory organs of the head and function as muscle attachment sites (mastication, facial expression, and eye muscles).
5. The mandible and maxillae possess alveolar processes with sockets for the attachment of the teeth.
6. The hyoid bone, which "floats" in the neck, is the attachment site for throat and tongue muscles.

Superior View of the Skull

The parietal bones are joined at the midline by the sagittal suture and are joined to the frontal bone by the coronal suture, to the occipital bone by the lambdoid suture, and to the temporal bone by the squamous suture.

Posterior View of the Skull

Nuchal lines are the points of attachment for neck muscles.

Lateral View of the Skull

1. The external auditory meatus transmits sound waves toward the eardrum.
2. Important neck muscles attach to the mastoid process.
3. The temporal lines are attachment points of the temporalis muscle.
4. The zygomatic arch, from the temporal and zygomatic bones, forms a bridge across the side of the skull.

Frontal View of the Skull

1. The orbits contain the eyes.
2. The nasal cavity is divided by the nasal septum, and the hard palate separates the nasal cavity from the oral cavity.
3. Sinuses within bone are air-filled cavities. The paranasal sinuses, which connect to the nasal cavity, are the frontal, ethmoidal, sphenoidal, and maxillary sinuses.
4. The mandible articulates with the temporal bone.

Interior of the Cranial Vault

1. The crista galli is a point of attachment for one of the meninges.
2. The olfactory nerves extend into the roof of the nasal cavity through the cribriform plate.
3. The sella turcica is occupied by the pituitary gland.
4. The spinal cord and brain are connected through the foramen magnum.

Inferior View of the Skull

1. Occipital condyles are points of articulation between the skull and the vertebral column.
2. Blood reaches the brain through the internal carotid arteries, which pass through the carotid canals, and the vertebral arteries, which pass through the foramen magnum.
3. Most blood leaves the brain through the internal jugular veins, which exit through the jugular foramina.
4. Styloid processes provide attachment points for three muscles involved in movement of the tongue, hyoid bone, and pharynx.
5. The hard palate forms the floor of the nasal cavity.

Vertebral Column

1. The vertebral column provides flexible support and protects the spinal cord.
2. The vertebral column has four major curvatures: cervical, thoracic, lumbar, and sacral/coccygeal. Abnormal curvatures are lordosis (lumbar), kyphosis (thoracic), and scoliosis (lateral).
3. A typical vertebra consists of a body, an arch, and various processes.
   - Adjacent bodies are separated by intervertebral disks. The disk has a fibrous outer covering (anulus fibrosus) surrounding a gelatinous interior (nucleus pulposus).
   - Vertebrae articulate with each other through the superior and inferior articular processes.
   - Part of the body and the arch (pedicle and lamina) form the vertebral foramen, which contains and protects the spinal cord.
   - Spinal nerves exit through the intervertebral foramina.
   - The transverse and spinous processes serve as points of muscle and ligament attachment.

4. Several different kinds of vertebrae can be recognized.
   - All seven cervical vertebrae have transverse foramina, and most have bifid spinous processes.
   - The 12 thoracic vertebrae are characterized by long, downward-pointing spinous processes and demi-facets.
   - The five lumbar vertebrae have thick, heavy bodies and processes.
   - The sacrum consists of five fused vertebrae and attaches to the pelvis.
   - The coccyx consists of four fused vertebrae attached to the sacrum.

Thoracic Cage

1. The thoracic cage, consisting of the ribs, their associated costal cartilages, and the sternum, functions to protect the thoracic organs and to prevent the collapse of the thorax during respiration.
2. Twelve pairs of ribs attach to the thoracic vertebrae. They are divided into seven pairs of true ribs and five pairs of false ribs. Two pairs of false ribs are floating ribs.
3. The sternum is composed of the manubrium, the body, and the xiphoid process.

APPENDICULAR SKELETON *p. 210*

The appendicular skeleton consists of the upper and lower limbs and the girdles that attach the limbs to the body.

Upper Limb

1. The upper limb is attached loosely and functions in grasping and manipulation.
2. The pectoral girdle consists of the scapula and the clavicle.
   - The scapula articulates with the humerus and the clavicle. It serves as an attachment site for shoulder, back, and arm muscles.
   - The clavicle holds the shoulder away from the body, permitting free movement of the arm.
3. The arm bone is the humerus.
   - The humerus articulates with the scapula (head), the radius (capitulum), and the ulna (trochlea).
   - Sites of muscle attachment are the greater and lesser tubercles, the deltoid tuberosity, and the epicondyles.
4. The forearm consists of the ulna and the radius.
   - The ulna and the radius articulate with each other, the humerus, and the wrist bones.
   - The wrist bones attach to the styloid processes of the radius and the ulna.
5. There are eight carpal, or wrist, bones arranged in two rows.
6. The hand consists of five metacarpal bones.
7. The phalanges are finger bones. Each finger has three phalanges, and the thumb has two phalanges.

Lower Limb

1. The lower limb is attached solidly to the coxa and functions in support and locomotion.
2. The pelvic girdle is formed by the coxae and the sacrum. Each coxa is formed by the fusion of the ilium, the ischium, and the pubis.
   - The coxae articulate with each other (symphysis pubis), the sacrum (sacroiliac joint), and the femur (acetabulum).
   - Important sites of muscle attachment are the iliac crest, the iliac spines, and the ischial tuberosity.
   - The female pelvis has a larger pelvic inlet and a larger pelvic outlet than the male pelvis.
3. The thigh bone is the femur.
   - The femur articulates with the coxa (head), the tibia (medial and lateral condyles), and the patella (patellar groove).
   - Sites of muscle attachment are the greater and lesser trochanters, the gluteal tuberosity, and the lateral and medial epicondyles.
4. The leg consists of the tibia and the fibula.
   - The tibia articulates with the femur, the fibula, and the talus. The fibula articulates with the tibia and the talus.
   - Knee ligaments attach to the intercondylar eminence, and tendons from the thigh muscles attach to the tibial tuberosity.
5. Seven tarsal bones form the ankle.
6. The foot consists of five metatarsal bones.
7. The toes have three phalanges each except for the big toe, which has two phalanges.

CONTENT REVIEW

1. Define axial skeleton and appendicular skeleton.
2. Name the bones of the cranial vault.
3. Name the bones of the face.
4. List the seven bones that form the orbit of the eye.
5. Name the eight bones that form the nasal cavity. Describe the bones and cartilage that form the nasal septum.
6. What is a sinus? What are the functions of the sinuses? Give the location of the paranasal sinuses.
7. Name the bones that form the hard palate. What is the function of the hard palate?
8. Through what foramen does the brainstem connect to the spinal cord? Name the foramina that contain nerves for the senses of vision (optic nerve), smell (olfactory nerve), and hearing (vestibulocochlear nerve).
9. Name the foramen through which the major blood vessels for the brain enter and exit the skull.
10. List the places where the following muscles attach to the skull: neck muscles, throat muscles, muscles of mastication, muscles of facial expression, and muscles that move the eyeballs.
11. Describe the four major curvatures of the vertebral column. Define lordosis, kyphosis, and scoliosis.
12. How do the vertebrae protect the spinal cord? Where do spinal nerves exit the vertebral column?
13. Name and give the number of each type of vertebra. Describe the characteristics that distinguish the different types from each other.
14. What is the function of the thoracic cage? Distinguish between true, false, and floating ribs.
15. Describe the different parts of the sternum.
16. Name the bones that comprise the pectoral girdle. What are the functions of these bones?
17. What are the functions of the acromion process and the coracoid process of the scapula?
18. Name the important sites of muscle attachment on the humerus.
19. Give the points of articulation between the scapula, humerus, ulna, radius, and wrist bones.
20. What is the function of the radial tuberosity?
21. List the eight carpal bones. What is the carpal tunnel?
22. What bones form the hand? The knuckles? How many phalanges are in each finger and in the thumb?
23. Define the term pelvis. What bones fuse to form each coxa? Where and with what bones do the coxae articulate?
24. Name the important sites of muscle attachment on the pelvis.
25. Distinguish between the true pelvis and the false pelvis.
26. Describe the difference between a male pelvis and a female pelvis.
27. What is the function of the greater trochanter and the lesser trochanter?
28. Name the bones of the leg.
29. Give the points of articulation between the pelvis, femur, leg, and ankle.
30. What is the function of the tibial tuberosity?
31. Name the seven tarsal bones. Which bones form the ankle joint? What bone forms the heel?
32. Describe the bones that constitute the foot.

CONCEPT QUESTIONS

1. A patient has an infection in the nasal cavity. Name seven places (routes) to which the infection could spread.
2. A patient is unconscious. X-ray films reveal that the superior articular process of the atlas has been fractured. Which of the following could have produced this condition: falling on the top of the head or being hit in the jaw with an uppercut? Explain.
3. If the vertebral column is forcefully rotated, what part of the vertebra most likely will be damaged? In what area of the vertebral column is such damage most likely?
4. An asymmetrical weakness of the back muscles can produce which of the following: scoliosis, kyphosis, or lordosis? Which could result from pregnancy? Explain.
5. What might be the consequences of a broken forearm involving both the ulna and radius when the ulna and radius fuse to each other during repair of the fracture?
6. Suppose you needed to compare the length of one lower limb to the other in an individual. Using bony landmarks, suggest an easy way to accomplish the measurements.
7. A paraplegic individual develops decubitus ulcers (pressure sores) on the buttocks from sitting in a wheelchair. Name the bony protuberance responsible.
8. Why are women knock-kneed more often than men?
9. Based on bone structure of the lower limb, explain why it is easier to turn the foot medially (sole of the foot facing toward the midline of the body) than laterally. Why is it easier to cock the wrist medially than laterally?
10. Justin Time leaped from his hotel room to avoid burning to death in a fire. If he landed on his heels, what bone was he likely to fracture? Unfortunately for Justin, a 240-pound fireman, Hefty Stomper, ran by and stepped heavily on the proximal part of Justin's foot (not the toes). What bones could now be broken?

NAMING OF JOINTS

Joints are commonly named according to the bones or portions of bones that are united at the joint such as the temporomandibular joint between the temporal bone and the mandible.

1

What would a joint between the metacarpals and the phalanges be called?

? ? ? ? ? ? ? ? ? ? ?

Some joints are given the name of only one of the articulating bones such as the humeral joint (shoulder) between the humerus and scapula. Still other joints are simply given the Greek or Latin equivalent of the common name such as **cubital** (ku'bĭ-tal) for the elbow joint.

CLASSES OF JOINTS

The three major classes of joints are classified structurally as fibrous, cartilaginous, and synovial (Table 8-1). In this classification scheme joints are classified according to the major connective tissue type that binds the bones together and whether or not there is a fluid-filled joint capsule. Another method of classifying joints is a functional classification that is based on the degree of motion at each joint and includes the terms synarthrosis (nonmovable joints), amphiarthrosis (slightly movable joints), and diarthrosis (freely movable joints). This functional classification is somewhat restrictive and will not be used in this text. The structural classification scheme with its various subclasses allows for a more precise classification and is the scheme used in this book.

Fibrous Joints

Fibrous joints consist of two bones that are united by fibrous tissue, have no joint cavity, and exhibit little or no movement. Joints in this group are classified further on the basis of structure as sutures, syndesmoses, or gomphoses.

Sutures

Sutures (su'churz; seams between flat bones) are only between some skull bones, and some sutures may be completely immovable in adults. Sutures are seldom smooth, and the opposing bones often interdigitate (interlocking of fingerlike processes). The tissue between the two bones is dense, regularly arranged fibrous connective tissue, and the periosteum on the inner and outer surfaces of the adjacent bones continues over the joint. The two layers of periosteum plus the dense fibrous connective tissue in between form the **sutural ligament.**

In a newborn the sutures are called **fontanels** (fon'tă-nels) and are fairly wide, allowing "give" in the skull during the birth process and growth of the head after birth (Figure 8-1).

The margins of bones within sutures are sites of continuous intramembranous bone growth, and many sutures eventually become ossified. For example, ossification of the suture between the two frontal bones occurs shortly after birth so that they usually form a single bone in the adult. In most normal adults the coronal, sagittal, and lambdoid sutures are not fused. However, in some very old adults even these sutures may become ossified. A **synostosis** (sin'os-to'sis) is the osseous union between the bones of a joint. This process occurs in certain other joints such as in the sternum in addition to sutures.

2

Predict the result of a sutural synostosis occurring prematurely in a child's skull before the brain has reached its full size.

? ? ? ? ? ? ? ? ? ? ?

Syndesmoses

A **syndesmosis** (sin'dez-mo'sis; to fasten or bind) is a type of fibrous joint in which the bones are separated by a greater distance than in a suture and are joined by ligaments. Some movement may occur at syndesmoses because of flexibility of the ligaments (e.g., the radioulnar syndesmosis, which binds the radius and ulna together; Figure 8-2).

Gomphoses

Gomphoses (gom-fo'sēs) are specialized joints consisting of pegs that fit into sockets and that are held in place by fine bundles of regular collagenous connective tissue. The joints between the teeth and the sockets (alveoli) along the alveolar processes of the mandible and maxillae are gomphoses. The connective tissue bundles between the teeth and their sockets are **periodontal** (pĕr'e-o-don'tal) **ligaments** (see Figure 24-6) and allow a slight amount of "give" to the teeth during mastication.

Periodontal disease, which is the leading cause of tooth loss in the United States, involves an accumulation of plaque and bacteria, which gradually destroys the periodontal ligaments and the bone. As a result, teeth fall out of their sockets. Proper brushing, flossing, and professional cleaning to remove plaque can prevent this type of tooth loss.

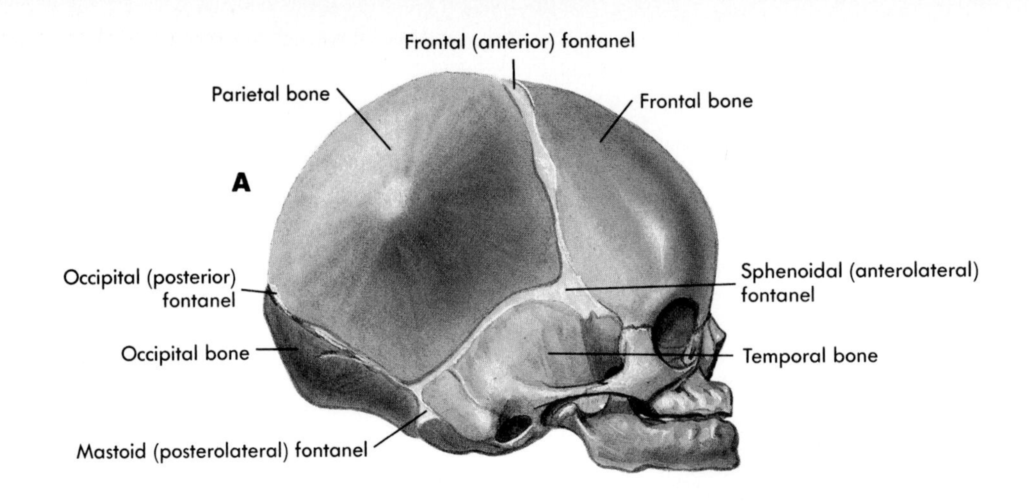

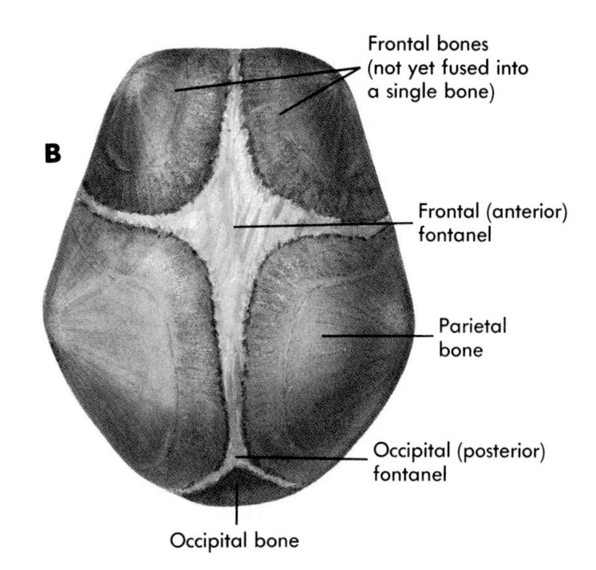

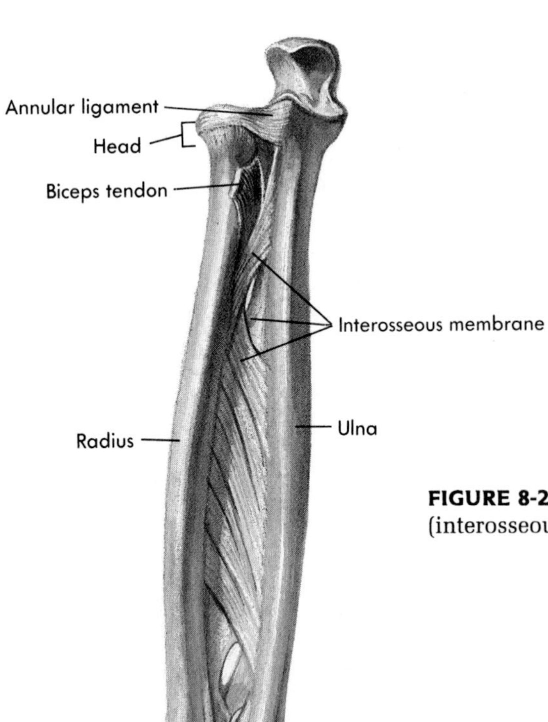

FIGURE 8-2 Radioulnar syndesmosis (interosseous membrane) of right forearm.

Cartilaginous Joints

Cartilaginous joints unite two bones by means of either hyaline cartilage or fibrocartilage. If the bones are joined by hyaline cartilage, they are called synchondroses; if they are joined by fibrocartilage, they are called symphyses.

Synchondroses

A **synchondrosis** (sin'kon-dro'sis; union through cartilage) consists of two bones joined by hyaline cartilage. The epiphyseal plates of growing bones are synchondroses in which no movement occurs. As a result, these synchondroses are temporary and are replaced by bone to form synostoses when growth ceases sometime before 25 years of age. On the other hand, costosternal synchondroses (Figure 8-3), which consist of the costal cartilage between most of the ribs and the sternum, persist throughout life. The costal cartilage has a fair amount of give, which allows the thorax to expand and contract during respiration.

Symphyses

A **symphysis** (sim'fă-sis; a growing together) consists of fibrocartilage uniting two bones. Symphyses include the junction between the manubrium and body of the sternum in adults, the symphysis pubis (the anterior junction between the pubic portions of the coxae; Figure 8-4), and the intervertebral disks. Some of these joints are slightly movable because of the somewhat flexible nature of fibrocartilage.

During pregnancy certain hormones (e.g., estrogen, progesterone, and relaxin) act on the connective tissue of the symphysis pubis, allowing it to stretch, loosening the joint. This change allows the pelvic opening to enlarge at the time of delivery. However, these same hormones may act on the connective tissue of other joints in the body such as the arches of the feet, causing them to relax. These hormones may also act on some of the baby's joints such as the hip, causing the joints to become more mobile than normal.

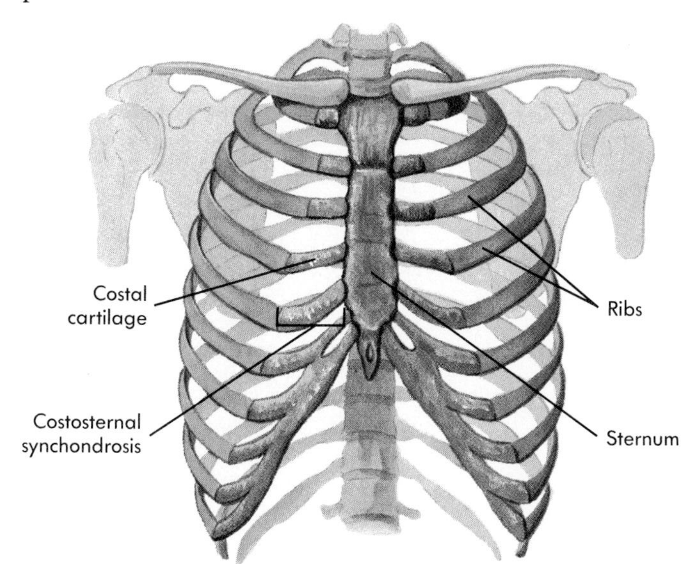

FIGURE 8-3 Costosternal synchondrosis.

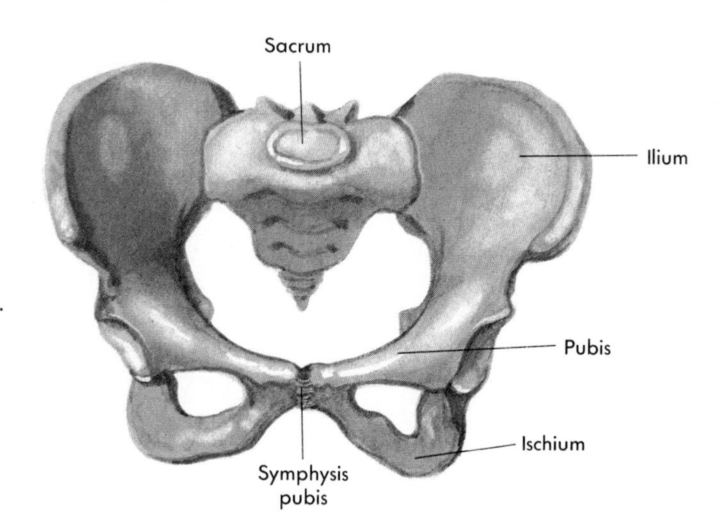

FIGURE 8-4 Symphysis pubis.

Joint Disorders

Arthritis

Arthritis, an inflammation of any joint, is the most common and best known of the joint disorders, affecting 10% of the world's population. There are at least 20 different types of arthritis. This classification scheme is based on the cause and progress of the arthritis. Causes include infectious agents, metabolic disorders, trauma, and immune disorders.

Suppurative (pus-forming) **arthritis** may result from a number of infectious agents. These joint infections may be transferred from some other infected site in the body or may be systemic (i.e., throughout the body). Usually only one joint (normally one of the larger joints) is affected, and the course of suppurative arthritis, if it is treated early, is transitory. However, with prolonged infection the articular surfaces may degenerate. **Tuberculous arthritis** can occur as a secondary infection from pulmonary tuberculosis and is more damaging than typical suppurative arthritis. It usually affects the spine or large joints and causes ulceration of the articular cartilages and even erosion of the underlying bone. Transient arthritis of multiple joints is a common symptom of **rheumatic fever,** but permanent damage seldom occurs in the joints with this disorder.

Hemophilia arthritis can result from bleeding into the joint cavity caused by hemophilia, a disease characterized by a deficient clotting mechanism in the blood. There is some evidence that the iron in the blood is toxic to the chondrocytes, resulting in degeneration of the articular cartilage.

Rheumatoid arthritis affects approximately 3% of all women and approximately 1% of all men in the United States. It is a general connective tissue disorder that affects the skin, vessels, lungs, and other organs, but it is most pronounced in the joints. It is severely disabling and most commonly destroys small joints such as those in the hands and feet (Figure 8-A). The initial cause is unknown but may involve a transient infection or an autoimmune disease (an immune reaction to one's own tissues; see Chapter 22) that develops against collagen. There may also be a genetic predisposition. Whatever the cause, the ultimate course apparently is immunological. All people with classic rheumatoid arthritis have a protein, **rheumatoid factor,** in their blood. In a patient with rheumatoid arthritis the synovia and associated connective tissue cells proliferate, forming a **pannus** (clothlike layer), which causes the joint capsule to thicken and which destroys the articular cartilage. In advanced stages opposing joint surfaces can become fused. **Juvenile rheumatoid arthritis** is similar to the adult type in many ways, but no rheumatoid factor is found in the serum.

Gout

Gout is a group of metabolic disorders in which joints are involved. These disorders are largely idiopathic (of unknown cause), although some cases of gout seem to be familial (occur in families and therefore probably are genetic). Gout is more common in males than in females. The ultimate problem in gout patients is an increase in uric acid in the body with precipitation of monosodium urate crystals in various tissues, including the kidneys and joint capsules.

Gout's earliest symptom is transient arthritis resulting from urate crystal accumulation in the synovia. This irritation can ultimately lead to an inflammatory response in the joints; both the crystal deposition and inflammation can become chronic. Normally, only one or two joints are affected. The most commonly affected joints (85% of the cases) are the base of the great toe and other foot and leg joints to a lesser extent. Ultimately any joint may be involved, and damage to the kidney from

A

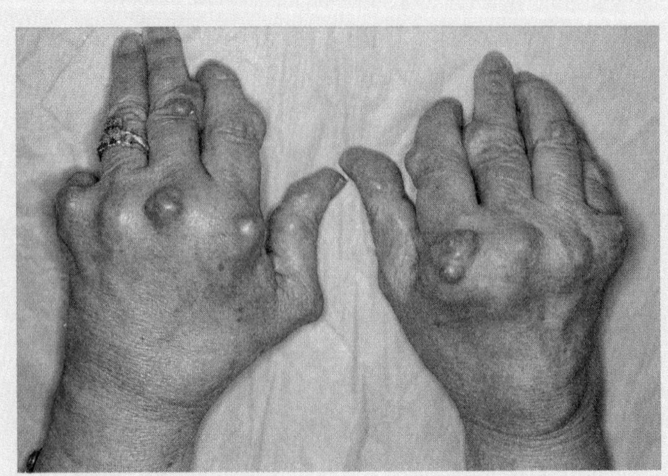

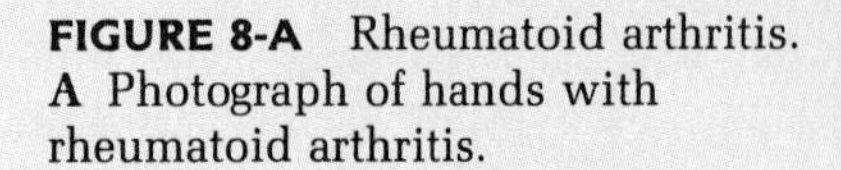

FIGURE 8-A Rheumatoid arthritis. **A** Photograph of hands with rheumatoid arthritis.

crystal formation occurs in almost all advanced cases.

Pseudogout is a disorder that causes pain and swelling similar to that seen in gout, but it is characterized by calcium hypophosphate crystal deposits rather than urate crystals in joints.

Hallux Valgus and Bunion

In people who wear pointed shoes the great toe can be deformed and displaced laterally, a condition called **hallux valgus.** Bunions are often associated with hallux valgus. A **bunion** is bursitis that develops over the first metatarsophalangeal joint because of pressure and rubbing by shoes.

Degenerative Joint Disease

Degenerative joint disease (DJD), also called osteoarthritis, consists of the gradual "wear and tear" of a joint that occurs with advancing age. It occurs as a result of accumulated minor trauma. Slowed metabolic rates with increased age also seem to contribute to DJD. This disorder is not actually an arthritis since no primary inflammation occurs and any associated inflammation is secondary to the disorder. It is very common in older individuals and affects 85% of all people in the United States over the age of 70. It tends to occur in the weight-bearing joints such as the knees and is more common in overweight individuals. Mild exercise retards joint degeneration and enhances mobility.

Joint Replacement

As a result of recent advancements in biomedical technology, many joints of the body can be replaced by artificial joints. Joint replacement, called **arthroplasty,** was first developed in the late 1950s. One of the major reasons for its use is to eliminate unbearable pain in patients who average 55 to 60 years of age and have joint disorders. Degenerative joint disease is the leading disease requiring joint replacement, accounting for two thirds of the patients. Rheumatoid arthritis accounts for more than half of the remaining cases.

The major objectives in the design of joint prostheses (artificial replacements) include the development of stable articulations, low friction, solid fixation to the bone, and normal range of motion. New synthetic replacement materials are being designed by biomedical engineers to accomplish these objectives. Prosthetic joints usually are composed of metal (e.g., stainless steel, titanium alloys, or cobalt-chrome alloys) in combination with modern plastics (e.g., high-density polyethylene, silastic, or elastomer). The bone of the articular area is removed on one side (hemireplacement) or both sides (total replacement) of the joint, and the artificial articular areas are glued to the bone with a synthetic adhesive (methylmethacrylate). The smooth metal surface rubbing against the smooth plastic surface provides a low-friction contact with a range of movement that depends on the design.

The success of joint replacement depends on the joint replaced, the age and condition of the patient, and the state of the technology. Most reports are based on examination of patients 2 to 10 years after joint replacement. The technology is improving constantly, so current reports do not adequately reflect the effect of the most recent improvements. Still, the current reports indicate a success rate of 80% to 90% in hip replacements and 60% or more in ankle and elbow replacements. The major reason for failure of prosthetic joints is loosening of the artificial joint from the bone to which it is attached. New prostheses with porous surfaces have been developed to overcome this problem.

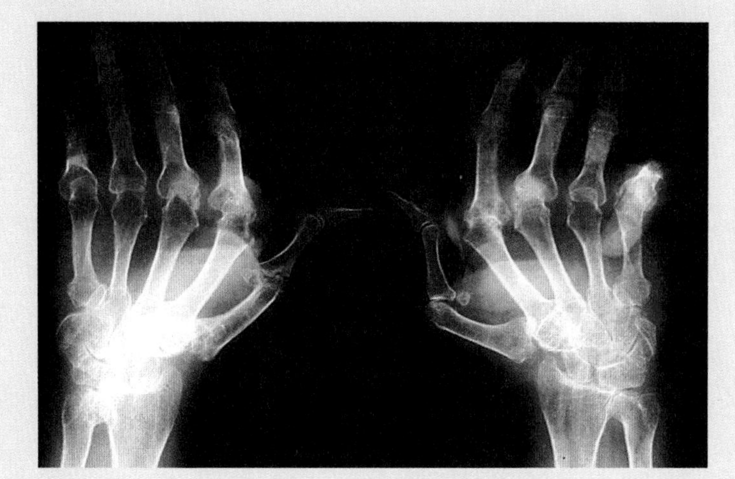

B

FIGURE 8-A, cont'd Rheumatoid arthritis. **B** X-ray of the same hands shown in **A.**

Synovial Joints

Synovial (sĭ-no've-al; joint fluid; *syn*, coming together, *ovia*, resembling egg albumin) **joints,** those containing **synovial fluid,** allow considerable movement between articulating bones (Figure 8-5). These joints are anatomically more complex than fibrous and cartilaginous joints. Most joints that unite the bones of the appendicular skeleton are synovial joints, reflecting the far greater mobility of the appendicular skeleton compared with the axial skeleton.

The articular surfaces of bones within synovial joints are covered with a thin layer of hyaline cartilage called **articular cartilage,** which provides a smooth surface where the bones meet. Additional fibrocartilage **articular disks** are associated with several synovial joints such as the knee and the temporomandibular joint. Articular disks provide extra strength and support to the joint and can increase the depth of the joint cavity.

The joint is enclosed by a **joint capsule,** which helps to hold the bones together and at the same time allows for movement. The joint capsule consists of two layers: an outer **fibrous capsule** and an inner **synovial membrane** (see Figure 8-5). The fibrous capsule is continuous with the fibrous layer of the periosteum that covers the bones united at the joint. Portions of the fibrous capsule may thicken to form ligaments. In addition, ligaments and tendons may be present outside the fibrous capsule, contributing to the strength of the joint.

A synovial membrane lines the joint everywhere except over the articular cartilage. It consists of a collection of modified connective tissue cells either intermixed with part of the fibrous capsule or separated from the fibrous capsule by a layer of areolar tissue or adipose tissue. The membrane produces synovial fluid, which consists of a serum (blood fluid) filtrate and secretions from the synovial cells. Synovial fluid is a complex mixture of polysaccharides, proteins, fat, and cells. The major polysaccharide is hyaluronic acid, which provides much of the slippery consistency of synovial fluid. Synovial fluid forms a thin lubricating film covering the surfaces of a joint.

In certain synovial joints the synovial membrane may extend as a pocket, or sac, called a **bursa** (bur'sah; pocket), for some distance away from the rest of the joint cavity (see Figure 8-5). Bursae contain synovial fluid and provide a fluid-filled cushion between structures that otherwise would rub against one another such as tendons rubbing on bones or other tendons. Some bursae are not associated with

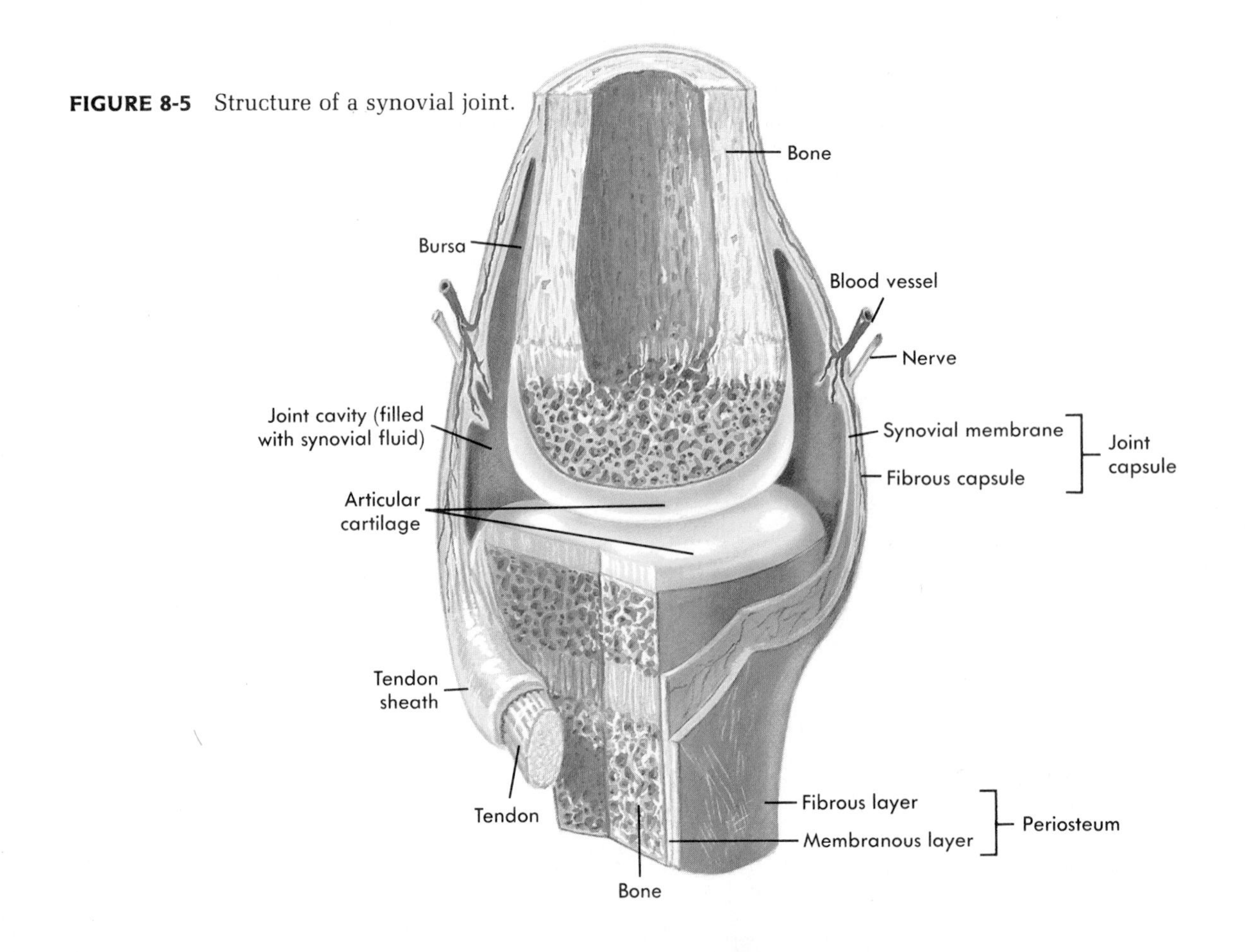

FIGURE 8-5 Structure of a synovial joint.

joints such as those located between the skin and underlying bony prominences where friction could damage the tissues. Other bursae extend along a tendon for some distance, forming a **tendon sheath. Bursitis** (bur-si′tis) is an inflammation of a bursa and may cause considerable pain around the joint and inhibition of movement.

At the peripheral margin of the articular cartilage blood vessels form a vascular circle that supplies the cartilage with nourishment, but no vessels penetrate the cartilage or enter the joint cavity. Additional nourishment to the articular cartilage comes from the underlying cancellous bone and from the synovial fluid covering the articular cartilage. Sensory nerves enter the fibrous capsule and, to a lesser extent, the synovial membrane. They not only supply information to the brain about pain in the joint but also furnish constant information to the brain about the joint's position and degree of movement (see Chapters 13 and 14). Nerves do not enter the cartilage or joint cavity.

Types of synovial joints

Synovial joints are classified according to the shape of the adjoining articular surfaces. The six types of synovial joints are plane, saddle, hinge, pivot, ball-and-socket, and ellipsoid. Movements at synovial joints can be described as **monoaxial** (occurring in one direction or plane), **biaxial** (occurring in two directions or planes), or **multiaxial** (occurring in many directions or planes).

Planes, or **gliding joints,** consist of two opposed flat surfaces approximately equal in size (Figure 8-6, *A*). These joints are multiaxial because the surfaces can slide over each other in many different directions. Some rotation of these joints is also possible but is limited by ligaments and adjacent bone. Examples of these joints are the articular processes between vertebrae.

Saddle joints consist of two saddle-shaped articulating surfaces oriented at right angles to one another so that complementary surfaces articulate with

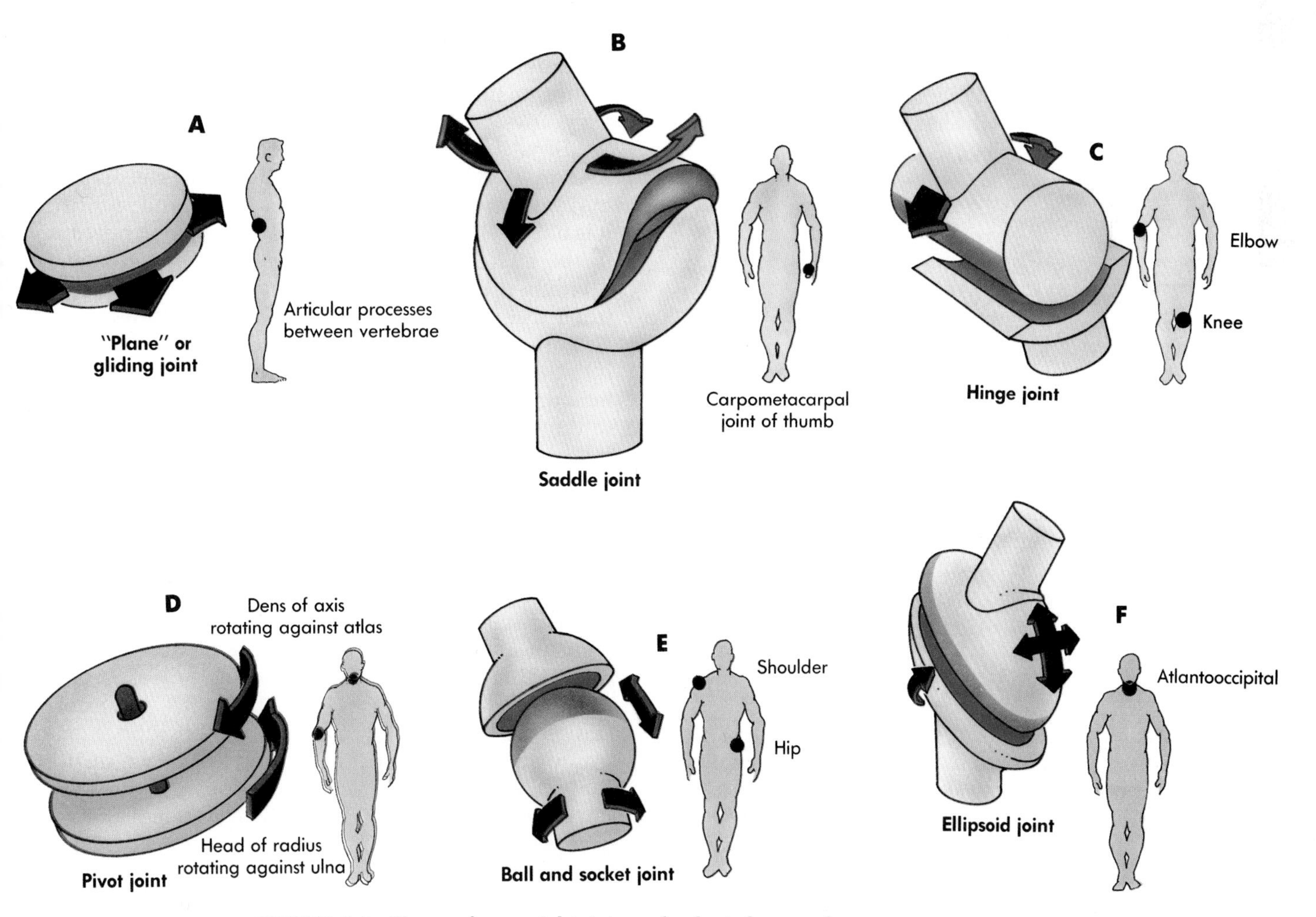

FIGURE 8-6 Types of synovial joints and selected examples.
A Plane. **B** Saddle. **C** Hinge. **D** Pivot. **E** Ball-and-socket. **F** Ellipsoid.

each other (Figure 8-6, *B*). Saddle joints are biaxial joints. The carpometacarpal joint of the thumb is an example.

Hinge joints are monoaxial joints (Figure 8-6, *C*). They consist of a convex cylinder in one bone applied to a corresponding concavity in the other bone. Examples include the elbow and knee joints.

Pivot joints are monoaxial joints, restricting movement to rotation around a single axis (Figure 8-6, *D*). Each pivot joint consists of a relatively cylindrical bony process that rotates within a ring composed partly of bone and partly of ligament. The articulation between the dens, a process on the axis (see Chapter 7), and the atlas is an example. The head of the radius articulating with the proximal end of the ulna is another example (see Figure 8-2).

Ball-and-socket joints consist of a ball (head) at the end of one bone and a socket in an adjacent bone into which a portion of the ball fits (Figure 8-6, *E*). This type of joint is multiaxial, allowing a wide range of movement in almost any direction. Examples are the shoulder and hip joints.

Ellipsoid joints (or condyloid joints) are modified ball-and-socket joints (Figure 8-6, *F*); the articular surfaces are ellipsoid in shape rather than spherical as in regular ball-and-socket joints. The shape of the joint limits its range of movement nearly to a hinge motion in two planes (biaxial) and restricts rotation. The atlantooccipital joint is an example.

TYPES OF MOVEMENT

The types of movement occurring at a given joint are related to the structure of that joint. Some joints are limited to only one type of movement; others can move in several directions. With few exceptions movement is best described in relation to the anatomical position: (1) movement away from the anatomical position and (2) movement returning a structure toward the anatomical position. Most movements are accompanied by other movements in the opposite direction and therefore are listed in pairs.

Angular Movements

Angular movements are those in which one part of a linear structure, such as the body as a whole or a limb, is bent relative to another part of the structure to change the angle between the two parts. Angular movements also involve the movement of a solid rod, such as a limb, that is attached at one end to the body so that the angle at which it meets the body is changed. The most common angular movements are flexion and extension, abduction and adduction.

Flexion and extension

Flexion means to bend (resulting in a decreased angle), and **extension** means to straighten (increasing the angle). These definitions can be easily understood in relation to hinge joints such as the elbow and knee (Figure 8-7, *A* and *B*). Flexion is bending of the elbow or knee; extension straightens it back to the anatomical position. Under normal conditions the elbow or knee cannot be extended beyond the anatomical position (abnormal extension beyond a joint's normal range of motion is called hyperextension). However, the concept of flexion as bending and extension as straightening can be confusing when applied to other joints that are capable of moving in a wider arc than are the elbow and knee. For example, the humoral head is the proximal end of a solid rod, the humerus, attached to the body by the shoulder joint. The shoulder joint allows the arm to move in an anterior direction (away from the anatomical position), back to the anatomical position, and in a posterior direction (again away from the anatomical position) (Figure 8-7, *C*). It is not clear in this movement when the shoulder is "bent" or "straightened," and as can be seen in Figure 8-7, *C*, the relative change in angle is the same in flexion and extension. A similar problem is encountered in using literal definitions to describe flexion and extension of the neck, trunk, hip, and wrist (Figure 8-7, *D* and *E*).

Because of the problems described above, applying the literal definition of flexion and extension to movements of the shoulder, neck, trunk, and so forth, it is helpful to define flexion and extension in relation to the anatomical position and to a coronal plane dividing the body into anterior and posterior halves (refer to Figure 8-7, *C* through *E*). **Flexion** moves a structure such as the upper limb (see Figure 8-7, *C*) into the space anterior to the coronal plane, and **extension** moves the structure into the space posterior to the coronal plane. Flexion can also return an extended structure back to the anatomical (neutral) position and vice versa. For example, moving the upper limb forward to grasp a doorknob involves flexion of the arm. Moving the arm posteriorly, as in inserting the upper limb into a coat sleeve, involves extension of the arm.

The knee, ankle, and toes are exceptions to this concept of the coronal plane. Knee flexion consists of bending the joint so that the leg moves in the posterior direction (see Figure 8-7, *B*); extension brings the leg back to the anatomical position. Flexion of the toes is toward the plantar surface of the foot (sole of the foot), and extension is toward the dorsal surface (Figure 8-7, *F*). These movements seem backward compared with other movements but can be understood when considering that the leg rotates 180 degrees during embryonic development so that the original anterior surface of the lower limb in the embryo becomes the posterior surface of the adult

Text continued on p. 240.

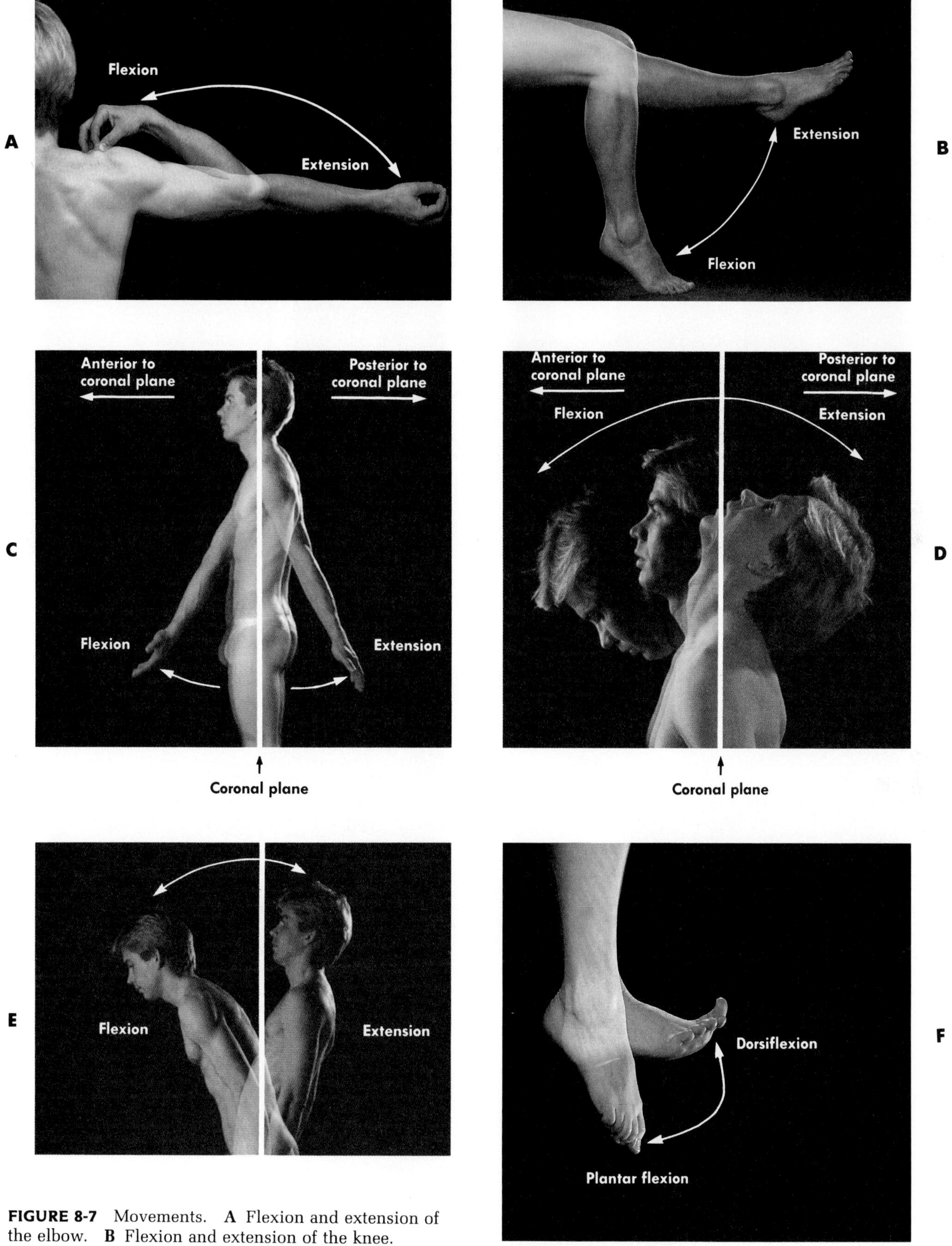

FIGURE 8-7 Movements. **A** Flexion and extension of the elbow. **B** Flexion and extension of the knee. **C** Flexion and extension of the shoulder. **D** Flexion and extension of the neck. **E** Flexion and extension of the trunk. **F** Plantar flexion and dorsiflexion of the foot.

Continued.

G

FIGURE 8-7, cont'd Movements. **G** Abduction and adduction of the upper limb. **H** Abduction and adduction of the fingers. **I** Medial and lateral rotation of the humerus. **J** Pronation and supination. **K** Circumduction of the shoulder. **L** Elevation and depression of the shoulder. **M** Protraction and retraction of the jaw. **N** Lateral excursion of the jaw.

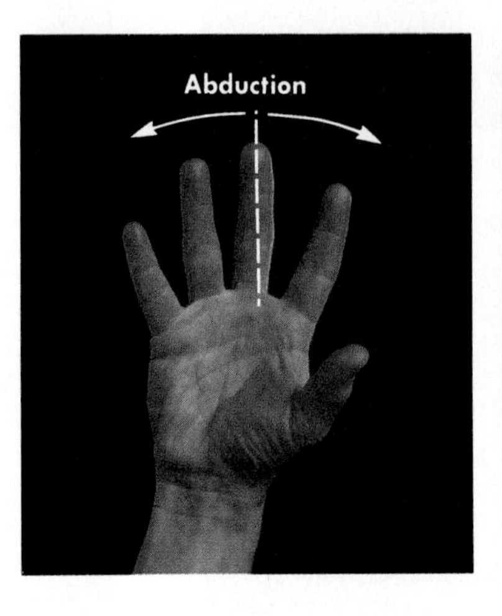

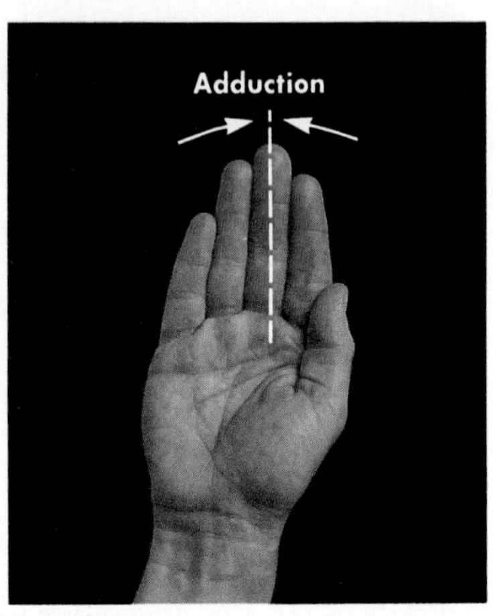

H

I

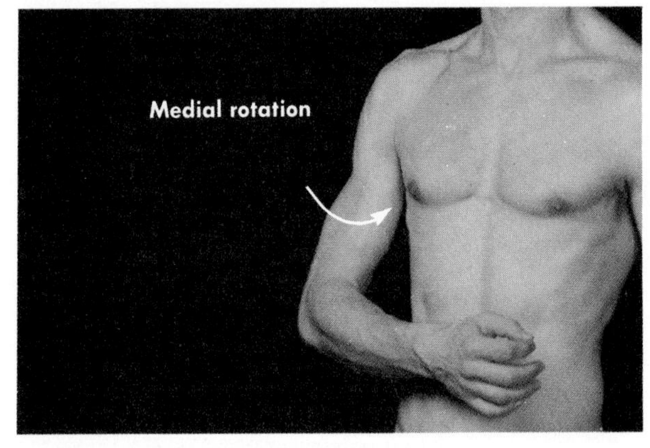

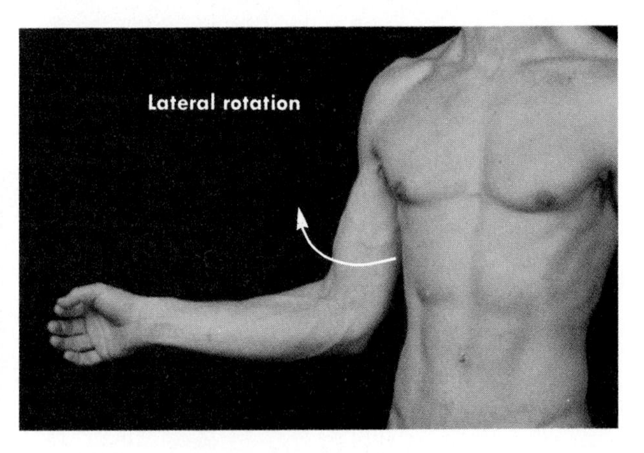

J

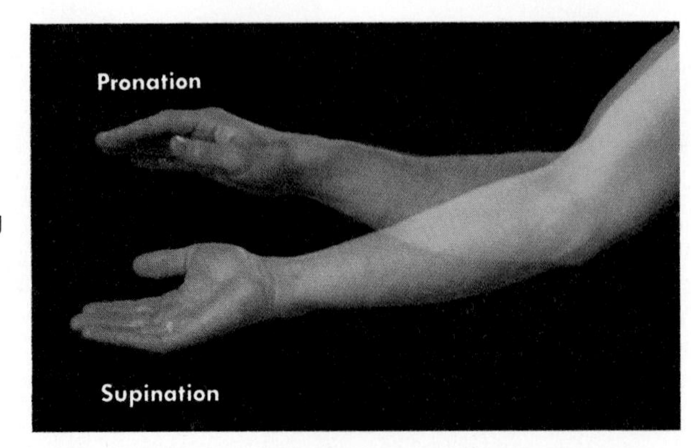

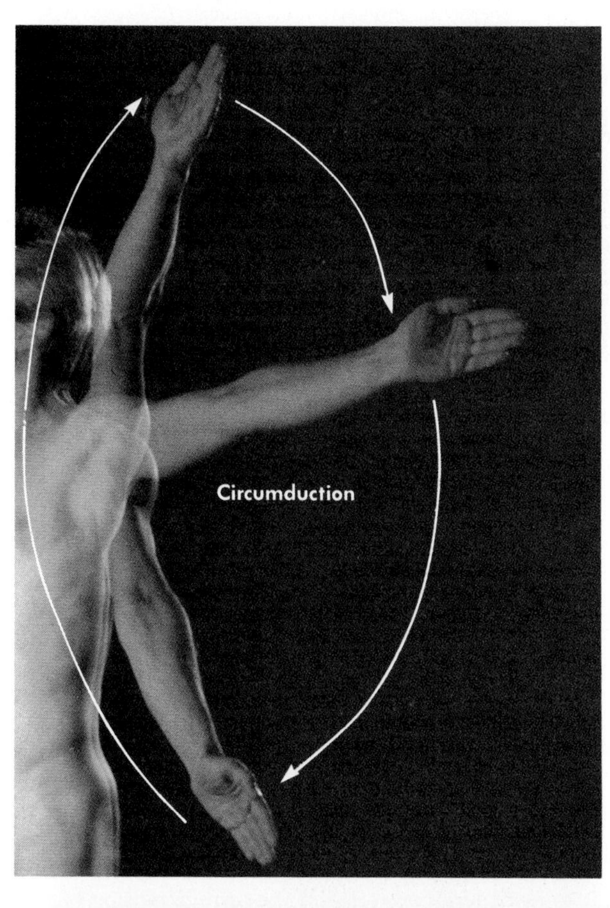

K

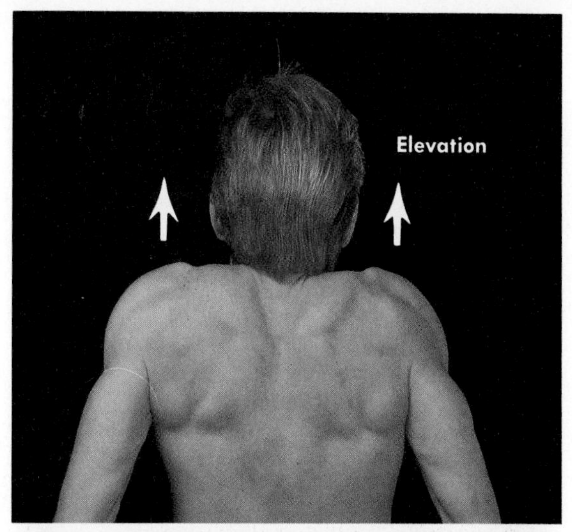

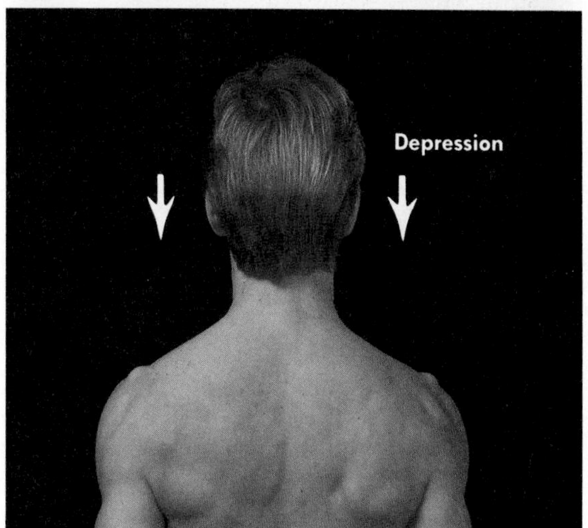

L

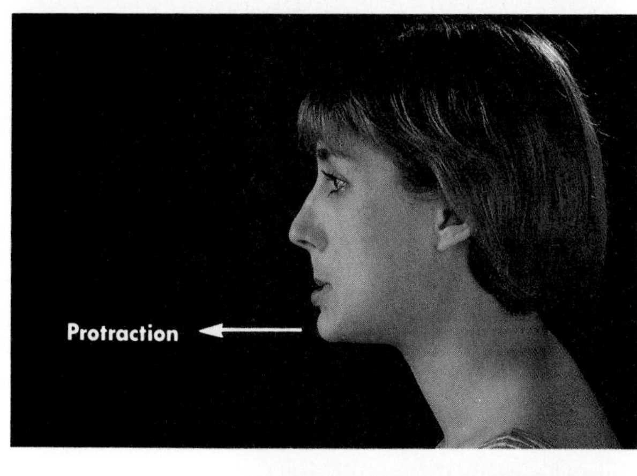

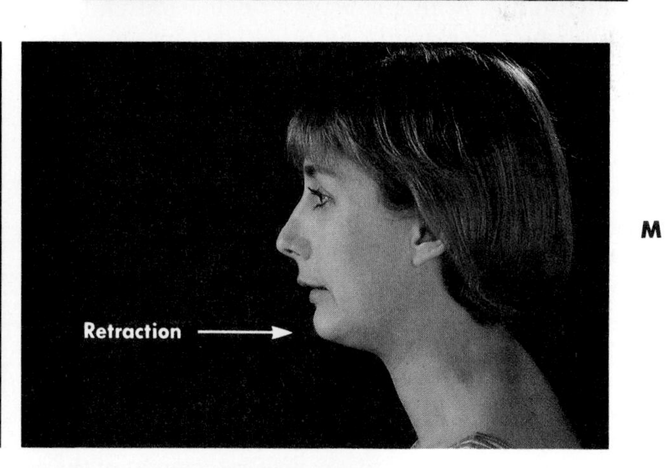

M

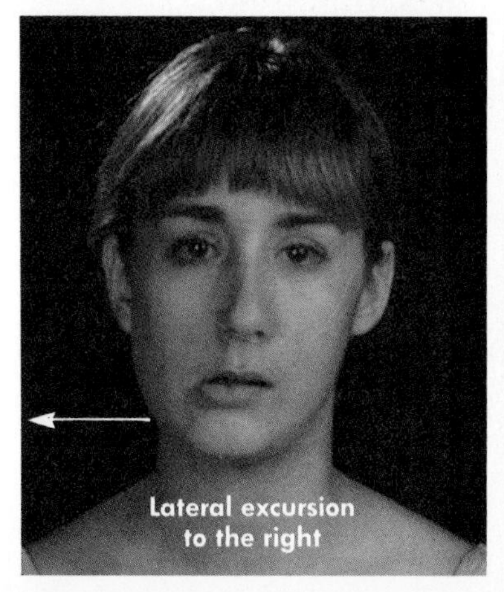

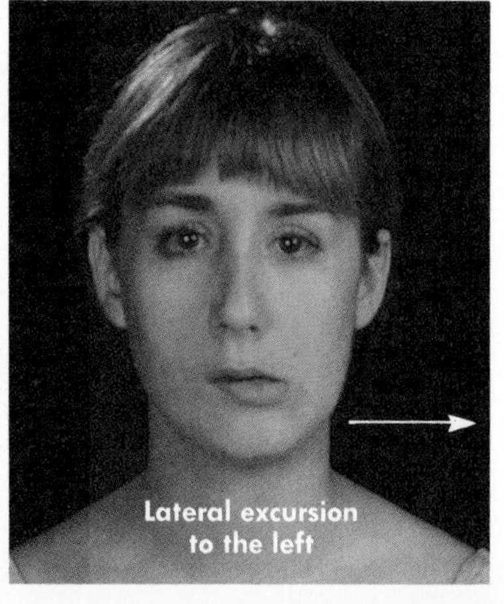

N

Continued.

O

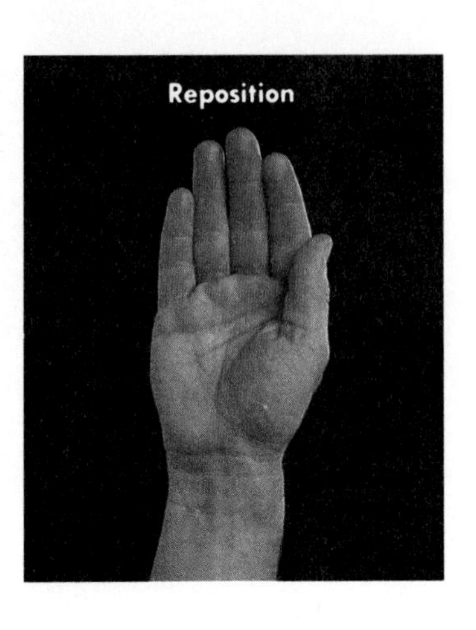

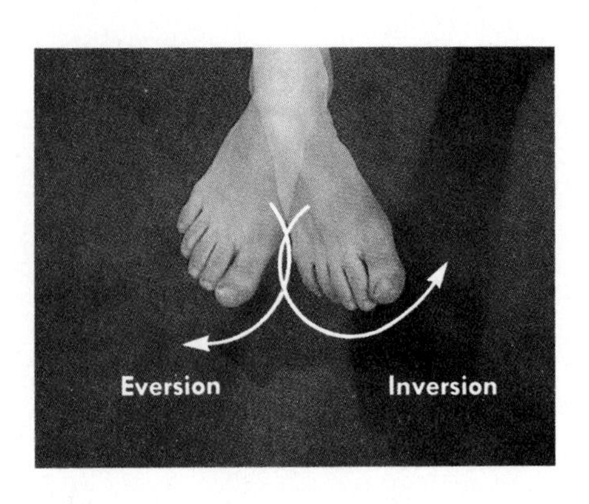

P

FIGURE 8-7, cont'd Movements. **O** Opposition and reposition of the thumb. **P** Inversion and eversion of the foot.

lower limb. As a result of this rotation, flexion and extension are reversed below midthigh in the adult.

Flexion and extension are not always used to describe ankle motion. Movement of the foot toward the plantar surface such as when standing on one's toes is often called **plantar flexion** (this is true flexion and is accomplished by flexor muscles; see Chapter 11). Movement of the foot toward the shin such as when walking on one's heels is called **dorsiflexion** (see Figure 8-7, *F;* this is actually extension and is accomplished by extensor muscles; see Chapter 11).

Abduction and adduction

Abduction (to take away) is movement away from the midline; **adduction** (to bring together) is movement toward the midline (Figure 8-7, *G*). Moving the lower limbs away from the midline of the body such as in the outward half of "jumping jacks" is abduction, and bringing the lower limbs back together is adduction. Abduction of the fingers involves spreading the fingers wide apart (away from the midline of the hand), and adduction is bringing them back together (Figure 8-7, *H*). Abduction (sometimes called radial deviation) of the wrist causes movement of the hand away from the midline of the body, whereas adduction (sometimes called ulnar deviation) of the wrist results in movement of the hand toward the midline of the body.

Hyperextension is an abnormal, forced extension of a joint beyond its normal range of motion. For example, if a person falls and attempts to break the fall by putting out his hand, the force of the fall directed into the hand and wrist may cause hyperextension of the wrist, which may result in sprained joints or broken bones.

Movement of a structure into the space posterior to the coronal plane sometimes has been referred to as hyperextension, and the term extension has been limited to movements ending at the (neutral) anatomical position. If used, this concept can confuse the issue of flexion and extension for two reasons. (1) the concept suggests that extension has two parts, one moving a structure to the anatomical position and the other moving it into the posterior space. By using two terms, there is an implication that the two parts of extension are in some way different. Based on this logic, it would seem appropriate to call flexion of the extended arm to the anatomical position flexion, and flexion of the arm into the space anterior to the coronal plane hyperflexion. (2) *Hyper-* is defined as excessive or above normal. Using the term hyperextension to describe the "normal" range of motion of a structure implies that the movement is abnormal. Conversely, if hyperextension is considered normal, then "excessive," "abnormal" extension resulting in injury, which is the true meaning of hyperextension, becomes confused with normal movement.

Circular Movements

Circular movements involve the rotation of a structure around an axis or movement of the structure in an arc.

Rotation

Rotation is the turning of a structure around its long axis (e.g., rotation of the head, the humerus, or the entire body; Figure 8-7, *I*). Medial rotation of the humerus with the forearm flexed brings the hand toward the body. Rotation of the humerus so that the hand moves away from the body is lateral rotation.

Pronation and supination

Pronation (pro-na'shun) and **supination** (su'pin-a'shun) refer to the unique rotation of the forearm (Figure 8-7, *J*). Prone means lying face down; supine means lying face up. Pronation is rotation of the palm so that it faces posteriorly (in relation to the anatomical position; facing down if the elbow is flexed); supination is rotation of the palm so that it faces anteriorly (facing up if the elbow is flexed). In pronation the radius and ulna cross; in supination they return to a parallel position. As described in Chapter 7, the head of the radius rotates against the radial notch of the ulna during supination and pronation.

Circumduction

Circumduction is a combination of flexion, extension, abduction, and adduction (Figure 8-7, *K*). It occurs at freely movable joints such as the shoulder. In circumduction the arm moves so that it describes a cone with the joint (e.g., the shoulder) at the apex.

Special Movements

Special movements are those movements unique to only one or two joints and do not fit neatly into one of the other categories.

Elevation and depression

Elevation moves a structure superiorly; **depression** moves it inferiorly (Figure 8-7, *L*). The mandible and scapulae are primary examples. Depression of the mandible opens the mouth, and elevation closes it. Shrugging the shoulders is an example of scapular elevation.

Protraction and retraction

Protraction consists of moving a structure in an anterior direction (Figure 8-7, *M*). **Retraction** moves the structure back to the anatomical position or even more posteriorly. As with elevation and depression, the mandible and scapulae are primary examples.

Excursion

Lateral excursion is essentially confined to the mandible and refers to moving the mandible to either the right or left of the midline (Figure 8-7, *N*) such as in grinding the teeth or chewing. **Medial excursion** returns the mandible to the neutral position.

Opposition and reposition

Opposition is a unique movement that is confined to the thumb and to the little finger (Figure 8-7, *O*). It occurs when these two digits are brought toward each other across the palm of the hand. The thumb can also oppose the other digits. **Reposition** is the movement returning the thumb and little finger to the neutral, anatomical position.

Inversion and eversion

Inversion and eversion are confined primarily to the ankle (Figure 8-7, *P*). **Inversion** consists of turning the ankle so that the plantar surface of the foot faces medially (toward the opposite foot). **Eversion** is turning the ankle so that the plantar surface faces laterally. Inversion of the foot is sometimes called supination, and eversion is called pronation.

Combination movements

Most movements that occur in the course of normal activities are combinations of the movements explained previously and are described by naming the individual movements involved in the combined movement. For example, if a person holds his hand straight out to his side at shoulder height and then brings it in front of him so that it is again at shoulder height, that movement could be considered a combination of adduction and flexion.

3

What combination of movements is required at the shoulder and elbow joints for a person to move his right upper limb from the anatomical position to touch the right side of his head with his fingertips?

? ? ? ? ? ? ? ? ? ? ?

DESCRIPTION OF SELECTED JOINTS

It is impossible in a limited space to describe all the joints of the body. Therefore only selected joints are described in this chapter, and they have been chosen because of their representative structure or important function.

Temporomandibular Joint

The mandible articulates with the temporal bone to form the **temporomandibular joint (TMJ).** The mandibular condyle fits into the mandibular fossa of

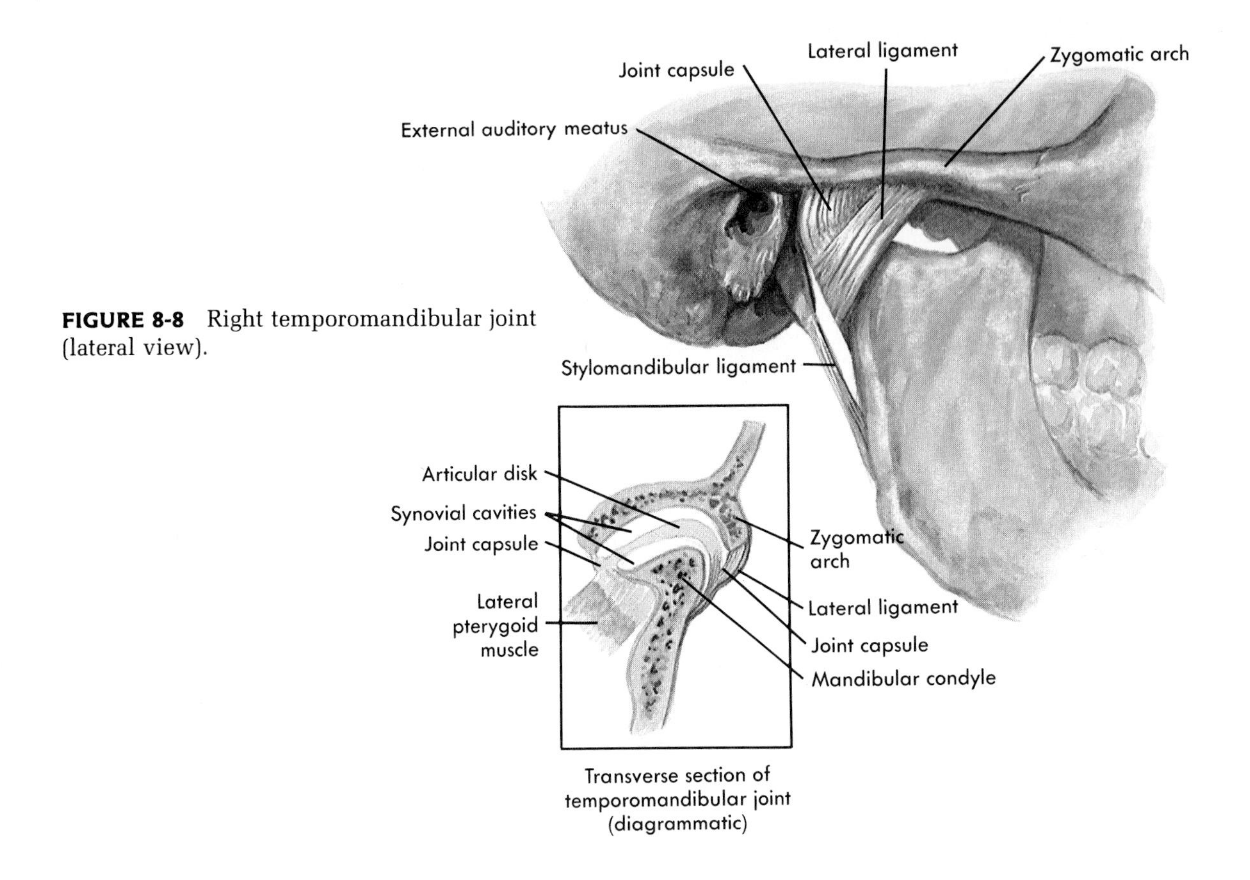

FIGURE 8-8 Right temporomandibular joint (lateral view).

the temporal bone. An articular disk is interposed between the mandible and the temporal bone, dividing the joint into superior and inferior cavities (Figure 8-8). The joint is surrounded by a fibrous capsule to which the articular disk is attached at its margin, and it is strengthened by lateral and accessory ligaments.

Temporomandibular joint syndrome involves pain around the joint and its associated muscles, clicking within the joint, and other features, such as severe headaches. This disorder is complex and is correlated with at least two separate factors: malocclusion (improper alignment of the teeth) and psychological tension, stress, or anxiety. The muscles apparently become painful because of increased tension and prolonged activity. The currently accepted procedure in treating temporomandibular joint syndrome is to reduce the psychological stress and anxiety of the patient and to assess the alignment of the teeth in the relaxed jaw. Mechanical devices can be applied to bring the teeth into alignment.

The temporomandibular joint is a combination plane and ellipsoid joint, with the ellipsoid portion predominating. Opening the mouth (depression of the mandible) involves an anterior gliding motion of the mandibular articular disk relative to the temporal bone (this is approximately the same motion that occurs in protraction of the mandible), followed by a hinge motion that occurs between the articular disk and the mandibular head. The mandibular condyle is also capable of slight mediolateral movement, allowing excursion of the mandible.

Shoulder Joint

The **shoulder,** or **humeral, joint** is a ball-and-socket joint in which stability is sacrificed somewhat for the sake of mobility (Figure 8-9). Flexion, extension, abduction, adduction, rotation, and circumduction can all occur at the shoulder joint. The rounded head of the humerus articulates with the shallow glenoid fossa of the scapula. The rim of the glenoid fossa is built up slightly by a fibrocartilage ring, the **glenoid labrum,** to which the joint capsule is attached. A **subscapular bursa** and a **subacromial bursa** open into the joint cavity.

The stability of the joint is maintained primarily by three sets of ligaments and four muscles. The ligaments are listed in Table 8-2. The four muscles, referred to collectively as the **rotator cuff,** pull the humeral head superiorly and medially toward the glenoid fossa. These muscles are discussed in more detail in Chapter 11. The head of the humerus is also supported against the glenoid fossa by the tendon from the biceps muscle of the anterior arm. This tendon is unusual in that it passes through the articular

Table 8-2

Ligaments of the shoulder joint (see Figure 8-9)

LIGAMENT	DESCRIPTION	LIGAMENT	DESCRIPTION
Glenohumeral (superior, middle, and inferior)	Three slightly thickened longitudinal sets of fibers on the anterior side of the capsule; extend from the humerus to the margin of the glenoid fossa	Coracohumeral	Crosses from the root of the coracoid process to the humeral neck
Transverse humeral	Lateral, transverse fibrous thickening of the joint capsule; crosses between the greater and lesser tubercles and holds down the tendon from the long head of the biceps muscle	Coracoacromial	Crosses above the joint between the coracoid process and the acromion process; an accessory ligament

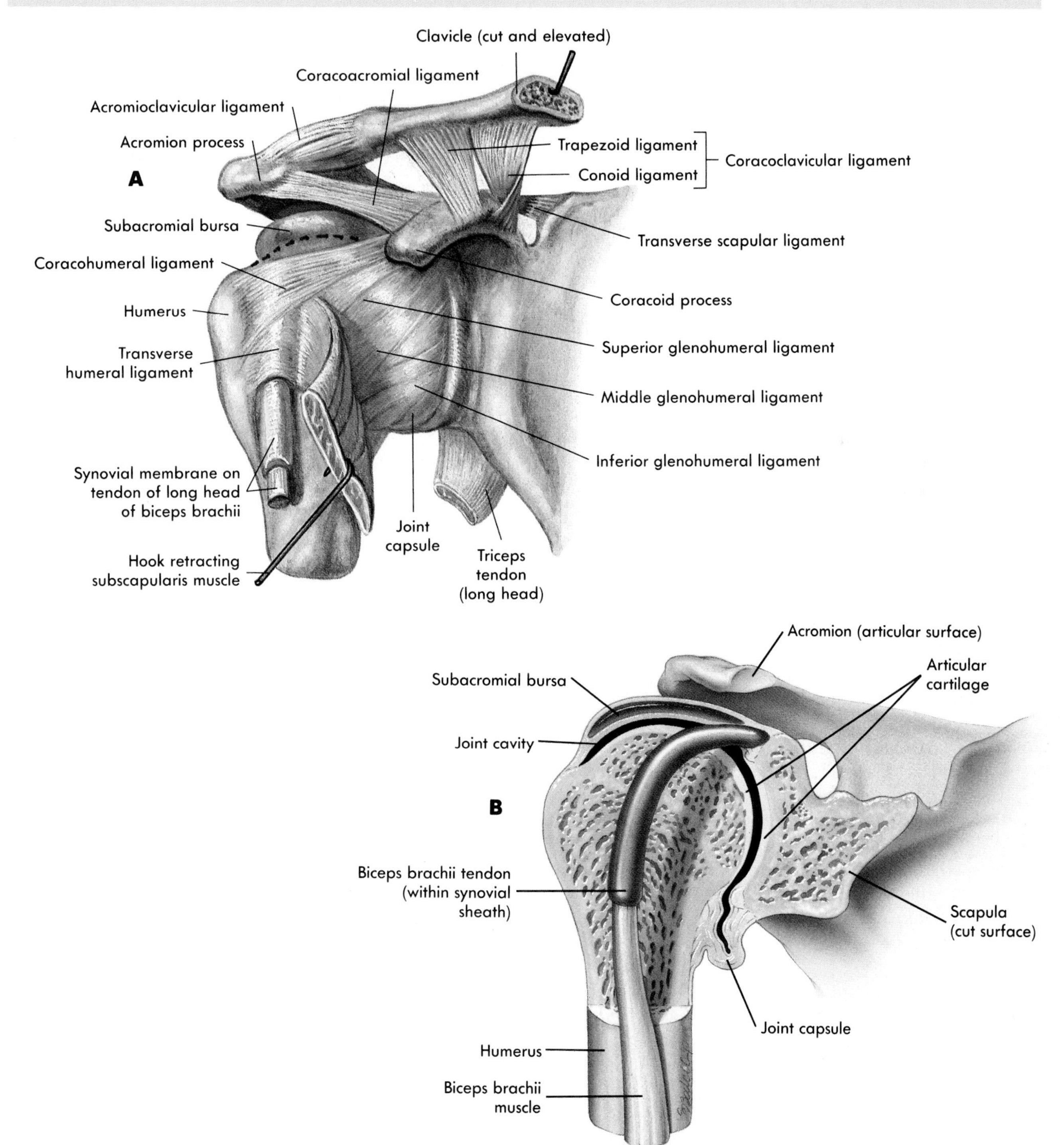

capsule of the shoulder joint before crossing the head of the humerus and attaching to the scapula (at the supraglenoid tubercle).

The most common traumatic shoulder disorders are dislocation and muscle or tendon tears. The major ligaments cross the superior portion of the shoulder joint, and no major ligaments or muscles are associated with the inferior side. As a result, dislocation of the humerus is most likely to occur inferiorly into the axilla. Because the axillae contain some very important nerves and arteries, severe and permanent damage may result from attempts to relocate a dislocated shoulder using inappropriate techniques (see Chapter 14). Chronic shoulder disorders include tendonitis, bursitis, and arthritis; they involve inflammation of tendons, bursae, or the joint, respectively. Bursitis of the subacromial bursa can become very painful when the large shoulder muscle (the deltoid muscle) compresses the bursa during shoulder movement.

4

Separation of the shoulder consists of stretching or tearing the ligaments of the acromioclavicular joint (acromioclavicular, or AC, separation). Using Figure 8-9, *B*, and your knowledge of the articulated skeleton for assistance, explain the nature of a shoulder separation and predict the problems that may follow a separation.

? ? ? ? ? ? ? ? ? ? ?

Hip Joint

The head of the femur is more nearly a complete ball than the articulating surface of any other bone of the body. The femoral head articulates with the relatively deep, concave acetabulum of the coxa to form the **coxageal,** or **hip, joint** (Figure 8-10). The acetabulum is deepened and strengthened by a lip of fibrocartilage, the **acetabular labrum,** which is incomplete inferiorly, and by a **transverse acetabular ligament,** which crosses the acetabular notch on the inferior edge of the acetabulum. The hip is capable of a wide range of movement, including flexion, extension, abduction, adduction, rotation, and circumduction.

An extremely strong articular capsule, reinforced by several ligaments, extends from the rim of the acetabulum to the neck of the femur (Table 8-3). The iliofemoral ligament is especially strong. When standing, most people tend to thrust the hips anteriorly. This position is relaxing because, in it, the iliofemoral ligament supports much of the body's weight. The ligament of the head of the femur (ligamentum teres) is located inside the hip joint and has very little function in strengthening the hip joint; however, it does carry a small nutrient artery to the head of the femur in approximately 80% of the population.

Dislocation of the hip may occur when the hip is flexed and the femur is driven posteriorly such as may occur when a person is sitting in an automobile and is involved in an accident. The head of the femur usually dislocates posterior to the acetabulum, tearing the labrum, the fibrous capsule, and the ligaments. Fracture of the femur and the coxa often accompany hip dislocation.

Knee Joint

The **knee** (genu) **joint** traditionally is classified as a hinge joint located between the femur and the tibia (Figure 8-11). Actually it is a complex ellipsoid joint that allows flexion, extension, and a small amount of rotation of the leg. The distal end of the femur has two large ellipsoid surfaces and a deep fossa between them. It articulates with the proximal end of the tibia, which is flattened and smooth laterally, with a crest called the intercondylar eminence in the center. The margins of the tibia are built up by thick fibrocartilage articular disks, **menisci** (mĕ-nis′si; crescent shaped; see Figure 8-11, *B* and *D*), that deepen the articular surface. The fibula does not articulate with the femur but articulates only with the lateral side of the tibia.

Two **cruciate** (kru′she-āt; crossed) **ligaments** extend between the intercondylar eminence of the tibia and the fossa of the femur (see Figure 8-11, *B*, *D*, and *E*). The anterior cruciate ligament prevents hyperextension (anterior movement) of the tibia, and the posterior cruciate ligament prevents posterior displacement of the tibia. The joint is also strengthened by collateral and popliteal ligaments and by the tendons of the thigh muscles, which extend around the knee (Table 8-4).

The knee is surrounded by a number of bursae (see Figure 8-11, *F*). The largest is the suprapatellar bursa, which is a superior extension of the joint capsule and allows movement of the anterior thigh muscles over the distal end of the femur. Other knee bursae include the popliteal bursa, gastrocnemius bursa, subcutaneous prepatellar bursa, subcutaneous infrapatellar bursa, and deep infrapatellar bursa.

Table 8-3

Ligaments of the hip joint (see Figure 8-10)

LIGAMENT	DESCRIPTION	LIGAMENT	DESCRIPTION
Transverse acetabular	Bridges gap in the inferior margin of the fibrocartilage acetabular labrum	Ischiofemoral	Bridges the ischial acetabular rim and the superior portion of the femoral neck; less well defined
Iliofemoral	Strong, thick band between the anterior inferior iliac spine and the intertrochanteric line of the femur	Ligamentum teres	Weak, flat band from the margin of the acetabular notch and the transverse ligament to the fovea capitis on the femoral head
Pubofemoral	Extends from the pubic portion of the acetabular rim to the inferior portion of the femoral neck		

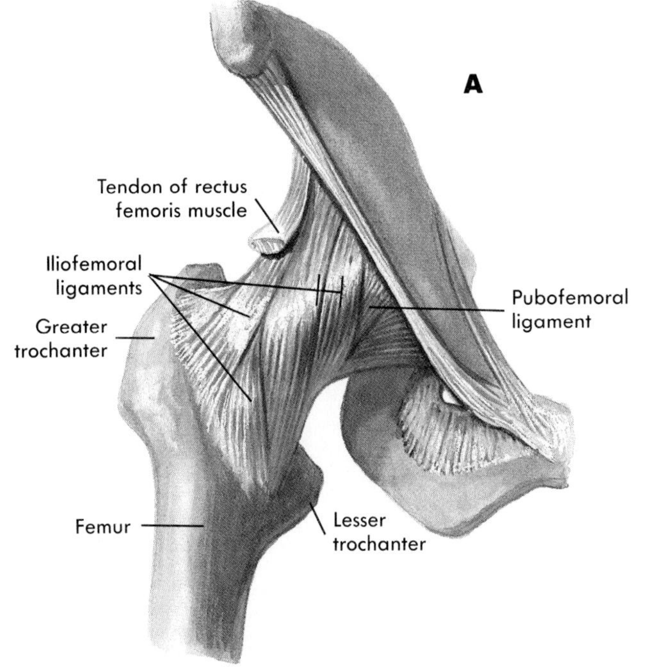

FIGURE 8-10 Right hip joint. **A** Anterior view. **B** Frontal section through the hip.

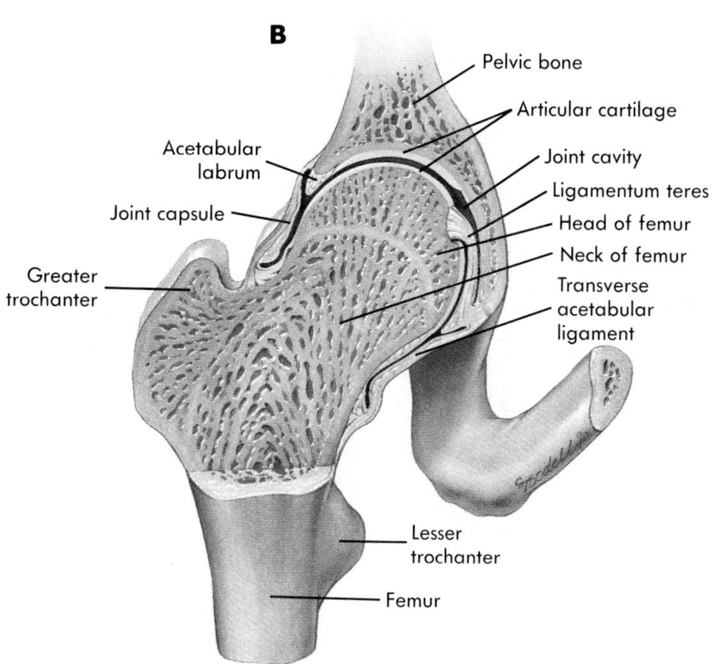

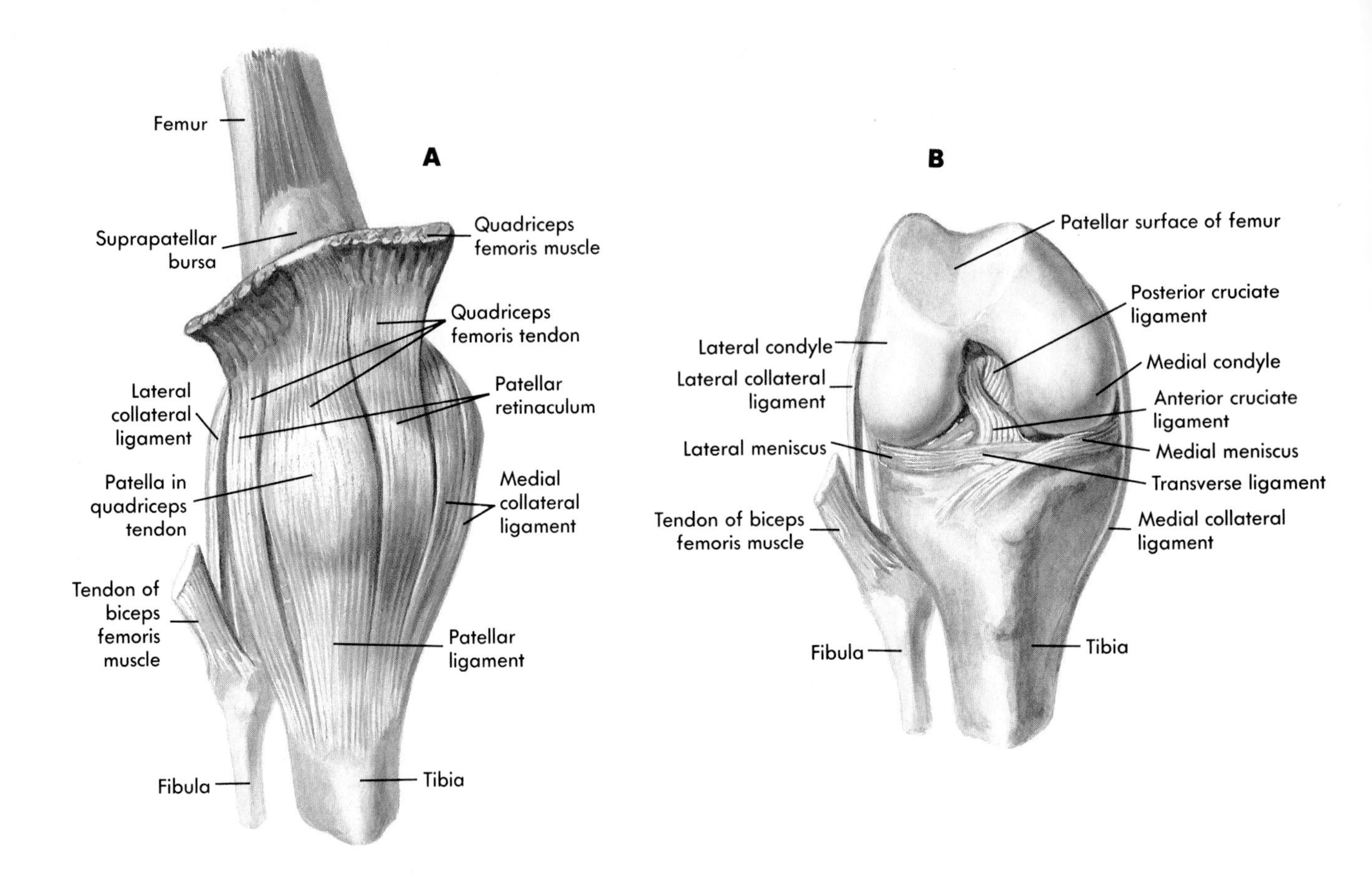

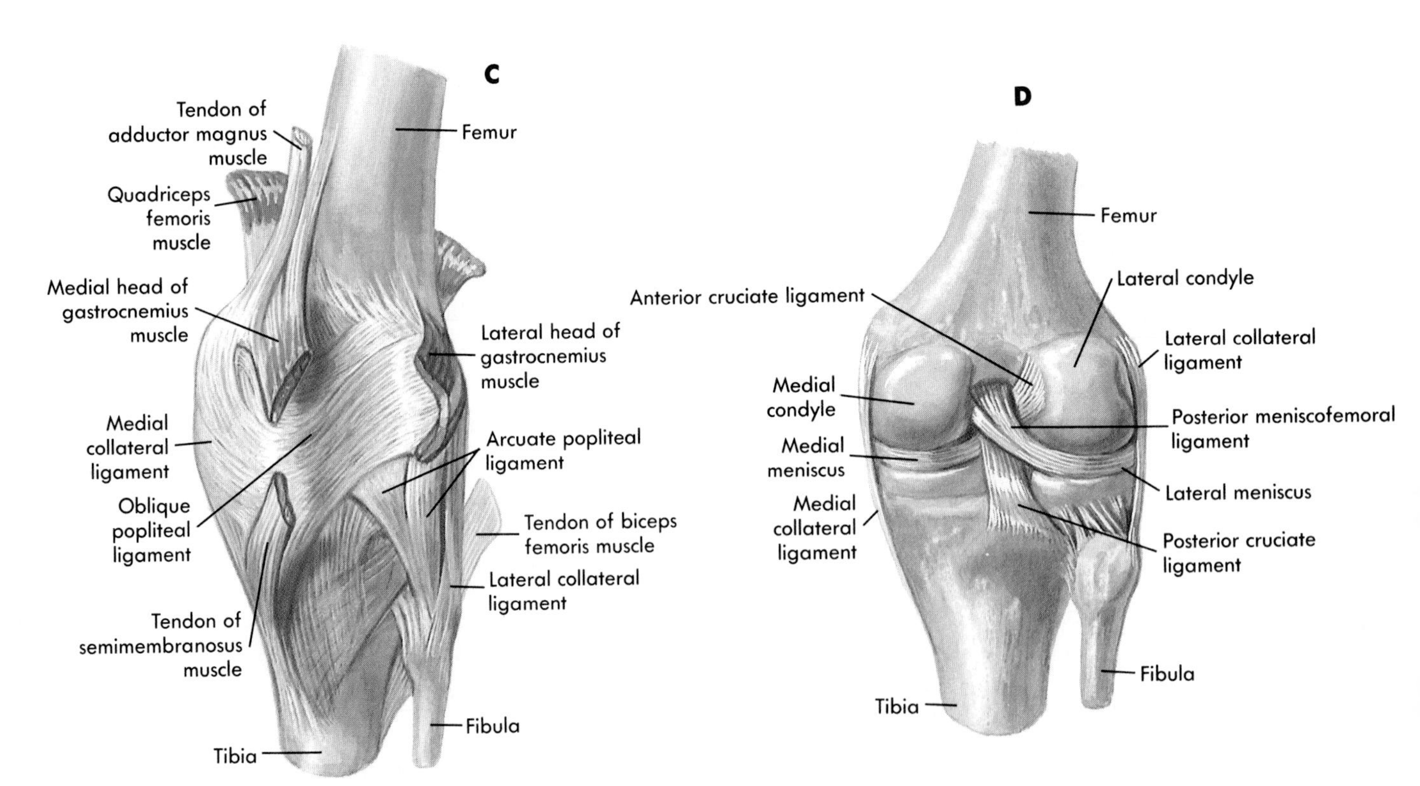

FIGURE 8-11 Right knee joint. **A** Anterior superficial view. **B** Anterior deep view (knee flexed). **C** Posterior superficial view. **D** Posterior deep view. **E** Photograph of anterior view. **F** Sagittal section.

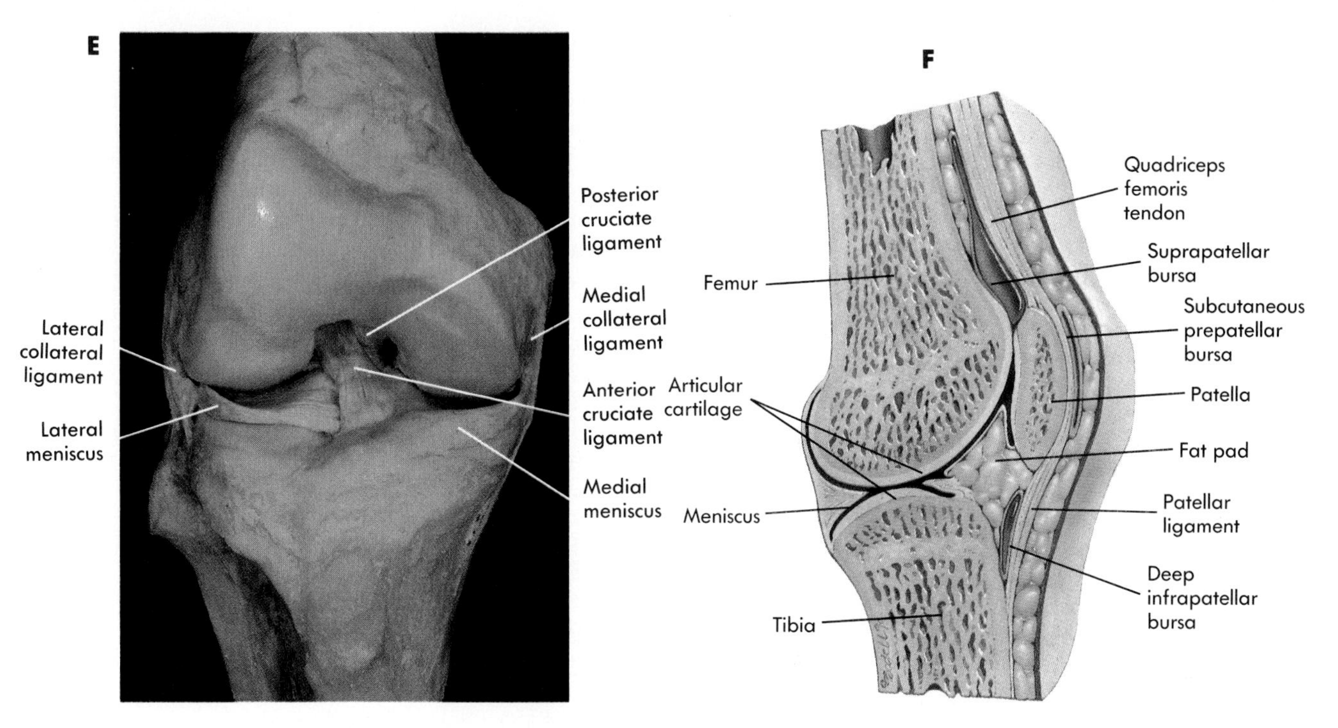

FIGURE 8-11, cont'd For legend see opposite page.

Table 8-4

Ligaments of the knee joint (see Figure 8-11)

LIGAMENT	DESCRIPTION
Patellar	Thick, heavy, fibrous band between the patella and the tibial tuberosity; actually part of the quadriceps tendon
Patellar retinaculum	Thin band from the margins of the patella to the sides of the tibial condyles
Oblique popliteal	Thickening of the posterior capsule; extension of the semimembranous tendon
Arcuate popliteal	Extends from the posterior fibular head to the posterior fibrous capsule
Medial collateral	Thickening of the lateral capsule from the medial epicondyle of the femur to the medial surface of the tibia; also called the tibial collateral ligament
Lateral collateral	Round ligament extending from the lateral femoral epicondyle to the head of the fibula; also called the fibular collateral ligament
Anterior cruciate	Extends obliquely, superiorly, and posteriorly from the anterior intercondylar eminence of the tibia to the medial side of the lateral femoral condyle
Posterior cruciate	Extends superiorly and anteriorly from the posterior intercondylar eminence to the lateral side of the medial condyle
Coronary (medial and lateral)	Attaches the menisci to the tibial condyles
Transverse	Connects the anterior portions of the medial and lateral menisci
Meniscofemoral (anterior and posterior)	Joins the posterior portion of the lateral menisci to the medial condyle of the femur, passing anterior and posterior to the posterior cruciate ligament

A common type of football injury results from a lateral blow (block or tackle) to the knee, which can cause the knee to bend inward, opening the medial side of the joint and tearing the medial collateral ligament (see Figure 8-11, *E*). Since this ligament is strongly attached to the medial meniscus, the medial meniscus often is torn as well. In severe injuries the anterior cruciate ligament, which is attached to the medial meniscus, also is damaged. The lateral collateral ligament strengthens the joint laterally and is not commonly torn because of its strength and because severe blows to the medial side of the knee are uncommon.

Bursitis in the subcutaneous prepatellar bursa, commonly called "housemaid's knees," may result from prolonged work performed while on the hands and knees. Another bursitis, "clergyman's knees," results from excessive kneeling and affects the subcutaneous infrapatellar bursa. This type of bursitis is common in carpet layers and roofers.

Other common knee problems include chondromalacia (softening of the cartilage), which results from abnormal movement of the patella within the patellar groove, and the "fat pad syndrome," which consists of an accumulation of fluid in the fat pad posterior to the patella. An acutely swollen knee appearing immediately after an injury is usually a sign of blood accumulation within the joint (hemarthrosis). A slower accumulation of fluid, "water on the knee," may be caused by bursitis.

Ankle Joint

The distal tibia and fibula form a highly modified hinge joint with the talus called the **talocrural** (tă'lo-kru'ral), or **ankle, joint** (Figure 8-12). The medial and lateral margins (medial and lateral malleoli of the tibia and fibula, respectively) are rather extensive, whereas the anterior and posterior margins are almost nonexistent. As a result, a hinge joint is created from a modified ball-and-socket arrangement. A fibrous capsule surrounds the joint, with the medial and lateral portions thickened to form ligaments. Other ligaments also help stabilize the joint (Table 8-5). Dorsiflexion, plantar flexion, and limited inversion and eversion can occur at this joint.

Table 8-5

Ligaments of the Ankle (see Figure 8-12)

LIGAMENT	DESCRIPTION
Medial	Thickening of the medial fibrous capsule that attaches the medial malleolus to the calcaneus, navicular, and talus; also called the deltoid ligament
Calcaneofibular	Extends from the lateral malleolus to the lateral surface of the calcaneus; separate from the capsule
Anterior talofibular	Extends from the lateral malleolus to the neck of the talus; fused with the joint capsule

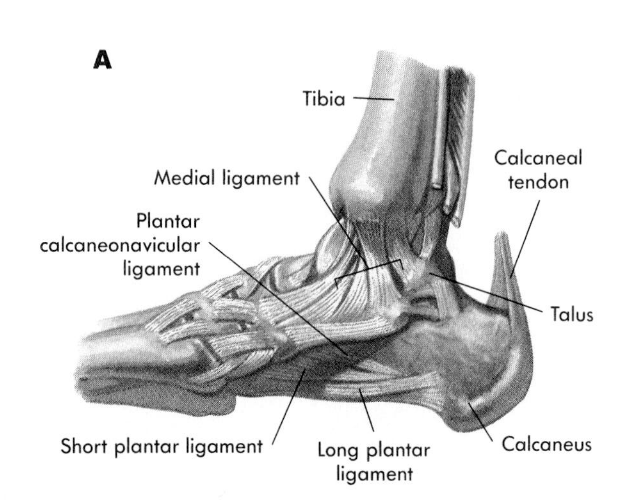

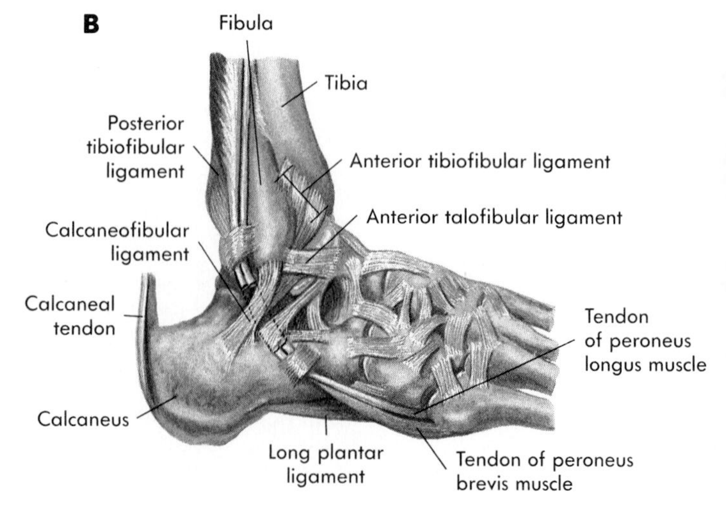

FIGURE 8-12 Ankle joint of the right foot.
A Medial view of ankle. **B** Lateral view of ankle.

The most common ankle injuries result from forceful inversion of the foot. A sprained ankle results when the ligaments or tendons of the ankle are torn partially or completely. The calcaneofibular ligament tears most often, followed in frequency by the anterior talofibular ligament. A fibular fracture can occur with severe inversion, for the talus can slide against the lateral malleolus and break it.

Arches of the Foot

The foot has three major arches that distribute the weight of the body between the heel and the ball of the foot during standing and walking (Figure 8-13). As the foot is placed on the ground, weight is transferred from the tibia and the fibula to the talus. From there, the weight is distributed first to the heel (calcaneus) and then through the arch system along the lateral side of the foot to the ball of the foot (head of the metatarsals). This effect can be observed when a person with wet feet walks across a dry surface; the print of the heel, the lateral border of the foot, and the ball of the foot can be seen, but the middle of the plantar surface and the medial border leave no impression. The medial side leaves no mark because the arches on this side of the foot are higher than those on the lateral side. The shape of the arches is maintained by the configuration of the bones, the ligaments connecting them, and the muscles acting on the foot (see Figure 8-12, A). The ligaments of the arch serve two major functions: to hold the bones in their proper relationship as segments of the arch and to provide ties across the arch somewhat like a bowstring. As weight is transferred through the arch system, some of the ligaments are stretched, resulting in mobility and allowing the foot to adjust to uneven surfaces. When weight is removed from the foot, the ligaments recoil and restore the arches to their unstressed shape.

The arches of the foot normally form early in fetal life. Failure to form results in congenital flat feet, or fallen arches, in which the arches (primarily the medial longitudinal arch) are depressed or collapsed. Flat feet may also occur when the muscles and ligaments supporting the arch tire and allow the arch to collapse.

Plantar fascitis, inflammation of the plantar fascia, which is a broad band of superficial connective tissue extending from the calcaneous to the ball of the foot, can be a problem for distance runners as a result of continuous stretching.

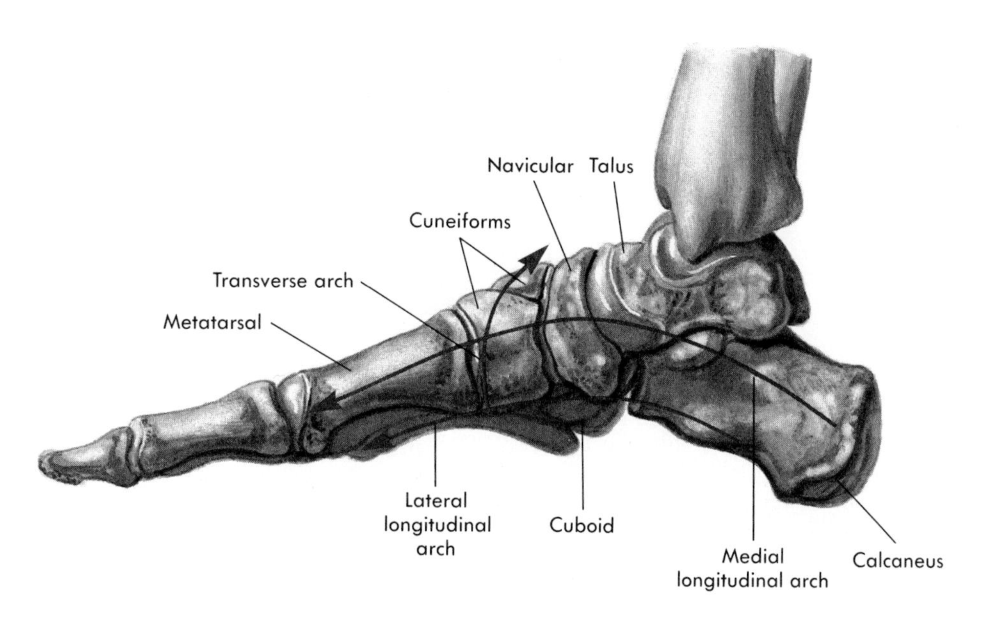

FIGURE 8-13 Arches *(arrows)* of the foot. The medial longitudinal arch is formed by the calcaneus, talus, navicular, the cuneiforms, and the three medial metatarsals. The lateral longitudinal arch is formed by the calcaneus, cuboid, and the two lateral metatarsals. The transverse arch is formed by the cuboid and cuneiforms.

SUMMARY

AN ARTICULATION, OR JOINT, IS A PLACE WHERE TWO BONES COME TOGETHER.

NAMING OF JOINTS *p. 229*

Joints are named according to the bones or parts of bones involved.

CLASSES OF JOINTS *p. 229*

Joints can be classified according to function or according to the type of connective tissue that binds them together and whether there is fluid between the bones.

Fibrous Joints

1. Fibrous joints are those in which bones are connected by fibrous tissue with no joint cavity. They are capable of little or no movement.
2. Sutures involve interdigitating bones held together by dense fibrous connective tissue. They occur between some skull bones.
3. Syndesmoses are joints consisting of fibrous ligaments.
4. Gomphoses are joints in which pegs fit into sockets and are held in place by periodontal ligaments (teeth in the jaws).
5. Some sutures and other joints can become ossified (synostosis).

Cartilaginous Joints

1. Synchondroses are immovable joints in which bones are joined by hyaline cartilage (epiphyses).
2. Symphyses are slightly movable joints made of fibrocartilage.

Synovial Joints

1. Synovial joints are capable of movement. They consist of the following:
   - Articular cartilage on the ends of bones, which provides a smooth surface for articulation. Articular disks can provide additional support.
   - A joint capsule of fibrous connective tissue (holds the bones together while permitting flexibility) and a synovial membrane (produces synovial fluid that lubricates the joint).
2. Bursae are extensions of synovial joints that protect skin, tendons, or bone from structures that could rub against them.
3. Synovial joints are classified according to the shape of the adjoining articular surfaces: plane (two flat surfaces), saddle (two saddle-shaped surfaces), hinge (concave and convex surfaces), pivot (cylindrical projection inside a ring), ball-and-socket (ball into a socket), and ellipsoid (ellipsoid concave and convex surfaces).

TYPES OF MOVEMENT *p. 236*

1. Angular movements include flexion and extension, abduction and adduction.
2. Circular movements include rotation, pronation and supination, and circumduction.
3. Special movements include elevation and depression, protraction and retraction, excursion, opposition and reposition, and inversion and eversion.
4. Combination movements involve two or more of the above listed movements.

DESCRIPTION OF SELECTED JOINTS *p. 241*

1. The temporomandibular joint is a complex hinge and gliding joint between the temporal and mandibular bones. It is capable of elevation/depression, protraction/retraction, and lateral/medial excursion movements.
2. The shoulder joint is a ball-and-socket joint between the head of the humerus and the glenoid fossa of the scapula that permits a wide range of movements. It is strengthened by ligaments and the rotator cuff. The tendon of the biceps brachii passes through the joint capsule. The shoulder joint is capable of flexion/extension, abduction/adduction, rotation, and circumduction.
3. The hip joint is a ball-and-socket joint between the head of the femur and the acetabulum of the coxa that is greatly strengthened by ligaments and that is capable of a wide range of movements.
4. The knee joint is a complex ellipsoid joint between the femur and the tibia that is supported by many ligaments. The joint allows flexion/extension and slight rotation of the leg.
5. The ankle joint is a special hinge joint of the tibia, fibula, and talus that allows dorsiflexion/plantar flexion and inversion/eversion.
6. The bony arches transfer weight from the heels to the toes and allow the foot to conform to many different positions.

CONTENT REVIEW

1. Define an articulation or joint.
2. On what criteria are joints named and classified? Name the three major classes of joints, based on their structure.
3. Define fibrous joints, describe the three different types, and give examples of each type.
4. Define cartilaginous joints, describe two different types, and give an example of each type.
5. Describe the structure of a synovial joint. How do the different parts of the joint function to permit joint movement?
6. Define bursa and tendon sheath. What is their function?
7. On what basis are synovial joints classified? Describe the different types of synovial joints, and give examples of each. What movements does each type of joint allow?
8. Define flexion and extension. How are they different for the upper and lower limbs? What is hyperextension?
9. Contrast abduction and adduction. Describe these movements for the upper limbs, lower limbs, fingers, and toes.
10. Distinguish among rotation, circumduction, pronation, and supination. Give an example of each.
11. Define the following jaw movements: protraction, retraction, lateral excursion, medial excursion, elevation, and depression.
12. Define opposition and reposition.
13. What terms are used for flexion and extension of the foot? For turning the sole of the foot medially or laterally?
14. For each of the following joints, name the bones of the joint, the specific part of the bones that form the joint, the type of joint(s) present, and the movement(s) that are possible at the joint: temporomandibular, shoulder, hip, knee, and ankle joint.

CONCEPT QUESTIONS

1. What would be the result if the costosternal synchondroses were to become a synostoses?
2. Using an articulated skeleton, examine the joints listed below. Describe the type of joint and the movement(s) possible.
   A The joint between the zygomatic bone and the maxilla.
   B The ligamentous connection between the coccyx and the sacrum.
   C The elbow joint.
3. For each muscle described below, describe the motion(s) produced when the muscle contracts. It may be helpful to use an articulated skeleton.
   A The biceps brachii muscle attaches to the coracoid process of the scapula (one head) and the radial tuberosity of the radius. Name two movements that the muscle accomplishes in the forearm.
   B The rectus femoris muscle attaches to the anterior superior iliac spine and the tibial tuberosity. How does contraction move the thigh? The leg?
   C The supraspinatus muscle is located in and is attached to the supraspinatus fossa of the scapula. Its tendon runs over the head of the humerus to the greater tubercle. When it contracts, how is the humerus moved?
   D The gastrocnemius muscle attaches to the medial and lateral condyles of the femur and to the calcaneus. What movement of the leg results when this muscle contracts? Of the foot?
4. Crash McBang hurt his knee in an auto accident by ramming the knee into the dashboard. The doctor tested the knee for ligament damage by having Crash sit on the edge of a table with his leg flexed at a 90-degree angle. The doctor attempted to pull the tibia in an anterior direction (the anterior drawer test) and then tried to push the tibia in a posterior direction (the posterior drawer test). There was no unusual movement of the tibia in the anterior drawer test, but there was during the posterior drawer test. Explain the purpose of each test, and tell Crash which ligament he has damaged.

ANSWERS TO PREDICT QUESTIONS

1 *p. 229.* Metacarpophalangeal joint.

2 *p. 229.* Premature sutural synostosis can interfere with normal brain growth and can result in brain damage if not corrected. It usually is corrected surgically by removing some of the bone around the suture and creating an artificial fontanel, which then undergoes normal synostosis.

3 *p. 241.* Abduction of the arm and flexion of the forearm.

4 *p. 244.* A shoulder separation involves stretching or tearing of the acromioclavicular ligament (and may involve tearing of the coracoclavicular ligament as well). Since the only bony attachment of the upper limb to the body is from the scapula through the clavicle to the sternum, separation of the acromioclavicular joint (the joint between the scapula and the clavicle) greatly reduces the stability of the shoulder. The scapula and humerus tend to be displaced inferiorly, and the proximal pivot point for the upper limb is destabilized.

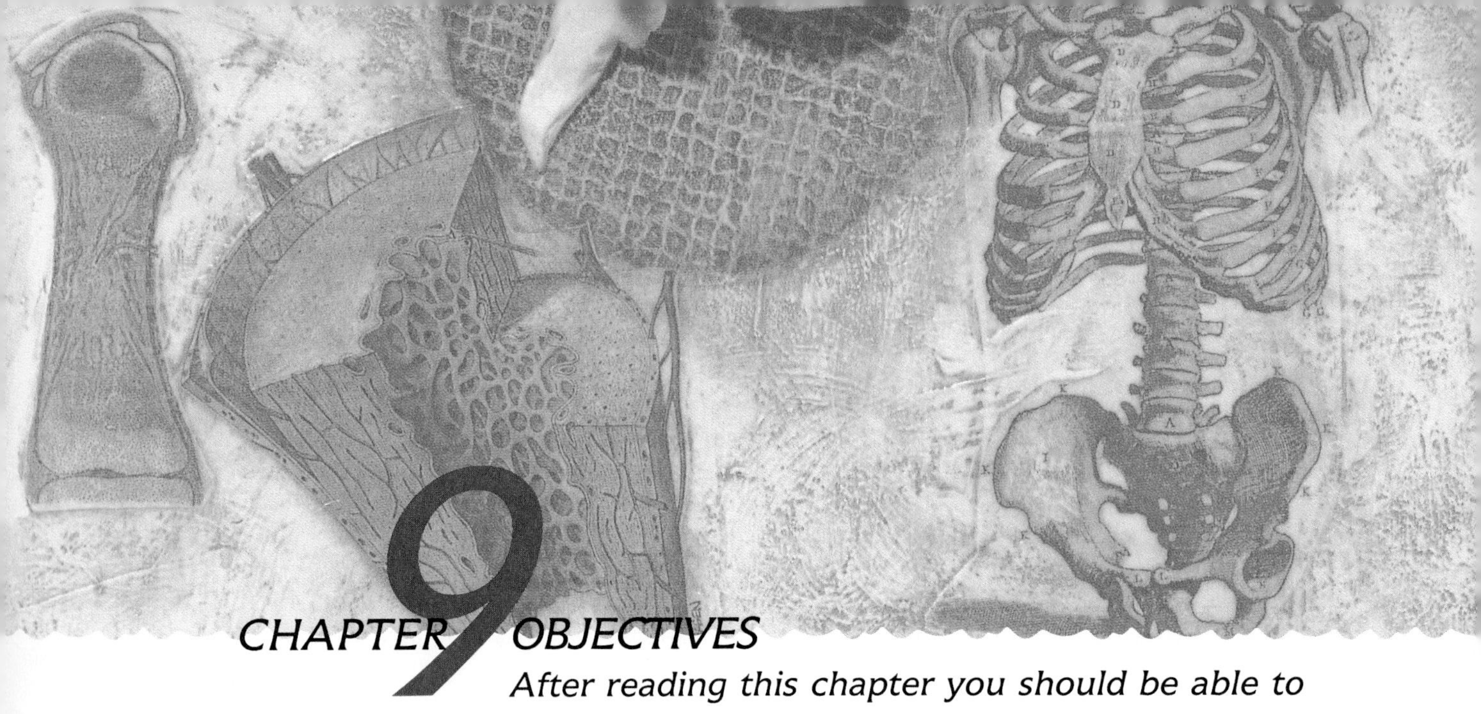

CHAPTER 9 OBJECTIVES

After reading this chapter you should be able to

1. Describe the concentration differences that exist between intracellular fluid and extracellular fluid.
2. Describe the factors that affect the concentration differences across the cell membrane for proteins and for potassium (K^+), sodium (Na^+), and chloride (Cl^-) ions.
3. Define resting membrane potential and explain how it is produced.
4. Predict and explain the changes that occur in the resting membrane potential as a result of changes in the K^+ ion concentration gradient across the cell membrane and do the same for changes in the permeability of the membrane to K^+ ions.
5. Explain the means by which ions cross the cell membrane.
6. List the characteristics of a local potential.
7. Explain how a local potential gives rise to an action potential.
8. Describe the phases of the action potential and the events that are responsible for each phase of the action potential.
9. Define the absolute and relative refractory periods and compare their effects on action potentials.
10. Describe how an action potential is propagated along a cell's membrane.
11. Define subthreshold, threshold, submaximal, maximal, and supramaximal stimuli.
12. Compare the effect of stimulus strength and stimulus duration on action potential frequency.

Membrane Potentials

KEY TERMS

absolute refractory period

accommodation

action potential

all-or-none

depolarization

hyperpolarization

ion channel

local potential

relative refractory period

resting membrane potential

sodium-potassium exchange pump

subthreshold stimulus

supramaximal stimulus

synapse

threshold potential

RELATED TOPICS

The following terms or concepts are important for a good understanding of this chapter. If you are not familiar with them, you should review them before proceeding.

Ions (Chapter 2)

Diffusion (Chapter 3)

Active transport (Chapter 3)

Cell membrane (Chapter 3)

ALL CELLS EXHIBIT VARIOUS ELECTRICAL PROPERTIES AND THE SPECIFIC ELECTRICAL PROPERTIES OF MANY CELL TYPES DRAMATICALLY INFLUENCE HOW THE BODY FUNCTIONS. For example, stimuli act on specialized cells in the eye, ear, mouth, and skin to produce electrical signals called **action potentials,** which are conducted from these cells to the spinal cord and brain. Within the brain the action potentials are interpreted and cause the sensations of vision, sound, taste, and touch. Action potentials originating within the brain and the spinal cord are conducted to muscles and certain glands to regulate their activities. Complex mental processes, including emotions and conscious thoughts, also depend on the electrical activity of cells within the brain.

This chapter introduces the electrical properties of cells under resting conditions and in response to stimuli. A basic knowledge of the electrical characteristics of cells is necessary to understand the normal functions and many pathologies of muscle and nervous tissues. The electrical properties of cells result from the ionic concentration differences across the cell membrane and from the cell membrane's permeability characteristics.

CONCENTRATION DIFFERENCES ACROSS THE CELL MEMBRANE

Table 9-1 lists the concentration differences for cations (positively charged ions) and anions (negatively charged ions) between the intracellular and extracellular fluids. The concentration of potassium (K^+) ions is much greater inside the cell than outside, and the concentration of sodium (Na^+) ions is much greater outside the cell than inside. Negatively charged proteins, other large anions, and a small concentration of chloride (Cl^-) ions are inside the cell. A smaller concentration of negatively charged proteins and a greater concentration of Cl^- ions are present outside the cell.

Differences in intracellular and extracellular concentrations are due to (1) the permeability characteristics of the cell membrane, (2) the presence of negatively charged proteins and other large anions within the cell, and (3) the sodium-potassium exchange pump.

As noted in Chapter 3, the cell membrane is selectively permeable, allowing some, but not all, substances to pass through the membrane. Negatively charged proteins are synthesized within the cell, and because of their large size and solubility characteristics, they cannot diffuse across the cell membrane. Negatively charged Cl^- ions are repelled by the negatively charged anions within the cell; as a result, Cl^- ions diffuse through the cell membrane and accumulate outside it. This results in a higher concentration of Cl^- ions outside the cell than inside. The negatively charged proteins and other large intracellular anions attract positively charged ions such as K^+ and Na^+ ions. However, K^+ ions are able to diffuse through the membrane more readily than Na^+ ions; as a consequence, only the concentration of K^+ ions increases within the cell (Figure 9-1).

The differences in K^+ and Na^+ ion concentrations across the cell membrane are also enhanced by the **sodium-potassium exchange pump.** Through active transport the sodium-potassium exchange pump moves K^+ and Na^+ ions through the cell membrane against their concentration gradients. K^+ ions are transported into the cell, increasing the concentration of K^+ ions inside the cell, and Na^+ ions are transported out of the cell, increasing the concentration of Na^+ ions outside the cell.

Although there are unequal concentrations of ions across the cell membrane, the extracellular and intracellular fluids are electrically neutral (i.e., in both the intracellular fluid and the extracellular fluid the cations equal the anions). However, a small electrical charge difference does exist between the immediate inside and the immediate outside of the cell membrane.

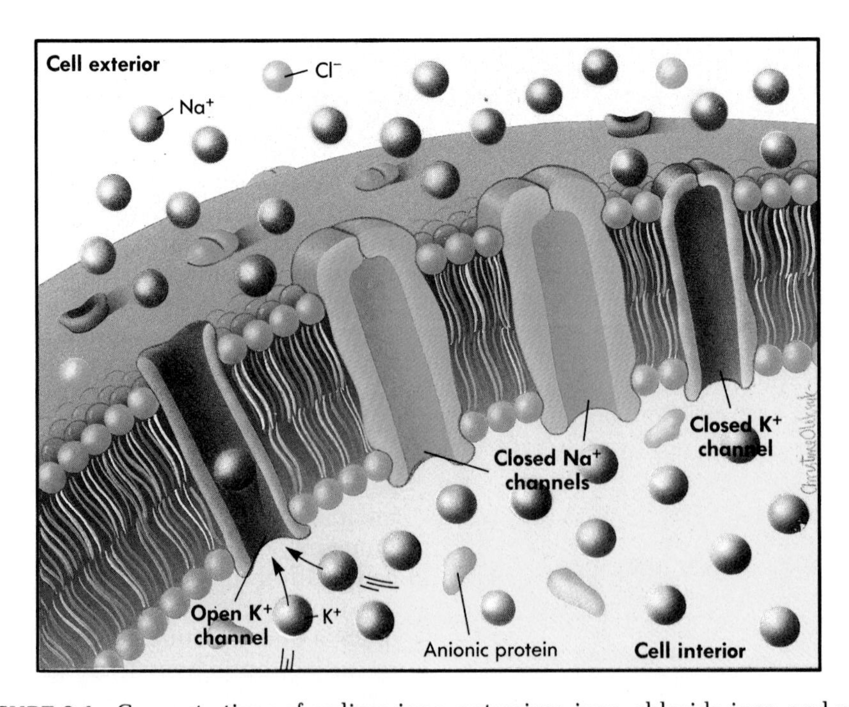

FIGURE 9-1 Concentrations of sodium ions, potassium ions, chloride ions, and negatively charged proteins across the cell membrane. The permeability of the membrane to potassium ions (note that some of the potassium ion channels are open) is greater than the permeability of the membrane to sodium ions (note the sodium ion channels are closed). The membrane is not permeable to the negatively charged proteins inside of the cell.

Table 9-1

Representative concentrations of the principal cations and anions in extracellular and intracellular fluids of vertebrates.

IONS	INTRACELLULAR FLUID (mEq/L)	EXTRACELLULAR FLUID (mEq/L)
Cations (Positive)		
Potassium (K^+)	148	5
Sodium (Na^+)	10	142
Calcium (Ca^{2+})	<1	5
Others	41	3
TOTAL	200	155
Anions (Negative)		
Proteins	56	16
Chloride (Cl^-)	4	103
Others	140	36
TOTAL	200	155

RESTING MEMBRANE POTENTIAL

An electrical charge, called a **potential difference,** can be measured between the inside and the outside of essentially all cells. By placing the tip of one microelectrode inside a cell and another microelectrode outside the cell and connecting the electrodes by wires to an appropriate measuring device (e.g., a voltmeter or an oscilloscope), the potential difference can be measured (Figure 9-2). The potential difference across the cell membranes of skeletal muscle fibers and nerve cells is 70 mV to 90mV (1 mV equals 0.001 volt). The inside of the cell membrane is negative when compared to the outside of the cell membrane. This potential difference is called the **resting membrane potential** and is reported as a negative number. The resting membrane potential exists when cells are in an unstimulated, or resting, state.

Unequal concentrations of charged molecules and ions that are separated by a selectively permeable cell membrane are essential to the development of the resting membrane potential. Because the cell membrane is somewhat permeable to K^+ ions, the K^+ ions tend to diffuse down their concentration gradient from inside to just outside of the cell membrane. Large negatively charged proteins and other molecules cannot diffuse through the cell membrane with the K^+ ions. Also, very few Na^+ ions can diffuse from the outside to the inside of the resting cell membrane. Since K^+ ions are positively charged, their

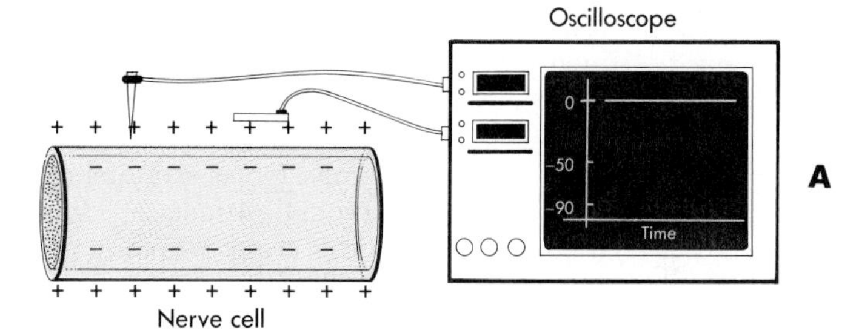

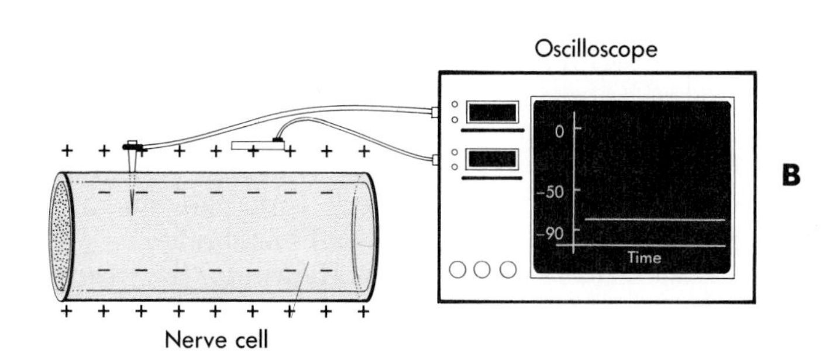

FIGURE 9-2 Recordings from a nerve cell. **A** Both recording (needle) and reference (block) electrodes are outside the cell, and no potential difference (0 mV) is recorded. **B** The recording electrode is inside the cell, the reference electrode is outside, and a potential difference (approximately −85 mV) is recorded, with the inside of the cell membrane negative with respect to the outside of the cell membrane.

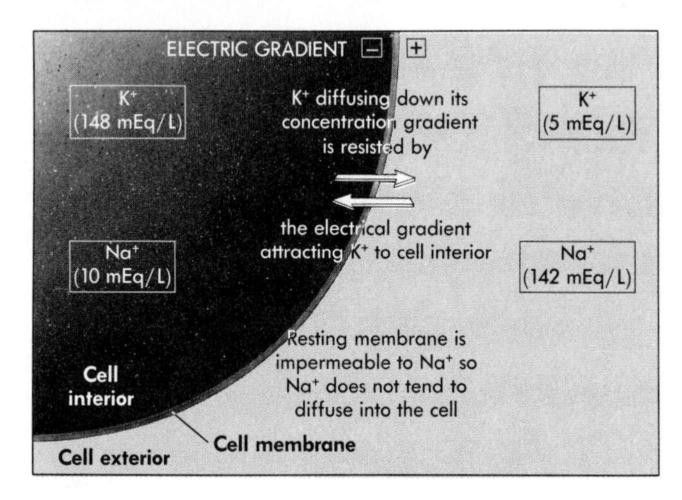

FIGURE 9-3 At equilibrium (resting conditions) the tendency for potassium ions to diffuse out of the cell is opposed by the potential difference (electrical gradient) across the cell membrane. As the resting membrane is not permeable to sodium ions, the sodium ions do not tend to diffuse into the cell.

movement causes the inside of the cell membrane to become negatively charged when compared to the outside of the cell membrane (Figure 9-3).

The K^+ ions do not diffuse out of the cell in large amounts. However, some K^+ ions do diffuse out of the cell, leaving negatively charged anions trapped inside of the cell until a negative charge develops on the inside of the cell membrane. Once established, the negative charge inside the cell attracts K^+ ions and prevents additional K^+ ions from diffusing out of the cell. Thus the K^+ ions only diffuse down their concentration gradient until the electrical charge across the cell membrane (the resting membrane potential) is just great enough to prevent any additional diffusion of K^+ ions out of the cell. The resting membrane potential is an equilibrium—the K^+ ion concentration gradient, which causes K^+ ions to diffuse out of the cell, is equal to the potential difference across the cell membrane, which opposes that movement. The resting membrane potential is proportional to the potential for K^+ ions to diffuse out of the cells but not to the actual rate of flow of K^+ ions. At equilibrium there is very little movement of charged particles across the cell membrane.

Factors of primary importance to the resting membrane potential are (1) the concentration difference of K^+ ions across the membrane and (2) the permeability of the membrane to K^+ ions. Thus it is possible to predict how the resting membrane potential will be affected by (1) alterations in the K^+ ion concentration on either side of the cell membrane and (2) changes in the permeability of the cell membrane to K^+ ions. In response to each of these conditions a new equilibrium is quickly established across the cell membrane.

For example, potassium succinate added to the extracellular fluid to increase the extracellular concentration of K^+ ions has a major effect on the resting membrane potential. The increased K^+ ion concentration outside the cell decreases the normal K^+ ion concentration gradient. As a consequence, there is a lesser tendency for K^+ ions to diffuse out of the cell, and a smaller negative charge inside of the cell is required to resist the diffusion of K^+ ions out of the cell. Once a new equilibrium is established, the charge difference across the cell membrane is decreased, and the resting membrane potential is less negative (Figure 9-4, *A*), a change that is called **depolarization** (the potential difference across the cell membrane becomes smaller or less polar).

A decrease in the extracellular concentration of K^+ ions (reducing the potassium chloride [KCl] in the extracellular fluid) also has an effect on the resting membrane potential. A decreased K^+ ion concentration outside the cell increases the K^+ ion concentration gradient from the inside to the outside of the cell and increases the tendency for K^+ ions to diffuse out of the cell. A greater negative charge inside of the cell membrane is required to resist the diffusion of the K^+ ions out of the cell. Thus the resting membrane potential becomes more negative (Figure 9-4, *B*), a change that is called **hyperpolarization** (the potential difference across the cell membrane becomes greater or more polar).

1

Predict the effect on the resting membrane potential if the intracellular concentration of potassium ions is increased.

? ? ? ? ? ? ? ? ? ? ?

A change in the permeability of the membrane to K^+ ions also affects the resting membrane potential. The resting membrane is not freely permeable to K^+ ions. With an increase in the permeability of the membrane to K^+ ions, a larger number of K^+ ions diffuse out of the cell, making the inside of the cell more negative when compared to the outside of the cell (hyperpolarization). A new equilibrium in which the greater negative charge inside the cell membrane resists the diffusion of K^+ ions out of the cell is established rapidly.

2

Explain the effect on the resting membrane potential of a reduced permeability of the cell membrane to K^+ ions.

? ? ? ? ? ? ? ? ? ? ?

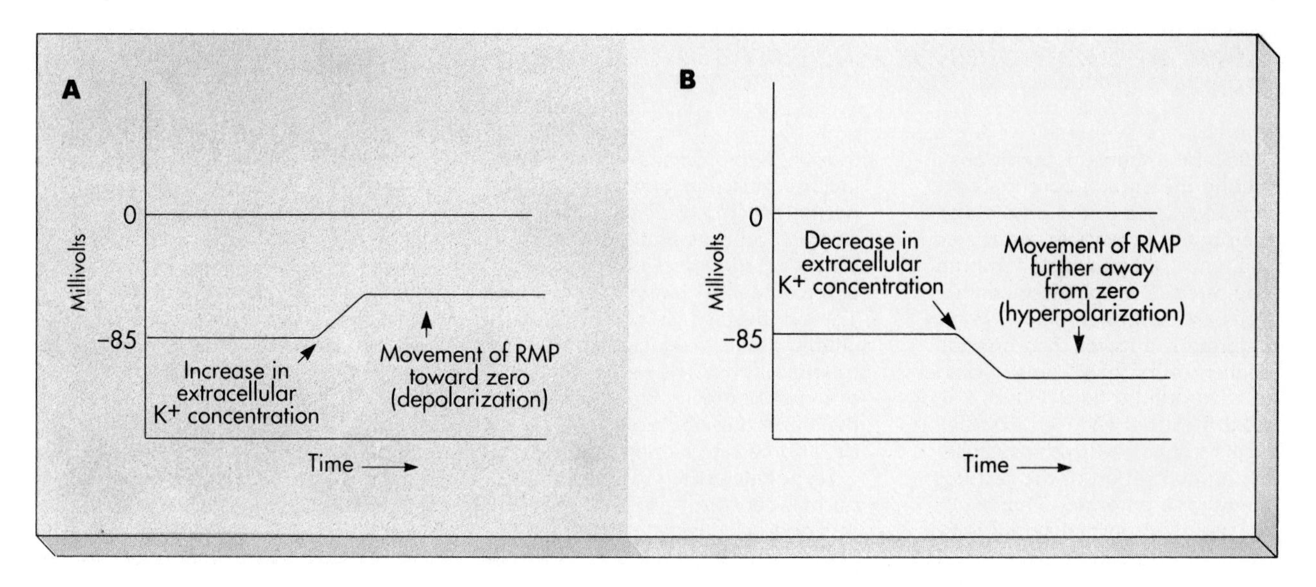

FIGURE 9-4 Changes in the resting membrane potential *(RMP)* caused by changes in extracellular potassium ion concentration. **A** Elevated extracellular potassium ion concentration causes depolarization. **B** Decreased extracellular potassium ion concentration causes hyperpolarization.

Table 9-2

Characteristics responsible for the resting membrane potential

1. There is an equal number of charged molecules and ions inside and outside of the cell.
2. There is a higher concentration of K^+ ions inside than outside of the cell, and there is a higher concentration of Na^+ ions outside than inside of the cell.
3. The cell membrane is 50 to 100 times more permeable to K^+ ions than to other positively charged ions such as Na^+ ions.
4. The cell membrane is impermeable to large intracellular anions such as proteins.
5. K^+ ions tend to diffuse across the cell membrane from the inside to the outside of the cell.
6. Since anions cannot follow the positively charged K^+ ions, a small negative charge develops just inside of the cell membrane.
7. The negative charge inside of the cell membrane attracts positively charged K^+ ions. When the negative charge inside of the cell membrane is great enough to prevent additional K^+ ions from diffusing out of the cell through the cell membrane, an equilibrium is established.
8. The negative charge inside of the cell at equilibrium is called the resting membrane potential.
9. The resting membrane potential is proportional to the potential for K^+ ions to diffuse out of the cell but not to the actual rate of flow of K^+ ions.
10. At equilibrium there is very little movement of K^+ or other ions across the cell membrane.

Changes in the concentration of Na^+ ions on either side of the cell membrane do not influence the resting membrane potential markedly because the membrane is 50 to 100 times less permeable to Na^+ ions than to K^+ ions. To have a significant effect on the resting membrane potential, large changes in the concentration gradient for sodium are required. However, if the permeability of the membrane to Na^+ ions changes, the resting membrane potential is affected dramatically. The concentration gradient for Na^+ ions is from the outside to the inside of the cell. If the permeability of the cell membrane to Na^+ ions increases, Na^+ ions diffuse down their concentration gradient into the cell, and the inside of the cell membrane becomes more positive, resulting in depolarization.

The characteristics responsible for a resting membrane potential are summarized in Table 9-2.

Clinical Examples of Abnormal Membrane Potentials

Several important conditions affecting membrane potentials provide examples of the physiology of membrane potentials and the consequence of abnormal membrane potentials. Two of these conditions are described below. **Hypokalemia** is a lower-than-normal concentration of K^+ ions in blood or extracellular fluid. Figure 9-4, *B*, illustrates that reduced extracellular K^+ ion concentrations cause hyperpolarization of the resting membrane potential. Thus a greater-than-normal stimulus is required to depolarize the membrane to its threshold level and to initiate action potentials in neurons, skeletal muscle, and cardiac muscle. Symptoms of hypokalemia include muscular weakness, an abnormal electrocardiogram, and sluggish reflexes and are consistent with the effect of a reduced extracellular K^+ ion concentration. The symptoms result from the excitable tissues' reduced sensitivity to stimulation. The several causes of hypokalemia include potassium depletion during starvation, alkalosis, and certain kidney diseases.

Hypocalcemia is a lower-than-normal concentration of Ca^{2+} ions in blood or extracellular fluid. Symptoms of hypocalcemia include nervousness, muscular spasms, and uncontrolled contraction of skeletal muscles **(tetany).** The symptoms are due to the increased membrane permeability to Na^+ ions that results because low blood levels of Ca^{2+} ions cause Na^+ channels in the membrane to open. Na^+ ions diffuse into the cell, cause depolarization of the cell membrane to threshold, and initiate action potentials. The tendency for action potentials to occur spontaneously in nerves and muscles accounts for the listed symptoms. A lack of dietary calcium, a lack of vitamin D, or a reduced secretion rate of a parathyroid gland hormone are examples of conditions that cause hypocalcemia.

MOVEMENT OF IONS THROUGH THE CELL MEMBRANE

There are two separate systems for the movement of ions through the cell membrane: (1) a series of ion channels that allow ions to diffuse through the membrane and (2) an active transport process called the sodium-potassium exchange pump that moves Na^+ ions out of the cell and K^+ ions into the cell.

Ion Channels

Ion channels, or pores, for K^+ and Na^+ ions exist in the cell membrane. They are formed from large protein molecules, and they are specific in that each pore type allows one type of ion, but not others, to pass through it.

The permeability characteristics of the channels are influenced by their electrical charge, their three-dimensional structure, their size, and the proteins, called gating proteins, that open and close the channels.

The molecular weight of potassium is greater than the molecular weight of sodium. However, the diameter of ions dissolved in water is determined by the size of the ions and the number of water molecules that surround them. Since Na^+ ions attract more water molecules than K^+ ions, the diameter of Na^+ ions is larger than the diameter of K^+ ions when they are dissolved in water.

The K^+ ion channels are smaller than the Na^+ ion channels, and their structure allows K^+ ions, but not Na^+ ions, to pass through them. Although K^+ ions are small enough to pass through Na^+ ion channels, the concentration gradient for Na^+ ions from the outside to the inside of the cell, the structure of the sodium channels, and the attraction of the negative charge inside of the cell favor the passage of Na^+ ions (not K^+ ions) through the Na^+ ion channels.

Gating proteins open and close the Na^+ ion channels, and their activity is influenced by factors such as changes in the potential difference across the cell membrane and the extracellular concentration of calcium (Ca^{2+}) ions. Decreases in the potential difference across the cell membrane (depolarization) can cause the sodium channels to open for a short time. For example, when the membrane depolarizes, the change in the membrane potential causes Na^+ ion channels to open, increasing the permeability of the cell membrane to Na^+ ions. The Na^+ ion channels remain open for a short period of time and then close (Figure 9-5, *A* and *B*).

Changes in the potential difference across the cell membrane cause a greater number of K^+ channels to open. Although the K^+ channels may begin to open at the same time as the sodium channels, they open more slowly in response to depolarization of the cell membrane, remain open for a short period of time, and then close.

The gating proteins associated with the Na^+ ion channels close the channels in the presence of high extracellular Ca^{2+} ion concentrations. On the other

A

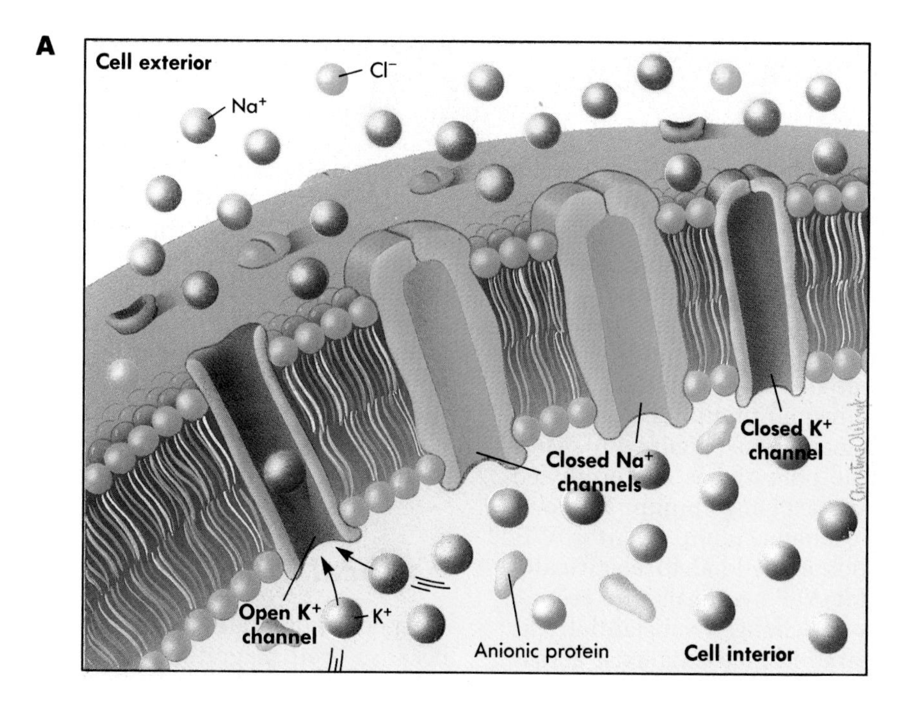

B

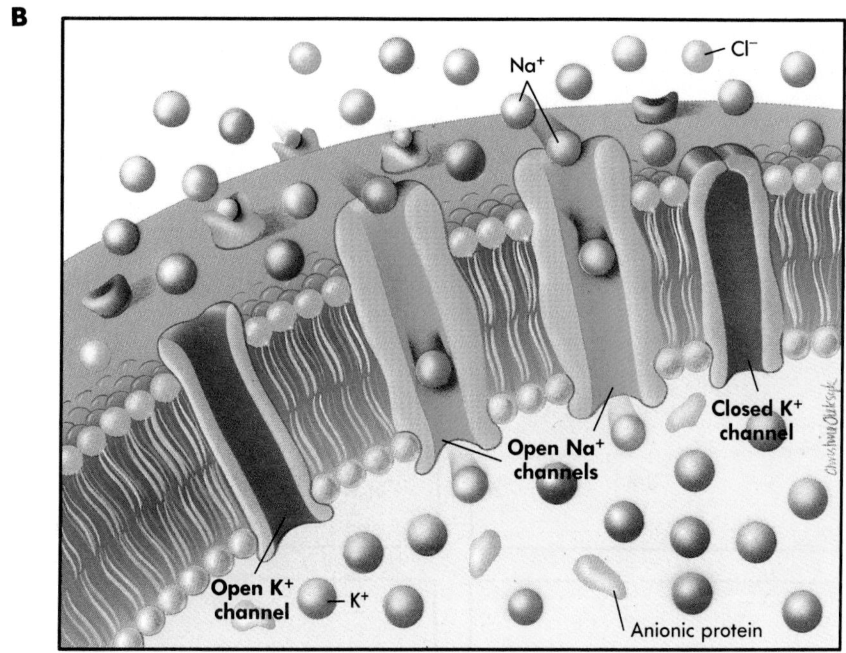

FIGURE 9-5 Effect of a stimulus that causes a voltage change across the cell membrane on the permeability of the cell membrane. **A** Sodium channels remain closed in a resting or unstimulated cell membrane. **B** Depolarization of the cell membrane causes sodium channels to open. Sodium ions then diffuse down their concentration gradient into the cell, causing depolarization of the cell membrane.

hand, very low extracellular Ca^{2+} ion concentrations cause the gating proteins to open the Na^+ ion channels, allowing Na^+ ions to diffuse into the cell. At the Ca^{2+} ion concentrations normally found in the extracellular fluid, only a small percentage of the Na^+ ion channels are open at any single moment.

3

Predict the effect on the resting membrane potential if the extracellular concentration of calcium ions decreases.

? ? ? ? ? ? ? ? ? ? ?

Sodium-Potassium Exchange Pump

The sodium-potassium exchange pump in the cell membrane actively transports Na^+ ions out of the cell and K^+ ions into the cell. The pump transports up to three Na^+ ions out of the cell for every two K^+ ions transported into the cell. Because more positively charged ions are pumped out of the cell than are pumped into the cell, a small portion of the resting membrane potential (usually less than 15 mV) is directly due to this unequal pumping of Na^+ and K^+ ions across the cell membrane (Figure 9-6). Most of the resting membrane potential is not produced by the sodium-potassium exchange pump. To work, the sodium-potassium exchange pump requires ATP. When metabolic poisons are added to electrically excitable cells to inhibit ATP synthesis, the resting membrane potential is not changed substantially as long as normal concentration gradients exist across the cell membrane. The major function of the pump is to maintain the normal concentration gradients for K^+ and Na^+ ions across the cell membrane.

ELECTRICALLY EXCITABLE CELLS

Nerve and muscle cells are electrically excitable; thus the application of a stimulus causes a series of characteristic changes in the resting membrane potential. A stimulus applied at one point on a cell normally causes a depolarization, called a **local potential,** that is confined to a small region of the cell membrane. If a local potential is large enough, it can produce an action potential, which is a larger depolarization and spreads or is propagated, without changing its magnitude, over the entire cell membrane.

Local Potential

Local potentials (Figure 9-7) are called **graded** because their magnitude is directly proportional to the stimulus strength; thus a weak stimulus causes a small depolarization, and a stronger stimulus causes a larger depolarization (Figure 9-7, *A*). In some cells the magnitude of a local potential may

Cell exterior
Na^+
ATP
Cell interior

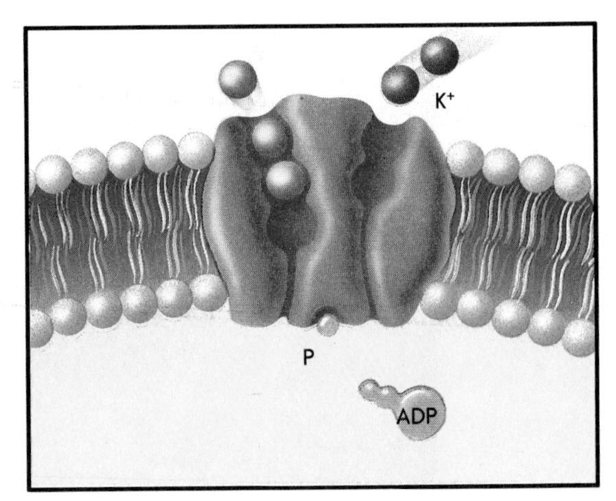

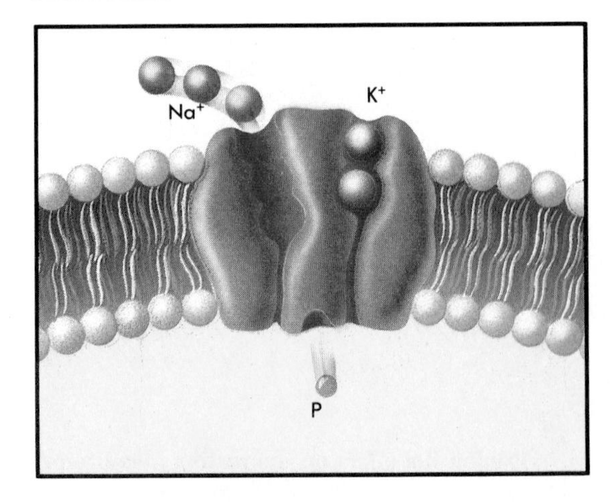

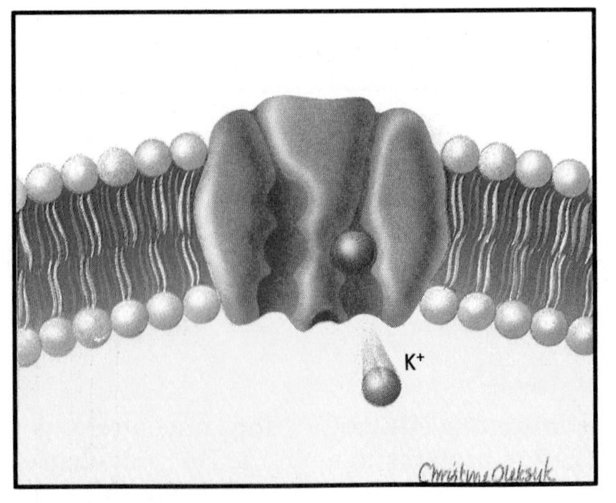

FIGURE 9-6 The sodium-potassium exchange pump actively transports sodium ions out of the cell across the cell membrane and potassium ions into the cell across the cell membrane. ATP is used as the energy source, and the pump can transport up to three sodium ions for every two potassium ions transported.

remain constant as long as the stimulus is applied (Figure 9-7, *B*), but in other cells the local potential becomes smaller, or accommodates, even though the stimulus still is being applied to the cell (Figure 9-7, *C*).

Local potentials are propagated in a decremental fashion, that is, they rapidly decrease in magnitude as they spread over the surface of the cell membrane. Normally the local potential cannot be detected more than a few millimeters from the site of stimulation. As a consequence, the local potential cannot transfer information over long distances from one area of the body to another. However, if the local potential is a depolarization of sufficient magnitude, it can trigger an action potential in electrically excitable cells (see Action Potential).

Some local potentials are hyperpolarizations instead of depolarizations. An increase in the permeability of the cell membrane to K^+ ions or Cl^- ions results in local potentials that are hyperpolarizations. In contrast to local depolarizations, the local hyperpolarizations do not cause action potentials, regardless of their magnitude.

The characteristics of local potentials are summarized in Table 9-3.

4

Given two cells that are identical in all ways except that the extracellular concentration for sodium ions is greater for cell A than for cell B, how would the magnitude of the local potential in cell A differ from that in cell B if stimuli of identical strength were applied to each cell?

? ? ? ? ? ? ? ? ? ? ?

Table 9-3

Characteristics of the local potential

1. A stimulus causes increased permeability of the membrane to Na^+ ions or increased permeability of the membrane to K^+ and Cl^- ions.
2. Depolarization is a result of increased permeability of the membrane to Na^+ ions; hyperpolarization is a result of increased permeability of the membrane to K^+ or Cl^- ions.
3. Depolarization and hyperpolarization are graded (proportional to the strength of the depolarizing or hyperpolarizing stimulus).
4. In some cell types the local potential may exist as long as the stimulus is applied; in other types it may become smaller (accommodate).
5. Local potentials are propagated in a decremental fashion—their magnitude decreases as they spread over the cell membrane.
6. The depolarizing local potential can cause an action potential.

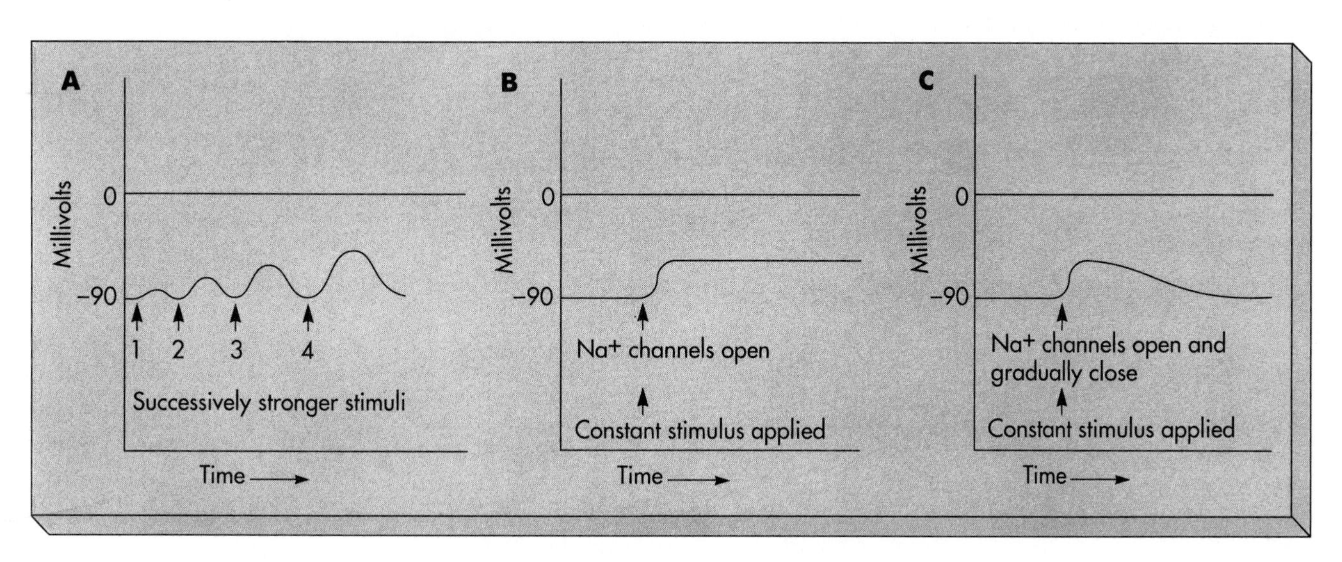

FIGURE 9-7 Local potentials. **A** A weak stimulus applied briefly causes a small depolarization, which quickly repolarizes *(1)*. Progressively stronger stimuli result in larger depolarizations *(2-4)*. **B** No accommodation occurs in some neurons when a constant stimulus is applied. **C** Accommodation occurs in some neurons when a constant stimulus is applied.

Action Potential

When the local potential causes depolarization of the cell membrane to a level called the **threshold potential,** a series of permeability changes occur that result in an **all-or-none action potential** (Figure 9-8, *A*). To understand what **all-or-none** implies, consider that when a depolarization reaches threshold, all the permeability changes responsible for an action potential proceed without stopping. If depolarization does not reach threshold, few of the permeability changes occur, and an action potential is not produced. Once threshold has been reached and the permeability changes begin, they proceed without stopping and are constant in magnitude (the "all" part). If the local potential does not reach threshold, the membrane potential returns to its resting level after a brief period of time without producing an action potential (the "none" part).

For a short period of time during the action potential the inside of the cell becomes positive. The action potential has a **depolarization phase** in which the membrane potential moves away from the resting membrane potential and becomes more positive and a **repolarization phase** in which the membrane potential returns toward the resting membrane state and becomes more negative. After the repolarization phase the cell membrane may be slightly hyperpolarized for a short period called the **afterpotential** (see Figure 9-8, *A*).

The change in charge across the cell membrane during a local potential causes increasing numbers of sodium channels to open for a brief period of time. In addition, potassium channels begin to open at the same time, but the potassium channels open more slowly. The increased movement of K^+ ions out of the cell through the open potassium channels partially counteracts the increased movement of Na^+ ions into the cell through the open sodium channels. As soon as threshold is reached, however, more Na^+ channels begin to open. Na^+ ions move rapidly into the cell, and the resulting change in the membrane potential causes additional sodium channels to open. As a consequence, more Na^+ ions rush into the cell, causing a greater depolarization in membrane potential, which, in turn, causes more Na^+ channels to open. This positive-feedback cycle continues until most of the sodium channels in the cell membrane are opened. The movement of Na^+ ions into the cell causes the depolarization phase of the action potential. Enough sodium ions move across the cell membrane to cause the membrane potential to become positive (the inside of the cell membrane becomes positive when compared to the outside of the cell membrane). Although sodium ions move through the Na^+ channel during the depolarization phase of the action potential, the total number of Na^+ ions crossing the cell membrane is small. Sodium ions move into the cell making the inside of the cell positive when compared to the outside of the cell. When the

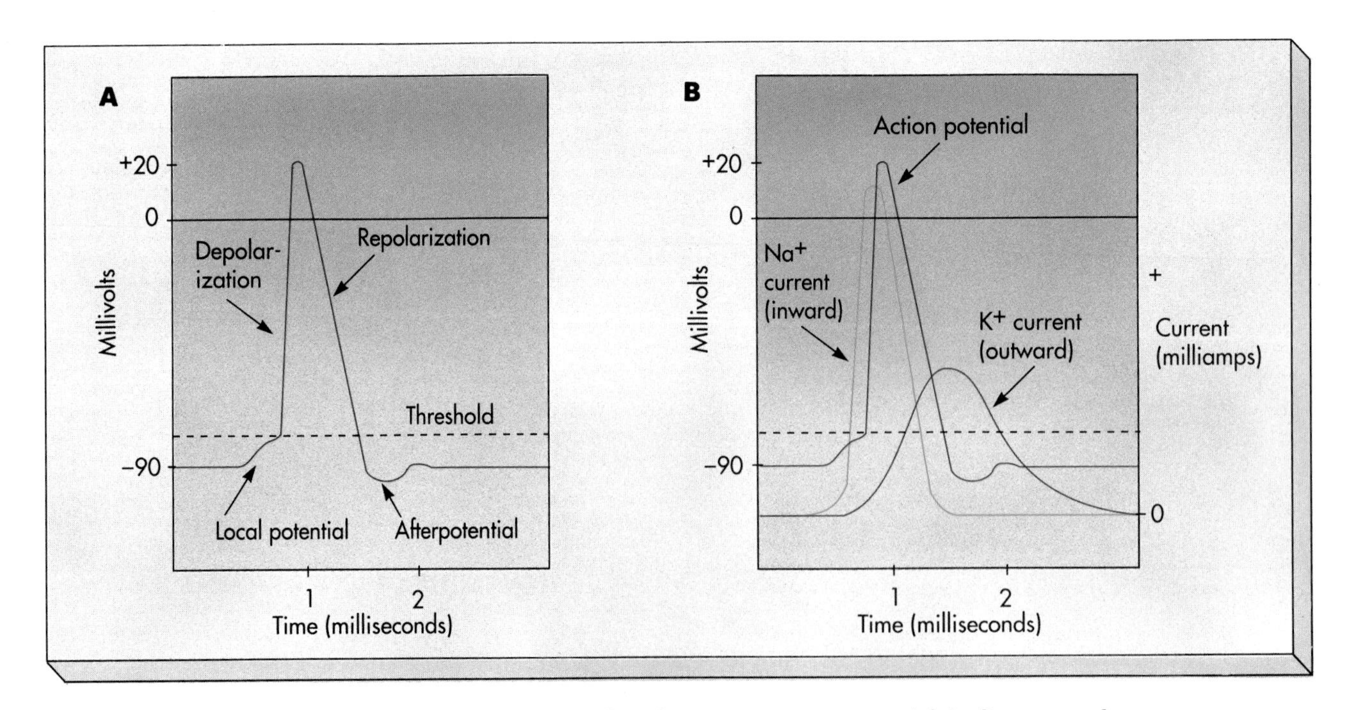

FIGURE 9-8 **A** An action potential *(red)*. **B** An action potential *(red)* compared to an inwardly directed sodium (Na^+) ion current during depolarization *(blue)* and an outwardly directed K^+ ion current during repolarization *(purple)*. An afterpotential occurs because of elevated potassium (K^+) ion permeability.

positive charge reaches approximately +20mV inside the cell, the negative charge outside of the cell is large enough to slow the movement of additional Na^+ ions across the cell membrane.

As the depolarization phase of the action potential approaches its peak, the sodium channels begin to close. At the same time the potassium channels are continuing to open. Consequently, the permeability of the membrane to Na^+ ions decreases, and the permeability to K^+ ions increases. The rapid influx of Na^+ ions slows, and movement of K^+ ions out of the cell increases, causing the inside of the cell membrane once again to become negative (Figure 9-8, *B*).

5

Predict the effect of a reduced extracellular concentration of sodium ions on the magnitude of the action potential in an electrically excitable tissue.

? ? ? ? ? ? ? ? ? ? ?

In many cells an afterpotential is observed after the action potential. The afterpotential occurs because the increased K^+ ion permeability during the repolarization phase of the action potential lasts slightly longer than the time required to bring the membrane potential back to its resting level. When the elevated K^+ ion permeability returns to normal, the membrane potential resumes its resting level.

As long as the Na^+ and K^+ ion concentrations remain unchanged across the cell membrane, all the action potentials produced by a cell are identical. They all take the same amount of time, and they all exhibit the same magnitude. These characteristics vary somewhat from one cell type to another, but it generally takes approximately 1 to 2 msec (1 msec equals 0.001 second) for an action potential to occur.

Once an action potential has been initiated at a given point on the cell's membrane, that area is insensitive to further stimulation during both the depolarization phase and most of the repolarization phase. The period of time of complete insensitivity to another stimulus is the **absolute refractory period.** During the last part of the repolarization phase of the action potential, the **relative refractory period,** a stimulus of greater-than-threshold strength can initiate another action potential (Figure 9-9). The existence of the absolute refractory period guarantees that once an action potential is begun, both the depolarization and nearly all of the repolarization phase will be completed before another action potential can be started and that a strong stimulus cannot lead to prolonged depolarization of the cell membrane.

Active transport of Na^+ and K^+ ions is not involved directly in the action potential, but it plays an important role in maintaining the concentration gradients across the cell membrane. The sodium-potassium exchange pump functions to transport Na^+ ions from the cell and K^+ ions into the cell to restore the normal concentrations of these ions on either side of the cell membrane after action potentials have occurred.

Propagation of Action Potentials

An action potential produced at any point on the cell membrane acts as a stimulus to adjacent regions of the cell membrane (Figure 9-10). The change in the membrane potential at the point where an action potential occurs causes sodium channels, which are sensitive to changes in the membrane potential, to open in the membrane adjacent to the site at which the action potential occurs. As a consequence an action potential is produced in the adjacent region.

Action potentials are propagated from their point of origin over the remainder of the cell. However, the absolute refractory period makes the membrane insensitive to restimulation long enough to prevent an action potential from reinitiating another action potential at that same point and keeps an action potential from reversing its direction of propagation.

The speed of propagation varies greatly from cell to cell. In general, neurons having large-diameter axons conduct action potentials faster than neurons

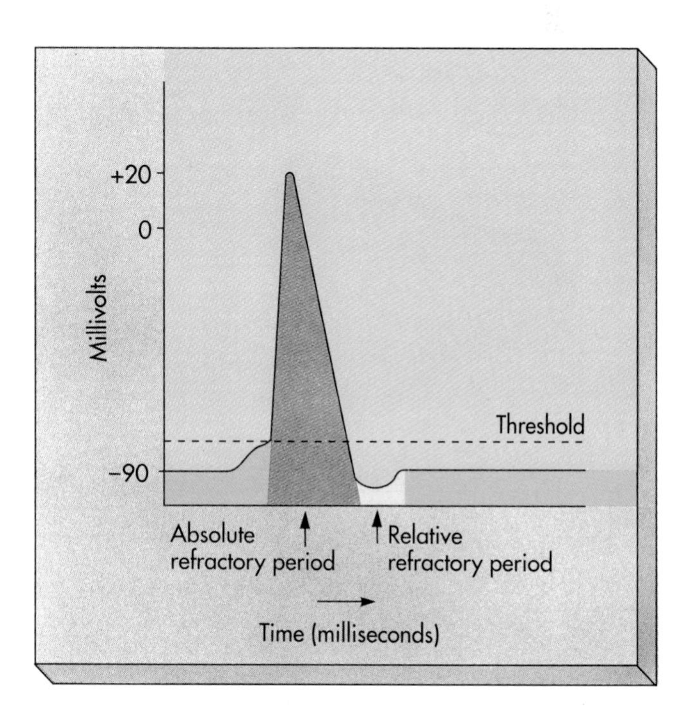

FIGURE 9-9 The absolute and relative refractory periods of an action potential.

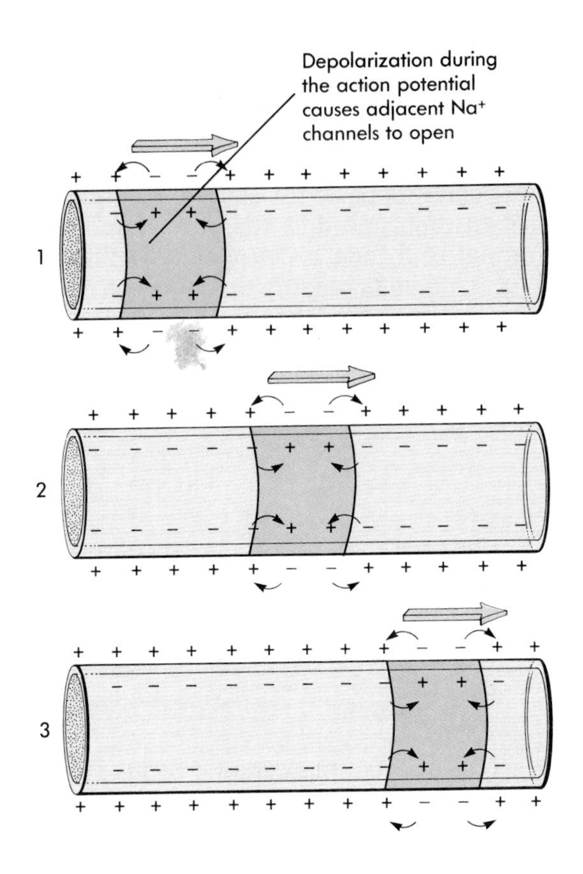

FIGURE 9-10 Propagation of an action potential along a cell membrane. An action potential initiated in one part of the cell membrane stimulates action potentials in adjacent parts of the membrane. The adjacent areas, now also producing an action potential, stimulate areas of the membrane adjacent to them so that the action potential spreads from the initial site. Because of the length of the refractory period relative to the time required to initiate an action potential, an area of the cell membrane that initiates an action potential in an adjacent area cannot in turn be stimulated to initiate another action potential by that adjacent area. As a result, action potentials are propagated away from the initial site of stimulation. In the figure action potentials in a local area of the cell membrane are represented by a change in charge across the cell membrane.

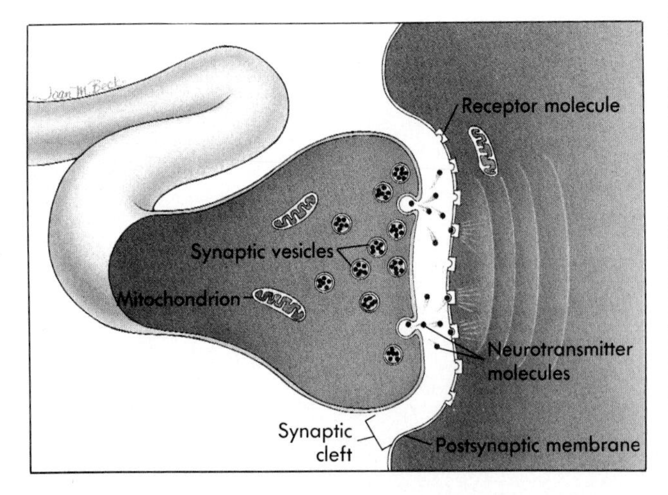

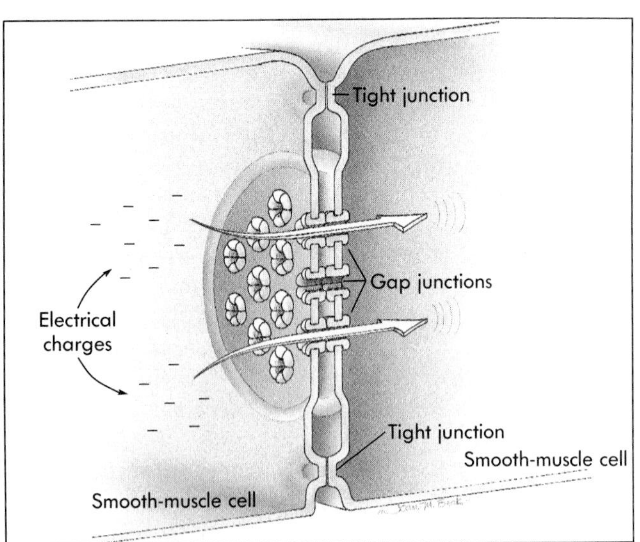

FIGURE 9-11 **A** Chemical synapse. In this synapse neurotransmitters that were synthesized in one cell are released by exocytosis into the space between the cells (the synaptic cleft). The neurotransmitters bind to receptors on the membrane of the receptor cell, causing a change in the permeability of the cell membrane. As the permeability of the cell membrane to sodium ions increases, depolarization results. If the depolarization reaches threshold, an action potential will be produced in the second (postsynaptic) cell. **B** Electrical synapse. In this synapse electrical charges in the form of action potentials can pass directly from one cell to another because gap junctions make an electrical connection between the cells.

having small-diameter axons. Nerve cells propagate action potentials more rapidly than do muscle cells, and specialized axons, which have insulating **myelin sheaths,** conduct action potentials more rapidly than those without myelin sheaths, even though the insulated neuron may have a smaller diameter (see Chapter 12).

Action potentials are not propagated from one cell to the next in the same way that they are propagated along the membrane of a single cell. Specialized structures called **synapses** are the sites at which action potentials from one cell are able to produce action potentials in an adjacent cell. Most synapses are chemical synapses in which an action potential causes the release of a chemical from the end of a nerve-cell process (Figure 9-11, *A*). The chemical, called a **neurotransmitter,** diffuses across the space within the synapse to bind with specific membrane receptors on the adjacent cell membrane, causing an increase in the permeability of that cell's membrane to Na^+ ions. As a consequence, Na^+ ions diffuse into that cell and cause a local potential; if the local potential exceeds threshold, an action potential results. In some synapses the neurotransmitter substance causes an increase in the permeability of the membrane to K^+ instead of Na^+. When that occurs, the membrane is hyperpolarized, and the cell, as a consequence, is less likely to produce an action potential. Some synapses are electrical synapses in which adjacent cell membranes, joined at **gap junctions,** allow depolarizations in one cell to spread to the adjacent cell as though the two cells were fused (Figure 9-11, *B*). Synapses are discussed further in Chapters 10 and 12.

Action Potential Frequency

The action potential frequency is directly proportional to the stimulus strength, whereas the length of time action potentials are produced is a measure of the duration of the stimulus. Most stimuli cause local depolarizations, and the size of the local potentials is proportional to the stimulus strength. A stimulus resulting in a local potential so small that it does not reach threshold, that is, a **subthreshold stimulus,** results in no action potential (Figure 9-12). A stimulus just strong enough to cause a local potential to reach threshold, a **threshold stimulus,** produces a single action potential. A stimulus strong enough to produce a maximum frequency of action potentials is a **maximal stimulus,** and a stronger stimulus is called a **supramaximal stimulus.** A **submaximal stimulus** includes stimuli between threshold and the maximal stimulus strength. For submaximal stimuli the action potential frequency increases in proportion to the strength of the stimulus.

The maximum frequency of action potentials in an excitable cell is determined by the duration of the absolute refractory period. If the duration is 1 msec for each action potential, the maximum frequency of action potentials for that cell is approximately 1000 action potentials per second.

The length of time action potentials are produced in a cell in response to a given stimulus reflects how long the stimulus is applied at a strength great enough to cause a local depolarization to exceed threshold (Figure 9-13, *A*). Two stimuli of identical strength applied to a cell for different lengths of time normally result in identical action potential frequencies in response to each stimulus, but the length of time the action potentials are produced will differ.

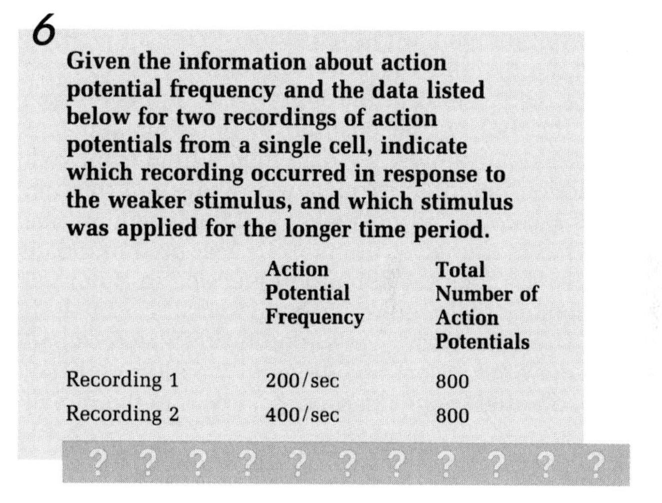

6

Given the information about action potential frequency and the data listed below for two recordings of action potentials from a single cell, indicate which recording occurred in response to the weaker stimulus, and which stimulus was applied for the longer time period.

	Action Potential Frequency	**Total Number of Action Potentials**
Recording 1	200/sec	800
Recording 2	400/sec	800

? ? ? ? ? ? ? ? ? ? ?

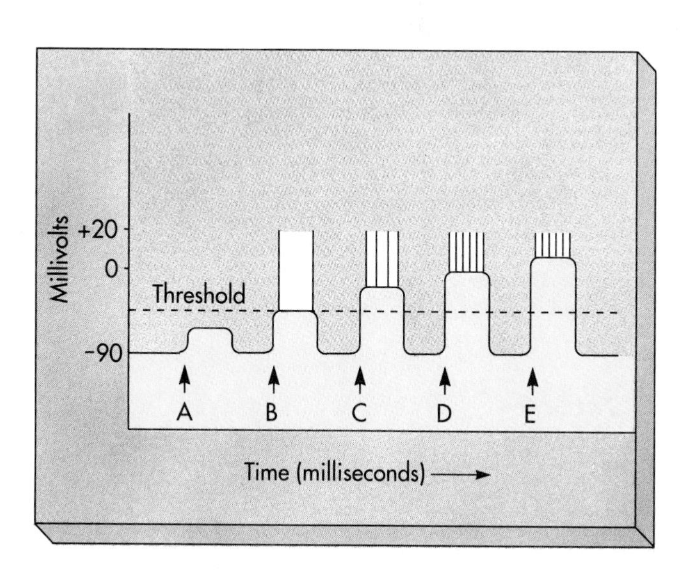

FIGURE 9-12 Relationship between the stimulus strength, the local potential, and the action potential frequency. Each stimulus is stronger than the previous one. Arrows indicate where a stimulus is applied. **A** Subthreshold stimulus. **B** Threshold stimulus. **C** Submaximal stimulus. **D** Maximal stimulus. **E** Supramaximal stimulus.)

In some cells the local potential is maintained at a constant magnitude as long as a stimulus of a given strength is applied to the cell; in other cells the local potential quickly returns to its resting membrane potential even though the stimulus is applied for a long period of time, an adjustment called **accommodation** (Figure 9-13, *B*). In cells that exhibit accommodation, a stimulus that is applied to the cell for a long duration will produce an action potential frequency proportional to the magnitude of the local potential. However, even though the stimulus remains constant, the local potential begins to decrease in magnitude, and the action potential frequency declines.

Cells that do not accommodate produce action potential frequencies proportional to the strength of the stimulus and the action potential frequency is maintained as long as the stimulus is applied. Nerve cells monitoring the position of the arms and legs accommodate slowly so that when a limb is moved, the action potential frequency provides information about the degree of movement and the length of time the limb is in that position. Cells that accommodate rapidly are adapted for producing action potential frequencies that are proportional to changes of stimulus strength. Nerve cells monitoring the acceleration of a limb accommodate rapidly and provide action potential frequencies proportional to the rate of change in the limb movement.

The characteristics of the action potential are summarized in Table 9-4.

7

Consider the following: A maximal stimulus is applied to a nerve cell and is maintained for a period of 10 seconds; the action potential frequency decreases from a maximum of 400/sec immediately after the stimulus to a frequency of 100/sec after 2 seconds and a frequency of 0/sec after 10 seconds. Explain the relationship between the stimulus' duration and the observed change in action potential frequency.

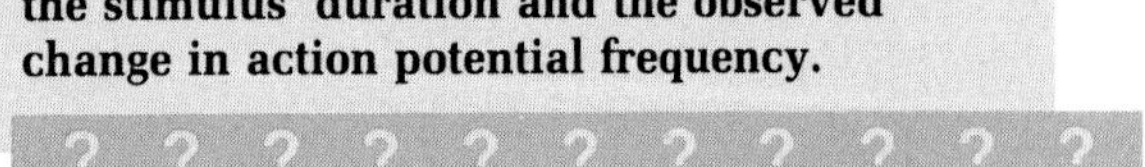

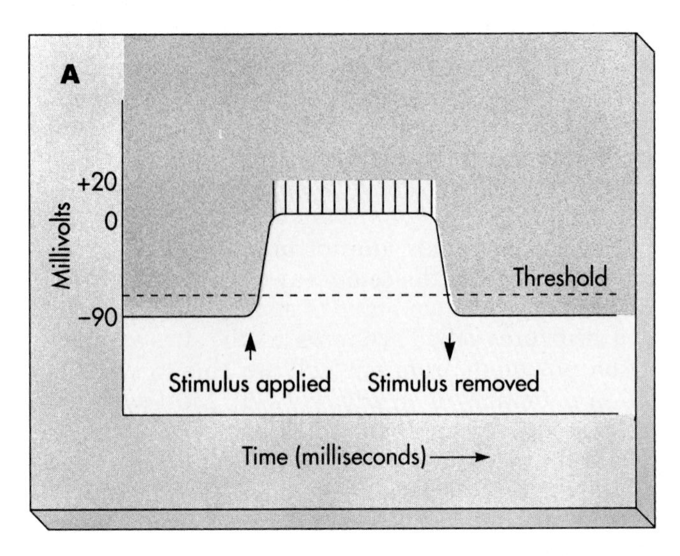

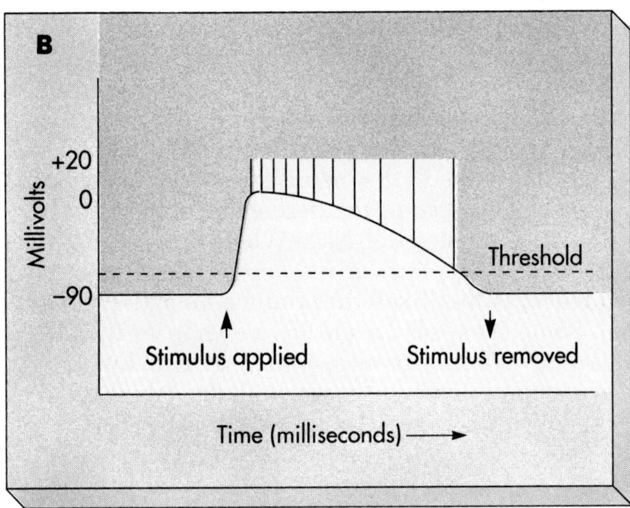

FIGURE 9-13 **A** When a stimulus is applied in some neurons, the local potential remains constant as long as the stimulus is applied so that the action potential frequency remains constant while the stimulus is applied. **B** In other neurons accommodation occurs when a constant stimulus is applied. The local potential becomes smaller, and the action potential frequency decreases. Action potentials still are produced (but at a decreasing frequency) until the local potential declines below threshold, at which time the action potentials cease.

Table 9-4

Characteristics of the action potential

1. Action potentials are produced when a local potential reaches threshold.
2. Action potentials are all-or-none.
3. Depolarization is a result of increased membrane permeability to Na^+ ions and movement of Na^+ ions into the cell. Repolarization is a result of decreased membrane permeability to Na^+ ions and increased membrane permeability to K^+ ions, which stop Na^+ ion movement into the cell and increase K^+ ion movement out of the cell.
4. No action potential is produced by a stimulus, no matter how strong, during the absolute refractory period. During the relative refractory period a stronger-than-threshold stimulus can produce an action potential.
5. Action potentials are propagated, and for a given axon or muscle fiber the magnitude of the action potential is constant.
6. Stimulus strength determines the frequency of action potentials. Unless accommodation occurs, stimulus duration determines how long action potentials are produced.

SUMMARY

CONCENTRATION DIFFERENCES ACROSS THE MEMBRANE p. 254

1. There is a higher concentration of K^+ ions inside the cell than outside and a higher concentration of Na^+ ions outside the cell than inside.
2. Negatively charged proteins and other anions are synthesized inside the cell and cannot diffuse out of it. They attract positively charged K^+ ions into the cell and repel negatively charged Cl^- ions.
3. The sodium-potassium exchange pump moves ions by active transport. K^+ ions are moved into the cell, and Na^+ ions are moved out of it.

RESTING MEMBRANE POTENTIAL p. 255

1. The resting membrane potential is a charge difference across the cell membrane when the cell is in an unstimulated condition. The inside of the cell membrane is negatively charged when compared to the outside of the cell membrane.
2. Depolarization is a decrease in the resting membrane potential and can result from an increase in extracellular K^+ ions or from a decrease in membrane permeability to K^+ ions.
3. Hyperpolarization is an increase in the resting membrane potential that can result from a decrease in extracellular K^+ ions or an increase in membrane permeability to K^+ ions.

MOVEMENT OF IONS THROUGH THE CELL MEMBRANE p. 258

Ion Channels

1. Na^+ and K^+ ions diffuse across the cell membrane through Na^+ and K^+ ion channels.
2. Changes in the charge across the cell membrane and changes in the extracellular Ca^{2+} ion concentration can open and close the Na^+ ion channels.

Sodium-Potassium Exchange Pump

The sodium-potassium exchange pump maintains K^+ and Na^+ ion concentrations across the cell membrane.

ELECTRICALLY EXCITABLE CELLS p. 260

1. A local potential, either a depolarization or a hyperpolarization, is a small change in the resting membrane potential that is confined to a small area of the cell membrane.
2. An action potential is a larger change in the resting membrane potential that spreads over the entire surface of the cell.

Local Potential

1. A local potential is termed graded because a stronger stimulus produces a greater potential change than a weaker stimulus.
2. A local potential decreases in magnitude as the distance from the stimulation increases.
3. A local potential can be produced in two ways.
   - An increase in membrane permeability to Na^+ ions can cause local depolarization.
   - An increase in membrane permeability to K^+ or Cl^- ions can result in local hyperpolarization.

Action Potential

1. The threshold potential is the membrane potential at which a local potential depolarizes the cell membrane sufficiently to produce an action potential.
2. Action potentials occur in an all-or-none fashion. If the action potential occurs at all, it is of the same magnitude and duration no matter how strong the stimulus.
3. Depolarization occurs as the inside of the membrane becomes more positive because of Na^+ ion movement into the cell. Repolarization is a return of the membrane potential toward the resting membrane potential because of K^+ ion movement out of the cell and because Na^+ ion movement into the cell slows to resting levels.
4. The afterpotential is a short period of hyperpolarization after the action potential.
5. The absolute refractory period is the time immediately after a stimulus during which a second stimulus, no matter how strong, cannot initiate another action potential.
6. The relative refractory period follows the absolute refractory period and is the time during which a stronger-than-threshold stimulus can evoke another action potential.

Propagation of Action Potentials

1. An action potential causes sodium channels in adjacent regions of the cell to open and to produce action potentials.
2. Reversal of the direction of action potential propagation is prevented by the absolute refractory period.
3. The conduction speed of action potentials is greatest in myelinated, large axons.
4. Action potentials can be transferred from cell to cell across a synapse. Transfer can be due to diffusion of a chemical (neurotransmitter) or can be electrical (gap junction).

Action Potential Frequency

1. Types of stimuli.
   - A subthreshold stimulus produces only a local potential.
   - A threshold stimulus causes a local potential that reaches threshold and results in a single action potential.
   - A submaximal stimulus is greater than a threshold stimulus and weaker than a maximal stimulus. The action potential frequency increases as the strength of the submaximal stimulus increases.
   - A maximal or a supramaximal stimulus produces a maximum frequency of action potentials.
2. The longer a threshold or greater stimulus is applied, the longer the action potentials will be generated unless accommodation occurs.
3. During accommodation, even though a long stimulus of constant strength is applied, the local potential returns to the resting membrane potential after a short time period, and no action potential is produced after the local potential decreases below the threshold value.

CONTENT REVIEW

1. Describe the concentration differences that exist across a cell membrane for K^+ ions, Na^+ ions, Cl^- ions, and proteins. Explain the cause of these differences.
2. Define the resting membrane potential. Is the outside of the cell membrane positively or negatively charged? The inside?
3. Explain the role of K^+ ions in establishing the resting membrane potential.
4. Define depolarization and hyperpolarization. How do changes in extracellular K^+ ion concentration or in membrane permeability to K^+ ions affect depolarization and hyperpolarization?
5. What effect does a change of extracellular sodium have on the resting membrane potential? Why is this so?
6. What are ion channels? What effect does a change in extracellular Ca^{2+} ion concentration and membrane polarity have on ion channels?
7. Describe the sodium-potassium exchange pump. What effect does it have on the resting membrane potential? On Na^+ and K^+ ion concentration gradients across the cell membrane?
8. Differentiate between a local potential and an action potential.
9. Describe two ways that a change in membrane permeability can produce a local potential.
10. What is meant by a graded, nonpropagated potential?
11. Define threshold potential. What happens to a local potential that reaches threshold?
12. Discuss the all-or-none production of an action potential. Are all action potentials the same?
13. Define the depolarization and repolarization phases of an action potential. Explain how changes in membrane permeability and movement of Na^+ and K^+ ions cause each phase.
14. Describe the afterpotential and its cause.
15. Distinguish between the absolute and relative refractory periods. Relate them to the depolarization and repolarization phases of the action potential.
16. What is the function of the sodium-potassium exchange pump after an action potential?
17. What causes the propagation of the action potential? What prevents the action potential from reversing its direction of propagation?
18. Describe two ways an action potential can pass from one cell to another cell.
19. Define a subthreshold, threshold, submaximal, maximal, and supramaximal stimulus.
20. What determines the maximum frequency of action potential generation?
21. How does stimulus strength affect the frequency of action potential production?
22. How does the length of stimulation affect the length of time that action potentials are produced? What is accommodation?

CONCEPT QUESTIONS

1. Predict the consequence of a reduced intracellular K^+ ion concentration on the resting membrane potential.
2. Predict the effect of an elevated extracellular potassium ion concentration on nerve and muscle tissue.
3. A child eats a whole bottle of salt (NaCl) tablets. What effect would this action have on action potentials?
4. Lithium ions reduce the permeability of cell membranes to sodium ions. Predict the effect lithium ions in the extracellular fluid would have on the response of a neuron to stimuli.
5. Predict the effect of an elevated extracellular calcium concentration on nerve and muscle tissue.
6. When severe burns occur, many cells are destroyed and release their contents into the blood. Assuming that shock caused by reduced blood volume and stress is under control, explain why many burn patients suffer from tachycardia (rapid heart rate).
7. Both smooth muscle and cardiac muscle have the ability to contract spontaneously (they will contract without external stimulation). They also contract rhythmically (contractions occur at regular intervals). Based on what you know about membrane potentials, propose an explanation for both the spontaneous and rhythmic characteristics of these muscle tissues. Assume that an action potential in a muscle cell will cause the muscle to contract.
8. Smooth muscle has some characteristics that differ from skeletal muscle or nerves. One characteristic involves the action potential. Although smooth-muscle action potentials are very similar to those in skeletal muscle, the following data suggest that some differences exist.
   - A chemical compound that specifically blocks the diffusion of sodium into the cell reduces the amplitude of the action potentials but does not eliminate them in smooth muscle.
   - The amplitude of smooth-muscle action potentials is reduced in a calcium-free medium.
   - Elevating the intracellular concentration of calcium reduces the amplitude of smooth-muscle action potentials.

   Based on the above information, which of the following is (are) most logical?

   A Calcium inhibits the entry of sodium into smooth-muscle cells.
   B Calcium regulates the permeability of smooth-muscle cell membranes to sodium.
   C Calcium participates with sodium in the depolarization phase of the action potential.
   D Calcium is responsible for the depolarization phase of the action potential.

? ?

ANSWERS TO PREDICT QUESTIONS

1 *p. 256.* If the intracellular concentration of potassium (K^+) ions is increased, the concentration gradient from the inside to the outside of the cell membrane is increased. This situation would be similar to decreasing the extracellular concentration of potassium ions. The greater concentration gradient for potassium ions would increase the tendency for potassium ions to diffuse out of the cell across the cell membrane. A greater negative charge then would develop inside of the cell membrane (hyperpolarization). At equilibrium the greater negative charge would be just enough to prevent the diffusion of additional potassium ions from the cell.

2 *p. 256.* A decrease in the cell membrane's permeability to K^+ ions results in depolarization of the cell membrane. When the permeability of the cell membrane to K^+ ions decreases, the tendency for K^+ ions to diffuse out of the cell decreases; fewer K^+ ions line up on the outside of the cell membrane, and a smaller negative charge is required inside the cell to prevent K^+ ions from leaving the cell. Thus a new equilibrium is established in which the membrane potential is less polar (is depolarized in comparison to the resting membrane potential), and fewer K^+ ions diffuse to the outside of the cell membrane.

3 *p. 259.* If the extracellular concentration of calcium ions is decreased, the resting membrane potential becomes depolarized. When extracellular calcium concentrations decline, sodium ion channels open. The open sodium ion channels allow sodium ions to diffuse into the cell and cause depolarization of the cell membrane. Calcium ions bind to gating proteins that regulate the sodium channels. Low concentrations of calcium cause the sodium channels to open, and high concentrations of calcium cause the sodium channels to close.

4 *p. 261.* If a cell is stimulated, usually an increase in the permeability of the cell membrane to Na^+ ions results, with the degree of permeability dependent on the strength of the stimulus. The greater the stimulus strength, the greater is the permeability of the membrane to Na^+ ions. Na^+ ions diffuse into the cell down their concentration gradient and cause depolarization of the cell membrane. If their concentration gradient is reduced, the tendency for Na^+ ions to diffuse into the cell is decreased in comparison to the normal condition. Thus two stimuli of the same strength will result in local potentials of differing magnitudes. In the cell with the reduced Na^+ ion concentration gradient, the local depolarization is of a smaller magnitude because fewer Na^+ ions are able to diffuse into the cell in response to the stimulus, even though the increase in the permeability of the cell membrane to Na^+ ions increases to the same value in both situations.

5 *p. 263.* If the extracellular concentration of sodium ions is decreased, the magnitude of the action potential will be reduced. For example, if depolarization occurs from a resting membrane potential of -80 mV to $+20$ mV during an action potential, the depolarization may be only from -80 mV to $+5$ or $+10$ mV if the extracellular concentration of sodium ions is reduced. The smaller concentration of sodium ions reduces the tendency for sodium ions to diffuse into the cell when the sodium channels are open during an action potential. Consequently, the inside of the cell membrane does not become as positive.

6 *p. 265.* The action potential frequency is proportional to the strength of the stimulus applied to a cell. Therefore when the cell produced action potentials at the higher frequency (400 per second), it was stimulated with the greater stimulus strength. Since the first recording produced a lower frequency of action potential but the same number of action potentials, it must have produced them for a longer period of time. For recording number 1 at 200 per second, the cell must have produced action potentials for 4 seconds, and for recording number 2 at 400 per second, it must have produced action potentials for 2 seconds.

7 *p. 266.* The data indicate that the neuron exhibited a marked decrease in action potential frequency even though a stimulus of constant strength was applied to the neuron for a long period of time. The data are consistent with accommodation. In neurons that exhibit accommodation the local potential declines in magnitude even though the stimulus is applied for a long period of time. As the local potential declines in magnitude, the action potential's frequency declines also. When the local potential declines below threshold, no more action potentials are produced.

10 CHAPTER OBJECTIVES

After reading this chapter you should be able to

1. List the major categories of muscles and describe their general characteristics.
2. Describe the structure of a muscle, including its connective tissue elements, blood vessels, and nerves.
3. Produce diagrams that illustrate the arrangement of myofilaments, myofibrils, sarcomeres, sarcoplasmic reticulum, and T tubules in a muscle fiber.
4. Explain the events that are responsible for the transmission of an action potential across the neuromuscular junction.
5. Describe the events that result in muscle contraction and relaxation in response to an action potential in a motor neuron.
6. Explain how muscle tone is maintained and how slow contraction and relaxation occur in skeletal muscle.
7. Describe how the length of a muscle influences the force of contraction.
8. Compare the mechanisms involved in psychological fatigue, muscular fatigue, and synaptic fatigue.
9. Explain the causes of physiological contracture and rigor mortis.
10. Describe the events that lead to an oxygen debt and recovery from it.
11. Distinguish between fast-twitch muscles and slow-twitch muscles and explain the functions for which each type is best adapted.
12. Predict the effects of both aerobic exercise and anaerobic exercise on the structure and function of skeletal muscle.
13. Explain the events responsible for the generation of heat produced by muscle before, during, and after exercise and when shivering.
14. List the types of smooth muscle and describe the characteristics of each.
15. Describe the relationship between the resting membrane potential, action potentials, and contraction in smooth muscle.
16. Compare the unique structural and functional characteristics of smooth muscle with those of skeletal muscle.

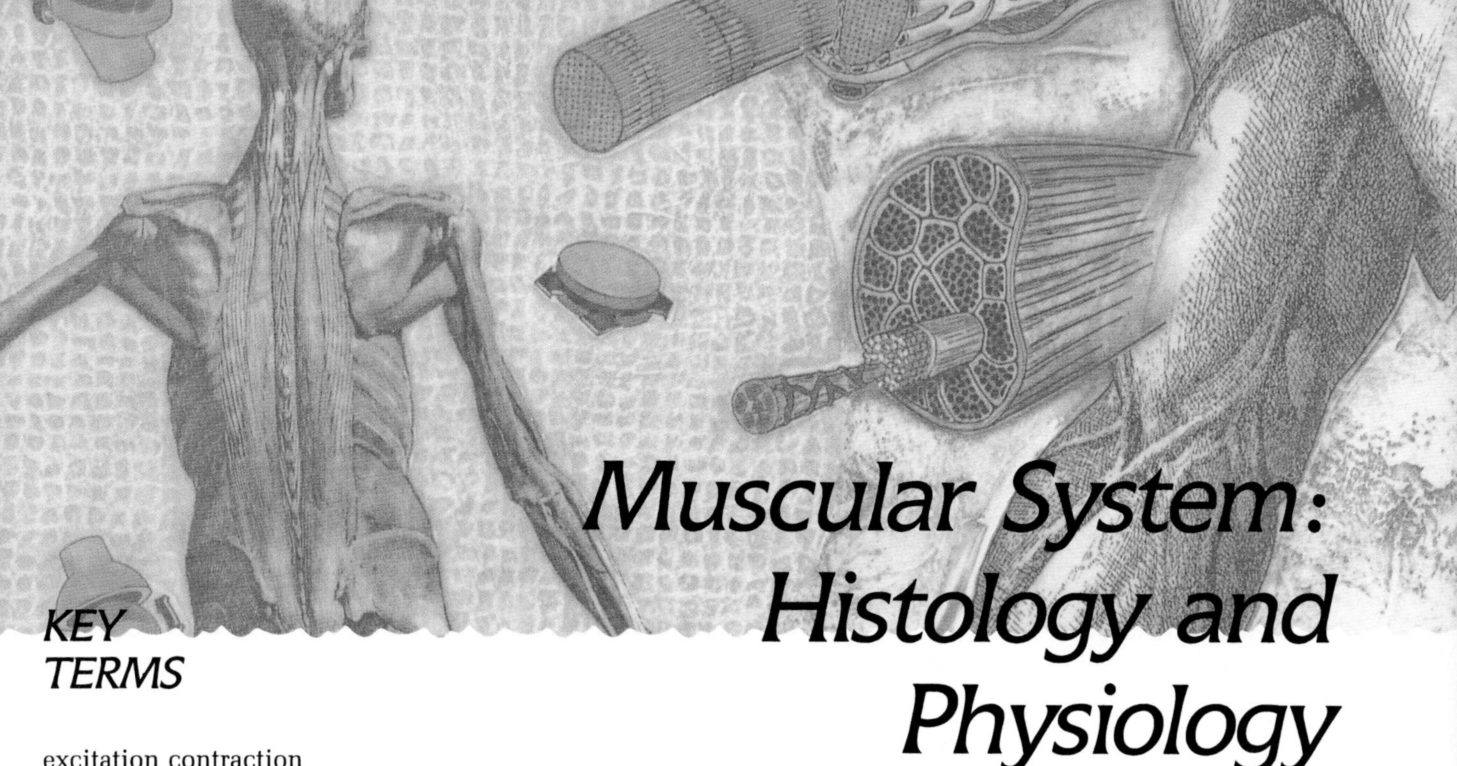

Muscular System: Histology and Physiology

KEY TERMS

excitation contraction coupling

fatigue

isometric contraction

isotonic contraction

motor unit

multiple motor unit summation

multiple wave summation

muscle twitch

myofibril

myofilament

neuromuscular junction

oxygen debt

sarcomere (sar′ko-mēr)

sliding filament mechanism

striated (stri′a-ted)

RELATED TOPICS

The following terms or concepts are important for a good understanding of this chapter. If you are not familiar with them, you should review them before proceeding.

Cellular organelles (Chapter 3)

Histology of muscle (Chapter 4)

Membrane potentials (Chapter 9)

MUSCLE TISSUE IS HIGHLY SPECIALIZED TO CONTRACT OR SHORTEN FORCEFULLY. WITH THE EXCEPTION OF MOVEMENTS PRODUCED BY CILIA AND FLAGELLA AND THE EFFECTS OF GRAVITY, MUSCLE IS RESPONSIBLE FOR THE MECHANICAL PROCESSES IN THE BODY. There are three types of muscle tissue: skeletal, cardiac, and smooth. Skeletal muscle moves the trunk and appendages, cardiac muscle propels blood through vessels, and smooth muscle forces food through the digestive system and moves glandular secretions through ducts. In addition, metabolism that occurs in the large mass of muscle tissue in the body produces heat essential to the maintenance of normal body temperature.

This chapter presents the basic structural and functional characteristics of muscle tissue. The interactions between structures and functions are emphasized to illustrate the mechanical properties of muscle, its energy requirements, and the relationship of these processes to the chemical and electrical events within muscle cells. Since skeletal muscle is more common than other types of muscle in the body and since more is known about it, skeletal muscle is examined in the greatest detail.

GENERAL FUNCTIONAL CHARACTERISTICS OF MUSCLE

Muscle has four major functional characteristics: contractility, excitability, extensibility, and elasticity. Contractility refers to the capacity of muscle to contract, or shorten, forcefully. Muscle is termed excitable because it responds to stimulation by nerves and hormones, making it possible for the nervous system and, in some muscle types, the endocrine system to regulate muscle activity. That muscles are extensible means they can be stretched. After a contraction muscles can be readily stretched to their normal resting length and beyond to a limited degree. If muscles are stretched, they recoil to their original resting length because of their elasticity.

Muscle contractions are responsible for the body's movements. Muscles shorten forcefully during contraction but lengthen passively. When they contract, muscles cause movement of structures to which they are attached, but the opposite movement requires an antagonistic force such as that produced by another muscle, gravity, or the force of fluid filling a hollow organ.

As stated previously, the major types of muscles are (1) skeletal, (2) cardiac, and (3) smooth (Table 10-1). Skeletal muscle with its associated connective tissue comprises approximately 40% of the body's weight and is responsible for locomotion, facial expressions, posture, and many other body movements (Figure 10-1). Its function, to a large degree, is under voluntary or conscious control by the nervous system. Smooth muscle, the most variable type of muscle in the body with respect to distribution and function, is in the walls of hollow organs and tubes, in the internal muscles of the eye, in walls of

Table 10-1

Comparison of muscle types

FEATURES	SKELETAL MUSCLE	CARDIAC MUSCLE	SMOOTH MUSCLE
Location	Attached to bones	Heart	Walls of hollow organs, blood vessels, eyes, glands, and skin
Cell shape	Very long and cylindrical (1-40 mm in length and may extend the entire length of short muscles; 10-100 μm in diameter)	Cylindrical and branched (100-500 μm in length; 100-200 μm in diameter)	Spindle shaped (15-200 μm in diameter)
Nucleus	Multiple, peripherally located	Single, centrally located	Single, centrally located
Special features	—	Intercalated disks join cells to each other	Gap junctions join some visceral smooth muscle cells together
Striations	Yes	Yes	No
Control	Voluntary	Involuntary	Involuntary
Capable of spontaneous contraction	No	Yes	Yes
Function	Body movement	Pumps blood	Food movement through the digestive tract, emptying of the urinary bladder, regulation of blood-vessel diameter, change in pupil size, contraction of many gland ducts, movement of hair, and many other functions

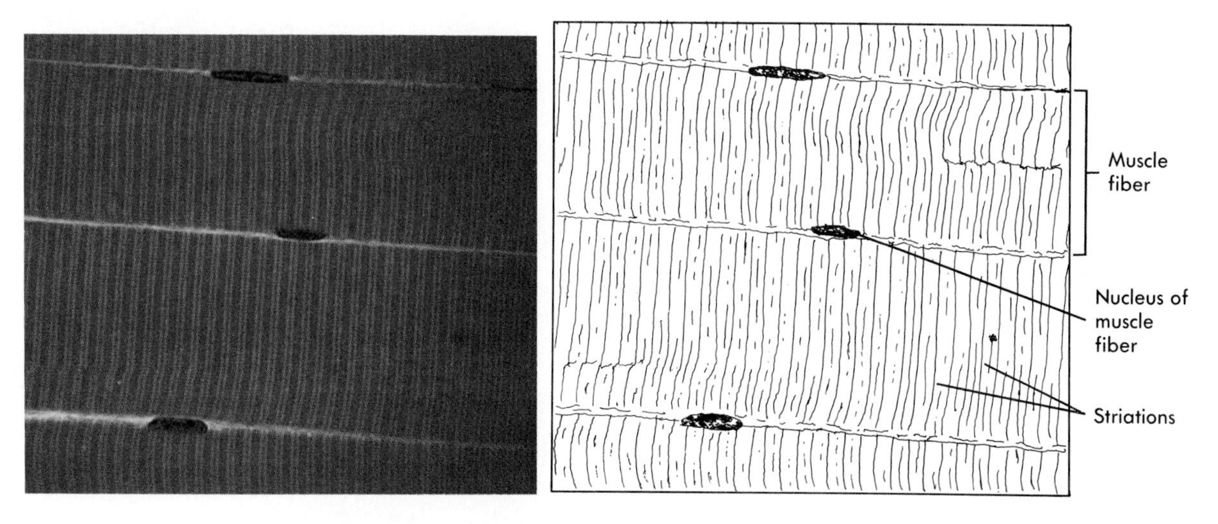

FIGURE 10-1 Skeletal muscle in longitudinal section.

blood vessels, and in other areas. Smooth muscle performs a variety of functions, including propelling urine through the urinary tract, mixing food in the stomach and intestine, dilating and constricting the pupil, and regulating the flow of blood through blood vessels. Cardiac muscle is found only in the heart, and its contractions provide the major force for propelling blood through the circulatory system. Unlike skeletal muscle, smooth muscle and cardiac muscle are autorhythmic (i.e., they contract spontaneously at somewhat regular intervals). Nervous or hormonal stimulation is not always required for them to contract. Furthermore, smooth muscle and cardiac muscle are not under direct conscious control but are innervated and in part regulated unconsciously or involuntarily by the autonomic nervous system and the endocrine system (see Chapters 16 and 18).

SKELETAL MUSCLE: STRUCTURE

Skeletal muscles are composed of skeletal **muscle fibers** associated with smaller amounts of connective tissue, blood vessels, and nerves. Skeletal muscle fibers are the equivalent of skeletal muscle cells. Each skeletal muscle fiber is a single cylindrical cell containing several nuclei located around the periphery of the fiber near the cell membrane. The muscle fibers develop from less mature multinucleated cells called **myoblasts** (mi'o-blasts). Their multiple nuclei result from the fusion of myoblast precursor cells and not from the division of nuclei within myoblasts. The myoblasts are converted to muscle fibers as contractile proteins accumulate within their cytoplasm. Shortly after the myoblasts form, nerves grow into the area and innervate the developing muscle fibers.

The number of skeletal muscle fibers remains relatively constant after birth. Therefore enlargement of muscles after birth is not due to a significant increase in the number of muscle fibers but to an increase in their size.

As seen in longitudinal section, alternating light and dark bands give the muscle fiber a **striated** (stri'a-ted; banded) appearance (see Figure 10-1). A single fiber may extend from one end of a small muscle to the other end, but several muscle fibers arranged end to end are required to extend the full length of most longer muscles. Muscle fibers range from 1 to 40 mm in length and from 10 to 100 μm in diameter. Large muscles contain large-diameter fibers, whereas small, delicate muscles have small-diameter fibers. All the muscle fibers in a given muscle have similar dimensions.

Connective Tissue

Surrounding each muscle fiber is a delicate **external lamina** composed primarily of reticular fibers. This external lamina is produced by the muscle fiber and when observed through the light microscope, cannot be distinguished from the muscle fiber's cell membrane, the **sarcolemma** (sar'ko-lem'ah; Figure 10-2). The **endomysium** (en'do-mīz'ĭ-um), a delicate network of loose connective tissue with numerous reticular fibers, surrounds each muscle fiber outside the external lamina. A bundle of muscle fibers with their endomysium is surrounded by another, heavier

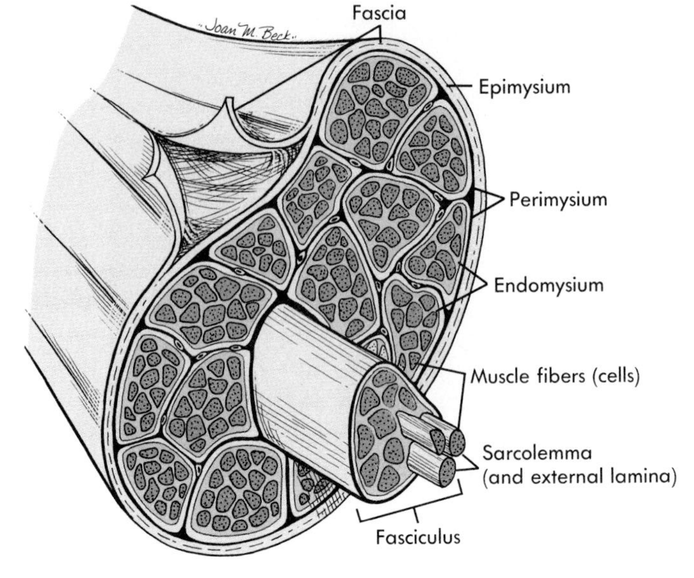

FIGURE 10-2 Relationship between muscle fibers, fasciculi, and associated connective tissue.

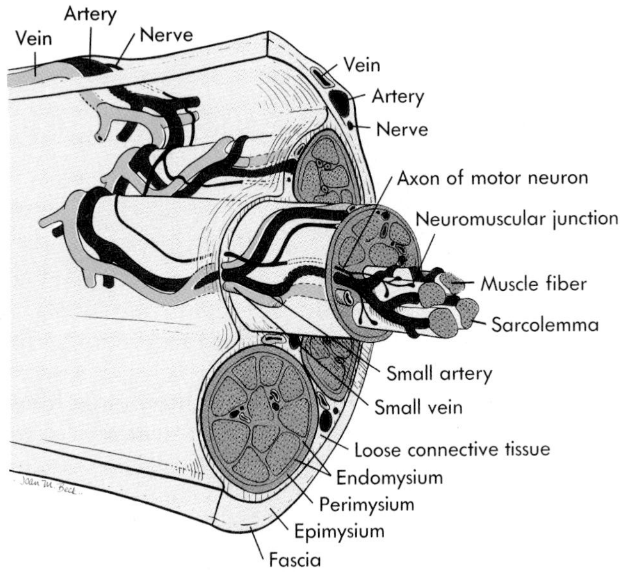

FIGURE 10-3 Innervation and blood supply of a muscle. Arteries, veins, and nerves course together through the connective tissue of muscle fibers. They branch frequently as they approach individual muscle fibers.

connective tissue layer called the **perimysium** (pĕr'ĭ-mīz'ĭ-um). Each bundle ensheathed by perimysium is a **muscle fasciculus** (fă-sik'u-lus). A muscle consists of many fasciculi grouped together and surrounded by a third and heavier layer, the **epimysium** (ep-ĭ-mīz'ĭ-um), which is composed of dense, fibrous, collagenous connective tissue and covers the entire surface of the muscle. The **fascia** (fash'e-ah), a layer of fibrous connective tissue outside the epimysium, separates individual muscles and in some cases surrounds muscle groups. The connective tissue components of muscles are continuous with each other. At the end of muscles the connective tissue components of muscle are continuous with the connective tissue of tendons or the connective tissue sheaths of bone (see Chapter 6). The connective tissue of muscle holds the muscle cells together, provides a passageway for blood vessels and nerves to reach the individual muscle cells, and attaches muscles to tendons or bones (Figure 10-3).

Muscle Fibers, Myofibrils, Sarcomeres, and Myofilaments

The many nuclei of each muscle fiber lie just inside the sarcolemma, whereas most of the fiber's interior is filled with myofibrils. The cytoplasm without the myofibrils is the **sarcoplasm** (sar'ko-plazm). Each **myofibril** is a threadlike structure approximately 1 to 3 μm in diameter that extends from one end of the muscle fiber to the other (Figure 10-4.) Myofibrils are composed of two kinds of protein filaments called **myofilaments**. **Actin myofilaments,** or thin myofilaments, are approximately 8 nm in diameter and 1000 nm in length, whereas **myosin myofilaments,** or thick myofilaments, are approximately 12 nm in diameter and 1800 nm in length. The actin and myosin myofilaments are organized in highly ordered units called **sarcomeres** (sar'ko-mērz), which are joined end to end to form the myofibrils (Figures 10-4 and 10-5).

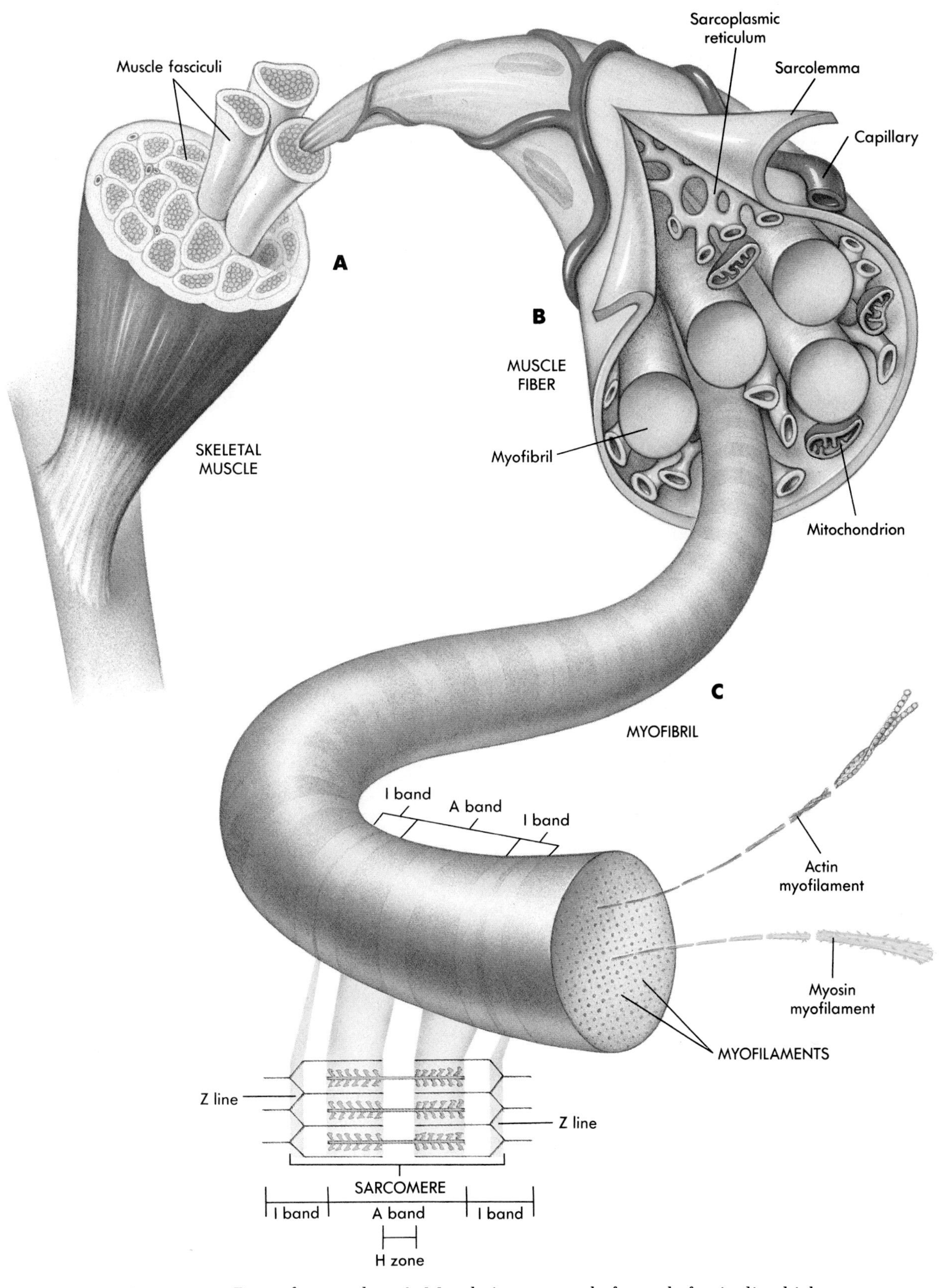

FIGURE 10-4 Parts of a muscle. **A** Muscle is composed of muscle fasciculi, which can be seen by the unaided eye in the muscle. The fasciculi are composed of bundles of individual muscle fibers (muscle cells). **B** Each muscle fiber contains myofibrils in which the banding patterns, called striations, of the sarcomeres are seen. **C** The myofibrils consist of units called sarcomeres. Each sarcomere is a highly organized structure consisting mainly of actin and myosin myofilaments. Each of the actin and myosin myofilaments are formed from thousands of actin and myosin molecules.

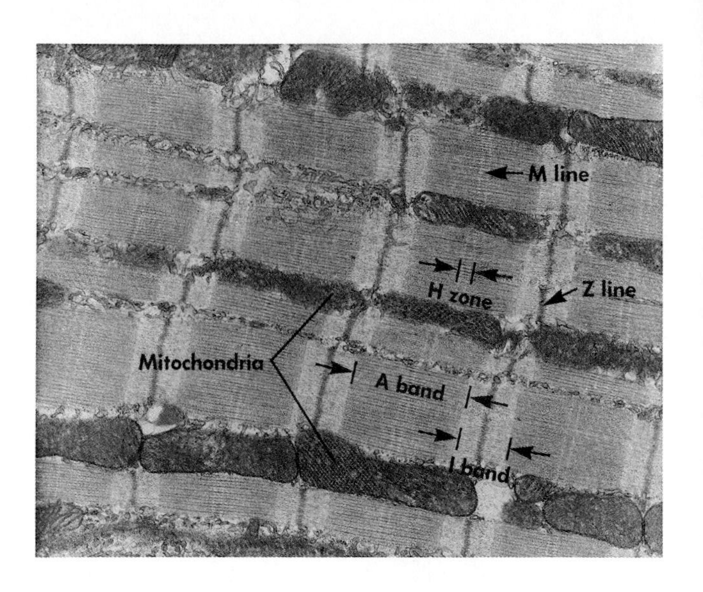

FIGURE 10-5 Electron micrograph of a section of a myofibril showing several sarcomeres.

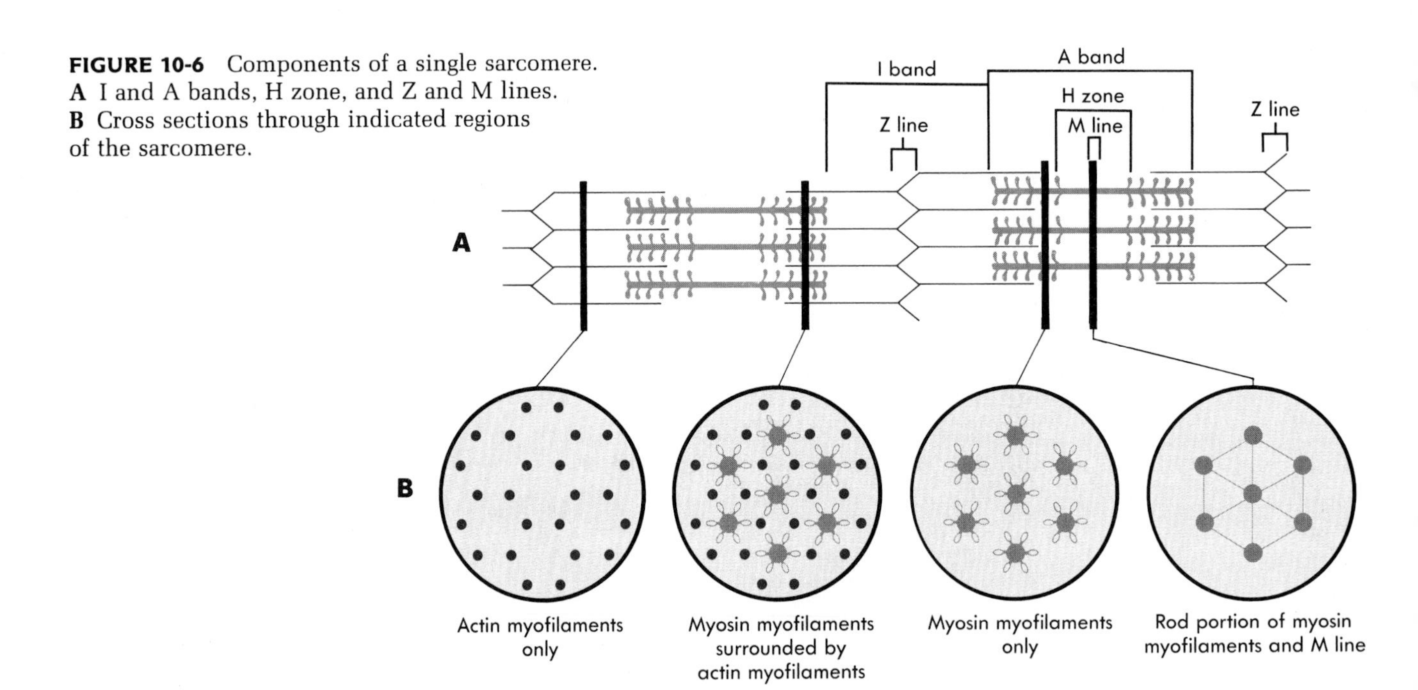

FIGURE 10-6 Components of a single sarcomere. **A** I and A bands, H zone, and Z and M lines. **B** Cross sections through indicated regions of the sarcomere.

Each sarcomere extends from one Z line to an adjacent Z line. A **Z line** is a filamentous network of protein forming a disklike structure for the attachment of actin myofilaments (Figure 10-6). The arrangement of the actin myofilaments and myosin myofilaments gives the myofibril a banded or striated appearance when viewed longitudinally. Each **isotropic** (light band), or **I, band** includes a Z line and extends from either side of the Z line to the ends of the myosin myofilaments. When seen in longitudinal and cross sections, the I band on either side of the Z line consists only of actin myofilaments. Each **anisotropic** (dark band), or **A, band** extends the length of the myosin myofilaments within a sarcomere. The actin and myosin myofilaments overlap for some distance at both ends of the A band. In a cross section of the A band in the area where actin and myosin myofilaments overlap, each myosin myofilament is visibly surrounded by six actin myofilaments. In the center of each A band is a smaller band called the **H zone** in which the actin and myosin myofilaments do not overlap and only myosin myofilaments are present. A dark band called the **M line** is in the middle of the H zone and consists of delicate filaments that attach to the center of the myosin myofilaments. The M line holds the myosin myofilaments in place similar to the way the Z line holds actin myofilaments in place.

Actin and myosin myofilaments

Each actin myofilament is composed of two strands of fibrous actin **(F-actin),** a series of **tropomyosin molecules,** and a series of **troponin molecules** (Figure 10-7, *A*). The two strands of F-actin are coiled to form a double helix that extends the length of the actin myofilament. Each F-actin strand is a polymer of approximately 200 small globular units called globular actin **(G-actin)** monomers. Each G-actin monomer has an active site to which myosin molecules can bind during muscle contraction. Tropomyosin is an elongated protein that winds along the groove of the F-actin double helix. Each tropomyosin molecule is sufficiently long to cover seven G-actin active sites. Troponin is composed of three subunits: one with a high affinity for actin, the second with a high affinity for tropomyosin, and the third with a high affinity for calcium ions. The troponin molecules are spaced between the ends of the tropomyosin molecules in the groove between the F-actin strands. The complex of tropomyosin and troponin regulates the interaction between active sites on G-actin and myosin.

Myosin myofilaments are composed of many elongated myosin molecules. Each myosin molecule is composed of two parts and is shaped like a golf club. The rodlike portions are wound together and lie parallel to the myosin myofilament with two heads extending laterally (Figure 10-7, *B*). Each myosin myofilament consists of approximately 100 myosin molecules arranged so that 50 myosin molecules have their heads projecting toward each end. The head is attached to the rodlike portion of the myosin molecule by a hingelike area that can bend and straighten during contraction. The heads of the myosin molecules contain ATPase, an enzyme that breaks down ATP, releasing energy, and a protein that binds the head of the myosin molecule to active sites on the actin molecules. The combination of the myosin heads with the active sites of actin molecules is called a **cross bridge.** The centers of the myosin myofilaments (the H zone) consist of only the rodlike portions of the myosin molecules and cannot form cross bridges (see Figure 10-6).

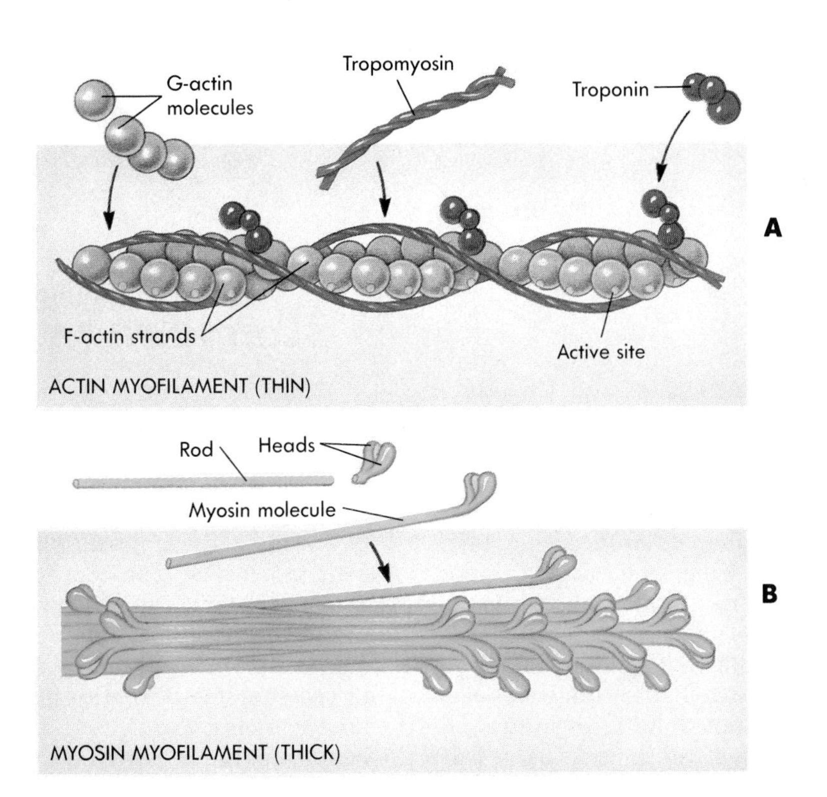

FIGURE 10-7 Fine structure of actin and myosin.
A Actin myofilaments are composed of individual globular-actin (G-actin) molecules *(purple spheres)*, filamentous tropomysin molecules *(blue strands)*, and globular troponin molecules *(red spheres)* assembled into a single filament. **B** Myosin myofilaments are composed of many individual golf-club–shaped myosin molecules, each of which has a rod portion and a globular head. The rod portions are in a parallel arrangement, with all the heads pointing in the same direction at one end and in the opposite direction at the other end of the myosin myofilament.

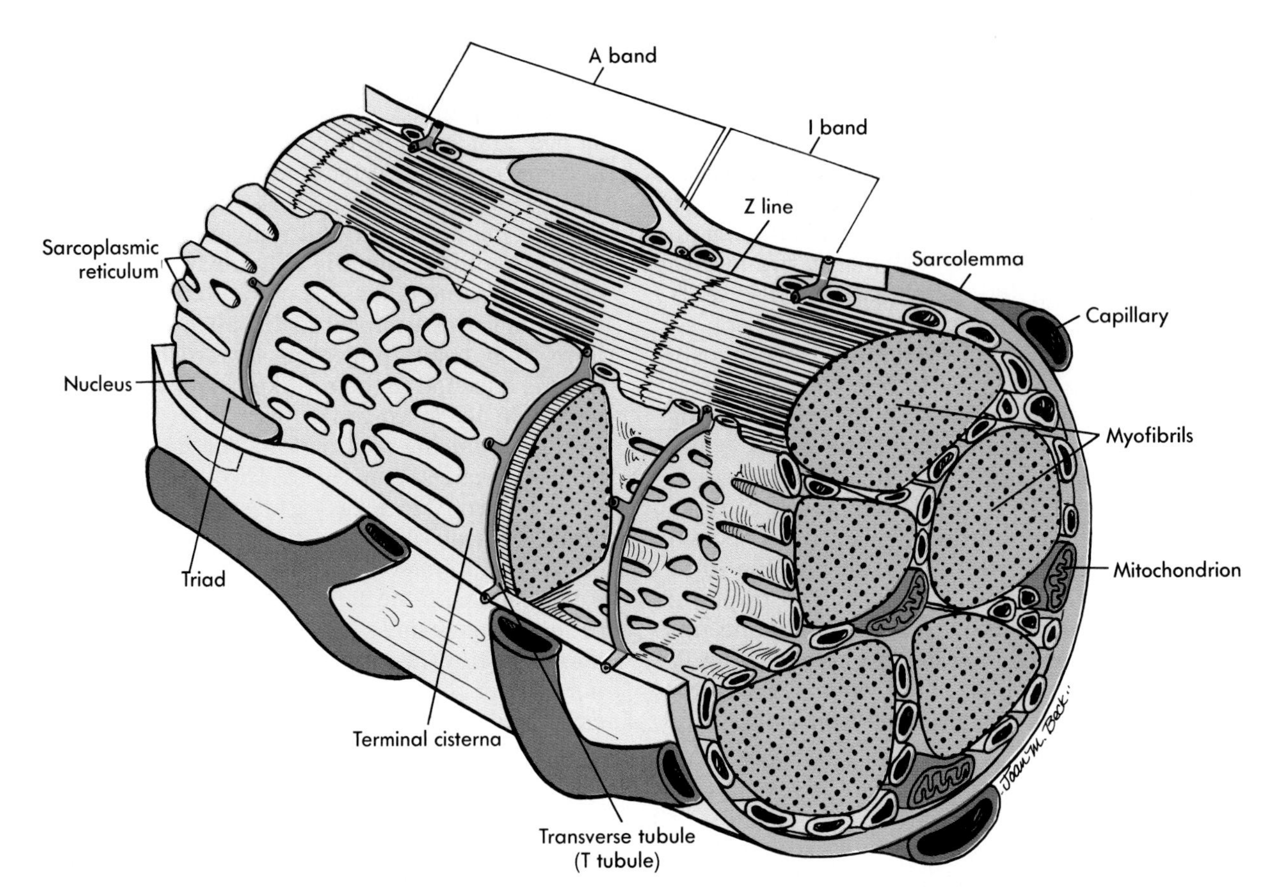

FIGURE 10-8 T tubules and sarcoplasmic reticulum of skeletal muscle.

Myofibrils and other organelles

The numerous myofibrils are oriented within each muscle fiber so that A bands and I bands of parallel myofibrils are aligned and thus produce the striated pattern seen through the microscope. Other organelles such as the numerous mitochondria and glycogen granules are packed between the myofibrils.

The sarcolemma has along its surface many tube-like invaginations called **transverse,** or **T, tubules,** which are regularly arranged and project into the muscle fiber and wrap around sarcomeres in the region in which actin myofilaments and myosin myofilaments overlap (Figure 10-8). The lumen of each T tubule is filled with extracellular fluid and is continuous with the exterior of the muscle fiber. Suspended in the sarcoplasm between the T tubules is a highly specialized, smooth endoplasmic reticulum called the **sarcoplasmic reticulum.** Near the T tubules the sarcoplasmic reticulum is enlarged to form **terminal cisternae.** A T tubule and the two adjacent terminal cisternae together are called a **triad.** The sarcoplasmic reticulum membrane actively transports calcium ions from the sarcoplasm into its lumen.

SLIDING FILAMENT THEORY

Actin and myosin myofilaments do not change length during contraction of skeletal muscle. Instead, the actin and myosin myofilaments slide past one another in a way that causes the sarcomeres to shorten. During contraction cross bridges form between the actin molecules and the heads of the myosin molecules. The cross bridges form, move, release, and then reform in a manner similar to the rowing motion of a boat. Movement of cross bridges forcefully causes the actin myofilaments at each end of the sarcomere to slide past the myosin myofilaments toward the H zone. As a consequence, the I bands and the H zones become more narrow, but the A bands remain constant in length (Figure 10-9). The H zone may disappear as the actin myofilaments overlap at the center of the sarcomere. As the actin myofilaments slide over the myosin myofilaments, the Z lines are brought closer together, and the sarcomere is shortened. The **sliding filament theory** of muscle contraction includes all of these events, which result in the movement of the myofilaments.

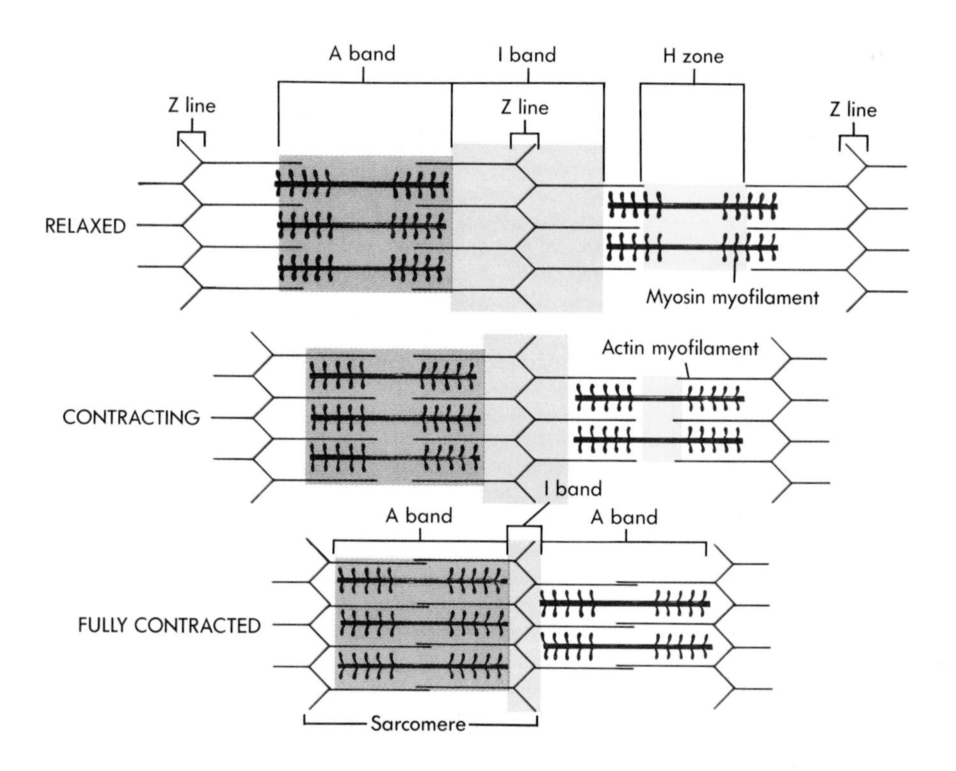

FIGURE 10-9 Sarcomere shortening in response to cross-bridge formation. Note that the I bands *(blue)* shorten, but the A bands *(pink)* do not. The H zone *(yellow)* narrows or even disappears as the actin myofilaments meet at the center of the sarcomere.

1

Although the following does not occur in muscle, which band would have to narrow if myosin fibers did shorten during muscle contraction?

? ? ? ? ? ? ? ? ? ? ?

Relaxation of muscle is more passive than contraction. The cross bridges are released, and the sarcomeres are allowed to lengthen passively. To do so, some force must be applied to the muscle to cause the sarcomeres to lengthen. That force is usually produced by an antagonistic muscle or by gravity.

PHYSIOLOGY OF SKELETAL MUSCLE FIBERS

Skeletal muscle contracts in response to electrochemical stimuli. Nerve cells regulate the function of skeletal muscle fibers by controlling the frequency of action potentials produced in the muscle cell membrane. The skeletal muscle action potentials then trigger a series of chemical events in the muscle fibers that results in the mechanical process of muscle contraction.

Neuromuscular Junction

Motor neurons are specialized nerve cells that propagate action potentials to skeletal muscle fibers at a relatively high velocity. Motor neuron axons enter the skeletal muscles along the same pathway as the arteries and veins. When the axons reach the level of the perimysium, they branch repeatedly, each branch projecting toward one muscle fiber and forming a **neuromuscular junction,** or **synapse** (sin'aps), near the center of the muscle fiber (see Figures 10-3 and 10-10). Thus each muscle fiber receives a branch of an axon, and each axon innervates more than one muscle fiber.

The neuromuscular junction is formed by an enlarged nerve terminal that rests in invaginations of the sarcolemma. An enlarged nerve terminal is the **presynaptic terminal;** the space between it and the muscle fiber is the **synaptic cleft;** and the muscle cell membrane in the area of the junction is the **motor endplate,** or the **postsynaptic terminal** (see Figure 10-10).

Each presynaptic terminal contains many small, spherical sacs approximately 45 μm in diameter, called **synaptic vesicles,** and numerous mitochondria. The vesicles contain **acetylcholine** (as'e-til-ko'lēn) (an organic molecule composed of acetic acid

FIGURE 10-10 Neuromuscular junction.

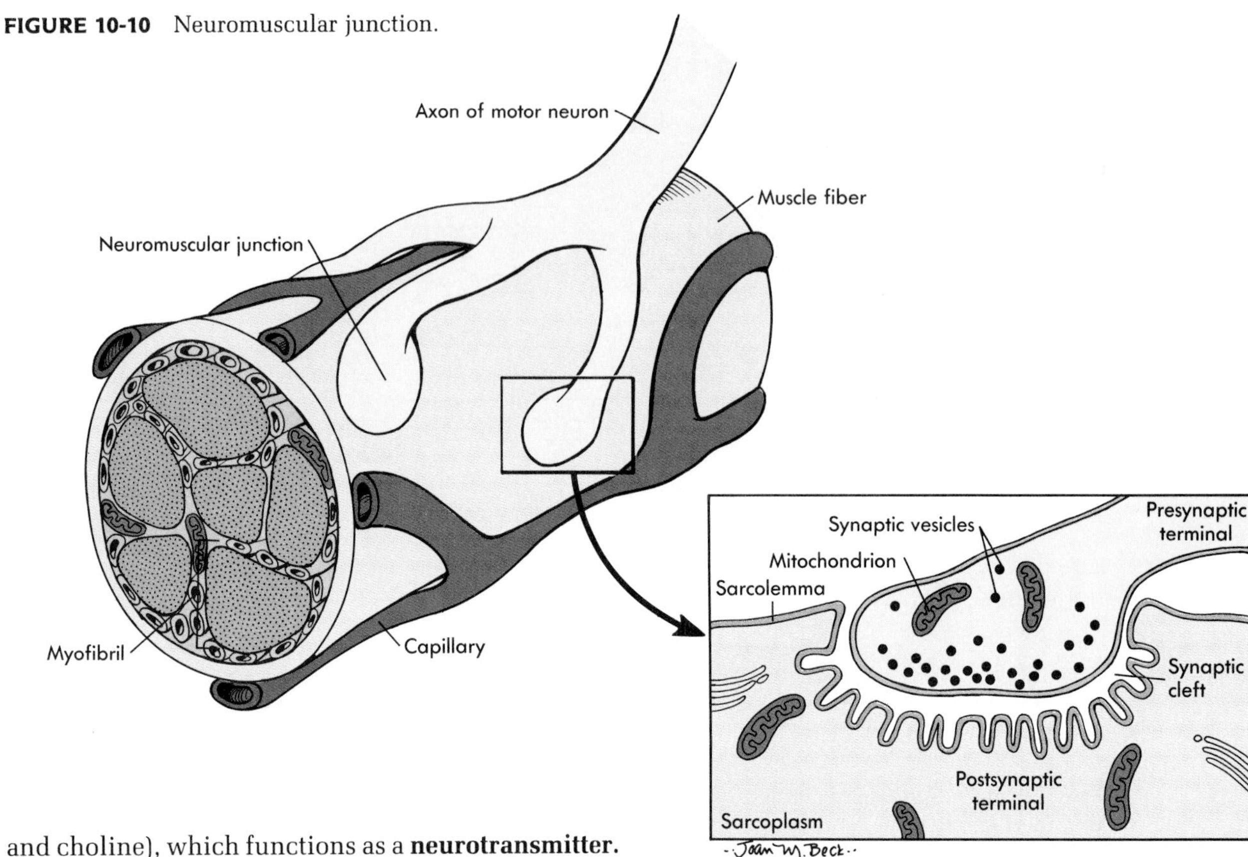

and choline), which functions as a **neurotransmitter.** A neurotransmitter is a substance released from a presynaptic terminal that diffuses across the synaptic cleft and stimulates (or inhibits) the production of an action potential in the postsynaptic terminal. Then, choline is combined with acetic acid to form acetylcholine.

When the action potential reaches the presynaptic terminal, it causes calcium (Ca^{2+}) channels in the axon's cell membrane to open, and as a result Ca^{2+} ions diffuse into the cell (Figure 10-11, *A*). Once inside the cell, the Ca^{2+} ions cause the contents of a few synaptic vesicles to be secreted by exocytosis from the presynaptic terminal into the synaptic cleft. The acetylcholine molecules then diffuse across the cleft and bind to receptor molecules located within the membrane of the postsynaptic terminal. This causes an increase in the permeability of the membrane that allows sodium (Na^+) ions to diffuse into the cell, producing a local potential. In skeletal muscle the local potential is great enough to exceed threshold, and an action potential in the muscle fiber is produced.

Acetylcholine released into the synaptic cleft is rapidly broken down to acetic acid and choline by **acetylcholinesterase** (as′e-til-ko-lin-es′ter-ās; Figure 10-11, *B*), an enzyme that prevents the accumulation of acetylcholine within the synaptic cleft where it would act as a constant stimulus at the postsynaptic terminal. The release of acetylcholine and its rapid degradation in the neuromuscular junction ensures that one presynaptic action potential yields only one postsynaptic action potential. Choline molecules are actively reabsorbed by the presynaptic terminal and then combined with the acetic acid produced within the cell to form acetylcholine. Recycling choline molecules requires less energy and is more rapid than completely synthesizing new acetylcholine molecules each time they are released from the presynaptic terminal. Acetic acid is an intermediate in the process of glucose metabolism (see Chapter 25) and can be taken up and used by a variety of cells after it diffuses away from the area of the neuromuscular junction.

2

Predict the consequence if presynaptic action potentials (in a nerve fiber) did not release sufficient acetylcholine to cause depolarization to threshold in a skeletal muscle fiber.

? ? ? ? ? ? ? ? ? ? ?

3

Predict the specific cause of death resulting from a lethal dose of (A) organophosphate poison and (B) curare.

? ? ? ? ? ? ? ? ? ? ?

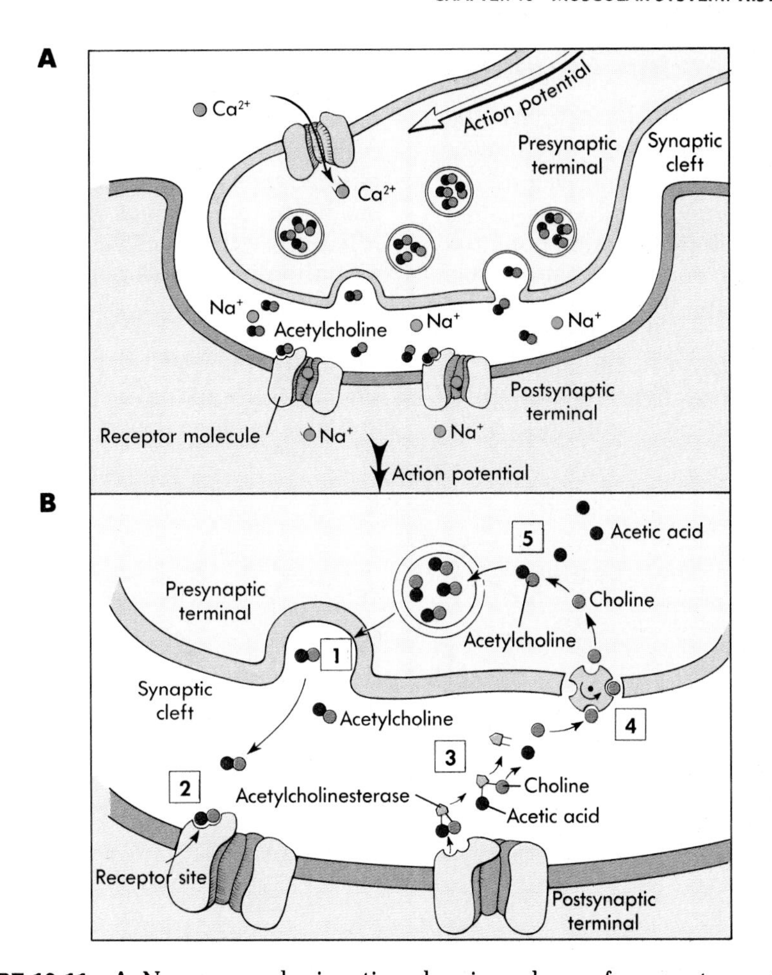

FIGURE 10-11 **A** Neuromuscular junction showing release of a neurotransmitter (acetylcholine, *ACh*) from the presynaptic terminal of a nerve fiber, diffusion across the synaptic cleft, and binding to acetylcholine receptors on the postsynaptic muscle fiber (postsynaptic terminal). These actions result in an increase in the permeability of the muscle fiber to sodium (Na^+) ions. **B** Acetylcholine is broken down in the synaptic cleft by acetylcholinesterase to acetic acid and choline. The choline is resorbed by the presynaptic terminal and is used to synthesize more acetylcholine.

Anything that affects the production, release, and degradation of acetylcholine or its ability to bind to its receptor molecule will also affect the transmission of action potentials across the neuromuscular junction. For example, some insecticides contain organophosphates that bind to and inhibit the function of acetylcholinesterase. As a result, acetylcholine is not degraded and accumulates in the synaptic cleft where it acts as a constant stimulus to the muscle fiber. Insects exposed to the insecticide die, partly because their muscles contract and cannot relax—a condition called spastic paralysis. Other organic poisons such as curare bind to the acetylcholine receptors, preventing acetylcholine from binding to them. Curare does not allow activation of the receptors; therefore the muscle is not capable of contracting in response to nervous stimulation—a condition called flaccid paralysis. **Myasthenia grăvis** (mi'as-the'ne-ah gră'vis) results from the production of antibodies that bind to acetylcholine receptors, eventually causing the destruction of the receptor and thus reducing the number of receptors. As a consequence, muscles exhibit a degree of flaccid paralysis or are extremely weak. A class of drugs that includes neostigmine partially blocks the action of acetylcholinesterase and sometimes is used to treat myasthenia gravis. The drugs cause acetylcholine levels to increase in the synaptic cleft and combine more effectively with the remaining acetylcholine receptor sites.

Excitation Contraction Coupling

Production of an action potential in a skeletal muscle fiber leads to contraction of the muscle fiber. The mechanism by which action potential production causes contraction of a muscle fiber is called **excitation contraction coupling.** The action potential is propagated along the sarcolemma of the muscle fiber and penetrates the T tubules. The T tubules carry the action potentials into the muscle fiber's interior. As action potentials reach the area of the triads, the membranes of the sarcoplasmic reticulum increase their permeability to Ca^{2+} ions. Since the sarcoplasmic reticulum actively transports Ca^{2+} ions into its lumen, the concentration of Ca^{2+} ions is approximately 2000 times higher within the sarcoplasmic reticulum than in the sarcoplasm of a resting muscle; thus when the sarcoplasmic reticulum's permeability to Ca^{2+} ions increases, the Ca^{2+} ions rapidly diffuse the short distance from the sarcoplasmic reticulum into the sarcoplasm surrounding the myofibrils (Figure 10-12, *A*).

The Ca^{2+} ions bind to troponin of the actin myofilaments (Figure 10-12, *B*). The combination of Ca^{2+} ions with troponin causes the troponin-tropomyosin complex to move deeper into the groove between the

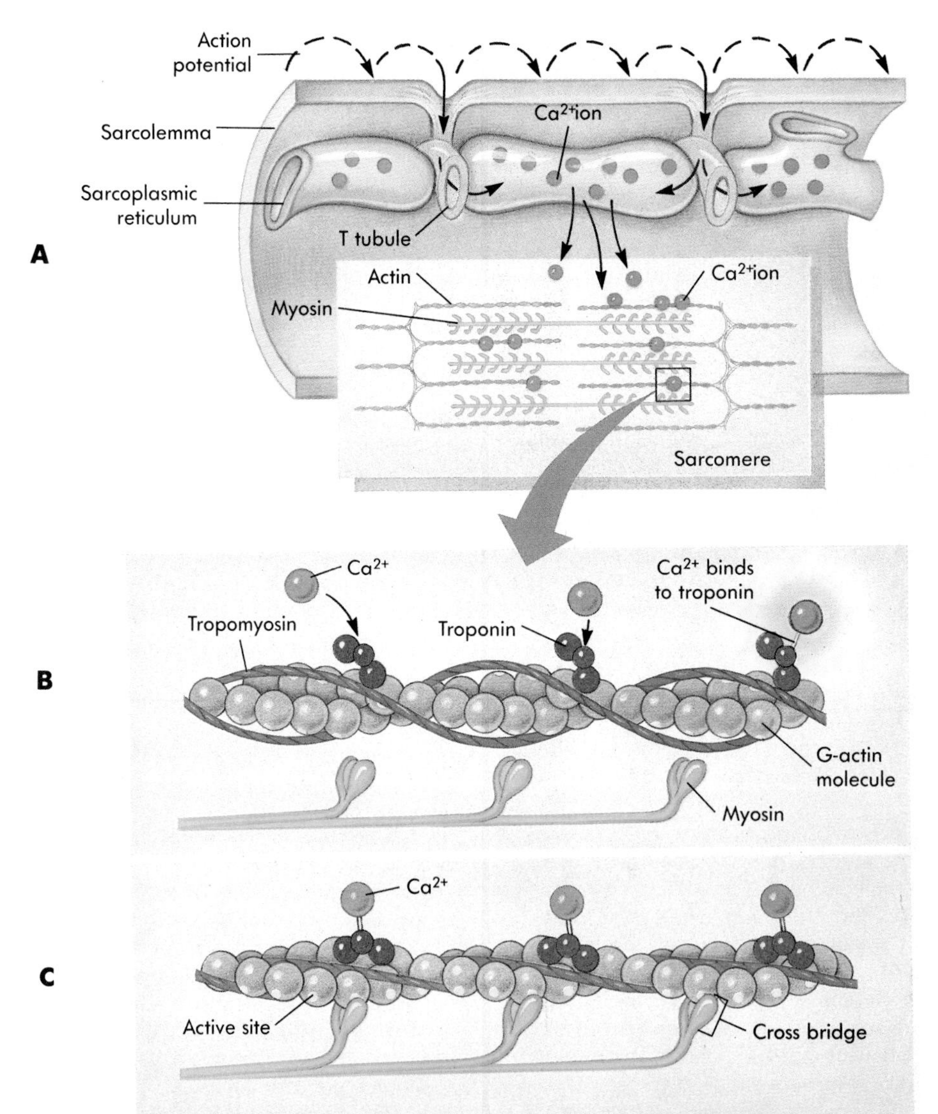

FIGURE 10-12 **A** Calcium (Ca^{2+}) ions are released from the sarcoplasmic reticulum in response to an action potential, which spreads from the sarcolemma to the T tubules and stimulates calcium release from the sarcoplasmic reticulum. The calcium release is accomplished by an increase in the permeability of the sarcoplasmic reticulum membrane. **B** Released Ca^{2+} ions bind to troponin, causing, **C,** the tropomyosin filament to move deeper into the groove along the actin myofilament and expose active sites on the actin molecules to which myosin can bind.

two F-actin molecules and thus expose active sites on the actin myofilaments. These exposed active sites bind to the heads of the myosin molecules to form cross bridges (Figure 10-12, *C*). When the heads of the myosin molecules bind to actin, a series of events resulting in contraction proceeds very rapidly. The heads move at their hinged area, forcing the actin myofilament, to which the heads of the myosin molecules are attached, to slide over the surface of the myosin myofilament. After movement, each myosin head releases from the actin and returns to its original position. It can then form another cross bridge.

Energy Requirements for Contraction

The energy from one ATP molecule is required for each cycle of cross-bridge formation, cross-bridge movement, and cross-bridge release. After a cross bridge has formed and movement has occurred, release of the myosin head from actin requires ATP to bind to the head of the myosin molecule. The ATP is broken down by ATPase in the head of the myosin myofilament and energy is stored in the head of the myosin molecule. The cross bridge is then released and the myosin head is restored to its original position (Figure 10-13, *A*). When the myosin molecule binds to actin to form another cross bridge, much of the stored energy is used for cross bridge formation and movement (Figure 10-13, *B* and *C*). Before the cross bridge can be released for another cycle, once again, an ATP molecule must bind to the head of the myosin molecule.

Movement of the myosin molecule while the cross bridge is attached is a power stroke, whereas return of the myosin head to its original position after cross-bridge release is a recovery stroke. Many cycles of power and recovery strokes occur during each muscle contraction. While muscle is relaxed, energy

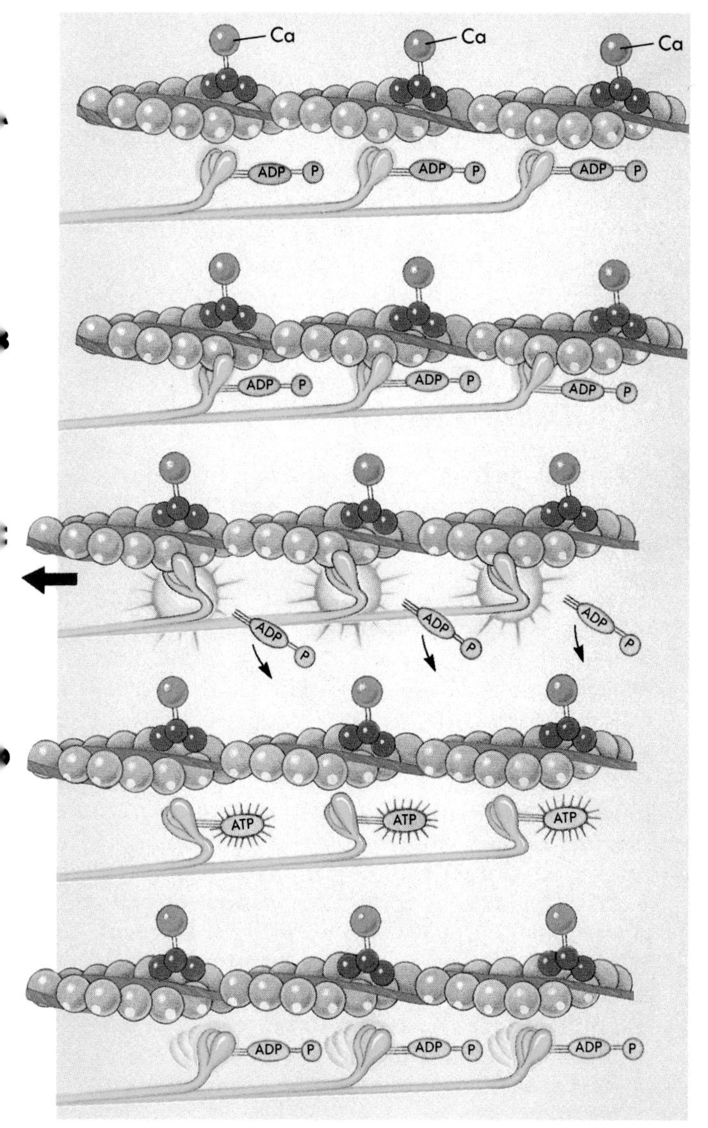

FIGURE 10-13 Breakdown of ATP and cross-bridge movement. At the end of a prior contraction, a molecule of ATP attaches to the globular head of a myosin molecule. The ATP is broken down to ADP and a phosphate molecule, and energy is stored in the myosin head. Then the myosin head is released from the active site on the actin myofilament and the head of the myosin molecule returns to its resting position. **A** During contraction calcium ions bind to troponin, causing exposure of cross bridges. **B** The myosin myofilament can then attach to the active sites on the actin myofilaments. **C** Energy stored in the head of the myosin myofilament is used to cause the actin myofilament to slide past the myosin myofilament. **D** Another ATP molecule then is able to bind to the myosin head, and the process of cross bridge release occurs. **E** The ATP is broken down to ADP, the head of the myosin molecule returns to its resting position, and energy is stored in the head of the myosin molecule. If Ca^{2+} ions are still attached to troponin, cross bridge formation and movement can occur many times during a muscle contraction.

Table 10-2

Summary of molecular events of skeletal muscle contraction

1. Before contraction ATP is broken down, and much of the energy is stored in the head of the myosin molecules.
2. Calcium (Ca^{2+}) ions are released from the sarcoplasmic reticulum in response to an action potential.
3. Ca^{2+} ions bind to troponin.
4. The troponin-tropomyosin complex moves to expose active sites on actin myofilament.
5. The myosin heads combine with active sites to form cross bridges, and the hinged areas of the myosin molecules move, causing the actin to slide past the myosin. The energy stored in the head of the myosin molecule is used for cross-bridge movement, and some is released in the form of heat.
6. ATP binds to the head of the myosin molecule and is broken down to ADP. A small amount of energy is used in the process of releasing actin from myosin, causing the hinged area of the myosin molecule to return to its original position. The remainder of the energy is stored in the head of the myosin molecule.
7. As long as actin-active sites are available, the process continues (go back to step 5), resulting in further contraction. If no additional action potentials are produced in the skeletal muscle fibers, Ca^{2+} ions are taken up by the sarcoplasmic reticulum, causing the Ca^{2+} ions to unbind from troponin; the troponin-tropomyosin complex covers the actin-active sites; and relaxation occurs.

stored in the heads of the myosin molecules is held in reserve until the next contraction. When calcium is released from the sarcoplasmic reticulum in response to an action potential, the cycle of cross-bridge formation and release, which results in contraction, begins (Table 10-2).

Muscle Relaxation

Relaxation occurs as a result of the active transport of Ca^{2+} ions back into the sarcoplasmic reticulum. As the Ca^{2+} ion concentration decreases in the sarcoplasm, Ca^{2+} ions diffuse away from the troponin molecules. The troponin-tropomyosin complex then reestablishes its position, which blocks the active sites on the actin molecules. As a consequence, cross bridges cannot reform once they have been released, and relaxation occurs.

4

Predict the consequences of having the following conditions develop in a muscle in response to a stimulus: (A) Na^+ ions cannot enter the skeletal muscle fiber; (B) inadequate ATP is present in the muscle fiber before a stimulus is applied; and (C) adequate ATP is present within the muscle fiber, but action potentials occur at a frequency so great that calcium is not transported back into the sarcoplasmic reticulum between individual action potentials.

? ? ? ? ? ? ? ? ? ? ?

In addition to the energy needed for muscle contraction, energy is needed for relaxation. The active transport of Ca^{2+} ions into the sarcoplasmic reticulum requires ATP. The active transport processes that maintain the normal concentrations of Na^+ and K^+ ions across the sarcolemma also require ATP.

PHYSIOLOGY OF SKELETAL MUSCLE

Muscle Twitch

A **muscle twitch** is contraction of a whole muscle in response to a stimulus that causes an action potential in one or more muscle fibers. Even though the normal function of muscles is more complex, an understanding of the muscle twitch makes the function of muscles in living organisms easier to comprehend.

A hypothetical contraction of a single muscle fiber in response to a single action potential is illustrated in Figure 10-14. The time period between application of the stimulus to the motor neuron and the beginning of contraction is the **lag,** or **latent, phase;** the time during which contraction occurs is the **contraction phase;** and the time during which relaxation occurs is the **relaxation phase** (Table 10-3). The action potential is an electrochemical event, but contraction is a mechanical event. The action potential is measured in millivolts (mV) and is completed in less than 2 msec, but muscle contraction is measured as a force (e.g., tension—the number of grams lifted) or the distance the muscle shortens, and requires up to 1 second to occur.

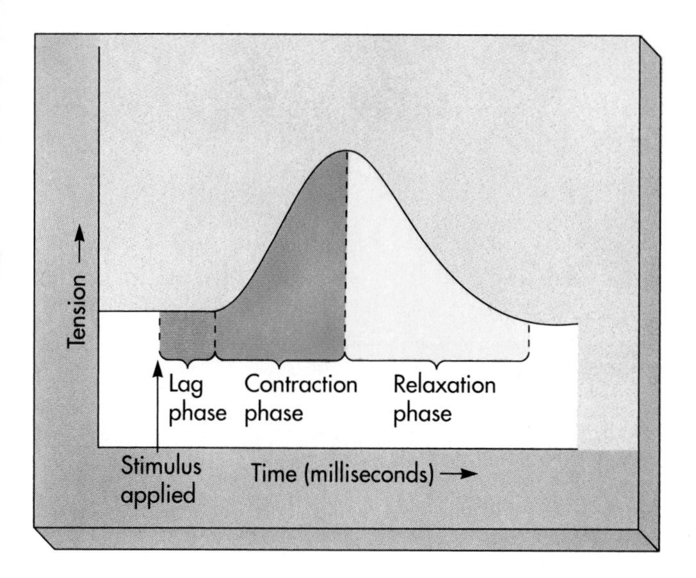

FIGURE 10-14 Muscle twitch in a single muscle fiber. There is a short lag phase after stimulus application, then a contraction phase and a relaxation phase.

Table 10-3

*Events that occur during each phase of a muscle twitch**

Lag Phase

An action potential is propagated to the presynaptic terminal of the motor neuron.
The action potential causes the permeability of the presynaptic terminal to increase.
Calcium ions diffuse into the presynaptic terminal, causing acetylcholine contained within several synaptic vesicles to be released by exocytosis into the synaptic cleft.
Acetylcholine released from the presynaptic terminal diffuses across the synaptic cleft and binds to acetylcholine-receptor molecules in the postsynaptic terminal of the sarcolemma.
The combination of acetylcholine and its receptor site causes the membrane of the postsynaptic terminal to become more permeable to sodium ions.
Sodium ions diffuse into the muscle fiber, causing a local depolarization that exceeds threshold and produces an action potential.
Acetylcholine is rapidly degraded in the synaptic cleft to acetic acid and choline, thus limiting the length of time acetylcholine is bound to its receptor site. The result is that one presynaptic action potential produces one postsynaptic action potential in the muscle fibers.
The action potential produced in the muscle fiber is propagated from the postsynaptic terminal near the middle of the fiber toward both ends and into the T tubules.
The electrical changes that occur in the T tubule in response to the action potential make the membrane of the sarcoplasmic reticulum very permeable to calcium ions.
Calcium ions diffuse from the sarcoplasmic reticulum into the sarcoplasm.
Calcium ions bind to troponin; the troponin-tropomyosin complex changes its position and exposes the active sites on the actin myofilaments.

Contraction Phase

Cross bridges between actin molecules and myosin molecules form, move, release, and reform many times, causing the sarcomeres to shorten. Energy stored in the head of the myosin molecule allows cross-bridge formation and movement. After cross-bridge movement has occurred, an ATP must bind to the myosin head. The ATP is broken down to ADP, and some of the energy is used to release the cross bridge and cause the head of the myosin molecule to move back to its resting position where it is ready to form another cross bridge. Most of the energy from the ATP is stored in the myosin head and is used for the next cross-bridge formation and movement. Some energy is released as heat.

Relaxation Phase

Calcium ions are actively transported into the sarcoplasmic reticulum.
The troponin-tropomyosin complexes inhibit cross-bridge formation.
The muscle fibers lengthen passively.

*Assuming that the process begins with a single action potential in the motor neuron.

Stimulus Strength and Muscle Contraction

In response to the appropriate stimulus, an isolated skeletal muscle fiber either contracts maximally or does not contract at all. This is the **all-or-none law of skeletal muscle contraction** and can be explained on the basis of action potential production in the skeletal muscle fiber. When brief electrical stimuli of increasing strength are applied to the muscle cell membrane, the following results occur: (1) a subthreshold stimulus does not produce an action potential, and no muscle contraction occurs; (2) a threshold stimulus produces an action potential and results in contraction of the muscle cell; or (3) a stronger-than-threshold stimulus produces an action potential of the same magnitude as the threshold stimulus and therefore produces an identical contraction. Thus for a given condition, once an action potential is generated, the skeletal muscle fiber contracts maximally. If internal conditions change, it is possible for the force of contraction to change as well. For example, increasing the amount of calcium available to the muscle cell results in a stronger force of contraction; conversely, muscle fatigue can result in a weaker force of contraction.

Within a skeletal muscle the individual skeletal muscle fibers are arranged into **motor units,** each of which consists of a single motor neuron and all of the muscle fibers it innervates (Figure 10-15). Like individual muscle fibers, motor units respond in an all-or-none fashion. All the muscle fibers of a motor unit contract maximally in response to a threshold stimulus because an action potential in a motor neuron initiates an action potential in all of the muscle fibers it innervates.

Whole muscles exhibit characteristics that are more complex than those of individual muscle fibers or motor units. A muscle is composed of many motor units, and the axons of the motor units combine to form a nerve. If brief electrical stimuli of increasing strength are applied to the nerve, the muscle responds in a graded fashion instead of an all-or-none fashion (Figure 10-16). A **subthreshold stimulus** is not strong enough to cause an action potential in any of the motor neuron axons and causes no contraction. As the stimulus increases in strength, however, it eventually becomes a threshold stimulus (i.e., a stimulus strong enough to produce an action potential in the axon of a single motor neuron), and all of the muscle fibers of that motor unit contract. Progressively stronger stimuli, **submaximal stimuli,** activate additional motor units until all of the motor units are activated by a **maximal stimulus,** at which point a greater stimulus strength, a **supramaximal stimulus,** has no additional effect. As the stimulus strength increases between threshold and maximum values, motor units are recruited, and the force of contraction produced by the muscle increases in a graded fashion. This relationship is called **multiple motor unit summation.** A whole muscle contracts with either a small force or a large force, depending on the number of motor units recruited, but each motor unit responds to an action potential either maximally or not at all.

Motor units in different muscles do not always contain the same number of muscle fibers. Muscles performing delicate and precise movements have

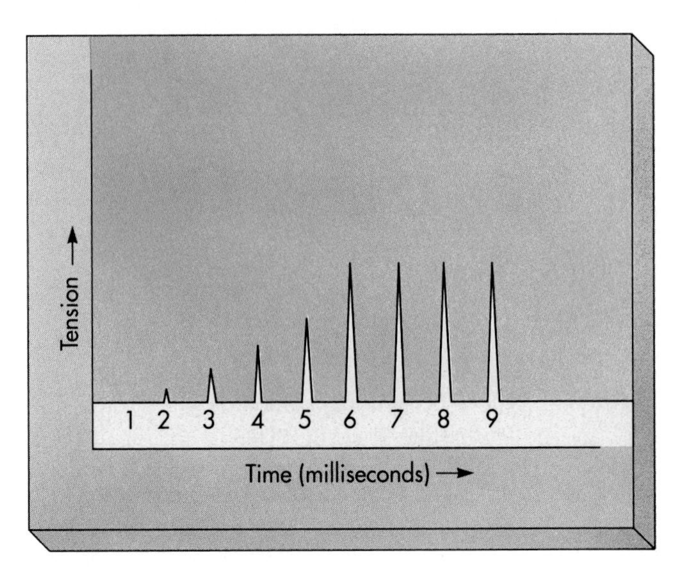

FIGURE 10-16 Muscle twitch—multiple motor unit summation caused by stimuli of increasing strength (*1 to 9*). The amount of tension (height of peaks) is influenced by the number of motor units responding: *1*, subthreshold stimulus (no motor units responding); *2*, threshold stimulus; *3 to 5*, submaximal stimulus; *6*, maximal stimulus (all motor units responding); and *7 to 9*, supramaximal stimuli.

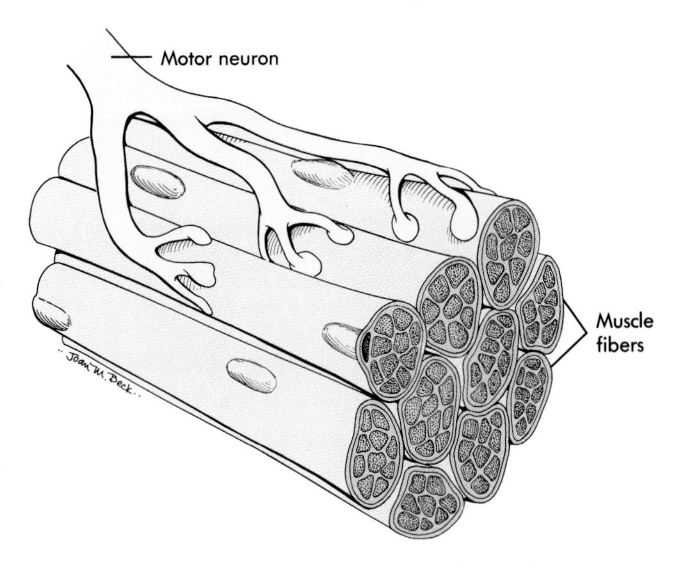

FIGURE 10-15 Motor unit consisting of a single neuron and all the muscle fibers it innervates.

motor units with a small number of muscle fibers, whereas muscles performing more powerful but less precise contractions have motor units with many motor fibers. For example, in very delicate muscles such as those that move the eye, the number of muscle fibers per motor unit ranges between one and ten, whereas in the heavy muscles of the leg the number may be several hundred.

5

In patients with poliomyelitis motor neurons are destroyed, causing loss of muscle function and even flaccid paralysis. Sometimes recovery occurs because of the formation of axon branches from the remaining motor neurons. These branches innervate the paralyzed muscle fibers to produce motor units with many more muscle fibers than usual, resulting in recovery of muscle function. What effect would this reinnervation of muscle fibers have on the degree of muscle control in a person who has recovered from poliomyelitis?

? ? ? ? ? ? ? ? ? ? ?

Stimulus Frequency and Muscle Contraction

A single muscle fiber contracts in response to an action potential, but unlike action potentials, the contractile mechanism exhibits no **refractory period** (see Chapter 9). Therefore relaxation is not required before a second action potential is able to stimulate a second contraction in a muscle. As the frequency of action potentials in a skeletal muscle fiber increases, the frequency of contraction increases until the muscle fiber remains contracted and does not relax (Figure 10-17). In **incomplete tetanus** muscle fibers partially relax between contractions, but in **complete tetanus** action potentials occur so rapidly there is no muscle relaxation between the action potentials.

The tension produced by a muscle increases as the stimulus frequency increases. The increased tension is called **multiple wave summation,** which is apparent when a muscle is exhibiting incomplete or complete tetanus.

Tetanus of a muscle caused by stimuli of increasing frequency can be explained by the effect of the action potentials on Ca^{2+} ion release from the sarcoplasmic reticulum. The first action potential causes Ca^{2+} ion release from the sarcoplasmic reticulum, the Ca^{2+} ions diffuse to the myofibrils, and contraction occurs. Relaxation begins as the Ca^{2+} ions are pumped back into the sarcoplasmic reticulum. However, if the next action potential occurs before relaxation is complete, two things happen. First, since there has not been enough time for the Ca^{2+} ions to reenter the sarcoplasmic reticulum, Ca^{2+} ion levels around the myofibrils remain elevated. Second, the next action potential causes the release of additional Ca^{2+} ions from the sarcoplasmic reticulum. Thus Ca^{2+} ion levels remain elevated in the sarcoplasm, producing continued contraction of the muscle fiber. Action potentials at a high frequency can increase Ca^{2+} ion concentrations in the sarcoplasm to an extent that the muscle fiber is contracted completely and does not relax at all.

At least two factors play a role in the increased tension observed during multiple wave summation. First, as the action potential frequency increases, the concentration of Ca^{2+} ions around the myofibrils becomes greater than during a single muscle twitch, causing a greater degree of contraction. The additional Ca^{2+} ions cause the exposure of a greater-than-normal number of active sites on the actin myofilaments. Second, the sarcoplasm and the connective tissue components of muscle have some elasticity. During each separate muscle twitch some tension produced by the contracting muscle fibers is used to stretch those elastic elements, and the remaining tension is applied to the load to be lifted. In a single muscle twitch relaxation begins before the elastic components are totally stretched; therefore the maximum tension produced during a single muscle twitch is not applied to the load to be lifted. When a muscle is stimulated at a high frequency, the elastic elements are stretched during the early part of the prolonged contraction, which subsequently allows

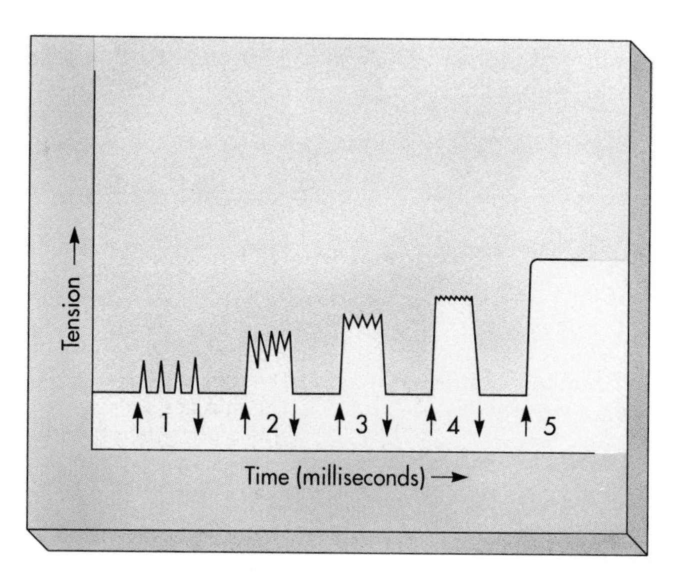

FIGURE 10-17 Muscle twitch—multiple wave summation caused by stimuli of increased frequency (*1 to 5*): complete relaxation between stimuli (*1*), incomplete tetanus—partial relaxation between stimuli. (*2 to 4*), and complete tetanus—no relaxation between stimuli (*5*).

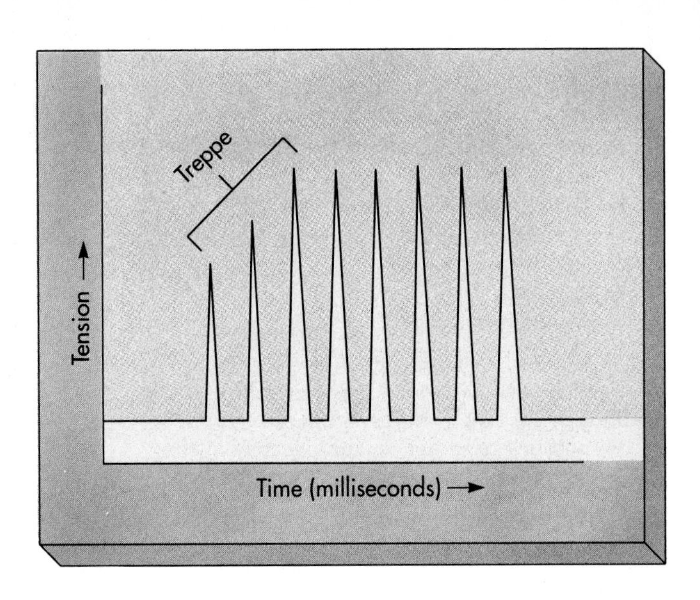

FIGURE 10-18 When a rested muscle is stimulated with maximal stimuli at a frequency that allows complete relaxation between stimuli, the second contraction produces a slightly greater tension than the first, and the third contraction produces greater tension than the second. After a few contractions, the tension produced by all contractions is equal.

Table 10-4

Types of muscle contractions

CONTRACTION TYPES	CHARACTERISTICS
Multiple motor unit summation	Each motor unit responds in an all-or-none fashion. A whole muscle is capable of producing an increasing amount of tension as the number of motor units that are stimulated increases.
Multiple wave summation	Summation results when many action potentials are produced in a muscle fiber. Contraction occurs in response to the first action potential, but there is not enough time for relaxation to occur between action potentials. Because each action potential causes the release of calcium ions from the sarcoplasmic reticulum, calcium ions remain elevated in the sarcoplasm to produce a tetanic contraction. Multiple wave summation can result in tetanic contractions. The tension produced as a result of multiple wave summation is greater than the tension produced by a single muscle twitch. The increased tension is due to the greater concentration of calcium in the sarcoplasm and to stretch of the elastic components of the muscle early in contraction.
Tetanic contractions	Tetanic contractions are due to multiple wave summation. Incomplete tetanus occurs when the action potential frequency is low enough to allow partial relaxation of the muscle fibers. Complete tetanus occurs when the action potential frequency is high enough so there is no relaxation of the muscle fibers.
Isotonic contraction	The muscle produces a constant tension during contraction. The muscles shorten during contraction. This type is characteristic of finger and hand movements.
Isometric contraction	The muscle produces an increasing tension during contraction. The length of the muscle remains constant during contraction. This type is characteristic of postural muscles that maintain a constant tension without changing their length.
Treppe	The tension produced increases for the first few contractions in response to a constant stimulus in a muscle that has been at rest for some time. The increased tension may be due to the accumulation of calcium in the sarcoplasm for the first few contractions or to the increased muscle temperature's increasing enzyme efficiency in a cold muscle.

all of the tension produced by the muscle to be applied to the load to be lifted; and the observed tension produced by the muscle increases.

Another example of a graded response is **treppe** (trep'eh; staircase), which occurs in muscle that has rested for a prolonged period of time (Figure 10-18). If the muscle is stimulated with a maximal stimulus at a frequency that allows complete relaxation between stimuli, the second contraction produces a slightly greater tension than the first, and the third produces greater tension than the second. After only a few stimuli, the tension produced by all the contractions is equal.

A possible explanation for treppe is an increase in Ca^{2+} ion levels around the myofibrils. The Ca^{2+} ions released in response to the first stimulus are not taken up completely by the sarcoplasmic reticulum before the second stimulus causes the release of additional Ca^{2+} ions. As a consequence, during the first few contractions of the muscle the Ca^{2+} ion concentration in the sarcoplasm increases, making contraction more efficient because of the increased number of Ca^{2+} ions available to bind to troponin. Treppe achieved during warm-up exercises may contribute to improved muscle efficiency during athletic events. Factors such as increased blood flow to the muscle and increased muscle temperature probably are involved as well. Increased muscle temperature causes the enzymes responsible for muscle contraction to function at a more rapid rate.

Types of Muscle Contractions

Muscle contractions are classified as either isometric or isotonic, depending on the type of contraction that predominates (Table 10-4). In **isometric contractions** the length of the muscle does not change, but the amount of tension does increase during the contraction process. Isometric contractions are responsible for the constant length of the postural muscles of the body. In **isotonic contractions** the amount of tension produced by the muscle is constant during contraction, but the length of the muscle changes. Movements of the arms or fingers are predominantly isotonic contractions. Most muscle contractions are a combination of isometric and isotonic contractions in which the muscles shorten some distance and the degree of tension increases. Although there are some mechanical differences, both types of contractions result from the same contractile process within muscle cells.

Movements of the body are usually smooth and occur at widely differing rates—some very slow and others quite rapid. All movements are produced by muscle contractions, but very few of the movements resemble the rapid contractions of individual muscle twitches. Smooth, slow contractions result from an increasing number of motor units contracting out of phase as the muscles shorten and from a decreasing number of motor units contracting out of phase as muscles lengthen. Each individual motor unit exhibits either incomplete or complete tetanus, but because the contractions are out of phase and because the number of motor units activated varies at each point in time, a smooth contraction results. Consequently, muscles are capable of contracting either slowly or rapidly, depending on the number of motor units stimulated and the rate at which that number increases or decreases.

Muscle tone refers to the constant tension produced by muscles of the body for long periods of time. Muscle tone is responsible for keeping the back and legs straight, the head held in an upright position, and the abdomen from bulging. Muscle tone depends on a small percentage of all the motor units contracting out of phase with each other at any point in time. However, the same motor units are not contracting all of the time. A small percentage of all motor units are stimulated with a frequency of nerve impulses that causes incomplete tetanus for short time periods. Those motor units that are contracting are stimulated in such as way that the tension produced by the whole muscle remains constant.

6

Is the contraction of muscle fibers that produces muscle tone more like isotonic or isometric contractions? Explain.

? ? ? ? ? ? ? ? ? ? ?

Length vs. Tension

When a muscle contracts, it applies **active tension** to an object to be lifted. The initial length of a muscle has a strong influence on the amount of active tension it produces. As the length of a muscle increases, its active tension also increases to a point. However, if the muscle is stretched farther than that optimum length, the active tension it produces begins to decline. The muscle length plotted against the tension produced by the muscle in response to maximal stimuli is the **active tension curve** (Figure 10-19). The length of the muscle affects the tension produced because of the effect of length on actin and myosin myofilament overlap. During normal body movements muscles are usually stretched near their optimum length before heavy objects are lifted.

Weight lifters and others who lift heavy objects frequently adjust their limbs so that the muscles are stretched close to the optimum length before lifting. Thus the maximum amount of active tension is produced by their muscles.

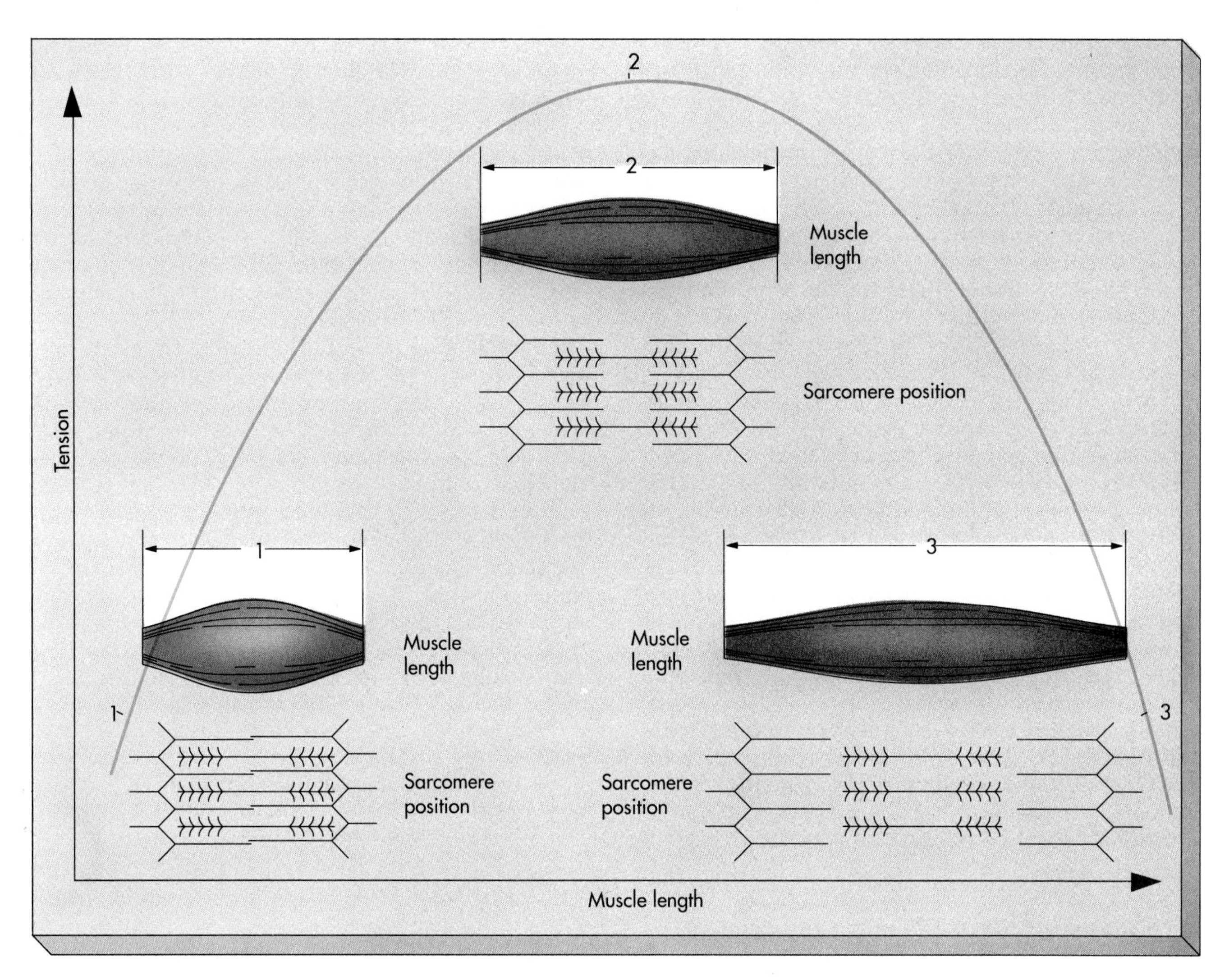

FIGURE 10-19 The tension produced by contractions of a muscle in response to a maximal stimulus is influenced by the length of the muscle (the degree to which it is stretched). At *1* the muscle is not stretched and the tension produced when the muscle contracts is small because the actin and myosin myofilaments are already overlapping. The sarcomere cannot shorten much more. At *2* the muscle is optimally stretched and the tension produced when the muscle contracts is maximal. At *3* the muscle is stretched severely and the tension produced is small because the actin and myosin myofilaments do not overlap. The number of cross bridges that can form is small.

When a muscle is stretched but not stimulated, the tension applied to the load is **passive tension** and is similar to the tension produced if the muscle were replaced with an elastic band. Passive tension exists because the muscle and its connective tissue have some elasticity. The sum of active and passive tension is **total tension.**

Fatigue

Fatigue is the decreased capacity to do work and the reduced efficiency of performance that normally follows a period of activity. The rate at which individuals develop fatigue is highly variable, but it is a phenomenon that everyone has experienced. Fatigue can develop at three possible sites: the nervous system, the muscles, and the neuromuscular junction.

Psychological fatigue, the most common type of fatigue, involves the central nervous system. The muscles are capable of functioning, but the individual "perceives" that additional muscular work is not possible. A burst of activity in a tired athlete in response to encouragement from spectators is an illustration of how psychological fatigue can be overcome. The onset and duration of psychological fatigue are highly variable and depend on the emotional state of the individual.

The second most common type of fatigue occurs in the muscle fiber. **Muscular fatigue** results from ATP depletion. Without adequate ATP levels in muscle fibers, cross bridges cannot function normally. As a consequence, the tension that a muscle is capable of producing declines.

The least common type of fatigue—**synaptic fatigue**—occurs in the neuromuscular junction. If the action potential frequency is great enough, the release of acetylcholine from the presynaptic terminals is greater than the rate of acetylcholine synthesis. As a result, the synaptic vesicles become depleted, and insufficient acetylcholine is released to stimulate the muscle fibers. Under normal physiological conditions, fatigue of neuromuscular junctions is rare; however, it may occur under conditions of extreme exertion.

Physiological Contracture and Rigor Mortis

Under conditions of extreme muscular fatigue, muscles occasionally become incapable of either contracting or relaxing—a condition called **physiological contracture,** which is caused by a lack of ATP within the muscle fibers. When ATP becomes limited while a muscle fiber is being stimulated, active transport of Ca^{2+} ions into the sarcoplasmic reticulum slows, Ca^{2+} ions accumulate within the sarcoplasm, and ATP is unavailable to bind to the myosin molecules that have formed cross bridges with the actin myofilaments. As a consequence, the previously formed cross bridges cannot release, resulting in physiological contracture. When muscle is stimulated repeatedly at a rapid rate, ATP is broken down rapidly. Under conditions of extreme exercise, the ATP levels may become so low that physiological contracture results.

Rigor mortis, the development of rigid muscles several hours after death, is similar to physiological contracture. ATP production stops shortly after death, and ATP levels within muscle fibers decline. As a result, Ca^{2+} ions leak from the sarcoplasmic reticulum into the sarcoplasm, and cross bridges form; however, too little ATP is available either to sustain the contractile process or to allow the cross bridges to release. As a consequence, the muscles remain stiff until tissue degeneration occurs.

Energy Sources

ATP, the immediate source of energy for muscular contraction, is produced by anaerobic or aerobic respiration. Anaerobic respiration and aerobic respiration are discussed in detail in Chapter 25; only the main points of these processes as they apply to muscle physiology are considered here.

Anaerobic respiration occurs in the absence of oxygen and results in the breakdown of glucose to yield ATP and lactic acid. **Aerobic respiration** requires oxygen and breaks down glucose to produce ATP, carbon dioxide, and water. Compared to anaerobic respiration, aerobic respiration is much more efficient—the metabolism of a glucose molecule by anaerobic respiration produces a net gain of 2 ATP, compared to approximately 38 ATP by aerobic respiration. In addition, aerobic respiration uses a greater variety of molecules as energy sources (e.g., fatty acids). Although anaerobic respiration is less efficient than aerobic respiration, it is faster, especially when oxygen availability limits aerobic respiration. By using many glucose molecules, anaerobic respiration rapidly produces ATP for a short time period.

During resting conditions only a small amount of ATP is present in muscle cells. Additional sources of energy are stored: glucose is converted to glycogen, which is stored in skeletal muscle cells; and ATP transfers a high-energy phosphate to creatine to form creatine phosphate. During exercise glycogen is broken down into glucose that can be used in aerobic or anaerobic respiration to produce ATP, and creatine phosphate combines with ADP to form ATP.

Resting muscles or muscles undergoing long-term exercise (e.g., long-distance running) depend primarily on aerobic respiration for ATP synthesis. Although some glucose is used as an energy source, fatty acids provide a more important source of energy during sustained exercise and during resting conditions. On the other hand, during short periods of intense exercise (e.g., sprinting) anaerobic respiration combined with the breakdown of creatine phosphate provides enough ATP to support intense muscle contraction for a short period of time (approximately 15 to 20 seconds). These processes are limited by depletion of creatine phosphate and glucose and the buildup of lactic acid within muscle fibers. However, lactic acid can diffuse out of the muscle cell into the blood, allowing anaerobic respiration to proceed longer.

Oxygen Debt

After intense exercise the rate of aerobic metabolism remains elevated for a period of time. The oxygen taken in by the body above that required for resting metabolism after exercise is called the **oxygen debt.** It represents the difference between the amount of oxygen needed for aerobic respiration during muscle activity and the amount that actually is used. ATP produced by anaerobic sources and used during muscle activity contributes to the oxygen debt. The increased aerobic metabolism after exercise rejuvenates depleted energy sources and reestablishes normal ATP and creatine phosphate levels in muscle fibers. Excess lactic acid is converted to glucose primarily in the liver, and glycogen levels are restored in muscle fibers.

During brief, but intense exercise such as during a sprint, much of the ATP used by exercising muscles comes from the conversion of creatine phosphate to creatine and from anaerobic sources. Glycogen is broken down to glucose and the glucose molecules are used in the process of glycolysis, which results in the buildup of lactic acid. Heavy breathing and elevated metabolism after the race represent the oxygen debt. The aerobic metabolism pays back the oxygen debt by converting creatine to creatine phosphate and converting the excess lactic acid to glucose, which is then stored as glycogen in muscles once again.

The magnitude of the oxygen debt depends on the severity of the exercise, the length of time it was sustained, and the physical condition of the individual. For those in poor physical condition, the capacity to perform aerobic metabolism is not as great as that of a well-trained athlete.

7

After a 1-mile run with a sprint at the end, a runner continues to breathe heavily for a period of time. Compare the function of the elevated metabolic processes during the run, near the end, and after the run.

? ? ? ? ? ? ? ? ? ? ?

Slow and Fast Fibers

Not all skeletal muscles have identical functional capabilities. Some muscle fibers contract quickly and fatigue quickly, whereas others contract more slowly and are more resistant to fatigue. The proportion of muscle fiber types differs within individual muscles.

Slow-twitch muscle fibers

Slow-twitch muscle fibers contract more slowly, are smaller in diameter, have a better developed blood supply, have more mitochondria, and are more fatigue resistant than fast-twitch muscle fibers. Slow-twitch muscle fibers respond to nervous stimulation and break down ATP at a limited rate within the heads of their myosin molecules. Aerobic metabolism is the primary source for ATP synthesis in slow-twitch muscles, and their capacity to perform aerobic metabolism is enhanced by a plentiful blood supply and the presence of numerous mitochondria. Slow-twitch fibers also contain large amounts of **myoglobin,** a dark pigment similar to hemoglobin, that binds oxygen and acts as a reservoir for oxygen when the blood does not supply an adequate amount. Myoglobin thus enhances the cell's capacity to perform aerobic metabolism.

Fast-twitch muscle fibers

Fast-twitch muscle fibers respond to nervous stimulation and contain myosin molecules that break down ATP more rapidly than slow-twitch muscle fibers, an ability associated with their cross bridges that form, release, and reform more rapidly than those in slow-twitch muscles. Muscles containing these fibers have a less well-developed blood supply than slow-twitch muscles. In addition, they have very little myoglobin and fewer and smaller mitochondria. Fast-twitch muscles have large deposits of glycogen and are well adapted to perform anaerobic metabolism. However, the anaerobic processes of fast-twitch muscles are not adapted for supplying large amounts of energy for a prolonged period. The muscles tend to contract rapidly for a shorter time and fatigue relatively quickly. Training causes fast-twitch muscles to improve their ability to carry out aerobic metabolism. Trained fast-twitch muscles are called **fatigue-resistant fast-twitch muscles.**

Distribution of fast-twitch and slow-twitch muscle fibers

The muscles of many animals are composed primarily of either fast-twitch or slow-twitch muscle fibers. The white meat of a chicken's or pheasant's breast, which is composed mainly of fast-twitch fibers, appears whitish because of its relatively poor blood supply and lack of myoglobin. The muscles are adapted to contract rapidly for a short time but fatigue quickly. The red, or dark, meat of a chicken's leg or of a duck's breast is composed of slow-twitch fibers and appears darker because of the relatively well-developed blood supply and a large amount of myoglobin. These muscles are adapted to contract slowly for a longer time period and to fatigue slowly. The distribution of slow-twitch and fast-twitch muscle fibers is consistent with the behavior of these animals. For example, pheasants can fly relatively fast for short distances, and ducks fly more slowly for long distances.

Humans exhibit no clear separation of slow-twitch and fast-twitch muscle fibers in individual muscles. Most muscles have both types of fibers, although the number of each varies in a given muscle. The large postural muscles contain more slow-twitch fibers, whereas muscles of the upper limbs contain more fast-twitch fibers. The distribution of the fibers in a given muscle is constant for each individual and apparently is established developmentally. People who are good sprinters have a greater percentage of fast-twitch muscle fibers, whereas good long-distance runners have a higher percentage of slow-twitch fibers in their leg muscles. Athletes who are able to perform a variety of anaerobic and aerobic exercises tend to have a more balanced mixture of fast-twitch and slow-twitch muscle fibers.

Effects of Exercise

Neither fast-twitch nor slow-twitch muscle fibers can be converted to muscle fibers of the other type. Nevertheless, training can increase the capacity of both types of muscle fibers to perform more efficiently. Intense exercise resulting in anaerobic metabolism increases muscular strength and mass and has a greater effect on fast-twitch muscle fibers. On the other hand, aerobic exercise increases the vascularity of muscle and causes enlargement of slow-twitch muscle fibers. Aerobic metabolism also can convert fast-twitch muscle fibers that fatigue readily to fast-twitch muscle fibers that resist fatigue by increasing the number of mitochondria in the muscle cells and increasing their blood supply. Thus through training a person with more fast-twitch muscle fibers can run long distances, and a person with more slow-twitch muscle fibers can increase the speed at which he runs.

8

What kind of exercise regimen is appropriate for people who are training to be weight lifters? What effect will the composition of their muscles, in terms of muscle fiber type, have on their ability to perform?

? ? ? ? ? ? ? ? ? ? ?

A muscle increases in size (i.e., it hypertrophies), strength, and endurance in response to exercise. Conversely, a muscle that is not used undergoes a decrease in size (i.e., it atrophies). The muscular atrophy that occurs in limbs placed in casts for several weeks is an example. Since muscle cell numbers do not change appreciably during a person's life, atrophy and hypertrophy of muscles result from changes in the size of individual muscle fibers. As fibers increase in size, the number of myofibrils and sarcomeres increases within each muscle fiber. Other elements (e.g., blood vessels, connective tissue, and mitochondria) also increase. Atrophy involves a decrease in these elements without a decrease in muscle fiber number. Severe atrophy such as occurs in the aged, however, does involve an irreversible decrease in the number of muscle fibers and may lead to paralysis.

The increased strength of trained muscle is greater than would be expected if it were based only on the change in muscle size. Part of the increased strength results from the nervous system's ability to recruit a large number of the motor units simultaneously in a trained person to perform movements with better neuromuscular coordination. In addition, trained muscles usually are restricted less by excess fat. Metabolic enzymes increase in hypertrophied muscle fibers, resulting in a greater capacity for ATP production. Improved endurance in trained muscles is in part a result of improved metabolism, increased circulation to the exercising muscles, an increased number of capillaries, more efficient respiration, and a greater capacity for the heart to pump blood.

Some people take synthetic hormones called **anabolic steroids** to increase the size and strength of their muscles. The anabolic steroids are related to testosterone, a reproductive hormone secreted by the testes, except that they have been altered so that the reproductive effects of these compounds are minimized but their effect on skeletal muscles is maintained. Testosterone and anabolic steroids cause skeletal muscle tissue to hypertrophy. People who take large doses of an anabolic steroid exhibit an increase in body weight and an increase in total skeletal muscle mass, and many athletes believe that anabolic steroids improve performance that depends on strength. Unfortunately, evidence indicates that harmful side effects are associated with taking large doses of anabolic steroids, including periods of irritability, testicular atrophy and sterility, cardiovascular diseases such as heart attack or stroke, and abnormal liver function. Most athletic organizations prohibit the use of anabolic steroids, and some athletic organizations analyze urine samples either randomly or periodically for evidence of anabolic steroid use. Penalties exist for athletes who have evidence of anabolic steroid metabolites in their urine.

Heat Production

The rate of metabolism in skeletal muscle differs before, during, and after exercise. As chemical reactions occur within cells, some energy is released in the form of heat. Normal body temperature is due, in large part, to this heat. Since the rate of chemical reactions increases in muscle fibers during contraction, the rate of heat production also increases, causing an increase in body temperature. After exercise elevated metabolism resulting from the oxygen debt helps keep the body temperature elevated.

When the body temperature declines below a certain level, the nervous system responds by inducing shivering, which involves rapid skeletal-muscle contractions that produce shaking rather than coordinated movements. The muscle movement increases heat production up to 18 times above resting levels, and the heat produced during shivering can exceed that produced during moderate exercise. The elevated heat production during shivering helps raise the body temperature to its normal range.

SMOOTH MUSCLE

Smooth muscle is distributed widely throughout the body and is more variable in function than other muscle types. Smooth-muscle cells (Figure 10-20) are smaller than skeletal muscle cells, ranging from 15 to 200 μm in length and from 5 to 10 μm in diameter. They are spindle shaped, with a single nucleus located in the middle of the cell. Compared to skeletal muscle, there are less actin and myosin myofilaments. Although the myofilaments approximate a longitudinal orientation within the smooth-muscle cell, they are not organized into sarcomeres; consequently, smooth muscle does not have a striated appearance. Sarcoplasmic reticulum is sparse in smooth-muscle cells, and there is no T-tubule system. Some shallow invaginated areas called **caveolae** (ka′ve-o-le) are along the surface of the cell membrane, and their function may be similar to both the T tubules and the sarcoplasmic reticulum of skeletal muscle. Because of the lack of an extensive sarcoplasmic reticulum, the Ca^{2+} ions required to initiate contractions usually enter the cell from the extracellular fluid. The distance that Ca^{2+} ions must diffuse, the rate at which action potentials are propagated between smooth-muscle cells, and the smaller number of actin and myosin filaments are responsible for the slower contraction of smooth muscle compared to skeletal muscle. Ca^{2+} ions bind to a protein called **calmodulin** (kal-mod′u-lin) in smooth-muscle cells. Calmodulin molecules with calcium ions bound to them initiate cross-bridge formation between actin and myosin in smooth-muscle cells.

Smooth-Muscle Types

There are two major types of smooth muscle: visceral and multiunit. Visceral, or unitary, smooth muscle is more common than multiunit smooth muscle. It normally occurs in sheets and includes smooth muscle of the digestive, reproductive, and urinary tracts. Visceral smooth muscle exhibits numerous gap junctions (see Chapter 4), which allow action potentials to pass directly from one cell to another. As a consequence, sheets of smooth-muscle cells function as a single unit, and a wave of contraction traverses the entire smooth-muscle sheet. Visceral smooth muscle is often autorhythmic, but some contracts only when stimulated. For example, visceral smooth muscles of the digestive tract contract spontaneously and at relatively regular intervals, whereas the visceral smooth muscle of the urinary bladder contracts when stimulated by the nervous system.

Multiunit smooth muscle occurs as sheets (e.g., the walls of blood vessels), in small bundles (e.g., arrector pili muscles and iris of the eye), or as single cells (e.g., the capsule of the spleen). They have few gap junctions, and each cell acts as an independent unit. Multiunit smooth muscle normally contracts only when stimulated by nerves or hormones.

Electrical Properties of Smooth Muscle

The resting membrane potential (RMP; see Chapter 9) of smooth-muscle cells ranges from −55 to −60 mV, in contrast to the RMP of −85 mV in skeletal muscle. Furthermore, the RMP fluctuates with slow depolarization and repolarization phases in many visceral smooth-muscle cells; these slow waves of depolarization and repolarization are propagated from cell to cell for short distances and cause contractions. More "classic" action potentials may be triggered by the slow waves of depolarization and usually are propagated for longer distances (Figure 10-21). The slow waves in the RMP result from a spontaneous and progressive increase in the permeability of the cell membrane to Na^{+} and Ca^{2+} ions. Both types of ions diffuse into the cell through their respective channels and produce the depolarization.

Smooth muscle does not respond in an all-or-none fashion to action potentials. A series of action potentials in smooth muscle may result in a single slow contraction followed by slow relaxation instead of individual contractions in response to each action potential as occurs in skeletal muscle. A slow wave of depolarization with one to several more classical-appearing action potentials superimposed on the slow depolarization wave is common in many types of smooth muscle. Subsequent to the wave of depolarization the smooth muscle undergoes a wave of contraction.

Spontaneously generated action potentials that lead to contractions are characteristic of visceral smooth muscle in the uterus, the ureter, and the digestive tract. Certain smooth-muscle cells in these organs function as pacemaker cells, which tend to develop action potentials more rapidly than other cells.

Functional Properties of Smooth Muscle

Smooth muscle has several functional properties not seen in skeletal muscle. An example is the autorhythmic contractions in some visceral smooth muscle. Smooth muscle tends to contract in response to a sudden stretch but not to a slow increase in length. Smooth muscle also exhibits a relatively constant tension over a long period of time and maintains that same tension in response to a gradual increase

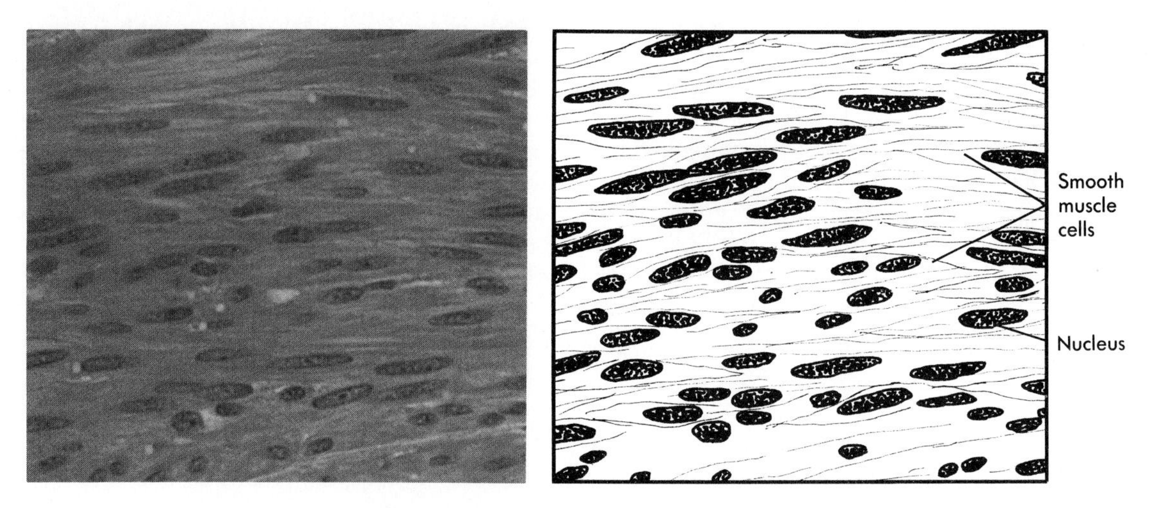

FIGURE 10-20 Smooth-muscle histology.

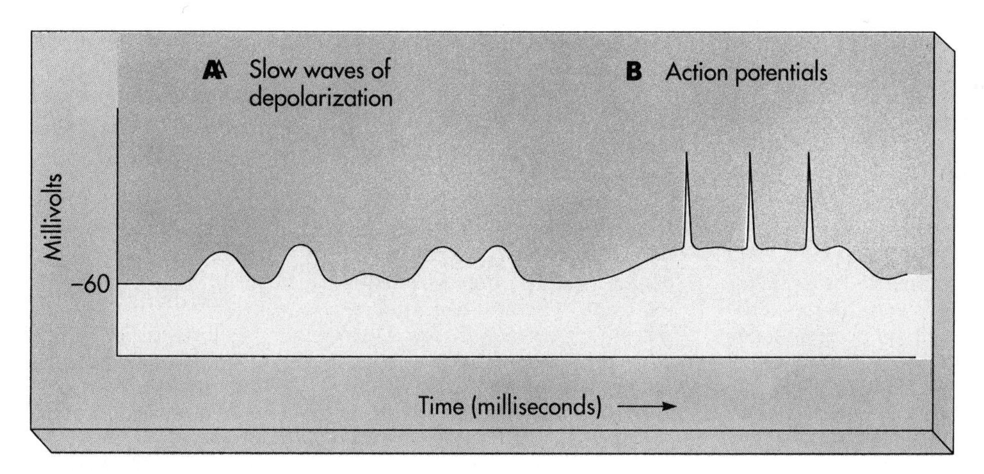

FIGURE 10-21 Membrane potential from a smooth muscle preparation. **A** Slow waves of depolarization; **B** Action potentials in smooth muscle superimposed on a slow wave of depolarization.

in the smooth muscle length. Therefore smooth muscle is well adapted for lining the walls of hollow organs such as the stomach and the urinary bladder. As the volume of the stomach or urinary bladder increases, only a small increase develops in the tension applied to their contents. The amplitude of contraction produced by smooth muscle also remains constant, although the muscle length varies. As the volume of the large and small intestines increases, the contractions that propel food through them do not change dramatically in amplitude.

Smooth muscle is innervated by nerves of the autonomic nervous system (see Chapter 16), whereas skeletal muscle is innervated by the somatomotor nervous system (see Chapter 14). Therefore the regulation of smooth muscle is involuntary, and the regulation of skeletal muscle is voluntary.

Hormones are also important in regulating smooth muscle. Epinephrine, a hormone from the adrenal medulla, stimulates some smooth muscles such as those in the blood vessels of the intestine and inhibits other smooth muscles such as those in the intestine. Oxytocin stimulates contractions of uterine smooth muscle, especially during delivery of a baby. These and other hormones are discussed more thoroughly in Chapters 17 and 18. Other chemical substances such as histamine and prostaglandins also influence smooth-muscle function.

The metabolism of smooth-muscle cells is similar to that of skeletal-muscle fibers. However, they are poorly adapted to perform anaerobic metabolism. An oxygen debt does not develop in smooth muscle, and fatigue occurs quickly in the absence of an adequate oxygen supply.

CARDIAC MUSCLE

Cardiac muscle is found only in the heart and is discussed in detail in Chapter 21. Cardiac muscle tissue is striated like skeletal muscle but each cell usually contains one nucleus located near the center. Adjacent cells join together to form branching fibers by specialized cell-to-cell attachments called **intercalated** (in-ter'kă-la-ted) **disks,** which have gap junctions that allow action potentials to pass from cell to cell. Cardiac muscle cells are autorhythmic; one portion of the heart normally acts as the pacemaker. The action potentials of cardiac muscle are similar to those in nerve and muscle but have a much longer duration and refractory period. The depolarization of cardiac muscle results from the influx of both sodium and calcium ions across the cell membrane.

Disorders of Muscle Tissue

Muscle disorders are caused by disruption of normal innervation, degeneration and replacement of muscle cells, injury, lack of use, or disease.

Atrophy

Exercise causes muscular hypertrophy. Conversely, disuse of muscle results in muscular atrophy. Extreme disuse of muscle results in muscular atrophy in which there is a permanent loss of skeletal muscle fibers and the replacement of those fibers by connective tissue. Immobility caused by damage to the nervous system or by old age may lead to permanent and severe muscular atrophy.

Denervation

When motor neurons innervating skeletal muscle fibers are severed, the result is flaccid paralysis. If the muscle is reinnervated, muscle function is restored, and atrophy is stopped. However, if skeletal muscle is permanently denervated, it atrophies and exhibits permanent flaccid paralysis. Muscles that have been denervated sometimes are stimulated electrically to prevent severe atrophy. The strategy is to slow the process of atrophy while motor neurons slowly grow toward the muscles and eventually reinnervate them. Neither cardiac muscle nor smooth muscle atrophies in response to denervation.

Muscular Dystrophy

Muscular dystrophy refers to a group of diseases called myopathies that destroy skeletal muscle tissue. Usually the diseases are inherited and are characterized by degeneration of muscle cells, leading to atrophy and eventual replacement by fatty tissue. Duchenne muscular dystrophy affects only males, and by early adolescence the individual is confined to a wheelchair. As the muscles atrophy, they shorten, causing conditions such as immobility of the joints and postural abnormalities such a scoliosis. Facioscapulohumoral (fa'sĭ-o-skap'u-lo-hu'mor-al) muscular dystrophy is generally less severe, and it affects both sexes later in life. The muscles of the face and shoulder girdle are primarily involved. Both types of muscular dystrophy are inherited and progressive, and no drugs prevent the progression of the disease. Therapy primarily involves exercises. Braces and corrective surgery sometimes help correct abnormal posture caused by the advanced disease.

Fibrosis

Fibrosis is the replacement of damaged cardiac muscle or skeletal muscle by connective tissue. Fibrosis, or scarring, is associated with severe trauma to skeletal muscle and with heart attack (myocardial infarction) in cardiac muscle.

Fibrositis

Fibrositis is an inflammation of fibrous connective tissue, resulting in stiffness, pain, or soreness. It is not progressive, nor does it lead to tissue destruction. Fibrositis may be caused by repeated muscular strain or prolonged muscular tension.

Cramps

Painful, spastic contractions of muscles (cramps) are usually due to an irritation within a muscle that causes a reflex contraction (see Chapter 13). Local inflammation resulting from a buildup of lactic acid and fibrositis causes reflex contraction of muscle fibers surrounding the irritated region.

SUMMARY

GENERAL FUNCTIONAL CHARACTERISTICS OF MUSCLE *p. 272*

1. Muscle exhibits contractility (shortens forcefully), excitability (responds to stimuli), extensibility (can be stretched), and elasticity (recoils to resting length).
2. The three types of muscle are skeletal, smooth, and cardiac.

SKELETAL MUSCLE: STRUCTURE *p. 273*

1. Muscle fibers are multinucleated and appear striated.
2. Muscle fibers are covered by the external lamina and the endomysium.
3. Muscle fasciculi, bundles of muscle fibers, are covered by the perimysium.
4. Muscle consisting of fasciculi is covered by the epimysium, which in turn is covered by fascia.
5. The connective tissue of muscle is bound firmly to the connective tissue of tendons and bone.

Connective Tissue

1. Endomysium surrounds each muscle fiber.
2. Perimysium surrounds each muscle fasciculus.
3. Epimysium surrounds each muscle.
4. Fascia surrounds muscles and muscle groups.

Muscle Fibers, Myofibrils, Sarcomeres, and Myofilaments

1. A muscle fiber is a single cell consisting of a cell membrane (sarcolemma), cytoplasm (sarcoplasm), several nuclei, and myofibrils.
2. Myofibrils are composed of many adjoining sarcomeres.
   - Sarcomeres are bound by Z lines that hold actin myofilaments.
   - Six actin myofilaments (thin filaments) surround a myosin myofilament (thick filament).
   - Myofibrils appear striated because of A bands and I bands.
3. Actin myofilaments consist of a double helix of F-actin (composed of G-actin monomers), tropomyosin, and troponin.
4. Myosin molecules, consisting of two globular heads and a rodlike portion, comprise myosin myofilaments.
5. A cross bridge is formed when the myosin binds to the actin.
6. Invaginations of the sarcolemma form T tubules that wrap around the sarcomeres.
7. A triad is a T tubule and two terminal cisternae (an enlarged area of sarcoplasmic reticulum).

SLIDING FILAMENT THEORY *p. 278*

1. Actin and myosin myofilaments do not change in length during contraction.
2. Actin and myosin myofilaments slide past one another in a way that causes sarcomeres to shorten.
3. The I band and H zones become more narrow during contraction, and the A band remains constant in length.

PHYSIOLOGY OF SKELETAL MUSCLE FIBERS *p. 279*

Neuromuscular Junction

1. The presynaptic terminal of the axon is separated from the postsynaptic terminal of the muscle fiber by the synaptic cleft.
2. Acetylcholine released from the presynaptic terminal binds to receptors of the postsynaptic terminal, thereby changing membrane permeability and producing an action potential.
3. After an action potential occurs, acetylcholinesterase splits acetylcholine into acetic acid and choline. Choline is reabsorbed into the presynaptic terminal to form acetylcholine.

Excitation Contraction Coupling

1. Action potentials move into the T-tubule system, causing the release of calcium from the sarcoplasmic reticulum.
2. Calcium diffuses to the myofilaments and binds to troponin, causing tropomyosin to move and expose actin to myosin.
3. Contraction occurs when actin and myosin bind, myosin changes shape, and actin is pulled past the myosin.
4. Relaxation occurs when calcium is taken up by the sarcoplasmic reticulum, ATP binds to myosin, and tropomyosin moves back between actin and myosin.

Energy Requirements for Contraction

1. One ATP molecule is required for each cycle of cross bridge formation, movement, and release.
2. ATP is also required to transport Ca^{2+} ions into the sarcoplasmic reticulum and to maintain normal concentrations across the cell membrane.

Muscle Relaxation

1. Ca^{2+} ions are transported into the sarcoplasmic reticulum.
2. Ca^{2+} ions diffuse away from troponin, preventing further cross bridge formation.

PHYSIOLOGY OF SKELETAL MUSCLE *p. 284*

Muscle Twitch

1. A muscle twitch is the contraction of a single muscle fiber or a whole muscle in response to a stimulus.
2. A muscle twitch has a lag, contraction, and relaxation phase.

Stimulus Strength and Muscle Contraction

1. For a given condition, a muscle fiber or motor unit will contract maximally or not at all (all-or-none law).
2. For a whole muscle, a stimulus of increasing magnitude results in a graded response of increased force of contraction as more motor units are recruited (multiple motor unit summation).

Stimulus Frequency and Muscle Contraction

1. A stimulus of increasing frequency increases the force of contraction (multiple wave summation).
2. Incomplete tetany is partial relaxation between contractions, and complete tetany is no relaxation between contractions.
3. The force of contraction of a whole muscle increases with increased frequency of stimulation because of an increasing concentration of Ca^{2+} ions around the myofibrils and because of complete stretching of muscle elastic elements.
4. Treppe is an increase in the force of contraction during the first few contractions of a rested muscle.

Types of Muscle Contraction

1. Isometric contractions cause a change in muscle tension but no change in muscle length.
2. Isotonic contractions cause a change in muscle length but no change in muscle tension.
3. Asynchronous contractions of motor units produce smooth, steady muscle contractions.
4. Muscle tone is maintenance of a steady tension for long periods of time.

Length vs. Tension

Muscle contracts with less-than-maximum force if its initial length is shorter or longer than optimum.

Fatigue

Fatigue is the decreased ability to do work and can be caused by the central nervous system, depletion of ATP in muscles, or depletion of acetylcholine in the neuromuscular synapse.

Physiological Contracture and Rigor Mortis

Physiological contracture (inability of muscles to contract or relax) and rigor mortis (stiff muscles after death) are due to inadequate ATP.

Energy Sources

1. Muscle cells produce ATP by aerobic respiration, and the energy in the ATP is transferred to creatine phosphate for storage.
2. During intense activity creatine phosphate reacts to form ATP, and additional ATP is formed by anaerobic respiration.
3. Lactic acid levels increase because of anaerobic respiration.

Oxygen Debt

After anaerobic respiration aerobic respiration is higher than normal, restoring creatine phosphate levels and converting lactic acid to glucose.

Slow and Fast Fibers

1. Slow-twitch fibers split ATP slowly and have a well-developed blood supply, many mitochondria, and myoglobin.
2. Fast-twitch fibers split ATP rapidly.
   - Fast-twitch, fatigue-resistant fibers have a well-developed blood supply, many mitochondria, and myoglobin.
   - Fast-twitch, fatigable fibers have large amounts of glycogen, a poor blood supply, fewer mitochondria, and little myoglobin.

Effects of Exercise

1. Muscle increase (hypertrophy) or decrease (atrophy) in size is due to a change in size of muscle fibers.
2. Anaerobic exercise develops fast-twitch, fatigable fibers. Aerobic exercise develops slow-twitch fibers and changes fast-twitch, fatigable fibers into fast-twitch, fatigue-resistant fibers.

Heat Production

1. Heat is produced as a by-product of chemical reactions in muscles.
2. Shivering produces heat to maintain body temperature.

SMOOTH MUSCLE *p. 294*

1. Smooth-muscle cells are spindle shaped with a single nucleus. They have actin myofilaments and myosin myofilaments but are not striated.
2. The sarcoplasmic reticulum is poorly developed, and caveolae may function as a T-tubule system.
3. Ca^{2+} ions enter the cell to initiate contraction; calmodulin may function like troponin in skeletal muscle.

Smooth-Muscle Types

1. Visceral smooth-muscle fibers contract slowly, have gap junctions (and thus function as a single unit), and may be autorhythmic.
2. Multiunit smooth-muscle fibers contract rapidly and function independently.

Electrical Properties of Smooth Muscle

1. Spontaneous contractions result from Na^+ ion leakage. Na^+ ion and Ca^{2+} ion movement into the cell is involved in depolarization.
2. The autonomic nervous system and hormones can inhibit or stimulate action potentials (and thus contractions).

Functional Properties of Smooth Muscle

1. Smooth muscle may contract autorhythmically in response to stretch or when stimulated by the autonomic nervous system or hormones.
2. Smooth muscle maintains a steady tension for long periods of time.
3. The force of smooth-muscle contraction remains nearly constant, despite changes in muscle length.
4. Smooth muscle does not develop an oxygen debt.

CARDIAC MUSCLE *p. 296*

Cardiac muscle fibers are striated, have a single nucleus, are connected by intercalated disks (thus function as a single unit), and are capable of autorhythmicity.

CONTENT REVIEW

1. Compare the structure, function, location, and control of the three major muscle types.
2. Name the connective tissue structures that surround muscle fibers, muscle fasciculi, and whole muscles.
3. Describe the blood and nerve supply of a muscle fiber.
4. Define sarcolemma, sarcoplasm, myofibril, and sarcomere.
5. What are Z lines and M lines, and what are their functions?
6. Explain how the arrangement of actin myofilaments and myosin myofilaments produce I bands, A bands, and H zones.
7. How do G-actin, tropomyosin, and troponin combine to form an actin myofilament?
8. What is the T-tubule system? What is a triad?
9. Describe the neuromuscular junction. How does an action potential in the neuron produce an action potential in the muscle cell?
10. How does an action potential produced in the postsynaptic terminal of the neuromuscular junction eventually result in contraction of the muscle fiber?
11. Where in the contraction and relaxation processes is ATP required?
12. Describe the phases of a muscle twitch and the events that occur in each phase.
13. Why does a single muscle fiber either not contract or contract with the same force in response to stimuli of different magnitudes?
14. How does increasing the magnitude of a stimulus cause a whole muscle to respond in a graded fashion?
15. Explain why increasing the frequency of stimulation increases the force of contraction of a single muscle fiber.
16. Define an isometric contraction and an isotonic contraction. What is muscle tone, and how is it maintained?

17. How are smooth contractions produced in muscles?
18. Draw an active tension curve. How does the overlap of actin and myosin explain the shape of the curve?
19. Define fatigue, and list three locations where fatigue can develop.
20. Define and explain the cause of physiological contracture and rigor mortis.
21. Contrast the efficiency of aerobic and anaerobic respiration. When is each type used by cells?
22. What is the function of creatine phosphate? When does lactic acid production increase in a muscle cell?
23. Contrast the structural and functional differences between slow and fast fibers.
24. What factors contribute to an increase in muscle strength and endurance? How does anaerobic vs. aerobic exercise affect muscles?
25. How do muscles contribute to the heat responsible for body temperature before, during, and after exercise? What is accomplished by shivering?
26. Describe a typical smooth muscle cell. How does it differ from a skeletal muscle cell or a cardiac muscle cell?
27. Compare visceral smooth muscle to multiunit smooth muscle. Explain why visceral smooth muscle contracts as a single unit.
28. How are sponatneous contractions produced in smooth muscle?
29. How do the nervous system and hormones regulate smooth-muscle activity?

CONCEPT REVIEW

1. Bob Canner improperly canned some homegrown vegetables. As a result, he contracted botulism poisoning after eating the vegetables. Symptoms included difficulty in swallowing and breathing. Eventually he died of respiratory failure (his respiratory muscles relaxed and would not contract). Assuming that botulism toxin affects the neuromuscular synapse, propose the ways in which botulism toxin could produce the observed symptoms.
2. A patient is suspected of suffering from either muscular dystrophy or myasthenia gravis. How would you distinguish between the two conditions?
3. Under certain circumstances the actin and myosin myofilaments can be extracted from muscle cells and placed in a beaker. They subsequently bind together to form long filaments of actin and myosin. Addition of what cell organelle or molecule to the beaker would make the actin and myosin myofilaments unbind?
4. Explain the effect of a lower-than-normal temperature on each of the processes that occur in the lag (latent) phase of muscle contraction.
5. Design an experiment to test the following hypothesis: muscle A has the same number of motor units as muscle B. Assume you could stimulate the muscles with an electronic stimulator and monitor the tension produced by the muscles.
6. Compare the differences in events that occur when a muscle (such as a biceps muscle) slowly lifts and lowers a weight to those that occur during a muscle twitch.
7. Predict the shape of an active tension curve for visceral smooth muscle. How does it differ from the active tension curve for skeletal muscle?
8. A researcher was investigating the composition of muscle tissue in the gastrocnemius muscles (in the calf of the leg) of athletes. A needle biopsy was taken from the muscle, and the concentration (or enzyme activity) of several substances was determined. Describe the major differences this researcher would see when comparing the muscles from athletes who performed in the following events: 100-meter dash, weight lifting, and 10,000-meter run.
9. Harvey Leche milked cows by hand each morning before school. One morning he slept later than usual and had to hurry to get to school on time. As he was milking the cows as fast as he could, his hands became very tired, and for a short time he could neither release his grip nor squeeze harder. Explain what happened.
10. If blood vessels that supply oxygen to smooth muscle undergo constriction, explain how that would affect the ability of smooth muscle to contract.
11. Shorty McFleet noticed that his rate of respiration was elevated after running a 100-meter race but was not as elevated after running slowly for a much longer distance. Since you had studied muscle physiology, he asked you for an explanation. What would you say?
12. It is known that high blood K^+ ion concentrations will cause depolarization of the resting membrane potential. Predict the effect of high blood K^+ ion levels on smooth muscle function. Explain.
13. If one could instantly increase the ATP concentration in a muscle that was exhibiting rigor mortis, predict and explain the response.

? ?

ANSWERS TO PREDICT QUESTIONS

1 *p. 279.* If myosin myofilaments shortened, the A band would have to shorten. The A band extends from one end of the myosin myofilaments in a sarcomere to the other end of the myosin myofilaments. Therefore if the myosin myofilaments shortened, the A band would have to shorten, also. Of course, the A band does not shorten during muscle contraction. That is one of the observations that gave rise to the sliding filament theory.

2 *p. 280.* If insufficient acetylcholine were released from the presynaptic terminal of a neuron to produce an action potential in a muscle fiber, the muscle could not contract. An action potential must be produced in the muscle fiber for contraction to occur. Several action potentials in neurons would have to occur to cause the neurons to release enough acetylcholine to produce an action potential in the muscle fibers.

3 *p. 280.*

A. Organophosphate poisons inhibit the activity of acetylcholinesterase, which breaks down acetylcholine at the neuromuscular junction and limits the length of time the acetylcholine stimulates the postsynaptic terminal of the muscle fiber. Consequently, acetylcholine accumulates in the synaptic cleft and continuously stimulates the muscle fiber. As a result, the muscle remains contracted until it fatigues. Death is caused by the inability of the victim to breathe. Either the respiratory muscles are in spastic paralysis or they are so depleted of ATP that they cannot contract at all.

B. Curare binds to acetylcholine receptors and thus prevents acetylcholine from binding to them. Because curare does not activate the receptors, the muscles do not respond to nervous stimulation. The person suffers from flaccid paralysis and dies from suffocation because the respiratory muscles are not able to contract.

4 *p. 284.*

A. If Na^+ ions cannot enter the muscle fiber, no action potentials are produced in the muscle fiber because the depolarization phase of the action potential is caused by the influx of Na^+ ions. Without action potentials the muscle fiber cannot contract at all. The result is flaccid paralysis.

B. If ATP levels are low in a muscle fiber before stimulation, the following will result. Energy from the breakdown of ATP already is stored in the heads of the myosin molecules. After stimulation cross bridges form. However, if there are not enough additional ATP molecules in the muscle cells to bind to the myosin molecules to allow cross-bridge release, the muscle becomes stiff without contracting or relaxing.

? ?

C. If ATP levels in the muscle fiber are adequate but the action potential frequency is so high that Ca^{2+} ions accumulate within the cell, the muscle contracts tetanically. As long as Ca^{2+} ions are numerous within the sarcoplasm in the area of the myofilaments, cross-bridge formation is possible. If ATP levels are adequate, cross-bridge formation, release, and formation can proceed again, resulting in a continuously contracting muscle.

5 *p. 287*. There is a decrease in muscle control when reinnervation of muscle fibers occurs after poliomyelitis because the number of motor units in the muscle is decreased. Reinnervation results in a greater number of muscle fibers per motor unit. Control is reduced because the number of motor units that can be recruited is decreased. The greater the number of motor units in a muscle, the greater is the ability to have fine gradations of muscle contraction as motor units are recruited. A smaller number of motor units means gradations of muscle contraction are not as fine.

6 *p. 289*. Contractions of muscle fibers that are responsible for muscle tone are isometric contractions. The length of the muscles does not change much, but the tension produced by the muscles is great enough to hold the body erect. In contrast, isotonic muscle contractions are those in which there is a change in the length of the muscle fibers but not a change in the amount of tension produced by the fibers.

7 *p. 292*. During a 1-mile run aerobic metabolism is the primary source of ATP production for muscle contraction. Anaerobic metabolism provides enough ATP for 15 to 20 seconds during vigorous exercise, but running a mile takes several minutes. However, if the runner sprints at the end of the mile run, anaerobic metabolism accounts for some of the energy production. After the run aerobic metabolism is elevated for a time to pay back the oxygen debt. Anaerobic metabolism near the end of the run produced lactic acid, which is converted back to glucose after the run, and ATP is required to restore the normal creatine phosphate levels in the muscle fibers.

8 *p. 293*. Weight lifters usually do not lift heavy objects for a prolonged period of time. Consequently, fast-twitch muscles function very well for weight lifting. In addition, exercise that causes the muscles to perform anaerobic metabolism is more effective than prolonged but less strenuous exercise. Therefore anaerobic exercise for people with a large percentage of fast-twitch muscle fibers is the best for weight lifting.

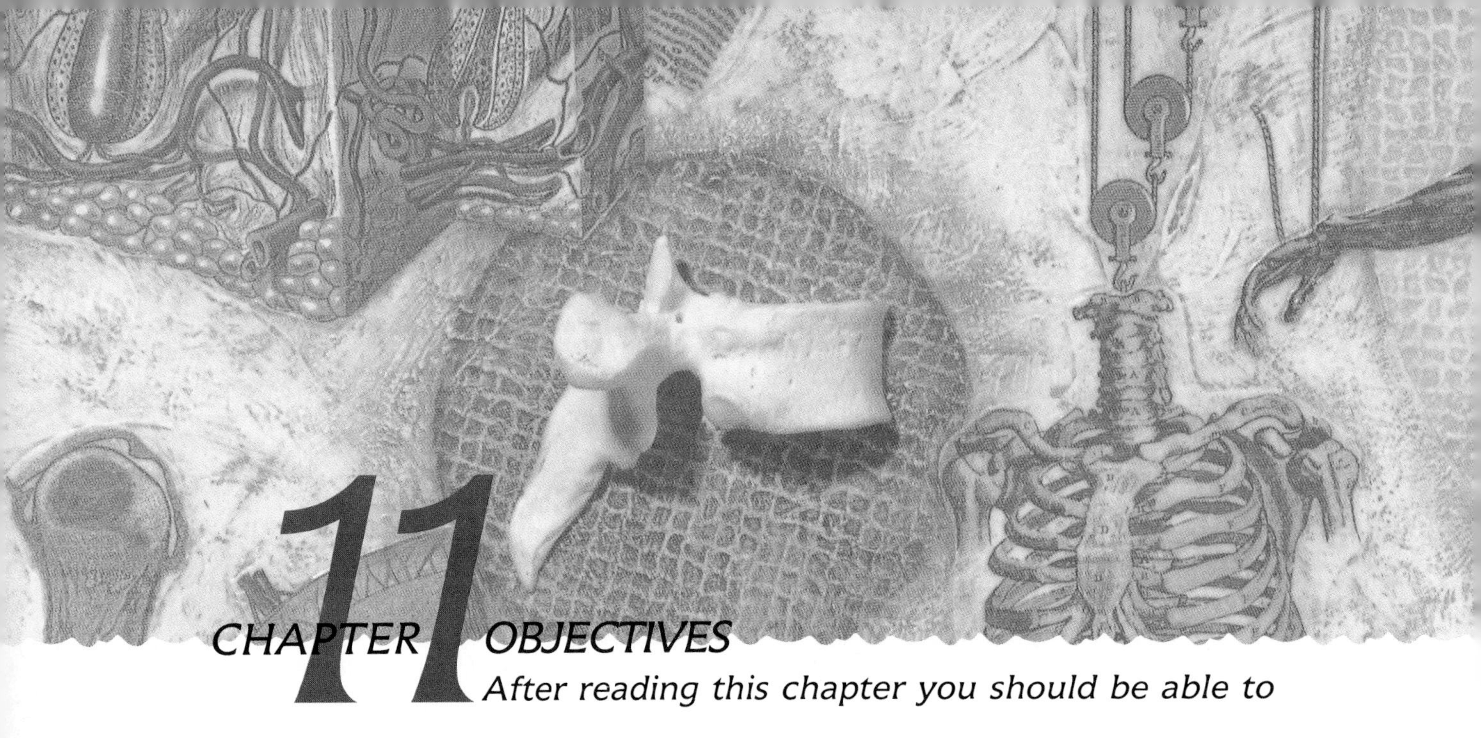

CHAPTER 11 OBJECTIVES

After reading this chapter you should be able to

1. Define the origin and insertion of a muscle.
2. Define the following terms and give an example of each: synergist, antagonist, prime mover, and fixator.
3. List the major muscle shapes and indicate how each relates to function.
4. List and describe the three lever classes and give an example of each.
5. Describe the major movements of the head and list the muscles involved in each movement.
6. Describe various facial expressions and list the major muscles causing them.
7. List the muscles of mastication and indicate the effect of each muscle on mandibular movement.
8. Explain the location and functional differences between extrinsic and intrinsic tongue muscles.
9. Describe the process of swallowing and explain the action of each muscle on this process.
10. Describe the muscles of the eye and how each affects eye movement.
11. Describe movements of the vertebral column and list the muscles involved.
12. Describe the placement, fascicular orientation, and function of the muscles of the thorax and abdominal wall.
13. Describe the pelvic floor and perineum and list the muscles forming them.
14. List the muscles forming the rotator cuff and describe its function.
15. Describe the movements of the arm and the muscles involved.
16. Describe the forearm muscles in terms of their functional groupings and the movements they produce.
17. Explain the difference between extrinsic and intrinsic hand muscles.
18. Describe the movements of the thigh and list the muscles involved in each movement.
19. Describe the leg in terms of compartments, list the muscles contained in each compartment, and indicate the function of each muscle.

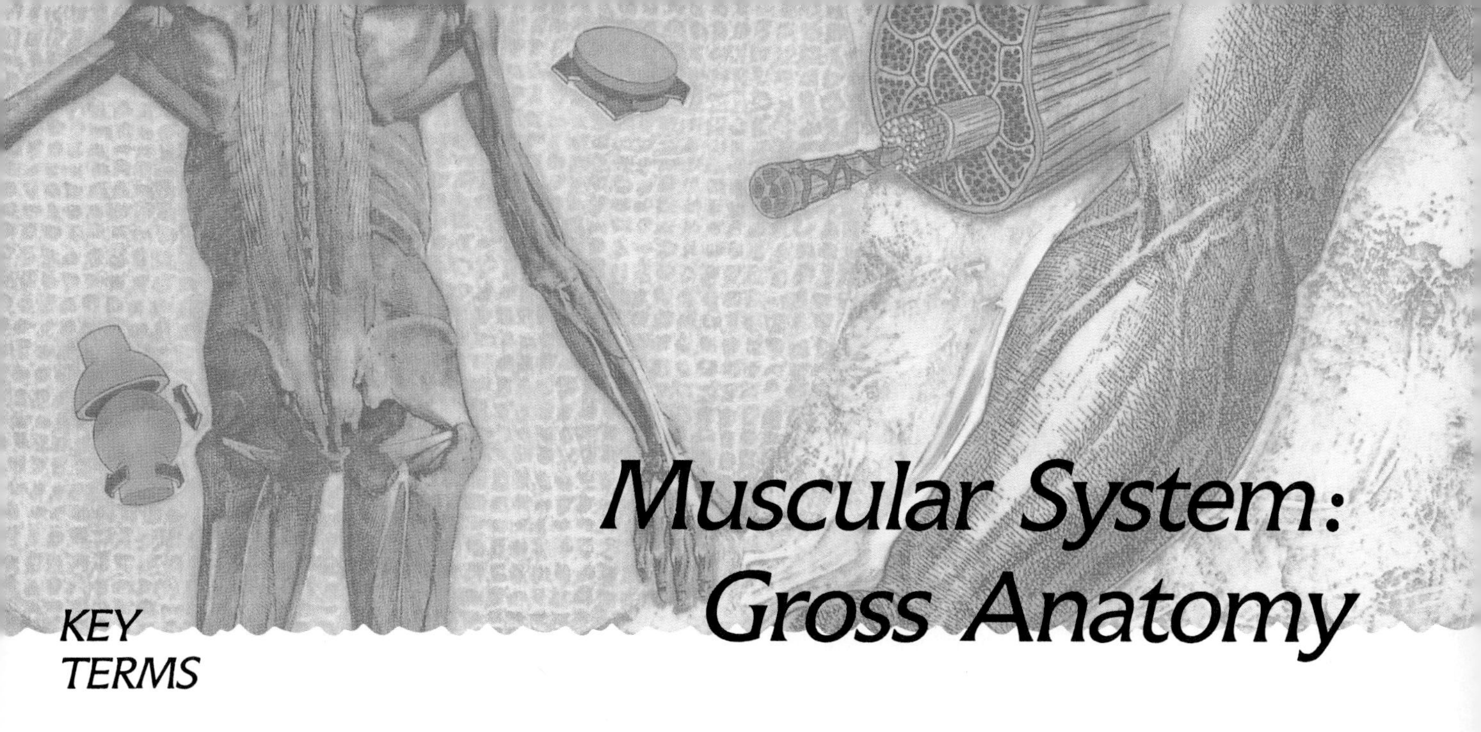

Muscular System: Gross Anatomy

KEY TERMS

antagonist

belly

extrinsic muscle

force

fulcrum

insertion

intrinsic muscle

lever

mastication (mas′tĭ-ka′shun)

origin

perineum (per′ĭ-ne′um)

prime mover

rotator cuff muscles

synergist (sin′er-jist)

thenar (the′nar)

RELATED TOPICS

The following terms or concepts are important for a good understanding of this chapter. If you are not familiar with them, you should review them before proceeding.

Directional terms and body regions (Chapter 1)

Bone anatomy (Chapter 6)

Joint anatomy (Chapter 8)

Movements (Chapter 8)

Muscle histology and function (Chapter 10)

BODY MOVEMENTS ARE ACCOMPLISHED BY THE CONTRACTION OF MUSCLES. This chapter is devoted entirely to the description of the major named skeletal muscles (Figure 11-1). The structure and function of cardiac and smooth muscle are considered in later chapters.

FIGURE 11-1 General overview of the body musculature.
A Anterior view.

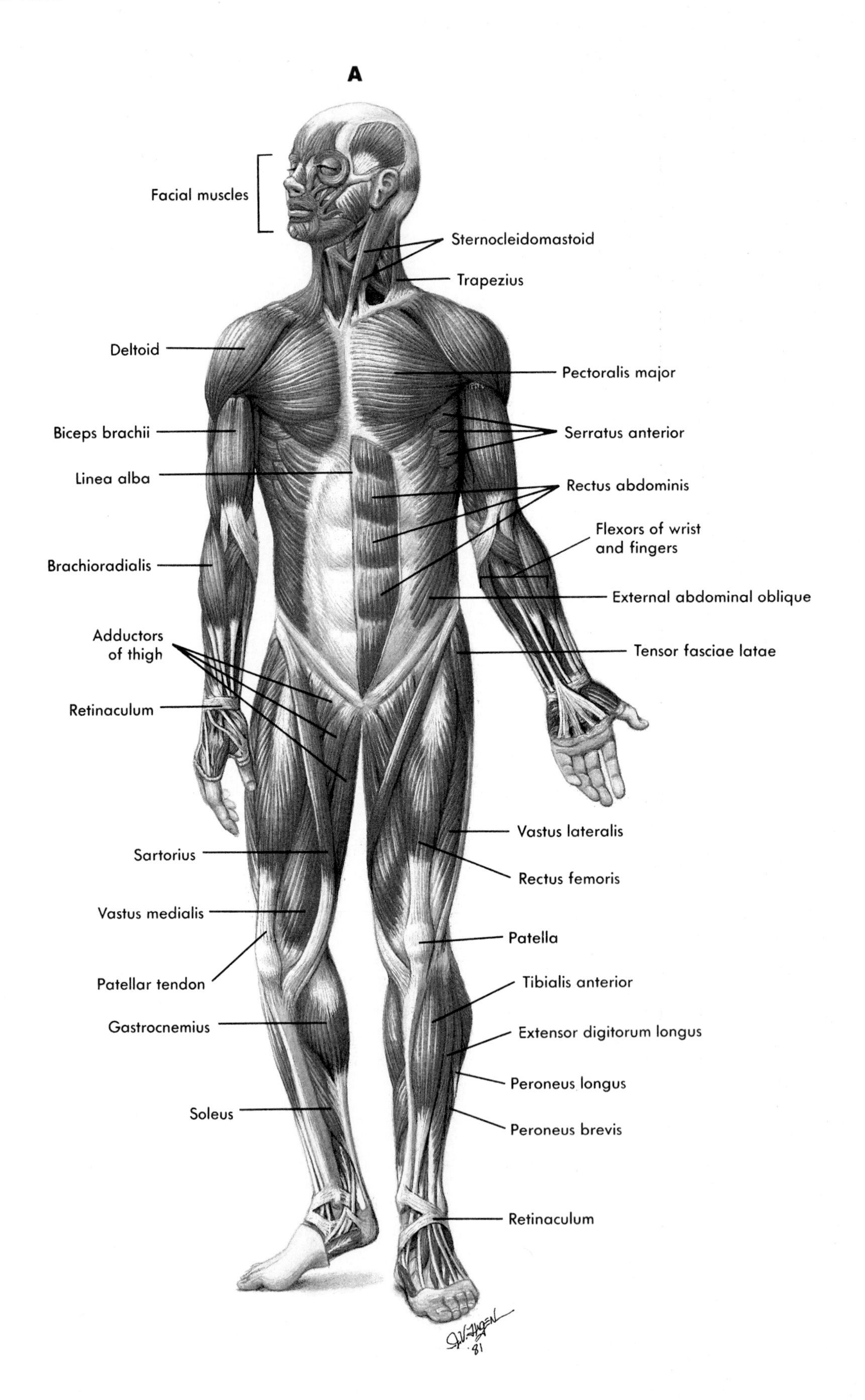

FIGURE 11-1, cont'd General overview of the body musculature.
B Posterior view.

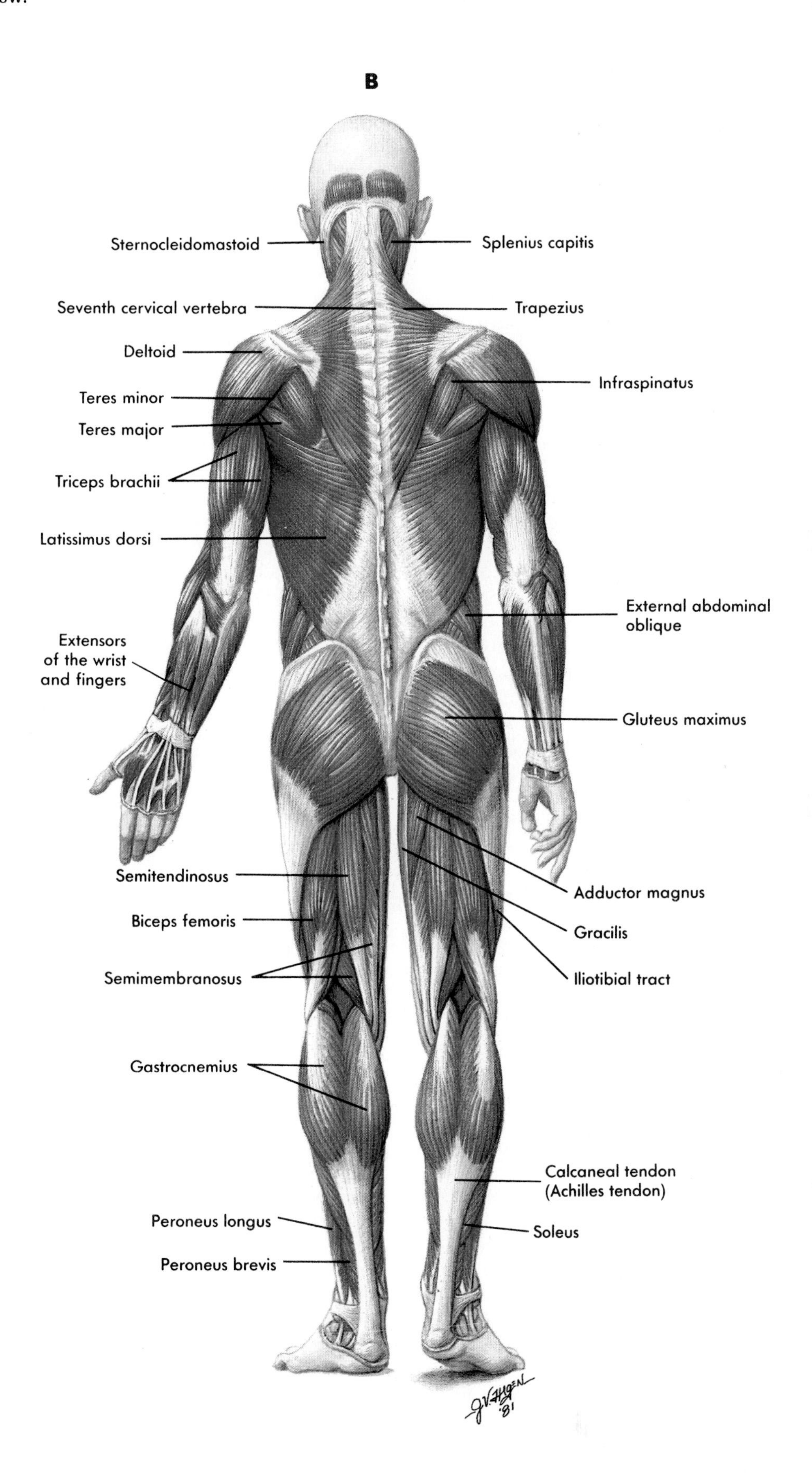

GENERAL PRINCIPLES

Most muscles extend from one bone to another and cross at least one joint. Muscle contraction causes most body movements by pulling one of the bones toward the other across a movable joint. Some muscles of the face, however, are not attached to bone at both ends but attach to the skin, which moves when the muscles contract.

The points of attachment for each muscle are the origin and insertion. The **origin,** also called the **head,** is normally that end of the muscle attached to the more stationary of the two bones, and the **insertion** is the end of the muscle attached to the bone undergoing the greatest movement. The largest portion of the muscle, between the origin and the insertion, is the **belly.** Some muscles have multiple origins and a common insertion and are said to have multiple heads (such as the biceps with two heads).

Most muscles function as members of a group to accomplish specific movements. Further, many muscles are members of more than one group, depending on the type of movement being considered. For example, the anterior part of the deltoid muscle functions with the flexors of the arm, whereas the posterior part functions with the extensors of the arm. Muscles that work together to cause movement are **synergists** (sin'er-jists), and a muscle working in opposition to another muscle, moving a structure in the opposite direction, is an **antagonist.** The brachialis and biceps brachii are synergists in flexing the forearm; the triceps brachii is the antagonist and extends the forearm. Among a group of synergists, if one muscle plays the major role in accomplishing the desired movement, it is the **prime mover.**

Other muscles, called **fixators** (fiks'a-ters), may stabilize one or more joints crossed by the prime mover. The extensor digitorum is the prime mover in finger extension. The flexor carpi radialis and flexor carpi ulnaris are the fixators that keep the wrist from extending as the fingers are extended.

Muscle Shapes

Muscles come in a wide variety of shapes; the shape of a given muscle determines the degree to which a muscle can contract and the amount of force it can generate. The large number of muscular shapes can be grouped into four classes according to the orientation of the muscle fasciculi: pennate, parallel, convergent, and circular. Some muscles have their fasciculi arranged like the barbs of a feather along a common tendon and are therefore called **pennate** (pen'āt; *penna* means feather) muscles. A muscle with fasciculi on one side of the tendon only is **unipennate** (Figure 11-2, *A*), one with fasciculi on both sides is **bipennate** (Figure 11-2, *B*), and a muscle with fasciculi arranged at many places around the central tendon is **multipennate** (Figure 11-2, *C*). The pennate arrangement allows a large number of fasciculi to attach to a single tendon with all the force of contraction concentrated at the tendon. The muscles that extend the leg are examples (see Table 11-19). In other muscles called **parallel** muscles, fasciculi are organized parallel to the long axis of the muscle. As a consequence, the muscles shorten to a greater degree than pennate muscles because the fasciculi are in direct line with the tendon; however, they contract with less force since fewer total fascicles are attached to the tendon. The infrahyoid muscles are an example (see Table 11-4). In **convergent** muscles such as the deltoid muscle, the base is much wider than the insertion, giving the muscle a triangular shape and allowing the muscle to contract with more force than could occur in a parallel muscle. **Circular** muscles such as the orbicularis oris and orbicularis oculi (see Table 11-2) have their fasciculi arranged in a circle around an opening and act as sphincters to close the opening.

The general shape of a muscle may be triangular (deltoid), quadrangular, trapezoid, rhomboid, or fusiform (Figure 11-2, *D* to *G*). Muscles also may have multiple components such as two bellies or two heads. A digastric muscle has two bellies separated by a tendon, whereas a bicipital muscle has two origins and a single insertion (Figure 11-2, *H* and *I*).

Nomenclature

Muscles are named according to their location, size, shape, orientation of fasciculi, origin and insertion, number of heads, or function. Recognizing the descriptive nature of muscle names makes learning those names much easier.

1. Location. Some muscles are named according to their location. For example, a *pectoralis* (chest) muscle is located in the chest, a *gluteus* (buttock) muscle is located in the buttock, and a *brachial* (arm) muscle is located in the arm.

2. Size. Muscle names may also refer to the relative size of the muscle. For example, the *gluteus maximus* (large) is the largest muscle of the buttock, and the *gluteus minimus* (small) is the smallest muscle of the gluteal group. A *longus* (long) muscle is longer than a *brevis* (short) muscle, which is shorter.

3. Shape. Some muscles are named according to their shape. The *deltoid* (triangular) muscle is triangular, a *quadratus* (quadrangular) muscle is rectangular, and a *teres* (round) muscle is round.

4. Orientation. Muscles are also named according to their fascicular orientation. A *rectus* (straight) muscle has muscle fasciculi running straight down the body, whereas the fasciculi of an *oblique* muscle lie oblique to the longitudinal axis of the body.

5. Origin and insertion. Muscles may be named according to the origin and insertion of the muscle.

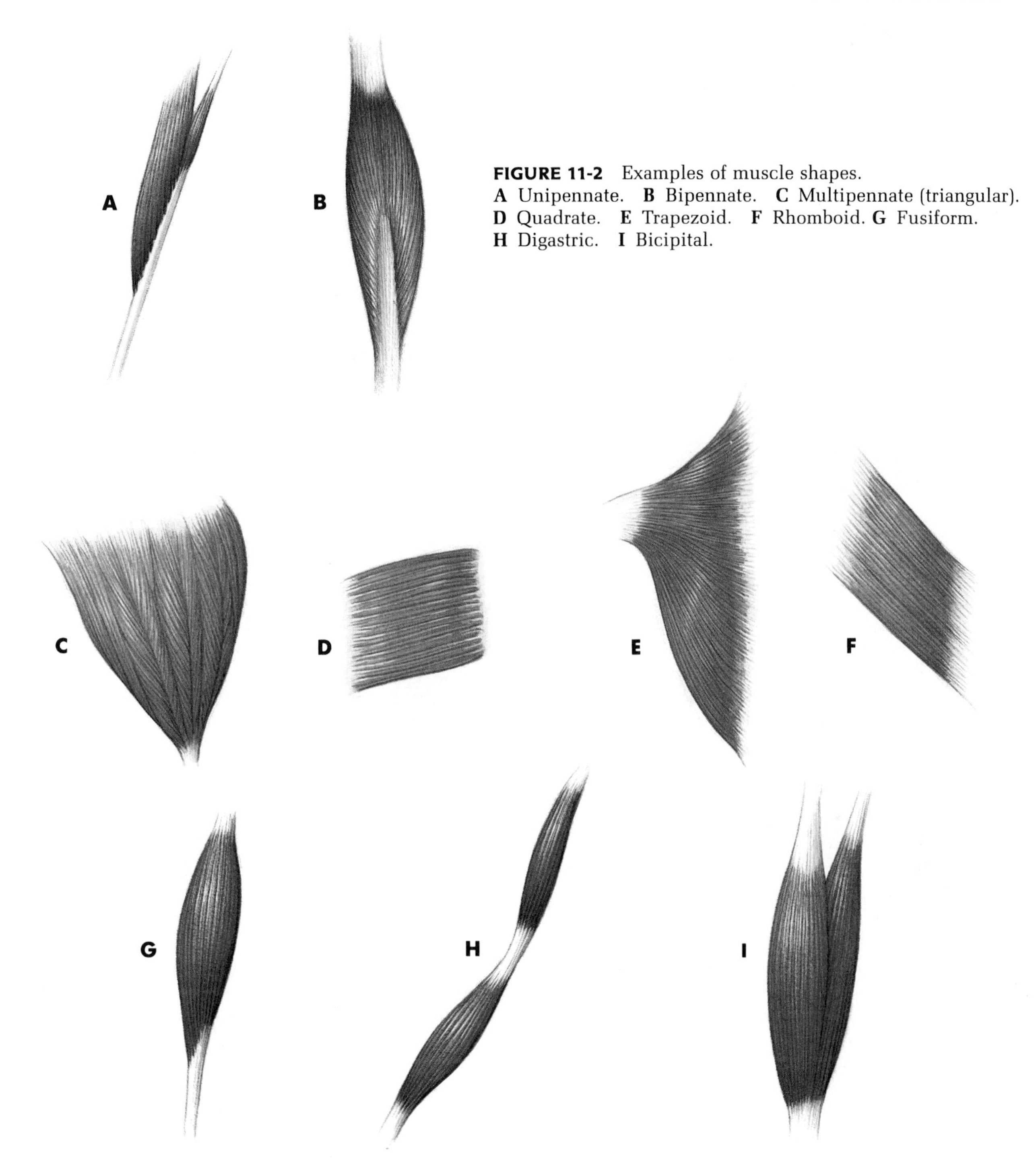

FIGURE 11-2 Examples of muscle shapes. **A** Unipennate. **B** Bipennate. **C** Multipennate (triangular). **D** Quadrate. **E** Trapezoid. **F** Rhomboid. **G** Fusiform. **H** Digastric. **I** Bicipital.

The *sternocleidomastoid* originates on the sternum and clavicle and inserts onto the mastoid process of the temporal bone. The *brachioradialis* originates in the arm (*brachium*) and inserts onto the radius.

6. Number of heads. The number of heads (origins) a muscle has may also be used in naming the muscle. A *biceps* muscle has two heads and a *triceps* muscle has three heads.

7. Function. Muscles are also named according to their function. An *abductor* moves a structure away from the midline, and an *adductor* moves a structure toward the midline. The *masseter* (a chewer) is the chewing muscle.

Movements Accomplished by Muscles

When muscles contract, **force** usually is applied to levers (such as bones), resulting in movement of the levers. A **lever** is a rigid shaft capable of turning about a pivot point, called a **fulcrum,** and transferring a force applied at one point along the lever to a weight placed at some other point along the lever. The joints function as fulcrums, the bones function as levers, and the muscles provide the force to move the levers. The relative positions of levers, weights, fulcrums, and forces comprise three classes of levers.

Class I lever

In a Class I lever system the fulcrum is located between the force and the weight (Figure 11-3, *A*). An example of this type of lever is a child's seesaw. The children on the seesaw alternate between being the weight and the force across a fulcrum in the center of the board. An example in the body is the head. The atlantooccipital joint is the fulcrum, the posterior neck muscles are the force depressing the back of the head, and the face, which is elevated, is the weight.

Class II lever

In a Class II lever system the weight is located between the fulcrum and the force (Figure 11-3, *B*). An example is a wheelbarrow in which the wheel is the fulcrum and the person lifting on the handles is the force. The weight, or load, carried in the wheelbarrow is placed between the wheel and the operator. In the body an example of a Class II lever is the foot when a person stands on his toes. The calf muscles (force) pulling on the calcaneus (end of the lever) elevate the foot and the weight of the entire body, with the ball of the foot acting as the fulcrum.

Class III lever

In a Class III lever system, the most common type in the body, the force is located between the fulcrum and the weight (Figure 11-3, *C*). An example is a person operating a shovel. The hand placed on the part of the handle closest to the blade acts as the force to lift the weight such as a shovel full of dirt, and the hand placed near the end of the handle acts as the fulcrum. In the body the action of the biceps brachii muscle (force) pulling on the radius (lever) to flex the elbow (fulcrum) and elevate the hand (weight) is an example of a Class III lever.

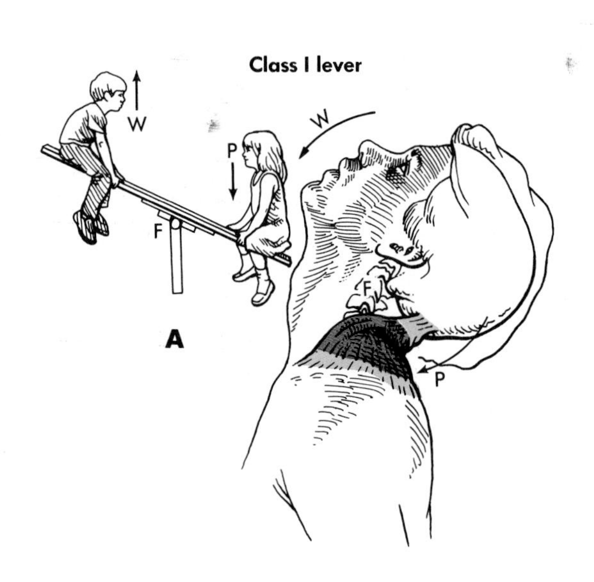

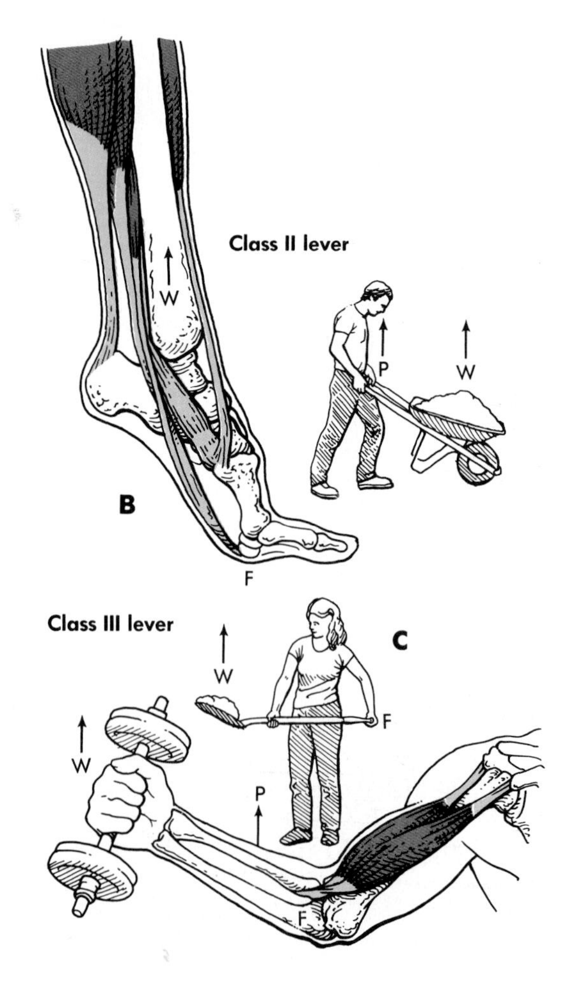

FIGURE 11-3 Lever classes. **A** Class I: fulcrum *(F)* between the weight *(W)* and force or pull *(P)*; **B** Class II: weight *(W)* between the fulcrum *(F)* and force or pull *(P)*; **C** Class III: force or pull *(P)* between the fulcrum *(F)* and the weight *(W)*.

HEAD MUSCLES

Head Movement

The flexors of the head and neck (Table 11-1 and Figure 11-4, *A*) lie deep within the neck along the anterior margins of the vertebral bodies. Extension of the head is accomplished by posterior neck muscles that attach to the occipital bone and function as the force of a Class I lever system (Figure 11-4, *B* and *C*).

The muscular ridge seen superficially in the posterior part of the neck and lateral to the midline is composed of the trapezius muscle overlying the splenius capitis (Figure 11-5, *A*). The trapezius muscles separate at the base of the neck, leaving a diamond-shaped area over the inferior cervical and superior thoracic vertebral spines.

Rotation and abduction of the head are accomplished by muscles of both the lateral and posterior groups (see Table 11-1). The **sternocleidomastoid** (ster′no-kli′do-mas′toyd) muscle, the prime mover of the lateral group, is very easily seen on the anterior and lateral sides of the neck, especially if the head is extended slightly and rotated to one side (see Figure 11-5, *B*). Adduction of the head (tilting the head to one side or the other) is accomplished by the abductors of the opposite side.

1

Shortening of the right sternocleidomastoid muscle would rotate the head in which direction?

? ? ? ? ? ? ? ? ? ? ?

Torticollis (meaning a twisted neck), or wry neck, may result from injury to one of the sternocleidomastoid muscles. It sometimes is caused by damage to an infant's neck muscles during a difficult birth and usually can be corrected by exercising the muscle.

Table 11-1

Muscles moving the head (see Figure 11-4)

MUSCLE	ORIGIN	INSERTION	NERVE	FUNCTION
Anterior				
Longus capitis	C4-C7	Occipital bone	Cervical plexus	Flexes neck
Rectus capitis anterior	Atlas	Occipital bone	C1-C2	Flexes neck
Posterior				
Longissimus capitis	Upper thoracic and lower cervical vertebrae	Mastoid process	Dorsal rami of cervical nerves	Extends head
Oblique capitis superior	Atlas	Occipital bone (inferior nuchal line)	Dorsal ramus of C1	Rotates head
Rectus capitis posterior	Axis, atlas	Occipital bone	Dorsal ramus of C1	Rotates and extends head
Semispinalis capitis	C4-T6	Occipital bone	Dorsal rami of cervical nerves	Rotates and extends head
Splenius capitis	C4-T6	Superior nuchal line and mastoid process	Dorsal rami of cervical nerves	Rotates and extends head
Trapezius	Occipital protuberance	Clavicle, acromion, and scapular spine	Accessory	Abducts and extends head
Lateral				
Rectus capitis lateralis	Atlas	Occipital bone	C1	Abducts head
Sternocleidomastoid	Manubrium and medial clavicle	Mastoid process and superior nuchal line	Accessory	Rotates and extends head; flexes neck

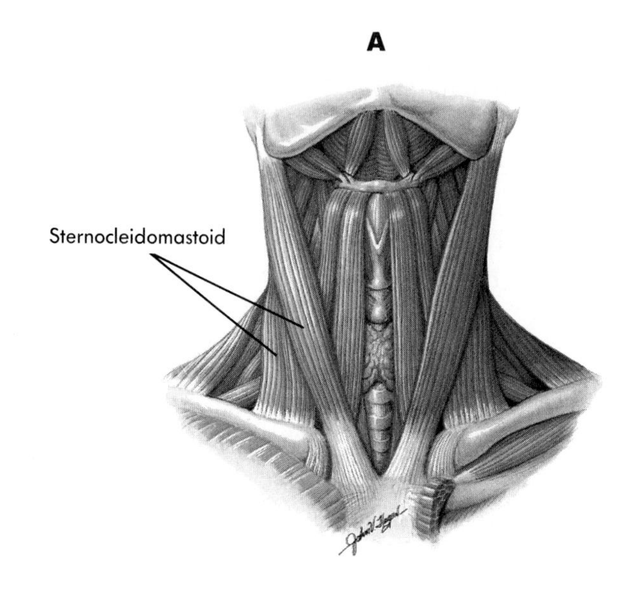

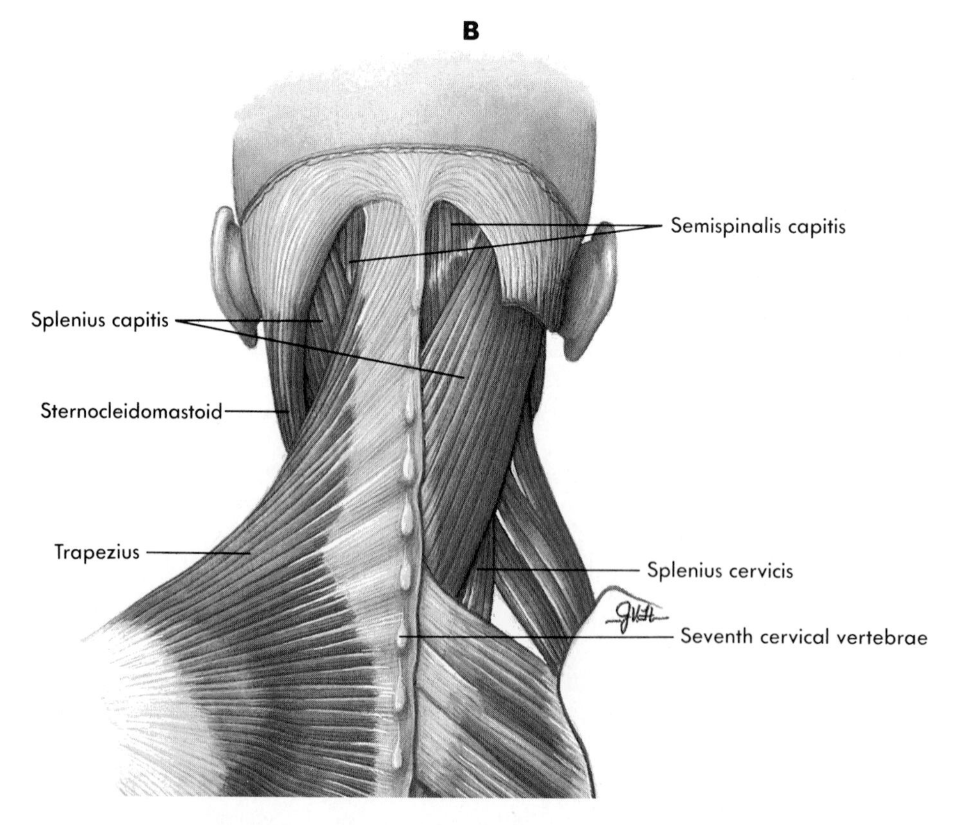

FIGURE 11-4 Muscles of the neck. **A** Anterior superficial. **B** Posterior superficial. **C** Posterior deep.

FIGURE 11-4, cont'd
For legend see opposite page.

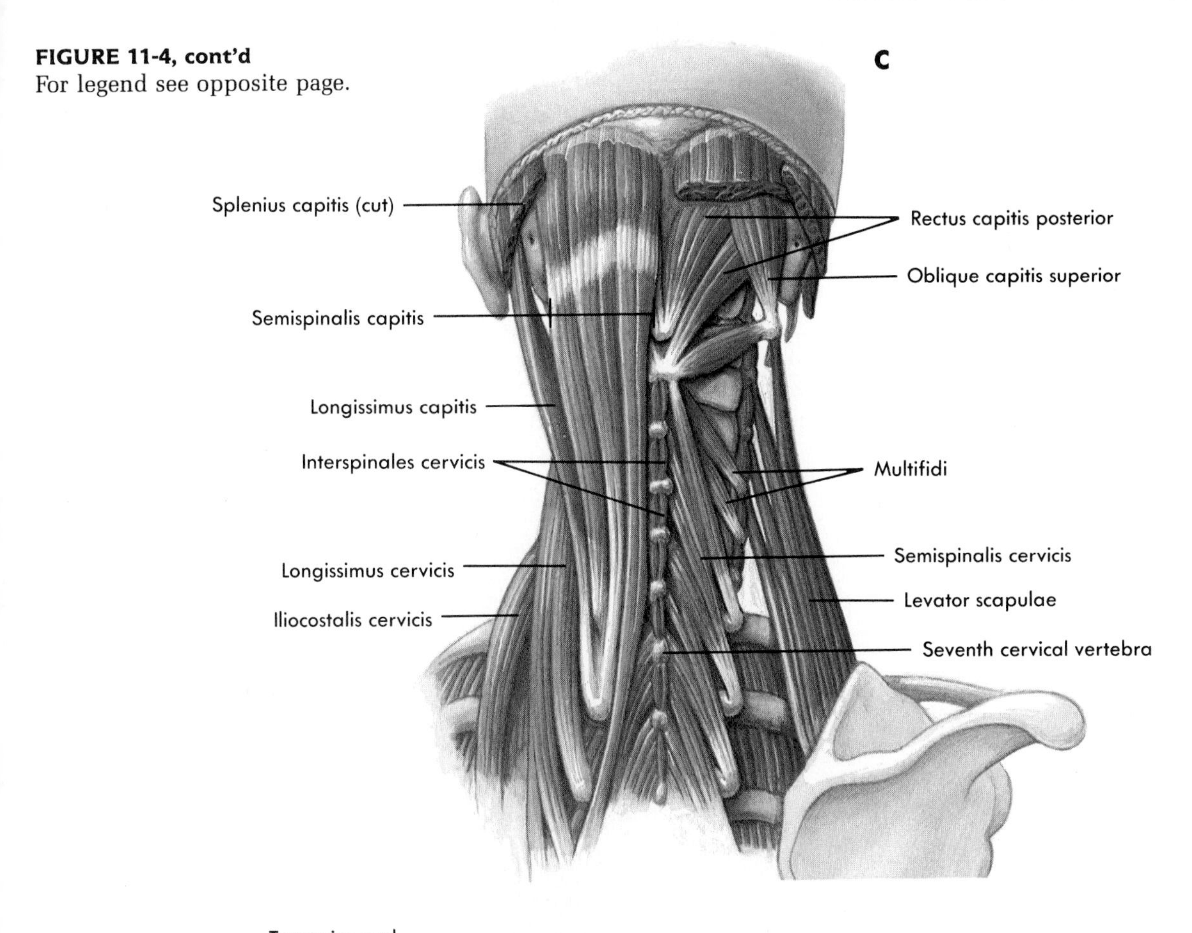

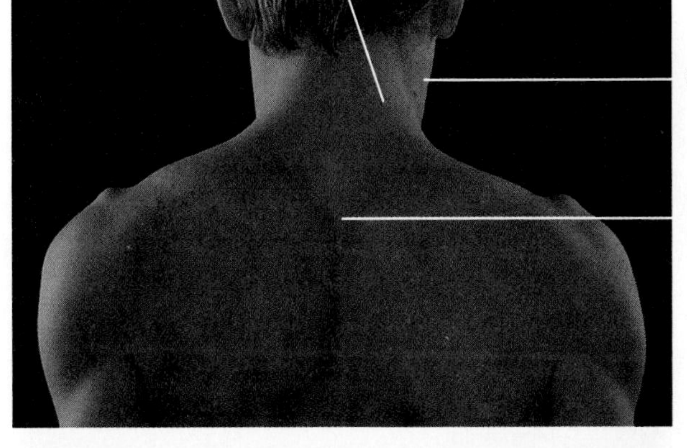

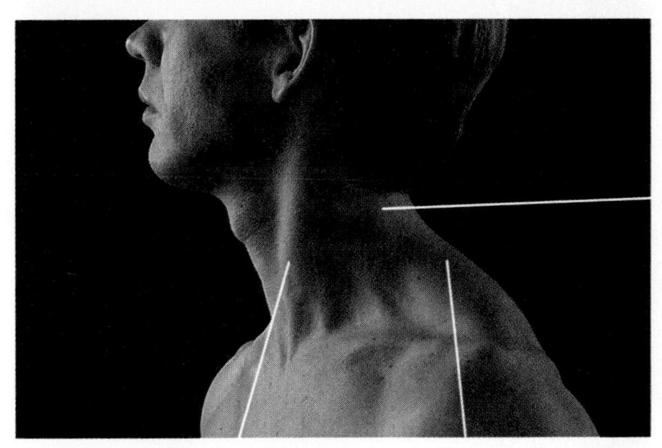

FIGURE 11-5 Surface anatomy, muscles of the neck. **A** Posterior view. **B** Lateral view.

Facial Expression

The skeletal muscles of the face (Table 11-2 and Figure 11-6) are cutaneous muscles attached to the skin. Many animals have cutaneous muscles over the trunk that allow the skin to twitch to remove irritants such as insects. In humans in whom facial expressions are important components of nonverbal communication, cutaneous muscles are confined primarily to the face and neck.

Several muscles act on the skin around the eyes and eyebrows (Figure 11-7). The **occipitofrontalis** raises the eyebrows and furrows the skin of the forehead. The **orbicularis oculi** (or-bik'u-lar'us ok'u-li) closes the eyelids and causes "crow's-feet" wrinkles in the skin at the lateral corners of the eyes. The **levator palpebrae** (pal-pe'bre; the palpebral fissure is the opening between the eyelids) **superioris** raises the upper lids (see Figures 11-6 and 11-7, *A*). A droopy eyelid on one side, called **ptosis** (to'sis), usually indicates that the nerve to the levator palpebrae superioris has been damaged. The **corrugator** (cor'ŭ-ga'tor) **supercilii** draws the eyebrows inferiorly and medially, producing vertical corrugations (furrows) in the skin between the eyes.

Several muscles function in moving the lips and

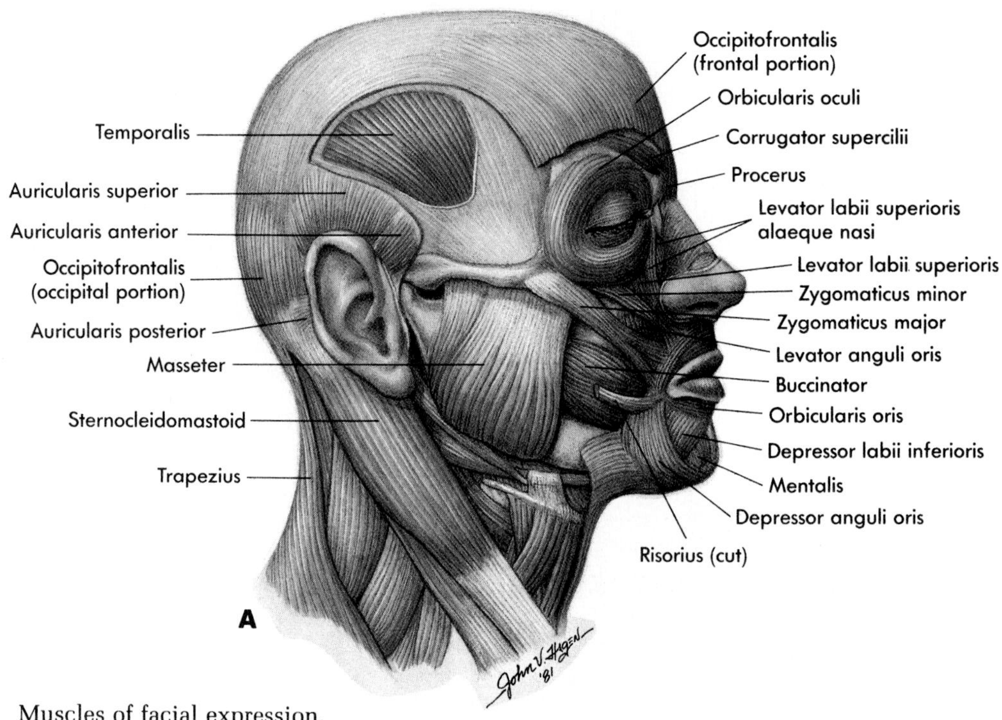

FIGURE 11-6 Muscles of facial expression. **A** Lateral view. **B** Anterior view.

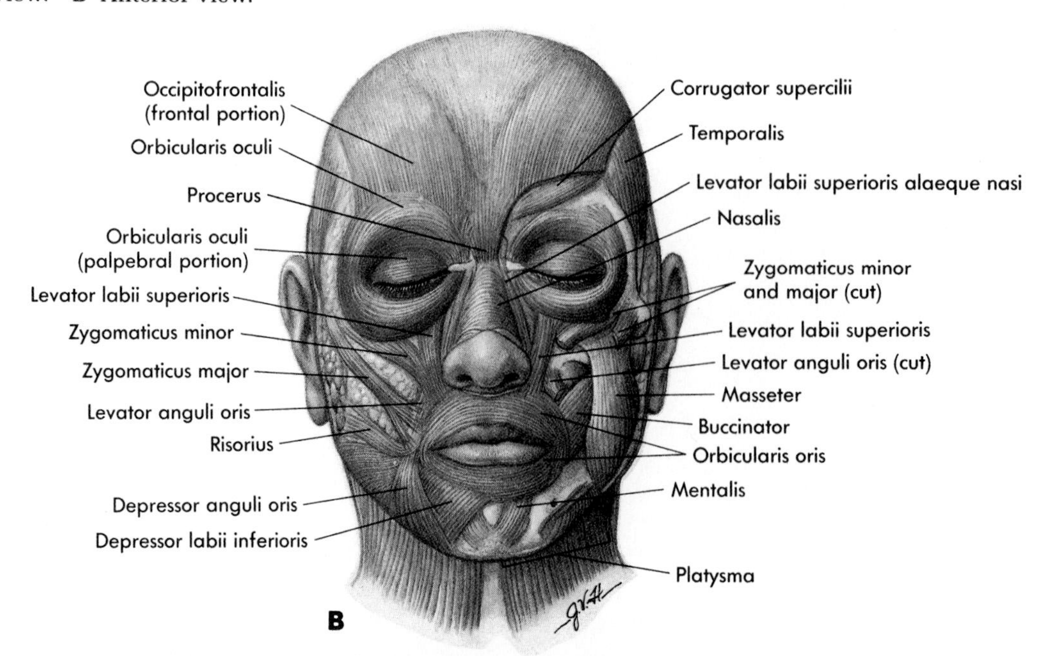

the skin surrounding the mouth. The **orbicularis oris** (or'us) and **buccinator** (buk'sĭ-na'tor), the kissing muscles, pucker the mouth. Smiling is accomplished by the **zygomaticus** (zi'go-mat'ĭ-kus) **major** and **minor,** the **levator anguli** (an'gu-li) **oris,** and the **risorius** (ri-so'rĭ-us). Sneering is accomplished by the **levator labii** (la'be-i) **superioris,** and frowning or pouting by the **depressor anguli oris, depressor labii inferioris,** and **mentalis** (men-tal'us). If the mentalis muscles are well developed on each side of the chin, a chin dimple may be located between the two muscles.

Table 11-2

Muscles of facial expression (see Figure 11-6)

MUSCLE	ORIGIN	INSERTION	NERVE	FUNCTION
Auricularis				
Anterior	Aponeurosis over head	Cartilage of auricle	Facial	Draws auricle superiorly and anteriorly
Posterior	Mastoid process	Posterior root of auricle	Facial	Draws auricle posteriorly
Superior	Aponeurosis over head	Cartilage of auricle	Facial	Draws auricle superiorly and posteriorly
Buccinator	Mandible and maxilla	Orbicularis oris at angle of mouth	Facial	Retracts angle of mouth; flattens cheek
Corrugator supercilii	Nasal bridge and orbicularis oculi	Skin of eyebrow	Facial	Depresses medial portion of eyebrow and draws eyebrows together as in frowning
Depressor anguli oris	Lower border of mandible	Lip near angle of mouth	Facial	Depresses angle of mouth
Depressor labii inferioris	Lower border of mandible	Skin of lower lip and orbicularis oris	Facial	Depresses lower lip
Levator anguli oris	Maxilla	Skin at angle of mouth and orbicularis oris	Facial	Elevates angle of mouth
Levator labii superioris	Maxilla	Skin and orbicularis oris of upper lip	Facial	Elevates upper lip
Levator labii superioris alaeque nasi	Maxilla	Ala of nose and upper lip	Facial	Elevates ala of nose and upper lip
Levator palpebrae superioris	Lesser wing of sphenoid	Skin of eyelid	Oculomotor	Elevates upper eyelid
Mentalis	Mandible	Skin of chin	Facial	Elevates and wrinkles skin over chin; elevates lower lip
Nasalis	Maxilla	Bridge and ala of nose	Facial	Dilates nostril
Occipitofrontalis	Occipital bone	Skin of eyebrow and nose	Facial	Moves scalp; elevates eyebrows
Orbicularis oculi	Maxilla and frontal bones	Circles orbit and inserts near origin	Facial	Closes eye
Orbicularis oris	Nasal septum, maxilla, and mandible	Fascia and other muscles of lips	Facial	Closes lips
Platysma	Fascia of deltoid and pectoralis major	Skin over inferior border of mandible	Facial	Depresses lower lip; wrinkles skin of neck and upper chest
Procerus	Bridge of nose	Frontalis	Facial	Creates horizontal wrinkle between eyes
Risorius	Platysma and masseter fascia	Orbicularis oris and skin at corner of mouth	Facial	Abducts angle of mouth
Zygomaticus major	Zygomatic bone	Angle of mouth	Facial	Elevates and abducts upper lip
Zygomaticus minor	Zygomatic bone	Orbicularis oris of upper lip	Facial	Elevates and abducts upper lip

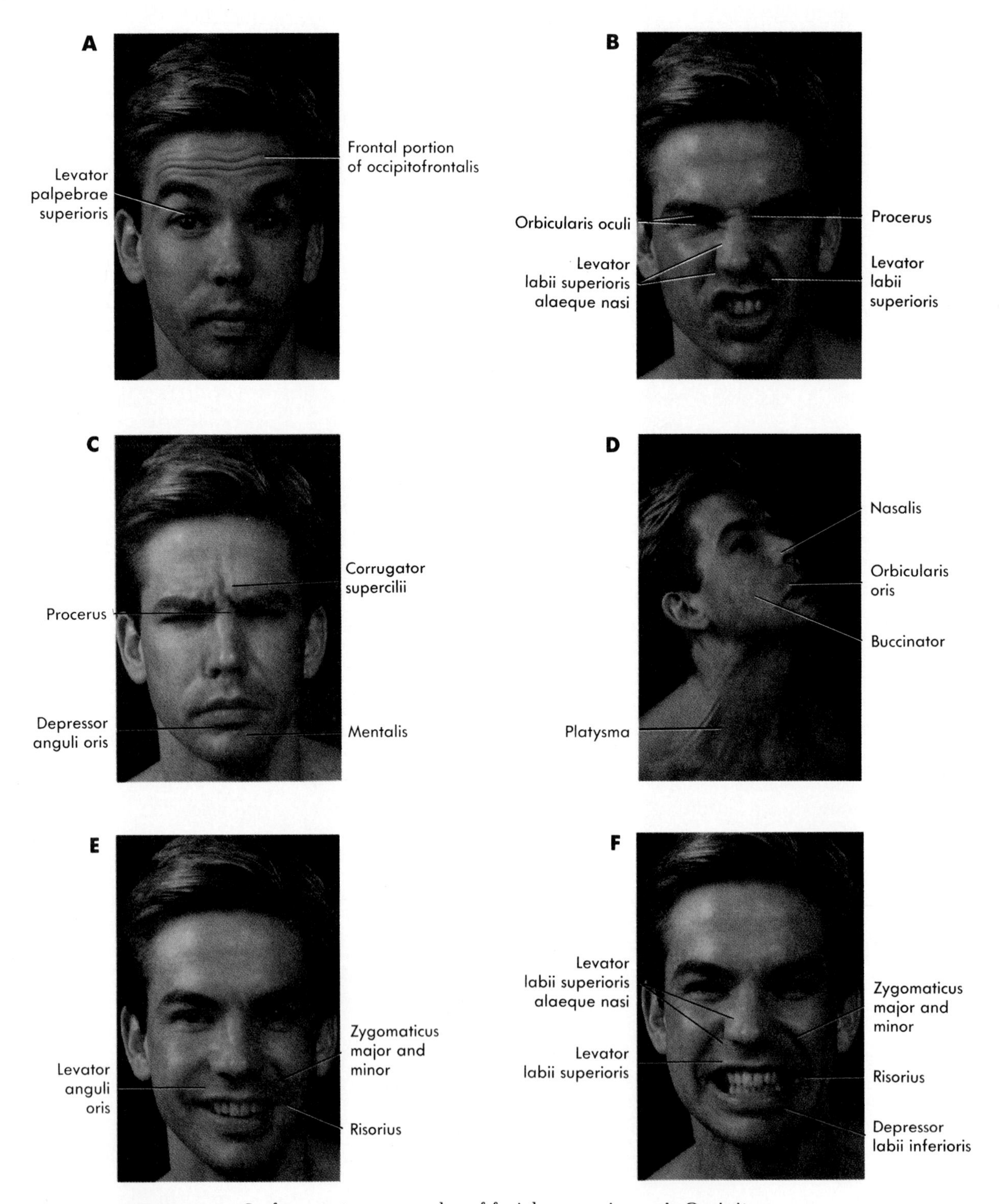

FIGURE 11-7 Surface anatomy, muscles of facial expression. **A** Occipitofrontalis. **B** Procerus and levator labii superioris alaeque nasi. **C** Corrugator supercilii. **D** Platysma. **E** Zygomaticus major and minor. **F** Multiple muscles around the mouth.

2

Harry Wolf, a notorious flirt, on seeing Sally Gorgeous raises his eyebrows, winks, whistles, and smiles. Name the facial muscles he used to carry out this communication. Sally, thoroughly displeased with this exhibition, frowns and flares her nostrils in disgust. What muscles did she use?

? ? ? ? ? ? ? ? ? ? ?

Mastication

Chewing, or **mastication** (mas'tĭ-ka'shun), involves forcefully closing the mouth (elevating the mandible) and grinding the food between the teeth (medial and lateral excursion of the mandible). The **muscles of mastication** and the **hyoid muscles** move the mandible (Tables 11-3 and 11-4; Figures 11-8 and 11-9). The elevators of the mandible are some of the strongest muscles of the body, bringing the mandib-

Table 11-3

Muscles of mastication (see Figures 11-6, A, and 11-8)

MUSCLE	ORIGIN	INSERTION	NERVE	FUNCTION
Temporalis	Temporal fossa	Anterior portion of mandibular ramus and coronoid process	Mandibular division of trigeminal	Elevates and retracts mandible; involved in excursion
Masseter	Zygomatic arch	Lateral side of mandibular ramus	Mandibular division of trigeminal	Elevates and protracts mandible; involved in excursion
Pterygoid				
Lateral	Pterygoid process and greater wing of sphenoid	Condylar process of mandible and articular disk	Mandibular division of trigeminal	Protracts and depresses mandible; involved in excursion
Medial	Pterygoid process of sphenoid and tuberosity of maxilla	Medial surface of mandible	Mandibular division of trigeminal	Protracts and elevates mandible; involved in excursion

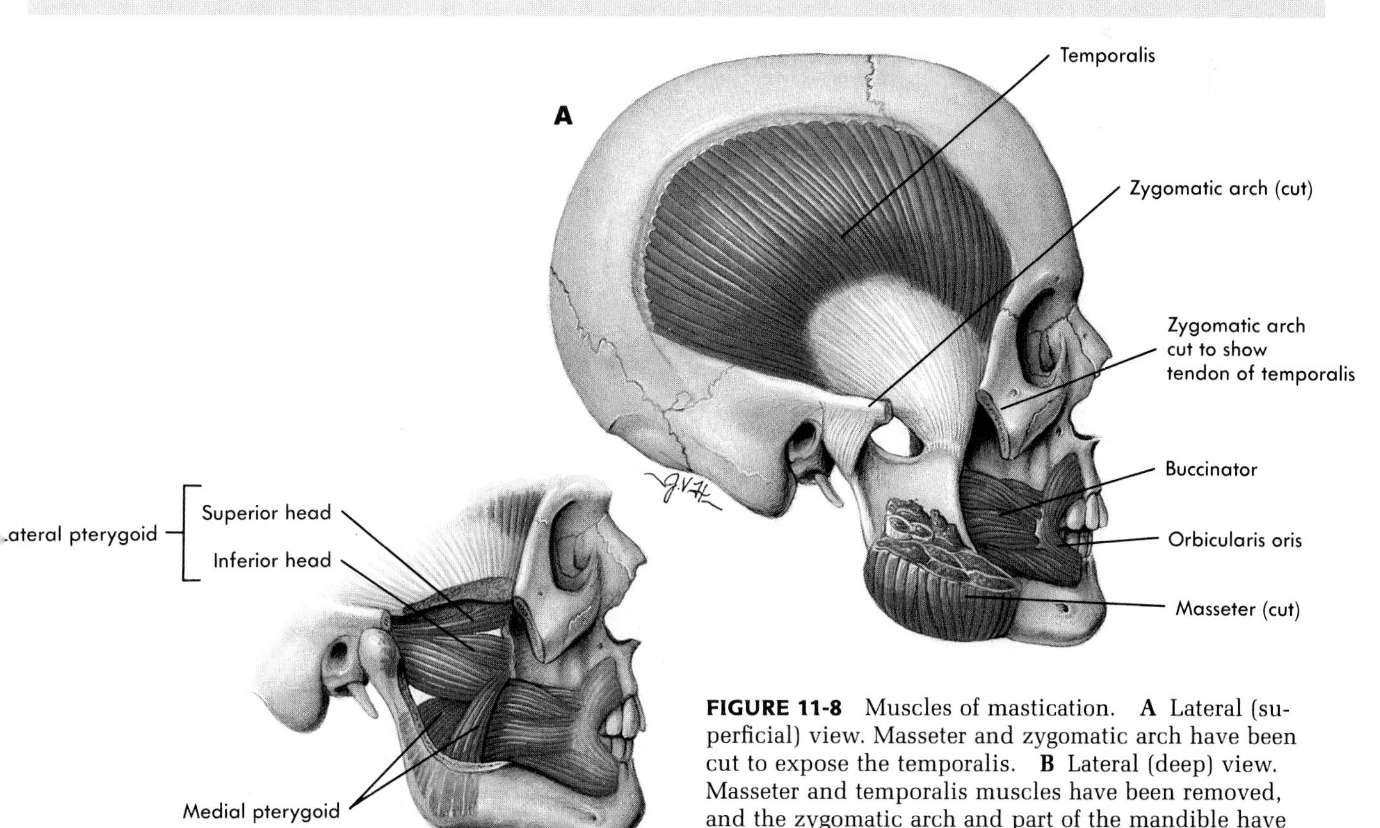

FIGURE 11-8 Muscles of mastication. **A** Lateral (superficial) view. Masseter and zygomatic arch have been cut to expose the temporalis. **B** Lateral (deep) view. Masseter and temporalis muscles have been removed, and the zygomatic arch and part of the mandible have been cut away to reveal the deeper muscles.

Table 11-4

Hyoid muscles (see Figures 11-9 and 11-10)

MUSCLE	ORIGIN	INSERTION	NERVE	FUNCTION
Suprahyoid Muscles				
Digastric	Mastoid process	Mandible near midline	Posterior—facial; anterior—mandibular of trigeminal	Elevates hyoid; depresses and retracts mandible
Geniohyoid	Genu of mandible	Body of hyoid	Fibers of C1 and C2 with hypoglossal	Protracts hyoid; depresses mandible
Mylohyoid	Body of mandible	Hyoid	Mandibular division of trigeminal	Elevates floor of mouth and tongue; depresses mandible when hyoid is fixed
Stylohyoid	Styloid process	Hyoid bone	Facial	Elevates hyoid
Infrahyoid Muscles				
Omohyoid	Superior border of scapula	Hyoid bone	Upper cervical through ansa cervicalis	Depresses hyoid; fixes hyoid in mandibular depression
Sternohyoid	Manubrium and costal cartilage 1	Hyoid	Upper cervical through ansa cervicalis	Depresses hyoid; fixes hyoid in mandibular depression
Sternothyroid	Manubrium and costal cartilage 1 or 2	Thyroid cartilage	Upper cervical through ansa cervicalis	Depresses larynx; fixes hyoid in mandibular depression
Thyrohyoid	Thyroid cartilage	Hyoid	Upper cervical, passing with hypoglossal	Depresses hyoid and elevates thyroid cartilage of larynx; fixes hyoid in mandibular depression

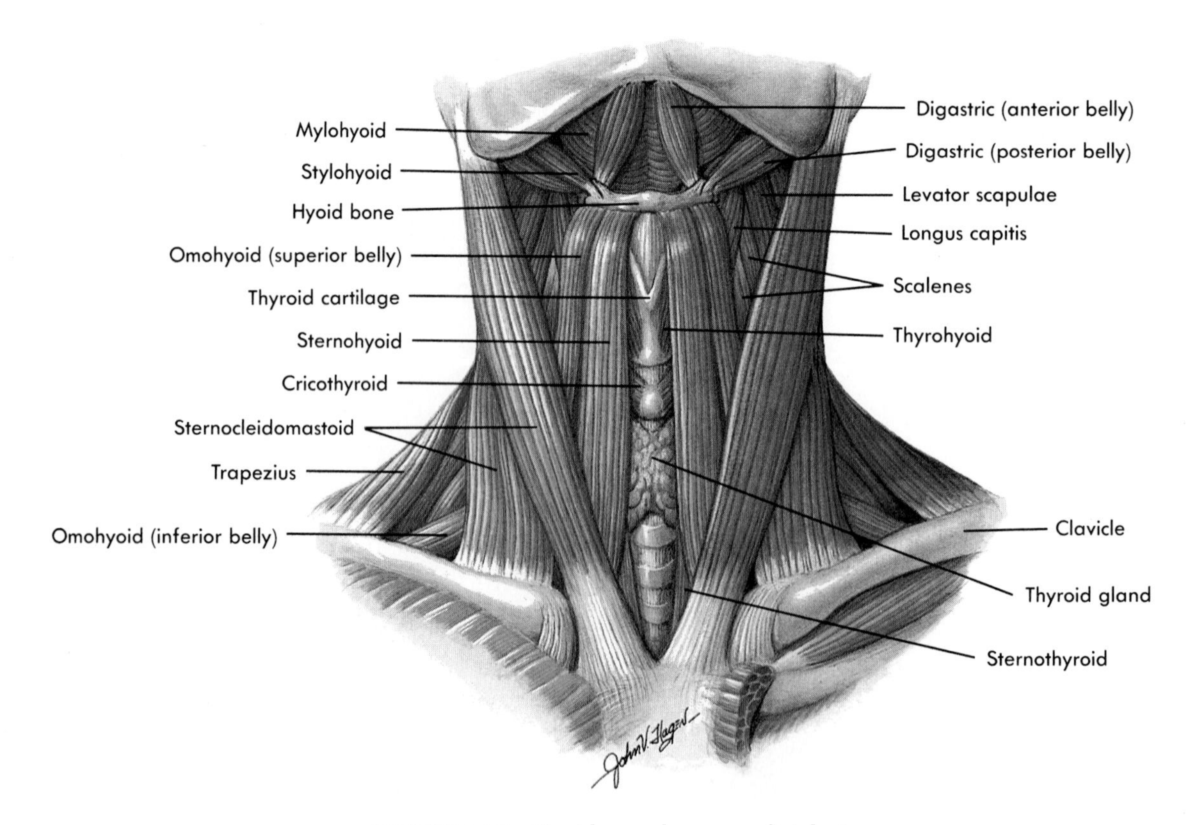

FIGURE 11-9 Hyoid muscles, superficial view.

ular teeth forcefully against the maxillary teeth to crush food. Slight mandibular depression involves relaxation of the mandibular elevators and the pull of gravity. Opening the mouth wide requires the action of the depressors of the mandible. The muscles of the tongue and the buccinator (see Table 11-2; Table 11-5), even though they are not involved in the actual process of chewing, help move the food in the mouth and hold it in place between the teeth.

Tongue Movements

The tongue is very important in mastication and speech: (1) it moves the food around in the mouth; (2) with the buccinator it holds the food in place while the teeth grind it; and (3) it pushes the food up to the palate and back toward the pharynx to initiate swallowing. The tongue consists of a mass of **intrinsic muscles** (entirely within the tongue), which are involved in changing the shape of the tongue, and **extrinsic muscles** (outside of the tongue but attached to it), which help change the shape and move the tongue (see Table 11-5; Figure 11-10).

Everyone can change the shape of the tongue, but not everyone can accomplish certain kinds of tongue movements. The ability to accomplish such movements apparently is controlled genetically. For example, some people are able to roll the tongue into a tube shape, and others cannot. It is not known exactly what tongue muscles are involved in these tongue movements, and no anatomical differences have been reported for tongue rollers vs. nonrollers.

Table 11-5

Tongue muscles (see Figure 11-10)

MUSCLE	ORIGIN	INSERTION	NERVE	FUNCTION
Intrinsic Muscles				
Longitudinal, transverse, and vertical	Within tongue	Within tongue	Hypoglossal	Changes tongue shape
Extrinsic Muscles				
Genioglossus	Genu of mandible	Tongue	Hypoglossal	Depresses and protrudes tongue
Hyoglossus	Hyoid	Side of tongue	Hypoglossal	Retracts and depresses side of tongue
Styloglossus	Styloid process of temporal bone	Tongue (lateral and inferior)	Hypoglossal	Retracts tongue
Palatoglossus	Soft palate	Tongue	Pharyngeal plexus	Elevates posterior tongue

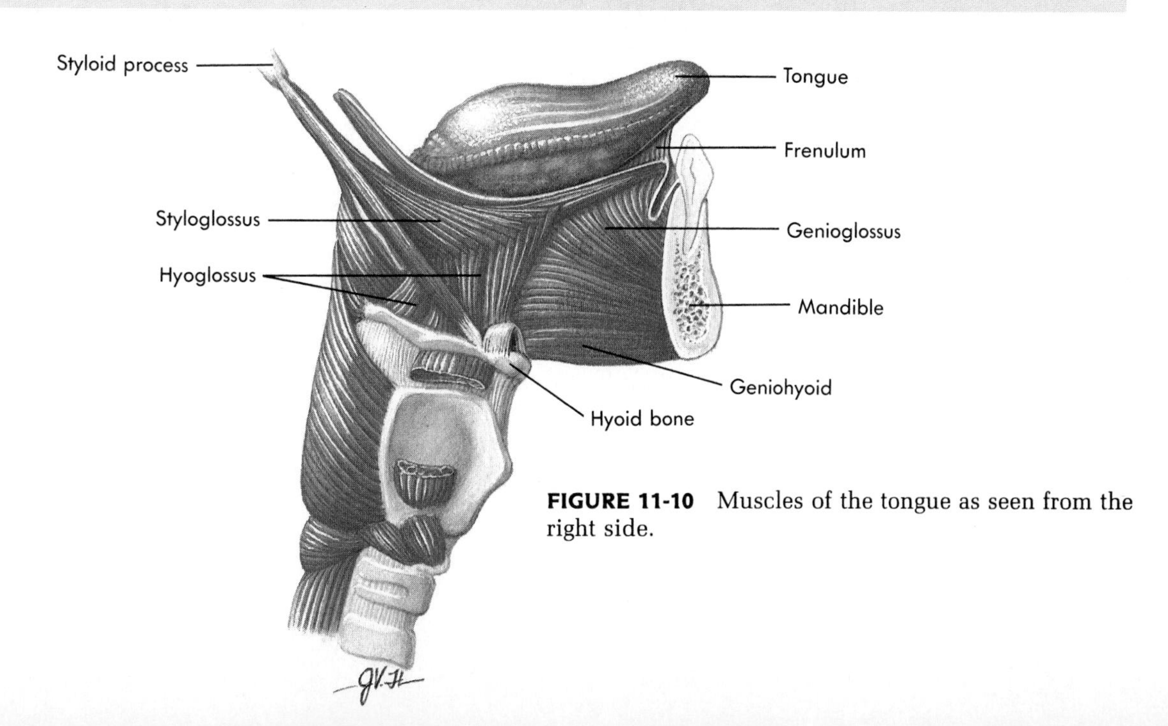

FIGURE 11-10 Muscles of the tongue as seen from the right side.

Swallowing and the Larynx

The hyoid muscles (see Table 11-4 and Figure 11-9) are divided into a suprahyoid group superior to the hyoid bone and an infrahyoid group inferior to the hyoid. When the hyoid bone is fixed by the infrahyoid muscles so that the bone is stabilized from below, the suprahyoid muscles can help depress the mandible. If the suprahyoid muscles fix the hyoid and thus stabilize it from above, the thyrohyoid muscle (an infrahyoid muscle) can elevate the larynx. To observe this effect, place your hand on your larynx (Adam's apple) and swallow.

The soft palate, pharynx, and larynx contain several muscles involved in swallowing and speech (Table 11-6 and Figure 11-11). The muscles of the soft palate close the posterior opening to the nasal cavity during swallowing.

Swallowing (see Chapter 24) is accomplished by elevation of the pharynx, which in turn is accomplished by elevation of the larynx, to which the pharynx is attached, and constriction of the **palatopharyngeus** (pal-at′o-far-in-je′us) and **salpingopharyngeus** (sal-ping′go-far-in-je′us; *salpingo* means trumpet and refers to the trumpet-shaped opening of the auditory, or eustachian, tube). The pharyngeal constrictor muscles then constrict from superior to inferior, forcing the food into the esophagus.

The salpingopharyngeus also opens the auditory tube, which connects the middle ear with the pharynx. Opening the auditory tube equalizes the pressure between the middle ear and the atmosphere and is why it is sometimes helpful to chew gum or swallow when ascending or descending a mountain in a car or when changing altitudes in an airplane.

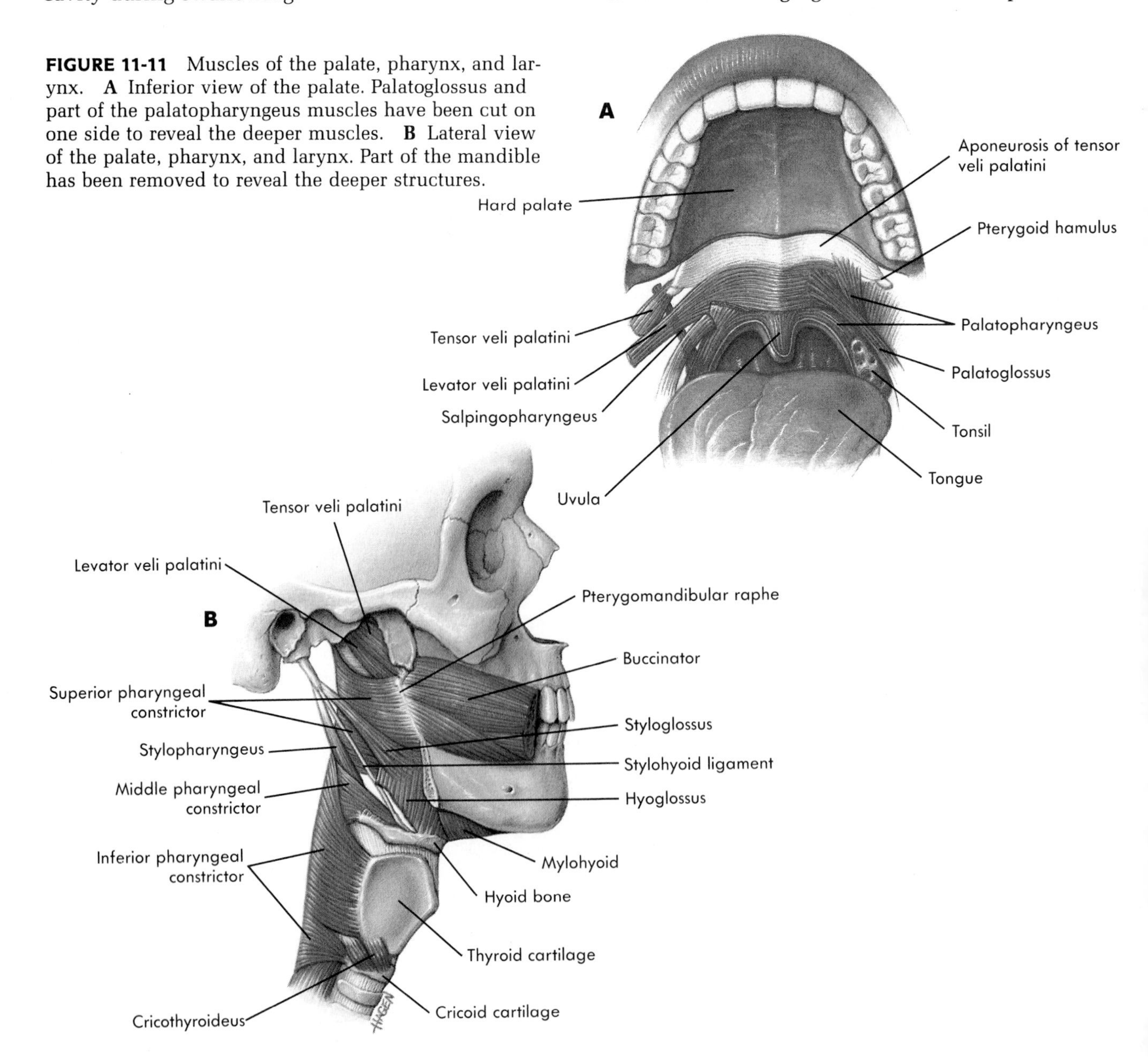

FIGURE 11-11 Muscles of the palate, pharynx, and larynx. **A** Inferior view of the palate. Palatoglossus and part of the palatopharyngeus muscles have been cut on one side to reveal the deeper muscles. **B** Lateral view of the palate, pharynx, and larynx. Part of the mandible has been removed to reveal the deeper structures.

Table 11-6

Muscles of swallowing and the larynx (see Figure 11-11)

MUSCLE	ORIGIN	INSERTION	NERVE	FUNCTION
Larynx (not illustrated)				
Arytenoideus				
Oblique	Arytenoid cartilage	Opposite arytenoid cartilage	Recurrent laryngeal	Narrows opening to larynx
Transverse	Arytenoid cartilage	Opposite arytenoid cartilage	Recurrent laryngeal	Narrows opening to larynx
Cricoarytenoideus				
Lateral	Lateral side of cricoid cartilage	Arytenoid cartilage	Recurrent laryngeal	Narrows opening to larynx
Posterior	Posterior side of cricoid cartilage	Arytenoid cartilage	Recurrent laryngeal	Widens opening to larynx
Cricothyroideus	Anterior cricoid cartilage	Thyroid cartilage	Superior laryngeal	Tenses vocal cords
Thyroarytenoideus	Thyroid cartilage	Arytenoid cartilage	Recurrent laryngeal	Shortens vocal cords
Vocalis	Thyroid cartilage	Arytenoid cartilage	Recurrent laryngeal	Shortens vocal cords
Soft Palate				
Levator veli palatini	Temporal bone and auditory tube	Soft palate	Pharyngeal plexus	Elevates soft palate
Palatoglossus	Soft palate	Tongue	Pharyngeal plexus	Narrows fauces; elevates posterior tongue
Palatopharyngeus	Soft palate	Pharynx	Pharyngeal plexus	Narrows fauces; depresses palate; elevates pharynx
Tensor veli palatini	Sphenoid and auditory tube	Soft palate	Mandibular division of trigeminal	Tenses soft palate; opens auditory tube
Uvulae	Posterior nasal spine	Uvula	Pharyngeal plexus	Elevates uvula
Pharynx				
Pharyngeal constrictor				
Inferior	Thyroid and cricoid cartilages	Pharyngeal raphe	Pharyngeal plexus and external laryngeal nerve	Narrows lower pharynx in swallowing
Middle	Styloid ligament and hyoid	Pharyngeal raphe	Pharyngeal plexus	Narrows pharynx in swallowing
Superior	Medial pterygoid plate, mandible, floor of mouth, and side of tongue	Pharyngeal raphe	Pharyngeal plexus	Narrows pharynx in swallowing
Salpingopharyngeus	Auditory tube	Pharynx	Pharyngeal plexus	Elevates pharynx; opens auditory tube in swallowing
Stylopharyngeus	Styloid process	Pharynx	Glossopharyngeus	Elevates pharynx

The muscles of the larynx are listed in Table 11-6 and are illustrated in Figure 11-11, *B*. Most of the laryngeal muscles help to narrow or close the laryngeal opening so food does not enter the larynx when a person swallows. The remaining muscles shorten the vocal cord to raise the pitch of the voice.

Snoring is a rough, raspy noise that can occur when a sleeping person inhales through the mouth and nose. The noise usually is made by vibration of the soft palate but also may occur as a result of vocal cord vibration.

Laryngospasm is a tetanic contraction of the muscles around the opening of the larynx. In severe cases the opening is closed completely, air no longer can pass through the larynx into the lungs, and the victim may die of asphyxiation. Laryngospasm can develop as a result of, for example, tetanus infections or hypocalcemia.

Movements of the Eyeball

The eyeball rotates within the orbital fossa, allowing vision in a wide range of directions. The movements of each eye are accomplished by six muscles named for the orientation of their fasciculi relative to the spherical eye (Figure 11-12).

Each rectus muscle (so named because the fibers are nearly straight with the axis of the eye) attaches to the globe of the eye anterior to the center of the sphere. The superior rectus rotates the anterior portion of the globe superiorly so that the pupil, and thus the gaze, are directed superiorly (looking up). The inferior rectus' depresses the gaze, the lateral rectus laterally deviates the gaze (looking to the side), and the medial rectus medially deviates the gaze (looking toward the nose). The superior rectus and inferior rectus are not completely straight in their orientation to the eye, so they also medially deviate the gaze as they contract.

The oblique muscles (so named because their fibers are oriented obliquely to the axis of the eye)

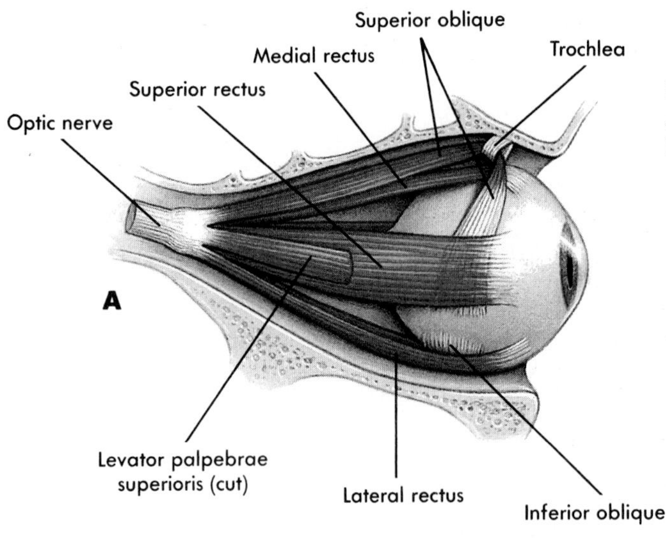

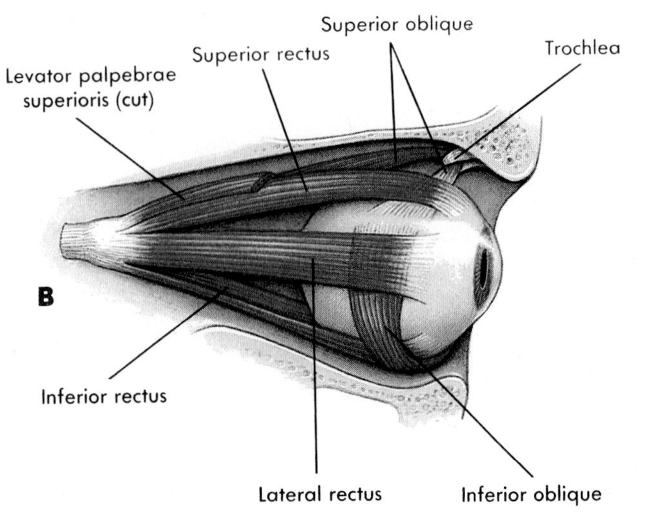

FIGURE 11-12 Muscles moving the eyeball.
A Superior view of right eyeball.
B Lateral view of right eyeball.

insert onto the posterolateral margin of the globe so that both muscles laterally deviate the gaze as they contract. The superior oblique elevates the posterior portion of the eye, thus directing the pupil inferiorly and depressing the gaze. The inferior oblique muscle elevates the gaze.

3

Strabismus (strā-biz'mus) is a condition in which one or both eyes deviate in a medial or lateral direction and in some cases may be caused by a weakness in either the medial or lateral rectus muscle. If the lateral rectus of the right eye were weak, in which direction would the eye deviate?

? ? ? ? ? ? ? ? ? ? ?

TRUNK MUSCLES

Muscles Moving the Vertebral Column

The muscles that extend the spine and abduct and rotate the vertebral column can be divided into deep and superficial groups (Table 11-7). In general, the muscles of the deep group extend from vertebra to vertebra, whereas the muscles of the superficial group extend from vertebra to ribs. In humans these back muscles are very strong to maintain erect posture, but comparable muscles in cattle are relatively delicate, constituting the area from which New York (top loin) steaks are cut. The **erector spinae** group of muscles on each side of the back consists of three subgroups, the **iliocostalis** (il'e-o-cos-tal'us), **longissimus** (lon-jis'ĭ-mus), and **spinalis** (spi-nal'us). The longissimus group accounts for most of the muscle mass in the lower back (Figure 11-13).

Text continued on p. 324.

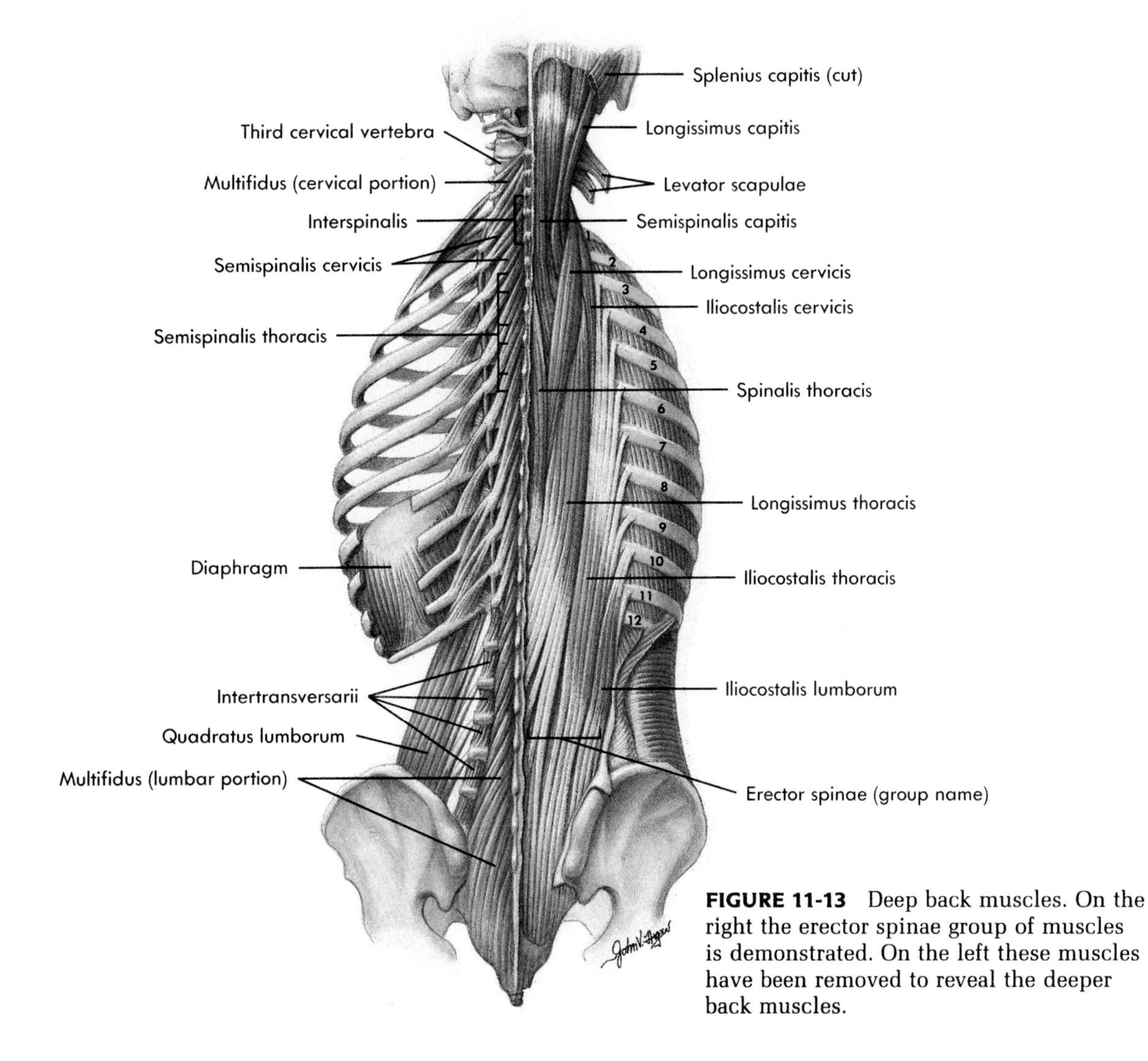

FIGURE 11-13 Deep back muscles. On the right the erector spinae group of muscles is demonstrated. On the left these muscles have been removed to reveal the deeper back muscles.

Table 11-7

Muscles acting on the vertebral column (see Figures 11-4 and 11-13)

MUSCLE	ORIGIN	INSERTION	NERVE	FUNCTION
Superficial				
Erector spinae (divides into three columns)	Sacrum, ilium, and lumbar spines	Ribs and vertebrae	Dorsal rami of spinal nerves	Extends vertebral column
Iliocostalis				
Cervicis	Superior six ribs	Middle cervical vertebrae	Dorsal rami	Extends, abducts, and rotates vertebral column
Thoracis	Inferior six ribs	Superior six ribs	Dorsal rami	Extends, abducts, and rotates vertebral column
Lumborum	Sacrum, ilium, and lumbar vertebrae	Inferior six ribs	Dorsal rami	Extends, abducts, and rotates vertebral column
Longissimus				
Capitis	Upper thoracic and lower cervical vertebrae	Mastoid process	Dorsal rami	Extends head
Cervicis	Upper thoracic vertebrae	Upper cervical vertebrae	Dorsal rami	Extends neck
Thoracis	Ribs and lower thoracic vertebrae	Upper lumbar vertebrae and ribs	Dorsal rami	Extends vertebral column
Spinalis				
Cervicis	C6-C7	C2-C3	Dorsal rami	Extends neck
Thoracis	T11-L2	Middle and upper thoracic vertebrae	Dorsal rami	Extends vertebral column
Longus colli	C3-T3	C1-C6	Ventral rami of cervical nerves	Rotates and flexes neck
Splenius cervicis	C3-C5	C1-C3	Dorsal rami	Rotates and extends neck
Deep				
Interspinales	Spinous processes of all vertebrae	Next superior spinous process	Dorsal rami	Extends back and neck
Intertransversarii	Transverse processes of all vertebrae	Next superior transverse process	Dorsal rami	Abducts vertebral column
Multifidus	Transverse processes of vertebrae, posterior surface of sacrum and ilium	Spinous processes of superior vertebrae	Dorsal rami	Extends and rotates vertebral column
Psoas minor	T12-L1	Near pubic crest	L1	Flexes vertebral column
Rotatores	Transverse processes of all vertebrae	Base of spinous process of superior vertebrae	Dorsal rami	Extends and rotates vertebral column
Semispinalis				
Cervicis	Transverse processes of T2-T5	Spinous processes of C2-C5	Dorsal rami	Extends neck
Thoracis	Transverse processes of T5-T11	Spinous processes of C5-T4	Dorsal rami	Extends vertebral column

Table 11-8

Muscles of the thorax (see Figures 11-4, B, and 11-14)

MUSCLE	ORIGIN	INSERTION	NERVE	FUNCTION
Diaphragm	Interior of ribs, sternum, and lumbar vertebrae	Central tendon of diaphragm	Phrenic	Inspiration; depresses floor of thorax
Intercostalis				
External	Inferior margin of each rib	Superior border of next rib below	Intercostal	Inspiration; elevates ribs
Internal	Superior margin of each rib	Inferior border of next rib above	Intercostal	Expiration; depresses ribs
Scalenus				
Anterior	C3-C6	First rib	Cervical plexus	Elevates first rib
Medial	C2-C6	First rib	Cervical plexus	Elevates first rib
Posterior	C4-C6	Second rib	Cervical and brachial plexuses	Elevates second rib
Serratus posterior				
Inferior	T11-L2	Inferior four ribs	Ninth to twelfth intercostals	Depresses inferior ribs and extends back
Superior	C6-T2	Second to fifth ribs	First to fourth intercostals	Elevates superior ribs
Transversus thoracis	Sternum and xiphoid process	Second to sixth costal cartilage	Intercostal	Narrows chest

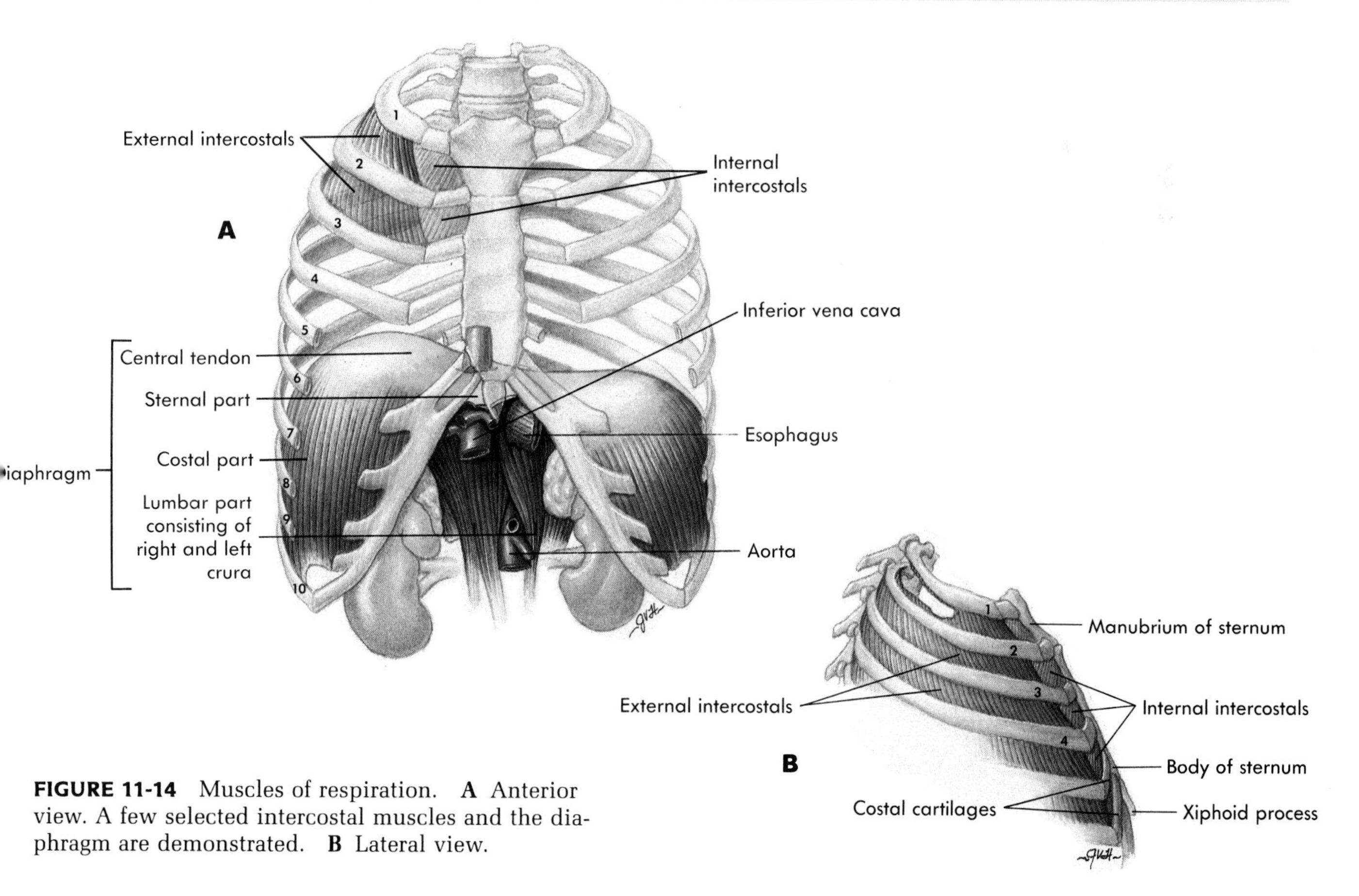

FIGURE 11-14 Muscles of respiration. **A** Anterior view. A few selected intercostal muscles and the diaphragm are demonstrated. **B** Lateral view.

Thoracic Muscles

The muscles of the thorax are involved almost entirely in the process of breathing (see Chapter 23). Four major groups of muscles are associated with the rib cage (Table 11-8 and Figure 11-14, p. 323). The **scalene** (ska'lēn) muscles elevate the first two ribs during inspiration. The **external intercostals** (in'ter-kos'tals) also elevate the ribs during inspiration. The **internal intercostals** and transverse thoracic muscles contract during forced expiration.

The major movement produced during quiet breathing, however, is accomplished by the **diaphragm** (di'ă-fram; see Figure 11-14, *A*). It is dome shaped when relaxed; when it contracts, the dome is flattened, causing the volume of the thoracic cavity to increase and resulting in inspiration. If this wall of skeletal muscle or the phrenic nerve supplying it is severely damaged, the amount of air exchanged in the lungs may be so small that the individual is likely to die unless connected to an artificial respirator.

Abdominal Wall

The muscles of the anterior abdominal wall (Table 11-9) flex and rotate the vertebral column. Contraction of the abdominal muscles when the vertebral column is fixed decreases the volume of the abdominal cavity and the thoracic cavity and can aid in such functions as forced expiration, vomiting, defecation, urination, and childbirth. The crossing pattern of the abdominal muscles creates a strong anterior wall that holds in and protects the abdominal viscera.

In a relatively muscular person with little fat, a vertical line is visible, extending from the area of the xiphoid process of the sternum through the navel to the pubis. This tendinous area of the abdominal wall devoid of muscle, the **linea alba** (lin'e-ah al'bah), or white line, is so named because it consists of white connective tissue rather than muscle (Figure 11-15). On each side of the linea alba is the **rectus abdominis** (Figures 11-16 and 11-17). **Tendinous inscriptions** (in-scrip'shuns) transect the rectus abdominis at three, or sometimes more, locations, causing the abdominal wall of a well-muscled person to appear segmented. Lateral to the rectus abdominis is the **linea semilunaris** (sem'i-lu-nar'us; crescent- or half moon–shaped line); lateral to it are three layers of muscle (see Figures 11-15 through 11-17). From superficial to deep, these muscles are the **external abdominal oblique, internal abdominal oblique,** and **transversus abdominis.**

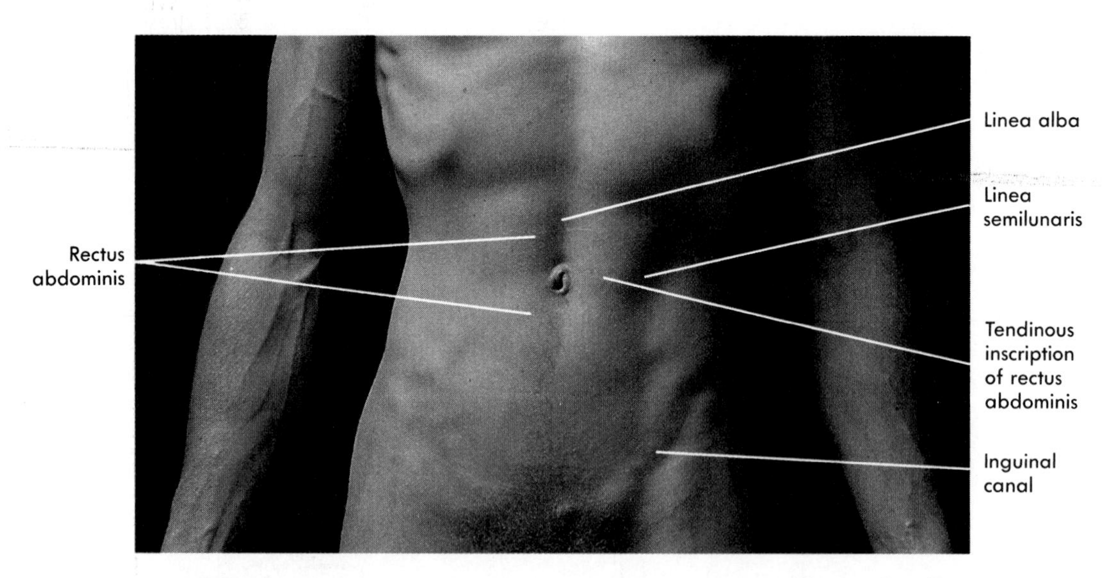

FIGURE 11-15 Surface anatomy, muscles of the anterior abdominal wall.

Table 11-9

Muscles of the abdominal wall (see Figures 11-13, 11-16, and 11-17)

MUSCLE	ORIGIN	INSERTION	NERVE	FUNCTION
Rectus abdominis	Pubic crest and symphysis	Xiphoid process and inferior ribs	Branches of lower thoracic	Flexes vertebral column; compresses abdomen
External abdominal oblique	Fifth to twelfth ribs	Iliac crest, inguinal ligament, and rectus sheath	Branches of lower thoracic	Flexes and rotates vertebral column; compresses abdomen; depresses thorax
Internal abdominal oblique	Iliac crest, inguinal ligament, and lumbar fascia	Tenth to twelfth ribs and rectus sheath	Lower thoracic	Flexes and rotates vertebral column; compresses abdomen; depresses thorax
Transversus abdominis	Seventh to twelfth costal cartilages, lumbar fascia, iliac crest, and inguinal ligament	Xiphoid process, linea alba, and pubic tubercle	Lower thoracic	Compresses abdomen
Quadratus lumborum	Iliac crest and lower lumbar vertebrae	Twelfth rib and upper lumbar vertebrae	Upper lumbar	Abducts vertebral column and depresses twelfth rib

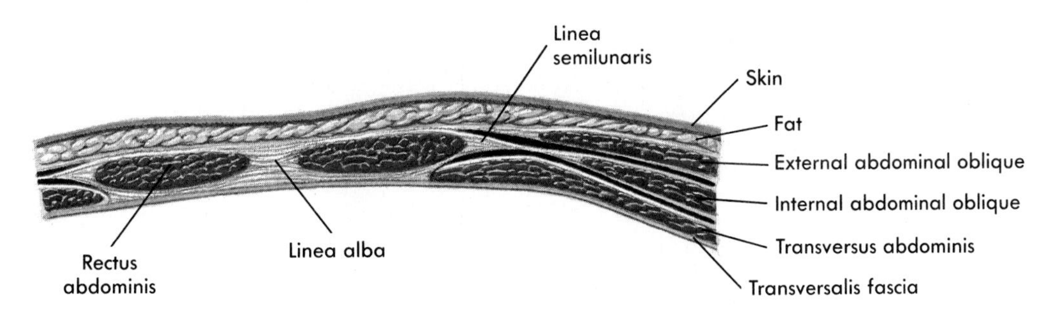

FIGURE 11-16 Muscles of the anterior abdominal wall. Cross section superior to the umbilicus.

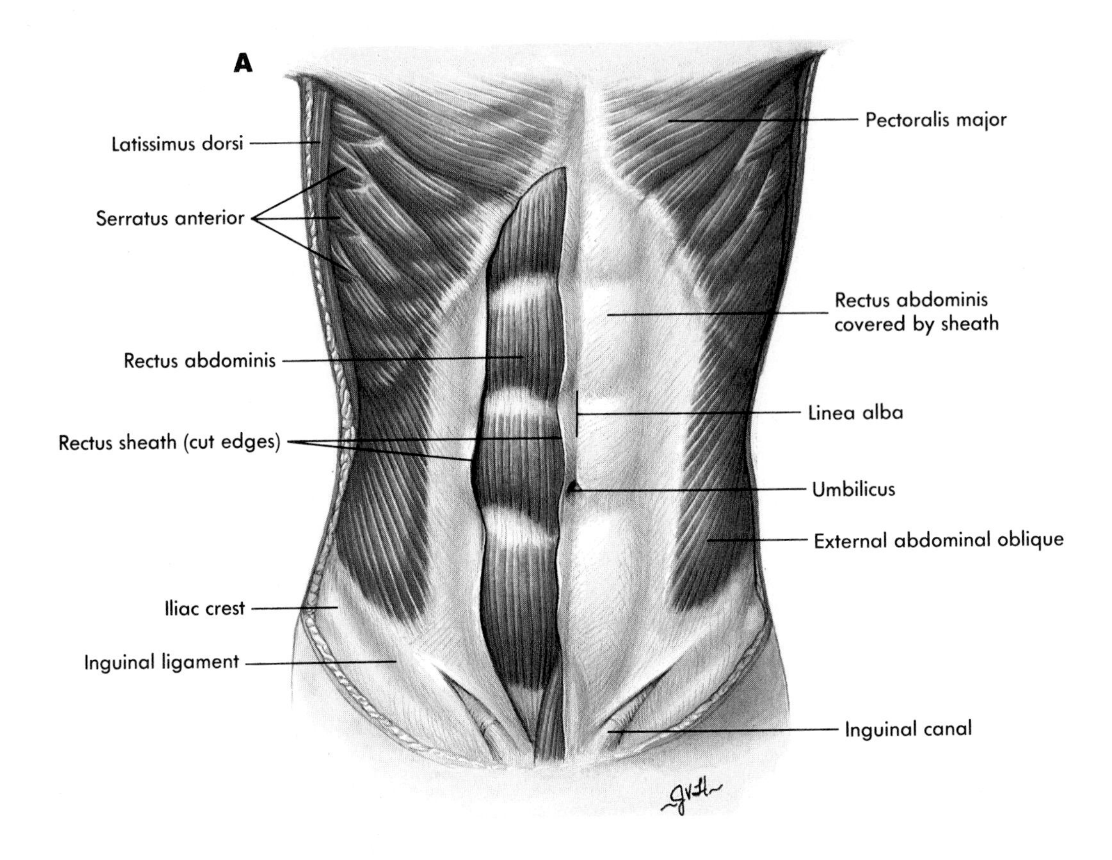

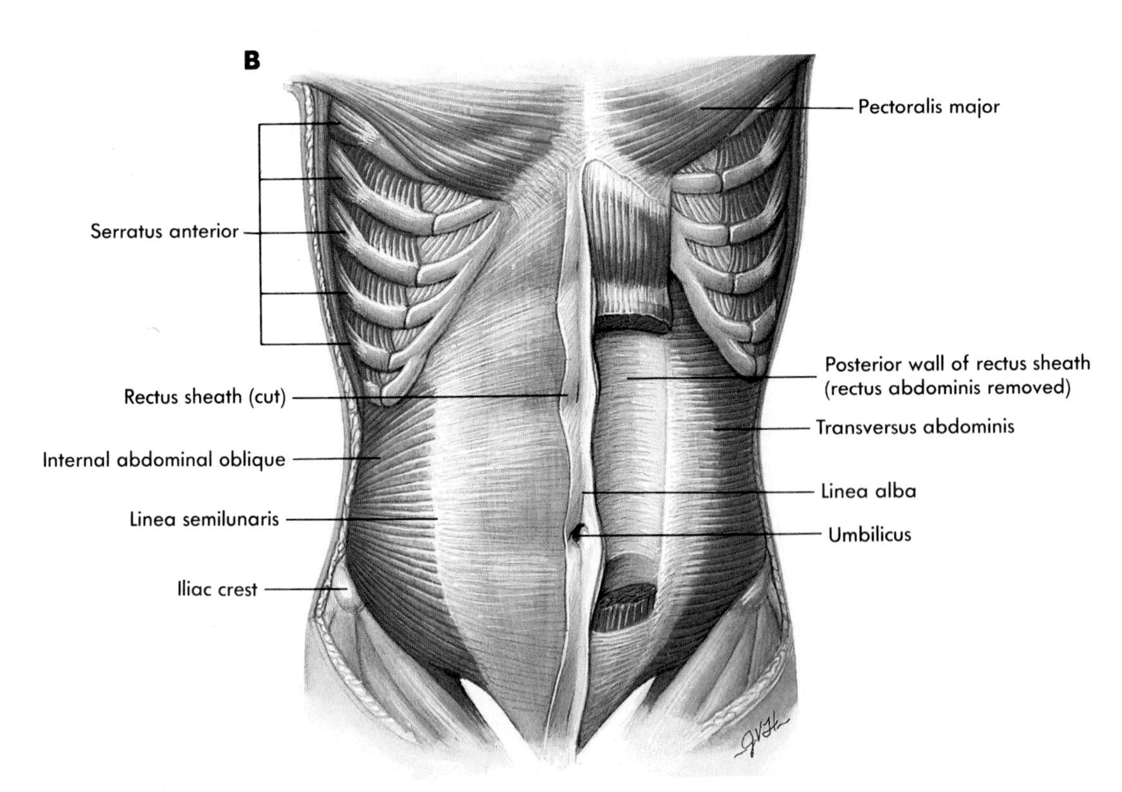

FIGURE 11-17 Muscles of the anterior abdominal wall. **A** Superficial. Rectus sheath has been removed on the right to reveal the rectus abdominis muscle. **B** Deep. External oblique has been removed on the right to reveal the internal oblique. The external and internal obliques have been removed on the left to reveal the transversus abdominis. The rectus abdominis has been cut to reveal the posterior rectus sheath.

Pelvic Floor and Perineum

The pelvis is a ring of bone (see Chapter 7) with an inferior opening that is closed by a muscular wall through which the anus and the urogenital openings penetrate (Table 11-10). Most of the pelvic floor is formed by the **coccygeus** muscle and the **levator ani** (a'ne) muscle, referred to jointly as the **pelvic diaphragm.** The area inferior to the pelvic floor is the **perineum** (per'ĭ-ne'um), which is somewhat diamond shaped (Figure 11-18). The anterior half of the diamond is the urogenital triangle, and the posterior half is the anal triangle (see Chapter 28). The urogenital triangle contains the **urogenital diaphragm,** which forms a "subfloor" to the pelvis in that area and consists of the **deep transverse perineus** muscle and the **sphincter urethrae** muscle. During pregnancy the muscles of the pelvic diaphragm and urogenital diaphragm may be stretched by the extra weight of the fetus, and specific exercises are designed to strengthen them.

Table 11-10

Muscles of the pelvic floor and perineum (see Figure 11-18)

MUSCLE	ORIGIN	INSERTION	NERVE	FUNCTION
Bulbospongiosus	Male—perineum and bulb of penis	Central tendon of perineum and bulb of penis	Pudendal	Constricts urethra; erects penis
	Female—central tendon of perineum	Base of clitoris	Pudendal	Erects clitoris
Coccygeus	Ischial spine	Coccyx	S3 and S4	Elevates and supports pelvic floor
Ischiocavernosus	Ischial ramus	Corpus cavernosum	Perineal	Compresses base of penis or clitoris
Levator ani	Posterior pubis and ischial spine	Sacrum and coccyx	Fourth sacral	Elevates anus; supports pelvic viscera
Sphincter ani externus	Coccyx	Central tendon of perineum	Fourth sacral and pudendal	Keeps orifice of anal canal closed
Sphincter urethrae	Pubic ramus	Median raphe	Pudendal	Constricts urethra
Transverse perinei				
Deep	Ischial ramus	Median raphe	Pudendal	Supports pelvic floor
Superficial	Ischial ramus	Central perineal tendon	Pudendal	Fixes central tendon

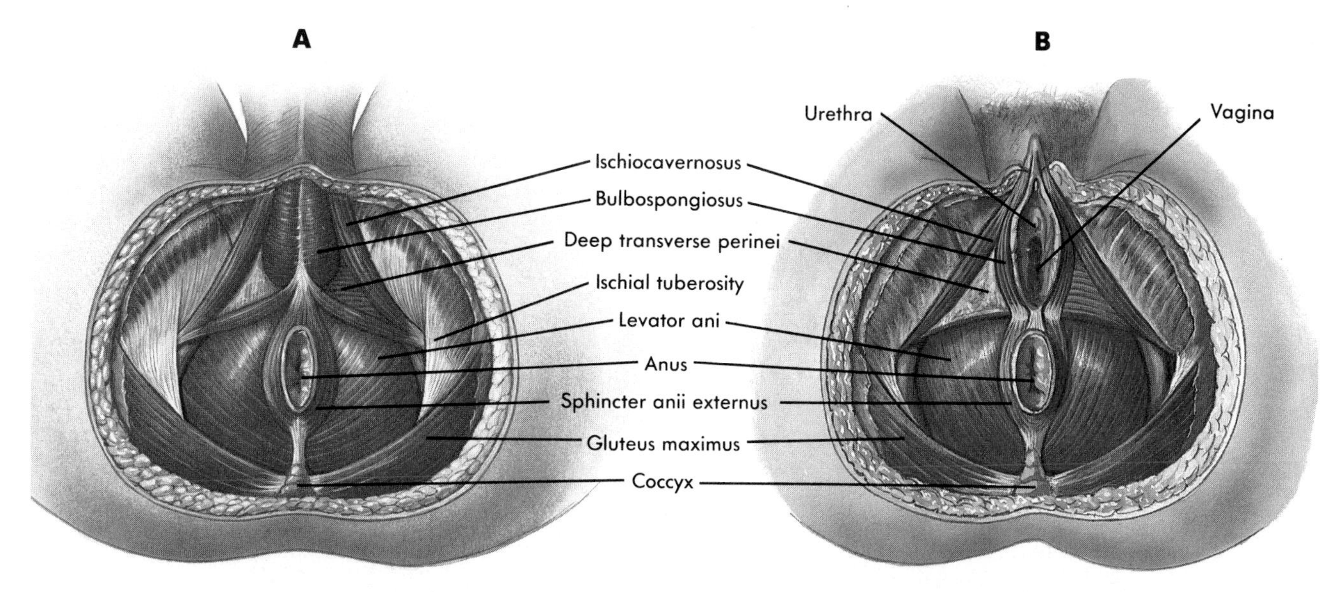

FIGURE 11-18 Muscles of the pelvic floor and perineum seen from an inferior view. **A** Male. **B** Female.

UPPER LIMB MUSCLES

The muscles of the upper limb include the ones that attach the limb and girdle to the body and that are in the arm, forearm, and hand.

Scapular Movements

The major connection of the upper limb to the body is accomplished by muscles (Table 11-11 and Figure 11-19). The muscles attaching the scapula to the thorax include the **trapezius, levator scapulae** (skap'u-le), **rhomboideus** (rom-boy'de-us) **major** and **minor, serratus** (ser-a'tus) **anterior,** and **pectoralis minor.** These muscles move the scapula, permitting a wide range of movements of the upper limb, or act as fixators to hold the scapula firmly in position when the muscles of the arm contract. The superficial muscles that act on the scapula can be easily seen on a living person (see Figure 11-22): the trapezius forms the upper line from each shoulder to the neck, and the origin of the serratus anterior from the first eight or nine ribs can be seen along the lateral thorax.

Arm Movements

The arm is attached to the thorax by the **pectoralis major** and **latissimus dorsi** (lah-tis'ĭ-mus dor'se) muscles (Table 11-12; Figures 11-19, *B* and 11-20). Notice that the pectoralis major muscle is listed in Table 11-12 as both a flexor and extensor. The muscle flexes the extended arm and extends the flexed arm. Try these movements yourself and notice the position and action of the muscle. The **deltoid** (deltoideus) muscle also is listed in Table 11-12 as a flexor and extendor. The deltoid muscle is like three muscles in one: the anterior fibers flex the arm; the lateral fibers abduct the arm; and the posterior fibers extend the arm. The deltoid muscle is part of the group of muscles that binds the humerus to the scapula. Four of these muscles are called the **rotator cuff muscles** (listed separately in Table 11-12) because they form a cuff or cap over the proximal humerus (Figure 11-21). A rotator cuff injury involves damage to one or more of these muscles or their tendons. The muscles moving the arm are involved in flexion, extension, abduction, adduction, rotation, and circumduction (Table 11-13).

4

A tennis player complains of pain in the shoulder when attempting to serve or when attempting an overhead volley (extreme abduction). What rotator cuff muscle probably is damaged?

? ? ? ? ? ? ? ? ? ? ?

Several muscles acting on the arm can be seen very clearly in the living individual (Figure 11-22). The pectoralis major forms the upper chest, and the deltoids are prominent over the shoulders. The deltoid is a common site for administering injections.

Text continued on p. 334.

Table 11-11

Muscles acting on the scapula (see Figure 11-19)

MUSCLE	ORIGIN	INSERTION	NERVE	FUNCTION
Levator scapulae	C1-C4	Superior angle of scapula	Dorsal scapular	Elevates and retracts scapula; abducts neck
Pectoralis minor	Third to fifth ribs	Coracoid process of scapula	Anterior thoracic	Depresses scapula or elevates ribs
Rhomboideus				
Major	T1-T4	Medial border of scapula	Dorsal scapular	Retracts, rotates, and fixes scapula
Minor	C6-C7	Medial border of scapula	Dorsal scapular	Retracts, slightly elevates, rotates, and fixes scapula
Serratus anterior	First to ninth ribs	Medial border of scapula	Long thoracic	Rotates and protracts scapula; elevates ribs
Subclavius	First rib	Clavicle	Subclavian	Fixes clavicle or elevates first rib
Trapezius	External occipital protuberance, ligamentum nuchae, and C7-T12	Clavicle, acromion, and scapular spine	Accessory and cervical plexus	Elevates, depresses, retracts, rotates, and fixes scapula; extends neck

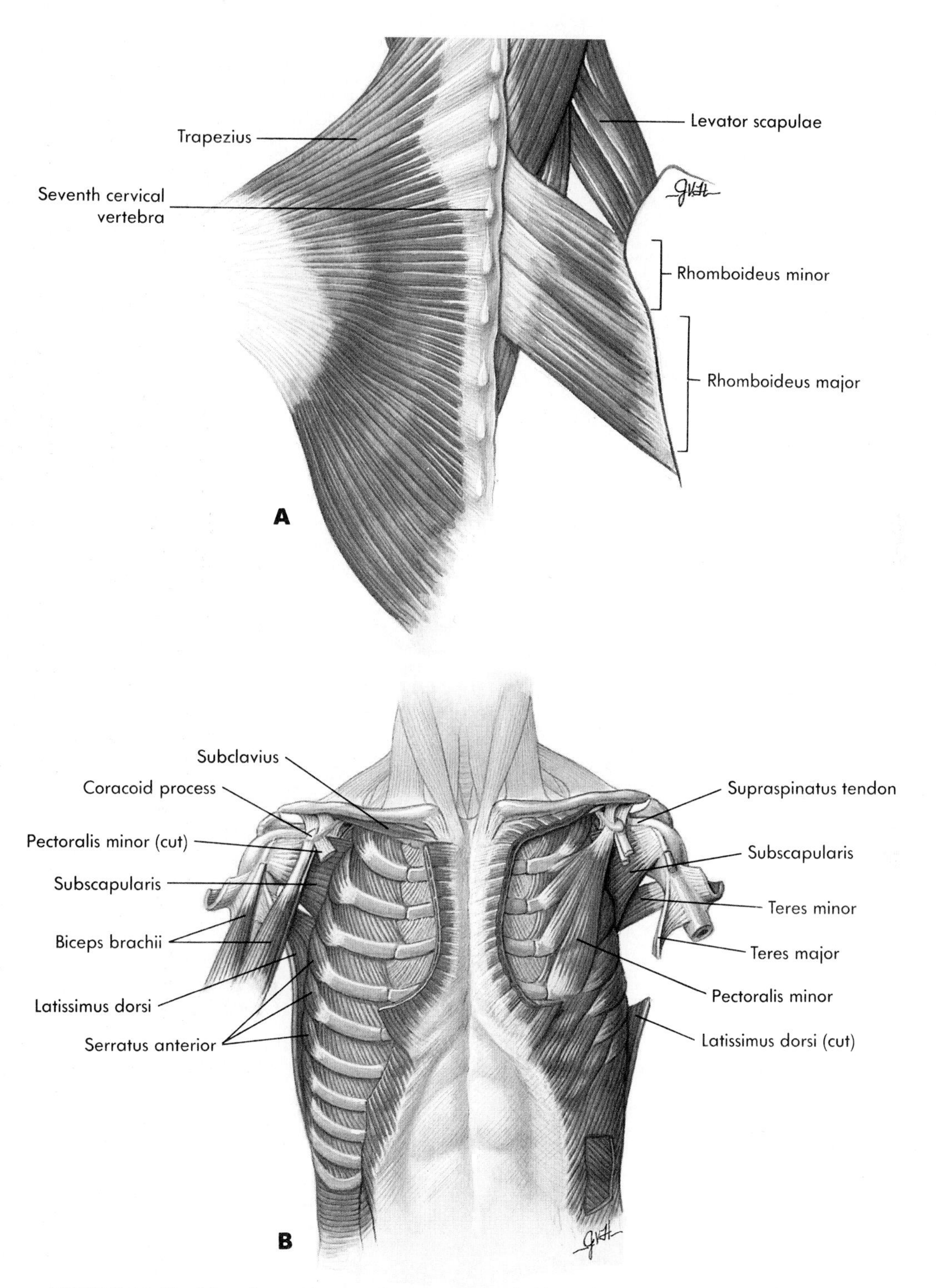

FIGURE 11-19 Muscles acting on the scapula. **A** Posterior view. Trapezius has been removed on the right to reveal the deeper muscles. **B** Anterior view. Pectoralis major has been removed on both sides. The pectoralis minor also has been removed on the right side.

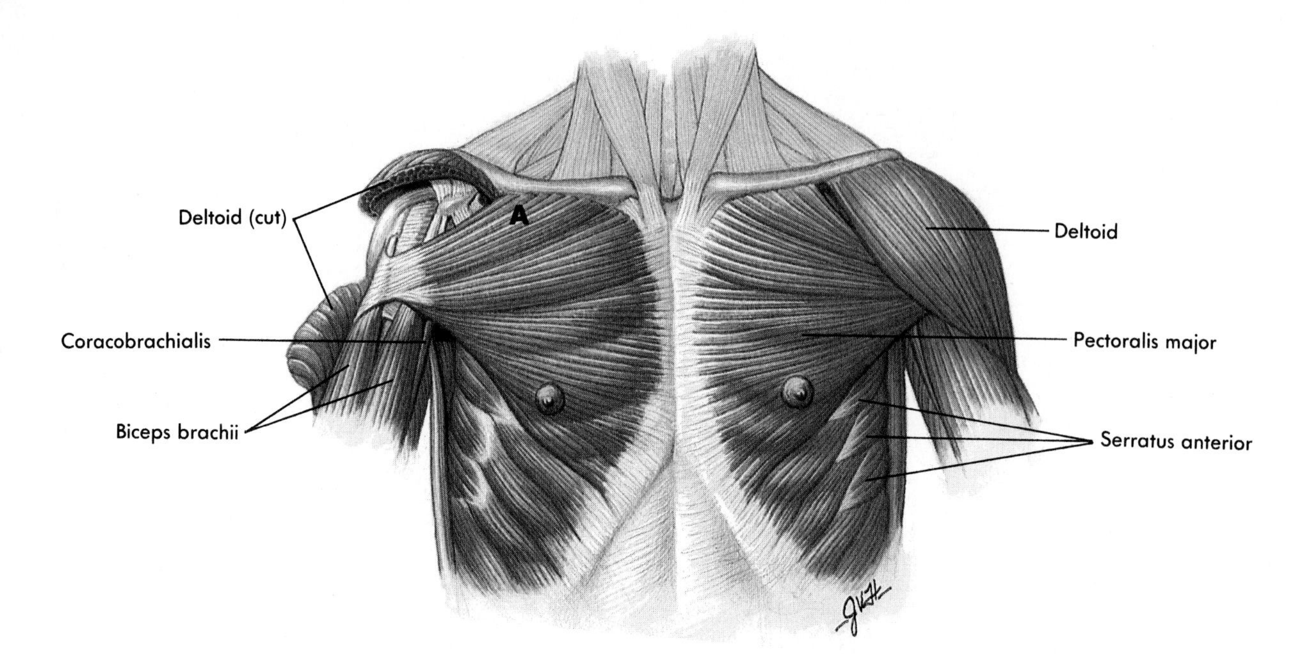

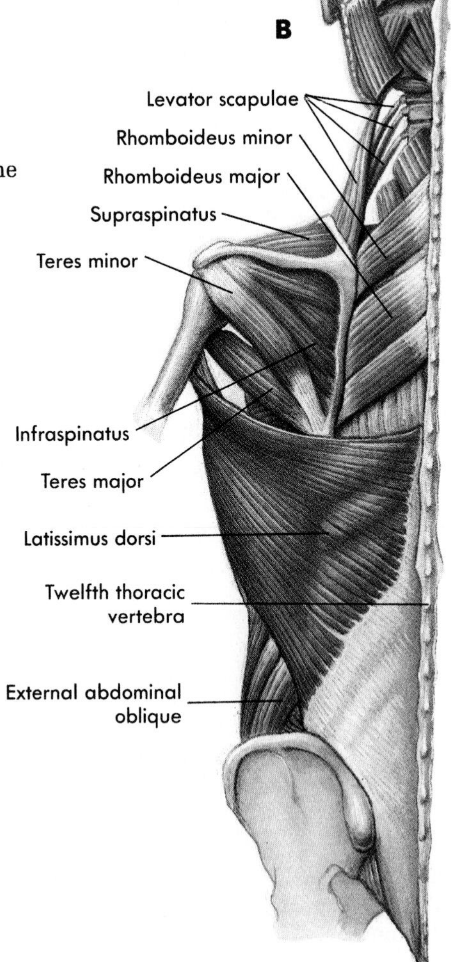

FIGURE 11-20 Muscles attaching the upper limb to the body. **A** Anterior view. **B** Posterior view.

Table 11-12

Muscles acting on the arm (see Figures 11-19, 11-20, 11-21, and 11-24)

MUSCLE	ORIGIN	INSERTION	NERVE	FUNCTION
Coracobrachialis	Coracoid process of scapula	Midshaft of humerus	Musculocutaneous	Adducts and flexes arm
Deltoid	Clavicle, acromion, and scapular spine	Deltoid tuberosity	Axillary	Abducts, flexes, extends, and rotates arm
Latissimus dorsi	T7-L5, sacrum and iliac crest	Lesser tubercle of humerus	Thoracodorsal	Adducts, medially rotates, and extends arm
Pectoralis major	Clavicle, sternum, and abdominal aponeurosis	Greater tubercle of humerus	Anterior thoracic	Adducts, flexes, extends, and medially rotates arm
Teres major	Lateral border of scapula	Lesser tubercle of humerus	Subscapular C5 and C6	Adducts, extends, and medially rotates arm
Rotator Cuff				
Infraspinatus	Inraspinous fossa of scapula	Greater tubercle of humerus	Suprascapular C5 and C6	Extends and laterally rotates arm
Subscapularis	Subscapular fossa	Lesser tubercle of humerus	Subscapular C5 and C6	Extends and medially rotates arm
Supraspinatus	Supraspinous fossa	Greater tubercle of humerus	Suprascapular C5 and C6	Abducts arm
Teres minor	Lateral border of scapula	Greater tubercle of humerus	Axillary C5 and C6	Adducts and laterally rotates arm

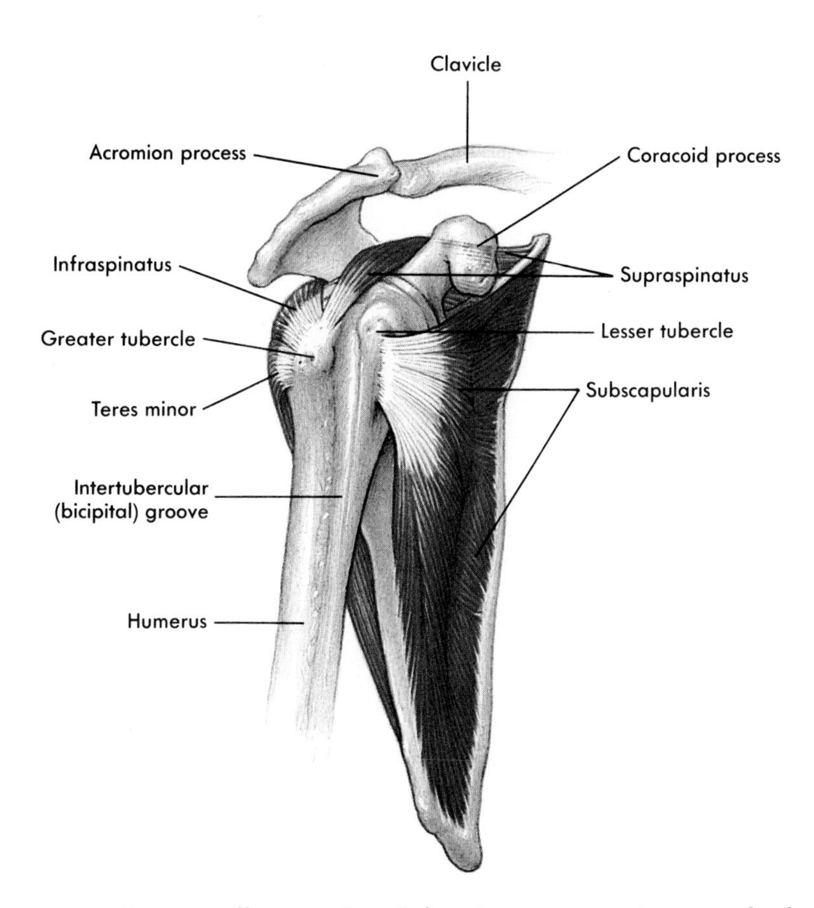

FIGURE 11-21 Rotator cuff: teres minor, infraspinatus, supraspinatus, and subscapularis muscles.

Table 11-13

Summary of muscle actions on the arm (see Figures 11-19 to 11-24)

FLEXION	EXTENSION	ABDUCTION	ADDUCTION	MEDIAL ROTATION	LATERAL ROTATION
Deltoid	Deltoid	Deltoid	Pectoralis major	Pectoralis major	Deltoid
Pectoralis major	Teres major	Supraspinatus	Latissimus dorsi	Teres major	Infraspinatus
Coracobrachialis	Latissimus dorsi	Biceps brachii	Teres major	Latissimus dorsi	Teres minor
Biceps brachii	Pectoralis major		Triceps brachii	Deltoid	
	Triceps brachii		Coracobrachialis	Subscapularis	

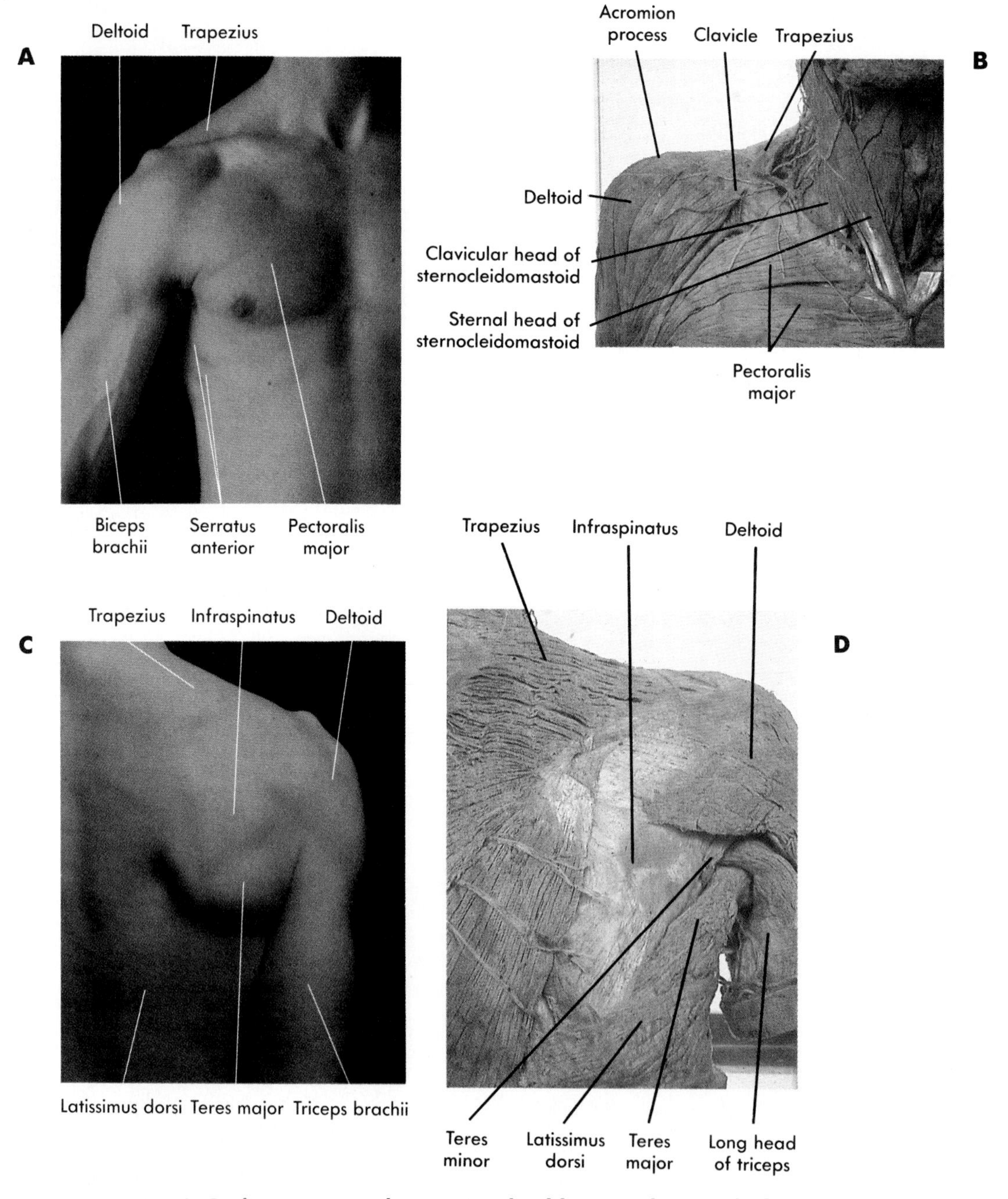

FIGURE 11-22 **A** Surface anatomy, the anterior shoulder. **B** Photograph showing a dissection of the anterior shoulder. **C** Surface anatomy, the posterior shoulder. **D** Photograph showing a dissection of the posterior shoulder.

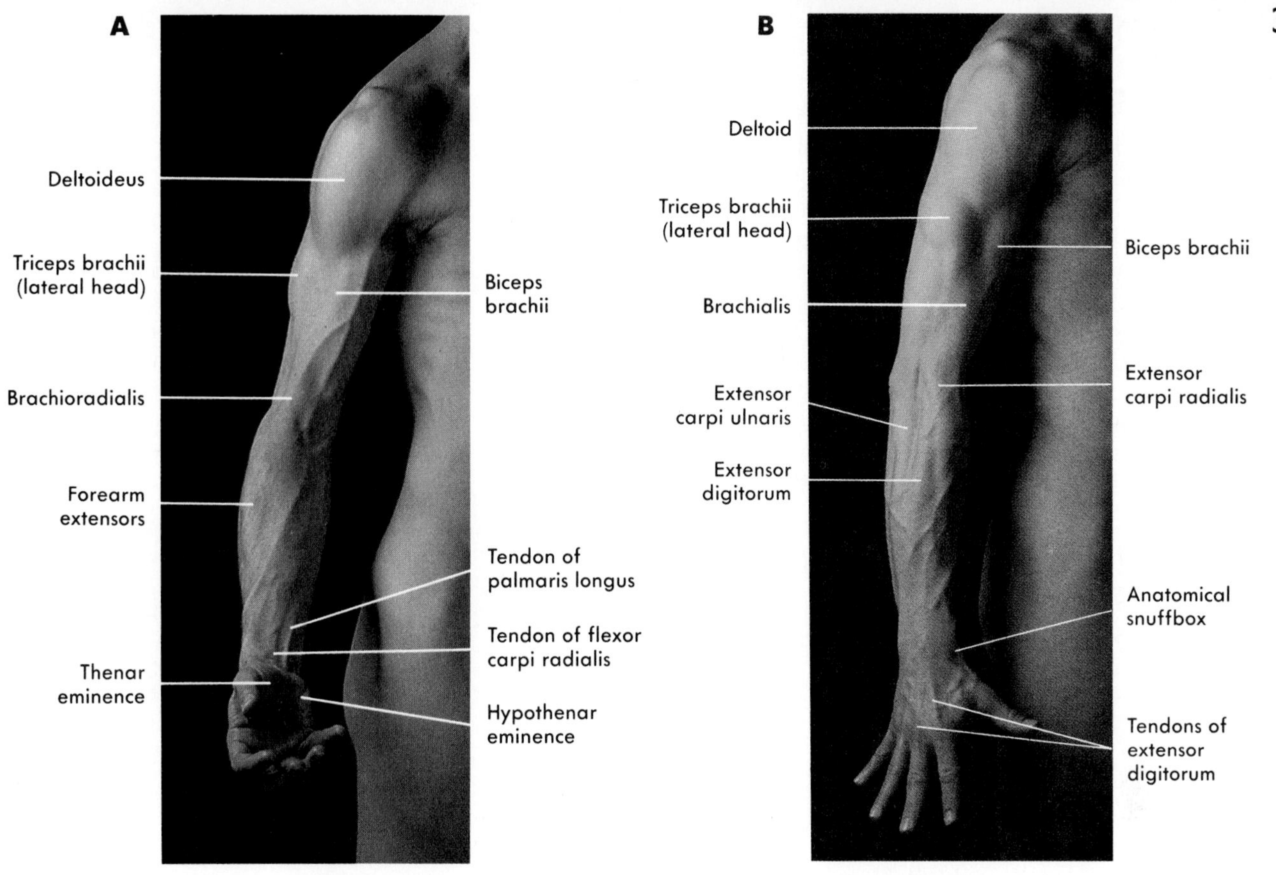

FIGURE 11-23 Surface anatomy, muscles of the upper limb.
A Anterior view. **B** Lateral view.

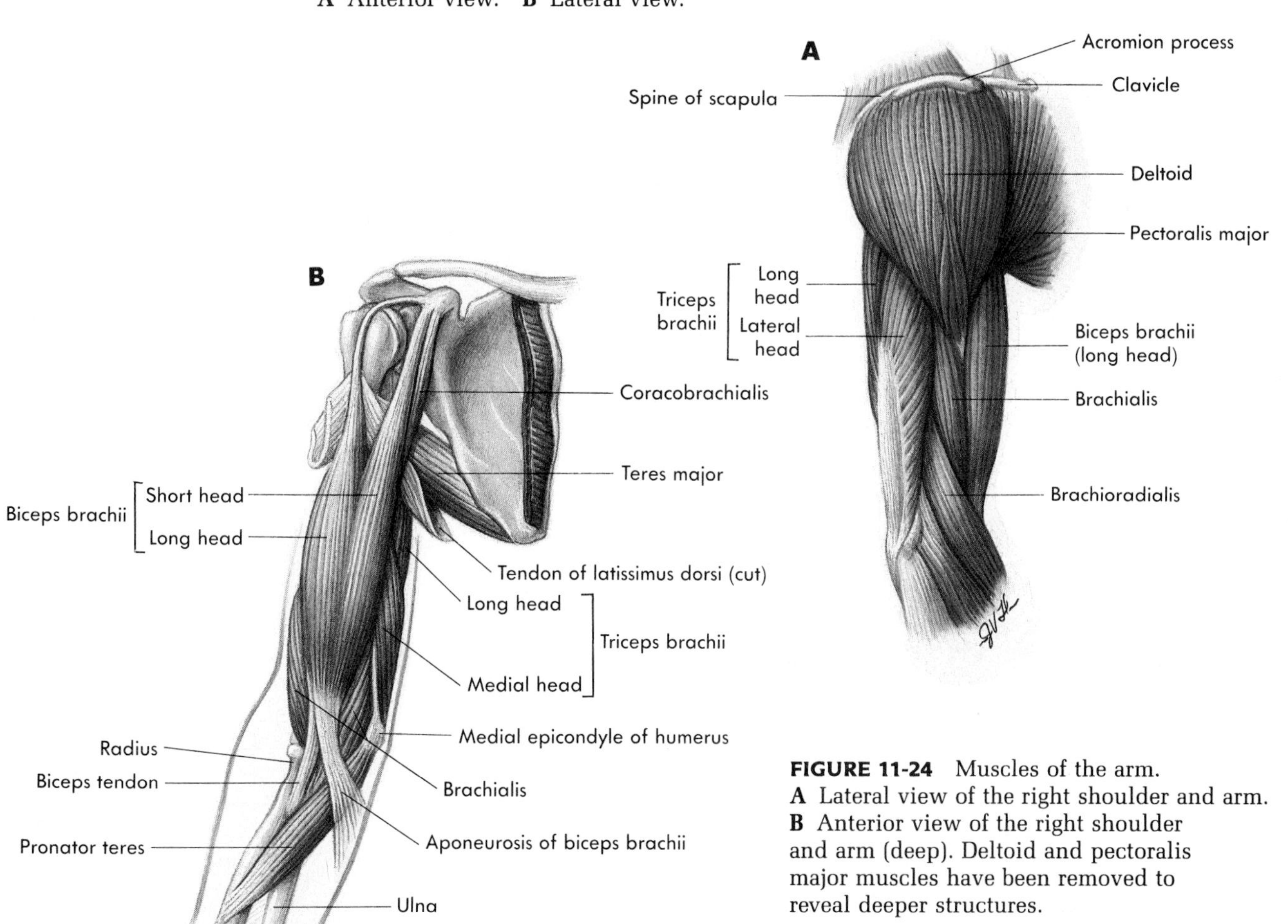

FIGURE 11-24 Muscles of the arm.
A Lateral view of the right shoulder and arm.
B Anterior view of the right shoulder and arm (deep). Deltoid and pectoralis major muscles have been removed to reveal deeper structures.

Forearm Movements

The surface anatomy of the arm muscles is illustrated in Figure 11-23 on p. 333. The triceps constitute the main mass visible on the posterior aspect of the arm. The biceps brachii is readily visible on the anterior aspect of the arm. The brachialis lies deep to the biceps and can be seen only as a small mass on the medial and lateral sides of the arm. The brachioradialis forms a bulge on the anterolateral side of the forearm just distal to the elbow. If the elbow is forcefully flexed in the midprone position (midway between pronation and supination), the brachioradialis stands out clearly on the forearm.

Flexion and extension of the forearm

Extension of the forearm is accomplished by the **triceps brachii** (bra'ke-i) and **anconeus** (an-kōn'e-us); flexion of the forearm is accomplished by the **brachialis** (bra'ke-al'us), **biceps brachii,** and **brachioradialis** (bra'ke-o-ra'de-al'us; Table 11-14; Figure 11-24).

Supination and pronation

Supination of the forearm is accomplished by the **supinator** and the **biceps brachii** (Figures 11-24, *B*, and 11-25, *D*). Pronation is a function of the **pronator quadratus** (kwad-ra'tus) and **pronator teres** (Figure 11-25, *A* and *C*).

5

Explain the difference between doing chin-ups with the forearm supinated vs. pronated. Which muscle or muscles would be used in each type of chin-up? Which type is easier? Why?

? ? ? ? ? ? ? ? ? ? ?

Table 11-14

Muscles acting on the forearm (see Figures 11-24 and 11-25)

MUSCLE (by location)	ORIGIN	INSERTION	NERVE	FUNCTION
Arm				
Biceps brachii	Long head—supraglenoid tubercle; short head—coracoid process	Radial tuberosity	Musculocutaneous	Flexes and supinates forearm; flexes and abducts arm
Brachialis	Humerus	Coronoid process of ulna	Musculocutaneous and radial	Flexes forearm
Triceps brachii	Long head—lateral border of scapula; lateral head—lateral and posterior surface of humerus; medial head—posterior humerus	Olecranon process of ulna	Radial	Extends forearm; extends and abducts arm
Forearm				
Anconeus	Lateral epicondyle of humerus	Olecranon process and posterior ulna	Radial	Extends forearm
Brachioradialis	Lateral supracondylar ridge of humerus	Styloid process of radius	Radial	Flexes forearm
Pronator quadratus	Distal ulna	Distal radius	Anterior interosseous	Pronates forearm
Pronator teres	Medial epicondyle of humerus and coronoid process of ulna	Radius	Median	Pronates forearm
Supinator	Lateral epicondyle of humerus and ulna	Radius	Radial	Supinates forearm

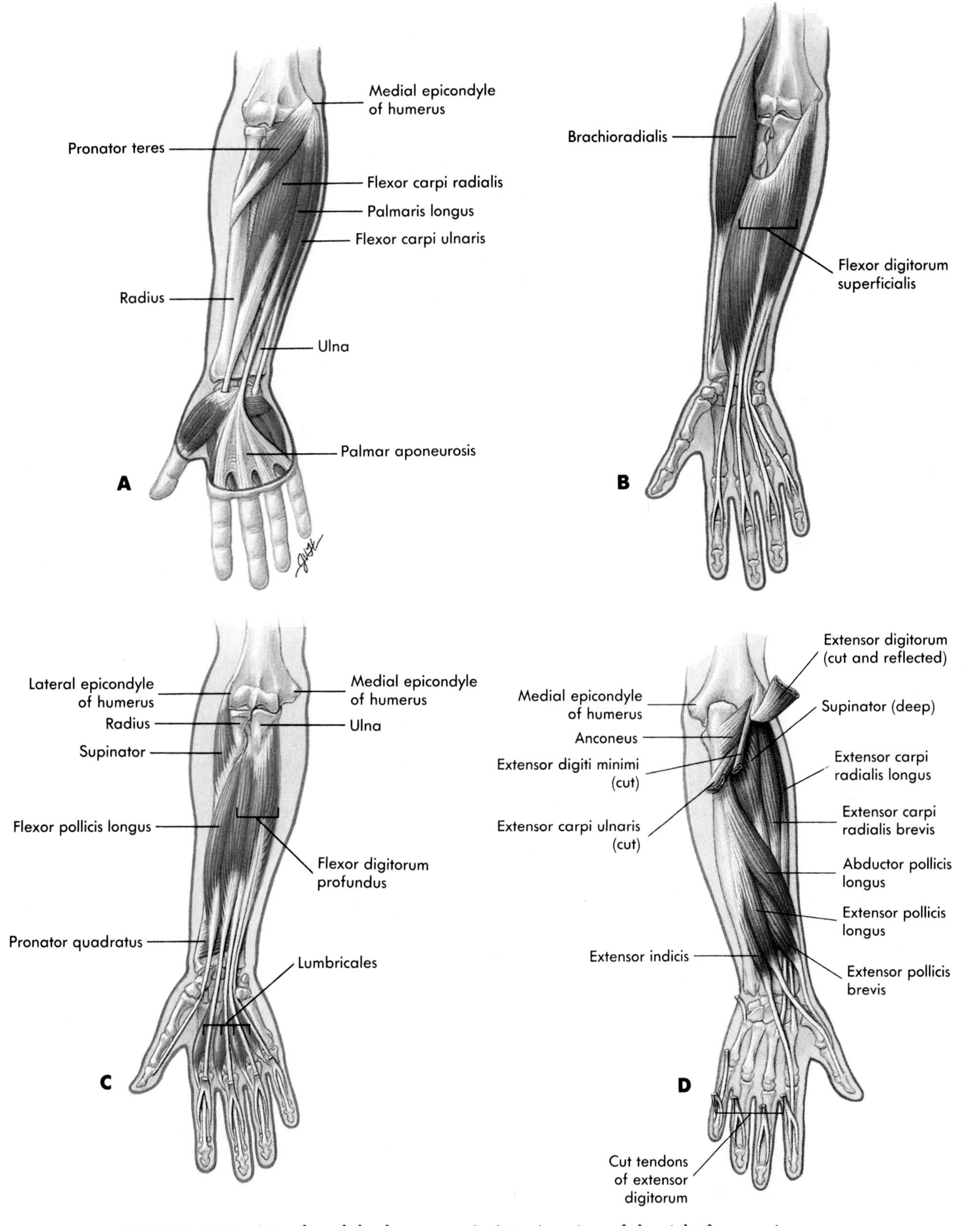

FIGURE 11-25 Muscles of the forearm. **A** Anterior view of the right forearm (superficial). Brachioradialis muscle has been removed. **B** Anterior view of the right forearm (deeper than **A**). Pronator teres, flexor carpi radialis and ulnaris, and palmaris longus muscles have been removed. **C** Anterior view of the right forearm (deeper than **A** or **B**). Brachioradialis, pronator teres, flexor carpi radialis and ulnaris, palmaris longus, and flexor digitorum superficialis muscles have been removed. **D** Deep muscles of the right posterior forearm. Extensor digitorum, extensor digiti minimi, and extensor carpi ulnaris muscles have been cut to reveal deeper muscles.

Table 11-15

Muscles of the forearm acting on the wrist, hand, and fingers (see Figure 11-25)

MUSCLE (by location)	ORIGIN	INSERTION	NERVE	FUNCTION
Anterior Forearm				
Flexor carpi radialis	Medial epicondyle of humerus	First and second metacarpals	Median	Flexes and abducts wrist
Flexor carpi ulnaris	Medial epicondyle of humerus and ulna	Pisiform	Ulnar	Flexes and adducts wrist
Flexor digitorum profundus	Ulna	Distal phalanges of digits two through five	Ulnar and median	Flexes fingers and wrist
Flexor digitorum superficialis	Medial epicondyle of humerus, coronoid process, and radius	Middle phalanges of digits two through five	Median	Flexes fingers and wrist
Flexor pollicis longus	Radius	Distal phalanx of thumb	Median	Flexes thumb and wrist
Palmaris longus	Medial epicondyle of humerus	Palmar fascia	Median	Tenses palmar fascia; flexes wrist
Posterior Forearm				
Abductor pollicis longus	Posterior ulna and radius and interosseous membrane	Base of first metacarpal	Radial	Abducts and extends thumb; abducts wrist
Extensor carpi radialis brevis	Lateral epicondyle of humerus	Base of third metacarpal	Radial	Extends and abducts wrist
Extensor carpi radialis longus	Lateral supracondylar ridge of humerus	Base of second metacarpal	Radial	Extends and abducts wrist
Extensor carpi ulnaris	Lateral epicondyle of humerus and ulna	Base of fifth metacarpal	Radial	Extends and adducts wrist
Extensor digiti minimi	Lateral epicondyle of humerus	Phalanges of fifth digit	Radial	Extends little finger and wrist
Extensor digitorum	Lateral epicondyle of humerus	Bases of phalanges of digits two through five	Radial	Extends fingers and wrist
Extensor indicis	Ulna	Second digit	Radial	Extends forefinger and wrist
Extensor pollicis brevis	Radius	Proximal phalanx of thumb	Radial	Extends and abducts thumb; abducts wrist
Extensor pollicis longus	Ulna	Distal phalanx of thumb	Radial	Extends thumb

Wrist, Hand, and Finger Movements

The forearm muscles can be divided into anterior and posterior groups (Table 11-15; see Figure 11-25). Most of the anterior forearm muscles are responsible for flexion of the wrist and fingers. Most of the posterior forearm muscles cause extension of the wrist and fingers.

Extrinsic hand muscles

The **extrinsic hand muscles** are in the forearm but have tendons that extend into the hand. A strong band of fibrous connective tissue, the **retinaculum** (ret′ĭ-nak′u-lum; bracelet), covers the flexor and extensor tendons and holds them in place around the wrist so that they do not "bowstring" during muscle contraction.

Two major anterior muscles, the **flexor carpi radialis** and the **flexor carpi ulnaris,** flex the wrist; and three posterior muscles, the **extensor carpi radialis longus,** the **extensor carpi radialis brevis,** and the **extensor carpi ulnaris,** extend the wrist. The wrist flexors and extensors are visible on the anterior and posterior surfaces of the forearm. The tendon of the flexor carpi radialis is an important landmark because the radial pulse can be felt just lateral to the tendon (see Figure 11-25, *A*).

Table 11-16

Intrinsic hand muscles (see Figure 11-26)

MUSCLE	ORIGIN	INSERTION	NERVE	FUNCTION
Interossei				
Dorsales	Sides of metacarpal bones	Proximal phalanges of second, third, and fourth digits	Ulnar	Abducts second, third, and fourth digits
Palmares	Second, fourth, and fifth metacarpals	Second, fourth, and fifth digits	Ulnar	Adducts second, fourth, and fifth digits
Lumbricales	Tendons of flexor digitorum profundus	Second through fifth digits	Two on radial side—median; two on ulnar side—ulnar	Flexes proximal and extends middle and distal phalanges
Thenar Muscles				
Abductor pollicis brevis	Trapezium	Proximal phalanx of thumb	Median	Abducts thumb
Adductor pollicis	Third metacarpal, second metacarpal, trapezoid, and capitate	Proximal phalanx of thumb	Ulnar	Adducts thumb
Flexor pollicis brevis	Flexor retinaculum and first metacarpal	Proximal phalanx of thumb	Median and ulnar	Flexes thumb
Opponens pollicis	Trapezium and flexor retinaculum	First metacarpal	Median	Opposes thumb
Hypothenar Muscles				
Abductor digiti minimi	Pisiform	Base of fifth digit	Ulnar	Abducts and flexes little finger
Flexor digiti minimi brevis	Hamate	Middle and proximal phalanx of fifth digit	Ulnar	Flexes little finger
Opponens digiti minimi	Hamate and flexor retinaculum	Fifth metacarpal	Ulnar	Opposes little finger

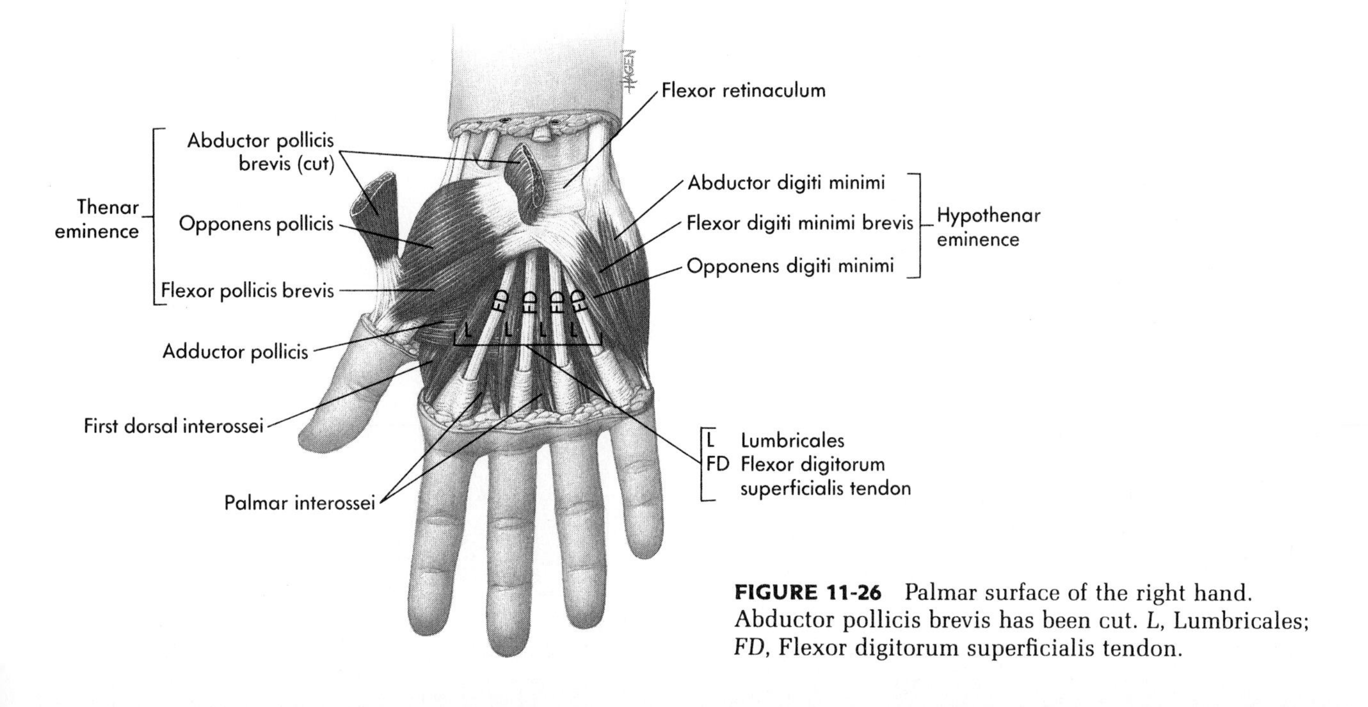

FIGURE 11-26 Palmar surface of the right hand. Abductor pollicis brevis has been cut. *L*, Lumbricales; *FD*, Flexor digitorum superficialis tendon.

Forceful extension of the wrist repeated over a period of time (e.g., movement with a tennis backhand) can result in inflammation and pain where the muscles attach to the lateral humeral epicondyle (the common point of origin for most extensors of the wrist and hand)—a condition referred to as "tennis elbow."

Flexion of the four medial digits is a function of the **flexor digitorum** (dij'ĭ-to'rum) **superficialis** and **flexor digitorum profundus** (pro-fun'dus; deep). Extension is accomplished by the **extensor digitorum.** The tendons of this muscle are very visible on the dorsum of the hand (see Figure 11-23, *B*). The little finger has an additional flexor and extensor, the **flexor digiti minimi** (dij'ĭ-te min'ĭ-me) and the **extensor digiti minimi.** The index finger also has an additional extensor, the **extensor indicis** (in'dĭ-sis).

Movement of the thumb is caused in part by the **abductor pollicis longus,** the **extensor pollicis longus,** and the **extensor pollicis brevis.** These tendons form the sides of a depression on the posterolateral side of the wrist called the "anatomical snuffbox" (see Figure 11-23, *B*). When snuff was in use, a small pinch could be placed into the anatomical snuffbox and inhaled through the nose.

Intrinsic hand muscles

The **intrinsic hand muscles** are entirely within the hand (Table 11-16 and Figure 11-26, p. 337). Abduction of the fingers is accomplished by the **interossei** (in'ter-os'e-i) **dorsales** and the **abductor digiti minimi,** whereas adduction is a function of the **interossei palmares.**

The **flexor pollicis brevis, abductor pollicis brevis,** and **opponens pollicis** form a fleshy prominence at the base of the thumb called the **thenar** (the'nar) **eminence** (see Figures 11-23, *A*, and 11-26). The **abductor digiti minimi, flexor digiti minimi brevis,** and **opponens digiti minimi** constitute the **hypothenar eminence** on the ulnar side of the hand. The thenar and hypothenar muscles are involved in the control of the thumb and little finger.

LOWER LIMB MUSCLES
Thigh Movements

Several hip muscles originate on the coxa and insert onto the femur (Table 11-17 and Figures 11-27 through 11-29). These muscles can be divided into three groups: anterior, posterolateral, and deep.

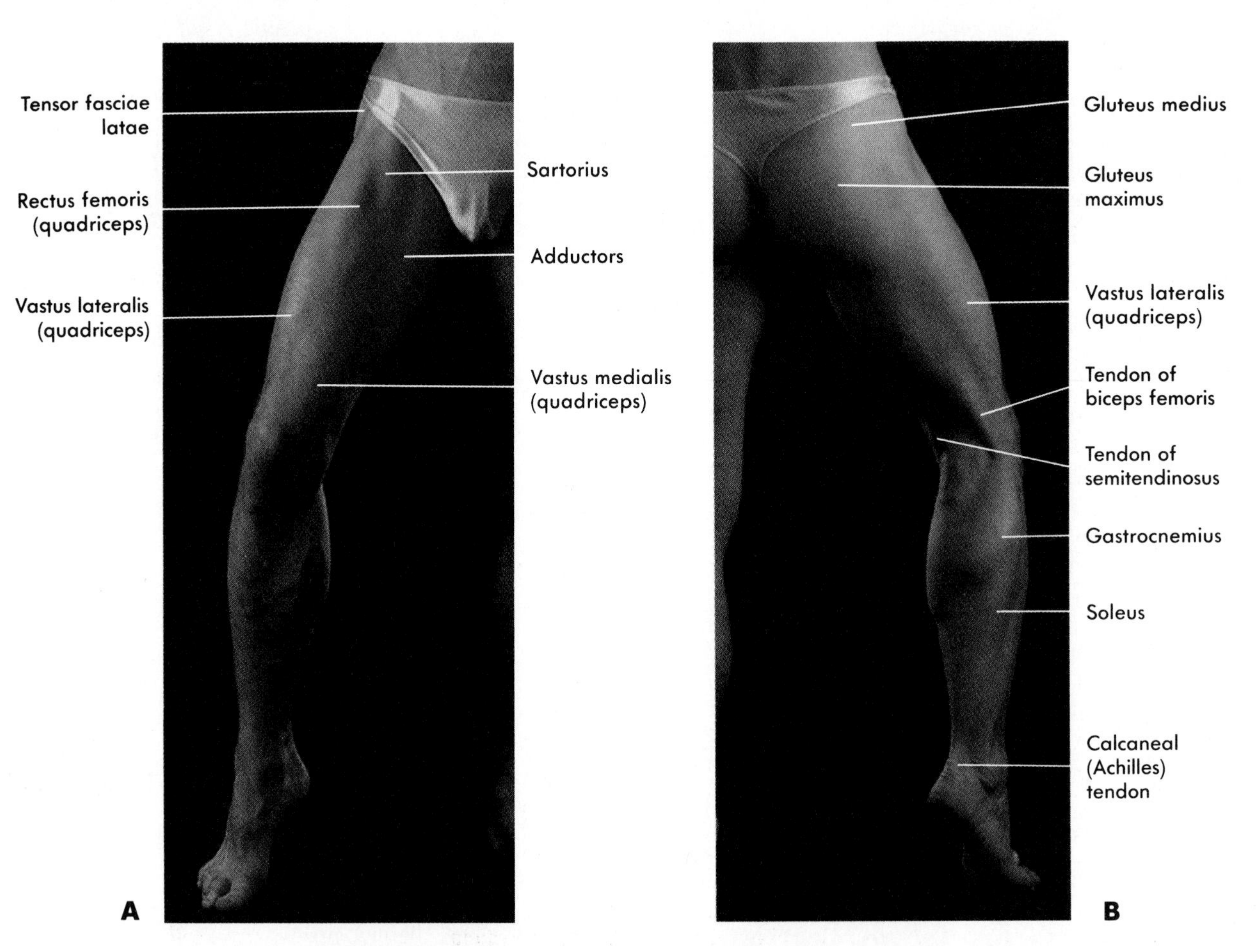

FIGURE 11-27 Surface anatomy, muscles of the lower limb. **A** Anterior view. **B** Posterior view.

Table 11-17

Muscles acting on the thigh (see Figures 11-28 and 11-29)

MUSCLE	ORIGIN	INSERTION	NERVE	FUNCTION
Anterior				
Iliopsoas				
Iliacus	Iliac fossa	Lesser trochanter of femur and capsule of hip joint	Lumbar plexus	Flexes and medially rotates thigh
Psoas major	T12-L5	Lesser trochanter of femur	Lumbar plexus	Flexes thigh
Posterior and Lateral				
Gluteus maximus	Ilium, sacrum, and coccyx	Gluteal tuberosity of femur and the fascia lata	Inferior gluteal	Extends, abducts, and laterally rotates thigh
Gluteus medius	Ilium	Greater trochanter of femur	Superior gluteal	Abducts and medially rotates thigh
Gluteus minimus	Ilium	Greater trochanter of femur	Superior gluteal	Abducts and medially rotates thigh
Tensor fasciae latae	Anterior superior iliac spine	Through iliotibial tract to lateral condyle of tibia	Superior gluteal	Tenses lateral fascia; flexes, abducts, and medially rotates thigh
Deep Thigh Rotators				
Gemellus				
Inferior	Ischial tuberosity	Obturator internus tendon	L5 and S1	Laterally rotates thigh
Superior	Ischial spine	Obturator internus tendon	L5 and S1	Laterally rotates thigh
Obturator				
Externus	Inferior margin of obturator foramen	Greater trochanter of femur	Obturator	Laterally rotates thigh
Internus	Margin of obturator foramen	Greater trochanter of femur	Sacral plexus	Laterally rotates and abducts thigh
Piriformis	Sacrum and ilium	Greater trochanter of femur	Sciatic plexus	Laterally rotates and abducts thigh
Quadratus femoris	Ischial tuberosity	Intertrochanteric ridge of femur	Sacral plexus	Laterally rotates and adducts thigh

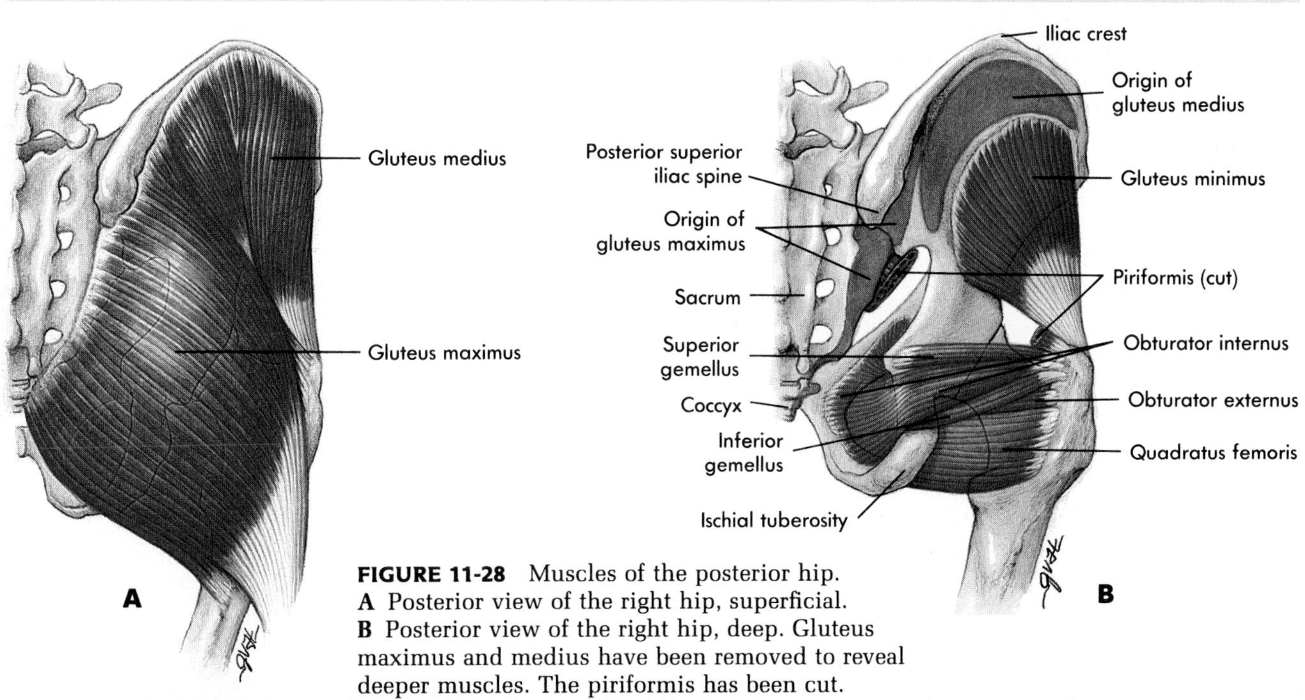

FIGURE 11-28 Muscles of the posterior hip. **A** Posterior view of the right hip, superficial. **B** Posterior view of the right hip, deep. Gluteus maximus and medius have been removed to reveal deeper muscles. The piriformis has been cut.

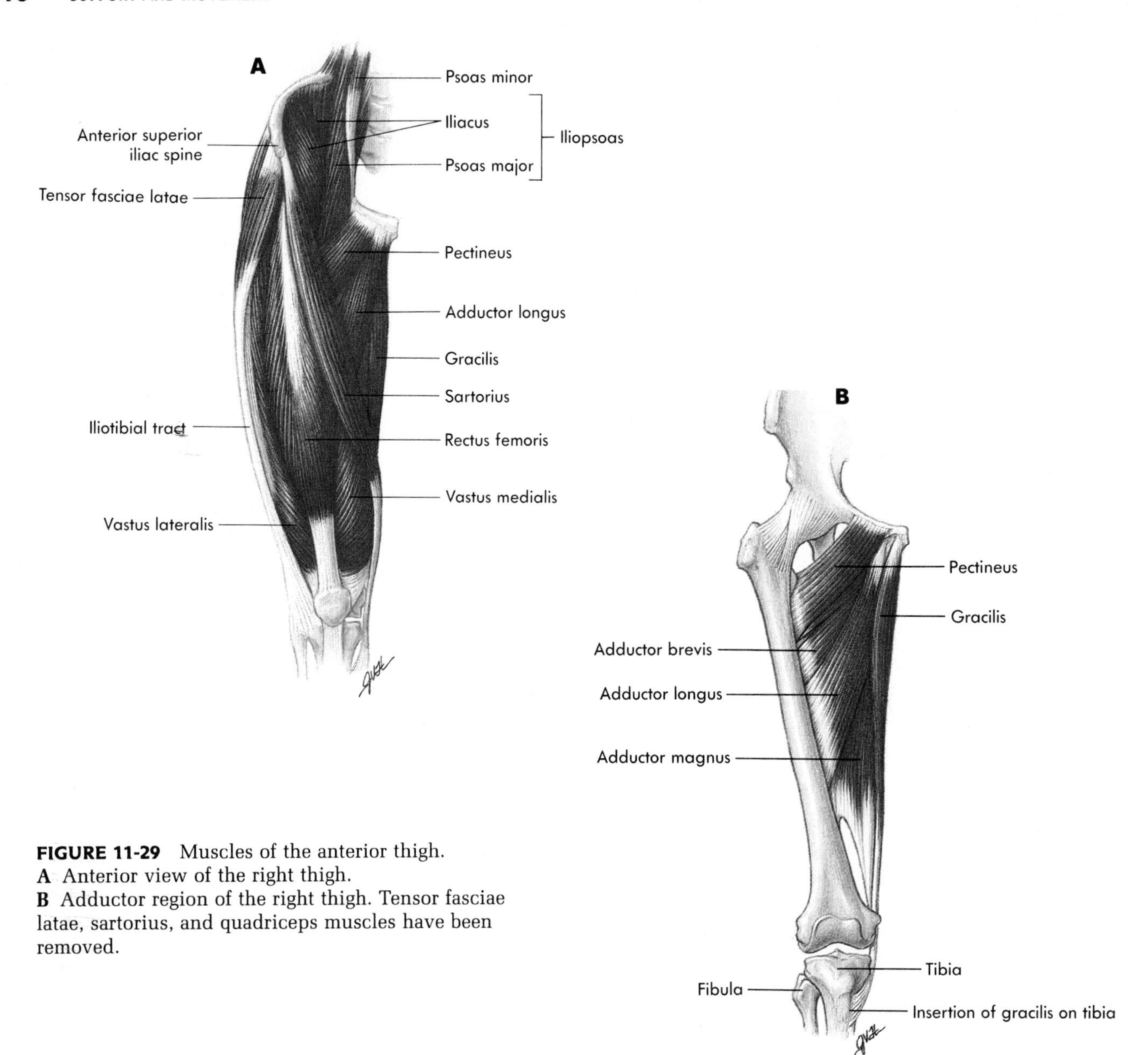

FIGURE 11-29 Muscles of the anterior thigh.
A Anterior view of the right thigh.
B Adductor region of the right thigh. Tensor fasciae latae, sartorius, and quadriceps muscles have been removed.

Table 11-18

Summary of muscle actions on the thigh (see Figures 11-28 and 11-29)

FLEXION	EXTENSION	ABDUCTION	ADDUCTION	MEDIAL ROTATION	LATERAL ROTATION
Iliopsoas	Gluteus maximus	Gluteus medius	Adductor magnus	Iliopsoas	Gluteus maximus
Tensor fasciae latae	Semitendinosus	Gluteus minimus	Adductor longus	Tensor fasciae latae	Obturator internus
Rectus femoris	Semimembranosus	Tensor fasciae latae	Adductor brevis	Gluteus medius	Obturator externus
Sartorius	Biceps femoris	Obturator internus	Pectineus	Gluteus minimus	Superior gemellus
Adductor longus	Adductor magnus	Piriformis	Gracilis		Inferior gemellus
Adductor brevis			Gluteus maximus		Quadratus femoris
Pectineus			Quadratus femoris		Piriformis
					Sartorius
					Adductor magnus
					Adductor longus
					Adductor brevis

The anterior muscles, the **iliacus** and **psoas major,** flex the thigh. Since these muscles share a common insertion and produce the same movement, they often are referred to as the **iliopsoas.** When the thigh is fixed, the iliopsoas flexes the trunk on the thigh. For example, the iliopsoas actually does most of the work in doing sit-ups.

The posterolateral hip muscles consist of the gluteal muscles and the **tensor fasciae latae.** The **gluteus maximus** contributes most of the mass that can be seen as the buttocks, and the **gluteus medius,** a common site for injections, creates a smaller mass just superior and lateral to the maximus. The gluteus maximus functions at its maximum force in extension of the thigh when the hip is flexed at a 45-degree angle so that the muscle is optimally stretched. That information accounts for both the sprinter's stance and the bicycle racing posture.

The deep hip muscles function as lateral thigh rotators (see Table 11-17).

In addition to the hip muscles, some of the muscles located in the thigh originate on to the coxa and can cause movement of the thigh (Table 11-18; see Table 11-19). There are three groups of thigh muscles, based on their location in the thigh: the anterior, which flex; the posterior, which extend; and the medial, which adduct the thigh.

Leg movements

The anterior thigh muscles are the **quadriceps femoris** and the **sartorius** (Table 11-19 and Figure 11-29, *A*). The quadriceps femoris is actually four muscles: the **rectus femoris,** the **vastus lateralis,** the **vastus medialis,** and the **vastus intermedius.** The vastus lateralis sometimes is used as an injection site,

Table 11-19

Muscles of the thigh (see Figures 11-29 and 11-30)

MUSCLE	ORIGIN	INSERTION	NERVE	FUNCTION
Anterior Compartment				
Quadriceps femoris	Rectus femoris—anterior inferior iliac spine; vastus lateralis—femur; vastus intermedius—femur; vastus medialis—linea aspera	Patella and onto tibial tuberosity through patellar ligament	Femoral	Extends leg; rectus femoris also flexes thigh
Sartorius	Anterior superior iliac spine	Medial side of tibial tuberosity	Femoral	Flexes thigh and leg; rotates leg medially and thigh laterally
Medial Compartment				
Adductor brevis	Pubis	Femur	Obturator	Adducts, flexes, and laterally rotates thigh
Adductor longus	Pubis	Femur	Obturator	Adducts, flexes, and laterally rotates thigh
Adductor magnus	Ischium	Femur	Obturator and tibial	Adducts, extends, and laterally rotates thigh
Gracilis	Pubis near symphysis	Tibia	Obturator	Adducts thigh; flexes leg
Pectineus	Pubic crest	Pectineal line of femur	Femoral and obturator	Adducts and flexes thigh
Posterior Compartment				
Biceps femoris	Long head—ischial tuberosity; short head—femur	Head of fibula	Long head—tibial; short head—peroneal	Flexes and laterally rotates leg; extends thigh
Semimembranosus	Ischial tuberosity	Medial condyle of tibia, collateral ligament, and initial epicondyle femur	Tibial	Flexes and medially rotates leg; tenses capsule of knee joint; extends thigh
Semitendinosus	Ischial tuberosity	Tibia	Tibial	Flexes and medially rotates leg; extends thigh

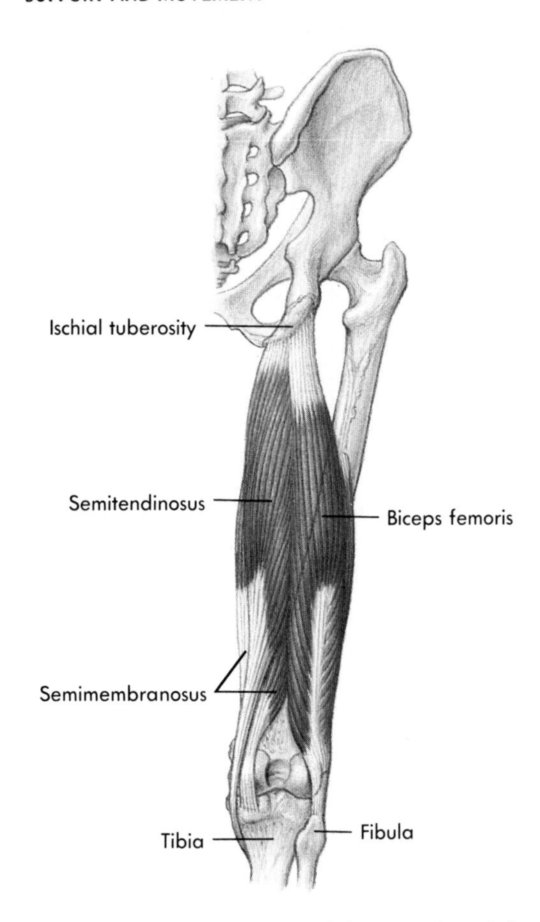

FIGURE 11-30 Posterior view of the muscles of the right thigh. Hip muscles have been removed.

especially in infants who may not have well-developed deltoid or gluteal muscles. The muscles of the quadriceps femoris have a common insertion, the patellar tendon, on and around the patella. The patellar ligament is an extension of the patellar tendon onto the tibial tuberosity. The patellar ligament is the point that is tapped with a rubber hammer when testing the knee-jerk reflex in a physical examination.

The sartorius is a very interesting muscle. It is the longest muscle of the body, crossing from the lateral side of the hip to the medial side of the knee. As the muscle contracts, it flexes the thigh and leg and laterally rotates the thigh. This movement is the action required for crossing the legs.

The medial group of muscles is involved primarily in adduction of the thigh (Figure 11-29, *B*).

The term sartorius means tailor. The sartorius muscle is so named because its action is to cross the legs, a common position preferred by tailors because they can hold their sewing in their lap as they sit and sew by hand.

The posterior thigh muscles are collectively called the hamstring muscles and consist of the **biceps femoris, semimembranosus,** and **semitendinosus** (see Table 11-19 and Figure 11-30). Their tendons are easily felt and seen on the medial and lateral posterior aspect of a slightly bent knee (see Figure 11-27).

The hamstrings are so named because in pigs these tendons can be used to suspend hams during curing. Some animals such as wolves often bring down their prey by biting through the hamstrings. Therefore "to hamstring" someone is to render him helpless. A "pulled hamstring" results from tearing one or more of these muscles or their tendons, usually near the origin of the muscle.

Ankle, Foot, and Toe Movements

Muscles of the leg that move the ankle and the foot are listed in Table 11-20 and are illustrated in Figure 11-31. These **extrinsic foot muscles** can be divided into three groups, each located within a separate compartment of the leg (Figure 11-32): anterior, posterior, and lateral. The anterior muscles are extensor muscles involved in dorsiflexion of the foot and extension of the toes.

The term shinsplints is a catchall term involving any one of the following four conditions associated with pain in the anterior portion of the leg.

- Excessive stress on the tibialis posterior resulting in pain along the origin of the muscle.
- Tibial periostitis, an inflammation of the tibial periosteum.
- Anterior compartment syndrome. During hard exercise the anterior compartment muscles may swell with blood. The overlying fascia is very tough and does not expand; thus the nerves and vessels are compressed, causing pain.
- Stress fracture of the tibia 2 to 5 cm distal to the knee.

The best treatment for any of these types of shinsplints is to rest the leg for 1 to 4 weeks, depending on the type of splint.

The superficial muscles of the posterior compartment, the **gastrocnemius** and **soleus,** form the bulge of the calf (posterior leg)(see Figure 11-27). They join with the small **plantaris** muscle to form the common **calcaneal** (kal-ka'ne-al), or **Achilles, tendon** (Figure 11-31, C). These muscles are flexors and are involved in plantar flexion of the foot. The deep muscles of the posterior compartment plantar flex and invert the foot and flex the toes.

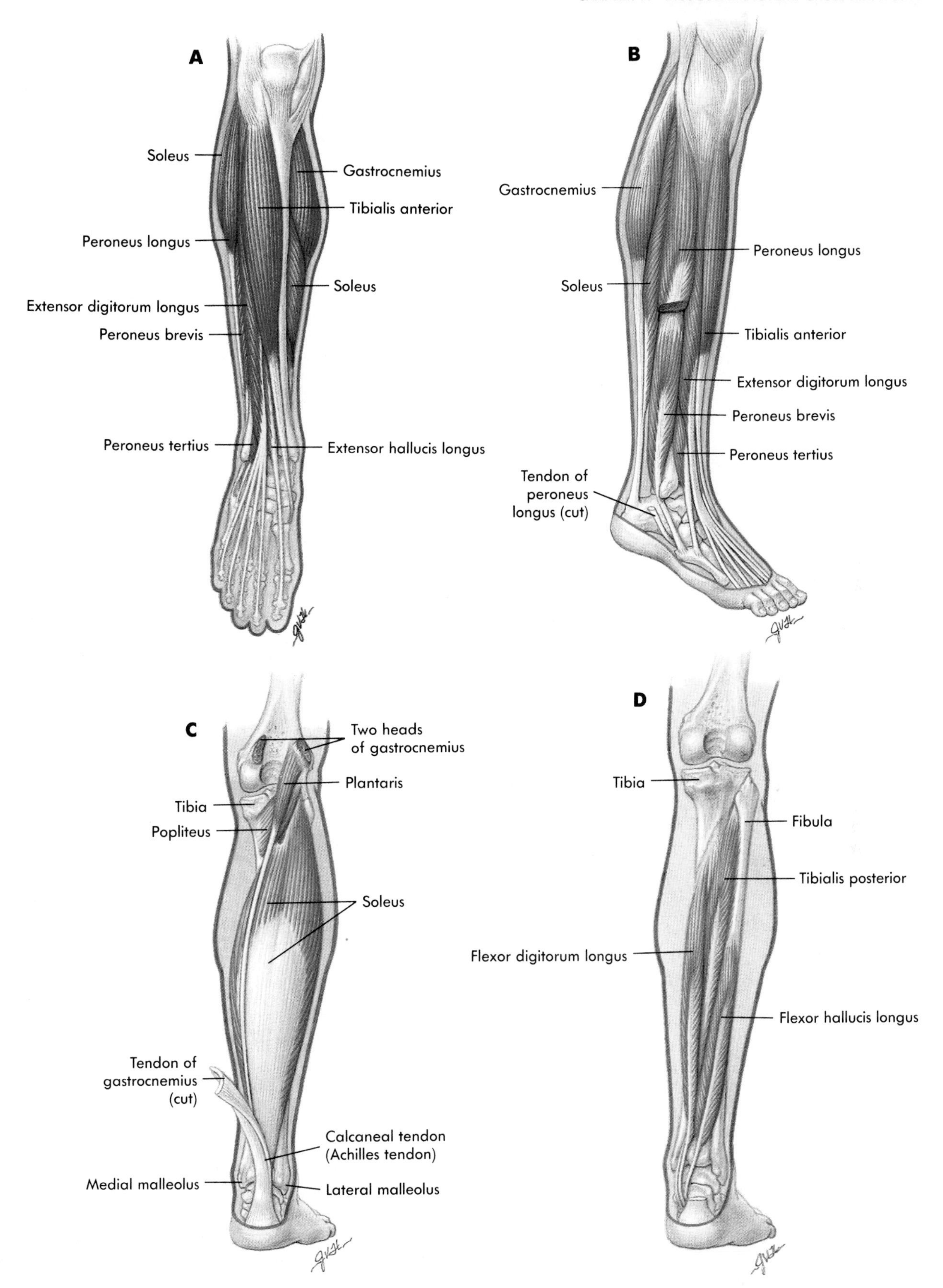

FIGURE 11-31 Muscles of the leg. **A** Anterior view of the right leg. **B** Lateral view of the right leg. **C** Posterior view of the right calf, superficial. Gastrocnemius has been removed. **D** Posterior view of the right calf, deep. Gastrocnemius, plantaris, and soleus muscles have been removed.

The lateral muscles are primarily everters of the foot, but they also aid plantar flexion.

Intrinsic foot muscles, located within the foot itself (Table 11-21; Figure 11-33), flex, extend, abduct, and adduct the toes. They are arranged in a manner similar to the intrinsic muscles of the hand.

The Achilles tendon derives its name from a hero of Greek mythology. As a baby, Achilles was dipped into magic water, which made him invulnerable to harm everywhere the water touched his skin. His mother, however, holding him by the heel, failed to submerge this part of his body under the water. Consequently, his heel was vulnerable and proved to be his undoing; he was shot in the heel with an arrow at the battle of Troy and died. Thus saying that someone has an "Achilles heel" means he has a weak spot that can be attacked.

Table 11-20

Muscles of the leg acting on the leg, ankle, and foot (see Figures 11-31 and 11-32)

MUSCLE	ORIGIN	INSERTION	NERVE	FUNCTION
Anterior Compartment				
Extensor digitorum longus	Lateral condyle of tibia and fibula	Four tendons to phalanges of four lateral toes	Deep peroneal	Extends four lateral toes; dorsiflexes and everts foot
Extensor hallicus longus	Lateral tibia and interosseous membrane	Distal phalanx of great toe	Anterior tibial	Extends great toe; dorsiflexes and inverts foot
Tibialis anterior	Tibia and interosseous membrane	Medial cuneiform and first metatarsal	Deep peroneal	Dorsiflexes and inverts foot
Peroneus tertius	Fibula and interosseous membrane	Fifth metatarsal	Deep branch of peroneal	Dorsiflexes and everts foot
Posterior Compartment				
Superficial				
Gastrocnemius	Medial and lateral epicondyles of femur	Through calcaneal (Achilles) tendon to calcaneus	Tibial	Plantar flexes foot; flexes leg
Plantaris	Femur	Through calcaneal tendon to calcaneus	Tibial	Plantar flexes foot; flexes leg
Soleus	Fibula and tibia	Calcaneal tendon to calcaneus	Tibial	Plantar flexes foot
Deep				
Flexor digitorum longus	Tibia	Four tendons to distal phalanges of four lateral toes	Tibial	Flexes four lateral toes; plantar flexes and inverts foot
Flexor hallucis longus	Fibula	Distal phalanx of great toe	Medial plantar	Flexes great toe; plantar flexes and inverts foot
Popliteus	Lateral femoral condyle	Posterior tibia	Tibial	Flexes and medially rotates leg
Tibialis posterior	Tibia, interosseous membrane, and fibula	Navicular, cuneiforms, cuboid, and second through fourth metatarsals	Tibial	Plantar flexes and inverts foot
Lateral Compartment				
Peroneus brevis	Fibula	Fifth metatarsal	Peroneal	Everts and plantar flexes foot
Peroneus longus	Fibula and lateral condyle of tibia	Medial cuneiform and first metatarsal	Peroneal	Everts and plantar flexes foot

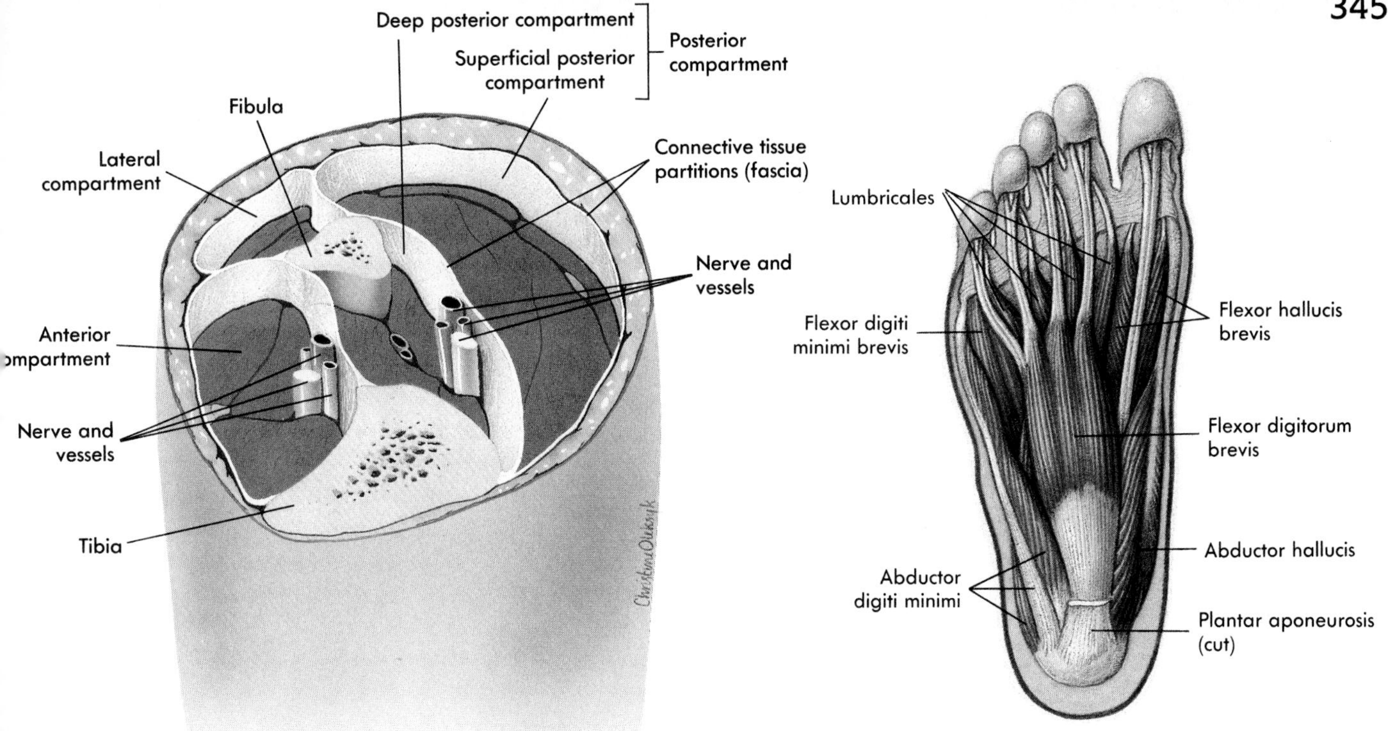

FIGURE 11-32 Cross section through the leg demonstrating the muscular compartments.

FIGURE 11-33 Muscles of the foot—plantar surface of the right foot.

Table 11-21

Intrinsic muscles of the foot (see Figure 11-33)

MUSCLE	ORIGIN	INSERTION	NERVE	FUNCTION
Abductor digiti minimi	Calcaneus	Proximal phalanx of fifth toe	Lateral plantar	Abducts and flexes little toe
Abductor hallucis	Calcaneus	Great toe	Medial plantar	Abducts great toe
Adductor hallucis	Lateral four metatarsals	Proximal phalanx of great toe	Lateral plantar	Adducts great toe
Extensor digitorum brevis	Calcaneus	Four tendons fused with tendons of extensor digitorum longus	Deep peroneal	Extends toes
Flexor digiti minimi brevis	Fifth metatarsal	Proximal phalanx of fifth digit	Lateral plantar	Flexes little toe (proximal phalanx)
Flexor digitorum brevis	Calcaneus and plantar fascia	Four tendons to middle phalanges of four lateral toes	Medial plantar	Flexes lateral four toes
Flexor hallucis brevis	Cuboid; medial and lateral cuneiform	Two tendons to proximal phalanx of great toe	Medial and lateral plantar	Flexes great toe
Dorsal interossei	Metatarsal bones	Proximal phalanges of second, third, and fourth digits	Lateral plantar	Abduct second, third, and fourth toes; adduct second toe
Plantar interossei	Third, fourth, and fifth metatarsals	Proximal phalanges of third, fourth, and fifth digits	Lateral plantar	Adducts third, fourth, and fifth toes
Lumbricales	Tendons of flexor digitorum longus	Second through fifth digits	Lateral and medial plantar	Flexes proximal and extends middle and distal phalanges
Quadratus plantae	Calcaneus	Tendons of flexor digitorum longus	Lateral plantar	Flexes toes

Bodybuilding

Bodybuilding is a growing sport worldwide. Once considered only for men, it currently is enjoyed by thousands of women as well. Participants in this sport combine diet and specific weight training to develop maximum muscle mass and minimum body fat, with their major goal a well-balanced, complete physique. Thus all skeletal muscles must be developed to their maximum. It is relatively simple for an uninformed, untrained muscle builder to build some muscles and ignore others; the result is a disproportioned body. Skill, training, and concentration are required to build all the muscles, to know which exercises build a large number of muscles and which are specialized to build certain parts of the body, and, above all, to know how to build a proportioned body.

Bodybuilding has its own language in which it is improper to refer to a muscle by its full name. Bodybuilders refer to the "lats," "traps," and "delts" rather than the lattissimus dorsi, trapezius, and deltoids. The exercises also have special names such as "lat pulldowns," "preacher curls," and "triceps extensions."

Bodybuilders concentrate on increasing skeletal muscle mass. Endurance tests conducted several years ago demonstrated that the cardiovascular and respiratory abilities of bodybuilders were similar to those abilities in normal, healthy persons untrained in a sport. However, more recent studies indicate that the cardiorespiratory fitness of bodybuilders is similar to that of other well-trained athletes. The difference between the new studies and the older studies is attributed to modern bodybuilding techniques that include aerobic exercise and running along with "pumping iron."

Photographs of bodybuilders are very useful in the study of anatomy to identify easily the surface anatomy of muscles that cannot be seen easily in untrained people (Figure 11-A).

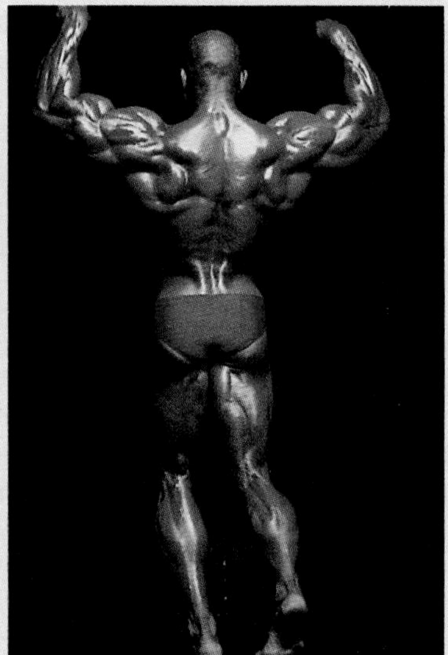

FIGURE 11-A
Bodybuilders.

SUMMARY

BODY MOVEMENTS RESULT FROM THE CONTRACTION OF SKELETAL MUSCLES.

GENERAL PRINCIPLES *p. 306*

1. The less movable end of muscle attachment is the origin; the more movable end is the insertion.
2. Synergists are muscles that function together to produce movement. Antagonists oppose or reverse the movement of another muscle.
3. Prime movers are mainly responsible for a movement. Fixators stabilize the action of prime movers.

Muscle Shapes

Muscle shape is determined primarily by the arrangement of muscle fasciculi.

Nomenclature

Muscles are named according to their shape, origin and insertion, location, size, orientation of fasciculi, or function.

Movements Accomplished by Muscles

Contracting muscles generate a force that acts on bones (levers) across joints (fulcrums) to create movement. There are three classes of levers.

HEAD MUSCLES *p. 309*

Head Movement

Origins of these muscles are mainly on the cervical vertebrae (except for the sternocleidomastoid); insertions are on the occipital bone or mastoid process. They

The Muscular System

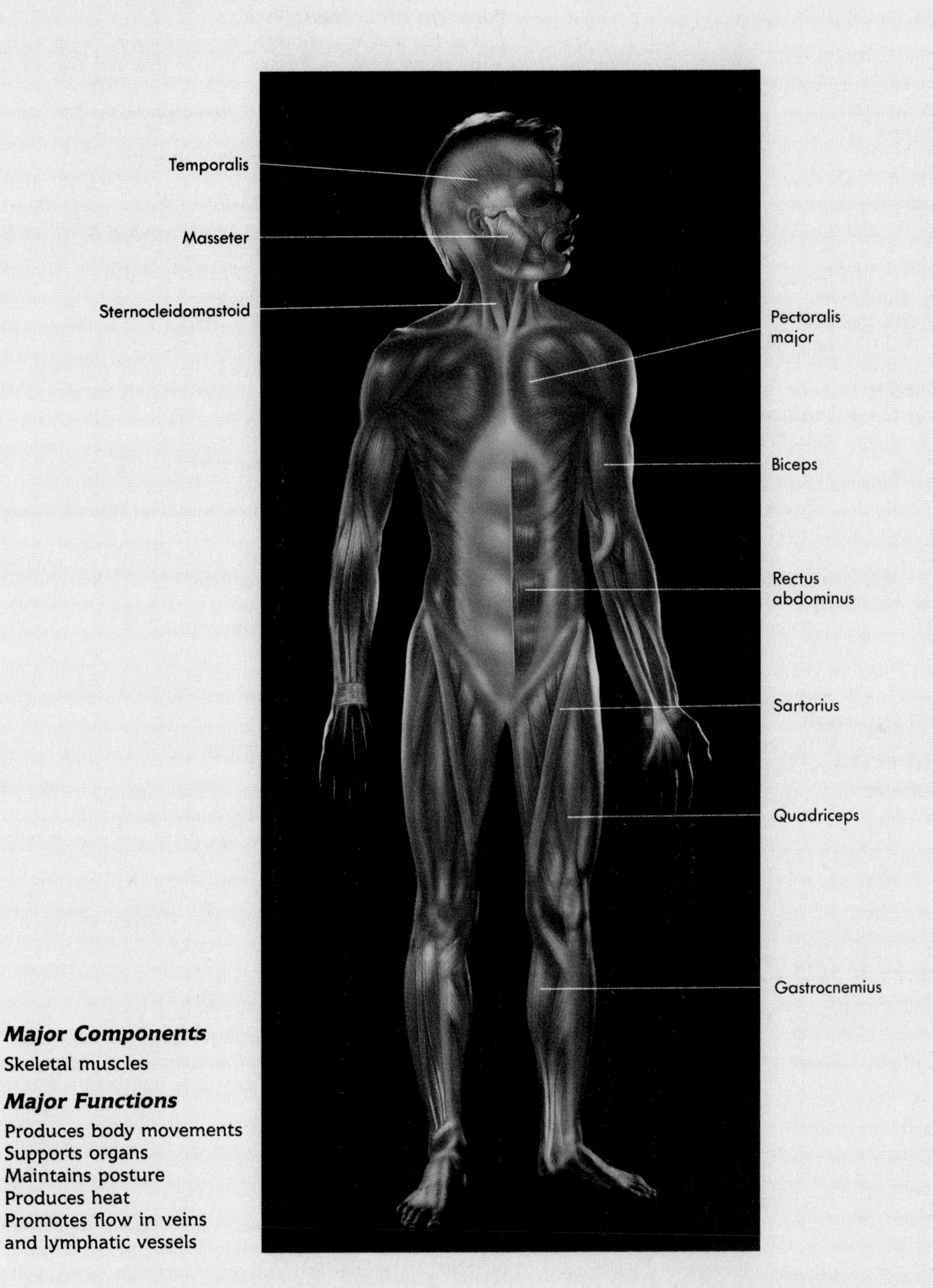

Major Components

Skeletal muscles

Major Functions

Produces body movements
Supports organs
Maintains posture
Produces heat
Promotes flow in veins and lymphatic vessels

cause flexion, extension, rotation, abduction, and adduction of the head.

Facial Expression

Origins of facial muscles are on skull bones or fascia; insertions are into the skin, causing movement of the facial skin, lips, and eyelids.

Mastication

Three pairs of muscles close the jaw; gravity opens the jaw. Forced opening is caused by the lateral pterygoids and the hyoid muscles.

Tongue Movement

Intrinsic tongue muscles change the shape of the tongue; extrinsic tongue muscles move the tongue.

Swallowing and the Larynx

1. Hyoid muscles can depress the jaw and assist in swallowing.
2. Muscles open and close the openings to the nasal cavity, auditory tubes, and larynx.

Movements of the Eyeball

Six muscles with their origins on the orbital bones insert on the eyeball and cause it to move within the orbit.

TRUNK MUSCLES *p. 321*

Muscles Moving the Vertebral Column

1. These muscles extend, abduct, rotate, or flex the vertebral column.
2. A deep group of muscles connects adjacent vertebrae.
3. A more superficial group of muscles runs from the pelvis to the skull, extending from the vertebrae to the ribs.

Thoracic Muscles

1. Most respiratory movement is caused by the diaphragm.
2. Muscles attached to the ribs aid in respiration.

Abdominal Wall

Abdominal wall muscles hold and protect abdominal organs and cause flexion, rotation, and abduction of the vertebral column.

Pelvic Floor and Perineum

These muscles support the abdominal organs inferiorly.

UPPER LIMB MUSCLES *p. 328*

Scapular Movements

Six muscles attach the scapula to the trunk, enabling the scapula to function as an anchor point for the muscles and bone of the arm.

Arm Movements

Seven muscles attach the humerus to the scapula. Two additional muscles attach the humerus to the trunk. These muscles cause flexion, extension, abduction, adduction, rotation, and circumduction of the arm.

Forearm Movements

1. Flexion and extension of the forearm are accomplished by three muscles located in the arm and two in the forearm.
2. Supination and pronation are accomplished primarily by forearm muscles.

Wrist, Hand, and Finger Movements

1. Forearm muscles that originate on the medial epicondyle are responsible for flexion of the wrist and fingers. Muscles extending the wrist and fingers originate on the lateral epicondyle.
2. Extrinsic hand muscles are in the forearm. Intrinsic hand muscles are in the hand.

LOWER LIMB MUSCLES *p. 338*

Thigh Movements

1. Anterior pelvic muscles cause flexion of the thigh.
2. Muscles of the buttocks are responsible for extension, abduction, and rotation of the thigh.
3. The thigh can be divided into three compartments.
   - The medial compartment muscles adduct the thigh.
   - The anterior compartment muscles flex the thigh.
   - The posterior compartment muscles extend the thigh.

Leg Movements

Some muscles of the thigh also act on the leg. The anterior thigh muscles extend the leg and the posterior thigh muscles flex the leg.

Ankle, Foot, and Toe Movements

1. The leg can be divided into three compartments.
   - Muscles in the anterior compartment cause dorsiflexion, inversion, or eversion of the foot, and extension of the toes.
   - Muscles of the lateral compartment plantar flex and evert the foot.
   - Muscles of the posterior compartment flex the leg, plantar flex and invert the foot, and flex the toes.
2. Intrinsic foot muscles flex or extend and abduct or adduct the toes.

CONTENT REVIEW

1. Define origin and insertion, synergist and antagonist, prime mover and fixator.
2. Describe the different shapes of muscles. How are the shapes related to the force of contraction of the muscle and the range of movement the contraction produces?
3. List the different criteria used to name muscles, and give an example of each.
4. Explain, using the terms fulcrum, lever, and force, how contraction of a muscle results in movement. Define the three classes of levers, and give an example of each in the body.
5. How does the sternocleidomastoid muscle differ from other muscles that move the head? What is wry neck?
6. What is unusual about the insertion (and sometimes origin) of facial muscles?
7. Which muscles are responsible for moving the ears, the eyebrows, the eyelids, and the nose? For puckering the lips, smiling, sneering, and frowning?
8. What causes a dimple on the chin?
9. Name the muscles responsible for closing and opening the jaw and for lateral and medial excursion of the jaw.
10. Contrast the movements produced by the extrinsic and intrinsic tongue muscles.
11. Which muscles open and close the openings to the auditory tube and larynx?
12. Describe the muscles of the eye and the movements that they cause.
13. Describe the group of muscles that attaches to the vertebrae and/or ribs and causes movement of the spine.
14. Name the muscle that is mainly responsible for respiratory movements. How do other muscles aid this movement?
15. Explain the anatomical basis for the lines seen on a well-muscled individual's abdomen. What are the functions of the abdominal muscles?
16. What is the function of the pelvic floor muscles?

17. What two muscles attach the humerus directly to the trunk? Name seven muscles that attach the humerus to the scapula.
18. What muscles cause extension and flexion of the arm? Abduction and adduction? Rotation?
19. List the muscles that cause flexion and extension of the forearm. Where are these muscles located?
20. Supination and pronation of the forearm are produced by what muscles? Where are these muscles located?
21. Describe the muscles that cause flexion and extension of the wrist.
22. Contrast the location and actions of the extrinsic and intrinsic hand muscles. What is the thenar and hypothenar eminence?
23. Describe the muscles that move the thumb. The tendons of what muscles form the anatomical snuffbox?
24. What muscle is the prime mover for flexion of the thigh? What muscles act as synergists to this muscle?
25. Describe the movements produced by the buttock muscles.
26. Name the muscle compartments of the thigh and the thigh movements produced by the muscles of each compartment. List the muscles of each compartment and the individual function of each muscle.
27. How is it possible for thigh muscles to move both the thigh and the leg? Name at least four muscles that can do this.
28. What movements are produced by the three muscle compartments of the leg? Name the muscles of each compartment, and describe the movements for which each muscle is responsible.
29. What movement do the peroneus muscles have in common? The tibialis muscles?
30. Name the leg muscles that flex the leg. Which of them can also flex the foot?
31. Describe the intrinsic foot muscles and their functions.

CONCEPT QUESTIONS

1. For each of the following muscles, (1) describe the movement that each muscle produces and (2) name the muscles that act as synergist and antagonist for them: sternocleidomastoid, depressor labii inferioris, masseter, erector spinae, rhomboideus major, coracobrachialis, brachialis, iliopsoas, rectus femoris, and soleus.
2. Propose an exercise that would benefit each of the following muscles specifically: biceps brachii, triceps brachii, deltoid, pectoralis major, rectus abdominis, quadriceps femoris, and gastrocnemius.
3. Consider only the effect of the brachioradialis muscle for this question. If a weight is held in the hand and the forearm is flexed, what type of lever system is in action? If the weight is placed on the forearm?
4. A patient was involved in an automobile accident in which the car was "rear-ended," resulting in whiplash injury of the head (hyperextension). What neck muscles would be injured in this type of accident?
5. During surgery a branch of the patient's facial nerve was accidentally cut. As a result, after the operation the skin of the patient's neck tended to droop and form wrinkles. What muscle was affected?
6. When a person becomes unconscious, the tongue muscles relax, and there is a tendency for the tongue to retract or fall back and obstruct the airway. Which tongue muscle is responsible?
7. In the active swinging of the arm, name the only muscle that is involved in all aspects of the arm's movement.
8. The support of the head of the humerus in the glenoid fossa is weakest in the inferior direction. What muscles help prevent dislocation of the shoulder when a heavy weight such as a suitcase is carried?
9. If the quadriceps femoris of the left leg were paralyzed, how would a person's ability to walk be affected?
10. Speedy Sprinter started a 200-meter dash and fell to the ground in pain. Examination of her right leg revealed the following symptoms: inability to plantar flex the foot against resistance, normal ability to evert the foot, dorsiflexion of the foot more than normal, and abnormal bulging of the calf muscles. Explain the nature of her injury.

? ?

ANSWERS TO PREDICT QUESTIONS

1 *p. 309.* Shortening the right sternocleidomastoid muscle would rotate the head to the left. It would also slightly elevate the chin.

2 *p. 315.* Raising eyebrows—occipitofrontalis; winking—orbicularis oculi; whistling—orbicularis oris and buccinator; smiling—levator anguli oris, risorius, zygomaticus major, and zygomaticus minor; frowning—corrugator supercilii, depressor anguli oris, depressor labii inferioris and mentalis; flaring nostrils—levator labii superioris alaque nasi and nasalis.

3 *p. 321.* Weakness of the lateral rectus would allow the eye to deviate medially.

4 *p. 328.* Pain in one of the four rotator cuff muscles associated with abduction would involve the supraspinatus.

5 *p. 334.* Two arm muscles are involved in flexion of the forearm—the brachialis and the biceps brachii. The brachialis only flexes, whereas the biceps brachii both flexes and supinates the forearm. With the forearm supinated, both muscles can flex the forearm optimally; when pronated, the biceps brachii does less to flex the forearm. Therefore chin-ups with the forearm supinated is easier since both muscles flex the forearm optimally in this position. Bodybuilders who wish to build up the brachialis muscle perform chin-ups with the forearms pronated.

PART III

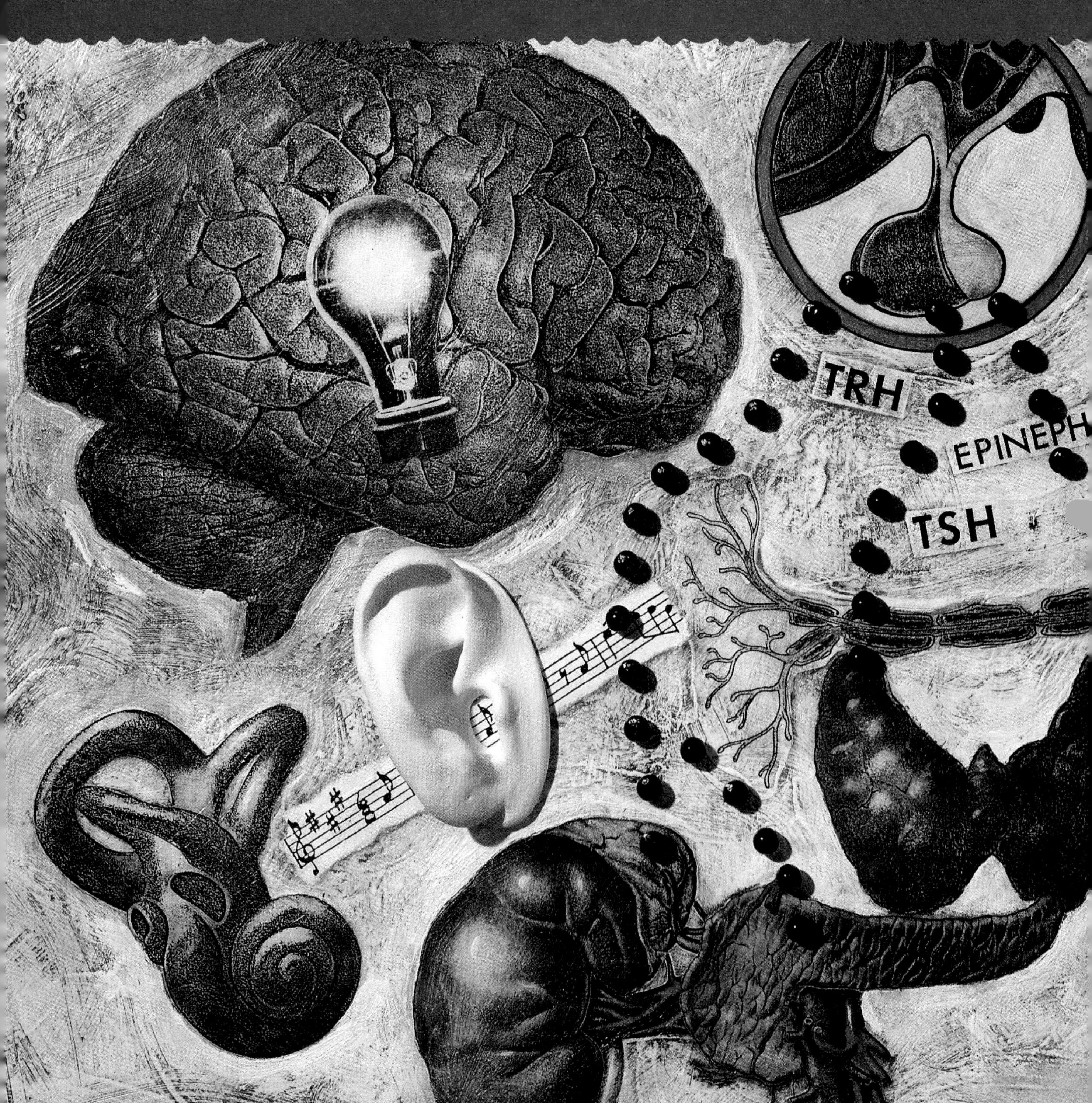

INTEGRATION AND CONTROL SYSTEMS

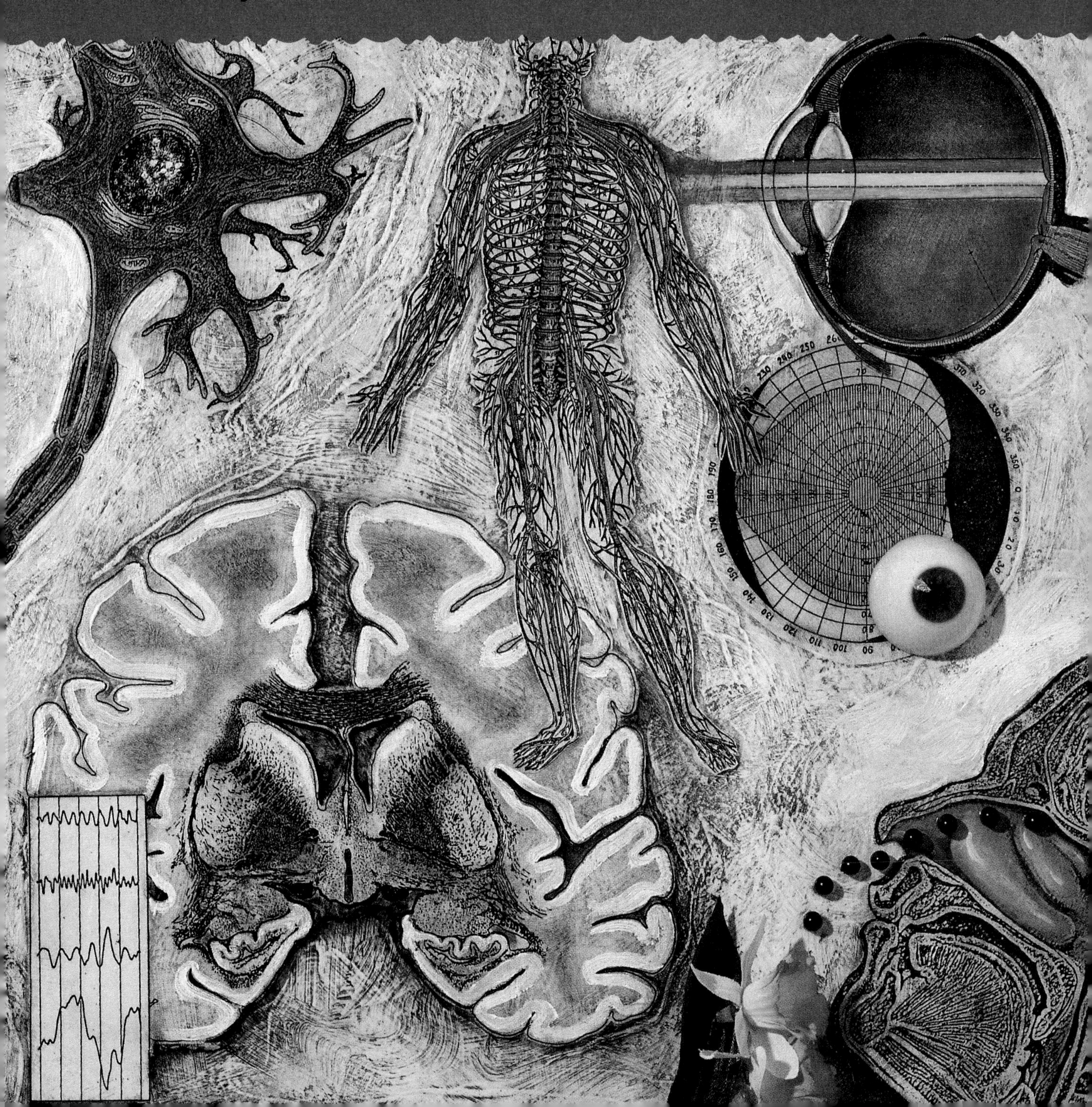

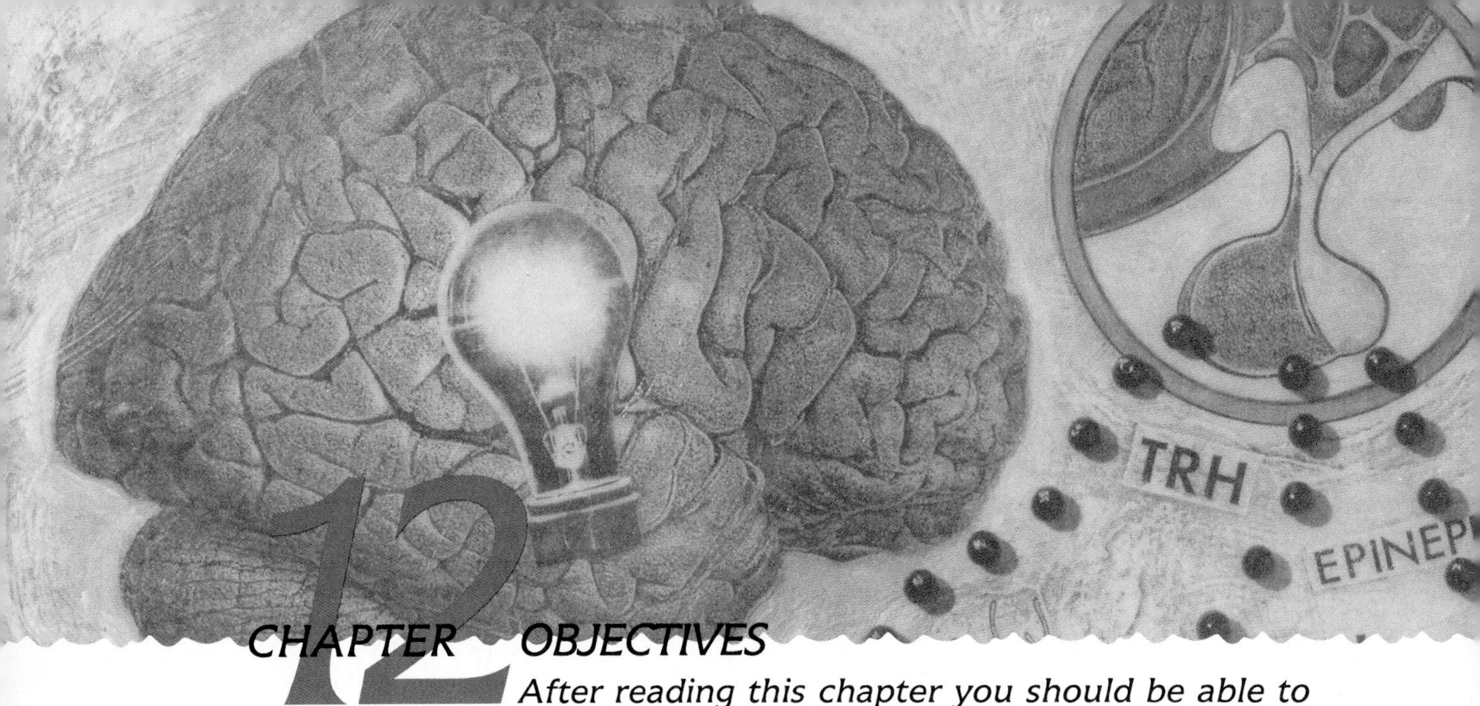

12 CHAPTER OBJECTIVES

After reading this chapter you should be able to

1. List the divisions of the nervous system and describe the characteristics of each.
2. Describe the structure of neurons and the function of their components.
3. Describe the three basic neuron shapes and give an example of where each one is found.
4. Describe the location, relative number, structure, and general function of neuroglia cells.
5. Explain the role of the myelin sheath in saltatory conduction of action potentials.
6. Define nerve, nerve tract, nucleus, and ganglion.
7. Describe the structure of a nerve, including its connective tissue components.
8. Describe the structure and function of the synapse.
9. Describe the release and metabolism of acetylcholine and norepinephrine in the synaptic cleft.
10. Explain how changes in the permeability of the neuron membrane result in the development of IPSPs and EPSPs.
11. Describe presynaptic inhibition and facilitation.
12. Define temporal and spatial summation.
13. Describe the role of the synapse in eliminating or enhancing information.
14. List the components and characteristics of a reflex.
15. Diagram a converging, a diverging, and an oscillating circuit and describe what is accomplished in each circuit type.

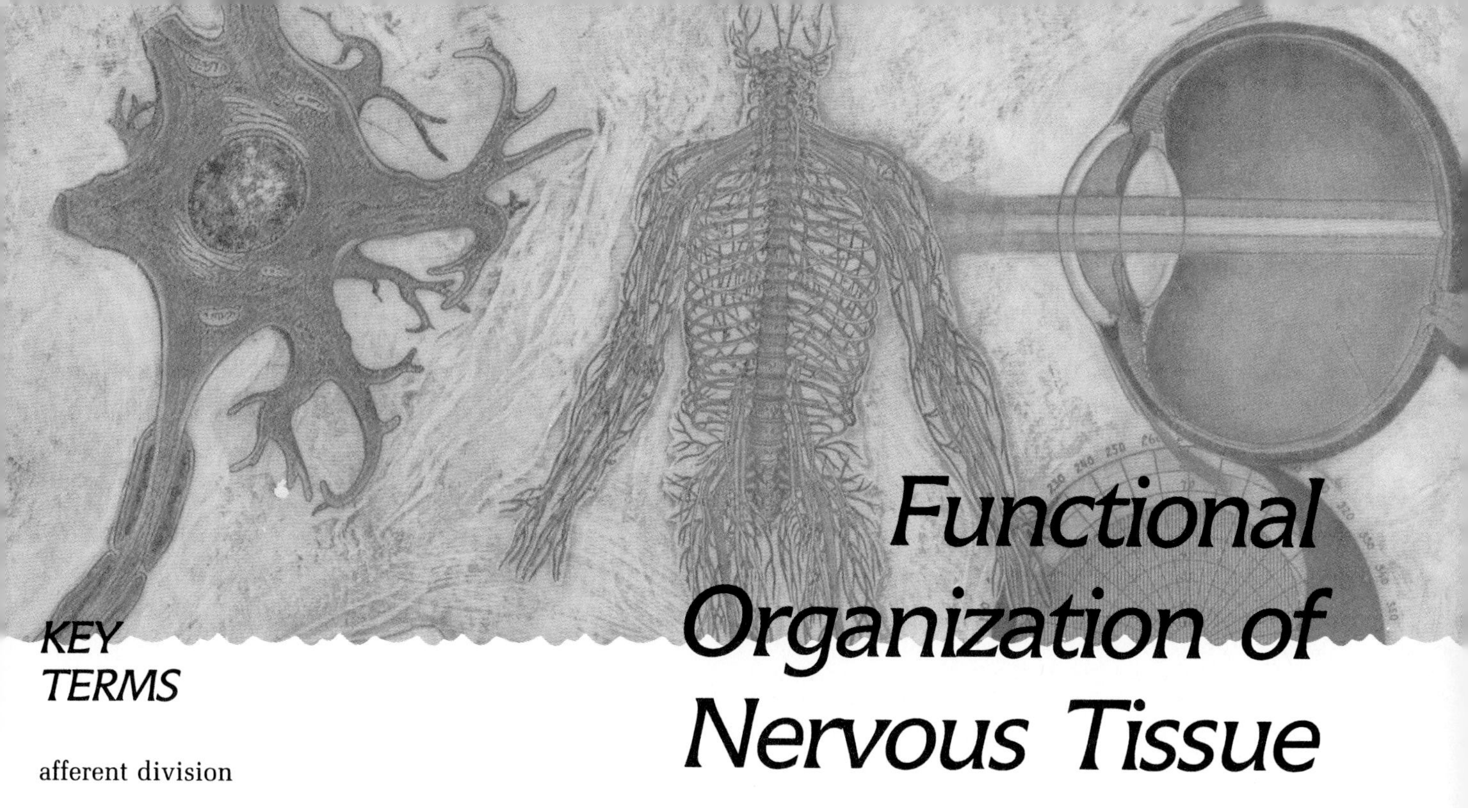

Functional Organization of Nervous Tissue

KEY TERMS

afferent division

central nervous system

efferent division

excitatory postsynaptic potential

gray matter

inhibitory postsynaptic potential

neuroglia

neuron

neurotransmitter

peripheral nervous system

reflex

saltatory conduction

spatial summation

temporal summation

white matter

RELATED TOPICS

The following terms or concepts are important for a good understanding of this chapter. If you are not familiar with them, you should review them before proceeding.

Membrane transport (Chapter 3)

Histology of nervous tissue (Chapter 4)

Membrane potential (Chapter 9)

THE NERVOUS SYSTEM AND THE ENDOCRINE SYSTEM ARE THE BODY'S MAJOR REGULATORY AND COORDINATING SYSTEMS. THE NERVOUS SYSTEM IS ALSO THE SEAT OF ALL MENTAL ACTIVITY, INCLUDING CONSCIOUSNESS, MEMORY, AND THINKING. The ability to solve problems, communicate using symbols, develop values, and exhibit a wide variety of emotions cannot occur without the nervous system.

Homeostasis is maintained to a large degree by the nervous system's regulatory and coordinating activities, which depend on its ability to detect, interpret, and respond to changes. Sensory organs monitor blood pressure, pH of the body fluids, temperature, taste, smell, touch, sound, light, the relative position of body parts, and other conditions. This information is transmitted from the sensory organs to centers of integration in the brain and the spinal cord in the form of action potentials. Based on this information, the brain and the spinal cord assess conditions within and outside of the body. The information may produce a response, may be stored for later use, or may be ignored.

DIVISIONS OF THE NERVOUS SYSTEM

There is only one nervous system, even though some of its subdivisions are referred to as separate systems. Thus the central nervous system and the peripheral nervous system are subdivisions of the nervous system instead of separate organ systems as their names suggest (Figure 12-1). Each subdivision has structural and functional features that separate it from the other subdivisions.

The **central nervous system (CNS)** consists of the brain and the spinal cord, both of which are encased in, and protected by, bone. The brain is located within the cranial vault of the skull, and the spinal cord is located within the vertebral canal formed by the vertebrae (see Chapter 7). The brain and spinal cord are continuous with each other at the foramen magnum.

The **peripheral nervous system (PNS)** consists of nerves and ganglia. **Nerves** are bundles of axons and their sheaths that extend from the CNS to peripheral structures such as muscles and glands and from sensory organs to the CNS. Forty-three pairs of nerves originate from the CNS to form the PNS. Twelve pairs, the **cranial nerves,** originate from the brain, and the remaining 31 pairs of nerves, the **spinal nerves,** originate from the spinal cord. **Ganglia** (gan'gle-ah; knots) are collections of nerve cell bodies located outside the CNS.

Two subdivisions comprise the PNS: the afferent, or sensory, division and the efferent, or motor, division (Figure 12-2). The **afferent division** transmits action potentials from the sensory organs to the CNS. The cell bodies of these neurons are located in ganglia near the spinal cord or near the origin of certain cranial nerves (Figure 12-3, *A*). The **efferent division** transmits action potentials from the CNS to effector organs such as muscles and glands.

The efferent division of the nervous system is divided into two subdivisions: the **somatic nervous system** and the **autonomic nervous system (ANS).** The somatic nervous system transmits action potentials from the CNS to skeletal muscle. Its nerve cell bodies are located within the CNS, and their axons extend through nerves to neuromuscular junctions (Figure 12-3, *B*), which are the only somatic nervous system synapses outside of the CNS. The ANS transmits action potentials from the CNS to smooth muscle, cardiac muscle, and certain glands. It sometimes is called the involuntary nervous system because control of its target tissues occurs subconsciously.

The ANS has two sets of neurons that exist in a series between the CNS and the effector organs. Cell bodies of the first neurons are within the CNS and send their axons to autonomic ganglia in which nerve cell bodies of the second neurons are located. Synapses exist between the first and second neurons within the autonomic ganglia, and the axons of the second neurons extend from the autonomic ganglia to the effector organs (Figure 12-3, *C*). The ANS is subdivided into the sympathetic and the parasympathetic divisions. In general the sympathetic division prepares the body for physical activity when activated, whereas the parasympathetic division regulates resting or vegetative functions such as digestion of food or emptying of the urinary bladder.

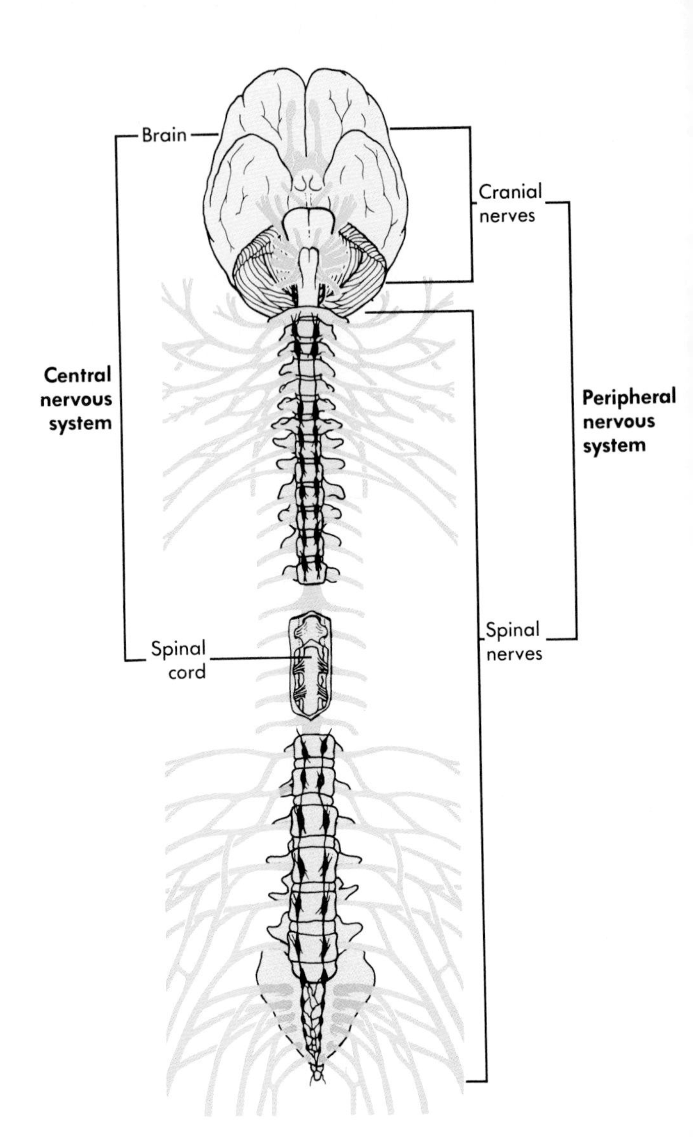

FIGURE 12-1 The central nervous system consists of the brain and spinal cord (some thoracic vertebrae have been removed, and the covering of the spinal cord has been cut to reveal the spinal cord). The peripheral nervous system consists of cranial nerves, which arise from the brain, and spinal nerves, which arise from the spinal cord.

The CNS is the major site for processing information, initiating responses, and integrating mental processes. It is analogous to a highly sophisticated

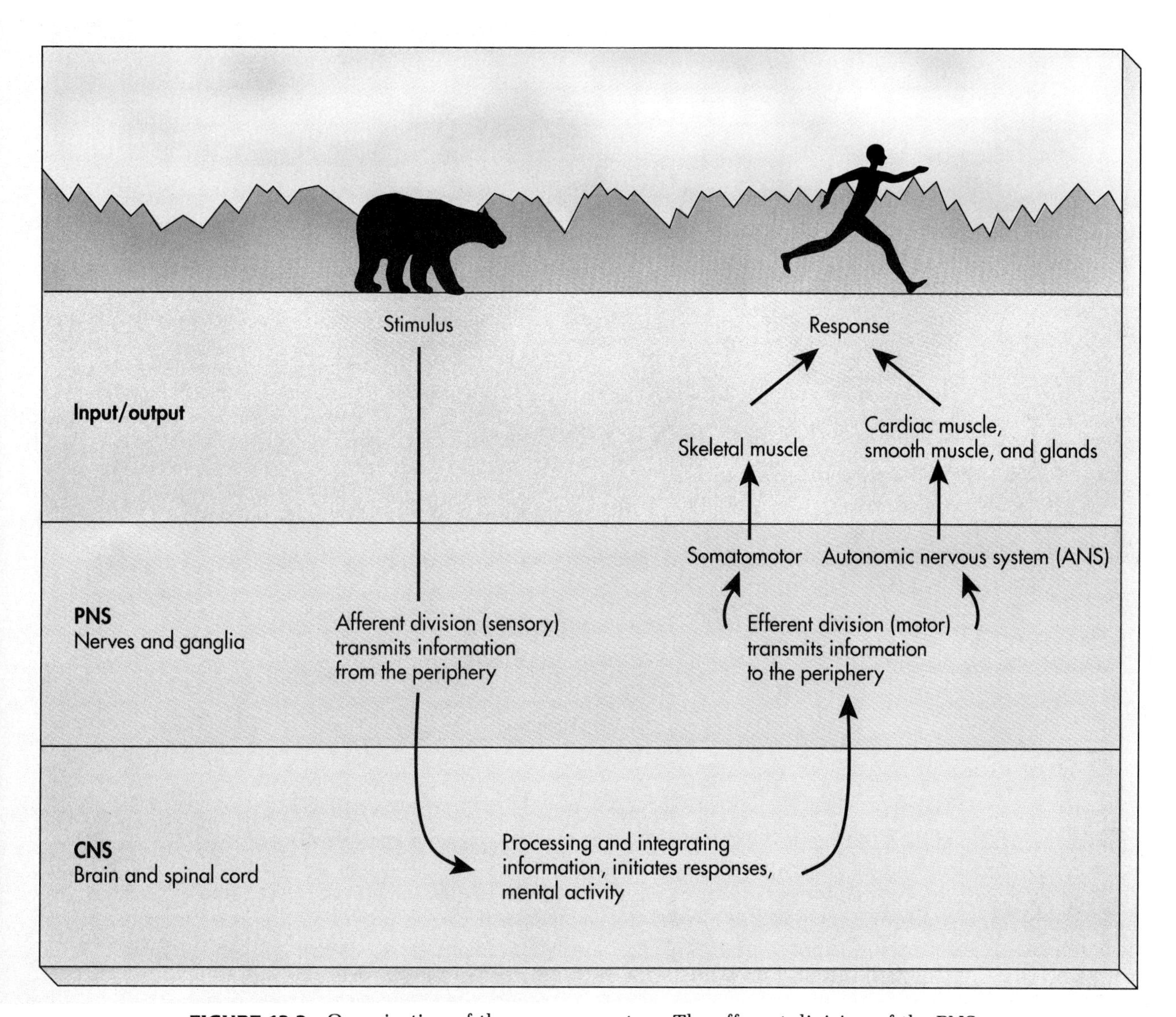

FIGURE 12-2 Organization of the nervous system. The afferent division of the PNS detects stimuli and conveys action potentials to the CNS. The CNS interprets incoming information and initiates action potentials that are transmitted through the efferent division to produce a response. The efferent division is divided into the somatic (somatomotor) nervous system and autonomic nervous system.

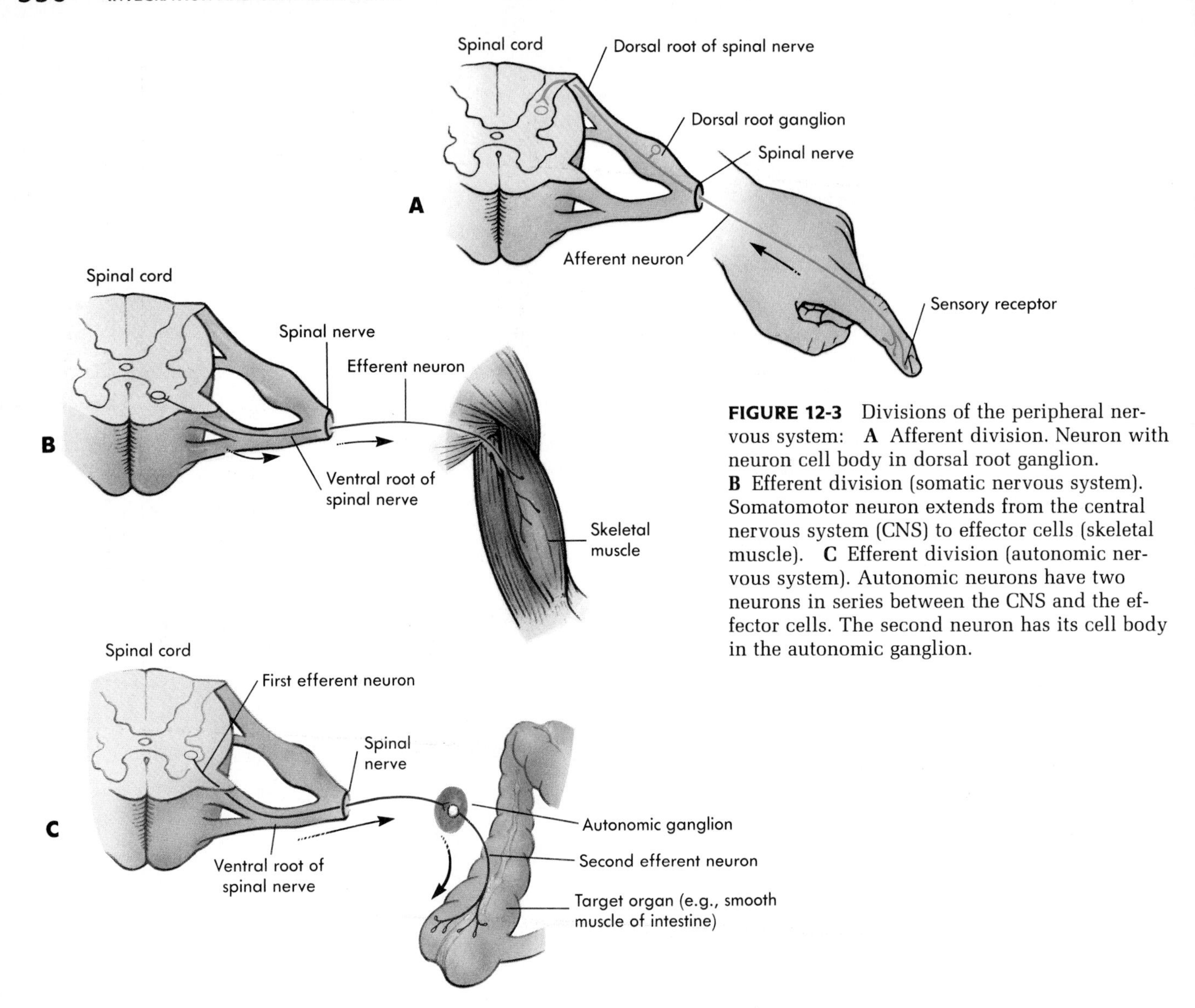

FIGURE 12-3 Divisions of the peripheral nervous system: **A** Afferent division. Neuron with neuron cell body in dorsal root ganglion. **B** Efferent division (somatic nervous system). Somatomotor neuron extends from the central nervous system (CNS) to effector cells (skeletal muscle). **C** Efferent division (autonomic nervous system). Autonomic neurons have two neurons in series between the CNS and the effector cells. The second neuron has its cell body in the autonomic ganglion.

computer with the ability to receive input, process and store information and to generate responses. It also has the ability to produce ideas, emotions, and other mental processes that are not the automatic consequences of information input. The PNS functions primarily to detect stimuli and transmit information in the form of action potentials to and from the CNS. However, the PNS does perform some limited integration at the sensory receptors and in some ganglia.

CELLS OF THE NERVOUS SYSTEM

Cells of the nervous system include nonneural cells and neurons. Nonneural cells are **neuroglia** (nu-rog'le-ah; nerve glue), or **glial cells,** which support and protect neurons and perform other functions, whereas neurons receive stimuli and conduct action potentials.

Neurons

Neurons, or **nerve cells,** receive stimuli and transmit action potentials to other neurons or to effector organs. They are organized to form complex networks that perform the functions of the nervous system. Each neuron consists of a cell body, called the **neuron cell body, nerve cell body,** or **soma** (so'mah; body), and two types of processes, **dendrites** (den'drītz; tree, referring to the branching organization of dendrites) and **axons** (ak'sonz; axle, referring to the straight alignment and uniform diameter of most axons) (Figure 12-4).

Neuron cell body

Each neuron cell body contains a single relatively large and centrally located nucleus with a prominent nucleolus. Extensive rough endoplasmic reticulum and Golgi apparatuses surround the nucleus, and a moderate number of mitochondria and other organelles are also present. Lipid droplets and melanin pigments accumulate in the cytoplasm of some neuron cell bodies as randomly arranged inclusions. The number of inclusions increases with age, but their functional significance is not known. Large numbers of intermediate filaments (neurofilaments) and microtubules form bundles that course through the cytoplasm in all directions. The neurofilaments separate areas of rough endoplasmic reticulum called **Nissl** (nis'l) **bodies,** which are apparent in the cytoplasm of the neuron cell body and the basal portion of the dendrites. The presence of organelles such as rough endoplasmic reticulum indicate that the neuron cell body is the primary site of protein synthesis within neurons.

1

Predict the effect on the portion of an axon that is severed so that it is no longer connected to its nerve cell body. Explain your prediction.

? ? ? ? ? ? ? ? ? ? ?

Dendrites

Dendrites are short, often highly branched cytoplasmic extensions that are tapered from their bases at the neuron cell body to their tips (see Figure 12-4). The surface of many dendrites has small extensions called **dendritic spines,** or **gemmules** (jem'ūls), with which axons of other neurons form synapses with the dendrites. Dendrites respond to **neurotransmitter** substances released from the axons of other neurons by producing local potentials. Most dendrites do not conduct action potentials. Functionally, dendrites have traditionally been classified as processes that conduct electrical signals toward the cell body. It now is known that this definition has exceptions.

Axons

In most neurons a single axon arises from an enlarged area, the **axon hillock,** of the neuron cell body. An axon may remain as a single structure or may branch to form collateral axons or side branches (see Figure 12-4). Each axon has a constant diameter and may vary in size from a few millimeters to more than 1 m in length. The axon's cytoplasm is called **axoplasm,** and its cell membrane is called the **axolemma** (*lemma* means sheath). Axons terminate by branching to form **telodendria** (tel-o-den'dre-ah) with enlarged ends called **terminal boutons** (bootonz') or **presynaptic terminals.** Numerous small vesicles that contain neurotransmitters are present in the presynaptic terminals. Functionally, axons conduct action potentials to the dendrites, cell bodies, or axons of other neurons.

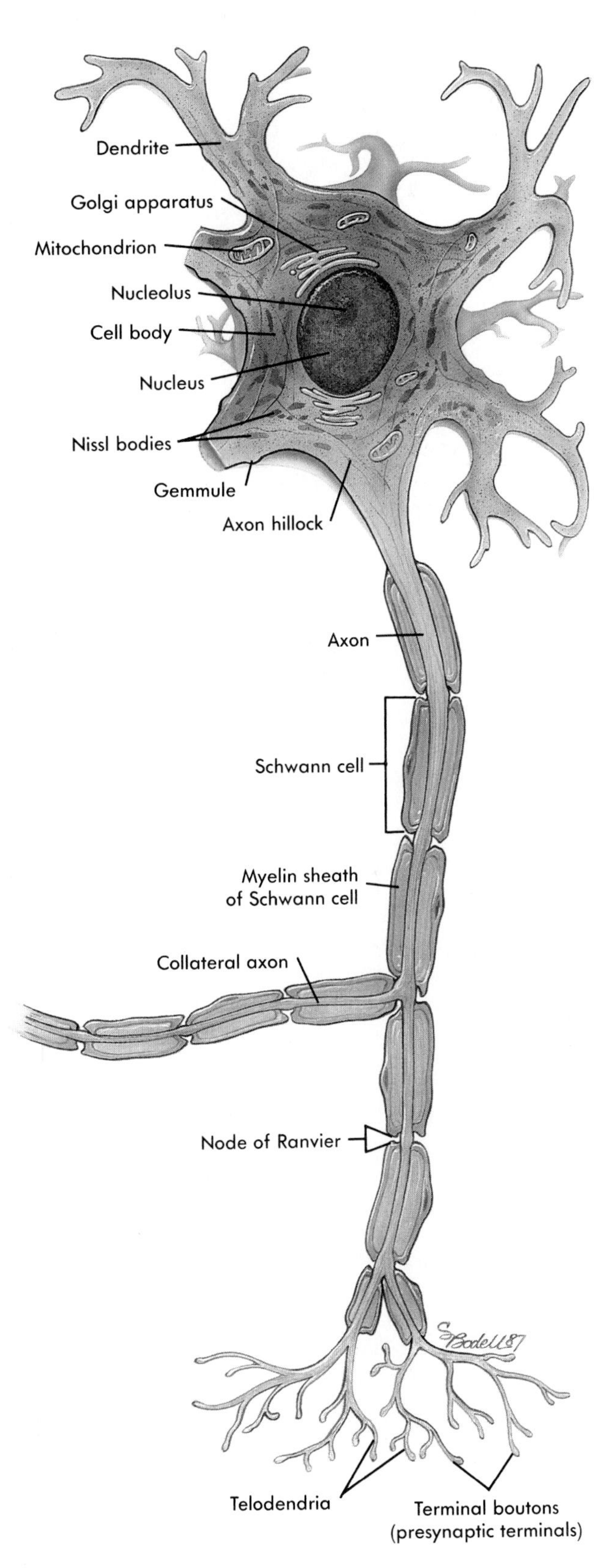

FIGURE 12-4 Structural features of neurons, including dendrites, cell body, and axon.

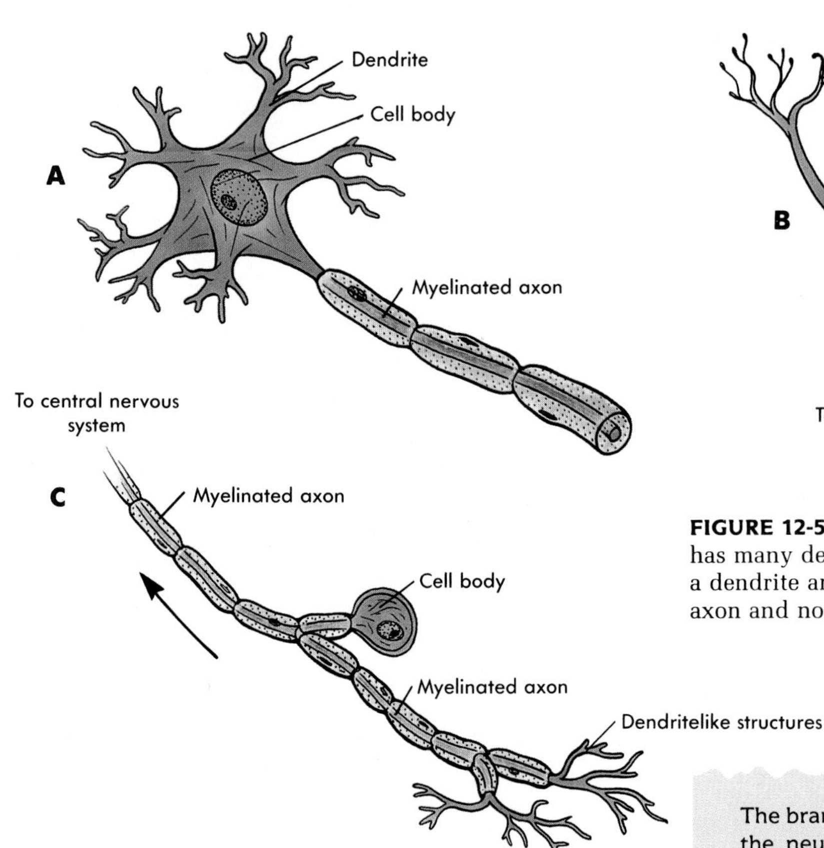

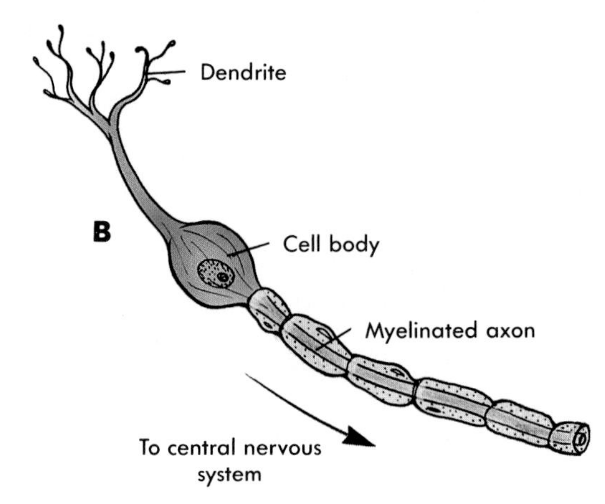

FIGURE 12-5 Types of neurons. **A** Multipolar neuron has many dendrites and an axon. **B** Bipolar neuron has a dendrite and an axon. **C** Unipolar neuron has an axon and no dendrites.

The branch of a unipolar neuron that extends from the neuron cell body to the periphery conducts action potentials toward the neuron cell body, and according to a functional definition of a dendrite, it could be classified as a dendrite. However, it often is referred to as an axon for two reasons: it cannot be distinguished from an axon based on its structure, and it conducts action potentials in the same fashion as an axon (see Axon Sheaths, p. 361).

Neuron Types

Three categories of neurons exist based on their shape: multipolar, bipolar, and unipolar. **Multipolar neurons,** the most numerous type, have several dendritic processes and a single axon. The dendritic processes vary in number and in their degree of branching (Figure 12-5, *A*). Most of the neurons within the CNS, including motor neurons, are multipolar.

Bipolar neurons consist of a single dendrite and a single axon (Figure 12-5, *B*). The dendrite often is specialized to receive the stimulus, and the axon conducts action potentials to the CNS. Bipolar neurons are components of some sensory organs (e.g., olfactory receptors of the nasal cavity and rods and cones of the eye).

Unipolar, or **pseudounipolar, neurons** have a single axon and no dendrites extending from the cell body (Figure 12-5, *C*). The term pseudounipolar (*pseudo* means false), which is used less commonly, applies because during development the axon and dendrite of a bipolar neuron come together and fuse to form a single process that is called the axon. Most sensory neurons are unipolar. The peripheral ends of their axons have dendritelike processes that respond to stimuli, producing action potentials that are conducted by the axon to the CNS.

Neuroglia

Neuroglia are far more numerous than neurons. They serve as the major supporting tissue in the CNS, participate in the formation of a permeability barrier between the blood and the nerve cells, produce cerebrospinal fluid, and form myelin sheaths around axons. There are five types of neuroglial cells, each with unique structural and functional characteristics.

Astrocytes

Astrocytes (as′tro-sītz) are neuroglia that are star shaped due to cytoplasmic processes that extend from the cell body. The astrocytes' processes extend to and cover the surfaces of blood vessels, neurons (Figure 12-6), and the pia mater (a membrane covering the outside of the brain and spinal cord). Astrocytes function as a nonrigid supporting matrix and help to regulate the composition of the extracellular fluid around neurons.

According to the concept of the **blood-brain barrier,** only certain substances are able to pass from the blood into the nervous tissue of the brain and spinal cord. The blood-brain barrier protects neurons from toxic substances in the blood, allows the exchange of nutrients and waste products between neurons and the blood, and prevents fluctuations in the composition of the blood from affecting the functioning of the brain. Endothelial cells of the blood vessels, which are joined by tight junctions (see Chapter 4), form the blood-brain barrier. Consequently, substances do not pass between the cells but must pass through the cells. Lipid-soluble substances such as nicotine, ethanol, and heroin can diffuse through the phospholipid membrane of the endothelial cells and enter the brain. Water-soluble molecules such as amino acids and glucose move across the blood-brain barrier by mediated transport (see Chapter 3).

Astrocytes play a role in regulating the extracellular composition of brain fluid by influencing the formation of the tight junctions and the types of molecules transported by the endothelial cells. In addition, after substances pass through the blood-brain barrier, astrocytes function to remove and/or process molecules and ions.

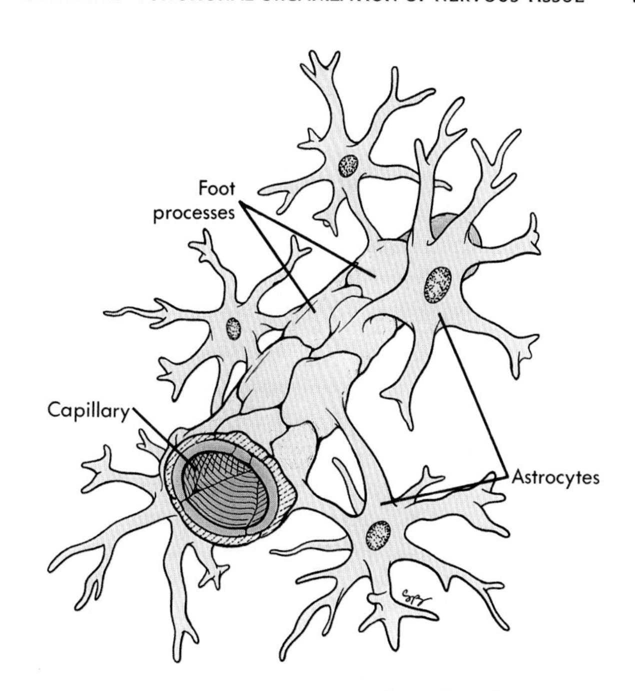

FIGURE 12-6 Astrocyte processes form feet that cover the surfaces of neurons and blood vessels. The astrocytes provide structural support and play a role in regulating which substances from the blood reach the neurons.

Pharmacologists must consider the permeability characteristics of the blood-brain barrier when developing drugs designed to affect neurons of the CNS. For example, levodopa (L-dopa), a precursor to a neurotransmitter called dopamine, is the substance sometimes used to treat patients with Parkinson's disease, a CNS disorder caused by the lack of dopamine, which normally is produced by certain neurons of the brain. This lack results in decreased muscle control and shaking movements. Levodopa, rather than dopamine, is administered because it can cross the blood-brain barrier and dopamine cannot. CNS neurons convert levodopa to dopamine, which reduces the symptoms of Parkinson's disease.

2

How can a drug that acts on CNS neurons but cannot cross the blood-brain barrier be administered to a person so that it is effective?

? ? ? ? ? ? ? ? ? ? ? ?

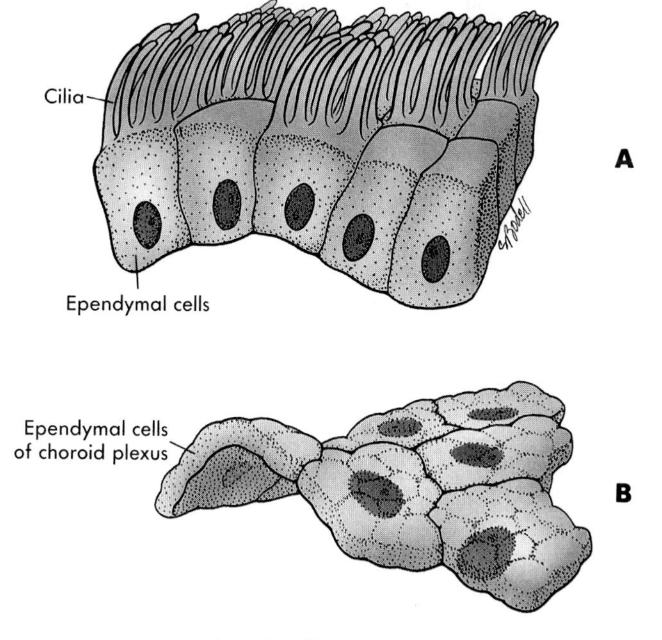

FIGURE 12-7 Ependymal cells. **A** Ciliated ependymal cells lining a ventricle of the brain help to move cerebrospinal fluid. **B** Ependymal cells on the surface of the choroid plexus secrete cerebrospinal fluid.

Ependymal cells

Ependymal (ep-en′dĭ-mal) cells line the ventricles (cavities) of the brain and the central canal of the spinal cord (Figure 12-7, *A*). Specialized ependymal cells within certain regions of the ventricles, the **choroid plexuses** (Figure 12-7, *B*), secrete the cerebrospinal fluid that circulates through the ventricles of the brain (see Chapter 13). The free surface of the ependymal cells frequently has patches of cilia that assist in moving cerebrospinal fluid through the cavities of the brain. Ependymal cells also have long processes at their basal surfaces that extend deep into the brain and the spinal cord and seem, in some cases, to perform astrocyte-like functions.

Numerous microglia migrate to areas damaged by infection, trauma, or stroke and perform phagocytosis. These damaged areas can be identified in the CNS by a pathologist because during an autopsy large numbers of microglia are found in them.

Microglia

Microglia (mi-krog′le-ah) originate from embryonic tissue different from that of the other neuroglia within the CNS (astrocytes and oligodendrocytes are derived from neuroectoderm, and microglia are modified macrophages derived from mesoderm; see Chapter 22), but they are included with the other neuroglia because of their location in the CNS and because it is traditional to do so. Microglia are small cells that become mobile and phagocytic in response to inflammation, phagocytizing necrotic tissue, microorganisms, and foreign substances that invade the CNS (Figure 12-8).

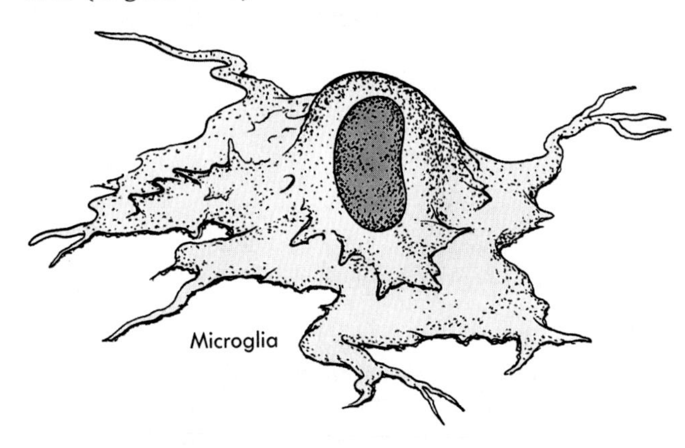

FIGURE 12-8 Microglia within the central nervous system are similar to macrophages.

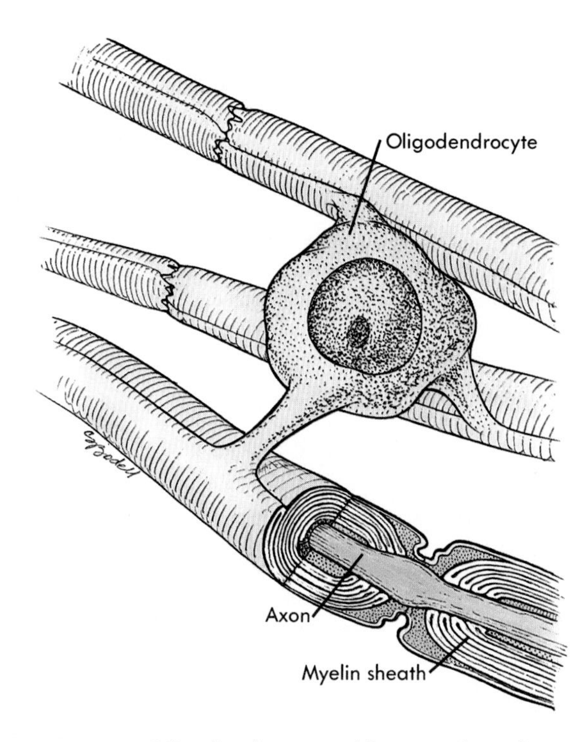

FIGURE 12-9 Oligodendrocyte with extensions that form the myelin sheaths of axons within the central nervous system.

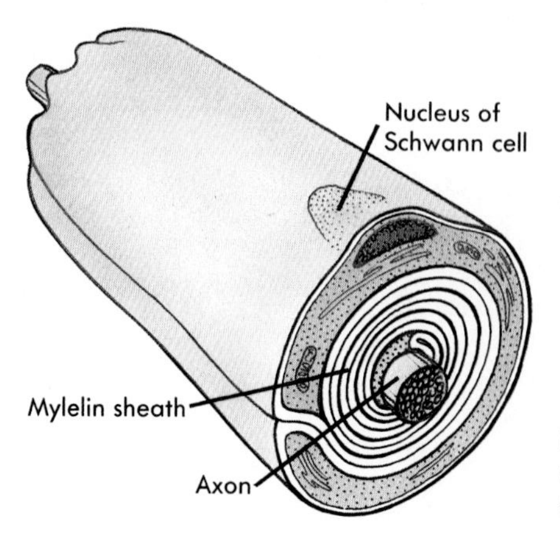

FIGURE 12-10 Schwann cell forming the myelin sheath of an axon within the peripheral nervous system.

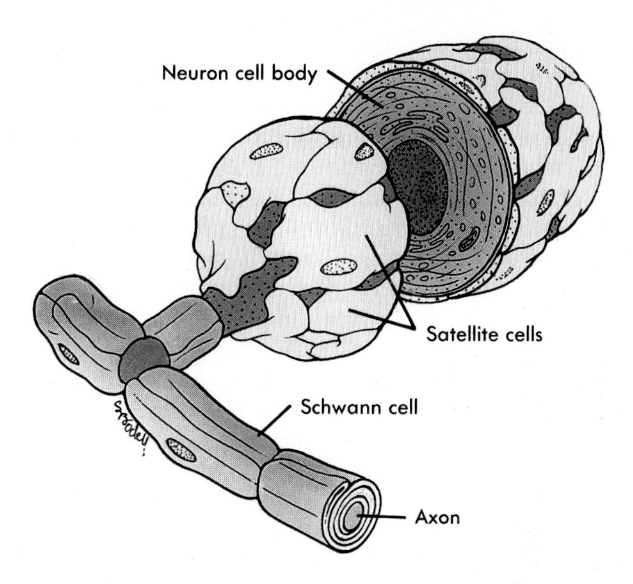

FIGURE 12-11 Satellite cells surrounding neuron cell bodies within ganglia.

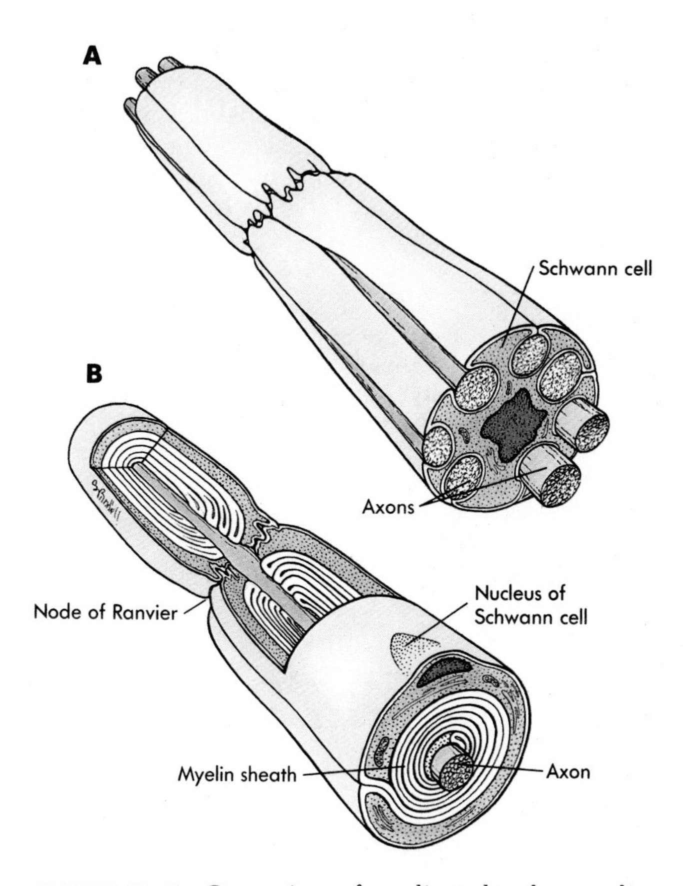

FIGURE 12-12 Comparison of myelinated and unmyelinated axons. **A** Unmyelinated axon with a series of Schwann cells surrounding several axons in parallel formation. Several axons are surrounded by the cytoplasm of a single Schwann cell. **B** Myelinated axon with a series of Schwann cells forming the myelin sheath around a single axon.

Oligodendrocytes

Oligodendrocytes (o-lig'o-den'dro-sītz; cells with few dendritic processes) have cytoplasmic extensions that form sheaths around axons in the CNS (Figure 12-9). A single oligodendrocyte may surround portions of several axons, but many oligodendrocyte processes form the sheath around a single axon. Oligodendrocyte processes are modified to form specialized sheaths, **myelin sheaths,** around many axons in the CNS. The normal rate of action potential propagation depends on these myelin sheaths.

Schwann cells

Schwann cells, or **neurolemmocytes,** are peripheral neuroglial cells that derive from neural crest tissue. They are in the PNS rather than the CNS, and like the oligodendrocytes of the CNS, they form the supportive and protective sheaths around axons. However, unlike oligodendrocytes, each Schwann cell forms a sheath around only one axon (Figure 12-10). Schwann cells form the specialized myelin sheaths of axons in the PNS and therefore are essential for the normal propagation of action potentials along these axons.

Satellite cells, which are specialized Schwann cells, surround nerve cell bodies in ganglia, provide support, and may provide nutrients to the nerve cell bodies (Figure 12-11).

Axon Sheaths

Cytoplasmic extensions of the oligodendrocytes in the CNS and of the Schwann cells in the PNS surround axons to form either unmyelinated or myelinated axon sheaths. The axon sheaths protect and electrically insulate the axons from each other. Additionally, action potentials are propagated along unmyelinated and myelinated axons in different ways and at different rates.

Unmyelinated axons

Unmyelinated axons rest in an invagination of the oligodendrocytes or the Schwann cells (Figure 12-12, *A*). The axons remain outside the cytoplasm of the cells that ensheathe them, but they are surrounded by the cells' cytoplasmic extensions. Thus each axon is ensheathed by a series of cells, and each cell may simultaneously ensheathe more than one unmyelinated axon. In unmyelinated axons action potentials are propagated along the entire axon membrane (see Figure 9-10).

Myelinated axons

In **myelinated axons** of the PNS each Schwann cell ensheathes only one axon (Figure 12-12, *B*) and repeatedly wraps around a segment of an axon to form a series of tightly wrapped membranes rich in phospholipids with little cytoplasm sandwiched between the membrane layers. The tightly wrapped membranes comprise the myelin sheath, giving myelinated axons a white appearance because of the high lipid concentration. When viewed longitudinally, gaps can be seen every 0.1 to 1.5 mm between the individual Schwann cells. These interruptions in the myelin sheath are the **nodes of Ranvier** (ron've-a), and the area between the nodes is called the **internode.** In the CNS there are also nodes of Ranvier between the myelin segments.

Myelinated axons conduct action potentials more rapidly than unmyelinated axons because action potentials appear to jump from one node of Ranvier to the next without propagation along the entire length of the axon. Conduction of action potentials along the axolemma in myelinated neurons is called **saltatory conduction** (L. *saltare,* to leap). The reversal of the electrical charge across the cell membrane, which occurs during an action potential at one node of Ranvier, causes electrical current to flow across the membrane at the adjacent node in an instantaneous fashion (Figure 12-13). The lipid within the membranes of the myelin sheath acts as a layer of insulation, forcing the electrical current to flow to the adjacent node and preventing flow across the membrane between the nodes. The flow of the electrical current acts as a stimulus and causes initiation of an action potential at the adjacent node; thus action potentials are conducted from node to node.

The speed of conduction of action potentials along an axon depends on the myelination and diameter of the axon. Action potentials are conducted more rapidly in myelinated than unmyelinated axons because action potentials are formed quickly at each successive node of Ranvier (saltatory conduction) instead of being propagated more slowly through every part of the axon's membrane. Large-diameter axons conduct action potentials more rapidly than small-diameter axons because large-diameter axons provide less resistance to action potential propagation.

Nerve fibers can be classified based on their size and myelination. Not surprisingly, the structure of nerve fibers reflects their functions. Type A fibers are large-diameter, myelinated axons that conduct action potentials at 15 to 120 m/sec. Motor neurons supplying skeletal muscles and most sensory neurons have type A fibers. Rapid response to the external environment is possible because of the rapid input of sensory information to the CNS and rapid output of action potentials to skeletal muscles.

Type B fibers are medium-diameter, lightly myelinated axons that conduct action potentials at 3 to 15 m/sec, and type C fibers are small-diameter, unmyelinated axons that conduct action potentials at 2 m/sec or less. Type B and C fibers are primarily part of the ANS, which supplies internal organs such as the stomach, intestines, and heart. The responses necessary to maintain internal homeostasis such as digestion need not be as rapid as responses to the external environment.

3

What is the advantage of having small-diameter axons that conduct action potentials rapidly? (HINT: Consider what an animal with only unmyelinated axons would be like.)

? ? ? ? ? ? ? ? ? ? ?

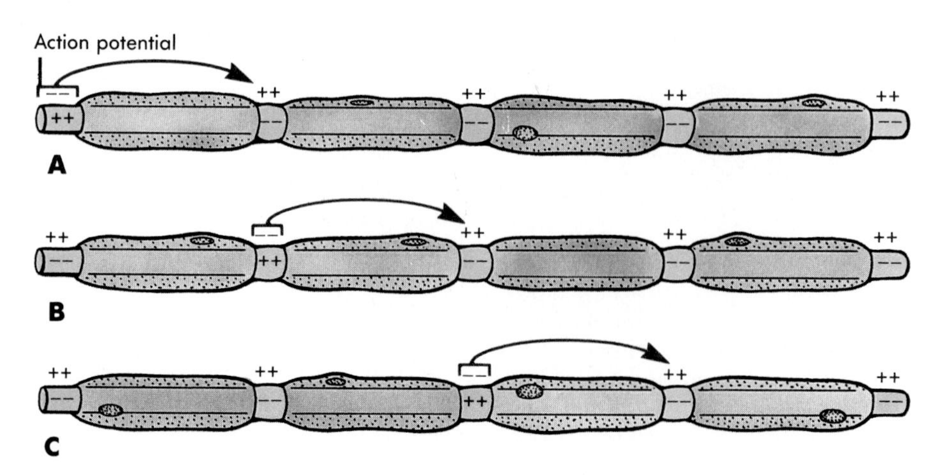

FIGURE 12-13 Flow of electrical charge from node to node during saltatory conduction produces action potentials at the nodes of Ranvier in myelinated neurons. **A** to **C** Conduction along an axon.

Myelin sheaths begin to form during the late part of fetal development. The process continues rapidly until the end of the first year after birth and continues thereafter at a slower rate. The development of myelin sheaths is associated with the infant's continuing development of more rapid and better coordinated responses.

ORGANIZATION OF NERVOUS TISSUE

Nervous tissue is organized so that axons form bundles and nerve cell bodies and their relatively short dendrites are organized into groups. Bundles of parallel axons with their associated sheaths are whitish in color and are called **white matter.** White matter propagates action potentials, and the axons comprising the white matter of the CNS form conduction pathways, **nerve tracts,** which propagate action potentials from one area in the CNS to another. In the PNS bundles of axons and their sheaths are called **nerves.** Collections of nerve cell bodies and their dendrites are more gray in color and are called **gray matter.** The terms white and gray matter normally are applied to the CNS, but the principle is appropriate for the PNS as well.

Gray matter is the site of integration within the nervous system. The outer surface of the cerebrum and the cerebellum (see Chapter 13) consists of gray matter called the cerebral cortex and cerebellar cortex, respectively. Within the brain are other collections of gray matter called **nuclei.** The nuclei of the CNS perform integrative functions or act as relay areas where axons of one nerve tract synapse with neurons of other nerve tracts. Collections of neuron cell bodies outside of the CNS are called **ganglia.**

Peripheral nerves consist of axon bundles (Figure 12-14). Each axon (also called a nerve fiber) and its Schwann cell sheath are surrounded by a delicate connective tissue layer, the **endoneurium** (en'do-nu're-um). Groups of axons are surrounded by a heavier connective tissue layer, the **perineurium** (pĕr'ĭ-nu're-um), to form **nerve fascicles** (fas'ĭ-klz). A third layer of connective tissue, the **epineurium** (ep'ĭ-nu're-um), binds the nerve fascicles together to form a nerve. The connective tissue of the epineurium merges with the loose connective tissue surrounding the nerve to make the PNS nerves tougher than the nerve tracts of the CNS.

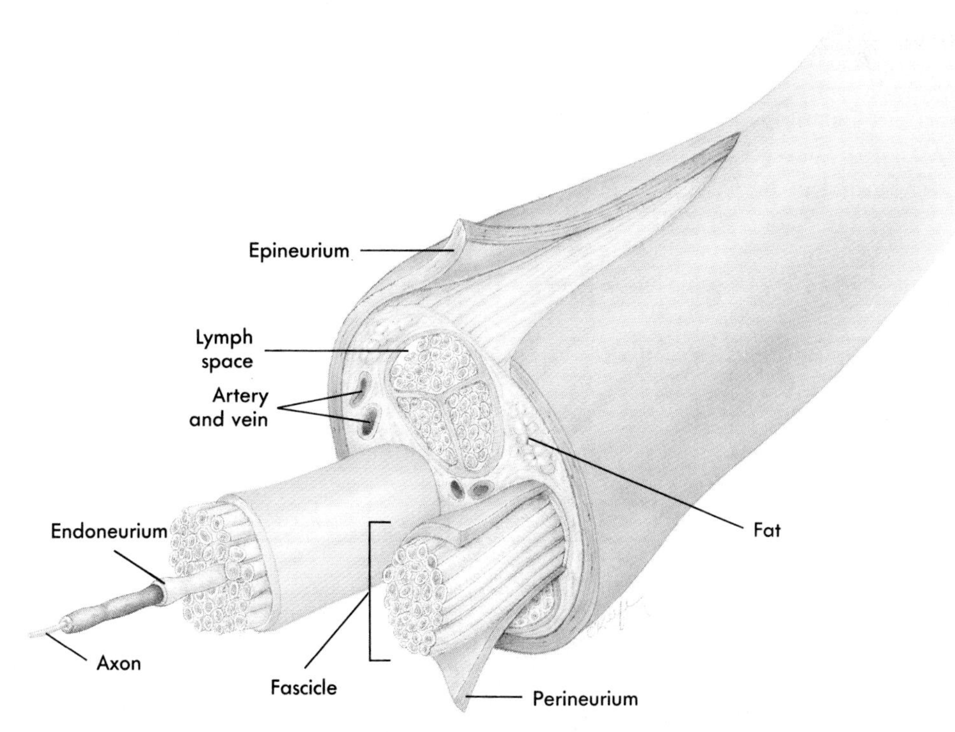

FIGURE 12-14 Nerve structure illustrating axons surrounded by various layers of connective tissue: epineurium around the whole nerve, perineurium around nerve fascicles, and endoneurium around Schwann cells and axons.

THE SYNAPSE

The **synapse** is the junction between one nerve cell and another nerve cell, an effector such as a muscle or gland, or a sensory receptor cell. Its essential components are the presynaptic terminal, the synaptic cleft, and the postsynaptic terminal (Figure 12-15, *A*). A close relationship exists between a **presynaptic terminal,** which is formed from the end of a nerve cell, and the cell with which it synapses. The space separating the two cells is the **synaptic cleft,** and the membrane opposed to the presynaptic terminal is the **postsynaptic terminal,** or postsynaptic membrane. An action potential in the presynaptic terminal can result in the production of an action potential in the postsynaptic terminal. The prefixes *pre-* and *post-* describe the direction of action potential propagation.

The size and degree of specialization of presynaptic terminals may vary from one axon to another; some axons branch to form an extensive series of enlarged presynaptic terminals, and others remain relatively simple. The major cytoplasmic organelles within presynaptic terminals are mitochondria and numerous membrane-bound **synaptic vesicles,** which contain neurotransmitter chemicals. The neurotransmitter can be released from the presynaptic

A
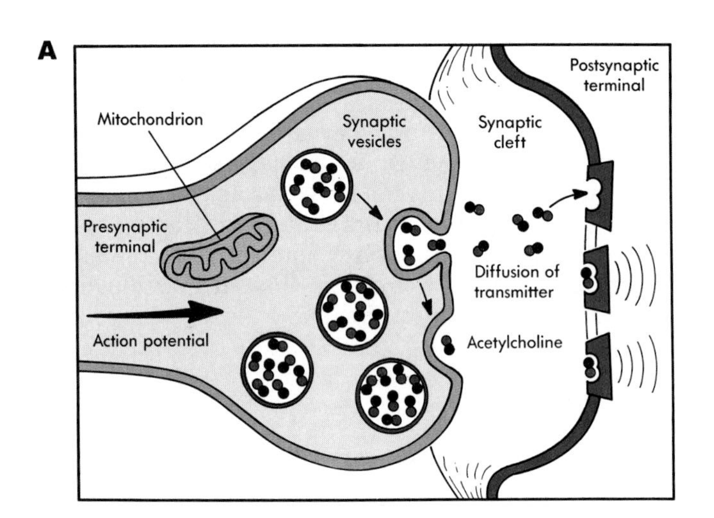

B
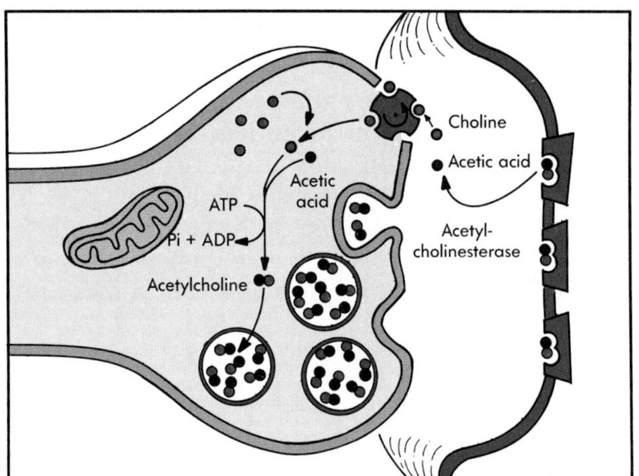

C
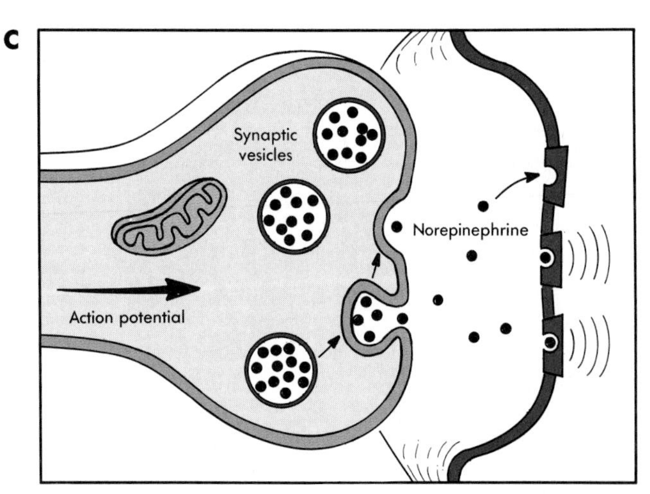

D
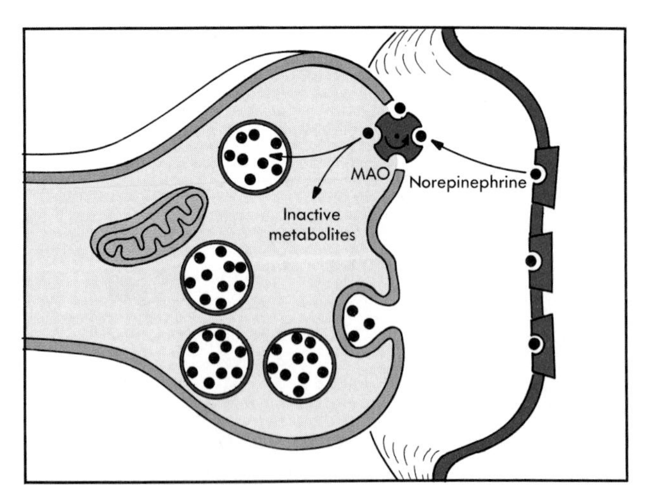

FIGURE 12-15 Synaptic transmission. The end of the axon, the presynaptic terminal, is separated by a space, the synaptic cleft, from the postsynaptic terminal.
A The neurotransmitter acetylcholine *(ACh)* diffuses from the presynaptic terminal across the synaptic cleft to the receptors in the postsynaptic terminal. **B** Acetylcholinesterase acts on the acetylcholine to break it down into acetic acid and choline. Choline is transported back to the presynaptic terminal where it is used to resynthesize acetylcholine. **C** Synaptic transmission in another type of synapse in which norepinephrine *(NE)* is the neurotransmitter. **D** Norepinephrine is transported back into the presynaptic terminal and packaged into synaptic vesicles for reuse, or the enzyme monoamine oxidase *(MAO)* alters its structure and inactivates it.

terminal to diffuse the short distance across the synaptic cleft to the postsynaptic terminal.

Postsynaptic terminals have specific receptors that can bind to the neurotransmitter released by the presynaptic terminal, and this binding causes a response in the postsynaptic terminal such as the production of an action potential. No synaptic vesicles are present in postsynaptic terminals; consequently, action potentials cannot be propagated from postsynaptic to presynaptic terminals. Synaptic conduction of action potentials occurs only from presynaptic to postsynaptic terminals.

Each action potential arriving at the presynaptic terminal initiates a series of specific events that results in the release of neurotransmitter substance. In response to an action potential calcium (Ca^{2+}) ion channels open, and Ca^{2+} ions diffuse into the presynaptic terminal. Ca^{2+} ions cause synaptic vesicles to fuse with the presynaptic membrane and release neurotransmitter by exocytosis into the synaptic cleft.

Experiments in rats indicate that the number of Ca^{2+} ion channels in the presynaptic terminals of neurons stimulating the heart decreases with age. Consequently, there is decreased movement of Ca^{2+} ions into the presynaptic terminals, causing a decreased release of neurotransmitter. The decreased amounts of neurotransmitter result in less stimulation of the heart and may explain the inability of the aged heart to pump faster and harder during exercise.

Once the neurotransmitter is released from the presynaptic terminal, it diffuses rapidly across the synaptic cleft, which is approximately 20 nm wide, and binds in a reversible fashion to specific receptors in the postsynaptic membrane. The interaction between the neurotransmitter substance and the receptor represents an equilibrium.

$$\text{Transmitter} + \text{Receptor} \leftrightarrow \text{Transmitter-receptor complex}$$

When the neurotransmitter concentration is high in the synaptic cleft, a large portion of the receptor molecules have neurotransmitter substance bound to them, and when the neurotransmitter concentration declines, the neurotransmitter diffuses away from the receptor molecules.

Neurotransmitter substances are rapidly removed from the synaptic clefts and therefore have short-term effects on postsynaptic terminals. The processes involved in this removal include catabolism, active transport, and diffusion. For example, in synapses in which acetylcholine is the neurotransmitter (see Figure 12-15, *A*) **acetylcholinesterase** (as′ē-til-ko-lin-es′ter-ās), an enzyme that breaks acetylcholine down to acetic acid and choline, is present. Choline is then transported back into the presynaptic terminal and is used to resynthesize acetylcholine (Figure 12-15, *B*). In another type of synapse norepinephrine is the neurotransmitter (Figure 12-15, *C*). Unlike acetylcholine, norepinephrine is actively transported back into the presynaptic terminal where most of the norepinephrine is repackaged into synaptic vesicles for reuse. The enzyme monoamine oxidase can alter the structure of some of the norepinephrine and make it incapable of binding to its receptors (Figure 12-15, *D*). Diffusion of neurotransmitter substances away from the synapse and into the extracellular fluid also limits the length of time the neurotransmitter substances remain bound to their receptor. Norepinephrine in the circulation is taken up primarily by liver and kidney cells in which monoamine oxidase and another enzyme, catechol-O-methyltransferase, convert it into inactive metabolites.

Receptor Molecules in Synapses

Receptor molecules have a high degree of specificity; consequently, only neurotransmitter molecules or very closely related substances normally bind to their receptors. For example, acetylcholine receptors bind to acetylcholine but not to other neurotransmitters. More than one type of receptor molecule exists for some neurotransmitters. Thus a single neurotransmitter may bind to one type of specific receptor to cause depolarization in one synapse and may bind to another type of specific receptor to cause hyperpolarization in another synapse. For example, norepinephrine is either inhibitory or stimulatory, depending on the type of norepinephrine receptor to which it binds and on the effect of that receptor on the permeability of the postsynaptic membrane.

Although neurotransmitter receptors are in greater concentrations on postsynaptic membranes, some receptors exist on presynaptic membranes. For example, norepinephrine released from the presynaptic membrane binds to receptors on both the presynaptic and postsynaptic membranes. Its binding to the presynaptic membrane decreases the release of additional synaptic vesicles. A high frequency of presynaptic action potentials results in the release of fewer synaptic vesicles in response to later action potentials; therefore norepinephrine may modify its own release by binding to presynaptic receptors.

Neurotransmitters and Neuromodulators

Several substances have been identified as neurotransmitters, and others are suspected neurotransmitters. It was thought that each neuron contains only one type of neurotransmitter; however, recent

Table 12-1

Substances that are neurotransmitters and/or neuromodulators

SUBSTANCE	LOCATION	EFFECT	CLINICAL EXAMPLE
Acetylcholine	Many nuclei scattered throughout the brain and spinal cord. Nerve tracts from the nuclei extend to many areas of the brain and spinal cord. Also found in the neuromuscular junction of skeletal muscle and many ANS synapses.	Excitatory or inhibitory	Alzheimer's disease (a type of senile dementia) is associated with a decrease in acetylcholine-secreting neurons. Myasthenia gravis (weakness of skeletal muscles) results from a reduction in acetylcholine receptors.
Monoamines			
Norepinephrine	A small number of small-sized nuclei in the brainstem. Nerve tracts extend from the nuclei to many areas of the brain and spinal cord. Also in some ANS synapses.	Excitatory or inhibitory	Cocaine and amphetamines,* resulting in overstimulation of postsynaptic neurons.
Serotonin	A small number of small-sized nuclei in the brainstem. Nerve tracts extend from the nuclei to many areas of the brain and spinal cord.	Generally inhibitory	Involved with mood, anxiety, and sleep induction. Levels of serotonin are elevated in schizophrenia (delusions, hallucinations, and withdrawal).
Dopamine	Confined to a small number of nuclei and nerve tracts. Distribution is more restricted than that of norepinephrine or serotonin. Also found in some ANS synapses.	Generally excitatory	Parkinson's disease (depression of voluntary motor control) results from destruction of dopamine-secreting neurons. Drugs used to increase dopamine production induce vomiting and schizophrenia.
Histamine	Hypothalamus, with nerve tracts to many parts of the brain and spinal cord.	Generally inhibitory	No clear indication of histamine-associated pathologies. Histamine apparently is involved with arousal, pituitary hormone secretion, control of cerebral circulation, and thermoregulation.

*Increase the release and block the re-uptake of norepinephrine.

evidence suggests that some neurons may secrete more than one type of neurotransmitter. If a neuron does produce more than one neurotransmitter, it secretes a mixture of neurotransmitters from all of its presynaptic terminals.

The physiological significance of presynaptic terminals that secrete more than one type of neurotransmitter has not been clearly established. However, some of these substances may not act as neurotransmitters but as neuromodulators. **Neuromodulators** can presynaptically or postsynaptically influence the likelihood that an action potential in the presynaptic terminal will result in the production of an action potential in the postsynaptic terminal. For example, a neuromodulator that decreases the release of an excitatory neurotransmitter from a presynaptic terminal would decrease the likelihood of action potential production in the postsynaptic terminal. A list of neurotransmitters and neuromodulators is presented in Table 12-1.

Table 12-1

Substances that are neurotransmitters and/or neuromodulators—cont'd

SUBSTANCE	LOCATION	EFFECT	CLINICAL EXAMPLE
Amino Acids			
Gamma-aminobutyric acid (GABA)	GABA-secreting neurons mostly control activities in their own area and are not usually involved with transmission from one part of the CNS to another. Most neurons of the CNS have GABA receptors.	Majority of postsynaptic inhibition in the brain; some presynaptic inhibition in the spinal cord	Drugs that increase GABA function have been used to treat epilepsy (excessive discharge of neurons).
Glycine	Spinal cord and brain. Like GABA, glycine predominantly produces local effects.	Most postsynaptic inhibition in the spinal cord	Glycine receptors are inhibited by strychnine.
Glutamate and aspartate	Widespread in the brain and spinal cord, especially in nerve tracts that ascend or descend the spinal cord or in tracts that project from one part of the brain to another.	Excitatory	Drugs that block glutamate or aspartate are under development. These drugs might prevent seizures and neural degeneration from overexcitation.
Neuropeptides			
Endorphins and enkephalins	Widely distributed in the CNS and PNS.	Generally inhibitory	The opiates morphine and heroin bind to endorphin and enkephalin receptors on presynaptic neurons and reduce pain by blocking the release of neurotransmitter.
Substance P	Spinal cord, brain, and sensory neurons associated with pain.	Generally excitatory	Substance P is a neurotransmitter in pain transmission pathways. Blocking the release of substance P by morphine reduces pain.

EPSPs AND IPSPs

The combination of neurotransmitters with their specific receptors causes either depolarization or hyperpolarization of the postsynaptic terminal. When depolarization occurs, the response is stimulatory, and the local depolarization is an **excitatory postsynaptic potential (EPSP)** (Figure 12-16, *A*). Neurons releasing neurotransmitter substances that cause EPSPs are **excitatory neurons.** In general, an EPSP occurs because of an increase in the permeability of the membrane to sodium (Na^+) ions (Figure 12-16, *B*). Because the concentration gradient is large for Na^+ ions and because the negative charge inside the cell attracts the positively charged Na^+ ions, Na^+ ions diffuse into the cells and cause depolarization. If depolarization reaches threshold, an action potential is produced.

When the combination of a neurotransmitter

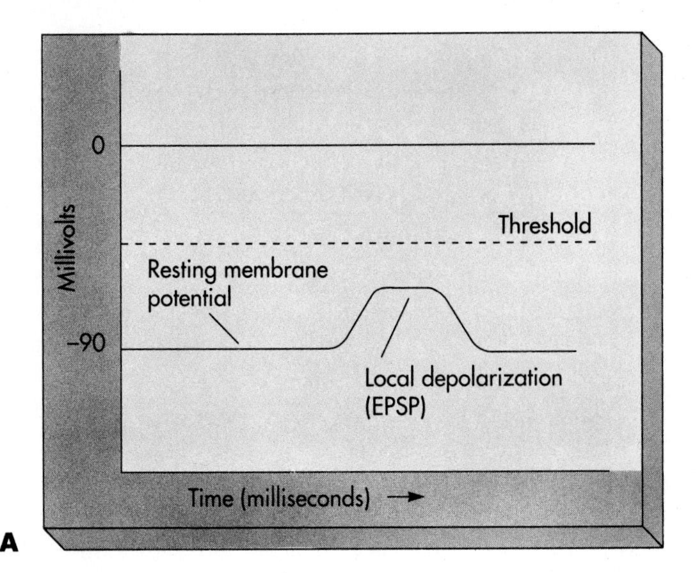

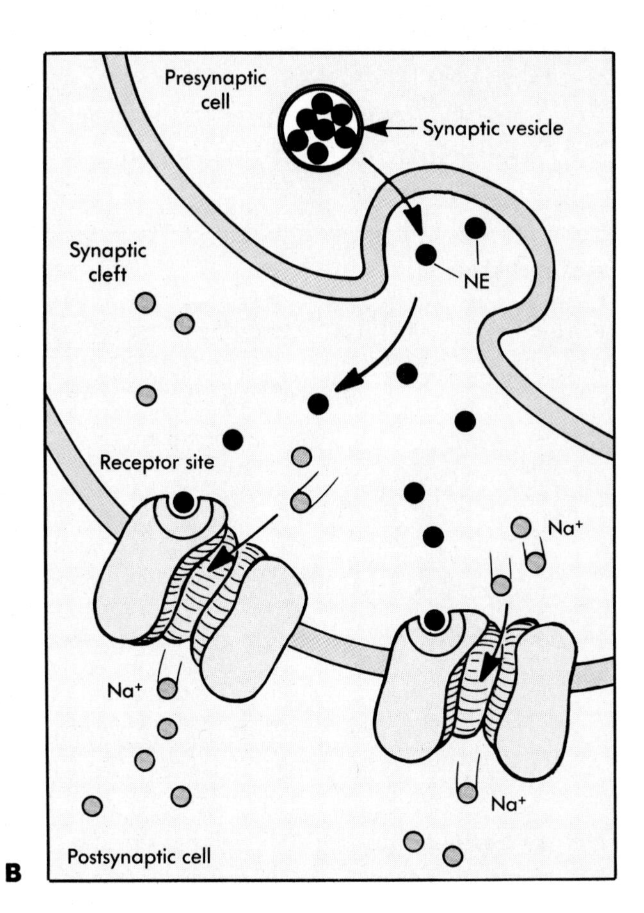

FIGURE 12-16 **A** Intracellular recording of an excitatory postsynaptic potential *(EPSP)*. **B** Norepinephrine *(NE)* produces an excitatory postsynaptic potential by allowing Na^+ ions to diffuse into the cell.

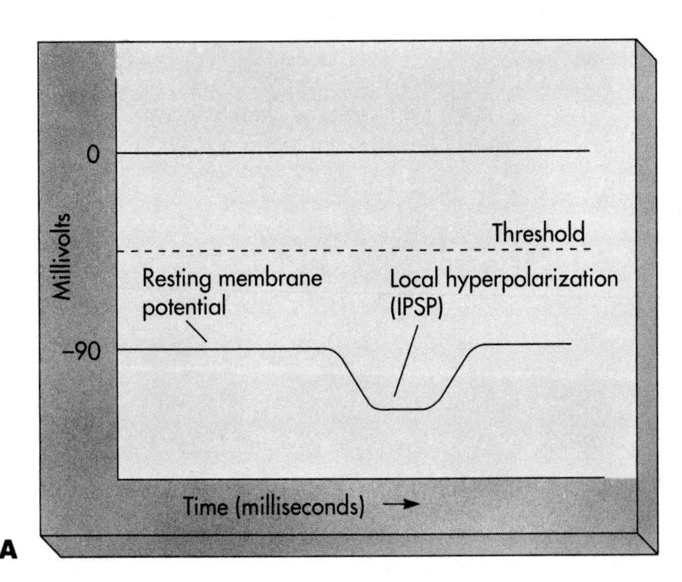

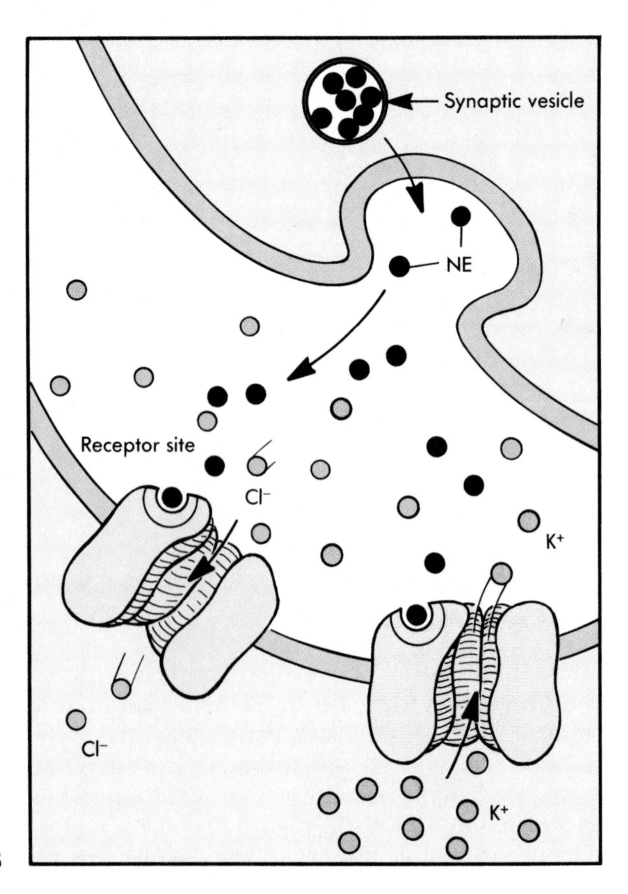

FIGURE 12-17 **A** Intracellular recording of an inhibitory postsynaptic potential *(IPSP)*. **B** Norepinephrine *(NE)* produces an inhibitory postsynaptic potential by allowing Cl^- ions to diffuse into the cell and K^+ ions to diffuse out of it.

with its receptor results in hyperpolarization of the postsynaptic membrane, the response is inhibitory, and the local hyperpolarization is an **inhibitory postsynaptic potential (IPSP)** (Figure 12-17, *A*). Neurons releasing neurotransmitter substances that cause IPSPs are **inhibitory neurons.** The IPSP is the result of an increase in the permeability of the cell membrane to chloride (Cl^-) or potassium (K^+) ions (Figure 12-17, *B*). Because Cl^- ions are more concentrated outside the cell than inside, when the permeability of the membrane to Cl^- ions increases, they diffuse into the cell, causing the inside of the cell to become more negative and resulting in hyperpolarization. The concentration of K^+ ions is greater inside the cell than outside, and increased permeability of the membrane to K^+ results in diffusion of K^+ ions out of the cell. Consequently, the outside of the cell becomes more positive than the inside, resulting in hyperpolarization.

Presynaptic Inhibition and Facilitation

Many of the synapses of the CNS are **axo-axonic synapses,** in which the axon of one neuron synapses with the presynaptic terminal (axon) of another neuron (Figure 12-18). The axo-axonic synapse does not initiate an action potential in the presynaptic terminal. When an action potential reaches the presynaptic terminal, however, neuromodulators released in the axo-axonic synapse can alter the amount of neurotransmitter released from the presynaptic terminal.

In **presynaptic inhibition** there is decreased neurotransmitter release from the presynaptic terminal, and in **presynaptic facilitation** there is increased neurotransmitter release from the presynaptic terminal (see Figure 12-18). The amount of neurotransmitter released by the presynaptic terminal affects the response (either an EPSP or IPSP) produced in the postsynaptic terminal. The greater the amount of neurotransmitter, the larger is the response. Consequently, presynaptic inhibition or facilitation can decrease or increase the likelihood of producing an action potential in the postsynaptic terminal. For example, enkephalins and endorphins in the brain and spinal cord produce presynaptic inhibition of neurons transmitting pain sensations. A reduction in neurotransmitter release can reduce or prevent the production of action potentials by postsynaptic neurons. This interference with the transmission of the pain signal results in reduction or elimination of the awareness of pain.

Spatial and Temporal Summation

Depolarizations produced in postsynaptic membranes are local depolarizations. Within the CNS and in many PNS synapses a single presynaptic action

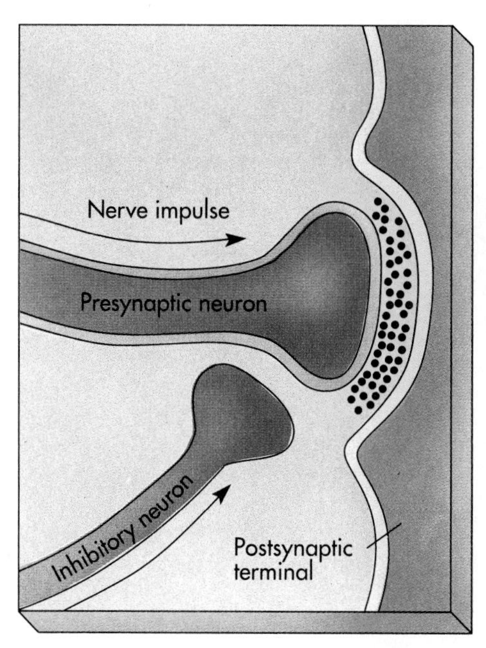

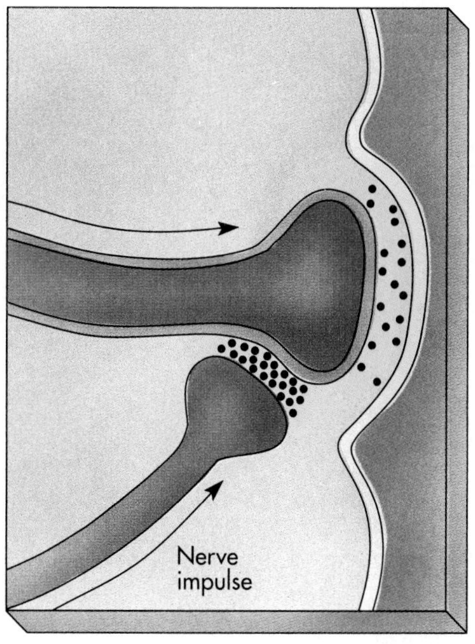

FIGURE 12-18 Presynaptic inhibition at an axo-axonic synapse. **A** The inhibitory neuron of the axo-axonic synapse is inactive and has no effect on the release of neurotransmitter from the presynaptic terminal. **B** Release of a neuromodulator from the inhibitory neuron of the axo-axonic synapse reduces the amount of neurotransmitter released from the presynaptic terminal.

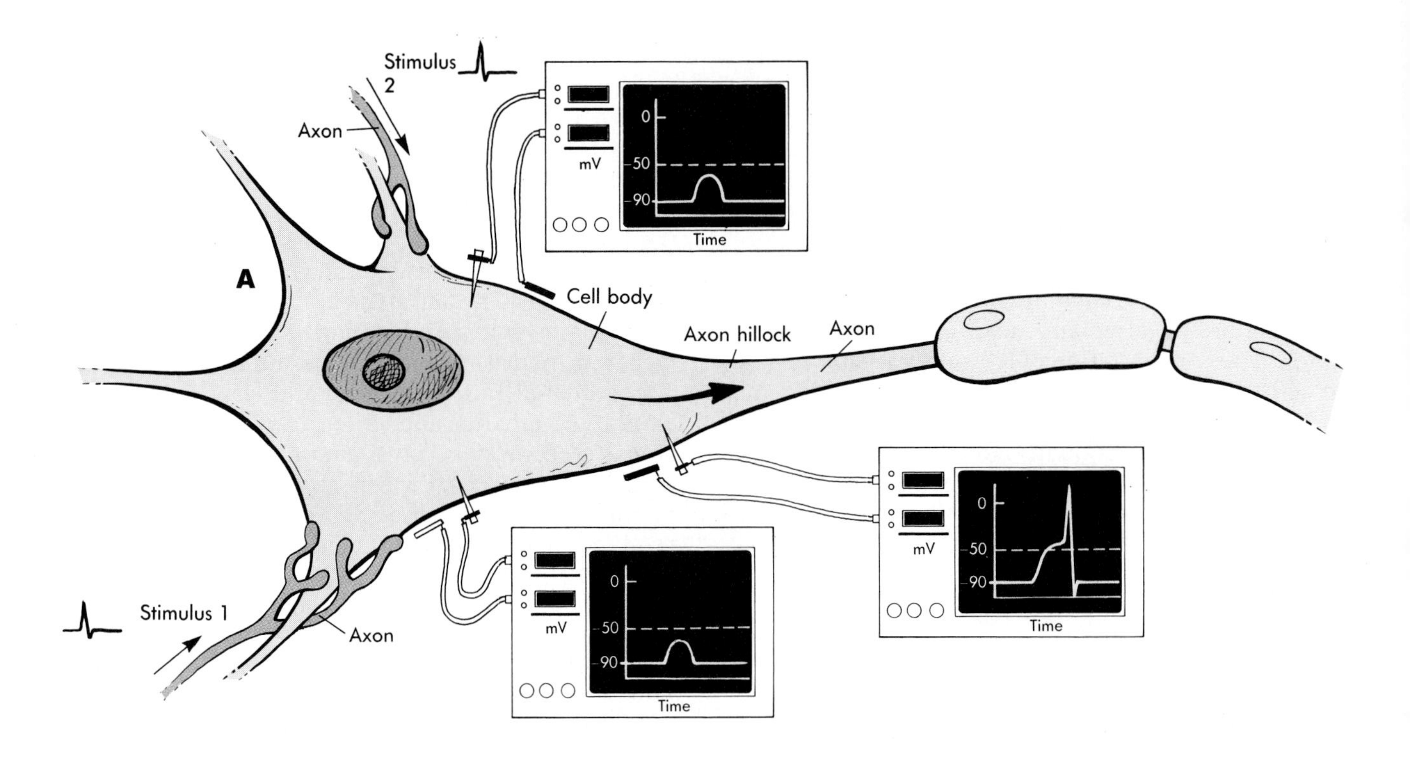

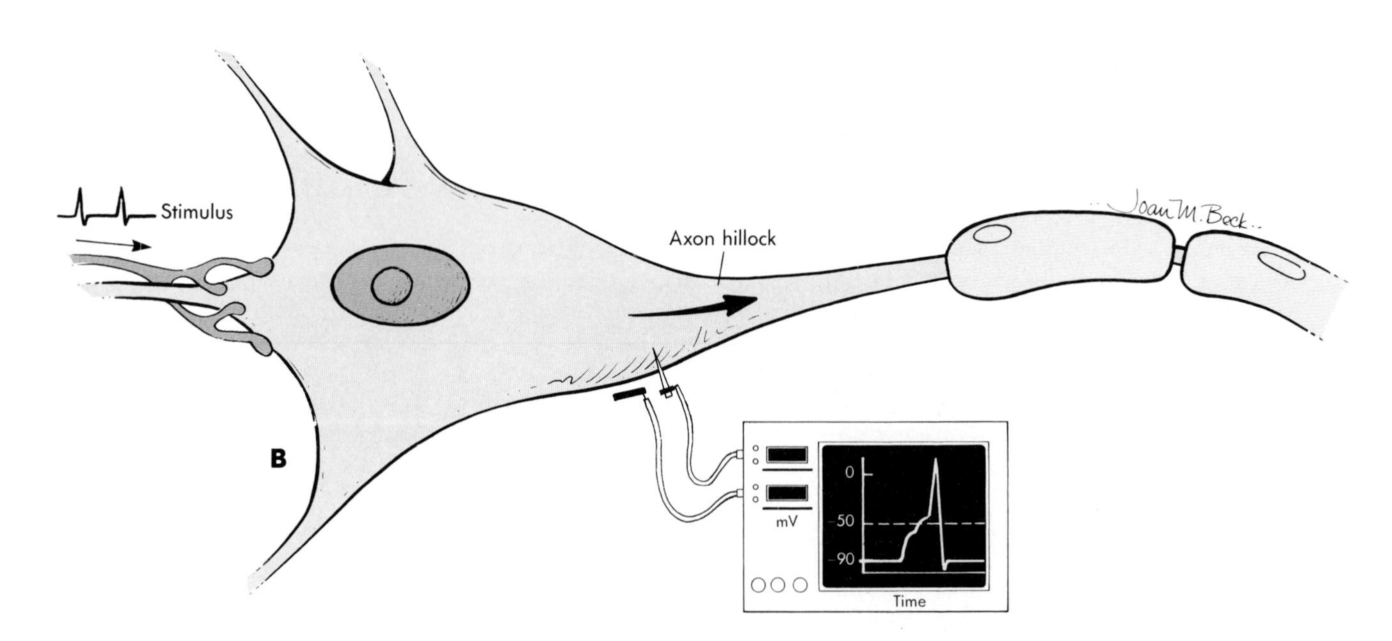

FIGURE 12-19 Summation. **A** Spatial summation. The two local depolarizations produced at 1 and 2 by action potentials that arrive simultaneously summate at the axon hillock to produce a local depolarization that exceeds threshold. **B** Temporal summation. Two action potentials arrive in close succession at the presynaptic membrane. Before the first local depolarization returns to threshold, the second is produced. They summate to exceed threshold to produce a postsynaptic action potential. **C** Combined spatial and temporal summation with both excitatory postsynaptic potentials and inhibitory postsynaptic potentials. The outcome, which is the product of summation, is determined by which influence is greater.

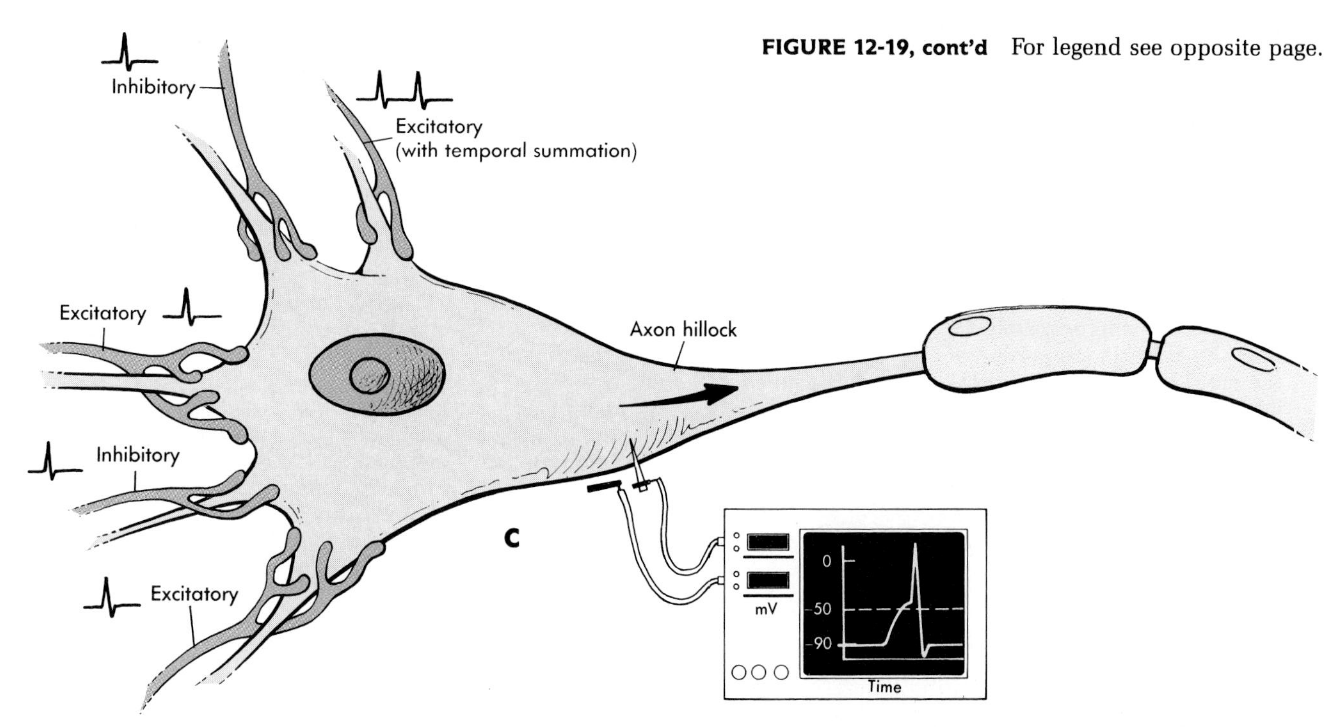

FIGURE 12-19, cont'd For legend see opposite page.

potential normally does not cause a local depolarization in the postsynaptic membrane sufficient to reach threshold and produce an action potential (see Chapter 9). Instead, a series of presynaptic action potentials causes a series of local potentials in the postsynaptic neuron. The local potentials combine in a process called **summation** at the axon hillock of the postsynaptic neuron, which is the normal site of action potential generation for most neurons. If summation results in a local potential that exceeds threshold at the axon hillock, an action potential is produced.

Two types of summation, spatial summation and temporal summation, are possible. The simplest type of **spatial summation** occurs when two action potentials arrive simultaneously at two different presynaptic terminals that synapse with the same postsynaptic neuron. In the postsynaptic neuron each action potential causes a local depolarization that undergoes summation at the axon hillock. If the summated depolarization reaches threshold, an action potential is produced (Figure 12-19, *A*).

4

Given two neurons: excitatory neuron A releases a neurotransmitter, and excitatory neuron B releases the same type and amount of neurotransmitter plus a neuromodulator that produces EPSPs in postsynaptic neurons. Will temporal summation of a postsynaptic neuron by neuron A produce the same number of action potentials as temporal summation of a postsynaptic neuron by neuron B? Explain.

? ? ? ? ? ? ? ? ? ? ?

Temporal summation results when two action potentials arrive in very close succession at a single presynaptic terminal. The first action potential causes a local depolarization in the postsynaptic membrane that remains for a few milliseconds before it disappears, although its magnitude decreases through time. Before the local depolarization caused by the first action potential repolarizes to its resting value, a second action potential initiates a second local depolarization. Temporal summation results when the second local depolarization summates with the remainder of the first local depolarization. If the summated local depolarization reaches threshold at the axon hillock, an action potential is produced in the postsynaptic neuron (Figure 12-19, *B*).

Excitatory and inhibitory neurons may synapse with a single postsynaptic neuron. Summation of EPSPs and IPSPs occurs in the postsynaptic neuron, and the IPSPs tend to cancel the EPSPs. Whether a postsynaptic action potential is initiated or not depends on which type of local potential has the greatest influence on the postsynaptic membrane potential (Figure 12-19, *C*).

The synapse is an essential structure for the process of integration carried out by the CNS. For example, action potentials propagated along axons from sensory organs to the CNS can produce a sensation, or they can be ignored. To produce a sensation, action potentials must be transmitted across synapses as they travel through the CNS to the cerebral cortex where information is interpreted. Stimuli that do not result in action potential transmission across synapses are ignored because information never reaches the cerebral cortex. The brain can ignore large amounts of sensory information as a result of complex integration.

Nervous Tissue: Response to Injury

When a nerve is cut, either healing or permanent interruption of the neural pathways occurs. The final outcome depends on the severity of the injury and on its treatment.

Several degenerative changes result when a nerve is cut (Figure 12-A). Within approximately 3 to 5 days the axons in the portion of the nerve distal to the cut break into irregular segments and degenerate. This occurs because the neuron cell body produces the substances essential for the maintenance of the axon, and these substances have no way of reaching parts of the axon distal to the point of damage. Eventually the distal part of the axon completely degenerates. At the same time the axons are degenerating, the myelin portion of the Schwann cells around them also degenerates, and macrophages invade the area to phagocytize the myelin. The Schwann cells then enlarge, undergo mitosis, and finally form a column of cells along the regions once occupied by the axons. The columns of Schwann cells are essential for the growth of new axons; if the ends of the axons proximal to the cut do not encounter the columns, they fail to reinnervate the peripheral structures. If the ends of the axons do encounter a Schwann cell column, their rate of growth increases, and reinnervation of peripheral structures is likely.

The portion of the axon proximal to the cut degenerates for a distance up to several Schwann cells in length and then begins regenerative processes that lead to growth from the end of the severed axons. The end of each axon forms bulbous enlargements and several axonal sprouts. It normally takes approximately 2 weeks for

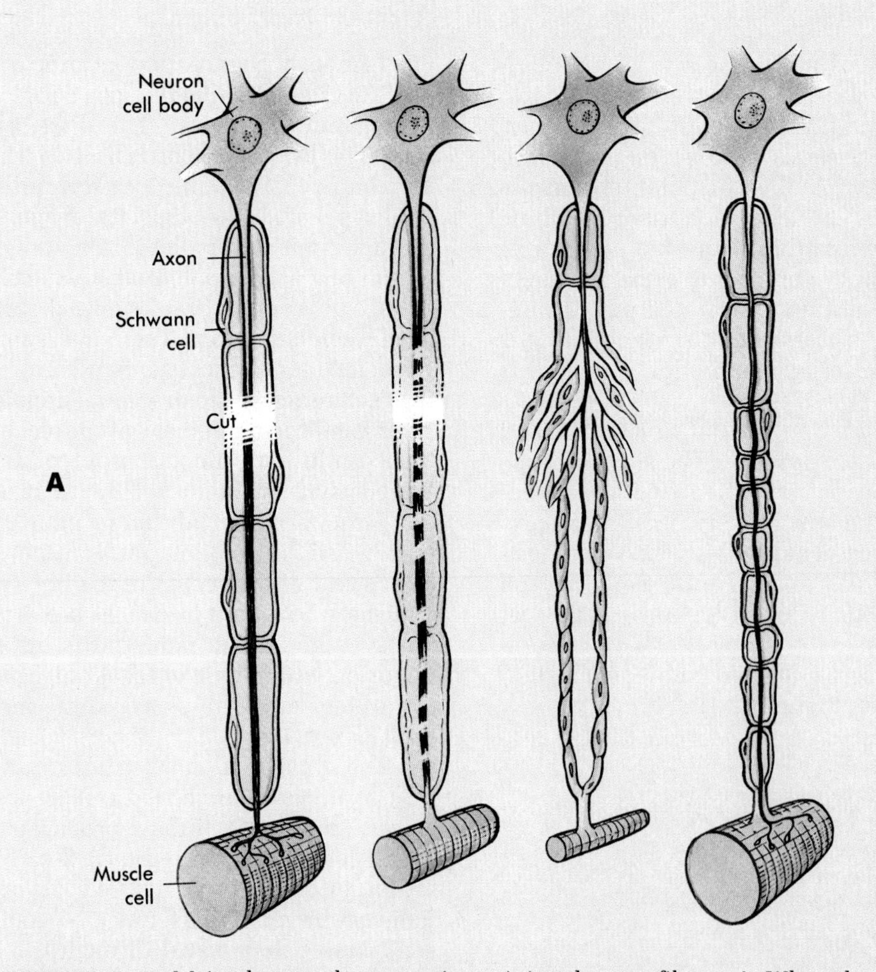

FIGURE 12-A Main changes that occur in an injured nerve fiber. **A** When the two ends of the injured nerve fiber are aligned in close proximity, healing and regeneration of the axon are likely to occur. Without stimulation from the nerve the muscle is paralyzed and atrophies (shrinks in size). After reinnervation the muscle can become functional and hypertrophy (increase in size).

the axonal sprouts to grow across the scar that develops in the area in which the nerve was cut and to enter the Schwann cell columns. However, only one of the sprouts from each severed neuron forms an axon. The other branches degenerate. After the axons grow through the Schwann cell columns, new myelin sheaths are formed, and the neurons reinnervate the structures they previously innervated.

If the ends of the neurons do not make contact with the Schwann cell columns, the Schwann cells eventually degenerate and are replaced by connective tissue. Treatment strategies that increase the probability of reinnervation include bringing the ends of the severed nerve close together surgically. In some cases in which sections of nerves are destroyed as a result of trauma, nerve transplants are performed to replace damaged segments. The transplanted nerve eventually degenerates, but it does provide Schwann cell columns through which axons can grow.

Regeneration of damaged nerve tracts within the CNS is very limited and is poor in comparison to regeneration of nerves in the PNS. In part the difference may be due to the oligodendrocytes, which are only in the CNS. Each oligodendrocyte has several processes, each of which forms part of a myelin sheath. The cell bodies of the oligodendrocytes are a short distance from the axons they ensheathe, and there are fewer oligodendrocytes than Schwann cells. Consequently, when the myelin degenerates after damage, no column of cells remains in the CNS to act as a guide for the growing axons.

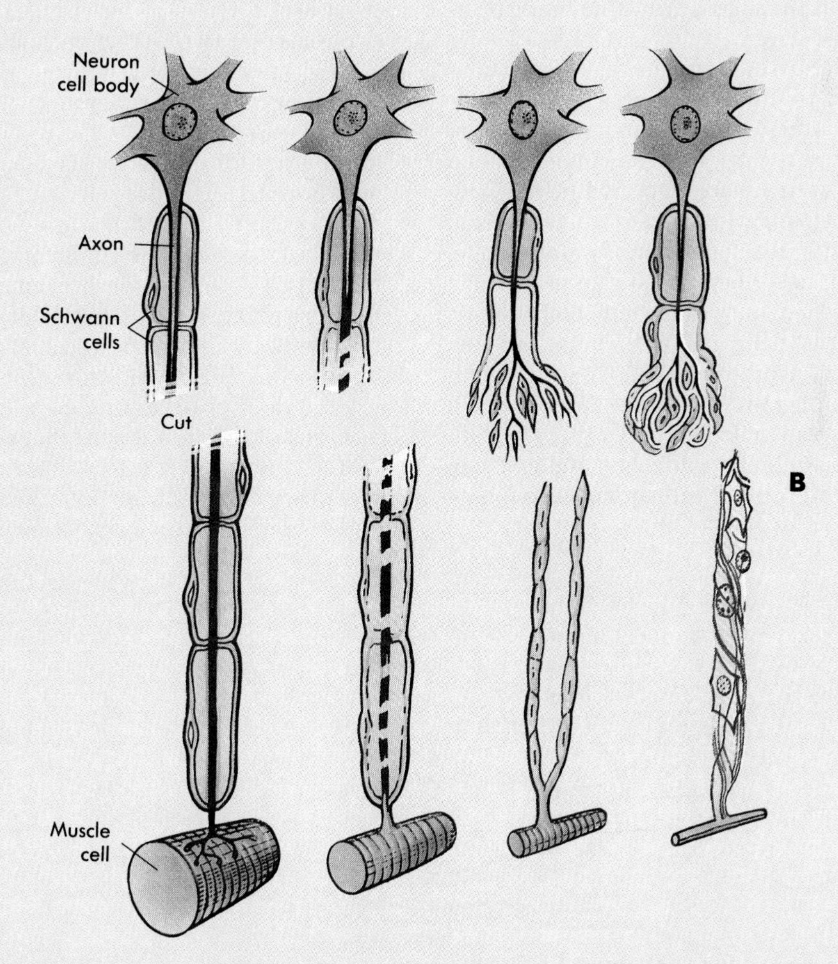

FIGURE 12-A, cont'd Main changes that occur in an injured nerve fiber. **B** When the two ends of the injured nerve fiber are not aligned in close proximity, regeneration is unlikely to occur. Without innervation from the nerve muscle function is completely lost, and the muscle remains atrophied.

REFLEXES

The basic structural unit of the nervous system is the neuron. The **reflex arc** is the basic functional unit of the nervous system and is the smallest, simplest portion capable of receiving a stimulus and yielding a response. It has several basic components: a sensory receptor, an afferent or sensory neuron, association neurons (also called interneurons or internuncial neurons), an efferent or motor neuron, and an effector organ (Figure 12-20).

Action potentials initiated in sensory receptors are propagated along afferent axons within the PNS to the CNS in which they usually synapse with association neurons. Association neurons synapse with efferent (motor) neurons, which send axons out of the spinal cord and through the PNS to muscles or glands where the action potentials of the efferent neurons cause effector organs to respond. The response produced by the reflex arc is called a **reflex.** It is an automatic response to a stimulus that occurs without conscious thought.

Reflexes are homeostatic. Some function to remove the body from painful stimuli that would cause tissue damage, and others function to keep the body from suddenly falling or moving because of external forces. A number of reflexes are responsible for maintaining relatively constant blood pressure, body fluid pH, blood carbon dioxide levels, and water intake. Specific reflexes are described in Chapter 13.

Individual reflexes vary in their complexity. Some involve simple neuronal pathways and few association neurons, whereas others involve complex circuits and integrative centers. Many are integrated within the spinal cord, and others are integrated within the brain. In addition, higher brain centers influence reflexes by either suppressing or exaggerating them.

Reflexes do not operate as isolated entities within the nervous system. Branches of the afferent neurons or association neurons send information along nerve tracts to the brain. A pain stimulus, for example, not only will initiate a response that removes the affected part of the body from the painful stimulus but will cause perception of the pain sensation as a result of the integration of action potentials sent to the brain.

NEURONAL CIRCUITS

Neurons are organized within the CNS to form circuits ranging from relatively simple to extremely complex. Although their complexity varies, three basic circuit patterns can be recognized: convergent, divergent, and oscillating. The functional characteristics of neuron circuits is determined to a large degree by these arrangements.

Convergent circuits have many neurons that converge and synapse with a smaller number of neurons (Figure 12-21, *A*). The simplest converging circuit occurs when two presynaptic neurons synapse with a single postsynaptic neuron, the activity of which is influenced by spatial summation. If action potentials in one presynaptic neuron cause a subthreshold depolarization in the postsynaptic neuron, no postsynaptic action potential will occur. However, that subthreshold depolarization will facilitate the response to action potentials from other presynaptic neurons. Also, if some presynaptic neurons are inhibitory and others are excitatory, the response of the postsynaptic neuron will depend on the summation of both the EPSPs and the IPSPs.

An example of a convergent pathway is the motor neurons of the spinal cord that control muscle movements (Figure 12-21, *B*). Afferent neurons from pain receptors carry action potentials to the spinal cord and synapse with association neurons, which, in

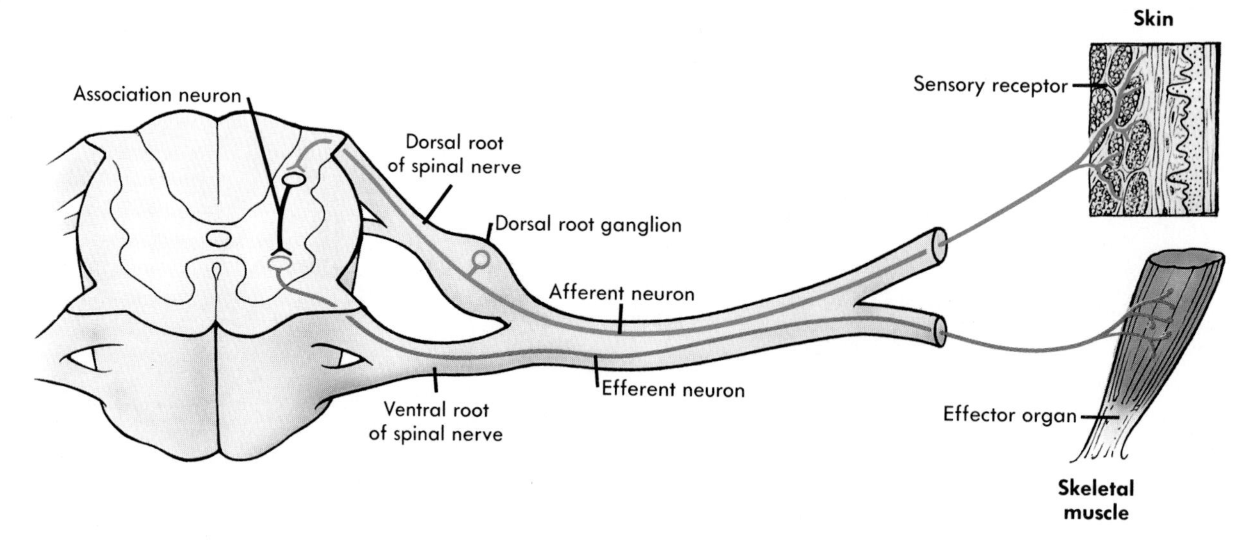

FIGURE 12-20 Basic diagram of a reflex arc, including the sensory receptor, afferent neuron, association neuron, efferent neuron, and effector organ.

turn, synapse with motor neurons. Stimulation of the pain receptors causes a reflex response that results in stimulation of the motor neurons. However, neurons with their cell bodies located within the cerebrum also synapse with the motor neurons. Conscious movements are controlled by the cerebrum by sending action potentials through nerve tracts that synapse with motor neurons in the spinal cord. Inhibitory axons also descend within the spinal cord and synapse either directly or through association neurons on the motor neurons. Thus the activity of motor neurons in the spinal cord depends on the activity in at least these three different types of presynaptic neurons.

In **divergent circuits** a smaller number of presynaptic neurons synapse with a larger number of postsynaptic neurons to allow information transmitted in one neuronal pathway to diverge into two or more pathways (Figure 12-22, *A*). The simplest diverging circuit occurs when a single presynaptic neuron branches to synapse with two postsynaptic neurons. An example of a divergent pathway is found within the spinal cord (Figure 12-22, *B*). Afferent neurons carrying action potentials from pain recep-

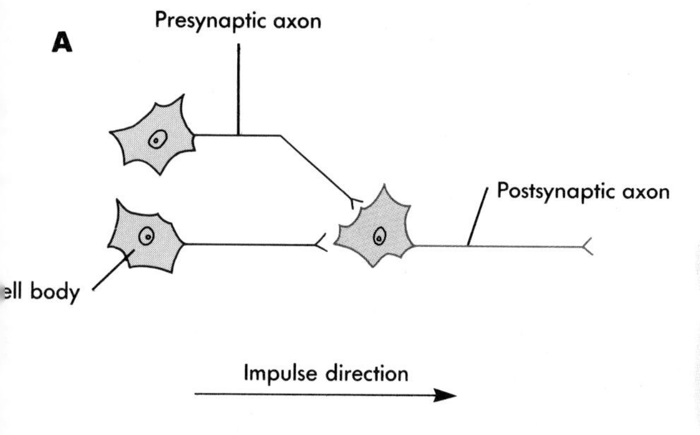

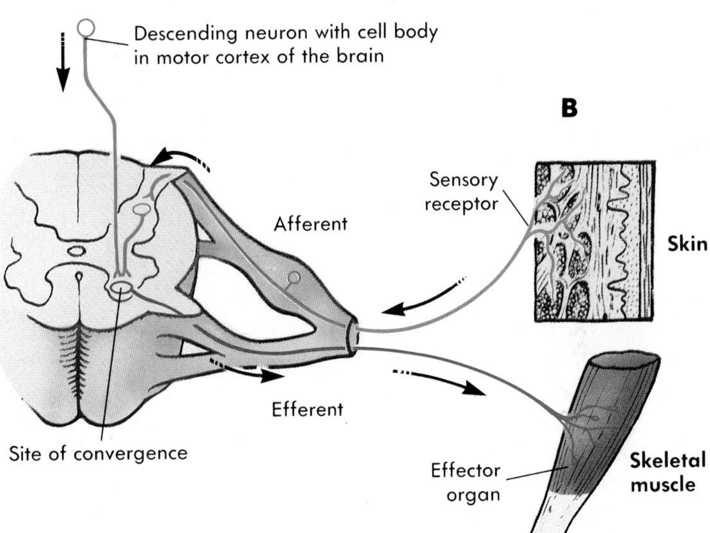

FIGURE 12-21 Convergent circuits. **A** General model. **B** Example of a convergent circuit in the spinal cord. Afferent neurons and descending neurons from the brain converge on a single motor neuron.

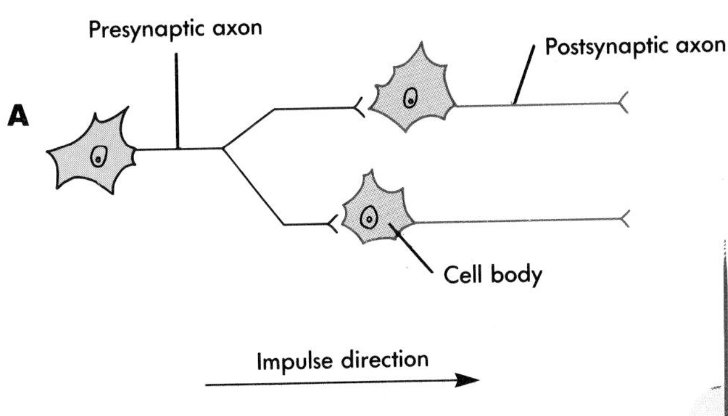

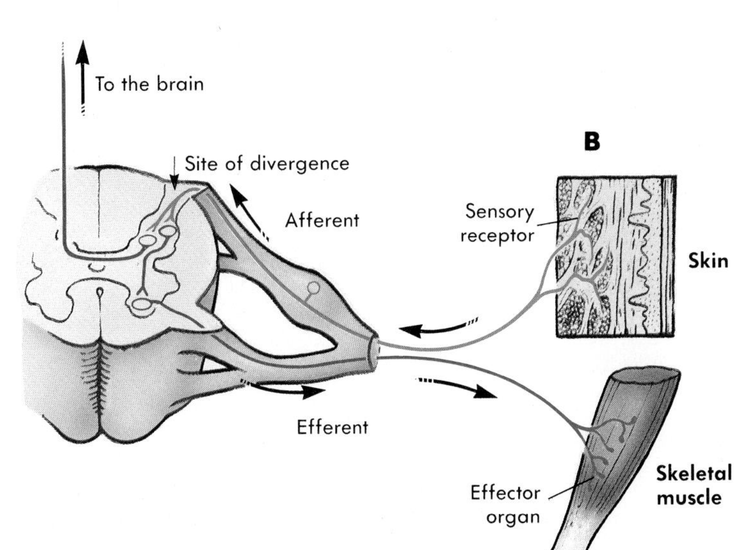

FIGURE 12-22 Divergent circuits. **A** General model. **B** Divergent circuit in the spinal cord. Afferent neurons send information to motor neurons in the spinal cord through an association neuron and to ascending neurons that send information to the brain.

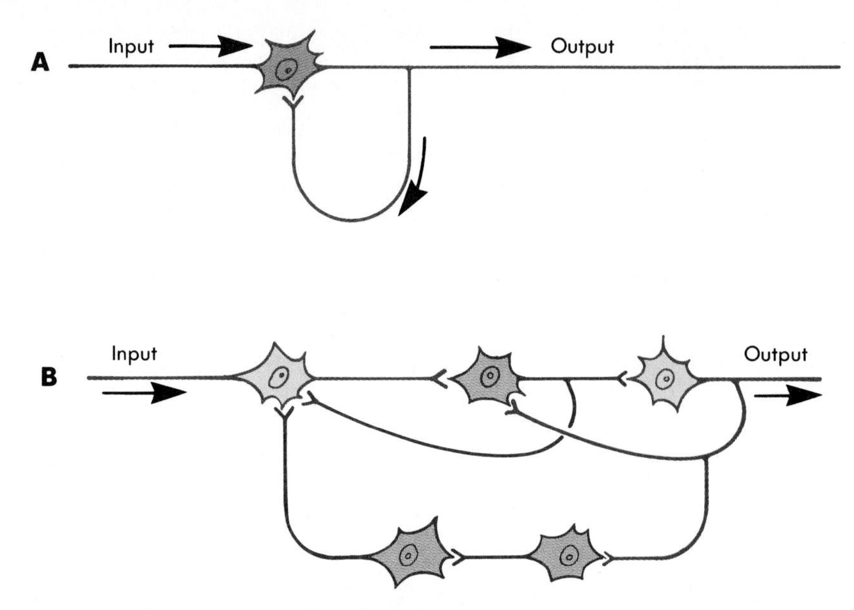

FIGURE 12-23 Oscillating circuits. **A** A single neuron stimulates itself. **B** A more complex oscillating circuit in which the input neuron is stimulated by two other neurons.

tors synapse within the spinal cord with association neurons that, in turn, induce a reflex response. In addition to synapsing with association neurons, collateral axons synapse with ascending neurons that carry action potentials toward the brain. The reflex response and the conscious sensation of pain are possible because of divergent pathways.

Oscillating circuits have neurons arranged in a circular fashion, which allows action potentials entering the circuit to cause a neuron further along in the circuit to produce an action potential more than once (Figure 12-23). This response is called **after discharge,** and its effect is to prolong the response to a stimulus. Oscillating circuits are similar to positive-feedback systems. Once an oscillating circuit is stimulated, it continues to discharge until the synapses involved become fatigued or until they are inhibited by other neurons. Figure 12-23, *A* illustrates a simple circuit in which a collateral axon stimulates its own cell body, and Figure 12-23, *B* a more complex circuit. Oscillating circuits play a role in neuronal circuits that are periodically active. Respiration may be controlled by an oscillating circuit that controls inspiration and another that controls expiration.

Neurons that spontaneously produce action potentials are common in the CNS and may activate oscillating circuits, which remain active for awhile. The cycle of wakefulness and sleep may involve circuits of this type. Spontaneously active neurons are also capable of influencing the activity of other circuit types. The complex functions carried out by the CNS are affected by the numerous circuits operating together and influencing the activity of one another.

SUMMARY

DIVISIONS OF THE NERVOUS SYSTEM p. 354

1. The nervous system has two anatomical divisions.
   - The central nervous system (CNS) consists of the brain and spinal cord and is encased in bone.
   - The peripheral nervous system (PNS), nervous tissue outside of the CNS, consists of nerves and ganglia.
2. The anatomical divisions perform different functions.
   - The CNS processes, integrates, stores, and responds to information from the PNS.
   - The PNS detects stimuli and transmits information to and receives information from the CNS.
3. The PNS has two divisions.
   - The afferent division transmits action potentials to the CNS and consists of single neurons that have their cell bodies in ganglia.
   - The efferent division carries action potentials away from the CNS in cranial or spinal nerves.
4. The efferent division has two subdivisions.
   - The somatic nervous system innervates skeletal muscle and is mostly under voluntary control. It consists of single neurons that have their cell bodies located within the CNS.
   - The autonomic nervous system (ANS) innervates cardiac muscle, smooth muscle, and glands. It has two neurons between the CNS and effector organs. The first has its cell bodies within the CNS, and the second has its cell bodies within autonomic ganglia.

CELLS OF THE NERVOUS SYSTEM p. 356

Neurons

1. Neurons receive stimuli and transmit acttion potentials.
2. Neurons have three components.
   - The cell body, or soma, contains the nucleus, nucleolus, and rough endoplasmic reticulum and is the primary site of protein synthesis.
   - Dendrites are short, branched cytoplasmic extensions of the cell body that usually receive action potentials from other neurons.
   - Axons are cytoplasmic extensions of the cell body that transmit action potentials to other cells.

Neuron Types

1. Multipolar neurons have several dendrites and a single axon. Most CNS and motor neurons are multipolar.
2. Bipolar neurons have a single axon and dendrite and are found as components of sensory organs.
3. Unipolar neurons have a single axon. Most sensory neurons are unipolar.

Neuroglia

1. Neuroglia are nonneural cells that support and aid the neurons of the CNS and PNS.
2. CNS neuroglia
   - Astrocytes provide structural support for neurons and blood vessels. The endothelium of blood vessels forms the blood-brain barrier, which regulates the movement of substances between the blood and the CNS. Astrocytes influence the functioning of the blood-brain barrier and process substances that pass through it.
   - Microglia are macrophages that phagocytize microorganisms, foreign substances, or necrotic tissue.
   - Ependymal cells line the ventricles and the central canal of the spinal cord. Some are specialized to produce cerebrospinal fluid.
   - Oligodendrocytes form myelin sheaths around the axons of neurons of the CNS.
3. Schwann cells (PNS neuroglia).
   - Schwann cells form myelin sheaths around the axons of neurons of the PNS.
   - Satellite cells support and nourish neuron cell bodies within ganglia.

Axon Sheaths

1. Unmyelinated axons rest in invaginations of oligodendrocytes (CNS) or Schwann cells (PNS). They conduct action potentials slowly.
2. Myelinated axons are wrapped by several layers of cell membrane from oligodendrocytes (CNS) or Schwann cells (PNS). Spaces between the wrappings are the nodes of Ranvier, and action potentials are conducted rapidly by saltatory conduction from one node of Ranvier to the next.

ORGANIZATION OF NERVOUS TISSUE p. 363

1. Nervous tissue can be grouped into white and gray matter.
   - White matter is myelinated axons and functions to propagate action potentials.
   - Gray matter is collections of neuron cell bodies. Axons synapse in the gray matter, which is functionally the site of integration in the nervous system.
   - White matter forms nerve tracts in the CNS and nerves in the PNS. Gray matter forms nuclei in the CNS and ganglia in the PNS.
2. In the PNS individual axons are surrounded by the endoneurium. Groups of axons, fascicles, are bound together by the perineurium. The fascicles form the nerve and are held together by the epineurium. Nerves are somewhat tough and elastic.

THE SYNAPSE p. 364

1. Anatomically the synapse has three components.
   - The enlarged ends of the axon are the presynaptic terminals containing synaptic vesicles.
   - The postsynaptic terminals contain receptors for the neurotransmitter.
   - The synaptic cleft, a space, separates the presynaptic and postsynaptic terminals.
2. An action potential arriving at the presynaptic terminal causes the release of a neurotransmitter, which diffuses across the synaptic cleft and binds to the receptors of the postsynaptic terminal.
3. The effect of the neurotransmitter on the postsynaptic terminal can be stopped in several ways.
   - The neurotransmitter is broken down by an enzyme.
   - The neurotransmitter is taken up by the presynaptic terminal.
   - The neurotransmitter diffuses out of the synaptic cleft.

Receptor Molecules in Synapses

1. Neurotransmitters are specific for their receptors.
2. A neurotransmitter can be stimulatory in one synapse and inhibitory in another, depending on the type of receptor present.
3. Some presynaptic terminals have receptors.

Neurotransmitters and Neuromodulators

Neuromodulators influence the likelihood that an action potential in a presynaptic terminal will result in an action potential in a postsynaptic terminal.

EPSPs AND IPSPs p. 367

1. Depolarization of the postsynaptic terminal caused by an increase in membrane permeability to sodium ions is an excitatory postsynaptic potential (EPSP).
2. Hyperpolarization of the postsynaptic terminal caused by an increase in membrane permeability to chloride ions or potassium ions is an inhibitory postsynaptic potential (IPSP).

Presynaptic Inhibition and Facilitation

1. Presynaptic inhibition decreases neurotransmitter release.
2. Presynaptic facilitation increases neurotransmitter release.

Spatial and Temporal Summation

1. Presynaptic action potentials through neurotransmitters produce local potentials in postsynaptic neurons. The local potential can summate to produce an action potential at the axon hillock.
2. Spatial summation occurs when two or more presynaptic terminals simultaneously stimulate a postsynaptic neuron.
3. Temporal summation occurs when two or more action potentials arrive in succession at a single presynaptic terminal.
4. Inhibitory and excitatory presynaptic neurons can converge on a postsynaptic neuron. The activity of the postsynaptic neuron is determined by the integration of the EPSPs and IPSPs produced in the postsynaptic neuron.

REFLEXES p. 374

1. A reflex arc is the functional unit of the nervous system.
   - Sensory receptors respond to stimuli and produce action potentials in afferent neurons.
   - Afferent neurons propagate action potentials to the CNS.
   - Association neurons in the CNS synapse with the afferent neuron and with the efferent neuron.
   - Efferent neurons carry action potentials from the CNS to effector organs.
   - Effector organs such as muscles or glands respond to the action potential.
2. Reflexes do not require conscious thought, and they produce a consistent and predictable result.
3. Reflexes are homeostatic.
4. Reflexes are integrated within the brain and spinal cord. Higher brain centers can suppress or exaggerate reflexes.

NEURONAL CIRCUITS p. 374

1. Convergent circuits have many neurons synapsing with a few neurons.
2. Divergent circuits have a few neurons synapsing with many neurons.
3. Oscillating circuits have a collateral branch of a postsynaptic neuron synapsing with a presynaptic neuron.

CONTENT REVIEW

1. Describe the CNS and PNS anatomically and functionally.
2. Define the afferent and efferent divisions of the PNS.
3. Contrast the afferent, somatic, and autonomic nervous systems in terms of the number of neurons, the location of neuron cell bodies, and the structures innervated.
4. What are the functions of neurons? Name the three parts of a neuron, and describe their functions.
5. Describe the three types of neurons based on their structure, and give an example of where each type is found.
6. Define neuroglia. Name and describe the functions of the different kinds of neuroglia.
7. What are the differences between unmyelinated and myelinated axons with regard to the arrangement of the cells that cover their axons? What are the nodes of Ranvier?
8. Do unmyelinated or myelinated neurons propagate action potentials more rapidly? Describe what occurs during saltatory conduction.
9. For nerve tract, nerve, nucleus, and ganglion, name the cells or parts of cells found in each, state if they are white or gray matter, and name the part (CNS or PNS) of the nervous system in which they are found.
10. Describe the layers of connective tissue found in nerves.
11. Describe the function of the synapse, starting with an action potential in the presynaptic neuron and ending with the generation of an action potential in the postsynaptic neuron.
12. Describe the specific, reversible reaction that occurs between a neurotransmitter and its receptor.
13. Name three ways that the effect of a neurotransmitter on the postsynaptic terminal can be stopped. Give an example of each way.
14. How can a neurotransmitter cause depolarization in one synapse but hyperpolarization in another synapse?
15. What is a neuromodulator?
16. Define and explain the production of EPSPs and IPSPs. Why are they important?
17. What are presynaptic inhibition and facilitation?
18. In what part of the postsynaptic neuron are local potentials produced? Where do they summate? What happens when they summate?
19. Contrast spatial and temporal summation. Give an example of each.
20. Explain how inhibitory and excitatory presynaptic neurons can influence the activity of a postsynaptic neuron.
21. At the cell level where does integration take place in the CNS?
22. Name the five components of a reflex arc. Describe the operation of a reflex arc, starting with a stimulus and ending with the reflex response.
23. Define a reflex. What is the relationship between a reflex response and awareness of the stimuli that caused the reflex? What effects can higher brain centers have on reflexes?
24. Define convergent, divergent, and oscillating circuits, and give an example of each.

CONCEPT QUESTIONS

1. Explain the consequences when an inhibitory neuromodulator is released from a presynaptic terminal and a stimulatory neurotransmitter is released from another presynaptic terminal, both of which synapse with the same neuron.
2. Students in a veterinary school were given the following hypothetical problem. A dog ingested organophosphate poison, and the students were responsible for saving the animal's life. Organophosphate poisons bind to and inhibit acetylcholinesterase. Several substances they could inject include the following: acetylcholine, curare (which blocks acetylcholine receptors), and potassium chloride. If you were a student in the class, what would you do to save the animal?
3. Design two oscillating neural circuits so that when the first is active, the second is inhibited, but when the second is active, the first is inhibited. Your design must result in each neural circuit's being active periodically. Note that respiration may be controlled in part by such a set of neural circuits.
4. Strychnine blocks receptor sites for inhibitory neurotransmitter substances in the central nervous system. Explain how strychnine could produce tetany in skeletal muscles.
5. Describe how a neuron that has its cell body in the cerebrum could, when stimulated, inhibit a reflex that is integrated within the spinal cord.

ANSWERS TO PREDICT QUESTIONS

1 p. 357. When the axon of a neuron is severed, the proximal portion of the axon remains attached to the nerve cell body; however, the distal portion of the axon is detached and has no way to replenish the enzymes and other proteins essential to its survival. Since the DNA in the nucleus provides the information that determines the structure of proteins by directing mRNA synthesis, the distal portion of the axon has no source of new proteins. Consequently, it will degenerate and die. On the other hand, the proximal portion of the axon is still attached to the nucleus and therefore has a source of new proteins. It remains alive and in many cases will grow to replace the severed distal axon.

2 p. 359. The ways the drug can gain access to the CNS neurons are very limited. One way is to bypass the blood-brain barrier mechanically by injecting the drug directly into the brain or at least into the cerebrospinal fluid. Such a technique would have a great risk of damage to the CNS neurons and possibly a greater risk of infection than if the drug could be taken orally. Another possibility is to change the permeability of the blood-brain barrier. Some preliminary data suggest that certain treatments (e.g., administration of a hypertonic glucose solution) can increase the permeability of the blood-brain barrier temporarily and may allow drugs to pass across the blood-brain barrier for a short period of time. However, the risks are much greater than if the drug could pass across the blood-brain barrier without assistance. Finally, an attempt could be made to change the permeability characteristics of the drug. Some success has been achieved by attaching the drugs to lipid-soluble molecules that diffuse across the blood-brain barrier, carrying the drug with them.

3 p. 362. Myelinated axons conduct action potentials much more rapidly than unmyelinated axons. In fact, the smallest-diameter myelinated axons conduct action potentials more rapidly than the largest diameter unmyelinated axons. An advantage of small-diameter myelinated axons over unmyelinated axons is that small-diameter myelinated axons take up little space (small diameter) but conduct action potentials rapidly (myelinated). For unmyelinated axons to conduct as rapidly, the diameters of unmyelinated axons would have to be so large that nerves and nerve tracts in the nervous system would have to be much larger than they are. The spinal cord, for example, would have to be many times larger in diameter than it is. Therefore animals with unmyelinated axons would have to have much larger nervous systems. Otherwise, they would move much more slowly, and their responses to stimuli would be much slower.

4 p. 371. Temporal summation by neuron B produces more action potentials in the postsynaptic neuron than temporal summation by neuron A. The neuromodulator produces EPSPs that depolarize the membrane potential of the postsynaptic neuron, bringing the membrane potential closer to threshold. Therefore a smaller amount of neurotransmitter is required to produce an action potential. Although neurons A and B release the same amount and type of neurotransmitter, the neuromodulator makes the neurotransmitter from neuron B more effective, resulting in more action potentials.

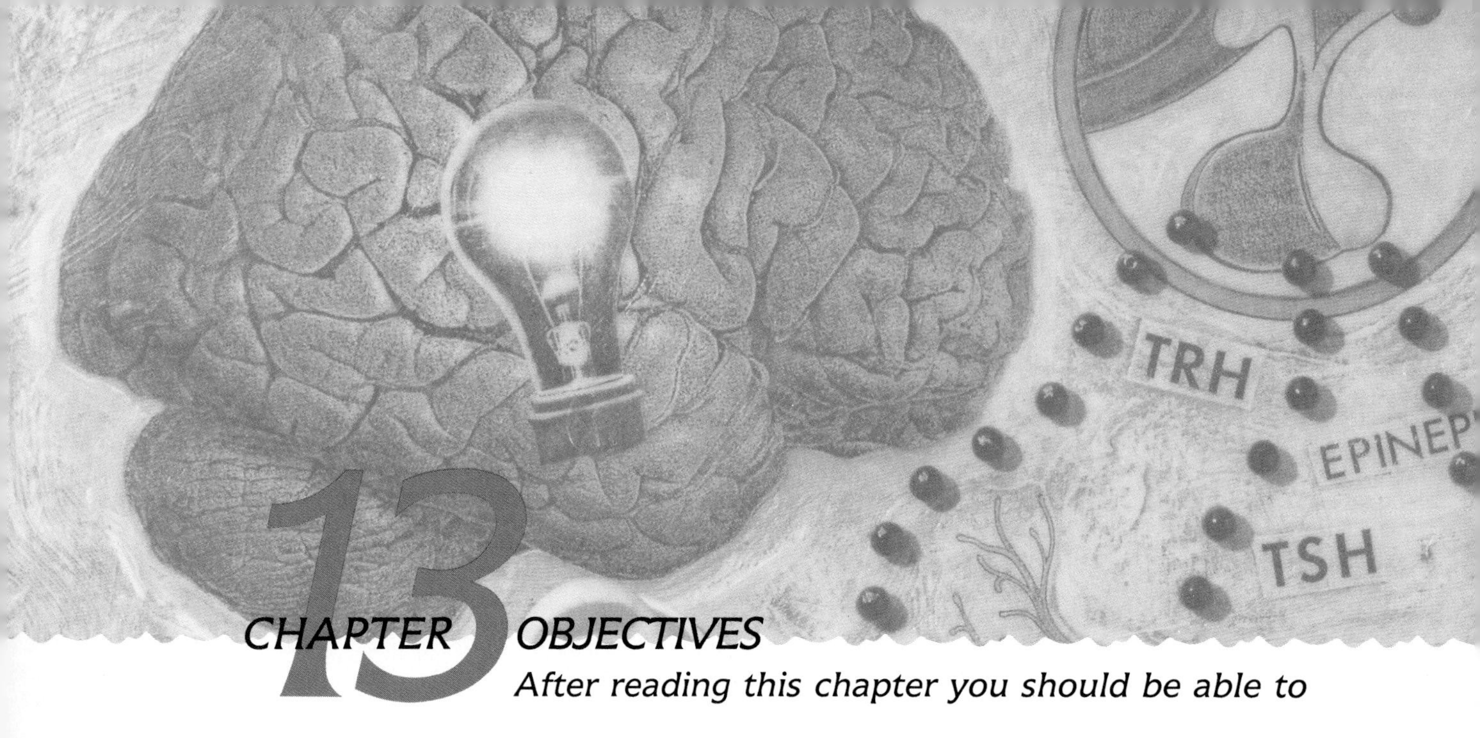

13 CHAPTER OBJECTIVES

After reading this chapter you should be able to

1. Describe the formation of the neural tube and list the structures that develop from the various portions of the neural tube.
2. List the parts of the brain.
3. Define what is included in the brainstem and describe its major features.
4. Describe the major function of the reticular formation.
5. List the regions of the diencephalon and indicate their major functions.
6. Describe the major functional areas of the cerebral cortex and explain their interactions.
7. Explain the pathway for speech.
8. Describe the basic brain waves and correlate them with brain function.
9. Explain how sensory, short-term, and long-term memory work.
10. Describe the major functions of the basal ganglia.
11. Describe the components and functions of the limbic system.
12. Describe the major functions of the cerebellum and explain its comparator function.
13. Describe the spinal cord in cross section, explaining the functions of each area.
14. Diagram the stretch reflex, Golgi tendon reflex, withdrawal reflex, reciprocal innervation, and the crossed extensor reflex. Explain the function of each of these reflexes.
15. Describe the course of the fibers associated with the spinothalamic and medial lemniscal systems.
16. Outline the course and describe the function of the corticospinal, corticobulbar, and extrapyramidal tracts.
17. Describe the three meningeal layers surrounding the central nervous system.
18. Name the four ventricles of the brain and describe their locations and the connections between them.
19. Describe the origin, composition, and circulation of the cerebrospinal fluid.

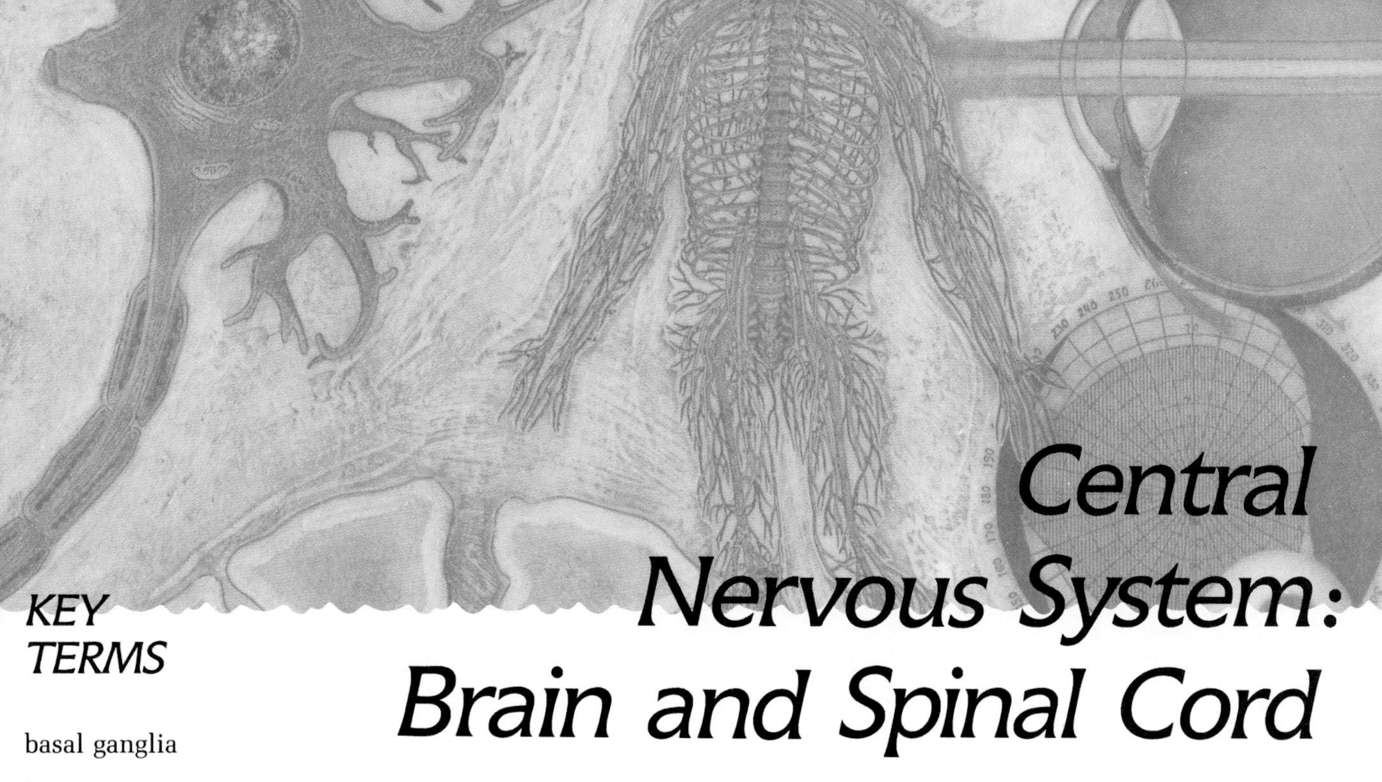

Central Nervous System: Brain and Spinal Cord

KEY TERMS

basal ganglia

brainstem

cauda equina (kaw′dah e-kwi′nah)

cerebrospinal (ser-e-bro-spi′nal) fluid

cerebrum

cortex

cerebellum (ser′ĕ-bel′um)

decussate (dĕ′kus-āt)

diencephalon (di-en-sef′ă-lon)

horn

limbic (lim′bik) system

meninges (mĕ-nin′jēz)

nerve tract

nucleus

ventricle (ven′trĭ-kul)

RELATED TOPICS

The following terms or concepts are important for a good understanding of this chapter. If you are not familiar with them, you should review them before proceeding.

Directional terms and body regions (Chapter 1)

Membrane potentials (Chapter 9)

Functional organization of the nervous system (Chapter 12)

THE CENTRAL NERVOUS SYSTEM (CNS) CONSISTS OF THE BRAIN AND THE SPINAL CORD (FIGURE 13-1), WITH THE DIVISION BETWEEN THESE TWO PORTIONS OF THE CNS PLACED SOMEWHAT ARBITRARILY AT THE LEVEL OF THE FORAMEN MAGNUM. The anatomical features and some basic functional features of the brain and spinal cord are presented in this chapter, followed by a description of the ascending and descending pathways. The pathways are described last because the basic anatomy and physiology of both the brain and spinal cord must be understood before the pathways can be fully comprehended.

DEVELOPMENT

The CNS develops from a flat plate of tissue, the **neural plate,** on the upper suface of the embryo (Figure 13-2, *A*). The lateral sides of the neural plate become elevated as waves called **neural crests** or **neural folds,** move toward each other in the midline, and fuse to create a **neural tube** (Figure 13-2, *B*). The cephalic portion of the neural tube develops into the brain, and the caudal portion develops into the spinal cord. **Neural crest cells** separate from the neural crests and give rise to part of the peripheral nervous system (see Chapter 14).

The portion of the neural tube that will become the brain forms a series of pouches (Table 13-1 and Figure 13-3). The pouch walls become the various portions of the adult brain, and the cavities become fluid-filled **ventricles.** The ventricles are continuous with the **central canal** of the spinal cord, which also derives from the hollow center of the neural tube. The neural tube also develops flexures that result in the brain's being oriented almost 90 degrees to the spinal cord.

Three brain regions can be identified in the early embryo (see Table 13-1), a forebrain, or **prosencephalon** (pro'sen-sef'ă-lon), a midbrain, or **mesencephalon,** and a hindbrain or **rhombencephalon.** Within a short time during development the forebrain di-

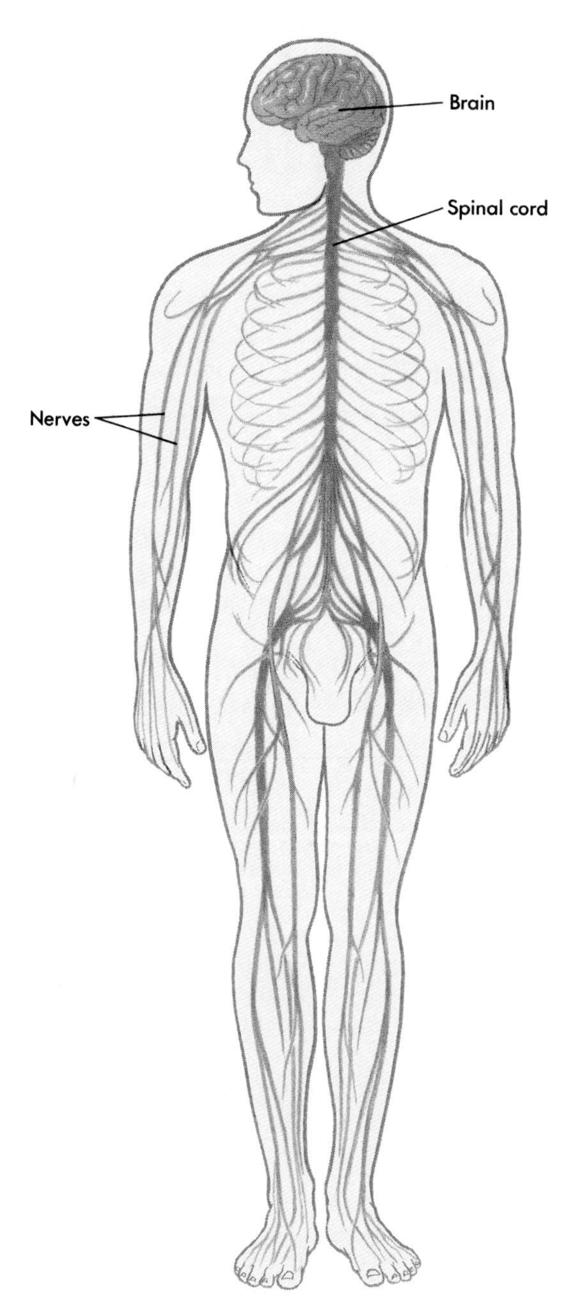

FIGURE 13-1 Brain, spinal cord, and peripheral nerves.

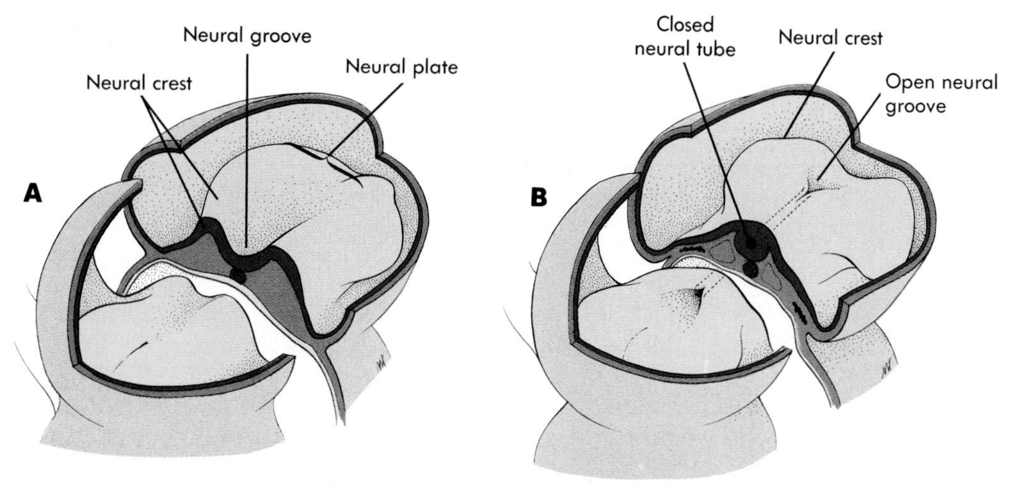

FIGURE 13-2 Formation of the neural tube. **A** Neural plate with the neural crests located laterally and the neural groove in the center. **B** The neural crests have come together in the center of the embryo to form the neural tube; open neural grooves can be seen at each end.

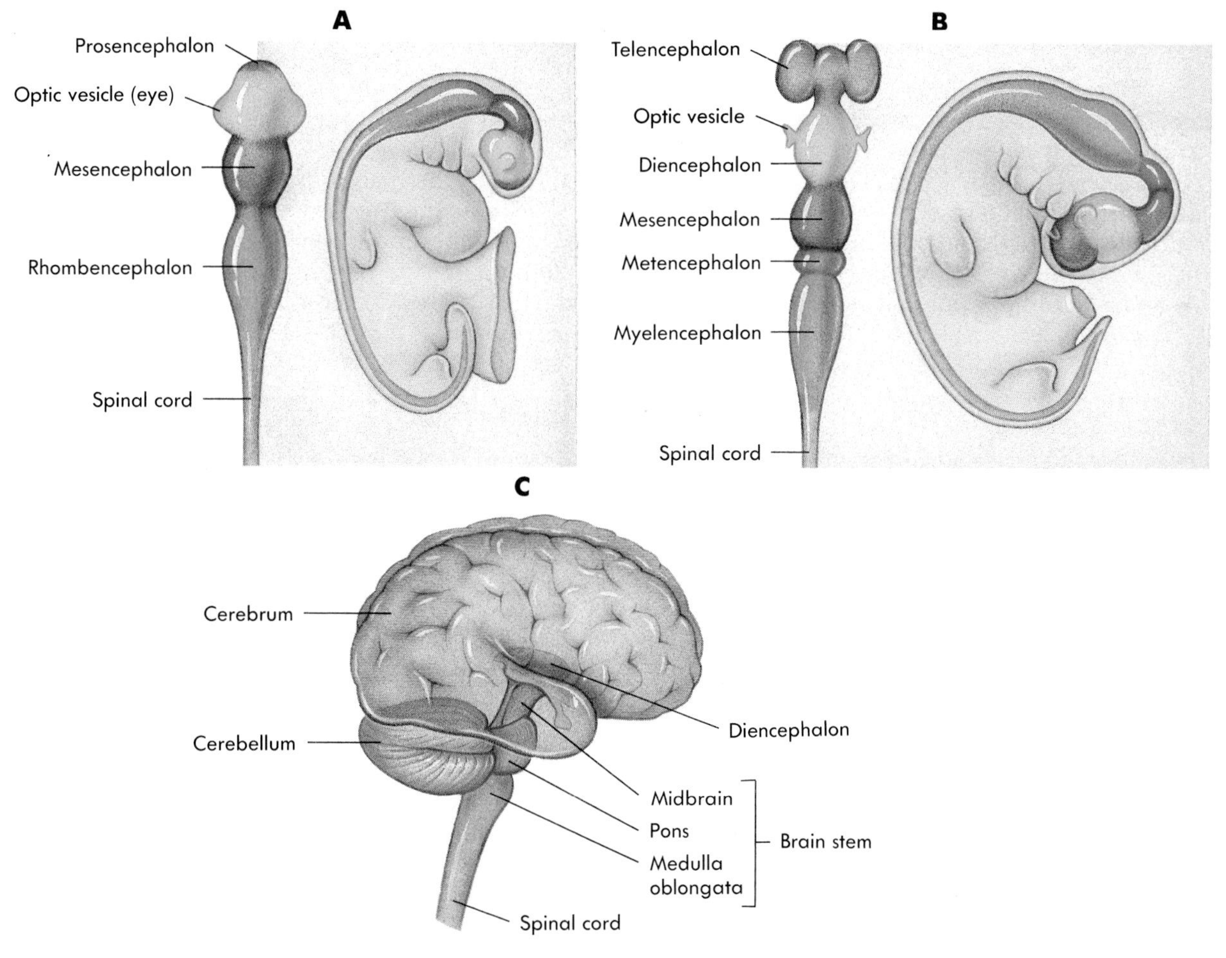

FIGURE 13-3 Development of the brain segments and ventricles. **A** Young embryo. **B** Older embryo. **C** Adult.

Table 13-1

Development of the central nervous system (see Figure 13-3)

EARLY EMBRYO	LATE EMBRYO	ADULT	CAVITY	FUNCTION
Prosencephalon (forebrain)	Telencephalon	Cerebrum	Lateral ventricles	Higher brain functions
	Diencephalon	Diencephalon (thalamus, subthalamus, epithalamus, hypothalamus)	Third ventricle	Relay center, autonomic nerve control, endocrine control
Mesencephalon (midbrain)	Mesencephalon	Mesencephalon (midbrain)	Cerebral aqueduct	Nerve pathways, reflex centers
Rhombencephalon (hindbrain)	Metencephalon	Pons and cerebellum	Fourth ventricle	Nerve pathways, reflex centers, muscle coordination, and balance
	Myelencephalon	Medulla oblongata	Central canal	Nerve pathways, reflex centers

vides into the **telencephalon,** which will become the cerebrum, and the **diencephalon.** The **mesencephalon** remains as a single structure, but the rhombencephalon divides into the **metencephalon,** which will become the pons and cerebellum, and the **myelencephalon,** which will become the medulla oblongata.

BRAINSTEM

The **brain** is that part of the CNS housed within the cranial vault. The major regions of the adult brain are the cerebrum, diencephalon (thalamus and hypothalamus), midbrain (mesencephalon), pons, medulla oblongata, and cerebellum (Table 13-2; Figure 13-4).

The medulla oblongata, pons, and midbrain constitute the **brainstem** (Figure 13-5). The brainstem connects the spinal cord to the remainder of the brain and is responsible for many essential functions. Damage to small brainstem areas often causes death, whereas relatively large areas of the cerebrum or cerebellum may be damaged without causing permanent symptoms. All but two of the 12 cranial nerves enter or exit the brain through the brainstem (see Chapter 14).

Medulla Oblongata

The **medulla oblongata** (ob'long-gah'tah), approximately 3 cm long, is the most inferior portion of the brainstem and is continuous inferiorly with the spinal cord. Superficially the spinal cord blends into the medulla, but internally there are several differences. Discrete **nuclei** (clusters of gray matter, composed mostly of cell bodies, surrounded by white matter) with specific functions are found within the medulla oblongata but not within the spinal cord. In addition, the spinal tracts that pass through the medulla do not have the same organization as the tracts of the spinal cord.

On the anterior surface two prominent enlargements, called **pyramids** because they are broader near the pons and taper toward the spinal cord, extend the length of the medulla (see Figure 13-5, *A*). The pyramids consist of descending nerve tracts involved in the conscious control of skeletal muscles. Near their inferior ends the descending nerve tracts cross to the opposite side, or **decussate** (dĕ'kus-āt; *decussis* means to form an X, as in the Roman numeral X). This decussation accounts, in part, for the fact that each half of the brain controls the opposite half of the body.

Two rounded, oval structures, called **olives,** protrude from the anterior surface of the medulla oblongata just lateral to the superior margins of the pyramids (see Figure 13-5, *A* and *B*). The olives consist of nuclei involved in functions such as balance, coordination, and modulation of sound impulses from the inner ear (see Chapter 15). The nuclei of cranial nerves IX (glossopharyngeal), X (vagus), XI (accessory), and XII (hypoglossal) also are located within the medulla.

Functionally the medulla oblongata acts as a conduction pathway for both ascending and descending nerve tracts. Its role as a conduction pathway is included as part of the description of ascending and descending nerve tracts, and the functional characteristics of the reticular system are described later in this chapter. Various medullary nuclei also function as centers for several reflexes (e.g., regulation of heart rate, blood vessel diameter, breathing, swallowing, vomiting, coughing, and sneezing).

Table 13-2

Divisions and functions of the central nervous system

BRAIN REGION	FUNCTION
Brainstem	Connects the spinal cord to the brain; several important functions (see below); location of cranial nerve nuclei
Medulla oblongata	Pathway for ascending and descending nerve tracts; center for several important reflexes (e.g., heart rate, breathing, swallowing, vomiting)
Pons	Contains ascending and descending nerve tracts; relay between cerebrum and cerebellum; reflex center
Midbrain	Contains ascending and descending nerve tracts; visual reflex center; part of auditory pathway
Reticular formation	Scattered throughout brainstem; controls cyclic activities such as the sleep-wake cycle
Diencephalon	See individual parts
Thalamus	Major sensory relay center; influences mood and movement
Diencephalon—cont'd	
Subthalamus	Contains nerve tracts and nuclei
Epithalamus	Contains nuclei responding to olfactory stimulation and contains pineal body
Hypothalamus	Major control center for maintaining homeostasis and regulating endocrine function
Cerebrum	Conscious perception, thought, and conscious motor activity; can override most other systems
Basal ganglia	Control of muscle activity and posture; largely inhibit unintentional movement
Limbic system	Autonomic response to smell, emotion, mood, and other such functions
Cerebellum	Control of muscle movement and tone; regulates extent of intentional movement

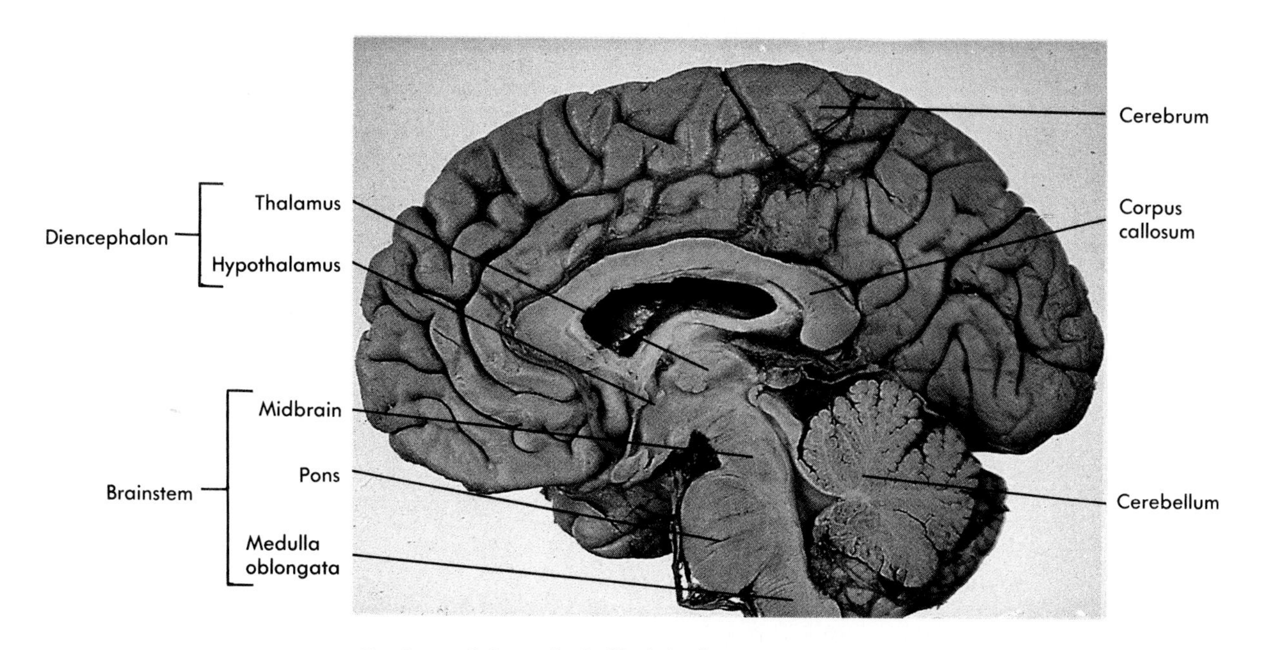

FIGURE 13-4 Regions of the right half of the brain as seen in a midsagittal section.

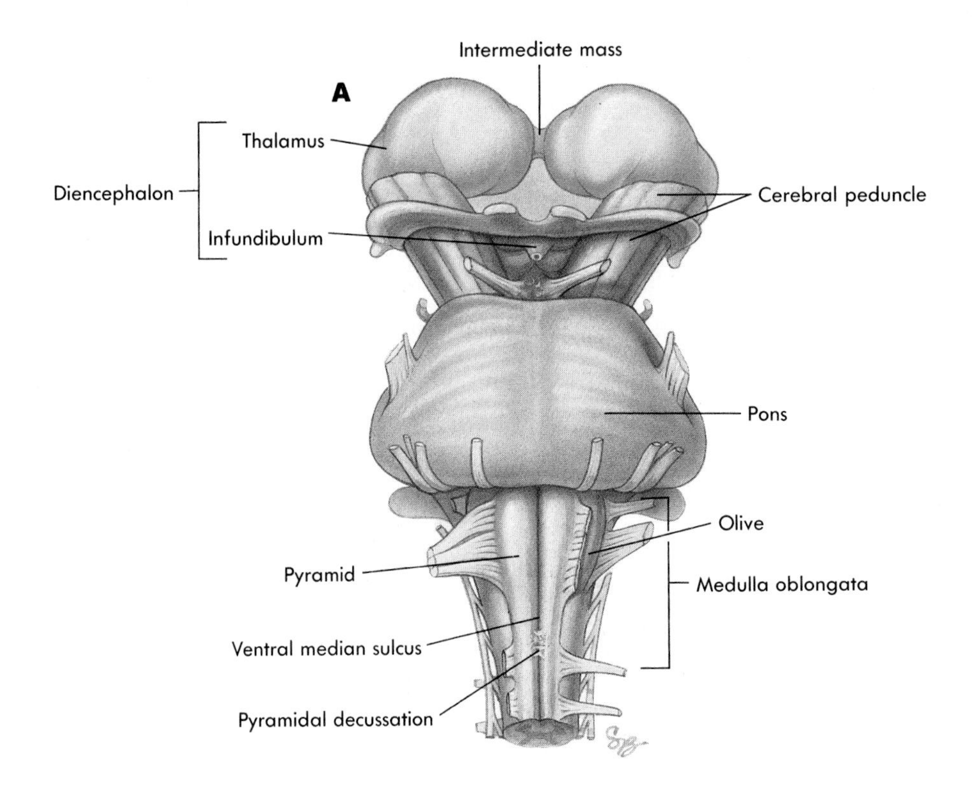

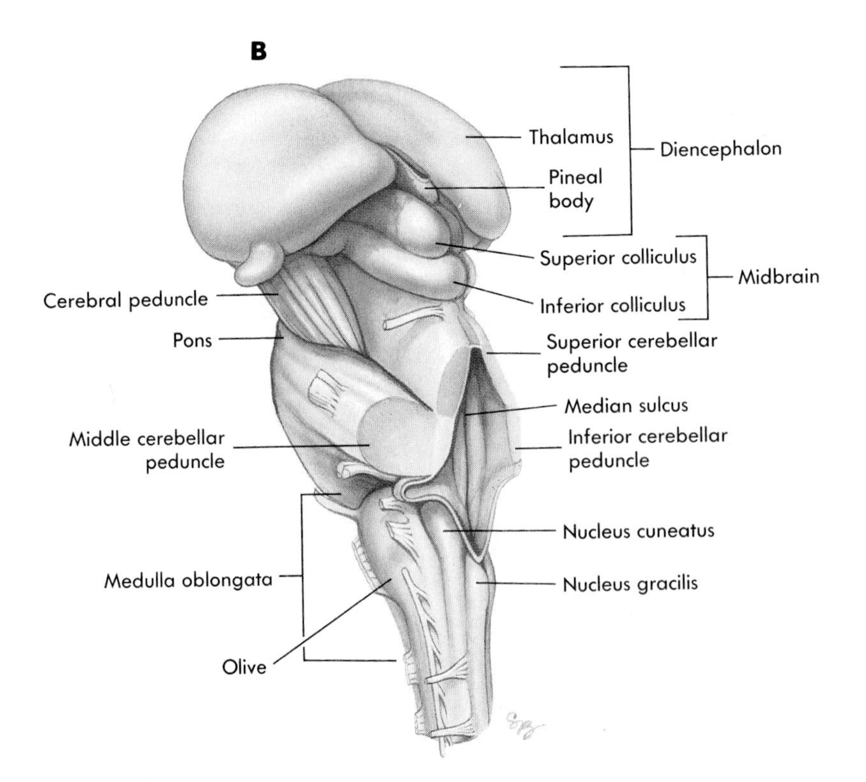

FIGURE 13-5 Brainstem. **A** Anterior view. **B** Posterolateral view.

Pons

The portion of the brainstem just superior to the medulla oblongata is the **pons** (see Figure 13-5, *A*), which contains ascending and descending nerve tracts and several nuclei. The pontine nuclei, located in the anterior portion of the pons, relay information from the cerebrum to the cerebellum.

The nuclei for cranial nerves V (trigeminal), VI (abducens), VII (facial), and VIII (vestibulocochlear) are contained within the posterior pons. Other important pontine areas include the pontine sleep center and the respiratory centers, which along with the medullary respiratory centers help control respiratory movements.

Midbrain

The **midbrain,** or mesencephalon, is the smallest region of the brainstem (see Figure 13-5, *B*). It is just superior to the pons and contains the nuclei of cranial nerves III (oculomotor) and IV (trochlear).

The **tectum** (tek'tum; roof) of the midbrain consists of four nuclei that form mounds on the dorsal surface, collectively called **corpora** (kōr'pōr-ah; bodies) **quadrigemina** (kwah'dră-jem'ĭ-nah; four twins). Each mound is called a **colliculus** (kol-lik'u-lus; hill); there are two **superior colliculi** and two **inferior colliculi.** The inferior colliculi are involved in hearing and are an integral portion of the auditory pathways in the CNS. Neurons conducting impulses from the structures of the inner ear (see Chapter 15) to the brain all synapse in the inferior colliculi. The superior colliculi are involved in visual reflexes, and they receive input from the eyes, the inferior colliculi, the skin, and the cerebrum. Fibers from the superior colliculi project to cranial nerve nuclei and to the superior cervical portion of the spinal cord where they stimulate motor neurons involved in turning the eyes (oculomotor, trochlear, and abducens cranial nerves) and the head (the accessory cranial nerve and superior cervical cord levels). Impulses reaching the superior colliculi from the cerebrum are involved in the visual tracking of moving objects (see Chapter 15).

1

The superior colliculi regulate the reflexive movement of the eyes and head in response to a number of different stimuli. What does a person do when a bright object suddenly appears in his range of vision? What does he do when he hears someone yell? What happens if someone else comes up quietly behind him, stands to his left side, and taps him on the right shoulder? Describe as much as possible the pathway for each of these reflexes.

? ? ? ? ? ? ? ? ? ? ?

The **tegmentum** (teg-men'tum; floor) of the midbrain largely consists of ascending tracts from the spinal cord to the brain and also contains the paired **red nuclei.** The red nuclei are so named because they have a pinkish color in fresh brain specimens because of an abundant blood supply. The red nuclei aid in the unconscious regulation and coordination of motor activities. **Cerebral peduncles** (pe-dun'klz) comprise that portion of the midbrain inferior to the tegmentum. They consist of descending tracts from the cerebrum to the spinal cord and constitute one of the major CNS motor pathways. The **substantia nigra** (ni'grah; black substance), a nuclear mass between the tegmentum and cerebral peduncles (Figure 13-6), is a pigmented region of the midbrain with cytoplasmic melanin granules that give it a dark gray-to-black color. The substantia nigra has interconnections with other basal ganglia nuclei of the cerebrum (described later in this chapter) and is involved in coordinating movement and muscle tone.

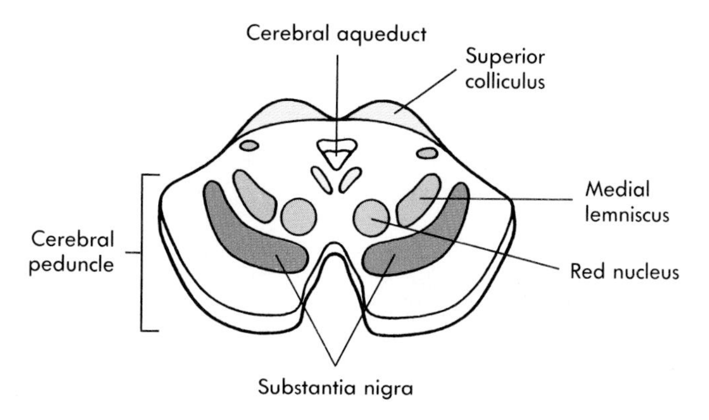

FIGURE 13-6 Cross section through the midbrain at the level of the superior colliculus.

Reticular Formation

Scattered like a cloud throughout most of the length of the brainstem is a group of nuclei collectively called the **reticular formation,** which receives afferent axons from a large number of sources and especially from nerves that innervate the face. These axons play an important role in arousing and maintaining consciousness. The reticular formation and its connections constitute a system, the **reticular activating system,** which is involved with the sleep/wake cycle.

Visual and acoustical stimuli and mental activities can stimulate the reticular activating system to maintain alertness and attention. Stimuli such as a ringing alarm clock, sudden bright lights, or cold water being splashed on the face can arouse consciousness. Conversely, removal of visual or auditory stimuli may lead to drowsiness or sleep (consider, for example, a monotonous lecture in a dark lecture hall). Damage to cells of the reticular formation can result in coma.

> The reticular activating system is relatively sensitive to certain drugs. General anesthetics function by suppressing this system. It may also be the target of many tranquilizers. On the other hand, ammonia (smelling salts) and other irritants stimulate trigeminal nerve endings in the nose, sending impulses to the reticular formation and the cerebral cortex to arouse an unconscious patient.

Descending fibers from the reticular formation constitute one of the most important motor pathways. Fibers from the reticular formation are critical in controlling respiratory and cardiac rhythms and other vital functions.

DIENCEPHALON

The **diencephalon** (di-en-sef'ă-lon) is the part of the brain between the brainstem and the cerebrum (see Figure 13-4; Figure 13-7, *A*). Its main components are the thalamus, subthalamus, hypothalamus, and epithalamus.

Thalamus

The **thalamus** (Figure 13-7, *A* and *B*) is by far the largest portion and constitutes approximately four fifths of the diencephalon's weight. The thalamus consists of a cluster of nuclei and is shaped somewhat like a yo-yo, with two large, lateral portions connected in the center by a small stalk called the **intermediate mass.** The space surrounding the intermediate mass and separating the two large portions of the thalamus is the third ventricle of the brain.

Most sensory input projects to the thalamus where afferent neurons synapse with thalamic neurons, which send projections from the thalamus to the cerebral cortex. Auditory impulses synapse in the **medial geniculate** (jĕ-nik'u-lat; *genu*, bent like a knee) **body** of the thalamus, visual impulses synapse in the **lateral geniculate body,** and most other sensory impulses synapse in the **ventral posterior nuclei.**

The thalamus also has other functions such as influencing mood and general body movements that are associated with strong emotions such as fear or rage.

Subthalamus

The **subthalamus** is a small area immediately inferior to the thalamus (see Figure 13-7, *A*). It contains several nerve tracts and the **subthalamic nuclei.** A small portion of the red nucleus and substantia nigra of the midbrain also extend into this area. The subthalamic nuclei are associated with the basal ganglia and are involved in controlling motor functions.

Epithalamus

The **epithalamus** is a small area superior and posterior to the thalamus (see Figure 13-7, *A*). It consists of habenular nuclei and the pineal body. The **habenular** (hă-ben'u-lar) **nuclei** are influenced by the sense of smell and are involved in emotional and visceral responses to odors. The **pineal** (pi'ne-al) **body,** or **epiphysis,** is shaped somewhat like a pinecone, from which the name pineal is derived. It plays a role in controlling the onset of puberty, but data are not conclusive, so active research continues in this field. The pineal body also may be involved in the sleep-wake cycle.

> The adult pineal body contains granules of calcium and magnesium salts called "brain sand." These granules can be seen on x-rays and are useful as a landmark in determining whether or not the pineal body has been displaced by some pathological enlargement of a portion of the brain such as a tumor or a hematoma.

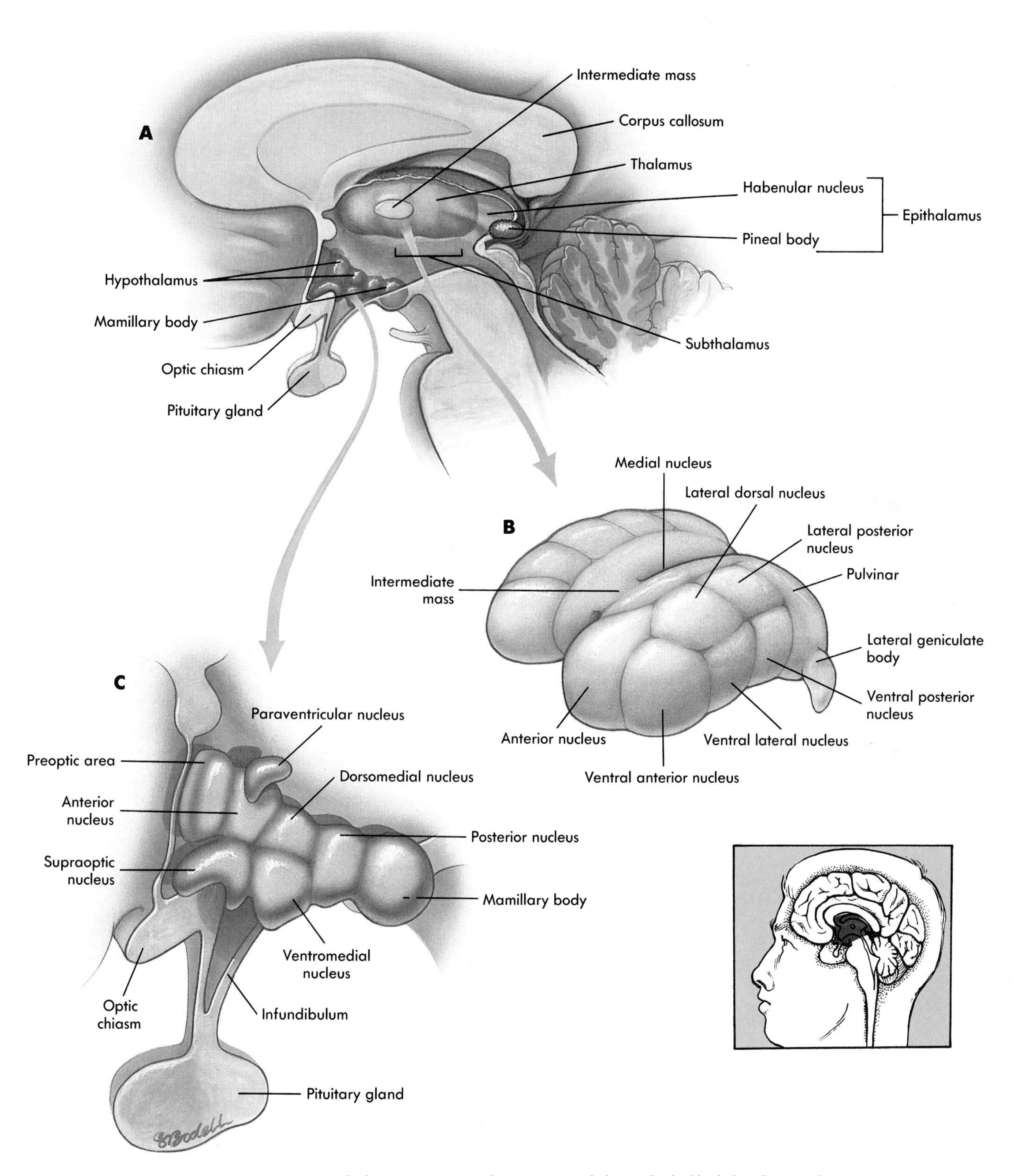

FIGURE 13-7 Diencephalon. **A** General overview of the right half of the diencephalon as seen in a midsagittal section. **B** Thalamus showing the nuclei. **C** Hypothalamus showing the nuclei and right half of the pituitary.

Hypothalamus

The **hypothalamus** is the most inferior portion of the diencephalon (Figure 13-7, *A* and C) and contains several small nuclei and nerve tracts. The most conspicuous nuclei, called the **mamillary bodies,** appear as bulges on the ventral surface of the diencephalon. They are involved in olfactory reflexes and emotional responses to odors. A funnel-shaped stalk, the **infundibulum,** extends from the floor of the hypothalamus, connecting it to the **posterior pituitary gland,** or **neurohypophysis.** The hypothalamus plays an important role in controlling the endocrine system because it regulates the pituitary gland's secretion of hormones, which influence functions as diverse as metabolism, reproduction, responses to stressful stimuli, and urine production (see Chapters 17 and 18).

Afferent fibers that terminate in the hypothalamus provide input from the following: (1) visceral organs; (2) taste receptors of the tongue; (3) the limbic system (involved in responses to smell); (4) specific cutaneous areas such as the nipples and external genitalia; and (5) the prefrontal cortex of the cerebrum carrying information relative to "mood" through the thalamus. Efferent fibers from the hypothalamus extend into the brainstem and the spinal cord where they synapse with neurons of the autonomic nervous system (see Chapter 16). Other fibers extend through the infundibulum to the posterior portion of the pituitary gland (see Chapter 17); some extend to trigeminal and facial nerve nuclei (see Chapter 14) to help control the head muscles that are involved in swallowing; and some extend to motor neurons of the spinal cord to stimulate shivering.

The hypothalamus is very important in a number of functions, all of which have emotional and mood relationships (Table 13-3). Sensations such as sexual pleasure, feeling relaxed and "good" after a meal, rage, and fear are related to hypothalamic functions.

Table 13-3

Hypothalamic functions

FUNCTION	DESCRIPTION
Autonomic	Helps control heart rate, urine release from the bladder, movement of food through the digestive tract, and blood vessel diameter
Endocrine	Helps regulate pituitary gland secretions and influences metabolism, ion balance, sexual development, and sexual functions
Muscle control	Controls muscles involved in swallowing and stimulates shivering in several muscles
Temperature regulation	Promotes heat loss when the hypothalamic temperature increases by increasing sweat production (anterior hypothalamus) and promotes heat production when the hypothalamic temperature decreases by promoting shivering (posterior hypothalamus)
Regulation of food and water intake	Hunger center promotes eating and satiety center inhibits eating; thirst center promotes water intake
Emotions	Large range of emotional influences over body functions; directly involved in stress-related and psychosomatic illnesses and with feelings of fear and rage
Regulation of the sleep-wake cycle	Coordinates responses to the sleep-wake cycle with other areas of the brain (e.g., the reticular activating system)

CEREBRUM

The cerebrum is the largest portion of the brain, weighing approximately 1200 g in females and 1400 g in males. Brain size is related to body size; large brains are associated with large bodies and not with greater intelligence. The cerebrum is what most people think of when the term brain is mentioned.

The cerebrum is divided into left and right hemispheres by a **longitudinal fissure** (Figure 13-8, *A*). The most conspicuous features on the surface of each hemisphere are numerous folds called **gyri** (ji'ri), which greatly increase the surface area of the cortex, and intervening grooves called **sulci** (sul'si; see Figure 13-8). The general pattern of the gyri is similar in all normal human brains, but some variation exists between individuals, even between the two hemispheres of the same cerebrum.

Each cerebral hemisphere is divided into lobes, which are named for the skull bones overlying each one (Figure 13-8, *B*). The **frontal lobe** is important in voluntary motor function, motivation, aggression, and mood. The **parietal lobe** is the major center for the reception and evaluation of most sensory information (excluding smell, hearing, and vision). The frontal and parietal lobes are separated by a prominent sulcus called the **central sulcus.** The **occipital lobe** functions in the reception and integration of visual input and is not distinctly separate from the other lobes. The **temporal lobe** receives and evaluates olfactory and auditory (hearing) input and plays an important role in memory. Its anterior and inferior portions are referred to as the "psychic cortex," and they are associated with brain function such as abstract thought and judgment. The temporal lobe is separated from the rest of the cerebrum by a **lateral fissure,** and deep within the fissure is the **insula** (in'su-lah; island), often referred to as a fifth lobe.

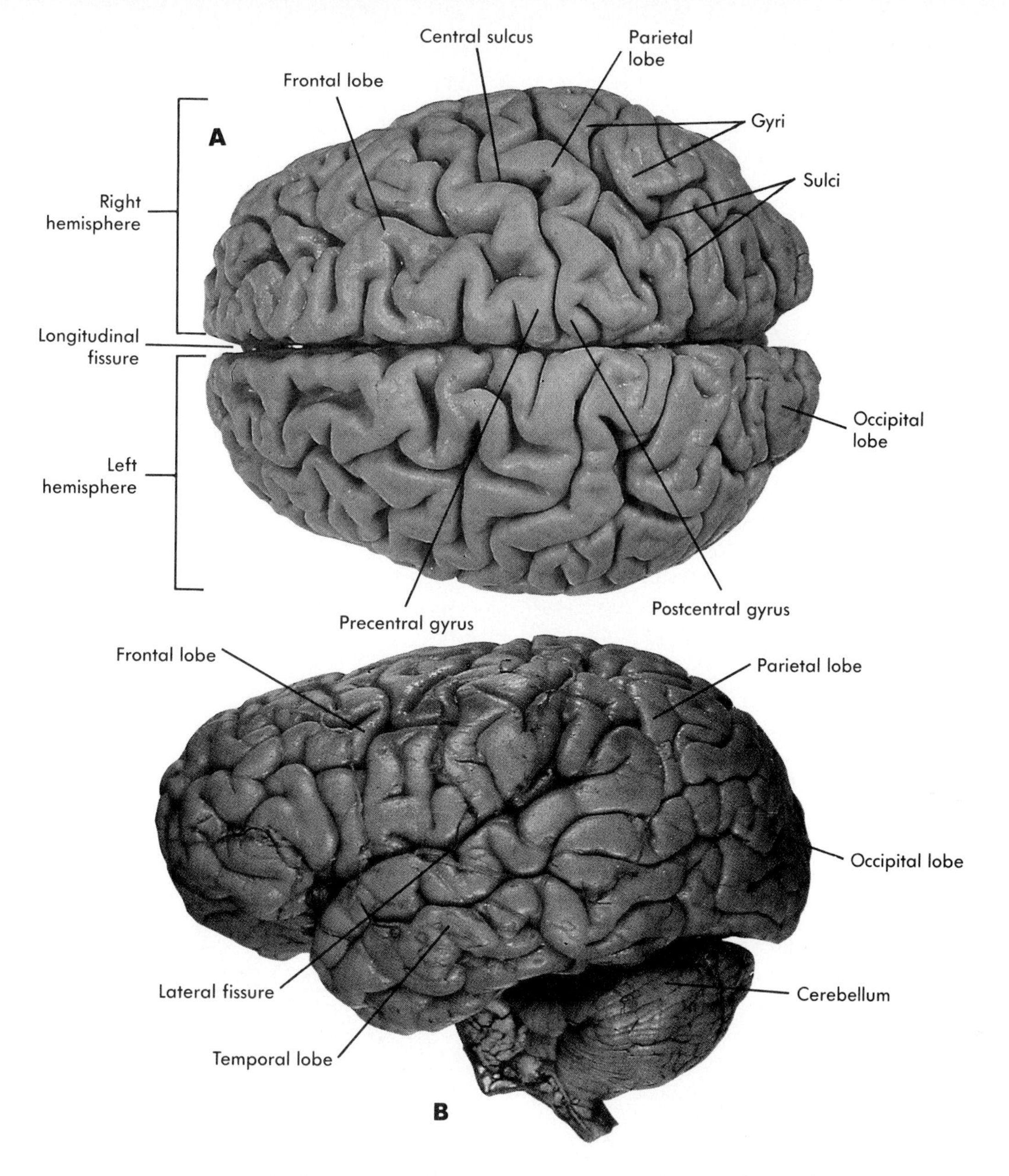

FIGURE 13-8 The brain. **A** Superior view. **B** Lateral view of the left cerebral hemisphere.

The gray matter on the outer surface of the cerebrum is the **cortex,** and clusters of gray matter deep inside the brain are **nuclei** (see Figure 13-9, *B*). The white matter of the brain between the cortex and nuclei is the **cerebral medulla** (not to be confused with the medulla oblongata; "medulla" is a general term meaning the center of a structure or bone marrow). The cerebral medulla consists of nerve tracts that connect the cerebral cortex to other areas of cortex or other parts of the CNS. These tracts fall into three main categories: (1) **association fibers,** which connect areas of the cerebral cortex within the same hemisphere; (2) **commissural fibers,** which connect one cerebral hemisphere to the other; and (3) **projection fibers,** which are between the cerebrum and other parts of the brain and spinal cord (Figure 13-9).

Cerebral Cortex

Figure 13-10 depicts a lateral view of the left cerebral cortex with some of the functional areas indicated. Sensory pathways project to specific regions of the cerebral cortex, called **primary sensory areas,** where those sensations are perceived.

Most of the **postcentral gyrus** (located posterior to the central sulcus) is called the **primary somesthetic** (so′mes-thet′ik) **area, (somesthetic cortex),** or **general sensory area.** Afferent fibers carrying general sensory input such as pain, pressure, and temperature synapse in the thalamus, and thalamic neurons relay the information to the somesthetic cortex.

The somesthetic cortex is organized topographically relative to the general plan of the body (Figure 13-11, *A*). Sensory impulses conducting input from

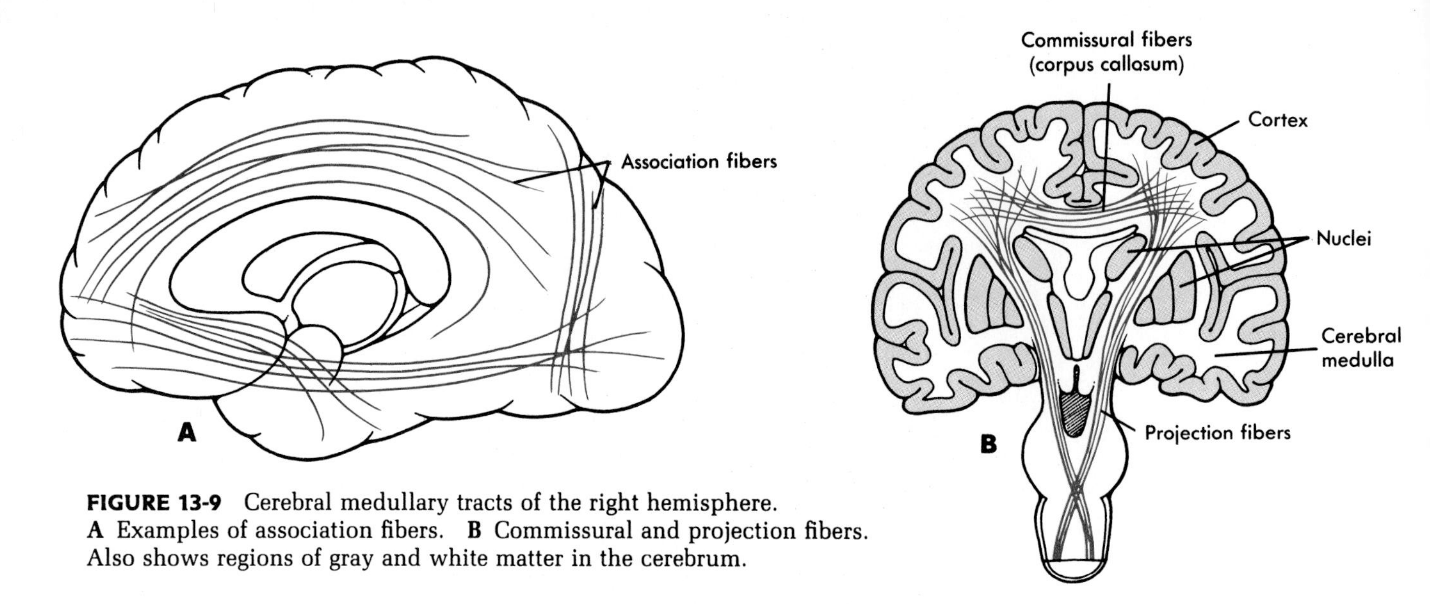

FIGURE 13-9 Cerebral medullary tracts of the right hemisphere. **A** Examples of association fibers. **B** Commissural and projection fibers. Also shows regions of gray and white matter in the cerebrum.

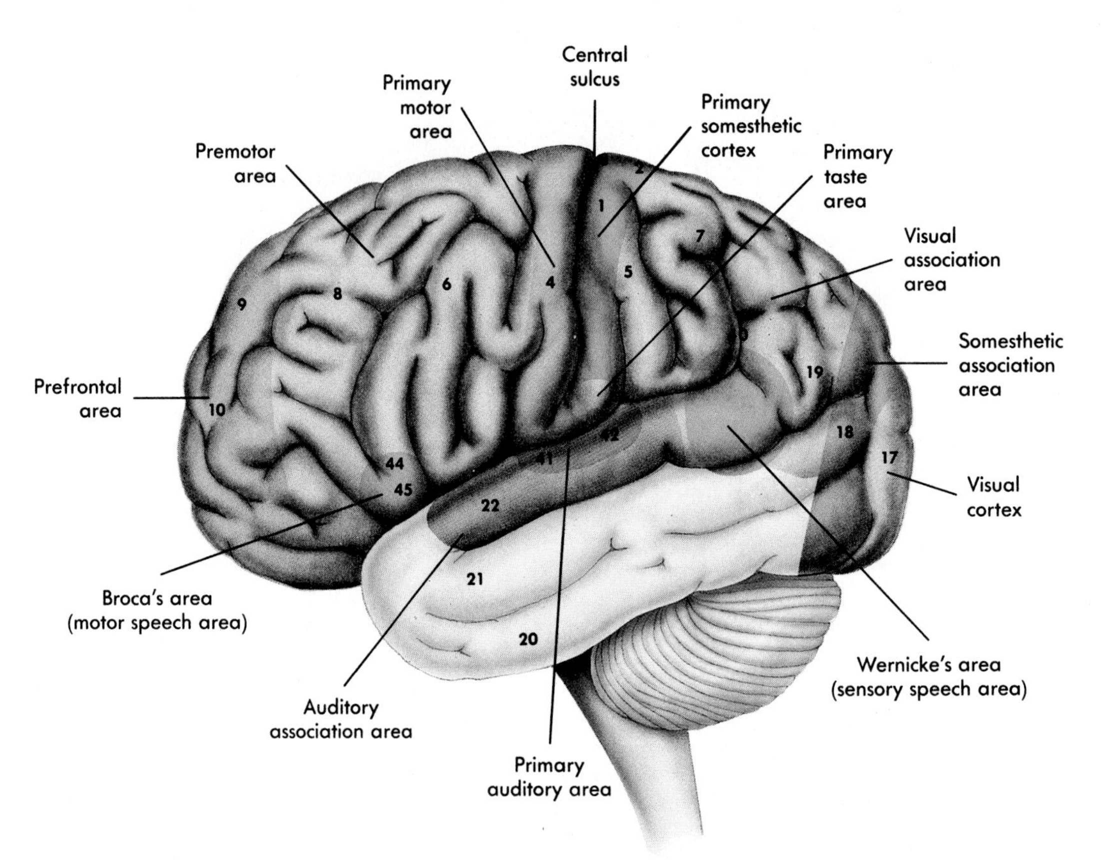

FIGURE 13-10 Some functional areas of the lateral side of the left cerebral cortex. The numbers shown are traditionally used to map the various areas of the cerebral cortex (referred to as Brodmann's areas).

the feet project to the most superior portion of the somesthetic cortex, and sensory impulses from the face project to the most inferior portion of the somesthetic cortex. The pattern of the somesthetic cortex in each hemisphere is arranged in the form of a half homunculus (a little human) representing the opposite side of the body, with the feet directed superiorly and the head directed inferiorly. In addition, the size of various regions of the somesthetic cortex is relative to the number of sensory receptors in the associated regions of the body. Many sensory receptors are in the face, but far fewer are in a comparably sized area of the legs; therefore a greater area of the somesthetic cortex contains sensory neurons associated with the face (the homunculus has a disproportionately large face).

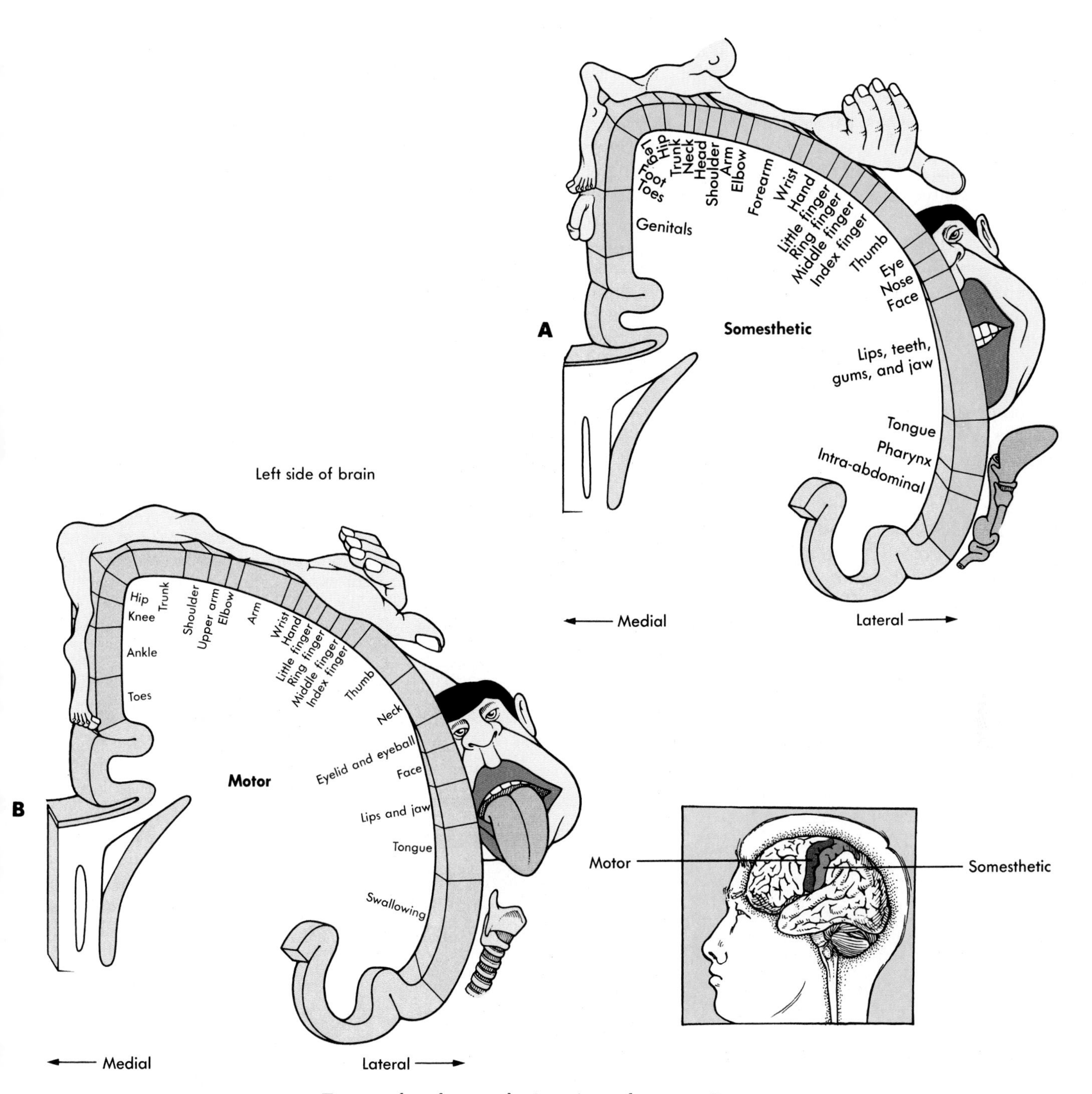

FIGURE 13-11 Topography of somesthetic, **A**, and motor, **B**, cortices seen in coronal section on the left side of the brain. The figure of the body (homunculus) depicts the relative nerve distributions; the size indicates relative innervation. Each cortex occurs on both sides of the brain but appears on only one side in this illustration.

The primary sensory areas of the cerebral cortex must be intact for conscious perception, localization, and identification of a stimulus. The conscious perceptions of cutaneous sensations, although integrated within the cerebrum, are perceived as though they were on the surface of the body. This is called **projection** and indicates that the brain refers a sensation to the superficial site where the stimulus interacts with the sensory receptors.

Cortical areas immediately adjacent to the primary sensory centers, **association areas,** are involved in the process of recognition. The **somesthetic association area** is posterior to the somesthetic area, and the **visual association area** is anterior to the primary visual area (visual cortex; see Figure 13-10). Afferent action potentials originating in the retina of the eye reach the visual cortex where the image is "seen." Action potentials then pass from the visual cortex to the visual association area where the present visual information is compared to past visual experience ("Have I seen this before?"). Based on this comparison, the visual association area "decides" whether or not the visual input is recognized and passes judgment concerning the significance of the input. You pay less attention, for example, to a person you have never seen before than to someone you know.

The visual association area, like other association areas of the cortex, has reciprocal connections with other parts of the cortex, and input from those parts influences decisions. These connections include input from the frontal lobe where emotional value is placed on the visual input. Because of these numerous connections, it is quite unlikely that visual information can pass beyond the visual association area without having several judgments made concerning the input and may be one of the reasons why two people may witness exactly the same event and if questioned immediately afterward, present somewhat different versions of what happened.

2

Using the visual association area as an example, explain the general functions of the association areas around the other primary cortical areas (see Figure 13-10).

? ? ? ? ? ? ? ? ? ? ?

The **precentral gyrus** (located anterior to the central sulcus) is also called the **primary motor area.** Efferent action potentials initiated in this region control many voluntary movements, especially the fine motor movements of the hands. Cortical neurons that control skeletal muscles are called **upper motor neurons** and are not confined to just the precentral gyrus; in fact, only approximately 30% of them are located there. Another 30% are in the **premotor area,** and the rest are in the somesthetic cortex.

The cortical functions of the precentral gyrus are arranged topographically according to the general plan of the body—similar to that of the postcentral gyrus (see Figure 13-11, *B*). The nerve cell bodies providing motor function to the feet are in the most superior and medial portions, whereas those for the face are in the inferior region. Muscle groups that have numerous motor units and therefore greater innervation are represented by a relatively larger area of the motor cortex. For example, muscles of the hands and mouth are represented by a larger area of the motor cortex than the muscles of the thighs and legs.

The premotor area (see Figure 13-10) is the staging area where motor functions are organized before they are initiated in the motor cortex. For example, if a person decides to take a step, the neurons of the premotor area are stimulated first, and the determination is made there as to which muscles must contract, in what order, and to what degree. Impulses are then passed to the upper motor neurons in the motor cortex, which actually initiate each planned movement.

The premotor area must be intact for a person to carry out complex, skilled, or learned movements, especially ones related to manual dexterity. Impairment in the performance of learned movements, called **apraxia** (a-prak'se-ah), can result from a lesion in the premotor area. Apraxia is characterized by hesitancy in performing these movements.

The motivation and the foresight to plan and initiate movements occur in the next most anterior portion of the brain, the **prefrontal area,** an area of association cortex that is well developed only in primates, especially in humans. It is involved in motivation and regulation of emotional behavior and mood. The large size of this area in humans may account for our relatively well-developed forethought and motivation and for our emotional complexity.

In relation to its involvement in motivation, the prefrontal area is the functional center for aggression. In the past one method used to eliminate uncontrollable aggression in mental hospital inmates was to surgically remove or destroy the prefrontal regions of the brain (prefrontal or frontal lobotomy). This operation was successful in eliminating aggression, but it also eliminated the motivation to do much else and dramatically altered the personality.

Speech

In the vast majority of people the speech area is in the left cortex. Two major cortical areas are involved in speech: **Wernicke's area** (sensory speech area), a portion of the parietal lobe, and **Broca's area** (motor speech area) in the inferior portion of the frontal lobe (see Figure 13-10). Wernicke's area is necessary for understanding and formulating coherent speech. Broca's area initiates the complex series of movements necessary for speech.

To repeat a word that one hears requires the functional integrity of the following pathway. Action potentials from the ear reach the primary auditory area (auditory cortex) where the word is heard; the word is recognized in the auditory association cortex and is comprehended in portions of Wernicke's area. Then, action potentials representing the word are conducted through association fibers that connect Wernicke's and Broca's areas. In Broca's area the word is formulated as it will be repeated; impulses then go to the premotor cortex where the movements are programmed and finally to the motor cortex where the proper movements are triggered.

Speaking a written word is somewhat similar. The information enters the visual cortex, passes to the visual association cortex where it is recognized, and continues to Wernicke's area where it is understood and formulated as it will be spoken. From Wernicke's area it follows the same route for repeating audibly received words.

3

Propose the pathway needed for a blindfolded person to name an object placed in her right hand.

? ? ? ? ? ? ? ? ? ? ?

Aphasia (a-fa'ze-ah), absent or defective speech or language comprehension, results from a lesion in the language areas of the cortex. The several types of aphasia depend on the site of the lesion. **Receptive aphasia** (Wernicke's aphasia), with defective auditory and visual comprehension of language and defective naming of objects and repetition of spoken sentences, is caused by a lesion in Wernicke's area. Both **jargon aphasia,** in which a person may speak fluently but unintelligibly, and **conduction aphasia,** in which a person has poor repetition but relatively good comprehension, can result from a lesion in the tracts between Wernicke's and Broca's areas. **Anomic aphasia,** caused by the isolation of Wernicke's area from the parietal or temporal association areas, is characterized by fluent but circular speech resulting from poor word-finding ability. **Expressive aphasia** (Broca's aphasia), caused by a lesion in Broca's area, is characterized by hesitant and distorted speech.

Brain waves

Electrodes placed on a person's scalp and attached to a recording device can record the brain's electrical activity, producing an **electroencephalogram (EEG)** (Figure 13-12). These electrodes are not sensitive enough to detect individual action potentials but can detect the simultaneous action potentials in large numbers of neurons. As a result, the EEG displays wavelike patterns known as **brain waves.** This electrical activity is constant, but the intensity and frequency of electrical discharge differ from time to time based on the state of brain activity.

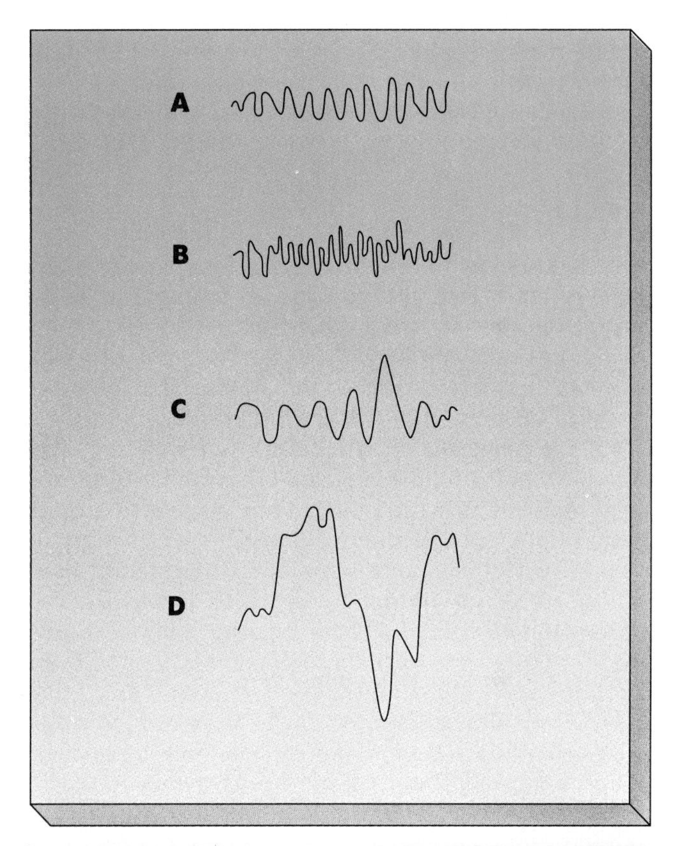

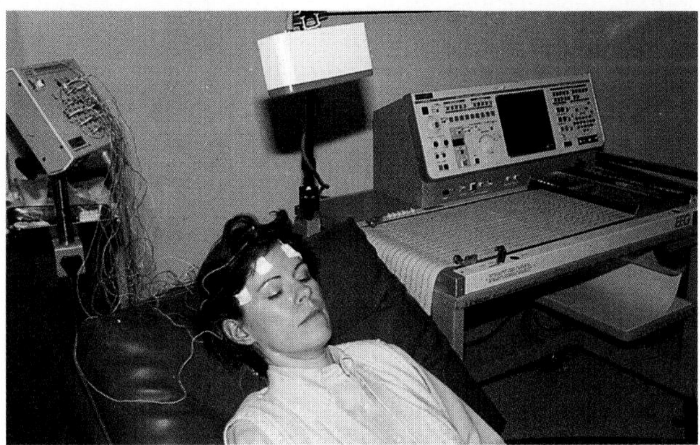

FIGURE 13-12 Electroencephalograms (EEGs) showing brain waves. **A** Alpha waves. **B** Beta waves; **C** Theta waves. **D** Delta waves. Photograph shows a person undergoing an electroencephalogram.

Most of the time EEG patterns from a given individual are irregular with no particular pattern because, although the normal brain is active, most of its electrical activity is not synchronous. At other times, however, specific patterns can be detected. These regular patterns are classified as alpha, beta, theta, or delta waves (see Figure 13-12). **Alpha waves** are observed in a normal person who is awake but in a quiet, resting state with the eyes closed. **Beta waves** have a higher frequency than alpha waves and occur during intense mental activity. **Theta waves** usually occur in children, but they can also occur in adults experiencing frustration or with certain brain disorders. **Delta waves** occur in deep sleep, in infancy, and in patients with severe brain disorders.

Distinct types of EEG patterns can be detected in patients with specific brain disorders such as epileptic seizures. Neurologists use these patterns to diagnose and determine the treatment for the disorders.

Memory

Memory can be divided into at least three types: sensory, short-term (or primary), and long-term. **Sensory memory** is the very short-term retention of sensory input received by the brain while something is scanned, evaluated, and acted upon. This type of memory lasts less than 1 second.

If a given piece of data held in sensory memory is considered valuable enough, it is moved into **short-term memory** in which information is retained for a few seconds to a few minutes. This memory is limited primarily by the number of bits of information that can be stored at any one time, which is usually approximately seven bits of information, although the amount varies from person to person. Have you ever wondered why telephone numbers are seven digits long? More bits can be stored when the numbers are grouped into specific segments separated by spaces such as when adding an area code. When new information is presented, old information previously stored in short-term memory is eliminated. Therefore if a person is given a second telephone number or if the person's attention is drawn to something else, the first number usually is forgotten.

How short-term memory works is not known, but it is known that part of the temporal lobe plays an important role—at least in the conversion of short-term to long-term memory. Patients with temporal lobe lesions can be invited into a room to meet a new person with whom they may talk for several minutes. If they are then escorted out of the room and immediately back in again, they will not remember ever being in the room or meeting the person to whom they were just talking.

Several physiological explanations have been proposed for short-term memory, most of which involve short-term changes in membrane potentials. The changes in membrane potentials are transitory and can be eliminated by new signals reaching the cells.

Certain pieces of information are transferred from short-term to **long-term memory,** which may involve a physical change in neuron shape, facilitating future transmission of impulses. Long-term memory storage in a single neuron involves a calcium influx into the cell that activates an enzyme called **calpain** (kal'pen). Calpain, in turn, partially degrades the dendritic cytoskeleton of the neuron, changing the shape of the dendrite. The change in shape is stabilized by the creation of a new cytoskeleton, and the memory becomes more-or-less permanent.

A whole series of neurons and their pattern of activity, called a **memory engram** or memory trace, probably are involved in the long-term retention of a given piece of information, thought, or idea. Rehearsal of information assists in the transfer of information from short-term to long-term memory.

There are two types of long-term memory: declarative and procedural. **Declarative memory** involves the retention of facts such as names, dates, and places. Declarative memory is localized in a part of the temporal lobe called the **hippocampus** (hip'po-kam'pus; shaped like a seahorse) and the **amygdaloid** (ă-mig'dă-loyd; almond shaped) **nucleus.** The hippocampus is involved in the actual memory (e.g., recall of a person's name), and the amygdala is involved in the emotional overtones of that memory (e.g., the feeling of like or dislike and good or bad memories associated with that person). Emotion and mood apparently serve as gates in the brain, determining what is or is not stored in long-term declarative memory.

Procedural memory involves the development of skills such as riding a bicycle or playing a piano. Procedural memory is stored primarily in the cerebellum and the premotor area of the cerebrum. Conditioned, or Pavlovian, reflexes are also procedural and can be eliminated in experimental animals by producing cerebellar lesions in the animals. The most famous example of a conditioned reflex is that of Pavlov's experiments with dogs. Each time he fed the dogs, a bell was rung; soon the dogs would salivate when the bell rang, even if no food was presented.

Right and left cortex

The cortex of the right cerebral hemisphere controls muscular activity in, and receives sensory input from, the left half of the body. The left cerebral hemisphere controls muscles and receives input from the right half of the body. Sensory information received by the cortex of one hemisphere is shared with the other through connections between the two hemispheres called **commissures** (kom'ĭ-shurz; a joining together). The largest of these commissures is the **corpus callosum** (kor'pus kah-lo'sum; callous body), a broad band of nerve tracts (white matter) at the base of the longitudinal fissure (see Figures 13-4 and 13-9).

Dominance for most functions is probably not very important in most people because the two hemispheres are in constant communication through the corpus callosum, literally allowing the right hand to know what the left hand is doing. Surgical cutting of the corpus callosum has been successful in treating epilepsy. However, under certain conditions interesting functional defects can be seen in people with the corpus callosum severed. For example, if a patient with a severed corpus callosum is asked to reach behind a screen to touch one of several items with one hand without being able to see it and then is asked to point out the same object with the other hand, he cannot do it. Tactile information from the left hand would enter the right somesthetic cortex but would not be transferred to the left hemisphere, which controls the right hand. As a result, the left hemisphere cannot direct the right hand to the correct object.

Language and perhaps other functions such as artistic activities are not shared equally between the cortices of the two hemispheres. The left hemisphere is believed to be the analytical hemisphere, involved in such skills as mathematics and speech. The right hemisphere is believed to be involved in functions such as three-dimensional or spatial perception and musical ability. Some people believe that the "dominant" left hemisphere represses the artistic abilities of the right hemisphere.

Basal Ganglia

The **basal ganglia** are a group of functionally related nuclei located bilaterally in the inferior cerebrum, diencephalon, and midbrain (Figure 13-13). The **subthalamic nucleus** is located in the diencephalon, and the **substantia nigra** is located in the midbrain. The nuclei in the cerebrum are collectively called the **corpus striatum** (kōr'pus stri-a'tum; striped body) and include the **caudate** (kaw'dāt; having a tail) **nucleus** and **lentiform** (lent'tĭ-form; lens shaped) **nucleus.** They are the largest nuclei of the brain and occupy a large portion of the cerebrum.

The basal ganglia play an important role in planning and coordinating motor movements and posture. Complex neural connections link the basal ganglia with the cerebral cortex. The major effect of the basal ganglia is to inhibit unwanted muscular activity, and disorders of the basal ganglia result in exaggerated, uncontrolled movements.

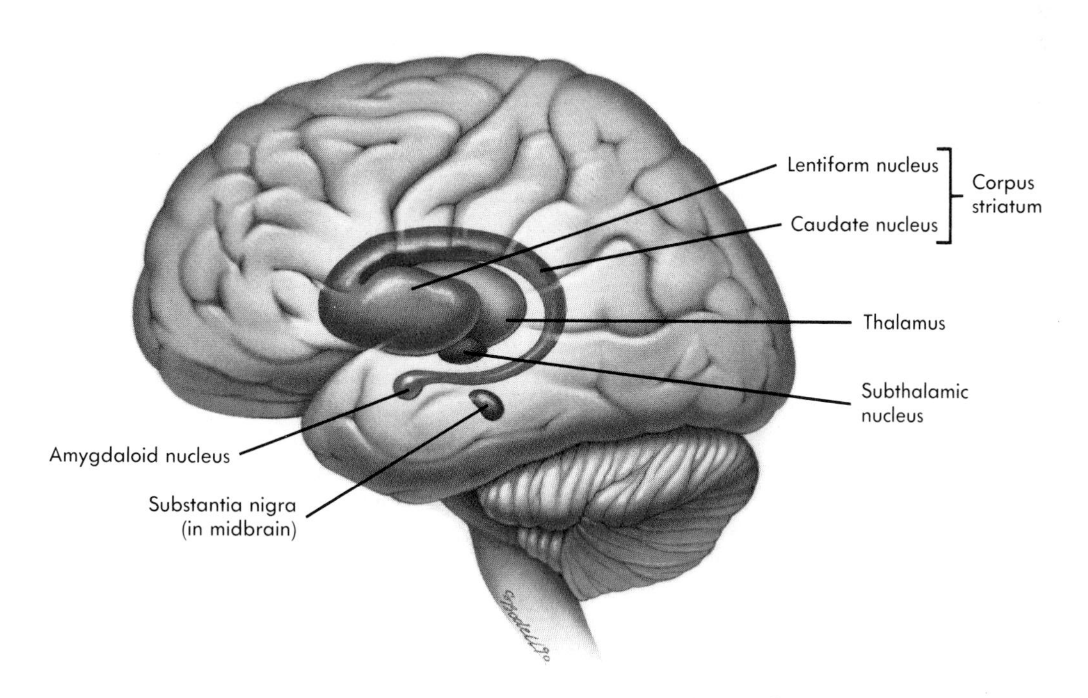

FIGURE 13-13 Basal ganglia of the left hemisphere.

Dyskinesias

Dyskinesias are a group of disorders often involving the basal ganglia in which unwanted, superfluous movements occur. Defects in the basal ganglia may result in brisk, jerky, purposeless movements resembling fragments of voluntary movements. **Sydenham's chorea** (St. Vitus' dance) is a disease usually associated with a toxic or infectious disorder that apparently causes temporary dysfunction of the corpus striatum and usually affects children. **Huntington's chorea** is a dominant hereditary disorder that begins in middle life and causes progressive degeneration of the corpus striatum in affected individuals.

Cerebral palsy is a general term referring to defects in motor functions or coordination resulting from several types of brain damage, which may be caused by abnormal brain development or birth-related injury. Some symptoms of cerebral palsy are related to basal ganglia dysfunction. **Athetosis,** often one of the features of cerebral palsy, is characterized by slow, sinuous, aimless movements. When the face, neck, and tongue muscles are involved, grimacing, protrusion and writhing of the tongue, and difficulty in speaking and swallowing are characteristics.

Damage to the subthalamic nucleus can result in **hemiballismus,** an uncontrolled, purposeless, and forceful throwing or flailing of the arm. Forceful twitching of the face and neck can also result from subthalamic nuclear damage.

Parkinson's disease, characterized by muscular rigidity, tremor, a slow, shuffling gait, and general lack of movement, is caused by a lesion in the substantia nigra. A resting tremor called "pill rolling" is characteristic of Parkinson's disease and consists of circular movement of the opposed thumb and index finger tips. The increased muscular rigidity in patients with Parkinson's disease is due to defective inhibition of some of the basal ganglia by the substantia nigra. In this disease dopamine, an inhibitory neurotransmitter substance, is deficient. The melanin-containing cells of the substantia nigra degenerate, resulting in a loss of pigment. Parkinson's disease can be treated with levodopa (L-dopa), a precursor to dopamine.

Cerebellar lesions result in a spectrum of characteristic functional disorders. Movements tend to be **ataxic** (jerky) and **dysmetric** (overshooting—e.g., pointing past or deviating from a mark that one tries to touch with the finger). Alternating movements such as supination and pronation are performed in a clumsy manner. **Nystagmus,** constant motion of the eyes, may also occur. A cerebellar tremor is an intention tremor, that is, the more carefully one tries to control a given movement, the greater the tremor becomes. For example, when a person with a cerebellar tremor attempts to drink a glass of water, the closer the glass comes to the mouth, the shakier the movement becomes. This type of tremor is in direct contrast to the basal ganglia tremors described previously in which the resting tremor largely or completely disappears during purposeful movement.

Limbic System

Portions of the cerebrum and diencephalon are grouped together under the title **limbic** (lim'bik) **system** (Figure 13-14). The term limbic is not precise and is used differently by various authors. *Limbus* means border, and the term limbic refers to the medial portion, or border, of the temporal lobe. Structurally the limbic system consists of (1) certain cerebral cortical areas, including the **cingulate** (sin'gu-lat; to surround) **gyrus,** located along the inner surface of the longitudinal fissure just above the corpus callosum, and the **hippocampus;** (2) various nuclei such as anterior nuclei of the thalamus and the **habenular nuclei** in the epithalamus; (3) parts of the **basal ganglia;** (4) the hypothalamus, especially the **mamillary bodies;** (5) the **olfactory cortex;** and (6) tracts connecting the various cortical areas and ganglia (the **fornix** is one such tract).

The limbic system influences emotions, the visceral responses to those emotions, motivation, mood, and sensations of pain and pleasure.

One of the major sources of sensory input into the limbic system is the olfactory nerves. The smell of food stimulates the hunger center in the hypothalamus. In animals such as dogs and cats olfactory detection of **pheromones** (fĕr'o-mōnz; molecules released into the air by one animal that attract another animal of the same species, usually of the opposite sex) are important in reproduction.

Lesions in the limbic system can result in voracious appetite, increased (often perverse) sexual activity, and docility (including loss of normal fear and anger responses). Since the hippocampus is part of the temporal lobe, damage to that portion of the limbic system can also result in a loss of memory. The hippocampus and the adjacent cortex are very important in the transition of information from short- to long-term memory, and the cells undergoing calcium-induced shape changes associated with long-term memory are localized in that region of the brain.

Cerebellum

The term **cerebellum** (ser'ĕ-bel'um; Figure 13-15) means little brain. It communicates with other regions of the CNS through three large nerve tracts, the

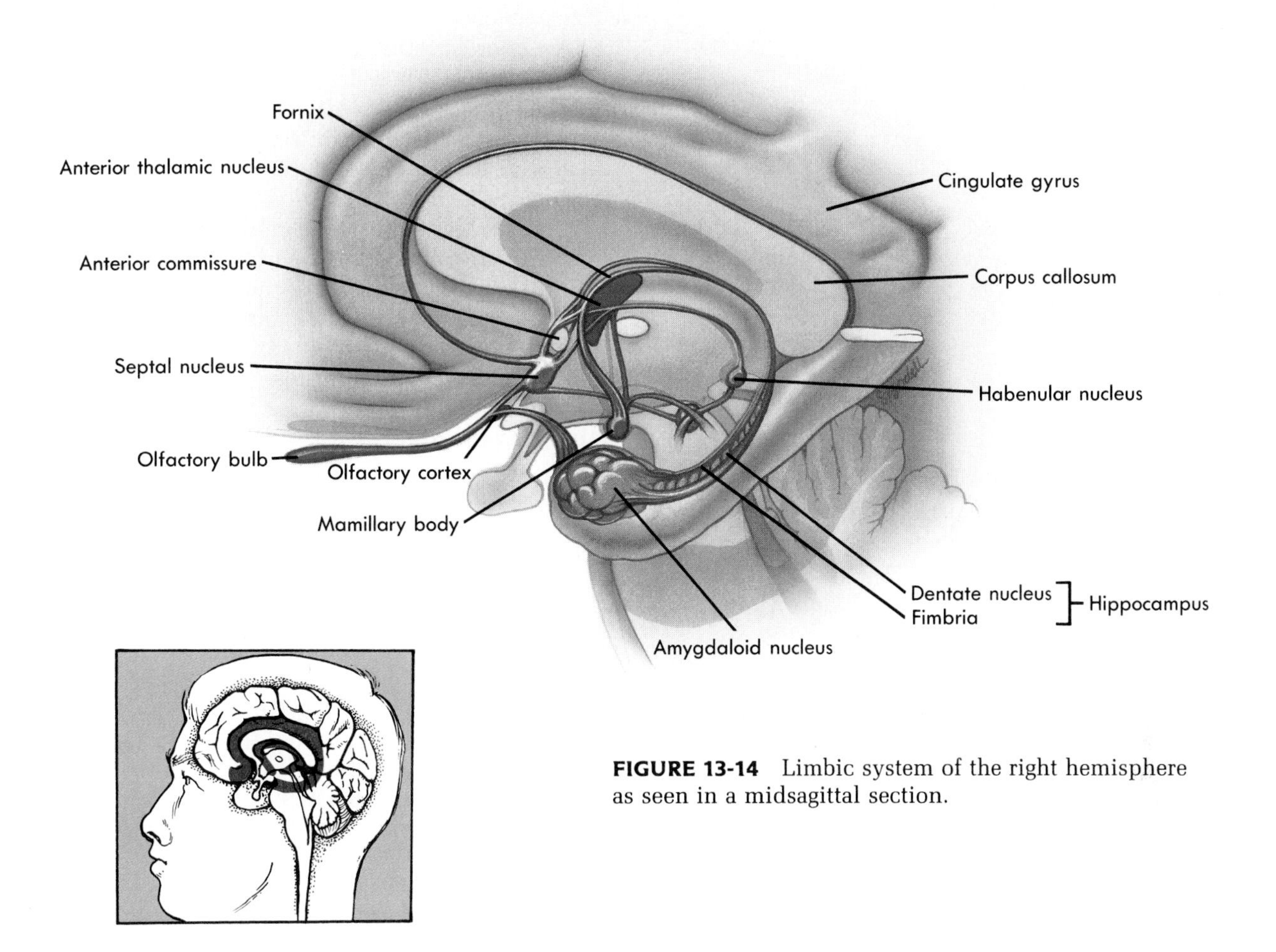

FIGURE 13-14 Limbic system of the right hemisphere as seen in a midsagittal section.

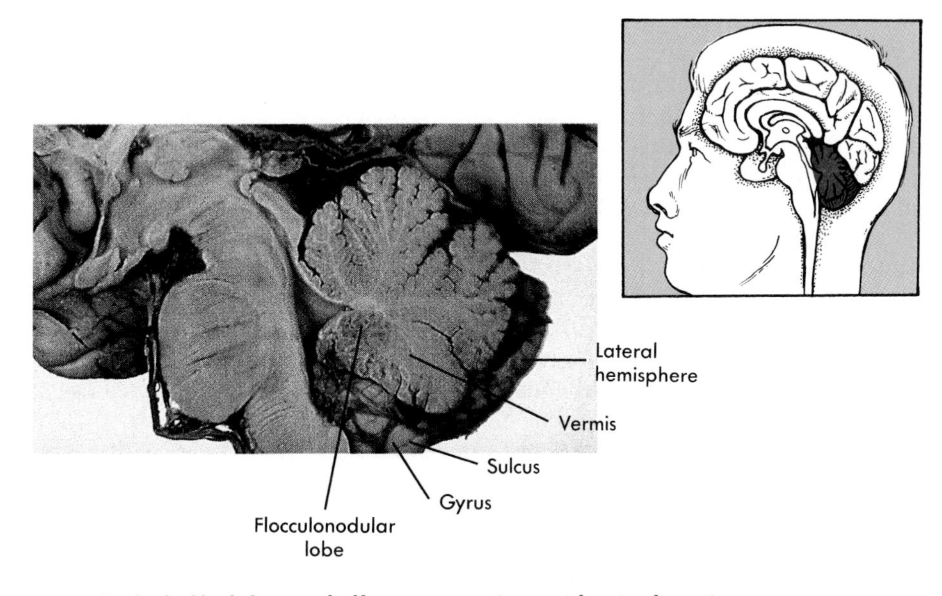

FIGURE 13-15 Right half of the cerebellum as seen in a midsagittal section.

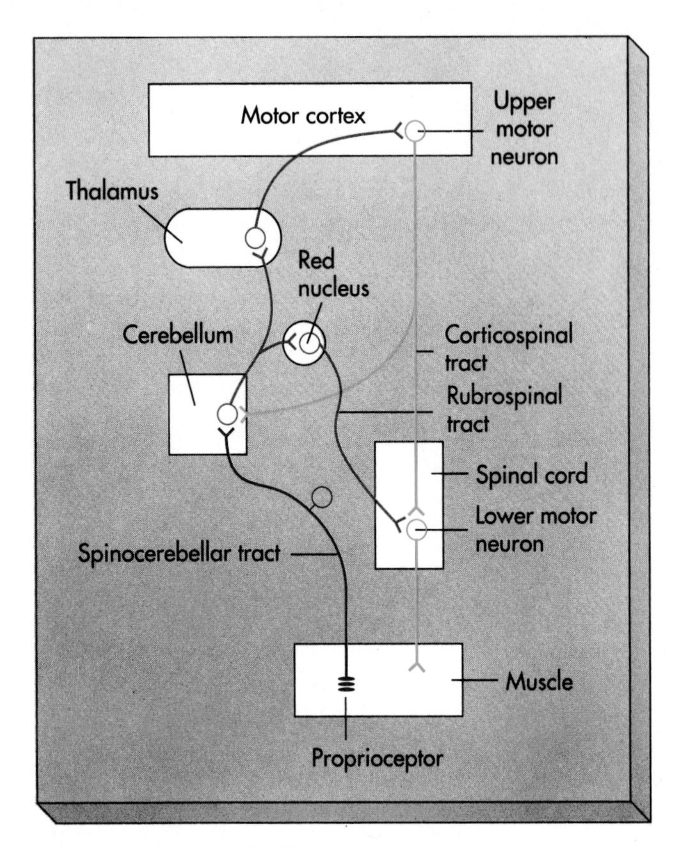

FIGURE 13-16 Cerebellar comparator function. Impulses from the motor cortex stimulate the spinal cord lower motor neurons, which supply innervation to skeletal muscles. At the same time impulses are sent from the motor cortex to the cerebellum to tell the cerebellum what movement is intended. The cerebellum also receives proprioceptive information from the muscle, telling it what movement actually is occurring. The cerebellum compares the intended movement with the actual movement and sends impulses to the motor cortex and lower motor neurons that adjust the movement toward what was intended.

superior, middle, and inferior **cerebellar peduncles** (pe-dun'klz).

The cerebellum is organized like the cerebrum, with gray matter both on the inside as nuclei and on the outside as cortex. The cerebellar cortex has gyri and sulci, but the gyri are much smaller than those of the cerebrum.

The cerebellum consists of three portions: a small anterior portion, the **flocculonodular** (flok'u-lo-nod'u-lar: floccular—a tuft of wool) **lobe;** a narrow central **vermis** (ver'mis; worm-shaped); and two large **lateral hemispheres** (see Figure 13-15). The flocculonodular lobe is the simplest portion of the cerebellum and is involved in balance. The anterior portion of the vermis is involved in gross motor coordination, and the posterior vermis and lateral hemispheres are involved in fine motor coordination, producing smooth, flowing movements.

A major function of the cerebellum is that of a **comparator.** Impulses from the motor cortex descend into the spinal cord to initiate voluntary movements, and at the same time impulses are sent from the motor cortex to the cerebellum, giving the cerebellar neurons information representing the intended movement (Figure 13-16). Simultaneously, impulses from the proprioceptive neurons (providing information about the position of the body or body parts) that innervate the joints and tendons of the structure being moved (e.g., the elbow or knee) reach the cerebellar cortex. These impulses give the cerebellar neurons information from the periphery about the actual movements. The cerebellum compares the impulses from the motor cortex with those from the moving structures (i.e., it compares the intended movement with the actual movement), and if a difference is detected, the cerebellum sends impulses to the motor cortex and the spinal cord to correct the discrepancy. The result is smooth and coordinated movements.

With training a person can develop highly skilled and rapid movements that are accomplished more rapidly than can be accounted for by the comparator function of the cerebellum. In these cases the cerebellum can "learn" highly specialized motor functions through specific, repeated comparator activities.

SPINAL CORD

The spinal cord is extremely important to the overall function of the nervous system. It is the communication link between the brain and the peripheral nervous system (PNS) inferior to the head, integrating incoming information and producing responses through reflex mechanisms.

General Structure

The **spinal cord** (Figure 13-17) extends from the foramen magnum to the level of the second lumbar vertebra. It is considerably shorter than the vertebral column because it does not grow as rapidly as the vertebral column during embryonic development. It is composed of cervical, thoracic, lumbar, and sacral segments, which are named according to the area of the vertebral column from which their nerves enter and exit. Because the spinal cord is shorter than the vertebral column, the nerves do not always exit the vertebral column at the same level that they exit the spinal cord. Thirty-one pairs of spinal nerves exit the spinal cord and pass out of the vertebral column through the intervertebral foramina (see Chapter 14).

The spinal cord is not uniform in diameter throughout its length. There is a general decrease in diameter superiorly to inferiorly, and there are two enlargements where nerves supplying the extremities enter and leave the cord (see Figure 13-17). The **cervical enlargement** in the inferior cervical region corresponds to the location at which nerves that supply the upper limbs enter or exit the cord, and the **lumbosacral enlargement** in the inferior thoracic and superior lumbar regions is the site at which the nerves that supply the lower limbs enter or exit.

4
Why is the cord enlarged in the cervical and lumbar areas?

? ? ? ? ? ? ? ? ? ? ?

Immediately inferior to the lumbar enlargement the spinal cord tapers to form a conelike region called the **conus medullaris.** Its tip is at the level of the second lumbar vertebra and is the inferior end of the spinal cord. A connective tissue filament, the **filum terminale** (fi'lum ter'mĭ-nal'e), extends inferiorly from the apex of the conus medullaris to the coccyx and functions to anchor the cord to the coccyx. The nerves supplying the legs and other inferior structures of the body (L2 to S5) exit the lumbar enlargement and conus medullaris, course inferiorly through the vertebral canal, and exit through the intervertebral foramina from L2 to S5. The conus medullaris and the numerous nerves extending inferiorly from it resemble a horse's tail and are therefore called the **cauda equina** (kaw'dah e-kwi'nah; see Figure 13-17).

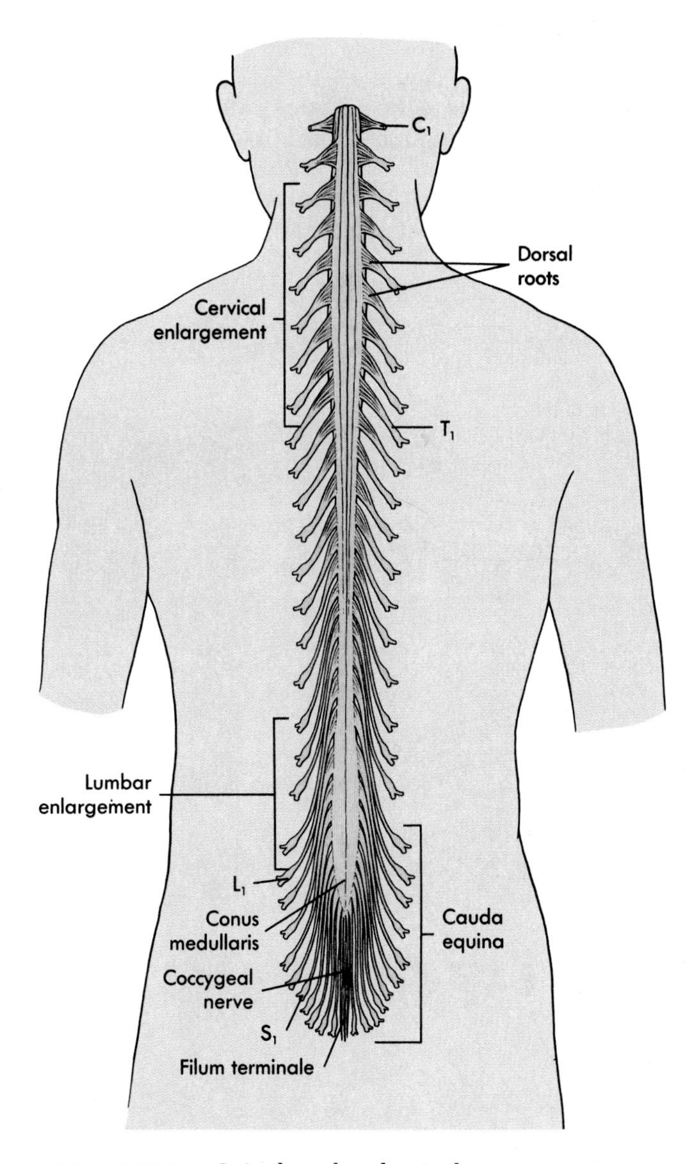

FIGURE 13-17 Spinal cord and spinal nerve roots.

Cross Section

A cross section of the spinal cord reveals that the cord consists of a central gray portion and a peripheral white portion (Figure 13-18). The white matter consists of nerve tracts, and the gray matter consists of nerve cell bodies and dendrites. An anterior median fissure and a posterior median sulcus are deep clefts partially separating the two halves of the cord. The white matter in each half of the spinal cord is organized into three columns, or **funiculi** (fu-nik'u-le), called the anterior (ventral), posterior (dorsal), and lateral funiculi. Each funiculus is subdivided into **fasciculi** (fă-sik'u-le), or tracts. Individual axons carrying impulses to (ascending) or from (descending) the brain are usually grouped together within the fasciculi, and axons within a given fasciculus carry basically the same type of information, although fasciculi may overlap to some extent.

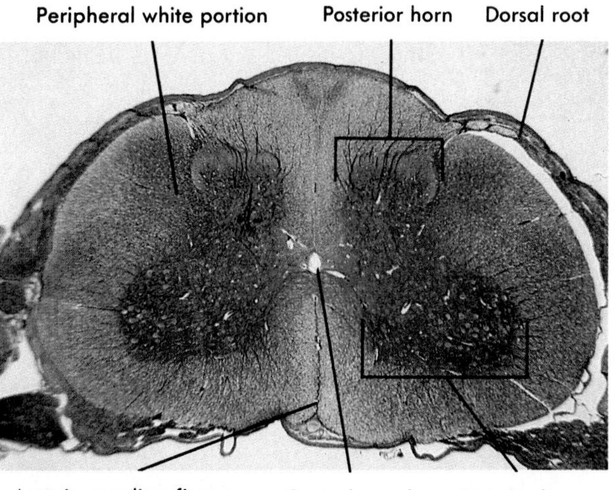

FIGURE 13-18 Cross section of the cat spinal cord. The lighter color represents tracts surrounding the darker central portion where cell bodies are located.

The central gray matter is organized into horns. Each half of the central gray matter of the spinal cord consists of a relatively thin **posterior** (dorsal) **horn** and a larger **anterior** (ventral) **horn.** Small **lateral horns** also exist in levels of the cord associated with the autonomic nervous system. The axons of sensory neurons synapse with cell bodies of neurons in the posterior horn; the cell bodies of motor neurons are in the anterior horn (also called the motor horn); and the cell bodies of autonomic neurons are in the lateral horns. The two halves of the spinal cord are connected by **gray** and **white commissures.** The central canal is in the center of the gray commissure.

Dorsal (posterior) and ventral (anterior) roots exit the spinal cord near the dorsal and ventral horns. The **dorsal root** conveys afferent nerve processes to the cord, and the **ventral root** conveys efferent processes away from the cord. The dorsal roots possess **dorsal root ganglia** (gang'gle-ah; a swelling or knot; also called spinal ganglia), which contain the cell bodies of sensory neurons. The axons of these neurons form the dorsal root and project into the posterior horn where they synapse with other neurons or ascend or descend in the spinal cord. The ventral root is formed by the axons of neurons in the anterior and lateral horns. The dorsal and ventral roots unite to form the spinal nerves.

5

Explain why the dorsal root ganglia are larger in diameter than the spinal nerves.

? ? ? ? ? ? ? ? ? ? ?

SPINAL REFLEXES

Automatic reactions to stimuli that occur without conscious thought are called **reflexes** (see Chapter 12). Because they occur without conscious thought, reflexes are considered involuntary, even though they often involve skeletal muscles.

Reflexes are integrated both in the brain and in the spinal cord. Many of the reflexes occurring in the brain are autonomic, or visceral, reflexes (see Chapter 16) and include such reactions as constriction of the pupil in response to increased light or an increased heart rate in response to reduced blood pressure. Major spinal cord reflexes include the stretch reflex, the Golgi tendon reflex, and the withdrawal reflex.

Stretch Reflex

The structurally simplest reflex is the **stretch,** or **myotonic, reflex** (Figure 13-19, *A*). The sensory receptor of the reflex is the **muscle spindle,** which consists of three to 10 small specialized muscle cells. The cells are contractile only at their ends, and the noncontractile centers of the muscle spindle cells are innervated by afferent neurons that carry impulses

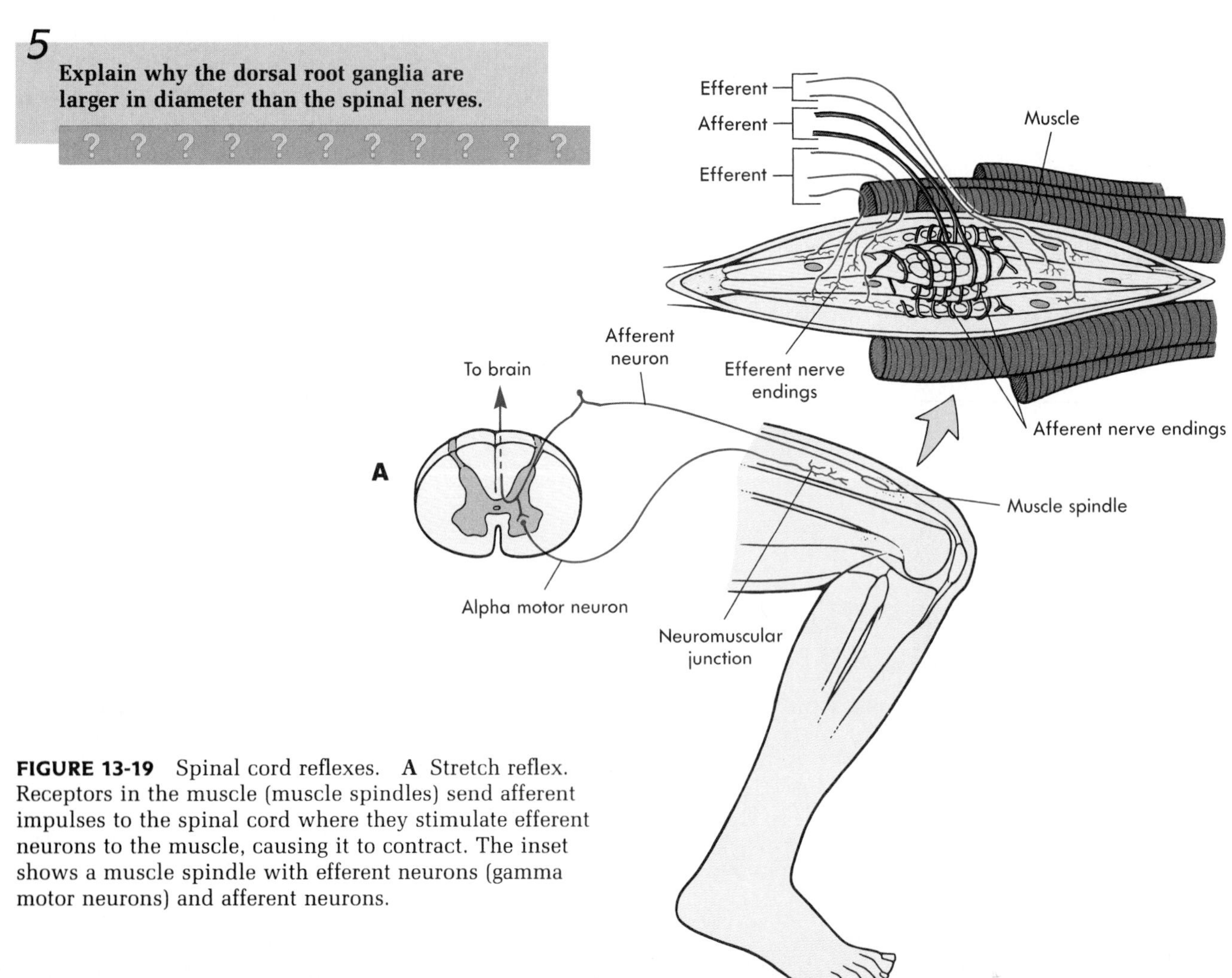

FIGURE 13-19 Spinal cord reflexes. **A** Stretch reflex. Receptors in the muscle (muscle spindles) send afferent impulses to the spinal cord where they stimulate efferent neurons to the muscle, causing it to contract. The inset shows a muscle spindle with efferent neurons (gamma motor neurons) and afferent neurons.

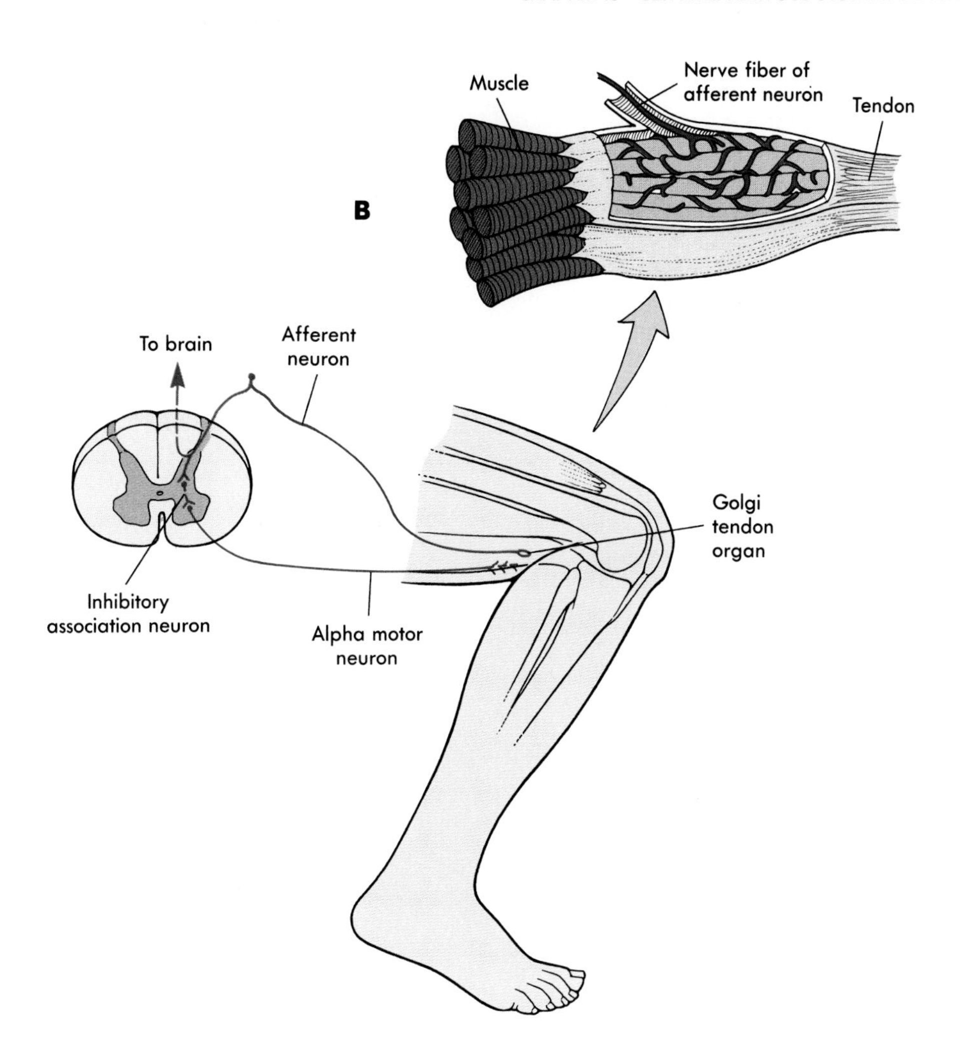

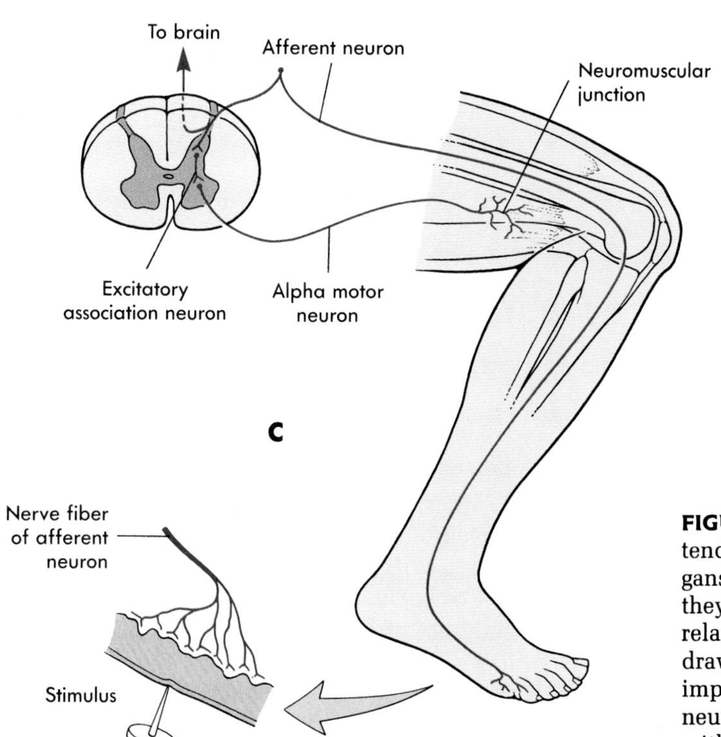

FIGURE 13-19, cont'd Spinal cord reflexes. **B** Golgi tendon reflex. Receptors in the tendon (Golgi tendon organs) send afferent impulses to the spinal cord where they inhibit efferent neurons to the muscle, causing it to relax and reduce the tension on the tendon. **C** Withdrawal reflex. Pain receptors in the skin send afferent impulses to the spinal cord where they stimulate efferent neurons to flexor muscles, causing them to contract and withdraw the limb from the painful stimulus.

Continued.

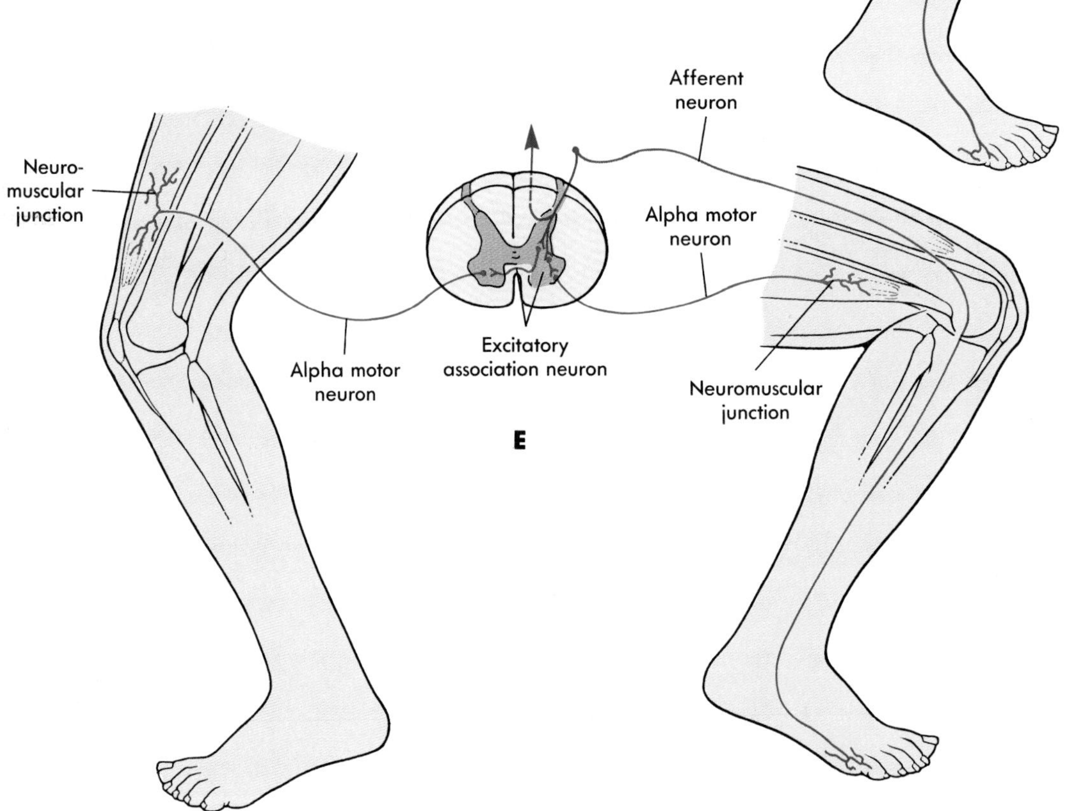

FIGURE 13-19, cont'd Spinal cord reflexes. **D** Reciprocal innervation. During the withdrawal reflex afferent branches to extensor muscles cause inhibition in those muscles, resulting in the muscles' relaxation and enhancing the withdrawal reflex. **E** Crossed extensor reflex. During the withdrawal reflex association neurons send collateral branches to the opposite side of the spinal cord and stimulate extensor muscles. This reflex is important in withdrawal of the lower limb in that extension of the opposite lower limb keeps the person from falling.

into the spinal cord where they synapse directly with motor neurons called **alpha motor neurons,** which in turn innervate the muscle in which the muscle spindle is embedded. The stretch reflex is unique in that it can occur without the involvement of association neurons between the afferent and efferent neurons.

Stretching a muscle also stretches the muscle spindle located among the muscle fibers and stimulates the afferent neurons that innervate the center of the muscle spindle. The increased frequency of nerve impulses in the afferent neurons stimulates the alpha motor neurons to cause a rapid contraction of the stretched muscle, thus resisting the stretch of the muscle. The postural muscles demonstrate this reflex's adaptive nature. If a person is standing upright and then bends slightly to one side, the postural muscles on the other side (e.g., those muscles associated with the vertebral column) are stretched. As a result, stretch reflexes will be initiated in those muscles to reestablish the normal posture.

Collateral axons from the afferent neurons of the muscle spindles also synapse with ascending nerve tracts, which enable the brain to perceive that a muscle has been stretched. Descending neurons within the spinal cord synapse with the neurons of the stretch reflex and modulate their activity. This activity is important in maintaining posture and in coordinating muscular activity.

Gamma motor neurons are responsible for regulating the sensitivity of the muscle spindle. Impulses carried by the gamma motor neurons stimulate the ends of the muscle spindle cells and cause them to contract, stimulating activity in the afferent neurons from the center of the muscle spindle cells and making the muscle spindle more sensitive to further stretch. Impulses from the brain that stimulate alpha motor neurons to the skeletal muscle and result in contraction of the muscle also stimulate gamma motor neurons to the muscle spindle, enhancing the activity of the muscle spindle, which in turn helps control and coordinate muscular activity (e.g., posture, muscle tension, and muscle length).

The knee-jerk reflex, or patellar reflex, is a classic example of the stretch reflex and is used by clinicians to determine if the higher CNS centers that normally influence this reflex are functional. When the patellar ligament is tapped, the quadriceps femoris muscle tendon and the muscles themselves are stretched. The muscle spindle fibers within these muscles are also stretched, and the stretch reflex is activated; contraction of the muscles extends the leg, producing the characteristic knee-jerk response. When the stretch reflex is greatly exaggerated, it indicates that the neurons within the brain that normally innervate the gamma motor neurons and enhance the stretch reflex are overly active. On the other hand, if the facilitatory impulses to the gamma motor neurons are depressed, the stretch reflex is greatly suppressed or absent.

Golgi Tendon Reflex

The Golgi tendon reflex prevents the production of excessive tension in a muscle. **Golgi tendon organs** are encapsulated nerve endings that have at their ends numerous terminal branches with small swellings that are associated with individual tendon fascicles. The organs lie within tendons near the muscle-tendon junction and are stimulated as the tendon is stretched during muscle contraction (Figure 13-19, *B*). As a muscle contracts, the attached tendons are stretched, resulting in increased tension in the tendon. The increased tension stimulates action potentials in the afferent fibers from the Golgi tendon organs.

The afferent neurons of the Golgi tendon organs pass through the dorsal root to the spinal cord and enter the posterior gray matter where they branch and synapse with inhibitory association neurons (see Chapter 12). The association neurons synapse with alpha motor neurons that innervate the muscle to which the Golgi tendon organ is attached. When a great amount of tension is applied to the tendon, this reflex inhibits the motor neurons of the associated muscle and causes it to relax, thus protecting muscles and tendons from damage caused by excessive tension. The sudden relaxation of the muscle reduces the tension applied to the muscle and tendons. A weight lifter who suddenly drops a heavy weight after straining to lift it does so, in part, because of the effect of the Golgi tendon reflex.

Tremendous amounts of tension can be applied to muscles and tendons in the legs. Frequently an athlete's Golgi tendon reflex is not adequate to protect muscles and tendons from excessive tension. The large muscles and sudden movements of football players and sprinters are correlated with relatively frequent hamstring pulls and Achilles (calcaneal) tendon injuries.

Withdrawal Reflex

The function of the **withdrawal reflex** (also called the flexor reflex) is to remove a limb or other body part from a painful stimulus. The sensory receptors are pain receptors (see Chapter 15). Impulses from painful stimuli are conducted by afferent neurons through the dorsal root to the spinal cord where they synapse with excitatory association neurons, which in turn synapse with alpha motor neurons (Figure 13-19, *C*). The alpha motor neurons stimulate muscles, usually flexor muscles, that remove the limb from the source of the painful stimulus. Collateral branches of the afferent neurons synapse with ascending fibers to the brain, providing conscious awareness of the painful stimuli.

Reciprocal innervation

Reciprocal innervation is associated with the withdrawal reflex and reinforces the efficiency of the reflex (Figure 13-19, *D*). Collateral axons of afferent neurons that carry impulses from pain receptors innervate inhibitory association neurons, which inhibit alpha motor neurons of the extensor (antagonist) muscles. When the withdrawal reflex is initiated, flexor muscles usually contract, and reciprocal innervation causes relaxation of extensor muscles, which reduces the resistance that extensor muscles otherwise would generate.

Reciprocal innervation is also involved in the stretch reflex, in which opposing muscles are inhibited. In the patellar reflex, for example, the leg flexors are inhibited when the leg extensors are stimulated.

Crossed extensor reflex

The **crossed extensor reflex** is another reflex associated with the withdrawal reflex (Figure 13-19, *E*). Association neurons that stimulate alpha motor neurons, resulting in withdrawal of a limb, send collateral axons through the white commissure to the opposite side of the spinal cord to stimulate alpha motor neurons that innervate extensor muscles in the opposite side of the body. If a withdrawal reflex is initiated in one leg, the crossed extensor reflex causes extension of the opposite leg.

6

Reciprocal innervation in the extended leg causes which muscles to relax, thus increasing the effectiveness of the crossed extensor reflex?

? ? ? ? ? ? ? ? ? ? ?

The crossed extensor reflex is adaptive in that the response prevents one from falling and is exemplified by a person's reaction to stepping on a sharp object. The withdrawal reflex occurs along

Pain

Pain is a sensation characterized by a group of unpleasant perceptual and emotional experiences that trigger autonomic, psychological, and somatomotor responses. Pain sensation consists of two portions: (1) rapidly conducted impulses resulting in sharp, well-localized, pricking, or cutting pain, followed by (2) more slowly propagated impulses resulting in diffuse, burning or aching pain. Research indicates that pain receptors have very uniform sensitivity that does not change dramatically from one point in time to another. Variations in pain sensation result from the differences in integration of impulses from the pain receptors and the mechanisms by which pain receptors are stimulated.

Although the medial lemniscal system contains no pain fibers, tactile and mechanoreceptors often are activated by the same stimuli that affect pain receptors. Impulses from the tactile receptors help localize the source of pain and monitor changes in the stimuli. Superficial pain is highly localized because of the simultaneous stimulation of pain receptors and mechanoreceptors in the skin. Deep or visceral pain is not highly localized because of the absence of numerous mechanoreceptors in the deeper structures, and it normally is perceived as a diffuse pain.

Medial lemniscal system neurons are part of the **gate-control theory** of pain control. Primary neurons of the medial lemniscal system send out collateral branches that synapse with association neurons in the posterior horn of the spinal cord. The association neurons have an inhibitory effect on the secondary neurons of the lateral spinothalamic tract. Thus pain impulses traveling through the lateral spinothalamic tract can be suppressed by impulses that originate in neurons of the medial lemniscal system. These neurons may act as a "gate" for pain impulses transmitted in the lateral spinothalamic tract. Increased activity in the medial lemniscal system tends to close the gate, reducing pain impulses transmitted in the lateral spinothalamic tract.

The gate-control theory may explain the physiological basis for the following methods that have been used to reduce the intensity of chronic pain: electrical stimulation of the medial lemniscal neurons, transcutaneous electrical stimulation (applying a weak electrical stimulation to the skin), acupuncture, massage, and exercise. The frequency of impulses that are transmitted in the medial lemniscal system is increased when the skin is rubbed vigorously and when the limbs are moved and may explain why vigorously rubbing a large area around a source of pricking pain tends to reduce the intensity of the painful sensation. Exercise normally decreases the sensation of pain, and exercise programs are important components in the clinical management of chronic pain not associated with illness. Impulses initiated by acupuncture procedures may also inhibit the impulses in neurons that transmit pain impulses upward in the spinal cord by influencing afferent cells of the posterior horn.

Referred Pain

Referred pain is a painful sensation in a region of the body that is not the source of the pain stimulus. Most commonly, referred pain is sensed in the skin or other superficial structures when internal organs are damaged or inflamed. This sensation usually occurs because both the area to which the pain is referred and the area in which the actual damage occurs are innervated by neurons from the same spinal segment.

Many cutaneous afferent neurons and visceral afferent neurons that transmit pain impulses converge on the same ascending neurons; however, the brain cannot distinguish between the two sources of painful stimuli, and the painful sensation is referred to the most superficial structures innervated by the converging neurons. This referral may be a result of the fact that the number of receptors is much greater in the periphery than in deep structures and the brain is more "accustomed" to dealing with superficial stimuli.

Referred pain is useful in diagnosing the actual cause of the painful stimulus. Heart attack victims often feel cutaneous pain radiating from the left shoulder down the arm. Figure 13-A shows other examples of referred pain.

with the painful stimulus in the affected leg, and the crossed extensor reflex occurs in the opposite leg. Initiating withdrawal reflex in both legs at the same time would cause one to fall.

SPINAL PATHWAYS

The names of most ascending and descending pathways in the CNS reflect their general function (Tables 13-4 and 13-5 and Figure 13-20). Each pathway usually is given a composite name in which the first half of the word indicates its origin and the second half indicates its termination. Ascending pathways therefore usually begin with *spino-*, indicating that they originate in the spinal cord. For example, a spinothalamic tract is one that originates in the spinal cord and terminates in the thalamus. Descending pathways usually begin with the term *cortico-* indicating that they begin in the cerebral cortex. The corticospinal tract is a descending tract that originates in the cerebral cortex and terminates in the spinal cord. The specific function of each ascending or descending tract, however, is not suggested by its name.

Phantom Pain

Phantom pain occurs in people who have had appendages amputated. Frequently these people perceive pain—sometimes intense pain—in the amputated structure as if it were still in place. If a neuron pathway that transmits action potentials is stimulated at any point along that pathway, action potentials may be initiated and propagated toward the CNS. Integration results in the perception of pain that is projected to the site of the sensory receptors, even if those sensory receptors are no longer present. A similar phenomenon can be demonstrated by bumping the ulnar nerve as it crosses the elbow (the funny bone). A sensation of pain is often felt in the fourth and fifth digits, even though the neurons were stimulated at the elbow.

A factor that may be very important in phantom pain results from the lack of touch, pressure, and proprioceptive impulses from the amputated limb. Those impulses suppress the transmission of pain impulses in the pain pathways. When a limb is amputated, the inhibitory effect of sensory information (normally transmitted through the medial lemniscal system) on the ascending pain pathways is removed. As a consequence, the intensity of phantom pain may be increased.

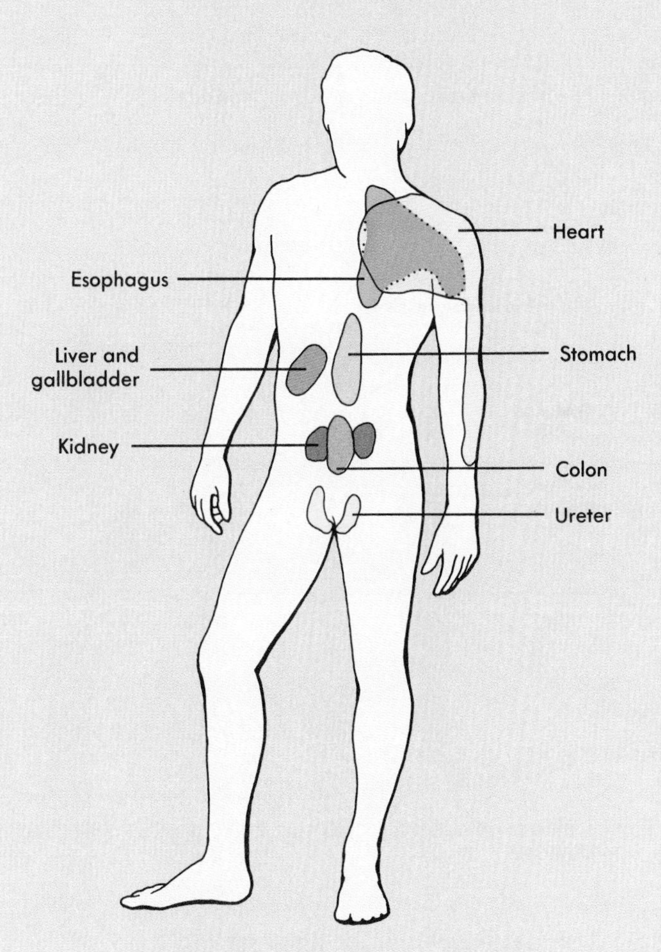

FIGURE 13-A Regions of referred pain.

Table 13-4

Ascending spinal pathways

PATHWAY	MODALITY (information transmitted)	ORIGIN	TERMINATION
Spinothalamic		Cutaneous receptors	Cerebral cortex
Lateral	Pain and temperature		
Anterior	Light touch, pressure, tickle, and itch sensation		
Medial lemniscal system	Proprioception, two-point discrimination, pressure, and vibration	Cutaneous receptors, joints	Cerebral cortex and cerebellum
Spinocerebellar	Proprioception to cerebellum	Joints, tendons	Cerebellum
Posterior			
Anterior			
Spino-olivary	Proprioception relating to balance	Joints, tendons	Accessory olivary nucleus, then to cerebellum
Spinotectal	Tactile stimulation causing visual reflexes	Cutaneous receptors	Superior colliculus
Spinoreticular	Tactile stimulation arousing consciousness	Cutaneous receptors	Reticular formation

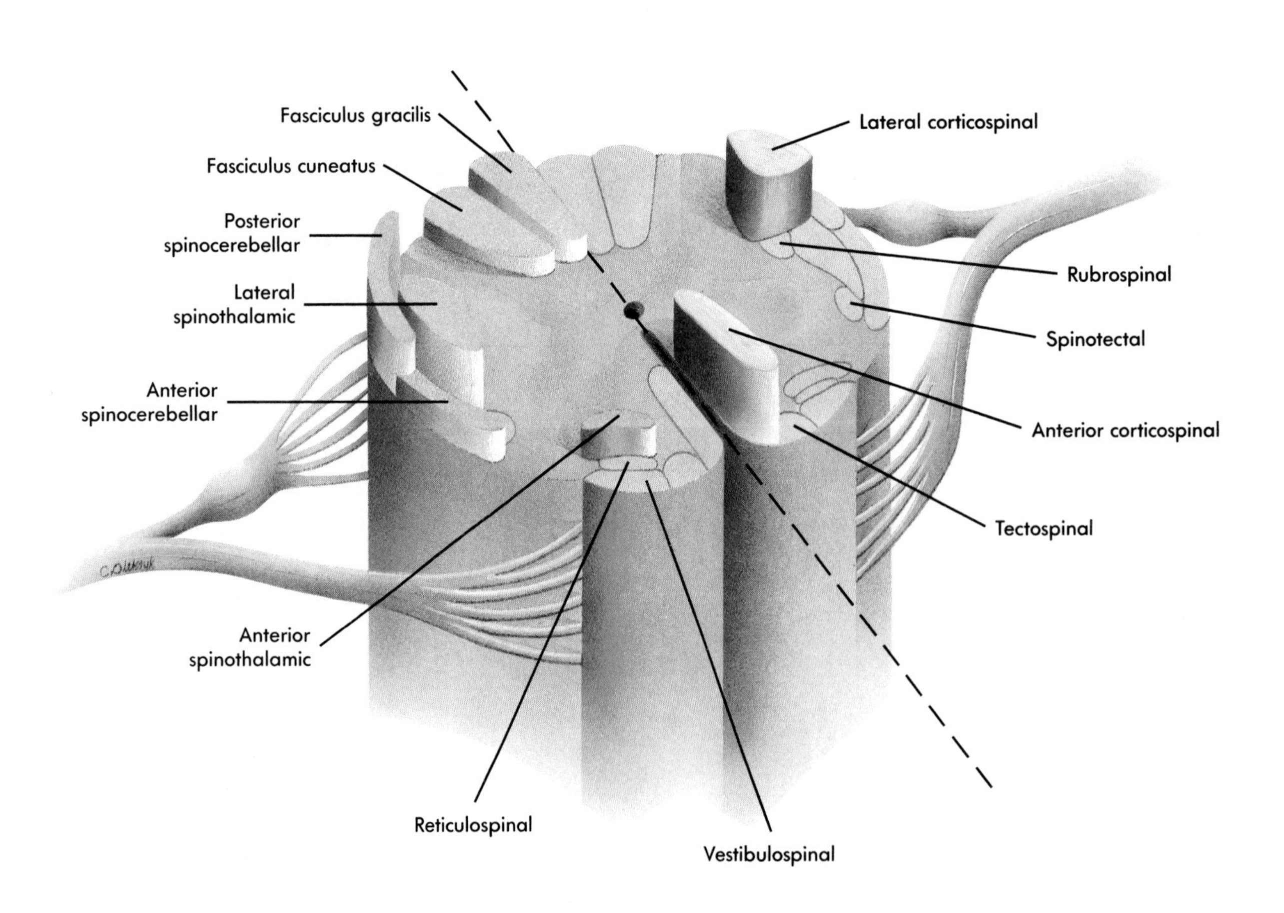

FIGURE 13-20 Cross section of the spinal cord depicting the pathways.

PRIMARY CELL BODY	SECONDARY CELL BODY	TERTIARY CELL BODY	CROSSOVER
Dorsal root ganglion	Posterior horn of spinal cord	Thalamus	
			Level at which primary neuron enters cord
			Eight to 10 segments from where primary neuron entered cord; many collaterals
Dorsal root ganglion	Medulla oblongata	Thalamus	Medulla oblongata
Dorsal root ganglion	Posterior horn of spinal cord	Cerebellum	
			Uncrossed
			Some uncrossed; some cross at point of origin and recross in cerebellum
Dorsal root ganglion	Posterior horn of spinal cord	Accessory olivary nucleus	At point of origin; recross to reach cerebellum
Dorsal root ganglion	Posterior horn of spinal cord	Superior colliculus	At point of origin
Dorsal root ganglion	Posterior horn of spinal cord	Reticular formation	Some uncrossed; some cross spinal cord at point of entry

Table 13-5

Descending spinal pathways

PATHWAY	FUNCTIONS CONTROLLED	ORIGIN	TERMINATION	CROSSOVER
Pyramidal	Muscle tone and conscious skilled movements, especially of the hands			
Corticospinal	Movements, especially of the hands	Cerebral cortex (upper motor neuron)	Anterior horn of spinal cord (lower motor neuron)	
Lateral				Inferior end of medulla oblongata
Anterior				At level of lower motor neuron
Corticobulbar	Facial and head movements	Cerebral cortex (upper motor neuron)	Cranial nerve nuclei in brainstem (lower motor neuron)	Varies for the various cranial nerves
Extrapyramidal	Unconscious movements			
Rubrospinal	Movement coordination	Red nucleus	Anterior horn of spinal cord	Midbrain
Vestibulospinal	Posture, balance	Vestibular nucleus	Anterior horn of spinal cord	Uncrossed
Reticulospinal	Control of vital functions such as respiration and heartbeat	Reticular formation	Anterior horn of spinal cord	Some uncrossed; some cross at level of termination
Tectospinal	Movement of head and neck in response to visual reflexes	Superior colliculus	Cranial nerve nucleus in medulla oblongata and anterior horn of upper levels of spinal cord (lower motor neurons that turn head and neck)	Midbrain

Ascending Pathways

The major ascending tracts involved in the conscious perception of external stimuli are the spinothalamic and medial lemniscal systems (see Table 13-4). The ones carrying sensations of which we are not consciously aware are the spinocerebellar, spino-olivary, spinotectal, and spinoreticular tracts.

Spinothalamic system

The spinothalamic system is the least discriminative of the two systems conveying cutaneous sensory information to the brain. Pain and temperature information are carried primarily by the **lateral spinothalamic tracts** (Figure 13-21, *A*). Light touch, pressure, tickle, and itch sensations are carried by the **anterior spinothalamic tracts** (Figure 13-21, *B*). Light touch is also called crude touch (poorly localized); although the receptors of these nerves respond to very light touch, the stimulus is not well localized.

Three neurons in sequence—the primary, secondary, and tertiary neurons—are involved in the pathway from the peripheral receptor to the cerebral cortex. The **primary neuron** cell bodies of the spinothalamic system are in the dorsal root ganglia. The primary neurons pick up sensory input from the periphery and relay it to the posterior horn of the spinal

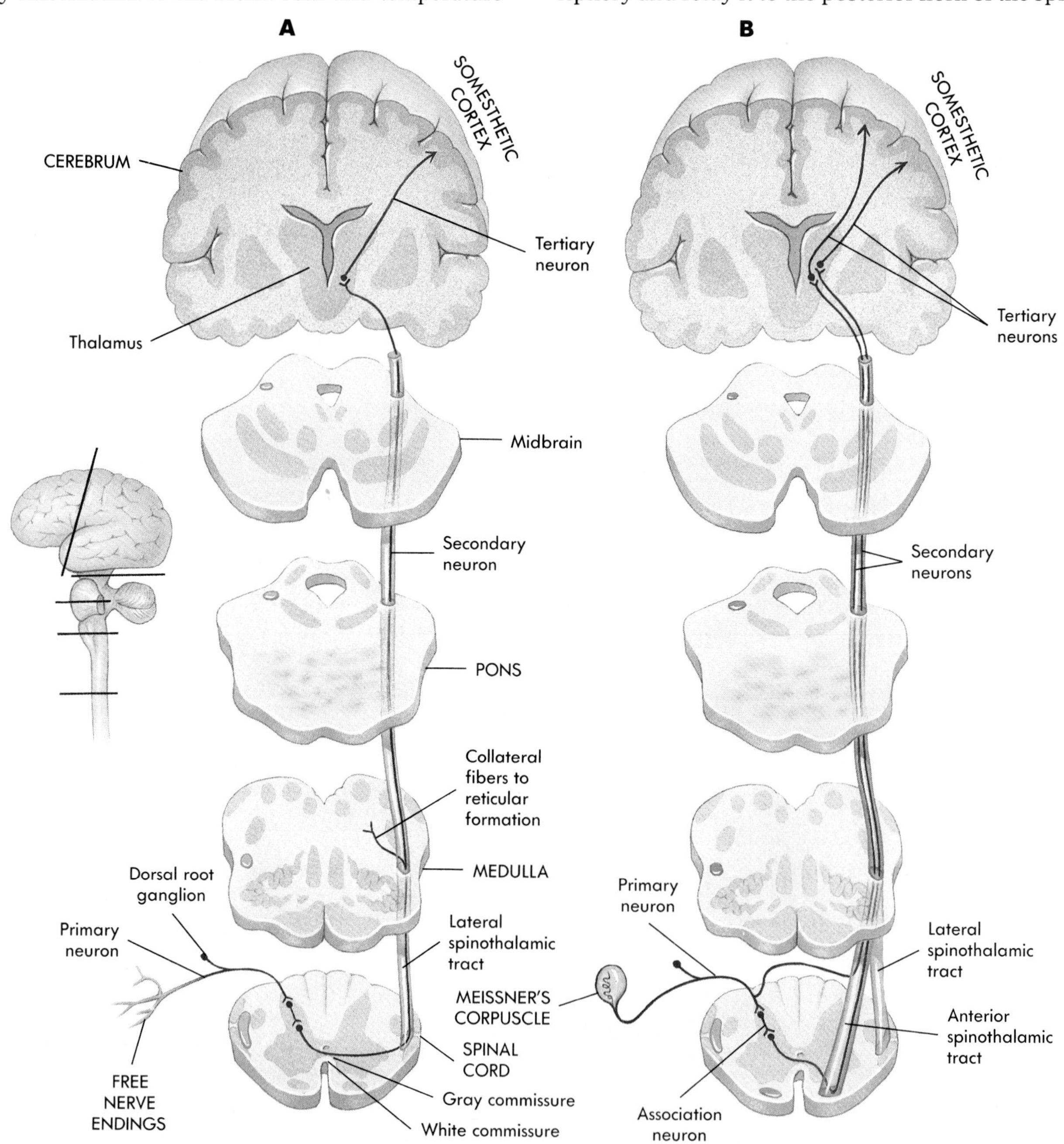

FIGURE 13-21 Spinothalamic system. **A** Lateral spinothalamic tract, which transmits impulses for pain and temperature. **B** Anterior spinothalamic tract, which transmits impulses for light touch.

cord where they synapse with association neurons. The association neurons, which are not specifically named in the three-neuron sequence, synapse with secondary neurons. Axons from the **secondary neurons** cross to the opposite side of the spinal cord through the anterior portion of the gray and white commissures and enter the spinothalamic tract where they ascend to the thalamus. As these fibers pass through the brainstem, they are joined by fibers of the **trigeminothalamic tract** (trigeminal nerve, or cranial nerve V; see Chapter 14), which carries pain and temperature impulses from the face and teeth. Collateral branches, especially from this tract, project to the reticular formation where they stimulate wakefulness and consciousness. The secondary neurons synapse with cell bodies of tertiary neurons in the thalamus. **Tertiary neurons** from the thalamus project to the somesthetic cortex.

Primary neurons contributing to the lateral spinothalamic tract (pain and temperature) ascend or descend only one or two segments before synapsing with secondary neurons, whereas those entering the anterior spinothalamic tract (light touch and pressure) may ascend or descend for eight to 10 segments before synapsing. Throughout this distance the primary neurons of the anterior spinothalamic system send out collateral branches that synapse with secondary neurons at several intermediate levels. Thus collateral branches from a number of sensory neurons, each conducting information from a different patch of skin, may converge on a single secondary neuron in the spinal cord.

7

Explain why light touch is very sensitive but not highly discriminative in relation to the exact point of stimulation.

? ? ? ? ? ? ? ? ? ? ?

Lesions on one side of the spinal cord that sever the lateral spinothalamic tract eliminate pain and temperature sensation below that level on the opposite side of the body. Lesions on one side of the spinal cord that sever the anterior spinothalamic tract, however, do not eliminate all of the light-touch and pressure sensations below the level of the lesion because of the large number of collateral branches crossing the cord at various levels.

Medial lemniscal system

The sensations of two-point discrimination (fine touch), proprioception (pro'pre-o-sep'shun; perception of position), pressure, and vibration are carried by the **medial lemniscal** (lem-nis'kal; ribbon; the fibers form a thin, ribbonlike pathway through the brainstem) **system.** This system is also called the **posterior column system** (located in the posterior column, which is also called the dorsal column or posterior funiculus of the spinal cord; Figure 13-22).

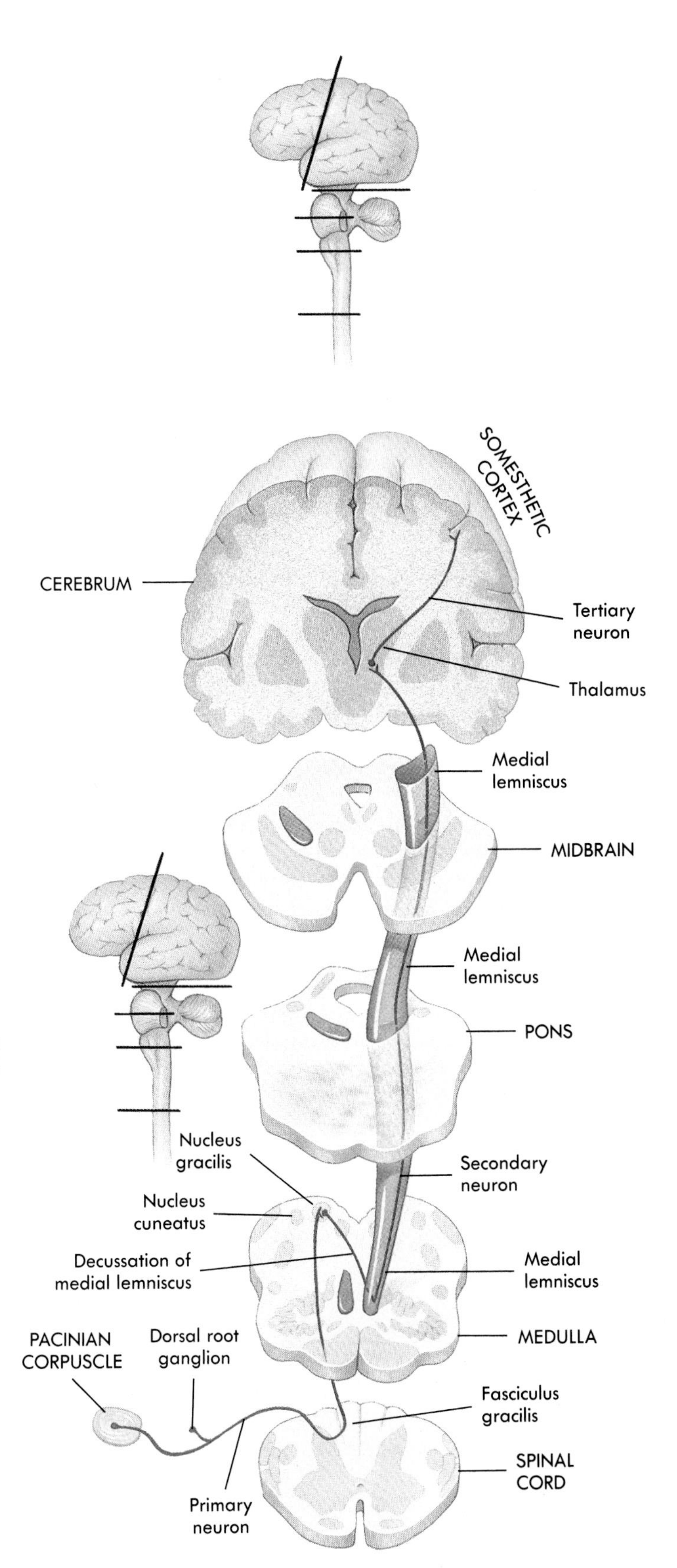

FIGURE 13-22 Medial lemniscal system. The fasciculus gracilis and fasciculus cuneatus convey proprioception and two-point discrimination. Only the fasciculus gracilis pathway is shown.

Two-point discrimination is the ability to detect simultaneous stimulation at two points on the skin. The distance between two points that a person can detect as separate points of stimulation differs for various regions of the body (Figure 13-23). This sensation is important in evaluating the texture of objects.

Proprioception provides information about the precise position and the rate of movement of various body parts, the weight of an object being held in the hand, and the range of movement of a joint. This information is involved in activities such as shooting a basketball, driving a car, eating, or writing.

The cell bodies of the afferent spinal neurons of the medial lemniscal system are the largest in the dorsal root ganglia, especially ones for discriminative touch. The primary neurons of the medial lemniscal system ascend the entire length of the spinal cord without crossing to its opposite side and synapse with secondary neurons located in the medulla oblongata.

In the spinal cord the medial lemniscal system can be divided into two separate tracts (see Figure 13-20) based on the source of the stimulus. The **fasciculus gracilis** (gras'ĭ-lis; thin) conveys sensations from nerve endings below the midthoracic level, and the **fasciculus cuneatus** (ku'ne-a'tus; wedge shaped) conveys impulses from nerve endings above the midthorax. The fasciculus gracilis terminates by synapsing with secondary neurons in the **nucleus gracilis** and with fibers of the posterior spinocerebellar tracts. The fasciculus cuneatus terminates by synapsing with secondary neurons in the **nucleus cuneatus.**

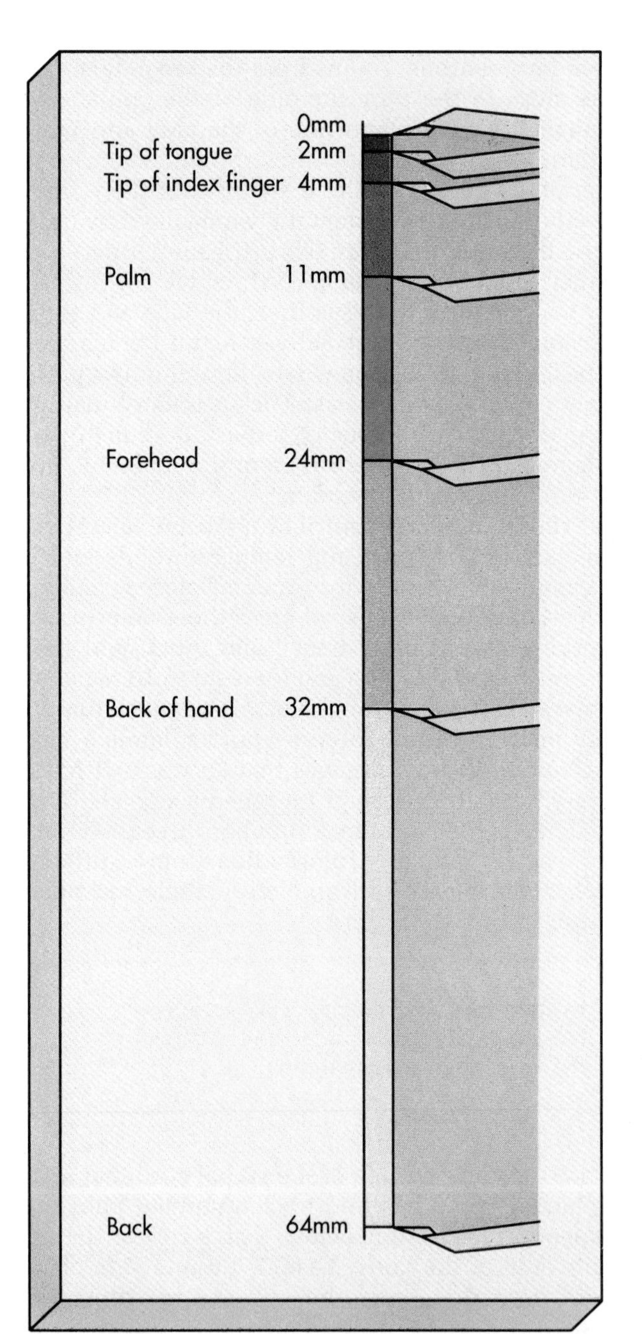

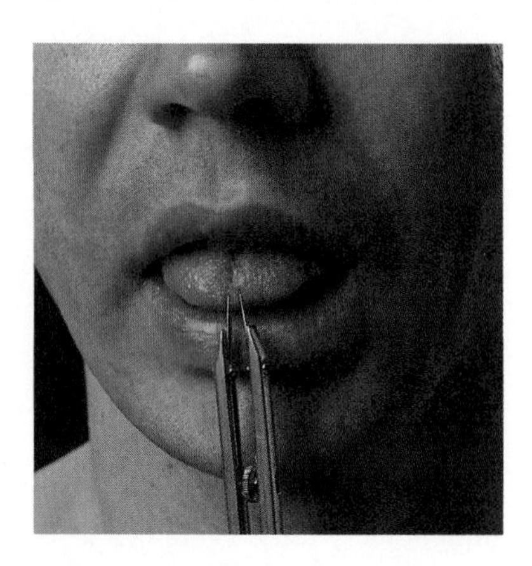

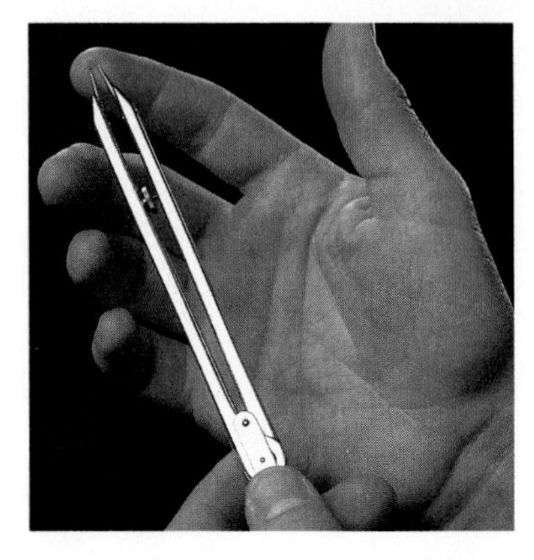

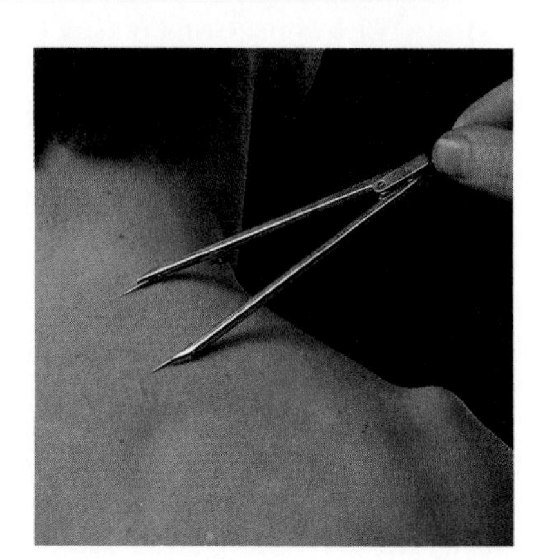

FIGURE 13-23 Two-point discrimination can be demonstrated by touching a person's skin with the two points of a compass. When the two points are close together, the individual perceives only one point. When the points of the compass are opened wider, the person becomes aware of two points. This awareness occurs when approximately 2 mm are between the points on the tip of the tongue, the most sensitive area of the body. The distance is approximately 4 mm for the finger tips and 64 mm for some areas of the back.

Both the nucleus gracilis and the nucleus cuneatus are in the medulla oblongata. The secondary neurons then exit the nucleus gracilis and the nucleus cuneatus, cross to the opposite side of the medulla (the decussations of the medial lemniscus), and ascend through the medial lemniscus to terminate in the thalamus. Tertiary neurons from the thalamus project to the somesthetic cortex.

8

Lesions of the medial lemniscal system in the spinal cord (posterior funiculus) cause loss of proprioception, fine touch, and vibration on the same side of the body below the level of the lesion. Lesions of the medial leminiscal system above the level of the medulla oblongata cause the same loss on the opposite side of the body below the level of the lesion. Explain why.

? ? ? ? ? ? ? ? ? ? ?

Spinocerebellar system and other tracts

The spinocerebellar tracts (see Figure 13-20) carry proprioceptive information to the cerebellum so that information concerning actual movements may be monitored and compared to cerebral information representing intended movements.

Two spinocerebellar tracts extend through the spinal cord: (1) the **posterior spinocerebellar tract** (Figure 13-24), which originates in the thoracic and upper-lumbar regions and contains uncrossed nerve fibers that enter the cerebellum through the inferior cerebellar peduncles; and (2) the **anterior spinocerebellar tract,** which carries information from the

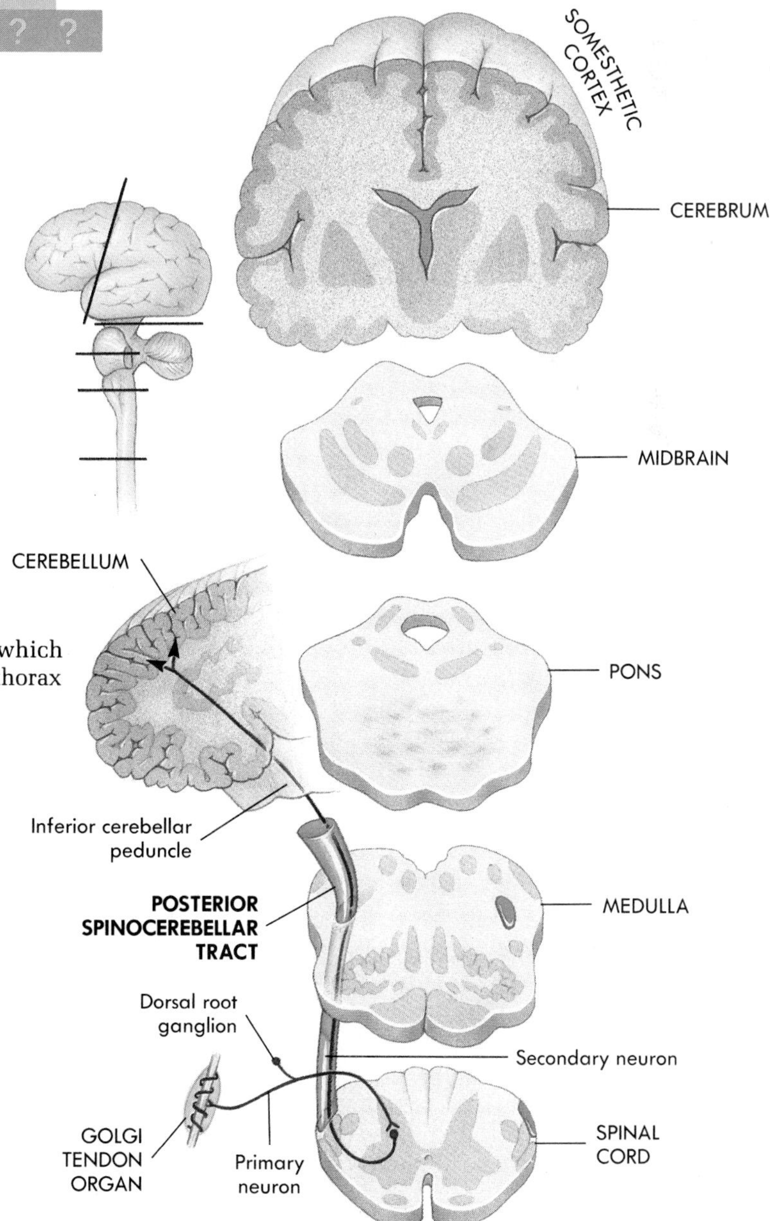

FIGURE 13-24 Posterior spinocerebellar tract, which transmits proprioceptive information from the thorax and upper limbs to the cerebellum.

lower trunk and lower limbs and contains both crossed and uncrossed nerve fibers that enter the cerebellum through the superior cerebellar peduncle. The crossed fibers recross in the cerebellum. Both spinocerebellar tracts transmit proprioceptive information to the cerebellum from the same side of the body as the cerebellar hemisphere to which they project. Why the anterior spinocerebellar tract crosses twice to accomplish this feat is unknown. Much of the proprioceptive information carried from the legs by the fasciculus gracilis of the medial lemniscal system is transferred by synapses in the inferior thorax to the spinocerebellar system and enters the cerebellum as unconscious proprioceptive information. In addition, the spinocerebellar tracts convey no information from the arms to the cerebellum. This input enters the cerebellum through the inferior peduncle from the cuneate nucleus of the medial lemniscal system. Therefore both the graciate and cuneate portions of the medial lemniscal system convey not only conscious awareness of proprioception to the cerebrum, but also unconscious awareness to the cerebellum.

9

Because most of the neurons from the fasciculus gracilis enter the spinocerebellar system and most of the neurons from the fasciculus cuneatus continue to the cerebrum, it can be deduced that most of the proprioception from the lower limbs is unconscious and most of the proprioception from the upper limbs is conscious. Explain why this difference in the two sets of limbs is of value.

? ? ? ? ? ? ? ? ? ? ?

The **spino-olivary tracts** project to the accessory olivary nucleus and to the cerebellum where their fibers contribute to coordination of movement associated primarily with balance. The **spinotectal tracts** end in the superior colliculi of the midbrain and are involved in reflexive turning of the head and eyes toward a point of cutaneous stimulation. The **spinoreticular tracts** are involved in arousing consciousness in the reticular activating system through cutaneous stimulation.

Descending Pathways

Most of the descending pathways are involved in the control of motor functions (see Table 13-5). However, some descending fibers are part of the sensory system and modulate the transmission of sensory information from the spinal cord to the somesthetic cortex (see discussion at the end of this section).

The voluntary motor system consists of two primary groups of neurons: upper motor neurons and lower motor neurons. **Upper motor neurons** originate in the cerebral cortex, cerebellum, and brainstem and modulate the activity of the lower motor neurons. Their fibers comprise the descending motor pathways. **Lower motor neurons** are either in the anterior horn of the spinal cord central gray matter or in the cranial nerve nuclei of the brainstem, and the axons of both groups extend through peripheral nerves to skeletal muscles.

The descending motor fibers are divided into two groups: the pyramidal system and the extrapyramidal system. The **pyramidal system** is involved in the maintenance of tone and in controlling the speed and precision of skilled movements, primarily fine movements involved in functions such as dexterity. The **extrapyramidal system** is involved in less precise control of motor functions, especially ones associated with overall body coordination and cerebellar function.

Pyramidal system

The **pyramidal system** is so named because the fibers of this group pass through the medullary **pyramids** (see Figure 13-5, *A*). It includes groups of nerve fibers arrayed into two tracts: the **corticospinal tract,** which is involved in direct cortical control of movements below the head (primarily the upper limbs), and the **corticobulbar tract,** which is involved in direct cortical control of head and neck movements.

Axons constituting the corticospinal tracts (Figure 13-25) originate from nerve cell bodies in the primary motor and premotor areas of the frontal lobes and from the somesthetic portion of the parietal lobe. They descend through the internal capsule (the major entrance and exit pathway of the cerebrum), the cerebral peduncles of the midbrain, and the pyramids of the medulla oblongata. At the inferior end of the medulla 75% to 85% (the number varies considerably from one person to the next) of the corticospinal fibers cross to the opposite side of the CNS through the **pyramidal decussation,** which is visible on the anterior surface of the inferior medulla (see Figure 13-5, *A*). These crossed fibers descend in the **lateral corticospinal tract** of the cord. The remaining 15% to 25% descend uncrossed in the **anterior corticospinal tract** and decussate near the level at which they synapse with lower motor neurons. The anterior corticospinal tracts supply the neck and upper limbs, and the lateral corticospinal tracts supply all levels of the body.

Most of the corticospinal fibers synapse with association neurons in the spinal cord's central gray matter. The association neurons, in turn, synapse with the lower motor neurons of the anterior horn.

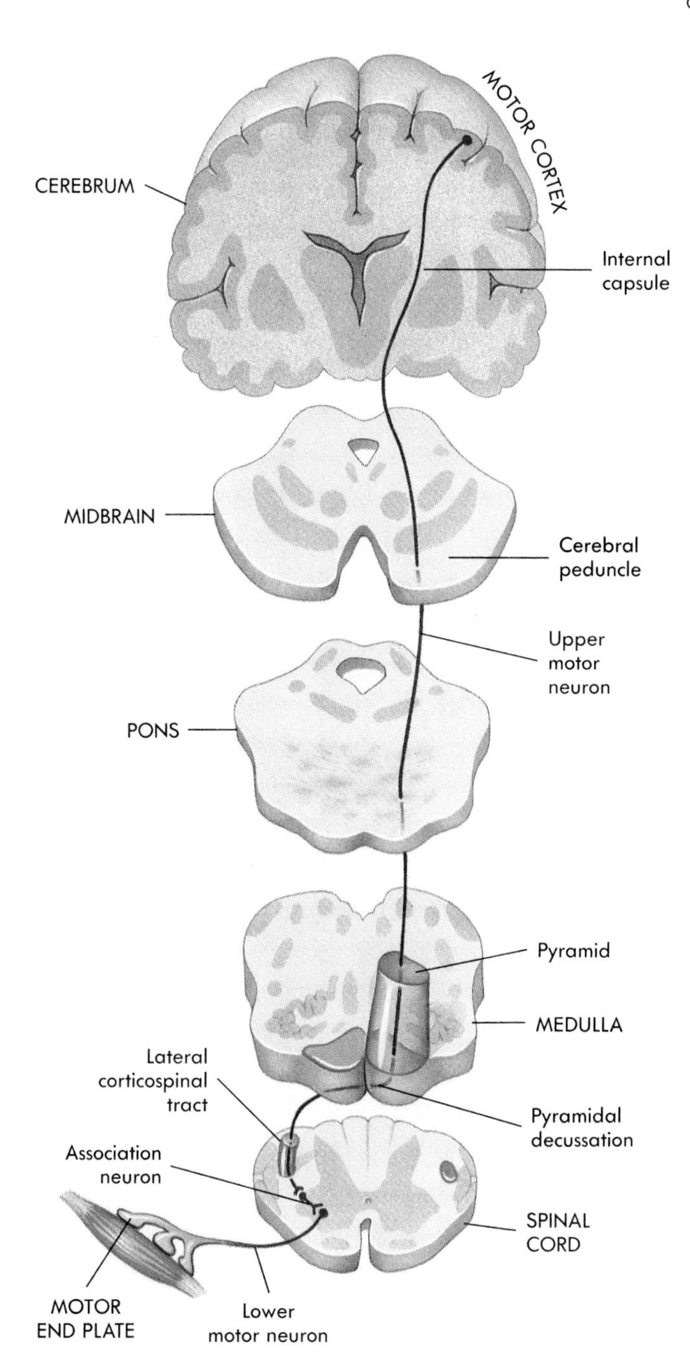

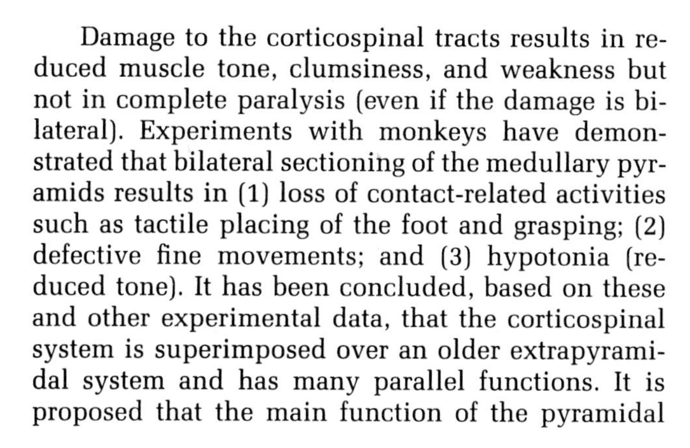

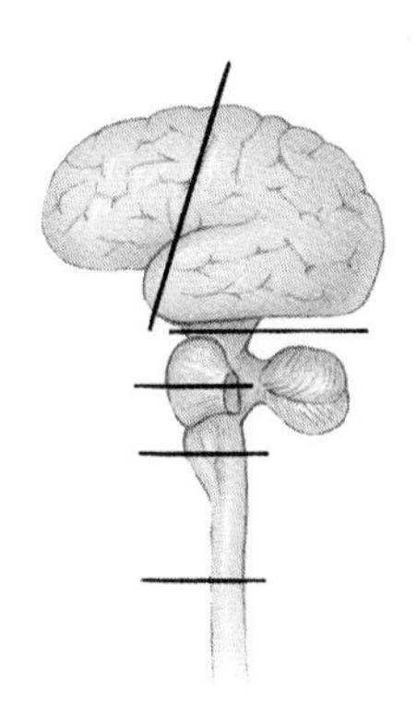

FIGURE 13-25 Pyramidal system. Lateral corticospinal tract, which is responsible for movement below the head.

Damage to the corticospinal tracts results in reduced muscle tone, clumsiness, and weakness but not in complete paralysis (even if the damage is bilateral). Experiments with monkeys have demonstrated that bilateral sectioning of the medullary pyramids results in (1) loss of contact-related activities such as tactile placing of the foot and grasping; (2) defective fine movements; and (3) hypotonia (reduced tone). It has been concluded, based on these and other experimental data, that the corticospinal system is superimposed over an older extrapyramidal system and has many parallel functions. It is proposed that the main function of the pyramidal system is to add speed and agility to conscious movements, especially of the hands, and to provide a high degree of fine motor control such as in movements of individual fingers. However, most spinal cord lesions affect both the pyramidal and extrapyramidal systems and result in complete paralysis.

The **corticobulbar tracts** are analogous to the corticospinal tracts. The former innervate the head, whereas the latter innervate the rest of the body. Cells that contribute to the corticobulbar tracts are in regions of the cortex similar to those of the corticospinal tracts except they are more laterally and inferiorly located on the cortex. Corticobulbar tracts follow the same basic route as the corticospinal system down to the level of the brainstem. At that point most corticobulbar fibers terminate in the reticular formation near the cranial nerve nuclei. Association neurons from the reticular formation then enter the **cranial nerve nuclei** in which they synapse with lower motor neurons. These nuclei give rise to nerves that control eye and tongue movements, mastication, facial expression, and palatine, pharyngeal, and laryngeal movements.

Extrapyramidal system

The extrapyramidal system (Figure 13-26) includes all of those descending motor fibers that do not pass through the pyramids or through the corticobulbar tracts. Its major tracts are the rubrospinal, vestibulospinal, and reticulospinal tracts. The **rubrospinal tract** begins in the red nucleus, decussates in the midbrain, and descends in the lateral funiculus. The red nucleus receives input from both the motor cortex and the cerebellum. Lesions in the red nucleus result in intention, or action, tremors similar to those seen in cerebellar lesions (see essay on dyskinesia, p. 398). Its function therefore is related closely to cerebellar function. Damage to the rubrospinal tract impairs distal arm and hand movements but does not greatly affect general body movements. The **vestibulospinal tracts** originate in the vestibular nuclei and descend in the anterior funiculus. Their fibers preferentially influence neurons innervating extensor muscles and are involved primarily in the maintenance of upright posture. The vestibular nuclei receive major input from the vestibular nerve (see Chapter 15) and the cerebellum. The **reticulospinal tract** originates in various regions of the reticular formation and descends in the anterior portion of the lateral funiculus. This tract's function is not well known, but it is assumed to mediate larger movements of the trunk and limbs that do not require balance or fine movement of the upper limbs. The basal ganglia also have a number of connections with the extrapyramidal system by which they modulate motor functions.

Descending pathways modulating sensation

The corticospinal and other descending pathways send collateral axons to the thalamus, reticular formation, trigeminal nuclei, and spinal cord, and the axons function to modulate pain impulses in ascending tracts. Through this route the cerebral cortex or other brain regions may reduce the conscious perception of sensation. The descending pathways reduce the transmission of pain impulses by secreting **endorphins** (natural analgesics).

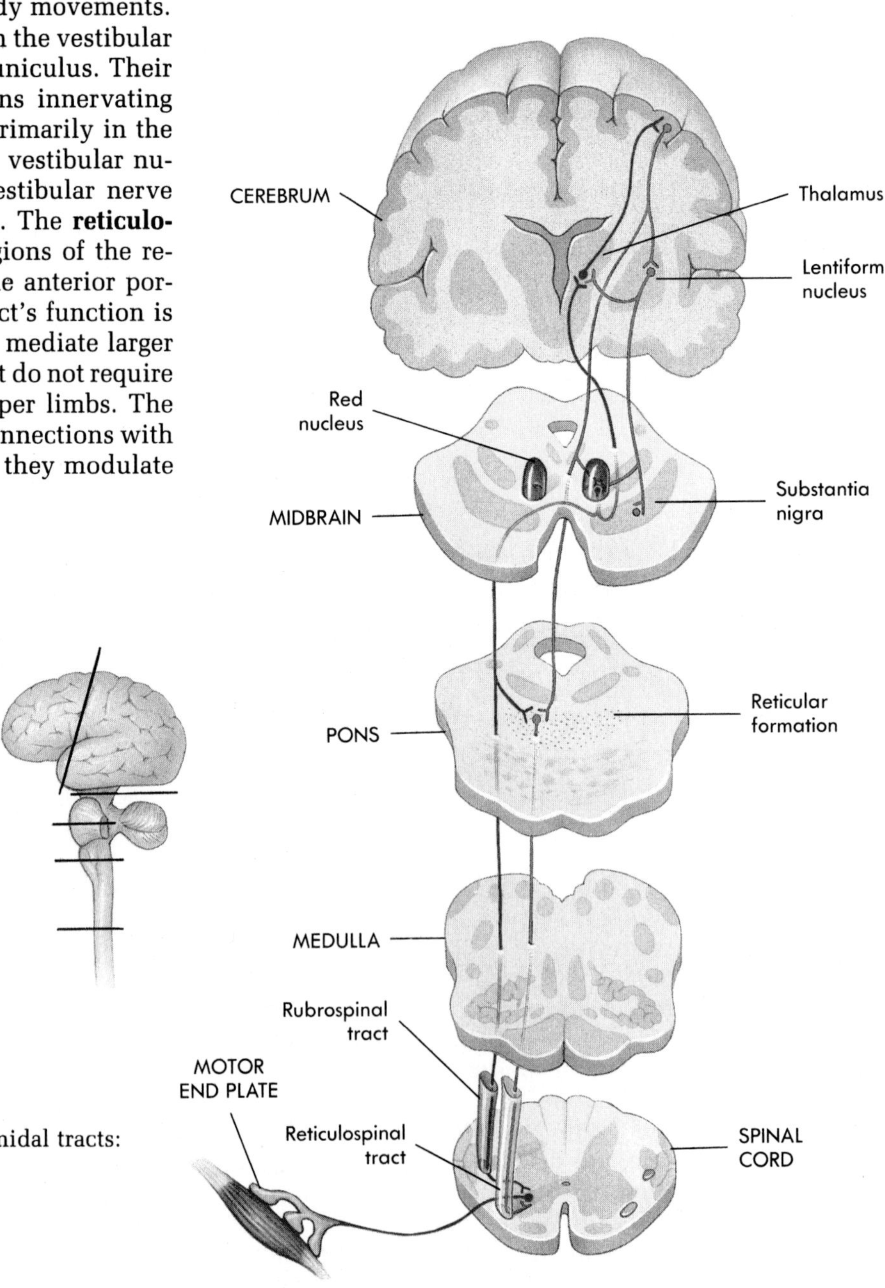

FIGURE 13-26 Examples of extrapyramidal tracts: rubrospinal and reticulospinal tracts.

General CNS Disorders

Infections

Encephalitis is an inflammation of the brain most often caused by a virus and less often by bacteria or other agents. A large variety of symptoms may result, including fever, paralysis, coma, or even death.

Myelitis is an inflammation of the spinal cord with causes and symptoms similar to those for encephalitis.

Meningitis is an inflammation of the meninges. It may be viral induced but is more often bacterial. Symptoms include neck stiffness, headache, and fever. In severe cases meningitis may cause paralysis, coma, or death.

Reye's syndrome may develop in children after an infection (especially influenza or chickenpox). The brain cells swell, and the liver and kidneys accumulate fat. Symptoms include vomiting, lethargy, and loss of consciousness and may progress to coma and death or to permanent brain damage.

Rabies is a viral disease transmitted by the bite of an infected mammal. The rabies virus infects the brain, salivary glands (through which it is transmitted), muscles, and connective tissue. When the patient attempts to swallow, the effort can produce pharyngeal muscle spasms; sometimes even the thought of swallowing water or the sight of water can induce pharyngeal spasms (thus the term hydrophobia, fear of water, is applied to the disease). The virus also infects the brain and results in abnormal excitability, aggression, and in later stages, paralysis and death.

Tabes dorsalis is a progressive disorder occurring as a result of an untreated syphilis infection. *Tabes* means wasting away, and dorsalis refers to degeneration of the dorsal roots and posterior columns of the spinal cord. The symptoms include ataxia (caused by lack of proprioceptive input), anesthesia (caused by dorsal root damage), and eventually paralysis as the infection spreads.

Multiple sclerosis (MS), although of unknown cause, is possibly viral in origin. It results in localized brain lesions and demyelination of the brain and spinal cord (the myelin sheaths become sclerotic, or hard, thus the name), resulting in poor conduction of action potentials. Its symptomatic periods are separated by periods of apparent remission. However, with each recurrence many neurons are permanently damaged so that the disease's progressive symptoms include exaggerated reflexes, tremor, nystagmus, and speech defects.

Other Disorders

Tumors of the brain develop from neuroglial cells. Symptoms vary widely, depending on the location of the tumor, but may include headaches, neuralgia (pain along the distribution of a peripheral nerve), paralysis, seizures, coma, and death.

Stroke is a term meaning a blow or sudden attack, suggesting the speed with which this type of defect can occur. It is also referred to as apoplexy or, clinically, as a **cerebrovascular accident** and is caused by hemorrhage, thrombosis, embolism, or vasospasm of the cerebral blood vessels. These causes result in an **infarct,** a local area of cell death caused by a lack of blood supply. Symptoms depend on the location but include anesthesia or paralysis on the side of the body opposite the cerebral infarct.

Senility once was thought a normal part of aging. The term *senile* is Latin and means old age. The several symptoms of senility—general intellectual deficiency and mental deterioration (called dementia), memory loss, short attention span, moodiness, and irritability—result from several specific disease states. **Alzheimer's disease** is a hereditary type of senility, often affecting people less than age 50. Apparently it is due to neuronal microfilament degeneration and results in general mental deterioration.

Tay-Sachs disease is a hereditary disorder of infants involving abnormal sphingolipid (lipids with long base chains) metabolism that results in severe brain dysfunction. Symptoms include paralysis, blindness, and death, usually before 5 years of age.

Mercury poisoning, when chronic, can cause brain disorders such as intention tremor, exaggerated reflexes, and emotional instability.

Epilepsy is actually a whole group of brain disorders that have seizure episodes in common. The seizure, a sudden massive neuronal discharge, can be either partial or complete, depending on the amount of brain involved and whether or not consciousness is impaired. Normally there is a balance between excitation and inhibition in the brain. When this balance is disrupted by increased excitation or decreased inhibition, a seizure may result. The neuronal discharges may stimulate muscles innervated by the nerves involved, resulting in involuntary muscle contractions, or convulsions.

Headaches have a variety of causes that can be grouped into two basic classes: extracranial and intracranial. Extracranial headaches can be caused by inflammation of the sinuses, dental irritations, ophthalmological disorders, or tension in the muscles moving the head and neck. Intracranial headaches may result from inflammation of the brain or meninges, vascular problems, mechanical damage, or tumors.

Alexia, loss of the ability to read, may result from a lesion in the visual association cortex. **Dyslexia** is a defect in which an individual's reading level is below that expected based on his overall intelligence. Most people with dyslexia have normal or above-normal intelligence quotients. The term means reading deficiency and is also called partial alexia. It is three times more common in males than females. As many as 10% of American males suffer from the disorder. The symptoms are quite variable from person to person and include transposition of letters in a word, confusion between *b* and *d,* and lack of orientation in three-dimensional space. The brains of dyslexics are physically different from other brains, with abnormal cellular arrangements, including cortical disorganization and the appearance of bits of gray matter in medullary areas. Dyslexia apparently results from abnormal brain development.

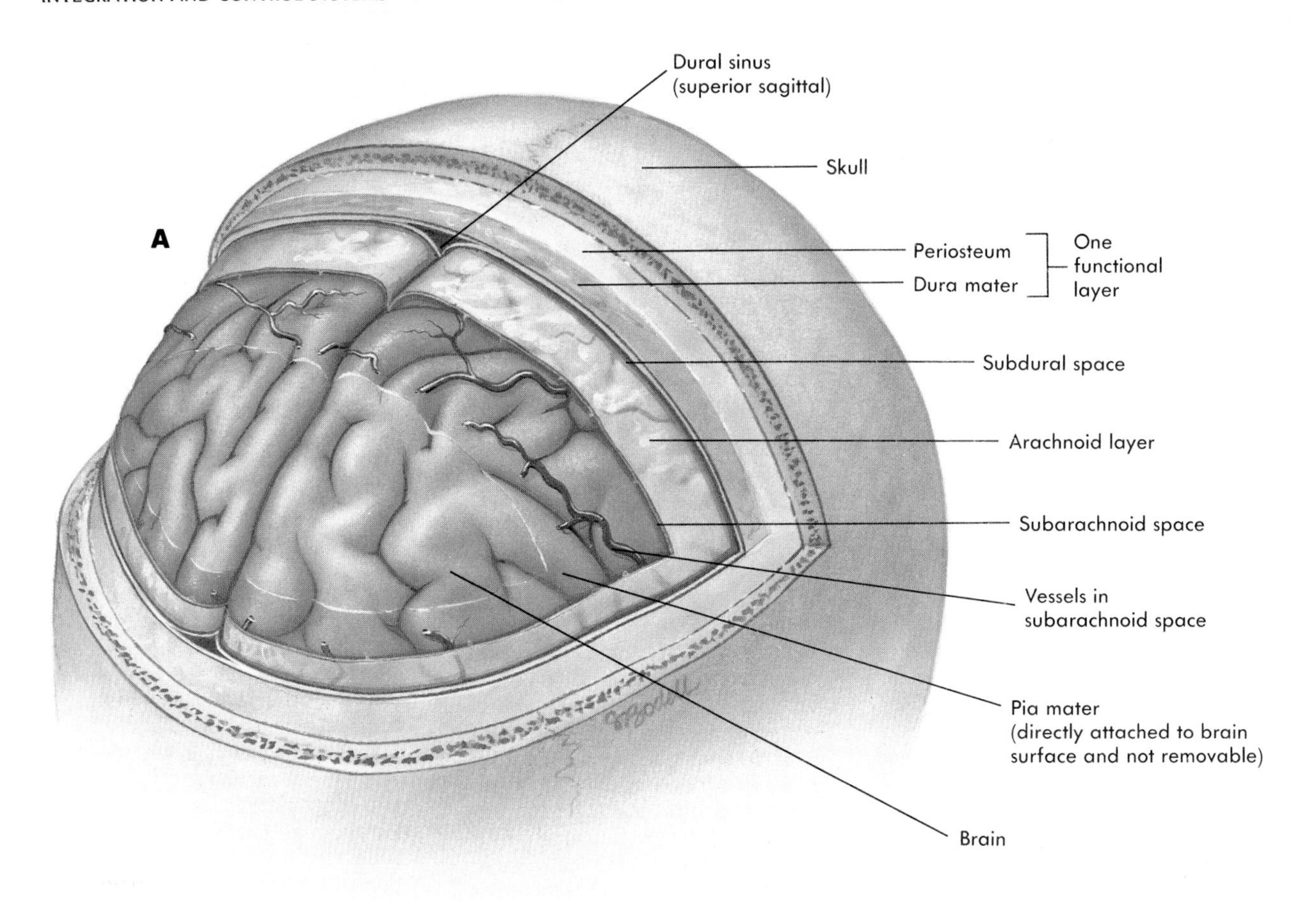

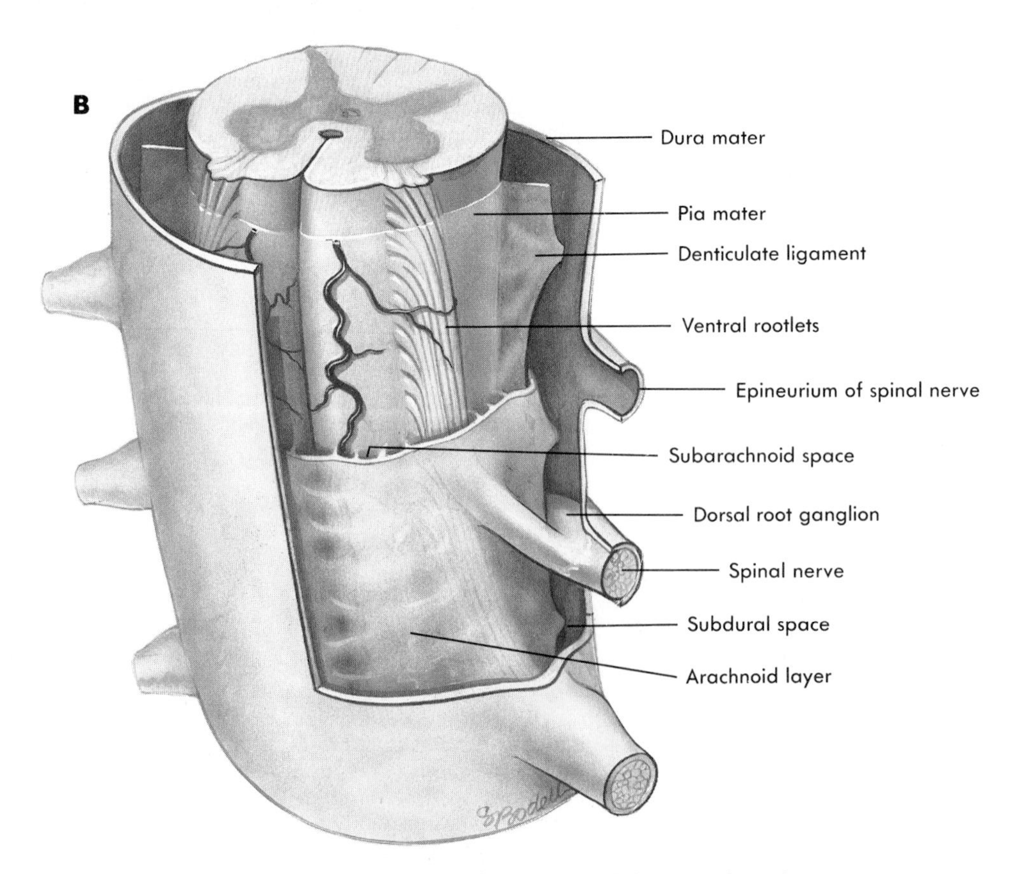

FIGURE 13-27 Meningeal coverings of the brain **A**, and the spinal cord, **B**.

MENINGES AND CEREBROSPINAL FLUID

Meninges

Three connective tissue layers, the **meninges** (mĕ-nin'jēz), surround and protect the brain and spinal cord (Figure 13-27). The most superficial and thickest layer is the **dura mater** (du'rah ma'ter; tough mother). Three dural folds, the **falx cerebri** (falks ser-e'bre; sickle shaped), **tentorium cerebelli** (ten-to'ri-um sĕr'ĕ-bel'e; tent), and the **falx cerebelli** extend into the major brain fissures. The falx cerebri is located between the two central hemispheres in the longitudinal fissure, the tentorium cerebelli is between the cerebrum and cerebellum, and the falx cerebelli lies between the two cerebellar hemispheres. The dura mater consists of two layers around the brain and one layer around the spinal cord. The two layers of the dura mater are fused around most of the brain but separate in several places, primarily at the bases of the three dural folds, to form venous **dural sinuses.** The dural sinuses collect most of the blood that returns from the brain and cerebrospinal fluid from around the brain (see next section). The sinuses then empty into the veins that exit the skull (see Chapter 21). The dura mater surrounding the brain is tightly attached to and is continuous with the periosteum of the cranial vault, forming a single functional layer, whereas the dura mater of the spinal cord is separated from the periosteum of the vertebral canal by the **epidural space.** It is a true space around the spinal cord and contains spinal nerves, blood vessels, areolar connective tissue, and fat. Epidural anesthesia of the spinal nerves is induced by injecting anesthetics into this space.

The next meningeal layer is a very thin, wispy **arachnoid** (ar-ak'noyd; spiderlike, i.e., cobwebs) **layer.** The space between this layer and the dura mater is the **subdural space** and contains only a very small amount of serous fluid. The third meningeal layer, the **pia** (pe'ah; affectionate mother) **mater,** is bound very tightly to the surface of the brain and spinal cord. Between the arachnoid layer and the pia mater is the **subarachnoid space,** which contains weblike strands of the arachnoid layer and blood vessels and is filled with cerebrospinal fluid.

The dura mater and dural folds help hold the brain in place within the skull and keep it from moving around too freely. The spinal cord is held in place within the vertebral canal by a series of connective tissue strands connecting the pia mater to the dura mater, which cause the arachnoid layer to form points between each of the nerves. Because of this "toothed" appearance, these attachments are called **denticulate ligaments** (see Figure 13-27, *B*).

Ventricles

As already stated, the CNS is formed as a hollow tube that may be quite reduced in some areas of the adult CNS and expanded in other areas (see Figure 13-3, *C*). It is lined by a single layer of epithelial cells called **ependymal cells** (ep-en'dĭ-mal; see Chapter 12). Each cerebral hemisphere contains a relatively large cavity, the **lateral ventricle** (Figure 13-28). The lateral ventricles are separated from each other by thin **septa pellucida** (sep'tah pel-lu'sid-ah; translucent walls), which lie in the midline just inferior to the corpus callosum, and usually are fused with each other. A smaller midline cavity, the **third ventricle,**

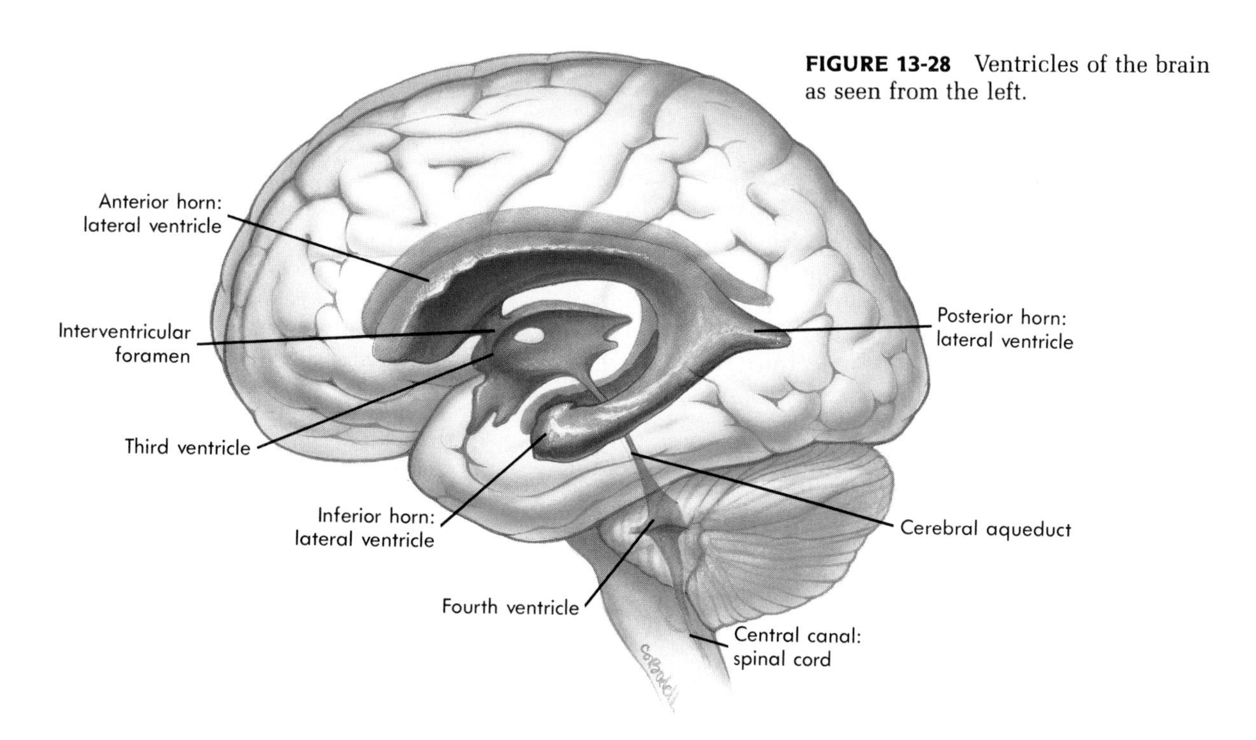

FIGURE 13-28 Ventricles of the brain as seen from the left.

is located in the center of the diencephalon between the two halves of the thalamus. The two lateral ventricles communicate with the third ventricle through two **interventricular foramina** (foramina of Monro). The lateral ventricles can be thought of as the first and second ventricles in the numbering scheme, but they are not designated as such. The **fourth ventricle** is in the inferior portion of the pontine region and the superior region of the medulla oblongata at the base of the cerebellum. The third ventricle communicates with the fourth ventricle through a narrow canal, the **cerebral aqueduct** (aqueduct of Sylvius), which passes through the midbrain. The fourth ventricle is continuous with the central canal of the spinal cord, which extends nearly the full length of the cord. The fourth ventricle is also continuous with the subarachnoid space through foramina in its walls and roof.

Because the spinal cord extends only to approximately level L2 of the vertebral column and the meninges extend to the end of the vertebral column, a needle can be introduced into the subarachnoid space inferior to the medullary cone to induce spinal anesthesia (spinal block) or to take a sample of cerebrospinal fluid (a spinal tap) without damaging the cord. The nerves of the cauda equina are pushed aside quite easily by the needle during these procedures and normally are not damaged. A spinal tap may be performed to examine the cerebrospinal fluid for infectious agents (meningitis), for the presence of blood (hemorrhage), or for CSF pressure. A radiopaque substance may also be injected into this area, and a myelogram (x-ray of the spinal cord) may be taken to visualize spinal cord defects or damage.

Cerebrospinal Fluid

Cerebrospinal (ser-e-bro-spi'nal) **fluid,** a fluid similar to plasma and interstitial fluid, bathes the brain and the spinal cord, providing a protective cushion around the CNS. It also provides some nutrients to the CNS tissues. Cerebrospinal fluid is produced by specialized ependymal cells within the lateral (approximately 80% to 90%), third, and fourth ventricles. These cells, their support tissue, and the associated blood vessels together are called **choroid** (ko'royd; lacy) **plexuses** (Figure 13-29). The choroid plexuses are formed by invaginations of the vascular pia mater into the ventricles, producing a vascular connective tissue core covered by ependymal cells.

Production of the cerebrospinal fluid by the choroid plexus is not fully understood. Some portions of the blood plasma cross the membranes of the plexus by diffusion (probably including both simple and facilitated), whereas other portions require active transport.

Cerebrospinal fluid fills the ventricles, the subarachnoid space of the brain and spinal cord, and the central canal of the spinal cord. Approximately 23 ml of fluid fills the ventricles, and 117 ml fills the subarachnoid space. The route taken by the cerebrospinal fluid from its origin in the choroid plexuses to its return to the circulation is depicted in Figure 13-29. The flow rate of cerebrospinal fluid from its origin to the point where it enters the bloodstream is approximately 0.4 ml per minute. Cerebrospinal fluid passes from the lateral ventricles through the interventricular foramina into the third ventricle and then through the cerebral aqueduct into the fourth ventricle. It can exit the interior of the brain only through the wall of the fourth ventricle. One **median foramen** (foramen of Magendie), which opens through the roof of the fourth ventricle, and two **lateral foramina** (foramina of Luschka) allow the cerebrospinal fluid to pass from the fourth ventricle to the subarachnoid space. Masses of arachnoid tissue, **arachnoid** (ar-ak'noyd) **granulations,** penetrate into the superior sagittal sinus (one of the dural sinuses), and cerebrospinal fluid passes into the venous sinuses through these granulations. The sinuses are blood filled, so it is within these dural sinuses that the cerebrospinal fluid reenters the bloodstream. From the venous sinuses the blood flows to veins of the general circulation.

Blockage of the foramina of the fourth ventricle or blockage of the cerebral aqueduct can result in accumulation of cerebrospinal fluid within the ventricles. This condition is called **internal hydrocephalus** and is the result of increased cerebrospinal fluid pressure. The production of cerebrospinal fluid continues even when the passages that normally allow it to exit the brain are blocked. Consequently, fluid continues to build inside the brain, causing pressure that compresses the nervous tissue and dilates the ventricles. Compression of the nervous tissue usually results in irreversible brain damage. If the skull bones are not completely ossified when the hydrocephalus occurs, the pressure may also cause severe enlargement of the head. Hydrocephalus can be successfully treated by placing a drainage tube (shunt) between the brain ventricles and one of the body cavities to eliminate the high internal pressures. However, these shunts may provide routes through which infection can be introduced into the brain, and the shunts must be replaced as the person grows. If cerebrospinal fluid accumulates in the subarachnoid space, the condition is called **external hydrocephalus.** Pressure is applied to the brain externally, compressing neural tissues and causing brain damage.

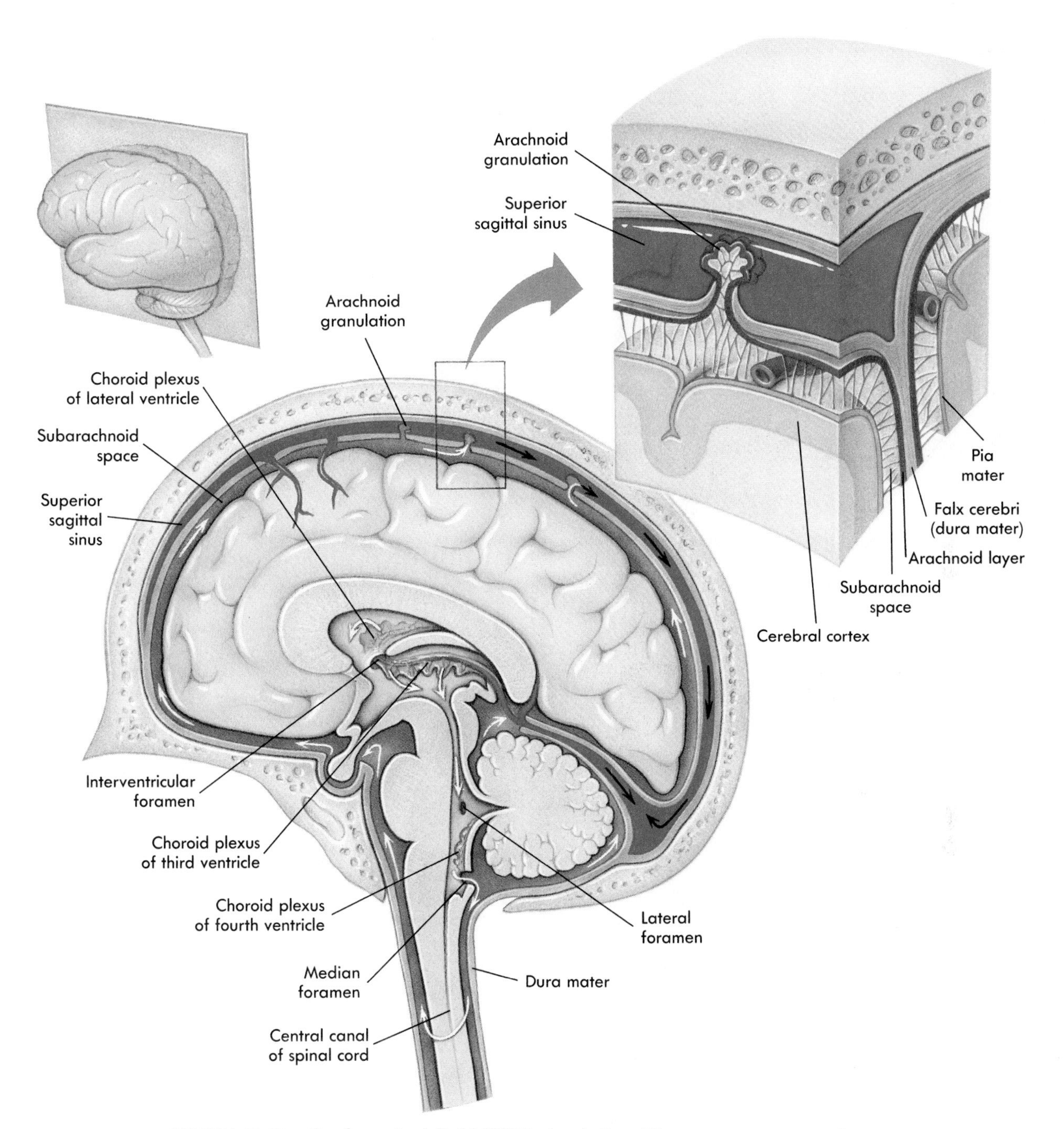

FIGURE 13-29 Cerebrospinal fluid (CSF) circulation. The arrows represent the route of the CSF. It is produced in the ventricles, exits the fourth ventricle, and returns to the venous circulation in the superior sagittal sinus. The inset depicts the arachnoid granulations in the superior sagittal sinus where the CSF enters the circulation.

The Nervous System

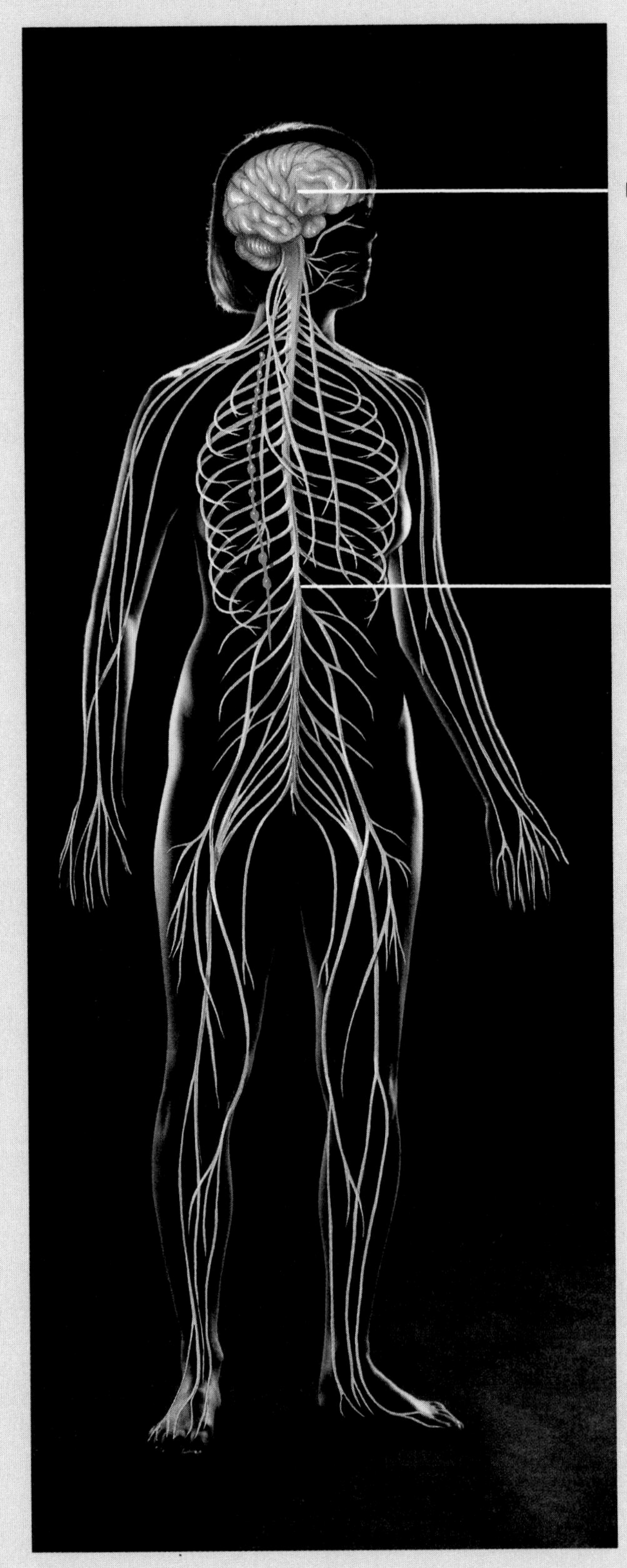

Major Components

Brain, spinal cord, nerves, and sensory receptors

Major Functions

Houses all mental activity such as consciousness, memory, and thought
Controls most physiological functions
Detects sensations
Interprets changes in internal and external environments
Produces movements

SUMMARY

DEVELOPMENT p. 382

The brain and spinal cord develop from the neural tube. The ventricles and central canal develop from the lumen of the neural tube.

BRAINSTEM p. 384

1. The medulla oblongata is continuous with the spinal cord and contains ascending and descending nerve tracts.
   - The pyramids are nerve tracts controlling voluntary muscle movement.
   - The olives are nuclei that function in equilibrium, coordination, and modulation of sound from the inner ear.
   - Medullary nuclei regulate the heart, blood vessels, respiration, swallowing, vomiting, coughing, sneezing, and hiccuping. The nuclei of cranial nerves IX through XII are in the medulla.
2. The pons is superior to the medulla.
   - Ascending and descending nerve tracts pass through the pons.
   - Pontine nuclei regulate sleep and respiration. The nuclei of cranial nerves V through VII are in the pons.
3. The midbrain is superior to the pons.
   - The midbrain contains the nuclei for cranial nerves III and IV.
   - The tectum consists of four colliculi. The inferior colliculi are involved in hearing and the superior colliculi in visual reflexes.
   - The tegmentum contains ascending tracts and the red nuclei, which are involved in motor activity.
   - The cerebral peduncles are the major descending motor pathway.
   - The substantia nigra connect to the basal ganglia and are involved with muscle tone and movement.
4. The reticular formation is nuclei scattered throughout the brainstem. The reticular activating system extends to the thalamus and cerebrum and maintains consciousness.

DIENCEPHALON p. 388

1. The diencephalon is located between the brainstem and the cerebrum.
2. The thalamus consists of two lobes connected by the intermediate mass. The thalamus functions as an integration center.
   - Most sensory input synapses in the thalamus.
   - The thalamus also has some motor functions.
3. The subthalamus is involved in motor function.
4. The epithalamus contains the habenular nuclei, which influence emotions through the sense of smell. The pineal body may play a role in the onset of puberty.
5. The hypothalamus contains several nuclei and tracts.
   - The mamillary bodies are reflex centers for olfaction.
   - The hypothalamus regulates many endocrine functions (e.g., metabolism, reproduction, response to stress, and urine production). The pituitary gland attaches to the hypothalamus.
   - The hypothalamus regulates body temperature, hunger, thirst, satiety, swallowing, and emotions.

CEREBRUM p. 390

1. The cortex of the cerebrum is folded into ridges called gyri and grooves called sulci or fissures. Nerve tracts connect areas of the cortex within the same hemisphere (association fibers), between different hemispheres (commissural fibers), and with other parts of the brain and the spinal cord (projection fibers).
2. The longitudinal fissure divides the cerebrum into left and right hemispheres. Each hemisphere has five lobes.
   - The frontal lobes are involved in voluntary motor function, motivation, aggression, and mood.
   - The parietal lobes contain the major sensory areas receiving general sensory input, taste, and balance.
   - The occipital lobes contain the visual centers.
   - The temporal lobes receive olfactory and auditory input and are involved in memory, abstract thought, and judgment.
3. Sensory pathways project to primary sensory areas in the cerebral cortex. Association areas interpret input from the primary sensory areas.
4. The frontal lobe contains areas dealing with voluntary muscle movement.
   - The primary motor area controls many muscle movements.
   - The premotor area is necessary for complex, skilled, and learned movements.
   - The prefrontal area is involved with the motivation and the foresight associated with movement.
5. Cortical functions are arranged topographically on the cortex.
6. Speech is located only in the left cortex in most people.
   - Wernicke's area comprehends and formulates speech.
   - Broca's area receives input from Wernicke's area and sends impulses to the premotor and motor areas, which cause the muscle movements required for speech.
7. Electroencephalograms (EEGs) record the electrical activity of the brain as alpha, beta, theta, and delta waves. Some brain disorders can be detected with EEGs.
8. There are at least three kinds of memory: sensory, short-term, and long-term.
9. Each cerebral hemisphere controls and receives input from the opposite side of the body.
   - The right and left hemispheres are connected by commissures. The largest commissure is the corpus callosum, which allows sharing of information between hemispheres.
   - In most people the left hemisphere is dominant, controlling speech and analytic skills. The right hemisphere controls spatial and musical abilities.

Basal Ganglia

1. Basal ganglia include the subthalamic nuclei, substantia nigra, and corpus striatum.
2. The basal ganglia are important in coordinating motor movements and posture. They mainly have an inhibitory effect.

Limbic System

1. The limbic system includes parts of the cerebral cortex, basal ganglia, thalamus, hypothalamus, and the olfactory cortex.
2. The limbic system controls visceral functions through the autonomic nervous system and the endocrine system and is also involved in emotions and memory.

Cerebellum

1. The cerebellum has three parts that control balance, gross motor coordination, and fine motor coordination.
2. The cerebellum functions to correct discrepancies between intended movements and actual movements.
3. The cerebellum can "learn" highly specific complex motor activities.

SPINAL CORD *p. 400*

General Structure

1. Thirty-one pairs of spinal nerves exit the spinal cord. The spinal cord has cervical and lumbar enlargements where nerves of the limbs enter and exit.
2. The spinal cord is shorter than the vertebral column. Nerves from the end of the spinal cord form the cauda equina.

Cross Section

1. The cord consists of peripheral white matter and central gray matter.
2. White matter is organized into funiculi, which are subdivided into fasciculi, or nerve tracts, which carry action potentials to and from the brain.
3. Gray matter is divided into horns.
   - The dorsal horns contain sensory axons that synapse with association neurons. The ventral horns contain the nerve cell bodies of somatic motor neurons, and the lateral horns contain the nerve cell bodies of autonomic neurons.
   - The gray and white commissures connect each half of the spinal cord.
4. The dorsal root conveys sensory input into the spinal cord, and the ventral root conveys motor output away from the spinal cord.

SPINAL REFLEXES *p. 402*

Stretch Reflex

Muscle spindles detect stretch of skeletal muscles and cause the muscle to shorten reflexively.

Golgi Tendon Reflex

Golgi tendon organs respond to increased tension within tendons and cause skeletal muscles to relax.

Withdrawal Reflex

1. Activation of pain receptors causes contraction of muscles and the removal of some part of the body from a painful stimulus.
2. Reciprocal innervation causes relaxation of muscles that would oppose the withdrawal movement.
3. In the crossed extensor reflex, during flexion of one limb caused by the withdrawal reflex, the opposite limb is stimulated to extend.

SPINAL PATHWAYS *p. 407*

Ascending Pathways

1. Ascending pathways carry conscious and unconscious sensations.
2. Spinothalamic System
   - The lateral spinothalamic tracts carry pain and temperature sensations. The anterior spinothalamic tracts carry light touch, pressure, tickle, and itch sensations.
   - Both tracts are formed by primary neurons that enter the spinal cord and synapse with secondary neurons. The secondary neurons cross the spinal cord and ascend to the thalamus where they synapse with tertiary neurons that project to the somesthetic cortex.
3. Medial Lemniscal System
   - The medial lemniscal system carries the sensations of two-point discrimination, proprioception, pressure and vibration.
   - Primary neurons enter the spinal cord and ascend to the medulla where they synapse with secondary neurons. The secondary neurons cross over and project to the thalamus where they synapse with tertiary neurons that extend to the somesthetic cortex.
4. Spinocerebellar System and Other Tracts
   - The spinocerebellar tracts carry unconscious proprioception to the cerebellum from the same side of the body.
   - Neurons of the medial lemniscal system synapse with the neurons that carry proprioception information to the cerebellum.
   - The spino-olivary tract contributes to coordination of movement, the spinotectal tract to eye reflexes, and the spinoreticular tract to arousing consciousness.

Descending Pathways

1. Upper motor neurons are located in the cerebral cortex, cerebellum, and brainstem. Lower motor neurons are found in the cranial nuclei or the ventral horn of the spinal cord gray matter.
2. The pyramidal system maintains muscle tone and controls fine, skilled movements. The extrapyramidal system controls conscious and unconscious muscle movements.
3. The corticospinal tracts control muscle movements below the head.
   - Approximately 75% to 85% of the upper motor neurons of the corticospinal tracts cross over in the medulla to form the lateral corticospinal tracts in the spinal cord.
   - The remaining upper motor neurons pass through the medulla to form the anterior corticospinal tracts, which cross over in the spinal cord.
   - The upper motor neurons of both tracts synapse with association neurons that then synapse with lower motor neurons in the spinal cord.
4. The corticobulbar tracts innervate the head muscles. Upper motor neurons synapse with association neurons in the reticular formation that, in turn, synapse with lower motor neurons in the cranial nerve nuclei.
5. The extrapyramidal system includes the rubrospinal, vestibulospinal, and reticulospinal tracts and fibers from the basal ganglia.
6. The extrapyramidal tracts are involved in conscious and unconscious muscle movements, posture, and balance.
7. Some axons from the somesthetic cortex synapse with secondary and tertiary neurons of the ascending sensory system and modify their activity.

MENINGES AND CEREBROSPINAL FLUID *p. 419*

Meninges

1. The brain and spinal cord are covered by the dura mater, arachnoid, and pia mater.
2. The dura mater attaches to the skull and has two layers that can separate to form dural sinuses.

3. Beneath the arachnoid layer the subarachnoid space contains cerebrospinal fluid that helps cushion the brain.
4. The pia mater attaches directly to the brain.

Ventricles

1. The lateral ventricles in the cerebrum are connected to the third ventricle in the diencephalon by the interventricular foramen.
2. The third ventricle is connected to the fourth ventricle in the pons by the cerebral aqueduct. The central canal of the spinal cord is connected to the fourth ventricle.

Cerebrospinal Fluid

1. Cerebrospinal fluid is produced from the blood in the choroid plexus of each ventricle. Cerebrospinal fluid moves from the lateral to the third and then to the fourth ventricle.
2. From the fourth ventricle cerebrospinal fluid enters the subarachnoid space through three foramina.
3. Cerebrospinal fluid leaves the subarachnoid space through arachnoid granulations and returns to the blood in the dural sinuses.

CONTENT REVIEW

1. Describe the formation of the neural tube. Name the five divisions of the neural tube and the parts of the brain that each division becomes.
2. Name the parts of the brainstem, and describe their functions.
3. What are the reticular formation and the reticular activating system?
4. Name the four main components of the diencephalon.
5. Describe the functions of the thalamus and the hypothalamus.
6. Name the five lobes of the cerebrum, and describe their location and functions.
7. Describe the cerebral cortex locations of the special and general senses and their association areas.
8. How do the primary and association areas interact to perceive a sensation?
9. Describe the topographical arrangement of the sensory and motor areas of the cerebral cortex.
10. Starting with hearing or reading a word, name the areas of the brain that are involved with receiving the word "stimulus" and the areas that eventually cause the word to be spoken aloud.
11. What is an EEG? What conditions produce alpha, beta, theta, and delta waves?
12. Name the three different types of memory, and describe the processes that eventually result in the ability to remember something for a long period of time.
13. Name the pathways that connect the right and left cerebral hemispheres. Which hemisphere is dominant in most people?
14. Name the basal ganglia, and state where they are located. What are their functions?
15. Name the parts of the limbic system. What does the limbic system do?
16. Describe the comparator activities of the cerebellum. Describe the role of the cerebellum in rapid and skilled motor movements such as playing the piano.
17. Define the cervical and lumbar enlargements of the spinal cord, the medullary cone, and the cauda equina.
18. Describe the spinal cord gray matter. Where are sensory, autonomic, and somatomotor neurons located in the gray matter?
19. Contrast and describe a stretch reflex and a Golgi tendon reflex.
20. What is a withdrawal reflex? How do reciprocal innervation and the crossed extensor reflex assist the withdrawal reflex?
21. Describe the operation of the gamma motor system. What does it accomplish?
22. What are the functions of the lateral and anterior spinothalamic tracts and the medial lemniscal system? Describe where the neurons of these tracts cross over and synapse.
23. What kind of information is carried in the spinocerebellar tracts?
24. What are the functions of the spino-olivary, spinotectal, and spinoreticular tracts?
25. Distinguish between upper and lower motor neurons.
26. What two tracts form the pyramidal system? What area of the body is supplied by each tract? Describe the location of the neurons in each tract, where they cross over, and where they synapse.
27. Name the tracts and structures that form the extrapyramidal system. What functions do they control? Contrast them with the functions of the pyramidal system.
28. Describe the three meninges that surround the CNS. What are the falx cerebri, tentorium cerebelli, and falx cerebelli?
29. What space between what dural layers contains the cerebrospinal fluid?
30. Describe the production and circulation of the cerebrospinal fluid. Where does the cerebrospinal fluid return to the blood?

CONCEPT REVIEW

1. A tumor or intracranial hemorrhage can cause increased intracranial pressure and can force the cerebellum downward through the foramen magnum, which in turn compresses the medulla and can lead to death. Give two likely causes of death, and explain why they would occur.
2. Woody Knothead was accidentally struck in the head with a baseball bat. He fell to the ground unconscious. Later, when he had regained consciousness, he was not able to remember any of the events that happened 10 minutes before the accident. Explain.
3. A patient suffered brain damage in an automobile accident. It was suspected that the cerebellum was the part of the brain that was affected. Based on what you know about cerebellar function, how could you determine that the cerebellum was involved?
4. If there were a decrease in the frequency of action potentials in gamma motor fibers to a muscle, would the muscle tend to relax or contract?
5. A patient is suffering from the loss of two-point discrimination and proprioceptive sensations on the right side of the body resulting from a lesion in the pons. What tract is affected, and which side of the pons is involved?
6. A patient suffered a lesion in the center of the spinal cord. It was suspected that the fibers that decussate and that are associated with the lateral spinothalamic tracts were affected in the area of the lesion. What observations would be consistent with that diagnosis?
7. A person in a car accident exhibited the following symptoms: extreme paresis on the right side, including the arm and leg, reduction of pain sensation on the left side, and normal tactile sensation on both sides. Which nerve tracts were damaged? Where did the patient suffer nerve tract damage?
8. If the right side of the spinal cord were completely transected, what symptoms would you expect to observe with regard to motor function, two-point discrimination, light touch, and pain perception?
9. A patient with a CNS lesion exhibited paralysis of the *left* hand, arm, and shoulder to the extent that gross postural movements were possible, but discrete, fine movements were not. Given that the lesion is on the *right* side of the body, is the problem in the cerebellar cortex, cerebral cortex, basal ganglia, or lateral corticospinal tract in the spinal cord? Explain why each choice is or is not the site of the lesion.
10. A baby is born with enlarged lateral and third ventricles but a normal fourth ventricle. Describe the defect and its location.
11. A blow to the head can rupture blood vessels of the meninges and cause bleeding. Name three places blood could accumulate in relation to the meninges.

? ?

ANSWERS TO PREDICT QUESTIONS

1 p. 387. When a bright object suddenly appears in a person's range of vision, his eyes reflexively turn to focus on the object. When a person hears a sudden, loud noise, his head and eyes reflexively turn toward the noise. When a part of the body such as the shoulder is touched, his head and eyes reflexively turn toward that part of the body. In each case the pathway involves the superior colliculus. Efferent fibers from the superior colliculus project to cranial nerve motor nuclei that initiate movement of the head and eyes.

Afferent fibers to the superior colliculus come from different sources for each of the three examples. Reflexes initiated by visual input (a bright object in the visual field) come to the superior colliculus as a side branch from the optic pathways. Visual reflexes initiated by auditory input (a sudden, loud noise) are stimulated by afferent impulses from the inferior colliculus to the superior colliculus. Visual reflexes initiated by tactile stimulation (touching someone on the shoulder) come from afferent impulses traveling to the superior colliculus from the skin by way of the spinotectal pathways. All these pathways are described in this chapter.

2 p. 394. In the visual cortex the brain "sees" an object. Without a functional visual cortex a person is blind. The visual association areas allow one to relate objects seen to previous experiences and to interpret what has been seen. Similarly, other association areas allow one to relate the sensory information integrated in the primary sensory areas with previous experiences and to make judgments about the information. The association cortices therefore function to interpret and recognize information coming into the primary sensory areas of the brain.

3 p. 395. If a person holds an object in her right hand, tactile sensations of various types (described in this and the next chapter) travel up the spinal cord to the brain where they reach the somesthetic cortex of the left hemisphere. The information is passed in the form of impulses to the somesthetic association cortex where the object is recognized. Impulses then travel to Wernicke's area (probably on both sides of the cerebrum) where the object is given a name. From there impulses travel to Broca's area where the spoken word is initiated. Impulses from Broca's area travel to the premotor and motor areas where impulses are initiated that stimulate the muscles necessary to form the word.

4 p. 401. The cord is enlarged in the inferior thoracic and superior lumbar regions because of the large numbers of nerve fibers exiting from the cord to the limbs and entering the cord from the limbs. Also more nerve cell bodies in the cord regions are associated with the increased numbers of afferent and efferent fibers.

5 p. 402. Dorsal root ganglia contain nerve cell bodies, which are larger in diameter than the axons of the spinal nerves.

6 p. 406. Reciprocal innervation in an extended leg would cause the flexor muscles to relax.

7 p. 411. Collateral branches result in increased light-touch sensitivity because collaterals from a number of sensory nerve endings can converge onto one ascending neuron and enhance its afferent conduction. As a result, light touch requires less peripheral stimulation to produce an action potential in the ascending pathway. However, collateral, converging pathways result in less discriminative information because sensory receptors from more than one point of the skin have input onto the same ascending neuron, and the neuron cannot distinguish one small area of skin from another within the zone in which its afferent receptors are located.

8 p. 413. Lesions of the posterior funiculus (medial lemniscal system) result in loss of proprioception, two-point discrimination, and vibration on the same side of the body below the lesion because the posterior funiculus terminates in the nucleus gracilis and cuneatus without crossing over to the other side of the spinal cord. However, fibers from the nucleus gracilis and cuneatus enter the medial lemniscus. Lesions of the medial lemniscus result in loss on the opposite side of the body because the nerve tracts cross to the opposite side in the medulla oblongata.

9 p. 414. Most proprioception from the lower limbs is unconscious, whereas that from the upper limbs is mostly conscious. This difference is valuable because walking and standing (balance) are not activities on which we want to focus our attention, whereas proprioceptive activities of the arms and hands are essential for gaining information about the environment.

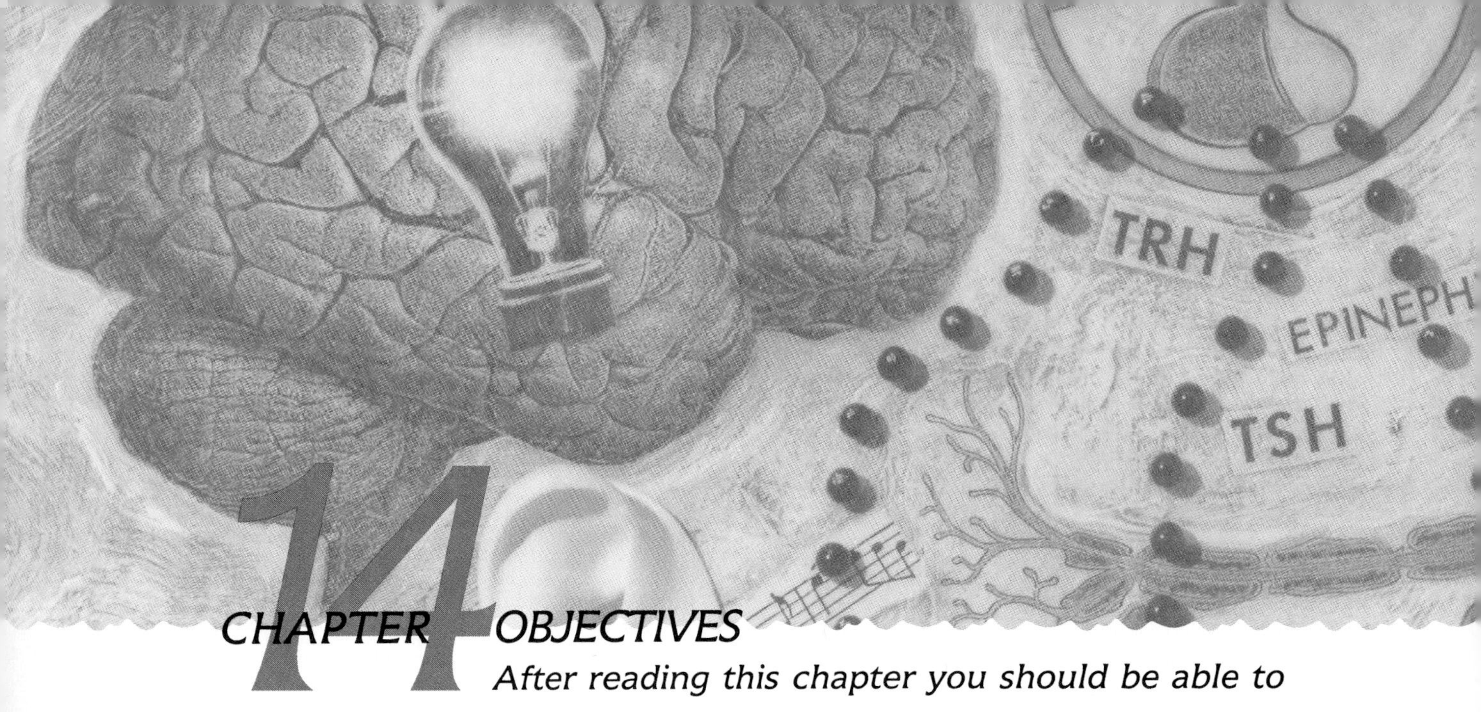

CHAPTER 14

OBJECTIVES

After reading this chapter you should be able to

1. Describe the distribution and function of each cranial nerve not involved in the special senses.
2. List the exclusively sensory cranial nerves and their functions.
3. List the somatomotor cranial nerves and their functions.
4. List the combination somatomotor and sensory cranial nerves and their functions.
5. List the combination somatomotor and parasympathetic cranial nerves and their functions.
6. List the combination somatomotor, sensory, and parasympathetic cranial nerves and their functions.
7. List, by letter and number, the spinal nerves existing at each vertebral level.
8. Explain what is meant by a dermatomal map.
9. Outline the pattern and distribution of the simplest spinal nerves, which do not participate in the formation of plexuses.
10. Explain the difference between ventral and dorsal rami.
11. Define the term plexus.
12. Describe the structure, distribution, and function of the cervical plexus.
13. Describe the general structure, distribution, and function of the nerves derived from the brachial plexus.
14. Explain the structural and functional basis of deficits resulting from damage to each upper-limb nerve.
15. Describe the structure, distribution, and function of the obturator, femoral, tibial, and common peroneal nerves.
16. Discuss the constitution of the sciatic nerve.
17. List the structures innervated by the coccygeal plexus.

Peripheral Nervous System: Cranial Nerves and Spinal Nerves

KEY TERMS

cranial nerve

dermatome (der′mă-tōm)

dorsal root

ganglion

parasympathetic

phrenic (fren′ik) nerve

plexus (plek′sus)

proprioception (pro′pre-o-sep′shun)

ramus (ra′mus)

sciatic (si-at′ik) nerve

somatomotor (so-mă′to-mo′tor)

spinal nerve

ventral root

RELATED TOPICS

The following terms or concepts are important for a good understanding of this chapter. If you are not familiar with them, you should review them before proceeding.

Directional terms and body regions (Chapter 1)

Membrane and action potentials (Chapter 9)

Muscle function and anatomy (Chapters 10 and 11)

Function of the nervous system (Chapter 12)

Central nervous system (Chapter 13)

THE PERIPHERAL NERVOUS SYSTEM (PNS) COLLECTS INFORMATION FROM NUMEROUS SOURCES BOTH INSIDE AND ON THE SURFACE OF THE INDIVIDUAL AND RELAYS IT BY WAY OF AFFERENT FIBERS TO THE CENTRAL NERVOUS SYSTEM (CNS), WHERE IT IS EVALUATED. EFFERENT FIBERS IN THE PNS RELAY INFORMATION FROM THE CNS TO VARIOUS PARTS OF THE BODY, PRIMARILY TO MUSCLES AND GLANDS, REGULATING ACTIVITY IN THOSE STRUCTURES.

The PNS can be divided into two parts: a cranial part, consisting of 12 pairs of nerves, and a spinal part, consisting of 31 pairs of nerves. The cranial portion of the PNS is discussed first in this chapter; discussion of the spinal nerves follows.

CRANIAL NERVES

The 12 **cranial nerves** are listed and illustrated in Table 14-1 and are illustrated in Figure 14-1. By convention, the cranial nerves are indicated by Roman numerals (I to XII) from anterior to posterior.

The three general categories of cranial nerve function are (1) sensory, (2) somatomotor and proprioception, and (3) parasympathetic. **Sensory** functions include the special senses such as vision and the more general senses such as touch and pain. **Somatomotor** (so-mă'to-mo'tor) functions refer to the control of skeletal muscles through motor neurons. **Proprioception** (pro'pre-o-sep'shun) provides the brain with information about the position of the body and its various parts, including joints and muscles.

Text continued on p. 436.

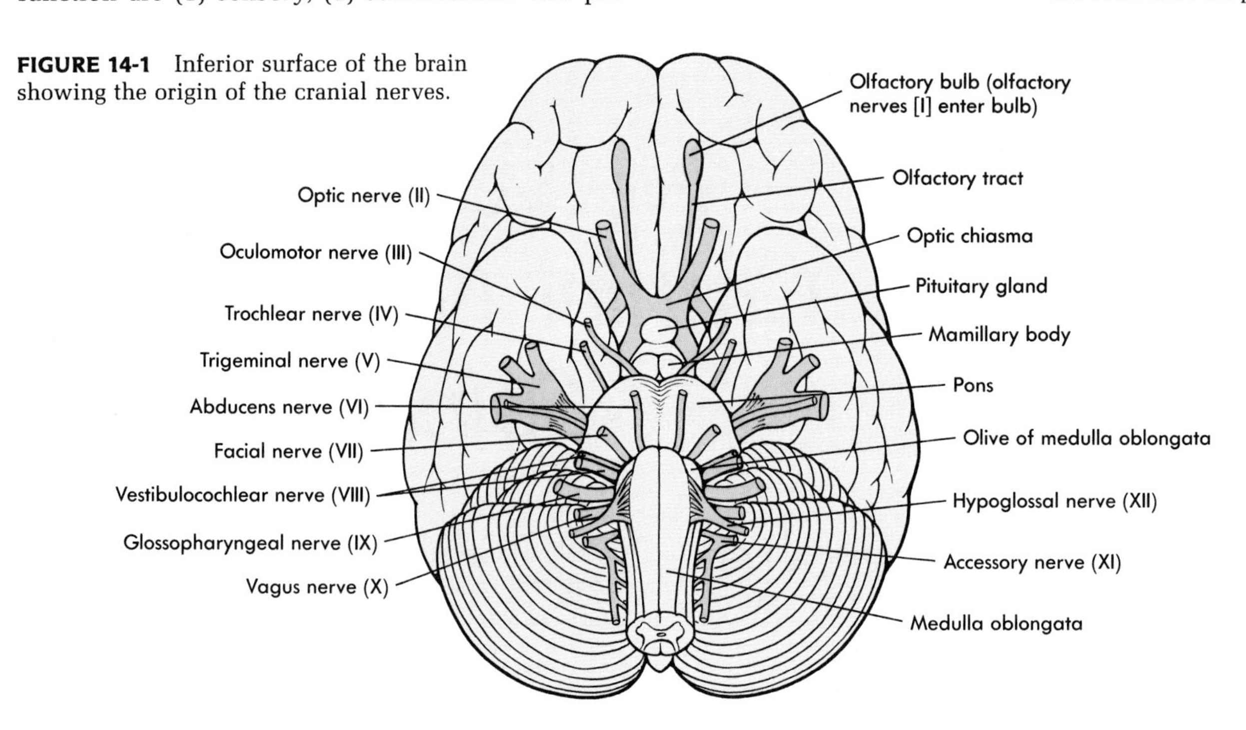

FIGURE 14-1 Inferior surface of the brain showing the origin of the cranial nerves.

Table 14-1

Cranial nerves and their functions

CRANIAL NERVE	FORAMEN OR FISSURE*	FUNCTION
I: Olfactory	Cribriform plate	Sensory Special sense of smell

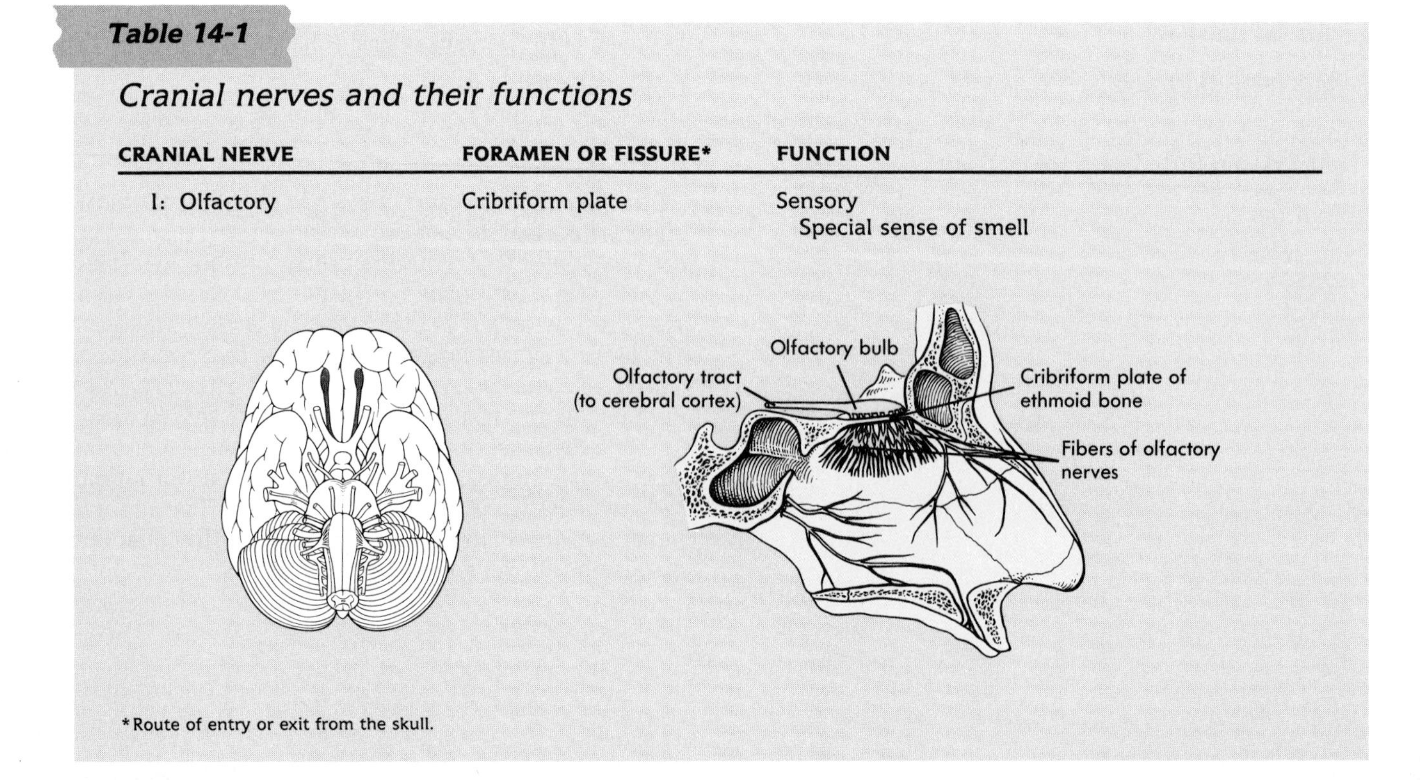

*Route of entry or exit from the skull.

Table 14-1

Cranial nerves and their functions—cont'd

CRANIAL NERVE	FORAMEN OR FISSURE	FUNCTION
II: Optic	Optic foramen	Sensory Special sense of vision
III: Oculomotor	Superior orbital fissure	Motor† and parasympathetic Motor to eye muscles (superior, medial, and inferior rectus; inferior oblique) and upper eyelid (levator palpebrae superioris) Proprioceptive to those muscles Parasympathetic to the sphincter of the pupil (causing constriction) and the ciliary muscle of the lens (causing accommodation)
IV: Trochlear	Superior orbital fissure	Motor Motor to one eye muscle (superior oblique) Proprioceptive to that muscle

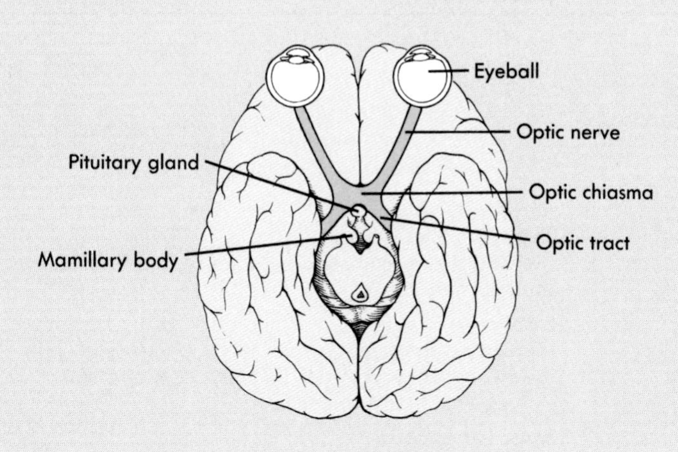

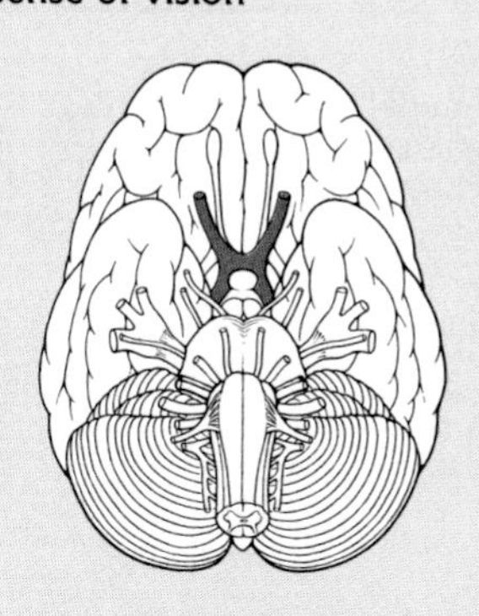

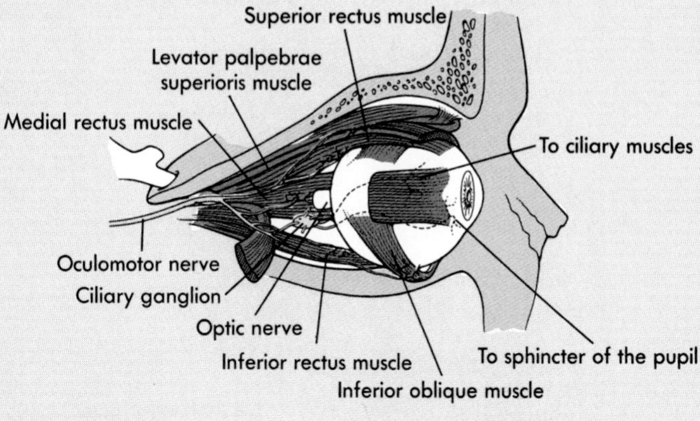

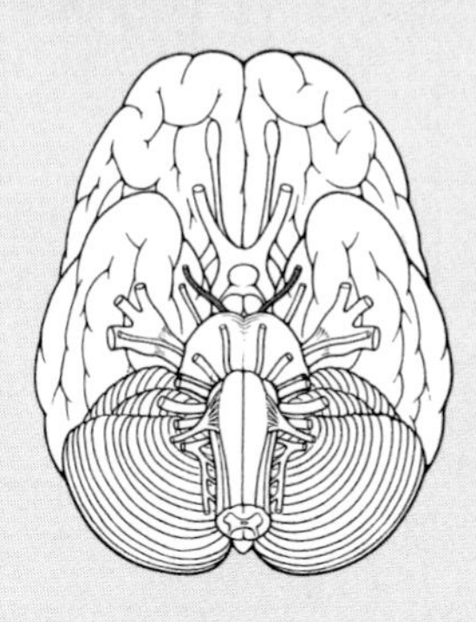

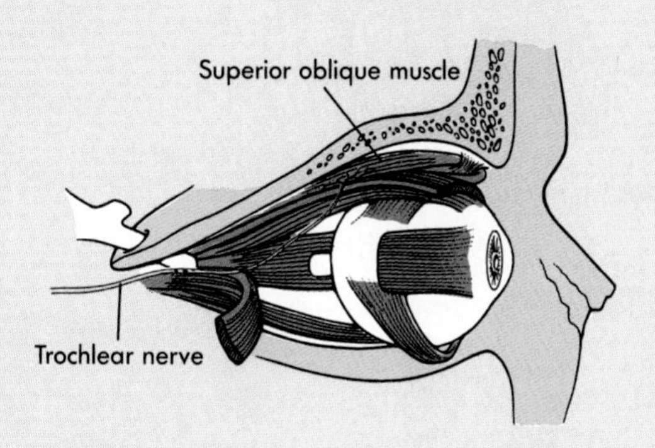

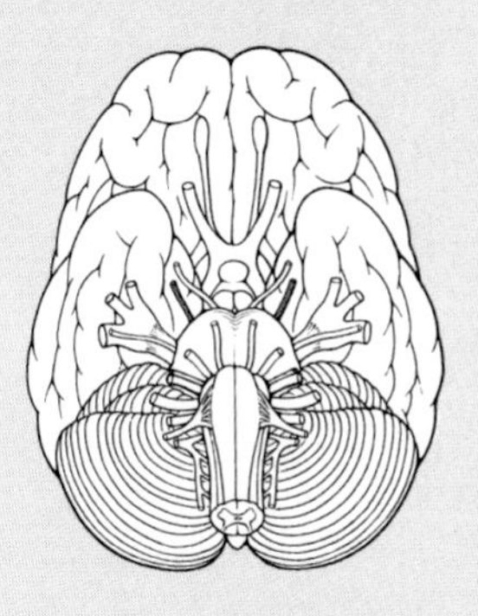

†Proprioception is a sensory function, not a motor function. However, motor nerves to muscles also conduct proprioceptive information from those muscles back to the brain. Since proprioception is the only sensory information carried by some cranial nerves, those nerves still are considered "motor."

Continued.

Table 14-1

Cranial nerves and their functions—cont'd

CRANIAL NERVE	FORAMEN OR FISSURE	FUNCTION
V: Trigeminal The trigeminal nerve is divided into three branches—the ophthalmic (V_1), the maxillary (V_2), and the mandibular (V_3)		
Ophthalmic branch (V_1)	Superior orbital fissure	Sensory Sensory from scalp, forehead, nose, upper eyelid, and cornea
Maxillary branch (V_2)	Foramen rotundum	Sensory Sensory from palate, upper jaw, upper teeth and gums, nasopharynx, nasal cavity, skin of cheek, lower eyelid, and upper lip
Mandibular branch (V_3)	Foramen ovale	Sensory and motor Sensory from lower jaw, lower teeth and gums, anterior two thirds of tongue, mucous membrane of cheek, lower lip, skin of chin, auricle, and temporal region Motor to muscles of mastication (masseter, temporalis, medial and lateral pterygoids), soft palate (tensor veli palatini), throat (anterior belly of digastric, mylohyoid), and middle ear (tensor tympani) Proprioceptive to those muscles

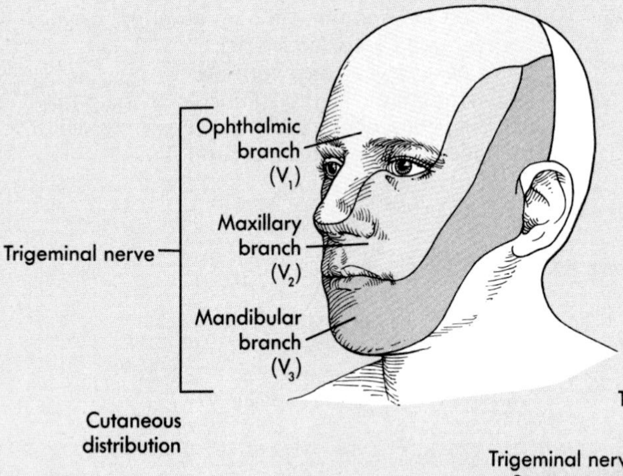

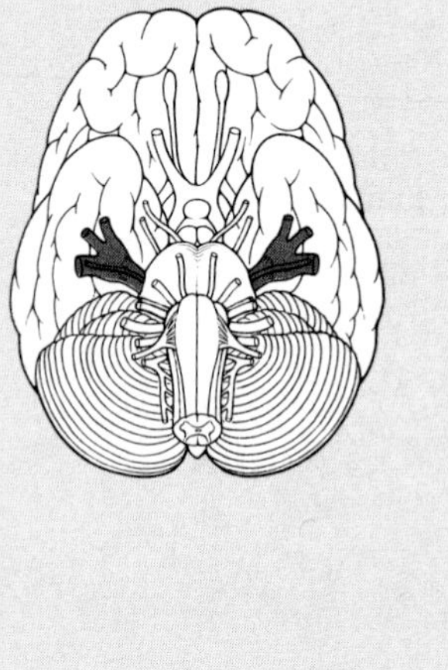

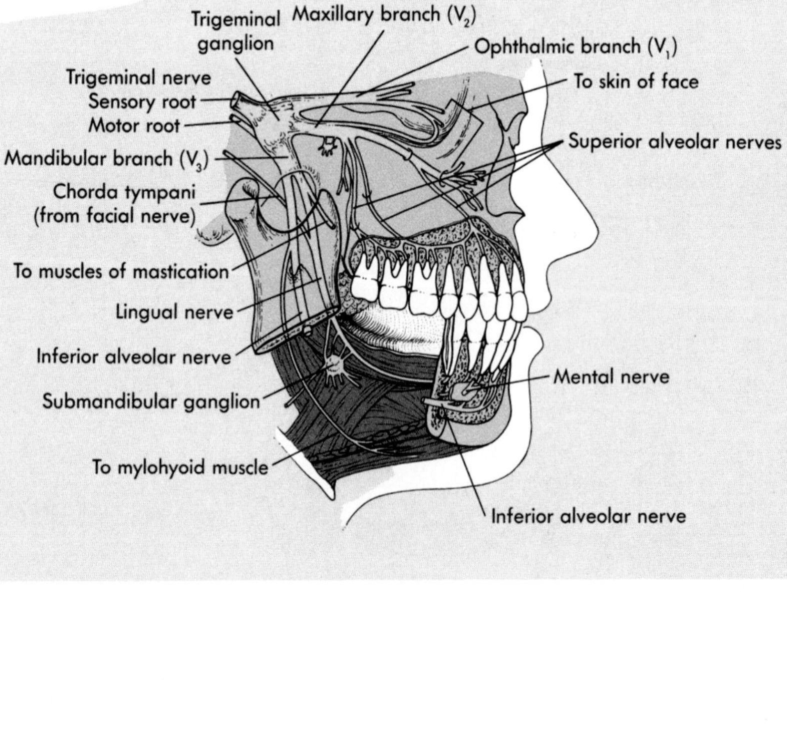

Table 14-1

Cranial nerves and their functions—cont'd

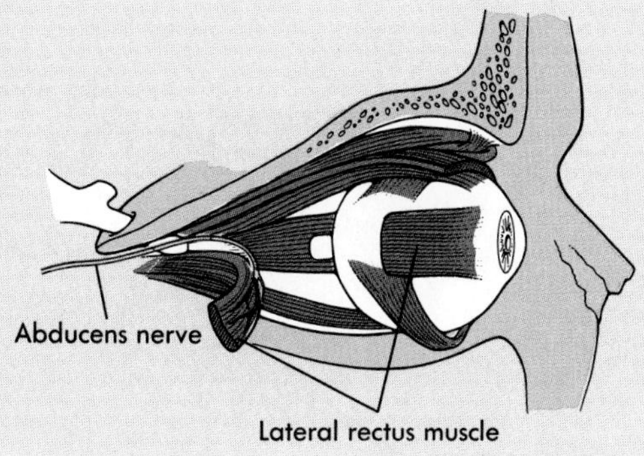

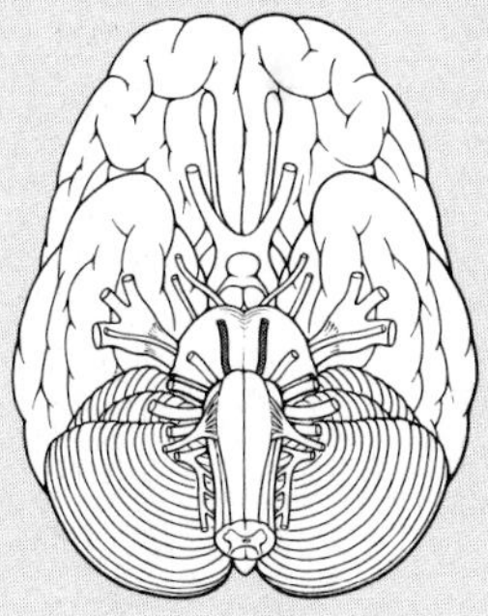

CRANIAL NERVE	FORAMEN OR FISSURE	FUNCTION
VI: Abducens	Superior orbital fissure	Motor Motor to one eye muscle (lateral rectus) Proprioceptive to that muscle
VII: Facial	Stylomastoid foramen	Sensory, motor, and parasympathetic Sense of taste from anterior two thirds of tongue, sensory from external ear and palate Motor to muscles of facial expression, throat (posterior belly of digastric, stylohyoid), and middle ear (stapedius) Proprioceptive to those muscles Parasympathetic to submandibular and sublingual salivary glands, lacrimal gland, glands of the nasal cavity and palate

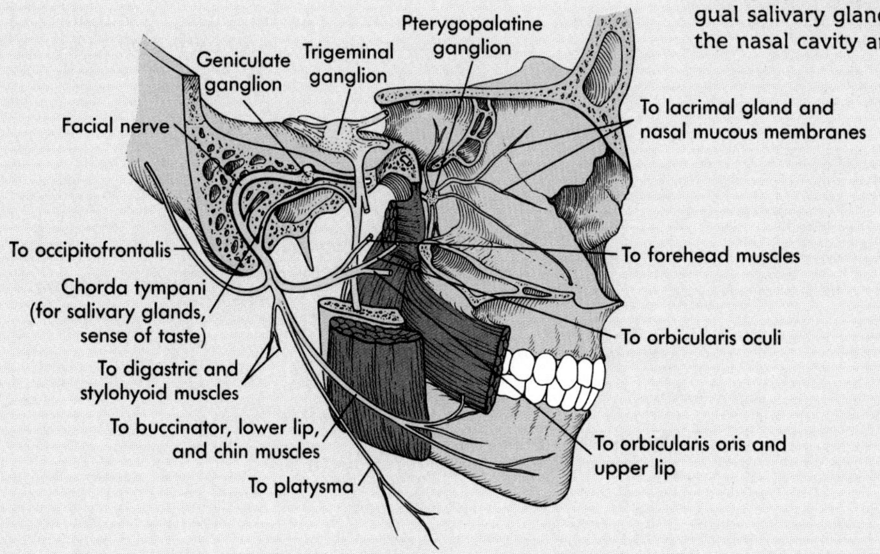

Abducens nerve
Lateral rectus muscle

Pterygopalatine ganglion
Geniculate ganglion
Trigeminal ganglion
Facial nerve
To lacrimal gland and nasal mucous membranes
To occipitofrontalis
To forehead muscles
Chorda tympani (for salivary glands, sense of taste)
To orbicularis oculi
To digastric and stylohyoid muscles
To buccinator, lower lip, and chin muscles
To orbicularis oris and upper lip
To platysma

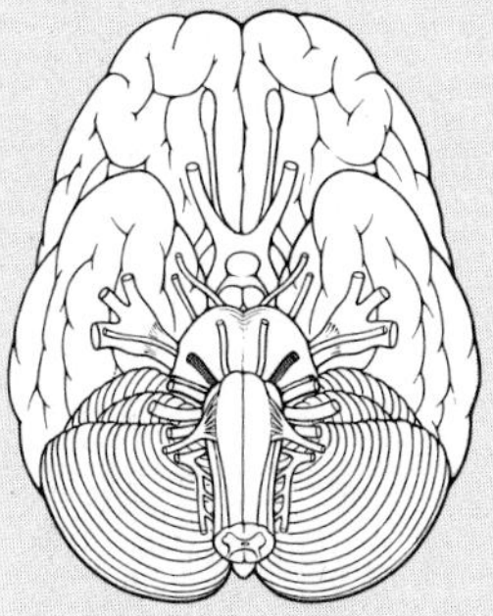

Continued.

Table 14-1

Cranial nerves and their functions—cont'd

CRANIAL NERVE	FORAMEN OR FISSURE	FUNCTION
VIII: Vestibulocochlear	Internal auditory meatus	Sensory Special senses of hearing and balance
IX: Glossopharyngeal	Jugular foramen	Sensory, motor, and parasympathetic Sense of taste from posterior one third of tongue, sensory from pharynx, palatine tonsils, posterior one third of tongue, middle ear, carotid sinus and carotid body Motor to pharyngeal muscle (stylopharyngeus) Proprioceptive to that muscle Parasympathetic to parotid salivary gland and the glands of the posterior one third of tongue

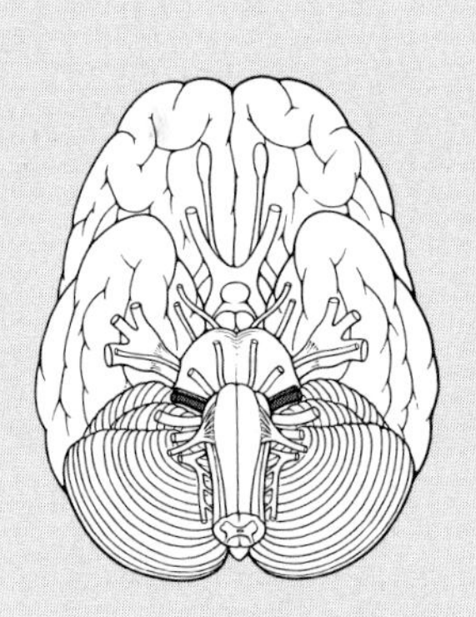

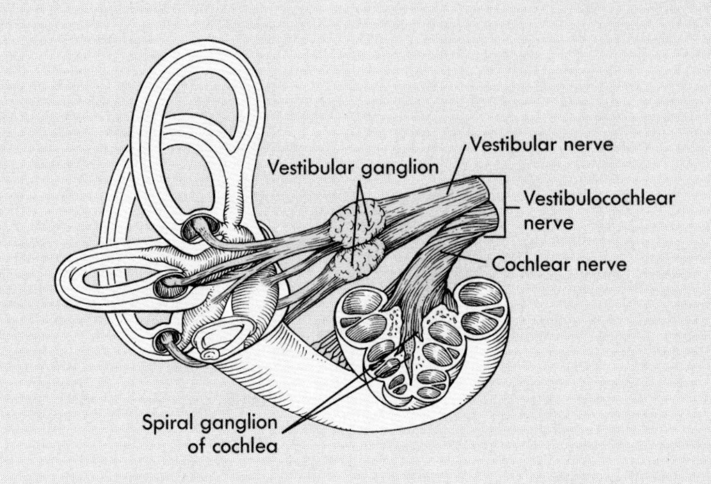

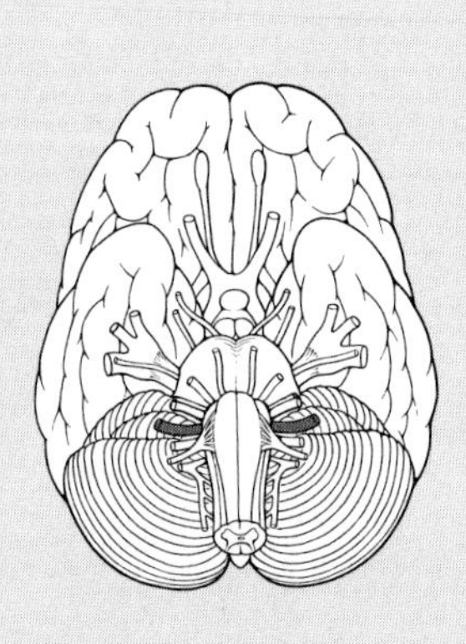

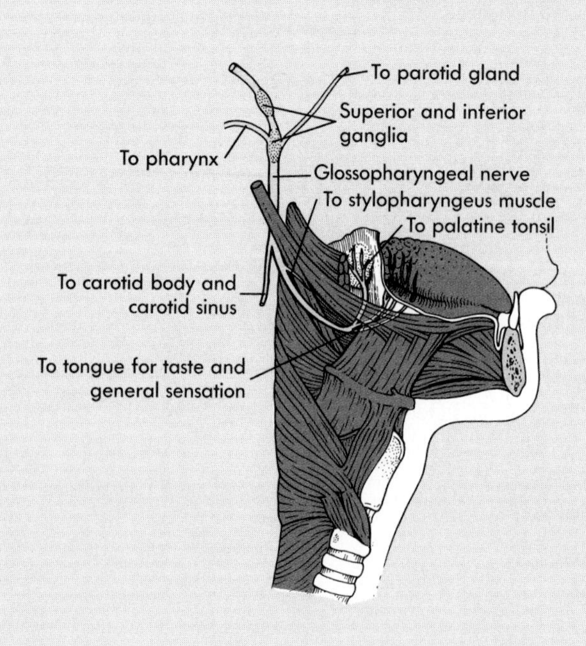

Table 14-1

Cranial nerves and their functions—cont'd

CRANIAL NERVE	FORAMEN OR FISSURE	FUNCTION
X: Vagus	Jugular foramen	Sensory, motor, and parasympathetic Sensory from inferior pharynx, larynx, thoracic and abdominal organs, sense of taste from posterior tongue Motor to soft palate, pharynx, intrinsic laryngeal muscles (voice production), and an extrinsic tongue muscle (palatoglossus) Proprioceptive to those muscles Parasympathetic to thoracic and abdominal viscera

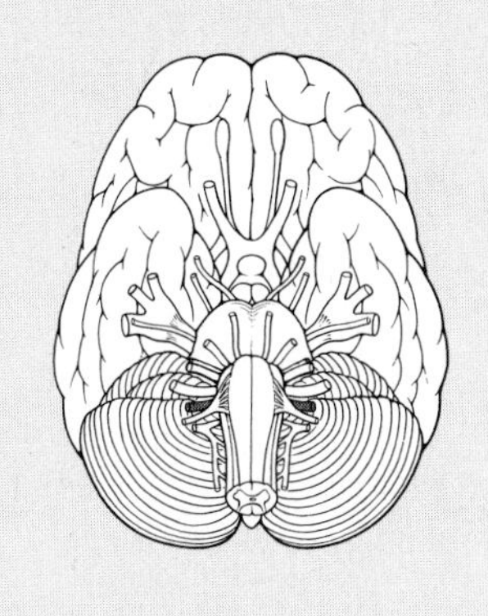

CRANIAL NERVE	FORAMEN OR FISSURE	FUNCTION
XI: Accessory	Jugular foramen	Motor Motor to soft palate, pharynx, sternocleidomastoid, and trapezius Proprioceptive to those muscles

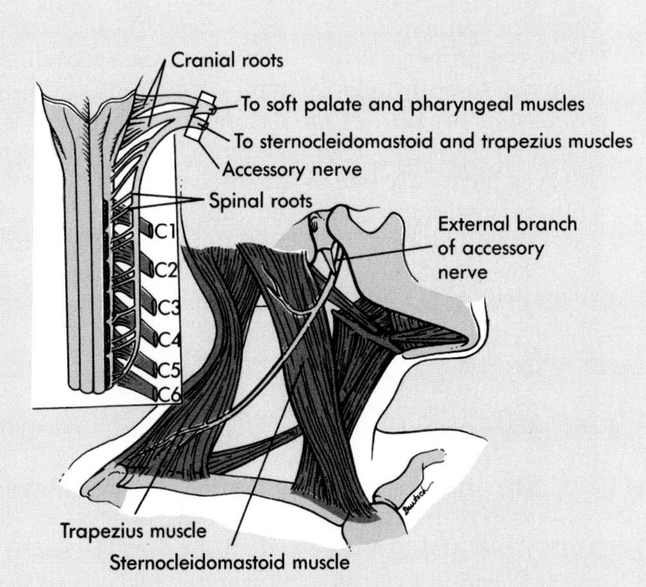

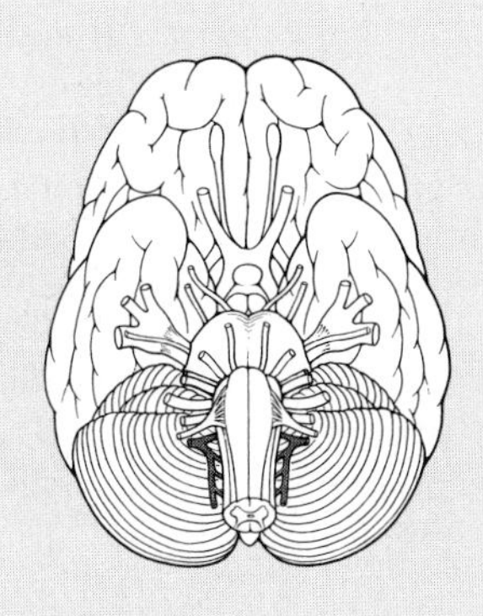

Continued.

Table 14-1

Cranial nerves and their functions—cont'd

CRANIAL NERVE	FORAMEN OR FISSURE	FUNCTION
XII: Hypoglossal	Hypoglossal canal	Motor Motor to intrinsic and extrinsic tongue muscles (styloglossus, hyoglossus, genioglossus), and throat muscles (thyrohyoid and geniohyoid) Proprioceptive to those muscles

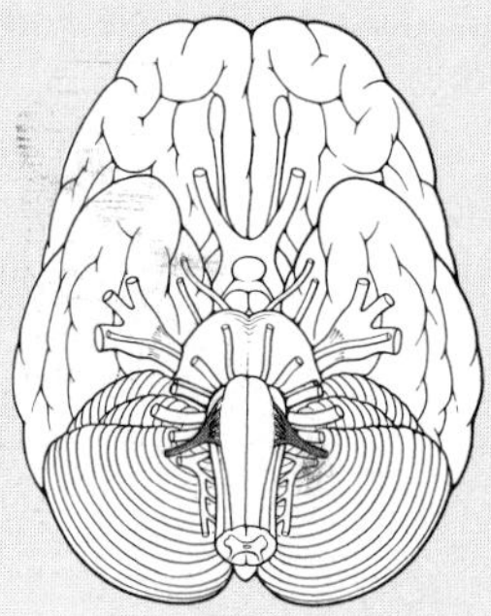

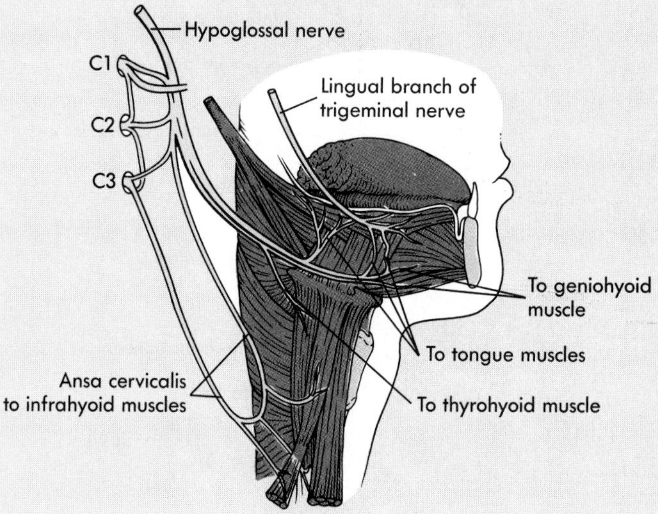

Table 14-2

Functional organization of the cranial nerves

NERVE FUNCTION	CRANIAL NERVE
Sensory	I Olfactory II Optic VIII Vestibulocochlear
Somatomotor/proprioception	IV Trochlear VI Abducens XI Accessory XII Hypoglossal
Somatomotor/proprioception and sensory	V Trigeminal
Somatomotor/proprioception and parasympathetic	III Oculomotor
Somatomotor/proprioception, sensory, and parasympathetic	VII Facial IX Glossopharyngeal X Vagus

Although proprioception involves sensory input to the CNS (see Chapter 13) concerning the position or movement of muscles and tendons, it is included with somatomotor function because the nerves supplying motor input to muscles also convey proprioceptive impulses to the CNS from those muscles. **Parasympathetic** function involves the regulation of glands, smooth muscles, and cardiac muscle (these functions are part of the autonomic nervous system and are discussed in Chapter 16). Some cranial nerves have only one of the three functions, whereas others have more than one. Table 14-2 lists the general organization of the cranial nerves by function.

Several of the cranial nerves have ganglia associated with them. These ganglia are of two types: parasympathetic and sensory.

Sensory

The **olfactory** (I), **optic** (II), and **vestibulocochlear** (ves-tib′u-lo-kōk′le-ar; VIII) nerves are sensory only and are involved in the special senses of smell, vision, hearing, and balance. These nerves are discussed in Chapter 15.

Somatomotor/Proprioception

The **trochlear** (trōk'le-ar; IV), **abducens** (ab-du'sens; VI), **accessory** (XI), and **hypoglossal** (XII) nerves are somatomotor and proprioceptive nerves. The trochlear and abducens nerves each control one of the six eye muscles responsible for moving the eyeball.

The accessory nerve has both a cranial and a spinal component. The cranial component joins the vagus nerve (hence the name accessory) and participates in its function. The spinal component of the accessory nerve provides the major innervation to the sternocleidomastoid and trapezius muscles of the neck and shoulder.

1

Injury to the spinal portion of the accessory nerve may result in sternocleidomastoid muscle dysfunction, a condition called wry neck. If the head of a person with wry neck were turned to the left, would this position indicate injury to the left or right spinal component of the accessory nerve?

? ? ? ? ? ? ? ? ? ? ?

The hypoglossal nerve supplies the intrinsic tongue muscles, three of the four extrinsic tongue muscles, and the thyrohyoid and the geniohyoid muscles.

2

Unilateral damage to the hypoglossal nerve results in loss of tongue movement on one side, which is most obvious when the tongue is protruded. If the tongue is deviated to the right, would the left or right hypoglossal nerve be damaged?

? ? ? ? ? ? ? ? ? ? ?

Somatomotor/Proprioception and Sensory

The **trigeminal** (tri-jem'ĭ-nal; V) **nerve** has somatomotor, proprioceptive, and cutaneous sensory functions. It supplies motor innervation to the muscles of mastication, one middle ear muscle, one palatine muscle, and two throat muscles. In addition to proprioception associated with its somatomotor functions, the trigeminal nerve also supplies proprioception to the temporomandibular joint. Damage to the trigeminal nerve may impede chewing.

The trigeminal nerve has the greatest general sensory function of all the cranial nerves and is the only cranial nerve involved in **sensory cutaneous innervation.** All other cutaneous innervation comes from spinal nerves (see Figure 14-4). Its sensory distribution in the face is divided into three regions (thus the name trigeminal), each supplied by a branch of the nerve. The three branches—ophthalmic, maxillary, and mandibular—arise directly from the trigeminal ganglion, which serves the same function as the dorsal root ganglia of the spinal nerves.

In addition to these cutaneous functions, the maxillary and mandibular branches are important in dentistry. The maxillary nerve supplies sensory innervation to the maxillary teeth, palate, and gingiva (jin'jĭ-vah; gum); the mandibular branch supplies sensory innervation to the mandibular teeth, tongue, and gingiva. The various nerves innervating the teeth are referred to as **alveolar** (al've-o'lar; refers to the sockets in which the teeth are located). Nerves to the maxillary teeth are derived from the maxillary branch of the trigeminal nerve and are called **superior alveolar;** ones to the mandibular teeth are derived from the mandibular branch of the trigeminal nerve and are called **inferior alveolar.**

Injections of anesthetic administered by a dentist are designed to block sensory transmission by the alveolar nerves. The superior alveolar nerves are not usually anesthetized directly because they are difficult to approach with a needle. For that reason the maxillary teeth usually are anesthetized locally by inserting the needle beneath the oral mucosa surrounding the teeth. The inferior alveolar nerve probably is anesthetized more often than any other nerve in the body. To anesthetize this nerve, the dentist inserts the needle somewhat posterior to the patient's last molar.

Several nondental nerves usually are anesthetized during an inferior alveolar block. The mental nerve, which is cutaneous to the anterior lip and chin, is a branch of the inferior alveolar nerve. Therefore when the inferior alveolar nerve is blocked, the mental nerve is blocked also, resulting in a numb lip and chin. Nerves lying near the point at which the inferior alveolar nerve enters the mandible often are also anesthetized during inferior alveolar anesthesia (e.g., the lingual nerve can be anesthetized to produce a numb tongue). The facial nerve lies some distance from the inferior alveolar nerve, but in rare cases anesthesia can travel far enough posteriorly to anesthetize that nerve. The result is a temporary facial palsy (with the injected side of the face drooping because of flaccid muscles), which will disappear when the anesthesia wears off. If the facial nerve is cut by an improperly inserted needle, permanent facial palsy may occur.

Somatomotor/Proprioception and Parasympathetic

The **oculomotor nerve** (III) innervates four of the six muscles that move the eyeball and the levator palpebrae superioris, which raises the superior eyelid. In addition, through the parasympathetic system the oculomotor nerve regulates the size of the pupil and the shape of the lens of the eye.

3

A drooping upper eyelid on one side of the face is a sign of possible oculomotor nerve damage. Describe how this possibility could be evaluated by examining other oculomotor nerve functions. Describe the movements of the eye that would distinguish between oculomotor, trochlear, and abducens nerve damage.

? ? ? ? ? ? ? ? ? ? ?

Somatomotor/Proprioception, Sensory, and Parasympathetic

The **facial** (VII), **glossopharyngeal** (IX), and **vagus** (X) nerves perform all three general functions listed for the cranial nerves. All three nerves have both sensory and parasympathetic ganglia associated with them. The facial nerve controls all the muscles of facial expression, a small muscle in the middle ear, and two throat muscles. The glossopharyngeal nerve innervates one muscle of the pharynx. Most muscles of the soft palate, pharynx, and larynx are innervated by the vagus nerve. Damage to the laryngeal branches of the vagus nerve can interfere with normal speech.

The facial, glossopharyngeal, and vagus nerves are involved with the sense of taste (see Chapter 15). The glossopharyngeal nerve also supplies tactile sensory innervation from the posterior tongue, middle ear, and pharynx. The vagus nerve is sensory for the inferior pharynx and the larynx. The glossopharyngeal and vagus nerves also transmit sensory stimulation from receptors in the carotid arteries and the aortic arch, which monitor blood pressure and blood carbon dioxide, oxygen, and pH levels (see Chapter 21). In addition, the vagus nerve conveys sensory information from the thoracic and abdominal organs.

The facial and glossopharyngeal nerves supply parasympathetic innervation to the salivary glands. The facial nerve also innervates the lacrimal glands. The parasympathetic portion of the vagus nerve is very important in regulating the functions of the thoracic and abdominal organs. It carries parasympathetic fibers to the heart and lungs in the thorax, to the digestive organs, and to the spleen and kidneys in the abdomen. The vagus nerve also carries afferent impulses from these organs back to the brain.

SPINAL NERVES

The **spinal nerves** arise through numerous rootlets along the dorsal and ventral surfaces of the spinal cord (Figure 14-2). Approximately six to eight of these rootlets combine to form a **ventral root** on the

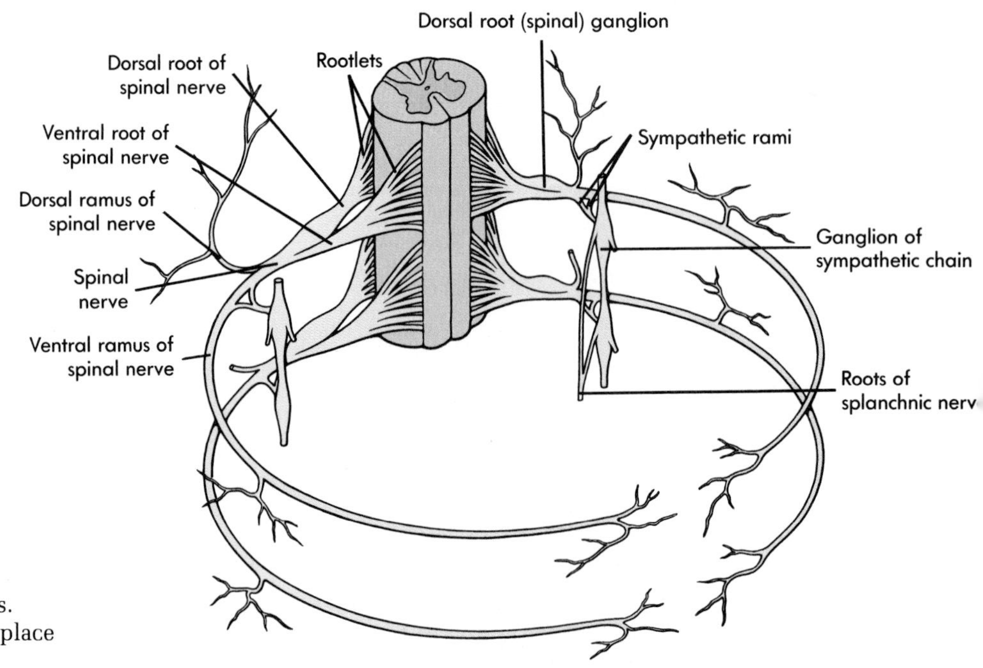

FIGURE 14-2
A Typical thoracic spinal nerves.
B Photograph of dorsal roots in place along the vertebral column.

ventral (anterior) side of the spinal cord, and six to eight rootlets form a **dorsal root** on the dorsal (posterior) side of the cord at each segment. The ventral root contains efferent (motor) fibers, and the dorsal root contains afferent (sensory) fibers. The dorsal and ventral roots join each other just lateral to the spinal cord to form the spinal nerve. The dorsal root contains a ganglion, called the **dorsal root ganglion,** or **spinal ganglion,** near where it joins the ventral root.

All of the 31 pairs of spinal nerves, except the first pair of spinal nerves and the spinal nerves in the sacrum, exit the vertebral column through an intervertebral foramen located between adjacent vertebrae. The first pair of spinal nerves exits between the skull and the first cervical vertebra. The nerves of the sacrum exit from this single bone through sacral foramina (see Chapter 7). Eight spinal nerve pairs exit the vertebral column in the cervical region, 12 in the thoracic region, five in the lumbar region, five in the sacral region, and one in the coccygeal region (Figure 14-3). For convenience, each of the spinal nerves is designated by a letter and number. The letter indicates the region of the vertebral column from which the nerve emerges: C, cervical; T, thoracic; L, lumbar; and S, sacral. The single coccygeal nerve is often not designated, but when it is, the symbol often used is Cx. The number indicates the location in each region where the nerve emerges from the vertebral column, with the smallest number always representing the most superior origin. For example, the most superior nerve exiting from the thoracic region of the vertebral column is designated T1. The cervical nerves are designated C1 to C8, the tho-

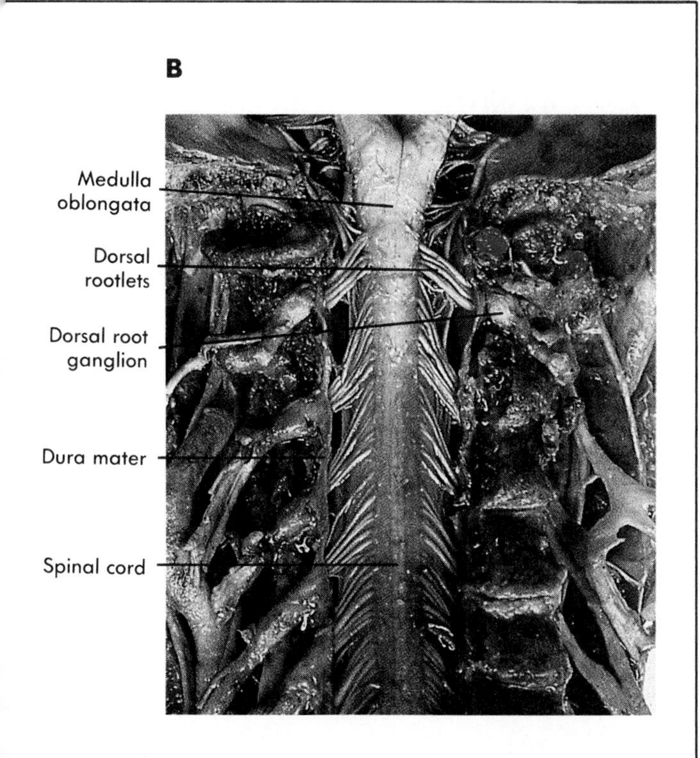

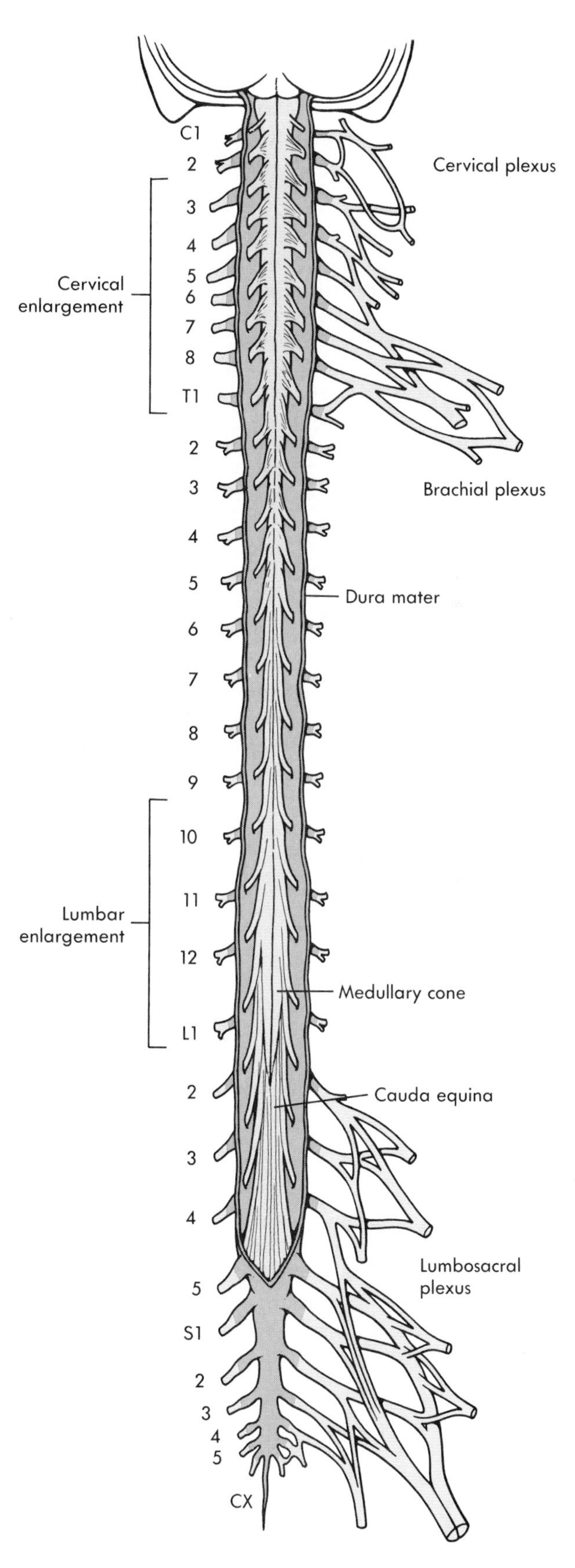

FIGURE 14-3 Spinal cord and spinal nerves.

racic nerves T1 to T12, the lumbar nerves L1 to L5, and the sacral nerves S1 to S5.

Each of the spinal nerves except C1 has a specific cutaneous sensory distribution. Figure 14-4 depicts the **dermatomal** (der-mă-to′mal) **map** for the sensory cutaneous distribution of the spinal nerves. A **dermatome** is the area of skin supplied by a given pair of spinal nerves.

4

The dermatomal map is important in clinical considerations of nerve damage. Loss of sensation in a dermatomal pattern can provide valuable information about the location of nerve damage. Predict the possible site of nerve damage for a patient who suffered whiplash in an automobile accident and subsequently developed anesthesia (no sensations) in the left arm, forearm, and hand (see Figure 14-4 for help).

? ? ? ? ? ? ? ? ? ? ?

Figure 14-2 depicts an idealized section through the trunk. Each spinal nerve has two major rami (ra′mi; branches), dorsal and ventral. Additional rami, called sympathetic rami, from the thoracic and upper lumbar spinal cord regions carry nerves associated with the sympathetic nervous system (see Chapter 16). The **dorsal rami** innervate most of the deep muscles of the dorsal trunk responsible for movement of the vertebral column. They also supply sensation to the connective tissue and skin near the midline of the back.

The **ventral rami** are distributed in two ways. In the thoracic region the ventral rami form **intercostal** (between ribs) **nerves** that extend along the inferior margin of each rib and innervate the intercostal muscles and the skin over the thorax. The ventral rami of the remaining spinal nerves are associated into five **plexuses** (plek′sus-ēz). The term plexus means "braid" and describes the organization produced by the intermingling of the nerves. Ventral rami of different spinal nerves join with each other to form a plexus. The axons from the spinal nerves mix, so the nerves that arise from plexuses usually have axons from more than one level of the spinal cord. The ventral rami of spinal nerves C1 to C4 form the cervical plexus; C5 to T1 form the brachial plexus; L1 to L4 form the lumbar plexus; L4 to S4 form the sacral plexus; and S4, S5, and the coccygeal nerve (Cx) form the coccygeal plexus.

Several smaller somatic plexuses are derived from more distal branches of the spinal nerves (an example is the pudendal plexus in the pelvis). Some of them are mentioned where appropriate in this chapter.

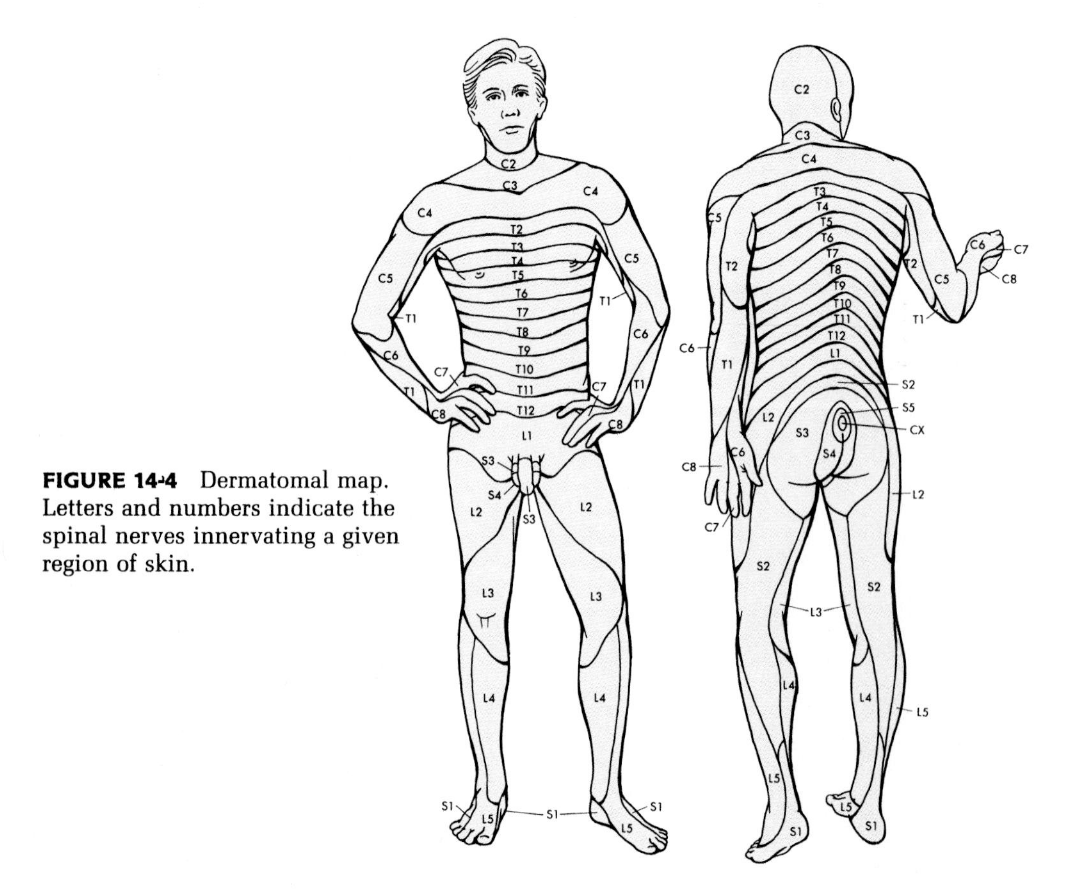

FIGURE 14-4 Dermatomal map. Letters and numbers indicate the spinal nerves innervating a given region of skin.

Cervical Plexus

The **cervical plexus** is a relatively small plexus originating from spinal nerves C1 to C4 (Figure 14-5). Branches derived from this plexus innervate superficial neck structures, including several of the muscles attached to the hyoid bone. Its cutaneous innervation is to the neck and posterior portion of the head (see Figure 14-4).

Perhaps the most important derivative of the cervical plexus, the **phrenic** (fren'ik) **nerve,** originates from spinal nerves C3 to C5 (derived from both the cervical and brachial plexus). The phrenic nerves descend along each side of the neck to enter the thorax. They descend along the side of the mediastinum (middle wall of the thorax) to reach the diaphragm, which they innervate. Contraction of the diaphragm is largely responsible for the ability to breathe.

5

The phrenic nerve may be damaged where it descends along the neck. Because the phrenic nerve descends along the mediastinum, it can also be damaged during open thoracic surgery or open heart surgery. Explain how damage to the right phrenic nerve would affect the diaphragm. Describe the effect on breathing of complete severance of the spinal cord in the lower- vs. the upper-cervical regions.

? ? ? ? ? ? ? ? ? ? ?

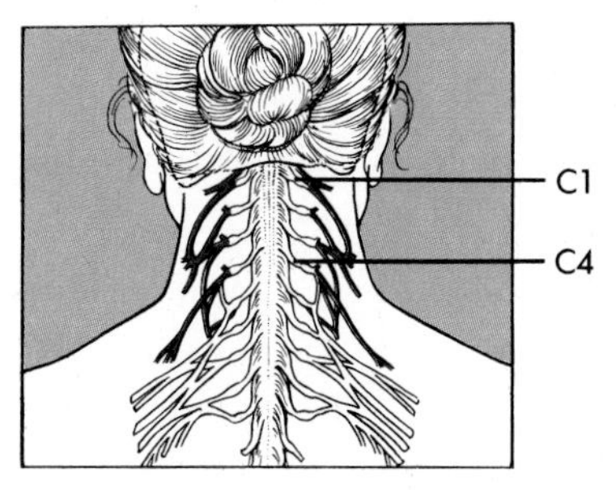

Cervical plexus

FIGURE 14-5 Cervical plexus. The roots of the plexus are formed by the ventral rami of the spinal nerves.

FIGURE 14-6 Brachial plexus. The roots (ventral rami) join to form an upper, middle, and lower trunk. Each trunk divides into anterior and posterior divisions. The divisions join together to form the posterior, lateral, and medial cords from which the major brachial plexus nerves arise.

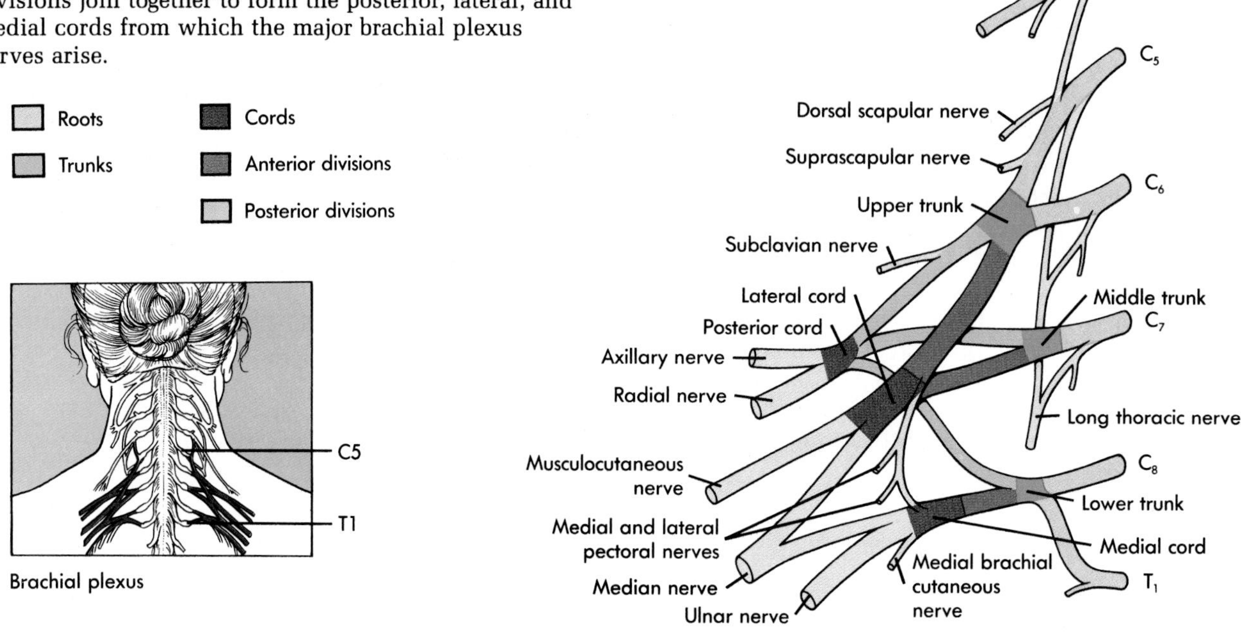

Brachial Plexus

The **brachial plexus** originates from spinal nerves C5 to T1 (Figure 14-6). The five ventral rami that comprise the brachial plexus join to form three **trunks,** which separate into six divisions, and then join again to create three **cords** (posterior, lateral, and medial) from which five **branches** or nerves of the upper limb emerge.

If a physician wishes to anesthetize the entire arm, anesthetic can be injected near the brachial plexus. The injection is made between the neck and the shoulder posterior to the clavicle and is called brachial anesthesia.

The five major nerves emerging from the brachial plexus to supply the upper limb are the axillary, radial, musculocutaneous, ulnar, and median nerves. The axillary nerve innervates part of the shoulder; the radial nerve innervates the posterior arm, forearm, and hand; the musculocutaneous nerve innervates the anterior arm; and the ulnar and median nerves innervate the anterior forearm and hand. Additional brachial plexus nerves innervate the shoulder and pectoral muscles.

Axillary nerve

The **axillary** (ak′sĭ-lăr′e) **nerve** innervates the deltoid and teres minor muscles (Figure 14-7). It also provides sensory innervation to the shoulder joint and to the skin over part of the shoulder (Table 14-3).

Radial nerve

The **radial nerve** emerges from the posterior cord of the brachial plexus and descends within the deep aspect of the posterior arm (Figure 14-8). Approximately midway down the shaft of the humerus it lies against the bone in the radial groove. The radial nerve innervates all of the extensor muscles of the upper limb, the supinator muscle, and two muscles that flex the forearm. Its cutaneous sensory distribution is to the posterior portion of the upper limb, including the posterior surface of the hand (Table 14-4).

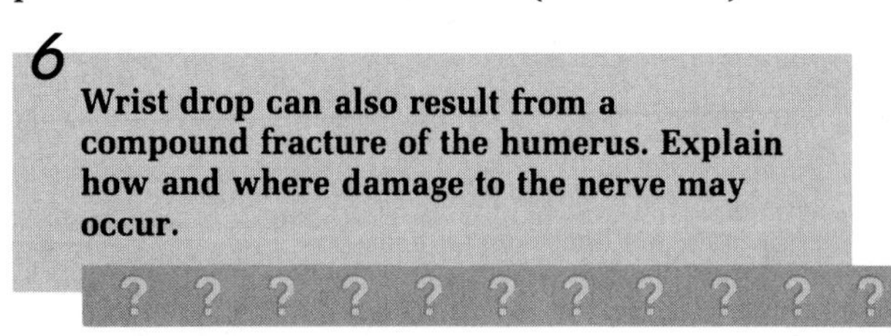

6

Wrist drop can also result from a compound fracture of the humerus. Explain how and where damage to the nerve may occur.

Table 14-3

Axillary nerve

Origin

Posterior cord of brachial plexus, C5-C6

Movements/Muscle Innervated

Abducts arm
 Deltoid

Laterally rotates arm
 Teres minor

Cutaneous Innervation

Inferior lateral shoulder

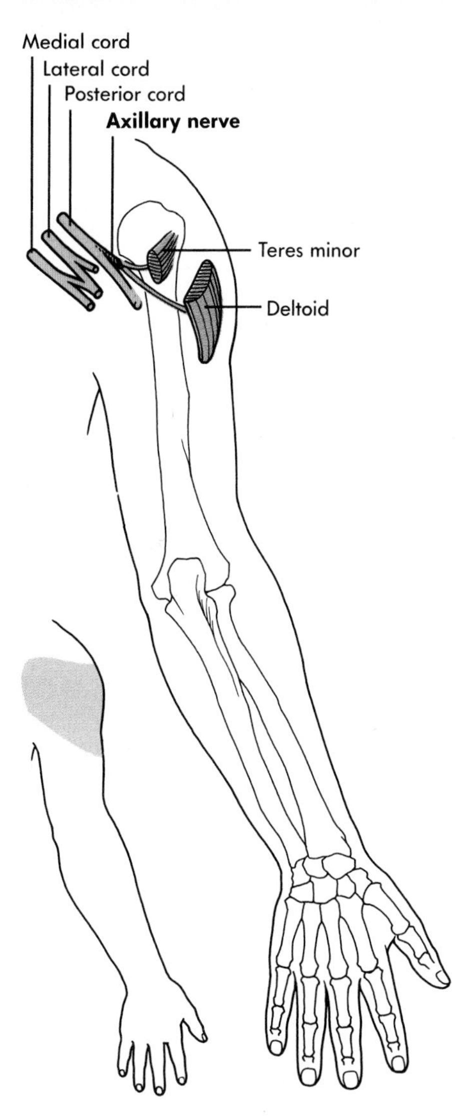

FIGURE 14-7 Route of the axillary nerve and the muscles it innervates. Inset depicts the cutaneous distribution of the axillary nerve *(shaded areas)*.

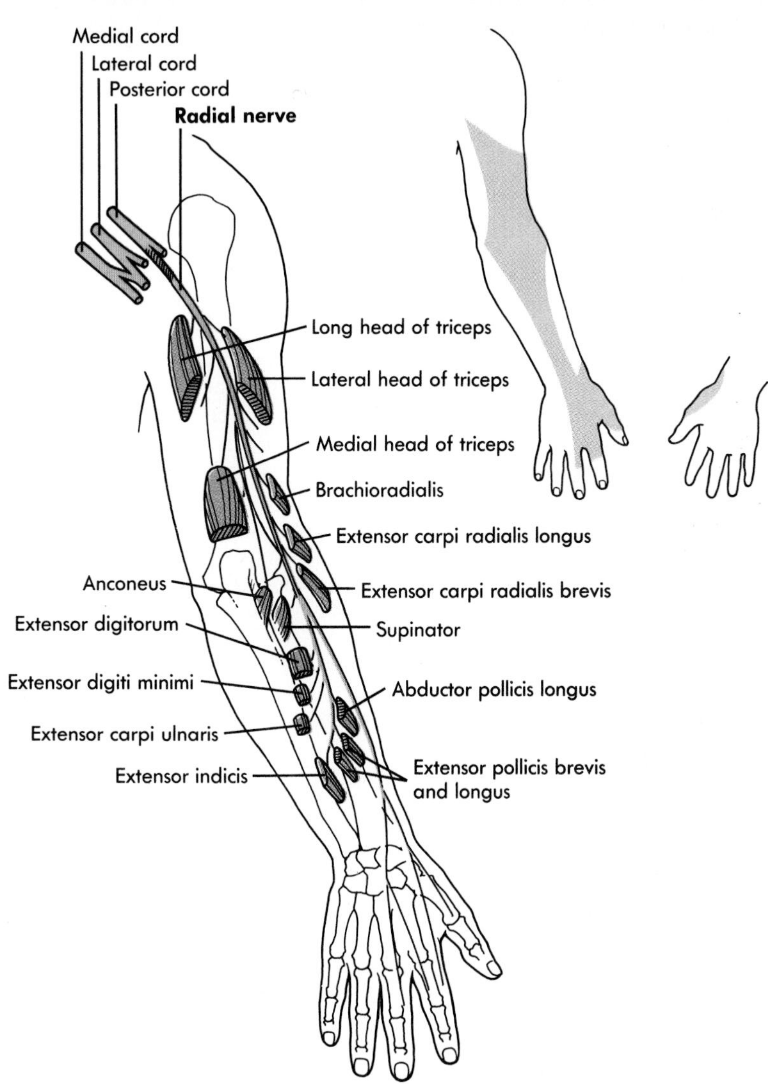

FIGURE 14-8 Route of the radial nerve and the muscles it innervates. Inset depicts the cutaneous distribution of the radial nerve *(shaded areas)*.

Table 14-4

Radial nerve

Origin

Posterior cord of brachial plexus, C5-T1

Movements/Muscle Innervated

Extends forearm
- Triceps brachii
- Anconeus

Flexes forearm
- Brachialis (part)
- Brachioradialis

Supinates forearm
- Supinator

Extends and abducts wrist
- Extensor carpi radialis longus
- Extensor carpi radialis brevis

Extends and adducts wrist
- Extensor carpi ulnaris

Extends fingers
- Extensor digitorum
- Extensor digiti minimi
- Extensor indicis

Extends thumb
- Extensor pollicis longus
- Extensor pollicis brevis

Abducts thumb
- Abductor pollicis longus

Cutaneous Innervation

Posterior surface of arm and forearm, lateral two thirds of dorsum of hand

Because the radial nerve lies near the humerus in the axilla, the radial nerve can be damaged if it is compressed against the humerus. Improper use of crutches in which the crutch is pushed tightly into the axilla can result in "crutch paralysis." In this disorder the radial nerve is compressed between the top of the crutch and the humerus. As a result, the radial nerve is damaged, and the muscles it innervates lose their function. The major symptom of crutch paralysis is "wrist drop" in which the extensor muscles of the wrist and fingers (innervated by the radial nerve) fail to function; as a result, the elbow, wrist, and fingers are constantly flexed.

The wrist drop caused by radial nerve damage can also occur as a result of improper attempts to reset a dislocated shoulder. Historically, one method of relocating the shoulder was to place a foot in the patient's axilla and pull on the arm. This method often led to radial nerve damage, and its use has been discontinued.

Musculocutaneous nerve

The **musculocutaneous** (mus′ku-lo-ku-ta′ne-us) **nerve** provides motor innervation to the anterior muscles of the arm. It also provides cutaneous sensory innervation for part of the forearm (Figure 14-9; Table 14-5).

Ulnar nerve

The **ulnar nerve** innervates two forearm muscles plus most of the intrinsic hand muscles (except some associated with the thumb). Its sensory distribution is to the ulnar side of the hand (Figure 14-10; Table 14-6).

Table 14-5

Musculocutaneous nerve

Origin

Lateral cord of brachial plexus, C5-C7

Movements/Muscles Innervated

Flexes arm
  Coracobrachialis

Flexes forearm
  Brachialis (also small amount of innervation from radial nerve)

Flexes and supinates forearm
  Biceps brachii (also flexes arm)

Cutaneous Innervation

Lateral surface of forearm

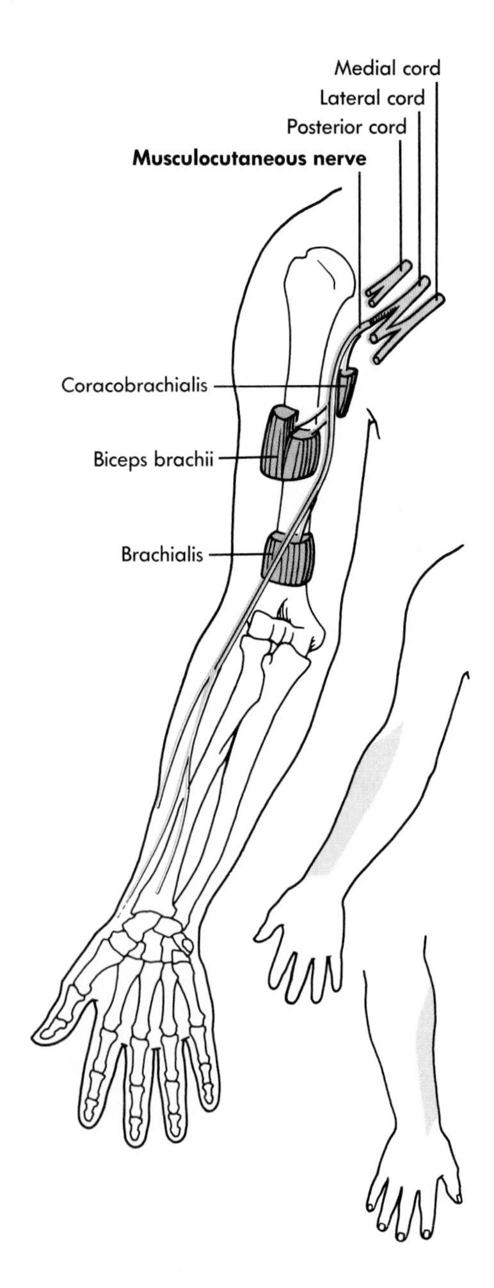

FIGURE 14-9 Route of the musculocutaneous nerve and the muscles it innervates. Inset depicts the cutaneous distribution of the musculocutaneous nerve *(shaded areas)*.

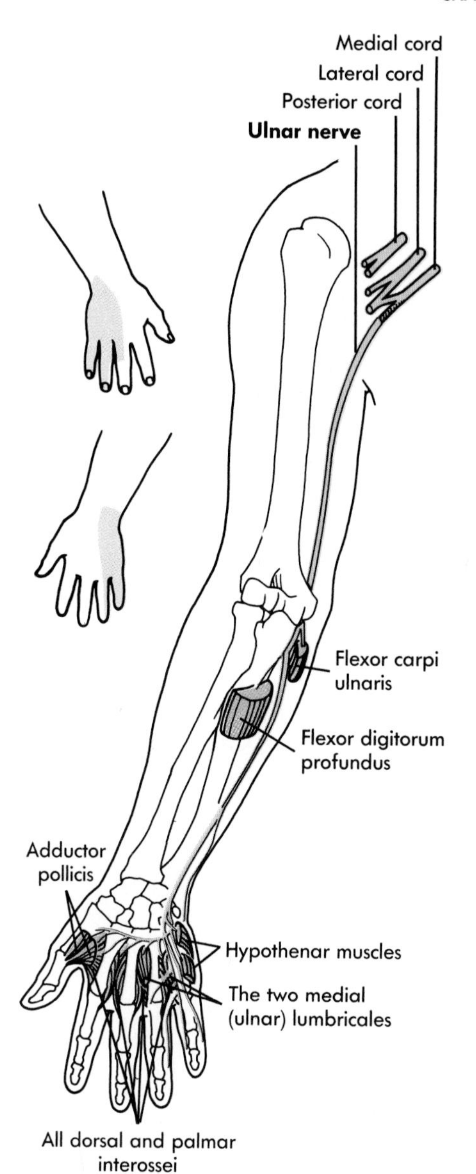

FIGURE 14-10 Route of the ulnar nerve and the muscles it innervates. Inset depicts the cutaneous distribution of the ulnar nerve *(shaded areas)*.

The ulnar nerve is the most easily damaged of all the peripheral nerves, but such damage is almost always temporary. Damage to the ulnar nerve most often occurs where it passes posterior to the medial epicondyle. The nerve can be felt just below the skin at this point, and if this region of the elbow is banged against a hard object, temporary ulnar nerve damage may occur. This damage results in painful tingling sensations radiating down the ulnar side of the forearm and hand. Because of this sensation, this area of the elbow is often called the "funny bone" or "crazy bone."

Table 14-6

Ulnar nerve

Origin

Medial cord of brachial plexus, C8-T1

Movements/Muscles Innervated

Flexes and adducts wrist
  Flexor carpi ulnaris

Flexes fingers
  Part of the flexor digitorum profundus controlling the distal phalanges of little and ring fingers

Abducts and adducts fingers
  Interossei

Adducts thumb
  Adductor pollicis

Controls hypothenar muscles
  Flexor digiti minimi brevis
  Abductor digiti minimi
  Opponens digiti minimi

Cutaneous Innervation

Medial one third of hand, little finger, and medial one half of ring finger

Nerve Replacement

Some patients paralyzed by strokes or spinal cord lesions are able to regain certain functions. Microcomputers are being perfected that will stimulate certain programmed activities such as grasping and walking. Fine wire leads convey the electrical impulse initiated by the microcomputer either to peripheral nerves or directly to the muscles responsible for the desired movement.

The program is initiated by the subtle movement of muscles not affected by the paralysis. Sensors connected to the microcomputer are attached to the skin overlying functional muscles and are able to detect electrical activity associated with movement of the underlying muscles. For example, a person with both legs paralyzed may have such a sensor attached to the abdomen. The abdominal muscles, normally involved in walking, are stimulated by descending tracts when walking is initiated by central nervous system centers. The resultant abdominal muscle activity is detected by the sensor, which activates the program that stimulates the appropriate sequence of muscles, and the paralyzed person walks. Similarly, a quadriplegic can initiate certain grasping actions by subtle movements of the shoulder, neck, or face onto which specific sensors can be placed.

Median nerve

The **median nerve** innervates all but one of the flexor muscles of the forearm and most of the hand muscles near the thumb (the thenar area of the hand). Its cutaneous sensory distribution is to the radial portion of the palm of the hand (Figure 14-11; Table 14-7).

Other brachial plexus nerves

Several nerves other than the five just described arise from the brachial plexus (see Figure 14-6). They supply most of the muscles acting on the scapula and arm and include the pectoral, long thoracic, thoracodorsal, subscapular, and suprascapular nerves. In addition, brachial plexus nerves supply the cutaneous innervation of the medial arm and forearm.

Damage to the median nerve occurs most commonly where it enters the wrist through the carpal tunnel. This tunnel is created by the concave organization of the carpal bones and the flexor retinaculum on the anterior surface of the wrist. None of the connective tissue components expands readily, and injury to the wrist can cause edema to develop within the carpal tunnel. Since the carpal tunnel cannot expand, the edema produces pressure within the tunnel, compressing the median nerve and resulting in numbness, tingling, and pain in the fingers. This condition is referred to as carpal tunnel syndrome. Treatment of this syndrome often requires surgery to relieve the pressure.

Table 14-7

Median nerve

Origin

Medial and lateral cords of brachial plexus, C5-T1

Movements/Muscles Innervated

Pronates forearm
- Pronator teres
- Pronator quadratus

Flexes wrist
- Palmaris longus

Flexes and abducts wrist
- Flexor carpi radialis

Flexes fingers
- Part of flexor digitorum profundus controlling the distal phalanx of the middle and index fingers
- Flexor digitorum superficialis

Controls thumb muscle
- Flexor pollicis longus

Controls thenar muscles
- Abductor pollicis brevis
- Opponens pollicis
- Flexor pollicis brevis

Cutaneous Innervation

Lateral two thirds of palm of hand, thumb, index and middle fingers, and the lateral half of ring finger and dorsal tips of the same fingers

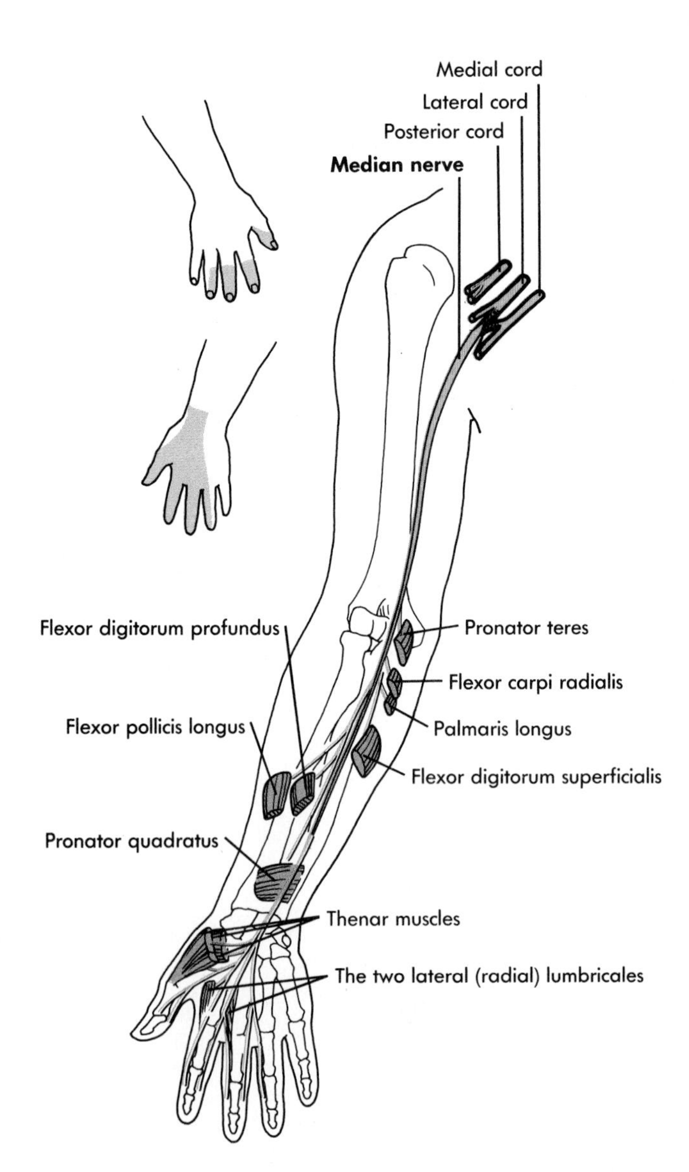

FIGURE 14-11 Route of the median nerve and the muscles it innervates. Inset depicts the cutaneous distribution of the median nerve *(shaded areas)*.

Lumbar and Sacral Plexuses

The **lumbar plexus** originates from the ventral rami of spinal nerves L1 to L4 and the **sacral plexus** from L4 to S4. However, because of their close, overlapping relationship and their similar distribution, the two plexuses often are considered together as a single **lumbosacral plexus** (L1 to S4; Figure 14-12). Four major nerves exit the lumbosacral plexus and enter the lower limb: the obturator, femoral, tibial, and common peroneal. The obturator nerve innervates the medial thigh; the femoral nerve innervates the anterior thigh; the tibial nerve innervates the posterior thigh and leg; and the common peroneal nerve innervates the posterior thigh, the anterior and lateral leg, and the foot. Other lumbosacral nerves supply the lower back, the hip, and the lower abdomen.

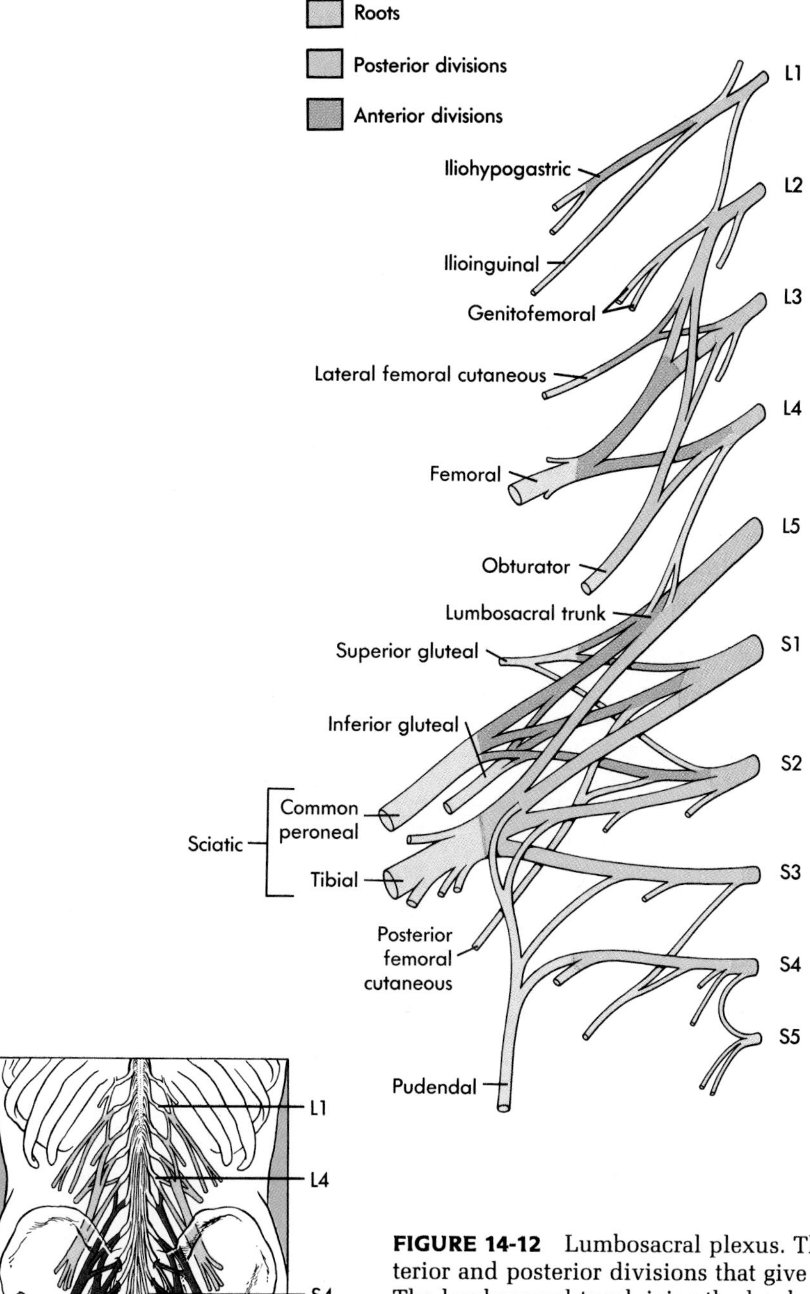

FIGURE 14-12 Lumbosacral plexus. The roots (ventral rami) form anterior and posterior divisions that give rise to the lumbosacral nerves. The lumbosacral trunk joins the lumbar and sacral plexuses.

Obturator nerve

The **obturator** (ob'tu-ra'tor) **nerve** supplies the muscles that adduct the thigh. Its cutaneous sensory distribution is to the medial side of the thigh (Figure 14-13; Table 14-8).

Femoral nerve

The **femoral nerve** innervates the iliopsoas and sartorius muscles and the quadriceps femoris group. Its cutaneous sensory distribution is the anterior and lateral thigh and the medial leg and foot (Figure 14-14; Table 14-9).

Tibial and common peroneal nerves

The **tibial** and **common peroneal** (pĕr'o-ne'al) **nerves** originate from spinal segments L4 to S3 and are bound together within a connective tissue sheath for the length of the thigh. These two nerves, combined within the same sheath, are referred to jointly as the **sciatic** (si-at'ik) **nerve.** The sciatic nerve, by far the largest peripheral nerve in the body, passes through the greater sciatic notch in the pelvis and descends in the posterior thigh to the popliteal fossa where the two portions of the sciatic nerve separate.

The **tibial nerve** innervates most of the posterior thigh and leg muscles. It branches in the foot to form the **medial** and **lateral plantar** (plan'tar) **nerves** that innervate the plantar muscles of the foot and the skin over the sole of the foot. Another branch, the **sural** (su'ral) **nerve,** supplies part of the cutaneous innervation over the calf of the leg and the plantar surface of the foot (Figure 14-15; Table 14-10).

The **common peroneal nerve** divides into the **deep** and **superficial peroneal nerves.** These branches innervate the anterior and lateral muscles of the leg and foot. The cutaneous distribution of the common peroneal nerve and its branches is the lateral and anterior leg and the dorsum of the foot (Figure 14-16; Table 14-11).

If a person sits on a hard surface for a considerable time, the sciatic nerve may be compressed against the ischial portion of the coxa. When the person stands up, tingling ("pins and needles") can be felt throughout the lower limb, and the limb is said to have "gone to sleep."

The sciatic nerve may be seriously injured in a number of ways. A ruptured disk or pressure from the uterus during pregnancy may compress the roots of the sciatic nerve. Other possibilities for causing sciatic nerve damage include hip injury or an improperly administered injection in the hip region.

Table 14-8

Obturator nerve

Origin

Lumbosacral plexus, L2-L4

Movements/Muscles Innervated

Adducts thigh
- Adductor magnus
- Adductor longus
- Adductor brevis

Adducts and flexes thigh
- Gracilis

Rotates thigh laterally
- Obturator externus

Cutaneous Innervation

Superior medial side of thigh

Table 14-9

Femoral nerve

Origin

Lumbosacral plexus, L2-L4

Movements/Muscles Innervated

Flexes thigh
- Iliacus
- Psoas major
- Pectineus

Flexes thigh and flexes leg
- Sartorius

Extends leg
- Vastus lateralis
- Vastus medialis
- Vastus intermedius

Extends leg and flexes thigh
- Rectus femoris

Cutaneous Innervation

Anterior and lateral branches supply the anterior and lateral thigh; saphenous branch supplies the medial leg and foot

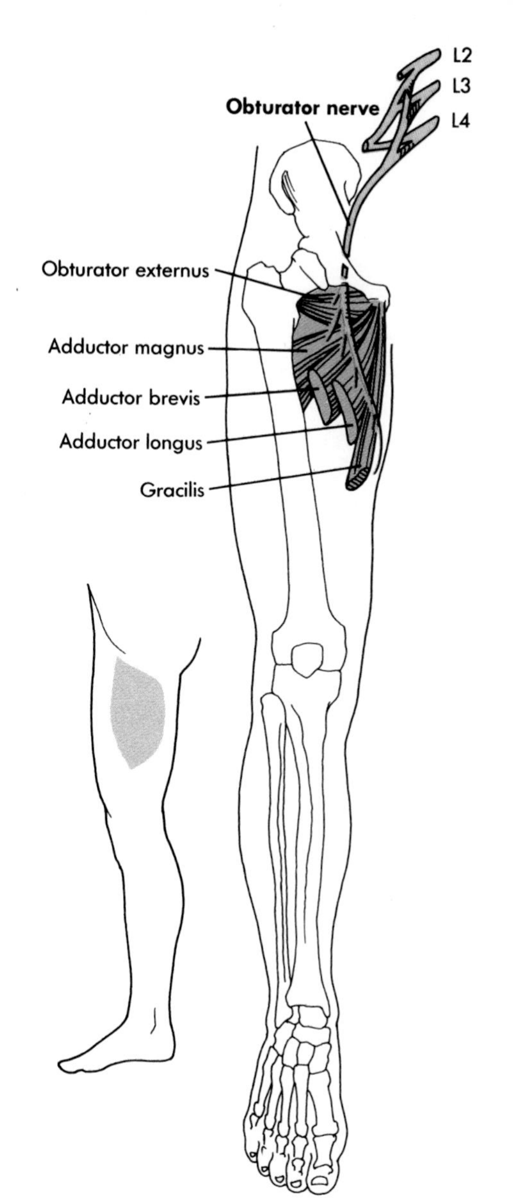

FIGURE 14-13 Route of the obturator nerve and the muscles it innervates. Inset depicts the cutaneous distribution of the obturator nerve *(shaded areas)*.

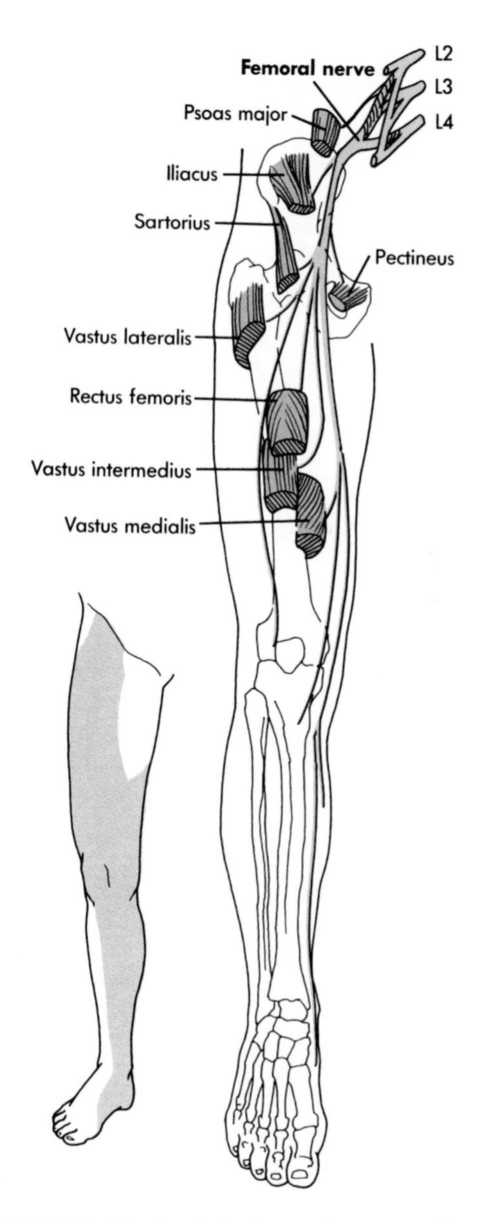

FIGURE 14-14 Route of the femoral nerve and the muscles it innervates. Inset depicts the cutaneous distribution of the femoral nerve *(shaded areas)*.

Table 14-10

Tibial, medial and lateral plantar, and sural nerves

TIBIAL NERVE	MEDIAL AND LATERAL PLANTAR NERVES	SURAL NERVE
Origin		
Lumbosacral plexus, L4-S3	Tibial nerve	Tibial nerve
Movements/Muscles Innervated		
Extends thigh and flexes leg Biceps femoris (long head) Semitendinosus Semimembranosus		
Flexes leg Popliteus	Plantar muscles of foot	None
Plantar flexes foot Gastrocnemius Soleus Plantaris Tibialis posterior		
Flexes toes Flexor hallucis longus Flexor digitorum longus		
Cutaneous Innervation		
None	Sole of foot	Lateral and posterior one third of leg and lateral side of foot

Table 14-11

Common, deep, and superficial peroneal nerves

COMMON PERONEAL NERVE	DEEP PERONEAL NERVE	SUPERFICIAL PERONEAL NERVE
Origin		
Lumbosacral plexus, L4-S2	Common peroneal nerve	Common peroneal nerve
Movements/Muscles Innervated		
Extends thigh and flexes leg Biceps femoris (short head)	Dorsiflexes foot Tibialis anterior Peroneus tertius	Plantar flexes and everts foot Peroneus longus Peroneus brevis
	Extends toes Extensor hallucis longus Extensor digitorum longus	Extends toes Extensor digitorum brevis
Cutaneous Innervation		
Lateral surface of knee	Great and second toe	Distal anterior third of leg and dorsum of foot

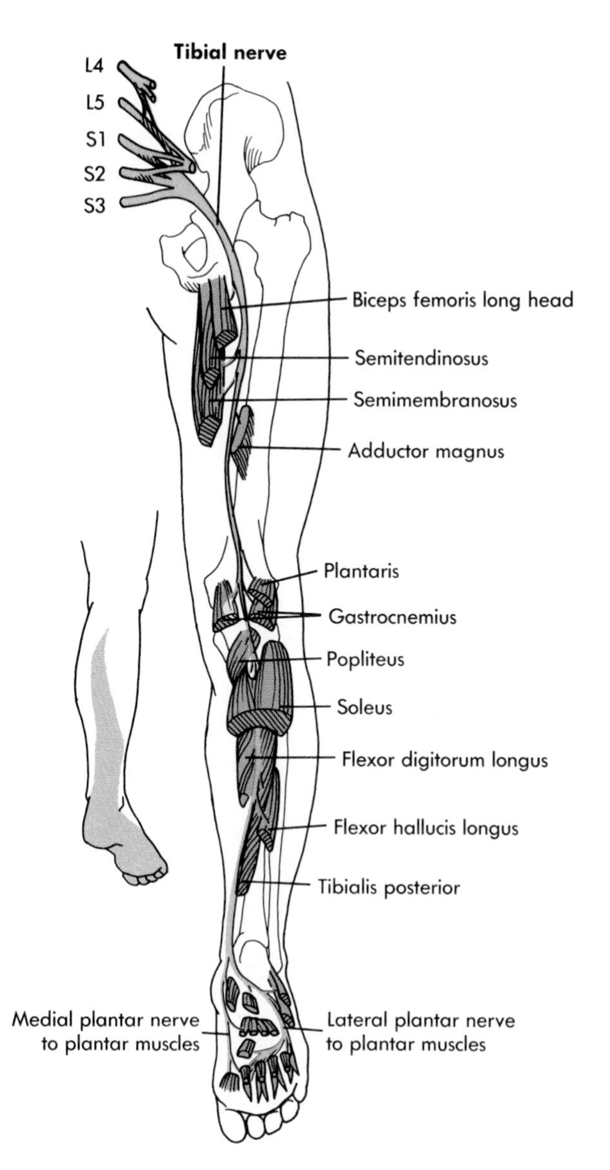

FIGURE 14-15 Route of the tibial nerve and the muscles it innervates. The tibial nerve divides to form the medial and lateral plantar nerves. Inset depicts the cutaneous distribution of these nerves and the sural nerve, a cutaneous branch of the tibial nerve *(shaded areas)*.

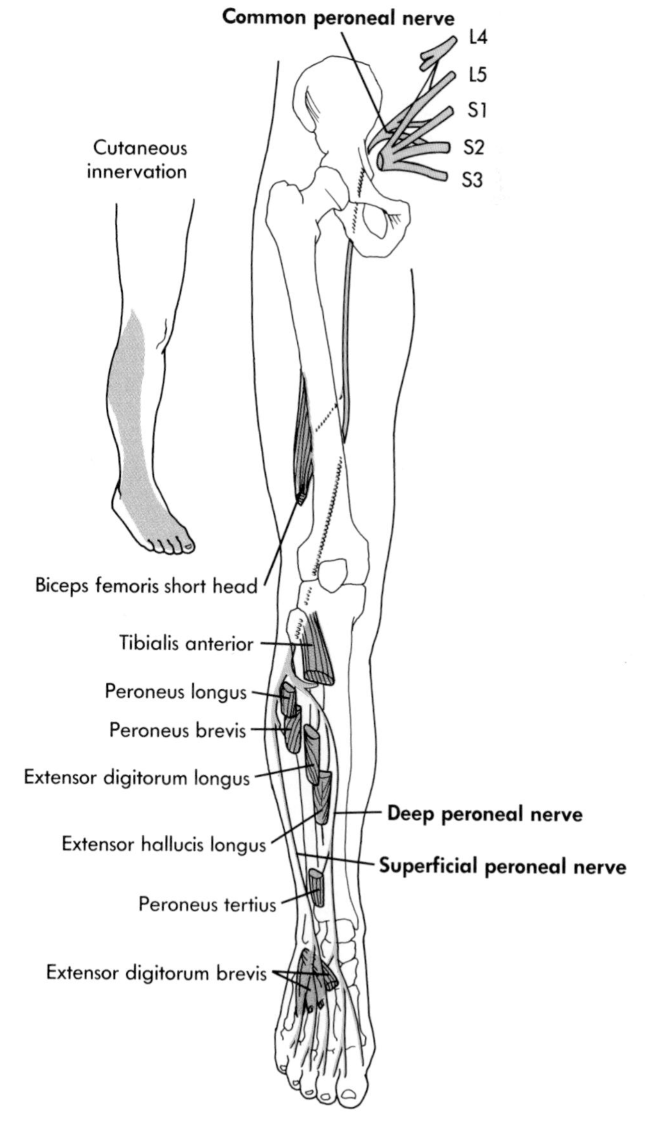

FIGURE 14-16 Route of the common peroneal nerve and the muscles it innervates. The common peroneal nerve divides to form the deep peroneal and superficial peroneal nerves. Inset depicts the cutaneous distribution of these nerves *(shaded areas)*.

Peripheral Nervous System (PNS) Disorders

General Types of PNS Disorders

Anesthesia is the loss of sensation (esthesis means sensation). It may be a pathological condition if it happens spontaneously, or it may be induced to facilitate surgery or some other medical action.

Hyperesthesia is an abnormal acuteness to sensation—an increased sensitivity.

Paresthesia is an abnormal spontaneous sensation such as tingling, prickling, or burning.

Neuralgia consists of severe spasms of throbbing or stabbing pain along the pathway of a nerve.

Neuritis is a general term referring to inflammation of a nerve resulting from a wide variety of causes, which include mechanical injury or pressure, viral or bacterial infection, poisoning by drugs or other chemicals, and vitamin deficiencies. Neuritis in sensory nerves is characterized by neuralgia or may result in anesthesia and loss of reflexes in the affected area. Neuritis in motor nerves results in loss of motor function.

Types of Neuralgia and Neuritis

Trigeminal neuralgia (also called tic douloureux) involves one or more of the trigeminal nerve branches and consists of sharp bursts of pain in the face. This disorder often has a trigger point in or around the mouth that, when touched, elicits the pain response in some other part of the face. The cause of trigeminal neuralgia is unknown.

Facial palsy (or Bell's palsy) is unilateral paralysis of the facial muscles. The affected side of the face droops because of the absence of muscle tone. Facial palsy involves the facial nerve and may result from facial nerve neuritis.

Sciatica is neuralgia of the sciatic nerve, with pain radiating down the back of the thigh and leg. The most common cause is a herniated lumbar disk, resulting in pressure on the spinal nerves contributing to the lumbar plexus. Sciatica also can be produced by sciatic neuritis resulting from a number of causes, including mechanical stretching during exertion, vitamin deficiency, or metabolic disorders (such as gout or diabetes).

Infections

Herpes is a family of diseases characterized by skin lesions, which are caused by a group of similar viruses. The term is derived from the Greek word *herpo,* meaning to creep, and indicates a spreading skin eruption. The viruses apparently reside in the ganglia of sensory nerves and cause lesions along the course of the nerve. **Herpes simplex I** is usually characterized by one or more lesions on the lips or nose. The virus apparently resides in the trigeminal ganglion. Eruptions are usually recurrent and often occur in times of reduced resistance such as during a cold episode. For this reason they are called cold sores or fever blisters. A different herpes virus, **herpes simplex II,** or genital herpes, is usually responsible for a venereal (sexually transmitted) disease causing lesions on the external genitalia. The **varicella,** or **herpes zoster,** virus causes chicken pox in children and shingles in older adults. Normally, this virus first enters the body in childhood to cause chicken pox. The virus then lies dormant in the spinal ganglia for many years and can become active during a time of reduced resistance to cause shingles, a unilateral patch of skin blisters and discoloration along the path of one or more spinal nerves, most commonly around the waist. The symptoms may persist for 3 to 6 months.

Poliomyelitis ("polio" or infantile paralysis) is a disease caused by *Poliovirus hominis.* It is actually a CNS infection, but its major effect is on the peripheral nerves and the muscles they supply. The virus infects the motor neurons in the anterior horn of the central gray matter of the spinal cord (the term polio means gray matter). The infection causes degeneration of the motor neurons and paralysis and atrophy of the muscles they innervate.

Anesthetic leprosy is a bacterial infection of the peripheral nerves caused by *Mycobacterium leprae.* The infection results in anesthesia, paralysis, ulceration, and gangrene.

Inborn, Genetic, and Autoimmune Disorders

Myotonic dystrophy is an autosomal dominant hereditary disease characterized by muscle weakness, dysfunction, and atrophy and by visual impairment as a result of nerve degeneration.

Myasthenia gravis is an autoimmune disorder resulting in the reduction of functional acetylcholine receptors at neuromuscular junctions. It results in muscular weakness and abnormal fatigability caused by neuromuscular dysfunction.

Neurofibromatosis is a genetic disorder in which small skin lesions appear in early childhood followed by the development of multiple subcutaneous neurofibromas (benign tumors resulting from Schwann cell proliferation). The neurofibromas may increase slowly in size and number over several years and cause extreme disfiguration. The most famous patient afflicted with this disorder was the so-called "Elephant Man."

Other lumbosacral plexus nerves

In addition to the nerves just described, the lumbosacral plexus gives rise to nerves that supply the lower abdominal muscles (iliohypogastric nerve), the hip muscles that act on the femur (gluteal nerves), and muscles of the abdominal floor (pudendal nerve; see Figure 14-12). The **iliohypogastric** (il′e-o-hi′po-gas′trik), **ilioinguinal** (il′e-o-in′gwĭ-nal), **genitofemoral** (jen′ĭ-to-fem′o-ral), **cutaneous femoral,** and **pudendal** (pu-den′dal) nerves innervate the skin of the suprapubic area, the external genitalia, the superior medial thigh, and the posterior thigh. The pudendal nerve plays a vital role in sexual stimulation and response.

Coccygeal Plexus

The **coccygeal plexus** is a very small plexus formed from the ventral rami of spinal nerves S4, S5, and the coccygeal nerve. This small plexus supplies motor innervation to muscles of the pelvic floor and sensory cutaneous innervation to the skin over the coccyx (although some skin over the coccyx is innervated by the dorsal rami of the coccygeal nerves).

Branches of the pudendal nerve are anesthetized before a doctor performs an episiotomy for childbirth. An episiotomy is a cut in the perineum that makes the opening of the birth canal larger.

SUMMARY

CRANIAL NERVES p. 430

Cranial nerves perform sensory, somatomotor, proprioceptive, and parasympathetic functions.

Sensory

The olfactory (I), optic (II), and vestibulocochlear (VIII) nerves are involved in the sense of smell, vision, hearing, and balance.

Somatomotor/Proprioception

1. The trochlear (IV) and abducens (VI) nerves each control an extrinsic eye muscle.
2. The accessory (XI) nerve has a cranial and a spinal component.
   - The cranial component joins the vagus nerve.
   - The spinal component supplies the sternocleidomastoid and trapezius muscles.
3. The hypoglossal (XII) nerve supplies the intrinsic tongue muscles, three of four extrinsic tongue muscles, and two throat muscles.

Somatomotor/Proprioception and Sensory

1. The trigeminal (V) nerve supplies the muscles of mastication, as well as a middle ear muscle, a palatine muscle, and two throat muscles.
2. The trigeminal nerve has the greatest cutaneous sensory distribution of any cranial nerve.
3. There are three branches of the trigeminal nerve.
4. Two of the three trigeminal nerve branches innervate the teeth.

Somatomotor/Proprioception and Parasympathetic

1. The oculomotor (III) nerve innervates four of six extrinsic eye muscles and the upper eyelid.
2. The oculomotor nerve provides the parasympathetic supply to the pupil and lens of the eye.

Somatomotor/Proprioception, Sensory, and Parasympathetic

1. The facial (VII), glossopharyngeal (IX), and vagus (X) nerves perform all three cranial nerve functions.
2. The facial nerve supplies the muscles of facial expression, an inner ear muscle, and two throat muscles. The glossopharyngeal nerve supplies a muscle in the pharynx, and the vagus nerve innervates the muscles of the pharynx, palate, and larynx.
3. All three nerves are involved in the sense of taste. The facial nerve is sensory for the external ear, tongue, and palate. The glossopharyngeal and vagus nerves are sensory for the palate, pharynx, and larynx and for receptors that monitor blood pressure and gas levels in the blood. The vagus nerve is sensory for thoracic and abdominal organs.
4. Parasympathetic activities of the facial and glossopharyngeal nerves regulate the salivary glands. The facial nerve also supplies the lacrimal glands. The vagus nerve provides parasympathetic innervation to the thoracic and abdominal organs.

SPINAL NERVES p. 438

1. Nerve rootlets from the spinal cord combine to form the ventral (efferent) and dorsal (afferent) roots, which join to form spinal nerves.
2. There are eight cervical, 12 thoracic, five lumbar, and five sacral pairs and one coccygeal pair of spinal nerves.
3. Spinal nerves have specific cutaneous distributions called dermatomes.
4. Spinal nerves branch to form rami.
   - The dorsal rami supply the muscles and skin near the midline of the back.
   - The ventral rami in the thoracic region form intercostal nerves that supply the thorax and upper abdomen. The remaining ventral rami join to form plexuses (see below).
   - Sympathetic rami supply autonomic nerves (see Chapter 16).

Cervical Plexus

Spinal nerves C1 to C4 form the cervical plexus that supplies some muscles and the skin of the neck and shoulder. The phrenic nerves innervate the diaphragm.

Brachial Plexus

1. Spinal nerves C5 to T1 form the brachial plexus, which supplies the upper limb.
2. The axillary nerve innervates the deltoid and teres minor muscles and the skin of the shoulder.
3. The radial nerve supplies the extensor muscles of the arm and forearm and the skin of the posterior surface of the arm, forearm, and hand.
4. The musculocutaneous nerve supplies the anterior arm muscles and the skin of the lateral surface of the forearm.
5. The ulnar nerve innervates most of the intrinsic hand muscles and the skin on the ulnar side of the hand.
6. The median nerve innervates the pronator and most of the flexor muscles of the forearm, most of the thenar muscles, and the skin of the radial side of the palm of the hand.
7. Other nerves supply most of the muscles that act on the arm, the scapula, and the skin of the medial arm and forearm.

Lumbar and Sacral Plexuses

1. Spinal nerves L1 to S4 form the lumbosacral plexus.
2. The obturator nerve supplies the muscles that adduct the thigh and the skin of the medial thigh.
3. The femoral nerve supplies the muscles that flex the thigh and extend the leg and the skin of the anterior and lateral thigh and the medial leg and foot.
4. The tibial nerve innervates the muscles that extend the thigh and flex the leg and the foot. It also supplies the plantar muscles and the skin of the posterior leg and the sole of the foot.
5. The common peroneal nerve supplies the short head of the biceps femoris, the muscles that dorsiflex and plantar flex the foot, and the skin of the lateral and anterior leg and the dorsum of the foot.
6. In the thigh the tibial nerve and the common peroneal nerve are combined as the sciatic nerve.
7. Other lumbosacral nerves supply the lower abdominal muscles, the hip muscles, and the skin of the suprapubic area, external genitalia, and upper medial thigh.

Coccygeal Plexus

Spinal nerves S4, S5, and Cx form the coccygeal plexus, which supplies the muscles of the pelvic floor and the skin over the coccyx.

CONTENT REVIEW

1. What are the three major functions of the cranial nerves?
2. Which cranial nerves are sensory only? With what sense is each of these nerves associated?
3. Name the cranial nerves that are somatomotor and proprioceptive only. What muscles does each nerve supply?
4. The sensory cutaneous innervation of the face is provided by what cranial nerve? How is this nerve important in dentistry? Name the muscles that would no longer function if this nerve were damaged.
5. Which four cranial nerves have a parasympathetic function? Describe the functions of each of these nerves.
6. Name the cranial nerves that control movement of the eyeball.
7. Which cranial nerves are involved in the sense of taste? What part of the tongue does each nerve supply?
8. Speech production involves what cranial nerves? Describe the branches of these nerves.
9. Differentiate between rootlet, dorsal root, ventral root, and spinal nerve. Which of them contain sensory fibers and/or motor fibers?
10. Describe all the spinal nerves by name and number. Where do they exit the vertebral column?
11. What is a dermatome? Why are dermatomes clinically important?
12. Contrast dorsal, ventral, and sympathetic rami of spinal nerves. What muscles do the dorsal rami innervate?
13. Describe the distribution of the ventral rami of the thoracic region.
14. What is a plexus? What happens to the axons of spinal nerves as they pass through a plexus?
15. Name the main spinal plexuses and the spinal nerves associated with each one.
16. Name the structures innervated by the cervical plexus. Describe the innervation of the phrenic nerve.
17. Name the five major nerves that emerge from the brachial plexus. List the muscles they innervate and the areas of the skin they supply. In addition to these five nerves, name the muscles and skin areas supplied by the remaining brachial plexus nerves.
18. Name the two principle nerves that arise from the lumbosacral plexus, and describe the muscles and skin areas they supply. Describe the structures innervated by the remaining lumbosacral nerves.
19. What structures are innervated by the coccygeal plexus?

CONCEPT REVIEW

1. Name the cranial nerve that, if injured, would produce the symptoms listed below:
   A A patient has strabismus, the left eye is turned inferiorly and laterally.
   B A patient is unable to move the eyeball medially.
   C The upper left eyelid is drooping (ptosis), and the left pupil is dilated.
2. What cranial nerve would be damaged to produce each of the following symptoms?
   A Vertigo (a balance disorder in which the patient feels as if he is spinning)
   B Tinnitus (a ringing sound in the ear)
   C Anosmia (loss of the sense of smell)
   D Blindness
3. Wendy Frost went cross-country skiing on a very cold day. Afterward she was unable to close her right eye or raise her right eyebrow. Although she could move her jaw, she had difficulty chewing because food would drool out of her mouth. What nerve was affected? Explain the observed symptoms.
4. Spot Blister has herpes zoster, a viral infection. The virus lies dormant in nervous tissue and sporadically becomes active, causing lesions in the skin supplied by the nerves that it infects. Spot exhibited lesions on the scalp, the forehead, and the cornea of the eye. Name the nerve that was infected by the herpes virus. Be as specific as you can.
5. The act of swallowing involves two components. The voluntary portion involves the movement of food to the superior part of the pharynx. There the food stimulates tactile receptors that initiate the second component, an involuntary swallowing reflex. Sensory impulses from the tactile receptors are transmitted to the medulla oblongata. From the medulla motor impulses are transmitted back to the muscles of the soft palate, pharynx, larynx, and throat, and the food is swallowed. Name the two cranial nerves that convey the sensory impulses and the five cranial nerves that carry the motor impulses.
6. A cancer patient has his left lung removed. To reduce the space remaining where the lung was removed, the diaphragm on the left side was paralyzed, allowing the abdominal viscera to push the diaphragm upward. What nerve would be cut? Where would be a good place to cut it?
7. Based on sensory response to pain in the skin of the hand, how could you distinguish between damage to the ulnar, median, and radial nerves?
8. During a difficult delivery the baby's arm delivered first. The attendant grasped the arm and forcefully pulled it. Later a nurse observed that the baby could not abduct or adduct the medial four fingers and flexion of the wrist was impaired. What nerve was damaged?
9. Two patients were admitted to the hospital. According to their charts, both had herniated disks that were placing pressure on the roots of the sciatic nerve. One patient had pain in the buttocks and the posterior aspect of the thigh. The other patient experienced pain in the posterior and lateral aspects of the leg and the lateral part of the ankle and foot. Explain how the same condition, a herniated disk, could produce such different symptoms.
10. In an automobile accident a woman suffered a crushing hip injury. For each of the conditions given below, state what nerve was damaged.
    A Unable to adduct the thigh
    B Unable to extend the leg
    C Unable to flex the leg
    D Loss of sensation from the skin of the anterior thigh
    E Loss of sensation from the skin of the medial thigh

ANSWERS TO PREDICT QUESTIONS

1 p. 437. The sternocleidomastoid muscle pulls the mastoid process (located behind the ear) toward the sternum, thus turning the face to the opposite side. If the innervation to one sternocleidomastoid muscle is eliminated (accessory nerve injury), the opposite muscle will be unopposed and will turn the face toward the side of injury. A person with wry neck, with the head turned to the left, most likely would have an injured left accessory nerve.

2 p. 437. The tongue is protruded by contraction of the geniohyoid muscle, which pulls the back of the tongue forward, pushing the muscle mass of the tongue forward. The situation is very similar to the previous predict question. With one side pushed forward and unopposed by muscles of the opposite side, the tongue will deviate toward the nonfunctional side. Therefore in the example the right hypoglossal nerve is damaged.

3 p. 438. The oculomotor nerve innervates four eye muscles and the levator palpebrae superioris muscle. These muscles move the eyeball so that the gaze is directed superiorly, inferiorly, medially, or superolaterally. Damage to this nerve can be tested by having the patient look in these directions. The abducens nerve directs the gaze laterally, and the trochlear nerve directs the gaze inferolaterally. If the patient can move his eyes in these directions, the associated nerves are intact.

4 p. 440. Nerves C5 to T1, which innervate the left arm, forearm, and hand.

5 p. 441. Damage to the right phrenic nerve would result in absence of muscular contraction in the right half of the diaphragm. Since the phrenic nerves originate from C3 to C5, damage to the upper-cervical region of the spinal cord would eliminate their function; damage in the lower cord below the point where the spinal nerves originated would not affect the nerves.

6 p. 442. The radial nerve lies along the shaft of the humerus approximately midway along its length. If the humerus is fractured, the radial nerve may be lacerated by bone fragments or, more commonly, pinched between two large fragments of bone, decreasing or eliminating the nerve's function.

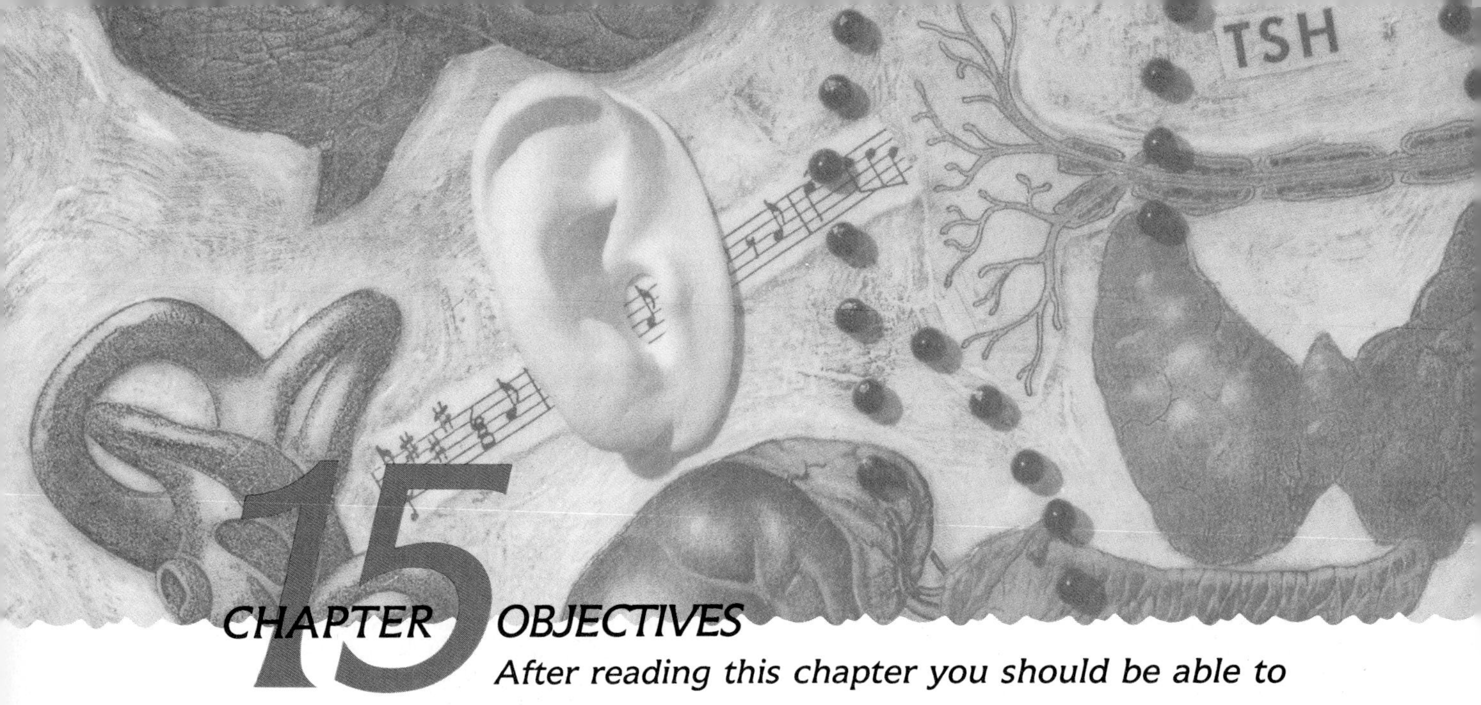

CHAPTER 15 OBJECTIVES

After reading this chapter you should be able to

1. Define sensation. Explain the differences between somatic, visceral, and special senses, and give examples of each.
2. Describe the major sensory nerve endings, their locations, and functions.
3. Describe the histological structure and function of the olfactory epithelium and the olfactory bulb.
4. Describe the central nervous system connections for smell and explain how these connections elicit various visceral and conscious responses to smell.
5. Explain adaptation to odor and describe various levels at which it can occur.
6. Describe the histology and function of a typical taste bud.
7. Describe the central nervous system pathways and cortical locations for taste.
8. List the accessory structures of the eye and explain their functions.
9. Describe the tunics of the eye and give the function of each.
10. Describe the internal structures of the eye, including the lens, ciliary body, iris, macula, and optic disc, and explain the function of each.
11. Describe the compartments and chambers of the eye and explain the function of the canal of Schlemm.
12. Explain light refraction and reflection and how they relate to eye function.
13. Name and describe the structure and function of the layers of the retina.
14. Describe the chemical reaction in rhodopsin as a result of light stimulation.
15. Outline the central nervous system pathway for visual input and describe what happens to images from each half of the visual field.
16. Describe the structures of the outer and middle ears and state their functions.
17. Describe the microanatomy of the cochlea and explain how sounds are detected.
18. Describe the central nervous system pathway for the appreciation of hearing, the pathway for controlling pitch, and the reflex pathway for dampening sound.
19. Explain how the static and kinetic labyrinths function in balance.
20. Describe the central nervous system pathways for balance.

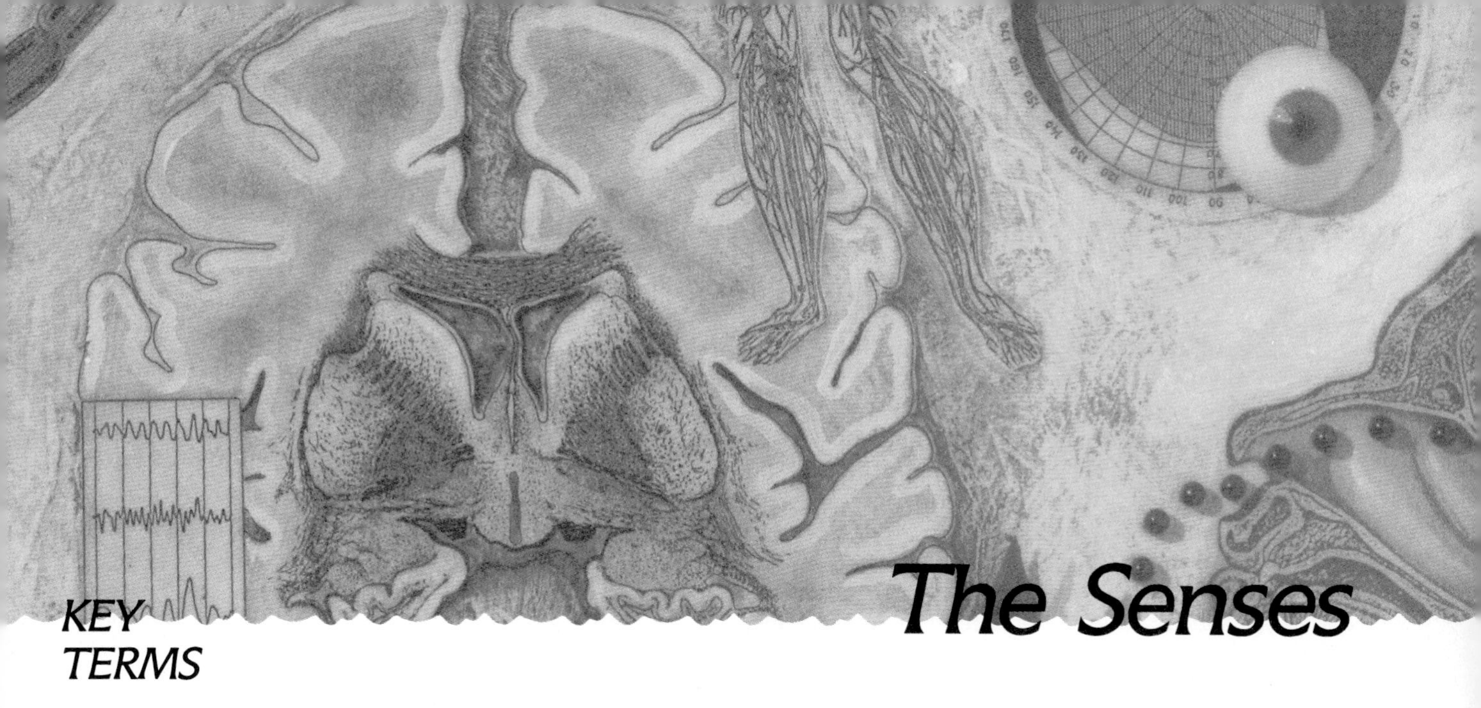

The Senses

KEY TERMS

chemoreceptor

cones

cupula (ku′pu-lah)

endolymph

exteroceptor

gustatory

labyrinth (lab′ĭ-rinth)

macula lutea (mak′u-lah lu′te-ah)

mechanoreceptor

nociceptor (no′sĭ-sep′tor)

olfactory

organ of Corti

perilymph

photoreceptor

rods

sensation

thermoreceptor

visceroreceptor

RELATED TOPICS

The following terms or concepts are important for a good understanding of this chapter. If you are not familiar with them, you should review them before proceeding.

Action potentials (Chapter 9)

Eye muscles (Chapter 11)

Nervous system function (Chapter 12)

Central nervous system anatomy (Chapter 13)

Cranial nerves (Chapter 14)

THE SENSES ARE THE MEANS BY WHICH THE BRAIN RECEIVES INFORMATION ABOUT THE "OUTSIDE WORLD." THEY CAN BE CHARACTERIZED AS EITHER SPECIAL OR GENERAL. Historically, five senses were recognized: smell, taste, sight, hearing, and touch. Today we recognize many more senses and classify them into two major groups. Senses produced by highly localized organs with very specialized sensory cells are referred to as special senses, and include the senses of smell, taste, sight, hearing and balance. General senses include touch (both "light" touch and the ability to discriminate between two points), pressure, pain, temperature, vibration, and proprioception (knowing where the body is located in space).

CLASSIFICATION OF THE SENSES

The **general senses** can be classified as **somatic,** those providing general sensory information about the body and the environment, or **visceral,** those providing information about various internal organs. The **special senses** are those with highly localized receptors providing specific information about the environment (Table 15-1). **Modalities** of sensation refer to the form of the sensation. For example, somatic modalities include touch, pressure, temperature, proprioception, and pain. Visceral modalities consist primarily of pain and pressure. Special modalities are smell, taste, sight, sound, and balance.

Usually receptors are quite specific and are most sensitive to only one type of stimulus. **Mechanoreceptors** respond to mechanical stimuli such as compression, bending, or stretching of cells. **Chemoreceptors** respond to chemicals that become attached to receptors on their membranes. **Photoreceptors** respond to light striking the receptor cells. **Thermoreceptors** respond to changes in temperature at the site of the receptor. **Nociceptors** (no′sĭ-sep′tors; *noceo* means hurt) respond to painful mechanical, chemical, or thermal stimuli.

The general senses of touch, pressure, and proprioception all depend on a variety of mechanoreceptors, as do the special senses of sound and balance. The special senses of smell and taste depend on chemoreceptors. The special sense of vision depends on photoreceptors, and the general sensation of temperature depends on thermoreceptors.

SENSATION

Sensation, or **perception,** is the conscious awareness of stimuli received by sensory receptors. The brain constantly receives a wide variety of stimuli from both inside and outside the body. To be perceived as a conscious sensation, action potentials generated by receptors must reach the cerebral cortex. Some action potentials project to other areas of the brain such as the cerebellum where they are not consciously perceived. In addition, the cerebral cortex screens much of what it receives, ignoring a large portion of the action potentials that reach it. Since we are not consciously aware of the information carried by these action potentials, they remain subconscious.

Sensation requires the following steps:

1. There must be a **stimulus** originating either inside or outside of the body.
2. There must be a **receptor** capable of detecting the stimulus and of converting the stimulus into an action potential.
3. The action potential generated in the receptor must be **conducted** to the central nervous system (CNS) through afferent nerves.

Table 15-1

Classification of the senses

TYPE OF SENSE	RECEPTOR TYPE	INITIATION OF RESPONSE
Somatic		
Touch	Mechanoreceptors	Compression of receptors
Pressure	Mechanoreceptors	Compression of receptors
Temperature	Thermoreceptors	Temperature around nerve endings
Proprioception	Mechanoreceptors	Compression of receptors
Pain	Nociceptors	Irritation of nerve endings (e.g., mechanical, chemical, or thermal)
Visceral		
Pain	Nociceptors	Irritation of nerve endings
Pressure	Mechanoreceptors	Compression of receptors
Special		
Smell	Chemoreceptors	Binding of molecules to membrane receptors
Taste	Chemoreceptors	Binding of molecules to membrane receptors
Sight	Photoreceptors	Chemical change in receptors initiated by light
Sound	Mechanoreceptors	Bending of microvilli on receptor cells
Balance	Mechanoreceptors	Bending of microvilli on receptor cells

4. The action potential must be **translated** within the CNS into information.
5. Information must be processed in the CNS so that the person is **aware** of the stimulus.

Because sensations are an awareness of a stimulus, they can occur only in the cerebral cortex where action potentials are translated. Receptors can only generate local potentials or action potentials in response to a stimulus. Action potentials for a given receptor are carried to a specific part of the cerebral cortex where they are interpreted. Sensation results from the interpretation of the action potential. In the case of touch sensation, the sensation is **projected** to the site of origin of the stimulus. Because of this projection, the sensation of touch is perceived to be at the site where the sensory receptor was stimulated rather than in the cerebral cortex.

The central nervous system is not capable of being conciously aware of all stimuli. Much information is processed at an unconcious level. For example, receptors that detect changes in blood pH or blood pressure do not produce a concious sensation.

In addition, we have the ability to focus attention on specific sensations. If we were simultaneously aware of all the stimuli with which the brain is constantly bombarded, it is unlikely that we would be able to function. Being aware of so many stimuli would require us to constantly make conscious decisions about which stimuli we should respond to. We would not be able to focus our attention on a given task and would be incapable of performing simple functions. Instead, we exhibit selective awareness. That is, we are more aware of sensations that are related to that on which we have our attention focused than on other sensations. When we switch our attention, our selective awareness changes also.

An example of projection is when your finger touches an object. Touch receptors are stimulated and action potentials are conducted to the primary somesthetic area of the cerebral cortex where the action potentials are interpreted as touch in the finger. Remember that the finger maps to a specific part of the somesthetic cortex (see Chapter 13). If an afferent neuron from a touch receptor were electrically stimulated anywhere along its length and action potentials were produced in that afferent neuron, the action potentials would be conducted to the finger region of the somesthetic cortex and would be interpreted as touch in the finger (also see the discussion of phantom pain in Chapter 13).

1

What would be the result of directly stimulating a neuron in the "finger region" of the somesthetic cortex?

? ? ? ? ? ? ? ? ? ? ?

Some sensations have the quality of **adaptation,** a decreased sensitivity to a continued stimulus. After exposure to a stimulus for a period of time, the response of the receptors and/or the afferent pathways to a certain stimulus level lessens from that when the stimulus was first applied. For example, when you first get dressed, tactile receptors and pathways relay information about the clothes to the brain; after a time the action potentials decrease, and the clothes are ignored.

Two types of proprioceptors are involved in providing positional information: tonic receptors and phasic receptors. **Tonic** receptors generate action potentials as long as a stimulus is applied. Information from tonic proprioceptors allows you to know, for example, where your little finger is at all times without your having to look for it. **Phasic** receptors, by contrast, are most sensitive to changes in stimuli. For example, information from phasic proprioceptors allows you to know where your hand is as it moves, allowing you to control the movement of your hand through space and to predict where it will be in the next moment.

We are usually not conscious of tonic or phasic input (selective awareness), but we can call up the information when we wish. For example, where is the thumb of your right hand at this moment? Were you aware of its position a few seconds ago?

TYPES OF AFFERENT NERVE ENDINGS

There are at least eight major types of sensory nerve endings: free nerve endings, Merkel's disks, hair follicle receptors, Pacinian corpuscles, Meissner's corpuscles, Ruffini's end-organs, Golgi tendon apparatuses, and muscle spindles (Table 15-2 and Figure 15-1). Many of these nerve endings are associated with the skin; others are associated with deeper structures such as tendons, ligaments, and muscles; and some can be found in both the skin and deeper structures. In general, sensory nerve endings are classified into three groups: **exteroreceptors, visceroreceptors,** or **proprioceptors.** Exteroreceptors (cutaneous receptors) are associated with the skin, visceroreceptors are associated with the viscera or organs, and proprioceptors are associated with joints, tendons, and other connective tissue. Exteroreceptors provide tactile information about the external environment, visceroreceptors provide information about the internal environment, and proprioceptors provide information about body position and movement.

The simplest and most common sensory nerve endings are the **free nerve endings** (see Figure 15-1), which are distributed throughout almost all parts of the body. Most visceroreceptors consist of free nerve endings. These nerve endings are responsible for a

Table 15-2

Afferent nerve endings

TYPE OF NERVE ENDING	STRUCTURE	FUNCTION
Free nerve endings	Branching, no capsule	Pain, itch, tickle, temperature, joint movement, and proprioception
Merkel's disks	Flattened expansions at the end of axons; each expansion associated with a Merkel cell	Light touch and superficial pressure
Hair follicle	Wrapped around hair follicles or extending along the hair axis, each axon supplies several hairs, and each hair receives branches from several neurons, resulting in considerable overlap	Light touch; responds to very slight bending of the hair
Pacinian corpuscle	Onion-shaped capsule of several cell layers with a single central nerve process	Deep cutaneous pressure, vibration, and proprioception
Meissner's corpuscles	Several branches of a single axon associated with wedge-shaped epithelioid cells and surrounded by a connective tissue capsule	Two-point discrimination
Ruffini's end organs	Branching axon with numerous small, terminal knobs surrounded by a connective tissue capsule	Continuous touch or pressure; respond to depression or stretch of the skin
Golgi tendon apparatus	Surrounds a bundle of tendon fascicles and is enclosed by a delicate connective tissue capsule; nerve terminations are branched with small swellings applied to individual tendon fascicles	Proprioception associated with tendon movement
Muscle spindle	Three to 10 striated muscle fibers enclosed by a loose connective tissue capsule, striated only at the ends, with sensory nerve endings in the center	Detects stretch in a muscle and helps control muscle tone

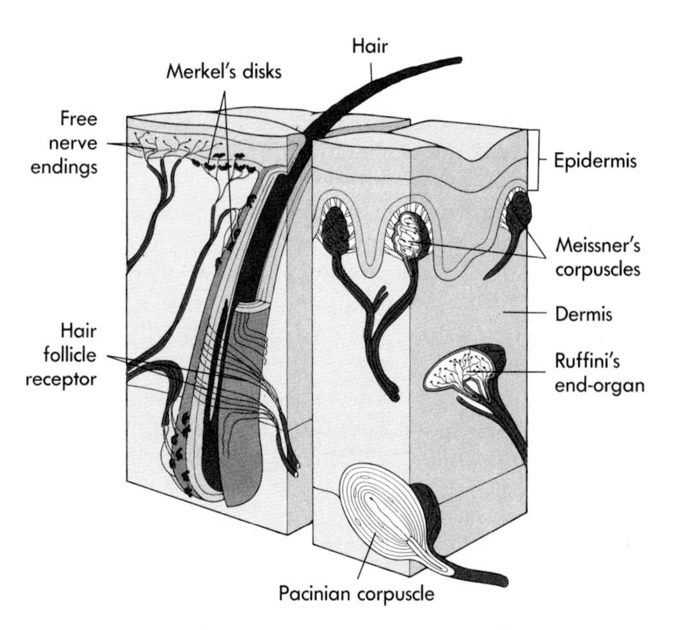

FIGURE 15-1 Sensory cutaneous nerve endings.

number of sensations, including pain, temperature, itch, and movement. The free nerve endings responsible for temperature detection are of three types. One type, the **cold receptors,** increases its rate of action potential firing as the skin is cooled. The second type, **warm receptors,** increases its rate of action potential firing as skin temperature increases. Both cold receptors and warm receptors are phasic receptors and, therefore, respond most strongly to changes in temperature. Cold receptors are 10 to 15 times more numerous in any given area of skin than warm receptors. The third type is a **pain receptor,** which is stimulated in extreme cold or heat. At very cold temperatures (0° to 12° C) only pain fibers are stimulated. The pain sensation ends as the temperature exceeds 15° C. Between 12° C and 35° C, cold fibers are stimulated. Nerve fibers from warm receptors are stimulated between 25° C and 47° C. Therefore "comfortable" temperatures, between 25° C and 35° C, stimulate both warm and cold receptors. Temperatures above 47° C no longer stimulate warm receptors but actually stimulate cold receptors and pain receptors.

2

Explain why very hot or cold objects placed into a blindfolded person's hand are often confused.

? ? ? ? ? ? ? ? ? ? ?

Merkel's (mer'kelz) **disks,** or **tactile disks,** are more complex than free nerve endings (see Figure 15-1) and consist of axonal branches that end as flattened expansions. They are distributed throughout the epidermis just superficial to the basement membrane and are associated with dome-shaped mounds of thickened epidermis in hairy skin. Merkel's disks are involved with sensation of light touch and superficial pressure.

Hair follicle receptors, or hair end organs, respond to very slight bending of the hair and are involved in light touch. Even though these nerve endings are very sensitive, requiring very little stimulation to elicit a response, they are not very discriminative (the sensation is not very well localized). The considerable overlap that exists in the sensory endings of afferent nerves helps explain why light touch is not very discriminative, but because of converging signals within the CNS, light touch is very sensitive (see Chapter 13).

Pacinian (pă-sĭ'ne-an), or **lamellated, corpuscles** are very complex nerve endings resembling an onion. They are located within the deep dermis or hypodermis where they are responsible for deep cutaneous pressure and vibration. Pacinian corpuscles associated with the joints help relay proprioceptive information about joint positions.

Meissner's (mīs'nerz), or **tactile, corpuscles** are distributed throughout the dermal papillae (see Chapter 5) and are involved in two-point, discrimination touch. Meissner's corpuscles are numerous and close together in the tongue (approximately 2 mm apart) and fingertips (approximately 4 mm apart) but are less numerous and more widely separated in other areas such as the back (approximately 64 mm apart).

Ruffini's (roo-fe'nēz) **end organs** are located in the dermis of the skin, primarily in the fingers. They respond to pressure on the skin directly superficial to the receptor and to stretch of adjacent skin. These nerve endings are important in responding to continuous touch or pressure.

Golgi tendon apparatuses or **organs** are proprioceptive nerve endings associated with the fibers of a tendon at the muscle-tendon junction (Figure 15-2, *A*). They are activated by an increase in tendon tension, whether it is caused by contraction of the muscle or by passive stretch of the tendon. When a great amount of tension is applied to the tendon, afferent fibers from the Golgi tendon apparatus inhibit the motor neuron of the associated muscle and cause it to relax (Golgi tendon reflex), preventing muscle and tendon damage caused by excessive tension.

Muscle spindles consist of three to 10 muscle fibers that are striated only on the ends so that only the ends of each cell can contract. Sensory nerve endings are wrapped around the center of the muscle fibers, and gamma motor neurons supply the striated ends (Figure 15-2, *B*). When the skeletal muscle is stretched or when the ends of the muscle spindle fibers contract, the center of the muscle spindle is stretched, stimulating the sensory neurons of the muscle spindle. The sensory neurons synapse with alpha motor neurons, which, when stimulated, cause a rapid contraction of the stretched muscle (the stretch reflex).

Muscle spindles are important to the control and tone of postural muscles. Brain centers act through descending tracts to either increase or decrease action potentials in gamma motor fibers. Stimulation of the gamma motor system activates the stretch reflex, which in turn increases the tone of the muscles involved.

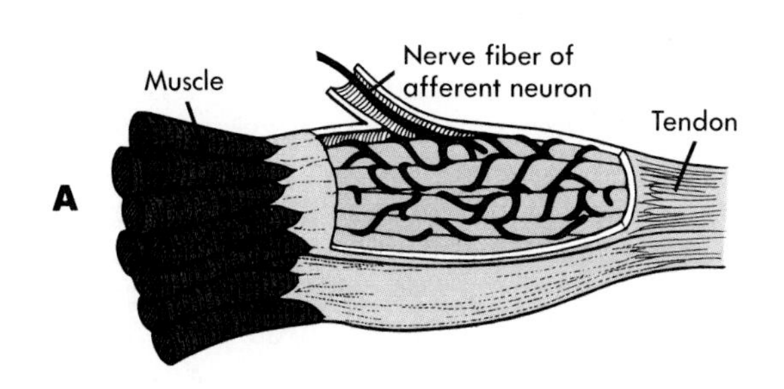

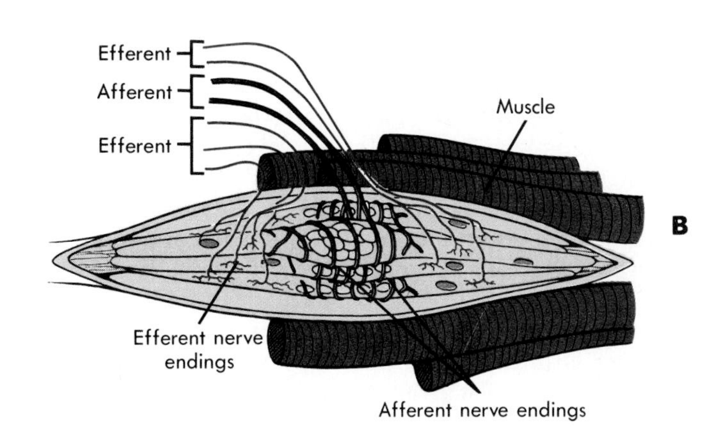

FIGURE 15-2 Nerve endings associated with muscles. **A** Golgi tendon apparatus. **B** Muscle spindle.

OLFACTION

Olfaction (ol-fak'shun), the sense of smell, occurs in response to odors that enter the extreme superior region of the nasal cavity, the **olfactory recess** (Figure 15-3). Most of the nasal cavity is involved in respiration, with only a small superior portion devoted to olfaction. During normal respiration air passes through the nasal cavity without much of it entering the olfactory recess. For this reason the major anatomical features of the nasal cavity are described in Chapter 23 in relation to respiration. The specialized nasal epithelium of the olfactory recess is called the **olfactory epithelium.**

3

Explain why it sometimes helps to inhale slowly and deeply through the nose when trying to identify an odor.

? ? ? ? ? ? ? ? ? ? ?

Olfactory Epithelium and Bulb

Olfactory neurons are bipolar neurons within the olfactory epithelium (Figure 15-4). Their axons project through numerous small foramina of the bony cribriform plate (see Chapter 7) to the **olfactory bulbs** (see Figures 15-3 and 15-4). **Olfactory tracts** project from the bulbs to the cerebral cortex.

The dendrites of olfactory neurons extend to the epithelial surface of the nasal cavity, and their ends are modified into bulbous enlargements called **olfactory vesicles** (see Figure 15-4). These vesicles possess extremely long (up to 100 mm) cilia called **olfactory hairs,** which lie in a thin mucous film on the epithelial surface.

Airborne molecules enter the nasal cavity and are dissolved in the fluid covering the olfactory epithelium. They interact with chemoreceptor molecules of the neuron ciliary membrane. Although the exact nature and site of this interaction are not yet fully understood, it may be presumed that, in general, chemoreceptors are membrane receptor molecules that somewhat specifically bind to certain molecules. Once the odor-producing molecule has become bound to the receptor, the cilia of the olfactory cells react by depolarizing and initiating action potentials in the olfactory neurons.

The threshold for the detection of odors is very low, so very few molecules are required. Apparently there is rather low specificity in the olfactory epithelium in that a given receptor may react with more than one type of airborne molecule.

The mechanism of olfactory discrimination is not known. Most physiologists believe that the wide variety (perhaps thousands) of detectable smells are actually combinations of a smaller number of primary odors. Seven primary classes of odors have been proposed: (1) camphoraceous, (2) musky, (3) floral, (4) pepperminty, (5) ethereal, (6) pungent, and (7) putrid. However, it is very unlikely that this list is an accurate representation of all primary odors. Indeed, recent studies point to the possibility of as many as 50 primary odors.

The primary olfactory neurons are constantly being replaced. Replacement is rare in other neurons. Olfactory neurons have the most exposed nerve endings; the entire olfactory epithelium, including the neurosensory cells, is lost approximately every 2 months as the olfactory epithelium degenerates and is lost from the surface. Lost olfactory cells are replaced by proliferation of **basal cells** in the olfactory epithelium.

Methylmercaptan, which has an odor similar to that of rotten cabbage, which can be quite nauseating, is added at a concentration of approximately 1 part per million to natural gas, which by itself is odorless. A person can detect the odor of approximately 1/25 billionth of a milligram of the substance and therefore is aware of the presence of the more dangerous but odorless natural gas.

Neuronal Pathways for Olfaction

Axons from the olfactory neurons (cranial nerve I) enter the olfactory bulb (see Figure 15-4) where they synapse with **mitral** (mi'tral; triangular cells; shaped like a bishop's miter or hat) **cells** or **tufted cells.** The mitral and tufted cells relay olfactory information to the brain through the olfactory tracts and synapse with **association neurons** in the olfactory bulb. Association neurons also receive input from nerve cell processes entering the olfactory bulb from the brain. As a result of input from both mitral cells and the brain, association neurons can modify olfactory information before it leaves the olfactory bulb.

Each olfactory tract terminates in an area of the brain called the **olfactory cortex.** The olfactory cortex is within the lateral fissure of the cerebrum and can be divided structurally and functionally into three areas: lateral, intermediate, and medial. The **lateral olfactory area** is involved in the conscious perception of smell. The **medial olfactory area** is responsible for the visceral and emotional reactions to odors. Axons extend from the **intermediate olfactory area** along the olfactory tract to the bulb, synapse with the association neurons, and thus constitute a major mechanism by which sensory information is modulated within the olfactory bulb.

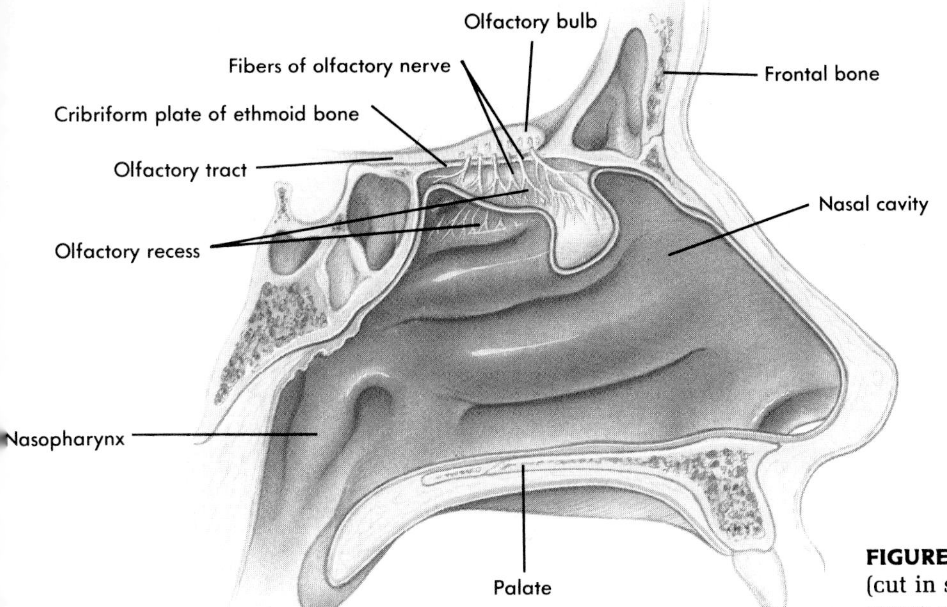

FIGURE 15-3 Lateral wall of the nasal cavity (cut in sagittal section) showing the olfactory recess and olfactory bulb.

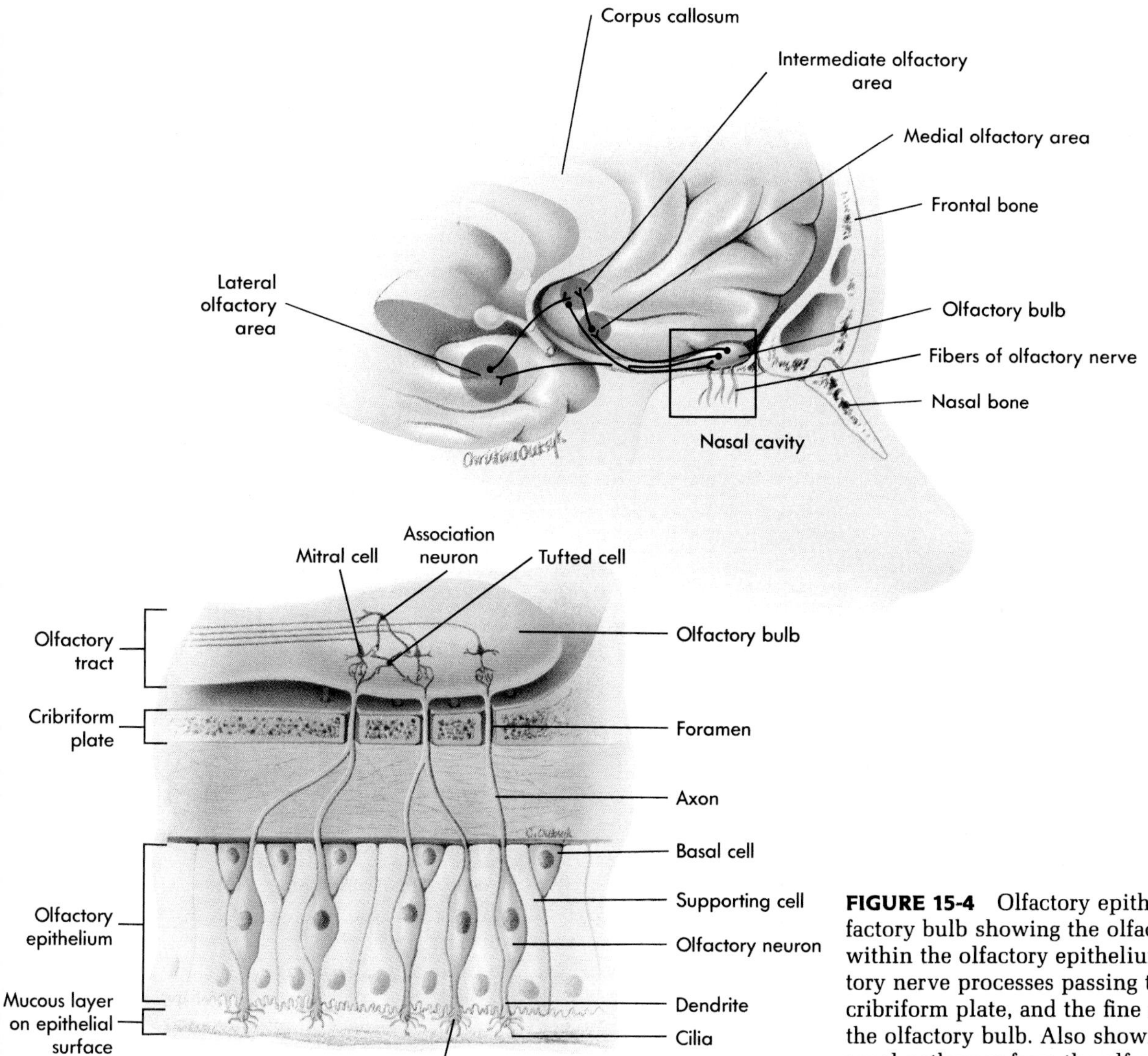

FIGURE 15-4 Olfactory epithelium and olfactory bulb showing the olfactory cells within the olfactory epithelium, the olfactory nerve processes passing through the cribriform plate, and the fine structure of the olfactory bulb. Also shown are the neuronal pathways from the olfactory bulb to the olfactory cortex of the brain.

4

The olfactory system quickly adapts to continued stimulation, and a particular odor becomes unnoticed before very long, even though the odor molecules are still present in the air. Describe as many olfactory sites as you can where such accommodation can occur.

? ? ? ? ? ? ? ? ? ? ?

TASTE

The sensory structures that detect **gustatory,** or **taste,** stimuli are the **taste buds.** Most taste buds are associated with specialized portions of the tongue called **papillae** (pă-pil′e). However, taste buds are also located on other areas of the tongue, the palate, and even the lips and throat, especially in children. There are four major types of papillae, named according to their shape (Figure 15-5): **circumvallate** (sur′kum-val′āt; surrounded by a groove or valley), **fungiform** (fun′jĭ-form; mushroom shaped), **foliate** (fo′le-āt; leaf shaped), and **filiform** (fil′ĭ-form; filament shaped). Taste buds (Figure 15-5, *F*) are associated with circumvallate, fungiform, and foliate papillae. Filiform papillae are the most numerous papillae on the surface of the tongue but have no taste buds.

Circumvallate papillae are the largest but least numerous of the papillae. Eight to 12 of these papillae form a V-shaped row along the border between the anterior and posterior parts of the tongue (see Figure 15-5, *A*). Fungiform papillae are scattered irregularly over the entire dorsal surface of the tongue and appear as small red dots interspersed among the far more numerous filiform papillae. Foliate papillae are distributed over the sides of the tongue and contain the most sensitive of the taste buds. They are most numerous in young children and decrease with age, becoming rare in adults.

Histology of Taste Buds

Taste buds are oval structures embedded in the epithelium of the tongue and mouth (see Figure 15-5, *F*). Each taste bud consists of two types of cells. Specialized epithelial cells form the exterior supporting capsule of the taste bud, and the interior of each bud consists of approximately 40 **gustatory,** or **taste, cells.** Like olfactory cells, taste bud cells are replaced continuously, each having a normal life span of approximately 10 days. Each gustatory cell has several microvilli, called **gustatory hairs,** extending from its apex into a tiny opening in the epithelium called the **gustatory,** or **taste, pore.**

Function of Taste

Substances dissolved in the saliva enter the taste pore and apparently become attached to chemoreceptor molecules on the cell membranes of the gustatory hair, causing a change in membrane permeability and subsequent depolarization of the taste cells. These cells have no axons and do not generate their own action potentials. Neurotransmitters apparently are released from the gustatory cells and stimulate action potentials in the axons of gustatory nerve cells associated with the taste cells.

Hot or cold food temperatures may interfere with the ability of the taste buds to function in tasting food. If a cold fluid is held in the mouth, the fluid will become warmed by the body, and the taste will become enhanced. On the other hand, adaptation is very rapid for taste. This adaptation apparently occurs both at the level of the taste bud and within the CNS. Adaptation may begin within 1 or 2 seconds after a taste sensation is perceived, and complete adaptation may occur within 5 minutes.

The basic tastes detected by the taste buds can be divided into four types: sour, salty, bitter, and sweet. Even though there are only four primary tastes, a fairly large number of different tastes can be perceived, presumably by combining the four basic taste sensations. As with olfaction, the specificity of the receptor molecules is not perfect. For example, artificial sweeteners have different chemical structures than the sugars they are designed to replace and are often many times more powerful than natural sugars in stimulating taste sensations.

Many of the sensations thought of as being taste are strongly influenced by olfactory sensations. This phenomenon can be demonstrated by pinching one's nose while trying to taste something. Much of the "taste" is lost by this action. Although all taste buds are able to detect all four of the basic tastes, each taste bud is usually most sensitive to one. The stimulus type to which each taste bud responds most strongly is related more to its position on the tongue than to the type of papilla with which it is associated (Figure 15-6). The tip of the tongue reacts more strongly to sweet and salty tastes, the back of the tongue to bitter taste, and the sides of the tongue to sour taste.

Thresholds vary for the four primary tastes. Sensitivity for bitter substances is the highest; sensitivities for sweet and salty tastes are the lowest. Sugars, some carbohydrates, and some proteins produce sweet tastes; acids produce sour tastes; metal ions tend to produce salty tastes; and alkaloids (bases) produce bitter tastes. Many alkaloids are poisonous, so the high sensitivity for bitter tastes may be protective. On the other hand, humans tend to crave sweet and salty tastes, perhaps in response to the body's need for sugars, carbohydrates, proteins, and minerals.

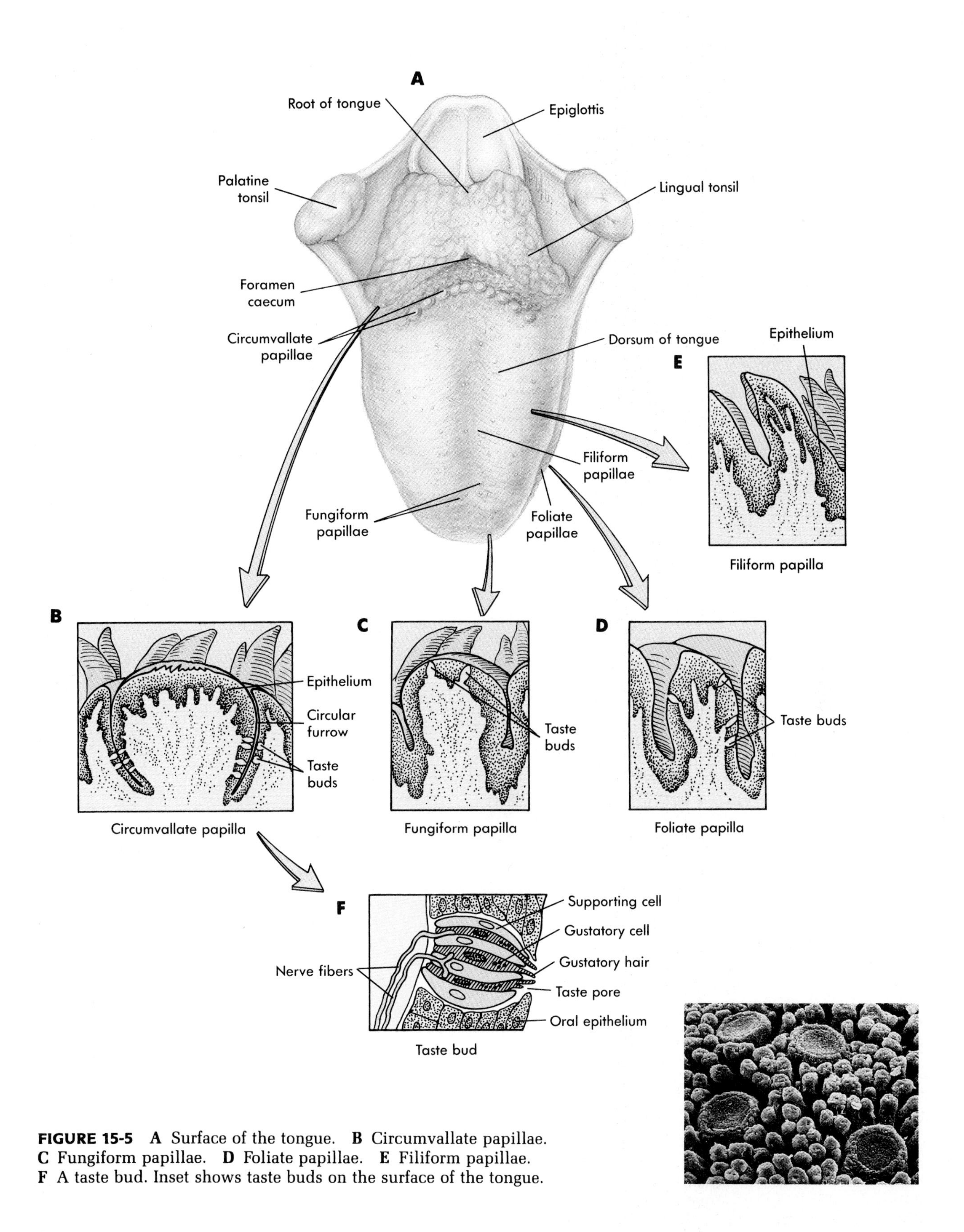

FIGURE 15-5 **A** Surface of the tongue. **B** Circumvallate papillae. **C** Fungiform papillae. **D** Foliate papillae. **E** Filiform papillae. **F** A taste bud. Inset shows taste buds on the surface of the tongue.

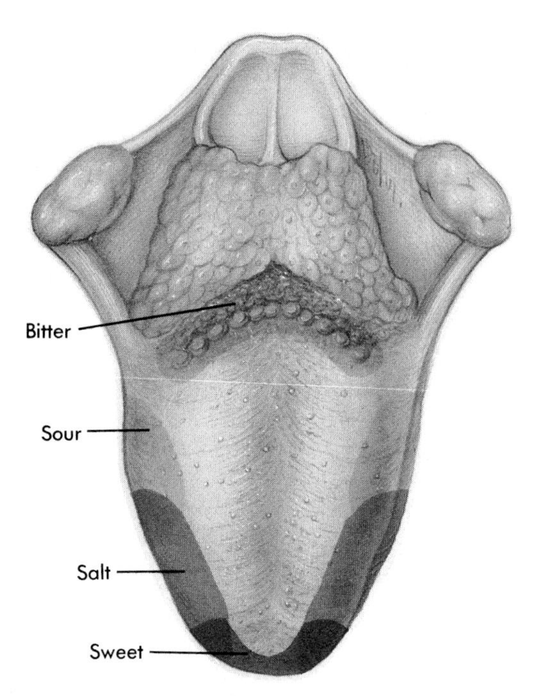

FIGURE 15-6 Regions of the tongue sensitive to various tastes.

Neuronal Pathways for Taste

Taste from the anterior two thirds of the tongue (except from the circumvallate papillae) is carried by means of a branch of the facial nerve (cranial nerve VII) called the **chorda tympani** (kor′dah tim′pah-ne; so named because it crosses over the surface of the tympanic membrane of the middle ear). Taste from the posterior one third of the tongue, the circumvallate papillae, and the superior pharynx is carried by means of the glossopharyngeal nerve (IX). In addition to these two major nerves, the vagus nerve (X) carries a few fibers for taste sensation from the epiglottis.

These nerves extend from the taste buds to the tractus solitarius of the medulla oblongata (Figure 15-7). Fibers from this nucleus decussate and extend to the thalamus. Neurons from the thalamus project to the taste area of the cortex, which is at the extreme inferior end of the postcentral gyrus.

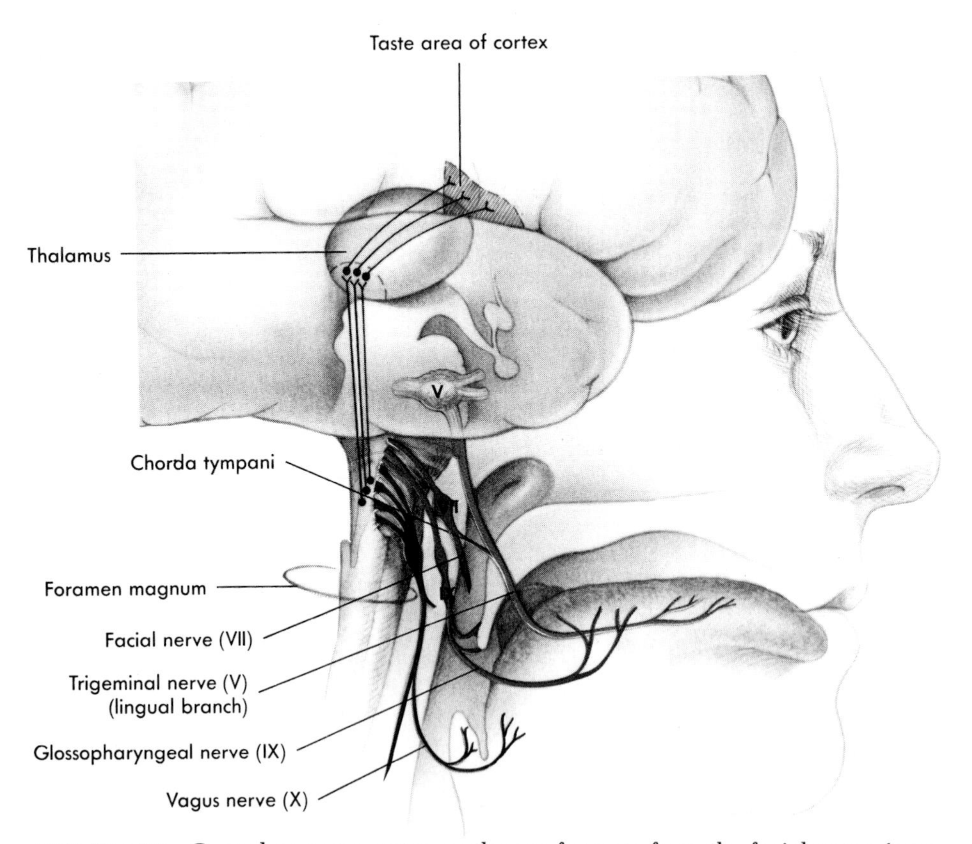

FIGURE 15-7 Central nervous system pathways for taste from the facial nerve (anterior two thirds of the tongue), glossopharyngeal nerve (posterior one third of the tongue), and vagus nerve (root of the tongue). The trigeminal nerve is also shown. It carries tactile sensations from the anterior two-thirds of the tongue. The chorda tympani from the facial nerve (carrying taste input) joins the trigeminal nerve. Nerves carrying taste impulses synapse in the ganglion of each nerve, in the nucleus of the tractus solitarius, and in the thalamus before terminating in the taste area of the cortex.

VISUAL SYSTEM

The visual system includes the eyes, the accessory structures, and the optic nerves, tracts, and pathways. The eyes respond to light and initiate afferent signals, which are transmitted from the eyes to the brain by the optic nerves and tracts. The accessory structures such as eyebrows, eyelids, eyelashes, and tear glands help protect the eyes from direct sunlight and damaging particles. Much of the information about the world around us is detected by the visual system. Our education is largely based on visual input and depends on our ability to read words and numbers. Visual input includes information about light and dark, color and hue.

Accessory Structures

Accessory structures protect, lubricate, move, and in other ways aid in the function of the eye (Figures 15-8 to 15-12). They include the eyebrows, eyelids, conjunctiva, lacrimal apparatus, and extrinsic eye muscles.

Eyebrows

The **eyebrows** protect the eyes by preventing perspiration, which can irritate the eyes, from running down the forehead and into the eyes, and they help shade the eyes from direct sunlight.

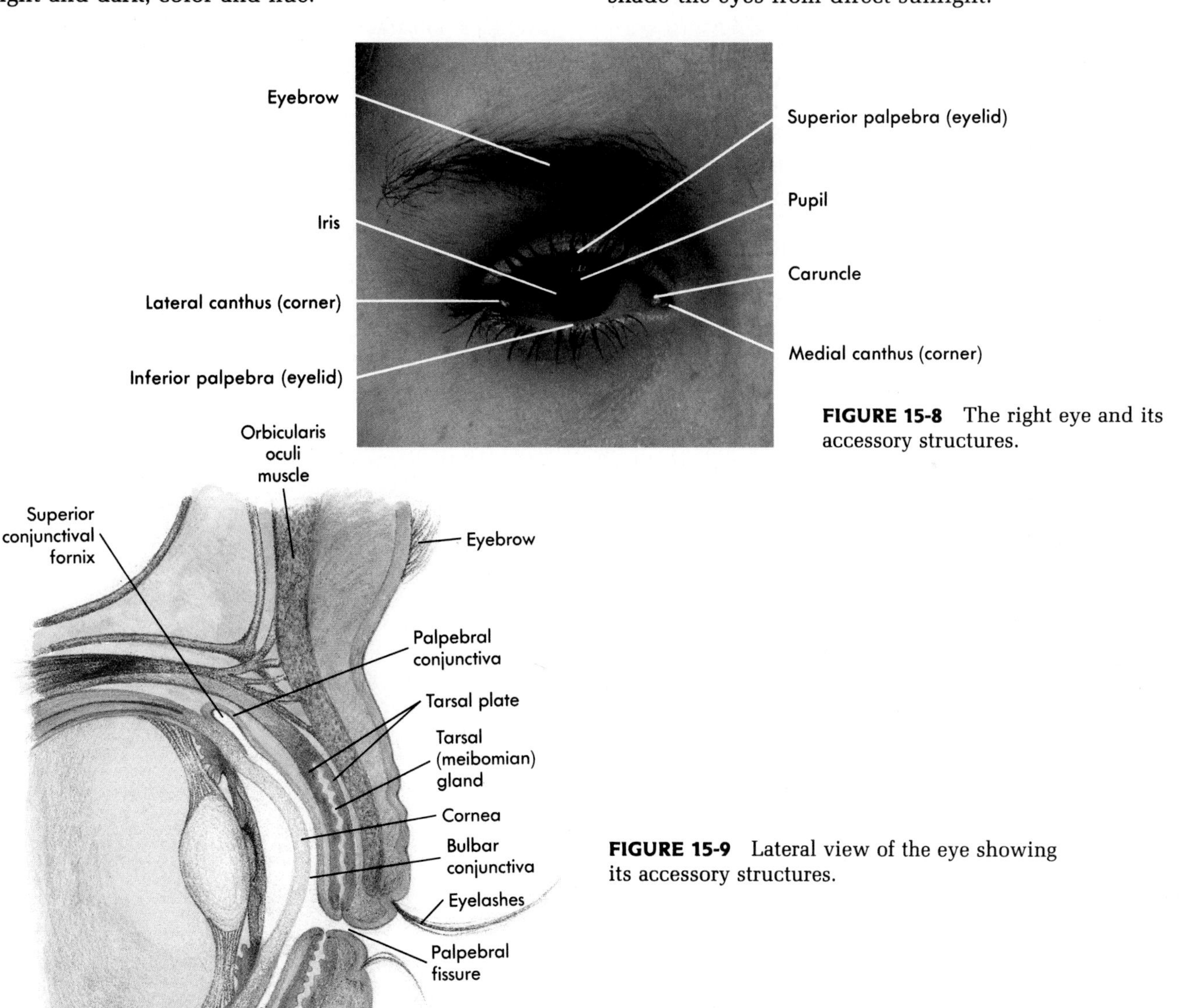

FIGURE 15-8 The right eye and its accessory structures.

FIGURE 15-9 Lateral view of the eye showing its accessory structures.

Eyelids

The **eyelids,** also called **palpebrae** (pal-pe'bre), with their associated lashes protect the eyes from foreign objects. The space between the two eyelids is called the **palpebral fissure,** and the angles where the eyelids join at the medial and lateral margins of the eye are called **canthi** (kan'thi; corners of the eye). The medial canthus contains a small reddish-pink mound called the **caruncle** (kar'ung-kl; a mound of tissue). This structure contains some modified sebaceous and sweat glands.

The eyelids consist of five layers of tissue, which, from the outer to the inner surface, are (1) a thin layer of integument on the external surface; (2) a thin layer of areolar connective tissue; (3) a layer of skeletal muscle (i.e., the orbicularis oculi and levator palpebrae superioris muscles); (4) a crescent-shaped layer of dense connective tissue called the **tarsal** (tar'sal) **plate,** which helps maintain the shape of the eyelid; and (5) the palpebral conjunctiva (described in the next section), which is the deepest portion of the lid, lining its inner surface and lying against the surface of the eyeball.

If an object suddenly approaches the eye, the eyelids protect the eye by closing and then opening quite rapidly (blink reflex). Blinking, which normally occurs approximately 25 times per minute, also helps keep the eye lubricated by spreading tears over the surface of the eye. Movements of the eyelids are a function of skeletal muscles; the orbicularis oculi muscle closes the lids, and the levator palpebrae superioris elevates the upper lid (see Chapter 11). The eyelids also help regulate the amount of light entering the eye.

Eyelashes (see Figure 15-9) are attached as a double or triple row of hairs to the free edges of the eyelids. **Ciliary glands** (modified sweat glands) open into the follicles of the eyelashes, keeping them lubricated. When one of these glands becomes inflamed, it is called a **sty. Meibomian** (mi-bo'me-an; also called tarsal) **glands** are sebaceous glands near the inner margins of the eyelids and produce **sebum** (se'bum; an oily semifluid substance) that lubricates the lids and restrains tears beneath them. An infection or blockage of a meibomian gland is called a **chalazion** (kal-a'ze-on), or **meibomian cyst.**

Conjunctiva

The **conjunctiva** (kon-junk-ti'vah) (see Figure 15-9) is a thin, transparent mucous membrane. The **palpebral conjunctiva** covers the inner surface of the eyelids and the **bulbar conjunctiva,** the anterior surface of the eye. The points where the palpebral and bulbar conjunctivae meet are the superior and inferior **conjunctival fornices.**

Conjunctivitis is an inflammation of the conjunctiva caused by infection or some other irritation. One example of conjunctivitis, which is caused by a bacterium, is **acute contagious conjunctivitis,** also called **pinkeye.**

Lacrimal apparatus

The **lacrimal** (lak'rĭ-mal) **apparatus** (Figure 15-10) consists of a lacrimal gland situated in the superolateral corner of the orbit and a nasolacrimal duct in the inferomedial corner of the orbit. The **lacrimal gland** is innervated by parasympathetic fibers from the facial nerve (cranial nerve VII). The gland produces lacrimal fluid (tears), which leaves the gland through several ducts and passes over the anterior surface of the eyeball. Tears are produced constantly by the gland at the rate of approximately 1 ml per day to moisten the surface of the eye, lubricate the eyelids, and wash away foreign objects. Tears also contain lysozyme, an enzyme that kills certain bacteria. Most of the fluid produced by the lacrimal glands evaporates from the surface of the eye, but excess tears are collected in the medial corner of the eye by the **lacrimal canaliculi.** The opening of each lacrimal canaliculus is called a **punctum** (punk'tum). The upper and lower eyelids each have a punctum near the medial canthus. Each punctum is located

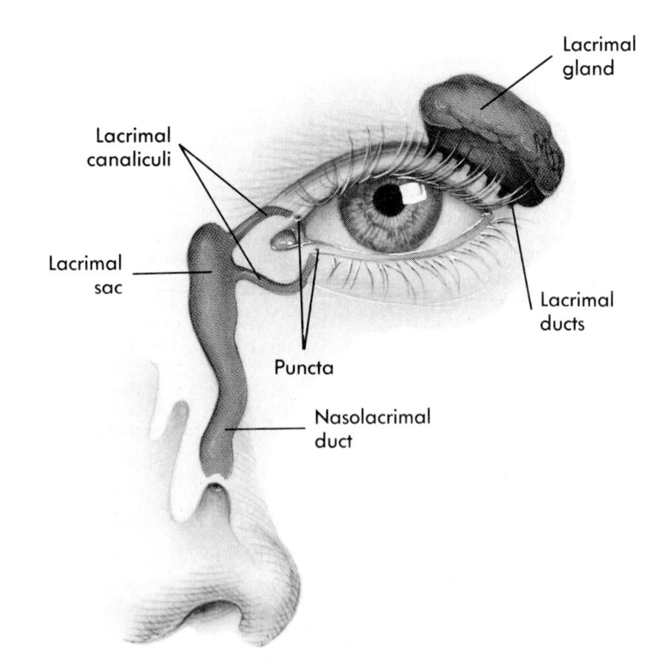

FIGURE 15-10 Lacrimal structures. Tears produced in the lacrimal gland pass over the surface of the eye and enter the lacrimal canaliculi. From there the tears are carried through the nasolacrimal duct to the nasal cavity.

on a small lump called the **lacrimal papilla.** The lacrimal canaliculi open into a **lacrimal sac,** which in turn continues into the **nasolacrimal duct** (see Figure 15-10). The nasolacrimal duct opens into the inferior meatus of the nasal cavity beneath the inferior nasal concha (see Chapter 23).

5

Explain why it is often possible to "taste" a medication such as eyedrops that has been placed into the eyes. Why does a person's nose "run" when he cries?

? ? ? ? ? ? ? ? ? ? ?

Extrinsic eye muscles

Movement of each eyeball is accomplished by six muscles, the extrinsic muscles of the eye (Figures 15-11 and 15-12; also see Chapter 11). Four of these muscles run more or less straight anteroposteriorly. They are the superior, inferior, medial, and lateral **rectus muscles.** Two muscles, the **oblique muscles** (superior and inferior), are placed at an angle to the globe of the eye.

The movements of the eye can be described by a figure resembling the letter "H." The clinical test for normal eye movement is therefore called the H test.

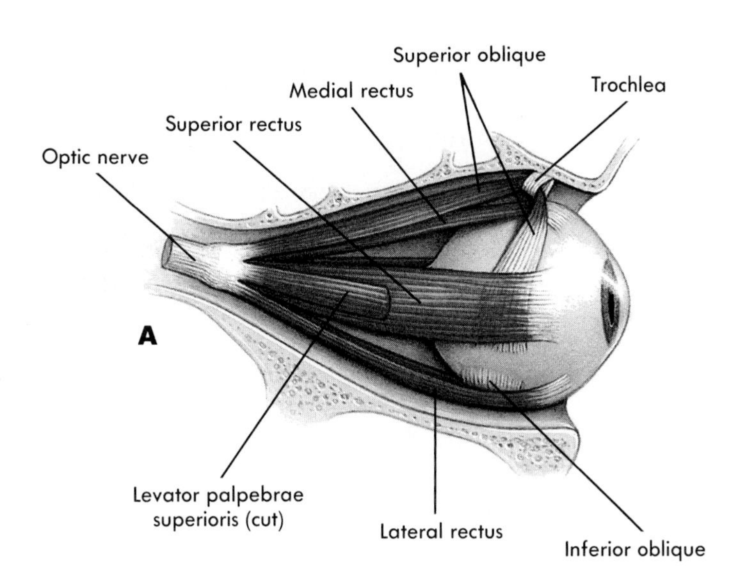

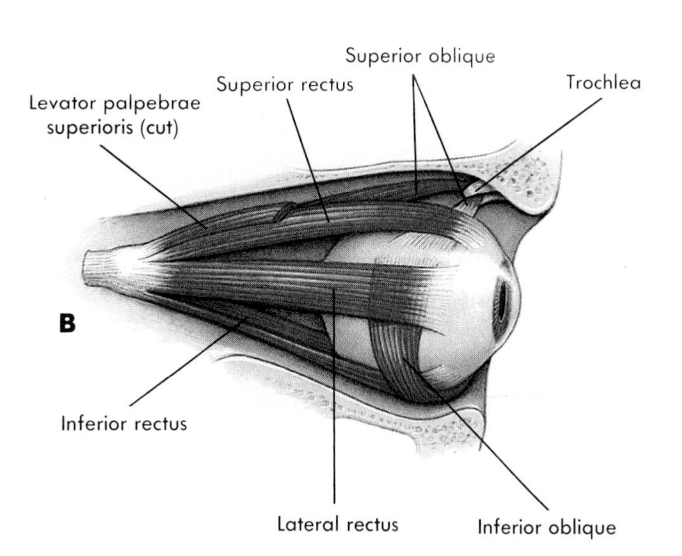

FIGURE 15-11 Extrinsic muscles of the eye. **A** Superior view. **B** Lateral view.

The superior oblique muscle is innervated by the trochlear nerve (cranial nerve IV). The nerve is so named because the superior oblique muscle goes around a little pulley, or trochlea, in the superomedial corner of the orbit. The lateral rectus muscle is innervated by the abducens nerve (cranial nerve VI), so named because the lateral rectus muscle abducts the eye. The other four extrinsic eye muscles are innervated by the oculomotor nerve (cranial nerve III).

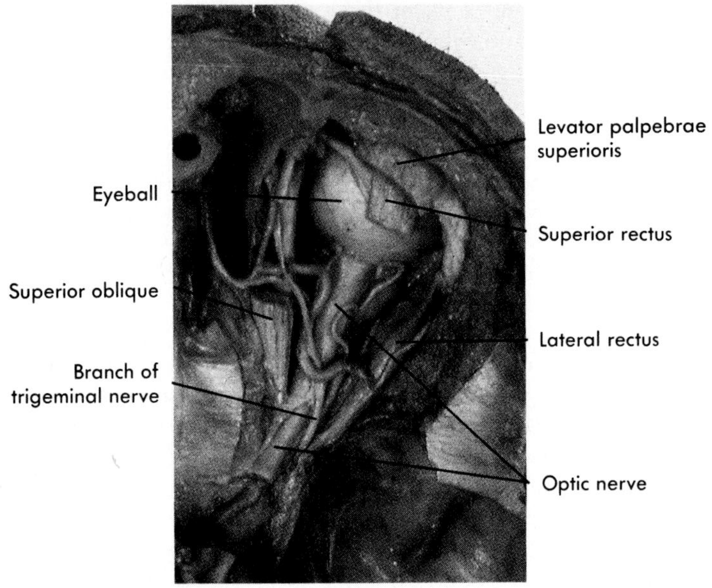

FIGURE 15-12 Superior photographic view of the eye and its associated structures.

Anatomy of the Eye

The eye is composed of three coats or tunics (Figure 15-13). The outer or **fibrous tunic** consists of the sclera and cornea; the middle or **vascular tunic** consists of the choroid, ciliary body, and iris; and the inner or **nervous tunic** consists of the retina.

Fibrous tunic

The **sclera** (skler'ah) is the firm, opaque, white, outer layer of the posterior five sixths of the eye. It consists of dense collagenous connective tissue with elastic fibers. The sclera helps maintain the shape of the eye, protects the internal structures of the eye, and provides an attachment point for the muscles that move the eye. A small portion of the sclera can be seen as the "white of the eye" when the eye and its surrounding structures are intact (see Figure 15-8).

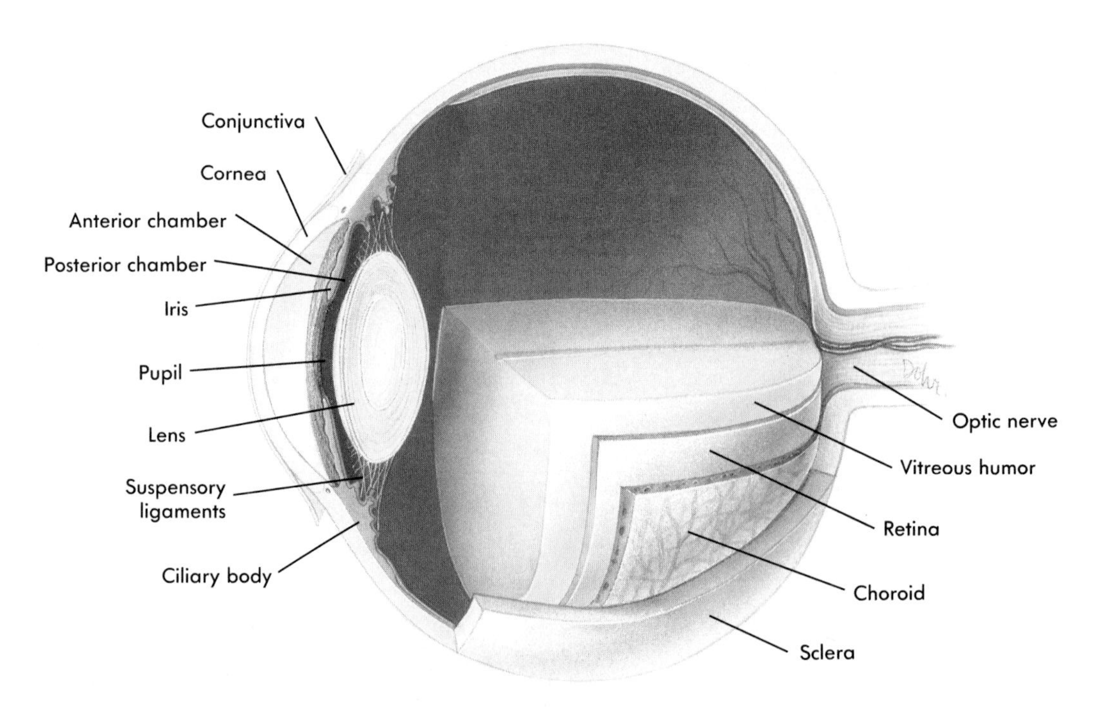

FIGURE 15-13 Sagittal section of the eye demonstrating its layers.

The sclera is continuous anteriorly with the cornea. The **cornea** (kor′ne-ah) is an avascular, transparent structure that permits light to enter the eye. As part of the focusing system of the eye, it bends or refracts the entering light. The cornea consists of a connective tissue matrix containing collagen, elastic fibers, and proteoglycans surrounded by a layer of stratified squamous epithelium. The transparency of the cornea is due, in part, to its low water content and the resultant change in the proteoglycans of the matrix. In the presence of water the proteoglycans are expanded and cause diffusion of light, whereas in the absence of water the proteoglycans decrease in size and do not interfere with light's passing through the matrix.

6

What effect would inflammation of the cornea have on vision?

? ? ? ? ? ? ? ? ? ? ?

The cornea was one of the first organs transplanted. Several characteristics make the cornea relatively easy to transplant: it is easily accessible and relatively easily removed; it is avascular and therefore does not require as extensive circulation as other tissues; and it is less immunologically active and less likely to be rejected than other tissues.

Vascular tunic

The middle tunic of the eyeball is called the vascular tunic because it is the layer containing most of the blood vessels of the eyeball (see Figure 15-13). The arteries of the vascular tunic are derived from a number of arteries called **short ciliary arteries,** which pierce the sclera in a circle around the optic nerve. These arteries are branches of the **ophthalmic artery,** which is a branch of the internal carotid artery. This layer also contains large numbers of melanin-containing pigment cells and appears black in color. The vascular tunic associated with the scleral portion of the eye is the **choroid** (ko′royd). The term means membrane and suggests that this layer is relatively thin (0.1 to 0.2 mm thick). Anteriorly the vascular tunic consists of the ciliary body and iris. The **ciliary** (sil′e-ăr-e) **body** is continuous with the choroid, and the **iris** is attached at its lateral margins to the ciliary body (Figure 15-14). The ciliary body consists of an outer **ciliary ring** and an inner group of **ciliary processes.** The ciliary ring contains smooth muscles called the **ciliary muscles** (the intrinsic eye muscles), which attach to the lens by **suspensory ligaments.** Contraction of the ciliary muscles can change the shape of the lens (this function is described in more detail later in this chapter). The ciliary processes are a complex of capillaries and cuboidal epithelium involved in the production of aqueous humor.

The iris is the "colored part" of the eye, and its color differs from person to person. Brown eyes have

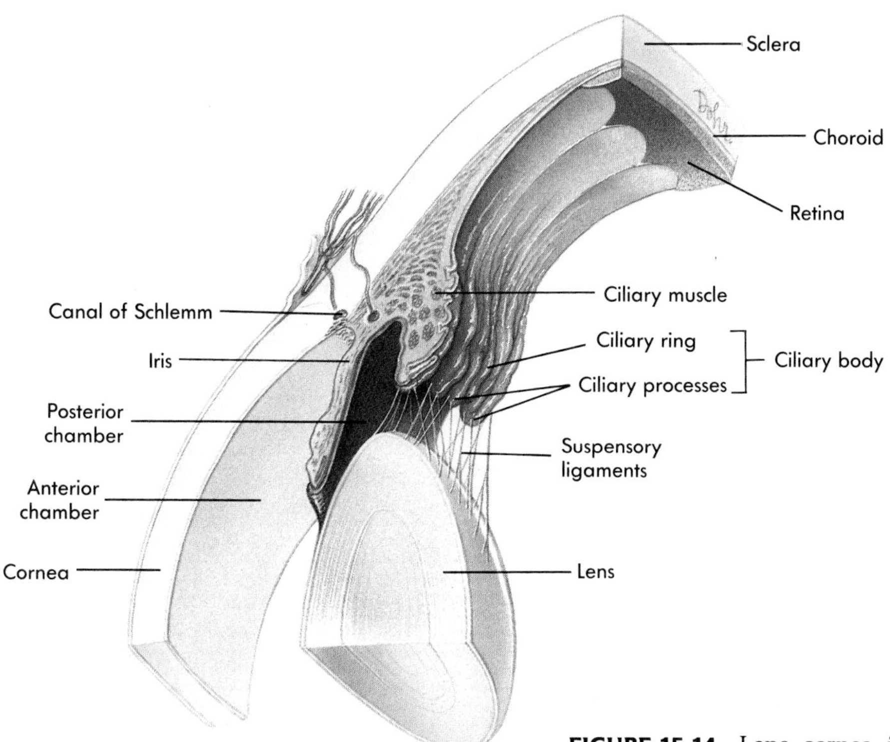

FIGURE 15-14 Lens, cornea, iris, and ciliary body.

brown pigment in the iris. Blue eyes are not caused by a blue pigment but result from the scattering of light by the tissue of the iris in a fashion similar to the scattering of light as it passes through the atmosphere to form the blue skies.

The iris is a contractile structure consisting mainly of smooth muscle and surrounding an opening called the **pupil.** Light enters the eye through the pupil, and the iris regulates the amount of light by controlling the size of the pupil. The iris contains two groups of muscles: a circular group called the **sphincter pupillae** (pu-pil'e); and a radial group called the **dilator pupillae.** The sphincter pupillae are innervated by parasympathetic fibers from the oculomotor nerve (cranial nerve III) and contract the iris, decreasing or constricting the size of the pupil. The dilator pupillae are innervated by sympathetic fibers and dilate the pupil.

Retina

The **retina** is the innermost tunic of the eye (see Figure 15-13). It consists of the outer **pigmented retina** (a pigmented simple cuboidal epithelium) and the inner **sensory retina,** which responds to light. The sensory retina contains photoreceptor cells called **rods** and **cones** and numerous relay neurons. The visual portion of the retina covers the inner surface of the eye posterior to the ciliary body. A more detailed description of the histology and function of the retina is presented later in this chapter.

The pupil appears black when you look into a person's eye because of the pigment in the choroid and the pigmented portion of the retina. The eye is a closed chamber, which allows light to enter only through the pupil. Light is absorbed by the pigmented inner lining of the eye, so looking into it is like looking into a dark room. If a bright light is directed into the pupil, however, the reflected light is red because of the blood vessels on the surface of the retina, which is why the pupils in the eyes of a person looking directly at a flash camera are red in a photograph. In a person with albinism (lacking melanin pigment) the pupil always appears red because there is no pigment to prevent light from entering the eye through the iris and less light is absorbed because of the lack of pigment in the inner lining of the eye. The diffusely lighted blood vessels in the interior of the eye contribute to the color of the pupil.

When the posterior region of the retina is examined with an ophthalmoscope (Figure 15-15), two interesting features can be observed. First, near the center of the posterior retina is a small yellow spot approximately 4 mm in diameter, the **macula lutea** (mak'u-lah lu'te-ah). In the center of the macula lutea is a small pit, the **fovea** (fo've-ah) **centralis.** The fovea is the portion of the retina with the greatest visual acuity (the ability to see fine images) and is normally the point where light is focused. Just medial to the macula lutea is a white spot, the **optic disc,** through which blood vessels enter the eye and spread over the surface of the retina. This is also the spot where nerve processes from the sensory retina meet, pass through the outer two tunics, and exit the eye as the optic nerve. The optic disc contains no photoreceptor cells and does not respond to light; therefore it is called the **blind spot** of the eye.

Compartments of the eye

Two major compartments exist within the eye, a large cavity posterior to the lens and a much smaller cavity anterior to the lens (see Figure 15-13). The anterior cavity is divided into two chambers; the **anterior chamber** lies between the cornea and iris, and a smaller **posterior chamber** lies between the iris and lens (see Figure 15-14). These two chambers are filled with **aqueous humor,** which helps maintain intraocular pressure (pressure within the eye that keeps the eye inflated), refracts light, and provides nutrition for the structures of the anterior chamber (such as the cornea, which has no blood vessels). Aqueous humor is produced by the ciliary processes as a blood filtrate and is returned to the circulation through a

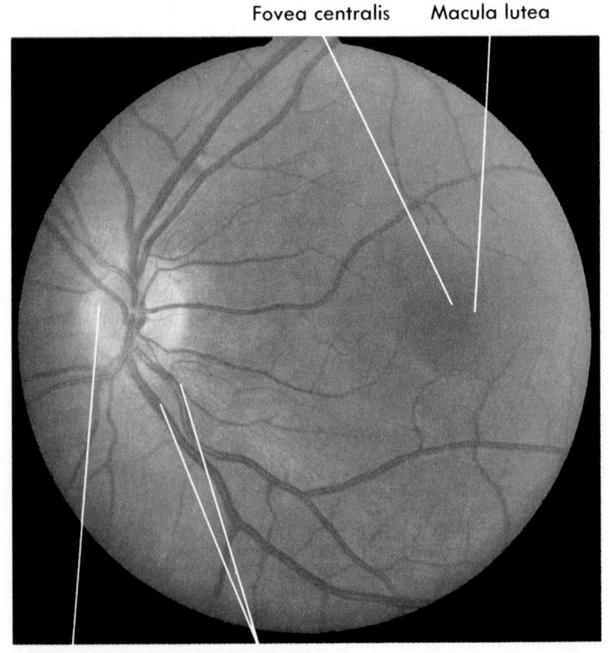

FIGURE 15-15 Ophthalmoscopic view of the retina showing the posterior wall of the retina as seen when looking through the pupil. Notice the vessels entering the eye through the optic disc (the optic nerve), and notice the macula lutea with the fovea (the part of the retina with the greatest visual acuity).

venous ring at the base of the cornea called the **canal of Schlemm.** The production and removal of aqueous humor results in "circulation" of aqueous humor similar to circulation of the cerebrospinal fluid. If circulation of the aqueous humor is inhibited, a defect called glaucoma can result (see essay).

The posterior cavity of the eye is much larger than the anterior cavity. It is surrounded almost completely by the retina and is filled with a transparent jellylike substance, the **vitreous** (vit're-us) **humor.** The vitreous humor is not produced on a regular basis as is the aqueous humor, and its turnover is extremely slow. The vitreous humor helps maintain intraocular pressure (thus maintaining the shape of the eyeball) and holds the lens and the retina in place. It also functions in the **refraction** (bending) of light in the eye.

Lens

The **lens** is an unusual biological structure. It is transparent and biconvex, with the greatest convexity on its posterior side. The lens consists of a layer of cuboidal epithelial cells on its anterior surface and a posterior region of very long columnar epithelial cells called **lens fibers.** Cells from the anterior epithelium proliferate and give rise to the lens fibers at the equator of the lens. The lens fibers lose their nuclei and other cellular organelles and accumulate a special set of proteins called **crystallines.** This crystalline lens is covered by a highly elastic transparent **capsule.**

The lens is suspended between the two eye compartments by the suspensory ligaments of the lens, which are connected to the lens capsule and to the ciliary body containing the smooth ciliary muscles.

Functions of the Complete Eye

The eye functions much like a camera. The iris allows light into the eye, and the light is focused by the lens, cornea, and humors onto the retina. The light striking the retina is converted into action potentials that are relayed to the brain.

Light

The electromagnetic spectrum is the entire range of wavelengths or frequencies of electromagnetic radiation from very short gamma waves at one end to the longest radio waves at the other end (Figure 15-16). **Visible light** is the portion of the electromagnetic spectrum that can be detected by the human eye. Light has characteristics of both particles (photons) and waves, with a wavelength between 400 and

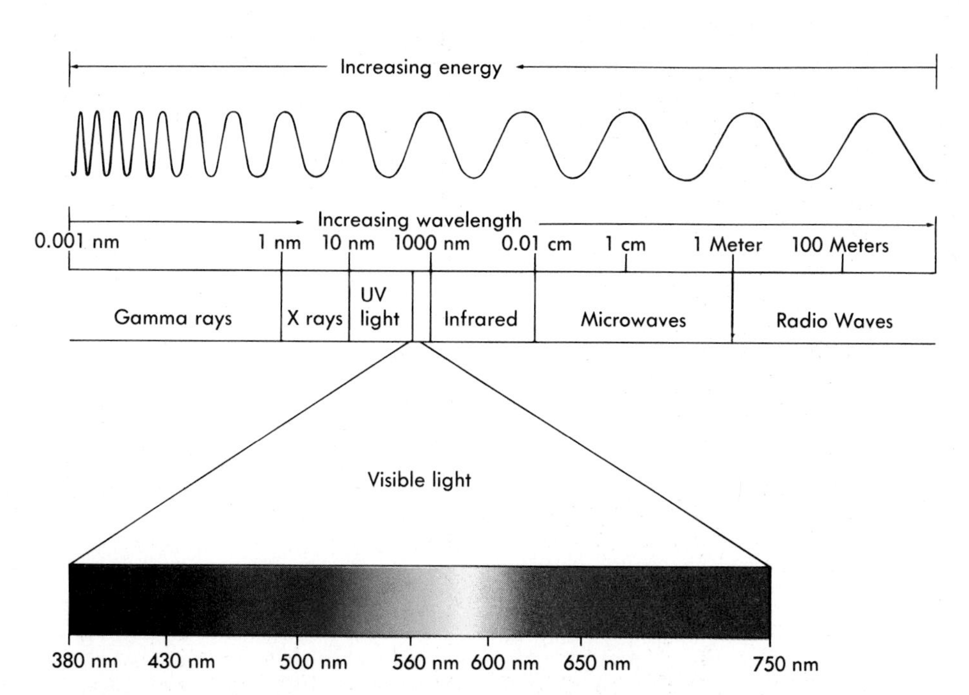

FIGURE 15-16 The electromagnetic spectrum, with the spectrum of visible light pulled out and expanded. The wavelengths of the various colors are also depicted.

700 nm. This range sometimes is called the range of visible light or, more correctly, the **visible spectrum.** Within the visible spectrum each color has a different wavelength.

Light refraction and reflection

An important characteristic of light is that it can be refracted (bent). As light passes from air to some other more dense substance such as glass or water, its speed is reduced. Furthermore, if the surface of that substance is at an angle other than 90 degrees to the direction the light rays are traveling, the rays are bent as a result of variation in the speed of light particles as they encounter the new medium. This is called **refraction.**

If the surface of a lens is concave (the lens is thinnest in the center), the light rays diverge, and the image is magnified; if the surface is convex (the lens is thickest in the center), the light rays tend to converge. As light rays converge, they finally reach a point where they cross. This point is called the **focal point,** and causing light to converge is called **focusing.** No image is formed exactly at the focal point, but a focused image forms just past the focal point. How far past the focal point the focused image forms depends on a number of factors. A biconvex lens causes light to focus closer to the lens than does a lens with a single convex surface. Furthermore, the more nearly spherical the lens, the closer to the lens the light will be focused; the more flattened the biconcave lens, the more distant will be the point where the light is focused.

If light rays strike an object that is not transparent, they bounce off the surface. This phenomenon is called **reflection.** If the surface is very smooth (as with a mirror), the light rays bounce off in a specific direction, depending on the angle of the mirror. If the surface is rough, the light rays are more diffuse as they bounce off the object. We can see most solid objects because of the light reflected off their surfaces.

Focusing of images on the retina

The function of the focusing system of the eye is to focus a clear image on the retina. The cornea is a convex structure; thus as light rays pass from the air through the cornea, they converge. Additional convergence occurs as light encounters the aqueous humor, lens, and vitreous humor. The greatest contrast in media density is between the air and the cornea; therefore the greatest amount of convergence occurs at that point. However, the shape of the cornea and its distance from the retina are fixed so that no adjustment in the location of the focal point can be made by the cornea. Fine adjustment in focal point location is accomplished by changing the shape of the lens. In general, focusing can be accomplished in two ways. One is to keep the shape of the lens constant and move it nearer or farther from the point at which the image will be focused such as occurs in a camera, microscope, or telescope. The second way is to keep the distance constant and to change the shape of the lens, which is the technique used in the eye.

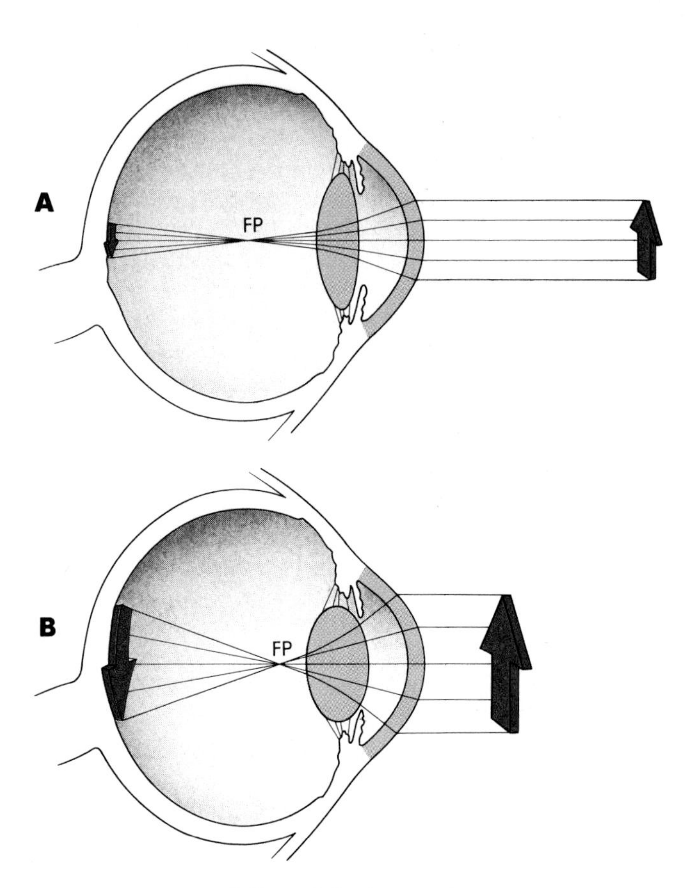

FIGURE 15-17 Ability of the lens to focus images on the retina. The focal point *(FP)* is where light rays cross. **A** Distant object: the lens is flattened, and the image is focused on the retina. **B** Close object: the lens is more rounded, and the image is focused on the retina.

As light rays enter the eye and are focused, the image formed just past the focal point is inverted (Figure 15-17). Action potentials that represent the inverted image are passed to the visual cortex of the cerebrum where they are interpreted by the brain as being right side up.

Because the visual image is inverted when it reaches the retina, the image of the world focused on the retina is upside down. The brain processes information from the retina so that the world is perceived the way "it really is." If a person wears glasses that invert the image entering the eye, he will see the world upside down for a few days, after which time the brain adjusts to the new input to set the world right side up again. If the glasses are then removed, another adjustment period is required before the world is made right by the brain.

When the ciliary muscles are relaxed, the suspensory ligament of the choroid maintains elastic pressure on the lens, keeping it relatively flat and allowing for distant vision (Figure 15-17, *A*). The suspensory ligament maintains tension on the lens because of its inherent elasticity (like a rubber band pulling on the lens). The condition in which the lens is flattened so that nearly parallel rays from a distant object are focused on the retina is referred to as **emmetropia** (em-ĕ-tro′pe-ah; measure) and is the normal resting condition of the lens. In this condition the focal point is located just slightly in front of the retina and the image is focused on the retina. The point at which the lens does not have to thicken for focusing to occur is called the **far point of vision** and normally is 20 feet or more from the eye.

When an object is brought closer than 20 feet to the eye, three events occur to bring the image into focus on the retina: accommodation by the lens, constriction of the pupil, and convergence of the eyes.

1. **Accommodation.** When focusing on a nearby object, the ciliary muscles contract as a result of parasympathetic stimulation from the oculomotor nerve (cranial nerve III). This contraction pulls the choroid toward the lens, reducing the tension on the suspensory ligaments of the lens and allowing the lens to assume a more spherical form because of its own elastic nature (Figure 15-17, *B*). The spherical lens then has a more convex surface, causing greater refraction of light. This process is called accommodation.

7

Explain how several hours of reading can cause eye strain or eye fatigue. Describe what structures are involved.

? ? ? ? ? ? ? ? ? ? ?

As light strikes a solid object, the rays are reflected in every direction from the surface of the object. A small portion of the light rays reflected from a solid object pass through the pupil and enter the eye (see Figure 15-17). An object that is far away from the eye appears small compared to an object that is nearby because the light rays that enter the eye from a distant object are rays that were reflected at a very oblique angle and the rays that enter the eye are nearly parallel (see Figure 15-17, *A*). For an object that is closer to the eye, rays leaving the object at a more acute angle enter the eye (see Figure 15-17, *B*), and the object appears larger. The nearly parallel light rays traveling to the eye from a distant object make the object appear smaller.

When rays from a distant object reach the lens, they do not have to be refracted to any great extent to be focused on the retina, and the lens can remain fairly flat. When an object is closer to the eye, the more obliquely directed rays must be refracted to a greater extent to be focused on the retina.

As an object is brought closer and closer to the eye, accommodation becomes more and more difficult because the lens cannot become any more convex. At some point the eye no longer can focus the object, and it is seen as a blur. The point at which this blurring occurs is called the **near point of vision,** which is approximately 2 to 3 inches from the eye for children, 4 to 6 inches for a young adult, 20 inches for a 45-year-old adult, and 60 inches for an 80-year-old adult. This increase in the near point of vision occurs because the lens becomes more rigid with increasing age, which is primarily why some older people say they could read with no problem if they only had longer arms.

When a person's vision is tested, a chart is placed 20 feet from the eye. and the person is asked to read a line that has been standardized for normal vision. If the person can read the line, the vision is considered to be 20/20, which means that the person can see at 20 feet what people with normal vision can see at 20 feet. If, on the other hand, the person can see words only at 20 feet that people with normal vision can see at 40 feet, the vision is considered 20/40.

2. **Pupil constriction.** Another factor involved in focusing is the **depth of focus,** which is the greatest distance through which an object can be moved and still remain in focus on the retina. The main factor affecting the depth of focus is the size of the pupil. If the pupillary diameter is small, the depth of focus is greater than if the pupillary diameter is large. Therefore with a smaller pupillary opening an object may be moved slightly nearer or farther from the eye without disturbing its focus. This is particularly important when viewing an object at close range because the interest in detail is much greater and therefore the acceptable margin for error is smaller. When the pupil is constricted the light entering the eye tends to pass more nearly through the center of the lens and is more accurately focused than is light passing through the edges of the lens. Pupillary diameter also regulates the amount of light entering the eye (i.e., the dimmer the light, the greater the pupil diameter must be). Therefore as the pupil constricts during close vision, more light is required on the object being observed.

3. **Convergence.** Because the light rays entering the eyes from a distant object are nearly parallel, both pupils can pick up the light rays when the eyes are directed more or less straight ahead. As an object moves closer, however, the eyes must be rotated medially so that the object is kept in view in each eye. This medial rotation of the eyes is accomplished by reflexive stimulation of the medial rectus muscle of each eye and is called convergence. Convergence can be easily observed. Have someone stand facing you.

Have him reach out one hand and extend his index finger as far in front of his face as he can. While he keeps his gaze fixed on his finger, have him slowly bring his finger in toward his nose until he finally touches it. Notice the movement of his pupils as he does this. What happens?

Structure and Function of the Retina

The retina consists of a pigmented retina and a sensory retina. The **sensory retina** contains three layers of neurons: photoreceptor, bipolar, and ganglionic. The cell bodies of these neurons form nuclear layers separated by plexiform layers where the neurons of adjacent layers synapse with each other (Figure 15-18). The outer plexiform layer is between the photoreceptor and bipolar cell layers. The inner plexiform layer is between the bipolar and ganglionic cell layers.

The **pigmented retina,** or pigmented epithelium, consists of a single layer of cells. This layer of cells is filled with melanin pigment and together with the pigment in the choroid, provides a black matrix, which enhances visual acuity by isolating individual photoreceptors and reducing light scattering. However, pigmentation is not strictly necessary for vision. People with albinism (lack of pigment) can see, although their visual acuity is reduced.

The layer of the sensory retina nearest the pigmented retina is the layer of rods and cones. The rods and cones are the photoreceptor cells, which are sensitive to stimulation from "visible" light. The light-sensitive portion of each photoreceptor cell is adjacent to the pigmented layer.

Rods

Rods are bipolar photoreceptor cells involved in noncolor vision and are responsible for vision under conditions of reduced light (Table 15-3). The modified dendritic light-sensitive part of rod cells is cylindrical, with no taper from base to apex (Figure 15-19, *A*). This rod-shaped photoreceptive part of the rod cell contains approximately 700 double-layered membranous discs. The discs contain **rhodopsin,** which consists of the protein **opsin** in loose chemical combination with a pigment called **retinal** (derived from vitamin A).

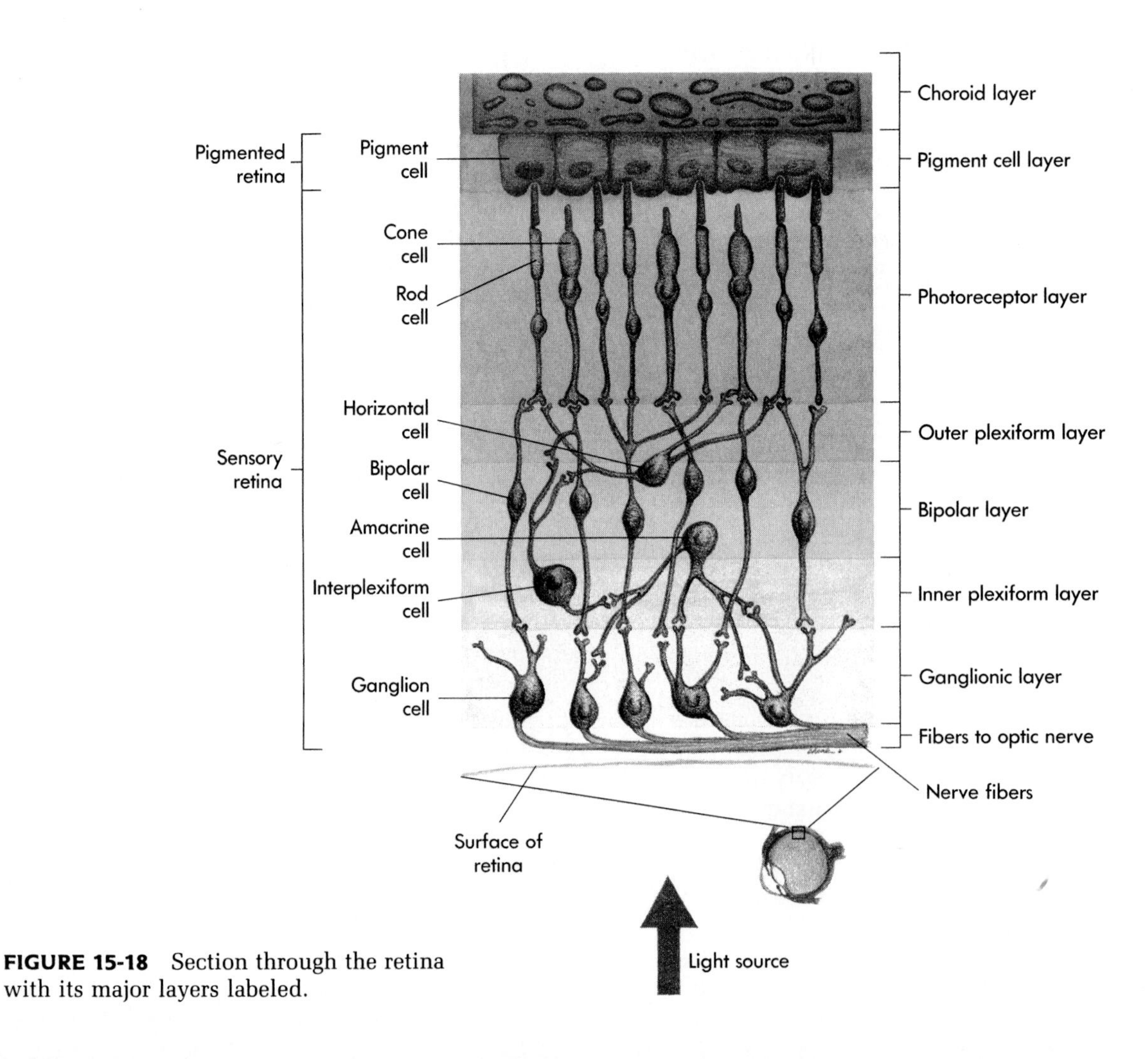

FIGURE 15-18 Section through the retina with its major layers labeled.

Table 15-3

Rods and cones

PHOTORECEPTIVE END	PHOTORECEPTIVE MOLECULE	FUNCTION	LOCATION
Rod			
Cylindrical	Rhodopsin	Noncolor vision; vision under conditions of low light	Over most of retina; none in fovea
Cone			
Conical	Iodopsin	Color vision; visual acuity	Numerous in fovea and macula lutea; sparse over rest of retina

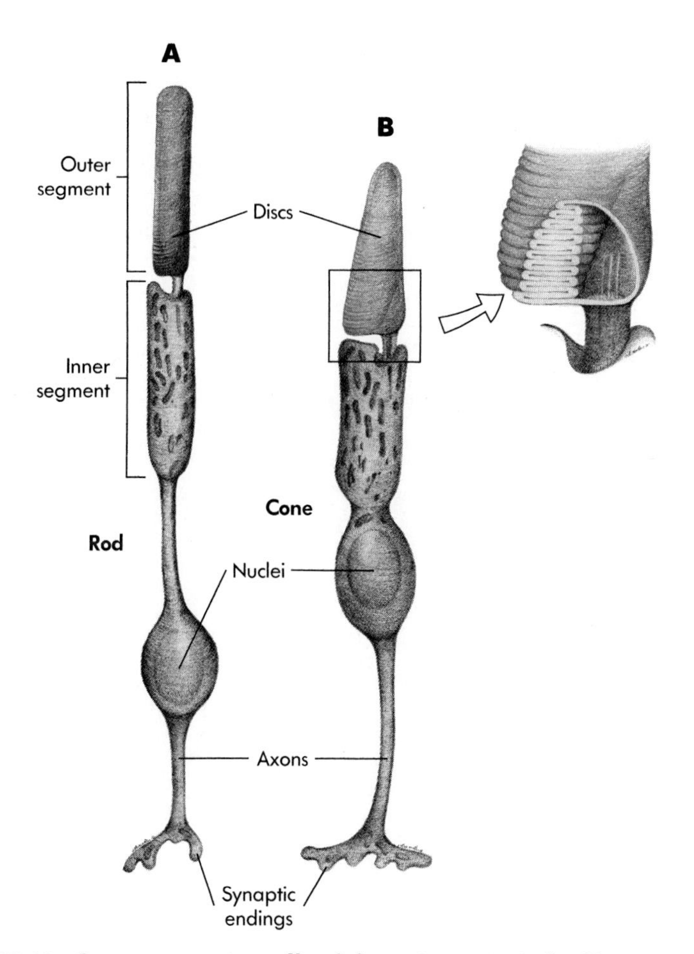

FIGURE 15-19 Sensory receptor cells of the retina. **A** Rod cell. **B** Cone cell.

Function of rhodopsin

Figure 15-20 depicts the changes that rhodopsin undergoes in response to light. In the resting state the shape of opsin and retinal keeps the retinal tightly bound to the opsin surface. In a process called **bleaching,** retinal separates from opsin when rhodopsin is exposed to light. As light is absorbed, retinal changes shape (becomes flattened rather than curved) and begins to lose its attachment to the opsin molecule. Because of the retinal detachment, opsin "opens up" with a release of energy. This reaction is somewhat like a spring (opsin) that is held by a trigger (retinal). Light simply activates the trigger, which, when released, allows the spring to forcefully uncoil. It is thought that the separation of opsin and retinal exposes some active sites that change the membrane potential of the rod cell.

At the final stage of this light-initiated reaction, retinal is completely released from the opsin. This free retinal may be converted back to vitamin A from which it was originally derived. The total vitamin A–retinal pool is in equilibrium so that under normal conditions the amount of free retinal is relatively constant. To create more rhodopsin, the altered retinal must be converted back to its original shape, a reaction that requires energy. Once the retinal resumes its original shape, its recombination with opsin is spontaneous, and the newly formed rhodopsin can again respond to light.

Adaptation to light or dark conditions such as coming out of a darkened building into the sunlight or vice versa is accomplished by changes in the amount of available rhodopsin. In bright light excess rhodopsin is broken down so that not as much is available to initiate action potentials, and the eyes become "adapted" to bright light. Conversely, in a dark room more rhodopsin is produced, making the retina more light sensitive.

8

If breakdown of rhodopsin occurs rapidly and production is slow, do eyes adapt more rapidly to light or dark conditions?

? ? ? ? ? ? ? ? ? ? ?

Light and dark adaptation (adjustment of the eyes to changes in light) also involves pupil reflexes (i.e., enlargement of the pupil in dim light and contraction in bright light), as well as decreased rod function and increased cone function in light conditions (and vice versa during dark conditions).

Cones

Color vision and visual acuity are functions of cone cells. Color is a function of the wavelength of light, and each color can be assigned a certain wavelength within the visible spectrum. Even though rods are very sensitive to light, they cannot detect color, and afferent signals that ultimately reach the brain from these cells are interpreted by the brain only as shades of gray. Cones require relatively bright light to function; as a result, as the light on a given object

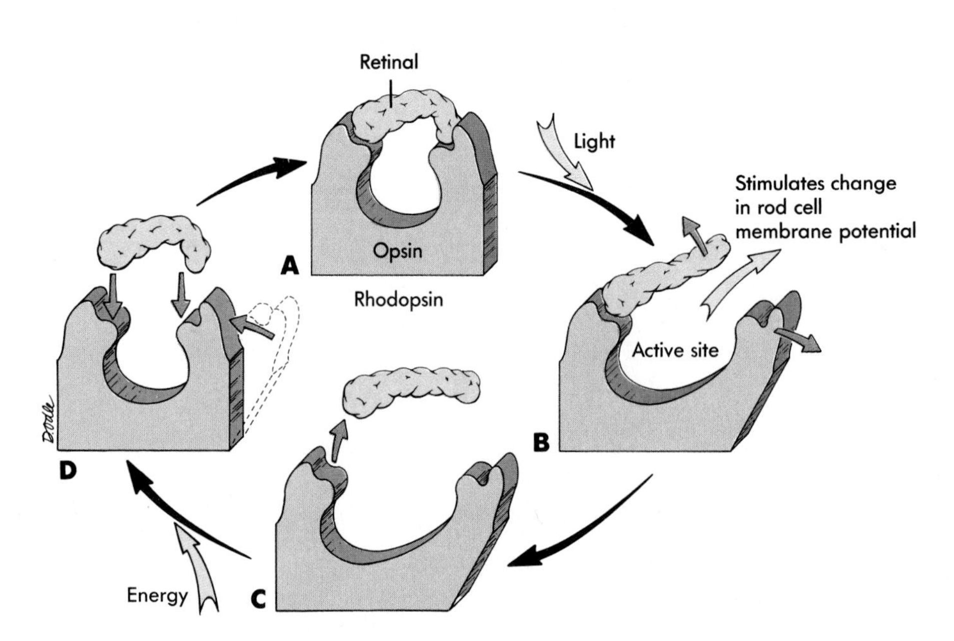

FIGURE 15-20 Rhodopsin cycle. **A** Retinal is attached to opsin. **B** Light causes retinal to change shape. Opsin molecule also changes shape (opens), exposing an active site, and stimulates a change in the rod cell membrane potential. **C** Retinal separates from opsin. **D** Energy is required to bring opsin back to its original form and to attach retinal to it.

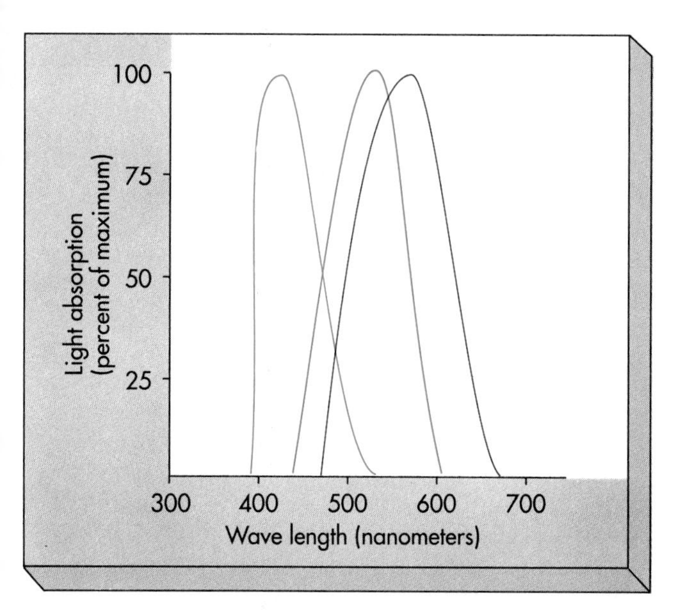

FIGURE 15-21 Wavelengths to which each of the three visual pigments are sensitive: blue, green, red, and their overlap.

decreases, so does the color that can be seen until, under conditions of very low illumination, the object appears gray.

Cones are bipolar photoreceptor cells with a conical light-sensitive part that tapers slightly from base to apex (Figure 15-19, *B*). The outer segments of the cone cells, like those of the rods, consist of double-layered discs. The discs are slightly more numerous and more closely stacked in the cones than in the rods. Cone cells contain a visual pigment, **iodopsin** (i'o-dop'sin), which consists of retinal combined with an opsin called **photopsin.**

There are actually three types of cones, each containing a different type of iodopsin: blue sensitive, red sensitive, and green sensitive. The functions of these pigments are not as well understood as those of rhodopsin, but it is assumed that they function in much the same manner. As can be seen in Figure 15-21, there is considerable overlap in the wavelength of light to which these pigments are sensitive. As light of a given wavelength, representing a certain color, strikes the retina, all cone cells capable of responding to that wavelength fire. Because of the overlap between the three types of cones, different proportions of cone cells respond to each wavelength, thus allowing color perception over a wide range. Color is interpreted in the visual cortex as combinations of afferent signals originating from cone cells. For example, when orange light strikes the retina, 99% of the red-sensitive cones fire, 42% of the green-sensitive cones fire, and no blue cones fire. The variety of combinations that can be created allows humans to distinguish numerous colors.

Distribution of rods and cones in the retina

In addition to their role in color vision, cones are involved in visual acuity. The fovea centralis is used when visual acuity is required (e.g., for focusing on the words of this page); it has a very large number of cones (approximately 35,000) and no rods. However, the 130 million rods are 20 times more plentiful than cones over most of the remaining retina. They are more highly concentrated away from the fovea and are therefore more important for vision in low-light conditions.

9

Explain why at night a person may notice a movement "out of the corner of his eye," but when he tries to focus on the area where the movement was noticed, it appears as though nothing is there.

? ? ? ? ? ? ? ? ? ? ?

Inner layers of the retina

The middle and inner nuclear layers of the retina consist of two major types of neurons: bipolar and ganglion cells. The rod and cone photoreceptor cells synapse with **bipolar cells,** which in turn synapse with **ganglion cells.** Axons from the ganglion cells pass over the inner surface of the retina (see Figure 15-18) except in the area of the fovea centralis, converge at the **optic disc,** and exit the eye as the **optic nerve.** The fovea centralis is devoid of ganglion cell processes, resulting in a small depression in this area (thus the name fovea, or small pit). As a result of the absence of ganglion cell processes (in addition to the concentration of cone cells mentioned previously), visual acuity is further enhanced in the fovea centralis since light rays do not have to pass through as many tissue layers before reaching the photoreceptor cells.

There is a difference between the way bipolar and ganglion cells receive input from rod and cone cells. One bipolar cell receives input from numerous rods, and one ganglion cell receives input from several bipolar cells so that spatial summation of the signal occurs and the signal is enhanced, allowing awareness of stimulus from very dim light sources but decreasing visual acuity in these cells. Cones, on the other hand, exhibit little or no convergence on bipolar cells so that one cone cell may synapse with only one bipolar cell. This system reduces light sensitivity but enhances visual acuity.

Within the inner layers of the retina, there are also **association neurons,** which modify the signals from the photoreceptor cells before the signal ever leaves the retina (see Figure 15-18). **Horizontal cells** are located in the outer plexiform layer and synapse with photoreceptor cells and bipolar cells. **Amacrine cells** are located in the inner plexiform layer and synapse with bipolar and ganglion cells. **Interplexiform cells** are located in the bipolar layer and synapse with amacrine, bipolar, and horizontal cells, forming a feedback loop. Association neurons may be either excitatory or inhibitory on the cells with which they synapse. These association cells function to enhance borders and contours, increasing the intensity at boundaries (e.g., at the edge of a dark object against a light background).

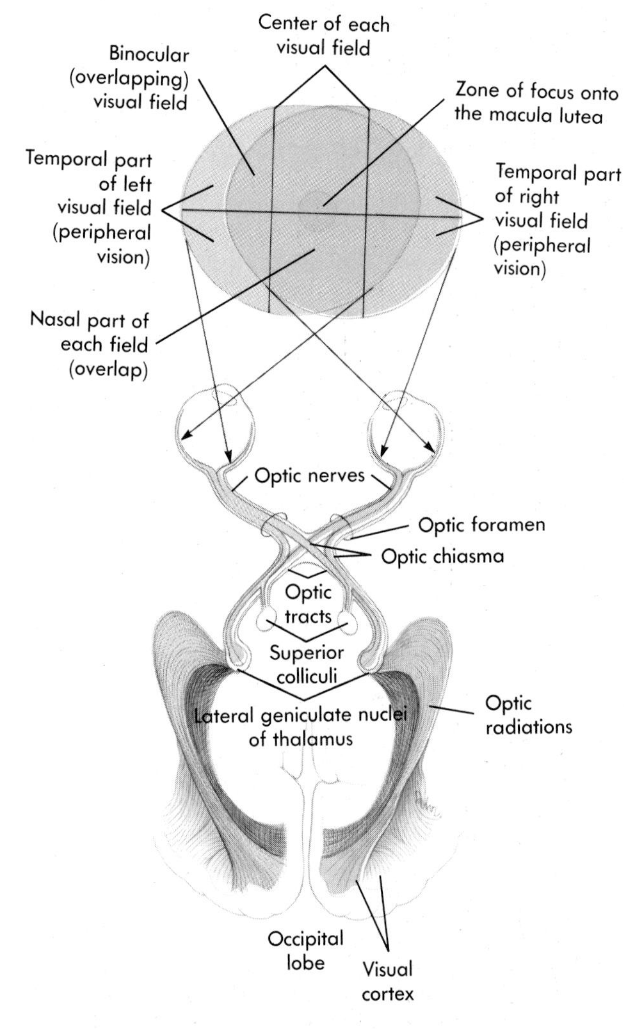

FIGURE 15-22 Right and left visual fields and the visual central nervous system pathways (superior view). Neuronal fibers from the eyes travel through the optic nerves to the optic chiasma where some of the fibers cross. The neurons synapse in the lateral geniculate nuclei of the thalamus, and neurons from there form the optic radiations, which project to the visual cortex.

Neuronal Pathways for Vision

The optic nerve (Figure 15-22) leaves the eye and exits the orbit through the **optic foramen** to enter the cranial vault. Just inside the vault and just anterior to the pituitary, the optic nerves are connected to each other at the **optic chiasma** (ki'az-mah). Ganglion cell axons from the medial portion of the retina (nasal retina) cross through the optic chiasma and project to the opposite side of the brain. Ganglion cell axons from the lateral portion of the retina (temporal retina) pass through the optic nerves and project to the brain on the same side of the body without crossing.

Beyond the optic chiasma the route of the ganglionic axons is called the **optic tract** (see Figure 15-22). Most of the optic tract axons terminate in the **lateral geniculate nucleus** of the thalamus. Some axons do not terminate in the thalamus but separate from the optic tract to terminate in the **superior colliculi,** the center for reflexes initiated by visual stimuli (see Chapter 13). Neurons of the lateral geniculate ganglion form the fibers of the **visual radiations,** which project to the **visual cortex** in the **occipital lobe** (see Figure 15-22). Neurons of the visual cortex integrate the messages coming from the retina into a single message, translate that message into a mental image, and then transfer that image to other parts of the brain where it is evaluated and either ignored or acted upon.

The projections of ganglion cells from the retina can be related to the **visual fields** (see Figure 15-22). The visual field of one eye can be evaluated by closing the other eye. Everything that can be seen with the one open eye is the visual field of that eye. The visual field of each eye can be divided into two parts, a temporal, or lateral, part and a nasal, or medial, part. In each eye the temporal half of the visual field projects onto the nasal portion of the retina, whereas the nasal portion of the visual field projects to the temporal portion of the retina. The projections and nerve pathways are arranged in such a way that images entering the eye from the right half of each visual field project to the left half of the brain. Conversely, the left half of each visual field projects to the right side of the brain.

10

Figure 15-A depicts examples of two lesions in the visual pathways. In the first example, *A*, the effect of a lesion in the optic radiations on the visual fields is depicted (with the right and left fields separated). The darkened areas indicate what parts of the visual fields are defective. In the second example, *B*, a lesion in the right optic nerve is depicted. Describe the effect the lesion would have on the visual fields.

? ? ? ? ? ? ? ? ? ? ? ?

The visual fields of the eyes partially overlap (see Figure 15-22). The region of overlap is the area of **binocular vision** (seen with two eyes), and it is responsible for **depth perception,** the ability to distinguish between near and far objects and to judge their distance. Because humans see the same object with both eyes, the image of the object reaches the retina of one eye at a slightly different angle from that of the other. With experience, the brain can interpret these differences in angle so that distance can be judged quite accurately.

Because the optic chiasma lies just anterior to the pituitary, a pituitary tumor can put pressure on the optic chiasma and may result in visual defects. Because the nerve fibers crossing in the optic chiasma are coming from the outsides of the visual fields, a person with optic chiasma damage cannot see objects in the outer portions of the visual fields, a condition called **tunnel vision.** An early sign of a pituitary tumor often is the development of tunnel vision.

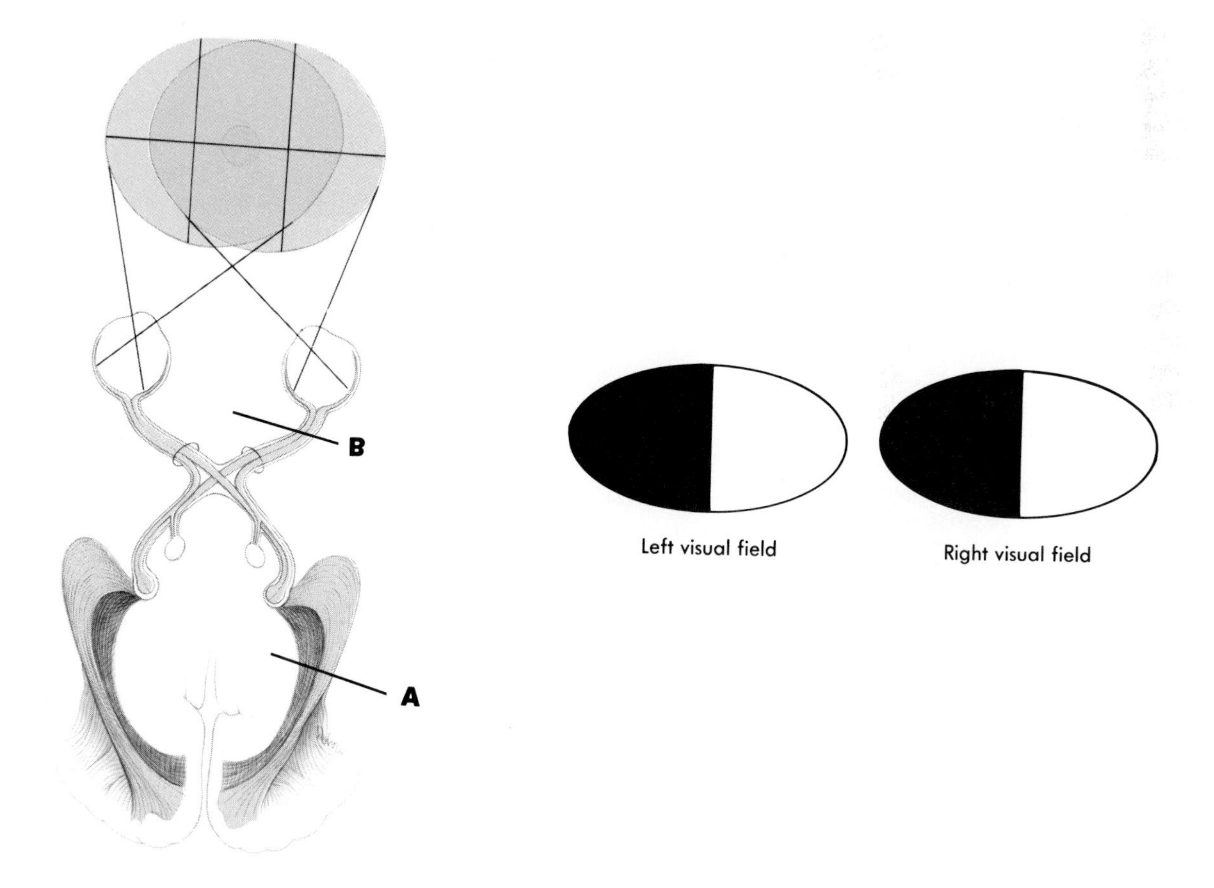

FIGURE 15-A Lesions of the visual pathways. The lines at *A* and *B* represent lesions. The oval insets depict the effects of the lesion at *A* on the visual fields. The darkened areas indicate lack of vision in the area.

Eye Disorders

Myopia

Myopia, or nearsightedness, is the ability to see close objects clearly, but distant objects appear blurry. Myopia is a defect of the eye in which the focusing system, the cornea and lens, is optically too powerful or the eyeball is too long (axial myopia). As a result, the focal point is too near the lens, and the image is focused in front of the retina (Figure 15-B, *A*).

Myopia is corrected by a concave lens that reduces the refractive power of the eye. Concave lenses spread out the light rays coming to the eye and are called "minus" lenses (Figure 15-B, *B*).

Hyperopia

Hyperopia, or farsightedness, is the ability to see distant objects clearly, but close objects appear blurry. Hyperopia is a disorder in which the cornea and lens system is optically too weak or the eyeball is too short. The image is focused behind the retina (Figure 15-B, *C*).

Hyperopia can be corrected by convex lenses that cause light rays to converge as they approach the eye (Figure 15-B, *D*). Such lenses are called "plus" lenses.

Presbyopia

Presbyopia is the normal, presently unavoidable, degeneration of the accommodation power of the eye that occurs as a consequence of aging. It occurs because the lens becomes sclerotic and less flexible. The eye is presbyopic when the near point of vision has increased beyond 9 inches. The average age for onset of presbyopia is the mid-forties. Avid readers or people engaged in fine, close work may develop the symptoms earlier.

Presbyopia can be corrected by the use of "reading glasses" that are worn only for close work and are removed when the person wants to see at a distance. However, it is sometimes an annoyance to keep removing and replacing glasses because reading glasses hamper vision of objects

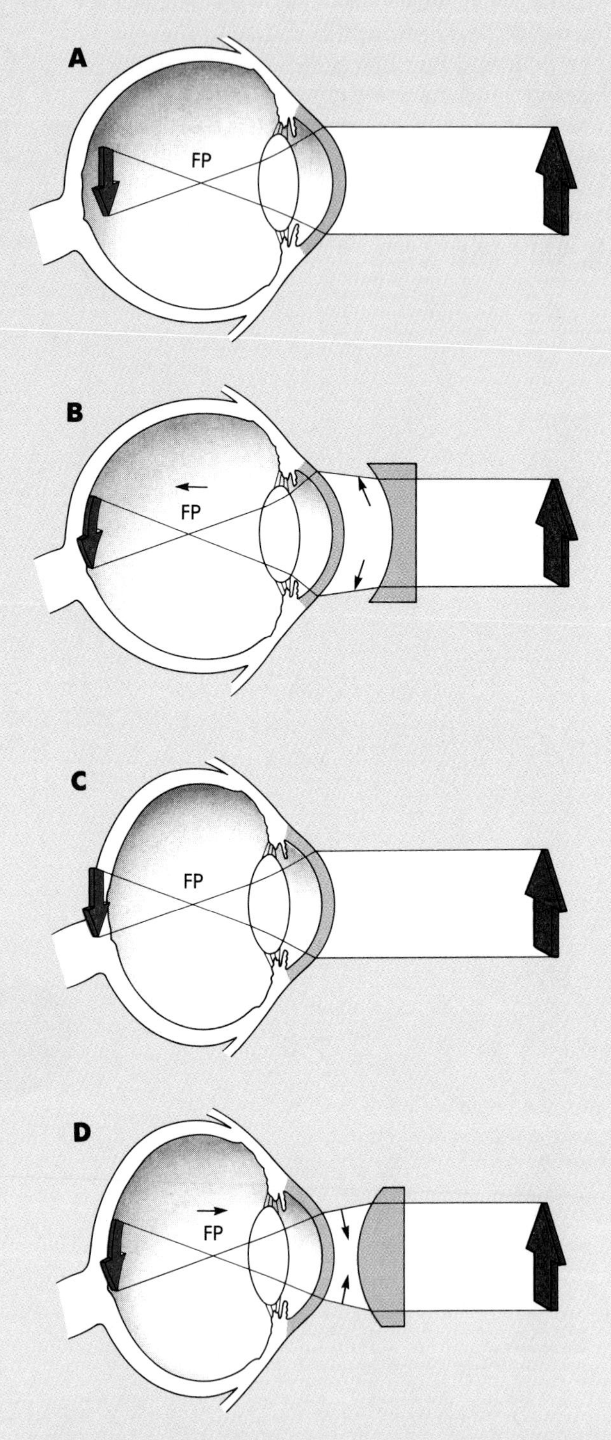

FIGURE 15-B Visual disorders and their correction by various lenses. *FP* is the focal point. **A** Myopia (nearsightedness). **B** Correction of myopia with a concave lens. **C** Hyperopia (farsightedness). **D** Correction of hyperopia with a convex lens.

only a few feet away. This problem can be corrected by the use of half glasses, or **bifocals,** which have different lens in the top and the bottom. Bifocals are particularly important for people who have myopia and develop presbyopia. The bottom half is convex to correct presbyopia when the person reads, and the top half is concave to correct the myopia when the person looks up.

Astigmatism

Astigmatism is a type of refractive error in which the quality of focus is affected. If the cornea or lens is not uniformly curved, the light rays will not focus at a single point but will fall as a blurred circle. Regular astigmatism can be corrected by glasses that are formed with the opposite curvature gradation. Irregular astigmatism is a situation in which the abnormal form of the cornea fits no specific pattern and is very difficult to correct with glasses.

Strabismus

Strabismus is a lack of parallelism of light paths through the eyes. Strabismus can involve only one eye or both eyes, and the eyes may turn in (convergent) or out (divergent). In **concomitant strabismus,** the most common congenital type, the angle between visual axes remains constant, regardless of the direction of the gaze. In **noncomitant strabismus** the angle varies, depending on the direction of the gaze, and deviates as the gaze changes.

In some cases the image that appears on the retina of one eye may be considerably different from that appearing on the other eye. This problem is called **diplopia** (double vision) and is often the result of weak or abnormal eye muscles.

Glaucoma

Glaucoma is a disease of the eye involving increased intraocular pressure caused by a buildup of aqueous humor. It usually results from blockage of the aqueous veins or the canal of Schlemm, restricting drainage of the aqueous humor, or from overproduction of aqueous humor. If untreated, glaucoma can lead to retinal, optic disc, and optic nerve damage. The damage results from the increased intraocular pressure, which is sufficient to close off the blood vessels, causing starvation and death of the retinal cells.

Retinal Detachment

Retinal detachment is a relatively common problem that can result in complete blindness. The integrity of the retina depends on the vitreous humor, which keeps the retina pushed against the other tunics of the eye. If a hole or tear occurs in the retina, fluid may accumulate between the sensory and pigmented retina. This separation may continue until the sensory retina is totally detached and is folded into a funnel-like form around the optic nerve. When the sensory retina becomes separated from its nutrient supply in the choroid, it degenerates, and blindness follows.

Color Blindness

Color blindness is the absence or deficiency of one or more of the cone pigments. It may be complete, involving the total absence of a pigment, or partial, involving only reduced levels of pigment. If one pigment is absent, the condition is called **dichromatism** (dichromatic is the presence of only two colors). An example of this condition is red-green color blindness (Figure 15-C). If a person lacks cones containing red-sensitive iodopsin, he cannot see red and therefore sees red and green as the same color. Individuals who lack cones containing green-sensitive iodopsin, on the other hand, cannot see green and will see red and green as the same color. In extremely rare cases a person may have what is called blue weakness, with decreased amounts of blue-sensitive iodopsin.

Most color blindness is a recessive X-linked inherited trait (i.e., the gene is on the X chromosome; see Chapter 29). In Western Europe approximately 8% of all males have some form of color blindness, whereas only approximately 1% of the females are color-blind.

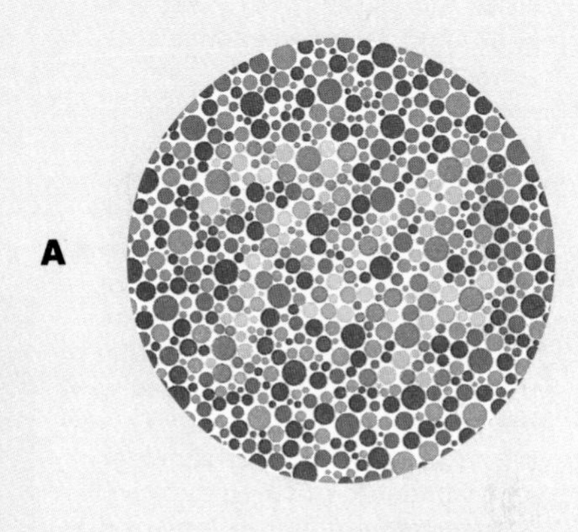

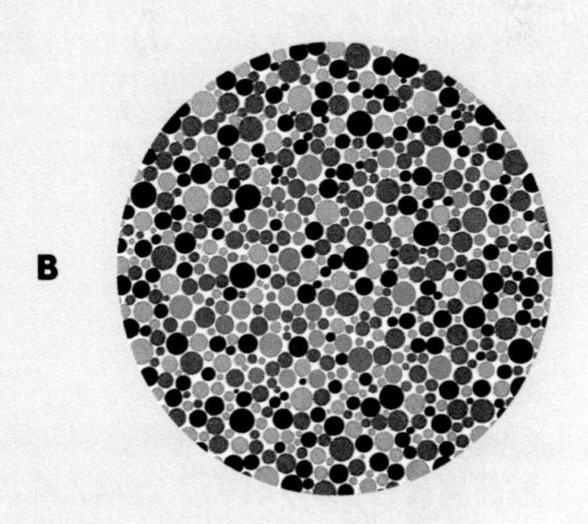

FIGURE 15-C Color blindness charts. **A** Person with normal color vision can see the number 74, whereas a person with red-green color blindness sees the number 21. **B** A person with normal color vision can see the number 42. A person with red color blindness sees the number 2, and a person with green color blindness sees the number 4.

Continued.

Eye Disorders—cont'd

Night Blindness

Everyone sees less keenly in the dark than in the light. However, a person with **night blindness** may not see well enough in a dimly lit environment to function adequately. **Progressive** night blindness results from general retinal degeneration. **Stationary** night blindness results from nonprogressive abnormal rod function. Temporary night blindness can result from a vitamin A deficiency.

Patients with night blindness can be helped with special electronic optical devices. They include monocular pocket scopes and binocular goggles that electronically amplify light.

Cataract

Cataract is a clouding of the lens (Figure 15-D) resulting from a buildup of proteins. The lens relies on the aqueous humor for its nutrition. Any loss of this nutrient source will lead to degeneration of the lens and, ultimately, opacity of the lens (i.e., a cataract). A cataract may occur with advancing age, infection, or trauma.

A certain amount of lens clouding occurs in 65% of patients over the age of 50 and 95% of patients over the age of 65. The decision of whether or not to remove the cataract depends on the extent to which light passage is blocked. More than 400,000 cataracts are removed in the United States each year. Surgery to remove a cataract is actually the removal of the lens. Although light convergence is accomplished by the cornea, with the lens gone the rays cannot be focused as well, and an artificial lens must be supplied to help accomplish focusing.

Macular Degeneration

Macular degeneration is very common in older people. It does not cause total blindness but results in the loss of acute vision. This degeneration has a variety of causes, including hereditary disorders, infections, trauma, tumor, or most often, poorly understood degeneration associated with aging. No satisfactory medical treatment has been developed; therefore optical aids such as magnifying glasses are used to improve visual function.

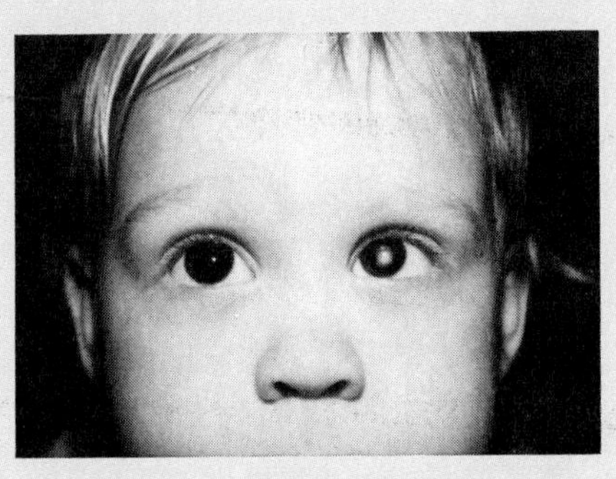

FIGURE 15-D Cataract. Notice how opaque the lens appears.

Diabetes

Loss of visual function is one of the most common consequences of diabetes since a major complication of the disease is dysfunction of the peripheral circulation. Defective circulation to the eye may result in retinal degeneration or detachment. Diabetic retinal degeneration is one of the leading causes of blindness in the United States.

Infections

Trachoma is the leading cause of blindness worldwide. It is caused by an intracellular microbial infection of the corneal epithelial cells that results in scar tissue formation in the cornea. The bacteria are spread from one eye to another eye by towels, fingers, and other objects.

Neonatal gonorrheal ophthalmia is a bacterial infection of the eye that causes blindness. If the mother has gonorrhea (a sexually transmitted disease of the reproductive tract), the bacteria can infect the newborn during delivery. The disease can be prevented by treating the infant's eyes with silver nitrate, tetracycline, or erythromycin drops.

HEARING AND BALANCE

The organs of hearing (auditory or acoustic organs) and balance can be divided into three portions: external, middle, and inner ears (Figure 15-23). The external and middle ears are involved in hearing only, whereas the inner ear functions in both hearing and balance.

The **external ear** includes the **auricle** (aw'rĭ-kl; ear) and the **external auditory meatus** (me-a'tus; passageway from the outside to the eardrum). The external ear terminates medially at the **eardrum,** or **tympanic** (tim-pan'ik) **membrane.** The **middle ear** is an air-filled space within the petrous portion of the temporal bone, which contains the **auditory ossicles.** The **inner ear** contains the sensory organs for hearing and balance. It consists of interconnecting fluid-filled tunnels and chambers within the petrous portion of the temporal bone.

Auditory Structures and Their Functions

External ear

The auricle, or the **pinna** (pin'ah), is the fleshy part of the external ear on the outside of the head and consists primarily of elastic cartilage covered with skin (Figure 15-24). Its shape helps to collect

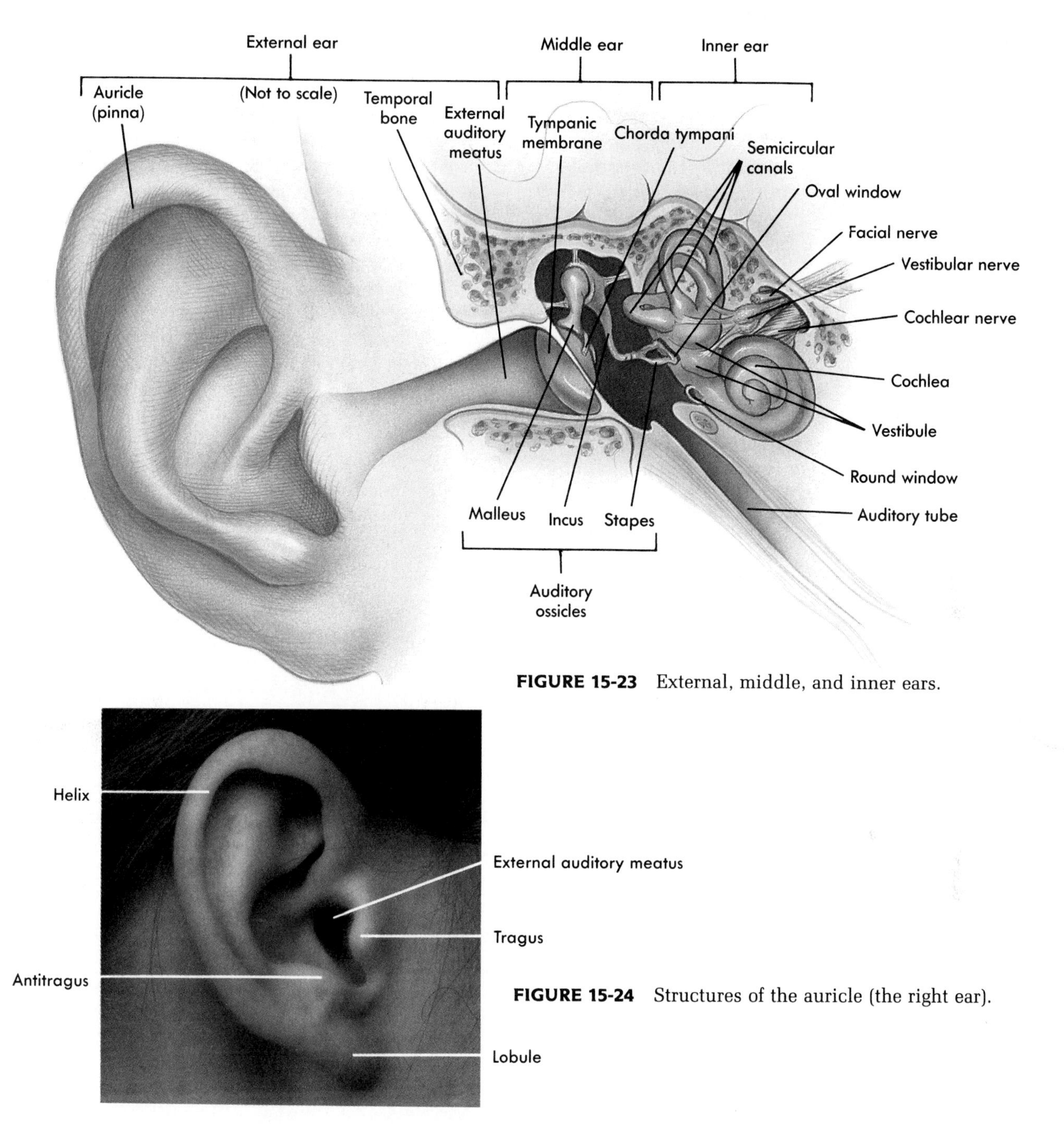

FIGURE 15-23 External, middle, and inner ears.

FIGURE 15-24 Structures of the auricle (the right ear).

sound waves and direct them toward the external auditory meatus. The external auditory meatus is lined by **hairs** and **ceruminous** (se-roo'mĭ-nus) **glands** that produce **cerumen,** a modified sebum commonly called earwax. The hairs and cerumen help prevent foreign objects from reaching the delicate eardrum.

The tympanic membrane, or eardrum, is a thin, semitransparent, nearly oval, three-layered membrane that separates the external ear from the middle ear. It consists of simple epithelium on the inner and outer surfaces with a layer of connective tissue between. Sound waves reaching the tympanic membrane through the external auditory meatus cause it to vibrate.

A somewhat surprising structure, the **chorda tympani,** a branch of the facial nerve carrying taste impulses from the anterior two thirds of the tongue, crosses over the inner surface of the tympanic membrane (see Figure 15-23). It has nothing to do with hearing but is just passing through. This nerve can be damaged during ear surgery or by a middle ear infection, resulting in loss of taste sensation carried by that nerve.

Middle ear

Medial to the tympanic membrane is the air-filled cavity of the middle ear (see Figure 15-23). Two covered openings, the round window and oval window, on the medial side of the middle ear separate it from the inner ear. Two openings provide air passages from the middle ear. One passage opens into the **mastoid air cells** in the mastoid process of the temporal bone posterior to the auricle; the other, the **auditory** or **eustachian tube,** opens into the pharynx and enables equalization of air pressure between the outside air and the middle ear cavity. Unequal pressure between the middle ear and the outside environment can distort the eardrum, dampen its vibrations, and make hearing difficult. Distortion of the eardrum also stimulates pain fibers associated with that structure. That distortion is why when a person changes altitude, sounds seem muffled and the eardrum may become painful. These symptoms can be relieved by opening the auditory tube, allowing air to pass through the auditory tube to equalize air pressure. Swallowing, yawning, chewing, and holding the nose and mouth shut while gently trying to force air out of the lungs are methods used to open the auditory tube.

The middle ear contains three auditory ossicles: the **malleus** (mal'e-us; hammer), **incus** (ing'kus; anvil), and **stapes** (sta'pēz; stirrup), which transmit vibrations from the tympanic membrane to the **oval window.** The handle of the malleus is attached to the inner surface of the tympanic membrane; vibration of the membrane causes the malleus to vibrate as well. The head of the malleus is attached by a very small synovial joint to the incus, which is in turn attached by a small synovial joint to the stapes. The foot plate of the stapes fits into the oval window and is held in place by a flexible **annular ligament.**

Inner ear

The bony tunnels inside the temporal bone are called the **bony labyrinth** (lab'ĭ-rinth; a maze). Inside the bony labyrinth is a similarly shaped but smaller set of membranous tunnels and chambers called the **membranous labyrinth.** The membranous labyrinth is filled with a clear fluid called **endolymph,** and the space between the membranous and bony labyrinth is filled with a fluid called **perilymph.** Perilymph is quite similar to cerebrospinal fluid (CSF), but endolymph has a high concentration of potassium and a low concentration of sodium, which is opposite to perilymph and CSF.

The bony labyrinth of the inner ear can be divided into three regions: cochlea, vestibule, and semicircular canals. The vestibule and semicircular canals are involved primarily in balance, and the cochlea is involved in hearing. The cochlea is divided into three parts: the scala vestibuli, the scala tympani, and the cochlear duct.

The oval window communicates with the vestibule of the inner ear, which in turn communicates with a cochlear chamber, the **scala** (ska'lah) **vestibuli** (Figure 15-25, *A*). The scala vestibuli extends from the membrane of the oval window to the **helicotrema** (hel'ĭ-ko-tre'mah; a hole at the end of a helix or spiral) at the apex of the cochlea; a second cochlear chamber, the **scala tympani,** extends from the helicotrema to the membrane of the **round window.**

The scala vestibuli and the scala tympani are the perilymph-filled spaces between the walls of the bony and membranous labyrinths. The bony walls of each of these chambers are covered by a layer of simple squamous epithelium that is attached to the periosteum of the bone. The wall of the membranous labyrinth that bounds the scala vestibuli is called the **vestibular membrane** (Reissner's membrane); the wall of the membranous labyrinth bordering the scala tympani is the **basilar membrane** (Figure 15-25, *B* and *C*). The space between the vestibular membrane and the basilar membrane, the interior of the membranous labyrinth, which is called the **cochlear duct** or **scala media,** is filled with endolymph.

The vestibular membrane, consisting of a double layer of squamous epithelium, is the simplest region of the membranous labyrinth. The vestibular membrane is so thin that it has little or no mechanical effect on the transmission of sound waves through the inner ear; therefore the perilymph and endolymph on the two sides of the vestibular membrane can be thought of mechanically as one fluid. The role of the vestibular membrane is to separate the two *chemically* different fluids. The basilar membrane is somewhat more complex and is of much greater physiological interest in relation to the mechanics of hearing. It consists of an acellular portion with collagen fibers, ground substance, and sparsely dispersed elastic fibers and a cellular portion with a thin layer of vascular connective tissue that is overlaid with simple squamous epithelium.

The basilar membrane is attached at one side to the osseous **spiral lamina,** which projects from the sides of the **modiolus** (mo'de-o'lus; bony core of the cochlea) like the threads of a screw, and at the other side to the lateral wall of the bony labyrinth by the **spiral ligament,** a local thickening of the periosteum. The distance between the spiral lamina and the spiral ligament (the width of the basilar membrane) increases from 0.04 mm near the oval window to 0.5 mm near the helicotrema. The collagen fibers of the basilar membrane are oriented across the membrane between the spiral lamina and the spiral ligament somewhat like the strings of a piano. The collagen fibers near the oval window are not only shorter than the ones near the helicotrema but are also thicker; the diameter of the collagen fibers in the membrane

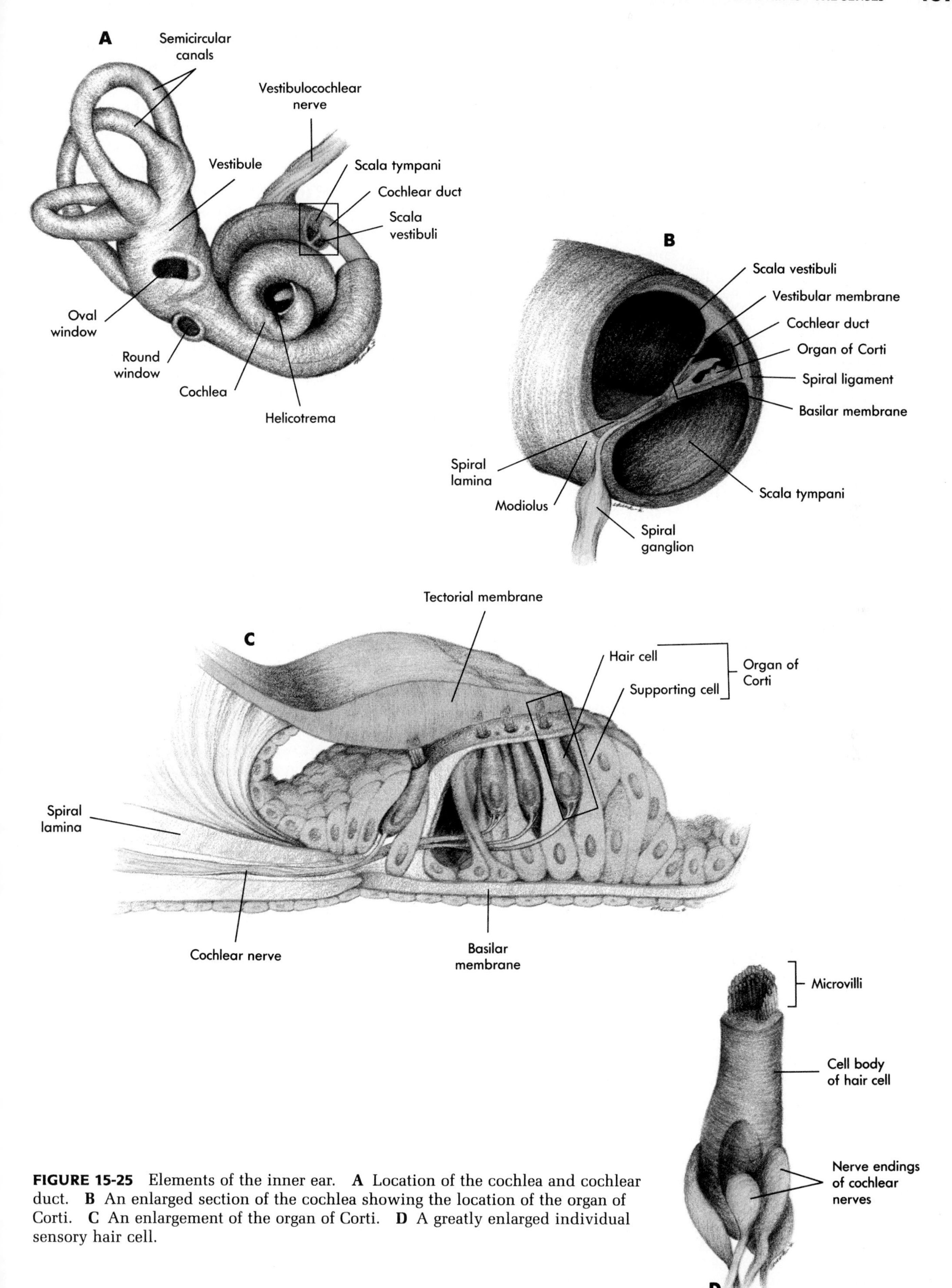

FIGURE 15-25 Elements of the inner ear. **A** Location of the cochlea and cochlear duct. **B** An enlarged section of the cochlea showing the location of the organ of Corti. **C** An enlargement of the organ of Corti. **D** A greatly enlarged individual sensory hair cell.

FIGURE 15-26 Scanning electron micrograph of cochlear hair cells.

decreases as the basilar membrane widens. As a result, the basilar membrane near the oval window is short and stiff and responds to high-frequency vibrations, whereas that part near the helicotrema is wide and limber and responds to low-frequency vibrations.

The cells inside the cochlear duct are highly modified to form a structure called the **organ of Corti** or the **spiral organ** (see Figure 15-25, *B* or *C*). The organ of Corti contains supporting epithelial cells and specialized sensory cells called **hair cells,** which have specialized hairlike projections at their apical ends. It once was thought that these projections were cilia, but it is now apparent that there are one cilium and a large number of microvilli at the apex of each hair cell (Figures 15-25, *D*, and 15-26). The hair cells are arranged in four long rows extending the length of the cochlear duct. The apical ends of the hairs are embedded within an acellular gelatinous shelf called the **tectorial** (tek-tōr′e-al) **membrane,** which is attached to the spiral lamina.

The hairs of the hair cells are bathed in endolymph. Because of the difference in the potassium and sodium ion concentrations between the perilymph and endolymph, there is approximately an 80 mV potential across the vestibular membrane between the two fluids. This is called the **endocochlear potential.** Because the hair cell hairs are surrounded by endolymph, the hairs have a greater electrical potential than if they were surrounded by perilymph. It is believed that this potential difference makes the hair cells much more sensitive to slight movement than they would be if surrounded by perilymph.

Hair cells have no axons, but the basilar regions of each hair cell are covered by synaptic terminals of sensory neurons, the cell bodies of which are located within the cochlear modiolus and are grouped into a **cochlear,** or **spiral, ganglion** (see Figures 15-25, *C*, and 15-31). Afferent fibers of these neurons join to form the **cochlear nerve.** This nerve then joins the vestibular nerve to become the **vestibulocochlear nerve** (cranial nerve VIII), which traverses the internal auditory meatus and enters the cranial vault.

Auditory Function

Sound is created by the vibration of matter such as air, water, or a solid material (there is no sound in a vacuum). When a person speaks, the vocal cords vibrate, causing the air passing out of the lungs to vibrate. The vibration consists of a band of compressed air followed by a band of less compressed air (Figure 15-27). These vibrations are propagated through the air as sound waves, somewhat like ripples are propagated over the surface of water. **Volume,** or loudness, is a function of wave amplitude (height); the greater the amplitude the louder is the sound. **Pitch** is a function of the wave frequency (the number of waves or cycles per second); the higher the frequency, the higher is the pitch. **Timbre** is the resonance quality, or overtones, of a sound. A smooth sigmoid curve is the image of a "pure" sound, but such a sound almost never exists in nature. The sounds made by musical instruments or the human voice are not smooth sigmoid curves but rather are rough, jagged curves. The roughness accounts for the timbre. Timbre allows one to distinguish between, for example, an oboe and a French horn playing the same note (pitch) at the same volume.

The steps involved in hearing are listed in Table 15-4 and are illustrated in Figure 15-28.

External ear

Sound waves are collected by the auricle and are conducted through the external auditory meatus toward the tympanic membrane. Sound travels relatively slowly in air, 332 meters per second, and there may be a significant time interval between the time that a given sound reaches one ear and the time that it reaches the other. The brain can interpret this interval to determine the direction from which a sound is coming.

Middle ear

Sound waves strike the tympanic membrane and cause it to vibrate. This vibration causes vibration of the three ossicles of the middle ear, and by this mechanical linkage vibration is transferred to the membrane of the oval window. More force is required to

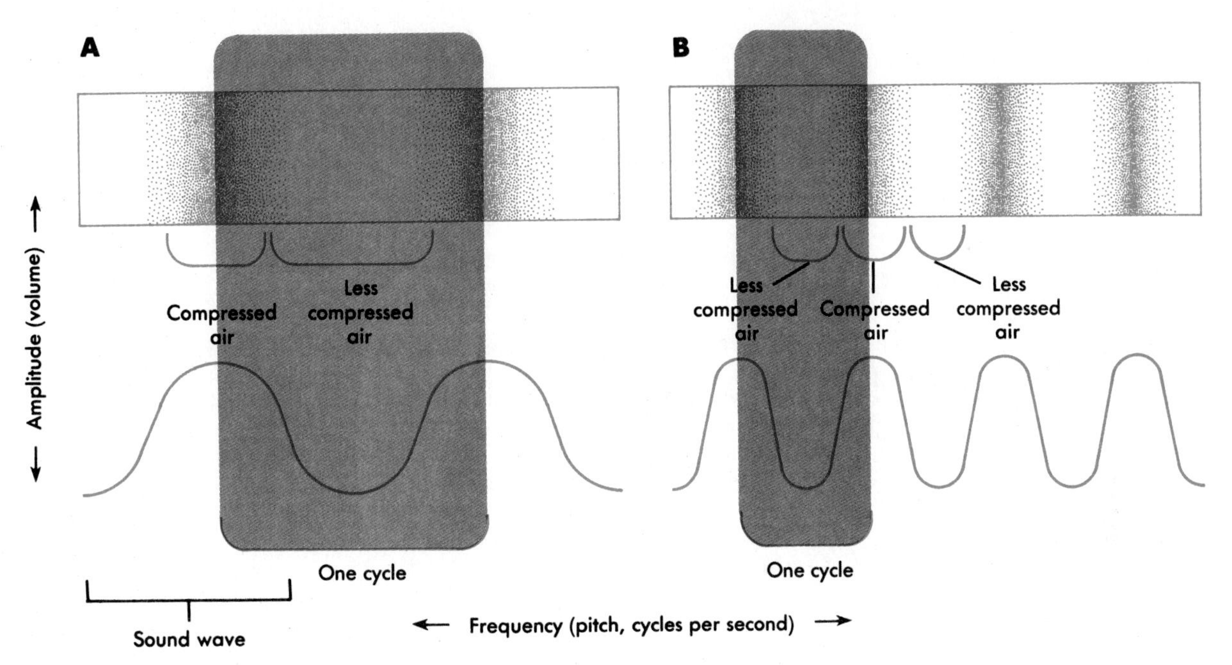

FIGURE 15-27 Each sound wave consists of a region of compressed air between two regions of less compressed air. The sigmoid waves depicted below correspond to the regions of more compressed air *(peaks)* and less compressed air *(troughs)*. The shadowed area represents the width of one cycle (distance between peaks). **A** Depicts a low-frequency sound (few waves per second). **B** Depicts a high-frequency sound (many waves per second).

Table 15-4

Steps involved in hearing

1. Sound waves are collected by the auricle and are conducted through the external auditory meatus to the tympanic membrane, causing it to vibrate.
2. The vibrating tympanic membrane causes the malleus, incus, and stapes to vibrate.
3. Vibration of the stapes produces vibration in the perilymph of the scala vestibuli.
4. The perilymph's vibration produces simultaneous vibration of the endolymph in the cochlear duct.
5. Vibration of the endolymph causes the basilar membrane to vibrate.
6. As the basilar membrane vibrates, the hair cells attached to the membrane move relative to the tectorial membrane, which remains stationary.
7. The hair cell microvilli, embedded in the tectorial membrane, become bent.
8. Bending of the microvilli causes depolarization of the hair cells.
9. The hair cells induce action potentials in the cochlear neurons.
10. The action potentials generated in the cochlear neurons are conducted to the CNS.
11. The action potentials are translated in the cerebral cortex and are perceived as sound.

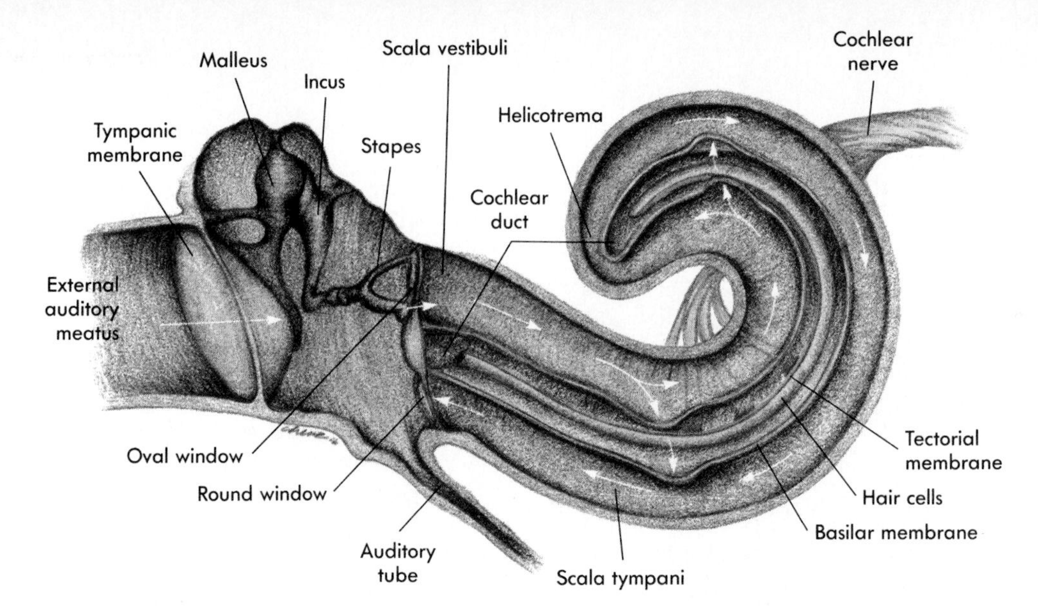

FIGURE 15-28 Effect of sound waves on cochlear structures. Sound waves strike the tympanic membrane and cause it to vibrate. This vibration causes the three bones of the middle ear to vibrate, causing the footplate of the stapes to vibrate in the oval window. This vibration causes the perilymph in the scala vestibuli to vibrate. Vibration of the perilymph causes simultaneous vibration of the endolymph in the cochlear duct, which causes the basilar membrane to vibrate. Short sound waves (high pitch) cause the basilar membrane near the oval window to vibrate, and longer sound waves (low pitch) cause the basilar membrane some distance from the oval window to vibrate. Sound is detected in the hair cells of the organ of Corti, which is attached to the basilar membrane. Vibrations are transferred to the perilymph of the scala tympani and to the round window where they are dampened.

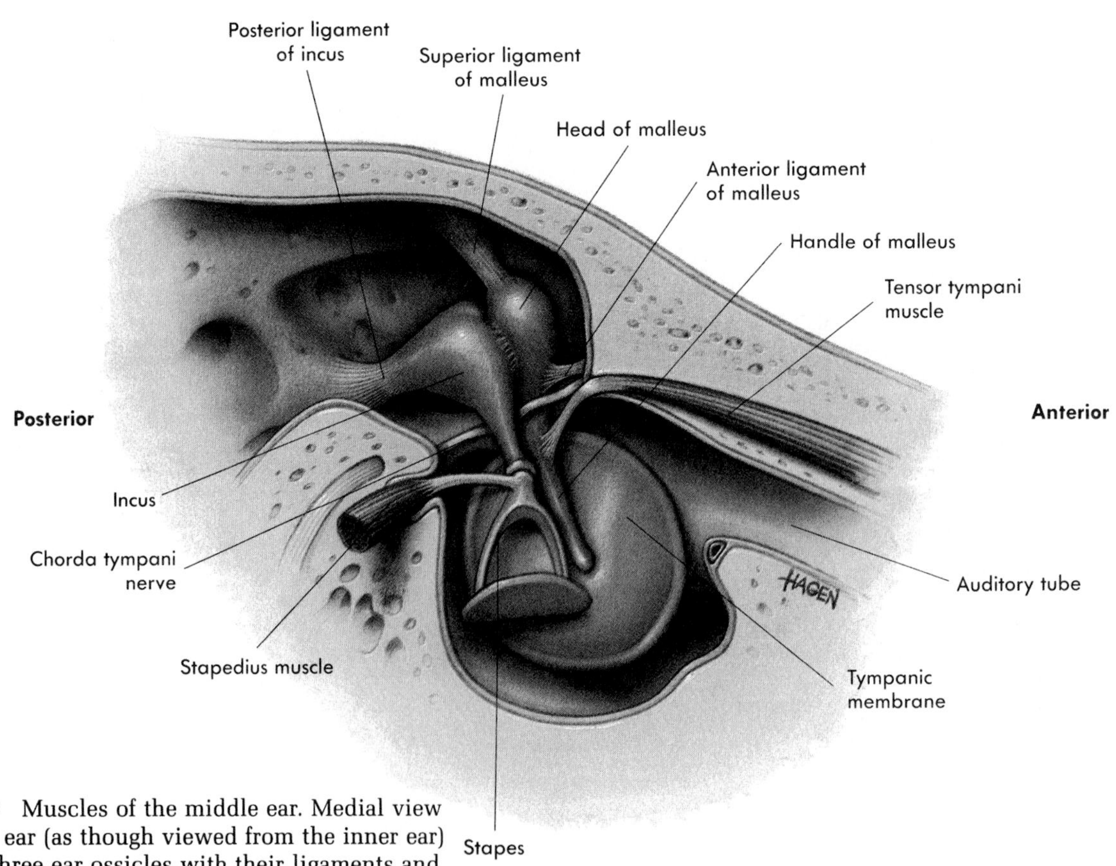

FIGURE 15-29 Muscles of the middle ear. Medial view of the middle ear (as though viewed from the inner ear) showing the three ear ossicles with their ligaments and the two muscles of the middle ear, the tensor tympani and the stapedius.

cause vibration in a liquid such as the perilymph of the inner ear than is required in air, so the vibrations reaching the perilymph must be amplified as they cross the middle ear. The footplate of the stapes and its annular ligament, which occupy the oval window are much smaller than the tympanic membrane. Because of this size difference, the mechanical force of vibration is amplified approximately twentyfold as it passes from the tympanic membrane, through the ossicles, and to the oval window.

Two small skeletal muscles are attached to the ear ossicles and reflexively dampen excessively loud sounds (Figure 15-29). This sound **attenuation reflex** protects the delicate ear structures from damage by loud noises. The **tensor tympani** muscle is attached to the malleus and is innervated by the trigeminal nerve (cranial nerve V). The **stapedius** muscle is attached to the stapes and is supplied by the facial nerve (cranial nerve VII).

11

Why do we have a system that amplifies sound (ossicles) and one that dampens it (muscles)? Would the sound attenuation reflex be more effective in protecting ear structures against sudden loud noises or against constant loud sounds?

? ? ? ? ? ? ? ? ? ? ?

Inner ear

As the stapes vibrates, it produces waves in the perilymph of the scala vestibuli (see Figure 15-28). Because the vestibular membrane is extremely thin and produces almost no mechanical impedance to the vibrating perilymph, vibrations of the perilymph cause simultaneous vibration of the endolymph. The mechanical effect is as though the perilymph and endolymph were a single fluid. Vibration of the endolymph causes vibration of the basilar membrane. Waves in the perilymph of the scala vestibuli are transmitted also through the helicotrema and into the scala tympani. However, because the helicotrema is very small, this transmitted vibration is probably of little consequence. Vibrations of the basilar membrane, together with weaker waves coming through the helicotrema, cause waves in the scala tympani perilymph and ultimately result in vibration of the membrane of the round window. Vibration of the round window membrane is important to hearing because it acts as a mechanical release for waves from within the cochlea. If this window were solid, it would reflect the waves, which would interfere with and dampen later sound waves. The round window also allows relief of pressure in the perilymph (since fluid is not compressible), preventing compression damage to the organ of Corti.

The vibration of the basilar membrane is most important to hearing. As this membrane vibrates, the hair cells resting on the basilar membrane move relative to the tectorial membrane, which remains stationary. The hair cell microvilli, which are embedded in the tectorial membrane, become bent, causing depolarization of the hair cells. The hair cells then induce action potentials in the cochlear neurons that synapse on the hair cells, apparently by direct electrical excitation through electrical synapses rather than by neurotransmitters.

The portion of the basilar membrane that vibrates as a result of endolymph vibration depends on the pitch of the sound that created the vibration and, as a result, on the vibration frequency within the endolymph. The width of the basilar membrane and the length and diameter of the collagen fibers stretching across the membrane at each level along the cochlear duct determine the optimum amount of basilar membrane vibration produced by a given pitch (Figure 15-30). As the basilar membrane vibrates, hair cells along a large portion of the basilar membrane are stimulated. In areas of minimum vibration the amount of stimulation may not reach threshold. In other areas a low-frequency afferent signal may be transmitted, whereas in the optimally vibrating regions of the basilar membrane a high-frequency impulse is initiated.

Afferent impulses conducted by cochlear nerve fibers from all along the spiral organ terminate in the **superior olivary nucleus** in the CNS (described in more detail in the section about neuronal pathways for hearing; see Figure 15-31 and Chapter 13). These impulses are compared to each other, and the strongest impulse, corresponding to the area of maximum basilar membrane vibration, is taken as standard. Efferent impulses then are sent from the superior oli-

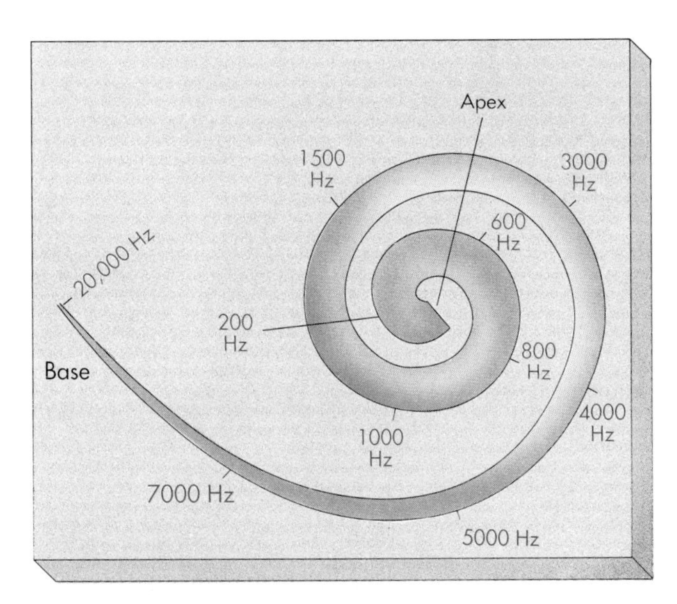

FIGURE 15-30 The basilar membrane and the points of maximum vibration resulting from stimulation by sounds of various frequencies.

vary nucleus back to the spiral organ to all regions where the maximum vibration did not occur. These impulses inhibit the hair cells from initiating additional action potentials in the afferent neurons. Thus only impulses from regions of maximum vibration are received by the cortex where they become consciously perceived.

By this process tones are localized along the cochlea; higher-pitched tones originate near the base, and lower-pitched tones originate near the apex. As a result of this localization, neurons along a given portion of the cochlea send action potentials only to the cerebral cortex in response to specific pitches; likewise, certain neurons in the auditory cortex, wired to those specific neurons in the cochlea, respond only to specific pitches.

Deafness and Functional Replacement of the Ear

Deafness can have many causes. In general, there are two categories of deafness: conduction and sensorineural (or nerve). Conduction deafness involves a mechanical deficiency in transmission of sound waves from the outer ear to the organ of Corti and can often be corrected surgically. Hearing aids help people with such hearing deficiencies by boosting the sound volume reaching the ear. Sensorineural deafness involves the organ of Corti or nerve pathways and is more difficult to correct.

Research is currently being conducted to develop ways of replacing the hearing pathways with electrical circuits. One approach involves the direct stimulation of nerves by electrical impulses, with considerable success achieved in the area of cochlear nerve stimulation. Certain types of sensorineural deafness in which the hair cells of the organ of Corti are impaired can now be partially corrected. In addition, prostheses are available that consist of a microphone for picking up the initial sound waves, a microelectronic processor for converting the sound into electrical signals, a transmission system for relaying the signals to the inner ear, and a long, slender electrode that is threaded into the cochlea. This electrode delivers electrical signals directly to the endings of the cochlear nerve (Figure 15-E). High-frequency sounds are picked up by the microphone and are transmitted through specific circuits to terminate near the oval window, whereas low-frequency sounds are transmitted farther up the cochlea to cochlear nerve endings near the helicotrema.

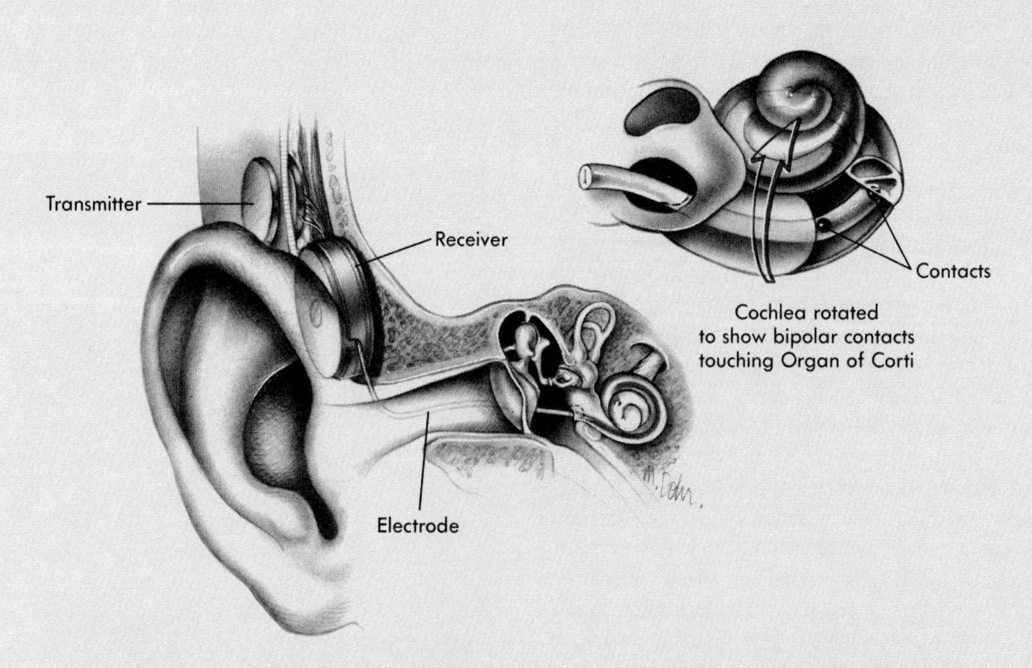

FIGURE 15-E Cochlear implant. A receiver and antenna are implanted under the skin near the auricle, and a small lead is fed through the external auditory meatus, eardrum, and middle ear into the cochlea where the cochlear nerve can be directly stimulated by electrical impulses from the receiver.

12

Suggest some possible sites and mechanisms to explain why certain people have "perfect pitch" and other people cannot seem to "carry a tune in a bucket."

? ? ? ? ? ? ? ? ? ? ?

Sound volume, or loudness, is a function of sound wave amplitude. As high-amplitude sound waves reach the ear, the perilymph, endolymph, and basilar membrane vibrate more intensely, and the hair cells are stimulated more strongly. As a result of the increased stimulation, more hair cells send action potentials at a higher frequency to the cerebral cortex where this information is perceived as a greater sound volume.

13

Explain why it is much easier to perceive subtle musical tones when music is played somewhat softly as opposed to very loudly.

? ? ? ? ? ? ? ? ? ? ?

Neuronal Pathways for Hearing

The special senses of hearing and balance are both transmitted by the vestibulocochlear (cranial nerve VIII) nerve. The term vestibular refers to the vestibule of the inner ear, which is involved in balance. The term cochlear refers to the cochlea and is that portion of the inner ear involved in hearing. The vestibulocochlear nerve functions as two separate nerves carrying information from two separate but closely related structures.

The auditory pathways within the CNS are very complex, with both crossed and uncrossed tracts (Figure 15-31). Therefore unilateral CNS damage usually has little impact on hearing. The neurons from the cochlear ganglion synapse with CNS neurons in the dorsal or ventral **cochlear nucleus** in the superior medulla near the inferior cerebellar peduncle. These neurons in turn either synapse in or pass through the superior olivary nucleus. Neurons terminating in this nucleus may synapse with efferent neurons returning to the cochlea (to modulate pitch perception). Nerve fibers from the superior olivary nucleus also project to the trigeminal (cranial nerve

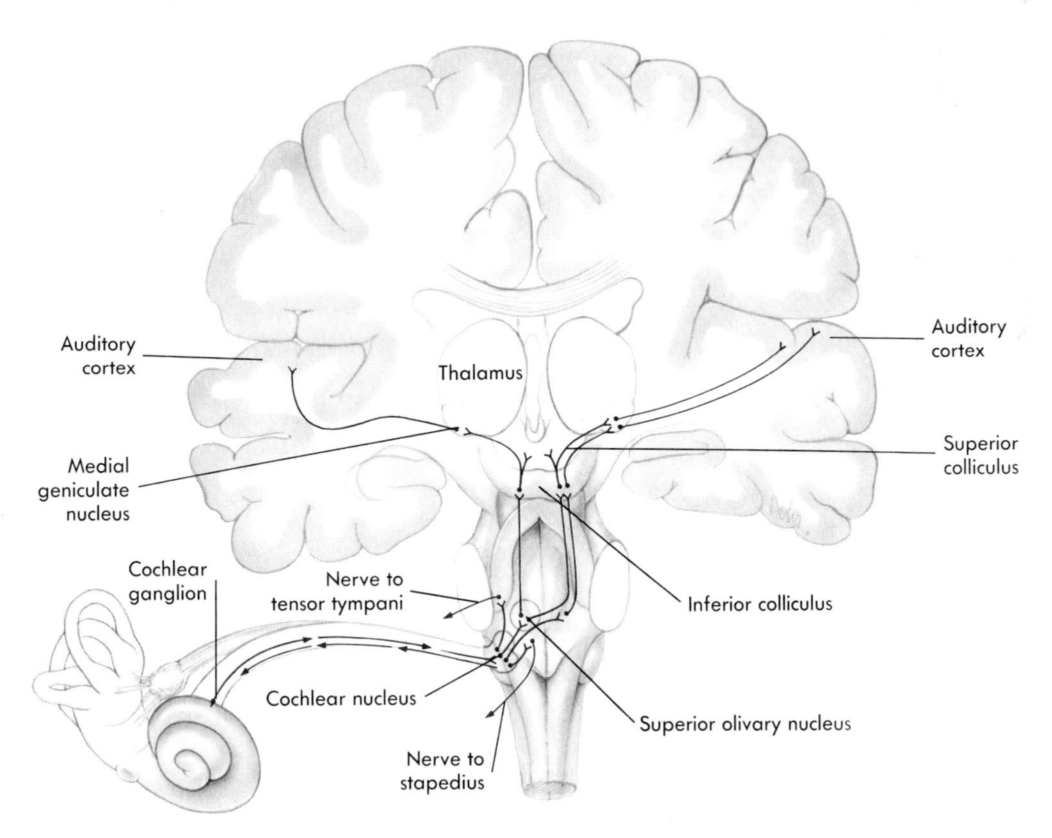

FIGURE 15-31 Central nervous system pathways for hearing. Afferent axons from the cochlear ganglion terminate in the cochlear nucleus in the brainstem. Axons from neurons in the cochlear nucleus project to the superior olivary nucleus or to the inferior colliculus. Axons from the inferior colliculus project to the medial geniculate nucleus of the thalamus, and thalamic neurons project to the auditory cortex. Neurons in the superior olivary nucleus send axons to the inferior colliculus, back to the inner ear, or to motor nuclei in the brainstem, which send efferent fibers to the middle ear muscles.

V) and facial (cranial nerve VII) nuclei, controlling the tensor tympani and stapedius muscles, respectively. This reflex pathway dampens loud sounds by initiating contractions of these muscles (the sound attenuation reflex described previously). Neurons synapsing in the superior olivary nucleus may also join other ascending neurons to the cerebral cortex.

Ascending neurons from the superior olivary nucleus travel in the **lateral lemniscus.** All ascending fibers synapse in the **inferior colliculi,** and neurons from there project to the **medial geniculate nucleus** of the **thalamus** where they synapse with neurons that project to the cortex. These neurons terminate in the **auditory cortex** in the dorsal portion of the temporal lobe within the lateral fissure and, to a lesser extent, on the superolateral surface of the temporal lobe (see Chapter 13). Neurons from the inferior colliculus also project to the **superior colliculus** and initiate reflexive turning of the head and eyes toward a loud sound.

Balance

The organs of balance can be divided structurally and functionally into two parts. The first, the **static labyrinth,** consists of the **utricle** and **saccule** of the vestibule and is involved in evaluating the position of the head relative to gravity or linear acceleration or deceleration (e.g., in a car that is increasing or decreasing speed). The second, the **kinetic labyrinth,** is associated with the semicircular canals and is involved in evaluating movements of the head.

Most of the utricular and saccular walls consist of simple cuboidal epithelium. However, the utricle and saccule each contain a specialized patch of epithelium approximately 2 to 3 mm in diameter called the **macula** (mak′u-lah; Figure 15-32, *A* and *B*). The macula of the utricle is oriented parallel to the base of the skull, and the macula of the saccule is perpendicular to the base of the skull.

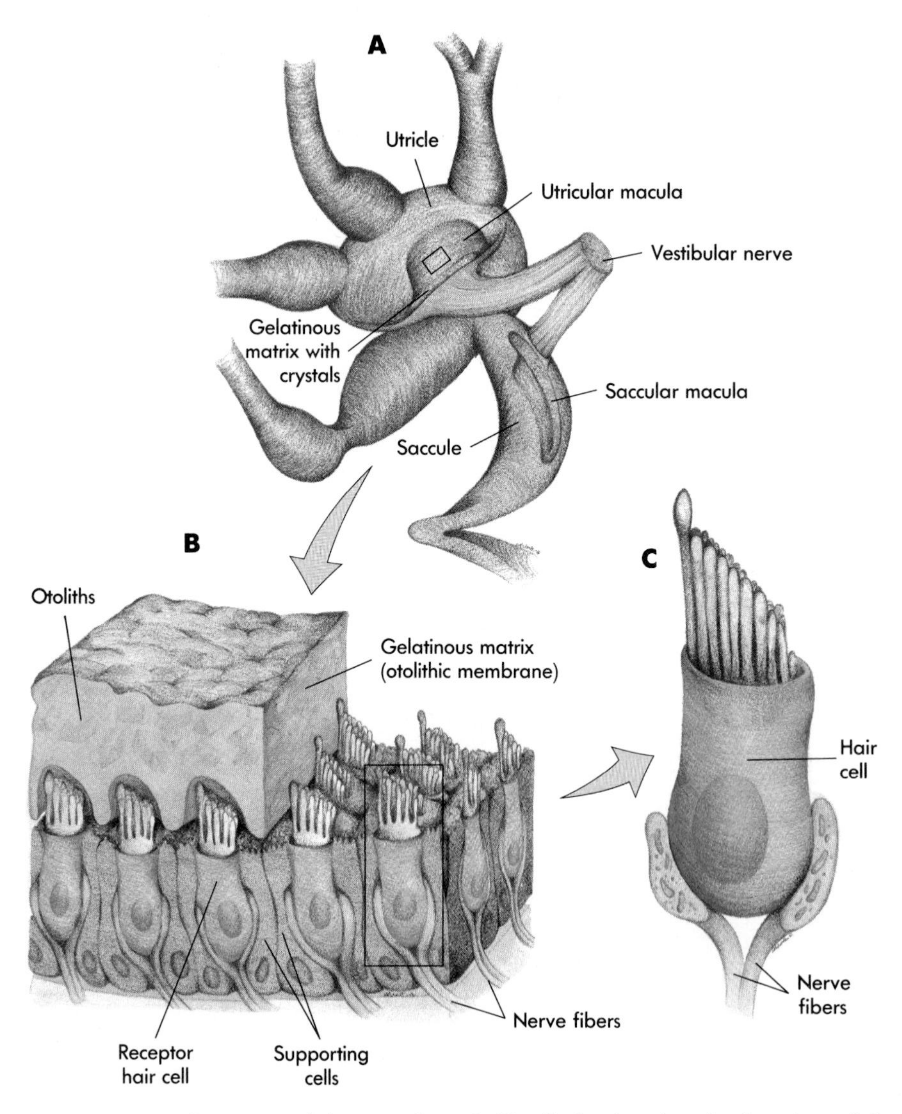

FIGURE 15-32 Structure of the macula. **A** Vestibule showing the location of the utricular and saccular maculae. **B** Enlargement of a section of the utricular macula showing otoliths in the macula. **C** Detail of a hair cell.

The maculae resemble the organ of Corti and consist of columnar supporting cells and hair cells. The "hairs" of these cells (microvilli as in the organ of Corti) are embedded into a gelatinous mass weighted by the presence of **otoliths** composed of protein and calcium carbonate (Figure 15-32, *C* and *D*). The gelatinous mass moves in response to gravity, acceleration, or deceleration, bending the hair cells and initiating action potentials in the associated neurons. If the head is tipped, otoliths move in response to gravity and stimulate certain hair cells (Figure 15-33). If the head accelerates, the weight of the otoliths causes them to stimulate hair cells. The hair cells are constantly being stimulated at a low level by the presence of the otolith-weighted covering of the macula; but as this covering moves in response to gravity or acceleration, the pattern of intensity of hair cell stimulation changes. This pattern of stimulation and the subsequent pattern of action potentials from the numerous hair cells of the maculae can be translated by the brain into specific information about head position or acceleration. Much of this information is not perceived consciously but is dealt with subconsciously. The body responds by making subtle tone adjustments in muscles of the back and neck, which are intended to restore the head to its proper neutral, balanced position.

The kinetic labyrinth (Figure 15-34) consists of three **semicircular canals** placed at nearly right angles to each other, one lying nearly in the transverse plane, one in the coronal plane, and one in the sagittal plane (see Chapter 1). The arrangement of the semicircular canals enables a person to detect movement in all directions. The base of each semicircular canal is expanded into an **ampulla** (Figure 15-34, *A* and *B*). Within each ampulla the epithelium is specialized to form a **crista ampullaris.** This specialized sensory epithelium is structurally and functionally very similar to that of the maculae. Each crista consists of a ridge or crest of epithelium with a curved gelatinous mass, the **cupula** (ku′pu-lah), suspended over the crest. The hairlike processes of the crista hair cells are embedded in the cupula (Figure 15-34, *C*). The cupula contains no otoliths and therefore does not respond to gravitational pull. Instead, the cupula is a float that is displaced by fluid movements within the semicircular canals. Endolymph movement within each semicircular canal moves the cupula, bends the hairs, and initiates action potentials (Figure 15-35).

As the head begins to move in a given direction (acceleration), the endolymph does not move at the same rate as the semicircular canals (Figure 15-35, *B*). This difference causes displacement of the cupula in a direction opposite to that of the movement of the head, resulting in relative movement between the cupula and the endolymph. As movement continues, the fluid of the semicircular canals begins to move

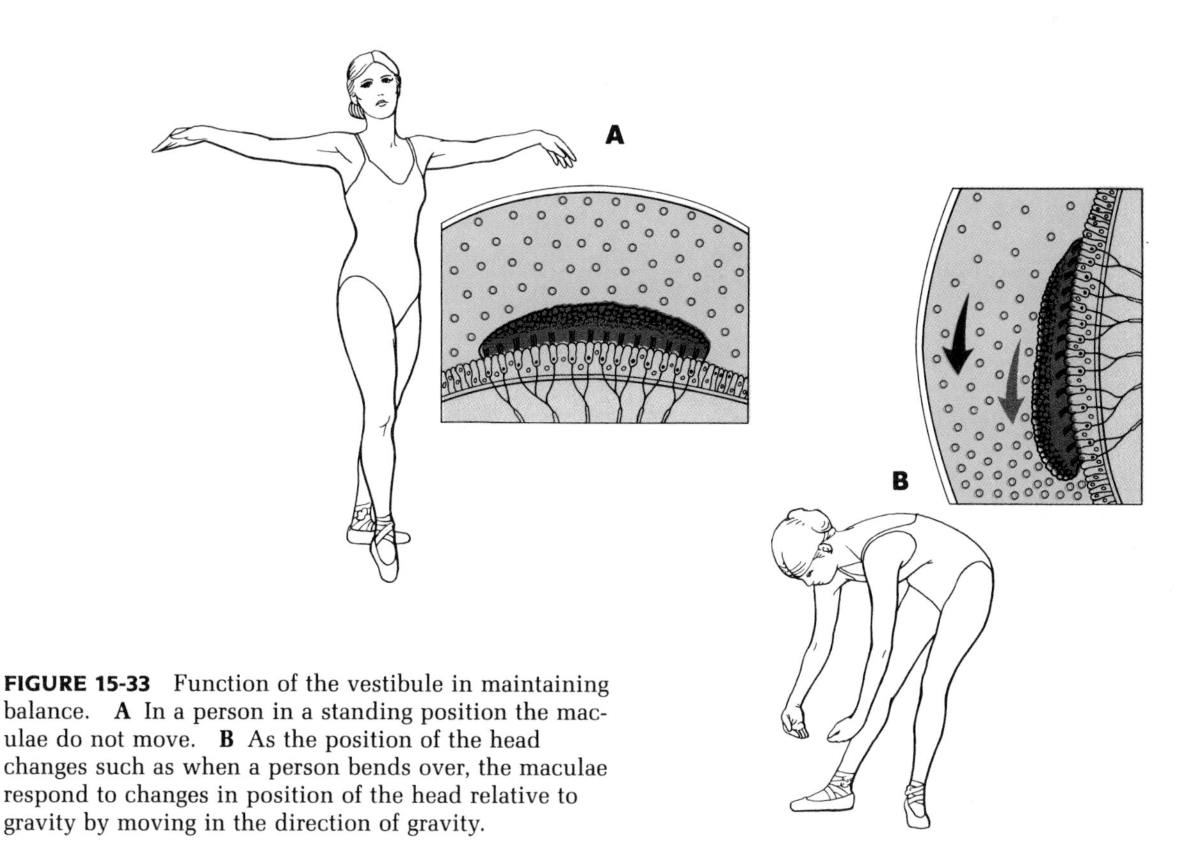

FIGURE 15-33 Function of the vestibule in maintaining balance. **A** In a person in a standing position the maculae do not move. **B** As the position of the head changes such as when a person bends over, the maculae respond to changes in position of the head relative to gravity by moving in the direction of gravity.

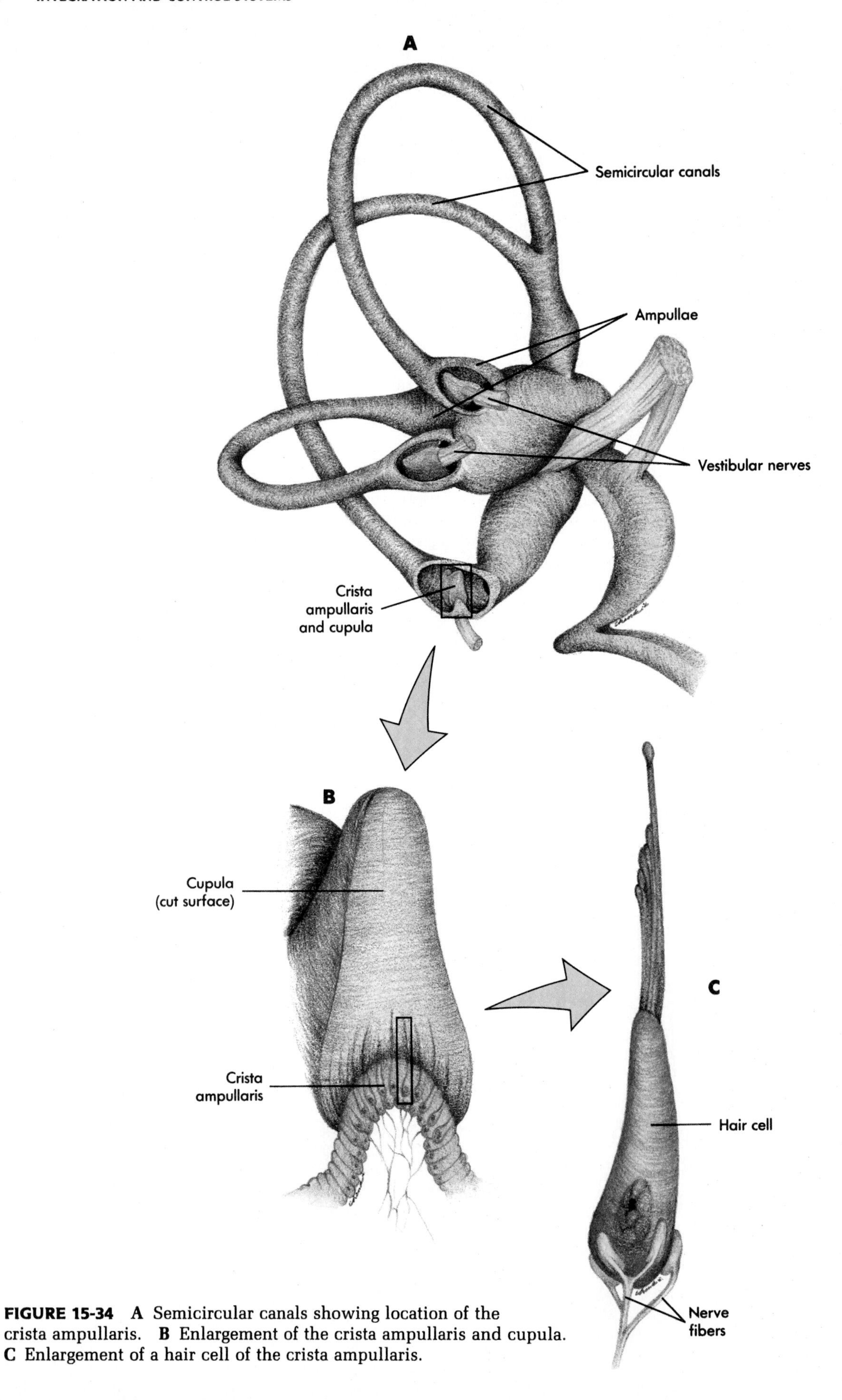

FIGURE 15-34 **A** Semicircular canals showing location of the crista ampullaris. **B** Enlargement of the crista ampullaris and cupula. **C** Enlargement of a hair cell of the crista ampullaris.

and "catches up" with the cupula, and stimulation is stopped. As movement of the head ceases, the endolymph continues to move because of its momentum, causing displacement of the cupula in the same direction as the head had been moving (Figure 15-35, C). Therefore this system detects changes in the rate of movement rather than movement alone. As with the static labyrinth, the information obtained by the brain from the kinetic labyrinth is largely subconscious.

Space sickness is a balance disorder occurring in zero gravity and resulting from unfamiliar sensory input to the brain. The brain must adjust to these unusual signals, or severe symptoms may result such as headaches and dizziness. Space sickness is unlike motion sickness in that motion sickness results from an excessive stimulation of the brain, whereas space sickness results from too little stimulation as a result of weightlessness.

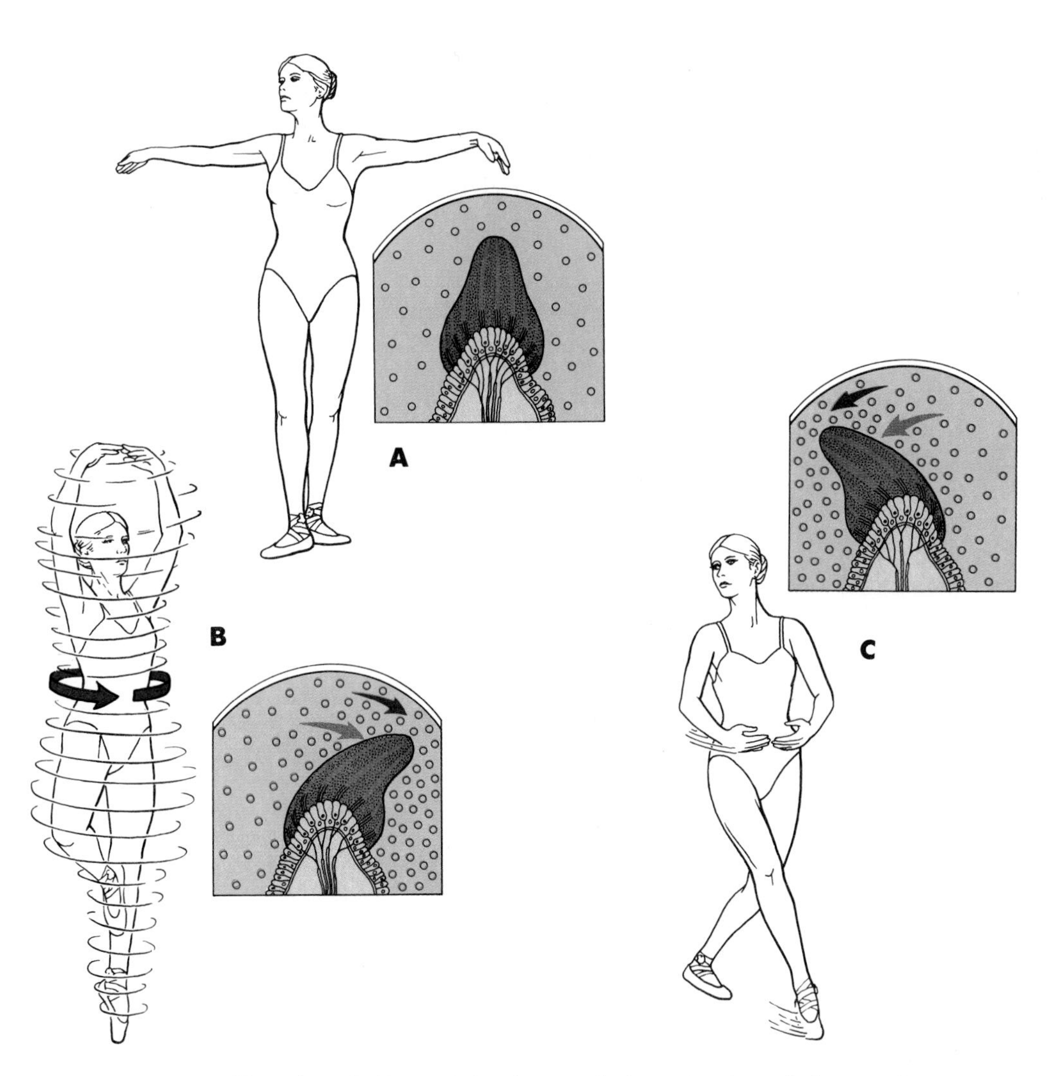

FIGURE 15-35 Function of the semicircular canals in maintaining balance. The crista ampullaris responds to changes in momentum. **A** When a person is at rest, the crista ampullaris does not move. **B** As a person begins to spin, the semicircular canals begin to move with the body (see inset), but the endolymph tends to remain stationary relative to the movement (*arrow* inside the semicircular canal pointing in the opposite direction of semicircular canal movement), and the crista ampullaris is displaced by the endolymph in a direction opposite to the direction of spin. **C** When the person stops spinning, the endolymph tends to continue moving, and the crista ampullaris is displaced by the endolymph in the same direction as the spin.

Neuronal Pathways for Balance

Neurons synapsing on the hair cells of the maculae and cristae ampullares converge into the **vestibular ganglion** where their cell bodies are located (Figure 15-36). Afferent fibers from these neurons join afferent fibers from the cochlear ganglion to form the vestibulocochlear nerve and terminate in the **vestibular nucleus** within the medulla oblongata. Fibers run from this nucleus to numerous areas of the CNS such as the spinal cord, cerebellum, cerebral cortex, and the nuclei controlling extrinsic eye muscles.

Balance is a complex process not simply confined to one type of input. In addition to vestibular sensory input, the vestibular nucleus receives input from proprioceptive nerves throughout the body and from the visual system. People are asked to close their eyes while balance is evaluated in a sobriety test because alcohol affects the proprioceptive and vestibular components of balance (cerebellar function) to a greater extent than it does the visual portion.

Reflex pathways exist between the kinetic portion of the vestibular system and the nuclei controlling the extrinsic eye muscles (oculomotor, trochlear, and abducens). A reflex pathway allows maintenance of visual fixation on an object while the head is in motion. This function can be demonstrated by spinning a person around several times (approximately 10 times in 20 seconds), stopping him, and observing his eye movements. The reaction is most pronounced if the individual's head is tilted forward approximately 30 degrees while he is spinning, thus bringing the lateral semicircular canals into the horizontal plane. There is slight oscillatory movement of the eyes. The eyes track in the direction of motion (slow component) and return with a rapid recovery movement before repeating the tracking motion. This oscillation of the eyes is called **nystagmus** (nis-tag'mus). If asked to walk in a straight line, the individual will deviate in the direction of rotation, and if he is asked to point to an object, his finger will deviate in the direction of rotation.

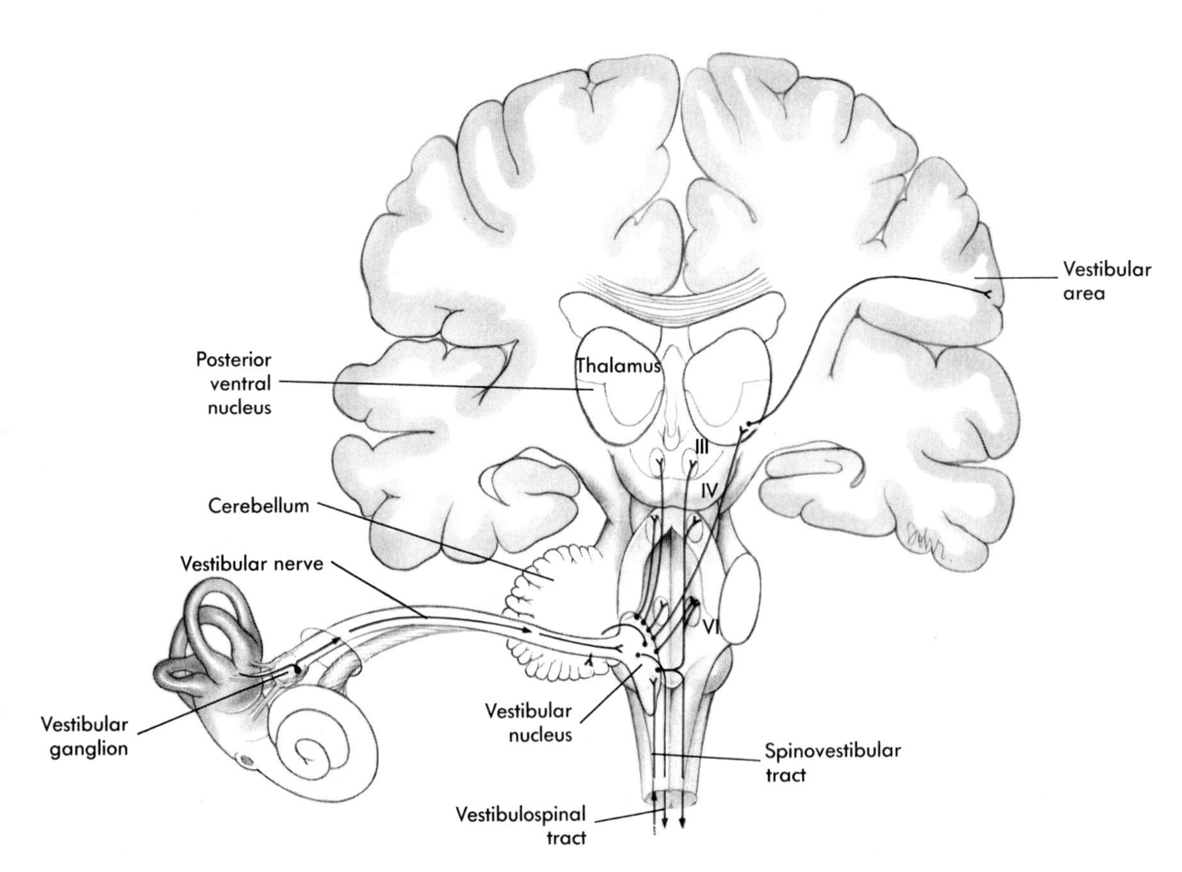

FIGURE 15-36 Central nervous system pathways for balance. Afferent axons from the vestibular ganglion pass through the vestibular nerve to the vestibular nucleus, which also receives input from several other sources such as proprioception from the legs. Vestibular neurons send projections to the cerebellum, controlling postural muscles, and to motor nuclei (oculomotor, trochlear, and abducens), controlling extrinsic eye muscles. Vestibular neurons also project to the posterior ventral nucleus of the thalamus and from there to the vestibular area of the cortex.

Ear Disorders

Otosclerosis

Otosclerosis is an ear disorder in which spongy bone grows over the oval window and immobilizes the stapes, leading to progressive loss of hearing. This disorder can be corrected surgically by breaking away the bony growth and the immobilized stapes. During surgery the oval window is covered by a fat pad or a synthetic membrane, and the stapes is replaced by a small rod connected to the fat or membrane over the oval window at one end and to the incus at the other.

Tinnitus

Tinnitus (tĭ-ni′tus) consists of noises (e.g., ringing, clicking, whistling, or booming) in the ears. These noises may occur as a result of disorders in the middle or inner ear or along the central neuronal pathways.

Motion Sickness

Motion sickness consists of nausea, weakness, and other dysfunctions caused by stimulation of the semicircular canals during motion (e.g., in a boat, automobile, airplane, swing, or amusement park ride). It may progress to vomiting and incapacitation.

Dimenhydrinate (Dramamine) or other motion sickness medications are taken to counter the labyrinthine stimulation of motion. Dimenhydrinate is primarily an antihistamine for which the mode of action is not exactly known.

Otitis Media

Infections of the middle ear **(otitis media)** are quite common in young children. These infections usually result from the spread of infection from the mucous membrane of the pharynx through the auditory tube to the mucous lining of the middle ear. The symptoms of otitis media, consisting of low-grade fever, lethargy, and irritability, are often not easily recognized by the parent as signs of middle ear infection. The infection can also cause a temporary decrease or loss of hearing because fluid buildup has dampened the tympanic membrane or ossicles.

SUMMARY

THE SENSES INCLUDE GENERAL SENSES AND SPECIAL SENSES.

CLASSIFICATION OF THE SENSES *p. 458*

1. The senses can be classified as somatic, visceral, and special.
2. Somatic modalities include touch, pressure, temperature, proprioception, and pain. Visceral modalities are primarily pain and pressure. Special modalities are smell, taste, sight, sound, and balance.
3. Receptors include mechanoreceptors, chemoreceptors, photoreceptors, thermoreceptors, and nociceptors.

SENSATION *p. 458*

1. Sensation is the conscious awareness of stimuli received by sensory receptors.
2. Sensation requires a stimulus, a receptor, conduction of an action potential to the central nervous system (CNS), translation of the action potential, and the person's awareness of the information obtained by the CNS.

TYPES OF AFFERENT NERVE ENDINGS *p. 459*

1. Free nerve endings detect light touch, pain, itch, tickle, and temperature.
2. Merkel's disks respond to light touch and superficial pressure.
3. Hair follicle receptors wrap around the hair follicle and are involved in the sensation of light touch when the hair is bent.
4. Pacinian corpuscles, located in the dermis and hypodermis, detect pressure. In joints they serve a proprioceptive function.
5. Meissner's corpuscles, located in the dermis, are responsible for two-point discriminative touch.
6. Ruffini's end organs are involved in continuous touch or pressure.
7. Golgi tendon apparatuses, embedded in tendons, respond to changes in tension.
8. Muscle spindles, located in skeletal muscle, are proprioceptors.

OLFACTION *p. 462*

Olfaction is the sense of smell.

Olfactory Epithelium and Bulb

1. Olfactory neurons in the olfactory epithelium are bipolar neurons. Their distal ends are enlarged as olfactory vesicles, which have long cilia. The cilia have receptors that respond to dissolved substances.
2. There are at least seven (perhaps 50) primary odors. The olfactory neurons have a very low threshold and accommodate rapidly.

Neuronal Pathways of Olfaction

1. Axons from the olfactory neurons extend as olfactory nerves to the olfactory bulb where they synapse with mitral and tufted cells. Axons from these cells form the olfactory tracts. Association neurons in the olfactory bulbs can modulate output to the olfactory tracts.
2. The olfactory tracts terminate in the olfactory cortex. The lateral olfactory area is involved in the conscious perception of smell, the intermediate area with modulating smell, and the medial area with visceral and emotional responses to smell.

TASTE p. 464

Taste buds usually are associated with papillae.

Histology of Taste Buds

1. The papillae are the circumvallate, fungiform, foliate, and filiform.
2. Taste buds consist of support and gustatory cells.
3. The gustatory cells have gustatory hairs that extend into taste pores.

Function of Taste

1. Receptors on the hairs detect dissolved substances.
2. There are four basic types of taste: sour, salty, bitter, and sweet.

Neuronal Pathways for Taste

1. The facial nerve carries taste sensations from the anterior two thirds of the tongue, the glossopharyngeal nerve from the posterior one third of the tongue, and the vagus nerve from the epiglottis.
2. The neural pathways for taste extend from the medulla oblongata to the thalamus, and to the cerebral cortex.

VISUAL SYSTEM p. 467

Accessory Structures

1. The eyebrows prevent perspiration from entering the eyes and help shade the eyes.
2. The eyelids consist of five tissue layers. They protect the eyes from foreign objects and help lubricate the eyes by spreading tears over their surface.
3. The conjunctiva covers the inner eyelid and the anterior part of the eye.
4. Lacrimal glands produce tears that flow across the surface of the eye. Excess tears enter the lacrimal canaliculi and reach the nasal cavity through the nasolacrimal canal. Tears lubricate and protect the eye.
5. The extrinsic eye muscles move the eyeball.

Anatomy of the Eye

1. The fibrous tunic is the outer layer of the eye. It consists of the sclera and cornea.
   - The sclera is the posterior four fifths of the eye. It is white connective tissue that maintains the shape of the eye and provides a site for muscle attachment.
   - The cornea is the anterior one fifth of the eye. It is transparent and refracts light that enters the eye.
2. The vascular tunic is the middle layer of the eye.
   - The iris is smooth muscle regulated by the autonomic nervous system. It controls the amount of light entering the pupil.
   - The ciliary muscles control the shape of the lens. They are smooth muscles regulated by the autonomic nervous system.
   - The ciliary process produces aqueous humor.
3. The retina is the inner layer of the eye and contains neurons sensitive to light.
4. The eye has two compartments.
   - The anterior compartment is filled with aqueous humor, which circulates and leaves by way of the canal of Schlemm.
   - The posterior compartment is filled with vitreous humor.
5. The lens is held in place by the suspensory ligaments, which are attached to the ciliary muscles.
6. The macula lutea (fovea centralis) is the area of greatest visual acuity.
7. The optic disc is the location through which nerves exit and blood vessels enter the eye. It has no photosensory cells and is therefore a blind spot in the eye.

Functions of the Complete Eye

1. Light is that portion of the electromagnetic spectrum that humans can see.
2. When light travels from one medium to another, it can bend or refract. Light striking a concave surface refracts inward (convergence). Light striking a convex surface refracts outward (divergence).
3. Converging light rays meet at the focal point and are said to be focused.
4. The cornea, aqueous humor, lens, and vitreous humor all refract light. The cornea is responsible for most of the convergence, whereas the lens can adjust the focal point by changing shape.
   - Relaxation of the ciliary muscles causes the lens to flatten, producing the emmetropic eye.
   - Contraction of the ciliary muscles causes the lens to become more spherical. This change in lens shape enables the eye to focus on objects that are less than 20 feet away, a process called accommodation.
5. The far point of vision is the distance at which the eye no longer has to change shape (20 feet or greater). The near point of vision is the closest an object can come to the eye and still be focused (2 to 4 inches).
6. The pupil becomes smaller during accommodation, increasing the depth of focus.

Structure and Function of the Retina

1. The pigmented retina provides a black backdrop for increasing visual acuity.
2. Rods are responsible for vision in low illumination (night vision).
   - A pigment, rhodopsin, is split by light into retinal and opsin, producing an action potential in the rod.
   - Light adaptation is caused by a reduction of rhodopsin; dark adaptation is caused by rhodopsin production.
3. Cones are responsible for color vision and visual acuity.
   - There are three types of cones, each with a different photopigment. The pigments are most sensitive to blue, red, and green lights.
   - Perception of many colors results from mixing the ratio of the different types of cones that are active at a given moment.
4. Most visual images are formed on the fovea centralis, which has a very high concentration of cones. Moving away from the fovea, there are fewer cones (the macula lutea); mostly rods are in the periphery of the retina.
5. The rods and the cones synapse with bipolar cells that in turn synapse with ganglion cells, which form the optic nerves.
6. Association neurons in the retina can modify information sent to the brain.

Neuronal Pathways for Vision

1. Ganglia cells extend to the lateral geniculate ganglion of the thalamus where they synapse. From there neurons form the visual radiations that project to the visual cortex.
2. Neurons from the nasal visual field (temporal retina) of one eye and the temporal visual field (nasal retina) of the opposite eye project to the same cerebral hemisphere. Some nerve fibers from the nasal portion of the retina cross in the optic chiasma, and fibers from the temporal portion of the retina remain uncrossed.
3. Depth perception is the ability to judge relative distances of an object from the eye and is a property of binocular vision. Binocular vision results because a slightly different image is seen by each eye.

HEARING AND BALANCE p. 484

The osseous labyrinth is a canal system within the temporal bone that contains perilymph and the membranous labyrinth. Endolymph is inside the membranous labyrinth.

Auditory Structures and Their Functions

1. The external ear consists of the auricle and external auditory meatus.
2. The middle ear connects the external and inner ears.
   - The tympanic membrane is stretched across the external auditory meatus.
   - The malleus, incus, and stapes connect the tympanic membrane to the oval window of the inner ear.
   - The auditory tube connects the middle ear to the pharynx and functions to equalize pressure.
   - The middle ear is connected to the mastoid air cells.
3. The inner ear has three parts: the semicircular canals; the vestibule, which contains the utricle and the saccule; and the cochlea.
4. The cochlea is a spiral-shaped canal within the temporal bone.
   - The canal is divided into three compartments by the vestibular and basilar membranes. The scala vestibuli and scala tympani contain perilymph. The cochlear duct contains endolymph and the spiral organ of Corti.
   - The spiral organ of Corti consists of hair cells that attach to the tectorial membrane.

Auditory Function

1. Sound waves are funneled by the auricle down the external auditory meatus, causing the tympanic membrane to vibrate.
2. The tympanic membrane vibrations are passed along the ossicles to the oval window of the inner ear.
3. Movement of stapes in the oval window causes the perilymph, vestibular membrane, and endolymph to vibrate, producing movement of the basilar membrane. Movement of the basilar membrane causes displacement of the hair cells in the spiral organ of Corti and the generation of action potentials, which travel along the vestibulocochlear nerve.
4. Some vestibulocochlear nerve axons synapse in the superior olivary nucleus. Efferent neurons from this nucleus project back to the cochlea where they regulate the perception of pitch.
5. The round window protects the inner ear from pressure buildup and dissipates sound waves.

Neuronal Pathways for Hearing

1. Axons from the vestibulocochlear nerve synapse in the medulla. Neurons from the medulla pass to the inferior colliculi where they synapse. Neurons from this point project to the thalamus and synapse. Thalamic neurons extend to the auditory cortex.
2. Efferent neurons project to cranial nerve nuclei responsible for controlling muscles that dampen sound in the middle ear.

Balance

1. Static balance evaluates the position of the head relative to gravity and detects linear acceleration and deceleration.
   - The utricle and saccule in the inner ear contain maculae. The maculae consist of hair cells with the hairs embedded in a gelatinous mass that contains otoliths.
   - The gelatinous mass moves in response to gravity.
2. Kinetic balance evaluates movements of the head.
   - There are three semicircular canals at right angles to each other in the inner ear. The ampulla of each semicircular canal contains the crista ampullaris, which has hair cells with hairs embedded in a gelatinous mass, the cupula.
   - The cupula is moved by endolymph within the semicircular canal when the head moves.
3. Balance also depends on proprioception and visual input.

Neuronal Pathways for Balance

Axons from the maculae and the cristae ampullares extend to the vestibular nucleus of the medulla. Fibers from the medulla run to the spinal cord, cerebellum, cortex, and nuclei that control the extrinsic eye muscles.

CONTENT REVIEW

1. Define somatic, visceral, and special sense.
2. List the types of receptors.
3. How is a stimulus perceived as a sensation?
4. Define adaptation.
5. List the eight major types of afferent nerve endings, where they are located, and the functions they perform.
6. Describe the initiation of an action potential in an olfactory neuron. Name all the structures and cells that the action potential would encounter on the way to the olfactory cortex.
7. Name the three areas of the olfactory cortex and give their functions.
8. How is the sense of smell modified in the olfactory bulb?
9. How is the sense of taste related to the sense of smell?
10. What is a primary odor? Name seven possible examples. How do the primary odors relate to our ability to smell many different odors?
11. Name and describe the four kinds of papillae found on the tongue. Which ones have taste buds associated with them?
12. Starting with a gustatory hair, name the structures and cells that an action potential would encounter on the way to the gustatory cortex.
13. What are the four primary tastes? Where are they concentrated on the tongue? How do they produce many different kinds of taste sensations?
14. Describe the following structures and state their functions: eyebrows, eyelids, conjunctiva, lacrimal apparatus, and extrinsic eye muscles.
15. Name the three layers (tunics) of the eye. For each layer describe the parts or structures it forms, and explain their functions.
16. What is the blind spot?

17. Name the two compartments of the eye and the substances that fill each compartment.
18. What is the function of the canal of Schlemm and the ciliary processes?
19. Describe the lens of the eye and how the lens is held in place.
20. How does the pupil constrict? How does it dilate?
21. What causes light to refract? What is a focal point?
22. Describe the changes that occur in the lens, pupil, and extrinsic eye muscles as an object moves from 25 feet away to 6 inches away. What is meant by the terms near and far points of vision?
23. Starting with a rod or a cone, name the cells or structures that an action potential would encounter while traveling to the visual cortex.
24. What is the function of the pigmented retina and of the choroid?
25. Describe the breakdown of rhodopsin by light. How does it reform?
26. Describe the arrangement of cones and rods in the fovea, the macula lutea, and the periphery of the eye.
27. What is a visual field? How do the visual fields project to the brain?
28. What is depth perception? How does it occur?
29. Name the three regions of the ear, and name each region's parts.
30. Describe the relationship between the tympanic membrane, the ear ossicles, and the oval window of the inner ear.
31. What is the function of the external auditory meatus and of the auditory tube?
32. Explain how the cochlear duct is divided into three compartments. What is found in each compartment?
33. Starting with the auricle, trace sound into the inner ear to the point where action potentials are generated in the vestibulocochlear nerve.
34. Describe the neural pathways for hearing from the vestibulocochlear nerve to the cerebral cortex.
35. What are the functions of the saccule and the utricle? Describe the macula and its function.
36. What is the function of the semicircular canals? Describe the crista ampullaris and its mode of operation.
37. Describe the neural pathways for balance.

CONCEPT QUESTIONS

1. Describe all the sensations involved in biting into an apple. Explain which of those sensations are special and which are general. What types of receptors are involved? Which parts of the taste of the apple are taste and which are olfaction?
2. An elderly man with normal vision developed cataracts. He was surgically treated by removing the lenses of his eyes. What kind of glasses would you recommend he wear to compensate for the removal of his lenses?
3. Some animals have a reflective area in the choroid called the tapetum lucidum. Light entering the eye is reflected back instead of being absorbed by the choroid. What would be the advantage of this arrangement? The disadvantage?
4. Perhaps you have heard someone say that eating carrots is good for the eyes. What is the basis for this claim?
5. On a camping trip Jean Tights ripped her pants. That evening she was going to repair the rip. As the sun went down, there was less and less light. When she tried to thread the needle, it was obvious that she was not looking directly at the needle but was looking a few inches to the side. Why did she do this?
6. A man stared at a black clock on a white wall for several minutes. Then he shifted his view and looked at only the blank white wall. Although he was no longer looking at the clock, he saw a light clock against a dark background. Explain what happened.
7. Persistent exposure to loud noise can cause loss of hearing, especially for high-frequency sounds. What part of the ear is probably damaged? Be as specific as possible.
8. A patient is suffering from paralysis of the facial nerve. One consequence of this condition is excessively acute hearing (hyperacusia). Explain how this occurs.
9. Professional divers are subject to increased pressure as they descend to the bottom of the ocean. Sometimes this pressure can lead to damage to the ear and loss of hearing. Describe the normal mechanisms that adjust for changes in pressure, suggest some conditions that might interfere with pressure adjustment, and explain how the increased pressure might cause loss of hearing.
10. If a vibrating tuning fork is placed against the mastoid process of the temporal bone, the vibrations will be perceived as sound, even if the external auditory meatus is plugged. Explain how this could happen.
11. Some student nurses are at a party. Since they love anatomy and physiology so much, they are discussing the phenomenon of accommodation of the special senses. They make the following observations:
   - **A** When entering a room, an odor such as brewing coffee is easily noticed. A few minutes later the odor might be barely, if at all, detectable, no matter how hard one tries to smell it.
   - **B** When entering a room, the sound of a ticking clock can be detected. Later the sound is not noticed until a conscious effort is made to hear it. Then it is easily heard.

   Explain the basis for each of the above observations.

ANSWERS TO PREDICT QUESTIONS

1 p. 459. Directly stimulating a neuron in the "finger region" of the somesthetic cortex would give the same result as stimulating the receptor or the nerve. The sensation would be projected to a given finger.

2 p. 461. Since hot and cold objects may not be perceived any differently for temperatures of 0° to 12° C or above 47° C (both temperature ranges stimulate pain fibers), the nervous system may not be able to discriminate between the two temperatures. At low temperatures both cold and pain receptors are stimulated; thus after the object has been in the hand for a very short time, discrimination usually can be made. However, if the central nervous system has been preprogrammed to think that the object that will be placed in the hand is hot, a cold object can elicit a rapid withdrawal reflex.

3 p. 462. Inhaling slowly and deeply allows a large amount of air to be drawn into the olfactory recess, whereas not as much air enters during normal breaths. Sniffing (rapid, repeated intake of air) is effective for the same reason.

4 p. 467. Adaptation can occur at several levels in the olfactory system. First, adaptation can occur at the receptor cell membrane where receptor sites become less sensitive to a specific odor. Second, association neurons within the olfactory bulb can modify sensitivity to an odor by inhibiting mitral cells or tufted cells. Third, neurons from the intermediate olfactory area of the cerebrum can send impulses to the association neurons in the olfactory bulb to inhibit further afferent impulses.

5 p. 469. Eyedrops placed into the eye tend to drain through the nasolacrimal duct into the nasal cavity. Recall that much of what is considered "taste" is actually smell. The medication is detected by the olfactory neurons and is interpreted by the brain as taste sensation. Crying produces extra tears, which are conducted to the nasal cavity, causing a "runny" nose.

6 p. 471. Inflammation of the cornea would involve edema, the accumulation of fluid. Fluid accumulation in the cornea would increase its water content. The water would cause the proteoglycans to expand, which would tend to decrease the transparency of the lens and interfere with normal vision.

7 p. 475. Eye strain or eye fatigue occurs primarily in the ciliary muscles. It occurs because for close vision the eyes must accommodate. Accommodation occurs as the ciliary muscles pull on the suspensory ligaments, releasing their tension and allowing the lens to become more rounded. Continued close vision requires maintenance of accommodation, which requires that the ciliary muscles remain contracted for a long period of time, resulting in fatigue of those muscles.

8 p. 478. Rhodopsin breakdown is associated with adaptation to bright light and occurs rapidly, whereas rhodopsin production occurs slowly and is associated with adaptation to conditions of little light. Therefore the eyes adapt rather quickly to bright light but quite slowly to very dim light.

9 p. 479. Rod cells distributed over most of the retina are involved in both peripheral vision (out of the corner of the eye) and vision under conditions of very dim light. However, when attempting to focus directly on an object, a person relies on the cones within the macula; although the cones are involved in visual acuity, they do not function well in dim light, so the object may not be seen at all.

10 p. 480. A lesion in the right optic nerve would result in loss of vision in the right visual field (see illustration below).

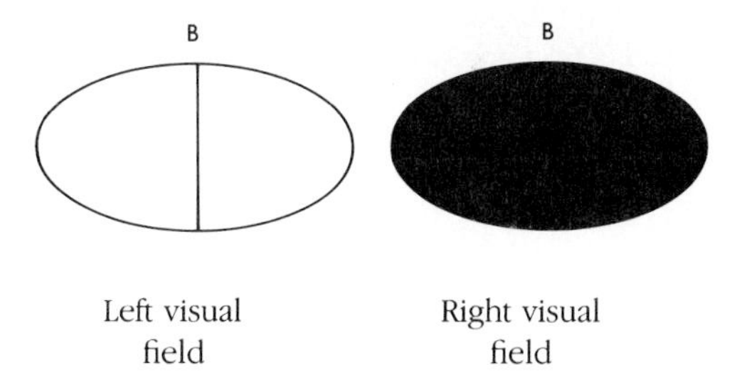

11 p. 491. Most sounds reaching the ear are not loud enough to directly stimulate the auditory neuron system, and amplification is necessary. However, some noises are very loud, and the sound waves may be very powerful and can damage the relatively delicate structures of the inner ear; thus it is helpful to dampen the sound to prevent such damage. However, the reflex, although rapid, does require a very brief passage of time so that the reflex is most effective against constant loud sounds, and a sudden loud sound is more damaging to the inner ear structures.

12 p. 493. Certain people seem to have "perfect pitch," and other people cannot seem to "carry a tune in a bucket" or are tone deaf. A condition such as tone deafness or at least a decreased ability to perceive tone differences could occur at a number of locations. The basilar membrane's structure may be such that tones are not adequately spaced along the cochlear duct to facilitate clear separation of tones. The reflex from the superior olive to the organ of Corti may not be functioning in some people as well as it does in others. The auditory cortex may not have the translation ability in some people for distinguishing differences in tones.

13 p. 493. It is much easier to perceive subtle musical tones when music is played somewhat softly as opposed to very loudly because loud sounds have sound waves with a greater amplitude, which causes the basilar membrane to vibrate more violently over a wider range. The spreading of the wave in the basilar membrane to some extent counteracts the reflex from the superior olive that is responsible for enabling a person to hear subtle tone differences.

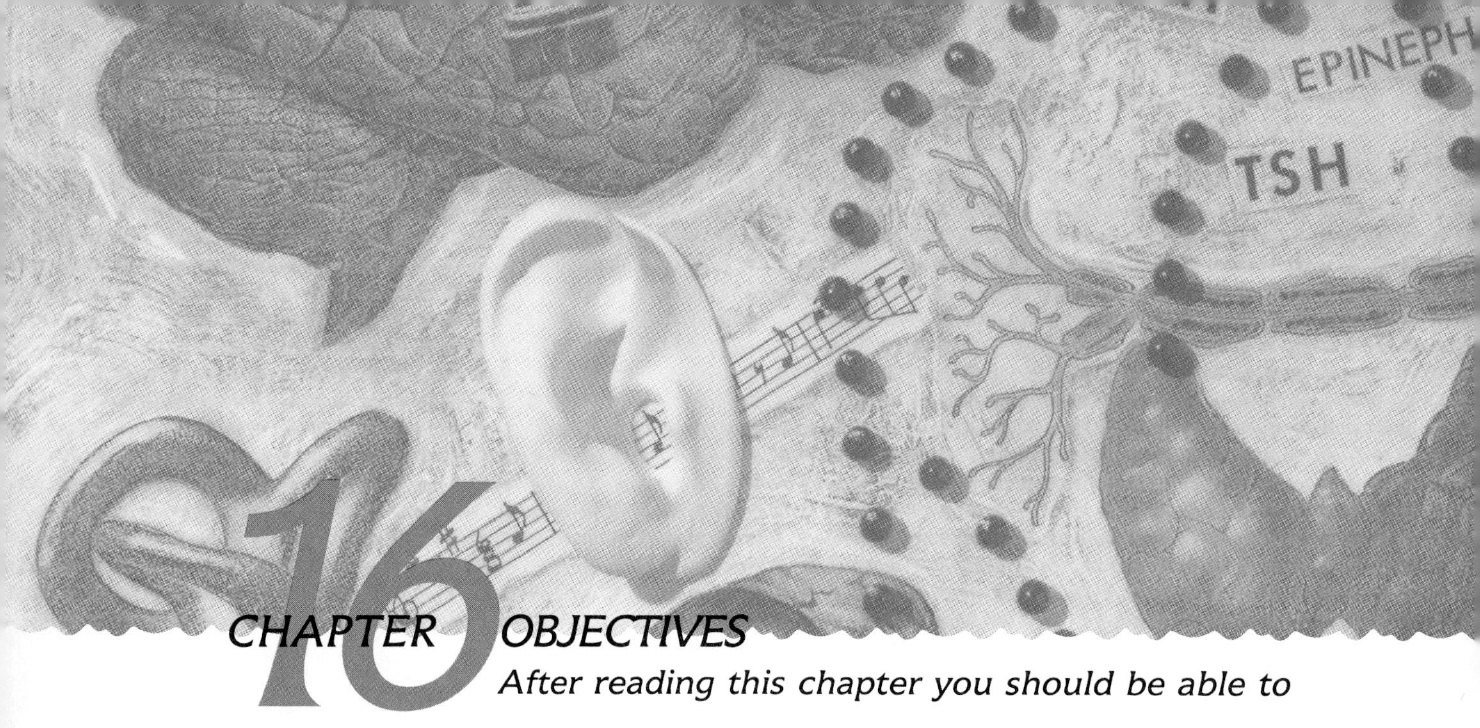

CHAPTER 16 OBJECTIVES

After reading this chapter you should be able to

1. Compare the structural differences of the autonomic nervous system and the somatomotor nervous system.
2. Define preganglionic neurons, postganglionic neurons, afferent neurons, somatomotor neurons, autonomic ganglia, and effector.
3. For both divisions of the autonomic nervous system, describe the location of the preganglionic and postganglionic neurons, the location of ganglia, the relative length of preganglionic and postganglionic axons, and the ratio of preganglionic to postganglionic neurons.
4. Describe the three pathways by which the sympathetic neurons extend from the sympathetic chain ganglia to target organs.
5. List the neurotransmitter substances for the preganglionic and postganglionic neurons for both the parasympathetic and sympathetic divisions.
6. Describe receptor types within autonomic synapses that respond to acetylcholine and describe their location.
7. Compare the autonomic nervous system's response to nicotine and muscarine.
8. Using examples, describe how autonomic reflexes help maintain homeostasis.
9. List the appropriate generalizations that can be made about the autonomic nervous system and describe the limitations of each generalization.
10. Give an example for each category of drugs that affect the autonomic nervous system and explain the general influence of the drug on the autonomic nervous system.

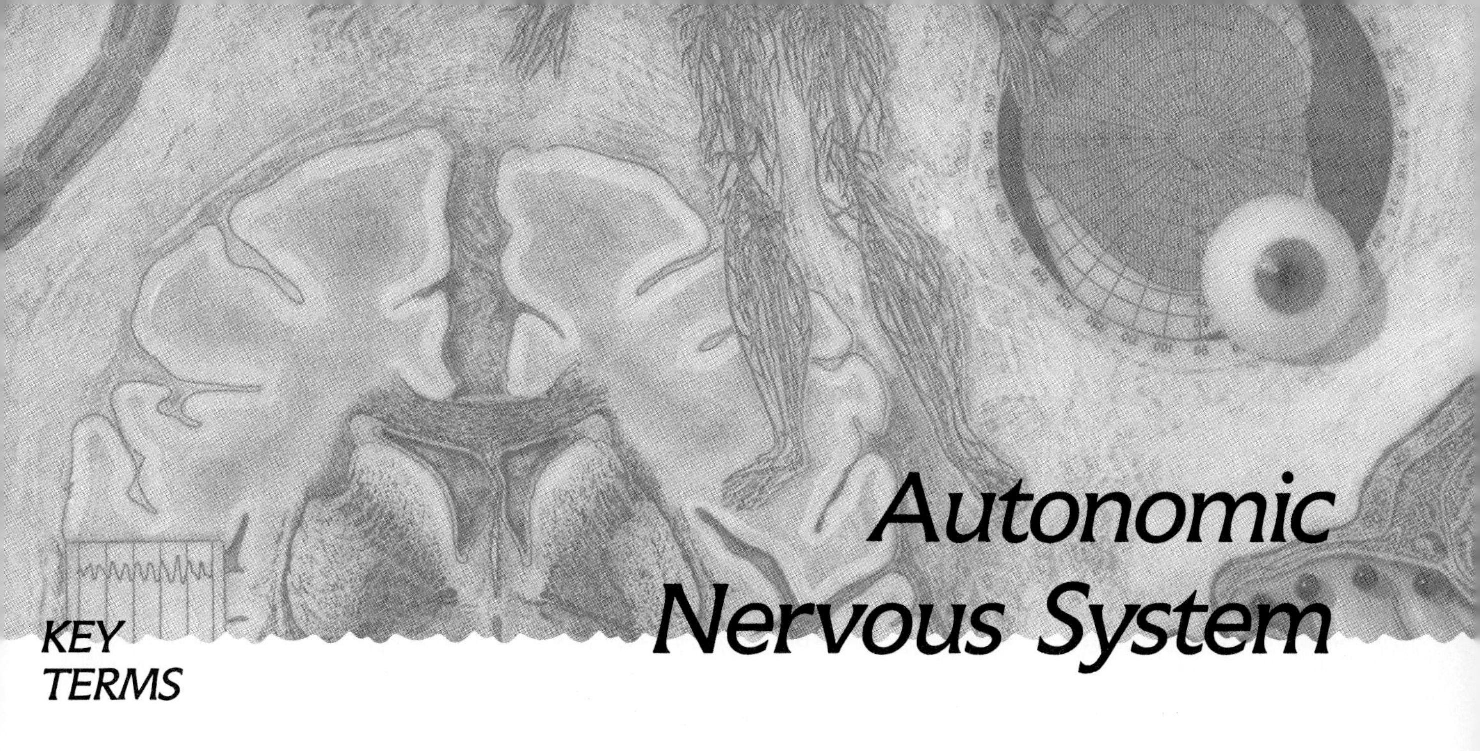

Autonomic Nervous System

KEY TERMS

adrenergic
(ă-drĕ-ner′jik) neuron

adrenergic receptor

cholinergic
(kol-in-er′jik) neuron

collateral ganglion

gray ramus communicans
(ra′mus kŏ-mu′nĭ-kans)

muscarinic
(mus′kar-in′ik) receptor

nicotinic (nik′o-tin′ik)
receptor

parasympathetic division

pelvic nerve

postganglionic neuron

preganglionic neuron

splanchnic (splangk′nik) nerve

sympathetic chain ganglion

sympathetic division

white ramus communicans
(ra′mus kŏ-mu′nĭ-kans)

RELATED TOPICS

The following terms or concepts are important for a good understanding of this chapter. If you are not familiar with them, you should review them before proceeding.

Action potentials (Chapter 9)

Histology of nervous tissue (Chapter 12)

Synaptic transmission (Chapter 12)

THE PERIPHERAL NERVOUS SYSTEM IS COMPOSED OF AFFERENT AND EFFERENT NEURONS THAT COURSE THROUGH THE SAME NERVES. AFFERENT NEURONS CARRY ACTION POTENTIALS FROM THE PERIPHERY TO THE CENTRAL NERVOUS SYSTEM (CNS), AND EFFERENT NEURONS CARRY ACTION POTENTIALS FROM THE CNS TO THE PERIPHERY. THE EFFERENT NEURONS BELONG TO EITHER THE SOMATOMOTOR NERVOUS SYSTEM, WHICH SUPPLIES SKELETAL MUSCLE, OR THE AUTONOMIC NERVOUS SYSTEM (ANS), WHICH SUPPLIES SMOOTH MUSCLE, CARDIAC MUSCLE, AND GLANDS.

The ANS regulates the activities of internal viscera such as the heart, blood vessels, digestive organs, and reproductive organs. The ability of our bodies to maintain homeostasis depends in large part on the ANS. For example, the delivery of blood to tissues is controlled by altering heart rate and blood vessel diameter.

The major anatomical and physiological characteristics of the ANS are described in this chapter. A functional knowledge of the ANS will enable you to predict general responses to a variety of stimuli, explain responses to changes in environmental conditions, comprehend symptoms that result from abnormal autonomic functions, and understand how drugs affect the ANS.

CONTRASTING THE SOMATOMOTOR AND AUTONOMIC NERVOUS SYSTEMS

Although axons of autonomic, somatomotor, and afferent neurons are found within the same nerves, the proportion varies from nerve to nerve. For example, nerves innervating smooth muscle, cardiac muscle, and glands consist primarily of autonomic neurons, and nerves innervating skeletal muscles consist primarily of somatomotor neurons. Other nerves such as the optic, vestibulocochlear, and trigeminal nerves are composed mainly or entirely of afferent neurons.

Unlike efferent neurons, afferent neurons are not divided into functional groups. Afferent neurons propagate action potentials from sensory receptors to the CNS and provide information for somatomotor reflexes, autonomic reflexes, and other functions performed by the CNS. For example, stimulation of pain receptors may initiate both somatomotor and autonomic reflexes such as the withdrawal reflex and an increase in heart rate, respectively. Some afferent neurons primarily affect autonomic functions, and others primarily influence somatomotor functions; however, the lack of a clear division and great functional overlap make classification of afferent neurons as autonomic or somatomotor misleading.

In contrast to afferent neurons, the efferent neurons are separated into somatomotor and autonomic divisions, which differ structurally and functionally. Single axons of somatomotor neurons extend from

Table 16-1

Comparison of the structural and functional features of the autonomic and somatomotor divisions of the peripheral nervous system

FEATURES	SOMATOMOTOR DIVISION	AUTONOMIC DIVISION
Neuron arrangement	One neuron extends from central nervous system (CNS) to effector organs	Two neurons in series extend from CNS to effector organs
Cell body location	Neuron cell bodies are in motor nuclei of the cranial nerves and in ventral horn of the spinal cord	Preganglionic neuron cell bodies are in autonomic nuclei of the cranial nerves and in the lateral horn of the spinal cord; postganglionic neuron cell bodies are in autonomic ganglia
Number of synapses	One synapse between the somatomotor neuron and the effector organ	Two synapses; first (ganglionic) occurs in the autonomic ganglia; second (neuroeffector) occurs at the target tissue
Axon sheaths	Myelinated	Preganglionic axons are myelinated; postganglionic axons are unmyelinated
Effector organs	Skeletal muscle	Smooth muscle, cardiac muscle, and glands
Neurotransmitter substance	Acetylcholine	Acetylcholine for ganglionic synapses and either acetylcholine or norepinephrine for neuroeffector synapses
Response to stimulation	Skeletal muscles contract	Target tissues are stimulated or inhibited
Receptor molecules	Receptor molecules for acetylcholine are muscarinic	Receptor molecules in ganglionic synapses are nicotinic; receptor molecules in neuroeffector synapses that are cholinergic are muscarinic and in neuroeffector synapses that are adrenergic are either alpha- or beta-adrenergic
Classes of responses	Controls all conscious and unconscious movements of skeletal muscle	Unconscious regulation, although influenced by conscious mental functions

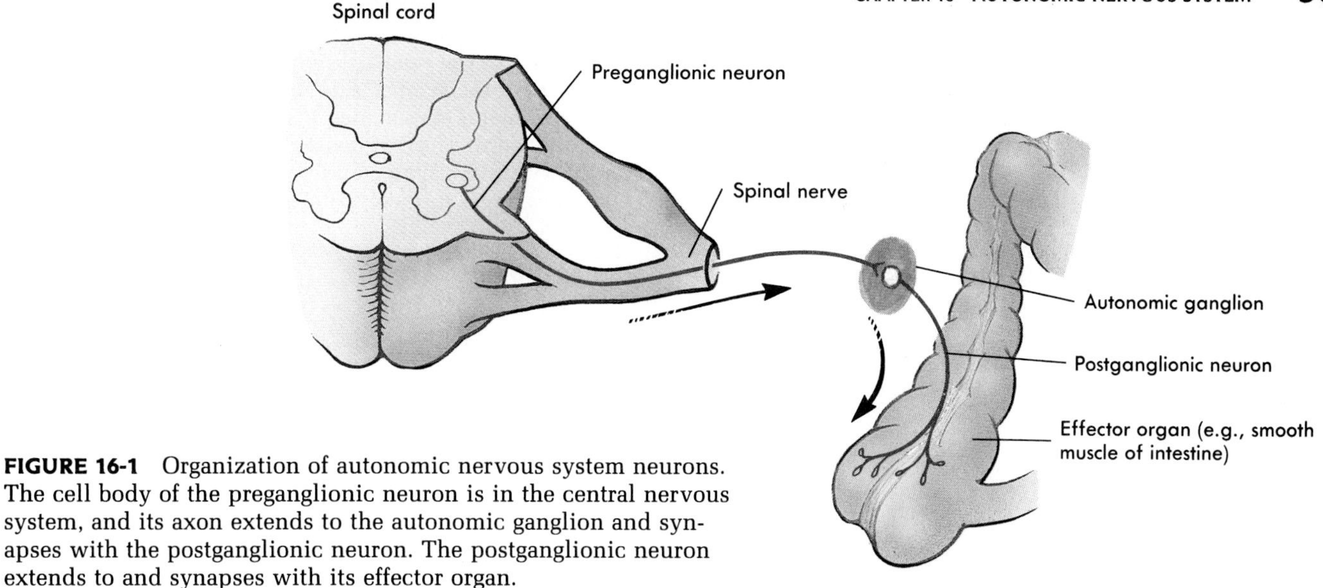

FIGURE 16-1 Organization of autonomic nervous system neurons. The cell body of the preganglionic neuron is in the central nervous system, and its axon extends to the autonomic ganglion and synapses with the postganglionic neuron. The postganglionic neuron extends to and synapses with its effector organ.

the CNS to skeletal muscle. The ANS, on the other hand, has two neurons in a series extending between the CNS and the organs innervated (Figure 16-1). The first neuron is called the **preganglionic neuron,** and the second neuron is called the **postganglionic neuron** because they both synapse in ganglia outside of the CNS. Cell bodies of the preganglionic neurons are within either the brainstem or the spinal cord, and their axons extend through nerves to the autonomic ganglia in which they synapse with postganglionic neurons (the **ganglionic synapse**). The cell bodies of the postganglionic neurons are in the **autonomic ganglia** and send their axons to effector organs, where they synapse with their target tissues (the **neuroeffector synapse**).

Somatomotor neurons innervate skeletal muscles and thus play an important role in locomotion, posture, and equilibrium. Many movements controlled by the somatomotor division are conscious. On the other hand, the ANS innervates smooth muscle, cardiac muscle, and glands, and ANS functions are unconsciously controlled. The effect of somatomotor neurons on skeletal muscle is always excitatory, but the effect of autonomic neurons on target tissues is either inhibitory or excitatory (Table 16-1).

DIVISIONS OF THE AUTONOMIC NERVOUS SYSTEM: STRUCTURAL FEATURES

The ANS is composed of **sympathetic** and **parasympathetic divisions,** each with unique structural and functional features. Structurally these divisions differ in the location of their preganglionic neuron cell bodies within the CNS, the location of their autonomic ganglia, the relative lengths of their preganglionic and postganglionic axons, and the ratio of preganglionic and postganglionic neurons.

Sympathetic Division

Cell bodies of sympathetic preganglionic neurons are in the lateral horns of the spinal cord gray matter between the first thoracic (T1) and the second lumbar (L2) segments (Figure 16-2). Because of the location of the preganglionic cell bodies, the sympathetic division is sometimes called the **thoracolumbar division.** The axons of the preganglionic neurons pass through the ventral roots of spinal nerves T1 to L2, course through the spinal nerves for a short distance, leave the spinal nerves, and project to autonomic ganglia. These ganglia, called **sympathetic chain ganglia,** are on either side of the vertebral column behind the epithelial linings of the pleural and peritoneal cavities. The ganglia are connected to each other and form a chain along both sides of the spinal cord. Although only ganglia from T1 to L2 receive preganglionic axons from the spinal cord, the sympathetic chain extends into the cervical and sacral regions so that one pair of ganglia is associated with nearly every pair of spinal nerves. In the cervical region the ganglia usually fuse during fetal development so there are only three pairs in the adult.

The axons of the preganglionic neurons are small in diameter and are myelinated. The short connection between a spinal nerve and a sympathetic chain ganglion through which the preganglionic axons pass is called the **white ramus communicans** (ra'mus kŏ-mu'nĭ-kans; pl. rami communicantes, ra'mi kŏ-mu-

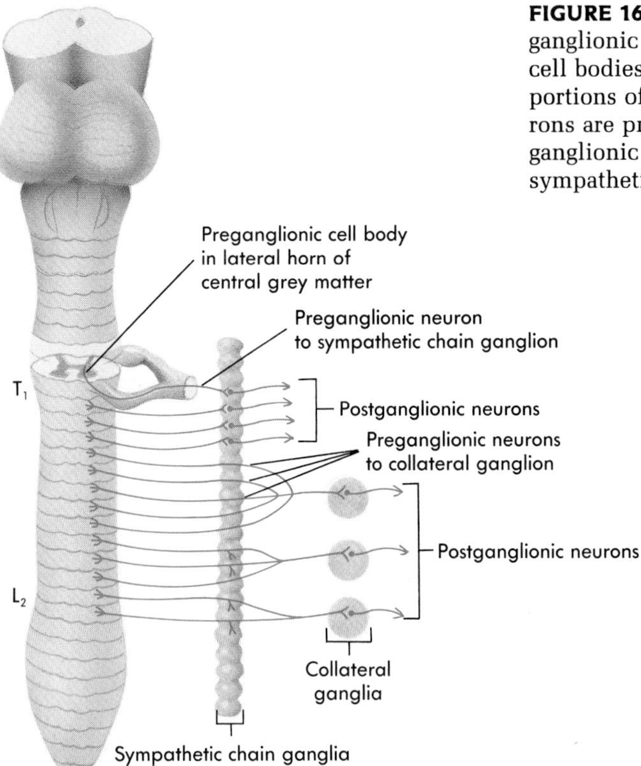

FIGURE 16-2 Location of the preganglionic cell bodies and the postganglionic cell bodies of the sympathetic neurons. The preganglionic cell bodies are in the lateral gray matter of the thoracic and lumbar portions of the spinal cord. The cell bodies of the postganglionic neurons are primarily within the sympathetic chain ganglia. Some postganglionic cell bodies are within collateral ganglia that lie outside the sympathetic chain ganglia.

nĭ-kan-tez) because of the whitish color of the myelinated axons (Figure 16-3).

Sympathetic axons exit the sympathetic chain ganglia by three different routes: the spinal, sympathetic, or splanchnic nerves.

1. **Spinal nerves** (see Figure 16-3, *A*). The preganglionic axons synapse with postganglionic neurons in a sympathetic chain ganglion at the same level the preganglionic axons enter the sympathetic chain. Or the preganglionic axons pass either superiorly or inferiorly through one or more ganglia and synapse with postganglionic neurons in a sympathetic chain ganglion at a different level from that at which the preganglionic axons entered the sympathetic chain. The axons of the postganglionic neurons pass through a **gray ramus communicans** and reenter a spinal nerve. The postganglionic axons are not myelinated, giving the gray ramus communicans a grayish color. The postganglionic axons then project through the spinal nerve to the organs they innervate. These organs include sweat glands in the skin, smooth muscle in skeletal and skin blood vessels, and the smooth muscle of the arrector pili.
2. **Sympathetic nerves** (see Figure 16-3, *B*). The preganglionic axons enter the sympathetic chain and synapse in a sympathetic chain ganglion at the same or at a different level with the postganglionic neuron. The postganglionic axons leave the sympathetic chain ganglion in a sympathetic nerve. Many of the sympathetic nerves supply the thoracic organs, including cardiac muscle of the heart, smooth muscle in thoracic blood vessels, and smooth muscle in the esophagus and lungs. In the cervical region sympathetic nerves form a plexus around the carotid artery and project to the organs of the head. These sympathetic axons supply areas of the head and neck not innervated through the spinal nerves. These areas include sweat glands in the skin, salivary glands in the mouth, and smooth muscle in blood vessels, the eye, and the arrector pili.
3. **Splanchnic** (splangk'nik) **nerves** (see Figure 16-3, *C*). The preganglionic axons that originate between T5 and T12 of the spinal cord enter the sympathetic chain ganglia and, without synapsing, exit at the same or at a different level. The preganglionic axons then pass through splanchnic nerves to **collateral ganglia** (also called **prevertebral ganglia**) in which they synapse with postganglionic neurons. Axons of the postganglionic neurons leave from the collateral ganglia through small nerves that extend to target organs. The collateral ganglia are in the abdomen close to sites where major arteries branch from the abdominal aorta and are named after the blood vessels near which they are located. There are three major collateral ganglia: celiac, superior mesenteric, and inferior mesenteric ganglia.

The celiac ganglia and associated nerves are referred to as the solar plexus. A sharp blow to the abdomen near the solar plexus disrupts normal autonomic nerve function, causing rapid dilation of abdominal blood vessels. Consequently, blood pools in the abdomen, and little blood flows back to the heart, resulting in a decreased amount of blood pumped by the heart to the brain. The brain thus suffers from a lack of oxygen, and dizziness or loss of consciousness (fainting) may occur.

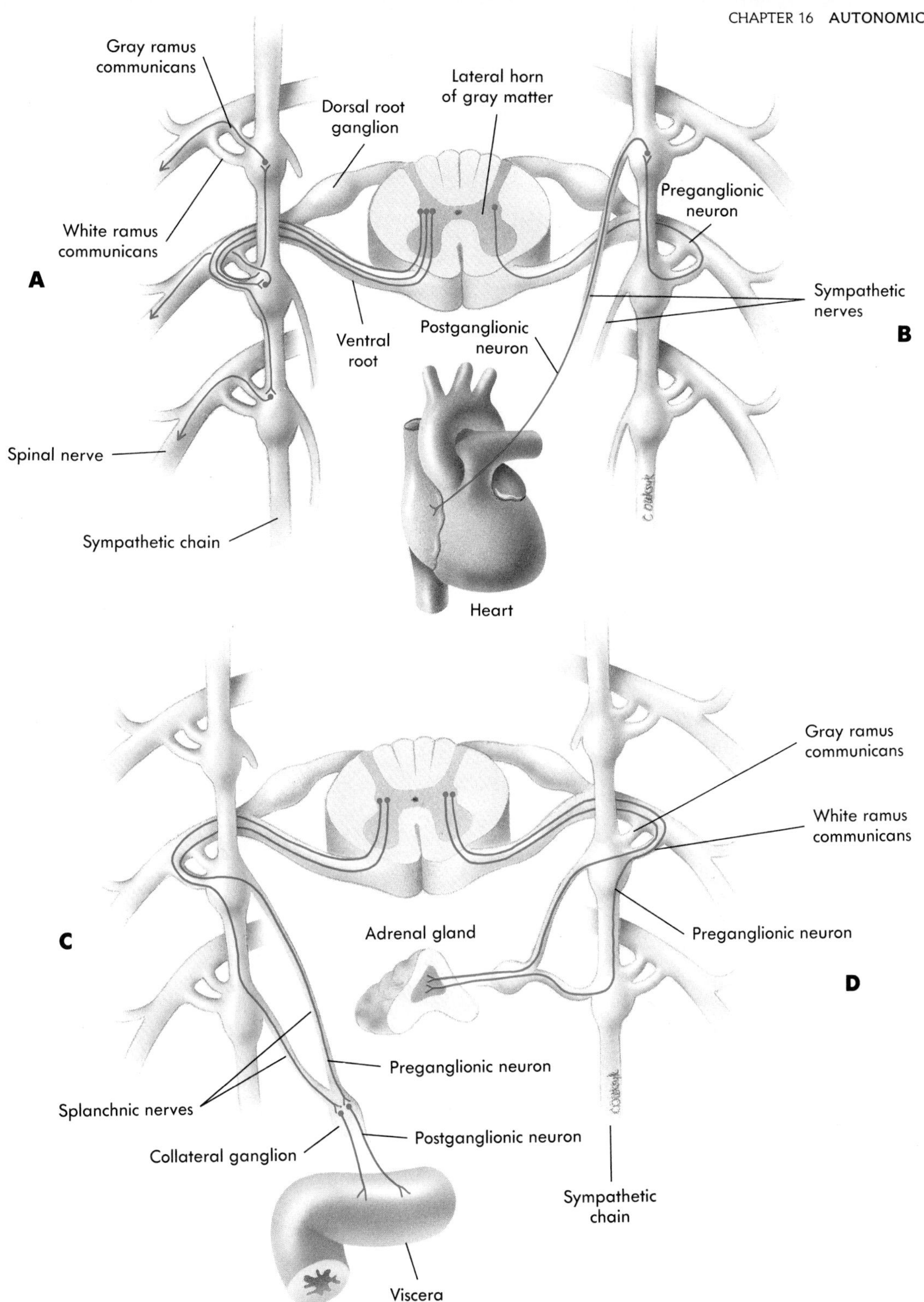

FIGURE 16-3 Different routes taken by sympathetic axons. **A** Preganglionic axons enter a sympathetic chain ganglion through a white ramus communicans. Some axons synapse with a postganglionic neuron at the level of entry; other axons ascend or descend to other levels before synapsing. Postganglionic axons exit the sympathetic chain ganglia through gray rami communicantes and enter spinal nerves. **B** Like **A** except a postganglionic axon exits through a sympathetic nerve (only an ascending axon is illustrated). **C** Preganglionic neurons do not synapse in the sympathetic chain ganglia but exit in splanchnic nerves and extend to collateral ganglia in which they synapse with postganglionic neurons. **D** Like **C** except the preganglionic axon extends to the adrenal medulla in which it synapses with specialized postganglionic cells.

The splanchnic nerves supply the abdominopelvic organs: smooth muscle in blood vessels or the walls of organs, and glands such as the pancreas, liver, prostate, and adrenal glands. The splanchnic nerve supply to the adrenal glands is different from other splanchnic nerves because the preganglionic axons do not synapse in collateral ganglia. Instead the preganglionic axons pass through or by the collateral ganglia and synapse with cells in the adrenal medulla (Figure 16-3, *D*). The adrenal medulla is the inner portion of the adrenal gland and consists of highly specialized postganglionic cells of the ANS. The postganglionic cells are polyhedral in shape, have no axons, and contain large amounts of epinephrine (adrenaline) and smaller amounts of norepinephrine (noradrenaline). Stimulation of these cells by the preganglionic axons causes the release of large amounts of epinephrine and some norepinephrine into the circulatory system. These substances circulate in the blood and affect all tissues having receptors to which they can bind. The general response to epinephrine and norepinephrine released from the adrenal medulla is to prepare the individual for physical activity. Secretions of the adrenal medulla are considered hormones because they are released into the general circulation and travel some distance to the tissues in which they have their effect (see Chapters 17 and 18).

The ratio between the number of postganglionic and preganglionic neurons is large for the sympathetic division of the ANS. A single preganglionic fiber synapses with many postganglionic neurons, and as a result, **divergence** is a characteristic of the sympathetic division. Functionally, this means that sympathetic activity tends to have a more generalized effect rather than a highly localized effect on specific organs.

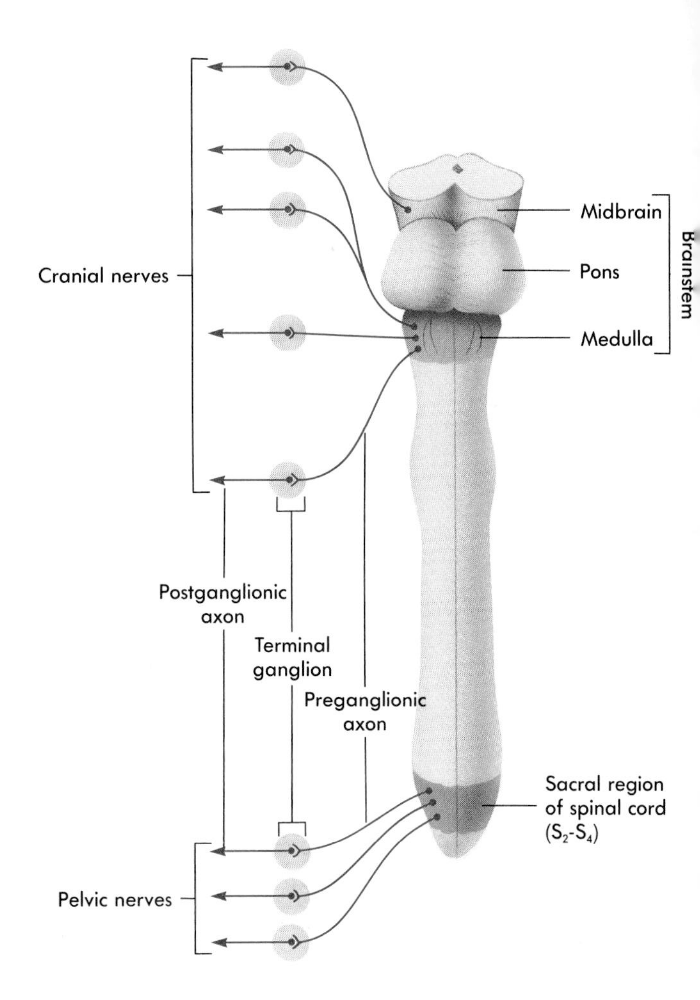

FIGURE 16-4 Location of the preganglionic and postganglionic nerve cell bodies of the parasympathetic division. The preganglionic nerve cell bodies are in the brainstem and the lateral gray matter of the sacral portion of the spinal cord, and the postganglionic nerve cell bodies are within terminal ganglia.

Parasympathetic Division

Preganglionic cell bodies of the parasympathetic division are either within brainstem nuclei or within the lateral horns of the gray matter in the sacral region of the spinal cord from S2 to S4 (Figure 16-4). For that reason the parasympathetic division is sometimes called the **craniosacral division.**

Axons of the preganglionic neurons course through cranial and pelvic nerves to **terminal ganglia** either near or embedded within the wall of the organ innervated by parasympathetic neurons. Consequently, the parasympathetic ganglia are not as structurally organized as the sympathetic ganglia. Many of the parasympathetic ganglia are small in size, but some such as those in the wall of the digestive tract are extensive. The axons of the postganglionic neurons extend the relatively short distance from the parasympathetic ganglia to the target organ.

Parasympathetic axons whose cell bodies are located within brainstem nuclei exit through the oculomotor (III), facial (VII), glossopharyngeal (IX), and vagus nerves (X). Specifically, the oculomotor nerves carry parasympathetic fibers that innervate smooth-muscle cells within the eyes, the facial and glossopharyngeal nerves carry parasympathetic fibers to the salivary glands, and the vagus nerves carry parasympathetic fibers to most thoracic and abdominal viscera, including the heart, lungs, esophagus, stomach, pancreas, liver, intestine, and upper colon. Approximately 75% of all parasympathetic neurons course through the vagus nerves. Parasympathetic axons in the head leave the oculomotor, facial, and glossopharyngeal nerves and travel with branches of the trigeminal nerve to their target organs.

Parasympathetic preganglionic axons whose cell bodies are in the sacral region of the spinal cord course through **pelvic nerves** that innervate the blad-

Table 16-2

Structural comparison of the sympathetic and parasympathetic divisions

FEATURE	SYMPATHETIC DIVISION	PARASYMPATHETIC DIVISION
Location of preganglionic cell body	Lateral horns of spinal cord gray matter (T1-L2)	Brainstem and lateral horns of spinal cord gray matter (S2-S4)
Outflow from central nervous system	Spinal nerves Sympathetic nerves Splanchnic nerves	Cranial nerves Pelvic nerves
Ganglia	Sympathetic chain ganglia along spinal cord for spinal and sympathetic nerves; collateral ganglia for splanchnic nerves	Terminal ganglia near or on effector organ
Number of postganglionic neurons for each preganglionic neuron	Many	Few
Relative length of neurons	Short preganglionic Long postganglionic	Long preganglionic Short postganglionic

The term parasympathetic implies that it surrounds the sympathetic division, and indeed, its preganglionic neuron cell bodies are found both superior and inferior to the area of the CNS in which preganglionic neuron cell bodies of the sympathetic division are found.

der, lower colon, rectum, and organs of the reproductive system.

The ratio of postganglionic to preganglionic neurons is much less for the parasympathetic division than for the sympathetic division. There are approximately two postganglionic neurons for each parasympathetic preganglionic neuron, whereas there are up to 17 postganglionic neurons for each sympathetic preganglionic neuron. For this reason the effects of the parasympathetic division are more specific and localized than the sympathetic division. For example, in contrast to the more general sympathetic response, the parasympathetic division can cause the urinary bladder to contract without activating other parasympathetic responses. Table 16-2 summarizes the structural differences between the sympathetic and parasympathetic divisions.

NEUROTRANSMITTER SUBSTANCES AND RECEPTORS

The sympathetic and parasympathetic nerve endings secrete one of two neurotransmitters. If the neuron secretes acetylcholine, it is a **cholinergic** (kol-in-er'jik) **neuron,** and, if it secretes norepinephrine, it is an **adrenergic** (ă-drĕ-ner'jik) **neuron.**

All preganglionic neurons of the sympathetic and parasympathetic divisions and all postganglionic neurons of the parasympathetic division are cholinergic. Most postganglionic neurons of the sympathetic division are adrenergic, but a few such as the postganglionic neurons that innervate sweat glands and a few blood vessels are cholinergic (Figure 16-5).

In cholinergic synapses acetylcholine molecules released from the cholinergic presynaptic terminals diffuse across the synaptic cleft and combine with receptor molecules within the postsynaptic membranes. **Cholinergic receptors** exist in two structurally different forms: nicotinic and muscarinic. Although acetylcholine binds to and activates both types of receptor molecules, nicotine, an alkaloid substance found in tobacco, specifically binds to and activates **nicotinic** (nik'o-tin'ik) **receptors** but not muscarinic receptors. Muscarine, an alkaloid extracted from some poisonous mushrooms, specifically binds to and activates **muscarinic** (mus'kar-in'ik) **receptors** but not nicotinic receptors. Although nicotine and muscarine are not naturally in the human body, they demonstrate differences in the two classes of cholinergic receptors.

Nicotinic receptors are in the membranes of all postganglionic neurons of autonomic ganglia, whereas muscarinic receptors are in the membranes of all effector cells that respond to acetylcholine, including skeletal muscle cells innervated by the somatomotor division of the peripheral nervous system.

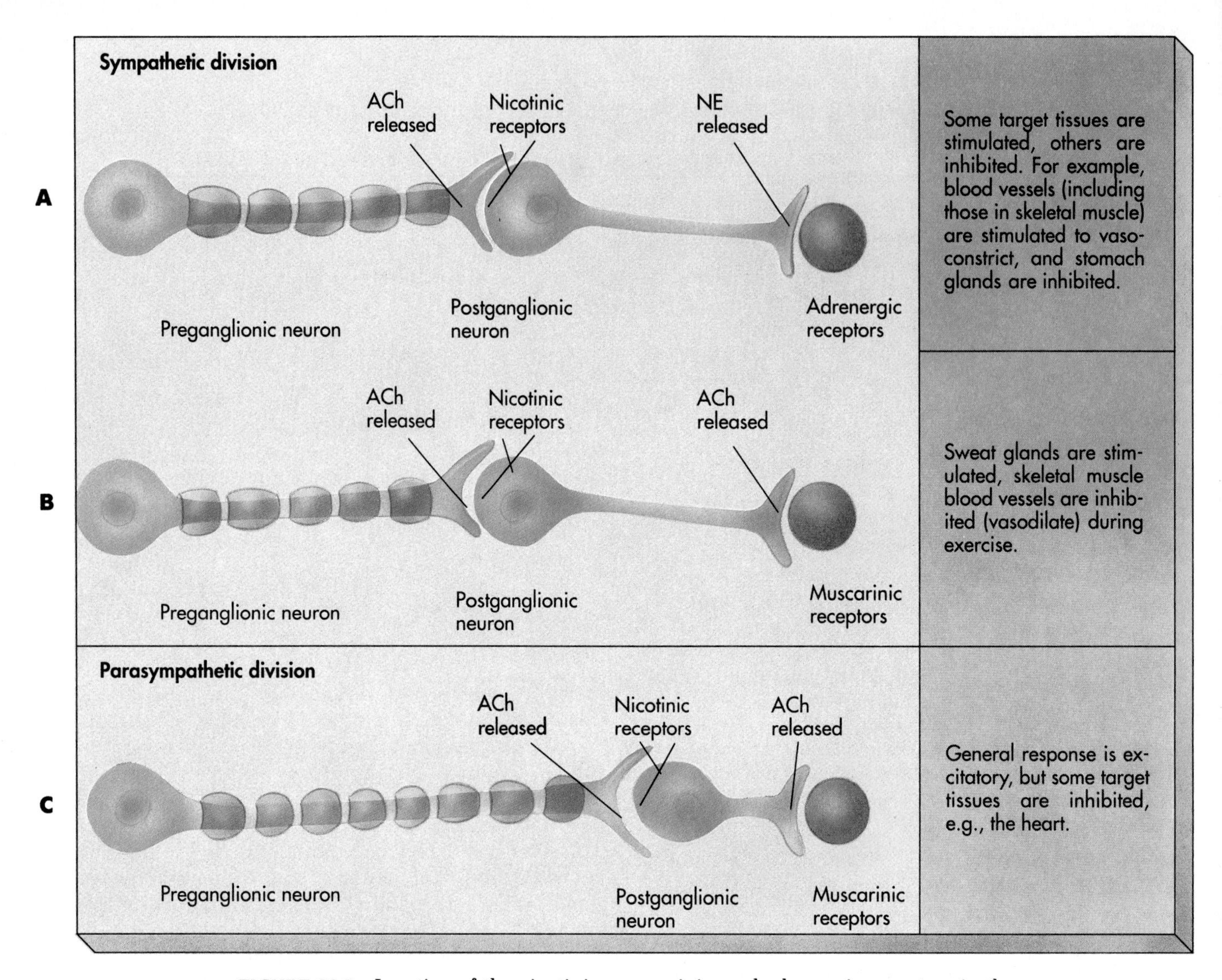

FIGURE 16-5 Location of the nicotinic, muscarinic, and adrenergic receptors in the autonomic nervous system. Nicotinic receptors are on the cell bodies of both sympathetic and parasympathetic postganglionic cells in the autonomic ganglia.
A Adrenergic receptors are in most target tissues innervated by the sympathetic division. **B** Some sympathetic target tissues have muscarinic receptors. **C** All parasympathetic target tissues have muscarinic receptors. *NE*, Norepinephrine; *ACh*, acetylcholine.

1

Would structures innervated by the sympathetic division or the parasympathetic division be stimulated after the consumption of nicotine? After the consumption of muscarine? Explain.

? ? ? ? ? ? ? ? ? ? ?

When acetylcholine binds to nicotinic receptors, the response of the postsynaptic membrane is excitatory. The combination of acetylcholine with muscarinic receptors, however, results in either excitation or inhibition, depending on the target tissue in which the receptors are found. For example, acetycholine binds to muscarinic receptors in cardiac muscle, causing a reduced heart rate; and acetylcholine binds to muscarinic receptors in most smooth-muscle cells of the gastrointestinal tract, causing an increased rate and amplitude of contraction.

Norepinephrine is released from adrenergic postganglionic neurons of the sympathetic division (see Figure 16-5), diffuses across the synapse, and binds to adrenergic receptor molecules within the membranes of the effector organ. The **adrenergic receptors** belong to two major structural categories: the **alpha- (α) adrenergic** and **beta- (β) adrenergic re-**

The Influence of Drugs on the ANS

Many drugs that directly influence the ANS either bind to specific receptors and activate them **(stimulating agents)** or bind to specific receptors and prevent their being activated **(blocking agents).** Other drugs stimulate the release of neurotransmitter substances and prevent either their metabolic breakdown, their biosynthesis, or their release. Some of these drugs have important therapeutic value, but others primarily are used to study the function of the ANS or are found in medically hazardous substances such as tobacco.

Drugs that Bind to Nicotinic Receptors

Drugs that bind to nicotinic receptors and activate them are **nicotinic agents.** Although these agents have little therapeutic value and are mainly of interest to researchers, nicotine is medically important because of its presence in tobacco. Nicotinic agents bind to the nicotinic receptors on all postganglionic neurons within the autonomic ganglia and produce stimulation. Responses to nicotine are variable and depend on the amount taken into the body. Since nicotine stimulates the postganglionic neurons of both the sympathetic and parasympathetic divisions, much of the variability of its effects is due to the opposing actions of these division. For example, in response to the nicotine contained in a cigarette, the heart rate may either increase or decrease, and its rhythm tends to become less regular as a result of the simultaneous actions of the sympathetic division, which increases the heart rate, and the parasympathetic division, which decreases the heart rate. Blood pressure tends to increase because the sympathetic neurons that constrict blood vessels are more numerous than either the sympathetic neurons or the parasympathetic neurons that dilate blood vessels. In addition to its influence on the ANS, nicotine also affects the CNS; therefore not all of its effects can be explained on the basis of action on the ANS. Nicotine is extremely toxic, and small amounts can be lethal.

Drugs that bind to and block nicotinic receptors are called **ganglionic blocking agents** because they block the effect of acetylcholine on both parasympathetic and sympathetic division. The effects of these substances on the sympathetic division, however, overshadows the effect on the parasympathetic division. Drugs that fall into this category include hexamethonium and tetraethylammonium ions, and they are used clinically to block sympathetic activity. Their primary use has been the reduction of blood pressure in some patients suffering from hypertension.

Drugs that Bind to Muscarinic Receptors

Drugs that bind to and activate muscarinic receptors are **muscarinic agents,** or **parasympathomimetic agents.** These drugs activate the muscarinic receptors at the neuroeffector synapses for both divisions of the ANS. Most observable responses are apparently parasympathetic because all structures innervated by the parasympathetic division have muscarinic receptors, whereas the majority of the structures innervated by the sympathetic division have adrenergic receptors. Muscarine causes increased sweating, increased secretion of glands in the digestive system, decreased heart rate, constriction of the pupil, and contraction of respiratory, digestive, and urinary system smooth muscles. In the past drugs related to muscarine, including pilocarpine, were used to treat conditions such as paroxysmal tachycardia (an increased heart rate of unknown cause) and to stimulate motility in the gastrointestinal tract or urinary bladder. These drugs have also been used to treat glaucoma (elevated pressure in the eyeball) because activation of muscarinic receptors dilates lymphatic vessels, allowing excess fluid in the eyeball to drain.

Drugs such as atropine that bind to and block the action of muscarinic receptors are **muscarinic blocking agents,** or **parasympathetic blocking agents.** These drugs cause the pupil of the eye to dilate and are used during eye examinations to allow the examiner to see the retina through the pupil. These drugs also decrease salivary secretion and are used during surgery to prevent patients from choking on excess saliva while they are anesthetized.

Drugs that Bind to Alpha- and Beta-Adrenergic Receptors

Drugs that activate adrenergic receptors are **adrenergic agents,** or **sympathomimetic agents.** Drugs such as phenylephrine stimulate alpha-adrenergic receptors, which are numerous in the smooth-muscle cells of certain blood vessels, especially in the digestive tract and the skin. These drugs increase blood pressure by causing vasoconstriction. On the other hand, isoproterenol is a drug that selectively activates beta-adrenergic receptors, which are found in cardiac muscle and bronchiolar smooth muscle. Beta-adrenergic—stimulating agents sometimes are used to dilate bronchioles in patients with respiratory disorders such as asthma and occasionally are used as cardiac stimulants.

Drugs such as phenoxybenzamine that bind to and block the action of alpha-adrenergic receptors are **alpha-adrenergic blocking agents.** The therapeutic uses of these drugs are limited, but they have been used to cause vasodilation in patients with diseases (e.g., Raynaud's disease and acrocyanosis) in which chronic vasoconstriction in the periphery is intense and must be reduced to prevent necrosis of tissues and the development of gangrene.

Propranolol is an example of a **beta-adrenergic blocking agent.** These drugs sometimes are used to treat hypertension (high blood pressure), some types of cardiac arrhythmias, and patients recovering from heart attacks. Blockade of the beta-adrenergic receptors within the heart prevents sudden increases in the heart rate and thus decreases the probability of arrhythmic contractions.

ceptors. Norepinephrine binds to and activates both types of receptor molecules, although it has a greater affinity for the alpha receptors. Epinephrine has nearly an equal affinity for both receptor types. In tissues containing adrenergic receptors, both alpha- and beta-adrenergic receptors are present, although one type or the other is more abundant. Several drugs specifically stimulate or inhibit alpha- or beta-adrenergic receptors. These drugs are important therapeutic agents (see boxed essay on p. 513).

Both alpha- and beta-adrenergic receptors can be excitatory or inhibitory. For example, beta-adrenergic receptors are stimulatory in cardiac muscle but inhibitory in intestinal smooth muscle. The number of receptors in a given tissue may also change. For example, beta-adrenergic receptors decrease in tissues when there is a lack of thyroid hormones secreted by the thyroid gland or a lack of cortisol secreted by the adrenal gland.

In recent years substances in addition to the regular neurotransmitters have been extracted from neurons of the autonomic ganglia. These substances include prostaglandins, gastrin, somatostatin, cholecystokinin, vasoactive intestinal peptide, enkephalins, and substance P. Most of these substances are peptides (prostaglandins are fatty acids) and are active in very low concentrations. The cells in which these substances are produced and the specific role that each of these compounds plays in the regulation of the ANS is not clear, but they may function as either neurotransmitters or neuromodulator substances (see Chapter 12).

REGULATION OF THE AUTONOMIC NERVOUS SYSTEM

Much of the regulation of structures by the ANS occurs through autonomic reflexes, but input from the cerebrum, hypothalamus, and other areas of the brain allows conscious thoughts and actions, emotions, and other CNS activities to influence autonomic functions. Without the regulatory activity of the ANS, an individual would have limited ability to maintain homeostasis and would readily succumb to sudden changes in environmental conditions.

Autonomic reflexes, like other reflexes, involve sensory receptors; afferent, association, and efferent neurons; and effector cells (Figure 16-6; see Chapter 12). For example, baroreceptors (stretch receptors) in the walls of large arteries near the heart detect changes in blood pressure, and afferent neurons transmit information from the baroreceptors through cranial nerves (primarily the glossopharyngeal and vagus nerves) to the medulla oblongata. In the medulla oblongata association neurons integrate the information, and action potentials are produced in autonomic neurons that extend to the heart. If baroreceptors detect a change in blood pressure, autonomic reflexes return blood pressure to normal by altering heart rate.

An increase in blood pressure initiates a parasympathetic reflex in which efferent neurons transmit action potentials, which inhibit cardiac muscle cells and reduce the heart rate, toward the heart, thus bringing blood pressure down toward its normal value. Conversely, a sudden decrease in blood pressure initiates a sympathetic reflex, which stimulates the heart to increase its rate and force of contraction, thus causing blood pressure to increase.

2

Sympathetic neurons stimulate sweat glands in the skin. Predict how they would function to control body temperature during exercise and during exposure to cold temperatures.

? ? ? ? ? ? ? ? ? ? ?

Other autonomic reflexes also participate in the regulation of blood pressure. For example, numerous sympathetic neurons transmit a low but relatively constant frequency of action potentials that innervate blood vessels throughout the body, keeping them partially constricted. If the vessels constrict further, blood pressure increases; and if they dilate, blood pressure decreases. Thus altering the frequency of action potentials delivered to blood vessels along sympathetic neurons can either raise or lower blood pressure.

3

How would sympathetic reflexes that control blood vessels respond to a sudden decrease in blood pressure and to a sudden increase in blood pressure?

? ? ? ? ? ? ? ? ? ? ?

Higher centers of the brain also affect autonomic functions. Emotions such as anger increase blood pressure by increasing heart rate and constricting blood vessels through sympathetic stimulation. Pleasant thoughts of a delicious banquet initiate increased secretion by salivary glands and by glands within the stomach and increased smooth-muscle contractions within the digestive system, all of which are controlled by parasympathetic neurons.

Specific areas of the CNS, including the spinal cord, medulla oblongata, and hypothalamus, integrate a large number of autonomic reflexes (Figure 16-7). However, areas in which autonomic reflexes are integrated are influenced by other areas of the CNS. For example, the hypothalamus integrates responses to temperature changes and coordinates responses to stress, rage, and other emotions. The limbic system, which plays an important role in emotions and their expression, and the cerebral cortex affect autonomic functions by influencing the hypothalamus.

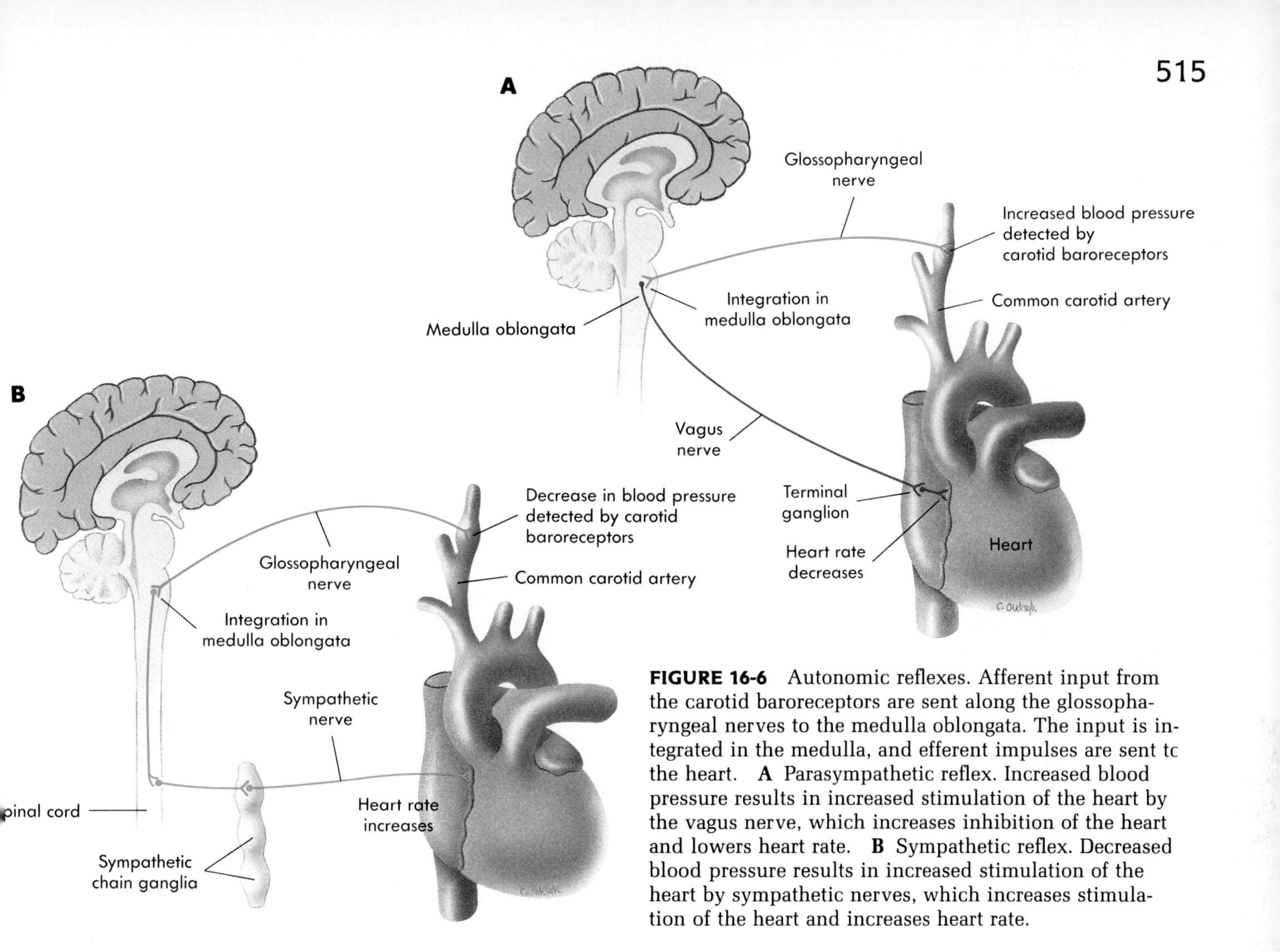

FIGURE 16-6 Autonomic reflexes. Afferent input from the carotid baroreceptors are sent along the glossopharyngeal nerves to the medulla oblongata. The input is integrated in the medulla, and efferent impulses are sent to the heart. **A** Parasympathetic reflex. Increased blood pressure results in increased stimulation of the heart by the vagus nerve, which increases inhibition of the heart and lowers heart rate. **B** Sympathetic reflex. Decreased blood pressure results in increased stimulation of the heart by sympathetic nerves, which increases stimulation of the heart and increases heart rate.

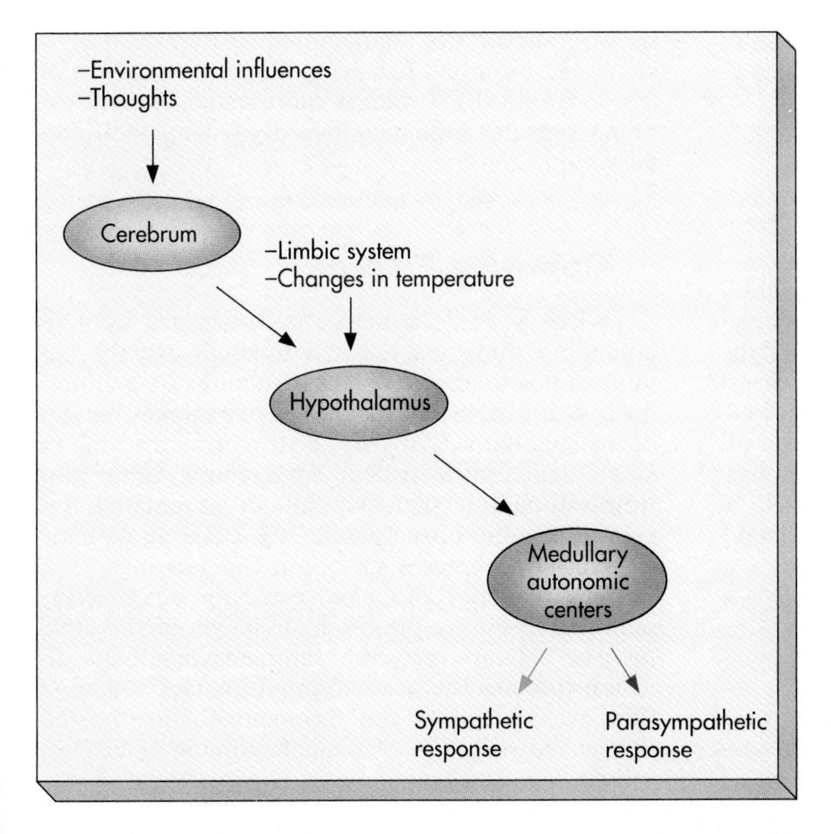

FIGURE 16-7 Influence of higher portions of the brain on autonomic functions, especially the influence of the hypothalamus and the cerebrum on the autonomic nervous system. Neural pathways extend from the cerebrum to the hypothalamus and from the hypothalamus to neurons of the autonomic nervous system.

Biofeedback, Meditation, and the Fight-or-Flight Response

Biofeedback takes advantage of electronic instruments or other techniques to monitor and change subconscious activities, many of which are regulated by the ANS. Skin temperature, heart rate, and nerve impulse patterns within the brain (brain waves) are monitored electronically. By watching the monitor and using biofeedback techniques, a person can learn how consciously to reduce his heart rate and blood pressure and regulate blood flow in the limbs. For example, it has been claimed that people can prevent the onset of migraine headaches or reduce their intensity by learning to dilate blood vessels in the skin of their arms and hands. Increased blood vessel dilation increases skin temperature, which is correlated with a decrease in the severity of the migraine.

Some people use biofeedback methods to relax by learning to reduce the heart rate or change the pattern of brain waves. The severity of stomach ulcers, high blood pressure, anxiety, and depression can be alleviated by using such biofeedback techniques.

Meditation is another technique that influences autonomic functions. Although numerous claims about the value of meditation include improving one's spiritual well-being, consciousness, and holistic view of the universe, it has been established that meditation does influence autonomic functions. Meditation techniques are useful in some people in reducing heart rate, blood pressure, severity of ulcers, and other symptoms frequently associated with stress.

The **fight-or-flight response** can occur when an individual is subjected to severe stress such as a threatening situation or a repugnant event. The response involves all parts of the nervous system, as well as the endocrine system, and may be consciously or unconsciously mediated. The autonomic portion of the fight-or-flight response results in a general increase in sympathetic activity, including heart rate, blood pressure, sweating, muscular strength, and other responses, that prepares the individual for physical activity. The flight-or-fight response is adaptive in that it usually removes the individual from the threatening situation or prepares him for optimum physical strength.

FUNCTIONAL GENERALIZATIONS ABOUT THE AUTONOMIC NERVOUS SYSTEM

Generalizations can be made about the function of the ANS on effector organs, but most have exceptions.

Stimulatory vs. Inhibitory Effects

Both divisions of the ANS produce stimulatory and inhibitory effects. For example, the parasympathetic division causes contraction (stimulation) of the urinary bladder and a decrease (inhibition) in heart rate, whereas the sympathetic division produces vasoconstriction (stimulation) of blood vessels and dilation (inhibition) of air passageways in the lungs. Thus it is *not* true that one division of the ANS is always stimulatory and the other division is always inhibitory.

Dual Innervation

Most organs that receive autonomic neurons are innervated by both the parasympathetic and sympathetic divisions (Figure 16-8). The gastrointestinal tract, heart, urinary bladder, and reproductive tract are examples (Table 16-3). However, dual innervation of organs by both divisions of the ANS is not universal. For example, sweat glands and blood vessels are innervated by sympathetic neurons almost exclusively. In addition, most structures receiving dual innervation are not regulated equally by both divisions. For example, parasympathetic innervation of the gastrointestinal tract is more extensive and exhibits a greater influence than does sympathetic innervation.

Opposite Effects

When a *single* structure is innervated by both autonomic divisions, the two divisions usually produce opposite effects on the structure. As a consequence, the ANS is capable of both increasing and decreasing the activity of the structure, resulting in an efficient control system. For example, in the gastrointestinal tract parasympathetic stimulation increases secretion from glands, whereas sympathetic stimulation decreases secretion. In a few instances, however, the effect of the two divisions is not clearly opposite. For example, both divisions of the ANS increase salivary secretion: the parasympathetic division initiates the production of a large volume of thin, watery saliva, and the sympathetic division causes the secretion of a small volume of viscous saliva.

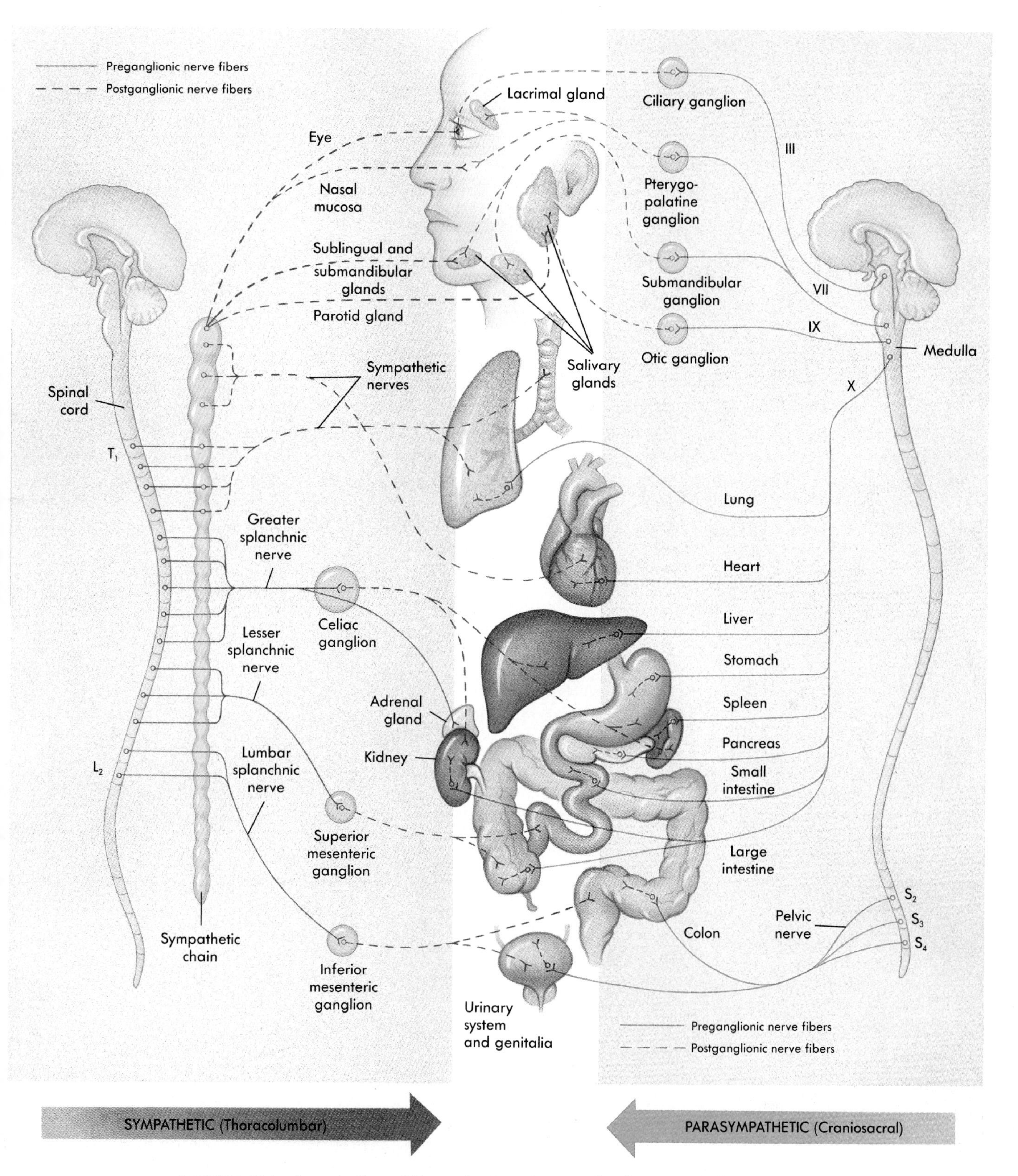

FIGURE 16-8 Innervation of the major target organs by the autonomic nervous system. Preganglionic fibers are indicated by solid lines, and postganglionic fibers are indicated by broken lines.

Table 16-3

Autonomic innervation of target tissues

ORGAN	EFFECT OF SYMPATHETIC STIMULATION	EFFECT OF PARASYMPATHETIC STIMULATION
Heart		
Muscle	Increased rate and force (b)*	Slowed rate (c)
Coronary arteries	Dilated (b)†, constricted (a)†	Dilated (c)
Systemic blood vessels		
Abdominal	Constricted (a)	None
Skin	Constricted (a)	None
Muscle	Dilated (b, c), constricted (a)	None
Lungs		
Bronchi	Dilated (b)	Constricted (c)
Liver	Glucose released into blood (b)	None
Skeletal muscles	Breakdown of glycogen to glucose (b)	None
Metabolism	Increased up to 100% (a, b)	None
Glands		
Adrenal	Release of epinephrine and norepinephrine (c)	None
Salivary	Constriction of blood vessels and slight production of a thick, viscous secretion (a)	Dilation of blood vessels and thin, copious secretion (c)
Gastric	Inhibition (a)	Stimulation (c)
Pancreas	Inhibition (a)	Stimulation (c)
Lacrimal	None	Secretion (c)
Sweat		
Merocrine	Copious, watery secretion (c)	None
Apocrine	Thick, organic secretion (c)	None
Gut		
Wall	Decreased tone (b)	Increased motility (c)
Sphincter	Increased tone (a)	Decreased tone (c)
Gallbladder and bile ducts	Relaxed (b)	Contracted (c)
Urinary bladder		
Wall	Relaxed (b)	Contracted (c)
Sphincter	Contracted (a)	Relaxed (c)
Eye		
Ciliary muscle	Relaxed for far vision (b)	Contracted for near vision (c)
Pupil	Dilated (a)	Constricted (c)
Arrector pili muscles	Contraction (a)	None
Blood	Increased coagulation (a)	None
Sex organs	Ejaculation (a)	Erection (c)

*a, Mediated by alpha-adrenergic receptors; b, mediated by beta-adrenergic receptors; c, mediated by cholinergic receptors.

†Normally there is increased blood flow through coronary arteries as a result of sympathetic stimulation of the heart because of increased demand by cardiac tissue for oxygen (local control of blood flow is discussed in Chapter 21). In experiments that isolate the coronary arteries, however, sympathetic nerve stimulation, acting through alpha-adrenergic receptors, causes vasoconstriction. The beta-adrenergic receptors are relatively insensitive to sympathetic nerve stimulation, but can be activated by drugs.

Cooperative Effects

One autonomic division alone or both divisions acting together can coordinate the activities of *different* structures. For example, the parasympathetic division stimulates the pancreas to release digestive enzymes into the small intestine. At the same time the parasympathetic division stimulates contractions of the small intestine to mix the digestive enzymes with food within the small intestine, resulting in increased digestion and absorption of the food.

Both divisions cooperate to achieve normal reproductive function. The parasympathetic division initiates erection of the penis, and the sympathetic division stimulates the release of secretions from male reproductive glands and initiates ejaculation in the male reproductive tract.

General vs. Localized Effects

The sympathetic division has a more general effect than does the parasympathetic division for two reasons. First, there is more divergence in the sympathetic than in the parasympathetic division. Each sympathetic preganglionic neuron synapses with many postganglionic neurons, resulting in the stimulation of many effector organs at the same time. The preganglionic neurons of the parasympathetic division, however, synapse with few postganglionic neurons. Consequently, the parasympathetic division has a more localized effect and can control effector organs independently of each other. Second, activation of the sympathetic system often causes secretion of both epinephrine and norepinephrine from the adrenal medulla. These hormones circulate in the

Diseases of the Autonomic Nervous System

Normal function of all components of the ANS is not required to maintain life as long as environmental conditions are constant and optimum. Abnormal autonomic functions, however, markedly affect the individual's ability to respond to changing conditions. For example, a complete sympathectomy (removal of sympathetic ganglia) in an animal makes it highly sensitive to heat, cold, or other forms of stress. In a hot environment the animal's ability to lose heat by increasing blood flow to the skin and by sweating is decreased. When exposed to the cold, the animal is less able to reduce blood flow to the skin and conserve heat. Sympathectomy initially results in hypotension (low blood pressure) caused by dilation of peripheral blood vessels and results in the inability to increase blood pressure during periods of physical activity.

Reduced vasomotor tone in humans causes **orthostatic hypotension,** a condition in which, when a person stands, blood pools in dilated blood vessels in the lower extremities, little blood returns to the heart, and as a consequence, the amount of blood the heart pumps decreases. Blood flow to the brain subsequently declines, causing fainting because of a lack of oxygen. Orthostatic hypotension sometimes is caused by a CNS disorder that decreases the frequency of impulses in sympathetic nerves innervating blood vessels.

Raynaud's disease involves the spasmodic contraction of blood vessels in the periphery, especially in the digits, and results in pale, cold hands that are prone to ulcerations and gangrene as a result of poor circulation. This condition may be caused by exaggerated sensitivity of blood vessels to sympathetic innervation. Preganglionic denervation (cutting the preganglionic neurons) occasionally is performed to alleviate the condition.

Hyperhidrosis (hi'per-hi-dro'sis), or excessive sweating, is caused by exaggerated sympathetic innervation of the sweat glands.

Achalasia (ă-kal-a'ze-ah) is characterized by difficulty in swallowing and in controlling contraction of the esophagus where it enters the stomach. Therefore normal peristaltic contractions of the esophagus are interrupted. The swallowing reflex is controlled partly by somatomotor reflexes and partly by parasympathetic reflexes. The cause of achalasia may be abnormal parasympathetic regulation of the swallowing reflex. The condition is aggravated by emotions.

Dysautonomia, (dis-aw-to-no'me-ah) an inherited condition involving an autosomal recessive gene, causes reduced lacrimation (tear gland secretion), poor vasomotor control, trouble in swallowing, and other symptoms. It is the result of poorly controlled autonomic reflexes.

Hirschsprung's disease, or **megacolon,** results in a functional obstruction in the lower colon and rectum. Ineffective parasympathetic innervation and a predominance of sympathetic innervation of the colon inhibit peristaltic contractions, causing feces to accumulate above the inhibited area. The resulting dilation of the colon may be so great that surgery is required to alleviate the condition.

blood and stimulate effector organs throughout the body. Because circulating epinephrine and norepinephrine can persist for a few minutes before being broken down, they can also produce an effect for a longer time than direct stimulation of effector organs by postganglionic sympathetic axons.

The sympathetic nervous system, however, does affect some specific structures, and the response to sympathetic innervation is not always general. For example, vasoconstriction of cutaneous blood vessels in a cold hand is not always associated with an increased heart rate or other responses controlled by the sympathetic division.

Functions at Rest vs. Activity

In cases in which both parasympathetic and sympathetic neurons innervate a single organ, the parasympathetic division tends to have a greater influence under resting conditions, whereas the sympathetic division has a major influence under conditions of physical activity or stress. Increased sympathetic activity results in increased nervous stimulation of effector organs and increased epinephrine and norepinephrine release from the adrenal medulla. Consequently, the pumping effectiveness of the heart increases, blood vessels in skeletal muscle dilate, and blood vessels in visceral structures and the skin constrict. There is also decreased activity of the gastrointestinal tract, increased glucose release from the liver, and a large increase in metabolism, especially in skeletal muscle. In general, the sympathetic division decreases the activity of organs not essential for the maintenance of physical activity and shunts blood and nutrients to structures that are active during physical exercise. This is sometimes referred to as the fight-or-flight response (see essay). The sympathetic division, however, also plays a major role during resting conditions by maintaining blood pressure and body temperature.

Increased activity of the parasympathetic division is generally consistent with resting conditions during which eating, digestion, urination, defecation, and other vegetative functions are emphasized. Many of the reflexes that regulate the digestive, urinary, and reproductive systems are mediated by the parasympathetic division (see Table 16-2).

4

Make a list of the responses controlled by the ANS in (a) a person who is extremely angry and (b) a person who has just finished eating and is relaxing.

? ? ? ? ? ? ? ? ? ? ?

SUMMARY

CONTRASTING THE SOMATOMOTOR AND AUTONOMIC NERVOUS SYSTEMS p. 506

1. The cell bodies of somatomotor neurons are located in the central nervous system (CNS), and their axons extend to skeletal muscles in which they have an excitatory effect that usually is controlled consciously.
2. The cell bodies of the preganglionic neurons of the autonomic nervous system (ANS) are located in the CNS and extend to ganglia in which they synapse with postganglionic neurons. The postganglionic axons extend to smooth muscle, cardiac muscle, or glands and have an excitatory or inhibitory effect that usually is controlled unconsciously.

DIVISIONS OF THE AUTONOMIC NERVOUS SYSTEM: STRUCTURAL FEATURES p. 507

Sympathetic Division

1. Preganglionic cell bodies are in the lateral horns of the spinal cord gray matter from T1 to L2.
2. Preganglionic axons pass through the ventral root to the white rami communicantes to the sympathetic chain ganglia. From there three courses are possible:
   - Preganglionic axons synapse (at the same or a different level) with postganglionic neurons, which exit the ganglia through the gray rami communicantes and enter spinal nerves.
   - Preganglionic axons synapse (at the same or a different level) with postganglionic neurons, which exit the ganglia through sympathetic nerves.
   - Preganglionic axons pass through the chain ganglia without synapsing to form splanchnic nerves. Preganglionic axons then synapse with postganglionic neurons in collateral ganglia. In the case of the adrenal gland, the preganglionic axons synapse with the adrenal medulla.

Parasympathetic Division

1. Preganglionic cell bodies are in nuclei in the brainstem or the lateral horns of the spinal cord gray matter from S2 to S4.
   - Preganglionic axons from the brain pass to ganglia through cranial nerves III, VII, IX, and X.
   - Preganglionic axons from the sacral region pass through ventral roots of the pelvic nerves to the ganglia.
2. Preganglionic axons pass to terminal ganglia within the wall of or near the organ that is innervated.

NEUROTRANSMITTER SUBSTANCES AND RECEPTORS p. 511

1. Acetylcholine is released by cholinergic neurons (all preganglionic, all parasympathetic, and some sympathetic postganglionic neurons). Norepinephrine is released by adrenergic neurons (most sympathetic postganglionic neurons).
2. Acetylcholine binds to nicotinic receptors (found in all postganglionic neurons) and muscarinic receptors (found in all parasympathetic and some sympathetic effector organs). Norepinephrine binds to alpha and beta receptors (found in most sympathetic effector organs).
3. Activation of nicotinic receptors is excitatory, whereas activation of the other receptors may be excitatory or inhibitory.

REGULATION OF THE AUTONOMIC NERVOUS SYSTEM p. 514

1. Autonomic reflexes control most of the activity of visceral organs, glands, and blood vessels.
2. Autonomic reflex activity can be influenced by the hypothalamus and higher brain centers.

FUNCTIONAL GENERALIZATIONS ABOUT THE AUTONOMIC NERVOUS SYSTEM p. 516

1. Both divisions of the ANS produce stimulatory and inhibitory effects.
2. Most organs are innervated by both divisions. Usually each division produces an opposite effect on a given organ.
3. Either division alone or both working together can coordinate the activities of different structures.
4. The sympathetic division produces more generalized effects than the parasympathetic division.
5. Sympathetic activity generally prepares the body for physical activity, whereas parasympathetic activity is more important for vegetative functions.

CONTENT REVIEW

1. Define preganglionic neuron, postganglionic neuron, autonomic ganglia, and effector organ.
2. Contrast the somatomotor nervous system and the autonomic nervous system (ANS) for each of the following:
   - **A** The number of neurons between the central nervous system (CNS) and the effector organ
   - **B** The location of neuron cell bodies
   - **C** The structures each innervates
   - **D** Inhibitory or excitatory effects
   - **E** Conscious or unconscious control
3. Contrast the sympathetic and parasympathetic with regard to the following:
   - **A** Location of preganglionic cell bodies
   - **B** Location of ganglia
   - **C** Length of preganglionic and postganglionic axon
   - **D** Number of preganglionic and postganglionic neurons
4. Describe three ways in which efferent neurons of the sympathetic system exit the CNS and extend to effector organs. Describe two ways for the parasympathetic system.
5. Why is the adrenal medulla considered a part of the sympathetic nervous system? What substances does it release, and what effects do these substances have? Are these substances neurotransmitters or hormones?
6. What kinds of neurons (preganglionic or postganglionic, myelinated or unmyelinated) are found in the following:
   - **A** Cranial nerves
   - **B** Spinal nerves
   - **C** Sympathetic nerves
   - **D** Splanchnic nerves
   - **E** Pelvic nerves
7. Describe three types (locations) of autonomic ganglia.
8. Generally speaking, what structures are innervated by autonomic fibers in cranial nerves, spinal nerves, sympathetic nerves, splanchnic nerves, and pelvic nerves?
9. What neurotransmitter is released by cholinergic neurons? Which neurons of the ANS are cholinergic?
10. What neurotransmitter is released by adrenergic neurons? Which neurons of the ANS are adrenergic?
11. With what type of receptors does acetylcholine bind? Where are these receptors found? Is the effect excitatory or inhibitory?
12. With what type of receptors does norepinephrine bind? Where are these receptors found? Is the effect excitatory or inhibitory?
13. Name the components of an autonomic reflex. Describe the autonomic reflex that maintains blood pressure by altering heart rate or the diameter of blood vessels.
14. In what area of the CNS are autonomic reflexes integrated? What role do other areas of the CNS have in modifying autonomic reflexes resulting from emotions or stress?
15. Most organs of the body are innervated by both divisions of the ANS. Give an exception.
16. For organs innervated by both the parasympathetic and sympathetic systems, list four organs for which the parasympathetic system has an excitatory effect and four organs for which the parasympathetic system has an inhibitory effect. For each set of organs listed, describe the effect of the sympathetic system on that set of organs. What conclusions can be made about the following?
    - **A** The ability of the parasympathetic or sympathetic system to have an excitatory or inhibitory effect
    - **B** The effect of the parasympathetic and sympathetic systems on an organ they both innervate
17. To help you remember whether or not the sympathetic or parasympathetic system has an excitatory or inhibitory effect on a particular organ, what generalization is useful?

CONCEPT REVIEW

1. When a person is startled or when he sees a "pleasurable" object, the pupils of the eyes may dilate. What division of the autonomic nervous system (ANS) is involved in this reaction? Describe the nerve pathway involved.
2. Reduced secretion from salivary and lacrimal glands could indicate damage to what nerve?
3. Johnny Uptight has a peptic ulcer that is treated surgically by resectioning the stomach (removal of the ulcerated area). At the same time, which of the following nerves should be cut to reduce stomach acid production: splanchnic, pelvic, or vagus?
4. In a patient with Raynaud's disease blood vessels in the skin of the hand can become chronically constricted, reducing blood flow and producing gangrene. These vessels are supplied by nerves that originate at levels T2 and T3 of the spinal cord and eventually exit through the first thoracic and inferior cervical sympathetic ganglia. Surgical treatment for Raynaud's disease severs this nerve supply. At which of the following locations would you recommend that the cut be made: white rami of T2 to T3, gray rami of T2 to T3, spinal nerves T2 to T3, or spinal nerves C1 to T1? Explain.
5. Patients with diabetes mellitus can develop autonomic neuropathy (i.e., damage to portions of the autonomic nerves). Given the following parts of the ANS—vagus nerve, splanchnic nerve, pelvic nerve, cranial nerve, outflow of gray ramus—match the part with the symptom it would produce if the part were damaged:
   A Impotence
   B Subnormal sweat production
   C Gastric atony and delayed emptying of the stomach
   D Diminished pupil reaction (constriction) to light
   E Bladder paralysis with urinary retention
6. Explain why methacholine, a drug that acts like acetylcholine, would be effective for treating tachycardia (heart rate faster than normal). Which of the following side effects would you predict: increased salivation, dilation of the pupils, sweating, or difficulty in breathing?
7. A patient has been exposed to the organophosphate malathion, which inactivates acetylcholinesterase. Which of the following symptoms would you predict: blurring of vision, excess tear formation, frequent or involuntary urination, pallor (pale skin), muscle twitching, or cramps? Would atropine be an effective drug to treat the symptoms? Explain.
8. Epinephrine or norepinephrine routinely is mixed with local anesthetic solutions. Why?
9. A drug blocks the effect of the sympathetic system on the heart. Careful investigation reveals that after administration of the drug, normal action potentials are produced in the sympathetic preganglionic and postganglionic neurons. Also, injection of norepinephrine produces a normal response in the heart. Explain in as many ways as you can the mode of action of the unknown drug.
10. Ergot alkaloids (basic substances derived from a fungus) cause a decrease in heart rate. After sympathectomy (cutting the white rami of T1 to T4) the drug still causes heart rate to decline. However, after vagotomy (cutting the vagal nerves) the drug no longer affects heart rate. Which division of the ANS does the drug affect, where does the drug have its effect (synapse between preganglionic and postganglionic neurons, synapse between postganglionic neurons and effector organs, or central nervous system), and is the effect excitatory or inhibitory?

? ?

ANSWERS TO PREDICT QUESTIONS

1 *p. 512*. Nicotinic receptors are located within the autonomic ganglia as components of the membranes of the postganglionic neurons of the sympathetic and parasympathetic divisions. Nicotine binds to the nicotinic receptors of the postganglionic neurons, resulting in action potentials. Consequently, the postganglionic neurons stimulate their effector organs. After the consumption of nicotine, structures innervated by both the sympathetic and parasympathetic divisions would be stimulated.

After the consumption of muscarine only the effector organs that respond to acetylcholine are affected. Because all of the postganglionic neurons of the parasympathetic division release acetylcholine at their synapses with effector organs, the structures innervated by the parasympathetic division would be affected to the greatest degree by muscarine. However, some of the postganglionic cells of the sympathetic division also release acetylcholine; therefore some structures innervated by the sympathetic division would be affected by the muscarine.

2 *p. 514*. The frequency of action potentials in sympathetic neurons to the sweat glands increases as the body temperature increases. The increasing body temperature is detected by the hypothalamus, which activates the sympathetic neurons. Sweating functions to cool the body by evaporation. As the body temperature declines, the frequency of action potentials in sympathetic neurons to the sweat glands decreases. During exposure to cold temperatures, a lack of sweating helps prevent heat loss from the body.

3 *p. 514*. In response to a sudden decrease in blood pressure, stretch receptors in major blood vessels detect the decrease in blood pressure. A decreased frequency of action potentials is sent along the vagus nerves to the medulla oblongata where the information is integrated. Consequently, there is a decrease in the frequency of action potentials delivered to the heart along the parasympathetic neurons of the vagus nerves. Because these action potentials inhibit the heart, less inhibition results in an increase in the heart rate.

In response to a sudden increase in blood pressure, the stretch receptors within the wall of major blood vessels detect the change. As a result, an increased frequency of action potentials is produced in nerves such as the vagus and is carried to the medulla oblongata. The cardioregulatory center within the medulla responds by decreasing the frequency of action potentials delivered to the heart along sympathetic neurons. Because these action potentials are stimulatory to the heart, less stimulation results in a decrease in the heart rate.

In response to an increase in blood pressure, information is transmitted in the form of action potentials along afferent neurons to the medulla oblongata. Within the medulla oblongata the frequency of action potentials delivered along sympathetic nerve fibers to blood vessels decreases. As a result, blood vessels dilate, causing the blood pressure to decrease.

4 *p. 520*. (a) Responses in a person who is extremely angry are primarily controlled by the sympathetic division of the autonomic nervous system. These responses include increased heart rate and blood pressure, decreased blood flow to the internal organs, increased blood flow to skeletal muscles, decreased contractions of the intestinal smooth muscle, flushed skin in the face and neck region, and dilation of the pupils of the eyes. (b) For a person who has just finished eating and is relaxing, the parasympathetic reflexes are more important than sympathetic reflexes. The blood pressure and heart rate are low, the blood flow to the internal organs is greater, contractions of smooth muscle in the intestines are greater, and secretions that achieve digestion are more active. If the urinary bladder or the colon becomes distended, autonomic reflexes that result in urination or defecation result. Blood flow to the skeletal muscles is reduced.

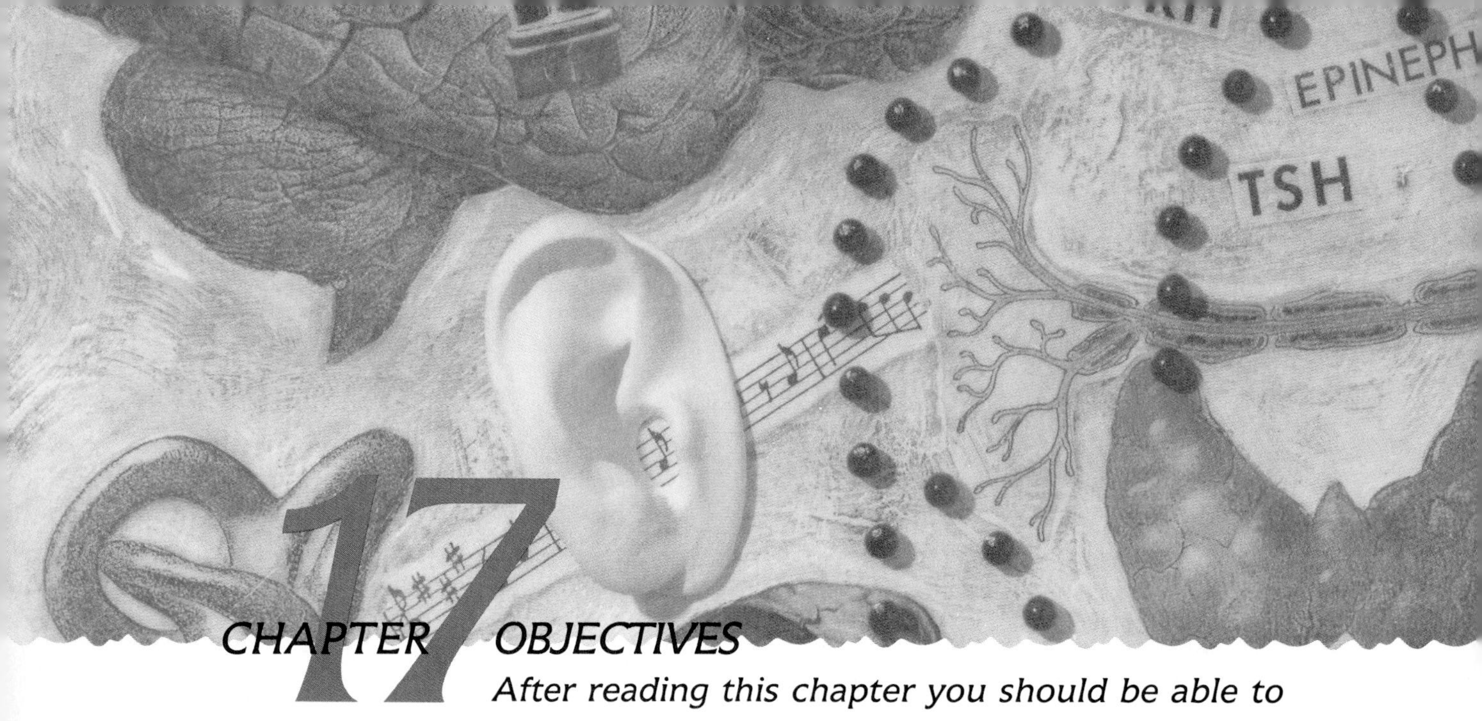

CHAPTER 17 OBJECTIVES

After reading this chapter you should be able to

1. Define endocrine gland, hormone, and endocrine system.
2. Explain why a simple definition for hormone is difficult to create.
3. Explain why the endocrine system is primarily an amplitude-modulated system.
4. Describe the functional relationship between the nervous system and the endocrine system.
5. Explain how the regulation of hormone secretion is achieved.
6. Define half-life and explain how the combination of hormones with plasma proteins affects their half-life.
7. Describe the means by which hormones are metabolized and excreted.
8. Explain how the sensitivity of target tissues to hormones can change.
9. Compare the relationship between hormones and their receptor molecules for membrane-bound and intracellular receptors.
10. List the responses that can occur following the combination of hormones with membrane-bound receptors.
11. Using diagrams, explain the second-messenger model of hormone action and describe the characteristics of the responses that are produced.

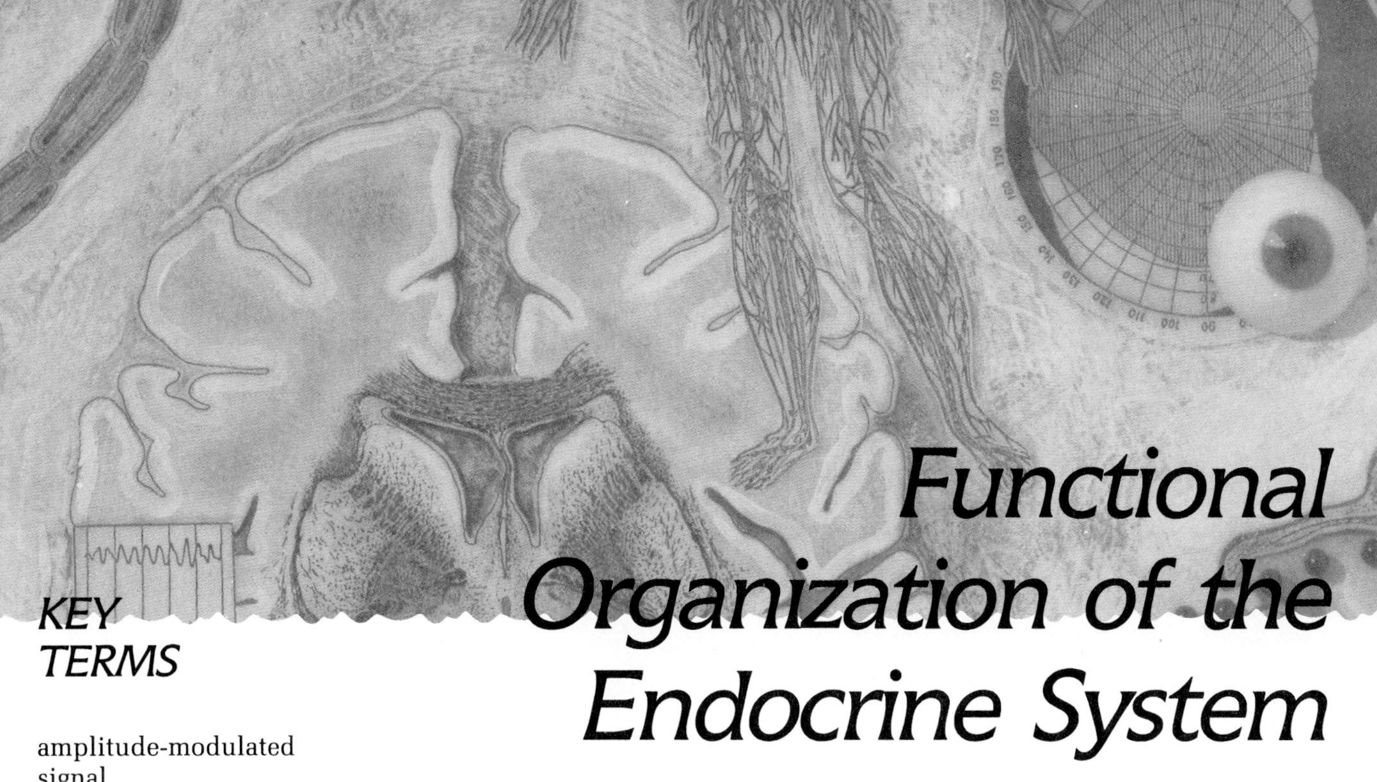

Functional Organization of the Endocrine System

KEY TERMS

amplitude-modulated signal

down regulation

endocrine

first messenger

frequency-modulated signal

half-life

hormone

intercellular chemical messenger

intracellular receptors

membrane-bound receptor

neurohormone

neuromodulator

second messenger

target tissue

up regulation

RELATED TOPICS

The following terms or concepts are important for a good understanding of this chapter. If you are not familiar with them, you should review them before proceeding.

Negative and positive feedback (Chapter 1)

Cellular organelles (Chapter 3)

Receptors (Chapter 3)

THE NERVOUS SYSTEM AND THE ENDOCRINE SYSTEM ARE THE TWO MAJOR REGULATORY SYSTEMS OF THE BODY, AND TOGETHER THEY REGULATE AND COORDINATE THE ACTIVITY OF ESSENTIALLY ALL OTHER BODY STRUCTURES. The basic features of the nervous system have been presented in Chapters 12 through 16. This chapter introduces the general characteristics of the endocrine system. It emphasizes the endocrine system's role in the maintenance of homeostasis and the means by which the endocrine system regulates body functions. The structure and function of each endocrine gland and its secretory products and the means by which its activity is regulated are described in Chapter 18.

GENERAL CHARACTERISTICS OF THE ENDOCRINE SYSTEM

The term **endocrine** is derived from the Greek words *endo*, meaning within, and *crino*, to separate. The term implies that endocrine glands secrete their products internally and influence tissues that are separated by some distance from the endocrine glands. The **endocrine system** is composed of glands that secrete their products into the circulatory system (Figure 17-1). The secretory products of endocrine glands are **hormones,** a term derived from the Greek word *hormon* meaning to set into motion. Traditionally a hormone is defined as a substance that (1) is produced in minute amounts by a collection of cells, (2) is secreted into the interstitial spaces, (3) enters the circulatory system in which it is transported some distance, and (4) acts on specific tissues called **target tissues** at another site in the body to influence the tissues' activity in a specific fashion. All hormones exhibit most components of this definition, but some components of this definition do not apply to every hormone.

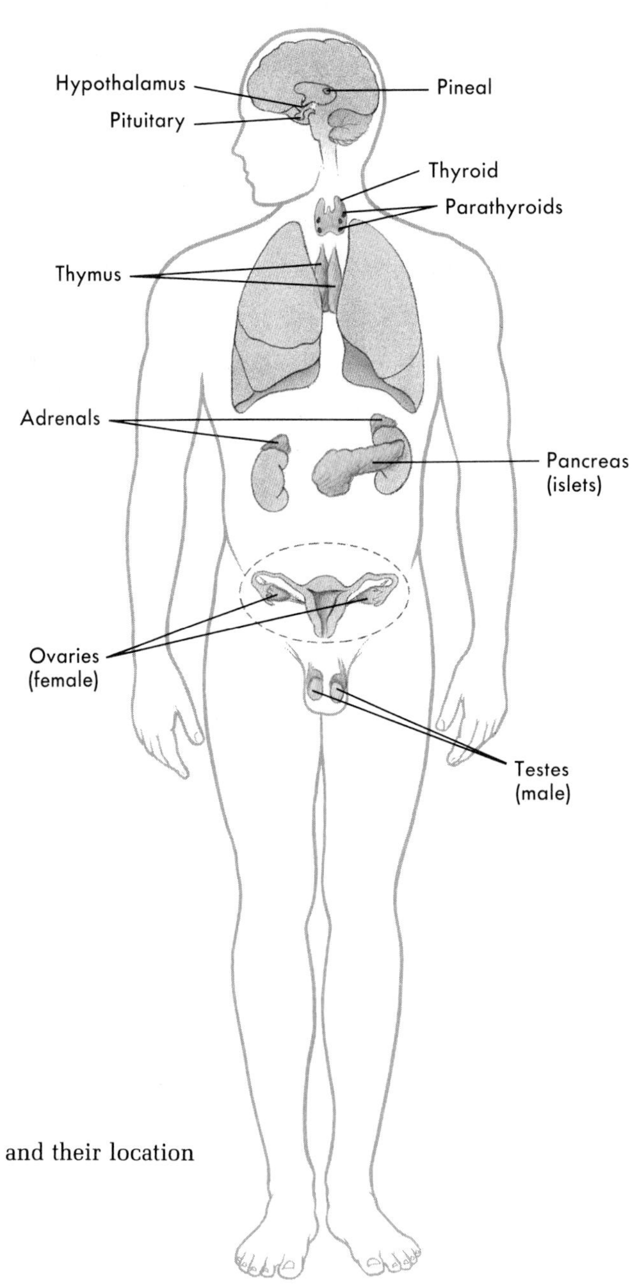

FIGURE 17-1 Endocrine glands and their location in the human body.

Although both the endocrine system and the nervous system regulate the activities of structures in the body, they do so in different ways. For example, most endocrine glands communicate with their target tissues using **amplitude-modulated signals,** which consist of increases or decreases in the concentration of a hormone in the body fluids (Figure 17-2, *A*). The effects produced by the hormones either increase or decrease responses as a function of the hormone concentration. On the other hand, neurons send **frequency-modulated signals** in the form of all-or-none action potentials (Figure 17-2, *B*), which vary in frequency but not in amplitude. Weak signals are represented by a lower frequency of action potentials, whereas strong signals are represented by a higher frequency of action potentials (see Chapter 9). The responses of the endocrine system are usually slower and of longer duration and can have effects that are more generally distributed than those of the nervous system.

Although the stated differences between the endocrine and nervous systems are generally true, exceptions do exist (e.g., some endocrine responses are more rapid than some neural responses, and some endocrine responses have a shorter duration than some neural responses). In fact, recent evidence suggests that some hormones may act as both amplitude- and frequency-modulated signals.

At one time the endocrine system was believed to be relatively independent and different from the nervous system, but an intimate relationship between these systems is now recognized. In fact, the two systems cannot be separated completely either anatomically or functionally. Some neurons secrete into the circulatory system regulatory chemicals called **neurohormones,** which function like hormones. Other neurons directly innervate endocrine glands and influence their secretory activity. Conversely, some hormones secreted by endocrine glands affect the nervous system and markedly influence its activity.

Several types of chemicals are produced by cells and act as chemical messengers, but not all of them are hormones. **Intercellular chemical messengers** act as signals that allow one cell type to communicate with other cell types. The signals coordinate and regulate the activities of the many cells that comprise the body. Terms such as hormones, neurohormones, neurotransmitters or neurohumors, neuromodulators, parahormones, and pheromones are used to classify these substances (Table 17-1). Although many intercellular chemical messengers consistently fit one specific definition, others do not. For example, norepinephrine functions both as a neurotransmitter substance and as a neurohormone, and prostaglandins function as neurotransmitters, neuromodulators, and parahormones. Therefore the schemes used to classify chemicals based on their functions are useful but are not without ambiguities and exceptions.

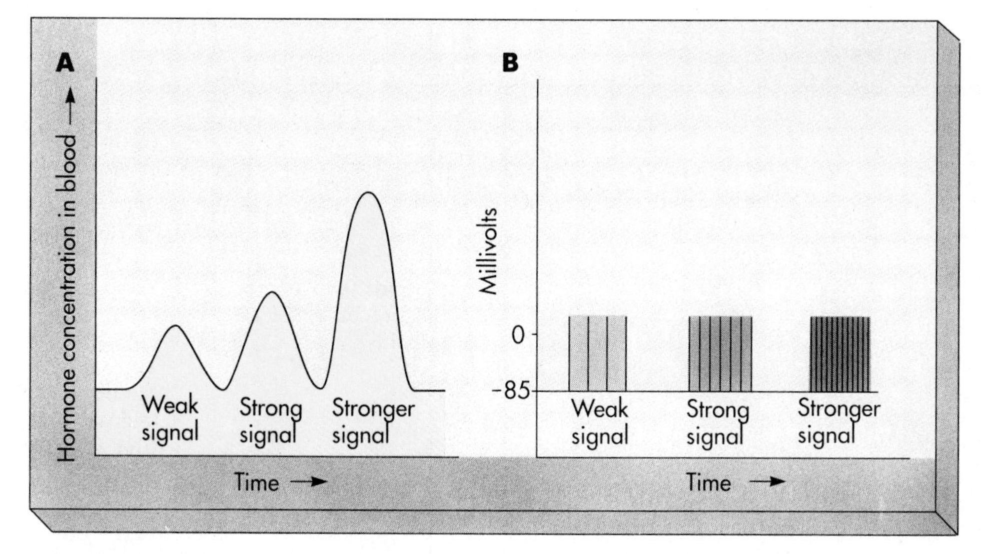

FIGURE 17-2 Regulatory systems. **A** Amplitude-modulated system. Hormones secreted into the circulatory system constitute signals, and their concentration determines the strength of the signal and the magnitude of the response. A small concentration represents a weak signal and produces a small response, whereas a larger concentration represents a stronger signal and results in a greater response. **B** Frequency-modulated system. The strength of the signal depends on the frequency, not the size, of the action potentials.

Table 17-1

Chemical messengers

INTERCELLULAR CHEMICAL MESSENGER	DESCRIPTION	EXAMPLE
Hormone	Secreted into the blood by specialized cells; travels some distance to target tissues; influences specific activities	Thyroxine
Neurohormone	Produced by neurons and functions like hormones	Oxytocin, antidiuretic hormone
Neurotransmitter or neurohumor	Produced by neurons and secreted into extracellular spaces by presynaptic nerve terminals; travels short distances; influences postsynaptic neurons	Acetylcholine, norepinephrine
Neuromodulator	Produced by neurons and released into the extracellular space; influences postsynaptic neurons; alters the sensitivity of postsynaptic neurons to neurotransmitters	Prostaglandins, endorphins
Parahormone	Produced by wide variety of tissues and secreted into tissue spaces; usually has a localized effect	Histamine, prostaglandins
Pheromone	Secreted into the environment; modifies physiology and behavior of other individuals	Sex pheromones released in the urine of many animals (e.g., dogs and cats)

Table 17-2

Structural categories of hormones

STRUCTURAL CATEGORY	EXAMPLES	STRUCTURAL CATEGORY	EXAMPLES
Proteins	Growth hormone Prolactin Insulin Parathyroid hormone	Amino acid derivatives	Epinephrine Norepinephrine Thyroxine (both T_4 and T_3) Melatonin
Glycoproteins	Follicle-stimulating hormone Luteinizing hormone Thyroid-stimulating hormone	Lipids	
Polypeptides	Thyrotropin-releasing hormone Oxytocin Antidiuretic hormone Calcitonin Glucagon Adrenocorticotropic hormone Endorphins Thymosin Melanocyte-stimulating hormone Hypothalamic hormones Lipotropins Somatostatin	Steroids (cholesterol is a precursor for all steroids)	Estrogens Progestins (progesterone) Testosterone Mineralocorticoids (aldosterone) Glucocorticoids (cortisol)
		Fatty acids	Prostaglandins Thromboxanes Prostacyclins Leukotrienes

CHEMICAL STRUCTURE OF HORMONES

Hormones, including neurohormones, are either proteins, short sequences of amino acids called polypeptides, derivatives of amino acids, or lipids. Some protein hormones are composed of more than one polypeptide chain and contain carbohydrate molecules as a component (glycoproteins). The lipid hormones are either steroids or derivatives of fatty acids (Table 17-2 and Figure 17-3).

CONTROL OF SECRETION RATE

Most hormones are not secreted at a constant rate; most endocrine glands increase and decrease their secretory activity dramatically over time. The specific mechanisms that regulate the secretion rates for each hormone are presented in Chapter 18, but the general patterns of regulation are introduced in this chapter. Negative-feedback mechanisms (see Chapter 1) play essential roles in maintaining hor-

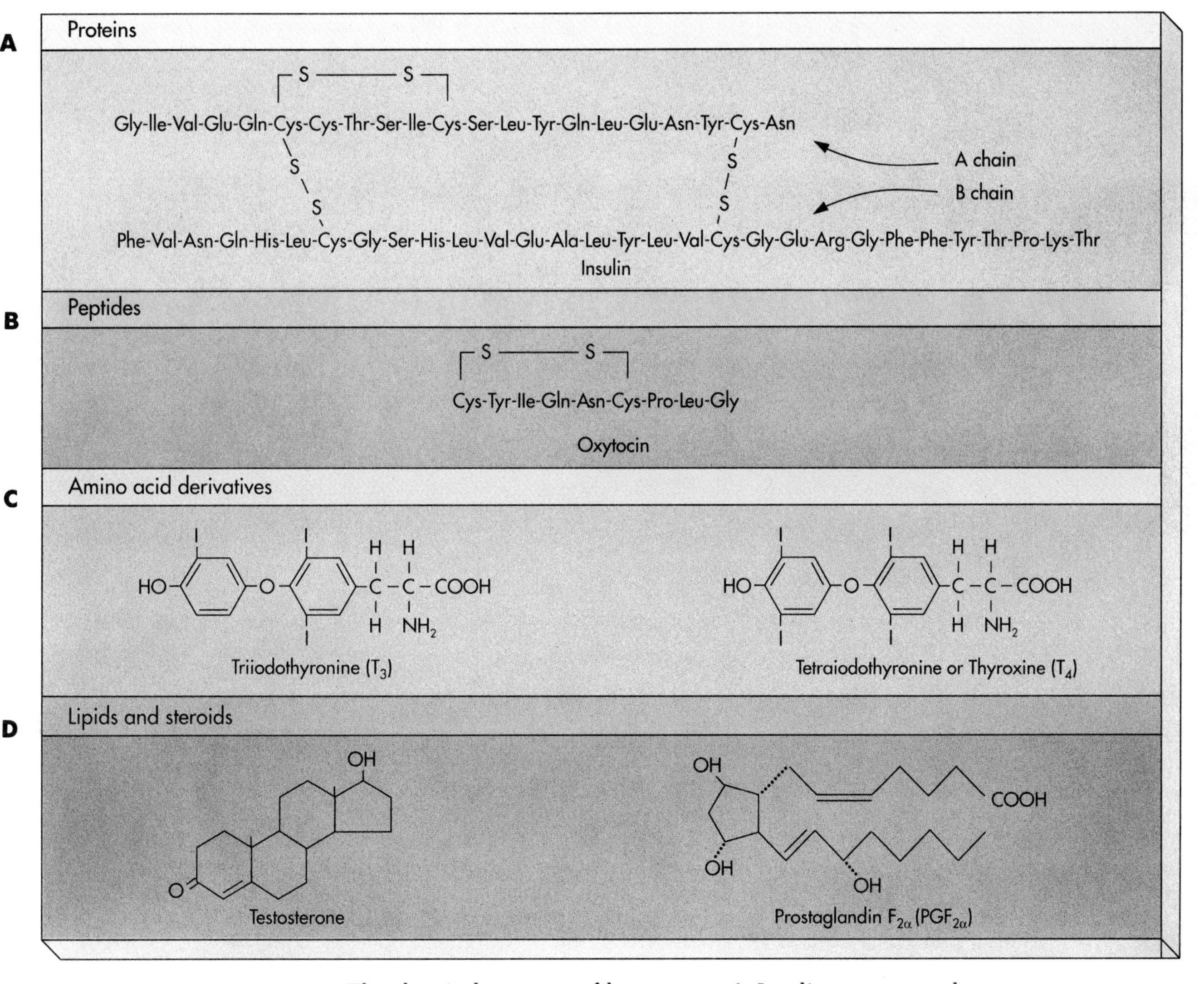

FIGURE 17-3 The chemical structure of hormones. **A** Insulin as an example of a protein hormone. **B** Oxytocin as an example of a peptide hormone. **C** Triiodothyronine and tetraiodothyronine as examples of modified amino acid hormones. **D** Testosterone and prostaglandin $F_{2\alpha}$ as examples of steroid and lipid hormones.

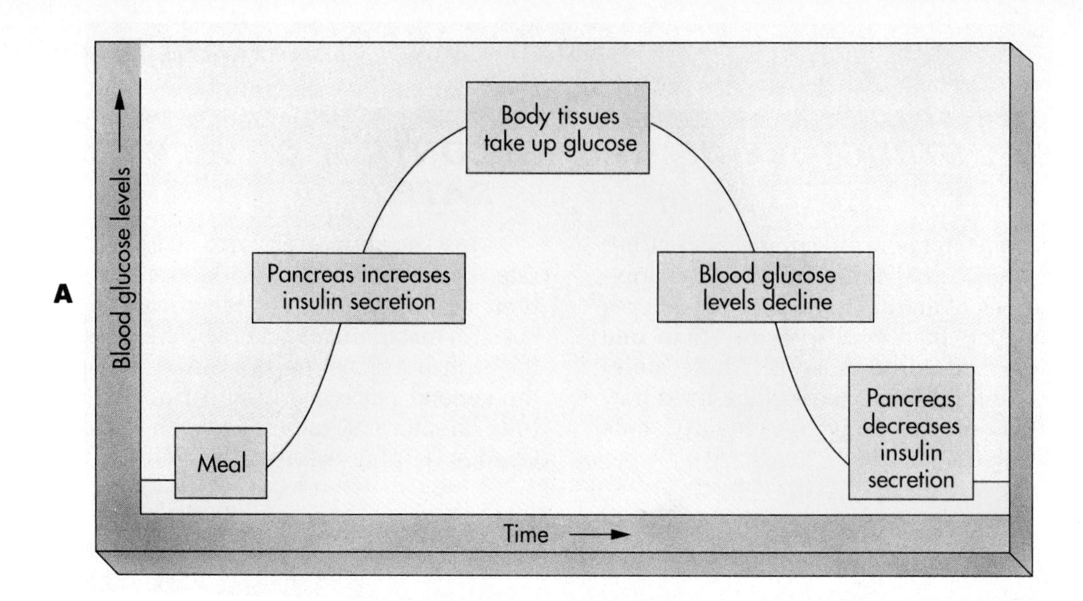

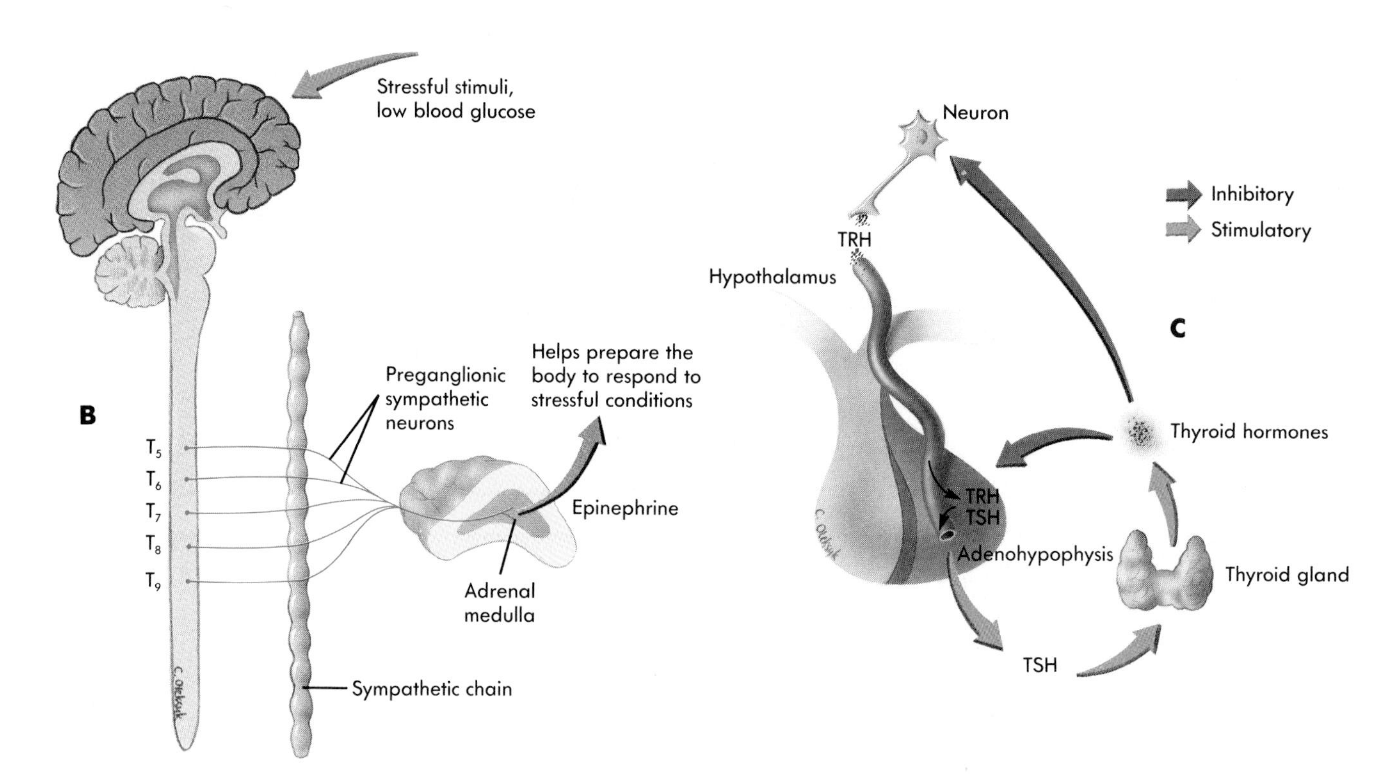

FIGURE 17-4 Control of hormone secretion.

A The concentration of glucose in the blood affects the rate of insulin secretion. After a meal increasing blood levels of glucose stimulate insulin secretion from the pancreas. Insulin acts on tissues to increase the rate at which they take up and use glucose. As glucose levels decline, the rate of insulin secretion also declines. When blood glucose levels are low, the resulting decrease in insulin secretion prevents the tissues from taking up too much glucose. This mechanism helps keep blood glucose levels within a normal range of values in the circulatory system.

B Epinephrine is released from the adrenal medulla in response to sympathetic stimulation. Stimuli such as stress activate the sympathetic division of the autonomic nervous system. Sympathetic neurons in turn stimulate the release from the adrenal medulla of epinephrine, which helps prepare the body to respond to stressful conditions. Once the stressful stimuli are removed, less epinephrine is released as a result of decreased stimulation from the autonomic nervous system.

C Hormones such as thyroid-releasing hormone *(TRH)* are released from neurons in the hypothalamus and in turn stimulate the release of other hormones such as thyroid-stimulating hormone *(TSH)* from the adenohypophysis (anterior pituitary). TSH then stimulates the secretion of thyroid hormones from the thyroid gland *(green arrows)*. These hormones act on tissues to produce the usual response to thyroid hormones and act on the hypothalamus and the adenohypophysis to inhibit both TRH secretion and TSH secretion *(red arrows)*. The thyroid hormones have a negative-feedback effect on their own secretion rate through this mechanism. If the thyroid hormones decrease in the circulatory system, the rate of both TRH and TSH secretion increases. Consequently, thyroid hormone concentrations fluctuate within a normal range of values.

mone levels within normal concentration ranges.

There are three major patterns of regulation for hormones. One method involves the action of a substance other than a hormone on the endocrine gland (Figure 17-4, *A*).

A second pattern of hormone regulation involves neural control of the endocrine gland (Figure 17-4, *B*). Neurons synapse with the cells that produce the hormone, and when action potentials result, the neurons release a neurotransmitter. The neurotransmitter in some cases is stimulatory and causes the cells to increase hormone secretion; in other cases the neurotransmitter is inhibitory and decreases hormone secretion. Thus sensory input and emotions acting through the nervous system can influence hormone secretion.

Nervous stimulation of the pancreatic islets can either increase or decrease insulin secretion. When action potentials in parasympathetic neurons that innervate the pancreatic islets increase, acetylcholine is released as the neurotransmitter substance. Acetylcholine causes depolarization of the pancreatic islet cells, and insulin is secreted. When action potentials in sympathetic neurons that innervate the pancreatic islets increase, norepinephrine is released as a neurotransmitter substance. Norepinephrine causes hyperpolarization of the pancreatic islet cells, and insulin secretion decreases.

A third pattern of hormone regulation involves the control of the secretory activity of one endocrine gland by a hormone or a neurohormone secreted by another endocrine gland (Figure 17-4, *C*).

1

Assuming that negative feedback is the major means by which a hormone's secretion rate is controlled and that the hormone causes the concentration of a substance called X to increase in the blood, predict the effect on the rate of hormone secretion if some abnormal condition causes the levels of X in the blood to remain very low.

? ? ? ? ? ? ? ? ? ? ?

One of the three major patterns by which hormone secretion is regulated applies to each hormone, but the picture is not quite so simple. The regulation of hormone secretion often involves more than one mechanism. In some cases the nervous system and hormones from other endocrine glands regulate the rate of hormone secretion.

There are a few examples of positive-feedback regulation in the endocrine system; however, in each instance negative-feedback mechanisms limit the positive-feedback process (Figure 17-5). The role of

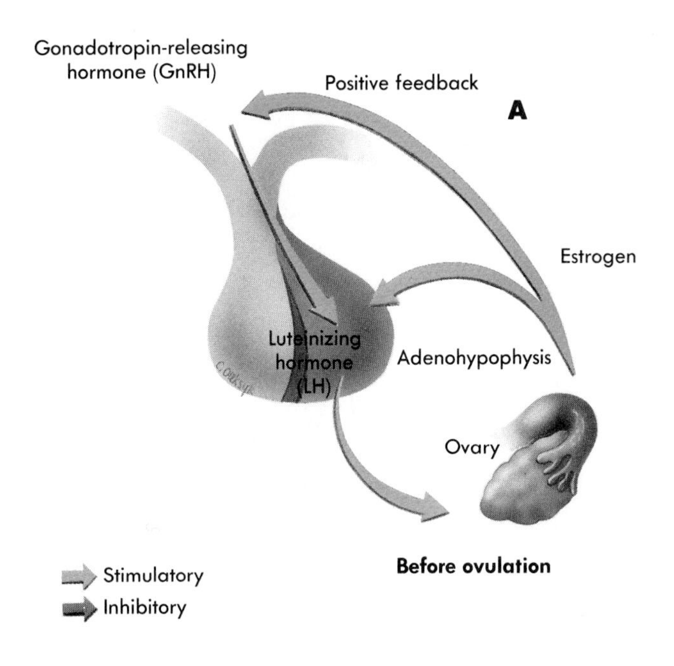

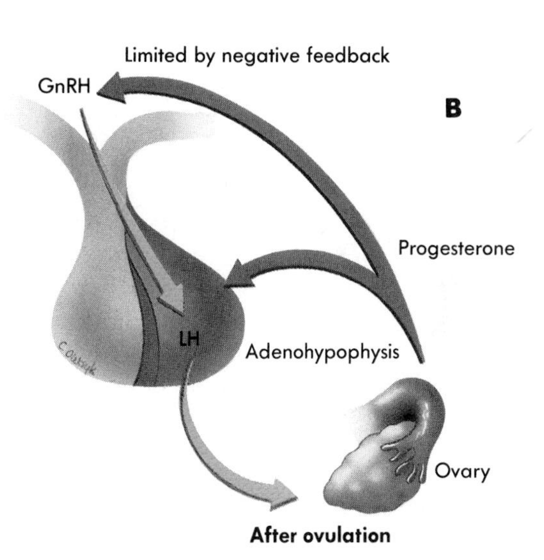

FIGURE 17-5 Positive and negative feedback.
A During the menstrual cycle of the human female small amounts of estrogen are secreted from the ovary and stimulate the release from the hypothalamus of gonadotropin-releasing hormone *(GnRH)*, a neurohormone. GnRH stimulates the release of luteinizing hormone (LH) from the adenohypophysis (anterior pituitary). LH causes the release of additional estrogen from the ovary, which stimulates additional GnRH release from the hypothalamus and also directly stimulates LH release from the adenohypophysis. Consequently, the blood levels of LH and estrogen increase because of this positive-feedback effect.
B Once ovulation occurs, the ovary begins to secrete progesterone, which has a negative-feedback effect on the release of GnRH from the hypothalamus and LH from the adenohypophysis. Therefore LH levels in the blood decrease and remain low until the next menstrual cycle.

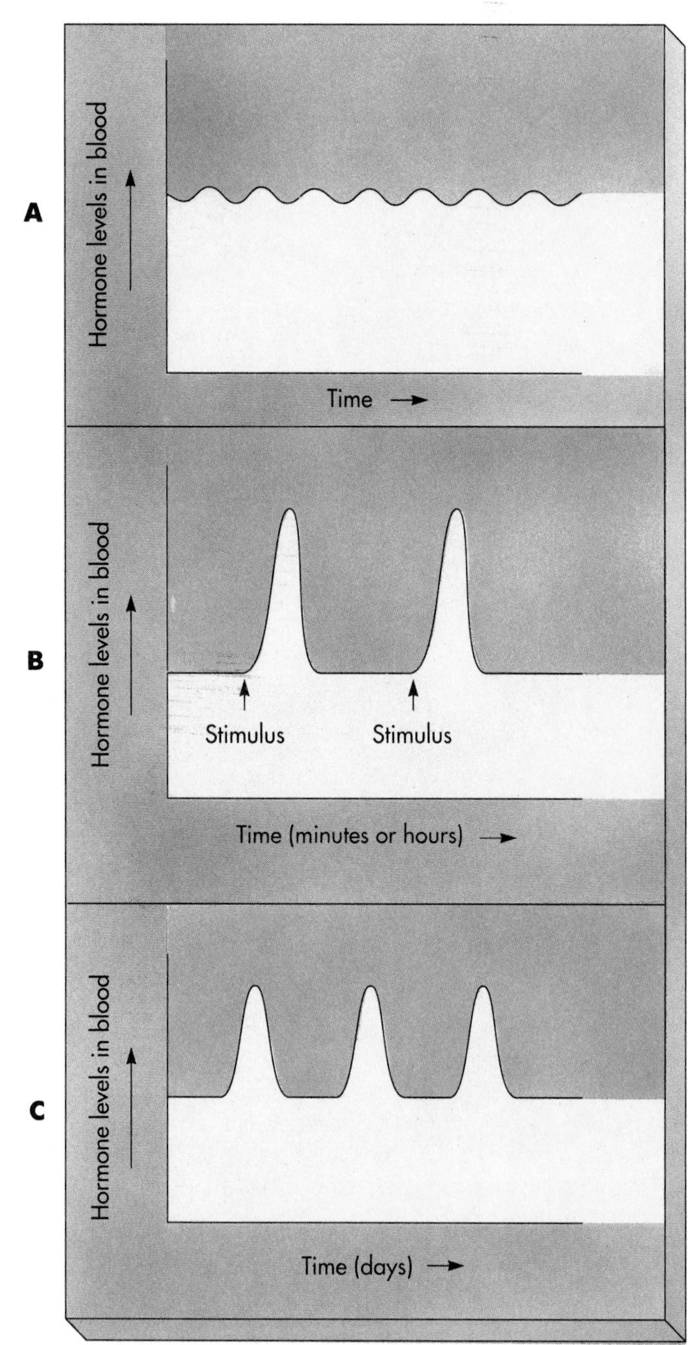

FIGURE 17-6 Changes in hormone secretion through time. At least three basic patterns of hormonal secretion exist. **A** Chronic hormone regulation—the maintenance of a relatively constant concentration of hormone in the circulating blood over a relatively long time period. **B** Acute hormone regulation—a hormone rapidly increases in the blood for a short time in response to a stimulus. **C** Cyclic hormone regulation—a hormone is regulated so that it increases in the blood at a relatively constant time and to roughly the same amount.

oxytocin in delivery (see Chapter 28) and the secretion of luteinizing hormone before ovulation are examples.

Some hormones are in the circulatory system at relatively constant levels; others change suddenly in response to certain stimuli; and others change in relatively constant cycles (Figure 17-6). For example, thyroid hormones in the blood vary within a small range of concentrations, so their concentration is regulated chronically. Epinephrine is released in large amounts in response to stress or physical exercise, so its concentration is regulated acutely. Reproductive hormones increase and decrease in a cyclic fashion in women during the reproductive years.

TRANSPORT AND DISTRIBUTION IN THE BODY

Hormones are dissolved in blood plasma and are transported either in a free form or bound to plasma proteins.

$$\underset{\text{Hormone}}{\text{H}} + \underset{\text{Binding protein}}{\text{BP}} \leftrightarrow \underset{\text{Hormone bound to binding protein}}{\text{HBP}}$$

Many hormones bind only to certain types of plasma proteins. For example, a specific plasma protein binds to thyroid hormones, and a different plasma protein binds to sex hormones such as testosterone. An equilibrium exists between the unbound hormone and the hormone bound to the plasma proteins. The equilibrium is important because only the free hormone is able to diffuse through capillary walls and bind to the target tissues. A large increase or decrease in plasma protein concentration can influence the concentration of free (unbound) hormone in the blood.

Because hormones circulate in the blood, they are distributed quickly throughout the body. They diffuse through the walls of capillaries and enter the interstitial spaces, although the rate at which this movement occurs varies from one hormone to the next. In general, the amount of hormone that reaches the target tissue is directly correlated with the concentration of the free hormone in the blood (Figure 17-7).

Lipid-soluble hormones readily diffuse through the walls of all capillaries. In contrast, water-soluble hormones such as proteins must pass through pores in the walls of capillaries. The capillaries of organs that are regulated by protein hormones have large pores in their walls.

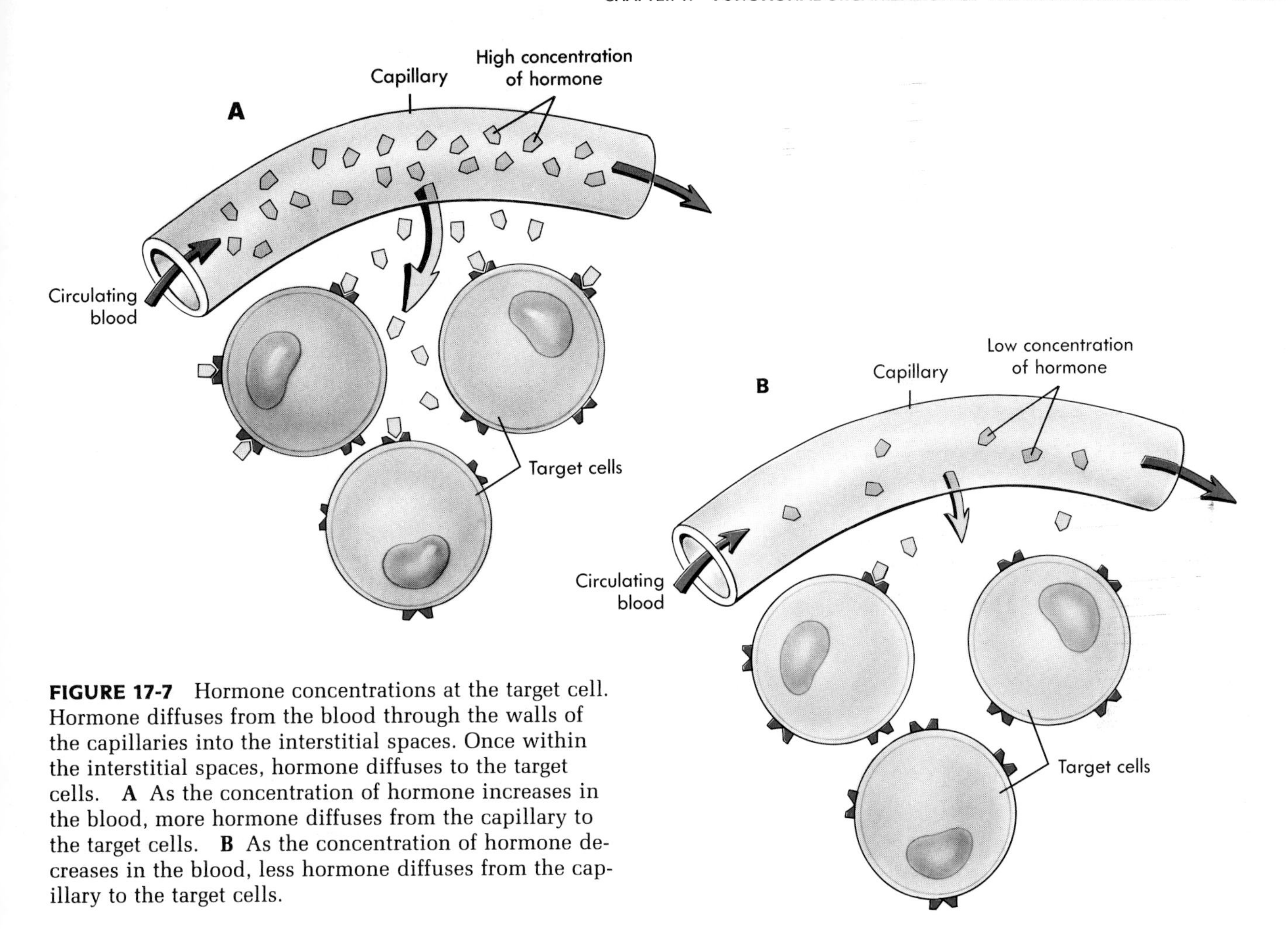

FIGURE 17-7 Hormone concentrations at the target cell. Hormone diffuses from the blood through the walls of the capillaries into the interstitial spaces. Once within the interstitial spaces, hormone diffuses to the target cells. **A** As the concentration of hormone increases in the blood, more hormone diffuses from the capillary to the target cells. **B** As the concentration of hormone decreases in the blood, less hormone diffuses from the capillary to the target cells.

METABOLISM AND EXCRETION

The destruction and elimination of hormones limit the length of time they are active, and regulation of body activities is more precise when hormones are secreted and remain active for only short time periods. The length of time it takes for elimination of half a dose of a substance from the circulatory system is called its **half-life.** The half-life of a hormone is a standard measurement used by endocrinologists because it allows them to predict the rate at which hormones are eliminated from the body. The length of time required for total removal of a hormone from the body is not as useful because that measurement is influenced dramatically by the starting concentration. Water-soluble hormones such as proteins, glycoproteins, epinephrine, and norepinephrine have relatively short half-lives because they are degraded rapidly by enzymes within the circulatory system or organs such as the kidneys, liver, or lungs.

Hormones with short half-lives normally have concentrations that increase and decrease rapidly within the blood. They generally regulate activities that have a rapid onset and a short duration.

Hormones that are lipid soluble such as steroids and thyroid hormones circulate in the blood in combination with the plasma proteins. The combination reduces the rate at which they diffuse through the wall of the blood vessels and increases their half-life. Hormones with a long half-life have blood levels that are maintained at a relatively constant level through time. Table 17-3 outlines the ways in which hormone half-life is shortened or lengthened.

Hormones are excreted by the kidney into the urine or are excreted by the liver into the bile. Some hormones are chemically modified by enzymes in the blood or in tissues such as the liver, kidney, lungs, or their target cells. The end products may be ex-

Table 17-3

Factors that influence the half-life of hormones

Means by which hormones are eliminated from the circulatory system

1. **Excretion**
   Hormones are excreted by the kidney into the urine or excreted by the liver into the bile.
2. **Metabolism**
   Hormones are enzymatically degraded in the blood, liver, kidney, lungs, or target tissues. End products of metabolism either are excreted in urine or bile or are used in other metabolic processes by cells in the body.
3. **Conjugation**
   Substances such as sulfate or glucuronic acid groups are attached to hormones primarily in the liver, normally making them less active as hormones and increasing the rate at which they are excreted in the urine or bile.
4. **Active transport**
   Some hormones are actively transported into cells and are used again as either hormones or neurotransmitter substances.

Means by which the half-life of hormones is prolonged

1. Some hormones are protected from rapid excretion or metabolism by binding reversibly with plasma proteins.
2. Some hormones are protected by their structure. The carbohydrate components of the glycoprotein hormones protect them from proteolytic enzymes in the circulatory system.

creted in the urine or bile, or they may be taken up by cells and used in metabolic processes. For example, epinephrine is modified enzymatically and then is excreted by the kidney. Protein hormones are broken down to their amino acid building blocks. The amino acids can then be taken up by cells and used to synthesize new proteins.

Some substances are conjugated by the liver. In the liver cells substances such as sulfate or glucuronic acid groups are attached to hormones. Once they are conjugated, hormones are excreted by the kidney and liver at a greater rate.

2

Explain the effect on a hormone's half-life if the concentration of the specific plasma protein to which that hormone binds decreases.

? ? ? ? ? ? ? ? ? ? ?

INTERACTION OF HORMONES WITH THEIR TARGET TISSUES

Hormones bind to receptors in their target tissues and alter the rate at which certain activities occur. Hormones do not cause cells to do new things, but they affect the rate at which target cells perform processes they can already do. Hormones may activate or inactivate enzymes that already exist in the cytoplasm of target cells, alter the rate at which specific molecules are synthesized within cells, or alter membrane permeability.

Hormone receptors are either protein or glycoprotein molecules that exist in specific three-dimensional shapes. Their unique shape and chemical composition allow receptors to be highly specific (i.e., each receptor type binds only to a single type of hormone or closely related substances). However, each hormone may bind to a number of different types of receptors. Hormones are secreted and distributed throughout the body by the circulatory system, but the presence or absence of specific receptor molecules in cells determines which cells will or will not respond to each hormone (Figure 17-8).

The response to a given concentration of a hormone is constant in some cases but variable in others. In some tissues the response rapidly decreases through time. Fatigue of the target tissues after prolonged stimulation explains some decreases in responsiveness. In many tissues the number of hormone receptors rapidly decreases after exposure to certain hormones—a phenomenon called **down regulation** (Figure 17-9, *A*). Two known mechanisms are responsible for the decrease in the number of receptor sites. First, the rate at which receptors are synthesized decreases in some tissues when the cells are exposed to a hormone; since most receptor molecules are degraded after a period of time, a decrease in synthesis rate reduces the total number of receptor molecules in a cell. Second, the combination of hormones and receptors may increase the rate at which receptor molecules are degraded.

Exposure of the pituitary gland to gonadotropin-releasing hormone (GnRH) causes the secretion of two hormones (luteinizing hormone and follicle-stimulating hormone). In addition, the number of receptor molecules in the pituitary gland dramatically decreases several hours after exposure to GnRH, causing the pituitary gland to become insensitive to additional GnRH. Therefore the normal response of the pituitary gland to GnRH depends on periodic exposure rather than constant exposure of the gland to the hormone.

Tissues that exhibit down regulation of receptor molecules are adapted to respond to short-term increases in hormone concentrations, and tissues that respond to hormones that are maintained at constant levels normally do not exhibit down regulation.

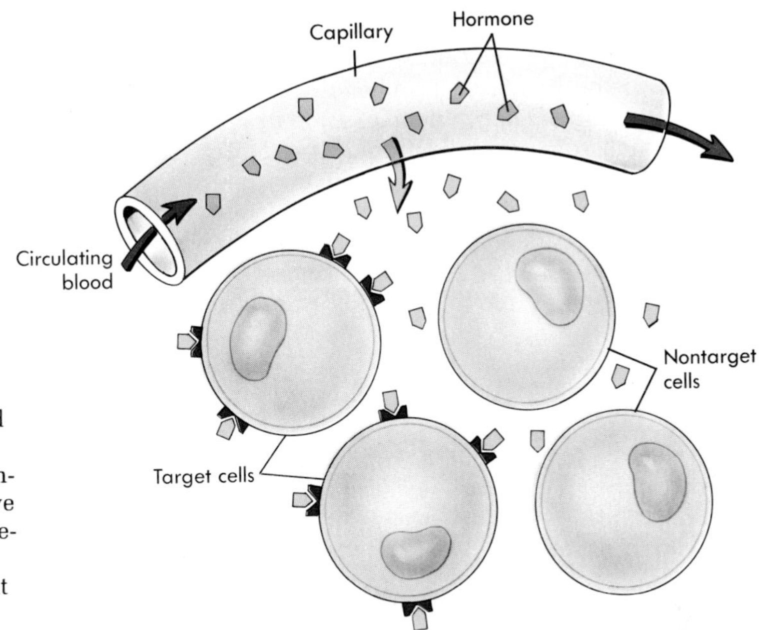

FIGURE 17-8 Target cell's response to hormones. Hormones are secreted into the blood and are distributed throughout the body in which they diffuse from the blood into the interstitial fluid. However, only target cells have receptors to which hormones can bind. Therefore, even though a hormone is distributed throughout the body, only target cells for that hormone can respond to it.

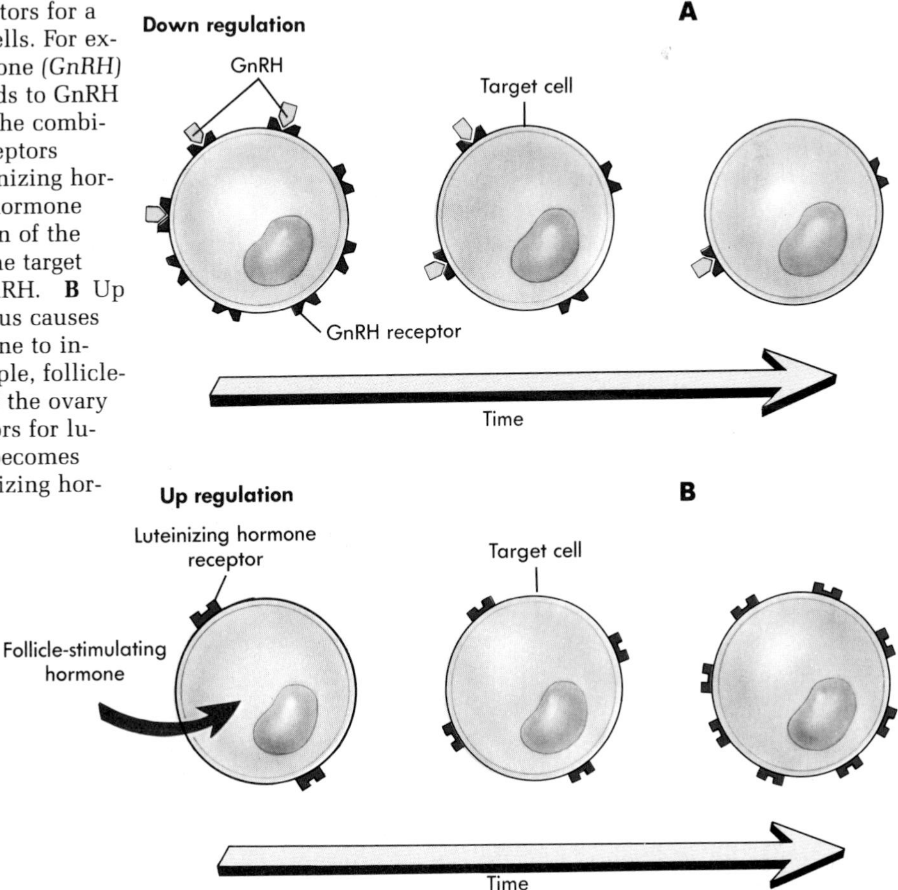

FIGURE 17-9 Down regulation and up regulation. **A** Down regulation occurs when some stimulus causes the number of receptors for a hormone to decrease within target cells. For example, gonadotropin-releasing hormone *(GnRH)* released from the hypothalamus binds to GnRH receptors in the adenohypophysis. The combination of the GnRH bound to its receptors causes the target cell to secrete luteinizing hormone (LH) and follicle-stimulating hormone (FSH); it also causes down regulation of the GnRH receptors so that eventually the target cells become less sensitive to the GnRH. **B** Up regulation occurs when some stimulus causes the number of receptors for a hormone to increase within a target cell. For example, follicle-stimulating hormone acts on cells of the ovary to up regulate the number of receptors for luteinizing hormone. Thus the ovary becomes more sensitive to the effect of luteinizing hormone.

In addition to down regulation, periodic increases in the sensitivity of some tissues to certain hormones also occur. This is called **up regulation,** and it results from an increase in the rate of receptor molecule synthesis (Figure 17-9, *B*). An example of up regulation is the increased number of receptor molecules for luteinizing hormone in ovarian tissues during each menstrual cycle. Follicle-stimulating hormone secreted by the pituitary gland increases the rate of luteinizing hormone receptor molecule synthesis. Thus exposure of a tissue to one hormone can increase its sensitivity to a second hormone by causing up regulation in the number of hormone receptor sites.

3

Consider that hormone A increases the concentration of receptors for hormone B in the target tissues for hormone B, and hormone A is secreted in large amounts before hormone B is secreted. Predict how the sensitivity of the target tissue for hormone B will change if hormone A is secreted in smaller-than-normal amounts.

? ? ? ? ? ? ? ? ? ? ?

CLASSES OF HORMONE RECEPTORS

There are two major classes of hormone receptors: **membrane-bound receptors** and **intracellular receptors.** Hormones that are water soluble or have a large molecular weight (and therefore cannot pass through the cell membrane) bind to membrane-bound receptor molecules. These hormones include proteins, glycoproteins, polypeptides, epinephrine, and norepinephrine. On the other hand, lipid-soluble hormones, including steroids and thyroid hormones, readily diffuse through plasma membranes, enter the cytoplasm, and bind to intracellular receptor molecules. In each case the combination of hormones with their receptors initiates a series of events that results in characteristic responses.

Membrane-Bound Receptors and the Second-Messenger Model

Hormones bind in a reversible fashion to that portion of membrane-bound receptor molecules exposed to the extracellular fluid. As a consequence, an equilibrium exists so that when the concentration of hormone at the target cell is high, the hormone receptors bind to the receptor molecules and when the concentration of the hormone declines, hormone molecules diffuse away from their receptors.

The combination of a hormone and its receptor molecule initiates a response. The occupied receptor molecule may cause membrane channels to open or close, resulting in a permeability change (Figure 17-10, *A*), which may lead to depolarization, hyperpolarization, or an influx of calcium ions, depending on the specific cell and the receptor molecule. For example, the combination of epinephrine with its epinephrine receptor molecules causes an increase in the permeability of the smooth-muscle cells in certain blood vessels to sodium and calcium ions. The influx of these ions initiates contraction of the smooth-muscle cells and, as a result, constriction of the blood vessels.

The combination of a hormone with its membrane-bound receptor molecule also may lead to the activation of an enzyme on the inner surface of the cell membrane. Within the cell the activated enzyme catalyzes the synthesis of a chemical that diffuses throughout the cytoplasm, binds to specific enzymes, and alters their activity. The hormone acts as the **first messenger,** which carries a signal to the cell membrane. The chemical produced at the membrane carries a signal from the cell membrane to intracellular structures and appropriately is called the **second messenger** in the sequence (Figure 17-10, *B*). Enzymes activated by the second messenger catalyze reactions that produce the characteristic response of the target cell to its hormone. For example, epinephrine binds to adrenergic receptors within the membrane of liver cells. The occupied receptors activate an enzyme on the inner surface of the cell membrane called adenylate cyclase, which catalyzes the following reaction:

$$\text{ATP} \xrightarrow[\text{cyclase}]{\text{Adenylate}} \text{Cylic AMP} + \text{PPi (two phosphates)}$$

The cyclic AMP (cAMP) functions as a second messenger. From the cell membrane cAMP diffuses throughout the cell's cytoplasm and binds to and activates an enzyme called protein kinase, which interacts with other enzymes and alters their activity. The final result is the breakdown by liver enzymes of glycogen to individual glucose molecules, which are released into the circulatory system. Finally, an enzyme called **phosphodiesterase** within the cell breaks down the cAMP molecules, and as cAMP levels decrease, enzyme activity returns to its previous rate. This process limits the length of time cAMP influences activities within cells.

4

Predict the effect of an inhibitor of the enzyme phosphodiesterase on a tissue's response to a hormone that has cAMP as a second messenger.

? ? ? ? ? ? ? ? ? ? ?

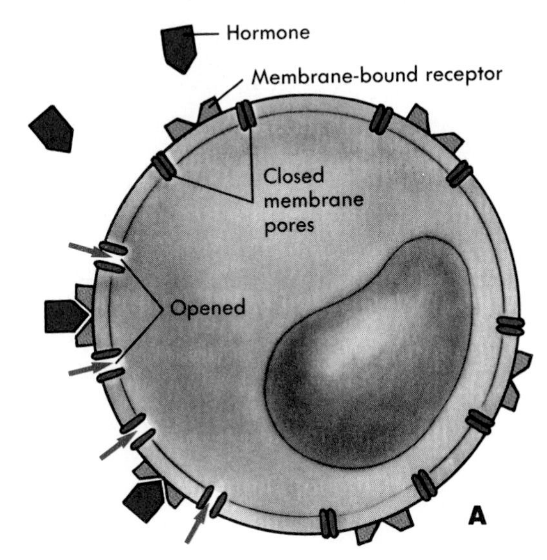

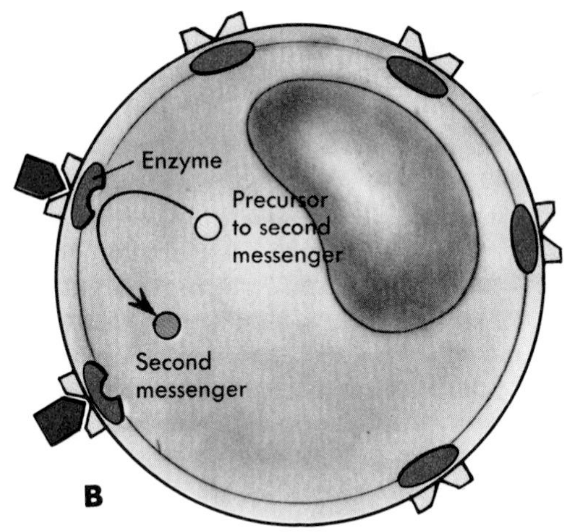

FIGURE 17-10 Membrane-bound receptor model. Some hormone receptors are within the cell membrane and are called membrane-bound receptors. Hormones in the interstitial fluid bind to the hormone receptor, and the combination of the hormone and receptor initiates a response in the target cells. **A** Some hormones bind to membrane-bound receptors, and the cell responds by altering the permeability of the cell membrane to ions such as sodium and calcium. The combination of the hormone with the receptor affects pores in the cell membrane and causes them to open *(arrows)*. **B** Some hormones bind to membrane-bound receptors, and the cell responds by activating a membrane-bound enzyme that synthesizes a "second messenger" inside the cell. This second messenger interacts with enzymes within the cell to alter their activity.

Hormones that stimulate the synthesis of a second-messenger molecule often produce rapid responses because the second messenger influences already existing enzymes and causes a **cascade effect,** which results when a few second-messenger molecules activate several enzymes and each of the activated enzymes in turn activates several other enzymes that produce the final response. Thus an amplification system exists in which a few cAMP molecules can affect the activity of many enzymes.

Second-messenger mechanisms apparently mediate the effect of numerous hormones. Examples include the gonadotropins, which control events in the ovary and testes, adrenocorticotropic hormone that regulates secretions from the adrenal cortex, and thyroid-stimulating hormone, which controls the rate of secretion from the thyroid glands. Cyclic AMP is the second-messenger molecule in many cells, and for each cell type, cAMP stimulates a different set of enzymes or other processes. Thus cAMP stimulates one type of response in liver cells but another type of response in the ovary. In addition, epinephrine binds to its receptor on liver cells and initiates cAMP synthesis, but there are no receptor sites for epinephrine in ovarian cells. Gonadotropins, on the other hand, bind to receptors on the ovary and stimulate cAMP synthesis, which initiates an increase in sex hormone synthesis in the ovary. These relationships allow cAMP to be a second-messenger molecule in more than one cell type in which responses differ.

Second-messenger molecules other than cAMP have also been discovered; cyclic guanosine monophosphate (cGMP) is an example. Modified second-messenger mechanisms may also exist. For example, some hormones bind to a membrane-bound receptor molecule that causes increased permeability of the membrane to calcium ions. Thus calcium ions enter the cell and bind to protein molecules called calmodulin. The calcium-calmodulin complex acts as an intracellular regulatory compound (Figure 17-11) that influences the activity of enzymes or other processes similar to the way cAMP regulates enzyme activity.

Intracellular Receptor Mechanism

Intracellular receptors are protein molecules inside the cell (Figure 17-12). Some of the receptor molecules float freely in the cytoplasm of the target cells whereas others are located in the nucleus. By the process of diffusion, lipid-soluble hormones cross the cell membrane into the cytoplasm. Some hormones then bind to the receptor molecules, and the receptor-hormone complex moves into the nucleus. Hormones in which there are no cytoplasmic receptors diffuse directly into the nucleus and bind to a receptor molecule, sometimes called an **acceptor molecule.** The acceptor molecule activates specific genes within the DNA of the nucleus to produce messenger RNA (mRNA). The mRNA moves to the cytoplasm and initiates the synthesis of new proteins at the ribosomes, which produce the cell response to the hormone. For example, the effect of the steroid aldosterone on its target cells within the kidney is to

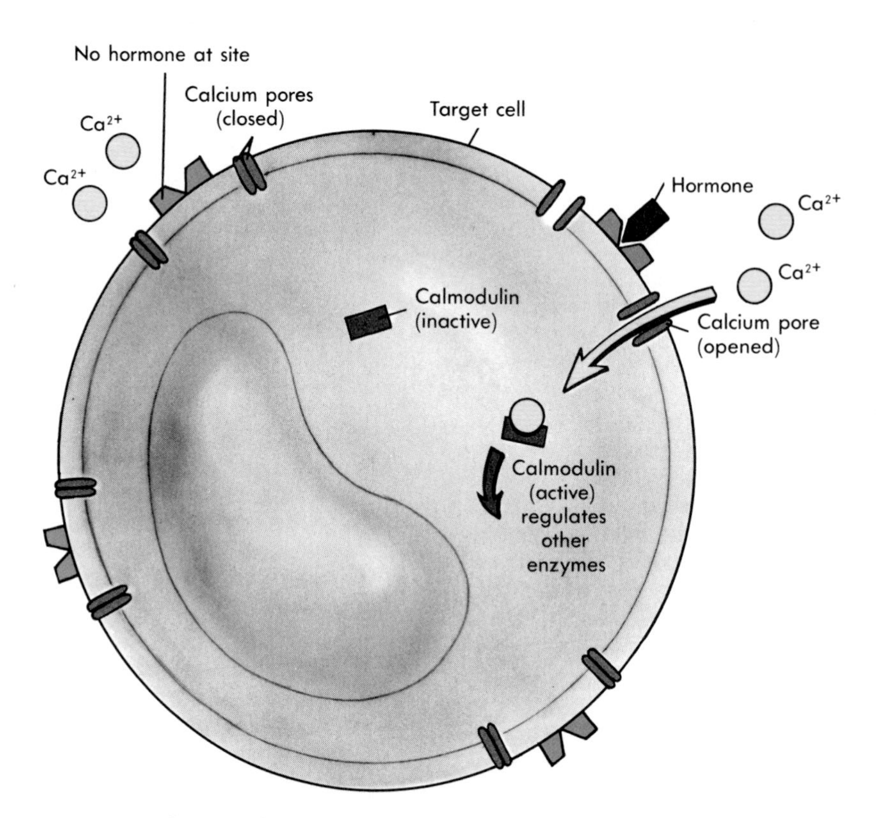

FIGURE 17-11 Calcium-calmodulin complex. Certain hormones bind to receptors on the cell membranes of target cells and cause an increase in permeability of the cell membrane to calcium (*Ca^{2+}; arrow*). Calcium diffuses into the cell because it exists in a higher concentration outside of the cell membrane than inside. Once the calcium enters the cell, it may bind to a protein called calmodulin. Once calcium is bound to it, calmodulin acts as a second messenger and alters the activity of other enzymes inside the cell. Calmodulin may be free within the cytoplasm of some cells or may be bound to the cell membrane.

increase the rate of sodium chloride transport. Newly synthesized proteins produced in the target cells are responsible for the increased rate of sodium chloride transport.

Cells that synthesize new protein molecules in response to a hormonal stimulus normally have a latent period of several hours between the time the hormone binds to its receptor and the time a response is observed. During this latent period mRNA and new proteins are synthesized. Receptor hormone complexes normally are degraded within the cell, limiting the length of time the hormone influences the cell's activity, and the cell slowly returns to its previous functional state.

5

Of membrane-bound receptors and intracellular receptors, which is better adapted for mediating a response that lasts a considerable length of time, and which is better for mediating a response with a rapid onset and a short duration? Explain why.

? ? ? ? ? ? ? ? ? ? ?

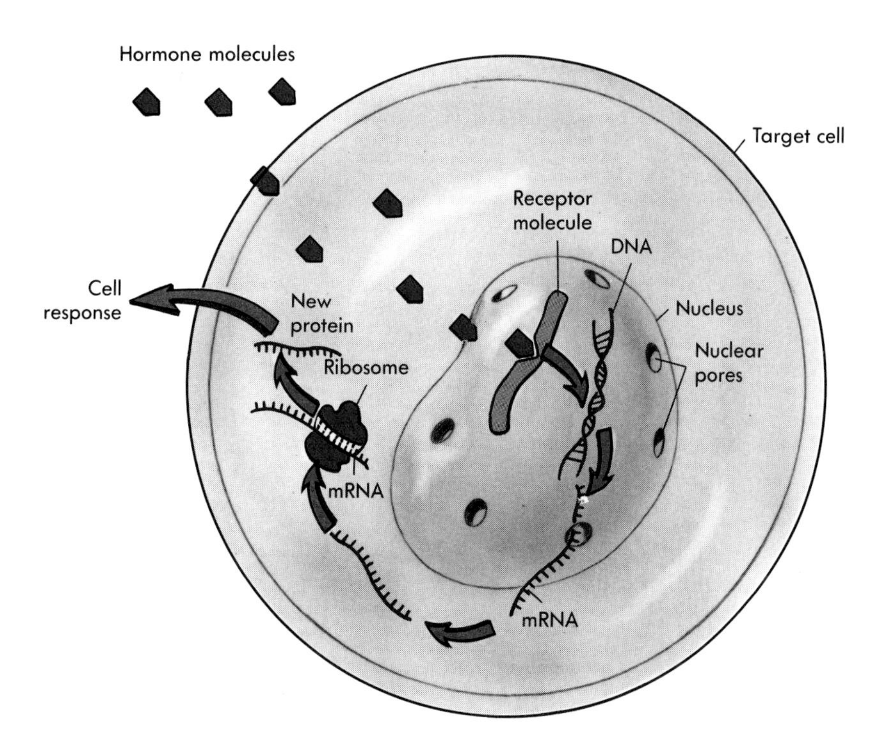

FIGURE 17-12 Intracellular receptor model and intracellular receptor mechanism. Some hormones are lipid soluble and can easily diffuse through the cell membrane and enter the cytoplasm of the cell. Once inside the cell, they bind to receptor molecules either in the cytoplasm or in the nucleus. The hormone receptor complex moves into the nucleus, and the combination of the hormone with the receptor molecules causes production of messenger RNA *(mRNA)* from specific regions of the chromosomes. The messenger RNA leaves the nucleus, passes into the cytoplasm of the cell, and binds to ribosomes, where it directs the synthesis of specific proteins. The protein synthesized on the ribosomes produces the cell's response to the hormone.

The Endocrine System

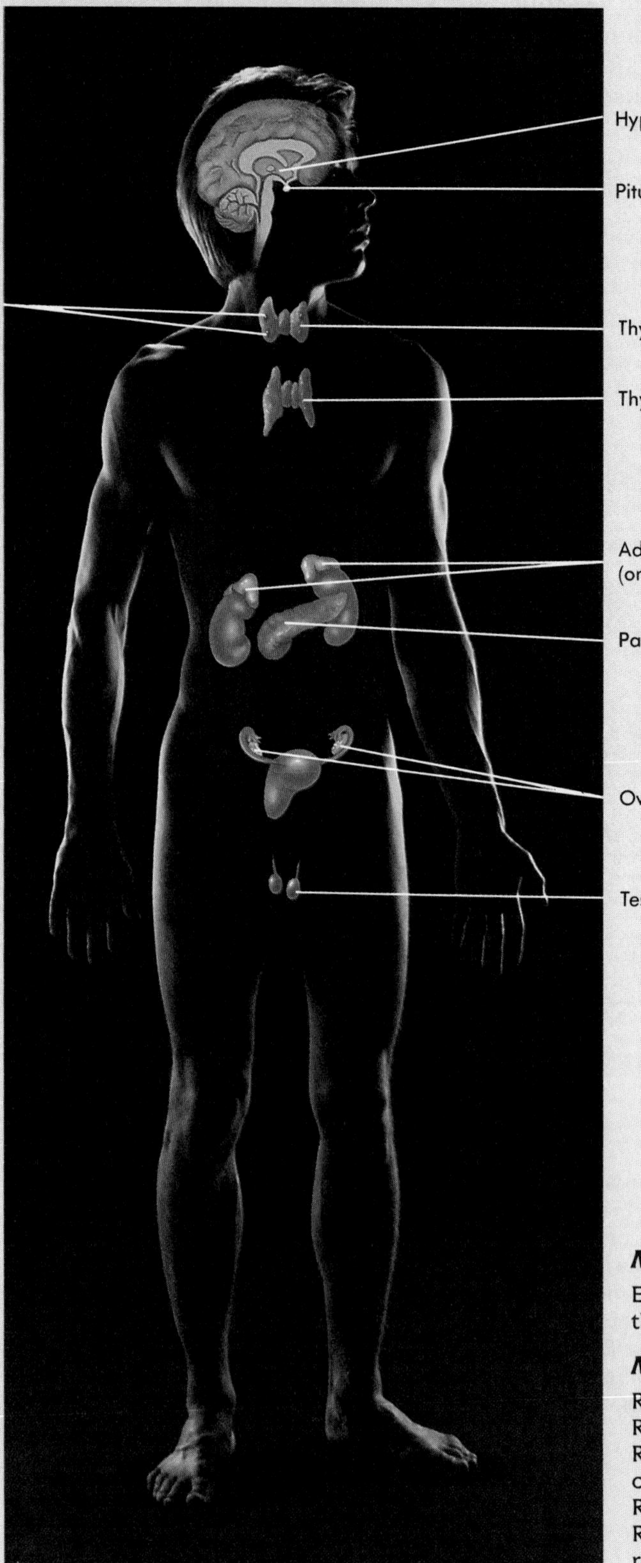

Major Components

Endocrine glands, such as the pituitary, thyroid, and adrenal glands

Major Functions

Regulates metabolism and growth
Regulates absorption of nutrients
Regulates fluid balance and ion concentration
Regulates the body's response to stress
Regulates sexual characteristics, reproduction, birth, and lactation

SUMMARY

GENERAL CHARACTERISTICS OF THE ENDOCRINE SYSTEM *p. 526*

1. Endocrine glands produce hormones that are released into the interstitial fluid, diffuse into the blood, and travel to target tissues in which they cause a specific response.
2. Other chemical messengers produced by endocrine glands include neurohormones, neurotransmitters, neuromodulators, parahormones, and pheromones.
3. Generalizations about the differences between the endocrine and nervous systems include (1) the endocrine system is amplitude modulated, whereas the nervous system is frequency modulated; and (2) the response of target tissues to hormones is usually slower and of longer duration than the response to neurons.

CHEMICAL STRUCTURE OF HORMONES *p. 529*

Hormones are either proteins, glycoproteins, polypeptides, derivatives of amino acids, or lipids (steroids or derivatives of fatty acids).

CONTROL OF SECRETION RATE *p. 529*

1. Most hormones are not secreted at a constant rate.
2. Most hormone secretion is controlled by negative-feedback mechanisms that function to maintain homeostasis.
3. Hormone secretion from an endocrine tissue is regulated by one or more of three mechanisms: (1) a nonhormone substance; (2) stimulation by the nervous system; or (3) a hormone from another endocrine tissue.

TRANSPORT AND DISTRIBUTION IN THE BODY *p. 532*

Hormones are dissolved in plasma or bind to plasma proteins. The blood quickly distributes hormones throughout the body.

METABOLISM AND EXCRETION *p. 533*

1. Nonpolar, readily diffusible hormones bind to plasma proteins and have an increased half-life.
2. Water-soluble hormones such as proteins, epinephrine, and norepinephrine do not bind to plasma proteins or readily diffuse out of the blood. Instead, they are broken down by enzymes or are taken up by tissues. They have a short half-life.
3. Hormones with a short half-life regulate activities that have a rapid onset and a short duration.
4. Hormones with a long half-life regulate activities that remain at a constant rate through time.
5. Hormones are eliminated from the blood by secretion from the kidneys and liver, enzymatic degradation, conjugation, or active transport.

INTERACTION OF HORMONES WITH THEIR TARGET TISSUES *p. 534*

1. Target tissues have receptor molecules that are specific for a particular hormone.
2. Hormones bound with receptors affect the rate at which already existing processes occur.
3. Down regulation is a decrease in the number of receptor molecules in a target tissue, and up regulation is an increase in the number of receptor molecules.

CLASSES OF HORMONE RECEPTORS *p. 536*

1. Membrane-bound receptors bind to water-soluble or large molecular weight hormones.
2. Intracellular receptors bind to lipid-soluble hormones.

Membrane-Bound Receptors and the Second-Messenger Model

1. Membrane-bound receptors are proteins or glycoproteins.
2. When a hormone binds to the membrane-bound receptor, the following can occur:
   - A change in membrane permeability produces depolarization, hyperpolarization, or an influx of calcium ions.
   - Activation of an enzyme within the membrane occurs. The activated enzyme catalyzes a reaction that produces a second messenger such as cAMP (the hormone is considered the first messenger). The second messenger then alters the actions of enzymes, producing the response of the target cells to the hormone.
3. Second-messenger mechanisms are rapid acting because they act on already existing enzymes and produce a cascade effect.
4. Second-messenger activated processes are limited by an enzyme that breaks down the second messenger.
5. A modified second-messenger mechanism involves calmodulin. As a result of the hormone's binding with the membrane-bound receptor, membrane permeability changes, and calcium ions move into the cell and bind with calmodulin. The calcium-calmodulin complex then alters the activities of enzymes.

Intracellular Receptor Mechanisms

1. Intracellular receptors are proteins in the cytoplasm or nucleus.
2. Hormones bind with the intracellular receptor, and the receptor-hormone complex activates DNA to produce mRNA. The mRNA initiates the production of a protein (enzyme) that produces the target cell's response to the hormone.
3. Intracellular receptor mechanisms are slow acting because time is required to produce the mRNA and the protein.
4. Intracellular receptor–activated processes are limited by the breakdown of the receptor-hormone complex.

CONTENT REVIEW

1. Define endocrine gland, endocrine system, and hormone.
2. Name and describe five chemical messengers, other than hormones, produced by endocrine glands.
3. Contrast the endocrine system and the nervous system for the following: (A) amplitude vs. frequency modulation and (B) speed and duration of target cell response.
4. List the categories of hormones based on chemical structure, and give an example of each.
5. Describe the ways in which hormone secretion is regulated. Give examples of three patterns of hormone secretion.
6. Define the half-life of a hormone. What happens to a hormone's half-life when a hormone binds to a plasma protein? What kinds of hormones bind to plasma proteins?
7. What kinds of activities do hormones with a short half-life regulate? With a long half-life?
8. List and describe four ways that hormones are eliminated from the blood.
9. Many different hormones circulate in the blood. What determines to which hormone a tissue will respond?
10. When a hormone combines with a receptor, does it cause the cell to do new things, alter already existing processes, or both?
11. Contrast membrane-bound receptors and intracellular receptors for the following: (A) their location in the cell and (B) the kind of hormone to which they bind.
12. What can happen to membrane permeability or enzyme activity within a cell when a hormone binds to a membrane-bound receptor?
13. Describe the second-messenger model of hormone action and the cascade effect. Does the second-messenger mechanism produce a slow or a rapid response?
14. How can the same second messenger such as cAMP produce different responses in different cells?
15. What finally limits the processes activated by the second messenger?
16. Describe the modified second-messenger mechanism that involves calcium and calmodulin.
17. Describe the intracellular model for hormone action. Compared to the second-messenger mechanism, is it fast or slow acting? Explain.
18. What finally limits the processes activated by the intracellular receptor mechanism?

CONCEPT REVIEW

1. Consider a hormone that is secreted in large amounts at a given interval and is modified chemically by the liver and excreted by the kidney at a rapid rate, making the half-life of the hormone in the circulatory system very short. Thus the hormone rapidly increases in the blood and then decreases rapidly. Predict the consequences of liver disease on the blood levels of that hormone.
2. Consider a hormone that controls the concentration of some substance in the circulatory system. If a tumor begins to produce that substance in large amounts in an uncontrolled fashion, predict the effect on the secretion rate for the hormone.
3. How could you determine whether or not a hormone-mediated response was due to the second-messenger mechanism or the intracellular receptor mechanism?
4. Prostaglandins are a group of hormones produced by nearly all cells of the body. Unlike other hormones, prostaglandins do not circulate but usually have their effect at or very near their site of production. Prostaglandins apparently affect many body functions, including blood pressure, inflammation, induction of labor, vomiting, fever, and inhibition of the clotting process. Prostaglandins also influence the formation of cyclic AMP. Explain how an inhibitor of prostaglandin synthesis could be used as a therapeutic agent. Inhibitors of prostaglandin synthesis may produce side effects. Why?
5. When an individual is confronted with a potentially harmful or dangerous situation, epinephrine (adrenaline) is released from the adrenal gland. Epinephrine prepares the body for action by increasing the heart rate and blood sugar levels. Explain the advantages or disadvantages that would be associated with a short half-life for epinephrine and with a long half-life.
6. Thyroid hormones are important in regulating the basal metabolic rate of the body. What would be the advantages or disadvantages of a long half-life for thyroid hormones and of a short half-life?
7. An increase in thyroid hormones causes an increase in metabolic rate. If a liver disease results in reduced production of the plasma proteins to which thyroid hormones normally bind, what would be the effect on metabolic rate? Explain.

? ?

ANSWERS TO PREDICT QUESTIONS

1 p. 531. If the concentration of substance X is lower than normal, the secretion of the hormone should increase. Normally the increase in hormone would cause an increase in substance X. As substance X increases, it would have a negative-feedback effect on the secretion of the hormone. The secretion rate for the hormone would then decrease (negative-feedback mechanism). As long as the abnormal condition keeps the level of substance X below normal, the rate of hormone secretion will remain high.

2 p. 534. A major function of plasma proteins to which hormones bind is to increase the half-life of the hormone. If the concentration of the plasma protein decreases, the half-life and, consequently, the concentration of the hormone in the circulatory system decrease also. Since the half-life of the hormone is decreased, the rate at which the hormone is removed from the circulatory system increases; and if the secretion rate for the hormone does not increase, its concentration in the blood will decline.

3 p. 536. If hormone A is secreted in smaller amounts than normal, the number of receptors in the target tissue for hormone B would be smaller than normal because it is hormone A that controls the concentration of the receptor molecules for hormone B. Consequently, the sensitivity of the target tissue for hormone B would be less than normal.

4 p. 536. An inhibitor of phosphodiesterase will slow the rate at which cAMP is broken down. The inhibitor will therefore increase the concentration of cAMP in the cell. Thus the inhibitor will have the same effect on the tissue as a hormone that increases the cAMP concentration.

5 p. 539. Intracellular receptor synthesis mechanisms result in new proteins that exist within the cell for a considerable amount of time. Therefore intracellular receptors are better adapted for mediating responses that last a relatively long time (i.e., for many minutes, hours, or longer). On the other hand, second messengers such as cAMP normally activate enzymes already existing in the cytoplasm of the cell for shorter periods of time. The synthesis of cAMP occurs quickly, but the duration is short because cAMP is broken down quickly and the activated enzymes are then deactivated. Therefore second-messenger mechanisms are better adapted to short-term and rapid responses.

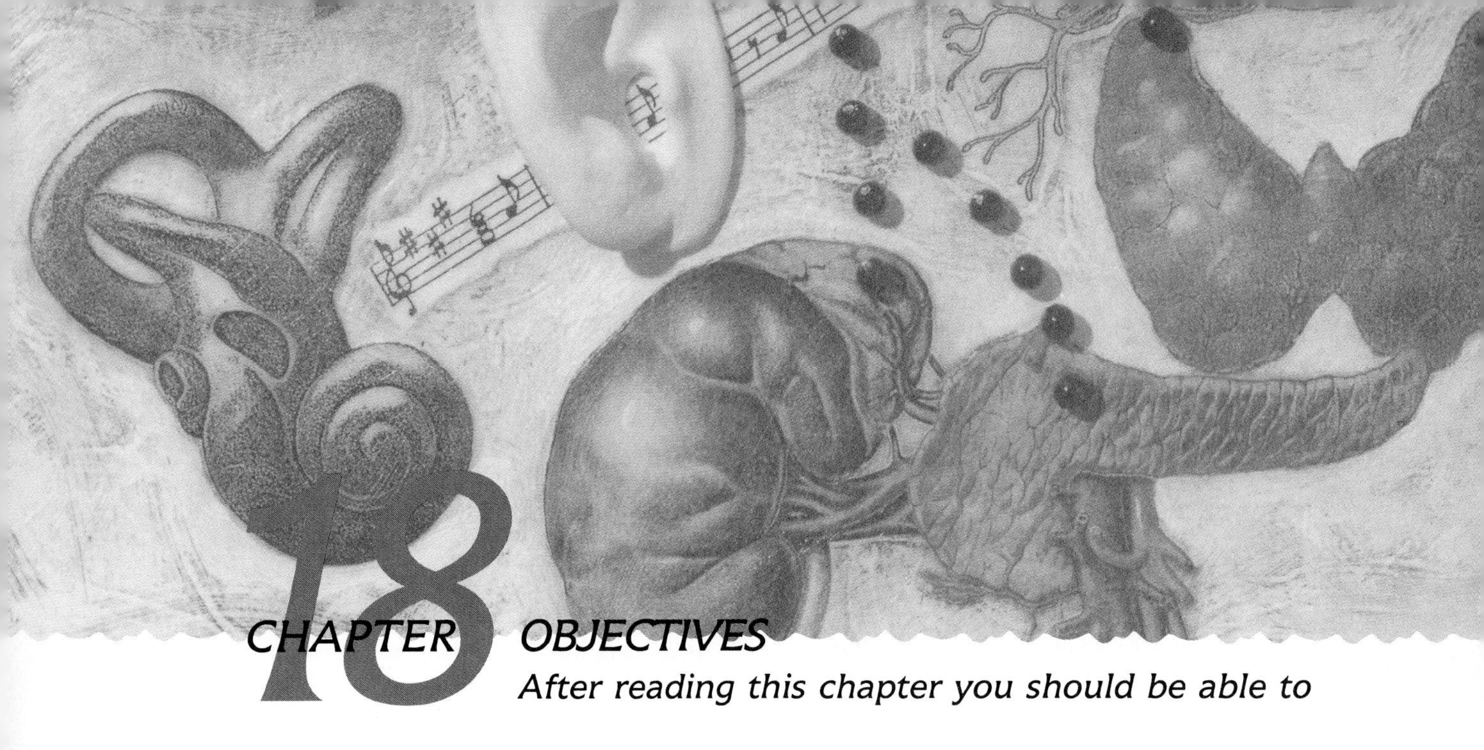

CHAPTER 18

OBJECTIVES

After reading this chapter you should be able to

1. List the most important information relating to endocrine glands and their secretions.
2. Describe the embryonic development, anatomy, and location of the pituitary gland and describe the functional and structural relationships between the hypothalamus of the brain and the pituitary gland.
3. Describe the secretory cells of the neurohypophysis and list the hormones secreted from the neurohypophysis.
4. Outline the means by which adenohypophyseal hormone secretion is regulated.
5. Describe the target tissues, regulation, and responses to each of the neurohypophyseal and adenohypophyseal hormones.
6. Describe the structure and location of the thyroid gland.
7. Describe the response of target tissues to thyroid hormones and outline the regulation of thyroid hormone secretion.
8. Describe the regulation of calcitonin secretion and describe its function.
9. Explain the function of parathyroid hormone and describe the means by which the secretion of parathyroid hormone is regulated.
10. Explain the relationship between parathyroid hormone and vitamin D.
11. Describe the structure and embryological development of the adrenal glands and describe the response of the target tissues to each of the adrenal hormones.
12. Describe the means by which the adrenal hormones are regulated.
13. Describe the position and structure of the pancreas and list the substances that are secreted by the pancreas and their functions.
14. Explain the regulation of insulin and glucagon secretion and describe how blood-nutrient levels are regulated by hormones after a meal and during exercise.
15. Describe the pineal body's structure, location, and secretory products and their functions.

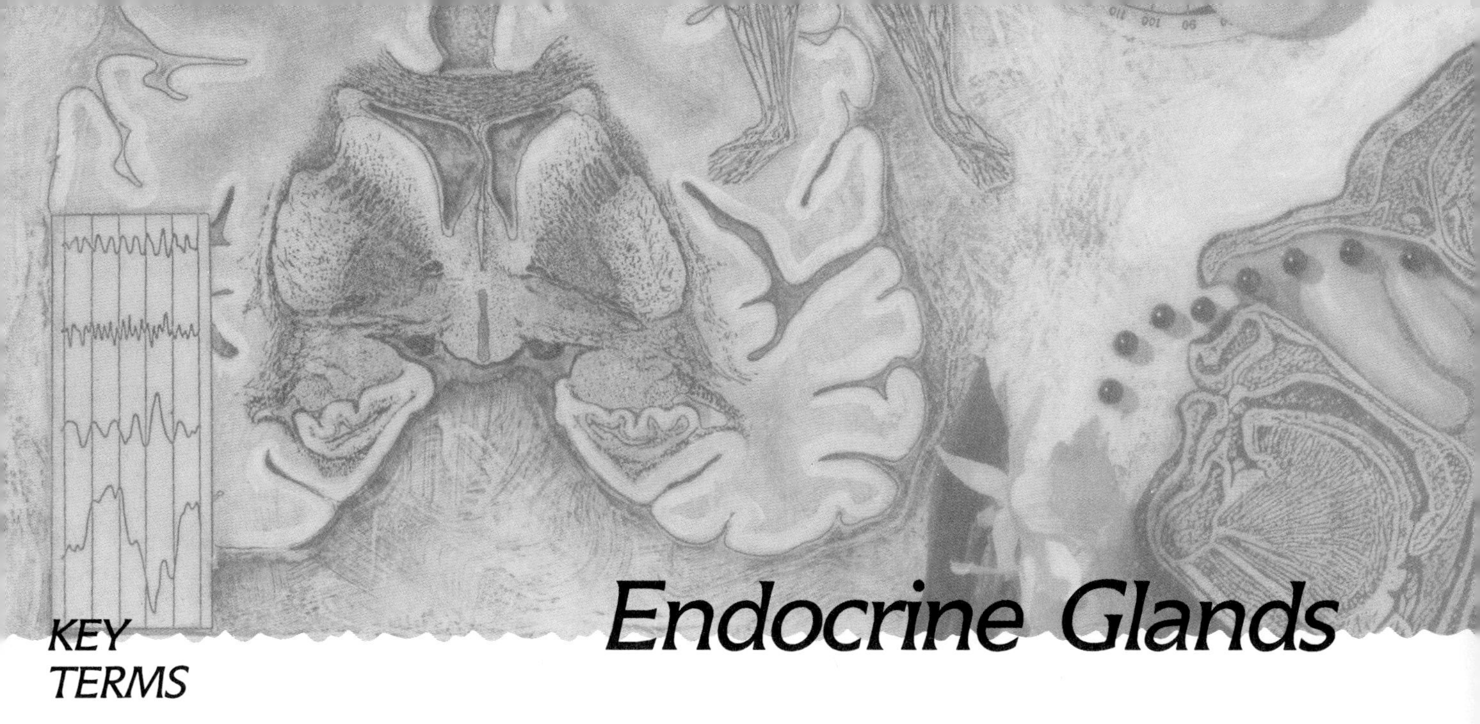

Endocrine Glands

KEY TERMS

adenohypophysis (ad′ĕ-no-hi-pof′ĭ-sis)

adrenal glands

endorphin (en′dor-fin)

hypothalamohypophyseal (hi′po-thal′ă-mo-hi′po-fiz′e-al) portal system

hypothalamohypophyseal tract

hypothalamus

neurohypophysis (nu′ro-hi-pof′ĭ-sis)

pancreas (pan′kre-us)

parafollicular (păr′ah-fŏ-lik′u-lar) cells

parathyroid gland

pineal (pi′ne-al) body

pituitary (pit-u′ĭ-tĕr-e)

prostaglandin (pros′tă-glan′din)

thymus gland

thyroid gland

RELATED TOPIC

The following term or concept is important for a good understanding of this chapter. If you are not familiar with it, you should review it before proceeding.

Organization of the endocrine system (Chapter 17)

SEVERAL PIECES OF INFORMATION ARE REQUIRED FOR A REASONABLE UNDERSTANDING OF THE ROLE ENDOCRINE GLANDS AND THEIR SECRETIONS PLAY IN THE BODY. THEY ARE AS FOLLOWS:

1. The anatomy of each gland and its location
2. The hormone secreted by each gland
3. The target tissues and the response of target tissues to each hormone
4. The means by which the secretion of each hormone is regulated
5. The consequences and causes, if known, of hypersecretion and hyposecretion of the hormone

This information is provided for each of the endocrine glands discussed in this chapter. Certain hormones such as those that regulate digestion and reproduction are mentioned only briefly, for they are explained more fully in later chapters.

PITUITARY GLAND AND HYPOTHALAMUS

The **pituitary** (pit-u'ĭ-tĕr-e) **gland,** or **hypophysis** (hi-pof'ĭ-sis; an undergrowth), secretes nine major hormones that directly regulate numerous body functions and the secretory activity of several other endocrine glands.

The **hypothalamus** of the brain regulates the secretory activity of the pituitary gland, and, in turn, the activity of the hypothalamus is influenced by hormones, by sensory information that enters the central nervous system, and by the emotional state of the individual. The hypothalamus and the pituitary are the major sites in which the two regulatory systems of the body (the nervous system and the endocrine system) interact (Figure 18-1). Indeed, a major portion of the pituitary gland (the posterior pituitary) is an extension of the hypothalamus.

Structure of the Pituitary Gland

The pituitary gland is roughly 1 cm in diameter, weighs 0.5 to 1 g, and rests in the sella turcica of the sphenoid bone (see Figure 18-1). The pituitary gland is located inferior to the hypothalamus and is connected to it by the **infundibulum** (in-fun-dib'u-lum).

The pituitary gland is divided functionally into two parts: the **neurohypophysis,** (nu'ro-hi-pof'ĭ-sis), or **posterior pituitary,** and the **adenohypophysis** (ad'ĕ-no-hi-pof'ĭ-sis), or **anterior pituitary.**

Neurohypophysis or posterior pituitary

The posterior pituitary is called the neurohypophysis because it is continuous with the brain (*neuro-* refers to the nervous system). It is formed during embryonic development from an outgrowth of the inferior portion of the brain in the area of the hypothalamus (see Chapter 29). The outgrowth of the brain forms a stalk of tissue, the infundibulum. The distal end of the infundibulum becomes enlarged to form the neurohypophysis (Figure 18-2). Secretions of the neurohypophysis are neurohormones because the neurohypophysis is an extension of the nervous system.

Adenohypophysis or anterior pituitary

The anterior pituitary, or adenohypophysis (*adeno-* means gland), arises as an outpocketing of the roof of the embryonic oral cavity called Rathke's pouch, which grows toward the neurohypophysis. As it nears the neurohypophysis, Rathke's pouch loses its connection with the oral cavity and becomes the adenohypophysis of the mature pituitary gland. The adenohypophysis is subdivided into three areas with indistinct boundaries: the pars tuberalis, the pars distalis, and the pars intermedia (see Figure 18-2). The hormones secreted from the adenohypophysis, in contrast to those from the neurohypophysis, are not neurohormones because the adenohypophysis is not derived from neural tissue.

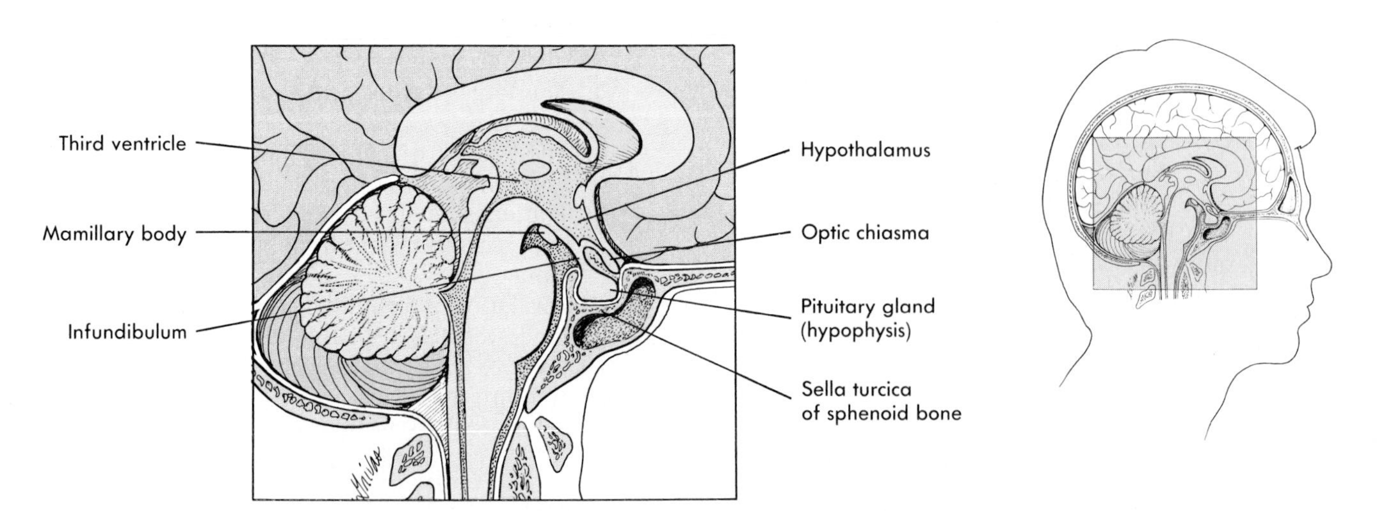

FIGURE 18-1 Midsagittal section of the head through the pituitary gland showing the anatomy of the pituitary and the hypothalamus. The pituitary gland is in a depression called the sella turcica in the floor of the skull. It is connected to the hypothalamus of the brain by the infundibulum.

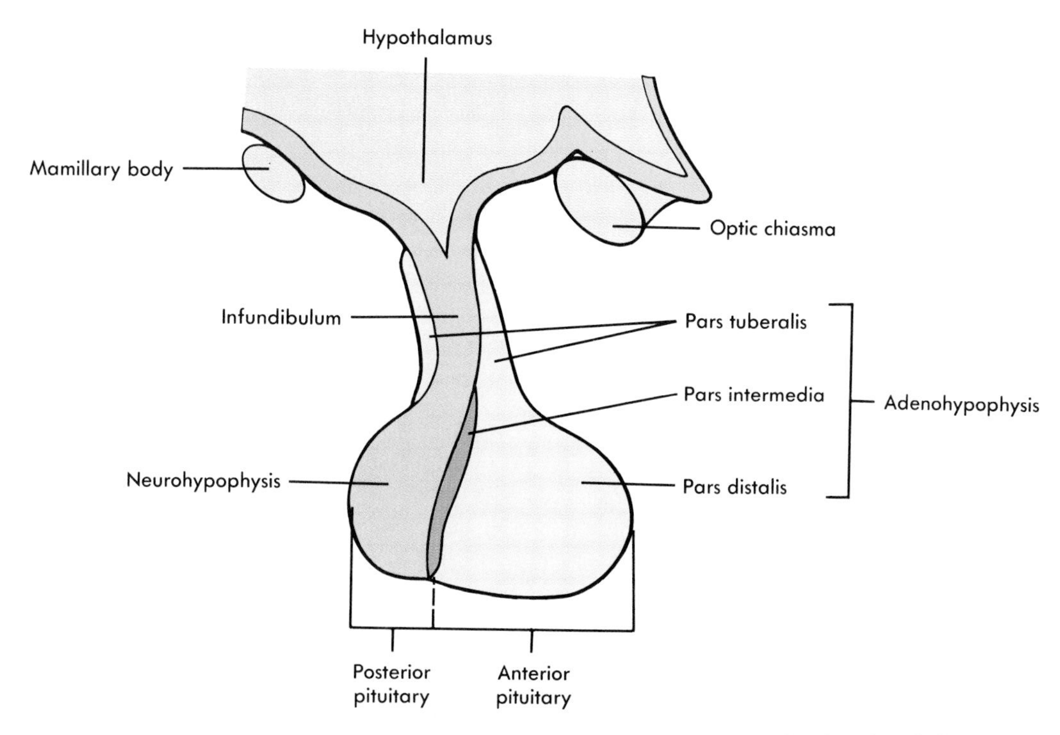

FIGURE 18-2 Subdivisions of the pituitary gland. The pituitary gland is divided into the anterior pituitary, or adenohypophysis, and the posterior pituitary, or neurohypophysis. The adenohypophysis is subdivided further into the pars distalis, pars tuberalis, and pars intermedia. The posterior pituitary consists of the enlarged distal end of the infundibulum called the neurohypophysis. The infundibulum connects the neurohypophysis to the hypothalamus.

Relationship of the Pituitary to the Brain

Portal vessels are blood vessels that begin and end in a capillary network. The **hypothalamohypophyseal** (hi′po-thal′ă-mo-hi′po-fiz′e-al) **portal system** extends from a portion of the hypothalamus to the adenohypophysis (Figure 18-3). The primary capillary plexus in the hypothalamus is supplied with blood from arteries that deliver blood to the hypothalamus. From the primary capillary plexus the hypothalamohypophyseal portal veins carry blood to a secondary capillary plexus in the adenohypophysis. Veins from the secondary capillary plexus eventually merge with the general circulation.

Neurohormones, produced and secreted by the hypothalamus, enter the primary capillary plexus and are carried to the secondary capillary plexus. There the neurohormones leave the blood and act on the cells of the adenohypophysis. Each neurohormone either inhibits or stimulates the production and secretion of a specific hormone by the adenohypophysis. In response to the hypothalamic neurohormones, adenohypophyseal cells secrete hormones that enter the secondary capillary plexus and are carried into the general circulation and hence to their target tissues. Thus the hypothalamohypophyseal portal system provides a means by which the hypothalamus, using neurohormones as chemical signals, can regulate the secretory activity of the adenohypophysis (Figure 18-4 and Table 18-1).

In contrast, there is no portal system that carries hypothalamic neurohormones to the neurohypophysis. Neurohormones released from the neurohypophysis are produced by neurosecretory cells with cell bodies in the hypothalamus. The axons of these cells extend from the hypothalamus through the infundibulum into the neurohypophysis and comprise a nerve tract called the **hypothalamohypophyseal tract** (see Figure 18-3). Neurohormones produced in the hypothalamus travel down these axons and are stored in secretory granules in the enlarged ends of the axons. When action potentials that originate in the neuron cell bodies in the hypothalamus are propagated along the axons to the axon terminals in the neurohypophysis, neurohormones are released from the axon terminals and enter the circulatory system (see Figure 18-4).

1

Surgical removal of the neurohypophysis in experimental animals results in marked symptoms, but the symptoms associated with hormone shortage are temporary. Explain these results.

? ? ? ? ? ? ? ? ? ? ?

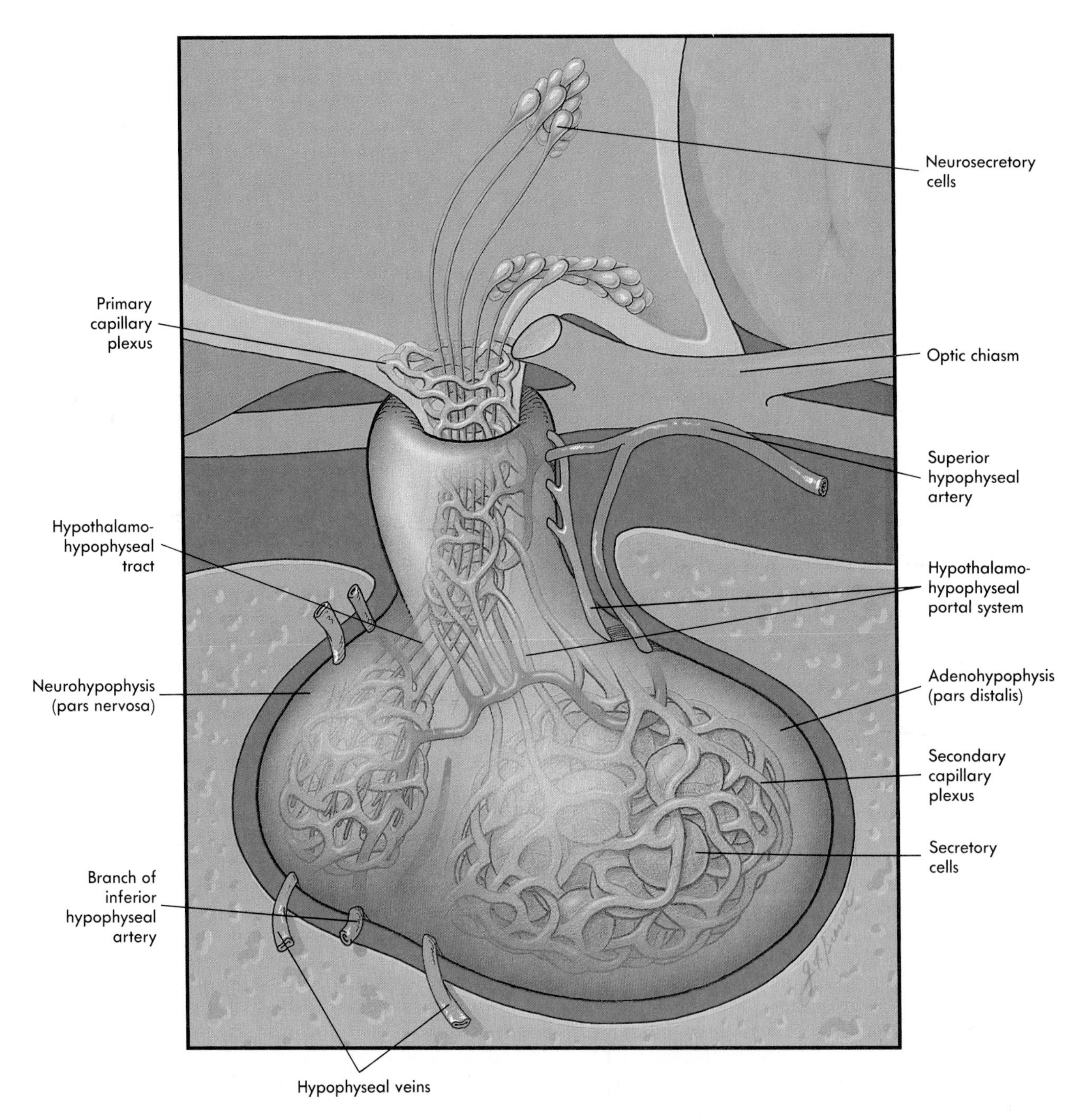

FIGURE 18-3 Hypothalamohypophyseal portal system and the hypothalamohypophyseal tract. The hypothalamohypophyseal portal system originates from a primary capillary plexus in the hypothalamus and extends to a secondary capillary plexus in the adenohypophysis. The hypothalamohypophyseal tract originates in the hypothalamus with the cell bodies of neurosecretory cells whose axons extend to the neurohypophysis.

Table 18-1

Hormones of the hypothalamus

HORMONE	STRUCTURE	TARGET TISSUE	RESPONSE
Growth hormone-releasing hormone (GH-RH)	Small peptide	Adenohypophyseal cells that secrete growth hormone	Increased growth hormone secretion
Growth hormone-inhibiting hormone (GH-IH) or somatostatin	Small peptide	Adenohypophyseal cells that secrete growth hormone	Decreased growth hormone secretion
Corticotropin-releasing hormone (CRH)	Peptide	Adenohypophyseal cells that secrete adrenocorticotropic hormone	Increased adrenocorticotropic hormone secretion
Gonadotropin-releasing hormone (GnRH)	Small peptide	Adenohypophyseal cells that secrete luteinizing hormone and follicle-stimulating hormone	Increased secretion of luteinizing hormone and follicle-stimulating hormone
Prolactin-inhibiting hormone (PIH)	Unknown (possibly dopamine)	Adenohypophyseal cells that secrete prolactin	Decreased prolactin secretion
Prolactin-releasing hormone (PRH)	Unknown	Adenohypophyseal cells that secrete prolactin	Increased prolactin secretion

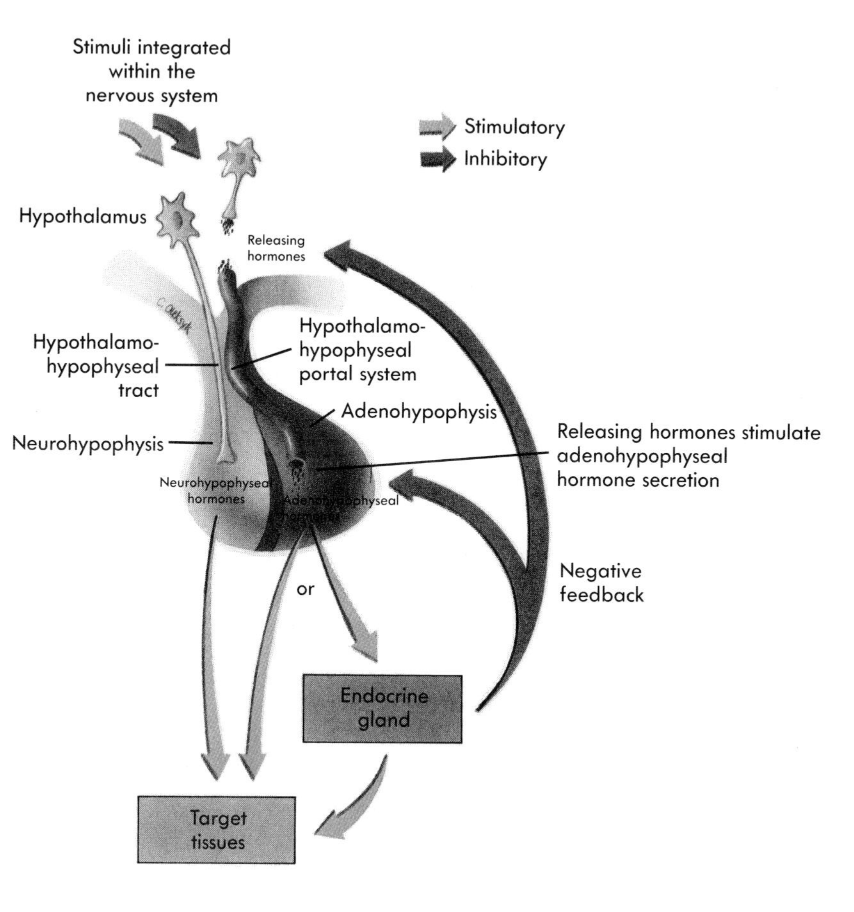

FIGURE 18-4 General relationship between the hypothalamus, the pituitary, and target tissues. Substances called releasing hormones or releasing factors are secreted from the hypothalamic neurons as a result of certain stimuli. They pass through the hypothalamo-hypophyseal portal system to the adenohypophysis. The releasing hormones either stimulate or inhibit the secretion of adenohypophyseal hormones. Secreted hormones from cells within the adenohypophysis pass through the blood and influence the activity of their target tissues. In response to stimulation of hypothalamic neurosecretory cells, action potentials pass along the axons of the neurosecretory cells to the neurohypophysis. The action potentials cause the release of neurohormones from the neurohypophysis and pass through the blood to target tissues.

HORMONES OF THE PITUITARY GLAND

This section describes the hormones secreted from the pituitary gland (Table 18-2), their effects on the body, and the mechanisms that regulate their secretion rate. Additionally, some major consequences of abnormal hormone secretion are stressed.

Neurohypophyseal Hormones

The neurohypophysis stores and secretes two polypeptide neurohormones: antidiuretic hormone and oxytocin. Each hormone is secreted by a separate population of cells.

Antidiuretic hormone

Antidiuretic hormone (ADH) is so named because it prevents *(anti-)* the output of large amounts of urine *(diuresis)*. ADH is also called **vasopressin** because it constricts blood vessels and raises blood pressure when present in high concentrations. ADH is synthesized in the hypothalamus and is transported within the axons of the hypothalamohypophyseal tract to the neurohypophysis where the ADH is stored in neuron terminal endings. It is released from these endings into the blood and is carried to its primary target tissue, the kidneys, where it acts to promote the retention of water and to reduce urine volume (see Chapter 26).

The secretion rate for ADH changes in response to alterations in blood osmolality and blood volume (Figure 18-5). Specialized neurons, osmoreceptors, synapse with the ADH neurosecretory cells in the hypothalamus. When blood osmolality increases, the frequency of action potentials in the osmoreceptors increases, causing an increase in action potential frequency in the neurosecretory cells. As a consequence, ADH secretion increases. Alternatively, the ADH neurosecretory cells may be stimulated directly by an increase in blood osmolality. Since ADH promotes water retention by the kidneys, it functions to reduce blood osmolality and resists any further increase in the osmolality of body fluids.

Sensory receptors that detect changes in blood pressure send action potentials through the vagus nerve along neurons that eventually synapse with the ADH neurosecretory cells. A drop in blood pressure, which normally accompanies a drop in blood volume, causes increased action potential frequency in the neurosecretory cells and increased ADH secretion. The ADH stimulates water retention by the kidneys and restores blood volume.

The inability to secrete ADH leads to the production of a large volume of dilute urine. This condition is called diabetes insipidus, and dehydration results unless water consumption is increased dramatically.

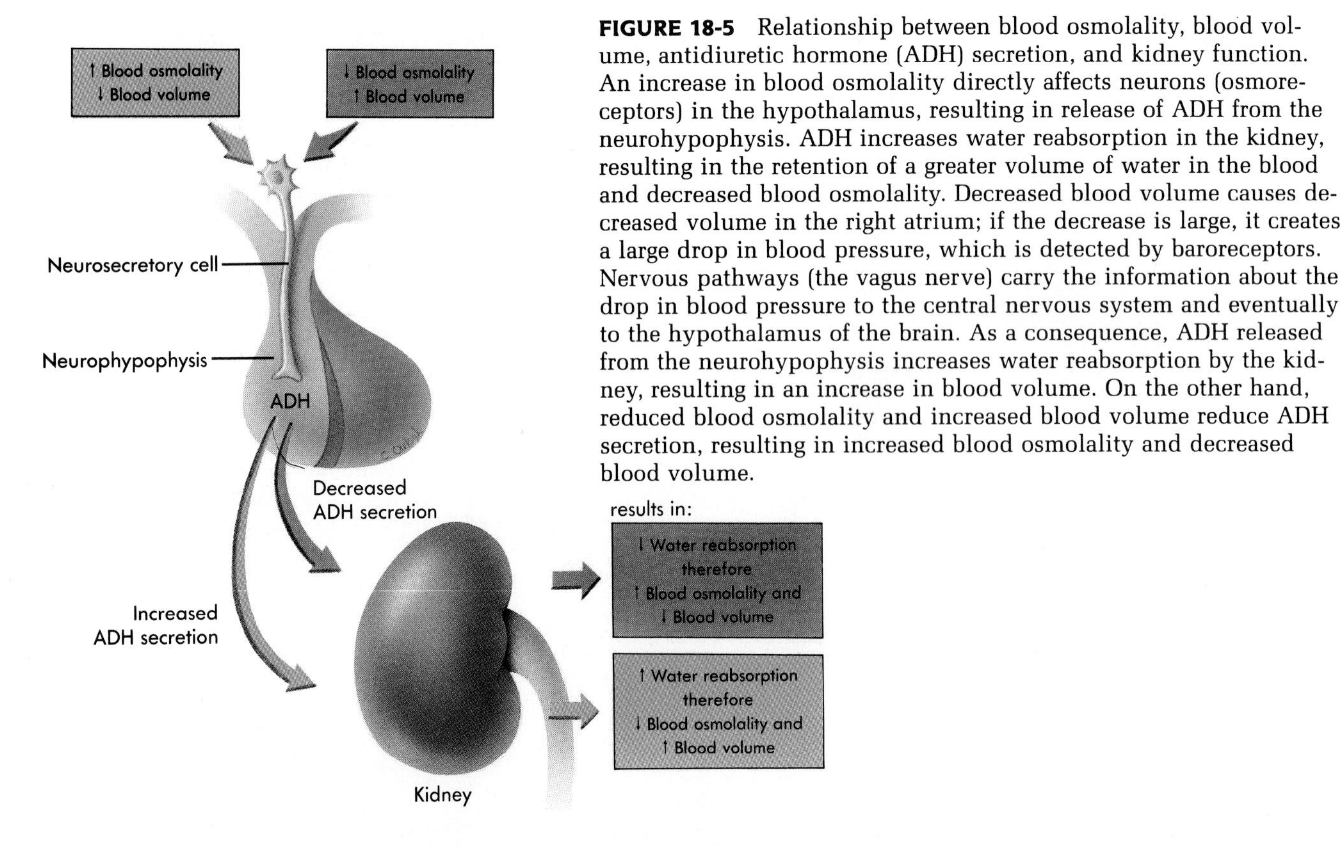

FIGURE 18-5 Relationship between blood osmolality, blood volume, antidiuretic hormone (ADH) secretion, and kidney function. An increase in blood osmolality directly affects neurons (osmoreceptors) in the hypothalamus, resulting in release of ADH from the neurohypophysis. ADH increases water reabsorption in the kidney, resulting in the retention of a greater volume of water in the blood and decreased blood osmolality. Decreased blood volume causes decreased volume in the right atrium; if the decrease is large, it creates a large drop in blood pressure, which is detected by baroreceptors. Nervous pathways (the vagus nerve) carry the information about the drop in blood pressure to the central nervous system and eventually to the hypothalamus of the brain. As a consequence, ADH released from the neurohypophysis increases water reabsorption by the kidney, resulting in an increase in blood volume. On the other hand, reduced blood osmolality and increased blood volume reduce ADH secretion, resulting in increased blood osmolality and decreased blood volume.

Oxytocin

Oxytocin is synthesized in the neuron cell bodies in the hypothalamus and then is transported through axons to the neurohypophysis, where it is stored within the axon terminals.

Oxytocin stimulates the smooth-muscle cells of the uterus. It causes contraction of the uterine smooth-muscle cells in nonpregnant women primarily during menses and during sexual intercourse. The uterine contractions may play a role in the expulsion of the uterine epithelium and small amounts of blood during menses and movement of spermatozoa through the uterus after sexual intercourse. Oxytocin plays an important role in the expulsion of the fetus from the uterus during delivery. It also pro-

Table 18-2

Hormones of the pituitary

HORMONE	STRUCTURE	TARGET TISSUE	RESPONSE
Neurohypophysis or Posterior Pituitary			
Antidiuretic hormone (ADH)	Small peptide	Kidney	Increased water reabsorption (less water is lost in the form of urine)
Oxytocin	Small peptide	Uterus; mammary glands	Increased uterine contractions; increased milk expulsion from mammary glands
Adenohypophysis or Anterior Pituitary			
Growth hormone (GH) or somatotropin	Protein	Most tissues	Increased growth in tissues; increased amino acid uptake and protein synthesis; increased breakdown of lipids and release of fatty acids from cells; increased glycogen synthesis and increased blood glucose levels; increased somatomedin production
Thyroid-stimulating hormone (TSH)	Glycoprotein	Thyroid gland	Increased thyroid hormone secretion
Adrenocorticotropic hormone (ACTH)	Peptide	Adrenal cortex	Increased glucocorticoid hormone secretion
Lipotropins	Peptides	Fat tissues	Increased fat breakdown
Beta endorphins	Peptides	Brain, but not all target tissues are known	Analgesia in the brain; inhibition of gonadotropin-releasing hormone secretion
Melanocyte-stimulating hormone (MSH)	Peptide	Melanocytes in the skin	Increased melanin production in melanocytes to make the skin darker in color
Luteinizing hormone (LH)	Glycoprotein	Ovary in females; testes in males	Ovulation and progesterone production in the ovary; testosterone synthesis and support for sperm production in the testes
Follicle-stimulating hormone (FSH)	Glycoprotein	Follicles in ovary in females and seminiferous tubes in males	Follicle maturation and estrogen secretion in ovary; spermatogenesis in testis
Prolactin	Protein	Ovary and mammary gland in females	Milk production in lactating women; increased response of follicle to LH and FSH; unclear function in males

motes contraction of smooth muscle-like cells surrounding the alveoli of the mammary gland to cause milk ejection in lactating females. Little is known about the effect of oxytocin in males.

Stretch of the uterus or mechanical stimulation of the cervix and stimulation of the nipples of the breast when a baby nurses cause a nervous reflex that stimulates oxytocin release. Action potentials are carried from the uterus and from the nipples to the spinal cord and then up the spinal cord to the hypothalamus of the brain. Action potentials in the oxytocin-secreting neurons pass along the axons to the neurohypophysis where they cause the release of oxytocin from the axon terminals.

The role of oxytocin in the reproductive system is described in greater detail in Chapter 28.

Adenohypophyseal Hormones

The adenohypophyseal secretions are influenced by neurohormones that pass from the hypothalamus through the hypothalamohypophyseal portal system to the adenohypophysis. The hypothalamic neurohormones act as releasing hormones that cause an increase in the secretion of adenohypophyseal hormones or as inhibiting hormones that cause a decrease in the secretion of adenohypophyseal hormones. For some adenohypophyseal hormones both releasing hormones and inhibiting hormones from the hypothalamus regulate their secretion (see Table 18-1).

All of the hormones released from the adenohypophysis are proteins, glycoproteins, or polypeptides. They are transported in the circulatory system without binding to specific plasma proteins, have a half-life measured in terms of minutes, and bind to membrane-bound receptor molecules on their target cells. For the most part each hormone is secreted by a separate cell type (adrenocorticotropic hormone and lipotropin are exceptions).

Growth hormone

Growth hormone (GH), sometimes called **somatotropin,** stimulates growth in most tissues and is one of the major regulators of metabolism. It increases the number of amino acids entering cells and favors their incorporation into proteins. GH increases the breakdown of lipids and the release of fatty acids from fat cells. The fatty acids then can be used as energy sources to drive chemical reactions, including anabolic reactions, by other cells. GH increases glycogen synthesis and storage in tissues, but the increased use of fats as an energy source spares glucose. GH plays an important role in regulating blood nutrient levels after a meal and during periods of fasting.

GH binds directly to target cells to mediate some of its effects, and it increases the production of a number of polypeptides by the liver, skeletal muscle, and possibly other tissues. These polypeptides, called **somatomedins** (so-mă'to-me'denz), circulate in the blood and affect target tissues. The best understood effect of the somatomedins is on the stimulation of cartilage and bone growth. The full effect of GH on target tissues requires the action of both GH and the somatomedins.

Two neurohormones released from the hypothalamus regulate the secretion of GH (Figure 18-6). One factor, **growth hormone-releasing hormone (GH-RH),** stimulates the secretion of GH, and the other, **growth hormone-inhibiting hormone (GH-IH),** or **somatostatin** (so'mă-to-stat'in), inhibits the secretion of GH. Note in Figure 18-6 that the stimuli that influence GH secretion act on the hypothalamus to increase or decrease the secretion of the releasing and inhibiting hormones. Secretion of GH is stimulated by low blood glucose levels and stress and is inhibited by high blood glucose levels. Rising blood levels of certain amino acids also increase GH secretion.

In most people a rhythm of GH secretion occurs with daily peak levels correlated with deep sleep. There is not a chronically elevated blood GH level during periods of rapid growth, although children

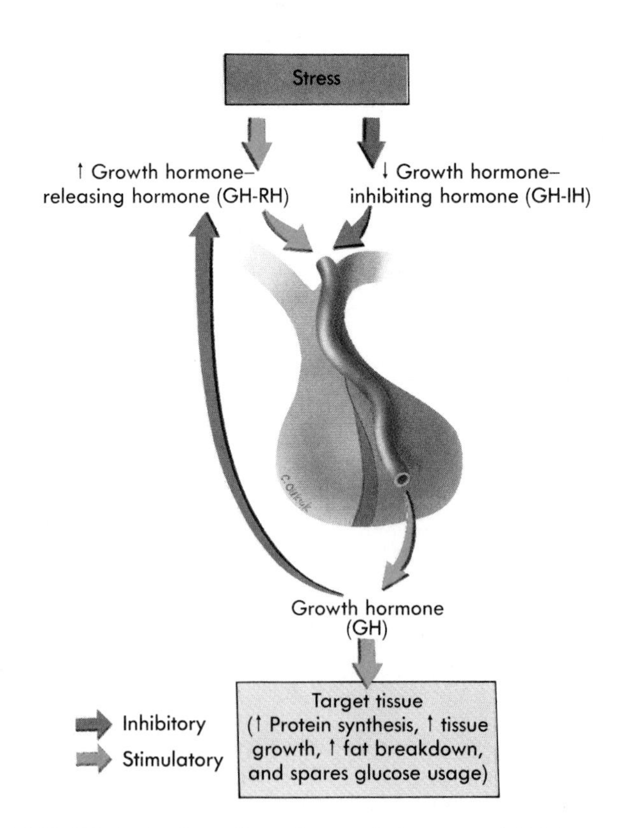

FIGURE 18-6 Secretion of growth hormone *(GH)* is controlled by two neurohormones released from the hypothalamus—growth hormone-releasing hormone (GH-RH), which stimulates GH secretion, and growth hormone-inhibiting hormone (GH-IH), which inhibits GH secretion. High levels of GH inhibit production of GH-RH by the hypothalamus.

Growth Hormone and Growth Disorders

Several pathological conditions are associated with abnormal GH secretion. In general, the causes for hypersecretion or hyposecretion of GH involve tumors in the hypothalamus or the pituitary, the synthesis of structurally abnormal GH, the inability of the liver to produce somatomedins, or the lack of receptor molecules in the target cells.

Chronic hyposecretion of GH in infants and children leads to dwarfism in which the stature is short because of delayed bone growth; however, the bones usually have a normal shape. In contrast to the dwarfism caused by hyposecretion of thyroid hormones, these dwarfs exhibit normal intelligence. Other symptoms that result from the lack of GH include mild obesity and retarded development of the adult reproductive functions. Two types of dwarfism result from a lack of GH secretion: (1) one that occurs in approximately two thirds of the cases in which GH and other anterior pituitary hormones are secreted in reduced amounts; and (2) one that occurs in approximately one third of the cases in which only a reduced amount of GH is observed and normal reproduction can occur. No obvious pathology is associated with hyposecretion of GH in adults.

The gene responsible for determining the structure of GH has been transferred successfully from human cells to bacterial cells, which produce GH that is identical to human GH. The GH produced in this fashion currently is available to treat patients who suffer from a lack of GH secretion.

Chronic hypersecretion of GH leads to one of two conditions (giantism and acromegaly), depending on whether the hypersecretion occurs before or after complete ossification of the epiphyseal plates in the skeletal system. Chronic hypersecretion of GH before the epiphyseal plates have ossified causes exaggerated and prolonged growth in long bones, resulting in giantism. Some individuals have grown to 8 feet tall or more.

In adults chronically elevated GH levels result in acromegaly; no height increase occurs because of the ossified epiphyseal plates. The condition does result in an increased diameter of fingers, toes, hands, and feet, the deposition of heavy bony ridges above the eyes, and a prominent jaw. The influence of GH on soft tissues results in a bulbous or broad nose, an enlarged tongue, thickened skin, and sparse subcutaneous adipose tissue. Nerves frequently are compressed as a result of the proliferation of connective tissue. Because GH spares glucose usage, chronic hyperglycemia results, frequently leading to diabetes mellitus and the development of severe atherosclerosis. Treatment for chronic hypersecretion of GH often involves surgical removal or irradiation of a GH-producing tumor.

tend to have somewhat higher blood levels of GH than adults. In addition to GH, factors such as genetics, nutrition, and sex hormones influence growth.

Thyroid-stimulating hormone

Hormones called tropic hormones are released from the pituitary gland and regulate the secretion of hormones from other glands. **Thyroid-stimulating hormone (TSH),** also called **thyrotropin,** is a tropic hormone that stimulates the synthesis and secretion of thyroid hormones from the thyroid gland. Other tropic hormones include luteinizing hormone, follicle-stimulating hormone, adrenocorticotropic hormone, and prolactin.

Adrenocorticotropic hormone and related substances

Adrenocorticotropic (ă-dre′no-kor′tĭ-ko-tro′pik) **hormone (ACTH)** is one of several adenohypophyseal hormones derived from the same large precursor molecule called **proopiomelanocortin.** In humans proopiomelanocortin's major products are ACTH, β-lipotropin, and to a lesser degree, γ-lipotropin, β-endorphin, and **melanocyte-stimulating hormone (MSH).**

ACTH functions to increase the secretion of hormones, primarily cortisol, from the adrenal cortex. ACTH also binds to melanocytes in the skin and increases skin pigmentation (see Chapter 5). In pathological conditions such as Addison's disease, blood levels of ACTH are chronically elevated, and the skin becomes markedly darker.

The **lipotropins** secreted from the anterior pituitary bind to membrane-bound receptor molecules on adipose tissue cells. They cause fat breakdown and the release of fatty acids into the circulatory system.

The **β-endorphins** have the same effects as opiate drugs such as morphine, and they may play a role in analgesia in response to stress and exercise. Other functions have been proposed for the β-endorphins, including regulation of body temperature, food intake, and water balance. Both ACTH and β-endorphin secretions increase in response to stress and exercise.

MSH binds to skin melanocytes and stimulates increased melanin deposition in the skin. The regulation of MSH secretion and its function in humans is not well understood, although it is an important regulator of skin pigmentation in some other vertebrates. ACTH also increases skin pigmentation because its structure is similar to that of MSH.

Gonadotropins and prolactin

Three adenohypophyseal hormones play important roles in regulating reproduction: **luteinizing** (lu'te-ĭ-nīz-ing) **hormone (LH), follicle-stimulating hormone (FSH),** and **prolactin.**

LH and FSH secreted into the blood bind to membrane-bound receptors and regulate the production of gametes (eggs and sperm) and reproductive hormones in the ovaries (estrogens and progesterones) of females and in the testes (testosterone) of males. LH and FSH are released from the adenohypophyseal cells under the influence of a single hypothalamic-releasing hormone called **gonadotropin-releasing hormone (GnRH).** GnRH is also called luteinizing hormone-releasing hormone (LH-RH).

Prolactin plays an important role in milk production in the mammary gland of lactating females. It also influences the number of receptor molecules for FSH and LH in the ovary, and it may play a role in enhancing progesterone secretion of the ovary after ovulation. No role for prolactin has been clearly established in males. The regulation of prolactin secretion is complex, and several compounds may be involved; however, hypothalamic neurohormones regulate its secretion. One neurohormone is **prolactin-releasing hormone (PRH),** and another is **prolactin-inhibiting hormone (PIH).** The regulation of gonadotropin and prolactin secretion and their specific effects are explained more fully in Chapter 28.

THYROID GLAND

The **thyroid gland** is composed of two lobes connected by a narrow band of thyroid tissue called the **isthmus.** The lobes are lateral to the upper portion of the trachea just inferior to the larynx, and the isthmus extends across the anterior aspect of the trachea (Figure 18-7, *A*). The thyroid gland is one of the largest endocrine glands with a weight of approximately 20 g. It is highly vascular and appears more red than its surrounding tissues.

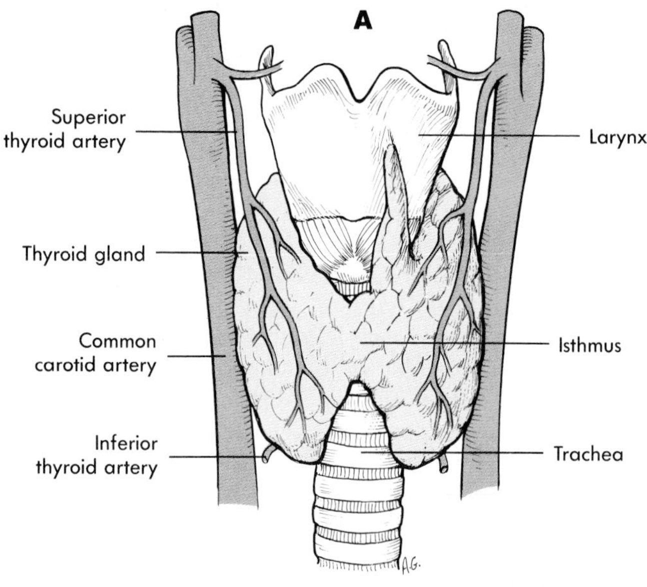

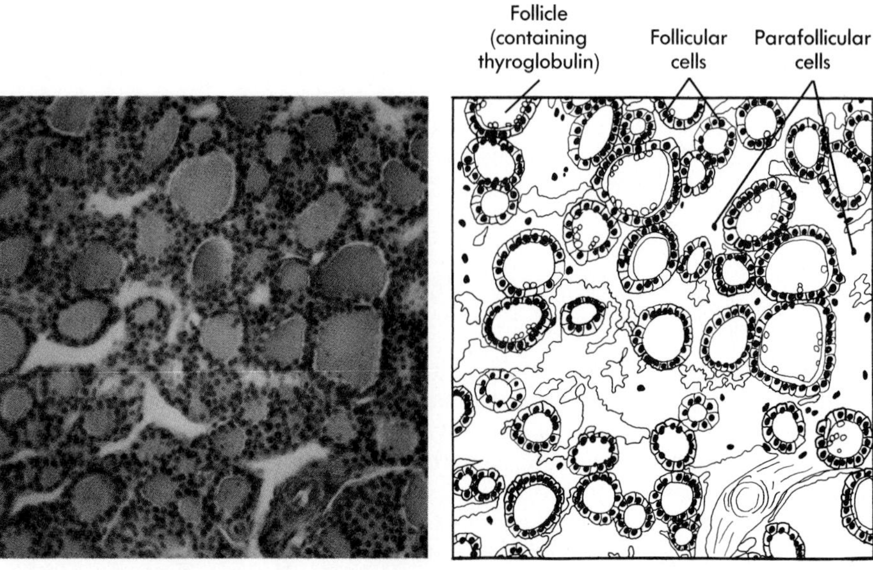

FIGURE 18-7 Anatomy and histology of the thyroid gland. **A** Frontal view of the thyroid gland. **B** Histology of the thyroid gland.

Histology

The thyroid gland contains numerous **follicles,** which are small spheres with their walls composed of a single layer of cuboidal epithelial cells (Figure 18-7, *B*). The center, or lumen, of each thyroid follicle is filled with a protein called **thyroglobulin** (thi′ro-glob′u-lin) to which thyroid hormones are bound. The thyroglobulin stores large amounts of thyroid hormone.

Between the follicles a delicate network of loose connective tissue contains numerous capillaries. Scattered **parafollicular** (păr′ah-fŏ-lik′u-lar) **cells** are found between the follicles and among the cells that comprise the wall of the follicle. **Calcitonin** (kal′sĭ-to′nin) is secreted from the parafollicular cells and plays a role in reducing the concentration of calcium in the body fluids when calcium levels become elevated.

Thyroid Hormones

The thyroid hormones include both **triiodothyronine** (tri-i′o-do-thi′ro-nēn; T_3) and **tetraiodothyronine** (T_4); T_4 is also called **thyroxine.** These substances constitute the major secretory products of the thyroid gland, with 10% T_3 and 90% T_4 (Table 18-3).

Thyroid hormone synthesis

The biosynthesis of the thyroid hormones is outlined in Figure 18-8. Thyroid-stimulating hormone (TSH) from the adenohypophysis must be present to maintain thyroid hormone synthesis and secretion. TSH binds to membrane-bound receptors on the cells of the thyroid follicles, and a second-messenger molecule mediates the response of thyroid cells to TSH. An adequate amount of iodine in the diet also is required. The following events in the thyroid follicles result in thyroid hormone synthesis and secretion:

1. Large proteins called thyroglobulins, which contain numerous tyrosine (an amino acid) molecules, are synthesized within the cells of the follicle.
2. Iodide (I^-) ions are actively absorbed by the cells of the thyroid follicle. The transport of the I^- ions is against a concentration gradient of approximately thirtyfold in healthy individuals.
3. Nearly simultaneously, the I^- ions are oxidized to form iodine (I) and then are bound chemically to tyrosine molecules of thyroglobulin. This binding occurs close to the time the thyroglobulin molecules are secreted by the process of exocytosis into the lumen of the follicle, and the secreted thyroglobulin contains the iodinated tyrosines. One or two I^- ions are bound to each tyrosine

Table 18-3

Hormones of the thyroid and parathyroid glands

HORMONE	STRUCTURE	TARGET TISSUE	RESPONSE
Thyroid			
Thyroid follicles			
Thyroid hormones (triiodothyronine and tetraiodothyronine)	Amino acid derivative	Most cells of the body	Increased metabolic rate; essential for normal process of growth and maturation
Parafollicular cells			
Calcitonin	Polypeptide	Bone	Decreased rate of breakdown of bone by osteoclasts; prevention of a large increase in blood calcium levels
Parathyroid			
Parathyroid hormone	Peptide	Bone; kidney; small intestine	Increased rate of breakdown of bone by osteoclasts; Increased reabsorption of calcium in kidneys, increased absorption of calcium from the small intestine, increased vitamin D synthesis, increases blood calcium levels

molecule to form either monoiodotyrosine or diiodotyrosine, respectively.

4. In the lumen of the follicle two diiodotyrosine molecules are combined to form tetraiodothyronine (T_4), and one monoiodotyrosine and one diiodotyrosine molecule are combined to form triiodothyronine (T_3). Large amounts of T_3 and T_4 are stored within the thyroid follicles as components of thyroglobulin. A reserve sufficient to supply thyroid hormones for approximately 2 weeks is stored in this form.
5. Thyroglobulin is taken into the thyroid cells by endocytosis in which lysosomes within the follicular cells fuse with the endocytotic vacuoles. Proteolytic enzymes break down thyroglobulin to release T_3 and T_4, which then diffuse from the follicular cells into the interstitial spaces and finally into the capillaries of the thyroid gland. The remaining amino acids of thyroglobulin are used to synthesize more thyroglobulin.

Transport in the blood

Thyroid hormones are transported in combination with plasma protein. Approximately 70% to 75% of the circulating T_3 and T_4 are bound to **thyroxine-binding globulin (TBG),** which is synthesized by the liver. Bound to plasma proteins, T_3 and T_4 form a large reservoir of circulating thyroid hormones, and the half-life of these hormones is increased greatly because of this binding. After thyroid gland removal in experimental animals, it takes approximately 1 week for the T_3 and T_4 levels in the blood to decrease by 50%. As free T_3 and T_4 levels decrease in the interstitial spaces, additional T_3 and T_4 dissociate from plasma proteins to maintain the levels in the tissue spaces. When sudden secretion of T_3 and T_4 occurs, excess T_3 and T_4 is taken up by the plasma proteins. As a consequence, the concentration of thyroid hormones in the tissue spaces fluctuates very little.

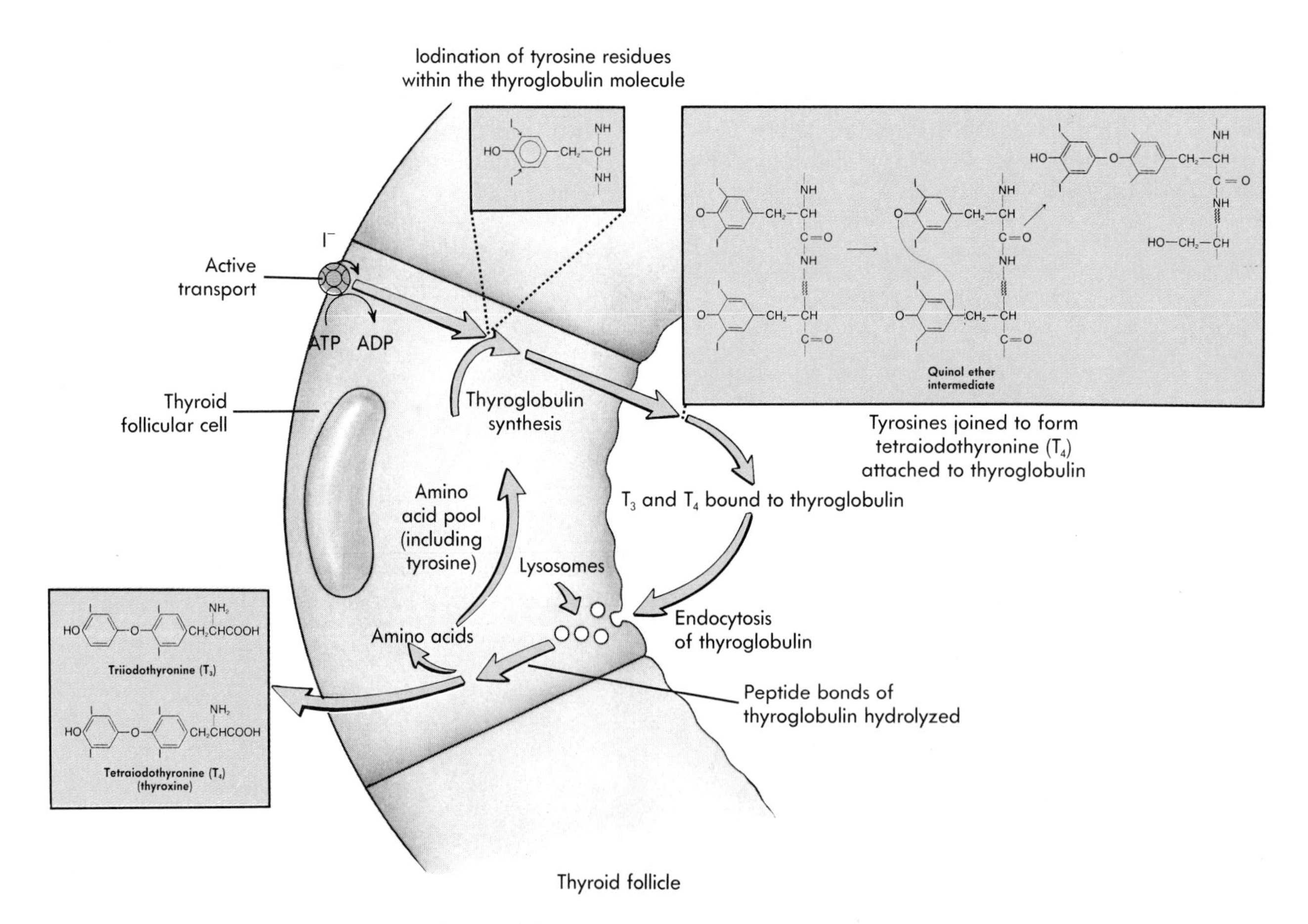

FIGURE 18-8 Biosynthesis of thyroid hormones. Thyroglobulin proteins are synthesized within the cells of the thyroid follicles, and iodide ions are actively transported into the thyroid follicle cells from the interstitial fluid. The tyrosine amino acids are iodinated within the thyroglobulin protein to form either triiodothyronine (T_3) or tetraiodothyronine (T_4). These hormones are secreted when the cells of the thyroid follicles take up the thyroglobulin and break the peptide bonds to release T_3 and T_4, which then diffuse into the blood. The hormones are transported by plasma proteins and enter target cells in which they bind to intracellular receptors.

Approximately 33% to 40% of the T_4 is converted to T_3 in the body tissues. This conversion may be important in the action of thyroid hormones on their target tissues since T_3 is the major hormone that interacts with the target cells. Much of the circulating T_4 is converted to tetraiodothyroacetic acid and is excreted in the urine or bile. In addition, a large amount is converted to an inactive form of T_3 and is rapidly metabolized and excreted.

Mechanism of action of thyroid hormones

Thyroid hormones interact with their target tissues in a fashion similar to that of the steroid hormones. Since they are nonpolar and lipid soluble, they readily diffuse through cell membranes into the cytoplasm of cells. Within the cell they bind to receptor molecules in the nuclei and initiate new protein synthesis. Thyroid hormones may also bind to and alter the function of mitochondria, resulting in greater ATP production and a greater rate of heat production. The newly synthesized proteins within the target cells mediate the response of the cell to the thyroid hormones. Approximately 1 week is needed for an observed response to occur after the administration of thyroid hormones, and new protein synthesis occupies much of that time.

Effects of thyroid hormones

Thyroid hormones affect nearly every tissue in the body, but not all tissues respond identically. Metabolism is primarily affected in some tissues, and growth and maturation are influenced in others.

The normal rate of metabolism for an individual depends on an adequate supply of thyroid hormone, which increases the rate of glucose, fat, and protein metabolism in many tissues, which in turn increases the body temperature. Low levels of thyroid hormones lead to the opposite effect. Normal body temperature depends on an adequate amount of thyroid hormone.

Normal growth and maturation of organs are also dependent on thyroid hormones. For example, bone, hair, teeth, connective tissue, and nervous tissue require thyroid hormone for normal growth and development. Also, T_3 and T_4 play a permissive role for growth hormone (GH), and GH does not have its normal effect on target tissues if the thyroid hormones are not present.

The specific effects of hyposecretion and hypersecretion of thyroid hormones are outlined in Table 18-4. Hypersecretion of thyroid hormone causes an increase in the rate of metabolism. High body temperature, weight loss, increased appetite, rapid heart rate, elevated blood pressure, and an enlarged thyroid gland are major symptoms.

Hyposecretion of thyroid hormone causes a decrease in the rate of metabolism. Low body temperature, weight gain, reduced appetite, reduced heart rate, reduced blood pressure, weak skeletal muscles, and apathy are major symptoms. If hyposecretion of thyroid hormones occurs in infants, in addition to a decreased rate of metabolism, abnormal nervous system development, abnormal growth, and abnormal

Table 18-4

Effects of hyposecretion and hypersecretion of thyroid hormones

HYPOTHYROIDISM	HYPERTHYROIDISM
Decreased metabolic rate, low body temperature, cold intolerance	Increased metabolic rate, high body temperature, heat intolerance
Weight gain, reduced appetite	Weight loss, increased appetite
Reduced activity of sweat and sebaceous glands, dry and cold skin	Copious sweating, warm and flushed skin
Reduced heart rate, reduced blood pressure, dilated and enlarged heart	Rapid heart rate, elevated blood pressure, abnormal electrocardiogram
Weak, flabby skeletal muscles, sluggish movements	Weak skeletal muscles that exhibit tremors, quick movements with exaggerated reflexes
Constipation	Bouts of diarrhea
Myxedema (swelling of the face and body) as a result of mucoprotein deposits	Exophthalmos (protruding of the eyes) as a result of mucoprotein and other deposits behind the eye
Apathetic, somnolent	Hyperactive, insomnia, restless, irritable, short attention span
Coarse hair, rough and dry skin	Soft, smooth hair and skin
Decreased iodide uptake	Increased iodide uptake
Possible goiter (enlargement of the thyroid gland)	Almost always develops goiter

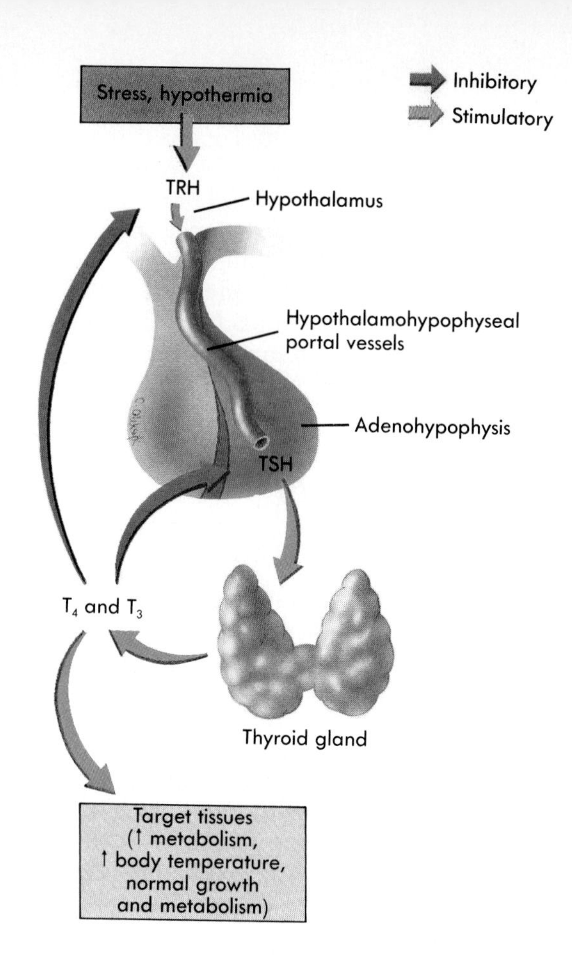

FIGURE 18-9 Regulation of thyroid hormone secretion. Thyrotropin-releasing hormone *(TRH)* is released from neurons within the hypothalamus and passes through the hypothalamohypophyseal portal blood vessels to the adenohypophysis, where it causes cells to secrete thyroid-stimulating hormone *(TSH)*. TSH passes through the general circulation to the thyroid gland, where it causes both increased synthesis and secretion of the thyroid hormones (T_3 and T_4). T_3 and T_4 have an inhibitory effect on the secretion of both TRH from the hypothalamus and TSH from the adenohypophysis.

Table 18-5

Abnormal thyroid conditions

CAUSE	DESCRIPTION
Hypothyroidism	
Iodine deficiency	Causes inadequate thyroid hormone synthesis, which results in elevated thyroid-stimulating hormone (TSH) secretion; thyroid gland enlarges (goiter) as a result of TSH stimulation; thyroid hormones frequently remain in the low-to-normal range
Goiterogenic substances	Found in certain drugs and in small amounts in certain plants such as cabbage; inhibit thyroid hormone synthesis
Cretinism	Caused by maternal iodine deficiency or congenital errors in thyroid hormone synthesis; results in mental retardation and a short, grotesque appearance
Lack of thyroid gland	Removed surgically or destroyed as a treatment for Grave's disease (hyperthyroidism)
Pituitary insufficiency	Results from lack of TSH secretion; often associated with inadequate secretion of other adenohypophyseal hormones
Hashimoto's disease	Autoimmune disease in which thyroid function is normal or depressed
Hyperthyroidism	
Grave's disease	Characterized by goiter and exophthalmos; apparently an autoimmune disease; most patients have long-acting thyroid stimulator, a TSH-like immune globulin, in their plasma
Tumors—benign adenoma or cancer	Result in either normal secretion or hypersecretion of thyroid hormones (rarely hyposecretion)
Thyroiditis—a viral infection	Produces painful swelling of the thyroid gland with normal or slightly increased thyroid hormone production
Elevated TSH levels	Result from a pituitary tumor
Thyroid storm	Sudden release of large amounts of thyroid hormones; caused by surgery, stress, infections, and unknown reasons

maturation of tissues occur. The consequence is a mentally retarded person of short stature and distinctive form called a **cretin.**

Regulation of thyroid hormone secretion

The most important regulator of thyroid hormone secretion is TSH. Small fluctuations occur in blood TSH levels on a daily basis, with a small nocturnal increase. The effects of increased TSH levels are increased synthesis and secretion of T_3 and T_4 and hypertrophy (increased cell size) and hyperplasia (increased cell number) of the thyroid gland. Decreased blood levels of TSH lead to decreased T_3 and T_4 secretion and thyroid gland atrophy. Figure 18-9 illustrates the regulation of thyroid hormone secretion by TSH and the negative-feedback mechanisms that influence TSH secretion. Note that T_3 and T_4 inhibit TSH secretion by acting on the adenohypophysis and hypothalamus. If the thyroid gland is removed or if T_3 and T_4 secretion declines, TSH levels increase dramatically.

The secretion of TSH is initiated by a hypothalamic-releasing hormone called **thyrotropin-releasing hormone (TRH).** Several phenomena affect TRH release from the hypothalamus. Exposure to cold and stress cause increased TRH secretion. As a consequence, TSH and thyroid hormone levels increase. In contrast, prolonged fasting decreases TRH secretion; subsequently, thyroid hormone levels decrease. The thyroid hormones play a role in acclimation to chronic exposure to cold, stress, and food deprivation. In all of these conditions the response is most important if the condition is relatively chronic since approximately 1 week is required before a measurable change in metabolism occurs in response to the thyroid hormones.

2

Explain why the thyroid gland enlarges in response to iodine deficiency in the diet.

? ? ? ? ? ? ? ? ? ? ? ?

Pathologies associated with abnormal thyroid hormone secretion are outlined in Table 18-5.

Calcitonin

The parafollicular cells of the thyroid gland, which secrete calcitonin, are dispersed between the thyroid follicles throughout the thyroid gland. The major stimulus for increased calcitonin secretion is an increase in calcium levels in the body fluids.

The primary target tissue for calcitonin is bone (see Chapter 6). Calcitonin decreases osteoclast activity, and it lengthens the life span of osteoblasts. The result is a decrease in blood calcium and phosphate levels caused by increased bone deposition.

The importance of calcitonin in the regulation of blood calcium levels is unclear. Its rate of secretion increases in response to elevated blood levels of calcium, and blood levels of calcitonin decrease with age to a greater extent in women than men (osteoporosis increases with age and occurs to a greater degree in women than men). However, complete thyroidectomy does not result in high blood levels of calcium. It is possible that the regulation of blood calcium levels by other hormones (e.g., parathyroid hormone and vitamin D) compensates for the loss of calcitonin in individuals who have undergone a thyroidectomy. No pathological condition is associated directly with a lack of calcitonin secretion.

PARATHYROID GLANDS

The **parathyroid glands** are usually embedded in the posterior portion of each lobe of the thyroid gland. There are usually four parathyroid glands with their cells organized in densely packed masses or cords rather than in follicles (Figure 18-10).

The parathyroid glands secrete **parathyroid hormone (PTH),** which is important in the regulation of calcium levels in body fluids (see Table 18-3). Bone, the kidneys, and the intestine are its major target tissues. Without functional parathyroid glands, the ability to adequately regulate blood calcium levels is lost.

PTH stimulates osteoclast activity in bone and may cause the number of osteoclasts to increase. The response to PTH includes osteoclast activity to release calcium and phosphate, causing an increase in blood calcium levels. PTH may also induce release of calcium from osteoblasts and shorten their life span.

PTH induces calcium reabsorption within the kidneys so that less calcium leaves the body in urine. PTH also increases the enzymatic formation of active vitamin D in the kidneys. Calcium is actively absorbed by the epithelial cells of the small intestine, and the synthesis of transport proteins in the intestinal cells requires active vitamin D. PTH increases the rate of active vitamin D synthesis, which, in turn, increases the rate of calcium and phosphate absorption in the intestine, elevating blood levels of calcium.

Although PTH increases the release of phosphate ions from bone and increases phosphate ion absorption in the gut, it increases phosphate ion excretion in the kidney. The overall effect of PTH is to decrease blood phosphate levels.

The regulation of PTH secretion is outlined in Figure 18-11. The primary stimulus for the secretion of PTH is a decrease in plasma calcium levels, whereas elevated plasma calcium levels inhibit PTH

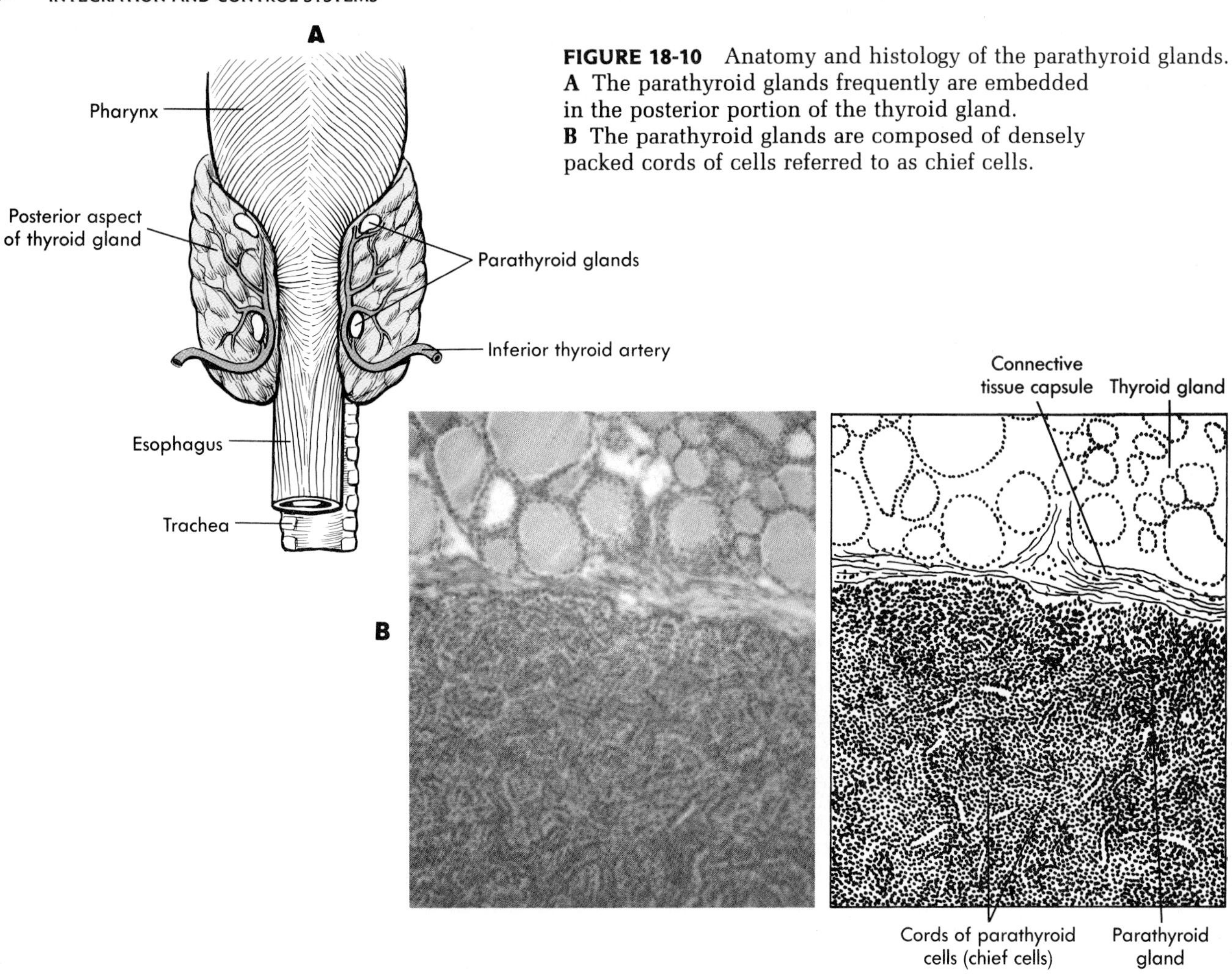

FIGURE 18-10 Anatomy and histology of the parathyroid glands. **A** The parathyroid glands frequently are embedded in the posterior portion of the thyroid gland. **B** The parathyroid glands are composed of densely packed cords of cells referred to as chief cells.

Table 18-6

Causes and symptoms of hypersecretion and hyposecretion of parathyroid hormone

HYPOPARATHYROIDISM	HYPERPARATHYROIDISM
Causes	
Accidental removal during thyroidectomy Idiopathic (unknown cause)	Primary hyperparathyroidism: a result of abnormal parathyroid function—adenomas of the parathyroid gland (90%), hyperplasia of parathyroid cells (9%), and carcinomas (1%) Secondary hyperparathyroidism: caused by conditions that reduce blood calcium levels such as inadequate calcium in the diet, inadequate levels of vitamin D, pregnancy, or lactation
Symptoms	
Hypocalcemia	Hypercalcemia or normal blood calcium levels; calcium carbonate salts may be deposited throughout the body, especially in the renal tubules (kidney stones), lungs, blood vessels, and gastric mucosa
Normal bone structure	Bones weak and eaten away as a result of resorption; some cases are first diagnosed when a broken bone is x-rayed
Increased neuromuscular excitability; tetany, laryngospasm, and death from asphyxiation can result	Neuromuscular system less excitable; muscular weakness may be present
Flaccid heart muscle; cardiac arrhythmia may develop	Increased force of contraction of cardiac muscle; at very high levels of calcium, cardiac arrest during contraction is possible
Diarrhea	Constipation

secretion. This regulation keeps calcium blood levels fluctuating within a normal range of values. Both hypersecretion and hyposecretion of PTH cause serious symptoms (Table 18-6).

3

Predict the effect of an inadequate dietary intake of calcium on PTH secretion and on PTH target tissues.

? ? ? ? ? ? ? ? ? ? ? ?

Inactive parathyroid glands result in hypocalcemia, which causes an increased permeability of cell membranes to sodium ions. As a consequence, sodium ions diffuse into cells and cause depolarization (see Chapter 9). Nervousness, muscle spasms, cardiac arrythmias, and convulsions in extreme cases are symptoms of hypocalcemia. Death may result because of tetany of the respiratory muscles.

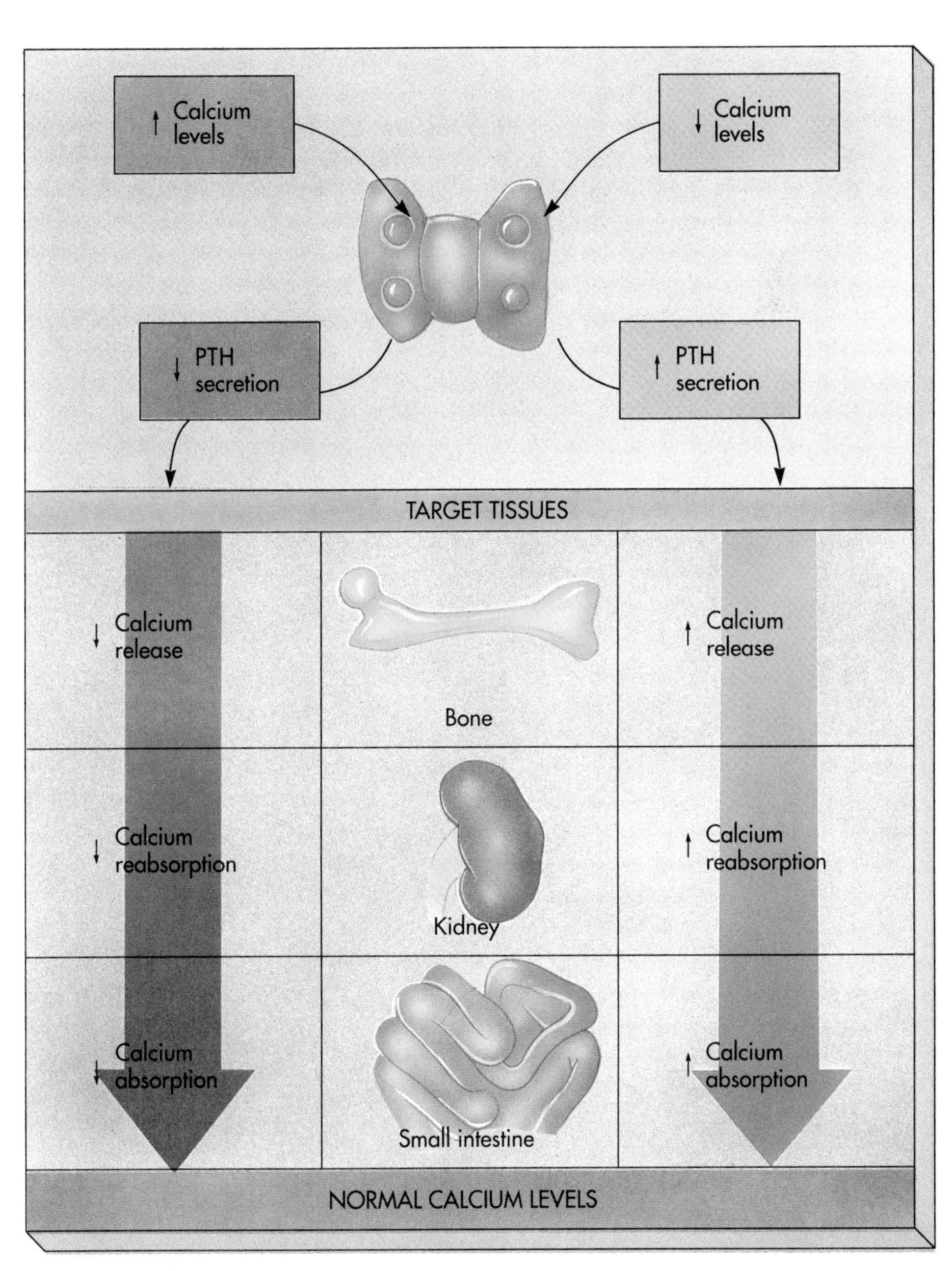

FIGURE 18-11 Regulation of parathyroid hormone *(PTH)* secretion and its effects on target tissues. Low blood levels of calcium ions stimulate PTH secretion, and high blood levels of calcium ions inhibit PTH secretion. PTH acts on target tissues, including bone, the kidneys, and the gastrointestinal tract, to increase the concentration of calcium in the blood. PTH also increases vitamin D synthesis, which is required for the gastrointestinal tract to absorb calcium ions.

4

A patient with a malignant tumor had his thyroid gland removed. What effect would this removal have on blood levels of thyroid hormone, thyrotropin-releasing hormone (TRH), TSH, and calcitonin? What would result if the parathyroid glands were inadvertently damaged during surgery?

? ? ? ? ? ? ? ? ? ? ?

ADRENAL GLANDS

The **adrenal glands,** also called the **suprarenal glands,** are near the superior pole of each kidney. Like the kidneys, they lie posterior to the parietal peritoneum and are surrounded by abundant adipose tissue. They are enclosed by a connective tissue capsule and receive a well-developed blood supply (Figure 18-12, *A*).

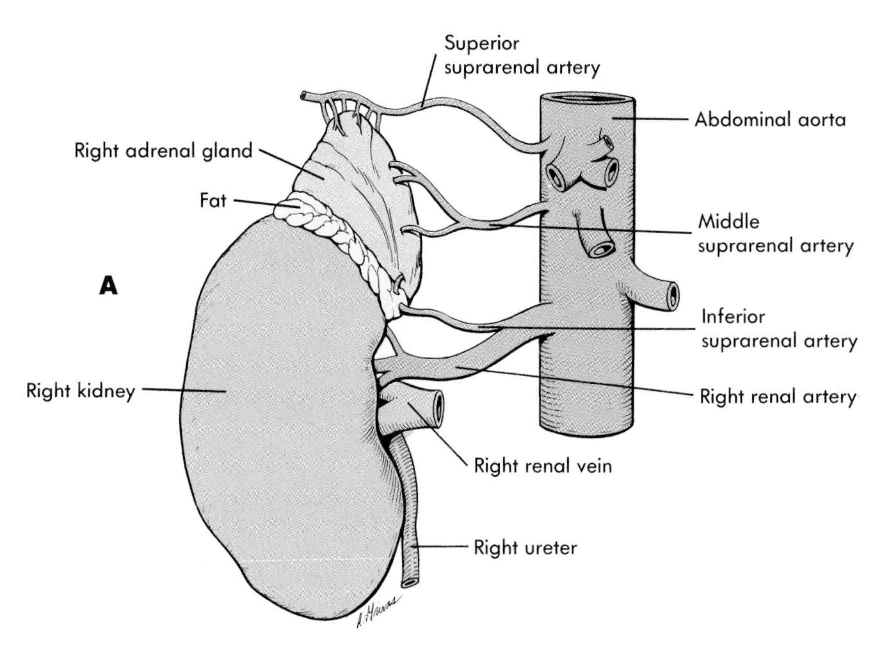

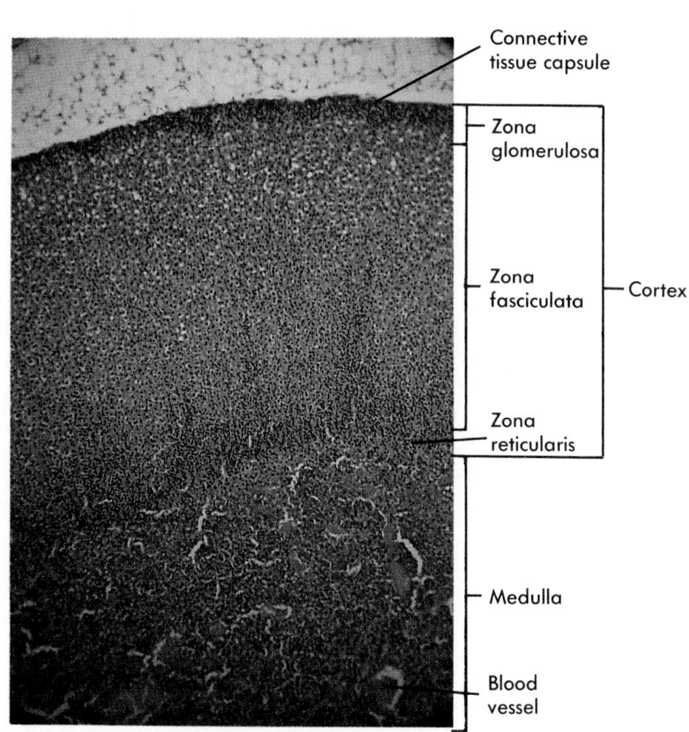

FIGURE 18-12 Anatomy and histology of the adrenal gland. **A** The adrenal glands are at the superior pole of each kidney. **B** The adrenal glands have an outer cortex and an inner medulla. The cortex consists of three layers: the zona glomerulosa, the zona fasciculata, and the zona reticularis.

The adrenal glands are composed of an inner **medulla** and an outer **cortex,** which are derived from two separate embryonic tissues. The adrenal medulla arises from neural crest cells and consists of specialized postganglionic neurons of the sympathetic division of the autonomic nervous system (see Chapter 16). Unlike most glands of the body, which develop from epithelial tissue, the adrenal cortex is derived from mesoderm.

Histology

Trabeculae of the connective tissue capsule penetrate into the adrenal gland in several locations, and numerous small blood vessels course with them to supply the gland. The medulla consists of closely packed polyhedral cells centrally located in the gland (see Figure 18-12, *B*). The cortex is comprised of smaller cells and forms three indistinct layers: the **zona glomerulosa, zona fasciculata,** and **zona reticularis.** These three layers are functionally and structurally specialized. The zona glomerulosa is immediately beneath the capsule and is composed of small clusters of cells. Beneath the zona glomerulosa is the thickest portion of the adrenal cortex, the zona fasciculata. In this layer the cells form long columns, or fascicles, of cells that extend from the surface toward the medulla of the gland. The deepest layer of the adrenal cortex is the zona reticularis, a thin layer of irregularly arranged cords of cells.

Hormones of the Adrenal Medulla

The adrenal medulla secretes two major hormones: **epinephrine (adrenaline),** 80%, and **norepinephrine (noradrenaline),** 20% (Table 18-7). The adrenal medulla is a component of the autonomic nervous system, and its secretory products are neurohormones.

Epinephrine increases blood levels of glucose. It combines with receptors in the liver cells and activates cyclic AMP synthesis within the cells. Cyclic AMP, in turn, activates enzymes that catalyze the breakdown of glycogen to glucose, causing its release into the blood. Epinephrine also increases glycogen breakdown, the intracellular metabolism of glucose in skeletal muscle cells, and the breakdown of fats in adipose tissue. Epinephrine and norepinephrine increase the heart rate and the force of contraction of the heart and cause blood vessels to constrict in the skin, kidneys, gastrointestinal tract, and other viscera; epinephrine causes dilation of blood vessels in skeletal muscles and cardiac muscle.

Secretion of adrenal medullary hormones prepares the individual for physical activity and is a major component of the fight-or-flight response (see Chapter 16). The response results in reduced activity in organs not essential for physical activity and in increased blood flow and metabolic activity in organs that participate in physical activity. Additionally, it

Table 18-7

Hormones of the adrenal gland

HORMONE	STRUCTURE	TARGET TISSUE	RESPONSE
Adrenal Medulla			
Epinephrine primarily; norepinephrine	Amino acid derivative	Heart, blood vessels, liver, fat cells	Increased cardiac output; increased blood flow to skeletal muscles and heart; increased release of glucose and fatty acids into blood; in general, preparation for physical activity
Adrenal Cortex			
Cortisol	Steroid	Most tissues	Increased protein and fat breakdown; increased fat production; inhibition of immune response
Aldosterone	Steroid	Kidney	Increased sodium ion reabsorption and potassium and hydrogen ion excretion
Sex steroids	Steroids	Many tissues	Minor importance in males; in females, development of some secondary sexual characteristics such as axillary and pubic hair

mobilizes nutrients that can be used to sustain physical exercise.

The effects of epinephrine and norepinephrine are short-lived because they are rapidly metabolized, excreted, or taken up by tissues. Their half-life in the circulatory system is measured in terms of minutes.

Regulation

The release of adrenal medullary hormones primarily occurs in response to stimulation by sympathetic neurons since the adrenal medulla is a specialized portion of the autonomic nervous system. Several conditions lead to the release of adrenal medullary neurohormones, including emotional excitement, injury, stress, exercise, and low blood glucose levels (Figure 18-13).

The two major disorders of the adrenal medulla are tumors: pheochromocytoma, a benign tumor, and neuroblastoma, a malignant tumor. Symptoms result from the release of large amounts of epinephrine and norepinephrine and include hypertension (high blood pressure), sweating, nervousness, pallor, and tachycardia (rapid heart rate). The high blood pressure is due to the effect of these hormones on the heart and blood vessels and is correlated with an increased chance of heart disease and stroke.

Hormones of the Adrenal Cortex

The adrenal cortex secretes three hormone types: **mineralocorticoids, glucocorticoids,** and **androgens** (see Table 18-7). All are similar in structure in that they are steroids, highly specialized lipids that are derived from cholesterol. Since they are lipid soluble, they are not stored in the adrenal gland cells but diffuse from the cells as they are synthesized. Adrenal cortical hormones are transported in the blood in combination with specific plasma proteins, and they are metabolized in the liver and are excreted in the bile and urine.

Mineralocorticoids

The major secretory products of the zona glomerulosa are the mineralocorticoids. **Aldosterone** (al-dos'ter-ōn) is produced in the greatest amounts, although other closely related mineralocorticoids are also secreted. Aldosterone increases the rate of sodium reabsorption by the kidneys, thereby increasing blood levels of sodium. Sodium reabsorption results in increased water reabsorption by the kidneys and an increase in blood volume. Aldosterone increases potassium excretion into the urine by the kidneys, decreasing blood levels of potassium. It also increases the rate of hydrogen ion excretion into the

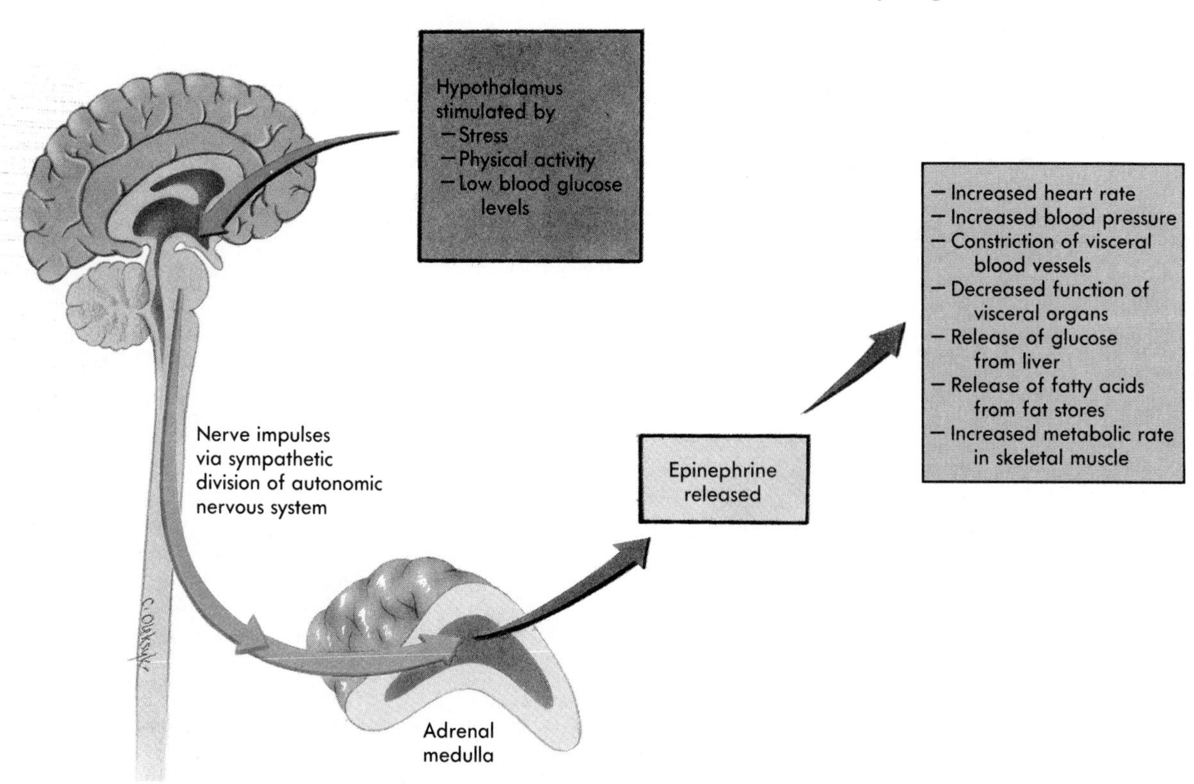

FIGURE 18-13 Regulation of adrenal medullary secretions. Sympathetic stimulation results in increased secretion of epinephrine and norepinephrine. Stress, physical exercise, and other conditions that cause increased activity of the sympathetic nervous system increase secretion from the adrenal medulla.

urine, and when present in high concentrations, aldosterone may result in alkalosis (elevated pH of body fluids). The details of aldosterone's effects and its regulation are discussed later in this text along with kidney function (Chapters 26 and 27) and the cardiovascular system (Chapter 21).

5

Alterations in blood levels of sodium and potassium have profound effects on the electrical properties of cells. Since high blood levels of aldosterone cause retention of sodium and excretion of potassium, predict and explain the effects of high aldosterone levels on nerve and muscle function. Conversely, since low blood levels of aldosterone cause low blood levels of sodium and elevated blood levels of potassium, predict the effects of low aldosterone levels on nerve and muscle function.

? ? ? ? ? ? ? ? ? ? ?

Glucocorticoids

The zona fasciculata primarily secretes glucocorticoid hormones; the major human glucocorticoid is **cortisol** (kor'tĭ-sol). The target tissues and responses to the glucocorticoids are numerous (Table 18-8). The responses are classified as metabolic, developmental, or anti-inflammatory. Glucocorticoids increase fat catabolism, decrease glucose and amino acid uptake in skeletal muscle, increase synthesis of glucose from amino acids in the liver, and increase protein degradation. Thus some major effects of glucocorticoids are to increase the metabolism of fats and proteins and to cause blood glucose levels and glycogen deposits in cells to increase. As a result, a reservoir of molecules that can be metabolized rapidly is available to cells. Glucocorticoids are also required for the maturation of tissues such as fetal lungs and for the development of receptor sites in target tissues for epinephrine and norepinephrine. Glucocorticoids decrease the intensity of the inflammatory response by decreasing both the number of white blood cells and the secretion of inflammatory chemicals from tissues. This anti-inflammatory effect is most important under conditions of stress when the rate of glucocorticoid secretion is relatively high.

Adrenocorticotropic hormone (ACTH) is required to maintain the secretory activity of the adrenal cortex, which rapidly atrophies without this hormone. The regulation of ACTH and cortisol secretion is outlined in Figure 18-14. **Corticotropin-releasing hormone (CRH)** is released from the hypothalamus and stimulates ACTH secretion from the adenohypophysis. ACTH and cortisol inhibit CRH secretion from the hypothalamus and thus constitute a negative-feedback influence on CRH secretion. In addition, high concentrations of cortisol in the blood inhibit, and low concentrations of cortisol stimulate, ACTH secretion from the adenohypophysis. This negative-feedback loop plays an important role in maintaining blood cortisol levels within a narrow range of concentrations. In response to stress or hypoglycemia, blood levels of cortisol increase rapidly because these stimuli trigger a large increase in CRH release from the hypothalamus. Table 18-9 outlines several abnormalities associated with hypersecretion and hyposecretion of adrenal hormones.

6

A drug similar to cortisol, cortisone, sometimes is given to people who have severe allergies. Taking this substance chronically may damage the adrenal cortex. Explain how that damage may occur.

? ? ? ? ? ? ? ? ? ? ?

Table 18-8

Target tissues and their responses to glucocorticoid hormones

TARGET TISSUES	RESPONSES
Peripheral tissues such as skeletal muscle, liver, and adipose tissue	Inhibits glucose use; stimulates amino acid uptake and formation of glucose from amino acids (gluconeogenesis), which results in elevated blood glucose levels; stimulates glycogen synthesis in cells; mobilizes fats by increasing lipolysis, which results in the release of fatty acids into the blood and in increased rate of fatty acid metabolism; increases protein catabolism
Immune tissues	Anti-inflammatory—depresses antibody production, white blood cell production, and the release of inflammatory components in response to injury
Target cells for epinephrine	Receptor molecules for epinephrine and norepinephrine decrease without adequate amounts of glucocorticoid hormone

Adrenal androgens

Some adrenal steroids, including **androstenedione** (an-dro-stēn′dī-ōn), are weak androgens. They are secreted by the zona reticularis and are converted by peripheral tissues to the more potent androgen, testosterone. Adrenal androgens stimulate pubic and axillary hair growth and sexual drive in females. Their effects in males are negligible in comparison to testosterone secreted by the testes. Additional information about androgens is presented in Chapter 28.

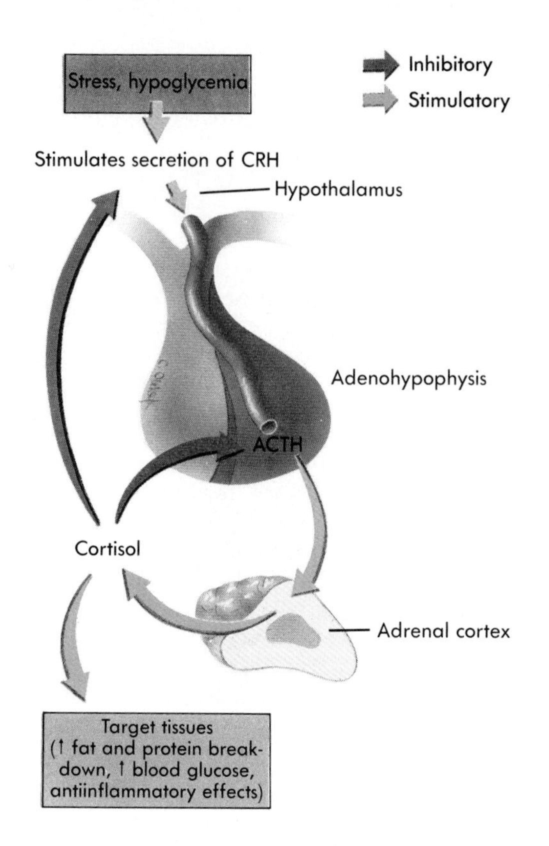

FIGURE 18-14 Regulation of cortisol secretion. Corticotropin-releasing hormone *(CRH)* is released from hypothalamic neurons and passes, by way of the hypothalamohypophyseal portal blood vessels, to the adenohypophysis, where it binds to and stimulates cells that secrete adrenocorticotropic hormone (ACTH). ACTH binds to membrane-bound receptors on cells of the adrenal cortex and stimulates the secretion of glucocorticoids, primarily cortisol, and small quantities of sex steroids. Cortisol, in turn, inhibits CRH and ACTH secretion.

Table 18-9

Symptoms of hyposecretion and hypersecretion of adrenal cortex hormones

HYPOSECRETION	HYPERSECRETION
Aldosterone	
Hyponatremia (low blood levels of sodium)	Slight hypernatremia (high blood levels of sodium)
Hyperkalemia (high blood levels of potassium)	Hypokalemia (low blood levels of potassium)
Acidosis	Alkalosis
Low blood pressure	High blood pressure
Tremors and tetany of skeletal muscles	Weakness of skeletal muscles
Polyuria	Acidic urine
Cortisol	
Hypoglycemia (low blood glucose levels)	Hyperglycemia (high blood glucose levels; adrenal diabetes)—leads to diabetes mellitus
Depressed immune system	Depressed immune system
Protein and fats from diet unused, resulting in weight loss	Destruction of tissue proteins, causing muscle atrophy and weakness, osteoporosis, weak capillaries (easily bruised), thin skin, and impaired wound healing; mobilization and redistribution of fats, causing depletion of fat from limbs and deposition in face (moon face), neck (buffalo hump), and abdomen
Loss of appetite, nausea, and vomiting	Emotional effects, including euphoria and depression
Increased skin pigmentation (caused by elevated ACTH)	Thin skin
Androgens	
In women reduction of pubic and axillary hair	In women hirsutism (excessive facial and body hair), acne, increased sex drive, regression of breast tissue, and loss of regular menses

Hormone Pathologies of the Adrenal Cortex

Several pathologies are associated with abnormal secretion of adrenal cortex hormones.

Addison's disease results from abnormally low levels of aldosterone and cortisol. The cause of many cases of Addison's disease is unknown (idiopathic), but it is a suspected autoimmune disease in which the body's defense mechanisms inappropriately destroy the adrenal cortex. Some cases of Addison's disease are caused by the destruction of the adrenal cortex by bacteria or cancer.

Aldosteronism is caused by excess production of aldosterone. Primary aldosteronism results from an adrenal cortex tumor, and secondary aldosteronism occurs when some extraneous factor (e.g., overproduction of renin, a substance produced by the kidney) increases aldosterone secretion.

Cushing's syndrome (Figure 18-A) is a disorder characterized by hypersecretion of cortisol and androgens; there may also be excess aldosterone production. The majority of cases are caused by excess ACTH production by nonpituitary tumors (usually a type of lung cancer) or by pituitary tumors. Sometimes adrenal tumors or unidentified causes may be responsible for hypersecretion of the adrenal cortex.

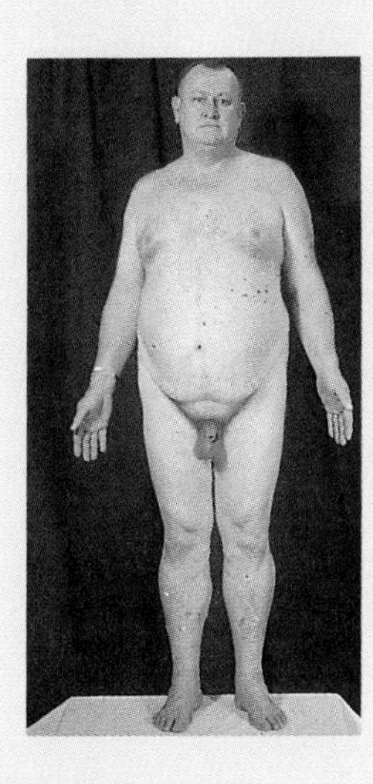

FIGURE 18-A
Patient with Cushing's syndrome

Hypersecretion of androgens from the adrenal cortex causes a condition called adrenogenital syndrome, in which early development of secondary sexual characteristics in male children and masculinization of female children occur. If the condition develops before birth in females, the external genitalia may be masculinized to the extent that the infant's reproductive structures may be neither clearly female nor male. Hypersecretion of adrenal androgens in male children before puberty results in rapid and early development of the reproductive system. If not treated, early sexual development and a short stature result. The short stature results from the effect of testosterone on skeletal growth. In adult females partial development of male secondary sexual characteristics (e.g., facial hair and a masculine voice) occurs.

PANCREAS

The **pancreas** (pan'kre-us) lies behind the peritoneum between the greater curvature of the stomach and the duodenum. It is an elongated structure approximately 15 cm long, weighing approximately 85 to 100 g. The head of the pancreas lies near the duodenum, and its body and tail extend toward the spleen.

Histology

The pancreas is both an exocrine gland and an endocrine gland. The exocrine portion consists of acini, which produce pancreatic juice, and a duct system, which carries pancreatic juice to the small intestine (see Chapter 24). The endocrine portion, consisting of pancreatic islets (islets of Langerhans), produces hormones that enter the circulatory system.

There are 500,000 to 1,000,000 pancreatic islets (Figure 18-15) dispersed among the ducts and acini of the pancreas. Each islet is composed of **alpha cells** (20%), which secrete glucagon, **beta cells** (75%), which secrete insulin, and other cell types (5%). The remaining cells are either immature cells of questionable function or **delta cells,** which secrete somatostatin. Nerves from both divisions of the autonomic nervous system innervate the pancreatic islets, and each islet is surrounded by a well-developed capillary network.

Effect of Insulin and Glucagon on Their Target Tissues

The pancreatic hormones play an important role in regulating the concentration of certain nutrients in the circulatory system, especially glucose (i.e., blood sugar) and amino acids (Table 18-10). Insulin's major target tissues are the liver, adipose tissue, muscles, and the satiety center within the hypothalamus of the brain. The satiety center is a collection of neurons in the hypothalamus that control appetite, but insulin does not directly affect most areas of the nervous system. The specific effects of insulin on these target tissues are listed in Table 18-11. Insulin binds to insulin receptors on the membranes of its target cells. Subsequently, the insulin and receptor mole-

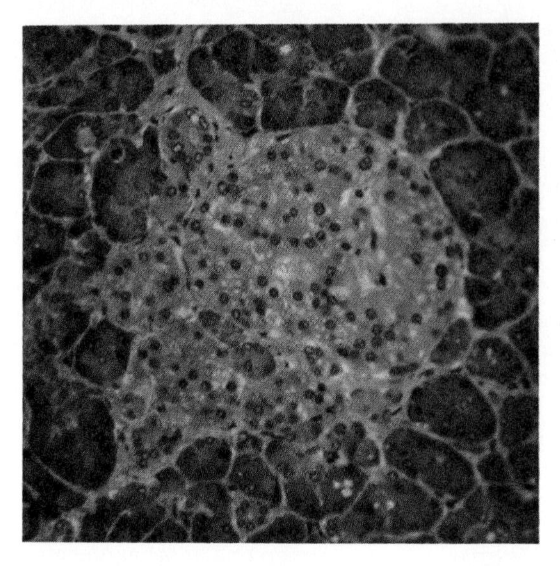

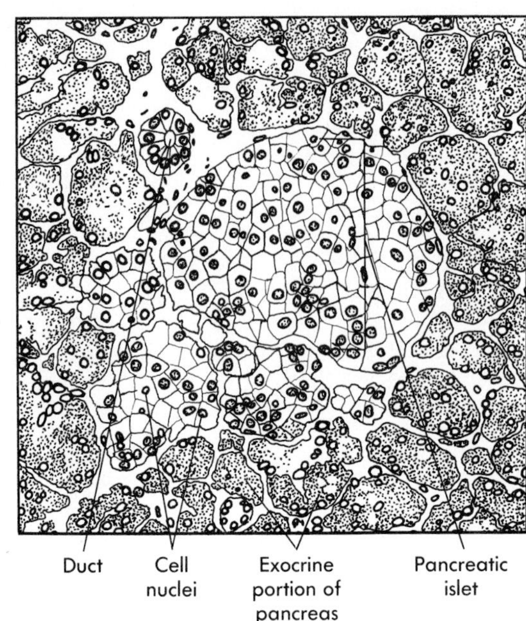

FIGURE 18-15 Histology of the pancreatic islets.

Table 18-10

Pancreatic hormones

CELLS IN ISLETS	HORMONE	STRUCTURE	TARGET TISSUE	RESPONSE
Beta	Insulin	Protein	Especially liver, skeletal muscle, fat tissue	Increased uptake and use of glucose and amino acids
Alpha	Glucagon	Polypeptide	Liver primarily	Increased breakdown of glycogen; release of glucose into the circulatory system
Delta	Somatostatin	Peptide	Alpha and beta cells (some somatostatin is produced in the hypothalamus)	Inhibition of insulin and glucagon secretion

cules are taken through endocytosis into the cell where insulin's effect may be mediated by a second messenger and by the effect of the insulin-receptor complex on the cell membrane. In general, insulin increases its target tissue's ability to take up and use glucose and amino acids. Glucose molecules that are not needed immediately as an energy source to maintain cell metabolism are stored as glycogen in skeletal muscle, liver, and other tissues and are converted to fat in adipose tissue. Amino acids can be broken down and used as an energy source or to synthesize glucose or can be converted to protein. Without insulin, the ability of these tissues to accept glucose and amino acids and use them is minimal.

In the presence of too much insulin, target tissues accept glucose rapidly from the circulatory system; as a consequence, blood levels of glucose decline to very low levels. Although most of the nervous system, with the exception of the satiety center, is not a target tissue for insulin, insulin plays an important role in regulating the concentration of blood glucose on which the nervous system depends. Since the nervous system is dependent on glucose as a nutrient, low blood levels of glucose cause the central nervous system to malfunction.

In the absence of insulin the uptake of glucose and amino acids declines dramatically, even though blood levels of these substances may be very high. To take up glucose the satiety center requires insulin. In the absence of insulin the satiety center is unable to detect the presence of glucose in the extracellular fluid even though it may be present at high levels. The result is an intense sensation of hunger in spite of the high blood glucose levels.

Glucagon primarily influences the liver, although it has some effect on skeletal muscle and adipose

Table 18-11

Effects of insulin and glucagon on target tissues

TARGET TISSUE	RESPONSE TO INSULIN	RESPONSE TO GLUCAGON
Skeletal muscle, cardiac muscle, cartilage, bone, fibroblasts, leukocytes, and mammary glands	Increased glucose uptake and glycogen synthesis; increased uptake of certain amino acids	Little effect
Liver	Increased glycogen synthesis; increased the use of glucose for energy (glycolysis)	Causes rapid increase in the breakdown of glycogen to glucose (glycogenolysis) and release of glucose into the blood Increased formation of glucose (gluconeogenesis) from amino acids and, to some degree, from fats Increased metabolism of fatty acids, resulting in increased ketones in the blood
Adipose cells	Increased glucose uptake, glycogen synthesis, fat synthesis, and fatty acid uptake; increased glycolysis	High concentrations cause breakdown of fats (lipolysis); probably unimportant under most conditions
Nervous system	Little effect except to increase glucose uptake in the satiety center	No effect

tissue (see Table 18-11). In general, glucagon causes the breakdown of glycogen and increased glucose synthesis in the liver. It also increases the breakdown of fats. The amount of glucose released from the liver into the blood increases dramatically after an increase in glucagon secretion. Because glucagon is secreted into the hepatic portal vein, which carries blood from the intestine and pancreas to the liver, glucagon is delivered in a relatively high concentration to the liver in which it is metabolized rapidly. Thus it has less effect on skeletal muscles and adipose tissue.

Regulation of Pancreatic Hormone Secretion

The secretion of insulin is under chemical, neural, and hormonal control. Hyperglycemia, or elevated blood levels of glucose, directly affects the beta cells and stimulates insulin secretion. Hypoglycemia, or low blood levels of glucose, directly inhibits insulin secretion. Thus blood glucose levels play a major role in the regulation of insulin secretion. Certain amino acids also stimulate insulin secretion by acting directly on the beta cells. After a meal when glucose and amino acid levels increase in the circulatory system, insulin secretion increases. During periods of fasting when blood glucose levels are low, the rate of insulin secretion declines (Figure 18-16).

The autonomic nervous system also controls insulin secretion. Since parasympathetic stimulation is associated with food intake, its stimulation acts with the elevated blood glucose levels to increase insulin secretion. Sympathetic innervation inhibits insulin secretion and helps prevent a rapid fall in blood glucose levels. Since most tissues, except nervous tissue, require insulin to take up glucose, sympathetic stimulation maintains blood glucose levels in a normal range during periods of physical activity or excitement. This response is important for the maintenance of normal nervous system function.

Gastrointestinal hormones involved with the regulation of digestion (e.g., gastrin, secretin, cholecystokinin; see Chapter 24) act to increase insulin secretion. **Somatostatin** inhibits insulin and glucagon secretion, but the factors that regulate somatostatin secretion are not clear. It may be released in response to food intake, in which case somatostatin may prevent over-secretion of insulin.

7

Explain why the increase in insulin secretion in response to parasympathetic stimulation and gastrointestinal hormones is consistent with the maintenance of homeostasis.

? ? ? ? ? ? ? ? ? ? ?

Low blood glucose levels stimulate, and high blood glucose levels inhibit, glucagon secretion. Certain amino acids and sympathetic stimulation also increase glucagon secretion. After a high-protein meal, amino acids increase both insulin and glucagon secretion. Insulin causes target tissues to accept the amino acids for protein synthesis, and glucagon increases the process of glucose synthesis from amino acids in the liver (gluconeogenesis). Both protein synthesis and the use of amino acids to maintain blood glucose levels result from the low, but simultaneous, secretion of insulin and glucagon induced by a high-protein intake.

8

Compare the regulation of glucagon and insulin secretion after a meal high in carbohydrates, after a meal low in carbohydrates but high in proteins, and during physical exercise.

? ? ? ? ? ? ? ? ? ? ?

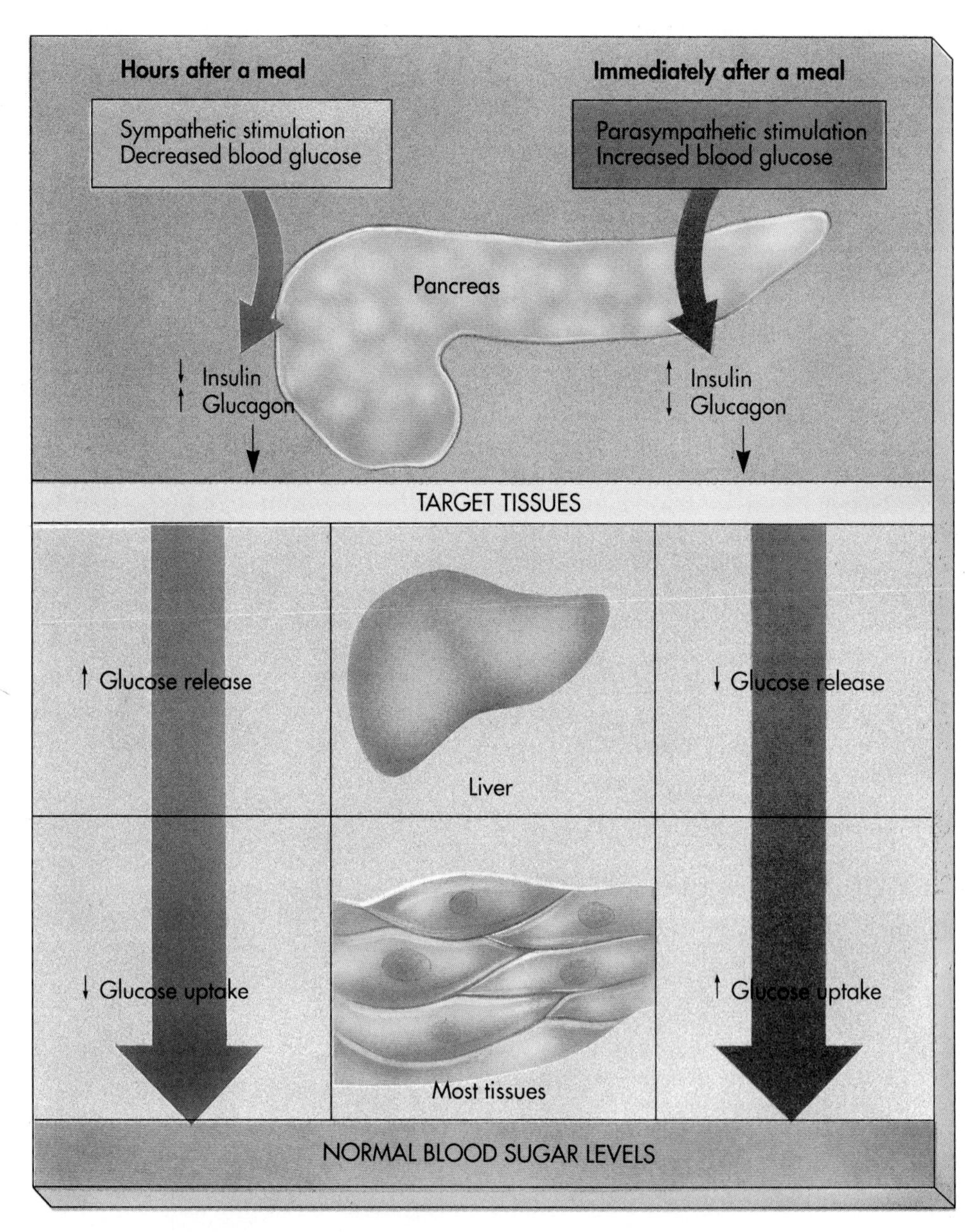

FIGURE 18-16 Regulation of insulin and glucagon secretion.
Sympathetic stimulation and decreasing concentrations of glucose increase the secretion of glucagon, which acts primarily on liver cells to increase the rate of glycogen breakdown and the secretion of glucose from the liver. The release of glucose from the liver helps maintain blood glucose levels. Increasing blood glucose levels have an inhibitory effect on glucagon secretion.
Increasing concentrations of blood glucose and amino acids stimulate the beta cells of the islets to secrete insulin. In addition, parasympathetic stimulation causes insulin secretion. Insulin acts on most tissues to increase the uptake and use of glucose and amino acids. As the blood levels of glucose and amino acids decrease, the rate of insulin secretion also decreases.

HORMONAL REGULATION OF NUTRIENTS

Two different situations— after a meal and during exercise—are presented to illustrate how several hormones function together to regulate blood nutrient levels.

After a meal and under resting conditions, secretion of glucagon, cortisol, growth hormone, and epinephrine is reduced (Figure 18-17, *A*). The high blood glucose levels and parasympathetic stimulation elevate insulin secretion, increasing the uptake of glucose, amino acids, and fats by target tissues. Substances not immediately used for cell metabolism are stored. Glucose is converted to glycogen in skel-

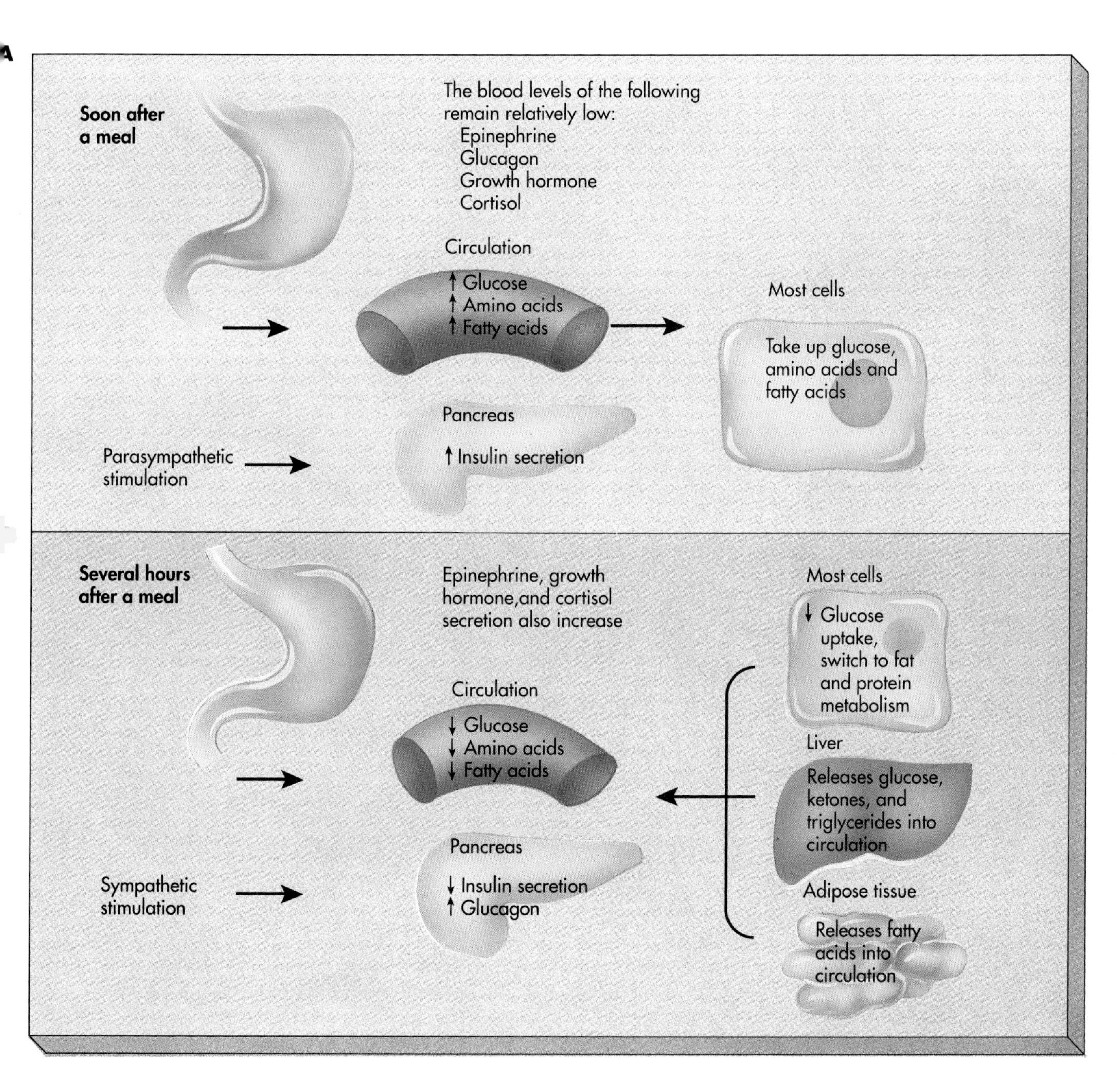

FIGURE 18-17 Regulation of blood nutrient levels after a meal.

A Soon after a meal glucose, amino acids, and fatty acids enter the bloodstream from the intestinal tract. Glucose and amino acids stimulate insulin secretion. In addition, parasympathetic stimulation increases insulin secretion. Cells take up the glucose and amino acids and use them in their metabolism.

B Several hours after a meal absorption from the intestinal tract decreases, and blood levels of glucose, amino acids, and fatty acids decrease. As a result, insulin secretion decreases, and glucagon, epinephrine, and growth hormone secretion increase. Cell uptake of glucose decreases, and usage of fats and proteins increases.

Diabetes Mellitus

Diabetes mellitus results from inadequate secretion of insulin, the secretion of abnormal insulin, or the inability of tissues to respond to insulin. Insulin hyposecretion is usually caused by degeneration of the beta cells in the pancreatic islets. Juvenile-onset diabetes apparently has an etiology different from maturity-onset diabetes, although both may result from diminished insulin secretion. Juvenile-onset diabetes (type I), as the name implies, usually develops in young people and usually is caused by diminished insulin secretion. It is not clear if heredity plays a major role in its onset, but viral infection of the pancreatic islets may be involved. Maturity-onset diabetes (type II) develops in older people and often does not result from a lack of insulin but from the inability of the tissues to respond to insulin. The age of onset may vary tremendously, and it apparently is, in part, hereditary.

The symptoms associated with diabetes mellitus are listed in Table 18-A. They are the consequence of the abnormal metabolism of nutrients, which is caused by diminished insulin secretion or a decreased number of insulin receptors. In patients with diabetes mellitus nutrients are absorbed from the intestine after a meal, but without insulin skeletal muscle, adipose tissue, the liver, and other target tissues do not readily take glucose into their cells. Consequently, blood levels of glucose increase dramatically.

Even though blood glucose levels increase, the glucose does not enter cells, and fat and protein catabolism increases to provide energy; thus wasting of body tissues is common in untreated diabetics in spite of a high food intake. Food intake may be excessive because without insulin glucose does not enter the satiety center, resulting in an increased appetite. If blood sugar levels are high enough, glucose is also excreted in the urine, a symptom used to confirm diabetes mellitus since urine normally does not contain glucose. Dehydration of cells and rapid urine production occur as results of the elevated concentration of blood glucose. Thirst is caused by the elevated osmotic concentration of blood and the rapid loss of water in the urine. The resultant ionic imbalances cause nerve cells to malfunction and result in diabetic coma in severe cases. Acidosis is caused by rapid fat catabolism that results in increased levels of acetoacetic acid, which is converted to acetone and β-hydroxybutyric acid. These three substances collectively are referred to as **ketone bodies.** The presence of excreted ketone bodies in urine and in expired air ("acetone breath") is used as a diagnostic test for diabetes mellitus.

Diabetes mellitus often is treated by the administration of insulin by injection. Insulin is extracted from sheep or pork pancreatic tissue and is purified for use by diabetic patients. Genetic engineering currently is used to synthesize human insulin. A gene for human insulin is placed in the bacterium *Escherichia coli* so that the *E. coli* synthesizes human insulin. The insulin is extracted from the bacterial cultures and is packaged for use by diabetic patients. In some cases diabetes mellitus can be treated by administering drugs that stimulate beta cells to secrete more insulin. This treatment is effective only if an adequate number of functional beta cells is present in the islets of Langerhans.

Too much insulin or too little food intake after an injection of insulin by a diabetic patient causes **insulin shock.** The high levels of insulin cause target tissues to take up glucose at a very high rate. As a result, blood glucose levels rapidly fall to a low level. Since the nervous system depends on glucose as its major source of energy, neurons malfunction because of a lack of metabolic energy. As the blood glucose levels decrease, the concentration of fatty acids increase in the blood, resulting in a decrease in blood pH, which also causes nerve cells to malfunction. The result is a series of nervous system malfunctions that include disorientation, confusion, and convulsions (see Table 18-A).

etal muscle and the liver and is used for fat synthesis in adipose tissue and the liver. The rapid uptake and storage of glucose prevents a too large increase in blood glucose levels. Amino acids are incorporated into protein, and fats that were ingested as part of the meal are stored in adipose tissue and the liver. If the meal is high in protein, a small amount of glucagon is secreted, increasing the rate at which the liver uses amino acids to form glucose.

Within 1 or 2 hours after the meal, absorption of digested materials from the gastrointestinal tract declines, and blood sugar levels decline (Figure 18-17, *B*). As a result, secretion of glucagon, cortisol, growth hormone, and epinephrine increases, stimulating the release of glucose from tissues. Insulin levels decrease, the rate of glucose entry into the target tissues for insulin decreases, and glycogen is converted back to glucose and released into the blood. The decreased uptake of glucose by most tissues, combined with its release from the liver, helps maintain blood glucose at levels necessary for normal brain function. Cells that use less glucose start using more fats and proteins. Adipose tissue releases fatty acids, and the liver releases triglycerides (in lipoproteins) and ketones into the blood. Thus fats are a major source of energy for most tissues when blood glucose levels are low.

The interactions of insulin, growth hormone, glu-

Table 18-A *Symptoms of diabetes mellitus and insulin shock*

Diabetes Mellitus

Associated with high glucose or low insulin
- Hyperglycemia (elevated blood glucose)
- Glucosuria (excess glucose in urine)
- Polyuria (copious urinary production)
- Polydipsia (thirst)
- Polyphagia (excessive eating)
- Dehydration

Associated with fat and protein metabolism
- Tissue wasting and weight loss caused by rapid fat and protein breakdown, especially in those with Type I (juvenile-onset) diabetes
- Ketosis (elevated ketones—acetoacetic acid, beta-hydroxybutyric acid, and acetone) in blood as a result of fat metabolism
- Ketonuria (ketones excreted in urine)
- Acidosis resulting from fat metabolism
- Shortness of breath caused by acidosis
- Acetone breath (ketones excreted in breath)
- Obesity—characteristic of Type II (maturity-onset) diabetes; fat cells apparently are more sensitive to insulin than are other tissues; fat cells take up glucose, leading to obesity in some individuals

Vascular complications—cause unknown
- Atherosclerosis
- Microangiopathy—can lead to blindness (diabetic retinopathy), kidney damage (diabetic nephropathy), and muscle damage
- Peripheral vascular disease—reduced blood flow, causing pain, numbness, and weakness, especially in the feet and legs

Nervous system complications
- Depressed reflexes, drowsiness, and possible coma
- Neuropathy—degeneration of the myelin sheath of peripheral nervous system nerves, causing pain, coldness, tingling, or burning; affects motor nerves, causing muscle weakness and paralysis
- Autonomic neuropathy—affects size of the pupil and emptying of gastrointestinal tract and bladder; impotence

Insulin Shock

Associated with low blood sugar and an inadequate supply of glucose in the nervous system; hypoglycemia
- Headache, drowsiness, and fatigue
- Behavioral changes and confusion
- Inability to focus eyes
- Stupor and coma
- Increased fat metabolism
- Convulsive seizures and death

Associated with epinephrine (as a result of low blood sugar, the adrenal gland secretes epinephrine)
- Sweating
- Pale skin
- Tachycardia (increased heart rate)
- Sensation of hunger and a "sinking" feeling in the stomach
- Restlessness and a feeling of anxiety

cagon, epinephrine, and cortisol are excellent examples of negative-feedback mechanisms. When blood sugar levels are high, these hormones cause rapid uptake and storage of glucose, amino acids, and fats. When blood sugar levels are low, they cause release of glucose and a switch to fat and protein metabolism as a source of energy for most tissues.

During exercise skeletal muscles require energy to support the contraction process (see Chapter 10). Although metabolism of intracellular nutrients may sustain muscle contraction for a short time, additional energy sources are required during prolonged activity. Sympathetic nervous system activity, which increases during exercise, stimulates the release of epinephrine from the adrenal medulla and of glucagon from the pancreas (Figure 18-18). These hormones induce the conversion of glycogen to glucose in the liver and the release of glucose into the blood, thus providing skeletal muscles with a source of energy. Since epinephrine and glucagon have short half-lives, they can rapidly adjust blood sugar levels for varying conditions of activity.

During sustained activity, glucose release from the liver and other tissues is not adequate to support muscle activity, and there is a danger that blood glucose levels will become too low to support brain function. A decrease in insulin prevents uptake of glucose by most tissues, thus conserving glucose for

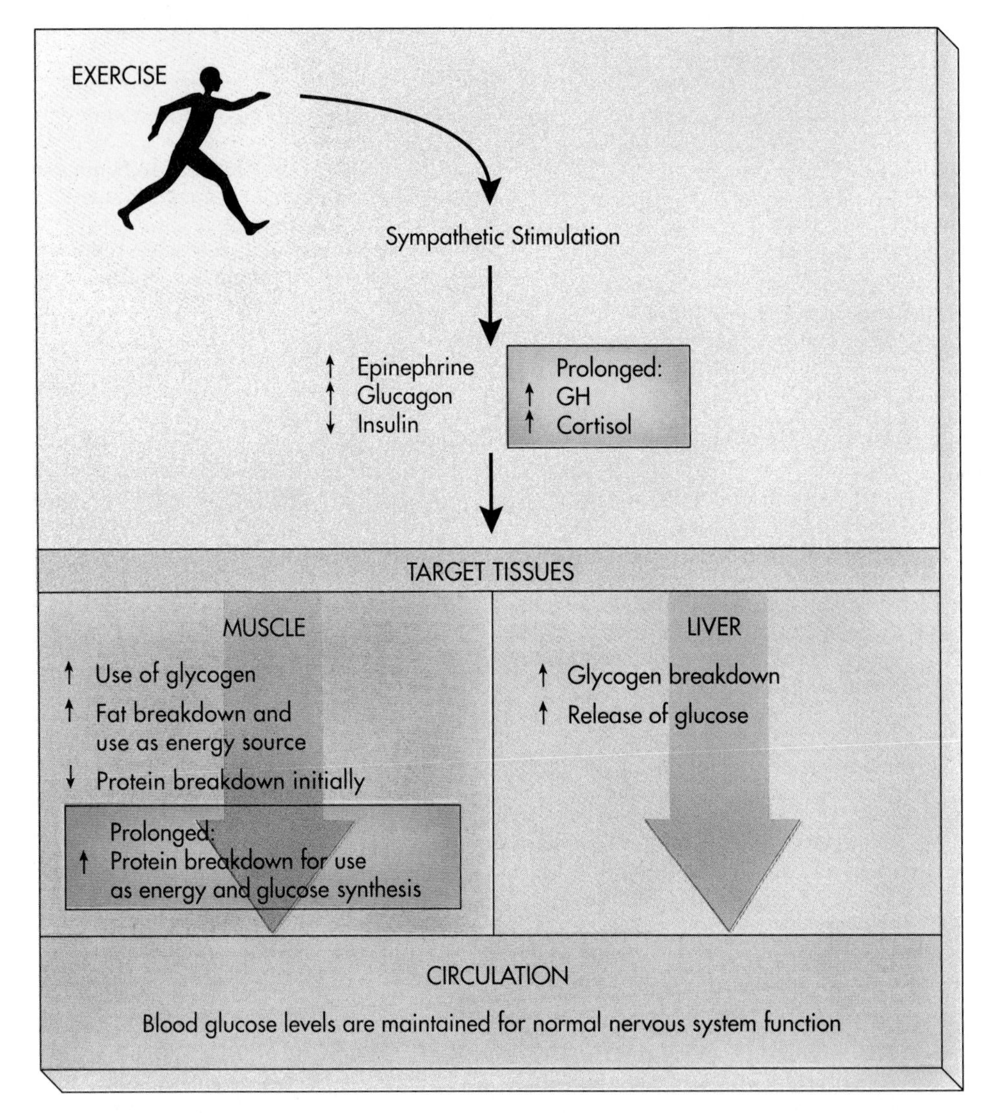

FIGURE 18-18 In response to exercise, sympathetic stimulation increases epinephrine and glucagon secretion and inhibits insulin secretion. If exercise is prolonged, both growth hormone and cortisol secretion increase. Epinephrine increases the rate at which glycogen in muscle cells is used so that the cells do not take up glucose from the blood. Epinephrine and glucagon increase glucagon breakdown in the liver, resulting in release of glucose into the circulatory system. Epinephrine and sympathetic stimulation also increase the breakdown of fat and the release of fatty acids from fat cells. During prolonged exercise, cortisol increases protein breakdown for use in glucose synthesis from amino acids and from some components of fat such as glycerol. Growth hormone slows the breakdown of proteins and conserves them. Thus glucose, glycogen, and fat are major sources of energy during exercise.

the brain. Epinephrine, glucagon, cortisol, and growth hormone cause an increase of fatty acids, triglycerides, and ketones in the blood. Growth hormone also inhibits the breakdown of proteins, preventing muscles from using themselves as an energy source. Consequently, there is a shift by skeletal muscles from glucose use to fat and glycogen metabolism. At the end of a long race, for example, muscles rely to a large extent on fat metabolism.

9

Explain why long-distance runners may not have much of a "kick" left when they try to sprint to the finish line.

? ? ? ? ? ? ? ? ? ? ?

REPRODUCTIVE HORMONES

Reproductive hormones are secreted primarily from the ovaries, testes, placenta, and pituitary gland (Table 18-12). These hormones are discussed in Chapter 28.

HORMONES OF THE PINEAL BODY, THYMUS GLAND, AND OTHERS

The **pineal** (pi′ne-al) **body** in the epithalamus of the brain functions as an endocrine gland that secretes hormones that act on the hypothalamus or the gonads to inhibit reproductive functions. Two substances have been proposed as secretory products: **melatonin** and **arginine vasotocin** (Table 18-13). Melatonin decreases gonadotropin-releasing hormone secretion from the hypothalamus and may inhibit reproductive functions through this mechanism.

In some animals, changes in the amount of daylight during each day (i.e., the photoperiod) regulate pineal secretions (Figure 18-19). For example, increased daylight initiates impulses in the retina of the eye that are propagated to the brain and cause a decrease in the action potentials sent first to the spinal cord and then through sympathetic neurons to the pineal body. Decreased pineal secretion results. In the dark, impulses delivered by sympathetic neurons to the pineal body increase, stimulating the secretion of pineal hormones. In animals that breed in the spring, the increased day length decreases pineal secretions. Since pineal secretions inhibit reproductive functions in these species, the increased day length results in hypertrophy of the reproductive structures.

The function of the pineal body in humans is not clear, but tumors that destroy the pineal body correlate with early sexual development, and tumors that result in pineal hormone secretion correlate with retarded development of the reproductive system. It is not clear, however, if the pineal body controls the onset of puberty.

Arginine vasotocin may function with melatonin to regulate the function of the reproductive system in some animals. The evidence for melatonin's role is more extensive, however.

The **thymus gland,** superior to the heart in the thorax and neck, secretes a hormone called **thymosin** (see Table 18-13). Both the thymus gland and thymosin play an important role in the development of the immune system and are discussed in Chapter 22.

Table 18-12

Hormones of the reproductive organs

HORMONE	STRUCTURE	TARGET TISSUE	RESPONSE
Testis			
Testosterone	Steroid	Most cells	Aids in spermatogenesis; maintenance of functional reproductive organs; secondary sexual characteristics; sexual behavior
Ovary			
Estrogens	Steroids	Most cells	Uterine and mammary gland development and function; external genitalia structure; secondary sexual characteristics; sexual behavior and menstrual cycle
Progesterone	Steroid	Most cells	Uterine and mammary gland development and function; external genitalia structure; secondary sexual characteristics; menstrual cycle

Table 18-13

Other hormones and hormonelike substances

HORMONE (?)	STRUCTURE	TARGET TISSUE	RESPONSE
Pineal Body			
Melatonin	Amino acid derivative	At least the hypothalamus	Inhibition of secretion of gonadotropin-releasing hormone, thereby inhibiting reproduction; significance is not clear in humans
Arginin vasotocin	Amino acid derivative	Possibly the hypothalamus	Possible inhibition of gonadotropin secretion
Thymus Gland			
Thymosin	Peptides	Immune tissues	Development and function of the immune system
Several Sources			
Prostaglandins	Modified fatty acids	Most tissues	Mediation of inflammatory responses; increased uterine contractions; ovulation; possible inhibition of progesterone synthesis; blood coagulation; others
Endorphins and enkephalins	Peptides	Nervous system	Reduction of pain sensation; possibly other functions

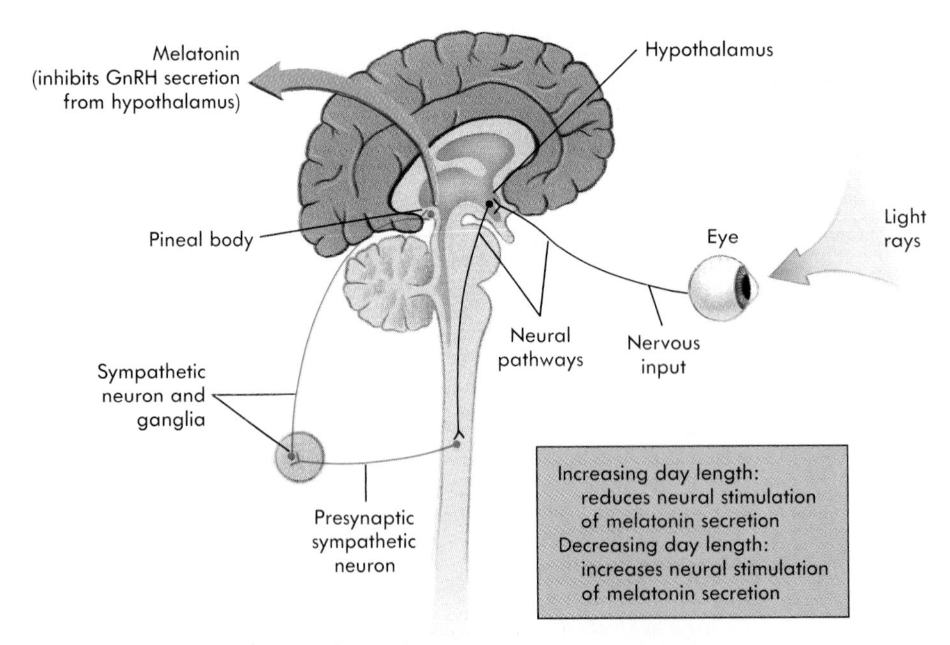

FIGURE 18-19 Regulation of pineal secretions. Since melatonin inhibits gonadotropin-releasing hormone (GnRH) secretion from the hypothalamus, reduced melatonin secretion results in an increase in GnRH secretion. As a consequence, luteinizing hormone (LH) and follicle-stimulating hormone (FSH) are secreted in greater and greater amounts as the daylight increases. LH and FSH stimulate ovulation in the female and sperm production in the male. Consequently, the photoperiod regulates the breeding season in some animals. The precise role that the pineal body plays in human reproduction is not known.

Several hormones are released from the gastrointestinal tract. They regulate digestive functions by influencing the activity of the stomach, intestines, liver, and pancreas. They are discussed in Chapter 24.

HORMONELIKE SUBSTANCES

Several substances have some characteristics of hormones. These substances include **prostaglandins,** (pros′tă-glan′dinz), **endorphins** (en′dor-finz), and **enkephalins** (en-kef′ă-linz) (see Table 18-13). These substances differ from hormones in that they are not secreted from discrete endocrine glands, have a local effect rather than a systemic one, or have functions that are not understood adequately to explain their role in the body.

Prostaglandins are released from injured cells and are responsible for initiating some of the symptoms of inflammation (see Chapter 22); in other cases they are released from healthy cells. Prostaglandins are involved in the regulation of uterine contractions during menstruation and childbirth, the process of ovulation, inhibition of progesterone synthesis by the corpus luteum, regulation of coagulation, kidney function, and modification of the effect of other hormones on their target tissues. They are "local" hormones because they are synthesized and secreted by the tissues on which they act. Once they enter the circulatory system, they are metabolized rapidly.

Pain receptors are stimulated directly by prostaglandins and other inflammatory compounds or, in the case of headaches, are stimulated by the dilation (caused by prostaglandin-like substances) of blood vessels. Anti-inflammatory drugs such as aspirin inhibit prostaglandin synthesis and, as a result, reduce inflammation and pain.

Endorphins and enkephalins bind to the same receptor molecules as morphine. They are produced in several sites in the body (e.g., parts of the brain, pituitary, spinal cord, and gut). These substances are endogenously produced analgesic substances and may affect both the central and peripheral nervous systems to moderate the sensation of pain (see Chapter 13). Decreased sensitivity to painful stimuli during exercise and stress may result from the increased secretion of these substances. The exact role of the endorphins and enkephalins is an area of intense research activity.

The number of hormonelike substances in the body is large. Chemical communication among cells in the body is complex, well developed, and necessary for maintenance of homeostasis. Investigations of chemical regulation increase our knowledge of body functions—knowledge that can be used in the development of techniques for the treatment of pathological conditions.

SUMMARY

PITUITARY GLAND AND HYPOTHALAMUS p. 546

1. The pituitary gland secretes at least nine hormones that regulate numerous body functions and other endocrine glands.
2. The hypothalamus regulates pituitary gland activity through neurohormones and action potentials.

Structure of the Pituitary Gland

1. The neurohypophysis develops from the floor of the brain and consists of the infundibulum and pars nervosa.
2. The adenohypophysis develops from the roof of the mouth and consists of the pars distalis, pars intermedia, and pars tuberalis.

Relationship of the Pituitary to the Brain

1. The hypothalamohypophyseal portal system connects the hypothalamus and the adenohypophysis.
   - Neurohormones are produced in hypothalamic neurons.
   - Through the portal system the neurohormones inhibit or stimulate hormone production in the adenohypophysis.
2. The hypothalamohypophyseal nerve tract connects the hypothalamus and the neurohypophysis.
   - Neurohormones are produced in hypothalamic neurons.
   - The neurohormones move down the axons of the nerve tract and are secreted from the neurohypophysis.

HORMONES OF THE PITUITARY GLAND p. 550

Neurohypophyseal Hormones

1. Antidiuretic hormone (ADH) promotes water retention by the kidneys.
2. Oxytocin promotes uterine contractions during delivery and causes milk ejection in lactating women.

Adenohypophyseal Hormones

1. Growth hormone (GH), or somatotropin.
   - GH stimulates the uptake of amino acids and their conversion into proteins and stimulates the breakdown of fats and glycogen.
   - GH stimulates the production of somatomedins; together they promote bone and cartilage growth.
   - GH secretion increases in response to an increase in blood amino acids, low blood glucose, or stress.
   - GH is regulated by growth hormone-releasing hormone (GH-RH) and growth hormone-inhibiting hormone (GH-IH), or somatostatin.
2. Thyroid-stimulating hormone (TSH), or thyrotropin, causes the release of thyroid hormones.
3. Adrenocorticotropic hormone (ACTH).
   - ACTH is derived from proopiomelanocortin.
   - ACTH stimulates cortisol secretion from the adrenal cortex and increases skin pigmentation.

4. Several hormones in addition to ACTH are derived from proopiomelanocortin.
   - Lipotropins cause fat breakdown.
   - Beta-endorphins play a role in analgesia.
   - Melanocyte-stimulating hormone (MSH) increases skin pigmentation.
5. Luteinizing hormone (LH) and follicle-stimulating hormone (FSH).
   - Both hormones regulate the production of gametes and reproductive hormones (testosterone in males; estrogen and progesterone in females).
   - Gonadotropin-releasing hormone (GnRH) from the hypothalamus stimulates LH and FSH secretion.
6. Prolactin.
   - Prolactin stimulates milk production in lactating females.
   - Prolactin-releasing hormone (PRH) and prolactin-inhibiting hormone (PIH) from the hypothalamus affect prolactin secretion.

THYROID GLAND *p. 554*

The thyroid gland is just inferior to the larynx.

Histology

1. The thyroid gland is composed of small, hollow balls of cells called follicles, which contain thyroglobulin.
2. Parafollicular cells are scattered throughout the thyroid gland.

Thyroid Hormones

1. Thyroid Hormone Synthesis
   - Iodide ions are taken into the follicles by active transport, are oxidized, and are bound to tyrosine molecules in thyroglobulin.
   - Thyroglobulin is secreted into the follicle lumen. Tyrosine molecules with iodine combine to form T_3 and T_4, thyroid hormones.
   - Thyroglobulin is taken into the follicular cells and is broken down; T_3 and T_4 diffuse from the follicles to the blood.
2. Transport in the Blood
   - T_3 and T_4 bind to thyroxine-binding globulin (TBG) and other plasma proteins.
   - The plasma proteins prolong the half-life of T_3 and T_4 and regulate the levels of T_3 and T_4 in the blood.
   - Approximately one third of the T_4 is converted into functional T_3.
3. Mechanism of Action of Thyroid Hormones
   - Thyroid hormones bind with intracellular receptor molecules and initiate new protein synthesis.
4. Effects of Thyroid Hormones
   - Thyroid hormones increase the rate of glucose, fat, and protein metabolism in many tissues, thus increasing body temperature.
   - Normal growth of many tissues is dependent on thyroid hormones.
5. Regulation of Thyroid Hormone Secretion
   - Increased thyroid-stimulating hormone (TSH) from the adenohypophysis increases thyroid hormone secretion.
   - Thyrotropin-releasing hormone (TRH) from the hypothalamus increases TSH secretion. TRH increases as a result of chronic exposure to cold, food deprivation, and stress.
   - T_3 and T_4 inhibit TSH and TRH secretion.

Calcitonin

1. The parafollicular cells secrete calcitonin.
2. An increase in blood calcium levels stimulates calcitonin secretion.
3. Calcitonin decreases blood calcium and phosphate levels by inhibiting osteoclasts.

PARATHYROID GLANDS *p. 559*

1. The parathyroid glands are embedded in the thyroid glands.
2. Parathyroid hormone (PTH) increases blood calcium levels.
   - PTH stimulates osteoclasts and inhibits osteoblasts.
   - PTH promotes calcium reabsorption by the kidneys and the formation of active vitamin D by the kidneys.
   - Active vitamin D increases calcium absorption by the intestine.
3. A decrease in blood calcium levels stimulates PTH secretion.

ADRENAL GLANDS *p. 562*

1. The adrenal glands are near the superior pole of each kidney.
2. The adrenal medulla arises from neural crest cells and functions as part of the sympathetic nervous system. The adrenal cortex is derived from mesoderm.

Histology

1. The medulla is composed of closely packed cells.
2. The cortex is divided into three layers: the zona glomerulosa, the zona fasciculata, and the zona reticularis.

Hormones of the Adrenal Medulla

1. Epinephrine accounts for 80% and norepinephrine for 20% of the adrenal medulla hormones.
   - Epinephrine increases blood glucose levels, use of glycogen and glucose by skeletal muscle, and heart rate and force of contraction and causes vasoconstriction in the skin and viscera and vasodilation in skeletal and cardiac muscle.
   - Norepinephrine stimulates cardiac muscle and causes constriction of most peripheral blood vessels.
2. The adrenal medulla hormones prepare the body for physical activity.
3. Release of adrenal medulla hormones is mediated by the sympathetic nervous system in response to emotions, injury, stress, exercise, and low blood glucose levels.

Hormones of the Adrenal Cortex

1. The zona glomerulosa secretes the mineralocorticoids, especially aldosterone. Aldosterone acts on the kidneys to increase sodium and to decrease potassium and hydrogen levels in the blood.
2. The zona fasciculata secretes glucocorticoids, especially cortisol.
   - Cortisol increases fat and protein breakdown, increases glucose synthesis from amino acids, decreases the inflammatory response, and is necessary for the development of some tissues.
   - Adrenocorticotropic hormone (ACTH) from the adenohypophysis stimulates cortisol secretion. Corticotropin-releasing hormone (CRH) from the hypothalamus stimulates ACTH release. Low blood sugar levels or stress stimulate CRH secretion.
3. The zona reticularis secretes androgens. In females androgens stimulate axillary and pubic hair growth and sexual drive.

PANCREAS *p. 567*

The pancreas is located along the small intestine and the stomach. It is both an exocrine gland and an endocrine gland.

Histology

1. The exocrine portion of the pancreas consists of a complex duct system that ends in small sacs called acini that produce pancreatic digestive juices.
2. The endocrine portion consists of the pancreatic islets. Each islet is composed of alpha cells that secrete glucagon, beta cells that secrete insulin, and delta cells that secrete somatostatin.

Effects of Insulin and Glucagon on Their Target Tissues

1. Insulin
   - Insulin's target tissues are the liver, adipose tissue, muscle, and the satiety center in the hypothalamus. The nervous system is not a target tissue, but it does rely on blood glucose levels maintained by insulin.
   - Insulin increases the uptake of glucose and amino acids by cells. Glucose is used for energy or is stored as glycogen. Amino acids are used for energy or are converted to glucose or proteins.
2. Glucagon
   - Glucagon's target tissue is mainly the liver.
   - Glucagon causes the breakdown of glycogen and fats for use as an energy source.

Regulation of Pancreatic Hormone Secretion

1. Insulin secretion increases because of elevated blood glucose levels, an increase in some amino acids, parasympathetic stimulation, and gastrointestinal hormones. Sympathetic stimulation decreases insulin secretion.
2. Glucagon secretion is stimulated by low blood glucose levels, certain amino acids, and sympathetic stimulation.
3. Somatostatin inhibits insulin and glucagon secretion.

HORMONAL REGULATION OF NUTRIENTS *p. 571*

1. After a meal the following events take place.
   - Glucagon, cortisol, growth hormone (GH), and epinephrine are inhibited by high blood glucose levels, reducing the release of glucose from tissues.
   - Insulin secretion increases as a result of the high blood glucose levels, increasing the uptake of glucose, amino acids, and fats, which are used for energy or are stored.
   - Sometime after the meal blood glucose levels drop. Glucagon, cortisol, GH, and epinephrine levels increase, insulin levels decrease, and glucose is released from tissues.
   - Adipose tissue releases fatty acids, triglycerides, and ketones, which are used for energy by most tissues.
2. During exercise the following events occur.
   - Sympathetic activity increases epinephrine and glucagon secretion, causing a release of glucose into the blood.
   - Low blood sugar levels, caused by uptake of glucose by skeletal muscles, stimulates epinephrine, glucagon, GH, and cortisol secretion, causing an increase in fatty acids, triglycerides, and ketones in the blood, all of which are used for energy.

REPRODUCTIVE HORMONES *p. 575*

Reproductive hormones are secreted by the ovaries, testes, placenta, and pituitary gland.

HORMONES OF THE PINEAL BODY, THYMUS GLAND, AND OTHERS *p. 575*

1. The pineal body produces melatonin and arginine vasotocin, which may inhibit reproductive maturation.
2. The thymus gland produces thymosin, which is involved in the development of the immune system.
3. Several hormones produced by the gastrointestinal tract regulate digestive functions.

HORMONELIKE SUBSTANCES *p. 577*

1. Prostaglandins are produced by most cells of the body and usually have a local effect. They affect many body functions.
2. Endorphins and enkephalins are analgesic substances.

CONTENT REVIEW

1. To understand the role of an endocrine gland and its secretions in the body, what five things should be kept in mind?
2. Where is the pituitary gland located? Contrast the embryonic origin of the neurohypophysis and the adenohypophysis.
3. Name the parts of the pituitary gland and the function of each part.
4. Describe the hypothalamohypophyseal portal system. How does the hypothalamus regulate the hormone secretion of the adenohypophysis?
5. Describe the production of a neurohormone in the hypothalamus and its secretion in the neurohypophysis.
6. Where is ADH produced, where is it secreted, and what is its target tissue? What happens when ADH levels increase?
7. Where is oxytocin produced and secreted, and what effects does it have on its target tissues?
8. Structurally, what kinds of hormones are released from the neurohypophysis and the adenohypophysis? Do these hormones bind to plasma proteins, how long is their half-life, and how do they activate their target tissues?
9. For each of the following hormones secreted by the adenohypophysis—GH, TSH, ACTH, LH, FSH, and prolactin—name their target tissues and the effect of the hormone on its target tissue.
10. What effect do amino acids, glucose, and stress have on GH secretion?
11. What stimulates somatomedin production, where is it produced, and what are its effects?
12. How are ACTH, MSH, lipotropins, and β-endorphins related? What are the functions of these hormones?
13. Where is the thyroid gland located? Describe the follicles and the parafollicular cells within the thyroid. What hormones do they produce?
14. Starting with the uptake of iodide by the follicles, describe the production and secretion of thyroid hormones.
15. How are the thyroid hormones transported in the blood? What effect does this transportation have on their half-life?
16. What are the target tissues of thyroid hormone? By what mechanism do thyroid hormones alter the activities of their target tissues? What effects are produced?

17. Starting in the hypothalamus, explain how chronic exposure to the cold, food deprivation, or stress can affect thyroid hormone production.
18. Diagram two negative-feedback mechanisms involving hormones that function to regulate the production of thyroid hormones.
19. What effect does calcitonin have on osteoclasts, osteoblasts, and blood calcium levels? What stimulus can cause an increase in calcitonin secretion?
20. Where are the parathyroid glands located, and what hormone do they produce?
21. What effect does PTH have on osteoclasts, osteoblasts, the kidneys, the small intestine, and blood calcium and phosphate levels? What stimulus can cause an increase in PTH secretion?
22. Where are the adrenal glands located? Describe the embryonic origin of the adrenal medulla and the adrenal cortex.
23. Name two hormones secreted by the adrenal medulla, and list the effects of these hormones.
24. List several conditions that can stimulate the production of adrenal medulla hormones. What role does the nervous system play in the release of adrenal medulla hormones? How does this role relate to the embryonic origin of the adrenal medulla?
25. Describe the three layers of the adrenal cortex, and name the hormones produced by each layer.
26. Name the target tissue of aldosterone, and list the effects of an increase in aldosterone secretion on the concentration of ions in the blood.
27. Describe the effects produced by an increase in cortisol secretion. Starting in the hypothalamus, describe how stress or low blood sugar levels can stimulate cortisol release.
28. What effects do adrenal androgens have on males and females?
29. Where is the pancreas located? Describe the exocrine and endocrine portion of this gland and the secretions produced by each portion.
30. Name the target tissues for insulin and glucagon, and list the effects they have on their target tissues.
31. How does insulin affect the nervous system in general and the satiety center in the hypothalamus in particular?
32. What effect do blood glucose levels, blood amino acid levels, the autonomic nervous system, and somatostatin have on insulin and glucagon secretion?
33. Describe the hormonal effects after a meal that result in the movement of nutrients into cells and their storage. Describe the hormonal effects that later cause the release of stored materials for use as energy.
34. During exercise how does sympathetic nervous system activity regulate blood sugar levels? Name five hormones that interact to ensure that both the brain and muscles have adequate energy sources.
35. Where is the pineal body located? Name the hormones it produces and their possible effects.
36. Where is the thymus gland located, what hormone does it produce, and what effects does the hormone have?
37. Describe the site of production and actions of prostaglandins, endorphins, and enkephalins.

CONCEPT REVIEW

1. The hypothalamohypophyseal portal system connects the hypothalamus with the adenohypophysis. Why is such a special circulatory system advantageous?
2. The secretion of ADH can be affected by exposure to hot or cold environmental temperatures. Predict the effect of a hot environment on ADH secretion, and explain why it is advantageous. Propose a mechanism by which temperature produces a change in ADH secretion.
3. A patient exhibited polydipsia (thirst), polyuria (excess urine production), and urine with a low specific gravity (contains few ions and no glucose). If you wanted to reverse the symptoms, would you administer insulin, glucagon, ADH, or aldosterone? Explain.
4. A patient complains of headaches and visual disturbances. A casual glance reveals that the patient's finger bones are enlarged in diameter, there is a heavy deposition of bone over the eyes, and the patient has a prominent jaw. The doctor tells you that the headaches and visual disturbances are due to increased pressure within the skull and that the patient is suffering from a pituitary tumor that is affecting hormone secretion. Name the hormone that is causing the problem, and explain why there is an increase in pressure within the skull.
5. Most laboratories have the ability to determine blood levels of TSH, T_3, and T_4. Given that ability, design a method of determining whether hyperthyroidism in a patient is due to a pituitary abnormality or to the production of a nonpituitary thyroid stimulatory substance.
6. An anatomy and physiology instructor asked two students to predict the response that would occur if an individual suffered from chronic vitamin D deficiency. One student claimed that the person would suffer from hypocalcemia and the symptoms would be associated with that condition. The other student claimed that calcium levels would remain within their normal range, although at the low end of the range, and bone resorption would occur to the point that advanced osteomalacia might be seen. With whom do you agree, and why?
7. Given the ability to measure blood glucose levels, design an experiment that distinguishes between a person with diabetes, a healthy person, and a person who has a pancreatic tumor that secretes large amounts of insulin.
8. A patient arrives in an unconscious condition. A medical emergency bracelet reveals that he is a diabetic. The patient may be in diabetic coma or insulin shock. How could you tell which, and what treatment would you recommend for each condition?
9. Diabetes mellitus can be due to a lack of insulin that results in hyperglycemia. Adrenal diabetes and pituitary diabetes also produce hyperglycemia. What hormones produce the last two conditions?

? ?

ANSWERS TO PREDICT QUESTIONS

1 p. 547. The cell bodies of the neurosecretory cells that produce ADH are in the hypothalamus, and their axons extend into the neurohypophysis where ADH is stored and secreted. Removing the neurohypophysis severs the axons, resulting in a temporary reduction in secretion. However, the cell bodies still produce ADH, and as ADH accumulates at the ends of severed axons, ADH secretion resumes.

2 p. 559. The thyroid gland enlarges in response to iodine deficiency because without iodine, thyroid hormones cannot be synthesized. Consequently, TSH levels in the circulatory system increase because of the lower-than-normal levels of thyroid hormones in the blood. Increased TSH levels cause the thyroid gland to enlarge because thyroglobulin is synthesized in large amounts and the thyroid follicles enlarge even though thyroid hormones cannot be produced.

3 p. 560. In response to a reduced dietary intake of calcium, the blood levels of calcium begin to decline. In response to the decline in blood levels of calcium, there is an increase of PTH from the parathyroid glands. The PTH functions to increase calcium resorption from bone. Consequently, blood levels of calcium are maintained within the normal range but at the same time bones are being decalcified. Severe dietary calcium deficiency will result in bones that become soft and eaten away because of the decrease in calcium content.

4 p. 562. Removal of the thyroid gland would remove the tissue responsible for thyroid hormone production (follicles), calcitonin (parafollicular cells), and PTH (parathyroid glands are embedded in the thyroid gland). Therefore thyroid hormones, calcitonin, and PTH would no longer be found in the blood. Without the negative-feedback effect of thyroid hormones TRH and TSH levels in the blood would increase. Without parathyroid hormone blood levels of calcium would fall. When the blood levels of calcium fall below normal, the permeability of nerve and muscle cells to sodium increases. As a consequence, spontaneous action potentials are produced that cause tetany of muscles. Death may result from tetany of respiratory muscles.

5 p. 565. High aldosterone levels in the blood lead to elevated sodium levels in the circulatory system and low blood levels of potassium. The effect of low blood levels of potassium would be hyperpolarization of muscle and nerve cells. The hyperpolarization results from the lower levels of potassium in the extracellular fluid and a greater tendency for potassium to diffuse from the cell. As a result, a greater-than-normal stimulus is required to cause the cells to depolarize to threshold and generate an action potential. Thus the symptoms include lethargy and muscle weakness. The elevated sodium concentrations would result in a greater-than-normal amount of water retention in the circulatory system, which may result in elevated blood pressure.

The major effect of a low rate of aldosterone secretion is elevated blood potassium levels. As a result, nerve and muscle cells depolarize. Because of their partial depolarization, they produce action potentials spontaneously or in response to very small stimuli. The result is muscle spasms or tetany.

6 p. 565. Large doses of cortisone may damage the adrenal cortex because cortisone inhibits ACTH secretion from the adenohypophysis. ACTH is required to keep the adrenal cortex from undergoing atrophy. Prolonged use of large doses of cortisone may cause the adrenal gland to atrophy to the point that it cannot recover if ACTH levels do increase again.

7 p. 569. An increase in insulin secretion in response to parasympathetic stimulation and gastrointestinal hormones is consistent with the maintenance of homeostasis because parasympathetic stimulation and increased gastrointestinal hormones result from conditions such as eating a meal. Therefore insulin levels increase just before the time large amounts of glucose and amino acids enter the circulatory system. The elevated insulin levels prevent a large increase in blood glucose and the loss of glucose in the urine.

8 p. 570. In response to a meal high in carbohydrates, insulin secretion is increased, and glucagon secretion is reduced. The stimulus for the insulin secretion comes from parasympathetic innervation and, more importantly, elevated blood levels of glucose. In response to a meal high in protein but low in carbohydrates, insulin secretion is increased slightly, and glucagon secretion is also increased. The stimulation for insulin secretion is parasympathetic stimulation and an increase in blood amino acid levels. Glucagon secretion is stimulated by low blood glucose levels and by some amino acids.

During periods of exercise sympathetic stimulation inhibits insulin secretion. As blood glucose levels decline, there is an increase of glucagon secretion.

9 p. 575. Sympathetic stimulation during exercise inhibits insulin secretion, and blood glucose levels are not high because of the rapid metabolism of the small amount of glucose that can enter the muscles. Much of the energy for muscle contraction depends on glucose stored in the form of glycogen in muscles, and during a long run glycogen levels are depleted. The "kick" at the end of the race is due to increased energy production through anaerobic respiration, which uses glucose or glycogen as an energy source. Since blood glucose levels and glycogen levels are low, there is an insufficient source of energy for greatly increased muscle activity.

PART IV

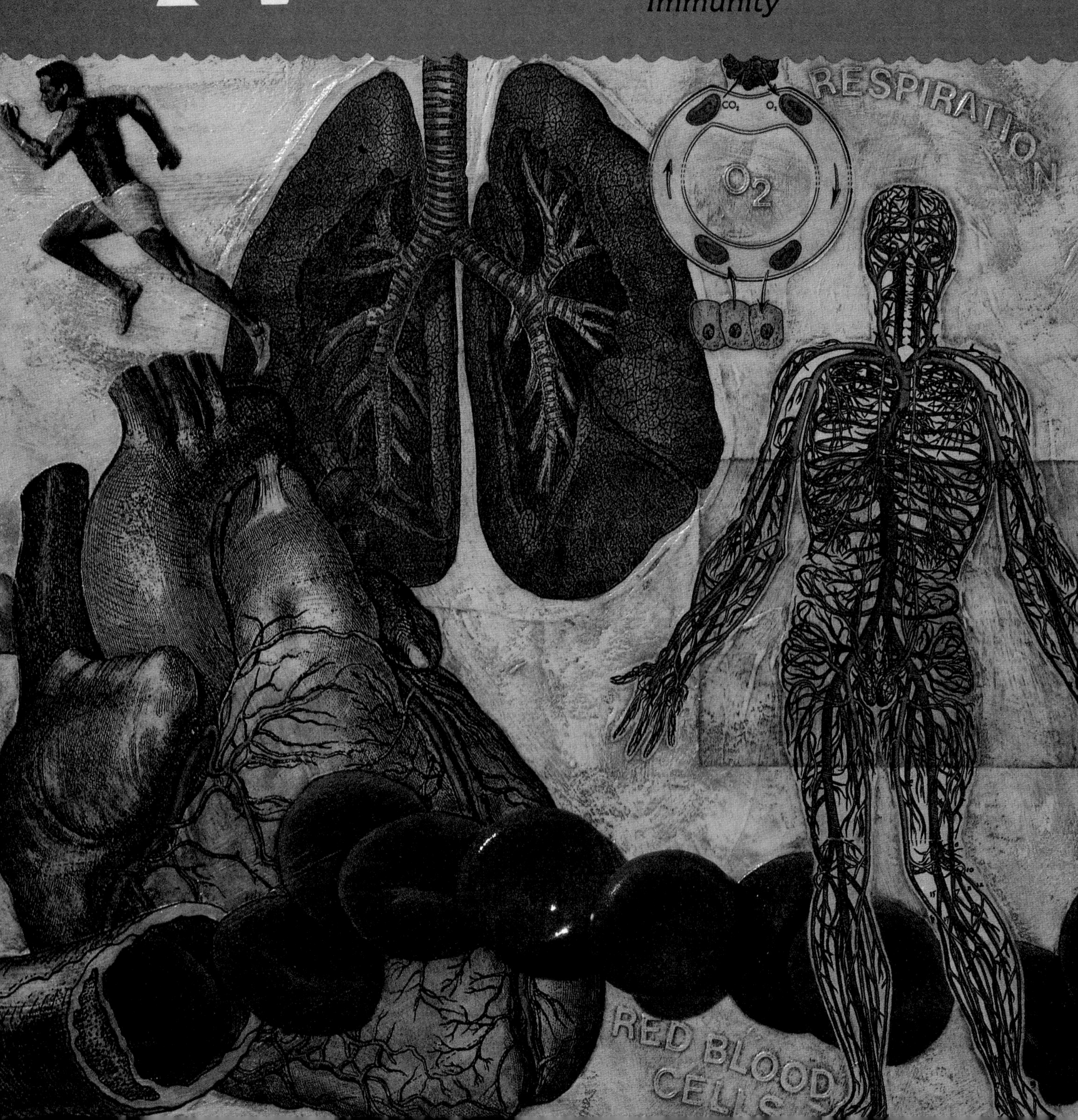

REGULATION AND MAINTENANCE

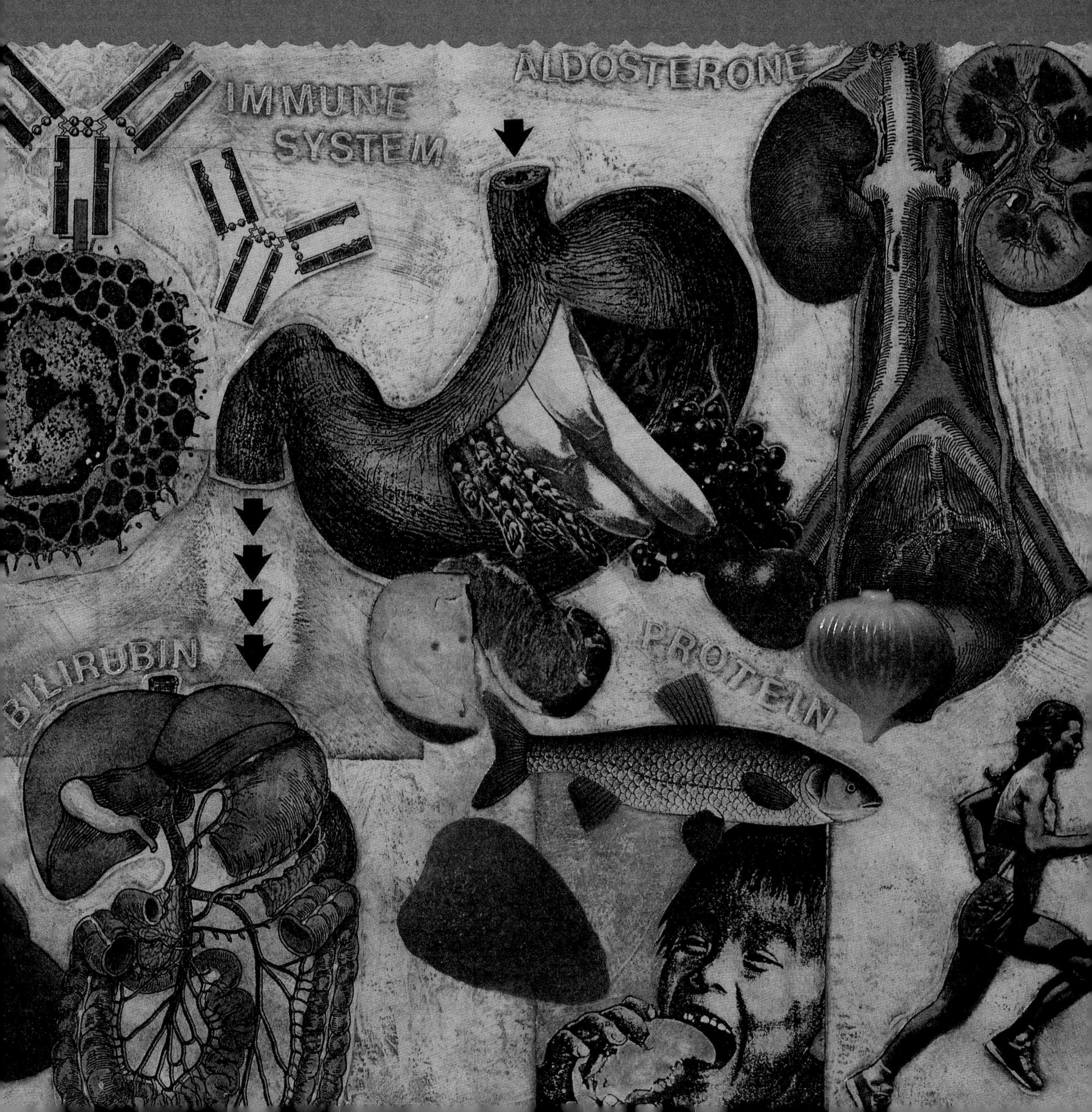

CHAPTER 19 OBJECTIVES

After reading this chapter you should be able to

1. List the functions of blood.
2. List the components of blood plasma and explain the difference between plasma and serum.
3. Define formed element and list the three types of formed elements.
4. Describe the origin and production of the formed elements.
5. Describe the structure and function of erythrocytes.
6. Explain the role of iron in blood.
7. Define erythropoiesis and name the cells produced during erythropoiesis.
8. Describe the removal of damaged or "worn-out" erythrocytes from the circulation and describe the production of bilirubin.
9. Describe the structures and functions of the five types of leukocytes.
10. Describe the structure, origin, and function of platelets.
11. Name and describe the three stages of hemostasis.
12. Compare the extrinsic and intrinsic pathways of coagulation.
13. Explain the importance of the balance between coagulation factors and anticoagulants.
14. Describe how a clot functions in wound healing and how the clot is removed.
15. Explain the basis of ABO and Rh incompatibilities.
16. Describe diagnostic blood tests and the normal values for the tests and give examples of disorders that produce abnormal test values.

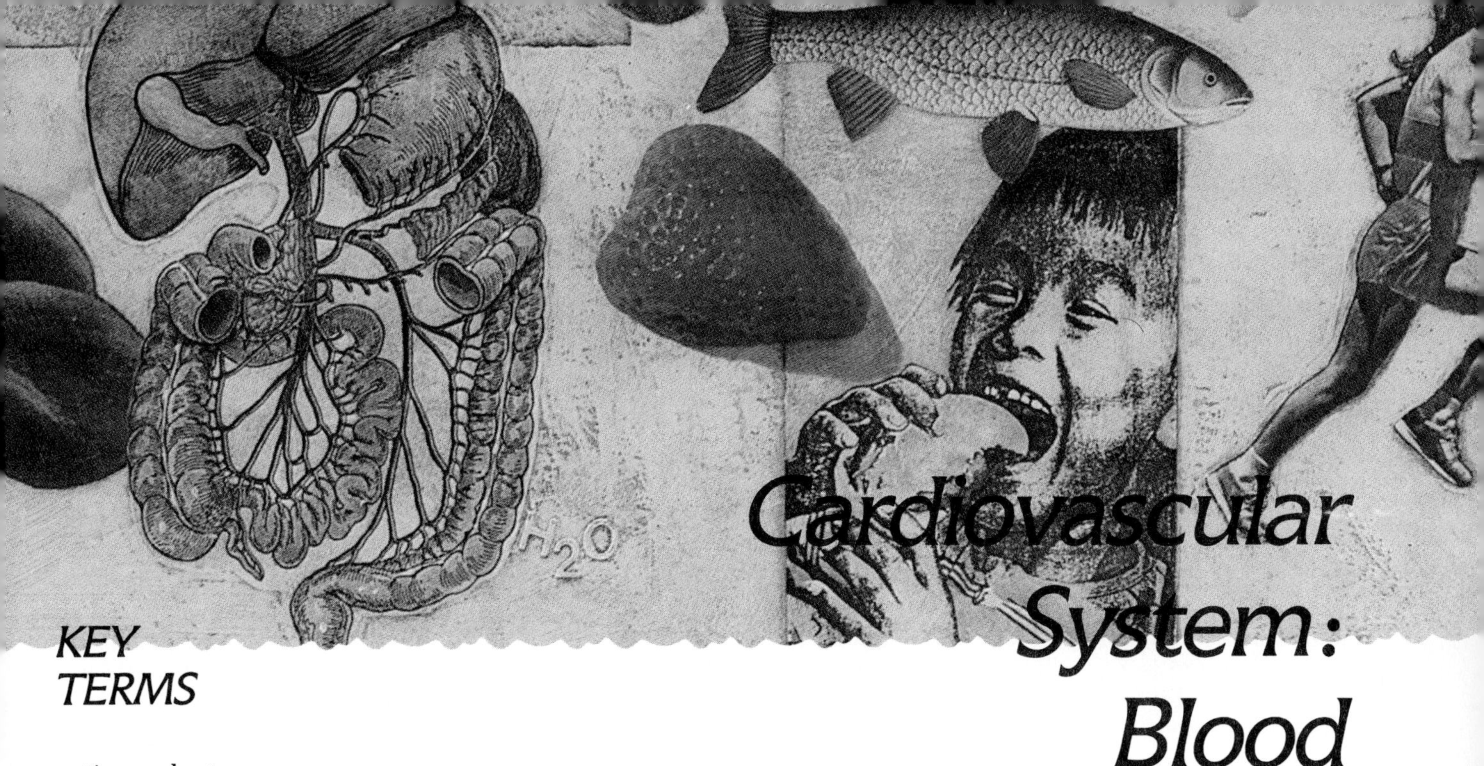

Cardiovascular System: Blood

KEY TERMS

anticoagulant
(an′tĭ-ko-ag′u-lant)

blood

blood group

clot retraction

erythrocyte
(ĕ-rith′ro-sīt)

erythropoietin
(ĕ-rith′ro-poy′e-tin)

extrinsic clotting pathway

formed elements

hemocytoblast
(he′mo-si′to-blast)

hemoglobin
(he′mo-glo′bin)

hemostasis
(he′mo-sta-sis)

intrinsic clotting pathway

leukocyte (lu′ko-sīt)

plasma

platelet plug

RELATED TOPIC

The following concept is important for a good understanding of this chapter. If you are not familiar with it, you should review it before proceeding.

Concept of connective tissues (Chapter 4)

CELLS ARE METABOLICALLY ACTIVE AND AS A RESULT REQUIRE CONSTANT NUTRITION AND WASTE REMOVAL. MOST CELLS, HOWEVER, ARE LOCATED SOME DISTANCE FROM NUTRIENT SOURCES SUCH AS THE DIGESTIVE TRACT AND SITES OF WASTE DISPOSAL SUCH AS THE KIDNEYS. THE CARDIOVASCULAR SYSTEM TAKES CARE OF THESE NEEDS BY PROVIDING A CONNECTION BETWEEN VARIOUS TISSUES. BLOOD VESSELS CARRY BLOOD TO AND FROM THE TISSUES. THE BLOOD PLAYS AN IMPORTANT ROLE IN MAINTAINING HOMEOSTASIS.

Blood is classified as a connective tissue, consisting of cells and cell fragments surrounded by a liquid matrix. The cells and cell fragments are the **formed elements,** and the fluid matrix is the **plasma.** The total blood volume in the average adult is approximately 4 to 5 L in females and 5 to 6 L in males. Blood makes up approximately 8% of the body's total weight.

FUNCTIONS

The functions of blood can be placed into the categories of transportation, maintenance, and protection. Many of the functions, however, could be placed in more than one category. For example, for blood cells to protect against microorganisms they must be transported to sites of infection.

Transportation

Blood is the primary transport medium of the body, and it participates in the movement of substances into and out of the body. Oxygen enters blood in the lungs and is carried to cells, and carbon dioxide, produced by cells, is carried to the lungs from which it is expelled. Ingested nutrients, electrolytes, and water are transported by the blood from the digestive tract to cells, and the waste products of cells are transported to the kidneys for elimination. In addition to the movement of substances into and out of the body, blood is also important for internal transport. For example, the precursor to vitamin D is produced in the skin (see Chapter 5) and is transported by the blood to the liver and then to kidneys for processing into active vitamin D. The active vitamin D is transported to the small intestines where it promotes the uptake of calcium. Another example is lactic acid produced by skeletal muscles during anaerobic respiration (see Chapter 10). The lactic acid is carried to the liver and is converted into glucose. Finally, many of the substances necessary for maintenance and protection must be transported throughout the body to perform their functions.

Maintenance

Blood plays a crucial role in maintaining homeostasis. Many of the hormones and enzymes that regulate body processes are found in blood, as are chemicals (buffers; see Chapter 2) that help keep the blood's pH within its normal limits of 7.35 to 7.45 (slightly alkaline). The osmotic composition of blood is also critical for maintaining the normal fluid and electrolyte balance. Because blood can hold heat, it is involved with temperature regulation, transporting heat from the interior to the surface of the body from which the heat is released. When tissues are damaged, the blood clot that forms is the first step in tissue repair and restoration of normal function.

Protection

Cells and chemicals of the blood comprise an important part of the immune system, protecting against foreign substances such as microorganisms and toxins. Blood clotting also provides protection against excessive fluid and cell loss when blood vessels are damaged.

PLASMA

Plasma is a pale yellow fluid accounting for slightly more than half the total blood volume and consisting of approximately 92% water and 8% dissolved or suspended molecules (Table 19-1). Plasma is considered a **colloidal** (ko-loy'dal) **solution** (fine particles suspended in a liquid and resistant to sedimentation or filtration). Plasma contains proteins such as **albumin, globulins,** and **fibrinogen.** When the proteins that produce clots are removed from plasma, the remaining fluid is called **serum.** In addition to the suspended molecules, plasma also contains a number of dissolved components such as salts, nutrients, gases, waste products, hormones, and enzymes.

Water enters the plasma from the digestive tract (ingested fluid), from interstitial fluids, and as a byproduct of metabolism. Excess water is removed from the plasma through the kidneys, lungs, intestinal tract, and skin. **Solutes** in the plasma come from several sources such as the liver, kidneys, intestines, endocrine glands, and immune tissues such as the spleen. Under normal conditions the intake of both water and solutes into the body equals the output so that the total volumes in the body are maintained within a narrow range.

FORMED ELEMENTS

The formed elements of the blood include several types of highly specialized cells and cell fragments. They are grouped into three major categories. Approximately 95% of the volume of the formed elements consists of **erythrocytes** (ĕ-rith'ro-sītz; red blood cells or corpuscles). The remaining 5% consists of **leukocytes** (lu'ko-sītz; white blood cells or corpuscles) and **platelets** (cell fragments), which are also called **thrombocytes** (throm'bo-sītz). The formed elements of the blood are outlined and illustrated in Table 19-2. Leukocytes are the only formed elements possessing nuclei in healthy adults, whereas erythrocytes and platelets have few organelles and lack nuclei.

Production of Formed Elements

The process of blood cell production, called **hematopoiesis** (hem'ă-to-poy-e'sis) or **hemopoiesis** (he'mo-poy-e'sis), occurs in the embryo and fetus in tissues such as the yolk sac, liver, thymus, spleen, lymph nodes, and red bone marrow. After birth, he-

Table 19-1

Composition of plasma

PLASMA COMPONENTS	PERCENT OF TOTAL PLASMA VOLUME	FUNCTION
Water	91.5	Acts as a solvent and suspending medium for blood components
Plasma protein	7.0	
Albumin (55% of total serum protein)		Partly responsible for blood viscosity and osmotic pressure; acts as a buffer
Globulins (38% of total serum protein)		Act as antibodies involved in immunity or as transport molecules
Fibrinogen (4% of total serum protein)		Functions in blood clotting
Complement (1% of total serum protein)		Part of the immune system
Ions	0.9	
Sodium, potassium, calcium, magnesium, chloride, iron, phosphate, hydrogen, hydroxide, bicarbonate		Involved in osmotic pressure, membrane potentials, and acid-base balance
Nutrients	0.3	
Glucose, amino acids, vitamins, cholesterol, triglycerides		Source of energy and basic "building blocks" of more complex molecules
Gases	0.1	
Carbon dioxide		Waste product of aerobic metabolism; as bicarbonate, helps buffer the blood
Oxygen		Necessary for aerobic metabolism; terminal electron receiver in electron transport chain
Nitrogen		Inert
Waste products	0.1	
Urea, uric acid, creatinine, ammonia salts		Breakdown products of protein metabolism; excreted by kidneys
Bilirubin		Breakdown product of red blood cells; excreted as part of the bile from the liver into the intestines
Lactic acid		End product of anaerobic metabolism; converted to glucose
Regulatory substances	0.1	
Enzymes		Catalyze numerous chemical reactions
Hormones		Stimulate or inhibit many body functions

matopoiesis is confined primarily to red bone marrow, with some lymphoid tissue helping in the production of lymphocytes (see Chapter 22). In young children nearly all bone marrow produces blood cells. However, in adults hematopoietic red marrow is confined to the skull, ribs, sternum, vertebrae, pelvis, proximal femur, and proximal humerus (see Chapter 6).

All the formed elements of the blood are derived from a single population of stem cells called **hemocytoblasts** (he′mo-si′to-blastz) (Figure 19-1). These stem cells give rise to the immediate progenitors of the various types of blood cells: **proerythroblasts** (pro′ĕ-rith′ro-blastz), from which erythrocytes develop; **myeloblasts,** from which granulocytes develop; **lymphoblasts,** from which lymphocytes develop; **monoblasts,** from which monocytes develop; and **megakaryoblasts** (meg′ă-kăr-ĭ-o-blastz), from which platelets develop. The development of each cell line is regulated by a specific growth factor. That is, the type of formed element of blood derived from the stem cells and how many formed elements are produced are determined by the growth factor.

Table 19-2

Formed elements of the blood

CELL TYPE	DESCRIPTION	FUNCTION
Erythrocyte	Biconcave disk; no nucleus; 7-8 μm in diameter	Transports oxygen and carbon dioxide
Leukocyte		
Neutrophil	Spherical cell; nucleus with two to four lobes connected by thin filaments; cytoplasmic granules stain a light pink or reddish-purple; 12-15 μm in diameter	Phagocytizes microorganisms
Basophil	Spherical cell; nucleus with two indistinct lobes; cytoplasmic granules stain blue-purple; 10-12 μm in diameter	Releases histamine, which promotes inflammation, and heparin, which prevents clot formation
Eosinophil	Spherical cell; nucleus often with two lobes; cytoplasmic granules stain orange-red or bright red; 10-12 μm in diameter	Releases chemicals that reduce inflammation; attacks certain worm parasites
Lymphocyte	Spherical cell with round nucleus; cytoplasm forms a thin ring around the nucleus; 6-8 μm in diameter	Produces antibodies and other chemicals responsible for destroying microorganisms; responsible for allergic reactions, graft rejection, tumor control, and regulation of the immune system
Monocyte	Spherical cell; nucleus round, kidney, or horse-shoe shaped; contains more cytoplasm than does lymphocyte; 10-15 μm in diameter	Phagocytic cell in the blood; leaves the blood and becomes a macrophage, which phagocytizes bacteria, dead cells, cell fragments, and debris within tissues
Platelet	Cell fragments surrounded by a cell membrane and containing granules; 2-5 μm in diameter	Forms platelet plugs; releases chemicals necessary for blood clotting

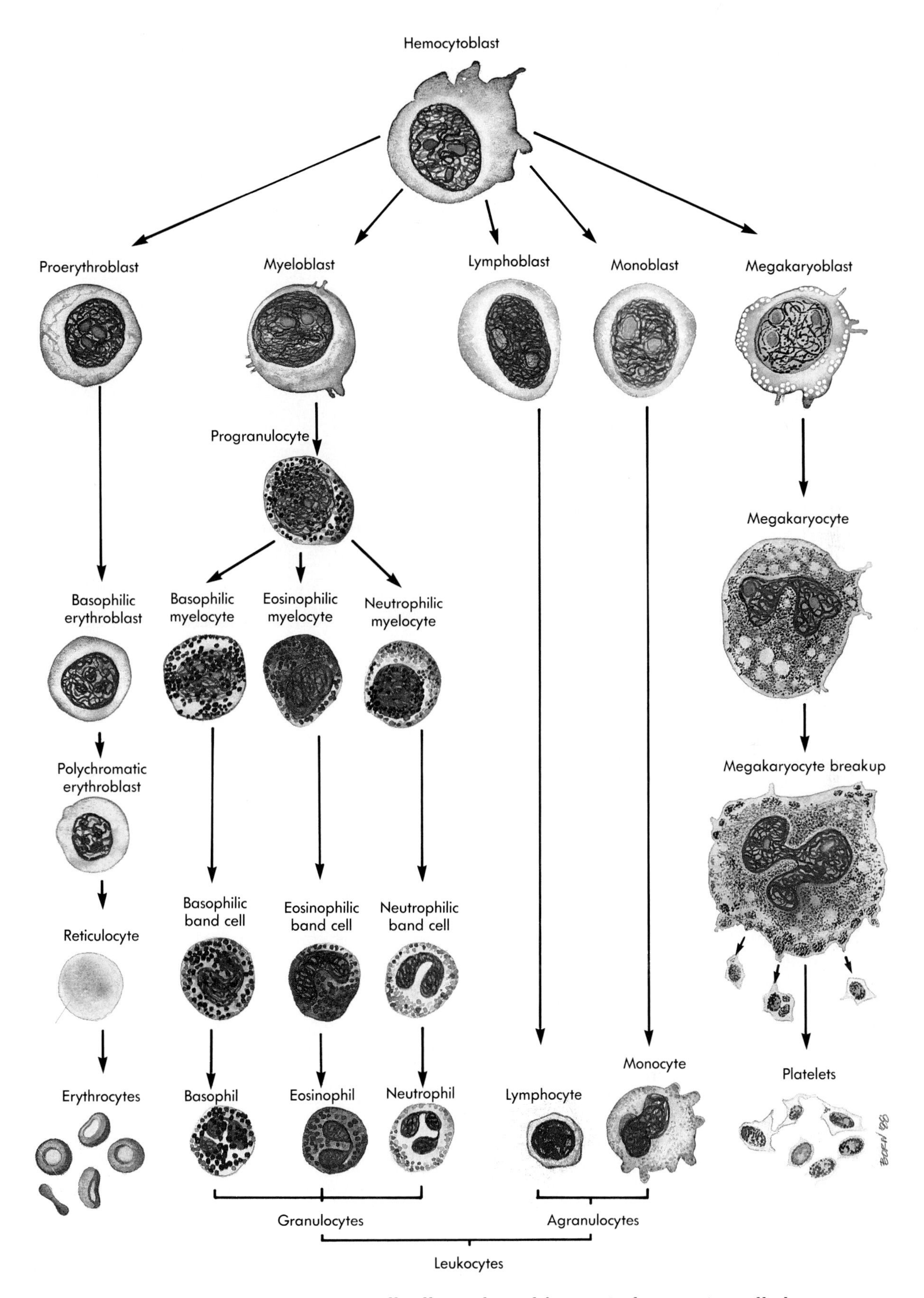

FIGURE 19-1 Hematopoiesis. All cells are derived from a single progenitor cell, the hemocytoblast, which gives rise to the progenitors of each cell type.

Many cancer therapies attack rapidly dividing cells such as those found in tumors. An undesirable side effect, however, can be the destruction of nontumor cells that divide rapidly, for example, the cells in red bone marrow. After treatment for cancer, growth factors are used to stimulate the rapid regeneration of the red bone marrow. Although not a cure for cancer, the use of growth factors can speed recovery from the cancer therapy.

Erythrocytes

Erythrocytes are the most numerous of the formed elements in the blood—approximately 700 times more numerous than leukocytes and 17 times more numerous than platelets. There are approximately 5.2 million erythrocytes per cubic millimeter (approximately one drop) of blood in males (range: 4.2 to 5.8 million) and approximately 4.5 million per cubic millimeter in females (range: 3.6 to 5.2 million). Erythrocytes cannot move of their own accord and therefore are passively propelled through the circulation by forces that cause the blood to flow.

Structure

Normal erythrocytes are biconcave disks approximately 7.7 μm in diameter, 1.9 μm thick at the margin, and less than 1 μm thick in the center (Figure 19-2). A sheet of paper is approximately 75 μm thick; therefore 10 erythrocytes could be aligned side by side across the edge of a sheet of paper. Compared to a flat disk of the same size, the biconcave shape increases the surface area of the erythrocyte. The greater surface area makes it easier for gases to move into and out of the erythrocyte. In addition, the erythrocyte can bend or fold around its thin center, decreasing the size of the erythrocyte and enabling it to pass more easily through small blood vessels.

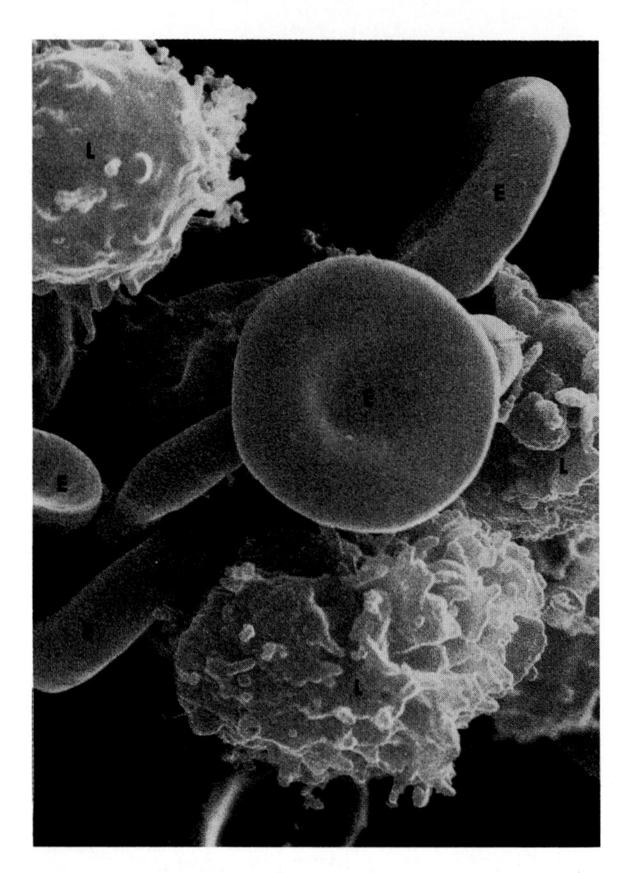

FIGURE 19-2 Scanning electron micrograph of formed elements: erythrocytes *(E)* and leukocytes *(L)*.

Erythrocytes are highly specialized cells that lose their nuclei and nearly all their cellular organelles during maturation. They consist of a relatively simple cell membrane surrounding an internal protein scaffolding, or stroma, and somewhat reduced cytoplasm. Some of the major erythrocyte contents include lipids, adenosine triphosphate (ATP), and the enzyme carbonic anhydrase. The main component of the erythrocyte is the pigmented protein **hemoglobin** (he'mo-glo'bin), which occupies approximately one third of the total cell volume and accounts for its red color.

Function

The primary functions of erythrocytes are to transport oxygen from the lungs to the various tissues of the body and to transport carbon dioxide from the tissues to the lungs. These functions are accomplished primarily by the hemoglobin contained within the erythrocytes. If erythrocytes are ruptured, the hemoglobin leaks out into the plasma and loses its normal configuration (becomes denatured) and most of its normal function. Erythrocyte rupture followed by hemoglobin release is called **hemolysis** (he-mol'ĭ-sis).

Another important function of erythrocytes occurs because of the presence of **carbonic anhydrase** in the cell. This enzyme catalyzes the reaction between carbon dioxide and water to form carbonic acid, which ionizes to form hydrogen and bicarbonate ions. The **bicarbonate ion** (HCO_3^-) is the major form of carbon dioxide transported in the blood (see Chapter 23); however, it is transported in the plasma rather than in the erythrocytes. Bicarbonate ions are produced in the cell and then pass by diffusion into the plasma. The production of bicarbonate ions is also important in the regulation of blood pH (see Chapters 23 and 27).

Hemoglobin

Hemoglobin consists of four protein chains and four heme groups. Each protein, called a **globin,** is bound to one **heme,** a red-pigment molecule. Each heme contains one **iron atom** (Figure 19-3).

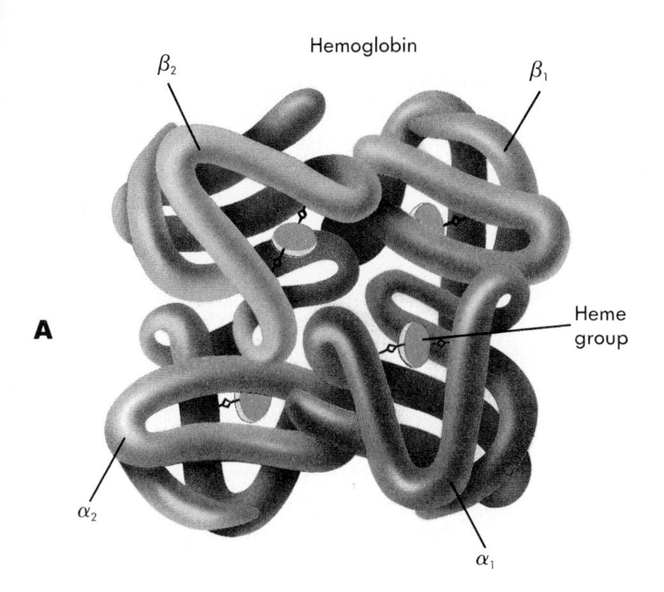

FIGURE 19-3 Hemoglobin. **A** Four protein chains, each with a heme group, form a hemoglobin molecule. **B** Each heme contains one iron atom.

A number of different types of globin exist, each having a slightly different amino acid composition. The four globins in normal adult hemoglobin consist of two alpha chains and two beta chains. Embryonic and fetal globins appear at different times during development and are replaced by adult globin near the time of birth. Embryonic and fetal hemoglobins are more effective at binding oxygen than is adult hemoglobin. Abnormal hemoglobins are less effective at attracting oxygen than is normal hemoglobin and may result in anemia (see essay).

1

What would happen to a fetus if maternal blood had an equal or greater affinity for oxygen than fetal blood?

? ? ? ? ? ? ? ? ? ? ?

Iron is necessary for the normal function of hemoglobin. Oxygen is transported in association with the iron atom, which is in the ferrous (bivalent) state. The adult human body normally contains approximately 4 g of iron, two thirds of which is associated with hemoglobin.

Current research is being conducted in an attempt to develop artificial hemoglobin. One chemical that has been used in clinical trials is Fluosol DA, a white liquid with a high oxygen affinity. Although the usefulness of hemoglobin substitutes is currently limited because artificial hemoglobin is destroyed fairly quickly in the body, future work may uncover more successful substitutes that can provide long-term relief for patients with blood disorders.

Various types of poisons affect the hemoglobin molecule. Carbon monoxide (CO) such as occurs in incomplete combustion of gasoline binds to the iron of hemoglobin, forming the relatively stable carboxyhemoglobin. As a result of the stable binding of carbon monoxide, hemoglobin cannot transport oxygen, and death may occur. Cigarette smoke also produces carbon monoxide, and the blood of smokers may contain 5% to 15% carboxyhemoglobin.

Small amounts of iron are regularly lost from the body in waste products such as urine and feces. Females lose additional iron as a result of menstrual bleeding and therefore require more dietary iron than do males. Dietary iron is absorbed into the circulation from the upper part of the intestinal tract. Acid from the stomach and ascorbic acid (vitamin C) in food increase the solubility of iron in the alkaline environment of the small intestine, thus facilitating the absorption of iron in the small intestine. Iron absorption is regulated according to need, and iron deficiency can result in anemia.

When hemoglobin is exposed to oxygen, one oxygen molecule may become associated with each heme group. This oxygenated form of hemoglobin is **oxyhemoglobin.** Hemoglobin containing no oxygen is **deoxyhemoglobin,** or **reduced hemoglobin** (see Chapter 2). Oxyhemoglobin is bright red, whereas deoxyhemoglobin has a darker color.

Hemoglobin also transports carbon dioxide, which does not combine with the iron atoms but is attached to amino groups of the globin molecule.

This hemoglobin form is **carbaminohemoglobin.** The transport of oxygen and carbon dioxide by the blood is discussed more fully in Chapter 23.

Life history of erythrocytes

Under normal conditions approximately 2.5 million erythrocytes are destroyed every second. This amount seems like a staggering loss of erythrocytes until it is realized that the loss represents only 0.00001% of the total 25 trillion erythrocytes contained in the normal adult circulation. Furthermore, those 2.5 million erythrocytes are being replaced by the production of an equal number of erythrocytes every second.

The process by which new erythrocytes are produced is called **erythropoiesis** (ĕ-rith′ro-poy-e′sis; see Figure 19-1), and the time required for the production of a single erythrocyte is approximately 4 days.

Proerythroblasts, the cells from which erythrocytes develop, are derived from hemocytoblasts, the stem cells from which all blood cells originate. After several mitotic divisions proerythroblasts become **basophilic erythroblasts,** so named because they attract basic dyes. Basophilic erythroblasts continue to undergo mitosis and begin to produce hemoglobin. **Polychromatic** (more than one color) **erythroblasts,** which develop from basophilic erythroblasts, have a rough endoplasmic reticulum, which attracts basic stains, in addition to their acidic-staining cytoplasmic granules. Polychromatic erythroblasts contain almost a complete complement of hemoglobin and during unusually rapid erythropoiesis are released into the circulation.

Polychromatic erythroblasts lose their nuclei by a process of extrusion, after which the cell is called a **reticulocyte** (rĕ-tik′u-lo-sīt). These cells still contain endoplasmic reticulum (hence the name), which can be stained with basic dyes, plus a few basophilic organelles and are naturally red because of the large amount of hemoglobin. Reticulocytes are released from the red bone marrow into the circulating blood, and within 1 or 2 days they lose their endoplasmic reticulum and become mature erythrocytes.

2

What does an elevated reticulocyte count indicate? Would a person's reticulocyte count change during the week after he had donated a unit (approximately 500 ml) of blood?

? ? ? ? ? ? ? ? ? ? ?

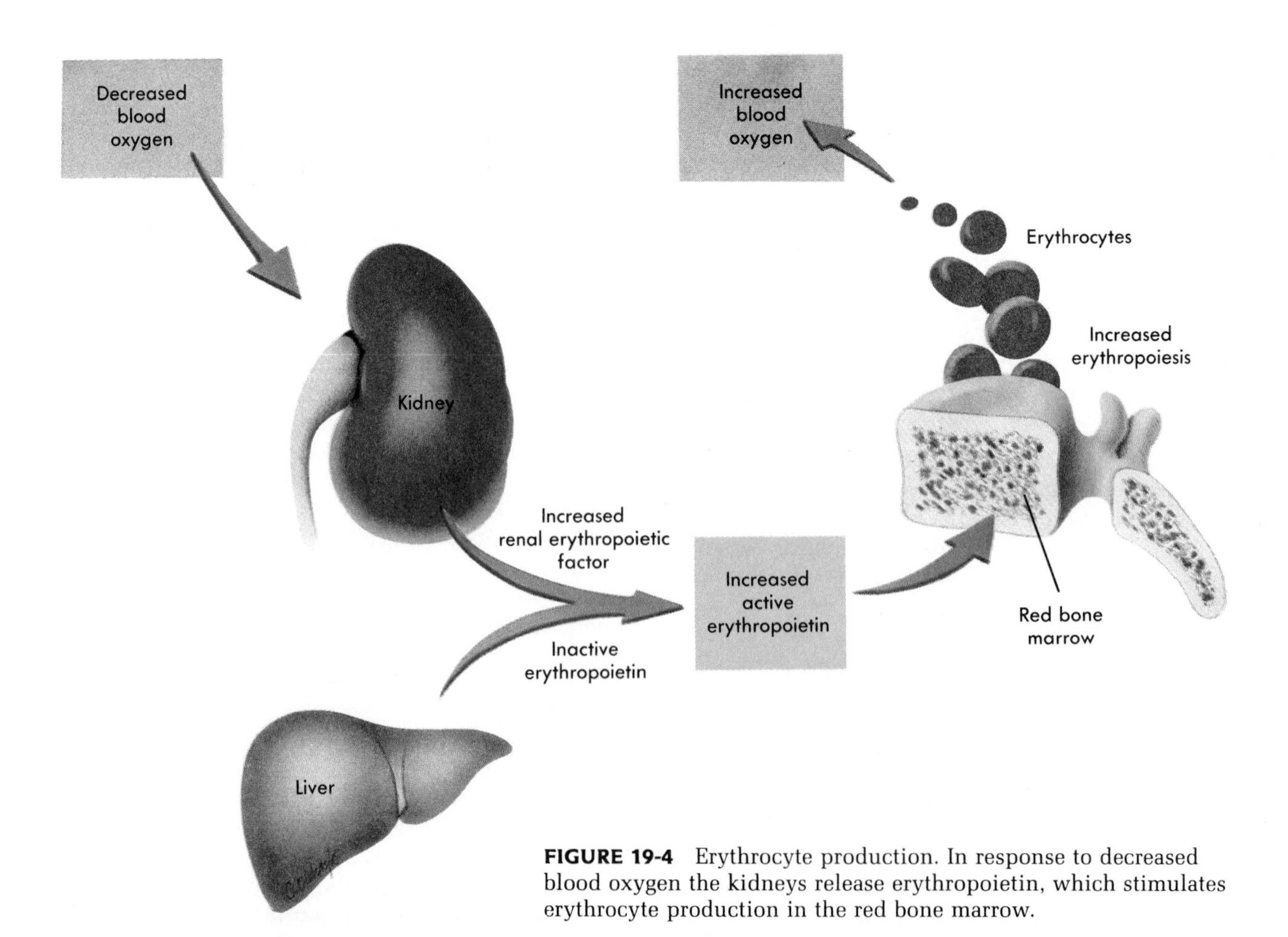

FIGURE 19-4 Erythrocyte production. In response to decreased blood oxygen the kidneys release erythropoietin, which stimulates erythrocyte production in the red bone marrow.

Erythropoiesis requires iron, vitamin B_{12}, folic acid, and erythropoietin. Iron is used in hemoglobin production, and vitamins B_{12} and folic acid are necessary for DNA synthesis. Without vitamins B_{12} and folic acid the series of cell divisions that produce erythrocytes is inhibited. **Erythropoietin** (ĕ-rith'ro-poy'e-tin) is a glycoprotein produced primarily by the kidneys, but also by the liver and other tissues. Erythropoietin stimulates hematopoietic bone marrow to increase the rate at which erythrocyte production occurs.

The production of erythropoietin is controlled by a negative-feedback mechanism involving oxygen (Figure 19-4). As the amount of oxygen reaching the kidneys decreases, the production of erythropoietin is increased and erythropoiesis is stimulated. On the other hand, increased oxygen levels result in decreased erythropoietin production and decreased erythropoiesis. Testosterone, a male sex hormone, also stimulates erythropoietin production.

Erythrocytes normally stay within the circulation for approximately 120 days in males and 110 days in females. These cells have no nuclei and therefore cannot produce new proteins. As their existing proteins, enzymes, cell membrane components, and other structures degenerate, the erythrocytes become old and abnormal in form and function. Erythrocytes may also be damaged in various ways while passing through the circulation.

Old, damaged, or defective erythrocytes are removed from the blood by **macrophages** (large "eating" cells) located in the spleen, liver, and other lymphatic tissues (Figure 19-5). Within the macrophage lysosomal enzymes break open erythrocytes and be-

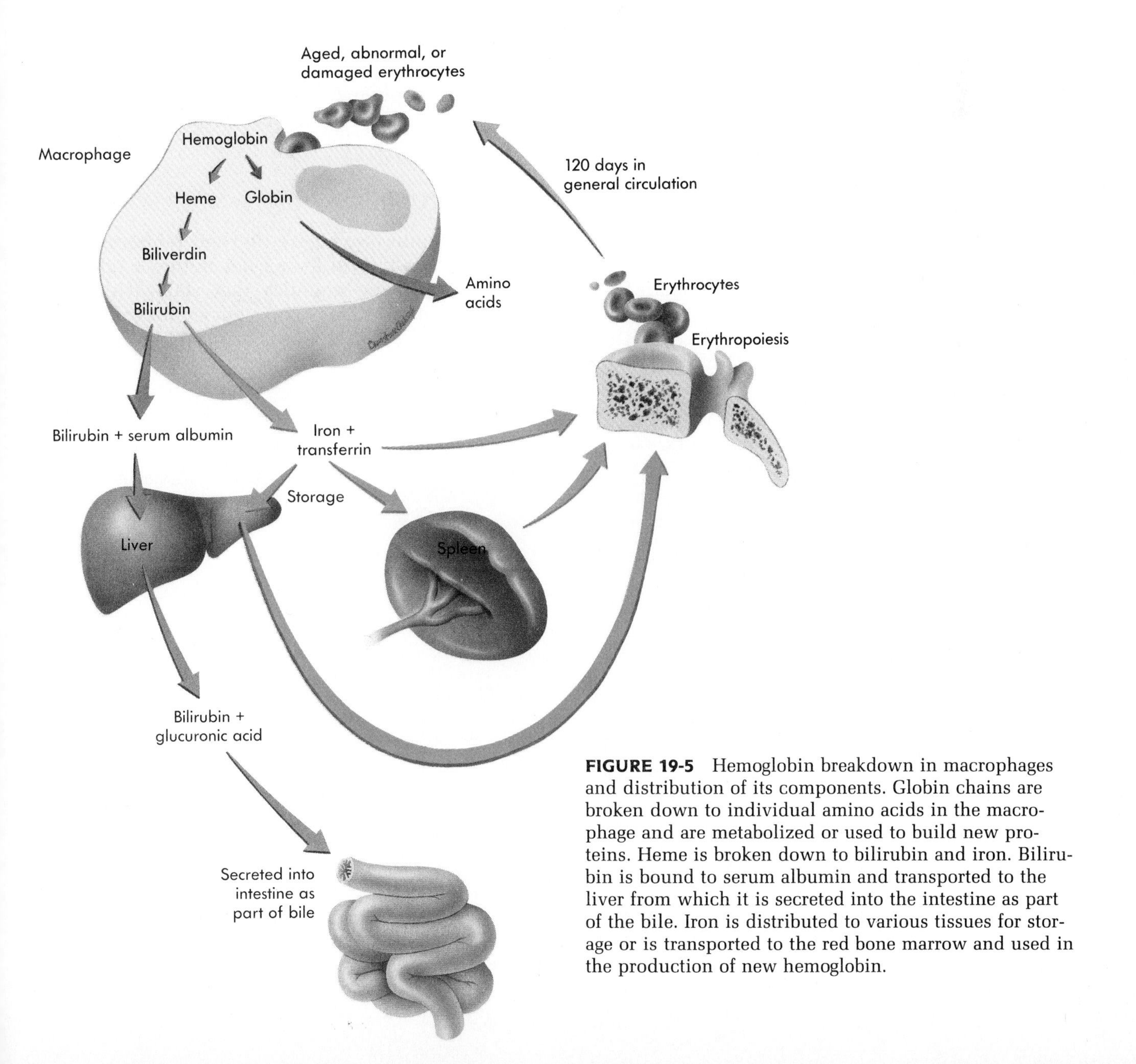

FIGURE 19-5 Hemoglobin breakdown in macrophages and distribution of its components. Globin chains are broken down to individual amino acids in the macrophage and are metabolized or used to build new proteins. Heme is broken down to bilirubin and iron. Bilirubin is bound to serum albumin and transported to the liver from which it is secreted into the intestine as part of the bile. Iron is distributed to various tissues for storage or is transported to the red bone marrow and used in the production of new hemoglobin.

gin to digest hemoglobin. Globin is broken down into its component amino acids, most of which are reused in the production of other proteins. Iron atoms also are released for recycling. The heme groups are converted to **biliverdin** (bil-ĭ-ver'din) and then to **bilirubin** (bil-ĭ-roo'bin), which is released into the plasma and becomes bound to serum albumin. Bound bilirubin is taken up by the liver, bound to glucuronic acid, and secreted in bile into the small intestine. In the intestines, bacteria convert bilirubin into the pigments that give the feces its characteristic yellow-brown color. Some of these pigments are absorbed from the intestine and are excreted in the urine, giving the urine its characteristic color. A yellowish staining of the integument and sclera by bile pigments, associated with a buildup of bilirubin in the circulation and interstitial spaces, is known as **jaundice** (jawn'dis).

Leukocytes

Leukocytes, or white blood cells, are nucleated blood cells that lack hemoglobin. They are clear or whitish in color and are larger than erythrocytes, ranging from 8 to 19 μm in diameter. In stained preparations leukocytes attract stain, whereas erythrocytes remain relatively unstained (Figure 19-6 and see Table 19-2).

Leukocytes protect the body against invading microorganisms and remove dead cells and debris from the body. Most leukocytes are motile, exhibiting ameboid movement (i.e., moving like an ameba by putting out irregular cytoplasmic projections). Leukocytes leave the circulation by the process of **diapedesis** (di'ă-pĕ-de'sis; movement through vessel walls) and move through the tissues in which they ingest foreign material or dead cells. Leukocytes find their way to such material by the process of **chemotaxis** (kem-o-tak'sis; attraction of cells to a chemical source). At the site of an infection leukocytes accumulate and phagocytize bacteria, dirt, and dead cells; then they die. This accumulation of dead leukocytes, along with fluid and cell debris, is called **pus.**

Leukocytes are named according to their appearance in stained preparations. Leukocytes containing large cytoplasmic granules are **granulocytes** (gran'u-lo-sītz), and those with very small granules that cannot be seen easily with the light microscope are **agranulocytes** (see Figure 19-1). The three types of granulocytes are named according to the staining characteristics of their cytoplasm: **neutrophils** (nu'tro-filz), **eosinophils** (e'o-sin'o-filz), and **basophils** (ba'so-filz). There are two types of agranulocytes: **monocytes** (mon'o-sītz), named according to nuclear morphology, and **lymphocytes** (lim'fosītz), named according to a major site of proliferation.

Neutrophils

Neutrophils (see Figure 19-6 and Table 19-2), the most common type of leukocytes in the blood, have small cytoplasmic granules that stain with both acidic and basic stains. Their nuclei are commonly trilobed, and for this reason neutrophils often are

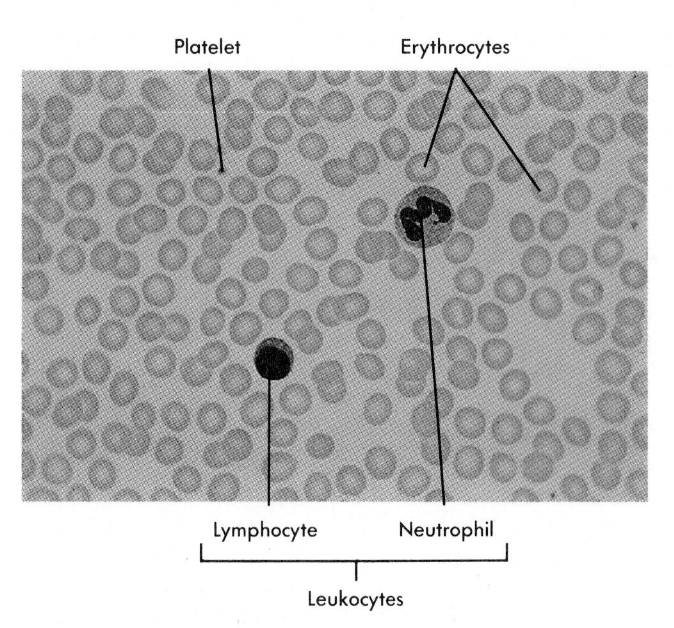

FIGURE 19-6 Standard blood smear. The erythrocytes are pink, and the leukocytes are more red (from stain) with a purple-stained nucleus.

called **polymorphonuclear** (pol'e-mor-fo-nu'kle-ar) **neutrophils,** or PMNs. Neutrophils usually remain in the circulation for only a short time (10 to 12 hours) and then move into other tissues in which they become motile and seek out and phagocytize bacteria, antigen-antibody complexes (antigens and antibodies bound together), and other foreign matter. Neutrophils also secrete a class of enzymes called **lysozymes** (li'so-zīmz), which are capable of destroying certain bacteria. Neutrophils usually survive for 1 or 2 days after leaving the circulation.

Eosinophils

Eosinophils (see Table 19-2) contain cytoplasmic granules that stain bright red with eosin, an acidic stain. They are motile cells that leave the circulation to enter the tissues during an inflammatory reaction. They are most common in tissues undergoing an allergic response, and their numbers are elevated in the blood of people with allergies or certain parasitic infections. Eosinophils apparently reduce the inflammatory response by producing enzymes that destroy inflammatory chemicals such as histamine. Eosinophils also phagocytize antigen-antibody complexes formed during the allergic response (although they are not as important in this function as neutrophils).

Basophils

Basophils (see Table 19-2), the least common of all leukocytes, contain large cytoplasmic granules that stain blue or purple with basic dyes. Basophils, like other granulocytes, leave the circulation and migrate through the tissues in which they play a role in both allergic and inflammatory reactions. Basophils contain large amounts of **histamine,** which they release within tissues to increase inflammation. They also release **heparin,** which inhibits blood clotting.

Lymphocytes

Lymphocytes (see Figure 19-6 and Table 19-2) are the smallest of all leukocytes. A lymphocyte is approximately 9 μm in diameter, which is not much larger than an erythrocyte. The lymphocyte's nucleus is smaller than the nuclei of other cells because the chromatin is tightly compacted. The lymphocytic cytoplasm consists of only a thin, sometimes imperceptible, ring around the nucleus. Lymphocytes are highly motile cells and actually are capable of migrating right through the cytoplasm of other cells. They play a major role in immunity, including antibody production.

Although lymphocytes originate in bone marrow, they migrate through the blood to lymphatic tissues in which they can proliferate and produce more lymphocytes. The majority of the total body's lymphocyte population is in the lymphatic tissues: the lymph nodes, spleen, tonsils, lymph nodules, and thymus. There are a number of different types of lymphocytes, which are described in Chapter 22.

Monocytes

Monocytes (see Table 19-2), measuring 19 μm in diameter, are by far the largest of the leukocytes. Monocytes normally remain in the circulation for approximately 3 days, leave the circulation, become transformed into macrophages, and migrate through the various tissues. They phagocytize bacteria, dead cells, cell fragments, and any other debris within the tissues. An increase in the number of monocytes is often associated with chronic infections.

Platelets

Platelets, or thrombocytes (Table 19-2), are minute fragments of cells consisting of a small amount of cytoplasm surrounded by a cell membrane. They are roughly disk shaped and average approximately 3 μm in diameter. Even though they are approximately 40 times more common in the blood than leukocytes (approximately 250,000 to 500,000 per cubic millimeter of blood), they often are not counted in typical blood smears because they tend to form platelet clumps and become difficult to distinguish.

The life expectancy of platelets is approximately 5 to 9 days. Platelets are produced within the marrow and are derived from **megakaryocytes** (meg'ă-kăr'e-o-sītz), which are extremely large cells with diameters up to 100 μm. Small fragments of these cells break off and enter the circulation as platelets.

Platelets play an important role in preventing blood loss. This prevention is accomplished in two ways: (1) the formation of platelet plugs, which seal holes in small vessels; and (2) the formation of clots, which help seal off larger wounds in the vessels.

HEMOSTASIS

Clot formation is very important to the maintenance of homeostasis. If not stopped, excessive bleeding from a cut or torn blood vessel can result in a positive-feedback pathway, consisting of ever-decreasing blood volume and blood pressure, leading away from homeostasis, and resulting in death. Fortunately, when a blood vessel is damaged, a number of events occur that help prevent excessive blood loss. **Hemostasis** (he'mo-sta-sis), the arrest of bleeding, can be divided into three stages: vascular spasm, platelet plug formation, and coagulation (clot formation).

Vascular Spasm

Vascular spasm is an immediate but temporary closure of a blood vessel resulting from contraction of smooth muscle within the blood vessel's wall. In small vessels the resulting constriction of the vessel can close the vessel completely and stop the flow of blood through the vessel. Vascular spasm is produced by nervous system reflexes and by chemicals released from platelets during the formation of a platelet plug.

Platelet Plug Formation

When a blood vessel is damaged, the endothelium becomes torn, and the underlying connective tissue is exposed. Platelets in the blood adhere to collagen fibers in the connective tissue and release adenosine diphosphate (ADP), thromboxane (a prostaglandin derivative), serotonin, and chemicals involved in coagulation. Thromboxane and serotonin stimulate vascular spasms. ADP makes the surface of platelets sticky, so additional platelets adhere to those already attached to the collagen fibers. Thromboxane induces the aggregating platelets to release ADP, thromboxane, and other chemicals. Thus platelets adhere to each other and release chemicals that cause other platelets to adhere and so on. The accumulating mass of platelets that results is a **platelet plug.**

Platelet plug formation is very important in maintaining the integrity of the circulatory system. Small tears occur in the smaller vessels and capillaries many times each day, and platelet plug formation quickly closes them. People who lack the normal number of platelets (a condition called thrombocytopenia) are prone to develop numerous small hemorrhages in their skin and internal organs.

The production of prostaglandin is very important to platelet plug formation as is demonstrated by the effect of aspirin, which is known to inhibit prostaglandin synthesis. If an expectant mother ingests aspirin near the end of pregnancy, prostaglandin synthesis is inhibited, having several effects. Two of these effects are (1) the mother experiences excessive postpartum hemorrhage because of decreased platelet function, and (2) the baby may exhibit numerous localized hemorrhages (called petechiae) over the surface of its body as a result of decreased platelet function. If the quantity of ingested aspirin is large, the infant, mother, or both may die as a result of hemorrhage. On the other hand, there are times when platelet plugs and clots (e.g., those resulting in strokes or heart attacks) form in vessels and threaten the life of the individual. Aspirin has been proposed as a useful substance in preventing such vascular problems.

Coagulation

Platelet plugs alone are not sufficient to close large tears or cuts; therefore when a blood vessel is severely damaged, a blood clot usually forms in the damaged area within 20 seconds. The formation of a blood clot is called **coagulation** (ko-ag′u-la-shun). A **blood clot** consists of a network of **fibrin** within which blood cells, platelets, and fluid become trapped.

The formation of a blood clot depends on a number of proteins found within plasma called **coagulation factors** (Table 19-3). Normally the coagulation factors are in an inactive state and do not cause clotting. After injury the clotting factors are activated to produce a clot. This activation is a complex process involving many chemical reactions, but it can be summarized in three main stages (Figure 19-7). **Stage 1** consists of the production of **prothrombin activator, Stage 2** consists of the conversion by prothrombin activator of **prothrombin** to **thrombin,** and **Stage 3** consists of the enzymatic conversion by thrombin of soluble **fibrinogen** to insoluble **fibrin.**

Depending on how prothrombin activator is formed in Stage 1, there are two separate pathways by which coagulation occurs: the **extrinsic clotting pathway** and the **intrinsic clotting pathway** (see Figure 19-7). Once prothrombin activator is formed, stages 2 and 3 are very similar for both pathways.

Extrinsic clotting pathway

The extrinsic clotting pathway is so named because the formation of prothrombin activator in stage 1 is initiated by chemicals from outside the blood (Figure 19-7, *A*). When a blood vessel is damaged, the damaged epithelial and connective tissue cells of that area release a number of chemicals into the blood. Among these chemicals is a complex mixture of lipoproteins known collectively as tissue thromboplastin (throm′bo-plas′tin), or factor III. Tissue thromboplastin, in the presence of calcium (Ca^{2+}) ions, forms a complex with factor VII, which activates factor X. Activated factor X combines with tissue thromboplastin, Ca^{2+} ions, and factor V to form prothrombin activator. In Stage 2 prothrombin activator converts the soluble plasma protein prothrombin to the enzyme thrombin. During Stage 3, thrombin, in the presence of Ca^{2+} ions, converts the soluble plasma protein fibrinogen to the insoluble protein fibrin, thus creating a clot. Factor XIII is necessary for the stabilization of the clot, and thrombin stimulates factor XIII activation.

Intrinsic clotting pathway

The intrinsic clotting pathway is so named because the formation of prothrombin activator in Stage 1 is initiated by chemicals within the blood (Figure 19-7, *B*). When plasma factor XII comes into contact

Table 19-3

Coagulation factors

FACTOR NUMBER	NAME (synonym)	DESCRIPTION AND FUNCTION
I	Fibrinogen	Plasma protein synthesized in liver; converted to fibrin in Stage 3
II	Prothrombin	Plasma protein synthesized in liver (requires vitamin K); converted to thrombin in Stage 2
III	Thromboplastin (tissue thromboplastin; tissue factor)	Mixture of lipoproteins released from damaged tissue; required in extrinsic Stage 1
IV	Calcium ion	Required throughout entire clotting sequence
V	Proaccelerin (labile factor)	Plasma protein synthesized in liver; activated form functions in Stages 1 and 2 of both intrinsic and extrinsic clotting pathways
VI	—	Once thought involved but no longer accepted as playing a role in coagulation; apparently the same as activated factor V
VII	Serum prothrombin conversion accelerator (stable factor, proconvertin)	Plasma protein synthesized in liver (requires vitamin K); functions in extrinsic Stage 1
VIII	Antihemophilic factor (antihemophilic globulin)	Plasma protein synthesized in liver; required for intrinsic Stage 1
IX	Plasma thromboplastin component	Plasma protein synthesized in liver (requires vitamin K); required for intrinsic Stage 1
X	Stuart factor (Stuart-Prower factor)	Plasma protein synthesized in liver (requires vitamin K); required in Stages 1 and 2 of both intrinsic and extrinsic clotting pathways
XI	Plasma thromboplastin antecedent	Plasma protein synthesized in liver; required for intrinsic Stage 1
XII	Hageman factor	Plasma protein required for intrinsic Stage 1
XIII	Fibrin-stabilizing factor	Protein found in plasma and platelets; required for Stage 3
Platelet Factors		
I	Platelet accelerator	Same as plasma factor V
II	Thrombin accelerator	Accelerates thrombin (intrinsic clotting pathway) and fibrin production
III	Platelet thromboplastic factor	Phospholipids necessary for the intrinsic and extrinsic clotting pathways
IV		Binds heparin, which prevents clot formation

Vitamin K deficiency may result in hemorrhages such as frequent nosebleeds. Vitamin K is produced by intestinal bacteria, and those bacteria may be killed by large doses of oral antibiotics, resulting in insufficient coagulation ability. Newborns lack these intestinal bacteria and therefore must obtain vitamin K from food such as milk. Because cow's milk contains more vitamin K than does human milk, breast-fed infants are more susceptible to hemorrhage than bottle-fed infants. Vitamin K is absorbed from the small intestine into the circulation and is necessary for the synthesis of several clotting factors by the liver. The absorption of vitamin K from the intestine requires the presence of bile; therefore certain disorders such as obstruction of bile flow to the intestine can interfere with vitamin K absorption and lead to insufficient clotting. Liver diseases that result in the decreased synthesis of clotting factors also may lead to insufficient clot formation.

with an area of damaged blood vessels, it becomes activated and stimulates factor XI, which in turn, in the presence of Ca^{2+} ions, activates factor IX. At the same time, platelet activation and aggregation result in the release of platelet chemicals, including platelet phospholipids (platelet factor III), which are necessary for the intrinsic clotting pathway. Activated factor IX, in the presence of platelet phospholipids, factor VIII, and Ca^{2+} ions, activates factor X. Activated factor X combines with Ca^{2+} ions, factor V, and platelet phospholipids to form prothrombin activator. Stages 2 and 3 then are activated, and a clot results.

Platelet factor III is released from activated platelets as they aggregate. Platelet factors I, II, and IV also are involved. Thrombin, which is produced during Stage 2, stimulates further platelet activation, thus

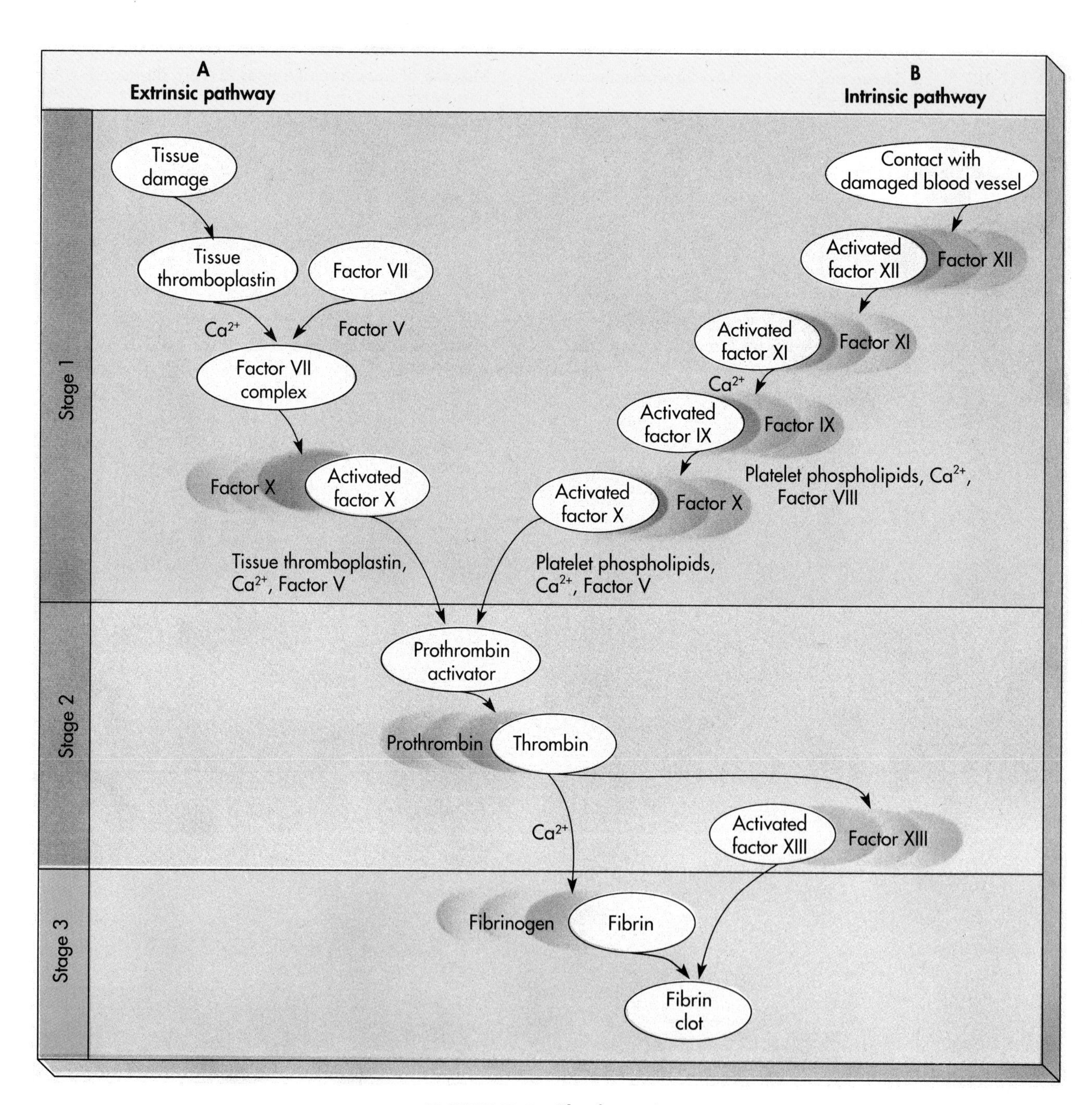

FIGURE 19-7 Clot formation.

A Extrinsic clotting pathway. Tissue damage releases tissue thromboplastin, which with factor VII and calcium ions activates factor X. Activated factor X, tissue thromboplastin, and calcium ions form prothrombin activator, which causes prothrombin to become thrombin. Thrombin causes fibrinogen to become fibrin and form a clot.

B Intrinsic clotting pathway. Damaged vessels cause activation of factor XII. Activated factor XII activates factor XI, which activates factor IX. Factor IX, along with factor VIII and platelet factors, activates factor X. From that point the intrinsic clotting pathway takes the same course as the extrinsic clotting pathway.

functioning as a positive-feedback loop in the intrinsic pathway. Thrombin also functions as a positive-feedback molecule by stimulating its own production from prothrombin. Many of the factors involved in clot formation require **vitamin K** for their production (see Table 19-3).

Control of Clot Formation

Without control coagulation would spread from the point of initiation to the entire circulatory system. Furthermore, vessels in a normal person contain rough areas that can stimulate clot formation; therefore small amounts of prothrombin are constantly being spontaneously converted into thrombin. To prevent unwanted clotting, the blood contains several **anticoagulants** (an'tĭ-ko-ag'u-lantz), which prevent coagulation factors from initiating clot formation. Only when coagulation factor concentrations exceed a given threshold does coagulation occur. At the site of injury so many coagulation factors are activated that the anticoagulants are unable to prevent clot formation. Away from the injury site, however, the activated coagulation factors are diluted in the blood, anticoagulants neutralize them, and clotting is prevented.

Examples of anticoagulants in the blood are antithrombin, heparin, and prostacyclin. **Antithrombin,** a plasma protein produced by the liver, slowly inactivates thrombin. **Heparin,** produced by basophils and endothelial cells, increases the effectiveness of antithrombin; heparin and antithrombin together rapidly inhibit thrombin activity. **Prostacyclin** (a prostaglandin derivative), produced by endothelial cells, counteracts the effects of thrombin by causing vasodilation and by inhibiting the release of coagulation factors from platelets.

A clot that forms when platelets encounter roughened plaque areas within diseased but otherwise undamaged blood vessels is called a **thrombus** (throm'bus). If a thrombus breaks loose and begins to float through the circulation, it is an **embolus** (em'bo-lus). Both thrombi and emboli can result in death if they occlude vessels that supply blood to essential organs such as the heart, brain, or lungs. Abnormal coagulation can be prevented or hindered by the injection of anticoagulants such as heparin, which acts rapidly, or warfarin, which acts more slowly than heparin. **Warfarin** (Coumadin) prevents clot formation by suppressing the production of vitamin K–dependent coagulation factors (II, VII, IX, and X) by the liver. Interestingly, warfarin was first used as a rat poison, causing rats to bleed to death. In small doses warfarin is a proven, effective anticoagulant in humans. However, with anticoagulant treatment there is a danger that the patient will either hemorrhage internally or bleed excessively when cut.

Anticoagulants are also important outside the body, preventing the clotting of blood used in transfusions and laboratory blood tests. Examples include heparin, **ethylenediaminetetraacetic acid (EDTA),** and **sodium citrate.** EDTA and sodium citrate prevent clot formation by binding to calcium ions, making them inaccessible for clotting reactions.

Clot Retraction and Dissolution

The fibrin meshwork constituting the clot adheres to the walls of the vessel. Once the clot has formed, it begins to condense into a denser, compact structure through a process known as **clot retraction.** Platelets, which contain the contractile protein actomyosin, pull on the fibrin in the clot and are responsible for clot retraction. As the clot condenses, serum (plasma without clotting factors) is squeezed out of the clot.

Consolidation of the clot pulls the edges of the damaged vessel together, which may help to stop the flow of blood, reduce infection, and enhance healing. The damaged vessel is repaired by the movement of fibroblasts into the damaged area and the formation of new connective tissue. In addition, epithelial cells around the wound proliferate and fill in the torn area.

The clot usually is dissolved within a few days after clot formation by a process called **fibrinolysis** (fi'brĭ-nol'ĭ-sis), which involves the activity of **plasmin** (plaz'min), an enzyme that hydrolyzes fibrin. Plasmin is formed from inactive plasminogen, which is a normal blood protein. It is activated by thrombin, factor XII, tissue plasminogen activator, and lysosomal enzymes released from damaged tissues (Figure 19-8). In disorders such as myocardial infarct (heart

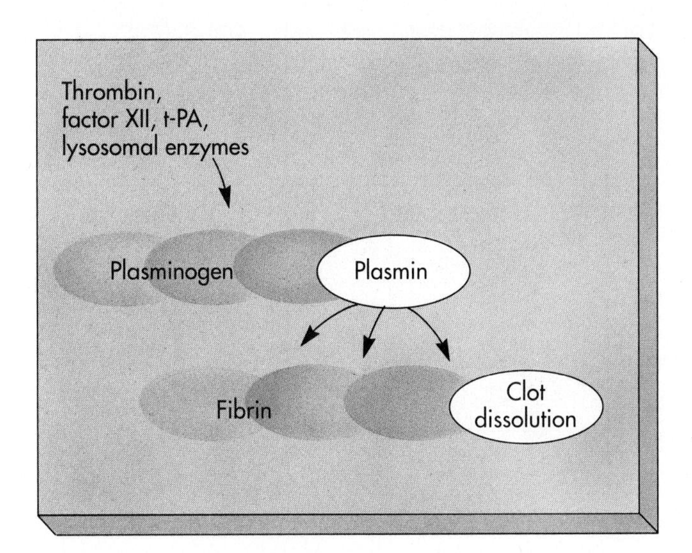

FIGURE 19-8 Fibrinolysis. Plasminogen is converted by thrombin, factor XII, t-PA (tissue plasminogen activator), or lysosomal enzymes to the active enzyme plasmin. Plasmin breaks the fibrin molecules and therefore the clot into smaller pieces, which are washed away in the blood or are phagocytized.

attack), which are caused by blockage of a vessel by a clot, the clot may be dissolved as a result of chemicals that stimulate plasmin production. For example, streptokinase, a bacterial enzyme, or tissue plasminogen activator produced by genetic engineering can be injected into the blood or introduced at the clot site by means of a catheter.

BLOOD GROUPING

If large quantities of blood are lost during surgery or in an accident, the blood volume must be increased, or the patient can go into shock and die. A **transfusion** is the transfer of blood or other solutions into the blood of the patient. In many cases the return of blood volume to normal levels is all that is necessary. This can be accomplished by the transfusion of plasma or prepared solutions that have the proper amounts of solutes. When large quantities of blood are lost, however, erythrocytes also must be replaced so that the oxygen-carrying capacity of the blood is restored.

Early attempts to transfuse blood from one person to another were unsuccessful. They often resulted in transfusion reactions, which included clotting within blood vessels, kidney damage, and death. It is now known that transfusion reactions are caused by interactions between antigens and antibodies (see Chapter 22). In brief, the surface of erythrocytes has molecules called **antigens** (an'tĭ-jenz), and in the plasma there are molecules called **antibodies.** An antibody is very specific, meaning it can combine only with a certain antigen. When the antibodies in the plasma bind to the antigens on the surface of the erythrocytes, they form molecular bridges that connect the erythrocytes together. As a result, **agglutination** (ă-glu'tĭ-na'shun), or clumping, of the cells occurs. The combination of the antibodies with the antigens can also initiate reactions that cause **hemolysis** (he-mol'ĭ-sis) or rupture of the erythrocytes. Because the antigen-antibody combination can cause agglutination, the antigens are often called **agglutinogens** (ă-glu-tin'o-jenz), and the antibodies are called **agglutinins** (ă-glu'tĭ-ninz).

The antigens on the surface of erythrocytes have been categorized into **blood groups,** and more than 35 blood groups, most of which are rare, have been identified. The ABO and Rh blood groups are among the most important. Other well-known groups include the Lewis, Duffy, MNSs, Kidd, Kell, and Lutheran groups.

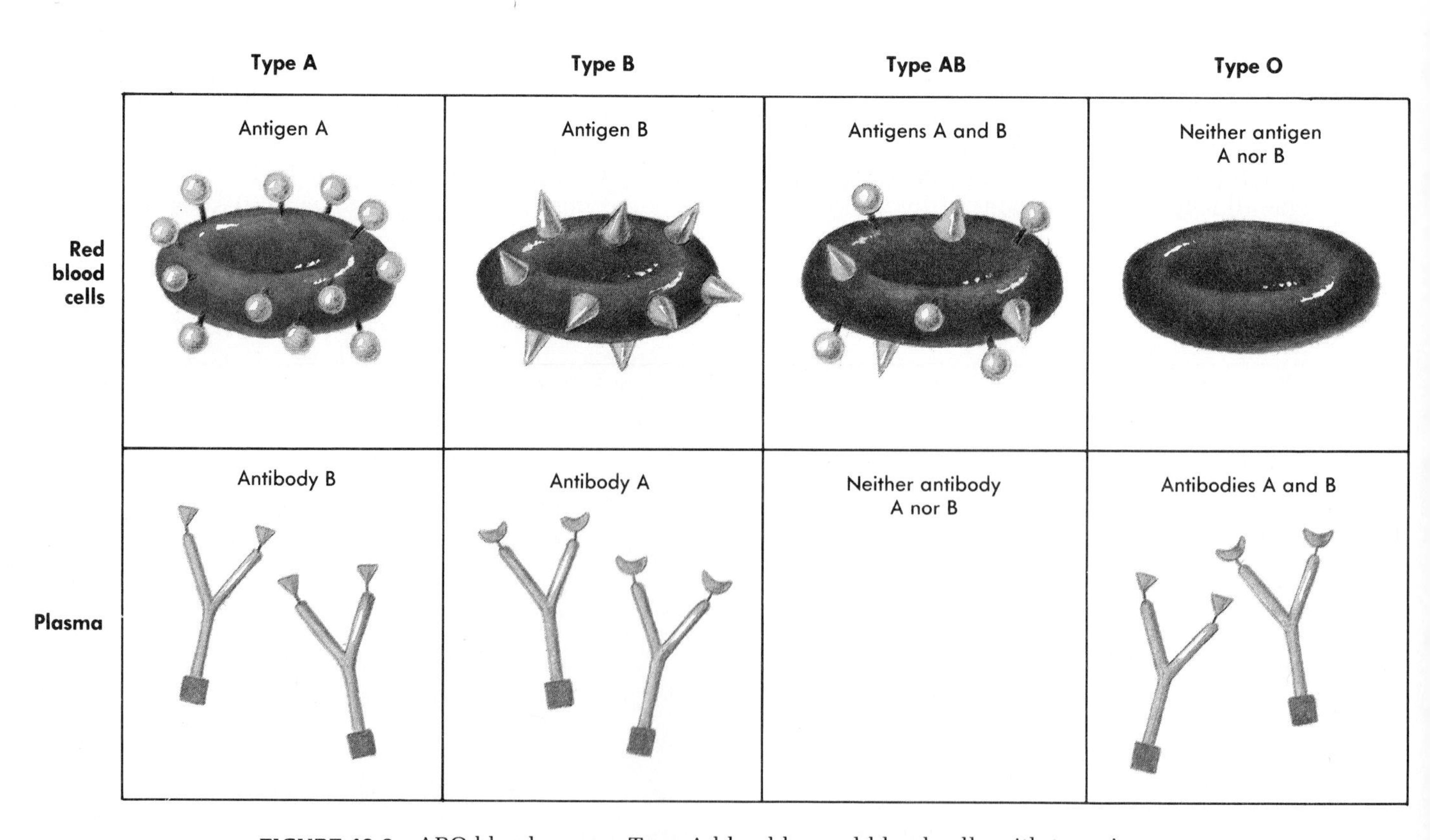

FIGURE 19-9 ABO blood groups. Type A blood has red blood cells with type A surface antigens and plasma with type B antibodies. Type B blood has type B surface antigens and plasma with type A antibodies. Type AB blood has both type A and type B surface antigens and no plasma antibodies. Type O blood has no ABO surface antigens but both A and B plasma antibodies.

ABO Blood Group

In the **ABO blood group** type A blood has type A antigens, type B blood has type B antigens, type AB blood has both types of antigens, and type O blood has neither A nor B antigens (Figure 19-9). In addition, plasma from type A blood contains antibodies against type B antigens, and plasma from type B blood contains antibodies against type A antigens. Type O blood has both A and B antibodies, and type AB blood has neither.

The ABO blood types are not found in equal numbers. In Caucasians in the United States the distribution is type O, 47%; type A, 41%; type B, 9%; and type AB, 3%. Among blacks in the United States the distribution is type O, 46%; type A, 27%; type B, 20%; and type AB, 7%.

The reason for the presence of A and B antibodies in blood is not clearly understood. Antibodies normally do not develop against an antigen unless the body is exposed to the antigen. This means, for example, that a person with type A blood should not have type B antibodies unless he has received a transfusion of type B blood, which contains type B antigens. Because people with type A blood (who never have received a transfusion of type B blood) do have type B antibodies, another explanation is needed. One possibility is that type A or B antigens on bacteria or food in the digestive tract stimulate the formation of antibodies against antigens that are different from one's own antigens. Thus a person with type A blood would produce type B antibodies against the B antigens on the bacteria or food. In support of this hypothesis is the observation that A and B antibodies are not found in the blood until 2 months after birth.

A **donor** is a person who gives blood, and a **recipient** is a person who receives blood. Usually a donor can give blood to a recipient if they both have the same blood type. For example, a person with type A blood could donate to another person with type A blood. There would be no ABO transfusion reaction because the recipient has no antibodies against the type A antigen. On the other hand, if type A blood were donated to a person with type B blood, there would be a transfusion reaction because the person with type B blood has antibodies against the type A antigen, and agglutination would result (Figure 19-10).

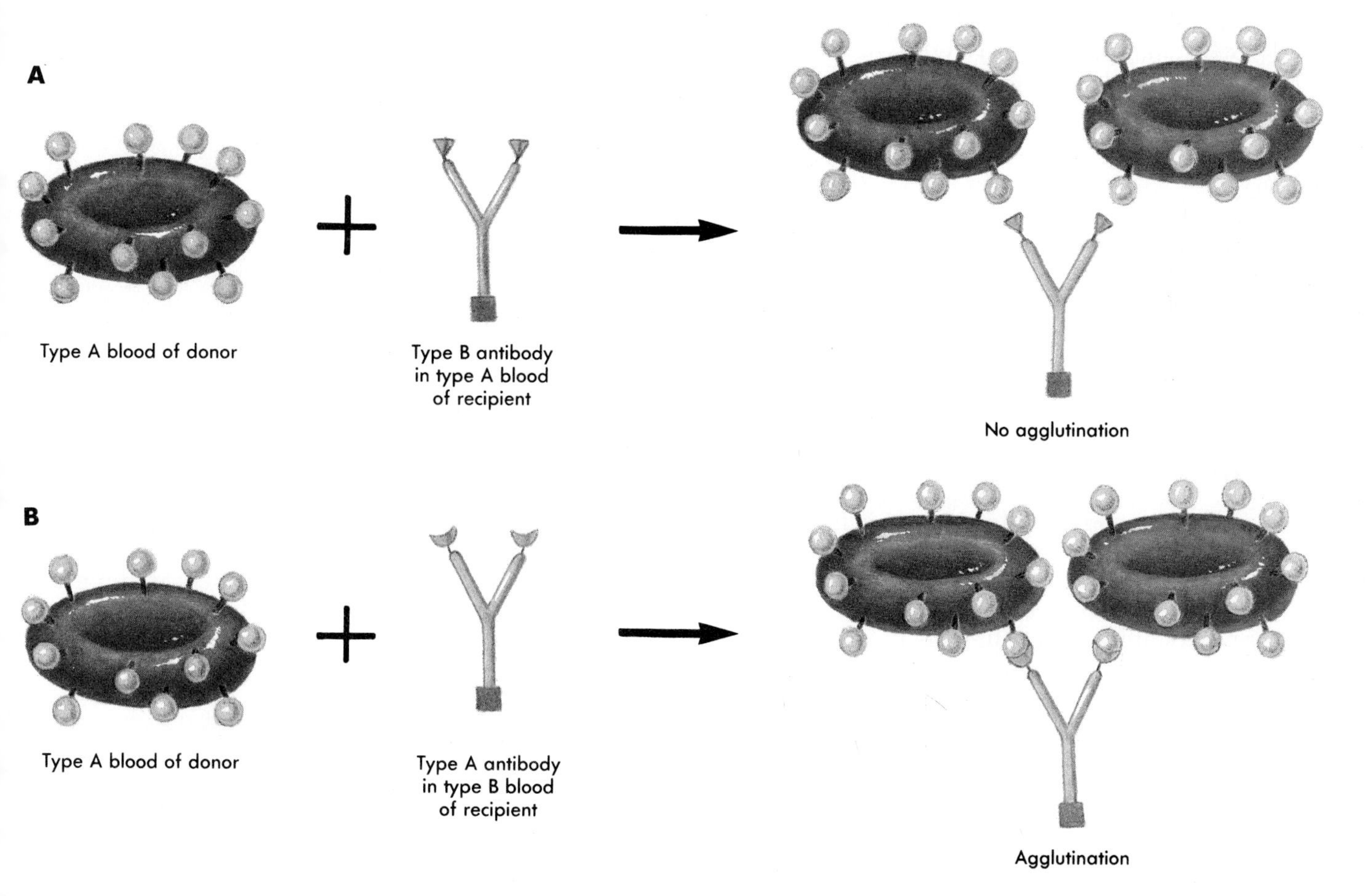

FIGURE 19-10 Agglutination reaction. **A** Type A blood donated to a type A recipient does not cause an agglutination reaction because the type B antibodies in the recipient do not combine with the type A antigens in the donated blood. **B** Type A blood donated to a type B recipient causes an agglutination reaction because the type A antibodies in the recipient combine with the type A antigens in the donated blood.

Historically, people with type O blood have been called universal donors because they usually can give blood to the other ABO blood types without causing an ABO transfusion reaction. Their erythrocytes have no ABO surface antigens and therefore do not react with the recipient's A or B antibodies. For example, if type O blood is given to a person with type A blood, the type O erythrocytes do not react with the type B antibodies in the recipient's blood. In a similar fashion, if type O blood is given to a person with type B blood, there would be no reaction with the recipient's type A antibodies.

The term universal donor is misleading. Transfusion of type O blood can produce a transfusion reaction for two reasons. First, there are other blood groups that can cause a transfusion reaction. To reduce the likelihood of a transfusion reaction, all the blood groups must be matched correctly. Second, antibodies in the blood of the donor can react with antigens in the blood of the recipient. For example, type O blood has type A and B antibodies. If type O blood is transfused into a person with type A blood, the A antibodies (in the type O blood) react against the A antigens (in the type A blood). Usually such reactions are not serious because the antibodies in the donor's blood are diluted in the blood of the recipient, and few reactions take place. Type O blood is given to a person with another blood type only in life-or-death emergency conditions because type O blood sometimes causes transfusion reactions in these situations.

3

Historically people with type AB blood were called universal recipients. What is the rationale for this term? Explain why the term is misleading.

? ? ? ? ? ? ? ? ? ? ?

Rh Blood Group

Another important blood group is the **Rh blood group,** so named because it was first studied in the rhesus monkey. People are Rh positive if they have a certain Rh antigen (the D antigen) on the surface of their erythrocytes, and they are Rh negative if they do not have this Rh antigen. Approximately 85% of white people and 88% of black people in the United States are Rh positive. The ABO blood type and the Rh blood type usually are designated together. For example, a person designated as A positive is type A in the ABO blood group and Rh positive. The rarest combination in the United States is AB negative, which occurs in less than 1% of all Americans.

Antibodies against the Rh antigen do not develop unless an Rh-negative person is exposed to Rh-positive blood. This can occur through a transfusion or by transfer of blood between a mother and her fetus across the placenta. When an Rh-negative person receives a transfusion of Rh-positive blood, the recipient becomes sensitized to the Rh antigen and produces Rh antibodies. If the Rh-negative person is unfortunate enough to receive a second transfusion of Rh-positive blood after becoming sensitized, a transfusion reaction results.

Rh incompatibility can pose a major problem in some pregnancies when the mother is Rh negative and the fetus is Rh positive (Figure 19-11). If fetal blood leaks through the placenta and mixes with the mother's blood, the mother becomes sensitized to the Rh antigen. The mother produces Rh antibodies that cross the placenta and cause agglutination and hemolysis of fetal erythrocytes. This disorder is called **erythroblastosis fetalis** (ĕ-rith′ro-blas-to′sis fe-tă′lis), and it can be fatal to the fetus. In the first pregnancy, however, there often is no problem. The leakage of fetal blood is usually the result of a tear in the placenta that takes place either late in the pregnancy or during delivery. Thus there is not enough time for the mother to produce enough Rh antibodies to harm the fetus. In later pregnancies, however, there can be a problem because the mother has been sensitized to the Rh antigen. Consequently, if the fetus is Rh positive and there is any leakage of fetal blood into the mother's blood, she rapidly produces large amounts of Rh antibodies, and erythroblastosis fetalis develops.

Prevention of erythroblastosis fetalis is possible if the Rh-negative woman is given an injection of a specific type of antibody preparation, called Rh_o (D) immune globulin (RhoGAM), before or immediately after each delivery or abortion. The injection contains antibodies against Rh antigens. The injected antibodies bind to the Rh antigens of any fetal erythrocytes that may have entered the mother's blood. This treatment inactivates the fetal Rh antigens and prevents sensitization of the mother.

If erythroblastosis fetalis develops, treatment consists of slowly removing the blood of the fetus or newborn and replacing it with Rh-negative blood. Exposure of the newborn to fluorescent light also is used because it helps to break down the large amounts of bilirubin formed as a result of erythrocyte destruction. High levels of bilirubin are toxic to the nervous system and can cause destruction of brain tissue.

DIAGNOSTIC BLOOD TESTS
Type and Cross Match

To prevent transfusion reactions the blood is typed, and a **crossmatch** is made. **Blood-typing** determines the ABO and Rh blood groups of the blood sample. Typically, the cells are separated from the serum. The cells are tested with known antibodies to determine the type of antigen on the cell surface.

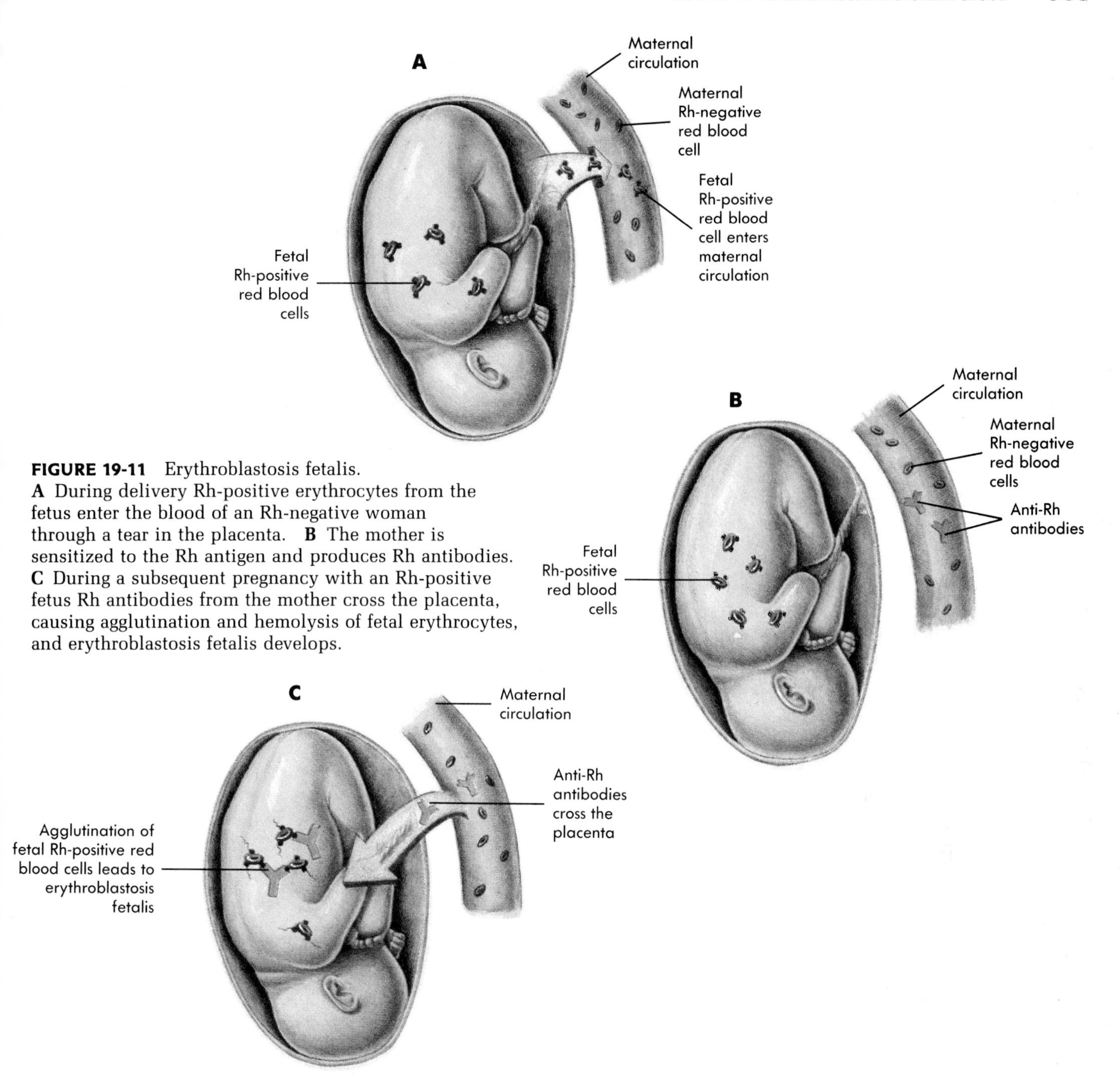

FIGURE 19-11 Erythroblastosis fetalis.
A During delivery Rh-positive erythrocytes from the fetus enter the blood of an Rh-negative woman through a tear in the placenta. **B** The mother is sensitized to the Rh antigen and produces Rh antibodies. **C** During a subsequent pregnancy with an Rh-positive fetus Rh antibodies from the mother cross the placenta, causing agglutination and hemolysis of fetal erythrocytes, and erythroblastosis fetalis develops.

For example, if a patient's blood cells agglutinate when mixed with type A antibodies but do not agglutinate when mixed with type B antibodies, it is concluded that the cells have type A antigen. In a similar fashion the serum is mixed with known cell types (antigens) to determine the type of antibodies in the serum.

Normally donor blood must match the ABO and Rh type of the recipient. Because other blood groups can also cause a transfusion reaction, however, a crossmatch is performed. In a crossmatch the donor's blood cells are mixed with the recipient's serum, and the donor's serum is mixed with the recipient's cells. The donor's blood is considered safe for transfusion only if there is no agglutination in either match.

Complete Blood Count

The **complete blood count (CBC)** is an analysis of the blood that provides much information. It consists of a red blood cell count, hemoglobin and hematocrit measurements, and a white blood cell count.

Red blood cell count

Blood cell counts usually are done electronically with a machine, but they can be done manually with a microscope. The normal range for a **red blood cell (RBC) count** is 4.2 to 5.8 million erythrocytes per cubic millimeter of blood for a male, and 3.6 to 5.2

million per cubic millimeter of blood for a female. **Polycythemia** (pol'ĭ-si-the'me-ah) is an overabundance of erythrocytes. It can result from a decreased oxygen supply, which stimulates erythropoietin production, or from red bone marrow tumors. Polycythemia makes it harder for the blood to flow through blood vessels and increases the work load of the heart. It can reduce blood flow through tissues and, if severe, can result in plugging of small blood vessels (capillaries).

Hemoglobin measurement

The **hemoglobin measurement** determines the amount of hemoglobin in a given volume of blood, usually expressed as grams of hemoglobin per 100 ml of blood. The normal hemoglobin count for a male is 14 to 18 g per 100 ml of blood, and for a female it is 12 to 16 g per 100 ml of blood. Abnormally low hemoglobin is an indication of **anemia** (ă-ne'me-ah), which is a reduced number of erythrocytes or a reduced amount of hemoglobin in each erythrocyte.

Hematocrit measurement

The percentage of total blood volume composed of erythrocytes is the **hematocrit** (hem'ă-to-krit). One way to determine hematocrit is to place blood in a tube and spin the tube in a centrifuge. The formed elements are heavier than the plasma and are forced to one end of the tube (Figure 19-12). The erythrocytes account for 44% to 54% of the total blood volume in males and 38% to 48% in females. Because the hematocrit measurement is based on volume, it is affected by the number and size of erythrocytes. For example, a decreased hematocrit can result from a decreased number of normal-sized erythrocytes or a normal number of small-sized erythrocytes. The average volume of an erythrocyte is calculated by dividing the hematocrit by the RBC count. A number of disorders result in smaller- or larger-than-normal erythrocytes. For example, inadequate iron in the diet can impair hemoglobin production. Consequently, during their formation erythrocytes do not fill with hemoglobin, and they remain smaller than normal.

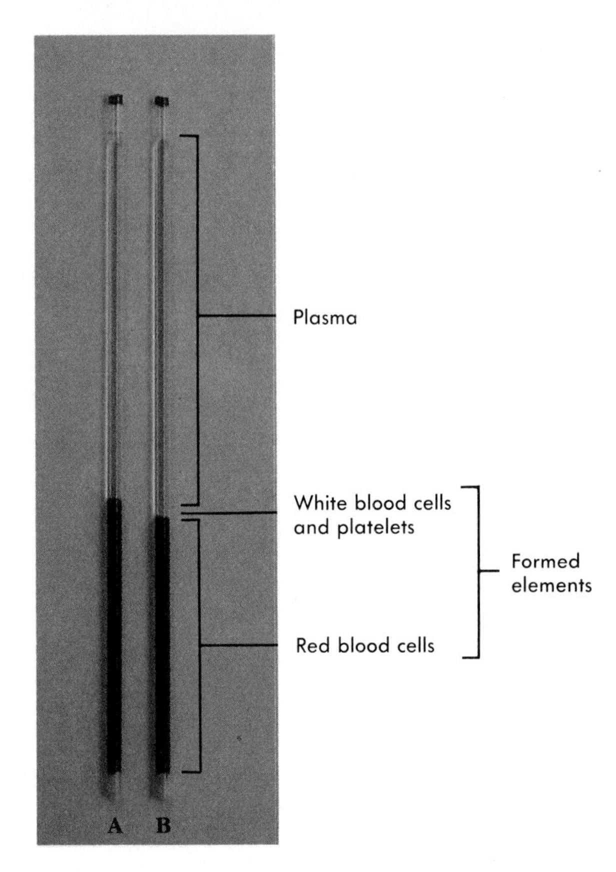

FIGURE 19-12 Normal hematocrits of, **A** male (44%) and, **B** female (40%). Blood is separated into plasma and formed elements. The relatively small amount of white blood cells (leukocytes) and platelets rests on the red blood cells (erythrocytes).

White blood cell count

A **white blood cell (WBC) count** measures the total number of leukocytes in the blood. There are normally 5000 to 10,000 leukocytes per cubic millimeter of blood. **Leukopenia** (lu-ko-pe'nĭ-ah) is a lower-than-normal WBC count and often indicates depression or destruction of the red marrow (e.g., by radiation, drugs, tumor, or a deficiency of vitamin B_{12} or folic acid). **Leukocytosis** (lu-ko-sī-to'sis) is an abnormally high WBC count. **Leukemia,** a tumor of the red marrow, and bacterial infections often cause leukocytosis.

White Blood Cell Differential Count

A **white blood cell differential count** determines the percentage of each of the five kinds of leukocytes in the WBC count. Normally neutrophils account for 60% to 70%; lymphocytes, 20% to 30%; monocytes, 2% to 8%; eosinophils, 1% to 4%; and basophils, 0.5% to 1%. Much insight about a patient's condition can be obtained from a WBC differential count. For example, in patients with bacterial infections the neutrophil count often is greatly increased, whereas in patients with allergic reactions the eosinophil and basophil counts are elevated.

Clotting

Two measurements that test the ability of the blood to clot are the platelet count and the prothrombin time.

Platelet count

A normal **platelet count** is 150,000 to 400,000 platelets per cubic millimeter of blood. **Thrombocytopenia** (throm-bo-si-to-pe'nĭ-ah) is a condition in which the platelet count is greatly reduced, resulting in chronic bleeding through small vessels and capillaries. It can be caused by decreased platelet production as a result of hereditary disorders, lack of vitamin B_{12} (pernicious anemia), drug therapy, or radiation therapy.

Prothrombin time measurement

Prothrombin time measurement is a measure of how long it takes for the blood to start clotting, which normally is 9 to 12 seconds. Because many clotting factors must be activated to form prothrombin, a deficiency of any one of them can cause an abnormal prothrombin time. Vitamin K deficiency, certain liver diseases, and drug therapy can cause an increased prothrombin time.

Blood Chemistry

The composition of materials dissolved or suspended in the plasma can be used to assess the functioning of many of the body's systems. For example, high blood glucose levels can indicate that the pancreas is not producing enough insulin; high blood urea nitrogen (BUN) is a sign of reduced kidney function; increased bilirubin can indicate liver dysfunction; and high cholesterol levels can indicate an increased risk of developing cardiovascular disease. A number of blood chemistry tests are routinely done when a blood sample is taken, and additional tests are available.

4

When a patient complains of acute pain in the abdomen, the physician suspects appendicitis, which is a bacterial infection of the appendix. What blood test should be done to support the diagnosis?

? ? ? ? ? ? ? ? ? ? ?

Disorders of the Blood

Polycythemia

Polycythemia (pol'ĭ-si-the'me-ah) is characterized by an overabundance of erythrocytes, resulting in increased viscosity of the blood, reduced flow rates, and, if severe, plugging of the capillaries. **Polycythemia vera** is the abnormally large production of erythrocytes (and other formed blood elements) and often results from red marrow tumors. **Secondary polycythemia** results from a decreased oxygen supply (e.g., in high altitudes or from increased production of erythropoietin).

Anemia

Anemia (ă-ne'me-ah) is a deficiency of hemoglobin in the blood. It can result from a decrease in the number of erythrocytes, a decrease in the amount of hemoglobin in each erythrocyte, or both. The decreased hemoglobin reduces the ability of the blood to transport oxygen. Anemic patients suffer from a lack of energy and feel excessively tired and listless. They may appear pale and quickly become short of breath with only slight exertion.

One general cause of anemia is insufficient production of erythrocytes. **Aplastic anemia** is caused by an inability of the red bone marrow to produce erythrocytes. It usually is acquired as a result of damage to the red marrow by chemicals (e.g., benzene), drugs (e.g., certain antibiotics and sedatives), or radiation.

Erythrocyte production also can be less than normal because of nutritional deficiencies. **Iron deficiency anemia** results from deficient intake or absorption of iron or from excessive iron loss. Consequently, not enough hemoglobin is produced, the number of erythrocytes decreases, and the erythrocytes that are manufactured are smaller than normal. Another type of nutritional anemia is **pernicious** (per-nish'us) **anemia,** which is caused by inadequate vitamin B_{12}. Because vitamin B_{12} is necessary for the cell divisions that result in erythrocyte formation, a shortage of vitamin B_{12} causes reduced erythrocyte production. Although inadequate levels of vitamin B_{12} in the diet can cause pernicious anemia, the usual cause is insufficient absorption of the vitamin. Normally the stomach produces intrinsic factor, a protein that binds to vitamin B_{12}. The combined molecules pass into the lower intestine in which intrinsic factor facilitates the absorption of the vitamin. Without adequate levels of intrinsic factor, insufficient vitamin B_{12} is absorbed, and pernicious anemia develops. Present evidence suggests that the inability to produce intrinsic factor is an autoimmune disease in which the body's immune system damages the cells in the stomach that produce intrinsic factor. **Folic acid deficiency** also can hinder cell division and cause anemia. Inadequate amounts of folic acid in the diet is

Continued.

Disorders of the Blood—cont'd

the usual cause, with the disorder developing most often in the poor, in pregnant women, and in chronic alcoholics.

Another general cause of anemia is loss or destruction of erythrocytes. **Hemorrhagic** (hem-ŏ-raj'ik) **anemia** results from a loss of blood such as can result from trauma, ulcers, or excessive menstrual bleeding. Chronic blood loss in which small amounts of blood are lost over a period of time can result in iron deficiency anemia. **Hemolytic** (he-mo-lit'ik) **anemia** is a disorder in which erythrocytes rupture or are destroyed at an excessive rate. It can be caused by inherited defects within the erythrocytes. For example, one kind of inherited hemolytic anemia is due to a defect in the cell membrane that causes erythrocytes to rupture easily. Many kinds of hemolytic anemia result from unusual damage to the erythrocytes by drugs, snake venom, artificial heart valves, autoimmune disease, or erythroblastosis fetalis.

Some anemias result from inadequate or defective hemoglobin production. **Thalassemia** (thal-ă-se'mĭ-ah) is a hereditary disease found in people of Mediterranean, Asian, and African ancestry. It is caused by insufficient production of the globin portion of the hemoglobin molecule. The major form of the disease results in death by age 20, and the minor form results in a mild anemia. **Sickle cell anemia,** a hereditary disease found mostly in blacks, results in the formation of abnormal hemoglobin. The erythrocytes assume a rigid, sickle shape and plug small blood vessels. They are also more fragile than normal. In its severe form sickle cell anemia is usually fatal before the patient is 30 years of age, whereas in its minor form, sickle cell trait, there are usually no symptoms.

Hemophilia

Hemophilia (he'mo-fil'ĭ-ah) is a genetic disorder in which clotting is abnormal or absent. Most often it is found in people from northern Europe and their descendants. Hemophilia is a sex-linked trait, meaning that it occurs almost exclusively in males. There are several types of hemophilia, each the result of a deficiency or dysfunction of a coagulation factor. **Hemophilia type A** (classic hemophilia; 83% of all hemophilias) results from a deficiency of plasma coagulation factor VIII. **Hemophilia type B** is caused by a deficiency in plasma factor IX. **Hemophilia type C** results from a deficiency in factor XI. Types A and B occur almost exclusively in males (X chromosome linked; see Chapter 29), whereas type C occurs in both males and females. Treatment of hemophilia involves injection of the missing clotting factor taken from donated blood.

Thrombocytopenia

Thrombocytopenia (throm-bo-si-to-pe'nĭ-ah) is a condition in which the platelet number is greatly reduced, resulting in chronic bleeding through small vessels and capillaries. There are several causes of thrombocytopenia, including increased platelet destruction or decreased platelet production caused by hereditary disorders, pernicious anemia, drug therapy, or radiation therapy.

Leukemia

Leukemia (lu-ke'me-ah) is a type of cancer in which abnormal production of one or more of the leukocyte types occurs. Because these cells are usually immature or abnormal and lack normal immunological functions, patients are very susceptible to infections. The excess production of leukocytes in the red marrow may also interfere with erythrocyte and platelet formation and thus lead to anemia and bleeding.

Infectious Diseases of the Blood

Normally blood is sterile. Microorganisms may be found in the blood for two reasons: transportation or multiplication. Microorganisms can gain entry to the body and be transported by the blood to the tissues they infect. For example, the poliomyelitis virus enters through the gastrointestinal tract and is carried to nervous tissue. After microorganisms are established at a site of infection, some can be picked up by the blood. These microorganisms can spread to other locations in the body, multiply within the blood, or be eliminated by the body's immune system.

Septicemia (sep'tĭ-se'mĭ-ah), or blood poisoning, is the multiplication of microorganisms in the blood. Often septicemia results from the introduction of microorganisms by a medical procedure such as the insertion of an intravenous tube into a blood vessel. The release of toxins by bacteria can cause **septic shock,** which is a decrease in blood pressure that can result in death.

There are a few diseases in which microorganisms actually multiply within blood cells. **Malaria** (mă-la'rĭ-ah) is caused by a protozoan that is introduced into the blood by the bite of the *Anopheles* mosquito. Part of the protozoan's development occurs inside erythrocytes. The symptoms of chills and fever are produced by toxins released when the protozoan causes the erythrocytes to rupture. **Infectious mononucleosis** (mon'o-nu'kle-o'sis) is caused by a virus that infects the salivary glands and lymphocytes. The lymphocytes are altered by the virus, and the immune system attacks and destroys the lymphocytes. The immune system response is believed to produce the symptoms of fever, sore throat, and swollen lymph nodes. The virus that causes **acquired immunodeficiency syndrome (AIDS)** also infects lymphocytes and causes suppression of the immune system (see Chapter 22).

The presence of microorganisms in blood is a concern when transfusions are performed because it is possible to infect the blood recipient. Blood is routinely tested in an effort to eliminate this risk, especially for AIDS and hepatitis. **Hepatitis** (hep'ă-ti'tis) is an infection of the liver caused by several different kinds of viruses. After recovering, hepatitis victims can become carriers. Although they show no signs of the disease, they release the virus into their blood or bile. To prevent infection of others, anyone who has had hepatitis is asked not to donate blood products.

SUMMARY

FUNCTIONS p. 586

1. Blood transports gases, nutrients, waste products, and hormones.
2. Blood is involved in the regulation of homeostasis and the maintenance of pH, body temperature, fluid balance, and electrolyte levels.
3. Blood protects against disease and blood loss.

PLASMA p. 586

1. Plasma is mostly water (92%) and contains proteins such as albumin (maintains osmotic pressure), globulins (function in transport and immunity), and fibrinogen (involved in clot formation) as well as hormones and enzymes (involved in regulation). Plasma also contains ions, nutrients, waste products, and gases.
2. When clot-forming proteins are removed from plasma, the remaining fluid is serum.

FORMED ELEMENTS p. 586

The formed elements include erythrocytes (red blood cells), leukocytes (white blood cells), and platelets (cell fragments).

Production of Formed Elements

1. In the embryo and fetus the formed elements are produced in a number of locations.
2. After birth, red bone marrow becomes the source of the formed elements.
3. All formed elements are derived from stem cells called hemocytoblasts.

Erythrocytes

1. Erythrocytes are biconcave disks containing hemoglobin and carbonic anhydrase.
   - Hemoglobin consists of heme and globin. The heme transports oxygen, and the globin transports carbon dioxide. Iron is required for oxygen transport.
   - Carbonic anhydrase is involved with the transport of carbon dioxide.
2. Erythropoiesis is the production of erythrocytes.
   - Stem cells in red bone marrow eventually give rise to polychromatic erythroblasts, which lose their nuclei and are released into the blood as reticulocytes. Loss of the endoplasmic reticulum by a reticulocyte produces an erythrocyte.
   - In response to low blood oxygen the kidneys produce erythropoietin, which stimulates erythropoiesis.
3. Worn out erythrocytes are phagocytized by macrophages. Hemoglobin is broken down, and heme becomes bilirubin, which is secreted in bile.

Leukocytes

1. Leukocytes protect the body against microorganisms and remove dead cells and debris.
2. Granulocytes are leukocytes that contain cytoplasmic granules.
   - Neutrophils are small phagocytic cells.
   - Eosinophils function to reduce inflammation.
   - Basophils release histamine and are involved with increasing the inflammatory response.
3. Agranulocytes are leukocytes in which cytoplasmic granules cannot be easily seen with a light microscope.
   - Lymphocytes are important in immunity, including the production of antibodies.
   - Monocytes leave the blood, enter tissues, and become large phagocytic cells called macrophages.

Platelets

Platelets, or thrombocytes, are cell fragments pinched off from megakaryocytes in the red bone marrow.

HEMOSTASIS p. 595

Vascular Spasm

Vasoconstriction of damaged blood vessels reduces blood loss.

Platelet Plug Formation

Platelets repair minor damage to blood vessels by forming platelet plugs. Platelets also release chemicals involved with coagulation.

Coagulation

1. Coagulation is the formation of a blood clot.
2. Coagulation consists of three stages.
   - Activation of prothrombin activator.
   - Conversion of prothrombin to thrombin by prothrombin activator.
   - Conversion of fibrinogen to fibrin by thrombin. The insoluble fibrin forms the clot.
3. The first stage of coagulation occurs through the extrinsic or intrinsic clotting pathway. Both pathways end with the production of prothrombin activator.
   - The extrinsic clotting pathway begins with the release of tissue thromboplastin from damaged tissues.
   - The intrinsic clotting pathway begins with the activation of factor XII and the release of platelet factor III.

Control of Clot Formation

1. Heparin and antithrombin III inhibit thrombin activity. Therefore fibrinogen is not converted to fibrin, and clot formation is inhibited.
2. Prostacyclin inhibits thromboxane.

Clot Retraction and Dissolution

1. Clot retraction results from the contraction of platelets, which pull the edges of damaged tissue closer together.
2. Factor XII, thrombin, and tissue plasminogen activator activate plasmin, which dissolves fibrin (the clot).

BLOOD GROUPING p. 600

1. Blood groups are determined by antigens on the surface of erythrocytes.
2. Antibodies can bind to erythrocyte antigens, resulting in agglutination or hemolysis of erythrocytes

ABO Blood Group

1. Type A blood has A antigens, type B blood has B antigens, type AB blood has A and B antigens, and type O blood does not have A or B antigens.
2. Type A blood has B antibodies, type B blood has A antibodies, type AB blood does not have A or B antibodies, and type O blood has both A and B antibodies.
3. Mismatching the ABO blood group is responsible for transfusion reactions.

Rh Blood Group

1. Rh-positive blood has a certain Rh antigen (the D antigen), whereas Rh-negative blood does not.
2. Antibodies against the Rh antigen are produced by an Rh-negative person when the person is exposed to Rh-positive blood.
3. The Rh blood group is responsible for erythroblastosis fetalis.

DIAGNOSTIC BLOOD TESTS *p. 602*

Type and Crossmatch

Blood typing determines the ABO and Rh blood groups of a blood sample. A crossmatch tests for agglutination reactions between donor and recipient blood.

Complete Blood Count

The complete blood count consists of the following: red blood cell count, hemoglobin measurement (grams of hemoglobin per 100 ml of blood), hematocrit measurement (percent volume of erythrocytes), and white blood cell count.

White Blood Cell Differential Count

The white blood cell differential count determines the percentage of each type of leukocyte.

Clotting

Platelet count and prothrombin time measure the ability of the blood to clot.

Blood Chemistry

The composition of materials dissolved or suspended in plasma (e.g., glucose, urea nitrogen, bilirubin, and cholesterol) can be used to assess the functioning and status of the body's systems.

CONTENT REVIEW

1. Describe the three major functions performed by blood, and give examples for each function.
2. Define plasma and serum. What is the function of albumin, globulins, and fibrinogen in plasma? What other substances are found in plasma?
3. Name the three general types of formed elements in blood.
4. Define hematopoiesis. What is a stem cell?
5. Describe the two basic parts of a hemoglobin molecule. Which part is associated with iron? What gases are transported by each part?
6. What is erythropoiesis? Where does it occur?
7. Describe the formation of red blood cells, starting with the stem cell in red bone marrow.
8. What is erythropoietin, where is it produced, what causes it to be produced, and what effect does it have on red blood cell production?
9. Where are red blood cells mainly broken down? List the three breakdown products of hemoglobin, and explain what happens to them.
10. What are the two major functions of leukocytes?
11. Name the three types of granulocytes and the two types of agranulocytes. Describe the morphology and function of each type.
12. Name the two leukocytes that function primarily as phagocytic cells.
13. Which leukocyte reduces the inflammatory response? Which leukocyte releases histamine and promotes inflammation?
14. What is the function of a platelet plug? Describe the process of platelet plug formation.
15. What is a clot? What is the function of a clot?
16. Clotting is divided into three stages. Describe the final event that occurs in each stage.
17. What is the difference between extrinsic and intrinsic activation of clotting?
18. What is the function of anticoagulants in blood? Name three anticoagulants in blood, and explain how they prevent clot formation.
19. Define a thrombus and an embolus, and explain why they are dangerous.
20. Describe clot retraction and clot dissolution. What do they accomplish?
21. What are blood groups, and how do they cause transfusion reactions? List the four ABO blood types. Why is type O blood considered a universal donor and type AB blood a universal recipient?
22. What is meant by the term Rh positive? How can Rh incompatibility affect a pregnancy?
23. For each of the following tests, define the test and give an example of a disorder that would cause an abnormal test result.
    - **A** Type and crossmatch
    - **B** Red blood cell count
    - **C** Hemoglobin measurement
    - **D** Hematocrit measurement
    - **E** White blood cell count
    - **F** White blood cell differential count
    - **G** Platelet count
    - **H** Prothrombin time measurement
    - **I** Blood chemistry tests

CONCEPT REVIEW

1. In hereditary hemolytic anemia there is massive destruction of red blood cells. Would you expect the reticulocyte count to be above or below normal? Explain why one of the symptoms of the disease is jaundice. In 1910 it was discovered that removal of the spleen cures hereditary hemolytic anemia. Explain why this cure occurs.
2. Red Packer, a physical education major, wanted to improve his performance in an upcoming marathon race. Approximately 6 weeks before the race 1 quart of blood was removed from his body, and the formed elements were separated from the plasma. The formed elements were frozen, and the plasma was reinfused into his body. Just before the competition, the formed elements were thawed and injected into his body. Explain why this procedure, called blood doping or blood boosting, would help Red's performance. Can you suggest any possible bad effects?
3. Chemicals such as benzene and chloramphenicol can destroy red bone marrow, causing aplastic anemia. What symptoms would you expect to develop as a result of the lack of (1) red blood cells, (2) platelets, and (3) leukocytes?
4. E.Z. Goen habitually used barbiturates to depress feelings of anxiety. Because barbiturates suppress the respiratory centers in the brain, they cause hypoventilation (i.e., slower-than-normal rate of breathing). What happens to the erythrocyte count of a habitual user of barbiturates? Explain.
5. What blood problems would you expect to observe in a patient after total gastrectomy (removal of the stomach)? Explain.
6. According to the old saying, "Good food makes good blood." Name three substances in the diet that are essential for "good blood." What blood disorders develop if these substances are absent from the diet?

ANSWERS TO PREDICT QUESTIONS

1 p. 591. The reason fetal hemoglobin must be more effective at binding oxygen than adult hemoglobin is so that the fetal circulation can draw the needed oxygen away from the maternal circulation. If maternal blood had an equal or greater oxygen affinity, the fetal blood would not be able to draw away the required oxygen, and the fetus would die.

2 p. 592. An elevated reticulocyte count indicates that erythropoiesis and the demand for erythrocytes are increased and that immature erythrocytes (reticulocytes) are entering the circulation in large numbers. An elevated reticulocyte count may occur for a number of reasons, including loss of blood; therefore after a person donates a unit of blood, his reticulocyte count increases.

3 p. 602. People with type AB blood were called universal recipients because they could receive type A, B, AB, or O blood with little likelihood of a transfusion reaction. Type AB blood does not have antibodies against type A or B antigens. Therefore transfusion of these antigens in type A, B, or AB blood does not cause a transfusion reaction in a person with type AB blood. The term is misleading, however, for two reasons. First, other blood groups can cause a transfusion reaction. Second, antibodies in the donor's blood can cause a transfusion reaction. For example, type O blood contains A and B antibodies that can react against the A and B antigens in type AB blood.

4 p. 605. A white blood cell differential count should be done. An increase in the number of neutrophils would support the diagnosis of a bacterial infection. Coupled with other symptoms, this result could mean appendicitis.

20 CHAPTER OBJECTIVES

After reading this chapter you should be able to

1. Describe the size, shape, and location of the heart.
2. Describe the structure and function of the pericardium.
3. Describe the external and internal anatomy of the heart.
4. Describe the histology of the three major layers of the heart and the histology of the conducting system.
5. Discuss the similarities and differences between cardiac muscle and skeletal muscle, particularly the autorhythmicity of cardiac muscle.
6. Describe the conducting system of the heart.
7. Explain why Purkinje fibers conduct impulses more rapidly than other cardiac muscle cells.
8. State the differences between slow and fast channels in cardiac muscle.
9. Describe depolarization in cardiac muscle.
10. Explain the importance of a refractory period in cardiac muscle.
11. Explain the various features of an electrocardiogram and the events that those features reflect.
12. Explain the various components of the cardiac cycle, including systole, diastole, and the heart sounds.
13. Describe the aortic pressure curve.
14. Explain the bases of the major heart sounds.
15. List the major factors involved in the intrinsic regulation of the heart and describe the major functions of each factor.
16. List the major factors involved in the extrinsic regulation of the heart and describe the major functions of each factor.
17. Describe the differences between sympathetic and parasympathetic stimulation on heart function.
18. Discuss the role of the heart in homeostasis and how changes in blood pressure, pH, oxygen, ions, and temperature affect the heart.

Cardiovascular System: The Heart

KEY TERMS

atrium (a′tre-um)

atrioventricular node

atrioventricular valve

baroreceptor (bar′o-re-sep′tor)

cardiac cycle

cardiac output

diastole (di-as′to-le)

heart skeleton

pericardium (pĕr-ĭ-kar′dĭ-um)

plateau phase of action potential

semilunar valve

sinoatrial node

starling's law of the heart

systole (sis′to-le)

ventricle (ven′trĭ-kul)

RELATED TOPICS

The following terms or concepts are important for a good understanding of this chapter. If you are not familiar with them, you should review them before proceeding.

Body cavities (Chapter 1)

Muscle histology (Chapters 4 and 10)

Membrane potentials (Chapter 9)

Autonomic nervous system function (Chapter 15)

Adrenal medullary stimulation of the heart (Chapter 18)

THE HEART IS RESPONSIBLE FOR THE CIRCULATION OF THE BLOOD. It contracts forcefully to pump blood through the vessels of the body. The heart is actually two pumps in one: one propels blood through the pulmonary circulation (to the lungs where the blood releases carbon dioxide and receives oxygen), and the other propels blood through the systemic circulation (to all remaining tissues of the body).

The heart of a healthy 70-kg person pumps approximately 7200 L of blood each day at a rate of 5 L per minute. For most people the heart continues to pump near that rate for more than 75 years. During short periods of vigorous exercise the amount of blood pumped per minute increases several times. If the heart loses its ability to pump blood for even a few minutes, the life of the individual is in danger.

The location and the anatomy of the heart are described first, followed by a discussion of its function and regulation. The structure and functional characteristics of the blood vessels and the integrated function of the heart and blood vessels are presented in Chapter 21.

SIZE, FORM, AND LOCATION OF THE HEART

The adult heart has the shape of a blunt cone and is approximately the size of a closed fist. It is in the thoracic cavity between the lungs in a space within the mediastinum called the **pericardial cavity** (see Chapter 1). The blunt, rounded point of the cone is the **apex,** and the larger, flat portion at the opposite end of the cone is the **base.**

The heart lies obliquely in the mediastinum with its base directed posteriorly and slightly superiorly and the apex directed anteriorly and slightly inferiorly. The apex is also directed to the left so that approximately two thirds of the heart's mass lies to the left of the midline of the sternum (Figure 20-1).

It is important for clinical reasons to know the exact location of the heart in the thoracic cavity. Positioning a stethoscope to hear the heart sounds and positioning electrodes to record an **electrocardiogram** (**ECG** or **EKG**) from chest leads depend on this knowledge. Effective cardiopulmonary resuscitation (CPR) also depends on a reasonable knowledge of the heart's position and form.

ANATOMY OF THE HEART

The heart is a muscular pump consisting of four chambers: two **atria** (a'tre-ah; entrance chamber) and two **ventricles** (ven'trĭ-kulz; belly) (Figure 20-2).

Pericardium

The **pericardium** (pĕr'ĭ-kar'dĭ-um), or **pericardial sac,** is a double-layered closed sac that surrounds the heart (Figure 20-3). It consists of a tough, fibrous connective tissue outer layer called the **fibrous pericardium** and a thin, transparent inner layer of simple squamous epithelium called the **serous pericardium.** The fibrous pericardium prevents over-distention of the heart and anchors the heart within the mediastinum. Superiorly the fibrous pericardium is continuous with the connective tissue coverings of the great vessels, and inferiorly it is attached to the surface of the diaphragm.

The portion of the serous pericardium lining the fibrous pericardium is the **parietal pericardium,** whereas that portion covering the heart surface is the **visceral pericardium,** or **epicardium** (see Figure 20-3). The parietal and visceral portions of the serous

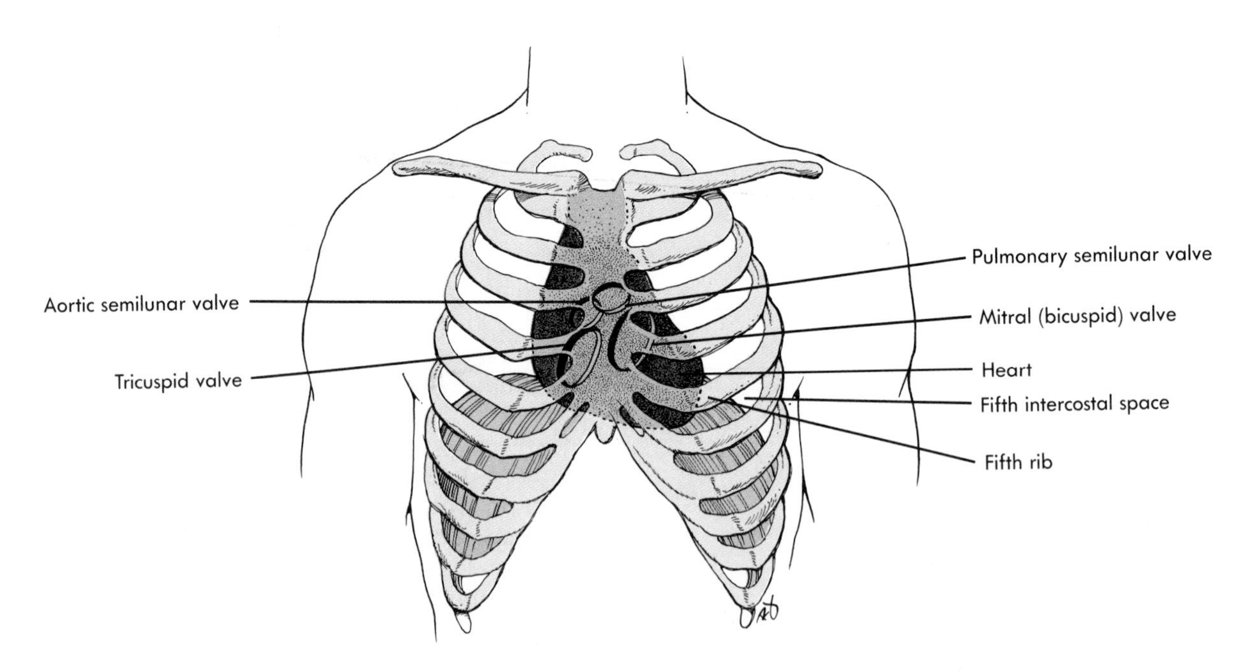

FIGURE 20-1 Location of the heart in the thorax. The heart lies deep and slightly to the left of the sternum. The base of the heart, located deep to the sternum, extends to the second intercostal space, and the apex of the heart is in the fifth intercostal space approximately 9 cm to the left of the midline.

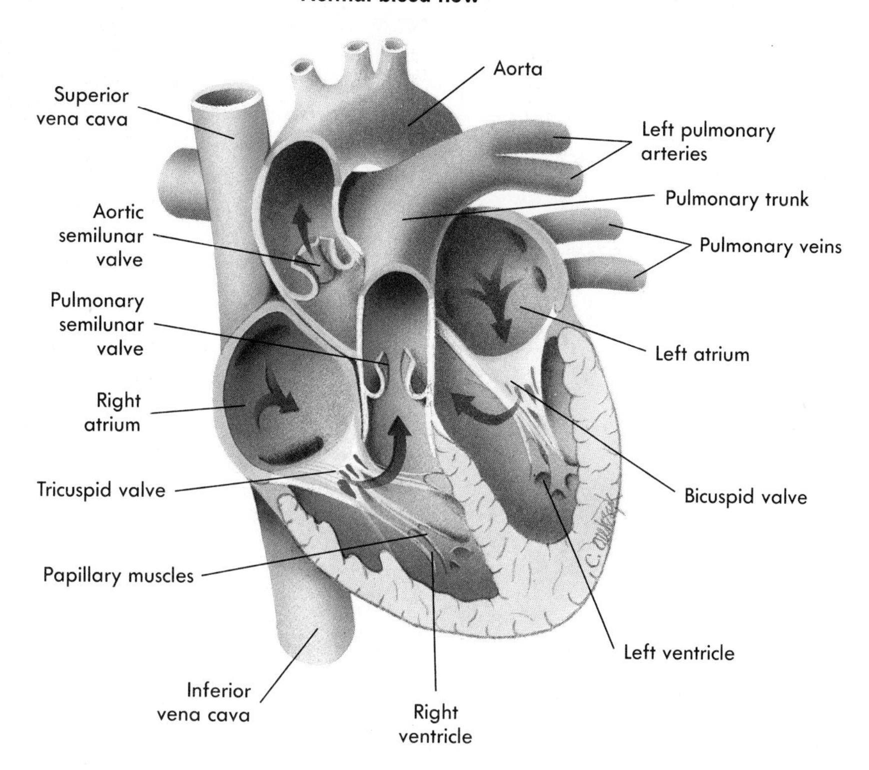

FIGURE 20-2 Frontal section of the heart revealing the four chambers and the direction of blood flow through the heart. The great vessels open into or out of the chambers.

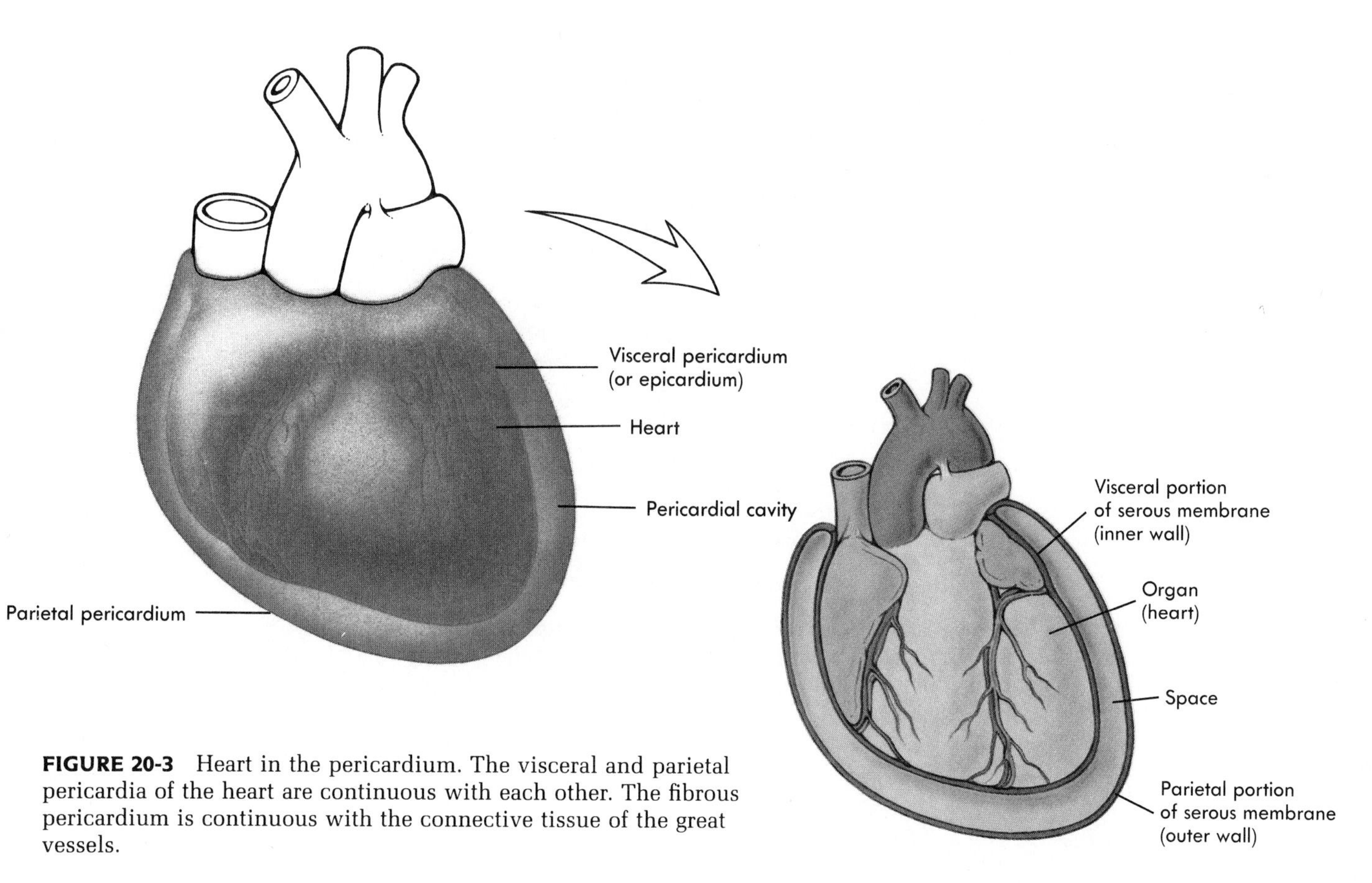

FIGURE 20-3 Heart in the pericardium. The visceral and parietal pericardia of the heart are continuous with each other. The fibrous pericardium is continuous with the connective tissue of the great vessels.

pericardium are continuous with each other where the great vessels enter or leave the heart. The **pericardial cavity** between the visceral and parietal pericardia is filled with a thin layer of serous **pericardial fluid** that helps reduce friction as the heart moves within the pericardial sac.

External Anatomy

The thin-walled atria form the superior and posterior portions of the heart, and the thick-walled ventricles form the anterior and inferior portions. Flaplike **auricles** (aw'rĭ-klz; ears) are extensions of the atria that can be seen anteriorly between each atrium and ventricle (Figure 20-4). The entire atrium used to be called the auricle, and some medical personnel still refer to it as such.

Seven large veins carry blood to the heart: four **pulmonary veins** carry blood from the lungs to the left atrium; the **superior** and **inferior venae cavae** carry blood from the body to the right atrium; and the **coronary sinus** carries blood from the walls of the heart to the right atrium. Two arteries, the **aorta** and the **pulmonary trunk,** exit the heart. The aorta carries blood from the left ventricle to the body, and the pulmonary trunk carries blood from the right ventricle to the lungs.

A large **coronary** (kor'o-năr-e; circling like a crown) **sulcus** (sul'kus; ditch) runs obliquely around the heart, separating the atria from the ventricles (see Figure 20-4). Two sulci extend inferiorly from the coronary sulcus, indicating the division between the right and left ventricles. The **anterior interventricular sulcus,** or **groove,** is on the anterior surface of the heart, and the **posterior interventricular sulcus,** or **groove,** is on the posterior surface of the heart. However, in the normal, intact heart the sulci are covered by fat, and only after this fat is removed can the actual sulci be seen.

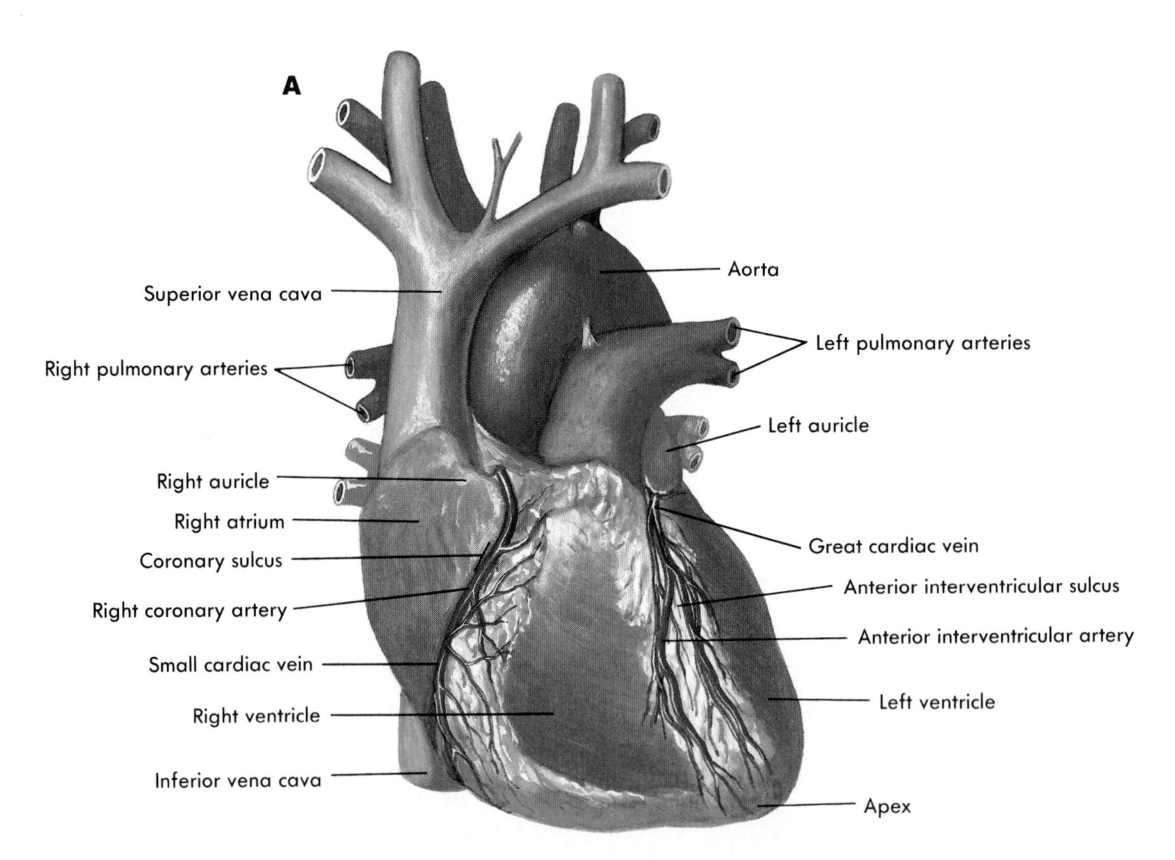

FIGURE 20-4 Surface of the heart. **A** View of the anterior (sternocostal) surface. **B** Photograph of the anterior surface. **C** View of the posterior (base) and inferior (diaphragmatic) surfaces of the heart.

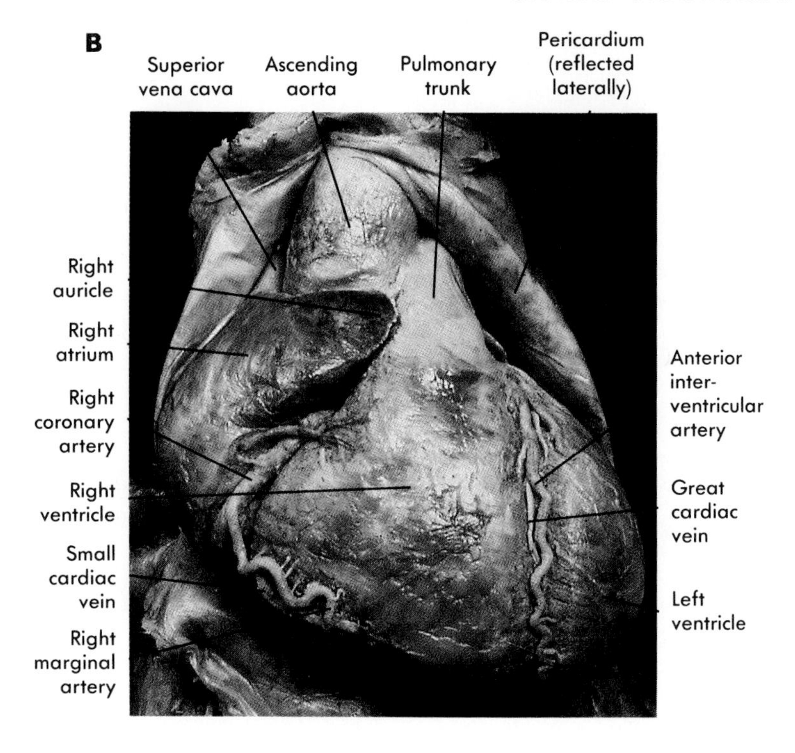

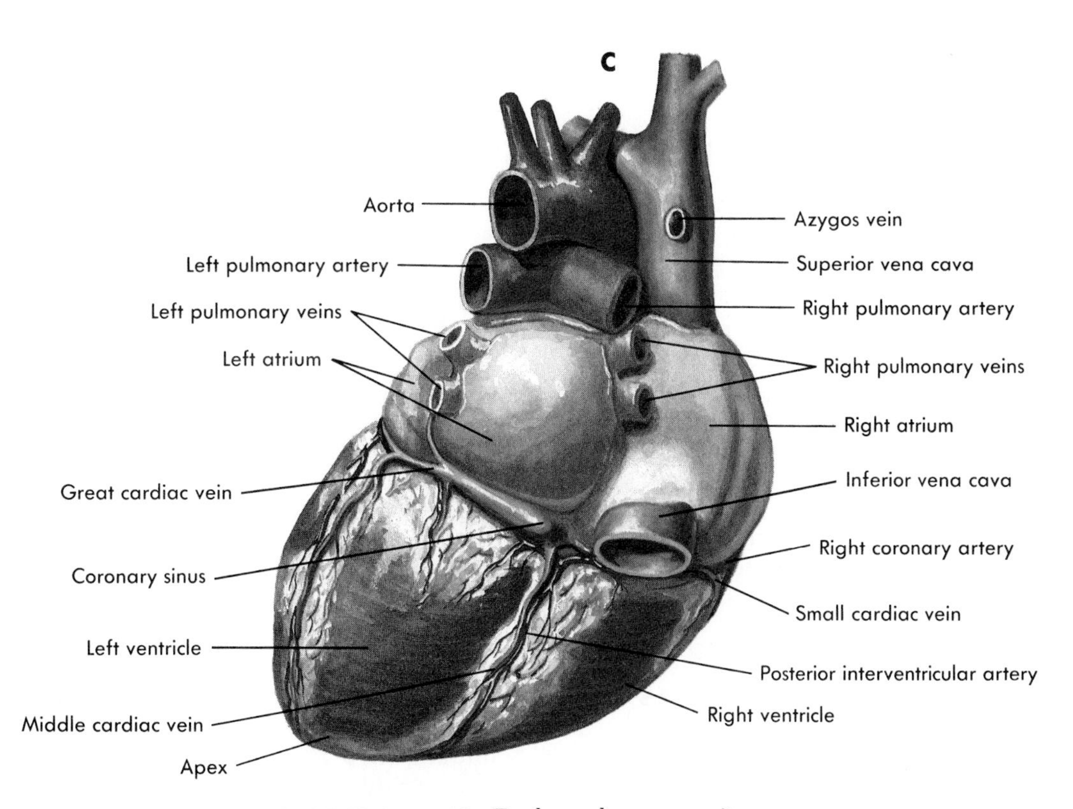

FIGURE 20-4, cont'd For legend see opposite page.

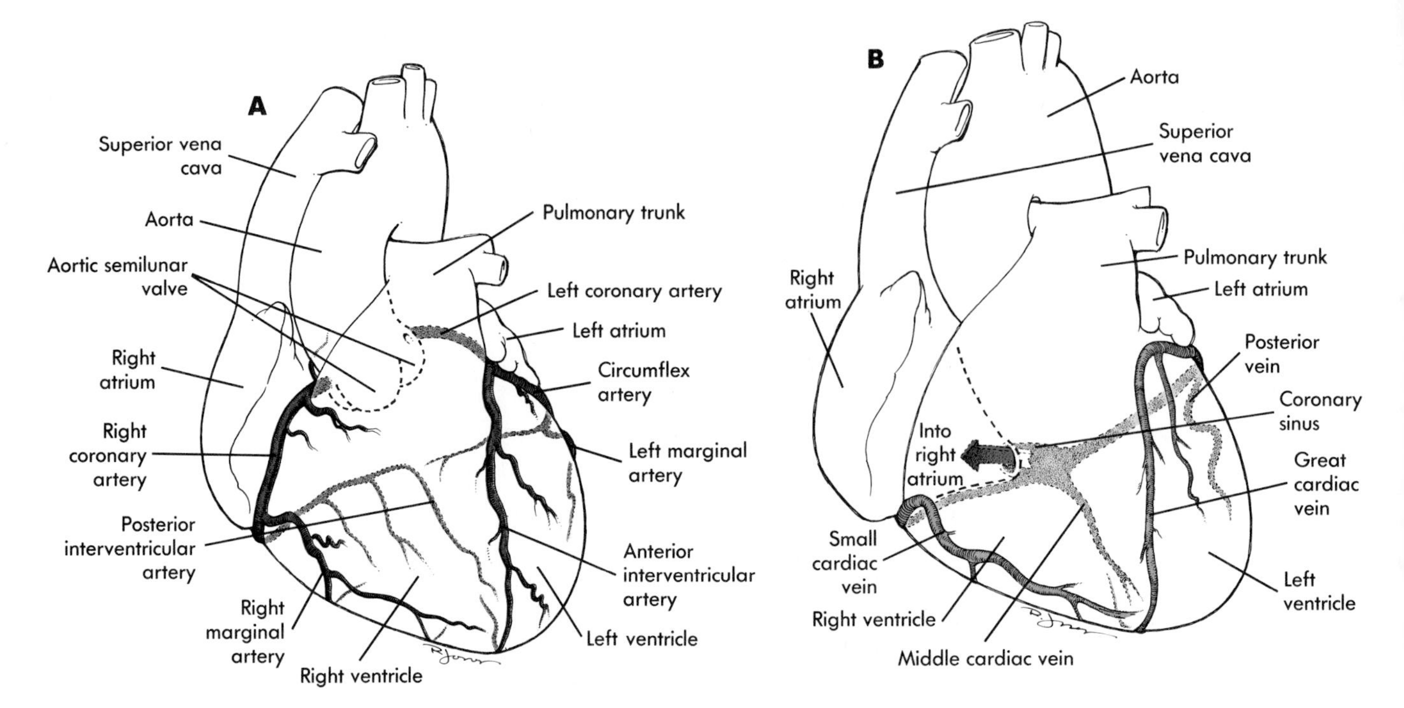

FIGURE 20-5 Blood vessels providing the circulation of the heart. **A** arteries and, **B** veins. The anterior surface of the heart is represented. The vessels of the anterior surface are seen directly and have a darker color, whereas the vessels of the posterior surface are seen through the heart and have a lighter color.

The major arteries supplying blood to the tissue of the heart lie within the coronary sulcus and interventricular grooves on the heart's surface. The right and left **coronary arteries** exit the aorta near the point at which the aorta leaves the heart and lie within the coronary sulcus (Figure 20-5, *A*).

> **1**
>
> **Predict the effect on the heart if blood flow through a coronary artery is restricted or completely blocked.**
>
> ? ? ? ? ? ? ? ? ? ? ?

The major vein draining the tissue on the left side of the heart is the **great cardiac vein,** and a **small cardiac vein** drains the right margin of the heart (Figure 20-5, *B*). These veins converge toward the posterior portion of the coronary sulcus and empty into a large venous cavity called the **coronary sinus,** which in turn empties into the right atrium. A number of smaller veins empty into either the cardiac veins, the coronary sinus, or directly into the right atrium.

Heart Chambers and Valves

Right and left atria

The right **atrium** has three major openings through which veins enter the heart from various parts of the body: the superior vena cava, the inferior vena cava, and the coronary sinus. The left atrium has four relatively uniform openings that receive the four pulmonary veins. The two atria are separated from each other by the **interatrial septum.** A slight oval depression, the **fossa ovalis** (o-vă'lis), on the right side of the septum marks the former location of the **foramen ovale** (o-vă'le), an opening between the right and left atria in the embryo and the fetus (see Chapter 29).

Right and left ventricles

The atria open into the ventricles through **atrioventricular canals** (Figure 20-6). Each ventricle has one large, superiorly placed outflow route near the midline of the heart. The right ventricle opens into the pulmonary trunk, and the left ventricle opens into the **aorta.** The two ventricles are separated from each other by the **interventricular septum,** which is thick (the muscular portion) toward the apex and very thin (the membranous portion) toward the atria.

Atrioventricular valves

An **atrioventricular valve** is on each atrioventricular canal and is composed of cusps, or flaps. These valves allow blood to flow from the atria into the ventricles but prevent blood from flowing back into the atria. The atrioventricular valve between the right atrium and the right ventricle has three cusps and is called the **tricuspid valve.** The atrioventricular valve between the left atrium and left ventricle has

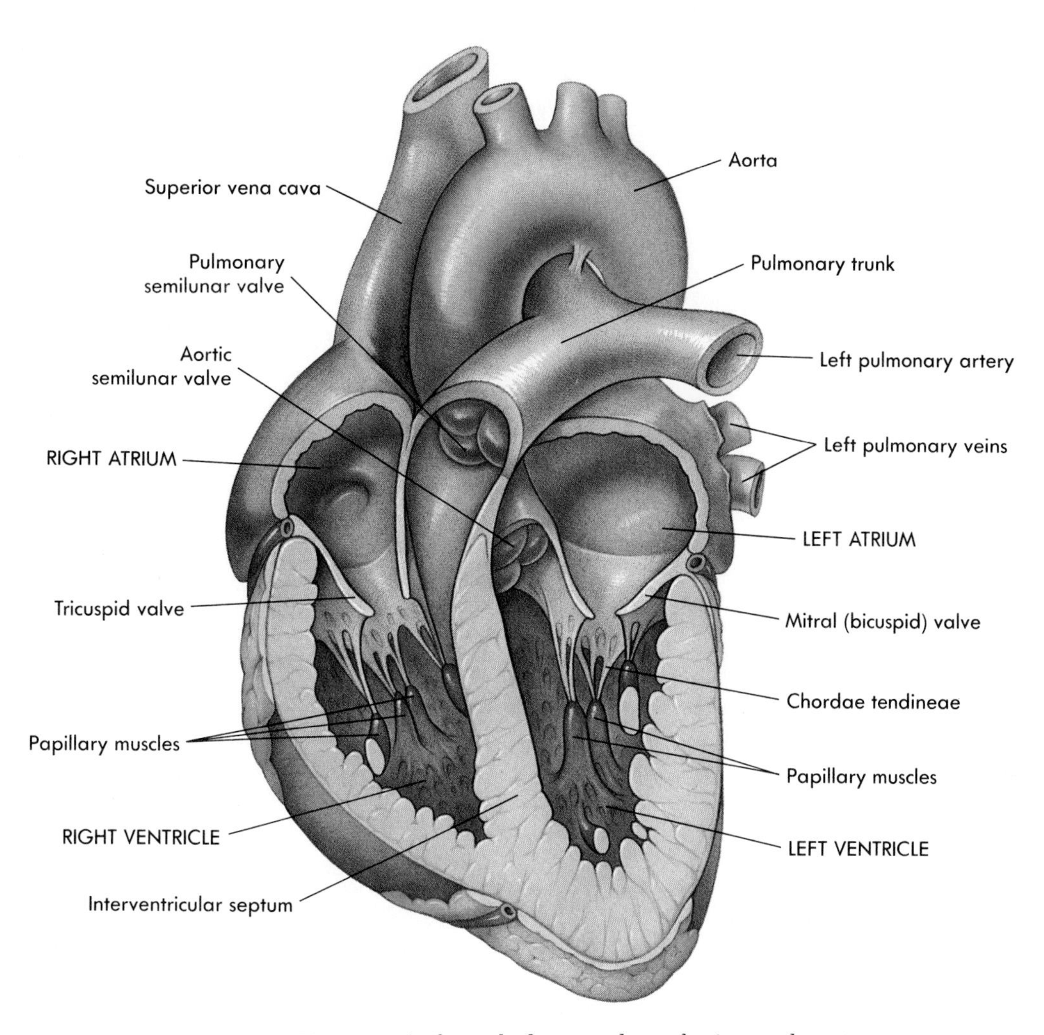

FIGURE 20-6 Heart cut in frontal plane to show the internal anatomy.

Angina pectoris is pain that results from a reduction in blood supply to cardiac muscle. The pain is temporary, and if blood flow is restored, no permanent change or damage results. Angina pectoris is characterized by chest discomfort deep to the sternum, often described as heaviness, pressure, or moderately severe pain, and is often mistaken for indigestion. The pain may also be referred to the neck, lower jaw, left arm, and left shoulder. Most often angina pectoris results from narrowed and hardened arterial walls. The reduced blood flow results in a reduced supply of oxygen needed by cardiac muscle cells. As a consequence, anaerobic metabolism results in a buildup of lactic acid and reduced pH in affected areas of the heart. Pain receptors are stimulated by the lactic acid. The pain is predictably associated with exercise because the increased pumping activity of the heart requires more oxygen and the narrowed blood vessels cannot supply it. Angina is frequently relieved by rest and by drugs such as nitroglycerin. Nitroglycerin causes blood vessel dilation in the heart and in peripheral blood vessels; consequently, it increases blood flow through coronary vessels to heart muscle and reduces the amount of work the heart has to perform by reducing the amount of blood that is returned to the heart from systemic vessels.

Myocardial infarction results from a prolonged lack of blood flow to a portion of the cardiac muscle resulting in a lack of oxygen and cellular death. Myocardial infarctions vary with the amount of cardiac muscle affected and the part of the heart that is affected. If blood supply to cardiac muscle is reestablished within 20 minutes, no permanent damage occurs. If the lack of oxygen lasts longer, cell death results. However, within 30 to 60 seconds after blockage of a coronary blood vessel, functional changes are obvious. The electrical properties of the cardiac muscle are altered, and the ability of the cardiac muscle to function properly is lost. The most common cause of myocardial infarctions apparently is the formation of a thrombus that blocks a coronary artery. Coronary arteries narrowed by atherosclerotic lesions provide one of the conditions that increase the chances for myocardial infarctions.

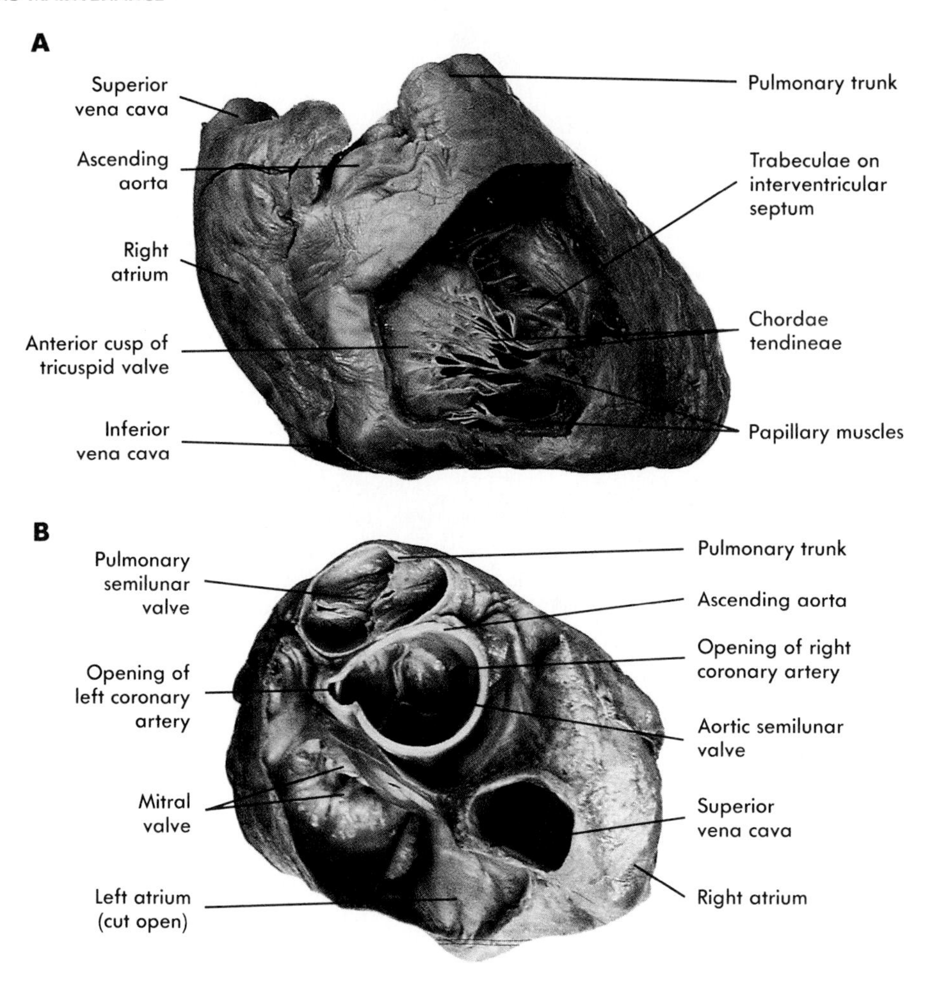

FIGURE 20-7 Photographs of the internal anatomy of the heart.
A Chordae tendineae of the right ventricle. **B** Superior view of heart valves.

two cusps and is called the **bicuspid,** or **mitral** (resembling a bishop's miter, a two-pointed hat), **valve.**

Each ventricle contains cone-shaped muscular pillars called **papillary** (pap'ĭ-lĕr'e; nipple or pimple shaped) **muscles.** These muscles are attached by thin, strong connective tissue strings called **chordae tendineae** (kor'de ten'dĭ-ne; heart strings) to the cusps of the atrioventricular valves (see Figure 20-6; Figure 20-7, *A*). The papillary muscles contract when the ventricles contract and prevent the valves from opening into the atria by pulling on the chordae tendineae attached to the valve cusps. Blood flowing from the atrium into the ventricle pushes the valve open into the ventricle, but when the ventricle contracts, blood pushes the valve back toward the atrium. The atrioventricular opening closes as the valve cusps meet.

Semilunar valves

The aorta and pulmonary trunk possess **aortic** and **pulmonary semilunar** (moon-shaped) **valves.** Each valve consists of three pocketlike semilunar cusps (see Figures 20-6 and 20-7, *B*), the free inner borders of which meet in the center of the artery to block blood flow. Blood flowing out of the ventricles pushes against each valve, forcing it open; but when blood flows back from the aorta or pulmonary trunk toward the ventricles, it enters the pockets of the cusps, causing them to meet in the center of the aorta or pulmonary trunk, thus closing them and keeping blood from flowing back into the ventricles.

ROUTE OF BLOOD FLOW THROUGH THE HEART

Blood flow through the heart is depicted in Figure 20-2. Even though it is more convenient to discuss blood flow through the heart one side at a time, it is important to understand that both atria contract at the same time and both ventricles contract at the same time. This concept is particularly important when the electrical activity, pressure changes, and heart sounds are considered.

Blood enters the right atrium from the systemic circulation, which supplies blood to all the tissues of the body, and from the coronary veins. Most of the blood in the right atrium then passes into the right ventricle as the ventricle relaxes after the previous contraction. When the right atrium contracts, the blood remaining in the atrium is pushed into the ventricle.

Contraction of the right ventricle pushes blood against the tricuspid valve, forcing it closed, and against the pulmonary semilunar valve, forcing it open, thus allowing blood to enter the pulmonary trunk.

The pulmonary trunk carries blood to the lungs, where carbon dioxide is released and oxygen is picked up (see Chapters 21 and 23). Blood returning from the lungs enters the left atrium through the four pulmonary veins. The blood passing from the left atrium to the left ventricle opens the bicuspid valve when the ventricles relax, and contraction of the left atrium completes left-ventricular filling.

Contraction of the left ventricle pushes blood against the bicuspid valve, closing it, and against the aortic semilunar valve, opening it and allowing blood to enter the aorta. Blood flowing through the aorta is distributed throughout all parts of the body except for the respiratory vessels in the lungs (Chapter 23).

HISTOLOGY

Heart Skeleton

The **skeleton of the heart** consists of a plate of fibrous connective tissue between the atria and ventricles. This connective tissue plate forms **fibrous rings** around the atrioventricular and semilunar valves and provides a solid support for them (Figure 20-8). The fibrous connective tissue plate also serves as electrical insulation between the atria and the ventricles and provides a rigid site of attachment for the cardiac muscles.

Heart Wall

The heart wall is composed of three layers of tissue: the epicardium, myocardium, and endocardium (Figure 20-9). The epicardium (visceral pericardium) is a thin serous membrane comprising the smooth outer surface of the heart. It consists of simple squamous epithelium overlaying a layer of connective tissue. The thick middle layer of the heart, the **myocardium,** is composed of cardiac muscle cells and is responsible for the ability of the heart to contract. The smooth inner surface of the heart chambers is the **endocardium,** which consists of simple squamous epithelium over a layer of connective tissue. The smooth inner surface allows blood to move easily through the heart. The heart valves are formed by a fold of the endocardium, making a double layer of endocardium with connective tissue in between.

The interior surfaces of the atria are mainly flat, but the interior of both auricles and a portion of the right atrial wall are modified by muscular ridges called **musculi pectinati** (pek'tĭ-nah'te; hair comb). The musculi pectinati of the right atrium are separated from the larger, smooth portions of the atrial wall by a ridge called the **crista terminalis** (terminal crest). The interior walls of the ventricles are modified by ridges and columns called **trabeculae** (trahbek'u-le; beams) **carneae** (kar'ne-e; meat).

Cardiac Muscle

Cardiac muscle cells are elongated, branching cells that contain one or occasionally two centrally located nuclei. Cardiac muscle cells also contain actin and myosin myofilaments organized to form sarcomeres and myofibrils, which are arranged so that the cardiac muscle fibers appear striated; but the striations are less regularly arranged and less nu-

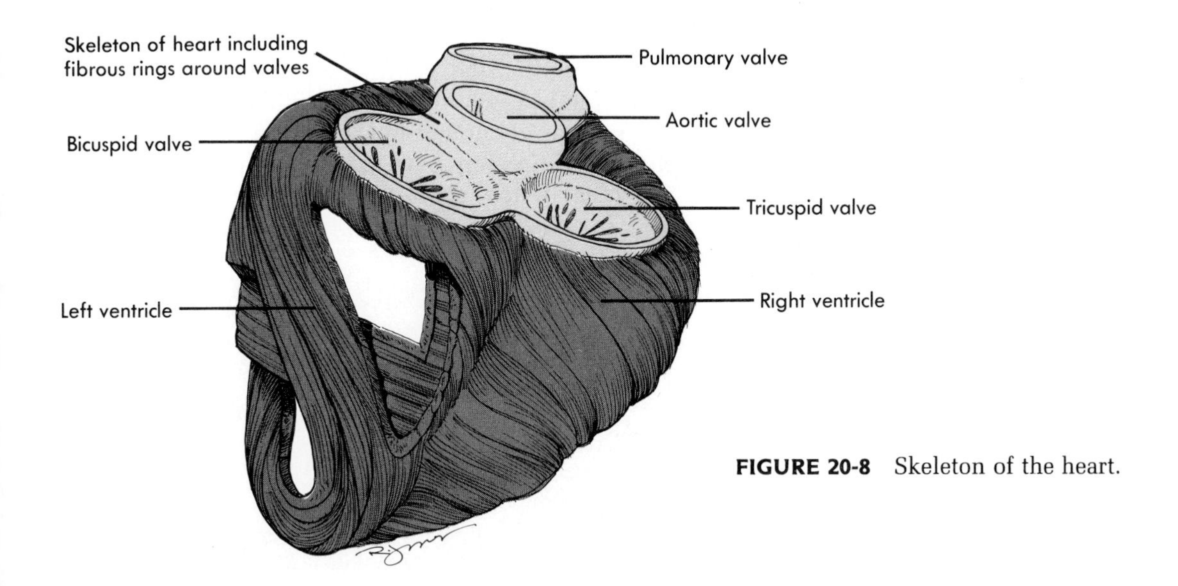

FIGURE 20-8 Skeleton of the heart.

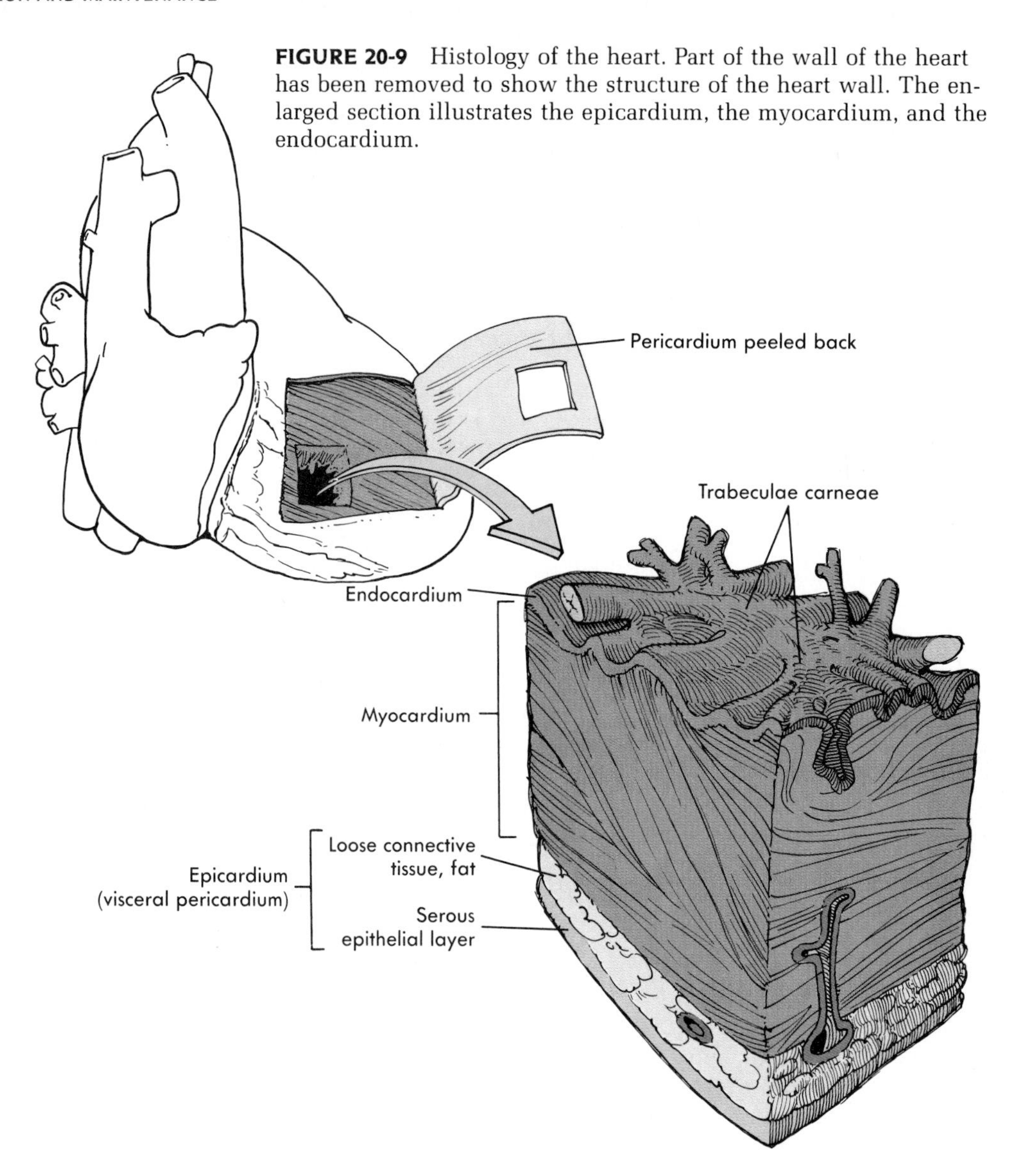

FIGURE 20-9 Histology of the heart. Part of the wall of the heart has been removed to show the structure of the heart wall. The enlarged section illustrates the epicardium, the myocardium, and the endocardium.

merous than those of skeletal muscle fibers (Figure 20-10, *A* and *B*).

Cardiac muscle has a smooth **sarcoplasmic reticulum,** but it is neither as regularly arranged nor as abundant as in skeletal muscle fibers, and there are no dilated cisternae as occurs in skeletal muscle. The sarcoplasmic reticulum comes into close association at various points with membranes of **transverse tubules (T tubules).** This loose association between the sarcoplasmic reticulum and the T tubules is partly responsible for the slow onset of contraction and the prolonged contraction phase in cardiac muscle. Action potentials are not carried from the surface of the cell to the sarcoplasmic reticulum as efficiently as they are in skeletal muscles, and calcium must diffuse a greater distance from the sarcoplasmic reticulum to the actin myofilaments.

Adenosine triphosphate (ATP) provides the energy for cardiac muscle contraction, and as in other tissues, ATP production depends on oxygen availability. However, cardiac muscle cannot develop a significant oxygen debt. The inability to develop an oxygen debt is consistent with the function of the heart since development of an oxygen debt could result in muscular fatigue and cessation of cardiac muscle contraction. Cardiac muscle cells are rich in mitochondria, which perform oxidative metabolism at a rate rapid enough to sustain normal myocardial energy requirements. An extensive capillary network provides an adequate oxygen supply to the cardiac muscle cells.

2

Under resting conditions most ATP produced in cardiac muscle is derived from the metabolism of fatty acids. However, during periods of heavy exercise cardiac muscle cells use lactic acid as an energy source. Explain why this arrangement is an advantage.

? ? ? ? ? ? ? ? ? ? ?

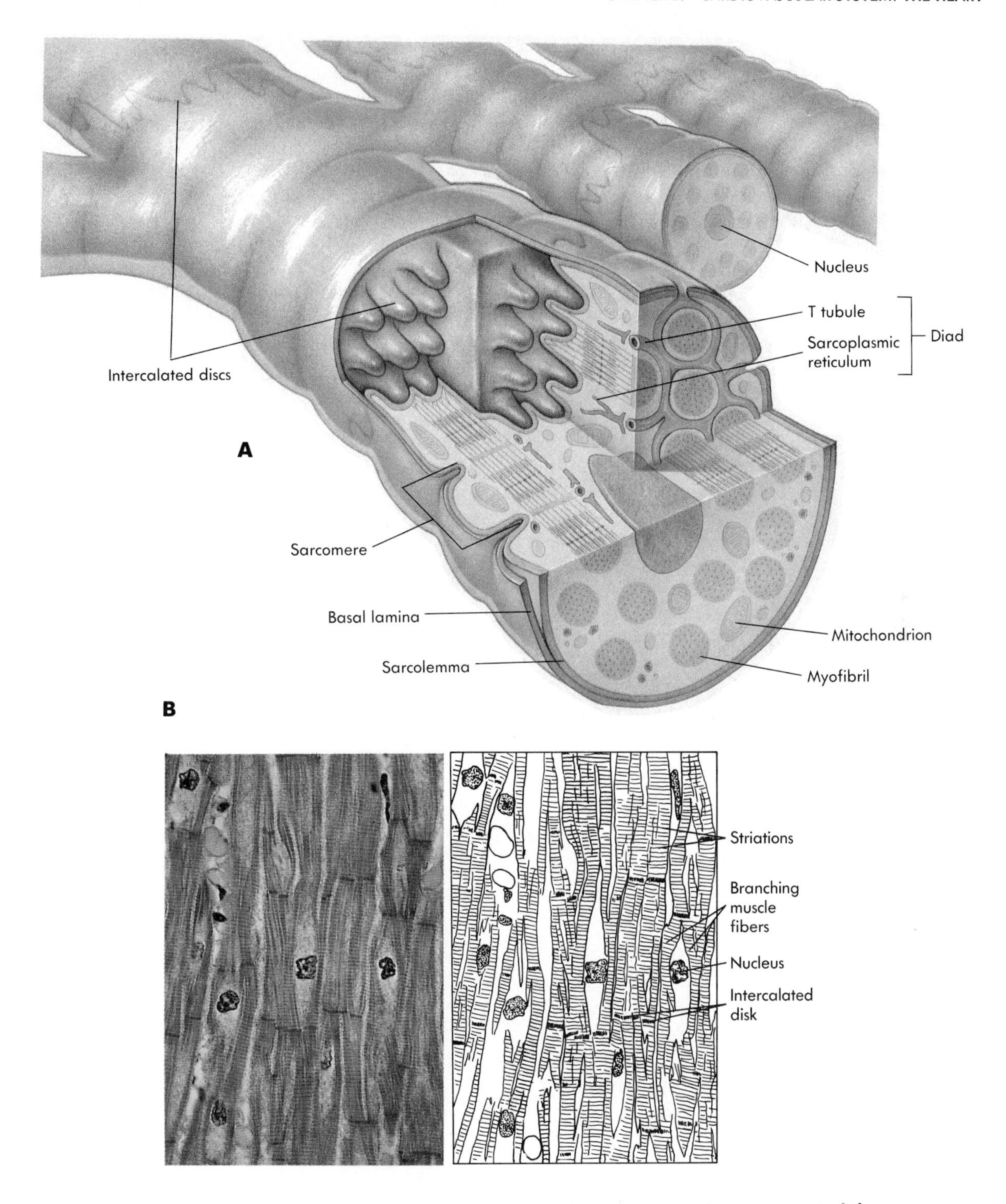

FIGURE 20-10 **A** Heart muscle demonstrating the structure and arrangement of the individual muscle fibers. **B** Histology of the heart muscle.

Cardiac muscle cells are organized in spiral bundles or sheets. The cells are bound end to end and laterally to adjacent cells by specialized cell-to-cell contacts called **intercalated disks** (see Figure 20-10). The membranes of the disks are greatly folded, and the adjacent cells fit together, greatly increasing contact between them. Specialized cell membrane structures called **desmosomes** hold the cells together, and **gap junctions** function as areas of low electrical resistance between the cells, allowing action potentials to pass from one cell to adjacent cells. Electrically the cardiac muscle cells behave as a single unit, and the highly coordinated contractions of the heart depend on this functional characteristic.

Conducting System

The conducting system of the heart, which relays electrical impulses through the heart, consists of modified cardiac muscle cells that form two **nodes** (a knot or lump) and a **conducting bundle** (Figure 20-11). The two nodes are contained within the walls of the right atrium and are named according to their position in the atrium. The **sinoatrial (SA) node** is medial to the opening of the superior vena cava, and the **atrioventricular (AV) node** is medial to the right atrioventricular valve. The AV node gives rise to a conducting bundle of the heart, the **atrioventricular bundle.** This bundle passes through a small opening in the fibrous skeleton to reach the interventricular septum where it divides to form the right and left **bundle branches,** which extend beneath the endocardium on either side of the interventricular septum to the apical portions of the right and left ventricles, respectively.

The inferior, terminal branches of the bundle branches become **Purkinje** (pur-kin′je) **fibers,** which are large-diameter cardiac muscle fibers. They have fewer myofibrils than most cardiac muscle cells and do not contract as forcefully. Intercalated disks are well developed between the Purkinje fibers and contain numerous gap junctions. As a result of these structural modifications, action potentials travel along the Purkinje fibers much more rapidly than through other cardiac muscle tissue.

All cardiac muscle cells generate spontaneous action potentials, but cells of the SA node do so at a greater frequency. As a result, the SA node is called the **pacemaker** of the heart. Once action potentials are produced, they spread from the SA node to adjacent cardiac muscle fibers of the atrium. Preferential pathways conduct action potentials from the SA node to the AV node at a greater velocity than they are transmitted in the remainder of the atrial muscle fibers, although such pathways cannot be distinguished structurally from the remainder of the atrium.

When the heart beats under resting conditions, approximately 0.04 second is required for action potentials to travel from the SA node to the AV node. Within the AV node action potentials are propagated slowly in comparison to the remainder of the conductile system. As a consequence, there is a delay of 0.11 second from the time action potentials reach the AV node until they pass to the AV bundle. The total delay of 0.15 second allows completion of the atrial

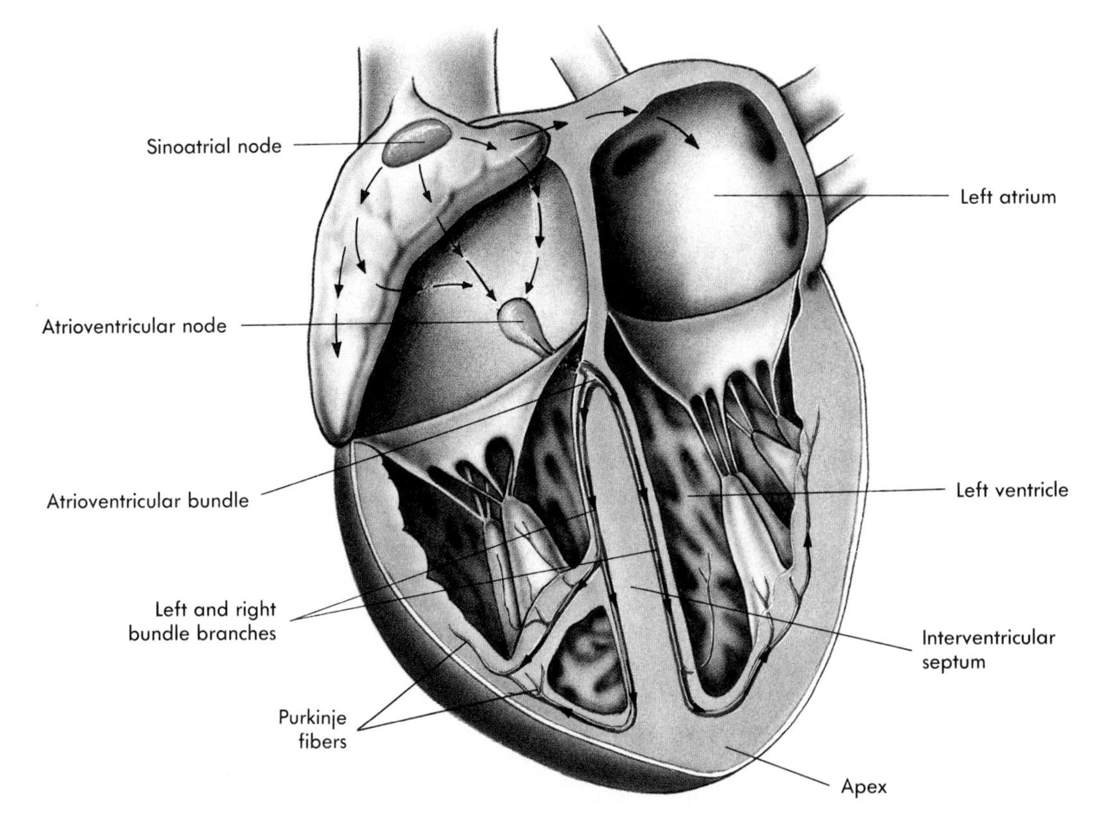

FIGURE 20-11 Conducting system of the heart. Impulses *(arrows)* travel across the wall of the right atrium from the SA node to the AV node. The atrioventricular bundle extends from the AV node, through the fibrous skeleton, and to the interventricular septum where it divides into right and left bundle branches. The bundle branches descend to the apex of each ventricle and then branch repeatedly for distribution throughout the ventricular walls.

contraction before ventricular contraction begins.

After action potentials pass from the AV node to the highly specialized conducting bundles, the velocity of conduction increases dramatically. The action potentials pass through the left and right bundle branches and through the individual Purkinje fibers that penetrate into the myocardium of the ventricles (see Figure 20-11).

Ventricular contraction begins at the apex and progresses throughout the ventricles. Because of the arrangement of the conducting system, the first portion of the myocardium that is stimulated is the inner wall of the ventricles near the apex. Once stimulated, the spiral arrangement of muscle layers in the wall of the heart results in a wringing action that proceeds from the apex toward the base of the heart. During the process the distance between the apex and the base of the heart decreases.

3

Explain why it is more efficient for contraction of the ventricles to begin at the apex of the heart than at the base.

? ? ? ? ? ? ? ? ? ? ?

ELECTRICAL PROPERTIES

Cardiac muscle cells have many characteristics in common with other electrically excitable cells such as neurons and skeletal muscle fibers. All these cells have a **resting membrane potential** (RMP), and when depolarized to their threshold level, action potentials result (see Chapter 9).

Action Potentials

In most respects the action potentials in skeletal muscle are similar to the action potentials in cardiac muscle, but the depolarization phase of cardiac muscle, which results from an increase in the permeability of cardiac muscle cell membranes to both sodium and calcium ions, is prolonged in comparison to that in skeletal muscle (Figure 20-12). The prolonged period of depolarization is the **plateau phase** of the action potential.

Action potentials in cardiac muscle are conducted from cell to cell, whereas action potentials in skeletal muscle fibers are conducted along the length of a single cell but not from cell to cell. Also, the rate of action potential propagation is slower in cardiac muscle than in skeletal muscle because cardiac muscle fibers are smaller in diameter and much shorter than skeletal muscle fibers. Even though the intercalated disks allow transfer of action potentials between cardiac muscle cells, they slow the rate of action potential conduction between cardiac muscle fibers.

In cardiac muscle there are two major channel systems in the cell membrane through which sodium and calcium ions enter the cell—**slow channels** and **fast channels.** Unstable slow channels open spontaneously, allowing sodium ions and calcium ions to diffuse into the cell, causing a slow depolarization. The fast channels are sensitive to small changes in membrane potential, and when the membrane begins to depolarize, the small change in electrical charge across the cell membrane causes fast channels to open, resulting in further depolarization. When the membrane potential is depolarized to approximately −60 mV (threshold), the remaining fast channels and the slow channels open, resulting in depolarization.

Various agents exist that specifically block either fast channels or slow channels. For example, tetrodotoxin blocks the fast channels, whereas manganese ions (Mn^{2+}) and verapamil block the slow channels. Slow channel–blocking agents prevent the movement of calcium ions into the cell and for that reason are called calcium channel blockers. Some calcium channel blockers now are widely used clinically in the treatment of various cardiac disorders, especially tachycardia and certain arrhythmias. On the other hand, catecholamines such as epinephrine and norepinephrine increase the heart rate and its force of contraction by opening the slow channels.

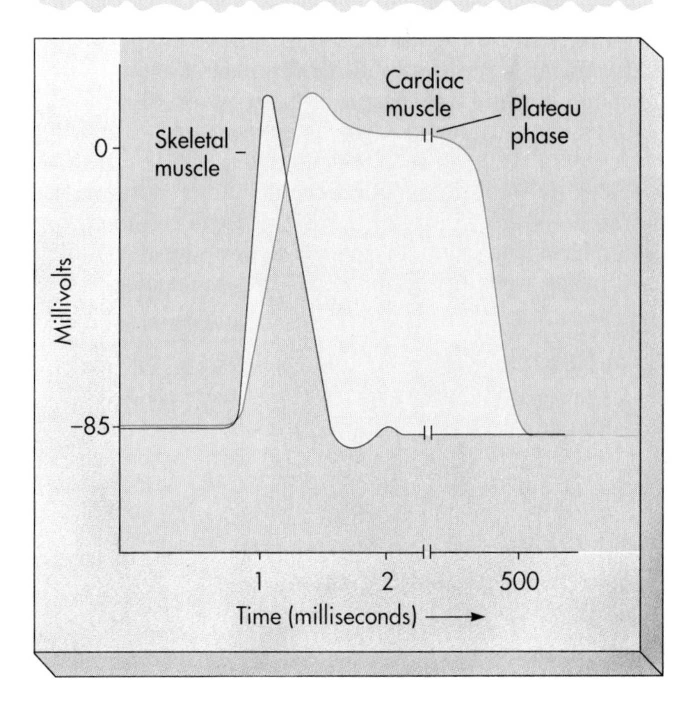

FIGURE 20-12 Action potentials compared for skeletal muscle *(red line)* and cardiac muscle *(blue line)*. Depolarization of cardiac muscle lasts considerably longer than that of skeletal muscle.

Autorhythmicity of Cardiac Muscle

After each action potential the membrane potential in cardiac muscle returns to its RMP, which has a value of approximately −85 to −95 mV in most cardiac muscle cells and −90 to −100 mV in Purkinje fibers. Immediately, some of the unstable slow channels open spontaneously, causing depolarization. The change in the membrane potential subsequently causes fast channels to open, causing further depolarization. When depolarization reaches threshold, an action potential is produced. Consequently, action potentials are produced in cardiac muscle without external stimulation.

Although all cardiac muscle cells can produce action potentials spontaneously, they are produced more rapidly in the SA node, the pacemaker, than in other areas of the heart. This is because that there are more slow channels in the SA node than in other areas of the heart.

Since action potentials are produced in the SA node spontaneously, automatic contractions result. Furthermore, the generation of action potentials occurs on a regular basis (i.e., rhythmically). Because the heart beats automatically and rhythmically, it is **autorhythmic.** If the heart is removed from the body and is maintained under physiological conditions with the proper nutrients and temperature, it will continue to beat regularly for a long period of time.

Even though cardiac muscle cells have the ability to contract autorhythmically, they normally function together in a coordinated fashion. Action potentials originate in the SA node and spread throughout the heart through the conductile system. However, under unusual circumstances other regions of the heart, called **ectopic pacemakers** or **ectopic foci,** can initiate beats. Ectopic foci appear when their own rate of action potential generation becomes enhanced, when the rhythmicity of the SA node is reduced, or when conduction pathways between the SA node and other areas of the heart become blocked.

4

Predict the consequences for the pumping effectiveness of the heart if numerous ectopic foci produce action potentials at the same time.

? ? ? ? ? ? ? ? ? ? ?

When the SA node is damaged or destroyed, either the AV node or other areas of the atria become the pacemaker. Cardiac muscle cells of the AV node usually have the next highest rate of action potential generation and therefore usually become the heart's pacemaker. The rhythmicity of the AV node, however, has a frequency of 40 to 60 beats per minute compared to the SA node's normal rate of 70 to 80 beats per minute.

Refractory Period of Cardiac Muscle

Cardiac muscle, like skeletal muscle, has a **refractory period** associated with the action potential. During the **absolute refractory period** the cardiac muscle cell is completely insensitive to further stimulation, and during the **relative refractory period** the cell exhibits reduced sensitivity to additional stimulation. Since the depolarization phase of the action potential in cardiac muscle is greatly prolonged, the refractory period is also prolonged. The long refractory period ensures that after contraction, relaxation is nearly complete before another action potential can be initiated, thus preventing tetanic contractions in cardiac muscle.

5

Predict the consequences if cardiac muscle could undergo tetanic contraction.

? ? ? ? ? ? ? ? ? ? ?

Electrocardiogram

The conduction of action potentials through the myocardium during the cardiac cycle produces electrical currents that can be measured at the surface of the body. Electrodes placed on the surface of the body and attached to an appropriate recording device can detect small voltage changes resulting from action potentials of the cardiac muscle. The electrodes detect a summation of all the action potentials (not individual action potentials) that are transmitted through the heart at a given time. The summated record of these electrical events is an **electrocardiogram** (**ECG** or **EKG**).

The ECG is not a direct measurement of mechanical events in the heart, and neither the force of contraction nor blood pressure can be determined from it. However, each deflection in the ECG record indicates an electrical event within the heart and correlates with a subsequent mechanical event. Consequently, it is an extremely valuable diagnostic tool in identifying a number of cardiac abnormalities (Table 20-1), particularly because it is painless, easy to record, and does not require surgical procedures. Abnormal heart rates or rhythms, abnormal conduction pathways, hypertrophy or atrophy of portions of the heart, and the approximate location of damaged cardiac muscle can be determined from analysis of an ECG.

The normal ECG consists of a P wave, a QRS complex, and a T wave (Figure 20-13). The **P wave,** which is the result of action potentials that cause depolarization of the atrial myocardium, signals the onset of atrial contraction. The **QRS complex** is composed of three individual waves—the Q, R, and S waves. The QRS complex results from ventricular depolarization and signals the onset of ventricular

Table 20-1

Major cardiac arrhythmias

CONDITION	SYMPTOMS	POSSIBLE CAUSES
Abnormal Heart Rhythms		
Tachycardia	Heart rate in excess of 100 beats/min	Elevated body temperature; excessive sympathetic stimulation; toxic conditions
Paroxysmal atrial tachycardia	Sudden increase in heart rate to 95-150 beats/min for a few seconds or even for several hours; P wave precedes every QT complex; P wave inverted and superimposed on T wave	Excessive sympathetic stimulation; abnormally elevated permeability of slow channels
Ventricular tachycardia	Frequently causes fibrillation	Often associated with damage to AV node or ventricular muscle
Abnormal Rhythms Resulting from Ectopic Action Potentials		
Atrial flutter	300 P waves/min; 125 QRS complexes/min resulting in two or three P waves (atrial contraction) for every QRS complex (ventricular contraction)	Ectopic action potentials in the atria
Atrial fibrillation	No P waves; normal QT complexes; irregular timing; ventricles constantly stimulated by atria; reduced pumping effectiveness and filling time	Ectopic action potentials in the atria
Ventricular fibrillation	No QT complexes; no rhythmic contraction of myocardium; many patches of asynchronously contracting ventricular muscle	Ectopic action potentials in the ventricles
Bradycardia	Heart rate less than 60 beats/min	Elevated stroke volume in athletes; excessive vagal stimulation; carotid sinus syndrome
Sinus Arrhythmia	Heart rate varies 5% during respiratory cycle and up to 30% during deep respiration	Cause not always known; occasionally caused by ischemia or inflammation or associated with cardiac failure
SA Node Block	Cessation of P wave; new low heart rate due to AV node acting as pacemaker; normal QT complex	Ischemia; tissue damage due to infarction; causes unknown
AV Node Block		
First degree	PR interval greater than 0.2 seconds	Inflammation of AV bundle
Second degree	PR interval 0.25 to 0.45; some P waves trigger QRS complexes and others do not; examples of 2:1, 3:1, and 3:2 P wave/QRS complex ratios	Excessive vagal stimulation
Complete heart block	P wave dissociated from QRS complex; atrial rhythm approximately 100 beats/min; ventricular rhythm less than 40 beats/min	Ischemia of AV nodal fibers or compression of AV bundle
Premature Atrial Contractions	Occasional shortened intervals between one contraction and the succeeding contraction; frequently occurs in healthy people P wave superimposed on QT complex	Excessive smoking; lack of sleep; too much coffee; alcoholism
Premature Ventricular Contractions (PVCs)	Prolonged QRS complex; exaggerated voltage because only one ventricle may depolarize; inverted T wave; increased probability of fibrillation	Ectopic foci in ventricles; lack of sleep; too much coffee; irritability; occasionally occurs with coronary thrombosis

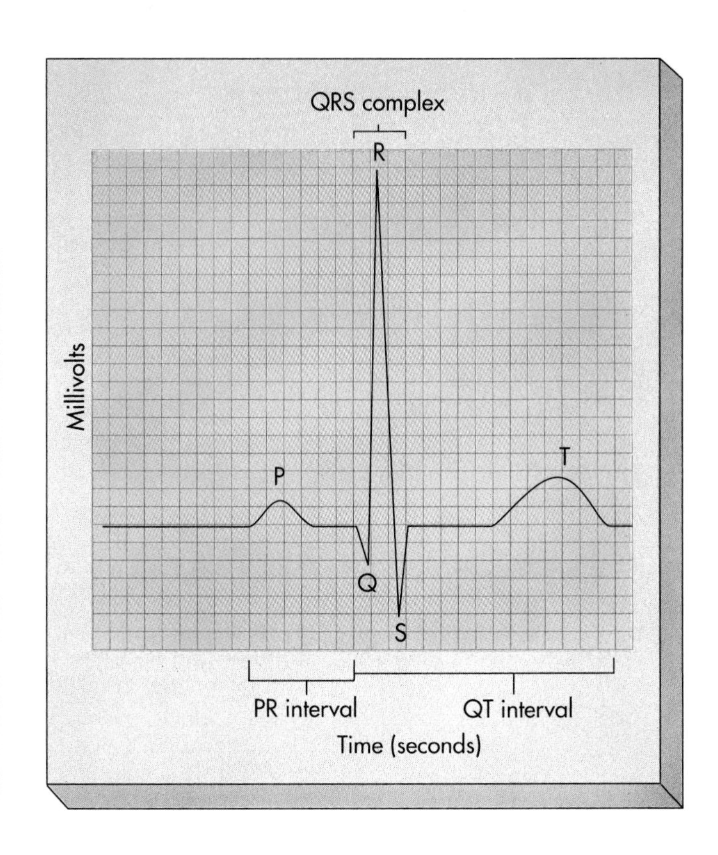

FIGURE 20-13 Electrocardiogram. The major waves and intervals are labeled.

Elongation of the PR interval may result from (1) a delay of action potential conduction through the atrial muscle because of damage such as that caused by ischemia (obstruction of the blood supply to the walls of the heart), (2) delay of action potential conduction through atrial muscle because of a dilated atrium, or (3) delay of action potential conduction through the AV node and bundle because of muscle damage, ischemia, or compression of the bundle. Common causes of a prolonged QRS complex include both ischemia and damaged bundle branches, which result in slow conduction of action potentials through the bundle branches. An unusually long QT interval reflects the abnormal conduction of action potentials through the ventricles, which may result from myocardial infarctions, or from an abnormally enlarged left or right ventricle.

contraction. The **T wave** represents repolarization and precedes ventricular relaxation. A wave representing repolarization of the atria cannot be seen because it occurs during the QRS complex.

The time between the beginning of the P wave and the beginning of the QRS complex is the **PQ interval,** commonly called the **PR interval** because the Q wave is often very small. During the PR interval, which lasts approximately 0.16 second, the atria contract and begin to relax. The ventricles begin to depolarize at the end of the PR interval. The **QT interval** extends from the beginning of the QRS complex to the end of the T wave, lasts approximately 0.3 second, and represents the approximate length of time required for the ventricles to contract and begin to relax.

CARDIAC CYCLE

The heart can be viewed as two separate pumps represented by the right and left halves of the heart. Each pump consists of a primer pump—the atrium—and a power pump—the ventricle. Both atrial primer pumps complete the filling of the ventricles with blood, and both ventricular power pumps produce the major force that causes blood to flow through the pulmonary and systemic arteries. The term **cardiac cycle** refers to the repetitive pumping process that begins with the onset of cardiac muscle contraction and ends with the beginning of the next contraction (Figures 20-14 and 20-15). Pressure changes produced within the heart chambers as a result of cardiac muscle contraction are responsible for blood movement since blood moves from areas of high pressure to areas of low pressure.

6

Explain why it is most important to replace the ventricles in artificial heart implantation.

? ? ? ? ? ? ? ? ? ? ?

The duration of the cardiac cycle varies considerably among humans and also varies during an individual's lifetime. It may be as short as 0.25 to 0.3 second in a newborn infant or as long as 1 or more seconds in a well-trained athlete. The normal cardiac cycle (0.7 to 0.8 second) depends on the capability of cardiac muscle to contract and on the functional integrity of the conducting system. Abnormalities of cardiac muscle, the valves, or the conducting system of the heart may alter the cardiac cycle and thus compromise the pumping effectiveness of the heart. In cases of severe dysfunction, repair is necessary and may involve techniques such as **angioplasty** in which a balloonlike structure on the end of a thin tube is inserted into the coronary arteries and is inflated to increase their diameter; bypassing an occluded coronary artery with a vein transplanted from

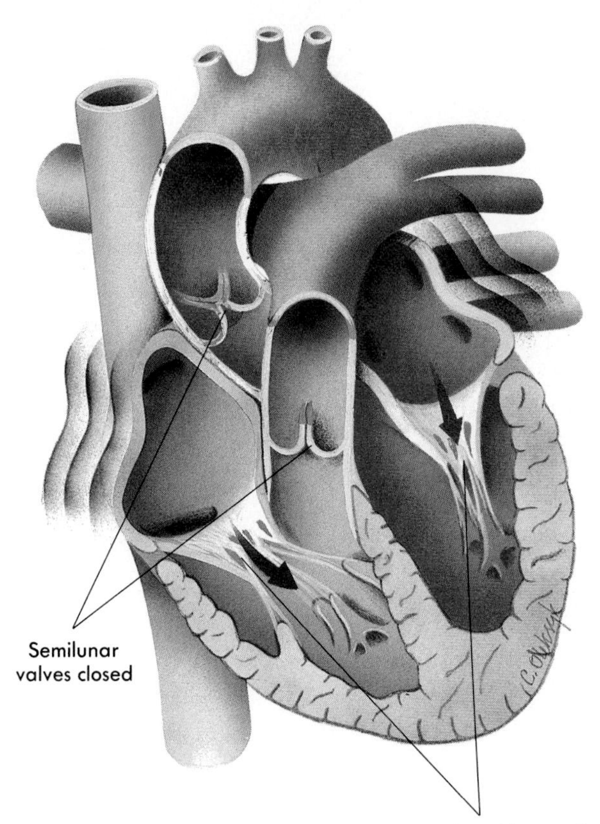

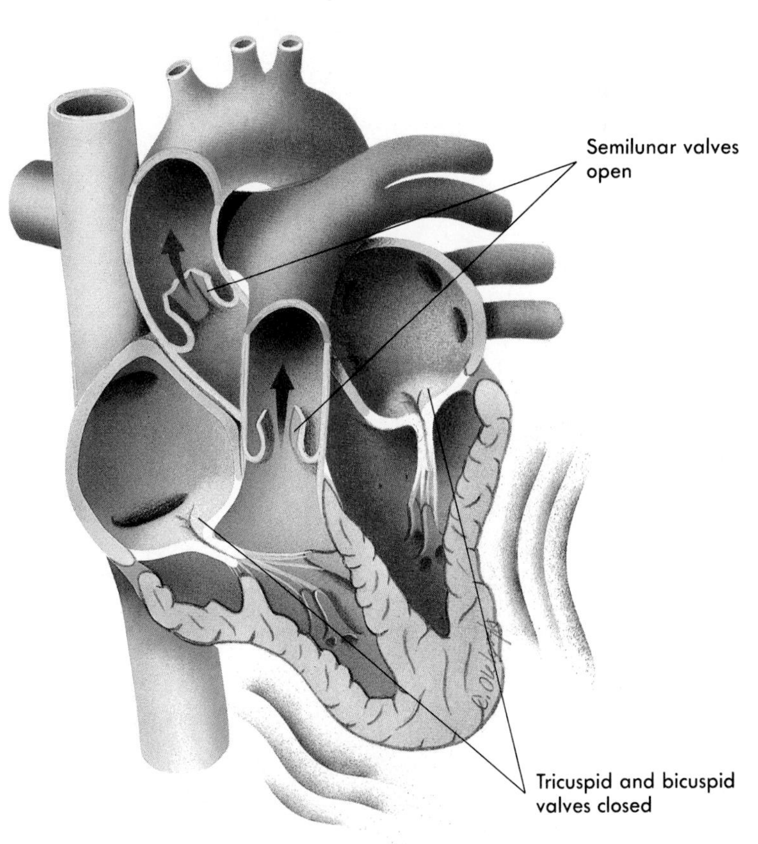

FIGURE 20-14 Events during atrial contraction (systole) **A** and ventricular contraction (systole) **B.**

FIGURE 20-15
For legend see opposite page.

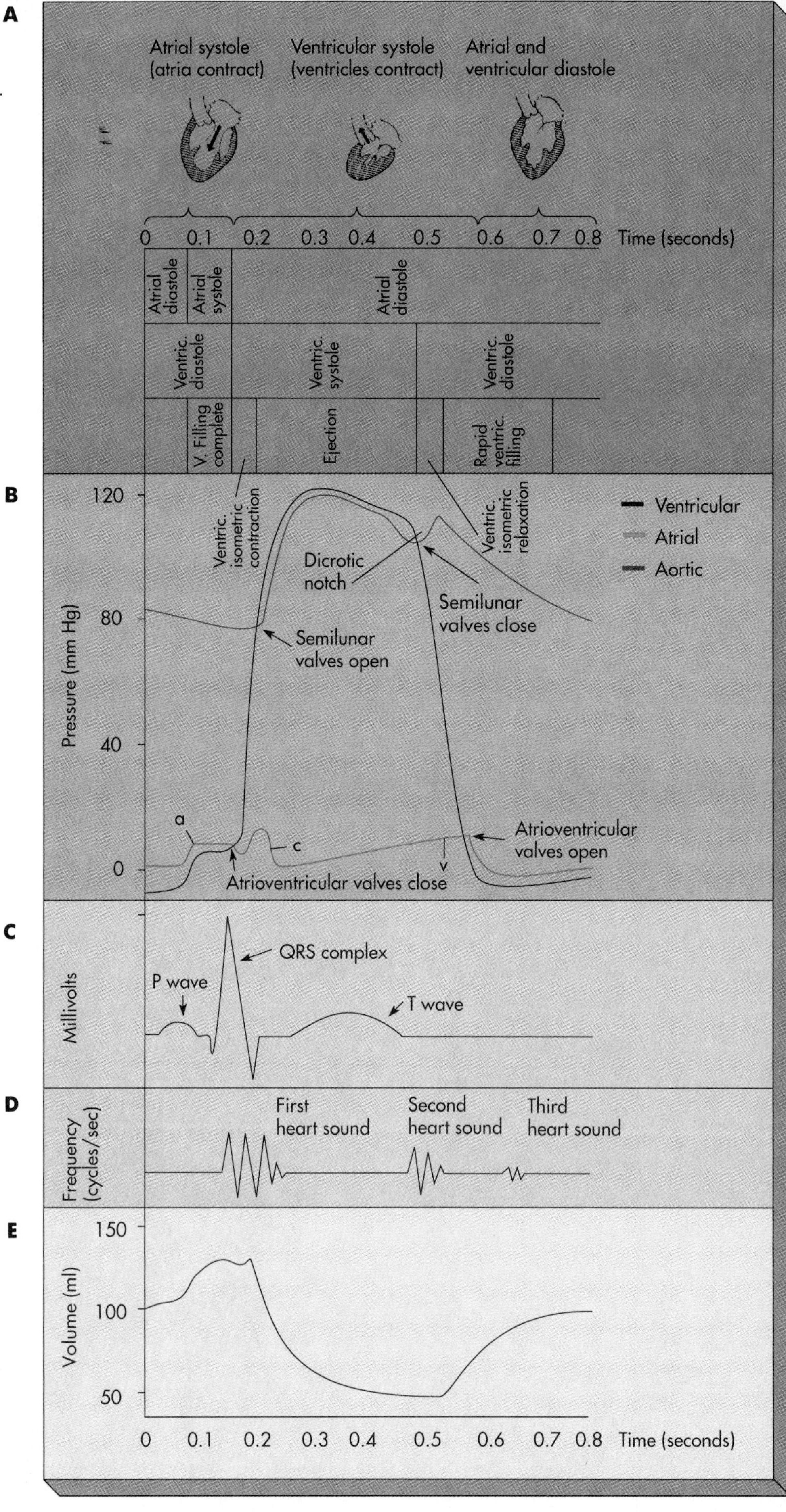

FIGURE 20-15 Events occurring during the cardiac cycle (elapsed time given in tenths of seconds). **A** Major atrial and ventricular events. **B** Pressure changes: *red line*, aortic pressure; *blue line*, atrial pressure; *black line*, ventricular pressure. **C** ECG tracing. The P wave represents depolarization of the atria, the QRS complex represents depolarization of the ventricle, and the T wave represents repolarization of the ventricles. **D** Heart sounds. The first heart sound corresponds with closure of the atrioventricular valves, and the second heart sound corresponds with closure of the semilunar valves. **E** Changes in ventricular volumes throughout the cardiac cycle.

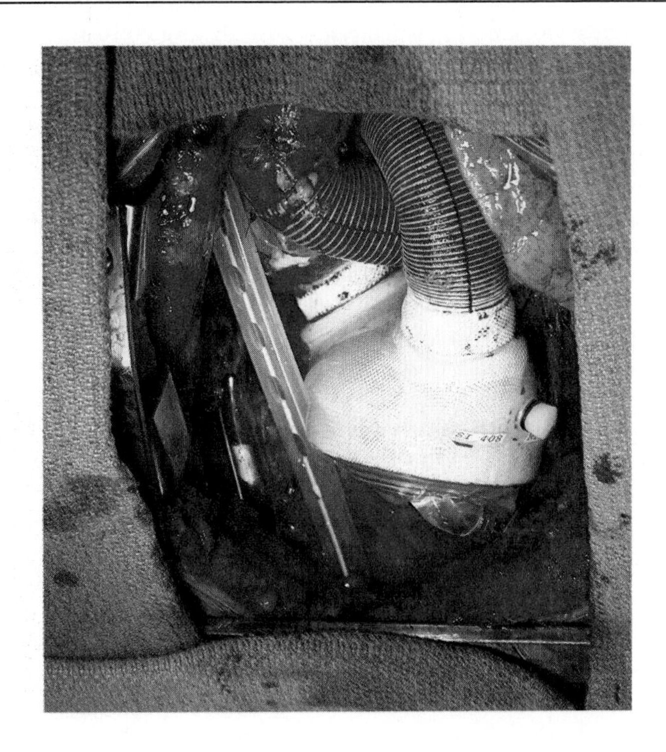

FIGURE 20-16 An artificial heart. This Jarvik-7 artificial heart is designed as a replacement for the ventricles and is grafted to the existing atria. The artificial pumps are connected to an air pump outside of the body that produces pulses of air that cause diaphragms in the artificial heart to force blood out of the heart and into the arteries.

another part of the body (coronary bypass); surgically implanting an electronic device to function as the pacemaker of the heart; or surgically replacing damaged valves with artificial valves. In cases in which the heart cannot be repaired, replacement of the heart with a heart transplant is possible if a compatible donor is available. Artificial hearts are still in experimental stages of development and cannot be viewed as permanent substitutes for the heart (Figure 20-16).

Systole and Diastole

The term **systole** (sis'to-le) means to contract, and **diastole** (di-as'to-le) means to dilate. **Atrial systole** is contraction of the atrial myocardium, and **atrial diastole** is relaxation of the atrial myocardium. Similarly, **ventricular systole** is contraction of the ventricular myocardium, and **ventricular diastole** is relaxation of the ventricular myocardium. However, when "systole" and "diastole" are used without reference to specific chambers, they mean ventricular systole or diastole.

Atrial systole and diastole

At the beginning of ventricular diastole the ventricular pressure falls below the pressure within the atria. As the atrioventricular valves open, blood that has accumulated in the atria flows into the ventricles. Approximately 70% of ventricular filling occurs during the first one third of ventricular diastole. During the second one third of ventricular diastole little ventricular filling occurs. Approximately two thirds of the way through ventricular diastole, the SA node depolarizes, and action potentials spread over the atria, producing the P wave and stimulating both atria to contract (atrial systole). The atria contract only during the last third of ventricular diastole and complete the last 30% of ventricular filling.

Under most conditions the atria function primarily as reservoirs, and the ventricles can pump sufficient blood to maintain homeostasis even if the atria do not contract at all. However, during exercise the heart pumps 300% to 400% more blood than during resting conditions. It is under these conditions that the pumping action of the atria becomes important in maintaining the pumping efficiency of the heart.

Three distinct changes, called the **a, c,** and **v waves,** can be detected in the left atrial pressure curve during the cardiac cycle (see Figure 20-15). The a wave is the result of atrial contraction and consists of a pressure increase of approximately 7 to 8 mm Hg as the atria contract during the last third of ventricular diastole. At the beginning of ventricular systole, the increasing ventricular pressure causes the atrioventricular valves to close and to bulge slightly into the atria. At the same time the contracting walls of the ventricles pull down on the atrial walls. These two events cause a slight increase in atrial pressure, which is the c wave. The v wave occurs near the end of ventricular systole and results from the continuous flow of blood into the atria, causing an increase in atrial blood volume while the atrioventricular valves are closed.

Ventricular systole and diastole

Ventricular contraction causes the ventricular pressure to increase rapidly, and as ventricular pressure exceeds atrial pressure, the atrioventricular valves close. As contraction proceeds, the pressure continues to rise; but no blood flows from the ventricles until the ventricular pressure exceeds that in the pulmonary artery on the right side or in the aorta on the left side of the heart (see Figure 20-15). That short period between atrioventricular valve closure and semilunar valve opening (during which ventricular contraction does not cause blood movement) is called the period of **isometric contraction.**

As soon as ventricular pressure exceeds that in the pulmonary trunk or aorta, the semilunar valves open, and blood flows from the ventricles into those arteries. The aortic semilunar valve opens at approximately 80 mm Hg ventricular pressure, whereas the pulmonary semilunar valve opens at approximately 25 mm Hg (both valves open at nearly the same time).

During the time that blood flows from the ventricles, the period of **ejection,** the left intraventricular pressure continues to climb from approximately 80 to approximately 120 mm Hg, and the right ventricular pressure increases from approximately 20 to 33 mm Hg. The pressure difference created by the two ventricles is reflected in the difference in their anatomy—the thickness of the left ventricular myocardium is three to four times that of the right. Even though the pressure generated by the left ventricle is much higher than that in the right ventricle, the amount of blood pumped by each is almost the same.

7

Why is it important for each ventricle to pump approximately the same volume of blood?

? ? ? ? ? ? ? ? ? ? ?

Ventricular volume decreases during the ejection period. During the last part of ventricular systole the ventricular volume becomes very low, and almost no blood flows from the ventricles into the aorta and the pulmonary trunk, even though ventricular contraction continues. Because no blood is being forced out of the ventricles, the ventricular pressure actually begins to decrease despite continued ventricular contraction.

Ventricular relaxation begins suddenly at the end of systole, causing the already decreasing ventricular pressure to fall very rapidly. When the ventricular pressure falls below the pressure in the aorta or the pulmonary trunk, the recoil of the elastic arterial walls, which were stretched during the period of ejection, forces the blood to flow back toward the ventricles, closing the semilunar valves. The aortic semilunar valves close when the ventricular pressure falls to approximately 95 mm Hg. As the ventricular pressure drops below the atrial pressure (see Figure 20-15), the atrioventricular valves open once more to allow blood to flow from the atria into the ventricles. The period between semilunar valve closure and atrioventricular valve opening, during which no blood flows from the atria into the ventricles, is called the period of **isometric relaxation.**

During diastole blood flows from the atria into the ventricles, and the volume of each ventricle, called the **end-diastolic volume,** normally increases to 120 to 130 ml. As the ventricles empty during systole, their volume decreases to 50 to 60 ml, the **end-systolic volume.** Therefore the volume of blood pumped during each cardiac cycle is approximately 70 ml (end-diastolic volume minus end-systolic volume) and is called the **stroke volume.** When the heart beats forcefully such as during exercise, the end-diastolic volume may be as high as 200 to 250 ml, and the end-systolic volume may fall to 10 to 30 ml. As a result, the stroke volume may be nearly 200 ml.

Under resting conditions the heart rate is approximately 72 beats per minute, and the stroke volume is approximately 70 ml for the average 70-kg male (this value may vary considerably from person to person). Therefore the total amount of blood pumped per minute, called the **cardiac output** or **minute volume,** is

$$70 \text{ ml/beat} \times 72 \text{ beats/min} = 5040 \text{ ml/min or approximately } 5 \text{ L/min}$$

During exercise the heart rate may increase to 120 beats per minute, and the stroke volume may increase to 200 ml or more. Consequently, cardiac output is

$$120 \text{ beats/min} \times 200 \text{ ml/beat} = 24{,}000 \text{ ml/min or approximately } 24 \text{ L/min}$$

The difference between the cardiac output when a person is at rest and the maximum cardiac output is called the **cardiac reserve.**

Cardiac output is a major factor in determining **blood pressure (BP),** which is responsible for blood movement and therefore is critical to the maintenance of homeostasis in the body. The average, or mean, BP in the aorta is proportional to the cardiac output (CO) times the peripheral resistance (PR): BP = CO × PR. The **peripheral resistance** is the total resistance against which blood must be pumped, and regulation of blood pressure involves mechanisms that change both the cardiac output and the peripheral resistance.

Aortic Pressure Curve

The elastic walls of the aorta are stretched as blood is ejected into the aorta from the left ventricle. The aortic pressure remains slightly below the ventricular pressure during this period of ejection. As the ventricular pressure drops below that in the aorta and the semilunar valve closes, a **dicrotic notch,** or **incisura** ("a cutting into"), occurs in the aortic pressure curve because of the momentum of blood flowing back toward the ventricle from the aorta. The aortic pressure then gradually falls throughout ventricular diastole as a result of the recoil of the elastic walls of the aorta, which maintains pressure in the aorta and forces blood to flow through the peripheral vessels. By the time the aortic pressure has fallen to approximately 80 mm Hg pressure, the ventricles again contract, forcing blood once more into the aorta.

Blood pressure measurements performed for clinical purposes reflect the pressure changes that occur in the aorta rather than in the left ventricle. Therefore the systolic pressure is approximately 120 mm Hg, and the diastolic pressure is approximately 80 mm Hg for the average young adult.

Heart Sounds

When a stethoscope is used to listen to the heart sounds, two distinct sounds, called the **first** and **second heart sounds,** normally are heard (see Figure 20-15). When the ventricles contract, both atrioventricular valves close nearly simultaneously, causing vibrations of the valves and the surrounding fluid and resulting in a low-pitched sound, which is the first heart sound (often described as a "lubb" sound). When the aortic and pulmonary semilunar valves close near the end of ventricular systole, they cause a higher-pitched sound, which is the second heart sound (described as "dupp"). Systole is approximately the time between the first and second heart sounds. Diastole, which lasts somewhat longer, is approximately between the second heart sound and the next first heart sound.

Occasionally a **third heart sound** can be detected near the end of the first third of diastole and is caused by blood flowing in a turbulent fashion into the ventricles. The third heart sound is normal, although faint, and is detected most easily in thin, young people.

Abnormal Heart Sounds

To clinicians heart sounds provide important information about the normal function of the heart and assist in diagnosing cardiac abnormalities. Abnormal heart sounds are called **murmurs,** and certain murmurs are important indicators of specific cardiac abnormalities. For example, an **incompetent valve** is a valve that leaks significantly. After closure of an incompetent valve, blood flows through it in a reverse direction (i.e., in a direction opposite to the normal flow of blood through the heart) and results in the turbulent flow of blood, which causes a gurgle or swish sound immediately after closure of the valve. An incompetent tricuspid valve or bicuspid valve exhibits a swish sound immediately after the first heart sound, and the first heart sound may be muffled. An incompetent aortic or pulmonary semilunar valve results in a swish sound immediately after the second heart sound. **Stenosed valves** have an abnormally narrow opening and also produce abnormal heart sounds. Blood flows through stenosed valves in a very turbulent fashion; consequently, a rushing sound precedes the valve closure. Therefore a stenosed atrioventricular valve results in a rushing sound immediately before the first heart sound, and a stenosed semilunar valve results in a rushing sound immediately before the second heart sound.

REGULATION OF THE HEART

The amount of blood pumped by the heart can vary dramatically. For example, during exercise the cardiac output can increase several times over resting values. The cardiac output is controlled by regulatory mechanisms that can be classified as either **intrinsic** or **extrinsic.** Intrinsic regulation results from the heart's normal functional characteristics and does not depend on either innervation or hormonal regulation. It functions when the heart is in place in the body or is removed and maintained outside the body under proper conditions. On the other hand, extrinsic factors involve neural and hormonal control. Neural regulation of the heart involves both sympathetic and parasympathetic reflexes, and the major hormonal regulation comes from epinephrine and norepinephrine secreted from the adrenal medulla.

Intrinsic Regulation

The amount of blood that flows into the right atrium from the veins during diastole is called the **venous return.** An increase in venous return causes an increase in cardiac output, and a decrease in venous return causes a decrease in cardiac output. If the heart rate is constant, an increase in venous return causes the ventricles to fill to a greater extent, resulting in an increased end-diastolic volume. Cardiac muscle exhibits a length vs. tension relationship similar to that of skeletal muscle. However, cardiac muscle fibers normally are not stretched to the point at which they contract with a maximal force (see Chapter 10). Therefore increased ventricular volume stretches the cardiac muscle fibers in the walls, and as a result, the cardiac muscle fibers contract with a greater force and produce a greater stroke volume. This relationship between venous return and cardiac output is commonly referred to as **Starling's law of the heart,** which describes the mechanism by which the heart's pumping effectiveness changes in response to alterations in venous return (Table 20-2). Venous return may decrease to a value as low as 2 L per minute or increase to as much as 24 L per minute.

In addition to Starling's law, venous return has another influence on cardiac output. Stretching of the right atrial wall in turn stretches the SA node, causing an increase in the rate of action potential generation in that node and an increase in heart rate by 10% to 30%. Stretching the node increases the permeability of the cell membranes in the SA node to sodium and calcium ions.

Although the pumping effectiveness of the heart is greatly influenced by relatively small changes in venous return, it is very insensitive to large changes in arterial blood pressure. Aortic blood pressure must increase to more than 170 mm Hg before it hampers the ability of the ventricles to pump blood. During ventricular diastole the semilunar valves are closed so that arterial pressure is isolated from ventricular pressure; and during ventricular systole the contracting ventricles can produce a pressure great enough to overcome the effect of the elevated arterial pressure.

During physical exercise muscle contractions repeatedly compress veins and cause an increased rate of blood flow toward the heart. Because of Starling's law of the heart, the elevated venous return increases the cardiac output and the volume of blood flow to the exercising muscles. When at rest, venous return to the heart decreases because muscular contractions are no longer repeatedly compressing the veins. As a result, the cardiac output declines, and the blood flow to the resting muscles decreases.

Extrinsic Regulation

The heart is innervated by both **parasympathetic** and **sympathetic** nerve fibers (Figure 20-17). They influence the pumping action of the heart by affecting both the heart rate and stroke volume. Hormones also play an important role in heart regulation.

Extrinsic regulation of the heart functions to keep blood pressure, blood oxygen levels, blood carbon dioxide levels, and blood pH within their normal ranges of values. For example, if the blood pressure suddenly decreases, extrinsic mechanisms detect the decrease and initiate responses that increase the cardiac output to bring blood pressure back to its normal range.

Parasympathetic control

Parasympathetic nerve fibers are carried to the heart through the **vagus nerves.** Preganglionic fibers of the vagus nerve extend to parasympathetic ganglia within the wall of the heart, and postganglionic fibers extend from the ganglia to the SA node, AV node, coronary vessels, and atrial myocardium.

Parasympathetic innervation has an inhibitory influence on the heart, primarily by decreasing heart rate. During resting conditions continuous parasympathetic innervation of the heart inhibits it to a small degree. Strong parasympathetic stimulation can decrease the heart rate 20 to 30 beats per minute. Parasympathetic stimulation has little impact on the stroke volume. In fact, if the venous return remains constant while the heart is inhibited by parasympathetic innervation, the stroke volume actually may increase because the longer time between heartbeats allows the heart to fill to a greater capacity, resulting in an increased stroke volume because of Starling's law of the heart.

Acetylcholine, the neurotransmitter produced by the postganglionic parasympathetic neurons, causes

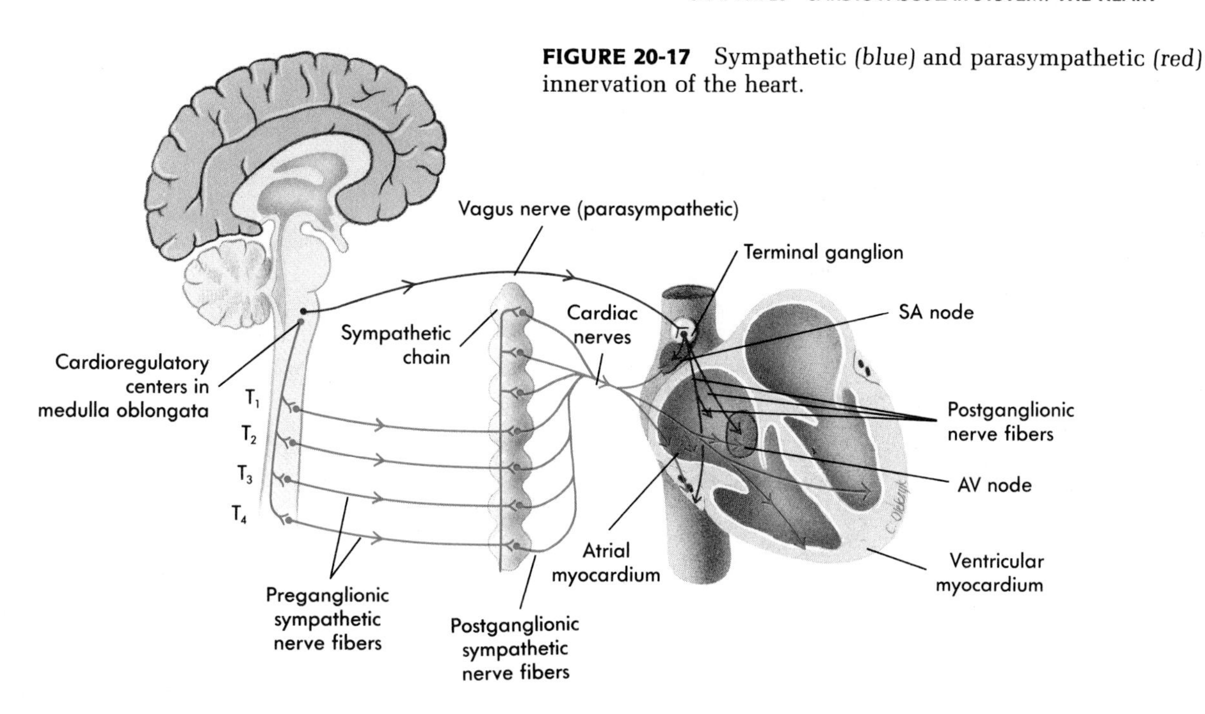

FIGURE 20-17 Sympathetic *(blue)* and parasympathetic *(red)* innervation of the heart.

Table 20-2

Regulation of the heart (see Figure 20-18)

STIMULUS	SENSORY RECEPTOR(s)	INTEGRATION	EFFECT ON HEART	RESPONSE
Intrinsic Regulation				
Starling's law of the heart				
Increase in venous return to the heart	None	Stretch of cardiac muscle	Increased force of contraction and slight increase in heart rate	Increased stroke volume; increased cardiac output
Decrease in venous return to the heart	None	Reduced stretch of cardiac muscle	Decreased force of contraction and slight decrease in heart rate	Decreased stroke volume; decreased cardiac output
Extrinsic Regulation of the Heart				
Baroreceptor reflex				
Sudden decrease in blood pressure	Baroreceptors in the carotid sinus and aortic arch	Reduced action potentials along afferent nerves to the cardioregulatory center	Increased sympathetic and decreased parasympathetic stimulation of the heart; increased release of epinephrine and norepinephrine from the adrenal medulla	Increased heart rate and increased stroke volume (if venous return is adequate) cause cardiac output to increase, resulting in increased blood pressure
Sudden increase in blood pressure	Baroreceptors in the carotid sinus and aortic arch	Increased action potentials along afferent nerves to the cardioregulatory center	Decreased sympathetic and increased parasympathetic stimulation of the heart; decreased release of epinephrine and norepinephrine from the adrenal medulla	Decreased heart rate and decreased stroke volume cause cardiac output to decrease, resulting in decreased blood pressure

Table 20-2

Regulation of the heart—cont'd

STIMULUS	SENSORY RECEPTOR(s)	INTEGRATION	EFFECT ON HEART	RESPONSE
Extrinsic Regulation of the Heart—cont'd				
Bainbridge reflex (of minor importance in humans)				
Increase in venous return to the heart	Increased stimulation of stretch receptors in the atria	Increased action potentials to the cardioregulatory center	Increased sympathetic stimulation of the heart	Slight increase in heart rate
Decrease in venous return to the heart	Decreased stimulation of stretch receptors in the atria	Decreased action potentials to the cardioregulatory center	Decreased sympathetic stimulation of the heart	Slight decrease in heart rate
*Chemoreceptor reflex: the cardioregulatory center**				
Increase in blood carbon dioxide (CO_2) levels and decreased blood pH	Chemoreceptors in the cardioregulatory center	Chemoreceptor reflex within the cardioregulatory center	Increased sympathetic and reduced parasympathetic stimulation of the heart; increased release of epinephrine and norepinephrine from the adrenal medulla	Increased stroke volume and heart rate increases cardiac output, resulting in increased blood pressure
Decrease in blood CO_2 levels and increased blood pH	Chemoreceptors in the cardioregulatory center	Chemoreceptor reflex within the cardioregulatory center	Decreased sympathetic and increased parasympathetic stimulation of the heart; decreased release of epinephrine and norepinephrine from the adrenal medulla	Decreased stroke volume and heart rate decreases cardiac output, resulting in decreased blood pressure
*Chemoreceptor reflex: the carotid and aortic bodies**				
Large decrease in blood oxygen (O_2) levels	Chemoreceptors in the carotid and aortic bodies	Increased action potentials sent to the cardioregulatory center	Increased parasympathetic stimulation of the heart	Decreased heart rate
Increase in blood O_2 levels to normal values	Chemoreceptors in the carotid and aortic bodies	Reduction in action potentials sent to the cardioregulatory center	Decreased parasympathetic stimulation of the heart	Increased heart rate

*The chemoreceptor reflexes influence the heart's function. However, they function primarily under emergency conditions. The chemoreceptor reflexes have more important influences on the peripheral blood vessels (peripheral resistance; see Chapter 21) and on the respiratory system (see Chapter 23).

the cardiac cell membranes to become more permeable to potassium ions. As a consequence, the resting membrane potential becomes more negative, the rate of depolarization is reduced, and heart rate is decreased.

Sympathetic control

Sympathetic nerve fibers originate in the thoracic region of the spinal cord as preganglionic neurons. These neurons synapse with postganglionic neurons of the **cervical sympathetic chain ganglia,** which project to the heart as **cardiac nerves** (see Figure 20-17 and Chapter 16). The postganglionic sympathetic nerve fibers innervate the SA and AV nodes, coronary vessels, and the atrial and ventricular myocardium.

Sympathetic innervation has a stimulatory influence on both the heart rate and the force of muscular contraction. In response to strong sympathetic stimulation the heart rate may increase to 250 or occasionally 300 beats per minute. If the heart fills with blood, the stroke volume also increases. The increased force of contraction resulting from sympathetic stimulation causes a lower end-systolic volume in the heart; therefore the heart empties to a greater extent. However, there are limitations to the relationship between increased heart rate and cardiac output. If the heart rate becomes too great, diastole is not long enough to allow complete ventricular filling, and the stroke volume actually decreases. In addition, if the heart rate increases beyond a critical level, the strength of contraction decreases, probably as a result of the accumulation of metabolites in the cardiac muscle cells. The limit of the heart's ability to pump blood is 170 to 250 beats per minute in response to intense sympathetic stimulation.

8

What effect does sympathetic stimulation have on stroke volume if the venous return remains constant? Sympathetic stimulation of the heart also causes dilation of the coronary blood vessels. Explain the functional advantage of that effect.

? ? ? ? ? ? ? ? ? ? ?

Sympathetic innervation of the ventricular myocardium plays a significant role in regulation of its contraction force during resting conditions. Sympathetic stimulation maintains the strength of ventricular contraction at a level approximately 20% greater than it would be with no sympathetic stimulation.

The influence of parasympathetic stimulation on the contractile strength of cardiac muscle is much less than that of sympathetic stimulation. Sympathetic stimulation can increase cardiac output by 50% to 100% over resting values, whereas parasympathetic stimulation can cause only a 10% to 20% decrease.

Norepinephrine, the postganglionic sympathetic neurotransmitter, increases the rate and degree of cardiac muscle depolarization so that both the frequency and the amplitude of the action potentials are increased. The effect of norepinephrine on the heart involves the association between norepinephrine and cell surface beta-adrenergic receptors. The combination of norepinephrine molecules with beta-adrenergic receptors causes increased synthesis and accumulation of cyclic AMP in the cytoplasm of cardiac muscle cells. The cyclic AMP increases the permeability of the cell membrane to sodium and calcium ions.

Hormonal control

Epinephrine and norepinephrine released from the adrenal medulla markedly influence the pumping effectiveness of the heart. Epinephrine has essentially the same effect on cardiac muscle as norepinephrine and therefore increases the rate and force of heart contractions. The secretion of epinephrine and norepinephrine from the adrenal medulla is controlled by sympathetic innervation and occurs in response to increased physical activity, emotional excitement, or stressful conditions. Many stimuli that increase sympathetic stimulation of the heart also increase release of epinephrine and norepinephrine from the adrenal gland (see Chapter 18). Epinephrine and norepinephrine are transported in the blood through the vessels of the heart to the cardiac muscle cells in which they bind to beta-adrenergic receptors and stimulate cyclic AMP synthesis. Epinephrine takes a longer time to act on the heart than sympathetic innervation does, but the effect lasts longer.

HEART AND HOMEOSTASIS

The heart's pumping efficiency plays an important role in the maintenance of homeostasis. Blood pressure in the systemic vessels must be maintained at a level at which blood flow is sufficient to achieve nutrient and waste product exchange across the walls of the capillaries at a rate that meets metabolic demands. Because the metabolic activities of the tissues change under conditions such as exercise and rest, the activity of the heart must be regulated.

Effect of Blood Pressure

The **baroreceptor reflexes** detect changes in blood pressure and result in changes in heart rate and in the force of contraction of the heart. Sensory receptors called **baroreceptors** (bar'o-re-sep'torz; stretch receptors) are in the walls of certain large arteries (i.e., the internal carotid arteries and the aorta) and function to measure blood pressure. The anatomy of these sensory structures and their afferent pathways are described in Chapter 21. Afferent neurons project primarily through the glossopharyngeal (IX) and vagus (X) nerves from these receptors to an area in the medulla oblongata called the **cardioregulatory center** where sensory impulses are integrated. The part of the cardioregulatory center that functions to increase heart rate is called the **cardioacceleratory center,** and the part that functions to decrease heart rate is called the **cardioinhibitory center.** Efferent impulses then are sent from the cardioregulatory center to the heart through both the sympathetic and parasympathetic divisions of the autonomic nervous system (Figure 20-18; see Table 20-2).

Increased blood pressure within the internal carotid arteries and aorta causes their walls to stretch, stimulating action potentials in the baroreceptors. At normal blood pressures (80 to 120 mm Hg) afferent impulses are sent from the baroreceptors to the medulla oblongata at a relatively constant frequency. When blood pressure increases, the arterial walls are stretched further, and the afferent impulse frequency increases. When blood pressure decreases, the arterial walls are stretched to a lesser extent, and the afferent impulse frequency decreases. The baroreceptor reflexes decrease sympathetic stimulation and increase parasympathetic stimulation of the heart in response to increased blood pressure, causing the heart rate to decrease. Decreased blood pressure causes increased sympathetic stimulation and decreased parasympathetic stimulation of the heart, resulting in an increased force of contraction and increased heart rate (see Figure 20-18). The baroreceptor reflexes are homeostatic in that they keep the blood pressure within a narrow range of values.

An increase in right atrial pressure also causes the cardioregulatory center to increase the heart rate through a reflex called the **Bainbridge reflex** (see Table 20-2). Stretch receptors within the wall of the right atrium send impulses to the medulla oblongata through the vagus nerves, and efferent impulses are

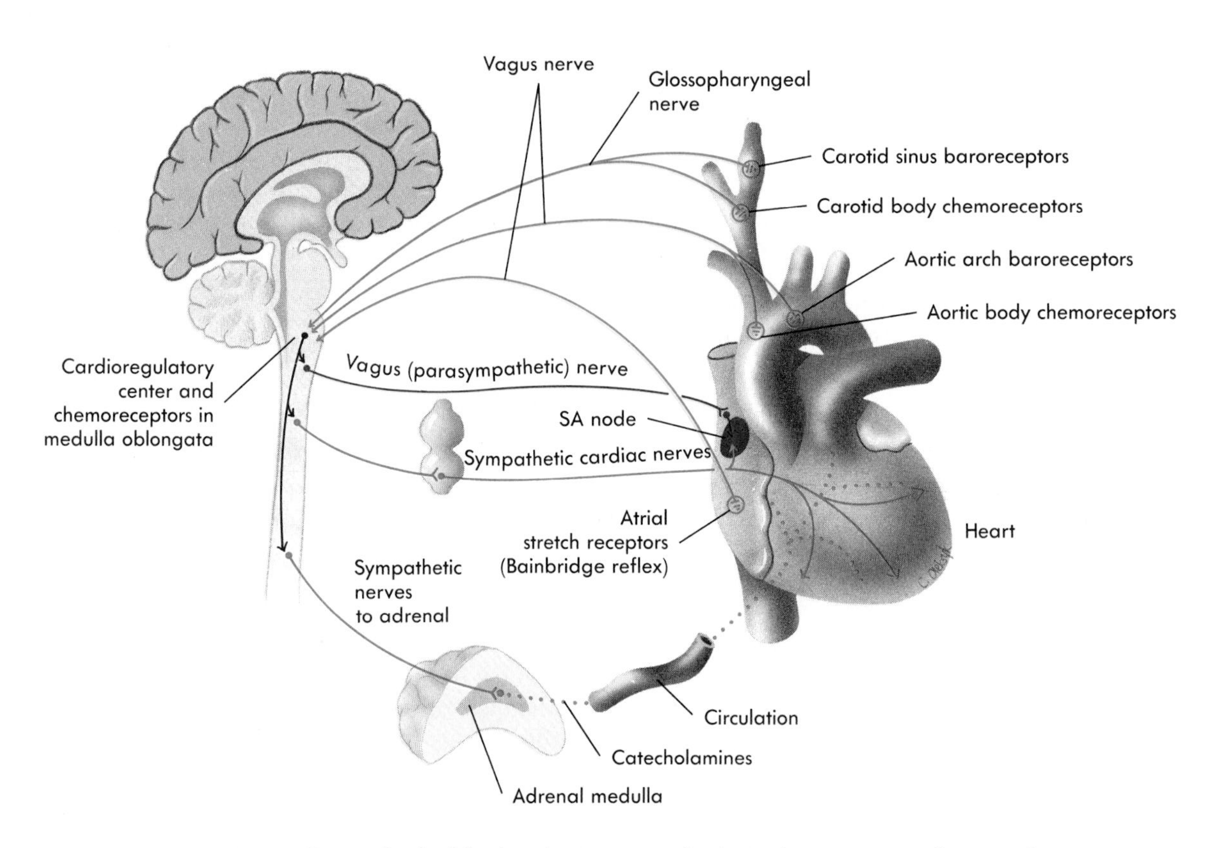

FIGURE 20-18 Sympathetic *(blue)* and parasympathetic *(red)* nerves exit the spinal cord or medulla and extend to the heart to regulate its function. Hormonal influences such as epinephrine from the adrenal gland also help regulate the heart's action.

sent through both sympathetic and parasympathetic neurons back to the heart. An increase in right atrial pressure causes increased sympathetic and decreased parasympathetic stimulation of the heart, resulting in an increased heart rate because of the Bainbridge reflex, which functions with Starling's law of the heart to increase cardiac output when venous return increases. The Bainbridge reflex is considered of minor importance in humans.

Effect of pH, Carbon Dioxide, and Oxygen

Chemoreceptor reflexes help regulate the activity of the heart. Chemoreceptors sensitive to changes in pH and carbon dioxide levels exist within the medulla oblongata. A decrease in pH and an increase in carbon dioxide increase sympathetic stimulation of the heart, whereas an increase in pH and a decrease in carbon dioxide decrease sympathetic stimulation. Very low levels of blood oxygen stimulate receptors near the carotid arteries. When they are stimulated alone, the result is increased parasympathetic stimulation of the heart which results in a decreased heart rate. When the chemoreceptors in both the medulla oblongata and near the carotid arteries are stimulated, the heart rate and stroke volume both increase.

Increased cardiac output in response to elevated blood carbon dioxide levels and decreased blood pH results in a greater blood flow through the lungs where carbon dioxide is eliminated from the body. The increased blood flow through the lungs, therefore, helps bring the blood carbon dioxide level down to its normal range of values and helps to increase the blood pH. Chemoreceptor reflexes also influence blood vessels and respiration (see Chapters 21 and 23).

Effect of Extracellular Ion Concentration

Ions that affect cardiac muscle function are the same ions, including potassium, calcium, and sodium, that influence membrane potentials in other electrically excitable tissues. However, there are some differences between the response of cardiac muscle and the response of nerve or muscle tissue to these ions, and extracellular levels of sodium rarely deviate enough from the normal value to affect the function of cardiac muscle significantly.

Excess potassium ions in cardiac tissue cause the heart rate and stroke volume to decrease. A twofold increase in extracellular potassium ions considerably slows the rate of action potential conduction through cardiac muscle and may result in heart blocks (i.e., loss of functional conduction of action potentials through the conducting system of the heart). The excess potassium in the extracellular fluid causes partial depolarization of the resting membrane potential, resulting in decreased amplitude and decreased rate at which action potentials are conducted along muscle fibers. As the conduction rates decrease, ectopic action potentials may occur. The reduced action potential amplitude also result in less calcium entering the sarcoplasm of the cell; thus the strength of cardiac muscle contraction decreases.

Although the extracellular concentration of potassium ions normally is small, a decrease in extracellular potassium results in a decrease in the heart rate because the resting membrane potential is hyperpolarized; as a consequence, it takes longer for the membrane to depolarize to threshold. The force of contraction, however, is not affected.

An increase in the extracellular concentration of calcium ions produces an increase in the force of cardiac contraction because of a greater influx of calcium into the sarcoplasm during action potential generation. Elevated plasma calcium levels have an indirect effect on heart rate because they reduce the frequency of action potentials in nerve fibers thus reducing sympathetic and parasympathetic stimulation of the heart (see Chapter 9). Generally, elevated blood calcium levels reduce the heart rate.

A low blood calcium level increases the heart rate, although the effect is imperceptible until blood calcium levels are reduced to approximately one tenth their normal value. The reduced extracellular calcium ion levels allow sodium ions to diffuse more readily into the cell, resulting in an elevated rate of action potential generation. However, reduced calcium levels usually will cause death as a result of tetany of skeletal muscles before they decrease enough to markedly influence the heart's function.

Effect of Body Temperature

Although under resting conditions the temperature of cardiac muscle normally does not change dramatically in humans, alterations in temperature influence the heart rate. Small increases in cardiac muscle temperature cause the heart rate to increase, and decreases in temperature cause the heart rate to decrease. For example, during exercise or fever, increased heart rate and force of contraction accompany temperature increases, but the heart rate decreases under conditions of hypothermia. During heart surgery the body temperature sometimes is reduced dramatically to slow the heart rate and other metabolic functions in the body.

The Cardiovascular System

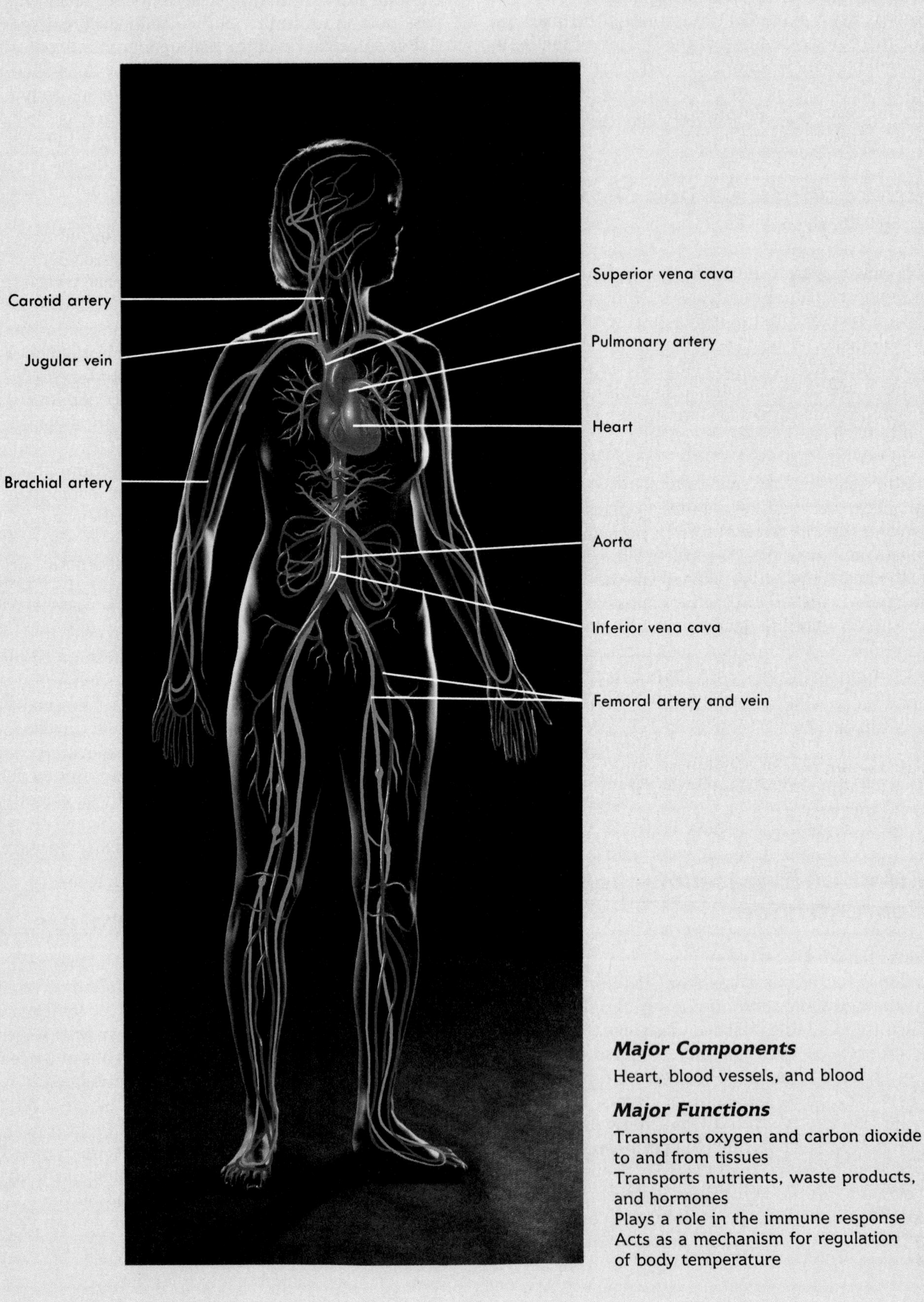

Major Components

Heart, blood vessels, and blood

Major Functions

Transports oxygen and carbon dioxide to and from tissues
Transports nutrients, waste products, and hormones
Plays a role in the immune response
Acts as a mechanism for regulation of body temperature

SUMMARY

THE HEART PRODUCES THE FORCE THAT CAUSES BLOOD CIRCULATION.

SIZE, FORM, AND LOCATION OF THE HEART *p. 612*

The heart is approximately the size of a closed fist and is shaped like a blunt cone. It is in the mediastinum.

ANATOMY OF THE HEART *p. 612*

The heart consists of two atria and two ventricles.

Pericardium

1. The pericardium is a sac that surrounds the heart and consists of the fibrous pericardium and the serous pericardium.
2. The fibrous pericardium helps hold the heart in place.
3. The serous pericardium reduces friction as the heart beats. It consists of the following parts:
   - The parietal pericardium, which lines the fibrous pericardium.
   - The visceral pericardium, which lines the exterior surface of the heart.
   - The pericardial cavity between the parietal and visceral pericardium, which is filled with pericardial fluid.

External Anatomy

1. Each atrium has a flap called the auricle.
2. The atria are separated from the ventricles by the coronary sulcus. The right and left ventricles are separated by the interventricular grooves.
3. The inferior and superior venae cavae and the coronary sinus enter the right atrium. The four pulmonary veins enter the left atrium.
4. The pulmonary trunk exits the right ventricle, and the aorta exits the left ventricle.
5. Coronary arteries branch off the aorta to supply the heart. Blood returns from the heart tissues to the right atrium through the coronary sinus and cardiac veins.

Heart Chambers and Valves

1. The atria are separated from each other by the interatrial septum, and the ventricles are separated by the interventricular septum.
2. The right atrium and ventricle are separated by the tricuspid valve. The left atrium and ventricle are separated by the bicuspid valve. The papillary muscles attach by the chordae tendineae to the atrioventricular valves.
3. The aorta and pulmonary trunk are separated from the ventricles by the semilunar valves.

ROUTE OF BLOOD FLOW THROUGH THE HEART *p. 618*

1. Blood from the body flows through the right atrium into the right ventricle and then to the lungs.
2. Blood returns from the lungs to the left atrium, enters the left ventricle, and is pumped back to the body.

HISTOLOGY *p. 619*

Heart Skeleton

The fibrous heart skeleton supports the openings of the heart, electrically insulates the atria from the ventricles, and provides a point of attachment for heart muscle.

Heart Wall

1. The heart wall has three layers.
   - The outer epicardium (visceral pericardium) provides protection against the friction of rubbing organs.
   - The middle myocardium is responsible for contraction.
   - The inner endocardium reduces the friction resulting from blood's passing through the heart.
2. The inner surfaces of the atria are mainly smooth. The auricles have raised areas called musculi pectinati.
3. The ventricles have ridges called trabeculae carneae.

Cardiac Muscle

1. Cardiac muscle cells are branched and have a centrally located nucleus. Actin and myosin are organized to form sarcomeres. The sarcoplasmic reticulum and T tubules are not as organized as in skeletal muscle.
2. Cardiac muscle cells are joined by intercalated disks, which allow action potentials to move from one cell to the next. Thus cardiac muscle cells function as a unit.
3. Cardiac muscle cells have a slow onset of contraction and a prolonged contraction time caused by the length of time required for calcium to move to and from the myofibrils.
4. Cardiac muscle is well supplied with blood vessels that support aerobic respiration.
5. Cardiac muscle aerobically uses glucose, fatty acids, and lactic acid to produce ATP for energy. Cardiac muscle does not develop a significant oxygen debt.

Conducting System

1. The sinoatrial (SA) node and the atrioventricular (AV) node are in the right atrium.
2. The AV node is connected to the bundle branches in the interventricular septum by the AV bundle.
3. The bundle branches give rise to Purkinje fibers, which supply the ventricles.
4. The SA node initiates action potentials, which spread across the atria and cause them to contract.
5. Action potentials are slowed in the AV node, allowing the atria to contract and blood to move into the ventricles. Then the action potentials travel through the AV bundles and bundle branches to the Purkinje fibers, causing the ventricles to contract, starting at the apex.

ELECTRICAL PROPERTIES *p. 623*

Action Potentials

1. Depolarization is prolonged (plateau phase) in cardiac muscle.
2. Spontaneous depolarization is caused by the movement of sodium and calcium ions through slow channels into the cell.
3. Voltage-dependent fast channels let sodium and calcium ions into the cell. When membrane potentials reach threshold, an action potential is produced.

Autorhythmicity of Cardiac Muscle

1. Cardiac muscle cells are autorhythmic.
2. Ectopic foci are areas of the heart that regulate heart rate under abnormal conditions.

Refractory Period of Cardiac Muscle

Cardiac muscle has a prolonged depolarization and thus a prolonged refractory period, which allows time for the cardiac muscle to relax before the next action potential causes a contraction.

Electrocardiogram

1. The electrocardiogram (ECG) records only the electrical activities of the heart.
   - Depolarization of the atria produces the P wave.
   - Depolarization of the ventricles produces the QRS complex. Repolarization of the atria occurs during the QRS complex.
   - Repolarization of the ventricles produces the T wave.
2. Based on the magnitude of the ECG waves and the time between waves, ECGs can be used to diagnose heart abnormalities.

CARDIAC CYCLE *p. 626*

1. The cardiac cycle is repetitive contraction and relaxation of the heart chambers.
2. Blood moves through the circulatory system from areas of higher pressure to areas of lower pressure. Contraction of the heart produces the pressure.

Systole and Diastole

1. Systole is contraction of a heart chamber, and diastole is relaxation of a heart chamber.
2. Atrial Systole and Diastole
   - Contraction of the atria is responsible for 30% of ventricular filling.
   - Pressure changes in the atria include the a wave (caused by atrial contraction), c wave (caused by ventricular contraction), and v wave (caused by blood's flowing into the atria).
3. Ventricular Systole and Diastole
   - Contraction of the ventricles causes blood to move to the lungs and to the body. Pressure generated by the right ventricle (blood to the lungs) is lower than pressure generated by the left ventricle (blood to the body).
   - During the period of isometric contraction ventricular contraction causes an increase in pressure within the ventricles but no movement of blood out of the ventricles.
   - Blood flows from the ventricles during the period of ejection.
   - The semilunar valves close, the ventricles relax, and pressure drops to zero during early stages of diastole, and the ventricles begin to fill with blood.
   - Stroke volume is the difference between end-diastolic volume and end-systolic volume (i.e., the amount of blood pumped by the heart per beat).
   - Cardiac output is stroke volume times heart rate.
   - Cardiac output and peripheral resistance determine blood pressure. Adequate blood pressure is necessary to ensure delivery of blood to the tissues.

Aortic Pressure Curve

1. Contraction of the ventricles forces blood into the aorta, thus producing the peak systolic pressure.
2. Blood pressure in the aorta falls to the diastolic level as blood flows out of the aorta.
3. Elastic recoil of the aorta maintains pressure in the aorta and produces the aortic incisura.

Heart Sounds

1. The first heart sound is produced by closure of the atrioventricular valves.
2. The second heart sound is produced by closure of the semilunar valves.

REGULATION OF THE HEART *p. 632*

Intrinsic Regulation

1. Venous return is the amount of blood that returns to the heart during each cardiac cycle.
2. Starling's law of the heart states that cardiac output is equal to venous return.
3. Venous return stretches the SA node and increases heart rate.

Extrinsic Regulation

1. The cardioregulatory center in the medulla oblongata regulates the parasympathetic and sympathetic nervous control of the heart.
2. Parasympathetic control.
   - Parasympathetic stimulation is supplied by the vagus nerve.
   - Parasympathetic stimulation decreases heart rate and can cause a small decrease in the force of contraction (stroke volume).
   - Postganglionic neurons secrete acetylcholine, which increases membrane permeability to potassium ions, producing hyperpolarization of the membrane.
3. Sympathetic control.
   - Sympathetic stimulation is supplied by the cardiac nerves.
   - Sympathetic stimulation increases heart rate and the force of contraction (stroke volume).
   - Postganglionic neurons secrete norepinephrine, which increases membrane permeability to sodium and calcium ions, producing hypopolarization of the membrane.
4. Epinephrine and norepinephrine are released into the blood from the adrenal medulla as a result of sympathetic stimulation.
   - The effects of epinephrine and norepinephrine on the heart are long lasting compared to the effects of neural stimulation.
   - Epinephrine and norepinephrine increase the rate and force of heart contraction.

HEART AND HOMEOSTASIS *p. 635*

Effect of Blood Pressure

1. Baroreceptors monitor blood pressure. In response to an increase in blood pressure, the baroreceptor reflexes decrease sympathetic stimulation and increase parasympathetic stimulation of the heart.
2. The Bainbridge reflex increases heart rate in response to stretching of the right atrial wall.

Effect of pH, Carbon Dioxide, and Oxygen

1. Chemoreceptors monitor blood carbon dioxide, pH, and oxygen levels.
2. In response to increased carbon dioxide, decreased pH, or decreased oxygen, autonomic nervous system reflexes increase sympathetic stimulation and decrease parasympathetic stimulation of the heart.

Effect of Extracellular Ion Concentration

1. An increase or decrease in extracellular potassium ions decreases heart rate.
2. Increased extracellular calcium ions increase the force of contraction of the heart and decrease the heart rate. Decreased calcium ion levels produce the opposite effect.

Effect of Body Temperature

Heart rate increases when body temperature increases, and it decreases when body temperature decreases.

CONTENT REVIEW

1. Give the approximate size and shape of the heart. Where is it located?
2. What is the pericardium? Name its parts and their functions.
3. Name the major blood vessels that enter and leave the heart. Which chambers of the heart do they enter or exit?
4. What structure separates the atria from each other? What structure separates the ventricles from each other?
5. Name the valves that separate the right atrium from the right ventricle and the left atrium from the left ventricle.
6. What are the functions of the papillary muscles and the chordae tendineae?
7. Describe the flow of blood through the heart.
8. What is the skeleton of the heart? Give three functions of the heart skeleton.
9. Describe the three layers of the heart and state their functions.
10. How does cardiac muscle differ from skeletal muscle?
11. Why does cardiac muscle have a slow onset of contraction and a prolonged contraction?
12. What anatomical features are responsible for the ability of cardiac muscle cells to contract as a unit?
13. What substances are used by cardiac muscle as an energy source? Do cardiac muscle cells develop an oxygen debt?
14. List the parts of the conducting system of the heart. Explain how the conducting system coordinates contraction of the atria and ventricles.
15. Describe ion movement during depolarization in cardiac muscle. What is the plateau phase?
16. Why is cardiac muscle referred to as autorhythmic? What are ectopic foci?
17. Why does cardiac muscle have a prolonged refractory period? What is the advantage of having a prolonged refractory period?
18. What does an ECG measure? Name the waves produced by an ECG, and state what events occur during each wave.
19. Define systole and diastole.
20. When the atria contract, what percentage of ventricular filling do they accomplish?
21. List the three pressure waves that occur in the atria, and explain what causes them.
22. Explain what happens in the ventricles during the period of isometric contraction, the period of ejection, and immediately following the period of ejection but before ventricular relaxation.
23. Define stroke volume, cardiac output, and peripheral resistance.
24. Explain the production in the aorta of systolic pressure, diastolic pressure, and the aortic incisura.
25. What produces the three heart sounds?
26. Define venous return. How does venous return affect cardiac output and heart rate? State Starling's law of the heart.
27. What part of the brain regulates the heart? Describe the autonomic nerve supply to the heart.
28. What effect do parasympathetic stimulation and sympathetic stimulation have on heart rate, force of contraction, and stroke volume?
29. What neurotransmitters are released by the parasympathetic and sympathetic postganglionic neurons of the heart? What effects do they produce on membrane permeability and excitability?
30. Name the two main hormones that affect the heart. Where are they produced, what causes their release, and what effects do they have on the heart?
31. How does the nervous system detect and respond to the following: (1) a decrease in blood pressure, (2) an increase in blood carbon dioxide, (3) a decrease in blood pH, and (4) a decrease in blood oxygen?
32. Describe the baroreceptor and Bainbridge reflexes.
33. What effect does an increase or decrease in extracellular potassium, calcium, and sodium ions have on heart rate and the force of contraction of the heart?
34. What effect does temperature have on heart rate?

CONCEPT REVIEW

1. Explain why the walls of the ventricles are thicker than the walls of the atria. Why does the left ventricle have a thicker wall than the right ventricle?
2. In most tissues peak blood flow occurs during systole and decreases during diastole. However, in heart tissue the opposite is true, and peak blood flow occurs during diastole. Explain why this difference occurs.
3. A patient has tachycardia. Would you recommend a drug that prolongs or shortens the plateau phase of cardiac muscle cell action potentials?
4. Endurance-trained athletes often have a decreased heart rate. Is this decrease caused by increased or decreased stimulation of the vagus nerve? Explain how this increased or decreased stimulation would decrease heart rate. Would a heart block be more or less likely to develop as a result of the effect of the vagus nerves?
5. A doctor lets you listen to a patient's heart with a stethoscope at the same time that you feel the patient's pulse. Every so often you hear two heartbeats very close together, but you feel only one pulse beat. Later the doctor tells you that the patient has an ectopic focus in the right atrium. Explain why you hear two heartbeats very close together. The doctor also tells you that the patient exhibits a pulse deficit (i.e., the number of pulse beats felt is less than the number of heartbeats heard). Explain why a pulse deficit occurs.
6. Heart rate and cardiac output were measured in a group of nonathletic students. After 2 months of aerobic exercise training their measurements were repeated. It was found that heart rate had decreased, but cardiac output remained the same for many activities. Explain these findings.
7. Explain why atrial fibrillation will not cause death (at least not immediately) but ventricular fibrillation will.

8. Mary Pinnacle records her resting heart rate at sea level. She then goes on a mountain climbing expedition. After reaching the top of the mountain (18,000 feet), she again records her resting heart rate. Predict what change, if any, would be observed, and explain the mechanisms involved.
9. During an experiment in a physiology laboratory a student named C. Saw was placed on a table that could be tilted. The instructor asked the students to predict what would happen to C. Saw's heart rate if the table were tilted so that her head were lower than her feet. Some students predicted an increase in heart rate, and others claimed it would decrease. Can you explain why both predictions might be true?
10. After C. Saw was tilted so that her head was lower than her feet for a few minutes, the table was tilted so that her head was higher than her feet. Predict the effect this change would have on C. Saw's heart rate.

? ?

ANSWERS TO PREDICT QUESTIONS

1 *p. 616.* The heart tissues supplied by the artery would lose oxygen and nutrient supply and would die. That part of the heart and possibly the entire heart would stop functioning. This condition is called a heart attack or coronary.

2 *p. 620.* The heart must continue to function under all conditions and requires energy in the form of ATP. During heavy exercise lactic acid is produced in skeletal muscle as a by-product of anaerobic metabolism. Heart tissue uses lactic acid as an additional energy source.

3 *p. 623.* Contraction of the ventricles, beginning at the apex and moving toward the base of the heart, forces blood out of the bottoms of the ventricles and toward their outflow vessels—the aorta and pulmonary trunk.

4 *p. 624.* Ectopic foci would cause various regions of the heart to contract at different times. As a result, pumping effectiveness would be reduced. Cardiac muscle contraction would not be coordinated, which would interrupt the cyclic filling and emptying of the ventricles.

5 *p. 624.* If cardiac muscle could undergo tetanic contraction, it would contract for a long time without relaxing. Its pumping action then would stop since that action requires alternating contraction and relaxation.

6 *p. 626.* In artificial heart implants it is most important to replace the ventricles because they are the major pumps of the heart. The heart can function fairly well without the pumping action of the atria.

7 *p. 630.* It is important for each ventricle to pump the same amount of blood because, with two connected circulation loops, the blood in one must equal the blood in the other so that one does not become overfilled with blood at the expense of the other.

8 *p. 635.* Sympathetic stimulation increases heart rate. If venous return remains constant, stroke volume will decrease as the number of beats per minute increases. Dilation of the coronary arteries is important because as the heart does more work, the cardiac tissue requires more energy and therefore a greater blood supply to carry more oxygen.

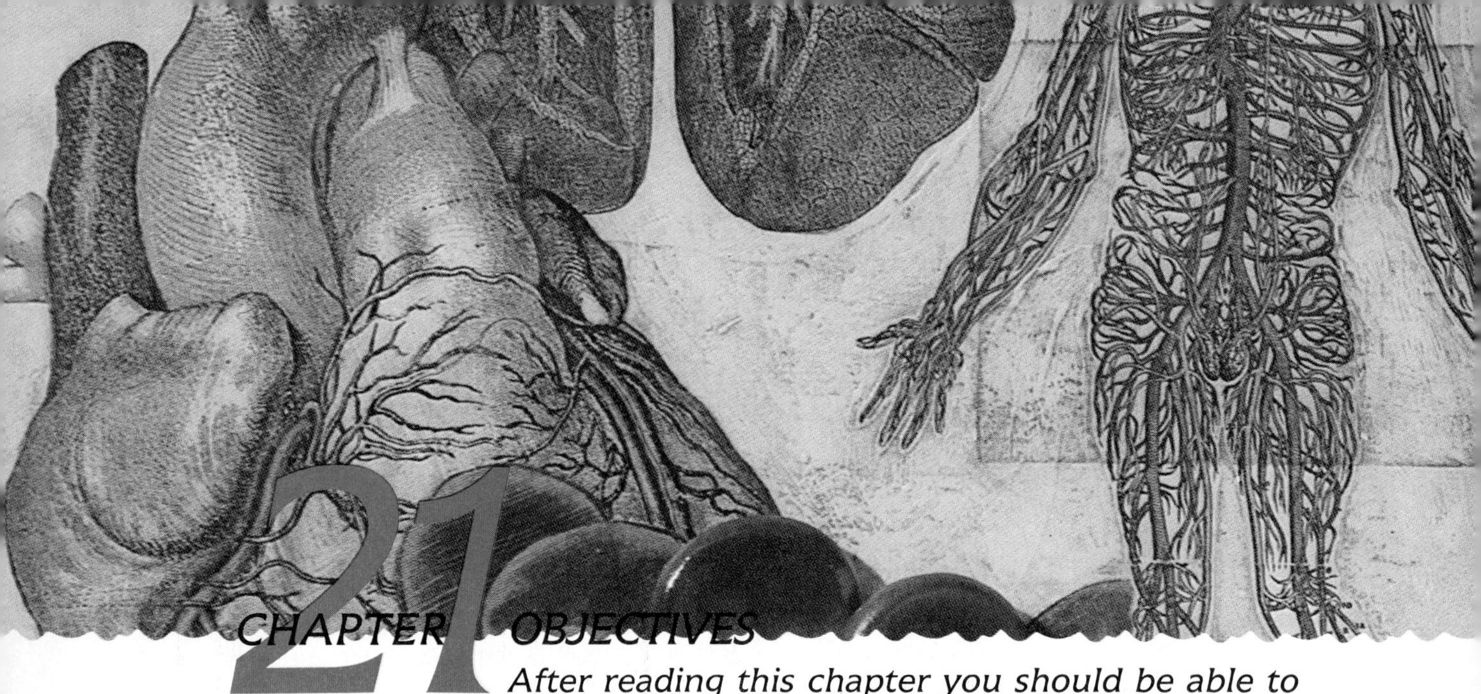

21 CHAPTER OBJECTIVES

After reading this chapter you should be able to

1. Describe the structure and function of capillaries, arteries, and veins.
2. Describe the structural and functional changes that occur in arteries as they age.
3. List the blood vessels of the pulmonary circulation and describe their function.
4. List the major arteries that supply each of the major body areas and describe their functions.
5. List the major veins that carry blood from each of the major body areas and describe their functions.
6. Explain how lymph is formed and transported.
7. List the major lymph vessels and describe their function.
8. Describe how each of the following affects blood flow in vessels: viscosity, laminar and turbulent flow, blood pressure, rate of blood flow, Poiseuille's law, critical closing pressure, law of LaPlace, and vascular compliance.
9. Explain how blood pressure can be measured.
10. Explain how the total cross-sectional area of blood vessels, blood pressure, and resistance to flow change as blood flows through the aorta, small arteries, arterioles, capillaries, venules, small veins, and venae cavae.
11. Describe how the exchange of materials across the capillary occurs and describe how edema can result from decreases in plasma protein and increases in the permeability of the capillary.
12. Describe the functional characteristics of veins.
13. Describe the mechanisms responsible for the local control of blood flow through tissues and explain under what conditions nervous control of blood flow through tissues is important.
14. Describe the short-term and long-term mechanisms that regulate the mean arterial pressure.
15. Define hypertension and explain its effect on the circulatory system.
16. Describe how the circulatory system responds to exercise and shock.

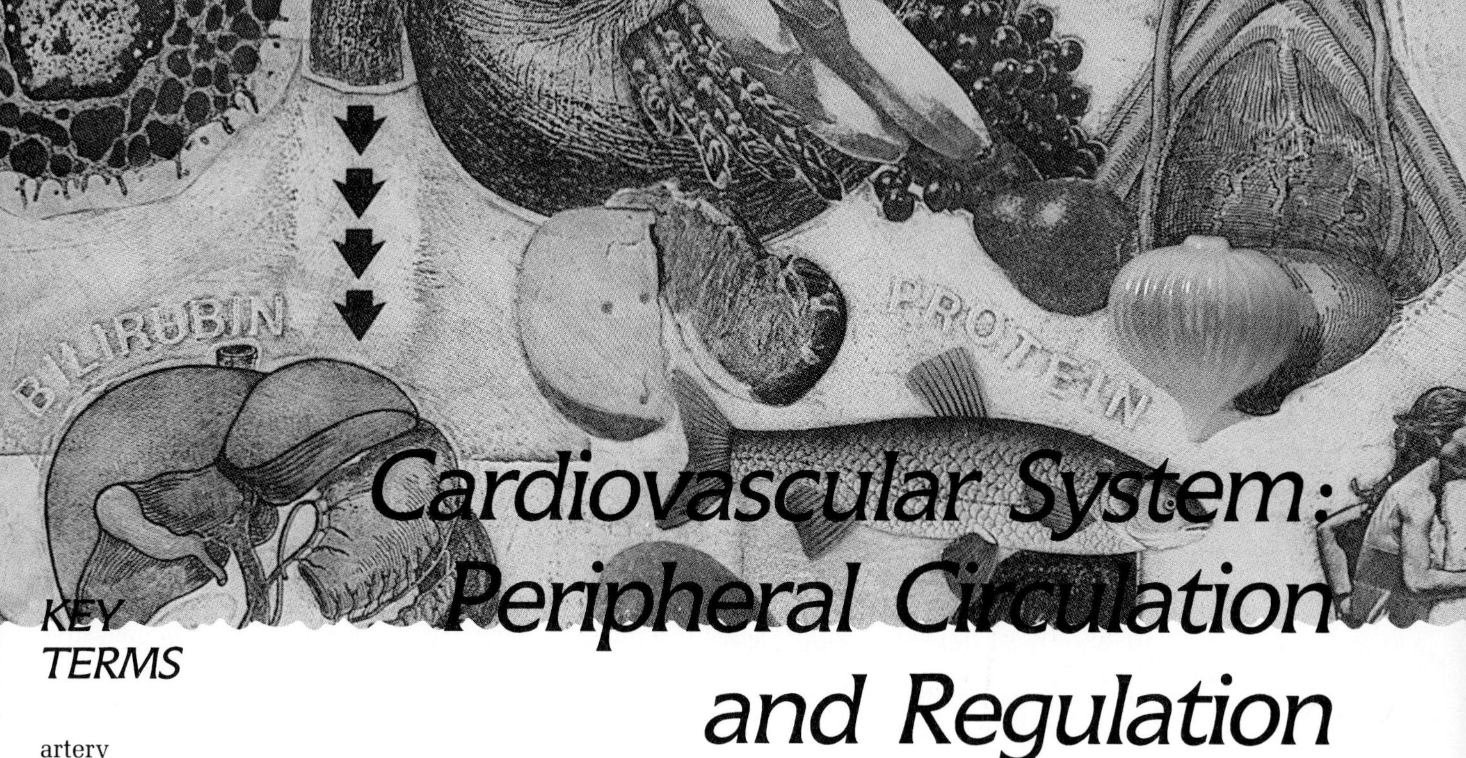

Cardiovascular System: Peripheral Circulation and Regulation

KEY TERMS

artery

autoregulation

baroreceptor reflex

blood pressure

capillary

central nervous system ischemic response

chemoreceptor reflex

Law of LaPlace

lymph

lymph vessels

peripheral resistance

Poiseuille's Law (puah-zuh'yez)

pulse pressure

vascular compliance

vasomotor tone

vein

viscosity

RELATED TOPICS

The following terms or concepts are important for a good understanding of this chapter. If you are not familiar with them, you should review them before proceeding.

Composition of blood (Chapter 19)

Circulation through the heart (Chapter 20)

THE PERIPHERAL CIRCULATORY SYSTEM CAN BE DIVIDED INTO SETS OF BLOOD VESSELS. THE SYSTEMIC VESSELS TRANSPORT BLOOD THROUGH ESSENTIALLY ALL PARTS OF THE BODY FROM THE LEFT VENTRICLE AND BACK TO THE RIGHT ATRIUM. THE PULMONARY VESSELS TRANSPORT BLOOD FROM THE RIGHT VENTRICLE THROUGH THE LUNGS AND BACK TO THE LEFT ATRIUM. Both the blood vessels and the heart are regulated to ensure that the blood pressure is high enough to cause blood flow in sufficient quantities to meet the metabolic needs of the tissues. The cardiovascular system ensures the survival of each tissue type in the body by supplying nutrients and removing waste products from tissues.

GENERAL FEATURES OF BLOOD VESSEL STRUCTURE

Blood is pumped from the ventricles of the heart into large elastic arteries that branch repeatedly to form many progressively smaller arteries. As they become smaller, the arteries undergo a gradual transition from having walls that contain a large amount of elastic tissue and a smaller amount of smooth muscle to having walls with a smaller amount of elastic tissue and a relatively large amount of smooth muscle. Although the arteries form a continuum from the largest to the smallest branches, they normally are classified as (1) elastic arteries, (2) muscular arteries, and (3) arterioles.

Blood flows from the arterioles into the **capillaries.** Most of the exchange that occurs between the interstitial spaces and the blood occurs across the walls of the capillaries. Their walls are the thinnest of all the blood vessels, blood flows through them slowly, and there is a greater number of them than any other blood vessel type.

From the capillaries blood flows into the venous system. When compared to arteries, the walls of the veins are thinner and contain less elastic tissue and fewer smooth-muscle cells. The veins increase in diameter and decrease in number, and their walls increase in thickness as they project toward the heart. They are classified as (1) venules, (2) small veins, and (3) medium-sized and large veins.

Capillaries

All blood vessels have an internal lining of simple squamous epithelial cells called the endothelium, which is continuous with the endocardium of the heart.

The capillary wall consists primarily of endothelial cells (Figure 21-1), which rest on a basement membrane. Outside the basement membrane is a delicate layer of loose connective tissue called the **adventitia** (ad-ven-tish'yah) that merges with the connective tissue surrounding the capillary.

Along the length of the capillary are some scattered cells that are closely associated with the endothelial cells. These scattered cells lie between the basement membrane and the endothelial cells and are called pericapillary cells. They apparently are fibroblasts, macrophages, or undifferentiated smooth muscle cells.

Most capillaries range from 7 to 9 μm in diameter, and they branch without a change in their diameter. Capillaries are variable in length, but in general, they are approximately 1.0 mm long. Red blood cells flow through most capillaries in a single file and frequently are folded as they pass through the smaller-diameter capillaries.

Types of capillaries

Capillaries can be classified as continuous, fenestrated, or sinusoidal, depending on their diameter and their permeability characteristics.

Continuous capillaries are approximately 7 to 9 μm in diameter, and their walls exhibit no gaps between the endothelial cells. Continuous capillaries are less permeable to large molecules than are other capillary types and are in muscle, nervous tissue, and many other locations.

In **fenestrated** (fen'es-trāt'ed) **capillaries** endothelial cells have numerous fenestrae. The fenestrae are areas approximately 70 to 100 nm in diameter in which the cytoplasm is absent and the cell membrane consists of a porous diaphragm that is thinner than the normal cell membrane. Fenestrated capillaries are in tissues in which capillaries are highly permeable (e.g., in the intestinal villi, ciliary process of the eye, choroid plexuses of the central nervous system, and glomeruli of the kidney).

Sinusoidal (si'nŭ-soy'dal) **capillaries** are larger in diameter than either continuous or fenestrated capillaries, and their basement membrane is less prominent. Their fenestrae are larger than ones in fenestrated capillaries. They are in places such as endocrine glands, where large molecules cross the wall of the capillary.

Sinusoids are large-diameter sinusoidal capillaries. Their basement membrane is sparse and often missing, and their structure suggests that large molecules and sometimes cells can move readily across their walls between the endothelial cells. Sinusoids

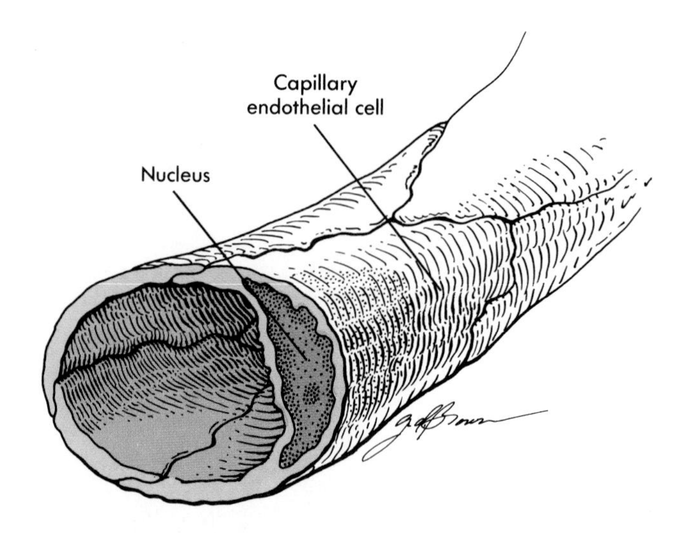

FIGURE 21-1 Section of a capillary showing that it is composed of flattened endothelial cells.

are common in the liver and the bone marrow. Macrophages are closely associated with the endothelial cells of the liver sinusoids. **Venous sinuses** are similar in structure to the sinusoidal capillaries, but they are even larger in diameter. They are primarily in the spleen, and they have large gaps between the endothelial cells that comprise their walls.

Substances cross capillary walls by diffusing through the endothelial cells, through fenestrae, and between the endothelial cells. Lipid-soluble substances such as oxygen and carbon dioxide and small water-soluble molecules readily diffuse through the cell membrane. Larger water-soluble substances must pass through the fenestrae or the gaps between the endothelial cells. In addition, transport by pinocytosis occurs, but little is known about its role in the capillaries. Because only certain substances can pass through the walls of capillaries, they are effective permeability barriers.

Capillary network

Arterioles supply blood to each capillary network (Figure 21-2). Blood then flows through the capillary network and into the venules. The ends of capillaries closest to the arterioles are **arterial capillaries,** and the ends closest to venules are **venous capillaries.**

Blood flows from arterioles through **metarterioles** (met'ar-tēr'e-ōlz), which have isolated smooth-muscle cells along their wall. From a metarteriole blood flows into a **thoroughfare channel** that extends in a relatively direct fashion from a metarteriole to a venule. Blood flow through thoroughfare channels is relatively continuous. Several capillaries branch from the thoroughfare channels, and in these branches blood flow is intermittent. Flow in these capillaries is regulated by smooth-muscle cells called **precapillary sphincters,** which are located at the origin of the branches (see Figure 21-2).

Capillary networks are more numerous and more extensive in highly metabolic tissues such as the lung, liver, kidney, skeletal muscle, and cardiac muscle. Capillary networks in the skin have many more thoroughfare channels than capillary networks in cardiac or skeletal muscle. Capillaries in the skin function in thermoregulation, and heat loss results from the flow of a large volume of blood through them. In muscle, however, nutrient and waste product exchange is the major function of the capillaries.

Structure of Arteries and Veins

General features

Except for the capillaries and the venules, the blood vessel walls consist of three relatively distinct layers, which are most apparent in the medium-sized or muscular arteries and are least apparent in the veins. From the lumen to the outer wall of the blood vessels the layers, or **tunics,** are (1) the tunica intima, (2) the tunica media, (3) and the tunica adventitia, or tunica externa (Figure 21-3).

The **tunica intima** consists of endothelium, a delicate connective tissue basement membrane, a lamina propria, and a fenestrated layer of elastic fibers

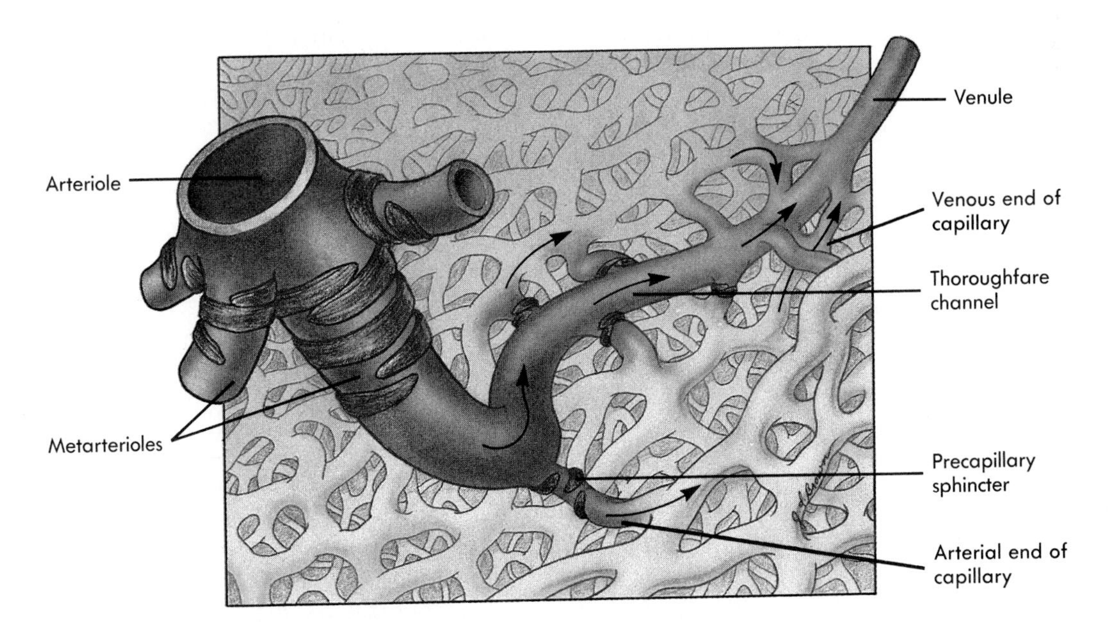

FIGURE 21-2 Capillary network. The metarteriole, giving rise to the network, feeds directly from an arteriole into the thoroughfare channel, which feeds into the venule. The network forms numerous branches that transport blood from the thoroughfare channel and may return to the thoroughfare channel.

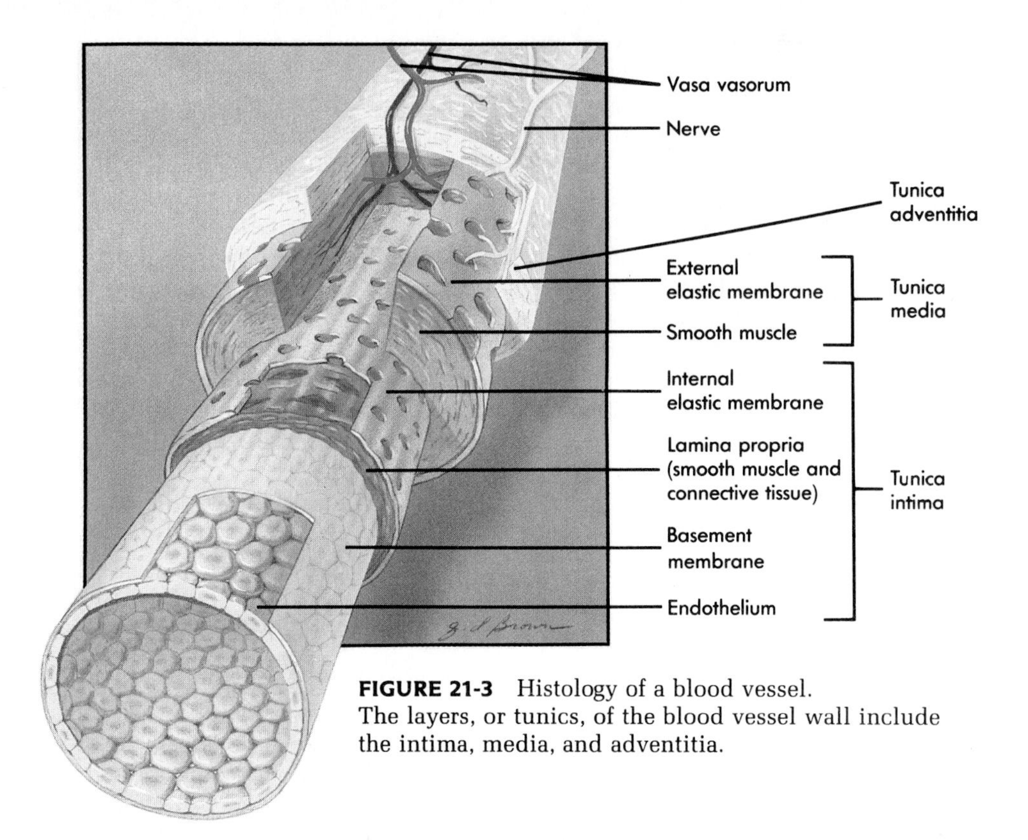

FIGURE 21-3 Histology of a blood vessel. The layers, or tunics, of the blood vessel wall include the intima, media, and adventitia.

called the **internal elastic membrane.** The internal elastic membrane separates the tunica intima from the next layer, the tunica media.

The tunica media, or middle layer, consists of smooth muscle cells arranged circularly around the blood vessel. It also contains variable amounts of elastic and collagen fibers, depending on the size of the vessel. At the outer border of the tunica media in certain arteries, an external elastic membrane, which separates the tunica media from the tunica adventitia, can be identified. A few longitudinally oriented smooth muscle cells are in some arteries near the tunica intima.

The **tunica adventitia** is composed of connective tissue, which varies from dense connective tissue that is near the tunica media and contains large amounts of collagen to loose connective tissue that merges with the connective tissue surrounding the blood vessels.

The relative thickness and composition of each layer varies with the diameter of the blood vessel and its type. The transition from one artery type or from one vein type to another is gradual, as are the structural changes.

Large elastic arteries

Elastic arteries are the largest diameter arteries (Figure 21-4, *A*) and often are called conducting arteries. The pressure is relatively high in these vessels, and it fluctuates between systolic and diastolic values. A greater amount of elastic tissue and a smaller amount of smooth muscle are in their walls when compared to other arteries. The elastic fibers are responsible for the elastic characteristics of the blood vessel wall, but collagenous connective tissue determines the degree to which the arterial wall can be stretched.

The tunica intima is relatively thick. The internal and external elastic membranes merge with the elastic fibers of the tunica media and are not recognizable as distinct layers. The tunica media consists of a meshwork of elastic fibers with interspersed circular smooth-muscle cells and some collagen fibers. The tunica adventitia is relatively thin.

Medium-sized and small arteries (muscular arteries)

The medium-sized arteries can be observed in a gross dissection, and they comprise most of the smaller arteries with names. Their walls are relatively thick when compared to their diameter mainly because of the smooth-muscle cells in the tunica media (Figure 21-4, *B*). The tunica intima has a well-developed internal elastic membrane, and the tunica media contains 25 to 40 layers of smooth muscle. Only the larger medium-sized arteries have an external elastic membrane. The tunica adventitia is composed of a relatively thick layer of collagenous connective tissue that blends with the surrounding connective tissue. Medium-sized arteries frequently are called **distributing arteries** because the smooth muscles allow these vessels to partially regulate

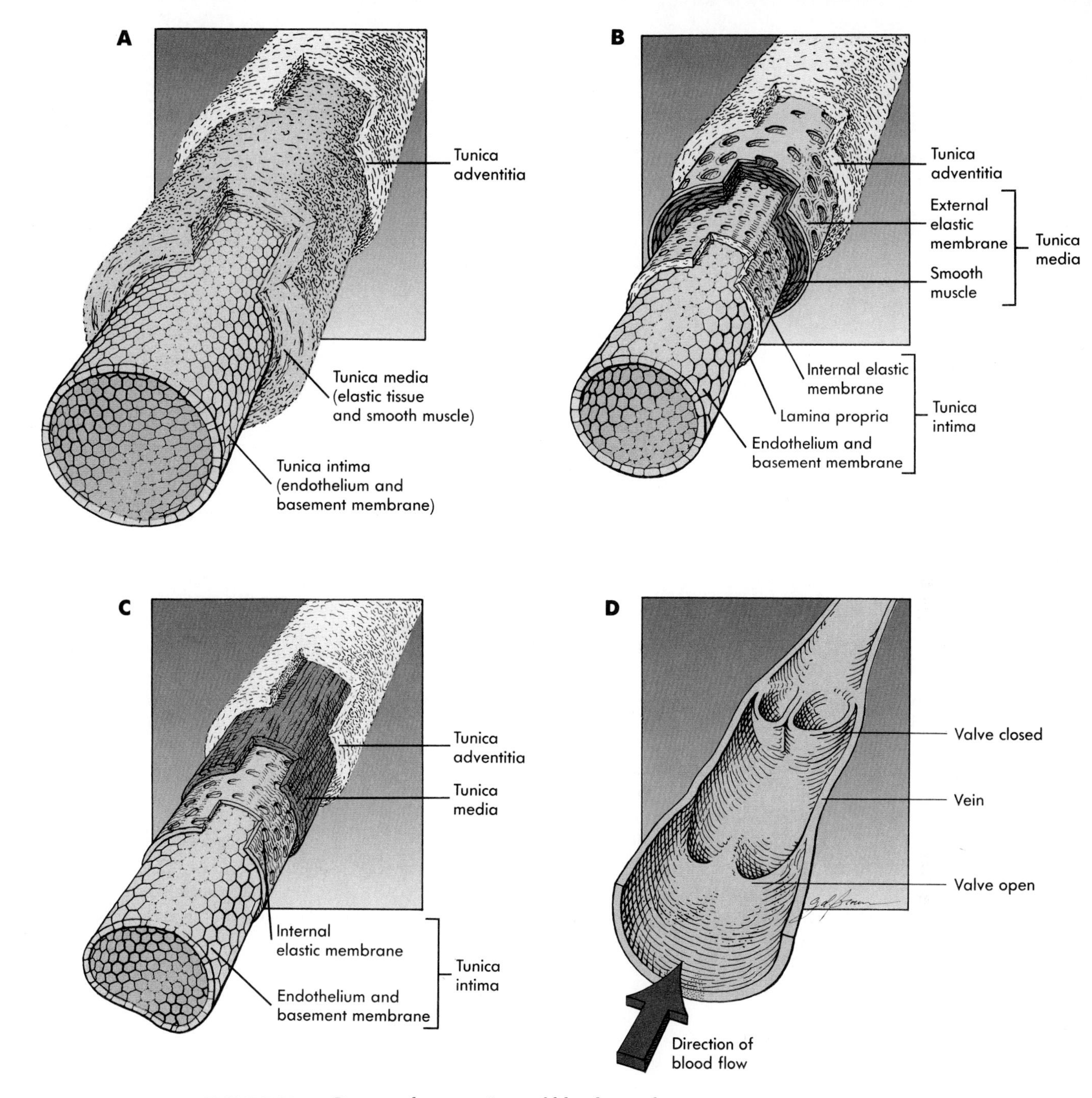

FIGURE 21-4 Structural comparison of blood vessel types.
A An elastic artery. **B** A medium-sized artery. **C** A medium-sized vein.
D Valves in veins.

blood supply to different regions of the body by either constricting or dilating.

The small arteries vary in diameter from approximately 40 to 300 μm. Arteries that are 40 μm in diameter have approximately three or four layers of smooth muscle in their tunica media, whereas arteries that are 300 μm have essentially the same structure as the medium-sized arteries. The small arteries are adapted for increasing (vasodilation) and decreasing (vasoconstriction) their diameter.

Arterioles

The **arterioles** transport blood from small arteries to capillaries and are the smallest arteries in which the three tunics can be identified. The tunica intima consists of endothelium and the underlying connective tissue, but there is no observable internal elastic membrane. The tunica media consists of one or two layers of circular smooth-muscle cells, and the tunica adventitia consists of collagenous connective tissue.

The arterioles, like the small arteries, are capable of vasodilation and vasoconstriction.

Venules and small veins

Venules, with a diameter of 40 to 50 μm, are tubes composed of endothelium resting on a delicate basement membrane. Their structure, except for their diameter, is very similar to that of capillaries. A few isolated smooth-muscle cells exist outside the endothelial cells, especially in the larger venules. As the vessels increase to 0.2 to 0.3 mm in diameter, the smooth-muscle cells form a continuous layer; the vessels then are called **small veins.** The small veins also have a tunica adventitia composed of collagenous connective tissue.

The venules collect blood from the capillaries and transport it to the small veins, which in turn transport it to the medium-sized veins. Nutrient exchange occurs across the walls of the venules, but as the walls of the small veins increase in thickness, the degree of nutrient exchange decreases.

Varicose veins result from incompetent valves that are caused by stretching of the veins in the legs. The veins become so dilated that the flaps of the venous valves no longer overlap or prevent the backflow of blood. As a consequence, the venous pressure is greater than normal in the veins of the legs, resulting in edema. Blood flow in the veins can become sufficiently stagnant that the blood clots. The condition can result in **phlebitis** (flĕ-bi'tis; inflammation of the veins), and if the condition becomes sufficiently severe, it can lead to **gangrene** (tissue death caused by a reduction or loss of blood supply).

Some people have a genetic propensity for the development of varicose veins. For women with that genetic propensity, conditions that increase the pressure in veins cause them to stretch, and varicose veins can develop. An example is pregnancy in which there is increased venous pressure in the veins that drain the lower limbs because of compression of the veins by the expanded uterus.

Medium-sized and large veins

Most of the veins observed in gross anatomical dissections, except for the large venous trunks, are **medium-sized veins.** They collect blood from small veins and deliver it to the large venous trunks. The **large veins** transport blood from the medium-sized veins to the heart. Their tunica intima is thin and consists of endothelial cells, a relatively thin layer of collagenous connective tissue, and a few scattered elastic fibers. There may be a poorly developed internal elastic membrane. The tunica media is also thin and is composed of a thin layer of circularly arranged smooth muscle cells and collagen fibers, plus a few sparsely distributed elastic fibers. The tunica adventitia is composed of collagenous connective tissue and is the predominant layer (Figure 21-4, *C*).

Valves

Veins having diameters greater than 2 mm contain **valves** that allow blood to flow toward the heart but not in the opposite direction (Fig. 21-4, *D*). The valves consist of folds in the tunica intima that form two flaps that are shaped like and function like the semilunar valves of the heart. The two folds overlap in the middle of the vein so that when blood attempts to flow in a reverse direction, they occlude the vessel. There are many valves in the medium-sized veins, and the number is greater in veins of the lower extremities than in veins of the upper extremities.

Vasa vasorum

For arteries and veins greater than 1 mm in diameter, nutrients cannot diffuse from the lumen of the vessel to all of the layers of the wall. Therefore nutrients are supplied to their walls by way of small blood vessels called **vasa vasorum,** which penetrate from the exterior of the vessel to form a capillary network in the tunica adventitia and the tunica media (see Fig. 21-3).

Arteriovenous anastomoses

Arteriovenous anastomoses (ah-nas'to-mo-sēz) allow blood to flow from arteries to veins without passing through capillaries. The arterioles directly enter the small veins without an intermediate capillary. A **glomus** (glo'mus) is an arteriovenous anastomosis that consists of arterioles arranged in a convoluted fashion surrounded by collagenous connective tissue.

Naturally occurring arteriovenous anastomoses are present in large numbers in the sole of the foot, the palm of the hand, the terminal phalanges, and the nail beds. They function in temperature regulation. Pathological arteriovenous anastomoses can result from injury or tumors. They cause the direct flow of blood from arteries to veins and can, if they are sufficiently severe, lead to heart failure because of the tremendous increase in venous return to the heart.

Nerves

The walls of most blood vessels are richly supplied by unmyelinated sympathetic nerve fibers (see Figure 21-3), although some blood vessels are innervated by parasympathetic fibers (such as those in the penis or clitoris). The nerve fibers project among the smooth-muscle cells of the tunica media and form synapses consisting of enlargements of the nerve fiber axons. The small arteries and arterioles are innervated to a greater extent than other blood vessel types. The response of the blood vessels to nervous stimulation is either vasoconstriction or vasodilation; the more common response is vasoconstriction.

The smooth-muscle cells of blood vessels act to some extent as a functional unit. There are frequent gap junctions between adjacent smooth-muscle cells; as a consequence, stimulation of a few smooth-muscle cells in the vessel wall results in constriction of a relatively large segment of the blood vessel.

A few myelinated sensory nerves also innervate some blood vessels. They are especially important in blood vessels that monitor stretch in the blood vessel wall and detect changes in blood pressure.

Aging of the Arteries

The walls of all arteries undergo changes as they age, although some arteries change more rapidly than others and some individuals are more susceptible to change than others. The most significant changes occur in the large elastic arteries (e.g., the aorta, large arteries that carry blood to the brain, and the coronary arteries). The age-related changes described here refer to these blood vessel types. Changes in muscular arteries do occur, but they are less dramatic and often do not result in disruption of normal blood vessel function.

Degenerative changes in arteries that make them less elastic are referred to collectively as **arteriosclerosis** (ar-tēr'ĭ-o-sklĕ-ro'sis; hardening of the arteries). These changes occur in nearly every individual, and they become more severe with advancing age. A related term, **atherosclerosis,** refers to the deposition of material in the walls of arteries to form plaques.

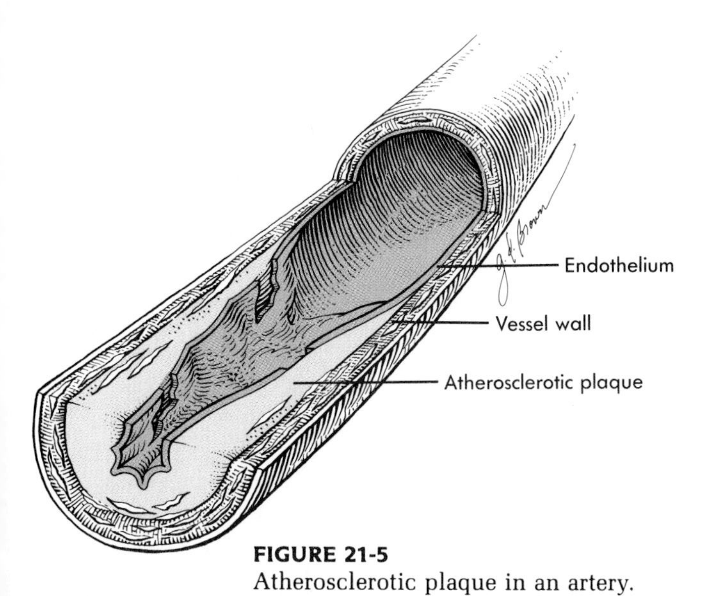

FIGURE 21-5
Atherosclerotic plaque in an artery.

The material is a fatlike substance containing cholesterol (Figure 21-5). The fatty material may be replaced later with dense connective tissue and calcium deposits. The initial signs of arteriosclerosis have been identified in the arteries of people in their teens, and it develops earlier and progresses more rapidly in some individuals than in others.

Arteriosclerosis is characterized by a thickening of the tunica intima and a chemical change in the elastic fibers of the tunica media, making the tunica media less elastic. Fat gradually accumulates between the elastic and collagen fibers to produce a lesion that protrudes into the lumen of the vessel and eventually may hamper normal blood flow. In advanced forms of arteriosclerosis calcium deposits, primarily in the form of calcium carbonate, accumulate in the walls of the blood vessels.

Arteriosclerosis greatly increases resistance to blood flow. Advanced arteriosclerosis, as a consequence, adversely affects the normal circulation of blood and greatly increases the work performed by the heart.

Some investigators think that arteriosclerosis may not be a pathological process. Instead, they think it may be simply an aging or wearing out process. Some recent evidence suggests that arteriosclerosis may be an autoimmune disease. In either case there are several factors that increase the rate at which it develops. Obesity, high dietary cholesterol and other fat consumption, and smoking are some of the factors correlated with the premature development of arteriosclerosis.

PULMONARY CIRCULATION

Blood from the right ventricle is pumped into the **pulmonary** (pul'mo-nĕr-e; relating to the lungs) **trunk**. This short vessel (5 cm long) bifurcates into the right and left **pulmonary arteries,** one transporting blood to each lung. Within the lungs gas exchange occurs between air in the lungs and blood. Two **pulmonary veins** exit each lung and enter the left atrium.

SYSTEMIC CIRCULATION: ARTERIES

Oxygenated blood entering the heart from the pulmonary veins passes through the left atrium into the left ventricle and from the left ventricle into the aorta. Blood is distributed from the aorta to all portions of the body (Figure 21-6).

Aorta

All **arteries** of the systemic circulation are derived either directly or indirectly from the **aorta** (a-or'tah), which usually is divided into three general

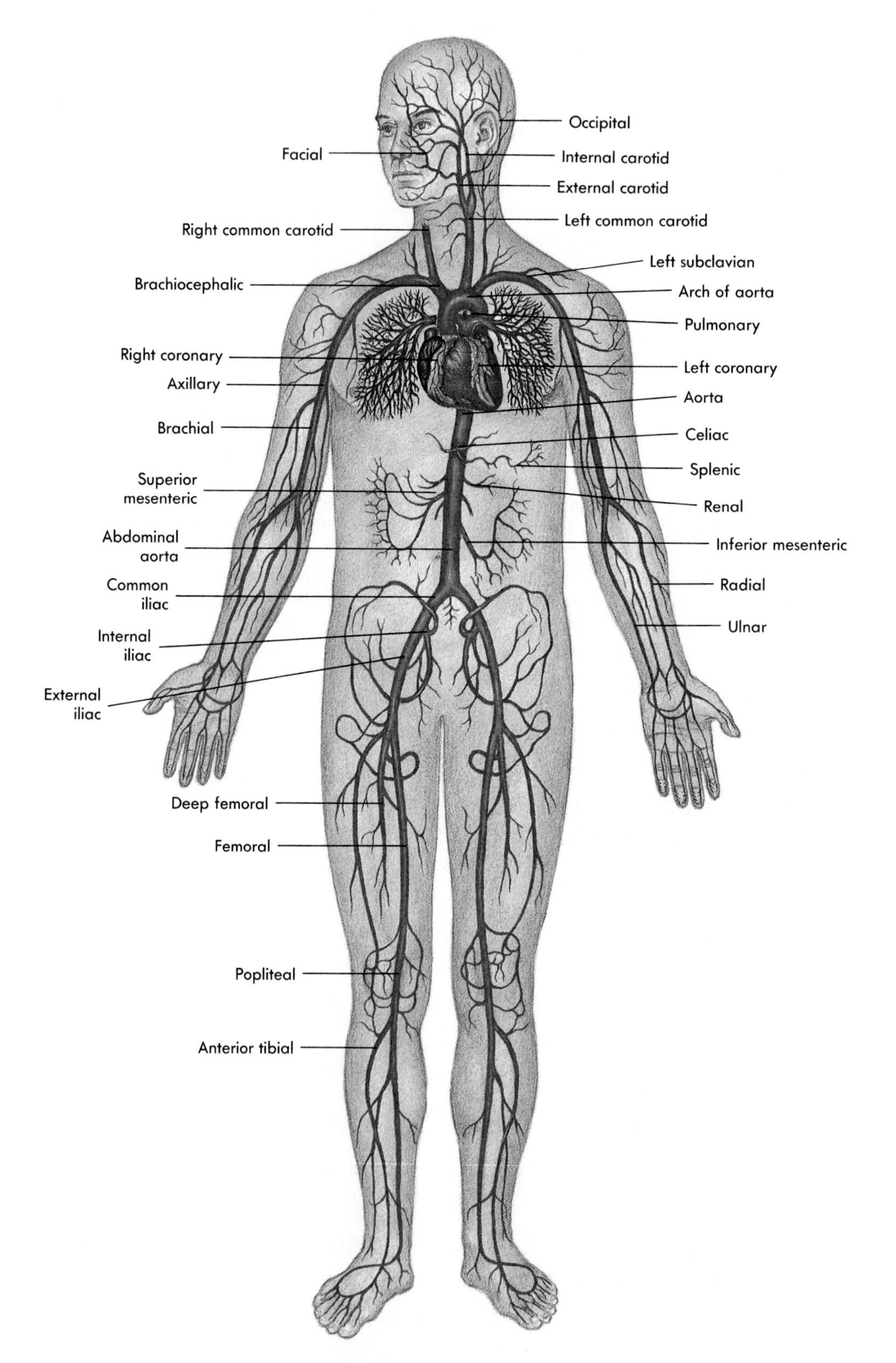

FIGURE 21-6 The major arteries. The arteries carry blood from the heart to the tissues of the body.

portions: the ascending aorta, the aortic arch, and the descending aorta. The descending aorta is divided further into a thoracic aorta and an abdominal aorta (see Figure 21-9).

At its origin from the left ventricle the aorta is approximately 2.8 cm in diameter. Because it passes superiorly from the heart, this portion is called the **ascending aorta.** It is approximately 5 cm long and has only two arteries branching from it, the right and left **coronary arteries,** which supply blood to the cardiac muscle.

The aorta then arches posteriorly and to the left as the **aortic arch.** Three major branches, which carry blood to the head and upper limbs, originate from the aortic arch: the brachiocephalic artery, the left common carotid artery, and the left subclavian artery.

Trauma that ruptures the aorta is almost immediately fatal. However, trauma may also lead to an **aneurysm** (a bulge caused by a weakened spot in the aortic wall) that leaks blood slowly into the thorax; this condition is not immediately fatal, but it must be corrected surgically. The majority of traumatic aortic arch ruptures occur in automobile accidents and result from the great force with which the body is thrown into the steering wheel, dashboard, or other objects. Waist-type safety belts alone do not prevent this type of injury as effectively as shoulder-type safety belts.

The next portion of the aorta is the **descending aorta.** It is the longest portion of the aorta and extends through the thorax on the left side of the mediastinum and through the abdomen to the superior margin of the pelvis. The thoracic aorta is that portion of the descending aorta located in the thorax. It has several branches that supply various structures between the aortic arch and the diaphragm. The abdominal aorta is that portion of the descending aorta between the diaphragm and the point at which the aorta ends by dividing into the two **common iliac** (il'e-ak; relating to the flank area) **arteries.** The abdominal aorta has several branches that supply the abdominal wall and organs. Its terminal branches, the common iliac arteries, supply blood to the pelvis and the lower limbs.

Coronary Arteries

The **coronary** (kor'o-năr-e; encircling the heart like a crown) **arteries,** which are the only branches of the ascending aorta, were described in Chapter 20.

Arteries to the Head and the Neck

The first vessel to branch from the aortic arch is the **brachiocephalic** (bra'ke-o-sĕ-fal'ik; vessel to the arm and head) **artery.** It is a very short artery and branches at the level of the clavicle to form the **right common carotid** (kah-rot'id) **artery,** which transports blood to the right side of the head and the neck, and the **right subclavian artery,** which transports blood to the right upper limb (Figure 21-7, *A*).

The second and third branches of the aortic arch are the **left common carotid artery,** which transports blood to the left side of the head and the neck, and the **left subclavian artery,** which transports blood to the left upper limb.

The common carotid arteries extend superiorly, without branching, along either side of the neck from their base to the inferior angle of the mandible at which each artery branches into **internal** and **external carotid arteries** (see Figure 21-7, *A*). At the point of bifurcation the common carotid arteries and the base of the internal carotid arteries are dilated slightly to form the **carotid sinuses,** structures important in monitoring blood pressure (baroreceptor reflex). The external carotid arteries have several branches that supply the structures of the face, nose, and mouth. The internal carotid arteries, together with the vertebral arteries, which are branches of the subclavian arteries, supply the brain (Table 21-1; see Figure 21-7, *A*).

1

The term carotid means to put to sleep, implying that if the carotid arteries are occluded for even a short time, the patient could lose consciousness (go to sleep). The blood supply to the brain is extremely important to its function. Elimination of this supply for even a relatively short time can result in permanent brain damage because the brain is dependent on oxidative metabolism and quickly malfunctions in the absence of oxygen. What is the physiological significance of arteriosclerosis, which slowly reduces blood flow through the carotids?

Branches of the subclavian arteries, the **left** and **right vertebral arteries,** enter the cranial vault through the foramen magnum, give off arteries to the cerebellum, and then unite to form a single, midline **basilar artery** (Figure 21-7, *B*). The basilar artery gives off branches to the pons and the cerebellum and then bifurcates to form the **posterior cerebral**

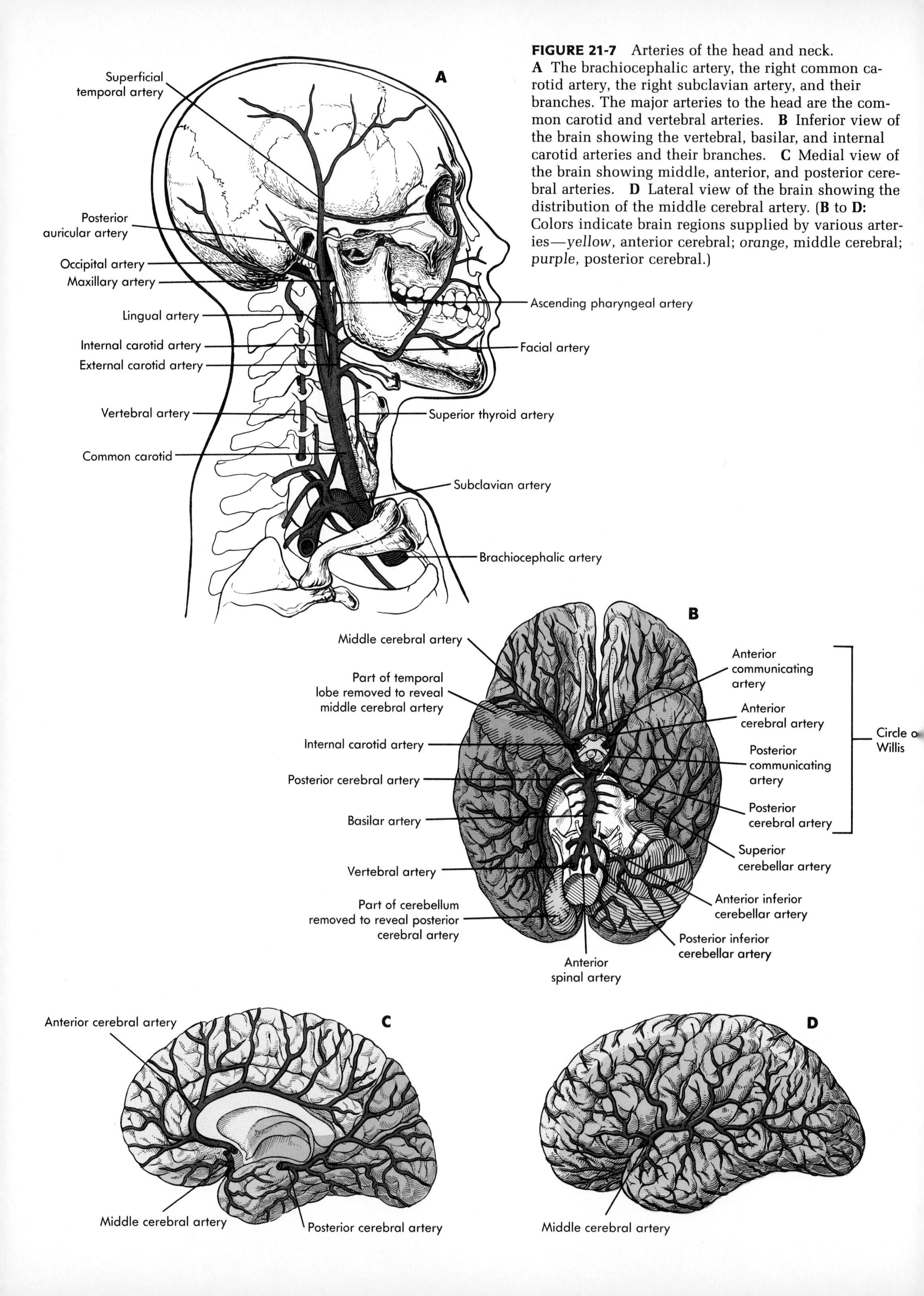

FIGURE 21-7 Arteries of the head and neck. **A** The brachiocephalic artery, the right common carotid artery, the right subclavian artery, and their branches. The major arteries to the head are the common carotid and vertebral arteries. **B** Inferior view of the brain showing the vertebral, basilar, and internal carotid arteries and their branches. **C** Medial view of the brain showing middle, anterior, and posterior cerebral arteries. **D** Lateral view of the brain showing the distribution of the middle cerebral artery. (**B** to **D**: Colors indicate brain regions supplied by various arteries—*yellow*, anterior cerebral; *orange*, middle cerebral; *purple*, posterior cerebral.)

arteries, which supply the posterior portion of the cerebrum (Figure 21-7, *C*).

The internal carotid arteries enter the cranial vault through the carotid canals and terminate by forming the **middle cerebral arteries,** which supply large portions of the lateral cerebral cortex (Figure 21-7, *D*). Posterior branches of these arteries, the **posterior communicating arteries,** unite with the posterior cerebral arteries; and anterior branches, the **anterior cerebral arteries,** supply blood to the frontal lobes of the brain. The anterior cerebral arteries are in turn connected by an **anterior communicating artery,** which completes a circle around the pituitary gland and base of the brain called the **circle of Willis** (see Figure 21-7, *B*).

A **stroke** is a sudden neurological disorder often caused by a decreased blood supply to a portion of the brain. It may occur as a result of a **thrombosis** (throm-bo'sis; a stationary clot), an **embolism** (em'bo-lizm; a floating clot that becomes lodged in smaller vessels), or **hemorrhage** (hem'ŏ-rij; rupture or leaking of blood from vessels). Any one of these conditions can result in a loss of blood supply or in trauma to a portion of the brain. As a result, the tissue normally supplied by the arteries becomes **necrotic** (nĕ-krot'ik; dead). The affected area is called an **infarct** (in'farkt; to stuff into, an area of cell death). The neurological results of a stroke are described in Chapter 13.

Table 21-1

Arteries of the head and neck (see Figure 21-7)

ARTERIES	TISSUES SUPPLIED
Common Carotid Arteries	Head and neck by branches listed below
External carotid	
Superior thyroid	Neck, larynx, and thyroid gland
Lingual	Tongue, mouth, and submandibular and sublingual glands
Facial	Mouth, pharynx, and face
Occipital	Posterior head and neck and meninges around posterior brain
Posterior auricular	Ear, inner ear, head, and neck
Ascending pharyngeal	Deep neck muscles, middle ear, pharynx, soft palate, and meninges around posterior brain
Superficial temporal	Temple, face, and anterior ear
Maxillary	Middle and inner ears, meninges, lower jaw and teeth, upper jaw and teeth, temple, external eye structures, face, palate, and nose
Internal carotid	
Posterior communicating	Joins the posterior cerebral artery
Anterior cerebral	Anterior portions of the brain and forms the anterior communicating arteries
Middle cerebral	Most of the lateral surface of the brain
Vertebral Arteries	
(branches of the subclavian arteries)	
Anterior spinal	Anterior spinal cord
Posterior inferior cerebellar	Cerebellum, fourth ventricle, and posterior plexus
Basilar Artery	
(formed by junction of vertebral arteries)	
Anterior inferior cerebellar	Cerebellum
Superior cerebellar	Cerebellum and midbrain
Posterior cerebral	Posterior portions of the brain

Arteries of the Upper Limb

The three major arteries of the upper limb, **subclavian** (sub-kla've-an; below the clavicle), **axillary** (ak'sĭ-lar'e; in the axilla), and **brachial** (bra'ke-al; in the arm) are a continuum rather than a branching system. The axillary artery is the continuation of the subclavian artery, and the brachial artery is a continuation of the axillary artery. The subclavian artery is located deep to the clavicle, the axillary artery is within the axilla, and the brachial artery lies within the arm itself (Table 21-2; Figure 21-8).

The brachial artery divides into **ulnar** and **radial arteries,** which form two arches within the palm of the hand referred to as the superficial and deep palmar arches. The **superficial palmar arch** is formed by the ulnar artery and is completed by anastomosing with the radial artery. The **deep palmar arch** is formed by the radial artery and is completed by anastomosing with the ulnar artery. This arch is not only deep to the superficial arch, but it is proximal as well.

Digital (dij'ĭ-tal; relating to the digits—the fingers and the thumb) **arteries** branch from each of the two palmar arches and unite to form single arteries on the medial and lateral sides of each digit.

Table 21-2

Arteries of the upper limbs (see Figure 21-8)

ARTERIES	TISSUES SUPPLIED
Subclavian Arteries	
(right subclavian originates from the brachiocephalic artery, and left subclavian originates directly from the aorta)	
Vertebral	Spinal cord and cerebellum (see Table 21-3)
Internal thoracic	Diaphragm, mediastinum, pericardium, anterior thoracic wall, and anterior abdominal wall
Thyrocervical trunk	Inferior neck and shoulder
Axillary Arteries	
(continuation of subclavian)	
Thoracoacromial	Pectoral region and shoulder
Lateral thoracic	Pectoral muscles, mammary gland, and axilla
Subscapular	Scapular muscles
Brachial Arteries	
(continuation of axillary arteries)	
Deep brachial	Arm and humerus
Radial	Forearm
Deep palmar arch	Hand and fingers
Digital arteries	Fingers
Ulnar	Forearm
Superficial palmar arch	Hand and fingers
Digital arteries	Fingers

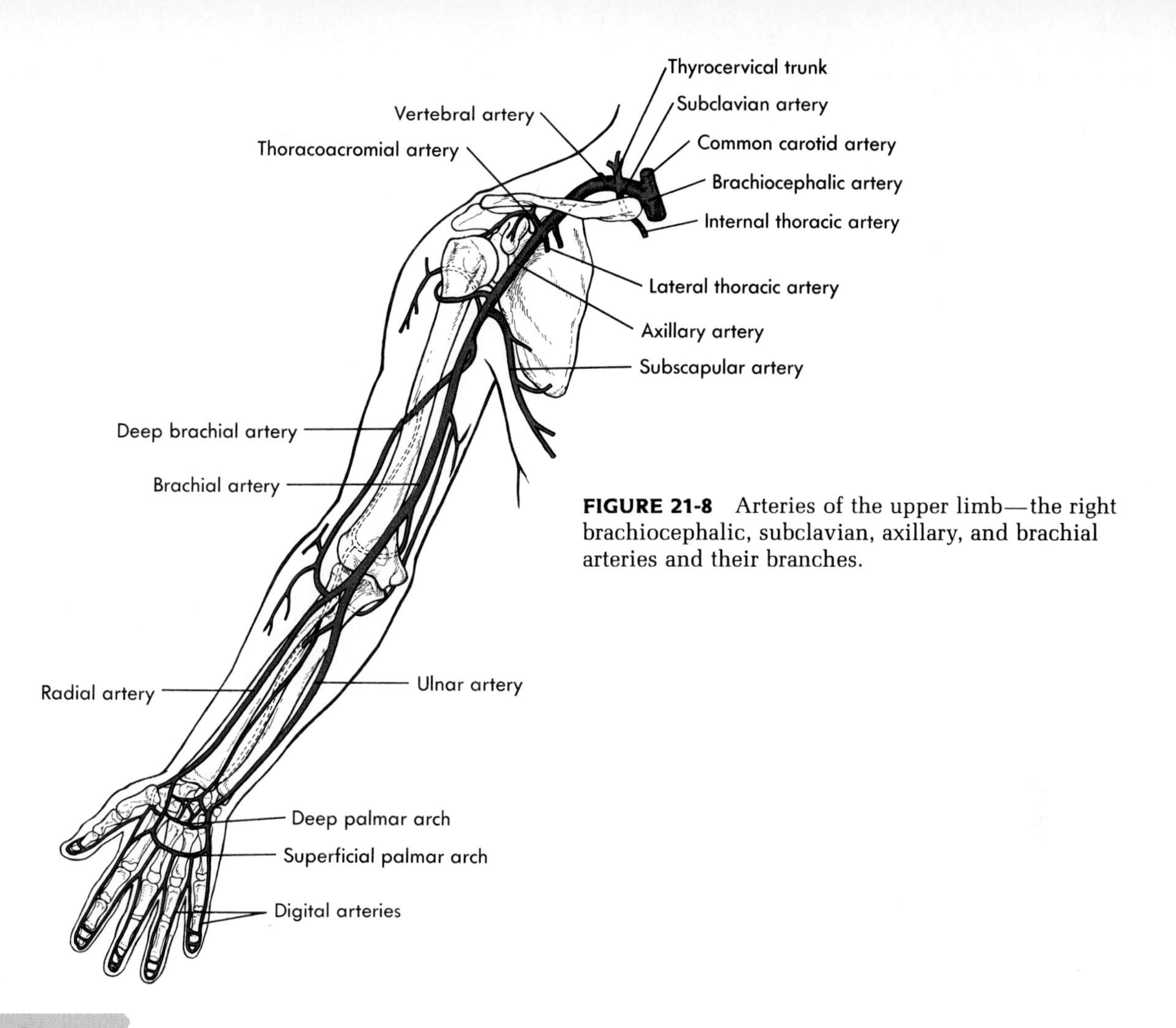

FIGURE 21-8 Arteries of the upper limb—the right brachiocephalic, subclavian, axillary, and brachial arteries and their branches.

Pulse

A pulse can be felt at locations where large arteries are close to the surface of the body. It is helpful to know where the major pulses can be detected because monitoring the pulse is important clinically. The heart rate, rhythmicity, and other characteristics can be determined by feeling the pulse.

A pulse can be felt at three major locations on each side of the head and neck. One site is the common carotid artery at the point where it divides into internal and external carotid arteries. A second is the superficial temporal artery immediately anterior to the ear. A third is in the facial artery at the point where it crosses the inferior border of the mandible approximately midway between the angle and the genu (Figure 21-A).

A pulse can be felt at three major points in the upper limb: in the axilla, in the brachial artery on the medial side of the arm slightly proximal to the elbow, and in the radial artery on the lateral side of the anterior forearm just proximal to the wrist. The radial artery is by tradition the most common site for detecting the pulse of a patient because it is the most easily accessible pulse in the body.

A pulse may be felt easily at the femoral artery in the groin, the popliteal artery just proximal to the knee, and the dorsalis pedis artery and the posterior tibial artery at the ankle (Figure 21-A).

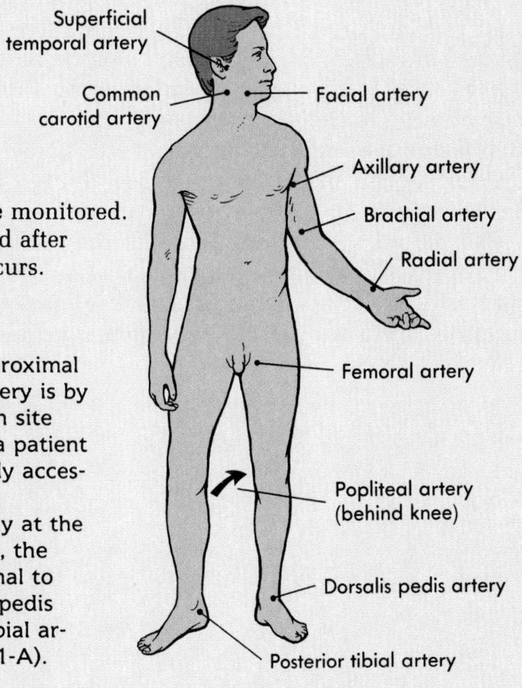

FIGURE 21-A
Location of major points at which the pulse can be monitored. Each pulse point is named after the artery on which it occurs.

Thoracic Aorta and Its Branches

The branches of the thoracic aorta can be divided into two groups: the **visceral arteries** supplying the thoracic organs and the **parietal arteries** supplying the thoracic wall (Table 21-3; Figure 21-9). The visceral branches supply the lungs, esophagus, and pericardial sac. Even though the lungs have a large quantity of blood flowing through them, the lung tissue requires a separate oxygenated blood supply from the left ventricle through small bronchial branches from the thoracic aorta.

The walls of the thorax are supplied with blood by the **intercostal** (in'ter-kos'tal; between the ribs) **arteries,** which consist of two sets, the anterior intercostals and the posterior intercostals. The **anterior intercostals** are derived from the **internal thoracic arteries,** which are branches of the **subclavian arteries,** and lie on the inner surface of the anterior thoracic wall (see Figure 21-9). The **posterior intercostals** are derived as bilateral branches directly from the descending aorta. The anterior and posterior intercostal arteries lie along the inferior margin of each rib and anastomose with each other approximately midway between the ends of the ribs. **Superior phrenic** (fren'ik; to the diaphragm) **arteries** supply blood to the diaphragm.

Abdominal Aorta and Its Branches

The branches of the abdominal aorta, like those of the thoracic aorta, can be divided into **visceral** and **parietal** portions (see Table 21-3 and Figure 21-9). The visceral arteries can in turn be divided into paired and unpaired branches. There are three major unpaired branches: the **celiac** (se'le-ak; belly) **trunk** (Figure 21-10, *A*), **superior mesenteric** (mes'enter'ik; relating to the mesenteries) (Figure 21-10, *B*), and **inferior mesenteric** (Figure 21-10, *C*) **arteries.** Each has several major branches supplying the abdominal organs.

The paired visceral branches of the abdominal aorta supply the kidneys, suprarenal gland, and gonads (testes or ovaries). The parietal arteries of the abdominal aorta supply the diaphragm and abdominal wall.

Table 21-3

Thoracic and abdominal aorta (see Figures 21-9 and 21-10)

ARTERIES	TISSUES SUPPLIED
Thoracic Aorta	
Visceral branches	
Bronchial	Lung tissue
Esophageal	Esophagus
Parietal branches	
Intercostal	Thoracic wall
Superior phrenic	Superior surface of diaphragm
Abdominal Aorta	
Visceral branches	
Unpaired	
Celiac trunk	
Left gastric	Stomach and esophagus
Common hepatic	Stomach, duodenum, and liver
Splenic	Spleen, pancreas, and stomach
Superior mesenteric	Pancreas, small intestine, and colon
Inferior mesenteric	Descending colon and rectum
Paired	
Suprarenal	Adrenal gland
Renal	Kidney
Gonadal	
Testicular (male)	Testis and ureter
Ovarian (female)	Ovary, ureter, and uterine tube
Parietal branches	
Inferior phrenic	Adrenal gland and inferior surface of diaphragm
Lumbar	Lumbar vertebrae and back muscles
Median sacral	Inferior vertebrae
Common iliac	
External iliac	Lower limb (see Table 21-5)
Internal iliac	Lower back, hip, pelvis, bladder, vagina, uterus, rectum, and external genitalia (see Table 21-4)

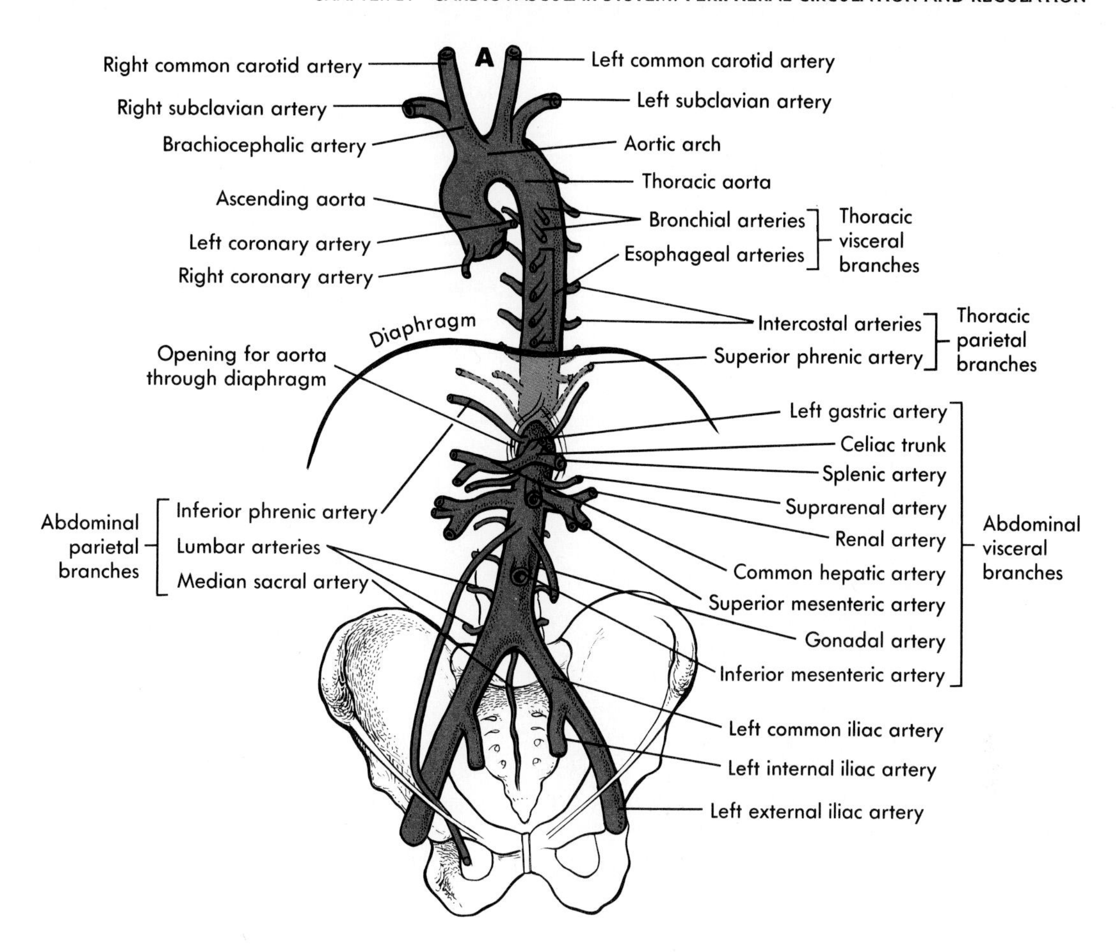

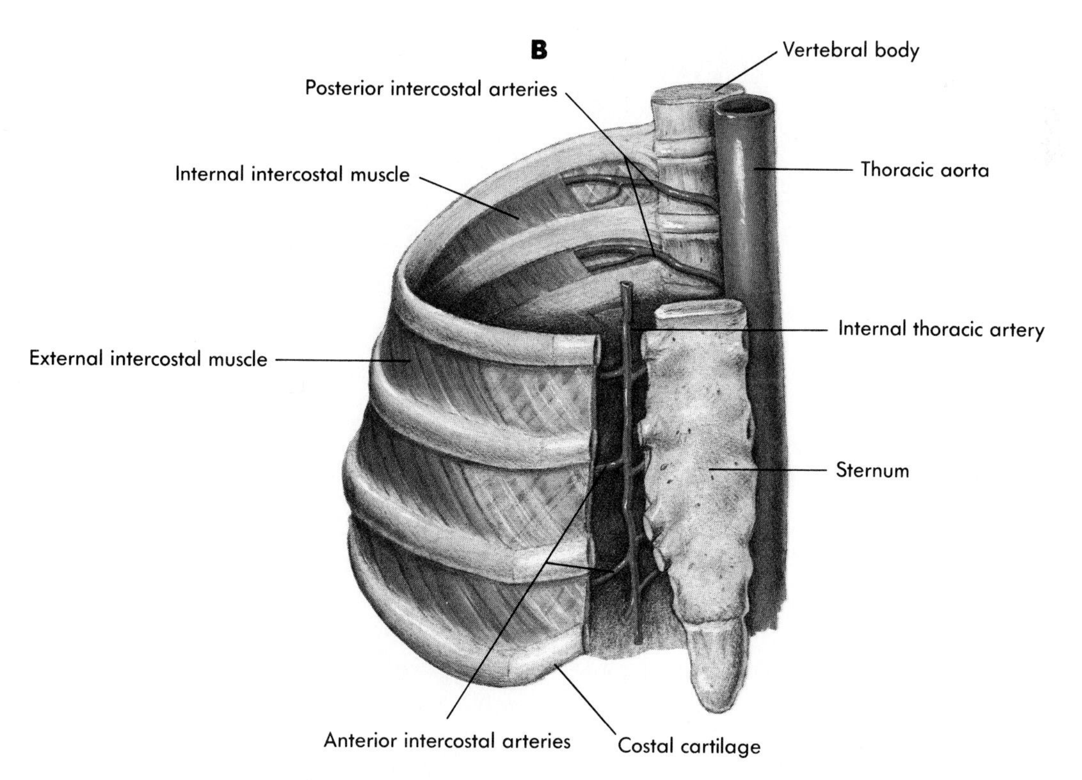

FIGURE 21-9 Branches of the aorta. **A** Aortic arch, thoracic aorta, and abdominal aorta and their branches. **B** Intercostal arteries.

FIGURE 21-10 Branches of the abdominal aorta.
A Celiac trunk and its branches.
B Superior mesenteric artery and its branches.
C Inferior mesenteric artery and its branches.

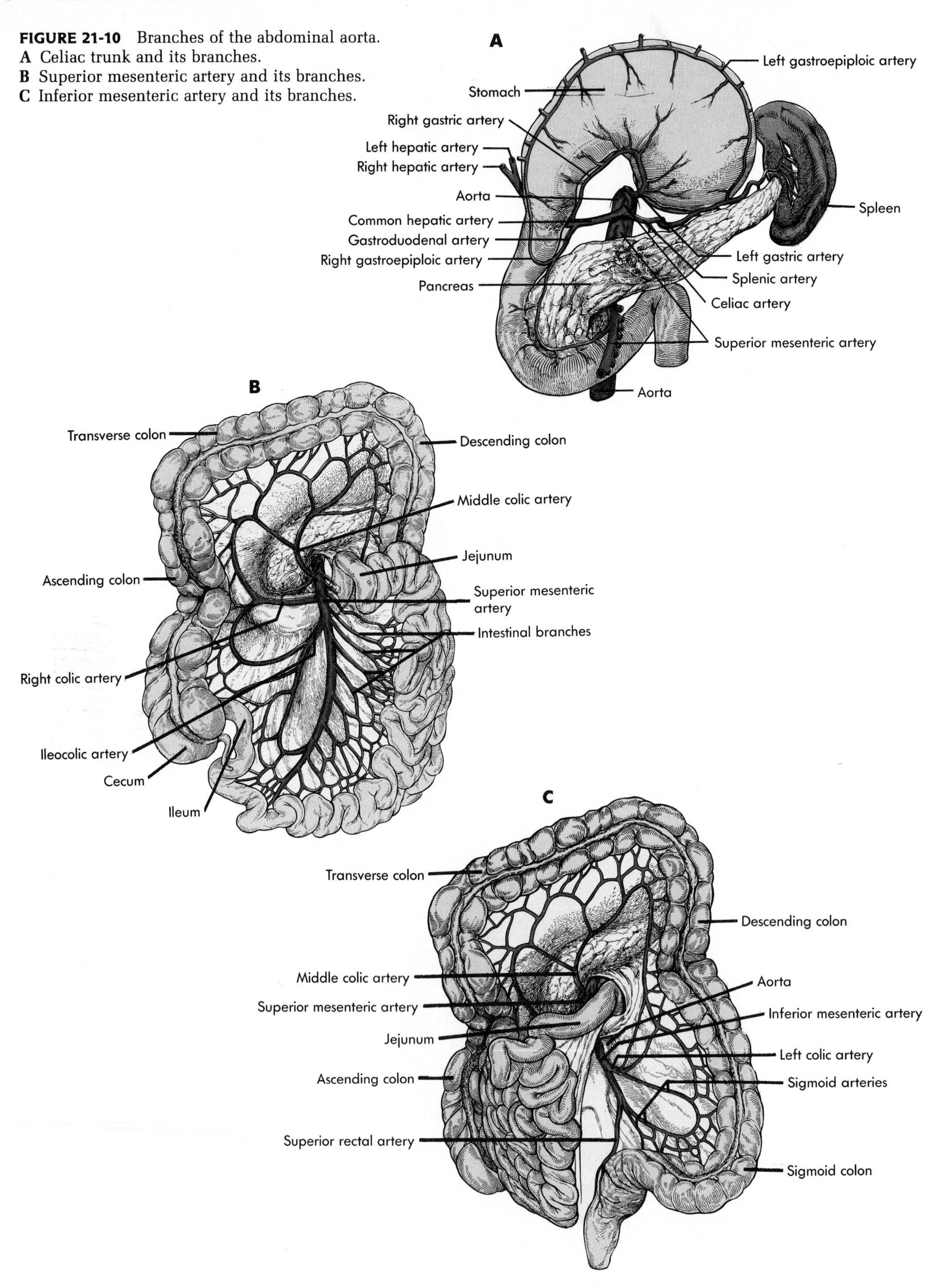

Arteries of the Pelvis

The abdominal aorta divides at the level of the fifth lumbar vertebra into two **common iliac arteries.** They divide to form the **external iliac arteries,** which enter the lower limbs, and the **internal iliac arteries,** which supply the pelvic area. Visceral branches supply the pelvic organs such as the urinary bladder, rectum, uterus, and vagina; parietal branches supply blood to the walls and floor of the pelvis, the lumbar, gluteal, and proximal thigh muscles, and the external genitalia (Table 21-4; Figure 21-11).

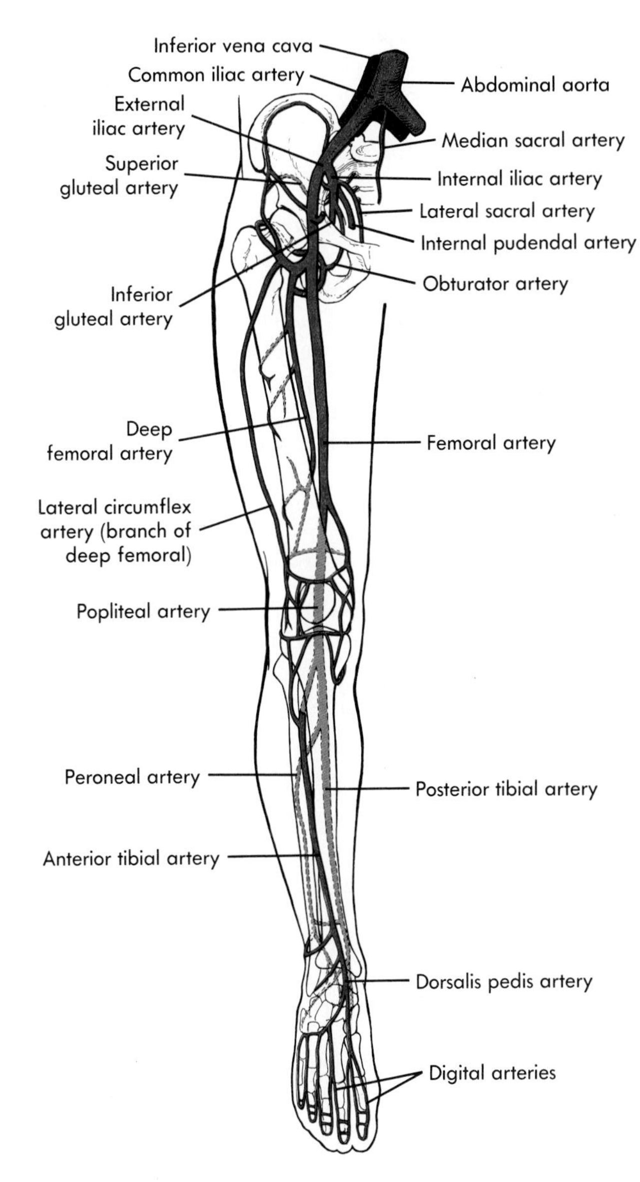

FIGURE 21-11 Arteries of the pelvis and lower limb—the internal and external iliac arteries and their branches. The internal iliac artery supplies the pelvis and hip, and the external iliac artery supplies the lower limb through the femoral artery.

Table 21-4

Arteries of the pelvis (see Figure 21-11)

ARTERIES	TISSUES SUPPLIED
Internal Iliac	Pelvis through the branches listed below
Visceral branches	
Middle rectal	Rectum
Vaginal	Vagina and uterus
Uterine	Uterus, vagina, uterine tube, and ovary
Parietal branches	
Lateral sacral	Sacrum
Superior gluteal	Muscles of the gluteal region
Obturator	Pubic region, deep groin muscles, and hip joint
Internal pudendal	Rectum, external genitalia, and floor of pelvis
Inferior gluteal	Inferior gluteal region, coccyx, and proximal thigh

Table 21-5

Arteries of the lower limb (see Figure 21-11)

ARTERIES	TISSUES SUPPLIED
Femoral	Thigh, external genitalia, anterior abdominal wall
Deep femoral	Thigh, knee, and femur
Popliteal (continuation of the femoral artery)	
Posterior tibial	Knee and leg
Peroneal	Calf and peroneal muscles and ankle
Medial plantar	Plantar region of foot
Digital arteries	Digits of foot
Lateral plantar	Plantar region of foot
Digital arteries	Digits of foot
Anterior tibial	Knee and leg
Dorsalis pedis	Dorsum of foot
Digital arteries	Digits of foot

Arteries of the Lower Limb

The arteries of the lower limb form a continuum similar to that of the arteries of the upper limb. The **external iliac arteries** become the **femoral** (fem'o-ral; relating to the thigh) **arteries** in the thigh, which become the **popliteal** (pop'lĭ-te-al means ham; the hamstring area posterior to the knee) **arteries** in the popliteal space. The popliteal arteries give off the **anterior tibial arteries** just inferior to the knee and then continue as the **posterior tibial arteries.** The anterior tibial artery becomes the **dorsalis pedis artery** at the foot. The posterior tibial artery gives off the **peroneal artery** and then gives rise to medial and lateral **plantar** (plan'tar; the sole of the foot) **arteries,** which in turn give off **digital branches** to the toes. The arteries of the lower limb are listed in Table 21-5 and are illustrated in Figure 21-11.

SYSTEMIC CIRCULATION: VEINS

Three major veins return blood from the body to the right atrium: the **coronary sinus,** returning blood from the walls of the heart; the **superior vena cava** (ve'nah ka'vah; venus cave), returning blood from the head, neck, thorax, and upper limbs; and the **inferior vena cava,** returning blood from the abdomen, pelvis, and lower limbs (Figure 21-12).

In a very general way, the smaller veins follow the same course as the arteries and often are given the same names. However, the veins are more numerous and more variable. The larger veins often follow a very different course and have names different from the arteries.

There are three major types of veins: superficial, deep, and sinuses. The superficial veins of the limbs are, in general, larger than the deep veins, whereas in the head and trunk the opposite is the case. Venous sinuses are primarily in the cranial vault and the heart.

Veins Draining the Heart

The **coronary veins,** which transport blood from the walls of the heart and return it to the right atrium, were described in Chapter 20.

Veins of the Head and Neck

The two pairs of major veins that drain blood from the head and neck are the **external** and **internal jugular** (jug'u-lar; jugular means neck) **veins.** The external jugular veins are the more superficial of the two sets, and they drain blood primarily from the posterior head and neck. The external jugular vein usually drains into the subclavian vein. The internal jugular veins are much larger and deeper than the external jugular veins. They drain blood from the cranial vault and the anterior head, face, and neck.

The internal jugular vein is formed primarily as the continuation of the **venous sinuses** of the cranial vault. The venous sinuses are actually spaces within the dura mater surrounding the brain (Chapter 13). They are depicted in Figure 21-13 and are listed in Table 21-6.

Because there is a venous communication between the facial veins and the cavernous sinuses through the ophthalmic veins, there is a potential for introducing infections into the cranial vault. A superficial infection of the face in the area on either side of the nose may enter the facial vein. The venous infection can then pass through the ophthalmic veins to the cavernous sinus and result in meningitis. For this reason people are warned not to aggravate pimples or boils in the face on either side of the nose.

Once the internal jugular veins exit the cranial vault, they receive several venous tributaries that drain the external head and face (Figure 21-14; Table 21-7). The **internal jugular veins** join the **subclavian veins** on each side of the body to form the **brachiocephalic veins.** The brachiocephalic veins also receive very small vertebral veins, which drain the deep neck muscles.

Veins of the Upper Limb

The **cephalic** (sĕ'fal'ik; toward the head) **basilic,** and **brachial veins** are responsible for draining most of the blood from the upper limb (Figure 21-15; Table 21-8). Many of the tributaries of the cephalic and basilic veins in the forearm and hand can be seen through the skin. Because of the considerable variation in the tributary veins of the forearm and hand, they often are left unnamed. The basilic vein of the arm becomes the **axillary vein** as it courses through the axillary region. The axillary vein then becomes the **subclavian vein** at the margin of the first rib. The cephalic vein enters the axillary vein but may have a branch that enters the external jugular vein.

The **median cubital** (ku'bĭ-tal; cubitus means elbow) **vein** is a variable vein that usually connects the cephalic vein or its tributaries with the basilic vein. In many people this vein is quite prominent on the anterior surface of the upper limb at the level of the elbow (cubital fossa), and it often is used as a site for drawing blood from a patient.

The deep veins draining the upper limb follow the same course as the arteries. The **radial** and **ulnar veins** therefore are named for the arteries they attend. They usually are paired, with one small vein lying

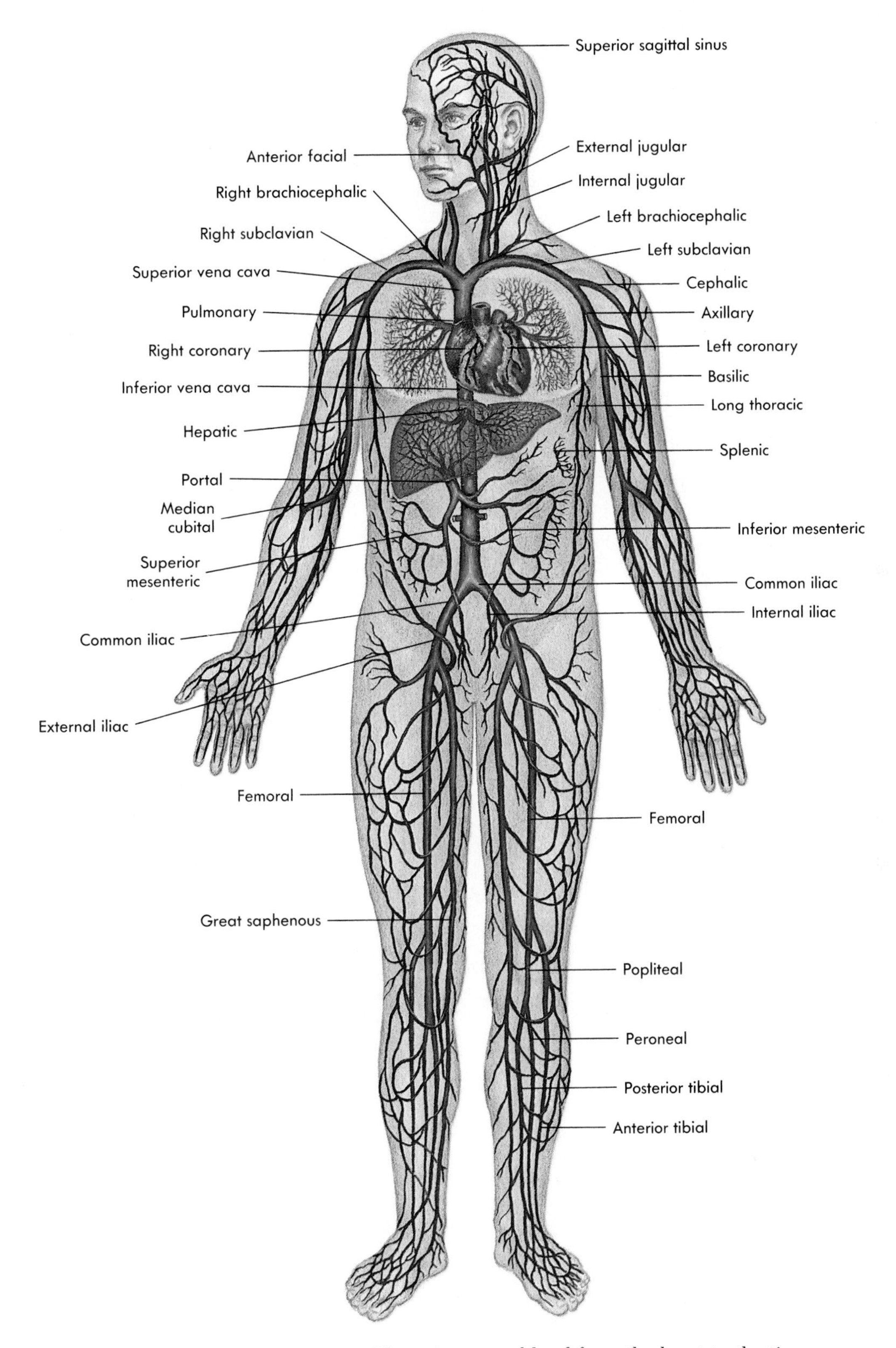

FIGURE 21-12 The major veins. The veins carry blood from the heart to the tissues of the body.

Table 21-6

Venous sinuses of the cranial vault (see Figure 21-13)

VEINS	TISSUES DRAINED
Internal Jugular Vein	
Sigmoid sinus	
Superior and inferior petrosal sinuses	Anterior portion of cranial vault
Cavernous sinus	
Ophthalmic veins	Orbit
Transverse sinus	
Occipital sinus	Central floor of posterior fossa of skull
Superior sagittal sinus	Superior portion of cranial vault and brain
Straight sinus	
Inferior sagittal sinus	Deep portion of longitudinal fissure

Table 21-7

Veins draining the head and neck (see Figure 21-14)

VEINS	TISSUES DRAINED
Brachiocephalic	
Internal jugular	Brain
Lingual	Tongue and mouth
Superior thyroid	Thyroid and deep posterior facial structures (also empties into external jugular)
Facial	Superficial and anterior facial structures
Lingual	Tongue
External Jugular	Superficial surface of posterior head and neck

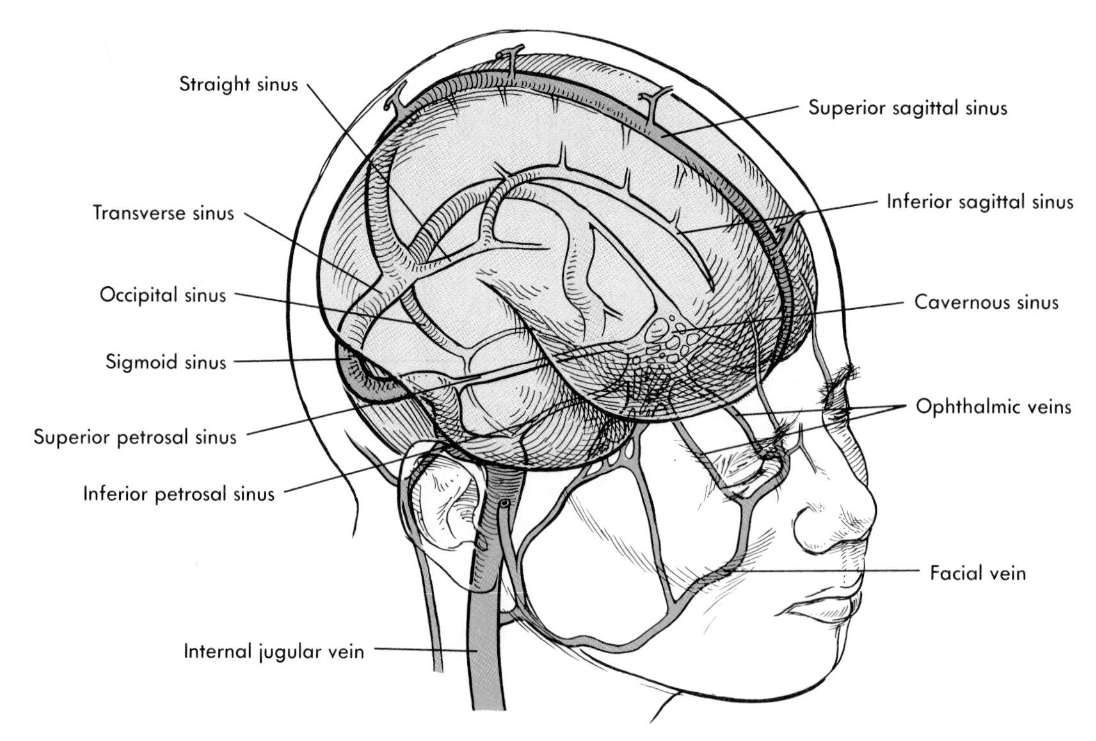

FIGURE 21-13 Venous sinuses associated with the brain.

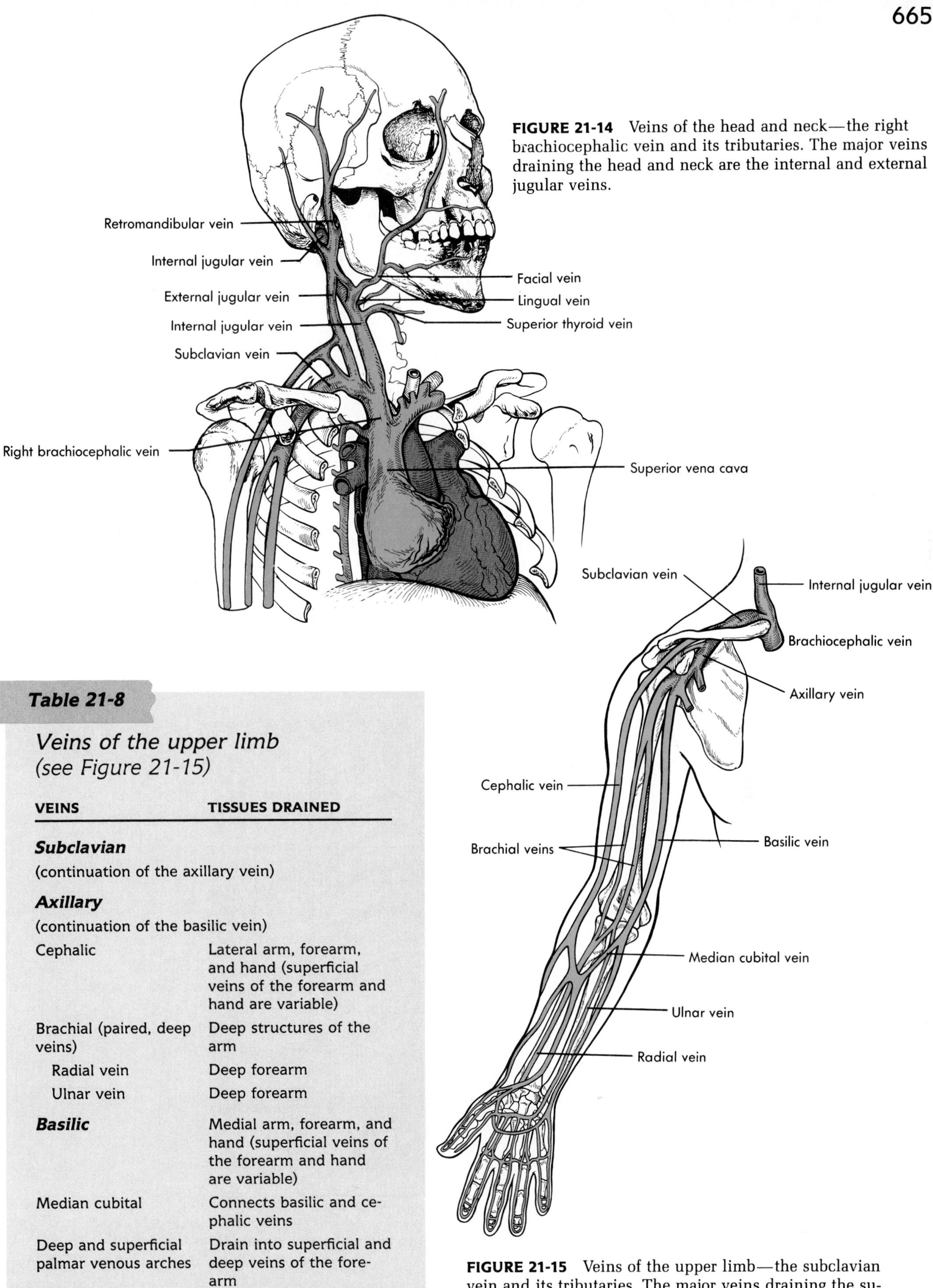

FIGURE 21-14 Veins of the head and neck—the right brachiocephalic vein and its tributaries. The major veins draining the head and neck are the internal and external jugular veins.

FIGURE 21-15 Veins of the upper limb—the subclavian vein and its tributaries. The major veins draining the superficial structures of the limb are the cephalic and basilic veins. The brachial veins drain the deep structures.

Table 21-8

Veins of the upper limb (see Figure 21-15)

VEINS	TISSUES DRAINED
Subclavian	
(continuation of the axillary vein)	
Axillary	
(continuation of the basilic vein)	
Cephalic	Lateral arm, forearm, and hand (superficial veins of the forearm and hand are variable)
Brachial (paired, deep veins)	Deep structures of the arm
Radial vein	Deep forearm
Ulnar vein	Deep forearm
Basilic	Medial arm, forearm, and hand (superficial veins of the forearm and hand are variable)
Median cubital	Connects basilic and cephalic veins
Deep and superficial palmar venous arches	Drain into superficial and deep veins of the forearm
Digital veins	Fingers

on each side of the artery, and they have numerous connections with each other and with the superficial veins. The radial and ulnar veins empty into the **brachial veins,** which accompany the brachial artery and empty into the axillary vein (see Figure 21-15).

Veins of the Thorax

Three major veins return blood from the thorax to the superior vena cava: the **right** and **left brachiocephalic veins** and the **azygos** (az′ĭ-gus; unpaired) **vein.** The thoracic drainage to the brachiocephalic veins is through the anterior thoracic wall by way of the **internal thoracic veins.** They receive blood from the **anterior intercostal veins.** Blood from the posterior thoracic wall is collected by **posterior intercostal veins** that drain into the azygos vein on the right and the **hemiazygos** or **accessory hemiazygos vein** on the left. The hemiazygos and accessory hemiazygos veins empty into the azygos vein, which drains into the superior vena cava. The thoracic veins are listed in Table 21-9 and are illustrated in Figure 21-16.

Table 21-9

Veins of the thorax (see Figure 21-16)

VEINS	TISSUES DRAINED
Superior Vena Cava	
Brachiocephalic	
Azygos vein	Right side, posterior thoracic wall and posterior abdominal wall; esophagus, bronchi, pericardium, and mediastinum
Hemiazygos	Left side, inferior posterior thoracic wall and posterior abdominal wall; esophagus and mediastinum
Accessory hemiazygos	Left side, superior posterior thoracic wall

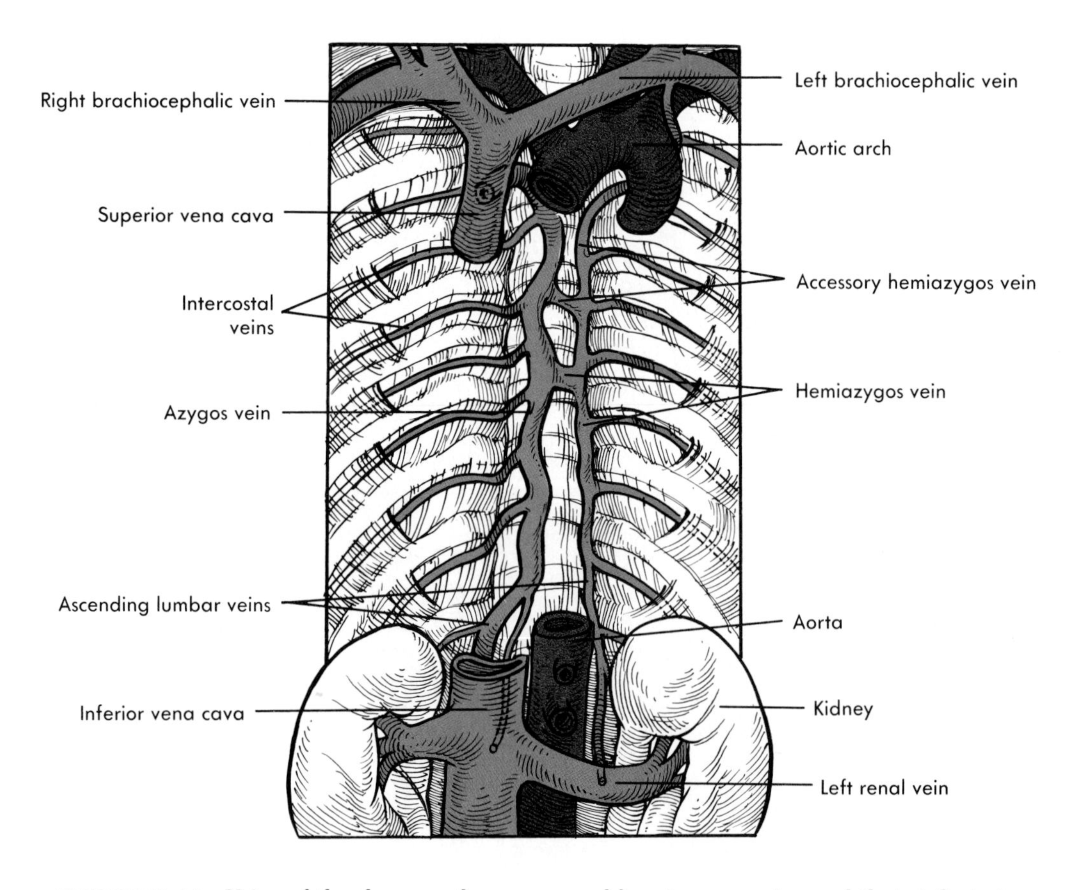

FIGURE 21-16 Veins of the thorax—the azygos and hemiazygos veins and their tributaries.

Veins of the Abdomen and Pelvis

Blood from the posterior abdominal wall drains into the **ascending lumbar veins.** These veins are continuous superiorly with the hemiazygos on the left and the azygos on the right. Blood from the rest of the abdomen, pelvis, and lower limbs returns to the heart through the inferior vena cava. The gonads (testes or ovaries), kidneys, and suprarenal glands are the only abdominal organs outside the pelvis that drain directly into the inferior vena cava. The **internal iliac veins** drain the pelvis and join the **external iliac veins** from the lower limbs to form the **common iliac veins,** which unite to form the inferior vena cava. The major abdominal and pelvic veins are listed in Table 21-10 and are illustrated in Figure 21-17.

Table 21-10

Veins draining the abdomen and pelvis (see Figure 21-17)

VEINS	TISSUES DRAINED
Inferior Vena Cava	
Hepatic veins	Liver (see hepatic portal system)
Common iliac	
External iliac	Lower limb (see Table 21-12)
Internal iliac	Pelvis and its viscera
Ascending lumbar	Posterior abdominal wall (empties into common iliac, azygos, and hemiazygos veins)
Renal	Kidney
Suprarenal	Adrenal gland
Gonadal	
Testicular (male)	Testis
Ovarian (female)	Ovary
Phrenic	Diaphragm

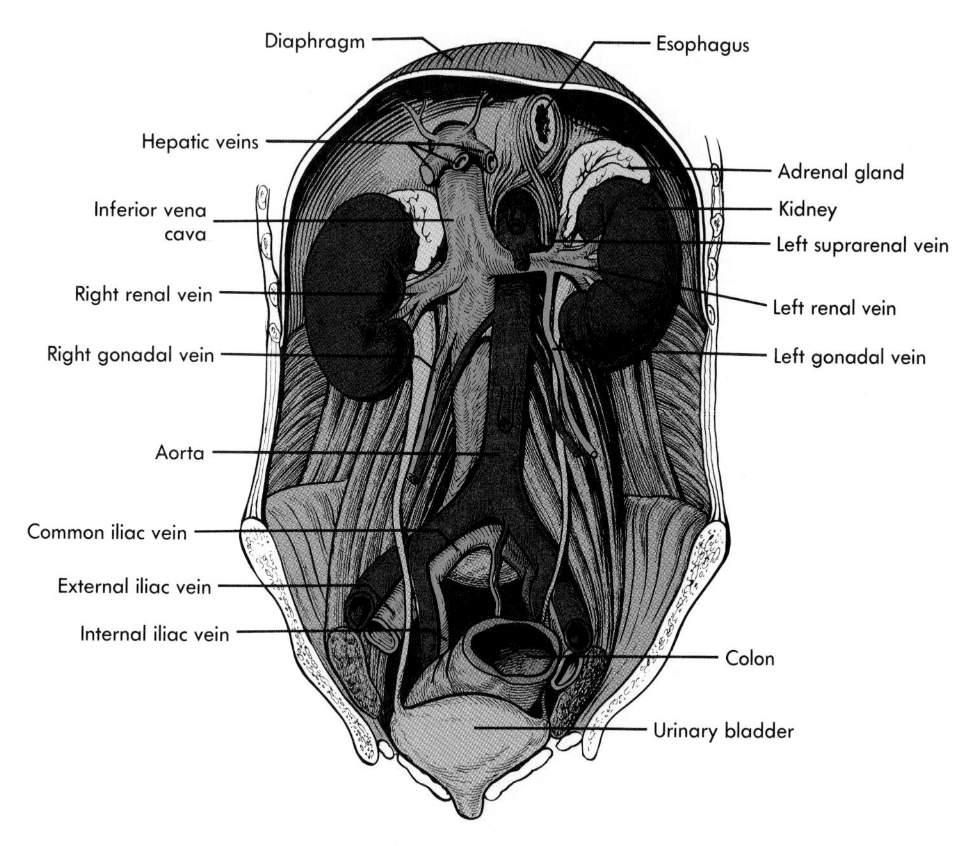

FIGURE 21-17 Inferior vena cava and its tributaries. The hepatic veins transport blood to the inferior vena cava from the hepatic portal system, which ends as a series of blood sinusoids in the liver (see Figure 21-18).

Hepatic portal system

Blood from the capillaries within most of the abdominal viscera such as the stomach, intestines, and spleen drains through a specialized system of blood vessels to the liver. Within the liver the blood flows through a series of dilated capillaries called **sinusoids.** A **portal** (pōr′tal; door) **system** is a vascular system that begins and ends with capillary beds and has no pumping mechanism such as the heart between the capillary beds. The portal system that begins with capillaries in the viscera and ends with the sinusoidal capillaries in the liver is the **hepatic** (hĕ-pat′ik; relating to the liver) **portal system** (Table 21-11 and Figure 21-18). The **hepatic portal vein,** the largest vein of the system, is formed by the union of the **superior mesenteric vein,** which drains the small intestine, and the **splenic vein,** which drains the spleen. The splenic vein receives the **inferior mesenteric** and **pancreatic veins,** which drain the large intestine and pancreas respectively. The hepatic portal vein also receives gastric and cystic veins before entering the liver.

Blood from the liver sinusoids is collected into **central veins,** which empty into **hepatic veins.** The hepatic veins join the inferior vena cava. The blood entering the liver through the hepatic portal vein is rich with nutrients collected from the intestines, but it also may contain a number of toxic substances harmful to the tissues of the body. Within the liver the nutrients are either taken up and stored or are modified chemically and used by other cells of the body (see Chapter 24). The cells of the liver also help remove toxic substances by altering their structure or by making them water soluble. The water-soluble substances can then be transported in the blood to the kidneys from which they are excreted in the urine (see Chapter 26).

Table 21-11

Hepatic portal system (see Figure 21-18)

VEINS	TISSUES DRAINED
Hepatic Portal	
Superior mesenteric	Small intestine and most of the colon
Splenic	Spleen
Inferior mesenteric	Descending colon and rectum
Pancreatic	Pancreas
Gastroepiploic	Stomach
Gastric	Stomach
Cystic	Gallbladder

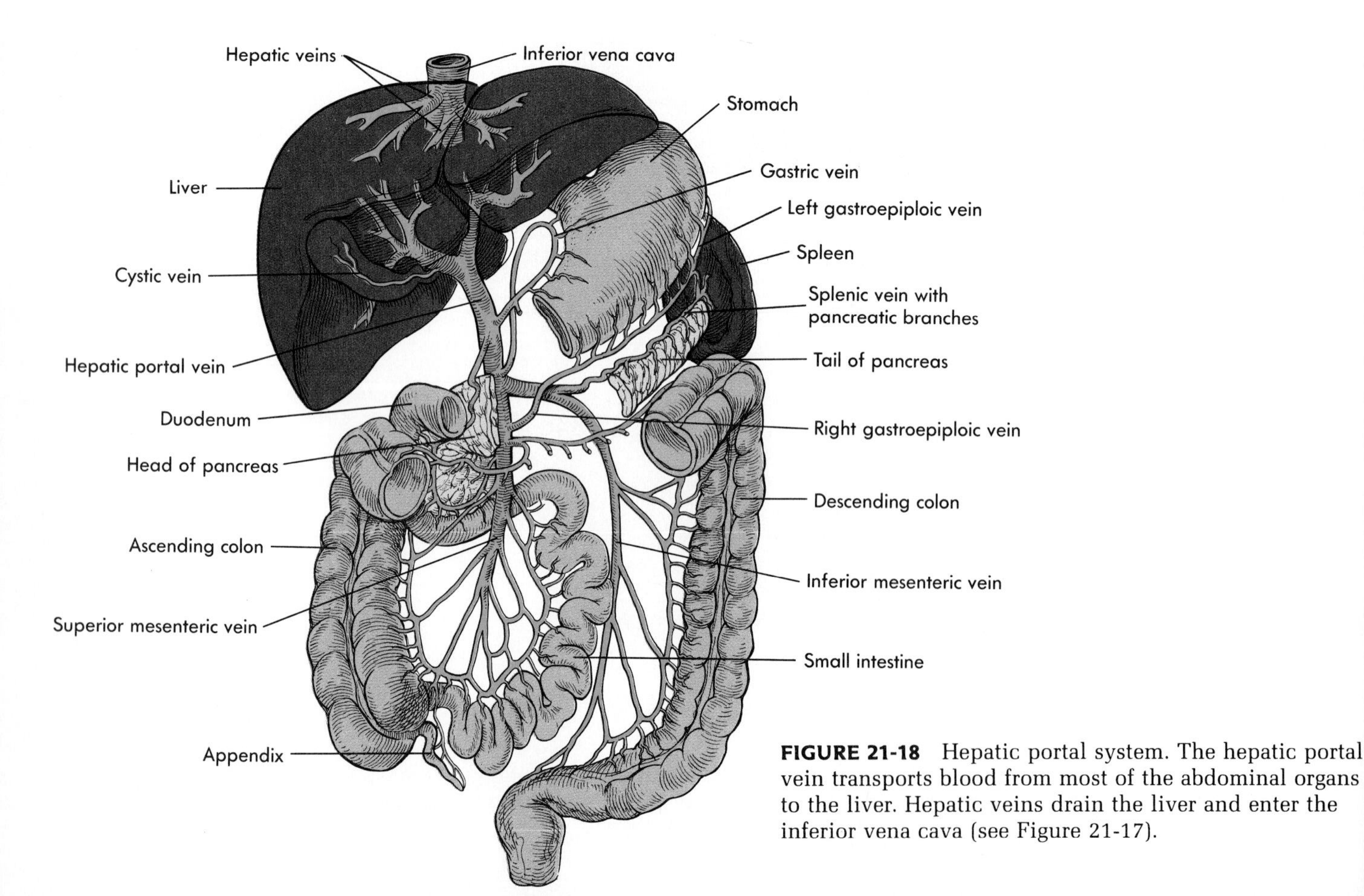

FIGURE 21-18 Hepatic portal system. The hepatic portal vein transports blood from most of the abdominal organs to the liver. Hepatic veins drain the liver and enter the inferior vena cava (see Figure 21-17).

Veins of the Lower Limb

The veins of the lower limb, like those of the upper limb, consist of superficial and deep groups. The distal deep veins are paired and follow the same path as the arteries, whereas the proximal deep veins are unpaired. The **anterior** and **posterior tibial veins** are paired and accompany the anterior and posterior tibial arteries. They unite just inferior to the knee to form the single **popliteal vein,** which ascends through the thigh and becomes the **femoral vein.** The femoral vein becomes the external iliac vein. **Peroneal veins** also are paired in each leg and accompany the peroneal arteries. They empty into the posterior tibial veins just before those veins contribute to the popliteal vein.

The superficial veins consist of the great and small **saphenous** (să-fe′nus; visible) **veins.** The **great saphenous vein,** the longest vein of the body, originates over the dorsal and medial side of the foot and ascends along the medial side of the leg and thigh to empty into the femoral vein. The **small saphenous vein** begins over the lateral side of the foot and ascends along the posterior leg to the popliteal space where it empties into the popliteal vein. The veins of the lower limb are illustrated in Figure 21-19 and are listed in Table 21-12.

Table 21-12

Veins of the lower limb (see Figure 21-19)

VEINS	TISSUES DRAINED
External Iliac Vein	
(continuation of the femoral vein)	
Femoral (continuation of the popliteal vein)	Thigh
Popliteal	
Anterior tibial	Deep anterior leg
Dorsal vein of foot	Dorsum of foot
Posterior tibial	Deep posterior leg
Plantar veins	Plantar region of foot
Peroneal	Deep lateral leg and foot
Small saphenous	Superficial posterior leg and lateral side of foot
Great saphenous	Superficial anterior and medial leg, thigh, and dorsum of foot
Dorsal vein of foot	Dorsum of foot
Dorsal venous arch	Foot
Digital veins	Toes

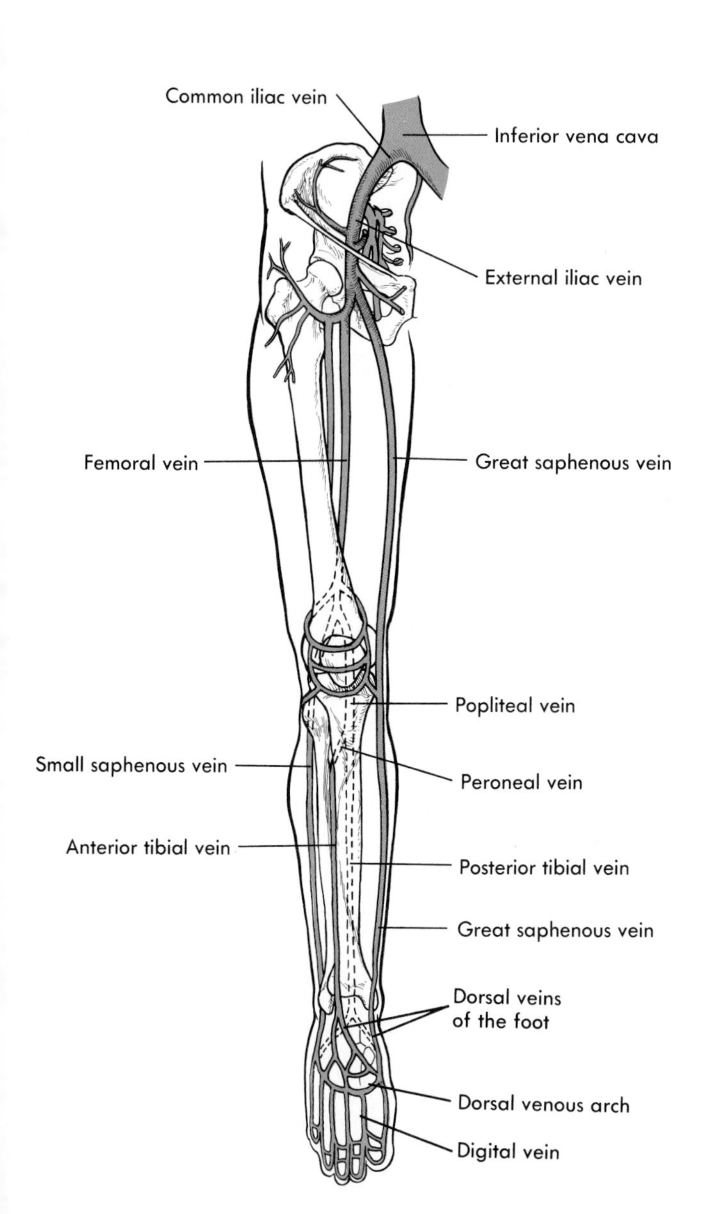

FIGURE 21-19 Veins of the pelvis and lower limb—the right common iliac vein and its tributaries.

LYMPH VESSELS

The lymphatic system (Figure 21-20), unlike the circulatory system, only carries fluid away from the tissues. The lymphatic system begins in the tissues as **lymph capillaries,** which differ from blood capillaries in that they lack a basement membrane and the cells of the simple squamous epithelium slightly overlap and are attached loosely to each other (Figure 21-21). Two things occur as a result of this structure. First, the lymph capillaries are far more permeable than blood capillaries, and nothing in the interstitial fluid is excluded from the lymph capillaries. Second, the lymph capillary epithelium functions as a series of one-way valves that allow fluid to enter the capillary but prevent fluid from passing back into the interstitial spaces.

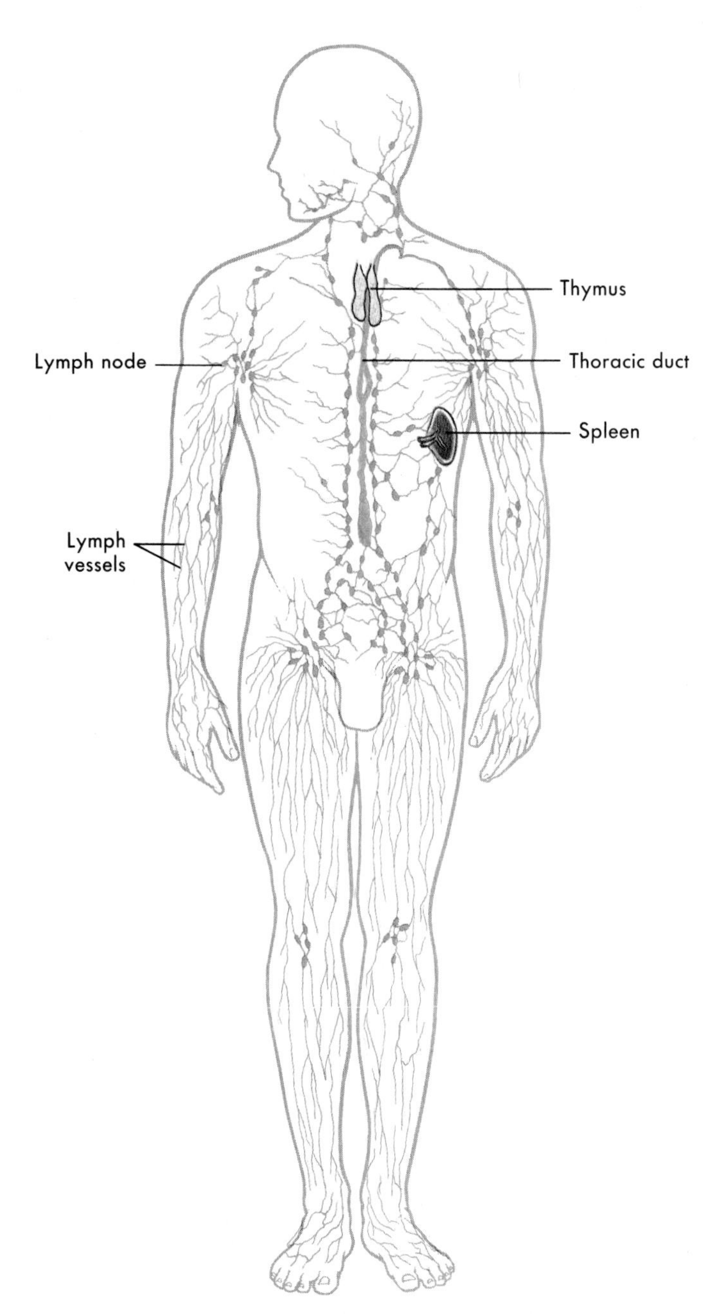

FIGURE 21-20 Lymphatic system showing the major lymphatic organs and vessels.

Lymph capillaries are in almost all tissues of the body with the exception of the central nervous system, bone marrow, and tissues without blood vessels (i.e., cartilage, epidermis, and cornea). A superficial group of lymph capillaries is in the dermis of the skin and in the hypodermis. A deep group of lymph capillaries drains the muscles, joints, viscera, and other deep structures. Fluids tend to move out of blood capillaries into tissue spaces and then out of the tissue spaces into lymph capillaries (Figure 21-22). The fluid moving into the lymph capillaries is called **lymph.**

The lymph capillaries join to form larger **lymph vessels** that resemble small veins. The inner layer of the lymph vessel consists of endothelium surrounded by an elastic membrane, the middle layer consists of smooth-muscle cells and elastic fibers, and the outer layer is a thin layer of fibrous connective tissue.

Small lymph vessels have a beaded appearance because of the presence of one-way valves along their lengths that are similar to the valves of veins (see Figure 21-21). When a lymph vessel is squeezed shut, backward movement of lymph is prevented by the valves; as a consequence, the lymph moves forward through the lymph vessel. Three factors are believed responsible for the compression of lymph vessels: (1) contraction of surrounding skeletal muscles during activity, (2) contraction of the smooth muscles in the lymph vessel wall, and (3) pressure changes in the thorax during respiration.

Lymph nodes (round, oval, or bean-shaped bodies) are distributed along the various lymph vessels. The lymph nodes function to filter lymph, and most lymph passes through at least one lymph node before entering the blood. See Chapter 22 for more information on lymph nodes.

After passing through superficial or deep lymph nodes, the lymph vessels converge toward either the right or the left subclavian vein. Vessels from the upper right limb and the right side of the head and the neck enter the right lymphatic duct. Lymph vessels from the rest of the body enter the larger thoracic duct (see Figure 21-20).

Thoracic Duct

The thoracic duct drains the lower limbs, abdomen, the left thorax, the left upper extremity, and the left side of the head and the neck (Figure 21-23). The duct ends by entering the left subclavian vein. Although the **thoracic duct** is the largest lymph vessel, it is still so small that it is difficult to see in cadavers. At the level of the superior abdominal cav-

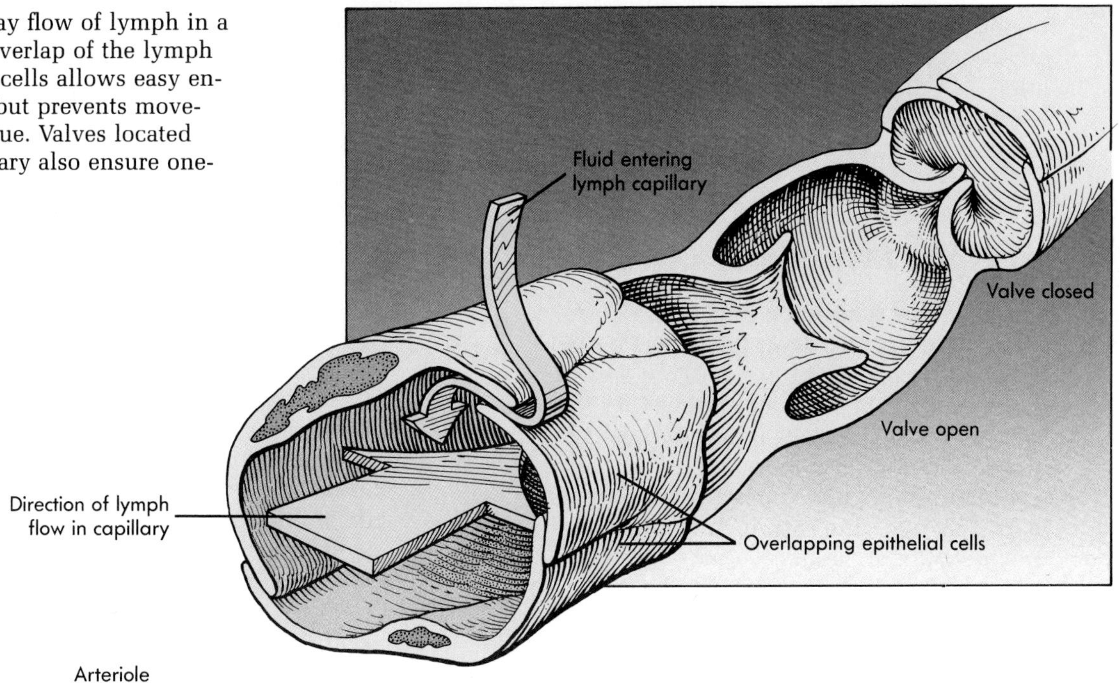

FIGURE 21-21 One-way flow of lymph in a lymph capillary. The overlap of the lymph capillary's endothelial cells allows easy entry of interstitial fluid but prevents movement back into the tissue. Valves located along the lymph capillary also ensure one-way flow of lymph.

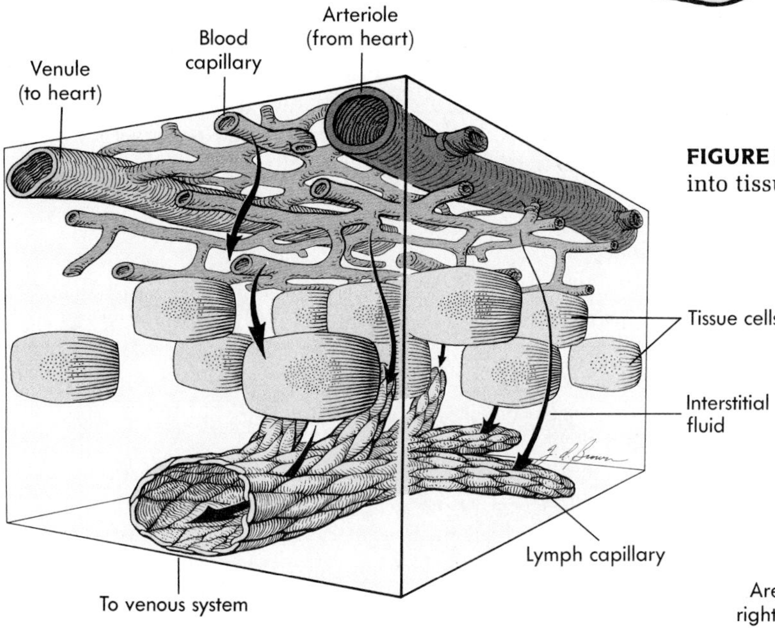

FIGURE 21-22 Movement of fluid from blood capillaries into tissues and from tissues into lymph capillaries.

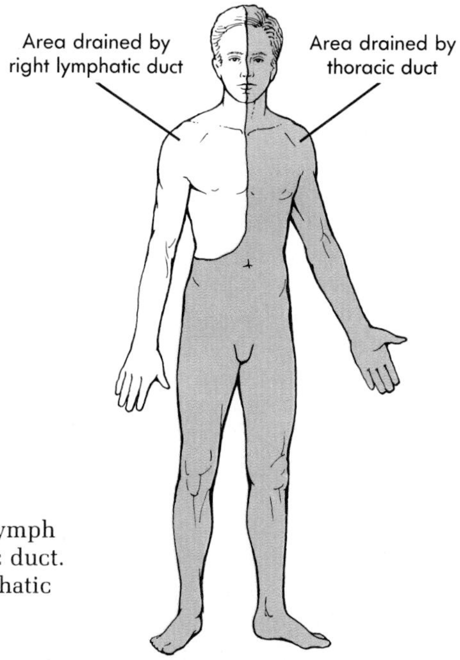

FIGURE 21-23 **A** Overall lymphatic drainage. Lymph from the colored area drains through the thoracic duct. Lymph from the white area enters the right lymphatic duct.

ity the thoracic duct is expanded to form the **cisterna chyli** (sis-ter'nah ki'le; cistern or tank that contains juice), which receives several lymph vessels from the lower limbs and from the abdomen, especially from the digestive tract.

Right Lymphatic Duct

The **right lymphatic duct** is much shorter and is smaller in diameter than the thoracic duct. It drains the right thorax, right upper limb, and right side of the head and the neck and opens into the right subclavian vein. The right lymphatic duct may consist of a single duct, but more commonly three separate right lymphatic ducts open into the right subclavian vein, the right internal jugular vein, and the right brachiocephalic vein.

2

During radical cancer surgery malignant lymph nodes are removed, and their vessels are tied off to prevent metastasis (spread) of the cancer. Predict the consequences of tying off the lymph vessels.

? ? ? ? ? ? ? ? ? ? ?

PHYSICS OF CIRCULATION

The basic physical characteristics of blood and the physical principles affecting the flow of liquids through vessels dramatically influence the circulation of blood. The interrelationships between pressure, flow, resistance, and the control mechanisms that regulate blood pressure and blood flow through vessels play a critical role in the function of the circulatory system.

Viscosity

Viscosity is a measure of the resistance of a liquid to flow. As the viscosity of a liquid increases, the pressure required to force it to flow increases. A common means for reporting the viscosity of liquids is to consider distilled water's viscosity as 1 and to compare the viscosity of other liquids to it. Using this procedure, whole blood has a viscosity of 3 to 4.5 (i.e., approximately three times as much pressure is required to force whole blood to flow through a given tube at the same rate as water).

The viscosity of blood is influenced largely by hematocrit (the percent of the total blood volume composed of blood cells); as the hematocrit increases, the viscosity of blood increases logarithmically. Blood with a hematocrit of 60% has a viscosity approximately seven to eight times that of water, whereas blood with a hematocrit of 45% has a viscosity only approximately three times that of water. Plasma proteins have only a minor effect on the viscosity of blood. Dehydration or uncontrolled production of red blood cells can increase hematocrit and the viscosity of blood substantially. Viscosity above its normal range of values increases the work load on the heart substantially, and if the work load on the heart is great enough, heart failure may result.

Laminar and Turbulent Flow in Vessels

Fluid, including blood, tends to flow through long, smooth-walled tubes in a streamlined fashion called **laminar flow** (Figure 21-24, *A*). Fluid behaves as if it is composed of a large number of concentric layers. The layer nearest the wall of the tube experiences the greatest resistance to flow since it moves against the stationary wall. The innermost layers slip over the surface of the outermost layers and experience less resistance to movement. Thus flow in a vessel consists of movement of concentric layers, with the outermost layer moving at the lowest velocity and the layer at the center moving at the greatest velocity.

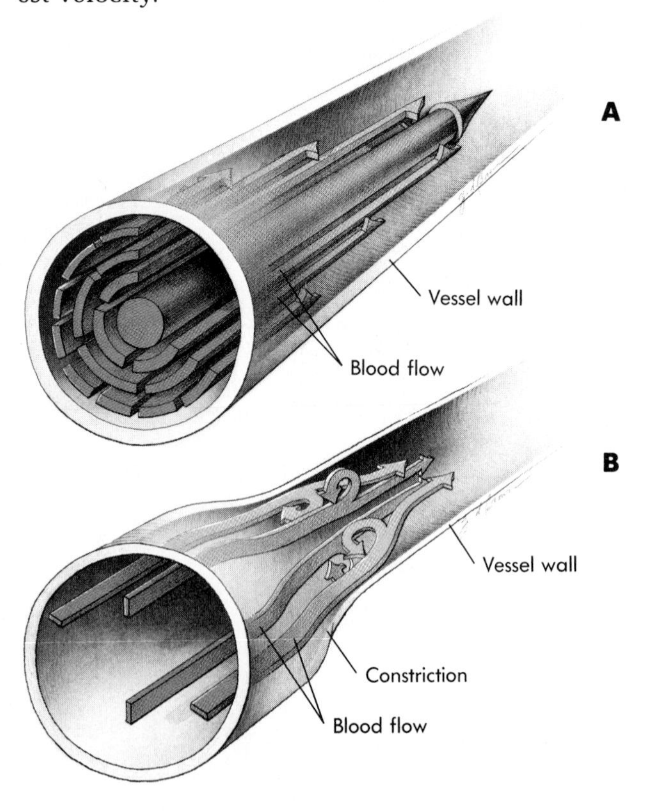

FIGURE 21-24 **A** Laminar flow. Fluid flows in long smooth-walled tubes as if it is composed of a large number of concentric layers. **B** Turbulent flow. Turbulent flow is caused by numerous small currents flowing crosswise or oblique to the long axis of the vessel, resulting in flowing whorls and eddy currents.

Laminar flow is interrupted and becomes **turbulent flow** when the rate of flow exceeds a critical velocity or when the fluid passes a constriction, a sharp turn, or a rough surface (Figure 21-24, *B*). Vibrations of the liquid and of the blood vessel walls during turbulent flow cause the sounds produced when blood pressure is measured using a blood pressure cuff. Turbulent flow is also common as blood flows past the valves in the heart and is partially responsible for the heart sounds.

Turbulent flow of blood through vessels occurs primarily in the heart and to a lesser extent where arteries branch. Sounds caused by turbulent blood flow in arteries are not normal and usually indicate that the blood vessel is constricted abnormally.

Blood Pressure

Blood pressure is a measure of the force blood exerts against the blood vessel walls. The standard reference for blood pressure is the mercury (Hg) manometer, and pressure is measured in terms of mm Hg. If the blood pressure is 100 mm Hg, the pressure is great enough to lift a column of mercury 100 mm.

Blood pressure can be measured directly by inserting a **cannula** (or tube) into a blood vessel and connecting a manometer or an electronic pressure transducer to it. Electronic transducers are better than the mercury manometer for measuring rapid blood pressure fluctuations up to approximately 60 cycles per second.

Placing catheters in blood vessels or in chambers of the heart to monitor pressure changes is possible, but these procedures are not appropriate for routine clinical determinations of systemic blood pressure. The **auscultatory** (aws-kul'tah-to're) method of determining blood pressure can be used to measure blood pressure without requiring surgical procedures and is used under most clinical conditions. A blood pressure cuff connected to a manometer, the **sphygmomanometer,** is placed around the patient's upper arm, and a stethoscope is placed over the brachial artery (Figure 21-25). The blood pressure cuff is inflated until the brachial artery is occluded completely. Since no blood flows through the constricted area, no sounds can be heard at this point. The pressure in the cuff is gradually lowered. As soon as the pressure in the cuff declines below the systolic pressure, blood flows through the constricted area during systole. The blood flow is turbulent and produces vibrations in the blood and surrounding tissues that can be heard through the stethoscope. These sounds are called **Korotkoff sounds,** and the pressure at which the first Korotkoff sound is heard represents the **systolic pressure.**

As the pressure in the blood pressure cuff is lowered still more, the Korotkoff sounds change tone and loudness. When the pressure has dropped until the sound disappears completely, continuous laminar blood flow is reestablished. The pressure at which continuous laminar flow is reestablished is the **diastolic pressure.** This method for determining systolic and diastolic pressures is not entirely accurate, but its results are within 10% of methods that are more direct.

Rate of Blood Flow

The **rate** at which blood or any other liquid flows through a tube is expressed as the volume that passes a specific point per unit of time. Blood flow usually is reported as either milliliters per minute or liters per minute. For example, when a person is resting, the **cardiac output** of the heart is approximately 5 L per minute; thus blood flow through the aorta is approximately 5 L per minute.

Blood flow in a vessel is proportional to the pressure gradient in that vessel. For example, if the pressures at point 1 (P_1) and point 2 (P_2) in a vessel are the same, no flow will occur. However, if the pressure at P_1 is greater than the pressure at P_2, flow will proceed from P_1 toward P_2, and the greater the pressure difference, the greater will be the rate of flow. If P_2 is greater than P_1, flow will proceed from P_2 toward P_1. Flow always occurs from a higher to a lower pressure.

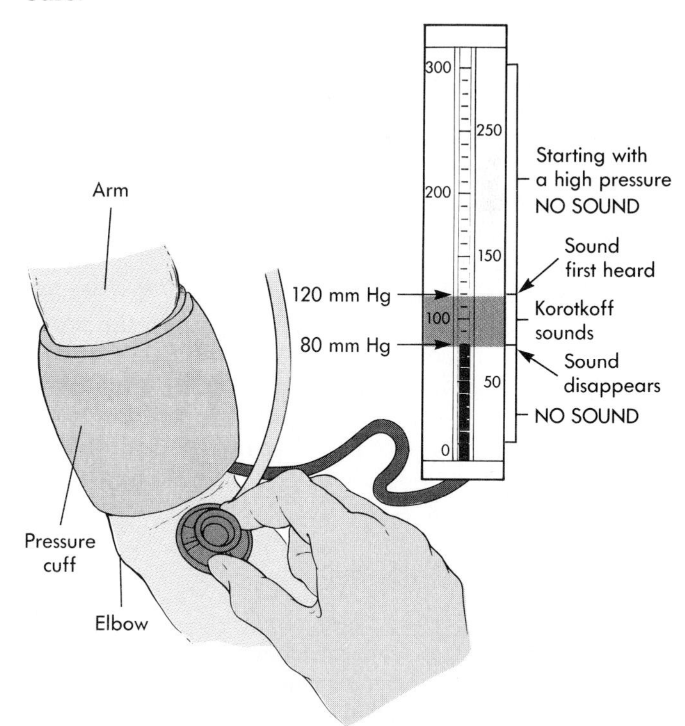

FIGURE 21-25 Blood pressure measurement using a sphygmomanometer. The blood pressure cuff is inflated to a high pressure, and the pressure is decreased slowly. The pressure at which turbulent blood flow is first heard is the systolic blood pressure. The pressure at which sounds disappear is the diastolic blood pressure.

The movement of blood as the result of a pressure difference is opposed by a resistance (R) to blood flow. As the resistance increases, blood flow decreases, and as the resistance decreases, blood flow increases. The effect of pressure differences and resistance to blood flow can be expressed mathematically:

$$\text{Flow} = \frac{P_1 - P_2}{R}$$

where *R* equals resistance.

Poiseuille's Law

Several factors affect resistance to blood flow and are expressed individually in **Poiseuille's** (puah-zuh'yez) **law:**

$$\text{Flow} = \frac{(P_1 - P_2)}{8\ vl/r^4} \quad \text{or} \quad \text{Flow} = \frac{(P_1 - P_2)r^4}{8\ vl}$$

where: $8\ vl/r^4$ = Resistance
v = Viscosity of blood
$P_1 - P_2$ = Pressure gradient
l = Length of vessel
r = Radius of the blood vessel

According to Poiseuille's law, flow is increased dramatically when the radius of the blood vessel is increased (resistance is decreased) because flow is proportional to the fourth power of the blood vessel's radius. Reciprocally, a small decrease in the blood vessel's radius (increase in resistance) results in a dramatic decrease in flow. In addition, either an increase in blood viscosity or an increase in blood vessel length reduces flow.

During exercise the heart contracts with a greater force, and the blood pressure increases in the aorta. In addition, blood vessels in skeletal muscles dilate, making their radii larger, which decreases the resistance to blood flow. As a consequence, the rate of flow may increase from 5 L per minute in the aorta to several times that value.

3

Compare the effects of the following on blood flow: (1) vasoconstriction of blood vessels in the skin in response to cold exposure; (2) vasodilation of the blood vessels in the skin in response to an elevated body temperature; (3) a decrease in the rate of red blood cell synthesis; and (4) polycythemia vera, which results in a greatly increased hematocrit.

? ? ? ? ? ? ? ? ? ? ?

Critical Closing Pressure and the Law of LaPlace

Each blood vessel exhibits a **critical closing pressure** (i.e., when the pressure decreases below some critical point, the vessel will collapse, and blood flow through the blood vessel stops). Under conditions of shock blood pressure may decrease below the critical closing pressure in vessels (see essay on shock). As a consequence, the blood vessels collapse, and flow ceases. Tissues supplied by those vessels may become necrotic because of the lack of blood supply.

The **law of LaPlace** helps to explain the critical closing pressure. It states that the force that stretches the vascular wall is proportional to the diameter of the vessel times the blood pressure:

$$F = D \times P$$

where *F* is force, *D* is vessel diameter, and *P* is pressure.

As the pressure in a vessel decreases, the force that stretches the vessel wall also decreases. Some minimum force is required to keep the vessel open; if the pressure decreases so that the force is below that minimum requirement, the vessel will close.

According to the law of LaPlace, as the diameter of the vessel increases, the force that is applied to the vessel wall also increases, even if the pressure remains constant. If a portion of an arterial wall becomes weakened so that a bulge forms in it, the force applied to the weakened part is greater than at other points along the blood vessel because its diameter is greater. The greater force causes the weakened vessel wall to bulge even more, further increasing the force applied to it. This series of events may proceed until the vessel finally ruptures. The bulges in weakened blood vessel walls are called **aneurysms** (an'u-riz-umz). As aneurysms enlarge, the danger that they will rupture increases. Ruptured aneurysms in the blood vessels of the brain or in the aorta often result in death of the individual.

Vascular Compliance

Compliance is the tendency for blood vessel volume to increase as the blood pressure increases. The more easily the vessel wall stretches, the greater is its compliance, which is expressed as a ratio:

$$\text{Compliance} = \frac{\text{Increase in volume (ml)}}{\text{Increase in pressure (mm Hg)}}$$

Vessels with a large compliance exhibit a large increase in volume when the pressure increases a small amount. Vessels with a small compliance do not show a large increase in volume when the pressure increases.

Venous compliance is approximately 24 times greater than the compliance of arteries. As venous pressure increases, the volume of the veins increases greatly. Consequently, veins act as storage areas for blood because their large compliance allows them to hold much more blood than other areas of the vascular system (Table 21-13).

Table 21-13

Distribution of blood volume in blood vessels

VESSELS		TOTAL BLOOD VOLUME (%)
Systemic		
Veins		64
Large veins	(39%)	
Small veins	(25%)	
Arteries		15
Large arteries	(8%)	
Small arteries	(5%)	
Arterioles	(2%)	
Capillaries		5
TOTAL IN SYSTEMIC VESSELS		84
Pulmonary Vessels		9
Heart		7
TOTAL BLOOD VOLUME		100

PHYSIOLOGY OF SYSTEMIC CIRCULATION

The anatomy of the circulatory system, the physics of blood flow, and the regulatory mechanisms that control the heart and blood vessels determine the physiological characteristics of the circulatory system. The entire circulatory system functions to maintain adequate blood flow to all tissues.

Approximately 84% of the total blood volume is contained in the systemic circulatory system. Most of the blood volume is contained in the veins (the vessels with the greatest compliance). Smaller volumes of blood are contained in the arteries and capillaries (see Table 21-13).

Cross-sectional Area of Blood Vessels

If the cross-sectional area of each blood vessel type is determined and is multiplied by the number of each type of blood vessel, the result is the total cross-sectional area for each blood vessel type. For example, there is only one aorta, which has a cross-sectional area of 5 cm^2. On the other hand, there are millions of capillaries, and each has a very small cross-sectional area; however, their total cross-sectional area (2500 cm^2) is much greater than the cross-sectional area of the aorta (Figure 21-26).

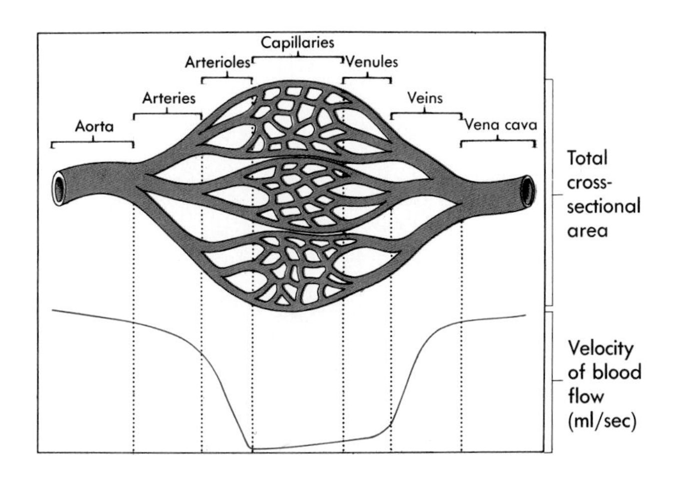

FIGURE 21-26 Total cross-sectional area and velocity of blood flow for each of the major blood vessel types. The line at the bottom shows that blood velocity drops dramatically in arterioles, capillaries, and venules because of the greatly increased cross-sectional area there.

The velocity of blood flow is greatest in the aorta, but the total cross-sectional area is small. In contrast, the total cross-sectional area for the capillaries is large, but the velocity of blood flow is low. As the veins become larger in diameter, their total cross-sectional area decreases, and the velocity of blood flow increases.

Pressure and Resistance

The left ventricle of the heart forcefully ejects blood from the heart into the aorta. Because the pumping action of the heart is pulsatile, the aortic pressure fluctuates between a systolic pressure of 120 mm Hg and a diastolic pressure of 80 mm Hg (Table 21-14; Figure 21-27). As blood flows from arteries through the capillaries and the veins, the pressure falls progressively to approximately 0 mm Hg or even slightly lower by the time it is returned to the right atrium.

The decrease in arterial pressure in each part of the systemic circulation is directly proportional to the resistance to blood flow. There is little resistance in the aorta so that the average pressure at the end of the aorta is nearly 100 mm Hg. The resistance in medium-sized arteries, which are as small as 3 mm in diameter, is also small so that their average pressure is still near 95 mm Hg. In the smaller arteries, however, the resistance to blood flow is greater; thus by the time blood reaches the arterioles, the mean pressure is approximately 85 mm Hg. Within the arterioles the resistance to flow is higher than in any other portion of the systemic circulation, and at their ends the mean pressure is only approximately 30 mm Hg. The resistance is also fairly high in the capillaries. The blood pressure at the arterial end of the capillaries is approximately 30 mm Hg, and it decreases to approximately 10 mm Hg at the venous end. Resistance to blood flow in the veins is low because of their relatively large diameter; by the time the blood reaches the right atrium in the venous system, the mean pressure has decreased from 10 mm Hg to approximately 0 mm Hg.

The muscular arteries and arterioles are capable of constricting or dilating in response to autonomic and hormonal stimulation. If constriction occurs, the resistance to blood flow increases, less blood flows through the constricted blood vessels, and blood is shunted to other, nonconstricted areas of the body.

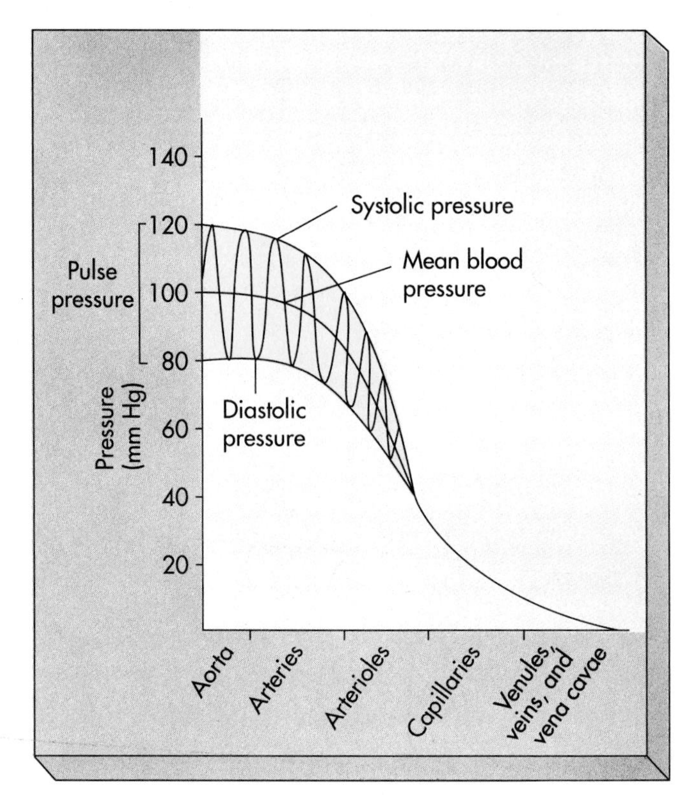

FIGURE 21-27 Blood pressure in each of the major blood vessel types. Blood pressure fluctuations between systole and diastole are damped in small arteries and arterioles. There are no large fluctuations in blood pressure in capillaries and veins.

Table 21-14

Mean systolic and diastolic blood pressures (with standard deviations) in healthy people

	MALES		FEMALES	
AGE (years)	SYSTOLIC	DIASTOLIC	SYSTOLIC	DIASTOLIC
20-24	123 ± 13.7	76 ± 9.9	116 ± 11.8	72 ± 9.7
30-34	126 ± 13.6	79 ± 9.7	120 ± 14.0	75 ± 10.8
40-45	129 ± 15.1	81 ± 9.5	127 ± 17.1	80 ± 10.6
50-54	135 ± 19.2	83 ± 11.3	137 ± 21.3	84 ± 12.4
60-64	142 ± 21.1	85 ± 12.4	144 ± 22.3	85 ± 13.0
70-74	145 ± 26.3	82 ± 15.3	159 ± 25.8	85 ± 15.3
80-84	145 ± 25.6	82 ± 9.9	157 ± 28.0	83 ± 13.1

Muscular arteries help control the amount of blood flowing to each region of the body, and arterioles regulate blood flow through specific tissues. Constriction of an arteriole decreases blood flow through the local area it supplies, and vasodilation increases the blood flow.

Pulse Pressure

The difference between the systolic and diastolic pressures is called the **pulse pressure** (see Figure 21-27). In a healthy young adult at rest the systolic pressure is approximately 120 mm Hg, and the diastolic pressure is approximately 80 mm Hg; thus the pulse pressure is approximately 40 mm Hg. Two major factors influence the pulse pressure: stroke volume of the heart and vascular compliance. When the stroke volume decreases, the pulse pressure also decreases, and when the stroke volume increases, the pulse pressure increases. The compliance of blood vessels decreases as arteries age. Arteries in older people become less elastic (arteriosclerosis), and the resulting decrease in compliance causes the pressure in the aorta to rise more rapidly and to a greater degree during systole and to fall to a lower level during diastole. Thus for a given stroke volume, the pulse pressure is increased as the vascular compliance decreases.

4

Predict the effect of arteriosclerosis on an aortic aneurysm.

? ? ? ? ? ? ? ? ? ? ?

The pulse pressure caused by the ejection of blood from the left ventricle into the aorta produces a pressure wave, or pulse, that travels rapidly along the arteries. Its rate of transmission is approximately 15 times greater in the aorta (7 to 10 m per second) and 100 times greater (15 to 35 m per second) in the distal arteries than the velocity of blood flow. The pressure wave is monitored frequently, especially in the radial artery (the **radial pulse**), to determine heart rate and rhythm. Weak pulses usually indicate a decreased stroke volume or increased constriction of the arteries as a result of intense sympathetic stimulation of the arteries. By the time the pulse pressure passes through the smallest arteries and arterioles, it has been gradually damped so that it is almost absent in the capillaries (see Figure 21-27).

5

Weak pulses also occur in response to ectopic and premature beats of the heart. Explain the cause of the weak pulses under these conditions.

? ? ? ? ? ? ? ? ? ? ?

Hypertension

Hypertension, or high blood pressure, affects approximately 20% of the human population at some time in their lives. Generally a person is considered hypertensive if his systolic blood pressure is greater than 150 mm Hg and his diastolic blood pressure is greater than 90 mm Hg. However, since normal blood pressure is age dependent, classification of an individual as hypertensive depends on his age.

Chronic hypertension has an adverse effect on the function of both the heart and the blood vessels. Hypertension requires the heart to perform a greater-than-normal amount of work. This extra work leads to hypertrophy of the cardiac muscle, especially in the left ventricle, and can lead to heart failure. Hypertension also increases the rate at which arteriosclerosis develops; arteriosclerosis, in turn, increases the probability that blood clots or thromboemboli may form and increases the probability that blood vessels will rupture. Common conditions associated with hypertension are cerebral hemorrhage, coronary infarction, hemorrhage of renal blood vessels, and poor vision caused by burst blood vessels in the retina.

Some conditions leading to hypertension include a decrease in functional kidney mass, excess aldosterone or angiotensin production, and increased resistance to blood flow in the renal arteries. All of these conditions cause an increase in the total blood volume, which causes the cardiac output to increase. The increased cardiac output forces blood to flow through tissue capillaries, causing the precapillary sphincters to constrict. Thus increased blood volume increases the cardiac output and the peripheral resistance, both of which result in a greater blood pressure.

Although these conditions result in hypertension, roughly 90% of the diagnosed cases of hypertension are called either idiopathic, or essential, hypertension—the cause of the condition is not known. Treatments that dilate blood vessels (vasodilators), increase the rate of urine production (diuretics), or decrease cardiac output normally are used to treat essential hypertension. The vasodilator drugs increase the rate of blood flow through the kidneys and thus increase urine production, and the diuretics prevent the resorption of salt and water once they have entered the urine. Substances that decrease cardiac output (such as beta-blocking agents) decrease the heart rate and force of contraction. In addition to these treatments, low-salt diets normally are recommended to reduce the amount of sodium chloride and water absorbed from the intestine into the bloodstream.

Capillary Exchange

There are approximately 10 billion capillaries in the body. The major means by which nutrients and waste products are exchanged is by the process of diffusion. Nutrients diffuse from the capillary into the interstitial spaces, and waste products diffuse in the opposite direction. In addition, fluid is forced out of the capillary at the arteriolar end, and most, but not all, of that fluid reenters the capillary at its venous end.

At the arterial end of the capillary the forces moving fluid out of the capillary are greater than the forces attracting fluids into it, but at the venous end the forces are reversed so that more fluid moves into the capillary. Blood pressure and a small negative pressure in the interstitial spaces force fluid from the capillary. The small negative pressure exists in the interstitial spaces because fluid moves into lymph capillaries when tissues are compressed during movement. One-way valves prevent the fluid from passing from the lymph capillaries into the interstitial spaces when the tissues are no longer compressed.

At the same time water moves by osmosis into the capillary (Figure 21-28). The concentration of small solute molecules is approximately the same in blood and in the interstitial spaces. However, the concentration of proteins in the interstitial spaces is much lower than in the plasma, and the proteins in the plasma are too large to pass through the wall of the capillary. The osmotic pressure caused by the plasma proteins is called the **blood colloid osmotic pressure,** and it is much higher than the osmotic pressure of the interstitial fluid.

Thus at the arteriolar end of the capillary the sum of the forces moving fluid from the capillary minus the forces moving fluid into the capillary results in movement of fluid from the capillaries into the tissue.

At the venous end of the capillary the blood pressure has decreased from approximately 30 to 10 mm Hg. The concentration of proteins within the capillary has increased slightly because of the movement of fluid out of the arteriolar end of the capillary, resulting in greater plasma protein concentration and greater blood colloid osmotic pressure. As a consequence, the force moving fluid from the capillaries is reduced, and the force moving fluid into the capillary is increased slightly. The sum of those forces results in a net inwardly directed movement of fluid at the venous end of the capillary. As a result, nine tenths of the fluid that leaves the capillary at its arteriolar end reenters the capillary at its venous end. The remaining one tenth enters the lymphatic capillaries and eventually is returned to the general circulation.

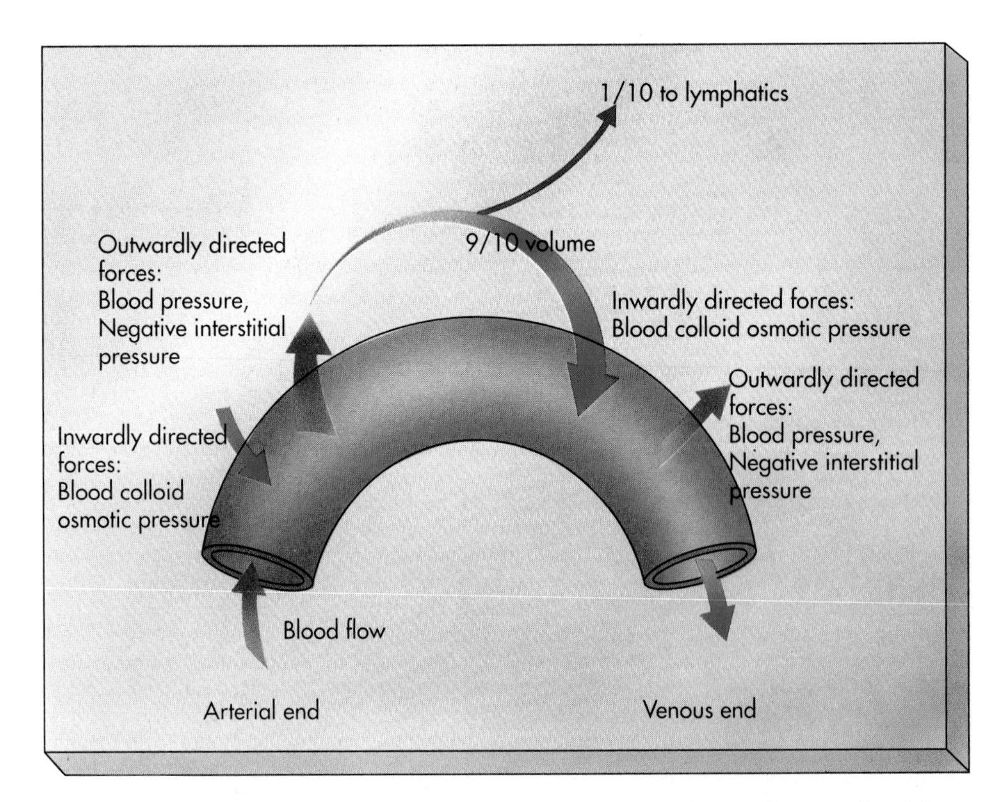

FIGURE 21-28 Total pressure differences between the inside and the outside of the capillary at its arteriolar and venous ends. At the arteriolar end, the sum of the forces causes fluid to move from the capillaries into the tissue. At the venous end, the sum of the forces attracts fluid into the capillary.

Circulatory Changes During Exercise

Blood pressure is regulated within a range of normal values, and blood flow through tissues is matched with the tissues' metabolic needs. During exercise blood flow through tissues is changed dramatically. Its rate of flow through exercising skeletal muscles may be 15 to 20 times greater than through resting muscles. The increased blood flow is the product of both local and systemic regulatory mechanisms. At rest at any point in time, only 20% to 25% of the capillaries in skeletal muscles are open. During exercise this amount increases to 100%. Low oxygen tensions resulting from greatly increased muscular activity or the release of vasodilators such as lactic acid, carbon dioxide, and potassium ions causes dilation of capillaries. Increased sympathetic innervation causes some vasoconstriction in the blood vessels of the skin and viscera, whereas blood vessels of skeletal muscles dilate. Consequently, blood is shunted from the viscera and the skin to the skeletal muscles. Sympathetic stimulation of the heart results in elevated cardiac output, and the blood pressure increases by 20 to 60 mm Hg. In addition, the movement of skeletal muscles and the constriction of veins greatly increase the venous return to the heart. Vasoconstriction in the spleen, liver, and large visceral veins causes them to contract, further increasing the venous return.

Because blood flow through the skin is reduced dramatically in response to sympathetic stimulation at the beginning of exercise, the skin often acquires a whitish color. However, as the body temperature increases in response to the increased muscular activity, temperature receptors in the hypothalamus are stimulated. As a result, sympathetic stimulation of blood vessels in the skin decreases, resulting in vasodilation. As a consequence, the skin turns a red or pinkish color. The skin is an important thermoregulatory organ, and a great deal of excess heat is lost from it as blood flows through its vessels.

Exchange of fluid across the capillary wall also may result from the cyclic dilation and constriction of the precapillary sphincter. When the precapillary sphincter dilates, the pressure rises in the capillary, forcing fluid to move into the interstitial spaces. When the precapillary sphincter constricts, the pressure in the capillary drops, and fluid moves into the capillary.

6

Edema, or swelling, often results from a disruption in the normal inwardly and outwardly directed pressures across the capillary wall. Based on what you know about fluid movement across the wall of the capillary, explain the following (see Figure 21-28): (1) edema as a result of decreased plasma protein concentration; (2) edema as a result of increased capillary permeability to the point that plasma proteins leak into the interstitial spaces; and (3) edema as a result of increased blood pressure within the capillaries.

? ? ? ? ? ? ? ? ? ? ?

Functional Characteristics of Veins

Since cardiac output depends on the volume of blood that enters the heart from the veins (Starling's law of the heart), the factors that affect flow in the veins are of great importance to the overall function of the cardiovascular system. If the blood volume is increased because of rapid transfusion, the amount of blood flow to the heart through the veins increases, and cardiac output increases because of Starling's law. On the other hand, a rapid loss of a large blood volume decreases venous return to the heart; consequently, cardiac output decreases. Increased venous tone (constriction of veins) caused by sympathetic stimulation greatly increases the venous return to the heart by forcing the large venous volume to flow toward the heart. Decreased venous tone decreases venous return to the heart.

The periodic muscular compression of veins forces blood to flow through them toward the heart more rapidly. The valves in the veins prevent flow away from the heart so that when veins are compressed, blood is forced to flow toward the heart. The combination of arteriolar dilation and compression of the veins by muscular movements during exercise causes blood to return to the heart more rapidly than under conditions of rest.

Hydrostatic Pressure and the Effect of Gravity

Blood pressure is approximately 0 mm Hg in the right atrium, and it averages approximately 100 mm Hg pressure in the aorta. However, the pressure in vessels above and below the heart is affected by gravity. While a person is standing, the venous pressure in the feet is influenced greatly by the force of gravity.

Instead of its usual 10 mm Hg pressure at the venules, the pressure may be as much as 90 mm Hg. The arterial pressure is influenced by gravity to the same degree, so the arteriolar end of the capillary may have a pressure of 110 mm Hg rather than 30 mm Hg. The normal pressure difference between the arterial and the venous ends of capillaries still remains the same so that flow continues through the capillary. The major effect of the high pressure in the feet and legs when a person stands for a prolonged period of time without moving is edema. Without muscular movement the pressure at the venous end of the capillaries increases. Up to 15% to 20% of the total blood volume can pass through the walls of the capillaries into the interstitial spaces of the legs during 15 minutes of standing still.

7

While an individual is in a standing position, the large veins of the brain have a negative pressure, as low as −10 mm Hg, and the veins of the neck are normally in a state of partial collapse. Explain the negative pressure in the veins of the head.

? ? ? ? ? ? ? ? ? ? ?

CONTROL OF BLOOD FLOW IN TISSUES

Blood flow provided to the tissues by the cardiovascular system is highly controlled. The blood flow is matched closely to the metabolic needs of tissues. Mechanisms that control blood flow through tissues are classified as (1) local control and (2) nervous control.

Local Control of Blood Flow by the Tissues

Blood flow is much greater in some organs than in others. For example, blood flow through the brain, kidneys, and liver is relatively high. The muscle mass of the body is large so that flow through resting skeletal muscles, although it is not high, is greater than that through other tissue types because skeletal muscle comprises 35% to 40% of the total body mass. However, flow increases through exercising skeletal muscles (up to a twentyfold increase), and the blood flow through the viscera, including the kidneys and liver, either remains the same or decreases.

In most tissues blood flow is proportional to the metabolic needs of the tissue (i.e., blood flow is proportional to the need for oxygen and other nutrients). Blood flow also may increase in response to a buildup of metabolic end products.

Metarteriole

Smooth muscle precapillary sphincter

Blood flow

FIGURE 21-29 Control of local blood flow through capillary beds. **A** Vasodilator theory. Substances such as carbon dioxide are released and cause vasodilation. **B** Nutrient demand theory. Lack of oxygen and other nutrients causes vasodilation.

A VASODILATOR THEORY

Smooth muscles relax due to an increase in vasodilator substances:

CO_2, lactic acid, adenosine, adenosine monophosphate, K^+, H^+

B NUTRIENT DEMAND THEORY

Smooth muscles relax in response to lack of oxygen and other nutrients

In some tissues, however, blood flow serves purposes other than the delivery of nutrients and the removal of waste products. In the skin, blood flow dissipates heat from the body. In the kidneys it eliminates metabolic waste products, regulates water balance, and controls the pH of body fluids. Blood flow through the liver, among other functions, delivers nutrients that have entered the blood from the small intestine enroute to the liver for processing.

Functional characteristics of the capillary bed

The innervation of the metarterioles and precapillary sphincters in capillary beds is sparse (see Figure 21-2; Table 21-15). These structures are regulated primarily by local factors. As the rate of metabolism increases in a tissue, blood flow through its capillaries increases. The precapillary sphincters relax, allowing blood to flow into the local capillary bed. Blood flow may increase sevenfold to eightfold as a result of vasodilation of the metarterioles and the precapillary sphincters in response to an increased rate of metabolism.

Blood flow through capillaries is not continuous but is cyclic. The cyclic fluctuation is the result of periodic contraction and relaxation of the precapillary sphincters called **vasomotion.**

Vasodilator substances are produced as the rate of metabolism increases (Figure 21-29, *A*). The vasodilator substances then diffuse from the tissues supplied by the capillary to the area of the precap-

Table 21-15

*Local control of blood flow**

STIMULUS	RESPONSE
Regulation by Metabolic Need of Tissues	
Increased vasodilator substances (e.g., carbon dioxide, lactic acid, decreased pH) or decreased nutrients (e.g., oxygen, glucose, amino acids) as a result of increased metabolism	Relaxation of precapillary sphincters and subsequent increase in blood flow through capillaries
Decrease in vasodilator substances (e.g., carbon dioxide, lactic acid or increased pH) or increased nutrients (e.g., oxygen, glucose, amino acids, fatty acids)	Contraction of precapillary sphincters and subsequent decrease in blood flow through capillaries
Regulation by Nervous Mechanisms	
Increased physical activity or increased sympathetic activity	Constriction of blood vessels in skin and viscera; dilation of blood vessels in skeletal muscle and coronary vessels
Increased body temperature detected by neurons of the hypothalamus	Dilation of blood vessels in skin (see Chapter 5)
Decreased body temperature detected by neurons of the hypothalamus	Constriction of blood vessels in skin (see Chapter 5)
Decrease in skin temperature below a critical value	Dilation of blood vessels in skin (protects skin from extreme cold)
Anger or embarrassment	Dilation of blood vessels in skin of face and upper thorax
Regulation by Hormonal Mechanisms	
(reinforces increased activity of the sympathetic nervous system)	
Increased physical activity and increased sympathetic activity, causing release of epinephrine and small amounts of norepinephrine from the adrenal medulla	Constriction of blood vessels in skin and viscera; dilation of blood vessels in skeletal and cardiac muscle
Autoregulation	
Increased blood pressure	Contraction of precapillary sphincters to maintain constant capillary blood flow
Decreased blood pressure	Relaxation of precapillary sphinters to maintain constant capillary blood flow
Long-Term Local Blood Flow	
Increased metabolic activity of tissues over a long time period	Increased diameter and number of capillaries
Decreased metabolic activity of tissues over a long time period	Decreased diameter and number of capillaries

*These mechanisms operate when the systemic blood pressure is maintained within a normal range of values.

illary sphincter, the metarterioles, and arterioles to cause vasodilation. Several chemicals—carbon dioxide, lactic acid, adenosine, adenosine monophosphate, adenosine diphosphate, potassium ions, and hydrogen ions—cause vasodilation, and they increase in concentration in the extracellular fluid as the rate of metabolism in tissues increases.

Lack of nutrients may also be important in regulating local blood flow (Figure 21-29, *B*). For example, oxygen and other nutrients are required to maintain vascular smooth-muscle contraction. An increased rate of metabolism decreases the amount of oxygen and other nutrients in the tissues. Smooth-muscle cells of the precapillary sphincter relax in response to a lack of oxygen and other nutrients, resulting in vasodilation.

Autoregulation of blood flow

Arterial pressure can change over a wide range, whereas blood flow through the tissues remains relatively normal. The maintenance of blood flow by tissues is called **autoregulation.** Between arterial pressures of approximately 75 mm Hg and 175 mm Hg, blood flow through tissues remains within 10% to 15% of its normal value. Thus local control mechanisms responsible for autoregulation play a fundamental role in controlling flow through tissues, even though the systemic blood pressure may vary.

8

When blood flow to a tissue has been blocked for a short time, the blood flow through that tissue increases to as much as five times its normal value after the removal of the blockade. The response is called reactive hyperemia. Create a reasonable explanation for that phenomenon based on what you know about the local control of blood flow.

? ? ? ? ? ? ? ? ? ? ?

Long-term local blood flow

The long-term regulation of blood flow through tissues is matched closely to the metabolic requirements of the tissue. If the metabolic activity of a tissue increases and remains elevated, the diameter and the number of capillaries in the tissue will increase, and local blood flow will increase. The increased density of capillaries in the well-trained skeletal muscles of athletes as compared to that in poorly trained skeletal muscles is an example.

The availability of oxygen to a tissue may be a major factor in determining the adjustment of the vascularity of a tissue to its long-term metabolic needs. If there is a lack of oxygen, capillaries increase in diameter and in number, and if the oxygen levels remain elevated in a tissue, the vascularity decreases.

Blockage, or occlusion, of a blood vessel leads to an increase in the diameter of smaller blood vessels that bypass the occluded vessel. In many cases the development of these collateral vessels is marked. For example, if a vessel such as the femoral artery becomes occluded, the small vessels that bypass the occluded vessel become greatly enlarged, and the leg often reestablishes an adequate blood supply over a period of weeks. If the occlusion is sudden and so complete that tissues supplied by a blood vessel suffer from ischemia (lack of blood flow), cell death may occur. In this instance collateral circulation does not have a chance to develop before necrosis occurs.

Nervous Regulation of Local Circulation

Nervous control of arterial blood pressure is important in minute-to-minute regulation and during exercise or shock (extremely low blood pressure). During exercise increased arterial blood pressure is needed to cause blood to flow through the capillaries of skeletal muscles at a rate great enough to supply their oxygen need. Capillaries in skeletal muscle also dilate, resulting in a dramatically increased blood flow through exercising skeletal muscles (see Table 21-15).

Nervous regulation also provides a means by which blood can be shunted from one large area of the peripheral circulatory system to another. For example, in response to blood loss, blood flow to the viscera and the skin is reduced dramatically. This helps maintain the arterial blood pressure within a range sufficient to allow adequate blood flow through the capillaries of the brain and cardiac muscle.

Nervous regulation can function rapidly (within 1 to 30 seconds), and it is controlled by the autonomic nervous system. The most important part of the autonomic nervous system for this regulation is the sympathetic division. Sympathetic vasomotor fibers innervate all blood vessels of the body except the capillaries, precapillary sphincters, and most metarterioles (Figure 21-30). The innervation of the small arteries and arterioles allows the sympathetic nervous system to increase or decrease resistance to blood flow.

9

Innervation of large vessels, especially the veins, makes it possible for the volume of these vessels to change. Predict the consequence on blood pressure if the volume of the veins either increases or decreases. Explain.

? ? ? ? ? ? ? ? ? ? ?

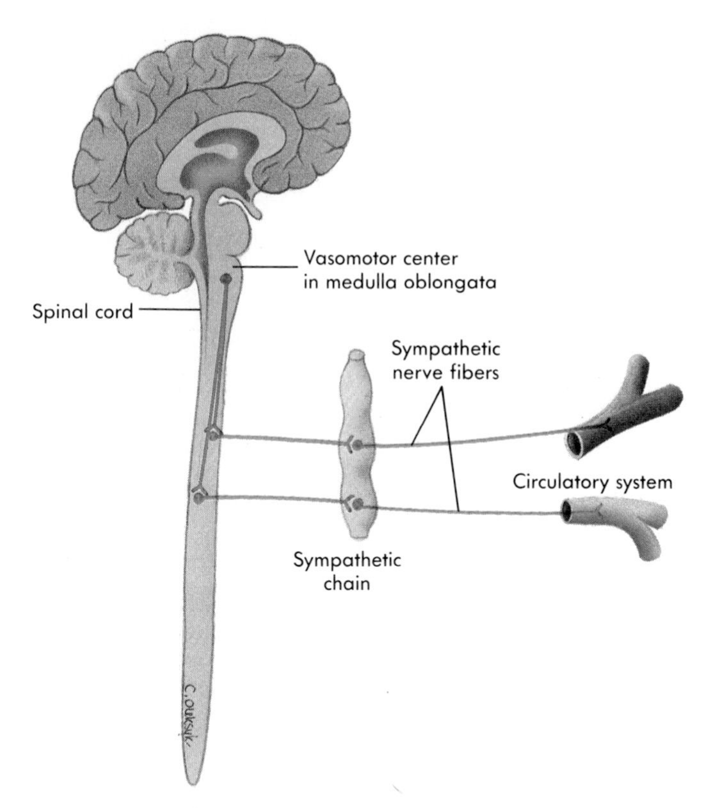

FIGURE 21-30 Nervous regulation of blood vessels. Most blood vessels are innervated by sympathetic nerve fibers. The vasomotor center within the medulla oblongata plays a major role in regulating the frequency of impulses in nerve fibers that innervate blood vessels.

Sympathetic innervation of blood vessels includes both vasoconstrictor and vasodilator fibers, but the sympathetic vasoconstrictor fibers are far more important. Vasoconstrictor fibers are distributed to most parts of the circulatory system; however, they are less prominent in skeletal muscle, cardiac muscle, and the brain and are more prominent in the kidneys, gut, spleen, and skin.

An area of the lower pons and upper medulla oblongata, called the **vasomotor center** (see Figure 21-30), is tonically active. Impulses are transmitted continually to maintain a low frequency of impulses in the sympathetic vasoconstrictor fibers. As a consequence, the peripheral blood vessels are partially constricted, a condition called **vasomotor tone.**

Part of the vasomotor center inhibits vasomotor tone. Thus the vasomotor center consists of an excitatory portion, which is tonically active, and an inhibitory portion, which induces vasodilation when it is active.

Areas throughout the pons, mesencephalon, and diencephalon can either stimulate or inhibit the vasomotor center. For example, the hypothalamus can exert either strong excitatory or inhibitory effects on the vasomotor center. Increased body temperature detected by temperature receptors in the hypothalamus causes vasodilation of blood vessels in the skin (see Chapter 5). The cerebral cortex also can either excite or inhibit the vasomotor center. For example, impulses that originate in the cerebral cortex during exercise or during periods of emotional excitement activate hypothalamic centers, which in turn increase vasomotor tone (see Table 21-15).

The neurotransmitter for the vasoconstrictor fibers is norepinephrine, which binds to alpha-adrenergic receptors on vascular smooth-muscle cells to cause vasoconstriction. Sympathetic impulses also cause the release of epinephrine and norepinephrine into the blood from the adrenal medulla. These hormones are transported in the blood to all parts of the body. In most vessels they cause vasoconstriction, but in some vessels, especially those in skeletal muscle, epinephrine binds to beta-adrenergic receptors, which are present in large numbers, and causes the skeletal muscle blood vessels to dilate.

In addition to vasoconstrictor fibers, sympathetic nerves to skeletal muscles and to the skin carry sympathetic vasodilator fibers, which usually have acetylcholine as their neurotransmitter substance, although some may have epinephrine. The major role for the vasodilator fibers is to cause vasodilation of skeletal muscle blood vessels, allowing an anticipatory increase in blood flow in muscles before they require increased nutrients during exercise.

REGULATION OF MEAN ARTERIAL PRESSURE

The range of normal blood pressures (systolic and diastolic) for people from birth to approximately 80 years of age is presented in Table 21-14. The mean arterial pressure is slightly less than the average of the systolic and diastolic pressure because diastole lasts for a longer time period than systole. The mean arterial pressure is approximately 70 mm Hg at birth, maintains at approximately 100 mm Hg from adolescence to middle age, and reaches 110 mm Hg in the healthy older person, but it may be as high as 130 mm Hg.

Blood flow through the entire circulatory system is determined by the cardiac output (CO), which is equal to the heart rate (HR) times the stroke volume (SV) (i.e., $CO = HR \times SV$), and by resistance to blood flow in all of the blood vessels, which is called **peripheral resistance** (PR). The **mean arterial pressure (MAP)** in the body is proportional to the cardiac output times the peripheral resistance:

$$MAP = CO \times PR \qquad \textit{or} \qquad MAP = HR \times SV \times PR$$

This equation expresses the effect of heart rate, stroke volume, and peripheral resistance on blood pressure. An increase in any one of them results in an increase in blood pressure. Conversely, a decrease in any one

Table 21-16

Control of mean arterial pressure

STIMULUS	SENSORY RECEPTORS	SITE OF INTEGRATION	RESPONSE	EFFECT ON CIRCULATORY SYSTEM
Short-Term (Rapidly Acting) Mechanisms (*respond in seconds to minutes*)				
Baroreceptor reflex				
Decrease in blood pressure	Baroreceptors in carotid sinus and aortic arch	Cardioregulatory and vasomotor centers	Increased sympathetic stimulation of heart and blood vessels; decreased parasympathetic stimulation of heart	Increased heart rate, stroke volume, and vasoconstriction cause an increase in blood pressure.
Increase in blood pressure	Baroreceptors in carotid sinus and aortic arch	Cardioregulatory and vasomotor centers	Decreased sympathetic innervation of heart and blood vessels; increased parasympathetic stimulation of heart	Decreased heart rate and vasodilation cause a decrease in blood pressure.
Chemoreceptor Reflexes				
Decrease in blood oxygen tension or increase in blood carbon dioxide tension	Carotid and aortic body chemoreceptors; medullary chemoreceptors	Cardioregulatory and vasomotor centers	Increased sympathetic stimulation of heart and blood vessels; decreased parasympathetic stimulation of heart	Increased heart rate, stroke volume, and vasoconstriction cause an increase in blood pressure.
Increase in blood oxygen tension and decrease in blood carbon dioxide tension	Carotid and aortic body chemoreceptors; medullary chemoreceptors	Cardioregulatory and vasomotor centers	Decreased sympathetic stimulation of heart and blood vessels; increased parasympathetic stimulation of heart	Decreased heart rate and vasodilation of blood vessels cause a decrease in blood pressure.
Central nervous system ischemic response				
Large decrease in pH, and blood oxygen; large increase in blood carbon dioxide	Chemoreceptors in medulla oblongata	Cardioregulatory and vasomotor centers	Increased sympathetic stimulation of heart and blood vessels; decreased parasympathetic stimulation of heart	Increased heart rate, stroke volume, and vasoconstriction of blood vessels cause an increase in blood pressure.
Adrenal medullary mechanism				
Decrease in blood pressure, excitement, and flight-or-fight response	Baroreceptors in carotid sinus and aortic arch; neurons of cerebrum	Medulla oblongata	Sympathetic stimulation of adrenal medulla causing increased release of epinephrine and norepinephrine	Increased heart rate, stroke volume, and vasoconstriction cause an increase in blood pressure.

of them produces a decrease in blood pressure. The mechanisms that control blood pressure do so by changing peripheral resistance (vasodilation or vasoconstriction), heart rate, or stroke volume. Since stroke volume depends on the amount of blood entering the heart (Starling's law), regulatory mechanisms that control blood volume also affect blood pressure (e.g., an increase in blood volume increases venous return and stroke volume).

When blood pressure suddenly drops because of hemorrhage or some other cause, the control systems respond by increasing the blood pressure to a value consistent with life and by increasing the blood volume to its normal value. Two major types of control systems operate to achieve these responses: (1) systems that respond acutely and (2) systems that respond on a long-term basis (Table 21-16).

Table 21-16

Control of mean arterial pressure—cont'd

STIMULUS	SENSORY RECEPTORS	SITE OF INTEGRATION	RESPONSE	EFFECT ON CIRCULATORY SYSTEM
Long-Term (Slow Acting) Mechanisms (respond in minutes to hours)				
Renin-angiotensin-aldosterone mechanism				
Decrease in blood pressure	Juxtaglomerular apparatus of kidneys	Kidney	Release of renin from juxtaglomerular cells, causing conversion of angiotensinogen to angiotensin I, which is converted to angiotensin II by angiotensin converting enzyme; angiotensin II stimulates the secretion of aldosterone from adrenal cortex	Aldosterone causes reabsorption of sodium ions and water from the tubules of the kidney, which reduces water loss in urine; therefore blood volume increases. Angiotensin II causes vasoconstriction. These responses cause blood pressure to increase.
Vasopressin mechanism				
Decrease in blood pressure or increase in osmolality of blood	Baroreceptors in carotid sinus and aortic arch; neurons in hypothalamus	Hypothalamus	Release of vasopressin (or antidiuretic hormone [ADH]) from neurohypophysis	Vasoconstriction of blood vessels results, and kidneys conserve water, which increases blood volume. Both responses function to increase blood pressure.
Atrial natriuretic mechanism				
Increase in atrial blood pressure	Stretch of cells in atria of the heart that produce atrial natriuretic factor	Atria of heart	Release of atrial natriuretic factor from cells of atria	Atrial natriuretic factor inhibits ADH secretion and acts on the kidney to increase urine volume, which decreases blood volume. Decreased blood volume decreases the blood pressure.
Fluid shift mechanism				
Increase in blood pressure	Increase in pressure in capillaries	Capillary	Movement of fluid out of the capillary across its wall	Decreased blood volume results in decreased blood pressure.
Decrease in blood pressure	Decrease in pressure in capillaries	Capillary	Movement of fluid into the capillary across its wall	Increased blood volume results in decreased blood pressure.
Stress-relaxation response				
Increase in blood pressure	Increased stretch of blood vessel walls	Smooth muscle of blood vessels	Relaxtion of smooth muscle	Vasodilation increases blood vessel volume, causing blood pressure to decrease.
Decrease in blood pressure	Reduced stretch of blood vessel walls	Smooth muscle of blood vessels	Increased contraction of smooth muscle	Vasoconstriction decreases blood vessel volume, resisting a further decrease in blood pressure.

The regulatory mechanisms that control pressure on a short-term basis begin to lose their capacity to regulate blood pressure a few hours to a few days after blood pressure is maintained at high or low values. Short-term regulation after hours or days occurs because sensory receptors adapt to the altered pressures. Long-term regulation of blood pressure is controlled primarily by mechanisms that influence kidney function and do not adapt rapidly to altered blood pressures.

Short-Term Regulation of Blood Pressure

Baroreceptor reflexes

Baroreceptors, or **pressoreceptors,** are sensory receptors that are sensitive to stretch. They are scattered along the walls of most of the large arteries of the neck and the thorax and are very numerous in the area of the carotid sinus at the base of the internal carotid artery (carotid sinus reflex) and in the walls of the aortic arch (aortic arch reflex; Figure 21-31, *A*). Action potentials are transmitted from the carotid sinus baroreceptors through the glossopharyngeal nerves to the cardioregulatory and vasomotor centers in the medulla oblongata and are transmitted from the aortic arch through the vagus nerves to the medulla oblongata.

Increased pressure in blood vessels stretches the vessel walls and results in an increased frequency of action potentials that are generated by the baroreceptors. Conversely, a decrease in blood pressure reduces the stretch of the arterial wall and results in a decreased frequency of action potentials. Normal blood pressure partially stretches the arterial wall so that a constant, but low, frequency of action potentials is produced by the baroreceptors.

The increased frequency of action potentials produced in the baroreceptors by an increase in blood pressure stimulates the cardioregulatory and vasomotor centers of the medulla oblongata. In response, the vasomotor center responds by causing vasodilation of blood vessels, and the cardioregulatory center increases parasympathetic innervation of the heart. As a result, increased systemic blood pressure causes both dilation of peripheral blood vessels and a decreased heart rate, resulting in decreased blood pressure.

Sudden decreases in arterial pressure result in a decreased frequency of impulses generated by the baroreceptors. As a consequence, the vasomotor center responds by increasing peripheral vasoconstriction. In addition, an increase in the sympathetic impulses delivered to the heart from the cardioregulatory center causes the heart rate and stroke volume to increase. This increase is accompanied by a decrease in parasympathetic impulses delivered to the heart (see Figure 21-31, *A*).

The **baroreceptor reflexes** are important in regulating blood pressure on a moment-to-moment basis. When a person rises rapidly from a sitting or lying position to a standing position, a dramatic drop in blood pressure in the neck and thoracic regions occurs because of the pull of gravity on the blood. This reduction may be so great that blood flow to the brain becomes sufficiently sluggish to cause dizziness or loss of consciousness. However, the falling blood pressure initiates the baroreceptor reflexes, which reestablish the normal blood pressure within a few seconds. In a healthy person a temporary sensation of dizziness is all that may be experienced.

10

Explain how the baroreceptor reflex would respond if a person did a headstand.

? ? ? ? ? ? ? ? ? ? ?

The baroreceptors do not change the average blood pressure on a long-term basis. In addition, the baroreceptors adapt within 1 to 3 days to any new blood pressure to which they are exposed. If the blood pressure is elevated for more than a few days, the baroreceptors adapt to the elevated pressure and do not reduce blood pressure to its original value. This adaptation is common in people who have hypertension (chronically elevated blood pressure).

Occasionally the application of pressure to the carotid arteries in the upper neck results in a dramatic decrease in blood pressure. This condition, called the **carotid sinus syndrome,** is most common in patients in whom arteriosclerosis of the carotid artery is advanced. In such patients a tight collar may apply enough pressure to the region of the carotid sinuses to stimulate the baroreceptors. The increased action potentials from the baroreceptors initiate reflexes that result in a decrease in vasomotor tone and an increase in parasympathetic impulses to the heart. As a result of the decreased peripheral resistance and heart rate, the blood pressure decreases dramatically. As a consequence, blood flow to the brain decreases to such a low level that the person either becomes dizzy or faints. People suffering from this condition must avoid applying external pressure to the neck region. If the carotid sinus becomes too sensitive, a treatment for this condition is surgical destruction of the innervation to the carotid sinuses.

Chemoreceptor reflexes

The **chemoreceptor reflexes** help maintain homeostasis when the oxygen tension in the blood decreases or when carbon dioxide and hydrogen ion concentrations increase.

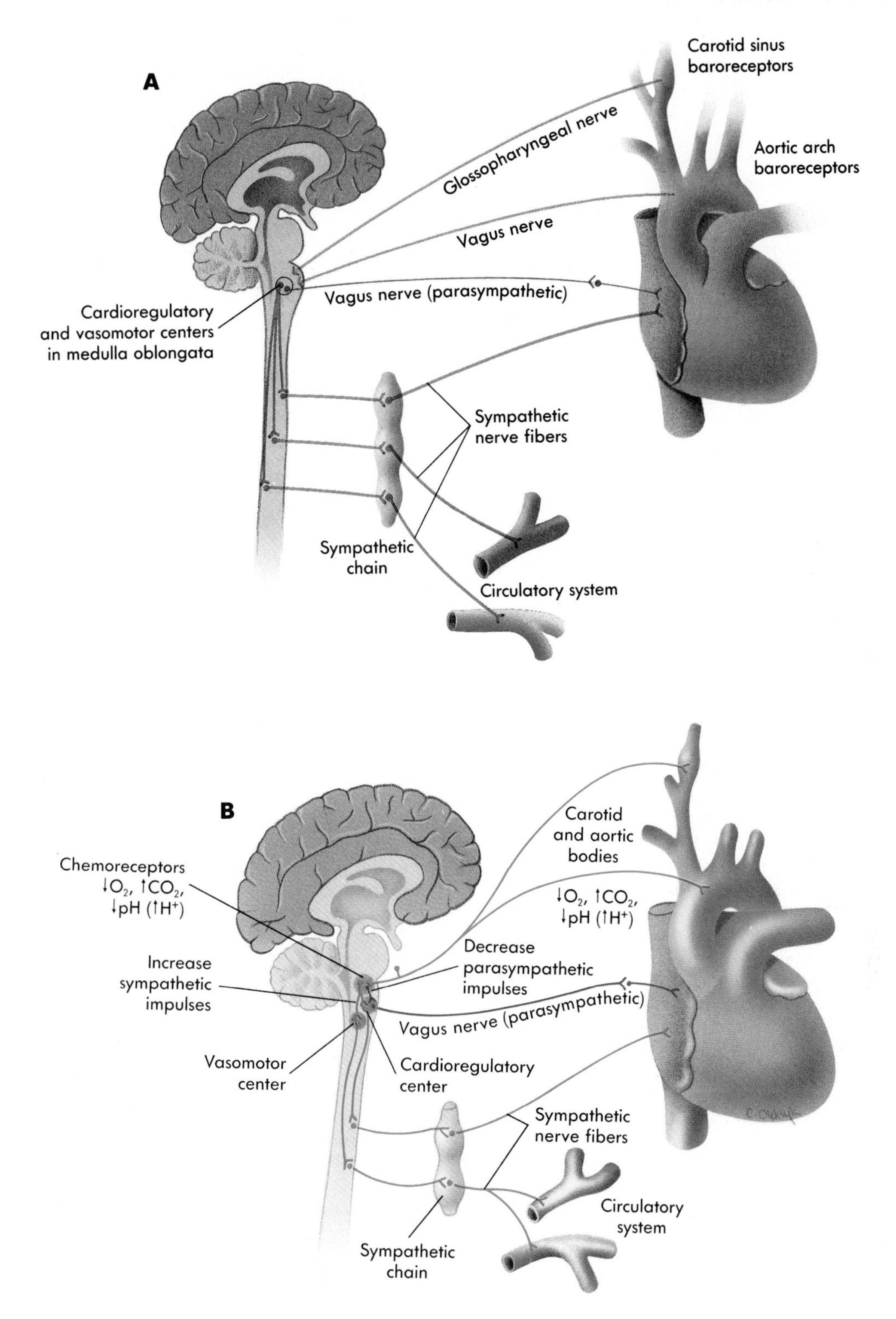

FIGURE 21-31 **A** Baroreceptor reflexes. Baroreceptors located in the carotid sinuses and aortic arch detect changes in blood pressure. Impulses are conducted to the cardioregulatory and vasomotor centers. The heart rate can be decreased by the parasympathetic system; the heart rate and stroke volume can be increased by the sympathetic system. The sympathetic system also can constrict or dilate blood vessels. **B** Chemoreceptor reflexes. Chemoreceptors located in the medulla oblongata and in the carotid and aortic bodies detect changes in blood oxygen, carbon dioxide, or pH. Impulses are conducted to the medulla oblongata. In response, the vasomotor center can cause vasoconstriction or dilation of blood vessels by the sympathetic system, and the cardioregulatory center can cause changes in the pumping activity of the heart through the parasympathetic and sympathetic system.

Shock

Circulatory shock is defined as an inadequate blood flow throughout the body. As a consequence, tissues may suffer from damage due to the delivery of too little oxygen to the cells. Severe shock may damage vital body tissues to the extent that death of the individual occurs.

Depending on its severity, shock can be divided into three separate stages: (1) the nonprogressive or compensated stage, (2) the progressive stage, and (3) the irreversible stage. All types of circulatory shock exhibit one or more of these stages regardless of their causes. There are several causes of shock, but hemorrhagic, or hypovolemic, shock is used to illustrate the characteristics of each stage.

If shock caused by hemorrhage is not severe, the blood pressure may decrease only a moderate amount. Under these conditions the mechanisms that normally regulate blood pressure function to reestablish normal blood pressure and blood flow. The baroreceptor reflexes, chemoreceptor reflexes, and ischemia within the medulla oblongata initiate strong sympathetic responses that result in intense vasoconstriction and increased heart rate. As the blood volume decreases, the stress-relaxation response of blood vessels causes the blood vessels to contract and helps sustain blood pressure. As a result of the reduced blood flow through the kidneys, increased amounts of renin are released. The elevated renin release results in a greater rate of angiotensin II formation, causing vasoconstriction and increased aldosterone release from the adrenal cortex. The aldosterone, in turn, promotes water and salt retention by the kidneys, conserving water. In response to reduced blood pressure, ADH is released from the posterior pituitary gland, and ADH enhances the retention of water by the kidneys. Water also moves from the interstitial spaces and the intestinal lumen to restore the normal blood volume. An intense sensation of thirst increases water intake, also helping to elevate normal blood volume.

In mild cases of shock the baroreceptor reflexes may be adequate to compensate for blood loss until the blood volume is restored, but in more severe cases all of the mechanisms are required to compensate for the blood loss.

In progressive shock the compensatory mechanisms are not adequate to compensate for its effects. As a consequence, a positive-feedback cycle begins to develop in which the blood pressure regulatory mechanisms lose their ability to compensate for shock, making the condition worse. As shock becomes worse, the effectiveness of the regulatory mechanisms deteriorates further. The cycle proceeds until the next stage of shock is reached or until treatment is applied that terminates the cycle.

During progressive shock the blood pressure declines to a very low level that is not adequate to maintain blood flow to the cardiac muscle; thus the heart begins to deteriorate. Substances that are toxic to the heart are released from tissues that suffer from severe ischemia. When the blood pressure declines to a very low level, blood begins to clot in the small vessels, and eventually, blood vessel dilation begins as a result of decreased sympathetic activity and due to the lack of oxygen in capillary beds. Capillary permeability increases under ischemic conditions allowing fluid to leave the blood vessels and enter the interstitial spaces; finally, intense tissue deterioration begins in response to inadequate blood flow.

Without medical intervention progressive shock leads to irreversible shock, which leads to death regardless of the amount or type of medical treatment applied. In this stage of shock, the damage is so extensive that the patient is destined to die.

Patients suffering from shock are normally placed in a horizontal plane, usually with the head slightly lower than the feet, and oxygen is often supplied. Replacement therapy (transfusions of whole blood, plasma, artificial solutions called plasma substitutes, and physiological saline solutions) is administered to increase blood volume. In some circumstances, drugs that enhance vasoconstriction are also administered. Occasionally (especially in patients in anaphylactic shock) antiinflammatory substances such as glucocorticoids and antihistamines are administered. The basic objective in treating shock is to reverse the condition to the point that progressive shock is arrested and to prevent it from progressing to the irreversible stage.

Several types of shock are classified below by the cause of the condition:

Hemorrhagic shock—reduce blood volume caused by either external or internal bleeding.

Plasma loss shock—reduced blood volume due to a loss of plasma into the interstitial spaces and greatly increased blood viscosity.

- *Intestinal obstruction*—results in the movement of large amount of plasma from the blood into the intestine.
- *Severe burns*—loss of large amounts of plasma from the burned surface.
- *Dehydration*—due to a severe and prolonged shortage of water intake.
- *Severe diarrhea or vomiting*—loss of plasma through the intestinal wall.

Neurogenic shock—rapid loss of vasomotor tone that leads to vasodilation to the extent that a severe decrease in blood pressure results.

- *Anesthesia*—deep general anesthesia or spinal anesthesia that decreases the activity of the medullary vasomotor center or the sympathetic nerve fibers.
- *Brain damage*—leads to an ineffective medullary vasomotor function.
- *Emotional shock (vasovagal syncope)*—results from emotions that cause strong parasympathetic stimulation of the heart and results in vasodilation in skeletal muscles and in the viscera.

Anaphylactic shock—due to an allergic response that results in the release of inflammatory substances that cause vasodilation and an increase in capillary permeability.

Septic shock or "blood poisoning"—results from peritoneal, systemic, and gangrenous infections that cause the release of toxic substances into the circulatory system, depressing the activity of the heart, leading to vasodilation, and increasing capillary permeability.

Carotid bodies, small organs approximately 1 to 2 mm in diameter, lie near the carotid sinuses, and several **aortic bodies** lie adjacent to the aorta. Afferent nerve fibers pass to the medulla oblongata through the glossopharyngeal nerve from the carotid bodies and through the vagus nerve from the aortic bodies. Within the carotid and aortic bodies are chemoreceptor cells.

The chemoreceptor cells receive an abundant blood supply. When oxygen availability decreases in the chemoreceptor cells, the frequency of action potentials increases and stimulates the vasomotor center, resulting in increased vasomotor tone (Figure 21-31, *B*). Thus a decrease in blood pressure that results in a lack of oxygen supply to the aortic and carotid bodies or simply a decrease in the oxygen tension in the blood stimulates the chemoreceptor cells.

The chemoreceptors act under emergency conditions and do not regulate the cardiovascular system under resting conditions. They normally do not respond strongly unless the mean blood pressure falls below 80 mm Hg or unless oxygen tension in the blood decreases markedly. The chemoreceptor cells are also stimulated by increased carbon dioxide and hydrogen ions. The response to these substances is similar to the response to low oxygen levels in the blood—the blood pressure increases as a result of vasoconstriction. Increased blood flow to the lungs helps eliminate excess carbon dioxide and hydrogen ions from the body and increases oxygen uptake.

Central nervous system ischemic response

When blood flow to the vasomotor center decreases enough, the neurons within the medulla are strongly excited by a buildup of carbon dioxide, an increase in hydrogen ions, and a decrease in oxygen. As a result, vasoconstriction occurs, and the systemic blood pressure rises dramatically. The elevation in blood pressure in response to a lack of blood flow to the medulla is called the **central nervous system (CNS) ischemic response.**

The CNS ischemic response is important only if the blood pressure falls below 50 mm Hg. Therefore it does not play an important role in regulating blood pressure under normal conditions and functions primarily in response to emergency situations in which blood flow to the brain is severely restricted.

If ischemia lasts longer than a few minutes, the vasomotor center becomes inactive, and extensive vasodilation occurs in the periphery. Prolonged ischemia of the medulla oblongata leads to a massive decline in blood pressure and death.

Hormonal mechanisms

In addition to the rapidly acting nervous mechanisms that regulate arterial pressure, four important hormonal mechanisms control arterial pressure: (1) the adrenal medullary mechanism, (2) the renin-angiotensin-aldosterone mechanism, (3) the vasopressin mechanism, and (4) the atrial natriuretic mechanism.

1. **Adrenal Medullary Mechanism**
   Stimuli that result in increased sympathetic stimulation of the heart and the blood vessels also cause increased stimulation of the adrenal medulla, which results in increased secretion of epinephrine and some norepinephrine from the adrenal medulla. These hormones affect the cardiovascular system in a fashion similar to direct sympathetic stimulation, causing increased heart rate, increased stroke volume, vasoconstriction in blood vessels to the skin and viscera, and vasodilation in blood vessels to skeletal and cardiac muscle (Figure 21-32, *A*).
2. **Renin-Angiotensin-Aldosterone Mechanism**
   The kidneys release an enzyme called **renin** into the circulatory system (see Chapter 26) from specialized structures called the **juxtaglomerular** (juks'tă-glo-mĕr'u-lar) **apparatuses.** Renin acts on plasma proteins called renin substrate, or **angiotensinogen,** to split a fragment off one end. The fragment, called **angiotensin I**, contains 10 amino acid molecules (a decapeptide). Another enzyme called **angiotensin converting enzyme,** found primarily in small blood vessels of the lung, cleaves two additional amino acid molecules from angiotensin I to produce an octapeptide (consisting of eight amino acids) called **angiotensin II,** or **active angiotensin** (Figure 21-32, *B*).

   Angiotensin II causes vasoconstriction in arterioles and to some degree in veins. As a result, it increases peripheral resistance and venous return to the heart, both of which function to raise the blood pressure. Angiotensin II also stimulates aldosterone secretion from the adrenal cortex. **Aldosterone** acts on the kidneys to decrease the production of urine, preventing reduced blood volume caused by the formation of urine (see Chapter 26). Angiotensin II also stimulates the sensation of thirst, increased salt appetite, and antidiuretic hormone (ADH) secretion. Stimuli that increase the rate of renin secretion include decreased blood pressure, elevated plasma concentration of potassium ions, and reduced plasma concentration of sodium ions, of which fluctuations in blood pressure and potassium ion concentration are the most important. Decreased blood pressure and elevated potassium ion concentration occur during plasma loss and dehydration and in response to tissue damage such as burns or crushing injuries.

   The **renin-angiotensin-aldosterone mechanism** is important in maintaining blood pressure under conditions of circulatory shock. The renin-angiotensin-aldosterone system requires approximately 20 minutes to become maximally effective in response to hemorrhagic shock. Its onset of ac-

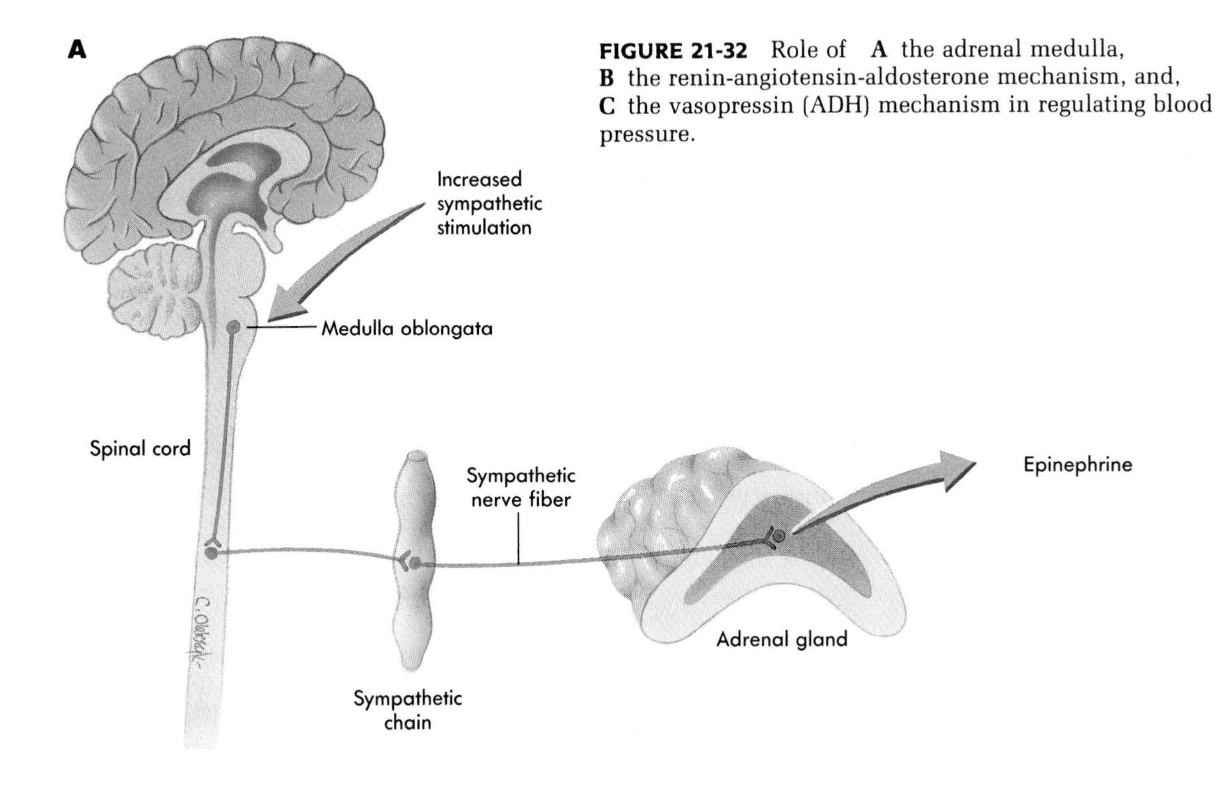

FIGURE 21-32 Role of **A** the adrenal medulla, **B** the renin-angiotensin-aldosterone mechanism, and, **C** the vasopressin (ADH) mechanism in regulating blood pressure.

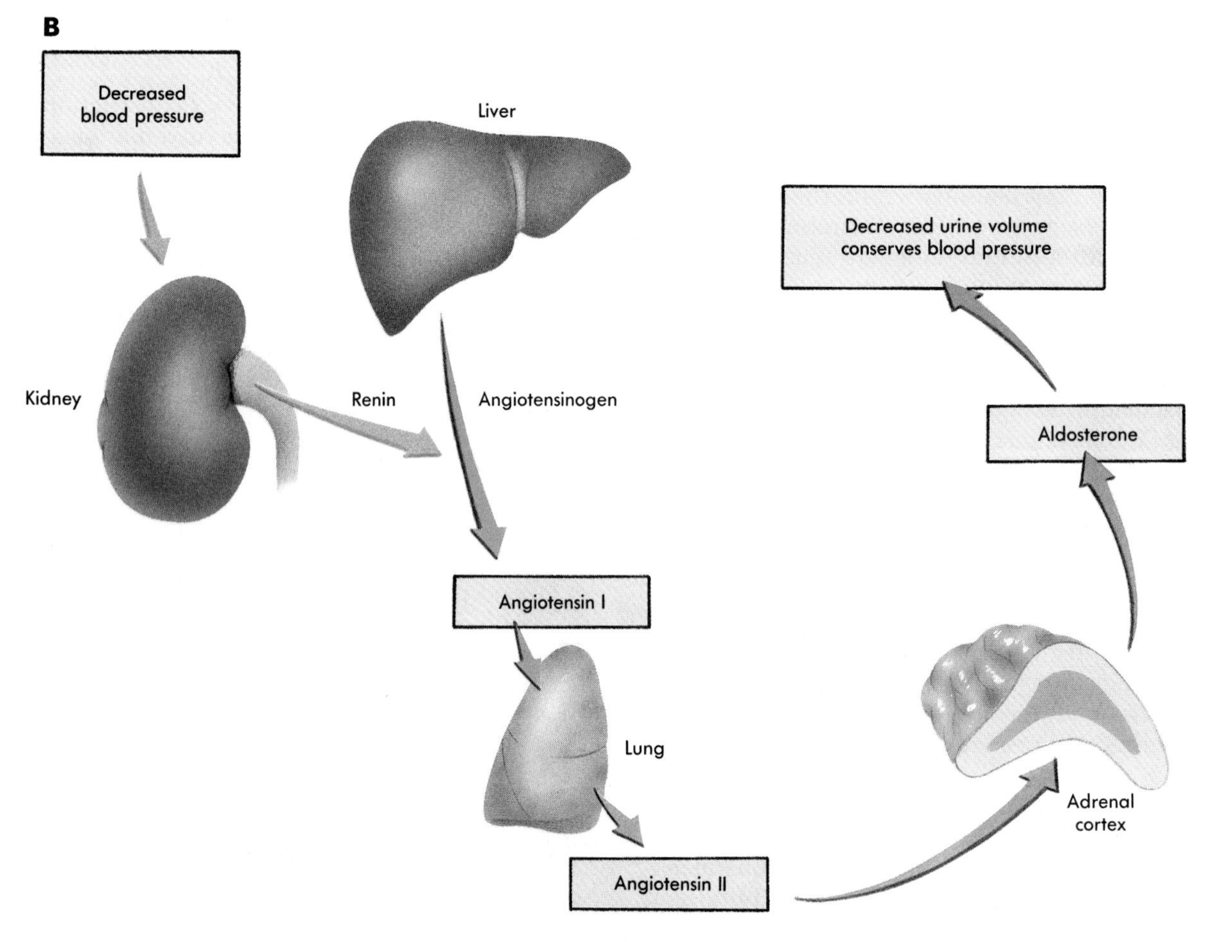

tion is not as fast as nervous reflexes or the adrenal medullary response, but its duration of action is longer. Once renin is secreted, it remains active for approximately 1 hour.

3. **Vasopressin Mechanism**
When blood pressure drops or the concentration of solutes in the plasma increases, hypothalamic neurons increase the frequency of impulses transmitted to the neurohypophysis and increase the secretion of vasopressin, or ADH. Changes in blood pressure affect the frequency of afferent impulses from baroreceptors, which influence the activity of the hypothalamic neurons (Figure 21-32, C). Increases in the plasma concentration of solutes directly affect hypothalamic neurons that stimulate ADH secretion. Under normal conditions small changes in the concentration of solutes in the plasma play a greater role in regulating the rate of ADH secretion than do changes in blood pressure. Under conditions of shock, however, the large decrease in blood pressure results in greatly elevated ADH secretion.
ADH acts directly on blood vessels to cause vasoconstriction, although ADH is not as potent as other vasoconstrictor agents. Evidence indicates that within minutes after a rapid decline in blood pressure, ADH is released in sufficient quantities to affect the reestablishment of normal blood pressures. ADH also decreases the rate of urine production by the kidneys, helping to maintain blood volume and blood pressure.

4. **Atrial Natriuretic Mechanism**
A polypeptide substance called **atrial natriuretic factor** is released from cells in the atria of the heart. A major stimulus for its release is elevated atrial blood pressure. Atrial natriuretic factor increases the rate of urine production by acting on the kidneys and possibly by inhibiting ADH secretion. Loss of water in the urine causes the blood volume to decrease, thus decreasing the blood pressure.

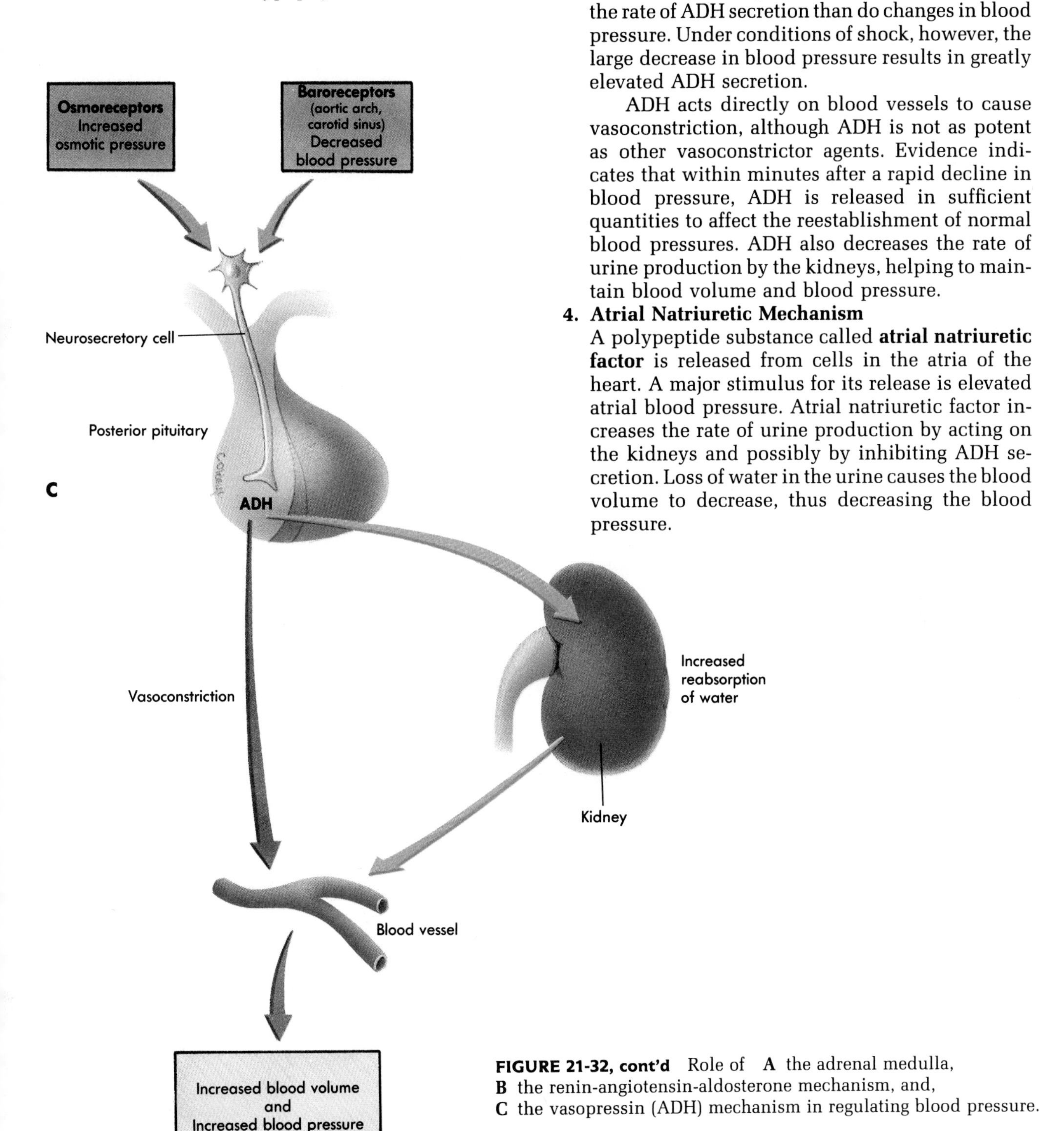

FIGURE 21-32, cont'd Role of **A** the adrenal medulla, **B** the renin-angiotensin-aldosterone mechanism, and, **C** the vasopressin (ADH) mechanism in regulating blood pressure.

Fluid shift mechanism and the stress-relaxation response

Two mechanisms in addition to the nervous and hormonal mechanisms help to regulate systemic blood pressure. The fluid shift and the stress-relaxation mechanisms begin to act within a few minutes and achieve their full functional capacity within a few hours.

The **fluid shift mechanism** operates in response to changes in the pressures across the capillary walls. If blood pressure increases too much, the increase forces some fluid from the blood vessels into the interstitial spaces. The movement of fluid into the interstitial spaces prevents the development of very high blood pressures. If the blood pressure falls to a level that is too low, the decreased blood pressure results in interstitial fluids passing into the capillaries to resist a further decline in the blood pressure. The fluid shift mechanism is a powerful method through which blood pressure is maintained because the interstitial fluid volume is large and acts as a reservoir.

A **stress-relaxation response** is characteristic of smooth-muscle cells (see Chapter 10). When blood volume suddenly declines, blood pressure also decreases, causing a reduction in the force applied to smooth-muscle cells in the blood vessel walls. As a result, during the next few minutes to an hour the smooth-muscle cells contract, reducing the volume of the blood vessels and thus preventing a further decline in blood pressure. Conversely, when the blood volume increases rapidly such as during a transfusion, the blood pressure increases, and the smooth-muscle cells of the blood vessel walls relax, resulting in a more gradual increase in blood pressure. The stress-relaxation mechanism is most effective when changes in blood pressure occur over a period of many minutes.

Long-Term Regulation of Blood Pressure

The mechanisms described previously are extremely important in regulating arterial blood pressure on a short-term basis. Some of the mechanisms also function on a long-term basis to maintain blood pressure in that they operate on a day-to-day and year-to-year basis, are influenced by relatively small disturbances in blood pressure, and respond by bringing the blood pressure back to its normal value.

One long-term mechanism involves the response of the kidneys to variations in blood pressure and includes (1) the renin-angiotensin-aldosterone system and (2) the atrial natriuretic factor.

When the blood pressure drops below 50 mm Hg, the volume of urine produced by the kidneys is close to zero. At 200 mm Hg the volume produced is approximately six to eight times greater than normal. Because of that relationship, the kidney is an important regulator of blood volume.

The kidney is also sensitive to small changes in blood volume. An acute increase of a few hundred milliliters increases the blood pressure significantly. Within several hours the small increase in blood volume is eliminated as urine because the elevated pressure results in increased urinary production (see Chapter 26).

When the intake of water and salt increases, the blood volume and blood pressure increase, and the amount of renin secreted by the kidney decreases. The decreased renin secretion results in a reduced rate at which angiotensin substrate is converted to angiotensin II. As a consequence, vasodilation occurs, causing a reduction in peripheral resistance and blood pressure but also allowing increased blood filtration through the kidneys to increase the urine volume. The decrease in renin secretion also causes a reduction in aldosterone secretion by the adrenal cortex. Since aldosterone promotes sodium and water retention by the kidney, its decrease results in a greater-than-normal excretion of sodium and water by the kidneys and a reduction in blood volume, causing the blood pressure to return to normal.

An increase in blood volume leads to a small increase in blood pressure within the atria of the heart, causing an elevated natriuretic factor secretion. Natriuretic factor causes increased urine production by acting on the kidneys and by inhibiting ADH secretion.

11

The response of long-term mechanisms that regulate blood pressure was just described for an increase in water and salt intake. Describe their response to a severe blood pressure decrease that is maintained for several hours.

? ? ? ? ? ? ? ? ? ? ?

SUMMARY

GENERAL FEATURES OF BLOOD VESSEL STRUCTURE *p. 646*

1. Blood flows from the heart through elastic arteries, muscular arteries, and arterioles to the capillaries.
2. Blood returns to the heart from the capillaries through venules, small veins, and large veins.

Capillaries

1. The entire circulatory system is lined with simple squamous epithelium called endothelium. Capillaries consist only of endothelium.
2. Capillaries are surrounded by loose connective tissue, the adventitia, which contains pericapillary cells.
3. There are three types of capillaries.
   - Fenestrated capillaries have pores called fenestrae that extend completely through the cell.
   - Sinusoidal capillaries are large-diameter capillaries with large fenestrae.
   - Continuous capillaries do not have fenestrae.
4. Materials pass through the capillaries in several ways—between the endothelial cells, through the fenestrae, and through the cell membrane.
5. Blood flows from arterioles through metarterioles and then through the capillary network. Venules drain the capillary network.
   - Smooth muscle in the arterioles, metarterioles, and precapillary sphincters regulates blood flow into the capillaries.
   - Blood can pass rapidly through the thoroughfare channel.

Structure of Arteries and Veins

1. Except for capillaries and venules, blood vessels have three layers.
   - The inner tunica intima consists of endothelium, basement membrane, and internal elastic lamina.
   - The tunica media, the middle layer, contains circular smooth muscle and elastic fibers.
   - The outer tunica adventitia is connective tissue.
2. The thickness and the composition of the layers vary with blood vessel type and diameter.
   - Large elastic arteries are thin walled with large diameters. The tunica media has many elastic fibers and little smooth muscle.
   - Muscular arteries are thick walled with small diameters. The tunica media has lots of smooth muscle and some elastic fibers.
   - Arterioles are the smallest arteries. The tunica media consists of smooth-muscle cells and a few elastic fibers.
   - Venules are endothelium surrounded by a few smooth-muscle cells.
   - Small veins are venules covered with a layer of smooth muscle.
   - Medium-sized veins and large veins contain less smooth muscle and fewer elastic fibers than arteries of the same size.
3. Valves prevent the back flow of blood in the veins.
4. Vasa vasorum are blood vessels that supply the tunica adventitia and tunica media.
5. Arteriovenous anastamoses allow blood to flow from arteries to veins without passing through the capillaries. They function in temperature regulation.

Nerves

The smooth muscle of the tunica media is supplied by sympathetic nerve fibers.

Aging of the Arteries

Arteriosclerosis results from a loss of elasticity in the aorta, large arteries, and coronary arteries.

PULMONARY CIRCULATION *p. 651*

The pulmonary circulation moves blood to and from the lungs. The pulmonary trunk arises from the right ventricle and divides to form the pulmonary arteries, which project to the lungs. From the lungs the pulmonary veins return to the left atrium.

SYSTEMIC CIRCULATION: ARTERIES *p. 651*

Aorta

The aorta leaves the left ventricle to form the ascending aorta, aortic arch, and descending aorta (consisting of the thoracic and abdominal aortae).

Coronary Arteries

Coronary arteries supply the heart.

Arteries to the Head and the Neck

1. The brachiocephalic, left common carotid, and left subclavian arteries branch from the aortic arch to supply the head and the upper limbs. The brachiocephalic artery divides to form the right common carotid and the right subclavian arteries. The vertebral arteries branch from the subclavian arteries.
2. The common carotid arteries and the vertebral arteries supply the head.
   - The common carotid arteries divide to form the external carotids, which supply the face and mouth, and the internal carotids, which supply the brain.
   - The vertebral arteries join within the cranial vault to form the basilar artery, which supplies the brain.

Arteries of the Upper Limbs

1. The subclavian artery continues (without branching) as the axillary artery and then as the brachial artery. The brachial artery divides into the radial and ulnar arteries.
2. The radial artery supplies the deep palmar arch, and the ulnar artery supplies the superficial palmar arch. Both arches give rise to the digital arteries.

Thoracic Aorta and Its Branches

The thoracic aorta has visceral branches that supply the thoracic organs and parietal branches that supply the thoracic wall.

Abdominal Aorta and Its Branches

1. The abdominal aorta has visceral branches that supply the abdominal organs and parietal branches that supply the abdominal wall.
2. The visceral branches are paired and unpaired.
   - The paired arteries supply the kidneys, suprarenal glands, and gonads.
   - The unpaired arteries supply the stomach, spleen, and liver (celiac trunk), the small intestine and upper part of the large intestine (superior mesenteric), and the lower part of the large intestine (inferior mesenteric).

Arteries of the Pelvis

1. The common iliac arteries arise from the abdominal aorta, and the internal iliac arteries branch from the common iliac arteries.
2. The visceral branches of the internal iliac arteries supply pelvic organs, and the parietal branches supply the pelvic wall and floor and the external genitalia.

Arteries of the Lower Limb

1. The external iliac arteries branch from the common iliac arteries.
2. The external iliac artery continues (without branching) as the femoral artery and then as the popliteal artery. The popliteal artery divides to form the anterior and posterior tibial arteries.
3. The posterior tibial artery gives rise to the peroneal and plantar arteries. The plantar arteries form the plantar arch from which the digital arteries arise.

SYSTEMIC CIRCULATION: VEINS *p. 662*

1. The three major veins returning blood to the heart are the superior vena cava (head, neck, thorax, and upper limbs), inferior vena cava (abdomen, pelvis, and lower limbs), and the coronary sinus (heart).
2. Veins are of three types: superficial, deep, and sinuses.

Veins Draining the Heart

Coronary veins enter the coronary sinus or the right atrium.

Veins of the Head and the Neck

1. The internal jugular veins drain the venous sinuses of the anterior head and neck.
2. The external jugular veins and the vertebral veins drain the posterior head and neck.

Veins of the Upper Limb

1. The deep veins are the small ulnar and radial veins of the forearm, which join the brachial vein of the arm. The brachial vein drains into the axillary vein.
2. The superficial veins are the basilic, cephalic, and median cubital. The basilic vein becomes the axillary vein, which then becomes the subclavian vein. The cephalic vein drains into the axillary vein and into the external jugular veins.

Veins of the Thorax

The left and right brachiocephalic veins and the azygos veins return blood to the superior vena cava.

Veins of the Abdomen and Pelvis

1. Abdominal veins join the azygos and hemiazygos veins.
2. Vessels from the kidneys, suprarenal gland, and gonads directly enter the inferior vena cava.
3. Vessels from the stomach, intestines, spleen, and pancreas connect with the hepatic portal vein. The hepatic portal vein transports blood to the liver for processing. Hepatic veins from the liver join the inferior vena cava.

Veins of the Lower Limb

1. The deep veins are the peroneal, anterior and posterior tibialis, popliteal, femoral, and external iliac.
2. The superficial veins are the small and great saphenous veins.

LYMPH VESSELS *p. 670*

1. Lymph vessels carry lymph away from tissues.
2. Lymph capillaries lack a basement membrane and have loosely overlapping epithelial cells. Fluids and other substances easily enter the capillary.
3. Lymph vessels are formed by the joining of lymph capillaries.
   - Lymph vessels have valves that ensure one-way flow of lymph.
   - Skeletal muscle action, contraction of lymph vessel smooth muscle, and thoracic pressure changes move the lymph.
4. Lymph nodes are along the lymph vessels from the abdomen and lower limbs, the left thorax, the upper-left limb, and the left side of the head and the neck. The expanded end of the thoracic duct is the cisterna chyli. The thoracic duct empties into the left subclavian vein.
5. The right lymph duct receives lymph vessels from the right thorax, the upper-right limb, and the right side of the head and the neck. The right lymph duct empties into the right subclavian vein.

PHYSICS OF CIRCULATION *p. 672*

Viscosity

1. Viscosity is the resistance of a liquid to flow. Most of the viscosity of blood is due to red blood cells.
2. The viscosity of blood increases when the hematocrit increases.

Laminar and Turbulent Flow in Vessels

Blood flow through vessels normally is streamlined or laminar. Turbulent flow is disruption of laminar flow.

Blood Pressure

1. Blood pressure is a measure of the force exerted by blood against the blood vessel wall. Blood moves through vessels because of blood pressure.
2. Blood pressure can be measured by listening for Korotkoff sounds produced by turbulent flow in arteries as pressure is released from a blood pressure cuff.

Rate of Blood Flow

Blood flow is the amount of blood that moves through a vessel in a given period of time. Blood flow is directly proportional to pressure differences and is inversely proportional to resistance.

Poiseuille's Law

Resistance is all the factors that inhibit blood flow. Resistance increases when viscosity increases and when blood vessels become smaller in diameter or longer in length.

Critical Closing Pressure and the Law of LaPlace

1. As pressure in a vessel decreases, the force holding it open decreases, and the vessel tends to collapse. The critical closing pressure is the pressure at which a blood vessel will close.
2. The law of LaPlace states that the force acting on the wall of a blood vessel is proportional to the diameter of the vessel times the blood pressure.

Vascular Compliance

Vascular compliance is a measure of the change in volume of blood vessels produced by a change in pressure. The venous system has a large compliance and acts as a blood reservoir.

PHYSIOLOGY OF SYSTEMIC CIRCULATION *p. 675*

The greatest volume of blood is contained in the veins. The smallest volume is in the arterioles.

Cross-Sectional Area of Blood Vessels

As the diameter of vessels decreases, their total cross-sectional area increases, and the velocity of blood flow through them decreases.

Pressure and Resistance

Blood pressure averages 100 mm Hg in the aorta and drops to 0 mm Hg in the right atrium. The greatest drop occurs in the arterioles, which regulate blood flow through tissues.

Pulse Pressure

1. Pulse pressure is the difference between systolic and diastolic pressures. Pulse pressure increases when stroke volume increases or vascular compliance decreases.

2. Pulse pressure waves travel through the vascular system faster than the blood flows. Pulse pressure can be used to take the pulse.

Capillary Exchange

1. Hydrostatic pressure, capillary permeability, and osmosis affect movement of fluid from the capillaries.
2. There is a net movement of fluid from the blood into the tissues. The fluid gained by the tissues is removed by the lymphatic system.

Functional Characteristics of Veins

Venous return to the heart increases because of an increase in blood volume, venous tone, and arteriole dilation.

Hydrostatic Pressure and the Effect of Gravity

In a standing person hydrostatic pressure caused by gravity increases blood pressure below the heart and decreases pressure above the heart.

CONTROL OF BLOOD FLOW IN TISSUES *p. 680*

Local Control of Blood Flow by the Tissues

1. Blood flow through a tissue is usually proportional to the metabolic needs of the tissue. Exceptions are tissues that perform functions that require additional blood.
2. Control of blood flow by the metarterioles and precapillary sphincters may be regulated by vasodilator substances or by lack of nutrients.
3. Only large changes in blood pressure have an effect on blood flow through tissues.
4. If the metabolic activity of a tissue increases, the number and the diameter of capillaries in the tissue will increase.

Nervous Regulation of Local Circulation

1. The sympathetic nervous system (vasomotor center in the medulla) controls blood vessel diameter. Other brain areas can excite or inhibit the vasomotor center.
2. Vasomotor tone is a state of partial contraction of blood vessels.
3. The nervous system is responsible for routing the flow of blood and maintaining blood pressure.

REGULATION OF MEAN ARTERIAL PRESSURE *p. 683*

Mean blood pressure is proportional to cardiac output times the peripheral resistance.

Short-Term Regulation of Blood Pressure

1. Baroreceptors are sensory receptors that are sensitive to stretch.
   - Baroreceptors are located in the carotid sinuses and the aortic arch.
   - The baroreceptor reflex changes peripheral resistance, heart rate, and stroke volume in response to changes in blood pressure.
2. Chemoreceptors are sensory receptors sensitive to oxygen, carbon dioxide, and pH levels in the blood.
   - Chemoreceptors are located in the carotid bodies and the aortic bodies.
   - The chemoreceptor reflex changes peripheral resistance, usually in response to low oxygen levels.
3. The central nervous system (CNS) ischemic response results from high carbon dioxide or low pH levels in the medulla and increases peripheral resistance.
4. Epinephrine and norepinephrine are released from the adrenal medulla as a result of sympathetic stimulation. They increase heart rate, stroke volume, and vasoconstriction.
5. Renin is released by the kidneys in response to low blood pressure. Renin promotes the production of angiotensin II, which causes vasoconstriction and an increase in aldosterone secretion.
6. Antidiuretic hormone (ADH) released from the posterior pituitary causes vasoconstriction.
7. Atrial natriuretic factor is released from the heart when atrial blood pressure increases. It stimulates an increase in urinary production, causing a decrease in blood volume and blood pressure.
8. Fluid shift is a movement of fluid from the interstitial spaces to maintain blood volume.
9. Stress-relaxation response is an adjustment of blood vessels' smooth muscles in response to a change in blood volume.

Long-Term Regulation of Blood Pressure

1. The kidneys regulate blood pressure by controlling blood volume.
2. In response to an increase in blood volume, the kidneys produce more urine and decrease blood volume. Renin, angiotensin II, aldosterone, atrial natriuretic factor, and sympathetic stimulation play a role in controlling urinary volume.

CONTENT REVIEW

1. Name, in order, all the types of blood vessels, starting at the heart, going to the tissues, and returning to the heart.
2. Describe the three types of capillaries. Explain the ways that materials pass through the capillary wall.
3. Describe a capillary network. Where is the smooth muscle that regulates blood flow into and through the capillary network located? What is the function of the thoroughfare channel?
4. Name the three layers of a blood vessel. What kinds of tissue are in each layer?
5. For the different types of arterial and venous blood vessels, compare the amount of elastic fibers and smooth muscle in each.
6. What is the function of valves in blood vessels? In which blood vessels are they found?
7. Define vasa vasorum and arteriovenous anastomoses, and give their function.
8. Name the different parts of the aorta. Name the major arteries that branch from the aorta to supply the heart, the head and upper limbs, and the lower limbs.
9. What areas of the body are supplied by the paired arteries that branch from the abdominal aorta? The unpaired arteries?
10. Name the three major vessels that return blood to the heart. What areas of the body do they drain?
11. Name the three major veins that return blood to the superior vena cava.
12. Explain the three ways that blood from the abdomen returns to the heart.

13. List the major deep and superficial veins of the upper and lower limbs.
14. Describe the structure of a lymph capillary. Explain why the structure makes it easy for fluid and other substances to enter the capillary.
15. What is the function of the valves in lymph vessels? What causes lymph to move through the lymph vessels?
16. What parts of the body are drained by the thoracic duct and right lymphatic duct? What is the cisterna chyli?
17. Define viscosity, and state the effect of hematocrit on viscosity. Define laminar flow and turbulent flow.
18. Define blood pressure, blood flow, and resistance. How can each be determined?
19. State Poiseuille's law. What effect do viscosity, blood vessel diameter, and blood vessel length have on resistance? On blood flow?
20. State the law of LaPlace. How does it explain critical closing pressure and aneurysms?
21. Define vascular compliance. Do veins or arteries have the greater compliance?
22. Describe the distribution of blood volumes throughout the circulatory system.
23. What is the relationship between blood vessel diameter, total cross-sectional area, and blood flow velocity?
24. Describe the changes in blood pressure, starting in the aorta, moving through the vascular system, and returning to the right atrium.
25. What is pulse pressure? How do stroke volume and vascular compliance affect pulse pressure?
26. Describe the factors that influence the movement of fluid from capillaries into the tissues. What happens to the fluid in the tissues? What is edema?
27. How do blood volume and tone in large blood vessels and arterioles affect cardiac output?
28. What effect does standing have on blood pressure in the feet and the head? Explain why this effect occurs.
29. Explain how vasodilator substances and nutrients are involved with local control of blood flow. What is autoregulation of local blood flow? How is long-term regulation of blood flow through tissues accomplished?
30. Describe nervous control of blood flow. Define vasomotor tone.
31. Where are baroreceptors located? Describe the response of the baroreceptor reflex when blood pressure increases and decreases.
32. Where are the chemoreceptors located? Describe what happens when oxygen levels in the blood decrease.
33. Describe the central nervous system ischemic response.
34. For each of the following hormones—epinephrine, norepinephrine, renin, angiotensin, aldosterone, antidiuretic hormone, and atrial natriuretic factor—state where the hormone is produced and what effects it has on the circulatory system.
35. What is fluid shift, and what does it accomplish? Describe the stretch-relaxation response of a blood vessel.
36. Discuss two ways that the kidneys are involved in the long-term regulation of blood pressure.

CONCEPT REVIEW

1. For each of the following destinations, name all the arteries that a red blood cell would encounter if it started its journey in the left ventricle.
   A Posterior interventricular groove of the heart
   B Anterior neck to the brain (give two ways)
   C Posterior neck to the brain (give two ways)
   D External skull
   E Tip of the fingers of the left arm (What other blood vessel would be encountered if the trip were through the right arm?)
   F Anterior compartment of the leg
   G Liver
   H Small intestine
   I Urinary bladder
2. For each of the following starting places, name all the veins that a red blood cell would encounter on its way back to the right atrium.
   A Anterior interventricular groove of the heart (give two ways)
   B Venous sinus in the brain
   C External posterior of skull
   D Hand (return deep and superficial)
   E Foot (return deep and superficial)
   F Stomach
   G Kidney
   H Left inferior wall of the thorax
3. In a study of heart valve functions it was necessary to inject a dye into the right atrium of the heart by inserting a catheter into a blood vessel and moving the catheter into the right atrium. What route would you suggest? If you wanted to do this procedure into the left atrium, what would you do differently?
4. In endurance-trained athletes the hematocrit may be lower than normal because plasma volume increases more than red blood cell numbers increase. Explain why this condition would be beneficial.
5. All the blood that passes through the aorta (past the point where the coronary arteries branch) returns to the heart through the venae cavae. (HINT: The radius of the aorta is 13 mm, and the radius of a vena cava is 16 mm.) Explain why the resistance to blood flow in the aorta is greater than the resistance to blood flow in the venae cavae. Since the resistances are different, explain why blood flow can be the same.
6. As blood vessels increase in diameter, the amount of smooth muscle decreases, and the amount of connective tissue increases. Explain why. (HINT: Law of LaPlace.)
7. A patient is suffering from edema in the lower-right limb. Explain why massage would help remove the excess fluid.
8. A very short nursing student was asked to measure the blood pressure of a very tall person. She decided to measure the blood pressure at the level of his foot while the tall person was standing. What artery did she use? After taking the blood pressure, she decided that the tall person was suffering from hypertension because the systolic pressure was 200 mm Hg. Is her diagnosis correct?
9. During hyperventilation carbon dioxide is "blown off," and carbon dioxide levels in the blood decrease. What effect would this decrease have on blood pressure? Explain. What symptoms would you expect to see as a result?
10. Epinephrine causes vasodilation of blood vessels in cardiac muscle but vasoconstriction of blood vessels in the skin. Explain why this is a beneficial arrangement.
11. One cool evening Skinny Dip jumped into a hot Jacuzzi. Predict what happened to Skinny's heart rate.

? ?

ANSWERS TO PREDICT QUESTIONS

1 p. 653. Arteriosclerosis slowly reduces blood flow through the carotid arteries and therefore the amount of blood that flows to the brain. As the resistance to flow increases in the carotid arteries during the late stages of arteriosclerosis, the blood flow to the brain is compromised, and confusion, loss of memory, and loss of the ability to perform other normal brain functions occur.

2 p. 672. Cutting and tying off the lymph vessels would prevent the movement of fluid from the affected tissue. The result would be edema.

3 p. 674. (1) Vasoconstriction of blood vessels in the skin in response to exposure to cold results in a decreased flow of blood through the skin and in a dramatic increase in resistance (see Poiseuille's law). (2) Vasodilation of blood vessels in the skin results in increased blood flow through the skin. Vasoconstriction makes the skin appear pale, and vasodilation makes the skin appear flushed or red in color. (3) A decrease in the rate of red blood cell synthesis will reduce the hematocrit and decrease the viscosity of blood. If blood's viscosity decreases and all other factors remain constant, blood flows through the vessels at a greater rate (see Poiseuille's law). (4) In the case of a patient with polycythemia vera the hematocrit is increased dramatically. As a result, the resistance to flow increases, flow decreases, or a greater pressure is needed to maintain the same flow.

4 p. 677. An aneurysm in the aorta is a major problem because the tension applied to the aneurysm becomes greater as the size of the aneurysm increases (see law of LaPlace). The aneurysm usually develops because of a weakness in the wall of the aorta. Arteriosclerosis complicates the matter by making the wall of the artery less elastic and by increasing the systolic blood pressure. The decreased elasticity and the increased blood pressure increase the probability that the aneurysm will rupture.

5 p. 677. Premature beats of the heart and ectopic beats result in contraction of the heart muscle before the heart has had time to fill to its normal capacity. Consequently, there is a reduced stroke volume, which results in a weak pulse.

6 p. 679. (1) Decreased plasma protein concentration reduces the colloid osmotic pressure of the blood. Edema results because less fluid reenters the venous end of the capillary and more fluid remains in the interstitial spaces. (2) The increased permeability allows plasma protein to leak into the interstitial spaces, causing an increase in the colloid osmotic pressure in the interstitial spaces. Consequently, the inwardly directed osmotic force that moves fluid from the interstitial spaces into the capillaries is reduced. More fluid leaves the arterial end of the capillary and less fluid enters the venous end of the capillary causing a buildup of interstitial fluid (i.e., edema). (3) Increased blood pressure within the capillary increases the amount of fluid that is forced from the arteriolar end of the capillary and reduces the amount of the fluid reentering the capillary at its venous end.

7 p. 680. The negative pressure within the veins of the head is due to the effect of gravity on blood flow. The pressure in the veins in the head increases if the head is lower than the heart.

8 p. 682. Reactive hyperemia can be explained based on any of the theories for the local control of blood pressure. When a blood vessel is occluded, nutrients are depleted, and waste products accumulate in tissue that is suffering from a lack of adequate blood supply. Both of these effects cause vasodilation and a greatly increased blood flow through the area after the occlusion has been removed.

9 p. 682. Sudden vasodilation of the veins in the abdomen causes blood flow return to the heart to decrease dramatically. The veins have a large capacity and can hold a large volume of blood. If the blood flow back to the heart decreases, the cardiac output and the systemic blood pressure also decrease. If the veins suddenly constrict, the blood flow into the heart increases, causing an increase in the cardiac output of the heart and an increase in blood pressure.

10 p. 686. During a headstand gravity acting on the blood would cause the blood pressure in the area of the aortic arch and carotid sinus baroreceptors to increase. The increased pressure would activate the baroreceptor reflexes, increasing parasympathetic stimulation of the heart and decreasing sympathetic stimulation. Thus the heart rate would decrease. Because standing on one's head also causes blood from the periphery to run downhill to the heart, the venous return would increase, causing the stroke volume to increase because of Starling's law of the heart. Some peripheral vasodilation also may occur because of the elevated baroreceptor pressure's causing a decrease in vasomotor tone.

11 p. 692. A decrease in blood pressure maintained over a period of several hours results in the following responses:

- Baroreceptor reflexes increase the heart rate.
- Baroreceptor reflexes increase the peripheral resistance.
- Baroreceptor reflexes increase epinephrine secretion and antidiuretic hormone secretion.
- Increased renin secretion causes angiotensin II activation, leading to vasoconstriction and aldosterone secretion.
- Decreased secretion of atrial natriuretic factor results.
- Lack of blood flow to tissues results in ischemia and deterioration of tissues, resulting in the release of substances from the tissues. These compounds are toxic to the heart and cause vasodilation and increased blood vessel permeability. If these events occur, shock becomes irreversible, and the victim dies (see essay on shock).

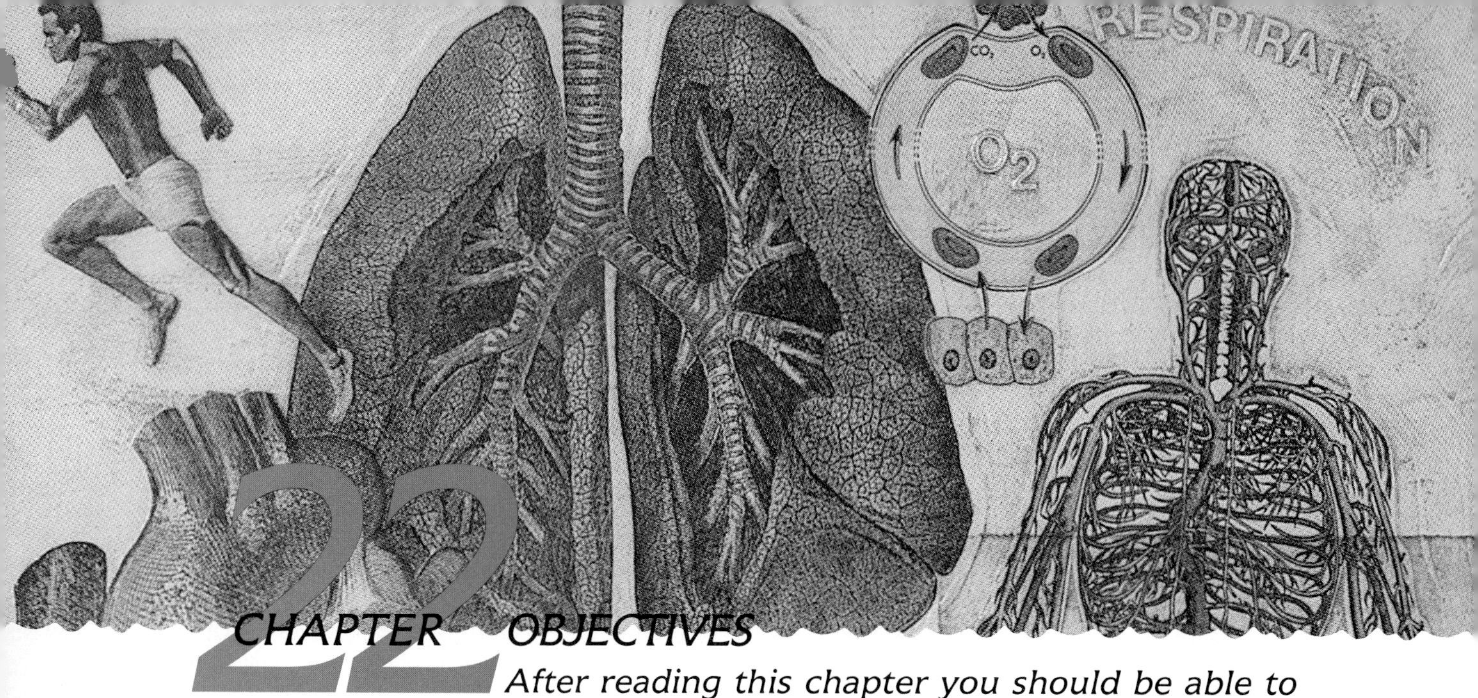

22 CHAPTER OBJECTIVES

After reading this chapter you should be able to

1. Describe the functions of the lymphatic system.
2. Describe the structure and functions of diffuse lymphatic tissue, lymph nodules, lymph nodes, tonsils, spleen, and thymus.
3. Define nonspecific resistance and describe the cells and chemicals involved.
4. Name the phagocytic cells of the immune system and describe their location and activities.
5. List the events that occur during an inflammatory response and explain their significance.
6. Define the term antigen.
7. Describe the origin, development, activation, and regulation of lymphocytes.
8. Define and explain the importance of the major histocompatibility complex.
9. Define antibody-mediated immunity and cell-mediated immunity and name the cells responsible for each.
10. Describe the structure of an antibody, list the different types of antibodies, and describe the effects produced by antibodies.
11. Discuss the primary and secondary response to an antigen. Explain the basis for long-lasting immunity.
12. Describe the functions of T cells.
13. Explain four ways by which specific immunity can be acquired.

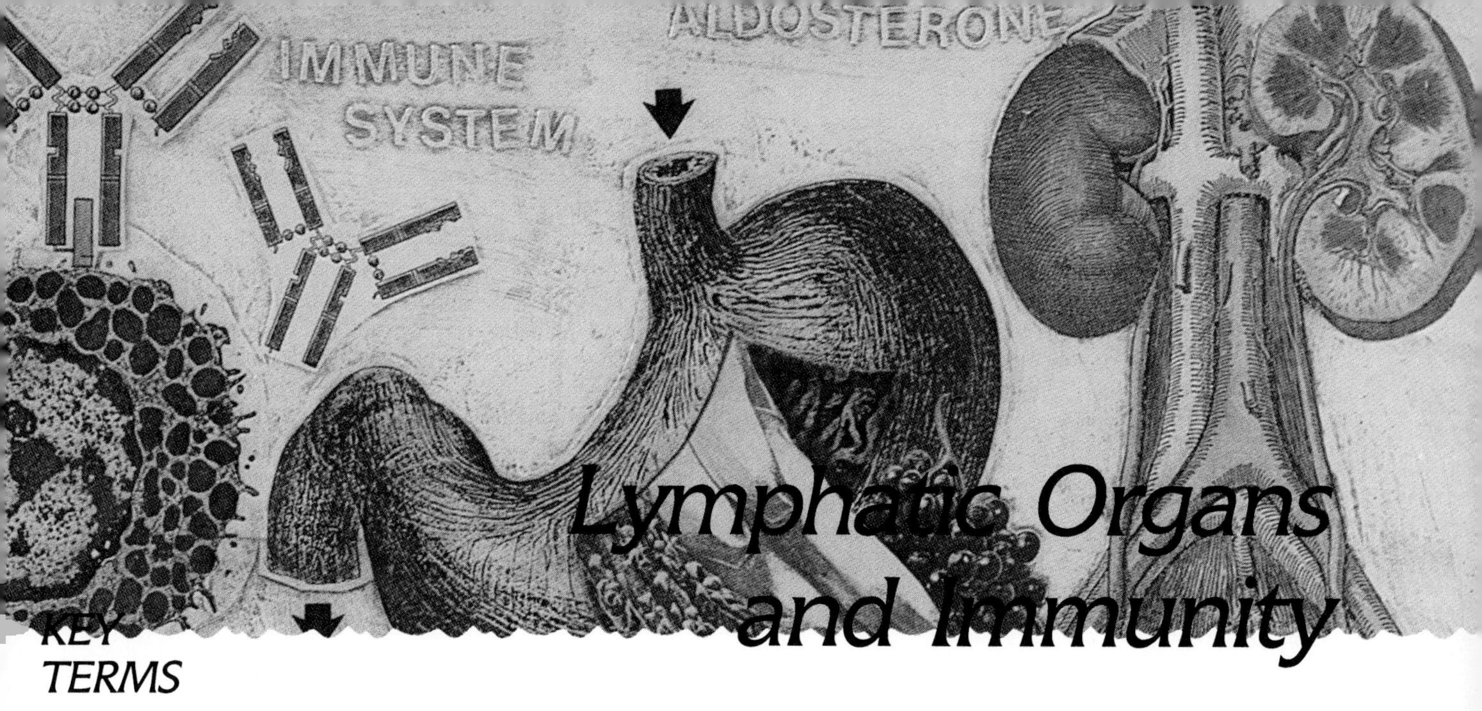

Lymphatic Organs and Immunity

KEY TERMS

antibody

antibody-mediated immunity

antigen (an′tĭ-jen)

cell-mediated immunity

complement

inflammatory response

interferon (in′ter-fēr′on)

lymph (limf)

lymph node

lymphokine (lim′fo-kīn)

major histocompatibility complex

memory cell

mononuclear phagocytic system

nonspecific resistance

specific immunity

RELATED TOPICS

The following terms or concepts are important for a good understanding of this chapter. If you are not familiar with them, you should review them before proceeding.

Lock-and-key model for enzyme function (Chapter 2)

Protein structure (Chapter 2)

Phagocytosis (Chapter 3)

Inflammation (Chapter 4)

Leukocyte structure and function (Chapter 19)

Lymph vessels (Chapter 21)

THE **LYMPHATIC** (LIM-FAT′IK) **SYSTEM** INCLUDES LYMPH, LYMPHOCYTES, LYMPH VESSELS, LYMPH NODES, TONSILS, THE SPLEEN, AND THE THYMUS GLAND (FIGURE 22-1). The lymphatic system performs three basic functions. First, it helps maintain fluid balance in the tissues (see Chapter 21). Approximately 30 L of fluid pass from the blood capillaries into the interstitial spaces each day, whereas only 27 L pass from the interstitial spaces back into the blood capillaries. If the extra 3 L of interstitial fluid were to remain in the interstitial spaces, edema would result, causing tissue damage and eventual death. These 3 L of fluid enter the lymph capillaries, where it is called **lymph** (limf, clear spring water). The lymph then passes through the lymph vessels to return to the blood. Lymph is similar in composition to plasma (see Chapter 19). In addition to water, lymph contains solutes derived from two sources: (1) substances in plasma such as ions, nutrients, gases, and some proteins pass from blood capillaries into the interstitial spaces to become part of the lymph; and (2) substances derived from cells within the tissues such as hormones, enzymes, and waste products are also found in lymph.

The lymphatic system's second basic function is to absorb fats and other substances from the digestive tract (see Chapter 24). Special lymph vessels called **lacteals** (lak′te-als) are in the lining of the small intestine. Fats enter the lacteals and pass through the lymph vessels to the venous circulation. The lymph passing through these capillaries has a milky appearance because of its fat content, and it is called **chyle** (kīl).

Third, the lymphatic system is part of the body's defense system. The lymph nodes filter lymph, and the spleen filters blood, removing microorganisms and other foreign substances. In addition, lymph tissue contains lymphocytes and other cells that are capable of destroying microorganisms and foreign substances. The emphasis of this chapter is on the major lymph organs and their role in protecting us against disease.

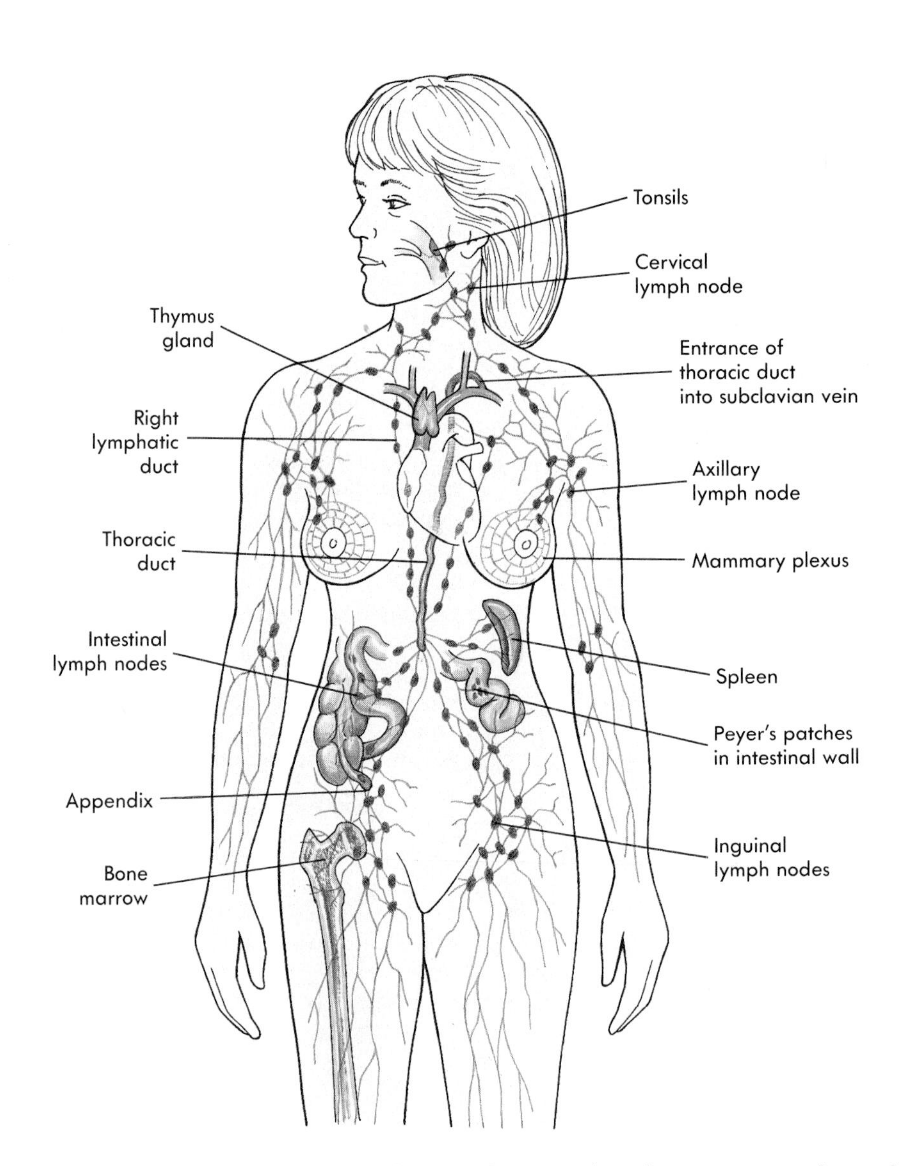

FIGURE 22-1 Lymphatic system showing the major lymphatic organs and vessels.

LYMPHATIC ORGANS

Lymphatic organs contain **lymphatic tissue,** which consists of many lymphocytes and other cells. The lymphocytes originate from red bone marrow (see Chapter 19) and are carried by the blood to lymph organs. When the body is exposed to microorganisms or foreign substances, the lymphocytes divide and increase in numbers. The lymphocytes are part of the immune system response that results in the destruction of microorganisms and foreign substances. In addition to cells, lymphatic tissue has very fine collagen fibers. These fibers form an interlaced network that holds the lymphocytes and other cells in place. When lymph or blood filters through lymph organs, the fiber network also traps microorganisms and other items in the fluid.

Diffuse Lymphatic Tissue and Lymph Nodules

Diffuse lymphatic tissue contains dispersed lymphocytes and other cells, has no clear boundary, and blends with surrounding tissues (Figure 22-2). It is under mucous membranes, around lymph nodules, and within the lymph nodes and spleen.

Lymph nodules are a denser arrangement of lymphoid tissue organized into compact, somewhat spherical structures, ranging in size from a few hundred microns to a few millimeters or more in diameter (see Figure 22-2). Lymph nodules are in the loose connective tissue of the digestive, respiratory, and urinary systems. In lymph nodes and the spleen, lymph nodules are usually referred to as lymph follicles. In some locations many lymph nodules join together to form larger structures. For example, **Peyer's patches** are aggregations of lymph nodules found in the lower half of the small intestine and the appendix.

Tonsils

Tonsils are unusually large groups of lymph nodules located beneath the mucous membranes within the oral cavity and the nasopharynx (back of the throat) (Figure 22-3; see Figure 23-2). They form a protective ring of lymphatic tissue around the openings between the nasal and oral cavities and the pharynx and provide protection against bacteria and other potentially harmful material in the nose and mouth. In adults the tonsils decrease in size and eventually may disappear.

There are three groups of tonsils: the palatine, the pharyngeal, and the lingual. The **palatine tonsils** usually are referred to as "the tonsils," and they are relatively large, oval lymphoid masses on each side of the posterior opening of the oral cavity. The **pharyngeal tonsil** is a collection of somewhat closely aggregated lymph nodules near the internal opening of the nasal cavity. An enlarged pharyngeal tonsil is called the **adenoids** (ad'noydz), and it can interfere with normal breathing. The **lingual tonsil** is a loosely associated collection of lymph nodules on the posterior margin of the tongue.

Sometimes the tonsils or adenoids become chronically infected and must be removed. The lingual tonsils become infected less often than the other tonsils, and they are more difficult to remove.

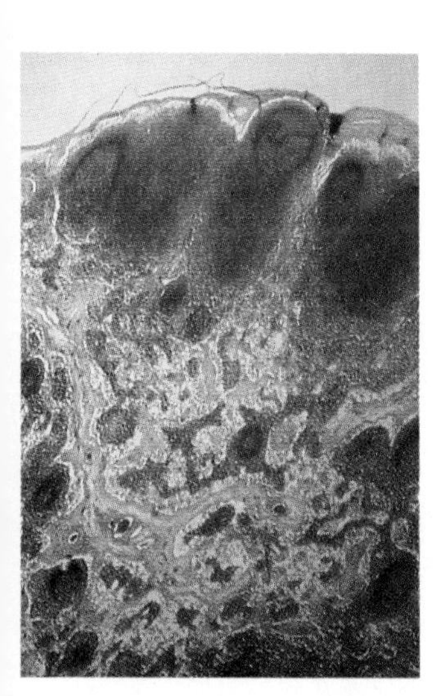

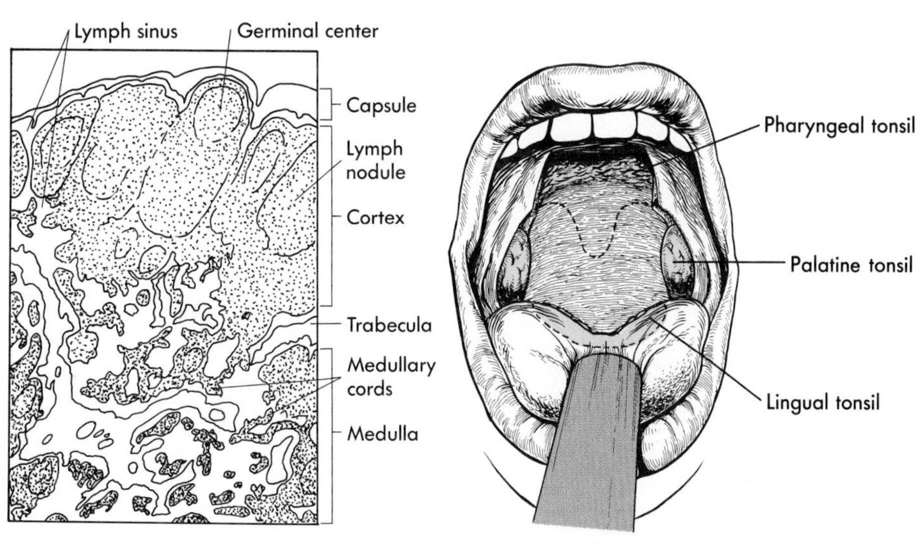

FIGURE 22-2 Histology of a lymph nodule surrounded by diffuse lymphatic tissue.

FIGURE 22-3 Location of the tonsils. Anterior view of the oral cavity with part of the hard and soft palates removed *(dotted line)* to show the pharyngeal tonsils.

Lymph Nodes

Lymph nodes are small, round, or bean-shaped structures, ranging in size from 1 to 25 mm long, and are distributed along the course of the lymph vessels (Figure 22-4; see Figure 22-1). They filter the lymph (removing bacteria and other materials) and provide a source of lymphocytes. Lymph nodes are found throughout the body. There are three superficial aggregations of lymph nodes on each side of the body: the inguinal nodes in the groin, the axillary nodes in the axillary (armpit) region, and the cervical nodes of the neck. Deep lymph nodes are located in the connective tissue around the intestines and along large blood vessels in the thorax and abdomen.

Lymph nodes are surrounded by a dense con-

FIGURE 22-4 Lymph node. **A** Arrows indicate direction of lymph flow. As lymph moves through the sinuses, phagocytic cells remove foreign substances. The germinal centers are sites of lymphocyte production. **B** Histology of a lymph node.

B

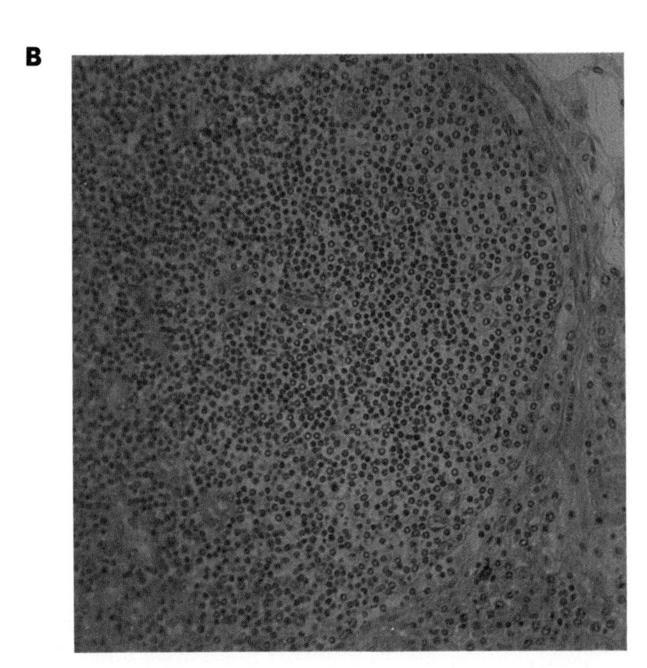

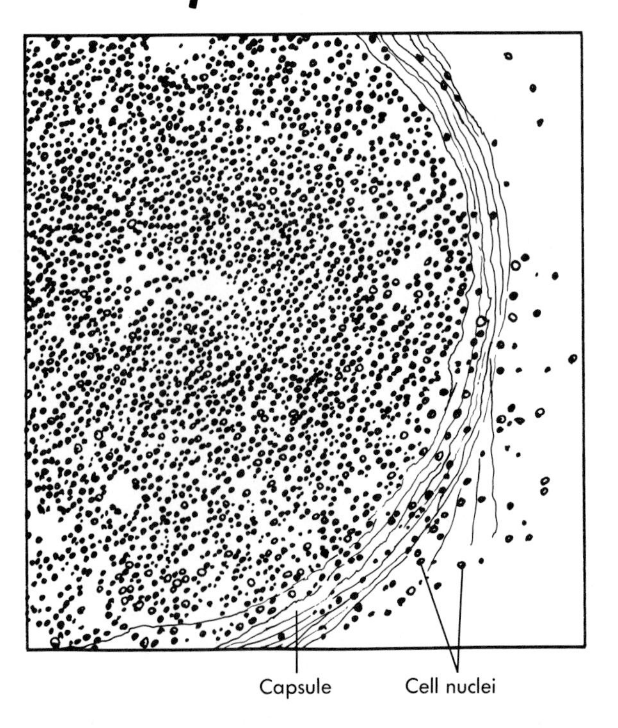

nective tissue **capsule.** Extensions of the capsule, called **trabeculae,** form a delicate internal skeleton in the lymph node. Reticular fibers, produced by the **reticular cells** that wrap around the individual fibers, extend from the capsule and trabeculae to form a fibrous network throughout the entire node. In some areas of the lymph node lymphocytes and macrophages are packed around the reticular fibers to form **lymphatic tissue,** and in other areas the reticular fibers extend across an open space called a **lymph sinus.** The lymphatic tissue and sinuses within the node are arranged into two somewhat indistinct layers. The outer **cortex** consists of lymph nodules separated by diffuse lymphatic tissue, trabeculae, and lymph sinuses. The inner **medulla** is organized into branching, irregular strands of diffuse lymphatic tissue, the **medullary cords,** separated by sinuses.

Lymph nodes are the only structures to filter lymph and to have both efferent and afferent lymph vessels. One to several **afferent lymph vessels** enter the lymph node through the cortex, and **efferent lymph vessels** exit from the opposite side. The sinuses are lined with phagocytic cells that remove bacteria and other foreign material from the lymph as it moves to the efferent vessels. The efferent vessels may become the afferent vessels of another node or may converge toward the thoracic duct or right lymphatic duct (see Chapter 21).

Cells of the lymph nodes consist primarily of lymphocytes, macrophages, and reticular cells. Microorganisms or other foreign substances in the lymph can stimulate lymphocytes throughout the lymph node to undergo cell division, with proliferation especially evident in the lymph nodules of the cortex. These areas of rapid lymphocyte division are **germinal centers.** The newly produced lymphocytes are released into the lymph and eventually reach the bloodstream where they circulate. Subsequently, the lymphocytes can leave the blood and enter other lymphatic tissues.

Cancer cells can spread from a tumor site to other areas of the body through the lymphatic system. At first, however, as the cancer cells pass through the lymphatic system, they are trapped in the lymph nodes, which filter the lymph. During cancer surgery malignant (cancerous) lymph nodes are often removed, and their vessels are cut and tied off to prevent the spread of the cancer.

Spleen

The **spleen** is roughly the size of a clenched fist, and it is located on the left side in the extreme superior, posterior corner of the abdominal cavity (see Figure 22-1). It has a fibrous **capsule** with **trabeculae** extending from the capsule into the tissue of the spleen, and it has an internal network of reticular fibers (Figure 22-5). The spleen contains two types of lymphatic tissue, **red pulp** and **white pulp.** White pulp is associated with the arterial supply to the spleen, and red pulp is associated with the venous supply.

The splenic arteries enter the spleen at the **hilum,** and their branches follow the various trabeculae into the spleen. Once an artery leaves the trabeculae, it is surrounded by dense accumulations of lymphocytes (the white pulp), which form a sheath of diffuse lymphatic tissue called the **periarterial sheath.** At various locations the sheath expands to form lymph nodules resembling ones in the cortex of lymph nodes. The arteries branch to form arterioles and eventually capillaries that supply the red pulp, a network of reticular fibers filled with blood cells that have come from the capillaries of the white pulp. The red pulp is divided into pulp cords by enlarged blood vessels called **venous sinuses.** The venous sinuses unite to form veins that eventually return to the trabeculae and leave the spleen as splenic veins.

The spleen detects and responds to foreign substances in the blood, destroys worn out red blood cells, and acts as a blood reservoir. As blood passes through the white pulp, lymphocytes in the periarterial sheath or the lymph nodules can be stimulated in the same manner as in lymph nodes. Before blood leaves the spleen through veins, it passes into the red pulp. Macrophages in the red pulp remove foreign substances and worn out red blood cells through phagocytosis. In emergency situations such as hemorrhage, smooth muscle in splenic blood vessels and in the splenic capsule can contract. The result is the movement of a small amount of blood into the general circulation.

Although the spleen is protected by the ribs, it often is ruptured in traumatic abdominal injuries. Injury to the spleen can cause severe bleeding, shock, and possible death. A splenectomy, removal of the spleen, is performed to stop the bleeding. Loss of the spleen's functions is compensated for by the liver and by other lymphatic tissues.

Thymus

The **thymus** is a bilobed gland roughly triangular in shape (Figure 22-6). It is primarily in the superior mediastinum, deep to the manubrium of the sternum. The size of the thymus differs markedly, depending on the age of the individual. In a newborn the thymus may extend halfway down the length of the thorax. The thymus continues to grow until puberty, although not as rapidly as other structures of the body.

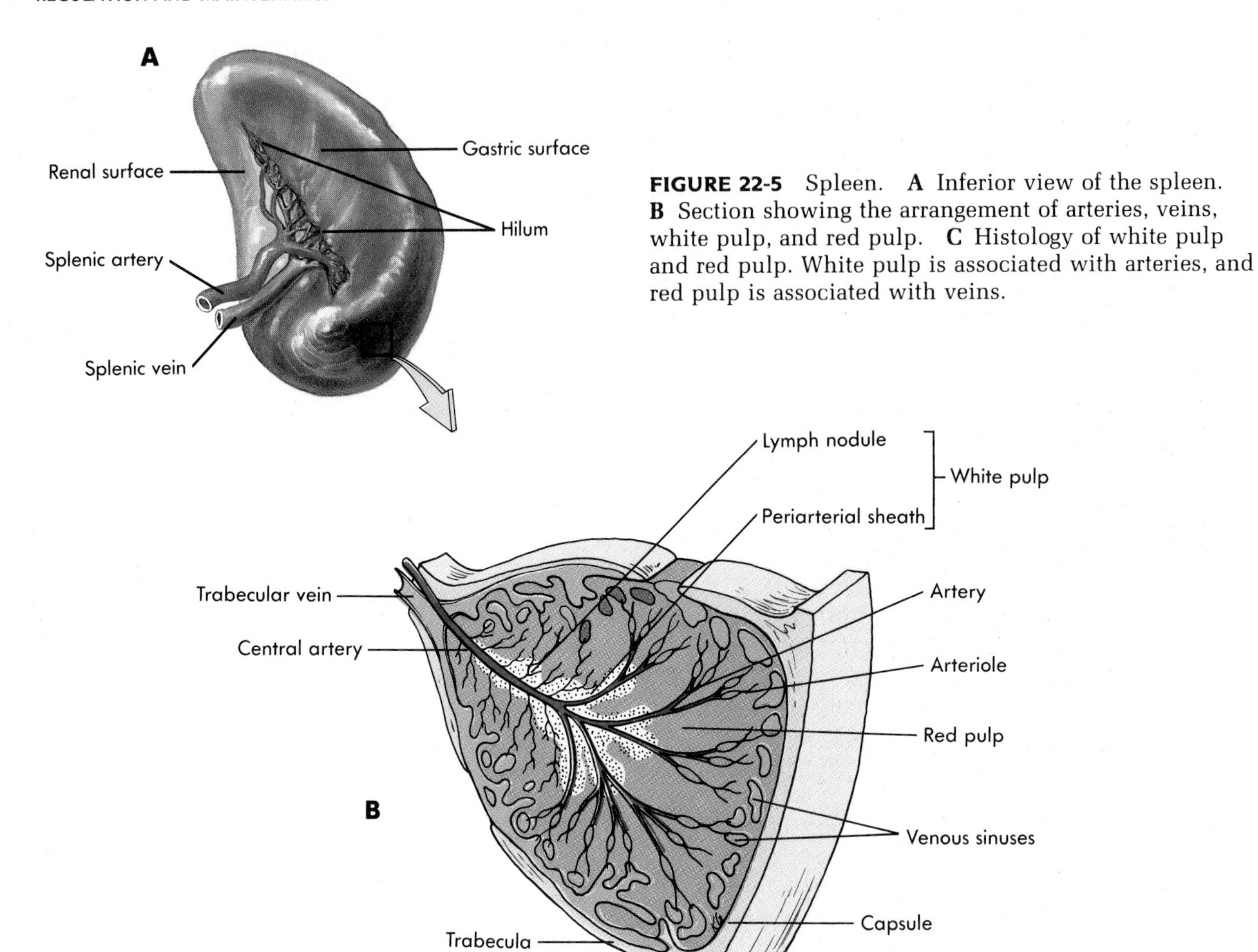

FIGURE 22-5 Spleen. **A** Inferior view of the spleen. **B** Section showing the arrangement of arteries, veins, white pulp, and red pulp. **C** Histology of white pulp and red pulp. White pulp is associated with arteries, and red pulp is associated with veins.

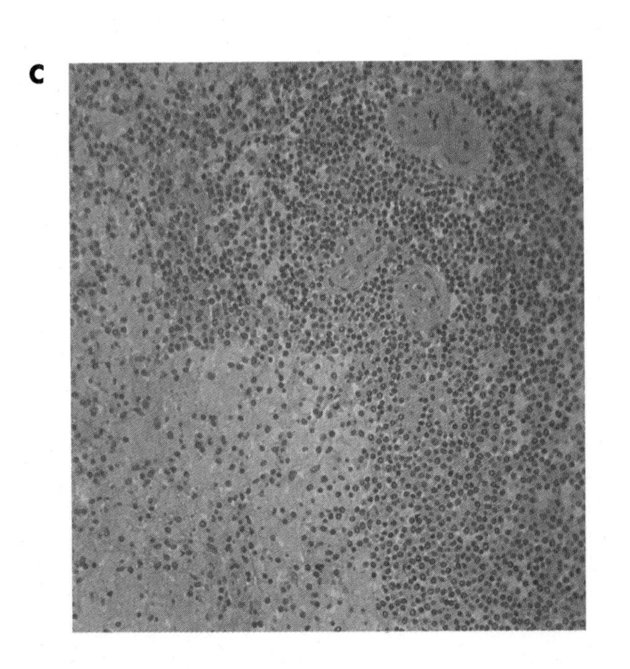

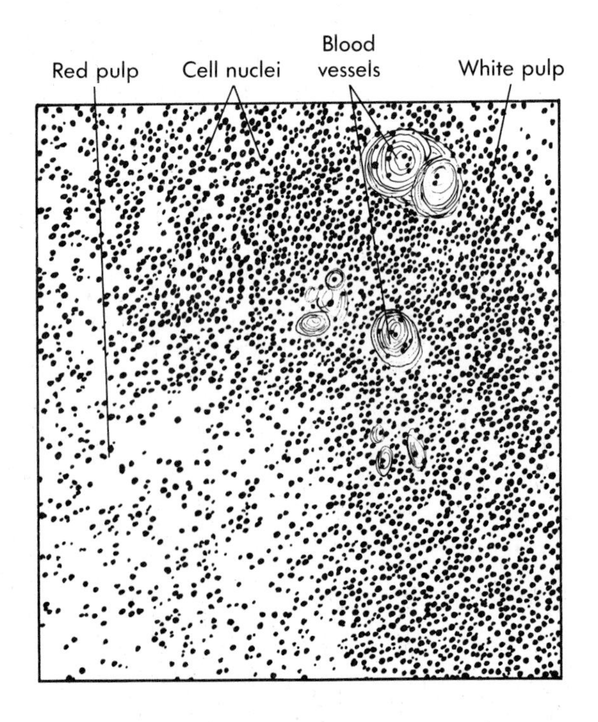

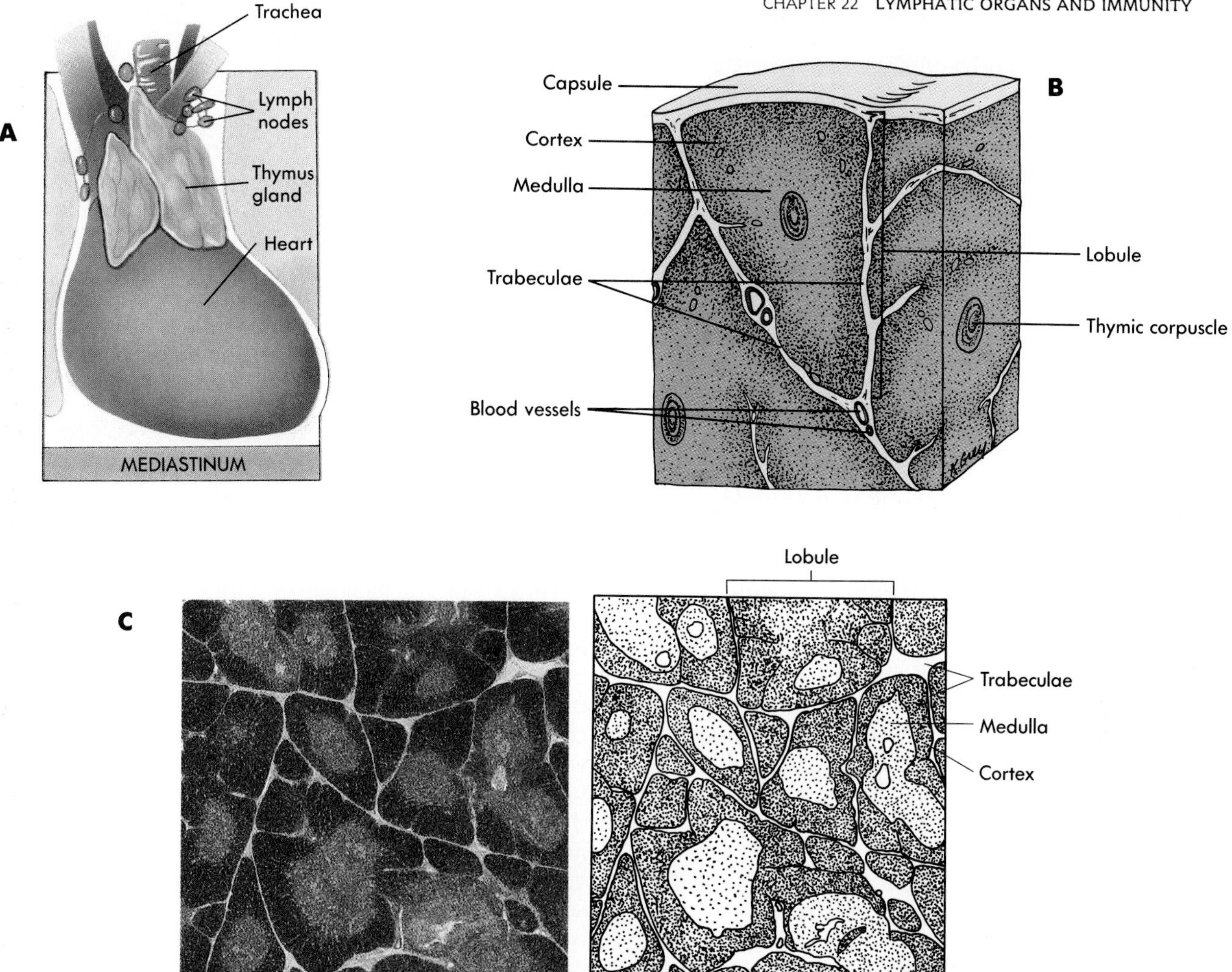

FIGURE 22-6 Thymus. **A** Location and shape of thymus. **B** Section showing a thymic lobule. **C** Histology of the thymus showing outer cortex and inner medulla.

After puberty it gradually decreases in size, and in older adults the thymus may be so small that it is difficult to find during dissection.

Each lobe of the thymus is surrounded by a thin connective tissue **capsule. Trabeculae** extend from the capsule into the substance of the gland, dividing it into **lobules.** Lymphocytes are concentrated near the capsule or trabeculae of each lobule and comprise the cortex. The relatively lymphocyte-free core of each lobule is the medulla. The medulla contains rounded epithelial structures, **thymic corpuscles** (Hassall's bodies), whose function is unknown.

Unlike other lymphatic tissues, few reticular fibers are in the thymus. Instead, reticular cells have long, branching processes that join to form an interconnected network of cells. In the cortex the reticular cells surround capillaries to form a **blood-thymic barrier,** which prevents large molecules from leaving the blood and entering the cortex.

The function of the thymus is to produce lymphocytes, which then move to other lymphatic tissues where they can respond to foreign substances. Lymphocytes within the thymus do not respond to foreign substances because the blood-thymic barrier prevents the entry of foreign substances. Large numbers of lymphocytes are produced in the thymus, but most degenerate. The lymphocytes that survive are capable of reacting to foreign substances, but they do not react to and destroy normal body cells. These surviving thymic lymphocytes migrate to the medulla, enter the blood, and travel to other lymphatic tissues.

IMMUNITY

Immunity is the ability to resist damage from foreign substances such as microorganisms and harmful chemicals (e.g., toxins released by microorganisms). Immunity is categorized as **nonspecific resistance** or **specific immunity.** The distinction between nonspecific resistance and specific immunity involves the concepts of memory and specificity. In nonspecific resistance each time the body is exposed to a substance, the response is the same. For example, each time a bacterial cell is introduced into the body, it is phagocytized with the same speed and efficiency. In specific immunity the response during second exposure to the same substance is faster and stronger than the response to the first exposure. For example, after the initial exposure to a virus the body may take many days to destroy the virus, allowing enough time for tissue damage and production of disease symptoms. After the second exposure to the same virus the response is so rapid and effective that the virus is destroyed before any symptoms develop, and the person is **immune.** This immune system response is possible because of specificity and memory, the ability of the immune system to recognize and remember a particular substance.

NONSPECIFIC RESISTANCE

The main components of nonspecific resistance include (1) mechanical mechanisms that prevent the entry of microbes into the body or that physically remove them from body surfaces; (2) chemicals that act directly against microorganisms or that activate other mechanisms, leading to the destruction of the microorganisms; (3) cells involved in phagocytosis and the production of chemicals that participate in the immune system's response; and (4) inflammation that mobilizes the immune system and isolates microorganisms until they can be destroyed.

Mechanical Mechanisms

Mechanical mechanisms such as the skin and mucous membranes form barriers that prevent the entry of microorganisms and chemicals into the tissues of the body. They also remove microorganisms and other substances from the surface of the body in several ways. The substances are washed from the eyes by tears, from the mouth by saliva, and from the urinary tract by urine. In the respiratory tract mucous membranes are ciliated, and microbes trapped in the mucus are swept to the back of the throat where they are swallowed. Coughing and sneezing also remove microorganisms from the respiratory tract. Microorganisms cannot cause disease if they cannot get into the body.

Chemicals

A variety of chemicals are involved in nonspecific resistance (Table 22-1). Some chemicals (e.g., lysozyme, sebum, and mucus) that are found on the surface of cells kill microorganisms or prevent their entry into the cells. Many other chemicals (e.g., histamine, complement, prostaglandins, and leukotrienes) promote inflammation by causing vasodilation, increasing vascular permeability, attracting leukocytes, and stimulating phagocytosis. In addition, the protein interferon protects cells against viral infections.

Complement

Complement is a group of at least 11 proteins that comprise approximately 10% of the globulin portion of serum. Normally complement proteins circulate in the blood in an inactive, nonfunctional form. Once activated, however, complement promotes inflammation and phagocytosis and can directly lyse bacterial cells. It becomes activated in the **complement cascade,** a series of reactions in which each component of the series activates the next component. The complement cascade begins through either the alternate pathway or the classical pathway (Figure 22-7). The **alternate pathway** is part of nonspecific resistance and is initiated when one of the complement proteins (C3) becomes spontaneously active. Normally the active complement protein is quickly inactivated, preventing the complement cascade. However, if the activated complement combines with some foreign substances (e.g., parts of a bacterial cell or virus) and another plasma protein (properdin), it becomes stabilized and causes activation of the complement cascade. The **classical pathway** is part of the specific immune system discussed later in this chapter.

Interferon

Interferon (in′ter-fēr′on) is a protein that protects the body against viral infection and perhaps some forms of cancer. When a virus infects a cell, viral nucleic acids take control of directing the cell's activities. The cell produces new nucleic acids and proteins, which are assembled into new virus particles, and the new viruses are released from the infected cell to infect other cells. Because infected cells usually stop their normal functions or die during viral replication, viral infections are clearly harmful to the body. Fortunately, viruses and other substances can also stimulate infected cells to produce interferon. Interferon neither protects the cell that produces it nor acts directly against viruses. Instead, interferon binds to the surface of neighboring cells where it stimulates them to produce antiviral proteins. These antiviral proteins stop viral reproduction in the neighboring cells by preventing the pro-

Table 22-1

Chemicals of nonspecific resistance and their functions

CHEMICAL	DESCRIPTION
Surface chemicals	Lysozymes (in tears, saliva, nasal secretions, and sweat) lyse cells; acid secretions (sebum in the skin and hydrochloric acid in the stomach) prevent microbial growth or kill microorganisms; mucus on the mucous membranes traps microorganisms until they can be destroyed
Histamine	An amine released from mast cells, basophils, and platelets; histamine causes vasodilation, increases vascular permeability, stimulates gland secretions (especially mucus and tear production), causes smooth muscle contraction of airway passages (bronchioles) in the lungs, and attracts eosinophils
Kinins	A polypeptide derived from plasma proteins. Kinins cause vasodilation, increase vascular permeability, stimulate pain receptors, and attract neutrophils
Interferon	A protein, produced by most cells, that interferes with virus production and infection
Complement	A group of plasma proteins that increase vascular permeability, stimulate the release of histamine, activate kinins, lyse cells, promote phagocytosis, and attract neutrophils, monocytes, macrophages, and eosinophils
Prostaglandins	A group of lipids (PGEs, PGFs, thromboxanes, and prostacyclins), produced by most cells, that cause smooth muscle relaxation and vasodilation, increase vascular permeability, and stimulate pain receptors
Leukotrienes	A group of lipids, produced primarily by mast cells and basophils, that cause prolonged smooth muscle contraction (especially in the lung bronchioles), increase vascular permeability, and attract neutrophils and eosinophils
Pyrogens	Chemicals, released by neutrophils, monocytes, and other cells, that stimulate fever production

duction of new viral nucleic acids and proteins. Interferon can also cause the production of defective viruses that are incapable of infecting other cells, and it can prevent the release of viruses from the infected cell. Thus interferon activates mechanisms that interfere with normal viral production and infection. Interferon viral resistance is not specific, and the same interferon acts against many different viruses. Infection by one kind of virus actually can produce protection against infection by other kinds of viruses.

Because some cancers are induced by viruses, interferon may play a role in controlling cancers. Interferon activates macrophages and natural killer cells (a type of lymphocyte) that attack tumor cells. Through genetic engineering interferon currently is produced in sufficient quantities for clinical use and, along with other therapies, has been effective in treating some cancers and viral infections.

Cells

Leukocytes and the cells derived from leukocytes (see Chapter 19) are the most important cellular components of the immune system (Table 22-2). Leukocytes are produced in red bone marrow and lymphatic tissue and are released into the blood where they are transported throughout the body. To be effective, leukocytes must move into the tissues where they are needed. **Chemotactic** (ke′mo-tak′tik) **factors** are parts of microbes or chemicals that are released by tissue cells and act as chemical signals to attract leukocytes. Important chemotactic factors include complement, leukotrienes, kinins, and histamine. They diffuse from the area where they are released. Leukocytes can detect small differences in chemotactic factor concentration and move from areas of lower chemotactic factor concentration to areas of higher concentration; thus they move toward the source of these substances, an ability called **chemotaxis.** Leukocytes can move by ameboid movement over the surface of cells, can squeeze between

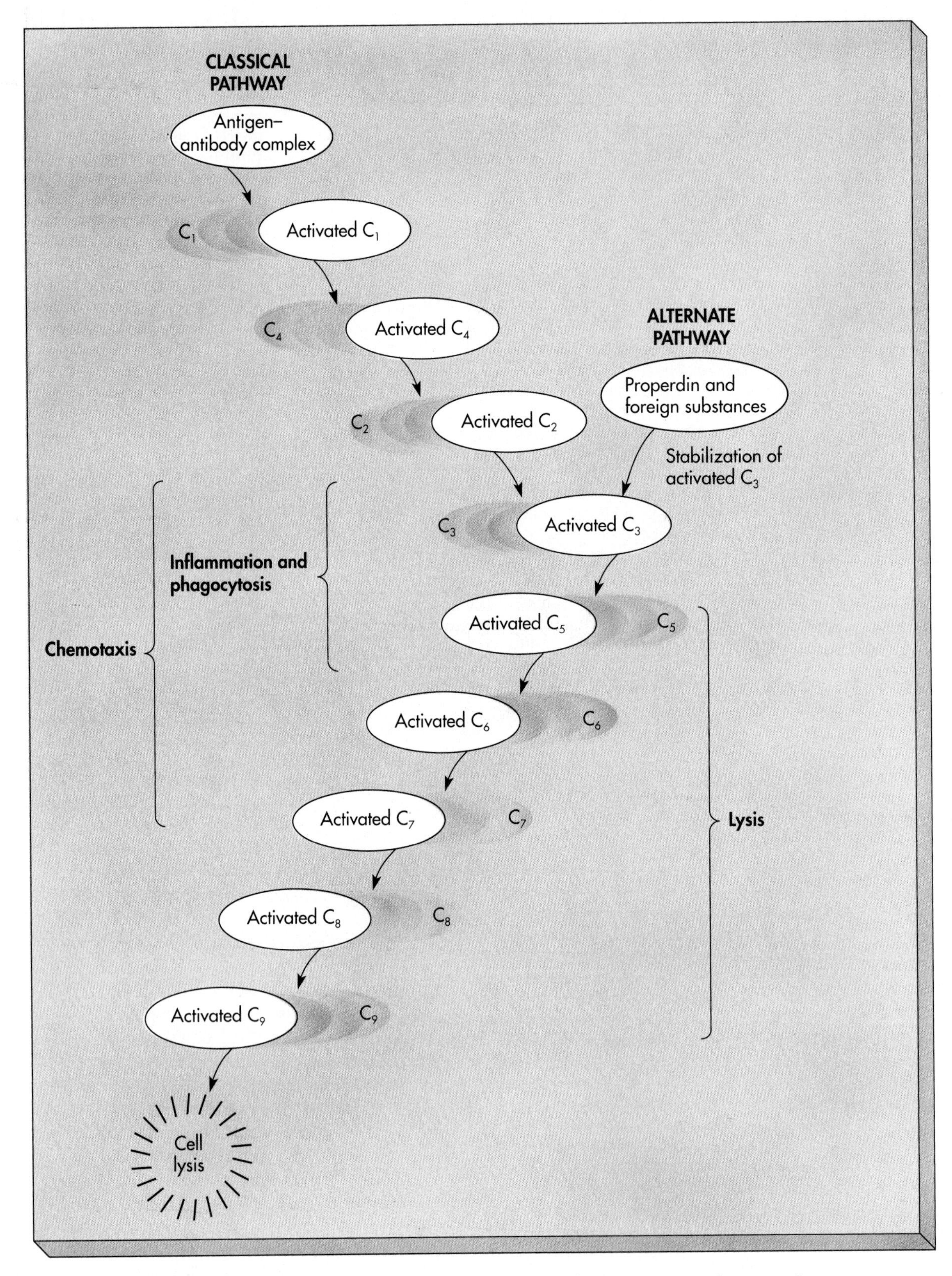

FIGURE 22-7 Complement cascade. The alternate pathway is activated at complement protein C3 and involves properdin. The classical pathway is activated at C1 and requires antibodies. Some effects of the activated complement proteins are shown. Complement proteins C5 to C9 combine to form a hole in the cell membrane of target cells, causing the cells to lyse.

Table 22-2

Summary of immune system cells and their primary functions

CELL	PRIMARY FUNCTION	CELL	PRIMARY FUNCTION
Nonspecific Resistance		***Specific Immunity***	
Neutrophil	Phagocytosis and inflammation; usually the first cell to leave the blood and enter infected tissues	B cell	After activation, differentiates to become plasma cell or memory B cell
Monocyte	Leaves the blood and enters tissues to become a macrophage	Plasma cell	Produces antibodies that are directly or indirectly responsible for the destruction of the antigen
Macrophage	Most effective phagocyte; important in later stages of infection and in tissue repair; located throughout the body to "intercept" foreign substances; processes antigens; involved in the activation of B and T cells	Memory B cell	Quick and effective response to an antigen against which the immune system has previously reacted; responsible for immunity
Basophil	Motile cell that leaves the blood, enters tissues, and releases chemicals that promote inflammation	Cytotoxic T cell	Once activated, responsible for the destruction of an antigen by lysis or by the production of lymphokines
Mast cell	Nonmotile cell in connective tissues that promotes inflammation through the release of chemicals	Delayed hypersensitivity T cell	Produces lymphokines that promote inflammation
		Helper T cell	Activates B and effector T cells
		Suppressor T cell	Inhibits B and effector T cells
Eosinophil	Enters tissues from the blood and releases chemicals that inhibit inflammation	Memory T cell	Quick and effective response to an antigen against which the immune system has previously reacted; responsible for immunity
Natural killer cell	Lyses tumor and virus-infected cells	Dendritic cell	Processes antigen and involved in the activation of B and T cells

cells, and in some instances can pass directly through other cells.

Phagocytosis (fag′o-si-to′sis) is the endocytosis and destruction of particles by cells called **phagocytes** (see Chapter 3). The particles can be microorganisms or their parts, foreign substances, or dead cells from the individual's body. The important phagocytic cells are neutrophils and macrophages.

Neutrophils

Neutrophils are small phagocytic cells that are produced in large numbers in red bone marrow and are released into the blood where they circulate for a few hours. Approximately 126 billion neutrophils per day leave the blood and pass through the wall of the gastrointestinal tract where they provide phagocytic protection. The neutrophils are then eliminated as part of the feces. Neutrophils are usually the first cells to enter infected tissues. However, neutrophils often die after a single phagocytic event; **pus** is primarily an accumulation of dead neutrophils. In addition to phagocytosis, neutrophils release lysosomal enzymes that kill microorganisms and also cause tissue damage and inflammation.

Macrophages

Macrophages are monocytes that leave the blood, enter tissues, enlarge approximately fivefold, and increase their number of lysosomes and mitochondria. They are large phagocytic cells that outlive neutrophils, and they can ingest more and larger phagocytic particles than neutrophils. Macrophages usually appear in tissues after neutrophils and are responsible for most of the phagocytic activity in the late stages of an infection, including the cleanup of dead neutrophils and other cellular debris. In addition to their phagocytic role, macrophages produce a variety of chemicals (e.g., interferon, prostaglandins, and complement) that enhance the immune system response.

Macrophages are beneath the free surfaces of the

body where they trap and destroy microbes entering the tissues, providing protection for the skin (dermis), hypodermis, mucous membranes, serous membranes, bladder, and uterus and around blood and lymph vessels.

If microbes do gain entry to the blood or lymphatic system, macrophages are waiting to filter them. As blood vessels pass through the spleen, bone marrow, and liver, they enlarge to form spaces called sinuses. Similarly, lymph vessels enlarge to form sinuses within lymph nodes. Within the sinuses reticular cells produce a fine network of reticular fibers that slows the flow of blood or lymph and provides a large surface area for the attachment of macrophages. In addition, macrophages are on the endothelial lining of the sinuses.

Because macrophages on the reticular fibers and endothelial lining of the sinuses were among the first macrophages studied, these cells were referred to as the **reticuloendothelial system.** It is now recognized that macrophages are derived from monocytes and are in locations other than the sinuses. Because monocytes and macrophages have a single, unlobed nucleus, they are now called the **mononuclear phagocytic system.** Sometimes macrophages are given specific names such as dust cells in the lungs, Kupffer cells in the liver, and microglia in the central nervous system.

Basophils, mast cells, and eosinophils

Basophils, which are derived from red bone marrow, are motile white blood cells that can leave the blood and enter infected tissues. **Mast cells,** which are also derived from red bone marrow, are nonmotile cells in connective tissue, especially near capillaries. Like macrophages, mast cells are located at potential points of entry of microorganisms into the body (e.g., the skin, lungs, gastrointestinal tract, and urogenital tract).

Basophils and mast cells can be activated through nonspecific resistance (e.g., by complement) or through specific immunity (see section on antibodies). When activated, they release chemicals (e.g., histamine and leukotrienes) that produce an inflammatory response or activate other mechanisms such as smooth muscle contraction in the lungs.

Eosinophils are produced in red bone marrow, enter the blood, and within a few minutes enter tissues. Enzymes released by eosinophils break down chemicals released by basophils and mast cells. Thus at the same time that inflammation is initiated, mechanisms are activated that will contain and reduce the inflammatory response. This process is similar to the blood clotting system in which clot prevention and removal mechanisms are activated while the clot is being formed (see Chapter 19). In patients with parasitic infections or allergic reactions with much inflammation, eosinophil numbers greatly increase. Eosinophils also secrete enzymes that effectively kill some parasites.

Natural killer cells

Natural killer (NK) cells are a type of lymphocyte produced in red bone marrow, and they account for 1% to 3% of all lymphocytes. NK cells kill certain tumor and virus-infected cells. Because NK cells recognize a general class of cells (e.g., tumor cells) rather than a specific cell (e.g., a specific type of tumor cell), NK cells are classified as part of nonspecific resistance. In addition, NK cells do not exhibit a memory response. NK cells use a variety of methods to kill their target cells, including the release of chemicals that damage cell membranes, causing the cells to lyse.

Inflammatory Response

The **inflammatory response** is a complex sequence of events involving many of the chemicals and cells previously discussed. Although they may vary in some details, depending on the events producing them or on the sites of inflammation, most inflammatory responses are strikingly similar. A bacterial infection can be used to illustrate the events in an inflammatory response (Figure 22-8). The microbe itself or damage to tissues causes the release or activation of chemical mediators—histamine, prostaglandins, leukotrienes, complement, kinins, and others. The mediators cause vasodilation, and increased blood flow brings phagocytes and other leukocytes to the area. Some of the mediators are chemotactic factors that stimulate phagocytes to leave the blood and enter the tissues. Mediators also increase vascular permeability, allowing fibrin, complement, and kinins to enter the tissue. Fibrin prevents the spread of the infection by walling off the infected area. The additional complement and kinins further enhance the inflammatory response and attract additional phagocytes. This process of increasing the numbers of mediators and phagocytes continues until the microorganisms are destroyed. Finally, phagocytes (mainly macrophages) clean the site of the infection, and the damaged tissues are repaired.

Inflammation can be localized or systemic. **Local inflammation** is an inflammatory response confined to a specific area of the body. Symptoms of local inflammation include redness, heat, swelling, pain, and loss of function. Redness, heat, and swelling result from increased blood flow and increased vascular permeability. Pain is caused by swelling and by chemicals acting on nerve receptors; loss of function results from tissue destruction, swelling, and pain.

Systemic inflammation is an inflammatory response that occurs in many parts of the body. In addition to the local symptoms at the sites of inflammation, three additional features can be present. First, red bone marrow produces and releases large numbers of neutrophils that promote phagocytosis. Second, **pyrogens** (pi'ro-jens), chemicals released by microorganisms, macrophages, neutrophils, and other cells, stimulate fever production. Pyrogens affect the body's temperature-regulating mechanism in the hypothalamus, heat is conserved, and body temperature increases. Fever promotes the activities of the immune system such as phagocytosis and inhibits the growth of some microorganisms. Third, in severe cases of systemic inflammation vascular permeability can increase so much that large amounts of fluid are lost from the blood into the tissues. The decreased blood volume can cause shock and death.

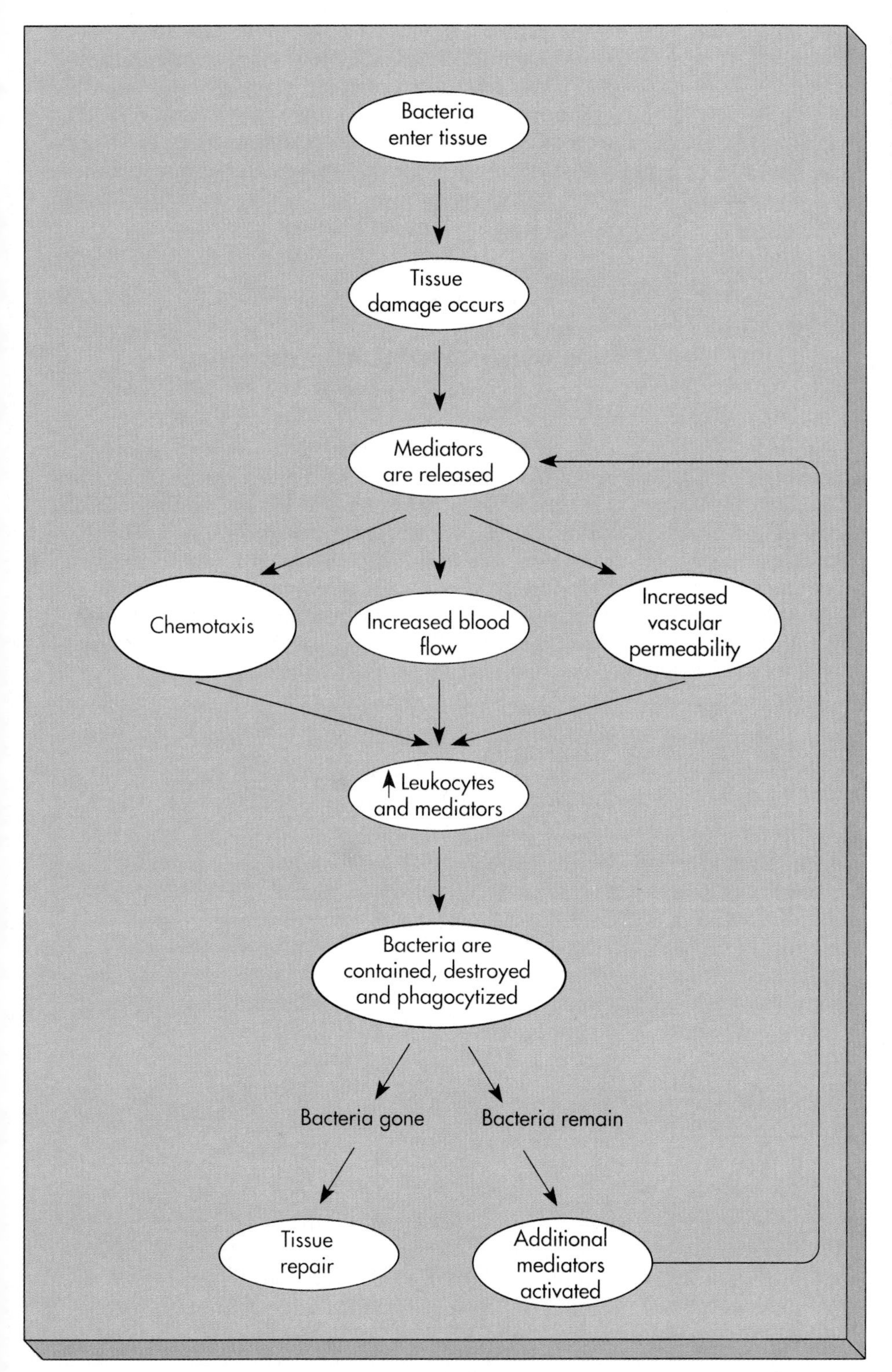

FIGURE 22-8
Flow diagram of the inflammatory response. Bacteria cause tissue damage and release of chemical mediators that initiate inflammation, resulting in the destruction of the bacteria.

SPECIFIC IMMUNITY

Specific immunity involves the ability to recognize, respond to, and remember a particular substance. Substances that stimulate specific immunity are **antigens** (an'tĭ-jenz), which usually are large molecules with a molecular weight of 10,000 or more. **Haptens** (hap'tenz) are small molecules (low molecular weight) capable of combining with larger molecules such as blood proteins to stimulate a specific immune system response.

Penicillin is an example of a hapten of clinical importance. It is a small molecule that does not evoke an immune system response. However, penicillin can break down and bind to serum proteins to form a combined molecule that can produce an allergic reaction. If the reaction is severe, death can result.

Antigens can be divided into two groups: foreign antigens and self-antigens. **Foreign antigens** are not produced by the body but are introduced from outside it. Components of bacteria, viruses, and other microorganisms are examples of foreign antigens that cause disease. Pollen, animal hairs, foods, and drugs are also foreign antigens and can trigger an overreaction of the immune system and produce allergies. Transplanted tissues and organs that contain foreign antigens result in the rejection of the transplant. **Self-antigens** are molecules produced by the body that stimulate a specific immune system response; the response to self-antigens can be beneficial or harmful (e.g., recognition of tumor antigens can result in tumor destruction, whereas **autoimmune disease** can result when self-antigens stimulate unwanted tissue destruction).

Specific immunity historically was divided into two parts: **humoral immunity** and **cell-mediated immunity.** Early investigators of the immune system found that when plasma from an immune animal was injected into the blood of a nonimmune animal, the nonimmune animal became immune. Because this process involved body fluids (humors), it was called humoral immunity. It was also discovered that blood cells transferred from an immune animal could be responsible for immunity; this process was called cell-mediated immunity.

Both types of immunity result from the activities of lymphocytes called B cells and T cells (see Table 22-2). **B cells** and their progeny produce proteins called **antibodies,** which are found in the plasma and are responsible for humoral immunity. Because antibodies are responsible, humoral immunity is now called **antibody-mediated immunity.** T cells can be divided into two groups: effector T cells and regulatory T cells. **Effector T cells** such as **cytotoxic T cells** and **delayed hypersensitivity T cells** are responsible for producing the effects of cell-mediated immunity. **Regulatory T cells** such as **helper T cells** and **suppressor T cells** function to control the activities of both antibody-mediated immunity and cell-mediated immunity.

Table 22-3 summarizes and contrasts the main features of nonspecific resistance, antibody-mediated immunity, and cell-mediated immunity. Although the immune system is divided into these three categories, this is an artificial division that is used to emphasize particular aspects of immunity. Actually, immune system responses often involve components of more than one type of immunity. For example, although the specific immune system has the ability to recognize and remember specific antigens, once recognition has occurred, many of the events that lead to the destruction of the antigen are nonspecific resistance activities such as inflammation and phagocytosis.

Origin and Development of Lymphocytes

All blood cells, including lymphocytes, are derived from stem cells in red bone marrow (see Chapter 19). This process begins during embryonic development and continues throughout life. Some stem cells give rise to pre-T cells that migrate through the blood to the thymus gland where they divide and are processed into T cells. Hormones (e.g., thymosin and thymopoietins) produced by the thymus stimulate T cell maturation, and the different types of T cells are formed. Other stem cells also produce pre-B cells, which are processed in the red bone marrow into B cells (Figure 22-9).

B cells are released from red bone marrow, T cells are released from the thymus, and both types of cells move through the blood to lymphatic tissue. Normally there are approximately five T cells for every B cell in the blood. These lymphocytes live for a few months to many years and continually circulate between the blood and the lymphatic tissues. The constant movement of lymphocytes serves two functions: it increases the likelihood lymphocytes will encounter antigens, and it enables lymphocytes to aggregate at sites of inflammation.

Evidence suggests that small groups of identical lymphocytes, called **clones,** are formed during embryonic development. Although each clone can respond only to a particular antigen, there is such a large number of clones that the immune system can react to most molecules. However, among the molecules to which the clones can respond are self-antigens. Because this response could destroy self-cells, clones acting against self-antigens are eliminated or suppressed. Most of this process occurs during de-

Table 22-3

Comparison of nonspecific resistance, antibody-mediated immunity, and cell-mediated immunity

CHARACTERISTICS	NONSPECIFIC RESISTANCE	ANTIBODY-MEDIATED IMMUNITY	CELL-MEDIATED IMMUNITY
Primary cells	Neutrophils, eosinophils, basophils, mast cells, monocytes, and macrophages	B cells	T cells
Origin of cells	Red bone marrow	Red bone marrow	Red bone marrow
Site of maturation	Red bone marrow (neutrophils, eosinophils, basophils, monocytes) and tissues (mast cells and macrophages)	Red bone marrow	Thymus
Location of mature cells	Blood, connective tissue, and lymphatic tissue	Blood and lymphatic tissue	Blood and lymphatic tissue
Primary secretory products	Histamine, kinins, complement, prostaglandins, leukotrienes, and interferon	Antibodies	Lymphokines
Primary actions	Inflammatory response and phagocytosis	Protection against extracellular antigens (bacteria, toxins, parasites, and viruses outside of cells)	Protection against intracellular antigens (viruses, intracellular bacteria, and intracellular fungi) and tumors: regulates antibody-mediated immunity and cell-mediated immunity responses (helper T and suppressor T cells)
Hypersensitivity reactions	None	Immediate hypersensitivity (atopy, anaphylaxis, cytotoxic reactions, and immune complex disease)	Delayed hypersensitivity (allergy of infection and contact hypersensitivity)

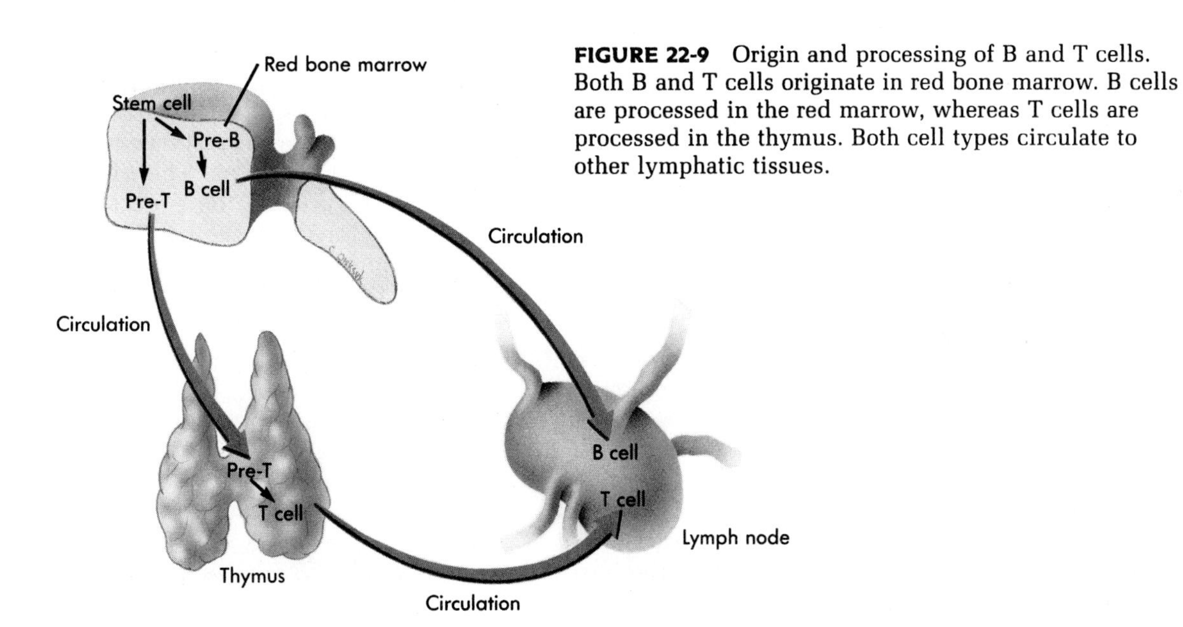

FIGURE 22-9 Origin and processing of B and T cells. Both B and T cells originate in red bone marrow. B cells are processed in the red marrow, whereas T cells are processed in the thymus. Both cell types circulate to other lymphatic tissues.

velopment, but it does continue after birth and throughout life.

The recognition of self-cells is determined by cell-surface glycoproteins that serve as self-identifying markers. The production of these glycoproteins is regulated by a group of genes called the **major histocompatibility complex (MHC),** and the glycoproteins are referred to as **MHC proteins.** During their processing B and T cell clones that could react *against* one's MHC proteins are eliminated or suppressed. The remaining B and T cells *use* the MHC proteins during their activation by antigens (see Antigen recognition below).

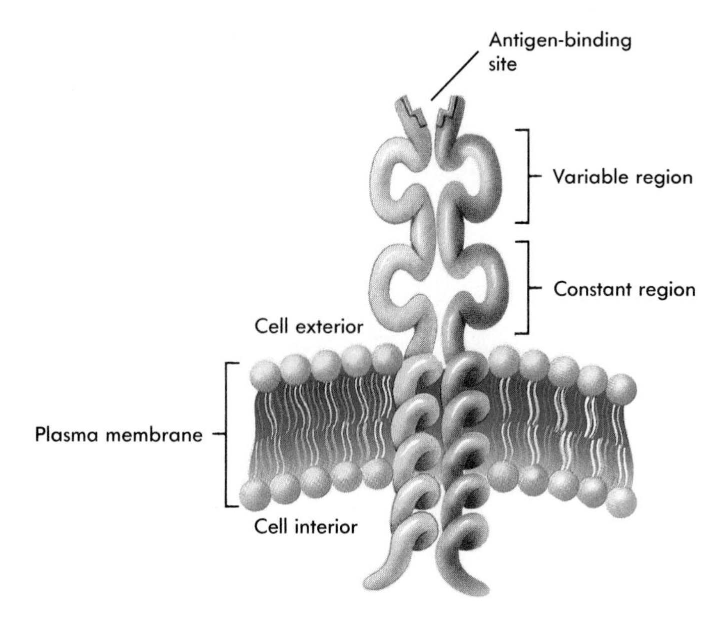

FIGURE 22-10 The T cell receptor consists of two polypeptide chains. The variable region of each type of T cell receptor is specific for a given antigen. The constant region attaches the T cell receptor to the plasma membrane.

Activation of Lymphocytes

The activation of lymphocytes by antigens can occur in different ways, depending on the type of lymphocyte and the type of antigen involved. Despite these differences, however, there are general principles of lymphocyte activation: lymphocytes must be able to recognize the antigen; and after recognition the lymphocytes must increase in number to effectively destroy the antigen.

Antigen recognition

To have a specific immune system response, lymphocytes must recognize and be activated by an antigen. Lymphocytes, however, do not interact with an entire antigen. Instead, **antigenic determinants,** or **epitopes,** are specific regions of a given antigen that activate a lymphocyte, and each antigen has many different antigenic determinants. All the lymphocytes of a clone have on their surface identical proteins, called **antigen-binding receptors,** which combine with antigenic determinants. The immune system response to an antigen with a particular antigenic determinant is similar to the lock-and-key model for enzymes (see Chapter 3), and any given antigenic determinant can combine only with a specific antigen-binding receptor on the lymphocyte. For example, the **T cell receptor** consists of two polypeptide chains, which are subdivided into a variable and a constant region (Figure 22-10). The variable region can bind to an antigen, and different T cell receptors are specific for given antigens because they have different variable regions. The **B cell receptor** consists of four polypeptide chains with two identical variable regions. It is a type of antibody and is considered in greater detail later in this chapter.

Although some antigens bind to their receptors and directly activate B cells and some T cells, most lymphocyte activation involves helper T cells, which are unable to respond to an antigen until it has been processed. Antigens are processed and presented to helper T cells by **antigen-presenting cells,** which include B cells, macrophages, and dendritic cells. **Dendritic cells** are large, motile cells with long cytoplasmic extensions, and they are scattered throughout most tissues (except the brain), with their highest concentrations in lymphatic tissues and in the skin. Dendritic cells in the skin are often called **Langerhans cells.**

Antigen-presenting cells take in antigens (Figure 22-11). An antigen binds to the B-cell receptor and is transported into the B cell by receptor-mediated endocytosis (see Chapter 3). Macrophages and dendritic cells are less specific and are capable of taking in almost any antigen. Once inside the antigen-presenting cell, the antigen is broken down, and fragments of the antigen are returned to the cell's surface bound to MHC proteins. The function of the MHC protein is to present the processed antigen to the T-cell receptor of the helper T cell for recognition. Because both the antigen and the MHC protein are required, this process is said to be **MHC restricted.**

1

In mouse A the helper T cells are able to respond to antigen X. If these helper T cells are transferred to mouse B and exposed to antigen X, will they respond to the antigen? Explain.

? ? ? ? ? ? ? ? ? ? ?

A

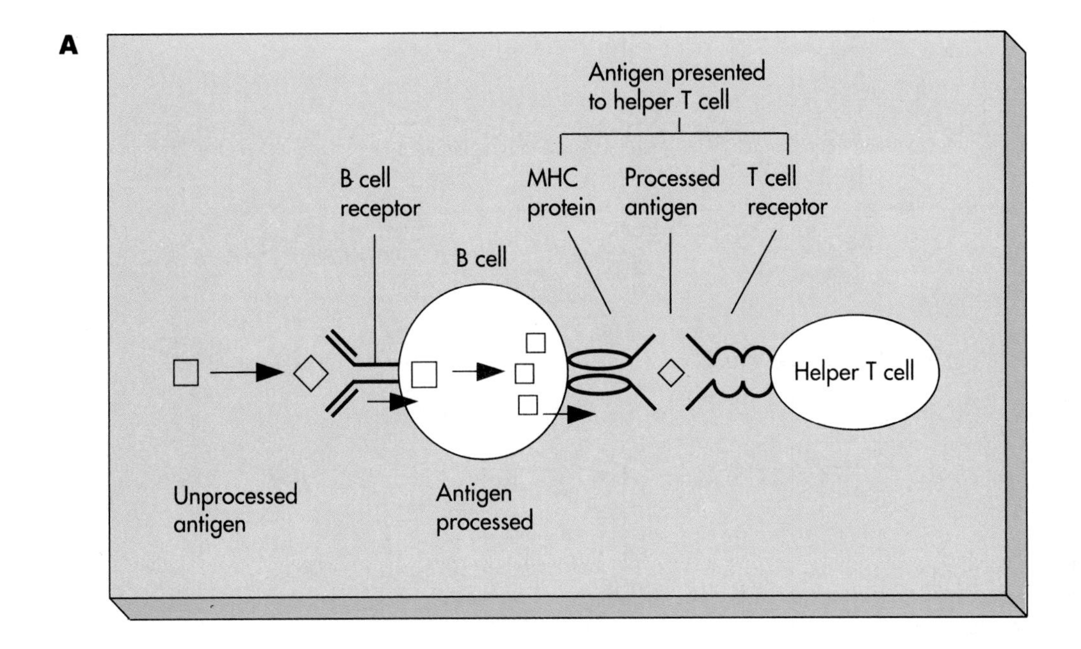

B

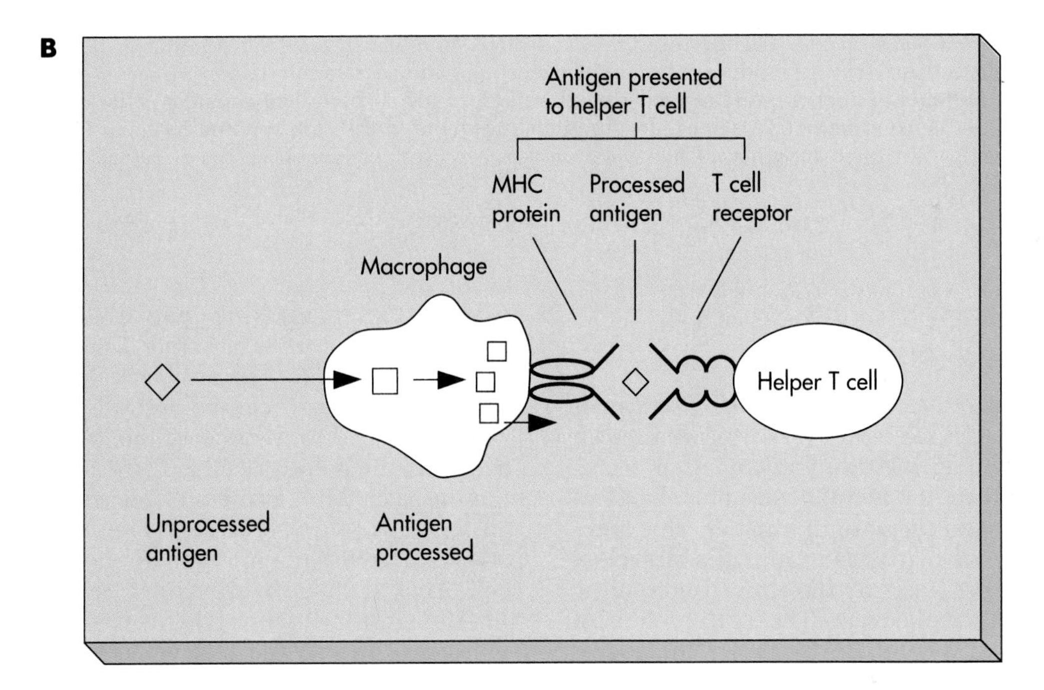

FIGURE 22-11 Antigen-presenting cells. **A** A B cell takes in the antigen using receptor-mediated endocytosis. The antigen is processed, and a fragment of the antigen, in combination with a MHC protein, moves to the cell surface. A helper T cell is activated when its T-cell receptor combines with the processed antigen and the MHC protein. **B** A macrophage or dendritic cell takes in the antigen using pinocytosis or phagocytosis. The antigen is processed and presented to the helper T cell.

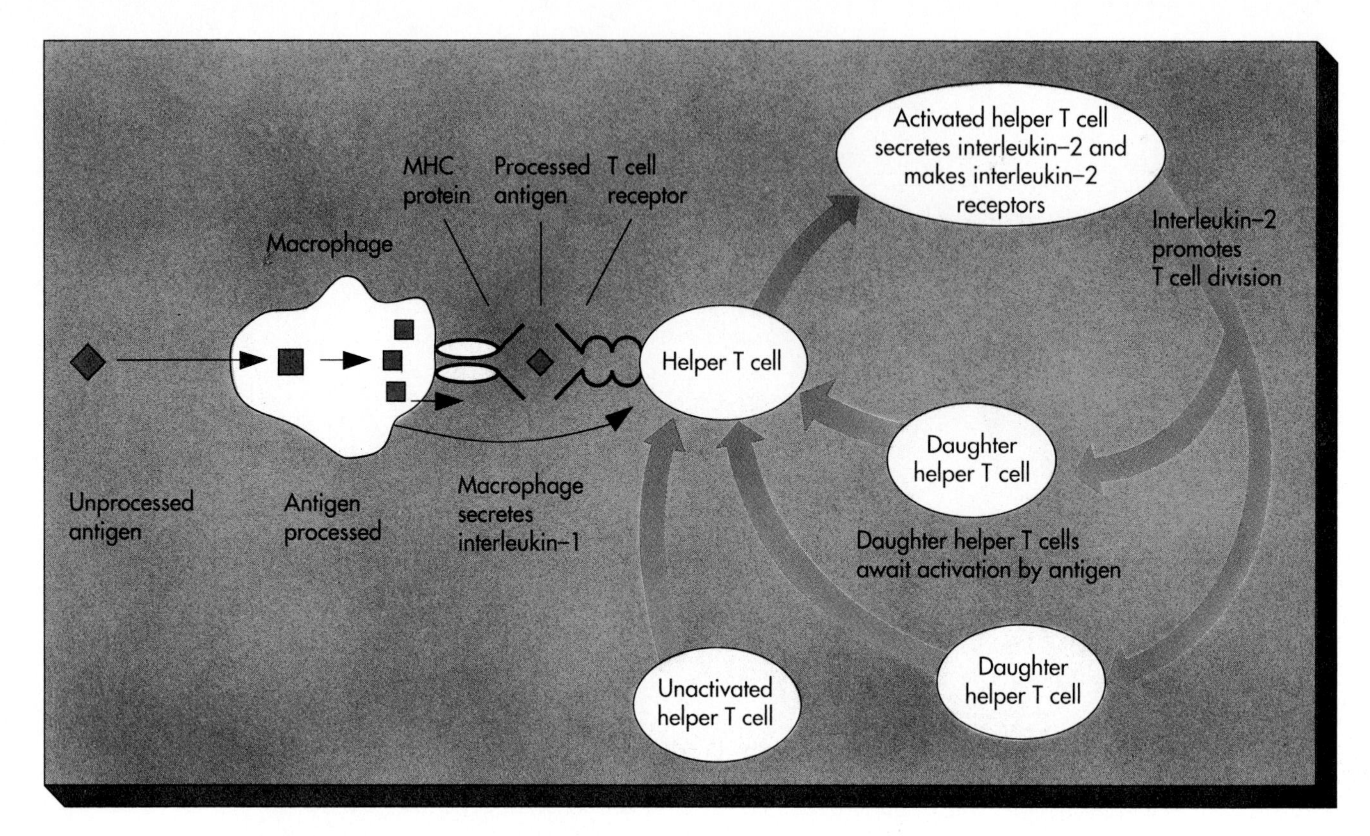

FIGURE 22-12 Macrophage (antigen-presenting cell) processes antigen and presents it to a helper T cell. The macrophage also secretes interleukin-1, which stimulates the helper T cell to produce interleukin-2 receptors and secrete interleukin-2. The interleukin-2 then stimulates the helper T cell to divide and produce daughter cells. The daughter helper T cells can be stimulated to divide if they are exposed to the antigen by the macrophage. This cyclic process can produce large numbers of helper T cells.

Lymphocyte proliferation

Before exposure to an antigen, the number of lymphocytes in a clone is too small to produce an effective response against the antigen. After recognition of an antigen, the lymphocytes must be stimulated to divide and increase in number. This stimulation generally involves chemicals called **interleukins,** which are produced by the antigen-presenting cells or by the lymphocytes. The proliferation of helper T cells and B cells illustrates this process.

A macrophage secretes interleukin-1 when the processed antigen bound to an MHC protein on the macrophage comes into contact with a helper T cell (Figure 22-12). **Interleukin-1** stimulates the helper T cell to secrete **interleukin-2** and to produce interleukin-2 receptors. The helper T cell then stimulates itself to divide when the interleukin-2 binds to the interleukin-2 receptors. The "daughter" helper T cells resulting from this division can be stimulated to divide again if they are exposed to the same antigen that stimulated the "parent" helper T cell; or the increased number of helper T cells can facilitate the activation of B cells or effector T cells.

Before a B cell can be activated by a helper T cell, the B cell must process the same antigen that activated the helper T cell (Figure 22-13). The B cell then uses an MHC protein to present the processed antigen to the helper T cell. The helper T cell responds by releasing interleukins that activate the B cell. They include interleukin-4, interleukin-5, and interleukin-6, which together cause the B cell to divide and differentiate into cells that produce antibodies. The antibodies are part of the antibody-mediated immune system response that eliminates the antigen.

2

How does elimination of the antigen stop the production of antibodies?

? ? ? ? ? ? ? ? ? ? ?

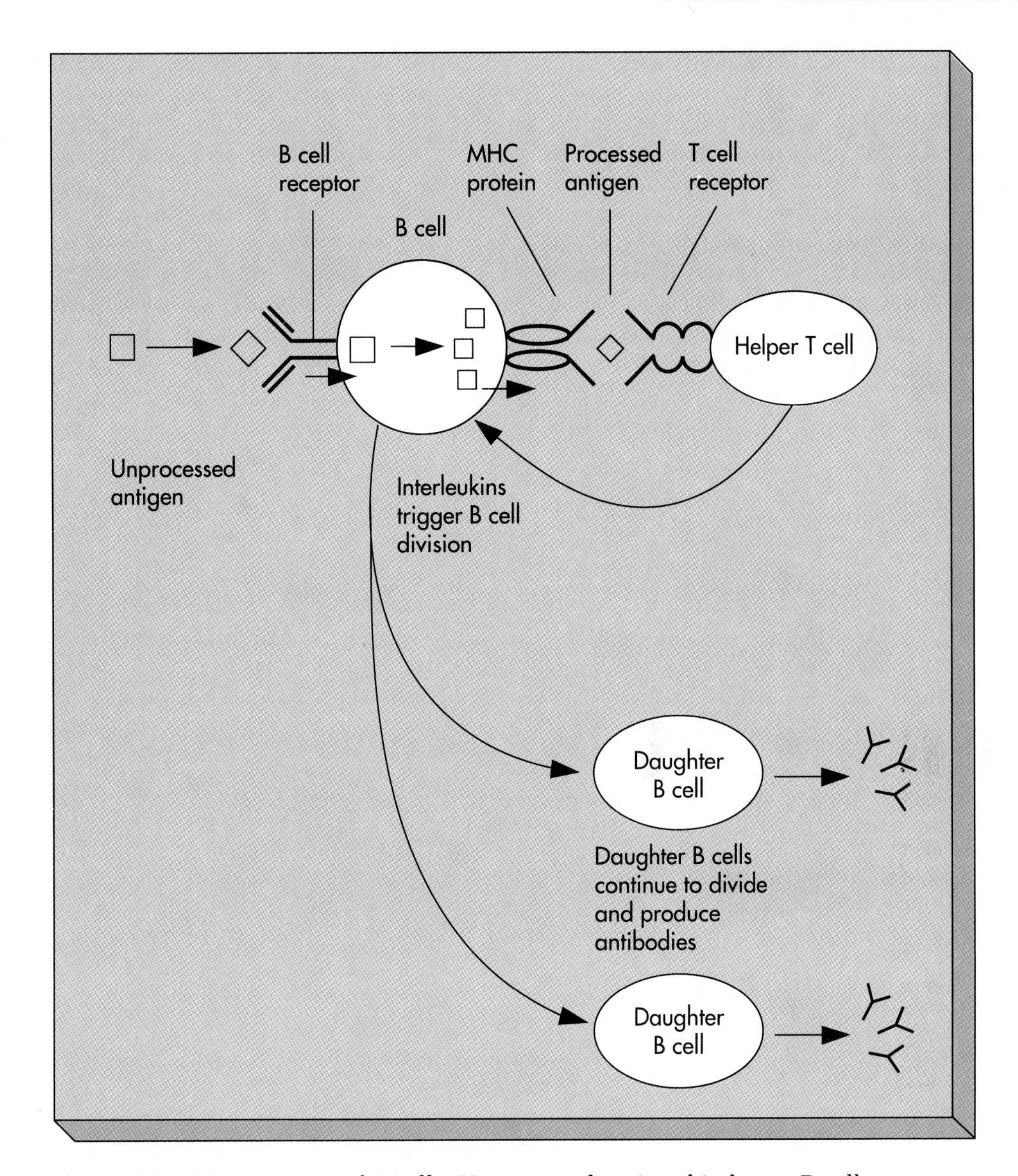

FIGURE 22-13 Activation of B cells. Unprocessed antigen binds to a B-cell receptor on a B cell. The antigen is taken into the B cell, processed, and presented by an MHC protein to a helper T cell. The helper T cell produces interleukins that stimulate the B cell to divide and produce antibodies.

Currently, researchers are using genetically engineered interleukins to stimulate the immune system. For example, interleukin-2, which increases the proliferation of helper T cells and effector T cells, may promote the destruction of cancer cells or boost the effectiveness of vaccinations. Conversely, decreasing the production or activity of interleukin-2 can suppress the immune system. For example, cyclosporine, a drug used to prevent the rejection of transplanted organs, inhibits the production of interleukin-2.

Inhibition of Lymphocytes

Inhibition of lymphocytes is an important part of the immune system's response. It contributes to the homeostatic control of the immune system by limiting lymphocyte proliferation and responses. Inhibition can also completely prevent an immune system response to certain antigens such as self-antigens or antigens ingested in food.

One way that the specific immune system response is controlled is through **suppressor T cells,** a subset of the T-cell population (Figure 22-14, *A*). Antigen exposure can activate suppressor T cells in addition to helper T, effector T, and B cells. Suppressor T cells can prevent the activity of helper T cells, and since helper T cells are involved with the activation of B cells and effector T cells, they can inhibit both antibody-mediated and cell-mediated responses. In addition, suppressor T cells can directly prevent B-cell and effector T-cell activity. Thus suppressor T-cell activation is a mechanism for dampening the specific immune system response. A balance between the activity of suppressor T cells and other lymphocytes results in a response of the appropriate magnitude.

Another example of control of the specific immune system is **tolerance** in which there is no response to the antigen. Apparently tolerance can be induced in many different ways. One mechanism involves suppressor T cells. If the suppressor T-cell response is strong enough, it is possible that there will be no detectable immune system response to a given antigen. Other mechanisms block, alter, or remove the antigen-binding receptors on the surface of B or T cells, thus preventing the activation of these lymphocytes (Figure 22-14, *B*). Finally, lymphocytes capable of acting against self-antigens are simply eliminated, especially during embryonic development (Figure 22-14, *C*).

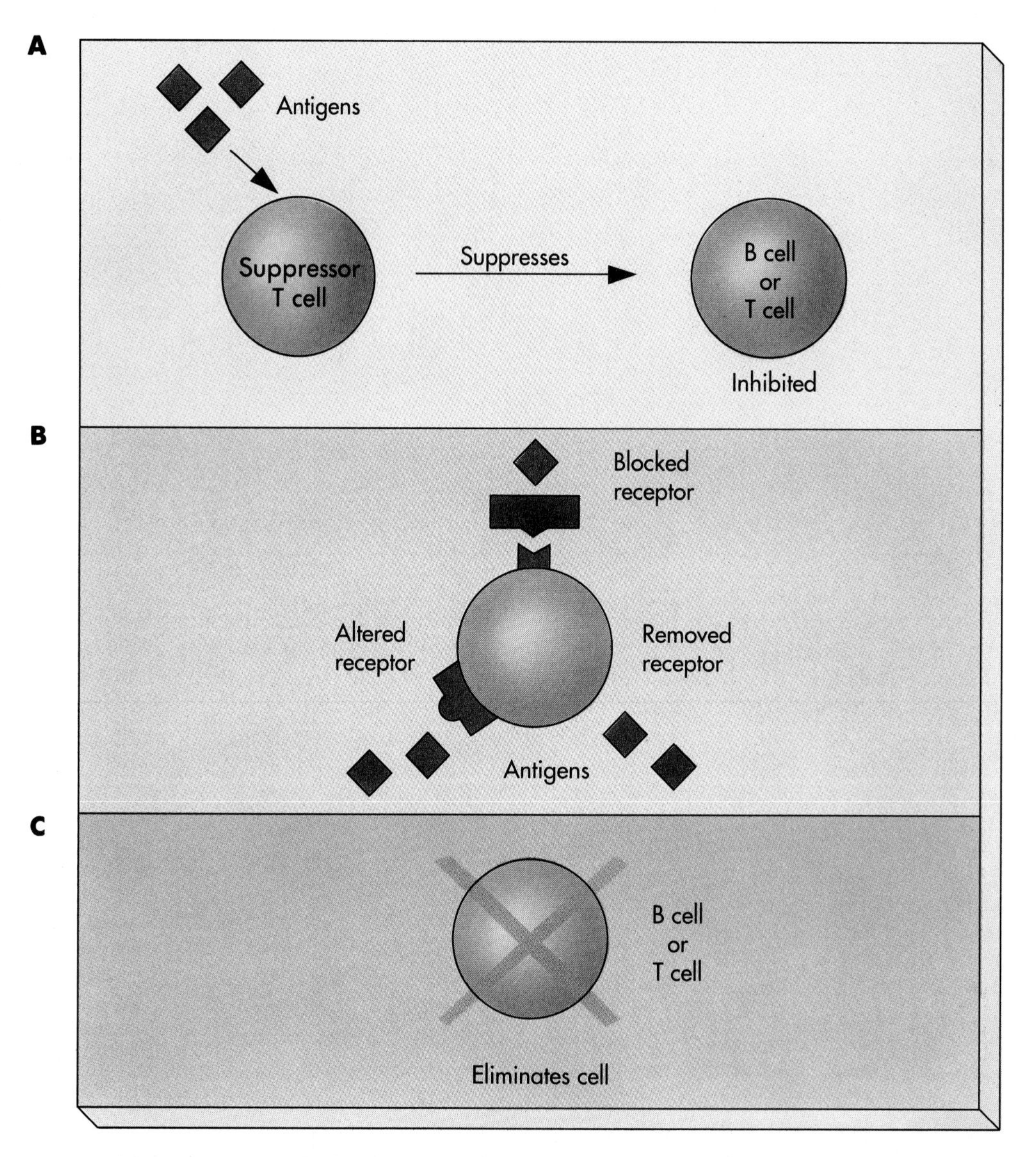

FIGURE 22-14 Methods of suppressing the activity of B or T cells. **A** Antigen activates suppressor T cells that inhibit the B or T cells. **B** Antigen-binding site on the B or T cells is blocked, altered, or removed. **C** B or T cells are destroyed.

The need to maintain tolerance and to avoid the development of autoimmune disease is obvious; without tolerance a woman would recognize the antigens of her developing fetus as foreign, and spontaneous abortion would result. Stimulation of tolerance could be used to treat allergies and to reduce organ transplant rejection, and the ability to decrease tolerance could be used to produce effective, long-lasting immunizations and to treat cancer.

Antibody-Mediated Immunity

Exposure of the body to an antigen can lead to activation of B cells and to production of antibodies, which are responsible for destruction of the antigen. Because antibodies are in body fluids, **antibody-mediated immunity** is effective against extracellular antigens such as bacteria, viruses (when they are not inside cells), toxins, and parasites and can cause immediate hypersensitivity reactions (see the box on pp. 727-728).

Antibodies

Antibodies are proteins produced in response to an antigen. Large amounts of antibodies are in plasma, although plasma also contains other proteins. Based on protein type and associated lipids, the plasma can be separated into albumin and alpha, beta, and gamma globulin portions. Antibodies are called **gamma globulins** because they are found (mostly) in the gamma globulin portion of plasma, and they are called **immunoglobulins** (Ig) because they are globulin proteins involved in immunity.

The five general classes of immunoglobulins are denoted IgG, IgM, IgA, IgE, and IgD (Table 22-4). All classes of antibodies have a similar structure, consisting of four polypeptide chains (Figure 22-15): two heavy chains that are identical to each other and two light chains that also are identical to each other. Each light chain is attached to a heavy chain, and the ends of the combined heavy and light chains form the antibody's **variable region,** which is the part that combines with the antigenic determinant of the antigen. Antibodies are specific for a given antigen because they have different variable regions. The rest of the antibody is the **constant region,** which is responsible for activities of antibodies such as the ability to activate complement or to attach the antibody to cells such as macrophages, basophils, mast cells, and eosinophils (see Figure 22-15).

Effects of antibodies

Antibodies can directly affect antigens in two ways (Figure 22-16). The antibody can bind to the antigenic determinant of an antigen and interfere with the ability of the antigen to function. Alternatively, the antibody can combine with an antigenic determinant on two different antigens, rendering the antigens ineffective and making them more susceptible to phagocytosis. The ability of antibodies to join antigens together is the basis for many clinical tests (e.g., blood typing) because when enough antigens are bound together, they become visible as a clump, or precipitate.

Although antibodies can directly affect antigens, most of the effectiveness of antibodies results from

Table 22-4

Classes of antibodies and their functions

ANTIBODY	TOTAL SERUM ANTIBODY (%)	DESCRIPTION
IgG	80-85	Activates complement and functions as an opsonin to increase phagocytosis; can cross the placenta and provide immune protection to the fetus and newborn; responsible for Rh reactions such as erythroblastosis fetalis
IgM	5-10	Activates complement and acts as an antigen-binding receptor on the surface of B cells; responsible for transfusion reactions in the ABO blood system; often the first antibody produced in response to an antigen
IgA	15	Secreted into saliva, tears, and onto mucous membranes to provide protection on body surfaces; found in colostrum and milk to provide immune protection to the newborn
IgE	0.002	Binds to mast cells and basophils and stimulates the inflammatory response
IgD	0.2	Functions as antigen-binding receptors on B cells

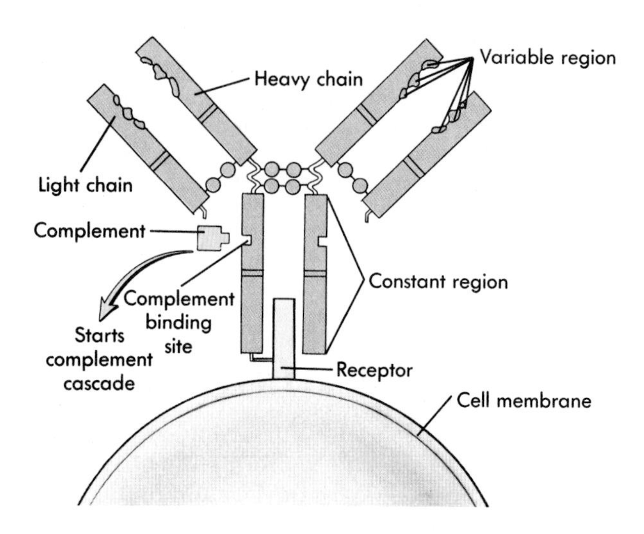

FIGURE 22-15 Structure of an antibody. Antibodies consist of two heavy and two light polypeptide chains. The variable region of the antibody binds to the antigen. The constant region of the antibody can activate the classical pathway of the complement cascade. The constant region of the antibody can also attach the antibody to the cell membrane of cells such as macrophages, basophils, or mast cells.

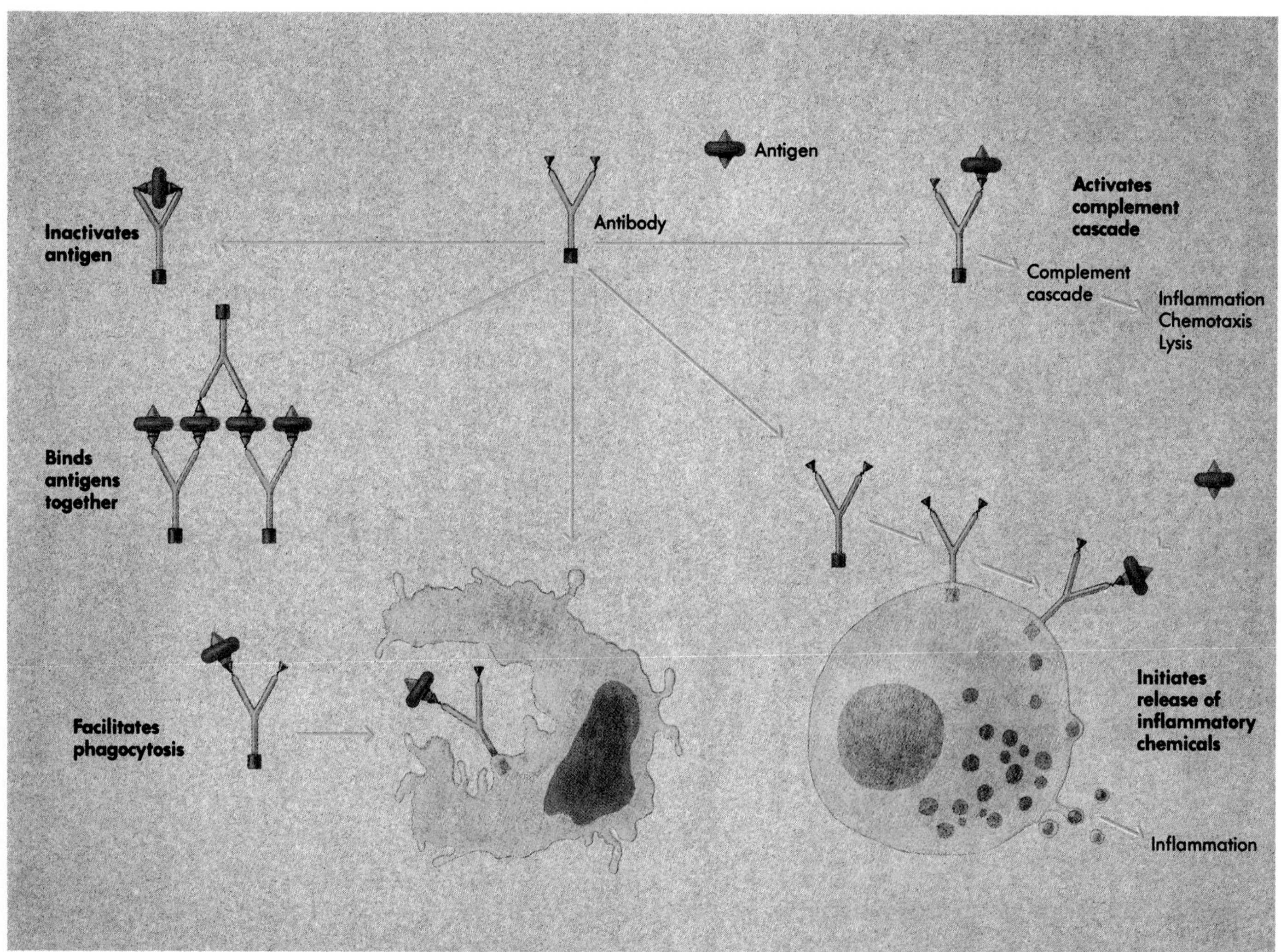

FIGURE 22-16 Actions of antibodies. Antibodies can inactivate antigen, promote phagocytosis (binding antigens together or opsonization), and cause inflammation (release of chemicals from mast cells or basophils and activation of the complement cascade).

other mechanisms such as opsonization, activation of complement, and stimulation of the inflammatory response (see Figure 22-16).

Opsonins (op′so-ninz) are substances that make an antigen more susceptible to phagocytosis. Antibodies (IgG) act as opsonins by connecting to the antigen through the antibody's variable region and to the macrophage through its constant region. The macrophage then phagocytizes the antigen and the antibody.

When an antibody (IgG or IgM) combines with an antigen through the variable region, the constant region can activate the complement cascade through the classical pathway (see Figure 22-10). Activated complement stimulates inflammation, attracts neutrophils, monocytes, macrophages, and eosinophils to sites of infection, and kills bacteria by lysis.

Antibodies (IgE) can initiate an inflammatory response. The antibodies attach to mast cells or basophils through their constant region. When antigens combine with the variable region of the antibodies, the mast cells or basophils release chemicals through exocytosis, and inflammation results.

Monoclonal antibodies are a pure antibody preparation that is specific for only one antigen. The antigen, injected into a laboratory animal, activates a B-cell clone against the antigen. The B cells are removed from the animal and are fused with tumor cells. The resulting hybridoma cells have two ideal characteristics: they divide to form large numbers of cells, and the cells of a given clone produce only one kind of antibody.

Monoclonal antibodies have many applications. They are used for determining pregnancy and for diagnosing diseases such as gonorrhea, syphilis, hepatitis, rabies, and cancer. These tests are specific and rapid because the monoclonal antibodies bind only to the antigen being tested. Monoclonal antibodies may also be used to treat cancer. Anticancer drugs are attached to monoclonal antibodies that bind to cancer cells. The drug then kills the cancer cell. This approach has the advantage of selectively destroying cancer cells while sparing normal, healthy cells.

Disorders of the Lymphatic System

It is not surprising that many infectious diseases produce symptoms associated with the lymphatic system because the lymphatic system is involved with the production of lymphocytes that fight infectious diseases and because the lymphatic system filters blood and lymph to remove microorganisms. **Lymphadenitis** (lim-fad-ĕ-ni′tis) is an inflammation of the lymph nodes that causes them to enlarge and become tender. It is an indication that microorganisms are being trapped and destroyed within the lymph nodes. Sometimes the lymph vessels become inflamed to produce **lymphangitis** (lim-fan-ji′tis). This condition often results in visible red streaks in the skin that extend away from the site of infection. If the microorganisms pass through the lymph vessels and nodes to reach the blood, **septicemia,** or blood poisoning, can result (see Chapter 19).

Bubonic plague and elephantiasis are diseases of the lymphatic system. In the sixth, fourteenth, and nineteenth centuries the **bubonic plague** killed large numbers of people. Fortunately, there are relatively few cases today. Bubonic plague is caused by bacteria that are transferred to humans from rats by the bite of the rat flea. The bacteria localize in the lymph nodes, causing the lymph nodes to enlarge. The term bubonic is derived from a Greek word referring to the groin since the disease often causes the inguinal lymph nodes of the groin to swell. Without treatment the bacteria enter the blood, multiply, and infect tissues throughout the body, rapidly causing death in 70% to 90% of those infected. **Elephantiasis** (el-ĕ-fan-ti′ă-sis) is caused by long, slender roundworms. The adult worms lodge in the lymph vessels and cause a blockage of lymph flow. The resulting accumulation of fluid in the interstitial spaces and lymph vessels can cause permanent swelling and enlargement of a limb. The resemblance of the affected limb to that of an elephant's leg is the basis for the name of the disease. The offspring of the adult worms pass through the lymphatic system into the blood from which they can be transferred to another human by mosquitoes.

A **lymphoma** (lim-fo′mah) is a neoplasm (tumor) of lymph tissue. Lymphomas usually are divided into two groups: (1) Hodgkin's disease and (2) all other lymphomas, which are called non-Hodgkin's lymphomas. Typically, lymphomas begin as an enlarged, painless mass of lymph nodes. Enlargement of the lymph nodes, however, can compress surrounding structures and produce complications. The immune system is depressed, and the patient has an increased susceptibility to infections. Fortunately, treatment with drugs and radiation is effective for many people who suffer from lymphoma.

Antibody production

The production of antibodies after the first exposure to an antigen is different from that after a second or subsequent exposure. The **primary response** results from the first exposure of a B cell to an antigen for which it is specific and includes a series of cell divisions, cell differentiation, and antibody production. The B-cell receptors on the surface of B cells are antibodies, usually IgM and IgD. The receptors have the same specificity (i.e., the same variable region) as the antibodies that eventually will be produced by the B cell. Before stimulation by an antigen, B cells are small lymphocytes. After activation the B cells undergo a series of divisions to produce large lymphocytes. Some of these enlarged cells become **plasma cells,** which produce antibodies, and others revert back to small lymphocytes and become **memory B cells** (Figure 22-17). Usually IgM is the first antibody produced in response to an antigen, but later other classes of antibodies are produced as well. The primary response normally takes 3 to 14 days to produce enough antibodies to be effective against the antigen. In the meantime, the individual usually develops disease symptoms because the antigen has had time to cause tissue damage.

The **secondary,** or **memory, response** occurs when the immune system is exposed to an antigen against which it has already produced a primary response. The secondary response is due to memory B cells, which rapidly divide to produce plasma cells and large amounts of antibody when exposed to the antigen. The secondary response provides better protection than the primary response for two reasons. First, the time required to start producing antibodies is less (hours to a few days), and second, the amount of antibody produced is much larger. As a consequence, the antigen is quickly destroyed, no disease symptoms develop, and the person is immune.

The memory response also includes the formation of new memory B cells, which provide protection against additional exposures to the antigen. Memory B cells are the basis for specific immunity. After destruction of the antigen plasma cells die, the antibodies they released are degraded, and antibody levels decline to the point at which they can no lon-

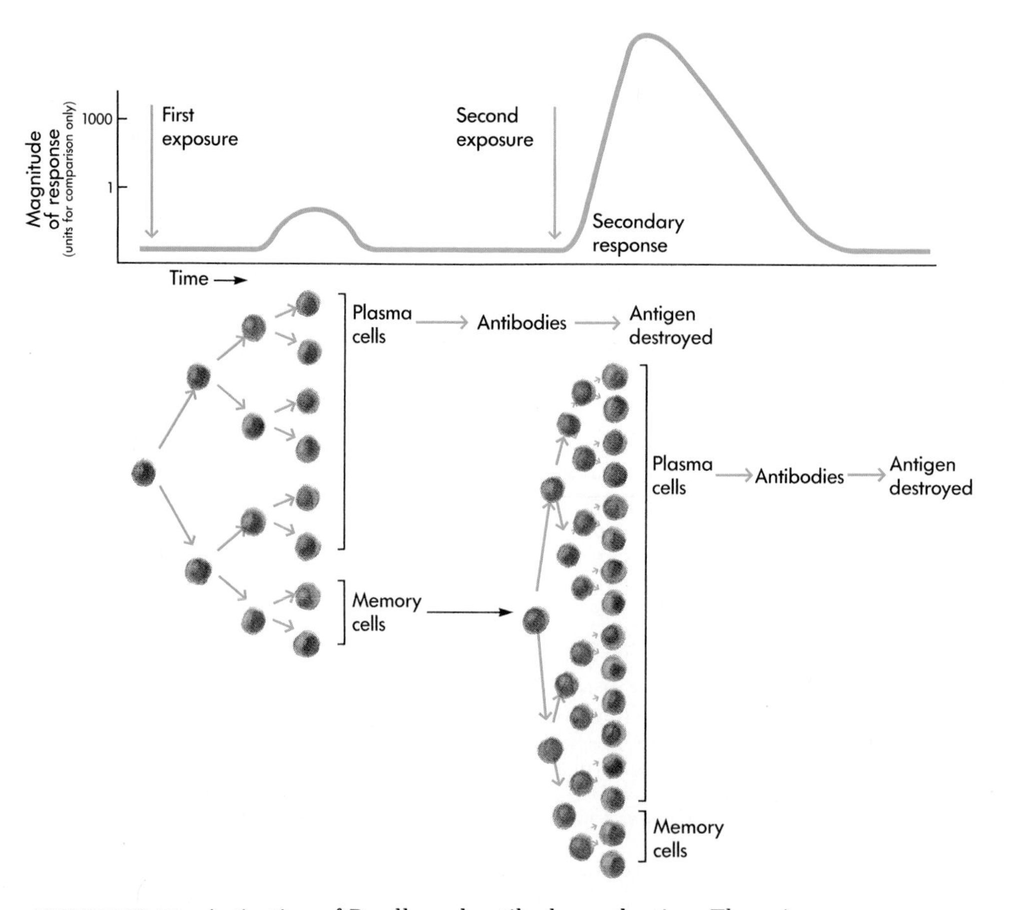

FIGURE 22-17 Activation of B cells and antibody production. The primary response occurs when a B cell is first activated by an antigen. The B cell proliferates to form plasma cells and memory cells. The plasma cells produce antibodies. The secondary response occurs when another exposure to the same antigen causes the memory cells to rapidly form plasma cells and additional memory cells. The secondary response is faster and produces more antibodies than the primary response.

ger provide adequate protection. However, memory B cells persist for many years and probably for life in some cases. On the other hand, if memory-cell production is not stimulated or if the memory B cells produced are short-lived, it is possible to have repeated infections of the same disease. For example, the same cold virus can cause the common cold more than once in the same person.

3

One theory for long-lasting immunity assumes that humans are continually exposed to the disease-causing agent. Explain how this exposure could produce lifelong immunity.

? ? ? ? ? ? ? ? ? ? ?

Cell-Mediated Immunity

Cell-mediated immunity is a function of T cells and is most effective against intracellular microorganisms such as viruses, fungi, intracellular bacteria, and parasites. Delayed hypersensitivity reactions and control of tumors also involve cell-mediated immunity (see the box on pp. 727-728).

Activation of T cells to antigens is regulated by antigen-presenting cells, helper T cells, and suppressor T cells. Once activated, T cells undergo a series of divisions and produce effector T cells and memory T cells (Figure 22-18). Effector T cells such as cytotoxic T cells and delayed hypersensitivity T cells are responsible for the cell-mediated immunity response. Memory T cells can provide a secondary response and long-lasting immunity in the same fashion as memory B cells.

Cytotoxic T cells

Cytotoxic T cells have two main effects: they lyse cells, and they produce lymphokines. Cytotoxic T cells can come into contact with other cells and cause them to lyse. Virus-infected cells have viral antigens, tumor cells have tumor antigens, and tissue transplants have foreign antigens on their surfaces that

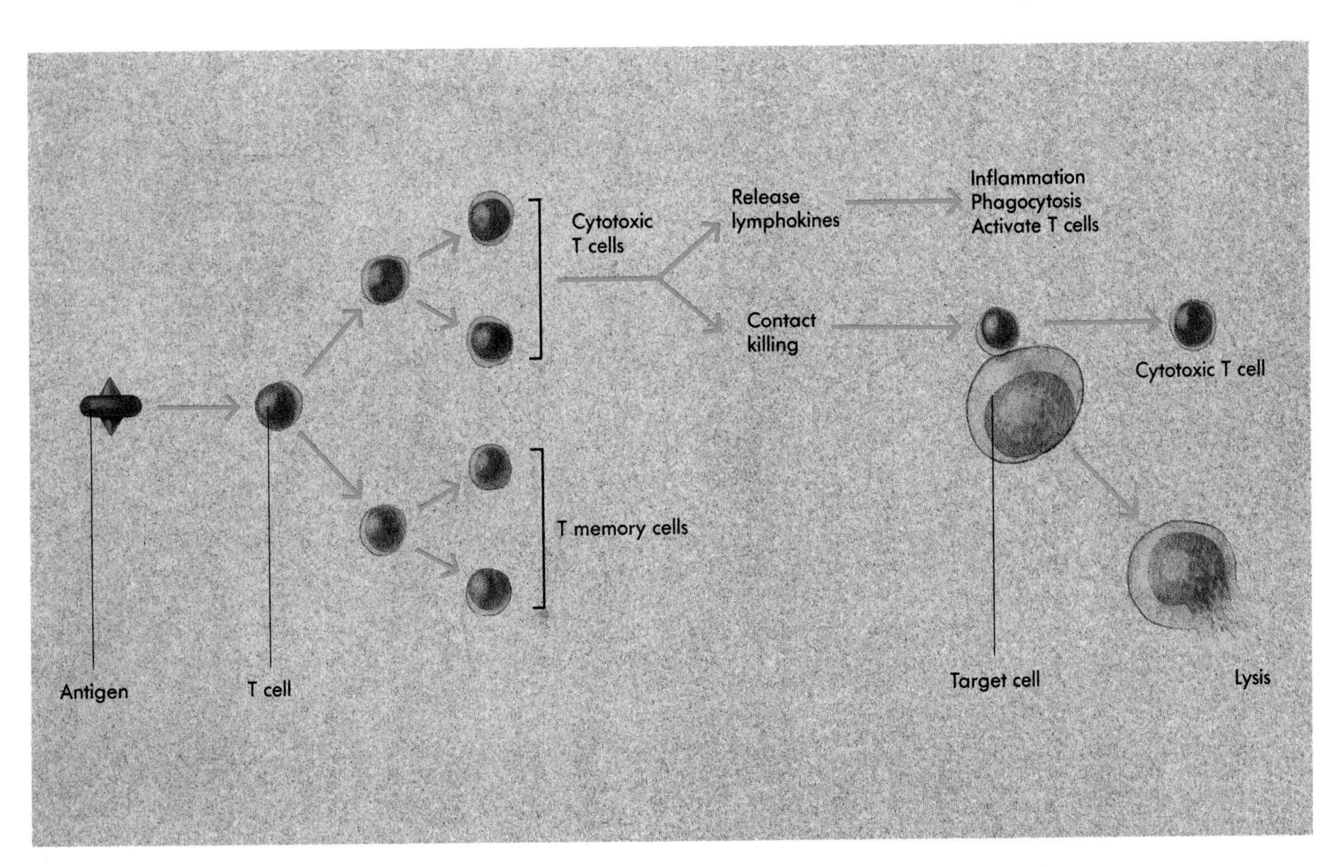

FIGURE 22-18 Stimulation of and effects of T cells. When activated, T cells form memory T and cytotoxic T cells. Memory T cells are responsible for the secondary response, and cytotoxic T cells cause contact killing or release lymphokines that promote the destruction of the antigen.

Table 22-5

Lymphokines and their functions

LYMPHOKINE	DESCRIPTION
Chemotactic factors	Attracts neutrophils, macrophages, eosinophils, and basophils
Macrophage-activating factor	Turns ordinary macrophages into "killer" macrophages that are larger, contain more lysosomes, and are capable of greater phagocytosis than normal macrophages
Lymphotoxins	Kills target cells
Interleukins	Activates additional T and B cells, amplifying the cell-mediated immune system response
Interferon	Prevents viral production and infection

can stimulate cytotoxic T-cell activity. The cytotoxic T cell binds to the target cell and releases chemicals that cause the target cell to lyse. In one method of lysis the chemicals, like complement, form a pore in the target cell's membrane. The cytotoxic T cell then moves on to destroy additional target cells.

In addition to lysing cells, cytotoxic T cells release glycoproteins called **lymphokines** (lim'fo-kīnz), which activate additional components of the immune system. One important function of lymphokines, for example, is the recruitment of nonspecific resistance cells, especially macrophages. These cells are then responsible for phagocytosis and inflammation. Table 22-5 lists some important lymphokines and their functions.

4

In patients with acquired immune deficiency syndrome (AIDS) helper T cells are destroyed by a viral infection. The patients usually die of pneumonia (caused by an intracellular protozoan) or Kaposi's sarcoma (tumorous growths in the skin and lymph nodes). Explain what is happening.

? ? ? ? ? ? ? ? ? ? ?

Delayed hypersensitivity T cells

Delayed hypersensitivity T cells respond to antigens by releasing lymphokines. Consequently, they promote phagocytosis and inflammation, especially in allergic reactions (see box on pp. 727-728). For example, poison ivy antigens can be processed by Langerhans cells in the skin, which present the antigen to delayed hypersensitivity T cells, resulting in an intense inflammatory response.

ACQUIRED IMMUNITY

There are four ways to acquire specific immunity: active natural, active artificial, passive natural, and passive artificial. "Natural" and "artificial" refer to the method of exposure. Natural exposure implies that contact with an antigen or antibody occurred as part of everyday living and was not deliberate. Artificial exposure, also called **immunization,** is a deliberate introduction of an antigen or antibody into the body.

"Active" and "passive" describe whose immune system is responding to the antigen. When an individual is exposed to an antigen (either naturally or artificially), there can be a specific immune system response, which is called **active immunity** because the individual's own immune system is the cause of the immunity. **Passive immunity** occurs when another person or animal develops immunity and the immunity is transferred to a nonimmune individual.

Durability (i.e., how long the immunity lasts) differs for active and passive immunity. The durability of active immunity varies from a few weeks (common cold) to lifelong (whooping cough and chickenpox). Immunity can be long lasting if enough **memory cells** (B or T) are produced and persist to respond to later antigen exposure. Passive immunity is not long lasting because the individual does not produce his own memory cells. Because active immunity can last longer than passive immunity, it is the preferred method. However, passive immunity is preferred in some situations in which immediate protection is needed.

Active Natural Immunity

Natural exposure to an antigen such as a disease-causing microorganism causes an individual's immune system to mount a specific immune system response against the antigen. Because the individual is not immune during the first exposure, he usually develops the symptoms of the disease. Interestingly, exposure to an antigen does not always produce symptoms. Many people, if exposed to the poliomyelitis virus at an early age, will have an immune system response (produce poliomyelitis antibodies) without any symptoms.

Active Artificial Immunity

In active artificial immunity an antigen is deliberately introduced into an individual to stimulate his immune system. This process is **vaccination,** and the

Acquired Immunodeficiency Syndrome (AIDS)

Acquired immunodeficiency syndrome (AIDS) is a life-threatening disease caused by the **human immunodeficiency virus (HIV).** AIDS was first reported in 1981. Since then more than 200,000 cases of AIDS have been confirmed, and it is estimated that more than 1 million people in the United States are infected with HIV, although they presently exhibit no symptoms. Evidence suggests that 54% of those infected will develop symptoms within 10 years, and that all who are infected will eventually develop the disease if they do not die of some other cause.

HIV is transmitted from an infected to a noninfected person by transfer of body fluids (e.g., blood, semen, vaginal secretions, breast milk) containing the virus. The major methods of transmission are intimate sexual contact, contaminated needles used by intravenous drug users, and blood products. Present evidence indicates that household, school, or work contacts do not result in transmission.

The pattern of AIDS cases in the United States can be summarized as follows: 71% are homosexual or bisexual men; 17% are heterosexual men or women who are intravenous drug users; 4% are heterosexual men or women who had sexual contact with an infected person; 2% are recipients of transfusions; 1% are hemophiliacs who received contaminated clotting factors; 1% are children who were infected before birth, during delivery, or by infected breast milk after birth; and a few are health care workers accidentally exposed to HIV-infected blood or body fluids. In other countries the pattern of AIDS cases may be different from that in the United States. For example, in Haiti and central Africa heterosexual transmission is the major route of spread of HIV.

Preventing transmission of HIV is presently the only way to prevent AIDS. The risk of transmission can be reduced by educating the public about safer sexual practices such as reducing one's number of sexual partners, avoiding anal intercourse, and using a condom. Public education also includes warnings to intravenous drug users about the dangers of using contaminated needles. Unfortunately, in many parts of the country there has been inadequate public education about the nature and dangers of AIDS.

Ensuring the safety of the blood supply is another important preventive measure. In April 1985, a test for HIV antibodies in blood became available. Transmission in blood is still possible, however, and the risk of transmission is estimated as 1 in 153,000 per unit of blood. Heat treatment of clotting factors taken from blood has also been effective in preventing transmission of HIV to hemophiliacs.

HIV infection begins when a viral protein called gp120 binds to a surface molecule of cells known as CD4. The CD4 molecule is found primarily on helper T cells, and it normally enables helper T cells to adhere to other lymphocytes (e.g., during the process of antigen presentation). Certain monocytes, macrophages, neurons, and neuroglial cells also have CD4 molecules. Once attached to the CD4 molecules, the virus injects its genetic material (RNA) and enzymes into the cell. Using complementary base pairing (see Chapter 2), the RNA and enzymes produce DNA that can direct the formation of new HIV RNA and proteins (i.e., additional viruses that can infect other cells). The virus eventually kills the helper T cell. In addition, other mechanisms can result in the death of helper T cells. The production of gp120 by the infected cell leads to the appearance of gp120 on the cell's surface and in the blood. The gp120 of infected cells can interact with the CD4 molecules of noninfected cells, and the cells fuse together into a mass of nonfunctional cells. Alternatively, the free gp120 in the blood can circulate and bind to uninfected cells. The immune system recognizes the gp120 on the surface of infected and uninfected cells as a foreign antigen, and the cells are destroyed.

Virtually all the manifestations of AIDS can be explained by the loss of helper T-cell functions or the infection of other cells with CD4 molecules. Without helper T cells cytotoxic T-cell and B-cell activation is impaired, and specific immunity is suppressed. Although macrophage functions are reduced, macrophages are more resistant to destruction than helper T cells. Macrophages serve as a reservoir for HIV and as means of carrying the virus into the central nervous system where the virus can infect and damage neural cells.

After their infection by HIV some patients develop an acute (sudden) mononucleosis-like syndrome that can last up to 14 days. Symptoms include fever, sweats, fatigue, muscle and joint aches, headache, sore throat, diarrhea, rash, and lymphadenopathy (swollen lymph nodes). More commonly there is a persistent version of the syndrome that lasts for several months and includes lymphadenopathy, fever, and fatigue. This condition has been called AIDS-related complex (ARC). During this time the patient becomes positive for HIV antibodies, and within a year many patients develop AIDS.

The most common clinical manifestations of AIDS are opportunistic infections and Kaposi's sarcoma. Opportunistic infections involve organisms that normally do not cause disease but can do so when the immune system is depressed. Examples include *Pneumocystis carinii* pneumonia (caused by an intracellular protozoan); cryptococcal meningitis (caused by a fungus); toxoplasmosis (a parasitic infection of the brain); candidiasis (a yeast infection of the mouth); protozoans that cause severe, persistent diarrhea; and many viral and bacterial agents. Kaposi's sarcoma is a type of cancer that produces lesions in the skin, lymph nodes, and visceral organs. Also associated with AIDS are symptoms resulting from the effects of HIV on the nervous system, including motor retardation, behavioral changes, progressive dementia, and possibly psychosis.

There currently is no cure for

Continued.

Acquired Immunodeficiency Syndrome (AIDS)—cont'd

AIDS. Management of AIDS can be divided into two categories: (1) management of secondary infections or malignancies associated with AIDS and (2) treatment of HIV infection itself. Management of secondary infections uses three approaches: prevention of infection, treatment of active infection, and prevention of recurrence of infection. Pentamidine and trimethoprim-sulfamethoxazole have proven effective against *P. carinii*, and specific treatments are available for many other secondary infections or malignancies. Zidovudine (also known as AZT) inhibits the replication of HIV but can produce serious side effects such as anemia or even total bone marrow failure. Zidovudine can increase the survival time of AIDS patients and provides clear benefits when given to asymptomatic persons infected with HIV. Resistance to zidovudine, however, often develops after 6 to 18 months of treatment, and there is clearly a need for alternative treatments. Another drug, dideoxyinosine (ddI), which also prevents HIV replication, recently has been approved for use in the United States. A different approach uses CD4 molecules produced through genetic engineering. Infection of cells may be prevented if the HIV binds to these "decoys" instead of the CD4 molecules on the helper T cells.

The ultimate goal is to produce a vaccine that prevents AIDS, and many different strategies are under development. Simian immunodeficiency virus (SIV) causes a type of AIDS in monkeys. A vaccine using killed SIV produced a protective antibody response in rhesus macaques, demonstrating that such a vaccine might work in humans. Because there is always a danger that not all of the viruses in a vaccine have been killed, another approach is to use just part of the virus, which could not cause an infection. The antigenic determinant in the gp120 molecule has been isolated and used to stimulate a protective antibody response in chimpanzees. Although antibody production may prevent infection, another strategy is to stimulate cell-mediated immunity to kill any cells that are infected. By removing the sites of HIV replication, infections could be prevented and possibly cured. A vaccine using the enzymes injected by HIV into cells may stimulate such a cell-mediated response. Hopefully, an effective vaccine using these or other strategies will soon become available.

introduced antigen is a **vaccine.** Injection of the vaccine is the usual mode of administration (tetanus toxoid, diphtheria, and whooping cough), although ingestion (Sabin poliomyelitis vaccine) is sometimes used. The vaccine usually consists of some part of the microbe, a dead microbe, or a live, altered microbe. The antigen has been changed so that it will stimulate the immune system but will not cause the symptoms of the disease. Because active artificial immunity produces long-lasting immunity without disease symptoms, it is the preferred method of acquiring specific immunity.

Memory response is the basis for many immunization schemes. The first injection of the antigen stimulates a primary response, and the booster shot causes a memory response, which produces high levels of antibody, many memory cells, and long-lasting protection.

Passive Natural Immunity

Passive natural immunity results from the transfer of antibodies from a mother to her child. During her life the mother has been exposed to many antigens, either naturally or artificially, and she therefore has antibodies against many of these antigens. These antibodies protect both the mother and the developing fetus against disease. The antibody IgG can cross the placenta and enter the fetal circulation. After birth the antibodies provide protection for the first few months of the infant's life. Eventually the antibodies are broken down, and the infant must rely on its own immune system. If the mother nurses her child, IgA in the mother's milk may also provide some protection for the infant.

Passive Artificial Immunity

Achieving passive artificial immunity usually begins with vaccinating an animal such as a horse. After the animal's immune system responds to the antigen, antibodies (sometimes T cells) are removed from the animal and are injected into the individual requiring immunity. In some cases a human that has developed immunity is used as a source. Passive artificial immunity provides immediate protection for the individual receiving the antibodies and is therefore the preferred treatment when there may not be time for the individual to develop his own immunity. However, this technique provides only temporary immunity since the antibodies are used or eliminated by the recipient.

Immune System Problems of Clinical Significance

Hypersensitivity Reactions

Immune and hypersensitivity (allergy) reactions involve the same mechanisms, and the differences between them are not clear. Both require exposure to an antigen and subsequent stimulation of antibody-mediated immunity and/or cell-mediated immunity. If immunity to the antigen is established, later exposure to the antigen results in an immune system response that eliminates the antigen, and no symptoms appear. In **hypersensitivity reactions** the antigen is called the allergen, and later exposure to the allergen stimulates much the same process that occurs during the normal immune system response. However, the processes that eliminate the allergen also produce undesirable side effects such as a very strong inflammatory reaction. This immune system response can be more harmful than beneficial and can produce many unpleasant symptoms. Hypersensitivity reactions are categorized as immediate or delayed.

Immediate Hypersensitivities

Immediate hypersensitivities are caused by antibodies interacting with the allergen, and symptoms appear within a few minutes of exposure to the allergen. Immediate hypersensitivity reactions include atopy, anaphylaxis, cytotoxic reactions, and immune complex disease.

Atopy (at'o-pe) is a localized IgE-mediated hypersensitivity reaction. For example, in patients with hay fever the allergens, usually plant pollens, are inhaled and absorbed through the respiratory mucosa. The resulting localized inflammatory response produces swelling and excess mucus production. In asthma patients the allergen stimulates the release of leukotrienes and histamine in the bronchioles of the lung, causing constriction of the smooth muscles of the bronchioles and difficulty in breathing. Hives (urticaria) is an allergic reaction that results in a skin rash or localized swellings and is usually caused by an ingested allergen.

Anaphylaxis (an-ă-fi-lak'sis) is a systemic IgE-mediated reaction that can be life threatening. Introduction of allergens such as drugs (e.g., penicillin) and insect stings is the most common cause. The chemicals released from mast cells and basophils cause systemic vasodilation, a drop in blood pressure, and cardiac failure. Symptoms of hay fever, asthma, and hives may also be observed.

In **cytotoxic reactions** IgG or IgM combines with the antigen on the surface of a cell, resulting in the activation of complement and subsequent lysis of the cell. Transfusion reactions caused by incompatible blood types, erythroblastosis fetalis (see Chapter 19), and some types of autoimmune disease are examples.

Immune complex disease occurs when too many immune complexes, which are combinations of soluble antigens (as opposed to cells in cytotoxic reactions) and IgG or IgM, are formed. When there are too many immune complexes, too much complement is activated, and an acute inflammatory response develops. Complement attracts neutrophils to the area of inflammation and stimulates the release of lysosomal enzymes. This release causes tissue damage, especially in small blood vessels where the immune complexes tend to lodge, and lack of blood supply causes tissue necrosis. Arthus reactions, serum sickness, some autoimmune diseases, and chronic graft rejection are examples of immune complex diseases.

An **Arthus reaction** is a localized immune complex reaction. For example, if an individual has been sensitized to antigens in the tetanus toxoid vaccine because of repeated vaccinations and if that individual were vaccinated again, at the injection site there would be large amounts of antigen with which the antibody could complex, causing a localized inflammatory response, neutrophil infiltration, and tissue necrosis.

Serum sickness is a systemic Arthus reaction in which the antibody-antigen complexes circulate and lodge in many different tissues. Serum sickness can develop from prolonged exposure to an antigen that provides enough time for an antibody response and the formation of many immune complexes. Examples of antigens include long-lasting drugs and passive artificial immunity. Symptoms include fever, swollen lymph nodes and spleen, and arthritis. Symptoms of anaphylaxis such as hives may also be present since IgE involvement is a part of serum sickness. If large numbers of the circulating antibody-antigen complexes are removed from the blood by the kidney, immune complex glomerulonephritis can develop in which kidney blood vessels are destroyed and the kidneys fail to function.

Delayed Hypersensitivity

Delayed hypersensitivity is mediated by T cells, and symptoms usually take several hours or days to develop. Like immediate hypersensitivity, delayed hypersensitivity is an acute extension of the normal operation of the immune system. Exposure to the allergen causes activation of effector T cells and the production of lymphokines. The lymphokines attract basophils and monocytes, which differentiate into macrophages. The activities of these cells result in progressive tissue destruction, loss of function, and scarring.

Delayed hypersensitivity can develop as allergy of infection and contact hypersensitivity. **Allergy of infection** is a side effect of cell-mediated efforts to eliminate intracellular microorganisms, and the amount of tissue destroyed is determined by the persistence and distribution of the antigen. The minor rash of measles is due to tissue damage as cell-mediated immunity destroys virus-infected cells. In patients with chronic infections with long-term antigenic stimulation, the allergy-of-infection response can cause extensive tissue damage. The destruction of lung tissue in tuberculosis is an example.

Contact hypersensitivity is a delayed hypersensitivity reaction

Continued.

Immune System Problems of Clinical Significance—cont'd

Delayed Hypersensitivity—cont'd

to allergens that contact the skin or mucous membranes. Poison ivy, poison oak, soaps, cosmetics, drugs, and a variety of chemicals can induce contact hypersensitivity, usually after prolonged exposure. The allergen is absorbed by epithelial cells, and T cells invade the affected area, causing inflammation and tissue destruction. Although itching may be intense, scratching is harmful because it damages tissues and causes additional inflammation.

Autoimmune Diseases

In an **autoimmune disease** the immune system fails to differentiate between self-antigens and foreign antigens. Consequently, an immune system response is produced, resulting in tissue destruction. In many instances autoimmunity is probably due to a breakdown of tolerance (e.g., suppressor T cells no longer prevent an immune system response to self-antigens). In molecular mimicry a foreign antigen that is very similar to a self-antigen stimulates an immune system response. After the foreign antigen is eliminated, the immune system continues to act against the self-antigen. It is hypothesized that type I diabetes (see Chapter 18) develops in this fashion. In susceptible people a foreign antigen may stimulate specific immunity, especially cell-mediated immunity, which destroys the insulin-producing beta cells of the pancreas. Other autoimmune diseases that involve antibodies are rheumatoid arthritis, rheumatic fever, Grave's disease, systemic lupus erythematosus, and myasthenia gravis.

Immunodeficiencies

Immunodeficiency is a failure of some part of the immune system to function properly. A deficient immune system is not uncommon because it can have many causes. Inadequate protein in the diet inhibits protein synthesis, and antibody levels will decrease. Stress can depress the immune system; fighting an infection can deplete lymphocyte and granulocyte reserves, making a person more susceptible to further infection. Diseases that cause proliferation of lymphocytes (e.g., mononucleosis, leukemias, and myelomas) can result in an abundance of lymphocytes that do not function properly. Finally, the immune system may purposefully be suppressed by drugs to prevent graft rejection.

Congenital (present at birth) immunodeficiencies may involve inadequate B-cell formation, inadequate T-cell formation, or both. **Severe combined immunodeficiency disease (SCID)** in which both B cells and T cells fail to differentiate is probably the best known. Unless the person suffering from SCID is kept in a sterile environment or is provided with a compatible bone marrow transplant, death from infection results.

Tumor Control

Tumor cells have tumor antigens that distinguish them from normal cells. T cells, NK cells, activated macrophages, antibodies, and complement are all involved in the destruction of tumor cells. According to the concept of **immune surveillance,** the immune system detects tumor cells and destroys them before a tumor can form. There is evidence that immune surveillance may exist for some forms of cancer. For example, people with deficient or depressed immune systems have a higher incidence of lymphosarcomas and leukemias than do people with normal immune systems. Apparently several different components of the immune system are involved in immune surveillance: T cells are effective against tumors caused by viruses, and NK cells prevent the formation and spread of tumors.

Transplantation

In humans the major histocompatibility complex genes are often referred to as **human leukocyte antigen (HLA) genes** because they were first identified in leukocytes. There are millions of different possible combinations of the HLA genes, and it is very rare for two individuals (except identical twins) to have the same set of HLA genes. Because they are genetically determined, however, the closer the relationship between two individuals, the greater is the likelihood they will share the same HLA genes.

The HLA genes control the production of HLAs (human leukocyte antigens, i.e., MHC proteins), which are inserted onto the surface of cells. The immune system can distinguish between self- and foreign cells because they both are marked with HLAs. Rejection of a transplanted tissue is caused by a normal immune system response to foreign antigens (i.e., to foreign HLAs).

Acute rejection of a graft occurs several weeks after transplantation and is due to a delayed hypersensitivity reaction and cell lysis. Lymphocytes and macrophages infiltrate the area, a strong inflammatory response occurs, and the foreign tissue is destroyed. If acute rejection does not develop, **chronic rejection** may occur at a later time; immune complexes form in the arteries supplying the graft, blood supply fails, and the graft is rejected.

Graft rejection can occur in two different directions. In **host vs. graft rejection** the recipient's immune system recognizes the donor's tissue as foreign and rejects the transplant. In a **graft vs. host rejection** the donor tissue recognizes the recipient's tissue as foreign, and the transplant rejects the recipient, causing destruction of the recipient's tissue and death.

To reduce graft rejection, a tissue match is performed. Only tissues with HLAs similar to the recipient's have a chance of acceptance. Because an exact match is possible only for a graft from one part to another part of the same person's body or between identical twins, in other graft situations immunosuppressive drugs must be administered throughout the patient's life to prevent rejection.

HLAs are important in ways in addition to organ transplants. Because they are genetically determined, HLAs help resolve paternity suits. In forensic medicine the HLAs in blood, semen, and other tissues help identify the person from whom the tissue came.

The Lymphatic/Immune System

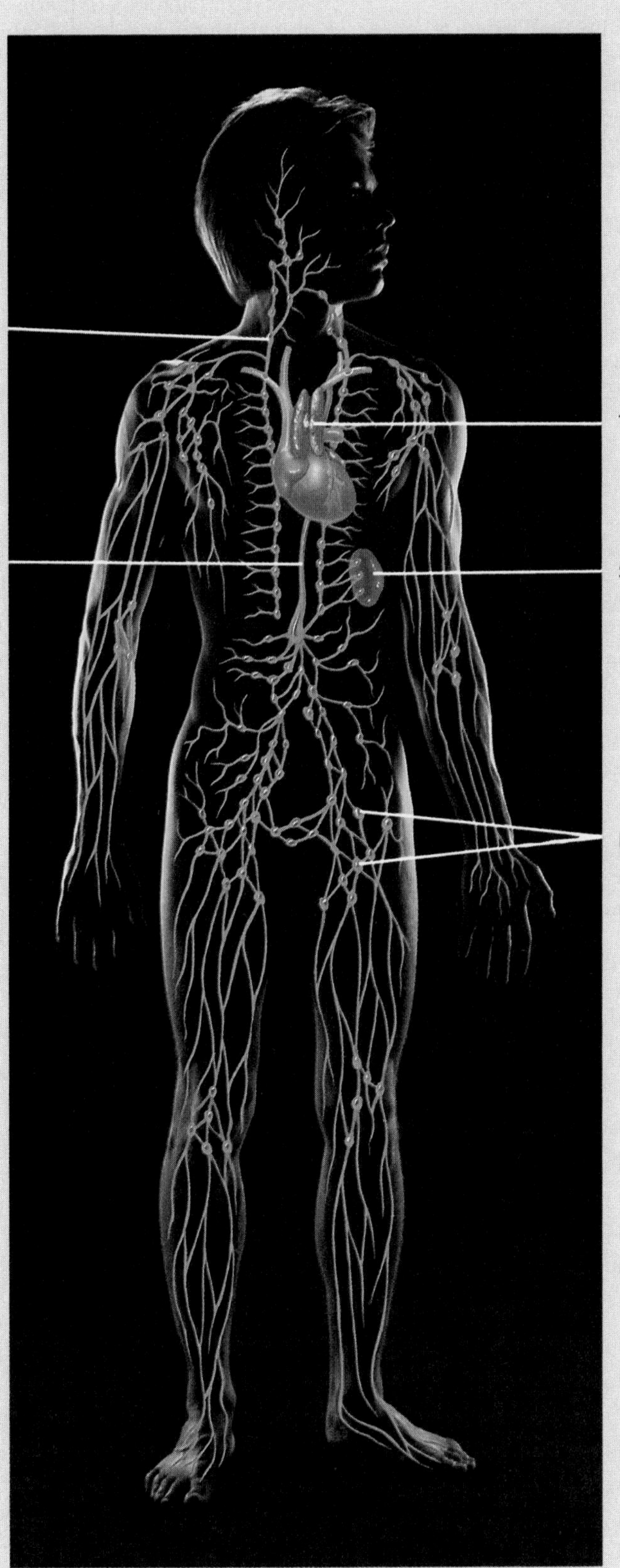

Major Components

Lymph vessels, lymph nodes, tonsils, spleen, thymus, and leukocytes

Major Functions

Lymph vessels absorb fats and helps maintain tissue fluid balance
Removes foreign substances from blood and lymph
Leukocytes destroy invading pathogens and produce antibodies against foreign substances
Produces the inflammatory response
Combats disease

SUMMARY

THE LYMPHATIC SYSTEM CONSISTS OF LYMPH, LYMPH VESSELS, LYMPHOCYTES, LYMPH NODULES, LYMPH NODES, TONSILS, SPLEEN, AND THYMUS. The lymphatic system maintains fluid balance in tissues, absorbs fats and other substances from the intestines, and defends against microorganisms and other foreign substances.

LYMPHATIC ORGANS *p. 701*

Lymphatic tissue is reticular connective tissue that contains lymphocytes and other cells.

Diffuse Lymphatic Tissue and Lymph Nodules

1. Diffuse lymphatic tissue consists of dispersed lymphocytes and has no clear boundaries.
2. Lymph nodules are small aggregates of lymphatic tissue (e.g., Peyer's patches in the small intestines).

Tonsils

1. Tonsils are large groups of lymph nodules in the oral cavity and nasopharynx.
2. The three groups of tonsils are the palatine, pharyngeal, and lingual tonsils.

Lymph Nodes

1. Lymphatic tissue in the node is organized into the cortex and the medulla. Lymph sinuses extend through the lymphatic tissue.
2. Substances in lymph are removed by phagocytosis, and/or they stimulate lymphocytes.
3. Lymphocytes leave the lymph node and circulate to other tissues.

Spleen

1. The spleen is in the left superior side of the abdomen.
2. Foreign substances stimulate lymphocytes in the white pulp.
3. Foreign substances and defective red blood cells are removed from the blood by phagocytes in the red pulp.

Thymus

1. The thymus gland is in the superior mediastinum and is divided into a cortex and a medulla.
2. Lymphocytes in the cortex are separated from the blood by reticular cells.
3. Lymphocytes produced in the cortex migrate through the medulla, enter the blood, and travel to other lymph tissues where they can proliferate.

IMMUNITY *p. 706*

Immunity is the ability to resist the harmful effects of microorganisms and other foreign substances.

NONSPECIFIC RESISTANCE *p. 706*

Mechanical Mechanisms

Mechanical mechanisms prevent the entry of microbes (skin and mucous membranes) or remove them (tears, saliva, and mucus).

Chemicals

1. Chemicals promote phagocytosis and inflammation.
2. Complement, activated by properdin (alternate pathway) or antibodies (classical pathway), increase phagocytosis and promote inflammation.
3. Interferon prevents viral replication. Interferon is produced by virally infected cells and moves to other cells, which are then protected.

Cells

1. Chemotactic factors are parts of microorganisms or chemicals that are released by damaged tissues. Chemotaxis is the ability of leukocytes to move to tissues that release chemotactic factors.
2. Phagocytosis is the ingestion and destruction of materials.
3. Neutrophils are small phagocytic cells.
4. Macrophages are large phagocytic cells.
   - Macrophages can engulf more than neutrophils can.
   - Macrophages in connective tissue protect the body at locations where microbes are likely to enter the body, and macrophages clean blood and lymph.
5. Basophils and mast cells release chemicals that promote inflammation.
6. Eosinophils release enzymes that reduce inflammation.
7. Natural killer cells lyse tumor cells and virus-infected cells.

Inflammatory Response

1. The inflammatory response can be initiated in many ways.
   - Chemical mediators cause vasodilation and increase vascular permeability, allowing the entry of other chemical mediators.
   - Chemotactic factors attract phagocytes.
   - The amount of chemical mediators and phagocytes increases until the cause of the inflammation is destroyed. Then the tissue undergoes repair.
2. Local inflammation produces the symptoms of redness, heat, swelling, pain, and loss of function. Symptoms of systemic inflammation include an increase in neutrophil numbers, fever, and shock.

SPECIFIC IMMUNITY *p. 712*

1. Antigens are large molecules that stimulate a specific immune system response. Haptens are small molecules that combine with large molecules to stimulate a specific immune system response.
2. B cells are responsible for humoral, or antibody-mediated, immunity. T cells are involved with cell-mediated immunity.

Origin and Development of Lymphocytes

1. B cells and T cells originate in red bone marrow. T cells are processed in the thymus, and B cells are processed in bone marrow.
2. B cells and T cells move to lymphatic tissue from their processing sites. They continually circulate from one lymphatic tissue to another.
3. A clone is a group of identical lymphocytes that can respond to a specific antigen. Clones that can react against self-cells are normally eliminated or suppressed.
4. The MHC genes produce MHC proteins, which are inserted on the surface of cells. The MHC proteins function as self-markers.

Activation of Lymphocytes

1. The antigenic determinant is the specific part of the antigen to which the lymphocyte responds. The antigen-binding receptor (T-cell receptor or B-cell receptor) on the surface of lymphocytes combines with the antigenic determinant.

2. Antigen-presenting cells use MHC proteins to present antigens to helper T cells.
3. Macrophages release interleukin-1 that stimulates helper T cells to release interleukin-2. The interleukin-2 stimulates the helper T cells to proliferate.

Inhibition of Lymphocytes

1. Suppressor T cells can block the activity of helper T, effector T, and B cells.
2. Tolerance is suppression of the immune system's response to an antigen.

Antibody-Mediated Immunity

1. Antibodies are proteins.
   - The variable region of an antibody combines with the antigen. The constant region activates complement or binds to cells.
   - There are five classes of antibodies: IgG, IgM, IgA, IgE, and IgD.
2. Antibodies affect the antigen in many ways.
   - Antibodies bind to the antigen and interfere with antigen activity or bind the antigens together.
   - Antibodies act as opsonins (a substance that increases phagocytosis) by binding to the antigen and to macrophages.
   - Antibodies can activate complement through the classical pathway.
   - Antibodies attach to mast cells or basophils and cause the release of inflammatory chemicals when the antibody combines with the antigen.
3. The primary response results from the first exposure to an antigen. B cells form plasma cells, which produce antibodies, and memory cells.
4. The secondary response results from exposure to an antigen after a primary response, and memory B cells quickly form plasma cells and additional memory cells.

Cell-Mediated Immunity

1. Antigen activates effector T cells and produces memory T cells.
2. Cytotoxic T cells lyse virus-infected cells, tumor cells, and tissue transplants.
3. Cytotoxic T cells produce lymphokines, which promote phagocytosis and inflammation.
4. Delayed hypersensitivity T cells release lymphokines and are involved in allergic reactions.

ACQUIRED IMMUNITY *p. 724*

1. Active natural immunity results from natural exposure to an antigen.
2. Active artificial immunity results from deliberate exposure to an antigen.
3. Passive natural immunity results from transfer of antibodies from a mother to her fetus or baby.
4. Passive artificial immunity results from transfer of antibodies (or cells) from an immune animal to a nonimmune animal.

CONTENT REVIEW

1. List the parts of the lymphatic system, and describe the three main functions of the lymphatic system.
2. What are diffuse lymphatic tissue and lymph nodules, and where are they found?
3. Describe the structure, location, and function of the Peyer's patches and the tonsils.
4. Describe the structure of a lymph node. What happens to substances in the lymph as lymph passes through the lymph node? What effect can foreign materials in lymph have on lymphocytes?
5. Describe the structure and location of the spleen. What are its functions?
6. Where is the thymus gland located? Describe its structure and function.
7. What is the difference between nonspecific resistance and specific immunity?
8. List some mechanical barriers and chemicals, and explain how they provide protection against microorganisms.
9. Define chemotactic factors and chemotaxis.
10. What are the functions of neutrophils and macrophages?
11. What effects do the chemicals released by basophils, mast cells, and eosinophils produce?
12. Describe the function of natural killer cells.
13. Describe the events that take place during an inflammatory response. What are the symptoms of local and systemic inflammation?
14. Define antigen and hapten. Distinguish between a foreign antigen and a self-antigen.
15. Define an antigenic determinant and an antigen-binding receptor. Describe a T-cell receptor.
16. Describe the origin and development of B and T cells. Define MHC proteins, and explain their function.
17. Describe the role of antigen-presenting cells and helper T cells in the activation of B cells.
18. What is the role of suppressor T cells in regulating lymphocyte activity? What is tolerance, and how is it accomplished?
19. What are the functions of the variable and constant regions of an antibody? List the five classes of antibodies and state their functions.
20. Describe the different ways that antibodies participate in the destruction of antigens.
21. What are plasma cells and memory cells, and what are their functions?
22. What are the primary and secondary antibody responses?
23. What are the functions of cytotoxic T cells, delayed hypersensitivity T cells, and memory T cells?
24. What are lymphokines? Describe some actions of lymphokines.
25. State four general ways of acquiring specific immunity. Which two provide the longest lasting immunity?

CONCEPT REVIEW

1. A patient had many allergic reactions (see boxed essay on pp. 727-728). As part of the treatment scheme, it was decided to try to identify the allergen that stimulated the immune system's response. A series of solutions, each containing an allergen that commonly causes a reaction, was composed. Each solution was injected into the skin at different locations on the patient's back. The following results were obtained: (1) at one location within a few minutes the injection site became red and swollen; (2) at another injection site swelling and redness did not appear until 2 days later; and (3) no redness or swelling developed at the other sites. Explain what happened for each observation (i.e., what part of the immune system was involved and what caused the redness and swelling).
2. Suppose you want to test a woman's serum to see if she has antibodies against the measles virus. Assume that you (1) can take a sample of the woman's serum, (2) have a supply of the measles virus, and (3) have test cells that can be grown in a test tube. When the test cells are infected by the measles virus, they lyse, a reaction that is easily observed. Design a test procedure that will verify or refute the presence of the measles virus in the woman's serum.
3. If the thymus of an experimental animal is removed immediately after its birth, the animal exhibits the following characteristics: (1) it is more susceptible to infections; (2) it has decreased numbers of lymphocytes in lymphatic tissue; and (3) its ability to reject grafts is greatly decreased. Explain these observations.
4. If the thymus of an adult experimental animal is removed, the following observations can be made: (1) there is no immediate effect; and (2) after 1 year the number of lymphocytes in the blood decreases, the ability to reject grafts decreases, and the ability to produce antibodies decreases. Explain these observations.
5. Adjuvants are substances that slow but do not stop the release of an antigen from an injection site into the blood. Suppose injection A of a given amount of antigen were given without an adjuvant and injection B of the same amount of antigen were given with an adjuvant that caused the release of antigen over a period of 2 to 3 weeks. Would injection A or B result in the greater amount of antibody production? Explain.
6. A researcher obtained two samples of blood from a heart attack victim. Sample A was taken the day after the heart attack, and sample B was taken several weeks later. The researcher tested the blood for the presence of antibodies that bind to mitochondria. She found no evidence for the antibodies in sample A but did find the antibodies in sample B. Explain these observations.

? ?

ANSWERS TO PREDICT QUESTIONS

1 *p. 714.* The helper T cells transferred to mouse B will not respond to the antigen. The helper T cells are MHC restricted and must have both the MHC proteins of mouse A and the antigen X to respond.

2 *p. 716.* Proliferation of helper T cells and B cells continues as long as they are exposed to the antigen for which they are specific. After the antigen is eliminated, helper T cells are not stimulated to divide, and B cells do not process and present antigen to the helper T cells. Consequently, the B cells are not stimulated, and the production of antibodies stops.

3 *p. 723.* The first exposure to the disease-causing agent (antigen) would evoke a primary immune system response. Gradually, however, antibodies would degrade, and memory cells would die. If before all the memory cells were eliminated, a second exposure to the antigen occurred, a secondary immune system response would result. The memory cells produced could provide immunity until the next exposure to the antigen.

4 *p. 724.* With depression of helper T-cell activity the ability of antigens to activate effector T cells would be greatly decreased. Depression of cell-mediated immunity results in an inability to deal with intracellular microorganisms and cancer.

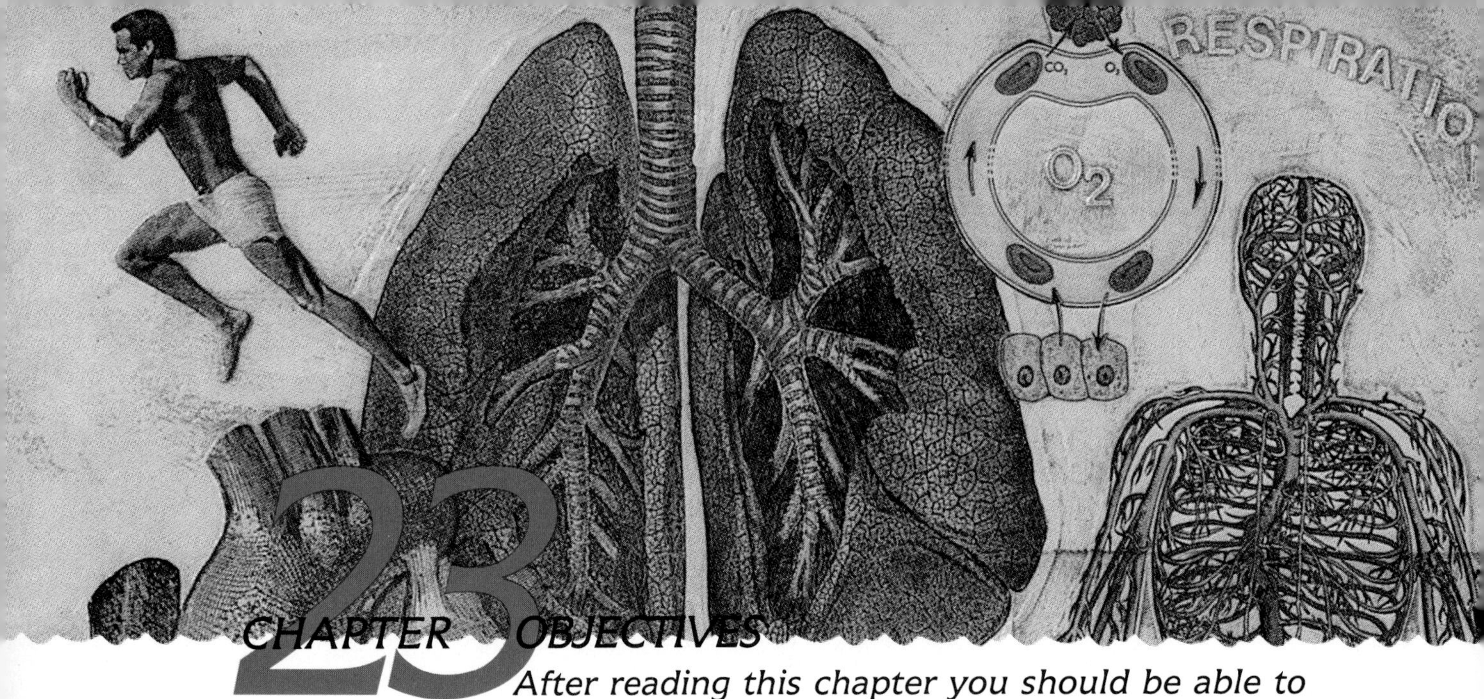

23 CHAPTER OBJECTIVES

After reading this chapter you should be able to

1. Describe the anatomy and histology of the respiratory passages.
2. Distinguish between vestibular cords and vocal cords, and explain how sounds of different loudness and pitch are produced.
3. Describe the lungs, their membranes, and the cavities in which they lie.
4. Explain how contraction of the muscles of respiration causes air to flow into and out of the lungs during quiet respiration and forced respiration.
5. Describe the factors that affect the flow of air through a tube and the factors that determine the pressure of a gas.
6. Name the factors that cause the lungs to collapse.
7. Define compliance and give its significance.
8. List the pulmonary volumes and capacities and define each of them.
9. Explain the significance of the following measurements: forced expiratory vital capacity, minute respiratory volume, and alveolar ventilation rate.
10. Define water vapor pressure and partial pressure of a gas.
11. Explain why the partial pressures of oxygen and of carbon dioxide are different for dry atmospheric air, alveolar air, and expired air.
12. Describe the factors affecting movement of a gas into and through a liquid.
13. List the components of the respiratory membrane and explain the factors that affect gas movement through it.
14. Explain the significance of the oxygen-hemoglobin dissociation curve and illustrate how it is influenced by exercise.
15. Describe how carbon dioxide is transported in blood and discuss the chloride shift and how respiration can affect blood pH.
16. Discuss the brain centers for controlling respiration.
17. Explain how alterations in blood P_{CO_2}, pH, and P_{O_2} influence the respiratory movements.
18. List the stimuli in addition to blood gases and pH that influence the respiratory movements and explain how they affect the respiratory movements.

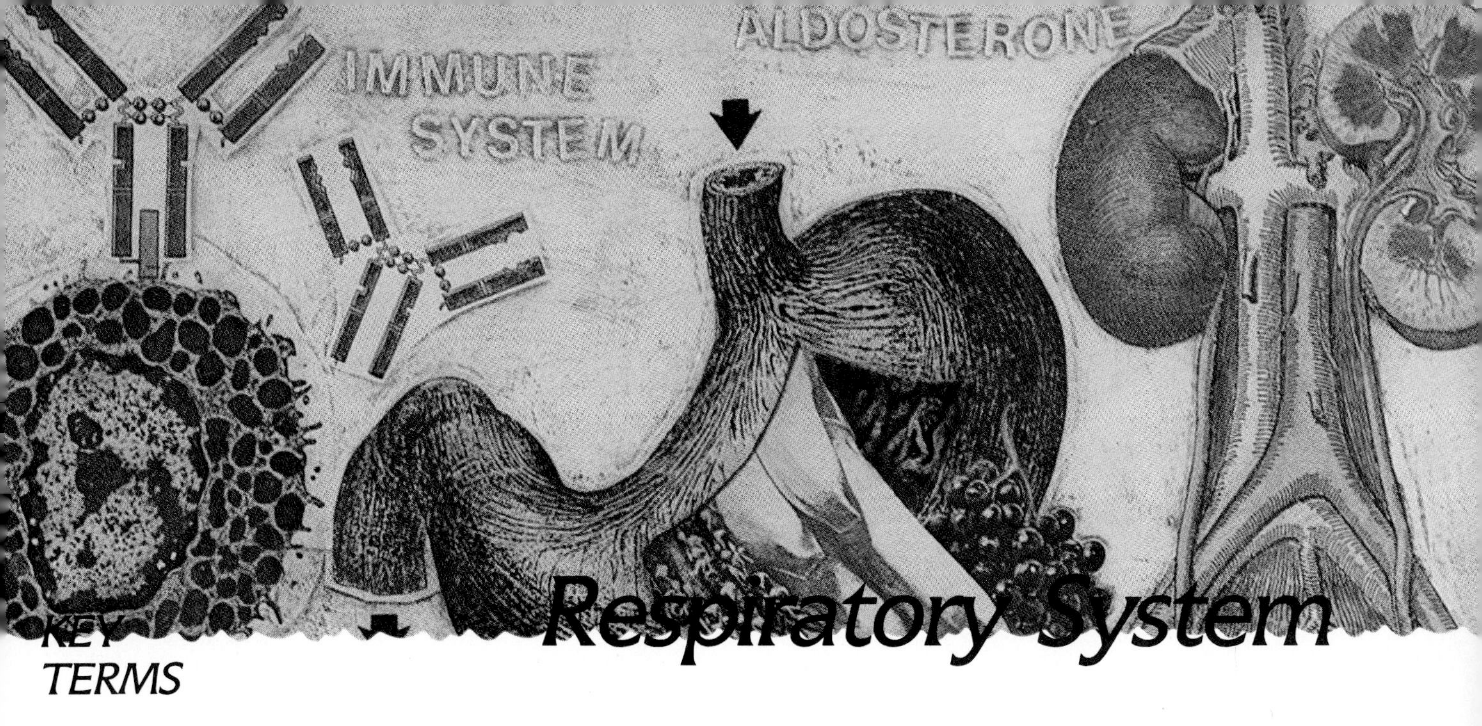

Respiratory System

KEY TERMS

alveolar ventilation rate

alveolus (al-ve′o-lus)

Bohr effect

bronchus (brong′kus)

chloride shift

compliance

Hering-Breuer (her′ing broy′er) reflex

intrapleural pressure

intrapulmonary pressure

larynx (lăr′ingks)

minute respiratory volume

partial pressure

pulmonary capacity

respiratory center

respiratory membrane

RELATED TOPICS

The following terms or concepts are important for a good understanding of this chapter. If you are not familiar with them, you should review them before proceeding.

Reflexes (Chapter 13)

Blood and hemoglobin (Chapter 19)

Pulmonary circulation (Chapter 21)

ALL CELLS OF THE BODY PERFORM AEROBIC METABOLISM FOR WHICH OXYGEN IS ESSENTIAL AND FROM WHICH CARBON DIOXIDE IS A MAJOR WASTE PRODUCT. BOTH OXYGEN AND CARBON DIOXIDE ARE GASES AT THE TEMPERATURES AND PRESSURES AT WHICH LIFE IS POSSIBLE. THE RESPIRATORY SYSTEM AND THE CARDIOVASCULAR SYSTEM TAKE OXYGEN FROM THE AIR AND TRANSPORT IT TO INDIVIDUAL CELLS. THEY THEN TRANSPORT CARBON DIOXIDE FROM CELLS AND RELEASE IT FROM THE BODY INTO THE AIR. THE RESPIRATORY SYSTEM ALSO PLAYS AN IMPORTANT ROLE IN REGULATING THE PH OF THE BODY FLUIDS.

Respiration is the exchange of oxygen and carbon dioxide between the atmosphere and body cells. Respiration involves several important processes: (1) ventilation, the movement of air into and out of the respiratory passages and the lungs; (2) gas exchange between the air in the lungs and the blood; (3) transport of oxygen and carbon dioxide in the blood; and (4) gas exchange between the blood and the tissues. The term respiration is also used in reference to cell metabolism. In aerobic respiration, for example, cells use oxygen and produce carbon dioxide. Cellular respiration is considered in Chapter 25.

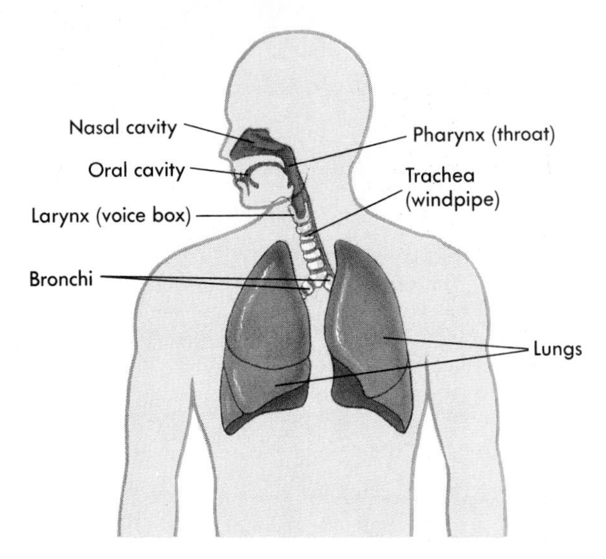

FIGURE 23-1 Major elements of the respiratory system: nasal cavity, pharynx, larynx, trachea, bronchi, and lungs.

The lungs and their accessory structures comprise the respiratory system (Figure 23-1). Gas exchange between the air and the blood requires a moist membrane with a large surface area so that gases can dissolve in the liquid film covering the membrane and diffuse through the membrane. The lungs contain air sacs composed of moist membranes and a system of ducts that transport air from the exterior of the body to the membranes.

This chapter describes the anatomy of the respiratory system, the physiology of gas exchange, the transport of gases in the body, and the mechanisms that regulate the respiratory system. Pathological conditions are described to illustrate the functional characteristics of the process of respiration.

ANATOMY AND HISTOLOGY

The respiratory system consists of the nasal cavity, pharynx, larynx, trachea, bronchi, and lungs. The **upper respiratory tract** refers to the nasal cavity, pharynx, and associated structures, and the **lower respiratory tract** includes the larynx, trachea, bronchi, and lungs. Respiratory movements are accomplished by the diaphragm and the muscles of the thoracic wall.

Nose and Nasal Cavity

The term **nose,** or **nasus** (nāz′us), usually refers to the visible structure that forms a prominent feature of the face (Figure 23-2) and can also refer to the internal nasal cavity. The bridge of the nose consists of the nasal bones plus extensions of the frontal and maxillary bones. The largest portion of the external nose is composed of cartilage plates (see Figure 7-10).

The **nasal cavity** is located inside the external nose and joins the pharynx. The external openings to the nasal cavity are the **external nares** (nă′rēz), or **nostrils,** and the posterior openings from the nasal cavity into the pharynx are the **internal nares,** or **choanae** (ko-a′ne). The anterior portion of the nasal cavity, just inside the external nares, is the **vestibule** (ves′tĭ-būl; entry room). The **nasal septum** separates the nasal cavity into two parts. The posterior half of this septum is bone (the vomer and the perpendicular plate of the ethmoid), and the anterior half is cartilage.

The floor of the nasal cavity is composed of the **hard palate,** a bony plate covered by mucosa. The lateral wall of the nasal cavity is modified by the presence of three bony ridges called **conchae** (kon′ke; resembling a conch shell; see Figure 23-2). Deep to each concha is a passageway called a **meatus** (me-a′tus; a tunnel or passageway). Within the superior and middle meatus are openings from the various **paranasal sinuses** (see Figure 7-11), and the opening of the **nasolacrimal** (na′zo-lak′rĭ-mal) **duct** is within the inferior meatus (see Chapter 15).

The vestibule is lined with stratified squamous epithelium that is continuous with the stratified squamous epithelium of the skin. The mucous membrane that lines the nasal cavity consists of pseudostratified ciliated columnar epithelium with goblet cells that secrete a thick layer of mucus. In the most superior part of the nasal cavity is the olfactory epithelium, which functions as the sensory organ for smell (see Chapter 15).

Air enters the nasal cavity through the external nares, and the vestibule is lined with hairs that trap some of the large particles of dust in the air. The mucus also traps debris in the air, and the cilia on the surface of the mucous membrane sweep the mucus posteriorly to the pharynx, where it is swallowed and eliminated by the digestive system. The air is also humidified by the addition of moisture from the mucous epithelium and is warmed within the nasal cavity before it passes into the pharynx, preventing damage to the more delicate linings in the rest of the respiratory passages.

1

Explain what happens to your throat when you sleep with your mouth open, especially when your nasal passages are plugged as a result of having a cold. Explain what may happen to your lungs when you run a long way in very cold weather while breathing rapidly through your mouth.

? ? ? ? ? ? ? ? ? ? ? ?

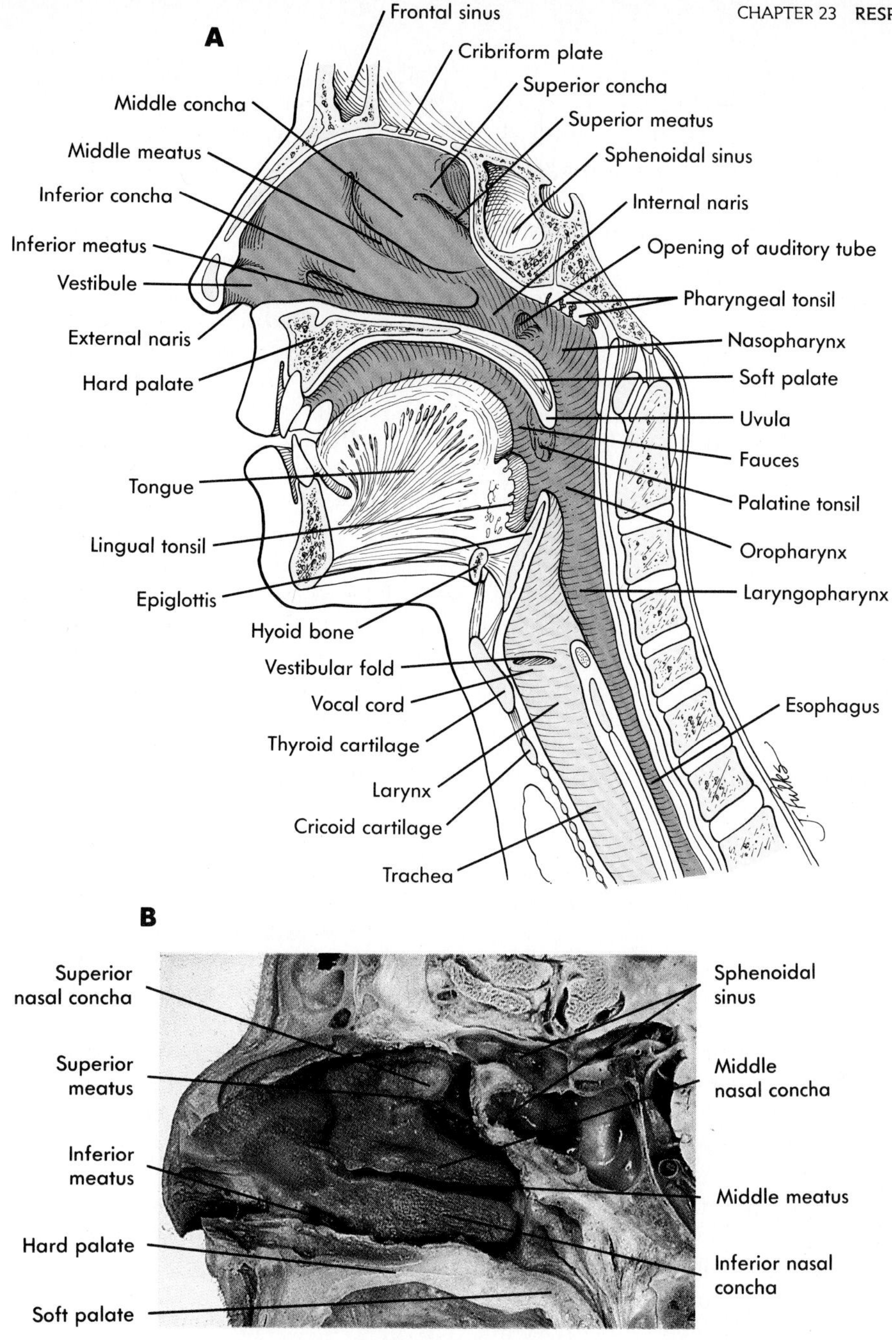

FIGURE 23-2 **A** Sagittal section through the nasal cavity and pharynx viewed from the medial side. **B** Photograph of sagittal section of nasal cavity.

Pharynx

The **pharynx** (făr′ingks; throat) is the common opening of both the digestive and respiratory systems. It receives air from the nasal cavity and air, food, and water from the mouth. Inferiorly, the pharynx leads to the separate openings of the respiratory system (opening into the larynx) and the digestive system (i.e., the esophagus). The pharynx can be divided into three regions, the nasopharynx, the oropharynx, and the laryngopharynx (see Figure 23-2).

The **nasopharynx** (na′zo-făr′ingks) is the superior portion of the pharynx and extends from the internal nares to the level of the **uvula** (u′vu-lah; a grape), a soft process that extends from the posterior edge of the soft palate. The nasopharynx is lined with a mucous membrane similar to that of the nasal cavity. Auditory tubes open into the nasopharynx (see Chapter 15), and the posterior surface of the nasopharynx contains the pharyngeal tonsil, which aids in defending the body against infection (see Chapter 22).

The **oropharynx** (o′ro-făr′ingks) extends from the uvula to the epiglottis. The oral cavity opens into the oropharynx through the **fauces** (faw′sēz). Thus food, drink, and air all pass through the oropharynx. The oropharynx is lined by stratified squamous epithelium that provides protection against abrasion. Two sets of tonsils—the palatine tonsils and the lingual tonsil—are located near the fauces (see Chapter 22).

The **laryngopharynx** (lă-ring′go-făr′ingks) extends from the tip of the epiglottis to the openings of the larynx and the esophagus. The laryngopharynx, like the oropharynx, is lined with stratified squamous epithelium.

Larynx

The **larynx** (lăr′ingks) consists of an outer casing of nine cartilages that are connected to each other by muscles and ligaments (Figure 23-3). Six of the nine cartilages are paired, and three are unpaired. The largest and most superior of the cartilages is the unpaired **thyroid cartilage** (the term means shield and refers to the shape of the cartilage), or Adam's apple.

The most inferior cartilage of the larynx is the unpaired **cricoid** (kri′koyd; ring shaped) **cartilage,** which forms the base of the larynx on which the other cartilages rest.

The third unpaired cartilage is the **epiglottis** (ep′ĭ-glot′is; on the glottis). It differs from the other cartilages in that it consists of elastic cartilage rather than hyaline cartilage. Its inferior margin is attached to the superior margin of the thyroid cartilage, and the superior part of the epiglottis projects as a free flap toward the tongue. During swallowing the epiglottis covers the opening of the larynx and prevents materials from entering the larynx.

The six paired cartilages are stacked in two pillars between the cricoid cartilage and the thyroid cartilage (Figure 23-3, *B*). The largest, most inferior cartilages are the **arytenoid** (ăr-ĭ-te′noyd; ladle shaped) **cartilages.** The middle pair is the **corniculate** (kōr-nik′u-lāt; horn shaped) **cartilages,** and the smallest, most superior cartilages are the **cuneiform** (ku′ne-ĭ-form; wedge shaped) **cartilages.**

Two pairs of ligaments extend from the anterior surface of the arytenoid cartilages to the posterior surface of the thyroid cartilage. The superior pair forms the **vestibular folds,** or **false vocal cords** (Figure 23-4). When the vestibular folds come together, they prevent air from leaving the lungs (e.g., when a person holds his breath) and prevent food and liquids from entering the larynx.

The inferior pair of ligaments composes the **vocal cords (folds),** or **true vocal cords.** The true vocal cords and the opening between them is called the **glottis** (glot′is). The vestibular folds and the vocal cords are lined with stratified squamous epithelium. The remainder of the larynx is lined with pseudostratified ciliated columnar epithelium. An inflammation of the mucosal epithelium of the vocal cords is called **laryngitis.**

When speech is produced, air moving past the vocal cords causes them to vibrate, producing sound. The greater the amplitude of the vibration, the louder the sound will be. Pitch is controlled by the frequency of vibrations. Variations in the length of the vibrating segments of the vocal cords affect the frequency of the vibrations. The cricoid cartilage and the arytenoid cartilages can be moved by various muscles to change the length of the vocal cords. Higher-pitched tones are produced when only the anterior portions of the cords vibrate, and progressively lower tones result when longer sections of the cords vibrate. Because males usually have longer vocal cords than females, men usually have lower-pitched voices than women. The sound produced by the vibrating vocal cords is modified by the tongue, lips, teeth, and other structures to form words. People with the larynx removed because of carcinoma of the larynx can produce sound by swallowing air and causing the esophagus to vibrate.

Trachea

The **trachea** (tra′ke-ah), or windpipe, is a membranous tube that consists of dense, regular connective tissue and smooth muscle reinforced with 15 to 20 C-shaped pieces of cartilage. The cartilages form the anterior and lateral sides of the trachea, and they protect the trachea and maintain an open passageway for air. The posterior wall of the trachea is devoid of cartilage and consists of a ligamentous membrane and smooth muscle, which can alter the diameter of the trachea (see Figure 23-3). The esophagus lies immediately posterior to the cartilage-free posterior wall of the trachea.

2

Explain what would happen if the cartilage supports of the trachea were solid rings all the way around the trachea and a person swallowed a large mouthful of food.

? ? ? ? ? ? ? ? ? ? ?

The trachea is lined with pseudostratified ciliated columnar epithelium that contains numerous goblet cells. The cilia propel mucus and foreign particles toward the larynx where they can enter the esophagus and be swallowed. Constant irritation to the trachea such as occurs in people who smoke cigarettes may cause the tracheal epithelium to become moist, stratified squamous epithelium that lacks cilia and goblet cells and thus also lacks normal function.

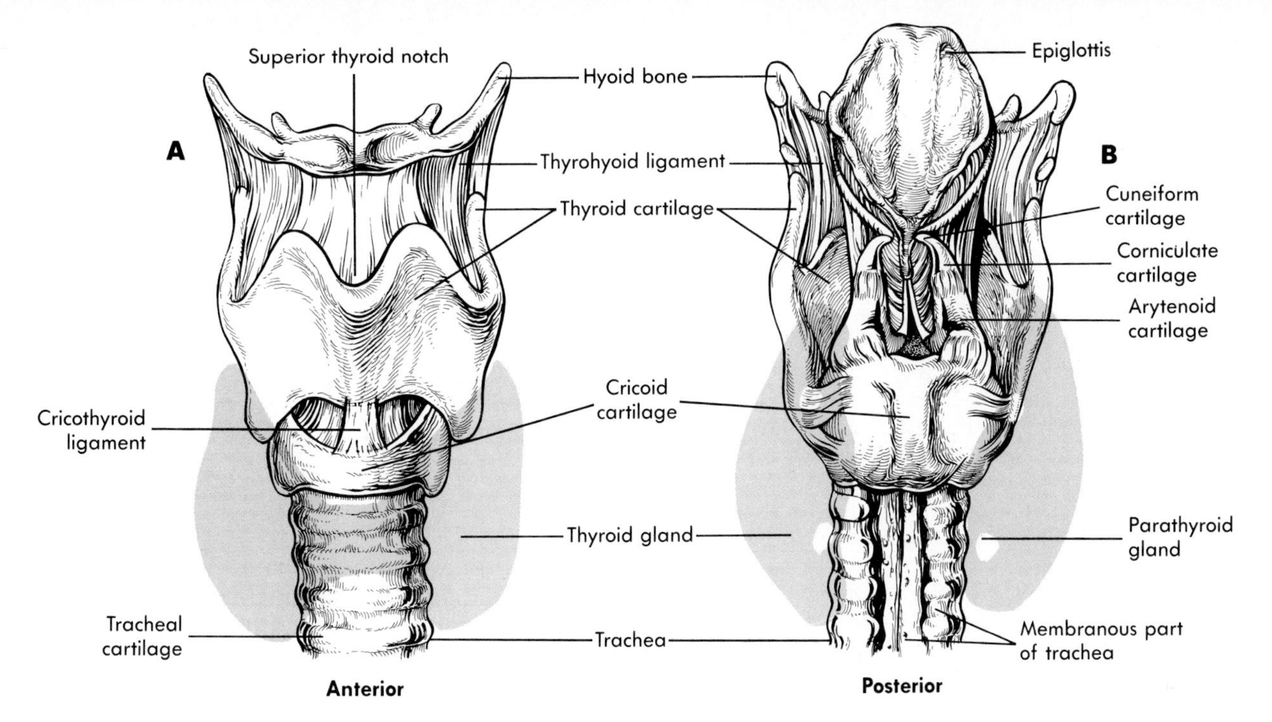

FIGURE 23-3 Anatomy of the larynx. **A** Anterior view. **B** Posterior view.

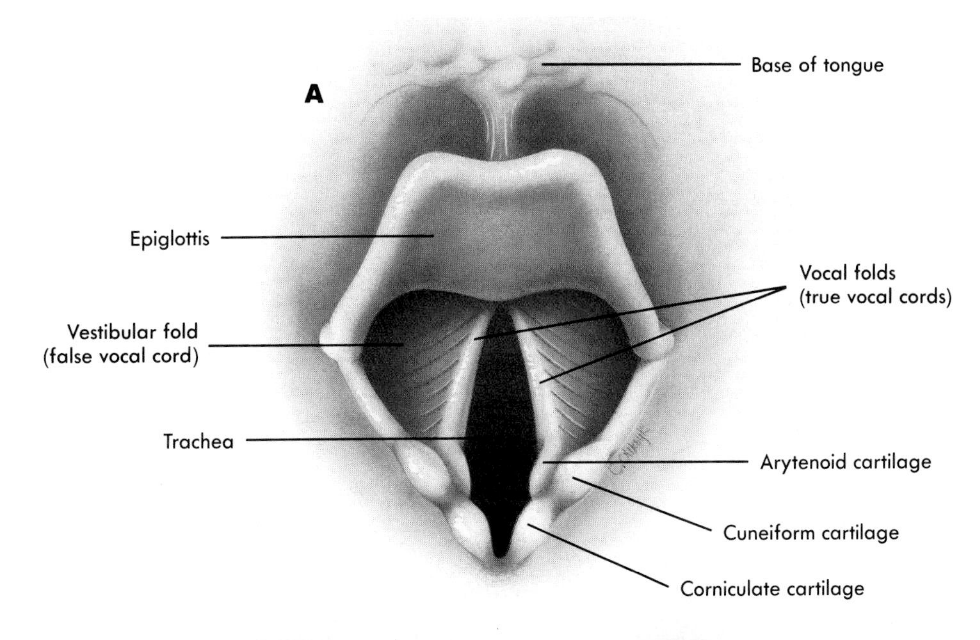

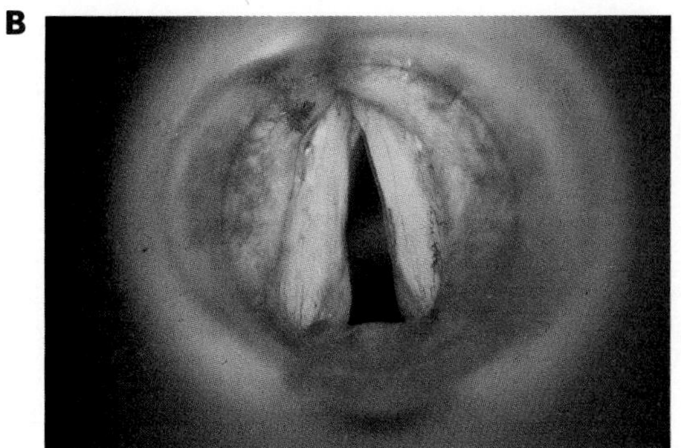

FIGURE 23-4
A Vocal cords viewed from above, showing their relationship to the paired cartilages of the larynx and the epiglottis. **B** Endoscopic view of the vocal cords.

In cases of extreme emergency when the upper air passageway is blocked by a foreign object to the extent that the victim cannot breathe, quick reaction is required to save the person's life. The **Heimlich maneuver** is designed to force such an object out of the air passage by the sudden application of pressure to the abdomen, forcing air up the trachea to dislodge the obstruction. The person who performs the maneuver stands behind the victim with his arms under the victim's arms and his hands over the victim's abdomen between the navel and the rib cage. With one hand formed into a fist, the other hand suddenly pulls the fist toward the abdomen with an accompanying upward motion. This maneuver, if done properly, will dislodge most foreign objects.

In rare cases when the obstruction cannot be removed using the Heimlich maneuver, it may be necessary to form an artificial opening in the victim's air passageway to save his life. The preferred point of entry is through the membrane between the cricoid and thyroid cartilages, a procedure referred to as a **cricothyrotomy.** A **tracheostomy** (tra'ke-os'to-mĭ) is an incision in the trachea with insertion of a tube to facilitate the passage of air. However, because the trachea has several structures overlying its anterior surface (e.g., arteries, nerves, and the thyroid gland), it is not advisable to enter the air passage through the trachea in emergency cases.

Bronchi

The trachea divides into the left and right **primary bronchi** (brong'ki; windpipe). The right primary bronchus is shorter and wider and is more vertical than the left primary bronchus (Figure 23-5).

3

In which lung would a foreign object that is small enough to pass into a primary bronchus become lodged?

? ? ? ? ? ? ? ? ? ? ?

The primary bronchi extend from the mediastinum to the lungs. Like the trachea, the primary bronchi are lined with pseudostratified ciliated columnar epithelium and are supported by C-shaped cartilage rings.

Lungs

The **lungs** are the principal organs of respiration, and on a volume basis they are one of the largest organs of the body. Each lung is conical in shape, with its base resting on the diaphragm and its apex extending superiorly to a point approximately 2.5 cm superior to each clavicle. The right lung is larger than

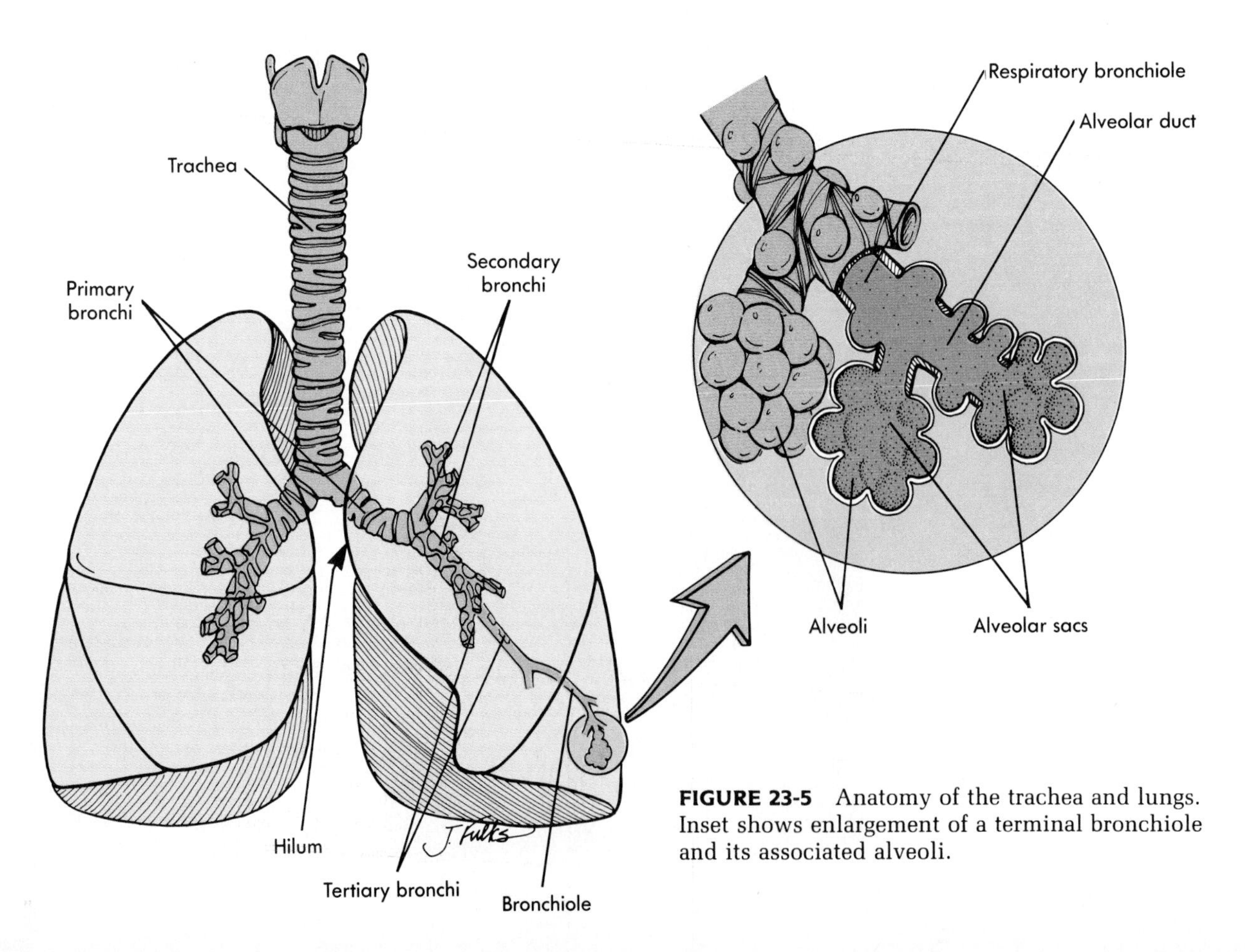

FIGURE 23-5 Anatomy of the trachea and lungs. Inset shows enlargement of a terminal bronchiole and its associated alveoli.

the left and weighs an average of 620 g, whereas the left lung weighs 560 g.

The right lung has three **lobes,** and the left lung has two (see Figure 23-5; Figure 23-6). The lobes are separated by deep, prominent fissures on the surface of the lung. Each lobe is divided into **lobules** that are separated from each other by connective tissue, but the separations are not visible as surface fissures. Because major blood vessels and bronchi do not cross the connective tissues, individual diseased lobules can be surgically removed, leaving the rest of the lung relatively intact. There are nine lobules in the left lung and 10 lobules in the right lung (see Figure 23-6).

The primary bronchi divide into **secondary bronchi** as they enter their respective lungs. The point of entry for the bronchi, vessels, and nerves in each lung is called the **hilum** (hi'lum), or root, of the lung. The secondary bronchi, two in the left lung and three in the right lung, conduct air to each lobe. The secondary bronchi, in turn, give rise to **tertiary bronchi,** which extend to the lobules. The bronchial tree continues to branch several times, finally giving rise to **bronchioles** (brong'kī-ōlz). The bronchioles also subdivide numerous times to become **terminal bronchioles,** which then divide into **respiratory bronchioles.** Each respiratory bronchiole divides to form **alveolar ducts** that end as clusters of air sacs called **alveoli** (al've-o'li; hollow sacs). An **alveolar sac** is composed of two or more alveoli that share a common opening (see Figure 23-5).

The bronchi are lined with pseudostratified ciliated columnar epithelium. The bronchi, other than the primary bronchi, are supported by numerous small cartilage plates embedded in their walls rather than by C-shaped rings. Farther into the respiratory tree the cartilage becomes more sparse and smaller, and smooth muscle becomes more abundant.

The bronchioles, devoid of cartilage in their walls, are very small tubes 1 mm or less in diameter. They are lined with ciliated columnar epithelium that undergoes a transition to ciliated simple cuboidal epithelium in the smaller branches; ultimately, they undergo a transition to simple squamous epithelium. Because the bronchioles have much smooth muscle and no cartilage in their walls, they can constrict if the smooth muscle contracts forcefully, which happens during an **asthma attack.**

All the walls of the alveolar ducts and the alveoli consist of thin, simple squamous epithelium that facilitates the diffusion of gas through the epithelial layer. The alveolar ducts also have scattered smooth

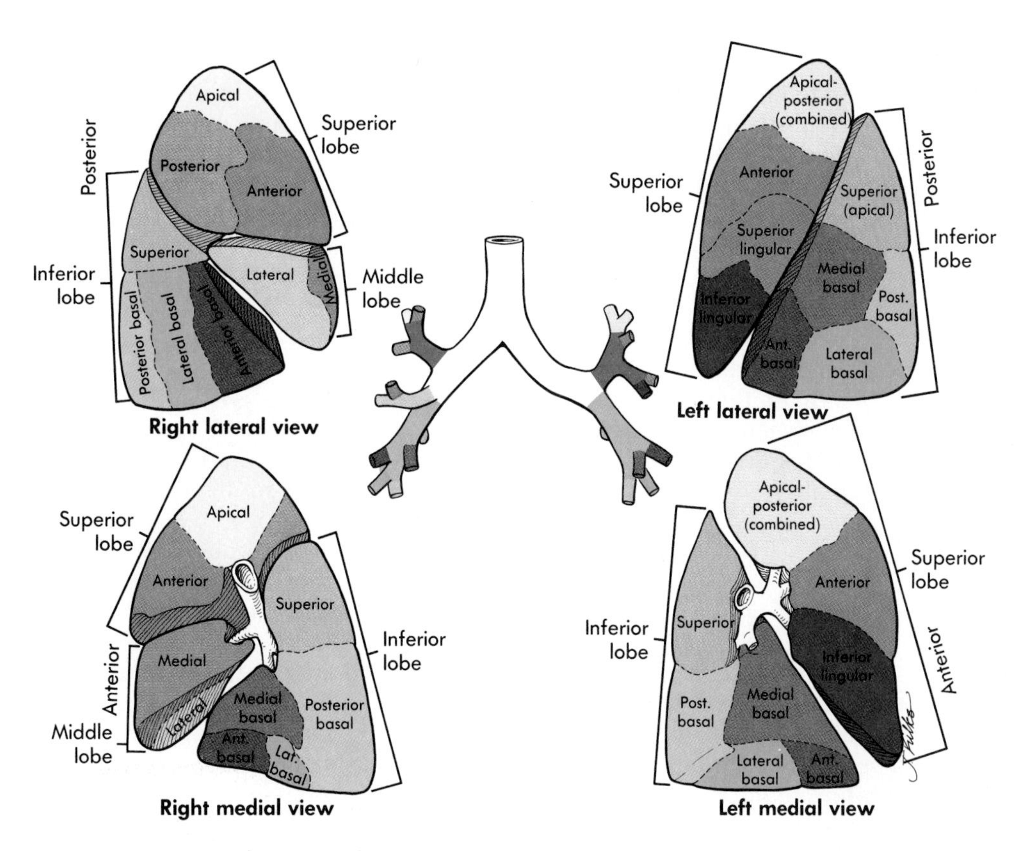

FIGURE 23-6 Lobes and lobules of the lungs. The trachea, primary bronchi, secondary bronchi, and tertiary bronchi are in the center of the figure surrounded by two views of each lung, showing the lobules. In general, each lobule is supplied by a tertiary bronchus (color coded to match the lobule it supplies).

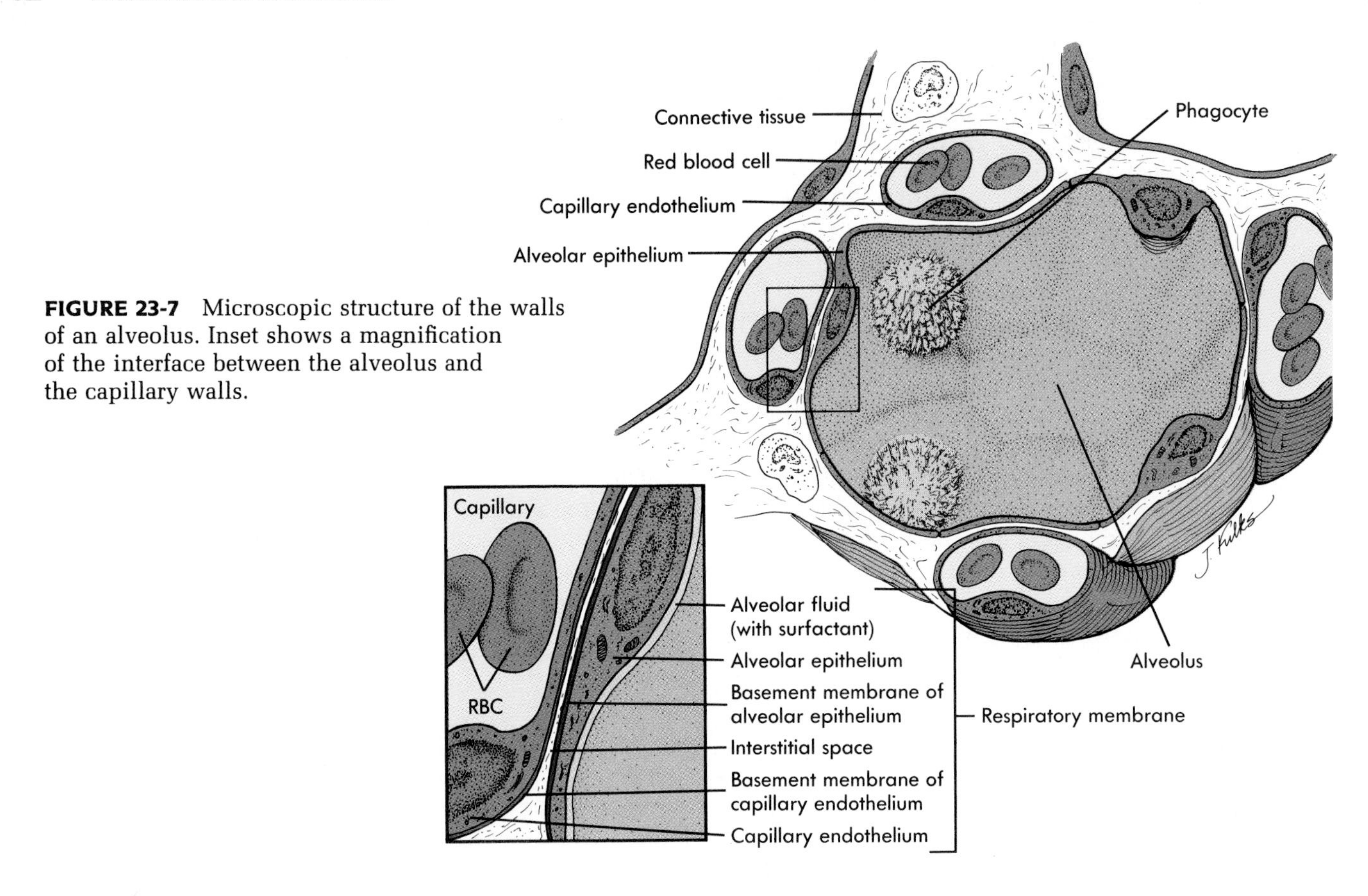

FIGURE 23-7 Microscopic structure of the walls of an alveolus. Inset shows a magnification of the interface between the alveolus and the capillary walls.

muscle cells associated with them, but the alveoli do not. The walls of the alveoli mostly consist of simple squamous epithelium supported by elastic connective tissue within the thin interstitial space (Figure 23-7). In addition to the simple squamous epithelium, secretory cells form part of the alveolar wall, and macrophages are on the surface of the alveolar epithelium.

The lungs are very elastic; when inflated artificially outside the thoracic cavity, they are capable of expelling the air and returning to their original, uninflated state. However, even when not inflated, the lungs retain some air in the respiratory tree, which gives the lungs a spongy quality.

Pleura

The lungs are contained within the **thoracic cavity,** and each lung is surrounded by a separate **pleural** (ploor′al; relating to the ribs) **cavity,** attached only along its medial border at the hilum. Each pleural cavity is lined with a serous membrane called the **parietal pleura.** At the hilum the parietal pleura becomes continuous with a serous membrane, the **visceral pleura,** which covers the surface of the lung (Figure 23-8, *A*).

The pleural cavity is filled with pleural fluid that is produced by the pleural membranes. The pleural fluid performs two functions: it acts as a lubricant, allowing the pleural membranes to slide past each other as the lungs and the thorax change shape during respiration; and it helps hold the pleural membranes together. The pleural fluid is analogous to a thin film of water between two sheets of glass (the visceral and parietal pleurae); the glass sheets can easily slide over each other, but it is difficult to separate them.

Blood Supply

There are two important blood flow routes to the lungs. The major route brings deoxygenated blood to the lungs, where it is oxygenated (see Chapter 21). The blood flows through the pulmonary arteries to the alveolar capillaries, becomes oxygenated, and returns to the heart through the pulmonary veins. The other route supplies the tissues of the bronchi with oxygenated blood through the bronchial arteries, which branch off the thoracic aorta. Deoxygenated blood from the proximal part of the major bronchi returns to the heart through the bronchial veins and azygous system (see Chapter 21); more distally, the venous drainage from the bronchi enters the pulmonary veins. Thus the oxygenated blood returning from the alveoli in the pulmonary veins is mixed with a small amount of deoxygenated blood returning from the bronchi.

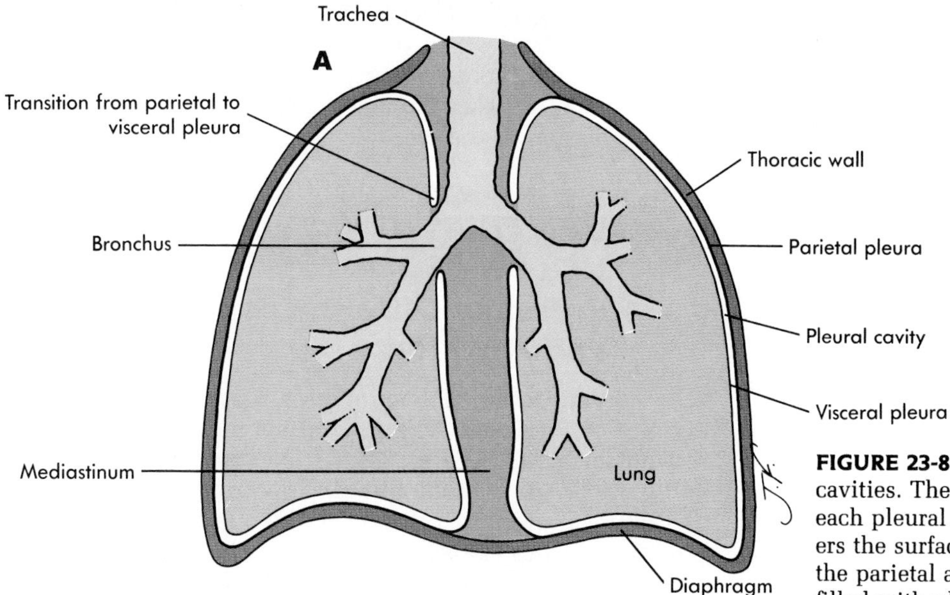

FIGURE 23-8 A Lungs surrounded by pleural cavities. The parietal pleura lines the wall of each pleural cavity, and the visceral pleura covers the surface of the lungs. The space between the parietal and visceral pleurae is small and is filled with pleural fluid. B Superior view of the diaphragm.

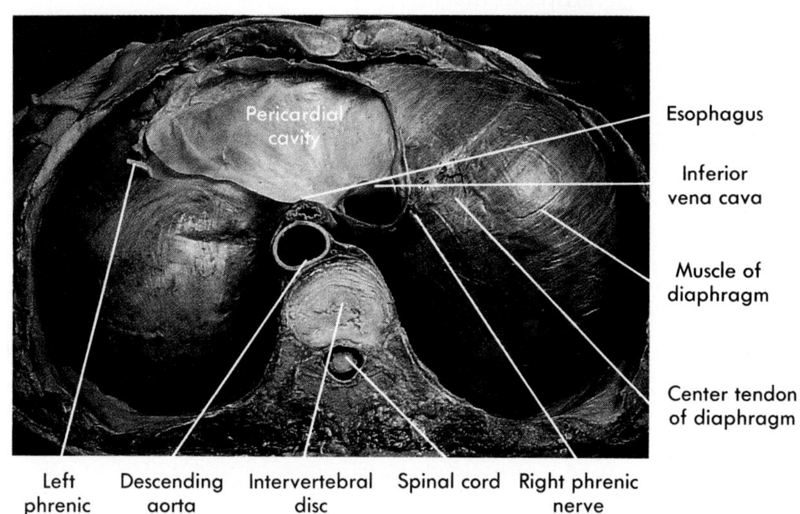

Cough and Sneeze Reflexes

The function of both the cough reflex and the sneeze reflex is to dislodge foreign matter or irritating material from respiratory passages. The bronchi and the trachea contain sensory receptors that are sensitive to foreign particles and irritating substances. To initiate the cough reflex, the sensory receptors detect these substances and initiate nerve impulses that pass along the vagus nerves to the medulla oblongata where the cough reflex is triggered.

The movements resulting in a cough occur in the following way: approximately 2.5 L of air are inspired, the epiglottis closes, and the vestibular folds and vocal cords close tightly to trap the inspired air in the lung; the abdominal muscles contract to force the abdominal contents up against the diaphragm; and the muscles of expiration contract forcefully. As a consequence, the pressure in the lungs increases to approximately 100 mm Hg. Then the vestibular folds and vocal cords and the epiglottis open suddenly, and the air rushes from the lungs at a high velocity, carrying foreign particles with it.

The sneeze reflex is similar to the cough reflex but differs in several ways. The source of irritation that initiates the sneeze reflex is in the nasal passages instead of in the trachea and bronchi, and the afferent impulses are conducted along the trigeminal nerves to the medulla where the reflex is triggered. During the sneeze reflex the uvula and the soft palate are depressed so that air is directed primarily through the nasal passages, although a considerable amount passes through the oral cavity. The rapidly flowing air dislodges particulate matter from the nasal passages and propels it a considerable distance from the nose.

Muscles of Respiration

The **diaphragm** (di'ă-fram; partition) is a large dome of skeletal muscles that separates the thoracic cavity from the abdominal cavity (Figure 23-8, *B*; see Figure 11-15, *A*). The muscles originate from the anterior, lateral, and posterior walls of the entire circumference of the body cavity and insert onto a tendon in the center of the diaphragm called the **central tendon.** When the muscles of the diaphragm contract, the dome is flattened, thus increasing the volume of the thorax and the pleural cavities.

Contraction of the muscles of inspiration (Table 23-1) results in expansion of the thorax, producing an increase in the thoracic cavity volume necessary for inspiration. Expiration occurs when the muscles of inspiration relax and the elastic properties of the thorax and lungs cause a passive decrease in thoracic volume. During labored breathing the inspiratory muscles contract more forcefully, causing a greater increase in thoracic volume. The muscles of expiration contract, producing a more rapid and greater decrease in thoracic volume than would be produced by the passive recoil of the thorax and lungs.

Thoracic Wall

The thoracic wall consists of the thoracic vertebrae, ribs, costal cartilages, sternum, and associated muscles (see Chapters 7 and 11). The ribs slope inferiorly from the vertebrae to the sternum, and elevation of the ribs can increase the anteroposterior dimension of the thoracic cavity. The costal cartilages allow lateral rib movement and lateral expansion of the thoracic cavity. The superior-inferior dimension of the thoracic cavity is expanded by depression of the diaphragm.

Table 23-1

Respiratory muscles

MUSCLES	ACTION
Of Inspiration	
Diaphragm	Depresses floor of thorax
External intercostals	Elevates ribs
Scalene	Elevates first two ribs
Serratus posterior superior	Elevates upper ribs
Quadratus lumborum	Depresses twelfth ribs
Of Expiration	
Internal intercostals	Depresses ribs
Transverse thoracic	Depresses ribs
Serratus posterior inferior	Depresses lower ribs
Rectus abdominis	Depresses thorax and compresses abdomen

VENTILATION AND LUNG VOLUMES

Pressure Differences and Air Flow

Ventilation is the process of moving air into and out of the lungs. The flow of air into the lungs requires a pressure gradient from the outside of the body to the alveoli, and air flow from the lungs requires a pressure gradient in the opposite direction. The physics of air flow in tubes such as the ones that comprise the respiratory passages is similar to the flow of fluids in tubes. Thus the following relationships hold:

$$F = \frac{P_1 - P_2}{R}$$

where:

F = Air flow (milliliters per minute) in a tube
P_1 = Pressure at a point called P_1
P_2 = Pressure at another point called P_2
R = Resistance to air flow

When P_1 is greater than P_2, gas will flow from P_1 to P_2 at a rate that is proportional to the pressure difference. During respiration the pressure in the alveoli changes so that there is a pressure difference between the atmosphere and the alveoli, resulting in air flow. The pressure in the alveoli can be described according to the **general gas law:**

$$P = \frac{nRT}{V}$$

where:

P = Pressure
n = Number of gram moles of gas (a measure of the number of molecules present)
R = Gas constant
T = Absolute temperature
V = Volume

The value of R is a constant, and the values of n and T (body temperature) are considered constants in humans. Thus the general gas law reveals that air pressure in alveoli, called the **intrapulmonary pressure,** is inversely proportional to the volume of the alveoli. As alveolar volume increases, intrapulmonary pressure decreases, and as alveolar volume decreases, intrapulmonary pressure increases (Table 23-2).

Movement of air into and out of the lungs is due to changes in thoracic volume, which cause changes in alveolar volume. The changes in alveolar volume produce changes in intrapulmonary pressure; the pressure differences between intrapulmonary pressure and atmospheric pressure result in air movement.

Contraction of inspiratory muscles increases tho-

Table 23-2

Gas laws

DESCRIPTION	IMPORTANCE
General Gas Law	
The pressure of a gas is inversely proportional to its volume (at a constant temperature).	Air flows from areas of higher to lower pressure. Inspiration results when alveolar volume increases, causing intrapulmonary pressure to decrease below atmospheric pressure. Expiration results when alveolar volume decreases, causing intrapulmonary pressure to increase above atmospheric pressure.
Dalton's Law	
The partial pressure of a gas in a mixture of gases is the percentage of the gas in the mixture times the total pressure of the mixture of gases.	Gases move from areas of higher to areas of lower partial pressures. The greater the difference in partial pressure between two points, the greater the rate of gas movement. Maintaining partial pressure differences ensures gas movements.
Henry's Law	
The concentration of a gas dissolved in a liquid is equal to the partial pressure of the gas over the liquid times the solubility coefficient of the gas.	Only a small amount of the gases in air dissolves in the fluid lining the alveoli. Carbon dioxide, however, is 24 times more soluble than oxygen; therefore carbon dioxide passes through the respiratory membrane more readily than oxygen.

racic volume during inspiration. This increased volume causes the lungs to expand because of negative intrapleural pressure in the pleural cavities (see below) and because the lungs adhere to the parietal pleura. As the lungs expand, the alveolar volume increases, resulting in an approximate 1 mm Hg decrease in intrapulmonary pressure below atmospheric pressure (Figure 23-9, *A* and *B*). Because of this pressure difference, air flows into the lungs. At the end of inspiration the thorax stops expanding, the alveoli stop expanding, intrapulmonary pressure becomes equal to atmospheric pressure, and air no longer moves into the lungs (Figure 23-9, *C*).

During expiration the thoracic volume decreases, producing a decrease in alveolar volume and an approximate 1 mm Hg increase in intrapulmonary pressure over atmospheric pressure (Figure 23-9, *D*). Consequently, air flows out of the lungs. At the end of expiration the decrease in thoracic volume stops, the alveolar volume no longer decreases, intrapulmonary pressure becomes equal to atmospheric pressure, and air movement out of the lungs ceases.

The pressure gradients that cause gas to flow into and out of the alveoli are not large. Nevertheless, air readily flows through the respiratory passages because the resistance to gas flow is small. One of the major factors affecting resistance to air flow in the lungs is the diameter of the respiratory passages; as diameter decreases, resistance to air flow increases. Consequently, constriction of the bronchioles during an asthma attack or a buildup of mucus caused by an infection can make breathing more difficult. Air flow can be maintained despite increased resistance by increasing the pressure difference between alveoli and the atmosphere. Within limits, this can be accomplished by increased contraction of the muscles of respiration.

Collapse of the Lungs

The lungs tend to collapse for two reasons: (1) elastic recoil caused by the elastic fibers in the alveolar walls and (2) surface tension of the film of fluid that lines the alveoli. Molecules of water attract each other, tending to form a droplet. Because the water molecules of the alveolar fluid are also attracted to the surface of the alveoli, formation of a droplet causes the alveoli to collapse, producing fluid-filled alveoli with smaller volumes than air-filled alveoli.

Normally the force produced by elastic recoil and surface tension is opposed, and the lungs remain inflated. Two factors keep the lungs from collapsing: (1) surfactant and (2) intrapleural pressure. **Surfactant** (sur-fak'tant) is a mixture of lipoprotein molecules produced by the secretory cells of the alveolar epithelium. The surfactant molecules form a monomolecular layer over the surface of the fluid within the alveoli to reduce the surface tension. With surfactant the force produced by surface tension is approximately 4 mm Hg; without surfactant the force can be as high as 20 to 30 mm Hg. Thus surfactant greatly reduces the tendency of the lungs to collapse.

In premature infants **hyaline membrane disease,** or **respiratory distress syndrome,** is common, especially for infants with a gestation age of less than 7 months. This occurs because surfactant is not produced in adequate quantities until approximately 7 months of development. Thereafter the amount produced increases as the fetus matures. If insufficient surfactant is produced, the lungs tend to collapse. Thus a great deal of energy must be exerted by the muscles of respiration to keep the lungs inflated and even then inadequate ventilation occurs. Without specialized treatment, most babies with this disease die soon after birth as a result of inadequate ventilation of the lungs and fatigue of the respiratory muscles.

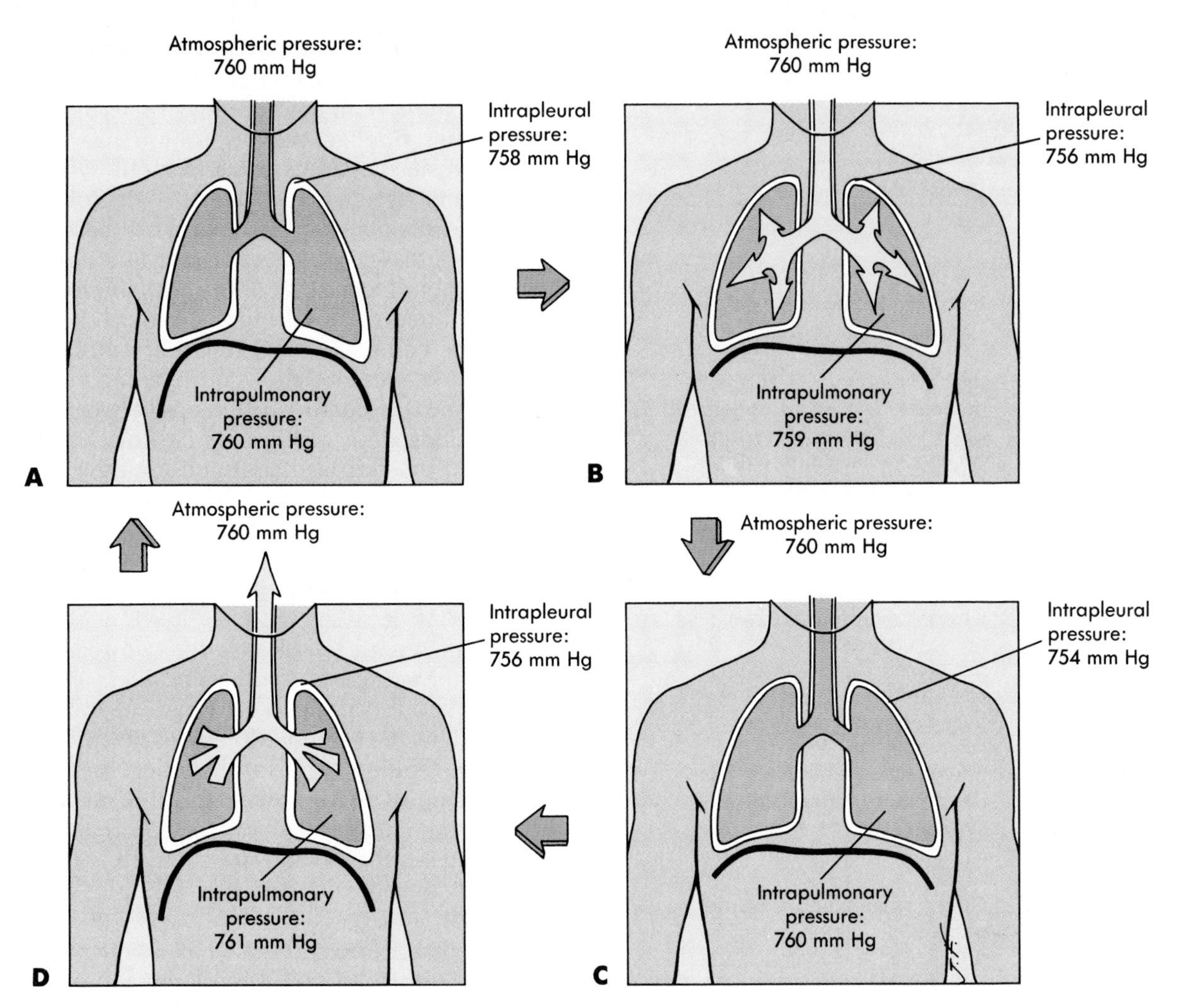

FIGURE 23-9 Pressure changes during inspiration and expiration. **A** At the end of expiration intrapulmonary pressure equals atmospheric pressure, and there is no movement of air. **B** During inspiration the volume of the pleural spaces increases, causing the pressure in the intrapulmonary spaces (alveoli) to decrease. Air then flows from outside the body where the pressure is greater (760 mm Hg) into the alveoli where the pressure is lower (759 mm Hg). **C** At the end of inspiration intrapulmonary pressure again equals atmospheric pressure, and there is no movement of air. **D** During expiration the volume of the pleural spaces decreases, causing the intrapulmonary pressure to increase. Because the intrapulmonary pressure exceeds the atmospheric pressure, air flows out of the body.

The second factor that prevents the lungs from collapsing, **intrapleural pressure,** is the pressure within the pleural cavity. At the end of a normal expiration it is approximately 2 mm Hg less than atmospheric pressure. If atmospheric pressure is 760 mm Hg (usually considered sea level atmospheric pressure), intrapleural pressure is 758 mm Hg (see Figure 23-9, *A*). As the lungs start to recoil from the thoracic wall because of their elasticity, a negative pressure is produced, much like the pressure produced by pulling on a syringe. The lungs stop recoiling when their tendency to collapse is counterbalanced by the negative intrapleural pressure. Negative intrapleural pressure is caused by the same factors that create a negative interstitial pressure (see Chapter 21) and by the tendency for the lungs to recoil from the thoracic wall. The lungs normally adhere to the thoracic wall because of the pleural fluid between the visceral and parietal pleurae.

A **pneumothorax** is the introduction of air into the pleural cavity. Air can enter by an external route when a sharp object such as a bullet or a broken rib penetrates the thoracic wall, or air can enter by an internal route if alveoli at the lung surface rupture such as may occur in a patient with emphysema. Such openings result in equalization of intrapleural and atmospheric pressures. The loss of the subatmospheric intrapleural pressure, which normally opposes the tendency of the lung to collapse, results in deflation of the lung. Pneumothorax can occur in one lung while the lung on the opposite side remains inflated because the two pleural cavities are separated by the mediastinum.

Compliance of the Lungs and the Thorax

The volume by which the lungs and the thorax increase for each unit of pressure change in the intrapulmonary pressure is called the **compliance** of the lungs and thorax. Compliance is usually expressed in liters (volume of air) per centimeter of water (pressure), and for the normal person the compliance of the lungs and thorax is 0.13 L per centimeter of water. That is, for every 1 cm of water change in the intrapulmonary pressure, the volume changes by 0.13 L.

Compliance is a measure of the expansibility of the lungs and the thorax. The greater the compliance, the easier it is for a change in pressure to cause expansion of the lungs and the thorax. For example, one possible result of emphysema is the destruction of elastic lung tissue. This reduces the elastic recoil force of the lungs, making expansion of the lungs easier and resulting in a higher-than-normal compliance. A lower-than-normal compliance means that it is harder to expand the lungs and the thorax. Conditions that decrease compliance include deposition of inelastic fibers in lung tissue (pulmonary fibrosis), collapse of the alveoli (respiratory distress syndrome and pulmonary edema), airway obstruction (asthma, bronchitis, and lung cancer), and deformities of the thoracic wall (kyphosis and scoliosis).

Pulmonary diseases markedly affect the total amount of energy required to perform ventilation, as well as the percentage of the total amount of energy expended by the body. Diseases that decrease compliance can increase the energy that is required for breathing up to 30% of the total energy expended by the body. Diseases such as respiratory distress syndrome and asthma can become sufficiently severe that the excess work load to perform ventilation causes fatigue of the respiratory muscles and possibly death.

Pulmonary Volumes and Capacities

Spirometry (spĭ-rom′ĕ-tre) is the process of measuring volumes of air that move into and out of the respiratory system, and the **spirometer** is the device that is used to measure these pulmonary volumes (Figure 23-10, *A*). The four pulmonary volumes and representative values (Figure 23-10, *B*) for a young adult male follow:

1. **Tidal volume:** the volume of air inspired or expired during a normal inspiration or expiration, respectively (approximately 500 ml)
2. **Inspiratory reserve volume:** the amount of air that can be inspired forcefully after inspiration of the normal tidal volume (approximately 3000 ml)
3. **Expiratory reserve volume:** the amount of air that can be forcefully expired after expiration of the normal tidal volume (approximately 1100 ml)
4. **Residual volume:** the volume of air still remaining in the respiratory passages and lungs after the most forceful expiration (approximately 1200 ml)

Pulmonary capacities are the sum of two or more pulmonary volumes (see Figure 23-10, *B*). Some pulmonary capacities follow:

1. **Inspiratory capacity:** the tidal volume plus the inspiratory reserve volume—the amount of air that a person can inspire maximally after a normal expiration (approximately 3500 ml)
2. **Functional residual capacity:** the expiratory reserve volume plus the residual volume—the amount of air remaining in the lungs at the end of a normal expiration (approximately 2300 ml)
3. **Vital capacity:** the sum of the inspiratory reserve volume, the tidal volume, and the expiratory reserve volume—the maximum volume of air that a person can expel from his respiratory tract after a maximum inspiration (approximately 4600 ml)
4. **Total lung capacity:** the sum of the inspiratory and expiratory reserve volumes plus the tidal volume and the residual volume (approximately 5800 ml)

Factors such as sex, age, body size, and physical conditioning cause variations in respiratory volumes and capacities from one individual to another. For example, the vital capacity of adult females is usually 20% to 25% less than that of adult males. The vital capacity reaches its maximum amount in the young adult, and it gradually decreases in the elderly. Tall people usually have a greater vital capacity than short people, and thin people have a greater vital capacity than obese people. Well-trained athletes can have a vital capacity 30% to 40% above normal (i.e., 6 to 7 L instead of the normal 4.6 L). In patients with paralysis of their respiratory muscles caused by spinal cord injury or paralyzing diseases such as poliomyelitis or muscular dystrophy, the vital capacity may be reduced to values not consistent with survival (less than 500 to 1000 ml). Factors that reduce compliance also reduce the vital capacity.

Determining the **forced expiratory vital capacity**, the rate at which lung volume changes during direct measurement of the vital capacity, is a simple and clinically important pulmonary test. The individual inspires maximally and then exhales into a spirometer maximally and as rapidly as possible. The spirometer records the volume of air that enters it per second. In some conditions the volume to which the lungs are inflated may not be dramatically affected, but the rate at which air can be expired may be greatly decreased. Abnormalities that influence the expiratory flow rate include ones that reduce the ability of the lungs to deflate (e.g., pulmonary fibro-

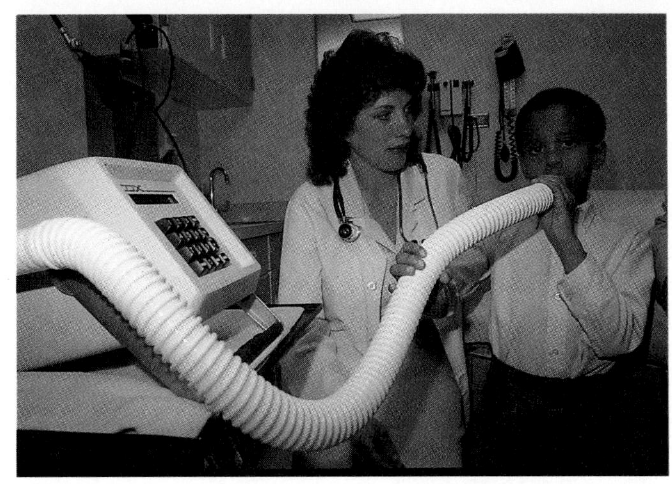

FIGURE 23-10 A Spirometer used to measure lung volumes and capacities. **B** Lung volumes and capacities. The tidal volume in the figure is the tidal volume during resting conditions.

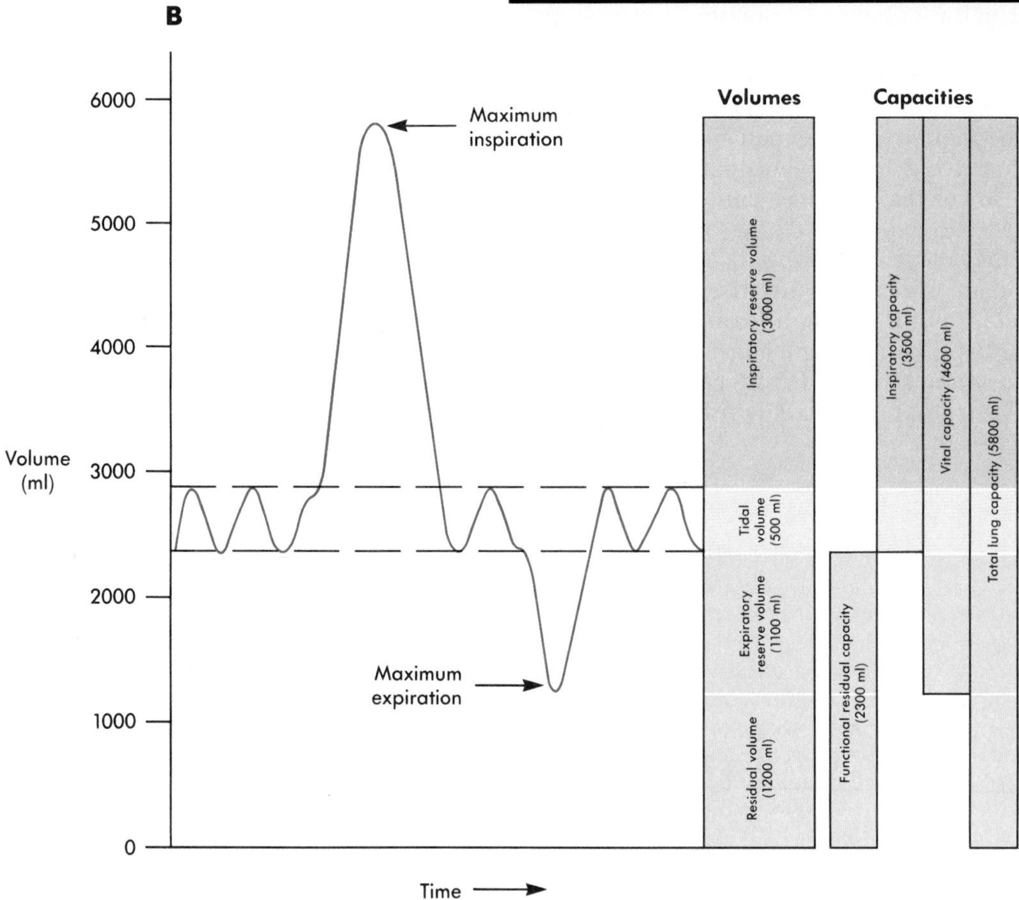

sis, silicosis, kyphosis, and scoliosis) and airway obstruction (e.g., asthma, collapse of bronchi in emphysema, and tumor).

Minute Respiratory Volume and Alveolar Ventilation Rate

The **minute respiratory volume** is the total amount of air moved into and out of the respiratory system each minute, and it is equal to the tidal volume times the respiratory rate. Because resting tidal volume is approximately 500 ml and respiratory rate is approximately 12 breaths per minute, the minute respiratory volume averages approximately 6 L per minute.

Although the minute respiratory volume measures the amount of air moving into and out of the lungs per minute, it is not a measure of the amount of air available for gas exchange because gas exchange takes place in the alveoli and to a lesser extent in the alveolar ducts and the respiratory bronchioles. The part of the respiratory system in which gas exchange does not take place is called the **dead air space.** A distinction can be made between anatomical and physiological dead air space. The **anatomical dead air space,** 150 ml, is formed by the nasal cavity, pharynx, larynx, trachea, bronchi, bronchioles, and terminal bronchioles. The **physiological dead air space** is the anatomical dead air space plus the volume of any nonfunctional alveoli. Normally the anatomical and physiological dead air spaces are nearly the same, meaning there are few nonfunctional alveoli.

In patients with emphysema many alveolar walls degenerate, and, although the remaining chambers are still ventilated, most of the ventilation is wasted because there is inadequate blood flow and inadequate surface area to complete gas exchange. The destruction of the alveoli results in an increase in the physiological dead air space.

During inspiration much of the inspired air fills the dead air space first before reaching the alveoli and thus is unavailable for gas exchange. The volume of air that is available for gas exchange per minute is called the **alveolar ventilation rate,** and it is calculated as follows:

$$AVR = RR\ (TV - DAS)$$

where:

AVR = Alveolar ventilation rate (milliliters per minute)
RR = Respiratory rate (respirations per minute)
TV = Tidal volume (milliliters per respiration)
DAS = Dead air space (milliliters per respiration)

4

If a resting person had a tidal volume of 500 ml per inspiration, a dead air space of 150 ml per inspiration, and a respiratory rate of 12 inspirations per minute, what would his alveolar ventilation rate be? Explain why increasing the respiration rate to 24 breaths per minute and tidal volume to 4000 ml per breath during exercise would be beneficial.

? ? ? ? ? ? ? ? ? ? ?

PHYSICAL PRINCIPLES OF GAS EXCHANGE

Ventilation supplies atmospheric air to the alveoli. The next step in the process of respiration is the diffusion of gases between the alveoli and the blood in the pulmonary capillaries. The molecules of gas move randomly, and if a gas is in a higher concentration at one point than at another, random motion will ensure that the net movement of gas will be from the higher concentration toward the lower concentration until a homogeneous mixture of gases is achieved. One measurement of the concentration of gases is partial pressure.

Partial Pressure

At sea level the atmospheric pressure is close to 760 mm Hg (i.e., the mixture of gases that comprise atmospheric air exerts a total pressure of 760 mm Hg). The major components of dry air are nitrogen (approximately 79%) and oxygen (approximately 21%). According to **Dalton's law,** in a mixture of gases the portion of the total pressure resulting from each type of gas is determined by the percentage of the total volume represented by each gas type (see Table 23-2). The pressure exerted by each type of gas in a mixture is referred to as the **partial pressure** of that gas. Because nitrogen comprises 79% of the volume of atmospheric air, the partial pressure resulting from nitrogen is 0.79 times 760 mm Hg, which equals 600.2 mm Hg. Because oxygen comprises approximately 21% of the volume of atmospheric air, the partial pressure resulting from oxygen is 0.21 times 760 mm Hg, which equals 159.5 mm Hg. It is traditional to designate the partial pressure of individual gases in a mixture as P_{N_2}, P_{O_2}, or P_{CO_2}, for example.

When air comes into contact with water, some of the water turns into a gas and evaporates into the air. Water molecules in the gaseous form also exert a partial pressure. This partial pressure (P_{H_2O}) is sometimes referred to as the **vapor pressure** of water. The composition of dry, humidified, alveolar, and expired air is presented in Table 23-3. The composition of

Table 23-3

Partial pressures of gases at sea level

	DRY AIR		HUMIDIFIED AIR		ALVEOLAR AIR		EXPIRED AIR	
GASES	mm Hg	%	mm Hg	%	mm Hg	%	mm Hg	%
Nitrogen	600.2	78.98	563.4	74.09	569.0	74.9	566.0	74.5
Oxygen	159.5	20.98	149.3	19.67	104.0	13.6	120.0	15.7
Carbon dioxide	0.3	0.04	0.3	0.04	40.0	5.3	27.0	3.6
Water vapor	0.0	0.0	47.0	6.20	47.0	6.2	47.0	6.2

alveolar air and of expired air is not identical to the composition of dry atmospheric air for several reasons. First, air entering the respiratory system during inspiration is humidified; second, oxygen diffuses from the alveoli into the blood, and carbon dioxide diffuses from the alveolar capillaries into the alveoli; and third, the air within the alveoli is only partially replaced with atmospheric air during each inspiration.

Diffusion of Gases Through Liquids

When a gas comes into contact with a liquid such as water, there is a tendency for the gas to dissolve in the liquid. At equilibrium the concentration of a gas in the liquid is determined by its partial pressure in the gas and by its solubility in the liquid (Figure 23-11). This relationship is described by **Henry's law** (see Table 23-2):

$$\text{Concentration of dissolved gas} = \text{Partial pressure of gas} \times \text{Solubility coefficient}$$

The solubility coefficient is a measure of how easily the gas dissolves in the liquid. In water the solubility coefficient for oxygen is 0.024, and for carbon dioxide it is 0.57. Thus carbon dioxide is approximately 24 times as soluble in water as oxygen.

Gases do not actually produce partial pressure in a liquid as they do when in the gaseous state. However, knowing the concentration of the gas in liquid, it is possible to determine mathematically (general gas law) its partial pressure as if it were in a gaseous state. Because the partial pressure thus calculated is a measure of concentration, it can be used to determine the direction of diffusion of gas through a liquid: gases move from areas of higher to areas of lower partial pressure.

5

As a scuba diver descends, the pressure of the water on his body prevents normal expansion of the lungs. To compensate, the diver breathes pressurized air, which has a greater pressure than sea level air pressure. What effect would the increased pressure have on the amount of gas dissolved in the diver's body? A diver who suddenly ascends to the surface from a great depth can develop decompression sickness (the bends) in which bubbles of nitrogen gas form. The bubbles can expand, causing damage to tissues. Explain the development of the bubbles.

? ? ? ? ? ? ? ? ? ? ?

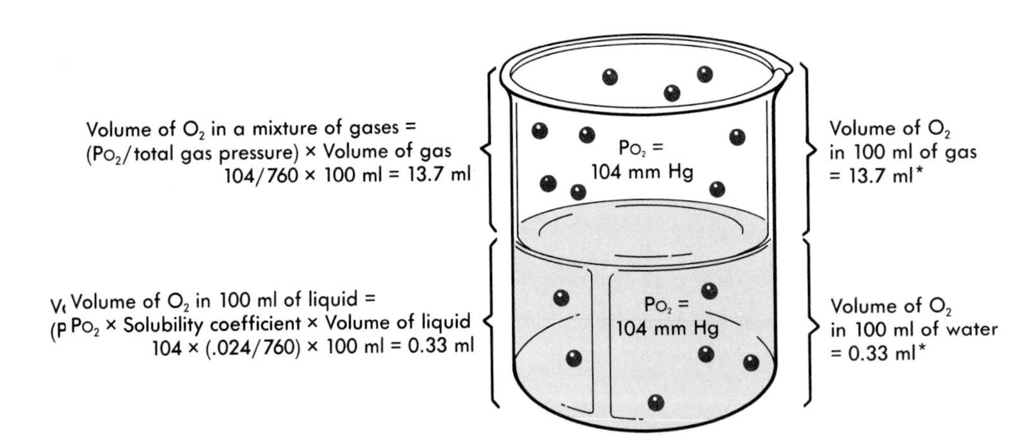

FIGURE 23-11 Amount of a gas dissolved in a liquid is determined by its partial pressure and by its solubility in the liquid. When a gas and a liquid come into contact, some of the gas will dissolve in the liquid until the partial pressure of the gas is equal to that in both the liquid and the gas. However, the volume of gas dissolved in a given volume of liquid may be very small if the gas is not soluble in the liquid or may be large if the gas is soluble in the liquid. Therefore the volume of a gas dissolved in a liquid is affected by the partial pressure of the gas and the degree to which that gas is soluble in the liquid (Henry's law). The solubility of a gas in a liquid is expressed by its solubility coefficient (units are milliliters of gas/milliliter of fluid/760 mm Hg). When oxygen has a partial pressure of 104 mm Hg and the atmospheric pressure is 760 mm Hg such as in the alveoli, 13.7 ml of oxygen is in 100 ml of alveolar air. If alveolar air is exposed to water, only 0.33 ml of oxygen will dissolve in 100 ml of water.

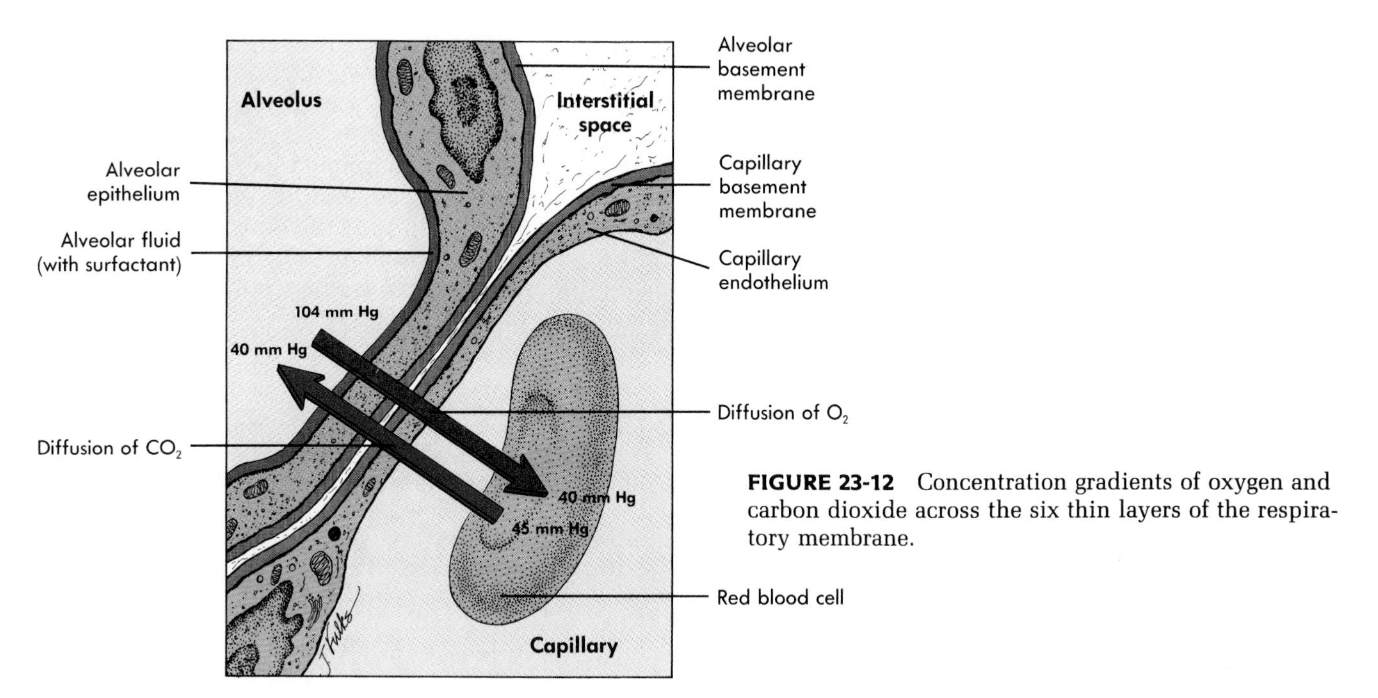

FIGURE 23-12 Concentration gradients of oxygen and carbon dioxide across the six thin layers of the respiratory membrane.

Diffusion of Gases Through the Respiratory Membrane

The **respiratory membranes** of the lungs are in the respiratory bronchioles, alveolar ducts, and alveoli. Approximately 300 million of these units are in the two lungs. The average diameter of each alveolus is approximately 0.25 mm, and its walls are extremely thin. Surrounding each alveolus is a network of capillaries arranged so that air within the alveoli is separated by a thin respiratory membrane from the blood contained within the alveolar capillaries.

The respiratory membrane (see Figure 23-7; Figure 23-12) consists of (1) a thin layer of fluid lining the alveolus, (2) the alveolar epithelium comprised of simple squamous epithelium, (3) the basement membrane of the alveolar epithelium, (4) a thin interstitial space, (5) the basement membrane of the capillary endothelium, and (6) the capillary endothelium comprised of simple squamous epithelium.

The factors that influence rate of gas diffusion across the respiratory membrane include (1) the thickness of the membrane, (2) the diffusion coefficient of the gas in the substance of the membrane, which is approximately the same as the diffusion coefficient for gas through water, (3) the surface area of membrane, and (4) the partial pressure difference of the gas between the two sides of the membrane.

Respiratory membrane thickness

Increasing the thickness of the respiratory membrane decreases the rate of diffusion. The thickness of the respiratory membrane normally averages 0.5 μm, but the thickness can be increased by respiratory diseases. For example, in patients with pulmonary edema fluid accumulates in the alveoli, and gases must diffuse through a thicker-than-normal layer of fluid. If the thickness of the respiratory membrane is increased two or three times, the rate of gas exchange is markedly decreased.

Diffusion coefficient

The **diffusion coefficient** is a measure of how easily a gas will diffuse through a liquid or tissue, taking into account the solubility of the gas in the liquid and the size of the gas molecule (molecular weight). If the diffusion coefficient of oxygen is assigned a value of 1, then the relative diffusion coefficient of carbon dioxide is 20 (i.e., carbon dioxide will diffuse through the respiratory membrane 20 times more rapidly than oxygen).

When the respiratory membrane becomes progressively damaged as a result of disease, its capacity for allowing the movement of oxygen into the blood is often impaired enough to cause death from oxygen deprivation before the diffusion of carbon dioxide is dramatically reduced. However, if life is being maintained by extensive oxygen therapy, which increases the concentration of oxygen in the lung alveoli, the reduced capacity for the diffusion of carbon dioxide across the respiratory membrane can result in substantial increases of carbon dioxide in the blood.

Surface area

The total surface area of the respiratory membrane is approximately 70 m^2 (approximately the area of one half of a tennis court) in the normal adult. The surface area of the respiratory membrane is decreased by several respiratory diseases, including emphysema and lung cancer. Even small decreases in this surface area adversely affect the respiratory exchange of gases during strenuous exercise. When the total surface area of the respiratory membrane is decreased to one third or one fourth of normal, the exchange of gases is significantly restricted even under resting conditions.

Partial pressure difference

The partial pressure difference of a gas across the respiratory membrane is the difference between the partial pressure of the gas in the alveoli and the partial pressure of the gas in the blood of the alveolar capillaries. When the partial pressure of a gas is greater on one side of the respiratory membrane than on the other side, net diffusion occurs from the higher to the lower pressure. Normally the partial pressure of oxygen (PO_2) is greater in the alveoli than in the blood of the alveolar capillaries, and the partial pressure of carbon dioxide (PCO_2) is greater in the blood than in the alveolar air.

The partial pressure difference for oxygen and carbon dioxide can be increased by increasing the alveolar ventilation rate. The greater volume of atmospheric air exchanged with the residual volume raises alveolar PO_2, lowers alveolar PCO_2, and thus promotes gas exchange. Conversely, inadequate ventilation causes a lower-than-normal partial pressure difference for oxygen and carbon dioxide, resulting in inadequate gas exchange.

Relationship Between Ventilation and Capillary Blood Flow

During conditions of normal ventilation and normal blood flow, exchange of oxygen and carbon dioxide through the respiratory membrane is approximately optimum. Oxygen diffuses from the alveolar air so that the alveolar PO_2 rises to a level between that of inspired air and that of the venous blood. Diffusion of both oxygen and carbon dioxide is continuous and is not intermittent between inspiration and expiration, and the alveolar PO_2 and PCO_2 average 104 mm Hg and 40 mm Hg, respectively (see Figure 23-12).

There are two ways that the optimum relationship between ventilation and blood flow can be affected. One way occurs when ventilation exceeds the ability of the blood to pick up oxygen, which could happen because of inadequate cardiac output after a heart attack. Another way occurs when ventilation is not great enough to provide the oxygen needed to oxygenate the blood flowing through the alveolar capillaries. At rest, the lungs are not fully expanded, a few alveoli have a lower-than-normal PO_2, and blood flowing through the capillaries of these alveoli does not become completely oxygenated. Blood that is not completely oxygenated is called **shunted blood.** The **physiological shunt** is the deoxygenated blood from the alveoli plus the deoxygenated blood returning from the bronchi and bronchioles. Normally, the physiological shunt accounts for approximately 2% of cardiac output.

In people suffering from obstruction of the bronchioles (e.g., in patients with asthma) the alveoli beyond the obstructed areas are not ventilated, causing a large increase in shunted blood because the blood flowing through the pulmonary capillaries in the obstructed area remains unoxygenated.

Normally local direct control of blood flow by the precapillary sphincters helps to match ventilation with blood flow. Local control in the lungs works in a fashion opposite to that in other tissues so that an increase in PO_2 or a decrease in PCO_2 results in relaxation of the precapillary sphincters and an increased blood flow. For example, during exercise alveolar ventilation increases, causing an increase in PO_2 and a decrease in PCO_2. Consequently, the precapillary sphincters relax, blood flow increases, and there is an increased exchange of gases.

6

People, even those in "good shape," can have trouble breathing at high altitudes. Explain how that difficulty can happen even when ventilation of the lungs increases.

? ? ? ? ? ? ? ? ? ? ?

OXYGEN AND CARBON DIOXIDE TRANSPORT IN THE BLOOD

Once oxygen diffuses across the respiratory membrane into the blood, most of the oxygen combines reversibly with hemoglobin, and a smaller amount remains dissolved in the plasma. Hemoglobin transports oxygen from the alveolar capillaries through the blood vessels to the tissue capillaries where it is released; there it diffuses from the tissue capillaries to the cells. Oxygen is used by the cells, and carbon dioxide is produced during aerobic metabolism. The carbon dioxide diffuses from the cells into the tissue capillaries. Once it enters the blood, carbon dioxide is transported dissolved in the plasma, in combination with hemoglobin, or in the form of bicarbonate ions.

Oxygen Diffusion Gradients

The partial pressure of oxygen (PO_2) within the alveoli averages approximately 104 mm Hg, and as blood flows into the alveolar capillary, it has a PO_2 of approximately 40 mm Hg (Figure 23-13). Consequently, oxygen diffuses from the alveoli into the pulmonary capillary blood because the PO_2 is greater in the alveoli than the PO_2 in the capillary blood. By the time blood flows through the first third of the pulmonary capillary, an equilibrium is achieved, and the PO_2 in the blood is 104 mm Hg and is equivalent to the PO_2 in the alveoli. During periods of exercise the rate that the body uses oxygen may increase fifteenfold. Even with the greater velocity of blood flow associated with exercise, by the time blood reaches the venous end of the alveolar capillaries, the PO_2 in the capillary has achieved the same value as in the alveoli.

Blood that leaves the pulmonary capillaries has a PO_2 of 104 mm Hg. This blood then mixes with shunted (deoxygenated) blood; as a consequence, blood leaving the lungs in the pulmonary veins has a PO_2 of approximately 95 mm Hg.

The blood that enters the arterial end of the tissue capillaries has a PO_2 of approximately 95 mm Hg. The PO_2 of the interstitial spaces, on the other hand, is close to 40 mm Hg and is probably near 20 mm Hg in the individual cells. Oxygen diffuses from the tissue capillaries to the interstitial fluid and from the interstitial fluid into the cells of the body in which it is used in aerobic metabolism. Because oxygen is continuously used by cells, a constant diffusion gradient for oxygen from the tissue capillaries to the cells exists.

Carbon Dioxide Diffusion Gradients

Carbon dioxide is continually produced as a by-product of cellular respiration, and a diffusion gradient is established from tissue cells to the blood within the tissue capillaries. The intracellular PCO_2 is approximately 46 mm Hg, whereas that in the interstitial fluid is approximately 45 mm Hg. At the arteriolar end of the tissue capillaries the PCO_2 is close to 40 mm Hg. As blood flows through the tissue capillaries, equilibrium in PCO_2 is established, and the blood at the venous end of the capillary has a PCO_2 of 45 mm Hg (see Figure 23-13).

After blood leaves the venous end of the capillaries, it is transported through the cardiovascular system to the lungs. At the arteriolar end of the pulmonary capillaries the PCO_2 is 45 mm Hg. Because the PCO_2 is approximately 40 mm Hg in the alveoli, carbon dioxide diffuses from the pulmonary capillaries into the alveoli. At the venous end of the pulmonary capillaries the PCO_2 has again decreased to 40 mm Hg.

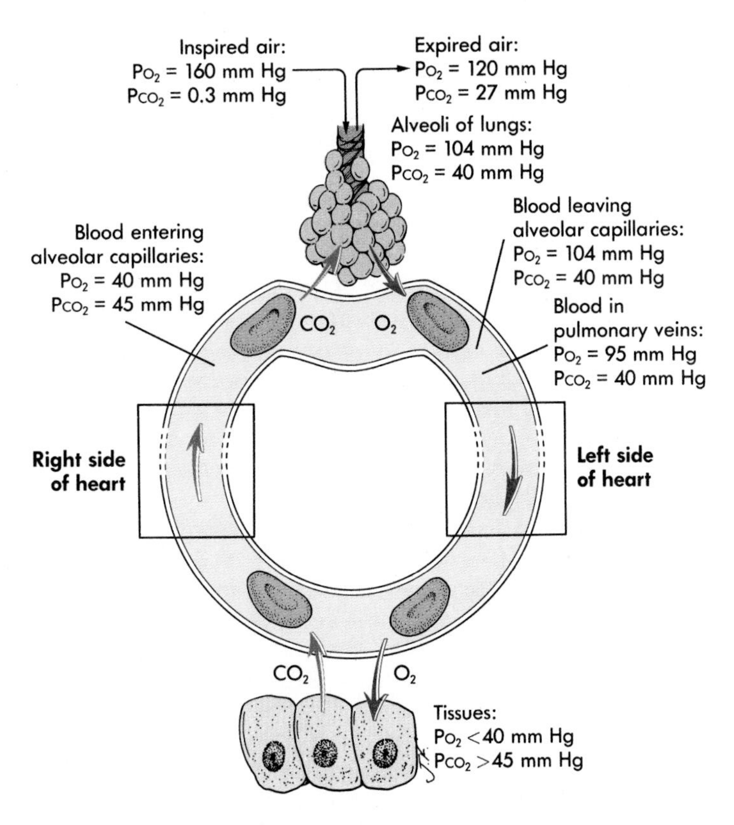

FIGURE 23-13 Oxygen and carbon dioxide diffusion gradients between the alveoli and the alveolar capillaries and between the tissues and the tissue capillaries.

Hemoglobin and Oxygen Transport

Approximately 97% of the oxygen transported in the blood from the lungs to the tissues is transported in combination with the hemoglobin in the red blood cells, and the remaining 3% is dissolved in the water portion of the plasma. The combination of oxygen with hemoglobin is reversible. In the pulmonary capillaries oxygen binds to hemoglobin, and in the tissue spaces oxygen diffuses away from the hemoglobin molecule and enters the tissues.

Effect of Po_2

The **oxygen-hemoglobin dissociation curve** describes the percentage of hemoglobin saturated with oxygen at any given Po_2. Hemoglobin is saturated when an oxygen molecule is bound to each of its four heme groups (see Chapter 19). At any Po_2 above 80 mm Hg, nearly 100% of the hemoglobin is saturated with oxygen (Figure 23-14). Because the Po_2 is well above 80 mm Hg in the alveolar capillaries, the hemoglobin is readily saturated.

At a Po_2 of 40 mm Hg, which is the normal Po_2

FIGURE 23-14
For legend see opposite page.

A

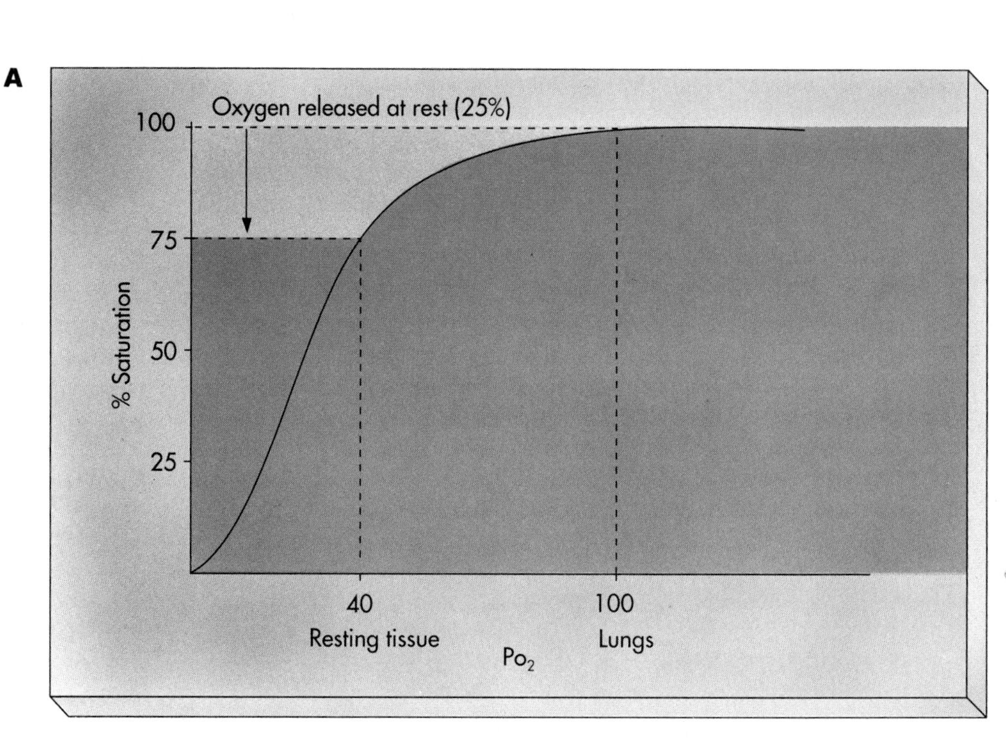

B

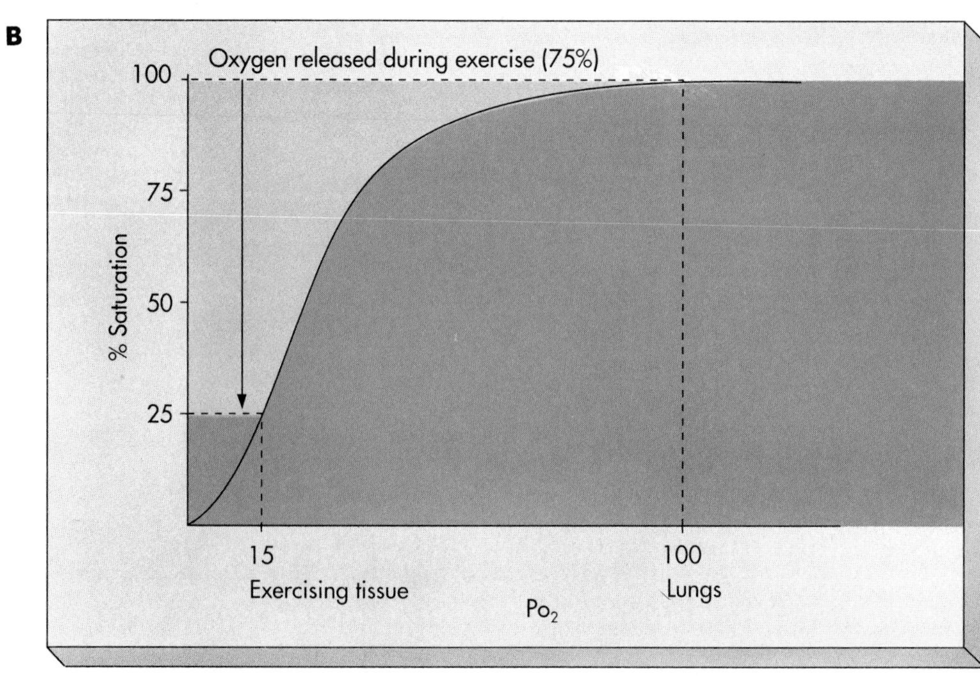

in the tissue capillaries, hemoglobin is only approximately 75% saturated (i.e., approximately 25% of the oxygen bound to hemoglobin is released into the blood and can diffuse into the tissue spaces). During conditions of vigorous exercise the PO_2 in the tissue spaces may decline to levels as low as 15 mm Hg. At a PO_2 of 15 mm Hg only approximately 25% of the hemoglobin is saturated with oxygen, and the hemoglobin releases 75% of the bound oxygen. Variations in the tissue metabolic rate cause the PO_2 to vary, thus controlling the amount of oxygen released from hemoglobin into the tissue capillaries.

Effect of pH, PCO_2, and temperature

In addition to PO_2, other factors influence the degree to which oxygen binds to hemoglobin. As the pH of the blood declines, the amount of oxygen bound to hemoglobin at any given PO_2 also declines. This occurs because a decrease in pH results from an increase in hydrogen ions, and the hydrogen ions combine with the protein portion of the hemoglobin molecule and change its three-dimensional structure, causing a decrease in the affinity of hemoglobin for oxygen. Conversely, an increase in blood pH results in an increased affinity of hemoglobin for oxygen. The effect of pH (hydrogen ions) on the oxygen-hemoglobin dissociation curve is called the **Bohr effect** after its discoverer.

An increase in PCO_2 also decreases the affinity of hemoglobin for oxygen because of the effect of carbon dioxide on pH. Within red blood cells an enzyme called **carbonic anhydrase** catalyzes this reversible reaction.

$$\underset{\text{Carbon dioxide}}{CO_2} + \underset{\text{Water}}{H_2O} \xrightleftharpoons{\text{Carbonic anhydrase}} \underset{\text{Carbonic acid}}{H_2CO_3} \rightleftharpoons \underset{\text{Hydrogen ion}}{H^+} + \underset{\text{Bicarbonate ion}}{HCO_3^-}$$

As the carbon dioxide levels increase, more hydrogen ions are produced, and the pH declines. As the carbon dioxide levels decline, the reaction proceeds in the opposite direction (i.e., the hydrogen ion concentration declines, and the pH increases).

Because carbon dioxide is present in greater concentrations in the tissue capillaries, hemoglobin has less affinity for oxygen in the tissue capillaries, and a greater amount of oxygen is released in the tissue capillaries than would be released if carbon dioxide were not present. When blood is returned to the lungs and passes through the pulmonary capillaries, carbon dioxide leaves the capillaries and enters the alveoli. As a consequence, carbon dioxide levels in the capillaries are reduced, and the affinity of hemoglobin for oxygen increases.

An increase in temperature also decreases the tendency for oxygen to remain bound to hemoglobin. Therefore elevated temperatures resulting from increased metabolism increase the amount of oxygen released into the tissues by hemoglobin. In less metabolically active tissues in which the temperature is lower, less oxygen is released from hemoglobin.

When the affinity of hemoglobin for oxygen decreases, the oxygen-hemoglobin dissociation curve is shifted to the right, and hemoglobin releases more oxygen (Figure 23-15, *A*). During exercise when carbon dioxide and acidic substances such as lactic acid accumulate and the temperature increases in the tissue spaces, the oxygen-hemoglobin curve shifts to the right. Under these conditions as much as 75% to 85% of the oxygen is released from the hemoglobin. In the lungs, however, the curve shifts to the left because of the lower carbon dioxide levels, lower temperature, and lower lactic acid levels. Therefore the affinity of hemoglobin for oxygen increases, and it becomes easily saturated (Figure 23-15, *B*).

During resting conditions approximately 5 ml of oxygen is transported to the tissues in each 100 ml of blood, and the cardiac output is approximately 5000 ml per minute. Consequently, 250 ml of oxygen is delivered to the tissues each minute. During conditions of exercise this value may increase up to 15 times. The oxygen transport can be increased threefold because of a greater degree of oxygen release from hemoglobin in the tissue spaces, and the rate of oxygen transport is increased another five times because of the increase in cardiac output. Consequently, the volume of oxygen delivered to the tissues may be as high as 3750 ml per minute (15×250 ml per minute). Highly trained athletes may increase this volume to as high as 5000 ml per minute.

FIGURE 23-14 Oxygen-hemoglobin dissociation curves. The graphs indicate the percentage of the hemoglobin saturated with oxygen as the partial pressure of oxygen increases. **A** At the PO_2 in the lungs hemoglobin is 100% saturated. At the PO_2 of resting tissues hemoglobin is 75% saturated. Consequently, 25% of the oxygen picked up in the lungs is released to the tissues. **B** In exercising tissues the percent saturation of hemoglobin can decrease to 25%, resulting in a release of 75% of the transported oxygen.

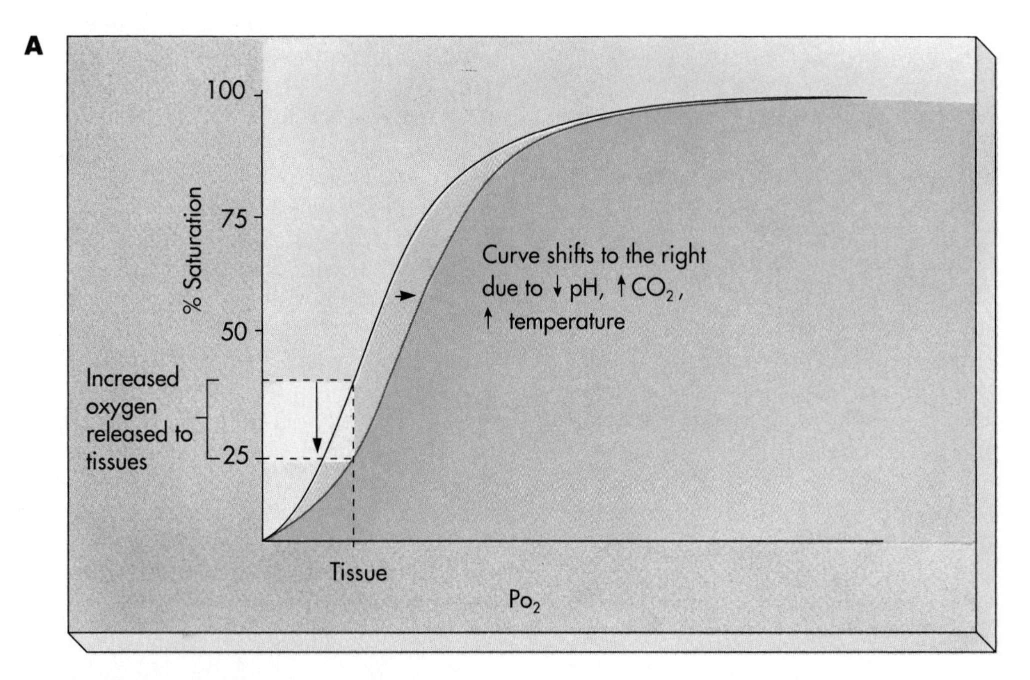

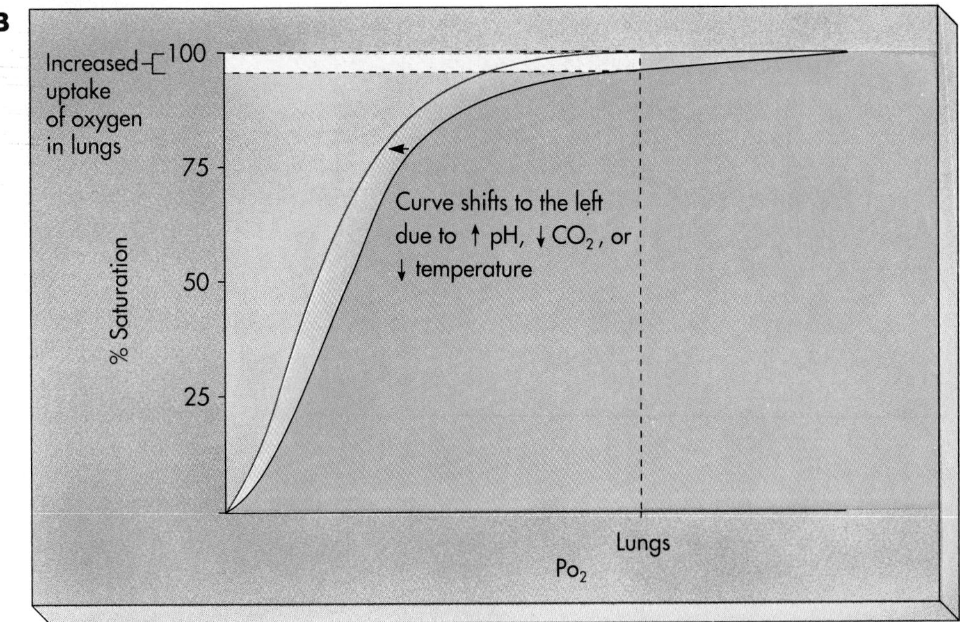

FIGURE 23-15 Effects of shifting the oxygen-hemoglobin dissociation curve. **A** As a result of decreased pH, increased P_{CO_2}, or increased temperature in the tissues, the curve shifts to the right, resulting in an increased release of oxygen. **B** As a result of increased pH, decreased P_{CO_2}, or decreased temperature in the lungs, the curve shifts to the left, resulting in an increased ability of hemoglobin to pick up oxygen.

Transport of Carbon Dioxide

Carbon dioxide is transported in the blood in three major ways: approximately 8% is transported as carbon dioxide dissolved in the plasma, approximately 20% is transported in combination with blood proteins (including hemoglobin), and 72% is transported in the form of bicarbonate ions.

Blood proteins that bind carbon dioxide are called **carbamino** (kar'bah-me'no) **compounds.** The most abundant protein to which carbon dioxide binds in the blood is hemoglobin, and when carbon dioxide is bound to hemoglobin, the combination is called **carbaminohemoglobin.** Carbon dioxide binds in a reversible fashion to the globin portion of the hemoglobin molecule, and many carbon dioxide molecules can combine to a single hemoglobin molecule.

Hemoglobin that has released its oxygen binds more readily to carbon dioxide than hemoglobin that has oxygen bound to it. This is called the **Haldane effect.** In tissues, after hemoglobin has released oxygen, the hemoglobin has an increased ability to pick up carbon dioxide; in the lungs, as hemoglobin binds to oxygen, the hemoglobin more readily releases carbon dioxide.

Carbon dioxide diffuses into red blood cells where some of the carbon dioxide binds to hemoglobin, but most of the carbon dioxide reacts with water to form carbonic acid, a reaction that is catalyzed by the carbonic anhydrase inside the red blood cell. The carbonic acid then dissociates to form bicarbonate ions and hydrogen ions. As a result of these reactions, a higher concentration of bicarbonate ions is inside the cells than outside, and the bicarbonate ions readily diffuse out of the red blood cells into the plasma. In response to this movement of negatively charged ions out of the red blood cells, negatively charged chloride ions move from the plasma into the red blood cells, maintaining electrical balance inside and outside them. The exchange of chloride ions for the bicarbonate ions across the red blood cells' membranes is called the **chloride shift** (Figure 23-16). The hydrogen ions formed by the dissociation of carbonic

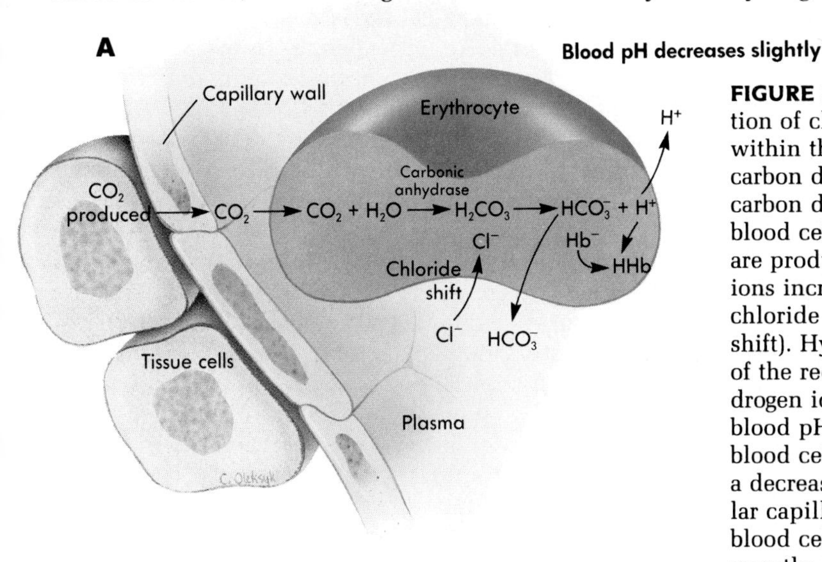

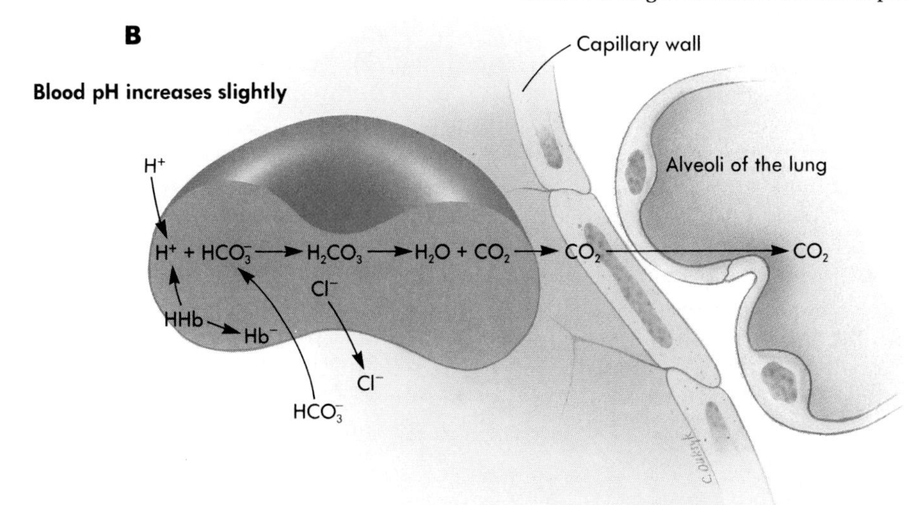

FIGURE 23-16 Chloride movement. **A** The concentration of chloride ions (Cl^-) in red blood cells increases within the tissue capillaries as a result of an increase in carbon dioxide concentration in the tissue capillaries. As carbon dioxide levels increase, carbon dioxide enters red blood cells, and more carbonic acid and bicarbonate ions are produced. As the concentration of the bicarbonate ions increases, they diffuse from the red blood cells, and chloride ions enter the red blood cells (the chloride shift). Hydrogen ions bind to hemoglobin or diffuse out of the red blood cells into the plasma. Movement of hydrogen ions into the plasma causes a slight decrease in blood pH. **B** The concentration of chloride ions in red blood cells decreases in alveolar capillaries as a result of a decrease in carbon dioxide concentration in the alveolar capillaries. As carbon dioxide levels decrease in red blood cells, the bicarbonate ions also decrease. Consequently, bicarbonate ions diffuse into, and chloride ions diffuse out of, the red blood cells. Hydrogen ions unbind from hemoglobin or diffuse from the plasma into the red blood cell. Movement of hydrogen ions out of the plasma causes a slight increase in blood pH.

acid bind to hemoglobin within the red blood cells. This binding prevents the hydrogen ions' leaving the cells and increasing the concentration of hydrogen ions in the plasma (i.e., hemoglobin acts as a buffer to prevent a decrease in blood pH). When the ability of hemoglobin to bind hydrogen ions is exceeded, the hydrogen ions diffuse out of the red blood cells. Despite the movement of hydrogen ions into the plasma, pH normally decreases very little because of buffering systems within the plasma (see Chapter 27).

At the pulmonary alveoli the reverse of the previous events occurs. Carbon dioxide diffuses from the pulmonary capillaries into the alveoli, causing the P_{CO_2} to decrease in the pulmonary capillaries. The carbon dioxide bound to the blood proteins dissociates and diffuses toward the alveoli. As the P_{CO_2} in the alveolar capillaries decreases, bicarbonate ions join with hydrogen ions to form carbonic acid. The carbonic acid then dissociates to form carbon dioxide and water molecules, and the carbon dioxide diffuses into the pulmonary alveoli. As bicarbonate ions and hydrogen ions combine to form carbonic acid, bicarbonate ions move into red blood cells, and chloride ions move out. Hydrogen ions also move into red blood cells or are released from hemoglobin. The movement of the hydrogen ions into the red blood cells lowers the concentration of hydrogen ions in the plasma, causing blood pH to increase slightly.

The transport of carbon dioxide as bicarbonate ions results in changes in blood pH (see Chapter 27). As blood carbon dioxide levels increase, blood pH decreases, and as blood carbon dioxide levels decrease, blood pH increases. Thus regulating blood carbon dioxide levels is a means of regulating blood pH and can be accomplished by varying the rate and depth of respiration.

7

What effect would hyperventilation and holding one's breath have on blood pH? Explain.

? ? ? ? ? ? ? ? ? ? ?

CONTROL OF RESPIRATION

The depth and rate of respiration are controlled by neurons within the medulla that stimulate the muscles of respiration. Increased frequency of stimulation of the respiratory muscles results in stronger contractions of the muscles and an increased depth of respiration. The rate of respiration is determined by the length of time respiratory muscles are stimulated. Although the medullary neurons establish the basic rhythm of breathing, their activities can be influenced by input from other parts of the brain and by input from peripherally located receptors (see Figure 23-18).

Nervous Control of Rhythmic Ventilation

Centers in the medulla

Nerve impulses responsible for controlling the respiratory muscles originate within neurons of the medulla oblongata. The **respiratory center** is composed of two groups of neurons called the inspiratory and expiratory centers (Figure 23-17). The **inspiratory center** is in the reticular formation in the ventral portion of the medulla oblongata. Its neurons are spontaneously active and exhibit rhythmicity. As a result, they exhibit a cycle of activity that arises spontaneously every few seconds and establishes the basic rhythm of the respiratory movements. The inspiratory neurons function as an oscillating circuit that becomes spontaneously active, fatigues, and is again spontaneously active. When the inspiratory neurons become active, impulses are sent along the reticulospinal tracts in the spinal cord, out of the spinal cord, and along the phrenic and intercostal nerves, which stimulate the muscles of inspiration.

The **expiratory center** is located bilaterally and ventrally along the entire length of the medulla oblongata. The expiratory center remains inactive during quiet respiration. There is little evidence to suggest that it plays a role either in the basic rhythm of respiration or in quiet breathing. When the rate and the depth of ventilation are increased, however, the expiratory center becomes active and increases the frequency of action potentials in the nerve fibers that innervate the muscles of expiration. Although the exact neural mechanisms that control the activity of the expiratory center are not known, during heavy breathing the increased activity of the inspiratory center, after a short delay, apparently activates the expiratory center. Increased neural activity in the expiratory center inhibits the inspiratory center and stimulates the muscles of expiration, and expiration proceeds forcefully. The inspiratory and expiratory centers are active in a reciprocal fashion during heavy breathing, with forceful expirations alternating with forceful inspirations.

Centers in the pons

In the pons, located superior to the respiratory center of the medulla oblongata, is the **pneumotaxic** (nu'mo-tak'sik) **center.** The pneumotaxic center is composed of a group of neurons that have an inhibitory effect on the inspiratory center. Removing the influence of the pneumotaxic center allows the inspiratory center to remain active for a much longer period of time than normal, resulting in prolonged and deep inspirations and brief and limited expirations. Increased activity of the pneumotaxic center, on the other hand, results in shallow inspirations (see Figure 23-17).

In the lower portion of the pons is a group of

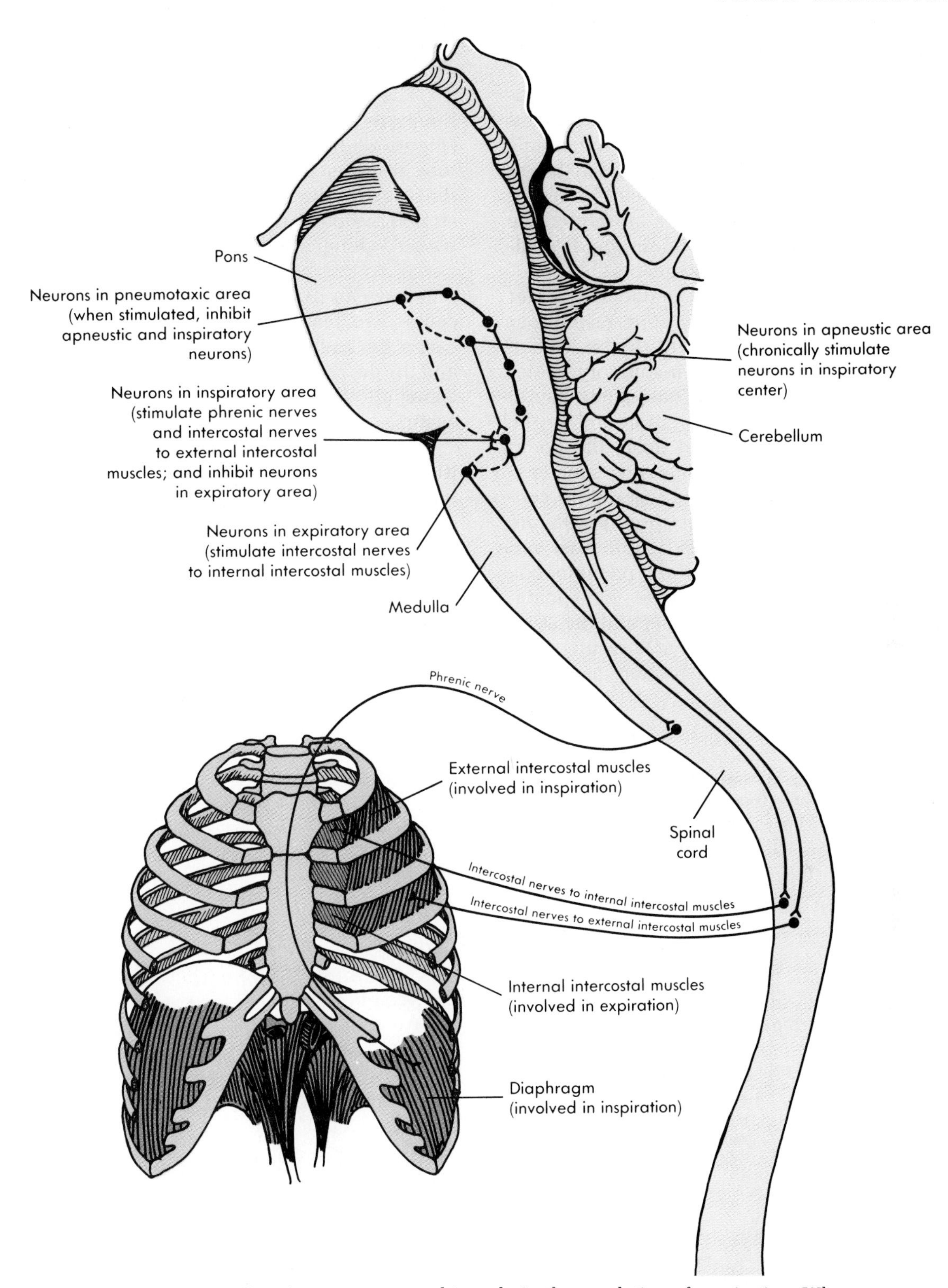

FIGURE 23-17 Respiratory center and its role in the regulation of respiration. When neurons in the inspiratory center are active, they stimulate the muscles of inspiration; when these neurons are inactive, the muscles of inspiration relax. Neurons of the apneustic center stimulate the inspiratory center, whereas neurons of the pneumotaxic center inhibit the inspiratory and apneustic centers. After each inspiration the pneumotaxic center is stimulated; the pneumotaxic center, in turn, inhibits the apneustic and inspiratory centers, resulting in expiration. The pneumotaxic center is activated, after a short delay, by neurons from the inspiratory center. The expiratory center inhibits the inspiratory center but plays a role in respiration only during heavy breathing.

scattered neurons called the **apneustic** (ap-nu'stik) **center.** Nerve impulses from the apneustic center stimulate the inspiratory center. The apneustic neurons are continuously active, but when the pneumotaxic center is active, the pneumotaxic center's influence on the inspiratory center overrides that of the apneustic center. Without the constant influence of the apneustic center on the inspiratory center, the cyclic activity of the inspiratory center yields shallow and irregular respiratory movements.

The apneustic center and the pneumotaxic center function together to ensure a rhythmic respiratory cycle, which reinforces the inherent rhythmicity of the inspiratory center. When the inspiratory center becomes active, action potentials pass along neural pathways to the muscles of inspiration and to the pneumotaxic center. After a short delay the pneumotaxic center is stimulated. The pneumotaxic center then inhibits the apneustic and inspiratory centers, causing the initiation of expiration (see Figure 23-17). After the inspiratory center is inhibited and its activity declines, the activity of the pneumotaxic center also declines. As a consequence, the spontaneous activity of the inspiratory center and the stimulatory effect of the apneustic center result, once again, in full activation of the inspiratory center.

Hering-Breuer reflex

The **Hering-Breuer** (her'ing broy'er) **reflex** also functions to ensure rhythmic respiratory movements (Figure 23-18). This reflex depends on stretch receptors in the walls of the bronchi and bronchioles of the lung. Action potentials are initiated in these stretch receptors when the lungs are inflated and are passed along afferent neurons within the vagus nerves to the medulla oblongata. The action potentials have an inhibitory influence on the inspiratory center and result in expiration. As expiration proceeds, the stretch receptors are no longer stimulated, and the decreased inhibitory effect on the inspiratory center allows the inspiratory center to become active again.

Thus the Hering-Breuer reflex plays a role in limiting the degree to which inspiration proceeds, and it prevents overinflation of the lungs.

Other modifications of rhythmic ventilation

Through the cerebral cortex it is possible to consciously increase or decrease the rate and depth of the respiratory movements (see Figure 23-18). Vol-

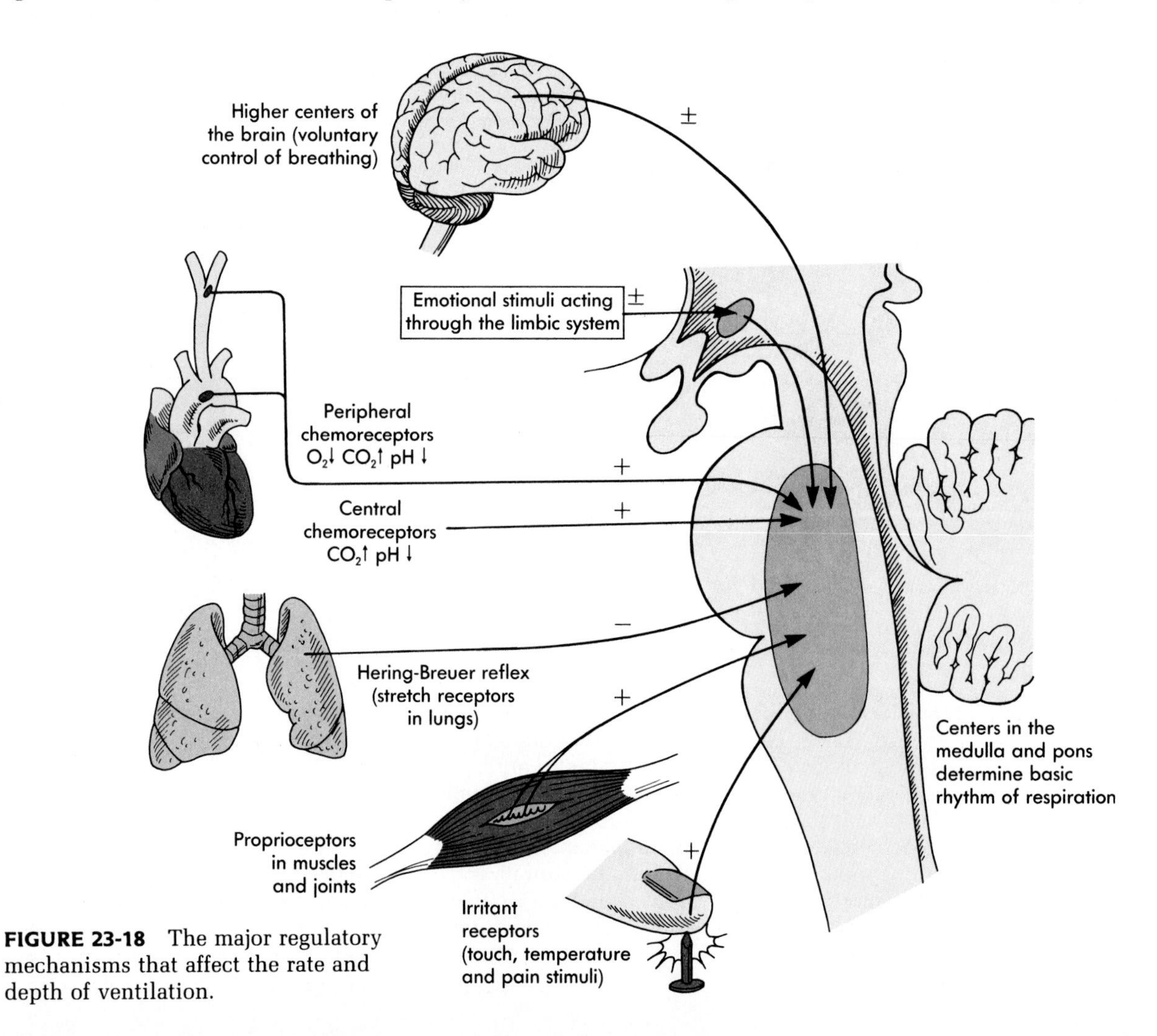

FIGURE 23-18 The major regulatory mechanisms that affect the rate and depth of ventilation.

untary hyperventilation can decrease the blood P_{CO_2} levels sufficiently to cause vasodilation of the peripheral blood vessels and a decrease in blood pressure. Dizziness or a giddy feeling may result because of a decreased rate of blood flow to the brain after the blood pressure drops. A person may also stop breathing voluntarily. As the period of voluntary apnea increases, a greater and greater urge to breathe develops. That urge is associated with increasing P_{CO_2} and decreasing P_{O_2} in the arterial blood. Finally the P_{CO_2} and the P_{O_2} reach levels that cause the respiratory center to override the conscious influence from the cerebrum. Occasionally people are able to hold their breath until the blood P_{O_2} declines to a level low enough that they lose consciousness. After consciousness is lost, the respiratory center resumes its normal function in automatically controlling respiration.

Emotions, acting through the limbic system of the brain, can also affect the respiratory center (see Figure 23-18). For example, strong emotions can produce the sobs and gasps of crying. In addition, the activation of touch, thermal, and pain receptors can also stimulate the respiratory center. For example, irritants in the nasal cavity can initiate a sneeze reflex, and irritants in the lungs can stimulate a cough reflex.

8

Describe the respiratory response when cold water is splashed onto a person. It was once a common practice to swat a newborn baby on the buttocks. Explain the rationale for this procedure.

? ? ? ? ? ? ? ? ? ? ?

Chemical Control of Respiration

The respiratory system maintains concentrations of oxygen and carbon dioxide and maintains the pH of the body fluids within a normal range of values. A deviation by any of these parameters from their normal range has a marked influence on respiratory movements. The effect of changes in oxygen and carbon dioxide concentrations and in pH is superimposed on the neural mechanisms that establish rhythmic respiratory movements.

Chemoreceptors

Chemoreceptors are specialized neurons that respond to changes in chemicals in solution. The chemoreceptors involved with the regulation of respiration respond to changes in hydrogen ion concentrations and/or changes in P_{O_2} (see Figure 23-18; Table 23-4). **Central chemoreceptors** are located bilaterally and ventrally in the **chemosensitive area** of the medulla oblongata, and they connect to the respiratory center. **Peripheral chemoreceptors** are found in the carotid and aortic bodies. These structures are small vascular sensory organs, which are encapsulated in connective tissue and are located near the carotid sinuses and the aortic arch (see Chapter 21). The respiratory center is connected to the carotid body chemoreceptors through the glossopharyngeal nerve and to the aortic body chemoreceptors by the vagus nerve.

Effect of pH

The chemosensitive area is bathed by cerebrospinal fluid and is sensitive to changes in pH of the fluid. Because the blood-brain barrier separates the chemosensitive area from the blood, the chemosensitive area does not directly detect changes in blood pH. Changes in blood pH can alter cerebrospinal fluid pH, however, so the chemosensitive area responds indirectly to changes in blood pH. On the other hand, the carotid and aortic bodies have a rich vascular supply and are directly sensitive to changes in blood pH.

Maintaining body fluid pH levels within normal parameters is necessary for the proper functioning of cells. Because changes in carbon dioxide levels can change pH, the respiratory system plays an important role in acid-base balance. For example, if pH decreases, the respiratory center is stimulated, resulting in elimination of carbon dioxide and an increase in pH back to normal levels. Conversely, if pH increases, the respiratory rate decreases, and carbon dioxide levels increase, causing pH to decrease back to normal levels. The role of the respiratory system in maintaining pH is considered in greater detail in Chapter 27.

Effect of carbon dioxide

Carbon dioxide levels are a major regulator of respiration during resting conditions and conditions when the carbon dioxide levels are elevated (e.g., during intense exercise). Even a small increase in carbon dioxide in the circulatory system triggers a large increase in the rate and the depth of respiration. An increase in P_{CO_2} of 5 mm Hg, for example, causes an increase in ventilation of 100%. A greater-than-normal amount of carbon dioxide in the blood is called **hypercapnia** (hi′per-kap′nĭ-ah). Conversely, lower-than-normal carbon dioxide levels, a condition called **hypocapnia** (hi′po-kap′nĭ-ah), result in periods in which respiratory movements do not occur.

Carbon dioxide apparently does not directly affect the chemosensitive area. Instead, carbon dioxide exerts its effect by changing hydrogen ion levels, which can affect the chemosensitive area. For example, if blood carbon dioxide levels increase, the carbon dioxide diffuses across the blood-brain barrier into the cerebrospinal fluid. The carbon dioxide combines with water to form carbonic acid, which dissociates into hydrogen ions and bicarbonate ions.

Table 23-4

Summary of chemical regulation of respiration

STIMULUS	MECHANISM	RESPONSE	IMPORTANCE
Central Chemoreceptors			
Decrease in pH detected by chemosensitive area	Increased stimulation of respiratory center	Increased ventilation results in decreased Pco_2 that causes pH to increase	Responsible for 85% to 90% of the response to changes in pH; helps to maintain acid-base balance of the body; indirectly but precisely regulates Pco_2 because changes in carbon dioxide levels result in changes in pH
Increase in pH detected by chemosensitive area	Decreased stimulation of respiratory center	Decreased ventilation results in increased Pco_2 that causes pH to decrease	
Peripheral Chemoreceptors			
Decrease in pH detected by carotid and aortic body chemoreceptors	Increased stimulation of respiratory center	Increased ventilation results in decreased Pco_2 that causes pH to increase	Responsible for 10% to 15% of the response to changes in pH; helps to maintain acid-base balance of the body; indirectly but precisely regulates Pco_2 because changes in carbon dioxide levels result in changes in pH
Increase in pH detected by carotid and aortic body chemoreceptors	Decreased stimulation of respiratory center	Decreased ventilation results in increased Pco_2 that causes pH to decrease	
Decrease in Po_2 detected by carotid and aortic body chemoreceptors	Increased stimulation of respiratory center	Increased ventilation results in increased Po_2	Normally not important until Po_2 levels significantly decrease (e.g., at high altitudes)

The increased levels of hydrogen ions stimulate the chemosensitive area, which then stimulates the respiratory center, resulting in a greater rate and depth of breathing. Consequently, carbon dioxide levels decrease as carbon dioxide is eliminated from the body.

9

Explain why a person who breathes rapidly and deeply (hyperventilates) for several seconds experiences a short period of time in which respiration does not occur (apnea) before normal breathing resumes.

? ? ? ? ? ? ? ? ? ? ?

The chemoreceptors in the carotid and aortic bodies also respond to changes in carbon dioxide because of the effects of carbon dioxide on blood pH. The carotid and aortic bodies, however, are responsible for, at most, 15% to 20% of the total response to changes in Pco_2 or pH. The chemosensitive area is far more important for the regulation of Pco_2 and pH than are the carotid and aortic bodies.

Effect of oxygen

Although Pco_2 levels detected by the chemosensitive area are responsible for most changes in respiration, changes in Po_2 can also affect respiration. A decrease in oxygen levels below normal values is called **hypoxia** (hi-pox'se-ah). If the Po_2 levels in the arterial blood are markedly reduced while the pH and Pco_2 are held constant, an increase in ventilation occurs. However, within a normal range of Po_2 levels, the effect of oxygen on the regulation of respiration is small. Only after arterial Po_2 has decreased to approximately 50% of its normal value does it begin to have a large stimulatory effect on respiratory movements.

At first it is somewhat surprising that small changes in Po_2 do not cause changes in respiratory rates. Consideration of the oxygen-hemoglobin dis-

Abnormalities That Reduce Alveolar Ventilation

Paralysis of the Respiratory Muscles

Paralysis of the respiratory muscles can result from poliomyelitis, which causes damage to the respiratory center or to lower motor neurons that stimulate the muscles of inspiration. Another cause of this paralysis is transection of the spinal cord in the cervical or thoracic regions, interrupting nerve impulses that normally pass to the muscles of inspiration. Finally, anesthetics or central nervous system depressants can depress the function of the respiratory center if taken or administered in large enough doses.

Increased Resistance in the Respiratory Passages

Asthma results in the release into the circulatory system of inflammatory chemicals (e.g., histamine) that result in severe constriction of the bronchioles. Emphysema results in increased airway resistance because the bronchioles are obstructed as a result of inflammation and because damaged bronchioles collapse during expiration, trapping air within the alveolar sacs. Cancer may also occlude respiratory passages as it replaces the lung tissue with the tumor.

Decreased Compliance of the Lungs and the Thoracic Wall

Conditions that either replace lung tissue with fibrous connective tissue or result in the encapsulation of materials or organisms in the lung reduce its elasticity and therefore reduce its compliance. For example, conditions such as silicosis and asbestosis in which silicone and asbestos particles are encapsulated by connective tissue elements reduce compliance. Tuberculosis and pneumonia are infections that result in pulmonary inflammation and edema, both of which also reduce compliance. Cancer involves the replacement of lung tissue with a very inelastic tumor, resulting in reduced compliance.

Decreased compliance of the chest wall may be caused by severe arthritis, scoliosis, and kyphosis. These conditions reduce the ability of the thoracic wall to increase its volume when the muscles of inspiration contract, thereby increasing the muscular effort required for inspiration.

Decreased Surface Area for Gas Exchange

A decreased surface area for gas exchange results from the surgical removal of lung tissue, the destruction of lung tissue by cancer, the degeneration of the alveolar walls by emphysema, or the replacement of lung tissue by connective tissue caused by tuberculosis. More acute conditions (e.g., pneumonia, pulmonary edema caused by failure of the left ventricle, and atelectasis [at′e-lik′tă-sis′; collapse of the lung]) that cause the alveoli to fill with fluid also reduce the surface area for gas exchange.

Increased Thickness of the Respiratory Membrane

Pulmonary edema caused by failure of the left side of the heart is the most common cause of an increase in the thickness of the respiratory membrane. The increased venous pressure in the pulmonary capillaries results in the accumulation of fluid in the alveoli. As a consequence, the efficiency of gas diffusion across the alveolar membrane is decreased. Conditions (e.g., tuberculosis, pneumonia, and silicosis) that result in inflammation of the lung tissues also cause pulmonary edema.

Ventilation and the Perfusion of Lung Tissue with Blood

In some alveoli there is too little ventilation to allow blood that flows through the alveolar capillaries to become saturated with oxygen. Diseases that reduce ventilation include asthma, which causes excess resistance of pulmonary airways, and emphysema, which damages lung tissues. In other alveoli the ventilation may be adequate, but blood flow through the alveolar capillaries may be inadequate. Disorders that reduce perfusion of lung tissue with blood include thrombosis of pulmonary arteries and reduced cardiac output that results from heart attack and shock.

Reduced Capacity of the Blood to Transport Oxygen

Anemias that result in a reduction of the total amount of hemoglobin available to transport oxygen also reduce the capacity of blood to transport oxygen. Carbon monoxide binds irreversibly to the heme portion of the hemoglobin molecule and makes it unavailable for oxygen transport. Thus carbon monoxide poisoning decreases the ability of hemoglobin to transport oxygen, even though it does not affect the total hemoglobin concentration in the blood (see Chapter 21).

sociation curve, however, provides an answer. Because of the S-shape of the curve, at any Po_2 above 80 mm Hg nearly 100% of the hemoglobin is saturated with oxygen. Consequently, until Po_2 levels change significantly, the oxygen-carrying capacity of the blood is not affected.

The carotid and aortic body chemoreceptors respond to decreased Po_2 by increasing their stimulation of the respiratory center. Normally, stimulation by the carotid and aortic bodies keeps the respiratory center active despite decreasing oxygen levels. If Po_2 decreases sufficiently, however, the respiratory center can fail to function, resulting in death.

Carbon dioxide is much more important than oxygen as a regulator of normal alveolar ventilation, but under certain circumstances a reduced Po_2 in the arterial blood does play an important stimulatory role. During conditions of shock in which the blood pressure is very low, the Po_2 in the arterial blood may decrease to levels sufficiently low to stimulate strongly the carotid and aortic body sensory receptors. At high altitudes where the barometric pressure is low, the Po_2 in the arterial blood may also decrease to levels sufficiently low to stimulate the carotid and aortic bodies. Although the Po_2 levels in the blood are reduced, the ability of the respiratory system to eliminate carbon dioxide is not greatly affected by the low barometric pressure. Thus the blood carbon dioxide levels become lower than normal because of the increased alveolar ventilation initiated in response to the low Po_2.

A similar situation exists in people who have emphysema. Because carbon dioxide diffuses across the respiratory membrane more readily than oxygen, the decreased surface area of the respiratory membrane (caused by the disease) results in low arterial Po_2 without elevated arterial Pco_2. The elevated rate and depth of respiration are due, to a large degree, to the stimulatory effect of low arterial Po_2 levels on the carotid and aortic bodies. More severe emphysema, in which the surface area of the respiratory membrane is reduced to a minimum, can also result in elevated Pco_2 levels in the arterial blood.

Effect of Exercise on Respiratory Movements

Only during very heavy exercise do the Po_2, Pco_2, and pH of the blood differ very much from their normal values. Thus there must be some other influence on the respiratory center that increases its activity during exercise. As impulses pass from the motor cortex of the cerebrum through the motor pathways, numerous collateral fibers project into the reticular formation of the brain. These fibers stimulate the respiratory center during conditions of exercise (see Figure 23-18).

Furthermore, during exercise body movements stimulate proprioceptors in the joints of the limbs. Nerve impulses from the proprioceptors pass along afferent nerve fibers to the spinal cord and along ascending nerve tracts (the medial lemniscal system) of the spinal cord to the brain. Collateral fibers project from these ascending pathways to the respiratory center in the medulla. Movement of the limbs has a strong stimulatory influence on the respiratory center (see Figure 23-18).

The neural impulses from the cerebral cortex and from the proprioceptors act as the major regulators of respiration during exercise. However, the system is fine-tuned by the chemosensitive receptor mechanisms that are sensitive to carbon dioxide. If the alveolar ventilation is too great, blood Pco_2 decreases, causing alveolar ventilation to decrease. On the other hand, if alveolar ventilation is too low, Pco_2 increases, stimulating alveolar ventilation.

The brain "learns" after a period of training to match the rate of respiration with the intensity of the exercise. Well-trained athletes match their respiratory movements more efficiently with their level of physical activity than do untrained individuals. Thus centers of the brain involved in learning have an indirect influence on the respiratory center, but the exact mechanism for this kind of regulation is not clear.

The Respiratory System

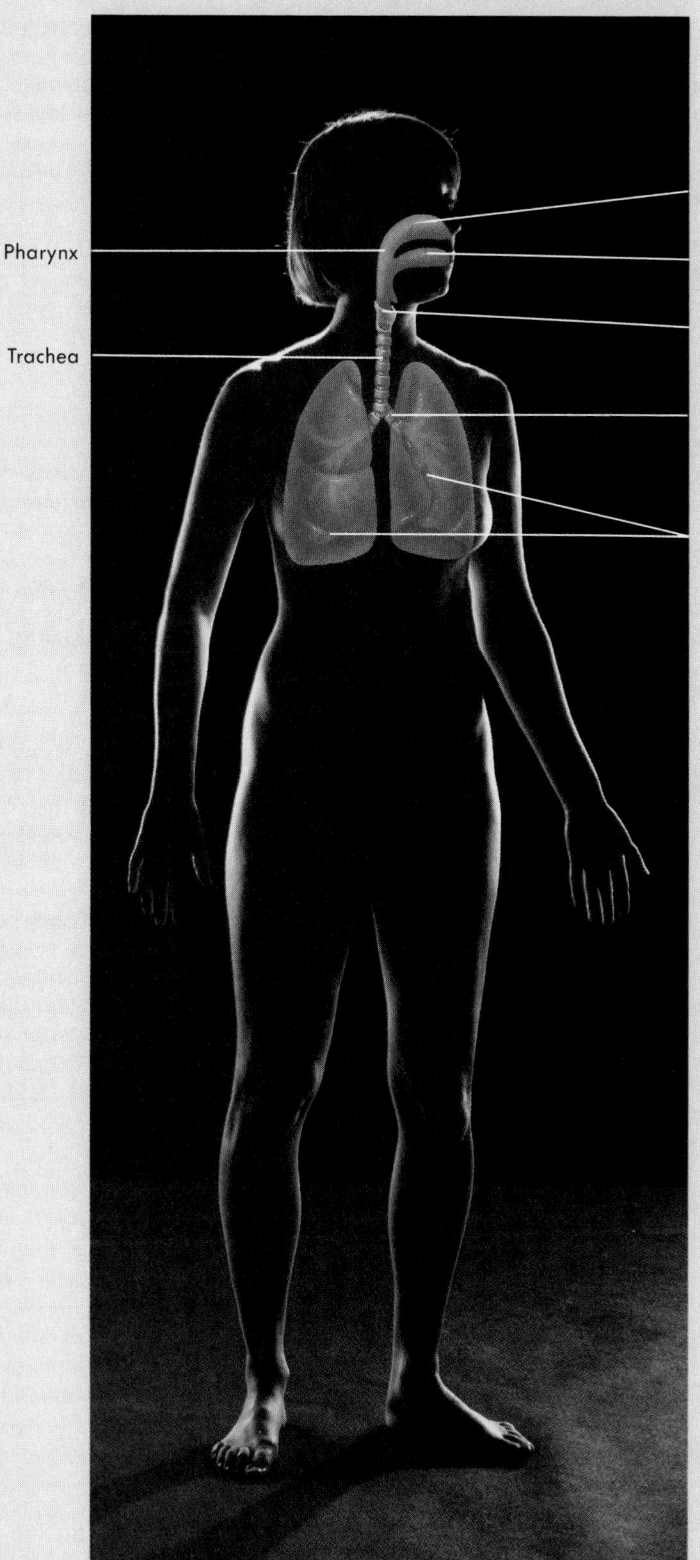

Major Components

Lungs and the respiratory passages

Major Functions

Transports air into and out of the lungs
Exchanges gases (oxygen and carbon dioxide) between the air and blood
Regulates blood pH

SUMMARY

RESPIRATION INCLUDES THE MOVEMENT OF AIR INTO AND OUT OF THE LUNGS, THE EXCHANGE OF GASES BETWEEN THE AIR AND THE BLOOD, THE TRANSPORT OF GASES IN THE BLOOD, AND THE EXCHANGE OF GASES BETWEEN THE BLOOD AND TISSUES.

ANATOMY AND HISTOLOGY p. 736

Nose and Nasal Cavity

1. The bridge of the nose is bone, and most of the external nose is cartilage.
2. Openings of the nasal cavity.
   - The external nares open to the outside, and the internal nares lead to the pharynx.
   - The paranasal sinuses and the nasolacrimal duct open into the nasal cavity.
3. Divisions of the nasal cavity.
   - The nasal cavity is divided by the nasal septum.
   - The anterior vestibule contains hairs that trap debris.
   - The nasal cavity is lined with pseudostratified ciliated epithelium that traps debris and moves it to the pharynx.
   - The superior part of the nasal cavity contains the olfactory epithelium.

Pharynx

1. The nasopharynx joins the nasal cavity through the internal nares and contains the openings to the auditory tube and the pharyngeal tonsils.
2. The oropharynx joins the oral cavity and contains the palatine and lingual tonsils.
3. The laryngopharynx opens into the larynx and the esophagus.

Larynx

1. Cartilage.
   - There are three unpaired cartilages. The thyroid cartilage and cricoid cartilage form most of the larynx. The epiglottis covers the opening of the larynx during swallowing.
   - There are six paired cartilages. The vocal cords attach to the arytenoid cartilages.
2. Sounds are produced as the vocal cords vibrate when air passes through the larynx. Tightening the cords produces sounds of different pitch by controlling the length of the cord that is allowed to vibrate.

Trachea

The trachea connects the larynx to the primary bronchi.

Bronchi

The primary bronchi go to each lung.

Lungs

1. There are two lungs.
2. The airway passages of the lungs branch and decrease in size.
   - The primary bronchi form the secondary bronchi, which go to each lobe of the lungs.
   - The secondary bronchi form the tertiary bronchi, which go to each lobule of the lungs.
   - The tertiary bronchi branch many times to form the terminal bronchioles.
   - The terminal bronchioles become the respiratory bronchioles from which the alveoli branch.
3. Important features of the tube system.
   - The area from the trachea to the terminal bronchioles is ciliated to facilitate removal of debris.
   - Cartilage helps to hold the tube system open (from the trachea to the bronchioles).
   - Smooth muscle controls the diameter of the tubes (terminal bronchioles).
   - The alveoli are sacs formed by simple squamous epithelium, and they facilitate diffusion of gases.

Pleura

The pleural membranes surround the lungs and provide protection against friction.

Blood Supply

Deoxygenated blood is transported to the lungs through the pulmonary arteries, and oxygenated blood leaves through the pulmonary veins. Oxygenated blood is mixed with a small amount of deoxygenated blood from the bronchi.

Muscles of Respiration

1. Contraction of the diaphragm increases thoracic volume.
2. Muscles can elevate the ribs and increase thoracic volume or can depress the ribs and decrease thoracic volume.

Thoracic Wall

The thoracic wall consists of vertebrae, ribs, sternum, and muscles that allow expansion of the thoracic cavity.

VENTILATION AND LUNG VOLUMES p. 744

Pressure Differences and Air Flow

1. Ventilation is the movement of air into and out of the lungs.
2. Air moves from an area of higher pressure to an area of lower pressure.
3. Pressure in the lungs decreases as the volume of the lungs increases, and pressure increases as lung volume decreases (general gas law).

Collapse of the Lungs

1. Recoil of elastic fibers and water surface tension make the lungs collapse.
2. Surfactant reduces water surface tension.
3. Intrapleural pressure prevents collapse. Pneumothorax is an opening between the pleural space and the air that causes a loss of intrapleural pressure.

Compliance of the Lungs and the Thorax

1. Compliance is a measure of lung expansion caused by intrapulmonary pressure.
2. Reduced compliance means that it is more difficult than normal to expand the lungs.

Pulmonary Volumes and Capacities

1. There are four pulmonary volumes: tidal volume, inspiratory reserve, expiratory reserve, and residual volume.
2. Pulmonary capacities are the sum of two or more pulmonary volumes and include inspiratory capacity, functional residual capacity, vital capacity, and total lung capacity.
3. Vital capacity is reduced by any disorder that reduces pulmonary compliance.
4. The forced expiratory vital capacity measures the rate at which air can be expelled from the lungs.

Minute Respiratory Volume and Alveolar Ventilation Rate

1. The minute respiratory volume is the total amount of air moved in and out of the respiratory system per minute.
2. Dead air space is the part of the respiratory system in which gas exchange does not take place.

3. Alveolar ventilation rate is how much air per minute enters the parts of the respiratory system in which gas exchange takes place.

PHYSICAL PRINCIPLES OF GAS EXCHANGE *p. 749*

Partial Pressure

1. Partial pressure is the contribution of a gas to the total pressure of a mixture of gases (Dalton's law).
2. Vapor pressure is the partial pressure produced by water.
3. Atmospheric air, alveolar air, and expired air have different compositions.

Diffusion of Gases Through Liquids

The concentration of a gas in a liquid is determined by its partial pressure and by its solubility coefficient (Henry's law).

Diffusion of Gases Through the Respiratory Membrane

1. The respiratory membranes are thin and have a large surface area that facilitates gas exchange.
2. The components of the respiratory membrane include a film of water, the walls of the alveolus and the capillary, and an interstitial space.
3. The rate of diffusion depends on the thickness of the respiratory membrane, the diffusion coefficient of the gas, the surface area of the membrane, and the partial pressure of the gases in the alveoli and the blood.

Relationship Between Ventilation and Capillary Blood Flow

1. Increased ventilation and/or increased capillary blood flow increases gas exchange.
2. The physiological shunt results from deoxygenated blood returning from the lungs.

OXYGEN AND CARBON DIOXIDE TRANSPORT IN THE BLOOD *p. 752*

Oxygen Diffusion Gradients

1. Oxygen moves from the alveoli (Po_2 equals 104 mm Hg) into the blood (Po_2 equals 40 mm Hg). Blood is saturated with oxygen when it leaves the capillary.
2. The partial pressure of oxygen in the blood decreases (Po_2 equals 95 mm Hg) because of mixing with deoxygenated blood.
3. Oxygen moves from the tissue capillaries (Po_2 equals 95 mm Hg) into the tissues (Po_2 equals 40 mm Hg).

Carbon Dioxide Diffusion Gradients

1. Carbon dioxide moves from the tissues (Pco_2 equals 45 mm Hg) into tissue capillaries (Pco_2 equals 40 mm Hg).
2. Carbon dioxide moves from the alveolar capillaries (Pco_2 equals 45 mm Hg) into the alveoli (Pco_2 equals 40 mm Hg).

Hemoglobin and Oxygen Transport

1. Oxygen is transported by hemoglobin (97%) and is dissolved in plasma (3%).
2. The oxygen-hemoglobin dissociation curve shows that hemoglobin is saturated when Po_2 equals 80 mm Hg or above. At lower partial pressures the hemoglobin releases oxygen.
3. A shift of the oxygen-hemoglobin dissociation curve to the right because of a decrease in pH (Bohr effect), an increase in carbon dioxide, or an increase in temperature results in a decrease in the ability of hemoglobin to hold oxygen.
4. A shift of the oxygen-hemoglobin dissociation curve to the left because of an increase in pH (Bohr effect), a decrease in carbon dioxide, or a decrease in temperature results in an increase in the ability of hemoglobin to hold oxygen.

Transport of Carbon Dioxide

1. Carbon dioxide is transported as bicarbonate ions (72%), in combination with blood proteins (20%), and in solution in plasma (8%).
2. Hemoglobin that has released carbon dioxide binds more readily to carbon dioxide than hemoglobin that has oxygen bound to it (Haldane effect).
3. In tissue capillaries carbon dioxide combines with water inside the red blood cells to form carbonic acid that dissociates to form bicarbonate ions and hydrogen ions.
   - The chloride shift is the movement of chloride ions into red blood cells as bicarbonate ions move out.
   - Hydrogen ions diffuse out of red blood cells and change blood pH. This effect is buffered.
4. In lung capillaries bicarbonate ions and hydrogen ions move into red blood cells, and chloride ions move out. Bicarbonate ions combine with hydrogen ions to form carbonic acid. The carbonic acid dissociates to form carbon dioxide that diffuses out of the red blood cell.

CONTROL OF RESPIRATION *p. 758*

Nervous Control of Rhythmic Ventilation

1. The inspiratory center stimulates the muscles of inspiration to contract. Inactivity of the inspiratory center causes passive expiration.
2. The expiratory center stimulates the muscles of forced expiration.
3. The apneustic center continuously activates the inspiratory center.
4. The pneumotaxic center inhibits the inspiratory and apneustic centers.
5. The Hering-Breuer reflex inhibits the inspiratory center when the lungs are stretched during inspiration.
6. It is possible to consciously control ventilation. Emotions, touch, temperature, and pain can also alter ventilation.

Chemical Control of Respiration

1. Carbon dioxide is the major regulator of respiration. An increase in carbon dioxide or a decrease in pH can stimulate the chemosensitive area, causing greater rate and depth of respiration.
2. Oxygen levels in the blood affect respiration when there is a 50% or greater decrease from normal levels. Decreased oxygen is detected by receptors in the carotid and aortic bodies, which then stimulate the respiratory center.

Effect of Exercise on Respiratory Movements

1. Blood levels of carbon dioxide, pH, and oxygen change very little during exercise.
2. Collateral fibers from motor neurons and from proprioceptors stimulate the respiratory centers.
3. Chemosensitive mechanisms and learning fine-tune the effects produced through the motor neurons and proprioceptors.

CONTENT REVIEW

1. What are the functions of the respiratory system?
2. Define respiration.
3. Describe the structure of the respiratory passages.
4. Name the three parts of the pharynx. With what structures does each part communicate?
5. Name and describe the three unpaired cartilages of the larynx. How are sounds of different pitch produced by the vocal cords?
6. What is the function of the C-shaped cartilages in the trachea? Why are they C-shaped? What happens to the amount of cartilage in the tube system of the respiratory system as the tubes become smaller? Explain why breathing becomes more difficult during an asthma attack.
7. What is the function of ciliated epithelium in the trachea, primary bronchi, and lungs?
8. Starting with the trachea, name all the structures a molecule of oxygen would encounter on its way to the alveoli.
9. Distinguish between the lungs, a lobe of the lung, and a lobule.
10. Describe the pleurae of the lungs. What is their function?
11. Describe the two major routes of blood flow to and from the lungs.
12. How does movement of the ribs and diaphragm affect thoracic volume?
13. Define ventilation. Describe the pressure changes that cause air to move into and out of the lungs. How do these pressure changes develop?
14. Name two factors that cause the lungs to collapse. How is this collapse prevented? Define a pneumothorax, and explain how it causes the lungs to collapse.
15. Define compliance. What is the effect on lung expansion when compliance is increased or decreased?
16. Define tidal volume, inspiratory reserve, expiratory reserve, and residual volume. Define inspiratory capacity, functional residual capacity, vital capacity, and total lung capacity.
17. Define minute respiratory volume and alveolar ventilation rate.
18. What is dead air space? What is the difference between anatomical and physiological dead air space? What conditions increase dead air space?
19. What is the partial pressure of a gas? What is vapor pressure?
20. Describe the factors that make inspired air, air in the alveoli, and expired air have different gas concentrations.
21. How do the partial pressure and the solubility of a gas affect the concentration of the gas in a liquid? How do they affect the rate of diffusion of the gas through the liquid?
22. List the components of the respiratory membrane. Describe the factors that affect the diffusion of gases across the respiratory membrane. Give some examples of diseases that decrease diffusion by altering these factors.
23. What effect do ventilation and alveolar capillary blood flow have on gas exchange? What is the physiological shunt?
24. Describe the partial pressures for oxygen and carbon dioxide in the alveoli, lung capillaries, tissue capillaries, and tissues. How do these partial pressures account for the movement of oxygen and carbon dioxide?
25. List two ways that oxygen is transported in the blood.
26. What is the oxygen-hemoglobin dissociation curve? How does it explain the release and uptake of oxygen? How does a shift of the curve to the left or right affect the ability of hemoglobin to bind to oxygen? Where do such shifts take place in the body?
27. List three ways that carbon dioxide is transported in the blood. What is the chloride shift? Where and why does it take place?
28. How can changes in respiration affect blood pH?
29. Describe the control of respiration by the inspiratory, expiratory, apneustic, and pneumotaxic centers.
30. Describe the Hering-Breuer reflex and its function.
31. How do carbon dioxide, pH, and oxygen influence the regulation of respiration?
32. During exercise how is respiration regulated?

CONCEPT REVIEW

1. What effect does rapid (respiration rate equals 24 breaths per minute), shallow (tidal volume equals 250 ml per breath) breathing have on minute respiratory volume, the alveolar ventilation rate, and the alveolar P_{O_2} and P_{CO_2}?
2. A person's vital capacity was measured while he was standing and while he was lying down. What difference, if any, in the measurement would you predict and why?
3. Ima Diver wanted to do some underwater exploration. However, she did not want to buy expensive scuba equipment. Instead, she bought a long hose and an inner tube. She attached one end of the hose to the inner tube so that the end was always out of the water, and she inserted the other end of the hose in her mouth and went diving. What would happen to her alveolar ventilation rate and why? How would she compensate for this change? How would diving affect lung compliance and the work of ventilation?
4. One technique for artificial respiration is the back-pressure–arm-lift method, which is performed with the victim lying face down. The rescuer presses firmly on the base of the scapulae for several seconds, then grasps the arms and lifts them. The sequence is then repeated. Explain why this procedure results in ventilation of the lung.
5. Another technique for artificial respiration is mouth-to-mouth resuscitation. The rescuer takes a deep breath, blows air into the victim's mouth, and lets air flow out of the victim. The process is repeated. Explain the following: (1) Why do the victim's lungs expand; (2) why does air move out of the victim's lungs; and (3) what effect do the P_{O_2} and the P_{CO_2} of the rescuer's air have on the victim?
6. During normal quiet respiration when does the maximum rate of diffusion of oxygen in the alveolar capillaries occur? The maximum rate of diffusion of carbon dioxide?

7. Would the oxygen-hemoglobin dissociation curve in humans who always live at high altitudes be to the left or to the right of a person who normally lives at low altitudes?
8. Predict what would happen to tidal volume if the vagus nerves were cut. The phrenic nerves? The intercostal nerves?
9. You and your physiology instructor are trapped in an overturned ship. To escape you must swim underwater a long distance. You tell your instructor it would be a good idea to hyperventilate before making the escape attempt. Your instructor calmly replies, "What good would that do since your alveolar capillaries are already 100% saturated with oxygen?" What would you do and why?

? ?

ANSWERS TO PREDICT QUESTIONS

1 p. 736. Air moving through the mouth is not as efficiently warmed and moistened as air moving through the nasal cavity, and the throat or lung tissue may become damaged.

2 p. 738. When food moves down the esophagus, the normally collapsed esophagus expands. If the cartilage rings were solid, expansion of the esophagus, and therefore swallowing, would be more difficult.

3 p. 740. The right primary bronchus.

4 p. 749. The alveolar ventilation rate would be 4200 ml per minute ($12 \times [500 - 150]$). Increasing the respiration rate to 24 breaths per minute and the tidal volume to 4000 ml per minute would increase the alveolar ventilation rate to 92,400 ml per minute ($24 \times [4000 - 150]$), a twenty-twofold increase. This increase would increase P_{O_2} and decrease P_{CO_2} in the alveoli, thus increasing gas exchange between the alveoli and the blood.

5 p. 750. The air the diver is breathing has a greater total pressure than sea level atmospheric pressure. Consequently, the partial pressure of each gas in the air is increased. According to Henry's law, as the partial pressure of a gas increases, the greater is the amount (concentration) of gas that is dissolved in the liquid (e.g., body fluids) with which the gas is in contact. When the diver suddenly ascends, the partial pressure of gases in the body returns toward sea level atmospheric pressure. As a result, the amount (concentration) of gas that can be dissolved in body fluids suddenly decreases. When the fluids can no longer hold all the gas, gas bubbles form.

6 p. 752. At high altitudes the atmospheric P_{O_2} decreases because of a decrease in atmospheric pressure. The decreased atmospheric P_{O_2} results in a decrease in alveolar P_{O_2} and less oxygen diffusion into lung tissue. If the person's local control mechanism is especially sensitive to the decreased oxygen levels, the lung's precapillary sphincters contract, blood flow through the lungs decreases, and there is a decreased ability to get adequate amounts of oxygen. Note that local control of blood flow in the lungs differs from that in most tissues. In the lungs decreased oxygen tension causes a reduction in blood flow through the capillaries.

7 p. 758. Hyperventilation would decrease P_{CO_2} in the blood, causing an increase in blood pH. Holding one's breath would increase P_{CO_2} in the blood and decrease blood pH.

8 p. 761. Through thermal or pain receptors the respiratory center can be stimulated to cause a sudden inspiration of air.

9 p. 762. When a person hyperventilates, P_{CO_2} in the blood decreases. Consequently, carbon dioxide moves out of cerebrospinal fluid into the blood. As carbon dioxide levels in cerebrospinal fluid decrease, hydrogen ions and bicarbonate ions combine to form carbonic acid that dissociates to form carbon dioxide. The result is a decrease in hydrogen ion concentration in cerebrospinal fluid and decreased stimulation of the respiratory center by the chemosensitive area. Until blood P_{CO_2} levels increase, the chemosensitive area is not stimulated, and apnea results.

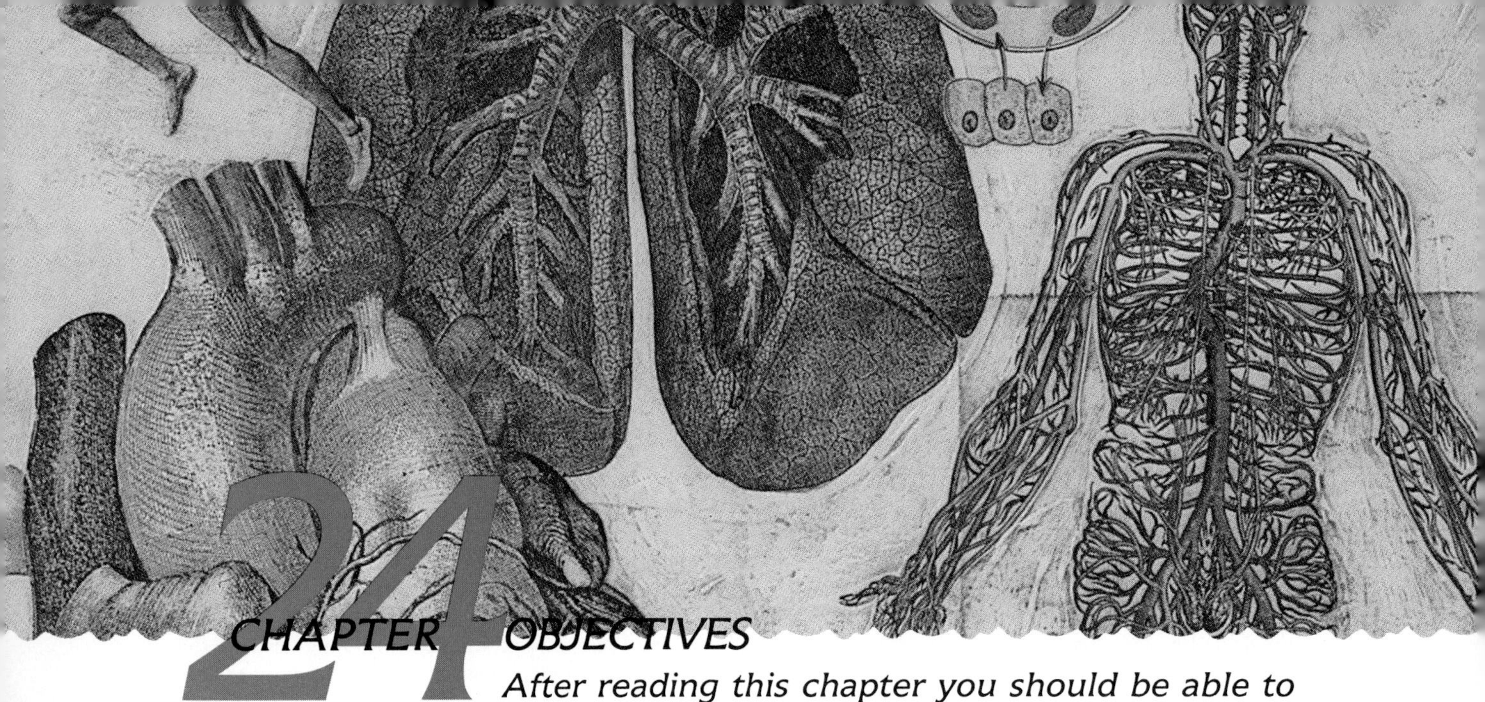

24 CHAPTER OBJECTIVES

After reading this chapter you should be able to

1. Describe the general anatomical features of the digestive tract.
2. Outline the basic histological characteristics of the digestive tract.
3. List and describe the major functions of the digestive tract.
4. Describe the anatomy of the oral cavity.
5. List the major types of teeth and describe the structure of an individual tooth.
6. Describe the functional differences between the major salivary glands.
7. Describe the anatomy of the esophagus.
8. List the stomach's anatomical and physiological characteristics that are most important to its function.
9. List the anatomical and histological characteristics of the small intestine that account for its large surface area.
10. Describe the structure of the liver, the gallbladder, and the pancreas.
11. Describe the anatomy of the large intestine.
12. Describe the peritoneum and the mesenteries.
13. Explain the functions of the oral cavity.
14. Describe mastication and deglutition.
15. Describe the stomach secretions and their functions and explain how they are regulated.
16. Describe gastric movements, stomach emptying, and their regulation.
17. Explain the functions of the small intestine's secretions and describe their regulation.
18. List the major functions of the pancreas, the liver, and the gallbladder and explain how they are regulated.
19. Describe the process of digestion for carbohydrates, lipids, and proteins and list the products of digestion for each.
20. Describe the functions of the large intestine.

Digestive System

KEY TERMS

bile

chyme (kīm)

defecation reflex

deglutition (dĕ′glu-tish′un)

digestive tract

gastric gland

intestinal gland

intramural plexus (in′trah-mu′ral plek′sus)

lacteal

mastication reflex

mucosa (mu-ko′sah)

plicae (pli′se) circulares

portal triad

ruga (ru′gah)

salivary gland

RELATED TOPICS

The following terms or concepts are important for a good understanding of this chapter. If you are not familiar with them, you should review them before proceeding.

Mesentery (Chapter 1)

Epithelial tissue types, loose connective tissue, and smooth muscle (Chapter 4)

Smooth muscle function (Chapter 10)

Reflexes (Chapter 13)

Autonomic control of the digestive system (Chapter 16)

THE PRIMARY FUNCTIONS OF THE DIGESTIVE SYSTEM ARE TO PROVIDE THE BODY WITH WATER, ELECTROLYTES, AND OTHER NUTRIENTS. To accomplish these functions the digestive system is specialized to ingest food, propel the food through the digestive tract, digest the food, and absorb water, electrolytes, and other nutrients from the lumen of the gastrointestinal tract. Once these useful substances are absorbed, they are transported through the circulatory system to cells where they are used. Undigested matter is moved through the digestive tract and is eliminated through the anus.

This chapter presents a general overview of the digestive system, describes the anatomy and histology of each section of the digestive tract in detail, and describes the physiology of each section.

GENERAL OVERVIEW

The **digestive system** (Figure 24-1) consists of the **digestive tract,** a tube extending from the mouth to the anus, and its associated **accessory organs** (primarily glands), which secrete fluids into the digestive tract. The digestive tract is also called the alimentary tract or alimentary canal. The term **gastrointestinal (GI;** gas′tro-in-tes′tĭ-nal) **tract** technically only refers to the stomach and intestines but is often used as a synonym for the digestive tract.

ANATOMY OVERVIEW

The first section of the digestive tract is the mouth, or **oral cavity.** It is surrounded by the lips, cheeks, teeth, and palate, and it contains the tongue. The **salivary glands** and **tonsils** are accessory organs of the oral cavity.

The oral cavity opens posteriorly into the **pharynx,** which, in turn, continues inferiorly into the **esophagus** (e-sof′ă-gus). The major accessory structures are single-celled or small, simple tubular mucous glands distributed the length of the pharynx and the esophagus.

The esophagus opens inferiorly into the **stomach.** The stomach wall contains many tubelike glands from which acid and enzymes are released into the stomach and are mixed with ingested food.

The stomach opens inferiorly into the **small intestine.** The first segment of the small intestine is the **duodenum** (du-o-de′num; or 12 inches long). The major accessory structures in this segment of the digestive tract are the **liver,** the **gallbladder,** and the **pancreas.** The next segment of the small intestine is the **jejunum** (jĕ-ju′num; empty). Small glands exist along its length, and it is the major site of absorption. The last segment of the small intestine is the **ileum** (il′e-um; twisted), which is similar to the jejunum except that fewer digestive enzymes and more mucus are secreted and less absorption occurs in the ileum.

The last section of the digestive tract is the **large intestine.** Its major accessory glands secrete mucus. It absorbs water and salts and concentrates undigested food into feces. The first segment is the **cecum** (se′kum; blind), with the attached **vermiform** (ver′mĭ-form) **appendix.** The cecum is followed by the ascending, transverse, descending, and sigmoid **colons** (ko′lonz) and the **rectum** (rek′tum; straight). The rectum joins the **anal canal,** which ends at the **anus,** the inferior termination of the digestive tract.

HISTOLOGY OVERVIEW

Figure 24-2 depicts a generalized view of the digestive tract histology. The digestive tube consists

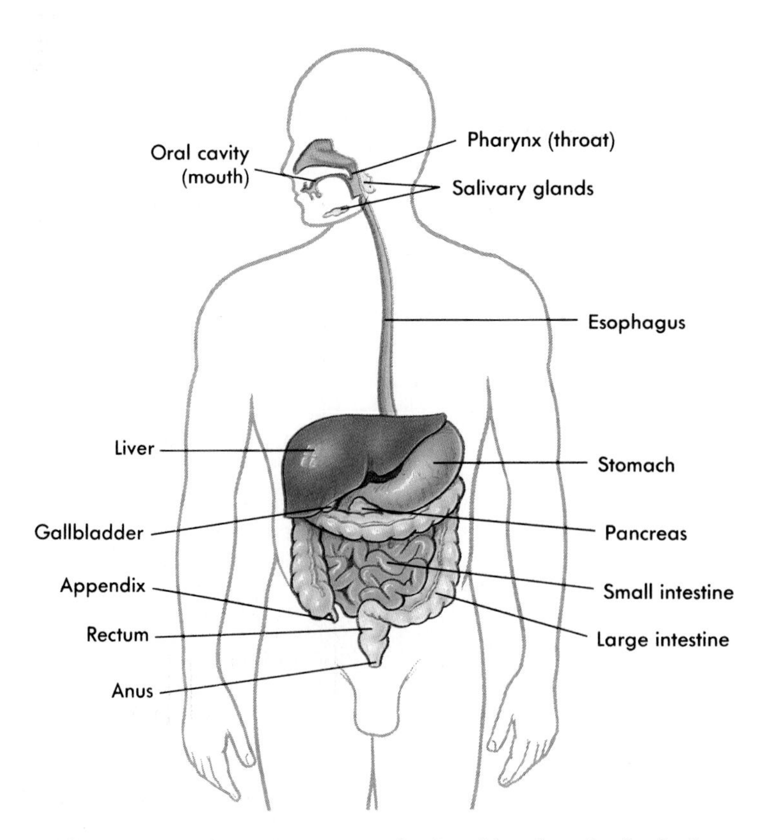

FIGURE 24-1 Digestive system depicted in place in the body.

of four layers or tunics: an internal mucosa and an external serosa with a submucosa and muscularis in between. These four tunics are present in all areas of the digestive tract from the esophagus to the anus. Three major types of glands are associated with the intestinal tract: (1) unicellular mucous glands in the mucosa, (2) multicellular glands in the mucosa and submucosa, and (3) multicellular glands (accessory glands) outside the digestive tract.

Mucosa

The innermost tunic, the **mucosa** (mu-ko'sah), consists of three layers: (1) the **mucous epithelium,** which is moist, stratified squamous epithelium in the mouth, oropharynx, esophagus, and anal canal and simple columnar epithelium in the remainder of the digestive tract; (2) a loose connective tissue called the **lamina propria** (lam'ĭ-nah pro'pre-ah); and (3) a thin smooth muscle layer, the **muscularis mucosa.**

Submucosa

The **submucosa** is a thick connective tissue layer containing nerves, blood vessels, and small glands, and lies beneath the mucosa. The nerves of the submucosa form the **submucosal plexus** (plek'sus; Meissner's plexus), a parasympathetic ganglionic plexus.

Muscularis

The next tunic is the **muscularis,** which consists of an inner layer of circular smooth muscle and an outer layer of longitudinal smooth muscle (except in the upper esophagus where the muscles are striated and in the stomach where there are three layers of smooth muscle). Another nerve plexus, the **myenteric plexus** (mi'en-tĕr'ik; Auerbach's plexus), which consists of nerve fibers and parasympathetic cell bodies, is between these two muscle layers (see Figure 24-2). Together, the submucosal and myenteric plexuses comprise the **intramural** (in'trah-mu'ral; within the walls) **plexus.** This plexus is extremely important in the control of movement and secretion.

Serosa or Adventitia

The fourth layer of the digestive tract is a connective tissue layer called either the **serosa** or the **adventitia** (ad'ven-tish'yah; foreign; coming from outside), depending on the structure of the layer. Portions of the digestive tract that protrude into the peritoneal cavity have a serosa as the outermost layer. This serosa consists of a connective tissue layer that is covered by a serous simple squamous epithelium, the visceral peritoneum. When the outer layer of the digestive tract is derived from adjacent connective tissue, the tunic is called the adventitia and consists

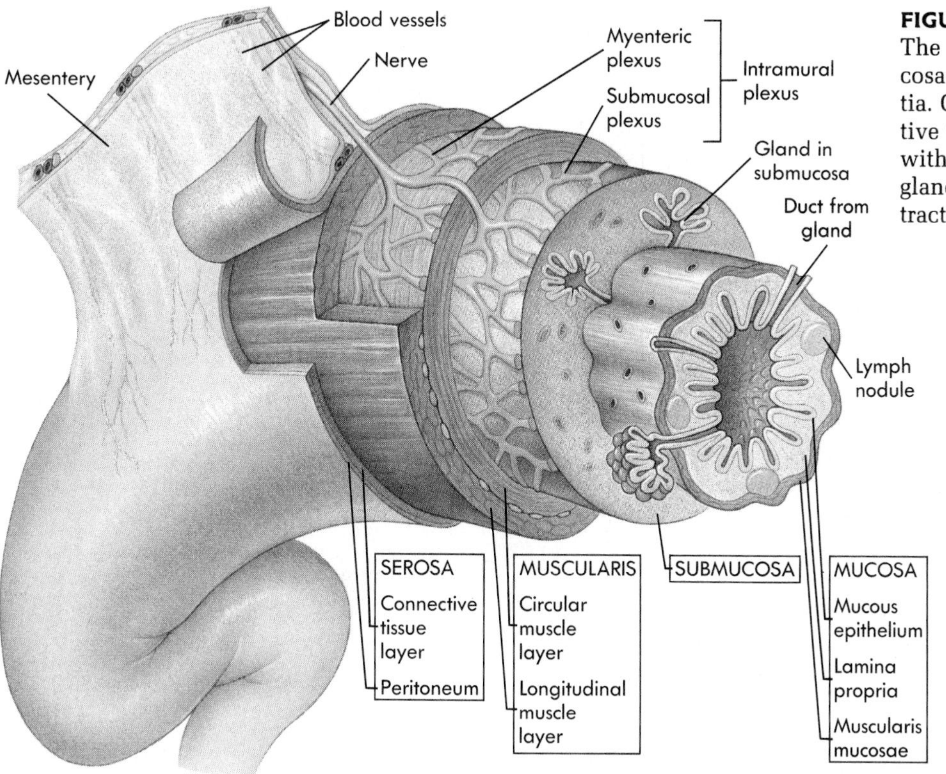

FIGURE 24-2 Digestive tract histology. The four tunics are the mucosa, submucosa, muscularis, and serosa or adventitia. Glands may exist along the digestive tract as part of the epithelium, within the submucosa, or as large glands that are outside the digestive tract.

Table 24-1

Functions of the digestive organs

ORGAN	FUNCTION	SECRETIONS
Oral Cavity		
Teeth	Mastication (cutting and grinding of food); communication	None
Lips and cheeks	Manipulation of food; hold food in position between the teeth; communication	Saliva from buccal glands (mucus only)
Tongue	Manipulation of food; holds food in position between the teeth; cleaning teeth; taste; communication	Some mucus; small amount of serous fluid
Salivary Glands		
Parotid gland	Secretion of saliva through ducts to superior and posterior portions of oral cavity	Serous saliva only, with amylase
Submandibular glands	Secretion of saliva in floor of oral cavity	Serous saliva, with amylase; mucous saliva
Sublingual glands	Secretion of saliva in floor of oral cavity	Mucous saliva only
Pharynx	Deglutition (movement of food from oral cavity to esophagus); breathing	Some mucus
Esophagus	Movement of food by peristalsis from pharynx to stomach	Mucus
Stomach	Mechanical mixing of food; enzymatic digestion; storage; absorption	
Mucous cells	Protection of stomach wall by mucus production	Mucus
Parietal cells	Decrease in stomach pH, vitamin B_{12} absorption	Hydrochloric acid, intrinsic factor
Chief cells	Protein digestion	Pepsin
Endocrine cells	Regulation of secretion and motility	Gastrin
Accessory Glands		
Liver	Secretion of bile into duodenum	Bile
Gallbladder	Bile storage; absorbs water and electrolytes to concentrate bile	No secretions of its own, stores bile
Pancreas	Secretion of several digestive enzymes and bicarbonate ions into duodenum	Trypsin, chymotrypsin, carboxypeptidase, pancreatic amylase, pancreatic lipase, ribonuclease, deoxyribonuclease, cholesterol esterase, bicarbonate ions
Small Intestine		
Duodenal glands	Protection	Mucus
Goblet cells	Protection	Mucus
Absorptive cells	Secretion of digestive enzymes and absorption of digested materials	Enterokinase, amylase, peptidases, sucrase, maltase, isomaltase, lactase, lipase
Endocrine cells	Regulation of secretion and motility	Gastrin, secretin, cholecystokinin, gastric inhibitory peptide
Large Intestine	Absorption, storage, and food movement	
Goblet cells	Protection	Mucus

of a connective-tissue covering that blends with the surrounding connective tissue. These areas include the esophagus and the retroperitoneal organs (discussed in relation to the peritoneum).

PHYSIOLOGY OVERVIEW

The functions of the digestive system are to **ingest** food; **masticate** the food; **propel** the food through the digestive tract, **mixing** it as it moves along; **add secretions** to the food to lubricate, liquefy, and **digest** the food; and **absorb** water, electrolytes, and other nutrients from the digested food (Table 24-1). Once these useful substances are absorbed, they are **transported** through the circulatory system to cells where they are used. Undigested matter is moved out of the digestive tract and is **excreted** through the anus. The processes of propulsion, secretion, and absorption are **regulated** by elaborate nervous and hormonal mechanisms.

Ingestion

Ingestion (a pouring in) is the introduction of solid or liquid food into the stomach. The normal route of ingestion is through the oral cavity, but food can be introduced directly into the stomach by a stomach tube.

Mastication

Mastication (chewing) is the process by which food taken into the mouth is chewed by the teeth. Digestive enzymes cannot easily penetrate solid food particles and can only work effectively on the surfaces of the particles. Therefore it is vital to normal digestive function that solid foods be mechanically broken down into small particles. Mastication breaks large food particles into smaller food particles, which have a much larger surface area than would a few large particles.

Propulsion

Propulsion in the digestive tract is the movement of food from one end of the digestive tract to the other. The total time that it takes food to travel the length of the digestive tract is approximately 24 to 36 hours (Table 24-2). Each segment of the digestive tract is specialized to assist in moving its contents from the oral end to the anal end. **Deglutition,** or swallowing, moves food from the oral cavity into the esophagus. **Peristalsis** (pĕr'ĭ-stal'sis; Figure 24-3) is responsible for moving food through the rest of the digestive tract. Muscular contractions occur in peristaltic waves, a wave of relaxation of the circular muscles and contraction of the longitudinal muscles, followed by a wave of strong contraction of the circular muscles accompanied by relaxation of the longitudinal muscles.

Table 24-2

Time food spends in each part of the digestive tract

REGION	TIME SPENT
Oral cavity	10-20 sec
Pharynx	1-2 sec
Esophagus	5-8 sec
Stomach	
Liquid	1.3-2.5 hr
Solid	3-4 hr
Small intestine (pyloric valve to ileocecal valve; proximal end most rapid movement)	3-5 hr
Large intestine	
Ileocecal valve to transverse colon	8-15 hr
Entire length (ileocecal valve to anus)	18-24 hr

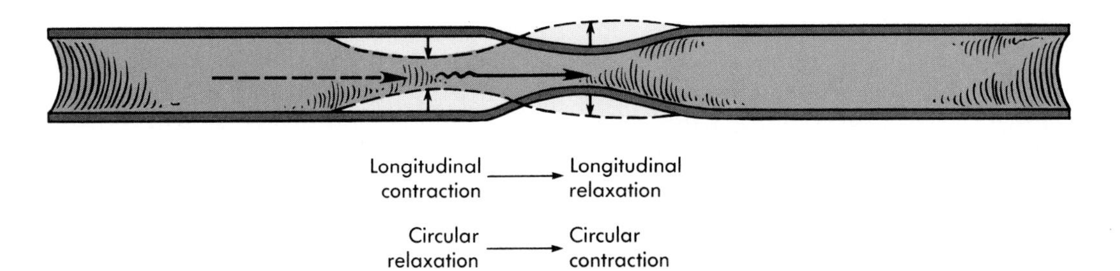

FIGURE 24-3 Peristalsis. A wave of relaxation of the circular muscles and contraction of the longitudinal muscles is followed by a wave of strong contraction of the circular muscles accompanied by relaxation of the longitudinal muscles.

Mixing

Some contractions do not propel food from one end of the digestive tract to the other but rather move the food back and forth within the digestive tract to **mix** the food with digestive secretions and to help break the food into smaller pieces.

Secretion

As food moves through the digestive tract, **secretions** are added to lubricate, liquefy, and digest the food. **Mucus,** secreted along the entire digestive tract, lubricates the food and the lining of the digestive tract. The mucus coats and protects the epithelial cells of the digestive tract from mechanical abrasion, from the damaging effect of acid in the stomach, and from the digestive enzymes of the digestive tract. The secretions also contain large amounts of **water,** which liquefies the food, making it easier to digest and absorb. **Enzymes,** secreted by the oral cavity, stomach, intestine, liver, and pancreas, break food down into small molecules that can be absorbed into the intestinal wall.

Digestion

Digestion is the breakdown of organic molecules into their component parts: carbohydrates into monosaccharides, proteins into amino acids, and triglycerides into fatty acids and glycerol. Digestion consists of **mechanical digestion,** which involves mastication and mixing of food, and **chemical digestion,** which is accomplished by digestive enzymes that are secreted along the digestive tract. Digestion of large molecules into their component parts must be accomplished before they can be absorbed by the digestive tract. Molecules such as vitamins, minerals, and water are not broken down before being absorbed. They are already small enough to be absorbed without digestion and would lose their function if they were digested (this is especially important in the case of vitamins).

Absorption

Absorption is the means by which molecules are moved out of the digestive tract and into the circulation or into the extracellular spaces. The mechanism by which absorption occurs depends on the type of molecule involved. Some molecules pass out of the digestive tract by diffusion, by facilitated diffusion, or by active transport (see Chapter 3).

Transportation

Transportation is the means by which molecules moved out of the digestive tract by absorption are distributed throughout the body. This distribution can occur either directly by way of the circulation or indirectly by first entering the lymphatic system and then passing to the circulatory system.

Excretion

Excretion is the process by which wastes are eliminated from the body. In specific relation to digestion, excretion is the means by which the waste products of digestion are removed from the body. During this process, occurring primarily in the large intestine, water and salts are absorbed, changing the material in the digestive tract from a liquefied state to a semisolid state. These semisolid waste products are then eliminated from the digestive tract by the process of **defecation.**

Regulation

The processes of propulsion, secretion, absorption, and excretion are **regulated** by elaborate **nervous** and **hormonal** mechanisms. Some of the nervous control is local, occurring as the result of **local reflexes** within the intramural plexus, and some is more general, mediated largely by the **vagus nerve.**

ANATOMY AND HISTOLOGY OF THE DIGESTIVE TRACT

Oral Cavity

The **oral cavity,** or mouth, is that portion of the digestive tract that is bounded by the **lips** anteriorly, the **fauces** (faw'sēz; throat; opening into the pharynx) posteriorly, the **cheeks** laterally, the **palate** superiorly, and a muscular floor inferiorly. The oral cavity can be divided into two regions: (1) the **vestibule** (ves'tĭ-būl; entry), which is the space between the lips or cheeks and the alveolar processes, which contain the teeth; and (2) the **oral cavity proper,** which lies medial to the alveolar processes. The oral cavity is lined with moist stratified epithelium, which provides protection against abrasion.

Lips and cheeks

The **lips,** or **labia** (la'be-ah) (Figure 24-4), are muscular folds covered internally by mucosa and externally by stratified squamous epithelium. The epithelial covering of the lips is relatively thin and is not as highly keratinized as the epithelium of the skin (see Chapter 5); consequently, it is more transparent than the epithelium over the rest of the body surface. The color from the underlying blood vessels can be seen through the transparent epithelium, giving the lips a reddish-pink appearance.

The **cheeks** form the lateral walls of the oral cavity. They consist of an interior lining of moist, stratified squamous epithelium and an exterior covering of skin. The substance of the cheek is contributed by the **buccinator muscle** (see Chapter 11), which flattens the cheek against the teeth, and the **buccal fat pad,** which rounds out the profile on the side of the face.

The lips and cheeks are important in the processes of **mastication** (mas'tĭ-ka'shun; chewing food) and **speech.** They help manipulate the food within the mouth and hold the food in place while the teeth crush or tear it. They also help form words during the speech process. A large number of the muscles of facial expression are involved in movement of the lips. They are listed in Chapter 11.

Tongue

The **tongue** is a large, muscular organ that occupies most of the oral cavity proper when the mouth is closed. Its major attachment in the oral cavity is through its posterior portion. The anterior portion of the tongue is relatively free and is attached to the floor of the mouth by a thin fold of tissue called the **frenulum** (fren'u-lum; bridle; see Figure 24-4). The muscles associated with the tongue are divided into two categories: **intrinsic muscles,** which are within the tongue itself; and **extrinsic muscles,** which are outside the tongue but are attached to it. The intrinsic muscles are largely responsible for changing the tongue's shape (e.g., flattening and elevating the tongue during drinking and swallowing). The extrinsic tongue muscles protrude and retract the tongue, move it from side to side, and change its shape (see Chapter 11).

A person is "tongue-tied" if the frenulum extends too far toward the tip of the tongue, inhibiting normal movement of the tongue and interfering with normal speech. Surgically cutting the frenulum can correct the condition.

The tongue is divided into two portions by a groove called the **terminal sulcus.** The portion anterior to the terminal sulcus accounts for approximately two thirds of the surface area and is covered by papillae, some of which contain taste buds (see Chapter 15). The posterior one third of the tongue is devoid of papillae and has only a few scattered taste buds. It has, instead, a few small glands and a large amount of lymphoid tissue, the **lingual tonsil** (see Chapter 22). The tongue is covered by moist, stratified squamous epithelium.

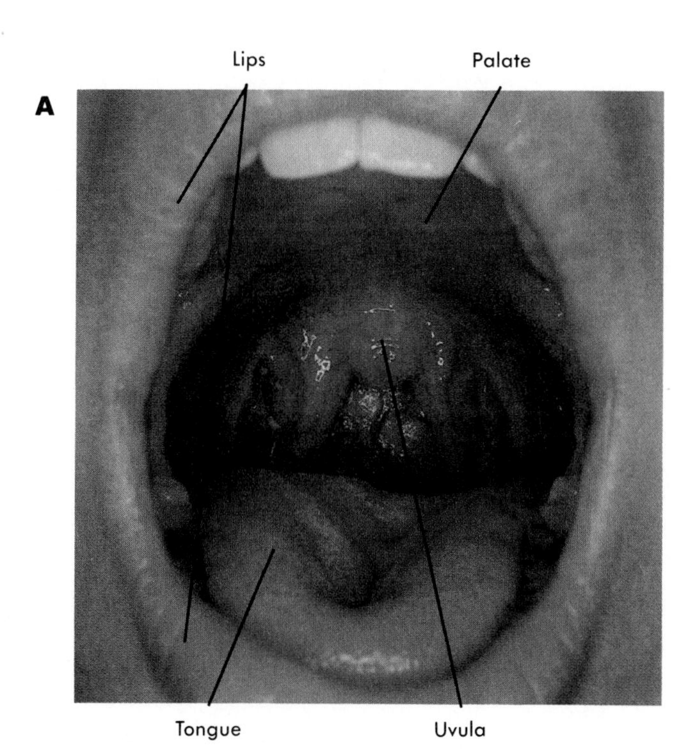

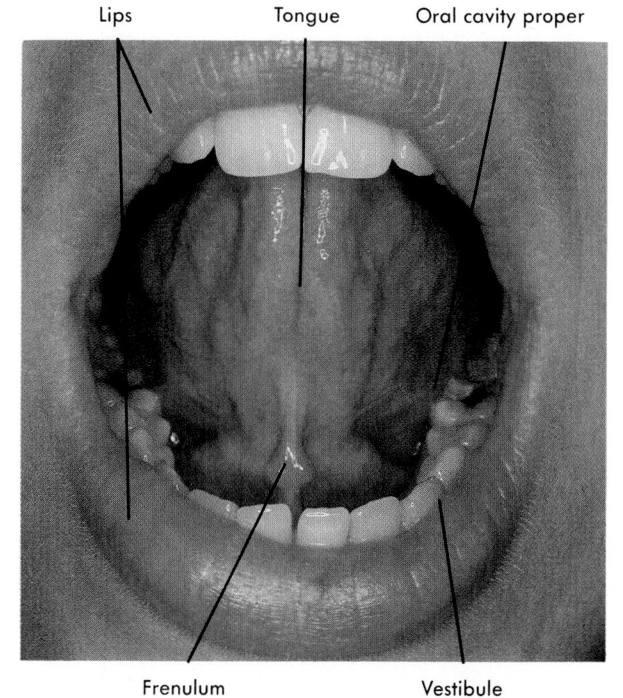

FIGURE 24-4 Oral cavity **A** with the tongue depressed and **B** with the tongue elevated.

The tongue moves food in the mouth and, in cooperation with the lips and gums, holds the food in place during mastication. It also plays a major role in the mechanism of swallowing (discussed later in this chapter). The tongue is a major sensory organ for taste (see Chapter 15), and it is one of the major organs of speech.

Patients who have undergone glossectomies (tongue removal) as a result of glossal carcinoma can compensate for loss of the tongue's function in speech, and they can learn to speak fairly well. However, these patients have a major problem, which they cannot entirely overcome, with chewing and swallowing food.

Teeth

There are 32 **teeth** in the normal adult mouth, which are distributed in two **dental arches,** one maxillary and one mandibular. The teeth in the right and left halves of each dental arch are roughly mirror images of each other. As a result, the teeth can be divided into four quadrants: right upper, left upper, right lower, and left lower. The teeth in each quadrant include one central and one lateral **incisor;** one **canine;** first and second **premolars;** and first, second, and third **molars** (Figure 24-5, *A*). The third molars are referred to as **wisdom teeth** because they usually appear when the person is in his late teens or early twenties and has supposedly acquired a little wisdom.

The third molars frequently do not have room to erupt into the oral cavity and remain embedded within the jaw. These embedded teeth are **impacted,** and their surgical removal is often necessary.

The teeth of the adult mouth are **permanent,** or **secondary, teeth.** Most of them are replacements for **primary,** or **deciduous, teeth** (dĕ-sid′u-us; those that fall out; also called milk teeth) that are lost during childhood (Figure 24-5, *B*).

Each tooth consists of a **crown** with one or more cusps (points), a **neck,** and a **root** (Figure 24-6). The center of the tooth is a **pulp cavity** that is filled with blood vessels, nerves, and connective tissue called **pulp.** The pulp cavity within the root is called the **root canal.** The nerves and blood vessels of the tooth enter and exit the pulp through a hole at the point of each root called the **apical foramen.** The pulp cavity is surrounded by a living, cellular, and calcified tissue called **dentin.** The dentin of the tooth crown is covered by an extremely hard, nonliving, acellular substance called **enamel,** which protects the tooth against abrasion and acids produced by

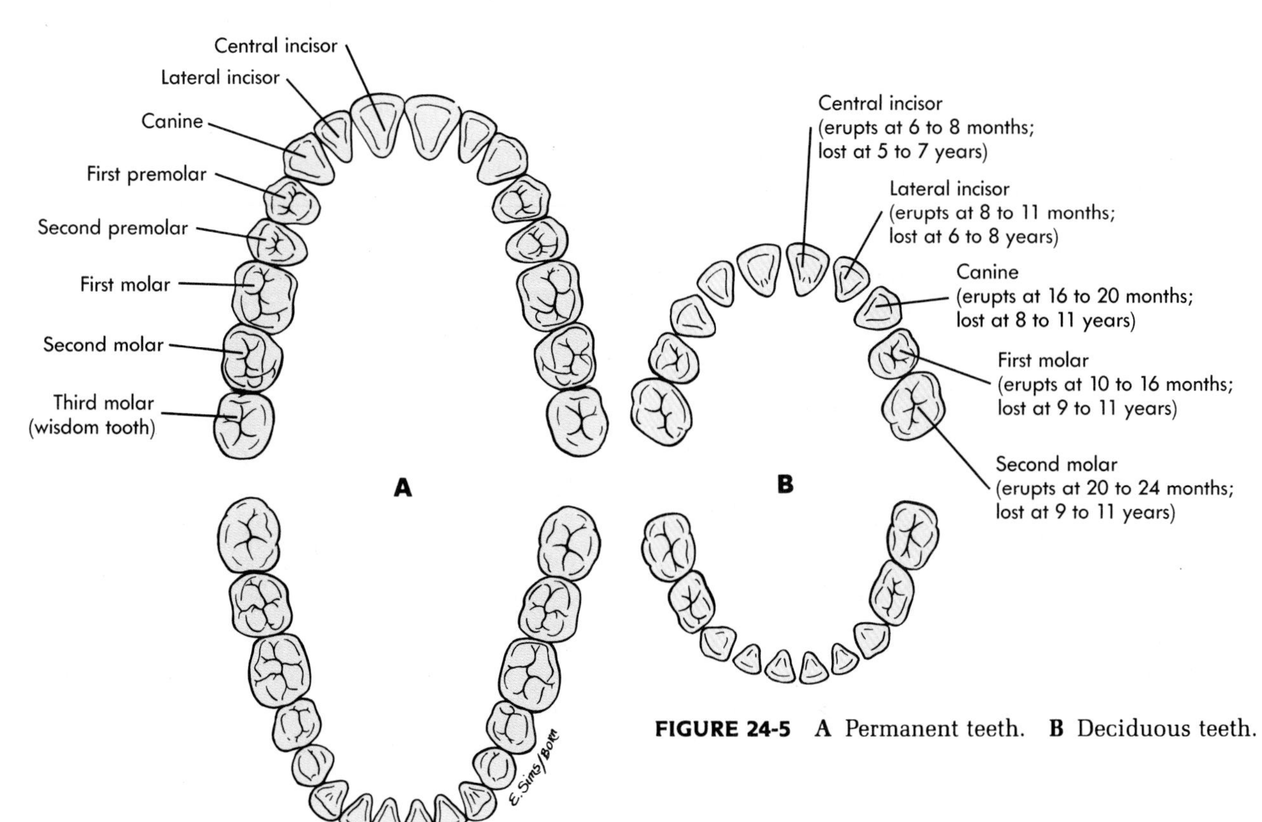

FIGURE 24-5 **A** Permanent teeth. **B** Deciduous teeth.

bacteria in the mouth. The surface of the dentin in the root is covered with **cementum,** which helps anchor the tooth in the jaw.

The teeth are set in **alveoli** (al-ve'o-li; sockets) along the alveolar ridges of the mandible and maxilla. The alveolar ridges are covered by dense, fibrous connective tissue and stratified squamous epithelium, referred to as the **gingiva** (jin'jĭ-vah; gums). The teeth are held in the alveoli by **periodontal** (per'e-o-don'tal; around the teeth) **ligaments,** and the alveolar walls are lined with a **periodontal membrane.**

The teeth play an important role in mastication and have a role in speech.

Dental caries or tooth decay is due to a breakdown of enamel by acids that are produced by bacteria on the tooth surface. Since the enamel is nonliving and cannot repair itself, a dental filling is necessary to prevent further damage.

Periodontal disease is the inflammation and degradation of the periodontal structures, gingiva, and alveolar bone. This disease is the most common cause of tooth loss in adults.

Muscles of mastication

Four pairs of muscles move the mandible during mastication: the **temporalis, masseter, medial pterygoid,** and **lateral pterygoid** (see Chapter 11 and Figure 11-8). The temporalis, masseter, and medial pterygoid muscles close the jaw, and the lateral pterygoid muscle opens it. Protrusion of the jaw and right and left excursion of the jaw are accomplished by the medial and lateral pterygoids and the masseter. Retraction of the jaw is accomplished by the temporalis. All these movements are involved in tearing, crushing, and grinding food.

Palate and palatine tonsils

The **palate** (see Figure 24-4, *A*) consists of two portions, an anterior bony portion, the **hard palate** (see Chapter 7), and a posterior, nonbony portion, the **soft palate,** which consists of skeletal muscle and connective tissue. The **uvula** (u'vu-lah; a grape) is the projection from the posterior edge of the soft palate. The palate is important in the swallowing process, preventing food from passing into the nasal cavity.

Palatine tonsils are located in the lateral wall of the fauces (see Chapter 22).

Salivary glands

A considerable number of **salivary glands** are scattered throughout the oral cavity. There are three pairs of large multicellular glands, the parotid (pă-rot'id; beside the ear), the submandibular (below the mandible), and the sublingual (below the tongue; Figure 24-7). In addition to these large consolidations of glandular tissue, numerous small, coiled tubular glands are in the tongue (lingual glands), palate (palatine glands), cheeks (buccal glands), and lips (labial glands). The secretions from all these glands help

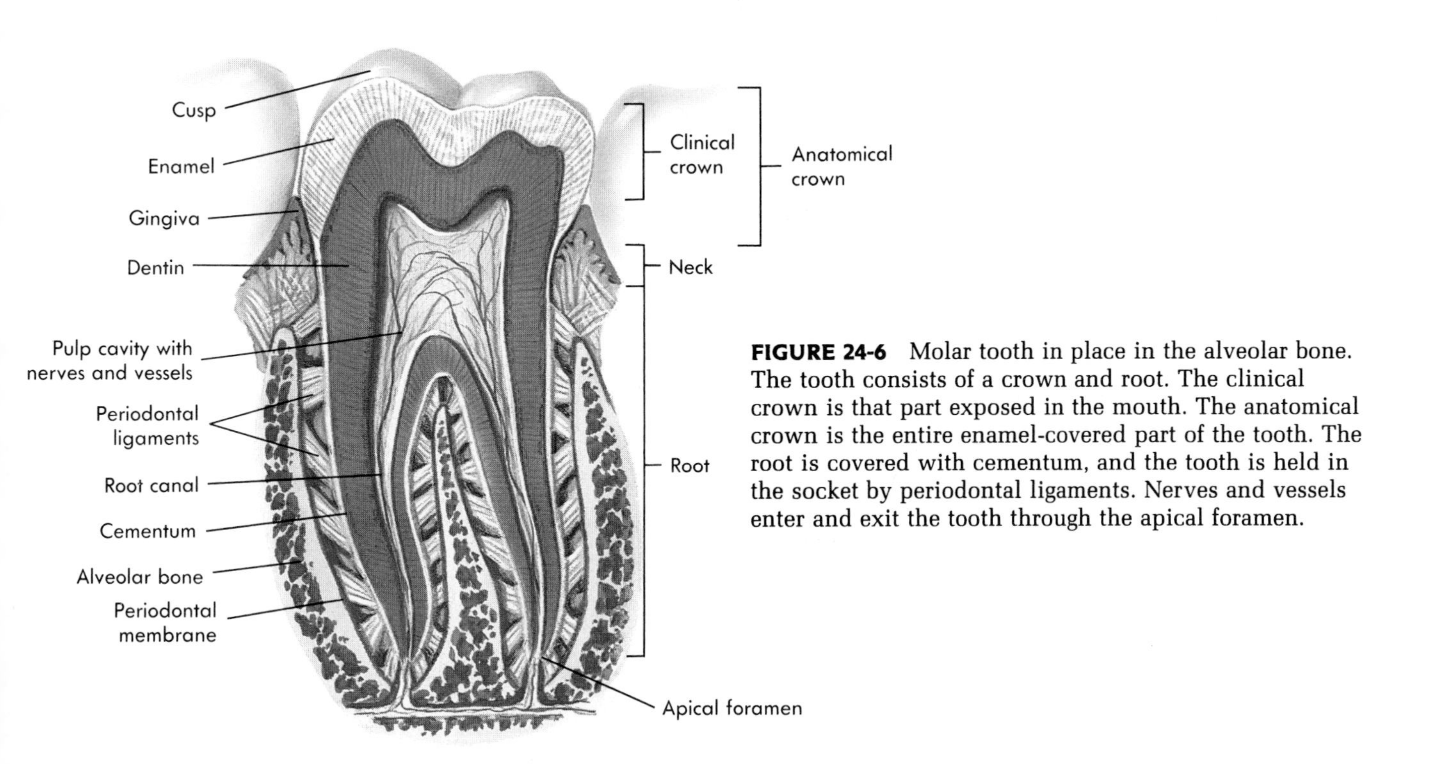

FIGURE 24-6 Molar tooth in place in the alveolar bone. The tooth consists of a crown and root. The clinical crown is that part exposed in the mouth. The anatomical crown is the entire enamel-covered part of the tooth. The root is covered with cementum, and the tooth is held in the socket by periodontal ligaments. Nerves and vessels enter and exit the tooth through the apical foramen.

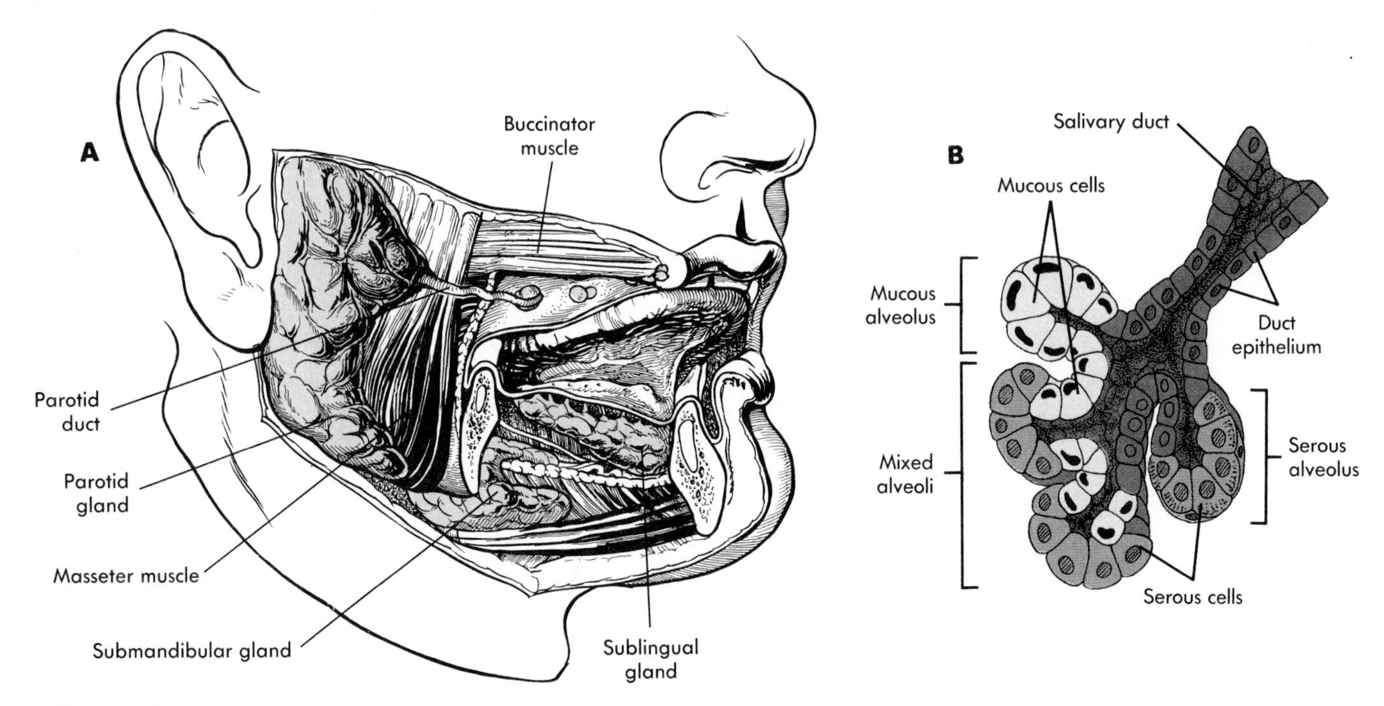

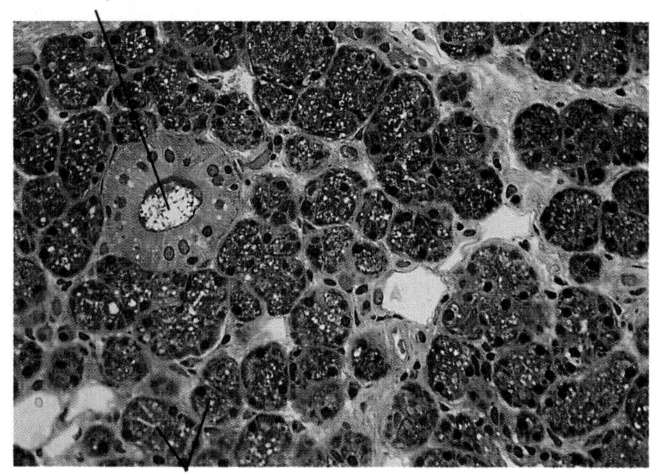

FIGURE 24-7 **A** Salivary glands. The large salivary glands are the parotid glands, the submandibular glands, and the sublingual glands. The minor salivary glands include the buccal and labial glands. The parotid duct extends anteriorly from the parotid gland. Inset shows a photomicrograph of the parotid gland. **B** An idealized schematic drawing of the histology of the larger glands. The figure is representative of all the glands and does not depict any one gland.

keep the oral cavity moist and begin the process of digestion.

All of the major large salivary glands are compound **alveolar glands** (branching glands with clusters of alveoli that resemble grapes; see Chapter 4). They produce thin serous secretions or thicker mucous secretions; thus saliva is a combination of serous fluids and mucus.

The largest salivary glands, the **parotid glands,** are serous glands (produce mostly watery saliva) and are located just anterior to the ear on each side of the head. The **parotid duct** exits the gland on its anterior margin, crosses the lateral surface of the masseter muscle, pierces the buccinator muscle, and enters the oral cavity adjacent to the second upper molar (see Figure 24-7).

Because the parotid secretions are released directly onto the surface of the second upper molar, it tends to have a considerable accumulation of mineral (secreted from the gland) on its surface.

Inflammation of the parotid gland is called parotiditis. The most common type of **parotiditis,** caused by a viral infection, is **mumps.**

The **submandibular glands** are mixed glands with more serous than mucous alveoli. Each gland can be felt as a soft lump along the inferior border of the posterior half of the mandible. The submandibular duct exits the gland, passes anteriorly deep to the mucous membrane on the floor of the oral cavity, and opens into the oral cavity on either side of the frenulum of the tongue (Figure 24-4, *B*). In certain people, if the mouth is opened and the tip of the tongue is elevated, saliva may squirt out of the mouth from the openings of these ducts.

The **sublingual glands,** the smallest of the three paired salivary glands, are mixed glands consisting primarily of mucous alveoli. They lie immediately below the mucous membrane in the floor of the mouth. These glands do not have single, well-defined ducts like those of the submandibular and parotid glands. Instead, each sublingual gland opens into the floor of the oral cavity through 10 to 12 small ducts.

Pharynx

The **pharynx** was described in detail in Chapter 23; thus only a brief description is provided in this chapter. The pharynx consists of three parts: the nasopharynx, the oropharynx, and the laryngopharynx. Normally, only the oropharynx and laryngopharynx transmit food. The **oropharynx** communicates with the nasopharynx superiorly, the larynx and laryngopharynx inferiorly, and the mouth anteriorly. The **laryngopharynx** extends from the oropharynx to the esophagus and is posterior to the larynx. The posterior walls of the oropharynx and laryngopharynx consist of three muscles, the superior, middle, and inferior pharyngeal constrictors, which are arranged like three stacked flower pots, one inside the other. The oropharynx and the laryngopharynx are lined with moist, stratified squamous epithelium, and the nasopharynx is lined with ciliated pseudostratified epithelium.

1

Explain the functional significance of the differences in epithelial types between the three pharyngeal regions.

? ? ? ? ? ? ? ? ? ? ?

Esophagus

The **esophagus** is that portion of the digestive tube that extends between the pharynx and the stomach. It is approximately 25 cm long and lies in the mediastinum, anterior to the vertebrae and posterior to the trachea. It passes through the esophageal hiatus (opening) of the diaphragm and ends at the stomach. The esophagus transports food from the pharynx to the stomach.

The esophagus has thick walls consisting of the four tunics common to the digestive tract: mucosa, submucosa, muscularis, and adventitia. The muscular tunic has an outer longitudinal layer and an inner circular layer, as is true of most parts of the digestive tract, but it differs from other parts of the digestive tract in that it consists of skeletal muscle in the superior portion of the esophagus and smooth muscle in the inferior portion. An **upper esophageal sphincter** and a **lower esophageal sphincter** regulate the movement of materials into and out of the esophagus. The esophagus' mucosal lining is moist, stratified squamous epithelium. Numerous mucous glands in the submucosal layer produce a thick, lubricating mucus that passes through ducts to the surface of the esophageal mucosa.

Stomach

The **stomach** is an enlarged segment of the digestive tract in the left superior portion of the abdomen (see Figure 24-1). Its shape and size vary from person to person; even within the same individual its size and shape change from time to time, depending on its food content and the posture of the body. Nonetheless, several general anatomical features can be described.

Stomach anatomy

The opening from the esophagus into the stomach is the **gastroesophageal,** or **cardiac** (located near the heart), **opening,** and the region of the stomach around the cardiac opening is the **cardiac region.** A portion of the stomach to the left of the cardiac region, the **fundus** (fun'dus; the bottom of a round-bottomed leather bottle), is actually superior to the cardiac opening. The largest portion of the stomach is the **body,** which turns to the right, thus creating a **greater curvature** and a **lesser curvature.** The body narrows to form the **pyloric** (pi-lōr'ik; gatekeeper) **region,** which joins the small intestine. The opening between the stomach and the small intestine is the **pyloric opening,** which is surrounded by a relatively thick ring of smooth muscle called the **pyloric sphincter.**

Stomach histology

The **serosa,** or visceral peritoneum, is the outermost layer of the stomach. It consists of an inner layer of connective tissue and an outer layer of simple squamous epithelium. The **muscularis** of the stomach consists of three layers: an outer longitudinal layer, a middle circular layer, and an inner oblique layer (Figure 24-8, *A*). Inside the muscular layer are the submucosa and the mucosa, which are thrown into large folds called **rugae** (ru'ge; wrinkles) when the stomach is empty (Figure 24-8, *A*). These folds allow the mucosa and submucosa to stretch, and the

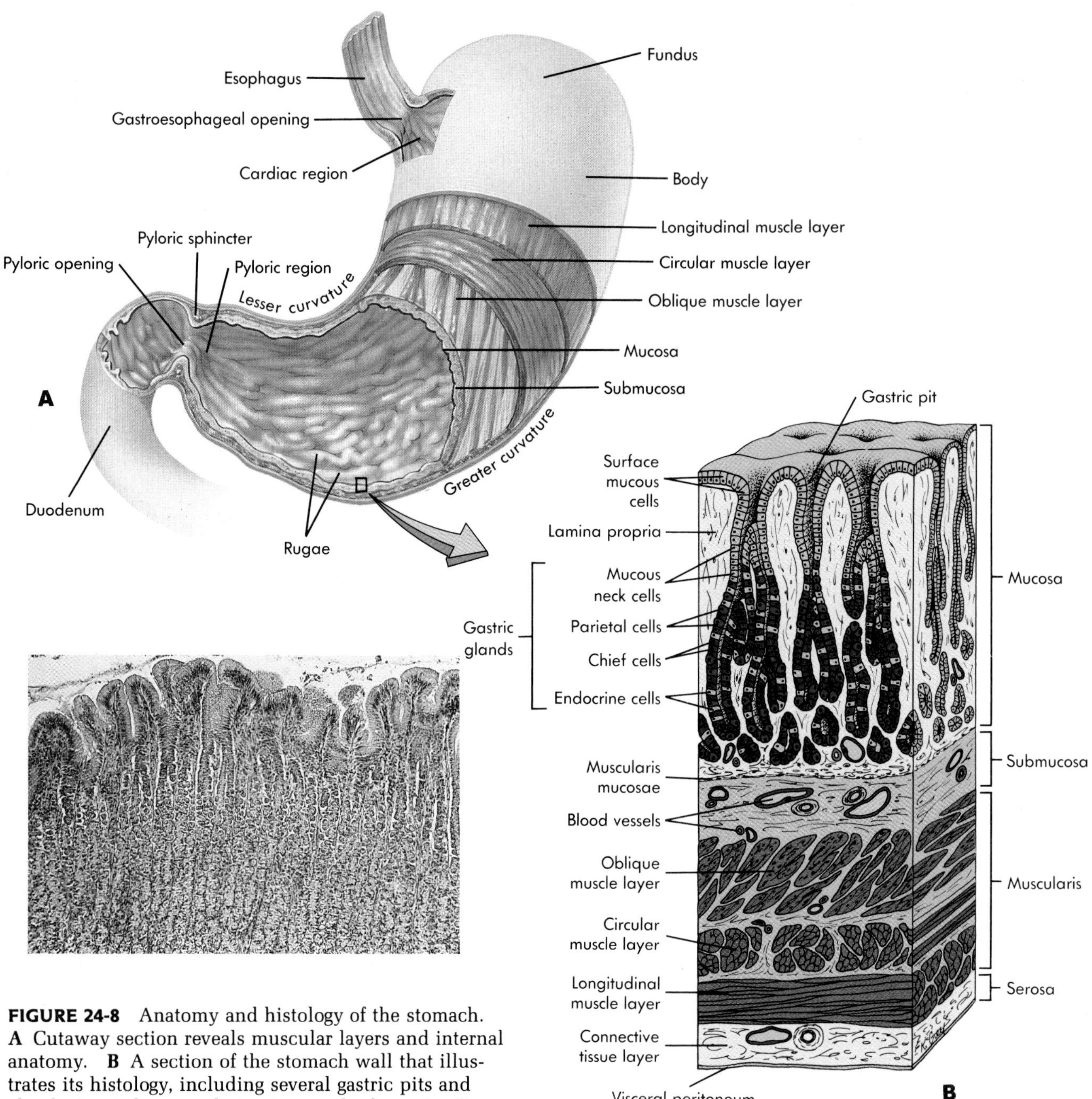

FIGURE 24-8 Anatomy and histology of the stomach. **A** Cutaway section reveals muscular layers and internal anatomy. **B** A section of the stomach wall that illustrates its histology, including several gastric pits and glands. Inset shows a photomicrograph of gastric pits.

folds disappear as the stomach is filled.

The stomach is lined with simple columnar epithelium. The mucosal surface forms numerous tubelike **gastric pits,** which are the openings for the **gastric glands** (Figure 24-8, *B*). The epithelial cells of the stomach can be divided into five groups. The first group, **surface mucous cells,** which produce mucus, is on the surface and lines the gastric pit. The remaining four cell types are in the gastric glands. They are **mucous neck cells,** which produce mucus; **parietal** (oxyntic) **cells,** which produce hydrochloric acid and intrinsic factor; **chief** (zymogenic) **cells,** which produce pepsinogen; and **endocrine cells,** which produce regulatory hormones.

Small intestine

The **small intestine** consists of three portions: the duodenum, the jejunum, and the ileum (Figure 24-9). The entire small intestine is approximately 6 m long (ranging from 4.6 m to 9 m); the duodenum is approximately 25 cm long (the term duodenum means 12, suggesting that it is 12 inches long); the jejunum, constituting approximately two fifths of the total length of the small intestine, is approximately 2.5 m long; and the ileum, constituting three fifths of the small intestine, is approximately 3.5 m long. Two major accessory glands, the liver and the pancreas, are associated with the duodenum.

FIGURE 24-9 Small intestine.

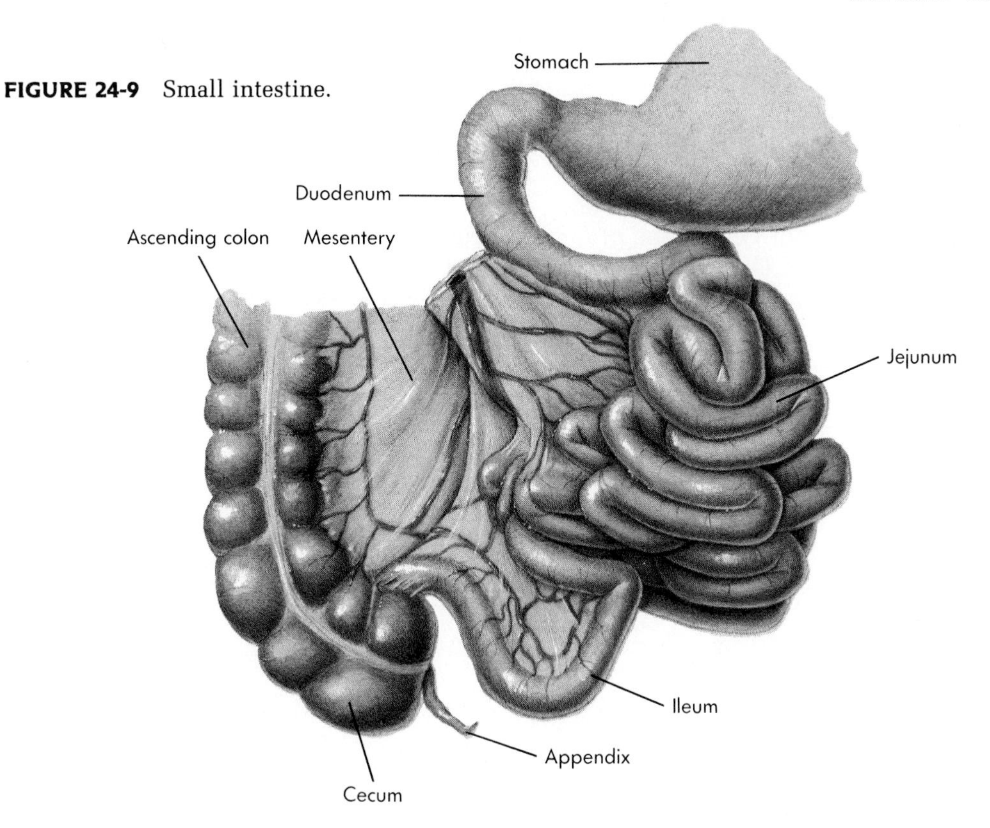

Duodenum

The **duodenum** nearly completes a 180-degree arc as it curves within the abdominal cavity (Figure 24-10), and the head of the pancreas lies within this arc. The duodenum begins with a short superior portion where it exits the pylorus of the stomach, and the duodenum ends in a sharp bend called the **duodenojejunal flexure** where the duodenum joins the jejunum.

Two small mounds are within the duodenum approximately two thirds of the way down the descending portion: the **greater duodenal papilla** and the **lesser duodenal papilla.** At the greater papilla the **common bile duct** and **pancreatic duct** join to form the **hepatopancreatic ampulla** (Vater's ampulla), which empties into the duodenum. The opening of the ampulla is usually kept closed by a smooth muscle sphincter, the **hepatopancreatic ampullar sphincter** (sphincter of Oddi). An accessory pancreatic duct, present in most people, opens at the apex of the lesser duodenal papilla.

The surface of the duodenum has several modifications that increase surface area approximately six-hundredfold, allowing more efficient digestion and absorption of food. The mucosa and submucosa form a series of folds called the **plicae** (pli'se; folds) **circulares,** or **circular folds** (Figure 24-10, *C*), which run perpendicular to the long axis of the digestive tract. Tiny fingerlike projections of the mucosa form numerous **villi** (vil'e; shaggy hair), which are 0.5 to 1.5 mm in length (Figure 24-10, *D*). Each villus is covered by simple columnar epithelium and contains a blood capillary network and a lymph capillary called a **lacteal** (lak'tēl) (Figure 24-10, *E*). Most of the cells comprising the surface of the villi have numerous cytoplasmic extensions (approximately 1 μm long) called microvilli, which further increase the surface area (Figure 24-10, *F*). The combined microvilli on the entire epithelial surface form the **brush border.** These various modifications greatly increase the number of capillaries near the intestinal surface and, as a result, greatly enhance absorption.

The mucosa of the duodenum is simple columnar epithelium with four major cell types: (1) **absorptive cells** (the cells with microvilli), which produce digestive enzymes and absorb digested food; (2) **goblet cells,** which produce a protective mucus; (3) **granular cells** (Paneth cells), which may help protect the intestinal epithelium from bacteria; and (4) **endocrine cells,** which produce regulatory hormones. The epithelial cells are produced within tubular invaginations of the mucosa, called **intestinal glands** (crypts of Lieberkühn), at the base of the villi. The absorptive and goblet cells migrate from the intestinal glands to cover the surface of the villi and eventually are shed from the tip of the villi. The granular and endocrine cells remain in the bottom of the glands. The submucosa of the duodenum contains mucous glands, called **duodenal glands** (Brunner's glands), which open into the base of the intestinal glands.

Jejunum and ileum

The **jejunum** and **ileum** are similar in structure to the duodenum except that there is a gradual decrease in the diameter of the small intestine, in the thickness of the intestinal wall, in the number of circular folds, and in the number of villi as one progresses through the small intestine. The jejunum and duodenum are the major sites of nutrient absorption, although some absorption occurs in the ileum. Lymph nodules called **Peyer's patches** are numerous in the ileum.

The junction between the ileum and the large intestine is the **ileocecal junction.** It has a ring of smooth muscle, the **ileocecal sphincter,** and a one-way **ileocecal valve** (see Figure 24-13).

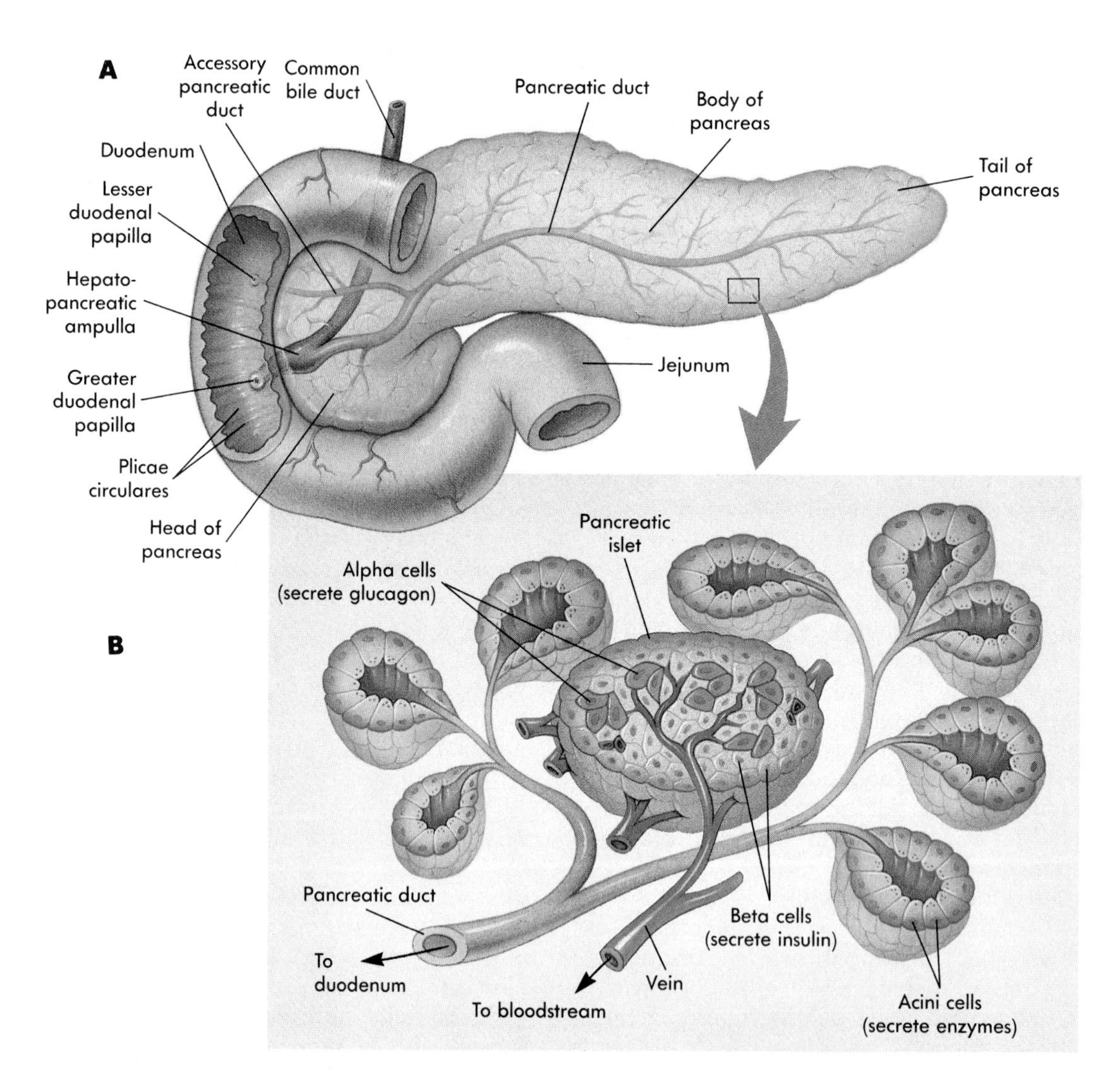

FIGURE 24-10 Anatomy and histology of the duodenum and pancreas. **A** The head of the pancreas lies within the duodenal curvature, with the pancreatic duct emptying into the duodenum. **B** Histology of the pancreas showing both the acini and the pancreatic duct.

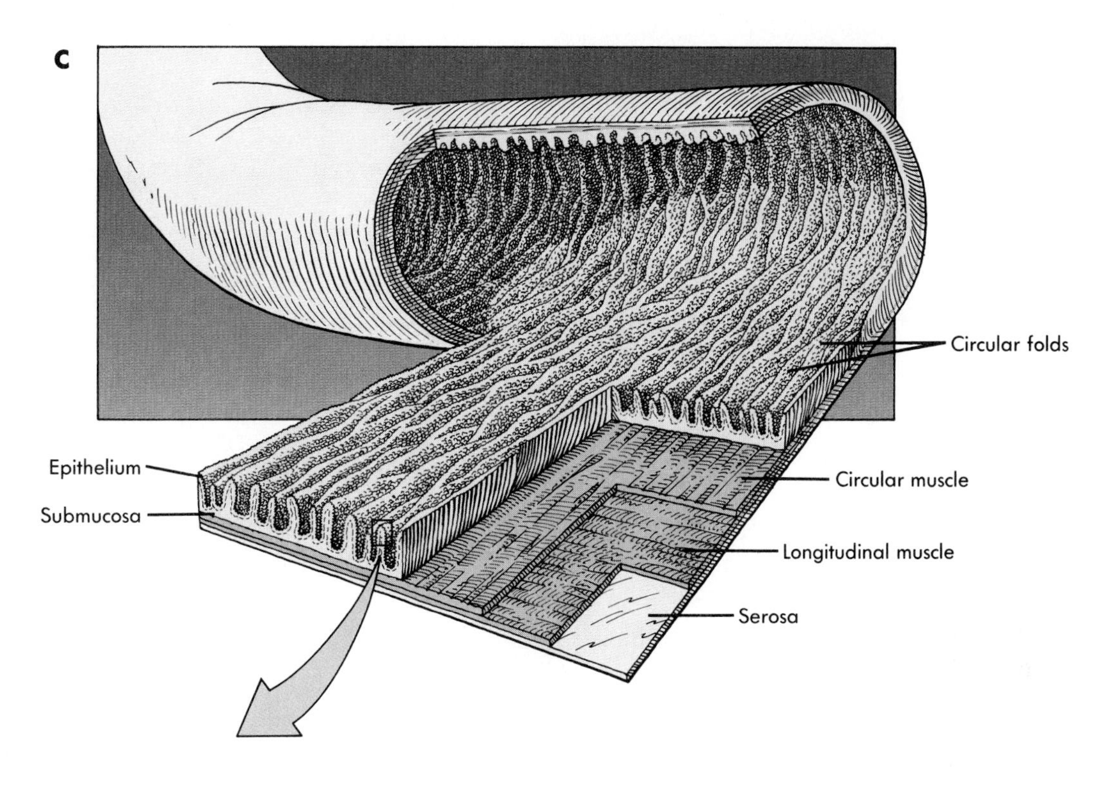

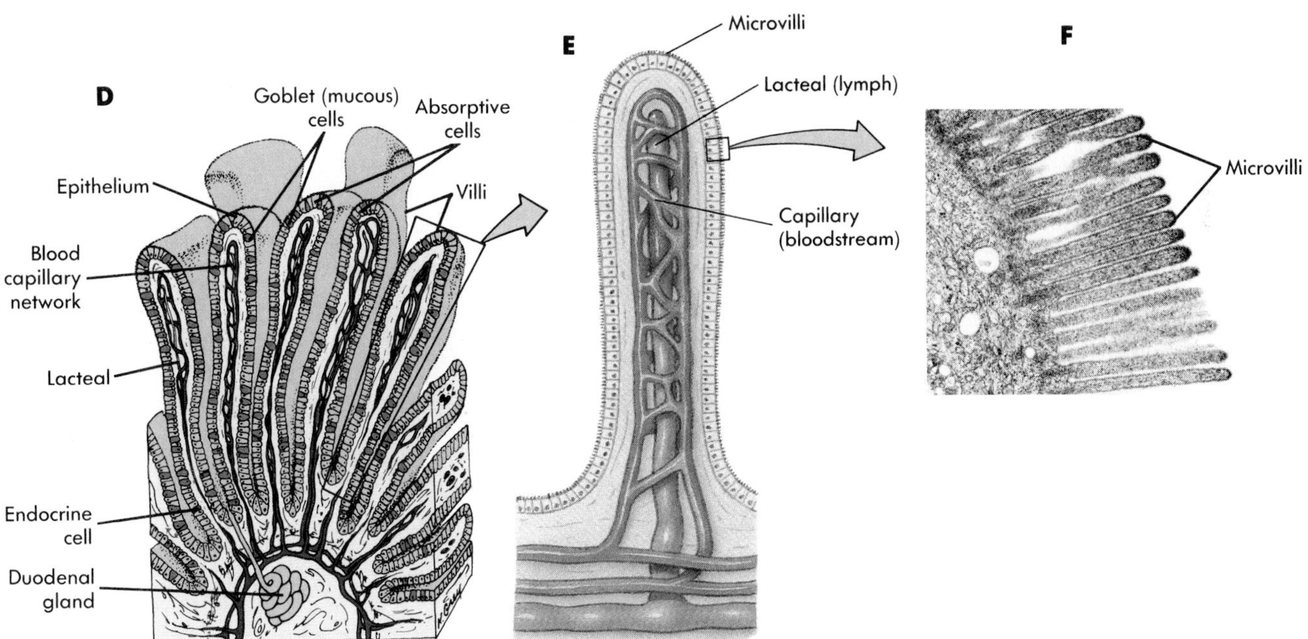

FIGURE 24-10, cont'd Anatomy and histology of the duodenum and pancreas. **C** Wall of the duodenum showing the circular folds. **D** the villi, and **E** a single villus showing the lacteal and capillary network. **F** Electron micrograph of the microvilli.

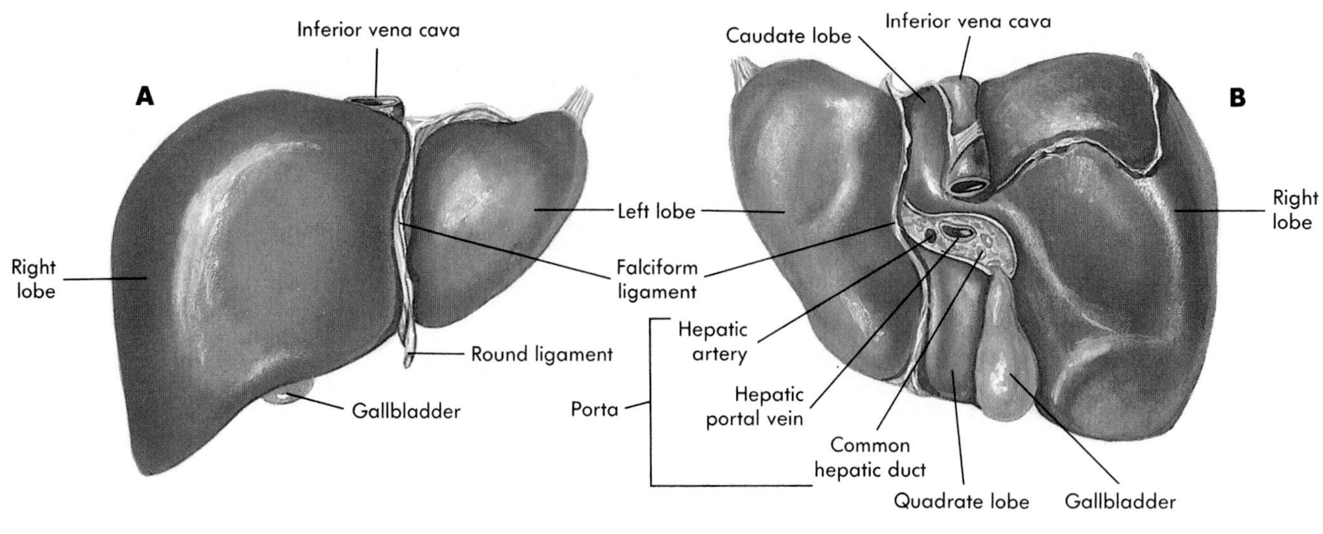

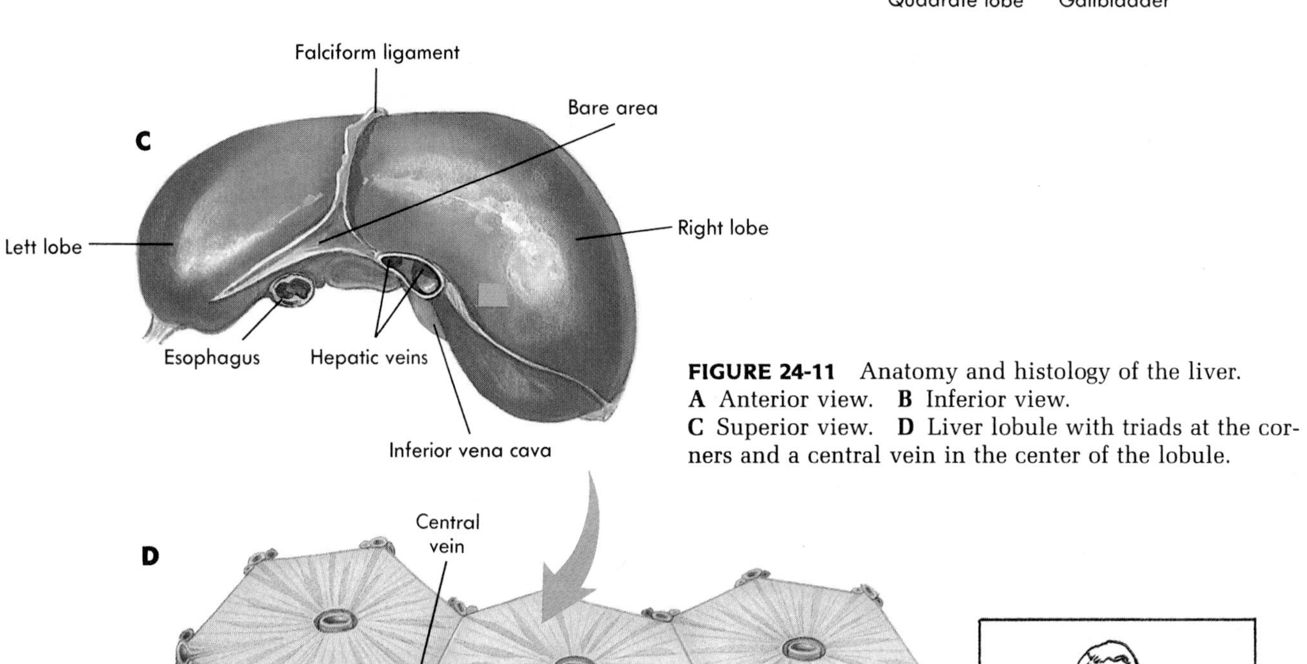

FIGURE 24-11 Anatomy and histology of the liver. **A** Anterior view. **B** Inferior view. **C** Superior view. **D** Liver lobule with triads at the corners and a central vein in the center of the lobule.

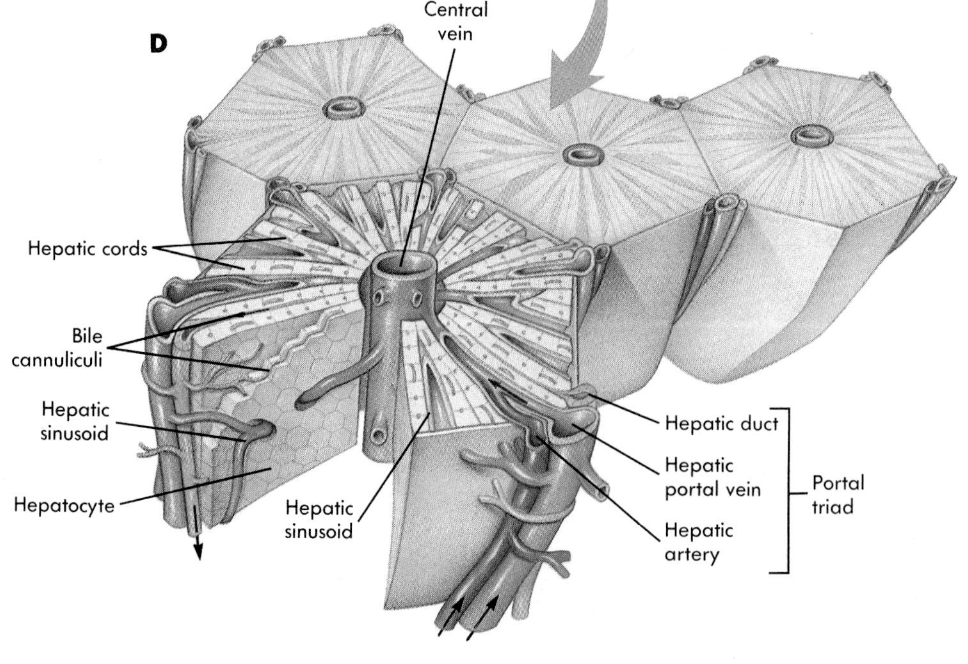

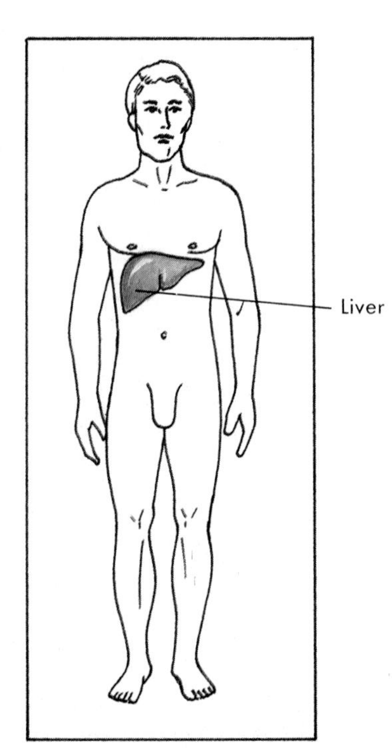

Liver

Liver anatomy

The **liver** is the largest internal organ of the body, weighing approximately 1.36 kg (3 pounds), and it is in the upper right quadrant of the abdomen, tucked against the inferior surface of the diaphragm (see Figure 24-1; Figure 24-11). The liver consists of two major **lobes, left** and **right,** and two minor lobes, **caudate** and **quadrate.**

A **porta** (gate) is on the inferior surface of the liver where the various vessels, ducts, and nerves enter and exit the liver. The **hepatic** (hĕ-pat'ik; associated with the liver) **portal vein,** the **hepatic artery,** and a small hepatic nerve plexus enter the liver through the porta. Lymphatic vessels and two hepatic ducts, one each from the right and left lobes, exit the liver at the porta. The hepatic ducts transport bile out of the liver. The right and left hepatic ducts unite to form a single **common hepatic duct.** The common hepatic duct is joined by the **cystic duct** from the gallbladder to form the **common bile duct,** which empties into the duodenum in union with the pancreatic duct (Figure 24-12). The gallbladder is a small sac on the inferior surface of the liver that stores bile.

Liver histology

The liver is covered by a connective tissue capsule and visceral peritoneum except for in a small area, the **bare area,** on the diaphragmatic surface (see Figure 24-11, C). At the porta the connective tissue capsule sends a branching network of septa (walls) into the substance of the liver to provide its main support. Vessels, nerves, and ducts follow the connective tissue branches throughout the liver.

The connective tissue septa divide the liver into hexagon-shaped **lobules** with a **portal triad** at each corner. The triads are so named because three vessels—the hepatic portal vein, hepatic artery, and hepatic duct—are commonly located in them (Figure 24-11, *D*). Hepatic nerves and lymph vessels, often too small to see easily in light micrographs, are also located in these areas. A **central vein** is in the center of each lobule. Central veins unite to form **hepatic veins,** which exit the liver on its posterior and superior surfaces and empty into the inferior vena cava.

Hepatic cords radiate out from the central vein of each lobule like the spokes of a wheel. The hepatic cords are composed of **hepatocytes,** the functional cells of the liver. The spaces between the hepatic cords are blood channels called **hepatic sinusoids.** The sinusoids are lined with a very thin, irregular squamous endothelium consisting of two cell populations: (1) extremely thin, sparse **endothelial** cells and (2) **hepatic phagocytic cells** (Kupffer cells). A cleftlike lumen, the **bile canaliculus** (kan'ă-lik'u-lus; little canal), lies between the cells within each cord (see Figure 24-11, *D*).

Hepatocytes have four major functions (described in more detail later in this chapter): (1) synthesis of bile, (2) storage, (3) biotransformation, and (4) synthesis of blood components. Nutrient-rich, oxygen-poor blood from the viscera enters the hepatic sinusoids from branches of the hepatic portal vein and mixes with oxygen-rich blood from the hepatic arteries. From the blood the hepatocytes can take up the oxygen and nutrients, which can be stored, detoxified, used for energy, or used to synthesize new molecules. Molecules produced by or modified in the hepatocytes are released into the hepatic sinusoids or into the bile canaliculi.

Mixed blood in the hepatic sinusoids flows to the central vein where it exits the lobule and then exits the liver through the hepatic veins. **Bile,** produced by the hepatocytes and consisting primarily

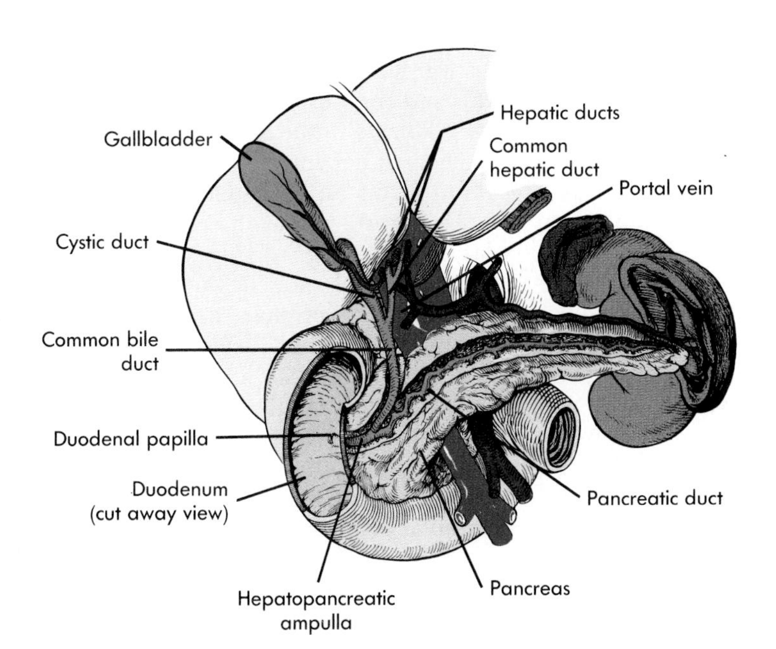

FIGURE 24-12 Duct system of the major abdominal digestive glands. The hepatic ducts from the liver lobes combine to form the common hepatic duct, which combines with the cystic duct to form the common bile duct. The common bile duct and pancreatic duct combine to form the short hepatopancreatic ampulla, which empties into the duodenum at the duodenal papilla.

of metabolic by-products, flows through the bile canaliculi toward the hepatic triad and exits the liver through the hepatic ducts. Therefore blood flows from the triad toward the center of each lobule, whereas bile flows away from the center of the lobule toward the triad.

In the fetus the flow of blood for the liver is different. The remnants of fetal blood vessels can be seen as the round ligament (ligamentum teres) and the ligamentum venosum (see Chapter 29).

Hepatitis is an inflammation of the liver caused by alcohol consumption or viral infection. If not corrected, liver cells can die and be replaced by scar tissue, resulting in loss of liver function. Death caused by liver failure can occur.

Gallbladder

The **gallbladder** is a saclike structure on the inferior surface of the liver that is approximately 8 cm long and 4 cm wide (see Figure 24-12). Three tunics form the gallbladder wall: (1) an inner mucosa folded into rugae that allow the gallbladder to expand; (2) a muscularis of smooth muscle that allows the gallbladder to contract; and (3) an outer covering of connective tissue. The gallbladder is connected to the common bile duct by the cystic duct.

Pancreas

The **pancreas** is a complex organ composed of both endocrine and exocrine tissues that perform several functions. The pancreas consists of a **head,** located within the curvature of the duodenum (see Figure 24-10, *A*), a **body,** and a **tail,** which extends to the spleen.

The endocrine portion of the pancreas consists of **pancreatic islets** (islets of Langerhans). The islet cells produce insulin and glucagon, which are very important in controlling blood levels of nutrients such as glucose and amino acids, and somatostatin, which regulates insulin secretion (see Chapter 18).

The exocrine portion of the pancreas consists of **acini** (as'ĭ-ne; grapes), which produce digestive enzymes. The acini form lobules that are separated by thin septa, and clusters of acini are connected by small **intercalated ducts** to **intralobular ducts,** which leave the lobules to join **interlobular ducts** between the lobules. The interlobular ducts attach to the main pancreatic duct, which joins the common bile duct at the hepatopancreatic ampulla (see Figures 24-10 and 24-12). The ducts are lined with simple cuboidal epithelium, and the epithelial cells of the acini are pyramid shaped.

Large Intestine

Cecum

The **cecum** (se'kum; blind) is the proximal end of the large intestine and is where the large and small intestines meet. The cecum extends inferiorly approximately 6 cm past the ileocecal junction in the form of a blind sac (Figure 24-13). Attached to the cecum is a small blind tube approximately 9 cm long called the **vermiform** (ver'mĭ-form; worm shaped) **appendix.** The walls of the appendix contain many lymph nodules.

Appendicitis is an inflammation of the vermiform appendix and usually occurs because of obstruction of the appendix. Secretions from the appendix cannot pass the obstruction and accumulate, causing enlargement and pain. Bacteria in the area cause infection of the appendix. If the appendix bursts, the infection can spread throughout the peritoneal cavity with life-threatening results. In the right inferior quadrant of the abdomen approximately midway along a line between the umbilicus and the right anterior superior iliac spine is an area on the body's surface called McBurney's point. This area becomes very tender in patients with acute appendicitis because of pain referred from the inflamed appendix to the body's surface.

Colon

The **colon** (ko'lon) is approximately 1.5 to 1.8 m long and consists of four portions: the ascending colon, transverse colon, descending colon, and sigmoid colon (see Figure 24-13). The **ascending colon** extends superiorly from the cecum and ends at the right colic flexure (hepatic flexure) near the right inferior margin of the liver. The **transverse colon** extends from the right colic flexure to the left colic flexure (splenic flexure), and the **descending colon** extends from the left colic flexure to the superior opening of the true pelvis where it becomes the **sigmoid colon.** The sigmoid colon forms an S-shaped tube that extends into the pelvis and ends at the rectum.

The circular muscle layer of the colon is complete, but the longitudinal muscle layer is incomplete. The longitudinal layer does not completely envelop the intestinal wall but forms three bands, called the **teniae coli** (te'ne-e ko'le; a band or tape along the colon), that run the length of the colon (Figure 24-14; see Figure 24-13). Contractions of the teniae coli cause pouches called **haustra** (haw'strah; to draw up) to form along the length of the colon, giving it a puckered appearance. Small, fat-filled connective tissue pouches called **epiploic** (ep'ĭ-plo'ik; related to the omentum) **appendages** are attached to the outer surface of the colon along its length.

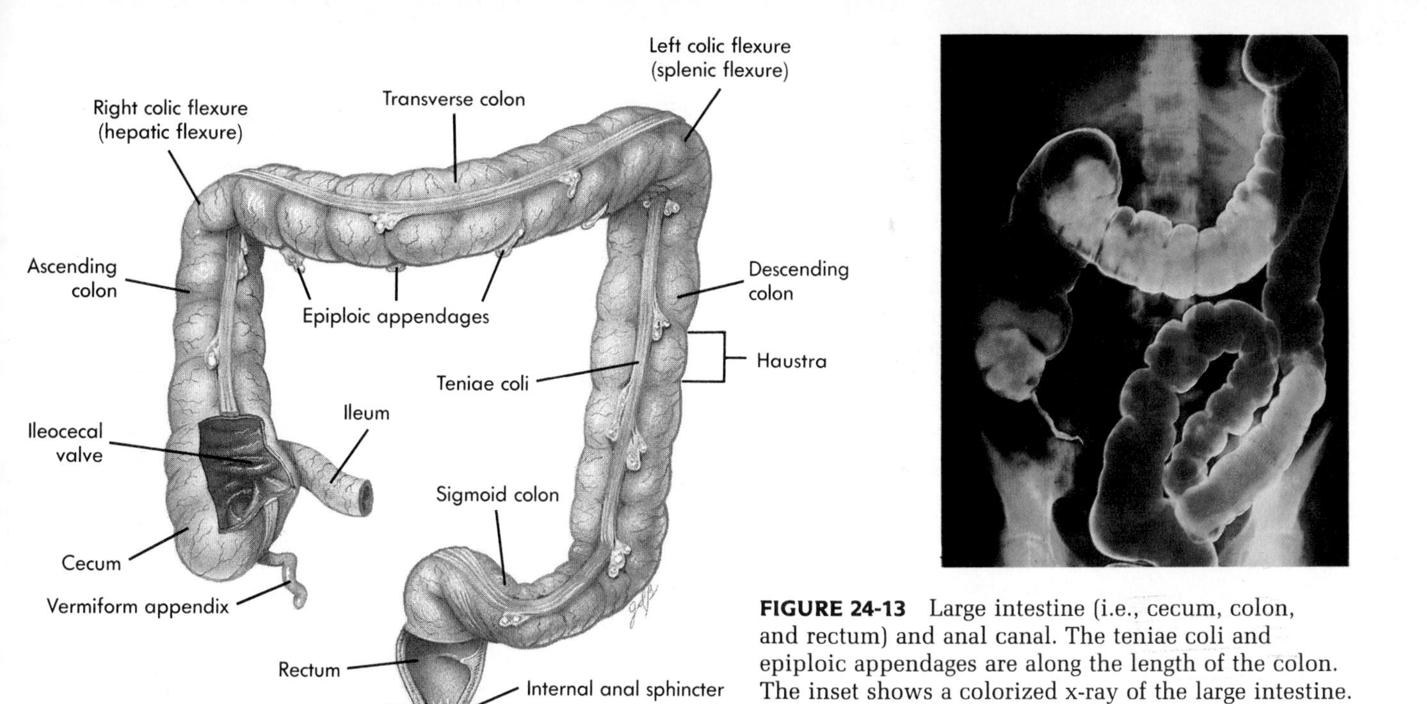

FIGURE 24-13 Large intestine (i.e., cecum, colon, and rectum) and anal canal. The teniae coli and epiploic appendages are along the length of the colon. The inset shows a colorized x-ray of the large intestine.

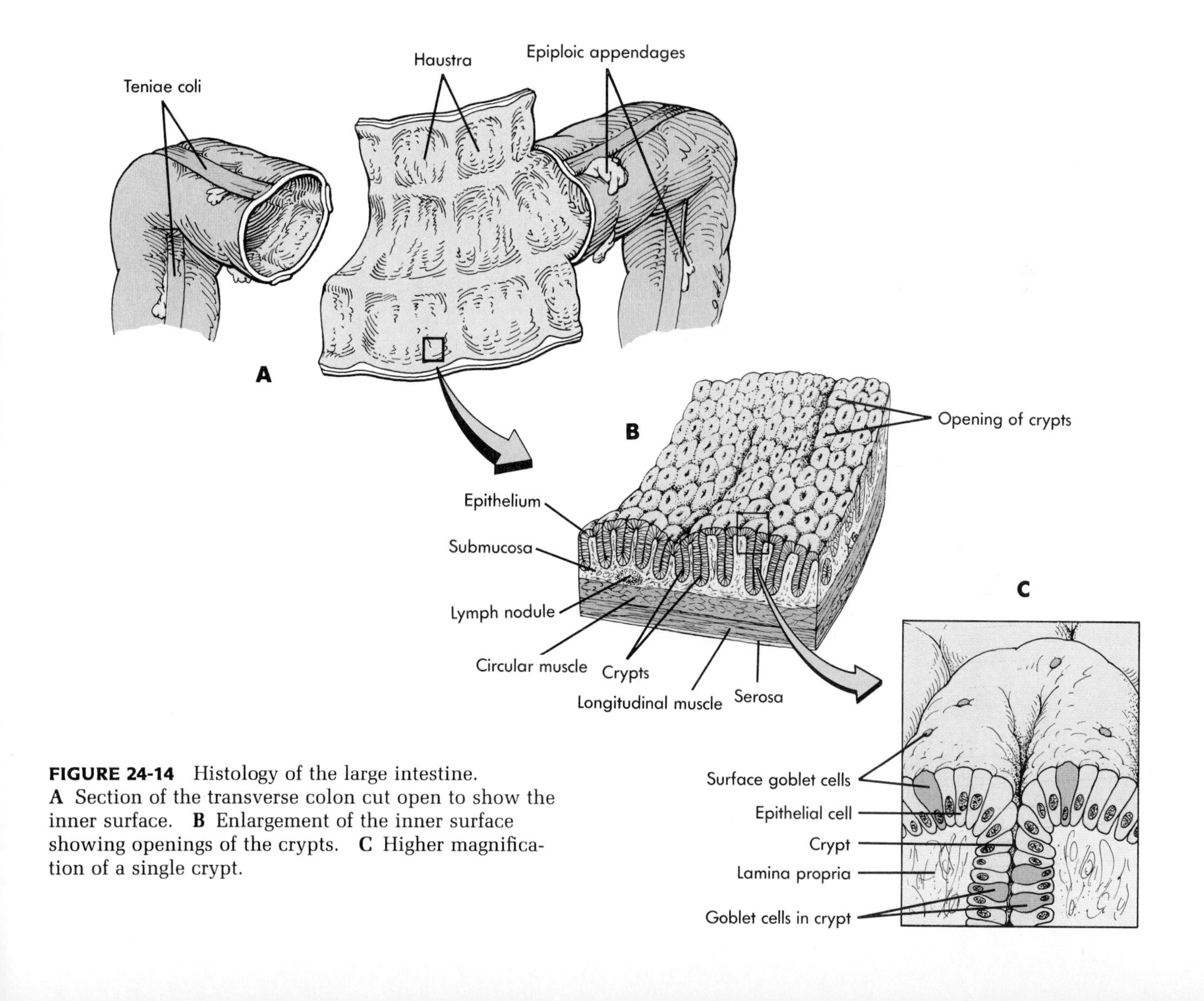

FIGURE 24-14 Histology of the large intestine. **A** Section of the transverse colon cut open to show the inner surface. **B** Enlargement of the inner surface showing openings of the crypts. **C** Higher magnification of a single crypt.

The mucosal lining of the large intestine consists of simple columnar epithelium. This epithelium is not formed into folds or villi like that of the small intestine but has numerous straight tubular glands called **crypts** (see Figure 24-14). The crypts are somewhat similar to the intestinal glands of the small intestine, with four cell types: absorptive, goblet, granular, and endocrine. The major difference is that in the large intestine goblet cells predominate and the other three cell types are greatly reduced in number.

Rectum

The **rectum** is a straight, muscular tube that begins at the termination of the sigmoid colon and ends at the anal canal (see Figure 24-13). The mucosal lining of the rectum is simple columnar epithelium, and the muscular tunic is relatively thick compared to that of the rest of the digestive tract.

Anal canal

The last 2 to 3 cm of the digestive tract is the **anal canal** (see Figure 24-13). It begins at the inferior end of the rectum and ends at the **anus** (external GI tract opening). The smooth muscle layer of the anal canal is even thicker than that of the rectum and forms the **internal anal sphincter** at the superior end of the anal canal. The **external anal sphincter** at the inferior end of the canal is formed by skeletal muscle. The epithelium of the superior portion of the anal canal is simple columnar, and that of the inferior portion is stratified squamous.

Hemorrhoids are the enlargement or inflammation of the hemorrhoidal veins (a condition also called varicose hemorrhoidal veins) that supply the anal canal.

Peritoneum

The body walls and organs of the abdominal cavity are lined with **serous membranes.** These membranes are very smooth and secrete a serous fluid that provides a lubricating film between the layers of membranes. These membranes and fluid allow the organs within the abdomen to move without producing friction. The serous membrane that covers the organs is the **visceral peritoneum** (per'ĭ-to-ne'um; to stretch over), and the one that covers the interior surface of the body wall is the **parietal peritoneum** (Figure 24-15).

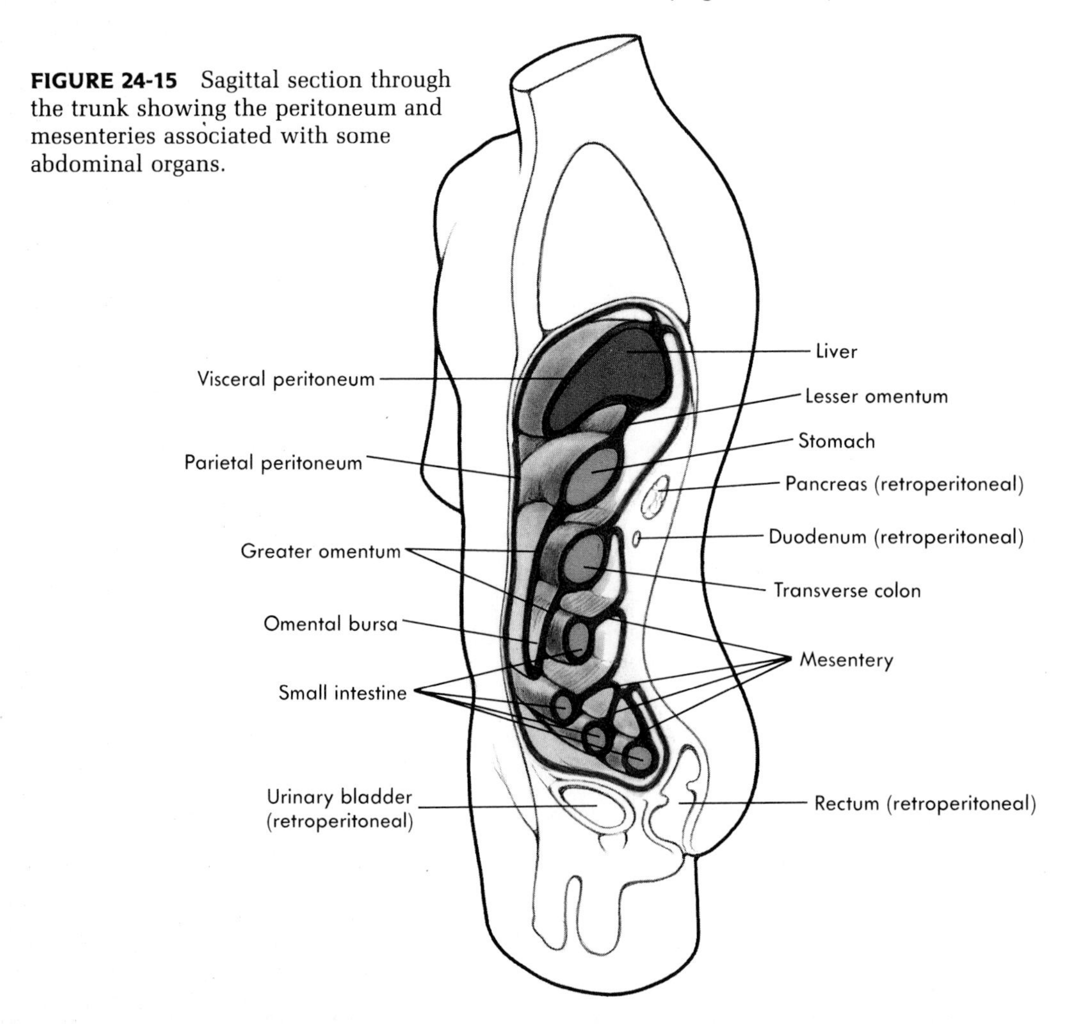

FIGURE 24-15 Sagittal section through the trunk showing the peritoneum and mesenteries associated with some abdominal organs.

Peritonitis is the inflammation of the peritoneal membranes. The inflammation may result from chemical irritation by substances such as bile that have escaped from the digestive tract; or it may result from infection, again originating in the digestive tract such as when the appendix ruptures. Peritonitis can be life threatening. An accumulation of excess serous fluid in the peritoneal cavity is called **ascites** (ă-si'tēz). Ascites may accompany peritonitis.

Many of the organs of the abdominal cavity are held in place by connective tissue sheets called **mesenteries** (mes'en-ter'ēz; middle intestine). The mesenteries consist of two layers of serous membranes with a thin layer of loose connective tissue between them. The mesenteries provide a route by which vessels and nerves can pass from the body wall to the organs. Other abdominal organs lie against the abdominal wall, have no mesenteries, and are referred to as **retroperitoneal** (rĕ'tro-pĕr'ĭ-to-ne'al; behind the peritoneum; see Chapter 1). The retroperitoneal organs include the duodenum, pancreas, ascending colon, descending colon, rectum, kidneys, adrenal glands, and urinary bladder.

Some mesenteries are given specific names. The mesentery connecting the lesser curvature of the stomach and the proximal end of the duodenum to the liver and diaphragm is called the **lesser omentum** (o-men'tum; membrane of the bowels), and the mesentery connecting the greater curvature to the transverse colon and posterior body wall is called the **greater omentum** (see Figure 24-15). However, the greater omentum does not attach directly to the body wall but forms a long, double fold of mesentery that extends inferiorly from the stomach over the small intestine. Because of this folding, a cavity or pocket called the **omental bursa** (bur'sah; pocket) is formed between the two layers of mesentery. A large amount of fat accumulates in the greater omentum, and it is sometimes referred to as the fatty apron.

2

If you placed a pin through the greater omentum, through how many layers of simple squamous epithelium would the pin pass?

? ? ? ? ? ? ? ? ? ? ?

The **coronary ligament** attaches the liver to the diaphragm. Unlike other mesenteries, the coronary ligament has a wide space in the center, the bare area of the liver. The **falciform ligament** attaches the liver to the anterior abdominal wall.

The mesenteries of the small intestine are called the **mesentery proper,** and the mesenteries of portions of the colon are the **transverse mesocolon** (which is actually a continuation of the posterior side of the greater omentum) and the **sigmoid mesocolon.** The vermiform appendix even has its own little mesentery called the **mesoappendix.**

FUNCTIONS OF THE DIGESTIVE SYSTEM

As food moves through the digestive tract, secretions are added to liquefy and digest the food and to provide lubrication (Table 24-3). Each segment of the digestive tract is specialized to assist in moving its contents from the oral end to the anal end. Portions of the digestive system are also specialized to transport molecules from the lumen of the digestive tract into the extracellular spaces. The processes of secretion, movement, and absorption are regulated by elaborate nervous and hormonal mechanisms.

Most of these mechanisms operate in a fashion already described (see Chapters 12, 16, 17, and 18). However, the digestive tract does have a unique regulatory mechanism, the **local reflex** of the intramural plexus, a nervous system response that does not involve the spinal cord or brain. Stimuli (e.g., distention of the digestive tract) activate receptors within the wall of the digestive tract, and action potentials are generated in the neurons of the intramural plexus. The action potentials travel up or down the intramural plexus and produce a response in an effector organ (e.g., in smooth muscle or a gland).

Functions of the Oral Cavity

Secretions of the oral cavity

Saliva is secreted at the rate of approximately 1 to 1.5 L per day. The serous portion of saliva contains a digestive enzyme called **salivary amylase** (am'ĭ-lās; starch-splitting enzyme), which breaks the covalent bonds between glucose molecules in starch and other polysaccharides to produce the disaccharides maltose and isomaltose (see Table 24-3). The release of maltose and isomaltose gives starches a sweet taste in the mouth. However, only approximately 3% to 5% of the total carbohydrates are digested in the mouth. Most of the starches are contained in cellulose-covered globules and are inaccessible to salivary amylase. Cooking and thorough chewing of food destroy the cellulose covering and increase the efficiency of the digestive process.

Saliva prevents bacterial infection in the mouth by washing the oral cavity, and it contains substances (e.g., lysozyme) with weak antibacterial action. The saliva also contains IgA, which helps prevent bacterial infection. The lack of salivary gland secretion increases the chance of ulceration and infection of the oral mucosa and of caries in the teeth.

The mucous secretions of the submandibular and

Table 24-3

Functions of various digestive secretions

FLUID OR ENZYME	SOURCE	FUNCTION	OPTIMUM pH FOR ENZYME ACTIVITY
Saliva			
Serous (watery)	Salivary glands	Moistens food and mucous membrane; lysozyme kills bacteria	—
Salivary amylase	Salivary glands	Starch digestion (conversion to maltose and isomaltose)	6.0-7.4
Mucus	Salivary glands	Lubricates food; protects gastrointestinal tract from digestion by enzymes	—
Gastric Secretions			
Hydrochloric acid	Parietal cells	Decreases stomach pH to activate pepsinogen	—
Pepsinogen	Chief cells	Active form (pepsin) digests protein into smaller peptide chains	2.0-3.0
Mucus	Mucous cells	Protects stomach lining from digestion	—
Liver			
Bile Sodium glycocholate (bile salt) Sodium taurocholate (bile salt) Cholesterol Biliverdin Bilirubin Mucus Fat Lecithin Cells and cell debris	Liver	Bile salts emulsify fats, making them available to intestinal lipases; help make end products soluble and available for absorption by the intestinal mucosa; aid peristalsis	—
Pancreas			
Trypsin	Pancreas	Digests proteins (breaks polypeptide chains at arginine or lysine residues)	8.0
Chymotrypsin	Pancreas	Digests proteins (cleaves carboxyl links of hydrophobic amino acids)	8.0
Carboxypeptidase	Pancreas	Digests proteins (removes amino acids from carboxyl end of peptide chains)	8.0
Pancreatic amylase	Pancreas	Digests carbohydrates (hydrolyzes starches and glycogen to form maltose and isomaltose)	8.0
Pancreatic lipas	Pancreas	Digests fat (hydrolyzes fats—mostly triglycerides—into glycerol and fatty acids)	8.0
Ribonuclease	Pancreas	Digests ribonucleic acid	8.0
Deoxyribonuclease	Pancreas	Digests deoxyribonucleic acid (hydrolyzes phosphodiester bonds)	8.0
Cholesterol esterase	Pancreas	Hydrolyzes cholesterol esters to form cholesterol and free fatty acids	8.0
Bicarbonate ions	Pancreas	Provides appropriate pH for pancreatic enzymes	—

Table 24-3

Functions of various digestive secretions—cont'd

FLUID OR ENZYME	SOURCE	FUNCTION	OPTIMUM pH FOR ENZYME ACTIVITY
Small Intestine Secretions			
Mucus	Duodenal glands and goblet cells	Protects duodenum from stomach acid, gastric enzymes, and intestinal enzymes; provides adhesion for fecal matter; protects intestinal wall from bacterial action and acid produced in the feces	—
Aminopeptidase	Small intestine epithelium	Splits polypeptides into amino acids (from amino end of chain)	6.5-7.5
Peptidase	Small intestine epithelium	Splits amino acids from polypeptides	6.5-7.5
Enterokinase	Small intestine epithelium	Activates trypsin from trypsinogen	6.5-7.5
Amylase	Small intestine epithelium	Digests carbohydrates	6.5-7.5
Sucrase	Small intestine epithelium	Splits sucrose into glucose and fructose	6.5-7.5
Maltase	Small intestine epithelium	Splits maltose into two glucose molecules	6.5-7.5
Isomaltase	Small intestine epithelium	Splits isomaltose into two glucose molecules	6.5-7.5
Lactase	Small intestine epithelium	Splits lactose into glucose and galactose	6.5-7.5
Lipase	Small intestine epithelium	Splits fats (monoglycerides) into glycerol and fatty acids	6.5-7.5

sublingual glands contain a large amount of **mucin** (mu'sin), a proteoglycan that gives a lubricating quality to the secretions of the salivary glands.

Salivary gland secretion is stimulated by the parasympathetic and sympathetic nervous systems, with the parasympathetic system being more important. Salivary nuclei in the brainstem increase salivary secretions through parasympathetic fibers of the facial (VII) and glossopharyngeal (IX) cranial nerves in response to a variety of stimuli (e.g., tactile stimulation in the oral cavity and certain tastes, especially sour). Higher centers of the brain also affect the activity of the salivary glands. Odors that trigger thoughts of food or the sensation of hunger can increase salivary secretions.

Mastication

Food taken into the mouth is **chewed,** or **masticated,** by the teeth. The anterior teeth, the incisors, and the canines primarily cut and tear food, whereas the premolars and molars primarily crush and grind food. Mastication breaks large food particles into smaller food particles, which have a much larger surface area than would a few large particles. Since digestive enzymes digest food molecules only at the surface of the particles, mastication increases the efficiency of digestion.

Chewing is controlled primarily by a reflex integrated in the medulla oblongata called the **chewing,** or **mastication, reflex.** The presence of food in the mouth stimulates sensory receptors, causing the muscles of mastication to relax. The muscles are stretched as the mandible is lowered, stimulating reflexive contraction of the muscles of mastication. Once the mouth is closed, the food again stimulates the muscles of mastication to relax, and the cycle is repeated. The cerebrum can influence the activity of the mastication reflex so that chewing can be initiated or stopped consciously.

Deglutition

Deglutition (dĕ'glu-tish'un), or **swallowing,** can be divided into three separate phases: voluntary, pha-

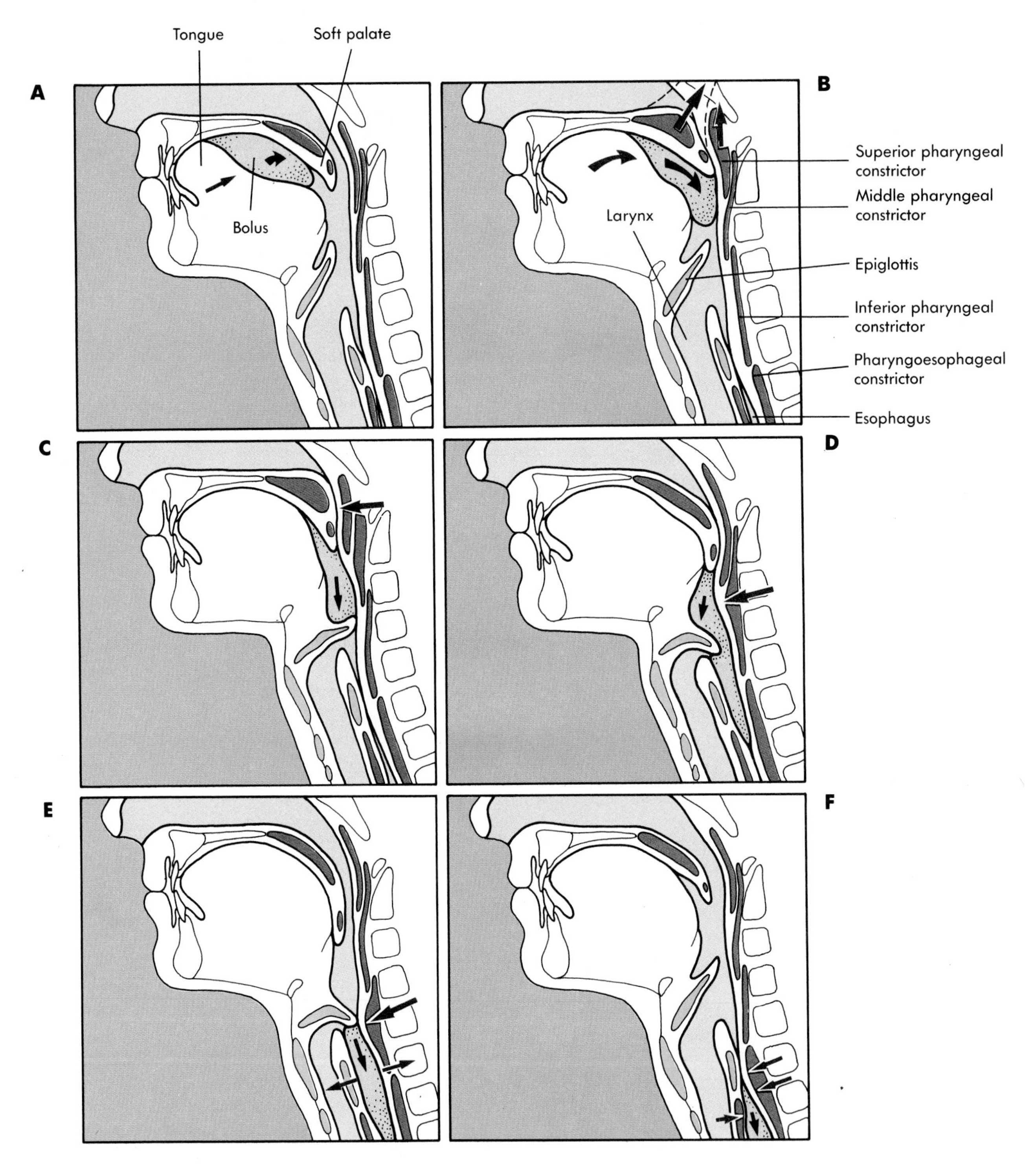

FIGURE 24-16 Three phases of swallowing (deglutition). (1) During the voluntary phase **A** a bolus of food *(yellow)* is pushed by the tongue against the hard palate and posteriorly toward the oropharynx (black arrow indicates movement of the bolus). (2) During the pharyngeal phase, **B** to **E,** the soft palate is elevated, closing off the nasopharynx, and the pharynx is elevated by the palatopharyngeal and salpingopharyngeal muscles **B** (red arrows indicate muscle movement). Successive constriction of the pharyngeal constrictors from superior to inferior, **C, D,** and **E** *(red arrows)*, forces the bolus through the pharynx and into the esophagus. As this occurs, the epiglottis is bent down over the opening of the larynx largely by the force of the bolus pressing against it, and **E** the pharyngoesophageal constrictor relaxes *(outwardly directed red arrows)*, allowing the bolus to enter the esophagus. (3) During the esophageal phase, **F** the bolus is moved by successive constrictions of the esophagus toward the stomach.

ryngeal, and esophageal. During the **voluntary phase** (Figure 24-16, *A*) a bolus of food is formed in the mouth and is pushed by the tongue against the hard palate, forcing the bolus toward the posterior portion of the mouth and into the oropharynx.

The **pharyngeal phase** (Figure 24-16, *B* to *E*) of swallowing is a reflex that is initiated by stimulation of tactile receptors in the area of the oropharynx. Afferent impulses travel through the trigeminal and glossopharyngeal nerves to the **swallowing center** in the medulla oblongata where they initiate motor stimuli that pass through the trigeminal, glossopharyngeal, vagus, and accessory nerves to the soft palate and pharynx. This phase of swallowing begins with the elevation of the soft palate, which closes the passage between the nasopharynx and oropharynx. The pharynx elevates to receive the bolus of food from the mouth and moves the bolus down the pharynx into the esophagus. The three **pharyngeal constrictor muscles** (superior, middle, and inferior) contract in succession, forcing the food through the pharynx. At the same time the upper-esophageal sphincter relaxes, the elevated pharynx opens the esophagus, and food is pushed into the esophagus. This phase of swallowing is unconscious and is controlled automatically even though the muscles involved are skeletal. The pharyngeal phase lasts approximately 1 to 2 seconds.

3

Why is it important to close off the nasopharynx during swallowing? What may happen if a person has an explosive burst of laughter while trying to swallow a liquid?

? ? ? ? ? ? ? ? ? ? ?

During the pharyngeal phase the vocal folds are moved medially; the **epiglottis** (ep′ĭ-glot′is; upon the glottis) is tipped posteriorly so that the epiglottic cartilage covers the opening into the larynx; and the larynx is elevated. These movements of the larynx prevent food from passing through the opening into the larynx.

4

What happens if you try to swallow and speak at the same time?

? ? ? ? ? ? ? ? ? ? ?

The **esophageal phase** (Figure 24-16, *F*) of swallowing takes approximately 5 to 8 seconds and is responsible for moving food from the pharynx to the stomach. Muscular contractions in the esophagus occur in **peristaltic** (pĕr′ĭ-stal′tik) **waves.**

The peristaltic contractions associated with swallowing cause relaxation of the lower-esophageal sphincter in the esophagus as the peristaltic waves approach the stomach. This sphincter is not anatomically distinct from the rest of the esophagus but can be identified physiologically because it remains tonically constricted to prevent the reflux of stomach contents into the lower portion of the esophagus; it relaxes when a bolus of food is moved down the esophagus by peristalsis.

The presence of food in the esophagus stimulates the intramural plexus, which initiates the peristaltic waves. The presence of food in the esophagus also stimulates tactile receptors, which send afferent impulses to the medulla oblongata through the vagus nerves. Motor impulses, in turn, pass along the vagal efferent fibers to the striated and smooth muscles within the esophagus, stimulating their contractions.

Gravity assists the movement of material through the esophagus, especially when liquids are swallowed. However, the peristaltic contractions that move material through the esophagus are sufficiently forceful to allow a person to swallow even while standing on his head.

Stomach Functions

Secretions of the stomach

The stomach functions primarily as a storage and mixing chamber for ingested food. Although some digestion and absorption occur in the stomach, they are not its major functions.

Stomach secretions include mucus, hydrochloric acid, gastrin, intrinsic factor, and pepsinogen, the inactive form of the protein-digesting enzyme pepsin.

The surface mucous cells and mucous neck cells secrete a viscous and alkaline **mucus** that covers the surface of the epithelial cells and forms a layer 1 to 1.5 mm thick. The thick layer of mucus lubricates and protects the epithelial cells of the stomach wall from the damaging effect of the acidic chyme and pepsin. Irritation of the stomach mucosa stimulates the secretion of a greater volume of mucus.

Parietal cells in the gastric glands of the pyloric region secrete intrinsic factor and a concentrated solution of hydrochloric acid. **Intrinsic factor** is a glycoprotein that binds with vitamin B_{12} and makes it more readily absorbed in the ileum. Vitamin B_{12} is important in DNA synthesis. **Hydrochloric acid** produces the low pH of the stomach, which is normally between 1 and 3. The hydrochloric acid secreted into the stomach has a minor digestive effect on ingested food. The low pH helps kill bacteria that are ingested with essentially everything humans put into their mouths. Pathogenic bacteria taken in through the mouth may avoid digestion in the stomach by having an outer coat that resists stomach acids. The low pH of the stomach also stops carbohydrate digestion by inactivating salivary amylase but provides the proper pH environment for the function of pepsin.

Hydrogen ions are derived from carbon dioxide and water, which enter the parietal cell from its serosal surface (the side opposite the lumen of the gastric pit) (Figure 24-17). Once inside the cell, carbonic anhydrase catalyzes the reaction between carbon dioxide and water to form carbonic acid. Some of the carbonic acid molecules then dissociate to form hydrogen ions and bicarbonate ions. The hydrogen ions are actively transported across the mucosal surface of the parietal cell into the lumen of the stomach, moving some potassium ions into the cell in exchange for the hydrogen ions. Although hydrogen ions are actively transported against a steep concentration gradient, chloride ions diffuse with the hydrogen ions through the cell membrane, reducing the positive charge outside the cell and thus reducing the amount of energy needed to transport hydrogen ions against both a concentration gradient and an electrical gradient. The bicarbonate ions are transported down a concentration gradient from the parietal cell into the extracellular fluid. This transport process exchanges bicarbonate ions for chloride ions.

5

Explain why a slight increase in the blood pH may occur following a heavy meal. The elevated pH of blood, especially in the veins that carry blood away from the stomach, is called "the postenteric alkaline tide."

? ? ? ? ? ? ? ? ? ? ?

Chief cells within the gastric glands secrete **pepsinogen** (pep-sin'o-jen). Pepsinogen is packaged in **zymogen** (zi'mo-jen; related to enzymes) **granules,** which are released by the process of exocytosis when pepsinogen secretion is stimulated. Once pepsinogen enters the lumen of the stomach, it is converted to **pepsin** by hydrochloric acid and by previously formed pepsin molecules. Pepsin exhibits optimum enzymatic activity at a pH of 3 or below. Pepsin catalyzes the cleavage of some covalent bonds in proteins, breaking the proteins into smaller peptide chains.

Regulation of stomach secretion

Approximately 2 to 3 L of gastric secretions (gastric juice) are produced each day. Diet dramatically affects the secretion amount; up to 700 ml are secreted as a result of a typical meal. Both nervous and hormonal mechanisms regulate gastric secretions. The neural mechanisms involve reflexes integrated within the medulla oblongata and local reflexes integrated within the intramural plexuses of the gastrointestinal tract. In addition, higher brain centers influence the reflexes. Hormones that regulate stomach secretions include gastrin, gastric-inhibitory polypeptide, and cholecystokinin (Table 24-4).

Heartburn, or **pyrosis,** is a painful or burning sensation in the chest usually associated with backflush of acidic chyme into the esophagus. The pain is usually short lived but may be confused with the pain of an ulcer or a heart attack.

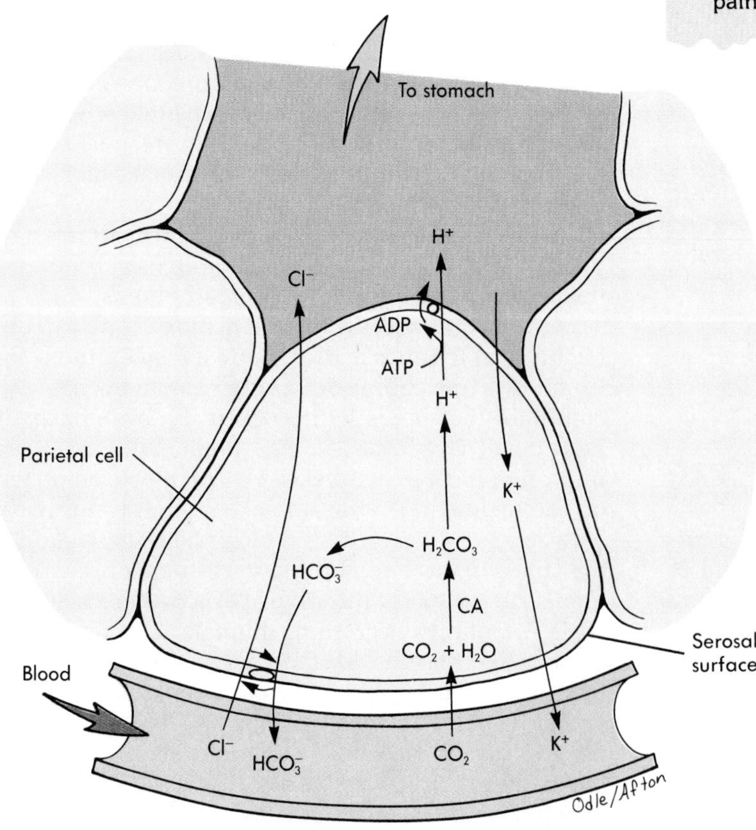

FIGURE 24-17
Hydrochloric acid production by parietal cells in the gastric glands of the stomach. Carbon dioxide (CO_2) is taken into the cell and is combined with water (H_2O) in an enzymatic reaction that is catalyzed by carbonic anhydrase (*CA*) to form carbonic acid (H_2CO_3) which becomes ionized. The bicarbonate ion ($HCO^-{}_3$) is transported back into the bloodstream in exchange for chloride (Cl^-), and the hydrogen ion (H^+) is pumped into the stomach. Chloride ions (Cl^-) diffuse with the charged hydrogen ions, and some potassium ions (K^+) are pumped into the cell in exchange for hydrogen ions.

Table 24-4

Functions of the gastrointestinal hormones

SITE OF PRODUCTION	METHOD OF STIMULATION	SECRETORY EFFECTS	MOTILITY EFFECTS
Gastrin			
Stomach and duodenum	Distention, partially digested proteins, autonomic stimulation, ingestion of alcohol or caffeine	Increases gastric secretion	Increases gastric emptying by increasing stomach motility and relaxing the pyloric sphincter
Secretin			
Duodenum	Acidity of chyme	Inhibits gastric secretion; stimulates pancreatic secretions high in bicarbonate ions; increases the rate of bile secretion; and increases intestinal mucus secretion	Inhibits gastric motility
Cholecystokinin			
Intestine	Fatty acids and other lipids	Slightly inhibits gastric secretion; stimulates pancreatic secretions high in digestive enzymes; causes contraction of the gallbladder and relaxation of the hepatopancreatic ampullar sphincter	Inhibits gastric motility
Gastric Inhibitory Peptide			
Duodenum and proximal jejunum	Fatty acids and other lipids	Inhibits gastric secretion	Inhibits gastric motility

Peptic Ulcer

Peptic ulcer is a condition in which the stomach acids digest the mucosal lining of the GI tract itself. The most common site of a peptic ulcer is near the pylorus, usually on the duodenal side (i.e., a duodenal ulcer). Ulcers occur less frequently along the lesser curvature or at the point where the esophagus enters the stomach. The most common cause of peptic ulcers is oversecretion of gastric juice relative to the degree of mucous and alkaline protection of the small intestine.

People experiencing severe anxiety for a long period of time are the most prone to develop duodenal ulcers. They often have a high rate of gastric secretion (as much as 15 times the normal amount) between meals. Cerebral impulses enhance vagal activity, which stimulates excessive gastric acid secretion. This secretion results in highly acidic chyme entering the duodenum. The duodenum is usually protected by sodium bicarbonate (secreted mainly by the pancreas), which neutralizes the chyme. However, when large amounts of acid enter the duodenum, the sodium bicarbonate is not adequate to neutralize it.

In patients with gastric ulcers the levels of gastric hydrochloric acid secretion are often normal or even low; however, the stomach has a reduced resistance to its own acid. Such inhibited resistance can result from excessive ingestion of alcohol or aspirin. Reflux of duodenal contents into the pylorus can also cause gastric ulcers. In this case, bile, which is present in the reflux, has a detergent effect that reduces gastric mucosal resistance to acid.

Regulation of stomach secretion can be divided into three phases: cephalic, gastric, and intestinal.

1. **Cephalic Phase.** In the cephalic phase of gastric regulation the sensations of the taste and the smell of food, stimulation of tactile receptors during the process of chewing and swallowing, and pleasant thoughts of food stimulate centers within the medulla that influence gastric secretions (Figure 24-18, *A*). Impulses are sent from the medulla along parasympathetic neurons within the vagus nerves to the stomach. Within the stomach wall the preganglionic neurons stimulate postganglionic neurons in the intramural plexuses. The postganglionic neurons, which are primarily cholinergic, stimulate secretory activity in the cells of the stomach mucosa.

   Acetylcholine stimulates the secretory activity of both the parietal and chief cells and stimulates the secretion of **gastrin** (gas'trin) from endocrine cells. Gastrin is released into the circulation and travels to the parietal cells where it stimulates additional hydrochloric acid secretion.
2. **Gastric Phase.** The greatest volume of gastric secretions is produced during the gastric phase of gastric regulation, which is initiated by the presence of food in the stomach (Figure 24-18, *B*). The primary stimuli are distention of the stomach and the presence of amino acids and peptides in the stomach.

   Distention of the stomach wall, especially in the body or fundus, results in the stimulation of mechanoreceptors. Impulses generated by these receptors initiate reflexes that involve both the central nervous system and the local intramural reflexes, resulting in secretion of mucus, hydrochloric acid, pepsinogen, intrinsic factor, and gastrin. Gastrin secretion is also stimulated by the presence of partially digested proteins or moderate amounts of alcohol or caffeine in the stomach.

   When the pH of the stomach contents falls below 2, the increased gastric secretion produced by distention of the stomach is blocked, an inhibitory effect that constitutes a negative-feedback mechanism that limits the secretion of gastric juice.

   Amino acids and peptides that result from the digestive action of pepsin on proteins directly stimulate parietal cells of the stomach to secrete hydrochloric acid. The mechanism by which this response is mediated is not clearly understood. It does not involve any known neurotransmitters, and when the pH drops below 2, this response also is inhibited.
3. **Intestinal Phase.** The intestinal phase of gastric regulation is controlled by the entrance of acidic chyme (stomach contents) into the duodenum of the small intestine (Figure 24-18, *C*). The presence of chyme in the duodenum initiates both neural and hormonal mechanisms, which first stimulate and later inhibit gastric secretions. When the pH of the chyme in the duodenum is 3 or above, the stimulatory response is greatest. When the pH of the chyme in the duodenum is 2 or below, the inhibitory influence prevails.

   Increased gastric secretion during the intestinal phase is due to the release of **gastrin** from the upper portion of the duodenum. The gastrin is carried in the blood to the stomach where it stimulates gastric secretions.

   Inhibition of gastric secretions is controlled by several hormonal mechanisms that are initiated within the duodenum and jejunum of the small intestine during the intestinal phase of gastric regulation. Acidic solutions in the duodenum cause the release of the hormone **secretin** (se-kre'tin) into the circulatory system. Secretin inhibits gastric secretion by inhibiting both parietal and chief cells. Acidic solutions also initiate a local nervous reflex that inhibits gastric secretions.

   Fatty acids and certain other lipids in the duodenum and the proximal jejunum initiate the release of two hormones: **gastric inhibitory polypeptide** and **cholecystokinin** (ko-le-sis-to-kīn'in). Gastric inhibitory polypeptide strongly inhibits gastric secretion, and cholecystokinin inhibits gastric secretions to a lesser degree. Hypertonic solutions in the duodenum and jejunum also inhibit gastric secretions. The mechanism appears to involve the secretion of a hormone, **enterogastrone,** but the actual existence of this hormone has never been established.

   Inhibition of gastric secretions is also under nervous control. Distention of the duodenal wall, the presence of irritating substances in the duodenum, reduced pH, and hypertonic or hypotonic solutions in the duodenum can activate the **enterogastric reflex** (composed of a local reflex and a reflex integrated within the medulla oblongata) and cause a reduction in gastric secretions.

Stomach filling

As food enters the stomach, the rugae flatten, and the stomach volume increases. Despite the increase in volume, the pressure within the stomach does not increase until the volume nears maximum capacity because smooth muscle can stretch without an increase in tension (see Chapter 10) and because of a reflex integrated within the medulla oblongata that inhibits muscle tone in the body of the stomach.

Mixing of stomach contents

Ingested food is thoroughly mixed with the secretions of the stomach glands to form a semifluid material called **chyme** (kīm; juice). This mixing is accomplished by gentle **mixing waves,** which are peristaltic-like contractions that occur approximately every 20 seconds and proceed from the body

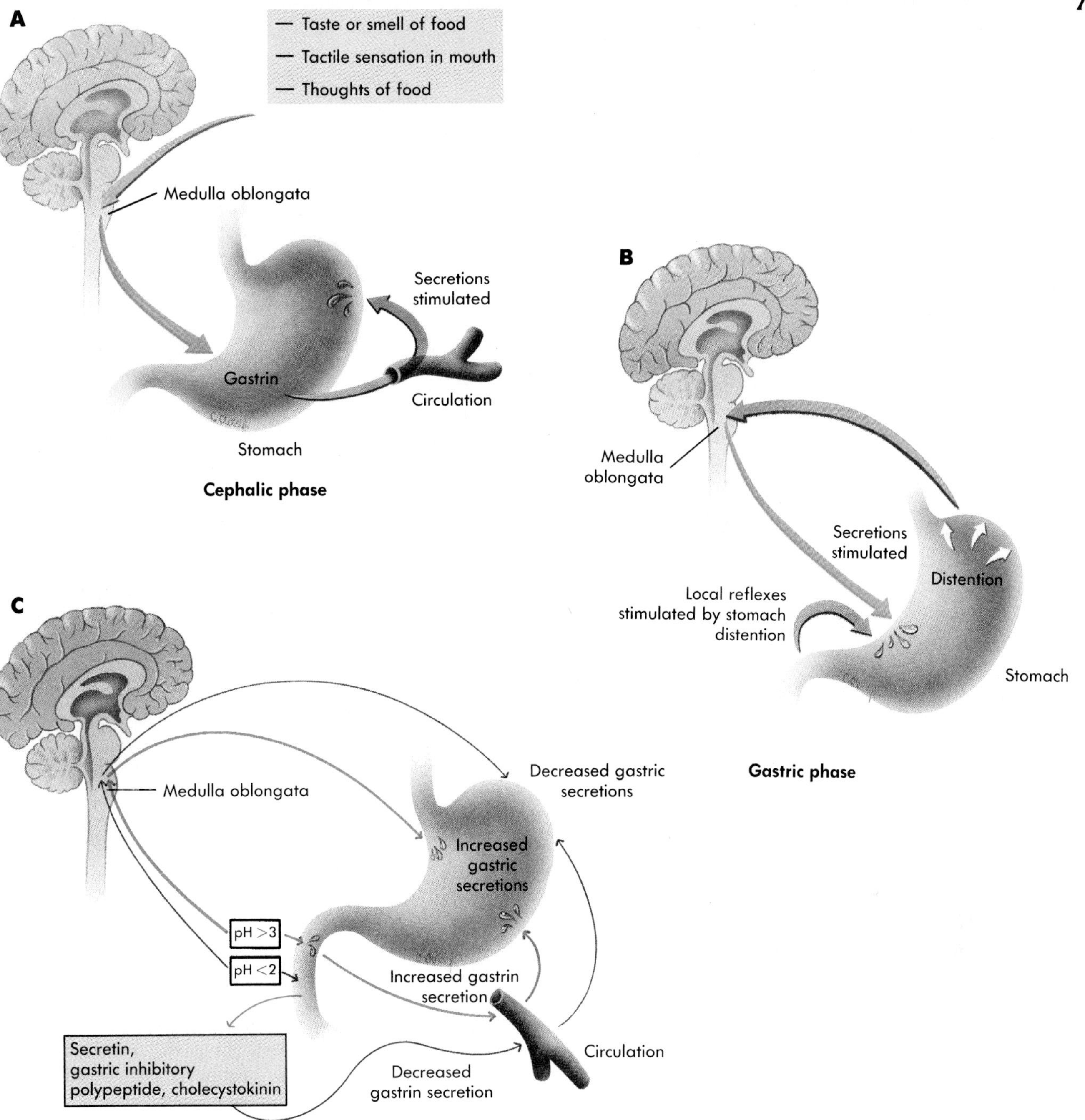

FIGURE 24-18 Three phases of gastric secretion. **A** Cephalic phase. Tastes or smells of food, tactile sensations of food in the mouth, or even thoughts of food stimulate vagal nuclei in the medulla, and parasympathetic sensations are carried by the vagus nerves to the stomach. Preganglionic parasympathetic vagus nerve fibers synapse with postganglionic neurons in the myenteric plexus of the stomach. Postganglionic fibers directly stimulate secretion by parietal and chief cells and stimulate gastrin secretion by endocrine cells. Gastrin is carried through the circulation back to other parts of the stomach where it also stimulates secretion by parietal and chief cells. **B** Gastric phase. Distention of the stomach by the presence of food stimulates local reflexes in the stomach and vagus nerve impulses to the medulla. Efferent vagal impulses and local impulses stimulate secretions by parietal, chief, and mucous cells. A low pH in the stomach can inhibit such secretions. **C** Intestinal phase. The presence in the duodenum of chyme with a pH greater than 3 or containing amino acids and peptides stimulates gastric secretions through vagus nerve pathways and through secretion of gastrin *(green lines)*. In the duodenum chyme with a pH less than 2 inhibits gastric secretions *(red lines)*. Secretin, gastric inhibitory polypeptide, and cholecystokinin also inhibit gastric secretions *(red lines)*.

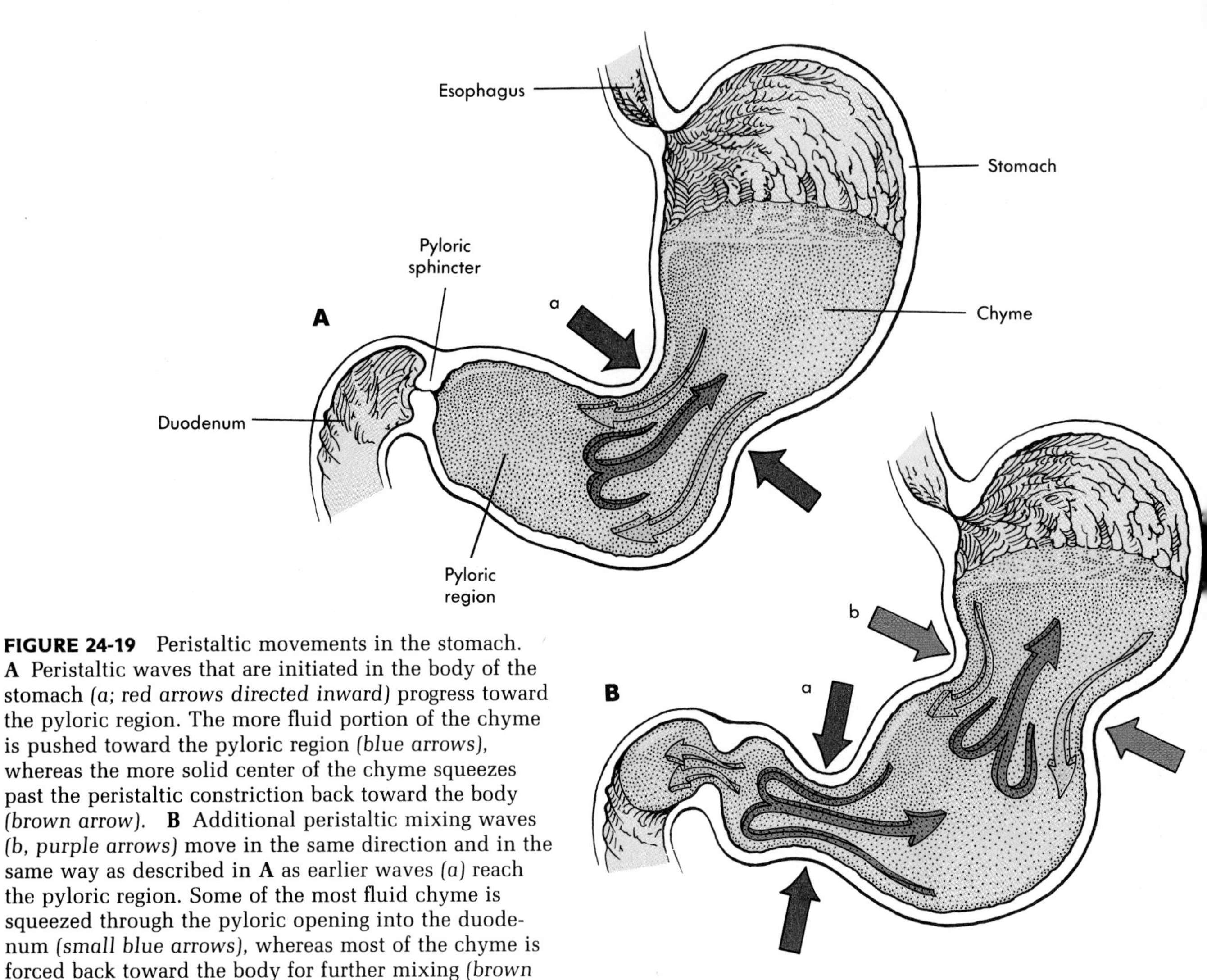

FIGURE 24-19 Peristaltic movements in the stomach. **A** Peristaltic waves that are initiated in the body of the stomach *(a; red arrows directed inward)* progress toward the pyloric region. The more fluid portion of the chyme is pushed toward the pyloric region *(blue arrows)*, whereas the more solid center of the chyme squeezes past the peristaltic constriction back toward the body *(brown arrow)*. **B** Additional peristaltic mixing waves *(b, purple arrows)* move in the same direction and in the same way as described in **A** as earlier waves *(a)* reach the pyloric region. Some of the most fluid chyme is squeezed through the pyloric opening into the duodenum *(small blue arrows)*, whereas most of the chyme is forced back toward the body for further mixing *(brown arrow)*.

toward the pyloric sphincter to mix the ingested material with the secretions of the stomach. **Peristaltic waves** occur less frequently, are significantly more powerful than mixing waves, and force the chyme near the periphery of the stomach toward the pyloric sphincter. The more solid material near the center of the stomach is pushed superiorly toward the cardiac region for further digestion (Figure 24-19). Roughly 80% of the contractions are mixing waves, and 20% are peristaltic waves.

Stomach emptying

The amount of time food remains in the stomach depends on a number of factors, including the type and volume of food. Liquids exit the stomach within 1½ hours to 2½ hours after ingestion. After a typical meal the stomach is usually empty within 3 to 4 hours. The pyloric sphincter usually remains partially closed because of mild tonic contraction. Each peristaltic contraction is sufficiently strong to force a small amount of chyme through the pyloric opening and into the duodenum. The peristaltic contractions responsible for movement of chyme through the partially closed pyloric opening are called the **pyloric pump.**

Hunger contractions are peristaltic contractions that approach tetany for periods of approximately 2 to 3 minutes. The contractions are increased by a decrease in blood glucose levels and are sufficiently strong to create an uncomfortable sensation often called a "hunger pang" or "hunger pain." Hunger pangs usually begin 12 to 24 hours after the previous meal. If nothing is ingested, they reach their maximum intensity within 3 or 4 days and then become progressively weaker.

Regulation of stomach movements

If the stomach empties too fast, the efficiency of digestion and absorption is reduced; and if the rate of emptying is too slow, the highly acidic contents of the stomach may damage the stomach wall and reduce the rate at which nutrients are digested and absorbed. Stomach emptying is regulated to prevent these two extremes. Many of the hormonal and neural mechanisms that stimulate stomach secretions also are involved with increasing stomach motility. For example, during the gastric phase of stomach secretion distention of the stomach stimulates local reflexes, central nervous system reflexes, and the release of gastrin, all of which increase stomach motility and cause relaxation of the pyloric sphincter. The result is an increase in stomach emptying. Conversely, the hormonal and neural mechanisms that decrease gastric secretions also inhibit gastric motility, increase constriction of the pyloric sphincter, and decrease the rate of stomach emptying.

Vomiting can result from irritation (e.g., overdistention or overexcitation) anywhere along the gastrointestinal tract. Impulses travel through the vagus nerve and spinal visceral afferent nerves to the vomiting center in the medulla. Once the vomiting center is stimulated and the reflex is initiated, the following events occur: (1) a deep breath is taken; (2) the hyoid bone and larynx are elevated, opening the upper-esophageal sphincter; (3) the opening of the larynx is closed; (4) the soft palate is elevated, closing the posterior nares; (5) the diaphragm and abdominal muscles are forcefully contracted, strongly compressing the stomach and increasing the intragastric pressure; (6) the lower-esophageal sphincter is relaxed, and (7) the gastric contents are forced out of the stomach, through the esophagus and oral cavity, to the outside.

Functions of the Small Intestine

The small intestine is the site where the greatest amount of digestion and absorption occurs.

Secretions of the small intestine

The mucosa of the small intestine produces secretions that primarily contain mucus, electrolytes, and water. Intestinal secretions lubricate and protect the intestinal wall from the acidic chyme and the action of digestive enzymes. They also keep the chyme in the small intestine in a liquid form to facilitate the digestive process (see Table 24-3). Most of the secretions that enter the small intestine are produced by the intestinal mucosa, but the secretions of the liver and the pancreas also enter the small intestine and play an important role in the process of digestion. Most of the digestive enzymes that enter the small intestine come from the pancreas. The intestinal mucosa also produces enzymes that remain associated with the intestinal epithelial surface.

Mucus is secreted in large amounts by duodenal glands, intestinal glands, and goblet cells. The mucus provides the wall of the intestine with protection against the irritating effects of acidic chyme and against the digestive enzymes that enter the duodenum from the pancreas. **Secretin** and **cholecystokinin** are released from the intestinal mucosa and stimulate hepatic and pancreatic secretions (Figure 24-20).

Secretion by duodenal glands is stimulated by the vagus nerve, secretin, and chemical or tactile irritation of the duodenal mucosa (see Figure 24-20). Goblet cells produce mucus in response to the tactile and chemical stimulation of the mucosa.

Duodenal gland secretion is inhibited by sympathetic nerve stimulation, thus reducing the duodenal wall's coating of mucus, which protects it against acid and gastric enzymes. Therefore if a person is highly stressed, elevated sympathetic activity may inhibit duodenal gland secretion and increase his susceptibility to a duodenal **ulcer.**

The membranes of the absorptive cell microvilli have enzymes bound to them. These surface-bound enzymes include **disaccharidases,** which break disaccharides down to monosaccharides, **peptidases,** which hydrolyze the peptide bonds between small amino acid chains, and **nucleases,** which break down nucleic acids (see Table 24-3). Although these enzymes are not secreted into the intestine, they influence the digestive process significantly, and the large surface area of the intestinal epithelium brings these enzymes into contact with the intestinal contents. Small molecules, which are breakdown products of digestion, are absorbed through the microvilli and enter the circulatory or lymphatic systems.

Movement in the small intestine

Mixing and propulsion of chyme are the primary mechanical events that occur in the small intestine. These functions are the result of two major types of contractions accomplished by the smooth muscle in the wall of the small intestine—segmental and peristaltic. **Segmental contractions** are propagated for only short distances and mix the intestinal contents.

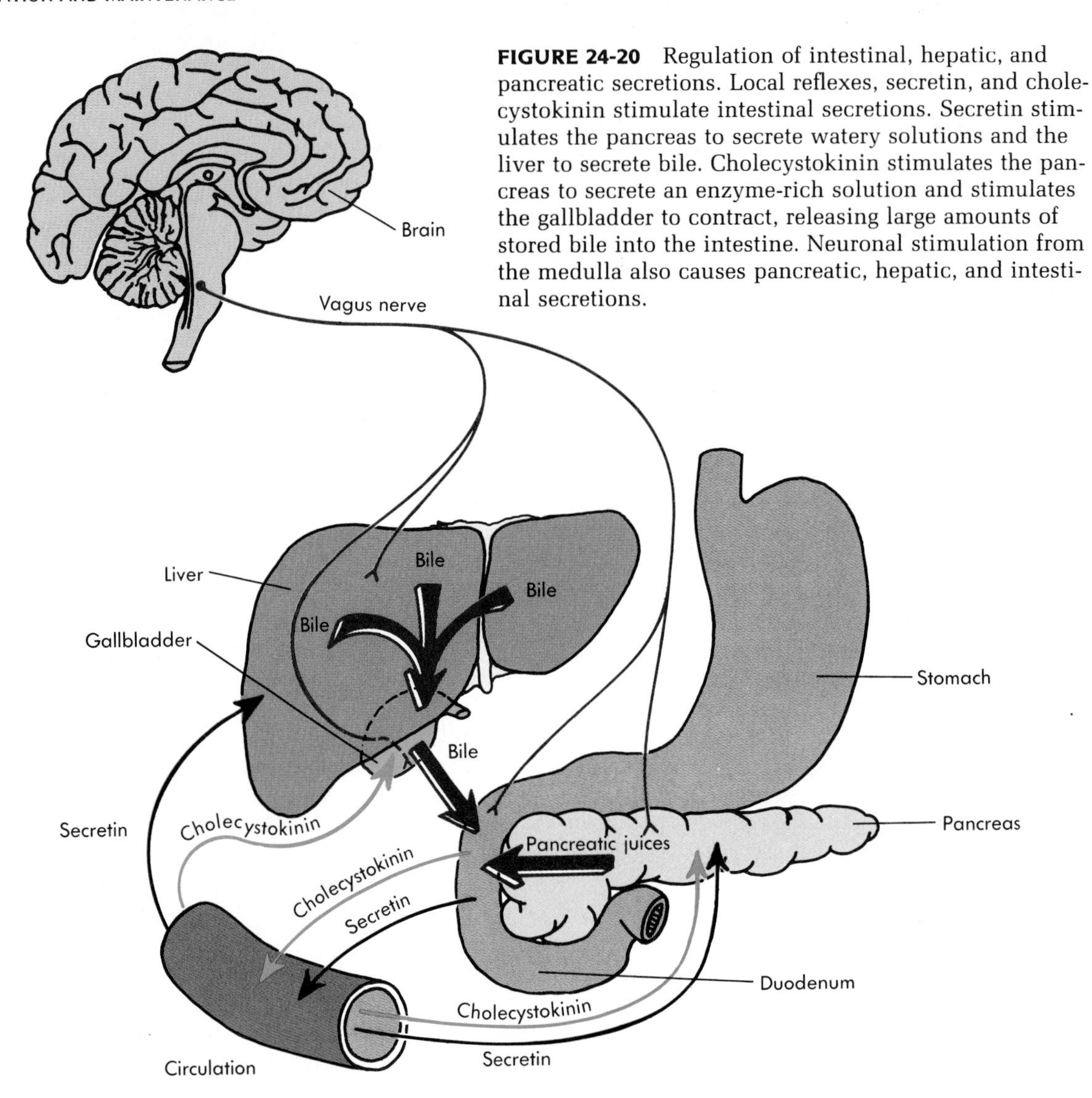

FIGURE 24-20 Regulation of intestinal, hepatic, and pancreatic secretions. Local reflexes, secretin, and cholecystokinin stimulate intestinal secretions. Secretin stimulates the pancreas to secrete watery solutions and the liver to secrete bile. Cholecystokinin stimulates the pancreas to secrete an enzyme-rich solution and stimulates the gallbladder to contract, releasing large amounts of stored bile into the intestine. Neuronal stimulation from the medulla also causes pancreatic, hepatic, and intestinal secretions.

In contrast, **peristaltic contractions** begin in the proximal portion of the small intestine and proceed along the length of the intestine for variable distances. Some peristaltic contractions proceed the entire length of the intestine, although most are propagated shorter distances. Frequently, intestinal peristaltic contractions are continuations of peristaltic contractions that begin in the stomach. These contractions both mix and propel substances through the small intestine as the wave of contraction proceeds. The contractions move at a rate of approximately 1 cm per minute. The movements are slightly faster at the proximal end of the small intestine and slightly slower at the distal end. It usually takes 3 to 5 hours for chyme to move from the pylorus to the ileocecal junction.

Local mechanical and chemical stimuli are especially important in regulating the motility of the small intestine. Smooth muscle contraction increases in response to distention of the intestinal wall. Solutions that are either hypertonic or hypotonic, solutions with a low pH, and certain products of digestion such as amino acids and peptides also stimulate contractions of the small intestine. Local reflexes, which are integrated within the intramural nervous plexuses of the small intestine, mediate the response of the small intestine to these mechanical and chemical stimuli. Stimulation through parasympathetic nerve fibers may also increase the motility of the small intestine, but the parasympathetic influences in the small intestine are not as important as those in the stomach.

The ileocecal sphincter at the juncture between the ileum and the large intestine remains mildly contracted most of the time, but peristaltic contractions reaching it from the small intestine cause it to relax and allow movement of chyme from the small intestine into the cecum. Cecal distention, however, initiates a local reflex that causes more intense constriction of the ileocecal sphincter. Closure of the sphincter facilitates digestion and absorption in the small intestine by slowing the rate of chyme movement from the small intestine into the large intestine and prevents material from returning to the ileum from the cecum.

Liver Functions

The liver performs important digestive and excretory functions, stores and processes nutrients, synthesizes new molecules, and detoxifies harmful chemicals.

Bile production

The liver produces and secretes approximately 600 to 1000 ml of **bile** each day (see Table 24-3). Although bile contains no digestive enzymes, it plays a role in digestion by diluting and neutralizing stomach acid. The pH of chyme as it leaves the stomach is too low for the normal function of pancreatic enzymes. Bile neutralizes the acidic chyme and brings the pH up to a level at which pancreatic enzymes can function. Bile salts **emulsify** fats by reducing surface tension on fat globules and by breaking the globules into smaller bits that are more easily digested and absorbed. Bile also contains excretory products such as bile pigments (i.e., bilirubin that results from the breakdown of hemoglobin), cholesterol, fats, fat-soluble hormones, and lecithin.

The rate of bile secretion is controlled in a variety of ways (see Figure 24-20). Secretin stimulates bile secretion, primarily by increasing the water and bicarbonate ion content of bile. Bile secretion is increased by parasympathetic stimulation through the vagus nerve and by increased blood flow through the liver. Bile salts also increase bile secretion through a positive-feedback system. Most bile salts are reabsorbed in the ileum and are carried in the blood back to the liver where they stimulate further bile secretion. This process continues until the duodenum empties and the release of bile into the duodenum stops. The loss of bile salts in the feces is reduced by this recycling process.

Storage

Hepatocytes can remove sugar from the blood and store it in the form of **glycogen.** They can also store fat, vitamins (A, B_{12}, D, E, and K), copper, and iron. This storage function is usually short-term, and the amount of stored material in the hepatocytes and thus the cell size fluctuate during a given day.

If a large amount of sugar were dumped into the general circulation after a meal, it would increase the osmolality of the blood and produce hyperglycemia. These results are prevented because the blood from the intestine passes through the hepatic portal vein to the liver where glucose and other substances are removed from the blood by hepatocytes, are stored, and are secreted back into the circulation when needed. By this means the hepatocytes can control blood sugar levels within very strict limits.

Nutrient interconversion

Another function that the liver performs is the interconversion of nutrients. Ingested foods are not always in the proportion that are needed by the tissues. If this is the case, the liver can convert some nutrients into others. If, for example, a person is on a fad diet that is very high in protein, an oversupply of protein and an undersupply of lipids and carbohydrates are delivered to the liver. The hepatocytes break down the amino acids and cycle many of them through metabolic pathways so they can be used to produce ATP, lipids, and glucose (see Chapter 25).

Hepatocytes also transform substances that cannot be used by the cells into more readily usable substances. Ingested fats, for example, are combined with choline and phosphorus in the liver to produce phospholipids, which are essential components of cell membranes. Vitamin D is converted to its active form necessary for calcium maintenance.

Detoxification

Many ingested substances are harmful to the cells of the body. In addition, the body itself produces many by-products of metabolism that, if accumulated, are toxic. The liver is one line of defense against many of those harmful substances. It detoxifies many substances by altering their structure, making them less toxic or making their excretion easier. Ammonia, for example, a by-product of amino acid metabolism, is toxic and is not readily removed from the circulation by the kidneys. Hepatocytes remove ammonia from the circulation and convert it to urea, which is secreted into the circulation and eliminated by the kidneys in the urine. Other substances are removed from the circulation and are excreted by the hepatocytes into the bile.

Phagocytosis

Hepatic phagocytic cells (Kupffer cells), which lie along the sinusoid walls of the liver, phagocytize worn out and dying red and white blood cells, some bacteria, and other debris that enters the liver through the circulation.

Synthesis

The liver can also produce its own unique new compounds. Many of the blood proteins (e.g., albumins, fibrinogen, globulins, heparin, and clotting factors) are produced by the liver and are released into the circulation (see Chapter 19).

Functions of the Gallbladder

Bile is continually secreted by the liver and is stored in the **gallbladder,** which can store 40 to 70 ml of bile. While the bile is in the gallbladder, water and electrolytes are absorbed, and bile salts and pigments become as much as five to 10 times more concentrated than they were when secreted by the liver. Shortly after a meal, the gallbladder contracts in response to stimulation by cholecystokinin and, to a lesser degree, in response to vagal stimulation, dumping large amounts of concentrated bile into the small intestine (see Figure 24-20).

Cholesterol, secreted by the liver, may precipitate in the gallbladder to produce **gallstones** (Figure 24-A). Occasionally a gallstone may pass out of the gallbladder and enter the cystic duct, blocking release of bile. Such a condition interferes with normal digestion, and the gallstone often must be removed surgically.

FIGURE 24-A Gallstones.

Functions of the Pancreas

The exocrine secretions of the pancreas are called **pancreatic juice** and have two major components—an aqueous component and an enzymatic component. Pancreatic juice is produced in the pancreas and is then delivered through the pancreatic ducts to the small intestine where it functions in digestion. The **aqueous component** is produced principally by columnar epithelial cells that line the smaller ducts of the pancreas. It contains sodium and potassium ions in approximately the same concentration found in extracellular fluid. Bicarbonate ions are a major part of the aqueous component, and they help neutralize the acidic chyme that enters the small intestine from the stomach. The increased pH stops pepsin digestion but provides the proper environment for the function of pancreatic enzymes. Bicarbonate ions are actively secreted by the duct epithelium, and water follows passively to make the pancreatic juice isotonic. The cellular mechanism that is responsible for the secretion of bicarbonate ions is diagrammed in Figure 24-21.

Pancreatic enzymes

The **enzymatic component** of the pancreatic juice is produced by the acinar cells of the pancreas and is important for the digestion of all major classes of food. Without the enzymes produced by the pancreas, lipids, proteins, and carbohydrates are not adequately digested (see Tables 24-1 and 24-3).

The proteolytic pancreatic enzymes, which digest proteins, are secreted in inactive forms, whereas many of the other enzymes are secreted in active form. The major proteolytic enzymes are **trypsin, chymotrypsin,** and **carboxypeptidase.** They are secreted in their inactive forms as trypsinogen, chymotrypsinogen, and procarboxypeptidase and are activated by the removal of certain peptides from the larger precursor proteins. If these were produced in their active forms, they would digest the tissues producing them. Trypsinogen is activated by the proteolytic enzyme **enterokinase** (en'ter-o-ki'nās; intestinal enzyme), which is secreted by the mucosa of the duodenum when chyme comes in contact with it. Trypsin then activates more trypsinogen, as well as chymotrypsinogen and procarboxypeptidase.

Pancreatic juice also contains **amylase,** which continues the polysaccharide digestion that was initiated in the oral cavity.

In addition, pancreatic juice contains a group of lipid-digesting enzymes called pancreatic **lipases,** which break down lipids into free fatty acids, glycerides, cholesterol, etc.

Enzymes that reduce DNA and RNA to their component nucleotides, **deoxyribonucleases** and **ribonucleases,** respectively, are also present in pancreatic juice.

Control of pancreatic secretion

The exocrine secretory activity of the pancreas is controlled by both hormonal and neural mechanisms (see Figure 24-20). Secretin stimulates the secretion of a watery solution that contains a large amount of bicarbonate ions from the pancreas. The primary stimulus for secretin release is the presence of acidic chyme in the duodenum.

6

Explain why secretin production in response to acidic chyme and its stimulation of bicarbonate ion secretion constitute a negative-feedback mechanism.

? ? ? ? ? ? ? ? ? ? ?

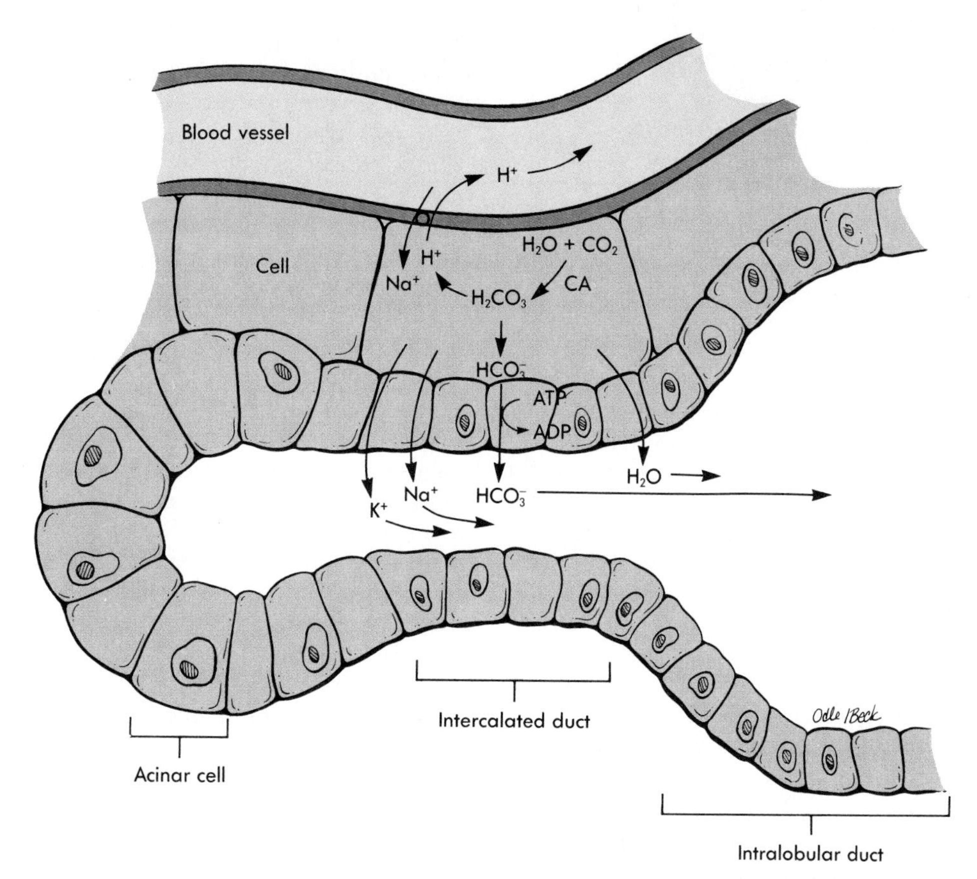

FIGURE 24-21 Bicarbonate ion production in the pancreas. Water (H_2O) and carbon dioxide (CO_2) combine under the influence of carbonic anhydrase (*CA*) to form carbonic acid (H_2CO_3), which dissociates to form hydrogen ions (H^+) and bicarbonate ions (HCO_3^-). The hydrogen ions are removed in the bloodstream, and the bicarbonate ions are secreted into the pancreatic ducts. Sodium, potassium, and water follow the bicarbonate ions into the duct.

Cholecystokinin stimulates the secretion of bile and the secretion of pancreatic juice rich in digestive enzymes. The major stimulus for the release of cholecystokinin is the presence of fatty acids and amino acids in the intestine.

Parasympathetic stimulation through the vagus nerves also stimulates the secretion of pancreatic juices rich in pancreatic enzymes, and sympathetic impulses inhibit secretion. The effect of vagal stimulation on pancreatic juice secretion is greatest during the cephalic and gastric phases of stomach secretion.

Functions of the Large Intestine

Normally 18 to 24 hours are required for material to pass through the large intestine in contrast to the 3 to 5 hours required for movement of chyme through the small intestine. Thus the movements of the colon are more sluggish than those of the small intestine. While in the colon, chyme is converted to **feces.** Absorption of water and salts, the secretion of mucus, and extensive action of microorganisms are involved in the formation of feces, which the colon stores until it is eliminated by the process of **defecation.** Approximately 1500 ml of chyme enter the cecum each day, but more than 90% of the volume is reabsorbed so that only 80 to 150 ml of feces are normally eliminated by defecation.

Secretions of the large intestine

The mucosa of the colon has numerous goblet cells that are scattered along its length and numerous crypts that are lined almost entirely with goblet cells. Little enzymatic activity is associated with secretions of the colon in which mucus is the major secretory product (see Table 24-3). Mucus lubricates the wall of the colon and helps the fecal matter stick together. Tactile stimuli and irritation of the wall of the colon trigger local intramural reflexes that increase mucous secretion. Parasympathetic stimulation also increases the secretory rate of the goblet cells.

When the large intestine is irritated and inflamed such as in patients with bacterial **enteritis** (inflamed intestine resulting from bacterial infection of the bowel), the intestinal mucosa secretes large amounts of water and electrolytes in addition to mucus. This condition is called **diarrhea,** and although it increases fluid and electrolyte loss, it also moves the infected feces out of the intestine more rapidly and speeds recovery from the disease.

Bicarbonate ions are actively secreted by epithelial cells of the colon in response to the acid produced by colic bacteria. Sodium ions are absorbed by active transport, and chloride ions follow sodium as the result of an electrical gradient. Water crosses the wall of the colon through osmosis after sodium chloride transport.

The feces that leave the digestive tract consist of water, solid substances (e.g., undigested food), microorganisms, and sloughed-off epithelial cells.

Numerous microorganisms inhabit the colon. They reproduce rapidly and ultimately comprise approximately 30% of the dry weight of the feces. Some bacteria in the intestine synthesize vitamin K, which is passively absorbed in the colon, and break down a small amount of cellulose to glucose.

Gases called **flatus** (fla′tus; blowing) are produced by bacterial actions in the colon. The amount of flatus depends partly on the bacterial population present in the colon and partly on the type of food consumed. Beans, for example, are well known for their flatus-producing effect.

Movement in the large intestine

Segmental mixing movements occur in the colon much less often than in the small intestine. Peristaltic waves are largely responsible for moving chyme along the ascending colon. At widely spaced intervals (normally three or four times each day), large portions of the transverse and descending colon undergo several strong peristaltic contractions, called **mass movements,** which propel their contents considerable distances toward the anus (Figure 24-22). Mass movements are very common after meals because the presence of food in the stomach initiates strong peristaltic contractions in the colon. Mass movements are most common approximately 15 min-

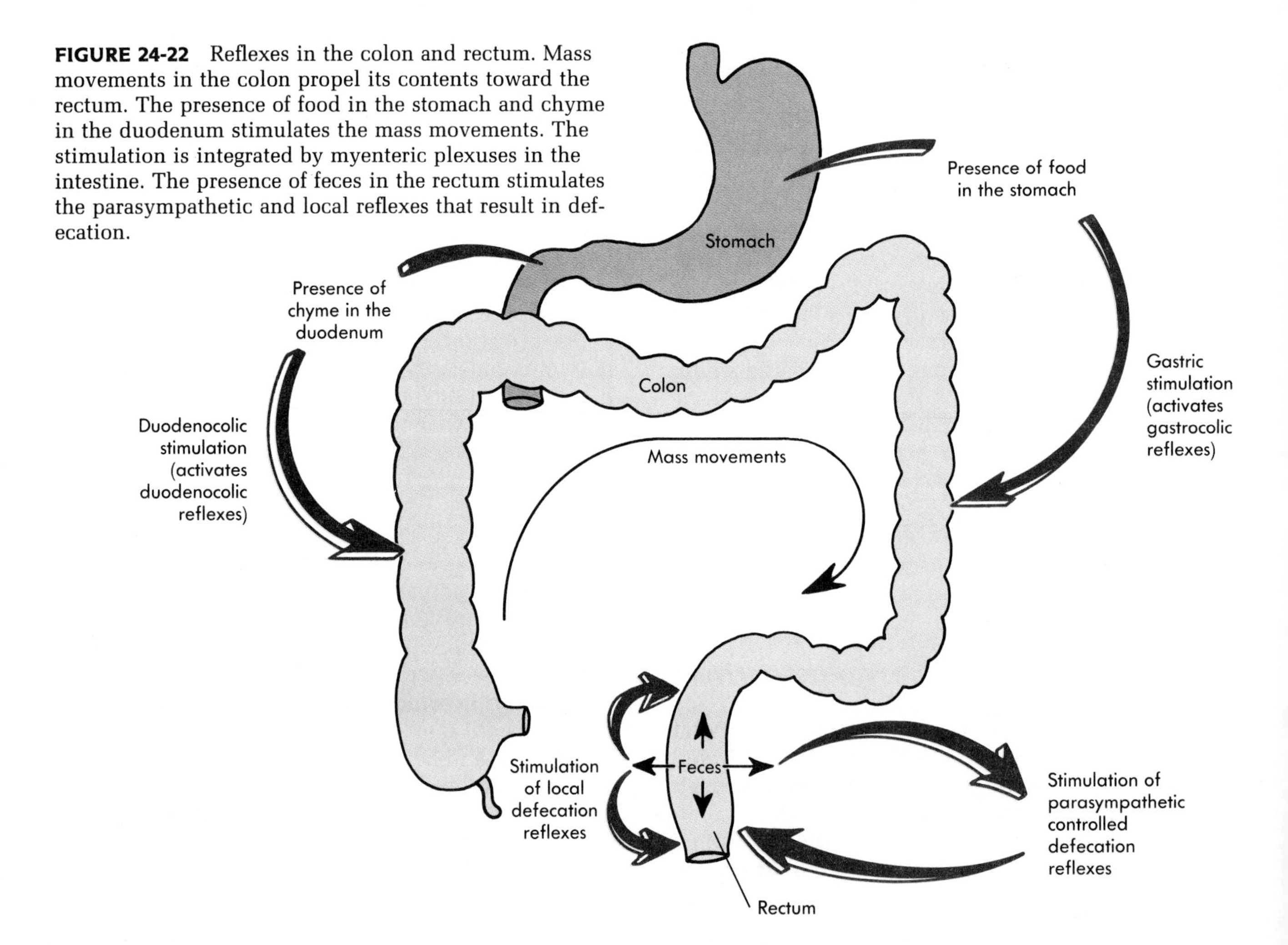

FIGURE 24-22 Reflexes in the colon and rectum. Mass movements in the colon propel its contents toward the rectum. The presence of food in the stomach and chyme in the duodenum stimulates the mass movements. The stimulation is integrated by myenteric plexuses in the intestine. The presence of feces in the rectum stimulates the parasympathetic and local reflexes that result in defecation.

utes after breakfast. They usually persist for 10 to 30 minutes and then stop for perhaps half a day. The peristaltic contractions of mass movements are integrated by local reflexes in the wall of the gastrointestinal tract, and they are called **gastrocolic reflexes** if initiated by the stomach or **duodenocolic reflexes** if initiated by the duodenum (see Figure 24-22).

Distention of the rectal wall by feces acts as a stimulus that initiates the **defecation reflex.** Local reflexes cause weak contractions and relaxation of the internal anal sphincter. Parasympathetic reflexes cause strong contractions and are normally responsible for most of the defecation reflex. Impulses produced in response to the distention travel along afferent nerve fibers to the sacral region of the spinal cord where efferent impulses are initiated that reinforce peristaltic contractions in the lower colon and rectum. Efferent impulses also cause the internal anal sphincter to relax. The external anal sphincter, which is composed of skeletal muscle and is under conscious cerebral control, prevents the movement of feces out of the rectum and through the anal opening. If this sphincter is relaxed voluntarily, feces are expelled. The defecation reflex persists for only a few minutes and quickly dies. Generally the reflex is reinitiated after a period of time that may be as long as several hours. Mass movements in the colon are usually the reason for the reinitiation of the defecation reflex.

Defecation is usually accompanied by voluntary movements that support the expulsion of feces. These voluntary movements include a large inspiration of air followed by closure of the larynx and forceful contraction of the abdominal muscles. As a consequence, the pressure in the abdominal cavity increases and forces the contents of the colon through the anal canal and out of the anus.

DIGESTION, ABSORPTION, AND TRANSPORT

Digestion is the breakdown of organic molecules into their component parts: carbohydrates into monosaccharides, proteins into amino acids, and fats into fatty acids and glycerol. Absorption and transport are the means by which molecules are moved out of the digestive tract and into the circulation for distribution throughout the body. Not all molecules (e.g., vitamins, minerals, and water) are broken down before being absorbed. Digestion begins in the oral cavity and continues in the stomach, but most digestion occurs in the proximal end of the small intestine, especially in the duodenum.

Absorption begins in the stomach where some very small molecules (e.g., alcohol and aspirin) can pass through the stomach epithelium into the circulation. Most absorption occurs in the duodenum and jejunum, although some absorption occurs in the ileum.

Once the digestive products have been absorbed, they are transported to other parts of the body by two different routes. Water, ions, and water-soluble digestion products such as glucose and amino acids enter the **hepatic portal system** (see Chapter 21) and are transported to the liver. The products of lipid metabolism are coated with proteins and are transported into **lacteals.** The lacteals are connected by lymph vessels to the thoracic duct (see Chapter 21), which empties into the left subclavian vein. The protein-coated lipid products then travel in the circulation to adipose tissue or to the liver.

Carbohydrates

Ingested **carbohydrates** consist primarily of polysaccharides such as starches and glycogen, disaccharides such as sucrose (table sugar) and lactose (milk sugar), and monosaccharides such as glucose and fructose (found in many fruits). During the digestion process polysaccharides are broken down into smaller chains and finally into disaccharides and monosaccharides. Disaccharides are broken down into monosaccharides. Carbohydrate digestion begins in the oral cavity with the partial digestion of starches by **salivary amylase** (am′ĭ-lās) and is completed in the intestine by **pancreatic amylase** (Table 24-5). The digestion of disaccharides into monosaccharides is accomplished by a series of **disaccharidases** that are bound to the microvilli of the intestinal epithelium.

Monosaccharides are taken up by intestinal epithelial cells by active transport (e.g., glucose and galactose) or by facilitated diffusion (e.g., fructose). The monosaccharides are transferred to the capillaries of the intestinal villi and are carried by the hepatic portal system to the liver where the nonglucose sugars are converted to glucose. Glucose is transported by the circulation to the cells that require energy. Glucose enters the cells through facilitated diffusion. The rate of glucose transport into most types of cells is greatly influenced by **insulin** and may increase tenfold in its presence.

In patients with **diabetes mellitus** insulin is lacking, and insufficient glucose is transported into the cells of the body. As a result, the cells do not have enough energy for normal function, blood glucose levels become significantly elevated, and abnormal amounts of glucose are released into the urine.

Table 24-5

Factors involved in the digestion of the three major food types

ENZYME	SITE OF PRODUCTION	ACTION
Carbohydrates		
Amylase	Salivary glands, pancreas, and lining of small intestine	Breaks long-chain starches into maltose and isomaltose
Disaccharidases	Intestine	Break disaccharides into monosaccharides
Sucrase		Splits sucrose into glucose and fructose
Maltase		Splits maltose into two glucose molecules
Isomaltase		Splits isomaltose into two glucose molecules
Lactase		Splits lactose into glucose and galactose
Lipids		
Lipase	Pancreas	Splits triglycerides into monoglycerides and fatty acids
Lipase	Intestine	Splits monoglycerides into glycerol and fatty acid
Esterase	Pancreas	Splits cholesterol esters into cholesterol and fatty acids
Proteins		
Pepsin	Stomach	Breaks proteins into smaller peptide chains
Trypsin	Pancreas	Breaks proteins and peptide chains into smaller peptide chains
Chymotrypsin	Pancreas	Breaks proteins and peptide chains into smaller peptide chains
Carboxypeptidase	Pancreas	Removes amino acid from the end of peptide chains
Aminopeptidase	Intestine	Removes amino acid from the end of peptide chains
Peptidase	Intestine	Completes the breakdown of small peptide chains
Tetrapeptidase	Intestine	Splits tetrapeptides into tripeptides plus one amino acid
Tripeptidase	Intestine	Splits tripeptides into dipeptides plus one amino acid
Dipeptidase	Intestine	Splits dipeptides into two amino acids

Lipids

Lipids are molecules that are insoluble or only slightly soluble in water. They include triglycerides, phospholipids, steroids, and fat-soluble vitamins. **Triglycerides** (tri-glis'er-īdz) consist of three fatty acids and one glycerol molecule covalently bound together. The first step in lipid digestion is **emulsification** (e-mul'sĭ-fĭ-ka'shun), which is the transformation of large lipid droplets into much smaller droplets. The enzymes that digest lipids are soluble in water and can digest the lipids only by acting at the surface of the droplets. The emulsification process increases the surface area of the lipid exposed to the digestive enzymes by decreasing the droplet size. Emulsification is accomplished by **bile salts** secreted by the liver and stored in the gallbladder.

Lipase (li'pās) secreted by the pancreas digests lipid molecules (see Table 24-5). The primary products of this digestive process are free fatty acids, glycerol, and monoglycerides. Cholesterol and phospholipids also comprise part of the lipid digestion products.

Cystic fibrosis is a hereditary disorder that results, in addition to other symptoms, in blockage of the pancreatic ducts so that the pancreatic digestive enzymes are prevented from reaching the duodenum; thus fats, which can only be digested by these enzymes, are not digested. As a result, fats and fat-soluble vitamins are not absorbed, and the patient suffers from vitamin A, D, E, and K deficiencies. These deficiencies result in conditions such as night blindness, skin disorders, rickets, and excessive bleeding. Therapy includes administering the missing vitamins to the patient and reducing his dietary fat intake.

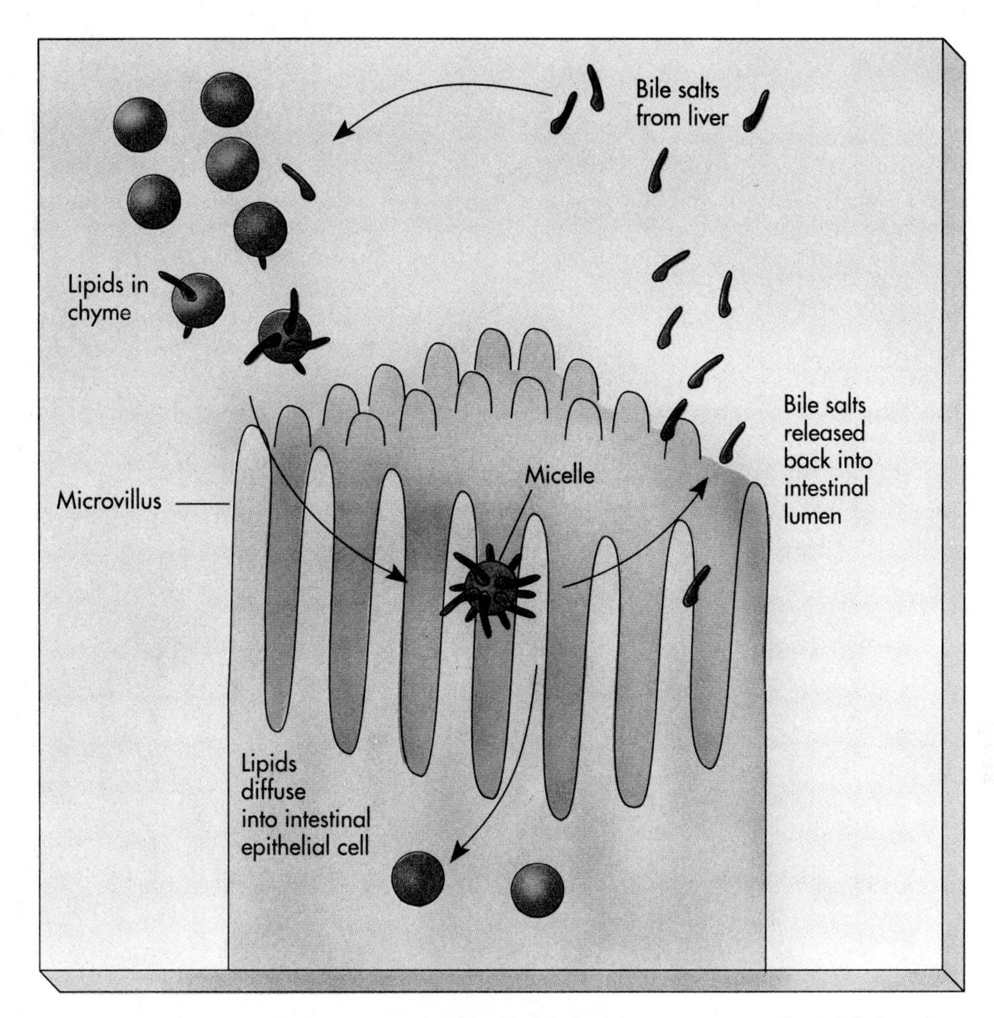

FIGURE 24-23 Lipid absorption. Lipids *(red spheres)* are associated with bile salts *(purple rods)* to form a micelle. The hydrophobic ends of the bile salts are directed inward toward the lipid core, and the hydrophilic ends are directed outward toward the watery environment of the digestive tract. When a micelle contacts the microvilli of an epithelial cell, the lipid diffuses into the cell, and the bile salts are released.

Once lipids are digested in the intestine, bile salts aggregate around the small droplets to form **micelles** (mi-sēlz′; a small morsel; Figure 24-23). The hydrophobic ends of the bile salts are directed toward the free fatty acids, cholesterol, and monoglycerides at the center of the micelle, and the hydrophilic ends are directed outward toward the water environment. When a micelle comes into contact with the epithelial cells of the small intestine, the micelle's contents pass by means of simple diffusion through the lipid cell membrane of the epithelial cells.

Within the smooth endoplasmic reticulum of the intestinal epithelial cells, free fatty acids are combined with glycerol molecules to form triglycerides. Proteins synthesized in the epithelial cells coat droplets of triglycerides, phospholipids, and cholesterol to form **chylomicrons** (ki-lo-mi′kronz; small particles in the chyle or fat-filled lymph). The chylomicrons leave the epithelial cells and enter the lacteals of the lymphatic system within the villi. They are carried through the lymphatic system to the bloodstream. The chylomicrons are carried by the blood to adipose tissue where triglycerides are converted to fat, which is stored until an energy source is needed elsewhere in the body. In the liver the chylomicron lipids are stored, converted into other molecules, or used as energy.

Proteins

Proteins are taken into the body from a number of dietary sources. **Pepsin** secreted by the stomach (see Table 24-5) catalyzes the cleavage of covalent bonds in proteins, producing smaller polypeptide chains (Figure 24-24). As much as 10% to 20% of the total ingested protein is digested by gastric pepsin. Once the proteins and polypeptide chains leave the stomach, proteolytic enzymes produced in the pancreas continue the digestive process, producing small peptide chains. These are broken down into amino acids by **peptidases** bound to the microvilli of the small intestine. Each peptidase is specific for a certain peptide chain length or for a certain peptide bond.

Absorption of individual amino acids occurs through intestinal epithelial cells by active transport, which requires the simultaneous transport of sodium. Dipeptides and tripeptides are also taken up by intestinal epithelial cells, probably through pinocytosis. Once inside the cells, dipeptidases and tripeptidases split the dipeptides and tripeptides into their component amino acids. Individual amino acids then leave the epithelial cells and enter the hepatic portal system, which transports them to the liver. The amino acids may be modified in the liver or released into the bloodstream and distributed throughout the body.

Amino acids are actively transported into the various cells of the body. This transport is stimulated by growth hormone and insulin. Most amino acids are used as building blocks to form new proteins (see Chapter 3), but some amino acids may be used for energy.

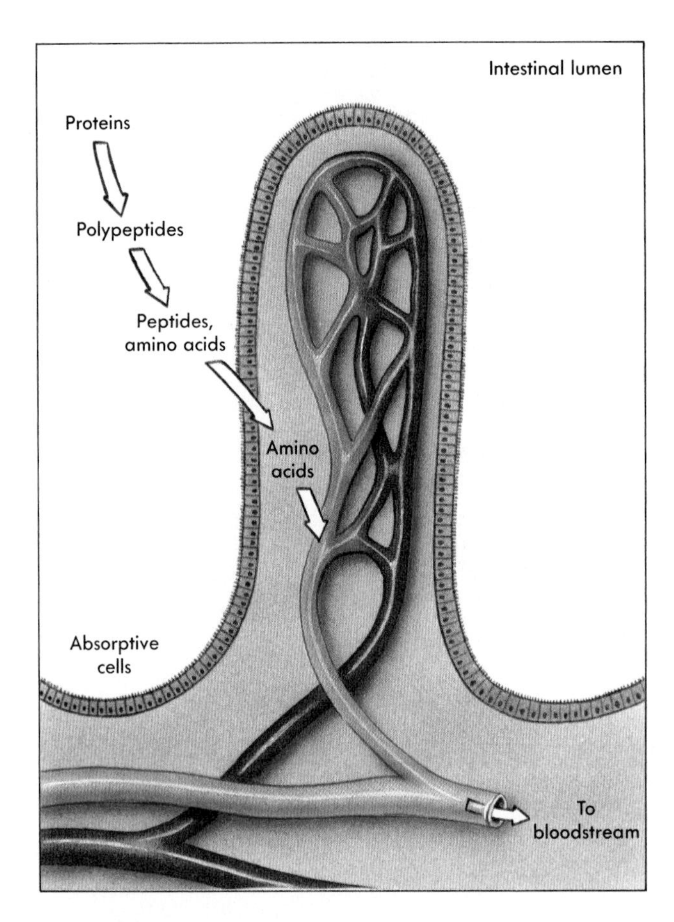

FIGURE 24-24 Digestion of protein molecules.

Water

Water can move in either direction across the wall of the small intestine. The direction of its diffusion is determined by osmotic gradients across the epithelium. When the chyme is dilute, water is absorbed by osmosis across the intestinal wall into the blood. When the chyme is very concentrated and contains very little water, water moves by osmosis into the lumen of the small intestine. As nutrients are absorbed in the small intestine, its osmotic pressure decreases; as a consequence, water moves from the intestine into the surrounding extracellular fluid. Water in the extracellular fluid can then enter the circulation. Because of the amount of nutrients that are absorbed in the small intestine and because of the osmotic gradient this absorption produces, nearly 90% of the water that enters the small intestine by way of the stomach or intestinal secretions is reabsorbed.

Ions

Active transport mechanisms for **sodium** ions are present within the epithelial cells of the small intestine. **Potassium, calcium, magnesium,** and **phosphate** are also actively transported. **Chloride** ions move passively through the intestinal wall of the duodenum and the jejunum following the positively charged sodium ions, but chloride ions are actively transported from the ileum. Although calcium ions are actively transported along the entire length of the small intestine, vitamin D is required for that transport process. The absorption of calcium is under hormonal control as is its excretion and storage. Parathyroid hormones, calcitonin, and vitamin D all play a role in regulating blood levels of calcium in the circulatory system (see Chapters 6, 18, and 27).

Intestinal Disorders

Malabsorption Syndrome

Malabsorption syndrome (sprue) is a spectrum of disorders of the small intestine that result in abnormal nutrient absorption. One type of malabsorption results from the toxic effects of gluten (present in certain types of grains) and involves the destruction of newly formed epithelial cells in the intestinal glands. Those cells fail to migrate to the villi surface, the villi become blunted, and the surface area decreases. As a result, the intestinal epithelium is less capable of absorbing nutrients. Sprue also results in inflammation of the digestive tract. Another type of malabsorption (called tropical) is apparently caused by bacteria, although no specific bacterium has been identified.

Enteritis

Enteritis is any inflammation of the intestines. It can result from an infection, chemical irritation, or from some unknown cause. **Regional enteritis,** or Crohn's disease, is a local enteritis of unknown cause characterized by patchy, deep ulcers developing in the intestinal wall, usually in the distal end of the ileum. The disease results in overproliferation of connective tissue and invasion of lymphatic tissue into the involved area, with subsequent thickening of the intestinal wall and narrowing of the lumen. **Colitis** is an inflammation of the colon.

Constipation

Constipation is the slow movement of feces through the large intestine. The feces often become dry and hard because of the increased fluid absorption during the extended time they are retained in the large intestine. Constipation often results from irregular defecation patterns that develop after a prolonged time of inhibiting normal defecation reflexes. Spasms of the sigmoid colon resulting from irritation can also result in slow feces movement and constipation.

The Digestive System

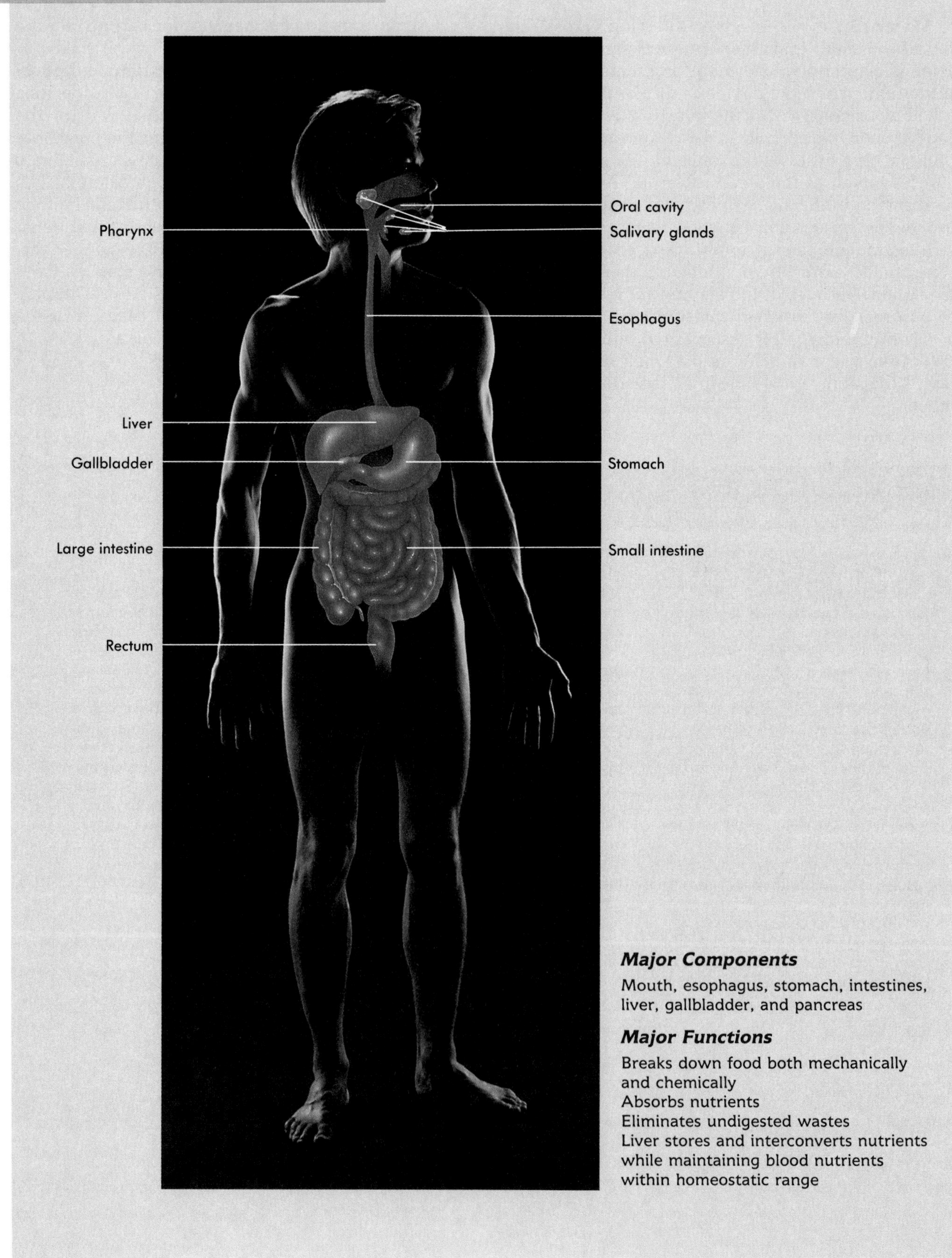

Major Components

Mouth, esophagus, stomach, intestines, liver, gallbladder, and pancreas

Major Functions

Breaks down food both mechanically and chemically
Absorbs nutrients
Eliminates undigested wastes
Liver stores and interconverts nutrients while maintaining blood nutrients within homeostatic range

SUMMARY

THE DIGESTIVE SYSTEM PROVIDES THE BODY WITH WATER, ELECTROCYTES, AND OTHER NUTRIENTS.

GENERAL OVERVIEW p. 722

The digestive system consists of a digestive tube and its associated accessory organs.

ANATOMY OVERVIEW p. 722

1. The digestive system consists of the oral cavity, pharynx, esophagus, stomach, small intestine, large intestine, and anus.
2. Accessory organs such as the salivary glands, liver, gallbladder, and pancreas are located along the digestive tract.

HISTOLOGY OVERVIEW p. 722

The digestive tract is composed of four tunics: mucosa, submucosa, muscularis, and serosa or adventitia.

Mucosa

The mucosa consists of a mucous epithelium, a lamina propria, and a muscularis mucosae.

Submucosa

The submucosa is a connective tissue layer containing nerves, blood vessels, and small glands.

Muscularis

1. The muscularis consists of an inner layer of circular smooth muscle and an outer layer of longitudinal smooth muscle.
2. The myenteric plexus is between the two muscle layers.

Serosa or Adventitia

The serosa or adventitia forms the outermost layer of the digestive tract.

PHYSIOLOGY OVERVIEW p. 775

1. The functions of the digestive system are ingestion, mastication, propulsion, mixing, secretion, digestion, absorption, transportation, excretion, and regulation.
2. The functions the digestive system are regulated by elaborate nervous and hormonal mechanisms.

Ingestion

Ingestion is the introduction of food into the stomach.

Mastication

1. Mastication is the chewing of food and its mechanical digestion.
2. Additional mechanical digestion occurs as food is mixed in the stomach and intestines.

Propulsion

Propulsion is the movement of food from one end of the digestive tract to the other.

Mixing

Some peristaltic contractions mix food with digestive secretions and break it down.

Secretion

Secretions are added to lubricate, liquefy, and digest the food as it moves through the digestive tract.

Digestion

1. Digestion is the breakdown of organic molecules into their component parts.
2. Digestion consists of mechanical digestion and chemical digestion.

Absorption

Absorption is the means by which molecules are moved out of the digestive tract and into the circulation or extracellular spaces.

Transportation

Transportation is the means by which molecules are distributed throughout the body.

Excretion

Excretion is the means by which the waste products of digestion are removed from the body by defecation.

Regulation

1. The processes of propulsion, secretion, absorption, and excretion are regulated by nervous and hormonal mechanisms.
2. Nervous control occurs through local reflexes and through the vagus nerve.

ANATOMY AND HISTOLOGY OF THE DIGESTIVE TRACT p. 776

Oral Cavity

1. The lips and cheeks are involved in facial expression, mastication, and speech.
2. The tongue is involved in speech, taste, mastication, and swallowing.
   - The intrinsic tongue muscles change the shape of the tongue, and the extrinsic tongue muscles move the tongue.
   - The anterior two-thirds of the tongue is covered with papillae, the posterior one-third is devoid of papillae.
3. There are 20 deciduous teeth that are replaced by 32 permanent teeth.
   - The types of teeth are incisors, canines, premolars, and molars.
   - A tooth consists of a crown, a neck, and a root.
   - The root is composed of dentin. Within the dentin of the root is the pulp cavity, which is filled with pulp, blood vessels, and nerves. The crown is dentin covered by enamel.
   - Teeth are held in the alveoli by the periodontal ligaments.
4. The muscles of mastication are the masseter, temporalis, medial pterygoid, and lateral pterygoid.
5. The roof of the oral cavity is divided into the hard and soft palates.
6. Salivary glands produce serous and mucous secretions. The three pairs of large salivary glands are the parotid, submandibular, and sublingual.

Pharynx

The pharynx consists of the nasopharynx, oropharynx, and laryngopharynx.

Esophagus

1. The esophagus connects the pharynx to the stomach. The upper- and lower-esophageal sphincters regulate movement.
2. The esophagus consists of an outer adventitia, a muscular layer (longitudinal and circular), a submucosal layer (with mucous glands), and a stratified squamous epithelium.

Stomach

1. Structures of the stomach.
   - The openings of the stomach are the gastroesophageal (to the esophagus) and the pyloric (to the duodenum).
   - The wall of the stomach consists of an external serosa, a muscle layer (longitudinal, circular, and oblique), a

submucosa, and simple columnar epithelium (surface mucous cells).
- Rugae are the folds in the stomach when it is empty.

2. Gastric pits are the openings to the gastric glands that contain mucous neck cells, parietal cells, chief cells, and endocrine cells.

Small Intestine

1. The small intestine is divided into the duodenum, jejunum, and ileum.
2. The wall of the small intestine consists of an external serosa, muscles (longitudinal and circular), submucosa, and simple columnar epithelium.
3. Circular folds, villi, and microvilli greatly increase the surface area of the intestinal lining.
4. Absorptive, goblet, and endocrine cells develop in intestinal glands. Duodenal glands produce mucus.

Liver

1. The liver has four lobes: right, left, caudate, and quadrate.
2. The liver is divided into lobules.
   - The hepatic cords are composed of columns of hepatocytes that are separated by the bile canaliculi.
   - The sinusoids are enlarged spaces filled with blood and lined with endothelium and hepatic phagocytic cells.
3. The portal triads supply the lobules.
   - The hepatic arteries and the hepatic portal veins bring blood to the lobules and empty into the sinusoids.
   - The sinusoids empty into central veins, which join to form the hepatic veins, which leave the liver.
   - Bile canaliculi converge to form hepatic ducts, which leave the liver.
4. Bile leaves the liver through the hepatic duct system.
   - The hepatic ducts receive bile from the lobules.
   - The cystic duct from the gallbladder joins the hepatic duct to form the common bile duct.
   - The common bile duct joins the pancreatic duct at the point where it empties into the duodenum.

Gallbladder

The gallbladder is a small sac on the inferior surface of the liver.

Pancreas

1. The pancreas is an endocrine and an exocrine gland. Its exocrine function is the production of digestive enzymes.
2. The pancreas is divided into lobules which contain acini. The acini connect to a duct system that eventually forms the pancreatic duct, which empties into the duodenum.

Large Intestine

1. The cecum forms a blind sac at the junction of the small and large intestines. The vermiform appendix is a blind sac off the cecum.
2. The ascending colon extends from the cecum superiorly to the right colic flexure. The transverse colon extends from the right to the left colic flexures. The descending colon extends inferiorly to join the sigmoid colon.
3. The sigmoid colon is an S-shaped tube that ends at the rectum.
4. Longitudinal smooth muscles of the large intestine wall are arranged into bands called teniae coli that contract to produce pouches called haustra.
5. The mucosal lining of the large intestine is simple columnar epithelium with mucus-producing crypts.
6. The rectum is a straight tube that ends at the anus.
7. The anal canal is surrounded by an internal anal sphincter (smooth muscle) and an external anal sphincter (skeletal muscle).

Peritoneum

1. The peritoneum is a serous membrane that lines the abdominal cavity and organs.
2. Mesenteries are peritoneum that extends from the body wall to many of the abdominal organs.
3. Retroperitoneal organs are located behind the peritoneum.

FUNCTIONS OF THE DIGESTIVE SYSTEM *p. 791*

The digestive system is regulated by neural and hormonal mechanisms. Intramural plexuses are responsible for local reflexes.

Functions of the Oral Cavity

1. Amylase in saliva starts starch digestion. Mucin provides lubrication.
2. Chewing is primarily a reflex activity. The teeth cut, tear, and crush the food.

Deglutition

1. During the voluntary phase of deglutition, a bolus of food is moved by the tongue from the oral cavity to the pharynx.
2. The pharyngeal phase is a reflex caused by stimulation of stretch receptors in the pharynx.
   - The soft palate closes the nasopharynx, and the epiglottis closes the opening into the larynx.
   - Pharyngeal muscles move the bolus to the esophagus.
3. The esophageal phase is a reflex initiated by the stimulation of stretch receptors in the esophagus. A wave of contraction (peristalsis) moves the food to the stomach.

Stomach Functions

1. Stomach secretions.
   - Mucus protects the stomach lining.
   - Pepsinogen is converted to pepsin, which digests proteins.
   - Hydrochloric acid promotes pepsin activity and kills microorganisms.
   - Intrinsic factor is necessary for vitamin B_{12} absorption.
2. Regulation of stomach secretions
   - The cephalic phase is initiated by the sight, smell, taste, or thought of food. Nerve impulses from the medulla stimulate hydrochloric acid, pepsinogen, and gastrin secretion.
   - The gastric phase is initiated by distention of the stomach, which stimulates gastrin secretion and activates central nervous system and local reflexes that promote secretion.
   - The intestinal phase is initiated by acidic chyme, which enters the duodenum and stimulates neuronal reflexes and the secretion of hormones that induce and then inhibit gastric secretions.
3. Movement in the stomach.
   - The stomach stretches and relaxes to increase volume.
   - Mixing waves mix the stomach contents with stomach secretions to form chyme.
   - Peristaltic waves move the chyme into the duodenum.
4. Regulation of stomach emptying.
   - Gastrin and stretching of the stomach stimulate stomach emptying.
   - Chyme entering the duodenum inhibits movement through neuronal reflexes and the release of hormones.

Functions of the Small Intestine

1. Secretions of the small intestine.
   - Mucus protects against digestive enzymes and stomach acids.
   - Digestive enzymes (disaccharidases and peptidases) are bound to the intestinal wall.
   - Chemical or tactile irritation, vagal stimulation, and secretin stimulate intestinal secretion.
2. Movement in the small intestine.
   - Segmental contractions mix intestinal contents. Peristaltic contractions move materials distally.
   - Stretch of smooth muscles, local reflexes, and the parasympathetic nervous system stimulate contractions. Distention of the cecum initiates a reflex that inhibits peristalsis.

Liver Function

1. The liver produces bile, which contains bile salts that emulsify fats. Secretin and parasympathetic stimulation increase bile production.
2. The liver stores and processes nutrients, produces new molecules, and detoxifies molecules.
3. The liver produces blood components.

Functions of the Gallbladder

1. The gallbladder stores and concentrates bile.
2. Cholecystokinin stimulates gallbladder contraction.

Functions of the Pancreas

1. Secretin stimulates the release of a watery bicarbonate solution that neutralizes acidic chyme.
2. Cholecystokinin stimulates the release of digestive enzymes.

Functions of the Large Intestine

1. Secretion and absorption.
   - Mucus provides protection to the intestinal lining.
   - Bicarbonate ions are secreted by epithelial cells. Sodium is absorbed by active transport, and water is absorbed by osmosis.
2. Microorganisms are responsible for vitamin K production, gas production, and much of the bulk of feces.
3. Movement in the large intestine.
   - Segmental movements mix the colon's contents.
   - Mass movements are strong peristaltic contractions that occur three to four times a day.
   - Defecation is the elimination of feces. Reflex activity moves feces through the internal anal sphincter. Voluntary activity regulates movement through the external anal sphincter.

DIGESTION, ABSORPTION, AND TRANSPORT p. 807

1. Digestion is the breakdown of organic molecules into their component parts.
2. Absorption and transport are the means by which molecules are moved out of the digestive tract and are distributed throughout the body.
3. Transportation occurs by two different routes.
   - Water, ions, and water-soluble products of digestion are transported to the liver through the hepatic portal system.
   - The products of lipid metabolism are transported through the lymphatic system to the circulatory system.

Carbohydrates

1. Carbohydrates consist of starches, glycogen, sucrose, lactose, glucose, and fructose.
2. Polysaccharides are broken down into monosaccharides by a number of different enzymes.
3. Monosaccharides are taken up by intestinal epithelial cells by active transport or by facilitated diffusion.
4. The monosaccharides are carried to the liver where the nonglucose sugars are converted to glucose.
5. Glucose is transported to the cells that require energy.
6. Glucose enters the cells through facilitated diffusion.
7. The rate of transport is influenced by insulin.

Lipids

1. Lipids include triglycerides, phospholipids, steroids, and fat-soluble vitamins.
2. Emulsification is the transformation of large lipid droplets into smaller droplets and is accomplished by bile salts.
3. Lipase digests lipid molecules to form free fatty acids, glycerol, and monoglycerides.
4. Micelles form around lipid digestion products and move to epithelial cells of the small intestine where the products pass into the cells by simple diffusion.
5. Within the epithelial cells, free fatty acids are combined with glycerol to form triglycerides.
6. Proteins coat triglycerides, phospholipids, and cholesterol to form chylomicrons.
7. Chylomicrons enter lacteals within intestinal villi and are carried through the lymphatic system to the bloodstream.
8. Triglycerides are stored in fat; or they are stored, converted into other molecules, or used as energy.

Proteins

1. Pepsin in the stomach breaks proteins into smaller polypeptide chains.
2. Proteolytic enzymes from the pancreas produce small peptide chains.
3. Peptides are broken down into amino acids by peptidases bound to the microvilli of the small intestine.
4. Amino acids are absorbed by active transport, which requires transport of sodium.
5. Amino acids are transported to the liver where the amino acids can be modified or released into the bloodstream.
6. Amino acids are actively transported into cells under the stimulation of growth hormone and insulin.
7. Amino acids are used as building blocks or for energy.

Water

Water can move in either direction across the wall of the small intestine, depending on the osmotic gradients across the epithelium.

Ions

1. Sodium, potassium, calcium, magnesium, and phosphate are actively transported.
2. Chloride ions move passively through the wall of the duodenum and jejunum but are actively transported from the ileum.
3. Calcium ions are actively transported, but vitamin D is required for transport, and the transport is under hormonal control.

CONTENT REVIEW

1. List the major digestive organs.
2. What are the major layers of the digestive tract? How do the serosa and the adventitia differ?
3. What are the general functions of the digestive system?
4. What are the functions of the lips and cheeks?
5. List the functions of the tongue. Distinguish between intrinsic and extrinsic tongue muscles.
6. What are deciduous and permanent teeth? Name the different kinds of teeth.
7. Describe the parts of a tooth. What are dentin, enamel, cementum, and pulp?
8. List the muscles of mastication and the actions they produce.
9. What are the hard and the soft palates?
10. Name and give the location of the three largest salivary glands. Name the other kinds of salivary glands. What is the difference between serous and mucous saliva?
11. Name the three portions of the pharynx.
12. Where is the esophagus located? Describe the layers of the esophageal wall and the esophageal sphincters.
13. Describe the parts of the stomach. List the layers of the stomach wall. How is the stomach different from the esophagus?
14. What are gastric pits and gastric glands? Name the different cell types in the stomach and the secretions they produce.
15. Name and describe the three parts of the small intestine. What are the greater and lesser duodenal papilla?
16. What are circular folds, villi, and microvilli in the small intestine? What are their functions?
17. What is the function of the ileocecal sphincter?
18. What are the hepatic cords and the sinusoids?
19. Describe the flow of blood to and through the liver. Describe the flow of bile away from the liver.
20. What kind of gland is the pancreas? Describe the acini and the duct system of the pancreas.
21. Describe the parts of the large intestine. What are teniae coli, haustra, and crypts?
22. What are the peritoneum, the mesenteries, and retroperitoneal organs?
23. What are the functions of saliva? How is salivary secretion regulated?
24. Describe the mastication reflex and the three stages of swallowing.
25. List five stomach secretions, and give their functions.
26. Name the three stages of stimulation of gastric secretions, and discuss the cause and result of each stage.
27. How are gastric secretions inhibited? Why is this inhibition necessary?
28. Why does pressure in the stomach not greatly increase as the stomach fills?
29. What are the two kinds of stomach movements? How are stomach movements regulated by hormones and nervous control?
30. List the enzymes of the small intestine wall, and state their functions.
31. What are the two kinds of movements of the small intestine? How are they regulated?
32. Describe the functions of the liver and the gallbladder. What stimulates the release of bile from the liver and the gallbladder?
33. Name the two kinds of exocrine secretions that are produced by the pancreas. Where are they produced, what stimulates their production, and what is their function?
34. What kinds of movements occur in the colon? Describe the defecation reflex.
35. Name the substances secreted and absorbed by the colon. What is the role of microorganisms in the colon?
36. Describe the mechanism of absorption and the route of transport for water-soluble and lipid-soluble molecules.
37. Describe the enzymatic digestion of carbohydrates, lipids, and proteins, and list the breakdown products of each.
38. Describe the movement of water through the intestinal wall.
39. Where and how are the various ions absorbed?

CONCEPT REVIEW

1. While anesthetized, patients sometimes vomit. Given that the anesthetic eliminates the swallowing reflex, explain why it is dangerous for an anesthetized patient to vomit.
2. Achlorhydria is a condition in which the stomach stops producing hydrochloric acid and other secretions. What effect would achlorhydria have on the digestive process? On red blood cell count?
3. Victor Worrystudent developed a duodenal ulcer during final examination week. Describe the possible reasons. Explain what habits could have caused the ulcer and recommend a reasonable remedy.
4. Gallstones sometimes obstruct the common bile duct. What would be the consequences of such a blockage?
5. Sometimes a gallstone can move to the pancreatic duct and block or impair the flow of pancreatic juices. What symptoms would you expect if this blockage occurred?
6. A patient has a spinal cord injury at level L2 of the spinal cord. How will this injury affect his ability to defecate? What components of his defecation response are still present, and which are lost?
7. The bowel (colon) occasionally can become impacted. Given what you know about the functions of the colon and the factors that determine the movement of substances across the colon wall, predict the effect of the impaction on the contents of the colon above the point of impaction.

? ?

ANSWERS TO PREDICT QUESTIONS

1 *p. 781.* The moist, stratified squamous epithelium of the oropharynx and the laryngopharynx protects these regions from abrasive food when it is first swallowed. The ciliated pseudostratified epithelium of the nasopharynx helps move mucus produced in the nasal cavity and nasopharynx into the oropharynx and esophagus. It is not as necessary to protect the nasopharynx from abrasion since food does not normally pass through this cavity.

2 *p. 791.* Four layers. The greater omentum is actually a folded mesentery, with each part consisting of two layers of serous squamous epithelium.

3 *p. 795.* It is important for the nasopharynx to be closed during swallowing so that food will not reflux into it or the nasal cavity. An explosive burst of laughter can relax the soft palate, open the nasopharynx, and cause the liquid to enter the nasal cavity.

4 *p. 795.* If a person tries to swallow and speak at the same time, the epiglottis would be elevated, and food or liquid could enter the larynx, causing the person to choke.

5 *p. 796.* After a heavy meal, blood pH may increase because as bicarbonate ions pass from the cells of the stomach into the extracellular fluid, the extracellular fluid's pH increases. As the extracellular fluid exchanges ions with the blood, the blood pH also increases.

6 *p. 804.* Secretin production and its stimulation of bicarbonate ion secretion constitute a negative-feedback mechanism because as the pH of the chyme in the duodenum decreases as a result of the presence of acid, secretin causes an increase in bicarbonate ion secretion that increases the pH, restoring the proper pH balance in the duodenum.

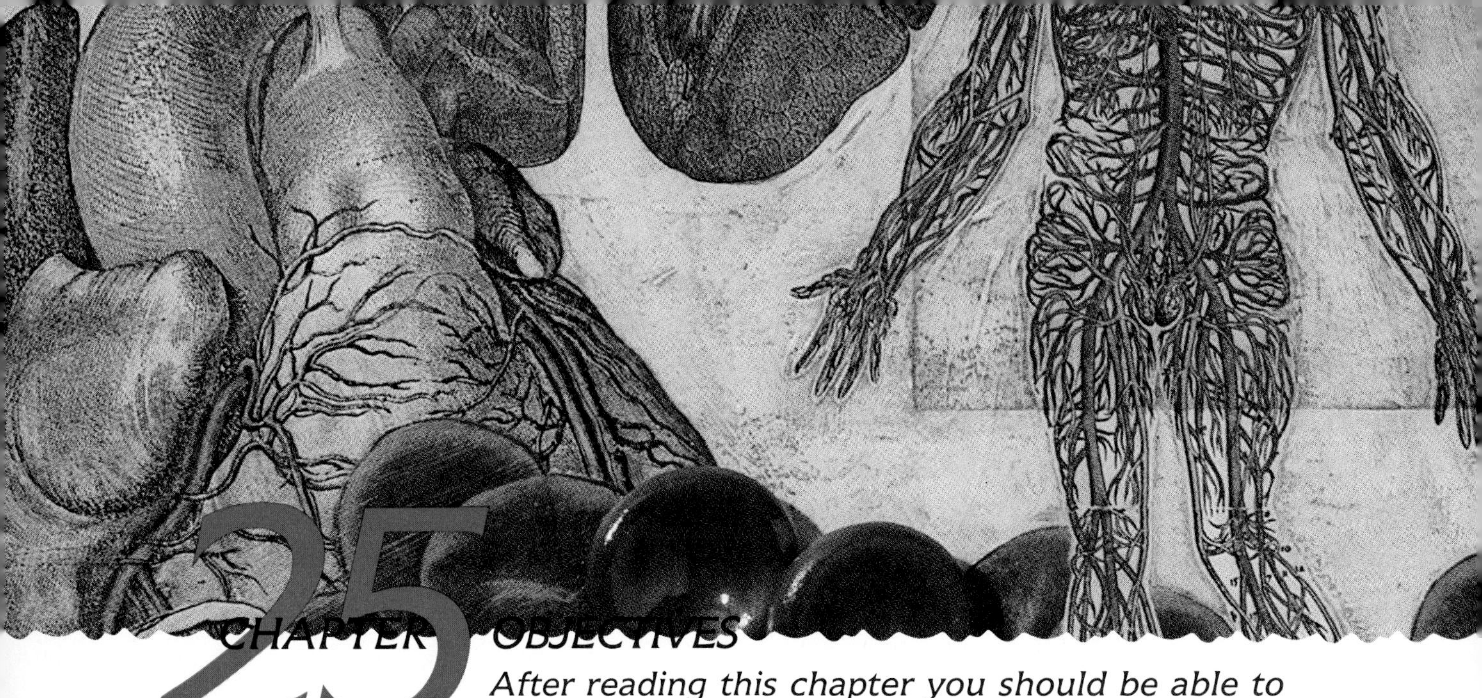

25 CHAPTER OBJECTIVES

After reading this chapter you should be able to

1. Define metabolism and nutrition.
2. Define a kilocalorie and list the kilocalories found in a gram of carbohydrate, lipids, and protein.
3. Describe for carbohydrates, lipids, and proteins their dietary sources, their uses in the body, and the daily recommended amounts of each in the diet.
4. List the common vitamins and indicate the function of each.
5. List the most common minerals and indicate the function of each.
6. Describe the basic steps in glycolysis and indicate its major products.
7. Describe the citric acid cycle and its major products.
8. Describe the electron-transport chain and how ATP is produced in the process.
9. Explain how two ATP molecules are produced in anaerobic respiration and 38 ATP molecules are produced in aerobic respiration from one molecule of glucose.
10. Describe the basic steps involved in using lipids as an energy source.
11. Explain how amino acids can be used for energy.
12. Describe the conversion of lipids and protein into glucose and the conversion of glucose into glycogen.
13. Differentiate between the absorptive and postabsorptive metabolic states.
14. Define metabolic rate.
15. Describe heat production and regulation in the body.

Nutrition and Metabolism

KEY TERMS

aerobic (ăr-o′bik) respiration

anaerobic (an′ăr-o′bik) respiration

beta-oxidation

citric acid cycle

electron-transport chain

gluconeogenesis (glu′ko-ne-o-jen′ĕ-sis)

glycogenesis (gli′ko-jen′ĕ-sis)

glycolysis (gli-kol′ĭ-sis)

kilocalorie (kil′o-kal-o-re)

metabolic rate

metabolism (mĕ-tab′o-lizm)

mineral

nutrition

oxidative deamination

vitamin (vi′tah-min)

RELATED TOPICS

The following terms or concepts are important for a good understanding of this chapter. If you are not familiar with them, you should review them before proceeding.

Oxidation-reduction reaction (Chapter 2)

Chemical characteristics of organic molecules (Chapter 2)

Structure and function of cellular organelles (Chapter 3)

Digestion and absorption (Chapter 24)

NUTRITION IS THE PROCESS BY WHICH FOOD ITEMS (NUTRIENTS) ARE OBTAINED AND USED BY THE BODY. THE PROCESS INCLUDES DIGESTION, ABSORPTION, TRANSPORT, AND CELL METABOLISM. NUTRITION CAN ALSO BE DEFINED AS THE EVALUATION OF FOOD AND DRINK REQUIREMENTS FOR NORMAL BODY FUNCTION.

Metabolism (mĕ-tab′o-lizm; change) is the total of all the chemical changes that occur in the body. It consists of **anabolism** (ah-nab′o-lizm; the building of molecules), the energy-requiring process by which small molecules are joined to form larger molecules, and **catabolism** (kah-tab′o-lizm; the breakdown of molecules), the energy-releasing process by which large molecules are broken down into smaller molecules. Anabolism occurs in all cells of the body as they divide to form new cells, maintain their own intracellular structure, and produce molecules such as hormones, neurotransmitters, or extracellular matrix molecules for export. Catabolism begins during the process of digestion and is concluded within individual cells where the energy released by the breaking of covalent bonds is used to produce adenosine triphosphate (ATP) and heat. The energy derived from catabolism is used to drive anabolic reactions.

The heat resulting from the chemical reactions of metabolism contribute to maintaining body temperature within a narrow range, which is essential for normal function. In addition, heat can be gained from or lost to the external environment. A constant body temperature can be maintained by regulating the internal production of heat and the exchange of heat with the external environment.

NUTRITION

Nutrients

Nutrients (food) are the chemicals taken into the body that are used to produce energy, provide building blocks for new molecules, or function in other chemical reactions. Some substances in food are not nutrients but provide bulk (fiber) in the diet. Nutrients can be divided into six major classes: carbohydrates, proteins, lipids, vitamins, minerals, and water. Carbohydrates, proteins, and lipids are the major organic nutrients and are broken down by enzymes into their individual components during digestion. Many of these subunits are broken down further to supply energy, whereas others are used as building blocks for other macromolecules. Vitamins, minerals, and water are taken into the body without being digested. Some nutrients are required in fairly substantial quantities, and others are required in only trace amounts (trace elements).

Essential nutrients are nutrients that must be ingested because the body cannot manufacture them or is unable to manufacture adequate amounts of them. The essential nutrients include certain amino acids, linoleic acid (a fatty acid), most vitamins, minerals, water, and a minimum amount of carbohydrates. The term essential does not mean, however, that only the essential nutrients are required by the body. Other nutrients are necessary, but if they are not ingested, they can be synthesized from the essential nutrients. Most of this synthesis takes place in the liver, which has a remarkable ability to transform and manufacture molecules.

Kilocalories

The energy stored within the chemical bonds of nutrients can be used by the body. A **calorie** (kal'o-re; cal) is the amount of energy (heat) necessary to raise the temperature of 1 g of water from 14° C to 15° C. A **kilocalorie** (kil'o-kal-o-re; kcal) is 1000 calories and is used to express the large amounts of energy available in foods and released through metabolism.

A kilocalorie is often called a Calorie (with a capital "C"). Unfortunately, this usage has resulted in confusion of the term calorie (with a lower case "c") with Calorie (with a capital "C"). It is common practice on food labels and in nutrition books to use calorie when Calorie (kilocalorie) is the proper term.

Almost all of the kilocalories found in food come from carbohydrates, proteins, or fats. For each gram of carbohydrate or protein metabolized by the body, approximately 4 kcal of energy are released. Fats contain more energy per unit of weight than carbohydrates and proteins and yield approximately 9 kcal per gram. Table 25-1 lists the kilocaloric content of some typical foods. A typical American diet consists of 50% to 60% carbohydrates, 35% to 45% fats, and 10% to 15% protein. Table 25-1 also lists the carbohydrate, fat, and protein composition of some foods.

Carbohydrates

Sources in the diet

Carbohydrates include monosaccharides, disaccharides, and polysaccharides (see Chapter 2). Most of the carbohydrates humans ingest come from plants. The exceptions are lactose (milk sugar), which is found in animal and human milk, and glycogen (animal starch), which is found in meat.

The most common monosaccharides in the diet are glucose and fructose. Plants capture the energy in sunlight and use the energy to produce glucose, which can be found in vegetables, fruits, molasses, honey, and syrup. Fructose (fruit sugar), an isomer of glucose (see Chapter 2), is most often derived from fruits and berries.

The disaccharide sucrose (table sugar) is what most people think of when they use the term sugar. Sucrose is a glucose and fructose molecule joined together, and its principal sources are sugar cane and sugar beets. Maltose (malt sugar), derived from germinating cereals, is a combination of two glucose molecules, and lactose (in milk) consists of a glucose and a galactose molecule.

The **complex carbohydrates** are the polysaccharides: starch, glycogen, and cellulose. Starch is an energy storage molecule in plants and is found pri-

Table 25-1

Food composition

FOOD	QUANTITY	FOOD ENERGY (kcal)	CARBOHYDRATE (g)	FAT (g)	PROTEIN (g)
Dairy Products					
Whole milk (3.3% fat)	1 cup	150	11	8	8
Low fat milk (2% fat)	1 cup	120	12	5	8
Butter	1 T	100	—	12	—
Grain					
Bread, white enriched	1 slice	75	24	1	2
Bread, whole wheat	1 slice	65	14	1	3
Fruit					
Apple	1	80	20	1	—
Banana	1	100	26	—	1
Orange	1	65	16	—	1
Vegetables					
Corn, canned	1 cup	140	33	1	4
Peas, canned	1 cup	150	29	1	8
Lettuce	1 cup	5	2	—	—
Celery	1 cup	20	5	—	1
Potato, baked	1 large	145	33	—	4
Meat, Fish, and Poultry					
Lean ground beef (10% fat)	3 oz	185	—	10	23
Shrimp, french fried	3 oz	190	9	9	17
Tuna, canned	3 oz	170	—	7	24
Chicken breast, fried	3 oz	160	1	5	26
Bacon	2 slices	85	—	8	4
Hot dog	1	170	1	15	7
Fast Foods					
McDonald's Egg McMuffin	1	327	31	15	19
McDonald's Big Mac	1	563	41	33	26
Taco Bell's beef burrito	1	466	37	21	30
Arby's roast beef	1	350	32	15	22
Pizza Hut Super Supreme	1 slice	260	23	13	15
Long John Silver's fish	2 pieces	366	21	22	22
McDonald's fish fillet	1	432	37	25	14
Dairy Queen malt, large	1	840	125	28	22
Desserts					
Cupcake with icing	1	130	21	5	2
Chocolate chip cookie	4	200	29	9	2
Apple pie	1 piece	345	51	15	3
Dairy Queen cone, large	1	340	52	10	10
Beverage					
Cola soft drink	12 oz	145	37	—	—
Beer	12 oz	150	14	—	1
Wine	3½ oz	85	4	—	—
Hard liquor (86 proof)	1½ oz	105	—	—	—
Miscellaneous					
Egg	1	80	1	6	6
Mayonnaise	1 T	100	—	11	—
Sugar	1 T	45	12	—	—

marily in vegetables, fruits, and grains; glycogen is an energy storage molecule in animals and is located in muscle and in the liver; and cellulose forms the cell walls of plants.

Uses in the body

During digestion, polysaccharides and disaccharides are split into monosaccharides, which are absorbed into the blood (see Chapter 24). Humans can break the bonds between the glucose molecules of starch and glycogen, but cellulose is not digestible by humans. Instead, cellulose provides fiber or "roughage," that increases the bulk of feces and promotes defecation.

Fructose and other monosaccharides absorbed into the blood are converted into glucose by the liver. Glucose, whether absorbed from the digestive tract or produced by the liver, is a primary energy source for most cells, which use it to produce ATP molecules (see anaerobic and aerobic respiration later in this chapter). Because the brain relies almost entirely on glucose for its energy, blood glucose levels are carefully regulated (see Chapter 18).

If excess amounts of glucose are present, the glucose is converted into glycogen that is stored in muscle and in the liver. The glycogen can be rapidly converted back to glucose when energy is needed. Because cells can store only a limited amount of glycogen, any additional glucose is converted into fat that is stored in adipose tissue.

In addition to being used as a source of energy, sugars have other functions. They form part of DNA, RNA, and ATP molecules, and they combine with proteins to form glycoprotein receptor molecules on the outer surface of the plasma membrane.

Recommended requirements

It is recommended that 125 to 175 g of carbohydrates be ingested every day. Although a minimum acceptable level of carbohydrate ingestion is not known, it is assumed that amounts of 100 g or less per day result in overuse of proteins and fats for energy sources. Because muscles are primarily protein, the use of proteins for energy can result in the breakdown of muscle tissue, and the use of fats can result in acidosis (see Chapter 27).

Complex carbohydrates are recommended because starchy foods often contain other valuable nutrients such as vitamins and minerals. Although foods such as soft drinks and candy are rich in carbohydrates, they are considered "empty calories." They are mostly sugar (e.g., a typical soft drink contains 9 teaspoons of sugar), which provides energy, but they may have little other nutritive value. In excess, the consumption of empty calories can result in obesity and tooth decay.

Lipids

Sources in the diet

Approximately 95% of the lipids in the human diet are **triglycerides,** which consist of three fatty acids attached to a glycerol molecule (see Chapter 2). Triglycerides are often referred to as fats, which can be divided into saturated and unsaturated fats. Fats are saturated if their fatty acids have only single covalent bonds between their carbon atoms, and they are unsaturated if they have one (monounsaturated) or more (polyunsaturated) double covalent bonds between their carbon atoms (see Figure 2-15). Saturated fats are found in the fats of meats (e.g., beef, pork), dairy products (e.g., whole milk, cheese, butter), eggs, nuts, coconut oil, and palm oil. Monounsaturated fats include olive and peanut oils, and polyunsaturated fats are in fish, safflower, sunflower, and corn oils.

The remaining 5% of lipids include cholesterol and phospholipids. Cholesterol is a steroid (see Chapter 2) found in high concentrations in brain, liver, and egg yolks, but it is also present in whole milk, cheese, butter, and meats. Cholesterol is not found in plants. Phospholipids are a major component of cell membranes, and they are found in a variety of foods.

Uses in the body

Triglycerides are an important source of energy that can be used to produce ATP molecules, and a gram of triglycerides delivers more than twice as many kilocalories as a gram of carbohydrates. Some cells (e.g., skeletal muscle cells) derive most of their energy from triglycerides.

After a meal excess triglycerides that are not immediately used are stored in adipose tissue or the liver. Later, when energy is required, the triglycerides are broken down, and their fatty acids are released into the blood where they can be taken up and used by various tissues. In addition to storing energy, adipose tissue surrounds and pads organs, and under the skin adipose tissue is an insulator, which prevents heat loss.

Cholesterol is an important molecule with many functions in the body. It is obtained in food, or it can be manufactured by the liver and most other tissues. Cholesterol is a component of the plasma membrane, and it can be modified to form other useful molecules such as bile salts and steroid hormones. Bile salts are necessary for fat digestion and absorption. Steroid hormones include the sex hormones estrogen, progesterone, and testosterone, which regulate the reproductive system, and prostaglandins, which are involved in inflammation, tissue repair, and smooth muscle contraction.

Phospholipids are part of the plasma membrane

and are used to construct the myelin sheath around the axons of nerve cells. Lecithin, a major component of plasma membranes, is found in many foods and is manufactured by the liver.

Recommended requirements

The American Heart Association recommends that fats account for 30% or less of the total kilocaloric intake. Furthermore, saturated fats should contribute no more than 10% of total fat intake, and cholesterol should be limited to 250 mg (the amount in an egg yolk) or less per day. These guidelines reflect the belief that excess amounts of fats, especially saturated fats and cholesterol, contribute to cardiovascular disease. The typical American diet derives 35% to 45% of its kilocalories from fats, indicating most Americans need to reduce fat consumption. On the other hand, fat intake can account for as few as 10% of the kilocalories in a healthy person's diet.

If insufficient amounts of fats are consumed, the body can synthesize fats from carbohydrates and proteins. However, linoleic acid, a fatty acid in many triglycerides, cannot be manufactured by the body. Therefore linoleic acid is an essential fatty acid, which must be ingested. It is found in plant oils and milk.

Proteins

Sources in the diet

Proteins are chains of amino acids (see Chapter 2). There are 20 amino acids found in food, and the adult human body can synthesize all but the eight essential amino acids (i.e., isoleucine, leucine, lysine, methionine, phenylalanine, threonine, tryptophan, and valine). If adequate amounts of the essential amino acids are ingested, they can be used to manufacture the nonessential amino acids, which are also necessary for good health. A complete protein food contains all eight essential amino acids, whereas an incomplete protein food does not have all of them. Examples of complete proteins are meat, fish, poultry, milk, cheese, and eggs; and examples of incomplete proteins are leafy green vegetables, grains, and legumes (peas and beans).

Uses in the body

Proteins perform numerous functions in the human body as the following examples illustrate. Collagen provides structural strength in connective tissue as does keratin in the skin, and the combination of actin and myosin makes muscle contraction possible. Enzymes are responsible for regulating the rate of chemical reactions, and protein hormones regulate many physiological processes (see Chapter 18). Proteins in the blood act as buffers to prevent changes in pH, and hemoglobin transports oxygen and carbon dioxide in the blood. Proteins also function as carrier molecules to move materials across plasma membranes, and other proteins in the plasma membrane function as receptor molecules and ion channels. Antibodies, lymphokines, and complement are part of the immune system response that protects against microorganisms and other foreign substances.

Proteins can also be used as a source of energy, yielding the same amount of energy as carbohydrates. If excess proteins are ingested, the energy in the proteins can be stored by converting their amino acids into glycogen or fats.

Recommended requirements

The recommended daily consumption of protein for an adult is 0.8 g per kilogram of body weight, or approximately 12% of total kilocalories. For a 58-kg (128 pounds) woman this is 46 g per day, and for a 70-kg man (154 pounds) it is 56 g per day. A cup of skim milk contains 8 g protein, one ounce of meat contains 7 g protein, and a slice of bread provides 2 g protein.

When protein intake is adequate, the synthesis and breakdown of proteins in a healthy adult occurs at the same rate. Because the amino acids of proteins contain nitrogen, when a person is in **nitrogen balance,** the nitrogen content of ingested protein is equal to the nitrogen excreted in urine or feces. A starving person is in negative nitrogen balance because the nitrogen gained in the diet is less than that lost by excretion. In other words, the rate at which proteins are broken down for energy results in greater nitrogen loss than is replaced in the diet. A growing child or a healthy pregnant woman, on the other hand, is in positive nitrogen balance because more nitrogen is going into the body to produce new tissues than is lost by excretion.

Vitamins

Vitamins (vi'tah-minz; life-giving chemicals) exist in minute quantities in food and are essential to normal metabolism (Table 25-2). Most vitamins cannot be produced by the body and must be obtained through the diet. Since no single food item or nutrient class provides all the essential vitamins, it is necessary to eat a variety of foods (maintain a balanced diet). The absence of a specific vitamin in the diet can result in a specific deficiency disease. A few vitamins (e.g., vitamin K) are produced by intestinal bacteria, and a few can be formed by the body from substances called **provitamins** (portions of vitamins that can be assembled or modified by the body into functional vitamins). Carotene is an example of a

Table 25-2

The principal vitamins

VITAMIN	FAT (F) OR WATER (W) SOLUBLE	SOURCE	FUNCTION	SYMPTOMS OF DEFICIENCY	MINIMUM DAILY REQUIREMENTS
A (retinol)	F	From provitamin carotene found in yellow and green vegetables: preformed in liver, egg yolk, butter, and milk	Necessary for rhodopsin synthesis, normal health of epithelial cells, and bone and tooth growth	Rhodopsin deficiency, night blindness, retarded growth, skin disorders, and increased infection risk	1000 μg (5000 IU) in males; 800 μg (4000 IU) in females*
B_1 (thiamine)	W	Yeast, grains, and milk	Involved in carbohydrate and amino acid metabolism; necessary for growth	Beriberi—muscle weakness (including cardiac muscle), neuritis, and paralysis	1.5 mg
B_2 (riboflavin)	W	Green vegetables, liver, wheat germ, milk, and eggs	Component of flavin adenine dinucleotide (FAD); involved in citric acid cycle	Eye disorders and skin cracking, especially at corners of the mouth	1.8 mg
Pantothenic acid (part of B_2 complex)	W	Liver, yeast, green vegetables, grains, and intestinal bacteria	Constituent of coenzyme A, glucose production from lipids and amino acids, and steroid hormone synthesis	Neuromuscular dysfunction and fatigue	10 mg
B_3 (niacin)	W	Fish, liver, red meat, yeast, grains, peas, beans, and nuts	Component of nicotinamide adenine dinucleotide (NAD); involved in glycolysis and citric acid cycle	Pellagra—diarrhea, dermatitis, and mental disturbance	20 mg
B_6 (pyridoxine)	W	Fish, liver, yeast, tomatoes, and intestinal bacteria	Involved in amino acid metabolism	Dermatitis, retarded growth, and nausea	2 mg
Folic acid	W	Liver, green leafy vegetables, and intestinal bacteria	Nucleic acid synthesis; hematopoiesis	Macrocytic anemia (enlarged red blood cells)	200 μg
B_{12} (cyanocobalamin)	W	Liver, red meat, milk and eggs	Necessary for erythrocyte production; some nucleic acid and amino acid metabolism	Pernicious anemia and nervous system disorders	2 μg
C (ascorbic acid)	W	Citrus fruit, tomatoes, and green vegetables	Collagen synthesis; general protein metabolism	Scurvy—defective bone formation and poor wound healing	60 mg
D (cholecalciferol, ergosterol)	F	Fish liver oil, enriched milk, and eggs; provitamin D converted by sunlight to cholecalciferol in the skin	Promotes calcium and phosphorus use; normal growth and bone and teeth formation	Rickets—poorly developed, weak bones; osteomalacia; bone reabsorption	10 μg (400 IU) for children or during pregnancy

*An international unit (IU) is the quantity of a substance that produces a particular biological effect agreed upon internationally.

Table 25-2

The principal vitamins—cont'd

VITAMIN	FAT (F) OR WATER (W) SOLUBLE	SOURCE	FUNCTION	SYMPTOMS OF DEFICIENCY	MINIMUM DAILY REQUIREMENTS
E (alpha-tocopherol)	F	Wheat germ; cottonseed, palm, and rice oils; grain, liver, and lettuce	Prevents catabolism of certain fatty acids; may prevent miscarriage	Muscular dystrophy and sterility	Unknown, 10 μg (40 IU) recommended
H (biotin), often considered part of the B-vitamin group	W	Liver, yeast, eggs, and intestinal bacteria	Fatty acid and purine synthesis; movement of pyruvic acid into citric acid cycle	Mental and muscle dysfunction, fatigue, and nausea	Unknown, 0.3 mg recommended
K (phylloquinone)	F	Alfalfa, liver, spinach, vegetable oils, cabbage, and intestinal bacteria	Required for synthesis of a number of clotting factors	Excessive bleeding due to retarded blood clotting	65-80 μg recommended

provitamin that can be modified by the body to form vitamin A. The other provitamins are 7-dehydrocholesterol, which can be converted to vitamin D, and tryptophan, which can be converted to niacin.

Vitamins are not broken down by catabolism but are used by the body in their original or slightly modified forms. Once the chemical structure of a vitamin is destroyed, its function is usually lost. The chemical structure of many vitamins is destroyed by heat (e.g., when food is overcooked). Vitamins function as coenzymes, parts of coenzymes, or as parts of enzymes in various metabolic reactions (Chapter 2). Many vitamins (e.g., riboflavin, pantothenic acid, niacin, and biotin) are critical to the production of energy, whereas others (e.g., folic acid and cyanocobalamin) are involved in nucleic acid synthesis. Retinol, thiamine, pyridoxine, cyanocobalamin, ascorbic acid (vitamin C), and vitamins D and E are necessary for general growth. Vitamin K is necessary for blood clotting.

There are two major classes of vitamins—**fat soluble** and **water soluble.** Fat-soluble vitamins such as vitamins A, D, E, and K are absorbed from the intestine along with lipids, and some of them can be stored in the body for long periods of time. Because they can be stored, it is possible to accumulate an overdose of these vitamins in the body (hypervitaminosis) to the point of toxicity. Water-soluble vitamins such as the B complex and C are absorbed with water from the intestinal tract and remain in the body only a short time before being excreted.

1

Predict what would happen if vitamins were broken down during the process of digestion rather than being absorbed intact into the circulation.

? ? ? ? ? ? ? ? ? ? ?

Minerals

A number of inorganic nutrients, **minerals,** are also necessary for normal metabolic functions. They comprise approximately 4% to 5% of the total body weight and are involved in a number of important functions such as adding mechanical strength to bones, combining with organic molecules, or acting as coenzymes, buffers, and osmotic regulators. Some of the important minerals and their functions are listed in Table 25-3.

Minerals are taken into the body by themselves or in combination with organic molecules. The foods with the highest mineral content include vegetables, legumes, and milk. Refined cereals and breads, fats, and sugar foods have hardly any minerals. A balanced diet can provide all the necessary minerals, with a few possible exceptions. For example, women who suffer from excessive menstrual bleeding may need an iron supplement.

Table 25-3

Important minerals

MINERAL	FUNCTION	SYMPTOMS OF DEFICIENCY	MINIMUM DAILY REQUIREMENTS
Calcium	Bone and teeth formation, blood clotting, muscle activity, and nerve function	Spontaneous nerve discharge and tetany	800-1200 mg
Chlorine	Blood acid-base balance; hydrochloric acid production in stomach	Acid-base imbalance	1.7-5.1 g
Chromium	Associated with enzymes in glucose metabolism	Unknown	0.05-0.2 mg
Cobalt	Component of vitamin B_{12}; erythrocyte production	Anemia	Unknown
Copper	Hemoglobin and melanin production; electron-transport system	Anemia and loss of energy	2.0-3.0 mg
Fluorine	Provides extra strength in teeth; prevents dental caries	No real pathology	1.5-4.0 mg
Iodine	Thyroid hormone production; maintenance of normal metabolic rate	Decrease of normal metabolism	150.0 µg
Iron	Component of hemoglobin; ATP production in electron-transport system	Anemia, decreased oxygen transport, and energy loss	10 mg in males; 15 mg in females
Magnesium	Coenzyme constituent; bone formation; muscle and nerve function	Increased nervous system irritability, vasodilation, and arrhythmias	280-350 mg
Manganese	Hemoglobin synthesis; growth; activation of several enzymes	Tremors and convulsions	2.5-5.0 mg
Molybdenum	Enzyme component	Unknown	0.15-0.5 mg
Phosphorus	Bone and teeth formation; important in energy transfer (ATP); component of nucleic acids	Loss of energy and cellular function	800-1200 mg
Potassium	Muscle and nerve function	Muscle weakness, abnormal electrocardiogram, and alkaline urine	1.8-5.6 g
Selenium	Component of many enzymes	Unknown	55-70 µg
Sodium	Osmotic pressure regulation; nerve and muscle function	Nausea, vomiting, exhaustion, and dizziness	1.1-3.3 g
Sulfur	Component of hormones, several vitamins, and proteins	Unknown	Unknown
Zinc	Component of several enzymes; carbon dioxide transport and metabolism; necessary for protein metabolism	Deficient carbon dioxide transport and deficient protein metabolism	15 mg

METABOLISM

Metabolism is the sum of all the chemical reactions taking place within the body, and two of the important components of metabolism are the chemical changes that occur during digestion and the metabolic processes that occur once the products of digestion are taken up by cells. The cellular metabolic processes are often referred to as **cellular metabolism** or **cellular respiration.** The digestive products of carbohydrates, proteins, and lipids taken into body cells are catabolized, and the released energy is used to combine **adenosine diphosphate (ADP)** and an inorganic phosphate group (P_i) to form **adenosine triphosphate (ATP).**

$$ADP + P_i + Energy \rightarrow ATP$$

ATP is often called the energy currency of the cell, and it is used to drive cell activities such as active transport and muscle contraction.

The chemical reactions responsible for the transfer of energy from the chemical bonds of nutrient molecules to ATP molecules involves oxidation-reduction reactions (see Chapter 2). A molecule is reduced when it gains electrons, and it is oxidized when it loses electrons. A nutrient molecule has many hydrogen atoms covalently bonded to the car-

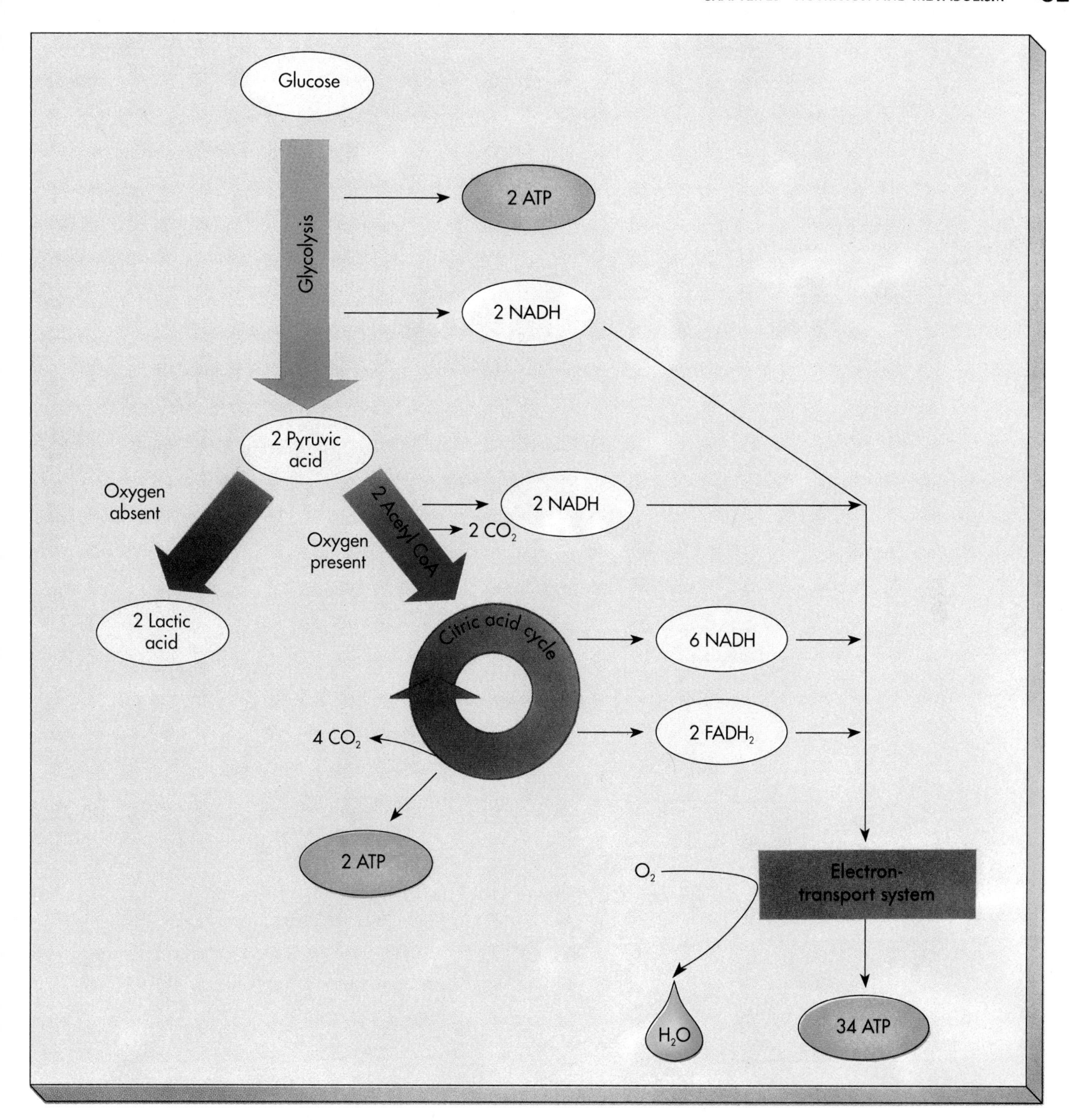

FIGURE 25-1 Overview of cellular metabolism, including glycolysis, citric acid cycle, and electron-transport system.

bon atoms that form the "backbone" of the molecule. Since a hydrogen atom is a hydrogen ion (proton) and an electron, the nutrient molecule has many electrons and is therefore highly reduced. When a hydrogen ion and an associated electron are lost from the nutrient molecule, the molecule loses energy and becomes oxidized. The energy in the electron is used to synthesize ATP. The major events of cellular metabolism are summarized in Figure 25-1.

CARBOHYDRATE METABOLISM

Glycolysis

Carbohydrate metabolism begins with **glycolysis** (gli-kol'ĭ-sis), which is a series of chemical reactions in the cytosol that result in the breakdown of glucose to two pyruvic acid molecules (Figure 25-2).

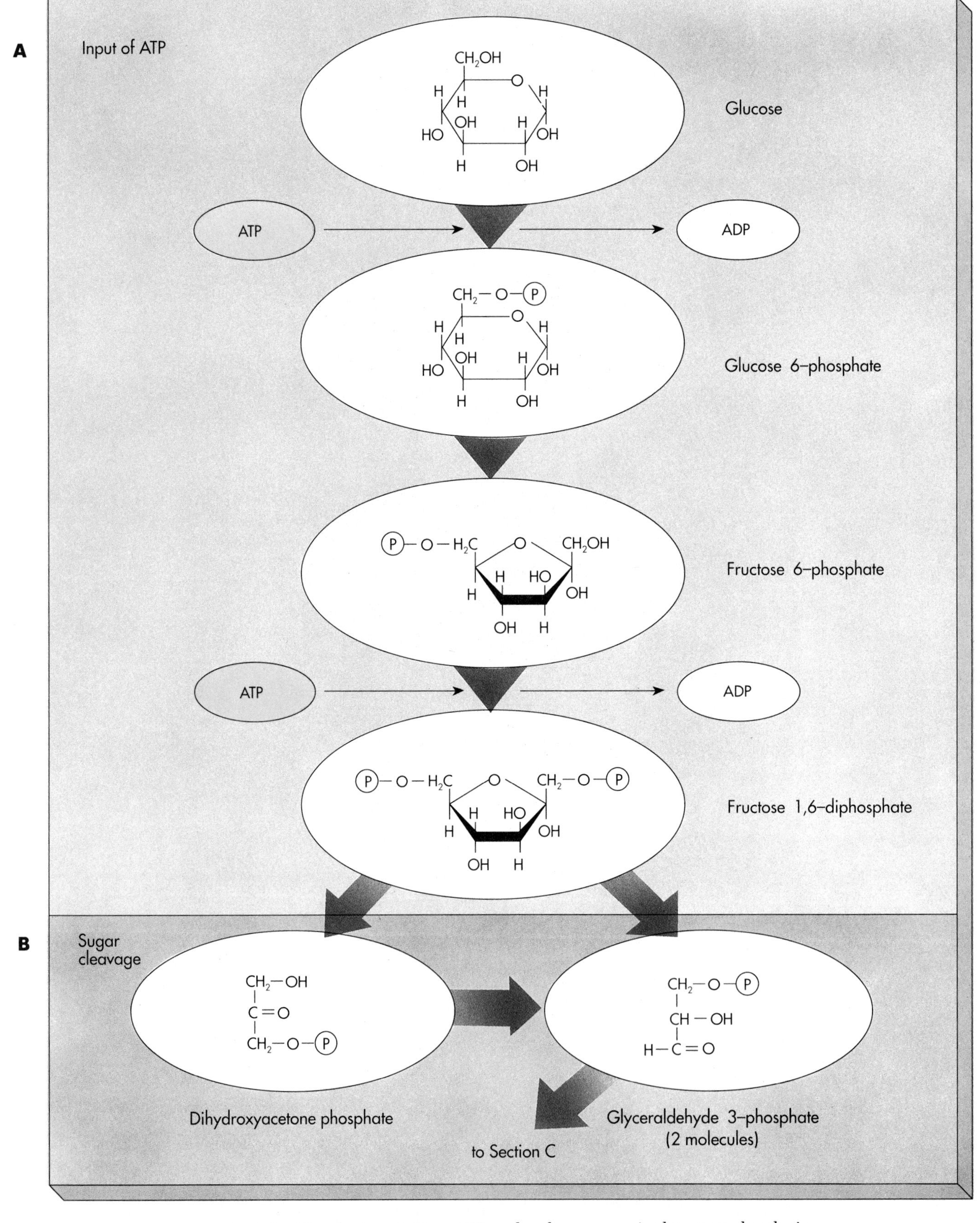

FIGURE 25-2 Glycolysis. **A** Two ATP molecules are required to start glycolysis, and fructose 1,6-diphosphate is formed. **B** Fructose 1,6-diphosphate is split to form two three-carbon glyceraldehyde 3-phosphate molecules.

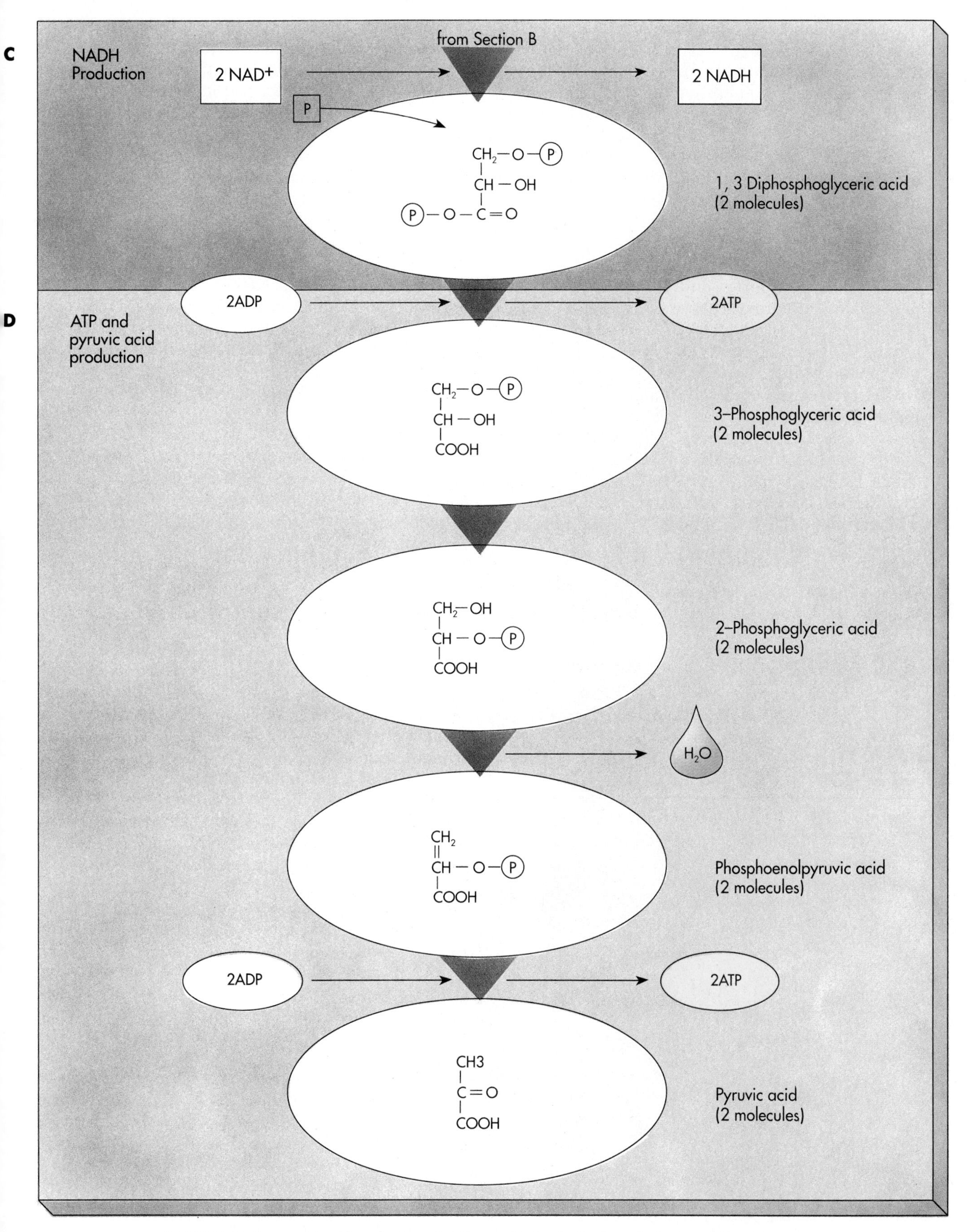

FIGURE 25-2, cont'd Glycolysis. **C** Glyceraldehyde 3-phosphate is oxidized to 1,3-diphosphoglyceric acid, and NAD^+ is reduced to NADH. **D** Two ATP molecules and a pyruvic acid molecule are produced for each glyceraldehyde 3-phosphate.

Glycolysis can be divided into four phases.

1. **Input of ATP.** The first steps in glycolysis require the input of energy in the form of two ATP molecules. A phosphate group is transferred from ATP to the glucose molecule, a process called **phosphorylation,** to form glucose 6-phosphate. The glucose 6-phosphate atoms are rearranged to form fructose 6-phosphate, which is then converted to fructose 1,6-diphosphate by the addition of another phosphate group from another ATP.
2. **Sugar cleavage.** Fructose 1,6-diphosphate is cleaved into two three-carbon molecules, glyceraldehyde 3-phosphate and dihydroxyacetone phosphate. Dihydroxyacetone phosphate is rearranged to form glyceraldehyde 3-phosphate; consequently, two molecules of glyceraldehyde 3-phosphate result.
3. **NADH production.** Each glyceraldehyde 3-phosphate molecule is oxidized (loses two electrons) to form 1,3-diphosphoglyceric acid, and **nicotinamide adenine dinucleotide (NAD^+)** is reduced (gains two electrons) to **NADH**. Glyceraldehyde 3-phosphate also loses two hydrogen ions, one of which binds to NAD^+.

$$NAD^+ + 2\,e^- + 2\,H^+ \rightarrow NADH + H^+$$

In the equation NAD^+ is the oxidized form of nicotinamide adenine dinucleotide, and *NADH* is the reduced form. NADH is a carrier molecule with two high energy electrons (e^-) that can be used to produce ATP molecules through the electron-transport chain (described later in this chapter).

4. **ATP and pyruvic acid production.** The last four steps of glycolysis produce two ATP molecules and one pyruvic acid molecule from each 1,3-diphosphoglyceric acid molecule.

The events of glycolysis are summarized in Table 25-4. Each glucose molecule that enters glycolysis forms two glyceraldehyde 3-phosphate molecules at the sugar cleavage phase. Each glyceraldehyde 3-phosphate molecule produces two ATP molecules, one NADH molecule, and one pyruvic acid molecule. Therefore each glucose molecule forms four ATP, two NADH, and two pyruvic acid molecules. Because the start of glycolysis requires the input of two ATP molecules, however, the final yield of each glucose molecule is two ATP, two NADH, and two pyruvic acid molecules (see Figure 25-1).

If the cell has adequate amounts of oxygen, the NADH and pyruvic acid molecules will be used in aerobic respiration to produce ATP. In the absence of sufficient oxygen they will be used in anaerobic respiration.

Table 25-4

ATP production from one glucose molecule

PROCESS	AMOUNT PRODUCED
Glycolysis	
Net (2 ATP required and 4 produced)	2
2 NADH	See electron-transport system (below)
Acetyl-CoA production 2 NADH	See electron-transport system (below)
Citric acid cycle	
2 ATP	2
Electron-transport system	
2 NADH (from glycolysis)	6 (or 4; see text)
2 NADH (from acetyl-CoA production)	6
6 NADH (from citric acid cycle)	18
2 $FADH_2$ (from citric acid cycle)	4
TOTAL	38 (or 36; see text)

Anaerobic Respiration

Anaerobic (an'ăr-o'bik) **respiration** is the breakdown of glucose in the absence of oxygen to produce two molecules of lactic acid and two ATP molecules (Figure 25-3). The ATP molecules are a source of energy during activities such as intense exercise when insufficient oxygen is delivered to tissues. The first phase of anaerobic respiration is glycolysis, and the second phase is the reduction of pyruvic acid to lactic acid. In this reaction the pyruvic acid gains electrons and hydrogen ions through the oxidation of the NADH molecules produced by glycolysis. Since there is a net gain of two ATP molecules during glycolysis, anaerobic respiration produces two ATP molecules for each molecule of glucose converted into two lactic acid molecules.

Lactic acid is released from the cells that produce it and is transported by the blood to the liver. If oxygen becomes available, the lactic acid in the liver can be converted through a series of chemical reactions into glucose. The glucose is released from the liver and is transported in the blood to cells that use the glucose as an energy source. This conversion of lactic acid to glucose is called the **Cori cycle.** Some of the reactions that are involved with converting lactic acid into glucose require the input of ATP (energy) produced by aerobic respiration. The oxygen necessary for the synthesis of the ATP is part of the **oxygen debt** (see Chapter 10).

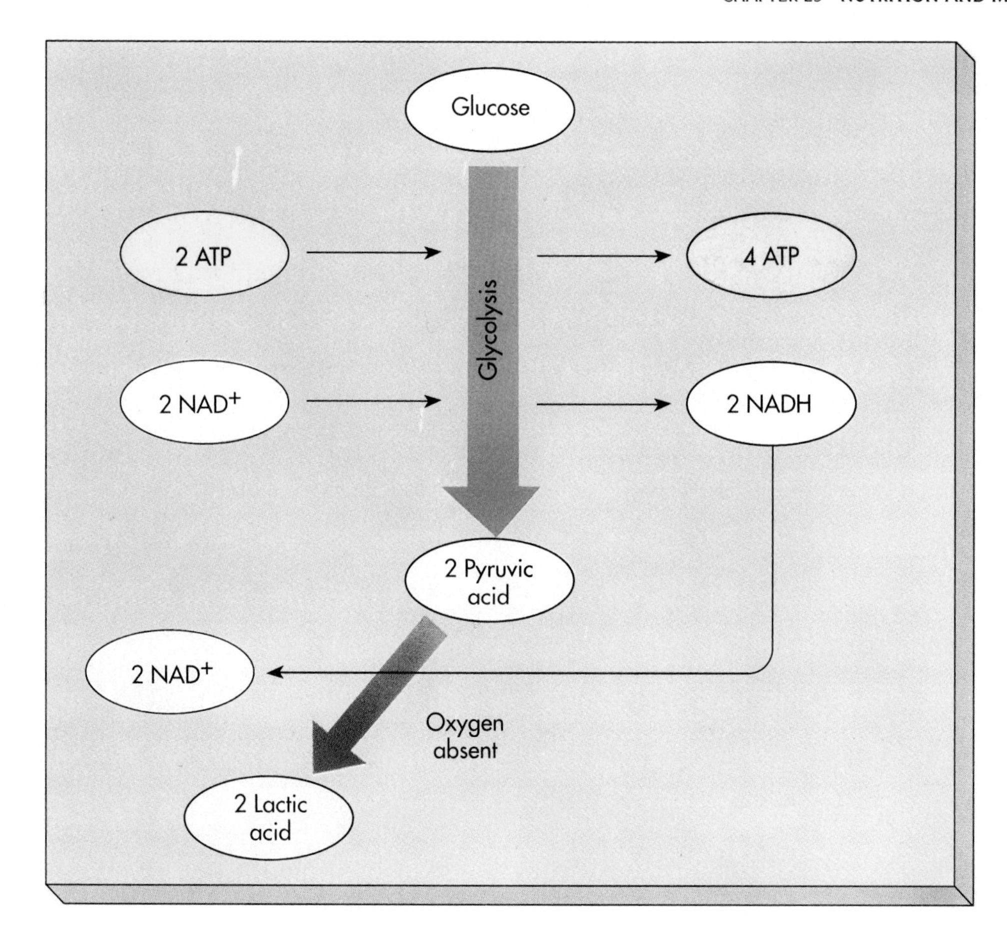

FIGURE 25-3 Anaerobic respiration. In the absence of oxygen the pyruvic acid produced in glycolysis is converted to lactic acid. The NADH produced in glycolysis is converted back to NAD^+. Two ATP molecules are used and four are produced, for a net gain of two ATP molecules.

Aerobic Respiration

Aerobic (ăr-o'bik) **respiration** is the breakdown of glucose in the presence of oxygen to produce carbon dioxide, water, and 38 ATP molecules. Aerobic respiration can be considered in four phases: glycolysis, acetyl-CoA formation, the citric acid cycle, and the electron-transport chain. The first phase of aerobic respiration, as in anaerobic respiration, is glycolysis in the cell's cytosol.

Acetyl-CoA formation

In the second phase of aerobic respiration pyruvic acid moves from the cytosol into a mitochondrion, which is separated into an inner and outer compartment by the inner mitochondrial membrane. Within the inner compartment, enzymes remove a carbon atom from the three-carbon pyruvic acid molecule to form carbon dioxide and a two-carbon acetyl group (Figure 25-4). Energy is released in the reaction and is used to reduce NAD^+ to NADH. The acetyl group combines with coenzyme A (CoA), derived from pantothenic acid (see Table 25-2), to form acetyl-CoA. For each two pyruvic acid molecules from glycolysis, two NADH and two carbon dioxide molecules are formed (see Figure 25-1).

Citric acid cycle

The third phase of aerobic respiration is the **citric acid cycle,** which is named after the six-carbon citric acid molecule formed in the first step of the cycle (see Figure 25-4). It is also called the Krebs cycle after its discoverer, the British biochemist Sir Hans Krebs. The citric acid cycle begins with the production of citric acid from the combination of acetyl-CoA and a four-carbon molecule called oxaloacetic acid. A series of reactions occurs, resulting in the formation of another oxaloacetic acid, which can

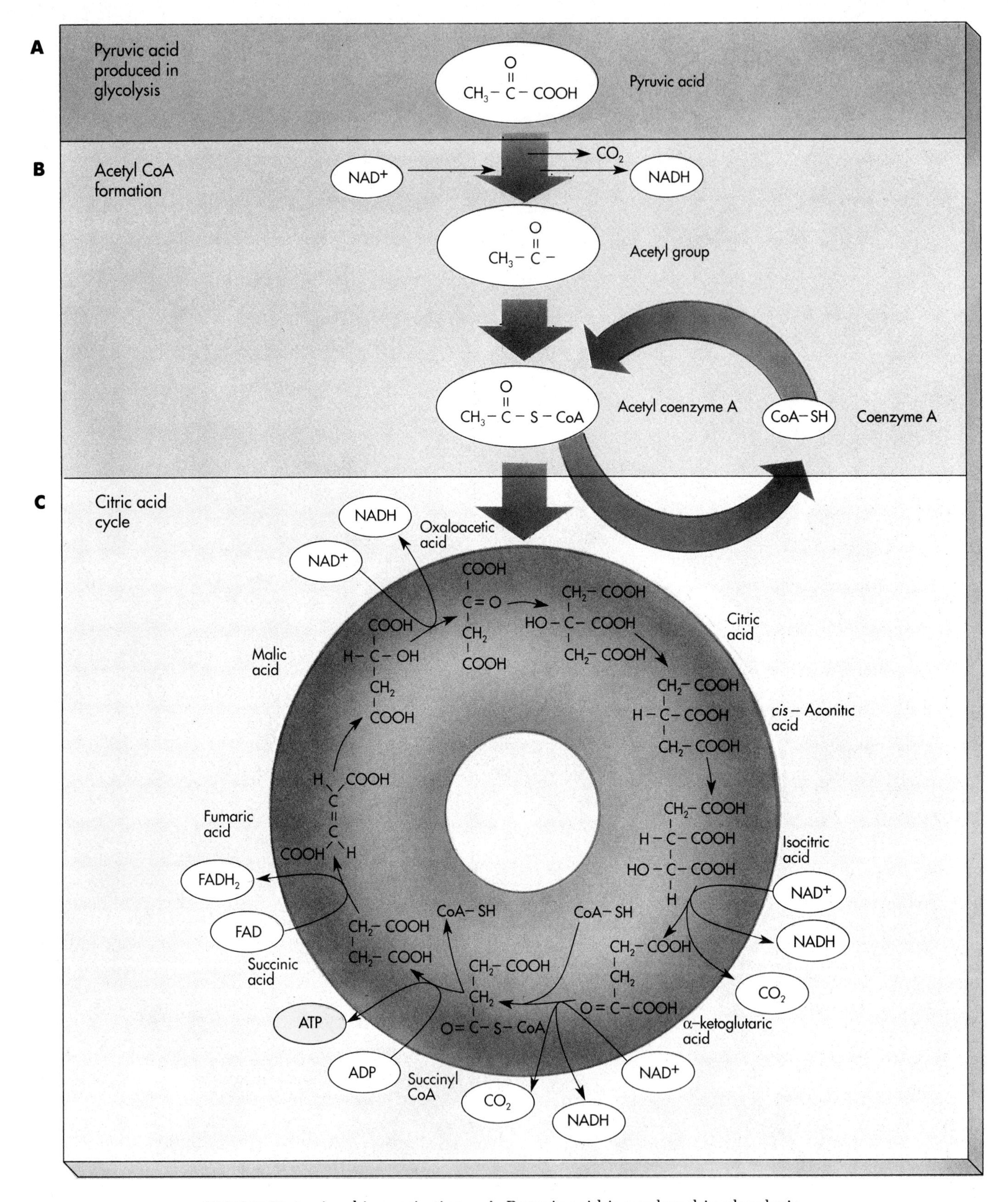

FIGURE 25-4 Aerobic respiration. **A** Pyruvic acid is produced in glycolysis. **B** In the presence of oxygen the pyruvic acid is converted to acetyl-CoA that enters the citric acid cycle. **C** Citric acid is converted through a series of reactions to oxaloacetic acid that can combine with acetyl-CoA to restart the cycle. In the process ATP, NADH, $FADH_2$, and carbon dioxide molecules are produced.

start the cycle again by combining with another acetyl-CoA. During the reactions of the citric acid cycle three important events occur.

1. **ATP production.** For each citric acid molecule one ATP is formed.
2. **NADH and $FADH_2$ production.** For each citric acid molecule three NAD^+ molecules are converted to NADH molecules, and one flavin adenine dinucleotide (FAD) molecule is converted to $FADH_2$. The NADH and $FADH_2$ molecules are electron carriers that enter the electron-transport chain and are used to produce ATP.
3. **Carbon dioxide production.** Each six-carbon citric acid molecule at the start of the cycle becomes a four-carbon oxaloacetic acid molecule at the end of the cycle. Two carbon atoms from the citric acid molecule are used to form two carbon dioxide molecules. Thus the carbon atoms that comprise food molecules such as glucose are eventually eliminated from the body as carbon dioxide. We literally breathe out part of the food we eat!

For each glucose molecule that begins aerobic respiration, two pyruvic acid molecules are produced in glycolysis, and they are converted into two acetyl-CoA molecules that enter the citric acid cycle. Therefore to determine the number of molecules produced from glucose by the citric acid cycle, two "turns" of the cycle must be counted; the results are two ATP, six NADH, two $FADH_2$, and four carbon dioxide molecules (see Figure 25-1).

Electron-transport chain

The fourth phase of aerobic respiration involves the **electron-transport chain** (Figure 25-5), which is a series of electron carriers in the inner mitochondrial membrane. Electrons are transferred from NADH and $FADH_2$ to the electron-transport carriers, and hydrogen ions are released from NADH and $FADH_2$. After the loss of the electrons and the hydrogen ions, the oxidized NAD^+ and FAD can be reused to transport additional electrons from the citric acid cycle to the electron-transport chain.

The electrons released from NADH and $FADH_2$

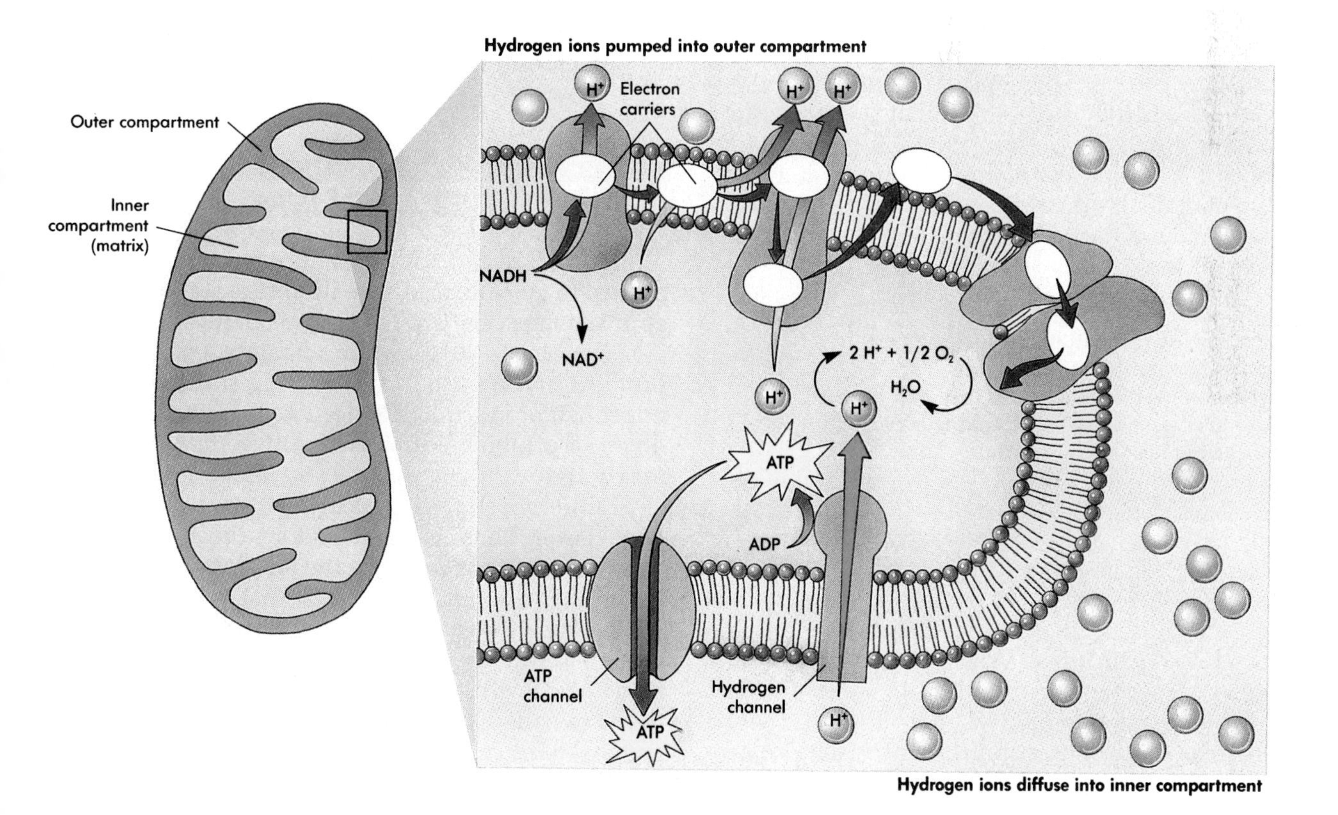

FIGURE 25-5 Electron-transport chain. NADH and $FADH_2$ (not shown) transfer their electrons (*red arrows*) to electron carriers located on the inner mitochondrial membrane. The electrons are transported along the electron carrier and some of the electron's energy is used to pump hydrogen ions across the inner mitochondrial membrane. A higher concentration of hydrogen ions in the outer compartment results, and the hydrogen ions diffuse back into the inner compartment through special channels that couple the hydrogen ion movement with the production of ATP, which moves out of the mitochondrion by facilitated diffusion. The electrons and hydrogen ions combine with oxygen to form water.

pass from one electron carrier to the next through a series of oxidation-reduction reactions. Three of the electron carriers also function as proton pumps that move the hydrogen ions across the inner mitochondrial membrane (from the inner compartment to the outer compartment of the mitochondrion). Each proton pump accepts an electron, uses some of the electron's energy to export a hydrogen ion, and passes the electron to the next electron carrier. The last electron carrier in the series collects four electrons and combines them with oxygen and four hydrogen ions to form water:

$$\frac{1}{2} O_2 + 2 H^+ + 2 e^- \rightarrow H_2O$$

Without oxygen to accept the electrons, the reactions of the electron-transport chain cease, effectively stopping aerobic respiration. Humans need oxygen because it is through aerobic respiration that most of the ATP necessary for life is produced.

The hydrogen ions released from NADH and $FADH_2$ are moved to the outer mitochondrial compartment by the proton pumps. As a result, the concentration of hydrogen ions in the outer compartment exceeds that of the inner compartment, and hydrogen ions diffuse back into the inner compartment. The hydrogen ions pass through special channels in the inner mitochondrial membrane that couple the movement of the hydrogen ions to ATP production. This process is called the **chemiosmotic model** because the chemical formation of ATP is coupled to a diffusion force similar to osmosis.

2

Many poisons function by blocking certain steps in the metabolic pathways. For example, cyanide blocks the last step in the electron-transport chain. Explain why this blockage would cause death.

? ? ? ? ? ? ? ? ? ? ?

Summary of ATP production

For each glucose molecule, aerobic respiration produces a net gain of 38 ATP molecules: 2 from glycolysis, 2 from the citric acid cycle, and 34 from the NADH molecules and $FADH_2$ molecules that pass through the electron-transport chain (see Table 25-4). For each NADH molecule formed, three ATP molecules are produced by the electron-transport chain, and for each $FADH_2$ molecule, two ATP molecules are produced.

The number of ATP molecules produced can also be reported as 36 ATP molecules. The two NADH molecules produced by glycolysis in the cytosol cannot cross the inner mitochondrial membrane, so their electrons are donated to a shuttle molecule that carries the electrons to the electron-transport chain. Depending on the shuttle molecule, each glycolytic NADH molecule can produce two or three ATP molecules. In skeletal muscle and the brain two ATP molecules are produced for each NADH molecule, resulting in a total number of 36 ATP molecules; but in the liver, kidneys, and heart three ATP molecules are produced for each NADH molecule, and the total number of ATP molecules formed is 38.

Six carbon dioxide molecules and six molecules of water are also produced in aerobic respiration. Thus aerobic respiration can be summarized as follows:

$$C_6H_{12}O_6 + 6 O_2 + 38 ADP + 38 P_i \rightarrow 6 CO_2 + 6 H_2O + 38 ATP$$

The number of ATP molecules produced per glucose molecule is a theoretical number that assumes two hydrogen ions are necessary for the formation of each ATP. If the number required is more than two, the efficiency of aerobic respiration decreases. In addition, it is now understood that it costs energy to get ADP and phosphates into the mitochondria and to get ATP out. Considering all these factors, it is currently estimated that each glucose molecule yields approximately 25 ATP molecules instead of 38 ATP molecules.

LIPID METABOLISM

Lipids are the body's main energy storage molecules. In a healthy person, lipids are responsible for approximately 99% of the body's energy storage, and glycogen accounts for approximately 1%. Although proteins can be used as an energy source, they are not considered storage molecules because the breakdown of proteins normally involves the loss of necessary tissue.

Lipids are stored primarily as triglycerides in adipose tissue. There is a constant synthesis and breakdown of triglycerides; thus the fat present in adipose tissue today is not the same fat that was there a few weeks ago. Between meals when triglycerides are broken down in adipose tissue, some of the fatty acids produced are released into the blood in which they are called **free fatty acids.** Other tissues, especially skeletal muscle and the liver, use the free fatty acids as a source of energy.

The metabolism of fatty acids occurs by **beta-oxidation,** a series of reactions in which two carbon atoms are removed from the end of a fatty acid chain to form acetyl-CoA. The process of beta-oxidation continues to remove two carbon atoms at a time until the entire fatty acid chain is converted into acetyl-CoA. Acetyl-CoA can enter the citric acid cycle and be used to generate ATP (Figure 25-6).

Acetyl-CoA can also be used in **ketogenesis** (ke-to-jen'ĕ-sis), the formation of ketone bodies. In the liver when large amounts of acetyl-CoA are pro-

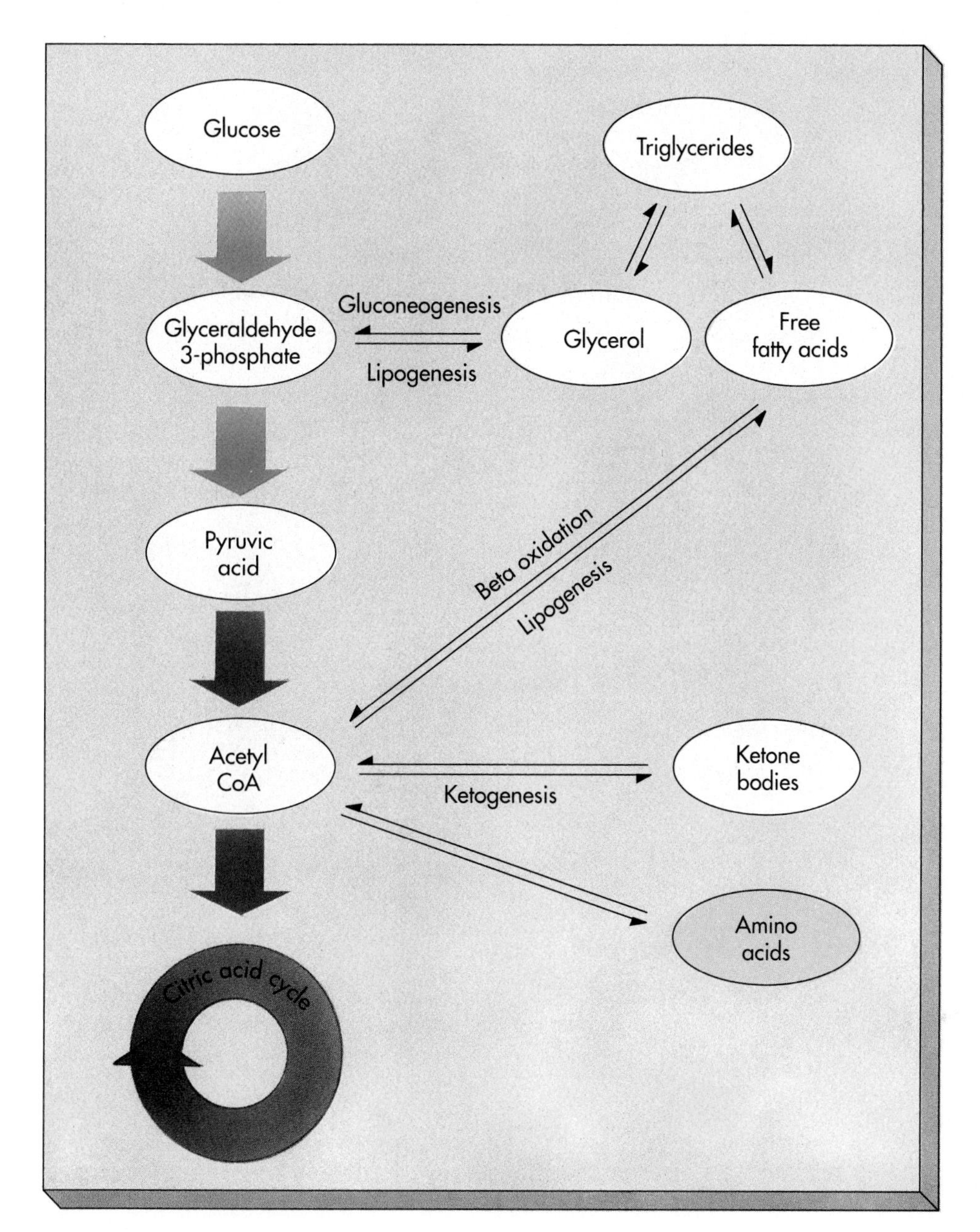

FIGURE 25-6 Lipid metabolism. Triglycerides are broken down into glycerol and fatty acids. Glycerol enters glycolysis to produce ATP. The fatty acids are broken down by beta-oxidation into acetyl-CoA, which enters the citric acid cycle to produce ATP. Acetyl-CoA can also be used to produce ketone bodies (ketogenesis). Lipogenesis is the production of lipids. Glucose is converted to glycerol, and amino acids are converted to acetyl-CoA molecules that combine to form fatty acids. Glycerol and fatty acids join to form triglycerides.

duced, not all of the acetyl-CoA enters the citric acid cycle. Instead, two acetyl-CoA molecules combine to form a molecule of acetoacetic acid, which is converted mainly into beta-hydroxybutyric acid and a smaller amount of acetone. Acetoacetic acid, beta-hydroxybutyric acid, and acetone are called **ketone bodies** and are released into the blood where they travel to other tissues, especially skeletal muscle. In these tissues the ketone bodies are converted back into acetyl-CoA that enters the citric acid cycle to produce ATP.

Normally the blood contains only small amounts of ketone bodies. However, during starvation (see essay) or in patients with diabetes mellitus the quantity of ketone bodies can increase to produce the condition ketosis. The increased number of ketone bodies can exceed the capacity of the body's buffering system, resulting in acidosis, a decrease in blood pH (see Chapter 27).

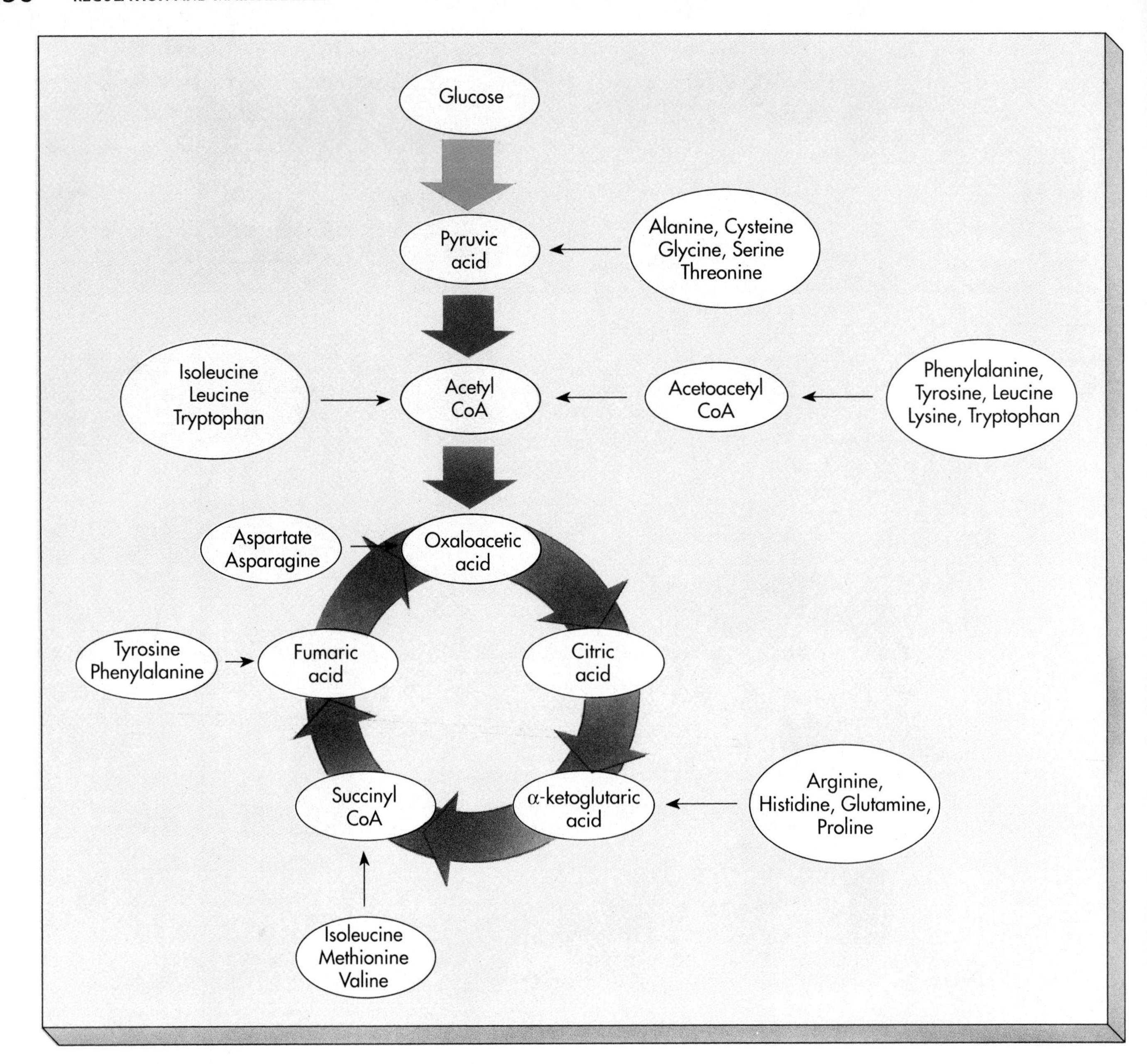

FIGURE 25-7 Various entry points for amino acids into carbohydrate metabolism.

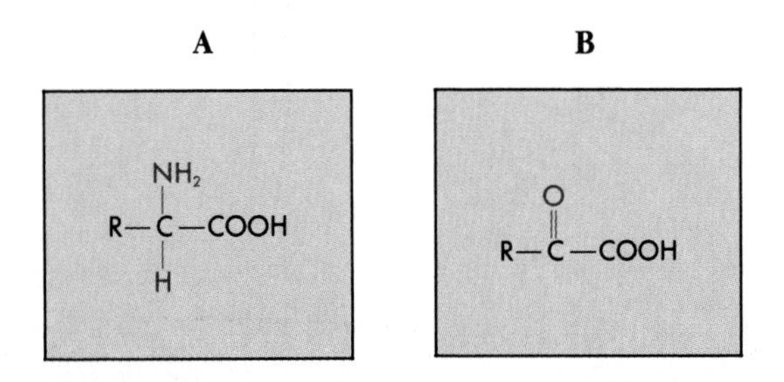

FIGURE 25-8 General formulas of an amino acid and a keto acid. **A** Amino acid with a carboxyl group (—COOH), an amine group (—NH_2), a hydrogen atom (H), and a group called "R" that represents the rest of the molecule. **B** Keto acid with a double-bonded oxygen replacing the amine group and the hydrogen atom of the amino acid.

PROTEIN METABOLISM

Once absorbed into the body, amino acids, the products of protein digestion, are quickly taken up by cells, especially in the liver. Amino acids can be used to synthesize needed proteins (see Chapter 3) or as a source of energy (Figure 25-7). Unlike glycogen and triglycerides, amino acids are not stored in the body.

Proteins in the human body are constructed of 20 amino acids (see Chapter 2), which can be divided into two groups. **Essential amino acids** cannot be

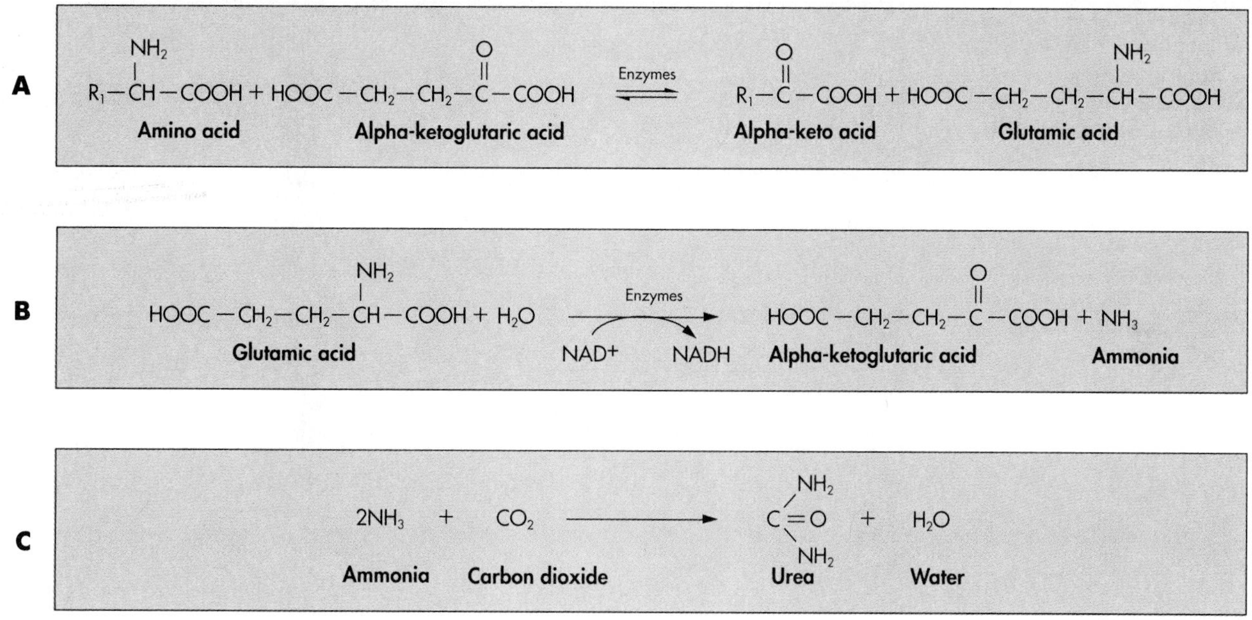

A

$$R_1{-}\underset{\text{NH}_2}{\text{CH}}{-}COOH + HOOC{-}CH_2{-}CH_2{-}\overset{\text{O}}{\overset{\|}{C}}{-}COOH \xrightleftharpoons{\text{Enzymes}} R_1{-}\overset{\text{O}}{\overset{\|}{C}}{-}COOH + HOOC{-}CH_2{-}CH_2{-}\underset{\text{NH}_2}{\text{CH}}{-}COOH$$

Amino acid + Alpha-ketoglutaric acid ⇌ Alpha-keto acid + Glutamic acid

B

$$HOOC{-}CH_2{-}CH_2{-}\underset{\text{NH}_2}{\text{CH}}{-}COOH + H_2O \xrightarrow[\text{NAD+} \rightarrow \text{NADH}]{\text{Enzymes}} HOOC{-}CH_2{-}CH_2{-}\overset{\text{O}}{\overset{\|}{C}}{-}COOH + NH_3$$

Glutamic acid → Alpha-ketoglutaric acid + Ammonia

C

$$2NH_3 + CO_2 \longrightarrow (NH_2)_2C{=}O + H_2O$$

Ammonia + Carbon dioxide → Urea + Water

FIGURE 25-9 Amino acid reactions. **A** Transamination reaction in which an amine group is transferred from an amino acid to a keto acid to form a different amino acid. **B** Oxidative deamination reaction in which an amino acid loses an amine group to become a keto acid and form ammonia. In the process NADH, which can be used to generate ATP, is formed. **C** Ammonia is converted to urea in the kidneys. The actual conversion of ammonia to urea is more complex, involving a number of intermediate reactions that comprise the urea cycle.

synthesized by the body and must be obtained in the diet. **Nonessential amino acids** can be produced by the body from other molecules.

The synthesis of nonessential amino acids usually begins with keto acids (Figure 25-8). A keto acid can be converted into an amino acid by replacing its oxygen with an amine group. Usually this conversion is accomplished by transferring an amine group from an amino acid to the keto acid, a reaction called **transamination.** For example, alpha-ketoglutaric acid (a keto acid) can react with an amino acid to form glutamic acid (an amino acid; Figure 25-9, *A*). *Most* amino acids can undergo transamination to produce glutamic acid. The glutamic acid can be used as a source of an amine group to construct most of the nonessential amino acids. A few nonessential amino acids are formed in other ways from the essential amino acids.

Amino acids can be used as a source of energy. In **oxidative deamination** an amine group is removed from an amino acid (usually glutamic acid), leaving ammonia and a keto acid (Figure 25-9, *B*). In the process NAD^+ is reduced to NADH that can enter the electron-transport chain to produce ATP. Ammonia is toxic to cells and is converted by the liver into urea, which is carried by the blood to the kidneys where the urea is eliminated (Figure 25-9, *C*; see Chapter 26).

Another way to produce energy from amino acids is to convert them into the intermediate molecules of carbohydrate metabolism (see Figure 25-7). These molecules then are metabolized to yield ATP. The conversion of the amino acid often begins with a transamination or oxidative-deamination reaction in which the amino acid is converted into a keto acid (see Figure 25-9). The keto acid can enter the citric acid cycle or be converted into pyruvic acid or acetyl-CoA.

INTERCONVERSION OF NUTRIENT MOLECULES

Blood glucose enters most cells by facilitated diffusion and is immediately converted to glucose 6-phosphate, which cannot recross the cell membrane (Figure 25-10). Glucose 6-phosphate can then continue through glycolysis to produce ATP. However, if there is excess glucose (e.g., after a meal), it can be used to form glycogen through a process called **glycogenesis** (gli′ko-jen′ĕ-sis). Most of the body's glycogen is in skeletal muscle and in the liver.

Once glycogen stores, which are quite limited, are filled, glucose and amino acids are used to synthesize lipids, a process called **lipogenesis** (lip′o-jen′ĕ-sis; see Figure 25-6). Glucose molecules can be

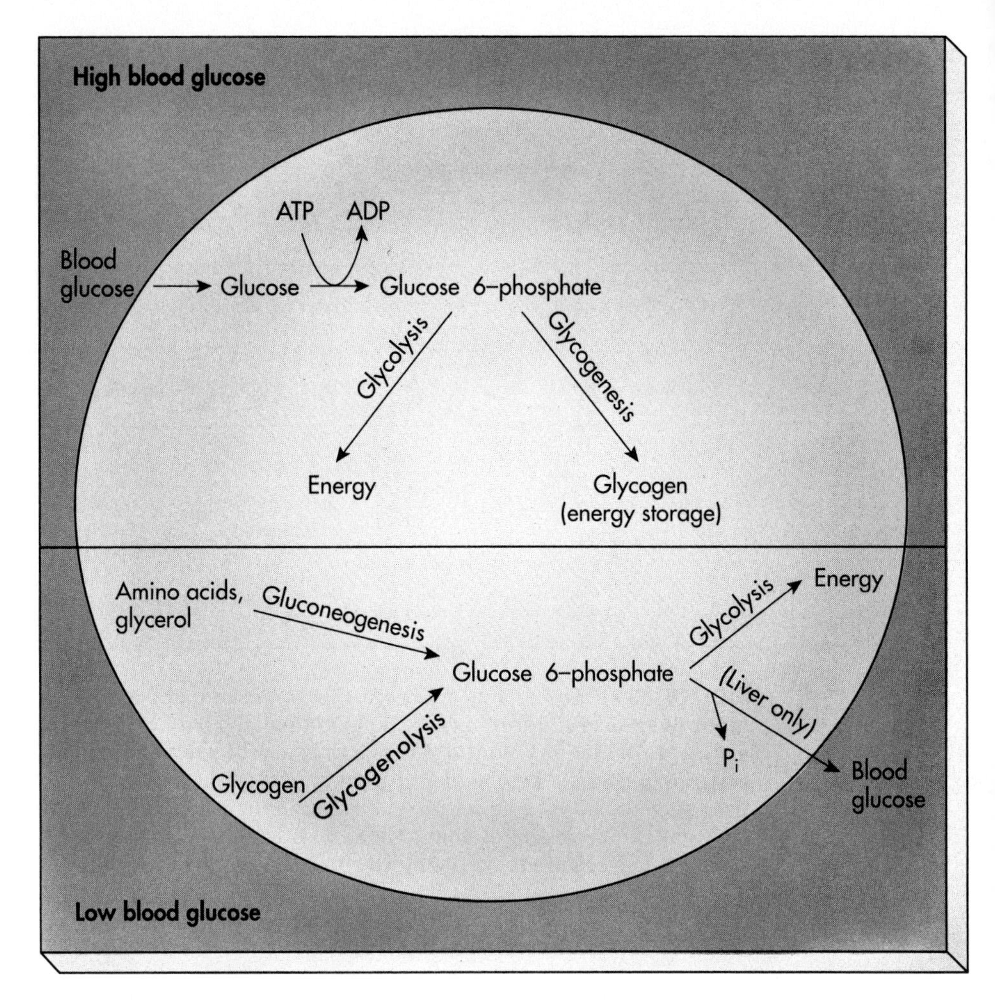

FIGURE 25-10 Interconversion of nutrient molecules. When blood glucose levels are high, glucose enters the cell and is phosphorylated to form glucose 6-phosphate, which can enter glycolysis or glycogenesis. When blood glucose levels drop, glucose 6-phosphate can be produced through glycogenolysis or gluconeogenesis. Glucose 6-phosphate can enter glycolysis, or the phosphate group can be removed and glucose released into the blood.

used to form glyceraldehyde 3-phosphate and acetyl-CoA. Amino acids can also be converted to acetyl-CoA. Glyceraldehyde 3-phosphate can be converted to glycerol, and the two-carbon acetyl-CoA molecules can be joined together to form fatty acid chains. Glycerol and fatty acids are then combined to form triglycerides.

Enzymes in the liver convert ethanol (beverage alcohol) into acetyl-CoA, and in the process two NADH molecules are produced. The NADH molecules enter the electron-transport chain and are used to produce ATP molecules. Each gram of ethanol provides 7 kcal of energy. Because of the high level of NADH in the cell that results from the metabolism of ethanol, the production of NADH by glycolysis and by the citric acid cycle is inhibited. Consequently, sugars and amino acids are not broken down but are converted into fats that accumulate in the liver. Death of fat-engorged cells, inflammation, and scar tissue formation result in cirrhosis of the liver, which can cause death because the liver is unable to carry out its normal functions.

When glucose is needed, glycogen can be broken down into glucose 6-phosphate through a set of reactions called **glycogenolysis** (gli′ko-jĕ-nol′ĭ-sis; see Figure 25-10). In skeletal muscle glucose 6-phosphate continues through glycolysis to produce ATP. The liver can use glucose 6-phosphate for energy or can convert it to glucose, which diffuses into the blood. The liver can release glucose, and skeletal muscle cannot because the liver has the necessary enzymes to convert glucose 6-phosphate into glucose.

Release of glucose from the liver is necessary to maintain blood glucose levels between meals. Maintaining these levels is especially important to the brain, which normally uses only glucose for an energy source and consumes approximately two thirds of the total glucose used each day. When liver glycogen levels are inadequate to supply glucose, amino acids from proteins and glycerol from triglycerides are used to produce glucose in a process called **gluconeogenesis** (glu′ko-ne-o-jen′ĕ-sis). Most amino acids can be converted into citric acid cycle molecules, acetyl-CoA, or pyruvic acid (see Figure 25-7). The citric acid molecules and acetyl-CoA can undergo reactions to become pyruvic acid, which can be converted into glucose. Glycerol can enter glycolysis by becoming glyceraldehyde 3-phosphate.

Starvation

Starvation, the inadequate intake of nutrients or the inability to metabolize or absorb nutrients, can have a number of causes (e.g., prolonged fasting, anorexia, deprivation, or disease). No matter what the cause, starvation takes approximately the same course and consists of three phases. The events of the first two phases occur even during relatively short periods of fasting or dieting, but the third phase occurs only during prolonged starvation and ends in death.

During the first phase of starvation blood glucose levels are maintained through the production of glucose from glycogen, proteins, and fats. At first glycogen is broken down into glucose. However, only enough glycogen is stored in the liver to last a few hours. Thereafter, blood glucose levels are maintained by the breakdown of proteins and fats. Fats are decomposed into fatty acids and glycerol. Fatty acids can be used as a source of energy, especially by skeletal muscle, thus decreasing the use of glucose by tissues other than the brain. Glycerol can be used to make a small amount of glucose, but most of the glucose is formed from the amino acids of proteins. In addition, some amino acids can be used directly for energy.

In the second stage, which can last for several weeks, fats are the primary energy source. The liver metabolizes fatty acids into ketone bodies that can be used as a source of energy. After approximately 1 week of fasting, the brain begins to use ketone bodies, as well as glucose, for energy. This usage decreases the demand for glucose, and the rate of protein breakdown diminishes but does not stop. In addition, there is a selective use of proteins (i.e., those proteins not essential for survival are used first).

The third stage of starvation begins when the fat reserves are depleted and there is a switch to proteins as the major energy source. Muscles, the largest source of protein in the body, are rapidly depleted. At the end of this stage, proteins essential for cellular functions are broken down, and cell function degenerates. Death can occur very rapidly, and the victim may just suddenly die.

METABOLIC STATES

There are two major metabolic states in the body. The first is the **absorptive state,** the period immediately after a meal when nutrients are being absorbed through the intestinal wall into the circulatory and lymphatic systems (Figure 25-11). The absorptive state usually lasts approximately 4 hours after each meal (a total of 12 hours a day if a person eats three meals). During the absorptive state most of the glucose that enters the circulation is used by cells to provide the energy they require. The remainder of the glucose is converted into glycogen or fats. Most of the absorbed fats are deposited in adipose tissue. Many of the absorbed amino acids are used by cells in protein synthesis, some are used for energy, and others enter the liver and are converted to fats or carbohydrates.

The second state, the **postabsorptive state,** occurs late in the morning, late in the afternoon, or during the night after each absorptive state is concluded (Figure 25-12). Normal blood glucose levels range between 70 and 110 mg/100 ml, and it is vital to the body's homeostasis that this range be maintained. During the postabsorptive state blood glucose levels are maintained by the conversion of other molecules to glucose. The first source of blood glucose during the postabsorptive state is the glycogen stored in the liver. However, this glycogen supply can provide glucose for only approximately 4 hours. The glycogen stored in skeletal muscles can also be used during times of vigorous exercise. Once the glycogen stores are depleted, fats are tapped as an energy source. The triglycerides are hydrolyzed to glycerol, which can be converted to glucose. The fatty acids from fat can be converted to acetyl-CoA, moved into the citric acid cycle, and used as a source of energy to produce ATP. In the liver, acetyl-CoA can be used to produce ketone bodies that other tissues can use for energy. The use of fatty acids as an energy source can partly eliminate the need to use glucose for energy, resulting in reduced need for glucose removal from the blood and maintaining blood glucose levels at homeostatic levels. Proteins can also be used as a source of glucose or can be directly used for energy production, again sparing blood glucose.

METABOLIC RATE

The **metabolic rate** is the total amount of energy produced and used by the body per unit of time. Any given molecule of ATP exists for less than 1 minute

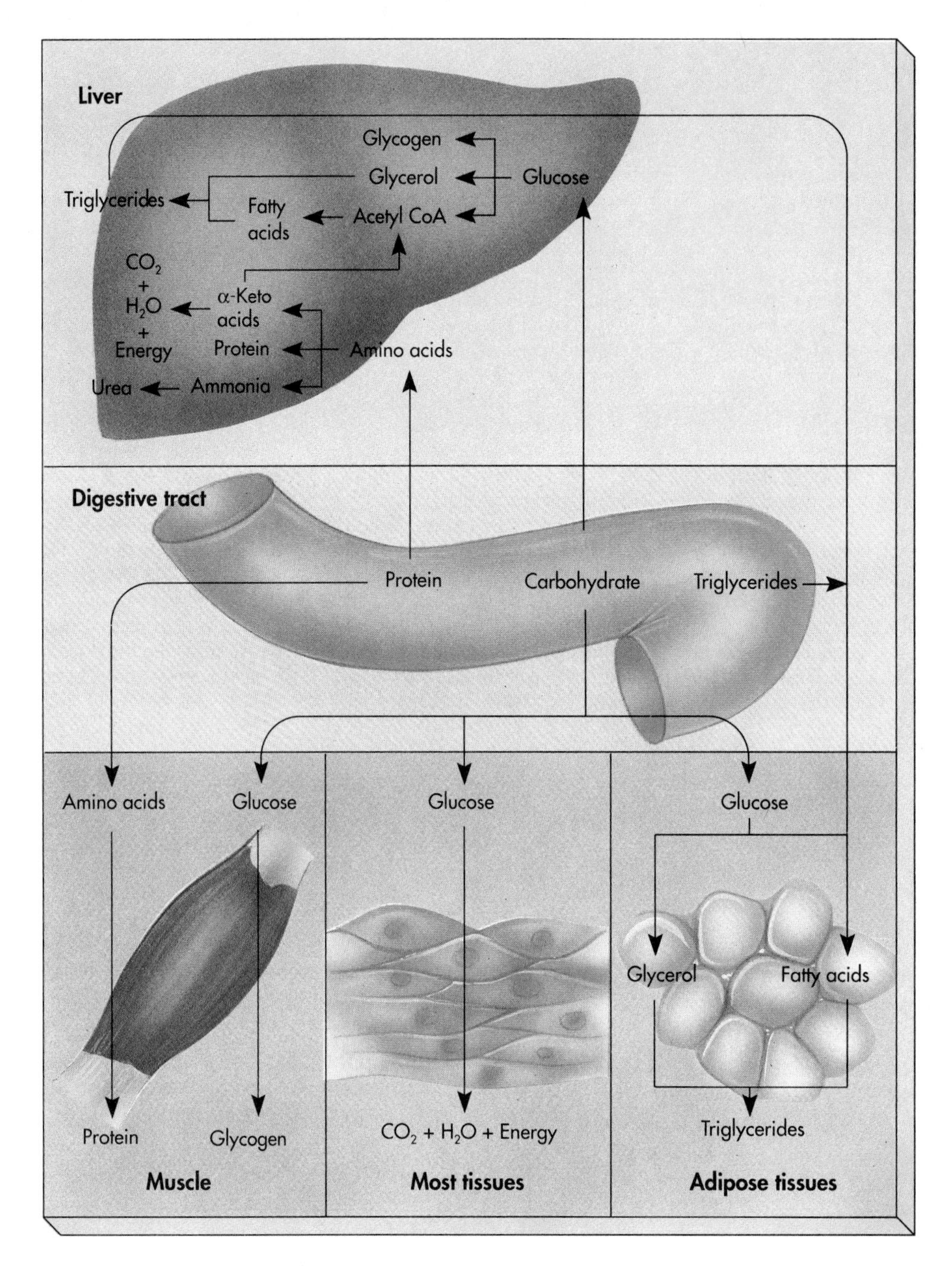

FIGURE 25-11 Events of the absorptive state.

before it is degraded back to ADP and phosphate. For this reason ATP is produced in cells at approximately the same rate as it is used. Thus in examining metabolic rate, ATP production and use can be roughly equated. Metabolic rate is usually estimated by measuring the amount of oxygen used per minute since most ATP production involves the use of oxygen. One liter of oxygen consumed by the body is assumed to produce 4.825 kcal of energy.

Metabolic energy can be used in three ways: for basal metabolism, for muscular activity, and for the assimilation of food. The **basal metabolic rate (BMR)** is the metabolic rate calculated in expended kilocalories per square meter of body surface area per hour. It is determined by measuring the oxygen consumption of a person who is awake but restful and has not eaten for 12 hours. The liters of oxygen consumed are then multiplied by 4.825 because each liter of oxygen used results in the production of 4.825 kcal of energy. A typical BMR for a 70-kg (154 pounds) man would be 38 kcal/m^2/hr.

BMR is the energy needed to keep the resting body functional. Active transport mechanisms, muscle tone, maintenance of body temperature, beating of the heart, and other activities are supported by basal metabolism. A number of factors can affect the BMR. Muscle tissue is metabolically more active than adipose tissue, even at rest. Younger people have a higher BMR than older people because of increased cell activity, especially during growth. Fever can in-

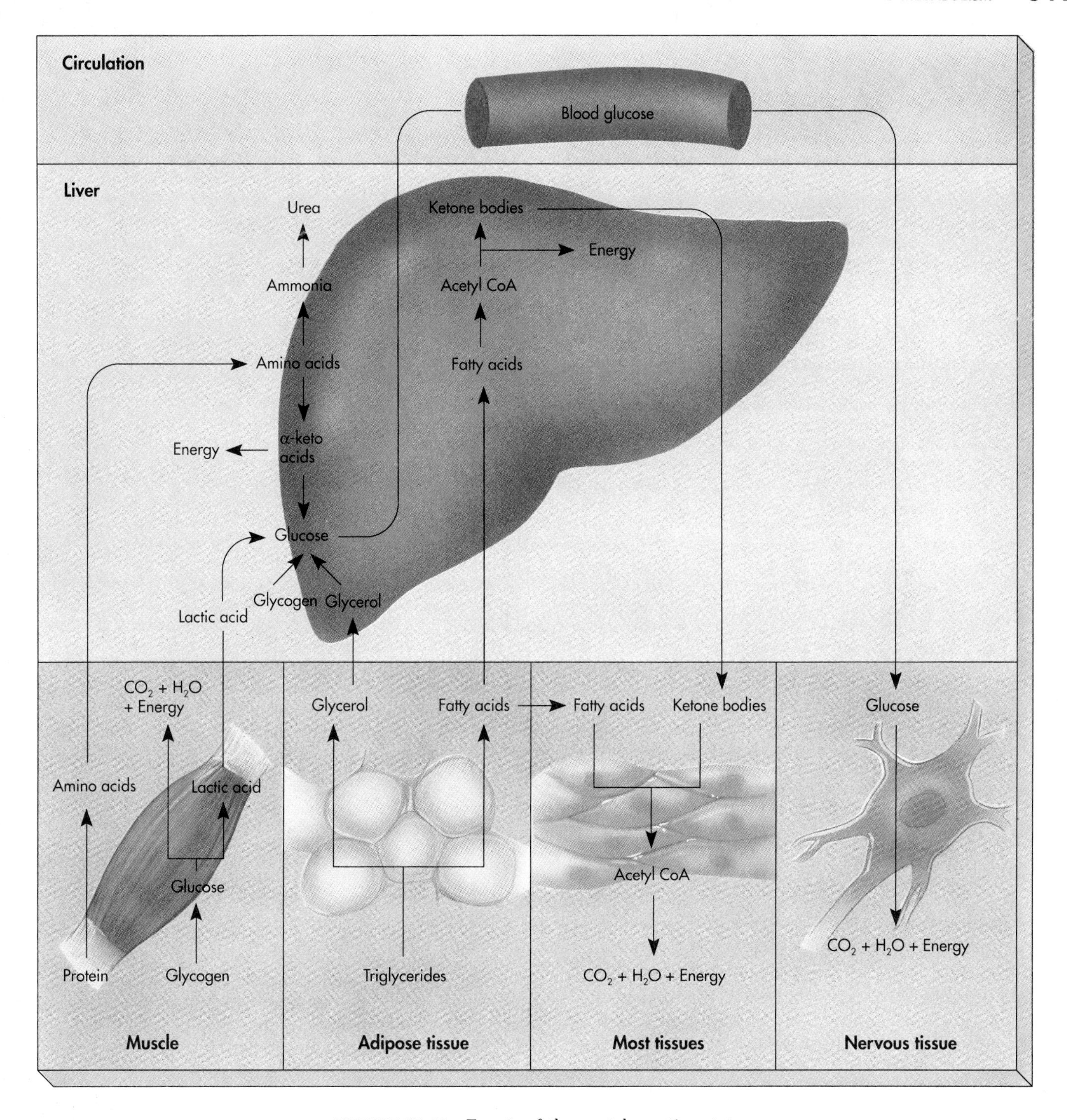

FIGURE 25-12 Events of the postabsorptive state.

crease BMR 7% for each degree Fahrenheit increase in body temperature. Greatly reduced kilocaloric input (i.e., dieting or fasting) depresses BMR, apparently a protective mechanism to prevent weight loss. BMR can be increased on a long-term basis by thyroid hormones and on a short-term basis by epinephrine (see Chapter 18). Males have a greater BMR than females because men have proportionately more muscle tissue and less adipose tissue than women. During pregnancy a woman's BMR may increase 20% because of the metabolic activity of the fetus.

Physical activity resulting from skeletal muscle movement requires the expenditure of energy. In addition, energy must be provided for increased contraction of the heart and of the muscles of respiration. The number of kilocalories used in an activity depends almost entirely on the amount of muscular work performed and on the length of the activity. Despite the fact that studying can make a person feel tired, intense mental concentration produces little change in the BMR.

The third component of metabolic energy con-

Obesity

Obesity is the storage of excess fat and it can be classified according to the number of fat cells and the size of fat cells. The greater the amount of lipid stored in the fat cells, the larger is their size. In an individual with **hyperplastic obesity** there is a greater-than-normal number of fat cells that are also larger than normal. This type of obesity is associated with massive obesity and begins at an early age. In nonobese children the number of fat cells triples or quadruples between birth and 2 years of age, then remains relatively stable until puberty when there is a further increase in number. In obese children, however, between 2 years of age and puberty there is also an increase in the number of fat cells. **Hypertrophic obesity** results from a normal number of fat cells that have increased in size. This type of obesity is more common, is associated with moderate obesity or being "overweight," and typically develops in adults. People who were thin or of average weight and quite active when young become less active as they become older. They begin to gain weight at age 20 to 40, and although they no longer use as many kilocalories, they still take in the same amount of food as when they were younger. The unused kilocalories are turned into fat, causing fat cells to increase in size. At one time it was believed that the number of fat cells did not increase after adulthood. It now is known that the number of fat cells can increase in adults. Apparently if all the existing fat cells are filled to capacity with lipids, new fat cells are formed to store the excess lipids. Once fat cells are formed, however, dieting and weight loss do not result in a decrease in the number of fat cells; instead, they become smaller in size as their lipid content decreases.

The distribution of fat in obese individuals can vary; it can be found mainly in the upper body such as the abdominal region, or it can be associated with the hips and buttocks. These distribution differences can be clinically significant because upper body obesity is associated with an increased likelihood of diabetes mellitus, cardiovascular disease, stroke, and death.

In some cases a specific cause of obesity can be identified. For example, a tumor in the hypothalamus can stimulate overeating. In most cases, however, no specific cause can be recognized. In fact, obesity can occur for many reasons, and obesity in an individual can result from more than one cause. There seems to be a genetic component for obesity, and if one or both parents are obese, their children also are more likely to be obese. Environmental factors such as eating habits, however, can play an important role. For example, adopted children can exhibit similarities in obesity to their adoptive parents. In addition, psychological factors such as overeating as a means for dealing with stress can contribute to obesity.

Regulation of body weight is actually a matter of regulating body fat because most changes in body weight reflect changes in the amount of fat in the body. According to the "set point" theory of weight control, each individual attempts to maintain a certain amount of body fat. If the amount of body fat decreases below or increases above this level, mechanisms are activated to return the amount of body fat to its normal value. In support of this hypothesis it is known that obese individuals have an increased amount of the enzyme lipoprotein lipase,

cerns the assimilation of food. When food is ingested, the accessory digestive organs and the intestinal lining produce secretions, the motility of the digestive tract increases, active transport increases, and the liver is involved in the synthesis of new molecules. These events are called the **thermic effect of feeding.**

The daily input of energy should equal the metabolic expenditure of energy; otherwise, a person will gain or lose weight. For a 23-year-old, 70-kg (154 pounds) man to maintain his weight the daily input should be 2700 kcal per day; for a 58-kg (128 pounds) woman of the same age 2000 kcal per day are necessary. For every 3500 kcal above the necessary energy requirement (not necessarily in 1 day), a pound of body fat can be gained, whereas for every 3500 kcal below the requirement (lost usually over several days), a pound of fat can be lost. Clearly, adjusting kilocaloric input is an important way to control body weight.

Not only the number of kilocalories ingested, but the proportion of fat in the diet has an effect on body weight. To convert dietary fat into body fat, 3% of the energy in the dietary fat is used, leaving 97% for storage as fat deposits. On the other hand, the conversion of dietary carbohydrate to fat requires 23% of the energy in the carbohydrate, leaving just 77% as body fat. If two people have the same kilocaloric intake, the one with the higher proportion of fat in his diet is more likely to gain weight because fewer kilocalories are used to convert the dietary fat into body fat.

The other way to control weight is through energy expenditure. In the average person basal metabolism accounts for approximately 60% of energy expenditure, muscular activity for 30%, and assimila-

which is responsible for the uptake and storage of triglycerides in fat cells. Furthermore, in obese individuals who have lost weight, the levels of lipoprotein lipase increases even more.

The two factors that affect the amount of adipose tissue are energy expenditure and energy intake. Energy expenditure occurs as a result of basal metabolic rate, exercise, and the thermic effect of feeding. At present there is no convincing evidence that these variables are significantly different in lean or obese individuals at their usual weights. This does not mean that expenditure of energy through exercise cannot result in weight loss. Instead, it means that the amount of energy lost through a specific amount of exercise is the same for lean and obese people.

The regulation of energy intake is poorly understood. Apparently appetite and food-seeking behaviors are continually and spontaneously stimulated by neurons originating in or passing through the hypothalamus. After food is consumed, several mechanisms may be responsible for decreasing further food intake. Neural mechanisms such as distension of the stomach are known to inhibit feeding, and a number of hormones released from the gastrointestinal tract or pancreas also decrease appetite. For example, somatostatin, cholecystokinin, glucagon, insulin, and other hormones have been shown to reduce food intake. The level of fatty acids, glucose, or amino acids in the blood may also provide the brain with information necessary to adjust appetite. Low levels of fatty acids, glucose, and amino acids stimulate appetite, whereas high levels of these substrates inhibit appetite.

It is a common belief that the main cause of obesity is overeating. Certainly for obesity to occur, at some time energy intake must have exceeded energy expenditure. A comparison of the kilocaloric intake of obese and lean individuals at their usual weights, however, reveals that on a per kilogram basis obese people consume fewer kilocalories than lean people. The feeding behavior of the obese at their usual weight is fairly normal.

When people lose a large amount of weight, their feeding behavior changes. They become hyper-responsive to external food cues, think of food often, and cannot get enough to eat without gaining weight. It is now understood that this behavior is typical of both lean and obese individuals who are below their relative set point for weight. Other changes such as a decrease in basal metabolic rate take place in a person who has lost a large amount of weight. Most of this decrease probably results from a decrease in muscle mass associated with weight loss. In addition, there is some evidence that energy lost through exercise and the thermic effect of feeding are also reduced. Thus a person who has lost a large amount of weight is a person with an increased appetite and a decreased ability to expend energy. It is no surprise that only a small percentage of obese people maintain weight loss on a long-term basis. Instead, the typical pattern is one of repeated cycles of weight loss followed by a rapid regain of the lost weight.

tion of food for approximately 10%. Of these amounts, energy loss through muscular activity is the only component that a person can reasonably control. A comparison of the number of kilocalories gained from food vs. the number of kilocalories lost in exercise reveals why weight loss can be a difficult task. For example, walking (3 mph) for 20 minutes would burn the kilocalories in one slice of bread, whereas jogging (5 mph) for the same time period would eliminate the kilocalories obtained from a soft drink or a beer (see Table 25-1).

BODY TEMPERATURE REGULATION

Free energy is the total amount of energy that can be liberated by the complete catabolism of food. It is usually expressed in terms of kilocalories (kcal) per mole of food consumed. For example, the complete catabolism of 180 g (1 mole; see Chapter 2) of glucose releases 686 kcal of free energy. Only a portion (approximately 43%) of the total energy released by catabolism is used to produce ATP and to accomplish biological work such as anabolism, muscular contraction, and other cellular activities. The remaining energy is lost as **heat.**

3

Explain why we become warm during exercise and why we shiver when it is cold.

? ? ? ? ? ? ? ? ? ? ?

Humans are **homeotherms** (ho′me-o-thermz; uniform warming), or **warm-blooded** animals, and can regulate body temperature rather than have it adjusted by the external environment. Maintenance of a constant body temperature is very important to homeostasis. Most enzymes are very temperature sensitive and function only in narrow temperature ranges. Environmental temperatures are too low for normal enzyme function, and the heat produced by metabolism and muscle contraction helps maintain the body temperature at a steady, elevated level that is high enough for normal enzyme function.

Normal body temperature is a range like any other homeostatically controlled condition in the body. The average normal temperature usually is considered 37° C (98.6° F) when it is measured orally and 37.6° C (99.7° F) when it is measured rectally. Rectal temperature comes closer to the true core body temperature, but an oral temperature is more easily obtained in older children and adults and therefore is the preferred measure.

Body temperature is maintained by balancing heat input with heat loss. Heat can be exchanged with the environment in a number of ways. **Radiation** is the loss of heat as infrared radiation, a type of electromagnetic radiation. For example, the coals in a fire give off radiant heat that can be felt some distance away from the fire. **Conduction** is the exchange of heat between objects in direct contact with each other (e.g., the bottom of the feet and the floor). **Convection** is a transfer of heat between the body and the air. A cool breeze results in movement of air over the body and loss of heat from the body. **Evaporation** is the loss of water from the body; the water carries heat away with it. The evaporation of 1 g of water results in the loss of 580 cal of heat.

The amount of heat exchanged between the environment and the body is determined by the difference in temperature between the body and the environment. The greater the temperature difference, the greater is the rate of heat exchange. Control of the temperature difference can be used to regulate body temperature. For example, if environmental temperature is very cold (e.g., a cold winter day), there is a large temperature difference between the body and the environment, and there is a large loss of heat. The loss of heat can be decreased by behaviorally selecting a warmer environment (e.g., going inside a heated house) or by insulating the exchange surface (e.g., putting on extra clothes). Physiologically, temperature difference can be controlled through dilation and constriction of blood vessels in the skin. When these blood vessels dilate, they bring warm blood to the surface of the body, raising skin temperature; on the other hand, vasoconstriction decreases blood flow and lowers skin temperature.

4

Explain why vasoconstriction of the skin's blood vessels on a cold winter day is beneficial.

? ? ? ? ? ? ? ? ? ? ?

When environmental temperature is greater than body temperature, vasodilation brings warm blood to the skin, causing an increase in skin temperature that decreases heat gain from the environment. At the same time, evaporation carries away excess heat to prevent heat gain and overheating.

Body temperature regulation is an example of a negative-feedback system that is controlled by a "set point." A small area in the anterior part of the hypothalamus can detect slight increases in body temperature through changes in blood temperature. As a result, mechanisms that cause heat loss (e.g., vasodilation and sweating) are activated, and body temperature decreases. A small area in the posterior hypothalamus can detect slight decreases in body temperature and can initiate heat gain by increasing muscular activity (shivering) and vasoconstriction.

Under some conditions the hypothalamus' set point is actually changed. For example, during a fever the set point is raised, heat-conserving and heat-producing mechanisms are stimulated, and body temperature increases. In recovery from a fever the set point is lowered to normal, heat-loss mechanisms are initiated, and body temperature decreases.

SUMMARY

1. Metabolism consists of anabolism and catabolism. Anabolism is the building up of molecules and requires energy. Catabolism is the breaking down of molecules and gives off energy.
2. Nutrition is the taking in and use of food.

NUTRITION p. 820

Nutrients

Nutrients are the chemicals used by the body and consist of carbohydrates, lipids, proteins, vitamins, minerals, and water.

Kilocalories

1. A kilocalorie (kcal) is 1000 calories. A calorie is the heat (energy) necessary to raise the temperature of 1 g of water from 14° C to 15° C.
2. A gram of carbohydrate or protein yields 4 kcal, and a gram of fat yields 9 kcal.

Carbohydrates

1. Carbohydrates are ingested as monosaccharides (glucose, fructose), disaccharides (sucrose, maltose, lactose), and polysaccharides (starch, glycogen, cellulose).

2. Polysaccharides and disaccharides are converted to glucose. Glucose can be used for energy or stored as glycogen or fats.
3. Approximately 125 to 175 g of carbohydrates should be ingested each day.

Lipids

1. Lipids are ingested as triglycerides (95%) or cholesterol and phospholipids (5%).
2. Triglycerides are used for energy or are stored in adipose tissue. Cholesterol forms other molecules such as steroid hormones. Cholesterol and phospholipids are part of the plasma membrane.
3. The daily diet should derive no more than 30% of its kilocalories from lipids, and no more than 250 mg should be in the form of cholesterol.

Proteins

1. Proteins are ingested and broken down into amino acids.
2. Proteins perform many functions: protection (antibodies), regulation (enzymes, hormones), structure (collagen), muscle contraction (actin and myosin), and transport (hemoglobin, carrier molecules, ion channels).
3. An adult should consume 0.8 g of protein per kg of body weight each day.

Vitamins

1. Vitamins function as coenzymes or as parts of coenzymes.
2. Most vitamins are not produced by the body and must be obtained in the diet. Some vitamins can be formed from provitamins.
3. Vitamins are classified as either fat soluble or water soluble.

Minerals

Minerals are necessary for normal metabolism, add mechanical strength to bones, function as buffers, and are involved in osmotic balance.

METABOLISM *p. 826*

The energy in carbohydrates, lipids, and proteins is used to produce ATP through oxidation-reduction reactions.

CARBOHYDRATE METABOLISM *p. 827*

1. Glycolysis is the breakdown of glucose to two pyruvic acid molecules. Also produced are two NADH and two ATP.
2. Anaerobic respiration is the breakdown of glucose in the absence of oxygen to two lactic acid and two ATP molecules.
3. Lactic acid can be converted to glucose (Cori cycle) using aerobically produced ATP (oxygen debt).
4. Aerobic respiration is the breakdown of glucose in the presence of oxygen to produce carbon dioxide, water, and 38 (or 36) ATP molecules.
   - The first phase is glycolysis, which produces two ATP, two NADH, and two pyruvic acid molecules.
   - The second phase is the conversion of the two pyruvic acid molecules to two molecules of acetyl-CoA. These reactions also produce two NADH and two carbon dioxide molecules.
   - The third phase is the citric acid cycle, which produces two ATP, six NADH, two $FADH_2$, and four carbon dioxide molecules.
   - The fourth phase is the electron-transport chain. The high-energy electrons in NADH and $FADH_2$ enter the electron-transport chain and are used in the synthesis of ATP and water.

LIPID METABOLISM *p. 834*

1. Adipose triglycerides are broken down and released as free fatty acids.
2. Free fatty acids are taken up by cells and broken down by beta-oxidation into acetyl-CoA.
   - Acetyl-CoA can enter the citric acid cycle.
   - Acetyl-CoA can be converted into ketone bodies.

PROTEIN METABOLISM *p. 836*

1. New amino acids are formed by transamination, the transfer of an amine group to a keto acid.
2. Amino acids are used for energy, and ammonia is produced as a by-product in oxidative deamination. Ammonia is converted to urea and is excreted.

INTERCONVERSION OF NUTRIENT MOLECULES *p. 837*

1. Glycogenesis is the formation of glycogen from glucose.
2. Lipogenesis is the formation of lipids from glucose and amino acids.
3. Glycogenolysis is the breakdown of glycogen to glucose.
4. Gluconeogenesis is the formation of glucose from amino acids and glycerol.

METABOLIC STATES *p. 839*

1. In the absorptive state nutrients are used as energy or are stored.
2. In the postabsorptive state stored nutrients are used for energy.

METABOLIC RATE *p. 839*

Metabolic rate is the total energy expenditure per unit of time, and it has three components.

- Basal metabolic rate is the energy used at rest.
- Muscular energy is used for muscle contraction.
- Assimilation energy is used to digest and absorb food.

BODY TEMPERATURE REGULATION *p. 843*

1. Body temperature is a balance between heat gain and heat loss.
   - Heat is produced through metabolism.
   - Heat is exchanged through radiation, conduction, convection, and evaporation.
2. The greater the temperature difference between the body and the environment, the greater is the rate of heat exchange.
3. Body temperature is regulated by a "set point" in the hypothalamus.

CONTENT REVIEW

1. Define metabolism, anabolism, and catabolism.
2. Define a nutrient, and list the six major classes of nutrients.
3. Define a kilocalorie, and state the number of kilocalories in a gram of carbohydrate, lipid, and protein.
4. List the dietary sources of carbohydrates. After they are converted to glucose, what happens to the glucose? What quantities of carbohydrate should be ingested daily?
5. List the dietary sources of lipids, explain how triglycerides, cholesterol, and phospholipids are used in the body, and describe the recommended dietary intake of lipids.
6. List the dietary sources of complete and incomplete protein foods. Describe some of the functions performed by proteins in the body. What is the recommended daily consumption of proteins?
7. What are a vitamin and a provitamin? Name the water-soluble vitamins and the fat-soluble vitamins. List some of the functions of vitamins.
8. List some of the minerals, and give their functions.
9. How does the removal of hydrogen atoms from nutrient molecules result in a loss of energy from the nutrient molecule?
10. Describe glycolysis. Although four ATP molecules are produced in glycolysis, explain why there is a net gain of only two ATP molecules.
11. What determines whether the pyruvic acid produced in glycolysis becomes lactic acid or acetyl-CoA?
12. Describe the two phases of anaerobic respiration. How many ATP molecules are produced? What happens to the lactic acid produced when oxygen becomes available?
13. Define aerobic respiration and the products produced by it. Name the four phases of aerobic respiration.
14. Why is the citric acid cycle a cycle? What molecules are produced as a result of the citric acid cycle?
15. What is the function of the electron-transport chain? Describe the chemiosmotic model of ATP production.
16. Define beta-oxidation, and explain how it results in ATP production.
17. What are ketone bodies, how are they produced, and for what are they used?
18. Distinguish between an essential and a nonessential amino acid.
19. Define transamination and oxidative deamination. How are proteins used to produce energy?
20. Define glycogenesis, lipogenesis, glycogenolysis, and gluconeogenesis.
21. Describe the events of the absorptive and the postabsorptive states. Why is it important to maintain blood glucose levels?
22. Define metabolic rate, and describe its three components.
23. How are kilocaloric input and output adjusted to maintain body weight?
24. Explain the ways that heat is produced and lost by the body. How does the hypothalamus regulate body temperature?

CONCEPT REVIEW

1. Why does a vegetarian usually have to be more careful about his diet than a person who includes meat in the diet?
2. Explain why a person suffering from copper deficiency would feel tired all the time.
3. Some people claim that fasting occasionally for short times may be beneficial. How can fasts be damaging?
4. Why can some people lose weight on a 1200 kilocalorie (Calories) per day diet and other people cannot?
5. Lotta Bulk, a muscle builder, wanted to increase her muscle mass. Knowing that proteins are the main components of muscle, she consumed large amounts of protein daily (high protein diet). Explain why this strategy will or will not work.

? ?

ANSWERS TO PREDICT QUESTIONS

1 p. 825. If vitamins were broken down during the process of digestion, their structures would be destroyed, and as a result, their ability to function would be lost.

2 p. 834. If the electron of the electron-transport chain cannot be donated to oxygen, the entire electron-transport chain stops, no ATP can be produced aerobically, and the patient dies because too little energy is available for the body to perform vital functions.

3 p. 843. When muscles contract, they must produce ATP. As a result, much heat is also produced. During exercise the large amounts of heat produced can raise body temperature, and we feel warm. Shivering consists of small, rapid muscle contractions that produce heat in an effort to prevent a decrease in body temperature in the cold.

4 p. 844. Vasoconstriction reduces blood flow to the skin, which cools as a result. As the difference in temperature between the skin and the environment decreases, there is less loss of heat.

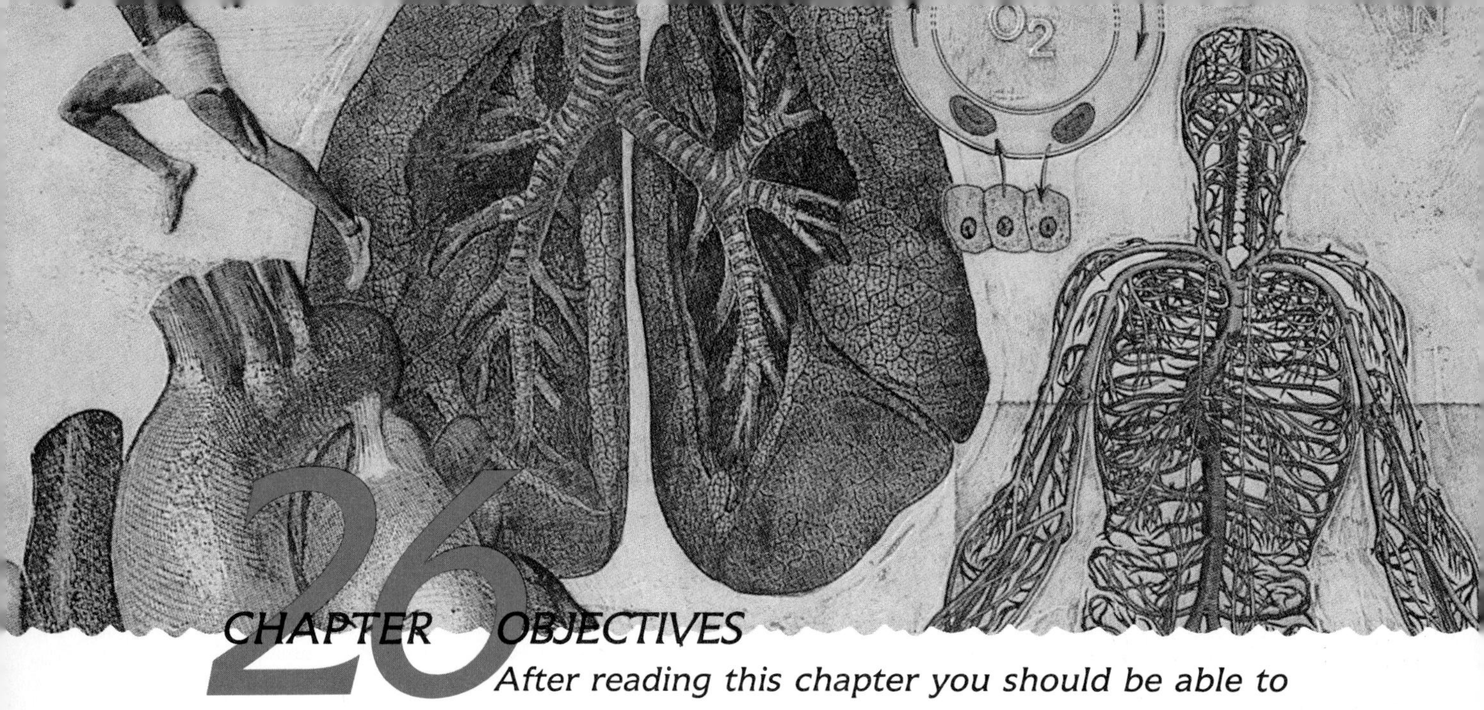

26 CHAPTER OBJECTIVES

After reading this chapter you should be able to

1. List the components of the urinary system and describe the overall functions the system performs.
2. Describe the location, size, shape, and internal anatomy of the kidneys.
3. Describe the structure of the nephron and the orientation of its parts within the kidney.
4. Describe the course of blood flow through the kidney and identify the blood volume that flows through the kidney.
5. List the components of the filtration barrier and describe its structure and the composition of the filtrate.
6. List factors that influence filtration pressure and the rate of filtrate formation.
7. Explain how tubular reabsorption in the proximal convoluted tubule is accomplished and how it influences filtrate composition.
8. Describe the permeability characteristics of the descending limb of the loop of Henle and discuss how the movement of substances across its wall influences the composition of the filtrate.
9. Describe permeability and transport characteristics of the ascending limb of the loop of Henle and explain how they influence filtrate composition.
10. Describe the permeability and transport characteristics of the distal convoluted tubule and the collecting duct.
11. Explain the function of the vasa recta.
12. Illustrate the major components of the countercurrent multiplier mechanism.
13. Demonstrate by using diagrams of the nephron how both increased and decreased antidiuretic hormone and aldosterone levels influence the volume and concentration of urine.
14. Define autoregulation and explain how it influences renal function.
15. Explain the effect that sympathetic stimulation has on the kidney during rest, exercise, and shock.
16. Define and explain tubular maximum and plasma clearance.
17. Describe the micturition reflex.

Urinary System

KEY TERMS

autoregulation

collecting duct

countercurrent multiplier mechanism

filtration

filtration membrane

filtration pressure

glomerular filtration rate

juxtaglomerular apparatus

micturition (mik-tu-rish′un) reflex

nephron (nef′ron)

plasma clearance

renal corpuscle

tubular reabsorption

tubular secretion

vasa recta

RELATED TOPICS

The following terms or concepts are important for a good understanding of this chapter. If you are not familiar with them, you should review them before proceeding.

Osmolality (Chapter 2)

Osmosis and active transport (Chapter 3)

Autonomic nervous control of blood flow to the kidney (Chapter 16)

Antidiuretic hormone and aldosterone (Chapter 18)

THE URINARY SYSTEM PARTICIPATES WITH OTHER ORGANS TO REGULATE THE VOLUME AND COMPOSITION OF THE INTERSTITIAL FLUID WITHIN A NARROW RANGE OF VALUES. Exchange across the walls of capillaries provides nutrients and removes waste products from the interstitial spaces. Exchange of gas in the lungs removes carbon dioxide from the blood and provides a supply of oxygen. The digestive system supplies nutrients to the blood, and the liver removes certain waste products. These organ systems function together to regulate the level of gases, nutrients, and some waste products in the blood. The kidneys remove waste products, many of which are toxic, from the blood and play a major role in controlling blood volume, the concentration of ions in the blood, and the pH of the blood. The kidneys are also involved in the control of red blood cell production and vitamin D metabolism. Although the kidneys are the major excretory organs in the body, the skin, liver, lungs, and intestines also eliminate wastes. However,

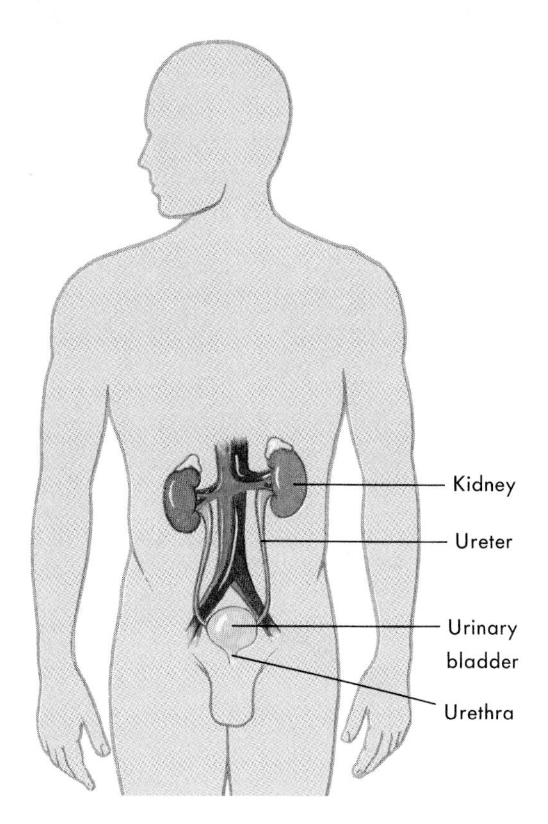

FIGURE 26-1 Anatomy of the urinary system. The urinary system consists of two kidneys, two ureters, a single urinary bladder, and a single urethra.

if the kidneys fail to function, other structures cannot adequately compensate to maintain a normal environment for the body cells.

The urinary system consists of the following organs: (1) two kidneys, (2) a single, midline urinary bladder, (3) two ureters, which carry urine from the kidneys to the urinary bladder, and (4) a single urethra, which carries urine from the bladder to the outside of the body (Figure 26-1).

URINARY SYSTEM

Kidneys

The kidneys are bean-shaped organs, each approximately the size of a tightly clenched fist. They lie on the posterior abdominal wall behind the peritoneum and on either side of the vertebral column near the lateral border of the psoas muscles (Figure 26-2). The superior pole of each kidney is protected by the rib cage, and the right kidney is slightly lower than the left because of the presence of the liver superior to it. Each kidney measures approximately 11 cm long, 5 cm wide, and 3 cm thick and weighs approximately 130 g. A fibrous connective tissue **renal capsule** surrounds each kidney, and around the renal capsule is a dense deposit of adipose tissue, the **renal fat pad,** which protects the kidney from mechanical shock. The kidneys and surrounding adipose tissue are anchored to the abdominal wall by a thin layer of connective tissue, the **renal fascia.**

On the medial side of each kidney is a relatively small area called the **hilum** (hi'lum; a small amount) where the renal artery and the nerves enter and the renal vein and the ureter exit. The hilum opens into a cavity called the **renal sinus,** which is filled with fat and connective tissue (Figure 26-3). In the center of the renal sinus the urinary channel is enlarged to form the **renal pelvis** (basin). Several large urinary tubes called **calyces** (kal'ĭ-sēz; cup of a flower; singular, **calyx**) extend to the renal pelvis from the kidney tissue. The calyces that open directly into the renal pelvis are called **major calyces,** and the smaller calyces that open into major calyces are called **minor calyces.** There are eight to 20 minor calyces and two or three major calyces per kidney. At the hilum the renal pelvis narrows to form the **ureter** (ur-re'ter).

The kidney is divided into an outer **cortex** and an inner **medulla.** The medulla consists of a number of **renal pyramids,** which are cone-shaped structures, although they appear triangular in shape when seen in a longitudinal section of the kidney. The base of each renal pyramid extends toward the cortex to form the **medullary rays.** The apex of each pyramid is called the **renal papilla,** and is surrounded by the opening of a minor calyx. Cortical tissue extends to the renal sinus between the pyramids, and these extensions are called the **renal columns.**

The basic histological and functional unit of the kidney is the **nephron** (nef'ron; Figure 26-4), which consists of an enlarged terminal end called a renal corpuscle, a proximal convoluted tubule, a nephric loop (the loop of Henle), and a distal convoluted tubule. Approximately one third of the 1,300,000 nephrons in each kidney must be functional to ensure survival. The distal convoluted tubule empties into a collecting duct, which carries the urine from the cortex of the kidney to the calyces. The renal corpuscle and both convoluted tubules are in the renal cortex. The collecting tubules and portions of the loops of Henle enter the medulla. Although most nephrons measure 50 to 55 mm in length, the nephrons with renal corpuscles located within the cortex near the medulla are longer than the nephrons with renal corpuscles in the cortex nearer to the exterior of the kidney. Nephrons that lie near the medulla are called **juxtamedullary** (juks'tă-med'u-lĕr-e; juxta means next to) **nephrons** and comprise approximately 15% of all the nephrons. The juxtamedullary nephrons have longer loops of Henle, which extend farther into the medulla than the loops of Henle of other neurons.

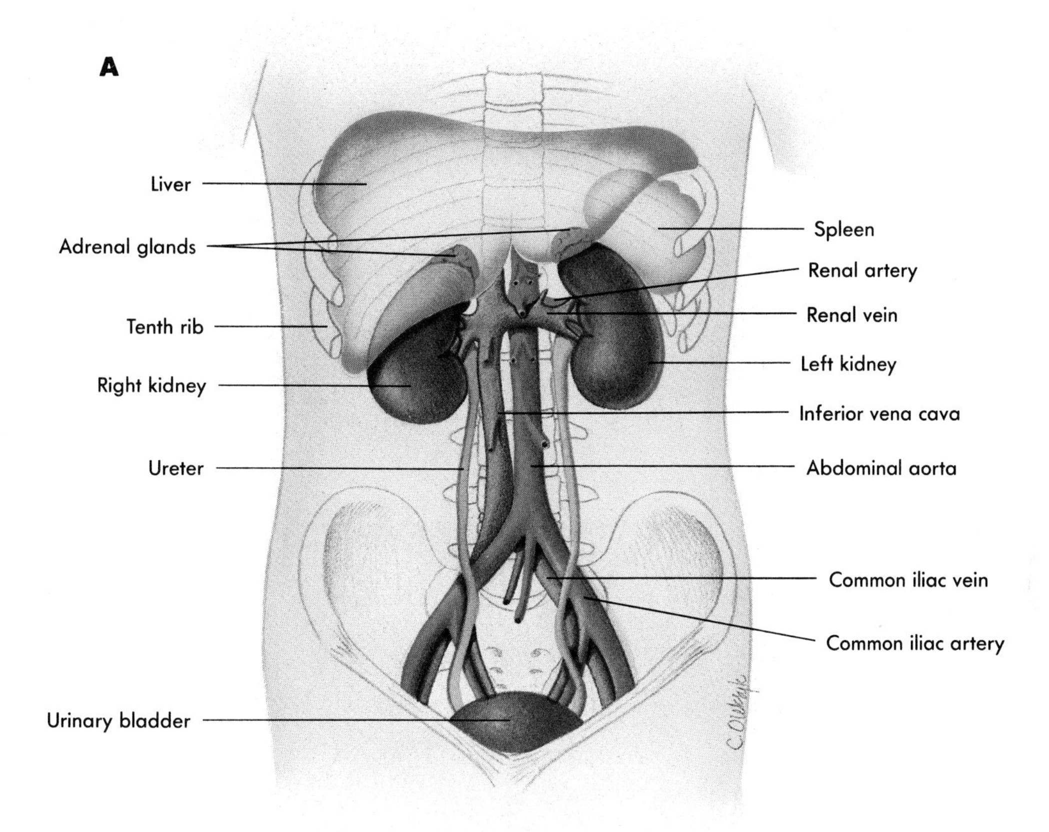

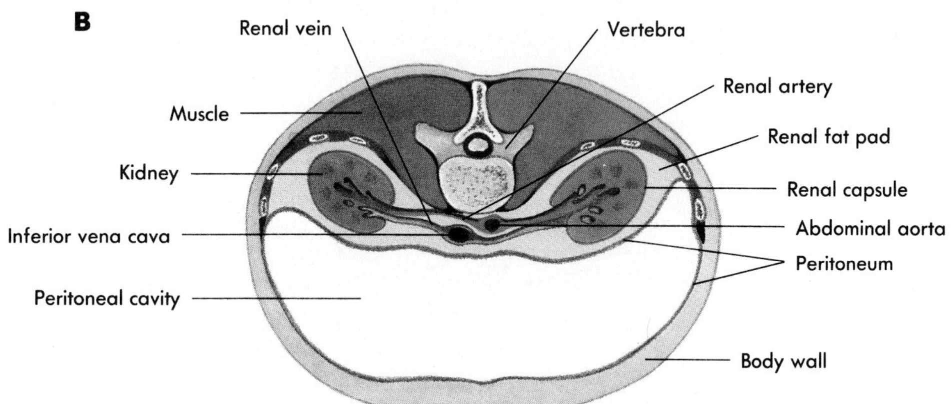

FIGURE 26-2 Anatomy of the kidney. **A** The kidneys are located in the abdominal cavity, with the right kidney just below the liver and the left kidney below the spleen. The ureters extend from the kidneys to the urinary bladder within the pelvic cavity. An adrenal gland is located at the superior pole of each kidney. **B** The kidneys are located behind the peritoneum. Surrounding each kidney is the renal fat pad. The renal arteries extend from the abdominal aorta to each kidney, and the renal veins extend from the kidneys to the inferior vena cava.

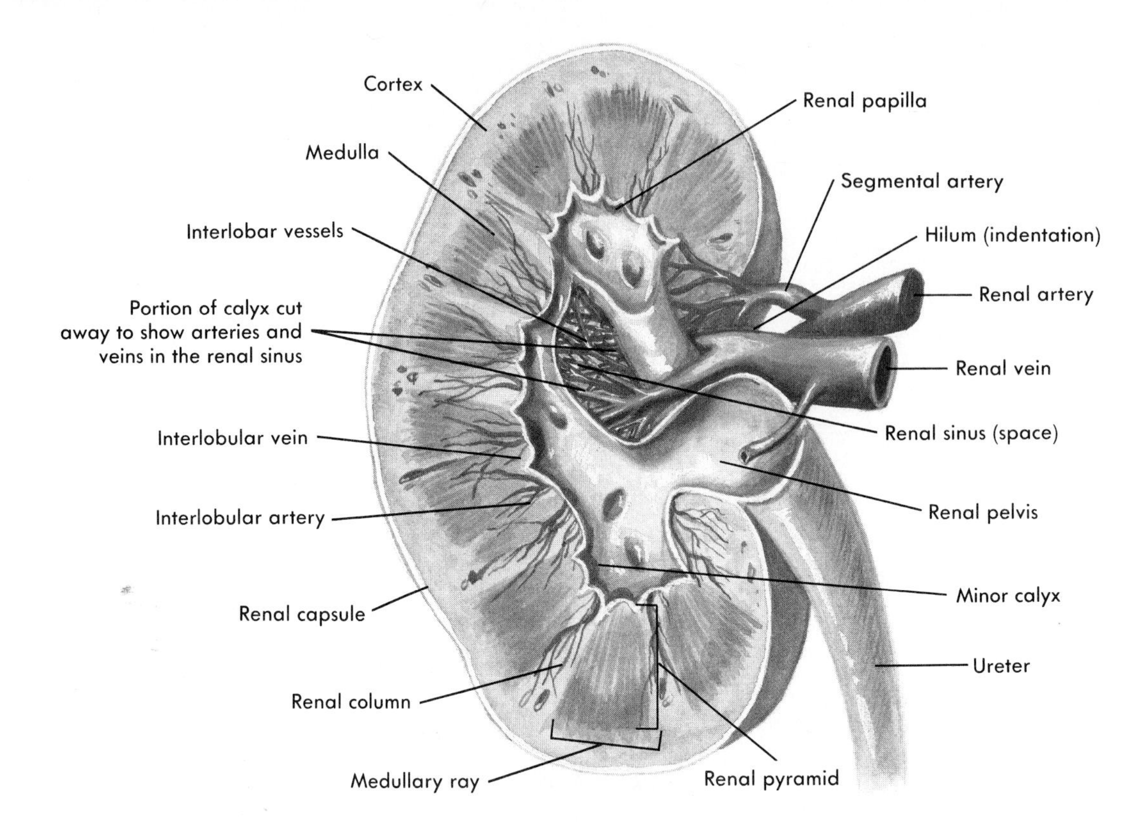

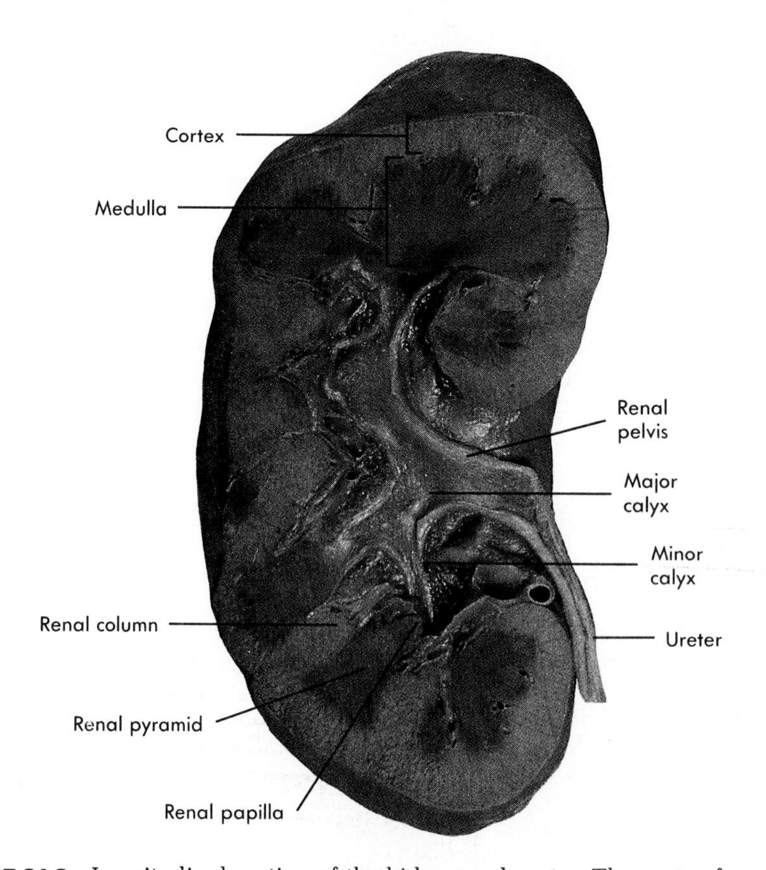

FIGURE 26-3 Longitudinal section of the kidney and ureter. The cortex forms the outer part of the kidney, and the medulla forms the inner part. A central cavity called the renal sinus contains the renal pelvis. The renal columns of the kidney project from the cortex into the medulla and are separated by the pyramids.

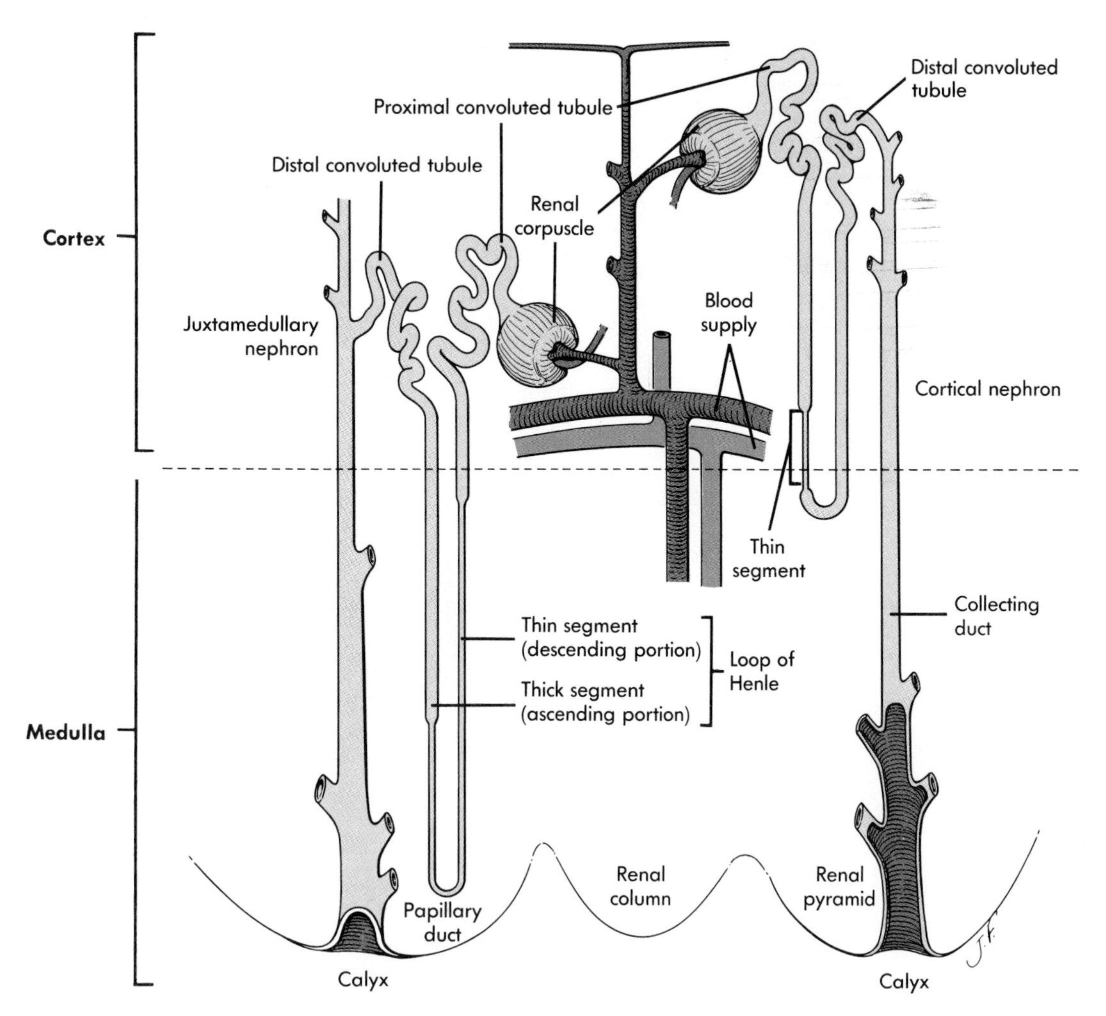

FIGURE 26-4 Functional unit of the kidney—the nephron. The juxtamedullary nephrons (those near the medulla of the kidney) have loops of Henle that extend deep into the medulla of the kidney, whereas other nephrons do not.

The proximal end of the nephron is enlarged to form **Bowman's capsule** (Figure 26-5, *A* and *B*). The wall of Bowman's capsule is indented to form a double-walled chamber. The indentation is occupied by a tuft of capillaries called the **glomerulus** (glo-mĕr′u-lus), which resembles a ball of yarn.

The glomerulus and Bowman's capsule together are called the **renal corpuscle.** The cavity of Bowman's capsule opens into a proximal convoluted tubule, which carries fluid away from the capsule (Figure 26-5, *B*). Surrounding the glomerulus is the inner layer of Bowman's capsule, which is called the **visceral layer.** It consists of specialized cells called **podocytes** (pōd′o-sītz). The outer **parietal layer** of Bowman's capsule is composed of simple squamous epithelium, which becomes cuboidal at the beginning of the proximal convoluted tubule.

The glomerular capillaries are fenestrated (i.e., have openings), and the podocyte processes surrounding the capillaries have gaps, the **filtration slits,** between them. A basement membrane is present between the glomerular capillary cells and the podocytes of Bowman's capsule. The capillary epithelium, basement membrane, and the podocytes constitute the **filtration membrane** (Figure 26-5, *C* and *D*). In the first step of urine formation, fluid passes from the glomerular capillaries into Bowman's capsule through the filtration membrane.

The glomerulus is supplied by an **afferent arteriole** and is drained by an **efferent arteriole.** The afferent and efferent arterioles both have a layer of smooth muscle. At the point where the afferent arteriole enters the renal corpuscle, the smooth muscle cells are modified to form a cuff around the arteriole. These modified cells are called **juxtaglomerular cells.** A portion of the distal convoluted tubule of the nephron lies adjacent to the renal corpuscle between the afferent and efferent arterioles. The specialized tubule cells in that area are collectively called the **macula densa.** The juxtaglomerular cells of the afferent arteriole and the macula densa cells are called the **juxtaglomerular apparatus** (see Figure 26-5, *B*).

The cells in this complex are in more intimate contact with each other than are the cells of other nephritic regions because of the absence of the internal elastic lamina of the afferent arterioles and because of the absence of the basement membrane in the macula densa.

The **proximal convoluted tubule** (proximal to Bowman's capsule) is approximately 14 mm long and 60 μm in diameter, and its wall is composed of simple cuboidal epithelium. The cells are broader at their base, which lies away from the lumen, than they are at the surface of the lumen (Figure 26-6, *A* and *B*), and they have microvilli at their luminal surface.

The **loops of Henle** are continuations of the proximal tubules. Each loop has a **descending limb** and an **ascending limb.** The first part of the descending

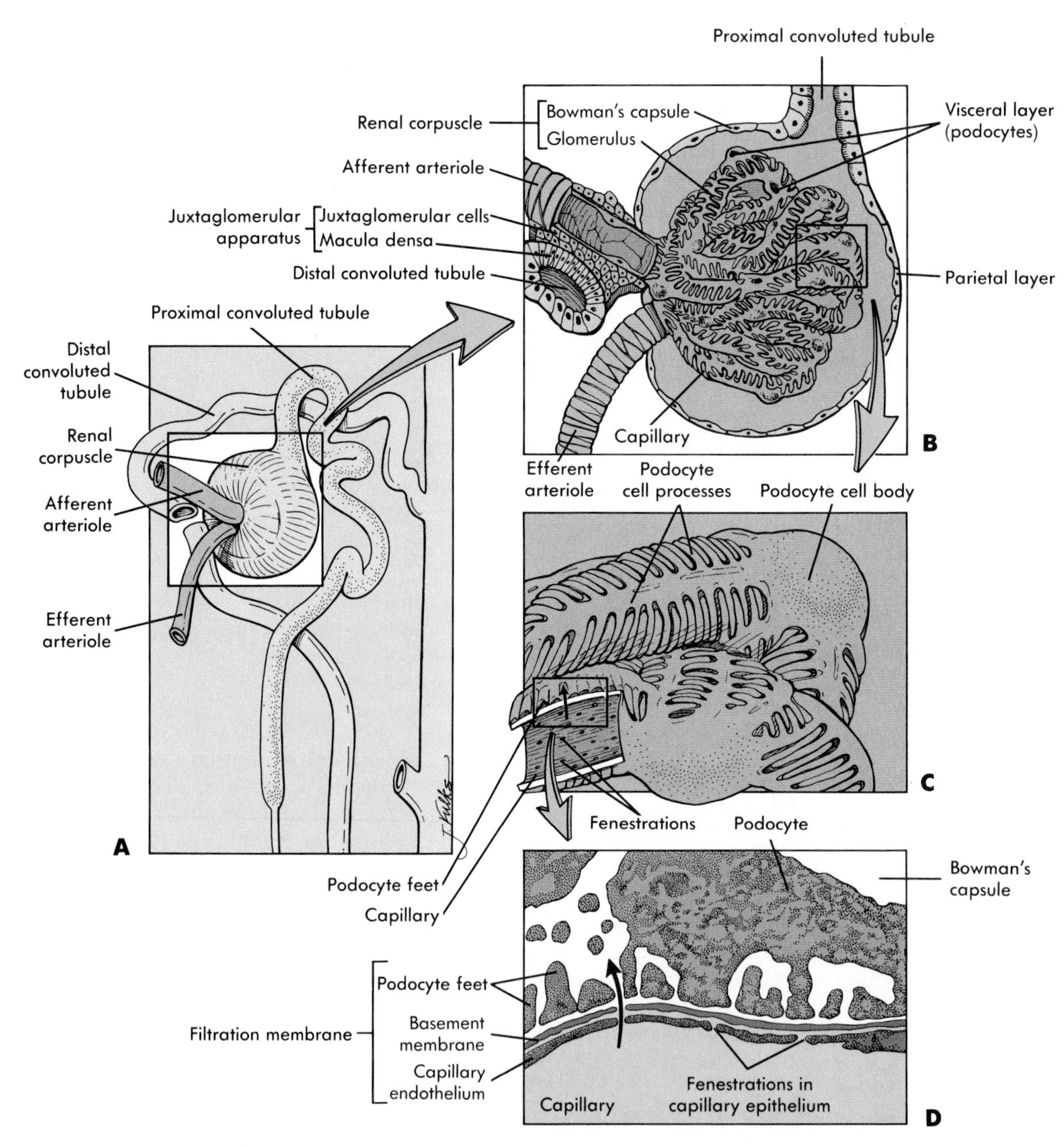

FIGURE 26-5 Renal corpuscle. **A** Bowman's capsule encloses the glomerulus. **B** Blood flows into the glomerulus through the afferent arterioles and leaves the glomerulus through the efferent arterioles. The proximal convoluted tubule exits Bowman's capsule. **C** Podocytes of Bowman's capsule surround the capillaries. Filtration slits between the podocytes allow fluid to pass into Bowman's capsule. The glomerulus is composed of capillary endothelium that is fenestrated. Surrounding the endothelial cells is a basement membrane. **D** Capillary endothelial cells, the basement membrane, and podocytes comprise the filtration membrane of the kidney.

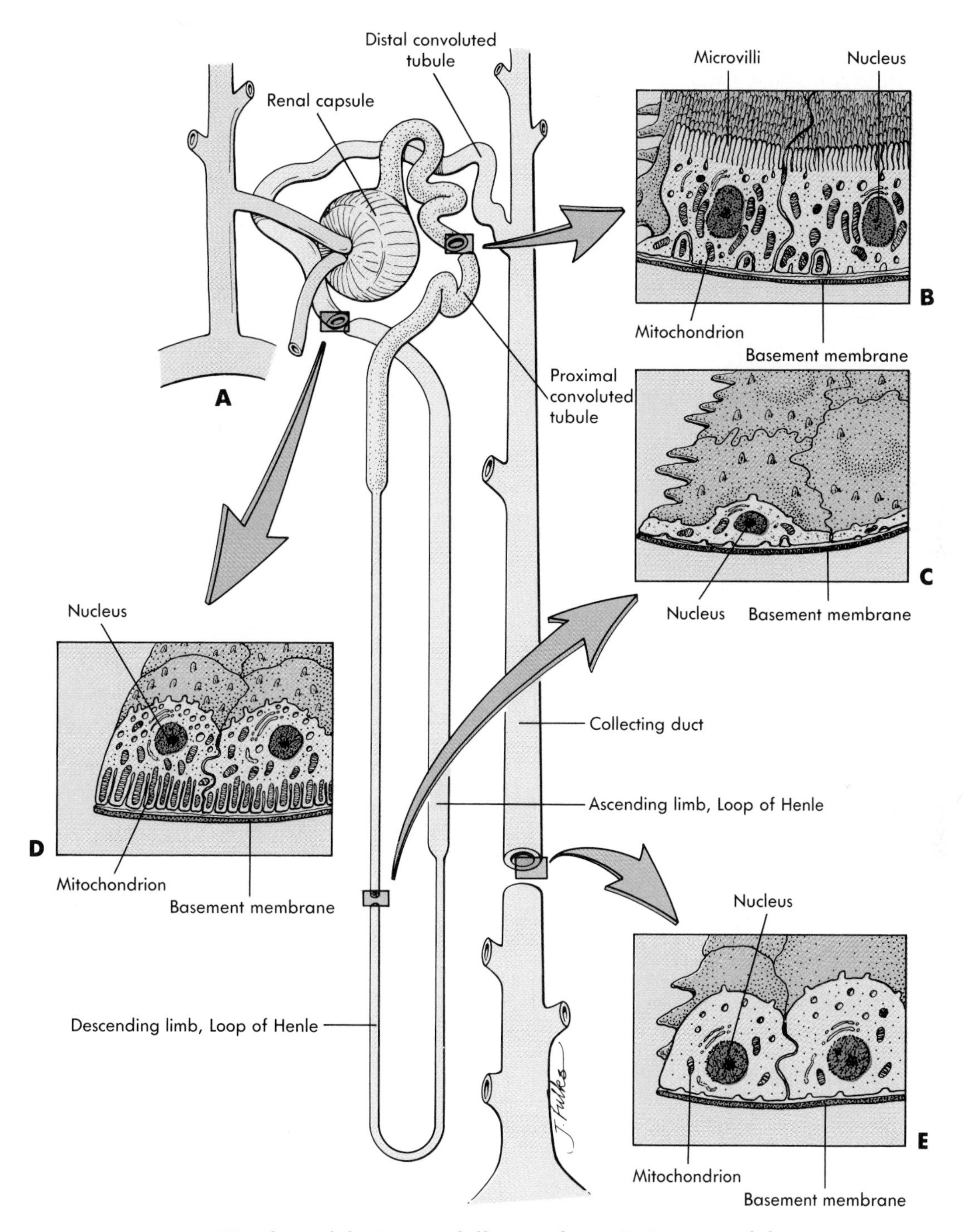

FIGURE 26-6 Histology of the juxtamedullary nephron. **A** Structure of the juxtamedullary nephron. **B** Structure of the cells of the proximal convoluted tubule. The luminal surface of the epithelial cells is lined with numerous microvilli. The basal surface of each cell rests on a basement membrane, and each cell is bound to the adjacent cells by cell-to-cell attachments. The basal margin of each epithelial cell has deep invaginations, and numerous mitochondria are adjacent to the basal cell membrane. **C** Structure of cells of the descending limb of the loop of Henle. The thin segment of the loop of Henle is composed of squamous epithelial cells that have microvilli and contain a relatively small number of mitochondria. **D** Structure of cells of the distal convoluted tubule. The cells have sparse microvilli and numerous mitochondria. **E** Structure of the collecting ducts.

limb is similar in structure to the proximal convoluted tubules, but near the end of the loop the epithelium becomes very thin (Figure 26-6, *C*). In the thin portion the lumen becomes narrow, and there is an abrupt transition from simple cuboidal epithelium to simple squamous epithelium. The first portion of the ascending limb is also very thin, but it soon becomes thicker and is again composed of simple cuboidal epithelium. The thick portion of the loop returns toward the glomerulus and ends by giving rise to the distal convoluted tubule near the macula densa. The **distal convoluted tubules** (distal to Bowman's capsule) are not as long as the proximal convoluted tubules. The epithelium is simple cuboidal, but the cells are smaller than the epithelial cells in the proximal tubules and do not possess a large number of microvilli (Figure 26-6, *D*). The **collecting ducts** are composed of simple cuboidal epithelium, are joined by the distal convoluted tubules of many nephrons, and are larger in diameter than segments of the nephron (Figure 26-6, *E*). The collecting ducts form much of the medullary rays, and they extend through the medulla to the tip of the renal pyramid.

Arteries and Veins

The **renal artery** branches off the abdominal aorta and enters the renal sinus of each kidney (Figure 26-7, *A*). **Segmental arteries** diverge to form **interlobar arteries,** which ascend within the renal columns toward the renal cortex. Branches from the interlobar arteries diverge near the base of each pyramid and arch over the base of the pyramids to form the **arcuate** (ar′ku-at) **arteries. Interlobular arteries** project from the arcuate arteries into the cortex, and the afferent arterioles are derived from the interlob-

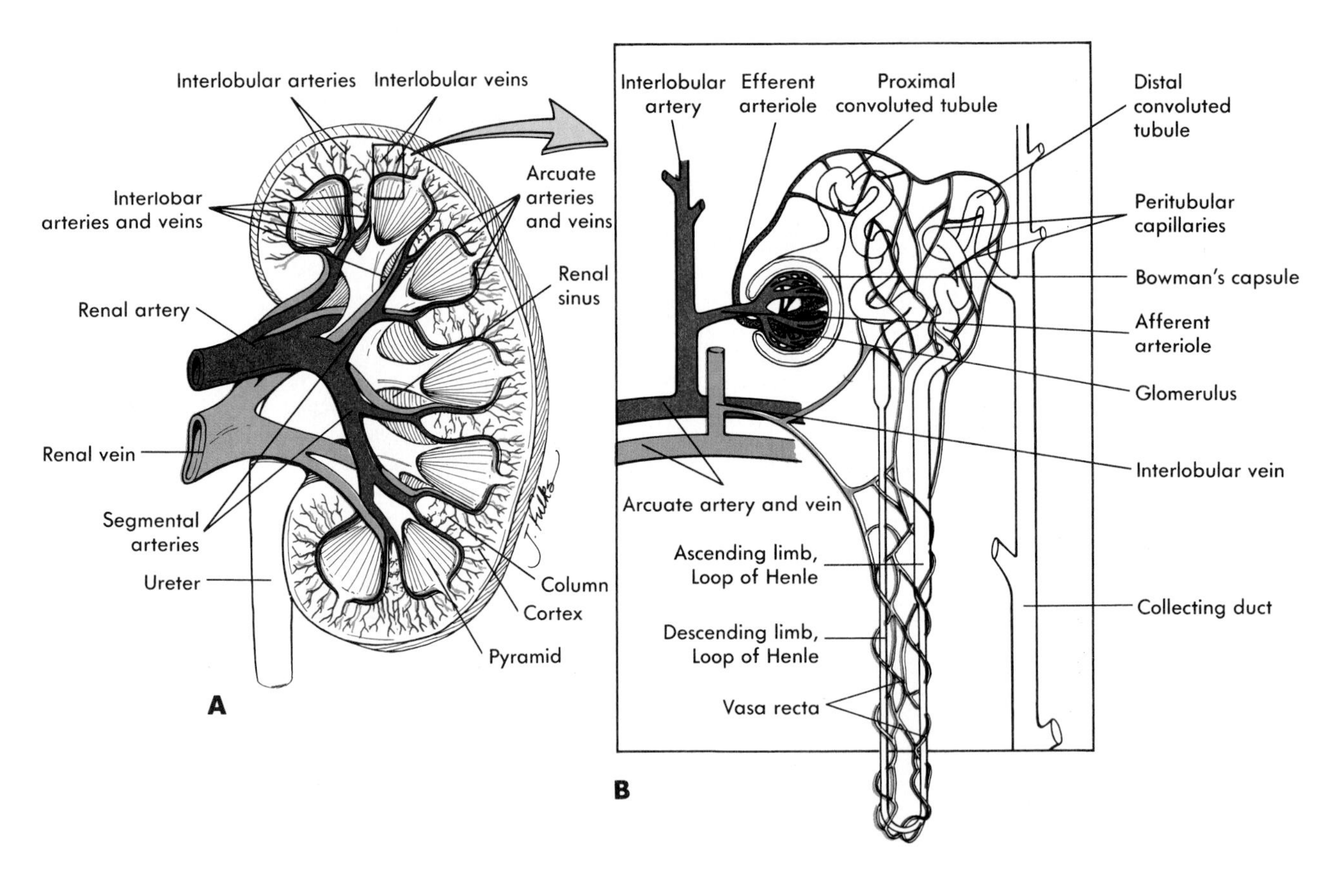

FIGURE 26-7 Blood flow through the kidney. **A** Renal arteries project to the renal sinus. The interlobar branches extend through the renal columns to the arcuate arteries near the border between the cortex and the medulla. **B** Interlobular arteries extend from the arcuate arteries toward the cortex. Branches of the interlobular arteries supply the afferent arterioles. Blood then flows into the glomeruli and from the glomeruli into the efferent arterioles. Efferent arterioles supply the peritubular capillaries. The vasa recta are specialized portions of the peritubular capillaries that extend deep into the medulla of the kidney. From the peritubular capillaries blood flows into the interlobular veins. From the interlobular veins blood flows through the arcuate veins, through the interlobar veins, and into the renal veins that extend to the inferior vena cava.

ular arteries or their branches. The afferent arterioles supply blood to the glomerular capillaries of the renal corpuscle. Efferent arterioles arise from the glomerulus and carry blood away from the glomerulus. After each efferent arteriole exits the glomerulus, it gives rise to a plexus of capillaries called the **peritubular capillaries** around the convoluted tubules. Specialized portions of the peritubular capillaries, called **vasa recta,** course into the medulla along with the loops of Henle (Figure 26-7, *B*). These capillaries drain into **interlobular veins,** which in turn drain into the **arcuate veins.** The arcuate veins empty into the **interlobar veins,** which drain into the renal vein. The renal vein exits the kidney and connects to the inferior vena cava.

Ureters and Urinary Bladder

The ureters extend inferiorly and medially from the renal pelvis at the renal hilum to reach the posterior and inferior surface of the urinary bladder (see Figure 26-1), which functions to store urine. The urinary bladder is a hollow muscular container that lies in the pelvic cavity just posterior to the symphysis pubis. In the male it is just anterior to the rectum, and in the female it is just anterior to the vagina and inferior and anterior to the uterus. The size of the bladder is dependent on the presence or absence of urine. The ureters enter the bladder inferiorly on its posterolateral surface, and the urethra exits the bladder inferiorly and anteriorly (Figure 26-8, *A*). The triangular area of the bladder wall between the two ureters posteriorly and the urethra anteriorly is called the **trigone** (tri'gōn). This region differs histologically from the rest of the bladder wall and does not expand during bladder filling.

The ureters and urinary bladder are lined with transitional epithelium, which is surrounded by a lamina propria, a muscular coat, and a fibrous adventitia (Figure 26-8, *B* and *C*). The wall of the bladder is much thicker than the wall of the ureter. This thickness is caused by the layers, composed primarily of smooth muscle, that are external to the epithelium. The epithelium itself ranges from four or five cells thick in the empty organ to two or three cells thick when the bladder is distended. Transitional epithelium is specialized so that the cells slide past one another, and the number of cell layers decreases as the volume of the urinary bladder in-

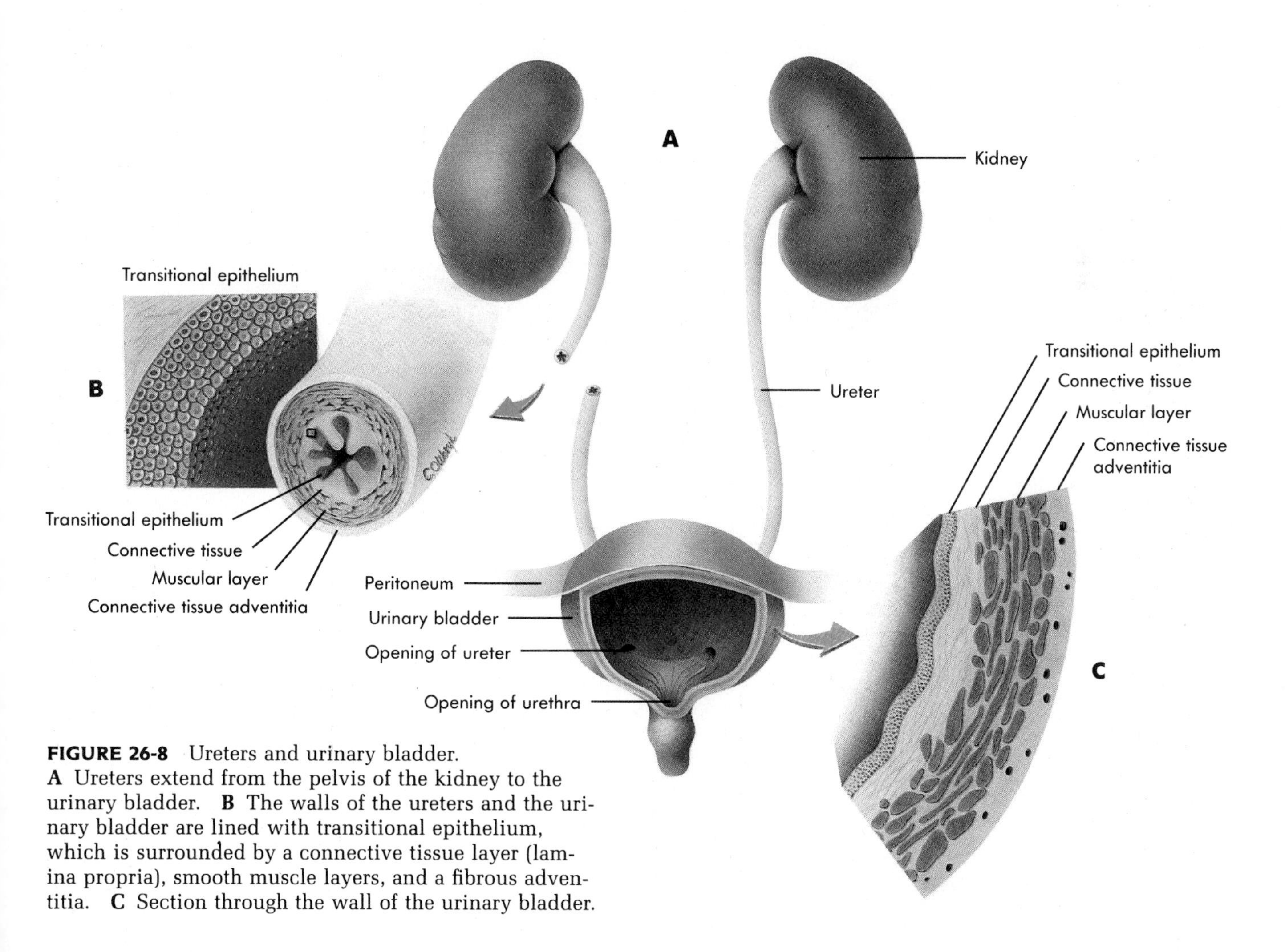

FIGURE 26-8 Ureters and urinary bladder.
A Ureters extend from the pelvis of the kidney to the urinary bladder. **B** The walls of the ureters and the urinary bladder are lined with transitional epithelium, which is surrounded by a connective tissue layer (lamina propria), smooth muscle layers, and a fibrous adventitia. **C** Section through the wall of the urinary bladder.

creases. The epithelium of the urethra is stratified or pseudostratified columnar epithelium.

At the junction of the urethra with the urinary bladder, smooth muscle of the bladder forms the **internal urinary sphincter.** The **external urinary sphincter** is skeletal muscle that surrounds the urethra as the urethra extends through the pelvic floor. The sphincters control the flow of urine through the urethra.

In the male the urethra extends to the end of the penis where it opens to the outside (see Chapter 28). The female urethra, approximately 3.8 cm long, is much shorter than the male urethra and opens into the vestibule anterior to the vaginal opening.

Since bladder infection (cystitis) often occurs when bacteria from outside the body enter the bladder, are males or females more prone to urinary bladder infection? Explain.

? ? ? ? ? ? ? ? ? ? ?

URINE PRODUCTION

Because nephrons are the smallest structural components capable of producing urine, they are called the functional units of the kidney. Filtration, reabsorption, and secretion are the three major processes critical to the formation of urine. **Filtration** is movement of plasma across the filtration membrane as a result of a pressure difference. The portion of the plasma entering the nephron becomes the **filtrate. Reabsorption** is the movement of substances from the filtrate back into the blood. In general, metabolic waste products are not reabsorbed, but useful substances are (Table 26-1). **Secretion** is the active transport of substances into the nephron. Urine produced by the nephrons consists of the constituents that are filtered and secreted into the nephron minus those substances that are reabsorbed.

Filtration

The portion of the total cardiac output that passes through the kidneys is called the **renal fraction.** Although the renal fraction varies from 12% to 30% of the cardiac output in healthy resting adults, it averages 21% to produce a **renal blood flow rate** of 1176 ml of blood per minute (see Table 26-2 for the calculation of renal blood flow rate and other kidney flow rates).

Approximately 19% of the plasma volume, the **filtration fraction,** is filtered through the filtration membrane in Bowman's capsule to become filtrate. Approximately 125 ml of filtrate are produced each minute (the **glomerular filtration rate**), which is equivalent to approximately 180 L of filtrate produced daily. Since only approximately 1 to 2 L of

Table 26-1

Concentrations of major solutes

SUBSTANCE	PLASMA	FILTRATE	NET MOVEMENT OF SOLUTE*	URINE	CONCENTRATION URINE/CONCENTRATION PLASMA
Water (L)	180	180	178.6	1.4	—
Organic molecules (mg/100 ml)					
Protein	3900-5000	6-11		0†	0
Glucose	100	100	−100.0	0	0
Urea	26	26	−11.4	1820	70
Uric acid	3	3	−2.7	42	14
Creatinine	1.1	1.1	0.5	196	140
Ions (mEq/L)					
Na^+	142	142	−141.0	128	0.9
K^+	5	5	−4.5	60	12.0
Cl^-	103	103	−101.9	134	1.3
HCO_3^-	28	28	−27.9	14	0.5

*In many cases there is movement of a solute into and out of the nephron. Figures indicate net movement. Negative numbers are net movement out of the filtrate, and positive numbers are net movement into the filtrate.

†Trace amounts of protein can be found in the urine. A value of zero is assumed here.

urine are produced each day by a healthy person, it is obvious that not all of the filtrate becomes urine. Approximately 99% of the filtrate volume is reabsorbed in the nephron, and less than 1% becomes urine.

2

If the filtration fraction increases from 19% to 22% and if 99.2% of the filtrate is reabsorbed, how much urine is produced in a normal person with a cardiac output of 5600 ml per minute?

? ? ? ? ? ? ? ? ? ? ?

Filtration barrier

The filtration membrane functions as a **filtration barrier,** which prevents the entry of blood cells and proteins into the nephron but allows other blood components to enter. The filtration barrier of the renal corpuscle is 100 to 1000 times more permeable than a typical capillary. Water and solutes of a small molecular diameter readily pass from the glomerular capillaries through the filtration barrier into Bowman's capsule, whereas larger molecules do not. The fenestrae of the glomerular capillary, the fused basement membrane of the glomerular endothelium and podocytes, and the podocyte cells (see Figure 26-5, D) prevent molecules larger than 7 nm in diameter or a molecular weight of 40,000 daltons from passing through. Since most plasma proteins are slightly larger than 7 nm in diameter, they are retained in the glomerular capillaries. Albumin, which has a diameter just slightly less than 7 nm, enters the filtrate in small amounts so that the filtrate contains no cells and approximately 0.03% protein. Protein hormones are also small enough to pass through the filtration

Table 26-2

Calculation of renal flow rates

SUBSTANCE	AMOUNT PER MINUTE (ml)	CALCULATION
Renal blood flow	1176	Amount of blood flowing through the kidneys per minute; equals cardiac output (5600 ml blood/min) times the percent (21%—renal fraction) of cardiac output that enters the kidneys. 5600 ml blood/min × 0.21 = 1176 ml blood/min
Renal plasma flow	650	Amount of plasma flowing through the kidneys per minute; equals renal blood flow times percent of the blood that is plasma. Since the hematocrit is the percent of the blood that is formed elements, the percent of the blood that is plasma is 100 minus the hematocrit. Assuming a hematocrit of 45, the percent of the blood that is plasma is 55% (100 − 45). Therefore renal plasma flow is 55% of renal blood flow. 1176 ml blood/min × 0.55 = 650 ml plasma/min
Glomerular filtration rate	125	Amount of plasma (filtrate) that enters Bowman's capsule per minute; equals renal plasma flow times the percent (19%—filtration fraction) of the plasma that enters the renal capsule. 650 ml plasma/min × 0.19 = 125 ml filtrate/min
Urine	1	Nonreabsorbed filtrate that leaves the kidneys per minute; equals glomerular filtration rate times the percent (0.8%) of the filtrate that is not reabsorbed into the blood. 125 ml filtrate/min × 0.008 = 1 ml urine/min Milliliters of urine per minute can be converted to liters of urine per day by multiplying by 1.44. 1 ml urine/min × 1.44 = 1.4 L/day

Kidney Dialysis

The artificial kidney (renal dialysis machine) is a machine used to treat patients who are experiencing renal failure. The use of this machine often allows people with severe acute renal failure to recover without developing the side effects of renal failure, and the machine can substitute for the kidneys for long periods of time in people suffering from chronic renal failure.

Renal dialysis is based on blood flow through tubes composed of a selectively permeable membrane. On the outside of the dialysis tubes is a fluid that contains the same concentration of solutes as the plasma except for the metabolic waste products. As a consequence, a diffusion gradient exists for the metabolic waste products from the blood to the dialysis fluid. The dialysis membrane has pores that are too small to allow the plasma proteins to pass through them. Since the dialysis fluid contains the same beneficial solutes as the plasma, the net movement of these substances is zero. The metabolic waste products, on the other hand, diffuse rapidly from the blood into the dialysis fluid.

Blood usually is taken from an artery, is passed through the tubes of the dialysis machine, and then is returned to a vein. The rate of blood flow is normally several hundred milliliters per minute, and the total surface area for exchange in the machine is close to 10,000 to 20,000 cm² (Figure 26-A).

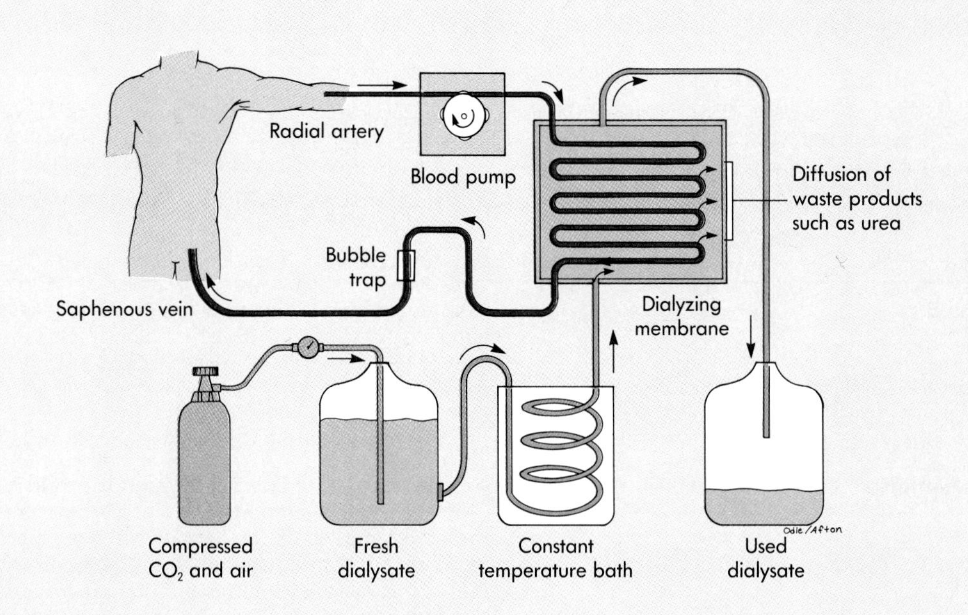

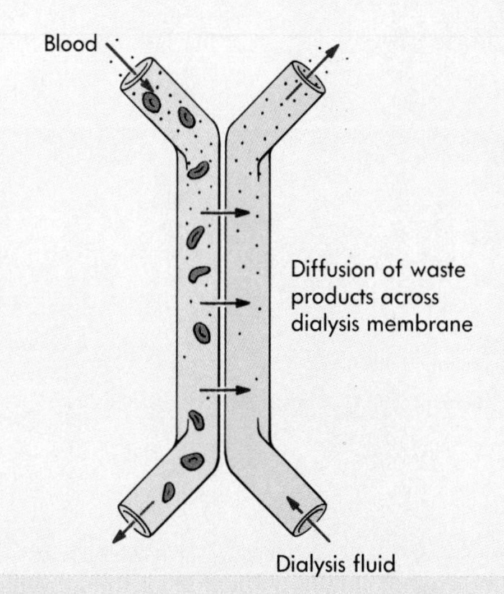

FIGURE 26-A Kidney dialysis. During kidney dialysis blood flows through a system of tubes composed of a selectively permeable membrane. Dialysis fluid, the composition of which is similar to that of blood except that the concentration of waste products is very low, flows in the opposite direction on the outside of the dialysis tubes. Consequently, waste products such as urea diffuse from the blood into the dialysis fluid. Other substances such as sodium, potassium, and glucose do not rapidly diffuse from the blood into the dialysis fluid because there is no concentration gradient since these substances are present in the dialysis fluid.

barrier. The protein is actively reabsorbed by endocytosis and is metabolized by the cells in the proximal convoluted tubule.

3

Hemoglobin has a smaller diameter than albumin, but very little hemoglobin passes from the blood into the filtrate. Explain why. Under what conditions would large amounts of hemoglobin enter the filtrate?

? ? ? ? ? ? ? ? ? ? ?

Filtration pressure

The formation of filtrate depends on a pressure gradient called the **filtration pressure,** which forces fluid from the glomerular capillary through the filtration membrane into Bowman's capsule. The filtration pressure results from forces that move fluid out of the capillary into Bowman's capsule and forces that move fluid out of Bowman's capsule into the capillary (Figure 26-9). The **glomerular capillary pressure** is the blood pressure within the capillary. It is approximately 60 mm Hg, which is two to three times higher than that in most capillaries, and it forces fluid out of the capillary through the filtration membrane. Opposing the movement of fluid into Bowman's capsule is the **capsule pressure,** which is approximately 18 mm Hg, caused by the pressure of filtrate already inside Bowman's capsule. The **colloid osmotic pressure,** which is caused by unfiltered plasma proteins remaining within the glomerular capillary, produces an osmotic force of approximately 32 mm Hg that causes fluid to move from the capsule into the capillary. Therefore the filtration pressure is approximately 10 mm Hg.

$$\underset{\text{Filtration pressure}}{10\text{ mm Hg}} = \underset{\text{Glomerular capillary pressure}}{60\text{ mm Hg}} - \underset{\text{Capsule pressure}}{18\text{ mm Hg}} - \underset{\text{Colloid osmotic pressure}}{32\text{ mm Hg}}$$

The high glomerular capillary pressure results from a low resistance in the afferent arterioles and a higher resistance in the efferent arterioles. As the diameter of a vessel decreases, the resistance to flow through the vessel increases (see Chapter 21). Pressure before the point of decreased vessel diameter is higher than pressure after the point of decreased diameter. For example, in the extreme case of a completely closed vessel, pressure would increase on the arterial side of the constriction and fall to zero after the constriction. Because the efferent arteriole has a small diameter, there is a high resistance to blood flow, and blood pressure within the glomerulus is increased. After the efferent arteriole, there is a large drop in pressure so that the pressure in the peritubular capillaries is very low. Consequently, filtrate is forced across the filtration membrane into Bowman's capsule, and the low pressure in the peritubular capillaries allows fluid to move into them from the interstitial fluid.

The afferent and efferent arterioles have smooth muscles in their walls that can alter the vessel diameter and affect glomerular filtration. For example, constriction of the efferent arteriole increases glomerular capillary pressure, causing an increase in filtration pressure and glomerular filtration.

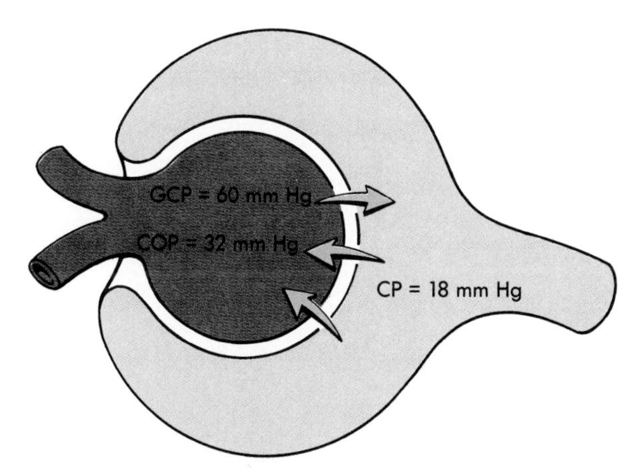

FIGURE 26-9 Filtration pressure across the filtration membrane of the kidney's glomeruli. Pressure across the filtration membrane favors the filtration of materials from the glomerulus into the Bowman's capsule.

4

What effect would constriction of the afferent arteriole have on the filtration pressure? What effect would a decrease in the concentration of plasma proteins have on the filtration pressure?

? ? ? ? ? ? ? ? ? ? ?

Tubular Reabsorption

The filtrate leaves the renal capsule and flows through the proximal convoluted tubule, the loop of Henle, and the distal convoluted tubule and then into the collecting ducts. As it passes through these structures, many of the substances in the filtrate undergo **tubular reabsorbtion.** Inorganic salts, organic molecules, and approximately 99% of the filtrate volume leave the nephron and enter the interstitial fluid. These substances ultimately enter the low-pressure peritubular capillaries and flow through the renal veins to enter the general circulation.

Substances transported from the lumen of the nephron to the interstitial spaces include protein, amino acids, glucose, and fructose, as well as sodium, potassium, calcium, bicarbonate, and chloride ions. Sodium ions are actively transported across the basal cell membrane from the cytoplasm to the interstitial fluid, creating a low concentration of sodium inside the cells (Table 26-3). There are carrier molecules in the cell membranes of the microvilli that line the lumen of the tubules. The carrier molecules transport sodium ions from the lumen of the nephron into the cell down their concentration gradient. Amino acids, glucose, fructose, and some inorganic ions are transported across the membrane of the microvilli with the sodium ions through a process called **cotransport** (Figure 26-10). Each molecule that is cotransported binds to its respective carrier molecule. Sodium ions bind to the same carrier molecules, and the sodium ion and the cotransported molecule move across the membrane into the cell. The concentration gradient for sodium is a source of energy for the cotransport of molecules into the cells against their concentration gradients (see Figure 26-10).

Table 26-3

Reabsorption of substances from the nephron

TRANSPORT PROCESS	SUBSTANCE TRANSPORTED
Proximal Convoluted Tubule	
Endocytosis	Protein
Active transport (or cotransport)	Sodium, chloride, and potassium ions; glucose, fructose, galactose; amino acids; lactate, succinate, citrate, bicarbonate, calcium, magnesium, and phosphate ions
Thick Ascending Limb	
Active transport (or cotransport)	Sodium, chloride, and potassium ions
Distal Convoluted Tubule	
Active transport (or cotransport)	Sodium, chloride, and calcium ions

The proximal convoluted tubule is permeable to water. Therefore as solute molecules are transported from the nephron to the interstitial spaces, water moves by osmosis in the same direction. By the time the filtrate has reached the end of the proximal convoluted tubule, its volume has been reduced by approximately 65%. The concentration of the filtrate in the nephron remains approximately the same as that of the interstitial fluid because the wall of the nephron is permeable to water.

The thin segment of the loop of Henle (see Figure 26-6, C) is highly permeable to water and moderately permeable to urea, sodium, and most other ions. It is adapted to allow passive movement of substances through its wall, although water passes through much more rapidly than solutes. As the filtrate passes through the thin segment of the loop of Henle, water moves out of the nephron by osmosis, and some solutes move into the nephron. These events occur be-

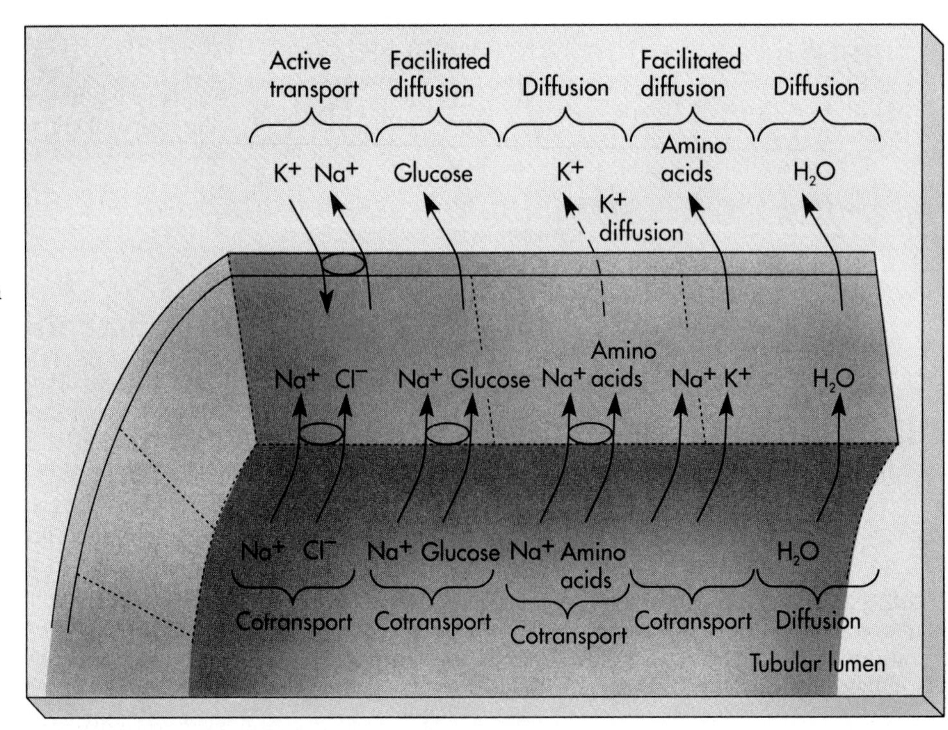

FIGURE 26-10
Transport of molecules across the epithelial lining of the proximal convoluted tubule. Amino acids, glucose, potassium ions, and chloride ions are transported with sodium ions through a process called cotransport.

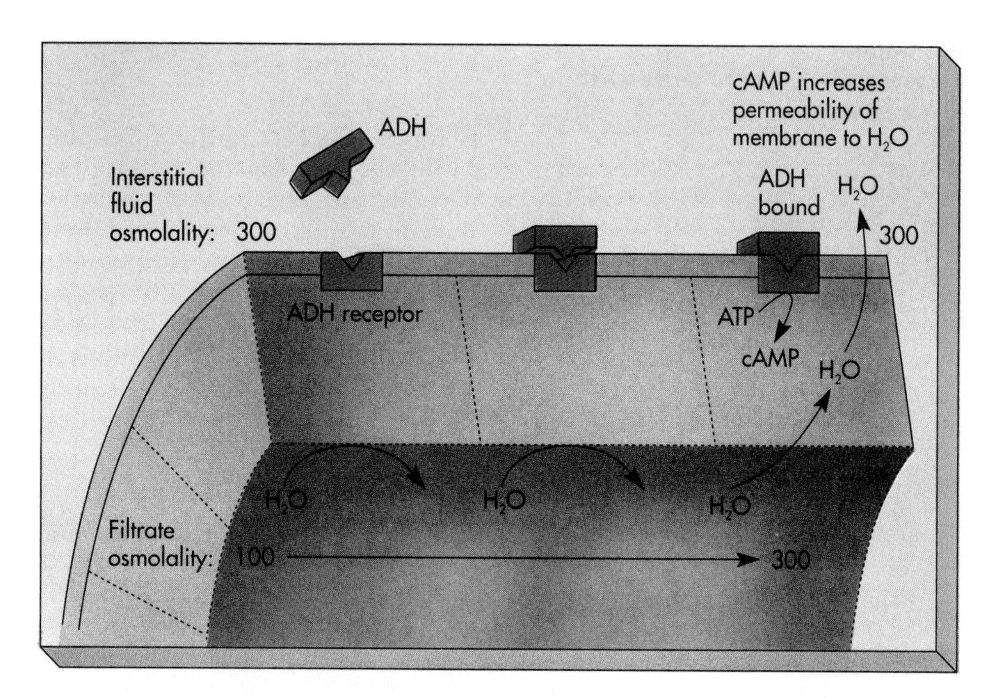

FIGURE 26-11 In the presence of antidiuretic hormone *(ADH)* the distal convoluted tubule and collecting ducts become more permeable to water. ADH binds to receptor molecules on the cell membrane. When ADH is bound to its receptor molecule, the cell membrane increases the rate of cyclic AMP *(cAMP)* synthesis; cyclic AMP, in turn, increases the permeability of the epithelial cells to water. Water then moves out of the tubule into the interstitial spaces by osmosis, both decreasing the volume of the filtrate and increasing its concentration.

cause the thin segment of the loop of Henle descends into the medulla of the kidney where the concentration of solutes in the interstitial fluid is very high. By the time the filtrate has reached the end of the thin segment of Henle's loop, the volume of the filtrate has been reduced by another 15%.

The ascending limb of Henle's loop is impermeable to water. However, solute molecules such as sodium, potassium, and chloride ions are transported from the nephron into the interstitial fluid. Transport of all solutes is tied to the active transport of sodium ions. The sodium ions are actively transported across the basal membrane of the cells of the nephron into the interstitial fluid, keeping the intracellular concentration of sodium ions very low. That process creates a substantial concentration gradient for sodium from the nephron into the cells of the ascending limb. A carrier protein allows the passive movement of sodium ions across the lumen membrane from the filtrate into the cells of the ascending limb. Chloride and potassium ions are cotransported along with the sodium ions (see Figure 26-10). The concentration gradient for sodium ions is responsible for the energy that moves potassium and chloride ions against their concentration gradients into the cells of the nephron. Once inside the cells of the ascending limb, chloride and potassium ions diffuse across the cell membrane into the interstitial fluid from a higher concentration inside the cells to a lower concentration outside the cells.

Because the ascending limb of Henle's loop is impermeable to water and because ions are transported out of the nephron, the concentration of solutes in the tubule is reduced to approximately 100 mOsm by the time the fluid reaches the distal convoluted tubule. In contrast, the concentration of the interstitial fluid is approximately 300 mOsm.

An osmole is a measure of the number of particles in solution. One osmole is the molecular weight, in grams, of a solute times the number of ions or particles into which it dissociates in 1 kg of solution. A milliosmole (mOsm) is 1/1000 of an osmole. The osmolality of a solution is the number of osmoles in a liter (1 kg) of solution. Water moves by osmosis from a solution with a low osmolality to a solution with a higher osmolality. Thus water moves by osmosis from a solution of 100 mOsm/L toward a solution of 300 mOsm/L.

Sodium ions and chloride ions are also actively transported across the wall of the distal convoluted tubule and collecting duct. The permeability of the distal convoluted tubule and collecting duct to water is controlled by hormones. **Antidiuretic hormone (ADH)** increases the permeability of the membrane to water, but the membrane is relatively impermeable to water in the absence of ADH (Figure 26-11). When ADH is present, water moves by osmosis out of the distal convoluted tubule and collecting duct, whereas in the absence of ADH, water remains within the nephron.

5

What effect would a lack of ADH secretion have on the volume and concentration of urine produced by the kidney?

? ? ? ? ? ? ? ? ? ? ?

Approximately 99% of the water that enters the filtrate is reabsorbed. Water moves by osmosis through the proximal convoluted tubule (65% of the filtrate volume) and through the thin segment of the loop of Henle (15% of the filtrate volume). Therefore approximately 80% of the volume of the filtrate is reabsorbed in these structures. Another 19% is reabsorbed (when ADH is present) in the distal convoluted tubules and collecting ducts.

Urea enters the glomerular filtrate and is present in the same concentration as it is in the plasma. As the volume of the filtrate decreases in the proximal convoluted tubule, the concentration of urea increases because renal tubules are not as permeable to urea as they are to water. Only 40% to 60% of the urea is passively reabsorbed in the nephron, although approximately 99% of the water is reabsorbed. In addition to urea, urate ions, creatinine, sulfates, phosphates, and nitrates are reabsorbed but not to the same extent as water. Since these substances are taken into the body in larger amounts than needed, their accumulation in the filtrate and elimination in the urine are adaptive.

Many drugs, environmental pollutants, and other foreign substances that gain access to the circulatory system are reabsorbed. These substances are usually lipid-soluble, nonpolar compounds. They enter the glomerular filtrate and are reabsorbed passively by a process similar to that by which urea is reabsorbed. Because these substances are passively resorbed within the nephron, they are not rapidly excreted. Some of these substances are bound to other molecules in the liver, a process called conjugation, to form more water-soluble molecules. These more water-soluble substances do not pass as readily through the wall of the nephron, are not reabsorbed from the renal tubules, and consequently are rapidly excreted in the urine. One of the important functions of the liver is to convert nonpolar toxic substances to more water-soluble forms, thus increasing the rate at which they are excreted in the urine.

Tubular Secretion

Some substances, including by-products of metabolism that become toxic in high concentrations and drugs or molecules not normally produced by the body, are secreted into the nephron (Table 26-4). As with tubular reabsorption, **tubular secretion** can be either active or passive. Ammonia is synthesized in the epithelial cells of the nephron and diffuses into the lumen of the nephron. Substances that are actively secreted into the nephron include hydrogen ions, potassium ions, penicillin, and paraaminohippuric acids. Hydrogen ions are secreted into the proximal tubules, distal tubules, and collecting ducts; and potassium ions are actively secreted into the distal convoluted tubules and collecting ducts (see Chapter 27). Penicillin and paraaminohippuric acid are examples of substances not normally produced by the body that are actively secreted into the proximal convoluted tubules.

URINE CONCENTRATION MECHANISM

When a large volume of water is consumed, it is necessary to eliminate the excess water without losing excessive electrolytes or other substances essential for the maintenance of a constant internal en-

Table 26-4

Secretion of substances into the nephron

TRANSPORT PROCESS	SUBSTANCE TRANSPORTED
Proximal Convoluted Tubule	
Active transport	Hydrogen ions
	Hydroxybenzoates
	para-Aminohippuric acid
	Neurotransmitters
	Dopamine
	Acetylcholine
	Epinephrine
	Bile pigments
	Uric acid
	Drugs and toxins
	Penicillin
	Atropine
	Morphine
	Saccharin
Passive transport	Ammonia
Distal Convoluted Tubule	
Active transport	Potassium ions
Passive transport	Potassium ions
	Hydrogen ions

vironment. Under this condition the body must eliminate a large volume of dilute urine. On the other hand, when drinking water is not available, the production of a large volume of dilute urine would lead to rapid dehydration. When water intake is restricted, the body must be able to produce a small volume of urine that contains sufficient waste products to prevent their accumulation in the circulatory system. The kidneys are able to produce urine with concentrations that vary between 65 and 1200 mOsm/L and maintain a plasma concentration of 300 mOsm.

Medullary concentration gradient

The ability of the kidney to concentrate urine depends on maintaining a medullary concentration gradient. The interstitial fluid concentration is approximately 300 mOsm/L in the cortical region of the kidney and becomes progressively higher in the medulla. The interstitial osmolality reaches approximately 1200 mOsm/L near the tips of the renal pyramids (Figure 26-12). As urine moves through the collecting ducts from the cortex to the medulla, water can leave the collecting ducts by osmosis. Consequently, urine concentration becomes essentially equal to interstitial fluid osmolality and a concentrated urine is produced.

The maintenance and production of the medullary concentration gradient depends upon the vasa recta, the loop of Henle, and the distribution of urea.

1. **Vasa recta.** As filtrate moves through the nephron and collecting duct, water and solutes are reabsorbed from the filtrate into the surrounding interstitial fluid. This movement of water and solutes would alter the medullary concentration gradient if they were not removed. The vasa recta can remove them without changing the high concentration of solutes in the interstitial fluid because the vasa recta functions as a countercurrent system.

A **countercurrent system** is one in which fluid flows in parallel tubes but in opposite directions, and heat or substances such as water or solutes diffuse from one tube to the next so that the fluid in both tubes has nearly the same composition. The vasa recta constitutes a countercurrent system because the blood flows through them to the medulla; after the vessels turn near the tip of the renal pyramid, the blood flows in the opposite direction (see Figure 26-12). As blood flows toward the medulla, water moves out of the vasa recta, and solutes diffuse into it. As blood flows back toward the cortex, water moves into the vasa recta, and solutes diffuse out of it. The composition of the blood at both ends of the vasa recta is nearly the same, with the volume and osmolality slightly greater as the blood once again reaches the cortex. The vasa recta supplies blood to the kidney's medulla without disturbing the high concentration of solutes in the interstitial fluid. In addition, the vasa recta carries away extra water and solutes that enter the interstitial fluid from the loop of Henle.

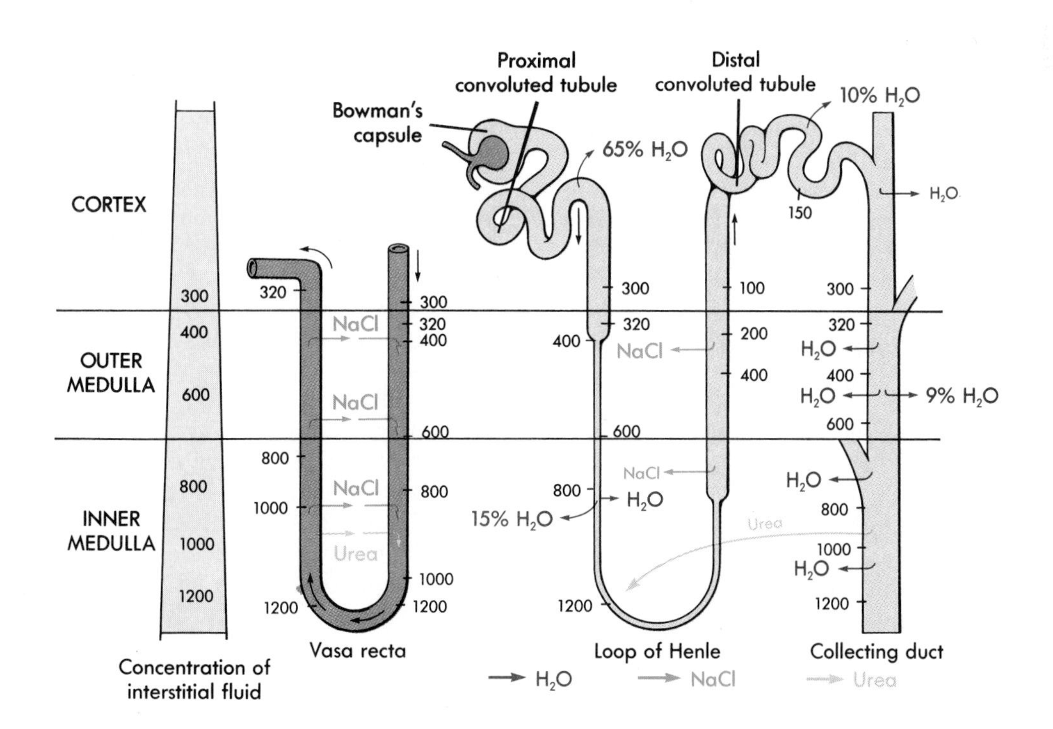

FIGURE 26-12 Urine concentrating mechanism (see text for details).

2. **Loop of Henle.** The loop of Henle, which is responsible for increasing the concentration of sodium ions within the medulla, is a countercurrent multiplier system. A **countercurrent multiplier system** is a countercurrent system that is assisted by active transport. Active transport mechanisms in the kidney transport solutes across the wall of the nephron and make the concentrating system of the kidney more efficient than it would be with a simple countercurrent system.

As filtrate descends into the medulla of the kidney through the descending limb of Henle's loop, water diffuses out of the nephron into the more concentrated interstitial fluid, and is removed by the vasa recta (see Figure 26-12). However, the wall of the ascending limb of Henle's loop is not permeable to water, and active transport mechanisms (including cotransport mechanisms) move large amounts of solutes such as sodium and chloride ions out of the filtrate and into the interstitial fluid. Although some of the sodium and chloride ions diffuse into the descending limb of the loop of Henle or the vasa recta, more sodium and chloride ions are transported from the ascending limb into the interstitial fluid of the medulla than enter the descending limb of the loop of Henle or the vasa recta. Thus, a high concentration of solutes is maintained in the interstitial fluid of the medulla.

3. **Urea.** Urea molecules are responsible for a substantial portion of the high osmolality in the medulla of the kidney (see Figure 26-12). Urea molecules diffuse into the descending limb of Henle's loop from the interstitial fluid. The ascending limb of Henle's loop and the distal convoluted tubule are not permeable to urea. However, the collecting ducts are permeable to urea, and urea diffuses out of the collecting ducts into the interstitial fluid of the medulla. The urea within the interstitial fluid of the medulla may cycle several times from the interstitial fluid into the descending limbs and from the collecting ducts back into the interstitial fluid. Consequently, a high urea concentration is maintained in the medulla of the kidney.

Changes in filtrate volume and concentration

Approximately 180 L of filtrate enters the proximal convoluted tubules daily in the average person. Most of the filtrate volume (65%) is reabsorbed in the proximal convoluted tubule. Substances (e.g., glucose, amino acids, sodium ions, calcium ions, potassium ions, and chloride ions; see Table 26-3) are actively transported, and water moves by osmosis from the lumen of the proximal convoluted tubules into the interstitial fluid. Consequently, the osmolality of both the interstitial fluid and the filtrate is maintained at approximately 300 mOsm/L as water and solutes move from the proximal convoluted tubule into the interstitial fluid. The excess solutes and water then enter the peritubular capillaries.

The filtrate then passes into the descending limb of the loop of Henle, which is highly permeable to water and solutes. As the descending limb penetrates deep into the medulla of the kidney, the surrounding interstitial fluid has a progressively greater osmolality. Water diffuses out of the nephron as solutes diffuse into the nephron. When the filtrate has reached the deepest portion of the loop of Henle, its osmolality has increased to approximately 1200 mOsm/L, and its volume has been reduced by an additional 15% to 20% of the original volume (see Figure 26-12). By the time the filtrate has reached the tip of the loop of Henle, 80% of the filtrate volume has been reabsorbed.

After passing through the descending limb of the loop of Henle, the filtrate enters the ascending limb, or the thick segment. The thick segment is not permeable to water, but sodium, chloride, and potassium ions are transported from the filtrate into the interstitial fluid (see Figures 26-10 and 26-12).

The movement of ions, but not water, across the wall of the ascending limb, causes the osmolality of the filtrate to decrease from 1200 to about 100 mOsm/L by the time the filtrate again reaches the cortex of the kidney. As a result, the content of the nephron is dilute when compared to the content of the surrounding interstitial fluid.

Filtrate reabsorption in the proximal convoluted tubule and the descending limb of the loop of Henle is obligatory (i.e., it is not under hormonal control and remains relatively constant). Reabsorption of filtrate in the remainder of the nephron and in the collecting duct is regulated by hormones and changes dramatically depending on the conditions to which the body is exposed. If it is necessary to eliminate a large volume of dilute urine, the filtrate can pass through the distal convoluted tubule and collecting duct with little change in concentration. On the other hand, if it is necessary to produce a small volume of concentrated urine, water is removed from the filtrate as it passes through the distal convoluted tubule and collecting duct.

In the presence of ADH, water diffuses passively across the wall of the distal convoluted tubule into the interstitial fluid (see Figure 26-11). Approximately 10% of the filtrate is reabsorbed in the distal convoluted tubules. The membrane of the distal convoluted tubules also actively transports some sodium ions. However, substances such as urea, creatinine, and other waste products do not readily cross the epithelial cells of the distal convoluted tubules. Consequently, as the filtrate volume decreases, concentration of these metabolic waste products increases in the filtrate. At the end of the distal convoluted tubule, the osmolality of the filtrate is once again equal to that of the interstitial fluid, and an additional 10% of the filtrate volume has been reabsorbed.

The filtrate flows from the distal convoluted tubule into the collecting duct, which penetrates the medulla to the tip of the renal pyramids (see Figure 26-4). The osmolality of the filtrate is approximately 320 mOsm/L as it enters the collecting duct. The collecting duct passes through the medulla where the interstitial fluid osmolality reaches 1200 mOsm/L at the tip of the renal pyramids. If ADH is present, the epithelial cells of the collecting duct are permeable to water, and an additional 9% (or slightly more) of the filtrate volume exits the collecting duct and the remaining urine is less than 1% of the original filtrate volume. In addition to the dramatic decrease in filtrate volume, however, there is a marked alteration in the filtrate composition. Waste products such as creatinine and urea as well as potassium, hydrogen, phosphate, and sulfate ions are present at a much higher concentration in urine than in the original filtrate because of the removal of water from the filtrate. Many substances are selectively reabsorbed from the nephron, and others are secreted into the nephron so that beneficial substances are retained in the body and toxic substances are eliminated.

Only the juxtamedullary nephrons descend deep into the medulla, but there are enough of them to maintain a high interstitial concentration of solutes in the interstitial fluid of the medulla. The juxtamedullary nephrons account for the reabsorption of approximately 15% of the filtrate volume. The remainder of urine concentration occurs as the filtrate passes through the collecting ducts. Not all of the nephrons have to have loops of Henle that descend into the medulla to have an effective urine concentrating mechanism. The cortical nephrons function like the juxtamedullary neurons with the exception that the loops of Henle are not as efficient at concentrating urine. However, filtrate from the cortical nephrons does pass through the collecting ducts where urine becomes concentrated. Animals that concentrate urine more effectively than humans have a greater percentage of nephrons that descend into the medulla of the kidney.

REGULATION OF URINE CONCENTRATION AND VOLUME

The volume and the composition of urine change, depending on conditions that exist in the body. Regulation of urine production involves hormonal mechanisms (Table 26-5), autoregulation, and sympathetic nervous system stimulation.

Hormonal Mechanisms

Aldosterone

Aldosterone, which is a steroid hormone secreted by the cortical cells of the adrenal gland (see Chapter 18), passes through the circulatory system from the adrenal gland to the kidney. Aldosterone affects the cells in the distal convoluted tubules, and the collecting ducts (Figure 26-13). The response of these segments of the nephron is to increase the rate of sodium and chloride ion transport.

Aldosterone diffuses through the cell membrane and binds to receptor molecules within the cell. The combination of the hormone with its receptor increases the synthesis of the protein molecules that are responsible for the active transport of sodium and chloride ions across the epithelial cells of the nephron.

Hyposecretion of aldosterone results in a decreased rate of sodium and chloride ion transport. As a consequence, the concentration of sodium and chloride ions in the distal convoluted tubules and the collecting ducts remains high. Since the concentration of the filtrate passing through the distal convoluted tubules and the collecting ducts has a greater-than-normal concentration of solutes, the capacity for water to move by osmosis from the distal convoluted tubules and the collecting ducts is diminished, the urine volume increases, and the urine has a greater-than-normal sodium and chloride ion concentration.

6

Drugs that increase the urine volume are called diuretics. Some diuretics inhibit the active transport of sodium and chloride ions in the nephron. Explain how diuretic drugs could cause increased urine volume.

? ? ? ? ? ? ? ? ? ? ?

Increased concentrations of potassium ions and decreased concentrations of sodium ions in the interstitial fluids act directly on the aldosterone-secreting cells of the adrenal cortex to increase the rate of aldosterone secretion. Conversely, decreases in blood levels of potassium ions and increases in blood levels of sodium ions depress aldosterone secretion (see Figure 26-13). **Angiotensin II** also acts on the cells of the adrenal cortex to stimulate aldosterone secretion (see Chapter 18). The influence of blood levels of potassium ions and angiotensin II is much more important than blood levels of sodium ions in regulating aldosterone secretion.

When a reduction in blood volume occurs, the plasma normally becomes more concentrated; consequently, the concentration of potassium ions in the plasma increases. Plasma ion concentrations also increase as a result of extensive tissue damage in which intracellular potassium ions are released into the extracellular fluid and during conditions of dehydration and anaphylactic shock, both of which result in the loss of plasma into the interstitial spaces. Elevated plasma levels of potassium ions stimulate the secretion of aldosterone, which acts on

Table 26-5

Hormonal regulation of urine volume and blood pressure

STIMULUS	INITIAL HORMONAL RESPONSE TO STIMULUS	EFFECT OF INITIAL HORMONAL RESPONSE	EFFECT ON URINARY VOLUME AND BLOOD PRESSURE
Renin-Angiotensin-Aldosterone			
Increased plasma concentration of potassium ions	Increased aldosterone secretion from adrenal cortex	Increased reabsorption of sodium and chloride ions, increased reabsorption of water and increased secretion of potassium and hydrogen ions	Reduced urinary volume and tendency to increase blood pressure
Decreased plasma concentration of sodium ions	Increased aldosterone secretion from adrenal cortex	Increased reabsorption of sodium and chloride ions, increased reabsorption of water and increased secretion of potassium ions and hydrogen ions	Reduced urinary volume and tendency to increase blood pressure
Decreased blood pressure	Increased renin secretion from kidney	Renin increases angiotensin secretion; angiotensin causes vasoconstriction and increased aldosterone secretion	Increased aldosterone reduces urinary volume and increases blood pressure by increasing sodium and chloride reabsorption in the distal nephron; vasoconstriction increases blood pressure
Decreased plasma concentration of potassium ions	Decreased aldosterone secretion from adrenal cortex	Decreased reabsorption of sodium and chloride ions and decreased secretion of potassium and hydrogen ions	Decreased reabsorption of water; increased urinary volume and tendency to decrease blood pressure
Increased plasma concentration of sodium ions	Decreased aldosterone secretion from adrenal cortex	Decreased reabsorption of sodium and chloride ions and decreased secretion of potassium ions and hydrogen ions	Decreased reabsorption of water; increased urinary volume and tendency to decrease blood pressure
Increased blood pressure	Decreased renin secretion from kidney	Reduced renin decreases angiotensin II production; reduced angiotensin causes vasodilation and decreased aldosterone secretion	Vasodilation reduces blood pressure; decreased aldosterone secretion increases urinary volume and decreases blood pressure by reducing sodium chloride reabsorption in the distal nephron
Antidiuretic Hormone			
Increased blood osmolality or decreased blood pressure	Increased antidiuretic hormone (ADH) secretion from posterior pituitary	Increased permeability of the distal convoluted tubule and collecting duct to water	Increased movement, by osmosis, of water from the distal convoluted tubule and collecting duct; reduced volume of concentrated urine; decreased blood osmolality and increased blood pressure
Decreased blood osmolality or increased blood pressure	Decreased ADH secretion from posterior pituitary	Decreased permeability of the distal convoluted tubule and collecting duct to water	Decreased movement, by osmosis, of water from the distal convoluted tubule and collecting duct; increased volume of dilute urine; increased blood osmolality and decreased blood pressure

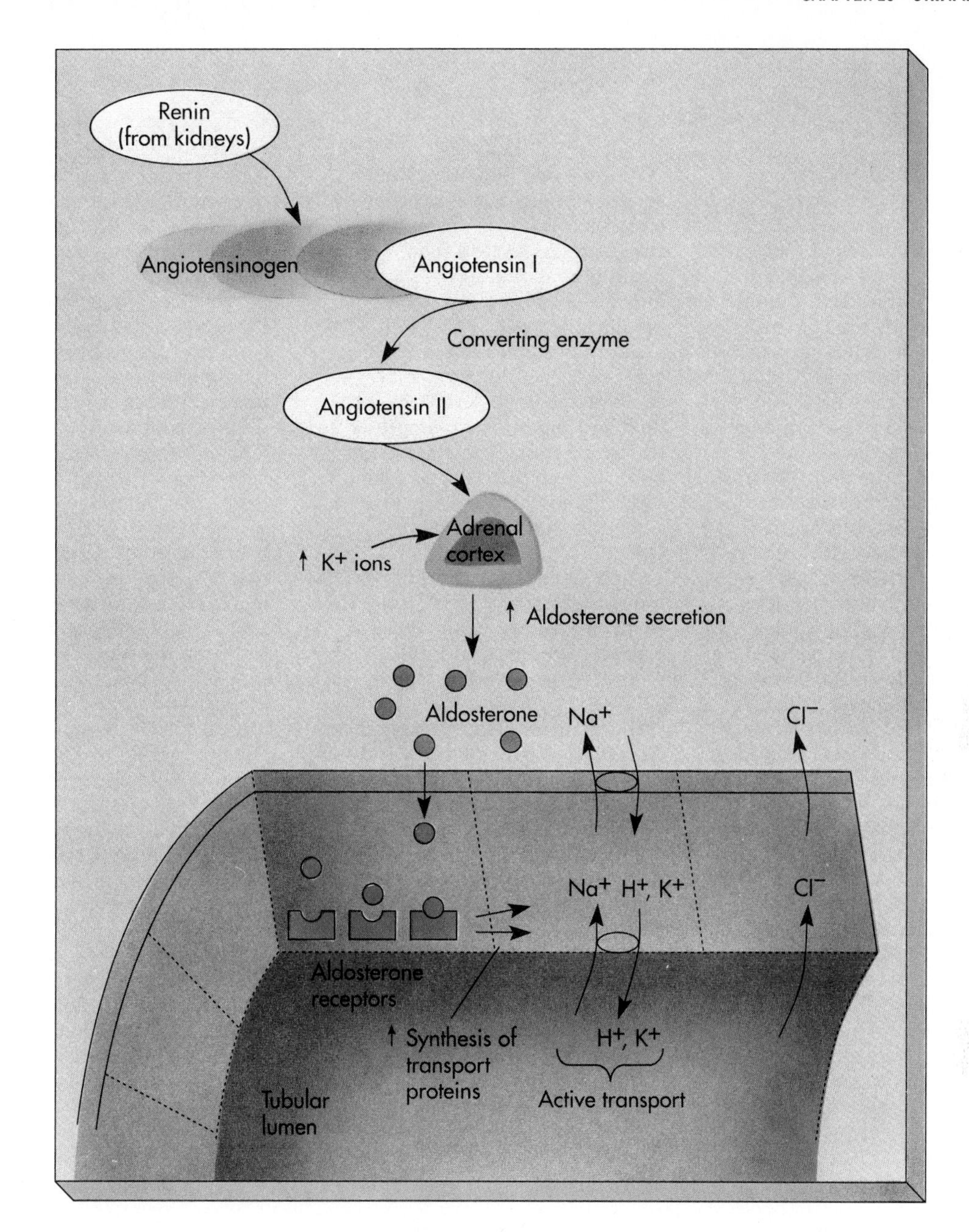

FIGURE 26-13 Effect of aldosterone on the distal convoluted tubule. Aldosterone increases the rate at which sodium ions are absorbed and the rate at which potassium and hydrogen ions are secreted. Chloride ions move with the sodium ions because they are attracted to the positive charge of the sodium ions.

the nephron to increase the rate of potassium ion secretion and sodium ion reabsorption. The increased sodium reabsorption causes the production of a concentrated urine that helps restore normal blood volume.

Renin-angiotensin

Elevated blood levels of angiotensin II result from the action of **renin** on angiotensinogen (see Chapter 18). Renin is secreted by cells of the juxtaglomerular apparatus. If blood pressure in the afferent arteriole decreases, or if sodium chloride concentration in the distal convoluted tubule decreases, the rate of renin secretion by the cells of the juxtaglomerular apparatus increases. Renin then enters the general circulation, acting on angiotensinogen and converting it to angiotensin I. Subsequently, a proteolytic enzyme called **angiotensin converting enzyme** converts angiotensin I to angiotensin II. Angiotensin II increases the systemic blood pressure in two ways. First, angiotensin II is a potent vasoconstrictor substance, and it increases the peripheral resistance, causing increased blood pressure. Second, it increases the rate of aldosterone secretion, which ultimately leads to an increased ability of the kidney

Diuretics

Diuretics are agents that increase the rate of urine formation. Although the definition is simple, a number of different physiological mechanisms may be involved.

Osmotic diuretics have the following characteristics: (1) they freely pass by filtration into the filtrate; and (2) they undergo limited reabsorption by the nephron. Osmotic diuretics increase urine volume by elevating the osmotic concentration of the nephron, thus reducing the amount of water moving by osmosis out of the nephron. Urea, mannitol, and potassium salts have been used as osmotic diuretics, although they are not commonly used clinically.

Acid-forming salts, which have a transient diuretic effect, usually act as osmotic diuretics. Acid-forming salts also cause slight acidosis, which results in increased secretion of ammonium and chloride ions and thus further increases the osmolality of the nephron contents. Ammonium chloride, ammonium nitrate, and calcium chloride have been used as acid-forming salts diuretics.

Mercurial diuretics and benzothiadiazides depress mechanisms responsible for the active reabsorption of sodium ions. Diuresis results from the increased sodium concentration in the nephron. Some mercurial diuretics are meralluride, mercurophylline, and mercumatilin.

Aldosterone inhibitors such as spironolactone prevent the increase in sodium reabsorption stimulated by aldosterone. Consequently, reabsorption of sodium ions is depressed, and urine volume is increased.

Xanthines and related substances act as diuretics partly because they increase renal blood flow and the rate of glomerular filtrate formation. They also influence the nephron by decreasing sodium and chloride reabsorption. Common xanthines include caffeine.

Alcohol acts as a diuretic, although it is not used clinically for that purpose. It inhibits ADH secretion from the posterior pituitary and results in increased urine volume.

Diuretics are used to treat disorders such as hypertension and several types of edema that are caused by conditions such as heart failure and cirrhosis of the liver. Complications, including dehydration and electrolyte imbalances, may arise because of the use of diuretics. In addition to the reduced rate of sodium and chloride ion reabsorption from the nephron, hyposecretion of aldosterone results in a decreased rate of potassium and hydrogen ion secretion. Therefore hydrogen ions and potassium ions accumulate in the body fluids. Symptoms of aldosterone hyposecretion include production of a large urine volume containing a high concentration of sodium chloride, increased thirst, reduced blood pressure, increased probability of dehydration, acidosis (i.e., high blood hydrogen ion concentration), and hyperkalemia (i.e., high blood potassium ion concentration). Three mechanisms are at work: concentration of potassium ions in the interstitial fluids, concentration of sodium ions in the interstitial fluid, and angiotensin II.

On the other hand, hypersecretion of aldosterone results in an increased rate of water, sodium ion, and chloride ion reabsorption from the nephron. In addition, the rate of potassium and hydrogen ion secretion into the nephron is increased. The symptoms include the production of a small volume of urine with a low sodium ion and high potassium ion concentration and with a lower-than-normal pH. High blood pressure and edema may occur along with symptoms associated with alkalosis and hypokalemia.

Hypersecretion and hyposecretion of aldosterone rarely occur alone; hypersecretion of aldosterone is associated with Cushing's syndrome, and hyposecretion of aldosterone is associated with Addison's disease (see Chapter 18).

to retain water and to produce a small volume of concentrated urine. If blood pressure in the afferent arteriole increases, or if the volume of the filtrate increases, there is an increased sodium chloride concentration passing through the juxtaglomerular apparatus in the distal convoluted tubule and the rate of renin secretion decreases.

Antidiuretic hormone

The distal convoluted tubules and the collecting ducts remain relatively impermeable to water in the absence of ADH (see Figure 26-11). As a result, a large part of the 19% of the filtrate that is normally reabsorbed in the distal convoluted tubules and the collecting ducts becomes part of the urine. People who suffer from a lack of ADH secretion often produce 10 to 20 L of urine per day and develop major problems such as dehydration and ion imbalances. A lack of ADH secretion results in a condition called **diabetes insipidus;** diabetes implies the production of a large volume of urine, and insipidus implies the production of a clear, tasteless, dilute urine. This condition is in contrast to **diabetes mellitus,** which implies the production of a large volume of urine that contains a high concentration of glucose (mellitus means honeyed or sweet).

ADH is secreted from the posterior pituitary, or neurohypophysis. Neurons with cell bodies primarily in the supraoptic nucleus of the hypothalamus have axons that course to the neurohypophysis. From these neuron terminals ADH is released into the circulatory system. Cells of the supraoptic nucleus are sensitive to changes in the osmolality of the interstitial fluid. If the osmolality of the blood and interstitial fluid increases, these cells stimulate the ADH-secreting neurons. Action potentials are then propagated along the axons of the ADH-secreting neurons to the neurohypophysis where ADH is released from the end of the axon. Reduced osmolality of the interstitial fluid within the supraoptic nucleus causes inhibition of ADH secretion.

Pressure receptors that monitor blood pressure, especially in the right atrium, also influence ADH secretion. Increased blood pressure causes action potentials to be sent along afferent neurons to the supraoptic region of the hypothalamus. These action potentials decrease ADH secretion.

When blood osmolality increases or when blood pressure declines, ADH secretion increases and acts on the kidney to increase the reabsorption of water. The retention of water by the kidney decreases blood osmolality and increases blood pressure. Conversely, when blood osmolality decreases or when blood pressure increases, ADH secretion declines. The reduced ADH levels cause the kidney to produce a larger volume of dilute urine, and the increased loss of water in the form of urine increases blood osmolality and decreases blood pressure.

Other hormones

A polypeptide hormone called **atrial natriuretic factor** is secreted from cells in the right atrium of the heart when blood pressure in the right atrium increases. The atrial natriuretic factor inhibits ADH secretion and reduces the ability of the kidney to concentrate urine, which leads to the production of a large volume of dilute urine. The resulting decrease in blood volume causes a decrease in blood pressure.

Two other substances, prostaglandins and kinins, are formed in the kidneys and affect kidney function. Their roles are not clear, but both substances influence the rate of filtrate formation and sodium ion reabsorption. The prostaglandins probably increase sensitivity of the renal blood vessels to neural stimuli and to angiotensin II.

7

Ethyl alcohol inhibits ADH secretion. Given this information, describe the mechanism by which alcoholic beverages affect urine production.

? ? ? ? ? ? ? ? ? ? ?

Autoregulation

Within the kidneys **autoregulation** is the maintenance of a relatively stable glomerular filtration rate over a wide range of systemic blood pressures. For example, when the arteriolar pressure increases to a value as high as 150 mm Hg, the pressure in the glomerular capillaries increases only a small amount. However, even a small increase in glomerular capillary pressure causes a substantial increase in the rate of filtrate formation. Consequently, large increases in the arterial blood pressure can increase the rate of urine production.

Autoregulation involves changes in the degree of constriction in both the afferent and efferent arterioles. The precise mechanism by which autoregulation is achieved is not clear, but as the systemic blood pressure increases, the afferent arterioles constrict and prevent an increase in renal blood flow and filtration pressure in the renal capsule. Conversely, a decrease in systemic blood pressure results in dilation of the afferent arteriole, thus preventing a decrease in the renal blood flow and filtration pressure in the renal capsule. The efferent arteriole also influences the filtration pressure. If the efferent arteriole constricts while the afferent arteriole remains dilated, the pressure in the glomerular capillary and the rate of filtrate formation increase. Conversely, dilation of the efferent arteriole while the afferent arteriole either constricts or remains unchanged results in decreased pressure in the glomerulus and a reduced rate of filtrate formation.

Effect of Sympathetic Innervation on Kidney Function

Sympathetic neurons that have norepinephrine as their neurotransmitter substance innervate the blood vessels of the kidney. Sympathetic stimulation constricts the small arteries and afferent arterioles, causing a decrease in renal blood flow and filtrate formation. Intense sympathetic stimulation (e.g., during shock or intense exercise) causes the rate of filtrate formation to decrease to only a few milliliters per minute.

In response to severe stress or circulatory shock the renal blood flow may be decreased to such low levels that the blood supply to the kidney is not adequate to maintain normal kidney metabolism. As a consequence, the kidney tissues may be damaged and thus be unable to perform their normal functions. One of the reasons that shock should be treated quickly is to prevent kidney damage.

CLEARANCE AND TUBULAR MAXIMUM

Plasma clearance is a calculated value representing the volume of plasma that is cleared of a specific substance each minute. For example, if the clearance value is 100 ml per minute for a substance, the substance is completely removed from 100 ml of plasma each minute.

The plasma clearance can be calculated for any substance that enters the circulatory system according to the following formula:

$$\text{Plasma clearance (ml/min)} = \text{Quantity of urine (ml/min)} \times \frac{\text{Concentration of substance in urine}}{\text{Concentration of substance in plasma}}$$

Plasma clearance can be used to estimate the glomerular filtration rate if the appropriate substance is monitored (see Table 26-2). Such a substance must have the following characteristics: (1) it must pass through the filtration membrane of the renal corpuscle as freely as water or other small molecules; (2) it must not be reabsorbed; and (3) it must not be secreted into the nephron. **Inulin** is a polysaccharide that has these characteristics. As filtrate is formed, it has the same concentration of inulin as plasma; but as the filtrate flows through the nephron, all of the inulin remains in the nephron to enter the urine. As a consequence, all of the volume of plasma that becomes filtrate is cleared of inulin, and the plasma clearance for inulin is equal to the rate of glomerular filtrate formation.

Plasma clearance can also be used to calculate the renal plasma flow (see Table 26-2). However, substances with the following characteristics must be used: (1) the substance must pass through the filtration membrane of the renal corpuscle, and (2) it must be secreted into the nephron at a sufficient rate so that very little of it remains in the blood as the blood leaves the kidney. **Para-aminohippuric acid (PAH)** meets these requirements. As blood flows through the kidney, essentially all of the PAH is either filtered or secreted into the nephron. Therefore the clearance calculation for PAH is a good estimate of the volume of plasma that flows through the kidney each minute. If the hematocrit is known, the total volume of blood that flows through the kidney each minute can be easily calculated.

The concept of plasma clearance can be used to make the measurements described previously, or it can be used to determine the means by which drugs or other substances are excreted by the kidney. A plasma clearance value greater than the inulin clearance value suggests that the substance is secreted by the nephron into the filtrate.

The **tubular load** of a substance is the total amount of the substance that filters through the filtration membrane into the nephrons each minute.

Normally glucose is almost completely reabsorbed from the nephron by the process of active transport. However, the capacity of the nephron to actively transport glucose across the epithelium of the nephron is limited. If the tubular load is greater than the nephron's capacity to reabsorb it, the excess amount of glucose remains in the urine. The maximum rate at which a substance can be actively reabsorbed is called the **tubular maximum** (Figure 26-14).

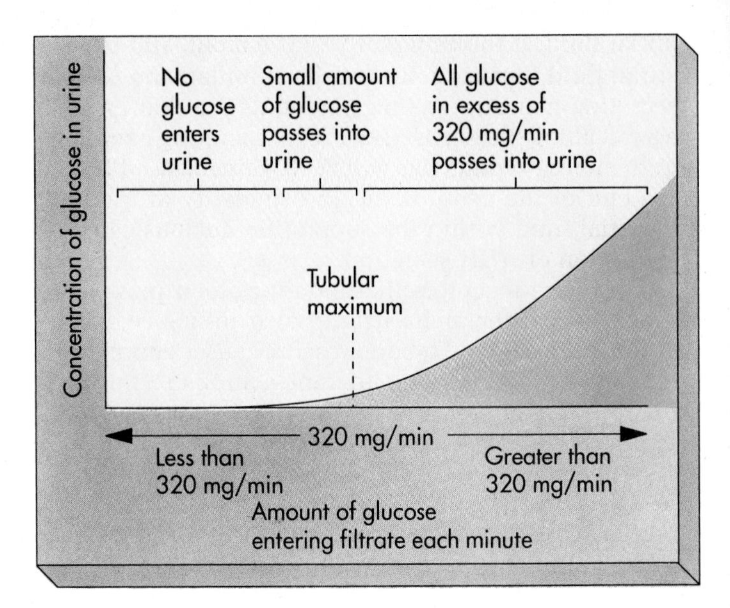

FIGURE 26-14 Tubular maximum for glucose. As the concentration of glucose increases in the filtrate, it reaches a point that exceeds the ability of the nephron to actively reabsorb it. That concentration is called the tubular maximum. Beyond that concentration the excess glucose enters the urine.

Each substance that is reabsorbed has its own tubular maximum, which is determined by the number of active transport carrier molecules and the rate at which they are able to transport the molecules of the substance. For example, in people suffering from diabetes mellitus the tubular load for glucose may exceed the tubular maximum by a substantial amount, and glucose appears in the urine. The urine volume is also greater than normal because the glucose molecules in the filtrate reduce the effectiveness of water reabsorption by osmosis in the distal convoluted tubules and the collecting ducts.

URINE MOVEMENT

Urine Flow Through the Nephron and the Ureters

The hydrostatic pressure averages 18 mm Hg in Bowman's capsule and nearly 0 mm Hg in the renal pelvis. The pressure gradient forces the filtrate to flow through the nephron into the renal pelvis. Since the pressure is 0 mm Hg in the renal pelvis, there is no pressure gradient to force urine to flow to the urinary bladder through the ureters. The walls of the ureters contain abundant smooth muscle arranged in a circular fashion. These smooth muscles exhibit peristaltic contractions that progress from the region of the renal pelvis to the urinary bladder and force the urine to flow through the ureters. The peristaltic waves occur from once every few seconds to once every 2 or 3 minutes. Parasympathetic stimulation increases their frequency, and sympathetic stimulation decreases it.

The peristaltic contractions of each ureter proceed at a velocity of approximately 3 cm per second and can generate pressures in excess of 50 mm Hg. At the point where the ureters penetrate the bladder, they course obliquely through the trigone. Pressure within the bladder compresses that part of the ureter, preventing backflow of urine.

When no urine is present in the urinary bladder, the internal pressure is approximately 0 mm Hg. When the volume is 100 ml of urine, the pressure is elevated to only 10 mm Hg. The pressure in the urinary bladder increases slowly as its volume increases to 400 to 500 ml, but above bladder volumes of 500 ml the pressure rises rapidly.

Kidney stones are hard objects that are usually found in the pelvis of the kidney. They are normally small (2 to 3 mm in diameter) with either a smooth or jagged surface; but occasionally a large branching kidney stone called a staghorn stone forms in the renal pelvis. Approximately 1% of all autopsies reveal the presence of kidney stones, and many of the stones occur without causing symptoms. The symptoms associated with kidney stones occur when a stone passes into the ureter, resulting in referred pain down the back, side, and groin area. The ureter contracts around the stone, causing the stone to irritate the epithelium and produce bleeding, which appears as blood in the urine (hematuria). In addition to causing intense pain, kidney stones can block the ureter, cause ulceration in the ureter, and increase the probability of bacterial infections.

Approximately 65% of all kidney stones are composed of calcium oxylate mixed with calcium phosphate, 15% are magnesium ammonium phosphate, and 10% are uric acid or cystine; in all cases approximately 2.5% of the kidney stone is composed of mucoprotein.

The cause of kidney stones is usually obscure. Predisposing conditions include a concentrated urine and an abnormally high calcium concentration in the urine, although the cause of the high calcium concentration is usually unknown. Magnesium ammonium phosphate stones are often found in people with recurrent kidney infections, and uric acid stones often occur in people suffering from gout. Severe kidney stones must be removed surgically. However, instruments that pulverize kidney stones with ultrasound have replaced most traditional surgical procedures.

Micturition reflex

The **micturition** (mik-tu-rish'un) **reflex** is initiated by stretching of the bladder wall, resulting in the elimination of urine from the bladder (micturition). As the bladder fills with urine, stretch receptors are stimulated. Afferent signals are conducted to the sacral segments of the spinal cord through the pelvic nerves. Integration of the reflex occurs in the spinal cord, and efferent signals are sent to the urinary bladder through parasympathetic fibers in the pelvic nerves (Figure 26-15). The efferent impulses cause the bladder to contract and the internal (smooth muscle) and external (skeletal muscle) urinary sphincters to relax. The micturition reflex normally produces a series of contractions of the urinary bladder.

The micturition reflex is an automatic reflex, but it can be either inhibited or stimulated by higher centers in the brain. The higher brain centers prevent micturition by sending impulses through the spinal cord to decrease the intensity of urinary bladder contractions and to stimulate efferent neurons that keep the external urinary sphincter tonically contracted. The ability to voluntarily inhibit micturition develops at the age of 2 to 3 years.

When the desire to urinate exists, the higher brain centers send impulses to the spinal cord to facilitate the micturition reflex and inhibit the external urinary sphincter. The desire to urinate is initiated because stretch of the urinary bladder stimulates ascending fibers in the spinal cord. The ascending fibers send impulses to the higher centers of the brain. Irritation of the urinary bladder or the urethra by bacterial infections or other conditions can also initiate the urge to urinate, even though the bladder may be empty.

If the spinal cord is damaged above the sacral region, a typical micturition reflex may still exist, but there is no conscious control over the onset or duration of the micturition reflex. Immediately after a spinal cord injury no reflex exists for a period of time; but if the bladder is emptied frequently, the micturition reflex eventually becomes adequate to cause the bladder to empty. This condition is called the automatic bladder.

A bladder that does not contract can result from damage to the sacral region of the spinal cord or to the nerves that carry impulses between the spinal cord and the urinary bladder. As a result, the micturition reflex cannot occur. The bladder fills to capacity, and urine is forced in a slow dribble through the urinary sphincters. In elderly people or in patients with damage to the brainstem or spinal cord, there can be a loss of inhibitory impulses to the sacral region of the spinal cord. Without inhibition, the sacral centers are overexcitable, and even a small amount of urine in the bladder can elicit an uncontrollable micturition reflex.

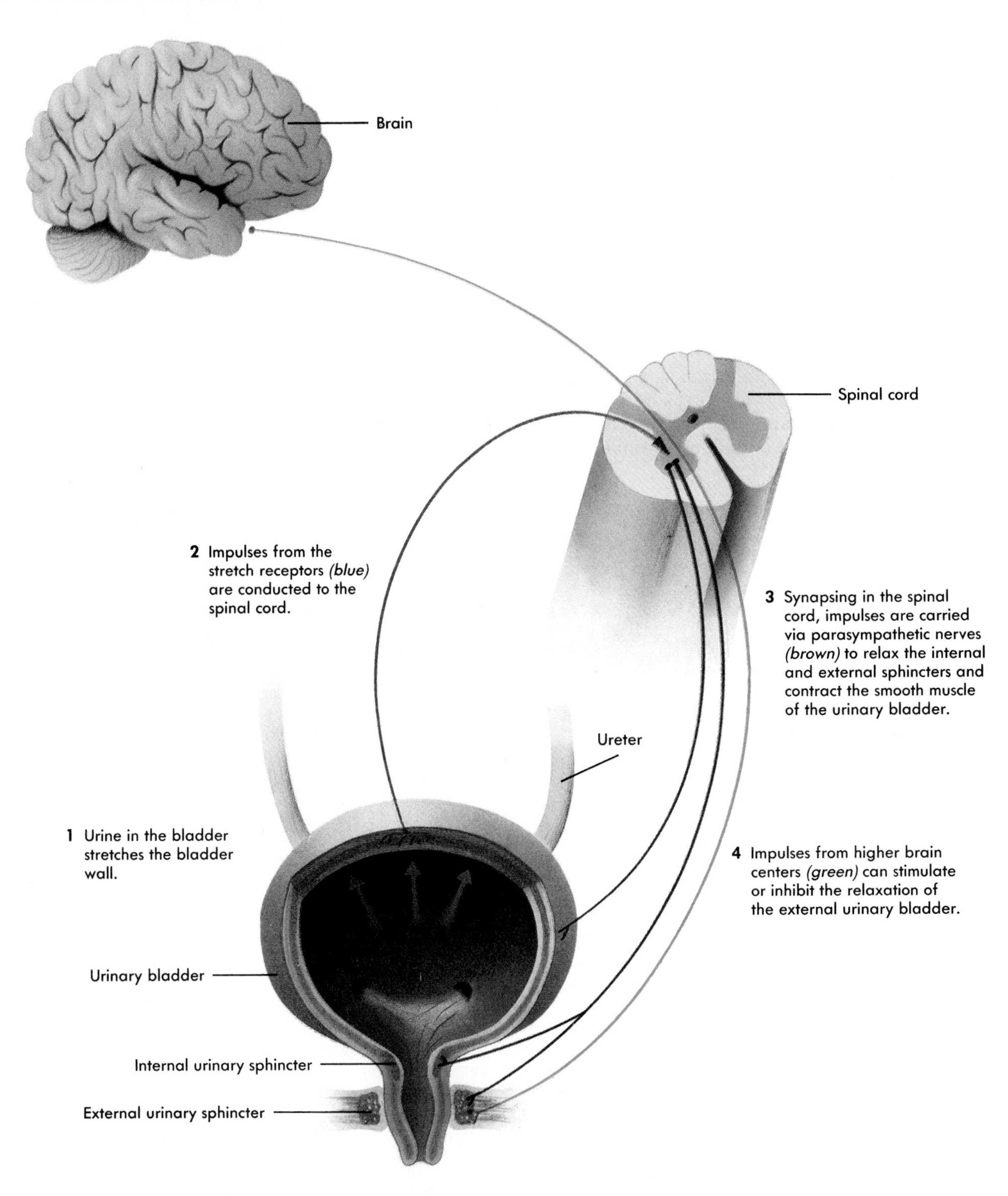

FIGURE 26-15 Micturition reflex. Stretch receptors in the wall of the urinary bladder are stimulated when the bladder volume increases. The action potentials are transported along afferent neurons to the spinal cord. In the spinal cord the micturition reflex is integrated. Efferent action potentials are transported to the urinary bladder and cause contraction of bladder smooth muscle. Relaxation of the internal and external urinary sphincter muscles occurs because of a decrease in efferent impulses to these structures. The brain has a major influence on the micturition reflex. Higher centers of the brain normally inhibit the micturition reflex and keep the external urinary sphincter contracted. Under appropriate conditions the inhibitory effect of the brain on the urinary reflex is reduced, and the reflex is allowed to initiate micturition.

Renal Pathologies

Glomerular nephritis results from inflammation of the filtration membrane within the renal corpuscle. It is characterized by an increased permeability of the filtration membrane and the accumulation of numerous white blood cells in the area of the filtration membrane. As a consequence, a high concentration of plasma proteins enters the urine along with numerous white blood cells. Plasma proteins in the filtrate increase the osmolality of the filtrate, causing a greater-than-normal urine volume.

Acute glomerular nephritis often occurs 1 to 3 weeks after a severe bacterial infection such as streptococcal sore throat or scarlet fever. Antigen-antibody complexes associated with the disease become deposited in the filtration membrane and cause its inflammation. This acute inflammation normally subsides after several days.

Chronic glomerular nephritis is long-term and usually progressive. The filtration membrane thickens and eventually is replaced by connective tissue. Although in the early stages chronic glomerular nephritis resembles the acute form, in the advanced stages many of the renal corpuscles are replaced by fibrous connective tissue, and the kidney eventually becomes nonfunctional.

Pyelonephritis is inflammation of the renal pelvis, medulla, and cortex. It often begins as a bacterial infection of the renal pelvis and then extends into the kidney itself. It can result from several types of bacteria, including *Escherichia coli.* Pyelonephritis may cause the destruction of nephrons and renal corpuscles, but because the infection starts in the pelvis of the kidney, it affects the medulla more than the cortex. As a consequence, the ability of the kidney to concentrate urine is dramatically affected.

Renal failure may result from any condition that interferes with kidney function. **Acute renal failure** occurs when damage to the kidney is extensive and leads to the accumulation of urea in the blood and to acidosis (see Chapter 27). In complete renal failure death may occur in 1 to 2 weeks. Acute renal failure may result from acute glomerular nephritis, or it may be caused by damage to or blockage of the renal tubules. Some poisons such as mercuric ions or carbon tetrachloride that are common to certain industrial processes cause necrosis of the nephron epithelium. If the damage does not interrupt the basement membrane surrounding the nephrons, extensive regeneration can occur within 2 to 3 weeks. Severe ischemia associated with circulatory shock caused by sympathetic vasoconstriction of the renal blood vessels can cause necrosis of the epithelial cells of the nephron.

Chronic renal failure is the result of permanent damage to so many nephrons that those nephrons remaining functional cannot adequately compensate. Chronic renal failure may result from chronic glomerular nephritis, trauma to the kidneys, absence of kidney tissue caused by congenital abnormalities, or tumors. Urinary tract obstruction by kidney stones, damage resulting from pyelonephritis, and severe arteriosclerosis of the renal arteries also cause degeneration of the kidney.

Chronic renal failure is characterized by the body's inability to excrete excess excretory products, including electrolytes and metabolic waste products. Water retention and edema result from the accumulation of solutes in the body fluids. Potassium levels in the extracellular fluid are elevated, and acidosis occurs because of the inability of the distal convoluted tubules to excrete sufficient quantities of potassium and hydrogen ions. Acidosis, elevated potassium levels in the body fluids, and the toxic effects of metabolic waste products cause mental confusion, coma, and finally death when chronic renal failure is severe.

The Urinary System

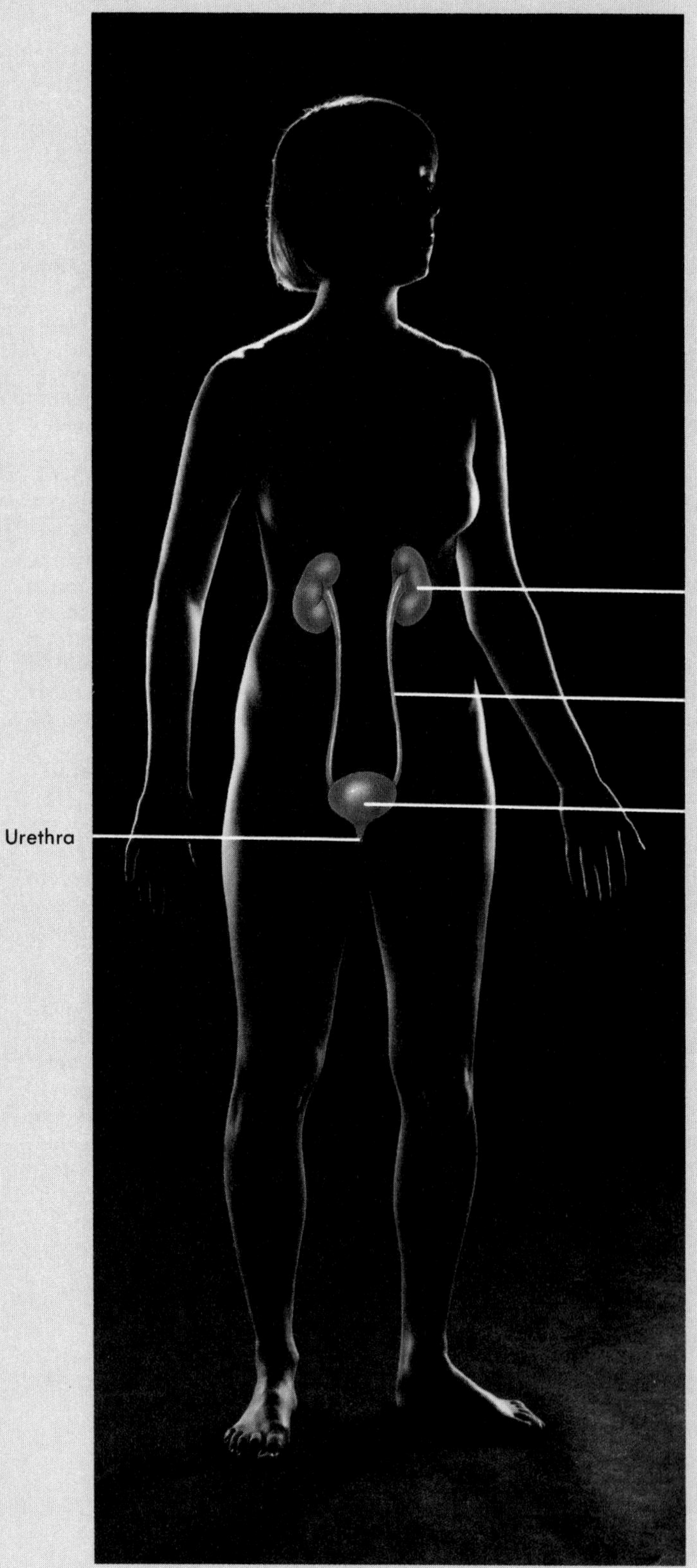

Major Components

Kidneys, ureters, urinary bladder, and urethra

Major Functions

Removes waste products from the circulatory system
Regulates blood pH, ion balance, and fluid balance
Assists in regulating blood pressure

SUMMARY

THE URINARY SYSTEM ELIMINATES WASTES, REGULATES BLOOD VOLUME, ION CONCENTRATION, AND PH, AND IS INVOLVED WITH RED BLOOD CELL AND VITAMIN D PRODUCTION. THE EXCRETORY SYSTEM CONSISTS OF THE KIDNEYS, URETERS, BLADDER, AND URETHRA.

URINARY SYSTEM *p. 850*

Kidneys

1. The kidney is surrounded by a renal capsule and a renal fat pad and is held in place by the renal fascia.
2. The two layers of the kidney are the cortex and the medulla.
   - The renal columns extend toward the medulla between the renal pyramids.
   - The renal pyramids of the medulla project to the minor calyces.
3. The minor calyces open into the major calyces, which open into the renal pelvis. The renal pelvis leads to the ureter.
4. The functional unit of the kidney is the nephron. The parts of a nephron are the renal corpuscle, the proximal convoluted tubule, the loop of Henle, and the distal convoluted tubule.
   - The renal corpuscle is Bowman's capsule and the glomerulus. Materials leave the blood in the glomerulus and enter Bowman's capsule through the filtration membrane.
   - The nephron empties through the distal convoluted tubule into a collecting duct.
5. The juxtaglomerular apparatus consists of the macula densa (part of the distal convoluted tubule) and the juxtaglomerular cells of the afferent arteriole.

Arteries and Veins

1. Arteries branch as follows: renal artery to segmental artery to interlobar artery to arcuate artery to interlobular artery to afferent arteriole.
2. Afferent arterioles supply the glomeruli.
3. Efferent arteries from the glomeruli supply the peritubular capillaries and vasa recta.
4. Veins form from the peritubular capillaries as follows: interlobular vein to arcuate vein to interlobar vein to renal vein.

Ureters and Urinary Bladder

1. Structure
   - The walls of the ureter and urinary bladder consist of epithelium, lamina propria, a muscular coat, and a fibrous adventitia.
   - The transitional epithelium permits changes in size.
2. Function
   - The ureters transport urine from the kidney to the urinary bladder.
   - The urinary bladder stores urine.

URINE PRODUCTION *p. 858*

Urine is produced by the processes of filtration, absorption, and secretion.

Filtration

1. The renal filtrate is plasma minus blood cells and blood proteins. Most (99%) of the filtrate is reabsorbed.
2. The filtration membrane is fenestrated endothelium, basement membrane, and the slitlike pores formed by podocytes.
3. Filtration pressure is responsible for filtrate formation.
   - Filtration pressure is glomerular capillary pressure minus capsule pressure minus colloid osmotic pressure.
   - Filtration pressure changes are primarily caused by changes in glomerular capillary pressure.

Tubular Reabsorption

1. Filtrate is reabsorbed by passive transport (simple diffusion and facilitated diffusion) or active transport (or cotransport) into the peritubular capillaries.
2. Specialization of tubule segments.
   - The thin segment of the loop of Henle is specialized for passive transport.
   - The rest of the nephron and collecting tubules perform active transport and passive transport.
3. Substances transported.
   - Active transport moves proteins, amino acids, glucose, fructose, and sodium, potassium, calcium, bicarbonate, and chloride ions.
   - Passive transport moves water, chloride and sodium ions, urea, and lipid-soluble, nonpolar compounds.

Tubular Secretion

1. Substances enter the proximal or distal convoluted tubules and the collecting ducts.
2. Hydrogen ions, potassium ions, and some substances not produced in the body are secreted.

URINE CONCENTRATION MECHANISM *p. 864*

1. Countercurrent systems (e.g., vasa recta and loop of Henle) and the distribution of urea are responsible for the concentration gradient in the medulla. The concentration gradient is necessary for the production of concentrated urine.
2. Production of urine.
   - In the proximal convoluted tubule sodium and other substances are removed by active transport. Water follows passively, filtrate volume is reduced 65%, and the filtrate concentration is 300 mOsm/L.
   - In the descending limb of the loop of Henle water exits passively, and solute enters. The filtrate volume is reduced 15%, and the filtrate concentration is 1200 mOsm/L.
   - In the ascending limb of the loop of Henle, sodium, chloride, and potassium ions are transported out of the filtrate, but water remains because this segment of the nephron is impermeable to water. The filtrate concentration is 100 mOsm/L.
   - In the distal convoluted tubules and collecting ducts, water movement out of them is regulated by antidiuretic hormone (ADH). If ADH is absent, water is not reabsorbed and a dilute urine is produced. If ADH is present, water moves out, and a concentrated urine is produced.

REGULATION OF URINE CONCENTRATION AND VOLUME *p. 867*

Hormonal Mechanisms

1. Aldosterone is produced in the adrenal cortex and affects sodium and chloride ion transport in the nephron and the collecting ducts.

- A decrease in aldosterone results in less sodium ion reabsorption and an increase in urine concentration and volume. An increase in aldosterone results in greater sodium ion reabsorption and a decrease in urine concentration and volume.
- Aldosterone production is stimulated by angiotensin II, increased blood potassium ion concentration, and decreased blood sodium ion concentration.

2. Renin, produced by the kidneys, causes the production of angiotensin II.
   - Angiotensin II acts as a vasoconstrictor and stimulates aldosterone secretion, causing a decrease in urine production and an increase in blood volume.
   - Decreased blood pressure or decreased sodium ion concentration stimulates renin production.
3. ADH is secreted by the posterior pituitary and increases water permeability in the distal convoluted tubules and the collecting ducts.
   - ADH decreases urine volume, increases blood volume, and thus increases blood pressure.
   - ADH release is stimulated by increased blood osmolality or a decrease in blood pressure.
4. Atrial natriuretic factor, produced by the heart when blood pressure increases, inhibits ADH production.

Autoregulation

Autoregulation dampens systemic blood pressure changes by altering afferent arteriole diameter.

Effects of Sympathetic Innervation on Kidney Function

Sympathetic stimulation decreases afferent arteriole diameter.

CLEARANCE AND TUBULAR MAXIMUM *p. 871*

1. Plasma clearance is the volume of plasma that is cleared of a specific substance each minute.
2. The tubular load is the total amount of substance that enters the nephron each minute.
3. Tubular maximum is the fastest rate at which a substance is reabsorbed from the nephron.

URINE MOVEMENT *p. 872*

Urine Flow Through the Nephron and Ureters

1. Hydrostatic pressure forces urine through the nephron.
2. Peristalsis moves urine through the ureters.

Micturition Reflex

1. Stretch of the urinary bladder stimulates a reflex that causes the bladder to contract and inhibits the urinary sphincters.
2. Higher brain centers can stimulate or inhibit the external urinary sphincter.

CONTENT REVIEW

1. What functions are performed by the urinary system? Name the structures that comprise the urinary system.
2. What structures surround the kidney?
3. Name the two layers of the kidney. What are the renal columns and the renal pyramids?
4. Describe the relationship between the calyces, renal pelvis, and ureter.
5. What is the functional unit of the kidney? Name its parts.
6. What is the juxtaglomerular complex?
7. Describe the type of epithelium found in the nephron and the collecting duct.
8. Describe the blood supply for the kidney.
9. What are the functions of the ureters and the urinary bladder? Describe their structure.
10. Name the three general processes that are involved in the production of urine.
11. Define renal blood flow, renal plasma flow, and glomerular filtration rate. How do they affect urine production?
12. Describe the filtration barrier. What substances do not pass through it?
13. What is filtration pressure? How does glomerular capillary pressure affect filtration pressure and the amount of urine produced?
14. How do systemic blood pressure and afferent arteriole diameter affect glomerular capillary pressure?
15. What happens to most of the filtrate that enters the nephron?
16. On what side of the nephron tubule cell does active transport take place during reabsorption and secretion of materials?
17. Describe how cotransport works in the nephron.
18. Name the substances that are moved by active and passive transport. In what part of the nephron does this movement take place?
19. Where does tubular secretion take place? What substances are secreted? Are these substances secreted by active or passive transport?
20. What is a countercurrent system? How does the vasa recta help maintain the concentration gradient in the medulla?
21. What is a countercurrent multiplier system? How do the loop of Henle and urea contribute to the production of a medullary concentration gradient?
22. Describe the net movement of solutes and water in the nephron and the collecting duct.
23. What are the effects of aldosterone on sodium and chloride ion transport? How does aldosterone affect urinary concentration, urinary volume, and blood pressure?
24. Where is aldosterone produced? What factors stimulate aldosterone secretion?
25. How is angiotensin II activated? What effects does it produce?
26. What factors cause an increase in renin production?
27. Where is antidiuretic hormone (ADH) produced? What factors stimulate an increase in ADH secretion?
28. What effect does ADH have on urine volume and concentration?
29. Where is atrial natriuretic factor produced, and what effect does it have on urine production?
30. Describe autoregulation.
31. How does sympathetic stimulation affect filtrate production?
32. Define plasma clearance, tubular load, and tubular maximum.
33. What is responsible for the movement of urine through the nephron and the ureter?
34. Describe the micturition reflex. How is voluntary control of micturition accomplished?

CONCEPT REVIEW

1. To relax after an anatomy and physiology examination, Mucho Gusto went to a local bistro and drank 2 quarts of low-sodium beer. What effect did this beer have on urine concentration and volume? Explain the mechanisms involved.
2. A man ate a full bag of salty potato chips. What effect did they have on urine concentration and volume? Explain the mechanisms involved.
3. During severe exertion in a hot environment a person can lose up to 4 L of hypoosmotic (less concentrated than plasma) sweat per hour. What effect would this loss have on urine concentration and volume? Explain the mechanisms involved.
4. Harry Macho was doing yard work one hot summer day and refused to drink anything until he was finished. He then drank glass after glass of plain water. Assume that he drank enough water to replace all the water he lost as sweat. How would this much water affect urine concentration and volume? Explain the mechanisms involved.
5. Which of the following symptoms are consistent with hyposecretion of aldosterone: polyuria (excessive urine production), low blood pressure, high plasma sodium levels, low plasma potassium levels, and muscle weakness?
6. A patient has the following symptoms: slight increase in extracellular fluid volume, large decrease in plasma sodium concentration, very concentrated urine, and cardiac fibrillation. An imbalance of what hormone is responsible for these symptoms? Are the symptoms caused by oversecretion or undersecretion of the hormone?
7. Propose as many ways as you can to decrease the glomerular filtration rate.
8. Design a kidney that can produce hypoosmotic (less concentrated than plasma) or hyperosmotic (more concentrated than plasma) urine by the active transport of water instead of chloride ions. Assume that the anatomical structure of the kidney is the same as that in humans. Feel free to change anything else you choose.
9. If the kidney had only a very small amount of urea present in the interstitial fluid instead of its normal concentration, what would be the effect on the kidney's ability to concentrate urine?

? ?

ANSWERS TO PREDICT QUESTIONS

1 p. 858. Because the urethra of females is much shorter than the urethra of males, the female urinary bladder is more accessible to bacteria from the exterior. This accessibility is one of the reasons that urinary bladder infection is more common in women than in men.

2 p. 859. If the cardiac output is 5600 ml of blood per minute and the hematocrit is 45, renal plasma flow is 650 ml of plasma per minute (see Table 26-2). If the filtration fraction increased from 19% to 22%, the glomerular filtration rate would be 143 ml of filtrate per minute (650 ml of plasma × 0.22). If 99.2% of the filtrate is reabsorbed, 0.8% becomes urine. Thus the urine produced is 1.14 ml of urine per minute (143 ml of filtrate × 0.008). Compared to the rate of urine production when the filtrate fraction was 19% (i.e., 1 ml per minute), the 3% increase in filtration fraction has caused a 14% increase in urine production. Converting 1.14 ml of urine per minute to liters of urine produced per day yields 1.64 L per day (1.14 × 1.44).

3 p. 861. Even though hemoglobin is a smaller molecule than albumin, it does not normally enter the filtrate because hemoglobin is contained within red blood cells. However, if red blood cells are ruptured (hemolysis), the hemoglobin is released into the plasma, and large amounts of hemoglobin enter the filtrate. Conditions that cause red blood cells to rupture in the circulatory system result in large amounts of hemoglobin entering the urine.

4 p. 861. A decrease in the concentration of plasma proteins would reduce the colloid osmotic pressure within the glomerular capillary. Since the total filtration pressure is determined by the glomerular blood pressure minus the colloid osmotic pressure minus the glomerular capsule pressure, a decrease in the colloid osmotic pressure would increase the total filtration pressure. As a result, the total volume of filtrate produced per minute would increase. Constriction of the afferent arteriole would decrease the blood pressure in the glomerulus. As a consequence, the total filtration pressure would decrease.

5 p. 864. Without ADH, the distal convoluted tubule and the collecting duct are impermeable to water. Consequently, water cannot move by osmosis from the nephron into the interstitial spaces and therefore remains in the nephron to become urine. Since approximately 19% of the filtrate volume leaves the nephron in the distal convoluted tubule and the collecting duct, much of that volume appears as urine. As a result, the urine volume increases, and the urine concentration decreases dramatically.

6 p. 867. Without the normal active transport of sodium and chloride ions, their concentration within the nephron remains elevated. Movement of water by osmosis out of the nephron into the interstitial spaces is decreased, resulting in an increased urine volume.

7 p. 871. Inhibition of ADH secretion is one of the numerous effects alcohol has on the body. The lack of ADH secretion causes the distal convoluted tubule and the collecting duct to be relatively impermeable to water. Therefore the water cannot move by osmosis from the nephron and remains in the nephron to become urine. In addition, since other fluids are normally consumed with the alcohol, the increased water intake also results in an increase in urine production.

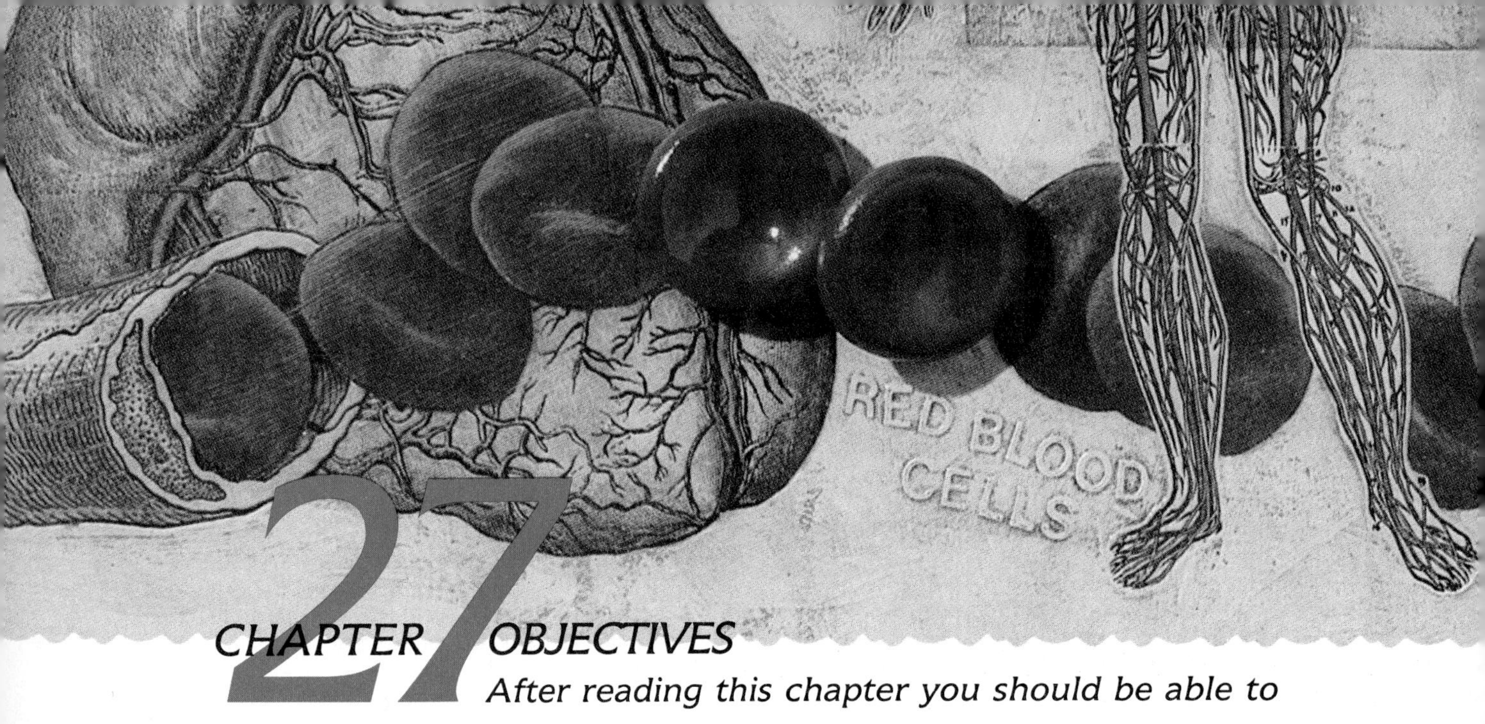

CHAPTER 27 OBJECTIVES

After reading this chapter you should be able to

1. List the major body fluid compartments and the approximate percent of body weight contributed by the fluid within each compartment and describe how the compartments are influenced by age and body fat.
2. Compare the composition of intracellular and extracellular fluids.
3. Diagram the mechanisms by which sodium, potassium, calcium, and chloride ions are regulated in the extracellular fluid.
4. Describe how the regulatory mechanisms respond to an increase or a decrease in extracellular sodium, potassium, chloride, calcium, or phosphate ion concentration.
5. Diagram the mechanisms by which the water content of the body fluids is regulated.
6. Describe how the regulatory mechanisms respond to either an increase or a decrease in the water content of body fluids.
7. Define acid, base, acidosis, alkalosis, and buffer.
8. Explain how buffers regulate the body fluid pH and list the major buffers that exist in the body fluids.
9. Diagram the mechanisms that regulate the body fluid pH and describe how they respond to either acidosis or alkalosis.
10. Describe how acidosis and alkalosis are classified and provide specific examples.

Water, Electrolytes, and Acid-Base Balance

KEY TERMS

acidosis

aldosterone

alkalosis

angiotensin II

antidiuretic hormone

atrial natriuretic factor

baroreceptor

buffer

electrolyte

extracellular

interstitial

intracellular

osmoreceptor cell

parathyroid hormone

pH

RELATED TOPICS

The following terms or concepts are important for a good understanding of this chapter. If you are not familiar with them, you should review them before proceeding.

Ions, acids, and bases (Chapter 2)

Endocrinology of the posterior pituitary, the adrenal cortex, and the right atrium of the heart (Chapters 18 and 20)

MAINTENANCE OF WATER VOLUME, PH, AND ELECTROLYTE CONCENTRATIONS WITHIN A NARROW RANGE OF VALUES IN THE BODY IS ESSENTIAL FOR SURVIVAL. The kidneys, along with the respiratory, integumentary, and gastrointestinal systems, regulate these parameters. The nervous and endocrine systems coordinate the activities of the systems.

BODY FLUIDS

The proportion of the body weight composed of water decreases from birth to old age, with the greatest decrease occurring during the first 10 years of life (Table 27-1). Because the water content of fat is relatively low, the fraction of the body weight composed of water decreases as fat content increases. The relatively low water content of adult females reflects the greater development of subcutaneous adipose tissue characteristic of women.

The two major fluid compartments in the body are the **intracellular** and **extracellular** fluid compartments. Each of the several trillion cells contains a small volume of intracellular fluid, which accounts for approximately 40% of the total body weight.

The extracellular fluid includes all of the fluid outside the cells, and it comprises nearly 20% of the total body weight. The extracellular fluid can be divided into several subcompartments such as lymph, cerebrospinal fluid, and synovial fluid, with interstitial fluid and plasma the major ones. **Interstitial fluid** occupies the extracellular spaces outside the blood vessels, and **plasma** occupies the extracellular space within blood vessels. All of the other subcompartments of the extracellular compartment comprise relatively small volumes.

The fluid contained in each subcompartment differs somewhat in composition from fluid in the other subcompartments, but there is continuous and ex-

Table 27-1

*Approximate volumes of body fluid compartments**

AGE OF PERSON	TOTAL BODY WATER	INTRACELLULAR FLUID	EXTRACELLULAR FLUID		
			PLASMA	INTERSTITIAL	TOTAL
Infants	75	45	4	26	30
Adult males	60	40	5	15	20
Adult females	50	35	5	10	15

*Expressed as percentage of body weight.

Table 27-2

*Approximate concentration of major solutes in body fluid compartments**

SOLUTE	PLASMA	INTERSTITIAL FLUID	INTRACELLULAR FLUID†
Cations			
Sodium (Na^+)	153.2	145.1	12.0
Potassium (K^+)	4.3	4.1	150.0
Calcium (Ca^{2+})	3.8	3.4	4.0
Magnesium (Mg^{2+})	1.4	1.3	34.0
TOTAL	162.7	153.9	200.0
Anions			
Chloride (Cl^-)	111.5	118.0	4.0
Bicarbonate (HCO_3^-)	25.7	27.0	12.0
Phosphate (HPO_4^{2-} plus HPO_4^-)	2.2	2.3	40.0
Protein	17.0	0.0	54.0
Other	6.3	6.6	90.0
TOTAL	162.7	153.9	200.0

*Expressed as milliequivalents per liter (mEq/L).
†Data are from skeletal muscle.

tensive exchange between the subcompartments. Water continuously diffuses from one subcompartment to another, and small molecules and ions are either transported or move freely between them. Large molecules such as proteins are much more restricted in their movement because of the permeability characteristics of the membranes that separate the fluid subcompartments (Table 27-2).

The osmotic concentration of most fluid compartments is approximately equal (e.g., the osmotic concentration of the hyaluronic acid of the synovial joints equals that of the proteins of the intraocular fluid).

REGULATION OF INTRACELLULAR FLUID COMPOSITION

The composition of intracellular fluid is substantially different from that of extracellular fluid. Cell membranes, which separate the two compartments, are selectively permeable, being relatively impermeable to proteins and other large molecules and having limited permeability to many smaller molecules and ions. Consequently, most large molecules (e.g., proteins) that are synthesized within the cell remain within the cell. However, some substances are actively transported across the cell membrane. The concentration of other substances (e.g., electrolytes) in the intracellular fluid is determined by the transport processes and by the electrical charge difference across the cell membrane (Figure 27-1).

Water movement across the cell membrane is influenced by the composition of the extracellular fluid. Water diffuses freely through cell membranes, and the net movement of water is affected by changes in the concentration of solutes in the extracellular fluid. For example, during conditions of dehydration the concentration of solutes in the extracellular fluid increases, resulting in the movement of water through the process of osmosis from the intracellular space into the extracellular space. If dehydration is severe enough, movement of water from the intracellular spaces causes the cells to function abnormally. When water intake increases after a period of dehydration, the concentration of solutes in the extracellular fluids decreases, resulting in the movement of water into cells through the process of osmosis.

REGULATION OF EXTRACELLULAR FLUID COMPOSITION

Homeostasis requires that the intake of substances such as water and **electrolytes** (molecules or ions with an electrical charge) must equal their elimination. Ingestion of water and electrolytes adds them to the body, whereas excretion by organs such as the kidneys and liver or, to a lesser degree, by the skin removes them from the body. Over a long period of time, the total amount of water and electrolytes in the body does not change unless the individual is growing, gaining weight, or losing weight. The regulation of water and electrolytes involves the coordinated participation of several organ systems.

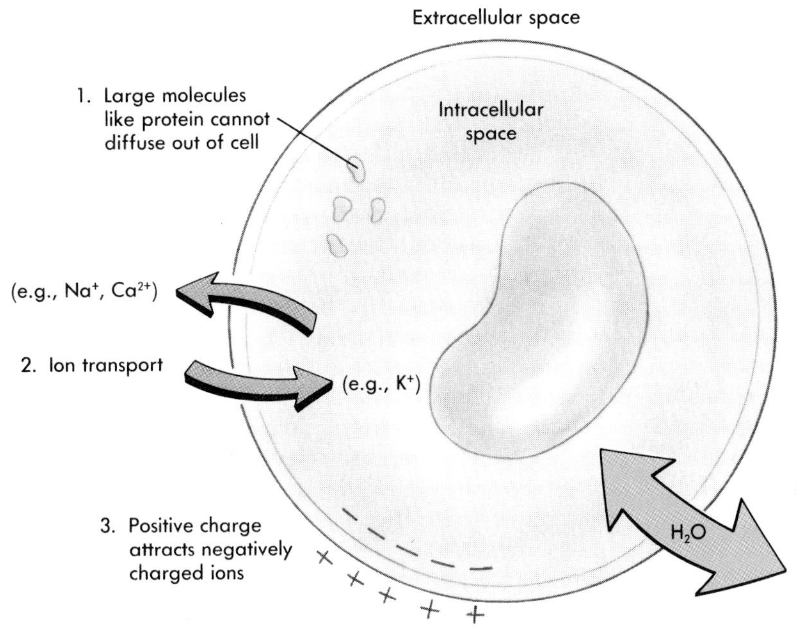

FIGURE 27-1 Mechanisms that influence the distribution of solutes and water between the intracellular and extracellular spaces. The distribution of the solutes is influenced by the large organic molecules such as proteins, which are synthesized within the cell and cannot diffuse out of it, by the transport of solute molecules across the cell membrane, and by the charge difference across the cell membrane. Since water readily moves across the cell membrane by osmosis, the distribution of solutes intracellularly and extracellularly determines the water volume on either side of the cell membrane.

REGULATION OF ION CONCENTRATIONS

Sodium Ions

Sodium ions are the dominant extracellular cations. Because of their abundance in the extracellular fluids, they exert substantial osmotic pressure. Approximately 90% to 95% of the osmotic pressure of the extracellular fluid is due to sodium ions and the negative ions associated with them.

In the United States the quantity of sodium ions ingested each day is 20 to 30 times the amount that is needed. Less than 0.5 g are required to maintain homeostasis, but approximately 10 to 15 g of sodium are ingested daily by the average individual. Regulation of the sodium ion content in the body, therefore, is primarily dependent on the excretion of excess quantities of sodium. However, the mechanisms for conserving sodium in the body are effective when the sodium ion intake is very low.

The kidneys provide the major route by which sodium ions are excreted. Sodium ions readily pass from the glomerulus into Bowman's capsule and are present in the same concentration in the filtrate as in the plasma. The concentration of sodium that is excreted in the urine is determined both by the amount of filtrate that enters the nephron and by the amount of sodium reabsorbed from filtrate in the nephron. If little sodium is reabsorbed from the nephron, a large amount is lost in the urine. On the other hand, if sodium is intensively reabsorbed by the nephron, little sodium is lost in the urine.

1

Indicate whether the rate at which sodium ions enter the nephron increases, decreases, or remains the same under the following conditions: (a) acute and severe hemorrhagic shock; and (b) slightly elevated mean arterial blood pressure.

? ? ? ? ? ? ? ? ? ? ?

The rate of sodium ion transport in the proximal convoluted tubule is relatively constant, but the sodium ion transport mechanisms of the distal convoluted tubule and the collecting duct are under hormonal control. When **aldosterone** is present, the reabsorption of sodium ions from the distal convoluted tubule and the collecting duct is very efficient. As little as 0.1 g of sodium is excreted in the urine each day in the presence of high blood levels of aldosterone. When aldosterone is absent, reabsorption of sodium in the nephron is greatly reduced, and as much as 30 to 40 g of sodium is lost in the urine daily.

Sodium ions are also excreted from the body in **sweat.** Normally only a small quantity of sodium is lost each day in the form of sweat, but the amount increases during conditions of heavy exercise in a warm environment. The amount of sodium excreted through the skin is controlled by the mechanisms that regulate sweating. As the body temperature increases, the thermoreceptor neurons within the hypothalamus cause the rate of sweat production to increase. As the rate of sweat production increases, the amount of sodium lost in the urine decreases to keep the extracellular concentration of sodium constant. The loss of sodium ions in sweat is rarely physiologically significant.

The primary mechanisms that regulate the sodium ion concentration in the extracellular fluid do not directly monitor sodium ion levels but are sensitive to changes in extracellular fluid osmolality or blood pressure (Figure 27-2 and Table 27-3). The amount of sodium in the body has a dramatic effect on the extracellular osmotic pressure. Consequently, if the sodium ion content increases, either the osmolality or the volume of the extracellular fluid increases. Increased **osmolality** of the extracellular fluids results in the production of a small volume of concentrated urine and an increased sensation of thirst. Decreased osmolality of the extracellular fluids results in the production of a large volume of dilute urine and a decreased sensation of thirst (see Chapter 26).

Elevated blood pressure under resting conditions results in increased sodium and water excretion because it stimulates baroreceptors, which, in turn, send signals to the hypothalamus of the brain to cause a reduction in **antidiuretic hormone (ADH)** secretion. At the same time, elevated blood pressure inhibits renin secretion from the juxtaglomerular apparatuses in the kidney. A reduced rate of renin secretion leads to a reduced rate of angiotensin II formation. In response to the lower levels of angiotensin II, aldosterone secretion declines, causing a reduction in the rate of sodium reabsorption from nephrons in the kidneys and allowing excretion of large quantities of sodium in the urine. The combined effect of these mechanisms is the regulation of the sodium ion concentration and the water content of the extracellular fluid.

If the blood pressure is low, the total sodium ion content of the body is usually also low. In response to low blood pressure, baroreceptors send fewer nerve impulses to the hypothalamus, resulting in an increased rate of ADH secretion. At the same time, the juxtaglomerular apparatuses of the kidney increase their rate of renin secretion.

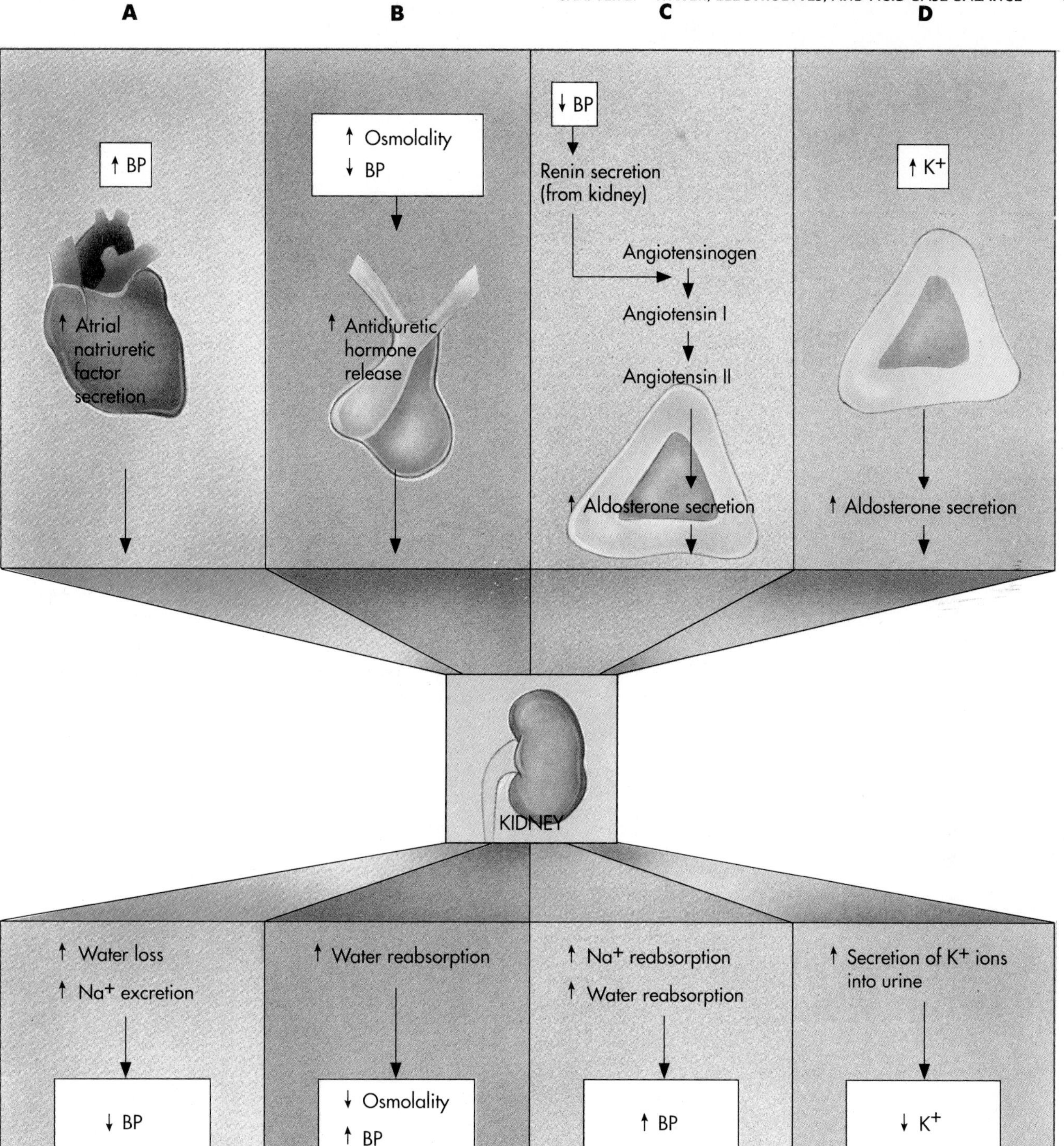

FIGURE 27-2 Major mechanisms that regulate the sodium ion levels in the extracellular fluids. **A** Increased blood pressure in the right atrium of the heart causes secretion of atrial natriuretic factor, which increases sodium ion excretion and water loss in the form of urine. **B** Increased blood osmolality affects hypothalamic neurons, and decreased blood pressure affects baroreceptors in the aortic arch, carotid sinuses, and atrium. As a result of these stimuli, an increased rate of antidiuretic hormone (ADH) secretion from the posterior pituitary results, which increases water reabsorption. **C** Low blood pressure stimulates renin release from the kidney. Renin stimulates the production of angiotensin II, which, in turn, stimulates aldosterone secretion from the adrenal cortex. Aldosterone stimulates sodium ion and water absorption in the kidney. **D** Elevated blood levels of potassium will stimulate aldosterone secretion by the adrenal cortex. Aldosterone, in turn, causes an increased rate of potassium secretion by the kidneys.

Table 27-3

Mechanisms regulating blood sodium

MECHANISM	STIMULUS	RESPONSE TO STIMULUS	EFFECT OF RESPONSE	RESULT
Responses to Changes in Blood Osmolality				
Antidiuretic hormone (ADH)—the most important regulator of blood osmolality	Increased blood osmolality (e.g., increased sodium ion concentration)	Increased ADH secretion from the posterior pituitary; mediated through osmoreceptors	Increased water reabsorption in the kidney; production of a small volume of concentrated urine	Decreased blood osmolality as reabsorbed water dilutes the blood
	Decreased blood osmolality (e.g., decreased sodium ion concentration)	Decreased ADH secretion from the posterior pituitary; mediated through osmoreceptors	Decreased water reabsorption in the kidney; production of a large volume of dilute urine	Increased blood osmolality as water is lost from the blood into the urine
Responses to Changes in Blood Pressure				
Renin-angiotensin-aldosterone	Decreased blood pressure in the kidney's afferent arterioles	Increased renin release from the juxtaglomerular apparatuses; renin initiates the conversion of angiotensinogen to angiotensin; angiotensin increases aldosterone secretion from the adrenal cortex	Increased sodium ion reabsorption in the kidney (because of increased aldosterone); increased water reabsorption as water follows the sodium ions; decreased urinary volume	Increased blood pressure as blood volume increases because of increased water reabsorption; blood osmolality is maintained because both sodium ions and water are reabsorbed*
	Increased blood pressure in the kidney's afferent arterioles	Decreased renin release from the juxtaglomerular apparatuses, resulting in reduced formation of angiotensin; reduced angiotensin causes a decrease in aldosterone secretion from the adrenal cortex	Decreased sodium ion reabsorption in the kidney (because of decreased aldosterone); decreased water reabsorption as fewer sodium ions are reabsorbed; increased urinary volume	Decreased blood pressure as blood volume decreases because water is lost in the urine; blood osmolality is maintained because both sodium ions and water are lost in the urine*
Atrial natriuretic factor	Decreased blood pressure in the atria of the heart	Decreased atrial natriuretic factor released from the atria	Increased sodium ion reabsorption in the kidney; increased water reabsorption as water follows the sodium ions; decreased urinary volume	Increased blood pressure as blood volume increases because of increased water reabsorption; blood osmolality is maintained because both sodium ions and water are reabsorbed*
	Increased blood pressure in the atria of the heart	Increased atrial natriuretic factor released from the atria	Decreased sodium ion reabsorption in the kidney; decreased water reabsorption as water is lost with sodium ions in the urine; increased urinary volume	Decreased blood pressure as blood volume increases because water is lost in the urine; blood osmolality is maintained because both sodium ions and water are lost in the urine*

*Assumes normal levels of ADH.

Table 27-3

Mechanisms regulating blood sodium—cont'd

MECHANISM	STIMULUS	RESPONSE TO STIMULUS	EFFECT OF RESPONSE	RESULT
Responses to Changes in Blood Pressure—cont'd				
ADH—activated by significant decreases in blood pressure; normally regulates blood osmolality (see above)	Decreased arterial blood pressure	Increased ADH secretion from the posterior pituitary; mediated through baroreceptors	Increased water reabsorption in the kidney; production of a small volume of concentrated urine	Increased arterial blood pressure resulting from increased blood volume; decreased blood osmolality
	Increased arterial blood pressure	Decreased ADH secretion from the posterior pituitary; mediated through baroreceptors	Decreased water reabsorption in the kidney; production of a large volume of dilute urine	Decreased arterial blood pressure resulting from decreased blood volume; increased blood osmolality

Table 27-4

Consequences of abnormal plasma levels of sodium ions

MAJOR CAUSES	SYMPTOMS
Hypernatremia (elevated blood levels of sodium ions)	
High dietary sodium rarely causes symptoms	Thirst, fever, dry mucous membranes, restlessness; most serious symptoms—convulsions and pulmonary edema
Administration of hypertonic saline solutions (e.g., sodium bicarbonate treatment for acidosis)	When occurs with an increased water volume—weight gain, edema, elevated blood pressure, and bounding pulse
Oversecretion of aldosterone (i.e., aldosteronism)	
Water loss (e.g., because of fever, respiratory infections, diabetes insipidus, diabetes mellitus, diarrhea)	
Hyponatremia (reduced blood levels of sodium ions)	
Inadequate dietary intake of sodium rarely causes symptoms—can occur in those on low-sodium diets and those taking diuretics	Lethargy, confusion, apprehension, seizures, and coma
Extrarenal losses—vomiting, prolonged diarrhea, gastrointestinal suctioning, burns	When accompanied by reduced blood volume—reduced blood pressure, tachycardia, and decreased urine output
Dilution—intake of large water volume after excessive sweating	When accompanied by increased blood volume—weight gain, edema, and distension of veins
Hyperglycemia, which attracts water into the circulatory system but reduces the concentration of sodium ions	

2

If the amount of sodium and water ingested in food exceeds the amount needed to maintain a constant extracellular fluid composition, explain the effect on (a) blood pressure, (b) urine volume, and (c) urine concentration.

? ? ? ? ? ? ? ? ? ? ?

Atrial natriuretic factor is synthesized by cells in the walls of the atria and is secreted in response to an elevation of blood pressure within the right atrium. Natriuretic hormone acts on the kidneys to increase urine production by inhibiting the reabsorption of sodium ions, inhibiting the effect of ADH on the distal convoluted tubules and collecting ducts, and inhibiting ADH secretion (see Chapter 26, Figure 27-2, *A* and Table 27-3).

Deviations from the normal concentration range for sodium ions in body fluids results in significant symptoms. Some major causes of elevated sodium ion concentrations (hypernatremia) and reduced sodium ion concentrations (hyponatremia) and the major symptoms of each are listed in Table 27-4.

Chloride Ions

The predominant anions in the extracellular fluid are **chloride ions.** The electrical attraction of anions and cations makes it expensive energy-wise to separate these charged particles. Consequently, the regulatory mechanisms that influence the concentration of cations in the extracellular fluid also influence the concentration of anions. The mechanisms that regulate sodium, potassium, and calcium ion levels in the body play a major role in influencing chloride ion levels.

Potassium Ions

The extracellular concentration of **potassium ions** must be maintained within a narrow range of concentrations. The concentration gradient of potassium ions across the cell membrane has a major influence on the resting membrane potential, and cells that are electrically excitable are highly sensitive to slight changes in that concentration gradient. An increase in the extracellular potassium ion concentration leads to depolarization, and a decrease in the extracellular potassium ion concentration leads to hyperpolarization of the resting membrane potential. Some major causes of elevated potassium ion levels (hyperkalemia) and reduced potassium ion levels (hypokalemia) in the body fluids and their symptoms are listed in Table 27-5.

Potassium ions pass freely through the filtration membrane of the renal corpuscle. They are actively reabsorbed in the proximal convoluted tubules and are actively secreted in the distal tubules. Potassium ion secretion into the distal convoluted tubule is highly regulated and is primarily responsible for controlling the extracellular concentration of potassium ions.

Aldosterone plays a major role in regulating the concentration of potassium ions in the extracellular fluid by increasing the rate of potassium ion secretion in the distal portion of the nephron. Aldosterone secretion from the adrenal cortex is stimulated by elevated potassium blood levels and angiotensin II (see Figure 27-2, *C* and *D* and Chapter 26). The elevated aldosterone concentrations in the circulatory system increase potassium ion secretion into the nephron, lowering the blood level of potassium. Circulatory system shock resulting from plasma loss, dehydration, and tissue damage (e.g., in burn patients) causes the extracellular potassium ions to be more concentrated than normal. Decreased blood pressure also stimulates renin secretion from the kidney. Renin causes the conversion of angiotensinogen to angiotensin I. Angiotensin converting enzyme present in the plasma and especially in the lungs converts angiotensin I to angiotensin II. Angiotensin II stimulates aldosterone secretion and vasoconstriction.

In response to an increase in the blood levels of potassium, the rate of aldosterone secretion increases. As a consequence, the rate at which potassium ions are secreted into the nephron increases, resulting in a lower blood potassium level.

Calcium Ions

The extracellular concentration of **calcium ions,** like that of potassium ions, is regulated within a narrow range. The normal concentration of calcium ions in plasma is 9.4 mg/100 ml. Minor symptoms of hypocalcemia develop when the concentration declines to 6 mg/100 ml, and major symptoms of hypercalcemia develop when concentrations reach 12 mg/100 ml. Increases and decreases in the extracellular concentration of calcium ions have marked effects on the electrical properties of excitable tissues. An elevated extracellular calcium ion concentration decreases the permeability of the cell membrane to sodium ions, thus preventing normal depolarization of nerve and muscle cells. High extracellular calcium ion levels also result in the deposition of calcium carbonate salts in soft tissues, resulting in irritation and inflammation of those tissues. Reduced extracellular calcium ion levels result in elevated permeability of cell membranes to sodium ions; as a result, nerve and muscle tissues undergo spontaneous action potential generation. Table 27-6 lists some of the major causes and symptoms of elevated blood levels of calcium (hypercalcemia) and reduced blood levels of calcium (hypocalcemia).

Table 27-5

Consequences of abnormal concentrations of potassium ions

COMMON CAUSES	SYMPTOMS
Hyperkalemia (elevated blood levels of potassium ions)	
Movement of potassium from intracellular to extracellular fluid occurs due to cell trauma (e.g., burns or crushing injuries) and alterations in cell membrane permeability (e.g., acidosis, insulin deficiency, and cell hypoxia)	Mild hyperkalemia (caused mainly by partial depolarization of cell membranes): Increased neuromuscular irritability, restlessness Intestinal cramping and diarrhea Electrocardiogram—alterations, including rapid repolarization with narrower and taller T waves and shortened QT intervals
Decreased renal excretion of potassium (e.g., from decreased secretion of aldosterone in persons with Addison's disease)	Severe hyperkalemia (caused mainly by partial depolarization of cell membranes severe enough to hamper action potential conduction): Muscle weakness, loss of muscle tone, and paralysis Electrocardiogram—alterations, including changes caused by reduced rate of action potential conduction (e.g., depressed ST segment, prolonged PR interval, wide QRS complex, arrhythmias, and cardiac arrest)
Hypokalemia (reduced blood levels of potassium ions)	
Alkalosis (potassium shifts into cell in exchange for hydrogen ions)	Symptoms are mainly due to hyperpolarization of membranes.
Insulin administration (promotes cellular uptake of potassium)	Decreased neuromuscular excitability—skeletal muscle weakness
Reduced potassium intake (especially with anorexia nervosa and alcoholism)	Decreased tone in smooth muscle
Increased renal loss (excessive aldosterone secretion, improper use of diuretics, kidney diseases that result in reduced ability to reabsorb sodium)	Cardiac muscle—delayed ventricular repolarization, bradycardia, and atrioventricular block

Table 27-6

Consequences of abnormal concentrations of calcium

COMMON CAUSES	SYMPTOMS
Hypercalcemia (elevated blood levels of calcium ions)	
Excessive parathyroid hormone secretion	Symptoms are mainly due to decreased permeability of cell membranes to sodium ions.
Excess vitamin D	Loss of membrane excitability—fatigue, weakness, lethargy, anorexia, nausea, and constipation
	Electrocardiogram—shorten QT segment and depressed T waves
	Kidney stones
Hypocalcemia (reduced blood levels of calcium ions)	
Nutritional deficiencies	Symptoms are mainly due to increased permeability of cell membranes to sodium ions.
Vitamin D deficiency	
Decreased parathyroid hormone secretion	Increase in neuromuscular excitability—confusion, muscle spasms, hyperreflexia, and intestinal cramping
Malabsorption of fats (reduced vitamin D absorption)	Severe neuromuscular excitability—convulsions, tetany, inadequate respiratory movements
Bone tumors that increase calcium deposition	Electrocardiogram—prolonged QT interval (prolonged ventricular depolarization)

The kidneys, intestinal tract, and bones play important roles in maintaining extracellular calcium ion levels (Figure 27-3). Close to 99% of total body calcium is contained in bone. Part of the extracellular calcium regulation involves the regulation of calcium ion deposition into and resorption from bone (see Chapter 6). However, the long-term regulation of calcium levels depends on maintaining a balance between calcium ion absorption across the wall of the intestinal tract and calcium ion excretion by the kidneys.

Parathyroid hormone, secreted by the parathyroid glands, increases extracellular calcium levels and reduces extracellular phosphate levels (see Figure 27-3). The rate of parathyroid hormone secretion is regulated by the extracellular calcium ion levels. Elevated calcium ion levels inhibit and reduced levels stimulate its secretion. Parathyroid hormone

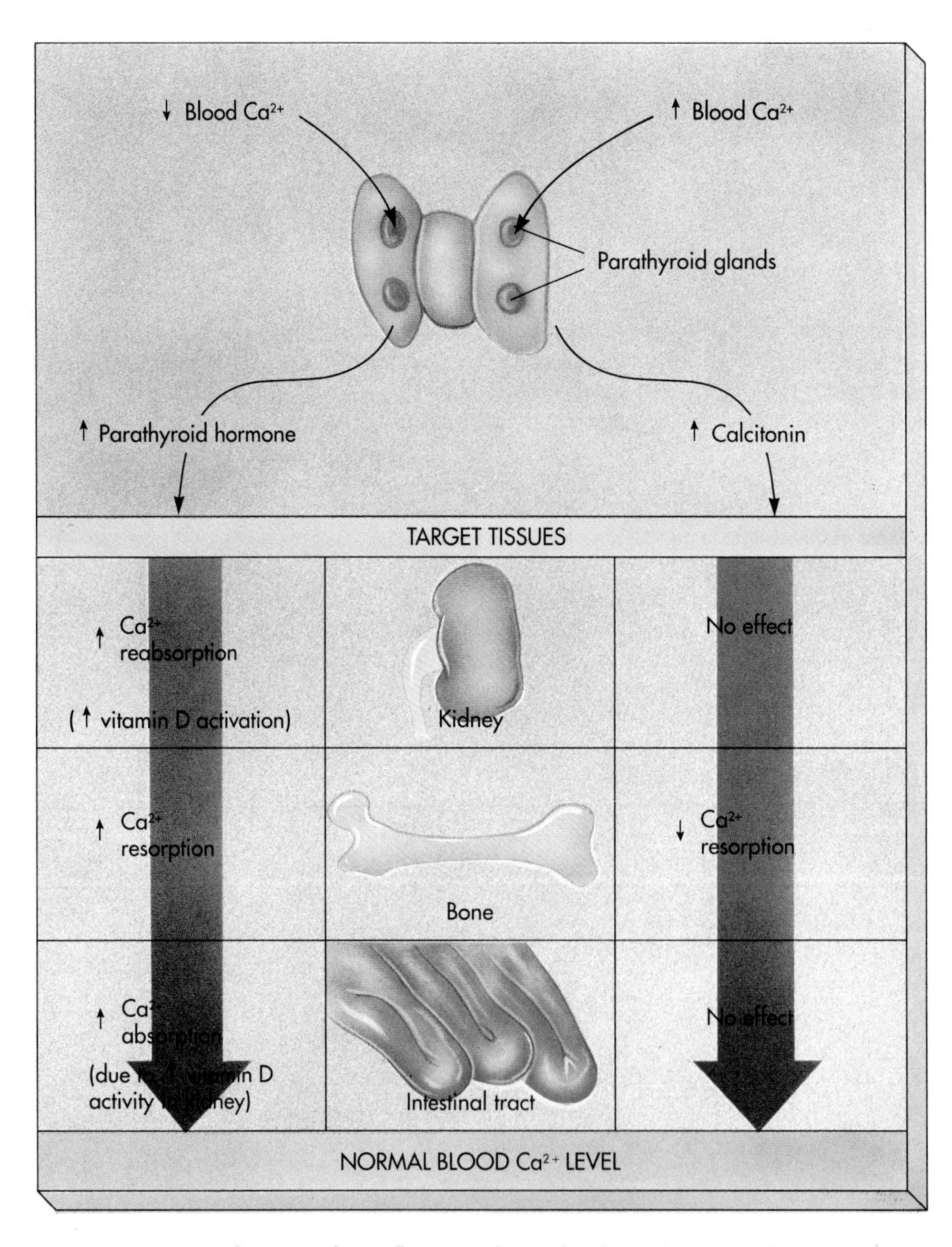

FIGURE 27-3 Mechanisms that influence calcium levels in the extracellular fluid. Decreased blood calcium levels stimulate parathyroid hormone secretion. Parathyroid hormone increases calcium absorption in the intestinal tract and reabsorption of calcium ions from bone and from filtrate in the kidney. Parathyroid hormone also increases the rate of active vitamin D synthesis in the kidney, which also stimulates calcium absorption in the gastrointestinal tract. Calcitonin, secreted by the thyroid gland, may also affect the blood calcium level through its action on bone.

causes osteoclasts to degrade bone and to release calcium and phosphate ions into the body fluids. It increases the rate of calcium ion reabsorption from nephrons in the kidneys and increases the concentration of phosphate in the urine. It also increases the rate at which vitamin D is converted to 1,25-hydroxycholecalciferol, or active vitamin D. Active vitamin D acts on the intestinal tract to increase calcium absorption across the intestinal mucosa. A lack of parathyroid hormone secretion results in a rapid decline in extracellular calcium ion concentration. This decline results from a reduction in the rate of absorption of calcium ions from the intestinal tract, increased calcium ion excretion by the kidneys, and reduced bone resorption. A lack of parathyroid hormone secretion can result in death because of tetany of the respiratory muscles caused by hypocalcemia.

The two sources of vitamin D are its consumption in food and vitamin D biosynthesis. Normally vitamin D biosynthesis is adequate, but prolonged lack of exposure to sunlight reduces the biosynthesis because ultraviolet light is required for one step in the process (see Chapter 5). The consumption of dietary vitamin D may involve the ingestion of active vitamin D or one of its precursors.

Without vitamin D the transport of calcium across the wall of the intestinal tract is negligible and leads to inadequate calcium intake, even though large amounts of calcium may be present in the diet. Thus calcium absorption depends on both the consumption of an adequate amount of calcium in food and the presence of an adequate amount of vitamin D.

Calcitonin, which is secreted by the parafollicular cells of the thyroid gland, causes a reduction in extracellular calcium levels. It is most effective when calcium levels are elevated, although greater-than-normal calcitonin levels in the blood are not consistently effective in causing blood levels of calcium to decline below normal values. The major effect of calcitonin is on bone. It inhibits bone demineralization by decreasing osteoclast activity. It may also increase the rate at which osteoblasts deposit calcium in the form of inorganic bone salts.

Elevated calcium levels stimulate calcitonin secretion, whereas reduced calcium levels inhibit calcitonin secretion from the parafollicular cells. Although calcitonin regulates blood calcium levels, calcitonin is not as important as parathyroid hormone in this regulation (see Figure 27-3).

Phosphate Ions

The nephrons of the kidney have a maximum rate at which phosphate ions are reabsorbed from the filtrate. Under normal conditions this transport mechanism is functioning near its maximum rate. If the concentration of phosphate ions increases in the plasma, it is lost in the urine. The additional phosphate ions enter the filtrate, and because the concentration exceeds the ability of the nephron to reabsorb phosphate, the additional phosphate is lost in the urine. Consequently, the major mechanism that controls the plasma levels of phosphate ions is the transport mechanisms for phosphate ions in the nephrons.

REGULATION OF WATER CONTENT

The body's water content is regulated so the total volume of water in the body remains constant. Thus the volume of water taken into the body is equal to the volume lost each day. Changes in the volume of water in body fluids alter both the osmolality and the total pressure (i.e., blood pressure and interstitial fluid pressure) of those fluids. The total volume of water that enters the body each day is 1500 to 3000 ml. Most of that volume comes from ingested fluids, some comes from food that is consumed, and a smaller amount is derived from water that is produced during cellular metabolism (Figure 27-4 and Table 27-7).

The movement of water across the wall of the gastrointestinal tract depends on osmosis, and the volume of water entering the body depends, to a large degree, on the volume of water consumed. If a large volume of dilute liquid is consumed, the rate at which water enters the body fluids increases. If a small volume of concentrated liquid is consumed, the rate decreases. At least one way in which the water content of the body can be altered, therefore, is by varying the volume of water ingested.

Although fluid consumption is heavily influenced by habit and by social phenomena, water ingestion does depend, at least in part, on regulatory mechanisms. The sensation of thirst results from an increase in the osmolality of the extracellular fluids and from a reduction in plasma volume. **Osmoreceptor cells** within the hypothalamus detect an increased extracellular fluid osmolality and initiate activity in neural circuits that results in a conscious sensation of thirst.

Baroreceptors may also influence the sensation of thirst. When they detect a decrease in blood pressure, action potentials are conducted to the brain along afferent pathways to influence the sensation of thirst. Low blood pressure associated with hemorrhagic shock, for example, is correlated with an intense sensation of thirst.

When **renin** is released from the juxtaglomerular apparatuses, it increases the formation of angiotensin II in the circulatory system (see Chapters 18 and 26). **Angiotensin II** stimulates the sensation of thirst by acting on the brain. Angiotensin II reverses decreases in blood pressure by stimulating the sensation of thirst in addition to increasing aldosterone secretion and initiating vasoconstriction.

Water gained or conserved		Water loss
↓ Blood pressure and ↑plasma osmolality cause ↑ADH secretion — results in ↑water reabsorption ↓ Blood pressure and ↑plasma K^+ levels cause ↑aldosterone secretion results in ↑water reabsorption	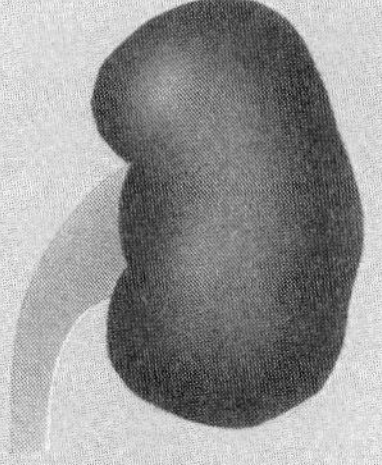 Kidney	↑ Blood pressure causes ↑atrial natriuretic hormone (ANH) secretion causing ↑water and Na^+ excretion ↑ Blood pressure and ↓plasma osmolality cause ↓ADH secretion — results in ↑water excretion ↑ Blood pressure and ↓K^+ levels cause ↓aldosterone secretion — results in ↑water and Na^+ excretion (plus ↓K^+ loss in urine) Water lost in urine as waste products are removed from the body
↑ Blood osmolality and large decreases in blood pressure affect hypothalamus — results in sensation of thirst	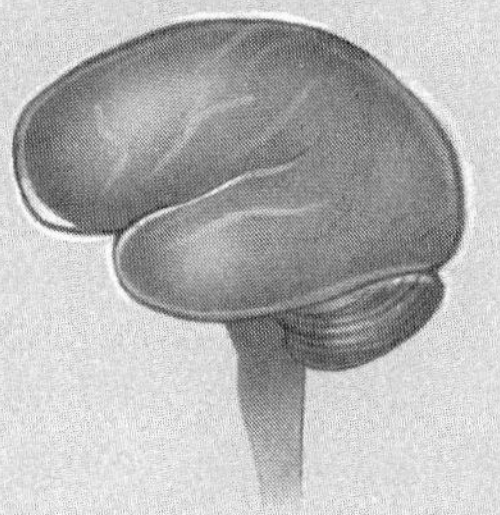 Central nervous system	↑ Body temperature, exercise, stress cause water loss by sweating and evaporation
Water absorbed following ingestion	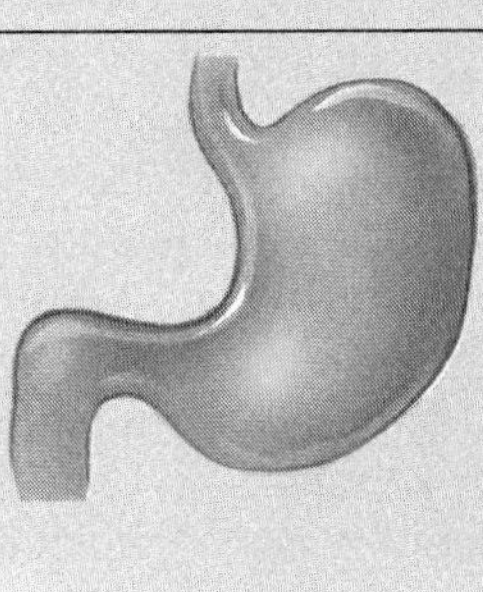 Digestive system	Water lost in feces
Water produced by cellular metabolism	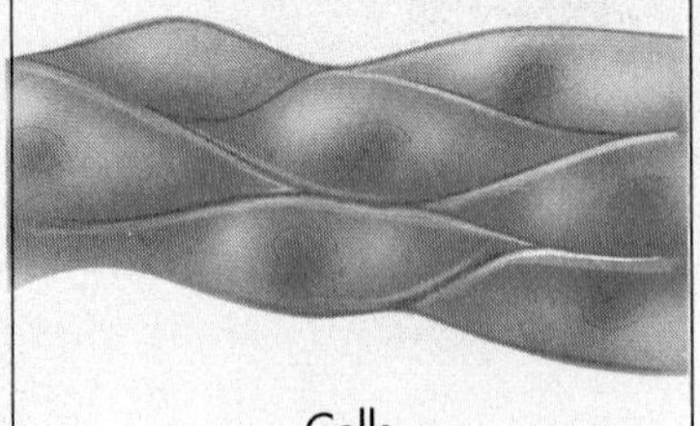 Cells	Water used in chemical reactions

FIGURE 27-4 Sources of water that contribute to the body's water content, routes through which water is lost from the body, and mechanisms that regulate water intake and water loss.

Table 27-7

Summary of water intake and loss

SOURCES OF WATER	ROUTES BY WHICH WATER IS LOST
Ingestion	Perspiration
Metabolism (small volume)	Insensible
	Sensible
	Urine
	Feces

Table 27-8

Composition of sweat

SOLUTE	CONCENTRATION (mM/L)
Sodium	9.8-77.2
Potassium	3.9-9.2
Chloride	5.2-65.1
Ammonia	1.7-5.6
Urea	6.2-12.1

When people who are dehydrated are allowed to drink water, they eventually consume a sufficient quantity to reduce the osmolality of the extracellular fluid to its normal value. However, they do not normally consume the water all at once but drink intermittently until the proper osmolality of the extracellular fluid is established. The thirst sensation is temporarily reduced after the ingestion of small amounts of liquid. At least two factors are responsible for this temporary interruption of the thirst sensation. First, when the oral mucosa becomes wet after it has been dry, inhibitory impulses are sent to the thirst center of the hypothalamus. Second, consumed fluid increases the gastrointestinal tract volume, and stretch of the gastrointestinal wall initiates afferent action potentials in stretch receptors. The action potentials are transmitted to the brain where they temporarily suppress the sensation of thirst.

Since the absorption of water from the gastrointestinal tract requires time, the mechanisms that temporarily suppress the sensation of thirst prevent the consumption of extreme volumes of fluid that would exceed the amount required to reduce the blood osmolality. Long-term suppression of the thirst sensation, however, requires that extracellular fluid osmolality and blood pressure come within their normal ranges.

Learned behavior may be very important in avoiding periodic dehydration through the consumption of fluids either with or without food even though blood osmolality is not reduced. The volume of fluid ingested by a healthy person usually exceeds the minimum volume required to maintain homeostasis, and the kidneys eliminate the excess water in the form of urine.

Water loss from the body occurs through three major routes (see Figure 27-4 and Table 27-7). The greater amount of water is lost through the urine and through evaporation, and a smaller volume is lost in the feces. Water lost through evaporation includes the volume lost from the respiratory passages and from the skin surface. The volume of water lost through the respiratory system depends on the temperature and humidity of the air, the body temperature, and the volume of air inspired.

The water lost through simple evaporation from the skin is called **insensible perspiration.** For each degree that the body temperature rises above normal, an increased volume of 100 to 150 ml of water is lost each day in the form of insensible perspiration.

Sweat, or **sensible perspiration,** is secreted by the sweat glands (see Chapter 5), and in contrast to insensible perspiration, it contains solutes. Sweat resembles extracellular fluid in its composition, with sodium chloride as the major component, but it does contain some potassium, ammonia, and urea (Table 27-8). The volume of sweat that is produced is determined primarily by neural mechanisms that regulate body temperature, although some sweat is produced as a result of sympathetic stimulation in response to stress. The volume of fluid lost as sweat is negligible for a person at rest in a cool environment. Under conditions of exercise, elevated environmental temperature, or fever, the volume increases substantially. Sweat losses of 8 to 10 L per day have been measured in outdoor workers in the summertime.

Evaporation of water is a major mechanism by which body temperature is regulated. Approximately one fourth of the total heat produced by metabolism is lost from the body through the evaporation of water. The rate of evaporation for a person exercising in a hot, dry environment is much greater than for a person at rest in a cool, humid environment.

Adequate fluid replacement during conditions of extensive sweating is important. Since sweat is usually hypotonic, the loss of a large volume of sweat causes hypertonicity in the body fluids. The loss of fluid volume is primarily from the extracellular space and leads to a reduction in plasma volume and an increase in hematocrit. The change may become sufficiently great during conditions of severe dehydration to cause blood viscosity to increase substantially. The increased work load created for the heart can result in heart failure.

3

List the mechanisms through which water loss changes during conditions of exercise.

? ? ? ? ? ? ? ? ? ? ?

The loss of water by way of the digestive tract is relatively small. The total volume of fluid secreted into the gastrointestinal tract is large, but nearly all of the fluid is reabsorbed under normal conditions (see Chapter 24). However, severe vomiting or diarrhea can result in a large volume of fluid loss from the gastrointestinal tract.

The kidneys are the primary organs that regulate the composition and the volume of body fluids by controlling the volume and the concentration of water excreted in the form of urine. Urine production varies greatly in response to mechanisms that regulate the body's water content. Reduced extracellular fluid osmolality and elevated blood pressure inhibit ADH secretion from the posterior pituitary, and elevated blood pressure also reduces renin secretion. The effect of these hormonal changes is the production of a large volume of dilute urine. If the osmolality of the blood increases, ADH and aldosterone levels will increase, producing a smaller volume of more concentrated urine. Reduced blood pressure also results in elevated ADH and aldosterone secretion (see Figure 27-4).

REGULATION OF ACID-BASE BALANCE

Hydrogen ions affect the activity of enzymes and interact with many electrically charged molecules. Consequently, most chemical reactions that occur within the body are highly sensitive to the hydrogen ion concentration of the fluid in which they occur, and maintenance of hydrogen ion concentration within a narrow range of values is essential for normal metabolic reactions. The two major components of the mechanism of pH regulation are the buffer systems and the regulatory mechanisms.

Acids and Bases

The acidity of a solution depends on its hydrogen ion concentration. The greater the hydrogen ion concentration, the greater is its acidity. Normally, the acidity of a solution is measured on the **pH scale** (Chapter 2 and Appendix). As the acidity of the solution becomes greater, the pH value becomes smaller, and as the solution becomes more basic, the pH value becomes larger. A solution is considered neutral at a pH of 7. Below 7 the pH is acidic, and above 7 it is basic.

For most purposes **acids** can be defined as substances that release hydrogen ions into a solution; **bases** bind to hydrogen ions and remove them from solution. Acids can be grouped as either strong or weak. Strong acids completely dissociate in solution so that all the hydrogen ions are released into the solution (Figure 27-5), whereas weak acids release hydrogen ions into the solution, but fewer of the acid molecules dissociate. Some of the acid molecules remain intact and do not release hydrogen ions into the solution. The proportion of weak acid molecules that do release hydrogen ions into solution is very predictable and is influenced by the pH of the solution into which the weak acid is placed. Weak acids are common in living systems and play an important role in preventing large changes in body fluid pH.

Buffer Systems

Buffers resist changes in the pH of a solution. Buffers within body fluids stabilize pH by chemically binding to excess hydrogen ions (H^+) when they are added to a solution or by releasing hydrogen ions when their concentration in a solution begins to fall.

A solution of carbonic acid (H_2CO_3) and sodium bicarbonate (HCO_3^-) is an example of a buffer system. Carbonic acid is a weak acid, and sodium bicarbonate is the salt of this weak acid. When these substances are dissolved in solution, an equilibrium is established:

$$H_2CO_3 \leftrightarrow HCO_3^- + H^+$$

Many carbonic acid molecules and many bicarbonate ions are in such a solution. When hydrogen ions are added, a large proportion of them bind to bicarbonate ions to form carbonic acid, and only a small percentage remain in the form of free hydrogen ions; thus a large decrease in pH is prevented when hydrogen ions are added to a solution. If a large number of hydrogen ions is removed from solution, many of the carbonic acid molecules dissociate to form bicarbonate and hydrogen ions; thus a large increase in pH is resisted by releasing hydrogen ions into solution.

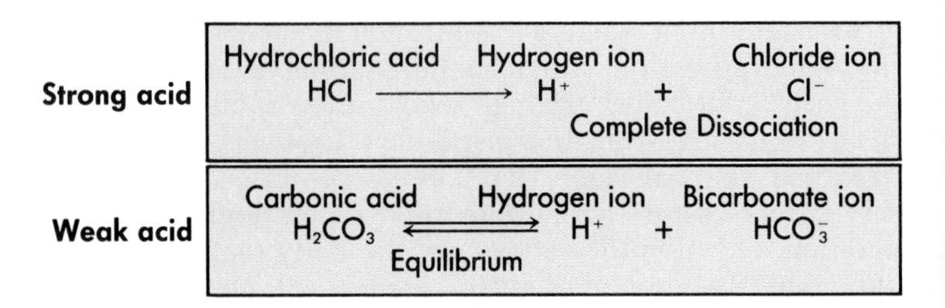

FIGURE 27-5 Comparison of strong and weak acids. Strong acids completely dissociate when dissolved in water, whereas weak acids do not completely dissociate. Weak acids partially dissociate so that an equilibrium is established between the acid and the ions (i.e., hydrogen ions plus anions) that are formed when the dissociation occurs.

Table 27-9

Buffer systems

Protein buffer system	Intracellular proteins and plasma proteins form a large pool of protein molecules that can act as buffer molecules. Because of their high concentration, they provide approximately three fourths of the buffer capacity of the body. Hemoglobin in red blood cells is an important intracellular protein. Other intracellular molecules such as histone proteins and nucleic acids also act as buffers.
Bicarbonate buffer system	Components of the bicarbonate buffer system are not present in high enough concentrations in the extracellular fluid to constitute a powerful buffer system. However, because the concentrations of the components of the buffer system are regulated, it plays an exceptionally important role in controlling the pH of extracellular fluid.
Phosphate buffer system	Concentration of the phosphate buffer components is low in the extracellular fluids in comparison to the other buffer systems, but it is an important intracellular buffer system.

Several important buffer systems in the body work together to resist changes in pH of body fluids (Table 27-9). The carbonic acid–bicarbonate buffer system, protein molecules such as hemoglobin and plasma protein, and phosphate compounds all act as buffers.

Mechanisms of Acid-Base Balance Regulation

The respiratory system and the urinary system play essential roles in regulation of acid-base balance. The **respiratory system** responds rapidly to a change in pH and functions to bring the pH of body fluids back toward its normal range. However, its capacity to regulate pH is not as great as the capacity of the **urinary system,** nor does it have the same ability to bring the pH back to its precise range of normal values. On the other hand, the urinary system responds more slowly than the respiratory system to a change in body fluid pH.

Respiratory regulation of acid-base balance

The ability of the respiratory system to regulate acid-base balance is dependent on the **carbonic acid–bicarbonate buffer system.** Carbon dioxide (CO_2) reacts with water (H_2O) to form carbonic acid (H_2CO_3) which, in turn, dissociates to form hydrogen ions (H^+) and bicarbonate ions (HCO_3^-) as follows:

$$H_2O + CO_2 \leftrightarrow H_2CO_3 \leftrightarrow H^+ + HCO_3^-$$

This reaction is in reversible equilibrium. The higher the concentration of carbon dioxide, the greater is the amount of carbonic acid that is formed, and the greater is the number of hydrogen ions and bicarbonate ions that are formed. On the other hand, if carbon dioxide levels decline, the equilibrium shifts in the opposite direction (i.e., hydrogen and bicarbonate ions combine to form carbonic acid, which then dissociates to form carbon dioxide and water).

The reaction between carbon dioxide and water is catalyzed by an enzyme, **carbonic anhydrase,** which is found in relatively high concentration within red blood cells (Figure 27-6). This enzyme does not influence equilibrium but accelerates the rate at which the reaction proceeds in either direction.

Increasing carbon dioxide levels and decreasing body fluid pH stimulate neurons in the medullary respiratory center of the brain and cause the rate and depth of ventilation to increase. In response to the increased rate and depth of ventilation, carbon dioxide is eliminated from the body through the lungs at a greater rate, and the concentration of carbon dioxide in the body fluids decreases. As carbon dioxide levels decline, the concentration of hydrogen ions and therefore the pH are returned to their normal range. The response of the respiratory system to changes in blood levels of carbon dioxide and pH are listed in Table 27-10.

If carbon dioxide levels become too low or the pH of the body fluids is elevated, the rate and the depth of respiration decline. As a consequence, the rate at which carbon dioxide is eliminated from the body is reduced. Carbon dioxide then accumulates in the body fluids because it is continually produced as a by-product of metabolism. As carbon dioxide increases in the body fluids, it reacts with water to form carbonic acid, which dissociates to form hydrogen and bicarbonate ions, and the pH is brought toward its normal value.

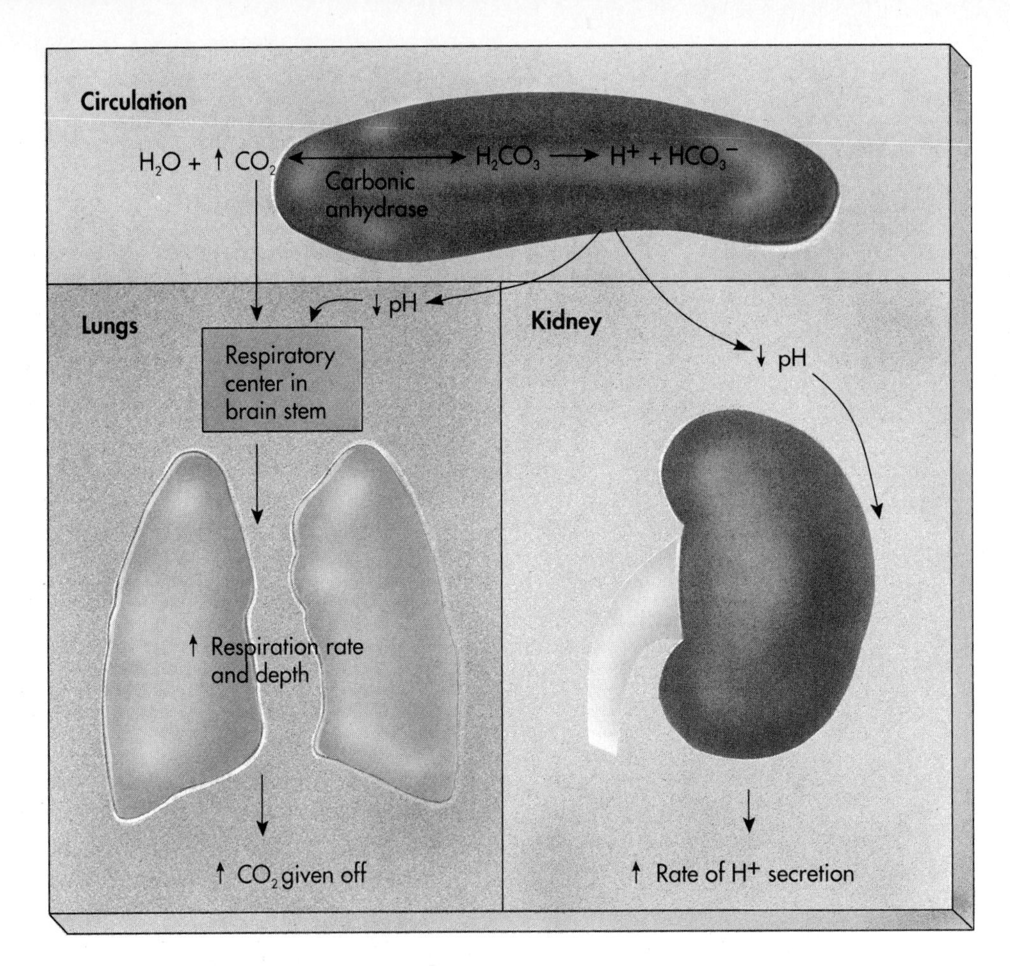

FIGURE 27-6 Equilibrium between carbon dioxide, carbonic acid, and bicarbonate and hydrogen ions within red blood cells. Carbon dioxide reacts with water to form carbonic acid. An enzyme, carbonic anhydrase, catalyzes the reaction. Carbonic acid then dissociates to form hydrogen ions and bicarbonate ions. An equilibrium exists so that an increase in carbon dioxide causes more carbonic acid formation and a decrease causes some of the carbonic acid to dissociate to form carbon dioxide and water. This reaction is an important mechanism through which the pH of the body fluids is regulated by the lungs and the kidneys.

Table 27-10

Respiratory regulation of acid-base balance

CHANGE IN PLASMA pH	RESPONSE TO CHANGE IN PLASMA pH	EFFECT ON PLASMA'S CARBON DIOXIDE LEVELS	RESPONSE
Increased pH (decreased hydrogen ion concentration); decreased level of plasma carbon dioxide	Affects neurons in the medullary respiratory center, resulting in decreased respiratory rate and depth	Accumulation of carbon dioxide, produced as a by-product of metabolism, in body fluids	Increased carbon dioxide reacts with water to produce carbonic acid, which lowers the plasma pH (increases hydrogen ion concentration)
Decreased pH (increased hydrogen ion concentration); increased level of plasma carbon dioxide	Affects neurons in the medullary respiratory center, resulting in increased respiratory rate and depth	Elimination of carbon dioxide through the lungs; reduces carbon dioxide levels in body fluids	Reduced carbon dioxide in body fluids increases the plasma pH (decreases hydrogen ion concentration in body fluids) because hydrogen ions combine with bicarbonate ions to form carbonic acid, which dissociates to form carbon dioxide

Renal regulation of acid-base balance

The kidneys can secrete hydrogen ions into the urine and therefore can directly regulate acid-base balance. Cells in the walls of the distal portion of the nephron are primarily responsible for the secretion of hydrogen ions. Carbonic anhydrase is present within these cells and catalyzes the formation of carbonic acid from carbon dioxide and water. The carbonic acid molecules dissociate to form hydrogen and bicarbonate ions, and the hydrogen ions are then transported into the lumen of the nephron by an active transport pump that exchanges sodium ions for the hydrogen ions. The bicarbonate ions and the sodium ions then pass from the nephron cells into the extracellular fluid. As a result, hydrogen ions are secreted into the nephron's lumen, and bicarbonate ions pass into the extracellular fluid where they combine with excess hydrogen ions. Both the secretion of hydrogen ions into the nephron and the movement of bicarbonate ions into the extracellular fluid raise the body fluid pH. The rate of this process increases as the pH of the body fluids decreases, and it slows as the pH of the body fluids increases (Figure 27-7).

Some of the hydrogen ions secreted into the filtrate combine with bicarbonate ions, which enter the filtrate through the renal corpuscle, to form carbonic acid. The carbonic acid molecules dissociate to form carbon dioxide and water. Carbon dioxide diffuses from the nephron into the tubular cells where it reacts with water to form carbonic acid, which subsequently dissociates to form hydrogen ions and bicarbonate ions. The hydrogen ions are actively transported into the lumen of the nephron, whereas bicarbonate ions reenter the extracellular fluid. As a result, many of the bicarbonate ions that enter the filtrate reenter the extracellular fluid. Normally the hydrogen ions secreted into the nephron exceed the amount of bicarbonate that is filtered so that almost all of the bicarbonate is reabsorbed. Bicarbonate ions are therefore not lost in the urine unless the pH of the body fluids is elevated above its normal value. Excess hydrogen ions remain in the urine, and the pH of the urine decreases.

If the pH of the body fluids increases, the rate of hydrogen ion secretion into the nephron decreases. As a result, the amount of bicarbonate filtered into the nephron exceeds the amount of secreted hydrogen ion, and the excess bicarbonate ions pass into

Text continued on p. 900.

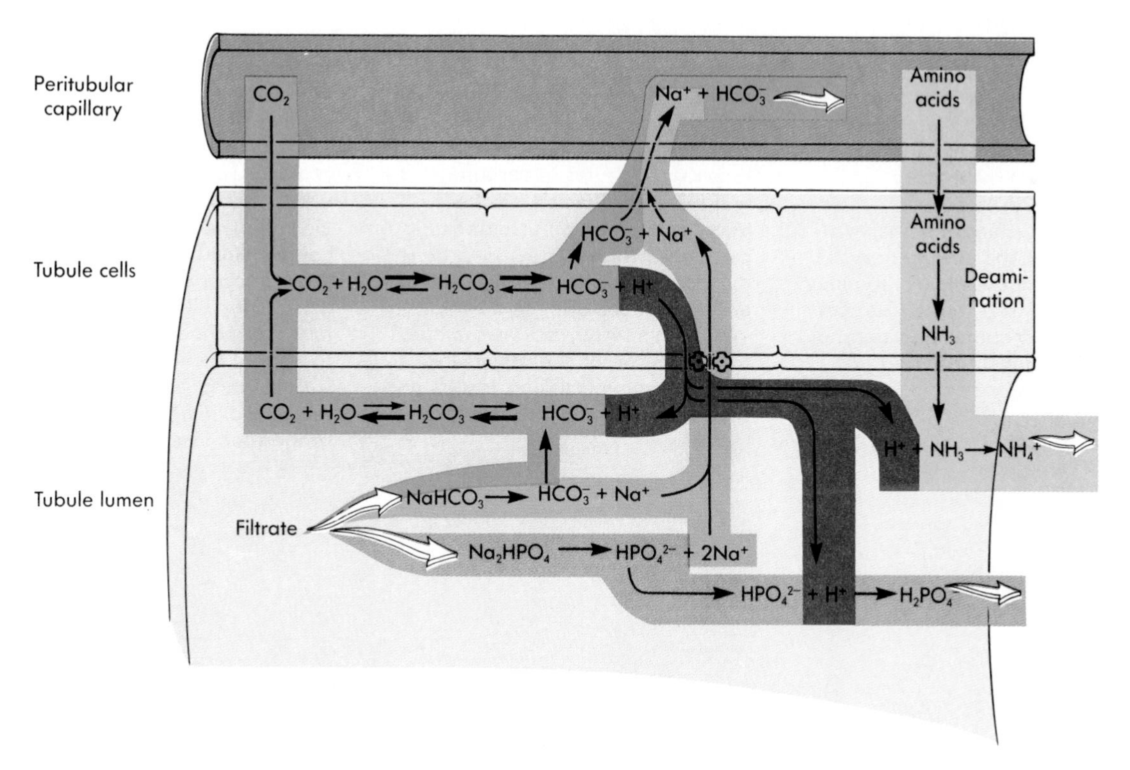

FIGURE 27-7 Mechanism through which the kidneys regulate the pH of body fluids. The epithelial cells of the nephrons transport hydrogen ions from the body fluids into the filtrate. The rate at which the hydrogen ions are transported is regulated so that a decrease in body fluid pH causes an increase in the rate of transport, whereas an increase in body fluid pH causes a decrease in the rate of transport. When hydrogen ions are transported into the lumen of the nephron, bicarbonate ions are transported into the body fluids. As hydrogen ions are transported into the nephron, they react with bicarbonate ions, phosphate ions, or ammonia. These cations act as buffers.

Acidosis and Alkalosis

When the pH value of the body fluids is below 7.35, the condition is referred to as **acidosis,** and when the pH value is above 7.45, the condition is called **alkalosis.**

Metabolism produces acidic products that lower the pH of the body fluids. Carbon dioxide is converted to carbonic acid, lactic acid is a product of anaerobic metabolism, protein metabolism produces phosphoric and sulfuric acids, and lipid metabolism produces fatty acids.

Acidosis and alkalosis are placed into two categories, depending on the cause of the condition. Respiratory acidosis or alkalosis results from abnormalities of the respiratory system. Metabolic acidosis or alkalosis results from causes other than abnormal respiratory functions (Table 27-A).

The major effect of acidosis is depression of the central nervous system. When the pH of the blood falls below 7, the central nervous system malfunctions, and the individual becomes disoriented and possibly comatose as the condition worsens.

A major effect of alkalosis is hyperexcitability of the nervous system. Peripheral nerves are affected first, resulting in spontaneous nervous stimulation of muscles. Spasms and tetanic contractions result, as may extreme nervousness or convulsions. Severe alkalosis can cause death as a result of tetany of the respiratory muscles.

Hypoventilation or hyperventilation can be observed in many acid-base disorders. In some cases the change in the respiratory rate is an attempt to correct the pH imbalance, and in some cases it is the cause of the pH imbalance. When respiratory acidosis is due to reduced elimination of carbon dioxide from the body, blood carbon dioxide levels increase, the respiratory center is stimulated, and hyperventilation results. Hyperventilation is an attempt to compensate for the acidosis by reducing blood carbon dioxide levels and thus increasing blood pH. When respiratory acidosis is due to depression of the respiratory center, hypoventilation results and causes buildup of carbon dioxide in the blood and decreased blood pH. During respiratory alkalosis hyperventilation causes a reduced blood carbon dioxide level and is responsible for the increased blood pH. When the cause of pH imbalance is metabolic, the changes in respiration rate are compensatory. Hypoventilation compensates for metabolic alkalosis, and hyperventilation compensates for metabolic acidosis.

Changes in urine pH also occur in acid-base disorders, and the changes in pH can be the cause of the disorder or an attempt to correct the disorder. If the disorder is caused by the production of a more-acid-than-normal urine, metabolic alkalosis results; if the disorder is caused by the production of a more-alkaline-than-normal urine, metabolic acidosis develops. On the other hand, an acidic urine compensates for respiratory acidosis and most cases of metabolic acidosis by eliminating hydrogen ions from the body. Also, an alkaline urine compensates for respiratory alkalosis and most cases of metabolic alkalosis by reducing hydrogen ion loss from the body.

Table 27-A
Acidosis and Alkalosis

CONDITION	CONSEQUENCE
ACIDOSIS	
Respiratory Acidosis Asphyxia Asthma Severe emphysema Hypoventilation (e.g., imparied respiratory center function due to trauma, tumor, shock, or heart failure)	Reduced elimination of carbon dioxide from the body fluids through the respiratory system, resulting in a higher-than-normal hydrogen ion concentration
Metabolic acidosis	
Severe diarrhea	Elimination of large amounts of bicarbonate due to mucous secretion in the colon
Vomiting of lower intestinal contents	Elimination of large amounts of bicarbonate due to production of mucus in the intestine
Ingestion of acidic drugs such as large doses of aspirin	Direct reduction of the body fluid pH
Untreated diabetes mellitus	Production of large amounts of fatty acids and other acidic metabolic end products (e.g., ketone bodies)
Lactic acid buildup (e.g., severe exercise, heart failure, and shock)	Inadequate oxygen delivery to tissue results in anaerobic respiration, increased lactic acid production, and acidosis
ALKALOSIS	
Respiratory alkalosis	
Emotions	Intense hyperventilation due to anxiety reduces carbon dioxide levels in the extracellular fluid, resulting in a lower-than-normal hydrogen ion concentration
High altitude	Decreased atmospheric pressure reduces the amount of oxygen transported in the blood; low blood oxygen stimulates chemoreceptor reflex, causing hyperventilation
Metabolic alkalosis	
Severe vomiting of stomach contents	Elimination of large amounts of acidic stomach contents
Ingestion of alkaline substances such as large amounts of sodium bicarbonate	Raised pH of the body fluids as bicarbonate ions are absorbed
Acidic urine (e.g., drugs such as most diuretics and aldosterone)	Higher-than-normal loss of hydrogen ions in the urine

the urine. Excretion of excess bicarbonate ions in the urine diminishes the amount of bicarbonate ions in the extracellular fluid, and as a consequence, the pH of the body fluids decreases.

Table 27-11 summarizes the response of the kidney to changes in body fluid pH, including changes in the rate of hydrogen ion secretion and bicarbonate ion reabsorption and changes in urine composition.

4

Predict the effect of aldosterone hyposecretion on body fluid pH.

? ? ? ? ? ? ? ? ? ? ?

Aldosterone increases the rate of sodium reabsorption and potassium secretion, but in high concentrations aldosterone also stimulates hydrogen ion secretion. Therefore elevated aldosterone levels such as occur in patients with Cushing's syndrome can cause an elevation of the body fluid pH above normal (alkalosis). The major factor that influences the rate of hydrogen ion secretion, however, is the pH of the body fluids.

The mechanisms that cause increased hydrogen ion secretion into the urine can achieve a urine pH of 4.5. Urine pH lower than 4.5 inhibits the secretion of additional hydrogen ions. The total amount of hydrogen ions that pass into the urine is greater than the quantity required to lower the pH of an unbuffered solution below 4.5. Consequently, buffers must be present in the urine to combine with hydrogen ions. Bicarbonate ions, phosphate ions (HPO_4^{2-}), and ammonia (NH_3) act as buffers in the urine. They enter the nephron through filtration to combine with secreted hydrogen ions (see Figure 27-7).

Ammonia is produced in the cells of the nephron when amino acids such as glycine are deaminated. The ammonia molecules then diffuse into the filtrate. The ammonia molecules subsequently combine with free hydrogen ions in the filtrate to form ammonium ions (NH_4^+), which are secreted in the urine in combination with chloride ions. The rate of ammonia production increases when the pH of the body fluids has been depressed for 2 or 3 days. The elevated ammonia production increases the buffering capacity of the filtrate, allowing secretion of additional hydrogen ions into the urine.

Although phosphate and ammonia constitute major buffers within the urine, other weak acids such as lactic acid also combine with hydrogen ions in the nephron and increase the amount of hydrogen ions that can be secreted.

Table 27-11

Renal regulation of acid-base balance

CHANGE IN PLASMA pH	RESPONSE TO CHANGE IN PLASMA pH	CHANGE IN URINARY COMPOSITION	RESPONSE
Increased pH (decreased hydrogen ion concentration)	Decreased hydrogen ion secretion into the nephron; decreased bicarbonate ion reabsorption	pH of urine increases; bicarbonate concentration in urine increases	Decreased plasma pH (increased hydrogen ion concentration) because fewer hydrogen ions are secreted into the nephron, allowing hydrogen ions to accumulate in the plasma, and fewer bicarbonate ions are present to act as a buffer
Decreased pH (increased hydrogen ion concentration)	Increased hydrogen ion secretion into the nephron; increased bicarbonate ion reabsorption	pH of urine decreases; bicarbonate concentration in urine decreases	Increased plasma pH (decreased hydrogen ion concentration) because more hydrogen ions are secreted into the nephron, reducing hydrogen ions in plasma, and more bicarbonate ions are present in the plasma to act as a buffer

SUMMARY

WATER, ACID, BASE, AND ELECTROLYTE LEVELS ARE MAINTAINED WITHIN A NARROW RANGE OF CONCENTRATIONS. THE URINARY, RESPIRATORY, GASTROINTESTINAL, INTEGUMENTARY, NERVOUS, AND ENDOCRINE SYSTEMS PLAY A ROLE IN MAINTAINING FLUID, ELECTROLYTE, AND PH BALANCE.

BODY FLUIDS p. 882

1. Intracellular fluid is inside cells.
2. Extracellular fluid is outside cells and includes interstitial fluid and plasma.

REGULATION OF INTRACELLULAR FLUID COMPOSITION p. 883

1. Intracellular fluid composition is determined by substances used or produced inside the cell and substances exchanged with the extracellular fluid.
2. Intracellular fluid is different from extracellular fluid because the cell membrane regulates the movement of materials.
3. Water movement is determined by the difference between intracellular and extracellular fluid concentrations.

REGULATION OF EXTRACELLULAR FLUID COMPOSITION p. 883

Extracellular fluid composition is determined by the intake and elimination of substances from the body and the exchange of substances between the extracellular and intracellular fluids.

REGULATION OF ION CONCENTRATION p. 884

Sodium Ions

1. Sodium is responsible for 90% to 95% of extracellular osmotic pressure.
2. The amount of sodium excreted in the kidneys is the difference between the amount of sodium that enters the nephron and the amount that is reabsorbed from the nephron.
   - Glomerular filtration rate determines the amount of sodium entering the nephron.
   - Aldosterone determines the amount of sodium reabsorbed.
3. Small quantities of sodium are lost in sweat.
4. Increased blood osmolality leads to the production of a small volume of concentrated urine and to thirst. Decreased blood osmolality leads to the production of a large volume of dilute urine and to decreased thirst.
5. Increased blood pressure increases water and salt loss.
   - Baroreceptor reflexes reduce antidiuretic hormone (ADH) secretion.
   - Renin secretion is inhibited, leading to reduced aldosterone production.

Chloride Ions

1. Chloride ions are the dominant negatively charged ions in extracellular fluid.

Potassium Ions

1. The extracellular concentration of potassium ions affects resting membrane potentials.
2. The amount of potassium excreted depends on the amount that enters with the glomerular filtrate, the amount actively reabsorbed by the nephron, and the amount secreted into the distal convoluted tubule.
3. Aldosterone increases the amount of potassium secreted.

Calcium Ions

1. Elevated extracellular calcium levels prevent membrane depolarization. Decreased levels lead to spontaneous action potential generation.
2. Parathyroid hormone increases extracellular calcium levels and decreases extracellular phosphate levels. It stimulates osteoclast activity, increases calcium reabsorption from the kidneys, and stimulates active vitamin D production.
3. Vitamin D stimulates calcium uptake in the intestines.
4. Calcitonin decreases extracellular calcium levels.

Phosphate Ions

1. Under normal conditions reabsorption of phosphate occurs at a maximum rate in the nephron.
2. An increase in plasma phosphate increases the amount of phosphate in the nephron beyond that which can be reabsorbed, and the excess is lost in the urine.

REGULATION OF WATER CONTENT p. 891

1. Water crosses the gastrointestinal tract through osmosis.
2. The sense of thirst is stimulated by an increase in extracellular osmolality or by a decrease in blood pressure.
3. Thirst is inhibited by wetting the oral mucosa or by stretch of the gastrointestinal tract.
4. Learned behavior plays a role in the amount of fluid ingested.
5. Routes of water loss.
   - Water is lost through evaporation from the respiratory system and the skin (insensible perspiration and sweat).
   - Water loss into the gastrointestinal tract normally is small. Vomiting or diarrhea can significantly increase this loss.
   - The kidneys are the primary regulator of water excretion. Urine output can vary from a small amount of concentrated urine to a large amount of dilute urine.

REGULATION OF ACID-BASE BALANCE p. 894

Acids and Bases

Acids release hydrogen ions into solution, and bases remove them.

Buffer Systems

1. A buffer resists changes in pH.
   - When hydrogen ions are added to a solution, the buffer removes them.
   - When hydrogen ions are removed from a solution, the buffer replaces them.
2. Proteins, carbonic acid–bicarbonate, and phosphate compounds are important buffers.

Mechanisms of Acid-Base Balance Regulation

1. Respiratory regulation of pH is achieved through the carbonic acid–bicarbonate buffer system.
   - As carbon dioxide levels increase, pH decreases.
   - As carbon dioxide levels decrease, pH increases.
   - Carbon dioxide levels and pH affect the respiratory centers. Hypoventilation increases blood carbon dioxide levels, and hyperventilation decreases blood carbon dioxide levels.

2. The loss of hydrogen ions into urine and the gain of bicarbonate ions into blood cause extracellular pH to increase.
   - Carbonic acid dissociates to form hydrogen ions and bicarbonate ions in nephron cells.
   - Active transport pumps hydrogen ions into the nephron lumen and sodium into the nephron cell.
   - Sodium and bicarbonate diffuse into the extracellular fluid.
3. Bicarbonate ions in the filtrate are reabsorbed.
   - Bicarbonate ions combine with hydrogen ions to form carbonic acid that dissociates to form carbon dioxide and water.
   - Carbon dioxide diffuses into nephron cells and forms carbonic acid, which dissociates to form bicarbonate ions and hydrogen ions.
   - Bicarbonate ions diffuse into the extracellular fluid, and hydrogen ions are pumped into the nephron lumen.
4. The rate of hydrogen ion secretion increases as body fluid pH decreases or as aldosterone levels increase.
5. Secretion of hydrogen ions is inhibited when urine pH falls below 4.5.
   - Ammonia and phosphate buffers in the urine resist a drop in pH.
   - As the buffers absorb hydrogen ions, more hydrogen ions are pumped into the urine.

CONTENT REVIEW

1. What systems are involved with the regulation of fluid, electrolyte, and pH balance?
2. Define intracellular fluid, extracellular fluid, interstitial fluid, and plasma.
3. What factors determine the composition of intracellular fluid and extracellular fluid?
4. Name the substance that is responsible for most of the osmotic pressure of extracellular fluid.
5. How do the glomerular filtration rate and aldosterone affect the amount of sodium in the urine?
6. What role does sweating play in sodium balance?
7. How does increased blood pressure lead to an increased loss of water and salt? What happens when blood pressure decreases?
8. What effect does atrial natriuretic factor have on sodium and water loss in urine?
9. How are chloride ion concentrations regulated?
10. What effect does an increase or a decrease in extracellular potassium concentration have on resting membrane potentials?
11. Where is potassium secreted in the nephron? How is its secretion regulated?
12. What effects are produced by an increase or a decrease in extracellular calcium concentration?
13. What effects on extracellular calcium concentrations do an increase or a decrease in parathyroid hormone have? What causes these effects?
14. What effect does calcitonin have on extracellular calcium levels?
15. How does an increase in extracellular osmolality or a decrease in blood pressure affect the sensation of thirst? Name two things that will inhibit the sense of thirst.
16. Explain how the kidney controls plasma levels of phosphate ions.
17. Describe three routes for the loss of water from the body.
18. Define an acid and a base. What is normal blood pH? Define acidosis and alkalosis.
19. Define a buffer. Describe how a buffer works when hydrogen ions are added to a solution or when hydrogen ions are removed from a solution. Name the three buffer systems of the body.
20. What happens to blood pH when blood carbon dioxide levels go up or down? What causes this change?
21. What effect do increased blood carbon dioxide levels or decreased pH have on respiration? How does this change in respiration affect blood pH?
22. Describe the process by which nephron cells move hydrogen ions into the nephron lumen and bicarbonate ions into the extracellular fluid.
23. Describe the process by which bicarbonate ions are reabsorbed from the nephron lumen.
24. Name the factors that can cause an increase in hydrogen ion secretion and the factors that can cause a decrease in hydrogen ion secretion.
25. What is the purpose of buffers in the urine? Describe how the ammonia buffer system operates.

CONCEPT REVIEW

1. In patients with diabetes mellitus not enough insulin is produced; as a consequence, blood glucose levels increase. If blood glucose levels rise high enough, the kidneys are unable to absorb the glucose from the glomerular filtrate, and glucose "spills over" into the urine. What effect would this glucose have on urine concentration and volume? How would the body adjust to the excess glucose in the urine?
2. A patient suffering from a tumor in the hypothalamus produces excessive amounts of antidiuretic hormone (ADH; inappropriate ADH syndrome). For this patient the excessive ADH production is chronic and has persisted for many months. A student nurse kept a fluid intake/output record about the patient. She was surprised to find that fluid intake and urinary output were normal. What effect was she expecting? Can you explain why urinary output was normal?
3. A patient exhibits the following symptoms: elevated urine ammonia and increased rate of respiration. Does the patient have metabolic acidosis or metabolic alkalosis?
4. Swifty Trotts has an enteropathogenic *Escherichia coli* infection that produces severe diarrhea. What would this diarrhea do to his blood pH, urine pH, and respiratory rate?
5. Acetazolamide is a diuretic that blocks the activity of the enzyme carbonic anhydrase inside kidney tubule cells. This blockage prevents the formation of carbonic acid from carbon dioxide and water. Normally carbonic acid dissociates to form hydrogen ions and bicarbonate ions, and the hydrogen ions are exchanged for sodium ions from the urine. Blocking the formation of hydrogen ions in the cells of the nephron tubule blocks sodium reabsorption, inhibiting water reabsorption and producing the diuretic effect. With this information in mind, what effect would acetazolamide have on blood pH, urine pH, and respiratory rate?
6. As part of a physiology experiment, a student was asked to breathe through a glass tube that was 3 feet long. What effect would this action have on his blood pH, urine pH, and respiratory rate?
7. A young boy is suspected of having epilepsy (i.e., he is prone to having convulsions). Based on your knowledge of acid-base balance and respiration, propose a test to determine if the boy is susceptible to convulsions.
8. Hardy Explorer climbed to the top of a very high mountain. To celebrate he drank a glass of whiskey (alcohol stimulates hydrochloric acid secretion in the stomach). What would you expect to happen to Hardy's respiratory rate and the pH of his urine?

? ?

ANSWERS TO PREDICT QUESTIONS

1 p. 884. A. During acute, severe hemorrhagic shock blood pressure is decreased, and visceral blood vessels are constricted (see Chapter 21). As a consequence, the blood flow to the kidneys and the blood pressure in the glomerulus are decreased dramatically. The total filtration pressure is decreased, and the amount of filtrate formed each minute is decreased. The rate at which sodium enters the nephron is therefore decreased.

B. When the arterial blood pressure is elevated, the pressure within the glomerulus also is elevated but not very much because of autoregulation. Therefore, as the systemic blood pressure increases, the rate at which filtrate is formed also increases. However, it increases by only a small amount. For a slightly elevated blood pressure the rate at which filtrate is formed and therefore the rate at which sodium ions enter the nephron may not increase significantly.

2 p. 888. A. If the amount of sodium and water ingested in food exceeds that needed to maintain a constant extracellular fluid composition, it increases the total blood volume and also increases the blood pressure.

B. Excessive sodium and water intake causes an increase in total blood volume and blood pressure. The elevated blood pressure causes a reflex response that results in decreased ADH secretion. The elevated pressure also causes reduced renin secretion from the kidneys, resulting in a reduction in the rate at which angiotensin II is formed. The reduced angiotensin II reduces the rate of aldosterone secretion. Together these changes cause increased loss of sodium in the urine and an increase in the volume of urine produced. Increased sodium ions and increased blood pressure cause the secretion of atrial natriuretic hormone, which also inhibits ADH secretion and sodium ion reabsorption in the nephron.

C. If the amount of water ingested is large, the urine concentration will be reduced, the urine volume will be increased, and the concentration of sodium ions in the urine will be low. If the amount of salt that is ingested is great, the concentration of the salt in the urine may be high, and the urine volume will be larger and will contain a substantial concentration of salt.

3 p. 893. During conditions of exercise the amount of water lost is increased because of increased evaporation from the respiratory system, in the form of insensible perspiration, and sweat. The amount of water lost in the form of sweat may increase substantially. The amount of urine formed usually decreases during conditions of exercise.

4 p. 900. Aldosterone hyposecretion results in acidosis. Aldosterone increases the rate at which sodium ions are reabsorbed, but it also increases the rate at which potassium and hydrogen ions are secreted. Hyposecretion of aldosterone decreases the rate at which hydrogen ions are secreted and therefore may result in acidosis.

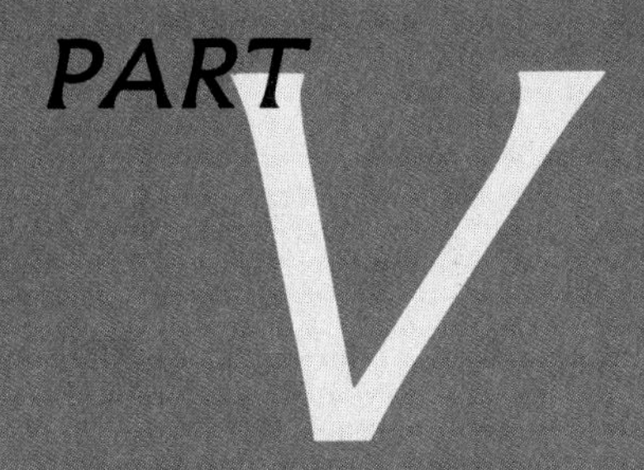
PART
V

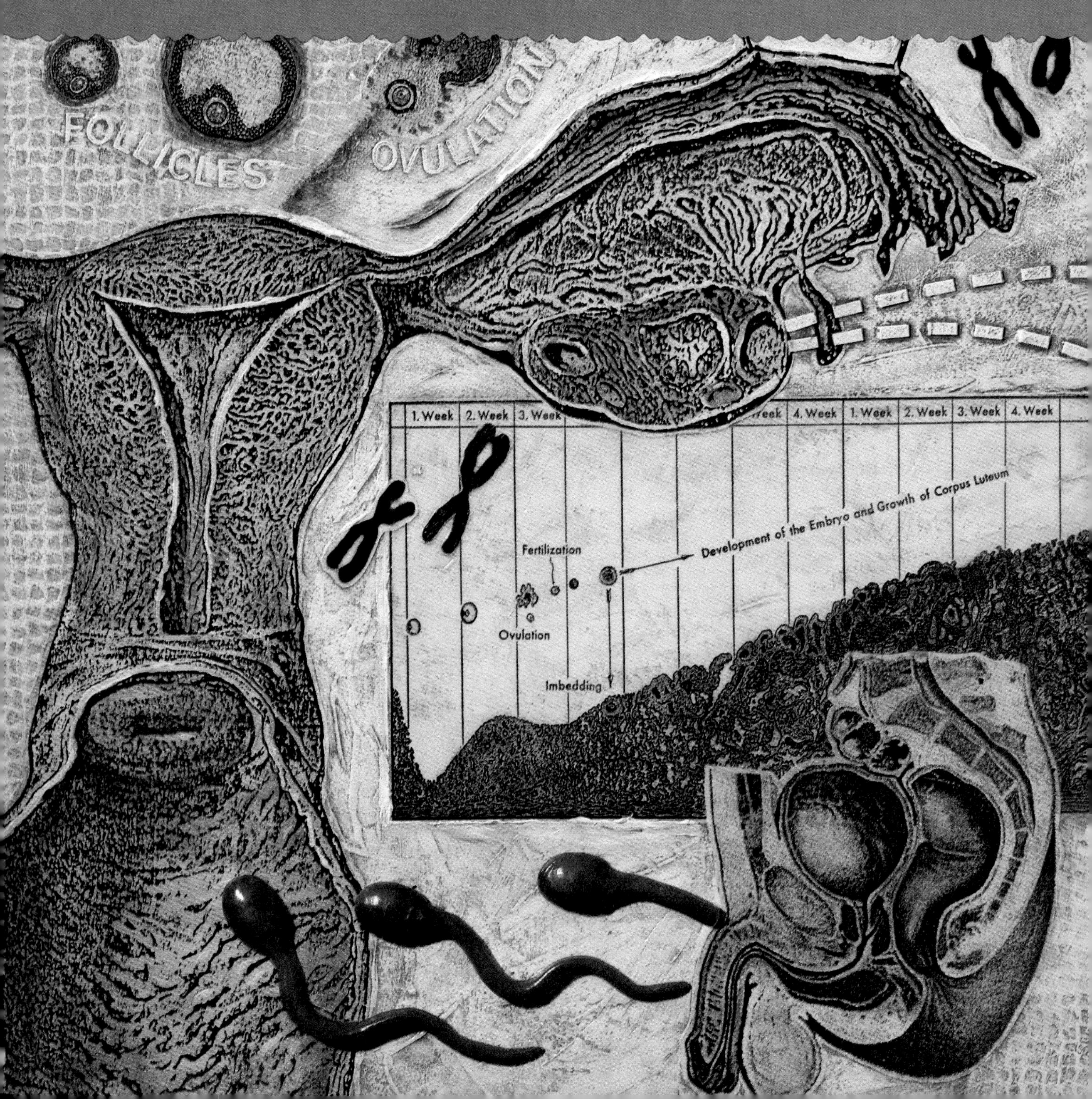
FOLLICLES
OVULATION
1. Week
2. Week
3. Week
eek
4. Week
1. Week
2. Week
3. Week
4. Week
Development of the Embryo and Growth of Corpus Luteum
Fertilization
Ovulation
Imbedding

REPRODUCTION AND DEVELOPMENT

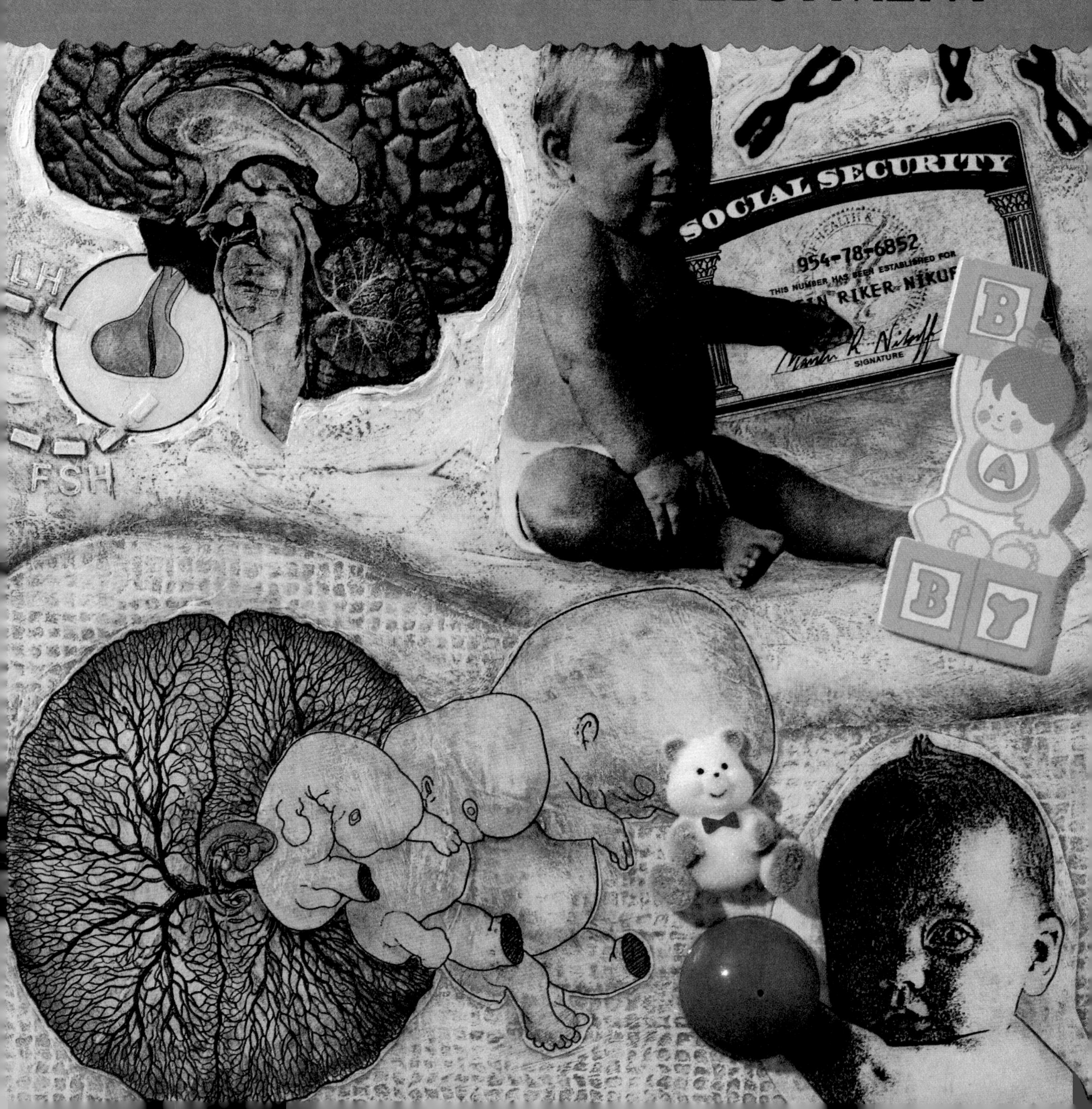

CHAPTER 28 OBJECTIVES

After reading this chapter you should be able to

1. Describe the scrotum and explain the role of the dartos and cremaster muscles in temperature regulation of the testes.
2. Describe the structure of the testes.
3. Describe the process of spermatogenesis.
4. Describe the route sperm follow from the site of their production to the outside of the body.
5. Name the parts of the spermatic cord.
6. Describe the parts of the penis.
7. Name the reproductive glands, state where they empty into the duct system, and describe their secretions.
8. List the hormones that influence the male reproductive system and explain how reproductive hormone secretions are regulated.
9. Explain the role of psychic stimulation, tactile stimulation, and the parasympathetic and sympathetic nervous systems in the male sex act.
10. Describe the anatomy and histology of the ovaries.
11. Discuss the development of the follicle and the oocyte, the process of ovulation, and fertilization.
12. Name and describe the parts of the uterine tube, uterus, vagina, external genitalia, and mammae.
13. Define the phases of the ovarian and uterine cycles.
14. List the hormones of the female reproductive system and explain how reproductive hormone secretions are regulated.
15. Discuss the effects of the ovarian hormones on the uterus.
16. Explain what happens to the ovaries and the uterus if fertilization occurs and if fertilization does not occur.
17. Describe the role of the nervous system in the female sex act.
18. Define menopause and describe the changes that occur because of it.

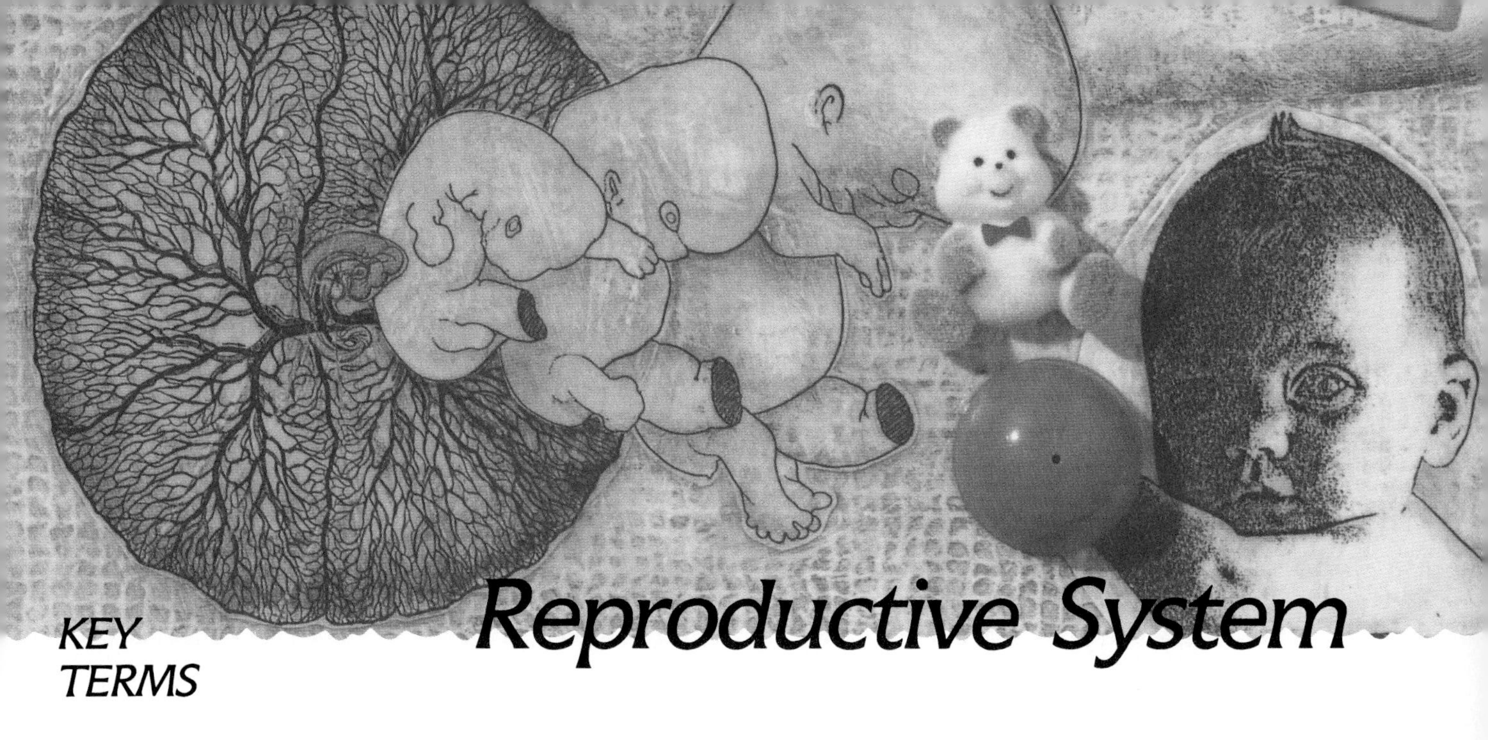

Reproductive System

KEY TERMS

follicle-stimulating hormone

gonadotropin-releasing hormone

luteinizing hormone

menopause

menstrual cycle

oogenesis (o-o-jen′ĕ-sis)

ovarian cycle

ovary

puberty

semen

spermatogenesis (sper′mă-to-jen′ĕ-sis)

testis (test′tis)

uterine cycle

RELATED TOPICS

The following terms or concepts are important for a good understanding of this chapter. If you are not familiar with them, you should review them before proceeding.

Autonomic regulation of the reproductive structures (Chapter 16)

Hormones from the adenohypophysis (Chapter 18)

Anatomy of the urethra (Chapter 26)

UNLIKE MOST ORGAN SYSTEMS OF THE BODY IN WHICH THERE IS LITTLE DIFFERENCE BETWEEN THE MALE AND THE FEMALE, THE REPRODUCTIVE SYSTEMS ARE VERY DIFFERENT. THE MALE REPRODUCTIVE SYSTEM PRODUCES SPERM CELLS AND TRANSFERS THEM TO THE FEMALE. THE FEMALE REPRODUCTIVE SYSTEM PRODUCES OOCYTES AND RECEIVES THE SPERM, ONE OF WHICH MAY UNITE WITH AN OOCYTE. THE FEMALE REPRODUCTIVE SYSTEM IS THEN INTIMATELY INVOLVED WITH NURTURING THE DEVELOPMENT OF THE INDIVIDUAL BEFORE AND AFTER BIRTH.

Although the male and female reproductive systems differ, there are a number of similarities since many of the reproductive organs are derived from the same embryological structures (see Chapter 29). In addition, many of the hormones are the same in the male and female, even though their functions may be quite different (Table 28-1).

Table 28-1

Major reproductive hormones in males and females

HORMONE	SOURCE	TARGET TISSUE	RESPONSE
Male Reproductive System			
Gonadotropin-releasing hormone (GnRH)	Hypothalamus	Adenohypophysis	Stimulates secretion of LH and FSH
Luteinizing hormone (LH) (also called interstitial cell-stimulating hormone [ICSH] in males)	Adenohypophysis	Leydig cells of the testes	Stimulates synthesis and secretion of testosterone
Follicle-stimulating hormone (FSH)	Adenohypophysis	Seminiferous tubules (Sertoli cells)	Supports spermatogenesis
Testosterone	Leydig cells of testes	Testes Body tissues	Supports spermatogenesis Development and maintenance of secondary sexual characteristics
		Adenohypophysis and hypothalamus	Inhibits GnRH, LH, and FSH secretion through negative feedback
Female Reproductive System			
Gonadotropin-releasing hormone (GnRH)	Hypothalamus	Adenohypophysis	Stimulates secretion of LH and FSH
Luteinizing hormone (LH)	Adenohypophysis	Ovaries	Causes follicles to complete maturation and undergo ovulation; causes ovulation; causes the ovulated follicle to become the corpus luteum
Follicle-stimulating hormone (FSH)	Adenohypophysis	Ovaries	Causes follicles to begin development
Estrogens	Follicles of ovaries	Uterus	Proliferation of endometrial cells
		Mammae	Development of the mammary glands (especially duct systems)
		Adenohypophysis and hypothalamus	Positive feedback before ovulation, resulting in increased LH and FSH secretion; negative feedback with progesterone on the hypothalamus and adenohypophysis after ovulation, resulting in decreased LH and FSH secretion
		Other tissues	Secondary sexual characteristics
Progesterone	Corpus luteum of ovaries	Uterus	Hypertrophy of endometrial cells and secretion of fluid from uterine glands

Table 28-1

Major reproductive hormones in males and females—cont'd

HORMONE	SOURCE	TARGET TISSUE	RESPONSE
Female Reproductive System—cont'd			
Progesterone—cont'd		Mammae	Development of the mammary glands (especially alveoli)
		Adenohypophysis	Negative feedback with estrogens on the hypothalamus and adenohypophysis after ovulation, resulting in decreased LH and FSH secretion
		Other tissues	Secondary sexual characteristics
Oxytocin*	Neurohypophysis	Uterus and mammary glands	Contraction of uterine smooth muscle and contraction of myoepithelial cells in the breast, resulting in milk letdown in lactating women
Human chorionic gonadotropin (HCG)	Placenta	Corpus luteum of ovaries	Maintains the corpus luteum and increases its rate of progesterone secretion during the first one third (first trimester) of pregnancy

*Covered in Chapter 29.

MALE REPRODUCTIVE SYSTEM

The male reproductive system consists of the testes (sing. testis), epididymides (sing. epididymis), ductus deferentia (sing. deferens), urethra, seminal vesicles, prostate gland, bulbourethral glands, scrotum, and penis (Figure 28-1, *A*). Sperm cells are very temperature sensitive and do not develop normally at usual body temperatures. The testes and epididymides, in which the sperm cells develop, are located outside the body cavity in the scrotum where the temperature is lower. The ductus deferentia lead from the testes into the pelvis where they join the ducts of the seminal vesicles to form the ampullae. Extensions of the ampullae, called the ejaculatory ducts, pass through the prostate and empty into the urethra within the prostate. The urethra, in turn, exits from the pelvis and passes through the penis to the outside of the body.

Scrotum

The **scrotum** contains the testes and is divided into two internal compartments by a connective tissue septum. Externally the scrotum is marked in the midline by an irregular ridge, the **raphe** (ra′fe; a seam), which continues posteriorly to the anus and anteriorly onto the inferior surface of the penis. The skin of the scrotum includes a layer of superficial fascia (loose connective tissue) and a layer of smooth muscle called the **dartos** (dar′tōs; to skin) **muscle.**

During cold weather the dartos muscle contracts, causing the skin of the scrotum to become firm and wrinkled and reducing the scrotum's overall size. At the same time the **cremaster muscles,** which are extensions of abdominal muscles into the scrotum, contract and help pull the testes nearer the body. During warm weather or exercise the dartos and cremaster muscles relax, and the skin of the scrotum becomes

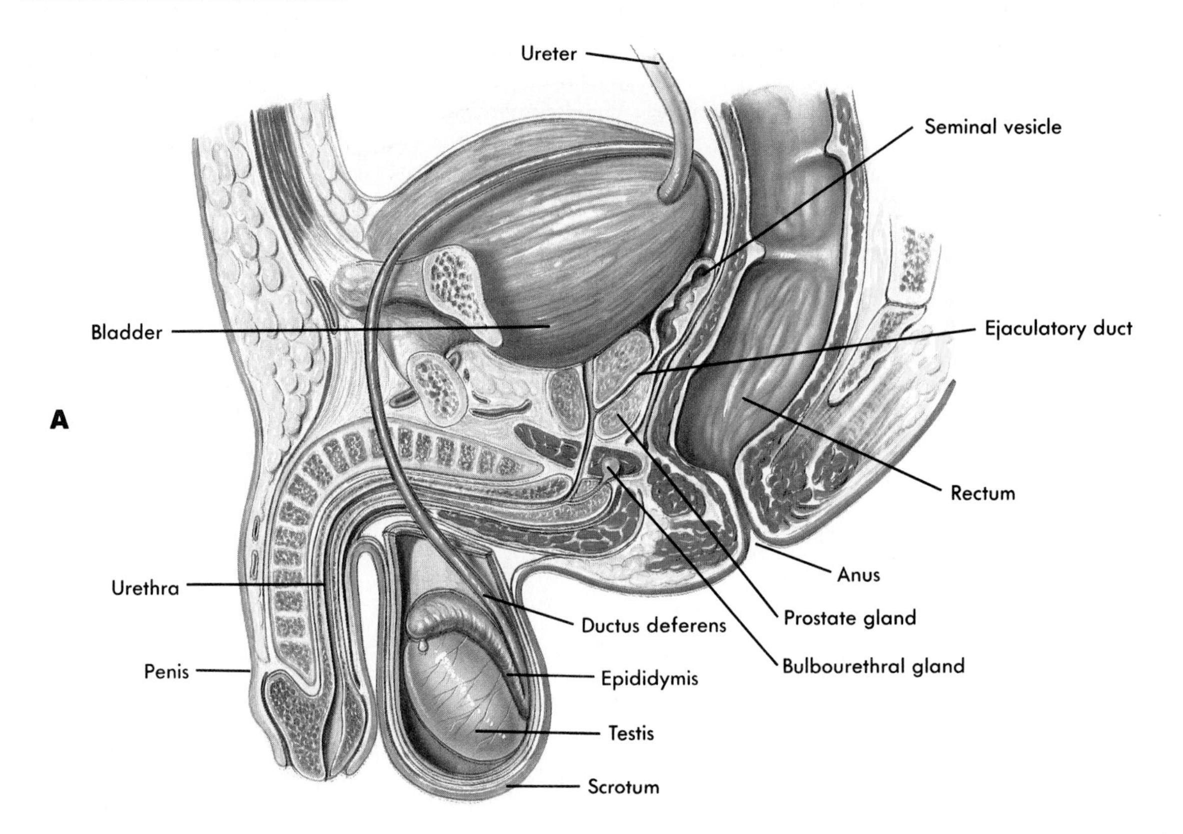

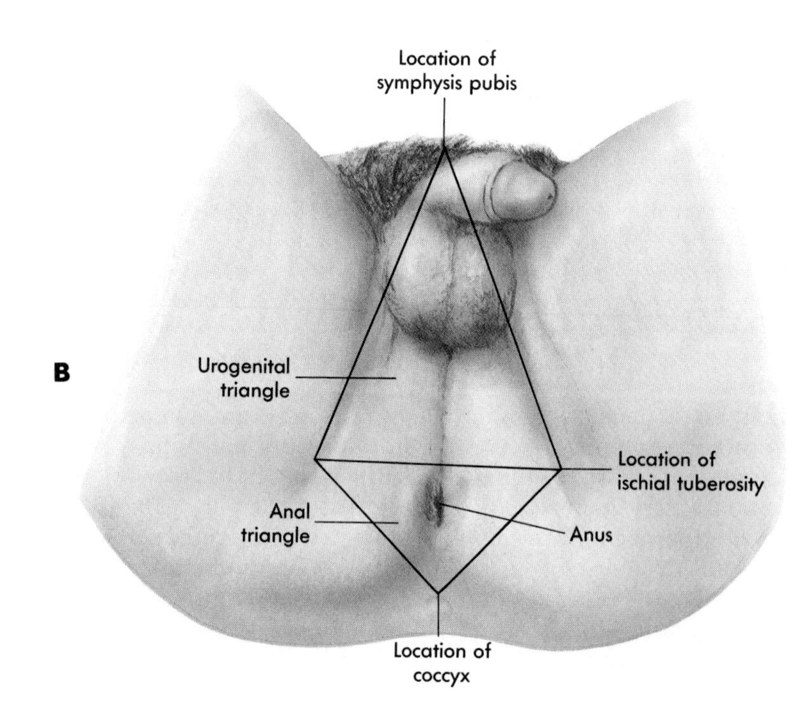

FIGURE 28-1 **A** Sagittal section of the male pelvis showing the male reproductive structures. **B** Inferior view of the male perineum.

loose and thin, allowing the testes to descend away from the body. The response of the dartos and cremaster muscles is important in the regulation of temperature in the testes. If the testes become too warm or too cold, normal spermatogenesis does not occur.

Perineum

The area between the thighs, which is bounded by the pubis anteriorly, the coccyx posteriorly, and the ischial tuberosities laterally, is called the **perineum** (pĕr′ĭ-ne′um). The perineum is divided into two triangles by a set of muscles, the superficial transverse and deep transverse perineal muscles, that runs transversely between the two ischial tuberosities. The anterior triangle, the **urogenital triangle,** contains the external genitalia (the base of the penis) and scrotum. The smaller, posterior triangle, the **anal triangle,** contains the anal opening (Figure 28-1, *B*).

Testes

Testicular histology

The testes are small ovoid organs, each approximately 4 to 5 cm long, within the scrotum (see Figure 28-1). The outer portion of each **testis** consists of a thick, white capsule called the **tunica albuginea** (al-bu-jin′e-ah; white). Connective tissue of the tunica albuginea enters the inferior part of the testis as incomplete **septa** (Figure 28-2, *A*). The septa divide each testis into approximately 300 to 400 cone-shaped **lobules.** The substance of the testis between the septa consists of two types of tissue—**seminiferous** (sem′ĭ-nif′er-us; seed carriers) **tubules** in which sperm cells develop and a loose connective tissue stroma that surrounds the tubules and contains clusters of endocrine cells called **interstitial cells,** or **cells of Leydig,** which secrete testosterone.

The combined length of the seminiferous tubules in both testes is nearly ½ mile. The seminiferous tubules empty into a set of short, straight tubules, which in turn empty into a tubular network called the **rete** (re′te; net) **testis.** The rete testis empties into 15 to 20 tubules called **efferent ductules.** They have a ciliated pseudostratified columnar epithelium that helps move sperm out of the testis. The efferent ductules pierce the tunica albuginea to exit the testis.

Descent of the testes

The testes develop as retroperitoneal organs in the abdominopelvic cavity and are connected to the scrotum by the **gubernaculum** (gu′ber-nak′u-lum), a fibromuscular cord (Figure 28-3, *A*; see Chapter 29). They move from the abdominal cavity through the **inguinal** (ing′gwi-nal) **canal** (Figure 28-3, *B*) to the scrotum (Figure 28-3, *C*). As the testes move into the scrotum, they are preceded by outpocketings of the peritoneum called the **process vaginalis** (vaj′ĭ-nă-lis). The superior portion of the process vaginalis usually becomes obliterated, and the inferior portion of the process vaginalis remains as a small, closed sac, the **tunica vaginalis,** which covers most of the testes.

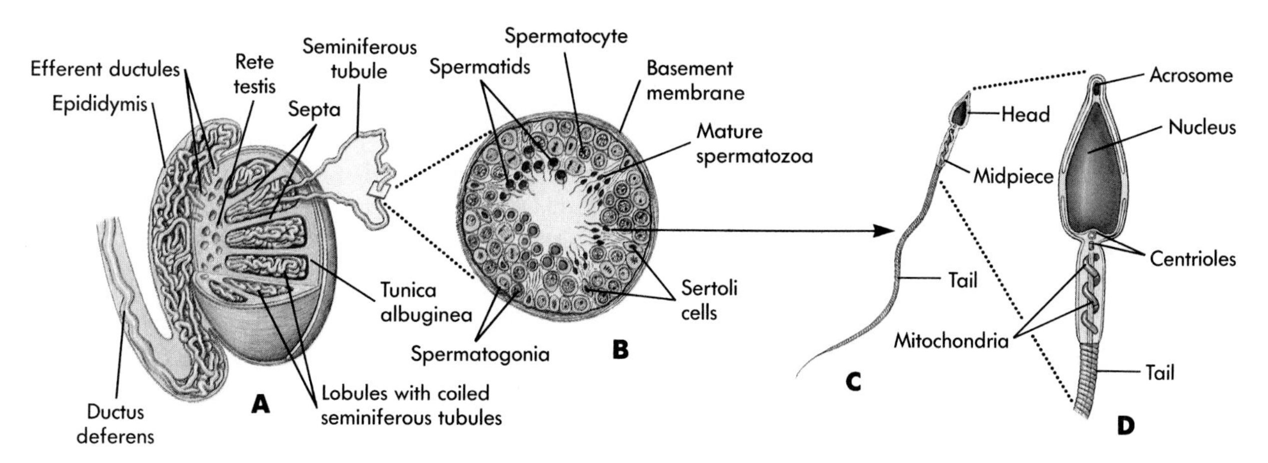

FIGURE 28-2 Histology of the testis. **A** Gross anatomy of the testis with a section cut away to reveal internal structures. **B** Cross section of a seminiferous tubule. Spermatogonia are near the periphery, and mature sperm cells are near the lumen of the seminiferous tubule. **C** Mature sperm cell. **D** Head of a mature sperm cell.

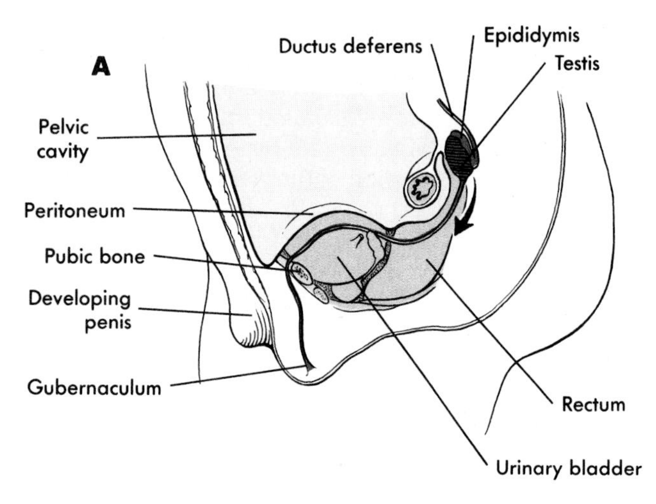

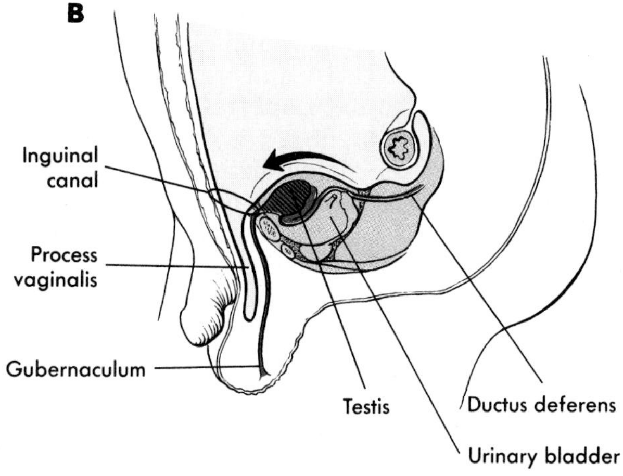

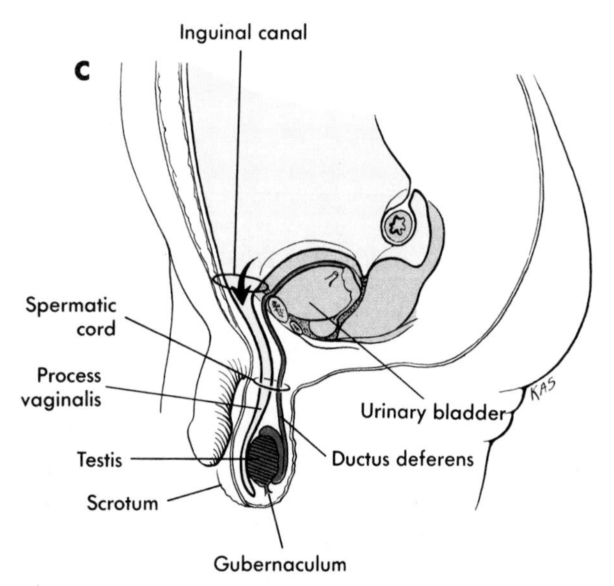

FIGURE 28-3 Descent of the testes. **A** Testes form as retroperitoneal structures near the level of the kidney. **B** Testes descend through the inguinal canals. **C** Testes descend into the scrotum. The testes are guided into the scrotum by a connective tissue strand called the gubernaculum and are preceded by an evagination of the peritoneum, the process vaginalis.

Normally the inguinal canal is closed, but it does represent a weak spot in the abdominal wall. If the inguinal canal weakens or ruptures, an **inguinal hernia** can result, and a loop of intestine can protrude into or even pass through the inguinal canal. This herniation can be quite painful and even very dangerous, especially if the inguinal canal compresses the intestine and cuts off its blood supply. Fortunately, inguinal hernias can be repaired surgically. Males are much more prone to inguinal hernias than females, apparently because a male's inguinal canal is weakened as the testis passes through it on its way into the scrotum.

Spermatogenesis

Before puberty the testes remain relatively simple and unchanged from the time of their initial development. The interstitial cells are not particularly prominent during this period, and the seminiferous tubules remain without a lumen and are not yet functional. At the time of puberty (12 to 14 years of age) the interstitial cells increase in number and size, a lumen develops in each seminiferous tubule, and sperm cell production begins.

A cross section of a mature seminiferous tubule reveals the various stages of sperm cell production, a process called **spermatogenesis** (sper'mă-to-jen'ĕ-sis; Figures 28-2, *B* and 28-4). The seminiferous tubules contain two types of cells, **germ cells** and **Sertoli** (ser-to'le; named for an Italian histologist) **cells.** Sertoli cells are also sometimes referred to as **sustentacular** (sus'ten-tak'u-lar) **cells** or **nurse cells.**

Sertoli cells are large cells that extend from the periphery to the lumen of the seminiferous tubule. They nourish the germ cells and probably produce, in cooperation with the Leydig cells, a number of hormones such as androgens, estrogens, and inhibins. In addition, the Sertoli cells join together to form a **blood-testes barrier,** which isolates the sperm cells from the immune system (see Figure 28-4). This barrier is necessary because as the sperm cells develop, they form surface antigens that could stimulate an immune response, resulting in destruction of the sperm cells.

Testosterone, produced by the cells of Leydig, passes into the Sertoli cells and binds to receptors. The combination of testosterone with the receptors is required for the Sertoli cells to function normally. Also testosterone is converted to two other steroids in the Sertoli cells: dihydrotestosterone and estradiol. The Sertoli cells also secrete a protein called **androgen-binding protein** into the seminiferous tubule. Testosterone and dihydrotestosterone bind to the androgen-binding protein and are carried along with other secretions of the seminiferous tubule to the epididymis. The estrogen and dihydrotestoster-

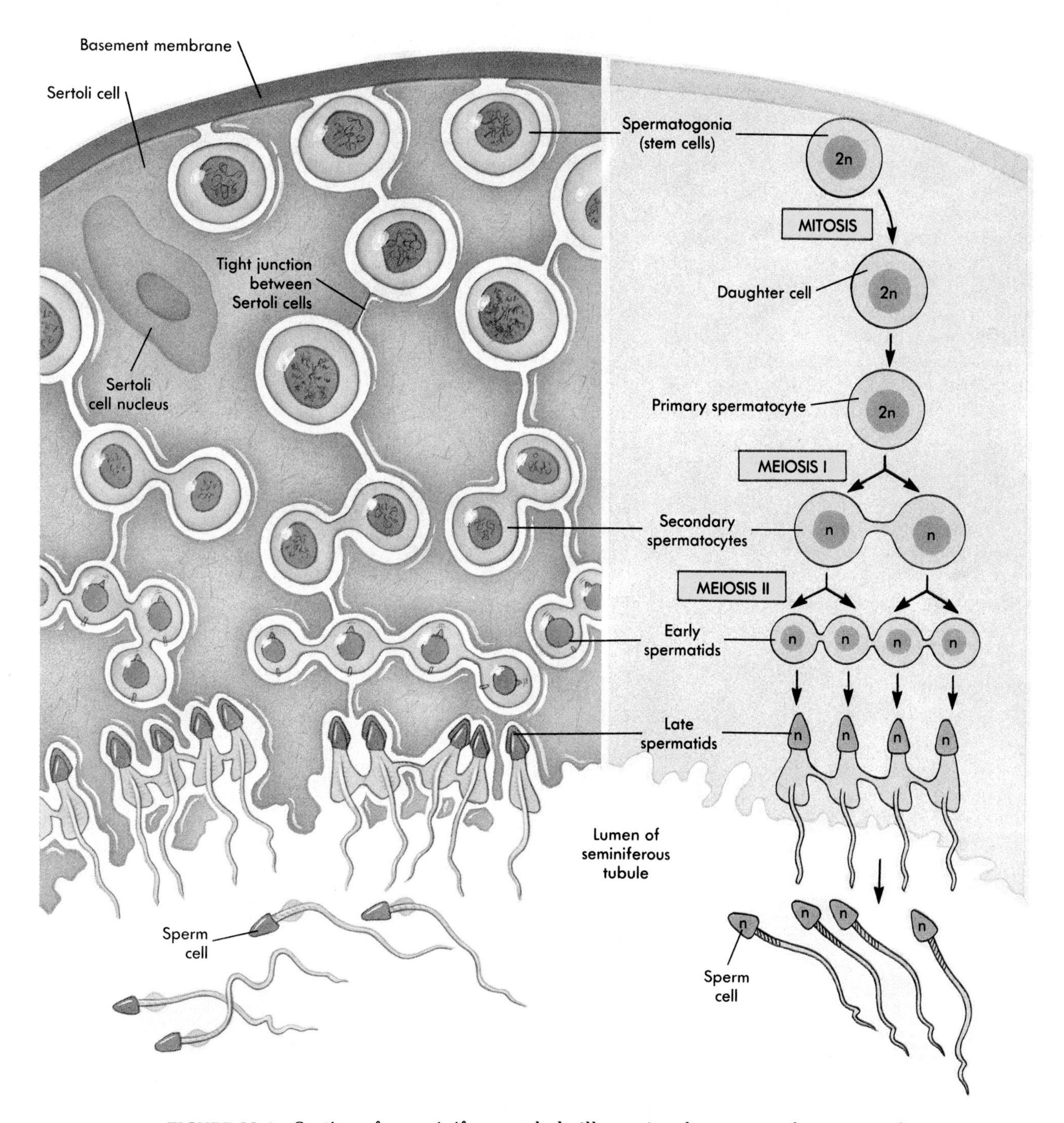

FIGURE 28-4 Section of a seminiferous tubule illustrating the process of meiosis and spermatogenesis.

one may be the active hormones that influence spermatogenesis.

Scattered between the Sertoli cells are the smaller germ cells from which sperm cells are derived. The germ cells are arranged according to maturity from the periphery to the lumen of the seminiferous tubules. The most peripheral cells, adjacent to the basement membrane of the seminiferous tubules, are **spermatogonia** (sper'mă-to-go'ne-ah), which divide by mitosis (see Figure 28-4). Some of the daughter cells produced from these mitotic divisions remain spermatogonia and continue to produce additional spermatogonia. The others divide through mitosis and differentiate to form **primary spermatocytes** (sper'mă-to-sītz).

Meiosis (see essay or Chapter 3) begins when the primary spermatocytes divide. Primary spermatocytes, which contain 23 pairs of chromosomes, pass through the first meiotic division to become **secondary spermatocytes,** which contain 23 chromosomes—one chromosome from each pair of chromosomes in the primary spermatocytes. Each secondary spermatocyte undergoes a second meiotic division to produce two even smaller cells called **spermatids** (sper'mă-tidz), which also have 23 chromosomes. Each spermatid then develops a head and flagellum (tail) to become a sperm cell or **spermatozoon** (sper'mă-to-zo'on; pl. spermatozoa) (see Figures 28-2, *C* and *D*, and 28-4). The head contains the chromosomes and at the leading end it has a cap, the **acrosome** (ak'ro-sōm), which contains enzymes necessary for penetration of the oocyte (female sex cell) by the sperm cell. The flagellum is composed of a middle piece and a tail. The flagellum is similar to a cilium (see Chapter 3), and movement of microtubules past each other causes the tail to move and propel the sperm cell forward. The middle piece has large numbers of mitochondria that produce the ATP necessary for microtubule movement.

The sperm cells gather around the lumen of the seminiferous tubules with their heads directed toward the surrounding Sertoli cells and their tails directed toward the center of the lumen (see Figures 28-2 and 28-4).

Ducts

After their production in the seminiferous tubules, sperm cells leave the testes through the efferent ductules and pass through a series of ducts to reach the exterior of the body.

Epididymis

The efferent ductules from each testis become extremely convoluted and form a comma-shaped structure on the posterior side of the testis called the **epididymis** (ep-ĭ-did'ĭ-mis; upon the twin; "twin" refers to the paired or twin testes). The final maturation of the sperm cells occurs within the ductules of the epididymis. Sperm cells taken directly from the testes in experimental animals are not capable of fertilizing ova, but after spending 1 to several days in the epididymis, the sperm cells develop the capacity to perform fertilization.

The epididymis consists of a head, a body, and a long tail (see Figure 28-2, *A*; Figure 28-5). The head

FIGURE 28-5 Testes, epididymis, ductus deferens, and glands of the male reproductive system.

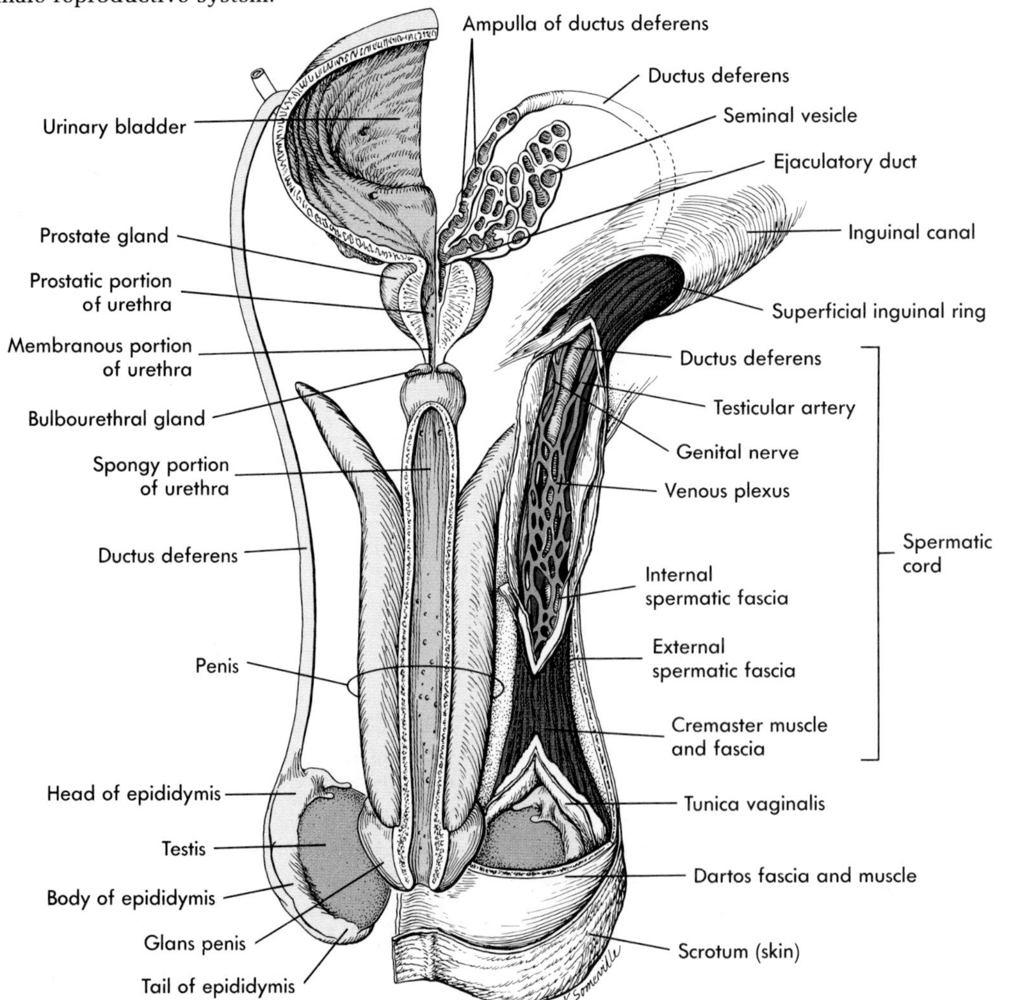

contains the convoluted efferent ductules, which empty into a single convoluted ductule, the **ductus epididymis,** located primarily within the body of the epididymis. This ductule alone, if unraveled, would extend for several meters. The epididymis has a pseudostratified columnar epithelium with microvilli called **stereocilia** (stĕr′e-o-sil′e-ah). The stereocilia function to increase the surface area of epithelial cells that absorb fluid from the ductus epididymis. The ductus epididymis ends at the tail of the epididymis, which is located at the inferior border of the testis.

Ductus deferens

The **ductus deferens,** or **vas deferens,** emerges (see Figures 28-1 and 28-5) from the tail of the epididymis and ascends along the posterior side of the testis medial to the epididymis and becomes associated with the blood vessels and nerves that supply the testis. These structures and their coverings constitute the **spermatic cord.** The spermatic cord consists of (1) the ductus deferens, (2) the testicular artery and venous plexus, (3) lymph vessels, (4) nerves, (5) fibrous remnants of the process vaginalis, and (6) three coats: the **external spermatic fascia;** the **cremaster muscle,** an extension of the muscle fibers of the internal oblique muscle of the abdomen; and the **internal spermatic fascia** (see Figure 28-5).

The spermatic cord passes obliquely through the inferior abdominal wall by way of the inguinal canal. The superficial opening of the inguinal canal, called the superficial inguinal ring, is medial, whereas the deep opening, the deep inguinal ring, is lateral.

The ductus deferens and the rest of the spermatic cord structures ascend and pass through the inguinal canal to enter the abdominal cavity (see Figures 28-1 and 28-5). The ductus deferens crosses the lateral wall of the cavity, travels over the ureter, and loops over the posterior surface of the urinary bladder to approach the prostate gland. The end of the ductus deferens enlarges to form the **ampulla.** The ductus deferens has a pseudostratified columnar epithelium and is surrounded by smooth muscle. Peristaltic contractions of these smooth muscles help propel the sperm cells through the ductus deferens.

Ejaculatory duct

Adjacent to the ampulla of each ductus deferens is a sac-shaped gland, which is called the **seminal vesicle.** A short duct from the seminal vesicle joins the ductus deferens to form the **ejaculatory duct.** The ejaculatory ducts are approximately 2.5 cm long. They project into the prostate gland, and end by opening into the urethra (see Figures 28-1 and 28-5).

Urethra

The male **urethra** (u-re′thrah) is approximately 20 cm long and extends from the urinary bladder to the distal end of the penis (see Figures 28-1 and 28-5; Figure 28-6). The urethra is a passageway for both urine and male reproductive fluids. The urethra can be divided into three portions: the prostatic portion, the membranous portion, and the spongy portion. The **prostatic portion** is closest to the bladder and passes through the prostate gland. Ducts from the prostate gland and the ejaculatory ducts empty into the prostatic urethra. The **membranous portion** of the urethra is the shortest and extends from the prostatic urethra through the urogenital diaphragm, which is part of the muscular floor of the pelvis. The **spongy portion,** by far the longest, extends from the membranous urethra through the length of the penis. Most of the urethra is lined by stratified columnar epithelium, but transitional epithelium is in the prostatic urethra near the bladder, and stratified squamous epithelium is near the opening of the spongy urethra. Several minute mucus-secreting **urethral glands** empty into the urethra.

Penis

The **penis** consists of three columns of erectile tissue (see Figure 28-6), and engorgement of this erectile tissue with blood causes the penis to enlarge and become firm, a process called **erection.** The penis is the male organ of copulation and functions in the transfer of sperm cells from the male to the female. Two of the erectile columns form the dorsum and sides of the penis and are called the **corpora cavernosa.** The **corpus spongiosum** expands to form a cap, the **glans penis,** over the distal end of the penis. The spongy urethra passes through the corpus spongiosum, penetrates the glans penis, and opens as the **external urethral orifice.** At the base of the penis the corpus spongiosum expands to form the **bulb of the penis,** and each corpus cavernosum expands to form a **crus of the penis.** Together these structures constitute the **root of the penis** and attach the penis to the coxae.

The shaft of the penis is covered by skin that is loosely attached to the connective tissue surrounding the penis. The skin is firmly attached at the base of the glans penis, and a thinner layer of skin tightly covers the glans penis. The skin of the penis, especially the glans penis, is well supplied with sensory receptors. A loose fold of skin called the **prepuce** (pre′pūs), or **foreskin,** covers the glans penis. **Circumcision** is accomplished by surgically removing the prepuce.

The primary nerves, arteries, and veins of the penis pass along its dorsal surface (see Figure 28-6). A single, midline dorsal vein lies in the middle,

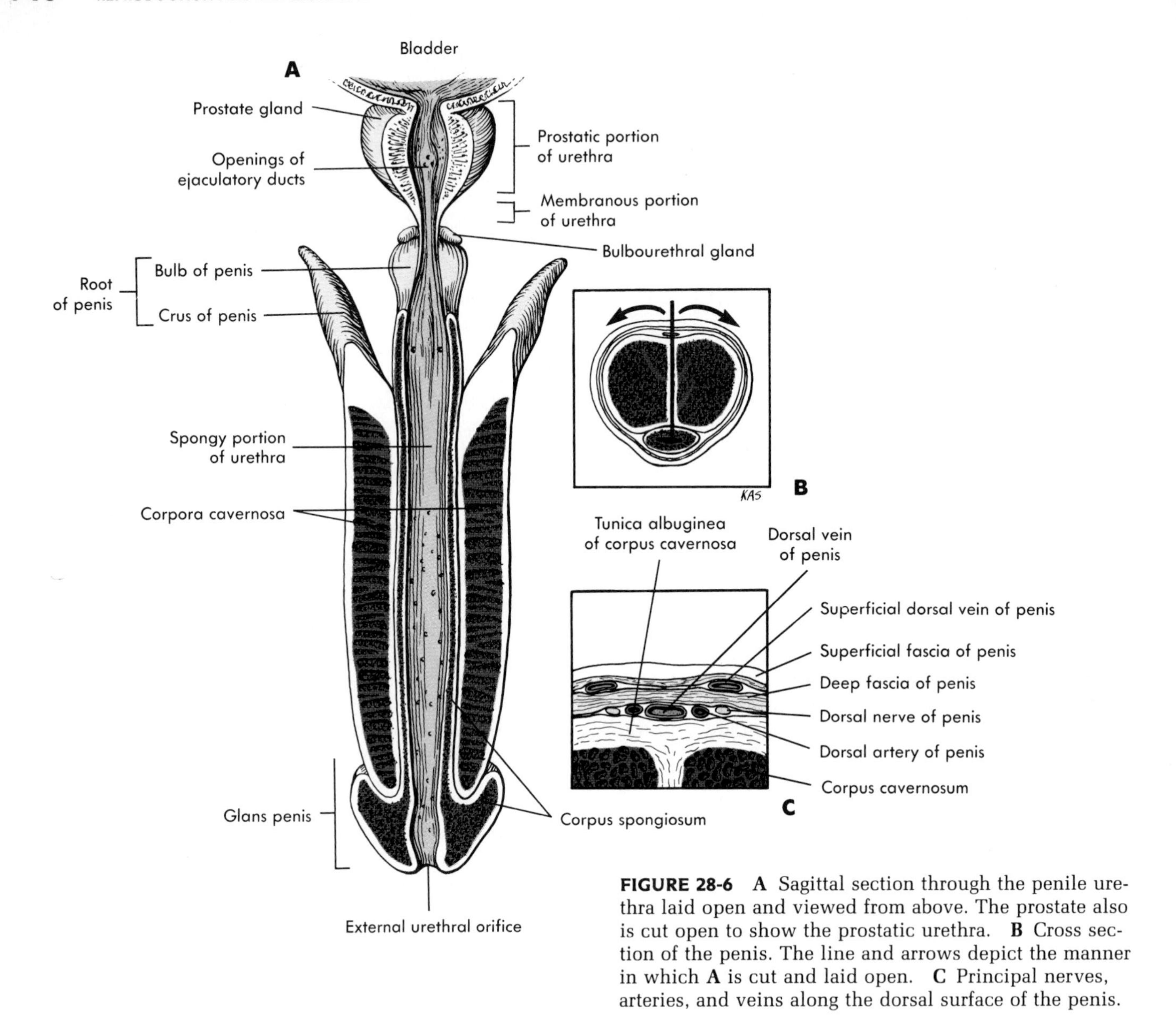

FIGURE 28-6 **A** Sagittal section through the penile urethra laid open and viewed from above. The prostate also is cut open to show the prostatic urethra. **B** Cross section of the penis. The line and arrows depict the manner in which **A** is cut and laid open. **C** Principal nerves, arteries, and veins along the dorsal surface of the penis.

Male Infertility

The most common cause of infertility in men is a low sperm cell count. Normal sperm cell counts in the semen range from 75-400,000,000 sperm cells. A normal ejaculation usually consists of approximately 2-5 ml of semen. Most of the sperm cells (millions) are expended in moving the general group of sperm cells through the female reproductive system. Enzymes carried in the acrosomal cap of each sperm cell help to digest a path through the mucoid fluids of the female reproductive tract, the cells surrounding the oocyte (cumulus cells), and the membrane that surrounds the oocyte (the zona pellucida). Once the acrosomal fluids are depleted, the sperm cell is no longer capable of fertilization. If the sperm cell count drops to 20,000,000 sperm cells per milliliter, the male is usually sterile.

Decreased sperm cell count can occur because of damage to the testes (e.g., because of mumps, radiation, or trauma), obstruction of the duct system, or inadequate hormone production. Fertility can sometimes be achieved by collecting several ejaculations, concentrating the sperm cells, and inserting the sperm cells into the female's reproductive tract (artificial insemination).

flanked on each side by dorsal arteries, with dorsal nerves lateral to them. Additional, deep arteries lie within the corpora cavernosa.

Accessory Glands

Seminal vesicles

The **seminal vesicles** are sac-shaped glands located next to the ampullae of the ductus deferentia (see Figure 28-4). Each gland is approximately 5 cm long and tapers into a short duct that joins the ductus deferens to form the ejaculatory duct.

Prostate gland

The **prostate** (pros′tāt; one standing before) **gland** consists of both glandular and muscular tissue and is near the size and shape of a walnut (i.e., approximately 4 cm long and 2 cm wide). The prostate gland is dorsal to the symphysis pubis at the base of the bladder where it surrounds the prostatic urethra and the two ejaculatory ducts (see Figure 28-1). The gland is composed of an indistinct smooth muscle capsule and numerous smooth muscle partitions that radiate inward toward the urethra. Covering these muscular partitions is a layer of columnar epithelial cells that form saccular dilations into which the cells secrete prostatic fluid. Twenty to 30 small prostatic ducts transport these secretions into the prostatic urethra.

1

The prostate gland can enlarge for several reasons, including infections, tumor, and old age. Cancer of the prostate is the second most common cause of male death from cancer in the United States (fewer deaths than from lung cancer and more than from colon cancer). The detection of enlargement or changes in the prostate is important. Suggest a way, without requiring surgery, that the prostate gland can be examined by palpation for any abnormal changes.

? ? ? ? ? ? ? ? ? ? ?

Bulbourethral glands

The **bulbourethral glands** are a pair of small glands located near the membranous portion of the urethra (see Figures 28-1 and 28-5). In young adults each is approximately the size of a pea, but they decrease in size with age and are almost impossible to see in old men. Each gland is a compound mucous gland (see Chapter 4). The small ducts of each gland unite to form a single duct. The duct from each gland then enters the spongy urethra at the base of the penis.

Secretions

Semen is a composite of sperm cells and secretions from the male reproductive glands. The seminal vesicles produce approximately 60% of the fluid, the prostate gland contributes approximately 30%, the testes contribute 5%, and the bulbourethral glands contribute 5%. **Emission** is the discharge of semen into the prostatic urethra. **Ejaculation** is the forceful expulsion of semen from the urethra caused by the contraction of the urethra, skeletal muscles in the floor of the pelvis, and muscles at the base of the penis.

The bulbourethral glands and the urethral mucous glands produce a mucous secretion some time (up to several minutes) before ejaculation. This mucus lubricates the urethra, neutralizes the contents of the normally acidic spongy urethra, provides a small amount of lubrication during intercourse, and helps to reduce acidity in the vagina.

Testicular secretions include sperm cells, a small amount of fluid, and metabolic by-products. The thick, mucoid secretions of the seminal vesicle contain large amounts of fructose and other nutrients that nourish the sperm cells. The seminal vesicle secretions also contain fibrinogen, which is involved in a weak coagulation reaction of the semen after ejaculation, and prostaglandins, which may cause uterine contractions.

The thin, milky secretions of the prostate have a rather high pH and, with secretions of the seminal vesicles, help to neutralize the acidic urethra, the acidic secretions of the testes, and the vagina. The prostatic secretions are also important in the transient coagulation of semen because they contain clotting factors that convert fibrinogen from the seminal vesicles to fibrin, resulting in coagulation. The coagulated material keeps the semen as a single, sticky mass for a few minutes after ejaculation; then fibrinolysin from the prostate causes the coagulum to dissolve, releasing the sperm cells to make their way up the female reproductive tract as somewhat free, motile cells.

Before ejaculation the ductus deferens begins to contract rhythmically, propelling sperm cells and testicular fluid from the tail of the epididymis to the ampulla of the ductus deferens. Contractions of the ampullae, seminal vesicles, and ejaculatory ducts cause the sperm, testicular secretions, and seminal fluid to move into the prostatic urethra where they mix with prostatic secretions released as a result of contractions of the prostate gland.

2

Explain a possible reason for having the coagulation reaction.

? ? ? ? ? ? ? ? ? ? ?

PHYSIOLOGY OF MALE REPRODUCTION

The male reproductive system depends on both hormonal and neural mechanisms to function normally. Hormones are primarily responsible for the following: development of reproductive structures and maintenance of their functional capacities; development of secondary sexual characteristics; control of spermatogenesis; and influence of sexual behavior. Neural mechanisms are primarily involved in controlling the sexual act and in the expression of sexual behavior.

Regulation of Sex Hormone Secretion

Hormonal mechanisms that influence the male reproductive system involve the hypothalamus, the pituitary gland, and the testes (Figure 28-7). A small peptide hormone called **gonadotropin-releasing hormone (GnRH)** or **luteinizing hormone-releasing hormone (LH-RH)** is released from neurons in the median eminence of the hypothalamus. GnRH passes through the hypothalamohypophyseal portal system to the adenohypophysis (see Chapter 18). In response to GnRH, cells within the adenohypophysis secrete two hormones referred to as **gonadotropins** because they influence the function of the **gonads** (the testes or ovaries).

The two gonadotropins are **luteinizing hormone (LH)** and **follicle-stimulating hormone (FSH);** although named for their functions in females, they also have important functions in males. LH in males is sometimes called **interstitial cell-stimulating hormone (ICSH).** LH binds to the cells of Leydig in the testes and causes the cells to increase their rate of testosterone synthesis and secretion. FSH binds primarily to Sertoli cells in the seminiferous tubules and promotes spermatogenesis. Both gonadotropins bind to specific receptor molecules on the membranes of the cells that they influence, and cyclic AMP may be an important second messenger in those cells.

For GnRH to stimulate large quantities of LH and FSH release, the adenohypophysis must be exposed to a series of brief increases and decreases in GnRH. Chronically elevated GnRH levels in the blood cause the adenohypophyseal cells to become insensitive to stimulation by GnRH molecules.

Testosterone is the major male hormone secreted by the testes. It is classified as an **androgen** (*andro* means male) because it encourages the development of male secondary sexual characteristics and it stimulates the accessory sex organs. Other androgens are secreted by the testes, but they are produced in smaller concentrations and are less potent than testosterone.

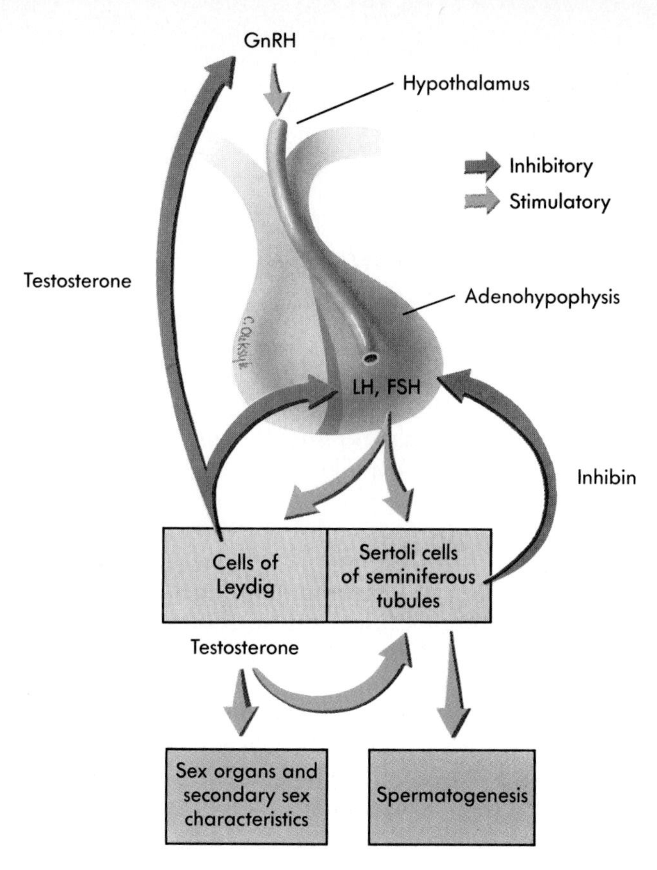

FIGURE 28-7 Regulation of reproductive hormone secretion in males. Gonadotropin-releasing hormone *(GnRH)* from the hypothalamus stimulates the secretion of luteinizing hormone *(LH)* and follicle-stimulating hormone *(FSH)* from the adenohypophysis. These hormones stimulate spermatogenesis, secretion of testosterone, and secretion of inhibin in the testes. Testosterone has a negative-feedback effect on the hypothalamus and pituitary to reduce LH and FSH secretion, whereas inhibin specifically inhibits FSH secretion. Testosterone has a stimulatory effect on the sex organs, secondary sex characteristics, and the Sertoli cells.

GnRH can be produced synthetically and is useful in treating people who are infertile if it is administered in small amounts in frequent pulses or surges. GnRH can also inhibit reproduction since chronic administration of GnRH can sufficiently reduce LH and FSH levels to prevent sperm production in males or ovulation in females.

Testosterone has a major influence on many tissues. It plays an essential role in the embryonic development of reproductive structures (see Chapter 29), in the further development of reproductive structures and secondary sex characteristics during puberty, in the maintenance of spermatogenesis and the functional integrity of accessory sexual organs, and in the regulation of gonadotropin secretion.

Another hormone, **inhibin,** is a polypeptide and is released from the testes. It inhibits FSH secretion from the adenohypophysis.

Puberty

A gonadotropin-like hormone called **human chorionic gonadotropin (HCG),** which is secreted by the maternal placenta, stimulates the synthesis and secretion of testosterone by the fetal testes before birth. After birth, however, no source of stimulation is present, and the testes of the newborn baby atrophy slightly and secrete only small amounts of testosterone until puberty, which normally begins when a boy is 12 to 14 years old.

Puberty is the age at which individuals become capable of sexual reproduction. Before puberty small amounts of testosterone and other androgens inhibit GnRH release from the hypothalamus. At puberty the hypothalamus becomes much less sensitive to the inhibitory effect of androgens, and the rate of GnRH secretion increases, leading to increased LH and FSH release. Elevated FSH levels promote spermatogenesis, and elevated LH levels cause the interstitial cells of Leydig to secrete larger amounts of testosterone. Testosterone still has a negative-feedback effect on GnRH secretion after puberty but is not capable of completely suppressing it.

Effects of Testosterone

Testosterone is by far the major androgen in males. Nearly all of the androgens, including testosterone, are produced by the cells of Leydig, with small amounts produced by the adrenal cortex and, possibly, by the Sertoli cells. Testosterone causes the enlargement and differentiation of the male genitals and reproductive duct system, is necessary for spermatogenesis, and is required for the descent of the testes near the end of fetal development. Testosterone stimulates hair growth in the following regions: (1) the pubic area and extending up the linea alba, (2) the legs, (3) the chest, (4) the axillary region, (5) the face, and (6) occasionally, the back.

Some men have a genetic tendency called male pattern baldness. When testosterone levels increase at puberty, the density of hair on the top of the head begins to decrease. Baldness usually reaches its maximum rate of development when the individual is in the third or fourth decade of life.

Testosterone also causes the texture of the skin and hair to become rougher or coarser. The quantity of melanin in the skin also increases, making the skin darker. Testosterone increases the rate of secretion from the sebaceous glands, especially in the region of the face, frequently resulting near the time of puberty in the development of acne. Beginning near the time of puberty, testosterone causes hypertrophy of the larynx. The structural changes may first result in a voice that is difficult to control, but ultimately the voice reaches its normal masculine quality.

Testosterone has a general stimulatory effect on metabolism so that males have a slightly higher metabolic rate than females. The red blood cell count is increased by nearly 20% as a result of the effects of testosterone on erythropoietin. Testosterone also has a minor mineralocorticoid-like effect, causing the retention of sodium in the body and, consequently, an increase in the volume of body fluids. Testosterone promotes protein synthesis in most tissues of the body; as a result, skeletal muscle mass increases at puberty. The average percentage of the body weight composed of skeletal muscle is greater for men than for women because of the effect of androgens.

Some athletes, especially weight lifters, ingest synthetic androgens in an attempt to increase muscle mass. The side effects of the large doses of androgens are often substantial and include testicular atrophy and kidney and liver damage. Administration of synthetic androgens is highly discouraged by the medical profession and is a violation of the rules of most athletic organizations.

Testosterone causes rapid bone growth and increases the deposition of calcium in bone, resulting in an increase in height. The growth in height is limited, however, because testosterone also causes early closure of the epiphyseal plates of long bones (see Chapter 6). Males who mature sexually at an earlier age grow rapidly and reach their maximum height earlier. Males who mature sexually at a later age do not exhibit a rapid period of growth, but they grow for a longer period of time and may become taller than men who mature sexually at an earlier age.

Male Sexual Behavior and the Male Sex Act

Testosterone is required to initiate and maintain normal male sexual behavior. Testosterone enters cells within the hypothalamus and the surrounding areas of the brain and influences the function of these cells, resulting in normal sexual behavior. However, male sexual behavior may depend, in part, on the conversion of testosterone to other substances in the cells of the brain.

The blood levels of testosterone remain relatively constant throughout the lifetime of a male from puberty until approximately 40 years of age. Thereafter, the levels slowly decline to approximately 20% of this value by 80 years of age, causing a slow decrease in sex drive and fertility.

The male sexual act is a complex series of reflexes that result in erection of the penis, secretion of mucus into the urethra, emission, and ejaculation. Sensations that are normally interpreted as pleasurable occur during the male sexual act and result in a climactic sensation, **orgasm,** associated with ejaculation. After ejaculation a phase called **resolution** occurs in which the penis becomes flaccid, an overall feeling of satisfaction exists, and the male is unable to achieve erection and a second ejaculation for many minutes to a few hours.

Afferent impulses and integration

Afferent impulses from the genitals are propagated through the pudendal nerve to the sacral region of the spinal cord where reflexes that result in the male sexual act are integrated. Impulses travel from the spinal cord to the cerebrum to produce the conscious sexual sensations.

Rhythmic massage of the penis, especially the glans, provides an extremely important source of afferent impulses that are required to initiate erection and ejaculation. Sensory impulses produced in surrounding tissues such as the scrotum and the anal, perineal, and pubic regions reinforce sexual sensations. Engorgement of the prostate and seminal vesicles with secretions and irritation of the urethra, urinary bladder, ductus deferens, and testes can also cause sexual sensations.

Although no drug treatment is consistently effective as an aphrodisiac (a substance that increases sexual excitement), substances that cause mild irritation of the urethra and urinary bladder may stimulate sexual sensation.

Psychic stimuli (e.g., sight, sound, odor, or thoughts) have a major effect on sexual reflexes. Thinking sexual thoughts or dreaming about erotic events tends to reinforce stimuli that trigger sexual reflexes such as erection and ejaculation. Ejaculation while sleeping (nocturnal emission) is a relatively common event in young males and is thought to be triggered by psychic stimuli associated with dreaming. Psychic stimuli can also inhibit the sexual act, and thoughts that are not sexual in nature tend to decrease the effectiveness of the male sexual act. The inability to concentrate on sexual sensations results in **impotence** (the inability to accomplish the male sexual act). Impotence can also be caused by physical factors such as inability of the erectile tissue to fill with blood.

Impulses from the cerebrum that reinforce the sacral reflexes are not absolutely required for the culmination of the male sexual act, and the male sexual act can occasionally be accomplished by males that have suffered spinal cord injuries superior to the sacral region.

Erection, emission, and ejaculation

When **erection** occurs the penis becomes enlarged and rigid. Erection is the first major component of the male sexual act. Nerve impulses from the spinal cord cause the arteries that supply blood to the erectile tissues to dilate. As a consequence, blood fills the sinusoids of the erectile tissue and compresses the veins. Since venous outflow is partially occluded, the blood pressure in the sinusoids causes inflation and rigidity of the erectile tissue. Nerve impulses that result in erection can come from parasympathetic centers (S2 to S4) or sympathetic centers (T2 to L1) in the spinal cord. Normally the parasympathetic centers are more important, but in cases of damage to the sacral region of the spinal cord it is possible for erection to occur through the sympathetic system.

Parasympathetic impulses also cause the mucous glands within the penile urethra and the bulbourethral glands at the base of the penis to secrete mucus.

Emission is the accumulation of sperm cells and secretions of the prostate gland and seminal vesicles in the urethra. Emission is controlled by sympathetic centers (T12 to L1) in the spinal cord, which are stimulated as the level of sexual tension increases. Efferent sympathetic impulses cause peristaltic contractions of the reproductive ducts and stimulate the seminal vesicles and the prostate gland to release their secretions. Consequently, sperm cells and secretions **(semen)** accumulate in the prostatic urethra, producing afferent impulses that pass through the pudendal nerves to the spinal cord. Integration of these impulses results in both sympathetic and somatic output. Efferent sympathetic impulses cause constriction of the internal sphincter of the urinary bladder so that semen and urine are not mixed. Efferent somatic impulses are sent to the skeletal muscles of the urogenital diaphragm and the base of the penis, causing several rhythmic contractions that force the semen out of the urethra. The movement of semen out of the urethra is called **ejaculation.** In addition, there is an increase in muscle tension throughout the body.

FEMALE REPRODUCTIVE SYSTEM

The female reproductive organs consist of the ovaries, uterine tubes, uterus, vagina, external genital organs, and mammary glands. The internal reproductive organs of the female (Figures 28-8 and 28-9) are within the pelvis between the urinary bladder and the rectum. The **uterus** (u'ter-us) and the **vagina** (vă-ji'nah) are in the midline, with the **ovaries** to each side of the uterus. The internal reproductive organs are held in place within the pelvis by a group

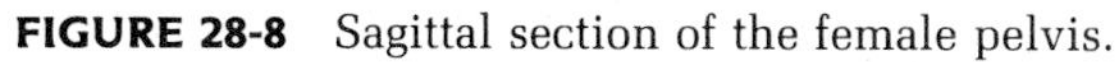

FIGURE 28-8 Sagittal section of the female pelvis.

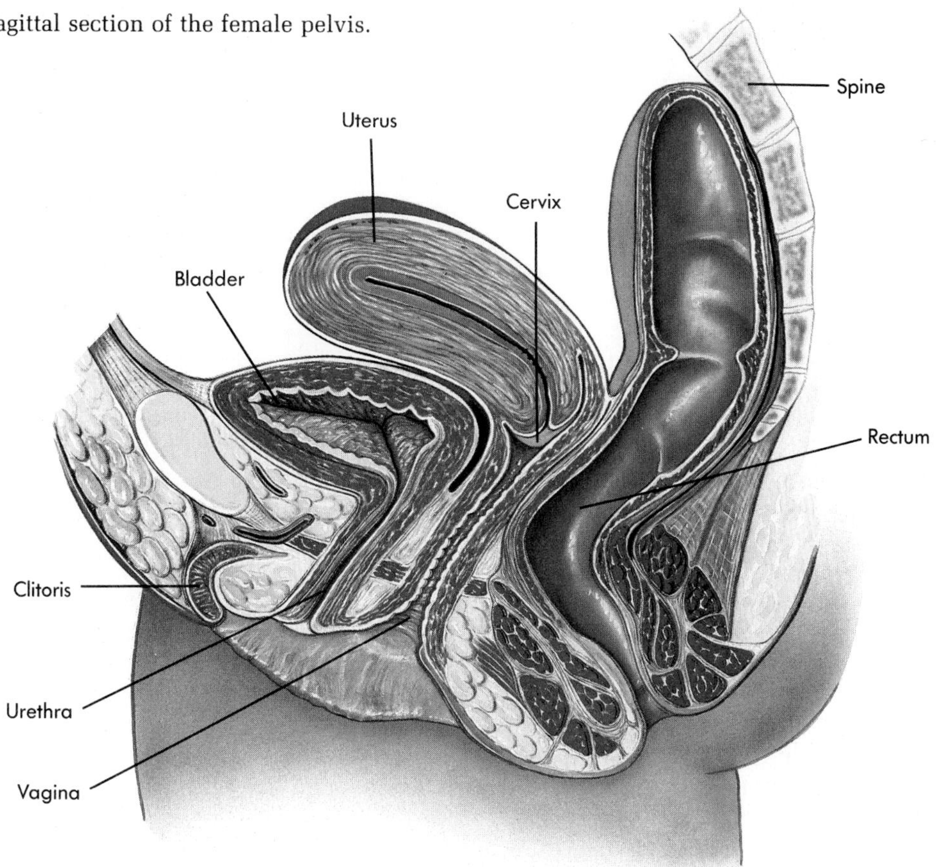

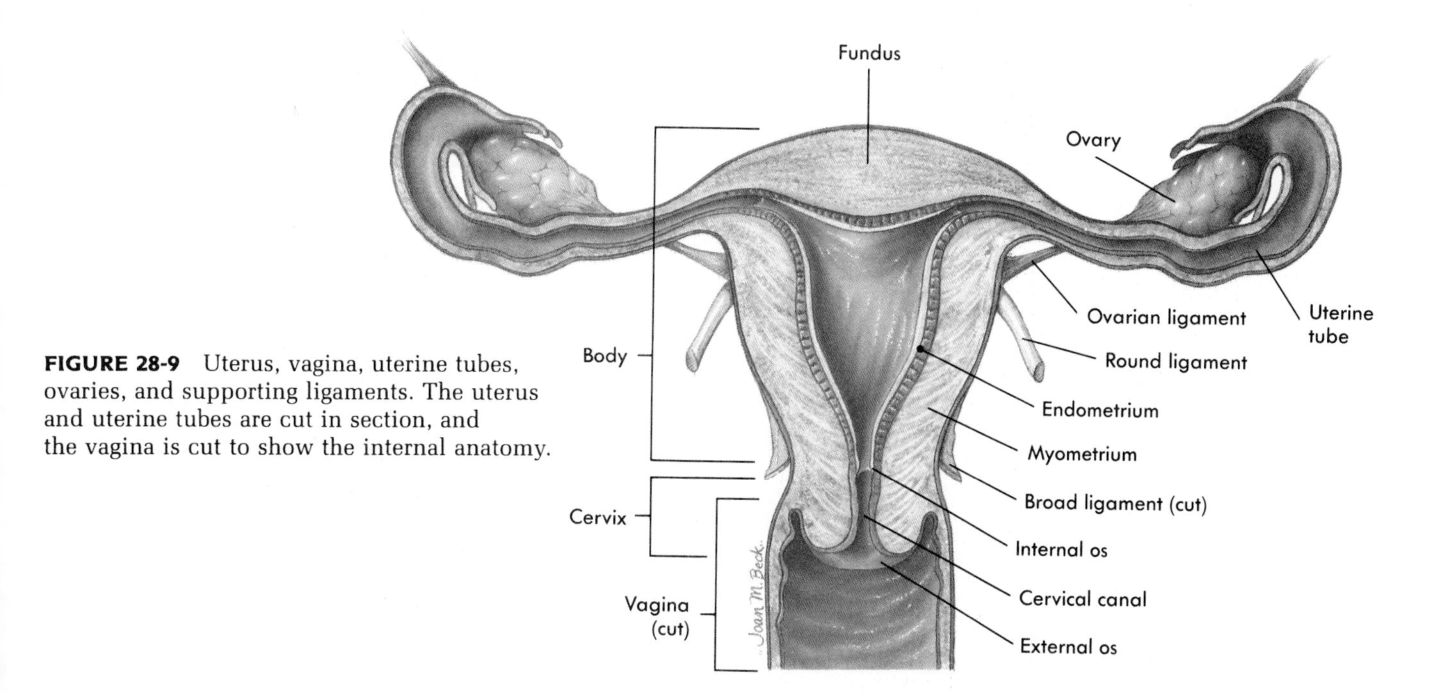

FIGURE 28-9 Uterus, vagina, uterine tubes, ovaries, and supporting ligaments. The uterus and uterine tubes are cut in section, and the vagina is cut to show the internal anatomy.

of ligaments. The most conspicuous is the broad ligament, an extension of the peritoneum that spreads out on both sides of the uterus and to which the ovaries and uterine tubes are attached.

Ovaries

The two ovaries are small organs approximately 2 to 3.5 cm long and 1 to 1.5 cm wide (see Figure 28-9). Each is attached to the posterior surface of the broad ligament by a peritoneal fold called the **mesovarium** (mes′o-vă′rĭ-um; mesentery of the ovary). Two other ligaments are associated with the ovary: the **suspensory ligament,** which extends from the mesovarium to the body wall, and the **ovarian ligament,** which attaches the ovary to the superior margin of the uterus. The ovarian arteries, veins, and nerves traverse the suspensory ligament and enter the ovary through the mesovarium.

Ovarian histology

The peritoneum covering the surface of the ovary is called the **ovarian epithelium** or the **germinal epithelium** because it was once thought to produce oocytes. Immediately below the epithelium a layer of dense, fibrous connective tissue, the **tunica albuginea,** surrounds the ovary. The ovary itself consists of a dense outer portion called the **cortex** and a looser inner portion called the **medulla** (Figure 28-10). Blood vessels, lymph vessels, and nerves from the mesovarium enter the medulla. Numerous small vesicles called **ovarian follicles,** each of which contains an **oocyte** (o′o-sīt), are distributed throughout the cortex.

Follicle and oocyte development

Oogenesis (o-o-jen′ĕ-sis) is the production of a secondary oocyte within the ovaries. By the fourth month of prenatal life, the ovaries may contain 5 million **oogonia** (o′o-go′nĭ-ah), the cells from which oocytes develop. By the time of birth the oogonia have degenerated or have begun meiosis. However, meiosis stops at prophase I of the first meiotic division (Figure 28-11). The cell at this stage is called a **primary oocyte,** and at birth there are approximately 2 million of them. From birth to puberty the number declines to 300,000 to 400,000; of these primary oocytes, only approximately 400 continue oogenesis and are released from the ovary. The pri-

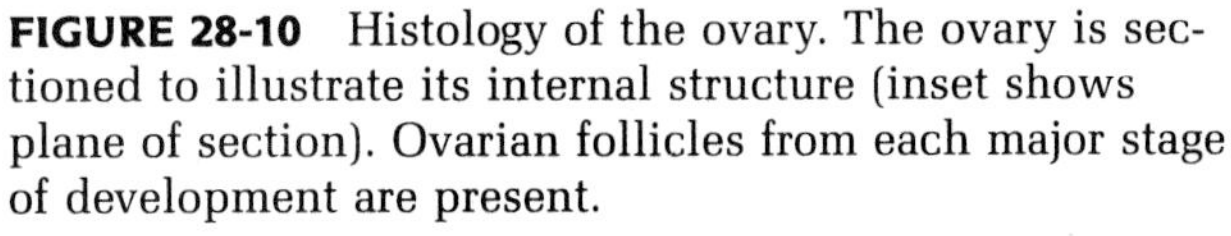

FIGURE 28-10 Histology of the ovary. The ovary is sectioned to illustrate its internal structure (inset shows plane of section). Ovarian follicles from each major stage of development are present.

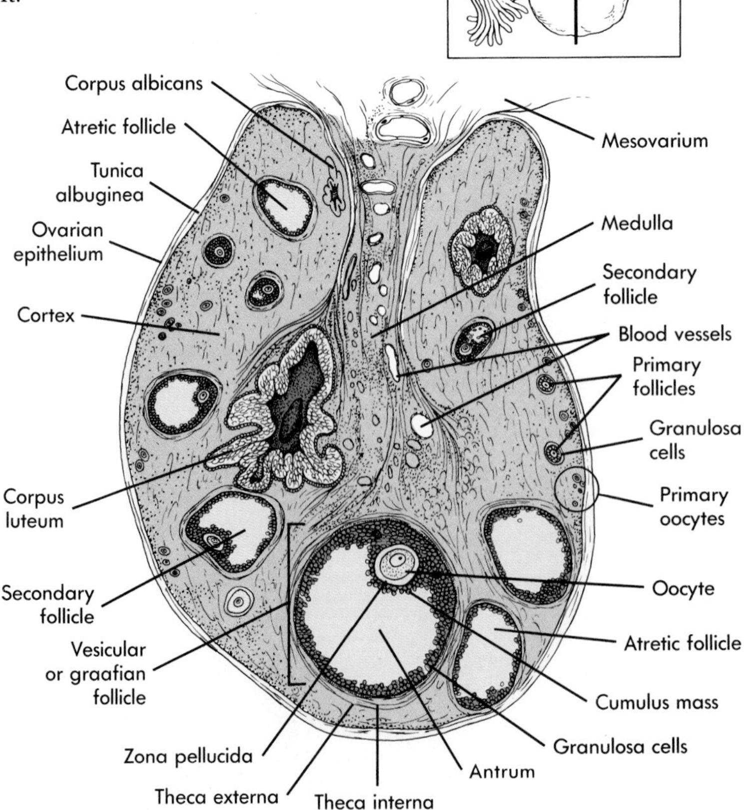

mary oocyte is surrounded by a layer of cells (the **granulosa cells**), and the entire structure is called a **primary follicle.**

Beginning during puberty, approximately every 28 days hormonal changes stimulate some of the primary follicles to continue development and become **secondary follicles.** The granulosa cells multiply and form an increasing number of layers around the oocyte. The center of the follicle becomes a chamber, the **antrum,** that is filled with fluid produced by the granulosa cells. The oocyte is pushed off to one side of the follicle and lies in a mass of follicular cells called the **cumulus mass,** or **cumulus oophorus** (o-of'or-us). The innermost cells of this mass resemble a crown radiating from the oocyte and are thus called the **corona radiata.**

The secondary follicle continues to enlarge, the antrum fills with additional fluid, and the follicle forms a lump on the surface of the ovary. This fully mature follicle is called the **vesicular,** or **graafian** (graf'ĭ-an), **follicle.** As the secondary follicle enlarges, surrounding cells are molded around it to form the **theca,** or capsule. Two layers of thecae can be recognized around the graafian follicle: the **theca interna** and the **theca externa** (see Figure 28-10). The theca interna is highly vascular, whereas the theca externa is mostly fibrous.

The primary oocyte enlarges because of an accumulation of yolk supplied to it by the follicular cells, and a layer of clear viscous fluid, the **zona pellucida** (pel-lu'cĭ-da), is deposited on the surface of the oocyte. Meiosis continues, and the first meiotic division is completed to produce a **secondary oocyte** and a **polar body** (Figure 28-12). Division of cytoplasm is unequal, and most of it is given to the secondary oocyte, whereas the polar body receives very little. The secondary oocyte begins the second meiotic division, which stops in metaphase II.

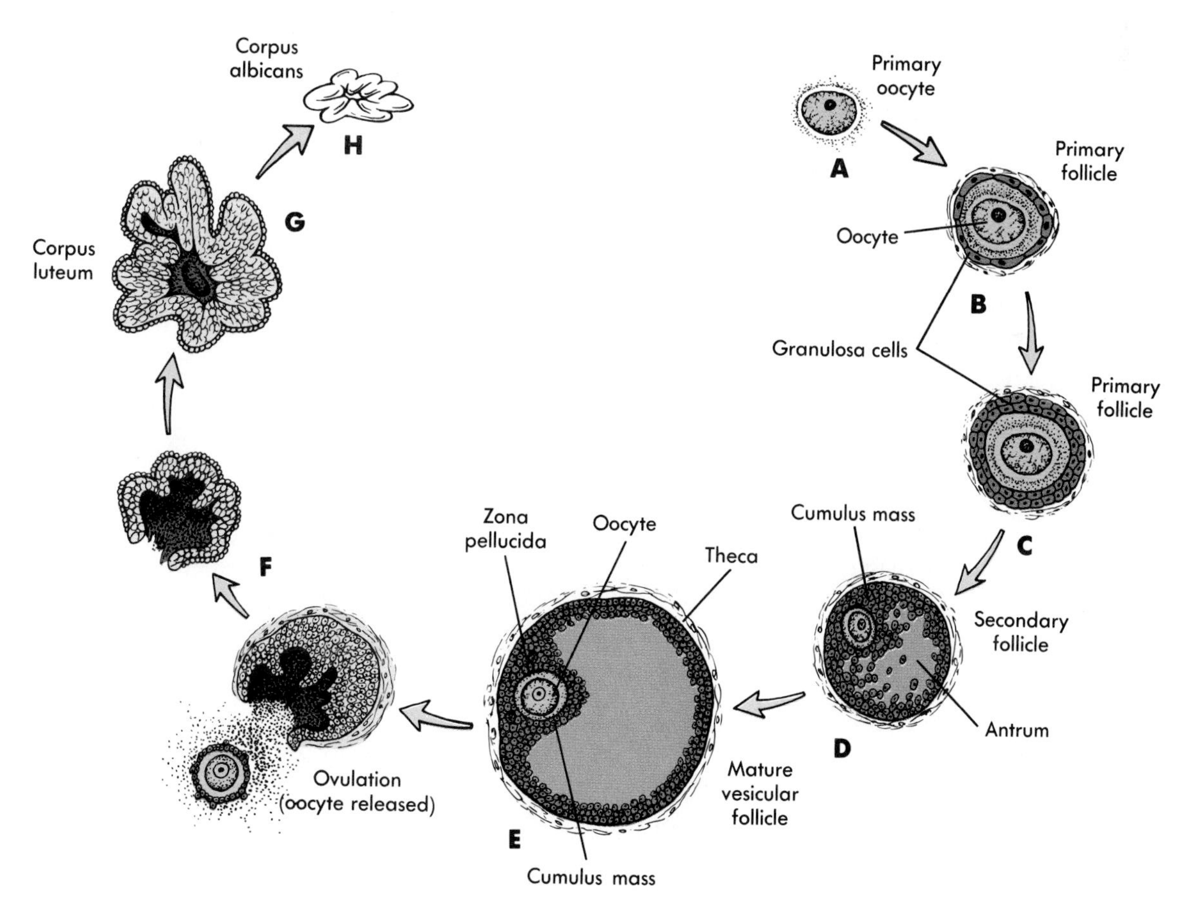

FIGURE 28-11 Maturation of the follicle and oocyte. **A** Primary oocytes begin to mature one or two menstrual cycles before they are ovulated. Several follicles begin to mature at the same time, but only one reaches the final stage of development and undergoes ovulation. **B** The primary follicles enlarge, and **C** granulosa cells form more than one layer. **D** An antrum begins to form and fill with fluid to form secondary follicles. **E** When a follicle becomes mature, it enlarges to its maximum size, and a large antrum is present. **F** During ovulation the oocyte is released from the follicle along with some surrounding granulosa cells. **G** The granulosa cells divide rapidly and enlarge to form the corpus luteum. **H** When the corpus luteum degenerates, it forms the corpus albicans.

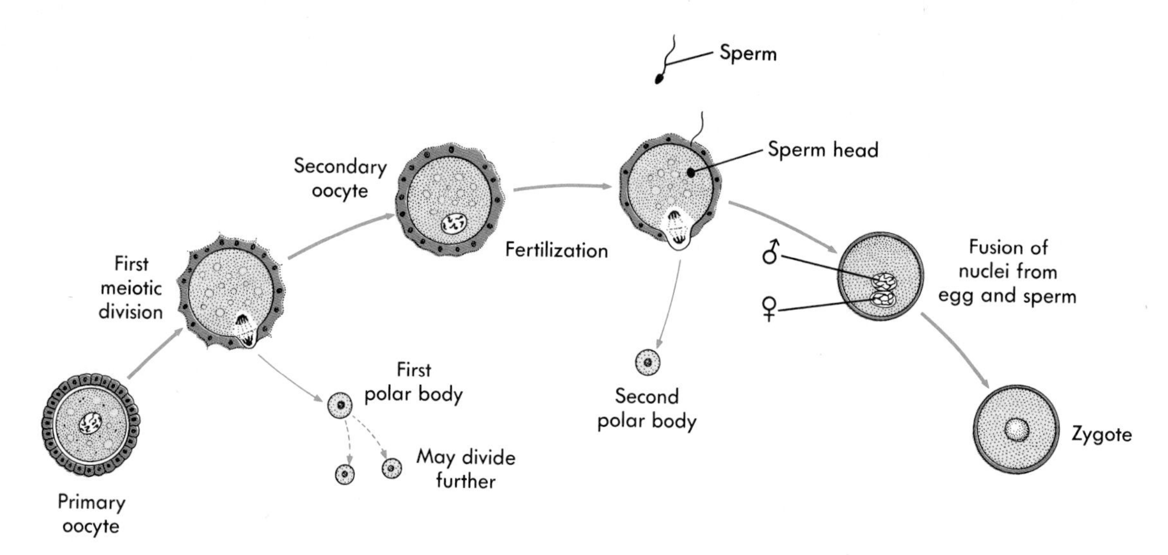

FIGURE 28-12 Maturation and fertilization of the oocyte. The primary oocyte undergoes meiosis and gives off the first polar body to become a secondary oocyte just before ovulation. Sperm penetration initiates the completion of the second meiotic division and the expulsion of a second polar body. Subsequently, the nuclei of the oocyte and the sperm unite. Fertilization results in the formation of a zygote.

Meiosis

Spermatogenesis and oogenesis involve meiosis (see Chapter 3). This kind of cell division occurs only in the gonads. It consists of two consecutive nuclear divisions without a second replication of the genetic material between the divisions. Four daughter cells are produced, and each has half as many chromosomes as the parent cell.

The normal chromosome number in human cells is 46. This number is called a diploid or a 2n number of chromosomes. The chromosomes consist of 23 pairs. Each pair of chromosomes is called a homologous pair. One chromosome of each homologous pair is from the male parent, and the other is from the female parent. The chromosomes of each homologous pair look alike, and they contain genes for the same traits.

In sperm cells and oocytes the number of chromosomes is 23. This number is called a haploid or a 1n number of chromosomes. Each gamete contains one chromosome from each of the homologous pairs. Reduction of the number of chromosomes in sperm cells or oocytes to a 1n number is important. When a sperm cell and an oocyte fuse to form a fertilized egg, each provides a 1n number of chromosomes, which reestablishes a 2n number of chromosomes. If meiosis did not occur, each time fertilization occurred the number of chromosomes in the fertilized oocyte would double. The extra chromosomal material would be lethal to the developing offspring.

The two divisions of meiosis are called meiosis I and meiosis II. The stages of meiosis are named the same as those that occur in mitosis (i.e., prophase, metaphase, anaphase, and telophase), but there are distinct differences between mitosis and meiosis.

Before meiosis begins, all the chromosomes are duplicated. At the beginning of meiosis each of the 46 chromosomes consists of two sister chromatids connected by a centromere (Figure 28-A). In prophase of meiosis I the chromosomes align with their homologous pairs near the middle of the cell. This process is called synapsis. Since each chromosome consists of two chromatids, the pairing of the homologous chromosomes brings two chromatids of each chromosome close together, an arrangement called a tetrad. Occasionally part of a chromatid of one homologous chromosome will break off and be exchanged with part of another chromatid from the other homologous chromosome of the tetrad. This exchange of genetic material is called crossing over. Crossing over allows the exchange of genetic material between maternal and paternal chromosomes.

During synapsis homologous pairs of chromosomes line up near the center of the cell undergoing meiosis. However, for each pair of homologous chromosomes, the side on which the maternal or paternal chromosome is located is random. The way the chromosomes align during synapsis results in the random assortment of maternal and paternal chromosomes in the daughter cells during meiosis. Crossing over and the random assortment of maternal and paternal chromosomes are responsible for the large degree of diversity in the genetic composition of sperm and oocyte produced by each individual.

During anaphase I the homolo-

Meiosis—cont'd

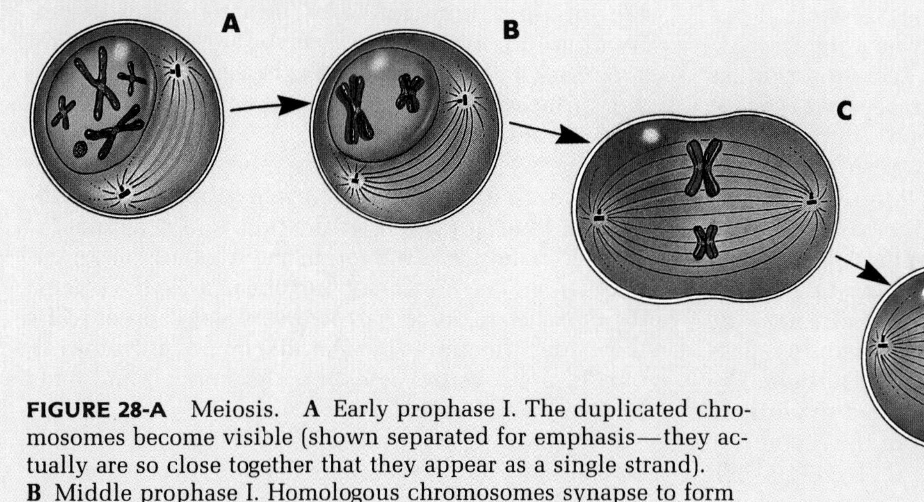

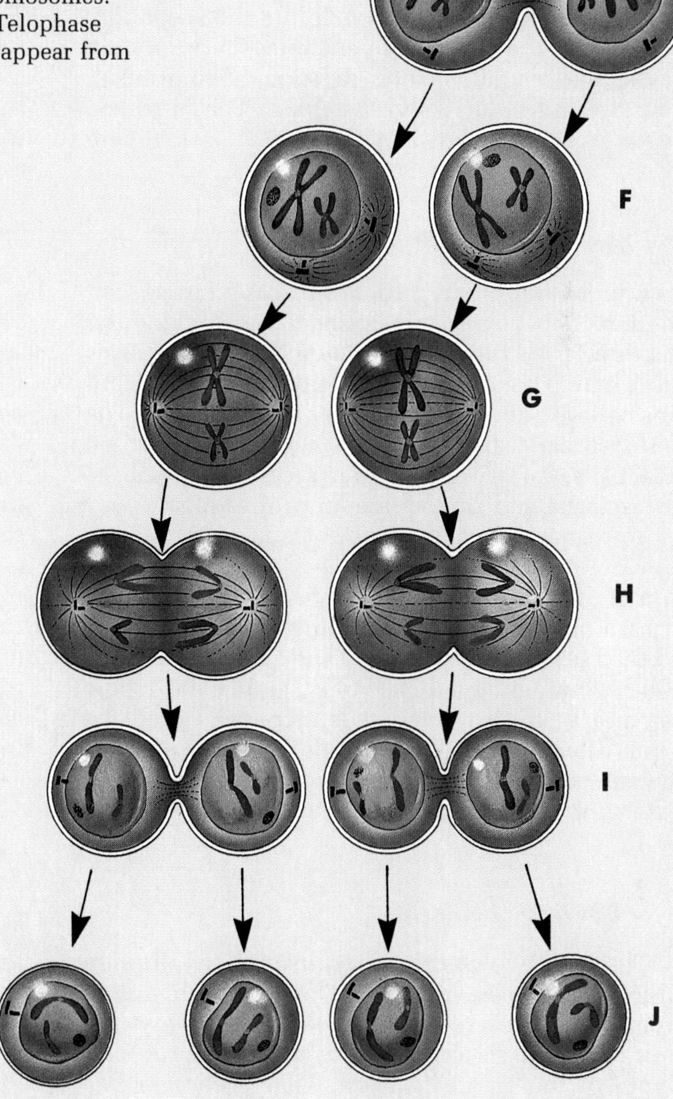

FIGURE 28-A Meiosis. **A** Early prophase I. The duplicated chromosomes become visible (shown separated for emphasis—they actually are so close together that they appear as a single strand). **B** Middle prophase I. Homologous chromosomes synapse to form tetrads. **C** Metaphase I. Tetrads align at the equatorial plane. **D** Anaphase I. Chromatids move apart to opposite sides of the cell. **E** Telophase I. New nuclei form, and the cells divide. During interkinesis (not shown) there is no duplication of chromosomes. **F** Prophase II. **G** Metaphase II. **H** Anaphase II. **I** Telophase II. **J** Haploid cells. The chromosomes are about to disappear from view.

gous pairs are separated to each side of the cell. As a consequence, when meiosis I is complete, each daughter cell has one chromosome from each of the homologous pairs. Each of the 23 chromosomes in each daughter cell consists of two chromatids joined by a centromere.

It is during the first meiotic division that the chromosome number is reduced from a 2n number (46 chromosomes—23 pairs) to a 1n number (23 chromosomes—one from each homologous pair). The first meiotic division is therefore called a reduction division.

The second meiotic division is similar to mitosis. The chromosomes, each consisting of two chromatids, line up near the middle of the cell (see Figure 28-A). Then the chromatids separate at the centromere, and each daughter cell receives one of the chromatids from each chromosome. When the centromere separates, each of the chromatids is called a chromosome. Consequently, each of the four daughter cells produced by meiosis contains 23 chromosomes.

Ovulation

As the graafian follicle continues to swell, it can be seen on the surface of the ovary as a tight, translucent blister. The follicular cells secrete a thinner fluid than previously and at an increased rate so that the follicle swells more rapidly than can be accommodated by follicular growth. As a result, the granulosa cells and theca become very thin over the area exposed to the ovarian surface.

The follicle expands and ruptures, forcing a small amount of blood and follicular fluid out of the vesicle. Shortly after this initial burst of fluid, the secondary oocyte, surrounded by the cumulus mass and the zona pellucida, escapes from the follicle. The release of the secondary oocyte is called **ovulation.**

During ovulation, development of the secondary oocyte has stopped at metaphase II. If sperm cell penetration does not occur, the secondary oocyte never completes this second division and simply degenerates and passes out of the system. Continuation of the second meiotic division is triggered by **fertilization,** the entry of a sperm cell into the secondary oocyte. Once the sperm cell penetrates the secondary oocyte, the second meiotic division is completed, and a second polar body is formed. The fertilized oocyte is now called a **zygote** (zi'gōt; see Figure 28-12).

Fate of the follicle

After ovulation the follicle still has an important function. It becomes transformed into a glandular structure called the **corpus luteum** (lu'te-um; yellow body), which has a convoluted appearance as a result of its collapse after ovulation (see Figure 28-11). The granulosa cells and the theca interna, now called luteal cells, enlarge and begin to secrete hormones—progesterone and smaller amounts of estrogen.

If pregnancy occurs, the corpus luteum enlarges and remains throughout pregnancy as the **corpus luteum of pregnancy.** If pregnancy does not occur, the corpus luteum lasts for approximately 10 to 12 days and then begins to degenerate. The connective tissue cells become enlarged and clear, giving the whole structure a whitish color; it is therefore called the **corpus albicans** (al'bĭ-kanz; white body). The corpus albicans continues to shrink and eventually disappears after several months or even years.

Uterine Tubes

There are two **uterine tubes,** also called **fallopian** (fal-lo'pĭ-an) **tubes** or **oviducts,** one on each side of the uterus and each associated with one ovary (see Figure 28-9). Each tube is located along the superior margin of the broad ligament. That portion of the broad ligament most directly associated with the tube is called the **mesosalpinx** (mez'o-sal'pinx; mesothelium of the trumpet-shaped uterine tube).

The uterine tube opens directly into the peritoneal cavity to receive the oocyte and expands to form the **infundibulum** (funnel). The opening of the infundibulum, the **ostium,** is surrounded by long, thin processes called **fimbriae** (fim'bre-e; fringe). The inner surfaces of the fimbriae consist of a ciliated mucous membrane.

The portion of the uterine tube that is nearest to the infundibulum is called the **ampulla.** It is the widest and longest portion of the tube and accounts for approximately 7.5 to 8 cm of the total 10 cm length of the tube. The portion of the tube nearest the uterus, the **isthmus,** is much narrower and has thinner walls than does the ampulla. The **uterine,** or **intramural, part** of the tube traverses the uterine wall and ends in a very small uterine opening.

The wall of each uterine tube consists of three layers. The outer **serosa** is formed by the peritoneum, the middle **muscular layer** consists of longitudinal and circular smooth muscle fibers, and the inner **mucosa** consists of a mucous membrane of simple ciliated columnar epithelium (see Figure 28-9). The mucosa is arranged into numerous longitudinal folds.

The mucosa of the uterine tubes provides nutrients for the ovum or developing embryo as long as it is traversing the uterine tubes. The ciliated epithelium helps move the small amount of fluid and the ovum through the uterine tubes.

Uterus

The **uterus** is the size and shape of a medium-sized pear and is approximately 7.5 cm long and 5 cm wide (see Figures 28-8 and 28-9). It is slightly flattened anteroposteriorly and is oriented in the pelvic cavity with the larger, rounded portion, the **fundus** (fun'dus; bottom of a rounded flask), directed superiorly and the narrower portion, the **cervix** (ser'viks; neck), directed inferiorly. The main portion of the uterus, the **body,** is between the fundus and the cervix. A slight constriction called the **isthmus** marks the junction of the cervix and the body. Internally, the **uterine cavity** continues as the **cervical canal,** which opens through the **ostium** (os'tĭ-um) into the vagina.

Cancer of the cervix is common in women and fortunately can be detected and treated. Early in the development of cervical cancer, the cells of the cervix change in a characteristic way. This change can be observed by taking a cell sample and examining the cells microscopically. The most common technique is to obtain a Papanicolaou (Pap) smear, which has a reliability of 90% for detecting cervical cancer.

The major ligaments holding the uterus in place are the **broad ligament, round ligaments,** and **uterosacral ligaments** (see Figure 28-9). The round ligaments extend from the uterus through the inguinal canals to the external genitalia (labia majora), and the uterosacral ligaments attach the uterus to the sacrum. Normally the uterus is anteverted, with the body of the uterus tipped slightly anteriorly. In some women the uterus may be retroverted, or tipped posteriorly. In addition to the ligaments, much support is provided inferiorly to the uterus by the skeletal muscles of the pelvic floor. If these muscles are weakened (e.g., in childbirth), the uterus can extend inferiorly into the vagina, a condition called a prolapsed uterus.

The uterine wall is composed of three layers: serous, muscular, and mucous (see Figure 28-9). The **perimetrium** (pĕr'ĭ-me'trĭ-um), or **serous coat,** of the uterus is the peritoneum. The next layer, just deep to the perimetrium, is the **myometrium** (mi'o-me'trĭ-um), or **muscular coat,** which consists of smooth muscle and is quite thick (in fact, it is the thickest area of smooth muscle in the body), accounting for the bulk of the uterine wall. In the cervix the muscular layer contains less muscle and more dense connective tissue; therefore the cervix is more rigid and less contractile than the rest of the uterus. The innermost layer of the uterus is the **endometrium** (en'do-me'trĭ-um), or **mucous membrane.** The endometrium consists of a simple columnar epithelial lining and a connective tissue lamina propria. Simple tubular glands are scattered about the lamina propria and open through the epithelium into the uterine cavity. The endometrium consists of two layers: a thin, deep **basal layer** (the deepest part of the lamina propria) is continuous with the myometrium, and a thicker, superficial **functional layer** (most of the lamina propria and the endothelium) lines the cavity itself. The functional layer is so named because it undergoes menstrual changes and sloughing during the female sex cycle.

The cervical canal is lined by columnar epithelial cells and contains cervical mucous glands. The consistency of the mucus changes during the female sex cycle. The mucus fills the cervical canal and acts as a barrier to substances that could pass from the vagina into the uterus. Near the time of ovulation the consistency of the mucus changes, making the passage of sperm cells from the vagina into the uterus easier.

Vagina

The **vagina** is a tube approximately 10 cm long that extends from the uterus to the outside of the body (see Figure 28-9). The vagina is the female organ of copulation and thus functions to receive the penis during intercourse, and it allows menstrual flow and childbirth. Longitudinal ridges called **columns** extend the length of the anterior and posterior vaginal walls, and several transverse ridges called **rugae** (ru'ge) extend between the anterior and posterior columns. The superior, domed portion of the vagina, the **fornix** (for'niks; domed), is attached to the sides of the cervix so that a portion of the cervix extends into the vagina.

The wall of the vagina consists of an outer muscular layer and an inner mucous membrane. The muscular layer is smooth muscle that allows the vagina to increase in size. Thus the vagina can accommodate the penis during intercourse and can stretch greatly during delivery. The mucous membrane is moist stratified squamous epithelium that forms a protective surface layer. Most of the lubricating secretions produced by the female during intercourse are released by the vaginal mucous membrane.

The **vaginal opening** or **orifice** is covered by a thin mucous membrane called the **hymen.** The hymen may completely close the vaginal opening (imperforate hymen), in which case it must be removed to allow menstrual flow, or, more commonly, the hymen may be perforated by one or several holes. The openings in the hymen are usually greatly enlarged during the first sexual intercourse, but this enlargement is not necessarily the case. In addition, the hymen may be perforated or torn at some earlier time in a young woman's life (e.g., during strenuous physical exercise). Thus the presence of a hymen is not a reliable indicator of virginity as was once thought.

External Genitalia

The external female genitalia, also referred to as the **vulva** (vul'vah) or **pudendum** (pu-den'dum), consist of the vestibule and its surrounding structures (Figure 28-13). The **vestibule** (ves'tĭ-būl) is the space into which the vagina (posteriorly) and urethra (anteriorly) open. It is bordered by a pair of thin, longitudinal skin folds called the **labia** (la'be-ah; lips) **minora** (sing. labium minus). A small erectile structure called the **clitoris** (klit'o-ris) is located in the anterior margin of the vestibule. Anteriorly, the two labia minora unite over the clitoris to form a fold of skin called the **prepuce.**

The clitoris is usually less than 2 cm in length and consists of a shaft and a distal glans. It is well supplied with sensory receptors and functions to initiate and intensify levels of sexual tension. The clitoris contains two erectile structures, the **corpora cavernosa,** each of which expands at the base end of the clitoris to form the crus of the clitoris and attaches the clitoris to the coxae. The corpora cavernosa of the clitoris are comparable to the corpora cavernosa of the penis, and they become engorged with blood as a result of sexual excitement. In most women this engorgement results in an increase in the diameter, but not the length, of the clitoris. With increased diameter, the clitoris makes better contact with the

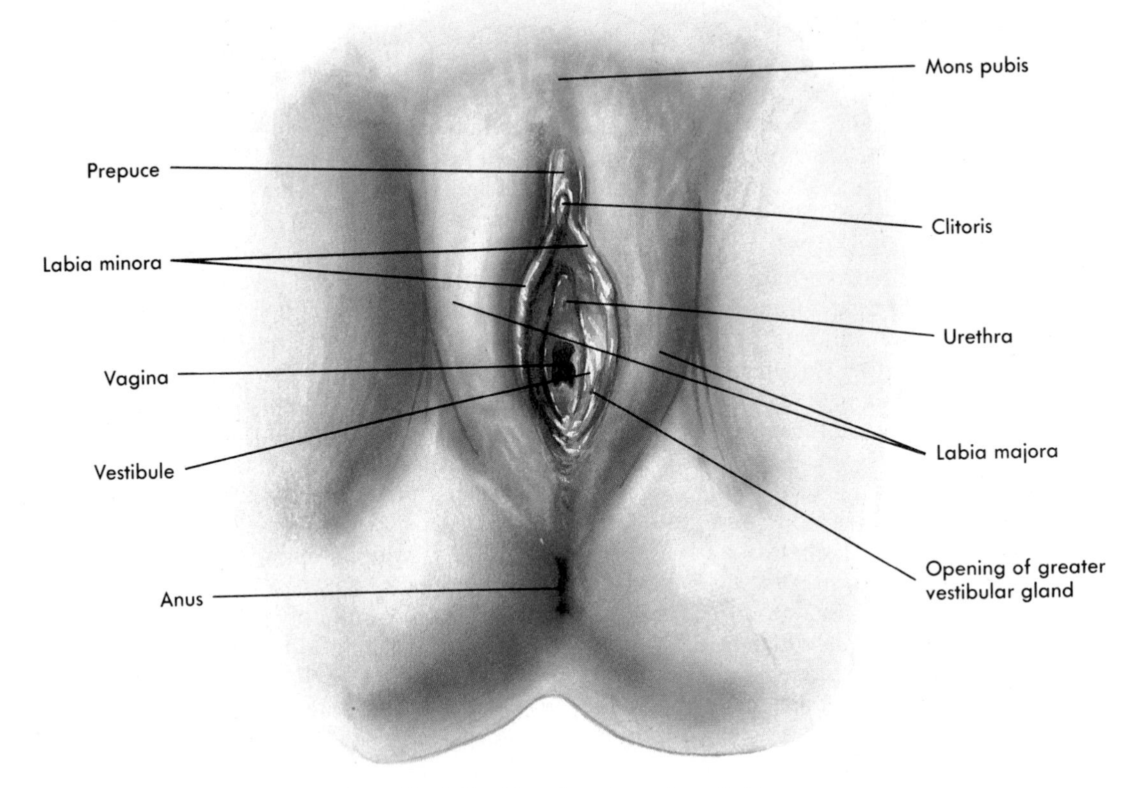

FIGURE 28-13 Female external genitalia.

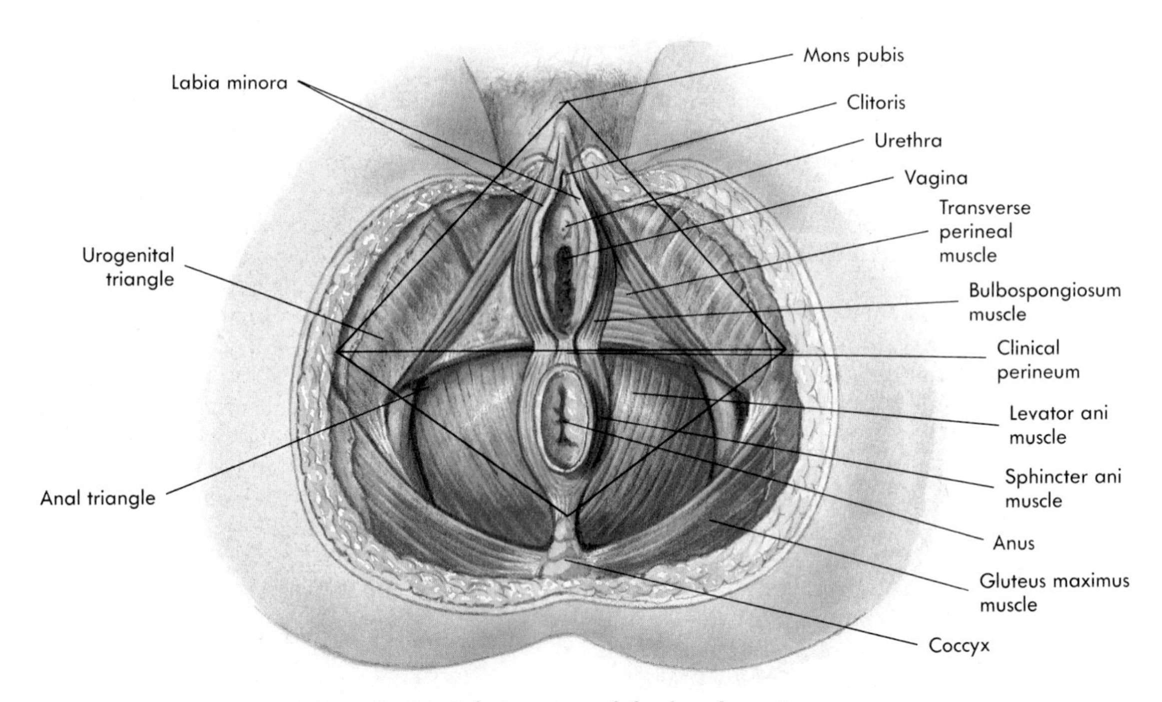

FIGURE 28-14 Inferior view of the female perineum.

prepuce and surrounding tissues and is more easily stimulated.

Erectile tissue that corresponds to the corpus spongiosum of the male lies deep to and on the lateral margins of the vestibular floor on either side of the vaginal orifice. This tissue is called the **bulbs of the vestibule,** and like other erectile tissue it becomes engorged with blood and is more sensitive during sexual arousal. Expansion of the bulb causes narrowing of the vaginal orifice, producing better contact of the vagina with the penis during intercourse.

On each side of the vestibule, between the vaginal opening and the labia minora, is an opening of the duct of the **greater vestibular gland.** Additional small mucous glands, the **lesser vestibular glands** or **paraurethral glands,** are located near the clitoris and urethral opening. They produce a lubricating fluid that helps to maintain the moistness of the vestibule.

Lateral to the labia minora are two prominent, rounded folds of skin called the **labia majora** (sing. labium majus). The prominence of the labia majora is primarily caused by the presence of subcutaneous fat within the labia. The two labia majora unite anteriorly in an elevation over the symphysis pubis called the **mons pubis.** The lateral surfaces of the labia majora and the surface of the mons pubis are covered with coarse hair. The medial surfaces are covered with numerous sebaceous and sweat glands. The space between the labia majora is called the **pudendal cleft.** Most of the time, the labia majora are in contact with each other across the midline, closing the pudendal cleft and concealing the deeper structures within the vestibule.

Perineum

The **perineum** (Figure 28-14), as in the male, is divided into two triangles by the superficial and deep transverse perineal muscles. The anterior, urogenital triangle contains the external genitalia, and the posterior, anal triangle contains the anal opening. The region between the vagina and the anus is the **clinical perineum.** The skin and muscle of this region may tear during childbirth. To prevent such tearing, an incision called an **episiotomy** sometimes is made in the clinical perineum. This clean, straight incision may be easier to repair than a tear would be. Alternatively, allowing the perineum to stretch slowly during the delivery may prevent tearing, making an episiotomy unnecessary.

Mammary Glands

The **mammary glands** are the organs of milk production and are located within the **mammae** (mam′e), or breasts (Figure 28-15). The mammary glands are modified sweat glands. Externally, the breasts of both males and females have a raised **nipple** surrounded by a circular, pigmented **areola** (ă-re′o-lah). The areolae normally have a slightly bumpy surface caused by the presence of rudimentary mammary glands, called **areolar glands,** just below the surface. Secretions from these glands protect the nipple and the areola from chafing during nursing.

In prepubescent children the general structure of the breasts is similar, and both males and females possess a rudimentary glandular system. The female breasts begin to enlarge during puberty primarily under the influence of estrogens and progesterone. This enlargement is often accompanied by increased sensitivity or pain in the breasts. Males often experience these same sensations during early puberty, and their breasts may even develop slight swellings; however, these symptoms usually disappear fairly quickly. If on rare occasions the breasts of a male become enlarged, this condition is called **gynecomastia** (gi′nĕ-ko-mas′tĭ-ah).

Each adult female mammary gland usually consists of 15 to 20 glandular **lobes** covered by a considerable amount of adipose tissue. It is primarily this superficial fat that gives the breast its form. The lobes of each mammary gland form a conical mass, the apex of which is located at the nipple. Each lobe possesses a single **lactiferous** (lak-tif′er-us) **duct,** which opens independently of other lactiferous ducts on the surface of the nipple. Just deep to the surface, each lactiferous duct enlarges to form a small, spindle-shaped **lactiferous sinus,** which ac-

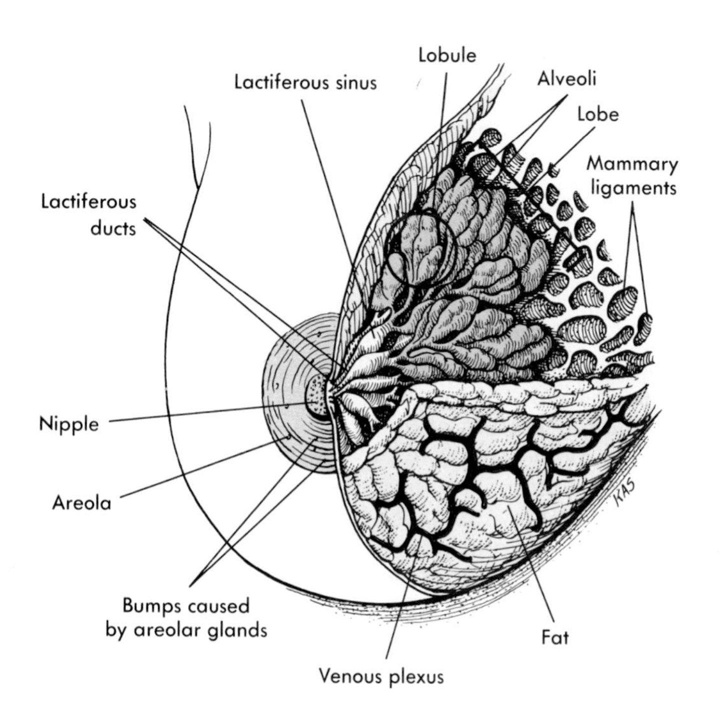

FIGURE 28-15 Right mamma. The section illustrates the blood supply, the mammary glands, and the duct system.

cumulates milk during milk production. The lactiferous duct supplying a lobe subdivides to form smaller ducts, each of which supplies a **lobule.** Within a lobule the ducts branch and become even smaller. In the milk-producing breast the ends of these small ducts expand to form secretory sacs called alveoli.

The mammae are supported and held in place by a group of **mammary,** or **Cooper's ligaments.** These ligaments extend from the fascia over the pectoralis major muscles to the skin over the mammary glands and prevent the mammary glands from excessive sagging. However, in older adults these ligaments weaken and elongate, allowing the breasts to sag to a greater extent than when the person was younger.

The nipples are very sensitive to tactile stimulation and contain smooth muscle that can contract, causing the nipple to become erect in response to stimulation. These smooth muscle fibers respond similarly to general sexual arousal.

Cancer of the breast is a serious, often fatal disease in women. The use of mammography and regular self-examination of the breast can lead to early detection of breast cancer and effective treatment.

PHYSIOLOGY OF FEMALE REPRODUCTION

As in the male, female reproduction is under the control of hormonal and nervous regulation. Development of the female reproductive organs and normal function depend on the relative levels of a number of hormones in the body.

Puberty

Puberty in females is marked by the first episode of menstrual bleeding, which is called **menarche** (me'nar'ke). During puberty the vagina, uterus, uterine tubes, and external genitalia begin to enlarge. Fat is deposited in the breasts and around the hips, causing them to enlarge and assume an adult form. The glandular portion of the breasts and the areolae develop, pubic and axillary hair grows, and the voice changes, although this is a more subtle change than in males. Development of a sexual drive is also associated with puberty.

The changes associated with puberty are due primarily to the elevated rate of estrogen and progesterone secretion by the ovaries. Before puberty estrogens and progesterone are secreted in very small amounts. Luteinizing hormone (LH) and follicle-stimulating hormone (FSH) levels also remain very low. The low secretory rates are due to a lack of gonadotropin-releasing hormone (GnRH) release from the hypothalamus. At puberty not only are GnRH, LH, and FSH secreted in greater quantities than before puberty, but the adult pattern is established in which a cyclic pattern of gonadotropin secretion occurs. This cyclic surge of LH and FSH triggers ovulation, the monthly changes in secretion of estrogens and progesterone, and the resultant changes in the uterus that characterize the menstrual cycle.

Menstrual Cycle

The term **menstrual cycle** technically refers to the series of changes that occur in sexually mature, nonpregnant females and culminate in menses. Typically, the menstrual cycle is approximately 28 days long, although it may be as short as 18 days in some women and as long as 40 days in others (Figure 28-16). **Menses** (derived from a Latin word meaning month) is a period of mild hemorrhage during which the uterine epithelium is sloughed and expelled from the uterus. Although the term menstrual cycle refers specifically to changes that occur in the uterus, several other cyclic changes are associated with it, and the term is often used to refer to all of the cyclic events that occur in the female reproductive system. These changes include cyclic changes in the ovary, in hormone secretion, and in the uterus.

The first day of menses is considered day 1 of the menstrual cycle. Menses typically lasts 4 or 5 days. Ovulation occurs on approximately day 14 of the menstrual cycle, although the timing of ovulation varies from individual to individual and varies within a single individual from one menstrual cycle to the next. The time between the ending of menses and ovulation is called the **follicular phase** (implies rapid development of ovarian follicles) or the **proliferative phase** (proliferation of the uterine mucosa). The period after ovulation and before the next menses is called the **luteal phase** (existence of the corpus luteum) or the **secretory phase** (maturation of and secretion by uterine glands). After approximately 28 days, menses again occurs, and a new menstrual cycle is initiated.

Ovarian cycle

The **ovarian cycle** specifically refers to the series of events that occur in a regular fashion in the ovaries of sexually mature, nonpregnant women. These events are controlled by hormones released from the hypothalamus and adenohypophysis. FSH is primarily responsible for initiating the development of the primary follicles, and as many as 25 begin to mature during each menstrual cycle. The follicles that start to develop in response to FSH are not ovulated during the same menstrual cycle in which they

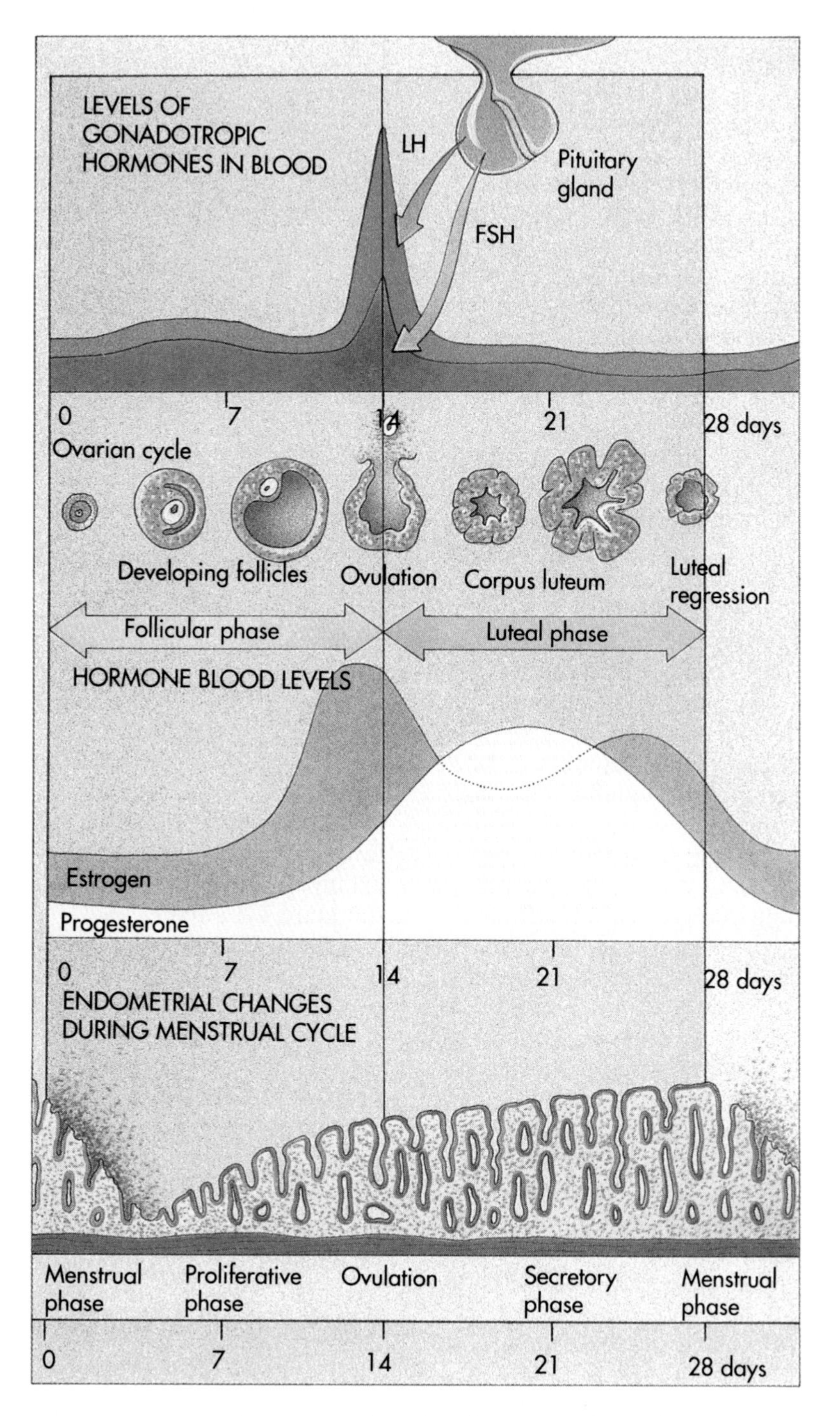

FIGURE 28-16 Events of the menstrual cycle. The various lines depict the changes in blood hormone levels, the development of the follicles, and the changes in the endometrium during the cycle.

begin to mature but are ovulated one or two cycles later.

Although several follicles begin to mature during each cycle, normally only one is ovulated. The remaining follicles degenerate. Larger and more mature follicles apparently secrete estrogen and other substances that have an inhibitory effect on other less mature follicles.

Gonadotropins are released from the adenohypophysis in large amounts just before ovulation (see Figure 28-16). An increase in blood levels of both LH, the **LH surge,** and FSH, the **FSH surge,** occurs. The LH surge occurs several hours earlier and to a greater degree than the FSH surge. The LH surge initiates ovulation and causes the ovulated follicle to become the corpus luteum. FSH may make the follicle more sensitive to the influence of LH by stimulating the synthesis of LH receptor molecules in the follicles.

The LH surge also causes the primary oocyte to complete the first meiotic division just before or during the process of ovulation. The LH surge triggers several events that are very much like inflammation in the mature follicle and that result in ovulation. The follicle becomes edematous, proteolytic enzymes cause the degeneration of the ovarian capsule, the follicle ruptures, and the oocyte and some surrounding cells are slowly extruded from the ovary.

After ovulation the granulosa cells enlarge and increase in number to become luteal cells. The cells of the corpus luteum secrete large amounts of progesterone and some estrogen throughout the luteal phase. Without the influence of additional LH-like hormones, the cells of the corpus luteum begin to atrophy after day 25 or 26, and the blood levels of estrogen and progesterone decrease rapidly. If fertilization of the ovulated oocyte takes place, the developing embryo begins to secrete an LH-like substance called **human chorionic gonadotropin (HCG),** which keeps the corpus luteum from degenerating. As a result, blood levels of estrogen and progesterone do not decrease, and menses does not occur.

The most obvious hormonal change that occurs during the menstrual cycle is the LH surge before ovulation. It is a major signal that sets into motion the cyclic events of the ovary, which in turn regulate the cyclic events of the uterus (see Figure 28-16). Before the preovulatory LH surge, LH and FSH are at very low levels. The first noticeable endocrine event before the LH surge is an increased blood level of estrogen secreted by the theca interna cells of the developing follicle. A positive-feedback system then develops in which estrogen stimulates GnRH secretion from the hypothalamus and GnRH triggers LH and FSH secretion from the adenohypophysis, which stimulates a greater rate of estrogen secretion from the ovary and an even greater level of GnRH secretion. LH positive-feedback loops produce a series of larger and larger surges occurring over approximately a 24-hour period and resulting in an increase of the total concentration of LH to a maximum level (see Figure 28-16). Ovulation occurs several hours after the LH maximum has been achieved.

Shortly after ovulation the production of estrogen by the follicle decreases, and the production of progesterone increases. After the corpus luteum forms, both estrogen and progesterone levels increase to much higher levels than before ovulation. This increase has a negative-feedback effect on GnRH release from the hypothalamus; as a result, LH and FSH release from the adenohypophysis decreases. Also, the anterior pituitary becomes less sensitive to GnRH. Because of the decreased secretion of GnRH and decreased sensitivity of the adenohypophysis to GnRH, the rate of LH and FSH secretion declines to very low levels after ovulation.

3

Predict the effect on the ovarian cycle of administering a relatively large amount of estrogen and progesterone just before the preovulatory LH surge. Also predict the consequences of continually administering high concentrations of GnRH.

? ? ? ? ? ? ? ? ? ? ?

FSH secretion follows a pattern similar to LH secretion except that the FSH surge is not as large as the LH surge, it occurs several hours later, and it lasts approximately twice as long as the LH surge (see Figure 28-16). During the remainder of the menstrual cycle, plasma levels of both FSH and LH are low. A few small surges of LH may occur just before menses occurs, but they are smaller and less consistent than the main LH surge.

Estrogen and progesterone levels decline after degeneration of the corpus luteum (beginning at approximately day 22 of the menstrual cycle) and reach a very low level in the circulatory system. The increase in estrogen and progesterone levels around the time of ovulation and the decrease in these two hormones after degeneration of the corpus luteum cause the cyclic changes that occur in the uterus during the menstrual cycle.

Uterine cycle

The term **uterine cycle** refers to changes that occur primarily in the endometrium of the uterus during the menstrual cycle (see Figure 28-16). Other, more subtle changes also occur in the vagina and other structures during the menstrual cycle. All these changes are caused primarily by the cyclic secretions of estrogen and progesterone.

The endometrium of the uterus begins to proliferate after menses. The epithelial cells of the basal layer rapidly divide and replace the cells of the functional layer that was sloughed during the last menses. A relatively uniform layer of low cuboidal endometrial cells is produced. It later becomes columnar and is thrown into folds to form **spiral tubular glands.** Blood vessels called **spiral arteries** project through the delicate connective tissue that separates the individual glands to supply nutrients to the endometrial cells. After ovulation the endometrium becomes thicker, and the spiral glands develop to a greater extent and begin to secrete small amounts of a fluid rich in glycogen. Approximately 7 days after ovulation (approximately day 21 of the menstrual cycle) the endometrium is prepared to receive the developing embryo if fertilization has occurred. If the developing embryo arrives in the uterus too early or too late, the endometrium does not provide a hospitable environment for implantation.

Estrogen causes proliferation of the endometrial cells and, to a lesser degree, of the myometrial cells. It also causes the uterine tissue to become more sensitive to progesterone by stimulating the synthesis of progesterone receptor molecules within the uterine cells. Progesterone then binds to the progesterone receptors, resulting in cellular hypertrophy in the endometrium and myometrium and causing the endometrial cells to become secretory. Progesterone also inhibits smooth muscle contractions.

4

Predict the effect on the endometrium of elevated progesterone levels in the circulatory system before the estrogen surge that occurs after menstruation.

? ? ? ? ? ? ? ? ? ? ?

Menstrual cramps are the result of strong myometrial contractions that occur before and during menstruation. The cramps may result from excessive prostaglandin secretion, which is inhibited by progesterone but is stimulated by estrogen. In some women menstrual cramps are extremely uncomfortable. Many women can alleviate painful menstruation by taking drugs (e.g., aspirin) that inhibit prostaglandin biosynthesis just before the onset of menstruation. These treatments, however, are not effective in treating all painful menstruation, especially when the causes of pain are different from the ones described above.

A topic of current research emphasis concerns a phenomenon called the **premenstrual syndrome (PMS).** Some women suffer from severe changes in mood that often result in aggression and other socially unacceptable behaviors just before menses. It has been hypothesized that hormonal changes associated with the menstrual cycle trigger these mood changes, and some women have been successfully treated with steroid hormones. It is unclear how many women are affected by this condition, and its precise cause and physiological mechanisms are unknown.

If pregnancy does not occur by day 24 or 25, progesterone and estrogen levels begin to decline as the corpus luteum degenerates. As a consequence, the uterine lining also begins to degenerate. The spiral arteries constrict in a rhythmic pattern for longer and longer time periods as progesterone levels fall. As a result, all but the basal portions of the spiral glands become ischemic and then necrotic. As the cells become necrotic, they slough into the uterine lumen. The necrotic endometrium, mucous secretions, and a small amount of blood released from the spiral arteries comprise the menstrual fluid. Decreases in progesterone levels and increases in inflammatory substances that stimulate myometrial smooth muscle cells cause uterine contractions that expel the menstrual fluid from the uterus through the cervix and into the vagina.

Female Sexual Behavior and the Female Sex Act

Sexual drive in females, like sexual drive in males, is dependent on hormones. Androgens and possibly estrogens affect cells (especially in the hypothalamic area) and influence sexual behavior. However, sexual drive cannot be influenced in a predictable fashion by injecting androgens into women. Androgens are produced in the adrenal gland and other tissues such as the liver by the conversion of other steroids (e.g., progesterone) to androgens. Psychic factors also play a role in sexual behavior. After ovariectomy (removal of the ovaries) or menopause many women report increased sex drive as a result of a lack of fear of pregnancy.

The female neural pathways, both afferent and efferent, involved in controlling sexual responses are the same as in the male. Afferent impulses are transported to the sacral region of the spinal cord where reflexes that govern sexual responses are integrated. Ascending pathways, primarily the spinothalamic tracts (see Chapter 13), transport sensory information through the spinal cord to the brain, and descending pathways transport impulses back to the sacrum. As a result, the sacral reflexes are modulated by cerebral influences. Motor impulses are transported from the spinal cord to the reproductive organs by both parasympathetic and sympathetic fibers.

During sexual excitement erectile tissue within the clitoris and around the vaginal opening become engorged with blood as a result of parasympathetic stimulation. The mucous glands within the vestibule, especially the vestibular glands, secrete small amounts of mucus. Large amounts of mucuslike fluid are also extruded into the vagina through its wall, although no well-developed mucous glands are within the vaginal wall. These secretions provide lubrication to allow easy entry of the penis into the vagina and easy movement of the penis during intercourse. The tactile stimulation of the female's genitals that occurs during sexual intercourse and psychological stimuli normally trigger an **orgasm,** or the female **climax.** The vaginal, uterine, and perineal muscles contract rhythmically, and there is an increase in muscle tension throughout much of the body. After the sexual act there is a period of **resolution** characterized by an overall sense of satisfaction and relaxation. The female can be receptive to further stimulation and can experience successive orgasms. Although orgasm is a pleasurable component of sexual intercourse, it is not necessary for females to experience an orgasm for fertilization to occur.

Female Fertility and Pregnancy

After the sperm cells are ejaculated into the vagina during sexual intercourse, they are transported through the cervix, the body of the uterus, and the uterine tubes to the ampulla (Figure 28-17). The forces responsible for the movement of sperm cells through the female reproductive tract involve the swimming ability of the sperm cells and the muscular contraction of the uterus and the uterine tubes. During sexual intercourse oxytocin is released from the posterior pituitary of the female, and the semen introduced into the vagina contains prostaglandins. Both of these hormones stimulate smooth muscle contractions in the uterus and the uterine tubes.

While passing through the vagina, uterus, and the uterine tubes, the sperm cells undergo **capacitation,** a process that enables the sperm cells to release acrosomal enzymes that allow penetration of the cervical mucus, cumulus mass cells, and the oocyte cell membrane.

The oocyte can be fertilized for up to 24 hours after ovulation, and some sperm cells remain viable in the female reproductive tract for up to 72 hours, although most of them have degenerated after 24 hours. Therefore for fertilization to occur successfully, sexual intercourse must occur approximately between 3 days before and 1 day after ovulation.

One sperm cell enters the secondary oocyte, and fertilization occurs (see Chapter 29). For the next several days a sequence of cell divisions occurs while the developing cells pass through the uterine tube to the uterus. By 7 or 8 days after ovulation (day 21 or 22 of the average menstrual cycle), the endometrium of the uterus is prepared for implantation. Estrogen and progesterone have caused it to reach its maximum thickness and secretory activity, and the developing cellular mass begins to implant. The outer layer of the developing embryonic mass, the trophoblast, secretes proteolytic enzymes that digest the cells of the thickened endometrium (see Chapter 29), and the mass digests its way into the endometrium.

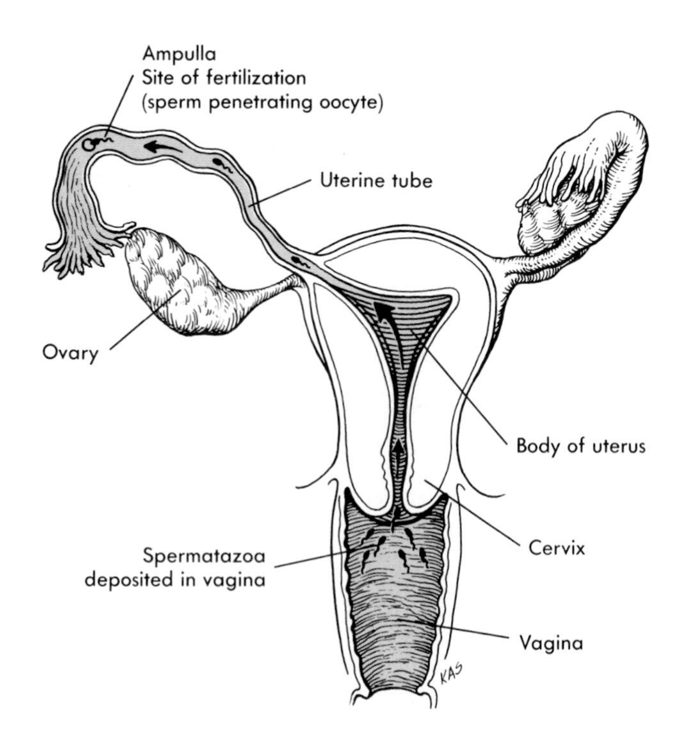

FIGURE 28-17 Movement of sperm cells from the vagina to the site of fertilization. Sperm cells are deposited in the vagina as part of the semen when the male ejaculates with the penis inside the vagina. Sperm cells pass through the cervix, the body of the uterus, and the uterine tube. Fertilization normally occurs when the oocyte is in the upper one third of the uterine tube (the ampulla).

An ectopic pregnancy results if implantation occurs anywhere other than in the uterine cavity. The most common site of ectopic pregnancy is the uterine tube. Implantation in the uterine tube eventually is fatal to the fetus and may cause the tube to rupture. In some cases implantation can occur in the mesenteries of the abdominal cavity, and the fetus can develop normally but must be delivered by cesarean section.

The trophoblast secretes human chorionic gonadotropin (HCG), which is transported in the blood to the ovary and causes the corpus luteum to remain functional. As a consequence, both estrogen and progesterone levels continue to increase rather than decrease. The secretion of HCG increases rapidly and reaches a peak approximately 8 or 9 weeks after fertilization. Subsequently, HCG levels in the circulatory system decline to a lower level by 16 weeks and remain at a relatively constant level throughout the remainder of pregnancy. Detection of HCG excreted in the urine is the basis for some pregnancy tests.

The estrogen and progesterone secreted by the corpus luteum are essential for the maintenance of pregnancy. However, after the **placenta** forms from the trophoblast, it also begins to secrete estrogens and progesterone. By the time the first 3 months of pregnancy are complete, the corpus luteum is no longer needed to maintain pregnancy; the placenta has become an endocrine gland that secretes sufficient quantities of estrogen and progesterone to maintain pregnancy. Estrogen and progesterone levels increase in the woman's blood throughout pregnancy (Figure 28-18).

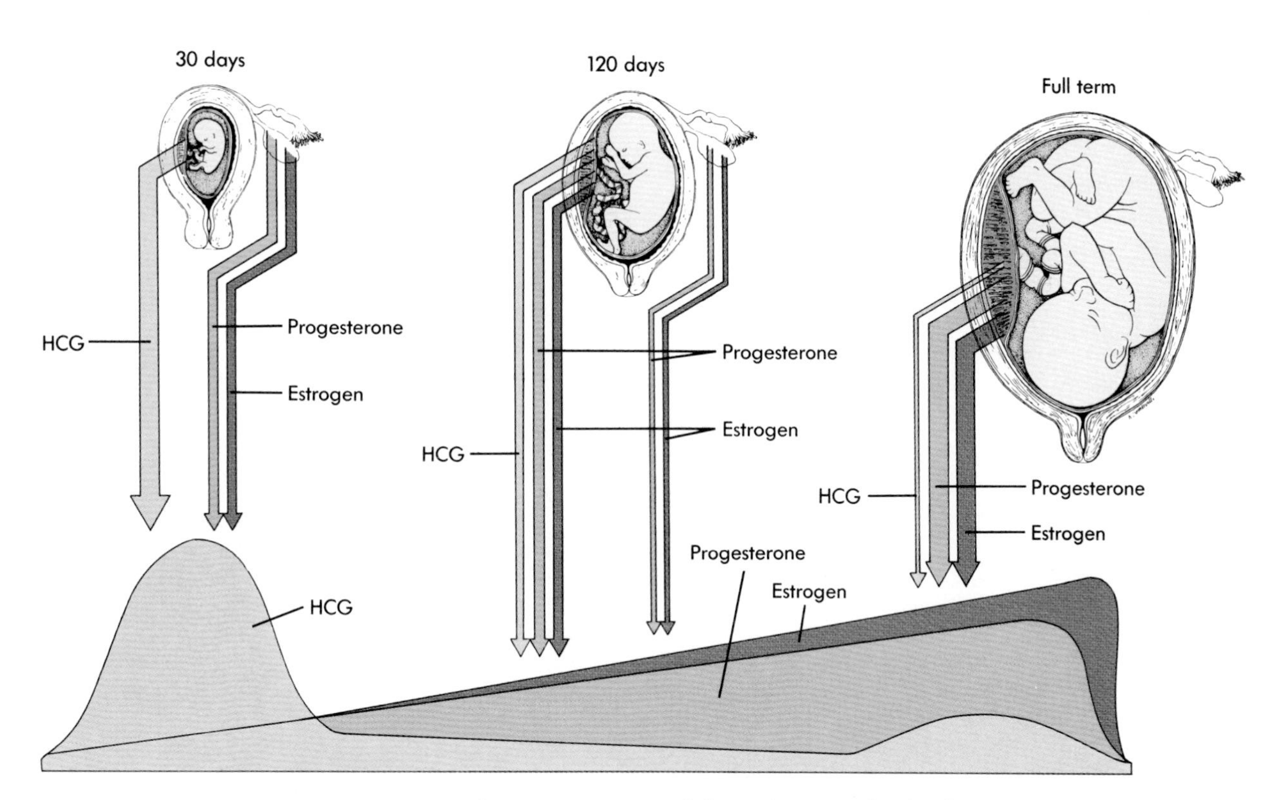

FIGURE 28-18 Changes in the concentration of the major reproductive hormones during pregnancy. Human chorionic gonadotropin *(HCG)*, progesterone, and estrogens are secreted from the placenta during pregnancy. The HCG increases until it reaches a maximum concentration near the end of the first trimester of pregnancy and then decreases to a low level thereafter. Progesterone continues to increase until it levels off near the end of pregnancy. Estrogen levels increase slowly throughout pregnancy, but they increase more rapidly as the end of pregnancy approaches. Early in pregnancy estrogen and progesterone are secreted by the ovary. During midpregnancy there is a shift toward estrogen and progesterone secretion by the placenta. Late in pregnancy these two hormones are secreted by the placenta.

Menopause

When a female is 40 to 50 years old, the menstrual cycles become less regular, and ovulation does not consistently occur during each cycle. Eventually the menstrual cycles stop completely. The cessation of menstrual cycles is called **menopause,** and the whole time period from the onset of irregular cycles to their complete cessation is called the **female climacteric.**

The major cause of menopause is age-related changes in the ovary. The number of follicles remaining in the ovaries of menopausal women is small. In addition, the follicles that remain become less sensitive to stimulation by LH and FSH, even though LH and FSH levels are elevated. As the ovaries become less responsive to stimulation by FSH and LH, fewer mature follicles and corpora lutea are produced. Gradual morphological changes occur in the female in response to the reduced amount of estrogen and progesterone produced by the ovaries (Table 28-2).

A variety of symptoms occur in some females during the climateric, including "hot flashes," irritability, fatigue, anxiety, and occasionally, severe emotional disturbances. Many of these symptoms can be treated successfully by administering small amounts of estrogen and then gradually decreasing the treatment over time or by providing psychological counseling. Although estrogen therapy has been successful, it prolongs the symptoms in many cases. Some potential side effects of estrogen therapy are of concern such as an increased possibility for the development of breast and uterine cancer.

Table 28-2

Possible changes caused by decreased ovarian hormone secretion in postmenopausal women

AFFECTED STRUCTURES/FUNCTIONS	CHANGES
Menstrual cycle	Five to 7 years before menopause the cycle becomes more irregular; finally the number of cycles in which ovulation does not occur increases, and corpora lutea do not develop
Oviduct	Little change
Uterus	Irregular menstruation gradually is followed by no menstruation; chance of cystic glandular hypertrophy of the endometrium increases; the endometrium finally atrophies, and the uterus becomes smaller
Vagina and external genitalia	Dermis and epithelial lining become thinner; vulva becomes thinner and less elastic; labia majora become smaller; pubic hair decreases; vaginal epithelium produces less glycogen; vaginal pH increases; reduced secretion leads to dryness; the vagina is more easily inflamed and infected
Skin	Epidermis becomes thinner; melanin synthesis increases
Cardiovascular system	Hypertension and atherosclerosis occur more frequently
Vasomotor instability	Hot flashes and increased sweating are correlated with vasodilation of cutaneous blood vessels; hot flashes are not due to abnormal FSH and LH secretion but are related to decreased estrogen levels
Libido	Temporary changes, usually a decrease, in libido are associated with the onset of menopause
Fertility	Fertility begins to decline approximately 10 years before the onset of menopause; by age 50 almost all germ cells and follicles are lost; loss is gradual, and no increased follicular degeneration is associated with the onset of menopause

Control of Pregnancy

Many methods are used to prevent or terminate pregnancy (Figure 28-B; Table 28-A), including methods that prevent fertilization (contraception), prevent implantation of the developing embryo (intrauterine devices [IUDs]), or remove the implanted embryo or fetus (abortion). Many of these techniques are quite effective when done properly. Often, however, effectiveness is reduced because of a human tendency to forget to use the technique or to ignore the correct procedure.

Behavioral Methods

Behavioral methods are those techniques that do not require drugs or an apparatus to perform. If practiced, abstinence is the surest way to prevent pregnancy. The rhythm method is to abstain from having sexual intercourse near the time of ovulation. A major factor in the success of this method is the ability to predict accurately the time of ovulation. Although the rhythm method provides some protection against becoming pregnant, it has a relatively high rate of failure that is a result of both the inability to predict the time of ovulation and the failure to abstain during the period of fertility.

One method for predicting ovulation is based on the length of the menstrual cycle. If the menstrual cycle is 28 days long, ovulation normally occurs within 1 day of the fourteenth day of the cycle. Sexual intercourse should be avoided 4 days before the expected day of ovulation and 3 days after it. For menstrual cycles of greater or shorter length, the expected day of ovulation must be estimated. For women with irregular menstrual cycles, the rhythm method is much less effective.

Changes in body temperature can be used to detect the time of ovulation. Body temperature increases by approximately 0.5° F in response to increasing levels of progesterone secreted by the corpus luteum. However, the techniques of detecting such temperature changes require charting the changes during the entire menstrual cycle and avoiding exercise or other activities that affect body temperature while the measurements are made. Even if the elevation in body temperature can be detected with certainty, it occurs several hours after ovulation has occurred.

Coitus interruptus is removal of the penis from the vagina just before ejaculation. This is a very unreliable method of preventing pregnancy since it requires perfect awareness and willingness on the male's part to withdraw the penis at the correct time. It also ignores the fact that some sperm cells are found in pre-ejaculatory emissions.

Barrier Methods

Barrier methods use a barrier that prevents sperm cells from reaching the ovum. A condom is a sheath of animal membrane, rubber, or plastic that is placed on the erect penis. The condom is a barrier device since the ejaculate is collected within the condom instead of within the vagina. Condoms also provide protection against sexually transmitted disease.

The diaphragm is a flexible plastic or rubber dome that is placed in the fornix of the vagina where it prevents passage of sperm cells from the vagina through the cervical canal of the uterus.

Spermicidal Agents

Spermicidal agents are substances that are toxic to sperm cells. The most commonly used spermicidal agents are foams or creams that are inserted within the vagina before sexual intercourse. They kill the sperm cells. After intercourse spermicidal douches, which remove and kill sperm cells, are sometimes used.

The vaginal sponge is either a natural or a synthetic sponge that is permeated with a spermicidal agent and is placed in the vagina where it acts as a barrier and kills the sperm cells. Spermicidal creams are also used with the diaphram to increases its effectiveness.

Continued.

Table 28-A
Effectiveness of various methods for preventing pregnancy

TECHNIQUE	EFFECTIVENESS WHEN USED PROPERLY (%)	ACTUAL EFFECTIVENESS (%)
Abortion	100	Unknown
Sterilization	100	99.9
Combination (estrogens and progesterones) pill	99.9	98
Intrauterine device	98	98
Mini pill (low dose of estrogens and progesterones)	99	97
Condom plus spermicide	99	96
Condom alone	97	90
Diaphragm plus spermicide	97	85
Foam	97	80
Rhythm	97	70

Control of Pregnancy—cont'd

Surgical Techniques

Vasectomy is a common method used to render males permanently incapable of fertilization without affecting the performance of the sexual act. Vasectomy is a surgical procedure used to sever and tie the ductus deferentia within the scrotal sac, preventing sperm cells from becoming part of the ejaculate. Since such a small volume of ejaculate comes from the testes and epididymides, vasectomy has little effect on the volume of the ejaculated semen. The sperm cells are destroyed in the epididymides.

A common method of permanent birth control in females is tubal ligation, a procedure in which the uterine tubules are tied and cut or clamped through an incision made through the wall of the abdomen. Laparoscopy is commonly used so that only a small opening is required to perform the operation.

Chemical Methods

The administration of hormones to suppress fertility is one of the most effective temporary methods for preventing pregnancy. Both estrogens and progesterone, if administered in sufficient amounts, inhibit ovulation. The estrogens and the progesterone used in birth control pills are synthetic steroids. They have a longer half-life in the circulatory system than the natural steroids, which are rapidly degraded by the liver. It is unclear precisely where estrogen and progesterone act, but the LH surge that precedes ovulation is reduced, presumably because estrogens and progesterone inhibit both GnRH and LH release from the hypothalamus and adenohypophysis, respectively. In women who take birth control pills with a very low dose of estrogen and progesterone ovulation occurs, but pregnancy does not occur perhaps because of an unusually rapid rate of oocyte transport through the uterine tube as a result of abnormal contractions.

Birth control pills usually consist of a mixture of estrogens and progesterone. The problem related to their development and effective use has been to discover the minimum dose of each substance that suppresses ovulation but does not cause unwanted side effects. Too much of either hormone causes abnormal menstrual bleeding. A greater risk of cardiovascular disease is associated with the use of birth control pills that contain higher doses of estrogen and progesterone, especially in women more than 35 years of age. A greater-than-normal risk of cardiovascular abnormalities such as stroke has been correlated with women who take birth control

A

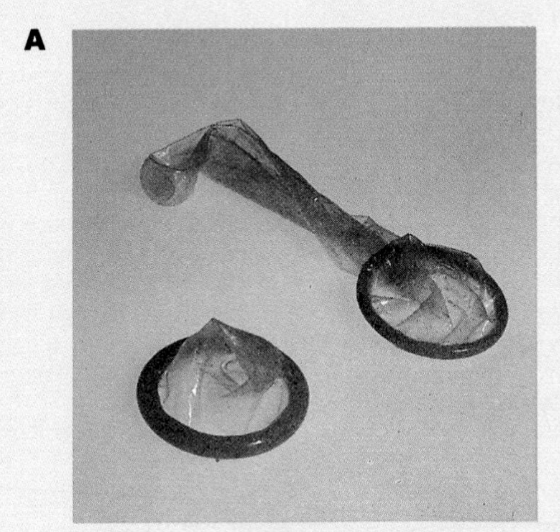

B

C

D

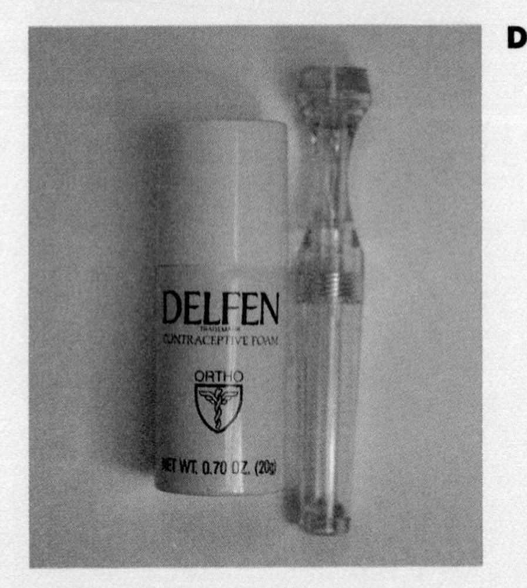

FIGURE 28-B Contraceptive devices and techniques.
A Condom; **B** Diaphragm with spermacidal; **C** Vaginal sponge; **D** Spermicidal foam; **E** Vasectomy; **F** Tubal ligation; and **G** Oral contraceptives.

pills and who smoke cigarettes or who have a history of high blood pressure. The long-term risks associated with birth control pill usage are not clear, but the pill is effective and has a minimum frequency of complications for women of reproductive age until at least age 35. After that age, there is an increased probability of stroke and cancer (especially breast and uterine cancer).

The normal procedure for ingestion of most birth control pills is to take a pill once each day for 21 days. Then the medication is stopped, or a placebo pill (one without estrogen or progesterone) is taken for the next 7 days. The resulting drop in hormones allows menstruation to occur. After the cessation of menstrual flow, the administration of the pills begins again.

Intrauterine Devices

Intrauterine devices (IUDs) are inserted into the uterus through the cervix. Precisely how the IUD works is not clear, but it successfully prevents implantation of the developing embryo within the endometrium of the uterus. The IUD is an effective means of preventing pregnancy with few complications. However, some IUDs have produced serious side effects such as perforation of the uterus; as a result of the controversy surrounding these IUDs, many of them have been removed from the market and are no longer in common use.

Lactation

Lactation prevents the ovarian cycle (and therefore the uterine cycle) for a few months after parturition in most women. Nerve impulses produced in the nipples by suckling travel to the hypothalamus and repress GnRH. Despite continual lactation, the ovarian and uterine cycles eventually resume. Since ovulation normally precedes menstruation, it is possible for pregnancy to occur before even one menstruation after delivery of an infant and lactation. Consequently, lactation is not a reliable method of birth control.

Interruption of Pregnancy

In some cases pregnancies are terminated by surgical procedures called abortions. The most common method for performing abortions is the insertion of an instrument through the cervix into the uterus. The instrument scrapes the endometrial surface, and at the same time a strong suction is applied. The embryo disintegrates and is suctioned out of the uterus. This technique is normally used only in pregancies that have progressed less than 3 months.

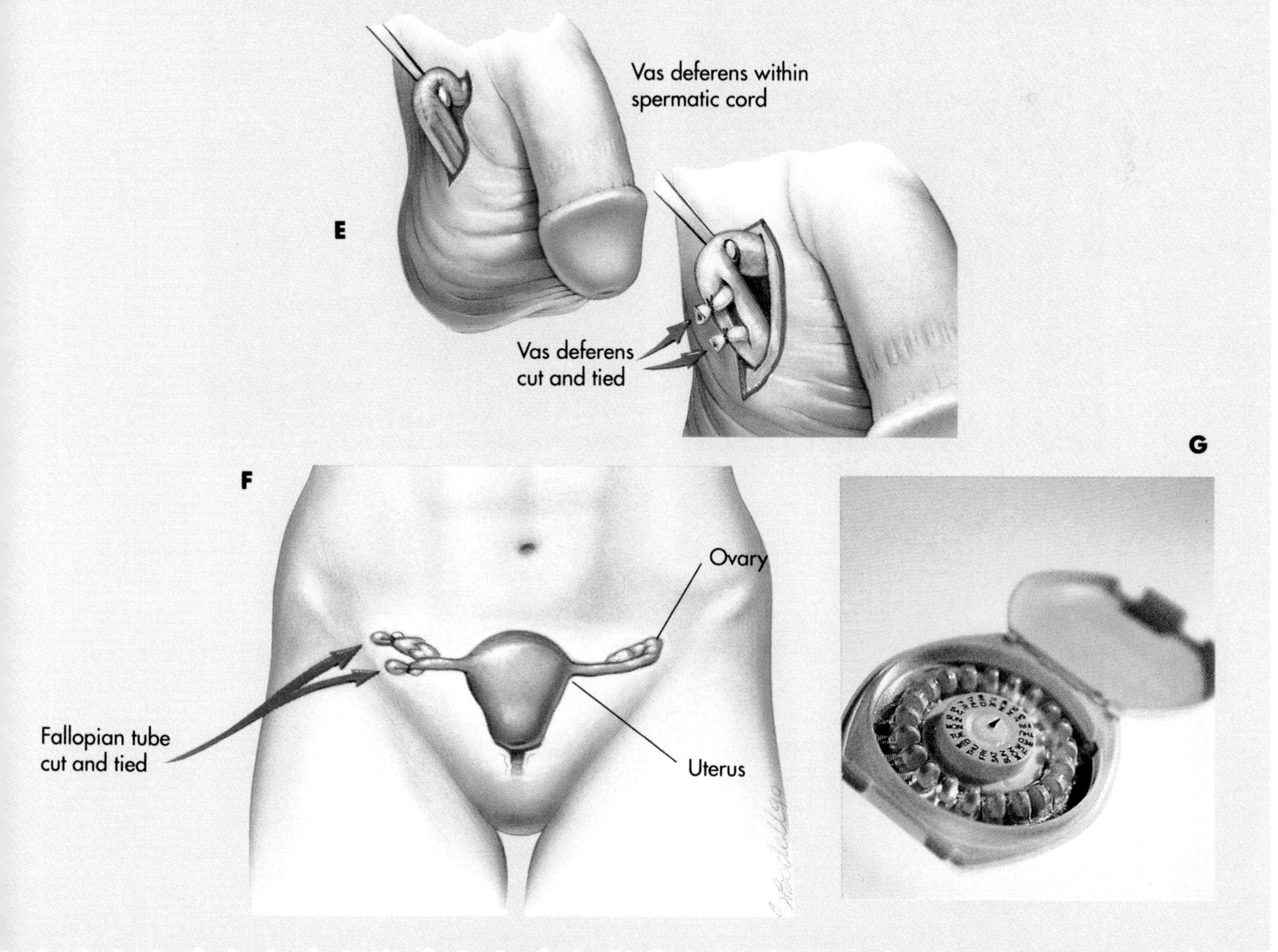

The Reproductive System

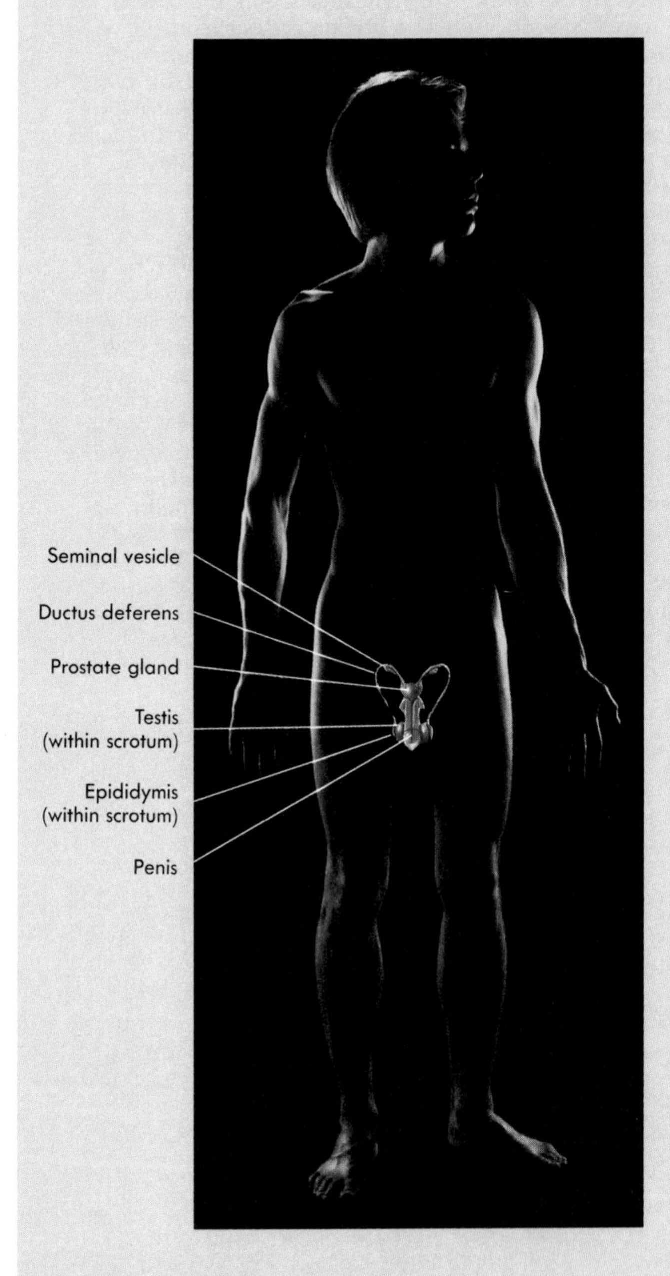

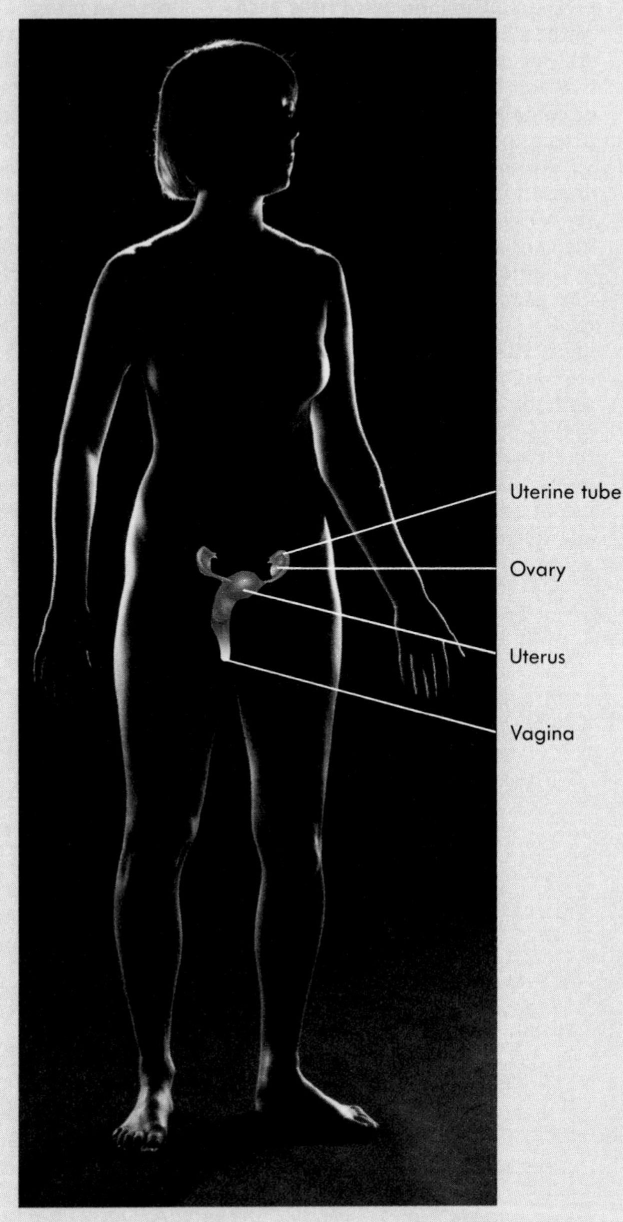

Major Components

Male

Penis, testis, epididymis, ductus deferens, seminal vesicle, and prostate gland

Female

Vagina, ovary, uterine tube, and uterus

Major Functions

Performs the processes of reproduction
Controls sexual functions and behavior
Produces hormones that play important roles in development, sexual behavior, and reproduction

SUMMARY

THE MALE REPRODUCTIVE SYSTEM PRODUCES SPERM CELLS AND TRANSFERS THEM TO THE FEMALE. THE FEMALE REPORUDCTIVE SYSTEM PRODUCES THE OOCYTE AND NURTURES THE DEVELOPING CHILD.

MALE REPRODUCTIVE SYSTEM *p. 909*

Scrotum

1. The scrotum is a two-chambered sac that contains the testes.
2. The dartos and cremaster muscles help to regulate testicular temperature.

Perineum

The perineum is the diamond-shaped area between the thighs and consists of a urogenital triangle and an anal triangle.

Testes

1. The tunica albuginea is the outer connective tissue capsule of the testes.
2. The testes are divided by septa into compartments that contain the seminiferous tubules and the cells of Leydig.
3. The seminiferous tubules empty into short ducts that lead to the rete testis. The rete testis opens into the efferent ductules of the epididymis.
4. During development the testes pass from the abdominal cavity through the inguinal canal to the scrotum.
5. Sperm cells (spermatazoa) are produced in the seminiferous tubules.
   - Spermatogonia divide (mitosis) to form primary spermatocytes.
   - Primary spermatocytes divide (first division of meiosis) to form secondary spermatocytes that divide (second division of meiosis) to form spermatids.
   - Spermatids develop an acrosome and a flagellum to become sperm cells.
   - Sertoli cells nourish the sperm cells, form a blood-testes barrier, and produce hormones.

Ducts

1. The epididymis is a coiled tube system located on the testis that is the site of sperm cell maturation.
2. The ductus deferens passes from the epididymis into the abdominal cavity.
3. The ejaculatory duct is formed by the joining of the ductus deferens and the duct from the seminal vesicle.
4. The prostatic urethra extends from the urinary bladder to join with the ejaculatory duct to form the membranous urethra.
5. The membranous urethra extends through the urogenital diaphragm and becomes the spongy urethra, which continues through the penis.
6. The spermatic cord consists of the ductus deferens, blood and lymph vessels, nerves, remnants of the process vaginalis, and tubular sheaths.
7. The spermatic cord passes through the inguinal canal into the abdominal cavity.

Penis

1. The penis consists of erectile tissue.
   - The two corpora cavernosa form the dorsum and the sides.
   - The corpus spongiosum forms the ventral portion and the glans penis.
2. The root of the penis attaches to the coxae.
3. The prepuce covers the glans penis.

Accessory Glands

1. The seminal vesicles empty into the ejaculatory ducts.
2. The prostate gland consists of glandular and muscular tissue and empties into the prostatic urethra.
3. The bulbourethral glands are compound mucous glands that empty into the spongy urethra.
4. Secretions
   - Semen is a mixture of gland secretions and sperm cells.
   - The bulbourethral glands and the urethral mucous glands produce mucus that neutralizes the acidic pH of the urethra.
   - The testicular secretions contain sperm cells.
   - The seminal vesicle fluid contains fructose and fibrinogen.
   - The prostate secretions neutralize the semen. Clotting factors activate fibrinogen, and fibinolysin breaks down fibrin.

PHYSIOLOGY OF MALE REPRODUCTION *p. 918*

Regulation of Sex Hormone Secretion

1. Gonadotropin-releasing hormone (GnRH) is produced in the hypothalamus and is released in surges.
2. GnRH stimulates luteinizing hormone (LH) and follicle-stimulating hormone (FSH) release from the anterior pituitary.
   - LH stimulates the cells of Leydig to produce testosterone.
   - FSH stimulates spermatogenesis.
3. Inhibin, produced by Sertoli cells, inhibits FSH secretion.

Puberty

1. Before puberty small amounts of testosterone inhibit GnRH release.
2. During puberty testosterone does not completely suppress GnRH release, resulting in increased production of FSH, LH, and testosterone.

Effects of Testosterone

1. Testosterone is produced by the cells of Leydig, the adrenal cortex, and possibly the Sertoli cells.
2. Testosterone causes the development of male sex organs in the embryo and stimulates the descent of the testes.
3. Testosterone causes enlargement of the genitals and is necessary for spermatogenesis.
4. Other effects of testosterone.
   - Hair growth stimulation (pubic area, axilla, and beard) and inhibition (male pattern baldness).
   - Enlargement of the larynx and deepening of the voice.
   - Increased skin thickness and melanin and sebum production.
   - Increased protein synthesis (muscle), bone growth, blood cell synthesis, and blood volume.
   - Increased metabolic rate.

Male Sexual Behavior and the Male Sex Act

1. Testosterone is required for normal sex drive.
2. Stimulation of the sexual act can be tactile or psychic.
3. Afferent impulses pass through the pudendal nerve to the sacral region of the spinal cord.
4. Parasympathetic stimulation.
   - Erection is due to vasodilation of the blood vessels that supply the erectile tissue.
   - Mucus is produced by the glands of the urethra and the bulbourethral glands.

5. Sympathetic stimulation causes erection, emission, and ejaculation.

FEMALE REPRODUCTIVE SYSTEM *p. 920*

Ovaries

1. The ovaries are held in place by the broad ligament, the mesovarium, the suspensory ligaments, and the ovarian ligaments.
2. The ovaries are covered by the peritoneum (ovarian epithelium) and the tunica albuginea.
3. The ovary is divided into a cortex (contains follicles) and a medulla (receives blood and lymph vessels and nerves).
4. Follicular development.
   - Oogonia proliferate and become primary oocytes that are in prophase I of meiosis.
   - Primary follicles are primary oocytes surrounded by granulosa cells.
   - During puberty primary follicles become secondary follicles.
   - The primary oocytes continue meiosis to metaphase II and become secondary oocytes surrounded by the zona pellucida. The center of the follicle fills with fluid to form the antrum, the granulosa cells increase in number, and theca cells form around the secondary follicle.
   - Graafian follicles are enlarged secondary follicles at the surface of the ovary.
5. Ovulation.
   - The follicle swells and ruptures, and the secondary oocyte is released from the ovary.
   - The second meiotic division is completed when the secondary oocyte unites with a sperm cell to form a zygote.
6. Fate of the follicle.
   - The graafian follicle becomes the corpus luteum.
   - If fertilization occurs, the corpus luteum persists. If there is no fertilization, it becomes the corpus albicans.

Uterine Tubes

1. The mesosalpinx holds the uterine tubes.
2. The uterine tubes transport the oocyte or zygote from the ovary to the uterus.
3. Structures.
   - The ovarian end of the uterine tube is expanded as the infundibulum. The opening of the infundibulum is the ostium, which is surrounded by fimbriae.
   - The infundibulum connects to the ampulla that narrows to become the isthmus. The isthmus becomes the uterine part of the uterine tube and passes through the uterus.
4. The uterine tube consists of an outer serosa, a middle muscular layer, and an inner mucosa with simple ciliated columnar epithelium.
5. Movement of the oocyte.
   - Cilia move the oocyte over the fimbriae surface into the infundibulum.
   - Peristaltic contractions and cilia move the oocyte within the uterine tube.
   - Fertilization occurs in the ampulla where the zygote remains for several days.

Uterus

1. The uterus consists of the body, the isthmus, and the cervix. The uterine cavity and the cervical canal are the spaces formed by the uterus.
2. The uterus is held in place by the broad, round, and uterosacral ligaments.
3. The wall of the uterus consists of the perimetrium (serous membrane), myometrium (smooth muscle), and endometrium (mucous membrane).

Vagina

1. The vagina connects the uterus (cervix) to the vestibule.
2. The vagina consists of a layer of smooth muscle and an inner lining of moist stratified squamous epithelium.
3. The vagina is folded into rugae and longitudinal folds.
4. The hymen covers the vestibular opening of the vagina.

External Genitalia

1. The vulva, or pudendum, is the external genitalia.
2. The vestibule is the space into which the vagina and the urethra open.
3. Erectile tissue.
   - The clitoris is formed by the two corpora cavernosa.
   - The bulb of the vestibule is formed by the corpus spongiosum.
4. The labia minora are folds that cover the vestibule and form the prepuce.
5. The greater and lesser vestibular glands produce a mucous fluid.
6. The labia majora cover the labia minora.
   - The pudendal cleft is a space between the labia majora.
   - The mons pubis is an elevated fat deposit superior to the labia majora.

Perineum

The clinical perineum is the region between the vagina and the anus.

Mammary Glands

1. The mammary glands are modified sweat glands.
   - The mammary glands consist of glandular lobes and adipose tissue.
   - The lobes consist of lobules that are divided into alveoli.
   - The lobes connect to the nipple through the lactiferous ducts.
   - The nipple is surrounded by the areola.
2. The breast is supported by Cooper's ligaments.

PHYSIOLOGY OF FEMALE REPRODUCTION *p. 930*

Puberty

1. Puberty begins with the first menstrual bleeding (menarche).
2. Puberty begins when GnRH levels increase.

Menstrual Cycle

1. Ovarian cycle.
   - FSH initiates development of the primary follicles.
   - The follicles secrete a substance that inhibits the development of other follicles.
   - LH stimulates ovulation and completion of the first meiotic division by the primary oocyte.
   - The LH surge stimulates the formation of the corpus luteum. If fertilization occurs, human chorionic gonadotropin (HCG) stimulates the corpus luteum to persist. If fertilization does not occur, the corpus luteum becomes the corpus albicans.
2. A positive-feedback mechanism causes FSH and LH levels to increase near the time of ovulation.
   - Estrogen produced by the theca cells of the follicle stimulates GnRH secretion.
   - GnRH stimulates FSH and LH, which stimulate more estrogen secretion, and so on.

- Inhibition of GnRH levels causes FSH and LH levels to decrease after ovulation. Inhibition is due to the high levels of estrogen and progesterone produced by the corpus luteum.

3. Uterine cycle.
    - Menses (day 1 to days 4 or 5). The spiral arteries constrict, and endometrial cells die. The menstrual fluid is composed of sloughed cells, secretions, and blood.
    - Proliferation phase (day 5 to day 14). Epithelial cells multiply and form glands, and the spiral arteries supply the glands.
    - Secretory phase (day 15 to day 28). The endometrium becomes thicker, and the endometrial glands secrete.
    - Estrogen stimulates proliferation of the endometrium and synthesis of progesterone receptors.
    - Increased progesterone levels cause hypertrophy of the endometrium, stimulate gland secretion, and inhibit uterine contractions. Decreased progesterone levels cause the spiral arteries to constrict and start menses.

Female Sexual Behavior and the Female Sex Act

1. Female sex drive is partially influenced by androgens (produced by the adrenal gland) and steroids (produced by the ovaries).
2. Parasympathetic effects.
    - The erectile tissue of the clitoris and the bulb of the vestibule become filled with blood.
    - The vestibular glands secrete mucus, and the vagina extrudes a mucuslike substance.

Female Fertility and Pregnancy

1. Intercourse must take place 3 days before to 1 day after ovulation if fertilization is to occur.
2. Sperm cells are transported to the ampulla through contractions of the uterus and the uterine tubes.
3. Implantation of the developing embryo into the uterine wall occurs when the uterus is most receptive.
4. Estrogen and progesterone secreted first by the corpus luteum and later by the placenta are essential for the maintenance of pregnancy.

Menopause

The female climacteric begins with irregular menstrual cycles and ends with menopause, the cessation of the menstrual cycle.

CONTENT REVIEW

1. What is the scrotum? Explain the function of the dartos and cremaster muscles.
2. Describe the covering and the structure of a testis.
3. When and how do the testes descend into the scrotum?
4. Where, specifically, are sperm cells produced in the testes? Describe the process of spermatogenesis.
5. Name all the ducts the sperm traverse to go from their site of production to the outside.
6. Where do sperm cells undergo maturation?
7. Distinguish between the prostatic, membranous, and spongy portions of the urethra.
8. Name the parts of the spermatic cord.
9. Describe the erectile tissue of the penis. Define glans penis, the crus, the bulb, and the prepuce.
10. State where the seminal vesicles, prostate gland, and bulbourethral glands empty into the male reproductive duct system.
11. Define emission and ejaculation.
12. Define semen. Describe the contribution to semen of the accessory sex glands. What is the function of each secretion?
13. Where are gonadotropin-releasing hormone (GnRH), follicle-stimulating hormone (FSH), luteinizing hormone (LH), and inhibin produced? What effects do they produce?
14. What changes in hormone production occur at puberty?
15. Where is testosterone produced? Describe the effects of testosterone on the embryo, during puberty, and on the adult male.
16. What effects does psychic, parasympathetic, and sympathetic stimulation have on the male sex act?
17. Name and describe the ligaments that hold the uterus, uterine tubes, and ovaries in place.
18. Describe the coverings and structure of the ovary.
19. Starting with the oogonia, describe the development and production of a graafian follicle that contains a secondary oocyte.
20. Describe the process of ovulation.
21. What is the corpus luteum? What happens to the corpus luteum if fertilization occurs? If fertilization does not occur?
22. Describe the structures of the uterine tube. How are they involved in moving the oocyte or the zygote?
23. Where does fertilization usually take place?
24. Name the parts of the uterus. Describe the layers of the uterine wall.
25. Where is the vagina located? Describe the layers of the vaginal wall. What are rugae and longitudinal folds?
26. What are the hymen, vulva, pudendum, and vestibule?
27. What erectile tissue is in the clitoris and the bulb of the vestibule? What is the function of the clitoris and the bulb of the vestibule?
28. Describe the labia minora, the prepuce, the labia majora, the pudendal cleft, and the mons pubis.
29. Where are the greater and lesser vestibular glands located? What is their function?
30. Define the perineum. What is the anterior clinical perineum? Define and give the purpose of an episiotomy.
31. Describe the route taken by a drop of milk from its site of production to the outside of the body. What are Cooper's ligaments?
32. Describe the events of the ovarian cycle. What role do FSH and LH play in the ovarian cycle? Where is human chorionic gonadotropin (HCG) produced, and what effect does it have on the ovary?
33. Describe how the cyclic increase and decrease in FSH and LH is produced.
34. Name the stages of the uterine cycle, and describe the events that take place in each stage. What are the effects of estrogen and progesterone on the uterus?
35. When must intercourse take place for fertilization to occur?
36. Define menopause and the female climacteric. What causes these changes and what symptoms commonly occur?

CONCEPT REVIEW

1. If an adult male were castrated, what would happen to the levels of gonadotropin-releasing hormone (GnRH), follicle-stimulating hormone (FSH), luteinizing hormone (LH), and testosterone in his blood? What effect would these hormonal changes have on sexual characteristics and behavior?
2. If a 9-year-old boy were castrated, what would happen to the levels of GnRH, FSH, LH, and testosterone in his blood? What effect would these hormonal changes have on sexual characteristics and behavior?
3. Suppose you wanted to produce a birth control pill for men. Based on what you know about the male hormone system, what would you want the pill to do? Discuss any possible side effects that could be produced by your pill.
4. If the ovaries were removed from a postmenopausal woman, what would happen to the levels of GnRH, FSH, LH, estrogen, and progesterone in her blood? What symptoms would you expect to observe?
5. If the ovaries were removed from a 20-year-old woman, what would happen to the levels of GnRH, FSH, LH, estrogen, and progesterone in her blood? What side effects would these hormonal changes have on her sexual characteristics and behavior?
6. Normal adult women were divided into two groups. Both groups were composed of females who had been married for at least 2 years and were not pregnant at the beginning of the experiment. They weighed approximately the same amount, and none smoked cigarettes, although some women did drink alcohol occasionally. Group A women received a placebo in the form of a sugar pill each morning during their menstrual cycles. Group B women received a pill containing estrogen and progesterone each morning of their menstrual cycles. Then plasma LH levels were measured before, during, and after ovulation. The results were as follows:

GROUP	4 DAYS BEFORE OVULATION	THE DAY OF OVULATION	4 DAYS AFTER OVULATION
A	18 mg/100 ml	300 mg/100 ml	17 mg/100 ml
B	21 mg/100 ml	157 mg/100 ml	15 mg/100 ml

The number of pregnancies in group A was 37 per 100 women per year. The number of pregnancies in group B was 1.5 per 100 women per year. What conclusion can you reach based on these data? Explain the mechanism involved.

7. A woman who was taking birth control pills that consisted of only progesterone experienced the hot flash symptoms of menopause. Explain why.
8. GnRH can be used to treat some women who want to have children but have not been able to get pregnant. Explain why it would be critical to administer the correct concentration of GnRH at the right time during the menstrual cycle.

? ?

ANSWERS TO PREDICT QUESTIONS

1 *p. 917.* The prostate gland is adjacent to the wall of the rectum. A finger inserted into the rectum can palpate the prostate gland through the rectal wall.

2 *p. 917.* Coagulation may help keep the sperm cells within the female reproductive tract, increasing the likelihood of fertilization.

3 *p. 932.* If administered before the preovulatory LH surge, estrogen would stimulate the hypothalamus to secrete GnRH. The GnRH, in turn, stimulates LH secretion from the adenohypophysis. Therefore a large amount of estrogen administered at this time should produce a large surge of LH. Continual administration of high levels of GnRH would cause adenohypophyseal cells to become insensitive to GnRH. Thus LH and FSH levels would remain low, and the ovarian cycle would stop.

4 *p. 933.* High progesterone levels after menses would inhibit GnRH secretion from the hypothalamus and therefore FSH and LH secretion from the adenohypophysis. Without FSH and LH, the events of the ovarian cycle, including estrogen production, would be inhibited. Since estrogen causes proliferation of the endometrium, thickening of the endometrium would not be expected. Also, estrogen increases the synthesis of uterine progesterone receptors, and without estrogen the secretory response of the endometrium to the elevated progesterone should be inhibited.

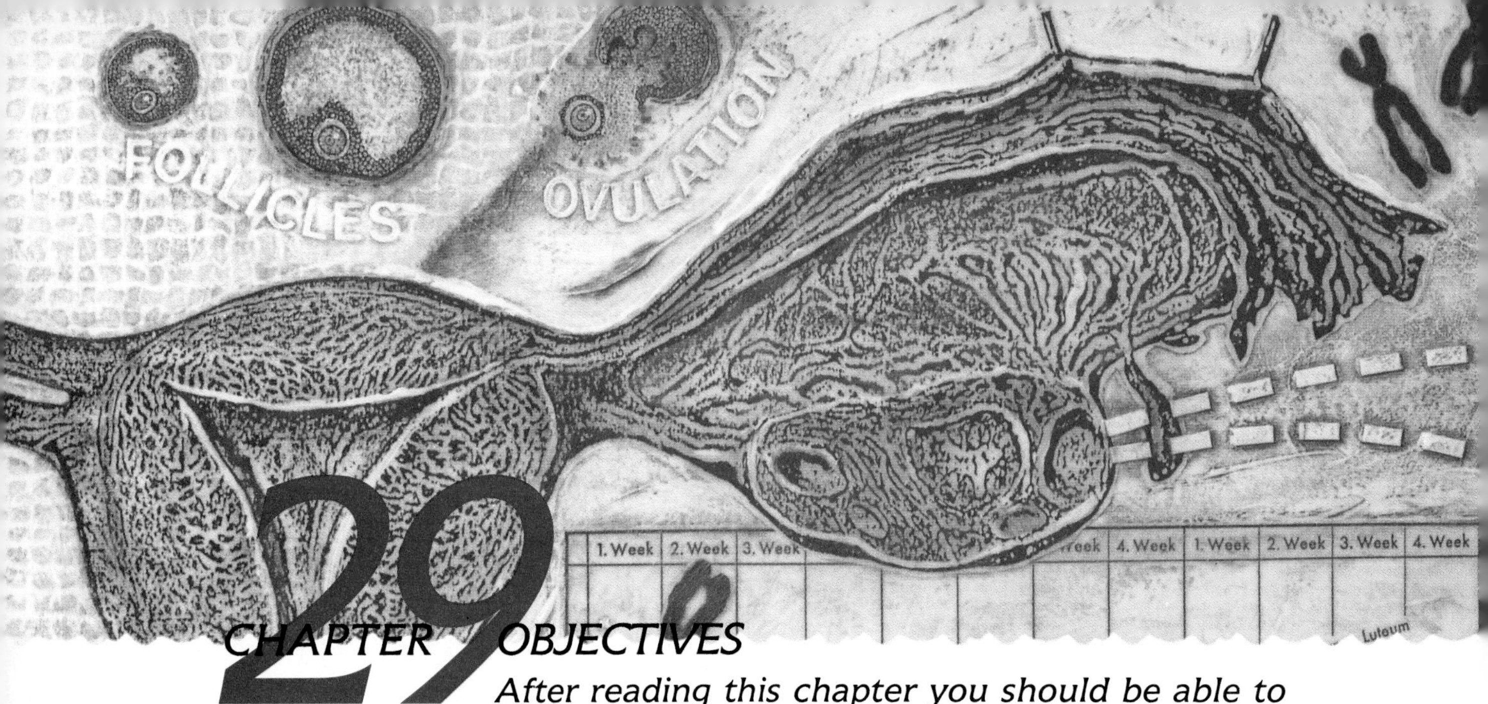

29 CHAPTER OBJECTIVES

After reading this chapter you should be able to

1. List the three prenatal periods and state major events associated with each.
2. Describe the events of fertilization.
3. Define the term pluripotent and explain what it means in terms of development.
4. Describe the morula and the blastocyst.
5. State the derivatives of the inner cell mass and the trophoblast.
6. Describe the process of implantation and placental formation.
7. List the three germ layers, describe their formation, and list the adult derivatives of the three germ layers.
8. Describe the formation of the neural tube and the somites.
9. Describe the formation of the gastrointestinal tract and the body cavities.
10. Describe the formation of the limbs, face, and palate.
11. Briefly describe the formation of the following major organ systems: integumentary, skeletal, muscular, nervous, endocrine, circulatory, respiratory, digestive, urinary, and reproductive.
12. Explain the process by which a one-chambered heart becomes a four-chambered heart.
13. Describe the effects of hormones on the development of the male and the female reproductive systems.
14. Explain the hormonal and nervous system factors responsible for parturition.
15. Discuss circulatory, digestive, and other changes that occur at the time of birth.
16. Explain the hormonal and nervous system factors responsible for lactation.
17. List the stages of life and describe the major events associated with each stage.
18. Describe the major changes associated with aging and death.
19. Explain the major concepts of genetics and the major inheritance patterns.

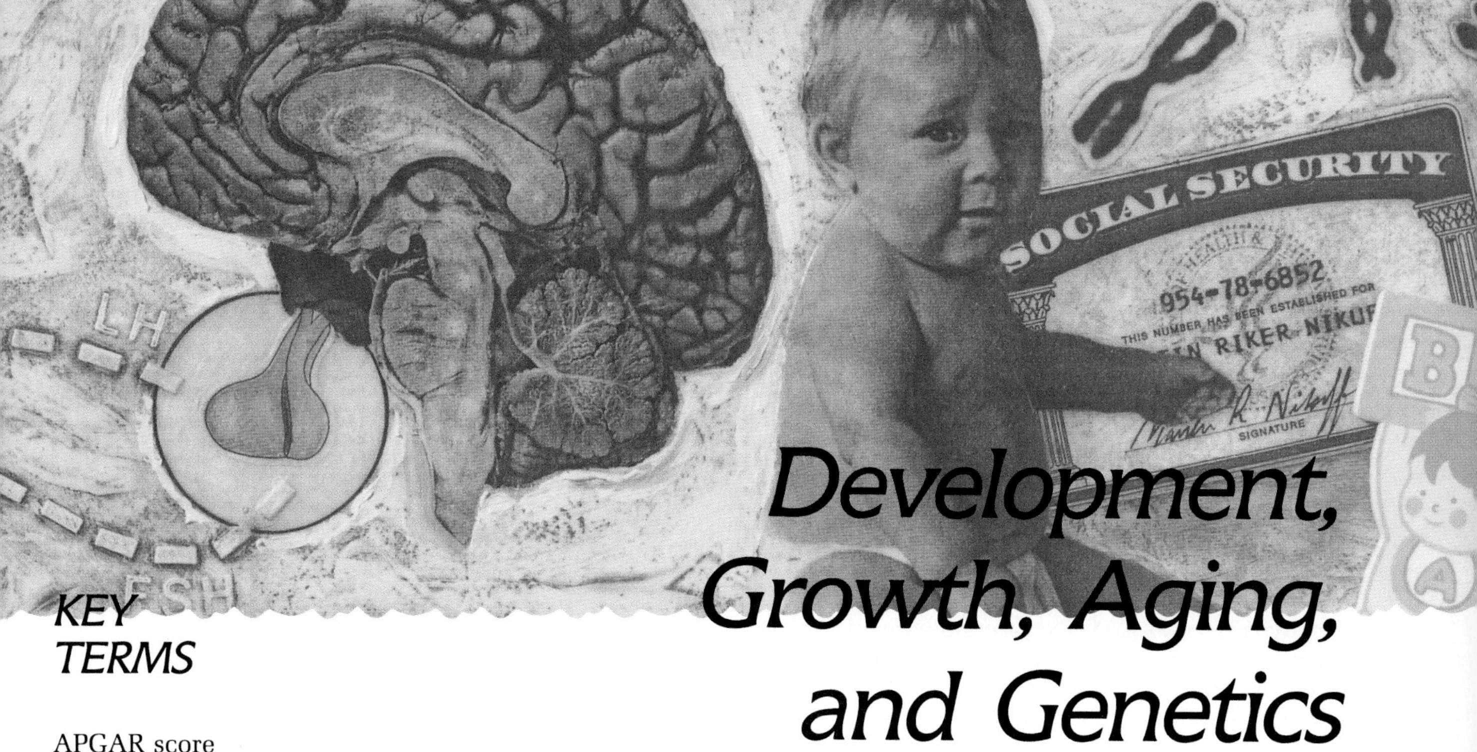

Development, Growth, Aging, and Genetics

KEY TERMS

APGAR score

blastocyst (blas′to-sist)

celom (se′lōm)

cerebral death

ectoderm (ek′to-derm)

embryo

endoderm (en′do-derm)

fetus

heterozygous (het′er-o-zi′gus)

homozygous (ho′mo-zi′gus)

lactation

mesoderm (mez′o-derm)

neural tube

parturition (par-tu-rish′un)

primitive streak

somite (so′mīt)

zygote (zi′gōt)

RELATED TOPICS

The following terms or concepts are important for a good understanding of this chapter. If you are not familiar with them, you should review them before proceeding.

Body cavities (Chapter 1)

Chromosome structure (Chapter 3)

General histology of the four tissue types (Chapter 4)

Hormones from the adenohypophysis (Chapter 18)

Arteriosclerosis (Chapter 21)

Ovulation, pregnancy, and birth (Chapter 28)

THE LIFE SPAN OF A PERSON IS USUALLY CONSIDERED THE TIME FROM BIRTH TO DEATH. HOWEVER, THE 9 MONTHS BEFORE BIRTH COMPRISE A CRITICAL PART OF AN INDIVIDUAL'S EXISTENCE, AND THE EVENTS THAT OCCUR DURING THAT PERIOD HAVE PROFOUND EFFECTS ON THE REST OF THAT PERSON'S LIFE. This chapter describes a number of events that occur during development. Genetics and the various life stages that occur after birth, including aging and death, are also discussed briefly.

Most people develop normally and are born without defects, but approximately 10 out of every 100 people are born with some type of birth defect, and 3 out of every 100 people are born with a birth defect so severe that it requires medical attention during the first year of life. Later in life many more people discover unknown congenital problems such as the tendency to develop asthma, certain brain disorders, or cancer.

PRENATAL DEVELOPMENT

The prenatal period (i.e., the period from conception until birth) can be divided into three portions: (1) the **germinal period**—approximately the first 2 weeks of development during which the primitive germ layers are formed; (2) the **embryonic period**—from approximately the second to the eighth week of development during which the major organ systems come into existence; and (3) the **fetal period**—the last 7 months of the prenatal period during which the organ systems grow and become more mature.

The medical community in general uses the **last menstrual period (LMP)** to calculate the **clinical age** of the unborn child. Most embryologists, on the other hand, use **postovulatory age** to describe the timing of developmental events. Since ovulation occurs approximately 14 days after LMP and fertilization occurs near the time of ovulation, it is assumed that postovulatory age is 14 days less than clinical age.

Fertilization

Thousands of sperm cells reach the oocyte and help digest a path through the cumulus mass cells and zona pellucida, but normally only one sperm cell penetrates the oocyte cell membrane and enters the cytoplasm in the process of **fertilization.** Entrance of a sperm cell into the oocyte stimulates the female nucleus to undergo the second meiotic division, and the second polar body is formed. The nucleus that results from the second meiotic division, called the **female pronucleus,** moves to the center of the cell where it meets the enlarged head of the sperm cell, the **male pronucleus.** Both the male and female pronuclei are haploid (each having one half of each chromosome pair; see Chapter 3), and their fusion, which completes the process of fertilization, restores the diploid number of chromosomes. The product of fertilization (Figure 29-1, *A*) is the **zygote** (zi'gōt).

The zygote may, in rare cases, split, resulting in **"identical"** or **monozygotic twins.** Occasionally a woman may ovulate two or more oocytes at the same time. Fertilization of these multiple oocytes by different sperm cells results in **"fraternal"** or **dizygotic twins.** Multiple ovulation can occur naturally or can be stimulated by injection of drugs that stimulate gonadotropin release and that are sometimes used to treat certain forms of infertility or for embryo transfer.

Early Cell Division

Approximately 18 to 39 hours after fertilization the zygote divides to form two cells. Those two cells divide to form four cells, which divide to form eight cells, and so on (Figure 29-1, *B* to *D*). The cells of this dividing embryonic mass are referred to as **pluripotent** (ploo-rĭ-po'tent; multiple powered; i.e., any cell has the ability to develop into a wide range of tissues). As a result, the total number of embryonic cells can be decreased, increased, or reorganized without affecting the normal development of the embryo.

Morula and Blastocyst

Once the dividing embryonic mass is a solid ball of 12 or more cells, it has the appearance of a sphere composed of numerous smaller spheres and is therefore called a **morula** (mor'u-lah; mulberry). Three or 4 days after ovulation the morula consists of approximately 32 cells. Near this time a fluid-filled cavity begins to appear in the midst of the cellular mass. The hollow sphere is then called a **blastocyst** (blas'to-sist) (Figure 29-2). The fluid-filled cavity, called the **blastocele** (blas'to-sēl), does not occupy the exact center of the blastocyst. A single layer of cells, the **trophoblast** (tro'fo-blast; feeding layer), surrounds most of the blastocele, but at one end of the blastocyst the cells are several layers thick. The thickened area is the **inner cell mass** and is the tissue from which the embryo will develop. The trophoblast forms the placenta and the membranes surrounding the embryo.

Implantation of the Blastocyst and Development of the Placenta

All of the events of the early germinal phase, including the first cell division through formation of the blastocele and the inner cell mass, occur as the embryonic mass moves from the site of fertilization in the ampulla of the uterine tube to the site of implantation in the uterus. Approximately 7 days after fertilization the blastocyst attaches itself to the uterine wall, usually in the area of the uterine fundus, and begins the process of **implantation** (i.e., burrowing into the uterine wall).

As the blastocyst invades the uterine wall, two populations of trophoblast cells develop and form the **placenta** (Figure 29-3), the organ of metabolic exchange between the fetus and the mother. The first is a proliferating population of individual trophoblast cells called the **cytotrophoblast** (si-to-tro'fo-

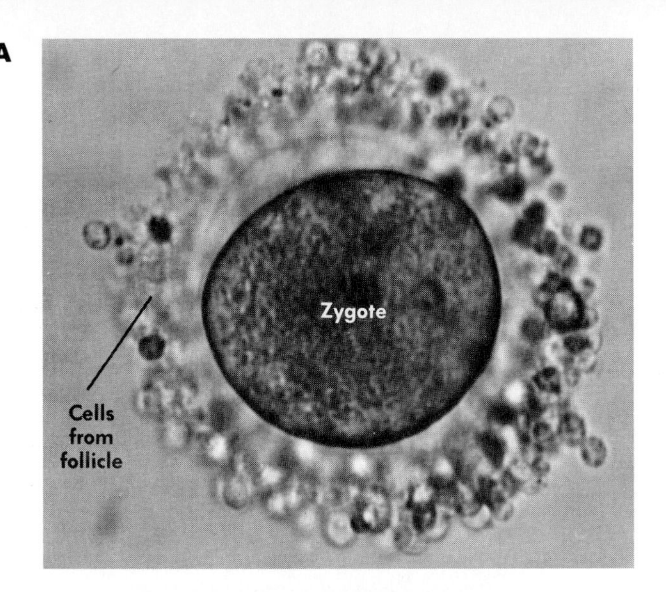

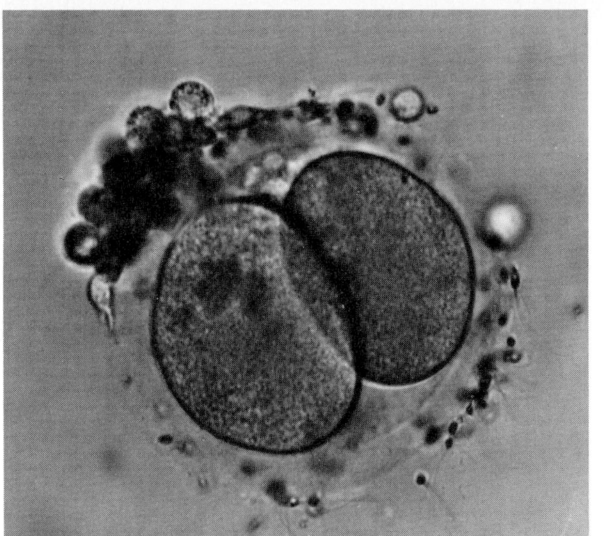

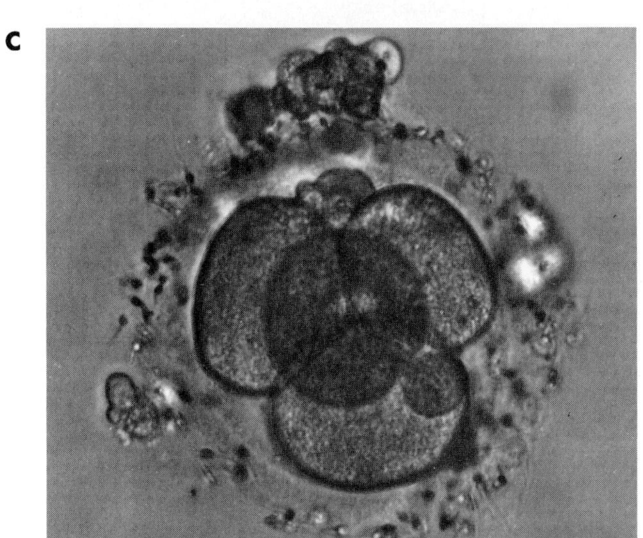

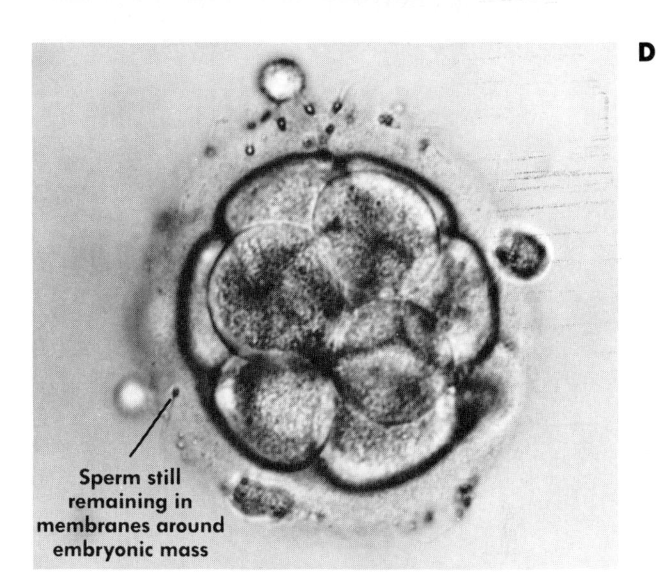

FIGURE 29-1 Early stages of human development. **A** Zygote (120 μm in diameter). **B** to **D** During the early cell divisions the zygote divides into more and more cells, but the total mass remains relatively constant—**B** 2 cells at approximately 18 to 36 hours after fertilization; **C** 4 cells at approximately 36 to 48 hours after fertilization; **D** 16 cells at approximately 48 to 72 hours after fertilization.

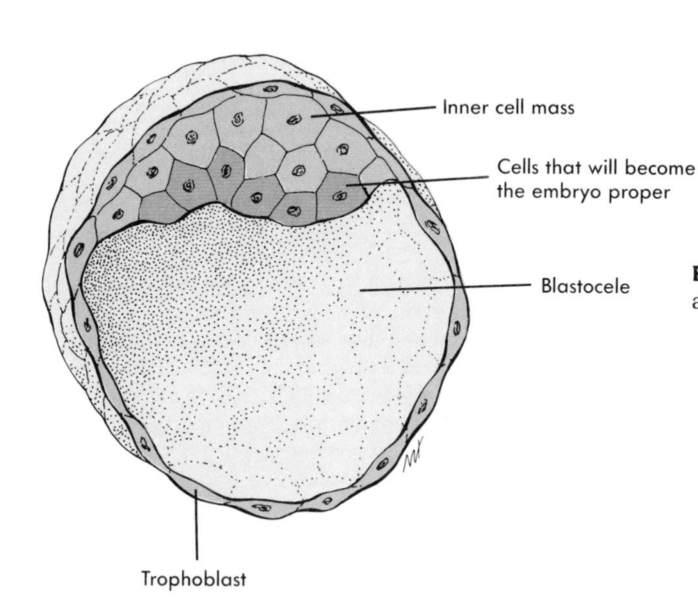

FIGURE 29-2 Blastocyst. Green cells are trophoblastic, and orange cells are embryonic.

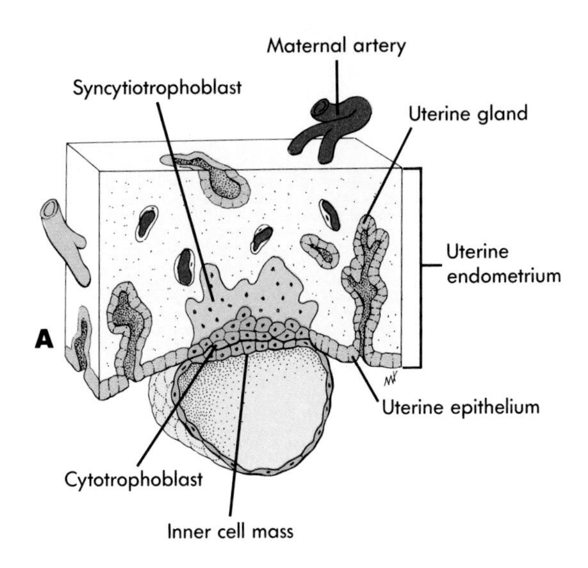

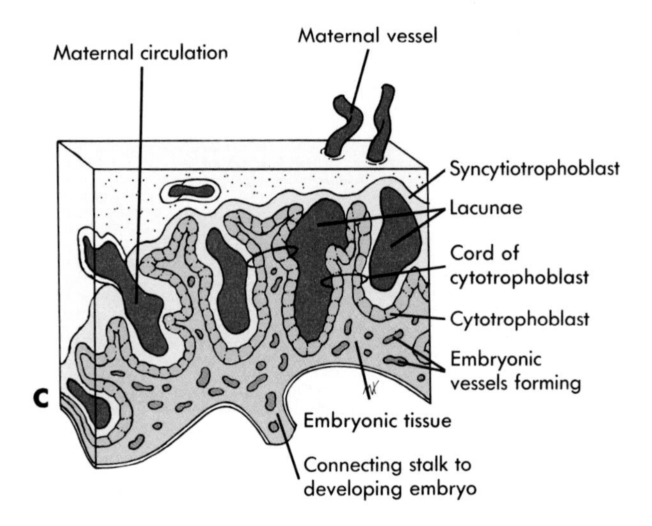

FIGURE 29-3 Implantation of the blastocyst and invasion of the trophoblast to form the placenta. **A** Implantation of the blastocyst with syncytiotrophoblast columns beginning to invade the uterine wall (at approximately 8 to 12 days). **B** Intermediate stage of placental formation (at approximately 14 to 20 days). As maternal blood vessels are encountered by the syncytiotrophoblast, lacunae are formed and are filled with maternal blood. **C** Cytotrophoblast cords surround the syncytiotrophoblast and lacunae, and embryonic tissue enters the cord (at approximately 1 month).

blast). The other is a nondividing syncytium (a multinucleated cell) called the **syncytiotrophoblast** (sin-sish'e-o-tro'fo-blast). The cytotrophoblast remains nearer the other embryonic tissues, whereas the syncytiotrophoblast invades the endometrium of the uterus. The syncytiotrophoblast is nonantigenic; therefore as it invades the maternal tissue, no immune reaction is triggered.

As the syncytiotrophoblast encounters maternal blood vessels, it surrounds them and digests the vessel wall, forming pools of maternal blood within cavities called **lacunae** (lă-ku'ne; Figure 29-3, *B*). The lacunae are still connected to intact maternal vessels so that blood circulates from the maternal vessels through the lacunae. Cords of cytotrophoblasts surround the syncytiotrophoblast and lacunae (Figure 29-3, *C*). Branches sprout from these cords and protrude into the lacunae-like fingers called **chorionic** (ko'rĭ-on'ik) **villi,** and the entire embryonic structure facing the maternal tissues is called the **chorion** (ko'rĭ-on). Embryonic blood vessels follow the cords into the lacunae. In the mature placenta (Figure 29-4) the cytotrophoblast disappears, leaving the embryonic blood supply separated from the maternal blood supply by only the embryonic capillary wall, a basement membrane, and a thin layer of syncytiotrophoblast.

If implantation of the embryo occurs near the cervix, a condition called **placenta previa** (pre've-ah) may occur. In this condition, as the placenta grows, it may extend partially or completely across the internal cervical opening. As the fetus and placenta continue to grow and the uterus stretches, the region of the placenta over the cervical opening may be torn, and hemorrhaging may occur. **Abruptio** (ab-rup'she-o) **placentae** is a tearing away of a normally positioned placenta from the uterine wall accompanied by hemorrhaging. Both of these conditions can result in miscarriage and may also be life threatening to the mother.

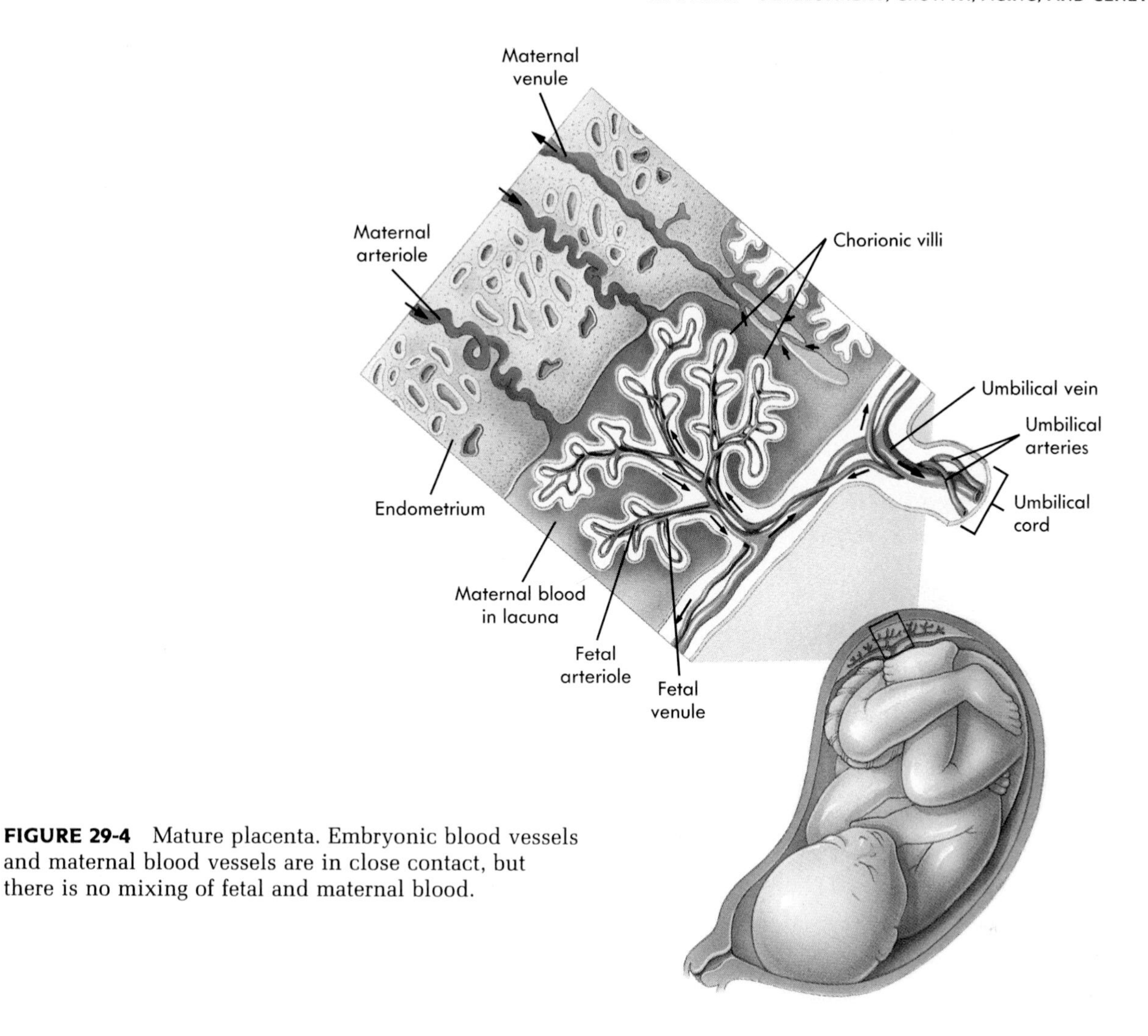

FIGURE 29-4 Mature placenta. Embryonic blood vessels and maternal blood vessels are in close contact, but there is no mixing of fetal and maternal blood.

In Vitro Fertilization and Embryo Transfer

In a small number of women, normal pregnancy is not possible because of some anatomical or physiological condition. In 87% of these cases the uterine tubes are incapable of transporting the zygote to the uterus or of allowing sperm to reach the oocyte. In vitro fertilization and embryo transfer have made pregnancy possible in hundreds of such women since 1978.

The woman is first induced to superovulate (causing ovulation of more than one oocyte at one time). Before the follicles rupture, the oocytes are removed from the ovary. The oocytes are incubated in a dish (thus the term *in vitro*, which means in glass) and maintained at body temperature for 6 hours; then sperm cells are added to the dish.

After 24 to 48 hours when the zygote has divided to a two- to eight-cell stage, several embryonic masses are transferred to the uterus (even in normal conditions half the zygotes do not implant). Implantation and subsequent development then proceed as they would for natural implantation. However, the woman is usually required to lie perfectly still for several hours after the embryonic mass has been introduced into the uterus to prevent possible expulsion before implantation can occur. It is not fully understood why such expulsion does not occur in natural fertilization and implantation.

The success rate of embryo transfer varies from clinic to clinic but is increasing steadily (the success rate at the best U.S. clinic in 1989 was 27%). Several multiple births have occurred after embryo transfer because of the practice of introducing more than one embryonic mass.

Formation of the Germ Layers

After implantation, a new cavity called the **amniotic** (am′nĭ-ot′ik) **cavity** forms inside the inner cell mass and is surrounded by a layer of cells called the **amnion** (am′nĭ-on). The formation of the amniotic cavity causes the part of the inner cell mass nearest the blastocele to separate as a flat disk of tissue called the **embryonic disk** (Figure 29-5). This embryonic disk is composed of two layers of cells: an **ectoderm** (ek′to-derm; outside layer) adjacent to the amniotic cavity and an **endoderm** (en′do-derm; inside layer) on the side of the disk opposite the amnion. A third cavity, the **yolk sac,** forms inside the blastocele from the endoderm. The amniotic cavity will eventually surround the developing embryo, providing it with a protective fluid environment, the "bag of waters," in which the embryo can form.

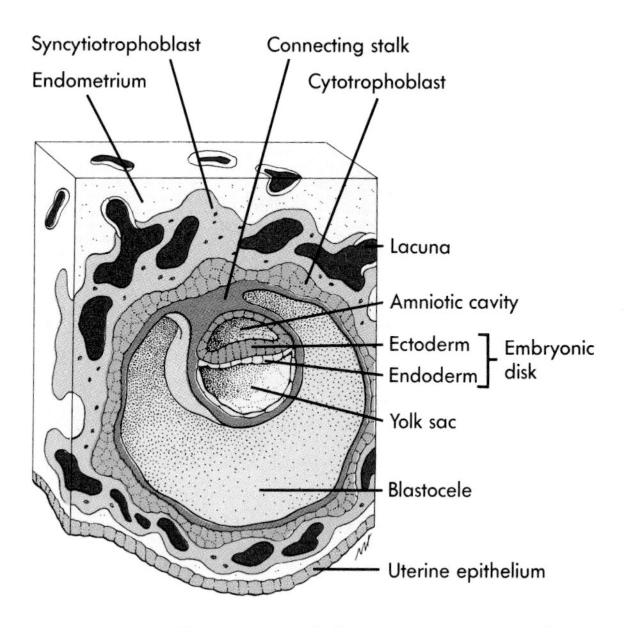

FIGURE 29-5 Embryonic disk consisting of ectoderm and endoderm, with the amniotic cavity and yolk sac.

Approximately 13 or 14 days after fertilization the embryonic disk has become a slightly elongated oval structure. Proliferating cells of the ectoderm migrate toward the center and the caudal end of the disk, forming a thickened line called the **primitive streak.** Some ectoderm cells leave the ectoderm, migrate through the primitive streak, and emerge between the ectoderm and endoderm as a new germ layer, the **mesoderm** (mez′o-derm; middle layer; Figure 29-6). The embryo is now three-layered, having ectoderm, mesoderm, and endoderm; all tissues of the adult can be traced to these three germ layers (Table 29-1). A cordlike structure called the **notochord** extends from the cephalic end of the primitive streak.

1

Predict the results of two primitive streaks forming in one embryonic disk. What if the two primitive streaks are touching each other?

? ? ? ? ? ? ? ? ? ? ?

During the first 2 weeks of development the embryo is quite resistant to outside influences that may cause malformations (although perhaps not safe from factors that could kill it). Between 2 weeks and the next 4 to 7 weeks (depending on the structure considered) the embryo is more sensitive to outside influences that cause malformations than at anv other time.

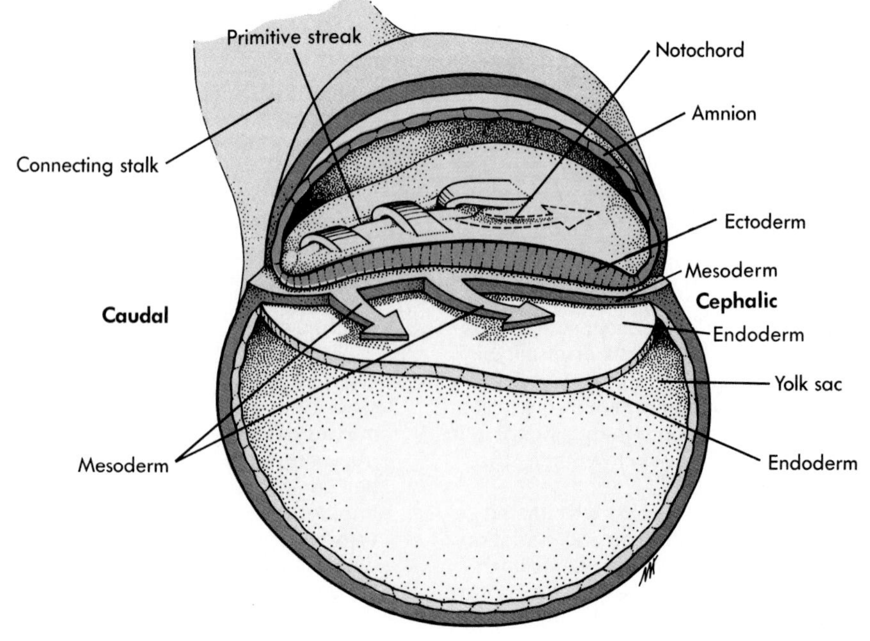

FIGURE 29-6 Embryonic disk with a primitive streak. The head of the embryo will develop over the notochord. Cells in the surface ectoderm *(blue)* move toward the primitive streak, fold into the streak *(blue arrow tails)*, and come out the other side of the streak as mesodermal cells *(orange arrows)*. The mesoderm *(orange)* lies between the ectoderm *(blue)* and the endoderm *(yellow)*.

Table 29-1

Germ layer derivatives

Ectoderm	***Endoderm***
Epidermis of skin	Lining of gastrointestinal tract
Tooth enamel	Lining of lungs
Lens and cornea of eye	Lining of hepatic, pancreatic, and other exocrine ducts
Outer ear	Kidney ducts and bladder
Nasal cavity	Adenohypophysis
Neuroectoderm	Thymus
Brain and spinal cord	Thyroid
Motor neurons	Parathyroid
Preganglionic autonomic neurons	Tonsils
Neuroglia cells (except microglia)	***Mesoderm***
Neural crest cells	Dermis of skin
Melanocytes	Circulatory system
Sensory neurons	Parenchyma of glands
Postganglionic autonomic neurons	Muscle
Adrenal medulla	Bones (except facial)
Facial bones	Microglia
Teeth: dentin, pulp, cementum, and gingiva	
Skeletal muscles in head	

Neural Tube and Neural Crest Formation

As a result of notochord formation, the overlying ectoderm is stimulated approximately 18 days after fertilization to form a thickened **neural plate.** The lateral edges of the plate begin to rise like two ocean waves coming together. These edges are called the **neural crests,** and a neural groove lies between them (Figure 29-7). The neural crests begin to meet in the midline and fuse into a **neural tube,** which is completely closed by 26 days. The neural tube becomes the brain and the spinal cord, and the cells of the neural tube are called **neuroectoderm** (see Table 29-1).

As the neural crests come together and fuse, a population of cells breaks away from the neuroectoderm all along the margins of the crests. These **neural crest cells** either migrate down along the side of the developing neural tube to become part of the peripheral nervous system or to form the adrenal

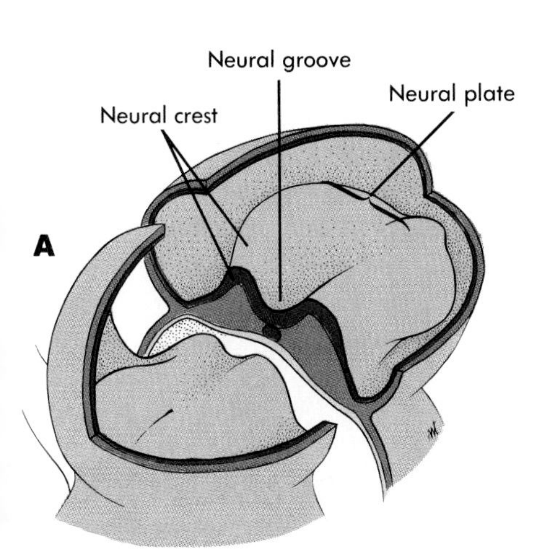

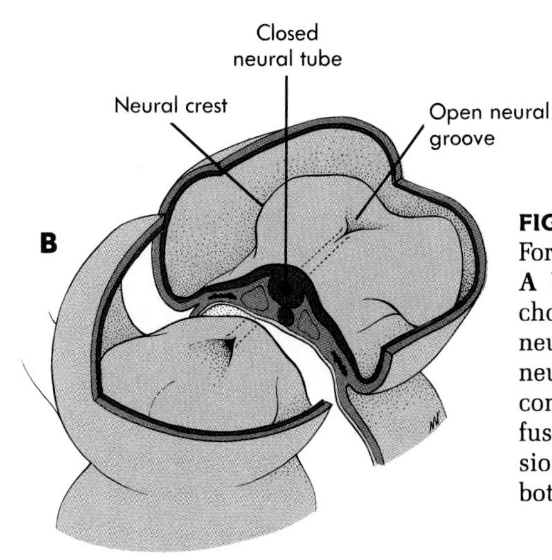

FIGURE 29-7
Formation of the neural tube.
A Under the influence of the notochord, the ectoderm thickens into a neural plate, whose edges rise up as neural folds. **B** The neural folds come together in the midline and fuse to form a neural tube. This fusion begins in the center and moves both cranially and caudally.

medulla or migrate laterally to just below the ectoderm to become melanocytes of the skin. In the head, neural crest cells perform additional functions; they contribute to the skull, the dentin of teeth, blood vessels, and general connective tissue.

Somite Formation

As the neural tube forms, the mesoderm immediately adjacent to the tube forms distinct segments called **somites** (so'mītz). In the head the first few somites never become clearly divided but develop into indistinct segmented structures called **somitomeres.** The somites and somitomeres eventually give rise to a portion of the skull, the vertebral column, and skeletal muscle.

Formation of the Gut and Body Cavities

At the same time the neural tube is forming, the **embryo** itself is becoming a tube along the upper portion of the yolk sac. The **foregut** and **hindgut** develop as the cephalic and caudal ends of the yolk sac are separated from the main yolk sac. This is the beginning of the digestive tract (Fig. 29-8, *A*). The developing digestive tract pinches off from the yolk sac as a tube, remaining attached in the center to the yolk sac by a yolk stalk.

The foregut and hindgut (Fig. 29-8, *B*) are in close relationship to the overlying ectoderm and form membranes called the **oropharyngeal membrane** and the **cloacal membrane,** respectively. The oropharyngeal membrane opens to form the mouth, and the

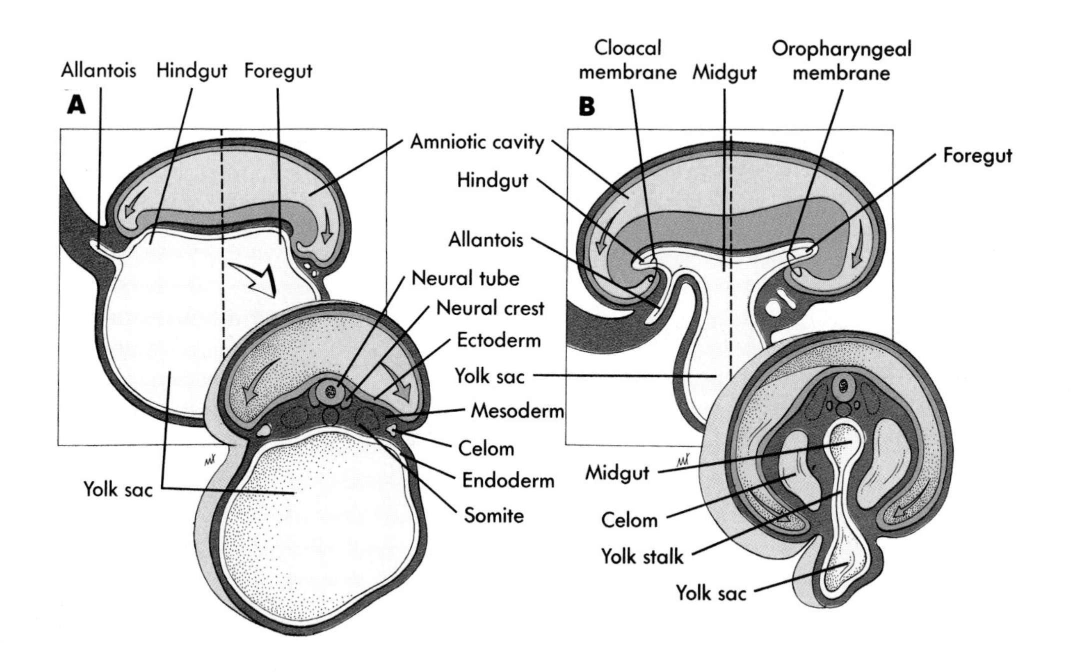

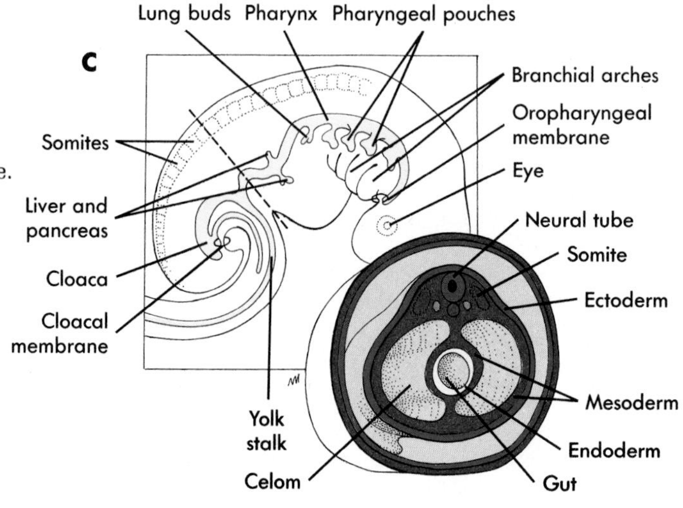

FIGURE 29-8 Formation of the digestive tract.
Arrows show the folding of the digestive tract into a tube. Dotted lines show plane of section from which insets were taken. **A** 20 days. **B** 25 days after fertilization. **C** 30 days after fertilization. Evaginations are identified along the pharynx and digestive tract.

cloacal membrane opens to form the urethra and anus. Thus the digestive tract becomes a tube that is open to the outside at both ends.

A considerable number of **evaginations** (e'vaj-ĭ-na'shunz; outpocketings) occur along the early digestive tract (Figure 29-8, C). They develop into structures such as the anterior pituitary, the thyroid gland, the lungs, the liver, the pancreas, and the urinary bladder. At the same time solid bars of tissue known as **branchial arches** (see Figures 29-8, C, and 29-9) form along the lateral sides of the head, and the sides of the foregut expand as pockets between the branchial arches. The central, expanded foregut is called the **pharynx,** and the pockets along both sides of the pharynx are called **pharyngeal pouches.** Adult derivatives of the pharyngeal pouches include the auditory tube, tonsils, thymus, and parathyroids.

At approximately the same time a series of isolated cavities starts to form within the embryo, thus beginning development of the **celom** (se'lōm; see Figure 29-8), or body cavities. The most cranial group of cavities enlarges and fuses to form the **pericardial cavity.** Shortly thereafter the celomic cavity extends

Fetal Monitoring

Amniocentesis (am'nĭ-o-sen-te'sis) is the removal of amniotic fluid from the amniotic cavity (Figure 29-A). As the fetus develops, molecules of various types and living cells are expelled into the amniotic fluid. These molecules and cells can be collected and analyzed. A number of normal conditions can be evaluated, and a number of metabolic disorders can be detected by analysis of the types of molecules expelled by the fetus. The cells collected by amniocentesis can be grown in culture, and additional metabolic disorders can be evaluated. Chromosome analysis, called a **karyotype** (kăr'ĭ-o-tīp), can also be accomplished from the cultured cells. Amniocentesis has been performed as early as 10 weeks after fertilization, but the success rate at that time is quite low. It is most commonly performed 13 to 14 weeks after fertilization.

Fetal tissue samples may also be obtained by **chorionic villi sampling** in which a probe is introduced into the uterine cavity through the cervix and a small piece of chorion is removed. This technique has an advantage over amniocentesis in that it can be used earlier in development—as early as the seventh to ninth week after fertilization.

One of the molecules normally produced by the fetus and released into the amniotic fluid is **alpha-fetoprotein.** If fetal tissues that are normally covered by skin such as nervous tissue (resulting from failure of the neural tube to close) or abdominal tissues (resulting from failure of the ab wall to fully form) are exposed to the amniotic fluid, an excessive amount of alpha-fetoprotein will be lost into the amniotic fluid.

Some of the metabolic by-products from the fetus such as alpha-fetoprotein and estriol (a weak form of estrogen produced in the placenta after 20 weeks of gestation) can enter the maternal blood and in some cases can be processed and passed to the maternal urine. The levels of those fetal products can then be measured in the mother's blood or urine.

The fetus can be seen within the uterus by **ultrasound,** which uses sound waves that are bounced off the fetus like sonar and then analyzed and enhanced by computer, or by **fetoscopy** wherein a fiberoptic probe is introduced into the amniotic cavity. Because of the constantly increasing resolution in ultrasound and because it is noninvasive compared to fetoscopy, the latter technique is not commonly used at present. The technique of ultrasound has not been found to pose any risk to the fetus or mother. Ultrasound can be accomplished by placing a transducer on the abdominal wall (transabdominal) or by inserting the transducer into the woman's vagina (transvaginal). The latter technique produces much higher resolution since there are fewer layers of tissue between the transducer and the uterine cavity. Transvaginal ultrasound can be used to identify the yolk sac of a developing embryo as early as 17 days after fertilization, and the embryo can be visualized by 25 days. Transabdominal ultrasound allows fetal monitoring by 6 to 8 weeks after fertilization.

Fetal heart rate can be detected with an ultrasound stethoscope by the tenth week after fertilization and with a conventional stethoscope by 20 weeks. The normal fetal heart rate is 140 beats per minute (normal range 110 to 160).

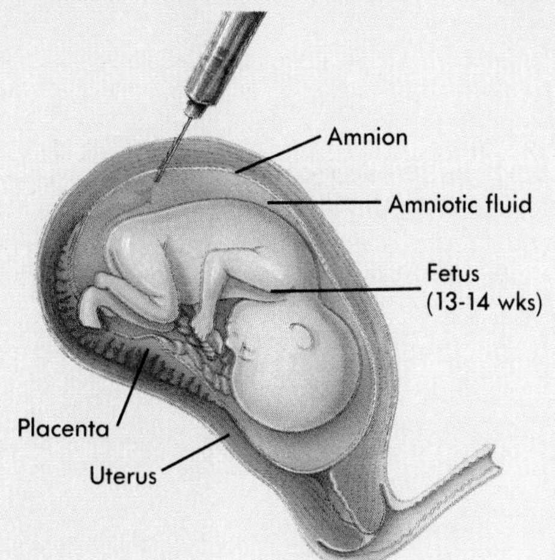

FIGURE 29-A
Removal of amniotic fluid for amniocentesis.

toward the caudal end of the embryo as the **pleural** and **peritoneal cavities.** Initially all three of these cavities are continuous, but they eventually separate into three distinct adult cavities (see Chapter 1).

Limb Bud Development

Arms and legs first appear as limb buds (Figure 29-9). The **apical ectodermal ridge,** a specialized thickening of the ectoderm, develops on the lateral margin of each limb bud and stimulates its outgrowth. As the buds elongate, limb tissues are laid down in a proximal-to-distal sequence. For example, in the upper limb the arm is formed before the forearm, which is formed before the hand.

Development of the Face

The face develops by fusion of five embryonic structures: the **frontonasal process,** which forms the forehead, nose, and midportion of the upper jaw and lip; two **maxillary processes,** which form the lateral portions of the upper jaw and lip; and two **mandibular processes,** which form the lower jaw and lip (Figure 29-10, *A*). **Nasal placodes** (plak′odz), which develop at the lateral margins of the frontonasal process, develop into the nose and the center of the upper jaw and lip (Figure 29-10, *B*).

As the brain enlarges and the face matures, the nasal placodes approach each other in the midline. The medial edges of the placodes fuse to form the midportion of the upper jaw and lip (Figure 29-10, *C* and *D*). This portion of the frontal process is between the two maxillary processes, which are expanding toward the midline, and fuses with them to form the upper jaw and lip, known as the **primary palate.**

A **cleft lip** results from failure of the frontonasal and two maxillary processes to fuse (see Figure 29-9, *F*). Because there are three structures—one midline and two lateral—involved in formation of the primary palate, cleft lips usually do not occur in the midline but to one side (or both sides) and extend from the mouth to the naris (nostril).

At approximately the same time the primary palate is forming, the lateral edges of the nasal placodes fuse with the maxillary processes to close off the groove extending from the mouth to the eye (Figure 29-10, *D* and *E*). On rare occasions these structures fail to meet, resulting in a facial cleft extending from the mouth to the eye.

The inferior margins of the maxillary processes fuse with the superior margins of the mandibular processes to decrease the size of the mouth.

All of the previously described fusions and the growth of the brain give the face a decidedly "human" appearance by approximately 50 days.

The roof of the mouth, known as the **secondary palate,** begins to form as vertical shelves, which swing to a horizontal position and begin to fuse with each other at approximately 56 days of development. Fusion of the entire palate is not completed until approximately 90 days. If the secondary palate does not fuse, a midline cleft in the roof of the mouth called a **cleft palate** results.

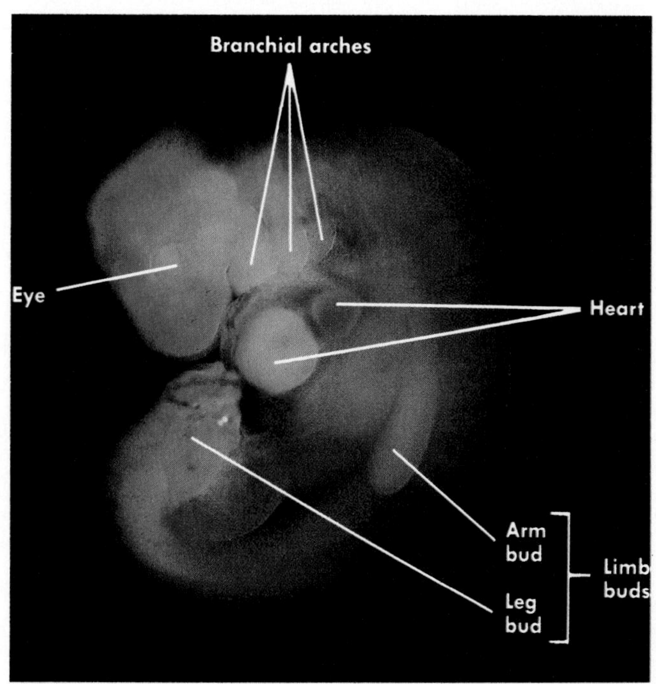

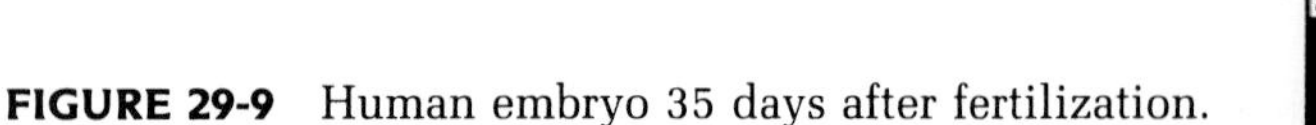
FIGURE 29-9 Human embryo 35 days after fertilization.

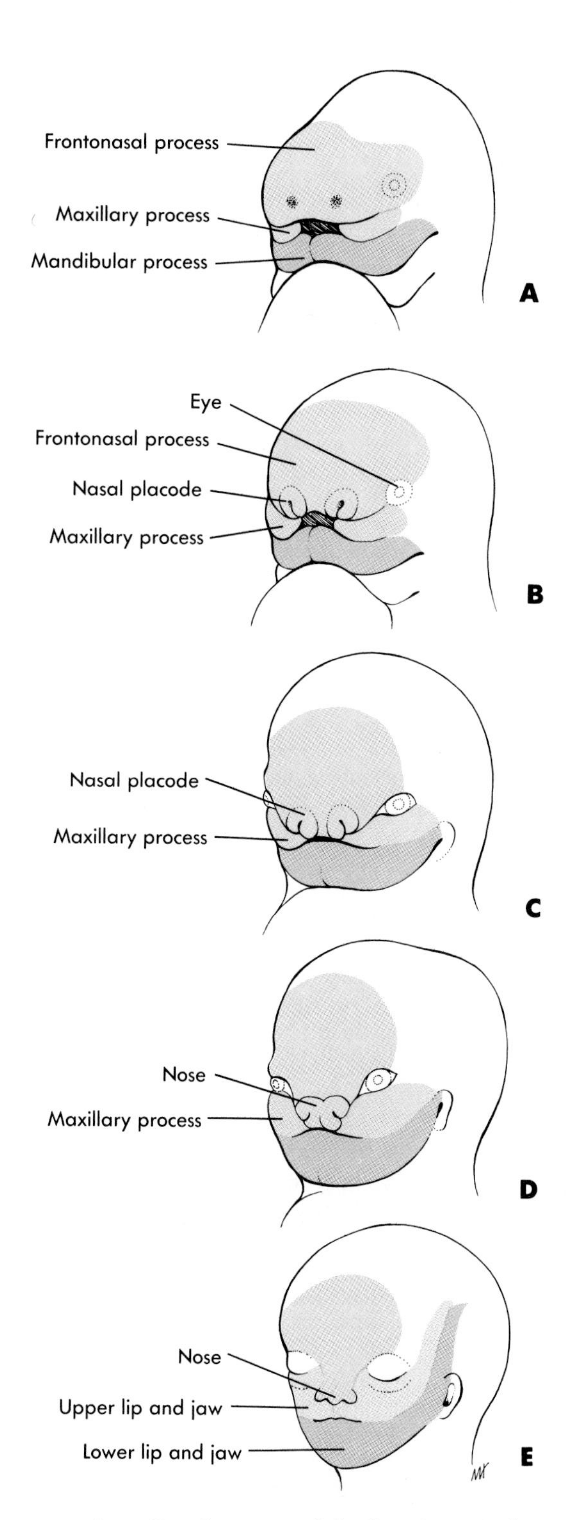

FIGURE 29-10 Development of the face (ages indicate postovulatory days). **A** 28 days. The face develops from five processes—frontonasal, two maxillary, and two mandibular (already fused). **B** 33 days. Nasal placodes appear on frontonasal process. **C** 40 days. Maxillary processes enlarge and move toward the midline. The nasal placodes also move toward the midline and fuse with the maxillary processes to form the upper jaw and lip. **D** 48 days. Continued growth brings structures more toward the midline. **E** 14-week fetus. Colors show the contributions of each process to the adult face.

Development of the Organ Systems

The major organ systems appear and begin to develop during the embryonic period. Therefore the period between 14 and 60 days is called the period of **organogenesis** (Table 29-2).

Skin

The **epidermis** of the skin is derived from ectoderm, and the **dermis** is derived from mesoderm. Nails, hair, and glands develop from the epidermis (see Chapter 5). Melanocytes and sensory receptors in the skin are derived from neural crest cells.

Skeleton

The skeleton develops from either mesoderm or the neural crest by intramembranous bone formation or endochondral bone formation (see Figure 6-12). The bones of the face develop from neural crest cells, whereas the rest of the skull, the vertebral column, and ribs develop from somite- or somitomere-derived mesoderm. The appendicular skeleton develops from limb bud mesoderm.

Muscle

Myoblasts (mi'o-blastz) are multinucleated cells that produce skeletal muscle fibers. Undifferentiated precursors of myoblasts migrate from somites or somitomeres to sites of future muscle development where they begin to fuse and form myoblasts. Shortly after myoblasts form, nerves grow into the area and innervate the developing muscle. After the basic form of each muscle is established, continued growth of the muscle occurs by an increase in the number of muscle fibers. The total number of muscle fibers is established before birth and remains relatively constant thereafter. Muscle enlargement after birth is due to an increase in the size of individual fibers.

Nervous system

The nervous system is derived from the neural tube and neural crest cells. Closure of the neural tube begins in the upper-cervical region and proceeds into the head and down the spinal cord. Soon after the neural tube has closed, the portion of the neural tube that will become the brain begins to expand and develops a series of pouches (see Figure 13-3). The central cavity of the neural tube becomes the ventricles of the brain and the central canal of the spinal cord.

The nerve cells that form the peripheral nervous system are located either within the neural tube (motor nerves and preganglionic neurons of the autonomic nervous system) or are derived from neural crest cells (sensory nerves and postganglionic neurons of the autonomic nervous system).

Table 29-2

Development of the organ systems

	AGE (days since fertilization)					
	1-5	6-10	11-15	16-20	21-25	26-30
General features	Fertilization Morula Blastocyst	Blastocyst implants	Primitive streak Three germ layers	Neural plate	Neural tube closed	Limb buds and other "buds" appear
Integumentary system			Ectoderm Mesoderm			Melanocytes from neural crest
Skeletal system			Mesoderm		Neural crest (will form facial bones)	Limb buds
Muscular system			Mesoderm	Somites begin to form		Somites all present
Nervous system			Ectoderm	Neural plate	Neural tube complete Neural crest Eyes and ears begin	Lens begins to form
Endocrine system			Ectoderm Mesoderm Endoderm	Thyroid begins to develop		Parathyroids appear
Cardiovascular system			Mesoderm	Blood islands form Two heart tubes	Single-tubed heart begins to beat	Interatrial septum begins to form
Lymphatic system			Mesoderm			Thymus appears
Respiratory system			Mesoderm Endoderm		Diaphragm begins to form	Trachea forms as single bud Lung buds (primary bronchi)
Digestive system			Endoderm		Foregut and hindgut form	Liver and pancreas appear as buds Tongue bud appears
Urinary system			Mesoderm Endoderm		Pronephros develops Allantois appears	Mesonephros appears
Reproductive system			Mesoderm Endoderm		Primordial germ cells on yolk sac	Mesonephros appears Genital tubercle forms

AGE (days since fertilization)					
31-35	36-40	41-45	46-50	51-55	56-60
Hand and foot plates on limbs	Fingers and toes appear Lips formed Embryo 15 mm	External ear forming Embryo 20 mm	Embryo 25 mm	Limbs elongate to a more adult relation Embryo 35 mm	Face is distinctly human in appearance
Sensory receptors appear in skin		Collagen fibers clearly present in skin		Extensive sensory endings in skin	
Mesoderm condensation in areas of future bone	Cartilage in site of future humerus	Cartilage in site of future ulna and radius	Cartilage in site of hand and fingers		Ossification begins in clavicle and then in other bones
Muscle precursor cells enter limb buds			Functional muscle		Nearly all muscles appear in adult form
Nerve processes enter limb buds		External ear forming Olfactory nerve begins to form		Semicircular canals in inner ear complete	Eyelids form Cochlea in inner ear complete
Hypophysis appears as evaginations from brain and mouth	Gonadal ridges form Adrenal glands forming		Pineal body appears	Thyroid gland in adult position and attachment to tongue lost	Adenohypophysis loses its connection to the mouth
Interventricular septum begins to form		Interventricular septum complete	Interatrial septum complete but still has opening until birth		
Large lymph vessels form in neck	Spleen appears			Adult lymph pattern formed	
Secondary bronchi to lobes form	Tertiary bronchi to lobules form		Tracheal cartilage begins to form		
Oropharyngeal membrane ruptures		Secondary palate begins to form Tooth buds begin to form			Secondary palate begins to fuse (fusion complete by 90 days)
Metanephros begins to develop				Mesonephros degenerates	Anal portion of cloacal membrane ruptures
	Gonadal ridges form	Primordial germ cells enter gonadal ridges	Paramesonephric ducts appear		Uterus forming Beginning of differentiation of external genitalia in male and female

Failure of the neural tube to close in the head results in **anencephaly.** Failure of the neural tube to close in the back results in a spectrum of defects referred to collectively as **spina bifida.**

Special senses

The **olfactory bulb** and **nerve** develop as an evagination from the telencephalon. The eyes develop as evaginations from the diencephalon. Each evagination elongates to form an **optic stalk,** and a bulb called the **optic vesicle** develops at its terminal end. The optic vesicle reaches the side of the head and stimulates the overlying ectoderm to thicken into a **lens.** The sensory portion of the ear first appears as an ectodermal thickening or placode, which invaginates and pinches off from the overlying ectoderm.

Endocrine system

The **neurohypophysis** of the pituitary is formed by an evagination from the floor of the diencephalon. The **adenohypophysis** develops from an evagination of ectoderm in the roof of the embryonic oral cavity and grows toward the floor of the brain. It eventually loses its connection with the oral cavity and becomes attached to the neurohypophysis (see Chapter 17).

The **thyroid gland** originates as an evagination from the floor of the pharynx in the region of the developing tongue and moves into the lower neck, eventually losing its connection with the pharynx. The **parathyroid glands,** which are derived from the third and fourth pharyngeal pouches, migrate inferiorly and become associated with the thyroid gland.

The **adrenal medulla** arises from neural crest cells and consists of specialized postganglionic neurons of the sympathetic division of the autonomic nervous system (see Chapter 16). The **adrenal cortex** is derived from mesoderm.

The **pancreas** originates as two evaginations from the duodenum, which come together to form a single gland (see Figure 29-8, C).

Circulatory system

The heart develops from two endothelial tubes (Figure 29-11, *A*), which fuse into a single, midline heart tube (Figure 29-11, *B*). Blood vessels form from **blood islands** (small masses of mesoderm that become blood vessels on the outside and blood cells on the inside) on the surface of the yolk sac and inside the embryo. These islands expand and fuse to form the circulatory system.

A series of dilations appears along the length of the primitive heart tube, and four major regions can be identified: the **sinus venosus,** the site where blood enters the heart; a single **atrium;** a single **ventricle;** and the **bulbus cordis,** where blood exits the heart (see Figure 29-11, *B*).

The elongating heart, confined within the pericardium, becomes bent into a loop, the apex of which is the ventricle (see Figure 29-9, *B*). The major chambers of the heart, the atrium and the ventricle, expand rapidly. The right portion of the sinus venosus becomes absorbed into the atrium, and the bulbus cordis is absorbed into the ventricle. The embryonic sinus venosus initiates contraction at one end of the tubular heart. Later in development, part of the sinus venosus becomes the sinoatrial (SA) node, which is the adult pacemaker.

2

What would happen if the sinus venosus did not contract before other areas of the primitive heart?

? ? ? ? ? ? ? ? ? ? ?

The single ventricle becomes divided into two chambers by the development of an **interventricular septum** (Figure 29-11, *C* to *E*). If the interventricular septum does not grow enough to completely separate the ventricles, a ventricular septal defect (VSD) results.

The **interatrial septum** (Figure 29-11, *C* to *E*), which separates the two atria in the adult heart, is formed from two portions—the **septum primum** (primary septum) and the **septum secundum** (secondary septum). An opening in the interatrial septum, called the **foramen ovale** (o-val'e), connects the two atria and allows blood to flow from the right to the left atrium in the embryo and fetus.

If the septum secundum fails to grow far enough or if the foramen secundum becomes too large, an interatrial septal defect (or atrial septal defect, ASD) occurs, allowing blood to flow from the left atrium to the right atrium in the newborn. Either an interatrial septal defect or a ventricular septal defect may result in a heart murmur (an abnormal heart sound).

Respiratory system

The lungs begin to develop as a single, midline evagination from the foregut in the region of the future esophagus. This evagination branches to form two **lung buds** (Figure 29-12, *A*). The lung buds elongate and branch, first forming the bronchi that project to the lobes of the lungs (Figure 29-12, *B*) and then the bronchi that project to the lobules of the lungs (Figure 29-12, *C*). This branching continues (Figure

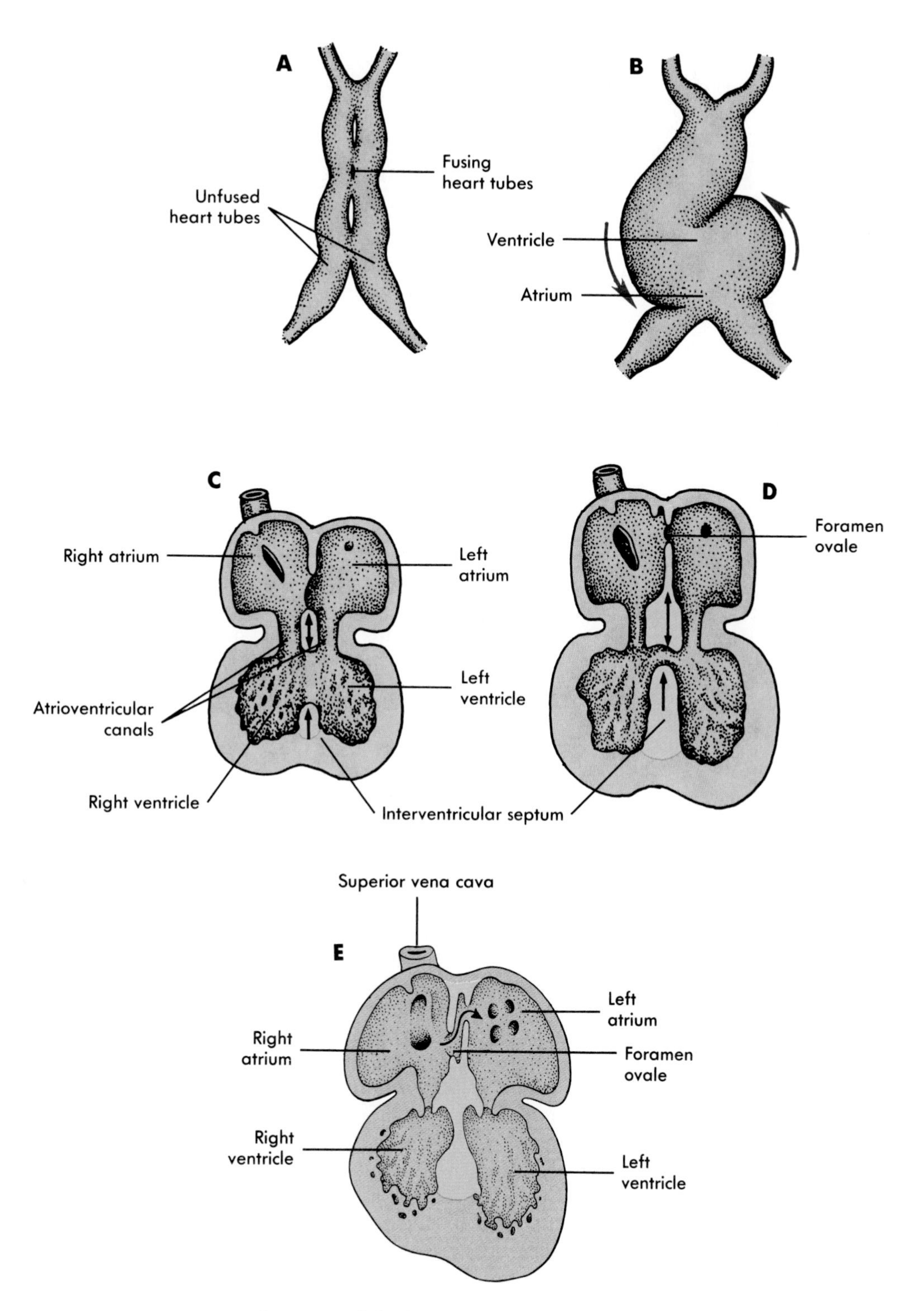

FIGURE 29-11 Development of the heart. **A** At 20 days after fertilization—two-tubed heart. **B** At 22 days—fused, bent heart tube resulting from elongation of the heart within the confined space of the pericardium. **C** At 31 days—the septum primum of the interatrial septum and the interventricular septum grow toward the center of the heart. **D** At 35 days—the septum primum is complete, a foramen opens in the septum, and the septum secundum begins to form. The interventricular septum is nearly complete. **E** The final embryonic condition of the interatrial septum. Blood from the right atrium can flow through the foramen ovale into the left atrium. As blood begins to flow in the other direction, the septum primum is forced against the septum secundum, closing the foramen ovale.

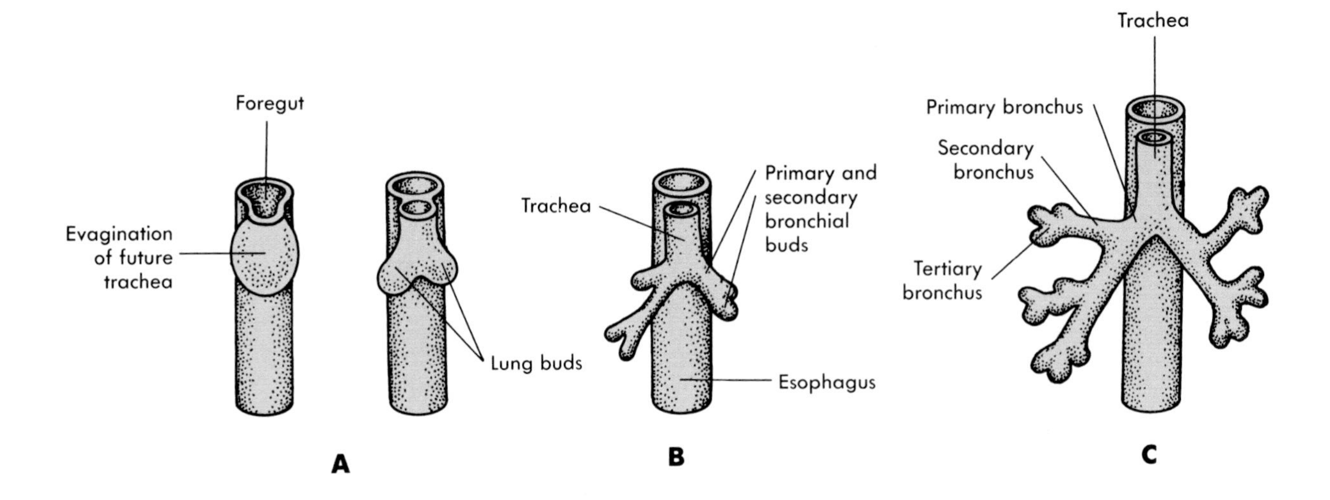

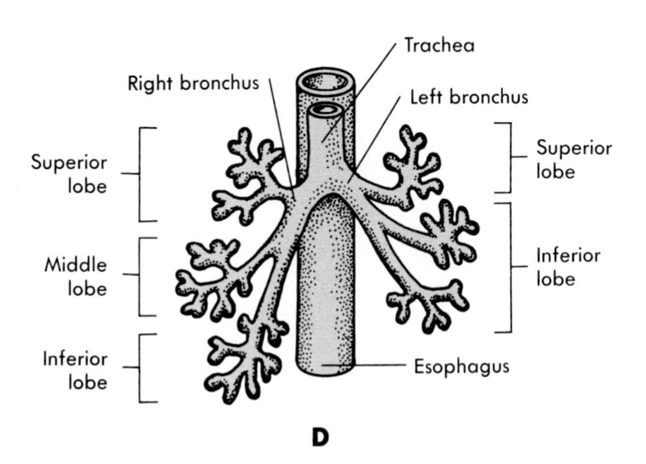

FIGURE 29-12 Development of the lung. **A** 28 days—a single lung bud forms and divides into two buds, forming primary bronchi. **B** 32 days—the secondary bronchi branch off the bronchial buds to form the lobes. **C** 35 days—tertiary bronchi branch to form lobules. **D** 50 days—continued branching.

29-12, *D*) until, by the end of the sixth month, approximately 17 generations of branching have occurred. Even after birth some branching continues as the lungs grow larger, and in the adult approximately 24 generations of branches have been established.

Urinary system

The kidneys develop from mesoderm located between the somites and the lateral portion of the embryo (Figure 29-13, *A*). Approximately 21 days after fertilization the mesoderm in the cervical region differentiates into a structure called the **pronephros** (the most forward or earliest kidney), which consists of a duct and simple tubules connecting the duct to the open celomic cavity. This type of kidney is the functional adult kidney in some lower chordates, but it is probably not functional in the human embryo and soon disappears.

The **mesonephros** (middle kidney) is a functional organ in the embryo. It consists of a duct, which is a caudal extension of the pronephric duct, and a number of minute tubules, which are smaller and more complex than those of the pronephros. One end of each tubule opens into the mesonephric duct, and the other end forms a glomerulus (see Chapter 26).

As the mesonephros is developing, the caudal end of the hindgut begins to enlarge to form the **cloaca** (klo-a'kah; sewer), the common junction of the digestive, urinary, and genital systems. The cloaca becomes divided by a urorectal septum into two portions: a digestive portion called the rectum and a urogenital portion called the urethra.

The cloaca has two tubes associated with it: the hindgut and the **allantois** (sausage), which is a blind tube extending into the umbilical cord (see Figures 29-8 and 29-13). The portion of the allantois nearest the cloaca enlarges to form the urinary bladder, and the remainder, from the bladder to the umbilicus, degenerates.

The mesonephric duct extends caudally as it develops and eventually joins the cloaca. At the point of junction another tube, the **ureter,** begins to form. Its distal end enlarges and branches to form the duct system of the adult kidney, called the **metanephros** (the last kidney; Figure 29-13, *B* and *C*), which takes over the function of the degenerating mesonephros. The mesonephric duct and a few tubules remain in the male as part of the reproductive system but almost completely disappear in the female (Figure 29-13, *D*).

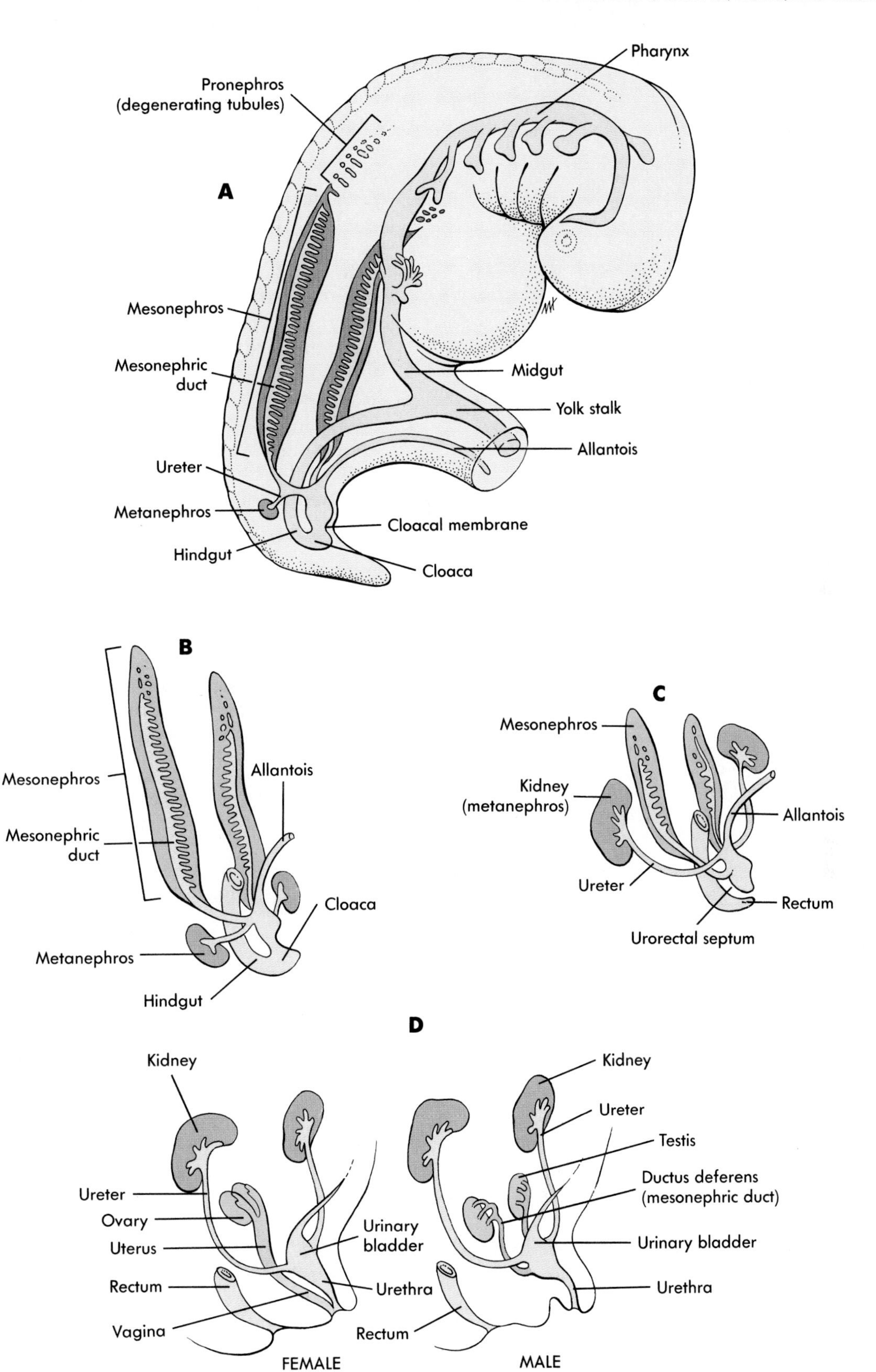

FIGURE 29-13 Development of the kidney and urinary bladder. **A** The three portions of the developing kidney—pronephros, mesonephros, and metanephros. **B** The metanephros (adult kidney) enlarges as the mesonephros degenerates. **C** The kidney continues to grow and develop. **D** Development of the male and female urogenital systems.

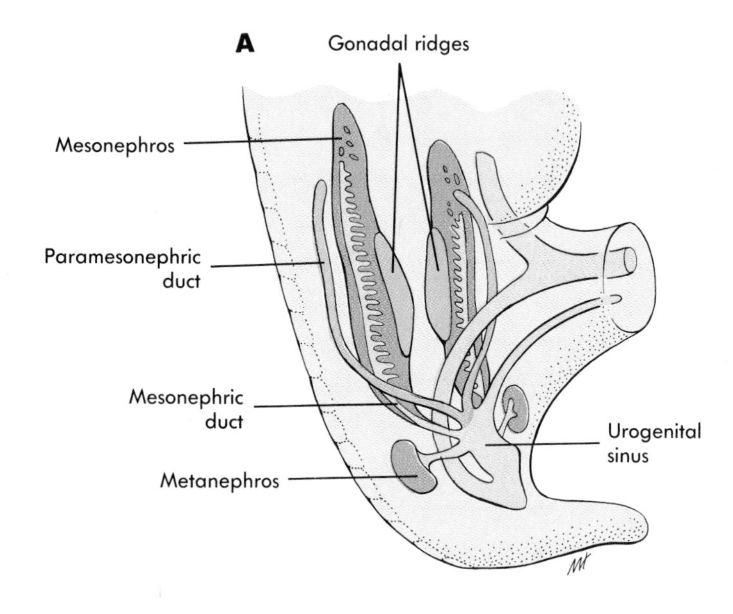

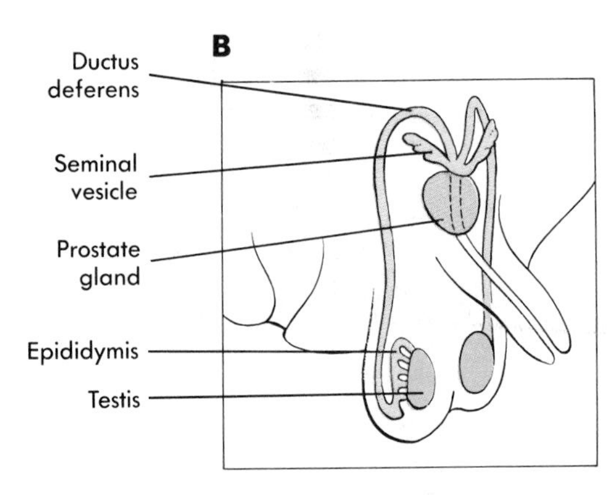

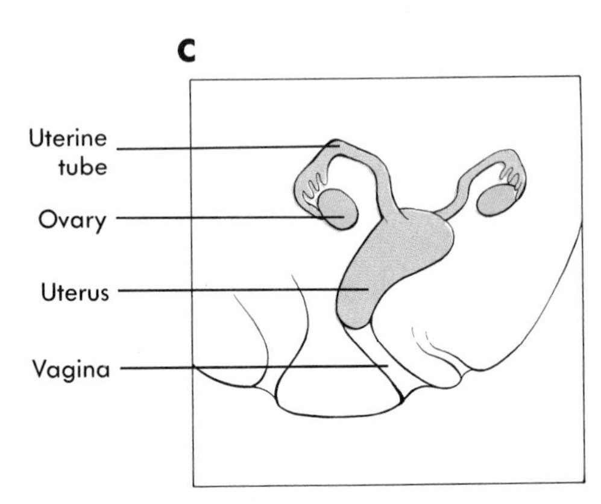

FIGURE 29-14 Development of the reproductive system. **A** Indifferent stage. **B** The male, under the influence of male hormones, develops a ductus deferens from the mesonephric duct, and the paramesonephric duct degenerates. **C** The female, without male hormones, develops uterine tubes from the paramesonephric duct, and the mesonephros disappears.

Reproductive system

The male and female gonads appear as **gonadal ridges** along the ventral border of each mesonephros (Fig. 29-14, *A*). **Primordial germ cells** (cells destined to become oocytes or sperm cells) form on the surface of the yolk sac, migrate into the embryo, and enter the gonadal ridge.

In the female the ovaries descend from their original position high in the abdomen to a position within the pelvis. The male testes descend even farther. As the testes reach the anteroinferior abdominal wall, a pair of tunnels called the **inguinal canals** form through the abdominal musculature. The testes pass through these canals, leaving the abdominal cavity and coming to lie within the **scrotum** (see Figure 28-3). Descent of the testes through the canals begins approximately 7 months after conception, and the testes enter the scrotum approximately 1 month before the infant is born.

In approximately 0.5% of male children one or both testes fail to enter the scrotum. This condition is called undescended testes or **cryptorchidism** (krip-tor′kĭ-dizm). Because testosterone is required for the testes to descend into the scrotum, cryptorchidism is often the result of inadequate testosterone secreted by the fetal testes. If neither testis descends and the defect is not corrected, the male will be infertile because the slightly higher temperature of the body cavity, in comparison to that of the scrotal sac, causes the spermatogonia to degenerate. Cryptorchidism is usually surgically corrected.

Paramesonephric ducts begin to develop just lateral to the mesonephric ducts and grow inferiorly to meet each other where they enter the cloaca as a single, midline tube.

Testosterone, secreted by the testes, causes the mesonephric duct system to enlarge and differentiate to form the ductus deferens, seminal vesicles, and prostate glands (Figure 29-14, *B*). Müllerian-inhibiting hormone, also secreted by the testes, causes the paramesonephric duct (also called the müllerian duct, which gives rise to the uterine tubes, the uterus, and part of the vagina in females) to degenerate. If neither testosterone nor müllerian-inhibiting hormone are secreted, the mesonephric duct system atrophies, and the paramesonephric duct system develops to form the internal female reproductive structures (Figure 29-14, *C*).

Like the other sexual organs, the external genitalia begin as the same structures in the male and female and then diverge. An enlargement called the **genital tubercle** develops in the groin of the embryo. **Urogenital folds** develop on each side of the urogenital opening, and **labioscrotal swellings** develop lateral to the folds. A **urethral groove** develops along the ventral surface of the genital tubercle.

In the male, under the influence of testosterone, the genital tubercle and the urogenital folds close over the urogenital opening and the urethral groove to form the penis. If this closure does not proceed all the way to the end of the penis, a defect known as **hypospadias** (hi-po-spa'dĭ-as) results. The testes move into the labioscrotal swellings, which become the scrotum of the male.

In the female, in the absence of testosterone, the genital tubercle becomes the clitoris. The urethral groove disappears, and the urogenital folds do not fuse. As a result, the urethra opens somewhat posterior to the clitoris but anterior to the vaginal opening. The unfused urogenital folds become the labia minora, and the labioscrotal folds become the labia majora.

Growth of the Fetus

The embryo becomes a **fetus** approximately 60 days after fertilization (Figure 29-15). The major difference between the embryo and the fetus is that in the embryo most of the organ systems are developing whereas in the fetus the organs are present. Most morphological changes occur in the embryonic phase of development, whereas the fetal period is primarily a "growing phase."

A

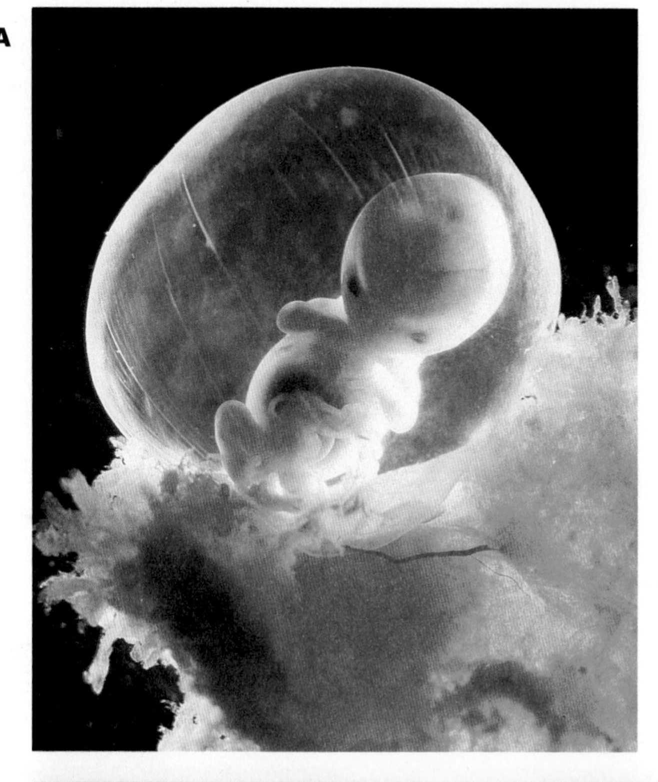

B

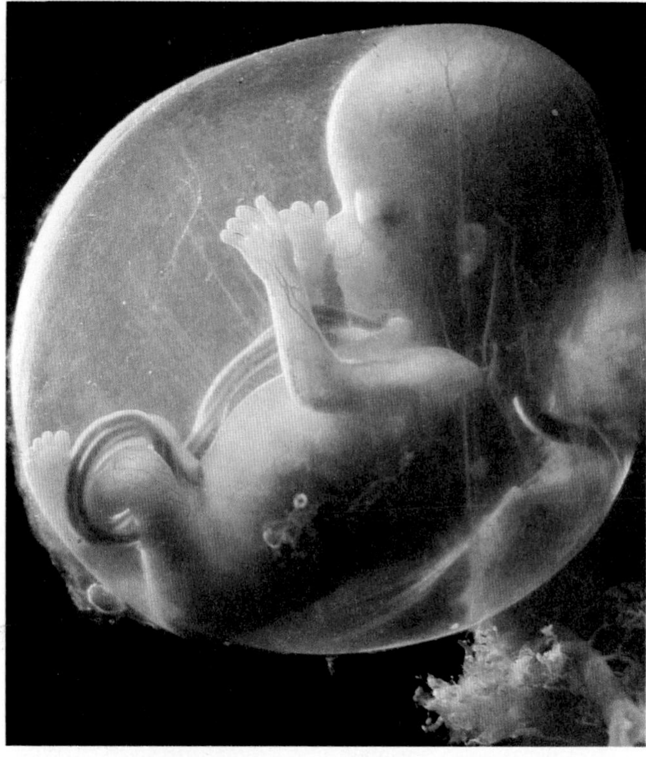

C

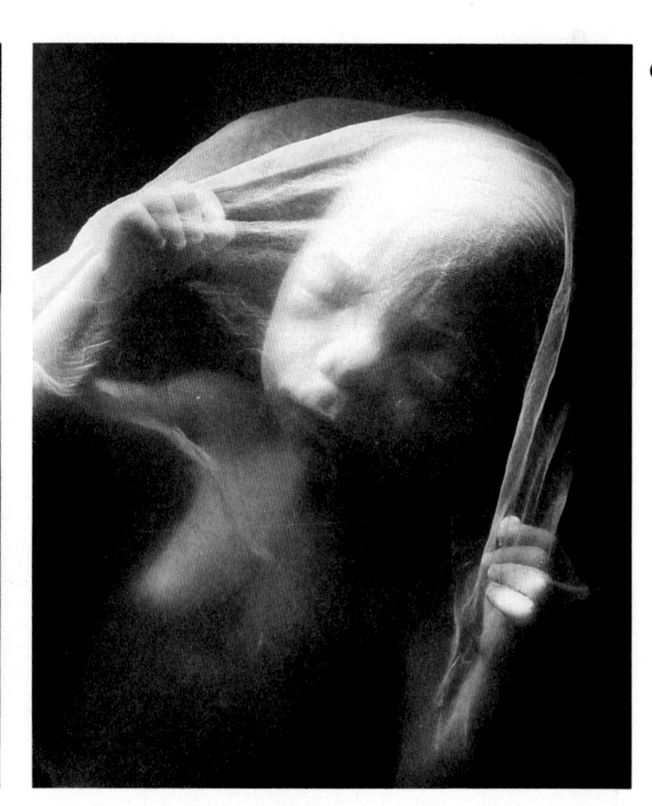

FIGURE 29-15 Embryos and fetuses at different ages. **A** 50 days; **B** 3 months; **C** 4 months.

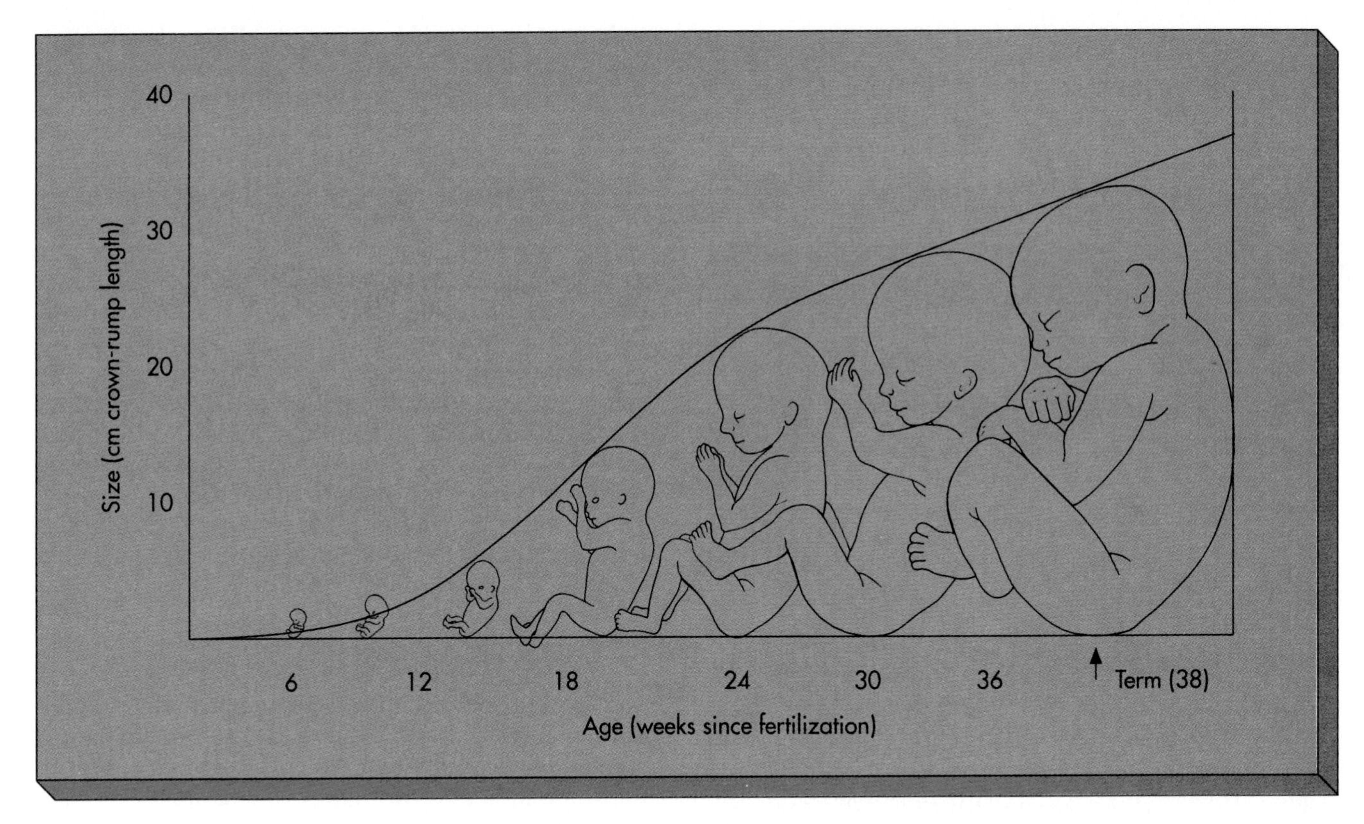

FIGURE 29-16 Growth of the fetus.

The fetus grows from approximately 3 cm and 2.5 g at 60 days to 50 cm and 3300 g at term—more than a 15-fold increase in length and a 1300-fold increase in weight (Figure 29-16). Although growth is certainly a major feature of the fetal period, it is not the only feature. Some of the major organ systems still continue to develop during the fetal period.

Fine, soft hair called **lanugo** (lă-nu′go) covers the fetus, and a waxy coat of sloughed epithelial cells called **vernix caseosa** protects the fetus from the somewhat toxic nature of the amniotic fluid formed by the accumulation of waste products from the fetus.

Subcutaneous fat that accumulates in the late fetus and newborn provides a nutrient reserve, helps insulate the, baby, and aids the baby in sucking by strengthening and supporting the cheeks so that negative pressure can be developed in the oral cavity.

Peak body growth occurs late in gestation, but as placental size and blood supply limits are approached, the growth rate slows. Growth of the placenta essentially stops at approximately 35 weeks, restricting further intrauterine growth.

At approximately 38 weeks of development the fetus has progressed to the point that it can survive outside the mother. The average weight at this point is 3250 g for a female fetus and 3300 g for a male fetus.

PARTURITION

Parturition (par-tu-rish′un) refers to the process by which the baby is born. Physicians usually calculate the gestation period (length of pregnancy) as 280 days (40 weeks or 10 lunar months) from the last menstrual period (LMP) to the date of confinement (i.e., the date of delivery of the infant).

3

How many days (developmental time) does it take an infant to develop from fertilization to parturition?

? ? ? ? ? ? ? ? ? ? ?

Occasionally the fetus is delivered before it has sufficiently matured. It is then considered **premature.** Prematurity is one of the most significant problems in pediatrics (the branch of medical science dealing with children) because of all the complications associated with prematurity.

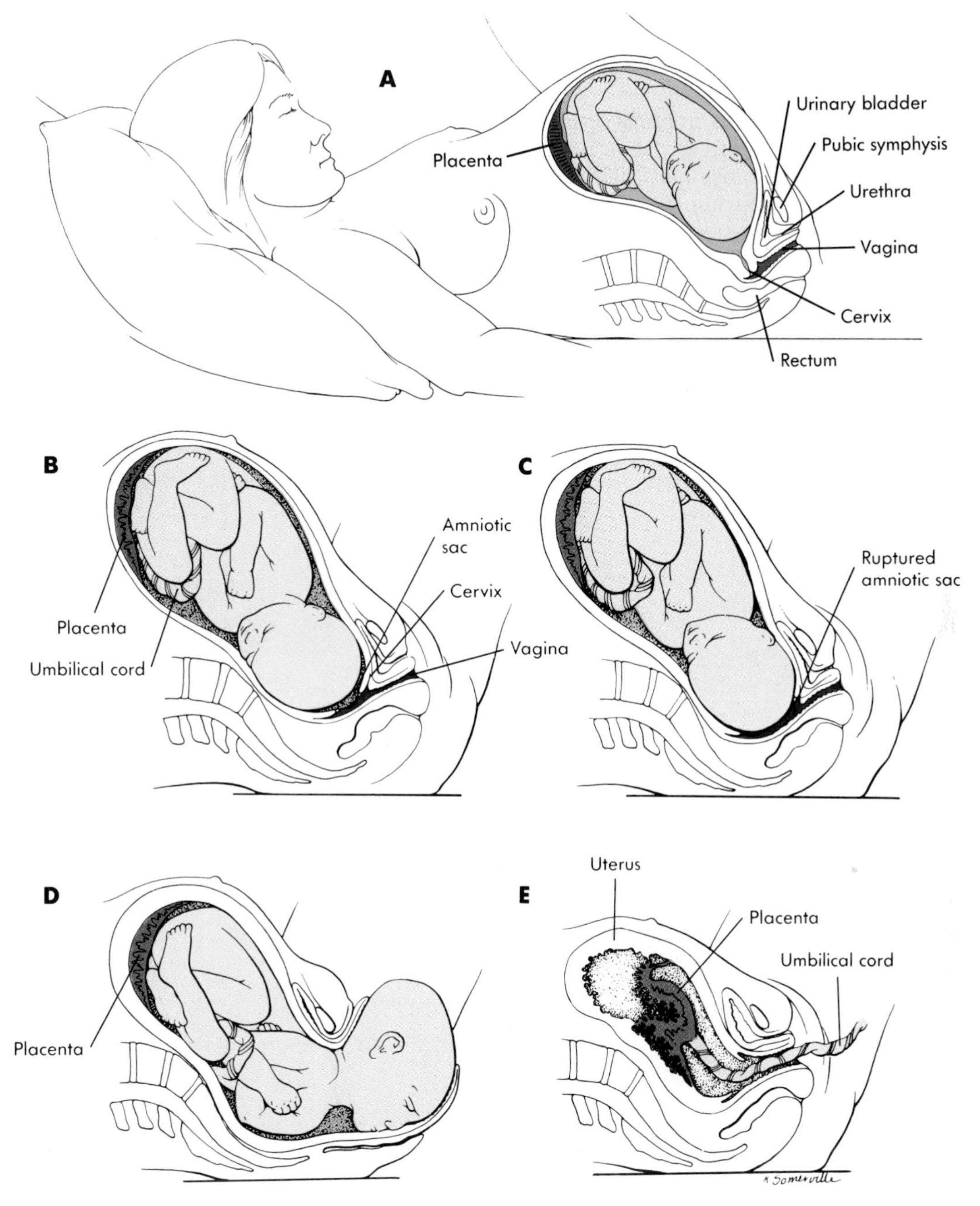

FIGURE 29-17 Process of parturition. **A** Fetal position before parturition. **B** Dilation of the cervix. **C** Rupture of the amniotic sac. **D** Expulsion of the fetus. **E** Expulsion of the placenta.

Near the end of pregnancy the uterus becomes progressively more irritable and usually exhibits occasional contractions that become stronger and more frequent until parturition is initiated. The cervix gradually dilates. Finally, strong uterine contractions complete cervical dilation and ultimately expel the fetus from the uterus through the vagina (Figure 29-17). Before expulsion of the fetus from the uterus, the amniotic sac ruptures, and amniotic fluid flows through the vagina to the exterior of the woman's body.

Labor is the period during which the contractions occur that result in expulsion of the fetus from the uterus. It occurs as three stages.

1. The **first stage** begins with the onset of regular uterine contractions and extends until the cervix dilates to a diameter approximately the size of the fetus' head. This stage of labor commonly lasts from 8 to 24 hours, but it may be as short as a few minutes in some women who have had more than one child. Normally (95% of the time) the head of the fetus is in an inferior position with-

in the woman's pelvis during labor. The head acts as a wedge, forcing the cervix and vagina to open as the uterine contractions push against the fetus.

2. The **second stage** of labor lasts from the time of maximum cervical dilation until the time that the baby exits the vagina. This stage may last from a minute to up to an hour. During this stage contractions of the abdominal muscles assist the uterine contractions. The contractions generate enough pressure to compress blood vessels in the placenta so that blood flow to the fetus is stopped. During periods of relaxation blood flow to the placenta resumes.

   Occasionally drugs such as oxytocin are administered to women during labor to increase the force of the uterine contractions. However, caution must be exercised so that tetanic-like contractions, which would drastically reduce the blood flow through the placenta, do not occur.
3. The **third stage** of labor involves the expulsion of the placenta from the uterus. Contractions of the uterus cause the placenta to tear away from the wall of the uterus. Some bleeding occurs because of the intimate contact between the placenta and the uterus. However, bleeding normally is restricted because uterine smooth muscle contractions compress the blood vessels to the placenta.

Blood levels of estrogen and progesterone fall dramatically after parturition. Once the placenta has been dislodged from the uterus, the source of these hormones is gone. In addition, during the 4 or 5 weeks after parturition the uterus becomes much smaller, but it remains somewhat larger than it was before pregnancy. The cells of the uterus become smaller, and many of them degenerate. A vaginal dis-

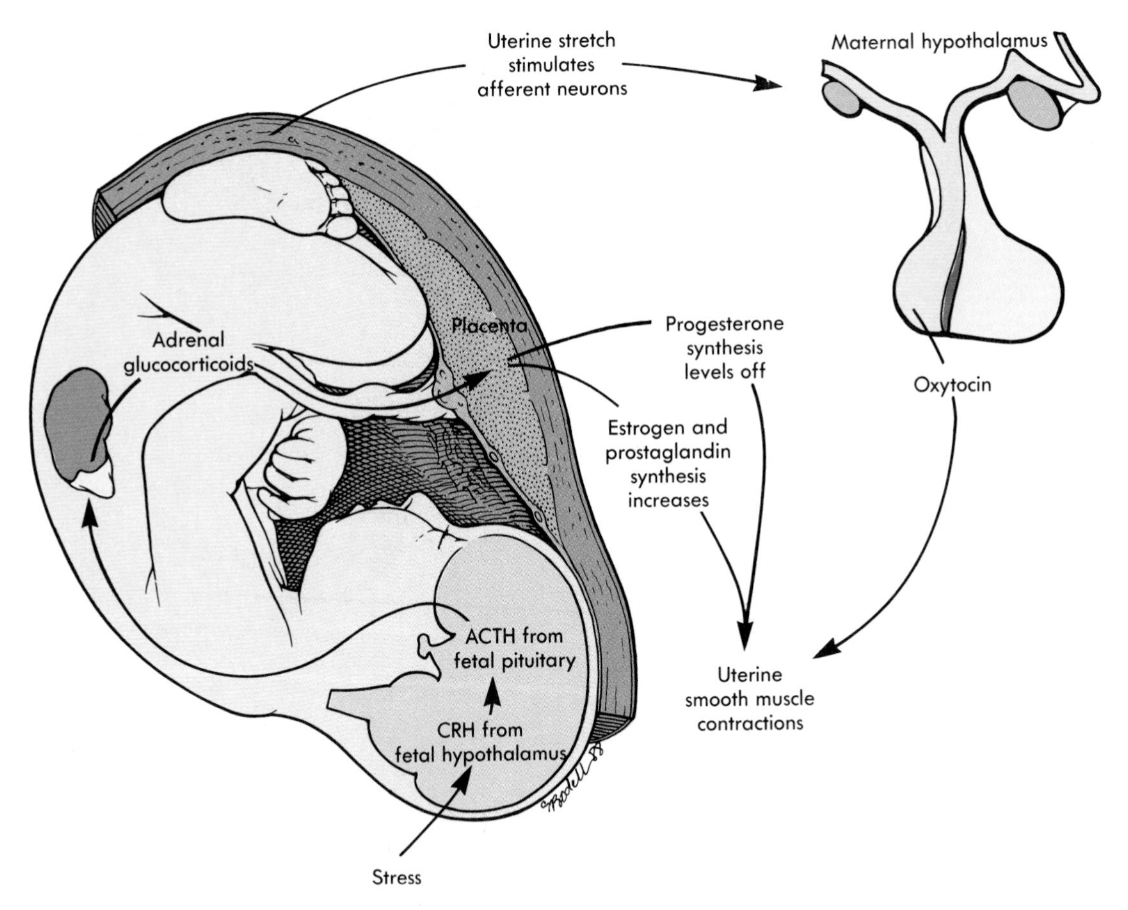

FIGURE 29-18 Factors that influence the process of parturition. The pituitary secretes adrenocorticotropic hormone *(ACTH)* in greater amounts near parturition. ACTH causes the fetal adrenal gland to secrete greater quantities of adrenal glucocorticoids, which travel in the umbilical blood to the placenta. There the adrenal glucocorticoids cause progesterone synthesis to level and estrogen and prostaglandin synthesis to increase, making the uterus more irritable. Stretch of the uterus causes neural impulses to the brain through ascending pathways and stimulates the secretion of oxytocin. Oxytocin also causes the uterine smooth muscle to contract. Although the precise control of parturition in humans is not known, these changes appear to play a role. (*CRH,* Corticotropin-releasing hormone.)

charge composed of small amounts of blood and degenerating endometrium persists for 1 week or more after parturition.

The precise signal that triggers parturition is not known, but many of the factors that support parturition have been identified (Figure 29-18). Before parturition the progesterone concentration in the maternal circulation is at its highest level (see Figure 28-16). Progesterone has an inhibitory effect on uterine smooth muscle cells. However, estrogen levels are rapidly increasing in the maternal circulation, and estrogens have an excitatory influence on uterine smooth muscle cells. As a result, the inhibitory influence of progesterone on smooth muscle cells is overcome by the stimulatory effect of estrogens near the end of pregnancy.

The adrenal gland of the fetus is greatly enlarged before parturition. Stress caused by the confined space of the uterus and the limited oxygen supply that results from a more rapid increase in the size of the fetus than in the size of the placenta cause an increased rate of adrenocorticotropic hormone (ACTH) secretion from the fetus' adenohypophysis. ACTH causes the fetal adrenal cortex to produce glucocorticoids, which travel to the placenta where they cause the rate of progesterone secretion to decrease and the rate of estrogen synthesis to increase. In addition, prostaglandin synthesis is initiated. Prostaglandins strongly stimulate uterine contractions.

During the process of parturition, oxytocin is released from the female's neurohypophysis as a result of nervous reflexes initiated by stretch of the uterine cervix. Oxytocin stimulates uterine contractions, which move the fetus farther into the cervix, causing further stretch. Thus a positive-feedback mechanism is established in which stretch stimulates oxytocin release and oxytocin causes further stretch. The positive-feedback system stops when the cervix is no longer stretched after delivery.

Progesterone inhibits oxytocin release, so decreased progesterone levels in the maternal circulation may support the increased secretion rate of oxytocin. In addition, estrogens make the uterus more sensitive to oxytocin stimulation by increasing the synthesis of receptor sites for oxytocin. Some evidence suggests that oxytocin also stimulates prostaglandin synthesis in the uterus. All of these events support the development of strong uterine contractions.

4

A woman is having an extremely prolonged labor. From her anatomy and physiology course she remembers the role of calcium in muscle contraction and asks the doctor to give her a calcium injection to speed the delivery. Explain why the doctor would or would not do as she requested.

? ? ? ? ? ? ? ? ? ? ? ?

THE NEWBORN

The newborn baby, or **neonate,** experiences several dramatic changes at the time of birth. The major and earliest changes deal with the separation of the infant from the maternal circulation and transfer from a fluid to a gaseous environment. The large, forced gasps of air that occur when the infant cries at the time of delivery help inflate the lungs.

Circulatory Changes

The initial inflation of the lungs causes important changes in the circulatory system (Figure 29-19). Expansion of the lungs reduces the resistance to blood flow through the lungs, resulting in increased blood flow through the pulmonary arteries. Consequently, an increased amount of blood flows from the right atrium to the right ventricle and into the pulmonary arteries, and less blood flows from the right atrium through the foramen ovale to the left atrium. In addition, an increased volume of blood returns from the lungs through the pulmonary veins to the left atrium, which increases the pressure in the left atrium. The increased left atrial pressure and decreased right atrial pressure (due to decreased pulmonary resistance) forces blood against the septum primum, causing the foramen ovale to close. This action functionally completes the separation of the heart into two pumps—the right side of the heart and the left side of the heart. The closed foramen ovale becomes the **fossa ovalis.**

The **ductus arteriosus,** which connects the pulmonary trunk to the aorta and allows blood to flow from the pulmonary trunk to the systemic circulation, closes off within 1 or 2 days after birth. This closure occurs because of the sphincterlike closure of the artery and is probably stimulated by local changes in blood pressure and blood oxygen content. Once closed, the ductus arteriosus is replaced by connective tissue and is known as the **ligamentum arteriosum.**

If the ductus arteriosus does not close completely, it is said to be patent. This is a serious birth defect, resulting in marked elevation in pulmonary blood pressure because blood flows from the left ventricle to the aorta, through the ductus arteriousus to the pulmonary arteries. If not corrected, it can lead to irreversible degenerative changes in the heart and lungs.

The fetal blood supply passes to the placenta through umbilical arteries from the internal iliac arteries and returns through an umbilical vein, which passes through the liver (ductus venosus) and joins the inferior vena cava. When the umbilical cord is

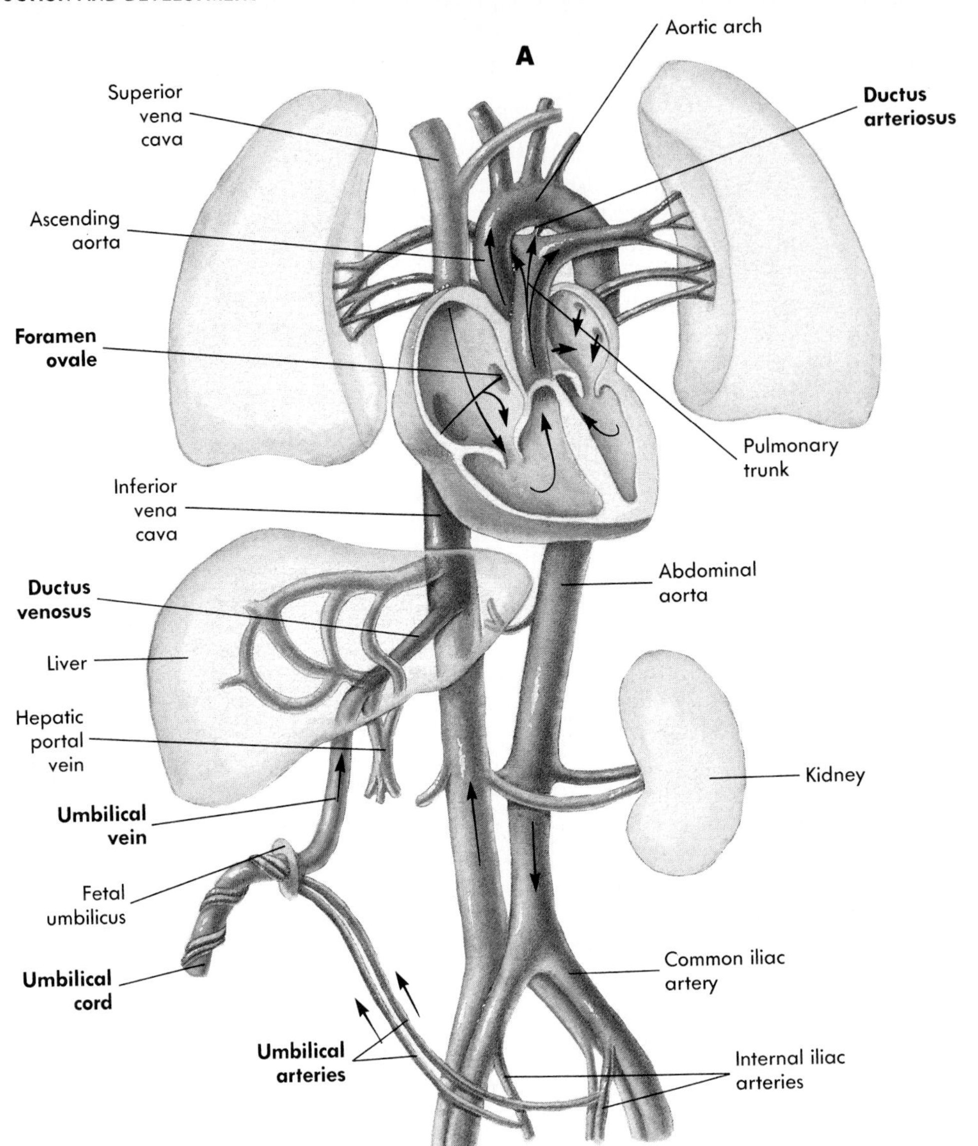

FIGURE 29-19 **A** Circulatory conditions in the fetus. Oxygen-rich blood is carried to the fetus from the placenta by the umbilical vein. Blood bypasses the liver sinusoids by flowing through the ductus venosus and bypasses the lungs by flowing through the foramen ovale and the ductus arteriosus. Oxygen-poor blood returns to the placenta through the umbilical arteries.

tied and cut, no more blood flows through the umbilical vein and arteries, and they degenerate. The remnant of the umbilical vein becomes the **round ligament of the liver,** and the ductus venosus becomes the **ligamentum venosum.**

Digestive Changes

When a baby is born, it is suddenly separated from its source of nutrients provided by the maternal circulation. Because of this separation and the shock of birth and new life, the neonate usually loses 5% to 10% of its total body weight during the first few days of life. Although the digestive system of the fetus becomes somewhat functional late in development, it is still very immature in comparison to that of the adult, and only a limited number of food types can be digested by the newborn.

The fetus swallows amniotic fluid from time to time late in gestation. Shortly after birth this swallowed fluid plus cells sloughed from the mucosal lining, mucus produced by intestinal mucous glands, and bile from the liver pass from the digestive tract as a greenish anal discharge called **meconium** (me-ko′nĭ-um).

The pH of the stomach at birth is nearly neutral because of the presence of swallowed alkaline am-

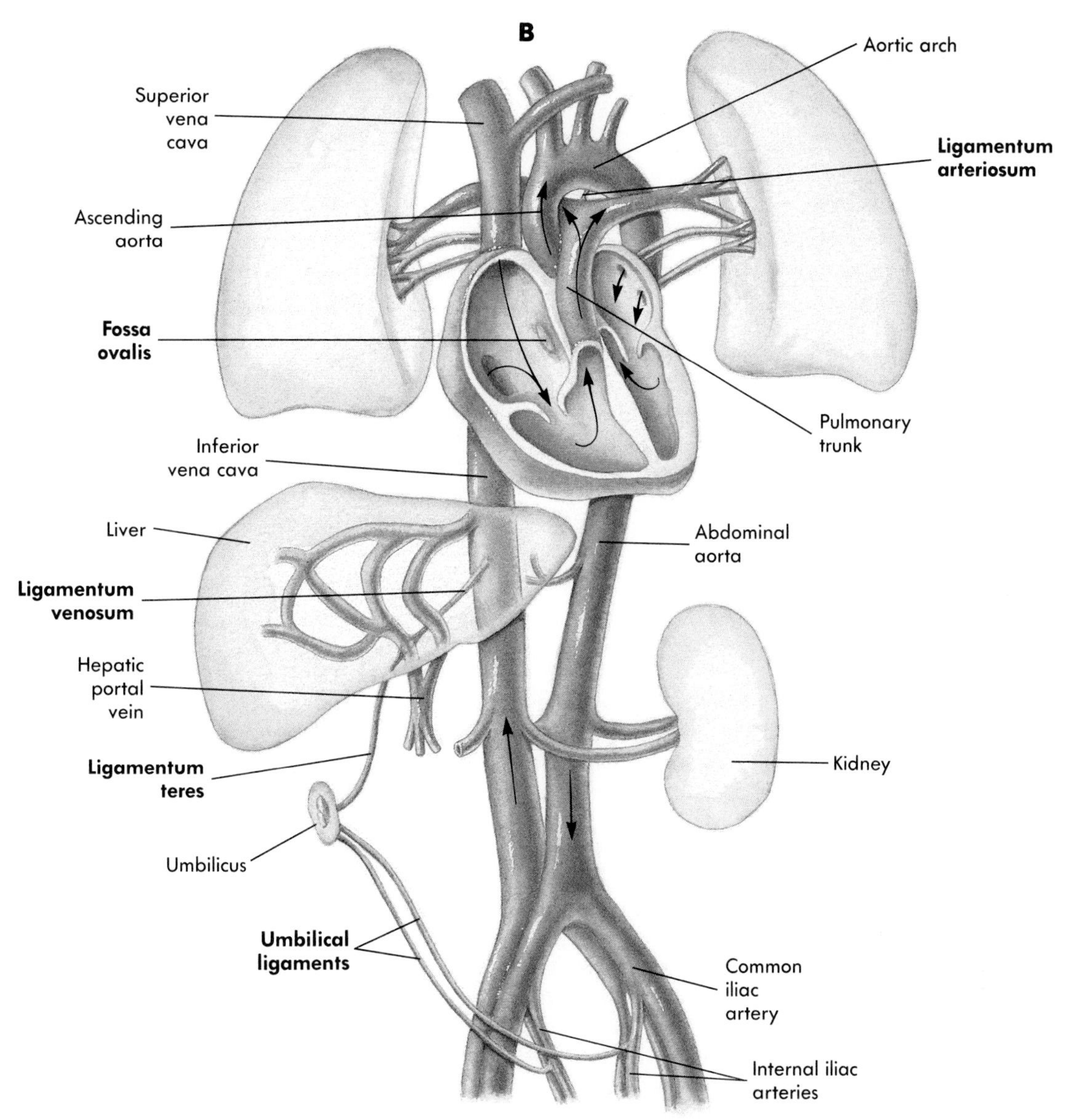

FIGURE 29-19, cont'd **B** Circulatory changes that occur at birth. The foramen ovale closes and becomes the fossa ovalis, the ductus arteriosus closes and becomes the ligamentum arteriosum, and the umbilical arteries and vein are cut so that the umbilical vein becomes the round ligament of the liver and the ductus venosus becomes the ligamentum venosum.

niotic fluid. Within the first 8 hours of life, a striking increase in gastric acid secretion occurs, causing the stomach pH to decrease. Maximum acidity is reached at 4 to 10 days, and the pH gradually increases for the next 10 to 30 days.

The neonatal liver is also functionally immature. It lacks adequate amounts of the enzyme required in the production of bilirubin; thus jaundice often occurs in premature babies. This enzyme system usually develops within 2 weeks after birth in a healthy neonate, but because this enzyme system is not fully developed at birth, some full-term babies may temporarily develop jaundice.

The newborn digestive system is capable of digesting lactose (milk sugar) from the time of birth. The pancreatic secretions are sufficiently mature for a milk diet, but the digestive system only gradually develops the ability to digest more solid foods over the first year or two. Therefore new foods should be introduced gradually during the first 2 years. It is also advised that only one new food at a time be introduced into the infant's diet so that if an allergic reaction occurs, the cause is more easily determined.

Amylase secretion by the salivary glands and the pancreas remains low until after the first year. Lactase activity in the small intestine is high at birth but declines during infancy, although the levels still exceed those in adults (activity is lost in many adults).

Table 29-3

Examples of APGAR rating scales

	0	1	2
Appearance (skin color)	White or blue	Limbs blue, body pink	Pink
Pulse (rate)	No pulse	100 beats/min	>100 beats/min
Grimace (reflexive grimace initiated by stimulating the plantar surface of the foot)	No response	Facial grimaces, slight body movement	Facial grimaces, extensive body movement
Activity (muscle tone)	No movement, muscles flaccid	Limbs partially flexed, little movement, poor muscle tone	Active movement, good muscle tone
Respiratory effort (amount of respiratory activity)	No respiration	Slow, irregular respiration	Good, regular respiration, strong cry

APGAR Scores

The newborn baby may be evaluated soon after birth to assess its physiological condition. This assessment is referred to as the **APGAR score.** APGAR is an acronym that stands for *a*ppearance, *p*ulse, *g*rimace, *a*ctivity, and *r*espiratory effort. Each of these characteristics is rated on a scale of 0 to 2 in which 2 denotes normal function, 1 denotes reduced function, and 0 denotes seriously impaired function. The total APGAR score is the sum of the scores from the five characteristics, ranging therefore from 0 to 10 (Table 29-3). A total APGAR score of 8 to 10 at 1 to 5 minutes after birth is considered normal. Other scoring systems to estimate normal growth and development, including general external appearance and neurological development, may also be applied to the neonate.

LACTATION

Lactation is the production of milk by the mother's breasts (mammary glands; Figure 29-20). It normally occurs in females after parturition and may continue for 2 or 3 years, provided suckling occurs often and regularly.

During pregnancy the high concentration and continuous presence of estrogens and progesterone cause expansion of the duct system and the secretory units of the breasts. The ducts grow and branch repeatedly to form an extensive network. Additional adipose tissue is deposited also, so the size of the breasts increases substantially throughout pregnancy. Estrogen is primarily responsible for breast growth during pregnancy, but normal development of the breast does not occur without the presence of several other hormones. Progesterone causes development of the breasts' secretory alveoli, which enlarge but do not secrete milk during pregnancy. The other hormones include growth hormone, prolactin, thyroid hormones, glucocorticoids, and insulin. A growth hormone–like substance (human somatotropin) and a prolactin-like substance (human placental lactogen) are secreted by the placenta, and these substances also help support the development of the breasts.

Prolactin, which is produced by the adenohypophysis, is the hormone responsible for milk production. Before parturition, high levels of estrogen stimulate an increase in prolactin production. However, milk production is inhibited during pregnancy because high levels of estrogen and progesterone inhibit the effect of prolactin on the mammary gland. After parturition, estrogen, progesterone, and prolactin levels decrease, and with lower estrogen and progesterone levels, prolactin can stimulate milk production. Despite a decrease in the basal levels of prolactin, a reflex response produces surges of prolactin release. During suckling, mechanical stimulation of the breasts initiates nerve impulses that reach the hypothalamus, causing the secretion of **prolactin-releasing factor (PRF)** and inhibiting the release of **prolactin-inhibiting factor (PIF).** Consequently, prolactin levels temporarily increase and stimulate milk production.

For the first few days after birth the mammary glands secrete **colostrum** (ko-los'trum), which contains little fat and less lactose than milk. Eventually, more nutritious milk is produced. Colostrum and milk not only provide nutrition, but they also contain antibodies (see Chapter 22), which help protect the nursing baby from infections.

Repeated stimulation of prolactin release makes nursing possible for several years. However, if nursing is stopped, within a few days the ability to pro-

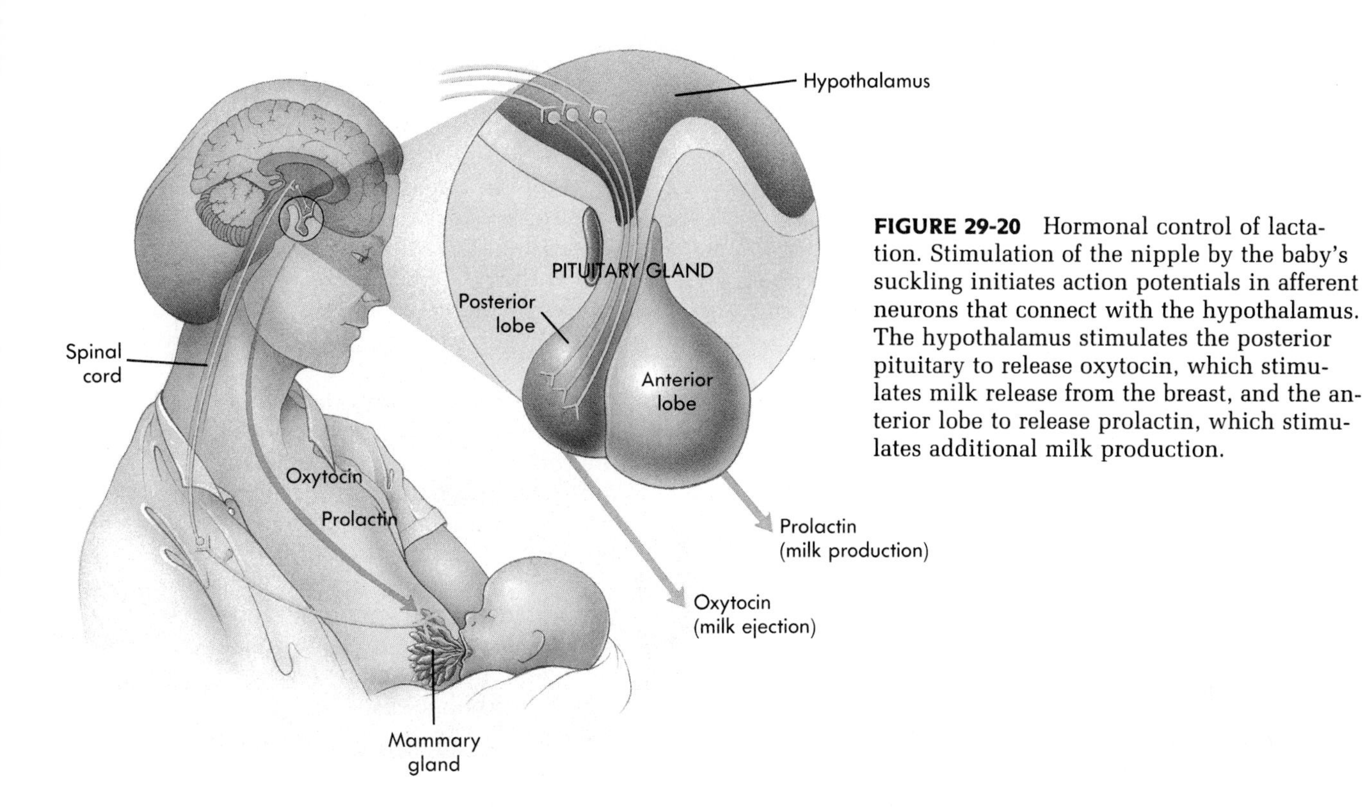

FIGURE 29-20 Hormonal control of lactation. Stimulation of the nipple by the baby's suckling initiates action potentials in afferent neurons that connect with the hypothalamus. The hypothalamus stimulates the posterior pituitary to release oxytocin, which stimulates milk release from the breast, and the anterior lobe to release prolactin, which stimulates additional milk production.

duce the prolactin increase ceases, and milk production stops.

Since it takes time to produce milk, an increase in prolactin results in the production of milk that will be used in the next nursing period. At the time of nursing, stored milk is released as a result of a reflex response. Mechanical stimulation of the breasts produces nerve impulses that cause the release of oxytocin, which stimulates cells surrounding the alveoli to contract; milk is then released from the breasts, a process that is called **milk letdown.** In addition, higher brain centers can stimulate oxytocin release, and such things as hearing an infant cry can result in milk letdown.

5

While nursing her baby, a woman noticed that she developed "stomach cramps." Explain what was happening.

? ? ? ? ? ? ? ? ? ? ?

FIRST YEAR AFTER BIRTH

A great number of changes occur in the life of the newborn from the time of birth until 1 year of age. The time when these changes occur may vary considerably from child to child, and the dates given are only rough estimates. The brain is still developing at this time, and much of what the neonate can accomplish depends on the amount of brain development achieved. It is estimated that the total adult number of neurons is present in the central nervous system at birth, but subsequent growth and maturation of the brain involve the addition of new neuroglial cells, some of which form new myelin sheaths, and the addition of new connections between neurons, which may continue throughout life.

By 6 weeks the infant is usually able to hold up its head when placed in a prone position and begins to smile in response to people or objects. At 3 months of age the infant's limbs are exercised aimlessly. However, the arms and hands are in enough control that voluntary thumb sucking can occur. The infant can follow a moving person with its eyes. At 4 months the infant begins to do push-ups (i.e., raises itself by its arms). It can begin to grasp things placed in its hand, coo and gurgle, roll from its back to its side, listen quietly when hearing a person's voice or music, hold its head erect, and play with its hands. At 5 months the infant can usually laugh out loud, reach for objects, turn its head to follow an object, lift its head and shoulders, sit with support, and roll over. At 8 months the infant can recognize familiar people, sit up without support, and reach for specific objects that it sees. At 12 months the infant may pull itself to a standing position and may be able to walk without support. The infant can pick up objects in its hands and examine them carefully. It can understand much of what is said to it and may say several words of its own.

LIFE STAGES

The stages of life previously described (i.e., prenatal and neonatal periods) are only a small portion of the total life span. The life stages from fertilization to death are as follows: (1) the germinal period—fertilization to 14 days; (2) the embryo—14 to 60 days after fertilization; (3) the fetus—60 days after fertilization to birth; (4) neonate—birth to 1 month after birth; (5) infant—1 month to 1 or 2 years (the end of infancy is sometimes set at the time that the child begins to walk); (6) child—1 or 2 years to puberty; (7) adolescent—puberty (age 11 to 14) to 20 years; and (8) adult—age 20 to death. Adulthood is sometimes divided into three periods: young adult, age 20 to 40; middle age, age 40 to 65; and older adult or senior citizen, age 65 to death (Figure 29-21). Much of this designation is associated more with social norms than with physiology.

During childhood the individual develops considerably. Many of the emotional characteristics that a person possesses throughout life are formed during early childhood.

Major physical and physiological changes occur during adolescence, and many of these changes also affect the emotions and behavior of the individual. Other emotional changes occur as the adolescent attempts to fit into an adult world.

Puberty usually occurs in females at 11 to 13 years, which is somewhat earlier than in males (approximately 12 to 14 years). The onset of puberty is usually accompanied by a growth spurt, followed by a period of slower growth. Full adult stature is usually achieved by age 17 or 18 in females and 19 or 20 years in males.

FIGURE 29-21 A conspicuous feature of the population of older adults is the range of variability. In some adults over 70 years, most systems are beginning to fail. Others can look forward to at least 10 more years of healthy living.

AGING

Development of a new and usually unique human being begins at fertilization, as does the process of aging. Cell proliferation occurs at an extremely rapid rate during early development and then begins to slow as various cells become committed to specific functions within the body.

Many cells of the body continue to proliferate throughout life, replacing dead or damaged tissue; but other cells such as the neurons in the CNS cease to proliferate once they have reached a certain number. Damage or death of these cells is irreversible. After the number of neurons reaches a peak (at approximately the time of birth), their numbers begin to decline. Neuronal loss is most rapid early in life and decreases to a slower, steadier rate.

The physical plasticity (i.e., the state of being soft and pliable) of young embryonic tissues is due largely to the presence of large amounts of hyaluronate and relatively small amounts of collagen. Furthermore, the collagen that is present is not highly cross-linked; thus the tissues are very flexible and elastic. However, many of the collagen fibers produced during development are permanent components of the individual; and as the individual ages, more and more cross-links form between the collagen molecules, rendering the tissues more rigid and less plastic.

The tissues with the highest collagen content and the greatest dependency on collagen for their function are the most severely affected by collagen cross-linking and tissue rigidity associated with aging. The lens of the eye is one of the first structures to exhibit pathological changes as a result of this increased rigidity. Vision of close objects becomes more difficult with advancing age until most middle-aged people require reading glasses (see Chapter 15). Loss of plasticity also affects other tissues, including the joints, kidneys, lungs, and heart, and greatly reduces the functional ability of these organs.

Like nervous tissue, muscle does not normally proliferate after terminal differentiation occurs before birth. As a result, the total number of skeletal and cardiac muscle fibers declines with age. The strength of skeletal muscle reaches a peak between 20 to 30 years of life and declines steadily thereafter. Furthermore, like the collagen of connective tissue, the macromolecules of muscle undergo biochemical changes during aging, rendering the muscle tissue less functional. However, a good exercise program can slow or even reverse this process.

The decline in muscular function also contributes to the decline in cardiac function with advancing age. The heart loses elastic recoil ability and muscular contractility. As a result, total cardiac output declines, and less oxygen and fewer nutrients reach cells such as neurons of the brain and cartilage cells of the joints, contributing to the decline in these tissues. Reduced cardiac function also may result in decreased blood flow to the kidneys, contributing to decreases in the kidneys' filtration ability. Degeneration of the connective tissues as a result of collagen cross-linking and other factors also decreases the filtration efficiency of the glomerular basement membrane.

Atherosclerosis (ath'er-o-sklĕ-ro'sis; the deposit and subsequent hardening of a soft, gruel-like material) is the deposit of lipid in the intima of large- and medium-sized arteries. These deposits then become fibrotic and calcified, resulting in **arteriosclerosis** (ar-ter-ĭ-o-sklĭ-ro-sis; hardening of the arteries). Arteriosclerosis interferes with normal blood flow and can result in a **thrombus,** which is a clot or plaque formed inside a vessel. A piece of the plaque, called an **embolus** (em'bo-lus) can break loose, float through the circulation, and lodge in smaller arteries to cause myocardial infarctions or strokes. Although atherosclerosis occurs to some extent in all middle-aged and elderly people and even may occur in certain young people, some people appear more at risk because of high blood cholesterol levels. This condition seems to have a heritable component, and blood tests are available to screen people for high blood cholesterol levels.

Many other organs such as the liver, pancreas, stomach, and colon undergo degenerative changes with age. The ingestion of harmful agents may accelerate such changes. Examples include the degenerative changes induced in the lungs (aside from lung cancer) by cigarette smoke and sclerotic changes in the liver as a result of alcohol consumption.

In addition to the previously described changes associated with aging, cellular wear and tear (or cytologic aging) is another factor that contributes to aging. Progressive damage from many sources such as radiation and toxic substances may result in irreversible cellular insults and may be one of the major factors leading to aging. It has been speculated that ingestion of vitamins C and E in combination may help slow this portion of aging by stimulating cell repair. Vitamin C also stimulates collagen production and may slow the loss of tissue plasticity associated with aging collagen.

Immune changes may also be a major factor contributing to aging. The aging immune system loses its ability to respond to outside antigens but becomes more sensitive to the body's own antigens. These autoimmune changes add to the degeneration of the tissues already described and may be responsible for such things as arthritic joint disorders, chronic glomerular nephritis, and hyperthyroidism. In addition, T lymphocytes tend to lose their functional capacity with aging and cannot destroy abnormal cells as efficiently. This change may be one reason that certain types of cancer occur more readily in older people.

Many changes associated with aging may be caused by genetic traits. As a general rule, animals with a very high metabolic rate have a shorter life span than those with a lower metabolic rate. In humans, a very small number of exceptional people have a slightly reduced normal body temperature, suggesting a lower metabolic rate. These same people often have an unusually long life span. This tendency appears to run in families and probably has some genetic basis. Studies of the general population suggest that if your parents and grandparents have lived long, so will you; if your parents and grandparents died young, you may expect the same.

Another piece of evidence suggesting that there is a strong genetic component to aging comes from a disorder called **progeria** (pro-jăr'ĭ-ah; premature aging). This apparent genetic trait causes the degenerative changes of aging to occur shortly after the first year, and the child may look like a very old person by age 7.

One of the greatest disadvantages of aging is the increasing lack of ability to adjust to stress. Older people have a far more precarious homeostatic balance than younger people, and eventually some stress is encountered that is so great that the body's ability to recover is surpassed and death results.

DEATH

Death is usually not attributed to old age. Some other problem such as heart failure, renal failure, or stroke is usually listed as the cause of death.

Death was once defined as the loss of heartbeat and respiration. However, in recent years more precise definitions of death have been developed since both the heart and the lungs can be kept working artificially and the heart can even be replaced by an artificial device. Modern definitions of death are based on the permanent cessation of life functions and the cessation of integrated tissue and organ function. The most widely accepted indication of death in humans is **brain death,** which is defined as irreparable brain damage manifested clinically by the absence of response to stimulation, the absence of spontaneous respiration and heart beat, and an isoelectric ("flat") electroencephalogram for at least 30 minutes (i.e., in the absence of known central nervous system poisoning or hypothermia).

GENETICS

Genetics is the study of heredity (i.e., those characteristics inherited by children from their parents). The functional unit of heredity is the **gene,** and **DNA** is the molecule responsible for heredity (see Chapter 2). Each gene is contained within a chromosome and consists of a certain portion of a DNA molecule but not necessarily a continuous stretch of DNA.

There are two major types of genes: structural and regulatory. **Structural genes** are those DNA sequences that code for specific amino acid sequences in proteins such as enzymes, hormones, or structural proteins (e.g., collagen). **Regulatory genes** are segments of DNA involved in controlling which structural genes are transcribed in a given tissue and are not understood as well as structural genes.

Human genetics is the study of inherited human traits. A major objective of geneticists is to prevent genetically caused birth defects by identifying people who are **carriers** of potentially harmful genes. **Genetic counseling** includes talking to parents of children with genetic disorders about the prognosis and possible treatment of the disorder, predicting the possible outcome of matings involving carriers of harmful genes, and giving counsel about available options. Genetic counselors often diagram a **pedigree** of the family to aid in making predictions (Figure 29-22). One common error about birth defects (made surprisingly often even among medical personnel) is the belief that all or most congenital disorders are genetic. **Congenital** means "occurring at birth" and does not imply a genetic component. In fact, only approximately 15% of all congenital disorders have a known genetic cause, and approximately 70% of all birth defects are of unknown cause.

Chromosomes

The DNA of each cell is packed into **chromosomes** (kro'mo-sōmz) within the nucleus (see Chapter 3). Each normal human somatic cell (i.e., all cells except certain reproductive cells) contains 23 pairs of chromosomes or 46 total chromosomes. Two of the 46 chromosomes are called **sex chromosomes,** and the remaining 44 chromosomes are called **autosomes.** A normal female has two **X** chromosomes (XX) in each somatic cell, whereas a normal male has one **X** and one **Y** chromosome (XY) in each somatic cell. For convenience, the autosomes are numbered in pairs from 1 through 22.

> There is a wide range of sex chromosome anomalies, suggesting that abnormalities in sex chromosome number are the most easily transmitted chromosomal anomalies still compatible with life. Maleness is determined by the presence of a Y chromosome, and femaleness is determined by its absence, regardless of the number of X chromosomes. Therefore XO (Turner's syndrome), XX, XXX, or XXXX results in females, and XY, XXY, XXXY, or XYY results in males.
>
> Secondary sexual characteristics are usually underdeveloped in both the XXX female and the XXY male (called Klinefelter's syndrome), and additional X chromosomes (XXXX or XXXY) are often associated with mental retardation.

Some genetic disorders involve abnormal chromosome numbers. For example, Down syndrome involves an extra chromosome 21, and Turner's syndrome involves a missing X chromosome (XO).

Each chromosome contains thousands of genes, and each gene occupies a specific **locus** on the chromosome. Both chromosomes of a given pair contain similar genes. The two chromosomes of a pair are homologous, and the genes occupying the same locus on homologous chromosomes are called **alleles** (ă-lēlz'). If the two allelic genes are identical, the person is **homozygous** (ho'mo-zi'gus) for the trait specified by that gene locus. If the two alleles are slightly different, the person is **heterozygous** (het'er-o-zi'gus) for the trait.

During reproduction the female contributes (through the oocyte) one set of 23 chromosomes to the new zygote, and the male contributes (through the sperm cell) one set of 23 chromosomes. Therefore, for any given pair of alleles controlling a given trait, the female contributes one allele, and the male contributes one allele. An exception occurs with the X and Y chromosomes—there are no alleles on the Y chromosome for most of the loci on the X chromosome.

Patterns of Inheritance

When people think of inheritance, they often think of such things as eye or hair color. These two traits are somewhat difficult to evaluate because they involve **complex,** or **polygenic,** inheritance patterns (i.e., more than one gene is involved). On the other hand, **simple** inheritance traits involve only one gene.

Considerations of inheritance patterns are based on **probability** predictions. The probability of the occurrence of a given genetic combination for a certain trait can be predicted for a large number of possible matings but cannot be precisely predicted for a given mating.

For certain simple genetic traits a given characteristic in one parent may also exist in the child without regard to the contrasting characteristic of the other parent. The trait that is inherited in this case is **dominant** over the opposite, **recessive** trait. For example, certain types of polydactyly (extra fingers and/or toes) are simple dominant traits. If a person with dominant polydactyly mates with a person without this trait, the probability is that either half or all of their children will also have polydactyly. If dominant polydactyly is designated *D* and the normal recessive condition is designated *d*, then the parent with polydactyly can be either homozygous (two copies of the same gene, *DD*) for the trait or heterozygous (different genes, *Dd*) for the trait (the normal parent will be homozygous recessive, *dd*). Polydactyly or normal is the **phenotype** (the trait that can be seen or tested for), and *DD*, *Dd*, and *dd* are the pos-

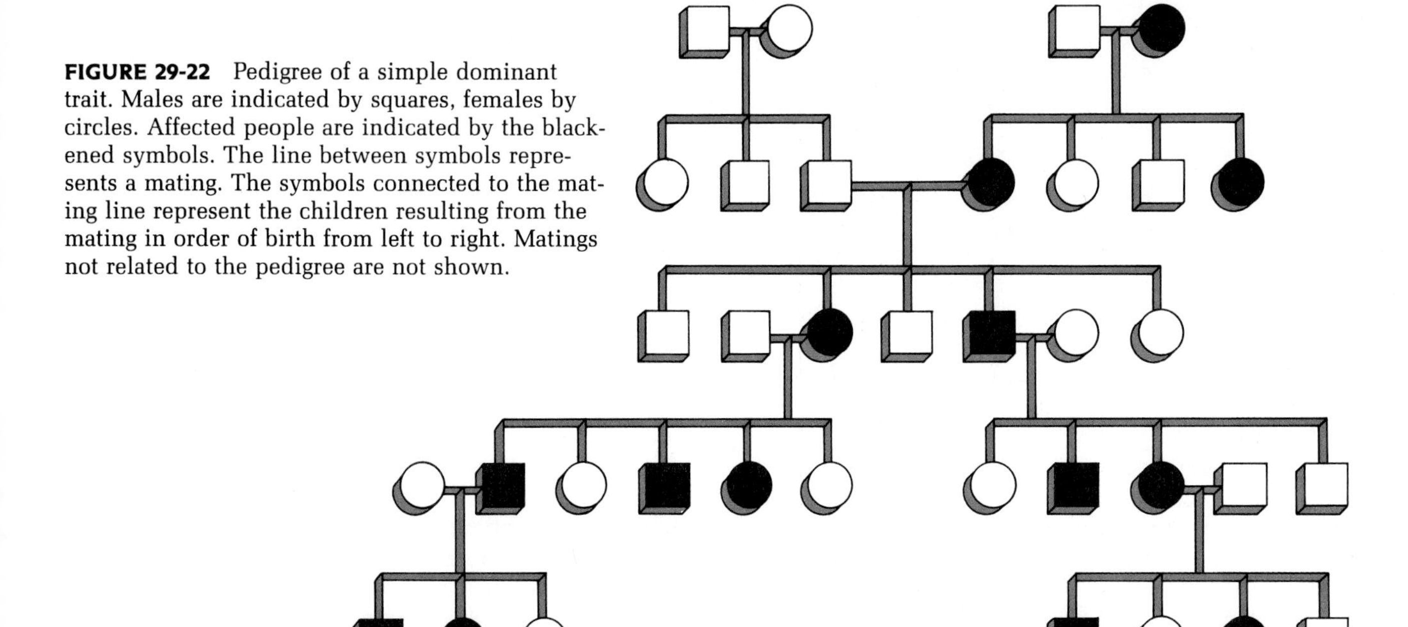

FIGURE 29-22 Pedigree of a simple dominant trait. Males are indicated by squares, females by circles. Affected people are indicated by the blackened symbols. The line between symbols represents a mating. The symbols connected to the mating line represent the children resulting from the mating in order of birth from left to right. Matings not related to the pedigree are not shown.

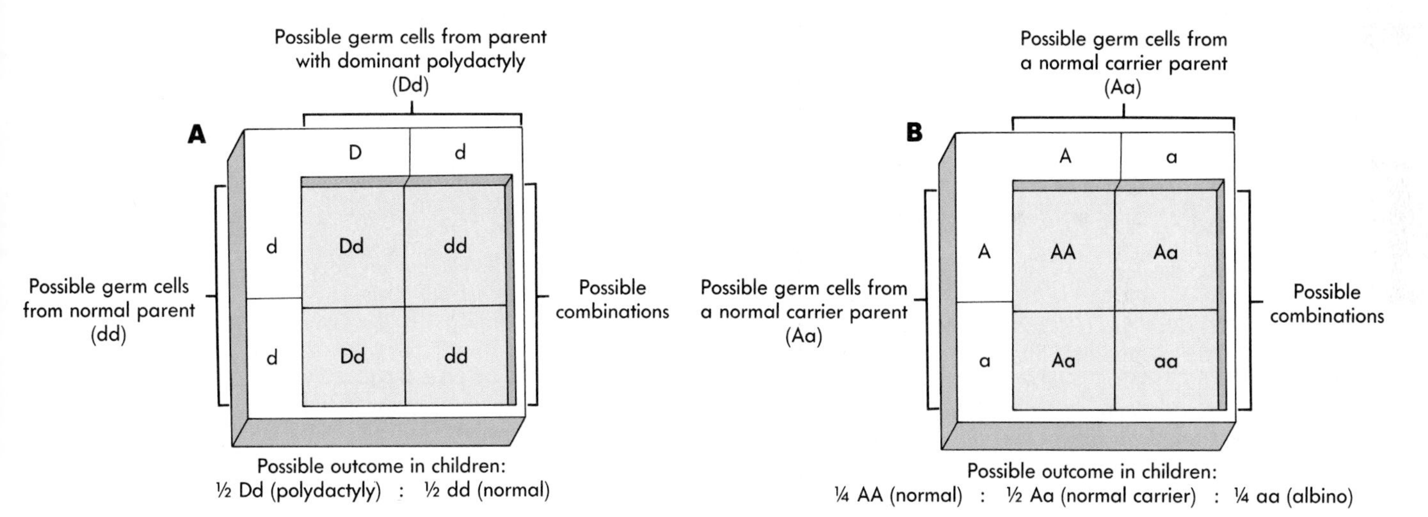

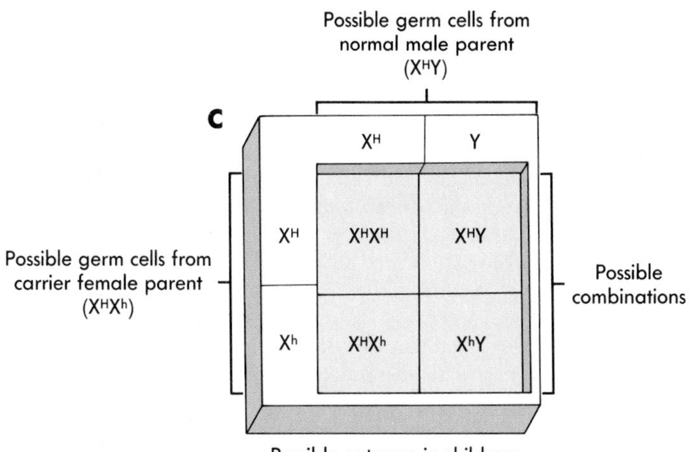

FIGURE 29-23 Examples of inheritance patterns. **A** A dominant trait (polydactyly; *D* represents the dominant condition with extra fingers or toes, and *d* represents the recessive normal condition). The figure represents a mating between a normal person and a person with dominant polydactyly. **B** A recessive trait (albinism; *A* represents the normal, pigmented condition, and *a* represents the recessive unpigmented condition). The figure represents a mating between two normal carriers. **C** An X-linked trait (hemophilia; X^H represents the normal X chromosome condition with all clotting factors, and X^h represents the X chromosome lacking a gene for one clotting factor). The figure represents a mating between a normal male and a normal carrier female.

sible **genotypes** (the actual genetic condition). Each parent has two alleles for polydactyly, and one of the two alleles is passed to the offspring. If the parent is homozygous, *D* is the only allele available, and all the children will have polydactyly, the same phenotype, and the same genotype—*Dd*—since *D* came from the parent with polydactyly and *d* came from the normal parent. However, if the parent with polydactyly is heterozygous, approximately half the reproductive cells from that person will carry the *D* allele and half will carry the *d* allele. As a result, the probability is that half the children will have polydactyly (*Dd*) and half will be normal (*dd*) since the normal parent contributed *d* in both cases (Figure 29-23, *A*).

A recessive condition such as albinism (absence of pigment) is somewhat more complex. A person with albinism has only one possible genotype (*aa*), whereas a normal person can be either homozygous (*AA*) or heterozygous (*Aa*). It is not possible with present technology to distinguish between the two genotypes *AA* or *Aa* because both present the same phenotype. If an albino person (*aa*) mates with a homozygous normal person (*AA*), all the children will be phenotypically normal with a heterozygous genotype (*Aa*). If an albino person (*aa*) mates with a heterozygous normal person (*Aa*), the probability is that approximately half the children will be albino (*aa*) and half will be normal heterozygous carriers (*Aa*). Albinism can also result from matings of two phenotypically normal individuals with genotypes *Aa*. Approximately half the children will be just like the parents (*Aa*), one fourth will be homozygous normal (*AA*) and will not be carriers of the trait, and one fourth will be albino (*aa*) (Figure 29-23, *B*).

Simple dominant and recessive traits are on the autosomes. Genes on the sex chromosomes are **X linked** (sex linked) or **Y linked** (holandric, male only). Y-linked traits are passed only from male to male and never appear in the female. In the case of an X-linked trait, females with two alleles function as though the gene were autosomal recessive or dominant. However, a heterozygous carrier female will have sons of whom approximately half do not have the trait and half do because the male receives only one X chromosome with no comparable allele on the Y chromosome. An example of an X-linked trait is hemophilia (lacking one of the clotting factors), where *H* represents the normal condition and *h* represents the recessive hemophilia (Figure 29-23, *C*). A heterozygous woman (X^HX^h) will have either normal sons (X^HY) or sons with hemophilia (X^hY).

SUMMARY

PRENATAL DEVELOPMENT IS AN IMPORTANT PART OF AN INDIVIDUAL'S LIFE.

PRENATAL DEVELOPMENT *p. 948*

1. Prenatal development is divided into the germinal, embryonic, and fetal periods.
2. Developmental age is 14 days less than clinical age.
3. Fertilization, the union of the oocyte and sperm, results in a zygote.
4. The product of fertilization undergoes divisions until it becomes a mass called a morula and then a hollow ball of cells called a blastocyst.
5. The cells of the morula are pluripotent (capable of making any cell of the body).
6. The blastocyst implants into the uterus approximately 7 days after fertilization. The placenta is derived from the trophoblast of the blastocyst.
7. All tissues of the body are derived from three primary germ layers: ectoderm, mesoderm, and endoderm.
8. The nervous system develops from a neural tube that forms in the ectodermal surface of the embryo and from neural crest cells derived from the developing neural tube.
9. Segments called somites that develop along the neural tube give rise to the musculature, vertebral column, and ribs.
10. The gastrointestinal (GI) tract develops as the developing embryo closes off part of the yolk sac.
11. The celom develops from small cavities that fuse within the embryo.
12. The limbs develop from proximal to distal as outgrowths called limb buds.
13. The face develops by the fusion of five major tissue processes.

Development of the Organ Systems

1. The skin develops from ectoderm (epithelium), mesoderm (dermis), and the neural crest (melanocytes).
2. The skeletal system develops from mesoderm or neural crest cells.
3. Muscle develops from myoblasts, which migrate from somites.
4. The brain and spinal cord develop from the neural tube, and the peripheral nervous system develops from the neural tube and the neural crest.
5. The special senses develop mainly as neural tube or neural crest derivatives.
6. The endocrine system develops mainly as outpocketings of the brain or digestive tract.
7. The heart develops as two tubes fuse into a single tube that bends and develops septa to form four chambers.
8. The peripheral circulation develops from mesoderm as blood islands become hollow and fuse to form a network.
9. The lungs form as evaginations of the digestive tract. These evaginations undergo repeated branching.
10. The urinary system develops in three stages—pronephros, mesonephros, and metanephros—from the head to the tail of the embryo. The ducts join the digestive tract.

11. The reproductive system develops in conjunction with the urinary system. Hormones are very important to sexual development.

Growth of the Fetus

1. The embryo becomes a fetus at 60 days.
2. The fetal period is from day 60 to birth. It is a time of rapid growth.

PARTURITION p. 966

1. The total length of gestation is 280 days (clinical age).
2. Uterine contractions force the baby out of the uterus during labor.
3. Increased estrogen levels and decreased progesterone levels help initiate parturition.
4. Fetal glucocorticoids act on the placenta to decrease progesterone synthesis and to increase estrogen and prostaglandin synthesis.
5. Stretching of the uterus and decreased progesterone levels stimulate oxytocin secretion, which stimulates uterine contraction.

THE NEWBORN p. 969

Circulatory Changes

1. The foramen ovale closes, separating the two atria.
2. The ductus arteriosus closes, and blood no longer flows between the pulmonary trunk and the aorta.
3. The umbilical vein and arteries degenerate.

Digestive Changes

1. Meconium is a mixture of amniotic fluid, bile, and mucus excreted by the newborn.
2. The stomach begins to secrete acid.
3. The liver does not form adult bilirubin for the first 2 weeks.
4. Lactose can be digested, but other foods must be gradually introduced.

APGAR Scores

1. APGAR represents *a*ppearance, *p*ulse, *g*rimace, *a*ctivity, and *r*espiratory effort.
2. APGAR and other methods are used to assess the physiological condition of the newborn.

LACTATION p. 972

1. Estrogen, progesterone, and other hormones stimulate the growth of the breasts during pregnancy.
2. Suckling stimulates prolactin and oxytocin synthesis. Prolactin stimulates milk production, and oxytocin stimulates milk letdown.

FIRST YEAR AFTER BIRTH p. 973

1. The number of neuron connections and glial cells increases.
2. Motor skills gradually develop, especially head, eye, and hand movements.

LIFE STAGES p. 974

The life stages include the following: germinal, embryo, fetus, neonate, infant, child, adolescent, and adult.

AGING p. 974

1. Loss of cells that are not replaced contributes to aging.
   - There is a loss of neurons.
   - Loss of muscle cells can affect skeletal and cardiac muscle function.
2. Loss of tissue plasticity results from cross-link formation between collagen molecules.
   - The lens of the eye loses the ability to accommodate.
   - Other organs such as the joints, kidneys, lungs, and heart also have reduced efficiency with advancing age.
3. The immune system loses the ability to act against foreign antigens and may attack self-antigens.
4. Many aging changes are probably genetic.

DEATH p. 975

Death is the loss of brain functions.

GENETICS p. 975

1. Genes are segments of DNA.
2. Human genetics is the study of inherited human traits.
3. Not all birth defects are genetic.

Chromosomes

1. There are 46 chromosomes in a normal somatic cell.
2. The chromosomes and thus the genes are paired. The male and the female each contribute one half the genetic material to the next generation.

Patterns of Inheritance

The main inheritance patterns are dominant, recessive, and X linked.

CONTENT REVIEW

1. Define clinical age and postovulatory age, and distinguish between the two.
2. What are the events during the first week after fertilization? Define zygote, morula, and blastocyst.
3. What is meant by the term pluripotent?
4. How does the placenta develop?
5. Describe the formation of the germ layers and the role of the primitive streak.
6. How are the neural tube and the neural crest formed? What do they become?
7. What is a somite?
8. Describe the formation of the gut and the body cavities.
9. What does proximal-to-distal growth mean in relation to the limbs?
10. Describe the processes involved in formation of the face. What clefts are closed by fusion of these processes?
11. Describe the formation of these major organs: skin, bones, skeletal muscles, eyes, pituitary gland, thyroid gland, pancreas, heart, lungs, kidneys, and gonads.
12. What major events distinguish embryonic and fetal development?
13. Describe the hormonal changes that take place before and during delivery. How is stretch of the cervix involved in delivery?
14. What changes take place in the newborn's circulatory system and digestive system shortly after birth?
15. Define meconium. Why does jaundice often develop after birth?
16. What is the APGAR score?
17. What hormones are involved in preparing the breast for lactation? Describe the events involved in milk production and milk letdown.

18. Describe the changes in motor and language skills that take place during the first year of life.
19. Define the different life stages, starting with the germinal stage and ending with the adult.
20. How does the loss of cells that are not replaced affect the aging process? Give examples.
21. How does loss of tissue plasticity affect the aging process? Give examples.
22. How does aging affect the immune system?
23. What role does genetics play in aging?
24. Define death.
25. How do genes relate to chromosomes?
26. Define homozygous, dominant, heterozygous, recessive, and X linked as used in the study of genetics.

CONCEPT REVIEW

1. A woman is told by her physician that she is 44 days past her last menstrual period (LMP). How many days has the embryo been developing, and what developmental events are occurring?
2. A high fever can prevent neural tube closure. If a woman has a high fever approximately 35 to 45 days after her LMP, what kinds of birth defects may be seen in the developing embryo?
3. If the apical ectodermal ridge were damaged during embryonic development when the limb bud was approximately half grown, what kinds of birth defects might be expected? Describe the anatomy of the affected structure.
4. What would be the results of exposing a female embryo to high levels of testosterone while she is developing?
5. Three minutes after birth a newborn has an APGAR score of five as follows: A, 0; P, 1; G, 1; A, 1; and R, 2. What are some of the possible causes for this low score? What might be done for this neonate?
6. When a woman nurses, it is possible for milk letdown to occur in the breast that is not being suckled. Explain how this response happens.
7. Dimpled cheeks are inherited as a dominant trait. Is it possible for two parents, each of whom have dimpled cheeks, to have a child without dimpled cheeks? Explain.
8. The ability to roll the tongue to form a "tube" is due to a dominant gene. Suppose that a woman and her son can both roll their tongues, but her husband cannot. Is it possible to determine if the husband is the father of her son?

ANSWERS TO PREDICT QUESTIONS

1 p. 952. Two primitive streaks on one embryonic disk result in the development of two embryos. If the two primitive streaks are touching or are very close to each other, the embryos may be joined, resulting in conjoined (or Siamese) twins.

2 p. 960. Since the early embryonic heart is a simple tube with no (or very primitive) valves, blood must be forced through the heart in almost a peristaltic fashion, and the contraction begins in the sinus venosus. If the sinus venosus did not contract first, blood could flow in the opposite direction.

3 p. 966. 266 days (i.e., 14 days less than clinical time [280 days minus 14 days equals 266 days]).

4 p. 969. Elevation of calcium levels might cause the uterine muscles to contract tetanically. This tetanic contraction could compress blood vessels and cut the blood supply to the fetus. Hypercalcemia can also result in arrythmias and muscle weakness (see Chapter 27).

5 p. 973. Nursing stimulates the release of oxytocin, which is responsible for milk letdown. Oxytocin can also cause uterine contractions and cramps.

APPENDIX A

MINI-ATLAS OF HUMAN ANATOMY

Most students who enroll in a course in human anatomy and physiology will never have the opportunity to actually dissect a human cadaver. Laboratory courses often include dissection of a cat or a fetal pig. These photographs of actual human dissections are intended to help you bridge the gaps between the idealized illustrations of human anatomy in the textbook and the actual anatomical structures.

Many photographs of human dissections are found within the text where particular systems or structures are discussed. The purpose of this atlas collection is to show the relationships among different systems in one illustration. For example, in the photographs of the head and neck regions, muscles, nerves, glands, and blood vessels are all labeled.

The photos are arranged so that you can study regions of the body from head to foot beginning with the superficial structures and progressing to the deeper structures. These should be used in conjunction with the drawings in your textbook to reinforce the anatomical relationships you will see in the laboratory.

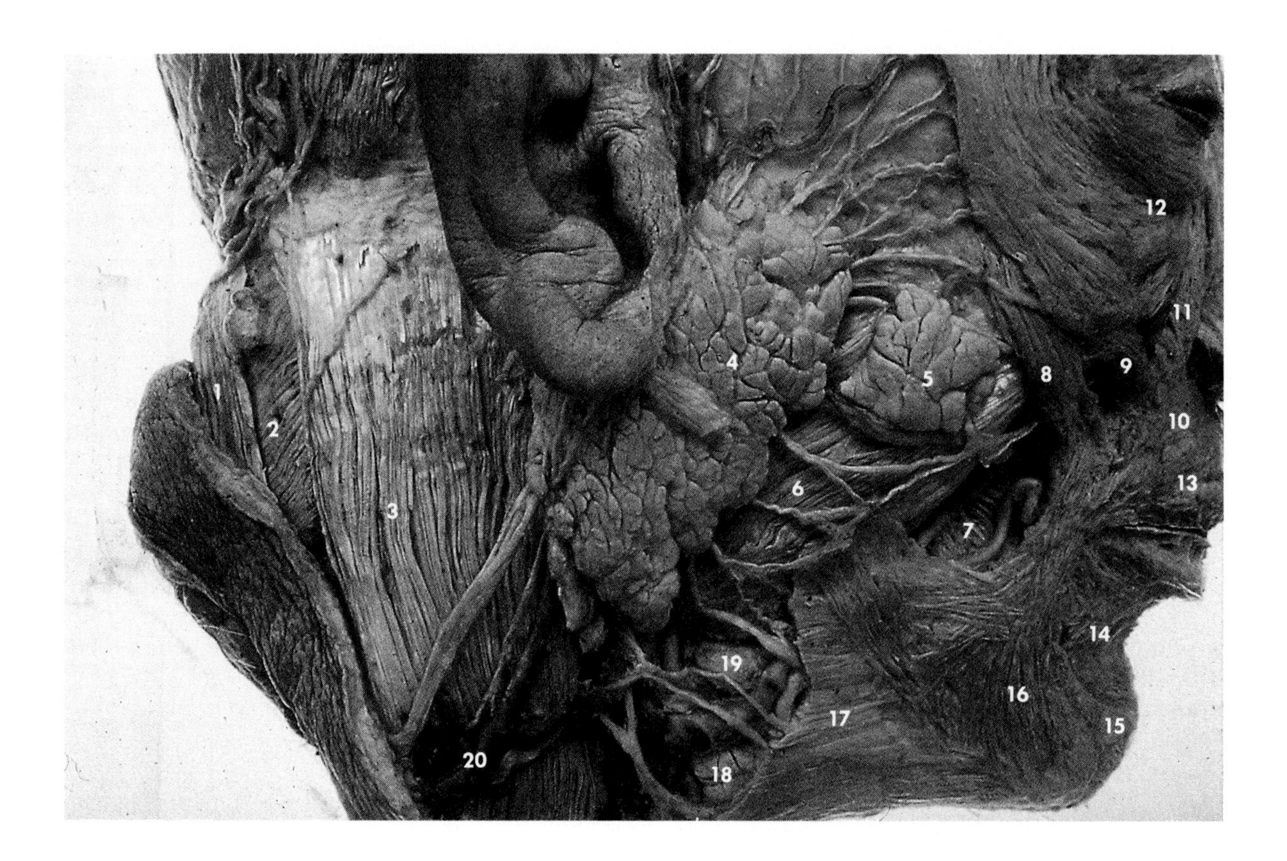

Right lower face and upper neck

1 Trapezius
2 Splenius capitis
3 Sternocleidomastoid
4 Parotid gland and facial nerve branches at anterior border
5 Accessory parotid gland
6 Masseter
7 Buccinator
8 Zygomaticus major
9 Zygomaticus minor
10 Levator labii superioris
11 Levator labii superioris alaeque nasi
12 Orbicularis oculi
13 Orbicularis oris
14 Depressor labii inferioris
15 Mentalis
16 Depressor anguli oris
17 Platysma
18 Submandibular gland
19 Submandibular lymph nodes
20 External jugular vein

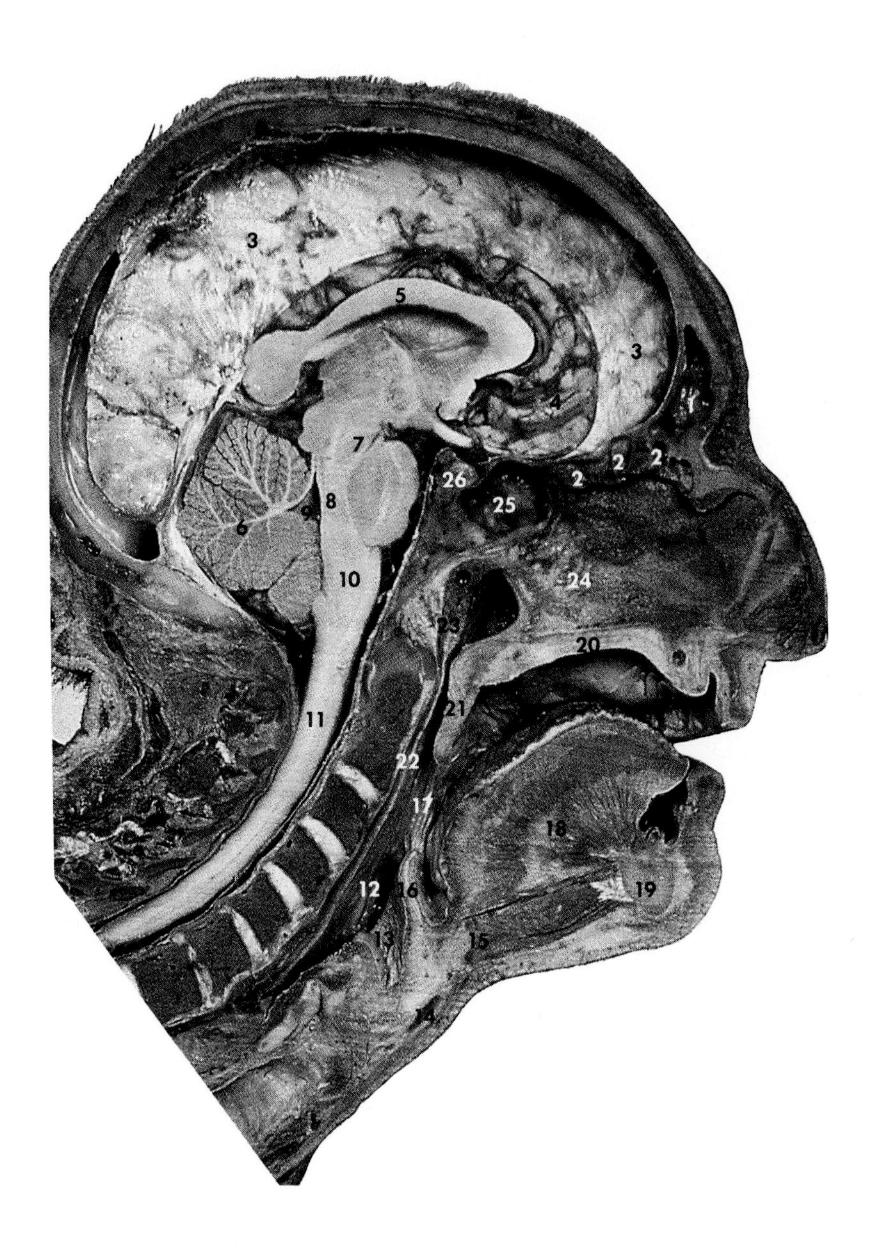

Right half of the head, in sagittal section

1 Frontal sinus
2 Ethmoidal air cells
3 Falx cerebri
4 Medial surface of cerebral hemisphere
5 Corpus callosum
6 Cerebellum
7 Midbrain
8 Pons
9 Fourth ventricle
10 Medulla oblongata
11 Spinal cord
12 Laryngopharynx
13 Opening into larynx
14 Thyroid cartilage
15 Hyoid bone
16 Epiglottis
17 Oropharynx
18 Tongue
19 Mandible
20 Hard palate
21 Soft palate
22 Nasopharynx
23 Pharyngeal tonsil
24 Nasal septum
25 Sphenoidal sinus
26 Pituitary gland

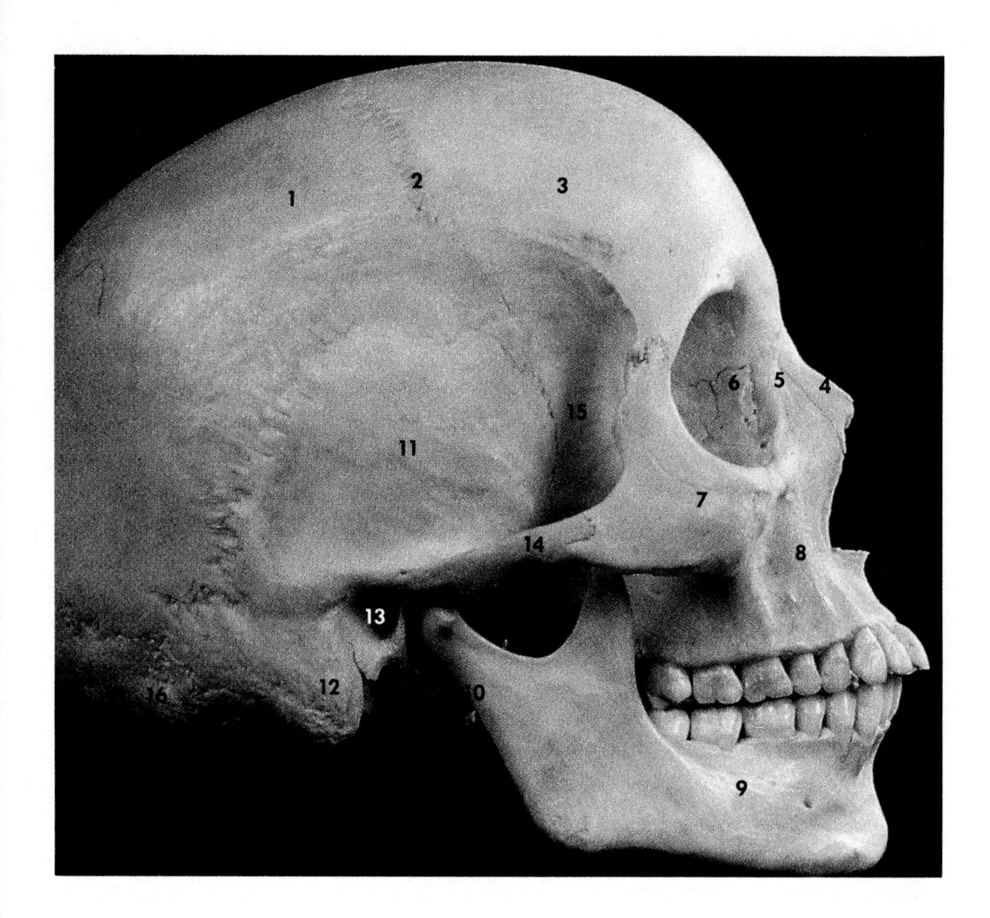

The skull, from the right

1 Parietal bone
2 Coronal suture
3 Frontal bone
4 Nasal bone
5 Frontal process of maxilla
6 Lacrimal bone
7 Zygomatic bone
8 Maxilla
9 Body of mandible
10 Styloid process
11 Temporal bone
12 Mastoid process
13 External auditory meatus
14 Zygomatic arch
15 Greater wing of sphenoid
16 Occipital bone

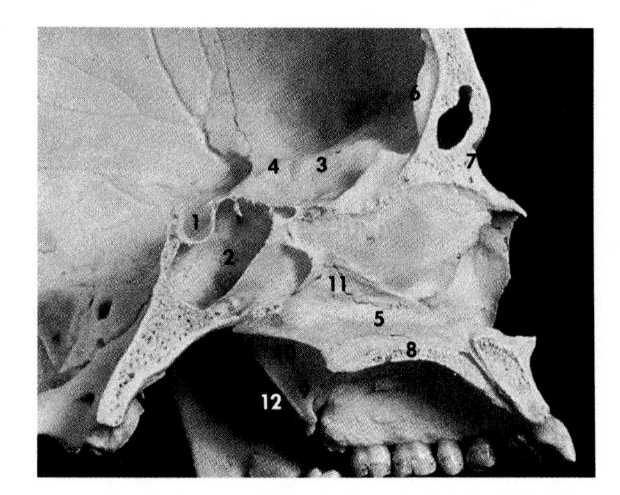

Lateral wall of the skull, midline sagittal section

1 Pituitary fossa (sella turcica)
2 Sphenoidal sinus
3 Cribriform plate of ethmoid bone
4 Air cells of ethmoidal sinus
5 Inferior nasal concha
6 Frontal sinus
7 Nasal bone
8 Palatine process of maxilla
9 Pterygoid hamulus
10 Medial pterygoid plate
11 Perpendicular plate of palatine bone
12 Lateral pterygoid plate

The brain, coronal section

1 Insula
2 Putamen
3 Globus pallidus
4 Body of caudate nucleus
5 Corpus callosum
6 Lateral ventricle
7 Fornix
8 Thalamus
9 Third ventricle
10 Hippocampus
11 Pons
12 Substantia nigra

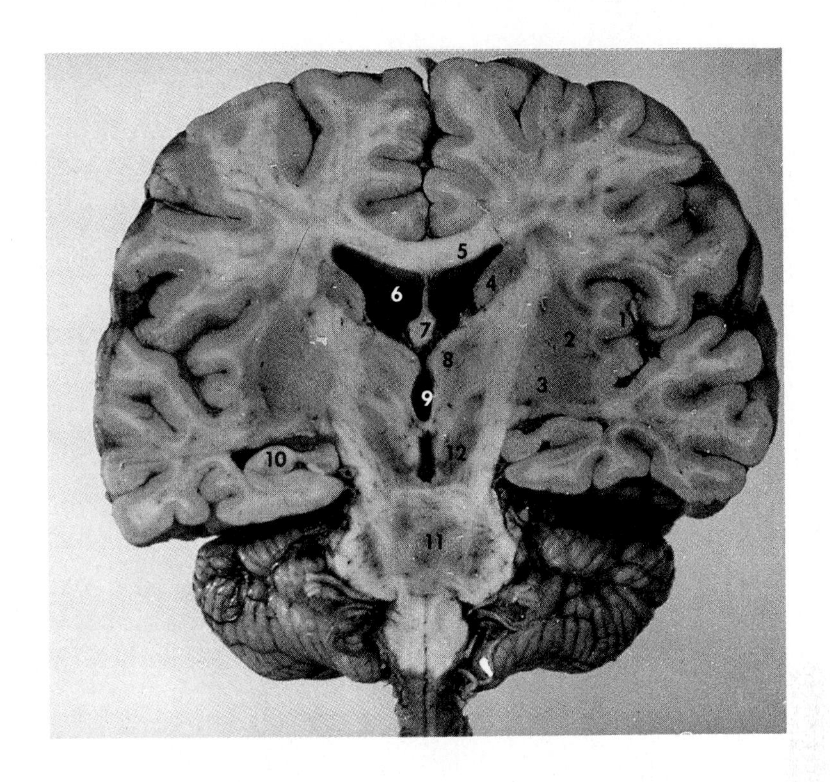

Front of the neck, superficial

1 Anterior belly of digastric
2 Hyoid bone
3 Laryngeal prominence (Adam's apple)
4 Sternohyoid
5 Superior belly of omohyoid
6 Sternocleidomastoid
7 Trapezius
8 Inferior belly of omohyoid
9 Cricoid cartilage
10 Isthmus of thyroid gland
11 Sternothyroid
12 Internal jugular vein
13 Common carotid artery

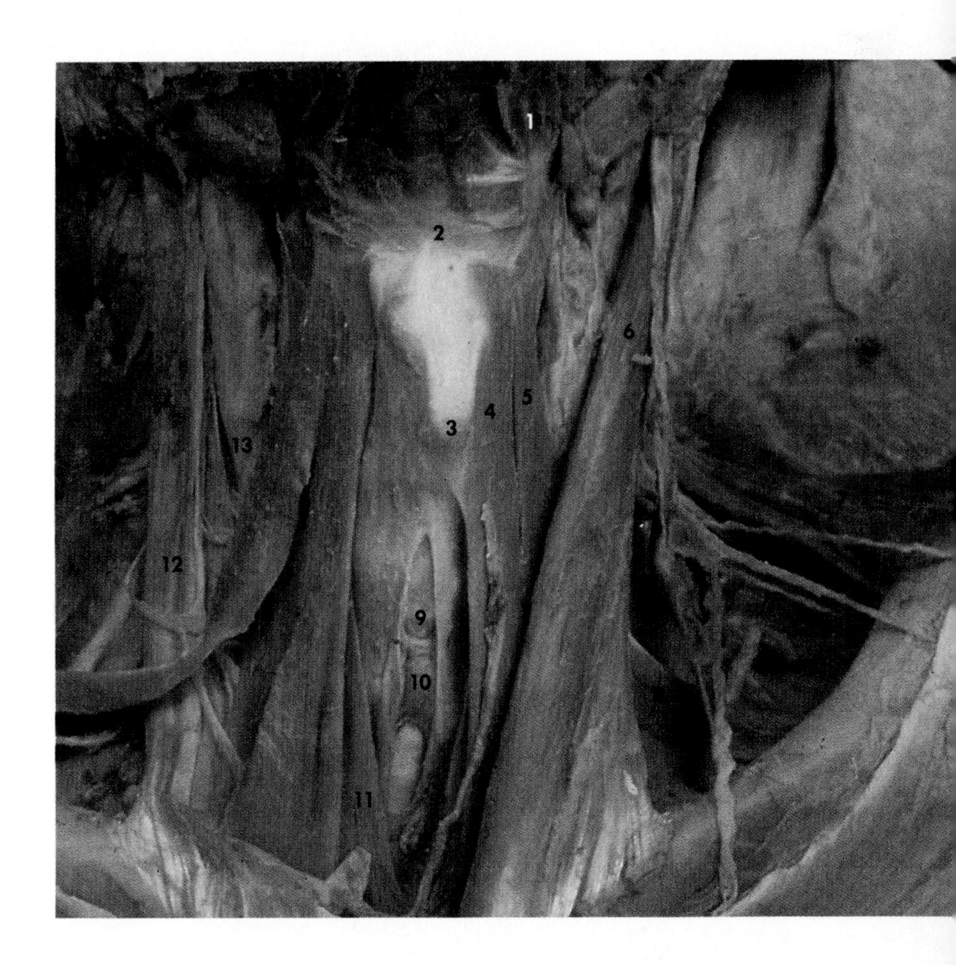

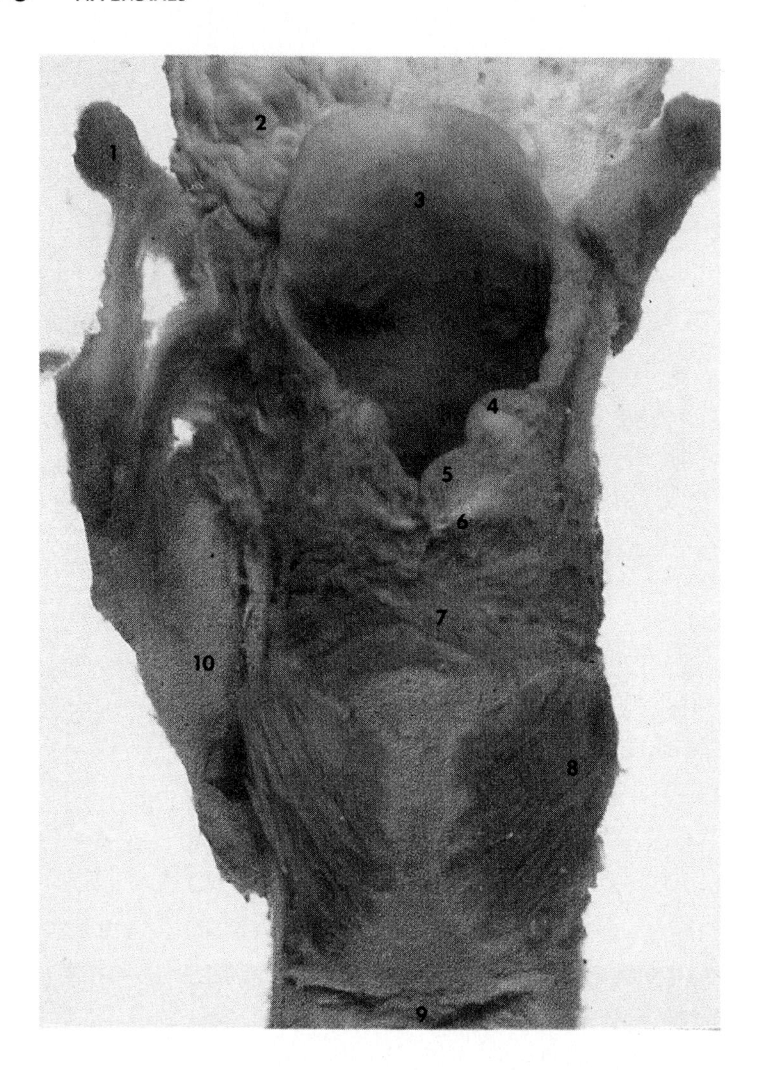

Larynx, posterior view

1 Greater horn of hyoid bone
2 Tongue
3 Epiglottis
4 Cuneiform cartilage
5 Corniculate cartilage
6 Transverse arytenoid muscle
7 Oblique arytenoid muscle
8 Posterior cricoarytenoid muscle
9 Trachea
10 Thyroid cartilage

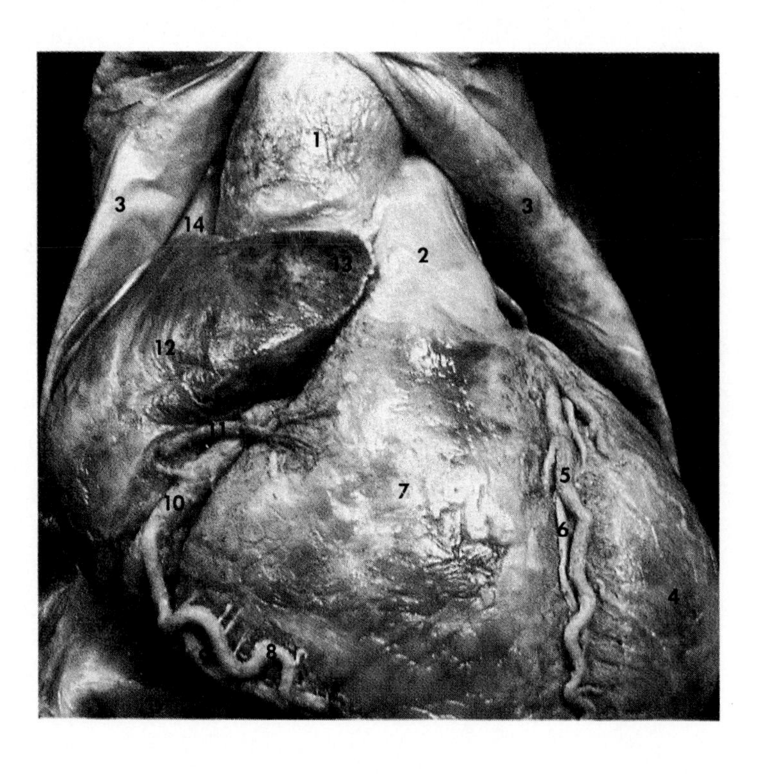

The heart

1 Ascending aorta
2 Pulmonary trunk
3 Pericardium (turned laterally)
4 Left ventricle
5 Anterior interventricular branch of left coronary artery
6 Great cardiac vein
7 Right ventricle
8 Marginal branch of right coronary artery
9 Small cardiac vein
10 Right coronary artery
11 Anterior cardiac vein
12 Right atrium
13 Auricle of right atrium
14 Superior vena cava

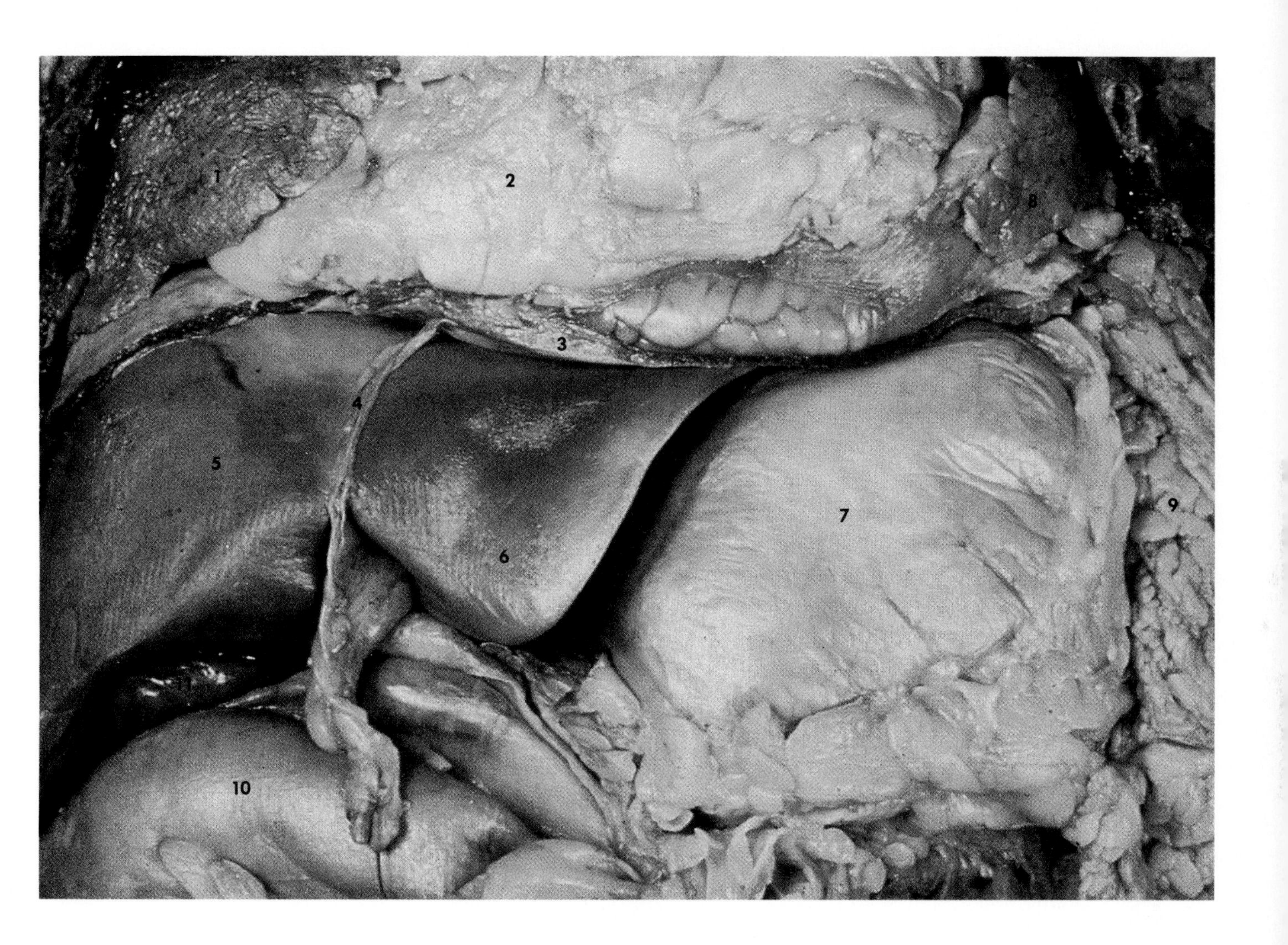

Upper abdomen, anterior view

1 Inferior lobe of right lung
2 Pericardial fat
3 Diaphragm
4 Falciform ligament
5 Right lobe of liver
6 Left lobe of liver
7 Stomach
8 Inferior lobe of left lung
9 Greater omentum
10 Transverse colon
11 Gallbladder

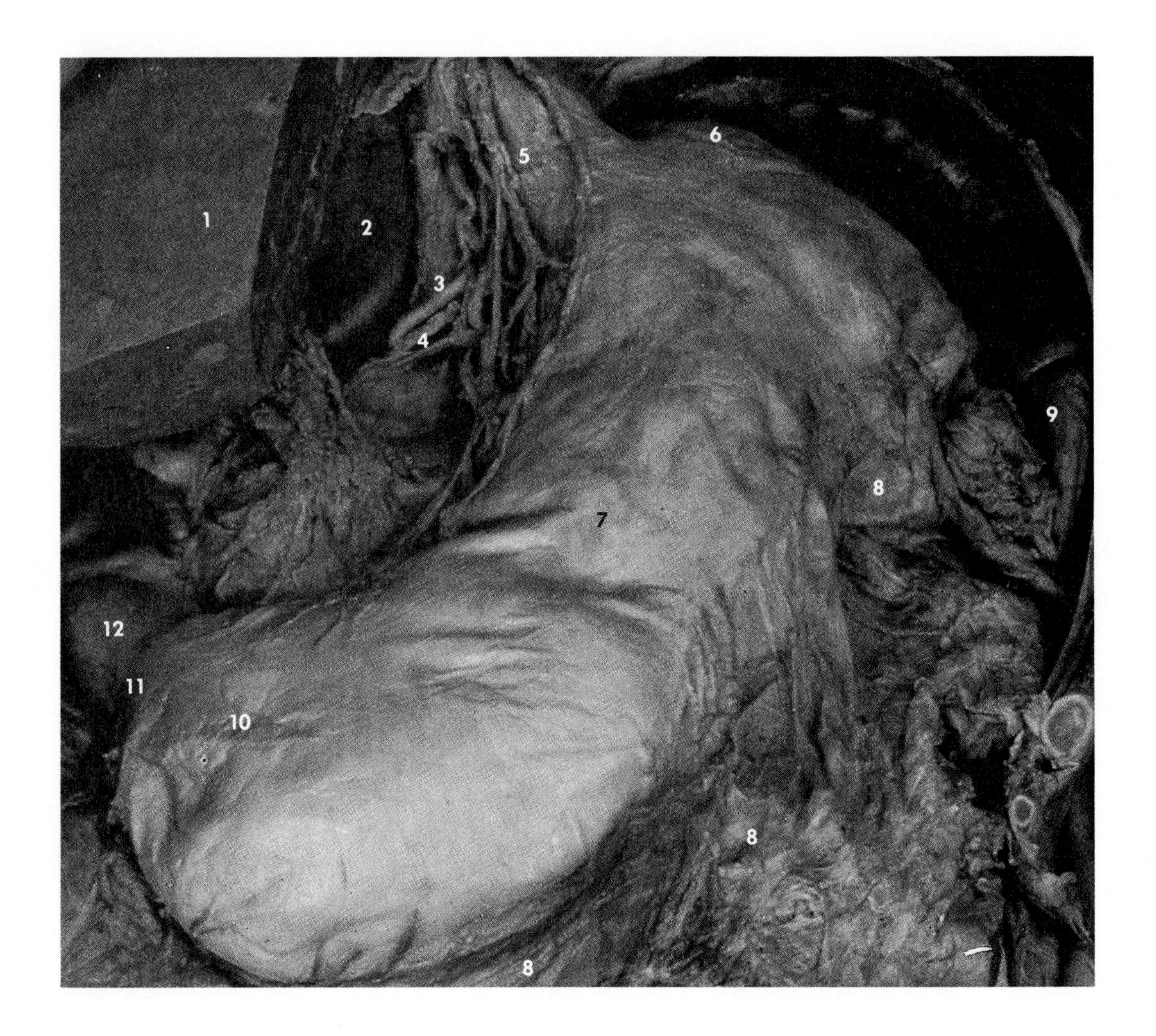

Stomach, anterior view

1 Right lobe of liver (cut)
2 Caudate lobe of liver
3 Left gastric artery
4 Left gastric vein
5 Esophagus
6 Fundus of stomach
7 Body of stomach
8 Greater omentum
9 Spleen
10 Pyloric part of stomach
11 Pylorus
12 Duodenum

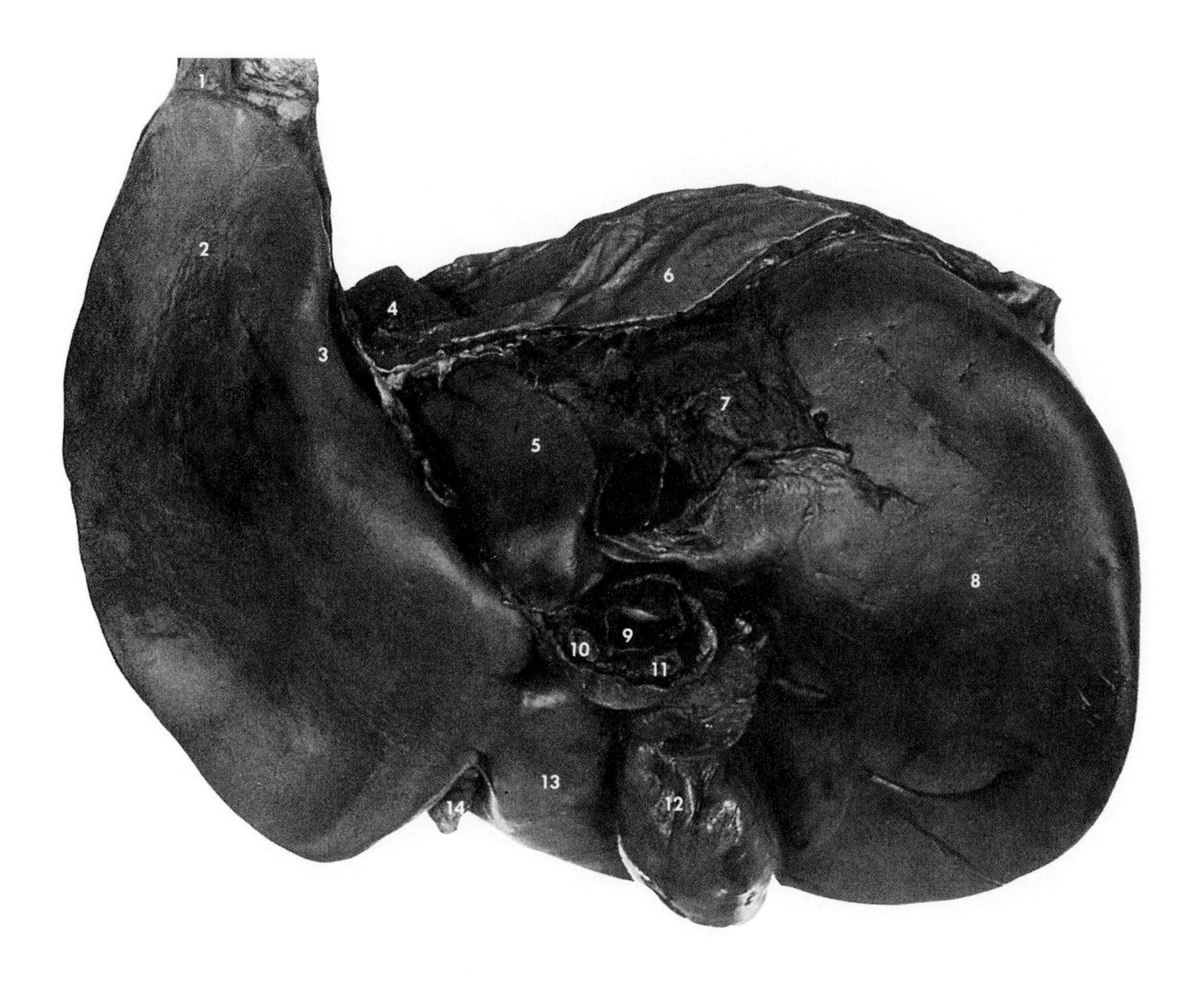

The liver, posterior view

1 Part of coronary ligament
2 Left lobe
3 Esophageal groove
4 Inferior vena cava
5 Caudate lobe
6 Diaphragm on part of bare area
7 Bare area
8 Right lobe
9 Hepatic portal vein
10 Hepatic artery
11 Common hepatic duct
12 Gallbladder
13 Quadrate lobe
14 Ligamentum teres and falciform ligament

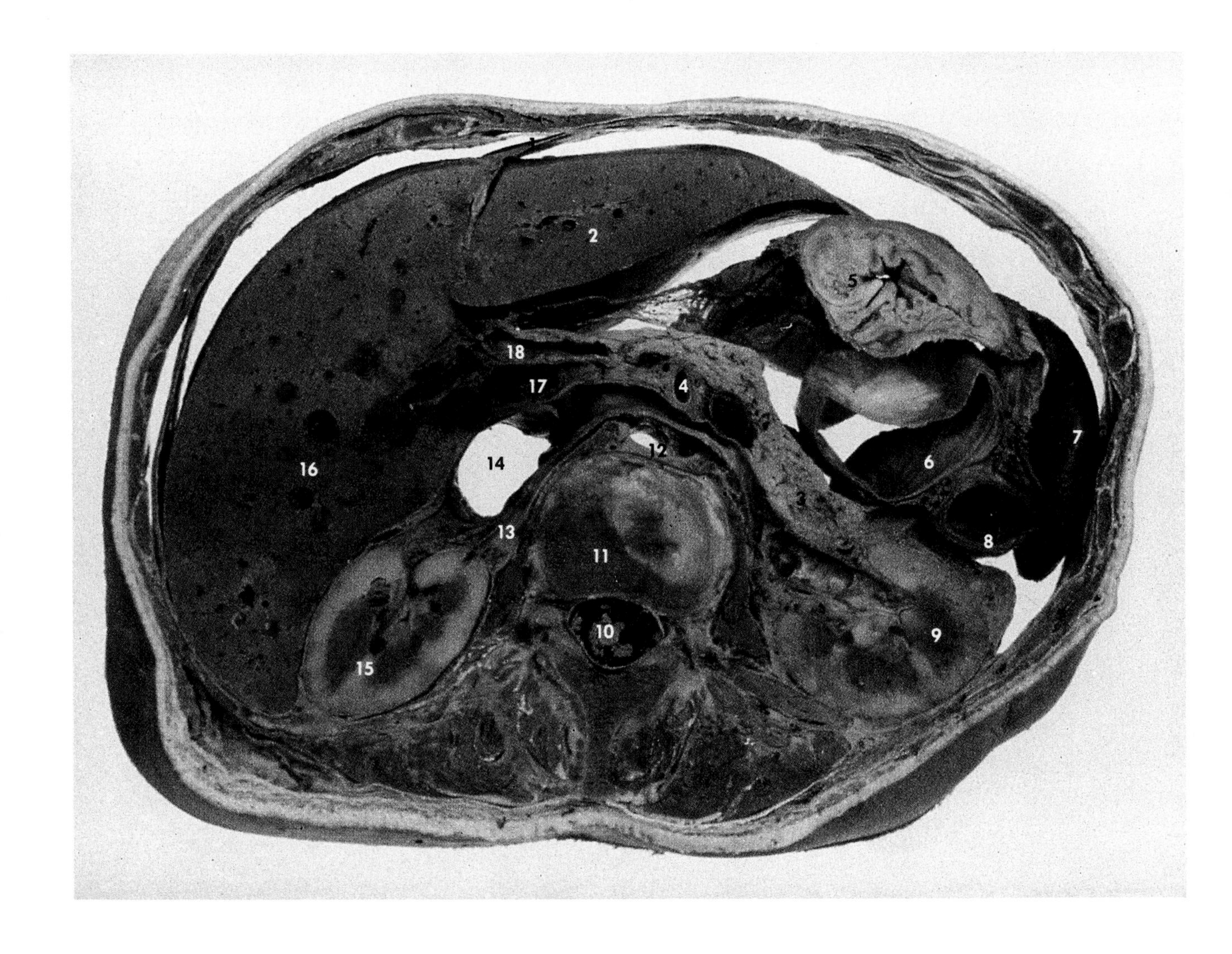

Upper abdomen, transverse section

1 Falciform ligament
2 Left lobe of liver
3 Pancreas
4 Superior mesenteric artery
5 Stomach
6 Transverse colon
7 Spleen
8 Descending colon
9 Left kidney
11 Body of first lumbar vertebra
12 Abdominal aorta
13 Right renal artery
14 Inferior vena cava
15 Right kidney
16 Right lobe of liver
17 Hepatic portal vein
18 Hepatic artery

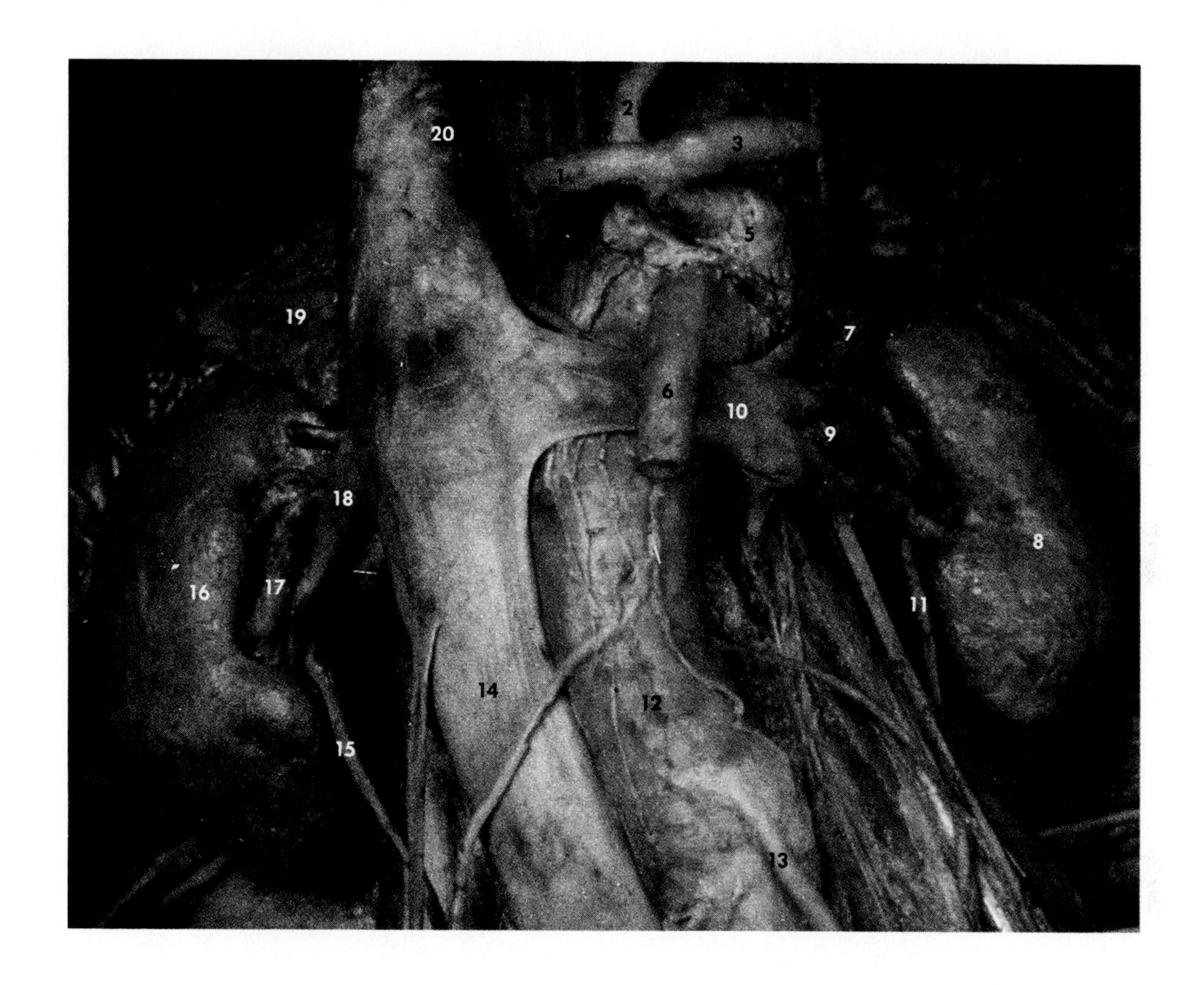

Kidneys and adrenal glands

1 Common hepatic artery
2 Left gastric artery
3 Splenic artery
4 Celiac trunk
5 Celiac ganglion
6 Superior mesenteric artery
7 Left adrenal gland
8 Left kidney
9 Left renal artery
10 Left renal vein
11 Left ureter
12 Abdominal aorta
13 Inferior mesenteric artery
14 Inferior vena cava
15 Right ureter
16 Right kidney
17 Right renal artery
18 Right renal vein
19 Right adrenal gland
20 A hepatic vein

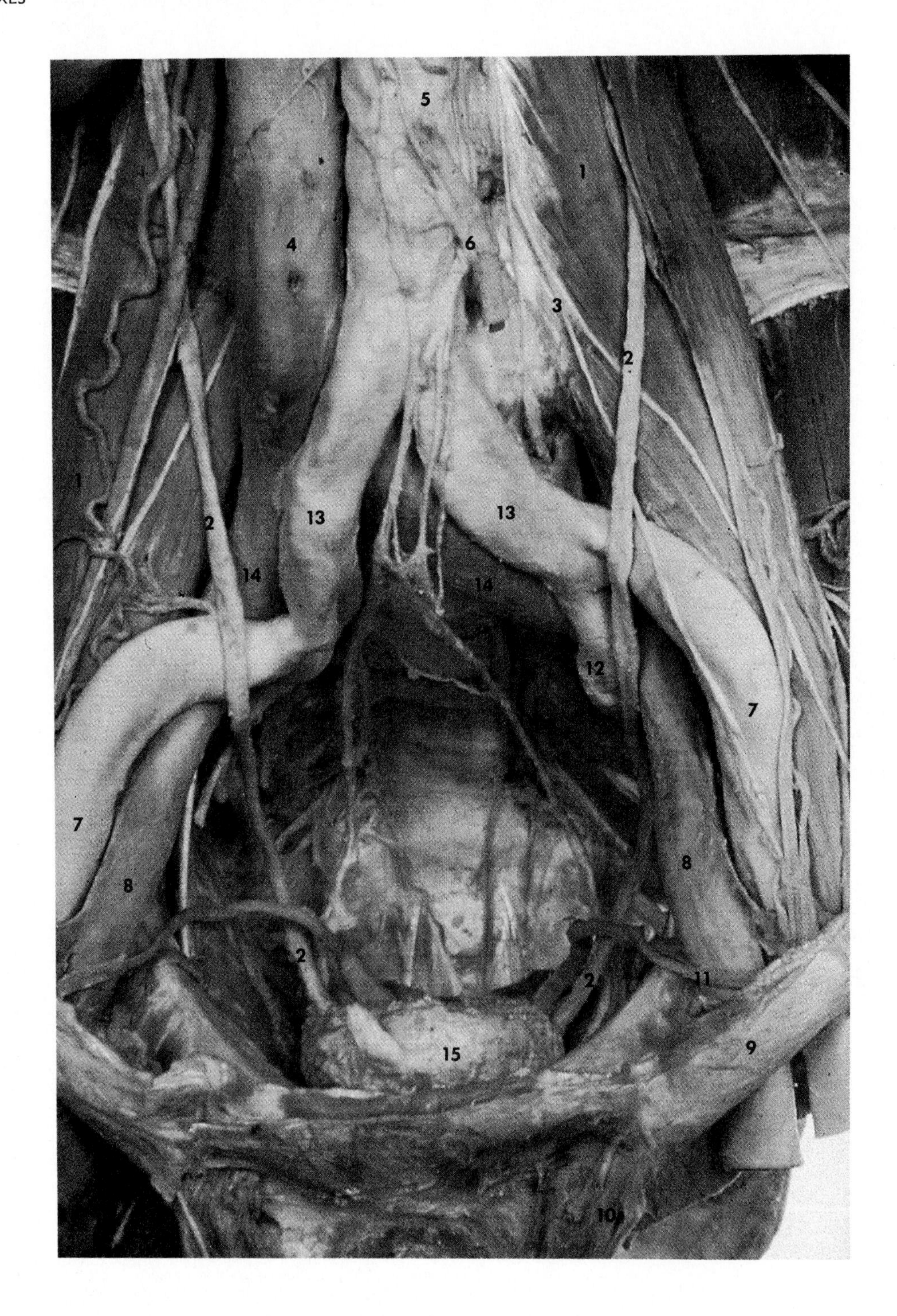

Posterior abdominal wall and pelvis

1 Psoas major
2 Ureter
3 Genitofemoral nerve
4 Inferior vena cava
5 Aorta
6 Inferior mesenteric artery
7 External iliac artery
8 External iliac vein
9 Inguinal ligament
10 Spermatic cord
11 Ductus deferens
12 Internal iliac artery
13 Common iliac artery
14 Common iliac vein
15 Urinary bladder

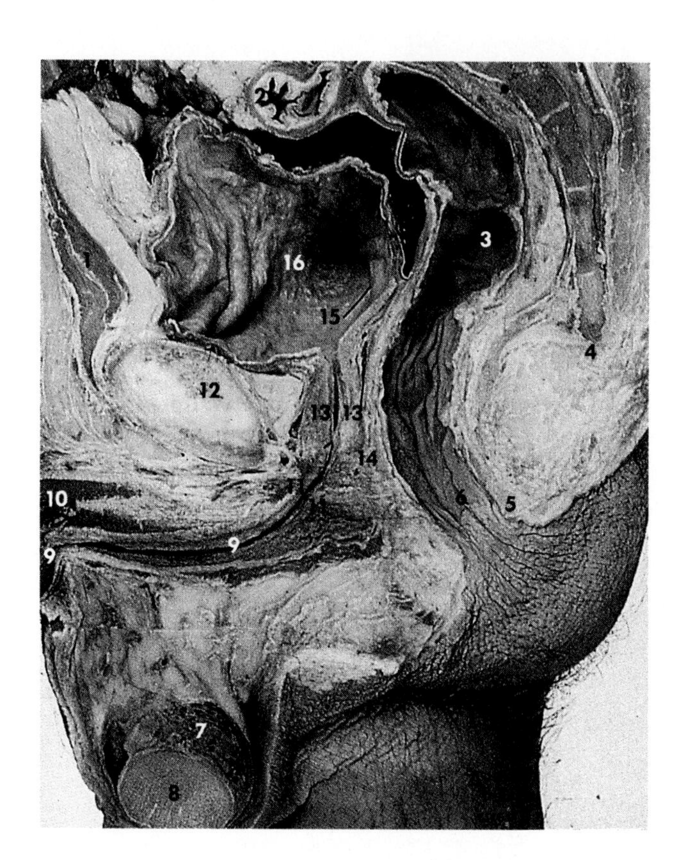

Male pelvis, midline sagittal section

1 Rectus abdominis
2 Sigmoid colon
3 Rectum
4 Coccyx
5 External anal sphincter
6 Anal canal
7 Epididymis
8 Testis
9 Spongy urethra and corpus spongiosum
10 Corpus cavernosum
11 Sphincter urethrae
12 Symphysis pubis
13 Prostate
14 Ejaculatory duct
15 Internal urethral orifice
16 Urinary bladder

Female pelvis, midline sagittal section

1 Ureter
2 Ovary
3 Uterine tube
4 Fundus of uterus
5 Body of uterus
6 Cervix of uterus
7 Urinary bladder
8 Urethra
9 Symphysis pubis
10 Labium minus
11 Vagina
12 Rectum
13 Anal canal
14 External anal sphincter

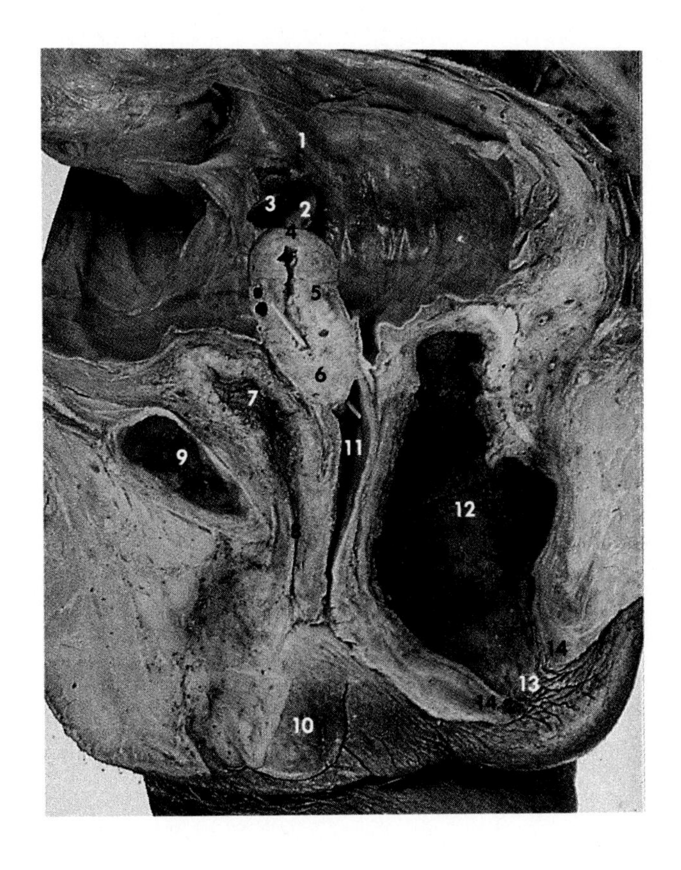

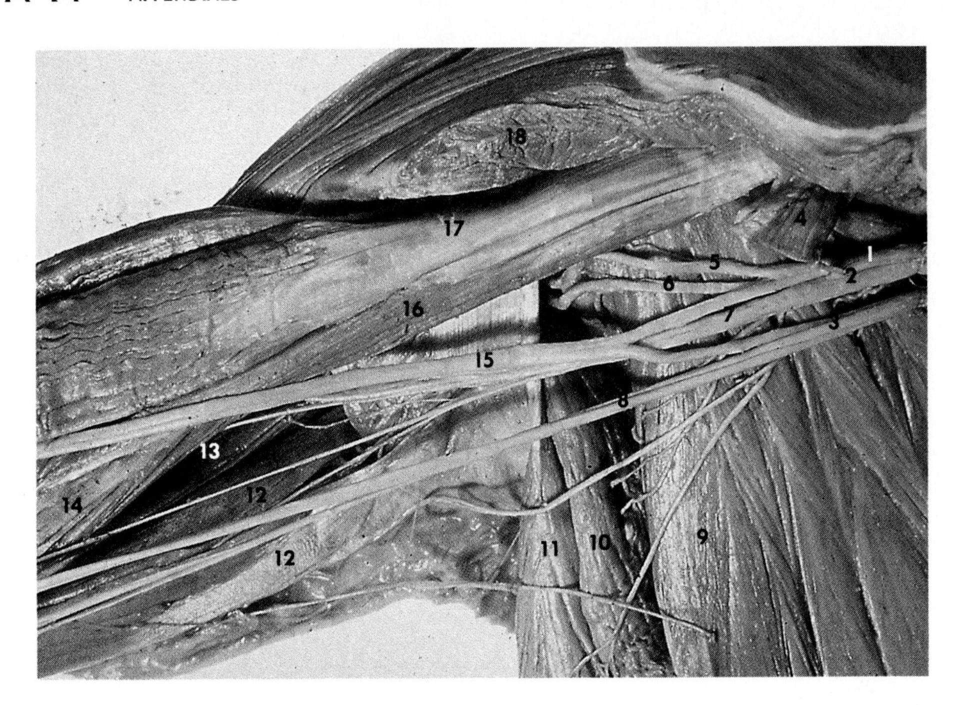

Right brachial plexus

1 Lateral cord
2 Posterior cord
3 Medial cord
4 Pectoralis minor (cut)
5 Musculocutaneous nerve
6 Axillary nerve
7 Radial nerve
8 Ulnar nerve
9 Subscapularis
10 Teres major
11 Latissimus dorsi
12 Long head of triceps
13 Lateral head of triceps
14 Medial head of triceps
15 Median nerve
16 Coracobrachialis
17 Biceps brachii
18 Deltoid (cut)

Forearm and hand, posterior view

1 Brachioradialis
2 Extensor carpi radialis longus
3 Extensor carpi radialis brevis
4 Abductor pollicis longus
5 Extensor pollicis brevis
6 Extensor pollicis longus
7 Extensor digitorum
8 Extensor digiti minimi
9 Extensor indicis
10 Extensor carpi ulnaris
11 Abductor digiti minimi
12 Extensor retinaculum

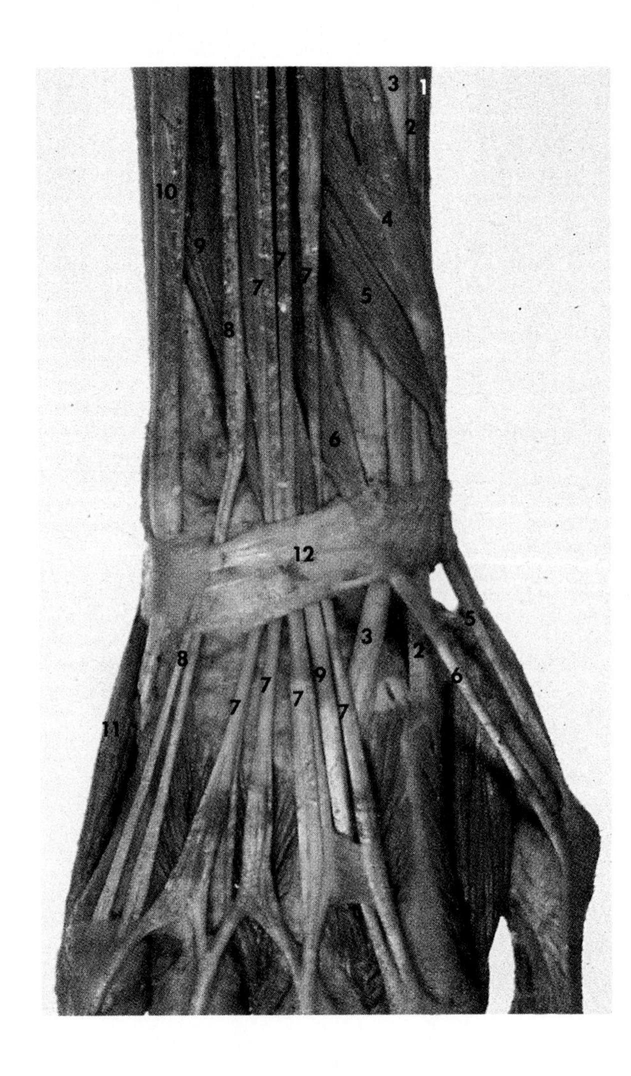

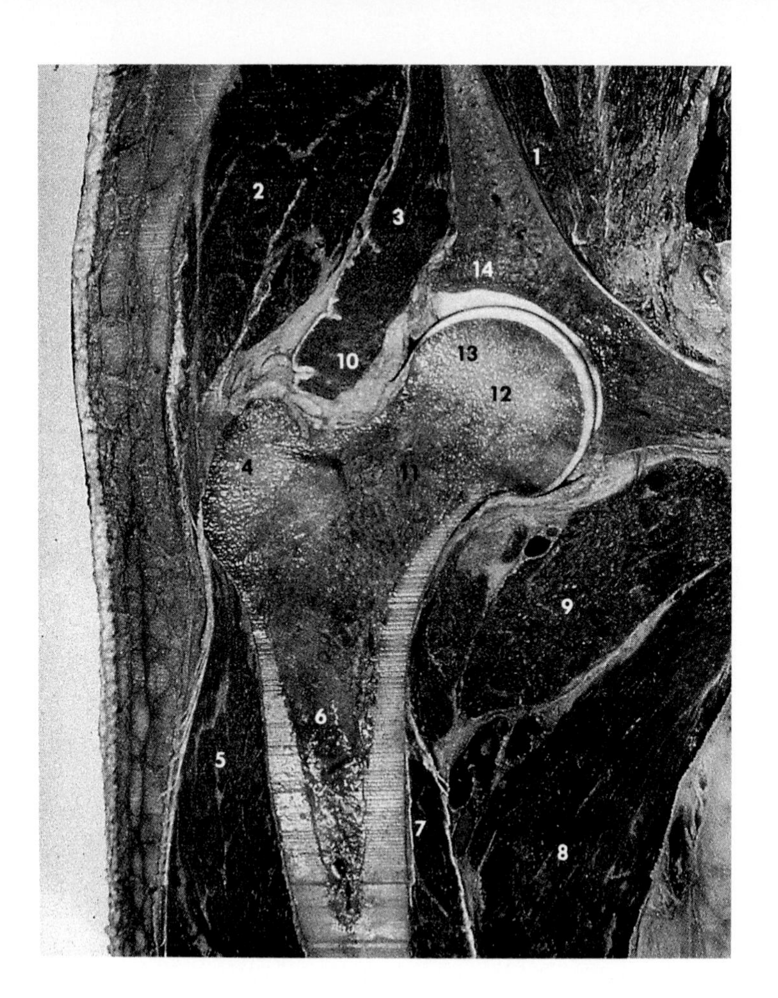

Hip joint, coronal section

1 Iliacus
2 Gluteus medius
3 Gluteus minimus
4 Greater trochanter
5 Vastus lateralis
6 Shaft of femur
7 Vastus medialis
8 Adductor longus
9 Pectineus
10 Capsule of hip joint
11 Neck of femur
12 Head of femur
13 Hyaline cartilage of head
14 Hyaline cartilage of acetabulum

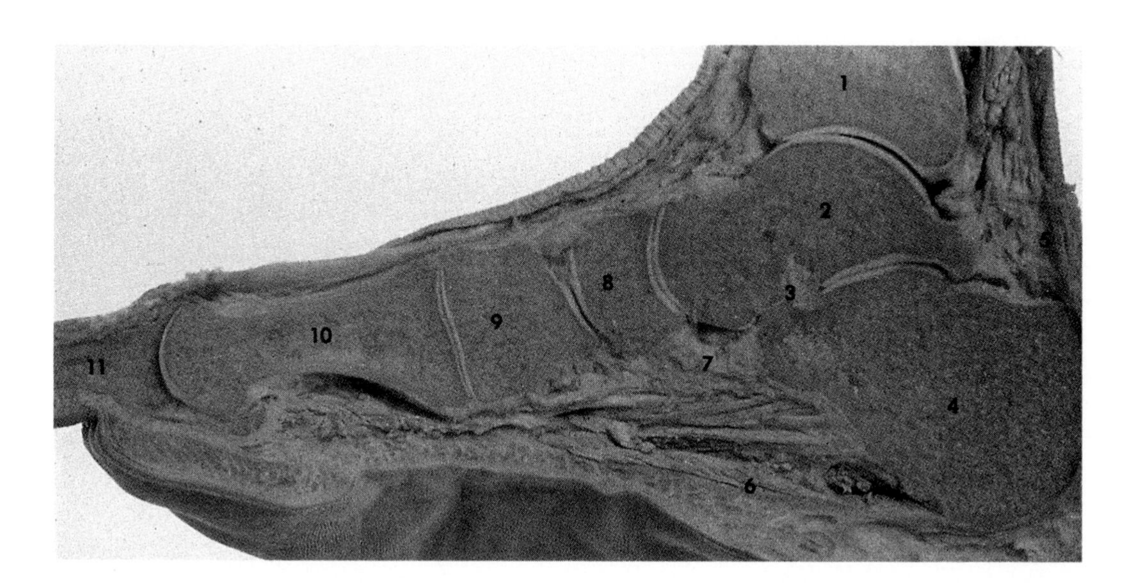

Foot, sagittal section

1 Tibia
2 Talus
3 Ligament
4 Calcaneus
5 Calcaneal tendon (Achilles' tendon)
6 Flexor digitorum brevis
7 Plantar calcaneonavicular ligament
8 Navicular
9 Medial cuneiform
10 First metatarsal
11 Proximal phalanx of great toe

TABLE OF MEASUREMENTS

UNIT	METRIC EQUIVALENT	SYMBOL	U.S. EQUIVALENT
Measures of Length			
1 kilometer	= 1000 meters	km	0.62137 mile
1 meter	= 10 decimeters or 100 centimeters	m	39.37 inches
1 decimeter	= 10 centimeters	dm	3.937 inches
1 centimeter	= 10 millimeters	cm	0.3937 inch
1 millimeter	= 1000 micrometers	mm	
1 micrometer	= 1/1000 millimeter or 1000 nanometers	µm	
1 nanometer	= 10 angstroms or 1000 picometers	nm	No U.S. equivalent
1 angstrom	= 1/10,000,000 millimeter	Å	
1 picometer	= 1/1,000,000,000 millimeter	pm	
Measures of Volume			
1 cubic meter	= 1000 cubic decimeters	m^3	1.308 cubic yards
1 cubic decimeter	= 1000 cubic centimeters	dm^3	0.03531 cubic foot
1 cubic centimeter	= 1000 cubic millimeters or 1 milliliter	cm^3 (cc)	0.06102 cubic inch
Measures of capacity			
1 kiloliter	= 1000 liters	kl	264.18 gallons
1 liter	= 10 deciliters	l	1.0567 quarts
1 deciliter	= 100 milliliters	dl	0.4227 cup
1 milliliter	= volume of 1 gram of water at standard temperature and pressure	ml	0.3381 ounces
Measures of mass			
1 kilogram	= 1000 grams	kg	2.2046 pounds
1 gram	= 100 centigrams or 1000 milligrams	g	0.0353 ounces
1 centigram	= 10 milligrams	cg	0.1543 grain
1 milligram	= 1/1000 gram	mg	

Note that a micrometer was formerly called a micron (µ), and a nanometer was formerly called a millimicron (mµ).

SCIENTIFIC NOTATION

Very large numbers with many zeros such as 1,000,000,000,000,000 or very small numbers such as 0.0000000000000001 are very cumbersome to work with. Consequently, the numbers are expressed in a kind of mathematical shorthand known as scientific notation. Scientific notation has the following form:

$$M \times 10^{n}$$

where *n* specifies how many times the number *M* is raised to the power of 10. The exponent n has two meanings, depending on its sign. If n is positive, M is multiplied by 10 n times. For example, if n = 2 and M = 1.2, then

$$1.2 \times 10^{2} = 1.2 \times 10 \times 10 = 120$$

In other words, if n is positive, the decimal point of M is moved to the right n times. In this case the decimal point of 1.2 is moved two places to the right.

1.20.

If n is negative, M is divided by 10 n times.

$$1.2 \times 10^{-2} = \frac{1.2}{(10 \times 10)} = \frac{1.2}{100} = 0.012$$

In other words, if n is negative, the decimal point of M is moved to the left n times. In this case the decimal point of 1.2 is moved two places to the left.

0.01.2

If M is the number 1.0, it often is not expressed in scientific notation. For example, 1.0×10^{2} is the same thing as 10^{2}, and 1.0×10^{-2} is the same thing as 10^{-2}.

Two common examples of the use of scientific notation in chemistry are Avogadro's number and pH. Avogadro's number, 6.023×10^{23}, is the number of atoms in 1 g molecular weight of an element. Thus

$$6.023 \times 10^{23} = 602{,}300{,}000{,}000{,}000{,}000{,}000{,}000$$

which is a very large number of atoms.

The pH scale is a measure of the concentration of hydrogen ions in a solution. A neutral solution has 10^{-7} moles of hydrogen ions per liter. In other words,

$$10^{-7} = 0.0000001$$

which is a very small amount (1 ten-millionth of a gram) of hydrogen ions.

SOLUTION CONCENTRATIONS

Physiologists often express solution concentration in terms of percent, molarity, molality, and equivalents.

Percent

The weight-volume method of expressing percent concentrations states the weight of a solute in a given volume of solvent. For example, to prepare a 10% solution of sodium chloride, 10 g of sodium chloride is dissolved in a small amount of water (solvent) to form a salt solution. Then additional water is added to the salt solution to form 100 ml of salt solution. Note that the sodium chloride was dissolved in water and then diluted to the required volume. The sodium chloride was not dissolved directly in 100 ml of water.

Molarity

Molarity determines the number of moles of solute dissolved in a given volume of solvent. A 1 molar (1 M) solution is made by dissolving 1 mole of a substance in enough water to make 1 L of solution. For example, 1 mole of sodium chloride solution is made by dissolving 58.5 g of sodium chloride in enough water to make 1 L of solution. One mole of glucose solution is made by dissolving 180 g of glucose in enough water to make 1 L of solution. Both solutions have the same number (Avogadro's number) of particles in solution.

Molality

Although 1-molar solutions have the same number of solute molecules, they do not have the same number of solvent (water) molecules. Because 58.5 g of sodium chloride occupies less volume than 180 g of glucose, the sodium chloride solution has more water molecules. **Molality** is a method of calculating concentrations that takes into account the number of solute and solvent molecules. A 1-molal solution (1 m) is 1 mole of a substance dissolved in 1 kg of water. Thus a 1-molal solution of sodium chloride and a 1-molal solution of glucose contain the same number of solute molecules dissolved in the same amount of water.

When sodium chloride is dissolved in a solvent, it dissociates, or separates, to form two ions, a sodium cation (Na^+) and a chloride anion (Cl^-). However, glucose does not dissociate when dissolved in a solvent. Although 1-molal solutions of sodium chloride and of glucose have the same number of molecules, because of dissociation the sodium chloride solution contains twice as many particles as the glucose solution (one Na^+ ion and one Cl^- ion for each glucose molecule). To report the concentration of these substances in a way that reflects the number of particles in a given volume of solution, the concept of **osmolality** is used. A 1 osmolal (Osm) solution is 1 mole of a solute times the number of particles into which the solute dissociates in 1 kg of solution. Thus 1 mole of sodium chloride in 1 kg of water is a 2-osmolal solution because sodium chloride dissociates to form two ions.

Osmolality of a solution is a reflection of the number, not the type, of particles in a solution. Thus a 1 osmolal solution contains 1 osmole of particles per kilogram of solution, but the particles may be all one type or a complex mixture of different types.

The concentration of particles in body fluids is so low that the measurement milliosmole (mOsm), 1/1000 of an osmole, is used. Most body fluids have an osmotic concentration of approximately 300 mOsm and consist of many different ions and molecules. The osmotic concentration of body fluids is important because it influences the movement of water into or out of cells (see Chapter 3).

Equivalents

Equivalents are a measure of the concentrations of ionized substances. One equivalent (Eq) is 1 mole of an ionized substance multiplied by the absolute value of its charge. For example, one mole of NaCl dissociates into 1 mole of Na^+ and 1 mole of Cl^-. Thus there is 1 Eq of Na^+ (1 mole $\times$ 1) and 1 Eq of Cl^- (1 mole $\times$ 1). One mole of $CaCl_2$ dissociates into 1 mole of Ca^{2+} and two moles of Cl^-. Thus there are 2 Eq of Ca^{2+} (1 mole $\times$ 2) and 2 Eq of Cl^- (2 moles $\times$ 1). In an electrically neutral solution the equivalent concentration of positively charged ions is equal to the equivalent concentration of the negatively charged ions. One milliequivalent (mEq) is 1/1000 of an equivalent.

pH

Pure water weakly dissociates to form small numbers of hydrogen and hydroxide ions:

$$H_2O \leftrightarrow H^+ + OH^-$$

At 25° C the concentration of both hydrogen ions and hydroxide ions is 10^{-7} moles/L. Any solution that has equal concentrations of hydrogen and hydroxide ions is considered **neutral.** A solution is an **acid** if it has a higher concentration of hydrogen ions than hydroxide ions, and a solution is a **base** if it has a lower concentration of hydrogen ions than hydroxide ions. In any aqueous solution the hydrogen ion concentration $[H^+]$ times the hydroxide ion concentration $[OH^-]$ is a constant that is equal to 10^{-14}.

$$[H^+] \times [OH^-] = 10^{-14}$$

Consequently, as the hydrogen ion concentration decreases, the hydroxide ion concentration increases, and vice versa. For example:

	$[H^+]$	$[OH^-]$
Acidic solution	10^{-3}	10^{-11}
Neutral solution	10^{-7}	10^{-7}
Basic solution	10^{-12}	10^{-2}

Although the acidity or basicity of a solution could be expressed in terms of either hydrogen or hydroxide ion concentration, it is customary to use hydrogen ion concentration. The pH of a solution is defined as

$$pH = -log_{10}(H^+)$$

Thus a neutral solution with 10^{-7} moles of hydrogen ions per liter has a pH of 7.

$$\begin{aligned} pH &= -log_{10}[H^+] \\ &= -log_{10}[10^{-7}] \\ &= -(-7) \\ &= 7 \end{aligned}$$

In simple terms, to convert the hydrogen ion concentration to the pH scale, the exponent of the concentration (e.g., -7) is used, and it is changed from a negative to a positive number. Thus an acidic solution with 10^{-3} moles of hydrogen ions/L has a pH of 3, whereas a basic solution with 10^{-12} hydrogen ions/L has a pH of 12.

SOME REFERENCE LABORATORY VALUES

Table F-1

Blood, plasma, or serum values

TEST	NORMAL VALUES	CLINICAL SIGNIFICANCE
Acetoacetate plus acetone	0.32-2 mg/100 ml	Values increase in diabetic acidosis, fasting, high fat diet, and toxemia of pregnancy
Ammonia	80-110 μg/100 ml	Values decrease with proteinuria and as a result of severe burns and increase in multiple myeloma
Amylase	4-25 U/ml*	Values increase in acute pancreatitis, intestinal obstruction, and mumps; values decrease in cirrhosis of the liver, toxemia of pregnancy, and chronic pancreatitis
Barbiturate	0	Coma level: phenobarbital, approximately 10 mg/100 ml; most other drugs, 1-3 mg/100 ml
Bilirubin	0.4 mg/100 ml	Values increase in conditions causing red blood cell destruction or biliary obstruction or liver inflammation
Blood volume	8.5%-9% of body weight in kilograms	
Calcium	8.5-10.5 mg/ml	Values increase in hyperparathyroidism, vitamin D hypervitaminosis; values decrease in hypoparathyroidism, malnutrition, and severe diarrhea
Carbon dioxide content	24-30 mEq/L 20-26 mEq/L in infants (as HCO_3^-)	Values increase in respiratory diseases, vomiting, and intestinal obstruction; they decrease in acidosis, nephritis, and diarrhea
Carbon monoxide	0	Symptoms with over 20% saturation
Chloride	100-106 mEq/L	Values increase in Cushing's syndrome, nephritis, and hyperventilation; they decrease in diabetic acidosis, Addison's disease, and diarrhea and after severe burns
Creatine phosphokinase (CPK)	Female 5-35 mU/ml Male 5-55 mU/ml	Values increase in myocardial infarction and skeletal muscle diseases such as muscular dystrophy
Creatinine	0.6-1.5 mg/100 ml	Values increase in certain kidney diseases
Ethanol	0	0.3%-0.4%, marked intoxication 0.4%-0.5%, alcoholic stupor 0.5% or over, alcoholic coma

*A unit *(U)* is the quantity of a substance that has a physiological effect.

Continued.

Table F-1

Blood, plasma, or serum values—cont'd

TEST	NORMAL VALUES	CLINICAL SIGNIFICANCE
Glucose	Fasting 70-110 mg/100 ml	Values increase in diabetes mellitus, liver diseases, nephritis, hyperthyroidism, and pregnancy; they decrease in hyperinsulinism, hypothyroidism, and Addison's disease
Iron	50-150 μg/100 ml	Values increase in various anemias and liver disease; they decrease in iron deficiency anemia
Lactic acid	0.6-1.8 mEq/L	Values increase with muscular activity and in congestive heart failure, severe hemorrhage, shock, and anaerobic exercise
Lactic dehydrogenase	60-120 U/ml	Values increase in pernicious anemia, myocardial infarction, liver diseases, acute leukemia, and widespread carcinoma
Lipids	Cholesterol 120-220 mg/100 ml Cholesterol esters 60%-75% of cholesterol Phospholipids 9-16 mg/100 ml as lipid phosphorus Total fatty acids 190-420 mg/100 ml Total lipids 450-1000 mg/100 ml Triglycerides 40-150 mg/100 ml	Increased values for cholesterol and triglycerides are connected with increased risk of cardiovascular disease, such as heart attack and stroke
Lithium	Toxic levels 2 mEq/L	
Osmolality	285-295 mOsm/kg water	
Oxygen saturation (arterial) see Po_2 values	96%-100%	
Pco_2	35-43 mm Hg	Values decrease in acidosis, nephritis, and diarrhea; they increase in respiratory diseases, intestinal obstruction, and vomiting
pH	7.35-7.45	Values decrease as a result of hypoventilation, severe diarrhea, Addison's disease, and diabetic acidosis; values increase due to hyperventilation, Cushing's syndrome, and vomiting
Po_2	75-100 mm Hg (breathing room air)	Values increase in polycythemia and decrease in anemia and obstructive pulmonary diseases
Phosphatase (acid)	Male: total 0.13-0.63 U/ml Female: total 0.01-0.56 U/ml	Values increase in cancer of the prostate gland, hyperparathyroidism, some liver diseases, myocardial infarction, and pulmonary embolism
Phosphatase (alkaline)	13-39 IU/L* (infants and adolescents up to 104 IU/L)	Values increase in hyperparathyroidism, some liver diseases, and pregnancy
Phosphorus (inorganic)	3-4.5 mg/100 ml (infants in first year up to 6 mg/100 ml)	Values increase in hypoparathyroidism, acromegaly, vitamin D hypervitaminosis, and kidney diseases; they decrease in hyperparathyroidism
Potassium	3.5-5 mEq/100 ml	
Protein	Total 6-8.4 g/100 ml Albumin 3.5-5 g/100 ml Globulin 2.3-3.5 g/100 ml	Total protein values increase in severe dehydration and shock; they decrease in severe malnutrition and hemorrhage

*A international unit *(IU)* is the quantity of a substance that produces a particular biological effect agreed upon internationally.

Table F-1

Blood, plasma, or serum values—cont'd

TEST	NORMAL VALUES	CLINICAL SIGNIFICANCE
Salicylate		
Therapeutic	0	20-25 mg/100 ml
Toxic		Over 30 mg/100 ml Over 20 mg/100 ml after age 60
Sodium	135-145 mEq/L	Values increase in nephritis and severe dehydration; they decrease in Addison's disease, myxedema, kidney disease, and diarrhea
Sulfonamide		
Therapeutic	0	5-15 mg/100 ml
Urea nitrogen	8-25 mg/100 ml	Values increase in response to increased dietary protein intake; values decrease in impaired renal function
Uric acid	3-7 mg/100 ml	Values increase in gout and toxemia of pregnancy and as a result of tissue damage

Table F-2

Blood count values

TEST	NORMAL VALUES	CLINICAL SIGNIFICANCE
Clotting (coagulation) time	5-10 minutes	Values increase in afibrinogenemia and hyperheparinemia, severe liver damage
Fetal hemoglobin	Newborns: 60%-90% Before age 2: 0%-4% Adults: 0%-2%	Values increase in thalassemia, sickle-celled anemia, and leakage of fetal blood into maternal bloodstream during pregnancy
Hemoglobin	Male: 14-16.5 g/100 ml Female: 12-15 g/100 ml Newborn: 14-20 g/100 ml	Values decrease in anemia, hyperthyroidism, cirrhosis of the liver, and severe hemorrhage; values increase in polycythemia, congestive heart failure, obstructive pulmonary disease, high altitudes
Hematocrit	Male: 40%-54% Female: 38%-47%	Values increase in polycythemia, severe dehydration, and shock; values decrease in anemia, leukemia, cirrhosis, and hyperthyroidism
Ketone bodies	0.3-2 mg/100 ml Toxic level: 20 mg/100 ml	Values increase in ketoacidosis, fever, anorexia, fasting, starvation, high fat diet
Platelet count	250,000-400,000/mm^3	Values decrease in anemias and allergic conditions and during cancer chemotherapy; values increase in cancer, trauma, heart disease, and cirrhosis
Prothrombin time	11-15 seconds	Values increase in prothrombin and vitamin deficiency, liver disease, and hypervitaminosis A
Red blood cell count	Males: 5.4 million/mm^3 Females: 4.8 million/mm^3	Values decrease in systemic lupus erythematosus, anemias, and Addison's disease; values increase in polycythemia and dehydration and following hemorrhage
Reticulocyte count	0.5%-1.5%	Values decrease in iron-deficiency and pernicious anemia and radiation therapy; values increase in hemolytic anemia, leukemia, and metastatic carcinoma
White blood cell count, differential	Neutrophils 60%-70% Eosinophils 2%-4% Basophils 0.5%-1% Lymphocytes 20%-25% Monocytes 3%-8%	Neutrophils increase in acute infections; eosinophils and basophils increase in allergic reactions; monocytes increase in chronic infections; lymphocytes increase during antigen-antibody reactions
White blood cell count, total	5000-9000/mm^3	Values decrease in diabetes mellitus, anemias, and following cancer chemotherapy; values increase in acute infections, trauma, some malignant diseases, and some cardiovascular diseases

Table F-3

Urine values

TEST	NORMAL VALUES	CLINICAL SIGNIFICANCE
Acetone and acetoacetate	0	Values increase in diabetic acidosis and during fasting
Albumin	0 to trace	Values increase in glomerular nephritis and hypertension
Ammonia	20-70 mEq/L	Values increase in diabetes mellitus and liver disease
Bacterial count	Under 10,000/ml	Values increase in urinary tract infection
Bile and bilirubin	0	Values increase in biliary tract obstruction
Calcium	Under 250 mg/24 h	Values increase in hyperparathyroidism and decrease in hypoparathyroidism
Chloride	110-254 mEq/24 h	Values decrease in pyloric obstruction, diarrhea; values increase in Addison's disease and dehydration
Potassium	25-100 mEq/L	Values decrease in diarrhea, malabsorption syndrome, and adrenal cortical insufficiency; values increase in chronic renal failure, dehydration, and Cushing's syndrome
Sodium	75-200 mg/24 h	Values decrease in diarrhea, acute renal failure, and Cushing's syndrome; values increase in dehydration, starvation, and diabetic acidosis
Creatinine clearance	100-140 ml/minute	Values increase in renal diseases
Creatinine	1-2 g/24 h	Values increase in infections and decrease in muscular atrophy, anemia, and certain kidney diseases
Glucose	0	Values increase in diabetes mellitus and certain pituitary gland disorders
Urea clearance	Over 40 ml of blood cleared of urea per minute	Values increase in certain kidney diseases
Urea	25-35 g/24 h	Values decrease in complete biliary obstruction and severe diarrhea; values increase in liver diseases and hemolytic anemia
Uric acid	0.6-1 g/24 h	Values increase in gout and decrease in certain kidney diseases
Casts		
Epithelial	Occasional	Increase in nephrosis and heavy metal poisoning
Granular	Occasional	Increase in nephritis and pyelonephritis
Hyaline	Occasional	Increase in glomerular membrane damage and fever
Red blood cell	Occasional	Values increase in pyelonephritis; blood cells appear in urine in response to kidney stones and cystitis
White blood cell	Occasional	Values increase in kidney infections
Color	Amber, straw, transparent yellow	Varies with hydration, diet, and disease states
Odor	Aromatic	Becomes acetone-like in diabetic ketosis
Osmolality	500-800 mOsm/kg water	Values decrease in aldosteronism and diabetes insipidus; values increase in high protein diets, heart failure, and dehydration
pH	4.6-8	Values decrease in acidosis, emphysema, starvation, and dehydration; values increase in urinary tract infections and severe alkalosis

Table F-4

Hormone levels

TEST	NORMAL VALUES
Steroid hormones	
Aldosterone	Excretion: 5-19 μg/24 h[1]
Fasting at rest, 210 mEq sodium diet	Supine: 48 ± 29 pg/ml[2] Upright: 65 ± 23 pg/ml
Fasting at rest, 10 mEq sodium diet	Supine: 175 ± 75 pg/ml Upright: 532 ± 228 pg/ml
Cortisol	
Fasting	8 a.m.: 5-25 μg/100 ml
At rest	8 p.m.: Below 10 μg/100 ml
Testosterone	Adult male: 300-1100 ng/100 ml[3] Adolescent male: over 100 ng/100 ml Female: 25-90 ng/100 ml
Peptide hormones	
Adrenocorticotropin (ACTH)	15-170 pg/ml
Calcitonin	Undetectable in normals
Growth hormone (GH)	
Fasting, at rest	Below 5 ng/ml
After exercise	Children: over 10 ng/ml Male: below 5 ng/ml Female: up to 30 ng/ml
Insulin	
Fasting	6-26 μU/ml
During hypoglycemia	Below 20 μU/ml
After glucose	Up to 150 μU/ml
Luteinizing hormone (LH)	Male: 6-18 mU/ml Preovulatory or postovulatory female: 5-22 mU/ml Midcycle peak 30-250 mU/ml
Parathyroid hormone	Less than 10 microl equiv/L
Prolactin	2-15 ng/ml
Renin activity	
Normal diet	
Supine	1.1 ± 0.8 ng/ml/h
Upright	1.9 ± 1.7 ng/ml/h
Low-sodium diet	
Supine	2.7 ± 1.8 ng/ml/h
Upright	6.6 ± 2.5 ng/ml/h
Thyroid-stimulating hormone (TSH)	0.5-3.5 μU/ml
Thyroxine-binding globulin	15.25 μg T_4/100 ml
Total thyroxine	4-12 μg/100 ml

[1]1 microgram (1 μg) is equal to 10^{-6} g.
[2]1 picogram (1 pg) is equal to 10^{-12} g.
[3]1 nanogram (1 ng) is equal to 10^{-9} g.

GLOSSARY

Many of the words in this glossary and throughout the text are followed by a simplified phonetic spelling showing pronunciation. The pronunciation key reflects standard clinical usage as presented in *Stedman's Medical Dictionary*, which has long been a leading reference volume in the health sciences.

As a general rule, vowels will be unmarked. If an unmarked vowel is at the end of a syllable, the sound is long (like the "a" in mate). If an unmarked vowel is followed by a consonant, the sound is short (like the "a" in mat). Accent marks follow stressed syllables.

Page numbers indicate where entries may be found in the text.

a at the end of a syllable, long, as in day (da), before a consonant (except r), short, as in hat (hat).
ă as in hat (flă-jel′ ah).
ā as in mate (māt).
ah as in father (fah′ ther).
ar as in far (far).
ăr as in fair (făr).
aw as in fall (fawl).
e at the end of a syllable, long, as in bee (be); before a consonant (except r), short, as in met.
ĕ as in met; (ap′ ĕ-tīt).
ē as in meet (mēt).
er as in term.
ĕr as in merry (mĕr′ e).
i at the end of a syllable, long, as in pie (pi); before a consonant (except r), short, as in pit.
ĭ as in pit; (kar′ tĭ-lij).
ī as in pine (pīn).
ir as in firm.
ĭr as in mirror (mĭr′ or).
īr as in fire (fīr).
o at the end of a syllable, long, as in no; before a consonant, short, as in not.
ŏ as in occult (ŏ-kult).
ō as in note (nōt).
ŏŏ as in food.
ŏr as in for.
ōr as in fore, four (fōr).
ow as in cow.
oy as in boy; as the oi in mastoid (mas′ toyd).
u at the end of a syllable, as in u′ nit; before a consonant, as in bud.
ŭ as in bud.
ū as in tune (tūn).
ur as in fur.
ūr as in pūre.

A band Length of the myosin myofilament in a sarcomere. (p. 276)

abdomen Belly, between the thorax and the pelvis. (p. 19)

abduction [L. *abductio,* take away] Movement away from the midline. (p. 240)

absolute refractory period Portion of the action potential during which the membrane is insensitive to all stimuli, regardless of their strength. (p. 263)

absorptive cell Cell on the surface of villi of the small intestines and the luminal surface of the large intestine that is characterized by having microvilli; secretes digestive enzymes and absorbs digested materials on its free surface. (p. 783)

absorptive state Immediately after a meal when nutrients are being absorbed from the intestine into the circulatory system. (p. 839)

acceptor molecule Receptor molecule to which a hormone binds and that is located in the nucleus and is associated with DNA (e.g., an estrogen molecule binds with an acceptor molecule in the nucleus of a cell and causes expression of specific genes in the chromosomes). (p. 538)

accommodation [L. *ac* + *commodo,* to adapt] Ability of electrically excitable tissues such as nerve or muscle cells to adjust to a constant stimulus so that the magnitude of the local potential decreases through time. (p. 475)

accommodation Increase in thickness and convexity of the lens to focus an external object on the retina as it moves closer to the eye. (p. 266)

acetabulum (a'sŭ-tab'u-lum) [L., shallow vinegar vessel or cup] Cup-shaped depression on the external surface of the coxa. (p. 215)

acetylcholine (as'ĕ-til-ko'lēn) Neurotransmitter substance released from motor neurons, all preganglionic neurons of the parasympathetic and sympathetic divisions, all postganglionic neurons of the parasympathetic division, some postganglionic neurons of the sympathetic division, and some central nervous system neurons. (p. 279)

acetylcholinesterase (as'ē-til-ko-lin-es'ter-ās) Enzyme found in the synaptic cleft that causes the breakdown of acetylcholine to acetic acid and choline, thus limiting the stimulatory effect of acetylcholine. (p. 280)

Achilles tendon See calcaneal tendon. (p. 342)

acid Molecule that is a proton donor; any substance that releases hydrogen ions (H^+). (p. 42)

acidic Solution containing more than 10^{-7} moles of hydrogen ions per liter; has a pH less than 7.

acidosis Condition characterized by blood pH of 7.35 or below. (p. 898)

acinus, pl. acini (as'ĭ-nus, as'ĭ-ne) [L., berry, grape] Grape-shaped secretory portion of a gland. The terms acinus and alveolus are sometimes used interchangeably. Some authorities differentiate the terms: acini have a constricted opening into the excretory duct, whereas alveoli have an enlarged opening. (p. 112)

acromegaly (ak'ro-meg'al-e) [Gr. *akron,* tip or extremity + *megas,* large] Disorder marked by progressive enlargement of the head and face, hands and feet, and thorax caused by excessive secretion of growth hormone by the adenohypophysis. (p. 176)

acromion (ak-ro'me-en) [Gr. *akron,* tip + *omos,* shoulder] Bone comprising the tip of the shoulder. (p. 211)

acrosome (ak'ro-sōm) [Gr. *akron,* extremity + *soma,* body] Cap on the head of the spermatozoon, with hydrolytic enzymes that help the spermatozoon to penetrate the ovum. (p. 914)

actin myofilament Thin myofilament within the sarcomere; composed of two F-actin molecules, tropomyosin, and troponin molecules. (p. 274)

action potential [L. *potentia,* power, potency] Change in membrane potential in an excitable tissue that acts as an electrical signal and is propagated in an all-or-none fashion. (p. 253)

activation energy Energy that must be added to molecules to initiate a reaction. (p. 51)

active site Portion of an enzyme in which reactants are brought into close proximity and that plays a role in reducing activation energy of the reaction. (p. 54)

active tension Tension produced by the contraction of a muscle. (p. 289)

active transport Carrier-mediated process that requires ATP and can move substances against a concentration gradient. (p. 70)

acute contagious conjunctivitis Acute inflammation of the conjunctiva. (p. 468)

acute rejection Type of immunologically mediated injury to a transplant manifested by delayed hypersensitivity reaction and cell lysis. (p. 728)

adduction [L. *adductus,* to bring toward] Movement toward the midline. (p. 240)

adenohypophysis (ad'ĕ-no-hi-pof'ĭ-sis) Portion of the hypophysis derived from the oral ectoderm; commonly called the anterior pituitary. (p. 546)

adenoid [Gr. *aden,* gland + *eidos,* appearance] Hypertrophy of the pharyngeal tonsil resulting from chronic inflammation. (p. 701)

adenosine diphosphate (ADP) Adenosine, an organic base, with two phosphate groups attached to it. Adenosine diphosphate combines with a phosphate group to form adenosine triphosphate. (p. 57)

adenosine triphosphate (ATP) Adenosine, an organic base, with three phosphate groups attached to it. Energy stored in ATP is used in nearly all of the endergonic reactions in cells. (p. 57)

adipose (ad'ĭ-pōs) [L. *adeps,* fat] Fat. (p. 120)

ADP See adenosine diphosphate.

adrenal gland [L. *ad,* to + *ren,* kidney] Also called the suprarenal gland. Located near the superior pole of each kidney, it is composed of a cortex and a medulla. The adrenal medulla is a highly modified sympathetic ganglion that secretes the hormones epinephrine and norepinephrine; the cortex secretes aldosterone and cortisol as its major secretory products. (p. 562)

adrenal medullary mechanism Increased release of epinephrine and norepinephrine from the adrenal medulla as a result of the same stimuli that increase sympathetic stimulation of the heart and blood vessels. (p. 690)

adrenaline Synonym for epinephrine. (p. 563)

adrenergic (ă-drĕ-ner'jik) Referring to nerve fibers of the autonomic nervous system that secrete norepinephrine or to drugs that mimic the actions of the sympathetic nervous system. (p. 511)

adrenergic receptor Receptor molecule that binds to adrenergic agents such as epinephrine and norepinephrine. (p. 512)

adrenocorticotropic hormone (ACTH) (ă-dre'no-kor'tĭ-ko-tró-pik) Hormone of the adenohypophysis that governs the nutrition and growth of the adrenal cortex, stimulates it to functional activity, and causes it to secrete cortisol. (p. 553)

adventitia (ad'ven-tish'yah) [L. *adventicius,* coming from abroad, foreign] Outermost covering of any organ or structure that is properly derived from without and does not form an integral part of the organ. (p. 646)

aerobic respiration (ăr-o'bik) Breakdown of glucose in the presence of oxygen to produce carbon dioxide, water, and approximately 38 ATPs; includes glycolysis, the citric acid cycle, and the electron transport chain. (p. 86)

afferent arteriole Branch of an interlobular artery of the kidney that conveys blood to the glomerulus. (p. 853)

afferent division Nerve fibers that send impulses from the periphery to the central nervous system. (p. 354)

after discharge Prolongation of response of neural elements after cessation of stimulation, commonly occurring in oscillating circuits. (p. 376)

after potential Small deviation in the membrane potential that is negative relative to the resting membrane potential; immediately follows an action potential. (p. 262)

agglutination (ă-glu′tĭ-na shun) [L. *ad*, to + *gluten*, glue] Process by which blood cells, bacteria, or other particles are caused to adhere to each other and form clumps. (p. 600)

agglutinin (ă-glu′tĭ-nin) Antibody that binds to an antigen and causes agglutination. (p. 600)

agglutinogen (ă-glu-tin′o-gen) Antigen on surface of erythrocytes that can stimulate the production of antibodies (agglutinins) that combine with the antigen and cause agglutination. (p. 600)

agranulocyte Nongranular leukocyte (monocyte or lymphocyte). (p. 594)

ala, pl. alae (ă′la, a′le) [L., a wing] Wing-shaped structure. (p. 208)

albinism (al′bĭ-nizm) [L. *albus*, white] Congenital inability to produce melanin, resulting in unpigmented skin and hair. (p. 141)

aldosterone (al-dos′ter-ōn) Steroid hormone produced by the zona glomerulosa of the adrenal cortex that facilitates potassium exchange for sodium in the distal renal tubule, causing sodium reabsorption and potassium and hydrogen secretion. (p. 564)

alkaline Solution containing less than 10^{-7} moles of hydrogen ions per liter; has a pH greater than 7.0. (p. 42)

alkalosis Condition characterized by blood pH of 7.45 or above. (p. 898)

allantois Tube extending from the hindgut into the umbilical cord; forms the urinary bladder. (p. 962)

allele [Gr. *allelon*, reciprocally] Any one of a series of two or more different genes that may occupy the same position or locus on a specific chromosome. (p. 976)

allergen Antigen that causes an allergic response. An allergic reaction, specifically one with strong familial tendencies, is caused by allergens, such as pollen, food, dander, and insect venoms and is associated with IgE antibodies. (p. 727)

allergy of infection Unavoidable inflammation produced in chronic infections by the cell-mediated immune system. (p. 727)

all-or-none When a stimulus is applied to a cell, an action potential is either produced or not. In muscle cells the cell either contracts to the maximum extent possible (for a given condition) or does not contract. (p. 286)

alternate pathway Part of the nonspecific immune system for activation of complement. (p. 706)

alveolar duct Part of the respiratory passages beyond a respiratory bronchiole; from it arise alveolar sacs and alveoli. (p. 741)

alveolar gland One in which the secretory unit has a saclike form and an obvious lumen. (p. 780)

alveolar sac Two or more alveoli that share a common opening. (p. 741)

alveolar ventilation rate Volume of air available for gas exchange per minute; equal to the respiratory rate times the tidal volume minus the dead air space. (p. 749)

alveolus, pl. alveoli (al-ve′o-lus, al-ve′o-li) Cavity. Examples include the sockets into which teeth fit, the endings of the respiratory system, and the terminal endings of secretory glands. (p. 112)

amino acid Class of organic acids that comprise the building blocks for proteins. (p. 49)

amniotic activity Fluid-filled cavity surrounding the developing embryo. (p. 952)

amplitude-modulated signal Signal that varies in magnitude or intensity such as with large vs. small concentrations of hormones. (p. 527)

ampulla (am-pul′lah) [L., two-handled bottle] Saclike dilatation of a semicircular canal; contains the crista ampullaris. Wide portion of the uterine tube between the infundibulum and the isthmus. (p. 495)

amylase One of a group of starch-splitting enzymes that cleave starch, glycogen, and related polysaccharides. (p. 804)

anabolism (ah-nab′o-lizm) [Gr. *anabole*, a raising up] All of the synthesis reactions that occur within the body; requires energy. (p. 37)

anaerobic respiration (an′ăr-o′bik) Breakdown of glucose in the absence of oxygen to produce lactic acid and two ATPs; consists of glycolysis and the reduction of pyruvic acid to lactic acid. (p. 86)

anal canal Terminal portion of the digestive tract. (p. 772)

anal triangle Posterior portion of the perineal region through which the anal canal opens. (p. 911)

anaphase Time during cell division when chromatids divide (or in the case of first meiosis, when the chromosome pairs divide). (p. 94)

anaphylaxis (an-ă-fi-lak′sis) [Gr. *ana*, away from, back from + *phylaxis*, protection] IgE-mediated allergic reaction in which released chemicals cause systemic vasodilation, a drop in blood pressure, and cardiac failure. (p. 727)

anatomical dead air space Volume of the conducting airways from the external environment down to the terminal bronchioles. (p. 748)

anatomical position Position in which a person is standing erect with the feet facing forward, arms hanging to the sides, and the palms of the hands facing forward with the thumbs to the outside. (p. 16)

anatomy [Gr. *ana*, up + *tome*, a cutting] Scientific discipline that investigates the structure of the body. (p. 4)

androstenedione (an-dro-stēn′di-ōn) Androgenic steroid of weaker potency than testosterone; secreted by the testis, ovary, and adrenal cortex. (p. 566)

anemia (ă-ne′me-ah) [Gr. *an* + *haima*, blood] Any condition in which the number of red blood cells per cubic millimeter, the amount of hemoglobin in 100 ml of blood, or the volume of packed red blood cells per 100 ml of blood is less than normal. (p. 604)

anencephaly [Gr. *an* + *enkephalos*, brain] Defective development of the brain and absence of the bones of the cranial vault. Only a rudimentary brainstem and some trace of basal ganglia are present. (p. 960)

aneurysm (an′u-riz-um) [Gr. *eurys*, wide] Dilated portion of an artery. (p. 674)

angina pectoris Severe constricting pain in the chest, often radiating to the left shoulder and down the arm; caused by ischemia of the heart muscle, usually because of coronary disease that occludes coronary vessels. (p. 617)

angioplasty (an′jĭ-o-plaz′te) Operation to enlarge a narrowed coronary arterial lumen by peripheral introduction of a balloon-tipped catheter and dilation of the lumen during withdrawal of the inflated catheter tip. (p. 626)

angiotensin I Peptide derived when renin acts on angiotensinogen. (p. 689)

angiotensin II Peptide derived from angiotensin I; stimulates vasoconstriction and aldosterone secretion. (p. 689)

angiotensinogen Precursor molecule to angiotensin I; produced by the liver. (p. 689)

anion (an′i-on) Ion carrying a negative charge. (p. 32)

antagonist Muscle that works in opposition to another muscle. (p. 306)

anterior That which goes first. In humans, toward the belly, or front. (p. 16)

anterior chamber of eye Chamber of the eye between the cornea and the iris. (p. 472)

anterior interventricular sulcus Groove on the anterior surface of the heart, marking the location of the septum between the two ventricles. (p. 614)

anterior pituitary See adenohypophysis. (p. 546)

antibody Protein found in the plasma that is responsible for humoral immunity; binds specifically to antigen. (p. 600)

antibody-mediated immunity Immunity due to B cells and the production of antibodies. (p. 712)

anticoagulant (an′tĭ-ko-ag′u-lant) Agent that prevents coagulation. (p. 599)

anticodon Sequence of three nucleotides in a tRNA molecule that binds to a corresponding codon in an mRNA molecule. (p. 89)

antidiuretic hormone (ADH) Hormone secreted from the neurohypophysis that acts on the kidney to reduce the output of urine; also called vasopressin because it causes vasoconstriction. (p. 550)

antigen (an'tĭ-jen) [anti(body) + Gr. *-gen,* producing] Any substance that induces a state of sensitivity and/or resistance to infection or toxic substances after a latent period; substance that stimulates the specific immune system. (p. 600)

antigen-binding receptor Receptor molecule on lymphocytes that specifically binds antigens. (p. 714)

antigenic determinant The specific part of an antigen that stimulates an immune system response by binding to receptors on the surface of lymphocytes. (p. 714)

antithrombin Any substance that inhibits or prevents the effects of thrombin so that blood does not coagulate. (p. 599)

antrum [Gr. *antron,* a cave] Cavity of an ovarian follicle filled with fluid containing estrogen. (p. 923)

anulus fibrosus (an'u-lus fi-bro'sus) [L., fibrous ring] Fibrous material forming the outer portion of an intervertebral disk. (p. 203)

anus Lower opening of the digestive tract through which fecal matter is extruded. (p. 790)

aorta (a-or'tah) [Gr. *aorte* from *aeiro,* to lift up] Large elastic artery that is the main trunk of the systemic arterial system; carries blood from the left ventricle of the heart and passes through the thorax and abdomen. (p. 614)

aortic arch [L., bow] Curve between the ascending and descending portions of the aorta. (p. 653)

aortic body One of the smallest bilateral structures, similar to the carotid bodies, attached to a small branch of the aorta near its arch; contains chemoreceptors that respond primarily to decreases in blood oxygen; less sensitive to decreases in blood pH or increases in carbon dioxide. (p. 689)

apex [L, summit or tip] Extremity of a conical or pyramidal structure. The apex of the heart is the rounded tip directed anteriorly and slightly inferiorly. (p. 612)

APGAR score Evaluation of a newborn infant's physical status by assigning numerical values to each of five criteria; *a*ppearance (skin color), *p*ulse (heart rate), *g*rimace (response to stimulation), *a*ctivity (muscle tone), and *r*espiratory effort; a score of 10 indicates the best possible condition. (p. 972)

aphasia (ă-fa'ze-ah) [Gr. *a,* unable + *phasis,* speech] Speechlessness; impaired or absent speech. (p. 395)

apical ectodermal ridge Layer of surface ectodermal cells at the lateral margin of the embryonic limb bud; they stimulate growth of the limb. (p. 956)

apical foramen (tooth) [L., aperture] Opening at the apex of the root of a tooth that gives passage to the nerve and blood vessels. (p. 778)

aplastic anemia Anemia characterized by greatly decreased formation of erythrocytes and hemoglobin as a result of hypoplastic or aplastic bone marrow. (p. 605)

apneustic center (ap-nus'tik) Group of neurons in the pons that have a stimulatory effect on the inspiratory center. (p. 760)

apocrine (ăp'o-krin) [Gr. *apo,* away from + *krino,* to separate] Gland whose cells contribute cytoplasm to its secretion (e.g., mammary glands). Sweat glands that produce organic secretions traditionally are called apocrine. These sweat glands, however, are actually merocrine glands; see merocrine and holocrine. (p. 113)

appendicular (ap'pen-dik'u-lar) [L. *appendo,* to hang something on] Relating to an appendix or appendage such as the limbs and their associated girdle. (p. 19)

appendicular skeleton The portion of the skeleton consisting of the upper limbs and the lower limbs and their girdles. (p. 183)

appositional growth [L. *ap* + *pono,* to put or place] To place one layer of bone, cartilage, or other connective tissue against an existing layer. (p. 158)

apraxia (ā-prak'se-ah) Disorder of voluntary movement characterized by partial or complete inability to perform purposeful movements. (p. 394)

aqueous humor Watery, clear solution that fills the anterior and posterior chambers of the eye. (p. 472)

arachnoid (ar-ak'noyd) [Gr. *arachne,* spider, cobweb] Thin, cobweb-appearing meningeal layer surrounding the brain; the middle of the three layers. (p. 419)

arcuate artery (ar'ku-āt) Originates from the interlobar arteries of the kidney and forms an arch between the cortex and medulla of the kidney. (p. 856)

areola [L., area] Circular pigmented area surrounding the nipple; its surface is dotted with little projections caused by the presence of the areolar glands beneath. (p. 929)

areolar (ah-re'o-lar) Relating to tissue with small areas or spaces within it. (p. 114)

areolar gland Gland forming small, rounded projections from the surface of the areola of the mamma. (p. 929)

arm That part of the upper limb between the shoulder and the elbow. (p. 19)

arrectores pilorum, pl. arrector pili (ah-rek'to-rez pi-lor'um, ah-rek'tor pĭ'le) [L., that which raises; hair] Smooth muscle attached to the hair follicle and dermis that raises the hair when it contracts. (p. 145)

arterial capillary Capillary opening from an arteriole or metarteriole. (p. 647)

arteriole Minute artery with all three tunics that transports blood to a capillary. (p. 649)

arteriosclerosis (ar-tēr-ĭ-o-sklĕ-ro'sis) [L. *arterio* + Gr. *sklerosis,* hardness] Hardening of the arteries. (p. 651)

arteriovenous anastomosis (ah-nas'to-mo-sis) Vessel through which blood is shunted from an arteriole to a venule without passing through the capillaries. (p. 650)

artery Blood vessel that carries blood away from the heart. (p. 651)

Arthus reaction Localized immune complex reaction that causes inflammation and tissue destruction. (p. 727)

articular cartilage Hyaline cartilage covering the ends of bones within a synovial joint. (p. 159)

articulation joint Place where two bones come together. (p. 227)

arytenoid cartilages (ăr-ĭ-te'noyd) Small pyramidal laryngeal cartilages that articulate with the cricoid cartilage. (p. 738)

ascending aorta Part of the aorta from which the coronary arteries arise. (p. 653)

ascending colon (kō'lon) Portion of the colon between the small intestine and the right colic flexure. (p. 788)

asthma Condition of the lungs in which there is widespread narrowing of airways caused by contraction of smooth muscle, edema of the mucosa, and mucus in the lumen of the bronchi and bronchioles. (p. 741)

astigmatism [Gr. *a* + *stigma,* a point] Condition of unequal curvatures of one or more of the refractive surfaces of the eye in which rays of light do not focus on a single point on the retina. (p. 483)

astrocyte (as'tro-sīt) [Gr. *astron,* star + *kytos,* a hollow, a cell] Star-shaped neuroglia cell involved with forming the blood-brain barrier. (p. 358)

atherosclerosis (ath'er-o-skle-ro'sis) Arteriosclerosis characterized by irregularly distributed lipid deposits in the intima of large and medium-sized arteries. (p. 651)

atom [Gr. *atomos,* indivisible, uncut] Smallest particles into which an element can be divided using conventional methods; composed of protons, neutrons, and electrons. (p. 30)

atomic number Equal to the number of protons in each type of atom. (p. 31)

atomic weight Weight, in grams, of Avogadro's number of atoms of that element, or the weight, in grams, of 1 mole of an element. (p. 32)

atopy (at'o-pe) [Gr. *atop*, out of place] Localized immediate hypersensitivity mediated by IgE; examples are asthma and hives. (p. 727)

ATP See adenosine triphosphate. (p. 830)

atrial diastole Dilation of the heart's atria. (p. 629)

atrial natriuretic factor Peptide released from the atria when atrial blood pressure is increased; acts to lower blood pressure by increasing the rate of urinary production, thus reducing blood volume. (p. 691)

atrial systole Contraction of the atria. (p. 629)

atrioventricular (AV) node Small node of specialized cardiac muscle fibers that gives rise to the atrioventricular bundle of the conduction system of the heart. (p. 622)

atrioventricular bundle Bundle of modified cardiac muscle fibers that projects from the AV node through the interventricular septum. (p. 616)

atrioventricular valve One of two valves closing the openings between the atria and ventricles. (p. 616)

atrium, pl. atria (a'tre-um) [L., entrance hall] One of two chambers of the heart into which veins carry blood. (p. 491)

atrophy (at'ro-fe) [Gr. *atrophia* + *trophi*, nourishment] Wasting of tissues as from decreased cell size or number. (p. 296)

attenuation reflex Reflex contraction of the tensor tympani and stapedius muscles attached to the ossicles in response to loud sounds; dampens loud sounds and protects inner ear structures. (p. 494)

auditory cortex Portion of the cerebral cortex that is responsible for the conscious sensation of sound; in the dorsal portion of the temporal lobe within the lateral fissure and on the superolateral surface of the temporal lobe. (p. 186)

auditory ossicle Bone of the middle ear: includes the malleus, incus, and stapes. (p. 484)

auricle (aw'rĭ-kl) [L. *auris*, ear] Part of the external ear that protrudes from the side of the head. Small pouch projecting from the superior, anterior portion of each atrium of the heart.

auricular (aw-rik'u-lar) [L. *auricularis*, ear] Relating to the ear. (p. 614)

autoimmune disease Disease resulting from a specific immune system reaction against self-antigens. (p. 712)

autonomic Refers to efferent neurons that innervate cardiac muscle, smooth muscle, and glands; characterized by having two neurons in series between the central nervous system and effector organs. (p. 354)

autonomic ganglia Ganglia containing the nerve cell bodies of the autoimmune division of the nervous system. (p. 507)

autonomic nervous system Composed of nerve fibers that send impulses from the central nervous system to smooth muscle, cardiac muscle, and glands. (p. 354)

autonomic reflex Reflexes in which the efferent neurons belong to the autonomic division of the nervous system. (p. 514)

autophagia (aw'to-fa'je-ah) [Gr. *auto*, self + *phagein*, to eat] Segregation and disposal of organelles within a cell. (p. 80)

autoregulation Maintenance of a relatively constant blood flow through a tissue despite relatively large changes in blood pressure; maintenance of a relatively constant glomerular filtration rate despite relatively large changes in blood pressure. (p. 871)

autorhythmic Spontaneous and periodic; e.g., in smooth muscle it implies spontaneous (without nervous or hormonal stimulation) and periodic contractions. (p. 624)

autosome [Gr. *auto*, self + *soma*, body] Any chromosome other than a sex chromosome; normally occur in pairs in somatic cells and singly in gametes. (p. 94)

axial [L. *axle*, axis] Head, neck, and trunk as distinguished from the extremities. (p. 19)

axial myopia Myopia or nearsightedness due to elongation of the globe of the eye. (p. 482)

axial skeleton Skull, vertebral column, and rib cage. (p. 183)

axilla (ak-sil'ah) Armpit. (p. 656)

axolemma [Gr. *axo* + *lemma*, husk] Cell membrane of the axon. (p. 357)

axon (ak'son) [Gr., axis] Main central process of a neuron that normally conducts action potentials away from the neuron cell body. (p. 123)

axon hillock Area of origin of the axon from the nerve cell body. (p. 357)

axoplasm Neuroplasm or cytoplasm of the axon. (p. 357)

B cell Type of lymphocyte responsible for antibody-mediated immunity. (p. 712)

Bainbridge reflex Increase in heart rate caused by a rise in pressure of the blood in the great veins and atria. (p. 636)

baroreceptor (bar'o-re-sep'tor) (pressoreceptor) Sensory nerve ending in the walls of the atria of the heart, venae cavae, aortic arch, and carotid sinuses; sensitive to stretching of the wall caused by increased blood pressure. (p. 635)

baroreceptor reflex Detects changes in blood pressure and produces changes in heart rate, heart force of contraction, and blood vessel diameter that return blood pressure to homeostatic levels. (p. 635)

basal ganglia Nuclei at the base of the cerebrum involved in controlling motor functions. (p. 397)

basal metabolic rate Metabolic rate of an individual at the lowest level of cell chemistry in the waking state; expressed as heat produced per unit of surface area over a specified time period. (p. 840)

base Molecule that is a proton acceptor; any substance that binds to hydrogen ions. (p. 894)

base [L. and Gr. *basis*] Lower part or bottom of a structure. The base of the heart is the flat portion directed posteriorly and superiorly. Veins and arteries project into and out of the base, respectively. (p. 612)

basement membrane Specialized extracellular material located at the base of epithelial cells and separating them from the underlying connective tissues. (p. 104)

basic See alkaline.

basilar membrane Wall of the membranous labyrinth bordering the scala tympani; supports the organ of Corti. (p. 486)

basophil [Gr. *basis*, base + *phileo*, to love] White blood cell with granules that stain specifically with basic dyes; promotes inflammation. (p. 594)

belly Largest portion of muscle between the origin and insertion. (p. 306)

benign (bĕ-nīn') [L. *benignus*, kind] Indicating the mild or harmless nature of an illness or tumor. (p. 127)

beta-oxidation Metabolism of fatty acids by removing a series of two carbon units to form acetyl CoA. (p. 834)

bicarbonate ion Anion (HCO_3^-) remaining after the dissociation of carbonic acid. (p. 590)

bicuspid (mitral) valve Valve closing the orifice between left atrium and left ventricle of the heart. (p. 618)

bifid (bi'fid) [L. *bifidus*, cleft in two parts] Separated into two parts. (p. 206)

bile Fluid secreted from the liver into the duodenum; consists of bile salts, bile pigments, bicarbonate ions, cholesterol, fats, fat-soluble hormones, and lecithin. (p. 787)

bile salt Organic salt secreted by the liver that functions as emulsifying agent. (p. 808)

biliary canaliculus (kan'ă-lik'u-lus) One of the intercellular channels approximately 1 μm or less in diameter that occurs between liver cells into which bile is secreted; empties into the hepatic ducts. (p. 787)

bilirubin (bil-ĭ-roo'bin) [L. *bili* + *ruber*, red] Bile pigment derived from hemoglobin during destruction of erythrocytes. (p. 594)

biliverdin (bil-ĭ-ver'din) Green bile pigment formed from the oxidation of bilirubin. (p. 594)

binocular vision [L. *bini*, paired + *oculus*, eye] Vision using two eyes at the same time; responsible for depth perception when the visual fields of each eyes overlap. (p. 481)

bipolar neuron One of the three categories of neurons consisting of a neuron with two processes—one dendrite and one axon—arising from opposite poles of the cell body. (p. 123)

blastocele (blas'to-sēl) [Gr. *blastos*, germ + *koilos*, hollow] Cavity in the blastocyst. (p. 948)

blastocyst (blas'to-sist) [Gr. *blastos*, germ + *kystis*, bladder] Stage of mammalian embryos that consists of the inner cell mass and a thin trophoblast layer enclosing the blastocele. (p. 948)

bleaching In response to light, retinal separates from opsin. (p. 478)

blind spot Point in the retina where the optic nerve penetrates the fibrous tunic; contains no rods or cones and therefore does not respond to light. (p. 472)

blocking agents Drugs that bind to but do not activate receptor sites. (p. 513)

blood [A.S. *blod*] Fluid and its suspended formed elements that are circulated through the heart, arteries, capillaries, and veins; means by which oxygen and nutritive materials are transported to the tissues and carbon dioxide and various metabolic products are removed for excretion. (p. 121)

blood-brain barrier Permeability barrier controlling the passage of most large-molecular compounds from the blood to the cerebrospinal fluid and brain tissue; consists of capillary endothelium and may include the astrocytes. (p. 359)

blood clot Coagulated phase of blood. (p. 596)

blood groups Classification of blood based on the type of antigen found on the surface of erythrocytes. (p. 600)

blood island Aggregation of mesodermal cells in the embryonic yolk sac that forms vascular endothelium and primitive blood cells. (p. 960)

blood pressure [L. *pressus*, to press] Tension of the blood within the blood vessels; commonly expressed in units of mm Hg. (p. 631)

blood-thymic barrier Layer of reticular cells that separates capillaries from thymic tissue in the cortex of the thymus gland; prevents large molecules from leaving the blood and entering the cortex. (p. 705)

Bohr effect Shift of the oxygen-hemoglobin dissociation curve to the right or left because of changes in blood pH. The definition sometimes is extended to include shifts caused by changes in blood carbon dioxide levels. (p. 755)

bony labyrinth (lab'ĭ-rinth) Part of the inner ear; contains the membranous labyrinth that forms the cochlea, vestibule, and semicircular canals. (p. 486)

Bowman's capsule Expanded beginning of a renal tubule; the visceral layer consists of podocytes that surround a tuft of capillaries, the glomerulus; the parietal layer forms the outside of the capsule. (p. 853)

brachial [L. *brachium*, arm] Relating to the arm. (p. 656)

brainstem Portion of the brain consisting of the midbrain, pons, and medulla oblongata. (p. 384)

branchial arch Typically, six arches in vertebrates; in the lower vertebrates they bear gills, but they appear transiently in the higher vertebrates and give rise to structures in the head and neck. (p. 955)

broad ligament Peritoneal fold passing from the lateral margin of the uterus to the wall of the pelvis on either side. (p. 927)

bronchiole One of the finer subdivisions of the bronchial tubes, less than 1 mm in diameter; has no cartilage in its wall but does have relatively more smooth muscle and elastic fibers. (p. 741)

bronchus, pl. bronchi (brong'kus, brong'ki) [Gr. *bronchos*, windpipe] Any one of the air ducts conducting air from the trachea to the bronchioles. (p. 740)

brush border Epithelial surface consisting of microvilli. (p. 783)

buffer Mixture of an acid and base that reduces any changes in pH that would otherwise occur in a solution when acid or base is added to the solution. (p. 43)

bulb of the penis Expanded posterior part of the corpus spongiosum penis. (p. 915)

bulb of the vestibule Mass of erectile tissue on either side of the vagina. (p. 929)

bulbar conjunctiva Conjunctiva that covers the surface of the eyeball. (p. 468)

bulbourethral gland One of two small compound glands that produce a mucoid secretion; it discharges through a small duct into the spongy urethra. (p. 917)

bulbus cordis [L., plant bulb] End of the embryonic cardiac tube where blood leaves the heart; becomes part of the ventricle. (p. 960)

bursa, pl. bursae (bur'sah, bur'se) [L., purse] Closed sac or pocket containing synovial fluid, usually found in areas where friction occurs. (p. 234)

bursitis (bur-si'tis) [L. *purse* + Gr. *ites*, inflammation] Inflammation of a bursa. (p. 235)

calcaneal tendon (kal-ka'ne-al) Common tendon of the gastrocnemius, soleus, and plantaris muscles that attaches to the calcaneus. (p. 342)

calcitonin (kal'sĭ-to'nin) Hormone released from parafollicular cells that acts on tissues to cause a decrease in blood levels of calcium ions. (p. 171)

callus (kal'us) [L., hard skin] Thickening of the stratum corneum of skin in response to pressure or friction. The hard bonelike substance that develops at the site of a broken bone. (p. 139)

calmodulin (kal-mod'u-lin) [calcium + modulate] Protein receptor for Ca^{2+} ions that plays a role in many Ca^{2+}-regulated processes such as smooth muscle contraction. (p. 294)

calorie [L. *calor*, heat] Unit of heat content or energy. The quantity of energy required to raise the temperature of 1 g of water 1° C. (p. 820)

calpain (kal'pen) Enzyme involved in changing the shape of dendrites; involved with long-term memory. (p. 396)

calyx, pl. calyces (ka'liks, kal'ĭ-sēz) [Gr., cup of a flower] Flower-shaped or funnel-shaped structure; specifically, one of the branches or recesses of a renal pelvis into which the tips of the renal pyramids project. (p. 850)

canal of Schlemm Series of veins at the base of the cornea that drain excess aqueous humor from the eye. (p. 473)

cannaliculus, pl. canaliculi (kan-ă-lik'u-lus, kan-ă-lik'u-li) [L. *canalis*, canal] Little canals, e.g., in bone containing osteocyte cell processes. (p. 162)

cancellous bone (kan'sĕ-lus) [L., grating or lattice] Bone with a latticelike appearance; spongy bone. (p. 121)

canine Referring to the cuspid tooth. (p. 159)

cannula [L. *canna*, reed] Tube; often inserted into an artery or vein. (p. 673)

canthus, pl. canthi (kan'thus, kan'thi) [Gr. *kanthos*, corner of the eye] Angle at the medial and lateral margin of the eye. (p. 468)

capacitation [L. *capax*, capable of] Process whereby spermatozoa acquire the ability to fertilize ova. This process occurs in the female genital tract. (p. 934)

capillary Minute blood vessel consisting of only simple squamous epithelium; major site for the exchange of substances between the blood and tissues. (p. 646)

capitulum (kă-pit'u-lum) [L. *caput*, head] Head-shaped structure. (p. 213)

carbamino compound (kar'bah-me'no) Blood proteins that bind carbon dioxide. (p. 757)

carbaminohemoglobin Carbon dioxide bound to hemoglobin by means of a reactive amino group on the hemoglobin. (p. 592)

carbohydrate Monosaccharide (simple sugar) or the organic molecules composed of monosaccharides bound together by chemical bonds, e.g., glycogen. For each carbon atom in the molecule there are typically one oxygen molecule and two hydrogen molecules. (p. 44)

carbonic acid–bicarbonate buffer system One of the major buffer systems in the body; major components are carbonic acid and bicarbonate ions. (p. 895)

carbonic anhydrase Enzyme that catalyzes the reaction between carbon dioxide and water to form carbonic acid. (p. 590)

carboxypeptidase Pancreatic enzyme that releases amino acids from the carboxyl end of peptide chains. (p. 804)

cardiac [Gr. *kardia*, heart] Related to the heart.

cardiac cycle [Gr. *kyklos*, circle] Complete round of cardiac systole and diastole. (p. 626)

cardiac nerve Nerve that extends from the sympathetic chain ganglia to the heart. (p. 635)

cardiac output (minute volume) Volume of blood pumped by the heart per minute. (p. 630)

cardiac region Region of the stomach near the opening of the esophagus. (p. 781)

cardiac reserve [L. *re* + *servo*, to keep back, reserve] Work that the heart is able to perform beyond that required during ordinary circumstances of daily life. (p. 631)

carotene [L. *carot*, a carrot] Type of yellow or orange plant pigment used as a source of vitamin A. Found in foods such as carrots, sweet potatoes, and egg yolks. (p. 141)

carotid body One of the small organs near the carotid sinuses; contains chemoreceptors that respond primarily to decreases in blood oxygen; less sensitive to decreases in blood pH or increases in carbon dioxide. (p. 689)

carotid sinus Enlargement of the internal carotid artery near the point where the internal carotid artery branches from the common carotid artery; contains baroreceptors. (p. 653)

carotid sinus syndrome Stimulation of a hyperactive carotid sinus, causing a marked fall in blood pressure due to vasodilation and cardiac slowing. (p. 686)

carpal (kar'pul) [Gr. *karpos*, wrist] Bone of the wrist. (p. 214)

carrier Person in apparent health whose chromosomes contain a pathologic mutant gene that may be transmitted to his or her children. (p. 976)

cartilage (kar'tĭ-lij) [L. *cartilage*, gristle] Firm, smooth, resilient, nonvascular connective tissue. (p. 120)

cartilaginous joint Bones connected by cartilage; includes synchondroses and symphyses. (p. 231)

caruncle (kăr'ung-kl) Raised mound of tissue at the medial angle or canthus of the eye. (p. 468)

catabolism (kah-tab'o-lizm) [Gr. *katabole*, a casting down] All of the decomposition reactions that occur in the body; releases energy. (p. 37)

catalyst (ka'tă-list) Substance that increases the rate at which a chemical reaction proceeds without being changed permanently. (p. 39)

cataract Loss of transparency of the lens or capsule of the eye due to a protein buildup. (p. 484)

cations [Gr. *kation*, going down] Ions carrying a positive charge. (p. 32)

cauda equina (kaw'dah e-kwi'nah) [L. *cauda*, tail + *equinus*, horse] Bundle of spinal nerves arising from the caudal end of the spinal cord and extending through the subarachnoid space within the vertebral canal below the first lumbar vertebra. (p. 401)

caveola, pl. caveolae (ka've-o-lah, ka've-o-le) [L., a small pocket] Shallow invagination in the membranes of smooth muscle cells that may perform a function similar to both the T tubules and sarcoplasmic reticulum of skeletal muscle. (p. 294)

cecum (se'kum) [L. *caecus*, blind] Cul-de-sac forming the first part of the large intestine. (p. 772)

cell [L. *cella*, storeroom, chamber] Basic living subunit of all plants and animals. (p. 5)

cell-mediated immunity Immunity due to the actions of T cells and null cells. (p. 712)

celom (se'lom) [Gr. *koilo* + *amma*, a hollow] Principal cavities of the trunk, e.g., the pericardial, pleural, and peritoneal cavities. Separate in the adult, they are continuous in the embryo. (p. 955)

cementum [L. *caementum*, rough quarry stone] Layer of modified bone covering the dentin of the root and neck of a tooth; blends with the fibers of the periodontal membrane. (p. 779)

central nervous system (CNS) Major subdivision of the nervous system consisting of the brain and spinal cord. (p. 354)

central nervous system ischemic response Increase in blood pressure due to vasoconstriction when oxygen levels are too low CO_2 levels are too high, or pH is too low in the medulla oblongata. (p. 689)

central tendon Three-lobed fibrous sheet occupying the center of the diaphragm. (p. 744)

central vein Terminal branches of the hepatic veins that lie centrally in the hepatic lobules and receive blood from the liver sinusoids. (p. 668)

centriole (sen'tre-ōl) [Gr. *kentron*, point or center] Usually paired organelles lying in the centrosome. A small cylindrical organelle composed of nine parallel sets of microtubules forming its wall. Each set is composed of three parallel microtubules joined together. (p. 80)

centrosome (sen'tro-sōm) Specialized zone of cytoplasm close to the nucleus and containing two centrioles. (p. 80)

cerebellum (ser'ĕ-bel'um) [L., little brain] Separate portion of the brain attached to the brainstem at the pons; important in maintaining muscle tone, balance, and coordination of movement. (p. 384)

cerebral death In the presence of cardiac activity the permanent loss of cerebral function, manifested clinically by absence of conscious responses to external stimuli, absence of cephalic reflexes, apnea, and an isoelectric electroencephalogram for at least 30 minutes in the absence of hypothermia and poisoning by central nervous system depressants. (p. 975)

cerebrospinal fluid (ser-e-bro-spi'nal) Fluid filling the ventricles and surrounding the brain and spinal cord. (p. 200)

cerebrum [L., brain] Portion of the brain derived from the telencephalon: the cerebral hemispheres, including cerebral cortex, cerebral medulla, and basal ganglia. (p. 384)

ceruminous glands (sĕ-roo'mĭ-nus) Modified sebaceous glands in the external auditory meatus that produce cerumen (earwax). (p. 485)

cervical canal Canal extending from the isthmus of the uterus to the opening of the uterus into the vagina. (p. 926)

cervix (ser'viks) [L., neck] Lower part of the uterus extending from the isthmus of the uterus into the vagina. (p. 926)

chalazion (kal-a'ze-on) [Gr. *chalaza*, a sty] Inflammation in the meibomian glands of the eyelid; also called a meibomian cyst. (p. 468)

cheek Side of the face forming the lateral wall of the mouth. (p. 776)

chemical bond Force holding two neighboring atoms in place and resisting their separation. (p. 32)

chemical energy Form of potential energy within the electrons of a chemical bond. (p. 40)

chemistry [Gr. *chemeia*, alchemy] Science dealing with the atomic composition of substances and the reactions they undergo. (p. 29)

chemoreceptor Sensory cell that is stimulated by a change in the concentration of chemicals to produce action potentials. Examples include taste receptors, olfactory receptors, and carotid bodies. (p. 458)

chemoreceptor reflex Chemoreceptors detect decrease in blood oxygen, increase in carbon dioxide, or decrease in pH and produce an increased rate and depth of respiration, and by means of the vasomotor center, vasoconstriction. (p. 686)

chemosensitive area Chemosensitive neurons in the medulla oblongata detect changes in blood, carbon dioxide, and pH. (p. 761)

chemotactic factor (ke'mo-tak'tik) Part of a microorganism or chemical released by tissues and cells that act as chemical signals to attract leukocytes. (p. 707)

chemotaxis (kem-o-tak'sis) [Gr. *chemo* + *taxis*, orderly arrangement] Attraction of living protoplasm (cells) to chemical stimuli. (p. 594)

chief cell Cell of the parathyroid gland that secretes parathyroid hormone. (p. 782) Cell of a gastric gland that secretes pepsinogen.

chloride Compound containing chlorine, e.g., salts of hydrochloric acid. (p. 35)

chloride shift Diffusion of chloride ions into red blood cells as bicarbonate ions diffuse out; maintains electrical neutrality inside and outside the red blood cells. (p. 757)

choana, pl. choanae (ko'an-ah, ko-a'ne) See internal naris.

cholecystokinin (ko-le-sis-to-kīn'in) hormone liberated by the upper intestinal mucosa on contact with gastric contents; stimulates the contraction of the gallbladder and the secretion of pancreatic juice high in digestive enzymes. (p. 798)

cholinergic (kol-in-er'jik) Referring to nerve fibers that secrete acetylcholine or to drugs that bind to and activate cholinergic receptor sites. (p. 511)

cholinergic neuron Refers to nerve fibers that secrete acetylcholine as a neurotransmitter substance. (p. 511)

chondroblast (kon'dro-blast) [Gr. *chondros*, cartilage + *blastos*, germ] Cartilage-producing cell. (p. 158)

chondrocyte (kon'dro-sīt) [Gr. *chondros*, gristle, cartilage + *kytos*, a cell] Mature cartilage cell. (p. 120)

chorda tympani (kor'dah tim'pah-ne) Branch of the facial nerve that conveys taste sensation from the front two thirds of the tongue. (p. 485)

chordae tendineae (kor'de ten'dī-ne-e) [L., cord] Tendinous strands running from the papillary muscles to the atrioventricular valves. (p. 618)

choroid (ko'royd) Portion of the vascular tunic associated with the sclera of the eye. (p. 471)

choroid plexus [Gr. *chorioeides*, membranelike] Specialized plexus located within the ventricles of the brain that secretes cerebrospinal fluid. (p. 359)

chromatid One half of a chromosome; separates from its partner during cell division. (p. 92)

chromatin (kro'mah-tin) Colored material; the genetic material in the nucleus. (p. 75)

chromosome Colored body in the nucleus, composed of DNA and proteins and containing the primary genetic information of the cell; 23 pairs in humans. (p. 94)

chronic rejection Rejection of a graft due to immune complexes forming in arteries supplying the graft, resulting in inadequate blood delivery. (p. 728)

chyle (kīl) [Gr. *chylos*, juice] Milky colored lymph with a high fat content. (p. 700)

chylomicron (ki-lo-mi'kron) [Gr. *chylos*, juice + *micros*, small] Microscopic particle of lipid surrounded by protein, occurring in chyle and in blood. (p. 809)

chyme (kīm) [Gr. *chymos*, juice] Semifluid mass of partly digested food passed from the stomach into the duodenum. (p. 798)

chymotrypsin Proteolytic enzyme formed in the small intestine from the pancreatic precursor chymotrypsinogen. (p. 804)

ciliary body (sil'e-ăr-e) Structure continuous with the choroid layer at its anterior margin that contains smooth muscle cells and functions in accommodation. (p. 471)

ciliary gland Modified sweat gland that opens into the follicle of an eyelash, keeping it lubricated. (p. 468)

ciliary muscle Smooth muscle in the ciliary body of the eye. (p. 471)

ciliary process Portion of the ciliary body of the eye that attaches by suspensory ligaments to the lens. (p. 471)

ciliary ring Portion of the ciliary body of the eye that contains smooth muscle cells. (p. 471)

cilium pl. cilia (sil'e-um, sil'e-ah) [L., eyelid] Motile extension of the cell surface containing nine longitudinal double microtubules arranged in a peripheral ring, together with a central pair. (p. 83)

circle of Willis "Circle" of interconnected blood vessels at the base of the brain. (p. 655)

circumcision [L. *circumcido*, to cut around] Operation in which part or all of the prepuce of the penis is removed. (p. 915)

circumduction [L., around + *ductus*, to draw] Movement in a circular motion. (p. 241)

circumferential lamellae Lamellae covering the surface of and extending around compact bone inside the periosteum. (p. 164)

circumvallate papilla (sur'kum-val'āt) Type of papilla on the surface of the tongue surrounded by a groove. (p. 464)

cisterna, pl. cisternae (sis-ter'nah, sis-ter'ne) Interior space of the endoplasmic reticulum. (p. 78)

cisterna chyli (ki'le) [L., tank; Gr. *chylos*, juice] Enlarged inferior end of the thoracic duct that receives chyle from the intestine. (p. 672)

citric acid cycle Series of chemical reactions in which citric acid is converted into oxaloacetic acid, carbon dioxide is formed, and energy is released. The oxaloacetic acid can combine with acetyl CoA to form citric acid and restart the cycle. The energy released is used to form NADH, FADH, and ATP. (p. 831)

classical pathway Part of the specific immune system for activation of complement. (p. 706)

cleavage furrow Inward pinching of the plasma membrane that divides a cell into two halves, which separate from each other to form two new cells. (p. 94)

cleft lip Failure of the frontonasal and maxillary processes to fuse during development, resulting in a gap to one side (or both sides) of the midline that extends from the mouth to the nostril. (p. 203)

cleft palate Failure of the embryonic palate to fuse along the midline, resulting in an opening through the roof of the mouth. (p. 203)

clinical age Age of the developing fetus from the time of the mother's last menstrual period before pregnancy. (p. 948)

clinical perineum Portion of the perineum between the vaginal and anal openings. (p. 929)

clitoris (klit'o-ris) Small cylindric, erectile body, rarely exceeding 2 cm in length, situated at the most anterior portion of the vulva and projecting beneath the prepuce. (p. 927)

cloaca (klo-a'kah) [L., sewer] In early embryos the endodermally lined chamber into which the hindgut and allantois empty. (p. 926)

clot retraction Condensation of the clot into a denser, compact structure; caused by the elastic nature of fibrin. (p. 599)

coagulation (ko-ag'u-la-shun) Process of changing from liquid to solid, especially of blood; formation of a blood clot. (p. 596)

cochlear duct Interior of the membranous labyrinth of the cochlea; cochlear canal. (p. 486)

cochlear nerve Nerve that carries sensory impulses from the organ of Corti to the vestibulocochlear nerve. (p. 488)

cochlear nucleus Neurons from the cochlear nerve synapse within the dorsal or ventral cochlear nucleus in the superior medulla oblongata. (p. 493)

codon Sequence of three nucleotides in mRNA or DNA that codes for a specific amino acid in a protein. (p. 89)

cofactor Nonprotein component of an enzyme such as coenzymes and inorganic ions essential for enzyme action. (p. 54)

collagen (kol'lă-jen) [Gr. *koila*, glue + *gen*, producing] Ropelike protein of the extracellular matrix. (p. 113)

collateral ganglia Sympathetic ganglia

that are found at the origin of large abdominal arteries; include the celiac, superior, and inferior mesenteric arteries.

collateral ganglion Collection of sympathetic postganglionic neurons within a splanchnic nerve. (p. 508)

collecting duct Straight tubule that extends from the cortex of the kidney to the tip of the renal pyramid. Filtrate from the distal convoluted tubes enters the collecting duct and is carried to the calyces. (p. 856)

colloid (kol'loyd) [Gr. *kolla*, glue + *eidos*, appearance] Atoms or molecules dispersed in a gaseous, liquid, or solid medium that resist separation from the liquid or gas. (p. 41)

colloid osmotic pressure Osmotic pressure due to the concentration difference of proteins across a membrane that does not allow passage of the proteins. (p. 861)

colloidal solution (ko-loy'del) Fine particles suspended in a liquid; particles are resistant to sedimentation or filtration. (p. 586)

colon (ko'lon) Division of the large intestine that extends from the cecum to the rectum. (p. 788)

color blindness Inability to distinguish between certain colors due to a deficiency of one or more visual pigments in the cones of the eye. (p. 483)

colostrum Thin, white fluid; the first milk secreted by the breast at the termination of pregnancy; contains less fat and lactose than the milk secreted later. (p. 972)

columnar Shaped like a column. (p. 104)

commissure (kom'ĭ-shur) [L. *commissura*, a joining together] Connection of nerve fibers between the cerebral hemispheres or from one side of the spinal cord to the other. (p. 397)

common bile duct Duct formed by the union of the common hepatic and cystic ducts; it empties into the small intestine. (p. 783)

common hepatic duct Part of the biliary duct system that is formed by the joining of the right and left hepatic ducts. (p. 787)

compact bone Bone that is more dense and has fewer spaces than cancellous bone. (p. 121)

competition Similar molecules binding to the same carrier molecule or receptor site. (p. 70)

complement Group of serum proteins that stimulates phagocytosis and inflammation. (p. 706)

complement cascade Series of reactions in which each component activates the next component, resulting in activation of complement proteins. (p. 706)

compliance Change in volume (e.g., in lungs or blood vessels) caused by a given change in pressure. (p. 674)

concentration gradient Concentration difference between two points in a solution divided by the distance between the points. (p. 67)

concha, pl. conchae (kon'kah, kon'ke) [L., shell] Structure comparable to a shell in shape; the three bony ridges on the lateral wall of the nasal cavity. (p. 736)

concomitant strabismus Same degree of strabismus in all directions of gaze. (p. 483)

condensation reaction [L. *con* + *denso*, to make thick, condense] Synthesis reactions in which water is a product; a dehydration reaction.

conduction [L. *con* + *ductus*, to lead, conduct] Transfer of energy such as heat from one point to another without evident movement in the conducting body. (p. 844)

condyle [Gr. *knodlyos*, knuckle] Rounded articulating surface of a joint. (p. 186)

cone Photoreceptor in the retina of the eye; responsible for color vision. (p. 472)

congenital [L. *congenitus*, born with] Occurring at birth; may be genetic or due to some influence (e.g., drugs) during development. (p. 976)

conjunctiva (kon-junk-ti'vah) [L. *conjungo*, to bind together] Mucous membrane covering the anterior surface of the eyeball and lining the lids. (p. 468)

conjunctival fornix Area in which the palpebral and bulbar conjunctiva meet. (p. 468)

connective tissue One of the four major tissue types, characterized by relatively more extracellular matrix than cells; the supporting or framework tissue of the body; derived from mesoderm. (p. 114)

constant region Portion of the antibody that does not combine with the antigen and is the same in different antibodies. (p. 719)

constipation Condition in which bowel movements are infrequent or incomplete. (p. 811)

contact hypersensitivity Delayed hypersensitivity reaction to antigens that contact the skin or mucous membranes. (p. 727)

continuous capillary [L. *capillaris*, relating to hair] Capillary in which pores are absent; less permeable to large molecules than other types of capillaries. (p. 646)

contraction phase One of the three phases of muscle contraction; the time during which tension is produced by the contraction of muscle. (p. 284)

convection [L. *con* + *vectus*, to carry or bring together] Transfer of heat in liquids or gases by movement of the heated particles. (p. 844)

convergent circuit Neuronal circuit in which two or more neurons converge on a smaller number of postsynaptic neurons. (p. 374)

coracoid (kor'ah-koyd) [Gr. *korakodes*, crow's beak] Resembling a crow's beak, e.g., a process on the scapula. (p. 212)

Cori cycle Lactic acid, produced by skeletal muscle, is carried in the blood to the liver, where it is aerobically converted into glucose. The glucose may return through the blood to skeletal muscle or may be stored as glycogen in the liver. (p. 830)

corn [L. *cornu*, horn] Thickening of the stratum corneum of the skin over a bony projection in response to friction or pressure. (p. 139)

cornea (kor'ne-ah) Transparent portion of the fibrous tunic that comprises the outer wall of the anterior portion of the eye. (p. 471)

corniculate cartilage (kōr-nik'u-lat) Conical nodule of elastic cartilage surmounting the apex of each arytenoid cartilage. (p. 738)

corona radiata [L., garland, crown] Innermost cells of the cumulus mass forming a radiating crown of cells around the oocyte. (p. 923)

coronal (ko-ro'nal) [Gr. *korone*, crown] Plane separating the body or any part of the body into anterior and posterior portions; frontal section. (p. 194)

coronary (kor'o-năr-e) [L. *coronarius*, a crown] Resembling a crown; encircling. (p. 614)

coronary artery One of two arteries that arise from the base of the aorta and carry blood to the muscle of the heart. (p. 616)

coronary ligament Peritoneal reflection from the liver to the diaphragm at the margins of the bare area of the liver. (p. 791)

coronary sinus Short trunk that receives most of the veins of the heart and empties into the right atrium. (p. 662)

coronary sulcus (sul'kus) [L., furrow or ditch] Groove on the outer surface of the heart marking the division between the atria and the ventricles. (p. 614)

coronoid (ko'ro-noyd) [Gr. *korone*, a crow] Shaped like a crow's beak, e.g., a process on the mandible. (p. 214)

corpus (kor'pus) [L. *body*] Any body or mass; the main part of an organ.

corpus albicans (al'bĭ-kanz) Atrophied corpus luteum leaving a connective tissue scar in the ovary. (p. 926)

corpus callosum (kah-lo'sum) [L., body; callous] Largest commissure of the brain, connecting the cerebral hemispheres. (p. 397)

corpus cavernosum, pl. corpora cavernosa One of two parallel columns of erectile tissue forming the dorsal part of the body of the penis or the body of the clitoris. (p. 915)

corpus luteum (lu'te-um) Yellow endocrine body formed in the ovary in the site of a ruptured vesicular follicle immediately after ovulation; secretes progesterone and estrogen. (p. 926)

corpus luteum of pregnancy Large corpus luteum in the ovary of a pregnant female; secretes large amounts of progesterone and estrogen. (p. 926)

corpus spongiosum Median column of erectile tissue located between and ventral to the two corpora cavernosa in the penis; posteriorly it forms the bulb of the penis, and anteriorly it terminates as the glans penis; it is traversed by the urethra. In the female it forms the bulb of the vestibule. (p. 915)

corpus striatum (stri-a'tum) [L. *corpus*, body + *striatus*, striated or furrowed] Collective term for the caudate nucleus, putamen, and globus pallidus; so named because of the striations caused by intermixing of gray and white matter that results from the number of tracts crossing the anterior portion of the corpus striatum. (p. 397)

cortex [L., bark] Outer portion of an organ (e.g., adrenal cortex or cortex of the kidney). (p. 144)

corticotropin releasing hormone Hormone from the hypothalamus that stimulates the adenohypophysis to release adrenocorticotropic hormone. (p. 565)

cortisol (kor'tĭ-sol) Steroid hormone released by the zona fasciculata of the adrenal cortex; increases blood glucose and inhibits inflammation. (p. 565)

costal Related to a rib. (p. 208)

cotransport Carrier-mediated simultaneous movement of two substances across a membrane in the same direction. (p. 862)

countercurrent multipler mechanism A U-tube arrangement in the kidney. Water diffuses from the tubule and solutes diffuse into the first part of the tube. In the other arm, solutes are actively transported from the arm. The volume of solute is reduced as fluid flows through the first part of the tube and the concentration is reduced as fluid flows through the second part of the tube. (p. 866)

covalent bond Chemical bond characterized by the sharing of electrons. (p. 35)

cranial nerve Nerve that originates from a nucleus within the brain; there are 12 pairs of cranial nerves. (p. 354)

cranial vault Eight skull bones that surround and protect the brain; brain case. (p. 186)

craniosacral division Synonym for the parasympathetic division of the autonomic nervous system. (p. 510)

cranium (kra'ne-um) [Gr. *kranion*, skull] Skull; in a more limited sense, the brain case. (p. 186)

cremaster muscle Extension of abdominal muscles originating from the internal oblique muscles; in the male, raises the testicles; in the female, envelops the round ligament of the uterus. (p. 909)

crenation (kre-na'shun) [L. *crena*, notched] Denoting the outline of a shrunken cell. (p. 68)

cricoid cartilage (kri'koyd) Most inferior laryngeal cartilage. (p. 738)

cricothyrotomy Incision through the skin and cricothyroid membrane for relief of respiratory obstruction.

crista ampullaris [L., crest] Elevation on the inner surface of the ampulla of each semicircular duct for dynamic or kinetic equilibrium. (p. 495)

cristae (kris'te) [L., crest] Shelflike infoldings of the inner membrane of a mitochondrion. (p. 80)

critical closing pressure Pressure in a blood vessel below which the vessel collapses, occluding the lumen and preventing blood flow. (p. 674)

crown (tooth) That part of a tooth that is covered with enamel. (p. 778)

cruciate (kru'she-āt) [L. *cruciatus*, cross] Resembling or shaped like a cross. (p. 244)

crus of the penis Posterior portion of the corpus cavernosum of the penis attached to the ischiopubic ramus. (p. 915)

cryptorchidism Failure of the testis to descend. (p. 964)

crystalline Protein that fills the epithelial cells of the lens in the eye. (p. 473)

cuboidal Something that resembles a cube. (p. 104)

cumulus mass See cumulus oophorus. (p. 923)

cumulus oophorus (o-of'or-us) [L., a heap] Mass of epithelial cells surrounding the oocyte; also called the cumulus mass. (p. 923)

cuneiform cartilage (ku'ne-ĭ-form) Small rod of elastic cartilage above the corniculate cartilages in the larynx. (p. 738)

cupula (ku'pu-lah) [L. *cupa*, a tub] Gelatinous mass that overlies the hair cells of the cristae ampullares of the semicircular ducts. (p. 495)

curare Plant extract that binds to acetylcholine receptors and prevents the normal function of acetylcholine.

cutaneous receptor Sensory receptor associated with the skin. (p. 145)

cuticle (ku'tĭ-kl) [L. *cutis*, skin] Outer thin layer, usually horny, e.g., the outer covering of hair or the growth of the stratum corneum onto the nail. (p. 144)

cyanosis (si-ă-no'sis) [Gr., dark blue color] Blue coloration of the skin and mucous membranes due to insufficient oxygenation of blood. (p. 141)

cystic duct Duct leading from the gallbladder; joins the common hepatic duct to form the common bile duct. (p. 787)

cytokinesis (si-to-kin-e'sis) [Gr. *cyto*, cell + *kinsis*, movement] Division of the cytoplasm during cell division. (p. 94)

cytology (si-tol'o-je) [Gr. *kytos*, a hollow (cell) + *logos*, study] Study of anatomy, physiology, pathology, and chemistry of the cell. (p. 4)

cytoplasm (si'to-plazm) Protoplasm of the cell surrounding the nucleus. (p. 63)

cytoplasmic inclusion Any foreign or other substance contained in the cytoplasm of a cell. (p. 76)

cytotoxic reaction [Gr. *cyto*, cell + L. *toxic*, poison] Antibodies (IgG or IgM) combine with cells and activate complement, and cell lysis occurs. (p. 727)

cytotrophoblast (si-to-tro'fo-blast) Inner layer of the trophoblast composed of individual cells. (p. 948)

Dalton's law In a mixture of gases the portion of the total pressure resulting from each type of gas is determined by the percentage of the total volume represented by each gas type. (p. 749)

dartos muscle (dar'tōs) Layer of smooth muscle in the skin of the scrotum; contracts in response to lower temperature and relaxes in response to higher temperature; raises and lowers testes in the scrotum. (p. 909)

dead air space Part of the respiratory system in which gas exchange does not take place. (p. 748)

deciduous tooth (dĕ-sid'u-us) Tooth of the first set of teeth; primary tooth. (p. 778)

decomposition reaction Disintegration of larger molecules into smaller molecules, ions, or atoms. (p. 37)

decubitus ulcer (de-ku'bi-tus) [L. *decumb*, to lie down] Destruction of tissue over a bony projection due to insufficient blood delivery or as a result of prolonged pressure on the tissue by objects such as a bed or cast. (p. 148)

decussate (dĕ'kūs-āt) [L. *decusso*, X shaped, from *decussis*, ten (X)] To cross. (p. 384)

deep inguinal ring Opening in the transverse fascia through which the spermatic cord (or round ligament in the female) enters the inguinal canal. (p. 915)

defecation [L. *defaeco*, to remove the dregs, purify] Discharge of feces from the rectum. (p. 776)

defecation reflex Combination of local and central nervous system reflexes initiated by distention of the rectum and resulting in movement of feces out of the lower colon. (p. 776)

deglutition (dĕ'glu-tish'un) [L. *de* + *glutio*, to swallow] Act of swallowing. (p. 775)

dehydration reaction [L. *de*, out + *hydr*, water] Synthesis reaction in which water is a product; a condensation reaction. (p. 38)

delayed hypersensitivity An overreaction of the cell-mediated immune system that produces harmful inflammation and tissue destruction within hours to days. (p. 712)

dendrite (den'drīt) [Gr. *dendrites*, tree] Branching processes of a neuron that receives stimuli and conducts potentials toward the cell body. (p. 123)

dendritic cell [Gr. *dendrites*, a tree] Any cell having a branching, treelike appearance. (p. 357)

dendritic cell Large cells with long cytoplasmic extensions that are capable of taking up and concentrating antigens leading to activation of B or T lymphocytes. (p. 120)

dendritic spine Extension of nerve cell dendrites where axons form synapses with the dendrites; also called gemmule. (p. 714)

dental arch [L. *arcus*, bow] Curved maxillary or mandibular arch in which the teeth are located. (p. 778)

dentin Bony material forming the mass of the tooth. (p. 778)

deoxyhemoglobin Hemoglobin without oxygen bound to it. (p. 591)

deoxyribonuclease Enzyme that splits DNA into its component nucleotides. (p. 804)

deoxyribonucleic acid Type of nucleic acid containing deoxyribose as the sugar component, found principally in the nuclei of cells; comprises the genetic material of cells. (p. 54)

depolarization Change in the electrical charge difference across the cell membrane that causes the difference to be smaller or closer to 0 mV; phase of the action potential in which the membrane potential moves toward zero, or becomes positive. (p. 256)

depression Movement of a structure in an inferior direction. (p. 241)

depth perception Ability to distinguish between near and far objects and to judge their distance. (p. 481)

dermatitis (der'mă-ti'tis) [Gr. *derma*, skin + *itis*, inflammation] Inflammation of the skin. (p. 149)

dermatome (der'mă-tōm) Area of skin supplied by a spinal nerve. (p. 440)

dermis (der'mis) [Gr. *derma*, skin] Dense, irregular connective tissue that forms the deep layer of the skin. (p. 135)

descending aorta Part of the aorta, further divided into the thoracic aorta and abdominal aorta. (p. 653)

descending colon Part of the colon extending from the left colonic flexure to the sigmoid colon. (p. 788)

desmosome (dez'mo-sōm) [Gr. *desmos*, a band + *soma*, body] Point of adhesion between cells. Each contains a dense plate at the point of adhesion and a cementing extracellular material between the cells. (p. 111)

desquamate (des'kuă-māt) [L. *desquamo*, to scale off] Peeling or scaling off of the superficial cells of the stratum corneum. (p. 136)

diabetes insipidus Chronic excretion of large amounts of urine of low specific gravity accompanied by extreme thirst; results from inadequate output of antidiuretic hormone. (p. 870)

diabetes mellitus Metabolic disease in which carbohydrate use is reduced and that of lipid and protein enhanced; caused by deficiency of insulin or an inability to respond to insulin and is characterized, in more severe cases, by hyperglycemia, glycosuria, water and electrolyte loss, ketoacidosis, and coma. (p. 870)

diad Transverse tubule and a cisterna in cardiac muscle fibers. (p. 273)

diapedesis (di'ă-pĕ-de'sis) [Gr. *dia*, through + *pedesis*, a leaping] Passage of blood or any of its formed elements through the intact walls of blood vessels. (p. 594)

diaphragm (di'ă-fram) Musculomembranous partition between the abdominal and thoracic cavities. (p. 324)

diaphysis (di-af'ĭ-sis) [Gr., growing between] Shaft of a long bone. (p. 159)

diarrhea [Gr. *dia*, through + *rhoia*, a flow, a flux] Abnormally frequent discharge of more or less fluid fecal matter from the bowel. (p. 806)

diastole (di-as'to-le) [Gr. *diastole*, dilation] Relaxation of the heart chambers during which they fill with blood; usually refers to ventricular relaxation. (p. 629)

dichromatism [Gr. *di*, two + *chroma*, color] Condition in which only two retinal cone pigments are present, resulting in color vision abnormality. (p. 483)

diencephalon (di-en-sef'ă-lon) [Gr. *dia*, through + *enkephalos*, brain] Second portion of the embryonic brain; in the inferior core of the adult cerebrum. (p. 388)

diffuse lymphatic tissue Dispersed lymphocytes and other cells with no clear boundary; found beneath mucous membranes, around lymph nodules, and within lymph nodes and spleen. (p. 701)

diffusion [L. *diffundo*, to pour in different directions] Tendency for solute molecules to move from an area of high concentration to an area of low concentration in solution; the product of the constant random motion of all atoms, molecules, or ions in a solution. (p. 66)

diffusion coefficient Measure of how easily a gas diffuses through a liquid or tissue. (p. 751)

digestive tract Mouth, oropharynx, esophagus, stomach, small intestine, and large intestine. (p. 772)

digit Finger, thumb, or toe. (p. 213)

dilator pupillae Radial smooth muscle cells of the iris diaphragm that cause the pupil of the eye to dilate. (p. 472)

diploid (dip'loyd) Normal number of chromosomes (in humans, 46 chromosomes) in somatic cells. (p. 94)

diplopia [Gr. *diplo* + *ops*, eye] Condition in which a single object is perceived as two objects. (p. 483)

disaccharide Condensation product of two monosaccharides by elimination of water. (p. 44)

dissociate [L. *dis* + *socio*, to disjoin, separate] Ionization in which ions are dissolved in water and the cations and anions are surrounded by water molecules. (p. 36)

distal convoluted tubule Convoluted tubule of the nephron that extends from the ascending limb of the loop of Henle and ends in a collecting duct. (p. 856)

distributing artery Medium-sized artery with a tunica media composed principally of smooth muscle; regulates blood flow to different regions of the body. (p. 648)

divergent circuit Neuronal circuit in which one or more neurons synapse with a greater number of postsynaptic neurons. (p. 374)

DNA Deoxyribonucleic acid; genetic material of a cell; the template from which mRNA (and other types of RNA) is made by transcription. (p. 75)

dominant [L. *dominus*, a master] In genetics a gene that is expressed phenotypically to the exclusion of a contrasting recessive gene. (p. 976)

dorsal root Sensory (afferent) root of a spinal nerve. (p. 402)

dorsal root ganglion (gang'gle-on) Collection of sensory neuron cell bodies within the dorsal root of a spinal nerve. (p. 402)

down regulation The decrease in the concentration of receptors in response to a signal. (p. 534)

ductus arteriosus Fetal vessel connecting the left pulmonary artery with the descending aorta. (p. 969)

ductus deferens Duct of the testicle, running from the epididymis to the ejaculatory duct; also called the vas deferens. (p. 915)

ductus venosus In the fetus, the continuation of the umbilical vein through the liver to the inferior vena cava. (p. 969)

duodenal gland Small gland that opens into the base of intestinal glands; secretes a mucoid alkaline substance. (p. 783)

duodenocolic reflex Local reflex resulting in a mass movement of the contents of the colon; produced by stimuli in the duodenum. (p. 807)

duodenum (du-o-de'num) [L. *duodeni*, twelve] First division of the small intestine; connects to the stomach. (p. 772)

dura mater (du'rah ma'ter) [L., hard

mother] Tough, fibrous membrane forming the outer covering of the brain and spinal cord. (p. 419)

eardrum Tympanic membrane; cellular membrane that separates the external from the middle ear; vibrates in response to sound waves. (p. 484)
eccrine (ek'rin) [Gr. *ek*, out + *krino*, to separate] Exocrine; refer to water-producing sweat glands; see merocrine. (p. 145)
ectoderm (ek'to-derm) Outermost of the three germ layers of an embryo. (p. 124)
ectopic pacemaker (ek-top'ik) Any pacemaker other than the sinus node of the heart; abnormal pacemaker; an ectopic focus. (p. 624)
eczema (ek'zĕ-mah) [Gr. *ekzeo*, to boil over] Inflammation of the skin, typically with vesicles that often break open. (p. 149)
edema [Gr. *oidema*, a swelling] Excessive accumulation of fluid, usually causing swelling. (p. 125)
efferent arteriole Vessel that carries blood from the glomerulus to the peritubular capillaries. (p. 853)
efferent division Nerve fibers that send impulses from the central nervous system to the periphery. (p. 354)
efferent ductule [L. *ductus*, duct] One of a number of small ducts leading from the testis to the head of the epididymis. (p. 911)
ejaculation Reflexive expulsion of semen from the penis. (p. 917)
ejaculatory duct Duct formed by the union of the ductus deferens and the excretory duct of the seminal vesicle; opens into the prostatic urethra. (p. 915)
ejection period Time in the cardiac cycle when the semilunar valves are open and blood is being ejected from the ventricles into the arterial system. (p. 630)
elastin [Gr. *elauno*, drive, push] Major connective tissue protein of elastic tissue that has a structure like a coiled spring. (p. 113)
electrocardiogram (ECG) [Gr. *elektron*, amber + *kardia*, heart + *gramma*, a drawing] Graphic record of the heart's electrical currents obtained with the electrocardiograph. (p. 612)
electrolyte [Gr. *electro* + *lytos*, soluble] Cation or anion in solution that conducts an electrical current. (p. 883)
electron Negatively charged subatomic particle in an atom. (p. 31)
electron's orbital Region around the nucleus of an atom, in which an electron orbits. (p. 31)
electron-transport chain Series of electron carriers in the inner mitochondrial membrane; they receive electrons from NADH and $FADH_2$, using the electrons in the formation of ATP and water. (p. 833)
element [L. *elementum*, a rudiment, beginning] Substance composed of atoms of only one kind. (p. 34)
elevation Movement of a structure in a superior direction. (p. 241)
embolism (em'bo-lizm) [Gr. *embolisma*, a piece of patch, literally something thrust in] Obstruction or occlusion of a vessel by a transported clot, a mass of bacteria, or other foreign material. (p. 655)
embolus (em'bo-lus) [Gr. *embolos*, plug, wedge, or stopper] Plug, composed of a detached clot, mass of bacteria, or other foreign body, occluding a blood vessel. (p. 579)
embryo Developing human from the second to the eighth week of development. (p. 954)
embryonic disk Point in the inner cell mass at which the embryo begins to be formed. (p. 952)
embryonic mass Group of cells formed during the first two weeks of development; includes the morula and blastocyst. (p. 948)
embryonic period From approximately the second to the eighth week of development, during which the major organ systems are organized. (p. 948)
emission [L. *emissio*, to send out] Discharge; accumulation of semen in the urethra prior to ejaculation. A nocturnal emission refers to a discharge of semen while asleep. (p. 917)
emmetropia (em-ĕ-tro'pe-ah) [Gr. *emmetros*, according to measure + *ops*, eye] In the eye the state of refraction in which parallel rays are focused exactly on the retina; no accommodation is necessary. (p. 475)
emulsify To form an emulsion. (p. 803)
emulsion Two liquids with one liquid dispersed through the other liquid or in very small globules; e.g., droplets of lipid suspended in an aqueous solution in the intestine.
enamel Hard substance covering the exposed portion of the tooth. (p. 778)
endergonic reaction (en-der-gon'ik) [L. *endo* + Gr. *ergon*, work] Reaction that results in absorption of energy from its surroundings. (p. 40)
endocardial cushion Pair of mounds of embryonic connective tissue covered by endothelium, bulging into the embryonic atrioventricular canal; they grow together and fuse, dividing the originally single canal into right and left atrioventricular orifices. (p. 952)
endocardium (en'do-kar'dĭ-um) Innermost layer of the heart, including endothelium and connective tissue. (p. 619)
endochondral ossification Bone formation that occurs by the formation and growth of a cartilage template, which is then replaced by bone. (p. 165)
endocrine (en'do-krin) [Gr. *endon*, inside + *krino*, to separate] Ductless gland that secretes a hormone internally, usually into the circulation. (p. 526)
endocytosis (en'do-si-to'sis) Bulk uptake of material through the cell membrane. (p. 72)
endoderm (en'do-derm) Innermost of the three germ layers of an embryo. (p. 124)
endolymph [Gr. *endo* + L. *lympha*, clear fluid] Fluid found within the membranous labyrinth of the inner ear. (p. 486)
endometrium (en'do-me'trĭ-um) Mucous membrane comprising the inner layer of the uterine wall; consists of a simple columnar epithelium and a lamina propria that contains simple tubular uterine glands. (p. 927)
endomysium (en'do-mĭz'ĭ-um) [Gr. *endo*, within + *mys*, muscle] Fine connective tissue sheath surrounding a muscle fiber. (p. 273)
endoneurium (en'do-nu're-um) [Gr. *endo*, within + *neuron*, nerve] Delicate connective tissue surrounding individual nerve fibers within a peripheral nerve. (p. 363)
endoplasmic reticulum (en'do-plaz'mik re-tik'u-lum) Double-walled membranous network inside the cytoplasm; rough has ribosomes attached to the surface; smooth does not have ribosomes attached. (p. 78)
endorphin (en'dor-fin) Opoiate-like polypeptide found in the brain and other parts of the body; binds in the brain to the same receptors that bind exogenous opiates. (p. 416)
endosteum (en-dos'te-um) [Gr. *endo*, within + *osteon*, bone] Membranous lining of the medullary cavity and the cavities of spongy bone. (p. 162)
endotendineum (en'do-ten-din'e-um) Loose connective tissue inside a tendon surrounding collagen fibers. (p. 114)
endothelium [Gr. *endo*, within + *thele*, nipple] Layer of flat cells lining blood and lymphatic vessels and the chambers of the heart. (p. 787)
enkephalin (en-kef'ă-lin) Pentapeptide found in the brain; binds to specific receptor sites, some of which may be pain-related opiate receptors. (p. 577)
enteritis Inflammation of the intestine, especially of the small intestine. (p. 806)
enterokinase (en'ter-o-ki'nās) Intestinal proteolytic enzyme that converts trypsinogen into trypsin. (p. 804)
enzyme (en'zīm) [Gr. *en*, in + *zyme*, leaven] Protein that acts as a catalyst. (p. 39)
eosinophil [Gr. *eos*, dawn + *philos*, fond] White blood cell that stains with

acidic dyes; inhibits inflammation. (p. 594)

epicardium (ep-ĭ-kar'dĭ-um) [Gr. *epi*, on + *kardia*, heart] Serous membrane covering the surface of the heart. Also called the visceral pericardium. (p. 612)

epicondyle [Gr. *epi*, on + *kondyles*, a knuckle] Projection on (usually to the side of) a condyle. (p. 186)

epidermis (ep'ĭ-der'mis) [Gr. *epi*, on + *derma*, skin] Outer portion of the skin formed of epithelial tissue that rests on or covers the dermis. (p. 135)

epididymis (ep-ĭ-did'ĭ-mis) [Gr. *epi*, on + *didymos*, twin] Elongated structure connected to the posterior surface of the testis, which consists of the head, body, and tail; site of storage and maturation of the spermatozoa. (p. 914)

epiglottis (ep'y-glot'is) [Gr. *epi*, on + *glottis*, the mouth of the windpipe] Plate of elastic cartilage covered with mucous membrane; serves as a valve over the glottis of the larynx during swallowing. (p. 738)

epimysium (ep-ĭ-mĭz'ĭ-um) [Gr. *epi*, on + *mys*, muscle] Fibrous envelope surrounding a skeletal muscle. (p. 274)

epinephrine Hormone (amino acid derivative) similar in structure to the neurotransmitter norepinephrine; major hormone released from the adrenal medulla; increases cardiac output and blood glucose levels. (p. 563)

epineurium (ep'ĭ-nu're-um) [Gr. *epi*, upon + *neuron*, nerve] Connective tissue sheath surrounding a nerve. (p. 363)

epiphyseal line (ep-ĭ-fiz'ĭ-al) Dense plate of bone in a bone that is no longer growing, indicating the former site of the epiphyseal plate. (p. 159)

epiphyseal plate Site at which bone growth in length occurs; located between the epiphysis and diaphysis of a long bone; area of hyaline cartilage where cartilage growth is followed by endochondral ossification; also called the metaphysis or growth plate. (p. 159)

epiphysis, pl. epiphyses (e-pif'ĭ-sis, e-pif'ĭ-sez) [Gr. *epi*, on + *physis*, growth] Portion of a bone developed from a secondary ossification center and separated from the remainder of the bone by the epiphyseal plate. (p. 159)

epiploic appendage (ep'ĭ-plo'ik) One of a number of little processes of peritoneum projecting from the serous coat of the large intestine except the rectum; they are generally distended with fat. (p. 788)

epitendineum Fibrous connective tissue surrounding a tendon. (p. 114)

epithelium, pl. epithelia [Gr. *epi*, on + *thele*, nipple] One of the four primary tissue types. "Nipples" refers to the tiny capillary-containing connective tissue in the lips, which is where the term was first used. The use of the term was later expanded to include all covering and lining surfaces of the body. (p. 104)

epitope [Gr. *epi*, upon + *top*, place] See antigenic determinant. (p. 714)

eponychium (ep-on-nik'e-um) [Gr. *epi*, on + *onyx*, nail] Outgrowth of the skin that covers the proximal and lateral borders of the nail. Cuticle. (p. 147)

equilibrium [L. *aequilibrium*, horizontal position] State created by two reactions proceeding in opposite directions at equal rates. (p. 38)

erection [L. *erectio*, to set up] Condition of erectile tissue when filled with blood; becomes hard and unyielding; especially refers to this state of the penis. (p. 920)

erythroblastosis fetalis (ĕ-rith'ro-blas-to'sis fe-tă'lis) [erythroblast + *osis*, condition] Destruction of erythrocytes in the fetus or newborn caused by antibodies produced in the Rh-negative mother acting on Rh-positive blood of the fetus or newborn. (p. 602)

erythrocyte (ĕ-rith'ro-sīt) [Gr. *erythros*, red + *kytos*, cell] Red blood cell; biconcave disk containing hemoglobin. (p. 586)

erythropoiesis (ĕ-rith'ro-poy-e'sis) [erythrocyte + Gr. *poiesis*, a making] Production of erythrocytes. (p. 592)

erythropoietin (ĕ-rith'ro-poy'ĕ-tin) Protein that enhances erythropoiesis by stimulating formation of proerythroblasts and release of reticulocytes from bone marrow. (p. 593)

esophageal sphincter Ring of muscle that regulates the passage of materials into or out of the esophagus. The upper esophageal sphincter is at the superior opening of the esophagus, and the lower esophageal sphincter is at the inferior end. (p. 781)

esophagus (e-sof'ă-gus) [Gr. *oisophagos*, gullet] Portion of the digestive tract between the pharynx and stomach. (p. 772)

essential amino acid Amino acid required by animals that must be supplied in the diet. (p. 836)

estrogen Steroid hormone secreted primarily by the ovaries; involved in maintenance and development of female reproductive organs, secondary sexual characteristics, and the menstrual cycle. (p. 575)

eustachian tube Auditory canal; extends from the middle ear to the nasopharynx. (p. 486)

evagination (e'vaj-ĭ-na'shun) [L. *e*, out + *vagina*, sheath] Protrusion of some part or organ from its normal position. (p. 955)

evaporation [L. *e*, out + *vaporare*, to emit vapor] Change from liquid to vapor form. (p. 844)

eversion [L. *everto*, to overturn] Turning outward. (p. 241)

exchange reaction Partly decomposition and partly synthesis in which part of a molecule is broken down and a portion of it is chemically bound to another molecule. (p. 38)

excitation contraction coupling Stimulation of a muscle fiber produces an action potential that results in contraction of the muscle fiber. (p. 282)

excitatory neuron Neuron that produces EPSPs and has a stimulatory influence. (p. 367)

excitatory postsynaptic potential (EPSP) Depolarization in the postsynaptic membrane that brings the membrane potential close to threshold. (p. 367)

excursion Movement of the mandible from side to side (lateral-medial). (p. 241)

exergonic reaction (ek'ser-gon'ik) [L. *exo* + Gr. *ergon*, work] Reaction resulting in a release of energy to its surroundings. (p. 40)

exocrine gland (ek'so-krin) [Gr. *exo*, outside + *krino*, to separate] Gland that secretes to a surface or outward through a duct. (p. 112)

exocytosis (eks-o-si-to'sis) Elimination of material from a cell through the formation of vacuoles. (p. 74)

expiratory center Region of the medulla oblongata that is electrically active during nonquiet expiration. (p. 758)

expiratory reserve volume Maximum volume of air that can be expelled from the lungs after a normal expiration. (p. 747)

extension [L. *extensio*, to stretch out] To stretch out. (p. 236)

external anal sphincter Ring of striated muscular fibers surrounding the anus. (p. 790)

external auditory meatus (me-a'tus) Short canal that opens to the exterior and terminates at the eardrum; part of the external ear. (p. 194)

external ear Portion of the ear that includes the auricle and external auditory meatus; terminates at the eardrum. (p. 484)

external naris, pl. nares Nostril; anterior or external opening of the nasal cavity. (p. 736)

external spermatic fascia Outer fascial covering of the spermatic cord. (p. 915)

external urethral orifice Slitlike opening of the urethra in the glans penis. (p. 915)

external urinary sphincter Sphincter skeletal muscle around the base of the urethra external to the internal urinary sphincter. (p. 858)

exteroceptor (eks'ter-o-sep'tor) [L. *exterus*, external + *receptor*, receiver] Sensory receptor in the skin or mucous membranes, which respond to stimulation by external agents or forces. (p. 458)

extracellular Outside the cell. (p. 65)
extracellular matrix Nonliving chemical substances located between connective tissue cells. (p. 113)
extrinsic clotting pathway Series of chemical reactions resulting in clot formation; begins with chemicals (e.g., tissue thromboplastin) found outside the blood. (p. 596)
extrinsic muscle Muscle located outside the structure being moved. (p. 317)
eyebrow Short hairs on the bony ridge above the eyes. (p. 467)
eyelash Hair at the margins of the eyelids. (p. 468)
eyelid Palpebra; Movable fold of skin in front of the eyeball. (p. 467)

F-actin Fibrous actin molecule that is composed of a series of globular actin molecules (G-actin). (p. 277)
facet (fas'et) "Little face." A small, smooth articular surface. (p. 186)
facilitated diffusion Carrier-mediated process that does not require ATP and moves substances into or out of cells from a high to a low concentration. (p. 70)
falciform ligament Fold of peritoneum extending to the surface of the liver from the diaphragm and anterior abdominal wall. (p. 791)
fallopian tube See uterine tube. (p. 926)
false or vertebrochondral rib (ver-te'bro-kon'dral) Does not attach directly to the sternum (attaches by means of a common cartilage to the cartilage of the seventh rib) or does not attach to the sternum at all. (p. 208)
false pelvis Portion of the pelvis superior to the pelvic brim; composed of the bone on the posterior and lateral sides and by muscle on the anterior side. (p. 218)
falx cerebelli (falks sĕr'ĕ-bel'e) Dural fold between the two cerebellar hemispheres. (p. 419)
falx cerebri (falks ser-e'bre) Dural fold between the two cerebral hemispheres. (p. 419)
far point of vision Distance from the eye where accommodation is not needed to have the image focused on the retina. (p. 475)
fascia (fash'e-ah) [L., band or fillet] Loose areolar connective tissue found beneath the skin (hypodermis) or dense connective tissue that encloses and separates muscles. (p. 274)
fascicle (fas'ĭ-kl) [L. *fascis*, bundle] Bundle of fibers in a tendon or muscle fibers in a muscle. (p. 273)
fasciculus (fă-sik'u-lus) [L. *fascis*, bundle] Band or bundle of nerve or muscle fibers bound together by connective tissue. (p. 401)
fat [A.S. *faet*] Greasy, soft-solid material found in animal tissues and many plants; composed of two types of molecules—glycerol and fatty acids. (p. 46)
fat-soluble vitamin Vitamin such as A, D, E, and K that is soluble in lipids and absorbed from the intestine along with lipids. (p. 49)
fatigue [L. *fatigo*, to tire] Period characterized by a reduced capacity to do work. (p. 290)
fatty acid Any organic acid composed of a long chain of carbon atoms with an acidic group at one end. (p. 46)
fauces (faw'sēz) [L., throat] Space between the cavity of the mouth and the pharynx. (p. 738)
feces Matter discharged from the bowel during defecation, consisting of the undigested residue of the food, epithelium, intestinal mucus, bacteria, and waste material. (p. 806)
female climacteric Period of life occurring in women, encompassing termination of the reproductive period and characterized by endocrine, somatic, and transitory psychologic changes and ultimately menopause. (p. 936)
female pronucleus Nuclear material of the ovum after the ovum has been penetrated by the spermatozoon. Each pronucleus carries the haploid number of chromosomes. (p. 948)
fertilization Process that begins with the penetration of the secondary oocyte by the spermatozoon and is completed with the fusion of the male and female pronuclei. (p. 926)
fetal period The last 7 months of development, during which the organ systems grow and become functionally mature. (p. 948)
fetus Developing human from the ninth week of development until birth. (p. 965)
fibroblast [L. *fibra*, fiber + Gr. *blastos*, germ] Spindle-shaped or stellate cells that form connective tissue. (p. 114)
fibrocyte Mature cell of fibrous connective tissue. (p. 158)
fibrous joint Bones connected by fibrous tissue with no joint cavity; includes sutures, syndesmoses, and gomphoses. (p. 229)
fibrous tunic Outer layer of the eye; composed of the sclera and the cornea. (p. 470)
filiform papilla (fil'ĭ-form) Filament-shaped papilla on the surface of the tongue. (p. 464)
filtrate Liquid that has passed through a filter; e.g., fluid that enters the nephron through the filtration membrane of the glomerulus. (p. 858)
filtration Movement, due to a pressure difference, of a liquid through a filter that prevents some or all of the substances in the liquid from passing through. (p. 69)
filtration fraction Fraction of the plasma entering the kidney that filters into Bowman's capsule. Normally it is around 19%. (p. 858)
filtration membrane Membrane formed by the glomerular capillary endothelium, the basement membrane, and the podocytes of Bowman's capsule. (p. 853)
filtration pressure Pressure gradient that forces fluid from the glomerular capillary through the filtration membrane into Bowman's capsule; glomerular capillary pressure minus glomerular capsule pressure minus colloid osmotic pressure. (p. 861)
filum terminale (fi'lum ter'mĭ-nal'e) [L., thread; terminal, end] Terminal thread; cord of pia mater tethering the end of the spinal cord to the end of the vertebral canal. (p. 401)
fimbria (fim'bre-ah) [L., fringe] Fringe-like structure located at the ostium of the uterine tube. (p. 926)
first messenger Molecule that acts as an intercellular messenger. (p. 536)
fixator (fiks'a-ter) Muscle that stabilizes the origin of a prime mover. (p. 306)
flaccid paralysis Inability to initiate contractions in muscle. The muscle remains without tone and does not contract. (p. 763)
flagellum, pl. flagella (flă-jel'ah) [L., whip] Whiplike locomotory organelle of constant structural arrangement consisting of double peripheral microtubules and two single central microtubules. (p. 83)
flatus (fla'tus) [L., a blowing] Gas or air in the gastrointestinal tract that may be expelled through the anus. (p. 806)
flexion [L. *flectus*] To bend. (p. 236)
focal point Point at which light rays cross after passing through a concave lens such as the lens of the eye. (p. 474)
foliate papilla (fo'le-āt) Leaf-shaped papilla on the lateral surface of the tongue. (p. 464)
follicle-stimulating hormone (FSH) Hormone of the adenohypophysis that, in females, stimulates the graafian follicles of the ovary and assists in follicular maturation and the secretion of estrogen; in males, stimulates the epithelium of the seminiferous tubules and is partially responsible for inducing spermatogenesis. (p. 554)
follicular phase Time between the end of menses and ovulation characterized by rapid division of endometrial cells and development of follicles in the ovary; the proliferative phase. (p. 930)
fontanel (fon'tă-nel') Membranous interval found between the cranial bones in the infant. (p. 229)
foramen, pl. foramina (fo-ra'men, fo-ram'i-nah) A hole. (p. 184)
foramen cecum Median pit on the dorsum of the posterior part of the tongue from which the limbs of a V-shaped furrow run forward and outward; the

site of the origin of the thyroid gland in the embryo. (p. 772)
foramen ovale (o-val′e) In the fetal heart the oval opening in the septum secundum; the persistent part of septum primum acts as a valve for this interatrial communication during fetal life; postnatally the septum primum becomes fused to the septum secundum to close the foramen ovale, forming the fossa ovale. (p. 616)
foramen primum [L., aperture] In the embryonic heart, the temporary opening between right and left atria. (p. 184)
foramen secundum Secondary opening appearing in the upper part of the septum primum between the atria of the heart in the sixth week of embryonic life just before the closure of the foramen primum. (p. 184)
force That which produces a motion in the body; power.
forearm Portion of the upper limb between the elbow and wrist. (p. 19)
foregut Cephalic portion of the primitive digestive tube in the embryo. (p. 954)
foreskin See prepuce. (p. 915)
formed elements Cells (i.e., erythrocytes and leukocytes) and cell fragments (i.e., platelets) of blood. (p. 585)
fornix [L. arch, vault] Recess at the cervical end of the vagina; also the recess deep to each eyelid where the palpebral and bulbar conjunctivae meet. (pp. 398, 927)
fovea centralis (fo′ve-ah) Depression in the middle of the macula where there are only cones and no blood vessels. (p. 472)
free energy Total amount of energy that can be liberated by the complete catabolism of food. (p. 843)
frenulum (fren′u-lum) [L. *frenum*, bridle] Fold extending from the floor of the mouth to the midline of the undersurface of the tongue. (p. 777)
frequency-modulated signals Signals, all of which are identical in amplitude, that differ in their frequency; e.g., strong stimuli may initiate a high frequency of action potentials and weak stimuli may initiate a low frequency of action potentials. (p. 527)
frontal See coronal. (p. 194)
fructose An isomer of glucose; fruit sugar. (p. 44)
FSH surge Increase in plasma follicle-stimulating hormone (FSH) levels before ovulation. (p. 932)
fulcrum Pivot point. (p. 307)
full-thickness burn Burn that destroys the epidermis and the dermis and sometimes the underlying tissue as well; also called a third degree burn. (p. 142)
fundus (fun′dus) [L., bottom] "Bottom" or rounded end of a hollow organ, e.g., the fundus of the stomach or uterus. (p. 181)
fungiform papilla (fun′jĭ-form) Mushroom-shaped papilla on the surface of the tongue. (p. 464)
funiculus, pl. funiculi (fu-nik′u-lus, fu-nik′u-le) [L. *funis*, cord] Small bundle of nerve fibers. (p. 401)

G-actin Globular protein molecules that, when bound together, form fibrous actin (F-actin). (p. 277)
galactose An isomer of glucose. (p. 44)
gallbladder Pear-shaped receptacle on the inferior surface of the liver; serves as a storage reservoir for bile. (p. 772)
gamete (gam′ēt) Ovum or spermatozoon. (p. 94)
gamma globulin [L. *globulus*, globule] Plasma proteins that include the antibodies. (p. 719)
ganglion, pl. ganglia (gan′gle-on, gan′gle-ah) [Gr., swelling or knot] Any group of nerve cell bodies in the peripheral nervous system. (p. 354)
gangrene (gang′grēn) [Gr. *gangraina*, an eating sore] Necrosis due to obstruction, loss, or diminution of blood supply. (p. 650)
gap junction Small channel between cells that allows the passage of ions and small molecules between cells; provides means of intercellular communication. (p. 111)
gastric gland Gland located in the mucosa of the fundus and body of the stomach. (p. 782)
gastric inhibitory polypeptide Hormone secreted by the duodenum that inhibits gastric acid secretion. (p. 798)
gastric pit Small pit in the mucous membrane of the stomach at the bottom of which are the mouths of the gastric glands that secrete mucus, hydrochloric acid, intrinsic factor, pepsinogen, and hormones. (p. 782)
gastrin (gas′trin) Hormone secreted in the mucosa of the stomach and duodenum that stimulates secretion of hydrochloric acid by the parietal cells of the gastric glands. (p. 798)
gastrocolic reflex Local reflex resulting in mass movement of the contents of the colon that occurs after the entrance of food into the stomach. (p. 807)
gastroenteric reflex Reflex initiated by stretch of the duodenal wall or the presence of irritating substances in the duodenum that causes a reduction in gastric secretions. (p. 797)
gastroesophageal (cardiac) opening Opening of the esophagus into the stomach. (p. 781)
gastrointestinal (gas′tro-in-tes′tĭ-nal) Referring to the stomach and intestines. (p. 772)
gating protein Protein that controls the rate at which ions move through an ion channel. (p. 258)
gemmule (jem′ūl) [L. *gemma*, bud] See dendritic spine. (p. 357)
gene [Gr. *genos*, birth] Functional unit of heredity. Each gene occupies a specific place or locus on a chromosome, is capable of reproducing itself exactly at each cell division, and often is capable of directing the formation of an enzyme or other protein. (p. 87)
general gas law The pressure of a gas is equal to the number of gram moles of the gas times the gas constant times the absolute temperature divided by the volume of the gas. Assuming a constant temperature, the pressure of a given amount of gas is inversely proportional to its volume. This relationship also is called Boyle's law. (p. 744)
genetics [Gr. *genesis*, origin or production] Branch of science that deals with heredity. (p. 975)
genital tubercle Median elevation just cephalic to the urogenital orifice of an embryo; gives rise to the penis of the male or the clitoris of the female. (p. 964)
genotype [Gr. *genos*, birth, descent + *typos*, type] Genetic makeup of an individual. (p. 978)
germ cell Spermatozoon or ovum. (p. 912)
germ layer One of three layers in the embryo (ectoderm, endoderm, or mesoderm) from which the four primary tissue types arise. (p. 124)
germinal center Lighter-staining center of a lymphatic nodule; area of rapid lymphocyte division. (p. 703)
germinal period Approximately the first 2 weeks of development. (p. 948)
giantism Abnormal growth in young people due to hypersecretion of growth hormone by the pituitary gland. (p. 553)
gingiva (jin′jĭ-vah) Dense fibrous tissue, covered by mucous membrane, that covers the alveolar processes of the upper and lower jaws and surrounds the necks of the teeth. (p. 437)
girdle Belt or zone; the bony region where the limbs attach to the body. (p. 19)
gland [L. *glans*, acorn] Secretory organ from which secretions may be released into the blood, a cavity, or onto a surface. (p. 112)
glans penis [L., acorn] Conical expansion of the corpus spongiosum that forms the head of the penis. (p. 915)
glaucoma [Gr. *glaukoma*, opacity of the crystalline lens, from *glaukos*, bluish green] Disease of the eye involving increased intraocular pressure caused by decreased outflow of the aqueous humor; results in degeneration of the optic disc, nerve fiber bundle damage, and defective field of vision. (p. 473)
glenoid (glen′oyd) [Gr. *glene*, socket of a joint] Socket of a joint. (p. 242)
globin Protein portion of hemoglobin. (p. 590)
glomerular capillary pressure Blood

pressure within the glomerulus. (p. 861)

glomerular capsule pressure Pressure of the fluid within Bowman's capsule. (p. 861)

glomerular filtration rate Amount of plasma (filtrate) that filters into Bowman's capsules per minute. (p. 858)

glomerulus (glo-mĕr'u-lus) [L. *glomus*, ball of yarn] Mass of capillary loops at the beginning of each nephron, nearly surrounded by Bowman's capsule. (p. 853)

glomus, pl. glomera (glo'mus) [L., ball] Highly organized arteriovenous anastomosis forming a tiny nodule. (p. 650)

glottis [Gr., aperture of the larynx] Vocal apparatus; includes vocal folds and the cleft between them. (p. 738)

glucagon Hormone secreted from the islets of Langerhans of the pancreas; acts primarily on the liver to release glucose into the circulatory system. (p. 567)

glucocorticoid Steroid hormone (e.g., cortisol) released by zonula fasciculata of the adrenal cortex; increases blood glucose and inhibits inflammation. (p. 564)

gluconeogenesis (glu'ko-ne-o-jen'ĕ-sis) [Gr. *glykys*, sweet + *neos*, new + *genesis*, production] Formation of glucose from noncarbohydrates such as proteins (amino acids) or lipids (glycerol). (p. 838)

glucose Six-carbon monosaccharide; dextrose or grape sugar. (p. 822)

glycogenesis (gli'ko-jen'e-sis) Formation of glycogen from glucose molecules. (p. 837)

glycogenolysis Hydrolysis of glycogen to glucose. (p. 838)

glycolysis (gli-kol'ĭ-sis) [Gr. *glykys*, sweet + *lysis*, a loosening] Anaerobic process during which glucose is converted to pyruvic acid; net of two ATP molecules is produced during glycolysis. (p. 85)

goblet cell Mucous-producing epithelial cell that has its apical end distended with mucin. (p. 783)

Golgi apparatus (gōl'je) Named for Camillo Golgi, Italian histologist and Nobel laureate, 1843-1926. Specialized endoplasmic reticulum that concentrates and packages materials for secretion from the cell. (p. 78)

Golgi tendon organ Proprioceptive nerve ending in a tendon. (p. 405)

gomphosis (gom-fo'sis) [Gr. *gomphos*, bolt, nail + *osis*, condition] Fibrous joint in which a peglike process fits into a hole. (p. 229)

gonad [Gr. *gone*, seed] Organ that produces sex cells; testis of a male or ovary of a female. (p. 918)

gonadal ridge Elevation on the embryonic mesonephros; primordial germ cells become embedded in it, establishing it as the testis or ovary. (p. 964)

gonadotropin Hormone capable of promoting gonadal growth and function. Two major gonadotropins are luteinizing hormone (LH) and follicle-stimulating hormone (FSH). (p. 554)

gonadotropin-releasing hormone (GnRH) Hypothalamic-releasing hormone that stimulates the secretion of gonadotropins (LH and FSH) from the adenohypophysis. (p. 554)

gout [L. *gutta*, drop, clot] Metabolic disease characterized by urate deposits or clots in the joints. (p. 232)

graafian follicle See vesicular follicle. (p. 923)

graft vs. host rejection The donor's tissues recognize the recipient's tissues as foreign, and the transplant rejects the recipient's tissues, causing destruction of the recipient's tissues and death. (p. 728)

granulocyte (gran'u-lo-sīt) Mature granular leukocyte (neutrophil, basophil, or eosinophil). (p. 394)

granulosa cell Cell in the layer surrounding the primary follicle. (p. 783)

gray matter Collections of nerve cell bodies, their dendritic processes, and associated neuroglial cells within the central nervous system. (p. 363)

gray ramus communicans, pl. rami communicantes (ra'mus kŏ-mu'nĭ-kans, rami kŏ-mu'nĭ-kon-tez) Connection between spinal nerves and sympathetic chain ganglia through which unmyelinated postganglionic axons project. (p. 508)

greater duodenal papilla Point of opening of the common bile duct and pancreatic duct into the duodenum. (p. 783)

greater omentum Peritoneal fold passing from the greater curvature of the stomach to the transverse colon, hanging like an apron in front of the intestines. (p. 791)

greater vestibular gland One of two mucus-secreting glands on either side of the lower part of the vagina. The equivalent of the bulbourethral glands in the male. (p. 929)

growth hormone Somatotropin; stimulates general growth of the individual; stimulates cellular amino acid uptake and protein synthesis. (p. 171)

gubernaculum (gu'ber-nak'u-lum) [L., helm] Column of tissue that connects the fetal testis to the developing scrotum; involved in testicular descent. (p. 911)

gustatory Associated with the sense of taste. (p. 464)

gustatory hair Microvillus of gustatory cell in a taste bud. (p. 464)

gynecomastia [Gr. *gyne*, woman + *mastos*, breast] Excessive development of the male mammary glands, which sometimes secrete milk. (p. 929)

gyrus, pl. gyri (ji'rus, ji'ri) [L. *gyros*, circle] Rounded elevation of the surface of the brain. (p. 390)

H zone Area in the center of the A band in which there are no actin myofilaments; contains only myosin. (p. 276)

hair [A.S., hear] Columns of dead keratinized epithelial cells. (p. 141)

hair follicle Invagination of the epidermis into the dermis; contains the root of the hair and receives the ducts of sebaceous and apocrine glands. (p. 144)

Haldane effect Hemoglobin that is not bound to carbon dioxide binds more readily to oxygen than hemoglobin that is bound to carbon dioxide. (p. 757)

half-life The time it takes for one half of an administered substance to be lost through biological processes. (p. 533)

haploid (hap'loyd) Having only one set of chromosomes, in contrast to diploid; characteristic of gametes. (p. 94)

hapten (hap'ten) [Gr. *hapto*, to fasten] Small molecule that binds to a large molecule; together they stimulate the specific immune system. (p. 712)

hard palate Floor of the nasal cavity that separates the nasal cavity from the oral cavity; compound of the palatine processes of the maxillary bones and the horizontal plates of the palatine bones. (p. 203)

haustra (haw'strah) [L., machine for drawing water] Sacs of the colon, caused by contraction of the taeniae coli, which are slightly shorter than the gut, so that the latter is thrown into pouches. (p. 788)

haversian canal (hă-ver'shan) Named for seventeenth century English anatomist, Clopton Havers. Canal containing blood vessels, nerves, and loose connective tissue and running parallel to the long axis of the bone. (p. 164)

haversian system See osteon. (p. 164)

heart skeleton Fibrous connective tissue that provides a point of attachment for cardiac muscle cells, electrically insulates the atria from the ventricles, and forms the fibrous rings around the valves.

heat The sensation produced by proximity to fire or an incandescent object as opposed to cold. The basis of heat is the kinetic energy of atoms, which becomes zero at absolute zero. (p. 843)

heat energy Energy that results from the random movement of atoms, ions, or molecules; the greater the amount of heat energy in an object, the higher is the object's temperature. (p. 41)

Heimlich maneuver Planned action designed to expel an obstructing bolus of food from the throat by standing behind the victim, wrapping one's arms around the victim, and suddenly thrusting the fist into the abdomen between the navel and the rib cage to force air up the trachea and dislodge the obstruction. (p. 740)

helicotrema (hel'ĭ-ko-tre'mah) [Gr. *helix*, spiral + *traema*, hole] Opening at the apex of the cochlea through which the scala vestibuli and the scala tympani of the cochlea connect. (p. 486)

hematocrit (hem-ă'to-krit) [Gr. *hemato*, blood + *krin*, to separate] Percentage of blood volume occupied by erythrocytes. (p. 604)

hematopoiesis (hem'ă-to-poy-e'sis) [Gr. *haima*, blood + *poiesis*, a making] Production of blood cells. (p. 586)

heme Oxygen-carrying, color-furnishing part of hemoglobin. (p. 590)

hemidesmosome Similar to half a desmosome, attaching epithelial cells to the basement membrane. (p. 111)

hemocytoblast (he'mo-si'to-blast) [Gr. *hemo* + *kytos*, cell + *blastos*, germ] Primitive blood cell from which the different lines of blood cells develop. (p. 587)

hemoglobin (he'mo-glo-bin) Red, respiratory protein of erythrocytes; consists of 6% heme and 94% globin; transports oxygen and carbon dioxide. (p. 590)

hemolysis (he-mol'ĭ-sis) [Gr. *hemo* + *lysis*, destruction] Destruction of red blood cells in such a manner that hemoglobin is released. (p. 590)

hemolytic anemia (he-mo-lit'ik) Any anemia resulting from abnormal destruction of erythrocytes in the body. (p. 606)

hemophilia Inherited blood disorder marked by a permanent tendency to hemorrhages due to a defect in the clotting mechanism. (p. 232)

hemopoiesis (he'mo-poy-e'sis) [Gr. *haima*, blood; *poiesis*, a making] Formation of the formed elements of blood, i.e., erythrocytes, leukocytes, and thrombocytes. (p. 586)

hemopoietic (he'mo-poy-et'ik) [Gr. *haima*, blood + *poiesis*, to make] Blood-forming tissue. (p. 586)

hemorrhage (hem'ŏ-rij) [Gr. *haima*, blood + *rhegnymi*, to burst forth] Loss of blood. (p. 655)

hemorrhagic anemia (hem-ŏ-raj'ik) Anemia resulting directly from loss of blood. (p. 606)

hemostasis (he'mo-sta-sis) Arrest of bleeding. (p. 595)

Henry's law The concentration of a gas dissolved in a liquid is equal to the partial pressure of the gas over the liquid times the solubility coefficient of the gas. (p. 750)

heparin Anticoagulant that prevents platelet agglutination and thus prevents thrombus formation. (p. 595)

hepatic artery Branch of the aorta that delivers blood to the liver. (p. 787)

hepatic cord Plate of liver cells that radiates away from the central vein of a liver lobule. (p. 787)

hepatic duct One of two ducts (left and right) that drain bile from the liver and join to form the common hepatic duct. (p. 787)

hepatic portal system (hĕ-pat'ik) System of portal veins that carry blood from the intestines, stomach, spleen, and pancreas to the liver. (p. 668)

hepatic portal vein (hĕ-pat'ik) Portal vein formed by the superior mesenteric and splenic veins and entering the liver. (p. 787)

hepatic sinusoid Terminal blood vessel having an irregular and larger caliber than an ordinary capillary within the liver lobule. (p. 787)

hepatic vein Vein that drains the liver into the inferior vena cava. (p. 668)

hepatocyte Liver cell. (p. 787)

hepatopancreatic ampulla Dilation within the major duodenal papilla that normally receives both the common bile duct and the main pancreatic duct. (p. 783)

hepatopancreatic ampullar sphincter Smooth muscle sphincter of the hepatopancreatic ampulla; sphincter of Oddi. (p. 783)

Hering-Breuer reflex (her'ing broy'er) Afferent impulses from stretch receptors in the lungs arrest inspiration; expiration then occurs. (p. 760)

heterozygous [Gr. *heteros*, other + *zygon*, yoke] State of having different allelic genes at one or more paired loci in homologous chromosomes. (p. 976)

hiatus (hi-a'tus) [L., aperture, to yawn] Opening.

hilum [L., small bit or trifle] Indented surface on many organs, serving as a point where nerves and vessels enter or leave. (p. 741)

hindgut Caudal or terminal part of the embryonic gut. (p. 954)

histamine Amine released by mast cells and basophils that promotes inflammation. (p. 595)

histology [Gr. *histo*, web (tissue) + *logos*, study] The science that deals with the microscopic structure of cells, tissues, and organs in relation to their function. (p. 4)

histone (his'tōn) Protein involved in the regulation of DNA function. (p. 75)

holocrine (hōl'o-krin) [Gr. *holos*, complete + *krino*, to separate] Gland whose secretion is formed by the disintegration of entire cells, (e.g., sebaceous gland; see apocrine and merocrine). (p. 113)

homeostasis (ho'me-o-sta'sis) [Gr. *homoio*, like + *stasis*, a standing] State of equilibrium in the body with respect to functions, composition of fluids and tissues. (p. 12)

homeotherm (ho'me-o-therm) (warm-blooded animals) [Gr. *homoiois*, like + *thermos*, warm] Any animal, including mammals and birds, that tends to maintain a constant body temperature. (p. 843)

homologous (ho-mol'o-gus) [Gr., ratio or relation] Alike in structure or origin. (p. 94)

homozygous [Gr. *homos*, the same + *zygon*, yoke] State of having identical allelic genes at one or more paired loci in homologous chromosomes. (p. 976)

hormone [Gr. *hormon*, to set into motion] Substance secreted by endocrine tissues into the blood that acts on a target tissue to produce a specific response. (p. 526)

hormone receptor Protein or glycoprotein molecule of cells that specifically binds to hormones and produces a response. (p. 534)

horn Subdivision of gray matter in the spinal cord. The axons of sensory neurons synapse with neurons in the posterior horn, the cell bodies of motor neurons are in the anterior horn, and the cell bodies of autonomic neurons are in the lateral horn.

host vs. graft rejection Recipient's immune system recognizes the donor's tissue as foreign and rejects the transplant. (p. 728)

human chorionic gonadotropin (HCG) Hormone produced by the placenta; stimulates secretion of testosterone by the fetus; during the first trimester stimulates ovarian secretion from the corpus luteum of the estrogen and progesterone required for the maintenance of the placenta. In a male fetus, stimulates secretion of testosterone by the fetal testis. (p. 919)

humoral immunity [L. *humor*, a fluid] Immunity due to antibodies in serum. (p. 712)

hyaline cartilage (hī'ă-lin) [Gr. *hyalos*, glass] Gelatinous, glossy cartilage tissue consisting of cartilage cells and their matrix; contains collagen, proteoglycans, and water. (p. 120)

hyaline membrane disease (respiratory distress syndrome) Disease seen especially in premature neonates with respiratory distress; associated with reduced amounts of lung surfactant. (p. 745)

hydrocephalus [Gr. *hydor*, water + *kephale*, head] Excessive fluid within the brain, swelling and thinning the cortex. (p. 420)

hydrochloric acid (HCl) Acid of gastric juice. (p. 795)

hydrogen bond Hydrogen atoms bound covalently to either N or O atoms have a small positive charge that is weakly attracted to the small negative charge of other atoms such as O or N; can occur within a molecule or between different molecules. (p. 36)

hydrolysis reaction [Gr. *hydor* + *lysis*, dissolution] Cleaving of a compound into two or more simpler compounds by the addition of H^+ and OH^- parts of a water molecule. (p. 38)

hydrophilic [Gr. *hydor*, water + *philos*, fond] Attracts water. (p. 65)

hydrophobic [Gr. *hydor*, water + *phobos*, fear] Repels water. (p. 65)

hydroxyapatite (hi-drok'se-ap'ĕ-tīt) Mineral with the empiric formula 3 $Ca_3(PO_4)_2 \cdot Ca(OH)_2$; the main mineral of bone and teeth. (pp. 120, 162)

hymen [Gr., membrane] Thin, membranous fold partly occluding the vaginal external orifice; normally disrupted by sexual intercourse or other mechanical phenomena. (p. 927)

hyoid (hi'oyd) [Gr. *hyoeides*, shaped like the Greek letter epsilon, ϵ] U-shaped bone between the mandible and larynx. (p. 186)

hypercalcemia Abnormally high levels of calcium in the blood. (p. 889)

hypercapnia (hi'per-kap'nĭ-ah) Higher-than-normal levels of carbon dioxide in the blood or tissues. (p. 761)

hyperopia [Gr. *hyper*, above + *ops*, eye] Condition due to an error in either refraction or shortening of the globe of the eye in which parallel rays of light are focused behind the retina; farsightedness. (p. 482)

hyperosmotic (hi'per-os-mot'ĭk) [Gr. *hyper*, above + *osmos*, an impulsion] Having a greater osmotic concentration or pressure than a reference solution. (p. 68)

hyperpolarization Increase in the charge difference across the cell membrane; causes the charge difference to move away from 0 mV. (p. 256)

hypertension [Gr. *hyper*, above + L. *tensio*, tension] High blood pressure. (p. 677)

hypertonic (hi'per-ton'ik) [Gr. *hyper*, above + *tonos*, tension] Solution that causes cells to shrink. (p. 68)

hypertrophy (hi-per'tro-fe) [Gr. *hyper*, above + *trophe*, nourishment] Increase in bulk or size; not due to an increase in number of individual elements. (p. 166)

hypocalcemia Abnormally low levels of calcium in the blood. (p. 258)

hypocapnia (hi'po-kap'nĭ-ah) Lower-than-normal levels of carbon dioxide in the blood or tissues. (p. 761)

hypochromic anemia Anemia characterized by a decrease in the ratio of the weight of hemoglobin to the volume of the erythrocyte. (p. 605)

hypodermis [Gr. *hypo*, under + *dermis*, skin] Loose areolar connective tissue found deep to the dermis that connects the skin to muscle or bone. (p. 136)

hypokalemia Abnormally small concentration of potassium ions in the blood. (p. 258)

hyponychium (hy-po-nik'e-um) [Gr. *hypo*, under + *onyx*, nail] Thickened portion of the stratum corneum under the free edge of the nail. (p. 147)

hypophysis (hi-pof'ĭ-sis) [Gr., an undergrowth] Endocrine gland attached to the hypothalamus by the infundibulum. Also called the pituitary gland. (p. 346)

hypopolarization Change in the electrical charge difference across the cell membrane that causes the charge difference to be smaller or move closer to 0 mV.

hyposmotic (hi'pos-mot'ik) [Gr. *hypo*, under + *osmos*, an impulsion] Having a lower osmotic concentration or pressure than a reference solution. (p. 68)

hypospadias [Gr., one having the orifice of the penis too low; *hypospao*, to draw away from under] Developmental anomaly in the wall of the urethra so that the canal is open for a greater or lesser distance on the undersurface of the penis; also a similar defect in the female in which the urethra opens into the vagina. (p. 965)

hypothalamohypophyseal portal system (hi'po-thal'ă-mo-hi'po-fiz'e-al) Series of blood vessels that carry blood from the area of the hypothalamus to the adenohypophysis; originate from capillary beds in the hypothalamus and terminate as a capillary bed in the adenohypophysis. (p. 547)

hypothalamohypophyseal tract Nerve tract, consisting of the axons of neurosecretory cells, extending from the hypothalamus into the neurohypophysis. Hormones produced in the neurosecretory cell bodies in the hypothalamus are transported through the hypothalamohypophyseal tract to the neurohypothesis where they are stored for later release. (p. 547)

hypothalamus (hi-po-thal'ă-mus) [Gr. *hypo*, under + *thalamus*, bedroom] Important autonomic and neuroendocrine control center beneath the thalamus. (p. 390)

hypothenar (hi'po-the'nar) [Gr. *hypo*, under + *thenar*, palm of the hand] Fleshy mass of tissue on the medial side of the palm; contains muscles responsible for moving the little finger. (p. 337)

hypotonic (hi'po-ton'ik) [Gr. *hypo*, under + *tonos*, tension] Solution that causes cells to swell. (p. 68)

hypoxia (hi-pox'se-ah) Lower-than-normal levels of oxygen in the blood or tissues. (p. 762)

I band Area between the ends of two adjacent myosin myofilaments within a myofibril; Z-line divides the I-band into two equal parts. (p. 276)

ileocecal sphincter Thickening of circular smooth muscle between the ileum and the cecum forming the ileocecal valve. (p. 784)

ileocecal valve Valve formed by the ileocecal sphincter between the ileum and the cecum. (p. 784)

ileum (il'e-um) [Gr. *eileo*, to roll up, twist] Third portion of the small intestine, extending from the jejunum to the ileocecal opening into the large intestine. (p. 772)

immediate hypersensitivity Overreaction of the antibody-mediated immune system that produces harmful inflammation and tissue destruction within a few minutes. (p. 727)

immune complex Antibody bound to an antigen. (p. 727)

immune surveillance Concept that the immune system recognizes and removes malignant cells as they arise. (p. 728)

immunity [L. *immunis*, free from service] Resistance to infectious disease and harmful substances. (p. 706)

immunization Process by which a subject is rendered immune by deliberately introducing an antigen or antibody into the subject. (p. 724)

immunodeficiency Failure of some component of the immune system to operate. (p. 728)

immunoglobulin Antibody found in the gamma globulin portion of plasma. (p. 719)

implantation Attachment of the blastocyst to the endometrium of the uterus; occurring 6 or 7 days after fertilization of the ovum. (p. 948)

impotence Inability to accomplish the male sexual act; caused by psychological or physical factors. (p. 920)

incisor [L. *incido*, to cut into] One of the anterior cutting teeth. (p. 778)

incisura (in'si-su'rah) [L., a cutting into] Notch or indentation at the edge of any structure. (p. 631)

incus (ing'kus) [L., anvil] Middle of the three ossicles in the middle ear. (p. 486)

infarct (in'farkt) [L. *in* + *fartus*, to stuff into] Area of necrosis resulting from a sudden insufficiency of arterial or venous blood supply. (p. 417)

infectious mononucleosis Viral infection that causes an increase in lymphocyte number, many of which resemble monocytes. (p. 606)

inferior Down, or lower, with reference to the anatomical position. (p. 16)

inferior colliculus (kol-lik'u-lus) [L. *collis*, hill] One of two rounded eminences of the midbrain; involved with hearing. (p. 387)

inferior vena cava Vein that returns blood from the lower limbs and the greater part of the pelvic and abdominal organs to the right atrium. (p. 614)

infertility [L. *in*, without + *fertilis*, fruitful] Inability to produce offspring; does not imply (either in the male or the female) an irreversible condition such as sterility. (p. 916)

inflammation (in-flă-ma'shun) [L. *in*, in + *flamma*, flame] Process consisting of reaction occurring in the blood vessels and tissues in the area of an injury or the response to an irritating physical, chemical, or biological agent. (p. 124)

inflammatory response Complex sequence of events involving chemicals and immune cells that results in the isolation and destruction of antigens and tissues near the antigens. See local and systemic inflammation. (p. 710)

infundibulum (in-fun-dib'u-lum) [L., funnel] Funnel-shaped structure or passage, e.g., the infundibulum that attaches the hypophysis to the hypothalamus or the funnel-like expansion of the uterine tube near the ovary. (p. 390)

inguinal canal Passage through the lower abdominal wall that transmits the spermatic cord in the male and the round ligament in the female. (p. 911)

inhibin Polypeptide secreted from the testes that inhibits FSH secretion. (p. 918)

inhibitory neuron Neuron that produces IPSPs and has an inhibitory influence. (p. 369)

inhibitory postsynaptic potential (IPSP) Hyperpolarization in the postsynaptic membrane that causes the membrane potential to move away from threshold. (p. 369)

inner cell mass Group of cells at one end of the blastocyst, part of which forms the body of the embryo. (p. 948)

inner ear Contains the sensory organs for hearing and balance; contains the bony and membranous labyrinth. (p. 484)

insensible perspiration [L. *per*, through + *spiro*, to breathe everywhere] Perspiration that evaporates before it is perceived as moisture on the skin; the term sometimes includes evaporation from the lungs. (p. 893)

insertion More movable attachment point of a muscle; usually the lateral or distal end of a muscle associated with the limbs. (p. 306)

inspiratory capacity Volume of air that can be inspired after a normal expiration; the sum of the tidal volume and the inspiratory reserve volume. (p. 747)

inspiratory center Region of the medulla oblongata that stimulates inspiration. (p. 758)

inspiratory reserve volume Maximum volume of air that can be inspired after a normal inspiration. (p. 747)

insula (in'su-lah) [L., island] Oval region of the cerebral cortex buried deep in the lateral fissure. (p. 390)

insulin Protein hormone secreted from the pancreas that increases the uptake of glucose and amino acids by most tissues. (p. 807)

interatrial septum [L. *saeptum*, a partition] Wall between the atria of the heart. (p. 616)

intercalated disk (in-ter'kă-la-ted) Cell-to-cell attachment with gap junctions between cardiac muscle cells. (p. 296)

intercalated duct Minute duct of glands such as the salivary gland and the pancreas; leads from the acini to the interlobular ducts. (p. 788)

intercellular Between cells. (p. 65)

intercellular chemical messenger Chemical that is released from cells and passes to other cells; acts as signal that allows cells to communicate with each other. (p. 527)

intercostal Between ribs. (p. 440)

interferon (in'ter-fēr'on) Protein that prevents viral replication. (p. 706)

interleukin Protein produced by macrophages and T cells; activates T and B cells. (p. 716)

interlobar artery Branch of the segmental arteries of the kidney; runs between the renal pyramids and gives rise to the arcuate arteries. (p. 856)

interlobular artery Artery that passes between lobules of an organ; branches of the interlobar arteries of the kidney pass outward through the cortex from the arcuate arteries and supply the afferent arterioles. (p. 856)

interlobular duct Any duct leading from a lobule of a gland and formed by the junction of the intercalated ducts draining the acini. (p. 788)

interlobular vein Parallels the interlobular arteries; in the kidney drains the peritubular capillary plexus, emptying into arcuate veins. (p. 857)

intermediate olfactory area Part of the olfactory cortex responsible for modulation of olfactory sensations. (p. 462)

internal anal sphincter [Gr. *sphinkter*, band or lace] Smooth muscle ring at the upper end of the anal canal. (p. 736)

internal naris, pl. nares Opening from the nasal cavity into the nasopharynx. (p. 736)

internal spermatic fascia Inner connective tissue covering of the spermatic cord. (p. 915)

internal urinary sphincter Traditionally recognized as a sphincter composed of a thickening of the middle smooth muscle layer of the bladder around the urethral opening. (p. 858)

internode Segment between two nodes of Ranvier. (p. 362)

interphase Period between active cell divisions when DNA replication occurs. (p. 91)

interstitial [L. *inter*, between + *sisto*, to stand] Space within tissue. Interstitial growth means growth from within. (p. 172)

interstitial cell (cell of Leydig) Cell between the seminiferous tubules of the testes; secretes testosterone. (p. 911)

interventricular septum Wall between the ventricles of the heart. (p. 960)

intestinal gland Tubular glands in the mucous membrane of the small and large intestines. (p. 783)

intracellular Inside a cell. (p. 65)

intracellular receptor Receptor molecule such as hormone receptor that is located in the cytoplasm or nucleus of the target cell. (p. 536)

intramembranous ossification Bone formation occurring within a connective tissue membrane. (p. 165)

intramural plexus (in'trah-mu'ral plek'sus) Combined submucosal and myenteric plexuses. (p. 773)

intrapleural pressure Pressure in the pleural cavity. (p. 746)

intrapulmonary pressure Air pressure in the alveoli. (p. 744)

intrinsic clotting pathway Series of chemical reactions resulting in clot formation that begins with chemicals (e.g., plasma factor XII) found within the blood. (p. 596)

intrinsic factor Factor secreted by the parietal cells of gastric glands and required for adequate absorption of vitamin B_{12}. (p. 795)

intrinsic muscles Muscles located within the structure being moved. (p. 317)

inulin Fructose polysaccharide used to determine the rate of glomerular filtration. (p. 872)

invagination Infolding or inpocketing.

inversion [L. *inverto*, to turn about] Turning inward. (p. 241)

iodopsin (i'o-dop'sin) Visual pigment found in the cones of the retina to which an opsin called photopsin binds. (p. 479)

ion channel Pore in the cell membrane through which ions, e.g., sodium and potassium, move. (p. 258)

ionic bond Chemical bond that is formed when one atom loses an electron and another accepts that electron. (p. 32)

ion (ī'on) [Gr. *ion*, going] Atom or group of atoms carrying a charge of electricity by virtue of having gained or lost one or more electrons. (p. 32)

iris Specialized portion of the vascular tunic; the "colored" portion of the eye that can be seen through the cornea. (p. 471)

iron deficiency anemia Anemia due to dietary lack of iron or iron loss as a result of chronic bleeding. (p. 605)

ischemia (is-ke'me-ah) [Gr. *ischo*, to keep back + *haima*, blood] Reduced blood supply to some area of the body. (p. 148)

islet of Langerhans See pancreatic islet.

isomaltose Isomer of maltose. (p. 793)

isomer [Gr. *isos*, equal + *meros*, part] Molecules having the same number and types of atoms but differing in their three-dimensional arrangement. (p. 44)

isometric contraction [Gr. *isos*, equal + *metron*, measure] Muscle contraction in which the length of the muscle does not change but the tension produced increases. (p. 289)

isosmotic (i'sos-mot'ik) [Gr. *isos*, equal + *osmos*, an impulsion] Having the same osmotic concentration or pressure as a reference solution. (p. 68)

isotonic contraction Muscle contraction in which the tension produced by the muscle stays the same but the muscle length becomes shorter. (p. 289)

isotonic solution (i'so-ton'ik) [Gr. *isos*, equal + *tonos*, tension] Solution that causes cells to neither shrink nor swell. (p. 68)

isotope [Gr. *isos*, equal + *topos*, part, place] Either of two or more atoms that have the same atomic number but a different number of neutrons. (p. 34)

isthmus Constriction connecting two larger parts of an organ, e.g., the constriction between the body and the cervix or the uterus, or the portion of the uterine tube between the ampulla and the uterus. (p. 554)

jaundice (jawn'dis) [Fr. *jaune*, yellow] Yellowish staining of the integument, sclerae, and other tissues with bile pigments. (p. 148)

jejunum (jĕ-ju'nam) [L. *jejunus*, empty] Second portion of the small intestine; located between the duodenum and the ileum. (p. 772)

jugular [L. *jugulum*, throat] Relating to the throat or neck. (p. 203)

juxtaglomerular apparatus (juks'tă-glo-mĕr'u-lar) Complex consisting of juxtaglomerular cells of the afferent arteriole and macular densa cells of the distal convoluted tubule near the renal corpuscle; secretes renin. (p. 689)

juxtaglomerular cell Modified smooth muscle cell of the afferent arteriole located at the renal corpuscle; a component of the juxtaglomerular apparatus. (p. 853)

juxtamedullary nephron (juks'tă-med'u-lĕr-e) Nephron located near the junction of the renal cortex and medulla. (p. 350)

keratin (kĕr'ah-tin) [Gr. *keras*, horn] Fibrous protein complex found in the stratum corneum, hair, and nails; provides protection against abrasion. (p. 136)

keratinization Production of keratin and changes in the chemical and structural character of epithelial cells as they move to the skin surface. (p. 136)

keratinized [Gr. *keras*, horn] Word means turned into a horn. In modern usage the term means to become a structure that contains keratin, a protein found in skin, hair, nails, and horns. (p. 138)

keratinocyte (kĕ-rat'ĭ-no-sīt) [Gr. *keras*, horn + *kytos*, cell] Epidermal cell that produces keratin. (p. 136)

keratohyalin (kĕr'ă-to-hi'ă-lin) Non-membrane-bound protein granules in the cytoplasm of stratum granulosum cells of the epidermis. (p. 138)

ketogenesis (ke-to-jen'ĕ-sis) Production of ketone bodies, e.g., from acetyl CoA. (p. 834)

ketone body One of a group of ketones, including acetoacetic acid, β-hydrobutyric acid, and acetone. (p. 572)

ketosis Condition characterized by the enhanced production of ketone bodies as in rapid lipid metabolism during diabetes mellitus. (p. 835)

kidney [A.S. *cwith*, womb, belly + *neere*, kidney] One of the two organs that excrete urine. The kidneys are bean-shaped organs approximately 11 cm long, 5 cm wide, and 3 cm thick lying on either side of the spinal column, posterior to the peritoneum, approximately opposite the twelfth thoracic and first three lumbar vertebrae. (p. 884)

kilocalorie (kil'o-kal-o-re) Quantity of energy required to raise the temperature of 1 kg of water 1° C; 1000 calories. (p. 820)

kinetic energy Motion energy or energy that can do work. (p. 39)

kinetic laryrinth Part of the membranous labyrinth composed of the semicircular canals; detects dynamic or kinetic equilibrium, i.e., movement of the head. (p. 494)

kinin Serum protein that causes vasodilation and increases vascular permeability. (p. 707)

Korotkoff sounds (kō-rot'kof) Sounds heard over an artery when blood pressure is determined by the auscultatory method; caused by turbulent flow of blood. (p. 673)

kyphosis (ki-fo'sis) [Gr., humpback] Abnormal anterior concavity of the vertebral column, especially of the thoracic region. (p. 203)

labium majus, pl. labia majora One of two rounded folds of skin surrounding the labia minora and vestibule; homologue of the scrotum in males. (p. 929)

labium minus, pl. labia minora One of two narrow longitudinal folds of mucous membrane enclosed by the labia majora and bounding the vestibule; anteriorly they unite to form the prepuce. (p. 927)

labor [L., toil, suffering] Process of expulsion of the fetus and the placenta from the uterus. (p. 967)

labyrinth (lab'ĭ-rinth) Intricate structure consisting of winding passageways, e.g., the bony and membranous labyrinths of the inner ear. (p. 484)

lacrimal apparatus (lak'rĭ-mal) Lacrimal or tear gland in the superolateral corner of the orbit of the eye and a duct system that extends from the eye to the nasal cavity. (p. 468)

lacrimal canal Canal that carries excess tears away from the eye; located in the medial canthus and opening on a small lump called the lacrimal papilla. (p. 468)

lacrimal gland Tear gland located in the superolateral corner of the orbit. (p. 468)

lacrimal papilla Small lump of tissue in the medial canthus or corner of the eye; the lacrimal canal opens within the lacrimal papilla. (p. 469)

lacrimal sac Enlargement in the lacrimal canal that leads into the nasolacrimal duct. (p. 469)

lactation [L. *lactatio*, suckle] Period after childbirth during which milk is formed in the breasts. (p. 972)

lacteal (lak'te-al) Lymphatic vessel in the wall of the small intestine that carries chyle from the intestine and absorbs fat. (p. 700)

lactic acid Three-carbon molecule derived from pyruvic acid as a product of anaerobic respiration. $NADH_2$ reacts with pyruvic acid to form lactic acid and NAD^+. (p. 42)

lactiferous ducts One of 15 or 20 ducts that drain the lobes of the mammary gland and open onto the surface of the nipple. (p. 929)

lactiferous sinus Dilation of the lactiferous duct just before it enters the nipple. (p. 929)

lactose Disaccharide present in mammalian milk and composed of glucose and galactose. (p. 44)

lacuna, pl. lacunae (lă-ku'nah, lă-ku'ne) [L. *lacus*, a hollow, a lake] Small space or cavity; potential space within the matrix of bone or cartilage normally occupied by a cell that can only be visualized when the cell shrinks away from the matrix during fixation; space containing maternal blood within the placenta. (p. 158)

lag phase One of the three phases of muscle contraction; time between the application of the stimulus and the beginning of muscular contraction. Also called the latent phase. (p. 284)

lamella, pl. lamellae (lă-mel'ah, lă-mel'e) [L. *lamina*, plate, leaf] Thin sheet or layer of bone. (p. 162)

lamellated corpuscle Pacinian corpuscle. Oval receptor found in the deep dermis or hypodermis (responsible for deep cutaneous pressure and vibration) and in tendons (responsible for proprioception). (p. 461)

lamina (lam'ĭ-nah) Thin plate, e.g., the thinner portion of the vertebral arch. (p. 206)

lamina propria (lam'ĭ-nah pro'pre-ah) Layer of connective tissue underlying the epithelium of a mucous membrane. (p. 124)

laminar flow Relative motion of layers

of a fluid along smooth parallel paths. (p. 672)

Langerhans cell Dendritic cell named after the German anatomist, Paul Langerhans; found in the skin. (p. 136)

lanugo (lă-nu'go) [L. *lana*, wool] Fine, soft, unpigmented fetal hair. (p. 966)

large intestine Portion of the digestive tract extending from the small intestine to the anus. (p. 772)

laryngitis Inflammation of the mucous membrane of the larynx. (p. 738)

laryngopharynx (lă-ring'go-făr'ingks) Part of the pharynx lying posterior to the larynx. (p. 738)

larynx (lăr'ingks) Organ of voice production located between the pharynx and the trachea; it consists of a framework of cartilages and elastic membranes housing the vocal folds and the muscles that control the position and tension of these elements. (p. 738)

last menstrual period (LMP) Beginning of the last menstruation before pregnancy; used clinically to time events during pregnancy. (p. 948)

latent phase See lag phase. (p. 284)

lateral geniculate nucleus Nucleus of the thalamus where fibers from the optic tract terminate. (p. 388)

lateral olfactory area Part of the olfactory cortex involved in the conscious perception of olfactory stimuli. (p. 462)

law of Laplace Force that stretches the wall of a blood vessel is proportional to the radius of the vessel times the blood pressure. (p. 674)

leg That part of the lower limb between the knee and ankle. (p. 19)

lens Transparent biconvex structure lying between the iris and the vitreous humor. (p. 473)

lens fiber Epithelial cell that comprises the lens of the eye. (p. 473)

lesser duodenal papilla Site of the opening of the accessory pancreatic duct into the duodenum. (p. 783)

lesser omentum (o-men'tum) [L., membrane that encloses the bowels] Peritoneal fold passing from the liver to the lesser curvature of the stomach and to the upper border of the duodenum for a distance of approximately 2 cm beyond the pylorus. (p. 791)

lesser vestibular gland Paraurethral gland. Number of minute mucous glands opening on the surface of the vestibule between the openings of the vagina and urethra. (p. 929)

leukocyte (lu'ko-sīt) White blood cell. (p. 586)

leukocytosis (lu-ko-si-to'sis) Abnormally large number of leukocytes in the blood. (p. 604)

leukopenia Lower-than-normal number of leukocytes in the blood. (p. 604)

leukotriene Specific class of physiologically active fatty acid derivatives present in many tissues. (p. 49)

lever Rigid shaft capable of turning about a fulcrum or pivot point. (p. 307)

Leydig cell See interstitial cell.

LH surge Increase in plasma luteinizing hormone (LH) levels before ovulation and responsible for initiating it. (p. 932)

ligament Band or sheet of dense connective tissue that connects two or more bones, cartilages, or other structures; a mesentery supporting an abdominal organ. (p. 158)

ligamentum arteriosum Remains of the ductus arteriosus. (p. 969)

ligamentum venosum Remnant of the ductus venosus. (p. 970)

limb Arm, forearm, wrist, and hand or thigh, leg, ankle, and foot considered as a whole.

limbic system (lim'bik) [L. *limbus*, border] Parts of the brain involved with emotions and olfaction; includes the cingulate gyrus, hippocampus, habenular nuclei, parts of the basal ganglia, the hypothalamus (especially the mammillary bodies, the olfactory cortex, and various nerve tracts (e.g., fornix). (p. 398)

lingual tonsil Collection of lymphoid tissue on the posterior portion of the dorsum of the tongue. (p. 701)

lip One of two muscular folds with an outer mucosa having a stratified squamous epithelial surface layer; the lips form the anterior border of the mouth. (p. 776)

lipase (li'pās) In general, any fat-splitting enzyme. (p. 54)

lipid [Gr. *lipos*, fat] Substance composed principally of carbon, oxygen, and hydrogen; contains a lower ratio of oxygen to carbon and is less polar than carbohydrates; generally soluble in nonpolar solvents. (p. 46)

lipid bilayer Double layer of lipid molecules forming the plasma membrane and other cellular membranes. (p. 65)

lipochrome (lip'o-krōm) Lipid-containing pigment that is metabolically inert. (p. 76)

lipogenesis Synthesis of lipids from glucose (carbohydrates) and amino acids (proteins). (p. 837)

lipotropin One of the peptide hormones released from the adenohypophysis; increases lipolysis in fat cells. (p. 553)

liver Largest gland of the body, lying in the upper-right quadrant of the abdomen just inferior to the diaphragm; secretes bile and is of great importance in carbohydrate and protein metabolism and in detoxifying chemicals. (p. 772)

lobe Rounded projecting part, e.g., the lobe of a lung, the liver, or a gland. (p. 787)

lobule Small lobe or a subdivision of a lobe, e.g., a lobule of the lung or a gland. (p. 705)

local inflammation Inflammation confined to a specific area of the body. Symptoms include redness, heat, swelling, pain, and loss of function. (p. 710)

local potential Depolarization that is not propagated and that is graded or proportional to the strength of the stimulus. (p. 260)

local reflex Reflex of the intramural plexus of the digestive tract that does not involve the brain or spinal cord. (p. 776)

locus Place; usually a specific site. (p. 976)

loop of Henle U-shaped part of the nephron extending from the proximal to the distal convoluted tubule and consisting of descending and ascending limbs. Some of the loops of Henle extend into the renal pyramids. (p. 854)

lordosis (lor-do'sis) [L., bending backward] Abnormal anterior convexity of the lumbar region of the vertebral column. (p. 203)

lower respiratory tract The larynx, trachea, and lungs. (p. 736)

lunula (lu'nu-lah) [L. *luna*, moon] White, crescent-shaped portion of the nail matrix visible through the proximal end of the nail. (p. 147)

luteal phase That portion of the menstrual cycle extending from the time of formation of the corpus luteum after ovulation to the time when menstrual flow begins; usually 14 days in length; the secretory phase. (p. 930)

luteinizing hormone (LH) (lu'te-ĭ-nīz-ing) In females, hormone stimulating the final maturation of the follicles and the secretion of progesterone by them, with their rupture releasing the ovum, and the conversion of the ruptured follicle into the corpus luteum; in males, stimulates the secretion of testosterone in the testes. (p. 554)

luteinizing hormone-releasing hormone (LH-RH) See gonadotropin-releasing hormone. (p. 918)

lymph (limf) [L. *lympha*, clear spring water] Clear or yellowish fluid derived from interstitial fluid and found in lymph vessels. (p. 699)

lymph capillary Beginning of the lymphatic system of vessels; lined with flattened endothelium lacking a basement membrane. (p. 670)

lymph node Encapsulated mass of lymph tissue found among lymph vessels. (p. 702)

lymph nodule Small accumulation of lymph tissue lacking a distinct boundary. (p. 701)

lymph sinus Channels in a lymph node crossed by a reticulum of cells and fibers. (p. 703)

lymph vessel One of the system of vessels carrying lymph from the lymph capillaries to the veins. (p. 670)

lymphatic tissue Network of reticular

fibers, lymphocytes, and other cells. (p. 701)

lymphoblast Cell that matures into a lymphocyte. (p. 587)

lymphocyte (lim'fo-sīt) Nongranulocytic white blood cell formed in lymphoid tissue. (p. 594)

lymphokine Chemical produced by lymphocytes that activates macrophages, attracts neutrophils, and promotes inflammation. (p. 724)

lysis [Gr. *lysis*, a loosening] Process by which a cell swells and ruptures. (p. 69)

lysosome (li'so-sōm) [Gr. *lysis*, loosening + *soma*, body] Membrane-bound vesicle containing hydrolytic enzymes that function as intracellular digestive enzymes. (p. 79)

lysozyme (li'so-zīm) Enzyme that is destructive to the cell walls of certain bacteria; present in tears and some other fluids of the body. (p. 595)

M line Line in the center of the H zone made of delicate filaments that holds the myosin myofilaments in place in the sarcomere of muscle fibers. (p. 276)

macrophage [Gr. *makros*, large + *phagein*, to eat] Any large mononuclear phagocytic cell. (p. 127)

macula (mak'u-lah) [L., a spot] Sensory structures in the utricle and saccule, consisting of hair cells and a gelatinous mass embedded with otoliths. (p. 494)

macula densa Cells of the distal convoluted tubule located at the renal corpuscle and forming part of the juxtaglomerular apparatus. (p. 853)

macula lutea (mak'u-lah lu'te-ah) [L. *macula*, a spot; *luteus*, yellow] Small spot different in color from surrounding tissue; spot in the retina directly behind the lens in which densely packed cones are located. (p. 472)

macular degeneration Partial degeneration of the macula lutea common in elderly people. (p. 484)

major calyx Primary subdivision of the renal pelvis, usually two or three in number. (p. 850)

major histocompatibility complex Group of genes that control the production of major histocompatibility complex proteins, which are glycoproteins found on the surfaces of cells. The major histocompatibility proteins serve as self-markers for the immune system and are used by antigen-presenting cells to present antigens to lymphocytes. (p. 714)

malabsorption syndrome State characterized by diverse features such as diarrhea, weakness, and edema; caused by ineffective absorption of nutrients. (p. 811)

male pronucleus Nuclear material of the sperm cell after the ovum has been penetrated by the sperm cell. (p. 948)

malleus (mal'e-us) [L., hammer] Largest of the three auditory ossicles; attached to the tympanic membrane. (p. 486)

maltose Disaccharide consisting of two glucose molecules bound together. (p. 44)

mamillary bodies [L., breast or nipple shaped] Nipple-shaped structures at the base of the hypothalamus. (p. 390)

mamma, pl. mammae (mam'ah, mam'e) Breast. The organ of milk secretion; one of two hemispheric projections of variable size situated in the subcutaneous layer over the pectoralis major muscle on either side of the chest; it is rudimentary in the male. (p. 929)

mammary ligaments Cooper's ligaments. Well-developed ligaments that extend from the overlying skin to the fibrous stroma of mammary gland. (p. 930)

manubrium (mă-nu'bre-um) [L., handle] Part of a bone representing the handle, e.g., the manubrium of the sternum representing the handle of a sword. (p. 210)

marrow Soft substance in the center of bones. (p. 161)

mass movement Forcible peristaltic movement of short duration, occurring only three or four times a day,which move the contents of the large intestine. (p. 806)

mass number Equal to the number of protons plus the number of neutrons in each atom. (p. 31)

mast cell Connective tissue cell that contains basic staining granules; promotes inflammation. (p. 710)

mastication (mas'tĭ-ka-shun) [L. *mastico*, to chew] Process of chewing. (p. 775)

mastication reflex Repetitive cycle of relaxation and contraction of the muscles of mastication that results in chewing of food. (p. 793)

mastoid (mas'toyd) [Gr. *mastos*, breast] Resembling a breast. (p. 196)

mastoid air cells Spaces within the mastoid process of the temporal bone connected to the middle ear by ducts. (p. 486)

matrix Noncellular substance surrounding the cells of connective tissue. (p. 80)

maximal stimulus Stimulus resulting in a local potential just large enough to produce the maximum frequency of action potentials. (p. 265)

meatus (me-a'tus) [L., to go, pass] Passageway or tunnel. (p. 736)

mechanoreceptor (mek'ă-no-re-sep'tor) A sensory receptor that has the role of responding to mechanical pressures. Examples are pressure receptors in the carotid sinus or touch receptors in the skin. (p. 458)

meconium (me-ko'nĭ-um) [Gr. *mekon*, poppy] First intestinal discharges of the newborn infant, greenish in color and consisting of epithelial cells, mucus, and bile. (p. 970)

medial olfactory area Part of the olfactory cortex responsible for the visceral and emotional reactions to odors. (p. 462)

mediastinum (me'de-as-ti'num) [L., middle septum] The middle wall of the thorax. (p. 20)

medulla oblongata (ob'long-gah'tah) Inferior portion of the brainstem that connects the spinal cord to the brain and contains autonomic centers controlling such functions as heart rate, respiration, and swallowing. (p. 384)

medullary cavity Large, marrow-filled cavity in the diaphysis of a long bone. (p. 161)

medullary ray Extension of the medulla into the cortex, consisting of collecting ducts and loops of Henle. (p. 850)

megakaryoblast (meg'ă-kăr-ĭ-o-blast) [Gr.*mega* + *karyon*, nut (nucleus) + *kytos*, hollow vessel cell)] Cell that gives rise to thrombocytes. (p. 587)

meibomian cyst See chalazion. (p. 468)

meibomian gland (mi-bo'me-an) Sebaceous gland near the inner margins of the eyelid; secretes sebum that lubricates the eyelid and retains tears. (p. 468)

meiosis (mi-o'sis) [Gr., a lessening] Process of cell division that results in the formation of gametes. Consists of two divisions that result in one (female) or four (male) gametes, each of which contains one half the number of chromosomes in the parent cell. (p. 94)

Meissner's corpuscle (mīs'nerz) Named for George Meissner, German histologist, 1829-1905. See tactile corpuscle. (p. 461)

melanin [Gr. *melas*, black] Brown to black pigment responsible for skin and hair color. (p. 139)

melanocyte (mel' ă-no-sīt) [Gr. *melas*, black + *kytos*, cell] Cell found mainly in the stratum basale that produces the brown or black pigment melanin. (p. 136)

melanocyte-stimulating hormone (MSH) Peptide hormone secreted by the adenohypophysis; increases melanin production by melanocytes, making the skin darker in color. (p. 553)

melanosome (mel'ă-no-sōm) [Gr. *melas*, black + *soma*, body] Membranous organelle containing the pigment melanin. (p. 141)

melatonin Hormone (amino acid derivative) secreted by the pineal body; inhibits secretion of gonadotropin-releasing hormone from the hypothalamus. (p. 575)

membrane-bound receptor Receptor molecule such as a hormone receptor that is bound to the cell membrane of the target cell. (p. 536)

membranous labyrinth Membranous

structure within the inner ear consisting of the cochlea, vestibule, and semicircular canals. (p. 486)

membranous urethra Portion of the male urethra, approximately 1 cm in length, extending from the prostate gland to the beginning of the penile urethra. (p. 915)

memory cell Small lymphocytes that are derived from B cells or T cells and that rapidly respond to a subsequent exposure to the same antigen. (p. 724)

menarche (me-nar'ke) [Gr. *mensis*, month + *arche*, beginning] Establishment of menstrual function; the time of the first menstrual period or flow. (p. 930)

meninx, pl. meninges (me'ningks) [Gr., membrane] Connective tissue membranes surrounding the brain. (p. 200)

meniscus, pl. menisci (mĕ-nis'kus) [Gr. *menisko*, crescent] Structure that is crescent shaped. (p. 244)

menopause [Gr. *mensis*, month + *pausis*, cessation] Permanent cessation of the menstrual cycle. (p. 936)

menses [L. *mensis*, month] Periodic hemorrhage from the uterine mucous membrane, occurring at approximately 28-day intervals. (p. 930)

menstrual cycle Series of changes that occur in sexually mature, nonpregnant women and result in menses. Specifically refers to the uterine cycle but is often used to include both the uterine and ovarian cycles. (p. 930)

Merkel's disk (mer'kelz) Named for Friedrick Merkel, German anatomist, 1845-1919. See tactile disk. (p. 461)

merocrine (mĕr'o-krin) [Gr. *meros*, part + *krino*, to separate] Gland that secretes products with no loss of cellular material; an example is water-producing sweat glands; see apocrine and holocrine. (p. 113)

mesencephalon (mes'en-sef'ă-lon) [Gr. *mesos*, middle + *enkephalos*, brain] Midbrain in both the embryo and adult; consists of the cerebral peduncle and the corpora quadrigemini. (p. 382)

mesenchymal cell Cell found between the ectoderm and endoderm of the early embryo. Usually derived from mesoderm or neural crest. (p. 952)

mesenchyme [Gr. *mesos*, middle + *enkyma*, infusion] Loose packing embryonic tissue consisting of mesenchymal cells, usually stellate in form, supported in a loose, fluid, homogenous ground substance. (p. 952)

mesentery (mes'en-tĕr'e) [Gr. *mesos*, middle + *enteron*, intestine] Double layer of peritoneum extending from the abdominal wall to the abdominal viscera, conveying to it its vessels and nerves. (p. 790)

mesoderm (mes'o-derm) Middle of the three germ layers of an embryo. (p. 124)

mesonephric duct One of two ducts in the embryo draining the mesonephric tubules; in the male it becomes the ductus deferens; in the female it becomes vestigial. (p. 960)

mesonephros One of three excretory organs appearing during embryonic development; forms caudal to the pronephros as the pronephros disappears. It is well developed and is functional for a time before the establishment of the metanephros, which gives rise to the kidney; undergoes regression as an excretory organ, but its duct system is retained in the male as the efferent ductule and epididymis. (p. 962)

mesosalpinx (mez'o-sal'pinx) [Gr. *mesos*, middle + *salpinx*, trumpet] Part of the broad ligament supporting the uterine tube. (p. 926)

mesovarium (mes'o-va'rī-um) Short peritoneal fold connecting the ovary with the broad ligament of the uterus. (p. 922)

messenger RNA (mRNA) Type of RNA that moves out of the nucleus and into the cytoplasm where it is used as a template to determine the structure of proteins. (pp. 75, 87)

metabolic acidosis Acidosis due to disorders other than respiratory that increases hydrogen ion concentration or decreases base concentration in the blood. (p. 898)

metabolic alkalosis Alkalosis due to disorders other than respiratory that decreases hydrogen ion concentration or increases base concentration in the blood. (p. 898)

metabolic rate Total amount of energy produced and used by the body per unit of time (e.g., 38 kcal/m²/hr). (p. 839)

metabolism (mĕ-tab'o-lizm) [Gr. *metabole*, change] Sum of all the chemical reactions that take place in the body, consisting of anabolism and catabolism. Cellular metabolism refers specifically to the chemical reactions within cells. (p. 37)

metacarpal Relating to the fine bones of the hand between the carpus (wrist) and the phalanges. (p. 215)

metanephros Most caudally located of the three excretory organs appearing during embryonic development; becomes the permanent kidney of mammals. In mammalian embryos it is formed caudal to the mesonephros and develops later as the mesonephros undergoes regression. (p. 962)

metaphase Time during cell division when the chromosomes line up along the equator of the cell. (p. 94)

metarteriole One of the small peripheral blood vessels that contain scattered groups of smooth muscle fibers in their walls; located between the arterioles and the true capillaries. (p. 647)

metastasis (mĕ-tas'tă-sis) [Gr. *meta*, a removing + *stasis*, a placing] Movement of a disease from one location in the body to another. (p. 127)

metatarsal [Gr. *meta*, after + *tarsos*, sole of the foot] Distal bone of the foot. (p. 220)

metencephalon (met'en-sef'ă-lon) [Gr. *meta*, after + *enkephalos*, brain] Second-most posterior division of the embryonic brain; becomes the pons and cerebellum in the adult. (p. 384)

micelle (mī-sel') [L. *micella*, small morsel] Droplets of lipid surrounded by bile salts in the small intestine. (p. 809)

microfilament Small fibril forming bundles, sheets, or networks in the cytoplasm of cells; provides structure to the cytoplasm and mechanical support for microvilli and stereocilia. (p. 76)

microglia (mi-krŏg'le-ah) [Gr. *micro* + *glia*, glue] Small neuroglial cells that become phagocytic and mobile in response to inflammation; considered to be macrophages within the central nervous system. (p. 360)

microphage [Gr. *micro*, small + *phag*, eat] Small mononuclear cells (neutrophils, eosinophils, basophils) that ingest and destroy antigens. (p. 588)

microtubule Hollow tube composed of tubulin, measuring approximately 25 nm in diameter and usually several μm long. Helps provide support to the cytoplasm of the cell and is a component of certain cell organelles such as centrioles, spindle fibers, cilia, and flagella. (p. 76)

microvillus, pl. microvilli (mi'kro-vil'us, mi'kro-vil'i) Minute projection of the cell membrane that greatly increases the surface area. (p. 83)

micturition reflex (mik-tu-rish'un) Contraction of the bladder stimulated by stretching of the bladder wall; results in emptying of the bladder. (p. 873)

middle ear Air-filled space within the temporal bone; contains auditory ossicles; between the external and internal ear. (p. 484)

midsagittal Sagittal plane dividing the body or any part of the body into equal left and right halves.

milk letdown Expulsion of milk from the alveoli of the mammary glands; stimulated by oxytocin. (p. 973)

mineral Inorganic nutrient necessary for normal metabolic functions. (p. 825)

mineralocorticoid Steroid hormone (e.g., aldosterone) produced by the zona glomerulosa of the adrenal cortex; facilitates exchange of potassium for sodium in the distal renal tubule, causing sodium reabsorption and potassium and hydrogen ion secretion. (p. 564)

minor calyx Subdivision of a major calyx that receives the renal papillae; the total number varies from 7 to 13. (p. 850)

minute respiratory volume Product of tidal volume times the respiratory rate. (p. 748)

minute volume Amount of blood pumped by either the left or right ventricle each minute. (p. 630)

mitochondrion, pl. mitochondria (mit'o-kon'dri-on, mi'to-kon'dre-ah) [Gr. *mitos*, thread + *chondros*, granule] Small, spherical, rod-shaped or thin filamentous structure in the cytoplasm of cells that is a site of ATP production. (p. 80)

mitosis (mi-to'sis) [Gr., thread] Cell division resulting in two daughter cells with exactly the same number and type of chromosomes as the mother cell. (p. 94)

modiolus (mo'de-o'lus) [L., nave of a wheel] Central core of spongy bone about which turns the spiral canal of the cochlea. (p. 486)

molar Tricuspid tooth; the three posterior teeth of each dental arch. (p. 778)

mole [A.S. *mael* + L. *moles*, mass] Amount of substance that contains 6.0225×10^{23} (Avogadro's number) atoms. (p. 32)

mole [L. *moles*, mass] Aggregation of melanocytes in the epidermis or dermis of the skin. (p. 149)

molecular formula Symbolic representation of composition of a molecule; elements are indicated by their symbols, and the number of atoms of each element is denoted by a subscript. (p. 32)

molecule Two or more atoms of the same or different type joined by a chemical bond. (p. 32)

monoblast Cell that matures into a monocyte. (p. 587)

monocyte (mon'o-sīt) Nongranulocytic, relatively large mononuclear leukocyte normally found in lymph nodes, spleen, bone marrow, and loose connective tissue. (p. 594)

mononuclear phagocytic system Phagocytic cells, each with a single nucleus; derived from monocytes. (p. 710)

monosaccharide Simple sugar carbohydrate that cannot form any simpler sugar by hydrolysis. (p. 44)

mons pubis [L., mountain] Prominence caused by a pad of fatty tissue over the symphysis pubis in the female. (p. 929)

morula (mor'u-lah) [L. *morus*, mulberry] Mass of 12 or more cells resulting from the early cleavage divisions of the zygote. (p. 948)

motor neuron Neuron that innervates skeletal, smooth, or cardiac muscle fibers. (p. 279)

motor unit Single neuron and the muscle fibers it innervates. (p. 279)

mucin (mu'sin) Secretion containing mucopolysaccharides produced by mucous glandular cells. (p. 793)

mucosa (mu-ko'sah) [L. *mucosus*, mucous] Mucous membrane consisting of epithelium and lamina propria. In the digestive tract there is also a layer of smooth muscle. (p. 773)

mucous membrane Thin sheet consisting of epithelium and connective tissue (lamina propria) that lines cavities that open to the outside of the body; many contain mucous glands that secrete mucus. (p. 124)

mucous neck cell One of the mucous-secreting cells in the neck of a gastric gland. (p. 782)

mucus Viscous secretion produced by and covering mucous membranes; lubricates mucous membranes and traps foreign substances. (p. 124)

multiple motor unit summation Increased force of contraction of a muscle due to recruitment of motor units. (p. 286)

multiple wave summation Increased force of contraction of a muscle due to increased frequency of stimulation. (p. 288)

multipolar neuron One of three categories of neurons consisting of a neuron cell body, an axon, and two or more dendrites. (p. 123)

mumps [a lump or bump] Inflammation of the parotid gland. (p. 780)

murmur Soft sound heard on auscultation of the heart, lungs, or blood vessels. (p. 631)

muscarine Alkaloid compound that is found in certain mushrooms; binds to and activates muscarinic receptors. (p. 513)

muscarinic receptor (mus'kar-in'ik) Class of cholinergic receptor that is specifically activated by muscarine in addition to acetylcholine. (p. 511)

muscle [L. *mus;* also Gr. *mys*, little mouse or muscle] Use of the term mouse to denote a muscle refers to the movement of the muscle beneath the skin as it contracts, looking like a "little animal" or "little mouse" moving beneath the skin. Muscle tissue is one of the four major tissue types, characterized by its contractile abilities. (p. 272)

muscle fasciculus (fă-sik'u-lus) [L. *fascis*, bundle] Bundle of muscle fibers surrounded by perimysium. (p. 274)

muscle fiber Muscle cell. (p. 273)

muscle spindle Three to 10 specialized muscle fibers supplied by gamma motor neurons and wrapped in sensory nerve endings; detects stretch of the muscle and is involved in maintaining muscle tone. (p. 402)

muscle tone Relatively constant tension produced by a muscle for long periods of time as a result of asynchronous contraction of motor units. (p. 289)

muscle twitch Contraction of a whole muscle in response to a stimulus that causes an action potential in one or more muscle fibers. (p. 284)

muscular fatigue Fatigue due to a depletion of ATP within the muscle fibers. (p. 291)

muscular tissue One of the four primary tissue types; characterized by its contractile abilities. (p. 121)

muscularis [Modern L., muscular] Muscular coat of a hollow organ or tubular structure. (p. 773)

muscularis mucosa Thin layer of smooth muscle found in most parts of the digestive tube; located outside the lamina propria and adjacent to the submucosa. (p. 773)

musculi pectinati (pek'tĭ-nah'te) Prominent ridges of atrial myocardium located on the inner surface of much of the right atrium and both auricles. (p. 619)

myasthenia gravis (mi'as-the'ne-ah grā'vis) [Gr. *mys*, muscle + *asthenia*, weakness] Muscular weakness and atrophy caused by destruction of acetylcholine receptors in the myoneural junction. (p. 281)

myelencephalon (mi'el-en-sef'ă-lon) [Gr. *myelos*, medulla, marrow + *enkephalos*, brain] Most caudal portion of the embryonic brain; medulla oblongata. (p. 384)

myelin [Gr. *myelos*, related to medulla, which means bone marrow] Lipoprotein surrounding the axons of some nerve fibers. The term demonstrates the Greek concept of the nervous system, i.e., that it was interconnected to all the bone marrow (and was made of the same material) as one great radiator or cooling system. The Greeks believed that thought and emotion originated in the heart and liver, not in the brain. (p. 363)

myelin sheath Envelope surrounding most axons; formed by Schwann cell membranes being wrapped around the axon. (p. 265)

myelinated axon Nerve fiber having a myelin sheath. (p. 362)

myeloblast Immature cell from which the different granulocytes develop. (p. 587)

myenteric plexus (mi'en-tĕr'ik) Plexus of unmyelinated fibers and postganglionic autonomic cell bodies lying in the muscular coat of the esophagus, stomach, and intestines; communicates with the submucosal plexuses. (p. 773)

myoblast (mi'o-blast) [Gr. *mys*, muscle + *blastos*, germ] Primitive multinucleated cell with the potential of developing into a muscle fiber. (p. 273)

myocardium (mi'o-kar'dĭ-um) Middle layer of the heart, consisting of cardiac muscle. (p. 619)

myofibril Fine longitudinal fibril within a muscle fiber; composed of thick and thin myofilaments. (p. 274)

myofilament Extremely fine molecular thread helping to form the myofibrils of muscle; thick myofilaments are formed of myosin, and thin myofilaments are formed of actin. (p. 274)

myometrium (mi'o-me'trĭ-um) Muscular wall of the uterus; composed of smooth muscle. (p. 927)

myopia [Gr. *myo*, to shut + *ops*, eye] Condition due to an error in either refraction or elongation of the globe of the eye in which parallel rays are focused in front of the retina; nearsightedness. (p. 482)

myosin myofilament Thick myofilament of muscle fibrils; composed of myosin molecules. (p. 274)

nail [A.S., naegel] Several layers of dead epithelial cells containing hard keratin on the ends of the digits. (p. 147)

nail matrix Portion of the nail bed from which the nail is formed. (p. 147)

nasal cavity Cavity between the external nares and the pharynx. It is divided into two chambers by the nasal septum and is bounded inferiorly by the hard and soft palates. (p. 198)

nasal septum Bony partition that separates the nasal cavity into left and right parts; composed of the vomer, the perpendicular plate of the ethmoid, and hyaline cartilage. (p. 198)

nasolacrimal duct Duct that leads from the lacrimal sac to the nasal cavity. (p. 469)

nasopharynx (na'zo-făr'ingks) Part of the pharynx that lies above the soft palate; anteriorly it opens into the nasal cavity. (p. 737)

near point of vision Closest point from the eye at which an object can be held without appearing blurred. (p. 475)

neck (tooth) Slightly constricted part of a tooth, between the crown and the root. (p. 778)

necrosis (nĕ-kro'sis) [Gr. *nekrosis*, death] Death of cells or tissues. (p. 148)

necrotic (nĕ-krot'ik) Pertaining to or affected by necrosis. (p. 655)

negative feedback Mechanism by which any deviation from an ideal normal value is resisted or negated.

neoplasm (ne'o-plazm) [Gr. *neos*, new + *plasis*, thing formed] Abnormal growth; benign or malignant tumor. (p. 127)

nephron (nef'ron) [Gr. *nephros*, kidney] Functional unit of the kidney, consisting of the renal corpuscle, the proximal convoluted tubule, the loop of Henle, and the distal convoluted tubule. (p. 850)

nerve Bundle of nerve fibers and accompanying connective tissue located outside of the central nervous system. (p. 354)

nerve tract Bundles of parallel axons with their associated sheaths in the central nervous system. (p. 363)

nervous tissue One of the four major tissue types; characterized by its conductile abilities. (p. 123)

neural [Gr. *neuron*, nerve] Relating to any structure composed of nerve cells. (p. 124)

neural crest Edge of the neural plate as it rises to meet at the midline to form the neural tube. (p. 382)

neural crest cells Cells derived from the crests of the forming neural tube in the embryo; together with the mesoderm, form the mesenchyme of the embryo; give rise to part of the skull, the teeth, melanocytes, sensory neurons, and autonomic neurons. (p. 124)

neural plate Region of the dorsal surface of the embryo that is transformed into the neural tube and neural crest. (p. 382)

neural tube Tube formed from the neuroectoderm by the closure of the neural groove. The neural tube develops into the spinal cord and brain. (p. 382)

neuroectoderm That part of the ectoderm of an embryo giving rise to the brain and spinal cord. (p. 124)

neuroglia (nu-rog'le-ah) [Gr. *neuro*, nerve + *glia*, glue] Cells in the nervous system other than the neurons; includes astrocytes, ependymal cells, microglia, oligodendrocytes, satellite cells, and Schwann cells. (p. 123)

neurohormone Hormone secreted by a neuron. (p. 527)

neurohypophysis (nu'ru-hi-pof'ĭ-sis) Portion of the hypophysis derived from the brain; commonly called the posterior pituitary. Major secretions include antidiuretic hormone and oxytocin. (p. 390)

neuromodulator Substance that influences the sensitivity of neurons to neurotransmitters but neither strongly stimulates nor strongly inhibits neurons by itself. (p. 366)

neuromuscular junction Specialized synapse between a motor neuron and a muscle fiber. (p. 279)

neuron [Gr., nerve] Morphologic and functional unit of the nervous system, consisting of the nerve cell body, the dendrites, and the axon. (p. 123)

neuron cell body Enlarged portion of the neuron containing the nucleus and other organelles; also called nerve cell body. (p. 356)

neurotransmitter [Gr. *neuro* + L. *transmitto*, to send across] Any specific chemical agent released by a presynaptic cell upon excitation that crosses the synaptic cleft and stimulates or inhibits the postsynaptic cell. (p. 280)

neutral solution Solution such as pure water that has 10^{-7} moles of hydrogen ions per liter and an equal concentration of hydroxide ions; has a pH of 7. (p. 42)

neutron [L. *neuter*, neither] Electrically neutral particle in the nuclei of atoms (except hydrogen). (p. 31)

neutrophil (nu'tro-fil) [L. *neuter*, neither + Gr. *philos*, fond] Type of white blood cell; small phagocytic white blood cell with a lobed nucleus and small granules in the cytoplasm. (p. 127)

nicotine Alkaloid compound found in tobacco that binds to and activates nicotinic receptors. (p. 513)

nicotinic receptor (nik'o-tin'ik) Class of cholinergic receptor molecule that is specifically activated by nicotine and by acetylcholine. (p. 511)

night blindness Inability to see in dim light; due to a lack of visual pigment in the rods of the eyes. (p. 484)

nipple Projection at the apex of the mamma, on the surface of which the lactiferous ducts open; surrounded by a circular pigmented area, the areola. (p. 929)

Nissl bodies (nis'l) Areas in the neuron cell body containing rough endoplasmic reticulum. (p. 357)

nociceptor (no'sĭ-sep'tor) [L. *noceo*, to injure + *capio*, to take] A sensory receptor that detects painful or injurious stimuli. (p. 458)

node of Ranvier (ron've-a) Short interval in the myelin sheath of a nerve fiber between adjacent Schwann cells. (p. 362)

noncomitant strabismus Lack of parallelism of the visual axis in which the amount of deviation in the two eyes differs. (p. 483)

nonessential amino acid Amino acid that may be synthesized by the organism and is not required in the diet. (p. 837)

nonspecific resistance Immune system response that is the same with each exposure to an antigen; there is no ability for the system to remember a previous exposure to the antigen. (p. 706)

norepinephrine Neurotransmitter substance released from most of the postganglionic neurons of the sympathetic division; hormone released from the adrenal cortex that increases cardiac output and blood glucose levels. (p. 563)

nose or nasus (nāz'us) Visible structure that forms a prominent feature of the face; can also refer to the nasal cavities. (p. 736)

notochord [Gr. *notor*, back + *chords*, cord] Small rod of tissue lying ventral to the neural tube. A characteristic of all vertebrates, in humans it becomes the nucleus pulposus of the intervertebral disks. (p. 952)

nuchal (nu'kal) Pertaining to the nape of the neck. (p. 119)

nuclear envelope Double membrane structure surrounding and enclosing the nucleus. (p. 75)

nuclear pores Porelike openings in the nuclear envelope where the inner and outer membranes fuse. (p. 75)
nucleic acid Polymer of nucleotides, consisting of DNA and RNA, forms a family of substances that comprise the genetic material of cells and control protein synthesis. (p. 54)
nucleolus, pl. nucleoli (nu-kle'o-lus) Somewhat rounded, dense, well-defined nuclear body with no surrounding membrane; contains ribosomal RNA and protein. (p. 76)
nucleotide Basic building block of nucleic acids consisting of a sugar (either ribose or deoxyribose) and one of several types of organic bases. (p. 54)
nucleus (nu'kle-us) [L., inside of a thing] Cell organelle containing most of the genetic material of the cell; collection of nerve cell bodies within the central nervous system; center of an atom consisting of protons and neutrons. (p. 31)
nucleus pulposus (pul-po'sus) [L., central pulp] Soft central portion of the intervertebral disk. (p. 203)
null cell A type of lymphocyte that acts against tumor- and virus-infected cells; neither a B cell nor a T cell. (p. 595)
nutrient [L. *nutriens*, to nourish] Chemicals taken into the body that are used to produce energy, provide building blocks for new molecules, or function in other chemical reactions. (p. 820)
nutrition Process by which nutrients are obtained and used by the body. (p. 820)
nystagmus (nis-tag'mus) [Gr. *nystagmos*, a nodding] Rhythmical oscillation of the eyeballs. (p. 398)

occipital Inferior, posterior portion of the head. (p. 203)
oculi (ok'u-li) [L. *oculus*, eye] Related to the eye.
olfaction (ōl-fak'shun) [L. *olfaçtus*, smell] Sense of smell. (p. 462)
olfactory bulb Ganglion-like enlargement at the rostral end of the olfactory tract that lies over the cribriform plate; receives the olfactory nerves from the nasal cavity. (p. 462)
olfactory cortex Termination of the olfactory tract in the cerebral cortex within the lateral fissure of the cerebrum. (p. 398)
olfactory epithelium Epithelium of the olfactory recess containing olfactory receptors. (p. 462)
olfactory receptor (ol-fak'tōr-e) [L., to smell] Sensory receptors that respond to chemicals suspended in air. Located in the superior portion of the nasal cavities. Responsible for the sense of smell.
olfactory recess Extreme superior region of the nasal cavity. (p. 462)
olfactory tract Nerve tract that projects from the olfactory bulb to the olfactory cortex. (p. 462)
oligodendrocyte (o-lig'o-den'dro-sīt) Neuroglial cell that has cytoplasmic extensions that form myelin sheaths around axons in the central nervous system. (p. 361)
omentum (o-men'tum) Fold of peritoneum passing from the stomach to another organ. (p. 791)
oncology (ong-kol'o-je) [Gr. *onkos*, mass + *logos*, study] Study and treatment of neoplasms; cancerology. (p. 127)
oocyte (o'o-sīt) [Gr. *oon*, egg + *kytos*, a hollow (cell)] Immature ovum.
oogenesis (o-o-jen'ĕ-sis) Formation and development of a secondary oocyte or ovum. (p. 922)
oogonium (o'o-go'nī-um) [Gr. *oon*, egg + *gone*, generation] Primitive cell from which oocytes are derived by meiosis. (p. 922)
opposition Movement of the thumb and little finger toward each other; movement of the thumb toward any of the fingers. (p. 241)
opsin Protein portion of the rhodopsin molecule. A class of proteins that bind to retinal to form the visual pigments of the rods and cones of the eye. (p. 476)
opsonin (op'so-nin) [Gr. *opsonein*, to prepare food] Substance such as antibody or complement that enhances phagocytosis. (p. 721)
optic Related to vision.
optic chiasma (ki'az-mah) [Gr., two crossing lines; *chi*, the letter X] Point of crossing of the optic tracts. (p. 480)
optic disc Point at which axons of ganglion cells of the retina converge to form the optic nerve, which then penetrates through the fibrous tunic of the eye. (p. 472)
optic nerve Nerve carrying visual signals from the eye to the optic chiasm. (p. 436)
optic stalk Constricted proximal portion of the optic vesicle in the embryo; develops into the optic nerve. (p. 960)
optic tract Tract that extends from the optic chiasm to the lateral geniculate nucleus of the thalamus. (p. 480)
optic vesicle One of the paired evaginations from the walls of the embryonic forebrain from which the retina develops. (p. 960)
oral cavity The mouth; consists of the space surrounded by the lips, cheeks, teeth, and palate; limited posteriorly by the fauces. (p. 772)
orbit Eye socket; formed by seven skull bones that surround and protect the eye. (p. 196)
organ [Gr. *organon*, tool] Part of the body composed of two or more tissue types and exercising one or more specific functions. (p. 5)
organ of Corti Spiral organ; rests on the basilar membrane and supports the hair cells that detect sounds. (p. 488)
organ system Group of organs classified as a unit because of a common function or set of functions. (p. 6)
organelle [Gr. *organon*, tool] Specialized part of a cell serving one or more specific individual functions. (p. 5)
organism Any living thing considered as a whole, whether composed of one cell or many. (p. 6)
organogenesis Formation of organs during development. (p. 957)
orgasm [Gr. *orgao*, to swell, be excited] Climax of the sexual act, associated with a pleasurable sensation. (p. 920)
origin Less movable attachment point of a muscle; usually the medial or proximal end of a muscle associated with the limbs. (p. 306)
oris (or'us) [L., mouth] Mouth. (p. 313)
oropharynx (o'ro-făr'ingks) Portion of the pharynx that lies posterior to the oral cavity; it is continuous above with the nasopharynx and below with the laryngopharynx. (p. 738)
oscillating circuit Neuronal circuit arranged in a circular fashion that allows action potentials produced in the circuit to keep stimulating the neurons of the circuit. (p. 376)
osmolality Osmotic concentration of a solution; the number of moles of solute in 1 kilogram of water times the number of particles into which the solute dissociates. (p. 42)
osmoreceptor cell [Gr. *osmos*, impulsion] Receptor in the central nervous system that responds to changes in the osmotic pressure of the blood. (p. 891)
osmosis (os-mo'sis) [Gr. *osmos*, thrusting or an impulsion] Diffusion of solvent (water) through a membrane from a less concentrated solution to a more concentrated solution. (p. 67)
osmotic pressure Force required to prevent the movement of water across a selectively permeable membrane. (p. 68)
ossicle One of the three small bones of the middle ear: the malleus, incus, or stapes. (p. 186)
ossification (os'ĭ-fĭ-ka'shun) [L. *os*, bone + *facio*, to make] Bone formation. (p. 165)
osteoblast (os'te-o-blast) [Gr. *osteon*, bone + *blastos*, germ] Bone-forming cell. (p. 161)
osteoclast (os'te-o-klast) [Gr. *osteon*, bone + *klastos*, broken] Large multinucleated cell that absorbs bone. (p. 161)
osteocyte (os'te-o-sīt) [Gr. *osteon*, bone + *kytos*, cell] Mature bone cell surrounded by bone matrix. (p. 120)
osteomalacia (os'te-o-mă-la'shī-ah) Softening of bones due to calcium depletion. Adult rickets. (p. 171)
osteon (os'te-on) Single haversian canal, with its contents, and the associated concentric lamellae and osteo-

cytes surrounding it; also called a haversian system. (p. 164)

osteoporosis (os'te-o-po-ro'sis) [Gr. *osteon*, bone + *poros*, pore + *osis*, condition] Reduction in quantity of bone, resulting in porous bone. (p. 178)

ostium [L., door, entrance, mouth] Small opening, e.g., the opening of the uterine tube near the ovary or the opening of the uterus into the vagina. (p. 926)

otitis media [Gr. *ot* + *itis*, inflammation] Inflammation of the middle ear. (p. 499)

otolith Crystalline particles of calcium carbonate and protein embedded in the maculae. (p. 495)

otosclerosis [Gr. *oto*, ear + *sklerosis*, hardening] Formation of spongy bone about the stapes and oval window, resulting in progressively increasing deafness. (p. 499)

oval window Membranous structure to which the stapes attaches; transmits vibrations to the inner ear. (p. 486)

ovarian cycle Series of events that occur in a regular fashion in the ovaries of sexually mature, nonpregnant females; results in ovulation and the production of the hormones estrogen and progesterone. (p. 930)

ovarian epithelium (germinal epithelium) Peritoneal covering of the ovary. (p. 922)

ovarian follicle [L. *folliculus*, a small sac] Spherical cell aggregation in the ovary containing an oocyte. (p. 922)

ovarian ligament Bundle of fibers passing to the uterus from the ovary. (p. 922)

ovary One of two female reproductive glands located in the pelvic cavity; produces the secondary oocyte, estrogen, and progesterone. (p. 920)

oviduct See uterine tube. (p. 926)

ovulation Release of an ovum, or secondary oocyte, from the vesicular follicle. (p. 926)

ovum Female gamete or sex cell that has completed both meiotic divisions; contains the genetic material transmitted from the female. (p. 94)

oxidation Loss of one or more electrons from a molecule. (p. 57)

oxidation-reduction reaction Reaction in which one molecule is oxidized and another is reduced. (p. 57)

oxidative deamination Removal of the amine group of an amino acid to form a keto acid, ammonia, and NADH. (p. 837)

oxygen debt Oxygen necessary for the synthesis of the ATP required to remove lactic acid produced by anaerobic respiration. (p. 291)

oxygen-hemoglobin dissociation curve Graph describing the relationship between the percentage of hemoglobin saturated with oxygen and a range of oxygen partial pressures. (p. 754)

oxyhemoglobin Oxygenated hemoglobin. (p. 591)

oxyntic cell See parietal cell.

oxytocin (ok-sī-to'sin) Peptide hormone secreted by the neurohypophysis; increases uterine contraction and stimulates milk ejection from the mammary glands. (p. 551)

P wave First complex of the electrocardiogram representing depolarization of the atria. (p. 624)

P-Q interval Time elapsing between the beginning of the P wave and the beginning of the QRS complex in the electrocardiogram; also called P-R interval. (p. 626)

P-R interval See P-Q interval. (p. 626)

pacemaker Region in the right atrium that establishes the rate of cardiac contractions. (p. 622)

pacinian corpuscle (pă-sī'ne-an) Named for Filippe Pacini, Italian anatomist, 1812-1883. See lamellated corpuscle. (p. 461)

palate [L. *palatum*, palate] Roof of the mouth. (p. 776)

palatine tonsil One of two large oval masses of lymphoid tissue embedded in the lateral wall of the oral pharynx. (p. 701)

palpebra, pl. palpebrae (pal-pe'brah, pal-pe'bre) [L., eyelid] An eyelid. (p. 468)

palpebral conjunctiva Conjunctiva that covers the inner surface of the eyelids. (p. 468)

palpebral fissure Space between the upper and lower eyelids. (p. 468)

pancreas (pan'kre-us) [Gr. *pankreas*, the sweetbread] Abdominal gland that secretes pancreatic juice into the intestine and insulin and glucagon from the pancreatic islets into the bloodstream. (p. 567)

pancreatic duct Excretory duct of the pancreas that extends through the gland from tail to head where it empties into the duodenum at the greater duodenal papilla. (p. 783)

pancreatic islet Islets of Langerhans; cellular mass varying from a few to hundreds of cells lying in the interstitial tissue of the pancreas; composed of different cell types that comprise the endocrine portion of the pancreas and are the source of insulin and glucagon. (p. 788)

pancreatic juice [L. *jus*, broth] External secretion of the pancreas; clear, alkaline fluid containing several enzymes. (p. 804)

pancreatic somatostatin Somatostatin released from the pancreas. (p. 514)

papilla (pă-pil'ah) [L., nipple] A small nipplelike process. Projection of the dermis, containing blood vessels and nerves, into the epidermis. Projections on the surface of the tongue. (p. 136)

papillary muscle (pap'ĭ-lĕr'e) Nipple-like conical projection of myocardium within the ventricle; the chordae tendineae are attached to the apex of the papillary muscle. (p. 618)

parafollicular cell (păr'ah-fŏ-lik'u-lar) Endocrine cell scattered throughout the thyroid gland; secretes the hormone calcitonin. (p. 555)

parahormone Substance secreted by a wide variety of tissues; usually has a localized effect. (p. 528)

paramesonephric duct One of two embryonic tubes extending along the mesonephros and emptying into the cloaca; in the female the duct forms the uterine tube, the uterus, and part of the vagina; in the male it degenerates. (p. 964)

paranasal sinus Air-filled cavity within a skull bone that connects to the nasal cavity; located in the frontal, maxillary, sphenoid, and ethmoid bones. (p. 200)

parasagittal Sagittal plane located to one side of the midline. (p. 19)

parasympathetic Subdivision of the autonomic nervous system; characterized by having the cell bodies of its preganglionic neurons located in the brainstem and the sacral region of the spinal cord (craniosacral division); usually involved in activating vegetative functions such as digestion, defecation, and urination. (p. 436)

parathyroid gland One of four glandular masses imbedded in the posterior surface of the thyroid gland; secretes parathyroid hormone. (p. 559)

parathyroid hormone Peptide hormone produced by the parathyroid gland; increases bone breakdown and blood calcium levels. (p. 171)

parietal (pa-ri'ĕ-tal) [L. *paries*, wall] Relating to the wall of any cavity. (p. 25)

parietal cell Gastric gland cell that secretes hydrochloric acid. (p. 782)

parietal pericardium Serous membrane lining the fibrous portion of the pericardial sac. (p. 25)

parietal peritoneum Layer of peritoneum lining the abdominal walls. (p. 25)

parietal pleura Serous membrane that lines the different parts of the wall of the pleural cavity. (p. 25)

parotid gland (pă-rot'id) Largest of the salivary glands; situated anterior to each ear. (p. 779)

partial pressure Pressure exerted by a single gas in a mixture of gases. (p. 749)

partial-thickness burn Burn that damages only the epidermis (first degree burn) or the epidermis and part of the dermis (second degree burn). (p. 142)

parturition [L. *parturio*, to be in labor] Childbirth. (p. 966)

passive tension Tension applied to a load by a muscle without contracting; produced when an external force stretches the muscle. (p. 290)

patella [L. *patina*, shallow disk] Kneecap. (p. 219)

pectoral (pek'to-ral) [L. *pectoralis*, breast bone] Relating to the chest.

pectoral girdle (pek'to-ral) Site of attachment of the upper limb to the trunk; consists of the scapula and the clavicle. (p. 19)

pedicle (ped'ĭ-kl) [L. *pes*, feet] Stalk or base of a structure, e.g., the pedicle of the vertebral arch. (p. 206)

pedigree [Old Fr. *pie de grue*, foot of crane] Ancestral line of descent, especially as diagrammed on a chart. (p. 976)

peduncle (pĕ-dun'kl) [L. *pedunculus*, foot] Stalk or stem connecting portions of the brainstem to the cerebrum or cerebellum. (p. 384)

pelvic (pel'vik) Relating to the pelvis.

pelvic brim Imaginary plane passing from the sacral promontory to the pubic crest. (p. 218)

pelvic girdle Site of attachment of the lower limb to the trunk; ring of bone formed by the sacrum and the coxae. (p. 19)

pelvic inlet Superior opening of the true pelvis. (p. 218)

pelvic nerve Parasympathetic nerve that arises from the sacral region of the spinal cord. (p. 510)

pelvic outlet Inferior opening of the true pelvis. (p. 218)

pelvis (pel'vis) [L., basin] Any basin-shaped structure; cup-shaped ring of bone at the lower end of the trunk, formed from the ossa coxae, sacrum, and coccyx. (p. 19)

pennate (pen'āt) [L. *penna*, feather] Muscles with fasciculi arranged like the barbs of a feather along a common tendon. (p. 306)

pepsin [Gr. *pepsis*, digestion] Principal digestive enzyme of the gastric juice, formed from pepsinogen; digests proteins into smaller peptide chains. (p. 796)

pepsinogen (pep-sin'o-jen) [pepsin + Gr. *gen*, producing] Proenzyme formed and secreted by the chief cells of the gastric mucosa; the acidity of the gastric juice and pepsin itself converts pepsinogen into pepsin. (p. 796)

peptidase An enzyme capable of hydrolyzing one of the peptide links of a peptide. (p. 801)

peptide bond Chemical bond between amino acids. (p. 49)

perforating canal Canal containing blood vessels and nerves and running through bone perpendicular to the haversian canals. (p. 164)

perforating fiber Connective tissue fiber from a tendon or ligament that penetrates the periosteum of a bone and anchors the tendon, ligament, or periosteum to the bone. (p. 162)

periarterial sheath Dense accumulations of lymphocytes (white pulp) surrounding arteries within the spleen. (p. 703)

pericapillary cell One of the slender connective tissue cells in close relationship to the outside of the capillary wall; relatively undifferentiated and may become a fibroblast, macrophage, or smooth muscle cell. (p. 113)

pericardial (pĕr-ĭ-kar'de-al) Around the heart.

pericardial cavity Space within the mediastinum in which the heart is located. (p. 25)

pericardial fluid Viscous fluid contained within the pericardial cavity between the visceral and parietal pericardium; functions as a lubricant. (p. 25)

pericardium (pĕr'ĭ-kar'dĭ-um) [Gr. *pericardion*, the membrane around the heart] Membrane covering the heart. (p. 612)

perichondrium (pĕr-e-kon'dre-um) [Gr. *peri*, around + *chondros*, cartilage] Double-layered connective tissue sheath surrounding cartilage. (p. 158)

perilymph [Gr. *peri*, around + L. *lympha*, a clear fluid (lymph)] Fluid contained within the bony labyrinth of the inner ear. (p. 486)

perimetrium (pĕr'ĭ-me'trĭ-um) Outer serous coat of the uterus. (p. 927)

perimysium (pĕr'ĭ-mĭz-ĭ-um) [Gr. *peri*, around + *mys*, muscle] Fibrous sheath enveloping a bundle of skeletal muscle fibers (muscle fascicle). (p. 279)

perineum (pĕr'ĭ-ne'um) Area inferior to the pelvic diaphragm between the thighs; extends from the coccyx to the pubis. (p. 911)

perineurium (pĕr'ĭ-nu're-um) [L. *peri*, around + Gr. *neuron*, nerve] Connective tissue sheath surrounding a nerve fascicle. (p. 363)

periodontal ligament (pĕr'e-o-don'tal) Connective tissue that surrounds the tooth root and attaches it to its bony socket. (p. 229)

periosteum (per'e-os'te-um) [Gr. *peri*, around + *osteon*, bone] Thick, double-layered connective tissue sheath covering the entire surface of a bone except the articular surface, which is covered with cartilage. (p. 161)

peripheral circulatory system All vessels of the circulatory system outside of the heart.

peripheral nervous system (PNS) Major subdivision of the nervous system consisting of nerves and ganglia. (p. 354)

peripheral resistance Resistance to blood flow in all the blood vessels. (p. 683)

peristaltic wave (pĕr'ĭ-stal'tik) Contraction in a tube such as the intestine characterized by a wave of contraction in smooth muscle preceded by a wave of relaxation that moves along the tube. (p. 795)

peritendineum (pĕr-ĭ-ten-din'e-um) Fibrous connective tissue surrounding bundles of collagen fibers in a tendon. (p. 119)

peritoneum (pĕr'ĭ-to-ne'um) [L. or Gr. to stretch over] Serous membrane that lines the peritoneal cavity and covers most of the viscera contained therein. (p. 790)

peritubular capillary The capillary network located in the cortex of the kidney; associated with the distal and proximal convoluted tubules. (p. 857)

permanent tooth One of the 32 teeth belonging to the second or permanent dentition. (p. 778)

pernicious anemia (per-nish'us) Anemia resulting from inadequate intake or absorption of vitamin B_{12}. (p. 605)

peroneal (pĕr'o-ne'al) [Gr. *perone*, fibula] Associated with the fibula. (p. 669)

peroxisome (per-oks'ĭ-som) Membrane-bound body similar to a lysosome in appearance but often smaller and irregular in shape; contains enzymes that either decompose or synthesize hydrogen peroxide. (p. 80)

Peyer's patch Lymph nodule found in the lower half of the small intestine and the appendix. (p. 701)

pH Symbol for logarithm of the reciprocal of the hydrogen ion concentration; a solution with pH 7.0 is neutral, with pH $>$7.0 is alkaline (basic), and with pH $<$7.0 is acidic. (p. 894)

phagocyte (fag'o-sīt) Cell possessing the property of ingesting bacteria, foreign particles, and other cells. (p. 709)

phagocytosis (fag'o-si-to'sis) [Gr. *phagein*, to eat + *kytos*, cell + *osis*, condition] Process of ingestion by cells of solid substances such as other cells, bacteria, bits of necrosed tissue, and foreign particles. (p. 72)

phalange (fa-lan'jē) [Gr. *phalanx*, line of soldiers] Bone of the fingers or toes. (p. 215)

pharyngeal pouch Paired evagination of embryonic pharyngeal endoderm between the brachial arches that gives rise to the thymus, thyroid gland, tonsils, and parathyroid glands. (p. 955)

pharyngeal tonsil One of two collections of aggregated lymphoid nodules on the posterior wall of the nasopharynx. (p. 701)

pharynx (făr'ingks) [Gr. *pharynx*, throat, the joint opening of the gullet and windpipe] Upper expanded portion of the digestive tube between the esophagus below and the oral and nasal cavities above and in front. (p. 737)

phenotype [Gr. *phaino*, to display, show forth + *typos*, model] Characteristic observed in an individual due to expression of his genotype. (p. 976)

phlebitis (flĕ-bi'tis) [Gr. *phelps*, vein + *itis*, inflammation] Inflammation of a vein. (p. 650)

phosphodiesterase (fos'fo-di-es'ter-ās)

Enzymes that split phosphodiester bonds, e.g., that break down cyclic AMP to AMP. (p. 536)

phospholipid Lipid with phosphorus, resulting in a molecule with a polar end and a nonpolar end; main component of the lipid bilayer. (p. 46)

phosphorylation Addition of phosphate to an organic compound. (p. 830)

photoreceptor (fo'to-re-sep'tor) [L. photo + L. *ceptus*, to receive] A sensory receptor that is sensitive to light. Examples are rods and cones of the retina. (p. 458)

phrenic nerve (fren'ik) Nerve derived from spinal nerves C3 to C5; supplies the diaphragm. (p. 441)

physiological contracture Temporary inability of a muscle to either contract or relax because of a depletion of ATP so that active transport of calcium ions into the sarcoplasmic reticulum cannot occur. (p. 291)

physiological dead air space Sum of anatomic dead air space plus the volume of any nonfunctional alveoli. (p. 749)

physiological shunt Deoxygenated blood from the alveoli plus deoxygenated blood from the bronchi and bronchioles. (p. 752)

physiology [Gr. *physis*, nature + *logos*, study] Scientific discipline that deals with the vital processes or functions of living things. (p. 4)

pia mater (pe'ah) [L., tender mother] Delicate membrane forming the inner covering of the brain and spinal cord. (p. 419)

pigmented retina Pigmented portion of the retina. (p. 472)

pineal body (pi'ne-al) [L. *pineus*, relating to pine trees] A small pine cone–shaped structure that projects from the epiphysis of the diencephalon; produces melatonin. (p. 388)

pinkeye Acute contagious conjunctivitis; inflammation of the conjunctiva. (p. 468)

pinna (pin'ah) [L. *pinna* or *penna*, feather, in plural wing] See auricle. (p. 484)

pinocytosis (pin'o-si-to'sis) [Gr. *pineo*, to drink + *kytos*, cell + *osis*, condition] Cell drinking; uptake of liquid by a cell. (p. 72)

pituitary gland (pit-u'ĭ-tĕr-e) See hypophysis. (p. 546)

plane (plān) [L. *planus*, flat] A flat surface. An imaginary surface formed by extension through any axis or two points. Examples include a midsagittal plane, a coronal plane, and a transverse plane. (p. 18)

plasma [Gr., something formed] Fluid portion of blood. (p. 585)

plasma cell Cell derived from B cells; produces antibodies. (p. 722)

plasma clearance Volume of plasma per minute from which a substance can be completely removed by the kidneys. (p. 871)

plasma membrane (plaz'mah) Cell membrane; outermost component of the cell, surrounding and binding the rest of the cell contents. (p. 63)

plasmin (plaz'min) Enzyme derived from plasminogen; dissolves clots by converting fibrin into soluble products. (p. 599)

plateau phase of action potential Prolongation of the depolarization phase of a cardiac muscle cell membrane; results in a prolonged refractory period. (p. 623)

platelet Irregularly shaped disk found in blood; contains granules in the central part and clear protoplasm peripherally but has no definite nucleus. (p. 586)

platelet plug Accumulation of platelets that stick to each other and to connective tissue; functions to prevent blood loss from damaged blood vessels. (p. 596)

pleura (ploor'ah) Serous membrane covering the lungs and lining the thoracic wall, diaphragm, and mediastinum. (p. 742)

pleural cavity (ploor'al) Potential space between the parietal and visceral layers of the pleura. (p. 25)

pleural fluid Serous fluid found in the pleural cavity; helps to reduce friction when the pleural membranes rub together. (p. 742)

plexus (plek'sus) [L., a braid] Intertwining of nerves or blood vessels. (p. 440)

plicae circulares (plī'se) (circular folds) Numerous folds of the mucous membrane of the small intestine. (p. 783)

pluripotent (ploo-rĭ-po'tent) [L. *pluris*, more + *potentia*, power] In development the term refers to a cell or group of cells that have not yet become fixed or determined as to what specific tissues they are going to become. (p. 948)

pneumotaxic center (nu'mo-tak'sik) Group of neurons in the pons that have an inhibitory effect on the inspiratory center. (p. 758)

pneumothorax Equalization of pressure in the pleural cavity and atmospheric air. (p. 748)

podocyte (pod'o-sīt) [Gr. *pous*, *podos*, foot + *kytos*, a hollow (cell)] Epithelial cell of Bowman's capsule attached to the outer surface of the glomerular capillary basement membrane by cytoplasmic foot processes. (p. 853)

Poiseuille's law (puah-zuh'yez) The volume of a fluid passing per unit of time through a tube is directly proportional to the pressure difference between its ends and to the fourth power of the internal radius of the tube and inversely proportional to the tube's length and the viscosity of the fluid. (p. 674)

polar body One of the two small cells formed during oogenesis because of unequal division of the cytoplasm. (p. 923)

polar covalent bond Covalent bond in which atoms do not share their electrons equally. (p. 36)

pollicis (pol-ĭ-sez) [L. *pollex*, thumb] Associated with the thumb.

polycythemia (pol'ĭ-si-the'me-ah) Increase in red blood cell number above the normal. (p. 604)

polygenic Relating to a hereditary disease or normal characteristic controlled by interaction of genes at more than one locus. (p. 976)

polysaccharide (pol-ĭ-sak'ărīd) Carbohydrate containing a large number of monosaccharide molecules. (p. 46)

polyunsaturated Fatty acid that contains two or more double covalent bonds between its carbon atoms. (p. 46)

pons [L., bridge] That portion of the brainstem between the medulla and midbrain. (p. 387)

popliteal (pop'lĭ-te-al) [L., ham] Posterior region of the knee. (p. 344)

porta [L., gate] Fissure on the inferior surface of the liver where the portal vein, hepatic artery, hepatic nerve plexus, hepatic ducts, and lymphatic vessels enter or exit the liver. (p. 787)

portal system (pōr'tal) System of vessels in which blood, after passing through one capillary bed, is conveyed through a second capillary network. (p. 668)

portal triad Branches of the portal vein, hepatic artery, and hepatic duct bound together in the connective tissue that divides the liver into lobules. (p. 787)

positive feedback Mechanism by which any deviation from a normal value is made greater. (p. 15)

postabsorptive state Following the absorptive state; blood glucose levels are maintained because of conversion of other molecules to glucose. (p. 839)

posterior That which follows. In humans, toward the back. (p. 16)

posterior chamber of the eye Chamber of the eye between the iris and the lens. (p. 472)

posterior interventricular sulcus Groove on the diaphragmatic surface of the heart, marking the location of the septum between the two ventricles. (p. 614)

posterior pituitary See neurohypophysis. (p. 390)

postganglionic neuron Autonomic neuron that has its cell body located within an autonomic ganglion and sends its axon to an effector organ. (p. 507)

postmenopausal period Relating to the period following the menopause. (p. 936)

postovulatory age Age of the developing fetus based on the assumption that fertilization occurs 14 days after the last menstrual period before the pregnancy. (p. 948)

postsynaptic Refers to the membrane of a nerve, muscle, or gland that is in close association with a presynaptic terminal. The postsynaptic membrane has receptor molecules within it that bind to neurotransmitter molecules. (p. 219)

potential difference Difference in electrical potential, measured as the charge difference across the cell membrane. (p. 255)

potential energy [Gr. *en*, in + *ergon*, work] Energy in a chemical bond that is not being exerted or used to do work. (p. 39)

precapillary sphincter Smooth muscle sphincter that regulates blood flow through a capillary. (p. 647)

preganglionic neuron Autonomic neuron that has its cell body located within the central nervous system and sends its axon through a nerve to an autonomic ganglion where it synapses with postganglionic neurons. (p. 507)

premenstrual syndrome (PMS) In some women of reproductive age, the regular monthly experience of physiological and emotional distress, usually during the several days preceding menses, typically involving fatigue, edema, irritability, tension, anxiety, and depression. (p. 933)

premolar Bicuspid tooth. (p. 778)

prepuce (pre′pus) In males, the free fold of skin that more-or-less completely covers the glans penis; the foreskin. In females, the external fold of the labia minora that covers the clitoris. (p. 915)

presbyopia [Gr. *presby* + *ops*, eye] Physiologic decrease in accommodation power of the eyes in advancing age as a result of the lens' becoming less elastic. (p. 482)

pressoreceptor See baroreceptor. (p. 686)

presynaptic Refers to the nerve terminal that contains neurotransmitter vesicles. Neurotransmitter substance is released from the presynaptic terminal, crosses the synaptic cleft, and affects the postsynaptic terminal. (p. 369)

presynaptic terminal Enlarged axon terminal or terminal bouton. (p. 279)

primary bronchus (brong′kus) One of two tubes arising at the inferior end of the trachea; each primary bronchus extends into one of the lungs. (p. 740)

primary follicle Ovarian follicle before the appearance of an antrum; contains the primary oocyte. (p. 923)

primary oocyte (o′o-sīt) Oocyte before completion of the first meiotic division (stops at prophase I). (p. 922)

primary palate In the early embryo, gives rise to the upper jaw and lips. (p. 956)

primary response Immune response that occurs as a result of the first exposure to an antigen. (p. 722)

primary spermatocyte (sper′mă-to-sīt) Spermatocyte arising by a growth phase from a spermatogonium; gives rise to secondary spermatocytes after the first meiotic division. (p. 913)

prime mover Muscle that plays a major role in accomplishing a movement. (p. 306)

primitive streak Ectodermal ridge in the midline of the embryonic disk from which arises the mesoderm by inward and then lateral migration of cells. (p. 952)

primordial germ cell Most primitive undifferentiated sex cell, found initially outside the gonad on the surface of the yolk sac. (p. 964)

process Projection on a bone. (p. 184)

processus vaginalis Peritoneal outpocketing in the embryonic lower anterior abdominal wall that traverses the inguinal canal; in the male it forms the tunica vaginalis testis and normally loses its connection with the peritoneal cavity. (p. 911)

product Substance produced in a chemical reaction. (p. 37)

progeria (pro-jer′ĭ-ah) [Gr. *pro*, before + *ge* + *amras*, old age] Severe retardation of growth after the first year accompanied by a senile appearance and death at an early age. (p. 975)

progesterone Steroid hormone secreted by the ovaries; necessary for uterine and mammary gland development and function. (p. 933)

prolactin Hormone of the adenohypophysis that stimulates the production of milk. (p. 554)

prolactin-inhibiting hormone (PIH) Neurohormone released from the hypothalamus that inhibits prolactin release from the adenohypophysis. (p. 554)

prolactin-releasing hormone (PRH) Neurohormone released from the hypothalamus that stimulates prolactin release from the adenohypophysis. (p. 554)

proliferative phase See follicular phase. (p. 930)

pronation (pro-na′shun) [L. *pronare*, to bend forward] Rotation of the forearm so that the anterior surface is down (prone). (p. 241)

prone Lying face down. (p. 18)

pronephros (pro-nef′ros) In the embryos of higher vertebrates, a series of tubules emptying into the celomic cavity. It is a temporary structure in the human embryo, followed by the mesonephros and still later by the metanephros, which gives rise to the kidney. (p. 962)

prophase First stage in cell division when chromatin strands condense to form chromosomes. (p. 94)

proprioception (pro′pre-o-sep′shun) [L. *proprius*, one's own + *capio*, to take] Information about the position of the body and its various parts. (p. 412)

proprioceptor (pro′pre-o-sep′tor) Sensory receptor associated with joints and tendons. (p. 459)

prostaglandin (pros′tă-glan′din) Class of physiologically active substances present in many tissues; among effects are those of vasodilation, stimulation and contraction of uterine smooth muscle, and promotion of inflammation and pain. (p. 49)

prostate gland (pros′tāt) [Gr. *prostates*, one standing before] Gland that surrounds the beginning of the urethra in the male. The secretion of the glands is a milky fluid that is discharged by 20 to 30 excretory ducts into the prostatic urethra as part of the semen. (p. 917)

prostatic urethra Part of the male urethra, approximately 2.5 cm in length, that passes through the prostate gland. (p. 915)

protease (pro′te-ās) Enzyme that breaks down proteins. (p. 54)

protein [Gr. *proteios*, primary] Macromolecule consisting of long sequences of amino acids linked together by peptide bonds. (p. 49)

proteoglycan (pro′te-o-gli′kan) Macromolecule consisting of numerous polysaccharides attached to a common protein core. (p. 113)

prothrombin Glycoprotein present in blood that, in the presence of prothrombin activator, is converted to thrombin. (p. 596)

proton [Gr. *protos*, first] Positively charged particle in the nuclei of atoms. (p. 31)

protoplasm (pro′to-plazm) [Gr. *proto*, first + *plasma*, a thing formed] Living matter of which cells are formed. (p. 65)

protraction [L. *protractus*, to draw forth] Movement forward or in the anterior direction. (p. 241)

provitamin Substance that may be converted into a vitamin. (p. 823)

proximal convoluted tubule Part of the nephron that extends from the glomerulus to the descending limb of the loop of Henle. (p. 854)

pseudostratified epithelium Epithelium consisting of a single layer of cells but having the appearance of multiple layers. (p. 104)

pseudounipolar neuron See unipolar neuron.

psoriasis (so-ri′ă-sis) [Gr. *psora*, the itch] Condition of unknown origin that produces large silvery scales over reddish bumps in the skin that bleed readily when scratched. (p. 148)

psychological fatigue Fatigue caused by the central nervous system. (p. 290)

pterygoid (tĕr′ĭ-goyd) [Gr. *pteryx*, wing] Wing-shaped structure.

ptosis (to′sis) [G. *ptosis*, a falling] Falling down of an organ, e.g., drooping of the upper eyelid. (p. 312)

puberty [L. *pubertas*, grown up] Series of events that transform a child into a sexually mature adult; involves an increase in the secretion of GnRH. (p. 919)

pudendal cleft (pu-den'dal) Cleft between the labia majora. (p. 929)

pudendum (pu-den'dum) See vulva. (p. 927)

pulmonary artery (pul'mo-nĕr-e) One of the arteries that extend from the pulmonary trunk to the right or left lungs. (p. 651)

pulmonary capacity Sum of two or more pulmonary volumes. (p. 747)

pulmonary trunk Large elastic artery that carries blood from the right ventricle of the heart to the right and left pulmonary arteries. (p. 614)

pulmonary vein One of the veins that carry blood from the lungs to the left atrium of the heart. (p. 614)

pulmonary volume Tidal volume, inspiratory reserve volume, expiratory reserve volume, or residual volume. (p. 747)

pulp (tooth) [L. *pulpa*, flesh] The soft tissue within the pulp cavity, consisting of connective tissue containing blood vessels, nerves, and lymphatics. (p. 778)

pulp cavity (tooth) Central hollow portion of a tooth consisting of the crown cavity and the root canal. (p. 778)

pulse (puls) Rhythmical dilation of an artery produced by the increased volume of blood ejected into the arteries by contraction of the left ventricle. (p. 657)

pulse pressure Difference between systolic and diastolic pressure. (p. 677)

punctum (punk'tum) [L. *pungo*, a prick, point] Small opening of the lacrimal canaliculus. (p. 468)

pupil Circular opening in the iris through which light enters the eye. (p. 472)

Purkinje fiber (pur-kin'je) Modified cardiac muscle cells found beneath the endocardium of the ventricles. Specialized to conduct action potentials. (p. 622)

pus Fluid product of inflammation; contains leukocytes, the debris of dead cells, and tissue elements liquefied by enzymes. (p. 127)

pyloric opening (pi-lōr'ik) Opening between the stomach and the superior part of the duodenum. (p. 781)

pyloric sphincter Thickening of the circular layer of the gastric musculature encircling the junction between the stomach and duodenum. (p. 781)

pyrogen (pi'ro-jen) Chemical released by microorganisms, neutrophils, monocytes, and other cells that stimulates fever production by acting on the hypothalamus. (p. 711)

pyruvic acid (pi-ru'vik) End product of glycolysis. Two three-carbon pyruvic acid molecules are produced from glucose as a result of glycolysis. (p. 830)

QRS complex Principle deflection in the electrocardiogram, representing ventricular depolarization, (p. 624)

Q-T interval Time elapsing from the beginning of the QRS complex to the end of the T wave, representing the total duration of electrical activity of the ventricles. (p. 626)

radial pulse Pulse detected in the radial artery. (p. 677)

radiation [L. *radius*, ray, beam] Radiant heat. (p. 844)

radioactive isotope Isotope with a nuclear composition that is unstable from which subatomic particles and electromagnetic waves are emitted. (p. 34)

ramus, pl. rami (ra'mus, ra'mi) [L., branch] One of the primary subdivisions of a nerve or blood vessel. The part of a bone that forms an angle with the main body of the bone. (p. 440)

raphe (ra'fe) [Gr., *rhaphe*, suture, seam] Central line running over the scrotum from the anus to the root of the penis. (p. 909)

Raynaud's disease Disease characterized by spasmodic contraction of blood vessels in the periphery, especially the digits; may be due to exaggerated sensitivity of blood vessels to sympathetic innervation. (p. 519)

reactant Substance taking part in a chemical reaction. (p. 37)

reactive hyperemia [Gr. *hyper* + *haima*, blood] Presence of an increased amount of blood flow following the arrest and subsequent restoration of the blood supply to a part. (p. 673)

recessive In genetics, a gene that may not be expressed because of suppression by a contrasting dominant gene. (p. 976)

rectum (rek'tum) [L. *rectus*, straight] Portion of the digestive tract that extends from the sigmoid colon to the anal canal. (p. 772)

red pulp [L. *pulpa*, flesh] Reddish-brown substance of the spleen consisting of venous sinuses and the tissues intervening between them called pulp cords. (p. 703)

reduced hemoglobin Form of hemoglobin in red blood cells after the oxygen of oxyhemoglobin is released in the tissues. (p. 591)

reduction Gain of one or more electrons by a molecule. (p. 57)

reflex Automatic response to a stimulus that occurs without conscious thought; produced by a reflex arc. (p. 374)

reflex arc Smallest portion of the nervous system that is capable of receiving a stimulus and producing a response; composed of a receptor, afferent neuron, association neuron, efferent neuron, and effector organ. (p. 374)

refraction Bending of a light ray when it passes from one medium into another of different density. (p. 474)

refractory period [Gr. *periodos*, a way around, a cycle] Period following effective stimulation during which excitable tissue such as heart muscle fails to respond to a stimulus of threshold intensity. (p. 624)

regulatory gene Gene involved with controlling the activity of structural genes. (p. 976)

relative refractory period Portion of the action potential following the absolute refractory period during which another action potential can be produced with a greater-than-threshold stimulus strength. (p. 263)

relaxation phase Phase of muscle contraction following the contraction phase; the time from maximal tension production until tension decreases to its resting level. (p. 284)

renal artery Originates from the aorta and delivers blood to the kidney. (p. 856)

renal blood flow rate Volume at which blood flows through the kidneys per minute; an average of approximately 1200 ml per minute. (p. 858)

renal column Cortical substance separating the renal pyramids. (p. 850)

renal corpuscle Glomerulus and Bowman's capsule that encloses it. (p. 853)

renal cortex Outer part of the kidney, consisting of the renal corpuscle and the proximal and distal convoluted tubules; also the renal columns, which are extensions inward between the pyramids. (p. 850)

renal fascia Connective tissue surrounding the kidney that forms a sheath or capsule for the organ. (p. 850)

renal fat pad Fat layer that surrounds the kidney and functions as a shock-absorbing material. (p. 850)

renal fraction Portion of the cardiac output that flows through the kidneys; averages 21%. (p. 858)

renal medulla Inner portion of the kidney, consisting of the renal pyramids and the medullary rays that extend into the cortex. (p. 850)

renal papillum Apex of a renal pyramid that projects into a minor calyx. (p. 850)

renal pelvis Funnel-shaped expansion of the upper end of the ureter that receives the calyces. (p. 850)

renal plasma flow Amount of plasma flowing through the kidneys per minute; approximately 650 ml per minute. (p. 872)

renal pyramid One of a number of pyramidal masses seen on longitudinal section of the kidney; they contain part of the loops of Henle and the collecting tubules. (p. 850)

renin Enzyme secreted by the juxtaglomerular apparatus that converts angiotensinogen to angiotensin I. (p. 689)

renin-angiotensin-aldosterone mechanism Renin, released from the kidneys in response to low blood pressure, converts angiotensinogen to angiotensin I. Angiotensin I is converted by angiotensin-converting enzyme to angiotensin II, which causes vasoconstriction, resulting in increased blood pressure. Angiotensin II also increases aldosterone secretion, which increases blood pressure by increasing blood volume. (p. 689)

replication Formation of new DNA from an existing DNA template. (p. 91)

repolarization Phase of the action potential in which the membrane potential moves from its maximum degree of depolarization toward the value of the resting membrane potential. (p. 262)

reposition Return of a structure to its original position. (p. 241)

residual volume Volume of air remaining in the lungs after a maximum expiratory effort. (p. 747)

resolution [L. *resolutio*, a slackening] Phase of the male sexual act after ejaculation during which the penis becomes flaccid; feeling of satisfaction; inability to achieve erection and second ejaculation. Last phase of the female sexual act, characterized by an overall sense of satisfaction and relaxation. (p. 920)

respiration [L. *respiratio*, to exhale, breathe] Process of life in which oxygen is used to oxidize organic fuel molecules, providing a source of energy, carbon dioxide, and water. (p. 735) Movement of air into and out of the lungs, the exchange of gases with blood, the transportation of gases in the blood, and gas exchange between the blood and the tissues. (p. 736)

respiratory acidosis Acidosis due to respiratory system disorders that result in increased blood carbon dioxide levels and decreased blood pH. (p. 898)

respiratory alkalosis Alkalosis due to respiratory system disorders that result in decreased blood carbon dioxide levels and increased blood pH. (p. 898)

respiratory bronchiole Smallest bronchiole (0.5 mm in diameter) that connects the terminal bronchiole to the alveolar duct. (p. 741)

respiratory center Inspiratory and expiratory centers together. (p. 758)

respiratory membrane Membrane in the lungs across which gas exchange occurs with blood. (p. 751)

resting membrane potential Electrical charge difference inside a cell membrane, measured relative to just outside the cell membrane. (p. 255)

rete testis (re'te tes'tis) Network of canals at the termination of the straight portion of the seminiferous tubules in the testes. (p. 911)

reticular (rĕ-tik'u-lar) [L. *rete*, net] Relating to a fine network of cells or collagen fibers. (p. 113)

reticular cell Cell with processes making contact with those of other similar cells to form a cellular network; along with the network of reticular fibers, the reticular cells form the framework of bone marrow and lymphatic tissues. (p. 120)

reticulocyte (rĕ-tik'u-lo-sīt) Young red blood cell with a network of basophilic endoplasmic reticulum occurring in larger numbers during the process of active red blood cell synthesis. (p. 592)

reticuloendothelial system See mononuclear phagocytic system. (p. 710)

retina Nervous tunic of the eyeball. (p. 472)

retinaculum (ret'ĭ-nak'u-lum) [L., band, halter, to hold back] Dense, regular connective tissue sheath holding down the tendons at the wrist, ankle, or other sites. (p. 115)

retinal Vitamin A derivative that binds to opsin to form rhodopsin. (p. 476)

retinal detachment Separation of the sensory retina from the pigmented retina; rods and cones in the sensory retina degenerate because of a lack of nutrition from the vascular choroid layer beneath the pigmented retina. (p. 483)

retraction [L., *retractio*, a drawing back] Movement in the posterior direction. (p. 241)

retroperitoneal (rĕ'tro-pĕr'ĭ-to-ne'al) Behind the peritoneum. (p. 25)

rheumatoid arthritis [Gr. *rheuma*, flux] Painful inflammation of the joints or bones. (p. 232)

rhodopsin (ro-dop'sin) Light-sensitive substance found in the rods of the retina; composed of opsin loosely bound to retinal. (p. 476)

ribonuclease Enzyme that splits RNA into its component nucleotides. (p. 804)

ribonucleic acid Nucleic acid containing ribose as the sugar component; found in all cells in both nuclei and cytoplasm; helps direct protein synthesis. (p. 54)

ribosomal RNA (rRNA) RNA that is associated with certain proteins to form ribosomes. (p. 77)

ribosome Small, spherical, cytoplasmic organelle where protein synthesis occurs. (p. 77)

right lymphatic duct Lymphatic duct that empties into the right subclavian vein; drains the right side of the head and neck, the right upper thorax, and the right upper limb. (p. 672)

rigor mortis Increased rigidity of muscle after death due to cross-bridge formation between actin and myosin as calcium ions leak from the sarcoplasmic reticulum. (p. 291)

RNA Ribonucleic acid; macromolecule formed by transcription from DNA and important in protein synthesis. See messenger RNA, ribosomal RNA, and transfer RNA. (p. 54)

rod Photoreceptor in the retina of the eye; responsible for noncolor vision in low-intensity light. (p. 472)

root of the penis Proximal attached part of the penis, including the two crura and the bulb. (p. 915)

root of the tooth That part below the neck of a tooth covered by cementum rather than enamel and attached by the periodontal ligament to the alveolar bone. (p. 778)

rotation Movement of a structure about its axis. (p. 240)

rotator cuff muscle One of four deep muscles that attach the humerus to the scapula. (p. 242)

round ligament Fibromuscular band that is attached to the uterus on either side in front of and below the opening of the uterine tube; it passes through the inguinal canal to the labium majus. (p. 970)

round ligament Remains of the umbilical vein. (p. 927)

round window Membranous structure separating the scala tympani of the inner ear from the middle ear.

Ruffini's end-organ (roo-fe'nēz) Named for Angelo Ruffini, Italian histologist, 1864-1929; receptor located deep in the dermis and responding to continuous touch or pressure. (p. 461)

ruga, pl. rugae (ru'gah, ru'ge) [L., a wrinkle] Fold or ridge; fold of the mucous membrane of the stomach when the organ is contracted; transverse ridge in the mucous membrane of the vagina. (p. 781)

saccule Part of the membranous labyrinth; contains sensory structure, the macula, that detects static equilibrium. (p. 494)

sagittal (saj'ĭ-tal) [L. *sagitta*, arrow] In the line of an arrow shot from a bow; plane running vertically through the body and dividing it into right and left portions. (p. 18)

salivary amylase (am'ĭ-lās) Enzyme secreted in the saliva that breaks down starch to maltose and isomaltose. (p. 791)

salivary gland Gland that produces and secretes saliva into the oral cavity. The three major pairs of salivary glands are the parotid, submandibular, and sublingual glands. (p. 772)

salt Molecule consisting of a cation

other than hydrogen and an anion other than hydroxide. (p. 42)

saltatory conduction Conduction in which action potentials jump from one node of Ranvier to the next node of Ranvier. (p. 362)

sarcolemma (sar'ko-lem'ah) [Gr. *sarco*, muscle + *lemma*, husk] Plasma membrane of a muscle fiber. (p. 273)

sarcomere (sar'ko-mēr) [Gr. *sarco*, muscle + *meros*, part] Part of a myofibril between adjacent Z lines. (p. 274)

sarcoplasm (sar'ko-plazm) [Gr. *sarco*, muscle + *plasma*, a thing formed] Cytoplasm of a muscle fiber, excluding the myofilaments. (p. 274)

sarcoplasmic reticulum [Gr. *sarco*, muscle + *plasma*, a thing formed + *reticulum*, net] Endoplasmic reticulum of muscle. (p. 278)

satellite cell Specialized cell that surrounds the cell bodies of neurons within ganglia. (p. 361)

satiety (sā-tī'ĕ-te) [L. *satie*, to fill, satisfy] Having hunger or thirst fulfilled.

saturated Fatty acid in which the carbon chain contains only single bonds between carbon atoms. (p. 46)

saturation Point when all carrier molecules or enzymes are attached to substrate molecules and no more molecules can be transported or reacted. (p. 70)

scala tympani (ska'lah) [L., stairway] Division of the spiral canal of the cochlea lying below the spiral lamina and basilar membrane. (p. 486)

scala vestibuli Division of the cochlea lying above the spiral lamina and vestibular membrane. (p. 486)

scar [Gr. *eschara*, scab] Fibrous tissue replacing normal tissue; cicatrix. (p. 127)

Schwann cell Cell that forms a myelin sheath around each nerve fiber of the peripheral nervous system. (p. 361)

sciatic (si-at'ik) [Gr., *ischiadikos*, the hipjoint] Relating to the hip or in the neighborhood of the hip. (p. 448)

sciatic nerve (si-at'ik) Tibial and common peroneal nerves bound together. (p. 448)

sclera (skler'ah) White of the eye; white, opaque portion of the fibrous tunic of the eye. (p. 470)

scoliosis (sko'le-o'sis) [Gr. *skoliōsis*, crookedness] Lateral curvature of the spine. (p. 203)

scrotum Musculocutaneous sac containing the testes. (p. 909)

sebaceous gland (se-ba'shus) [L. *sebum*, tallow] Gland of the skin, usually associated with a hair follicle, that produces sebum. (p. 145)

sebum (se'bum) [L., tallow] Oily, white, fatty substance produced by the sebaceous glands. (p. 145)

second heart sound Sound due to closure of the semilunar valves. (p. 631)

second messenger Molecule that is produced in a cell in which the first messenger interacts with a membrane-bound receptor molecule; the second messenger then acts as a signal and carries information to a site within the cell; e.g., cyclic AMP. (p. 536)

secondary bronchus (brong'kus) Branch from a primary bronchus that conducts air to each lobe of the lungs. There are two branches in the left lung and three branches from the primary bronchus in the right lung. (p. 741)

secondary follicle Follicle in which the secondary oocyte is surrounded by granulosa cells at the periphery of the fluid-filled antrum. (p. 923)

secondary (memory) response Immune response that occurs when the immune system is exposed to an antigen against which it has already produced a primary response. (p. 722)

secondary oocyte (o'o-sīt) Oocyte in which the second meiotic division stops at metaphase II unless fertilization occurs. (p. 923)

secondary palate Roof of the mouth in the early embryo that gives rise to the hard and the soft palates. (p. 956)

secondary spermatocyte (sper'mă-to-sīt) Spermatocyte derived from a primary spermatocyte by the firt meiotic division; each secondary spermatocyte gives rise by the second meiotic division to two spermatids. (p. 913)

secretin (se-kre'tin) Hormone formed by the epithelial cells of the duodenum; stimulates secretion of pancreatic juice high in bicarbonate ions. (p. 798)

secretion General term for a substance produced inside a cell and released from the cell. (p. 776)

secretory phase See luteal phase. (p. 930)

segmental artery One of five branches of the renal artery, each supplying a segment of the kidney. (p. 856)

self-antigen Antigen produced by the body that is capable of initiating an immune response. (p. 712)

semen [L., seed (of plants, men, animals)] Penile ejaculate; thick, yellowish-white, viscous fluid containing spermatozoa and secretions of the testes, seminal vesicles, prostate, and bulbourethral glands. (p. 917)

semicircular canal Canal in the petrous portion of the temporal bone that contains sensory organs that detect kinetic or dynamic equilibrium. Three semicircular canals are within each inner ear. (p. 495)

semilunar valve One of three semilunar segments serving as the three cusps of a valve; prevents regurgitation of blood at the beginning of the aorta or pulmonary trunk. (p. 618)

seminal fluid Semen.

seminal vesicle One of two glandular structures that empty into the ejaculatory ducts; its secretion is one of the components of semen. (p. 915)

seminiferous tubule (sem'ĭ-nif'er-us) Tubule in the testis in which spermatozoa develop. (p. 911)

sensible perspiration Perspiration excreted by the sweat glands that appears as moisture on the skin; produced in large quantity when there is much humidity in the atmosphere. (p. 893)

sensory ganglion Collection of sensory nerve cell bodies within a cranial nerve.

sensory retina Portion of the retina containing rods and cones. (p. 472)

septum (sep'tum) [L. *saeptum*, a partition] Thin wall dividing two cavities or masses of soft tissue. (p. 911)

septum pellucidum (sep'tum pel-lu'sid-um) One of two thin plates of brain tissue that separate the left and right ventricles. (p. 419)

septum primum First septum in the embryonic heart that arises on the wall of the originally single atrium of the heart and separates it into right and left chambers. (p. 960)

septum secundum Second of two major septal structures involved in the partitioning of the atrium, arising later than the septum primum and located to the right of it; it remains an incomplete partition until after birth, with its unclosed area constituting the foramen ovale. (p. 960)

serosa (se-ro'sah) [L. *serosus*, serous] Outermost covering of an organ or structure that lies in a body cavity; see adventitia. (p. 773)

serotonin (sēr-o-to'nin) Vasoconstrictor released by blood platelets. (p. 366)

serous (sēr'us) Relating to or producing a watery substance. (p. 927)

serous fluid Fluid similar to lymph that is produced by and covers serous membrane; lubricates the serous membrane. (p. 124)

serous membrane Thin sheet composed of epithelial and connective tissues; lines cavities that do not open to the outside of the body or contain glands but do secrete serous fluid. (p. 124)

serous pericardium Lining of the pericardial sac composed of a serous membrane. (p. 612)

Sertoli cell (ser-to'le) Elongated cell in the wall of the seminiferous tubules to which spermatids are attached during spermatogenesis. (p. 912)

serum [L., whey] Fluid portion of blood after the removal of fibrin and blood cells. (p. 586)

serum sickness Systemic immune complex reaction. (p. 727)

sesamoid bone (ses'ă-moyd) [Gr. *sesamoceies*, like a sesame seed] Bone found within a tendon; e.g., the patella. (p. 215)

sex chromosomes Pair of chromosomes responsible for sex determination; XX in female and XY in male. (p. 976)

Sharpey's fiber See perforating fiber.
shivering Rapid and rhythmic contractions of skeletal muscle that are involuntary; initiated by the nervous system as the body temperature falls below normal values. (p. 293)
shunted blood Blood that is not completely oxygenated. (p. 752)
sickle cell anemia Anemia characterized by the presence of crescent-shaped erythrocytes and excessive hemolysis; an inheritable condition. (p. 606)
sigmoid colon Part of the colon between the descending colon and the rectum. (p. 788)
sigmoid mesocolon Fold of peritoneum attaching the sigmoid colon to the posterior abdominal wall. (p. 791)
simple epithelium Epithelium consisting of a single layer of cells. (p. 104)
sinoatrial (SA) node Mass of specialized cardiac muscle fibers; acts as the "pacemaker" of the cardiac conduction system. (p. 622)
sinus [L., cavity] Hollow in a bone or other tissue; enlarged channel for blood or lymph. (p. 161)
sinus venosus End of the embryonic cardiac tube where blood enters the heart; becomes a portion of the right atrium, including the SA node. (p. 960)
sinusoid [sinus + Gr. *eidos*, resemblance] Terminal blood vessel having a larger diameter than an ordinary capillary. (p. 646)
sinusoidal capillary (si'nŭ-soy'dal) Capillary with caliber of from 10 to 20 μm or more; lined with a fenestrated type of endothelium. (p. 646)
sliding filament theory Mechanism by which actin and myosin myofilaments slide over one another during muscle contraction. (p. 278)
small intestine [L. *intestinus*, the entrails] Portion of the digestive tube between the stomach and the cecum; consists of the duodenum, jejunum, and ileum. (p. 772)
sodium-potassium exchange pump Biochemical mechanism that uses energy derived from ATP to achieve the active transport of potassium ions opposite to that of sodium ions. (p. 254)
soft palate Posterior muscular portion of the palate, forming an incomplete septum between the mouth and the oropharynx and between the oropharynx and the nasopharynx. (p. 779)
solubility coefficient Measure of how easily a gas dissolves in a liquid. (p. 750)
solute [L. *solutus*, dissolved] Dissolved substance in a solution. (p. 41)
solution [L. *solutio*] Homogenous mixture formed when a solute is dissolved in a solvent. (p. 41)
solvent [L. *solvens*, to dissolve] Liquid that holds another substance in solution. (p. 41)
soma (so'mah) [Gr., body] Neuron cell body or the enlarged portion of the neuron containing the nucleus and other organelles. (p. 356)
somatic (so-mat'ik) [Gr. *somatikos*, bodily] Relating to the body; the cells of the body except the reproductive cells. (p. 94)
somatic nervous system Composed of nerve fibers that send impulses from the central nervous system to skeletal muscle. (p. 354)
somatomedin (so-mă'to-me'den) Peptide synthesized in the liver capable of stimulating certain anabolic processes in bone and cartilage such as synthesis of DNA, RNA, and protein. (p. 552)
somatomotor (so-mă'to-mo'tor) [Gr. *soma*, body + motor] Motor nerves to the skeletal muscles. (p. 430)
somatostatin (so'mă-to-stat'in) Hypothalamic hormone capable of inhibiting the release of growth hormone by the adenohypophysis. (p. 552)
somatotropin (so'mă-to-tro'pin) Protein hormone of the adenohypophysis; it promotes body growth, fat mobilization, and inhibition of glucose utilization. (p. 552)
somesthetic (so'mes-thet'ik) [Gr. *somat*, body + *aisthesis*, sensation] Body sensations consciously perceived. (p. 391)
somite (so'mīt) [Gr. *soma*, body + *ite*] One of the paired segments consisting of cell masses formed in the early embryonic mesoderm on either side of the neural tube. (p. 954)
somitomere An indistinct somite in the head region of the embryo. (p. 954)
spatial summation Summation of the local potentials in which two or more action potentials arrive simultaneously at two or more presynaptic terminals that synapse with a single neuron. (p. 371)
specific dynamic activity Energy required to transport, digest, and absorb food. (p. 826)
specific heat Heat required to raise the temperature of any substance 1° C compared with the heat required to raise the same volume of water 1° C. (p. 41)
specific immunity Immune status in which there is an ability to recognize, remember, and destroy a specific antigen. (p. 706)
speech Use of the voice in conveying ideas. (p. 777)
spermatic cord Cord formed by the ductus deferens and its associated structures; extends through the inguinal canal into the scrotum. (p. 915)
spermatid (sper'mă-tid) [Gr. *sperma*, seed + *id*] Cell derived from the secondary spermatocyte; gives rise to a spermatozoon. (p. 914)
spermatocyte (sper'mă-to-sīt) Cell arising from a spermatogonium and destined to give rise to spermatozoa. (p. 913)
spermatogenesis Formation and development of the spermatozoon. (p. 912)
spermatogonium (sper'mă-to-go'ne-um) [Gr. *sperma*, seed + *gone*, generation] Cell that divides by mitosis to form primary spermatocytes. (p. 913)
spermatozoon, pl. spermatozoa (sper'mă-to-zo'on, sper'mă-to-zo'ah) [Gr. *sperma*, seed + *zoon*, animal] Sperm cell. Male gamete or sex cell, composed of a head and a tail. The spermatozoon contains the genetic information transmitted by the male. (p. 912)
sphenoid (sfe'noyd) [Gr. *shen*, wedge] Wedge shaped. (p. 196)
sphincter pupillae (pu-pil'e) Circular smooth muscle fibers of the iris' diaphragm that constrict the pupil of the eye. (p. 472)
sphygmomanometer (sfig'mo-mă-nom'ĕ-ter) [Gr. *sphygmos*, pulse + *manos*, thin, scanty + *metron*, measure] Instrument for measuring blood pressure. (p. 673)
spina bifida (spi'nah bif'ĭ-dah) [L., thorn, backbone, spine] Absence of the vertebral arches through which the spinal membranes, with or without spinal cord tissue, may protrude. (p. 205)
spinal nerve One of 31 pairs of nerves formed by the joining of the dorsal and ventral roots that arise from the spinal cord. (p. 354)
spindle fiber Specialized microtubule that develops from each centrosome and extends toward the chromosomes during cell division. (pp. 82, 92)
spiral artery One of the corkscrewlike arteries in premenstrual endometrium; most obvious during the secretory phase of the uterine cycle. (p. 933)
spiral ganglion Cell bodies of sensory neurons that innervate hair cells of the organ of Corti are located in the spiral ganglion. (p. 439)
spiral lamina Attached to the modiolus and supports the basilar and vestibular membranes. (p. 486)
spiral ligament Attachment of the basilar membrane to the lateral wall of the bony labyrinth. (p. 486)
spiral organ Organ of Corti; rests on the basilar membrane and consists of the hair cells that detect sound. (p. 488)
spiral tubular gland Well-developed simple or compound tubular glands that are spiral in shape within the endometrium of the uterus; prevalent in the secretory phase of the uterine cycle. (p. 933)
spirometer (spi-rom'ĕ-ter) [L. *spiro*, to breathe + Gr. *metron*, measure] Gasometer used for measuring the volume of respiratory gases; usually understood to consist of a counterbalanced cylindrical bell sealed by

dipping into a circular trough of water. (p. 747)

spirometry (spĭ-rom'ĕ-tre) Making pulmonary measurements with a spirometer. (p. 747)

splanchnic nerve (splangk'nik) Sympathetic nerve formed by preganglionic fibers that pass through the sympathetic chain ganglia without synapsing. (p. 508)

spleen Large lymphatic organ in the upper part of the abdominal cavity on the left side between the stomach and diaphragm, composed of white and red pulp. It responds to foreign substances in the blood, destroys worn out erythrocytes, and is a storage site for blood cells. (p. 703)

spongy urethra Portion of the male urethra, approximately 15 cm in length, that traverses the corpus spongiosum of the penis. (p. 915)

squamous (skwa'mus) [L. *squama,* a scale] Scalelike, flat. (p. 104)

stapedius Small skeletal muscles attached to the stapes. (p. 491)

stapes (sta'pēz) [L., stirrup] Smallest of the three auditory ossicles; attached to the oval window. (p. 486)

Starling's law of the heart Force of contraction of cardiac muscle is a function of the length of its muscle fibers at the end of diastole; the greater the ventricular filling, the greater is the stroke volume produced by the heart. (p. 632)

static labyrinth Part of the membranous labyrinth composed of the utricle and saccule involved in static equilibrium. (p. 494)

stereocilium (stĕr'e-o-sil'e-um) Elongated nonmobile microvillus. (p. 915)

sternum [L. *sternon,* chest] Breast bone. (p. 210)

steroid Large family of lipids, including some reproductive hormones, vitamins, and cholesterol. (p. 49)

stomach Large sac between the esophagus and the small intestine, lying just beneath the diaphragm. (p. 772)

strabismus (stră-biz'mus) [Gr. *strabismos,* a squinting] Lack of parallelism of the visual axes of the eyes. (p. 483)

stratified epithelium Epithelium consisting of more than one layer of cells. (p. 104)

stratum [L., bed cover, layer] Layer of tissue. (p. 138)

stratum basale (bah-sal'e) [L., layer; basal] Basal or deepest layer of the epidermis. (p. 138)

stratum corneum (kor'ne-um) [L., layer; *corneus,* horny] Most superficial layer of the epidermis consisting of flat, keratinized, dead cells. (p. 139)

stratum germinativum (jer'mĭ-na-tiv'um) [L., layer; *germen,* sprout, bud] Combined stratum basale and stratum spinosum; the layer of the epidermis where cells replicate by mitosis. (p. 138)

stratum granulosum (gran'u-lo'sum) [L., layer; granulum] Layer of cells in the epidermis filled with granules of keratohyalin. (p. 138)

stratum lucidum (lu'sĭ-dum) [L., layer; *lucidus,* clear] Clear layer of the epidermis found in thick skin between the stratum granulosum and the stratum corneum. (p. 139)

stratum spinosum (spi-no'sum) [L., layer; *spina,* spine] Layer of many-sided cells in the epidermis with intercellular connections (desmosomes) that give the cells a spiny appearance. (p. 138)

stress-relaxation response Change in blood vessel diameter in response to changes in blood pressure. The change in the volume of the blood vessel helps to maintain normal blood pressure. (p. 692)

stria, pl. striae (stri'ah, strī'e) [L., channel] Line or streak in the skin that is a different texture or color from the surrounding skin. Stretch mark. (p. 136)

striated (stri'a-ted) [L. *striatus,* furrowed] Striped; marked by stripes or bands. (p. 121)

stroke Lay term denoting a sudden neurological affliction, usually related to the cerebral blood supply. (p. 417)

stroke volume [L. *volumen,* something rolled up, scroll, from *volvo,* to roll] Volume of blood pumped out of one ventricle of the heart in a single beat. (p. 630)

structural gene Gene with the function of determining the structure of a specific protein or peptide. (p. 976)

sty Inflamed ciliary gland of the eye. (p. 468)

subcutaneous (sub'ku-ta'ne-us) [L. *sub,* under + *cutis,* skin] Under the skin; same tissue as the hypodermis. (p. 136)

sublingual gland One of two salivary glands in the floor of the mouth beneath the tongue. (p. 779)

submandibular gland One of two salivary glands in the neck, located in the space bounded by the two bellies of the digastric muscle and the angle of the mandible. (p. 779)

submucosa Layer of tissue beneath a mucous membrane. (p. 772)

submucosal plexus [L., a braid] Gangliated plexus of unmyelinated nerve fibers in the intestinal submucosa. (p. 772)

substantia nigra (ni'grah) [L., substance; black] Black nuclear mass in the midbrain; involved in coordinating movement and maintaining muscle tone. (p. 387)

subthreshold stimulus Stimulus resulting in a local potential so small that it does not reach threshold and produce an action potential. (p. 265)

sucrose Disaccharide composed of glucose and fructose; table sugar. (p. 44)

sulcus, pl. sulci (sul'sus, sul'si) [L., furrow or ditch] Furrow or groove on the surface of the brain between the gyri; may also refer to a fissure. (p. 390)

summation [Medieval L. *summatio,* to sum up] Summation of more than one local potential to produce a larger local potential; increased force of contraction of a muscle when stimulated in rapid succession. (p. 371)

superficial inguinal ring Slitlike opening in the aponeurosis of the external oblique muscle of the abdominal wall through which the spermatic cord (round ligament in the female) emerges from the inguinal canal. (p. 325)

superior Up, or higher, with reference to the anatomical position. (p. 16)

superior colliculus (kol-lik'u-lus) [L. *collis,* hill] One of two rounded eminences of the midbrain; aids in coordination of eye movements. (p. 387)

superior vena cava (ve'nah ka'vah) Vein that returns blood from the head and neck, upper limbs, and thorax to the right atrium. (p. 614)

supination (su'pĭn-a'shun) [L. *supino,* to bend backward, place on back] Rotation of the forearm (when the forearm is parallel to the ground) so that the anterior surface is up (supine). (p. 241)

supine Lying face up.

suppurative (su-pur'ah-tiv) [L. *sup* + *puro,* to form pus] Forming pus. (p. 232)

supramaximal stimulus Stimulus of greater magnitude than a maximal stimulus; however, the frequency of action potentials is not increased above that produced by a maximal stimulus. (p. 265)

surfactant (sur-fak'tant) Lipoproteins forming a monomolecular layer over pulmonary alveolar surfaces; stabilizes alveolar volume by reducing surface tension and the tendency for the alveoli to collapse. (p. 745)

suspension Liquid through which a solid is dispersed and from which the solid separates unless the liquid is kept in motion. (p. 41)

suspensory ligament Band of peritoneum that extends from the ovary to the body wall; contains the ovarian vessels and nerves. Small ligament attached to the margin of the lens in the eye and the ciliary body to hold the lens in place. (p. 471)

suture (su'chur) [L. *sutura,* a seam] Junction between flat bones of the skull. (p. 229)

sweat Perspiration; secretions produced by the sweat glands of the skin. See sensible and insensible perspiration. (p. 146)

sweat gland [A.S. *swat*] Usually means structures that produce a watery secretion called sweat. Some sweat glands, however, produce viscous organic secretions. (p. 145)

sympathetic chain ganglion Collection of sympathetic postganglionic neurons that are connected to each other to form a chain along both sides of the spinal cord. (p. 507)

sympathetic division Subdivision of the autonomic division of the nervous system characterized by having the cell bodies of its preganglionic neurons located in the thoracic and upper lumbar regions of the spinal cord (thoracolumbar division); usually involved in preparing the body for physical activity. (p. 507)

symphysis, pl. symphyses (sim'fă-sis, sim'fă-sez) [Gr., a growing together] Fibrocartilage joint between two bones. (p. 231)

synapse (sin'aps) [Gr. *syn*, together + *haptein*, to clasp] Functional membrane-to-membrane contact of a nerve cell with another nerve cell, muscle cell, gland cell, or sensory receptor; functions in the transmission of action potentials from one cell to another. (p. 94)

synaptic cleft Space between the presynaptic and the postsynaptic membranes. (p. 279)

synaptic fatigue Fatigue due to depletion of neurotransmitter vesicles in the presynaptic terminals. (p. 291)

synaptic vesicle Secretory vesicle in the presynaptic terminal containing neurotransmitter substances. (p. 279)

synchondrosis, pl. synchondroses (sin'kon-dro'sis, sin'kon-dro'sez) [Gr. *syn*, together + *chondros*, cartilage + *osis*, condition] Union between two bones formed by hyaline cartilage. (p. 231)

syncytiotrophoblast (sin-sish'e-o-tro'fo-blast) Outer layer of the trophoblast composed of multinucleated cells. (p. 950)

syncytium (sin-sish'-um) A true syncytium is produced when several cells fuse to form a single multinucleated structure. A functional syncytium behaves like a true syncytium, but the cells remain separate. Numerous gap junctions allow visceral smooth muscle cells and cardiac cells to function as a syncytium.

syndesmosis, pl. syndesmoses (sin'dez-mo'sis, sin'dez-mo'sez) [Gr. *syndeo*, to bind + *osis*, condition] Form of fibrous joint in which opposing surfaces that are some distance apart are united by ligaments.

synergist (sin'er-jist) Muscle that works with other muscles to cause a movement. (p. 306)

synostosis, pl. synostoses (sin'os-to'sis, sin'os'to'sez) [Gr. *syn*, together + *ostem*, bone + *osis*, condition] Bony union between the bones of a joint. (p. 229)

synovial (sĭ-no've-al) [Gr. *syn*, together + *oon*, egg] Relating to or containing synovia (a substance that serves as a lubricant in a joint, tendon sheath, or bursa). (p. 234)

synovial fluid (sĭ-no've-al) Slippery fluid found inside synovial joints and bursae; produced by the synovial membranes. (p. 234)

synovial joint (sĭ-no've-al) Bone joint in which the ends of the bones are covered with articular cartilage but are separated by a joint cavity filled with synovial fluid. The bones are held together by the joint capsule. (p. 234)

synthesis reaction [Gr. *syn*, together + *thesis*, a placing, arranging] Formation of larger molecules by the union of atoms or molecules. (p. 37)

system [Gr. *systema*, organized whole] Consistent and complex whole composed of correlated and semi-independent parts; in the case of organ systems, a complex of anatomically and functionally related organs. (p. 411)

systemic inflammation Inflammation that occurs in many areas of the body. In additionl to symptoms of local inflammation, increased neutrophil numbers in the blood, fever, and shock can occur. (p. 710)

systole (sis'to-le) [Gr. *systole*, a contracting] Contraction of the heart chambers during which blood leaves the chambers; usually refers to ventricular contraction. (p. 629)

T cell Thymus-derived lymphocyte of immunological importance; it is of long life and is responsible for cell-mediated immunity. (p. 714)

effector T Subset of T lymphocytes that is responsible for cell-mediated immunity. (p. 712)

helper T Subset of T lymphocytes that increases the activity of B cells and T cells. (p. 712)

suppressor T Subset of T lymphocytes that decreases the activity of B cells and T cells. (p. 712)

T tubule Tubelike invagination of the sarcolemma that conducts action potentials toward the center of the cylindrical muscle fibers. (p. 273)

T wave Deflection in the electrocardiogram following the QRS complex, representing ventricular repolarization. (p. 626)

tactile corpuscle Oval receptor found in the papillae of the dermis; responsible for fine, discriminative touch; Meissner's corpuscle. (p. 461)

tactile disk Cuplike receptor found in the epidermis; responsible for light touch and superficial pressure; Merkel's disk. (p. 461)

taenia colus, pl. taeniae coli (te'ne-ah ko'lus, te'ne-e ko'le) [Gr. *tainia*, band, tape, tapeworm] One of three bands in which the longitudinal muscular fibers of the large intestine, except the rectum, are collected. (p. 790)

talus (tal'us) [L., ankle bone, heel] Tarsal bone contributing to the ankle. (p. 220)

target tissue Tissue on which a hormone acts. (p. 526)

tarsal plate (tar'sal) Crescent-shaped layer of connective tissue that helps maintain the shape of the eyelid. (p. 468)

tarsal (tar'sal) [Gr. *tarsos*, sole of foot] One of seven ankle bones. (p. 220)

taste Sensations created when a chemical stimulus is applied to the taste receptors in the tongue. (p. 464)

taste bud Sensory structure, mostly on the tongue, that functions as a taste receptor. (p. 464)

taste pore Small opening in a taste bud. (p. 464)

tectorial membrane (tek-tōr'e-al) Attached to the spiral lamina and extends over the hair cells; cilia extend from the apical surface of the hair cells to the tectorial membrane. (p. 488)

tectum (tek'tum) Roof of the midbrain. (p. 387)

tegmentum (teg-men'tum) Floor of the midbrain. (p. 387)

telencephalon (tel-en-sef'ă-lon) [Gr. *telos*, end + *enkephalos*, brain] Anterior division of the embryonic brain from which the cerebral hemispheres develop. (p. 384)

telodendron, pl. telodendria (tel-o-den'dre-ah) [Gr. *telos*, end + *dendron*, tree] Terminal branch of an axon. (p. 357)

telophase Time during cell division when the chromosomes are pulled by spindle fibers away from the cell equator and into the two halves of the dividing cell. (p. 94)

temporal [L. *tempus*, time] Indicating the temple; the temple of the head is so named because it is there that the hair first begins turning white, indicating the passage of time. (p. 196)

temporal summation Summation of the local potential that results when two or more action potentials arrive at a single synapse in rapid succession. (p. 371)

tendon Band or cord of dense connective tissue that connects a muscle to a bone or other structure. (p. 158)

tensor tympani Small skeletal muscle attached to the malleus. (p. 491)

tentorium cerebelli (ten-to'rĭ-um sĕr'ĕ-bel'e) Dural folds between the cerebrum and the cerebellum. (p. 419)

teres [L. *tero*, to rub] Round, smooth, and tubular. (p. 331)

terminal bouton (boo-ton) [Fr., button] Enlarged axon terminal or presynaptic terminal. (p. 357)

terminal bronchiole End of the conducting airway; the lining is simple columnar or cuboidal epithelium without mucous goblet cells; most of the cells are ciliated, but a few nonciliated, serous-secreting cells occur. (p. 741)

terminal cisterna (sis-ter'nah) [L. *terminus*, limit + *cista*, box] Enlarged end of the sarcoplasmic reticulum in the area of the T tubules. (p. 278)

terminal hair [L. *terminus*, a boundary, limit] Long, coarse, usually pigmented hair found in the scalp, eyebrows, and eyelids and replacing vellus hair. (p. 141)

terminal sulcus [L., furrow or ditch] V-shaped groove on the surface of the tongue at the posterior margin. (p. 777)

tertiary bronchus (brong'kus) Extends from the secondary bronchus and conducts air to each lobule of the lungs. (p. 741)

testis, pl. testes (tes'tis, tes'tēz) One of two male reproductive glands located in the scrotum; produces spermatozoa, testosterone, and inhibin. (p. 911)

testosterone Steroid hormone secreted primarily by the testes; aids in spermatogenesis, maintenance and development of male reproductive organs, secondary sexual characteristics, and sexual behavior. (p. 918)

tetanus (tet'ă-nus) A smooth, sustained muscular contraction caused by a series of stimuli repeated so rapidly that individual muscular responses are fused; also refers to a disease marked by painful tonic muscular contractions. (p. 287)

tetany (tet'ă-ne) [Gr. *tetanos*, convulsive tension] Uncontrolled contraction of skeletal muscle; may be intermittent and accompanied by tremors in some cases.)

tetraiodothyronine (tet'ră-i'o-do-thi'ro-nēn) One of the iodine-containing thyroid hormones; also called thyroxine. (p. 555)

tetrodotoxin (tet'ro-do-tok'sin) Potent neurotoxin found in the liver and ovaries of the Japanese pufferfish and certain newts; produces axonal blocks of the preganglionic cholinergic fibers and the somatic motor nerves. (p. 623)

thalamus (thal'ă-mus) [Gr. *thalamos*, a bed, a bedroom] Large mass of gray matter that forms the larger dorsal subdivision of the diencephalon. (p. 388)

thalassemia (thal-ă-se'mĭ-ah) [Gr. *thalassa*, sea + *haima*, blood] Any of a group of inherited disorders of hemoglobin metabolism in which there is a decrease in synthesis of a particular globin chain without change in the structure of that chain. (p. 606)

theca [Gr. *theke*, a box] Sheath or capsule. (p. 923)

theca externa External fibrous layer of the theca or capsule of a vesicular follicle. (p. 923)

theca interna Inner vascular layer of the theca or capsule of the vesicular follicle; produces estrogen and contributes to the formation of the corpus luteum after ovulation. (p. 923)

thenar (the'nar) [Gr., palm of the hand] Fleshy mass of tissue at the base of the thumb; contains muscles responsible for thumb movements. (p. 337)

thick skin Found in the palms, soles, and tips of the digits and has all five epidermal strata. (p. 140)

thigh That part of the lower limb between the hip and knee. (p. 19)

thin skin Found over most of the body, usually without a stratum lucidum, and has fewer layers of cells than thick skin. (p. 140)

third heart sound Sound sometimes heard and corresponding with the first phase of rapid ventricular filling. (p. 631)

thoracic cavity Space within the thoracic walls, bounded below by the diaphragm and above by the neck. (p. 20)

thoracic duct Largest lymph vessel in the body, beginning at the cisterna chyli and emptying into the left subclavian vein; drains the left side of the head and neck, the left upper thorax, the left upper limb, and the inferior half of the body. (p. 670)

thoracolumbar division Synonym for the sympathetic division of the autonomic nervous system. (p. 507)

thorax [L. *thorax*, breastplate, chest] Chest; upper part of the trunk between the neck and abdomen. (p. 19)

thoroughfare channel Channel for blood through a capillary bed from an arteriole to a venule. (p. 647)

threshold potential Value of the membrane potential at which an action potential is produced as a result of depolarization in response to a stimulus. (p. 262)

threshold stimulus Stimulus resulting in a local potential just large enough to reach threshold and produce an action potential. (p. 265)

thrombin Enzyme, formed in blood, that converts fibrinogen into fibrin.

thrombocyte (throm'bo-sīt) Platelet. (p. 586)

thrombocytopenia (throm-bo-si-to-pe'nĭ-ah) [*thrombocyte* + Gr. *penia*, poverty] Condition in which there is an abnormally small number of platelets in the blood. (p. 605)

thrombosis (throm-bo'sis) [Gr. *thrombos*, a clot] Clotting within a blood vessel that may cause infarction of tissues supplied by the vessel. (p. 655)

thromboxane (throm'bok-zan) Specific class of physiologically active fatty acid derivatives present in many tissues. (p. 49)

thrombus (throm'bus) [Gr. *thrombos*, a clot] Clot in the cardiovascular system formed from constituents of blood; may be occlusive or attached to the vessel or heart wall without obstructing the lumen. (p. 599)

thymus gland [Gr. *thymos*, sweetbread] Bilobed lymph organ located in the inferior neck and superior mediastinum; secretes the hormone thymosin. (p. 575)

thyroglobulin (thi'ro-glob'u-lin) Thyroid hormone–containing protein, stored in the colloid within the thyroid follicles. (p. 555)

thyroid cartilage Largest laryngeal cartilage. It forms the laryngeal prominence, or Adam's apple. (p. 738)

thyroid gland [Gr. *thyreoeides*, shield] Endocrine gland located inferior to the larynx and consisting of two lobes connected by the isthmus; secretes the thyroid hormones triiodothyronine and tetraiodothyronine. (p. 554)

thyroid-stimulating hormone (TSH) Glycoprotein hormone released from the hypothalamus; stimulates thyroid hormone secretion from the thyroid gland. (p. 553)

thyrotropin See thyroid-stimulating hormone. (p. 553)

thyroxine See tetraiodothyronine. (p. 555)

tidal volume Volume of air that is inspired or expired in a single breath during regular, quiet breathing. (p. 747)

tinnitus (tĭ-ni'tus) [L. *tinnio*, to jingle, clink] Spontaneous sensation of noise without sound stimuli. (p. 499)

tissue [L. *texo*, to weave] Collection of similar cells and the substances between them. (p. 5)

tissue repair Substitution of viable cells for damaged or dead cells by regeneration or replacement. (p. 125)

tolerance Failure of the specific immune system to respond to an antigen.

tongue Muscular organ occupying most of the oral cavity when the mouth is closed; major attachment is through its posterior portion. (p. 777)

tonsil, pl. tonsils [L. *tonsilla*, stake] Any collection of lymphoid tissue; usually refers to large collections of lymphatic tissue beneath the mucous membrane of the oral cavity and pharynx; lingual, pharyngeal, and palatine tonsils. (p. 772)

tooth, pl. teeth Hard, conical structure set in the alveoli of the upper or lower jaws; used in mastication and assists in articulation. (p. 778)

total lung capacity Volume of air contained in the lungs at the end of a maximum inspiration; equals vital capacity plus residual volume. (p. 747)

total tension Sum of active and passive tension. (p. 290)

trabecula, pl. trabeculae (tră-bek'u-lah, tră-bek'u-le) [L. *trabs*, beam] One of the supporting bundles of fibers traversing the substance of a structure, usually derived from the capsule or one of the fibrous septa, e.g., trabeculae of lymph nodes, testes; a beam or plate of cancellous bone. (p. 121)

trabeculae carneae (trah-bek'u-le kar'ne-e) [L. *trabs*, beam] Muscular bundles lining the walls of the ventricles of the heart. (p. 619)

trachea (tra'ke-ah) [Gr. *tracheia arteria*, rough artery] Air tube extending from the larynx into the thorax where it divides to form the bronchi; composed of 16 to 20 rings of hyaline cartilage. (p. 738)

tracheostomy (tra'ke-os'to-mī) [tracheo + Gr. *stoma*, mouth] Incision into the trachea, creating an opening into which a tube can be inserted to facilitate the passage of air. (p. 740)

transamination Transfer of an amine group from an amino acid to a keto acid. (p. 837)

transcription Process of forming RNA from a DNA template. (p. 87)

transfer RNA (tRNA) RNA that attaches to individual amino acids and transports them to the ribosomes where they are connected to form a protein polypeptide chain. (p. 88)

transfusion [L. *trans*, across + *fundo*, to pour from one vessel to another] Transfer of blood from one person to another. (p. 600)

transitional epithelium Stratified epithelium that may be either cuboidal or squamouslike, depending on the presence or absence of fluid in the organ (as in the urinary bladder). (p. 104)

translation Synthesis of polypeptide chains at the ribosome in response to information contained in mRNA molecules. (p. 87)

transverse Plane separating the body or any part of the body into superior and inferior portions; a cross section.

transverse colon Part of the colon between the right and left colic flexures. (p. 788)

transverse mesocolon Fold of peritoneum attaching the transverse colon to the posterior abdominal wall. (p. 791)

transverse tubule [L. *tubus*, tube] Tubule that extends from the sarcolemma to a myofibril of striated muscles. (p. 620)

treppe (trep'eh) [Ger., staircase] Series of successively stronger contractions that occur when a rested muscle fiber receives closely spaced stimuli of the same strength but with a sufficient stimulus interval to allow complete relaxation of the fiber between stimuli. (p. 289)

triad Two terminal cisternae and a T tubule between them. (p. 278)

tricuspid valve Valve closing the orifice between the right atrium and the right ventricle of the heart. (p. 616)

triglyceride (tri-glis'er-īd) Glycerol with three attached fatty acids. (p. 808)

trigone (tri'gōn) [Gr. *trigonon*, triangle] Triangular smooth area at the base of the bladder between the openings of the two ureters and that of the urethra. (p. 857)

triiodothyronine (tri-i'o-do-thi'ro-nēn) One of the iodine-containing thyroid hormones. (p. 555)

trochlea (trok'le-ah) [L., pulley] Structure shaped like or serving as a pulley or spool. (p. 213)

trochlear nerve (trōk'le-ar) [L. *trochlea*, pulley] Cranial nerve IV, to the muscle (superior oblique) turning around a pulley. (p. 437)

trophoblast (tro'fo-blast) [Gr. *trophe*, nourishment + *blastos*, germ] Cell layer forming the outer layer of the blastocyst, which erodes the uterine mucosa during implantation; the trophoblast does not become part of the embryo but contributes to the formation of the placenta. (p. 948)

tropomyosin (tro'po-mī'o-sin) Fibrous protein found as a component of the actin myofilament. (p. 277)

troponin (tro'po-nin) Globular protein component of the actin myofilament. (p. 277)

true or vertebrosternal rib (ver-te'bro-ster'nal) Rib that attaches by an independent costal cartilage directly to the sternum. (p. 208)

true pelvis Portion of the pelvis inferior to the pelvic brim. (p. 218)

trypsin (trip'sin) Proteolytic enzyme formed in the small intestine from the inactive pancreatic precursor trypsinogen. (p. 804)

tubercle (tu'ber-kul) Lump on a bone. (p. 184)

tubular load Amount of a substance per minute that crosses the filtration membrane into Bowman's capsule. (p. 872)

tubular maximum Maximum rate of secretion or reabsorption of a substance by the renal tubules. (p. 872)

tubular reabsorption Movement of materials, by means of diffusion, active transport or cotransport, from the filtrate within a nephron to the blood. (p. 861)

tubular secretion Movement of materials, by means of active transport, from the blood into the filtrate of a nephron. (p. 864)

tunic [L., coat] One of the enveloping layers of a part; one of the coats of a blood vessel; one of the coats of the eye; one of the coats of the digestive tract. (p. 647)

tunica adventitia (ad-ven-tish'yah) Outermost fibrous coat of a vessel or an organ that is derived from the surrounding connective tissue. (p. 648)

tunica albuginea (al-bu-jin'e-ah) Dense, white, collagenous tunic surrounding a structure; e.g., the capsule around the testis. (p. 911)

tunica intima (in'ti-ma) Innermost coat of a blood vessel; consists of endothelium, a lamina propria, and an inner elastic membrane. (p. 647)

tunica media Middle, usually muscular, coat of an artery or other tubular structure. (p. 649)

tunica vaginalis Closed sac derived from the peritoneum that contains the testis and epididymis. It forms from an outpocket of the abdominal cavity, the processus vaginalis. (p. 911)

turbulent flow Flow characterized by eddy currents exhibiting nonparallel blood flow. (p. 673)

tympanic membrane (tim-pan'ik) Eardrum; cellular membrane that separates the external from the middle ear; vibrates in response to sound waves. (p. 484)

unipolar neuron One of the three categories of neurons consisting of a nerve cell body with a single axon projecting from it; also called a pseudounipolar neuron. (p. 123)

unmyelinated axon Nerve fibers lacking a myelin sheath. (p. 361)

unsaturated Carbon chain of a fatty acid that possesses one or more double or triple bonds. (p. 46)

up regulation An increase in the concentration of receptors in response to a signal. (p. 536)

upper respiratory tract The nasal cavity, pharynx, and associated structures. (p. 736)

ureter (ur-re'ter) [Gr. *oureter*, urinary canal] Tube conducting urine from the kidney to the urinary bladder. (p. 850)

urethra (u-re'thrah) Urogenital canal; canal leading from the bladder, discharging the urine externally. (p. 915)

urethral gland One of numerous mucous glands in the wall of the spongy urethra in the male. (p. 915)

urogenital fold Paired longitudinal ridges developing in the embryo on either side of the urogenital orifice. In the male they form part of the penis; in the female they form the labia minora. (p. 964)

urogenital triangle Anterior portion of the perineal region containing the openings of the urethra and vagina in the female and the urethra and root structures of the penis in the male. (p. 911)

uterine cavity Space within the uterus extending from the cervical canal to the openings of the uterine tubes. (p. 926)

uterine cycle Series of events that occur in a regular fashion in the uterus of sexually mature, nonpregnant females; prepares the uterine lining for implantation of the embryo. (p. 932)

uterine part Portion of the uterine tube that passes through the wall of the uterus. (p. 926)

uterine tube One of the tubes leading on either side from the uterus to the ovary; consists of the infundibulum, ampulla, isthmus, and uterine parts;

also called the fallopian tube or oviduct. (p. 926)

uterus Hollow muscular organ in which the fertilized ovum develops into a fetus. (p. 920)

utricle (u'trĭ-kul) Part of the membranous labyrinth; contains sensory structure, the macula, that detects static equilibrium. (p. 494)

uvula (u'vu-lah) [L. *uva*, grape] Small grapelike appendage at posterior margin of soft palate. (p. 737)

vaccination Deliberate introduction of an antigen into a subject to stimulate the immune system and produce immunity to the antigen. (p. 724)

vaccine [L. *vaccinus*, relating to a cow] Preparation of killed microbes, altered microbes, or derivatives of microbes or microbial products intended to produce immunity. The method of administration is usually inoculation, but ingestion is preferred in some instances, and nasal spray is used occasionally. (p. 726)

vagina [L., sheath] Genital canal in the female, extending from the uterus to the vulva. (p. 920)

vapor pressure Partial pressure exerted by water vapor. (p. 749)

variable region Part of the antibody that combines with the antigen. (p. 719)

varicose vein Permanent dilation and tortuosity of veins as a result of incompetent valves. (p. 650)

vas deferens See ductus deferens. (p. 915)

vasa recta Specialized capillary that extends from the cortex of the kidney into the medulla and then back to the cortex. (p. 857)

vasa vasorum [L., vessel, dish] Small vessels distributed to the outer and middle coats of larger blood vessels. (p. 650)

vascular compliance Tendency for blood vessel volume to change as a result of change in blood pressure (e.g., increased blood pressure results in increased blood vessel volume). (p. 674)

vascular tunic Middle layer of the eye; contains many blood vessels. (p. 470)

vasoconstriction Decreased diameter of blood vessels. (p. 844)

vasodilation Increased diameter of blood vessels. (p. 123)

vasomotion Periodic contraction and relaxation of the precapillary sphincter, resulting in cyclic blood flow through capillaries. (p. 125)

vasomotor center Area within the medulla oblongata that regulates the diameter of blood vessels by way of the sympathetic nervous system. (p. 681)

vasomotor tone Relatively constant frequency of sympathetic impulses that keep blood vessels partially constricted in the periphery. (p. 683)

vasopressin Hormone secreted from the neurohypophysis that causes vasoconstriction and acts on the kidney to reduce urinary volume; also called antidiuretic hormone. (p. 550)

vasopressin mechanism Increase in ADH secretion from the neurohypophysis when blood pressure drops or plasma osmolality increases. The ADH reduces urinary production and stimulates vasoconstriction. (p. 690)

vein Blood vessel that carries blood toward the heart; after birth, all veins except the pulmonary veins carry unoxygenated blood. (p. 679)

vellus (vel'us) [L., fleece] Short, fine, usually unpigmented hair that covers the body except for the scalp, eyebrows, and eyelids. Much of the vellus is replaced at puberty by terminal hairs. (p. 141)

venous capillary Capillary opening into a venule. (p. 647)

venous return Volume of blood returning to the heart. (p. 632)

venous sinus Endothelium-lined venous channel in the dura mater that receives cerebrospinal fluid from the arachnoid granulations. (p. 647)

ventilation [L. *ventus*, the wind] Movement of gases into and out of the lungs. (p. 744)

ventral root Motor (efferent) root of a spinal nerve. (p. 438)

ventricle (ven'trĭ-kul) [L. *venter*, belly] Chamber of the heart that pumps blood into arteries (i.e., the left and right ventricles). In the brain, a fluid-filled cavity. (p. 382)

ventricular diastole Dilation of the heart ventricles. (p. 629)

ventricular systole Contraction of the ventricles. (p. 629)

venule Minute vein, consisting of endothelium and a few scattered smooth muscles, that carries blood away from capillaries. (p. 650)

verapamil (ve-ra'pa-mil) Coronary vasodilator. (p. 623)

vermiform appendix (ver'mi-form) [L. *vermis*, worm + *forma*, form; appendage] Wormlike sac extending from the blind end of the cecum. (p. 772)

vernix caseosa [L., varnish + Gr. *easeus*, cheese] Fatty, cheeselike substance of sloughed epithelial cells and secretions that cover the fetal skin. (p. 966)

vertebral column The 26 vertebrae considered together; bears the weight of the trunk, protects the spinal cord, is the site of exit of the spinal nerves, and provides attachment sites for muscles. (p. 203)

vesicle (ves'ĭ-kl) [L. *vesica*, bladder] Small sac containing a liquid or gas, e.g., a blister in the skin or an intracellular, membrane-bound sac.

vesicular follicle Secondary follicle; the oocyte attains its full size and is surrounded by granulosa cells at the periphery of the fluid-filled atrium; the follicular cells proliferate; the theca develops into internal and external layers; a graafian follicle. (p. 923)

vestibular fold (false vocal cord) One of two folds of mucous membrane stretching across the laryngeal cavity from the angle of the thyroid cartilage to the arytenoid cartilage superior to the vocal cords; helps close the glottis; false vocal cord. (p. 738)

vestibular membrane Membrane separating the cochlear duct and the scala vestibuli. (p. 486)

vestibule (ves'ti-būl) [L., antechamber, entrance court] Anterior part of the nasal cavity just inside the external nares that is enclosed by cartilage; space between the lips and the alveolar processes and teeth; middle region of the inner ear containing the utricle and saccule; space behind the labia minora containing the openings of the vagina, urethra, and vestibular glands. (p. 736)

vestibulocochlear nerve Formed by the cochlear and vestibular nerves and extends to the brain. (p. 436)

villus, pl. villi (vil'us, vil'e) [L., shaggy hair (of beasts)] Projections of the mucous membrane of the intestine; they are leaf shaped in the duodenum and become shorter, more finger shaped, and sparser in the ileum. (p. 783)

viscera (vis'er-ah) [L. *viscus*, the soft parts, internal organs] Internal organs. (p. 6)

visceral (vis'er-al) Relating to the internal organs. (p. 25)

visceral pericardium (pĕr'ĭ-kar'dĭ-um) Serous membrane covering the surface of the heart. Also called the epicardium. (p. 25)

visceral peritoneum (pĕr'ĭ-to-ne'um) [Gr. *periteino*, to stretch over] Layer of peritoneum covering the abdominal organs. (p. 25)

visceral pleura (ploor'ah) Serous membrane investing the lungs and dipping into the fissures between the several lobes. (p. 25)

visceroreceptor Sensory receptor associated with the organs.

viscosity [L. *viscosus*, viscous] In general, the resistance to flow or alteration of shape by any substance as a result of molecular cohesion. (p. 672)

visual cortex Area in the occipital lobe of the cerebral cortex that integrates visual information and produces the sensation of vision. (p. 480)

visual field Area of vision for each eye. (p. 480)

visual radiation Nerve fibers that project from the lateral geniculate body to the visual cortex of the brain. (p. 480)

vital capacity Greatest volume of air

that can be exhaled from the lungs after a maximum inspiration. (p. 747)

vitamin (vi'tah-min) [L. *vita*, life + amine] One of a group of organic substances present in minute amounts in natural foodstuffs that are essential to normal metabolism; insufficient amounts in the diet may cause deficiency diseases. (p. 823)

vitamin D Fat-soluble vitamin produced from precursor molecules in skin exposed to ultraviolet light; increases calcium and phosphate uptake from the intestines. (pp. 150, 170)

vitamin K Fat-soluble compound essential for the information of normal clotting factors. (p. 599)

vitiligo (vit-ĭ-li'go) [L., skin eruption] White patches of skin caused by loss of melanin pigment. (p. 141)

vitreous humor (vit're-us) Transparent jellylike material that fills the space between the lens and the retina. (p. 473)

vocal cord One of two folds of elastic ligaments covered by mucous membrane stretching from the thyroid cartilage to the arytenoid cartilage; vibration of the vocal cords is responsible for voice production; true vocal cord. (p. 738)

Volkmann's canal Canal in bone containing blood vessels; not surrounded by lamellae; runs perpendicular to the long axis of the bone and the haversian canals, interconnecting the latter with each other and the exterior circulation. (p. 164)

vomiting [L. *vomitus*, to vomit] To eject matter from the stomach or small intestine through the mouth. (p. 801)

vulva (vul'vah) [L., wrapper or covering, seed covering, womb] External genitalia of the female composed of the mons pubis, the labia majora and minora, the clitoris, the vestibule of the vagina and its glands, and the opening of the urethra and of the vagina; the pudendum. (p. 927)

water-soluble vitamin Vitamin such as B complex and C that is absorbed with water from the intestinal tract. (p. 825)

white matter Bundles of parallel axons with their associated sheath in the central nervous system. (p. 363)

white pulp That part of the spleen consisting of lymphatic nodules and diffuse lymphatic tissue; associated with arteries. (p. 703)

white ramus communicans, pl. rami communicantes (ra'mus kō-mu-nĭ-kans, ra'mi kō-mu'nĭ-kan-tez) Connection between spinal nerves and sympathetic chain ganglia through which myelinated preganglionic axons project. (p. 507)

wisdom tooth Third molar tooth on each side in each jaw. (p. 778)

X linked Gene located on an X chromosome. (p. 978)

xiphoid (zĭf'oyd) [Gr. *xiphos*, sword] Sword shaped, with special reference to the sword tip; the inferior part of the sternum. (p. 210)

Y linked Gene located on a Y chromosome. (p. 978)

yolk sac Highly vascular layer surrounding the yolk of an embryo. (p. 952)

Z line Delicate membranelike structure found at either end of a sarcomere to which the actin myofilaments attach. (p. 276)

zona fasciculata (zo'nah fă-sik'u-la'ta) [L. *zone*, a girdle, one of the zones of the sphere] Middle layer of the adrenal cortex that secretes cortisol. (p. 563)

zona glomerulosa (zónah glo-mĕr-u-lo'sa) Outer layer of the adrenal cortex that secretes aldosterone. (p. 563)

zona pellucida (pel-lu'cĭ-da) Layer of viscous fluid surrounding the oocyte. (p. 923)

zona reticularis (zo'nah re-tik'u-lar'is) Inner layer of the adrenal cortex that secretes androgens and estrogens. (p. 563)

zonula adherens (zo'nu-lah ad-hĕ'renz) [L., a small zone; adhering] Small zone holding or adhering cells together. (p. 111)

zonula occludens (zo'nu-lah o-klood'enz) [L., occluding] Junction between cells in which the cell membranes may be fused; occludes or blocks off the space between the cells. (p. 111)

zygomatic (zi-go-mat'ik) [Gr. *zygon*, yoke] To yoke or join. Bony arch created by the junction of the zygomatic and temporal bones. (p. 196)

zygote (zi'gōt) [Gr. *zygotos*, yoked] Diploid cell resulting from the union of a sperm cell and an oocyte. (p. 926)

INDEX

Page numbers in *italics* indicate illustrations. Page numbers followed by *t* indicate tables.

B

F

G

I

J

O

P

U

V

CREDITS

Chapter 1

1-1 Christine Oleksyk/Tom Tracy, Photographic Resources. 1-3, 1-4 Christine Oleksyk. 1-8, 1-9A, 1-11, Table 1-2 Terry Cockerham, Synapse Media Production. 1-9B, 1-9C, 1-9D R.T. Hutchings. 1-10, 1-14, 1-15 Michael Schenk. 1-12, 1-13 Nadine Sokol. 1-A SIU Biomedical Communications/Photoresearchers, Inc. 1-B St. Bartholomew's Hospital/Science Photo Library. 1-C, 1-D, 1-E, 1-G Howard Sochurek. 1-F Phillippe Plailly/Science Photo Library. Table 1-1 Cynthia Turner Alexander/Terry Cockerham, Synapse Media Production/Christine Oleksyk.

Chapter 2

2-1, 2-6, 2-10, 2-21, 2-22, 2-A Ronald J. Ervin. 2-3B Nadine Sokol. 2-3C Michael Godomski/Tom Stack & Associates. 2-7 Nadine Sokol after William Ober. 2-13 Barry King/Tom Stack & Associates. 2-14B, 2-14C, 2-16 William Ober. 2-20 George Klatt after William Ober. 2-25 Christine Oleksyk.

Chapter 3

3-1, 3-2A, 3-7, 3-8, 3-9A, 3-10A, 3-11A, 3-15A, 3-16, 3-17A, 3-18A, 3-20A, 3-22, 3-23, 3-24, 3-25 Christine Oleksyk. 3-2 B J. David Robertson, from Charles Flickinger, *Medical Cell Biology*, Philadelphia, 1979, W.B. Saunders. 3-3, 3-21 Nadine Sokol. 3-4, 3-5, 3-27, 3-28 Barbara Cousins. 3-9B, 3-11B D.W. Fawcett/Photoresearchers, Inc. 3-10B Birgit Satir. 3-12 C.P. Morgan & R.A. Jersild, *Anatomical Record*, vol. 166, 1970. 3-13, 3-19 William Ober. 3-14, 3-26 Ronald J. Ervin. 3-15B Richard Rodewald. 3-17B Charles Flickinger, *Medical Cell Biology*, Philadelphia, 1979, W.B. Saunders. 3-18B Kent McDonald. 3-20B Susumu Ito. 3-26 Ed Reschke/Michael Abbey, Science Source.

Chapter 4

4-1A, 4-1B, 4-1D, 4-1E, 4-1F, 4-1G, 4-1H, 4-5A, 4-5D, 4-5F, 4-5G, 4-5H, 4-5I, 4-5J, 4-5K, 4-5N, 4-6A, 4-6B, 4-6C Ed Reschke. 4-1C, 4-5B, 4-5C, 4-5E, 4-5L, 4-5M, 4-7A, 4-7B, 4-8, Trent Stephens. 4-2 Christine Oleksyk. 4-3, 4-4, 4-9, 4-10 Michael Schenk.

Chapter 5

5-1, 5-A Ronald J. Ervin. 5-2, 5-7 John Cunningham/Visuals Unlimited. 5-3, 5-4, 5-5, 5-6 Christine Oleksyk. 5-B G. David Brown. 5-Unn Cynthia Turner Alexander/Terry Cockerham, Synapse Media Production.

Chapter 6

6-1 Ed Reschke. 6-2, 6-A David Mascaro & Associates. 6-3, 6-8, 6-9, 6-10A John V. Hagen. 6-4 Karen Waldo. 6-5, 6-13C, 6-15 Christine Oleksyk. 6-6, 6-10B Trent Stephens. 6-7A Fred E. Hossler/Visuals Unlimited. 6-7B Biophoto Associates/Photoresearchers, Inc. 6-11 Science VU/Visuals Unlimited. 6-12, 6-13A Joan Beck. 6-13B Ed Reschke. 6-14 J.M. Booher. 6-16, 6-17 Rusty Jones. 6-Ba Ewing Galloway. 6-Bb, 6-Bd N.E. Hilt, and S.B. Cogburn, *Manual of Orthopedics*. 6-Bc Robert Brashear, Jr. and Raney Beverly, Sr., *Handbook of Orthopedic Surgery*, ed. 10, 1986, St. Louis, Mosby–Year Book, Inc. 6-Be Custom Medical Stock Photo.

Chapter 7

7-1, 7-2A, 7-2B, 7-2C, 7-2D, 7-2E, 7-2F, 7-2H, 7-2I, 7-2J, 7-2M, 7-2N, 7-3, 7-4, 7-5, 7-7, 7-12, 7-13, 7-14, 7-18, 7-19, 7-21, 7-22, 7-23, 7-25, 7-26, 7-27, 7-29, 7-30, 7-31, 7-32, 7-34 David J. Mascaro & Associates. 7-2G, 7-2K, 7-2L, 7-9, 7-10 Christine Oleksyk. 7-6, 7-8, 7-17, 7-20, 7-24, 7-28, 7-33 Terry Cockerham/Synapse Media Production. 7-11, 7-16 Karen Waldo. 7-15 Robert Brashear, Jr. and Raney Beverly, Sr., *Handbook of Orthopedic Surgery*, ed. 10, 1986, St. Louis, Mosby–Year Book, Inc. 7-Unn Cynthia Turner Alexander/Terry Cockerham, Synapse Media Production.

Chapter 8

8-1, 8-2, 8-3, 8-4, 8-5, 8-8, 8-9A, 8-10A, 8-11A, 8-11B, 8-11C, 8-11D, 8-12, 8-13 David J. Mascaro & Associates. 8-6 Rusty Jones. 8-7 Terry Cockerham, Synapse Media Productions. 8-9B, 8-10B, 8-11F Scott Bodell. 8-11E R.T. Hutchings. 8-A Paul Manske.

Chapter 9

9-1, 9-5, 9-6 Christine Oleksyk. 9-2, 9-3, 9-10, 9-11 Joan Beck.

Chapter 10

10-1, 10-20 Ed Reschke. 10-2, 10-3, 10-6, 10-8 Joan Beck. 10-4, 10-7, 10-12, 10-13 Barbara Cousins. 10-5 Richard Rodewald. 10-9, 10-10, 10-11, 10-15 Joan Beck/Andrew Grivas.

Chapter 11

11-1, 11-2, 11-4, 11-6, 11-8, 11-9, 11-10, 11-11, 11-12, 11-13, 11-14, 11-16, 11-17, 11-18, 11-19, 11-20, 11-21, 11-24, 11-25, 11-26, 11-28, 11-29, 11-30, 11-31, 11-33 John V. Hagen. 11-3 G. David Brown. 11-5, 11-7, 11-15, 11-22A, 11-22C, 11-23, 11-27 Terry Cockerham, Synapse Media Production. 11-22B, 11-22D R.T. Hutchings. 11-32 Christine Oleksyk. 11-A Sylvan Legrand/Jean Marc Barey/Photoresearchers, Inc. 11-Unn Cynthia Turner Alexander/Terry Cockerham, Synapse Media Production.

Chapter 12

12-1 David J. Mascaro & Associates. 12-3, 12-4, 12-5, 12-6, 12-7, 12-8, 12-9, 12-10, 12-11, 12-12, 12-13, 12-20, 12-21, 12-22, 12-23, 12-A Scott Bodell. 12-14 John Daugherty. 12-16B, 12-17B, 12-19 Joan Beck.

Chapter 13

13-1 Joan Beck. 13-2 Marcia Hartsock. 13-3, 13-21, 13-22, 13-24, 13-25, 13-26, 13-29 Barbara Cousins. 13-4, 13-8, 13-15A R.T. Hutchings. 13-5, 13-7, 13-13, 13-14, 13-19, 13-27, 13-28 Scott Bodell. 13-9 Michael Schenk. 13-10 after William Ober. 13-12 Rebecca S. Montgomery. 13-17 Michael Schenk/Nadine Sokol. 13-18 Ed Reschke. 13-20 Christine Oleksyk. 13-23B Terry Cockerham, Synapse Media Production. 13-A G. David Brown. 13-Unn Cynthia Turner Alexander/Terry Cockerham, Synapse Media Production.

Chapter 14

14-1, 14-2A, 14-5, 14-6, 14-12, Table 14-1, III, IV, VI, X Michael Schenk. 14-2B R.T. Hutchings. 14-4 G. David Brown. 14-7, 14-8, 14-9, 14-10, 14-11, 14-13, 14-14, 14-15, 14-16 Janis K. Atlee. Table 14-1 I, II, V, VII, VIII, IX, XI, XIII David J. Mascaro & Associates. Table 14-1 V G. David Brown/Karen Waldo.

Chapter 15

15-1 Janis K. Atlee/Michael Schenk. 15-2 Michael Schenk. 15-3, 15-5, 15-6, 15-7, 15-10, 15-13, 15-14, 15-22, 15-23, 15-31, 15-36, 15-A Marsha J. Dohrmann. 15-4 Christine Oleksyk. 15-5 (inset) Omikron/Science Source. 15-8, 15-24 Terry Cockerham, Synapse Media Production. 15-9, 15-18, 15-19, 15-25, 15-28, 15-32, 15-34 Lisa Chuck/Michael Schenk. 15-11, 15-29 John V. Hagen. 15-12 R.T. Hutchings. 15-15 Trent Stephens. 15-17, 15-33, 15-B G. David Brown. 15-20 Donna Odle. 15-26 Dr. Tilney. 15-C S. Ishihara, Tests for Colour-Blindness, Tokyo, Japan; provided by Washington University Department of Ophthalmology. 15-D Eugene M. Helveston, and Forrest D. Ellis, *Pediatric Ophthalmology Practice*, 1984, St. Louis, Mosby–Year Book, Inc. 15-E Marsha J. Dohrman from materials provided by 3M Company.

Chapter 16

16-1 Scott Bodell. 16-2, 16-3, 16-4, 16-6 Christine Oleksyk. 16-8 Barbara Cousins.

Chapter 17

17-1 Joan Beck. 17-4B, 17-4C, 17-5 Christine Oleksyk. 17-7, 17-8, 17-9, 17-10, 17-11, 17-12 Andrew Grivas. 17-Unn Cynthia Turner Alexander/Terry Cockerham, Synapse Media Production.

Chapter 18

18-1, 18-2, 18-7A, 18-8, 18-10A, 18-12A Andrew Grivas. 18-3 G. David Brown. 18-4, 18-5, 18-6, 18-9, 18-13, 18-14, 18-19 Christine Oleksyk. 18-7B, 18-10B, 18-15 Trent Stephens. 18-12B Robert Callentine. 18-A Science VU/Visuals Unlimited.

Chapter 19

19-1 Kathryn A. Born. 19-2 David Phillips/Visuals Unlimited. 19-3, 19-4, 19-5 Christine Oleksyk. 19-6 Ed Reschke. 19-9, 19-10, 19-11 Molly Babich/John Daugherty. 19-12 Trent Stephens. Table 19-2 William Ober.

Chapter 20

20-1, 20-5, 20-8, 20-9 Rusty Jones. 20-2, 20-14, 20-17, 20-18 Christine Oleksyk. 20-3 Michael Schenk. 20-4A, 20-4C David J. Mascaro & Associates. 20-4B, 20-7 R.T. Hutchings. 20-6, 20-10A Barbara Cousins. 20-10B Ed Reschke. 20-11 Ronald J. Ervin. 20-16 the University Medical Center, Tucson, Arizona. 20-Unn Cynthia Turner Alexander/Terry Cockerham, Synapse Media Production.

Chapter 21

21-1, 21-2, 21-3, 21-4, 21-5, 21-24 Ronald J. Ervin. 21-6, 21-12 George J. Wassilchenko. 21-7A, 21-8, 21-9A, 21-10, 21-11, 21-14, 21-15, 21-17, 21-19 David J. Mascaro & Associates. 21-7B, 21-7C, 21-7D, 21-13, 21-16, 21-18 Karen Waldo. 21-9B John Daugherty. 21-20 Joan Beck. 21-21, 21-22, 21-23, 21-A G. David Brown. 21-24 Ronald J. Ervin. 21-25, 21-29 Joan Beck/Donna Odle. 21-30, 21-31. 21-32 Christine Oleksyk.

Chapter 22

22-1 Barbara Cousins. 22-2, 22-6C John Cunningham/Visuals Unlimited 22-3 David J. Mascaro & Associates. 22-4A, 22-5A Walter D. Glanze, editor, *Mosby's Medical and Nursing Dictionary*, ed. 2, 1986, Mosby–Year Book, Inc. 22-4B, 22-5C Trent Stephens 22-5B, 22-6B Kathy Mitchell Grey. 22-6A Nadine Sokol. 22-9, 22-10 Christine Oleksyk. 22-15 Michael Schenk. 22-16, 22-17, 22-18 John Daugherty. 22-Unn Cynthia Turner Alexander/Terry Cockerham, Synapse Media Production.

Chapter 23

23-1 Joan Beck. 23-2A, 23-5, 23-6, 23-7, 23-8A, 23-9, 23-11, 23-12 Jody L. Fulks. 23-2B, 23-8B R.T. Hutchings. 23-3 David J. Mascaro & Associates. 23-4A, 23-16 Christine Oleksyk. 23-4B, 23-10A Custom Medical Stock Photo. 23-10B John Daugherty. 23-13 Joan Beck/Donna Odle. 23-17 Trent Stephens. 23-18 Barbara Cousins. 23-Unn Cynthia Turner Alexander/Terry Cockerham, Synapse Media Production.

Chapter 24

24-1, 24-22 Joan Beck 24-2, 24-10A, 24-10B, 24-11D Barbara Cousins. 24-3 Barbara Stackhouse. 24-4, 24-20 Trent Stephens. 24-5 Kathryn A. Born. 24-6, 24-7A, 24-12 David J. Mascaro & Associates. 24-7B, 24-10C, 24-10D, 24-14B, 24-14C Kathy Mitchell Grey. 24-7 (inset) Ed Reschke. 24-8A, 24-13A, 24-24A, 24-16, 24-19 G. David Brown. 24-8 (inset) S. Elems/Visuals Unlimited. 24-9 John Daugherty. 24-11A, 24-11B, 24-11C Walter D. Glanze, editor, *Mosby's Medical and Nursing Dictionary*, ed. 2, 1986, St. Louis, Mosby–Year Book, Inc. 24-13 (inset) Science Photo Library/Photoresearchers, Inc. 24-15 Michael Schenk. 24-17, 24-21 Joan Beck/Donna Odle. 24-24 William Ober. 24-A SIU Biomedical Communications/Science Source/Photoresearchers, Inc. 24-Unn Cynthia Turner Alexander/Terry Cockerham, Synapse Media Production.

Chapter 25

25-5 Kevin Sommerville. 25-11, 25-12 Nadine Sokol.

Chapter 26

26-1 Joan Beck. 26-2, 26-8, 26-15 Christine Oleksyk. 26-3A, David J. Mascaro & Associates. 26-3B R.T. Hutchings. 26-4, 26-5, 26-6, 26-7 Jody L. Fulks. 26-9, 26-12 Michael Schenk. 26-A Joan Beck/Donna Odle. 26-Unn Cynthia Turner Alexander/Terry Cockerham, Synapse Media Production.

Chapter 27

27-1 Andrew Grivas. 27-2, 27-3, 27-4, 27-5 Nadine Sokol. 27-6 Joan Beck.

Chapter 28

28-1A, 28-8, 28-A, 28-Be, 28-Bf Ronald J. Ervin. 28-1B, 28-13, 28-14 David J. Mascaro & Associates. 28-2 William Ober 28-3, 28-5, 28-6, 28-9, 28-10, 28-11, 28-12, 28-15, 28-16, 28-17, 28-18 Kevin A. Sommerville. 28-4 Barbara Cousins. 28-7 Christine Oleksyk. 28-Ba, 28-Bc, 28-Bg Custom Medical Stock Photo. 28-Bb, 28-Bd Joel Gordon. 28-Unn Cynthia Turner Alexander/Terry Cockerham, Synapse Media Production.

Chapter 29

29-1 Lucinda L. Veeck, Jones Institute for Reproductive Medicine, Norfolk, Virginia. 29-2, 29-3, 29-5, 29-7, 29-8, 29-10, 29-13, 29-14 Marcia Hartsock. 29-4, 29-16, 29-17, 29-20, 29-A Kevin A. Sommerville. 29-6 Michael Schenk. 29-9, 29-15 Lennart Nilsson. 29-11, 29-12 David J. Mascaro & Associates. 29-18 Scott Bodell. 29-19 Molly Babich/John Daugherty. 29-21 Stephen Marks/The Image Bank.

Prefixes, Suffixes, And Combining Forms

Continued from inside front cover.

Term	Meaning	Example
intra-	Within	Intraocular (within the eye)
-ism	Condition, state of	Dimorphism (condition of two forms)
iso-	Equal	Isotonic (same tension)
-itis	Inflammation	Gastritis (inflammation of the stomach)
-ity	Expressing condition	Acidity (condition of acid)
kerato-	Cornea or horny tissue	Keratinization (formation of a hard tissue)
-kin-	Move	Kinesiology (study of movement)
leuko-	White	Leukocyte (white blood cell)
-liga-	Bind	Ligament (structure that binds bone to bone)
lip-	Fat	Lipolysis (breakdown of fats)
-logy	Study	Histology (study of tissue)
-lysis	Breaking up, dissolving	Glycolysis (breakdown of sugar)
macro-	Large	Macrophage (large phagocytic cell)
mal-	Bad	Malnutrition (bad nutrition)
malaco-	Soft	Osteomalacia (soft bone)
mast-	Breast	Mastectomy (excision of the breast)
mega-	Great	Megacolon (large colon)
melano-	Black	Melanocyte (black pigment producing skin cell)
meso-	Middle, mid	Mesoderm (middle skin)
meta-	Beyond, after, change	Metastasis (beyond original position)
micro-	Small	Microorganism (small organism)
mito-	Thread, filament	Mitosis (referring to threadlike chromosomes during cell division)
mono-	One, single	Monosaccharide (one sugar)
-morph-	Form	Morphology (study of form)
multi-	Many, much	Multinucleated (two or more nuclei)
myelo-	Marrow, spinal cord	Myeloid (derived from bone marrow)
myo-	Muscle	Myocardium (heart muscle)
narco-	Numbness	Narcotic (drug producing stupor or weakness)
neo-	New	Neonatal (first four weeks of life)
nephro-	Kidney	Nephrectomy (removal of a kidney)
neuro-	Nerve	Neuritis (inflammation of a nerve)
oculo-	Eye	Oculomotor (movement of the eye)
odonto-	Tooth or teeth	Odontomy (cutting a tooth)
-oid	Expressing resemblance	Epidermoid (resembling epidermis)
oligo-	Few, scanty, little	Oliguria (little urine)
-oma	Tumor	Carcinoma (cancerous tumor)
-op-	See	Myopia (nearsighted)
ophthalm-	Eye	Ophthalmology (study of the eye)
ortho-	Straight, normal	Orthodontics (discipline dealing with the straightening of teeth)
-ory	Referring to	Olfactory (relating to the sense of smell)
-ose	Full of	Adipose (full of fat)
-osis	A condition of	Osteoporosis (porous condition of bone)
osteo-	Bone	Osteocyte (bone cell)
oto-	Ear	Otolith (ear stone)
-ous	Expressing material	Serous (composed of serum)
para-	Beside, beyond, near to	Paranasal (near the nose)